NEW HORIZONS IN EARTH REINFORCEMENT

T0144228

BALKEMA – Proceedings and Monographs
in Engineering, Water and Earth Sciences

PROCEEDINGS OF THE 5th INTERNATIONAL SYMPOSIUM ON EARTH REINFORCEMENT (IS Kyushu '07) FUKUOKA, KYUSHU, JAPAN, 14–16 NOVEMBER 2007

New Horizons in Earth Reinforcement

Edited by

Jun Otani
Kumamoto University, Kumamoto, Japan

Yoshihisa Miyata
National Defense Academy, Yokosuka, Japan

Toshifumi Mukunoki
Kumamoto University, Kumamoto, Japan

Routledge
Taylor & Francis Group
LONDON AND NEW YORK

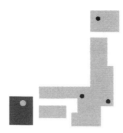

IS Kyushu '07

organized by
the Japanese Geotechnical Society,

under the auspices of the International Society for Soil Mechanics and Geotechnical Engineering,

with support of the International Geosynthetics Society

and the Japan Society of Civil Engineers

First published 2008 by Routledge

2 Park Square, Milton Park, Abingdon, Oxfordshire OX14 4RN
52 Vanderbilt Avenue, New York, NY 10017

Routledge is an imprint of the Taylor & Francis Group, an informa business

First issued in paperback 2019

Copyright © 2008 Taylor & Francis

All rights reserved. No part of this book may be reprinted or reproduced or utilized in any form or by any electronic, mechanical, or other means, now known or hereafter invented, including photocopying and recording, or in any information storage or retrieval system, without permission in writing from the publishers.

Notice:
Product or corporate names may be trademarks or registered trademarks, and are used only for identification and explanation without intent to infringe.

Although all care is taken to ensure integrity and the quality of this publication and the information herein, no responsibility is assumed by the publishers nor the author for any damage to the property or persons as a result of operation or use of this publication and/or the information contained herein.

Typeset by Charon Tec Ltd (A Macmillan Company), Chennai, India

ISBN: 978-0-415-45775-0 (hbk)
ISBN: 978-0-367-38849-2 (pbk)

New Horizons in Earth Reinforcement – Otani, Miyata & Mukunoki (eds)
© 2008 Taylor & Francis Group, London, ISBN 978-0-415-45775-0

Table of Contents

Materials and new testing

Advanced numerical modeling

Design and measurement on full-scale behavior of reinforced structure

Physical modeling

Combined technologies

Geo-hazards and mitigation

New Horizons in Earth Reinforcement – Otani, Miyata & Mukunoki (eds)
© 2008 Taylor & Francis Group, London, ISBN 978-0-415-45775-0

Preface

Earth reinforcement techniques are used worldwide and offer proven solutions to a wide range of geotechnical engineering problems. The International Symposia on Earth Reinforcement (IS Kyushu) held in the city of Fukuoka, Japan have played a key role in disseminating developments in earth reinforcement starting in 1988 and again in 1992, 1996, and 2001. The important role of this series of symposia in promoting these technologies has been recognized by the strong support offered by the International Society for Soil Mechanics and Geotechnical Engineering (ISSMGE) and the International Geosynthetics Society (IGS).

While earth reinforcement technologies are well-established, new materials, construction techniques, design and analysis methods continue to be developed. Furthermore, reinforced earth structures are now being used more and more as expedient and economical solutions in response to earthquake, heavy rain, tsunami, and other natural disasters.

Based on the success of previous IS Kyushu symposia and the continued importance of earth reinforcement technologies to geotechnical engineering, it was decided that the *5th International Symposium on Earth Reinforcement* (IS Kyushu'07) will be held on November 14–16, 2007 in Fukuoka, Japan. The theme subtitle of IS Kyushu'07 is "New Horizons in Earth Reinforcement". The main purpose of the symposium is to provide a forum for the exchange of technical information on current earth reinforcement techniques and design procedures around the world, introduce new reinforcement material, and present new and emerging applications with special emphasis on disaster mitigation and geoenvironmental issues. We encourage the new generation of engineers and researchers to attend this important technical event so that they are part of the future "New Horizons" in earth reinforcement.

A total of 175 abstracts from 32 countries were submitted to the conference; of these, 124 full papers have been accepted for publication in the conference proceedings and for presentation. Three distinguished engineers and researchers from around the world under the guidance of the Scientific Committee carefully reviewed each paper.

We believe that all the participants will benefit greatly from the presentations and discussions that will take place during the symposium. The conference proceedings will serve as a milestone not only for those involved in earth reinforcement technologies but also for the larger geotechnical engineering community.

Jun Otani
Chairperson of IS Kyushu'07

New Horizons in Earth Reinforcement – Otani, Miyata & Mukunoki (eds)
© 2008 Taylor & Francis Group, London, ISBN 978-0-415-45775-0

Organization

Steering Committee

Prof. Ochiai, H. (Chairperson)
Prof. Asaoka, A.
Dr. Fukuda, N.
Prof. Gibo, S.
Prof. Hyodo, M.
Prof. Iizuka, A.
Prof. Kamon, M.
Prof. Kimura, M.
Prof. Kitamura, R.
Prof. Koseki, J.
Prof. Kodaka, T.
Prof. Koumoto, T.
Prof. Kumagai, K.
Prof. Kusakabe, O.
Prof. Kuwano, J.
Prof. Kazama, M.
Prof. Makiuchi, K.
Prof. Mitachi, T.
Prof. Miura, N.
Prof. Murakami, A.
Prof. Murata, H.
Prof. Nakai, T.
Prof. Ohta, H.
Prof. Okuzono, S.
Prof. Otsuka, S.
Prof. Tanabashi, Y.
Dr. Tateyama, M.
Prof. Umezaki, T.
Prof. Yashima, A.
Prof. Yasuda, S.
Prof. Yasuhara, K.
Prof. Yokota, H.
Prof. Zen, K.

International Advisory Group

Dr. Garcia-Mina, J.
Prof. Bathurst, R.J.
Prof. Bergado, D.T.
Prof. Bekenov, T.N.
Prof. Bouassida, M.
Dr. Cazzuffi, D.
Prof. Chang, D.T.T.
Prof. Davies, M.
Prof. Ehrlich, M.
Prof. Gourc, J.P.
Dr. Guler, E.
Dr. Heerten, G.
Dr. Herle, V.
Dr. Jones, C.J.F.P.
Dr. Kitazume, M.
Prof. Kvasnicka, P.
Prof. Lam, A.
Dr. Lawson, C.R.
Prof. Leshchinsky, D.
Dr. Liausu, Ph.
Dr. Lopes, M.G.
Prof. Madav, M.R.
Dr. Miki, H.
Prof. Lo, R.
Prof. Mathys, M.
Prof. Martens, J.
Prof. Moraci, M.
Prof. Palmeira, E.M.
Prof. Rajagopal, K.
Dr. Rathmayer, H.
Prof. Rowe, R.K.
Prof. Scharle, P.
Dr. Varaksin, S.
Prof. Stanic, B.
Prof. Telekes, G.
Prof. Timofeeva, L.M.
Dr. Uriel, S.
Prof. Tatsuoka, F.
Prof. Tan, S.A.
Dr. Voskamp, W.
Dr. Yeo. K.
Prof. Yoo, C.
Prof. Zornberg, J.

Scientific Committee

Prof. Miyata, Y. (Chairperson)
Prof. Bathurst, R.J.
Dr. Cazzuffi, D.
Prof. Gourc, J.P.
Prof. Guler, E.
Dr. Hery, P.
Dr. Hirakawa, D.
Dr. Hironaka, J.
Dr. Izawa, J.
Dr. Kasama, K.
Dr. Kawamura, T.
Dr. Kobayashi, T.
Dr. Kojima, K.
Prof. Kohata, Y.
Prof. Kuwano, J.
Prof. Leshchinsky, D.
Prof. Lo, R.
Prof. Mukunoki, T.
Prof. Nakamura, T.
Prof. Nabeshima, Y.
Prof. Omine, K.
Prof. Otani, J.
Prof. Palmeira, E.M.
Prof. Rowe, R.K.
Prof. Sawada, K.
Dr. Shinoda, M.
Dr. Suetsugu, D.
Prof. Tan, S.A.
Prof. Uchimura, T.
Dr. Yamada, S.
Prof. Yoo, C.
Dr. Yamamoto, K.

Organizing Committee

Prof. Otani, J. (Chairperson)
Prof. Maeda, Y. (Vice-chairperson)
Prof. Mukunoki, T. (Secretary General)
Prof. Aramaki, S.
Prof. Chen, G.
Dr. Hirai, T.
Prof. Hirooka, A.
Dr. Irie, T.
Dr. Kasama, K.
Mr. Kidera, S.
Dr. Kobayashi, T.
Prof. Kudou, M.
Prof. Kondou, F.
Prof. Miyata, Y.
Mr. Matsugu, H.
Prof. Nagase, H.
Mr. Nakayama, H.
Dr. Nishida, K.
Prof. Omine, K.
Prof. Ohtsubo, M.
Dr. Sakata, T.
Mr. Sakimoto, S.
Prof. Sato, K.
Ms. Sato, T. (Secretary)
Dr. Suetsugu, D.
Dr. Sugimoto, T.
Prof. Sezaki, M.
Dr. Tajiri, M.
Dr. Tanoue, Y.
Prof. Tokashiki, N.
Mr. Ueyama, Y.
Prof. Yasufuku, N.
Mr. Yamauchi, Y.
Dr. Yamada, S.
Dr. Yamamoto, K.
Dr. Yakabe, H.
Dr. Wada, H.

International Review Panel

Prof. Alfaro, M.
Prof. Bathurst, R.J.
Prof. Bergado, D.T.
Prof. Bouassida, M.
Dr. Cazzuffi, D.
Dr. Chew, S.H.
Prof. Gourc, J.P.
Dr. Greenwood, J.H.
Prof. Guler, E.
Prof. Han, J.
Dr. Hatami, K.
Dr. Hery, P.
Dr. Hirai, T.
Dr. Hirakawa, D.
Dr. Hironaka, J.
Prof. Hirooka, A.
Prof. Huang, C.C.
Prof. Iizuka, A.
Dr. Izawa, J.
Dr. Jenner, C.
Dr. Jeon, H.Y.
Prof. Kaneko, K.
Dr. Kasama, K.
Prof. Kawabata, T.
Dr. Kawamura, T.
Prof. Kimura, M.
Dr. Kitazume, M.
Dr. Kobayashi, T.
Prof. Kodaka, T.
Prof. Kohata, Y.
Dr. Kojima, K.
Mr. Konami, T.
Prof. Koseki, J.
Dr. Kotake, N.
Prof. Kuwano, J.
Prof. Lam, A.
Prof. Lo, R.
Prof. Michalowsky, R.L.
Prof. Miyata, Y.
Prof. Moraci, N.
Prof. Mukunoki, T.
Prof. Nabeshima, Y.
Prof. Nakamura, T.
Prof. Nishigata, T.
Prof. Otsuka, S.
Prof. Okamura, M.
Prof. Omine, H.
Dr. Otani, Y.
Prof. Palmeira, E.M.
Dr. Sankey, J.
Prof. Sawada, K.
Prof. Shin, E.C.
Dr. Shinoda, M.
Dr. Suetsugu, D.
Prof. Tan, S.A.
Prof. Tsukamoto, Y.
Prof. Uchimura, T.
Dr. Watanabe, K.
Dr. Yamada, S.
Dr. Yamamoto, K.
Prof. Yasufuku, N.
Prof. Yoo, C.
Dr. Yoshida, T.

Special lectures

New Horizons in Earth Reinforcement – Otani, Miyata & Mukunoki (eds)
© 2008 Taylor & Francis Group, London, ISBN 978-0-415-45775-0

Earth reinforcement technique as a role of new geotechnical solutions – memory of IS Kyushu

H. Ochiai
Civil Engineering, Kyushu University, Japan
Immediate past chairperson of IS Kyushu Conferences

ABSTRACT: Earth reinforcing techniques are increasingly becoming a useful, powerful and economical solution to various problems encountered in geotechnical engineering practice. Expansion of the experiences and knowledge in this area has succeeded in developing new techniques and their applications to geotechnical engineering problems. The series of IS Kyushu Conferences and the activities of TC-9/ISSMGE have played an important role on the expansion of the latest experiences and knowledge, thus provided successful contributions for further development of the earth reinforcement practice. In this paper, the history of the IS Kyushu conferences was summarized with careful review of the activities of TC-9/ISSMGE, and finally the future prospects on earth reinforcement techniques were proposed.

1 INTRODUCTION

Studies with respect to modern earth reinforcing techniques and their applications to the construction activities have been started historically by Henry Vidal in France since 1966. In addition, the first stage of their remarkable development was mainly in Europe where the First International Conference on Geotextiles was held in Paris in 1977. In Japan, the fundamental studies and applications of earth reinforcement commenced in laboratory in the middle of the 1960s to meet the extensive needs of practicing engineers who encountered many geotechnical problems caused by widely distributed soft ground in Japan. Thereafter, they had been very active not only in carrying out the fundamental studies of earth reinforcement but also in applying them to geotechnical engineering practices.

Among various districts in Japan, the geotechnical group in Kyushu was one of pioneer in geotextile research, and in the beginning of the year 1985, an international symposium concerning the theory and practice of earth reinforcement was proposed by the Kyushu Chapter of the Japanese Society of Soil Mechanics and Foundation Engineering (JSSMFE), which is now the Japanese Geotechnical Society (JGS). This was the beginning of the IS Kyushu (the International Symposium on Earth Reinforcement). IS Kyushu have played a key role on development of this technique starting from 1988, and was held in 1992, 1996, and 2001, successfully. Figure 1 shows the proceedings of these four times IS Kyushu. In this paper, history of IS Kyushu is summarized (Ochiai, 1999) with careful review of international activity under TC9/ISSMGE (Asian TC, 1997; TC9, 2001; TC9, 2006). Finally, the future prospects on earth reinforcement techniques are proposed.

Figure 1. Proceedings of previous IS Kyushu.

2 THE INITIATION OF "IS KYUSHU"

2.1 *Preparations by Kyushu chapter of JGS*

2.1.1 *International committee*

Kyushu Chapter of the JGS celebrated 30th anniversary in 1979 and they started to discuss a new progress to the next generation. The Kyushu chapter, from 1960s, under Prof. Yamanouchi's leadership, has been made their effort to the effective development of the international exchanges by having the opportunities to have the international academic exchange events such as lectures and seminars performed by the distinguished foreign researchers who had a chance of visit in Japan. The importance and need of the international activities were discussed diligently and an establishment of the international committee in the Kyushu chapter was agreed. As a result, the international committee in the Kyushu chapter was formally approved at the officers' meeting of the chapter in August 1982. The statement of intent to establish this committee at that time is shown as follows:

"Recently, as internationalization and informatization in geotechnical engineering fields made further progress, the necessity of promotion of the international research activities is getting increase as one of the role of the Kyushu chapter of JGS. Now, for members of the Kyushu chapter, they have not only regional disadvantage but also not good access to transportation, and also, the international information tend to become individual belonging. The purpose of this committee is that to collect all the international information and not only share them to the members of this chapter, but also promote the international activities and research in Kyushu area and become the parent organization in order to enhance their relationship with the foreign researchers to be expected to increase." The objectives of this committee were stated and those are shown in Table 1.

The members of the committee were not only university professors but also the engineers who were in Kyushu and had a mind of international activities. And one of the final purposes was to hold an international conference in Kyushu in the future. This purpose led to the beginning of "International Symposium on Earth reinforcement (IS Kyushu)" and made IS Kyushu'88 possible after six years from that time.

2.1.2 *Research committee*

Moreover, at the same time of establishment of the international committee in the Kyushu chapter, the needs of a permanent research committee was also discussed on May 1984. And the topic of "earth reinforcement" was selected as a theme of the first research committee. The motivations of choosing this theme are summarized as follows:

1) New type of reinforcing materials such as geotextiles and geogrids were just developed in Europe and this could be the new wave of the construction technologies;
2) The topic of earth reinforcement has been relatively familiar with the researchers and engineers in Kyushu, which is not only the research activities but also those of construction sites, compared to the other areas of Japan; and
3) Earth reinforcement technique has a high potential for widely use in Asian countries. This could be taken advantage of establishing the regional community.

Table 1. Objectives of the international committee of Kyushu chapter of Japan Geotechnical Society (1982).

1) To collect the international information and transmitting them to the members
2) To suggest projects to promote the international research activities
3) To interact with the foreign researchers
4) To activate as the parent organization to preparation of the international conference

The purposes of this committee at that time are summarized as follows:

1) to collect the data and information on earth reinforcement, straighten and transmit them to the chapter members, not only utilize them to the design and application of geotechnical problems, but also aim to get the new guideline for the development of the reinforced soil works adapted to the ground in Kyushu; and
2) to be a parent research organization for international research meeting to be expected in the future.

The committee consisting about 30 members including engineers from the government and companies, and academic researchers in Kyushu were established and the first meeting was held on June 1985. At the discussion in this meeting, it was confirmed that the subject of earth reinforcement includes the case that the different materials except soil is installed or inserted in the ground but it dose not include the concept of the stabilization such as grouting and mixing. This idea was taken over as the concept of the "earth reinforcement" and later this was chosen as the theme of the "IS Kyushu".

After several discussions by the working groups consisting of several members, the research committee on earth reinforcement was successfully achieved and as a result, the international symposium was decided to be held in Kyushu chapter and following was the proposed contents:

1) On the purposes of promotion of the exchanges of academics and technologies among Asian countries, vitalization of Kyushu area and

(a) Current issues on earth reinforcement method (b) Case histories of earth reinforcement method

Figure 2. Committee Reports in Japanese by the Kyushu Chapter on the topic on earth reinforcement (1987).

establishment of the periodical international research meeting, this symposium with researchers and engineers from abroad and in Japan will be held in Fukuoka city in October 1988.

2) The name of this conference will be "International Geotechnical Symposium, Kyushu-1988" and its abbreviation will be "IGSK '88".
3) The theme of this conference will be "Theory and Practice for Earth Reinforcement". The expected number of participants will be around 100 (30 from overseas, 70 in Japan). Proceedings should be published worldwide.

Figure 2 shows the final reports by the research committee on earth reinforcement in the Kyushu chapter published in 1987.

2.2 *How to begin IS Kyushu*

As described above, the international committee and research committee in the Kyushu chapter played a key role on the preparation of "IS Kyushu".

After several discussions on the idea of IS Kyushu, it was decided at the officers' meeting of the Kyushu chapter on August 1985 that "This symposium will be hosted by the Japanese Geotechnical Society and the Kyushu chapter will be responsible for its organizing, preparations and managing". Then, this idea was approved by the officers' meeting of Japanese Geotechnical Society. This was final decision of starting IS Kyushu. In fact, this idea followed even at second time or later IS Kyushu; so that it was obvious that IS Kyushu became one of key activities for not only the Kyushu chapter but also the Japanese Geotechnical Society. At the same time, Prof. Norihiko Miura of Saga University accepted the position of the secretariat of this "International Symposium on the theory and practice of earth reinforcement". Since then, preparation of the symposium has been remarkably advanced.

After the secretariat was formed, the steering committee and the organizing committee were also organized, which was considered that its members should be gathered the researchers and engineers from the industrial, academic and governmental institutions. After that, because "earth reinforcement" was the theme of this symposium, the organizing committee of the symposium asked the International Geosynthetics Society (IGS) to support this symposium. Then, due to the abbreviation of the symposium "IGSK" that looked alike "IGS", the abbreviation name of the symposium was changed. The new abbreviation name of this symposium was decided to be "IS Kyushu" by the suggestion of the international committee of Japanese Geotechnical Society, and the word "Geotechnical" was omitted from "International Geotechnical Symposium, Kyushu (IGSK)". It is the name that is considered that other chapters of JGS might become the host in the future. And also it seemed that it was well sounding and accepted to the overseas. This derived the nickname "IS" which is familiar to the people in geotechnical engineering field world-wide. If the organizing committee had not asked IGS to support, the present success of IS Kyushu might not be happened.

Needless to say, Kyushu and Okinawa area is closest to the countries in East Asia compared with the other areas in Japan, so that more participants from these Asian countries were expected. The reputation of this symposium was extremely good overseas as well as in Japan, and the timely theme of the symposium and the friendship between the academic researchers and the engineers have been built up worldwide interest of this symposium. Furthermore, because the venue, the City of Fukuoka, is convenient by any transportation systems, especially aircrafts from the other parts of Japan and from abroad and the appropriate size of the city with a population of 11 million people, and the period of the symposium was also good season. The number of papers and participants resulted was beyond the expectations. As the press in Kyushu area reported at the day of the symposium, it was the exceptional one of the international scale events in Kyushu area. Figure 3 shows the symposium logo of the first time IS Kyushu which is IS Kyushu'88.

3 THE HISTORY OF "IS KYUSHU" AND THE TREND OF EARTH REINFORCEMENT TECHNIQUE

3.1 *IS Kyushu '88*

"IS Kyushu '88" was held at the hotel in Fukuoka city for 3 days from Oct. 5, 1988. The symposium "IS Kyushu '88" aimed at inviting more than two hundred participants from twenty countries to discuss various problems and/or topics with respect to earth reinforcement, for the benefit of collecting and exchanging a large amount of knowledge of the methods or techniques being developed recently and to spread these to all countries in the world for further development. As, 94 papers were submitted from 21 countries and there were 275 participations from 24 countries.

The symposium was sponsored by the Science Council of Japan. Also, the symposium received support from fourteen societies and/or associations including the International Geotextile Society which is now called International Geosynthetics Society (IGS), the Japan Chapter of the International Geotextile Society and the Japan Society of Civil Engineers. It should be especially noted that helpful subsidies are granted to the symposium from four societies and/or foundations including the Japan Society for the Promotion of Science.

Professor B.B. Broms and Professor M. Fukuoka, the present and past presidents of the International Society of Soil Mechanics and Foundation Engineering, were invited to give special lectures.

The first special lecture was "Fabric reinforcement retaining walls" by Prof. Broms. The lecture started from the introduction of the basic issues such as kind of reinforcement made of polymeric material, advantage and disadvantage of geosynthetics compared with steel reinforcement. The brief summary of this lecture was as follows; Many test methods, which were examined in order to find the tensile strength and the internal friction angle that were needed to know the mechanical characteristics of fabric such as short and long terms strength, axial strain of fabric, the friction and adhesion between ground and fabric, were introduced and those problems were indicated. Then the behavior of fabric reinforcement was indicated, based on some examples of measurement and calculation. That is, how sharing force, stress-strain curve and so on were influenced by number of fabric layers, its space and height of retaining wall were examined and the mechanism of fabric reinforcement for the reduction of earth pressure and stabilization was introduced comprehensively. Final subject was design method of the retaining wall as the application of fabric. Theoretical and practical calculation methods such as the distribution of Rankin earth pressure and fabric pressure, anchoring of fabric, design of folds, and length of fabric and strength of retaining wall when right concrete panel was used as a wall element were explained. Two impressive figures are shown in Figure 4 and Figure 5.

Continuously, next special lecture was "Earth reinforcement – West and east–" by Prof. Fukuoka. The brief summary of this lecture was as follows; the history of the earth reinforcement was old enough, and from the ancient, natural reinforcing materials such as woods, bamboo and plants have been used in the various countries. In Japan, earth reinforcement method, using woods and bamboo, for mainly flood prevention has been done after the 6th century. Modern methods of earth reinforcement were remarkably developed from 1950s. The reinforcements used for these modern methods are steel and geosynthetics. The former is steel net used for prevention of

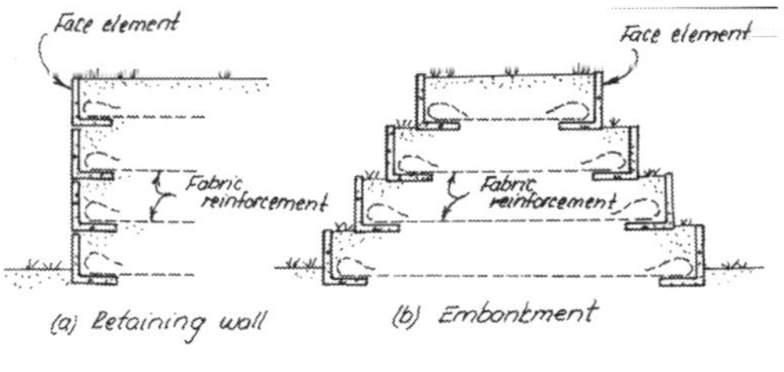

Figure 4. Applications of fabric as reinforcement in soil (Broms, 1988).

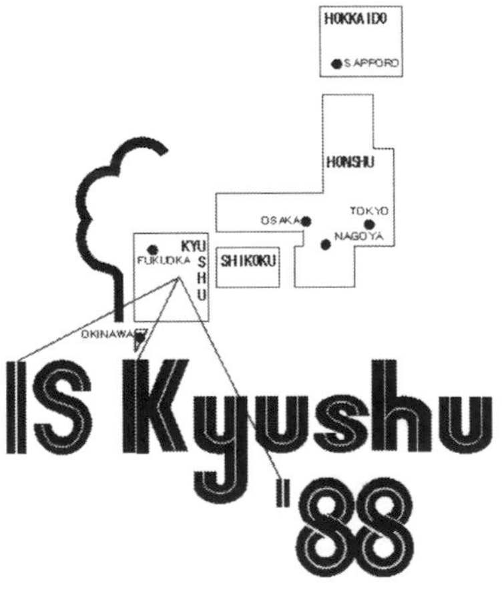

Figure 3. The symposium logo of the first time IS Kyushu (IS Kyushu'88).

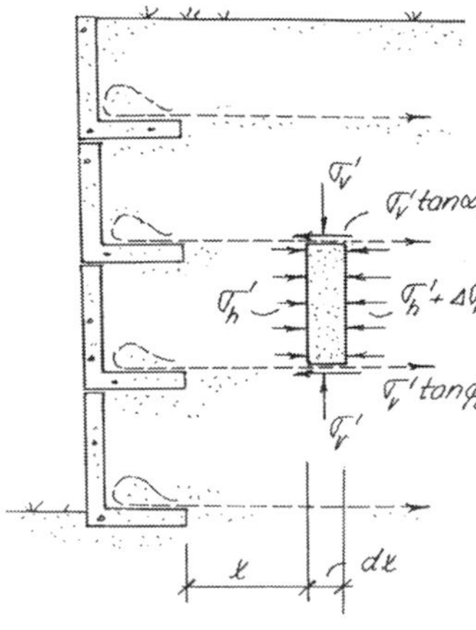

Figure 5. Stress distribution in fabric reinforced soil (Broms, 1988).

subsidence of soft ground, steel reinforcement materials typified by "Terre Armee" method, combination with steel bar and anchor plates, NATM which is the tunneling method for protection of the inside of wall of the tunnel, and reinforcement of slopes with steel bars so on. On the other hand, the latter is geotextiles, which are mainly used for separation and reinforcement of embankments, geonets and geogrids, which have interlocking effect. For each reinforcement method, many applications, tests on the site, and basic researches were introduced, and its progress, transmission and present situation of the techniques and researches of east and west were summarized widely.

However, techniques and researches of Japanese earth reinforcement method have not been introduced to the western countries, the present situation of these Japanese methods, which are characterized by the procedures for earthquakes and heavy rain falls were introduced widely in this lecture. Two impressive figures are shown in Figure 6 and Figure 7.

This times' feature were that the number of papers written by Japanese authors on the theme of

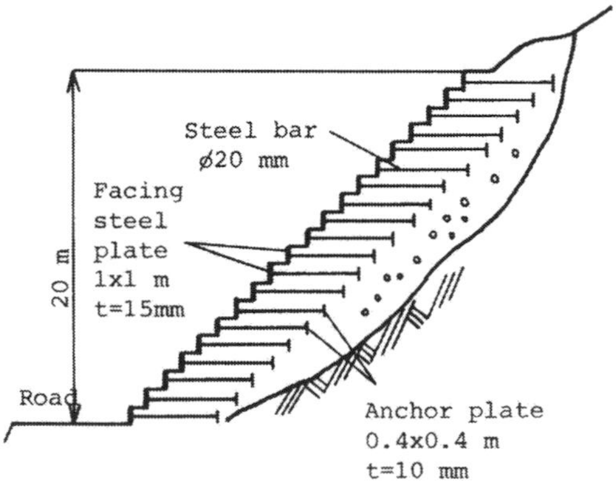

Figure 6. High road embankment reinforced by steel bars with concrete plates, 1963. (Fukuoka, 1988).

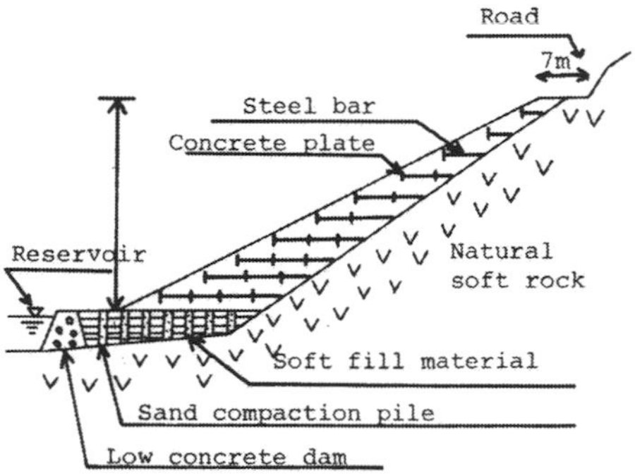

Figure 7. Slope reinforced by steel bars with anchor plates. (Fukuoka, 1988).

application of the slopes and excavation with the steel materials, or application of embankment with geosynthetics such as geonets, geogrids, woven and non-woven, was large. And also, the number of papers by Japanese authors on the basic researches such as test methods and materials was relatively large. When it comes to papers from abroad, there were many papers on case histories that geotextile were applied to the wall structure from UK, France and China. And various reinforcements were used widely in the papers from USA. On the other hand, we could learn that natural materials such as stones were used for the reinforcement of foundation from the papers of India, Sri Lanka and Thailand. In terms of the contents of papers, there was not much application to the natural ground such as slope, excavation and foundation compared to application to the artificial earth structure such as embankment and wall structure. It was considered to be caused by the difficulty of construction management and evaluation of reinforcing effects. In terms of the main aims and methods of research, it was common that there were many papers aimed at the application to the stability problem in earth reinforcement characteristics, but there are some papers that aimed the application to the procedure for subsidence and liquefaction caused by earthquakes. About the methods of research, most of the papers were written about theory, design, laboratory test and on-site test. That is attributed to the fact that its basic research was performed energetically in reflection of the situation that in earth reinforcement, more likely than not, application to the site have been ahead but the design method and evaluation of effect has not been established, but application to the site. The contents and themes of the symposium are shown in Table 2. The summaries of the technical sessions are shown in Table 3.

3.2 *IS Kyushu '92*

The second international symposium on Earth Reinforcement Practice, "IS Kyushu '92" was held at Fukuoka city from 11 to 13 November 1992. This symposium was a continuation and expansion of a previous international symposium entitled "Theory and Practice of Earth Reinforcement" which was held in 1988 at the same location. The symposium was being held under the auspices of the Japanese Society of Soil Mechanics and Foundation Engineering, and is being supported by the International Geotextile Society and the Japan Society of Civil Engineers.

In this period, earth reinforcing techniques had become a useful and economical solution to many problems in geotechnical engineering practice, such as improvement of soft ground, stabilization of slopes, reduction of earth pressure and others. Also, considerable interest in dais area had promoted both fundamental and practical studies as well as the

Table 2. Contents and themes of the IS Kyushu '88.

Themes	Contents
Tests and materials	Strength and durability tests, corrosion problems of materials, contrivance and characterization of new material, and the new application of conventional materials involving natural fibers.
Shallow and deep foundations	Basic theory, development of the design methods and their applications a soft grounds and in road constructions.
Slopes and excavations	Theory of slope reinforcement, reinforcing techniques of natural and cutoff slopes, application for slope protection, the ground reinforcement accompanied by excavation, and the temporary reinforcement method in excavation works in soft grounds.
Embankments	Guideline for design, the reinforcement method of embankment foundations, the reinforcement at the boundary between soft ground and embankment, the damage of reinforcement during construction.
Wall structures	Reinforcement theory of retaining walls, design methods for retaining wall reinforcement, the reinforcement of backfill materials, jointing techniques of reinforcement materials, and the monitoring system.

development of various types of reinforcing materials. It follows that new techniques for earth reinforcement and their applications to geotechnical engineering practice were developed rapidly.

For this symposium 161 abstracts were submitted from a total of 28 countries. Each of the abstracts and papers were reviewed by the Paper Review Committee. The committee finally selected 126 papers from 25 countries for presentation during the symposium, partly in technical sessions, partly in poster sessions. 77 participants from overseas 21 countries and 314 participants including 8 foreigners in Japan, and a total of 391 people participated into this conference and it was unprecedented great success as this kind of international symposium that the theme was limited.

The aim of the symposium was to discuss various problems and topics on earth reinforcement for the benefit of collecting and exchanging knowledge concerning recently developed techniques and to spread this knowledge to all the countries of the world for further development. For this purpose, six internationally distinguished scholars in this area had been invited to IS Kyushu '92 as one special lecture: Professor T. Yamanouchi and five keynote lecturers: Professor J.P. Gourc, Dr R.A. Jewell, Professor D. Leshchinsky, Professor R. K. Rowe and Professor F. Tatsuoka.

The brief summary of the special lecture "Historical Review of Geotextiles for Reinforcement of Earth Works in Asia" by Prof. Yamanouchi was as follows; It was lectured that usage of various geotextile made of natural material to polymeric material used from ancient to today and the research developments on earth reinforcement separated by the era and the volume with wide range of source. At first, it was emphasized that earth reinforcement has been developed by harmonizing reinforcement and nature, climate and soil property since ancient times. And the earth reinforcement using natural materials that has been performed in China and Japan since ancient times was introduced according to application. Then, earth reinforcement using plastic and polymeric materials, which have been remarkably used these 10 years, especially, method for very soft ground as Japanese specific method were explained. And also, comparison of each strengths and economic efficiency of polymeric materials, its characteristics, and development and problem of design method were referred. Finally, as the conclusion, actual condition and problem of the earth reinforcement in Asia, and also future prospective were referred. An impressive figure and a table in his paper are shown in Figure 8 and Table 4.

Then, before the technical sessions started, Keynote lectures were presented by the following five lecturers. The contents of the each lecture are indicated in Table 5.

The first volume of the proceedings contained these 126 papers. The papers were arranged under five categories which cover almost all aspects in the area of earth reinforcement. The second volume, which was published after the symposium, contained papers of the special and keynote lectures, reports of technical session chairmen and discussion leaders, records of discussions, concluding remarks, as well as the symposium program.

Compared with the distribution of submitted papers by county of IS Kyushu '88, it was feature that some papers were newly submitted from east Europe that progress the democratization, and Africa, and also number of papers from USA, UK, Germany and France where "earth reinforcement" were developed, were increased. When distribution of the accepted papers were looked at the viewpoint of the subject of research and structure, there were slightly large number of the papers on basic research such as test method and materials, but, the rate of papers classified "foundation", "slopes and ground", "embankment" and "retaining wall" were about the same if it was considered that the border line of reinforced embankment and reinforced wall as the subject of research tend not to be clear, and pavement added in reinforcements of foundation. And as for aims of research, papers of the

Table 3. Summary of each technical session in IS Kyushu '88.

Themes	Summary
Tests and materials	"Strength and durability tests, corrosion problems of materials, contrivance and characterization of new material, and the new application of conventional materials involving natural fibers"; it was discussed that the test method such as shear box test, pullout test, and triaxial compressive test are easy and convenient to perform as the method for evaluating mechanism of the reinforced soil and interaction between reinforcement and soil, but the mechanical properties of the reinforced soil cannot be completely clarified. For example, although the actual composite reinforcement is influenced force of tension, the test piece of the triaxial compressive test apparatus is in compression. Furthermore, the boundary condition of the actual reinforced soil is not uniform, but that of the test piece is uniform. Thus, how these test result, which are performed with the stress and boundary condition, which are different from the actual reinforced soil, should be reflected to design should be considered when these tests are preformed, and it becomes problem.
Shallow and deep foundations	"The basic theory, development of the design methods and their applications a soft grounds and in road constructions"; three themes were discussed. First theme is "model test", and many interesting results were indicated. However, because model test has its limitation, the actual sized test should be needed. Second theme is "asking for the theoretical solutions". Some of the papers appeared in the proceedings give the useful information for the actual design. We should focus on the interaction between reinforcement, membrane and soil and should progress its research. Third theme is "observation and investigation at the site" and the results were discussed.
Slopes and excavations	"The theory of slope reinforcement, reinforcing techniques of natural and cutoff slopes, application for slope protection, the ground reinforcement accompanied by excavation, and the temporary reinforcement method in excavation works in soft grounds"; it was focused that most of the method of the design for reinforcement are safe side. However, it will be improved as the elucidation of the theory progress. Most of theoretical prognosis seem to well consistent with a range of experimental results. However, the researcher who suggests the theories should verify each theory by the data of the experience that not only you done but also other researcher performed. At the practical level, more detailed investigation of the site should be needed in order to prevent the expensive repair work.
Embankments	Many theories and results of the experiments were presented at this session. The guideline for design, the reinforcement method of embankment foundations, the reinforcement at the boundary between soft ground and embankment, the damage of reinforcement during construction, and it was sufficient. It was regrettable that the number of the oral presentation was limited. There was the interesting thesis on reinforcement using bamboo in the proceedings besides the oral-presented papers. Though the actual sized test requires cost and time, its result is useful and sufficient. The mechanism of reinforcement by not only non-woven but also woven should be investigated. As for safety against rainfall, both woven and non-woven can be expected there drainage effect, and it is important.
Wall structures	"The reinforcement theory of retaining walls, design methods for retaining wall reinforcement, the reinforcement of backfill materials, jointing techniques of reinforcement materials, and the monitoring system"; many presentation of research on nonmetallic materials and research of this field should be needed to be continued. It seems that the limit equilibrium method be practically common method. FEM is not common but is method that can give the solution for the question that how the behavior of the element connect to the entire behavior. There were a lot of discussion on durability of the reinforcement associated with the phenomenon attributed by the chemical factor such as creep and this research should need to be continued.

case histories about reinforced structure for the permanent structure, and ones of experimental research about creep property and durability were increased. As one of the major movements, the papers subjected on seismic stability of reinforcement were seen. As for method and measure of research, creep test, durability test, and model test with centrifugal load tended to be increased. Reinforcement used for these tests have been got more and more widely variety such as composite materials and metal grids instead of natural materials which are getting decreased. In the subjected soil, sand and gravel constitute the considerable amount of them, but clay and waste soil became used from this period. And also it was feature that the technical papers on reinforced wall, which was one of the biggest points of discussion, were increased.

Figure 8. The use of fascines for a sea-side dyke construction, Kyushu, Japan, 1987. (Yamanouchi, 1993).

Table 4. Practical uses of various geotextiles in Japan. (Yamanouchi, 1993).

Material and uses	1960's	1970's	1980's	1990's
Woven or nonwoven fabrics				
Restraining	- - - -	→		
Reinforcement		- - -	- - - - - -	→
Asphalt surface		- - -	- - - - - -	→
Plastic nets				
Restraining	- - - - -	- - - - - -	- - - - - -	→
Polymer grids				
Restraining			- - -	- - - →
Reinforcement			- - -	- - - →
Geocell mattress			- - -	- - - →
Steel mesh				
Retaining wall		-	- - - - - - -	- - - - - →
Embankment reinforcement		-	- - - - - - -	- - - - - →
Mesh element stabilization			- -	- - →
Continuous yarn reinforcement			- -	- - →
Polymeric flood control works			——	——→

Solid line: fully in practice. Broken line: partially in practice.

Table 5. Lecturer and title of keynote lectures. (IS Kyushu '92).

Lecturer	Title
Dr. R.A. Jewell	Links between the testing, modeling, and design of reinforced soil
Prof. J.P. Gourc	Geosynthetics in embankments, review of theory and walls
Prof. R.K. Rowe	A review of the behaviour of reinforced soil walls
Prof. F. Tatsuoka	Roles of facing rigidity in soil reinforcing
Prof. D. Leshchinsky	Issues in geosynthetic reinforced soil

3.3 *IS Kyushu '96*

The third international conference on Earth Reinforcement Practice, "IS Kyushu '96" was held at Fukuoka city from 11 to 13 November 1996. In this period, about 30 years had passed successfully since the modern concept of earth reinforcement was proposed by Henry Vidal in the middle of the 1960s. The expansion of experiences arid knowledge in earth reinforcement practice had allowed the development of new technique and their application to geotechnical engineering problems. It was therefore of great advantage to scholars and engineers to share the latest information on earth reinforcement and to discuss about their own experience and knowledge. The symposium was being held under the auspices of the Japanese Geotechnical Society, and was being supported by the International Geosynthetics Society and the Japan Society of Civil Engineers.

The aim of the symposium "IS Kyushu '96" was to discuss various problems and topics on earth reinforcement for the benefit of collecting and exchanging knowledge concerning recently developed techniques and to spread this knowledge to all countries in the world for further development. This third symposium was a continuation and expansion of the two previous symposia. A total of 201 abstracts were submitted for this symposium from 33 countries. Both the abstracts and papers were carefully reviewed by the Paper Review Committee. The Committee finally selected 151 papers for presentation during the symposium, some in technical sessions, and some in poster sessions. 124 participants from overseas 32 countries and 321 participants in Japan, and a total of 445 people participated into this conference and this international conference was a great success. Those numbers were more than expected and it was confirmed that the IS Kyushu '96 had a great success.

In this symposium, one special lecture: Professor C.J.F.P. Jones and three keynote lectures: Dr. R.J. Bathurst, Dr. H. Miki and Prof. G. Gässler by internationally distinguished scholars had been delivered, as well as presentations of the papers accepted for the symposium. Special reports on the performance of earth reinforcement structures in two major earthquakes, The Northridge Earthquake in 1994 and The Great Hanshin Earthquake in 1995, had been presented, too. Summary discussion sessions on Testing

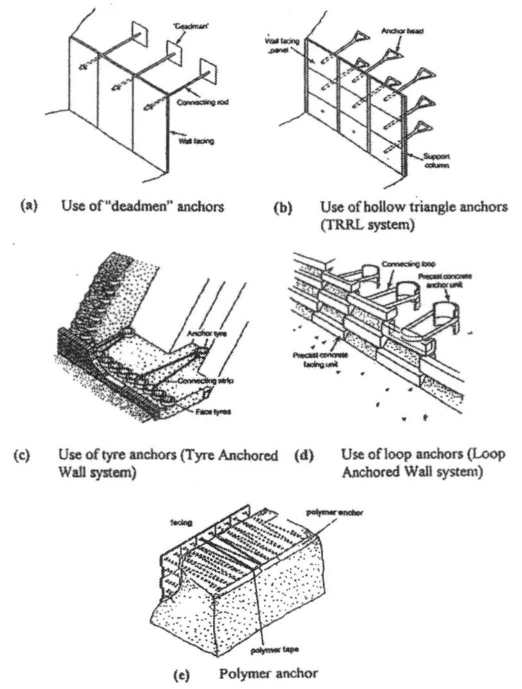

Figure 9. Anchored soil retaining wall system. (Jones & Hassan, 1992, Jones, 1997).

methods' and "Design methods" had also been organized to summarize results of research activities on the development of these areas in the last three decades.

The brief summary of the special lecture "Geosynthetic materials with improved reinforcement capabilities" by Prof. Jones was as follows: At first, development of the polymeric geosynthetics and history of its application for reinforcement were explained and the uses of not only sandy soil but also viscous soil were also explained. Then new geosynthetics which has electrical conducting property called "Electrokinetic geosynthetics (EKG)" were introduced, its new applicability to the technology of reinforcement which was positively combined technology of electrosmotic electrocataphoresis properties which are different from conventional concept of reinforcement was indicated with the lots of basic data. Two impressive figures in his paper are shown in Figure 9 and Figure 10.

Next, the keynote lectures were performed by the following three lecturers on the contents shown in Table 6. Moreover, the special reports were presented by the following two lecturers as shown in Table 7. One was the report on damage of the reinforced soil structures in the earthquake in Northridge, California, 1994, which was one year before this symposium, and another was the report on damage of the reinforced

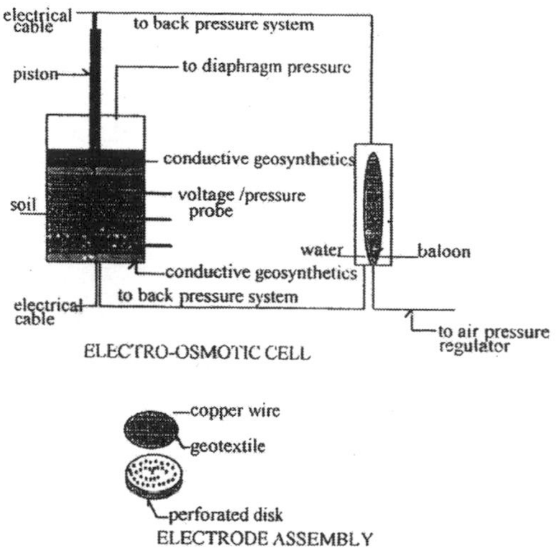

Figure 10. Schematic diagram of the electroosmotic cell and the electrode assembly. (Jones, 1997).

Table 6. Lecturer and title of keynote lectures. (IS Kyushu '96)

Lecturer	Title
Dr. R.J. Bathurst	Review of seismic design, analysis and performance of geosynthetic reinforced walls, slopes and embankment
Dr. H. Miki	Application of geosynthetics to embankment on soft ground and reclamation using soft soils
Prof. G. Gässler	Design of reinforced excavations and natural slopes using new European Codes

Table 7. Lecturer and title of special reports. (IS Kyushu '96)

Lecturer	Title
Mr. D.M. White	Performance of geosynthetic reinforced slopes and walls during the Northridge, California of January 17, 1994
Prof. F. Tatsuoka	Performance of reinforced soil structures during the 1995 Hyogorken Nanbu Earthquake

soil structures in Hyogo Nanbu earthquake in 1995. Both reports indicated that earthquake resistances of the reinforced soil structure have relatively good quality and the importance that ductility effect should be taken in the seismic design.

As the themes of the summary discussion sessions, "test method" and "design method" were set up in order to indicate analysis on the present research on the international perspective view and the direction of future researches on the basis of past two IS Kyushu's achievements.

At the session of "test method", it was reported from the panelists that the present status and what remains to be done of the geosynthetic test method in Japan, need of monitoring on durability of reinforcement, suggestion of the safety factor for lifetime of reinforced soil structures, property of stress and rupture on design of earth reinforcement. In the response this report, it was discussed sufficiently that 1) creep property, dynamic property of reinforcement, 2) interaction between reinforcement and soil, creep stress relaxation in the soil, influence of boundary friction of apparatus, 3) predictive accuracy of model test, 4) problems of real sized on site test and monitoring of behaviors. And what remains to be done of each issue were cleared.

At the session of "design method", it was reported by the panelists that 1) present status of the design method for slope and embankment on the soft ground, 2) actual behavior and reinforced mechanism of reinforced wall and slope 3) limit state design method and numerical analysis 4) new usage of reinforcement and its possibility.

At the technical sessions, oral presentations and discussions were done on 4 themes such as "reinforced embankment", "reinforced soil wall structure", "basic earth reinforcement", and "slope and excavation".

3.4 *IS Kyushu '01*

The fourth international conference on Earth Reinforcement Practice, "IS Kyushu '01" was held at Fukuoka city from 14 to 16 November 2001. In this period, earth reinforcing techniques were increasingly becoming a useful, powerful and economical solution to various problems encountered in geotechnical engineering practice. Expansion of the experiences and knowledge in this area had succeeded developing new techniques and their applications to geotechnical engineering problems. In order to discuss the latest experiences and knowledge, and with the purpose of spreading them all over the world for further development, the IS Kyushu conference series on the subject of earth reinforcement had been held in Fukuoka, Japan, every four years since 1988. The symposium were "Theory and Practice of Earth Reinforcement" for the first one in 1988, "Earth Reinforcement Practice" for the second one in 1992 and "Earth Reinforcement" for the third one in 1996. They had provided successful contributions towards the development of the earth reinforcement practice. This symposium was being held under the joint-auspices of the Japanese Geotechnical Society (JGS) and the TC-9 of ISSMGE, and was being supported by the International Geosynthetics Society (IGS) and the Japan Society of Civil Engineers (JSCE).

A total of 212 abstracts from 32 countries were submitted for this symposium. The Scientific Committee carefully reviewed both the abstracts and full papers, and finally selected 137 papers for presentation during the symposium, some in technical sessions for oral presentations, and the rests in poster sessions. They were included in Volume 1 of the proceedings. 99 participants from overseas 30 countries and 372 participants in Japan, and a total of 471 people participated into this conference and this international conference was another great success.

This fourth symposium, entitled "Landmarks in Earth Reinforcement", was a continuation of the series of IS Kyushu conferences, and also aimed at being one of the landmarks in the progress of the modern earth reinforcement practice. With this objective, one special lecture: Professor R.J. Bathurst and five keynote lectures: Professor R.K. Rowe, Mr. C.R. Lawson, Dr. O. Murata, Mr. G.R.A. Watts and Mr. M. Hirano were arranged to deliver, in addition to the presentations of the papers accepted for the symposium.

The brief summary of the special lecture "Full scale performance testing and numerical modeling of reinforced soil walls" by Prof. Bathurst was as follows; its deformation behavior property and the problem of the present analysis method were indicated with the latest full-scaled test result on geosynthetics and metallic reinforced soil wall. Two impressive figures in his paper are shown in Figure 11 and Figure 12.

Then keynote lectures on the contents, indicated on the following Table 8, were lectured by the following five lecturers. The special session on "Soil Nailing" and the summary discussion session on "Design Procedure" had also been organized to summarize the results of fundamental and practical aspects on these areas developed in the recent years. These valuable reports were included in Volume 2 of the proceedings to be published after the symposium.

As the themes of the summary discussion sessions, "design method" were set up in order to indicate analysis on the present research on the international perspective view and the direction of future researches on the basis of past two IS Kyushu's achievements. Prof. D. Leshchinsky and Prof. J.G. Zornberg were accepted to be the chairperson. And the special session mainly based on discussion was organized. "Reinforcement of slopes" was selected as the theme of this session on the basis of characteristic of this symposium. The chairperson of this session was Dr. R. Jewell.

The summery discussion "design method", how to progress this session and point of the discussion were explained by the chairperson and topics were provided by five panelists. The topics were 1) stability analysis method combined retained strength and peak strength, 2) international comparison of design method

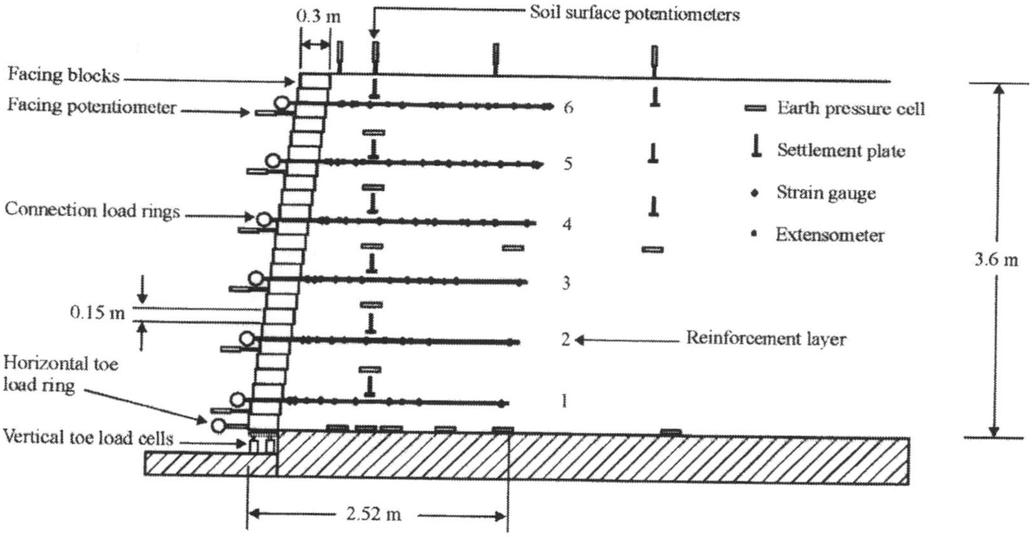

Figure 11. Typical instrumentation plan for segmental walls. (Bathurst, 2003).

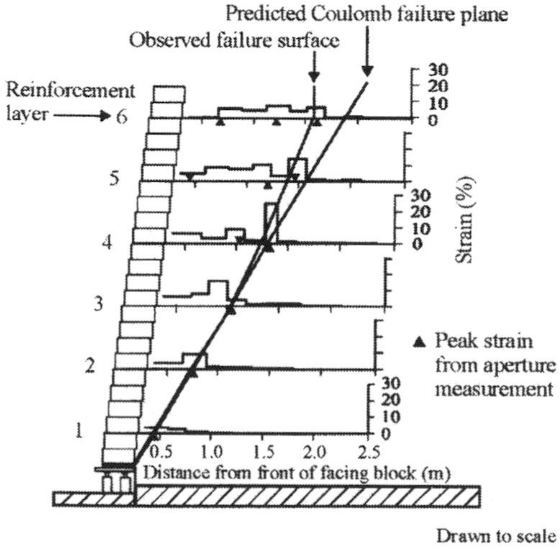

Figure 12. Location of peak reinforcement strain and internal failure surface for wall 2. (Bathurst, 2003).

for geosynthetics reinforced soil wall, 3) application limit of reinforced soil wall, 4) evaluation and design of material characteristics, 5) reinforced effect of geosynthetics which were not considered to design. In addition, the presentation about 1) evaluation method of design tensile strength of geosynthetics, 2) role of numeric analysis to design were performed. The discussion was progressed with the above topics and was lively as not to be answered all questions gathered. At last, what remain to be done about "earth reinforcement" were summarized by the chairman as time flow from IS Kyushu '88 to '01 was related to development of reinforcement technique.

Table 8. Lecturer and title of keynote lectures. (IS Kyushu '01).

Lecturer	Title
Prof. R.K. Rowe	Insights from case histories: Reinforced embankments and retaining walls
Mr. C.R. Lawson	Performance related issues affecting reinforced soil structures in Asia
Dr. O. Murata	An outlook on recent research and development concerning long-term performance and extreme loading
Mr. G.R.A. Watts	The durability of geosynthetics for retaining walls and slopes for long term performance
Mr. M. Hirano	Actual status of the application of the soil nailing to expressway cut-slope construction in Japan

At the special session "soil nailing", oral presentation, panelist report, and discussion were done. On the oral presentation, it was done that 1) case history of reinforced slope combined ground anchor and soil nailing, 2) composite structure of reinforced embankment and reinforced excavation used by soil nailing and elongated soil nail, 3) case history that high strength wire mesh applied to surface of slope and introduction of method of thinking about design, 4) report that reinforced slope used soil nailing effected brittleness than non-reinforced slope during earthquake, and stability analysis method of limit equilibrium method for reinforced slope considered progressive deformation and its verification analysis. On the panelist report, as the latest case histories in Malaysia, 1) application method that metal pipe was injected as the excavation of soft clay, 2) example of measurement of cutting reinforcement of steep slope using the entire welding nail 3) basic experiment and case history on the various condition such as tropical soil property were introduced. Then, the earth reinforcement method for steep slope which have been recently used in Germany, 1) long soil nailing used with the specialized machine, 2) soil nailing which were self punching 3) application example of reinforced slope with precast concrete panel were introduced. And also, it was summarized and reported about the important points to keep in mind of FEM analysis related soil nailing such as consideration of application process, modeling about 2 dimensional, and 3 dimensional problems. At the discussion, the following themes, which consist with the themes of the oral presentation and the panelists' reports, were arisen from the session floor, 1) materials and mechanical equipments 2) case histories 3) seismic adequacy 4) analysis. As related to the reinforcement, which have been used recently, reported by panelist, question-and-answer about definition of soil nailing, material characteristics and reason to be selected were done. And also, question-and-answer about modeling for FEM analysis and evaluation of seismic test result were done. Meanwhile, the application example of the reinforced excavation slope in India was introduced from audience of the session floor.

At the technical sessions, oral presentation and discussion were performed on 4 themes such as "reinforced soil wall structure", "test method and materials", "reinforced embankment", and "reinforced soil foundation". Before presentation, introduction of the contents of papers were introduced and questions were arose by the discussion leaders.

3.5 *History of "IS Kyushu" and changes of earth reinforcement technique*

About 20 years have past since IS Kyushu '88 was held in 1988. For this period, the earth reinforcement technology has been remarkably widespread and developed and the number of its application has been rapidly increased. As the typical example, Figure 13 shows the cumulative number of the applications of the typical reinforced earth wall (Metallic strips, Geosynthetics and Multi-anchor) in Japan. And Figure 14 shows the overview on the rate of the number of the applications of earth reinforcement in Japan. Because more than 60 percents of the total earth reinforcement

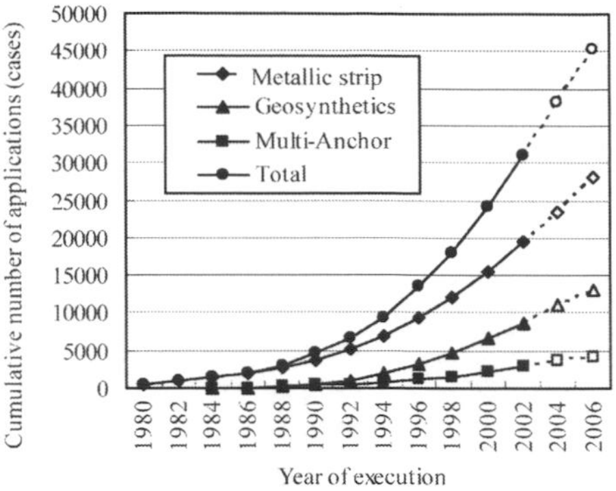

Figure 13. Cumulative number of typical reinforced earth wall in Japan.

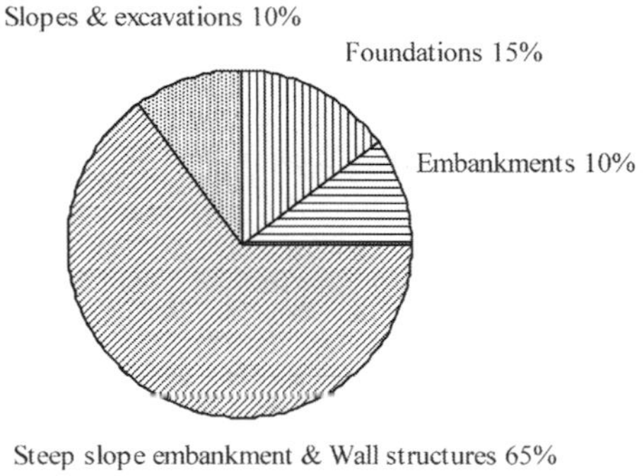

Figure 14. Overview on the rate of the number of the applications of earth reinforcement in Japan.

technique are the case of the wall structures, it can be said that the growth of earth reinforcement is near to the growth of the reinforced earth wall. Because its reliable statistic value counting has not been done since 2002, we can count only estimated amount, the number of applications of earth reinforcement has been steadily increased. It is estimated that more than 100 millions square meters of applications of reinforced earth wall at the wall area have been done in total when other earth reinforcement add to these three major reinforcements.

With those backgrounds of the development of the earth reinforcement, the number of participants and papers to IS Kyushu, as shown in Table 9 and Table 10, has been increased such as 262 participants in '88, 391 in '92, 445 in '96 and 471 in '01.

It should be noted that nearly 100 people participated from oversea countries in '92 and '96. Table 11 shows the changes of contents of the papers made presentations. It is indicated that the papers related on reinforced soil wall have been steadily increased; however the total number of submitted paper has varied year by year.

Table 9. Number of participants of IS Kyushu.

	Japan	Asia	Oceania	America	Europe	Africa	Total
IS Kyushu '88	205	26	2	7	21	1	262
IS Kyushu '92	314	26	2	17	31	1	391
IS Kyushu '96	321	62	3	24	32	3	445
IS Kyushu '01	372	42	3	16	36	2	471

Table 10. Number of papers.

	Japan	Asia	Oceania	America	Europe	Africa	Total
IS Kyushu '88	39	21	3	6	25	0	94
IS Kyushu '92	54	22	3	12	34	2	127
IS Kyushu '96	50	40	4	20	36	1	151
IS Kyushu '01	49	33	4	11	38	2	137

Table 11. Contents of papers.

	Testing and materials	Foundations	Slopes and excavations	Embankments	Wall structures	Total
IS Kyushu '88	21	14	18	21	20	94
IS Kyushu '92	36	23	24	22	22	127
IS Kyushu '96	32	27	24	20	48	151
IS Kyushu '01	32	24	22	23	36	137

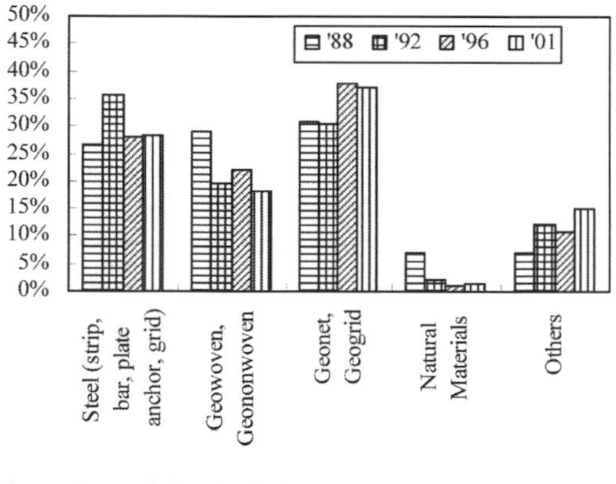

Figure 15. Kinds of reinforcements in papers cumulative number of amount of usage of geogrids in Japan.

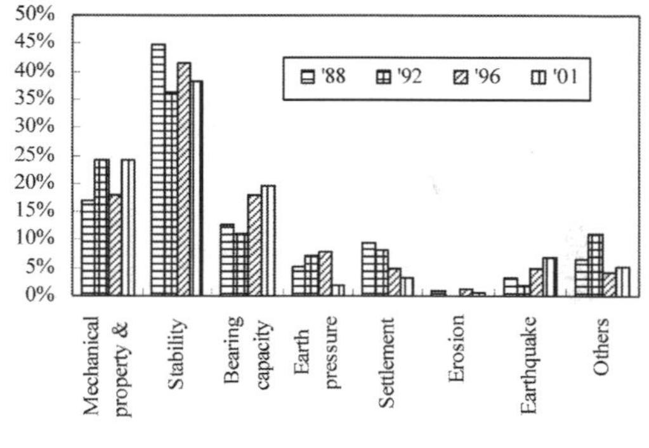

Figure 16. Kinds of subject of papers.

Figure 15 shows the rate and kinds of reinforcement subjected in the papers made presentations. The rate of the steel and geogrids are large and the rate of geogrids has been rapidly increased on reaching 1992. The rate of others also has been increased because composite materials appeared on 1992, fiber mixture, geo tube and geo cell appeared on 1996.

Figure 16 shows the kinds of subject of the papers made presentations. It is obvious that the number of papers on the stability problem is large, because the theme of this symposium is earth reinforcement, however, it is featured that the number of the papers on baring capacity and earthquakes has been increased each time.

Figure 17 shows the kinds of method used in the papers made presentation. It is indicated that the rate of the papers on laboratory test and full-scale test are remarkably large. It is the noteworthy point as one of the features of IS Kyushu. It is also remarkable that the number of papers about report on management and measurement, and those used numeral analysis has been increased gradually at the IS Kyushu '92 and '96.

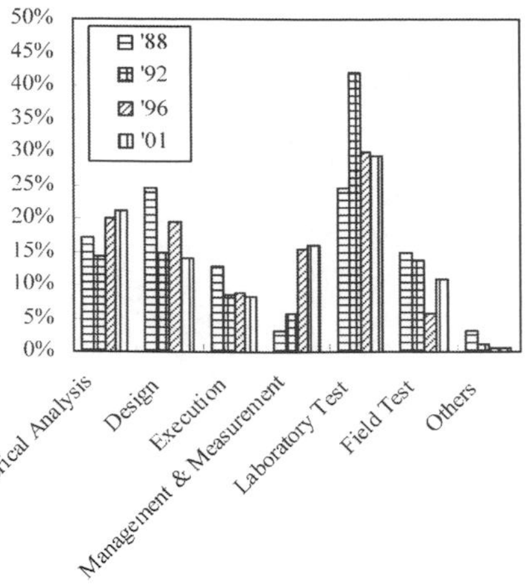

Figure 17. Kinds of method used in papers.

Table 12. Member list of Asian TC of ISSMGE. (1995–1996).

CHAIRMAN: Ochiai, H., Japan
SECRETARY: Otani, J., Japan
MEMBERS:

Bergado, D., Thailand	Cho, S.D., Korea
Fukuda, N., Japan	Guler, E.F., Turkey
Huang, C.C., Taiwan	Karunaratne, G.P., Singapore
Kabir, M.H., Bangladesh	Madhav, M., India
Yang, C.W., China	Yeo, K.C., Hong Kong
Wiseman, G., Israel	

4 THROUGH THE ACTIVITIES OF INTERNATIONAL COMMITTEES

4.1 *Technical committee of ISSMGE*

IS Kyushu '88, '92 was profitable for engineer and researcher to discuss on earth reinforcement technology inclusively. In those days, earth reinforcement technology was spreading rapidly in Asia. In Asia, there are many regional and difficult subsoils. Earth reinforcement technology was very effective for many projects. However this technology still was new. The international information exchange was needed to apply the technology correctly. Under these circumstances, new technical committee concerning earth reinforcement was set in the Asian regional society of ISSMGE. The objectives of this committee were 1) to continue in developing and spreading the knowledge of earth reinforcement practice, 2) to facilitate the activities of earth reinforcement research in Asia and 3) to support IS Kyushu '96. List of members are shown is the Table 12. In order to complete these activities, a supporting committee has been organized under the Japanese Geotechnical Society. In this committee, three working groups (WG) had been set as follows, 1) WG-1: Testing method, 2) WG-2: Design methods, 3) WG-3: Numerical analysis. The first technical meeting was opened at 10th Asian regional conference on SMFE at China, 1995. To exchange the information about each activity in earth reinforcement in Asia, Electronic newsletter had been published among this committee in every three or four month. This letter includes 1) member news, 2) diary event, 3) technical report, and 4) the others. The final report of this committee was published as a technical paper in the proceeding of IS Kyushu '96. The result greatly contributed to the success in IS Kyushu '96 (Asian TC, 1997).

In ISSMGE, there was a technical committee No.9 (TC-9) whose name was "Geotextile and Geosynthetics". This committee was established in 1986. Chairperson was Prof. J.P. Giroud (1986–1989), Prof. A. McGown (1990–1993), and Prof. J.P. Gourc (1994–1997). They did contribute the developing of earth reinforcement with geosynthetics. They published the book concerning case histories (Raymond and Giroud eds., 1993). In 1998, the name of TC-9 was changed to "Geosynthetics and Earth Reinforcement" in order to match the activities which are expected to develop in the future. This was based on the great success of IS Kyushu in 1988, 1992 and 1996. For this reason, the JGS supported the activities of TC-9 as a host member society, and a supporting committee was established in the framework of JGS. Terms of reference for the TC-9 are shown as follows; 1) to continue in developing and spreading the knowledge on geosynthetics and earth reinforcement, 2) to promote the activities for the use of geosynthetics in geotechnical engineering, 3) to participate actively in the technical program of the conferences and symposia related to the topic on geosynthetics and earth reinforcement, 4) to support International Symposium on Earth Reinforcement (IS Kyushu 2001), and 5) to cooperate with the International Geosynthetics Society. A well selected international group of experts became appointed as responsible members of this committee based on the candidates recommended by ISSMGE member societies, presided from 1998 to 2001. List of members are shown in the Table 13. The subcommittees (SC) have been set up in the TC-9 in order to enhance the activities.

The themes of each subcommittee with the head persons are shown as follows; 1) SC-1: Case history and data base, 2) SC-2: Long term performance under extreme loading, 3) SC-3: Design and parameter determination, 4) SC-4: Education, and 5) SC-5: New approach and applications. TC-9 has also set up the supporting committee in the framework of the Japanese Geotechnical Society (JGS) in order to enhance the activities of TC-9. TC-9 opened at technical meeting at 12th European regional conference

Table 13. Member list of TC9 of ISSMGE. (1998–2002).

CHAIRMAN: Ochiai, H., Japan	
SECRETARY: Otani, J., Japan	
CORE MEMBERS:	
Gourc, J.P., France	Kuwano, J., Japan
Lawson, C.R., Malaysia	Leshchinsky, D., USA
Rowe, R.K., Canada	
MEMBERS:	
Abramento, M., Brazil	Atmatzidis, D.K., Greece
Bouassida, M., Tunisia	Cancelli, A., Italy
Cazzuffi, D.A., Italy	Cho, S.D., Korea
Delmas, P., France	Dembicki, E., Poland
Didier, G., France	Floss, R., Germany
Gnanendran, C.T., Australia	Guler, E.F., Turkey
Hausmann, M.R., Australia	Heerten, G., Germany
Hermann, S., Norway	Jones, C.J.F.P., UK
Legrand, C., Belgium	Lopes, M.D.G.A., Portugal
Matys, M., Slovak Republic	Murata, O., Japan
Palmeira, E.M., Brazil	Paredes, L., Chile
Paskauskas, S., Lithuania	Rao, G.V., India
Rogbeck, Y., Sweden	Timofeeva, L.M., Russia
Uriel, S., Spain	Voskamp, W., Netherlands
Watn, A., Norway	Yeo, K.C., Hong Kong
Zornberg, J.G., USA	

Table 14. Member List of TC9 of ISSMGE. (2002–2006).

CHAIRMAN: Ochiai, H., Japan	
SECRETARY: Otani, J., Japan	
CORE MEMBERS:	
Gassler, G., Germany	Gourc, J.P., France
Madhav, M., India	Oden, K., Sweden
Palmeira, E.M., Brazil	Tan, S.A., Singapore
Zornberg, J., U.S.A.	
MEMBERS:	
Abramento, M., Brazil	Bouassida, M., Tunisia
Bekenov, T.N., Kazakhstan	Bathurst, R.J., Canada
Bergado, D., Thailand	Cazzuffi, D.A., Italy
Chang, D.T.T., Taiwan	Ehrlich, M., Brazil
Garcia-Mina, J., Spain	Guler, E.F., Turkey
Kvasnicka, P., Croatia	Herle, V., Czech Rep.
Heerten, G., Germany	Jones, C.J.F.P., UK
Kitazume, M., Japan	Lafleur, J., Canda
Lam, A., Hong Kong	Lawson, C.R., Malaysia
Liausu, Ph., France	Lopes, M.D.G.A., Portugal
Leshchinsky, D., U.S.A.	Lo, R., Australia
Matys, M., Slovak Rep.	Moraci, N., Italy
Rajagopal, K., India	Rathmayer, H., Finland
Rowe, R.K., Canada	Scharle, P., Hungary
Timofeeva, L.M., Russia	Uriel, S., Spain
Voskamp, W., Netherlands	Yeo, K.C., Hong Kong
Yoo, C., Korea	

on SMGE (1999), 11th Pan-American regional conference on SMGE (1999), 11th Asian regional conference on SMGE (1999), Geo Denver 2000, and EuroGeo 2 in IGS (2000). TC-9 member also exchanged their information by using mailing list in internet. The TC-9 organized a special session on the title of Earth Reinforcement Technique in Asia Current Topics in Asia and New Horizon for 21st Century during 11th Asian regional conference on SMGE, 1999. In this session, 6 panelists gave a presentation and discussion session was organized. The final reports of the technical committee were published as a technical paper in the proceeding of IS Kyushu '01. The result greatly contributed to the success in IS Kyushu '01 (TC9, 2001).

For the period 2002–2005, the ISSMGE president presented the new general objective for TC-9, by which the topics under TC-9 became wide range of "earth reinforcement" including the topic of "ground improvement" which was originally covered by TC-17. Terms of reference for the TC-9 are shown as follows; 1) to continue enhancement in developing and spreading the experiences and knowledge on earth reinforcement practice, along the lines of the objectives stipulated for this TC, 2) to promote research and development, along the lines of objectives stipulated for this TC, 3) to promote publication of education materials on earth reinforcement practice for engineers and students, 4) to contribute international symposia/conferences related to earth reinforcement; and 5) to collaborate with TC-17. A well-selected international group of experts were appointed as responsible members of this committee based on the candidates recommended by ISSMGE member's societies, presided from 2002 to 2005. List of members are shown in Table 14. TC-9 has also set up the supporting committee in the framework of the JGS in order to enhance the activities of TC-9. TC-9 opened at technical meeting at 7th International Geosynthetics Conference (2002), 12th Pan-American regional conference on SMGE (2003), 13th European regional conference on SMGE (2003), 3rd IGS Asian regional conference (2004) and Geo-Frontiers 2005. TC-9 member also exchanged their information by using mailing list in internet. The TC-9 organized six of special sessions at the conference related to earth reinforcement. Their titles are shown in Table 15. These sessions were very effective to promote the developing of earth reinforcement. The detailed report was included to proceeding of 16th ICSMGE (TC9, 2006a, 2006b).

4.2 *Activity results of technical committee*

In this section, activity results of ATC and TC9 concerning earth reinforcement are summarized briefly.

4.2.1 *Case histories and education*

Soil is complicated material and geotechnical problem has a regional dependency. An unexpected failure is caused when the technology is not applied appropriately. Earth reinforcement technology is still a

Table 15. TC9 Organizing sessions in the regional conferences and related conferences. (2002–2006).

Conference	Theme
TC-9 session during 12th Asia Conf. of ISSMGE (Singapore, 2003)	Combined technology in earth reinforcement
TC-9 and TC-17 Jointed Special Session during 13th Africa Conf. of ISSMGE (Marrakech, 2003)	Reinforcement and Improvement of Soil
Special Session at 3rd Asian Regional Conference on Geo-synthetics 3 (Seoul, 2004)	Role of numerical analysis on earth reinforcement
TC-9 and TC-17 Jointed Seminar in Hong Kong (2004)	One-Day Seminar on Ground Treatment
Co-sponsored session between TC-9 and ASCE during Geo-Frontiers (2005)	International perspectives on earth reinforcement applications
Technical session at ICSMGE (2006)	Reinforcement and stone columns

new technology. Collecting the case histories and summarizing it are important to develop the technology. In Asian TC concerning earth reinforcement and TC9 for two terms, case histories from over the world were collected and summarized. Lawson (2003a) reported the earth reinforcement technique in ASEAN. Its range of case histories is shown as Figure 18. Hirai et al. (2003) reported the case histories of earth reinforcement technique in Japan. An example is shown in Figure 19. A lot of valuable data has been obtained in Asia and Japan. Those were good opportunity to send the information to the world. Case histories are effective for the spread of new technologies. The limit state design of the reinforced soil structure is examined in a lot of areas and countries. Many data is needed to determine the safety factor in the limit state design. These attempts will contribute for improving new design method.

4.2.2 *Materials and testing*

In the construction project of reinforced soil structure, it is important to measure the material properties in high accuracy and determine the design value. In the Asian TC, current status and future task was reviewed in detail.

Hayashi et al. (1997) reported the examination results on the testing methods of reinforcing materials and interaction properties. In this report, the examination method concerning both the metal and geosynthetics was summarized. For the pullout test, testing and evaluation method was especially examined. Figure 20 shows that recommended analysis method for the pullout test. Current testing method depends on kind of reinforcing materials. In the Asian TC, the testing method was discussed without any distinction of reinforcing material. In the near future, it seems that a more inclusive examination method is examined. These achievements will contribute for future development of international standard of reinforcing material.

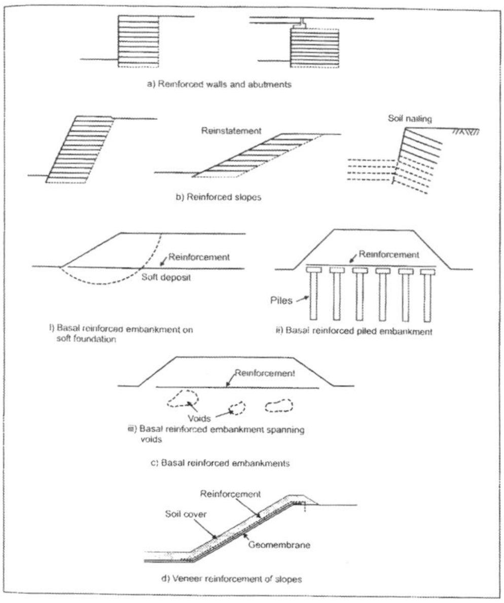

Figure 18. Range of applications where the reinforced soil technique is used (Lawson, 2003a).

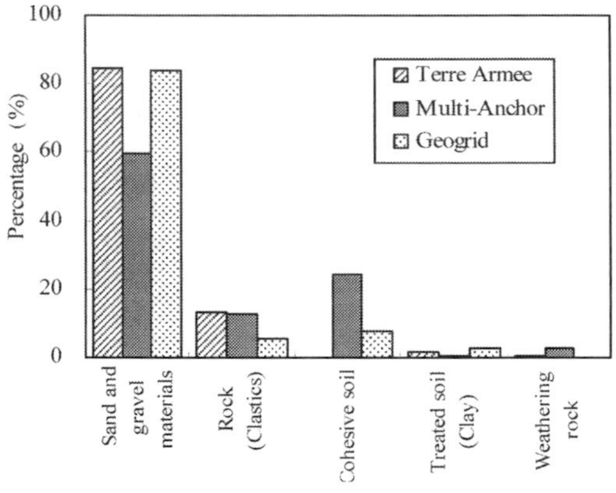

Figure 19. Case histories of earth reinforcement technique in Japan: Fill material of construction examples (Hirai, et al., 2003).

4.2.3 *Physical and numerical modeling*

Reinforcing effects depends on initial and boundary condition. This means that physical and numerical modeling is especially important in this research field.

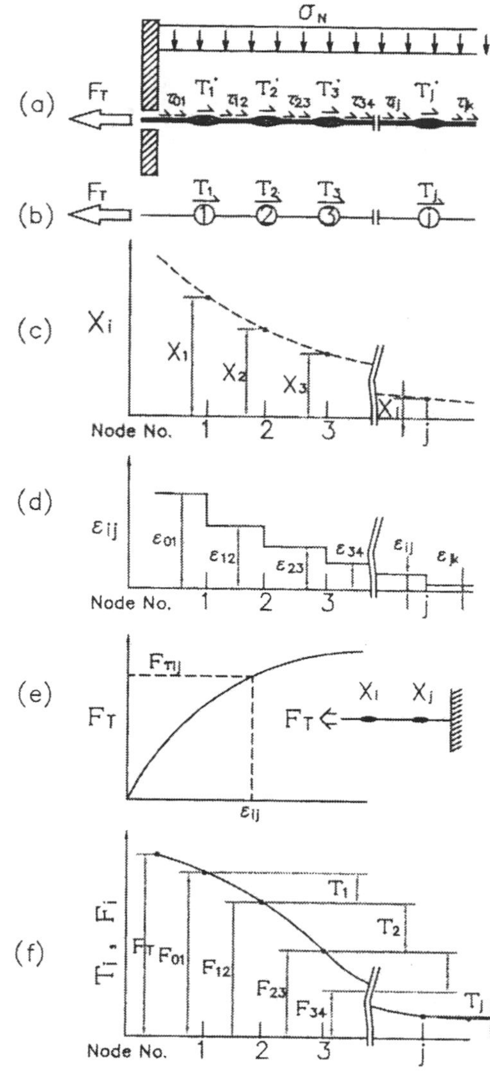

Figure 20. Analysis method of pullout test (Hayashi, et al., 1997).

Otani et al. (1997) reported the state of art concerning the numerical modeling of earth reinforcement. In the ATC, the following pointes were discussed as 1) what kind of numerical analysis is conducted, 2) what kind of models including is used for soil, reinforcing material and interaction as shown in Figure 21, 3) what in indispensable modeling at least; and 4) the task of numerical analysis to the design method. Ten years passed after the report by Otani et al. (1997) had been submitted. Earth reinforcement technique is applied on a more complex condition. Rule of numerical analysis has changed from the tool of research to the tool of the design. Such attempt should be done again.

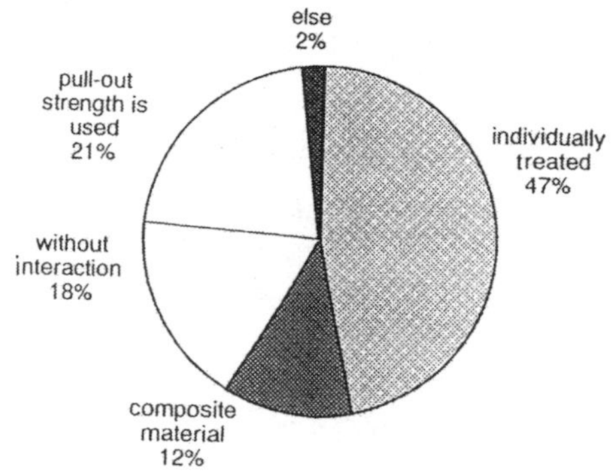

Figure 21. Interaction model used in analysis of reinforced soil structure. (Otani, et al., 1997).

Validity of numerical modeling should be verified by computational analysis of real structure and natural ground. However, there is a limit in the number of sites. Data of appropriate physical model experiment might be very useful for the verification of the numerical modeling. In the future TC, it is expected that how to obtain data with high accuracy in the physical model test is discussed.

4.2.4 *Design method*

Recently, a global standardization became a key issue for designs around the world. In the TC9, new design concepts such as limit state design and performance based design were discussed. Zornberg and Leshchinsky (2003) compared currently available criteria for the design of geosynthetic reinforced soil walls, slopes, embankments founded on soft soil deposit. Compared criteria are shown in Table 16. Compared item are performance criteria, earth reinforcement interaction, reinforced fill, geosynthetics reinforcements, and design methods/considerations. Detailed items are shown in Table 17. Miyata et al. (2003a) compared the design method of different reinforcing method used in Japan. They described that the more inclusive design method should be established based on this comparison work. These achievements will contribute for developing comprehensive and harmonized design code for earth reinforcement technology.

4.2.5 *Geo-hazards and mitigation*

Not a week goes by without news of a disaster, natural or man-made. Especially geo-hazards affect on huge losses on humans and the environment. High performance of reinforced soil structure against seismic was guaranteed through the experience of the earthquake, Northridge Earthquake (1994, USA), Hyogoken-Nanbu (Kobe) Earthquake (1995, Japan), Chi-chi Earthquake (1999, Taiwan), Kocaeli Earthquake (1999, Turkey), and El Salvador Earthquake

Table 16. Compared design manual for reinforces soil wall. (from the presentation by Zornberg and Leshchinsky (2003)).

Country	Agency	Reference
Australia	RTA, NSW DOT	RTA (1997)
	OMRD, QL DOT	QMRD (1997)
Brazil	GFSSCC	Geo Rio (1989)
Canada	CGS	CGS (1992)
Germany	GSMFE	EBGEO (1997)
Hong Kong	GEO	GEO (1989)
Italy	IMPW	IMPW (1988)
Japan	PWRC	PWRC (2000)
United Kingdom	BSI	BSI (1995)
United State	AASHTO/FHWA	FHWA (1997)
	NCMA	NCMA (1998)

Table 17. Considered item in the comparison of design manual. (from the presentation by Zornberg and Leshchinsky (2003)).

Performance criteria	Sliding, Overturning, Eccentricity of base, Bearing capacity, Compound and deep seated stability, Seismic stability, Pullout resistance
Earth reinforcement interaction	Default earth reinforcement interaction (static and dynamic)
Reinforced fill	Maximum cohesion, Default value ϕ, Peak or constant volume ϕ, Gradation requirements, Plasticity index, Soundness, PH
Geosynthetics reinforcements	Ultimate tensile strength, Material safety factors, Factor of safety in the design
Design methods/ considerations	External and internal stability, seismic stability, limitations for lay out of reinforcement etc.

(2001, Central America). Murata et al. (2003) summarized seismic stability of reinforced soil structure. In this report, Japanese case histories were focused, and 200 reports published mainly in Japan were summarized. Figure 22 is comparison of shaking table tests performed to four kinds of retaining walls. This information was valuable for not only researcher but also engineer in the countries developing the earthquake engineering.

4.2.6 *Combined technologies*

In the construction work, reduction of costs, realization of higher performance, and reduction of environmental impacts are demanded. In these circumstances, earth reinforcement is often applied with the other technology (Ochiai, et al. eds., 2003). Combined technology is mixing, piling, EPS, preloading-prestressed

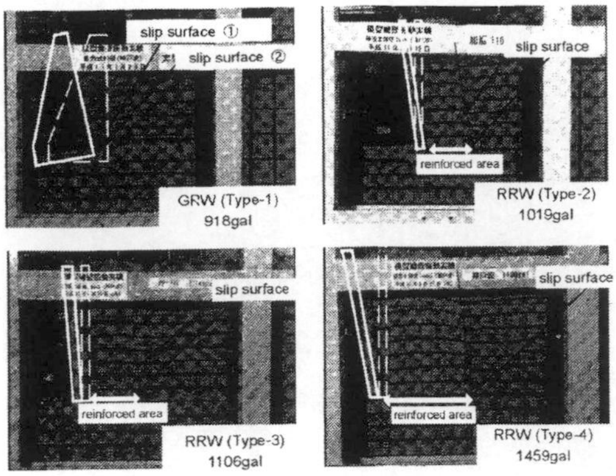

Figure 22. Deformation of retaining wall modes observed after in the shaking table test. (Murata, et al., 2003, original data was observed by Watanabe, Munaf, Koseki, Tateyama, and Kojima, 1999).

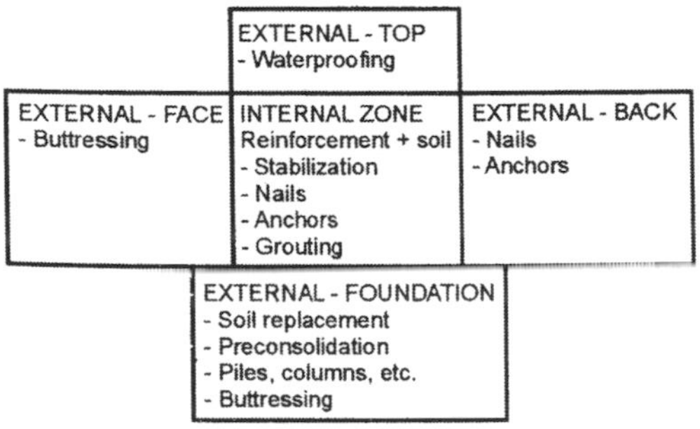

Figure 23. Range of ground improvement techniques used at different locations within and around a reinforced soil structure. (Lawson, 2003b).

and others. Lawson (2003b) summarized this combined technology in Southeast Asia. Figure 23 lists the range of ground improvement techniques used in conjunction with reinforced structure. Miyata et al. (2003b) summarized Japanese case histories shown in Table 18. Combined technology is still in the primary stage. There might be a more effective combining method. When they are systematically examined, these results might be useful.

5 FUTURE PROSPECTS ON EARTH REINFORCEMENT

As a recent trend of earth reinforcement technology, development of the global standard on testing should be paid to attention. In 2002, chairpersons of ISO/TC221 and ASTM/D35 signed a memorandum of understanding which stated that the two committees agreed to work together to avoid duplication of efforts in standards development. This meant that both

Table 18. Summarized case histories of combined technology in Japan. (Miyata et al., 2003b).

Combined method	Improvement of soil properties				
Application of reinforcement	Densification	Mixing & Grouting	Using of other member	PLPS
Retaining Wall	–	Ex. 1, Ex. 2, Ex. 3	Ex. 4 (EPS)	Ex.5
Embankment	Ex.6, EX.7	–	Ex. 8 (Impact Absorber)	Ex.7
Slope	–	Ex. 9, Ex. 10	Ex. 11 (Ring Reinforcement)	–
Foundation	–	Ex. 13	Ex. 12 (Pile)	–

committees would recognize the work being carried on within the other committee, and where possible, accept each other's standards. It is joyous that the integration of geosynthetics fields was attempted by this agreement. For the metallic reinforcement, many standards are being developed. Further cooperation of geosynthetics and a metallic field will develop this technology further.

ISO 2394; General principles on reliability for structures was established in 1998. Eurocode 7 was issued in 2005. Both of ISO 2394 and Eurocode 7 are based on the limit state philosophy. In these circumstances, developing and revising of design method of reinforced soil structure started in the world. Geoguide 6 (Hong Kong), Nordic Handbook was newly established in the last 5 years. AASHTO (USA) was revised at 2006. BS 8006 (UK) and PWRC (Japan) are under the revision work. EN 14445, whose subject has been execution, was issued in 2006. An applicable design method in both metallic and geosynthetics reinforcement has became mainstream. It seems to be very strong. However, we have not got adequate conclusion for the problem about the stability analysis and the partial safety factor. Can we use same analysis method for both geosynthetic and metallic reinforced soil structure? How can we decide the safe factor of a different reinforcement method with the correspondence? Development of good data base and opening it to public is needed. The development of the information technology enabled such an activity more low-cost.

In the construction of reinforced soil structure, the quality control is a problem in the state of insufficient examination. How to maintain the quality during the design life can be classed into same category problem. To solve these problems, application of Radio Frequency Identification (RFID) technology seems to be effective. Concept is shown as Figure 24. The RFID is attached to reinforcing material, wall facing, and so on. Each sensor perceives the change in the state of the fill material, reinforcement, and the wall material. Measured data will be sent to the receiver out side of reinforced soil structure. The engineer manages information that has been transmitted, and gives engineering judgment. This technology will be effective for quality control, maintenance support, and

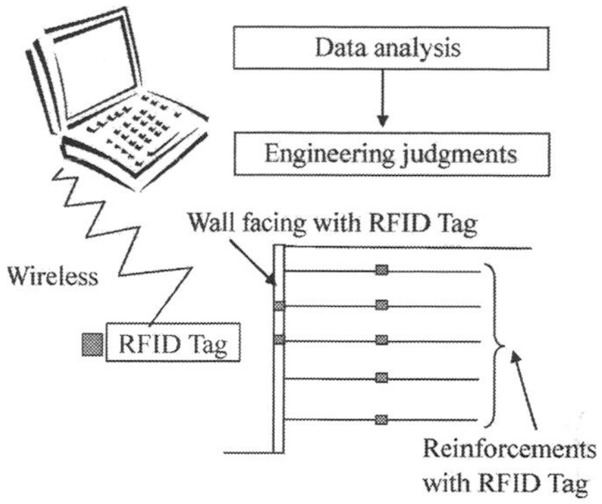

Figure 24. Concept of application of RDID tag to reinforced soil structure.

life-cycle management in re-construction. To develop this technique, we have to corporate with IT engineer, structural engineer, manager of the infrastructure.

In the recent construction work, engineer has to consider not only conventional structure evaluation items but also environmental evaluation. Applying this technology to an important structure became unusual. Higher performance is required than before. It is necessary to develop the technology to unite structural safety, low-cost, and a low-environmental impact. Development of the combined technology is one solution to the problem. As TC9 report has already clarified, combined technology has many new possibilities. However, it is difficult to apply the conventional design method to the reinforced soil structure that combined is applied. Quality control for the combined technology is more complicated than conventional one. We have to remember that there are positive and negative synergistic effects in the combined technology. Further examination is necessary.

6 SUMMARY

The International Symposium on Earth Reinforcement (IS Kyushu) have played a key role on the development of earth reinforcement technique starting from

1988, and was held in 1992, 1996, and 2001, successfully. And this time of IS Kyushu is the 5th one on November 2007. Under these circumstances, the authors who deeply involved the design and execution of the IS Kyushu tried to summarize the history of IS Kyushu, precisely. And the international activities under TC9/ISSMGE with the purpose of familiarizing the earth reinforcement technique worldwide were also introduced.

Earth reinforcement technique has become more and more complicate as time goes on. However, it may say that the technique itself has become a successful solution in geotechnical engineering nowadays. And the new direction on this technique should be expected in the 21st century. The author tried to propose the next key issues on this technique at the last part of this paper, which are quality control of the technique and expectation of the combination with other techniques or field of engineering.

I hope that the earth reinforcement technique will be long lasting and effective solution for any types of geotechnical engineering problems. And it is for sure that this technique will be used as not only the new construction techniques but also the rehabilitation techniques suffering from any natural disasters such as earthquakes, heavy rain, tsunami, and so on.

ACKNOWLEDGEMENTS

The author has had a large number of data including visual information and several statistics related to IS Kyushu and the activities of TC9/ISSMGE. The author would like to express his sincere gratitude to Prof. J. Otani of Kumamoto University, Prof. Y. Miyata of National Defense Academy, Prof. Omine of Kyushu University and Dr. T. Hirai of Mitsui Chemicals Industrial Product, Ltd., who have collaborated in providing the author with useful information and assistance. The author could not complete this paper without those valuable data. Those supports are highly acknowledged.

REFERENCES

Asian TC for earth reinforcement, ISSMGE (1995–1996) 1996. Asian technical committee report on earth reinforcement, *Proc. of 3rd International Symposium on Earth Reinforcement*, Fukuoka, Japan, Ochiai, Yasufuku & Omine (eds.), Vol.2, 1127–1170, Rotterdam: Balkema.

Bathurst, R.J., Walters, D.L., Hatami, K., Allen, T.M. 2003. Full-scale performance testing and numerical modeling of reinforced soil retaining walls. *Proc. of 4th International Symposium on Earth Reinforcement*, Fukuoka, Japan, Ochiai, Otani, Yasufuku & Omine (eds.), Vol.2, 777–799, Rotterdam: Balkema.

Broms, B.B. 1988. Fabric reinforced retaining Walls. *Proc. of International Symposium on Earth Reinforcement*, Fukuoka, Japan, Yamanouchi, Miura & Ochiai (eds.), Vol.2, 3–31, Rotterdam: Balkema.

Fukuoka, M. 1988. Earth Reinforcement-West and east. *Proc. of International Symposium on Earth Reinforcement*, Fukuoka, Japan, Yamanouchi, Miura & Ochiai (eds.), Vol.2, 33–47, Rotterdam: Balkema.

Hirai, T., Konami, T., Yokota, Y., Otani, Y., and Ogata, K. 2003. Case histories of earth reinforcement technique in Japan, *Proc. of 4th International Symposium on Earth Reinforcement*, Fukuoka, Japan, Ochiai, Otani, Yasufuku & Omine (eds.), Vol.2, 1009–1030, Rotterdam: Balkema.

Jones, C.J.F.P., Hassan, C.A. 1992. Reinforced Soil formed using polymeric anchors. *Proc. of 2nd International Symposium on Earth Reinforcement*, Fukuoka, Japan, Ochiai, Hayashi, & Otani (eds.), Vol.1, Fukuoka, Japan: 345–350. Rotterdam:
Balkema.

Jones, C.J.F.P., Fakher, A., Hamir, R. & Nettleton, I.M. 1997. Geosynthetic materials with improved reinforcement capabilities. *Proc. of 3rd International Symposium on Earth Reinforcement*, Fukuoka, Japan, Ochiai, Yasufuku & Omine (eds.), Vol.2, 865–883, Rotterdam: Balkema.

Lawson, C.R. 2003a. Earth reinforcement technique with geosynthetics in ASEAN region, *Proc. of 4th International Symposium on Earth Reinforcement*, Fukuoka, Japan, Ochiai, Otani, Yasufuku & Omine (eds.), Vol.2, 995–1008, Rotterdam: Balkema.

Lawson C.R. 2003b. Southeast Asian Practice of Soil Reinforcement in Combination with Other Soil Improvement Methods, *Proc. of 12th Asia Conference on SMFE*, Singapore, Vo.2, pp.1317–1322.

Miyata, Y., Fukuda, N., Kojima, N., Konami, T., and Otani, Y. 2003a. Design of reinforced soil wall – Overview of design manuals in Japan, *Proc. of 4th International Symposium on Earth Reinforcement*, Fukuoka, Japan, Ochiai, Otani, Yasufuku & Omine (eds.), Vol.2, 995–1008, Rotterdam: Balkema.

Miyata, Y., Ochiai, H. and Otani, J. 2003b. Recent case histories on combined technology with earth reinforcement in Japan, *Proc. of 12th Asia Conference on SMFE*, Singapore, Vo.2, pp.1317–1322.

Murata, O., Kojima, K., Uchimura, T., Nishimura, J., Ogata, K., Tayama, S., Hirata, M., Ogisako, E., and Miyatake, H. 2003. Long-term performance and seismic stability of reinforced soil structure reported in Japan, *Proc. of 4th International Symposium on Earth Reinforcement*, Fukuoka, Japan, Ochiai, Otani, Yasufuku & Omine (eds.), Vol.2, 1065–1091, Rotterdam: Balkema.

Hayashi, S., Hirai, T., Kuwano, J., and Yokota, Y. 1997. Testing method of reinforcements for use in reinforced soil structures, *Proc. of 3rd International Symposium on Earth Reinforcement*, Fukuoka, Japan, Ochiai, Yasufuku & Omine (eds.), Vol.2, 1131–1146, Rotterdam: Balkema.

Ochiai, H. 1999. Activities of Kyushu Branch in the Latest 20 years and the next expectation, *the report of 50th anniversary of the Kyushu branch of JGS*, Japan Geotechnical Society, 22–29.

Ochiai, H., Otani, J. & Miyata, Y. (eds.) 2003. Combined Technology in Earth Reinforcement, TC9 Sponsored session, Special print in 12th Asian Regional Conference on SMGE, Singapore.

Otani, J., Yamamoto, A. Kodaka, T., Yasufuku, N., and Yashima, A. 1997. Current state on numerical analysis of reinforced soil structures, *Proc. of 3rd International Symposium on Earth Reinforcement*, Fukuoka, Japan, Ochiai, Yasufuku & Omine (eds.), Vol.2, 1159–1170, Rotterdam: Balkema.

Raymond, G.P. and Giroud, J.P. (eds.) 1993. *Geosynthetics Case Histories*, ISSMGE Technical Committee TC9 "Geotextiles and Geosynthetics (1986–1994)", 300pp., ISBN No. 0-969-6924-0-4, BiTech Publishers Ltd.

TC9 of ISSMGE, Geosynthetics and Earth Reinforcement (1998–2001), 2003. Committee reports, *Proc. of 4th International Symposium on Earth Reinforcement*, Fukuoka, Japan, Ochiai, Otani, Yasufuku & Omine (eds.), Vol.2, 982–1125, Rotterdam: Balkema.

TC9 of ISSMGE, Earth Reinforcement (2002–2005), 2006a. Administrative reports, *Proc. of 16th International Conference on Soil Mechanics and Geotechnical Engineering*, Vol.5, 3483–3487, Rotterdam: Millpress.

TC9 of ISSMGE, Earth Reinforcement (2002–2005), 2006b. Extended reports, *Proc. of 16th International Conference on Soil Mechanics and Geotechnical Engineering*, Vol.5, 3591–3610, Rotterdam: Millpress.

Yamanouchi, T. 1993. Historical review of geotextiles for reinforcement of earth works in Asia. *Proc. of 2nd International Symposium on Earth Reinforcement*, Fukuoka, Japan, Ochiai, Hayashi, & Otani (eds.), Vol.2, 737–751, Rotterdam: Balkema.

Zornberg, J.G. and Leshchinsky, D. 2003. Comparison of International design criteria for geosynthetic-reinforced soil structures. *Proc. of 4th International Symposium on Earth Reinforcement*, Fukuoka, Japan, Ochiai, Otani, Yasufuku & Omine (eds.), Vol.2, 1095–1106, Rotterdam: Balkema.

New Horizons in Earth Reinforcement – Otani, Miyata & Mukunoki (eds)
© 2008 Taylor & Francis Group, London, ISBN 978-0-415-45775-0

New horizons in reinforced soil technology

J.G. Zornberg
The University of Texas at Austin

ABSTRACT: Traditional soil reinforcing techniques involve the use of continuous geosynthetic inclusions such as geogrids and geotextiles. The acceptance of geosynthetics in reinforced soil construction has been triggered by a number of factors, including aesthetics, reliability, simple construction techniques, good seismic performance, and the ability to tolerate large deformations without structural distress. Following an overview of conventional reinforced soil applications, this paper focuses on recent advances in reinforced soil technology. Examples include advances in reinforced soil design for conventional loading (e.g. validation of analysis tools), advances in design for unconventional loading (e.g., reinforced bridge abutments), and advances in reinforcement materials (e.g., polymeric fiber reinforcements).

1 INTRODUCTION

Geosynthetic inclusions within a soil mass can provide a reinforcement function by developing tensile forces which contribute to the stability of the geosynthetic-soil composite (a reinforced soil structure). Design and construction of stable slopes and retaining structures within space constrains are aspects of major economical significance in geotechnical engineering projects. For example, when geometry requirements dictate changes of elevation in a highway project, the engineer faces a variety of distinct alternatives for designing the required earth structures. Traditional solutions have been either a concrete retaining wall or a conventional, relatively flat, unreinforced slope. Although simple to design, concrete wall alternatives have generally led to elevated construction and material costs. On the other hand, the construction of unreinforced embankments with flat slope angles dictated by stability considerations is an alternative often precluded in projects where design is controlled by space constraints.

Geosynthetics are particularly suitable for soil reinforcement. Geosynthetic products typically used as reinforcement elements are nonwoven geotextiles, woven geotextiles, geogrids, and geocells. Reinforced soil vertical walls generally provide vertical grade separations at a lower cost than traditional concrete walls. Reinforced wall systems involve the use of shotcrete facing protection or of facing elements such as precast or cast-in-place concrete panels. Alternatively, steepened reinforced slopes may eliminate the use of facing elements, thus saving material costs and construction time in relation to vertical reinforced walls. A reinforced soil system generally provides an optimized alternative for the design of earth retaining structures.

A reduced scale geotextile-reinforced slope model built using dry sand as backfill material. The maximum slope inclination of an unreinforced sand under its own weight is the angle of repose of the sand, which is well below the inclination of the slope face of the model. Horizontal geotextile reinforcements placed within the backfill provided stability to the steep sand slope. In fact, not only the reinforced slope model did not fail under its own weight, but its failure only occurred after the unit weight of the backfill was increased 67 times by placing the model in a geotechnical centrifuge (Zornberg et al., 1998).

The use of inclusions to improve the mechanical properties of soils dates to ancient times. However, it is only within the last quarter of century or so (Vidal, 1969) that analytical and experimental studies have led to the contemporary soil reinforcement techniques. Soil reinforcement is now a highly attractive alternative for embankment and retaining wall projects because of the economic benefits it offers in relation to conventional retaining structures. Moreover, its acceptance has also been triggered by a number of technical factors, that include aesthetics, reliability, simple construction techniques, good seismic performance, and the ability to tolerate large deformations without structural distress. The design of reinforced soil slopes is based on the use of limit equilibrium methods to evaluate both external (global) and internal stability of the structure. The required tensile strength of the reinforcements is selected during design so that the margins of safety, considering an internal failure are adequate. Guidance in soil reinforcement design procedures is provided by Elias et al. (2001).

2 VALIDATION OF DESIGN TOOLS

2.1 *Overview*

The use *of* inclusions to improve the mechanical properties of soils dates to ancient times. However, it is only within the last three decades or so (Vidal 1969) that analytical and experimental studies have led to the contemporary soil reinforcement techniques. Soil reinforcement is now a highly attractive alternative for embankment and retaining wall projects because of the economic benefits it offers in relation to conventional retaining structures. Moreover, its acceptance has also been triggered by a number of technical factors, which include aesthetics, reliability, simple construction techniques, good seismic performance, and the ability to tolerate large deformations without structural distress. The design of reinforced soil slopes is based on the use of limit equilibrium methods to evaluate both external (global) and internal stability. After adopting the shear strength properties of the backfill material, the required tensile strength of the reinforcements can be defined in the design so that the margin of safety is adequate.

Geosynthetics are classified as extensible reinforcements. Consequently, the soil strength may be expected to mobilize rapidly, reaching its peak strength before the reinforcements achieve their ultimate strength. This rationale has led to some recommendations towards the adoption of the residual shear strength for the design of geosynthetic-reinforced slopes. This is the case of commonly used design methods such as those proposed by Jewell (1991) and Leshchinsky and Boedeker (1989). Several agencies have endorsed the use of residual shear strength parameters in the design of reinforced soil structures, as summarized in Table 1. Zornberg and Leshchinsky (2001) present a review of current design criteria used by different agencies for geosynthetic-reinforced walls, geosynthetic-reinforced slopes, and embankments over soft soils.

The use of the peak friction angle has been common practice in the US for the design of geosynthetic-reinforced slopes. Guidance in soil reinforcement design procedures has been compiled by several federal agencies in the US, including the American Association of State Highway and Transportation Officials (AASHTO 1996), and the Federal Highway Administration (Elias et al. 2001). Design guidance is also provided by the National Concrete Masonry Association (NCMA 1997), possibly the only industry manual of soil reinforcement practice. The above mentioned design guidance manuals recommend the use of the peak friction angle in the limit equilibrium analyses. Other agencies that have also endorsed the use of peak shear strength parameters in the design of reinforced soil structures are summarized in Table 1.

Table 1. Summary of Guidelines on Selection of Soil Shear Strength Parameters for Geosynthetic-Reinforced Soil Design.

Method/agency	Shear Strength parameters	Reference
Jewell's Method	Residual	Jewell (1991)
Leshchinsky and Boedeker's method	Residual	Leshchinsky and Boedeker (1989)
Queensland DOT, Australia	Residual	RTA (1997)
New South Wells, Australia	Residual	QMRD (1997)
Bureau National Sols-Routes (draft French Standard)	Residual	Gourc et al. (2001)
Federal Highway Administration (FHWA), AASHTO	Peak	Elias et al. (2001), AASHTO 1996
National Concrete Masonry Association	Peak	NCMA (1997, 1998)
GeoRio, Brazil	Peak	GeoRio (1989)
Canadian Geotechnical Society	Peak	Can. Geotechnical Society (1992)
German Society of Soil Mech. and Geot. Eng.	Peak	EBGEO (1997)
Geotechnical Engineering Office, Hong Kong	Peak	GCO (1989), GEO (1993)
Public Works Research Center, Japan	Peak	Public Works Res. Ctr. (2000)
British Standards, United Kingdom	Peak	British Standard Institution (1995)
Leshchinsky's hybrid method	Hybrid	Leshchinsky (2001)

A hybrid approach was recently proposed by Leshchinsky (2000, 2001). Central to his approach is the use of a design procedure in which peak soil shear strength properties would be used to locate the critical slip surface, while the residual soil shear strength properties would subsequently be used along the located slip surface to compute the reinforcement requirements.

In order to address the controversial issue regarding selection of shear strength properties in reinforced soil design, this paper presents experimental evidence on failed reinforced slopes. Specifically, the experimental information obtained from centrifuge modeling supports the use of peak shear strength parameters in the design of geosynthetic-reinforced soil structures. The perceived conservatism in design is also not supported by the generally observed good performance of monitored reinforced soil structures.

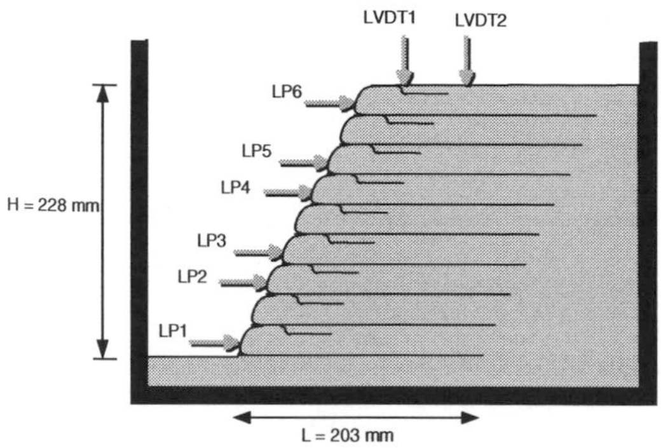

Figure 1. Typical centrifuge model.

2.2 *Centrifuge testing program*

Limit equilibrium analysis methods have been traditionally used to analyze the stability of slopes with and without reinforcements. However, to date, limit equilibrium predictions of the performance of geosynthetic-reinforced slopes have not been fully validated against monitored failures. This has led to a perceived overconservatism in their design. Consequently, an investigation was undertaken to evaluate design assumptions for geosynthetic-reinforced slopes (Zornberg et al. 1998a, 2000). The results of centrifuge tests provide an excellent opportunity to examine the validity of various assumptions typically made in the analysis and design of reinforced soil slopes. This paper presents the aspects of that study aimed at evaluating the shear strength properties governing failure of reinforced soil slopes.

All reinforced slope models in the experimental testing program had the same geometry and were built within the same strong box. A transparent Plexiglas plate was used on one side of the box to enable side view of the models during testing. The other walls of the box were aluminum plates lined with Teflon to minimize side friction. The overall dimensions of the geotextile-reinforced slope models are as shown in Figure 1 for a model with nine reinforcement layers. Displacement transducers are also indicated in the figure.

The number of reinforcement layers in the models ranged from six to eighteen, giving reinforcement spacing ranging from 37.5 mm to 12.5 mm. All models used the same reinforcement length of 203 mm. The use of a reasonably long reinforcement length was deliberate, since this study focused on the evaluation of internal stability against breakage of the geotextile reinforcements. In this way, external or compound failure surfaces were not expected to develop during testing. As shown in the figure, the geotextile layers were wrapped at the slope facing in all models. Green colored sand was placed along the Plexiglas wall at the level of each reinforcement in order to identify the failure surface. In addition, black colored sand markers were placed at a regular horizontal spacing (25 mm) in order to monitor lateral displacements within the backfill material.

The variables investigated in this study were selected so that they could be taken into account in a limit equilibrium framework. Accordingly, the selected variables were:

- Vertical spacing of the geotextile reinforcements: four different reinforcement spacings were adopted;
- soil shear strength parameters: the same sand at two different relative densities was used; and
- ultimate tensile strength of the reinforcements: two geotextiles with different ultimate tensile strength were selected.

Of particular relevance, for the purpose of the issues addressed in this paper, is the fact that that the same sand placed at two different relative densities was used as backfill material for the centrifuge models. The backfill material at these two relative densities has different peak shear strength values but the same residual shear strength.

The model slopes were built using Monterey No. 30 sand, which is a clean, uniformly graded sand classified as SP in the Unified Soil Classification System (Zornberg et al. 1998b). The particles are rounded to subrounded, consisting predominantly of quartz with a smaller amount of feldspars and other minerals. The average particle size for the material is 0.4 mm, the coefficient of uniformity is 1.3, and the coefficient of curvature is about 1.1. The maximum and minimum void ratios of the sand are 0.83 and 0.53, respectively. To obtain the target dry densities in the model slopes, the sand was pluviated through air at controlled combinations of sand discharge rate and discharge height. The unit weights for the Monterey No. 30 sand at the target relative densities of 55% and 75% are 15.64 kN/m^3 and 16.21 kN/m^3, respectively.

Two series of triaxial tests were performed to evaluate the friction angle for the Monterey No. 30 sand as a function of relative density and of confining pressure. The tests were performed using a modified form of the automated triaxial testing system developed by Li et al. (1988). The specimens had nominal dimensions of 70 mm in diameter and 150 mm in height and were prepared by dry tamping. Figure 2 shows the stress strain response obtained from the series of tests conducted to evaluate the behavior of Monterey No. 30 sand as a function of relative density. All tests shown in the figure were conducted using a confining pressure of 100 kPa. As can be observed in the figure, while the sand shows a different peak shear strength for different relative densities, the shear stress tends to a single residual shear strength for large strain conditions. Figure 3 shows the increase in peak friction

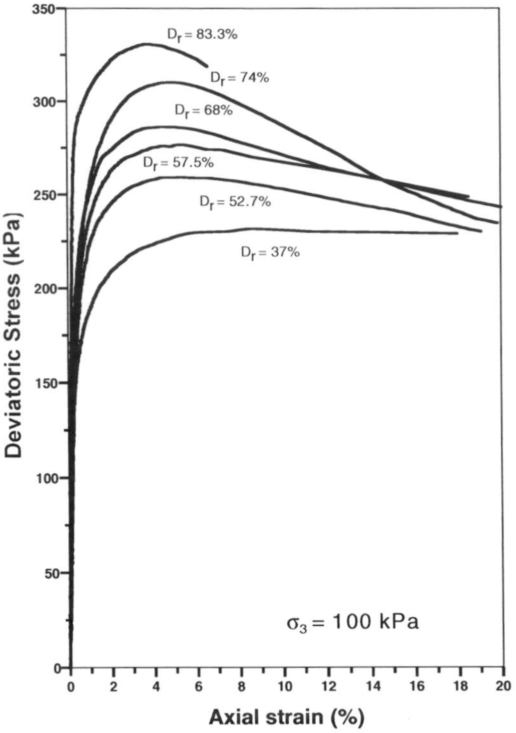

Figure 2. Stress strain behavior of Monterey No. 30 sand pluviated at different relative densities and tested in triaxial compression under the same confinement.

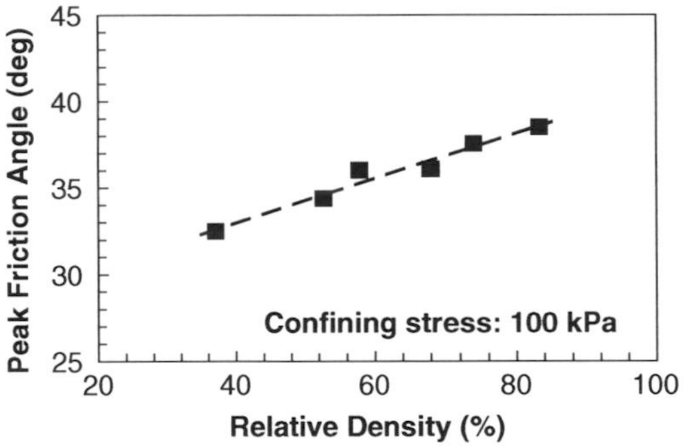

Figure 3. Friction angle for Monterey No. 30 sand obtained from triaxial testing at different relative densities.

angle with increasing relative density at a confining pressure of 100 kPa. Of particular interest are the friction angles obtained at relative densities of 55% and 75%, which correspond to the relative density of the backfill material in the models. The estimated triaxial compression friction angles (ϕ_{tc}) at these relative densities are 35° and 37.5°, respectively. Although the tests did not achieve strain values large enough to guarantee a critical state condition, the friction angles at large strains appear to converge to a residual value (ϕ_r) of approximately 32.5°. This value agrees with the critical state friction angle for Monterey No. 0 sand obtained by Riemer (1992). As the residual friction angle is mainly a function of mineralogy (Bolton 1986), Monterey No. 0 and Monterey No. 30 sands should show similar ϕ_r values. The effect of confining pressure on the frictional strength of the sand was also evaluated. The results showed that the friction angle of Monterey No. 30 decreases only slightly with increasing confinement. The fact that the friction angle of this sand does not exhibit normal stress dependency avoids additional complications in the interpretation of the centrifuge model tests.

Scale requirement for the reinforcing material establish that the reinforcement tensile strength in the models be reduced by N. That is, an Nth-scale reinforced slope model should be built using a planar reinforcement having $1/N$ the strength of the prototype reinforcement elements (Zornberg et al. 1998a). Two types of nonwoven interfacing fabrics, having mass per unit area of 24.5 g/m^2 and 28 g/m^2, were selected as reinforcement. Unconfined ultimate tensile strength values, measured from wide-width strip tensile tests ASTM D4595, were 0.063 kN/m and 0.119 kN/m for the weaker and stronger geotextiles, respectively. Confined tensile strength values, obtained from backcalculation of failure in the centrifuge slope models, were 0.123 kN/m and 0.183 kN/m for the weaker and stronger geotextiles, respectively (Zornberg et al. 1998b). Confined tensile strength values were used for estimating the factor of safety of the models analyzed in this study under increasing g-levels.

2.3 *Typical centrifuge test results*

The models were subjected to a progressively increasing centrifugal acceleration until failure occurred. A detailed description of the characteristics of the centrifuge testing program is presented by Zornberg et al. (1998a). The centrifuge tests can be grouped into three test series (B, D, or S). Accordingly, each reinforced slope model in this study was named using a letter that identifies the test series, followed by the number of reinforcement layers in the model. Each test series aimed at investigating the effect of one variable, as follows:

- Baseline, B-series: Performed to investigate the effect of the reinforcement vertical spacing.
- Denser soil, D-series: Performed to investigate the effect of the soil shear strength on the stability of geosynthetic-reinforced slopes. The models in this series were built with a denser backfill but the same reinforcement as in the B-series.

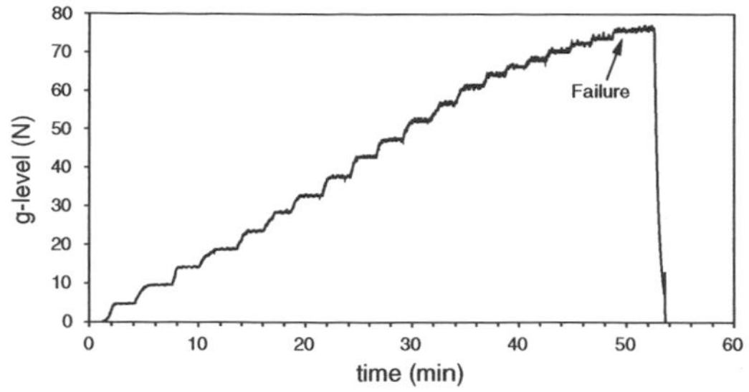

Figure 4. G-level (N) versus time during centrifuge testing.

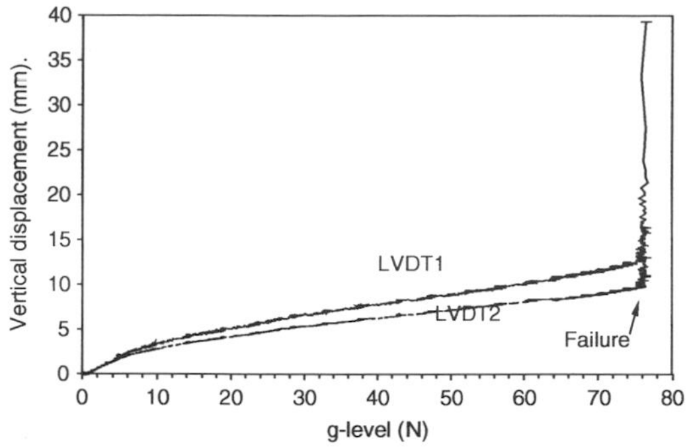

Figure 5. Settlements at the crest of a reinforced slope model.

- Stronger geotextile, S-series: Performed to investigate the effect of the reinforcement tensile strength on the performance of reinforced slopes. The models in this series were built using reinforcements with a higher tensile strength than in the B-series but with the same backfill density.

The history of centrifugal acceleration during centrifuge testing of one of the models is indicated in Figure 4. In this particular test, the acceleration was increased until sudden failure occurred after approximately 50 min of testing when the acceleration imparted to the model was 76.5 times the acceleration of gravity. Settlements at the crest of the slope, monitored by LVDTs, proved to be invaluable to accurately identify the moment of failure. Figure 5 shows the increasing settlements at the top of a reinforced slope model during centrifuge testing. The sudden increase in the monitored settlements indicates the moment of failure when the reinforced active wedge slid along the failure surface. Figure 6 shows a typical failure surface as developed in the centrifuge models. As can be seen, the failure surface is well defined and goes through the toe of the reinforced slope.

Following the test, each model was carefully disassembled in order to examine the tears in the geotextile layers. Figure 7 shows the geotextiles retrieved after

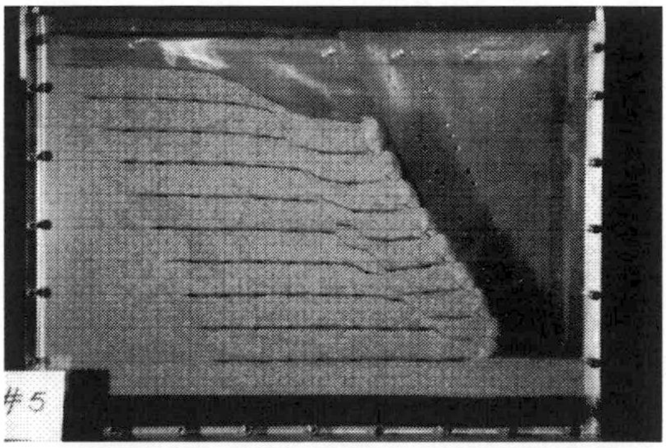

Figure 6. Failed geotextile-reinforced slope model.

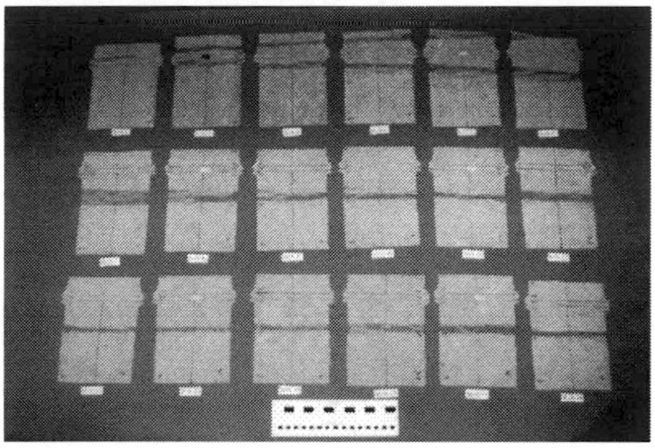

Figure 7. Geotextile reinforcements retrieved after testing.

centrifuge testing of a model reinforced with eighteen geotextile layers. The geotextile at the top left corner of the figure is the reinforcement layer retrieved from the base of the model. The geotextile at the bottom right corner is the reinforcement retrieved from the top of the model. All retrieved geotextiles show clear tears at the location of the failure surface. The pattern observed from the retrieved geotextiles shows that internal failure occurred when the tensile strength on the reinforcements was achieved. The geotextile layers located towards the base of the slope model also showed breakage of the geotextile overlaps, which clearly contributed to the stability of the slope. No evidence of pullout was observed, even on the short overlapping layers.

2.4 *Effect of backfill shear strength on the experimental results*

The criteria for characterizing reinforcements as extensible or inextensible has been established by comparing the horizontal strain in an element of reinforced backfill soil subjected to a given load, to the strain required to develop an active plastic state in an element of the same soil without reinforcement (Bonaparte

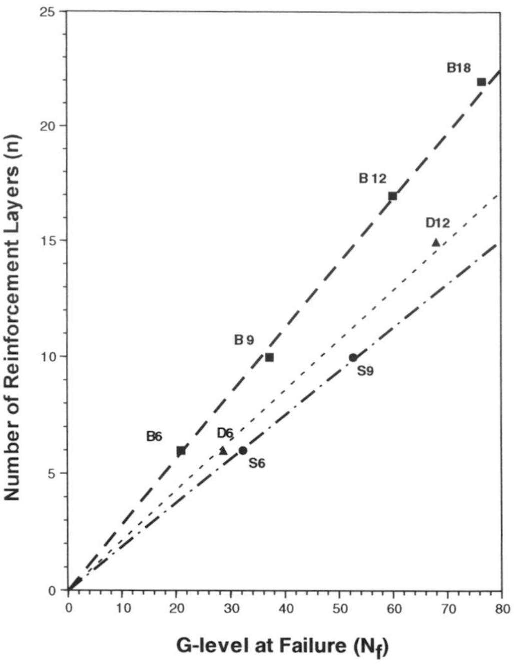

Figure 8. G-level at failure for the centrifuge models.

and Schmertmann 1987). Accordingly, reinforcements have been typically classified as:

- extensible, if the tensile strain at failure in the reinforcement exceeds the horizontal extension required to develop an active plastic state in the soil; or as
- inextensible, if the tensile strain at failure in the reinforcement is significantly less than the horizontal extension required to develop an active plastic state in the soil.

The geotextiles used to reinforce the centrifuge model slopes are extensible reinforcements. The effect of reinforcement spacing on the stability of the reinforced slope models, as indicated by the measured g-level at failure N_f, is shown in Figure 8. The number of reinforcement layers n in the figure includes the total number of model geotextiles intersected by the failure surface (i.e. primary reinforcements and overlaps intersected by the failure surface). The overlaps intersected by the failure surface developed tensile forces and eventually failed by breakage and not by pullout. The figure shows that a well-defined linear relationship can be established between the number of reinforcement layers and the g-level at failure. As the fitted lines for each test series passes through the origin, the results in each test series can be characterized by a single n/N_f ratio.

Models in the B- and D-series were reinforced using the same geotextile reinforcement, but using sand backfill placed at two different relative densities (55 and 75%). As mentioned, the Monterey sand at these two relative densities has the same soil residual friction angle (32.5°) but different peak friction angles (35° and 37.5°). As shown in Figure 8, models in the D-series failed at higher g-levels than models in the B-series built with the same reinforcement spacing and reinforcement type. Since the backfill soil in models from the D- and B-series have the same residual soil shear strength, the higher g-level at failure in the Dseries models is due to the higher peak soil shear strength in this test series.

Analysis of the data presented in this figure emphasizes that the use of a single residual shear strength value, common to the two backfill materials used in the test series, can not explain the experimental results. Instead, the experimental results can be explained by acknowledging that the stability models constructed with the same reinforcement layout and the same sand backfill, but placed at different densities, is governed by different shear strength values. Indeed, limit equilibrium analyses (Zornberg et al. 1998b) indicated that the shear strength value that should be used in the analysis of these slope failures is the plain strain peak shear strength of the backfill.

The experimental results indicate that the stability of structures with extensible reinforcements is governed by the peak shear strength and not by the residual shear strength of the backfill soil. A plausible explanation of these experimental results is that, although the soil shear strength may have been fully mobilized along certain active failure planes within the reinforced soil mass, shear displacements have not taken place along these failure surfaces. That is, although the soil may have reached active state due to large horizontal strains because of the extensible nature of the reinforcements, large shear displacements (and drop from peak to residual soil shear strength) only take place along the failure surface during final sliding of the active reinforced wedge (Zornberg et al. 1998b).

An additional way of evaluating these experimental results is by using dimensionless coefficients, which have been used in order to develop design charts for geosynthetic-reinforced soil slopes (Schmertmann et al. 1987, Leshchinsky and Boedeker 1989, Jewell 1991). The validity of the proposed normalization can be investigated from the centrifuge results of this study. For a reinforced slope model that failed at an acceleration equal to N_f times the acceleration of gravity, a dimensionless coefficient K can be estimated as follows:

$$K = n\,T_{ult} \cdot \left(\frac{2}{\gamma H^2} \right) \cdot \frac{1}{N_f} \qquad (1)$$

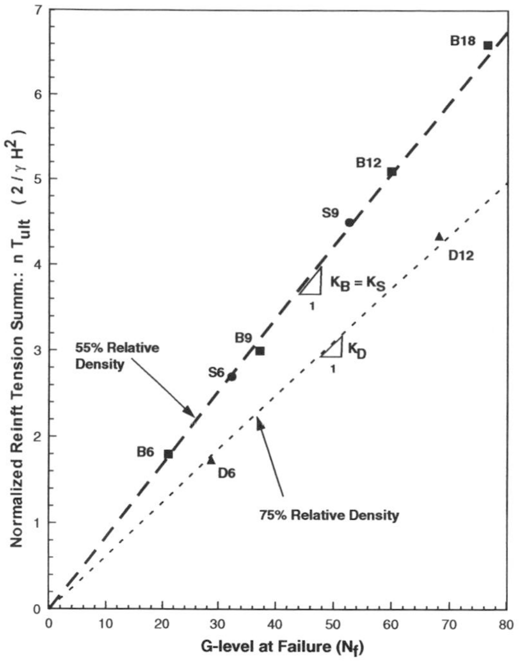

Figure 9. Normalized Reinforcement Tension Summation (RTS) values from centrifuge test results.

where n is the number of reinforcements, T_{ult} is the reinforcement tensile strength, H is the slope height, N_f is the g-level at failure from the centrifuge test, and γ is the sand unit weight. The value of n used in Equation 1 includes the number of overlaps that were intersected by the failure surface in the centrifuge slope models in addition to the number of primary reinforcement layers. The coefficient K is a function of the shear strength of the soil and of the slope inclination. [i.e. $K = K(\phi, \beta)$]. All centrifuge slope models were built with the same slope inclination β. Consequently, validation of the suggested normalization requires that a single coefficient $K(\phi, \beta)$ be obtained for all models built with the same backfill. If the soil shear strength governing failure of the models is the residual strength, a single coefficient $K(\phi, \beta)$ should be obtained for all models. On the other hand, if the soil shear strength governing failure is the peak shear strength, a single coefficient should be obtained for those models built with sand placed at the same relative density.

Figure 9 shows the centrifuge results in terms of $(n\ T_{ult})\ (2/\gamma H^2)$ versus the g-level at failure N_f. The results in the figure show that a linear relationship can be established for those models built with sand placed at the same relative density. As inferred from Equation 1, the slope of the fitted line corresponds to the dimensionless RTS coefficient $K = K(\phi, \beta)$. The results obtained using the centrifuge models from the B- and S-series, built using Monterey sand placed at 55% relative density, define a normalized coefficient $K(\phi, \beta) = K_B = K_S = 0.084$. Similarly, centrifuge results from the D-series models, built using Monterey sand at 75% relative density, define a normalized coefficient $K(\phi, \beta) = K_D = 0.062$. These results provide sound experimental evidence supporting the use of charts based on normalized coefficients for preliminary design of geosynthetic-reinforced slopes. If failure of reinforced soil slopes were governed by the residual soil shear strength, the results of all centrifuge tests should have defined a single line. However, as can be observed in the figure, different normalized coefficients are obtained for different soil densities. This confirms that the normalization should be based on the peak shear strength and not on the residual shear strength of the backfill material.

2.5 *Final remarks on validation of design tools*

The selection of the backfill shear strength properties in the design of geosynthetic-reinforced soil structures is an issue over which design guidelines disagree. The main debate has been over whether the peak or the residual shear strength of the backfill material should be adopted for design. The use of residual shear strength values in the design of geosynthetic reinforced slopes while still using peak shear strength in the design of unreinforced embankments could lead to illogical comparisons of alternatives for embankment design. For example, an unreinforced slope that satisfies stability criteria based on a factor of safety calculated using peak strength, would become unacceptable if reinforced using inclusions of small (or negligible, for the purposes of this example) tensile strength because stability would be evaluated in this case using residual soil shear strength values. The main purpose of this investigation was to provide experimental evidence addressing this currently unsettled issue.

The experimental results presented herein indicate that the soil shear strength governing the stability of geosynthetic-reinforced soil slopes is the peak shear strength. A centrifuge experimental testing program was undertaken which indicated that reinforced slopes constructed with the same reinforcement layout and the same backfill sand, but using different sand densities failed at different centrifuge accelerations. That is, nominally identical models built with backfill material having the same residual shear strength but different peak shear strength did not have the same factor of safety. Since the residual shear strength of the sand backfill is independent of the relative density, these results indicate that the soil shear strength governing stability is the peak shear strength of the backfill material.

Several design guidance manuals have implicitly recommended the selection of the peak shear strength for the design of reinforced soil slopes. Considering the current debate over the selection of the soil shear strength in design and the experimental results presented herein, design manuals should explicitly endorse selection of peak shear strength values for the design of reinforced soil structures. This approach would not only be consistent with the observed experimental centrifuge results, but also with the US practice of using peak shear strength in the design of unreinforced slopes.

3 GEOSYNTHETIC-REINFORCED BRIDGE ABUTMENTS

3.1 *Overview*

The technology of geosynthetic-reinforced soil (GRS) systems has been used extensively in transportation systems to support the self-weight of the backfill soil, roadway structures, and traffic loads. The increasing use and acceptance of soil reinforcement has been triggered by a number of factors, including cost savings, aesthetics, simple and fast construction techniques, good seismic performance, and the ability to tolerate large differential settlement without structural distress. A comparatively new use of this technology is the use of GRS systems as an integral structural component of bridge abutments and piers. Use of a reinforced soil system to directly support both the bridge (e.g. using a shallow foundation) and the approaching roadway structure has the potential of significantly reducing construction costs, decreasing construction time, and smoothing the ride for vehicular traffic by eliminating the "bump at the bridge" caused by differential settlements between bridge foundations and approaching roadway structures.

The most prominent GRS abutment for bridge support in the U.S. is the recently-opened-to-traffic Founders/Meadows Parkway bridge, which crosses I-25 approximately 20 miles south of downtown Denver, Colorado. Designed and constructed by the Colorado Department of Transportation (CDOT), this is the first major bridge in the United States to be built on footings supported by a geosynthetic-reinforced system, eliminating the use of traditional deep foundations (piles) altogether. Phased construction of the almost 9-m high, horseshoe-shaped abutments, located on each side of the highway, began July 1998 and was completed twelve months later. Significant previous research by FHWA and CDOT on GRS bridge abutments, which has demonstrated their excellent performance and high load-carrying capacity, led to the construction of this unique structure.

The performance of bridge structures supported by GRS abutments has not been tested under actual service conditions to merit acceptance without reservation in highway construction. Consequently, the Founders/Meadows structure was considered experimental and comprehensive material testing, instrumentation, and monitoring programs were incorporated into the construction operations. Design procedures, material characterization programs, and monitoring results from the preliminary (Phase I) instrumentation program are discussed by Abu-Hejleh et al. (2000). Large-size direct shear and triaxial tests were conducted to determine representative shear strength properties and constitutive relations of the gravelly backfill used for construction. Three sections of the GRS system were instrumented to provide information on the structure movements, soil stresses, geogrid strains, and moisture content during construction and after opening the structure to traffic.

3.2 *Past experiences in GRS bridge abutments*

Although the Founders/Meadows structure is a pioneer project in the U.S. involving permanent GRS bridge abutments for highway infrastructure, significant efforts have been undertaken in Japan, Europe and Australia regarding implementation of such systems in transportation projects. Japanese experience includes preloaded and prestressed bridge piers (Tatsuoka et al. 1997, Uchimura et al. 1998) and geosynthetic-reinforced wall systems with continuous rigid facing for railway infrastructure (Kanazawa et al. 1994, Tateyama et al. 1994). European experience includes vertically loaded, full-scale tests on geosynthetic reinforced walls constructed in France (Gotteland et al. 1997) and Germany (Brau and Floss 2000). Finally, Won et al. (1996) reported the use of three terraced geogrid-reinforced walls with segmental block facing to directly support end spans for a major bridge in Australia.

The experience in the U.S. regarding geosynthetic-reinforced bridge abutments for highway infrastructure includes full-scale demonstration tests conducted by the Federal Highway Administration (FHWA) (e.g. Adams 1997, 2000) and by CDOT (e.g. Ketchart and Wu 1997). In the CDOT demonstration project, the GRS abutment was constructed with roadbase backfill reinforced with layers of a woven polypropylene geotextile placed at a spacing of 0.2 m. Dry-stacked hollow-cored concrete blocks were used as facing. A vertical surcharge of 232 kPa was applied to the 7.6 m high abutment structure. The measured immediate maximum vertical and lateral displacements were 27.1 mm and 14.3 mm, respectively. The maximum vertical and lateral creep displacements after a sustained vertical surcharge pressure of 232 kPa, applied during 70 days, were 18.3 mm and 14.3 mm, respectively. The excellent performance and high loading capacity demonstrated by these geosynthetic-reinforced soil abutments with segmental block facing convinced CDOT design engineers to select GRS walls to support the bridge abutment at the Founders/Meadows structure.

3.3 *Description of the GRS bridge abutment*

The Founders/Meadows bridge is located 20 miles south of Denver, Colorado, near Castle Rock. The bridge carries Colorado State Highway 86, Founders/Meadows Parkway, over U.S. Interstate 25. This structure, completed by CDOT in July of 1999, replaced a deteriorated two-span bridge structure. In this project, both the bridge and the approaching roadway structures are supported by a system of geosynthetic-reinforced segmental retaining walls. The bridge superstructure is supported by the "front MSE wall," which extends around a 90-degree curve into a "lower MSE wall" supporting the "wing wall" and a second tier, "upper MSE wall".

Each span of the new bridge is 34.5 m long and 34.5 m wide, with 20 side-by-side prestressed box girders. The new bridge is 13 m longer and 25 m wider than the previous structure, accommodating six traffic lanes and sidewalks on both sides of the bridge. A typical monitored cross-section through the "front MSE wall" and "abutment wall" transmits the load through abutment walls to a shallow strip footing placed directly on the top of a geogrid-reinforced segmental retaining wall. The centerline of the bridge abutment wall and edge of the foundation are located 3.1 m and 1.35 m, respectively, from the facing of the front MSE wall. A short reinforced concrete abutment wall and two wing walls, resting on the spread foundation, confine the reinforced backfill soil behind the bridge abutment and support the bridge approach slab. The bridge is supported by central pier columns along the middle of the structure, which in turn are supported by a spread footing founded on bedrock at the median of U.S. Interstate 25.

When compared to typical systems involving the use of deep foundations to support bridge structures, the use of geosynthetic-reinforced systems to support both the bridge and the approaching roadway structures has the potential to alleviate the "bump at the bridge" problem caused by differential settlements between the bridge abutment and approaching roadway. In addition, this approach also allows for construction in stages and comparatively smaller construction working areas. Several of the common causes for development of bridge bumps were addressed in the design of the Founders/Meadows structure. The main cause of uneven settlements in typical systems is the use of different foundation types. That is, while the approaching roadway structure is typically constructed on compacted backfill soil (reinforced or not), the bridge abutment is typically supported on stronger soils by deep foundations. The roadway approach embankment and the bridge footing were integrated at the Founders/Meadows structure with an extended reinforced soil zone in order to minimize uneven settlements between the bridge abutment and approaching roadway. A second cause of differential settlements can be attributed to erosion of the fill material around the abutment wall induced by surface water runoff. Several measures were implemented in this project to prevent that surface water, as well as groundwater, reach the reinforced soil mass and the bedrock at the base of the fill (e.g. placement of impervious membranes with collector pipes). Finally, a third potential cause of differential settlements is the thermally induced movements, i.e., expansion and contraction of bridge girders strongly attached to the abutment wall (integral abutment). A compressible 75 mm low-density expanded polystyrene sheet was placed between the reinforced backfill and the abutment walls. It was expected that this system would accommodate the thermally induced movements of the bridge superstructure without affecting the retained backfill.

The backfill soil used in this project includes fractions of gravel (35%), sand (54.4%), and fine-grained soil (10.6%). The liquid limit and plasticity index for the fine fraction of the backfill are 25% and 4%, respectively. The backfill soil classifies as SW-SM per ASTM 2487, and as A-1-B (0) per AASHTO M 145. The backfill met the construction requirements for CDOT Class 1 backfill. A friction angle of 34° and zero cohesion were assumed in the design of the GRS walls. To evaluate the suitability of these design parameters, conventional direct shear tests and large size direct shear and triaxial tests were conducted. In the conventional tests, the 35% gravel portion was removed from the specimens, but in the large-size triaxial and direct shear tests, the backfill soil specimens included the gravel portion. The results of conventional direct shear tests and large size direct shear and triaxial tests indicate that assuming zero cohesion in the design procedure and removing the gravel portion from the test specimens lead to significant underestimation of the actual shear strength of the backfill.

The geogrid reinforcements used in this project were manufactured by the Tensar Corporation. Three types of geogrid reinforcements were used: UX 6 below the footing, and UX 3 and UX 2 behind the abutment wall. The long-term-design-strength (LTDS) of these reinforcements is 27 kN/m, 11 kN/m, and 6.8 kN/m, respectively. CDOT specifications imposed a global reduction factor of 5.82 to determine the long-term design strength (LTDS) of the geogrid reinforcements from their ultimate strength. This global reduction factor accounts for reinforcement tensile strength losses over the design life period due to creep, durability, and installation damage. It also includes a factor of safety to account for uncertainties.

3.4 *Performance*

The instrumentation program was conducted in two phases (Phases I and II), which correspond, respectively to the construction of two phases of the GRS bridge abutment structure. A pilot instrumentation plan was conducted during construction of the Phase I structure in order to obtain information that will tailor the design of a more comprehensive monitoring program to be implemented during Phase II. The Phase I instrumentation program included survey targets, pressure cells, jointmeters, and inclinometer. The more comprehensive Phase II instrumentation program included monitoring using survey targets, digital road profiler, pressure cells, strain gauges, moisture gauges, and temperature gauges. A view of the instrumentation plan for Phase II is shown in Figure 10. The figure shows the four critical locations that were instrumented in Phase II:

(i) Location A, close to the facing. Data collected at this location is particularly useful for guiding the structural design of the facing and of the connection between facing and reinforcements.
(ii) Locations B and C along the center and interior edge of the abutment foundation. Information collected at these locations is relevant for the design of the reinforcement elements.
(iii) Location D, behind the bridge foundation, and horizontal plane at the base of the fill. Data measured at these locations is useful to estimate the external forces acting behind and below the reinforced soil mass.

A comprehensive discussion of the instrumentation results, the collection and analysis of which is under progress, is beyond the scope of this paper. Results of

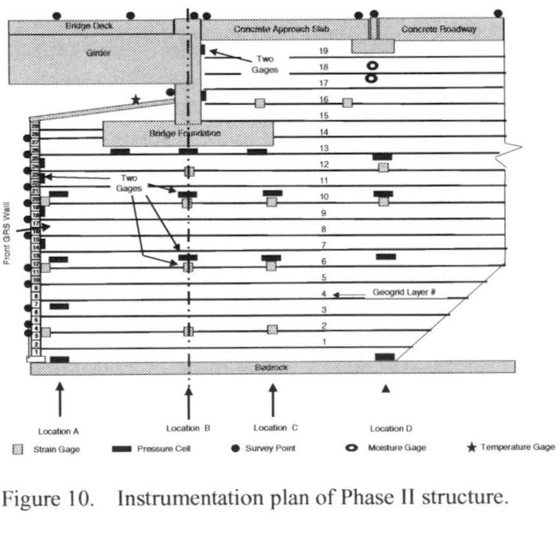

Figure 10. Instrumentation plan of Phase II structure.

the preliminary Phase I instrumentation program have been reported by Abu-Hejleh et al. (2000). Some of the relevant findings obtained based on the information collected so far are the following:

- The measured response from both the pressure cells and strain gauges correlates well with the applied loads during the construction stages.
- The maximum geogrid strains experienced during construction are comparatively very small (approximately 0.45%).
- Horizontal earth pressures collected at the facing and of the reinforcement maximum tensile strains are well below design values.
- Most of the straining of the geogrid reinforcements occurred during construction of the wall and not during placement of the bridge surcharge load. This can be explained by the effect of compaction operations and presence of slacks in the geogrid reinforcements. Strain gauge monitoring results collected so far suggest that approximately 50% of the total recorded strains occurred during placement and compaction of a few lifts of soil above the geogrid layers (e.g. approximately 2 m of soil or 40 kPa). The maximum measured front wall outward displacement induced by wall construction (before placement of the bridge superstructure) was 12 mm, which corresponds to 0.20% of the wall height.
- The maximum outward displacement induced by placement of the bridge superstructure was additional 10 mm, which corresponds to 0.17% of the wall height. The maximum settlement of the bridge footing due to placement of the bridge superstructure was 13 mm.
- The maximum outward displacements induced after opening the structure to traffic and until June 2000 (18 months) was 13 mm. These movements correspond to 0.22% of the wall height. The measured settlement of the leveling pad supporting the front wall facing was approximately 5 mm. However, it is important to emphasize that these movements took place only during the initial 12 months of service (until January 2000). Lateral and vertical movements have been negligible from January to June 2000.
- Elevation profiling and surveying results show no signs of development of the "bump at the bridge" problem.

Overall, the performance of the Founders/Meadows bridge structure, based on the monitored behavior recorded so far, showed excellent short- and long-term performance. Specifically, the monitored movements were significantly smaller than those expected in design or allowed by performance requirements, there were no signs for development of the "bump at the bridge" problem or any structural damage, and post-construction movements became negligible after an in-service period of 1 year.

4 ADVANCES IN FIBER-REINFORCED SOIL DESIGN

4.1 *Overview*

Fiber reinforcement has become a promising solution to the stabilization of thin soil veneers and localized repair of failed slopes. Randomly distributed fibers can maintain strength isotropy and avoid the existence of the potential planes of weakness that can develop parallel to continuous planar reinforcement elements. The design of fiber-reinforced soil slopes has typically been performed using composite approaches, where the fiber-reinforced soil is considered a single homogenized material. Accordingly, fiber-reinforced soil design has required non-conventional laboratory testing of composite fiber-reinforced soil specimens which has discouraged implementation of fiber-reinforcement in engineering practice.

Several composite models have been proposed to explain the behavior of randomly distributed fibers within a soil mass (Maher and Gray, 1990, Michalowski and Zhao, 1996, Ranjan et al., 1996). The mechanistic models proposed by Gray and Ohashi (1983) and Maher and Gray (1990) quantify the "equivalent shear strength" of the fiber-reinforced composite as a function of the thickness of the shear band that develops during failure. Information needed to characterize shear band development for these models is, however, difficult to quantify (Shewbridge and Sitar, 1990). Common findings from the various testing programs implemented to investigate composite models include: (i) randomly distributed fibers provide strength isotropy in a soil composite; (ii) fiber inclusions increase the "equivalent" shear strength within a reinforced soil mass; and (iii) the "equivalent" strength typically shows a bilinear behavior, which was experimentally observed by testing of comparatively weak fibers under a wide range of confining stresses.

A discrete approach for the design of fiber-reinforced soil slopes was recently proposed to characterize the contribution of randomly distributed fibers to stability (Zornberg, 2002). In this approach, fiber-reinforced soil is characterized as a two-component (soil and fibers) material. Fibers are treated as discrete elements that contribute to stability by mobilizing tensile stresses along the shear plane. Consequently, independent testing of soil specimens and of fiber specimens, but not of fiber-reinforced soil specimens, can be used to characterize fiber-reinforced soil performance.

This paper initially reviews the main concepts of the discrete approach and subsequently validates the framework for design purposes.

4.2 *Discrete frame work for fiber reinforcement*

The volumetric fiber content, χ, used in the proposed discrete framework is defined as:

$$\chi = \frac{V_f}{V} \tag{2}$$

where V_f is the volume of fibers and V is the control volume of fiber-reinforced soil.

The gravimetric fiber content, χ_w, typically used in construction specifications, is defined as:

$$\chi_w = \frac{W_f}{W_s} \tag{3}$$

where W_f is the weight of fibers and W_s is the dry weight of soil.

The dry unit weight of the fiber-reinforced soil composite, γ_d, is defined as:

$$\gamma_d = \frac{W_f + W_s}{V} \tag{4}$$

The contribution of fibers to stability leads to an increased shear strength of the "homogenized" composite reinforced mass. However, the reinforcing fibers actually work in tension and not in shear. A major objective of the discrete framework is to explicitly quantify the fiber-induced distributed tension, t, which is the tensile force per unit area induced in a soil mass by randomly distributed fibers.

Specifically, the magnitude of the fiber-induced distributed tension is defined as a function of properties of the individual fibers. In this way, as in analysis involving planar reinforcements, limit equilibrium analysis of fiber-reinforced soil can explicitly account for tensile forces.

The interface shear strength of individual fibers can be expressed as:

$$f_f = c_{i,c} \cdot c + c_{i,\phi} \cdot \tan\phi \cdot \sigma_{n,ave} \tag{5}$$

where c and ϕ are the cohesive and frictional components of the soil shear strength and $\sigma_{n,ave}$ is the average normal stress acting on the fibers. The interaction coefficients, $c_{i,c}$ and $c_{i,\phi}$, commonly used in soil reinforcement literature for continuous planar reinforcement, is adopted herein to relate the interface shear strength to the shear strength of the soil. The interaction coefficients are defined as:

$$c_{i,c} = \frac{a}{c} \tag{6}$$

$$c_{i,\phi} = \frac{\tan\delta}{\tan\phi} \tag{7}$$

where a is the adhesive component of the interface shear strength between soil and the polymeric fiber, tan δ is the skin-frictional component.

The pullout resistance of a fiber of length l_f should be estimated over the shortest side of the two portions of a fiber intercepted by the failure plane. The length of the shortest portion of a fiber intercepted by the failure plane varies from zero to l_f /2. Statistically, the average embedment length of randomly distributed fibers, $l_{e,ave}$, can be analytically defined by:

$$l_{e,ave} = \frac{l_f}{4} \tag{8}$$

where l_f is total length of the fibers.

The average pullout resistance can be quantified along the average embedment length, $l_{e,ave}$, of all individual fibers crossing a soil control surface A. The ratio between the total cross sectional area of the fibers A_f and the control surface A is assumed to be defined by the volumetric fiber content χ. That is:

$$\chi = \frac{A_f}{A} \tag{9}$$

When failure is governed by the pullout of the fibers, the fiber-induced distributed tension, t_p, is defined as the average of the tensile forces inside the fibers over the control area A. Consequently, t_p can be estimated as:

$$t_p = \chi \cdot \eta \cdot \left(c_{i,c} \cdot c + c_{i,\phi} \cdot \tan\phi \cdot \sigma_{n,ave}\right) \tag{10}$$

where η is the aspect ratio defined as:

$$\eta = \frac{l_f}{d_f} \tag{11}$$

where d_f is the equivalent diameter of the fiber.

When failure is governed by the yielding of the fibers, the distributed tension, t_t, is determined from the tensile strength of the fiber:

$$t_t = \chi \cdot \sigma_{f,ult} \tag{12}$$

where $\sigma_{f,ult}$ is the ultimate tensile strength of the individual fibers.

The fiber-induced distributed tension t to be used in the discrete approach to account for the tensile contribution of the fibers in limit equilibrium analysis is:

$$t = \min\left(t_p, t_t\right) \tag{13}$$

The critical normal stress, $\sigma_{n,crit}$, which defines the change in the governing failure mode, is the normal stress at which failure occurs simultaneously by pullout and tensile breakage of the fibers. That is, the following condition holds at the critical normal stress:

$$t_t = t_p \tag{14}$$

An analytical expression for the critical normal stress can be obtained as follows:

$$\sigma_{n,crit} = \frac{\sigma_{f,ult} - \eta \cdot c_{i,c} \cdot c}{\eta \cdot c_{i,\phi} \cdot \tan\phi} \tag{15}$$

As in analyses involving planar inclusions, the orientation of the fiber-induced distributed tension should also be identified or assumed. Specifically, the fiber-induced distributed tension can be assumed to act: a) along the failure surface so that the discrete fiber-induced tensile contribution can be directly "added" to the shear strength contribution of the soil in a limit equilibrium analysis; b) horizontally, which would be consistent with design assumptions for reinforced soil structures using planar reinforcements; and c) in a direction somewhere between the initial fiber orientation (which is random) and the orientation of the failure plane.

This equivalent shear strength of fiber-reinforced specimens can be defined as a function of the fiber-induced distributed tension t, and the shear strength of the unreinforced soil, S:

$$S_{eq} = S + \alpha \cdot t = c + \sigma_n \tan\phi + \alpha \cdot t \tag{16}$$

where α is an empirical coefficient that accounts for the orientation of fiber and the efficiency of the mixing of fibers. α is equal to 1, if the fibers are randomly distributed and working with 100% efficiency, otherwise α should be smaller than 1.

Depending on whether the mode of failure is fiber pullout or yielding, the equivalent shear strength can be derived by combining (9) or (11) with (15). It should be noted that the average normal stress acting on the fibers, $\sigma_{n,ave}$, does not necessarily equal the normal stress on the shear plane σ_n . For randomly distributed fibers, $\sigma_{n,ave}$ could be represented by the octahedral stress component. However, a sensitivity evaluation undertaken using typical ranges of shear strength parameters show that $\sigma_{n,ave}$ can be approximated by σ_n without introducing significant error.

Accordingly, the following expressions can be used to define the equivalent shear strength when failure is governed by fiber pullout:

$$S_{eq,p} = c_{eq,p} + (\tan\phi)_{eq,p} \cdot \sigma_n \tag{17}$$

$$c_{eq,p} = \left(1 + \alpha \cdot \eta \cdot \chi \cdot c_{i,c}\right) \cdot c \tag{18}$$

$$(\tan\phi)_{eq,p} = \left(1 + \alpha \cdot \eta \cdot \chi \cdot c_{i,\phi}\right) \cdot \tan\phi \tag{19}$$

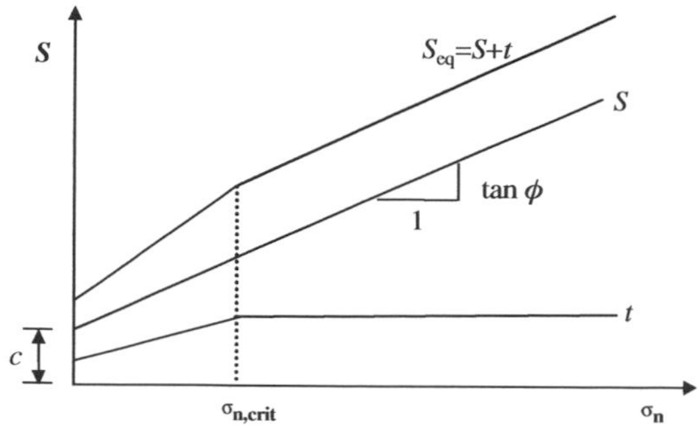

Figure 11. Representation of the equivalent shear strength according to the discrete approach.

Table 2. Summary of soil properties.

Soil type	Soil 1	Soil 2
USCS classification	SP	CL
LL %	–	49
PL %	–	24
IP %	–	25
% Fines	1.4	82.6

Equivalently, the following expressions can be obtained to define the equivalent shear strength when failure is governed by tensile breakage of the fibers:

$$S_{eq,t} = c_{eq,t} + (\tan\phi)_{eq,t} \cdot \sigma_n \quad (20)$$

$$c_{eq,t} = c + \alpha \cdot \chi \cdot \sigma_{f,ult} \quad (21)$$

$$(\tan\phi)_{eq,t} = \tan\phi \quad (22)$$

The above expressions yield a bilinear shear strength envelope, which is shown in Figure 11.

4.3 *Experimental validation*

A triaxial compression testing program on fiber-reinforced soil was implemented to validate the proposed discrete framework. Both cohesive and granular soils were used in the testing program, and the soil properties were summarized in Table 2.

The tests were conducted using commercially available polypropylene fibers, and the properties of fibers were summarized in Table 3. A series of tensile test were performed in general accordance with ASTM D2256-97 to evaluate the ultimate tensile strength of fibers. The average tensile strength of the fibers was approximately 425,000 kPa.

The triaxial testing program involved consolidated drained (CD) tests for SP soils and consolidated undrained (CU) tests for CL soils. The specimens have a diameter of 71 mm and a minimum length-to-diameter ratio of 2. The CU tests were performed in general accordance with ASTM D4767, and the specimens were back pressure saturated and the pore water pressure was measured. The unreinforced tests of SP soil yielded an effective shear strength envelope defined by cohesion of 6.1 kPa and friction angle of 34.3°, while the cohesion and friction angle of CL soil were 12.0 kPa and 31.0° respectively.

Table 3. Summary of fiber properties.

	SP tests	CL tests
Linear density (denier)	1000& 360	2610
Fiber content (%)	0.2 & 0.4	0.2 & 0.4
Length of fibers (mm)	25 & 51	25 & 51
Type of fiber	fibrillated & tape	fibrillated

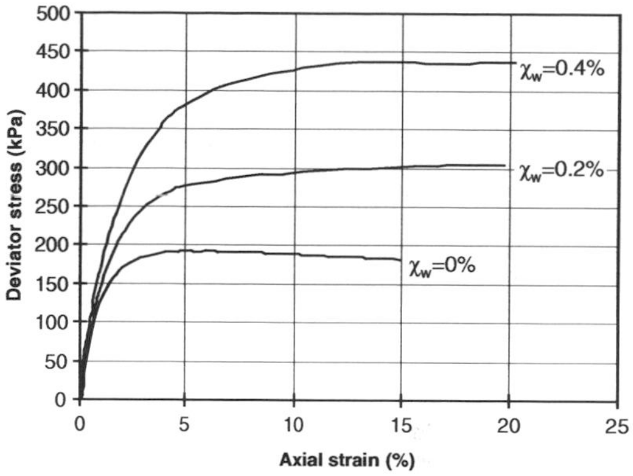

Figure 12. Stress-strain behavior of specimens prepared using $\chi_w = 0$, 0.2 and 0.4% with lf = 25 mm fibers (360 denier), $\sigma_3 = 70$ kPa, Soil 1.

The governing failure mode for the polymeric fibers used in this investigation is pullout because of the comparatively high tensile strength and short length of the fibers. Accordingly, the triaxial testing program conducted in this study focuses only on the first portion of the bilinear strength envelope shown in Figure 11.

Figure 12 shows the stress-strain behavior of SP soil specimens reinforced with 360 denier fibers, and placed at gravimetric fiber contents of 0, 0.2 and 0.4%. Specimens were tested under confining pressure of 70 kPa. The peak deviator stress increases approximately linearly with increasing fiber content, which is consistent with the discrete framework (see Equations 17–19). The post-peak shear strength loss is smaller in the reinforced specimens than in the unreinforced specimens. However, the initial portions of the stress-strain curves of the reinforced and unreinforced specimens are approximately similar. Accordingly, the

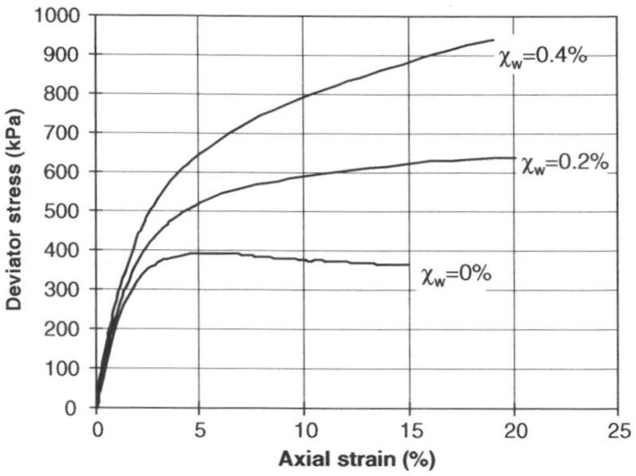

Figure 13. Stress-strain behavior of specimens prepared using $\chi_w = 0$, 0.2 and 0.4% with lf = 25 mm fibers (360 denier), $\sigma_3 = 140$ kPa, Soil 1.

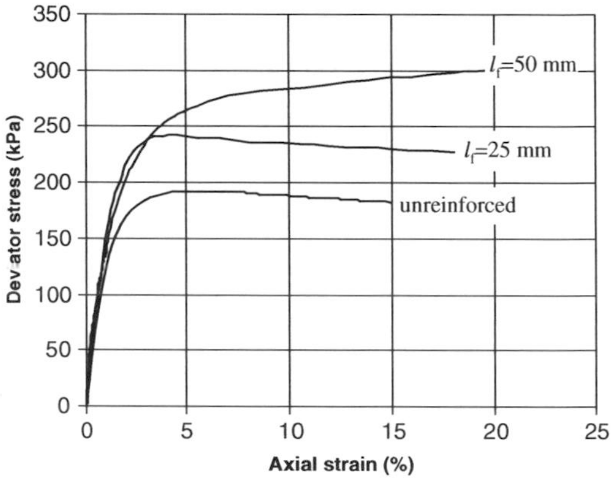

Figure 14. Stress-strain behavior of specimen prepared using $\chi_w = 0.2$ %, with $l_f = 25$ mm and 50 mm fibers (1000 denier), $\sigma_3 = 70$ kPa, Soil 1.

soil appears to take most of the applied load at small strain levels, while the load resisted by the fibers is more substantial at higher strain level. The larger strain corresponding to the peak deviator stress displayed by the fiber-reinforced specimens suggests that fibers increase the ductility of the reinforced soil specimen. These findings are confirmed in Figure 13, which shows the test results obtained under higher confining stress (140 kPa).

The effect of fiber length on the stress-strain behavior is shown in Figure 14. The specimens were prepared using fibers with a different fiber type (1000 denier) than that used in the tests shown in Figures 12 and 13. The specimens were prepared using the same gravimetric fiber content, but with varying fiber length. The specimens reinforced with longer (50 mm) fibers displayed higher shear strength. The peak deviator stress increases linearly with increasing aspect ratio,

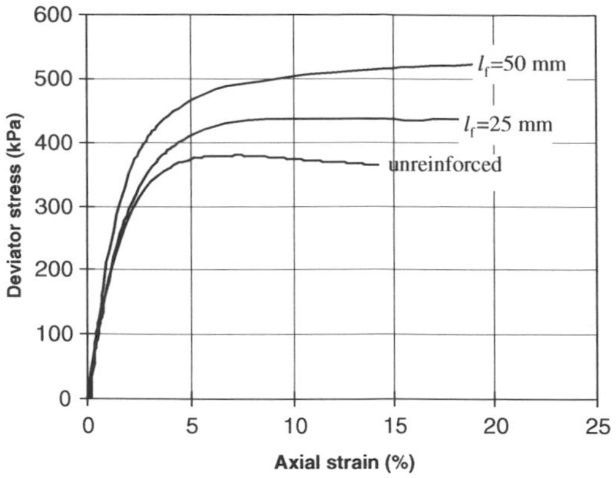

Figure 15. Stress-strain behavior of specimen prepared using $\chi_w = 0.2$ %, with $l_f = 25$ mm and 50 mm fibers (1000 denier), $\sigma_3 = 140$ kPa, Soil 1.

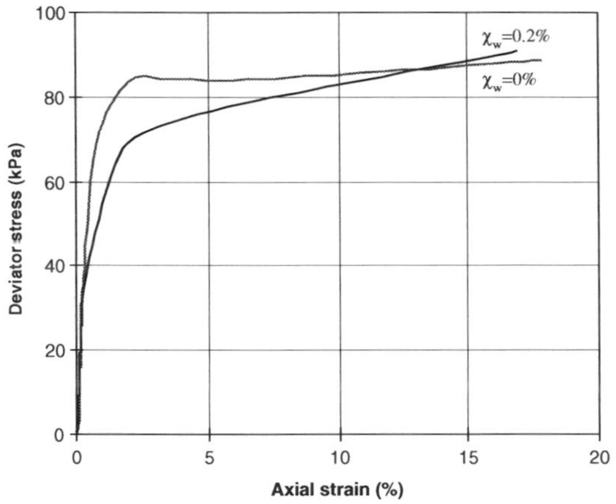

Figure 16. Stress-strain behavior of specimens prepared using $\chi_w = 0$, 0.2%, with $l_f = 50$ mm fibers (2610 denier), $\sigma_3 = 98$ kPa, Soil 2.

which is also consistent with the trend indicated by Equations 17–19. The strain corresponding to the peak strength increases with increasing fiber length. When the governing failure mode is pullout, the fiber-induced distributed tension reaches its peak when the pullout resistance is fully mobilized. For longer fibers, it usually requires a larger interface shear deformation to fully mobilize the interface strength. Consequently, the macroscopic axial strain at peak stress should be larger for specimen reinforced with longer fibers. Figure 15 shows a similar trend for the case of tests conducted under higher confining pressures.

Figure 16 compares the stress-strain behavior of both unreinforced and fiber-reinforced specimen using soil 2. The reinforced specimen were prepared at $\chi_w = 0.2\%$, using 2-inch long 2610 denier fibers.

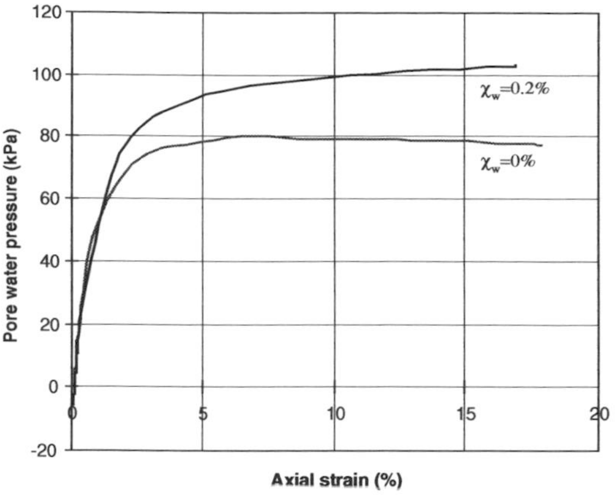

Figure 17. Excess pore water pressure of specimens prepared using $\chi_w = 0$, 0.2%, with $l_f = 50$ mm fibers (2610 denier), $\sigma_3 = 98$ kPa, Soil 2.

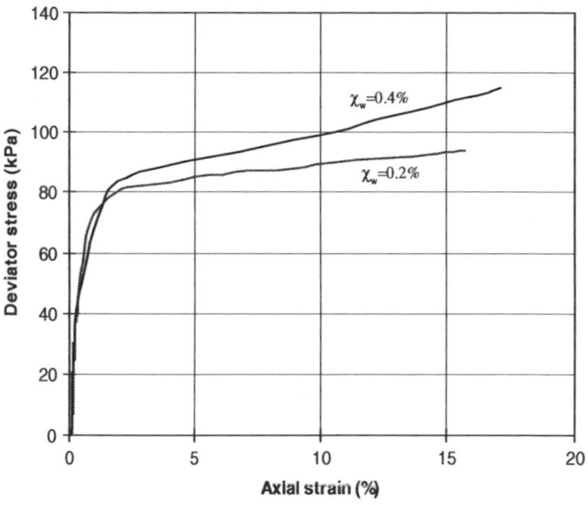

Figure 18. Stress-strain behavior of specimens prepared using $\chi_w = 0.2$, 0.4%, with $l_f = 25$ mm fibers (2610 denier), $\sigma_3 = 116$ kPa, Soil 2.

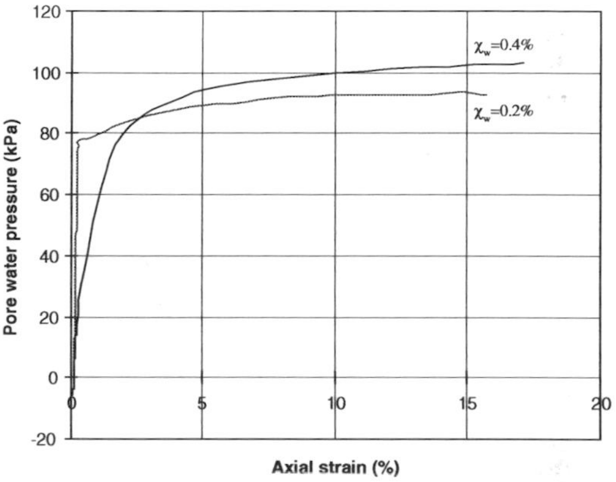

Figure 19. Excess pore water pressure of specimens prepared using $\chi_w = 0.2$, 0.4%, with $l_f = 25$ mm fibers (2610 denier), $\sigma_3 = 116$kPa, Soil 2.

Both specimens were compacted at optimum moisture content to 90% of the maximum dry density achieved in the standard Proctor test as specified in ASTM D 698, and tested under confining pressure $\sigma_3 = 98$ kPa. Due to the undrained test condition, the effective confining stress changes with the excess pore water pressure induced in the process of shearing. The peak shear strength was selected in terms of the maximum value of (σ_1'/σ_3'). The increment of deviator stress due to fiber addition is not as obvious as in the case of SP sand. However, the pore water pressure generated during shearing is larger for reinforced specimen than for unreinforced specimen (see Figure 17). Consequently the effective confining stress inside the reinforced specimen is smaller than that inside the unreinforced specimen. The fiber-reinforced specimen achieved equal or higher peak deviator stress than the unreinforced specimen under a lower effective confining stress. This shows that the addition of fibers increases the shear strength of the reinforced specimen. Since positive water pressure is associated with the tendency of volume shrinkage, this observation shows that fiber reinforcement restrains the dilatancy of the reinforced soil. Other researchers (Michalowski and Cermak, 2003, Consoli et al., 1998) reported that fiber-reinforced specimens displayed smaller volume dilatation than unreinforced specimen in consolidated drained (CD) test. This observation confirms their findings in a different test condition.

Similar observation can be made from Figures 18 and 19, which shows the stress-strain behavior and the pore water pressure evolution obtained using 25-mm fibers placed at 0.2% and 0.4% gravimetric fiber contents. As the fiber content increases, the pore water pressure generated during undrained shearing also increases.

Equations 17–19 were used to predict the equivalent shear strength for fiber-reinforced specimens. Interaction coefficients ($c_{i,c}$ and $c_{i,\phi}$) of 0.8 are assumed in the analyses conducted in this study. The interface shear strength obtained from pullout test results conducted on woven geotextiles was considered representative of the interface shear strength on individual fibers. For practical purposes, interaction coefficients can be selected from values reported in the literature for continuous planar reinforcements. This is because pullout tests conducted using a variety of soils and planar geosynthetics have been reported to render interaction coefficient values falling within a narrow range (Koutsourais et al., 1998, Michalowski and Cermak, 2003). α is assumed to be 1.0 for randomly distributed

Table 4. Summary of parameters used in the prediction.

	α	$\phi(°)$	c (kpa)	$c_{i,c}$	$c_{i,\phi}$
Soil 1	1.0	34.3	6.1	0.8	0.8
Soil 2	1.0	31.0	12.0	0.8	0.8

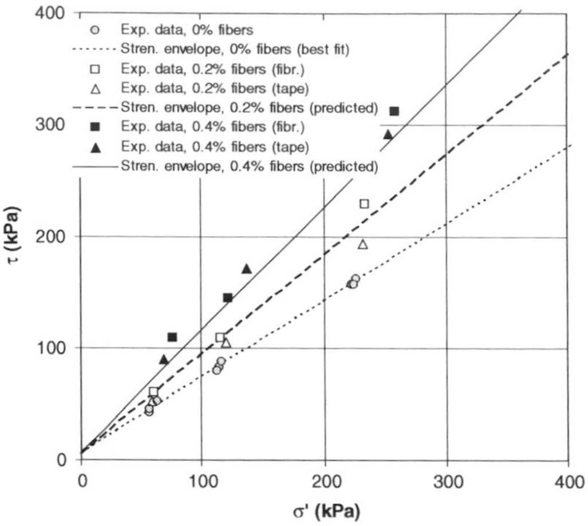

Figure 20. Comparison between predicted and experimental shear strength results for specimens reinforced at $\chi_w = 0$, 0.2%, 0.4% with 25 mm-long fibers (360 denier), Soil 1.

fibers. Table 4 summarized the values of parameters used in the analyses.

The effect of fiber content on shear strength is shown in Figure 20, which compares the experimental data and predicted shear strength envelopes obtained from Soil 1 using 25 mm fibers with linear density of 360 denier placed at fiber contents of 0.0%, 0.2%, and 0.4%. The experimental results show a clear increase in equivalent shear strength with increasing fiber content. No major influence of fibrillation is perceived in the results of the testing program. The shear strength envelope for the unreinforced specimens was defined by fitting the experimental data. However, the shear strength envelopes shown in the figure for the reinforced specimens were predicted analytically using the proposed discrete framework. A very good agreement is observed between experimental data points and predicted shear strength envelopes. As predicted by the discrete framework, the distributed fiber-induced tension increases linearly with the volumetric fiber content. Similar observation can be made in Figure 21, which shows the results obtained from Soil 2 using 50-mm-long fibers with linear density of 2610 denier.

The effect of fiber aspect ratio on shear strength is shown in Figure 22, which compares the experimental and predicted shear strength envelopes of specimens

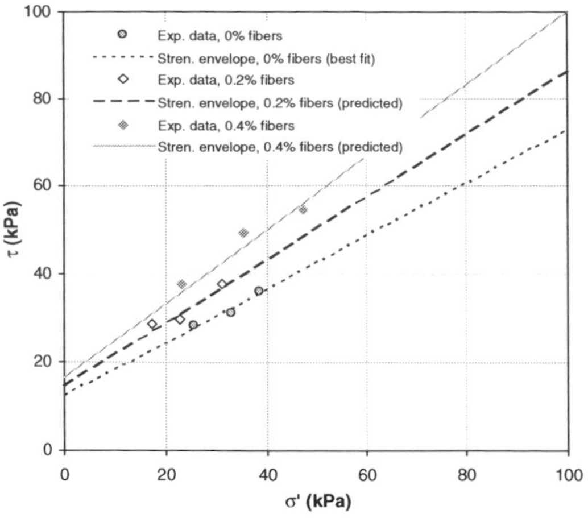

Figure 21. Comparison between predicted and experimental shear strength results for specimens reinforced at $\chi_w = 0$, 0.2%, 0.4% with 50 mm-long fibers (2610 denier), Soil 2.

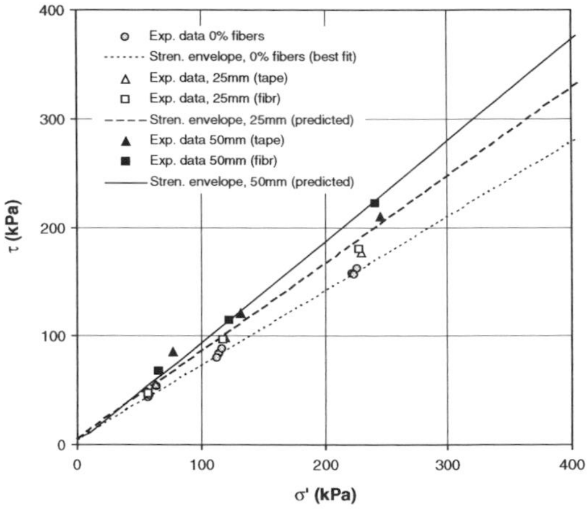

Figure 22. Comparison between predicted and experimental shear strength results for specimens reinforced at $\chi_w = 0.2\%$, with 25 mm-long and 50 mm-long fibers (1000 denier), Soil 1.

of Soil 1 placed at $\chi_w = 0.2\%$, with 25 and 50 mm-long fibers. As predicted by the discrete framework, increasing the fiber length increases the pullout resistance of individual fibers, and results in a higher fiber-induced distributed tension. Consequently, for the same fiber content, specimens reinforced with longer fibers will have higher equivalent shear strength. This trend agrees well with the experimental data. Similar observation can be made from Figure 23, which shows the results obtained from Soil 2 using 25 mm and 50 mm-long fibers and placed at $\chi_w = 0.4\%$.

Additional insight into the validity of the proposed discrete approach can be obtained by comparing the

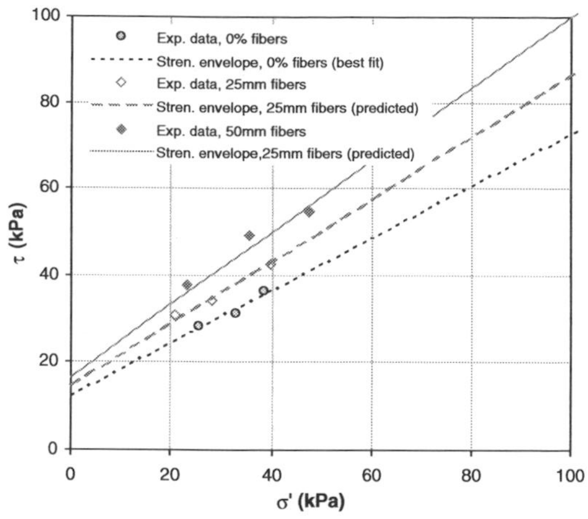

Figure 23. Comparison between predicted and experimental shear strength results for specimens reinforced at $\chi_w = 0.4\%$, with 25 mm-long and 50 mm-long fibers (2610 denier), Soil 2.

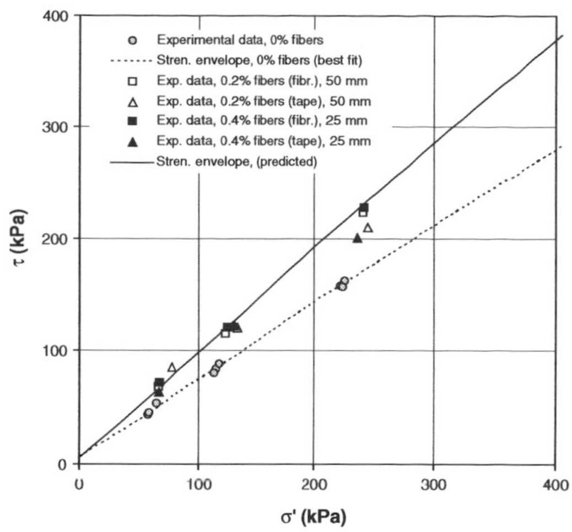

Figure 24. Consolidated shear strength results for specimen reinforced with 50 mm-long fibers (1000 denier) placed at $\chi_w = 0.2\%$ and 25 mm fibers placed at $\chi_w = 0.4\%$, Soil 1.

results obtained for specimens reinforced with 50 mm-long fibers placed at a fiber content of 0.2% with those obtained for specimens reinforced with 25 mm-long fibers placed at a fiber content of 0.4%. That is specimens with a constant value of $(\chi_w \cdot \eta)$. As inferred from inspection of Equations 17–19, the fiber-induced distributed tension is directly proportional to both the fiber content and the fiber aspect ratio. Consequently, the predicted equivalent shear strength parameters for the above combinations of fiber length and fiber content are the same. Figures 24 and 25 combine these experimental results.

The good agreement between experimental results and predicted values provides additional evidence of the suitability of the proposed discrete approach. From the practical standpoint, it should be noted that using 50 mm-long fibers placed at a fiber content of 0.2% corresponds to half the reinforcement material than using 25 mm-long fibers placed at a fiber content of 0.4%. That is, for the same target equivalent shear strength the first combination leads to half the material costs than the second one. It is anticipated, though, that difficulty in achieving good fiber mixing may compromise the validity of the relationships developed herein for comparatively high aspect ratios (i.e. comparatively long fibers) and for comparatively high fiber contents. The fiber content or fiber length at which the validity of these relationships is compromised should be further evaluated. Nonetheless, good mixing was achieved for the fiber contents and fiber lengths considered in this investigation, which were selected based on values typically used in geotechnical projects.

Figure 26 shows the stress-strain behavior of specimen reinforced with 50 mm fibers placed at $\chi_w = 0.2\%$

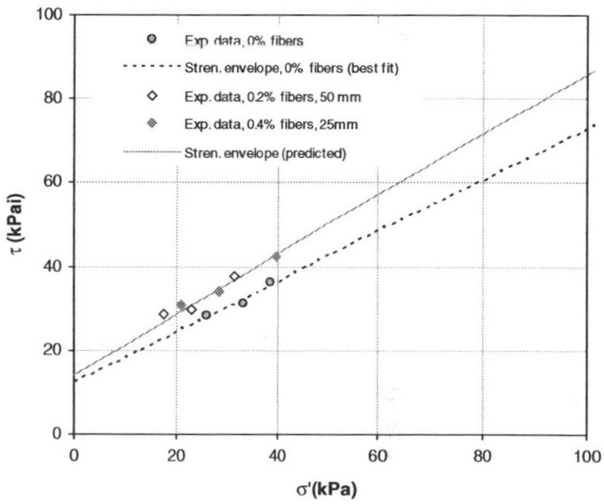

Figure 25. Consolidated shear strength results for specimen reinforced with 50 mm-long fibers (2610 denier) placed at $\chi_w = 0.2\%$ and 25 mm fibers placed at $\chi_w = 0.4\%$, Soil 2.

and 25 mm fibers placed at $\chi_w = 0.4\%$. While the discrete approach was developed only to predict the shear strength response, the results in the figure show that fiber-reinforced specimens prepared using a constant value ??display similar stress-strain behavior. This similar response is observed for both fibrillated and tape fibers, suggesting that the fibrillation procedure does not have a significant impact on the mechanical response of fiber-reinforced soil. The experimental results suggest that the proportionality of shear strength with the fiber content and fiber aspect ratio predicted by the discrete framework can

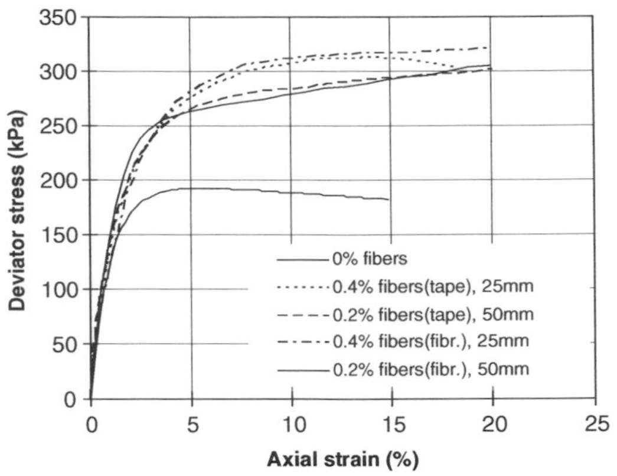

Figure 26. Comparison between stress-strain behavior for specimen reinforced with 50 mm fibers (1000 denier) placed at $\chi_w = 0.2\%$ and 25 mm fibers placed at $\chi_w = 0.4\%$, $\sigma_3 = 70$ kPa.

be extrapolated to the entire stress-strain response of fiber-reinforced specimens.

4.4 *Remarks on fiber reinforced soil*

The discrete approach for fiber-reinforced soil was validated in this investigation using experimental data from a triaxial testing of both sand and clay. The effect of fiber reinforcement on stress-strain behavior and shear strength was investigated and compared with the analytical results of the discrete approach. The main conclusions drawn from this investigation are:

- The addition of fibers can significantly increase the peak shear strength and limit the post peak strength loss of both cohesive and granular soil. An increase in fiber content leads to increasing strain at failure and, consequently, to a more ductile behavior.
- The fiber reinforcement tends to restrain the volume dilation of the soil in drained condition, or equivalently, increase the positive water pressure in undrained condition.
- The peak shear strength increases with increasing aspect ratio. The strain at peak deviator stress increases with increasing fiber aspect ratio.
- As predicted by the discrete framework, the experimental results confirmed that the fiber-induced distributed tension increases linearly with fiber content and fiber aspect ratio when failure is characterized by pullout of individual fibers.
- Experimental results conducted using specimens with a constant $(\chi_w \cdot \eta)$ value show not only the same shear strength but also display a similar stress-strain behavior.
- If good mixing can be achieved, fibers with comparatively high aspect ratio can lead to lower fiber contents while reaching the same target equivalent shear strength, resulting in savings of reinforcement material.
- Overall, for both sand and clay specimens, the discrete approach was shown to predict accurately the shear strength obtained experimentally using specimens reinforced with polymeric fibers tested under confining stresses typical of slope stabilization projects.

ACKNOWLEDGEMENTS

The author is indebt to Dr. Abu-Hejleh for his contributions to the geosynthetic-reinforced bridge abutment study and to Dr. Chunling Li, from GeoSyntec Consultants, for his contribution to the fiber-reinforced portion of this study.

REFERENCES

AASHTO, 1996, *Standard specifications for highway bridges*. American Association of State Highway and Transportation Officials, Washington, D.C., USA.

Abu-Hejleh, N. Wang, T., and Zornberg, J. G. (2000). "Performance of geosynthetic-reinforced walls supporting bridge and approaching roadway structures." *Advances in Transportation and Geoenvironmental Systems using Geosynthetics*, ASCE Geotechnical Special Publication No. 103, Zornberg, J.G. and Christopher, B.R. (Eds.), pp. 218–243.

Adams, M. (1997). "Performance of a prestrained geosynthetic reinforced soil bridge pier." *Int. Symp. On Mechanically Stabilized Backfill,* T.H. Wu, editor, A. A. Balkema, Denver, USA, 35–53.

Adams, M. (2000). "Reinforced soil technology at FHWA – Making old technology new." *Geotechnical Fabrics Report*, August 2000, pp. 34–37.

Bolton, M., 1986, "The Strength and Dilatancy of Sands." *Géotechnique*, Vol. 36, No. 1, pp. 65–78.

Bonaparte, R. and Schmertmann, G.R., 1987, "Reinforcement Extensibility in Reinforced Soil Wall Design." *The Application of Polymeric Reinforcement in Soil Retaining Structures*, pp. 409–457.

Brau, G. and Floss, R. (2000). "Geotextile structures used for the reconstruction of the motorway Munich-Salzburg." *Proceedings of Second European Geosynthetics Conference*, Bologna, Italy, Vol. 1, pp. 373–377.

British Standards Institution, 1995, BS 8006:1995, *Code of Practice for Strengthened/Reinforced Soil and Other Fills*, P.162, BSI, London.

Canadian Geotechnical Society, 1992, *Canadian Foundation Engineering Manual*. 3rd Edition.

Christopher, B., Bonczkiewicz, C., and Holtz, R., 1992, "Design, Construction and Monitoring of Full Scale Test of Reinforced Soil Walls and Slopes." *Recent Case Histories of Permanent Geosynthetic-Reinforced Soil Retaining Walls*, Tokyo, Japan: A.A. Balkema, pp. 45–60.

Consoli N.C., Prietto P.D.M., Ulbrich L.A. 1998, "Influence of Fiber and Cement Addition on Behavior of Sandy Soil", ASCE J. of Geotech. and Geoenviron. Engrg., 124(12): pp. 1211–1214.

EBGEO, 1997, *Empfehlungen fur Bewehrungen aus Geokunststoffen*, Ernst & Sohn Verlag.

Elias, V., Christopher, B.R., Berg, R.R., 2001, *Mechanically Stabilized Earth Walls and Reinforced Soil Slopes*. Publication Number FHWA NH-00-043, March 2001, NHI-FHWA.

GCO, 1989, *Model Specification for Reinforced Fill Structures*, Geospec 2, Geotechnical Engineering Office, Hong Kong.

GEO, 1993, *A Partial Factor Method for Reinforced Fill Slope Design*, GEO Report No. 34, Geotechnical Engineering Office, Hong Kong.

GeoRio, 1999, *Technical Manual for Slope Stabilization*. (in Portuguese). Foundation for Slope Stability Control in the City of Rio de Janeiro, Volumes 1–4, Rio de Janeiro, Brazil, 682 p.

Gotteland, Ph., Gourc, J.P., and Villard, P. (1997). "Geosynthetic reinforced structures as bridge abutment: Full Scale experimentation and comparison with modelisations." *Int. Symp. On Mechanically Stabilized Backfill,* T.H. Wu, editor, A. A. Balkema, Denver, USA, 25–34.

Gray, D.H. and Ohashi, H. (1983). "Mechanics of Fiber-reinforcement in Sand", ASCE J. Geotech. Engrg. 109(3): pp. 335–353.

Gourc, J.P., Arab, R., and Giraud, H., 2001, "Calibration and Validation of Design Methods for Geosynthetic-Reinforced Retaining Structures using Partial Factors," *Geosynthetics International*, Vol.8, No. 2, pp. 163–191.

Jewell, R. A., 1991, "Application of revised design charts for steep reinforced slopes." *Geotextiles and Geomembranes*, Vol. 10, pp. 203–233.

Ketchart, K. and Wu, J.T.H. (1997). "Performance of geosynthetic-reinforced soil bridge pier and abutment, Denver, Colorado, USA." International Symposium on Mechanically Stabilized Backfill, T.H. Wu (Ed.), Balkema, pp. 101–116.

Koutsourais, M., Sandri, D., and Swan, R. (1998). "Soil Interaction Characteristics of Geotextiles and Geogrids". Proc. 6th Int. Conf. Geosynthetics, Atlanta, Georgia, March 1998, pp. 739–744.

Leshchinsky, D., 2000, Discussion on "Performance of geosynthetic reinforced slopes at failure." By Zornberg, Sitar, and Mitchell, *Journal of Geotechnical and Geoenvironmental Engineering*, ASCE, Vol. 126, No. 3, pp. 281–283.

Leshchinsky, D., 2001, "Design Dilemma: Use peak or residual strength of soil." *Geotextiles and Geomembranes*, Vol. 19, pp. 111–125.

Leshchinsky, D. and Boedeker, R.H., 1989, "Geosynthetic Reinforced Soil Structures." *Journal of Geotechnical Engineering*, ASCE, Vol. 115, No. 10, pp. 1459–1478.

Li, X.S., Chan, C.K., and Shen, C.K., 1988, "An Automated Triaxial Testing System." *Advanced Triaxial Testing of Soil and Rock*, ASTM STP 977, pp. 95–106.

Maher, M.H. and Gray, D.H. (1990). "Static Response of Sand Reinforced With Randomly Distributed Fibers", ASCE J. Geotech. Engrg. 116(11): pp. 1661–1677.

McGown, A., Murray, R.T., and Jewell, R.A., 1989, "State-of-the-art report on reinforced soil." Proc., *12th Int. Conference Soil Mechanics and Foundation Engineering*, Rio de Janeiro, Vol. 4.

Michalowski R.L., Cermak, J. (2003), "Triaxial compression of sand reinforced with fibers", ASCE J. of Geotch. and Geoenviron. Engrg., 129(2): pp. 125–136.

Michalowski, R.L. and Zhao, A. (1996). "Failure of Fiber-Reinforced Granular Soils". ASCE J. Geotech. Engrg. 122(3): pp. 226–234.

National Concrete Masonry Association, 1997, *Design Manual for Segmental Retaining Walls*, Second edition, Second printing, J. Collin (Editor), Herndon, Virginia.

National Concrete Masonry Association, 1998, *Segmental Retaining Walls – Seismic Design Manual*, 1st Edition, Bathurst (Editor), Herndon, Virginia.

National Road Administration Publication 1992:10, 1992, *Soil reinforcement – Design tensile strength for synthetic materials* (In Swedish).

Public Works Research Center, 2000, *Design Manual for Geotextile Reinforced Soil* (in Japanese).

QMRD, 1997, *Reinforced Soil Structures*, Specification MRS11.06, Main Roads Department, Queensland, Australia.

Ranjan, G., Vassan, R.M. and Charan, H.D. (1996). "Probabilistic Analysis of Randomly Distributed Fiber-Reinforced Soil", ASCE J. Geotech. Engrg. 120(6): pp. 419–426.

Riemer, M.F., 1992, *The Effects of Testing Conditions on the Constitutive Behavior of Loose, Saturated Sand under Monotonic Loading*, Ph.D. Dissertation, Dept. of Civil Engineering, Univ. of California, Berkeley.

RTA, 1997, *Design of Reinforced Soil Walls*, QA Specification R57, Roads and Traffic Authority, New South Wales, Australia.

Schmertmann, G. R., Chouery–Curtis, V.E., Johnson, R.D., and Bonaparte, R., 1987, "Design Charts for Geogrid–Reinforced Soil Slopes." Proc., *Geosynthetics'87* Conference, New Orleans, pp. 108–120.

Shewbridge, S.E. and Sitar, N. (1990). "Deformation Based Model for Reinforced Sand", ASCE J. Geotechnical Engineering 116(7): pp. 1153-1170.

Tatsuoka, F., Uchimura, T., and Tateyama, M. (1997) "Preloaded and prestressed reinforced soil." *Soils and Foundations*, Vol. 37, No. 3, pp. 79–94.

Uchimura, T., Tatsuoka, F., Tateyama, M., and Koga, T. (1998). "Preloaded-prestressed geogrid-reinforced soil bridge pier." Proceedings of the *6th International Conference on Geosynthetics*, Atlanta, Georgia, March 1998, Vol. 2, pp. 565–572.

Vidal, H., 1969, "La Terre Armée." *Annales de l'Institut Technique du Bâtiment et des Travaux Publics*, Materials, Vol. 38, No. 259–260, pp. 1–59.

Won, G.W., Hull, T., and De Ambrosis, L. (1996). "Performance of a geosynthetic segmental block wall structure to support bridge abutments." *Earth Reinforcement*, Ochiai, Yasufuku and Omine (eds), Balkema, pp. 543–549.

Zornberg, J.G., Sitar, N., and Mitchell, J.K., 1998a, "Performance of Geosynthetic Reinforced Slopes at Failure." *Journal of Geotechnical and Geoenvironmental Engineering*, ASCE, Vol. 124, No.8, pp. 670–683.

Zornberg, J.G., Sitar, N., and Mitchell, J.K., 1998b, "Limit Equilibrium as a Basis for Design of Geosynthetic Reinforced Slopes." *Journal of Geotechnical and*

Geoenvironmental Engineering, ASCE, Vol. 124, No.8, pp. 684–698.
Zornberg, J.G., Sitar, N., and Mitchell, J.K., 2000, Closure on discussion to "Limit Equilibrium as a Basis for Design of Geosynthetic Reinforced Slopes." *Journal of Geotechnical and Geoenvironmental Engineering*, ASCE, Vol. 126, No. 3, pp. 286–288.
Zornberg, J.G., and Leshchinsky, D., 2001, "Comparison of International Design Criteria for Geosynthetic-Reinforced Soil Structures." *Geosynthetics and Earth Reinforcement*, H. Ochiai, J. Otani, and Y. Miyata (Editors), ISSMGE-TC9, pp. 106–117.
Zornberg, J.G. (2002) "Discrete framework for limit equilibrium analysis of fibre-reinforced soil", Géotechnique 52(8): pp. 593–604.

Keynote lectures

New Horizons in Earth Reinforcement – Otani, Miyata & Mukunoki (eds)
© 2008 Taylor & Francis Group, London, ISBN 978-0-415-45775-0

Newly standardized procedures for assessing the reduction factors for high strength geosynthetic reinforcements

S.R. Allen
TRI/Environmental, Inc. Geosynthetic Services, Austin, Texas, USA

ABSTRACT: Reduction factors to modify short term tensile properties and estimate allowable design strengths of geosynthetic reinforcements is well established, yet the procedures for determining these reduction factors has just recently been standardized. Significantly, these procedures employ the most recent testing technologies with associated criteria in place to qualify and properly employ resulting data. This paper updates a previously published paper providing summary explanations of testing technologies as well as context for their use within standardized product reduction factor qualification efforts. The contents are updated with a description of the Stepped Isothermal Method (SIM), a test procedure representing a significant change in the approach to accelerated creep testing of geosynthetic reinforcements. Detailed explanations of the SIM test procedure as well as underlying theory are presented.

1 DETERMINATION OF DESIGN STRENGTH

We begin with a review of the available design strengths of reinforcement geosynthetics. These reinforcement geosynthetics may include geogrids, woven geotextiles and strips, where the reinforcing component is made from polyester (polyethylene terephthalate), polypropylene, polyethylene, polyamides and polyvinyl alcohol.

As geosynthetic materials serve as reinforcement when they assume load, and knowing they experience elongation when loaded, the measurement of their strain behavior and its incorporation into design is required for successful design. In addition, many other in-situ conditions bear on the realized performance of the geosynthetic reinforcement including affects caused by installation damage, weathering and chemical degradation. The traditional approach to including these phenomenon in design is to place a multi-component reduction factor on short term measured tensile strengths. This allowable, or long-term strength, of a geosynthetic reinforcement may be determined in accordance with equation 1.

$$T_{al} = T_{ult}/RF \quad (1)$$

where:

$RF = RF_{ID} \times RF_{CR} \times RF_{D}$

T_{al} = allowable strength – The long term tensile strength that will not result in rupture of the reinforcement during the required design life, calculated on a load per unit of reinforcement width basis

T_{ult} = the ultimate tensile strength of the reinforcement determined from width tensile tests (ASTM D 6637 or ISO 10319)

RF = a combined reduction factor to account for potential long-term degradation due to installation damage, creep, and chemical/biological aging

RF_{ID} = a strength reduction factor to account for installation damage to the reinforcement

RF_{CR} = a strength reduction factor to prevent long-term creep rupture of the reinforcement

RF_{D} = a strength reduction factor to prevent rupture of the reinforcement due to chemical and biological degradation.

In addition to reduction factors, the allowable strength may also be reduced by a safety factor, f_s, accounting for the uncertainties of application that are not be quantified, and the quality of supporting data for the reduction factors used. It is this quality of reduction factor data, however, that will benefit significantly as a function of new testing technologies and associated standardization.

2 HISTORICAL CHALLENGES IN DETERMINING REDUCTION FACTORS

2.1 *Manufacturer driven*

Manufacturers of geosynthetic reinforcement products have historically supported the financial burden of tension creep and creep rupture testing. Due to the relative expense of this considerable product evaluation effort, it has been common practice among many manufacturers to characterize an initial product and publish resulting design guidance, or relevant reduction factors, based on these results. Future modifications via product improvements and raw material source changes are often not incorporated into revised or updated guidance. This has been enabled by the burdens of excessive re-characterization costs as well as the pace of raw material change exceeding the pace of testing. When a 10,000 hour creep test exceeds the duration of two or more formulation changes, unique product characterization is rarely considered feasible.

2.2 *Testing duration*

Prior to developments during the past few years, the readily available and practical testing tools responsive to the needs of real time design support have been few. In fact, specific to time-dependent creep testing, conventional measurement procedures have called for testing times up to 10,000 hours ($\approx$14 months) and longer. This burdensome time requirement has traditionally removed conventional creep testing from any context of construction quality assurance and site-specific or lot-specific approval processes. Additionally, the requirement for long testing times has sometimes stifled new product development or design specific performance investigations due the burden of expensive and long-term testing commitments that may not result in confirmation of desired performance.

2.3 *Data quantity, quality and use*

The establishment of geosynthetic reinforcement reduction factors has traditionally involved compiling the results of numerous tests to establish a well defined performance evaluation. However, the requirements for the quantity and validity of established data and, more importantly, the resulting reduction factors, have been variable at best and non-existent at worst. These missing criteria have often challenged the market place and overall acceptance of geosynthetic reinforcements.

While the manufacturing community has invested considerable resources in defining product reduction factors, they have often done so in accordance with market and regulatory driven requests. And while regulatory agencies have used this data they have often found fault with generated results and required augmenting additional product testing. This is often difficult if not impossible to resolve if product lines have undergone modifications with time, or complete replacement with improved or more economical materials.

Before more recent developments, several regulatory and standardization bodies had attempted to define standardized approaches with regard to requested reduction factors, limiting and allowable strengths, etc. These organizations have included: The British Board of Agrement (UK), Highway Innovative Technology Evaluation Center (USA), United States Department of Transportation (USA), The Geosynthetic Institute (USA), BAM Federal Institute for Materials and Testing (Germany), and many others throughout the world. These regulatory agencies reviewing reduction factor data for product approval or acceptance have often reviewed the data differently resulting in different conclusions regarding product performance and acceptability. This has caused great frustration on the part of manufacturers who are told their products are suitable for a given application and then unsuitable for the same application in a differently governed area.

Geosynthetic testing and research facilities have often been consulted to define the minimum requirements for reduction factor testing as they are seen to have experience in product characterization. That is, they know "what the last client did" and are sometimes familiar with local or regional data review practices. Still, laboratory personnel are not regulators and are not authorized to speak for regulatory or designer approval. Just because a testing program and data set has been recommended by a qualified laboratory and has been successful in review and approval for some applications, does not necessarily guarantee approval or suitability in another.

The overall quality of individual reduction factor test measurements have traditionally been unmeasured and undefined. That is, there have historically been no criteria for repeatability of individual tests, linearity of regressions, validity of shift factors or extrapolations, sameness of product data for combined product presentations, etc. Indeed, a great deal of effort is expended to compare competitive reduction factors with rare attention and criticism given to the responsible test results. This has provided ample room for debates regarding individual product merits based on marketing agendas rather than consideration of actual laboratory work.

3 STANDARDIZATION EFFORTS

Two significant efforts to respond to the challenges outlined above have been realized within the last few years. This work has been accomplished to provide consistent and well considered guidance on the specific test procedures and data evaluation techniques for the determination of geosynthetic reinforcement reduction factors. Two documents representing this work and establishing standardized procedures are as follows.

- Washington State Department of Transportation Standard Practice T 925, *Standard Practice for*

Determination of Long-Term Strength for Geosynthetic Reinforcement.
- ISO CD 20432, *Guide to the determination of long-term strength of geosynthetics for soil reinforcement.*

Significantly, the author of the first document served as a coauthor of the ISO document. This served the industry in that, while some differences are present, a single approach was encouraged throughout the work efforts. In both documents, a fully developed approach for the determination of geosynthetic reinforcement reduction factors is presented. The T 925 standard practice is currently in its third version and incorporates all the most recent testing and measurement technologies. The authors have drawn upon testing and review experience to standardize a single approach and implement state-of-the-art test procedures and evaluation matrices.

Based on these documents, the following sections summarize reduction factor determinations and provide details regarding requirements.

4 THE CREEP REDUCTION FACTOR

The creep reduction factor, RF_{CR}, and the testing involved, often represents the most significant reduction factor in the determination of allowable strength. When testing to predict the strengths intended to prevent the long-term rupture of a geosynthetic reinforcement, a testing laboratory is responding to **limit state** design requirements. However, creep testing is also performed to respond to specific job applications establishing limits on the total strain allowable over the service lifetime. In this case, a **strain limit** design and strength is desired to prevent excess strain over the intended service life.

Figure 1. Laboratory creep test.

A typical creep test involves the unconfined loading of a geosynthetic reinforcement with some percentage of its short term tensile strength. This loading of the specimen results in its elongating under load. The test is performed by monitoring this strain over time and reporting results accordingly.

Creep tests are performed in accordance with the following established procedures.

- ASTM D 5262, *Standard Test Method for Evaluating the Unconfined Tension Creep Behavior of Geosynthetics.*
- ISO 13431, *Geotextiles and geotextile-related products – Determination of tensile creep and creep rupture behaviour.*
- ASTM D 6992, *Standard Test Method for Accelerated Tensile Creep and Creep-Rupture of Geosynthetic Materials Based on Time-Temperature Superposition Using the Stepped Isothermal Method.*

5 THE SPECIAL CASE OF THE STEPPED ISOTHERMAL METHOD (SIM)

5.1 *Introduction*

The single most significant development in the measurement of geosynthetic allowable strength involves technology advancement with regard to creep and creep rupture characterization. In a relatively few short years, the long term test durations and excessive expense associated with time dependent strain characterization were largely made insignificant. First proposed by Thornton, et. al. (1997), the stepped isothermal method (SIM) for creep testing has been a significant contributor to more robust and frequent creep and stress rupture characterization of geosynthetic reinforcements. The procedure as applied to tension creep testing was standardized through ASTM International and is currently employed as a measurement technique in consensus based characterization guides. Detailed procedural descriptions follow.

5.2 *Time-temperature superposition*

To understand SIM and how it works, it is useful to review the basic principles and practices of elevated temperature creep testing and time-temperature superposition (TTS). The fundamental notion of all TTS techniques, including SIM, is that elevating temperature accelerates the response to mechanical load. Deformations, such as creep strain, occur relatively rapidly when load is first applied, but the rate of increase decreases with time. The total strain and strain rate associated with creep is the study of creep and creep rupture, or maximum strain, tests. Consequently, graphs presenting creep data often present time in a log scale along the abscissa.

The precise way that increasing temperature accelerates strain behavior governs how creep response can be "shifted" along a log time scale. A temperature dependent time factor a_T, relates the ratio of the time, t_T, for a viscoelastic process to proceed a given amount at an arbitrary temperature to the time, t_R, for the same process at a reference temperature. This is shown in equation 2.

$$a_T = (t_T)/(t_R) \tag{2}$$

At test temperatures greater than the reference temperature, test time is shortened, and a_T is less than 1. For this reason, some in the accelerated testing community refer to a_T, as an attenuation factor and A_t, defined as its reciprocal, as an acceleration factor. Both, collectively, are called shift factors. These shift factors are often used with creep test data developed for individual specimens, tested under various elevated temperature conditions, to shift all creep curves, or time dependent events defined by creep curves, to a standard reference temperature, usually 20 degree C.

5.3 *TTS theory*

One popular theory of TTS is demonstrated by the Williams-Landel-Ferry equation 3,

$$\log A_T = \frac{17.4(T - T_g)}{51.6 + T - T_g} \tag{3}$$

This is not the case for all materials and is not reflected in all TTS equations.

The rate theory for TTS is expressed in equation 4 and was applied by Zhurkov to creep rupture of sevmany rigid solid materials.

$$\log A_T = \frac{U_0 - \gamma\sigma}{2.3k}\left(\frac{1}{T_R} - \frac{1}{T}\right) \tag{4}$$

where A_T is the shift factor and T is the test temperature and T_g is the glass transition temperature. Note that this equation was developed to describe TTS for linear viscoelastic amorphous polymers in the transition temperature region from just below T_g to $T_g + 50°C$. Most significantly, in this equation, $\log A_T$ depends only on temperature. In this equation A_T is the shift factor, T is temperature, T_R is reference temperature, σ is stress, U_0 and γ are rate theory parameters (activation energy and volume) and k is Boltzman's constant.

This approach states that the energy barrier for creep-rupture is reduced by the applied stress. In other words, the creep rate is dependent not only on temperature, but also on stress.

The implications of these differing equations is that if A_T is a function of temperature only (as in WLF), creep-rupture curves obtained at different temperatures will always be parallel. But if A_T is a function of temperature and stress (as in Rate theory), creep-rupture curves obtained at different temperatures will not be parallel. Significantly, the TTS approach taken when evaluating data may impact the determination of creep or creep rupture reduction factors.

Baker & Thornton (2001) performed a series of creep rupture tests and developed associated rupture curves for a woven polypropylene geotextile. The results are presented in Figure 2. They also demonstrated in their work that the resulting shift factor was influenced by the level of stress applied, and that a specific TTS approach worked well for shifting test results.

Revisiting these results, one may demonstrate the sensitivity to shifting technique. Using the same set of test results one may determine resulting rupture limited 114 year design strengths using a variety of approaches. Table 1 demonstrates the differences achieved.

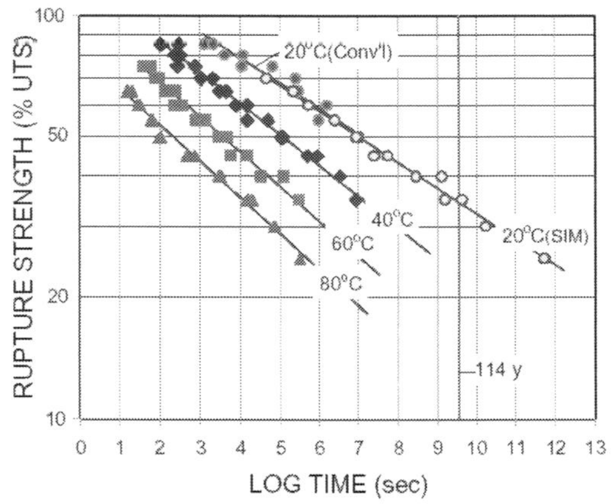

Figure 2. PP geotextile creep rupture test results (Tornton and Baker, 2001).

Table 1. Effect of shifting approach on 114 y strength.

Curve	R2 Rank	% UTS @ 114 y
SIM only	0.9903	34.42
Strain shift	0.9876	33.54
SIM + 20°C conventional	0.9833	34.44
Non-linear block shift	0.9756	28.00
Linear block shift	0.9727	29.12
Non-linear WLF shift	0.9509	36.91
Linear WLF shift	0.9129	36.91

While successful at showing the impact of various shifting and data evaluation techniques, it is not wise to draw conclusions with regard to their relative conservatism. Indeed the ranking presented in table 1 is valid for a finite set of test results only and may not apply to other results. What is significant is that robust use of TTS includes the evaluation of how best to shift a data set.

5.4 *TTS caution*

There are several other relevant issues with the practice of TTS, and when applied to new or untested materials, study is warranted with regard to testing and data shifting practices. Do test temperatures exceed the material's glass transition temperature and if so, change the mechanism of response? In a semi-crystalline polymer, does a greater percentage of the crystalline structure melt? Are other response mechanisms engaged that would affect the creep response?

With regard to acceleration factors used in shifting data to a reference temperature, is the creep rate of the tested material sensitive to applied stress in addition to elevated temperature? If so, does this dependence change as a function of temperature? Is there a physical time dependence that's not captured in accelerated testing, such as physical aging of the material or additive migration? These and other issues should be understood before putting too much emphasis on TTS predicted creep results. Of course the best measure of the accuracy of TTS predictions will always be direct comparisons to conventional tests performed at in-application test temperatures. In lieu of this long term data it is always wise to use multiple predictive tests and apply conservative design principles accordingly.

5.5 *Conventional accelerated testing*

In figures 3 and 4, an imaginary sample has been characterized for creep behavior under 40% of its measured ultimate tensile strength, at four test temperatures: 20, 30, 40 and 50 degrees C.

The resulting curves are then horizontally shifted and overlapped to define the master curve at 40% UTS at a 20 degree C reference temperature. This method of strain shifting is called classical shifting and is defined as the shifting of a continuous elevated temperature strain vs log time curve along the log time axis, to achieve overlapping superposition with another strain curve at a lower, usually the reference temperature.

Typically, the sample is characterized for creep behavior at a variety of load levels representing different percentages of its ultimate strength. By studying these multiple curves generated at different load levels, one may select features such as time to rupture or time to some arbitrary strain limit. This approach has been used historically in the geosynthetics industry to predict time dependent strain or rupture limits.

Figures 5 and 6 demonstrate the shifting of a time dependent 10% strain limited strength.

In these plots, the 10% strain-limited strength has been identified from each load-temperature creep test, and then plotted for each test temperature (figure 5). The individual strain-limited plots are then shifted

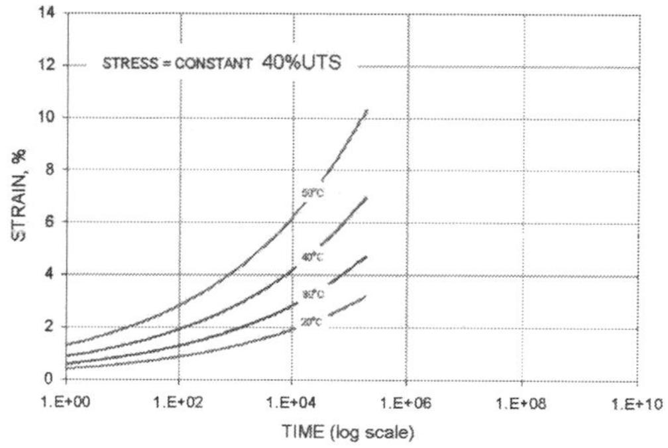

Figure 3. 40% UTS loaded creep tests at various temperatures.

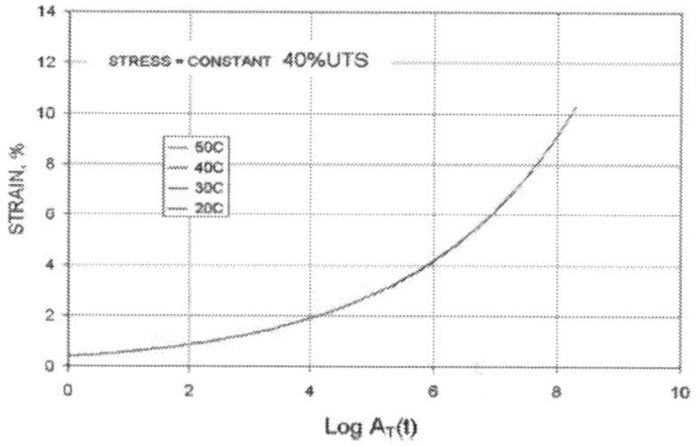

Figure 4. Figure 3 test results strain shifted to 20°C.

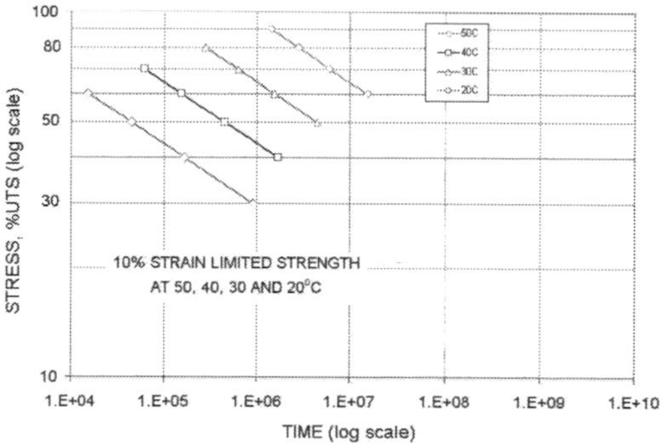

Figure 5. 10% limited strength curves at various temperatures.

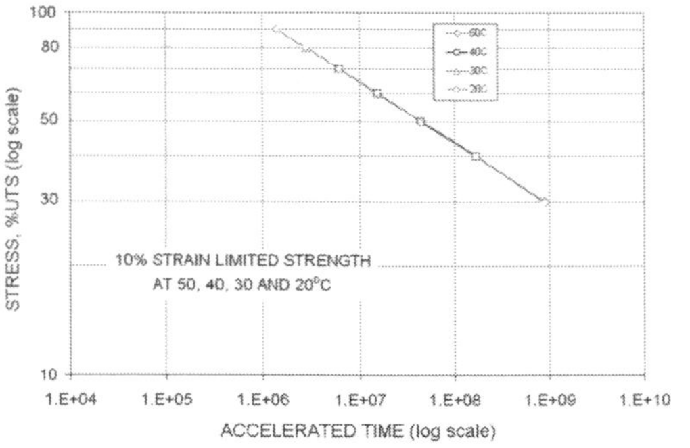

Figure 6. 10% limited strength block shifted to 20°C.

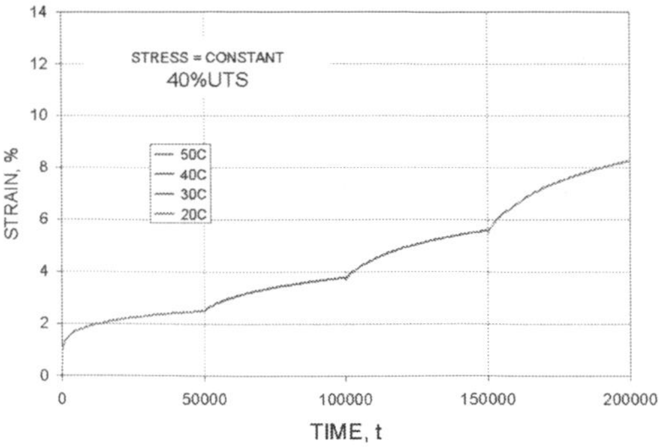

Figure 7. Example test results for a SIM test.

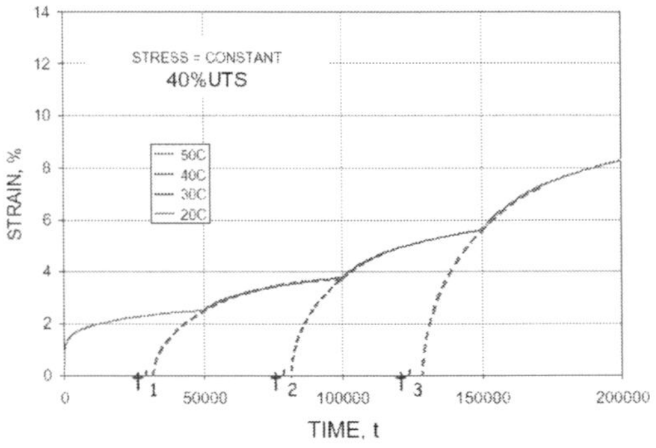

Figure 8. Demonstration of t' representing the beginning of each creep test.

again to a reference temperature of 20 degree C in order to predict at what strength the sample will achieve 10% strain at a specified time (figure 6). This movement of curves is called block shifting and is defined as the shifting of an event limited (strain or rupture), elevated temperature strength vs log time curve along the log time axis, to achieve overlapping superposition with another like strength curve at a lower, usually the reference, temperature.

5.6 *The SIM test*

SIM is a special case of classical TTS, and relies on strain measurements to employ strain shifting. SIM utilizes a single specimen instead of many to generate a master curve at each stress level. The immediate benefit of a single specimen test to achieve a master curve is the avoidance of specimen-to-specimen variability that often masks small differences in creep rate and confuses the TTS shifting process.

The SIM test consists of a series of timed isothermal creep tests performed at a sequence of increasing temperatures. As in an ordinary creep test the load is held constant and the creep strain in measured throughout the duration of the test. The number, heights and durations of temperature steps are designed to produce a master curve of creep compliance over a long term period defined by the test objective. For the geosynthetics community, long term design is generally 114 years.

Preparation for the SIM test involves mounting a single specimen into an environmental chamber equipped with specimen grips. The specimen is mounted in the grips, achieving temperature equilibrium and applying a small pre-load. Ramping the load to a predetermined stress level starts the test. This loading is performed rapidly to mimic the strain rate employed during short term tensile tests used to define ultimate tensile strength. Many have fund it useful to "start the clock" of a test immediately prior to ramp-up as this preserves the load up portion of the creep curve.

The first creep exposure of the SIM procedure is a normal creep test in the sense that the specimen does not have a history of creep loading. The second and subsequent creep exposures are complicated slightly by having the thermal histories of the previous temperature steps. This feature, however, represents the essence of the SIM method. Since no strain recovery is permitted during the time that the temperature step takes place, and since the temperature step is accomplished rapidly, typically within 1 to 2 minutes, the mechanical state of the sample is nearly the same after the temperature step as it was before the temperature step took place. Inherent in the SIM is the rigid necessity that the corresponding states at both the lower and higher temperature are stable ones that would be readily achievable under a more traditional TTS (isothermal) approach. That condition would not be achieved if, for example, the temperature step was so large as to negate sample equilibrium at the new temperature step.

An example of the data collected during a SIM test performed at 40% UTS loading is shown in Figure 7.

Each new temperature step may be seen as the initiation of a new creep test having a hypothetical start time t', as presented in Figure 8.

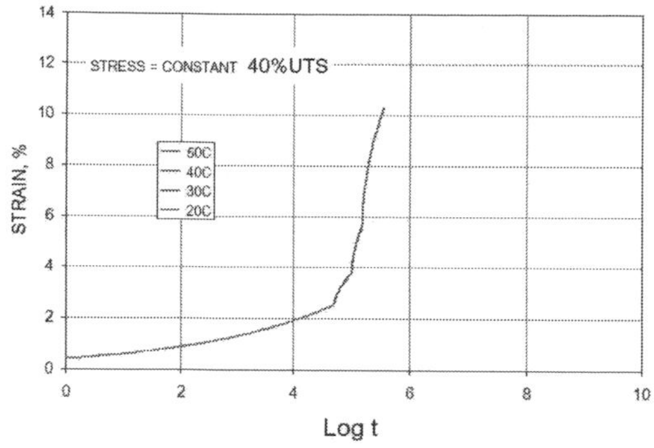

Figure 9. Test results plotted with log time abscissa.

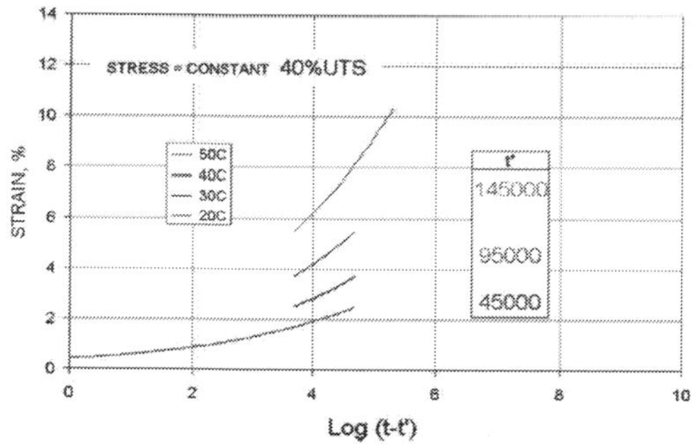

Figure 10. Horizontal shifting to define t-t'.

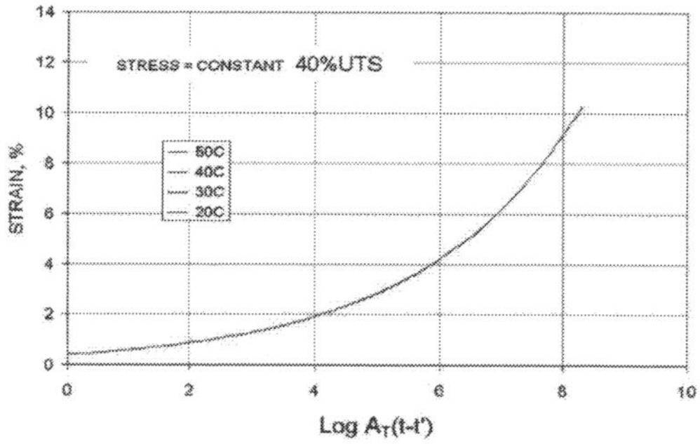

Figure 11. Master creep curve resulting from SIM test.

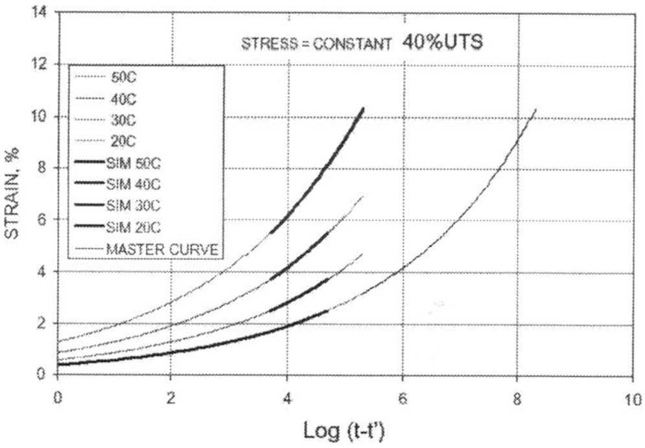

Figure 12. Conventional creep tests overlain with SIM segments.

It is assumed that with each temperature step, only the rate of creep has been changed, not the creep mechanism. Based on the notion that for a pair of corresponding states obtainable at two temperatures, the one at the higher temperature must have occurred at an earlier time than the one at the lower temperature, we begin the rescaling process to find an appropriate t'.

This is accomplished by first changing the abscissa of the strain vs linear time plot into log scale, as shown in Figure 9. Rescaling has a dramatic effect on the shapes of the curve segments. Part of this is due to the nature of logarithmic scales, but the initial slopes of the rescaled curves are influenced by the new starting times. The virtual starting time for each segment, t', is now selected based on the principle that the creep response at the end of any temperature step must be nearly equivalent to the creep response at the beginning of the next temperature step. This , involves the rescaling of the times associated with each individual creep curves.

Rescaling is performed by plotting the individual curves vs. the logarithm of the initial value of (t-t') where t' is adjusted to account for the strain history. The correct rescaling adjustment is achieved when the slope at the beginning of a new segment is exactly the same as the slope of the ending slope of the previous segment as shown in Figure 10.

The final phase of data processing is accomplished by shifting each segment horizontally to achieve juxtaposition of the rescaled individual creep curve segments. Figure 11 presents the master creep curve resulting from the horizontal shifts. The abscissa is now labeled LOG A_T (t-t') to indicate the data have been both rescaled and shifted.

SIM shifting is essentially equivalent to classical shifting of strain data, but achieves juxtaposition instead of overlapping superposition. What is accomplished in relatively short times (<24 hours) is a master creep curve for the creep response of a specimen under a specific load. One may think of a single SIM test as "visiting" a portion of each of the component creep tests if they would have been performed in their entirety under isothermal conditions. In other words, during the SIM test the material is experiencing a specific period of time during each of the tests outlined in Figure 1 and shown again in Figure 12.

The aftermath of a SIM test involves several "reasonability" checks to asses the relative quality of results. While the horizontal shift factors are determined empirically, their relative value may be compared to shift factors determined through conventional accelerated testing as well as shift factors published in the literature for like materials. Indeed, one of the many benefits of SIM is the robust practice of reporting all shift factors used in developing predicted results. It is during the shifting process and the development of associated shift factors that poor test parameters reveal themselves, such as temperature steps that are too large, prohibiting rapid specimen temperature and equilibrium. In addition, shift factors and associated creep rates may become exceedingly large. In this way, the shifting process itself serves to monitor the quality of the test.

While conventional time-temperature shifting of results obtained on specimens tested at elevated temperatures can provide improved predictions of longer-term performance at ambient temperatures, SIM has evolved this conventional approach to provide shorter term, and more numerous predictions of creep and creep rupture behavior. The SIM test is standardized in ASTM D 6992, and its use in development of creep reduction factors is well established and referenced in the two guidance documents.

5.7 *SIM use in RF_{CR} characterization*

Traditional limit state characterization has involved the development of a creep rupture curve specific to the geosynthetic and documenting the time to failure, or rupture, of the reinforcement when loaded to various stress levels. This requires the performance of creep tests which terminate in rupture (thus "creep-rupture" tests) caused by the relatively high loading of individual test specimens. A typical rupture curve is shown in Figure 13.

In order to develop a creep rupture curve, the following general requirements are established.

In addition to establishing the requirements above for minimum data quantity and qualification of SIM test results, the procedures also give extensive guidance regarding establishment of a reference temperature for all developed test results and the time-temperature data shifting of accelerated data.

Finally, while the standards realize a common approach for the determination of a creep and creep-rupture reduction factor, they also suggest the use of accelerated testing and short term conventional tests as a means to re-qualify and "check" products on a more routine basis. While not suitable for use alone for regulatory submittals or for formal establishment of creep reduction factors, the use of SIM has already realized increased creep and creep-rupture testing as new raw materials are characterized, new products are developed and site specific reduction factor estimates are desired. In the relatively short time since development in the late 90's, SIM has proven a powerful tool in a

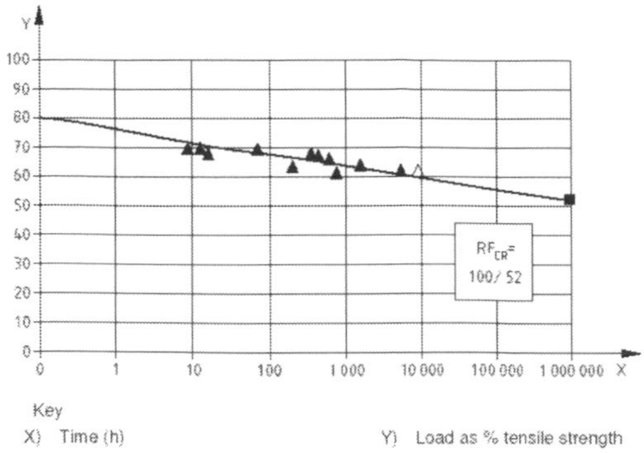

Figure 13. Example creep-rupture curve showing the results of individual creep-rupture tests (from ISO 20432).

Table 2. Example creep reduction factor test guidance from WSDOT T924 and ISO 20432.

Number of creep rupture points (creep-rupture tests) for creep rupture curve
Requirement: 12–18

Spacing of creep rupture points on creep rupture curves
Requirement: 3 between 10 and 100 hours (not accelerated by temperature); 4 between 100 and 1000 hours; 4 between 1000 and 10,000 hours; 1 at 10,000 or more hours

Use of SIM testing
Requirement: Must accompany conventional testing. Estimates of RF_{CR} calculated using accelerated tests and conventional tests must agree within specific tolerances.**

Combining creep data developed from "similar" geosynthetic reinforcements (manufactured from same product line)
Requirement: RFCR for "similar" product must show equivalent or better performance. Tests on new products should be taken to between 1000 and 2000 hours in length. If SIM has been established as viable with the product, 3 tests projecting performance out to between 100,000 and 10,000,000 hours is advised for maximum statistical efficiency.

** ISO criteria: 2000 hour RF_{CR} must differ by no more than 0.15. T925 criteria: RF_{CR} as determined at 1000 and 50,000 hours must agree using 90% confidence limit via Student's t test.

better understanding of creep behavior under accelerated conditions.

6 THE INSTALLATION DAMAGE REDUCTION FACTOR

The installation damage reduction factor, RF_{ID}, is born of an acknowledgement that when a reinforcement geosynthetic is installed it undergoes damage

Figure 14. Example of the installation damage exposure procedure.

that immediately reduces its long-term strength. The relevant standards and procedures for this test are as follows.

- ISO 13437, *Geotextiles and geotextile-related products – Method for installing and extracting samples in soil, and testing specimens in laboratory.* This procedure allows for site-specific installation damage testing involving numerous testing periods and longer-term evaluation.
- ASTM D 5818, *Standard Practice for Exposure and Retrieval of Samples to Evaluate Installation Damage of Geosynthetics.*
- ISO 10722-1, *Geotextiles and geotextile-related products. Procedure for simulating damage during installation. Installation in granular materials.* This procedure is used for index testing only; and is not used for the determination of RF_{ID}.

In general, these procedures all differ slightly in their approach but some common features enable testing and research laboratories to accommodate the requirements established by all.

During the last several years a number of RF_{ID} assessments have been performed by employing a test procedure initially based on a protocol developed by Watts and Brady of the Transport Research Laboratory (TRL) in the United Kingdom. This procedure has met the general criteria defined by ISO 13437 and ASTM D 5818.

In this approach, the entire process of geosynthetic exposure and exhumation is made controllable and repeatable by employing a small out-door laboratory approach. The tested samples are placed perpendicular to the running direction of the compaction equipment simulating most retaining wall and reinforced slope installation guidelines. A substratum of four steel plates, equipped with lifting chains, are incorporated into the exposure regime. The layers of compacted soil/aggregate are constructed on top of the plates with the geosynthetic installed accordingly. The geosynthetic is then exhumed via the raising one end of the steel plates with lifting chains to about a 45°. Soil located at the bottom of the plate is removed and, if necessary, the plate is struck with a sledgehammer to loosen fill. As the upper lift falls away, the geosynthetic is removed by "rolling" it away from the underlying lift. This procedure significantly minimizes exhumation damage unrelated to installation damage.

The exhumed samples are then cut and tested in their "soiled" condition. Other un-exposed samples are also cut and tested from the same roll of material to establish "baseline" tensile strength. The evaluation of RF_{ID} is based on the results of wide width tensile testing in accordance with various responsive test procedures. These include:

- ASTM D 4595, *Standard Test Method for Tensile Properties of Geotextiles by the Wide Width Strip Method*, for woven and non-woven geotextiles.
- ASTM D 6637, *Standard Test Method for Determining Tensile Properties of Geogrids by the Single or Multi-Rib Method, Method B*, for stiff and flexible geogrids, and
- ISO 10319, *Geosynthetics – Wide Width Tensile Test*, for European installation damage determinations.

Table 3. Example of RF_{ID} testing program where X denotes a test.

Product/d50 (mm)	0.03	0.8	10
Lightweight	X	X	
Midweight	X	X	X
Heavyweight	X	X	X

Finally, the RF_{ID} is determined as follows:

$$RFID = T_{lot} / T_{dam} \quad (5)$$

where, T_{lot} is the average lot specific tensile strength before exposure to installation, and T_{dam} is the average lot specific tensile strength after installation.

Most exposures are based upon the use of numerous aggregates and fills differing by features (d_{50} particle size, particle angularity, overall gradation and hardness (durability), etc.). A typical test matrix for the establishment of RF_{ID} may be include the weakest, mid-strength and strongest products in a product line as follows.

Because it is possible to curve fit the generated test results, it is also possible to reasonably extrapolate to other products in the product family or to other aggregates having differing d_{50} characterization.

It is important to note that through the wealth of installation damage testing experience, several observations have been recorded with respect to material performance. Many aspects of product exposure, other that the gradations of soils and aggregates encountered, contribute to observed strength loss. These include, coating thickness and stability of coated geogrids, the overall orientation and dimensional features of geogrid ribs, the weave patterns of woven geotextiles, and other product-specific characteristics. For these reasons manufacturers are considering product design features with new appreciation to product susceptibility to installation damage.

7 THE DURABILITY REDUCTION FACTOR

The durability reduction factor is in many ways the most complex with a host of possible measurement techniques based on geosynthetic polymer type and product application. Test procedures are established based upon the anticipated degradation mechanism of the tested material. Several types of polymer degradation mechanisms exist.

All geosynthetic polymers are susceptible to environmental degradation due to weathering, including exposure to ultra-violet light, and chemical attack. All three effects are further accelerated by elevated temperature and, for some polymers, by moisture availability and uptake. The durability of reinforcement geosynthetics is generally possible by their high degree of orientation and high molecular weights, while for polyolefins in particular, molecular stability (non-reactivity) is the principal reason for durability.

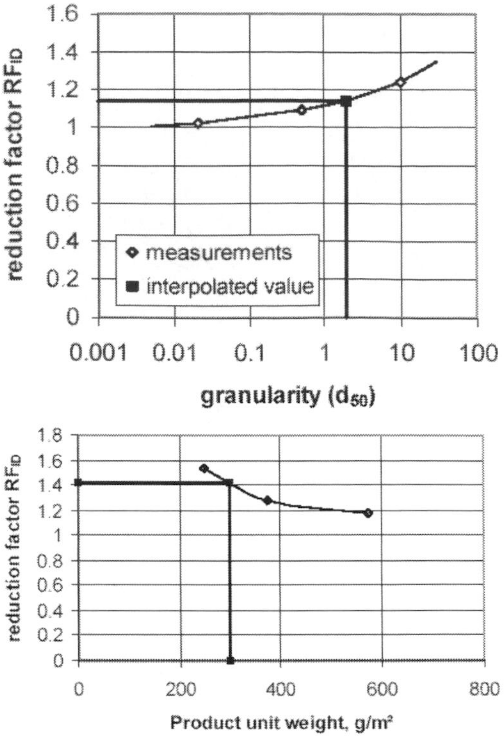

Figure 15. Example presentations of installation damage test results and extrapolation exercises (from ISO 20432).

7.1 *UV resistance*

For UV stability and resistance to weathering testing, the following two test procedures are used.

- ASTM D 4355, *Standard Test Method for Deterioration of Geotextiles by Exposure to Light, Moisture and Heat in a Xenon Arc Type Apparatus*.
- ISO 12959, *Test Method for Deterioration of Geotextiles from Exposure to Ultraviolet Light and Water (Xenon Arc Type Device*.

Unfortunately, there are enough small differences between these two standards that the results of one test do not always support the results of the other. Indeed, still other tests such as the European standard, ENV 12224, *Geotextiles and geotextile-related products – determination of the resistance to weathering, uses a fluorescent UV weatherometer to affect accelerated UV exposure*. This use of a significantly different UV source further prohibits straight-forward correlation between test standards. For these reasons, market and regulatory area requirements continue to govern which type of UV resistance testing approach is followed.

7.2 *Chemical resistance*

A number of test procedures and general experimental approaches exist for the establishment of chemical and biological resistance. Summaries of the technical and procedural considerations are available in the following documents.

- ASTM D 5819, *Standard Guide for Selecting Test Methods for Experimental Evaluation of Geosynthetic Durability*.
- ISO 13434, *Guidelines on durability of geotextiles and geo textile-related products*.

The principal cause of degradation of polyester (PET) geosynthetics is hydrolysis. The rate of hydrolysis will be less if the soil is partially saturated instead of fully saturated, but is not zero. Alkaline liquids with pH > 9 can erode the PET surfaces. Work has been performed to establish minimum criteria for the long-term chemical durability of polyesters in relatively neutral aqueous environments, that is environments having pH of 4.5 to 9, where the organic content is 1% or less, and the effective design temperature for site application is less than 30° C. For these conditions of use, two index tests may be performed to confirm the following.

- The polyester geosynthetics used for reinforcement, or the yarns from which they are made, should exhibit a carboxyl end group count less than 30 meq/g, when measured in accordance with GRI GG7, Carboxyl End Group Content of Polyethylene Terephthalate (PET) Yarns.
- The polyester geosynthetics used for reinforcement, or the yarns from which they are made, should exhibit a minimum number average molecular weight equal to or greater than 25,000 Mn, when measured in accordance with GRI GG8, Determination of the Number Average Molecular Weight of Polyethylene Terephthalate (PET) Yarns Based on a Relative Viscosity Value.

Both criteria should be satisfied.

The principal cause of degradation of polyolefins, polypropylene and polyethylene, is oxidation. Assessment of the rate of oxidation is complex and thus various test procedures have been attempted to model long-term oxidation with variable success. Currently, most oxidation resistance testing is performed by heating the geosynthetic in a forced air oven via ISO 13438, *Geotextiles and geotextile-related products – Screening test method for determining the resistance to oxidation*. WSDOT T925 assumes exposure to oxidation during the relatively aggressive UV exposure in ASTM D 4355 testing, which combines UV exposure with moisture and heat. In either case, before and after exposure testing is performed via tensile testing to determine the effect of exposure.

Chemical resistance may also be of concern when geosynthetic reinforcements are used in aggressive chemical environments. In these cases, resistance to chemical degradation may be studied via a range of test procedures, most involving incubation of the geosynthetics in chemical baths at a number of exposure temperatures. The change in tensile properties is measured with increasing exposure times and plotted to determine time to reach designated failure criteria.

A listing of test procedures used to assess durability of geosynthetic reinforcements follows.

7.2.1 *Biological resistance*

- ISO 12961, *Geotextile and geotextile related products – Method for determining the microbiological resistance by a soil burial test*.
- ASTM G160, *Standard Practice for Evaluating Microbial Susceptibility of Nonmetallic Materials by Laboratory Soil Burial*.

7.2.2 *UV and oxidation resistance*

- ISO 13438, *Geotextiles and geotextile-related products – Screening test method for determining the resistance to oxidation*.
- ASTM D 5721, *Standard Practice for Air-Oven Aging of Polyolefin Geomembranes*.

7.2.3 *Chemical resistance*

- GRI GG7, *Carboxyl End Group Content of Polyethylene Terephthalate (PET) Yarns*.
- GRI GG8, *Determination of the Number Average Molecular Weight of Polyethylene Terephthalate (PET) Yarns Based on a Relative Viscosity Value*.
- ISO 13439, *Geotextile and geotextile related products – Method for determining the resistance to hydrolysis*.
- ISO 12960, *Geotextiles and geotextile-related products – Screening test method for determining the resistance to liquids*.
- ASTM D 5322, *Standard Practice for Immersion Procedures for Evaluating the Chemical Resistance of Geosynthetics to Liquids*.
- ASTM D 5496, *Standard Practice for In Field Immersion Testing of Geosynthetics*.
- ASTM D 6389, *Standard Practice for Tests to Evaluate the Chemical Resistance of Geotextiles to Liquids*.
- ASTM D 6213, *Standard Practice for Tests to Evaluate the Chemical Resistance of Geogrids to Liquids*.

8 A STANDARDIZED APPROACH FOR PRODUCT CHARACTERIZATION

The establishment of a program to fully characterize design reduction factors, and maintain this characterization via periodic re-qualification, is one of the most significant developments in the geosynthetic reinforcement industry in many years. The program was developed by the American Association of State Highway and Transportation Officials (AASHTO) and is based on the WSDOT Test Method 925 introduced at the beginning of this paper. By virtue of the similarities in approach between T925 and the newer ISO 20432 document, the program responds successfully to both with a carefully considered testing plan.

The program of initial characterization and follow-up testing is explained in detail by the National Transportation Product Evaluation Program (NTPEP) – Geosynthetic Reinforcement Program (available at www.ntpep.org). The following summarizes specific features of the program.

- The program is based on the test procedures and data evaluation techniques subscribed by WSDOT T925.
- Initial product characterization is intended to fully develop design reduction factors for a geosynthetic reinforcement product family having the same raw materials and manufacturing regime.
- The ultimate tensile strength, specific product dimensions, RF_{ID}, RF_{CR} and RF_D are all tested and reported under the program.
- The products are also required to undergo quality assurance testing on a more limited basis via a reduced testing program, scheduled for three year intervals. This serves to document product consistency and reward those manufacturers who have realized product improvements or materials.
- The testing procedures involved are performed in accordance with existing ASTM, ISO and GRI standards and require specific implementation with regard to data evaluation. This is significant in that all products are evaluated in a technically equivalent manner, making biased manipulation of data or partial reporting of results a non-issue.
- Results are suitable for immediate use in design procedures and may be issued throughout the world as a comprehensive set of technical reports for presentation and full explanation of reported reduction factors.

9 CONCLUSION

The development of the ISO 20432 and WSDOT T925 guidance documents and the implementation realized by the NTPEP geosynthetic reinforcement program, establish a powerful and unique process for consistent characterization of existing geosynthetic reinforcement products as well as full investigation of newly developed products and product families.

Integral to each of these programs is the standardization of SIM and the acknowledgement of the SIM as a useful application of TTS with testing schedule benefits to the user. More significantly, SIM is beneficial the collection and review of creep and creep rupture test results. Employed correctly, the use of SIM can support a very responsive and technically valid prediction of long term creep and creep rupture reduction factors.

These standardized technical approaches have already served the manufacturer and design communities with significant characterization guidance and economy-of-testing benefits. It is hoped that these new developments will continue to support and grow a robust and successful application of geosynthetic reinforcement products.

REFERENCES

Washington Department of Transportation Test Method T 925, 1) Washington State Department of Transportation Standard Practice T 925, *Standard Practice for Determination of Long-Term Strength for Geosynthetic Reinforcement, 2006,* http://www.ntpep.org/ProfileCenter/Uploads/WSDOT_T925_11-23-05.pdf

ISO CD 20432, *Guide to the determination of long-term strength of geosynthetics for soil reinforcement*, ww.iso.org

Thornton, J.S., Allen, S.R., Thomas, R.W., Sandri, D., (1997), "Approaches for the Prediction of Long Term Viscolelastic Properties of Geosynthetics from Short-Term Tests," Geosynthetics '97, Vol. 1, Industrial Fabrics Association International, Long Beach, California

Baker, T.L., Thornton, J.S., (2001), Comparison of Results Using the Stepped Isothermal Method and Conventional Creep Tests on a Woven Polypropylene Geotextile, Geosynthetics '01, Portland, Oregon

Allen, S.R. (2003), Considerations on the Stepped Isothermal Method - A Breather on the Way to Widespread Use, GSI Conference Proceedings, Hot Topics in Geosynthetics - IV Waste Properties; Geotextile Tubes; Challenges Vol: 4 Pages: 272 to 291

Allen, S.R. (2005), Geosynthetic Installation Damage Testing – A Status Report, GSI-NAGS Conference Proceedings, Hot Topics in Geosynthetics - Vol: 6 CD

New Horizons in Earth Reinforcement – Otani, Miyata & Mukunoki (eds)
© 2008 Taylor & Francis Group, London, ISBN 978-0-415-45775-0

Model testing to evaluate the performance of soil nailed structures

M.C.R. Davies
University of Auckland, Auckland, New Zealand

ABSTRACT: Soil nailing was introduced in the United Kingdom in the mid 1980's. In situ strengthening of the ground has proved to be cost effective and applicable to stabilise new and existing embankments and slope cuttings, and strengthen existing retaining walls. At first the take up of the use of soil nailing in the UK was hindered by a lack of both a fundamental understanding of the behaviour of soil nailed systems and comprehensive design guidance. The publication of HA68/94 and BS 8006:1995 has led to a steady increase in utilisation since the early 1990's. Nevertheless, there remains a concern that these guidelines are not fully comprehensive for UK construction conditions and practice. To address this issue a study of mechanisms associates with soil nailed systems has been conducted that was centred on a number of series of centrifuge model tests. The major findings of these studies are presented.

1 INTRODUCTION

In recent decades soil nailing has become increasingly more popular worldwide as a technique for reinforcing soil to form retaining structures in cut and to create, steepen or strengthen existing slopes. However, as has been reported in the best practice guide for soil nailing recently published in the UK by CIRIA (2004), in the UK soil nailing has been adopted more slowly as a construction technique than elsewhere in the world; particularly in other comparable European markets (i.e. France and Germany) and in the USA. Statistics assembled during the writing of the CIRIA guide, Figure 1, indicate that. Although soil nailing was first used in the UK in the mid 1980s, the face area of soil nailing installed in the UK did not start to increase significantly until the latter half of the 1990s. The slow acceptance of the technique in the UK prior to this date is partly attributable to a debate amongst British academics and practitioners about the mechanisms of soil nail/soil interaction and concerns about the potential corrosion of steel nails, but more probably resulted from a lack of UK guidance until the publication of the design manual for the reinforcement of highways slopes, HA68/94, in 1994 and the British Standard for reinforced soil BS8006, in 1995. In contrast, major national research projects conducted in France and Germany – the Clouterre (1991) and Bodenvernagelun (Gassler & Gudehus 1981) projects, respectively – and, later, in the USA, FHWA (1996), provided designers with not only design guidelines but also confidence in the technique of soil nailing resulting from the experience obtained from well instrumented case histories.

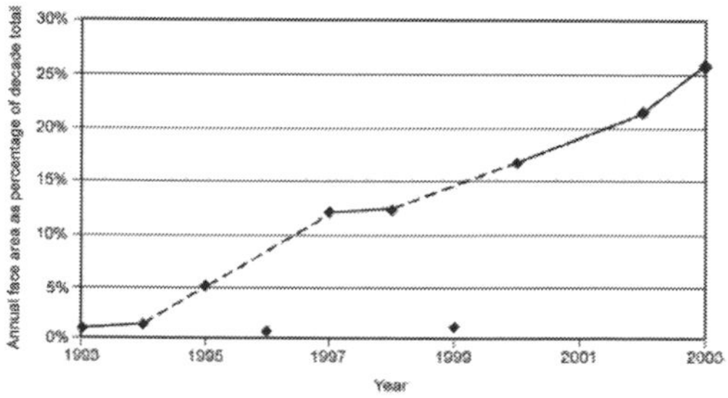

Figure 1. Growth of the UK soil nailing market in the decade up to 2003 (after CIRIA 2004).

As with other reinforcement techniques for retaining structures and slopes, the major components of soil nailed structures can have a number of different forms. Although most commonly formed from steel, soil nails can be made out of different materials (e.g. polymers), and they may be in direct contact with the soil or be surrounded by an annulus of grout. To prevent corrosion the surface of the nail might be treated with an epoxy or by galvanising, and there might be secondary protection in the form of plastic tubes (typically PVC or HDPE) surrounding the nail. Whilst the vast majority of nails are installed by being grouted into predrilled holes, different methods for placing nails exist; including self-drilled and driven installation techniques. The component of a soil nailed retaining structure that is most visible is the facing and these may be categorised into three major types: soft facings, flexible structural facings and hard structural facings. The materials used for these range from geofabrics to sprayed concrete reinforced with steel mesh depending on the requirement of the facing and the amount of facing deformation that can be permitted.

Although the technique of soil nailing is becoming more widely applied internationally there still remain aspects of the technique that are not fully defined and this lack of understanding has resulted in a number of both serviceability and ultimate limit state failures of soil nailed structures. In the survey of problems with soil nailing construction in the UK (CIRIA 2006) failure were attributed to two major causes. The first was construction details (such as inadequate control of construction operations) and the second to a lack of understanding of mechanisms – this included poor facing design (particularly flexible facings) and a lack of appreciation of the interrelationship between nail spacing and type of facing.

Whilst problems associated with construction details can be eliminated with tighter control of construction operations, improved understating of the mechanisms requires a systematic study of soil nailed structures. However, since such tests are expensive to conduct at prototype (i.e. full) scale, this has restricted the scope of such investigations. Monitoring of in-service structures may also be conducted but, clearly, it is not possible to investigate the performance of such structures under failure conditions, However, it is possible to gain a greater understanding into the mechanisms of soil nailed systems in carefully controlled, correctly scaled model tests which can be achieved using the technique of geotechnical centrifuge modelling. This paper describes results from a number of studies conducted by the author and his research students to gain a greater understanding of a number of the key mechanisms of soil nailed systems. These include the development of global deformations and stresses induced in the inclusions – both during construction of soil nailed structures and during subsequent loading of the structure – and the influence of geometric parameters (e.g. slope angle, nail orientation and nail spacing) on these. The paper also considers the effects on nail forces and global deformations of changing the effective stress in the retained soil.

In addition, this purpose of this paper is to illustrate more generally, using the example of soil nailing, how the technique of geotechnical centrifuge modelling may be used to systematically investigate mechanisms in a range of ground reinforcement applications (such as goesynthetic reinforced slopes, e.g. Zornberg et al 1998, Zornberg & Arriaga 2003).

2 GEOTECHNICAL CENTRIFUGE MODELLING

2.1 *Centrifuge scaling laws*

The stress/strain behaviour of granular soils is highly non-linear, stress level dependent and stress history dependent. Accurate scale modelling therefore requires both similitude between material properties in the prototype (i.e. the full scale structure) and the model and the correct stress distribution within the model. If a model is constructed at 1/N scale using soil from the prototype (i.e. providing similitude in soil properties), and accelerated in a centrifuge so that the model experiences an acceleration N times Earth's gravity, stress similitude in model and prototype may be demonstrated (e.g. Taylor 1995) as follows:

Prototype:

$$\sigma_p = h.\rho.g \quad (1)$$

Model:

$$\sigma_m = (h/N).\rho.(N.g) \quad = h.\rho.g \quad (2)$$

Therefore:

$$\sigma_p = \sigma_m \quad (3)$$

where σ_p and σ_m are self-weight stress in the prototype and model respectively, h is depth below the surface and g the gravitational field.

It can, similarly, be shown that the scaling law for force in a centrifuge model (such as the axial force in a soil nail) is given by

$$F_m = F_p/N^2 \quad (4)$$

A summary of centrifuge scaling laws appropriate to modelling soil nailed structures is given in Table 1.

Table 1. Scaling factors for centrifuge modelling.

Quantity	Full scale	Model scale
Linear Dimension	1	1/N
Stress	1	1
Strain	1	1
Density	1	1
Force	1	$1/N^2$

2.2 *Experimental procedure*

2.2.1 *Centrifuge model*

The results presented in this paper were obtained from studies conducted at the Geotechnical Centrifuge centres at Cardiff University (Gammage 1997, Aminfar 1998) and the University of Dundee (Morgan 2002), Figure 2. The two machines have similar platform dimensions, with both able to accommodate a model package of up to 1 m × 0.8 m × 1.0 m; permitting relatively large models to be tested. This is an important factor in the study of reinforced soils where accurate modelling of the inclusions is required. The model scale selected for the experiments described herein was 1:20 (i.e. at an elevated gravity 20.g). This permitted

Figure 2. The University of Dundee geotechnical centrifuge.

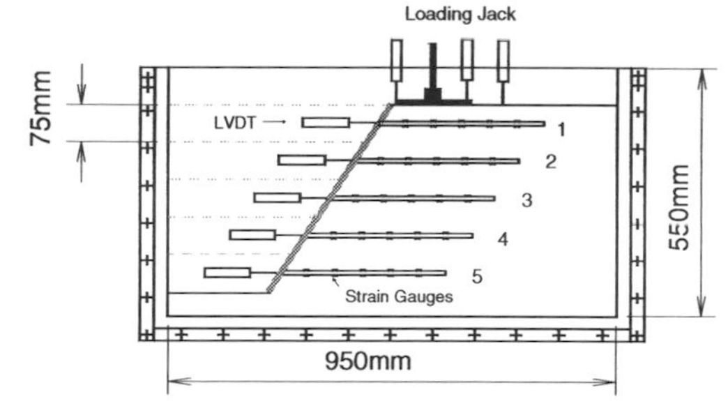

Figure 3. Centrifuge model – geometry and instrumentation (surcharge loading tests).

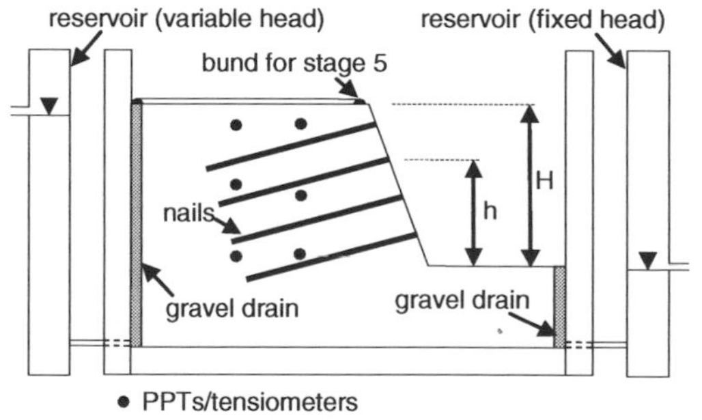

Figure 4. Centrifuge model – geometry and instrumentation (effective stress change tests).

prototype soil nail slopes and walls of up to 7.5 m high, and varying in slope angle between 45° and 90° to the horizontal, to be modelled whilst allowing dimensions and spacing of model nails to correctly replicate prototype conditions.

Each test consisted of two major phases. The nailed slope (or wall) was constructed in the first phase and then loading was applied in the second. During both stages instrumentation was monitored to provide information about the development of slope displacements and forces in the nails. The typical configuration of the models is shown in Figures 3 and 4. The models represent two different ways in which a soil nailed

Figure 5. Centrifuge model in strong box.

slope or wall was loaded following its construction. Figure 3 shows a model with loading in the form of a surcharge at the top of the slope (e.g. to represent construction at the top of the slope), whilst the loading in the model represented by Figure 4 was achieved by reducing the effective stress in the soil, by permitting water to flow through the slope. Figure 5 shows the photograph of a centrifuge strong box containing a model soil nailed slope with a flexible facing following an experiment (i.e. the "model package").

The soil used for the models was a fine sand, $D_{50} = 0.18$ mm, $c' = 0$ and $\phi' = 41.2°$. This was placed in the centrifuge strong box by pluviation to form a specimen with a uniform density typically of $\gamma = 17$ kN/m3. The front and rear faces of the strong boxes were fabricated from thick Perspex to permit observation of the model during testing.

2.2.2 *Soil nails*

It is experimentally very difficult to construct and reliably instrument a model nail which is an exact replica of the prototype. The nails to be modelled were 6.0 m long, 20 mm diameter steel bars grouted into a 160 mm diameter bore hole. This was replicated using an 8 mm diameter, 300 mm long acrylic bar designed to model the axial stiffness of the prototype nail (Gammage 1997, Morgan 2002). The model nails were coated with sand to replicate the soil/nail interface properties. The nails were instrumented with strain gauges on their upper and lower faces at five locations spaced at 50 mm intervals. These enabled both axial and bending strains to be measured; from which the axial force and bending moment distribution along the nail could be obtained.

2.2.3 *Facing*

A facing is required to prevent local failure occurring between reinforcements. To allow comparison between suites of experiments, the majority of the centrifuge models reported in this paper were constructed using a facing formed from a woven geofabric. The fabric used for the tests was typically a HF550 geotextile, manufactured by Don & Low. This has a tensile strength of 25 kN/m and 16 kN/m in the warp and weft directions, respectively. As in a prototype structure, the geotextile was held in place by the nail heads, to form a flexible facing.

2.2.4 *Simulation of construction*
When constructing a soil nailed retaining structure the normal procedure is first to excavate a bench and then install the nail before, finally, applying the facing (although in situations where the unsupported face will not hold for a sufficient period to permit nail installation the second and third stages are sometimes reversed). It is not practically possible to simulate the soil nailing process exactly in a centrifuge model. However, in order both to investigate the construction process and to ensure that the construction displacements are equivalent to field conditions, two techniques have been used successfully by the author to achieve this. The first is to build a slope with the nails and facing in place on the laboratory floor and then place a flexible bag containing a solution of Zinc Chloride in the volume to be excavated. This is a technique that has been used by both for simulating excavation for soil nailing (e.g. Bolton & Stewart 1990, Jones & Davies 2000) and in other centrifuge model applications (e.g. Powrie & Daly 2002). If the Zinc Chloride solution is mixed to the same mass density as the soil, then this results in a lateral earth pressure coefficient, K_0, of approximately unity. Excavation may be simulated during centrifuge operation by draining the Zinc Chloride solution from the bag, halting to simulate the different stages of construction as required. In the second technique the model is also prepared on the laboratory floor with both nails and facing in location but in this case the same soil as that forming the slope is placed – at the same density and using the same placement technique – in the volume in front of the slope. Construction is then modelled in stages in a number of sequential centrifuge runs. Each stage of construction is achieved by excavating a layer of soil whist the centrifuge is stationary and then accelerating the model in the centrifuge to replicate the self weight loading of the reinforced soil system.

3 EXPERIMENTAL FINDINGS

3.1 *Mechanisms during construction*

3.1.1 *Development of displacements*
Figure 6 shows the displacement profile of the face of a model slope that developed during the sequential centrifuge runs of the construction phase in one of the experiments. In this experiment the prototype slope modelled was 7.5 m high with a slope angle of 60°. The wall was constructed in five lifts using 6 m long grouted soil nails inserted at vertical and horizontal spacing of $S_v = 1.5$ m and $S_h = 2.33$ m, respectively,

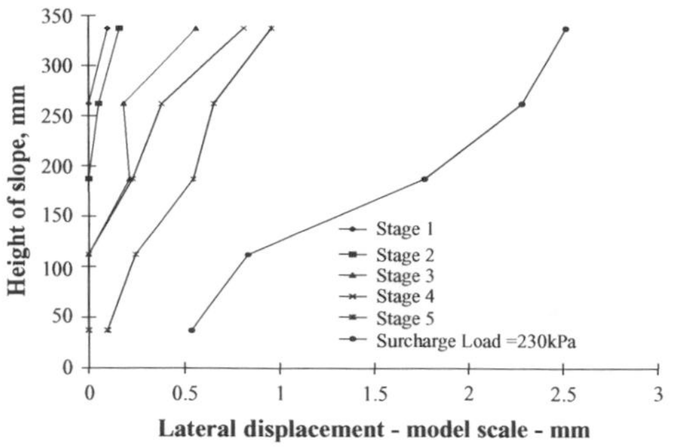

Figure 6. Development of lateral displacements (model scale).

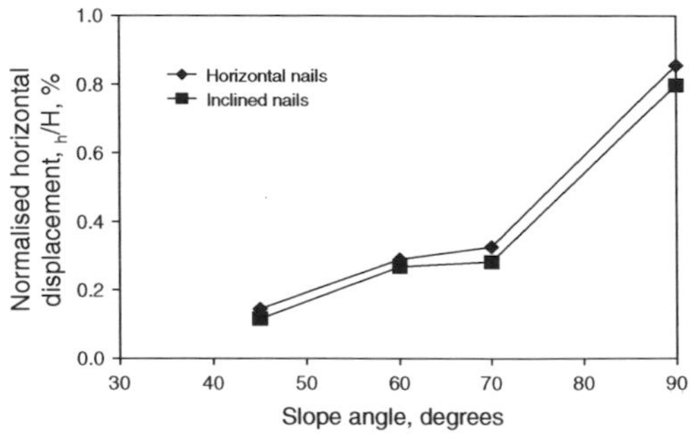

Figure 7. Variation in horizontal displacement at top of slope with slope angle.

at an angle of inclination of 15°. The figure indicates that there was a gradual development of lateral movement of the slope as excavation proceeded. The pattern of generally decreasing lateral displacement with depth conforms to observations of full scale soil nailed slopes. The maximum horizontal displacement of 0.95 mm is 0.25% of the slope height (i.e. 19 mm at prototype scale). Vertical deflections at the crest of the slope were of very similar magnitude. The displacement profile when a surcharge of 230 kPa was applied is also plotted in Figure 7. The figure shows that the front face of the wall bulged and the maximum lateral displacement was 2.50 mm; which corresponded to 0.67% of the slope height (i.e. 50 mm at prototype scale). In both phases of the tests the measured behaviour was in very good quantitative agreement with the results of the limited quantity of field measurements e.g. Gassler & Gudehus (1981), Clouterre (1991), Pedley & Pugh (1992) and Lazarte et al (2003).

The influence of both slope angle and nail inclination on horizontal displacement during the construction phase was investigated by conducting experiments on identical models in which these two variables were changed systematically. As can be seen from Figure 7, the maximum displacement at the top of the wall was reduced when the nail was inclined; but the reduction

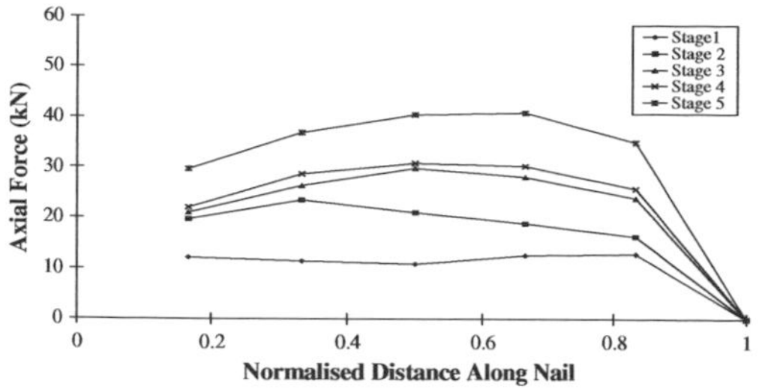

Figure 8. Axial forces in nail 2 during construction.

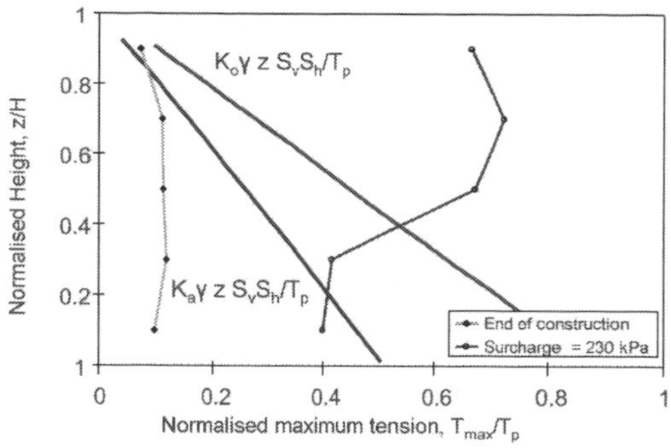

Figure 9. Normalised maximum axial force in nails following construction and surcharge of 320 kPa – 60° slope.

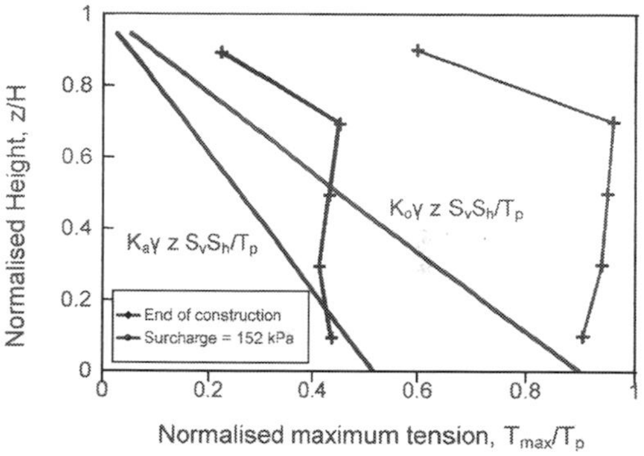

Figure 10. Normalised maximum axial force in nails following construction and surcharge of 152 kPa – vertical face wall.

in displacement was relatively minor in all cases. The results do, however, indicate very markedly that, as would be expected, the measured lateral displacement of the soil nailed system increases as the slope angle is increased. This figure indicates that there appears to be a transition in behaviour at slope angles between 70° and the vertical as the soil nailed retaining systems moves from a reinforced slope to a reinforced soil retaining wall.

3.2 *Development of nail forces*

Results are presented in Figure 8 for axial force development in the nail located at level 2 i.e. the nail installed in the second stage (see Figure 2). As indicated above, the technique adopted for building the model resulted in the nails being located in the soil during model preparation. This permitted measurements of strain development in the nails to be monitored prior to the head of the nail being exposed by excavation. During this stage the nails showed a development of tensile force; which is attributed to the reduction in lateral stress in the soil beneath the excavation. At the second excavation stage, when nail 2 was exposed and began to act as a reinforcing element in the slope, it can be seen that the maximum axial force developed at a normalised distance of approximately 0.3 (1.8 m) from the face. As construction proceeded the location of maximum force migrated to 0.6 (3.6 m) from the face; indicating that thc location of the potential failure surface moved, as would be expected, deeper into the slope. This observation agrees with the findings of centrifuge experiments of soil wall construction conducted by other workers e.g. Tei (1993).

The stress distribution in all the nails in the 60° slope may be observed by plotting the maximum axial force in each nail, T_{max}, (normalised by the tensile capacity of the prototype nail, T_p) against normalised height. The plot in Figure 9 shows that following construction the maximum stress recorded in the three nails located in the mid height of the slope were approximately 20% greater than that in the top and bottom nails. Lines representing the normalised active (K_a) and at rest (K_0) lateral earth pressure distributions are also plotted on Figure 9.

Comparison of the maximum force distribution in the nails with these lines indicates that with the exception of the top nail the force carried by the nails is lower than would be required to maintain active conditions. In order to examine the influence of changing the angle of the slope on the distribution of maximum axial force in the nails, the normalised axial forces, T_{max}/T_p, recorded following construction of a vertical soil nail wall is shown in Figure 10.

Comparison of the distributions at the end of construction in this figure with that in Figure 9 indicates that the forces monitored in the wall were, as would be expected, significantly greater than those in the 60° slope. It is interesting to note that, with the exception of the top nail, the distribution of maximum force with depth is very linear. However, unlike the slope, the forces in all except one of the nails were greater that required to maintain active conditions. At the top of the wall the nails indicate axial forces above the K_0 distribution line. This implies that the nails are not acting to resist active earth pressures – in a mechanism analogous to that used for the coherent gravity method of analysis used in the design of reinforced earth – but are involved in a more complex slip mechanism that is analogous to that used in the tie back method of reinforced soil design. This observation supports methods of analysis based on both wedges and slip circles that are widely used for the design of soil nailed systems, e.g. BS 8006 (1995). Estimates of the pullout capacity of the nails indicated that the top nail was very close to, or had reached, its maximum capacity and this explains why it contributed less to stabilising the wall than each of the other four nails.

3.3 *Mechanisms during loading*

3.3.1 *Surcharge loading of slopes and walls*

On the completion of construction a 150 mm wide rigid footing (the full width of the model), representing a 3.0 m wide strip load (this might represent, for example, the footing for a bridge abutment), was placed at the top of the slope; set back from the crest of the slope by 25 mm, Figure 2. Force was applied to the footing, via a 20 kN load cell, using displacement control.

The development of maximum axial force in each nail following application of a surcharge load may be investigated by comparing the distribution of axial force at the end of construction with that when a surcharge has been applied. For the case of the 60° slope the maximum axial forces in each nail when a surcharge of 230 kPa was applied to the top of the slope are plotted on Figure 9. During loading the maximum force was recorded in nail 2. This reflects a non-uniform distribution of shear stress along the potential shear plane; which was highest near the surface, where the load intensity due to the surcharge loading was at its greatest. Despite the very high surcharge loading the peak stress in all of the nails was significantly less than the tensile capacity of these nails, T_p. As has been described above, the surcharge loading resulted in the top of slope displacing horizontally by a prototype distance of 30 mm during this phase of the experiment.

Similarly, the maximum axial forces in each nail in the soil nailed wall following surcharging are plotted in Figure 10. In this case the test was stopped when the surcharge was 152 kPa. The test was stopped at this point because, as can be seen from Figure 10, a number of the nails were very close to rupture and therefore the wall was near to failure. Although the upper nail was at (or very close to) pullout following construction of the slope, application of the surcharge increased its pullout capacity allowing the nail to carry greater axial forces.

For the prototype slope and wall modelled in the experiments it would generally be expected that in the absence of a surcharge load the failure of nails by pull out would be more critical than by tendon rupture. A prediction of the factor of safety of the 60° slope at this end of the test was conduced using a circular slip analysis that incorporated the contribution of nails based on that proposed in BS8006 (1995) (Gammage 1997). The minimum calculated factor of safety was 1.6. A comparison of the predicted nail capacities with the peak measured axial forces is shown in Table 1. Although the predicted sum of the nail forces at the minimum factor of safety is very similar to measured value the distribution of forces in the nails was not predicted particularly accurately. The capacity of the top nail was greatly under predicted because the method of slices used in the analysis did not account for load spreading from the footing which increased the pull out capacity of the nail in the model. This indicates that the influence of surcharge has to be included when assessing the pull out capacity of a nail.

3.3.2 *Loading by changing the effective stress in the slope*

One of the applications of soil nailing is to stabilise slopes that are showing signs of impending failure or to strengthen slopes that have a calculated factor of safety that is considered to be too low. These conditions may be caused a variety of factors, including dissipation of negative excess pore water pressures generated during excavation and changes in hydrological conditions, e.g. Vaughan & Walbanche (1973). In addition to its application in stabilising existing slopes, soil nailing is also being used increasingly for the construction of new cuttings. Such new soil nailed slopes may also be subjected to changes in effective stress following construction – such as from increased rainfall intensity as a result of climate change, Hadley Centre (1998). There is a requirement, therefore, to establish the response of soil nailed systems to variations in effective stress in order to assess the long term serviceability of both existing slopes strengthened using soil nails and newly excavated soil nailed retaining structures. A study to investigate the mechanism resulting from variation in the effective stress of the soil forming a soil nailed slope has been reported by Davies & Morgan (2005) and the major findings reported in this paper are reproduced here.

The construction phase in these tests followed the same procedures as that described above for the surface loading tests. On completion of construction, the long term serviceability experiments were conducted by changing the drainage boundary conditions of the centrifuge models to examine the effects of both the increase and the cycling of pore water pressure in the slope. During and immediately following construction the water table was maintained at the level of the toe of the slope – stage 1. To reduce the effective stress in the model, in stages 2 and 4 of the test, the water level was raised in the gravel drain to the elevation of the top of the slope. In stage 3 the water level was returned to the same elevation as in stage 1. Finally, in stage 5 the model was subjected to a rise in ground water level combined with inundation from the top surface, achieved by forming a 1 cm high bund at the crown of the slope to impound water introduced to the

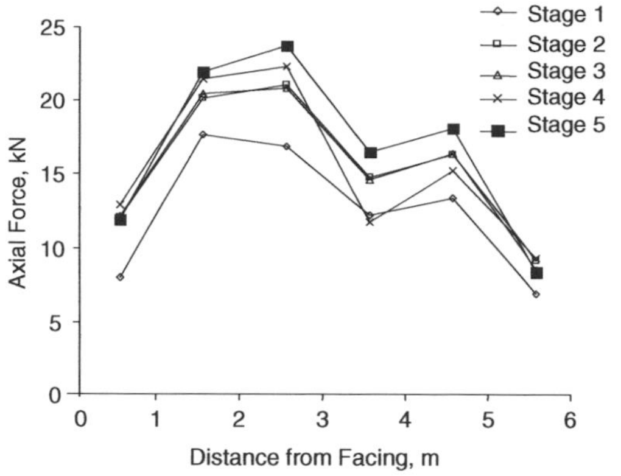

Figure 11. Axial forces along nail at elevation 0.375 at the end of each experimental stage.

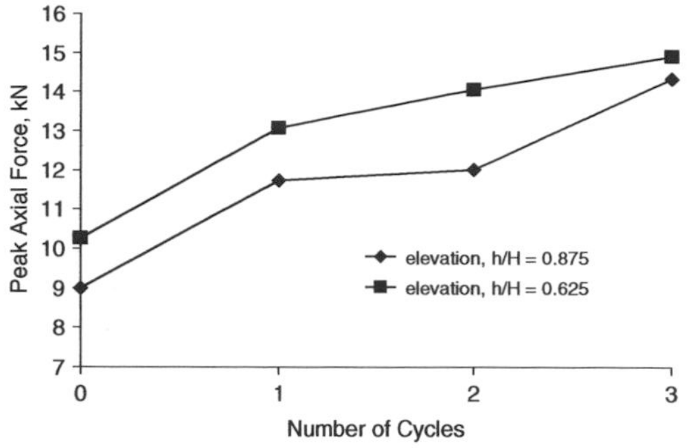

Figure 12. Peak axial forces (T) at each experimental stage (60° slope).

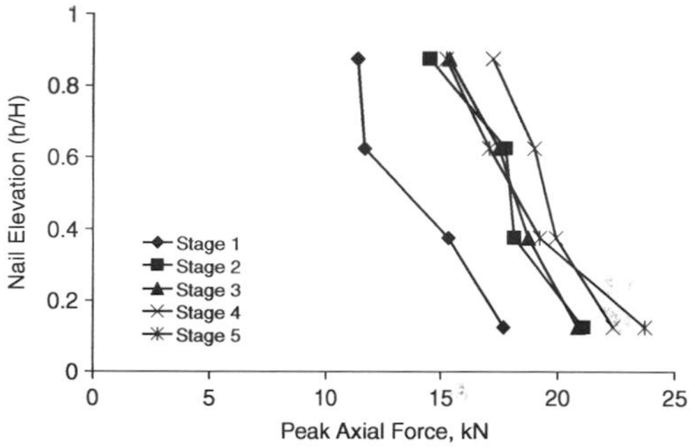

Figure 13. Peak axial forces in nails following cycles of water table rise in model (50° slope).

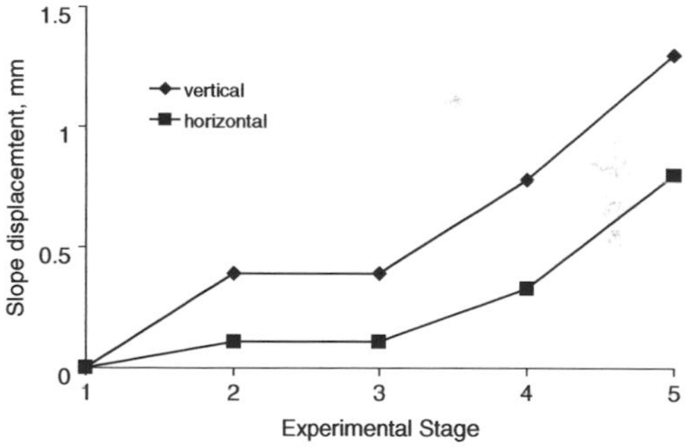

Figure 14. Vertical displacements at the top of the slope (0.7 m from the slope crest) and horizontal displacements measured adjacent to the head of a nail at a elevation of 0.625 (60° slope).

top of the slope (Fig. 3). Stages 1 to 5 were applied to all the models except one in which an extra cycle of water table fall and rise was applied (i.e. repeat of stages 3 and 4). This phase of the experiment was completed without stopping the centrifuge.

3.3.2.1 Nail forces

Typical variation in axial force distribution along a nail during the experiments is shown in Figure 11. Immediately after construction (stage 1) the peak axial force in the nail is 17.65 kN, representing 19% of the predicted pullout capacity of the length of the nail located in the resistive zone of the slope. With the rise in water table in the upslope gravel drain in the model in stage 2 of the experiment, resulting in a reduction in effective stress in the slope, the force in the nail rose by 19%. However, on lowering of the water table to its original level – stage 3 – there was hardly any change in the force distribution in the nail, implying negligible rebound or settlement of the soil in response to the increase in effective stress re-loading of the soil.

Repeating the rise in water table (stage 4) resulted in extra forces being generated in the nail – the maximum change during this stage being 7% of the original loading. Finally, inundation from the top surface of the slope resulted in a further decrease in effective stresses resulting in additional forces (8%) being taken by the nail. The change in peak axial force, T, in each of the four nails reinforcing a 60° slope during an experiment are shown in Figure 12, from which it can be seen that all the nails in the slope were subjected to significant increases in axial load when water was permitted to flow through the slope.

The development of axial load in two nails in a 50° slope that was subjected to three cycles of effective stress reduction and increase, shown in Figure 13, indicates an increase in axial force with number of cycles. In common with all the other tests in the programme of experiments, the largest percentage change (i.e. increase) in axial force was induced during the first decrease in effective stress. Subsequent effective stress cycles increased the axial force in the nails.

3.3.2.2 Slope displacements

Examples of surface displacements measured on a 60° slope developed following stage 1 of an experiment are shown in Figure 14. Both sets of data show the

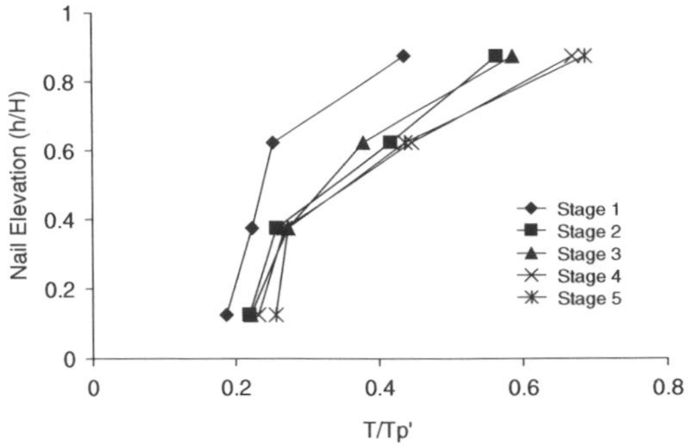

Figure 15. Comparison of measured peak axial forces (T) to theoretical pullout values (Tp').

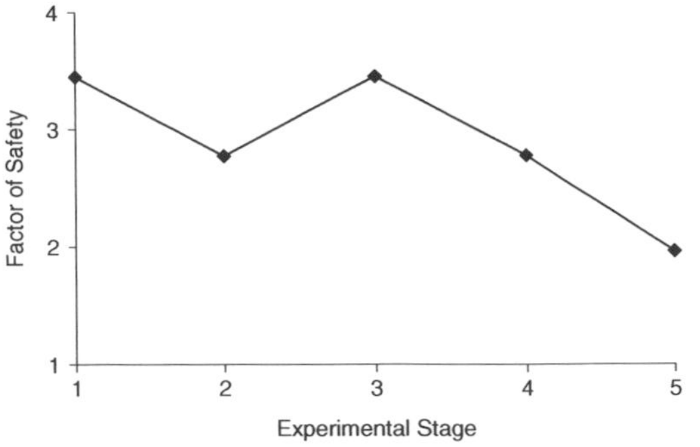

Figure 16. Predicted factor of safety (M_R/M_O) at each experimental stage (60° slope).

same trends. As can be seen in Figure 14, the slope displaced following the reduction of effective stress during stages 2, 4 and 5 of the experiment, but showed negligible displacement during the lowering of the phreatic surface in stage 3 – indicating a stiff response on re-loading. Since soil nails are passive inclusions, in order for forces to develop they have to be subjected to soil displacements and these measurements explain the sequence of axial force change in the nails presented above, i.e. when there are negligible displacements in the soil there is, similarly, negligible development of axial force in the nails. It is well established that cyclic shear loading of soils can result in the accumulation of deformation which can lead to serviceability limit state failure, such as in pavements, e.g. O'Reilly & Brown (1991). The results of the experiments indicated that there is the possibility that cyclic loading of soil nailed systems could lead to serviceability failure; however, this would depend on the number and amplitude of the cycles together with the geometry of and soil type forming the slope.

3.3.2.3 Nail pullout capacity

Since pore water pressures were measured in the experiments it is possible to assess the influence of change in effective stresses in the model on the pullout capacity of the nails. This may be achieved by comparing the measured maximum axial force in a nail, T, with the predicted pullout capacity of the length of nail located in the resistive zone of the slope, Tp'. Values of Tp' were calculated using values of pore water pressure interpolated from the spot values measured at the appropriate stage in the experiments. Values of T/Tp' for the four nails in the 60° slope at each stage of a test are shown in Figure 15, from which it may be seen that T/Tp' increased at all levels in the experiment during stages 2, 4 and 5. Figure 15 indicates that the ratio increased at all levels with each decrease in effective stress and that the axial force in the nail in the row nearest the top of the slope was approaching its pullout capacity.

3.3.3 *Stability analysis*

The effect of excess pore water pressure on the predicted stability of the model slopes was assessed in a limit equilibrium analysis, using a method of slices in which the critical slip surface was a log spiral. Pore pressures, measured at each stage of the experiments using miniature pore water pressure transducers, were applied at the base of the slices and the calculation was carried out to determine the out of balance moment, M_O. The resisting moment, M_R, was determined from the tension in the nails. The factor of safety is defined as M_R/M_O. The predicted factor of safety for the 60° slope is plotted for each experimental stage in Figure 16 from which it can be seen that, as should be expected, increase of pore water pressure in the slope leads to a reduction in the factor of safety; in this case by stage 5 the predicted factor of safety has almost halved. This analysis demonstrates the requirement to take into account in the design of soil nailed systems possible variations in pore water pressure within a slope, as monotonically increasing or cyclically changing pore water pressures might result in excessive deformations leading to serviceability failure.

3.4 *Nail spacing*

The effects of both nail spacing and facing stiffness on mechanisms associated with both walls and slopes is being investigated in the latest phase of the study. Aspects being investigated include both the forces developed in the soils nails and the stresses acting on the facing. The experimental procedure in these tests is the same as used for the investigation of the change in effective stress in the slopes, the only significant difference is that these slopes were built with four rows of nails as opposed to five in the experiments reported above. Tests were conduced on both walls and slopes with vertical nail spacing, S_v, of 1.5 m and with three horizontal spacings, S_h, of 1.4 m, 2.0 m and 3.4 m.

Figure 17 shows the measured maximum axial force measured in nails in a 70° slope for three horizontal

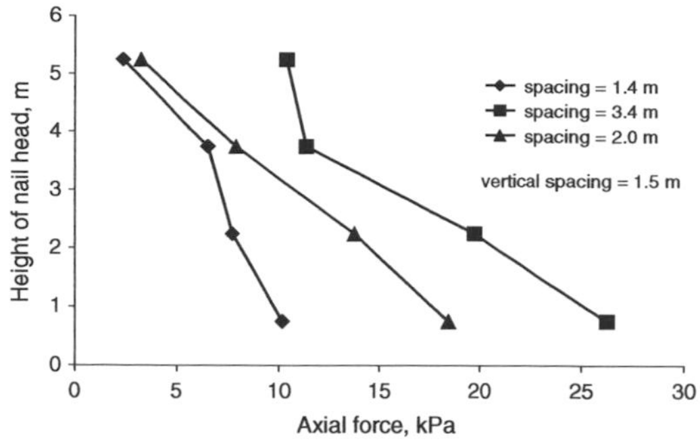

Figure 17. Distribution of maximum axial forces in nails at different horizontal spacings, S_h, (70° slope).

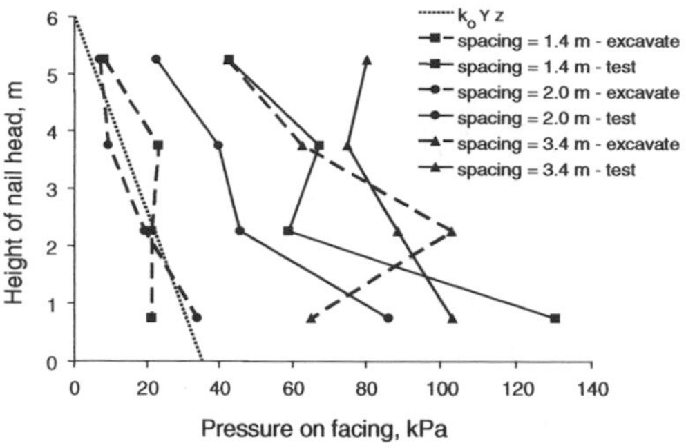

Figure 18. Distribution of measured pressure on facing at the end of construction and following reduction of effective stress for with different horizontal spacing of nails, S_h, (70° slope).

Table 2. Measured axial forces in nails at different horizontal spacings, S_h, (70° slope).

Height of nail head, m	Maximum axial force in nail, T_{max}, kN		
	$S_h = 1.4$ m	$S_h = 2$ m	$S_h = 3.4$ m
5.25	2.4	3.3	10.5
3.75	6.6	7.9	11.4
2.25	7.8	13.8	19.8
0.75	10.3	18.5	26.3
ΣT_{max}	27.0	43.5	68.0
T_{max}/S_h, kN/m	19.3	21.8	20.0

spacings following both construction and loading to stage 4 (i.e. water table is at the highest level within the slope). Although the configuration for this model is very similar to that for the other experiments presented herein, the facing was manufactured from horizontal strips of 2 mm thick aluminium. It can be seen that, as would be expected, the nails with the wider spacing carry higher axial load. The values of the maximum recorded axial forces are presented in Table 2 from which it can be seen that the sum of the nail forces in each of the models is almost identical. This result indicates that in the range of nail spacing investigated – which was typical of prototype conditions – there is no measurable interaction between the nails that might result in group action.

Miniature load cells attached to the rear of the facing permitted total stresses to be measured at different stages of the exponents. These are plotted in Figure 18 for two stages in each test, viz. immediately after construction ("excavate") and at loading stage 4 ("test"). These results show that immediately after construction of the walls with the two smaller spacings (1.4 m and 2.0 m) the stresses measured immediately behind the facing was very near to the active earth pressure; as indicated by the K_0 line which is also included on the diagram. However for the spacing of 3.4 m, the stresses are significantly higher that K_0 values. In all cases, Figure 18 shows that decreasing the effective stress in the slope results in an increase in the stress acting on the back of the facing to values above K_0.

4 CONCLUSIONS

The technique of soil nailing is now become established internationally as a technique for constructing earth retaining structures and for strengthening and steeping slopes. However, since the significant national studies conducted in the 1980s and 1990s there have been few opportunities for significant symmetric studies of the technique despite there remaining a number of unanswered questions about the mechanisms associated with aspects of the technique.

The technique of geotechnical centrifuge modelling allows studies of the mechanisms associated with nailed systems in experiments in which the major features of the prototype soil nailed systems are modelled correctly. The results of series of tests to investigate the performance of soil nailed systems during both construction and when subjected to different forms of load yielded the following conclusions:

1. The technique adopted for constructing the slope in a series of sequential centrifuge runs permitted the development of displacements of the slope and forces in the nails to be monitored following each stage of construction. Comparisons of the results with available data from field trials indicated that the models closely represent field conditions both during construction and the application of surcharge loading. In stability analyses the distribution of force in the nails was not well predicted.
2. Variations in effective stress within a soil nailed slope may result from dissipation of excess pore water pressures following excavation or through temporary or permanent – both natural and anthropogenic – changes in hydrological conditions. The results of the centrifuge model experiments

indicated that the axial force developed in soil nails may increase significantly to maintain stability of an excavation when effective stresses are decreased. This increase in nail load is caused by movement of the soil in the slope that could lead to serviceability conditions being exceeded or, in the extreme, ultimate limit state conditions being reached. The results indicate, therefore, that the level of possible reduced effective stresses within a soil nailed structure over time needs to be considered to assess any long-term effects on the structure, although initially a design may appear overly conservative.

3. Cycling of effective stresses within the model slopes led to an increase in axial forces in the nails resulting from accumulated deformations. The findings of the experiments indicate that repeated cycling of effective stress could lead to serviceability failure. In this experimental programme only a limited number of cycles of effective stress were applied to the model slopes and to assess the effects of a larger number of cycles and variations in boundary conditions this aspect is currently the subject of further investigation.
4. The results of the centrifuge tests indicate that the nail forces developed in a soil nailed system are influenced by the spacing of the nails. However, for the range of spacings examined to date – which are typical of dimensions used in practice – the axial force carried by the nails per unit width of the structure was found to be the same for all nail spacing. Increasing the spacing of the nails resulted in an increase in soil pressure acting at the rear of the facing.

Finally, the results of this study demonstrate the way in which the technique of geotechnical centrifuge modelling may be used to investigate a range of complex mechanisms associated with a particular construction technique. There are many other applications in the field of soil reinforcement where this technique may be applied to gain a greater insight into complex soil structure interaction mechanisms and provide quantitative data to validate analytical methods used in engineering design.

ACKNOWLEDGEMENTS

The experimental work described in this paper was conducted mainly by a number of the author's past and current research students; Dr Paul Gammage, Dr Mohamed Aminfa, Dr Neil Morgan and Ms Tanja Waaser. Their invaluable contributions to this programme of research are most gratefully acknowledged.

REFERENCES

Aminfar, M.H. 1998. *Centrifuge modelling soil nailed slopes*. PhD Thesis, University of Wales, Cardiff.

Bolton, M.D. & Stewart, D.I. 1990. The response of nailed walls to the elimination of suction in clay. Proceedings of the International Reinforced Soil Conference, Glasgow, United Kingdom.

BS 8006 1995. *Strengthened/ reinforced soils and other fills*. British Standards Institution, London, United Kingdom.

Clouterre 1993. Recommendations CLOUTERRE 1991 – Soil Nailing Recommendations 1991, English Translation, Presses de l'Ecole Nationale des Ponts et Chaussées, Paris, France.

Davies, M.C.R. and Morgan, N. 2005. The Influence of the Variation of Effective Stress on the Serviceability of Soil Nailed Slopes, XVI International Conference of the International Society for Soil Mechanics and Geotechnical Engineering, Osaka, Japan, 2005, pp 1335–1338.

FHWA 1996. *Manual for design and construction of soil nail walls*. Soil nail walls -Demonstration project 103, Washington D.C.

Gammage, P.J. 1997. *Centrifuge Modelling of soil nailed walls*. PhD Thesis, University of Wales, Cardiff, United Kingdom.

Gässler, G. and Gudehus, G., 1981. Soil Nailing-Some Aspects of a New Technique, *In* Proceedings of the 10th International Conference on Soil Mechanics and Foundation Engineering, Vol. 3., Session 12, Stockholm, Sweden, pp. 665–670.

Hadley Centre 1998. Climate change scenarios for the United Kingdom. DEFRA/ Met Office.

Jones, A.M. and Davies, M.C.R. 2000. An investigation of long-term stability of a soil nailed excavation using centrifuge modelling, GeoEng 2000, Proceedings of the International Conference on Geotechnical and Geological Engineering, Melbourne Australia, November, 2000.

Lazarte, C.A., Elias, V., Espinoza, R.D. and Sabatini, P.J. 2003. Geotechnical Engineering Circular No. 7 – Soil Nail Walls Publication No.FHWA0-IF-03-017, Federal Highway Administration, Washington, D.C.

Morgan, N. 2002. The influence of variation in effective stress on the serviceability of soil nailed slopes. PhD Thesis, University of Dundee, United Kingdom.

O'Reilly, M.P. & Brown, S.F. (eds) 1991. *Cyclic loading of soils from theory to design*. Blackie, London.

Pedley, M.J. and Pugh, R.S. 1992. Soil nailing in the Hastings beds, Engineering Society Geological Special publication No. 10, 6–10 Sept. pp 361–368.

Powrie, W. & Daly M.P. 2002. Centrifuge model tests on embedded retaining walls supported by earth berms, Geotechnique 52, No. 2, 89–106.

Taylor, R.N. 1995. *Geotechnical Centrifuge Technology*. Blackie Academic and Professional.

Tei, K. 1993. *A study of soil nailing in sand*, PhD Thesis, University of Oxford, United Kingdom.

Vaughan, P.R. and Walbanche, H.J. 1997. Pore pressure changes and the delayed failure of cutting slopes in over-consolidated clay. *Géotechnique*, 23, No.4, 531–539.

Zornberg, J.G., and Arriaga, F. 2003. Strain Distribution within Geosynthetic-Reinforced Slopes. *Journal of Geotechnical and Geoenvironmental Engineering, ASCE*, Vol. 129, No. 1, pp. 32–45.

Zornberg, J.G., Sitar, N., and Mitchell, J.K. 1998. Performance of Geosynthetic Reinforced Slopes at Failure. *Journal of Geotechnical and Geoenvironmental Engineering, ASCE*, Vol. 124, No. 8, pp. 670–683.

New Horizons in Earth Reinforcement – Otani, Miyata & Mukunoki (eds)
© 2008 Taylor & Francis Group, London, ISBN 978-0-415-45775-0

Combined technology with other technique – Current innovation on earth reinforcement technique

D.T. Bergado, Y.P. Lai & T. Tanchaisawat
GTE Program, School of Engineering and Technology, Asian Institute of Technology, Bangkok, Thailand

S. Hayashi & Y.J. Du
Institute of Lowland Technology, Saga University, Saga, Japan

ABSTRACT: This paper contains current innovations in combining the earth reinforcing with other ground improvement technologies. Obviously, this paper cannot cover all aspects of combined technologies. Consequently, a few aspects of combined technologies are only covered here. First, the combination of earth reinforcement and deep mixing method with cement admixtures is described. Second, the combination between earth reinforcement and lightweight fill consisting of tire chips-sand mixtures is presented. Finally, the innovative applications of dual function geosynthetics combining reinforcement and drainage are discussed. The data and subsequent analyses cited in this paper are based on the actual laboratory and field tests performed on lowland environment of soft Bangkok clay in the Central Plain of Thailand.

1 INTRODUCTION

Combined technology involving earth reinforcement and other soil improvement techniques has been used in various researches and actual projects nowadays. The numerous schemes include, but not limited to such various combinations as follows: lightweight fill with reinforcement, dual function geosynthetics, above ground and below ground improvement, etc. This paper will present some combined technology involving soft Bangkok clay and the surrounding environment in Thailand.

2 REINFORCED EMBANKMENT ON DMM IMPROVED FOUNDATIONS

2.1 *Introduction*

Deep mixing method (DMM) with cement admixtures has been successfully applied in Thailand as foundation improvement for highway embankments, retaining structures for foundation works, etc. (Bergado et al, 1999; Petchgate et al, 2003). The methods of mixing generally used in DMM installation are either mechanical mixing or jet grouting/mixing (Kamon and Bergado, 1991; Porbaha, 1998). In the mechanical mixing, the cement admixtures are mixed into the clay by mixing blades. In jet grouting/mixing, the grouting/mixing is done by high pressure jets of cement slurry. A full scale study of soft clay foundation improved by deep mixing piles installed by jet grouting/mixing technique was performed. The improved ground was loaded with a 6 m high reinforced embankment and was instrumented and monitored. Back-analyses were performed in order to derive essential parameters for the analyses of deep mixing improved ground supporting a reinforced embankment.

2.2 *Project site and soil profile*

The site of the full scale study was located in Wangnoi, Ayuthaya, Thailand. The site is underlain by the well-known soft Bangkok clay deposit. The properties of the foundation soil are shown in Figure 1. The soft clay which is overlain by 1.0 m thick weathered crust layer and 1.5 m thick clay fill, is encountered from 2.5 to 9.0 m depth. The corrected undrained shear strength obtained from field vane test is approximately 15 kPa. Underlying the soft clay layer is the medium stiff to stiff clay layer, having undrained strength of more than 50 kPa. The soft clay layer has thickness of 6.5 m.

2.3 *Installation of deep mixing piles by jet grouting*

The soft foundation subsoil was improved with deep cement mixing (DMM) piles which were installed in-situ by the jet grouting/mixing method employing a jet pressure of 20 MPa (Bergado et al, 2004) The deep mixing piles were installed at 1.5 m center to center spacing in square pattern, except for the perimeter deep mixing piles which has 2.0 m center to center spacing (see Figs. 2, 3 and 4). The water to cement ratio (W/C) of the cement slurry and the cement content were 1.5 and 150 kg/m^3, respectively. Each deep mixing pile had diameter of 0.5 m and length of 9.0 m. The deep cement mixing piles were allowed to cure and the dissipation of excess pore pressure were monitored until about 80 days when the overlying embankment was constructed.

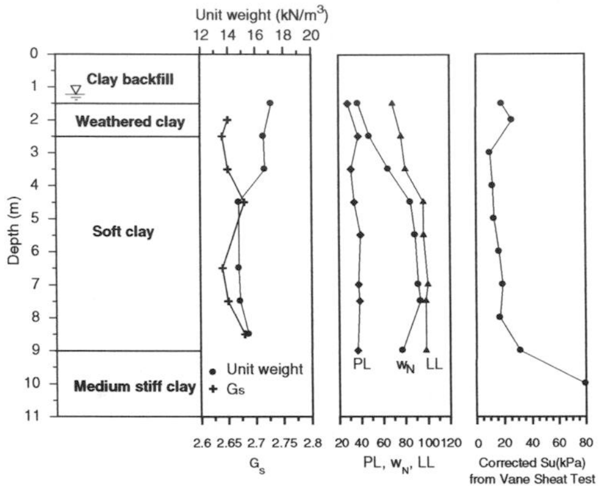

Figure 1. Soil profiles under the test embankment.

2.4 *Construction of reinforced embankment*

The full scale test embankment was constructed with well-compacted silty sand backfill reinforced with PVC-coated hexagonal wire mesh, (Lai et al, 2006). The backfill soil had compacted unit weight of 18.20 kN/m^3, cohesion of 7.70 kPa and angle of friction of 22 degrees with maximum dry density of 16.1 kN/m^3 at 15% optimum water content. To support the vertical side of the embankment, precast concrete facing with dimensions of 1.50 × 1.50 × 0.15 m were utilized. Each precast concrete block was attached to 2 hexagonal wire mesh reinforcements having a vertical reinforcement spacing of 0.75 m. All reinforcements were 4.0 m long and were arranged horizontally behind the precast concrete facing. The first layer of the reinforcement was located above the ground surface. The completed test embankment was 6.0 m high and its construction was finished within 15 days.

2.5 *Pullout resistance of reinforcement*

Pullout tests were conducted in the laboratory using a 1.27 × 0.76 × 0.51 m pullout box on the PVC-coated

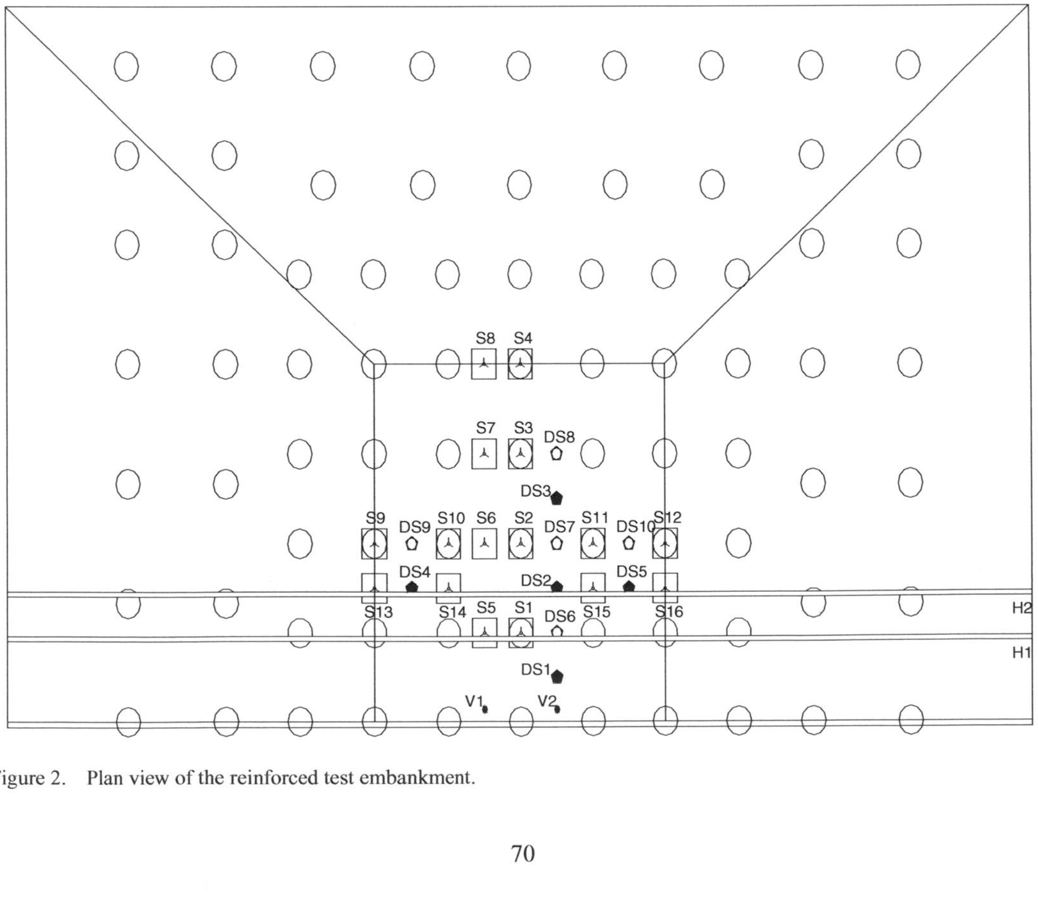

Figure 2. Plan view of the reinforced test embankment.

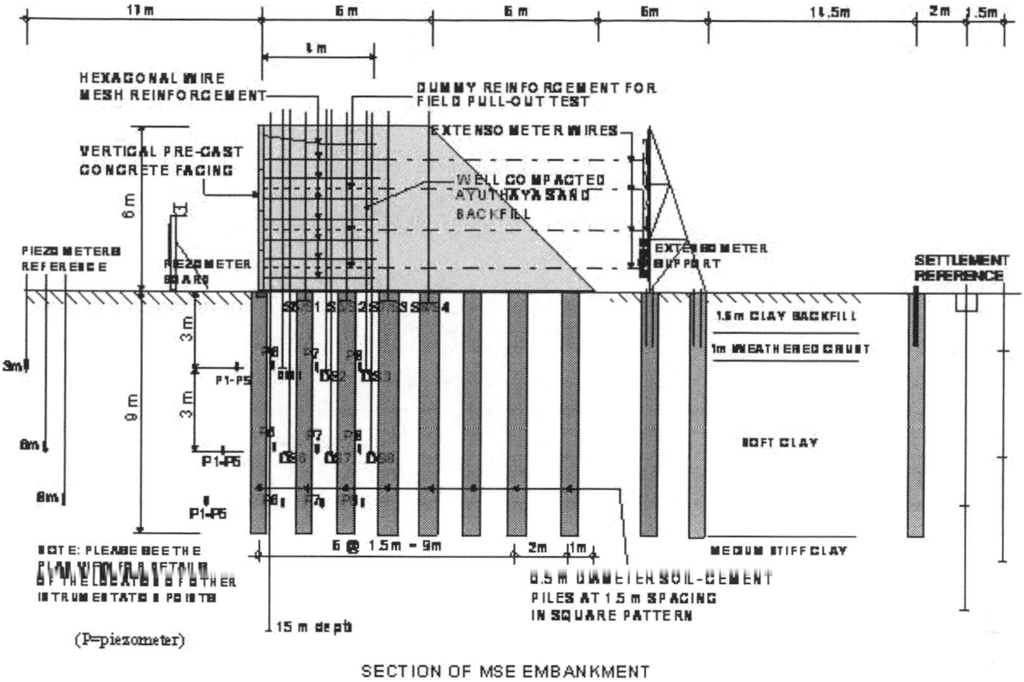

Figure 3. Longitudinal cross-section of the reinforced test embankment.

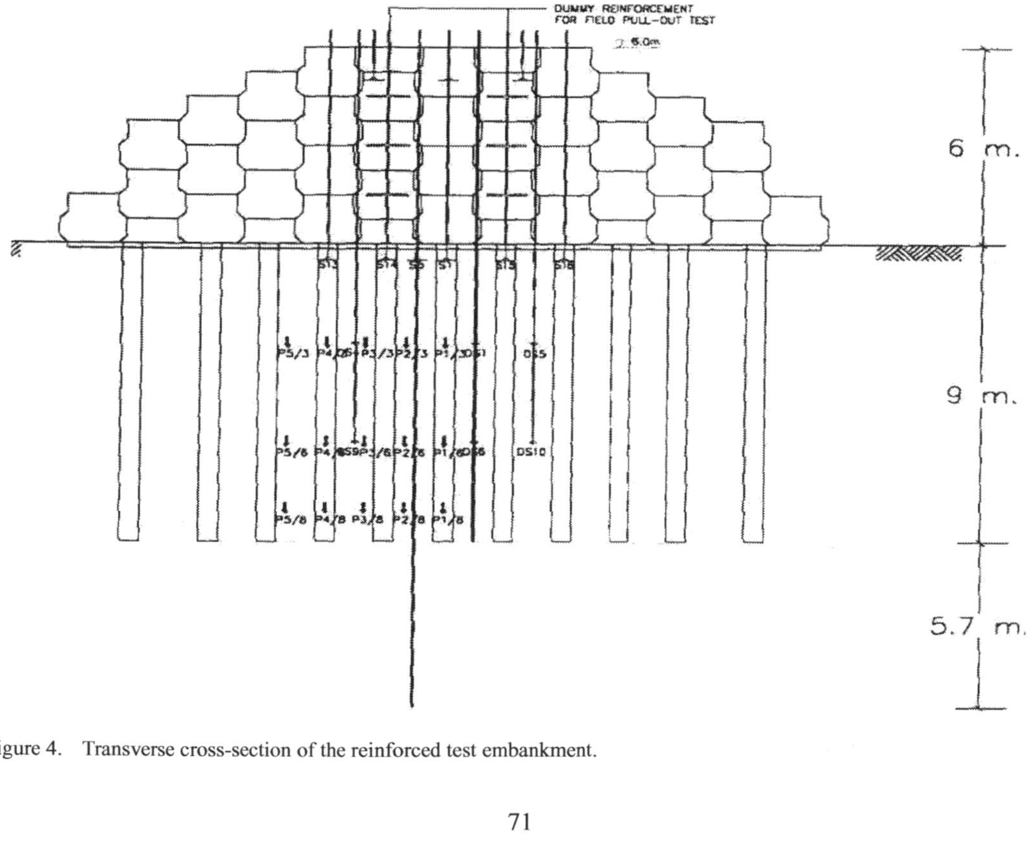

Figure 4. Transverse cross-section of the reinforced test embankment.

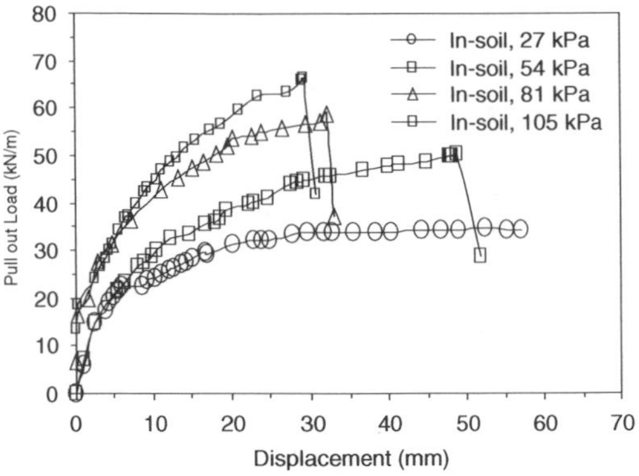

Figure 5. Laboratory pullout resistance of PVC-coated hexagonal wire reinforcement.

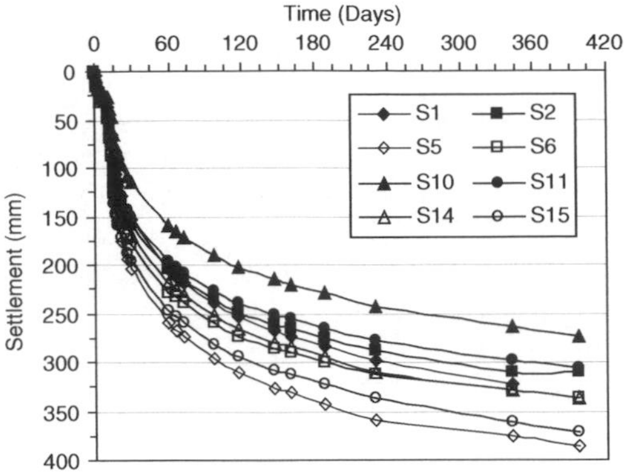

Figure 6. Surface settlement "on clay" and "on piles" near the center of test embankment (solid symbols = on piles; hollow symbols=on clay).

hexagonal wire reinforcement (Bergado et al, 2004). The pullout resistance plotted in Figure 5 correspond to the reinforcement embedment length of 0.90 m. As shown in Figure 5, the pullout resistance increased while the maximum pullout displacement decreased with increasing normal pressure.

2.6 *Instrumentation and monitoring*

The test embankment and the improved foundation were instrumented. Piezometers(P) which monitored the excess pore pressure in the foundation during and after deep mixing pile installation were installed at various points within and outside the embankment zone (Figs. 3 and 4). Surface settlement plates (S) were installed both "on pile" and "on clay" at the bottom of the embankment. Deep settlement plates (DS) were also installed at 3.0 and 6.0 m depths at few locations (Figs. 2 and 3). In addition, vertical and horizontal inclinometers were placed near the vertical side and at the bottom of the test embankment to monitor the lateral and settlement profiles, respectively. The monitoring program lasted for 12 months starting from the installation of the deep mixing cement piles.

2.7 *Construction and consolidation settlements*

Figure 6 shows the settlements on top of the deep mixing piles and on the surface of the surrounding clay in-between the piles during and after construction up to one year of embankment loading. The average settlements on deep mixing piles and on clay amounted to about 122 and 162 mm, respectively, after construction as well as 285 and 355 mm, respectively, one year after construction. From the method of Asaoka (1978), the average total settlements on deep mixing pile and on surrounding ground amounted to 340 and 440 mm, respectively. Thus, about 40% of the total settlements occurred during the construction of the embankment. Moreover, without the deep mixing piles, the total settlements could have been greater than 1.0 m (Lorenzo and Bergado, 2003), The deep mixing pile have, therefore, transferred the load down to their pile tips and, consequently, reduced the amount of settlements in the clay foundation by about 70%.

The deep mixing piles also resulted in faster rate of consolidation. The consolidation settlements of the improved foundation reached 90% degree of consolidation one year after construction. Otherwise, this degree of consolidation could have been attained 9 years after construction without ground improvement for 6.5 m thickness of soft clay layer using the vertical coefficient of consolidation, C_v, of 4 m^2/yr.

2.8 *Differential settlements on top and in-between deep mixing piles*

The differential settlements between the deep mixing pile and the ground surface in-between piles ranged from 25 to 60 mm (Fig. 6) which correspond from 8 to 20% of the average total settlements. However, the differential settlements were not visible at the top of the test embankment due to the combined effects of fill compaction as well as reinforcement stiffness and arching of the overlying reinforced soil in-between deep mixing piles. Significantly, the data in Figure 6 also demonstrated that the magnitudes of the differential settlements between the deep mixing piles and the surrounding ground have been fully attained one month after embankment construction. Consequently, for reinforced road embankments supported by deep mixing piles, differential settlements no longer occurred one month after its construction as evidenced by the parallel settlement curves in Figure 6.

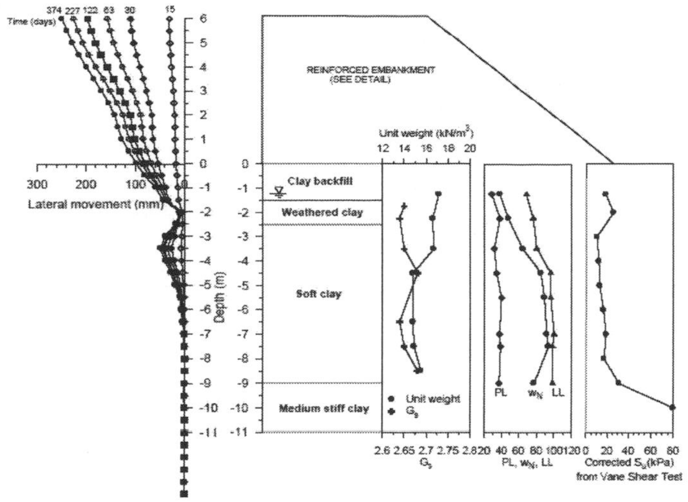

Figure 7. Lateral movement profiles of test embankment on deep mixing piles.

2.9 *Lateral movements*

The lateral movement profiles of the improved foundation as well as the vertical facing of the reinforced embankment are plotted together in Figure 7. The maximum measured lateral movements in the improved foundation amounted to 5 and 45 mm, respectively, just after and 7 months after embankment construction. The maximum lateral movement occurred at 3.5 m depth corresponding to the weakest part of the soft clay foundation. Since the corresponding amount of vertical settlements amounted to 162 and 325 mm, respectively, these magnitudes of lateral movements were only 3% and 14% of the corresponding vertical settlements of the embankment.

The top of the vertical facing of the reinforced embankment experienced a forward movement of only 30 mm just after embankment construction and increasing to 230 mm at 7 months after construction. These lateral movements can be attributed to the lateral movements of the improved foundation as well as the rigid body rotation of the reinforced embankment due to the unequal settlements of the improved foundation. At seven months after embankment construction, the translational movement at the bottom of the embankment amounted to 90 mm which is caused by the horizontal thrust from the sloping side of the embankment. Moreover, after 7 months, the embankment underwent rotation due to the uneven settlements of the improved foundation as shown in Figure 8. The gradient of the settlement profile at the bottom of the embankment after 7 months as can be interpreted from Figure 8 is 0.02 or 20 mm vertical per meter horizontal. Assuming the reinforced embankment behaved rigidly after mobilizing the interface resistances of the reinforcement, this level of rotation will mobilize a magnitude of 120 mm at the top of the 6.0 m high embankment. Thus, after 7 months, the lateral movement at the top of the embankment vertical facing can be composed of the following, namely: 14 mm elongation of the reinforcements, 90 mm translational movement of the embankment, and 120 mm due to the rotation of the embankment. Subsequently, the total horizontal movement at the top of the embankment can be 224 mm which is close to the measured value of 230 mm as shown in Figure 7. Therefore, the lateral movement of the vertical facing of the embankment was greatly affected by the unsymmetrical configuration and loading of the embankment and the consequent uneven settlements of the improved foundation.

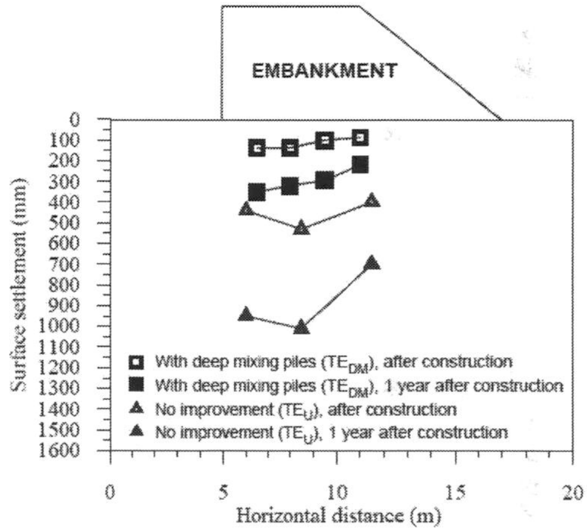

Figure 8. Comparison of settlement profiles between TE_u and TE_{DM}.

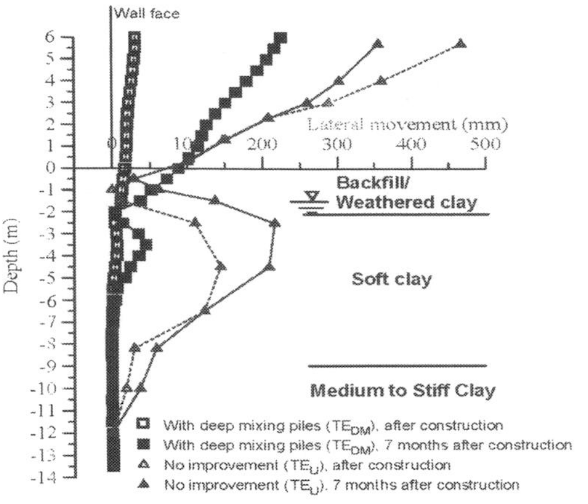

Figure 9. Comparison of lateral movement between TE_u and TE_{DM}.

2.10 *Effects of deep mixing piles on reinforced embankment*

The degree of improvement of the soft clay foundation improved by deep mixing cement piles can be assessed by comparing the settlements and lateral movements to the previously constructed reinforced embankment (TE_u) constructed on unimproved soft clay foundation in adjacent site with similar embankment configurations as shown in Figure 10 a, b. The TE_u embankment had 5.7 m height, reinforced with 6.1 mm by 5.4 mm diameters galvanized steel grids with 152 mm by 228 mm grid geometry in the longitudinal and transverse directions, respectively.

The reinforced embankment (TE_u) was constructed on the similar soil profile and properties on soft Bangkok clay as shown in Figure 11. The full scale reinforced embankment on improved foundation is designated as TE_{DM}. Both test embankments, TE_u and TE_{DM}, were reinforced embankments underlain by similar soil profiles and properties with both underlain by 6.5 m thick soft Bangkok clay layer.

The installation of deep mixing piles in the soft clay foundation yielded favorable effects on the reinforced embankment. The comparison of the settlements of TE_u on unimproved foundation and TE_{DM} on improved foundation is plotted in Figure 8. One year after construction of settlement of TE_u is 1.0 m while that of TE_{DM} is 0.31 m. Thus, the vertical compression of the improved foundation is reduced by as much as 70%.

The lateral deformation of the foundation is also reduced. At 7 months after construction, the lateral movement of the soft clay foundation under TE_u amounted to 220 mm white the corresponding value for TE_{DM} reached only 50 mm, a reduction of nearly 80%.

Moreover, the decrease in the lateral movement and decrease in compressibility of the improved foundation reduced the lateral movement of the vertical facing of the reinforced embankments. At 7 months after construction, the lateral movement at the top of the vertical facing of TE_u amounted to 450 mm compared to only 230 mm for TE_{DM}. Therefore, the installation of the deep mixing piles in the soft clay foundation has reduced the overall lateral movement of the reinforced embankment.

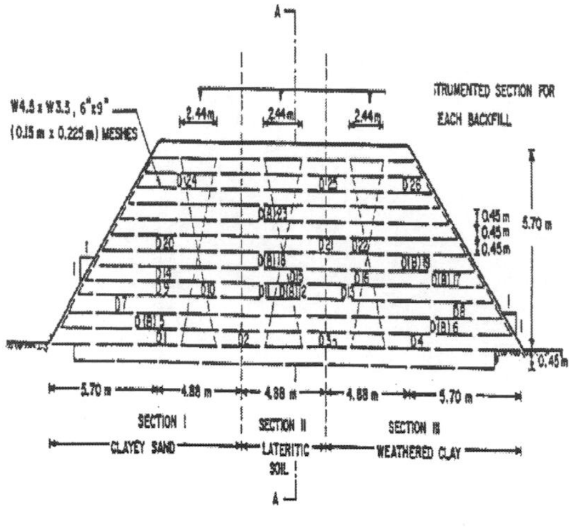

Figure 10a. Longitudinal section of test embankment TE_u.

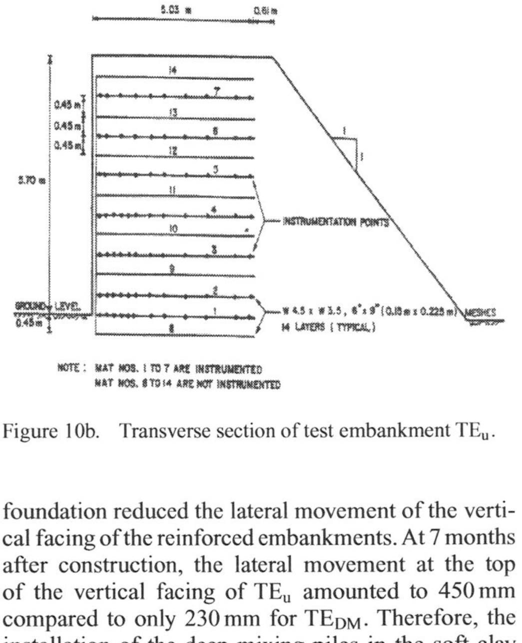

Figure 10b. Transverse section of test embankment TE_u.

2.11 *Analytical model for the rate of settlement of deep mixing improved ground*

To calculate the average degree of consolidation of the soil-cement pile improved ground, the following modified time factors obtained from the analytical model of Lorenzo and Bergado (2003) must be substituted to the standard solution to any approximate solutions

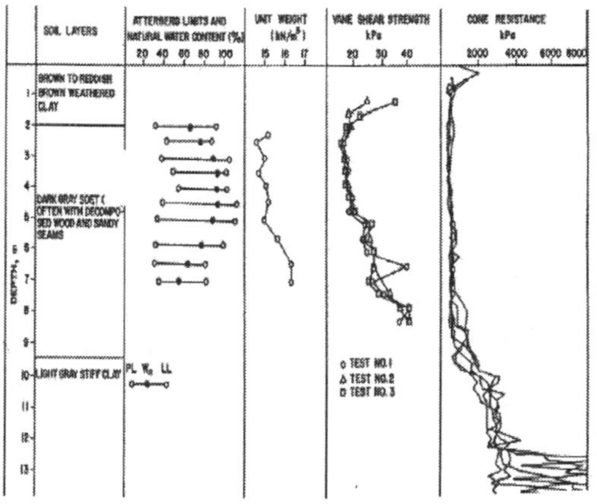

Figure 11. Subsoil at the site of test embankment TE_u.

(e.g., Sivaram and Swamee, 1977) of one-dimensional consolidation equation:

Equal stress condition between DMM pile and soil:

$$T_{v,\sigma} = \left(\frac{\left(\frac{m_{v,p}}{m_{v,c}}\right)}{\left(\frac{m_{v,p}}{m_{v,c}}\right) + (n^2 - 1)\left(\frac{C_c}{C_s}\right)_p} \right) \left(\frac{C_{v,p} t}{H_p^2} \right) \tag{1}$$

Equal strain condition between DMM pile and soil:

$$T_{v,\varepsilon} = \left(\frac{1}{1 + (n^2 - 1)\left(\frac{C_c}{C_s}\right)_p} \right) \left(\frac{C_{v,p} t}{H_p^2} \right) \tag{2}$$

where:$(C_c/C_s)_p$ is the ratio of the compression and swelling indices of the DMM pile at stress level corresponding to loading condition (if the DMM pile does not reach to its yield stress, this constant can be taken as unity); $C_{v,p}$ is coefficient of consolidation of the DMM pile material as obtained from oedometer test; $m_{v,p}/m_{v,c}$ is the ratio of the coefficient of volume change of the DMM pile and the surrounding clay; n equal to (D_e/D_p) is the ratio of the equivalent diameter of the unit cell to the diameter of the pile; where D_e is equal to 1.03S and 1.13S corresponding to triangular and square pattern of the piles, respectively, S is the center-to-center spacing of the piles, and D_p is the diameter of pile; H_p is the effective longest drainage path of the consolidating soil-cement pile; t is the time when a particular degree of consolidation is desired.

The actual load transfer mechanism is neither equal strain nor equal stress; however, it must fall within these two extreme conditions. The actual average degree of consolidation can be better estimated by applying appropriate weighting factor to each average degree of consolidation from the two extreme conditions, namely: equal stress and equal strain conditions, (Lai et al, 2006). Thus, the actual average degree of consolidation of the improved ground, $\overline{U}$, will be predicted using the following relationship:

$$\overline{U} = \alpha_\varepsilon\left(\overline{U}_{v,\varepsilon}\right) + \alpha_\sigma\left(\overline{U}_{v,\sigma}\right) \tag{3}$$

where α_ε and α_σ are the weighting factors of the average degree of consolidation corresponding to equal strain and equal stress conditions, respectively; $\overline{U}_{v,\varepsilon}$ is the average degree of consolidation under equal strain condition calculated using the standard solution or to any approximate solutions (e.g., Sivaram and Swamee, 1977) of one-dimensional consolidation equation with the time factor, $T_{v\varepsilon}$, given in Eq. (2); and $\overline{U}_{v,\sigma}$ is the average degree of consolidation under equal stress condition calculated using the time factor, $T_{v\sigma}$, given in Eq. (1). Obviously, the sum of these two weighting factors, α_ε and α_σ, must be equal to unity (Lai et al, 2006).

The consolidation parameters as well as the strength parameters of soil-cement piles used in the back-analyses were estimated based from the test piles, which were installed few meters away from the embankment. Petchgate *et al.* (2003) reported the following properties of the tested soil-cement piles: water content = 160%, γ_{wet} = 12.8 kN/m^3; q_u = 300~700 kPa and E_{50} = 60,000 ~ 120,000 kPa. From laboratory tests, the specific gravity of the cement-admixed clay composing the pile is about 2.65. Accordingly, the after-curing void ratio of cement-admixed clay composing the soil-cement pile can be obtained as 4.3, which is almost twice the void ratio of the natural clay. At this magnitude of after-curing void ratio of soil-cement piles, the coefficient of vertical permeability, $K_{v,p}$, ranges from 150 to 200 × 10^{-10} m/s and the corresponding coefficient of consolidation, $C_{v,p}$, ranges from 200 to 400 m^2/yr (Lorenzo and Bergado, 2003). In addition, for the surrounding clay, the coefficient of vertical permeability, $K_{v,c}$, ranges from 3 to 6 × 10^{-10} m/s and the corresponding coefficient of consolidation, $C_{v,c}$, ranges from 1 to 3 m^2/yr (Lorenzo and Bergado, 2003).

2.12 Results of analytical back-analysis

Figures 12 and 13 show the predicted settlement-time plots together with the corresponding measured settlement-time plots from settlement plates S1 vs. S5, and S11 vs. S15, respectively (refer to Fig. 2 for their locations). Two adjacent settlement plates are paired, one on DMM pile and other one on the adjacent clay in-between DMM. In the analysis, the immediate settlement and the consolidation settlement were first obtained by trial until the actual settlement just after embankment construction and one year after

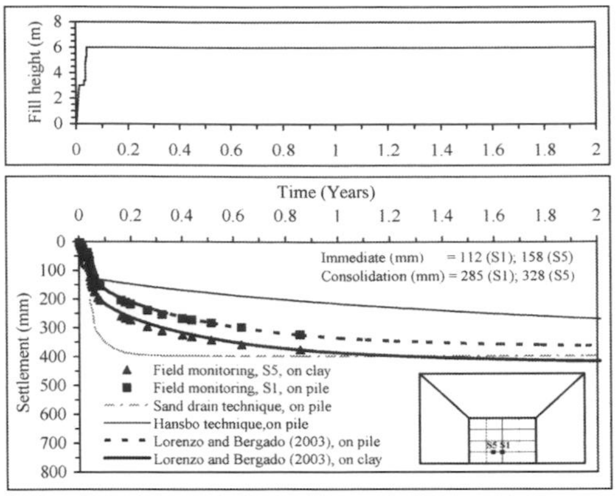

Figure 12. Comparison between field data and back-analyses (S1 vs S5).

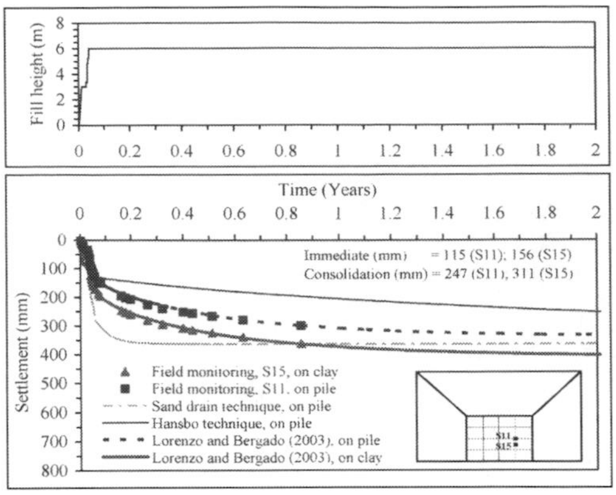

Figure 13. Comparison between field data and back-analyses (S11 vs S15).

construction (last observed data) agreed to the predicted or projected ones. The behavior of the settlement against time reflects the consolidation properties such as permeability ratio ($k_{v,p}/k_{v,c}$), compressibility ratio ($m_{v,p}/m_{v,c}$), and the coefficient of consolidation of the deep mixing pile (C_{vp}).

The good agreement between the measured and the predicted settlement-time plots shown in Figures 12 and 13 were obtained using the coefficient of consolidation of the pile (C_{vp}) of 800 m^2/yr and coefficient of consolidation of surrounding clay ($C_{v,c}$) of 2.0 m^2/yr. Bergado et al. (1999) and Lorenzo and Bergado (2003) also utilized coefficient of consolidation of surrounding clay ($C_{v,c}$) of 2.0 for the back-analysis of Bangna-Bangpakong Highway embankment which was also improved by deep mixing method. Moreover, the compressibility ratio ($m_{v,p}/m_{v,c}$) of 0.10 was used, which also confirmed to the back-analysis of another case study done previously by Lorenzo and Bergado (2003). Consequently, the permeability ratio ($k_{v,p}/k_{v,c}$) was derived as 40, which is twice to what was obtained by Bergado et al. (1999) and Lorenzo and Bergado (2003) for the Bangna-Bangpakong Highway embankment. The higher permeability ratio obtained from this study compared to the previous case of Bangna-Bangpakong Highway embankment could be attributed to the different method of mixing applied. The deep mixing piles supporting the reinforcement embankment mentioned in this paper were installed by jet mixing method with slurry of cement employing a jet pressure of 20 MPa, while those supporting the Bangna-Bangpakong Highway embankment were installed by mechanical mixing method with slurry of cement as reported by Bergado et al. (1999). It has been found from the laboratory investigation that the increase in mixing water content corresponds to an equivalent increase in the after-curing void ratio and, hence, the coefficient of permeability and consolidation of the resulting cement-admixed clay (Lorenzo and Bergado, 2003; Lorenzo and Bergado, 2004). Thus, in the case of the present test embankment, the higher water content deliberately added into the soil during jet mixing must have effected the higher coefficient of consolidation of the resulting soil-cement piles compared to those from Bangna-Bangpakong Highway embankment which were installed by mechanical mixing method.

Furthermore, the weighting factors of the average degree of consolidation, α_ε and α_σ, as mentioned in Eq. (3) corresponding to "equal strain" and "equal stress" conditions, respectively, that were utilized in the analysis and simulated closely to the actual rate of settlement of the improved ground are 80% for equal strain and 20% for equal stress. This means that at any time the overall degree of consolidation of the improved ground was taken equal to 80% of the average degree of consolidation under equal strain condition plus 20% of the average degree of consolidation under equal stress condition (Lai et al, 2006). The 80% for equal strain can be possible since the bottom layer of the reinforcement was not located in the ground surface but rather located above ground as shown in Figure 3.

Moreover, the corresponding settlement-time curves from the existing methods, such as Hansbo (1979) for prefabricated vertical drain and Barron (1948) for sand drains, of predicting the consolidation settlement using unit cell technique are also presented in Figures 12 and 13. Each of these figures shows that the sand drain technique overestimated the rate of settlement of soil-cement pile improved ground. It follows, therefore, that the soil-cement pile cannot behave exactly as sand drain. This is because, owing to the very high permeability of sand versus soil-cement pile, the consolidation process of sand can occur very quickly as compared to that of the soil-cement pile. On the other hand, the use of Hansbo's (1979) technique, which assumed that soil-cement pile could be converted to an equivalent vertical drain, underestimated the actual settlement. Lorenzo and Bergado (2003) presented the following reasons. First, due to the fact that the discharge capacity of the soil-cement pile is quite small compared to that of actual prefabricated vertical drain; thus, much time are required to discharge the excess pore water in the soil-cement pile. Second, the very different boundary conditions that must be met for soil-cement pile unit cells versus prefabricated vertical drain unit cells. While the prefabricated vertical drain is considered to be a drain within the unit cell, the pile in a soil-cement unit cell cannot be considered as a drain due to its lower permeability and lesser capacity to discharge water than the prefabricated vertical drain. This means that at any time the excess pore water pressure at any depth of soil-cement pile cannot be assumed as zero, which is otherwise assumed as zero in the prefabricated vertical drain analysis. Moreover, the consolidation process of the pile in a soil-cement pile unit cell is quite quantifiable and cannot be assumed to as "quick" as in the case of either sand drains or prefabricated vertical drains.

Significantly, the good agreement of the predicted settlement-time plots, predicted using the method of Lorenzo and Bergado (2003), against the measured ones does not only confirm to the previous finding of Lorenzo and Bergado (2003) but also indicate the suitability of the method of deep mixing improved ground to which it was designed for.

3 REINFORCED EMBANKMENT WITH LIGHTWEIGHT TIRE CHIPS-SAND FILL

3.1 *Introduction*

Generally, to improve the stability and performance of infrastructures on soft foundation, two alternative methods are available, namely: 1) improve the strength and deformation characteristics of the foundation, and 2) reduce the weight of the structure acting on the soft foundation. The former was discussed in Section 1 of this paper. The latter will be discussed in this section. The lightweight expanded polystyrene (EPS) was first used for road embankment fill in Oslo, Norway (Freudelund and Aaboe, 1993). Yasuhara (2002) summarized the utilization of lightweight geomaterials in Japan. Several materials and methods have been proposed to produce lightweight geomaterials and are classified into 3 categories as follows: 1) use of lightweight materials alone; 2) mix lightweight material with natural soil and/or cementing agents; and 3) add foam agent to reduce weight. The advantages of using lightweight geomaterials are not only the reduction of vertical pressures on foundation but also to decrease the lateral earth pressures as well as decrease the traffic induced vibration.

In recent years, there has been a growing emphasis on using industrial by-products and waste materials in construction. Used rubber tires, which are cut into small pieces called tire-chips, are waste materials that can be used as lightweight backfills to embankments. In order to reduce its compressibility and to prevent igniting fires due to its self-heating characteristics, these rubber chips are mixed with silty sand. To improve their stability, the tire chips-sand mixtures are reinforced with high strength geogrids. In this section, the performance of a full scale test embankment with tire chips-sand lightweight fill reinforced with geogrids is discussed. The data on settlements, excess pore pressures, and lateral movements are presented (Bergado, et al, 2007).

3.2 *Laboratory testing program*

The laboratory testing program consists of 4 phases. The first phase involves the determination of the physical properties for both the silty sand and tire chips. In the second phase, the compaction tests on tire chips-sand mixtures were performed on 30:70, 40:60 and 50:50 percent by weight. The third phase concerns with the in-air tensile test of the geogrid reinforcements. Finally, in the fourth phase, large scale direct shear and in-soil pullout tests were performed to investigate the interaction, between the geogrid reinforcement and the lightweight tire-chips-sand fill. Two types of geogrids were utilized in the laboratory tests, namely: a) Saint Gobain geogrids (DJG 120 × 120-1) denoted as Geogrid A, and, b) Polyfelt geogrids (GX 100/30) referred as Geogrid B (Fig. 14). Geogrid A is made of high tenacity polyester yarns knitted into mesh and coated with modified polymer mixture with high tensile strength, high modulus, and good creep resistance. Geogrid B consists of high molecular weight and high strength polyester yarns knitted into a stable grid with polymeric coating protection. Geogrids A and B have ultimate tensile strength of 120 kN/m and 100 kN/m, respectively.

The specific gravity tests of sand and tire chips were conducted according to ASTM D854-97 and ASTM C127-01, respectively. The grain size distributions of sand and tire chips were conducted according to ASTM D422-63. To obtain the optimum moisture content and maximum dry unit weight of the tire chip-sand mixtures, ASTM D689-91 test procedures were adopted. The pullout tests were done to investigate the interactions between the tire chip-sand mixtures and the 2 types geogrid reinforcements particularly the relationship between pullout force and pullout displacement. Four normal stresses, namely: 30, 60, 90, and 120 kPa were applied on the fill materials to

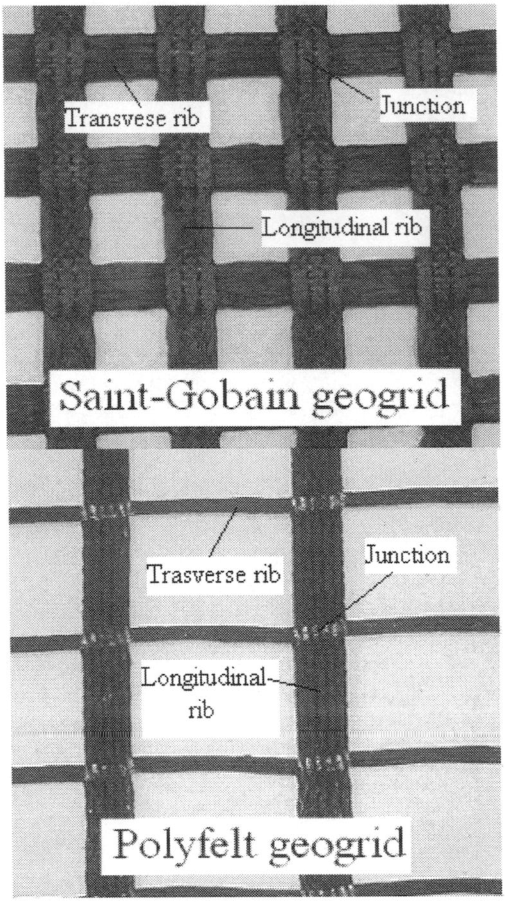

Figure 14. Direct shear interaction coefficient between Geogrid B and tire chip-sand mixture.

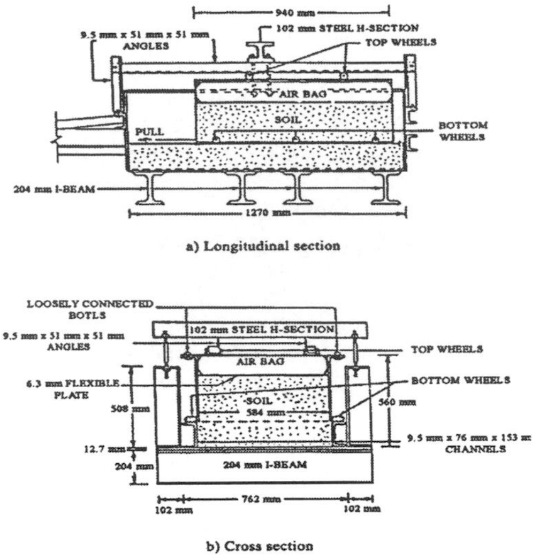

Figure 15. Schematic pullout test apparatus.

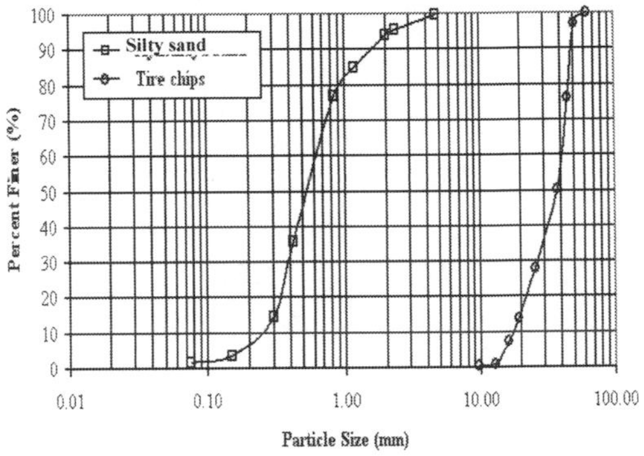

Figure 16. Particle size distribution of silty sand and tire chips.

cover the range of possible stress loads. The pullout machine used is shown in Figure 15 which is the same apparatus used to obtain the data in Figure 5. The pullout forces were generated by a 225kN capacity electro-hydraulic controlled jack. The normal stresses were applied by means of inflated air bag which was inserted between the flexible steel plate and the top cover of the pullout box. The load cell used to measure the pullout resistance and pullout displacement was connected to the electronic data logger. The pullout displacements of the geogrid reinforcements were monitored by using LVDTs connected to the inextensible wires mounted in the geogrid specimen at predetermined positions. The pullout displacements rate of 1mm/min was adopted during the tests. The geogrid specimens used in the pullout and direct shear test have dimensions of 500×700 mm. The large scale direct shear apparatus was adapted from the pullout machine with the same dimensions and the same rate of displacement.

3.3 *Laboratory test results*

The specific gravity of the silty sand was found to be 2.65 while the corresponding value for the tire chips was 1.12. The silty sand had 1.64 percent passing no. 200 standard sieve with D_{10} of 0.22 mm, D_{30} of 0.38 mm and $D_{60}$of 0.62 mm. Consequently, the uniformity coefficient was 2.82 and its gradation coefficient was 1.06. Thus, the silty sand was classified as poorly-graded sand (SP). For the tire chips most of the particle sizes ranged from 12 to 50 mm with irregular shapes due to the random cutting process. The grain size distributions of the silty sand and tire chips are plotted in Figure 16. The compaction test results from standard Proctor procedures of the tire chips-sand

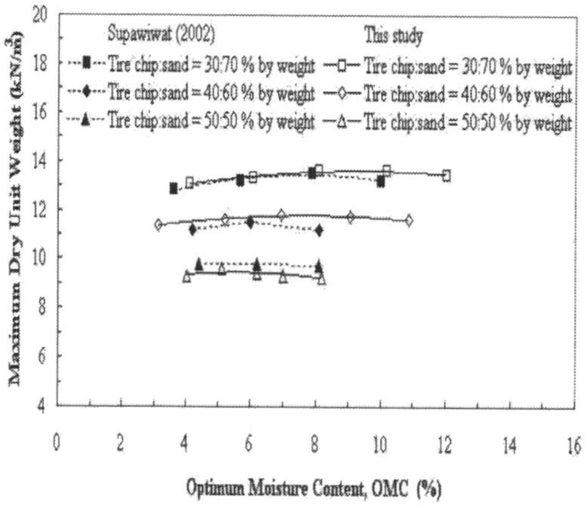

Figure 17. Compaction test results.

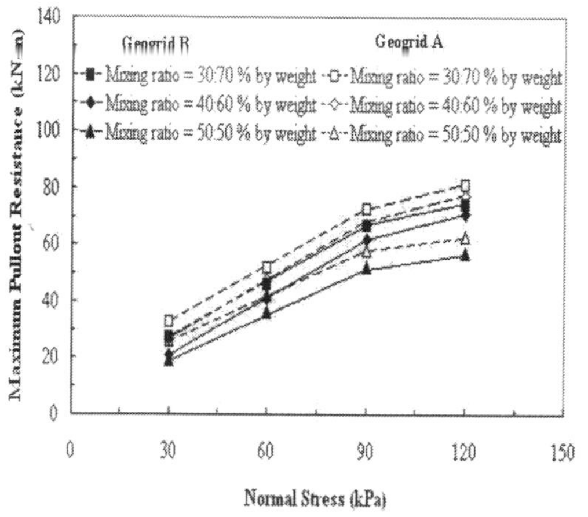

Figure 18. Maximum pullout resistance versus normal stress.

mixtures are plotted in Figure 17. The maximum dry unit weights and optimum moisture contents of the tire chips-sand mixtures vary from 9.5 to 13.6 kN/m^{3}and from 5.7 to 8.8, respectively.

The in-soil pullout test results as shown in Figure 18 revealed increasing pullout resistances with increased normal stresses while the displacement at maximum pullout resistances tended to decrease as the normal stresses increased. The mixing ratio of 30:70 percent tire chip-sand by weight yielded the highest pullout resistance for both types of geogrids. The frictional resistance influences more the pullout resistance than the bearing resistance. Hence, the sand content of the backfill directly affected the pullout resistance because of its higher frictional properties. The pullout resistances of both geogrids are compared

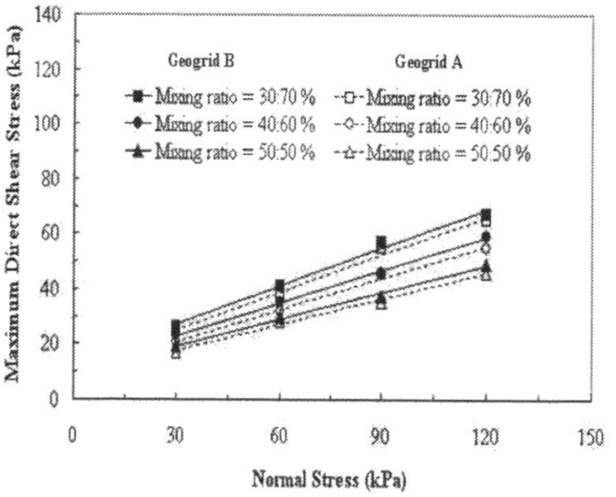

Figure 19. Maximum direct shear stress versus normal stress.

in Figure 18. The Geogrid A yielded slightly higher pullout resistance than Geogrid B.

As shown in Figure 19, at the same normal stresses and mixing ratios, the direct shear resistances of geogrid B were higher than those of geogrid A because the aperture sizes of the former is larger than the latter. At the same normal stresses and mixing ratios, the shear resistances of the tire chips-sand mixtures were higher than the corresponding values between both geogrids and the backfills.

3.4 *Full scale test embankment*

The test embankment was constructed in the campus of Asian Institute of Technology (AIT). The general soil profile consists of weathered crust layer of heavily overconsolidated reddish brown clay over the top 2 m. This layer is underlain by soft grayish clay down to about 8.0 in depth. The medium stiff clay with silt seams and fine sand lenses was found at the depth of 8.0 to 10.5 m depth. Below this layer is the stiff clay layer. Figure 11 summarizes the subsoil profile and relevant parameters which is the same locations as the previous study in Section 2. The geogrid reinforcement embankment was extensively instrumented both in the subsoil and within the embankment itself. Since the embankment was founded on a highly compressible and thick layer of soft, several field instruments were installed in the subsurface soils. The 3D illustration of full-scale field test embankment is shown in Figure 20. The instrumentation in the subsoil were installed prior to the construction of the geogrid reinforcement wall and consisted of the surface settlement plates, subsurface settlement gauges, temporary bench marks, open standpipe, groundwater table observation wells, inclinometers, dummy open standpipe, dummy surface settlement plates and dummy subsurface settlement gauges (Fig. 21). Six surface settlement plates were placed beneath the embankment at 0.45 m depth below the general ground surface. Settlements were measured by precise leveling with reference to a benchmark.

Twelve subsurface gages, six of which were installed at 6 m depth, the rest at 3 m depth below the general ground surface at different locations.Two dummy gages were also installed at depths of 3 m and 6 m. The pore water pressure was monitored by

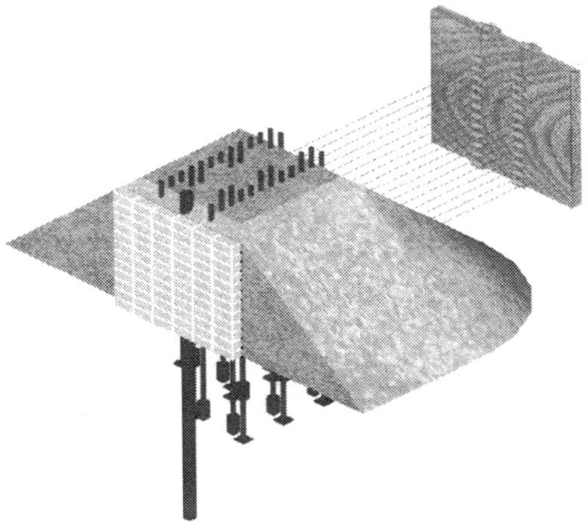

Figure 20. Embankment drawing in 3D.

the conventional open stand pipe piezometers. Six piezometers were installed in the soft clay subsoil at 3 m and 6 m depths from the ground level. Two of dummy open standpipes were installed at the area nearby temporary benchmarks. The lateral movements of the subsoil and the embankment were measured by using a digitilt inclinometer. The inclinometer was installed vertically near the face to measure the lateral movements of the vertical face of the wall and the subsoil. The inclinometers were installed down to 12 m below the original ground level. Wire extensometers were used to measure the displacements of the geogrid reinforcement and the surrounding soil as well. The extensometer consisted of a 2 mm diameter high-strength stainless cable inside a flexible PVC tube. One end of the wire was fixed at the measured point in the geogrid reinforcement and the other end was connected to a counterweight of about 0.8 kg through a pulley system at the readout board.

The construction of the reinforced embankment involved the precast concrete block facing unit connected to the geogrid reinforcement. The rubber tire chips were mixed with sand at the ratio of 30:70 by weight. The vertical spacing of the geogrid reinforcement was 0.60 m. The backfill was compacted in layers of 0.15 m thickness to density of about 95% of standard Proctor. The compactions were carried out with a roller compactor and with a hand compactor near the instrumentation such as settlement plates,

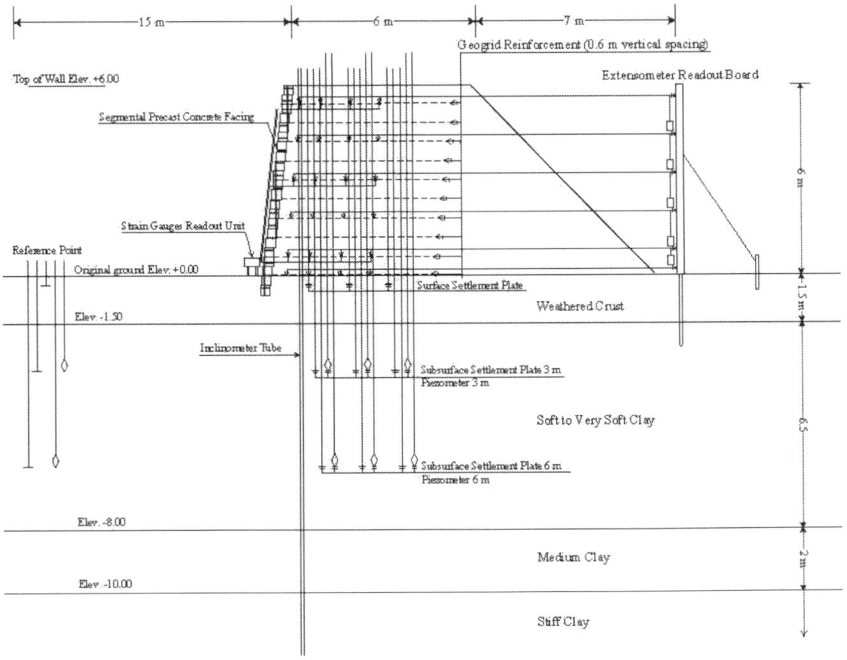

Figure 21. Section view of embankment with instrumentation.

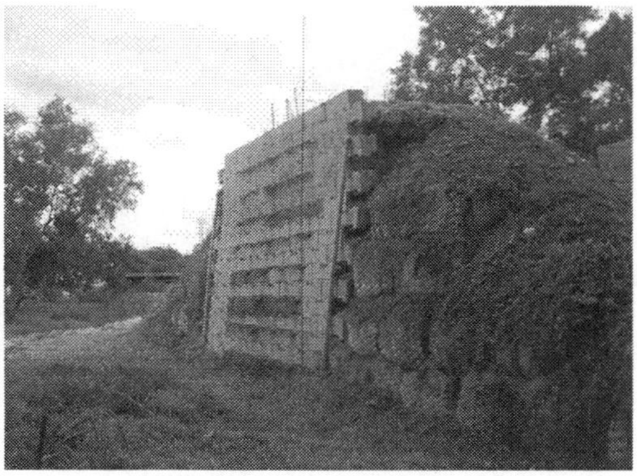

Figure 22. Completed full scale test embankment construction.

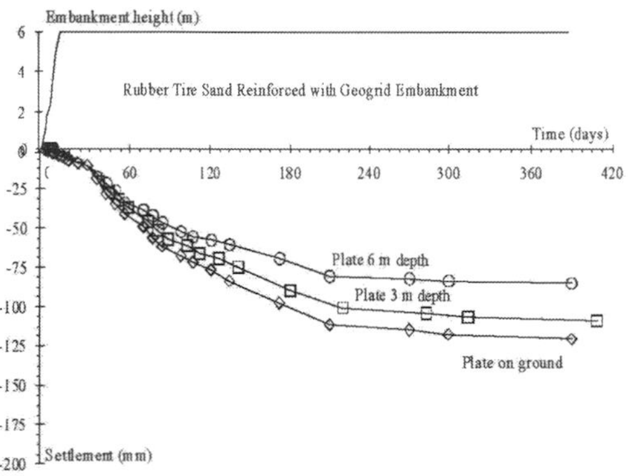

Figure 23. Observed average settlements at the different depths.

stand pipe piezometers and inclinometer. The degree of compaction and the moisture content were checked regularly at several points. The sand backfill was used as the surface cover for the rubber tire chips-sand with thickness of 0.6 m. Hexagonal wire gabions were used on both sides of the concrete facing at the front sloping side slopes. Figure 22 illustrates the completed embankment construction (Bergado et al, 2007).

3.5 *Observed behavior of test embankment on soft ground*

The observed surface and subsurface settlements of the test embankment are illustrated in Figure 23. During the construction period, immediate elastic settlements were observed. The rate of settlement was low in all the surface and subsurface settlement plates during the construction period. After the construction, the rate of settlement increased. After 210 days from the end of construction, the maximum settlement was 122 mm as recorded in surface settlement plates

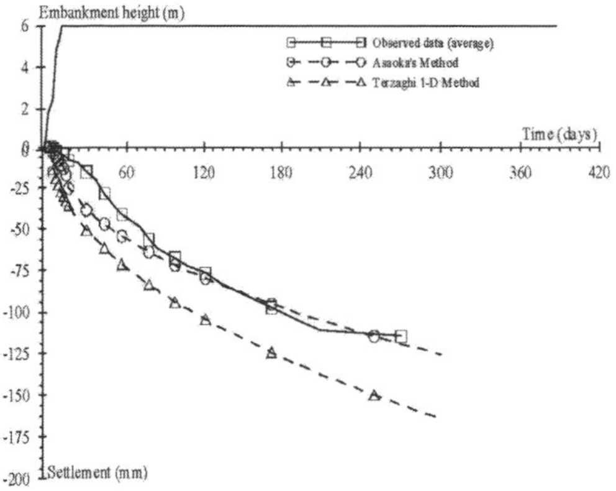

Figure 24. Observed and predicted surface settlement at original ground level.

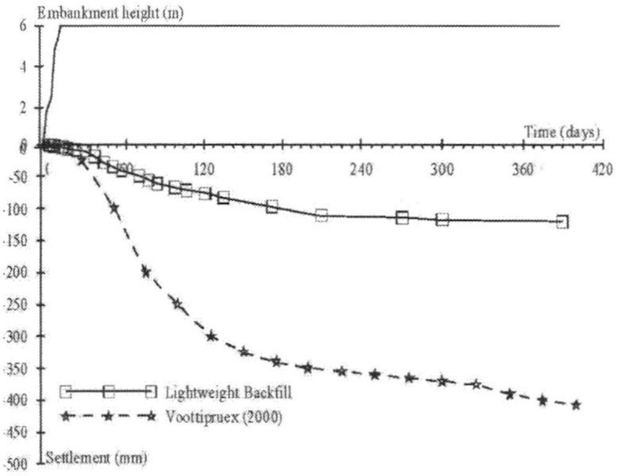

Figure 25. Comparison of settlement between conventional and lightweight backfill.

near the facing. This is because the weight of the concrete facing is more than the lightweight embankment and the forward tilting of the embankment. Along the cross-section of the embankment, settlement decreased from front (122 mm) middle (112 mm) and back (104 mm). The average surface settlement on the ground after 210 days from the end of construction is about 111 mm. The settlements at 3 m and 6 m depths were lower than at ground surface, as expected. The observed and predicted surface settlements of the test embankment are plotted together in Figure 24. As expected the predictions from Asaoka (1978) closely follow the observed data while the predictions from one-dimensional method overpredicted. Figure 25 demonstrates the comparison of the maximum settlements between conventional sand backfill reinforced embankment (Bergado & Voottipruex, 2000) and lightweight backfill embankment in this study. The maximum settlement of lightweight embankment

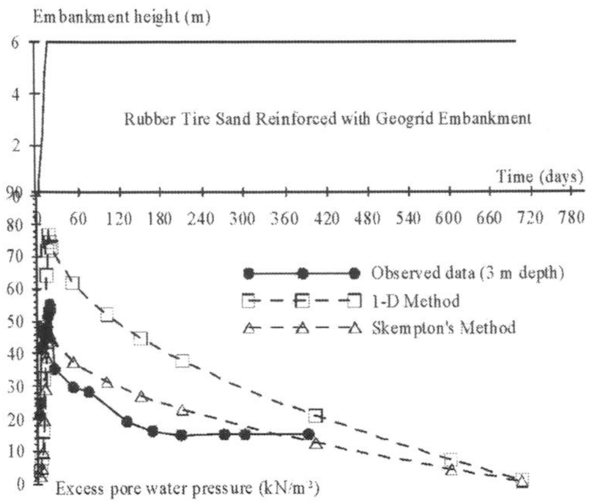

Figure 26. Observed and predicted excess pore water pressure at 3 m depth.

with unit weight of 13.6 kN/m^3 was 130 mm compared to 400 mm for conventional embankment with unit weight of 18 kN/m^3. The reduction of settlement amounted to 67.5%. The use of rubber tire chip-sand mixture as a lightweight geomaterials can alleviate the problem of settlement in soft ground area.

The excess pore water pressure below the lightweight embankment was obtained from open stand pipe piezometer. Figure 26 shows the excess pore water pressures during and after construction at 3 m depth. The maximum pore water pressure of 57 kN/m^3 occurred at 15 days after full height of embankment. The trend of excess pore water pressure dissipation is an indication of consolidation of soft foundation subsoil in the overconsolidation range when the load is below the maximum past pressure. After 50 days, the excess pore water pressure tends to dissipate very fast with time. The excess pore water pressure decreased to 18 kN/m^2 and 25 kN/m^2 at 3 m and 6 m depths, respectively. The excess pore water pressure becomes constant with time after 120 days from the end of construction. The observed and predicted excess pore water pressures below the embankment are also plotted in Figure 26. The 1-D method overpredicted excess pore water pressures while the predictions from the Skempton and Bjerrum (1957) method agrees well with the observed data

The lateral wall movement was measured by digitilt inclinometer which was located near the embankment facing. The plots of lateral wall movement with depth from top of embankment to 12 m depth below original ground are shown in Figure 27. The lateral wall movement was monitored once a week since end of construction and every month afterwards until 13 months. The lateral movement increased significantly until 4 months after construction and decreased to negligible amounts 13 months after embankment construction. The lateral movement occured in the short term after construction. The total wall movement is quite small from 95 to 100 mm (at the top of embankment). The maximum lateral movement in the soft clay subsoil occurred at about 3.0 to 5.0 m depth below the ground surface, corresponding to the weakest zone of the subsoil. The maximum lateral wall movement at the top was 100 mm which is less than the corresponding value of 350 mm as reported by Bergado and Voottipruex (2000) which used conventional backfill of silty sand reinforced with hexagonal wire. The lateral wall movement of this study is about 70 % less than that of the conventional backfill material. This indicates that the use of lightweight backfill significantly reduces the lateral movement of the embankment.

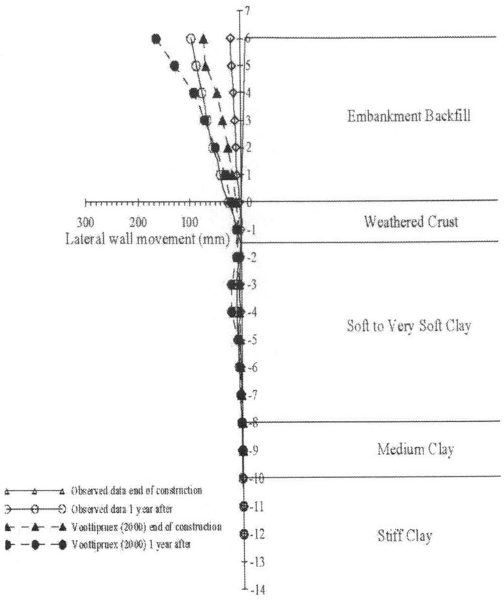

Figure 27. Observed lateral wall movement.

4 DUAL FUNCTION GEOSYNTHETICS

Recent innovation of soil: reinforcement for slope protection utilizes the versatility of dual function geosynthetic. Strictly, the term 'dual function' used herein includes only drainage and reinforcement functions. Moreover, the dual function geosynthetic studied in this paper refers to the high strength nonwoven/woven geotextile only.

Slope failures, caused by the drawdown of water level in an irrigation/drainage canal, occurred at Klong 15, 28 and 31 in Nakornnayok Province, Thailand. The deep mixing method (DMM) soil-cement columns had been implemented to solve such problem. However,

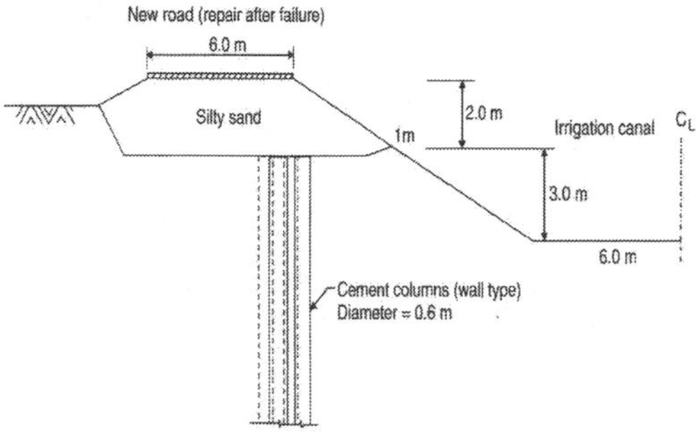

Figure 28. Use of DMM columns for the repair of irrigation canal at Klong 15, 29, 31.

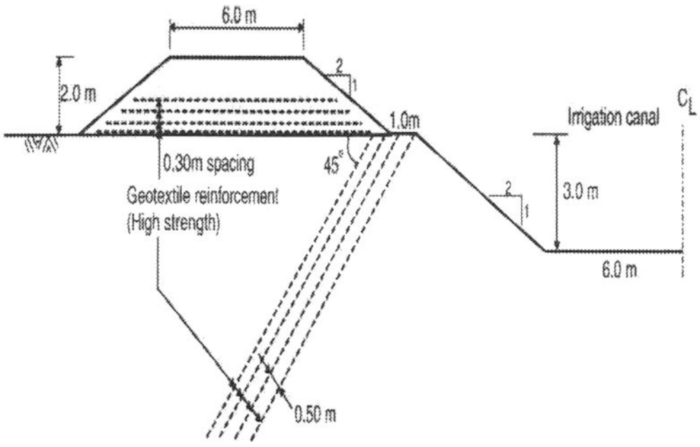

Figure 29. Slope reinforcement scheme using dual function high strength nonwoven/woven geotextile reinforcement.

due to its high-cost, the search for a cost-effective alternative has been undertaken. The possibility of using dual function geosynthetic has been studied (Lorenzo et al., 2004). The actual slope protection scheme using DMM soil-cement columns is shown in Figure 28, and the alternative scheme using dual function geosynthetic is given in Figure 29. The generalized soil profile at the site is summarized in Table 1. For the original scheme, soil-cement piles, with 0.60 m diameter and with 500 kPa UC strength, were installed down to 8 m depth. For the alternative scheme, dual function high strength nonwoven/woven geotextiles were utilized. In the latter scheme, two modes of slope instability were analyzed, namely, Case A with limitation of foundation deformation, and Case B with allowable large settlement. In Case A, the mobilized tensile force in the geotextile was calculated based on the critical strain (ε_c), which is about 3%, at the initiation of bearing failure of foundation soil. In Case B, the mobilized tensile force was calculated based on the critical strain (ε_c) plus the localized strain (ε_{lc}) associated with the development of slip surface at the onset of failure caused by the reorientation of reinforcement tension force (Bergado et al., 2003). Furthermore, the results of the stability analyses using SLOPE/W program are

Table 1. Parameters for limit equilibrium analysis.

Soil condition	Depth (m)	γ_t (kN/m^3)	S_u (kPa)	ϕ' (Degree)
Silty sand (embankment)		19.0		30
Weathered crust	0–2	17.5	25.0	–
Soft clay	2–16	15.0	12.5	–
Medium clay	16–22	16.0	25.0	–
Stiff clay	22–27	18.0	120.0	–

Table 2. Factor of safety calculated by SLOPE/W.

Description	PS (Case A)	FS (Case B)
Geotextile reinforcement		
(a) Inside embankment (2 layers)	0.983	1.022
(b) Inside embankment (4 layers)	1.022	1.260
(c) Inside embankment (2 layers) and soft clay (2 layers)	1.388	1.578
(d) Inside embankment (4 layers) and soft clay (2 layers)	1.662	2.020
(e) Inside embankment (4 layers) and soft clay (3 layers)	1.810	2.449
Soil-cement columns	FS = 1.972	

tabulated in Table 2. The factor of safety of 1.972 was obtained for the scheme using DMM columns. From Table 2, the scheme with four layers of reinforcements in the embankment and three layers in the soft clay (Figure 29) yielded the same stability performance as the DMM columns slope protection scheme. However, subsequent cost analysis revealed that utilizing geotextile reinforcements is cheaper by about half the cost of implementing DMM soil-cement columns (Lorenzo et al., 2004).

5 CONCLUSIONS

A few aspects of innovative combined earth reinforcement technology with other techniques have been presented and discussed including the combination with deep mixing method (DMM) using cement admixtures, the combination with lightweight tire chips-sand fill, and the combination with prefabricated geosynthetics for drainage.

The deep mixing improvement in the soft clay foundation has reduced the settlement of reinforced embankment by 70 percent. Just after construction, the settlement of the improved foundation already amounted to 40 percent of the total settlements. The 90 percent consolidation was achieved within one year implying that the DMM contributed to the partial drainage of the improved foundation. Furthermore, the weighing factors for "equal strain" and "equal stress" conditions were utilized and evaluated. From the back-analysis, the actual rate of settlements simulated that 80 percent corresponds to "equal strain" and only 20 percent belongs to "equal stress". This 80 percent equal strain is possible since the firs layer of reinforcement was not installed in the ground surface but rather located above ground

The percentage of sand mixed with the tire chips-sand mixture was most significant factor controlling its strength and compressibility characteristics. The tire chip-sand mixture with mixing ratio of 30:70 percent by weight yielded the highest pullout and direct shear resistances. Although Geogrid A has higher ultimate tensile strength than Geogrid B and slightly higher pullout resistance, Geogrid A was selected considering cost. The unit weight of the tire chips-sand lightweight fill at 13.6 kN/m^3 was 75% lighter than the conventional sand fill at 18.0 kN/m^3. The total settlement of 122 mm for the lightweight embankment was 67.5 percent less compared to the corresponding value using conventional sand fill.

The use of dual function geosynthetics performing both drainage and reinforcement functions, would lead to more economical design in mitigating excavation slope failures at irrigation/drainage canals.

REFERENCES

Asaoka, A. (1978), Observational procedure of settlement predictions. *Soils and Foundations* 18(4): 53–66.

Barron, R.A. (1948), Consolidation of fine-grained soils by sand drain wells. *Transactions of ASCE* 124: 709–739.

Bergado, D.T., Tawatchai, T., Voottipruex, P. & Kanjanak, T. (2007), Reinforced lightweight tire chips-sand mixtures for bridge approach utilization, *Proceedings International Workshop of Tire Derived Geomaterials (IW-TDGM), Yokosuka, Japan.*

Bergado, D.T., Lorenzo, G.A. & Duangchan T. (2004), Consolidation settlement of reinforced embankment on deep mixing cement piles. *Geotechnical Engineering Journal* 36 (1): 77–84.

Bergado, D.T., Lorenzo, G.A. & Long, P.V. (2003), LEM-back analysis of geotextile reinforced embankment on Soft Bangkok clay- a case study. *Geosynthetic International* 9 (3): 217–245.

Bergado, D.T. & Vootipruex, P. (2000), Interaction coefficient between silty sand backfill and various types of reinforcement. *Proceeding 2nd Asian Geosynthetic Conference, Kuala Lumpur, Malaysia* 1: 119–151.

Bergado, D.T., Ruenkrairergsa, T., Taesiri, Y. & Balasubramaniam, A.S. (1999), Deep soil mixing used to reduce embankment settlement. *Ground Improvement* 3: 145–162.

Freudelund, T.E. & Aaboe, R. (1993), Expan- polystylene-A light way across soft ground. *Proceeding 14th International Conference on Soil Mechanics and Geotechnical Engineering, New Delhi, India* 1: 119–151.

Hansbo, S. (1979), Consolidation of clay by bandshaped prefabricated drains. *Ground Engineering* 12 (5): 16–25.

Kamon, M. & Bergado, D.T. (1991), Ground improvement techniques. *Proceeding 9th Asian Regional Conference on Soil Mechanics and Geotechnical Engineering, Bangkok, Thailand* 2: 526–546.

Lai, Y.P., Bergado, D.T., Lorenzo, G.A. & Duangchan, T. (2006), Full scale reinforced embankment on deep jet mixing improved ground. *Ground Improvement* 10 (4): 153–164.

Lorenzo, G.A. & Bergado, D.T. (2003), New consolidation equation for soil-cement pile improved ground. *Canadian Geotechnical Journal* 40 (2): 265–275.

Lorenzo, G.A. & Bergado, D.T. (2004), Fundamental parameters of cement-admixed clay-new approach. *Journal of Geotechnical and Geoenvironmental Engineering, ASCE* 130 (10): 1042–1050.

Lorenzo, G.A., Bergado, D.T., Bunthai, W., Hormdee, D. & Phothiraksanon, P. (2004), Innovation and performance of PVD and dual function geosynthetic applications. *Geotextiles and Geomembranes* 22: 75–99.

Petchgate, K., Jongpradist, P. & Panmanajareonphol, S. (2003), Field pile load test of soil-cement column in soft clay. *Proceedings International Symposium on Soil/Ground Improvement and Geosynthetics in Waste Containment and Erosion Control Application, AIT, Bangkok, Thailand.*

Porbaha, A. (1998), State-of-the-art in deep mixing technology. Part I: Basic concept and overview. *Ground Improvement* 2: 81–92.

Skempton, A.W. & Bjerrum, L. (1957), A contribution to the settlement analysis of foundations on clays, *Geotechnique*, 7 (4): 168–178.

Sivaram, B. & Swamee, P. (1977), A computational method of consolidation coefficient. *Soils and Foundations* 17 (12): 48–52.

Yasuhara, K. (2002), Recent Japanese experiences with lightweight geomaterials. *Proceeding International Workshop on Lightweight Geomaterials, Tokyo, Japan*: 32–59.

New Horizons in Earth Reinforcement – Otani, Miyata & Mukunoki (eds)
© 2008 Taylor & Francis Group, London, *ISBN 978-0-415-45775-0*

New direction of earth reinforcement – disaster prevention

Y. Mohri & K. Matsushima
National Institute for Rural Engineering (NIRE)

S. Yamazaki
Mitsui Chemicals Industrial Products,Ltd

T.N. Lohani & A. Goran
JSPS Post-Doctoral Fellow, Kobe University & NIRE

U. Aqil
Associated Consulting Engineers, ACE (Pvt) Limited, Pakistan Formerly research Engineer, National Institute for Rural Engineering(NIRE)

ABSTRACT: Numerous authors have proposed and developed the geosynthetic reinforcing technology for embankments and retaining walls and have examined the behavior of those structures based on the laboratory and field tests, including the field case histories. Special emphasis is given to those seismic behaviors under earthquake in Japan. Regarding the earth-fill dam, there is presently both a need and opportunity to achieve significant improvements in construction technology used to increase earth fill-dam breach protection. This paper provide a review of the damage feature of small earth-fill dam during by earthquake and heavy rainfall using published investigation reports and personal experiences. The paper describes a new technology to stabilize the earth-fill dam by using the geosynthetic soil bags with extended tails allowing over flow and earthquake. Finally, some fundamental properties of reinforced embankment by soil bags and seismic behavior using shaking table is demonstrated.

1 INTRODUCTION

Earth-fill dams must withstand two extreme events: major floods due to heavy rainfall, and earthquakes. In Japan, there are more than 210,000 small earth-fill dams for agricultural irrigation, and almost all are located near cities. Among small earth-fill dams currently operated, 48,000 or more reservoirs were constructed before B.C. 1600. These reservoirs have been maintained by regional management organizations for some 200 years, but at least 20,000 sites are deteriorating with age such as leakage and sliding. The body of the dam consists of soil materials, so there is a possibility that slight erosion on the slope of the dam caused by rainfall may gradually develop into a large-scale collapse. In addition, it has been reported in many papers that a dam body damaged by an earthquake unexpectedly failed.

Figure 1 shows the cumulative and annual numbers of damaged small earth-fill dams in Japan, from 1976 to 2004. The figure shows that complete dam failure occurs several times a year, but most small earth-fill dams have been damaged by sliding and leakage. It is obvious that many dams were damaged by earthquakes in 1995 and in 2004, but most of the dams were adversely affected by embankment landslide and overflow caused by rainfall.

These small earth-fill dams damaged by earthquakes and heavy rainfall are usually reconstructed to the original condition and structural type, even if the reconstructed earth fill dams may suffer the same

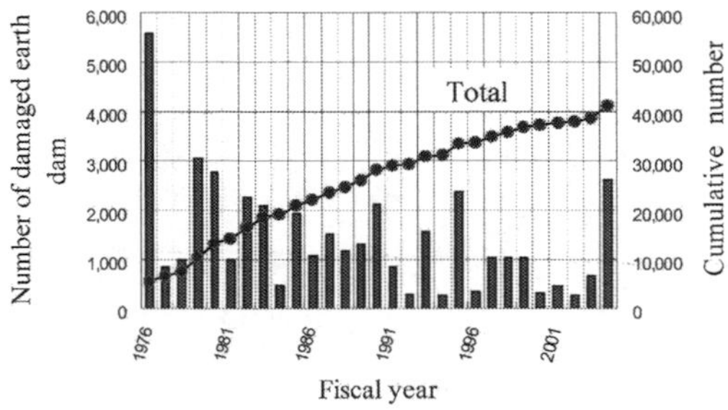

Figure 1. Accumulated and annual numbers of damaged small earth-fill dams in Japan.

damage again when subjected to an earthquake or rainfall of the same level.

However, considering that localized torrential rain exceeding 1,000 mm daily often occurs throughout Japan causing an unexpected flood, and that earthquakes occur frequently, it is essential to develop a small earth-fill dam of high durability.

Breaching issues related to earth-fill dams due to overflow by heavy rainfall have been addressed in many reports at previous symposiums of ICOLD and some hydraulic journals. A general report presented by J.P. Tournier (2006) provided an extensive review of dam safety, breaching, evaluation of risk, and rehabilitation.

There is a strong need to develop a technique to prevent frequent damage to earth-fill dams and avoid earth-fill dam-related disasters. Consequently, in this paper we collate past findings and cases concerning the destruction of dams and discuss the direction of technologies and research to be developed. In this report, several case histories in which earth-fill dams were serious damaged by earthquake and heavy rainfall are described in order to clarify the key causes. Next, we give examples of a technology for reinforcing small earth-fill dams which deliver high performance under seismic load and heavy rainfall. In the following section, we give the test results for breaching and seismic performance for the new technology.

2 CASE HISTORIES FOR SMALL EARTH-FILL DAMS

It is first necessary to understand the detailed destruction patterns of small earth-fill dams and their mechanisms in order to develop technologies for improving the durability of the dams in an innovative manner.

These dams have several destruction patterns due to heavy rainfall. Here, the main causes of destruction can be classified into three types as shown in Figure 2: failure by overflow, failure by sliding and failure by internal erosion. The combination of these causes and multiple conditions that destroy the body of the dams must also be considered, but it is reasonable to note individual damage cases when developing technologies to prevent destruction. Namely, the dams must be able to fully withstand individual destruction patterns in order to comprehensively enhance the safety of the dam body. These damage patterns can be classified as shown in Table 1.

A failure by overflow is an event in which a flood exceeding a spillway's capacity for discharging the flow in a small earth-fill dam occurs, and the dam's water flows over the crest of the body, erodes the crest surface and downstream slope, and thus destroys the embankment. A failure by sliding is an event in which the seepage of water and rainfall into the downstream slope increase the pore pressure within the embankment and decrease the strength of the body, causing the slope to slide. A failure by internal erosion is a generic name for destructive phenomena caused by "hydraulic fracturing", "piping" and other events.

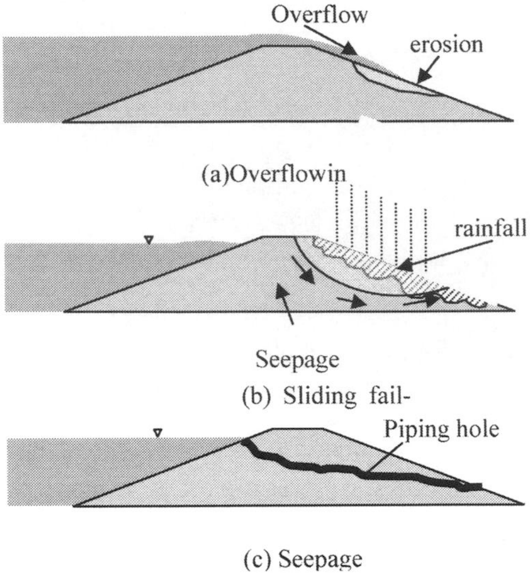

Figure 2. 3 Dam failure modes by heavy rainfall (Hori et al., 1997).

Table 1. Destruction patterns of the body of small earth-fill dams.

Category	Events or reasons
Overflow	Flood, Small embankment-height
Sliding	Increase upstream water level, Seepage to the downstream slope of rainfall
Int. erosion	Piping route, sink fall, decrease contact pressure around outlet structure

Int.erosion: Internal erosion.

Yasunaka's (Yasunaka et al., 1985) and Fujii's (Fujii et al., 1991) survey results indicated that 30% of small earth-fill dams damaged by heavy rainfall were destroyed by overflow. Furthermore, when failures caused by sliding on the downstream slope were included, this failure rate reached 60%.

2.1 *Damage by rainfall*

2.1.1 *Failure by overflow*

Water flowing over the body of a small-fill dam erodes the embankment soil with its streaming momentum and causes the embankment to collapse local erosion on the slope. Considering soil erosion caused by flowing water, a cohesive and clay slope is more resistant

Figure 3. Erosion of downstream slope by overflow and progress into failure of the slope.

to erosion than a sand slope. A larger overflow depth naturally increases the volume of embankment soil eroded. Conversely, a smaller overflow depth and a slower flow velocity decrease the erosion by water, so the embankment provides resistance to short-time overflow.

Figure 3 shows the state of a small earth-fill dam that was destroyed by overflow. Figure 3a shows the condition immediately after the water level rose to the crest surface of the dam, and that the erosion of the downstream slope continues to develop with the start of overflow. In a case where a large-scale flood occurs or where natural ground around a reservoir collapses and the water level is increases significantly, the whole embankment can collapse, or it may be partially destroyed when the overflow duration time is short. In any case, the body of the dam does not provide great resistance to an overflowing stream, so the erosion of the downstream slope proceeds, reduces the section of the embankment and finally destroys the body. The degree of failure of the embankment depends greatly on the soil materials of the embankment and the paving condition of the upper surface, as well as the water flow velocity and duration of the overflow.

2.1.2 *Failure by sliding*

An embankment of a small earth-fill dam is unsaturated before rainfall, and is under a suction and so is stable against sliding. As the embankment's water content increases, its suction and shear resistance decline, and its lowered resistance increases the risk of failure by sliding. Most sliding events on an embankment during heavy rainfall occur at the downstream slope. Continuous rainfall gradually infiltrates the embankment, thereby increasing the degree of ground saturation. At the same time, the ground water table generated by seepage into the body also increases in a staged manner. For this reason, as the pore pressure within the embankment increases, the downstream slope decrease strength, resulting in sliding collapse. The causes of slope failure by sliding resulting from this rainfall include:

(a) Reservoir water level conditions
(b) Embankment geometry (slope gradient)
(c) Embankment soil conditions (permeability coefficient, water retention properties, cohesion, internal friction angle, etc.)

When rehabilitating an existing old earth-fill dam, it is most effective to install a drain at the toe of the downstream slope. If this drain lowers the ground water table, it will prevent the reduction of strength of the sliding body, thus significantly increasing the safety factor. In addition, it is also effective to cover the surface of the downstream slope and directly reduce the infiltration of rainfall into the body.

2.1.3 *Key factors for rehabilitating existing old earth-fill dams*

2.1.3.1 For failure by overflow

It is necessary to develop a slope reinforcement method to avoid immediate destruction of the body of an existing old earth-fill dam even if the dam water overflows the embankment. It is also required to develop a method to evaluate the structural safety of the body that allows for overflow.

2.1.3.2 For failure by sliding

It is important to control rainfall seepage and take measures against erosion in order to maintain the mechanical safety of the downstream slope. In addition, a drain to maintain the embankment's saturation line at a safe level is very effective and economical.

2.2 *Damage by earthquake*

2.2.1 *Damage to dike*

In the 1995 Hyogoken-Nambu earthquake that struck the Kobe area of Japan, even agricultural infrastructure, including earth-fill dams, land reclamation embankments and farm roads were affected. For large-scale earth-fill dams designed using modern soil mechanics technology, even this large earthquake caused only very slight damage. Small earth-fill dams for irrigation suffered serious damage, including total failure and sliding. Concrete spillways, which are installed on the crest, also suffered major damage.

Figure 4 shows a damaged small earth-fill dam, in which the whole dam dike underwent large settlement due to widespread liquefaction of the foundation below the dike.

The crest of the dam settled about 5m during the earthquake and many soil boils were found around the dike. Figure 5 shows slip failure on the upstream slope of a small earth-fill dam, which consists of clay materials.

When a small earth-fill dam is hit by an earthquake, it greatly vibrates along the dam axis and in the

Figure 4. Total failure and large settlement caused by Kobe earthquake, due to liquefaction.

Figure 5. Upstream slope sliding by seismic loading (Hyogo Pref., 1996. Final Rept. for Earthquake).

perpendicular direction. As a result, tension cracking occurs near the crest surface of the embankment, causing shear failure within the body. If the embankment is greatly shaken, the deformation by these failures is cumulated, resulting in large sliding collapse. When the dam retains water and is struck by an earthquake, it may cause a great sliding collapse downstream under water pressure. If the foundation of the dam is soft or undergoes liquefaction, as reported by Tani, S. (1996), often the whole body settles down and completely collapses.

The factors that affect the destruction of the embankment in an earthquake include:

(a) Strength of embankment and soil conditions
(b) Properties of foundation
(c) Earth-fill dam geometry (slope gradient)
(d) Magnitude and duration of earthquake ground motion.

Due to the motion of the embankment during an earthquake, its crest surface is often cracked. As shown in Figure 6, several continuous cracking events in parallel with its axis can occur. These surface cracks may extend to the depth of the embankment even if the

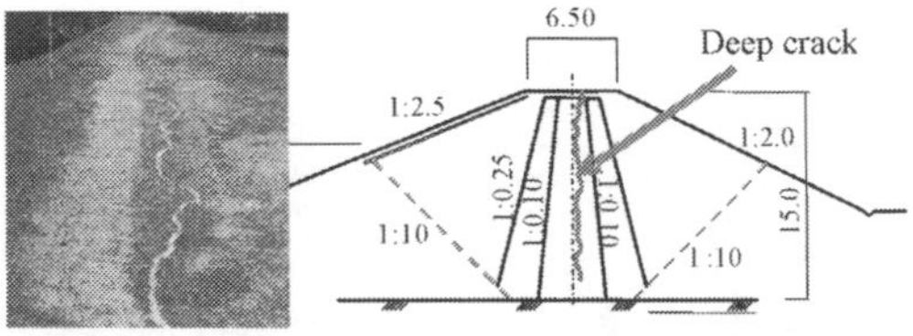

Figure 6. Open crack in the direction parallel to the dam axis and propagating to the bottom to the core zone from the crest. (Hyogo Pref., 1996. Final Rept. for Earthquake).

Figure 7. Damage to the spillway of Mukumoto Dam, overturning the side wall and causing slippage at the interface between the dike and spillway.

opening width of the surface is approximately 5 cm. In one small earth-fill dam hit by the 1995. Hyogoken-Nambu earthquake, cracks reached the bottom of the impermeable core layer, so the entire dam had to be excavated to reconstruct a new embankment. In a case where slope sliding as well as cracking is caused by an earthquake, larger cracking occurs on the crest surface, and such a damaged dam will require large-scale rehabilitation including the slope.

2.2.2 *Damage to spillway*

The example of Mukumoto Dam illustrates a complete failure of dam facilities. Figure 7 shows the result of total failure of the concrete spillway on the left side of the embankment. Almost no damage was seen in the embankment section itself. The embankment consists of a mixture of silt and fine sand with an N-value of 5. This concrete spillway suffered a slip surface between the side wall and the dike, and also rupture of the bottom concrete slab of the spillway.

In addition, there was a cavity in the interface between the channel side wall and the embankment, while a part of the side wall fell down as the bottom cracked. This heavy structure on the embankment's upper surface causes a difference phase behavior between the embankment and the structure in earthquake, so the side wall and the bottom concrete slab are easily detached from the embankment. Cracking and segregation at the interface with such a structure provide leakage routes to the dam itself, thereby causing loss of the embankment's soil and reduced safety of the embankment.

For this reason, a spillway channel should not be placed on the embankment. It is important to consider the installation position of the channel, with highest priority on locating the channel on natural ground with a large bearing capacity.

The factors that affect the destruction of a spillway channel caused by an earthquake include:

(a) Size of spillway channel (scale and weight)
(b) Strength of embankment
(c) Embankment geometry (height and slope gradient)
(d) Magnitude and duration of earthquake ground motion

3 FORMULATION OF A NEW CONSTRUCTION METHOD FOR OVERFLOWING TOLERANT AND EARTHQUAKE RESISTANT SMALL EARTH DAMS

Existing old earth-fill dams consist of an earth embankment and a concrete spillway. When such dams are designed, additional seismic force due to earthquakes is usually not taken into consideration. Also, because almost all old earth-fill dams were constructed before the design standard for spillway capacity was drawn up, many dam spillways have insufficient capacity to discharge the water of occasional floods.

A great number of small earth-fill dams for agricultural irrigation have been seriously damaged or completely failed by overflowing during typhoon-induced flooding which exceeds the drainage capacity of the spillway discharge system, or during earthquakes. This natural hazard of overflowing can totally destroy downstream city areas, so a new method for constructing small earth-fill dams which can tolerate overflowing due to flooding and the seismic forces of earthquakes is needed.

It is very expensive to increase the drainage capacity of a flood discharge spillway system of a reinforced concrete (RC) structure so that it can discharge the design flood that might take place once every 200 years. Moreover, a large spillway system on a small earth-fill dam requires reinforcing the dam dike itself in order to increase the stability of the dam. This rehabilitation approach is not cost effective and takes too long. Mohri et al. (2005) proposed protecting the downstream slope of such earth-fill dams by using soil bags anchored with geosynthetic reinforcement layers arranged inside the slope as shown in Figure 8. This is a realistic, cost-effective and quicker method to rehabilitate a great number of old earth-fill dams without increasing the capacity of an existing

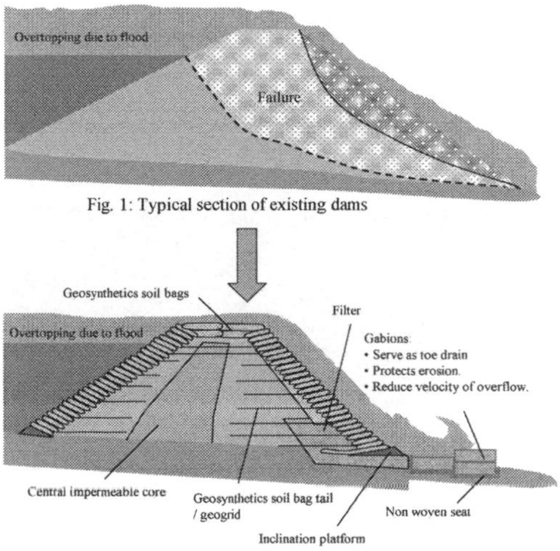

Figure 8. New technology to rehabilitate existing old earth-fill dams to have a high flood discharge capacity and a high seismic stability (Mohri et al., 2005; Matsushima et al., 2005a).

flood discharge RC structure. Moreover, the slope constructed or reconstructed by the new technology is more stable against seismic load.

In the new construction method, the dam section is a composite structure of earthwork, impermeable core zone, geosynthetic soil bags with extended tail(GSET) filled with appropriate backfill material and geosynthetic wing. The use of advanced soil bags not only increases stability against overflowing but also provides resistance against earthquake-induced forces.

Performance for GSET

(a) A dam with stacked soil bags can tolerate overflowing.
(b) The existing spillway can also be used as an emergency spillway during floods.
(c) High stability against seismic force due to geosynthetic reinforcement.

Efficient, cost-effective construction method for GSET

(a) Cost effective, as there is no need to build an additional spillway.
(b) No heavy equipment is needed so construction in remote locations is possible.
(c) Reduced base width and top crest require fewer earthworks.

3.1 *Geosynthetic soil bag system with extended tail for earth fill dam (GSET-dam)*

GSET is a method for piling up soil bags as shown in Figure 9 to construct an earth-fill dam that has higher

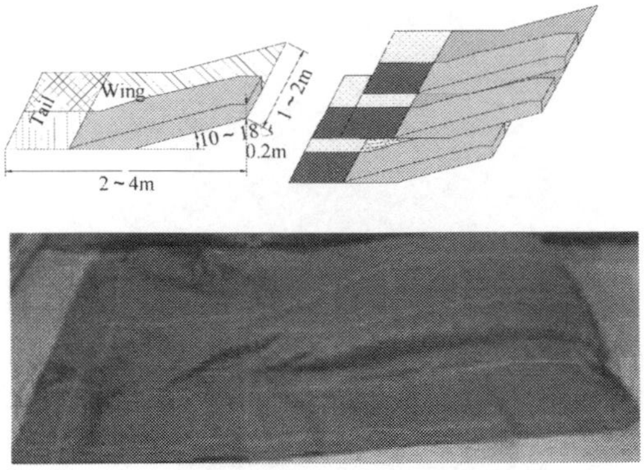

Figure 9. Geosynthetic soil bag with extended tail and inclined stacking system.

stability even when a flood overflows its embankment. This soil bag, which has a flat shape with an aspect ratio of 5 to 8, consists of a large soil bag with a weight of 200 kgf or more, and a tail and a wing that are connected to this soil bag. The wing is inserted between neighboring soil bags, thereby ensuring the strength of stacked soil bags serving as a potential wall. The tail is installed within the embankment, so it exerts a reinforcement effect in combination with stacked soil bags and embankment. Namely, the body of the soil bag system and the tail reinforce a wider area of the embankment, and increase the strength of the whole body as a flexible wall. Furthermore, as the materials within the soil bag are confined by the bag and exert a larger bearing capacity, on-site soil of low quality can be used. When high-permeability materials such as crushed stone are used for the embankment, its stacked layer can also serve as a drain. Therefore, these materials provide resistance to overflow and also improve measures against leakage and the safety of the dam during an earthquake.

It is important to make clear the soil bag strength and full use of the capability of the soil bag system in order to stabilize the structure of stacked soil bags. Compressive properties of soil bags were reported by Matsuoka et al. (2001), Lohani et al. (2004) and Matsushima et al. (2005). Regarding the shear resistance of stacked soil bags, Aqil et al. (2005,2006) and Matsushima et al. (2006) reported the results of a large-scale shearing test. It is difficult to ensure sufficient resistance by simple stacking because sliding friction between soil bag materials is small and its friction dominates the shear deformation of the whole stacked body. For this reason, Matsushima et al. (2006) suggested a method to increase the shear resistance of the body by installing soil bags at a inclined angle, and provided detailed data on the increased resistance.

3.2 *Compression strength for stacked soil bags (Lohani et al., 2005; Matsushima et al., 2005)*

The use of geosynthetics in the closed form, such as soil bags, is also becoming attractive (e.g., Tatsuoka et al., 1997; Matsuoka et al., 2001; Lohani et al., 2004; Matsushima et al., 2005). Stacking soil bags to form a structure is a practical and cost-effective solution when construction space is limited and also when rapid construction is required.

It is very important to ensure that the stacked soil bags used to stabilize an old earth-fill dam are high strength in order to increase the stability of the dam dike itself.

Recycled concrete aggregate has mainly been used as a material for filling soil bags but considering the cost-effective use of locally excavated material, sands mixed with different proportions of clay fraction were also tested as backfills. To evaluate all these issues and propose a rational method of applying soil bags in permanent civil engineering works, a series of unconfined compression tests was conducted on the stacks of geosynthetic soil bags filled with various geomaterials and height/width ratio greater than 2.

Polyethylene (PE) and polypropylene (PP) geosynthetics were used in these tests as soil bag materials and their tensile strength was 3.75 kN/m and 14.5 kN/m, respectively.

Along with REPA (Material A), other materials used to infill the soil bags were: Toyoura sand (Material B), Hokota sand (Material C), FC35 (Material D, Mixture of Hokota sand & Kanazawa clay in 7:3 proportion by weight) and FC50 (Material E, Mixture of Hokota sand & Kanazawa clay in 1:1 proportion by weight).

The methods used to prepare soil bags can be categorized into the following main subgroups:

(a) L-Series: The total weight of a filled bag was 19.5 kg/bag (REPA), which was about 50% of its full volume. The filled bags were placed under a vibrating rammer of weight 170 kg.
(b) F-Series: Soil bags were packed with 39 kg at the standard Proctor compaction energy level. Four different types of infill material, Materials A, C, D and E, were packed in the PP geosynthetic bags.

A series of unconfined compression tests (Table 2 and Figure 10) was conducted on the stacks of geosynthetic soil bags filled with various geomaterials. A constant strain rate of loading of 1.08 mm/min was maintained in all of the tests. The H/W ratio, which is expected to play a major role, was varied and for many of the tests it exceeded the minimum requirement of 2 in normal shear tests.

3.2.1 *Effect of initial compaction of soil bags*

The test pairs L4, L6 & L9 show the magnitude of difference resulting from the initial compaction condition. The soil bags for all these cases were prepared

Table 2. Soil bag test data of bag preparations. (Lohani et al., 2005, modified by Mohri)

Case	No*.	Type**	Density	Infill***.	H/W	Comp.
L4	4	PE	1.54	REPA	0.80	No
L6	4	PE	1.52	REPA	0.59	Yes
L9	4	PP	1.49	REPA	0.54	Yes
L10	14	PE	1.55	REPA	2.02	Yes
L11	14	PP	1.57	REPA	1.74	Yes
F19	9	PP	1.61	REPA	2.33	Yes
F29	9	PP	1.56	Hokota	2.43	Yes
F39	9	PP	1.37	FC35	2.32	Yes
F49	9	PP	1.38	FC50	2.19	Yes

No*: No. of bags, Type**: Bag material Infill***: Infill material for bag.

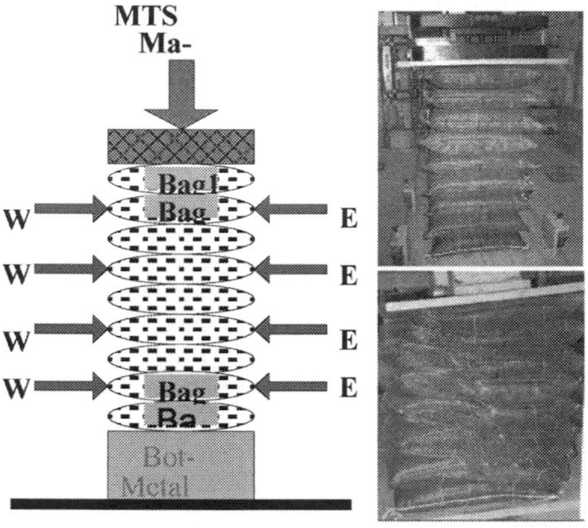

Figure 10. Schematic diagram of unconfined compression test and after test. (Lohani et al., 2004).

from PE type geotextiles. The identical difference among the stress-strain curves not only confirms the repeatability of the soil bag tests but also that the initial compaction improves the initial stiffness. Figure 11 shows this with relatively small creep strain for initially compacted samples. However, the range of strain is rather higher than can be allowed in regular and permanent civil engineering structures.

3.2.2 *Effect of geosynthetics used to prepare soil bags*

Figure 12 shows that the soil bags made from stronger geotextiles had increased strength and the L11 test using stronger PP type geotextiles showed larger strength than the test using weaker PE material. Figure 11 compares the creep magnitude of the soil bags prepared from PE and PP materials, clarifying the superiority of the stronger bag material. In this figure, the L9 test with PP bags yielded a smaller creep strain at similar loading levels than the L6 test.

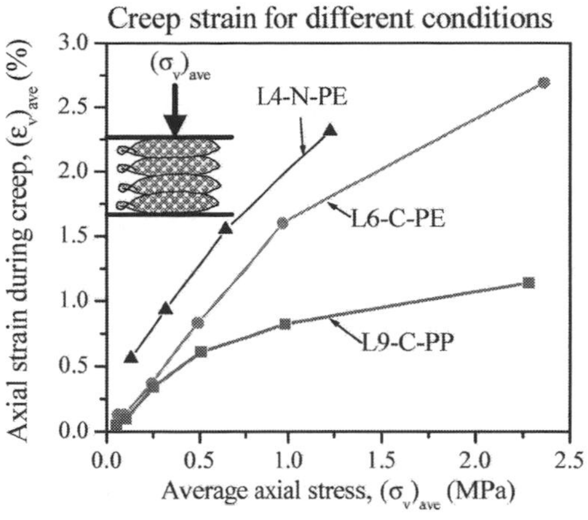

Figure 11. Effect of compaction and the material used for the soil bags prepared in L-series of tests.

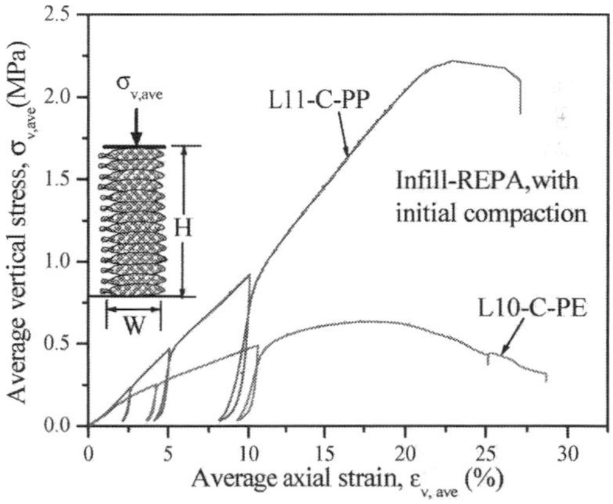

Figure 12. Stress-strain behavior of soil bags filled with REPA materials for L-Series samples.

3.2.3 *Effectiveness of preloading*

Figure 13, which will be described in subsequent paragraphs, shows the tests with sustained loading on soil bags with REPA and FC35 materials during normal loading and during reloading. The sustained loading was performed under 100 kN load for all the tests, and the small difference in stress is duly incorporated in the difference in x-sectional area. Under sustained loading for 8 hours, REPA material showed better performance with smaller creep strain of 0.6% compared to the nearly 2% strain in finer FC35 material during loading.

3.2.4 *Effect of backfill materials used in soil bags*

Figure 13 shows the average stress-strain response of the stacked soil bags in the F-Series tests. Unloading to near zero stress and then reloading were performed numerous times to evaluate the effectiveness

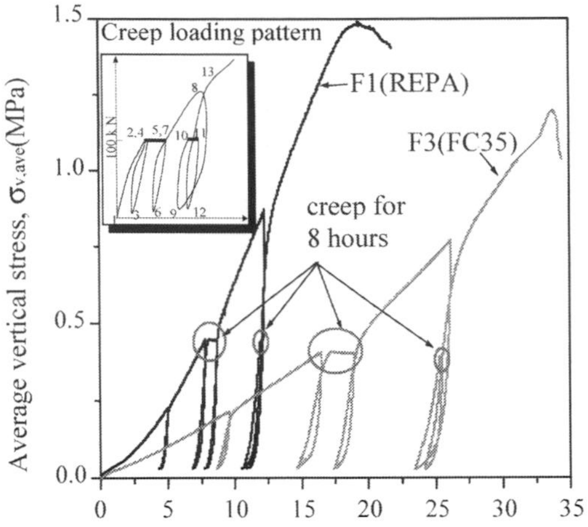

Figure 13. Stress-strain behavior of F-Series test specimens with different backfills.

of preloading as stated in the previous paragraphs. In the tests, very similar loads were repeatedly applied on the stacked soil bags by changing the infilling material. The general trend is that the REPA material, which is a well-graded gravel with higher angle of internal friction, is better than the finer soils: Hokota sand, FC35 and FC50 designated here as materials C, D & E, respectively. Although the infill material with the largest percentage of fines was expected to show the weakest behavior, the tests with FC50 infill differed slightly, the results of which may be summarized as follows:

(a) Both the strength and stiffness of soil bags are improved by using stronger geosynthetic bags.
(b) Well-graded REPA aggregate is better suited as a backfill material for soil bags than those containing a large proportion of fines.
(c) The soil bag system shows a clear increase in stiffness during the reloading phase.
(d) While the initial response is not so encouraging, the stronger stress-strain behavior and negligible creep strain during the reloading phase confirm that soil bags can be used cost-effectively in civil engineering works.

3.3 *Lateral shear tests for stacked soil bags*

Regarding the lateral deformation of a reinforced soil structure, although it is assumed to be a rigid body in many design codes, the results of physical model tests show that its reinforced area undergoes simple shear-type deformation under large earthquake loads as illustrated in Figure 14 (Tatsuoka et al., 1998; Bathurst and Hatami 1998).

During seismic events, stacked soil bag systems may be subjected to lateral shear loading, in which

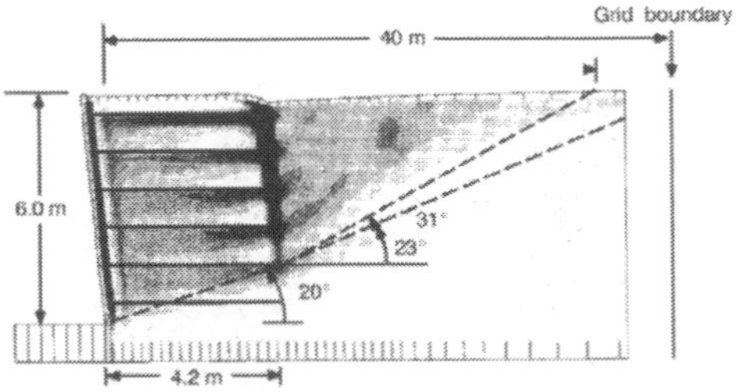

(a) FLAC model (Bathurst and Hatami 1998)

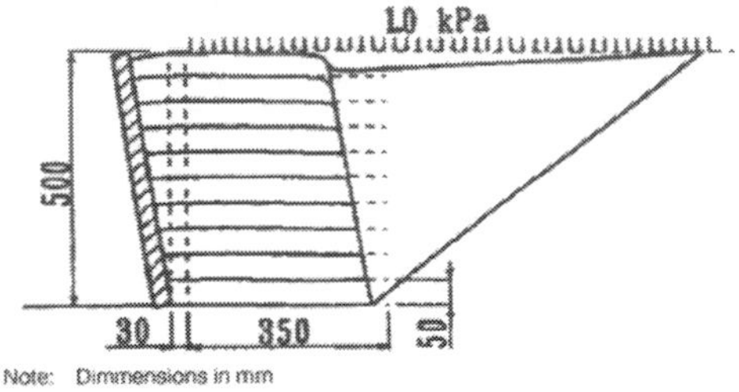

(b) Reduced-scale shaking table test (Tatsuoka et al. 1998)

Figure 14. Shear deformation of reinforced soil zone (Tatsuoka et al., 1998, Bathurst and Hatami 1998).

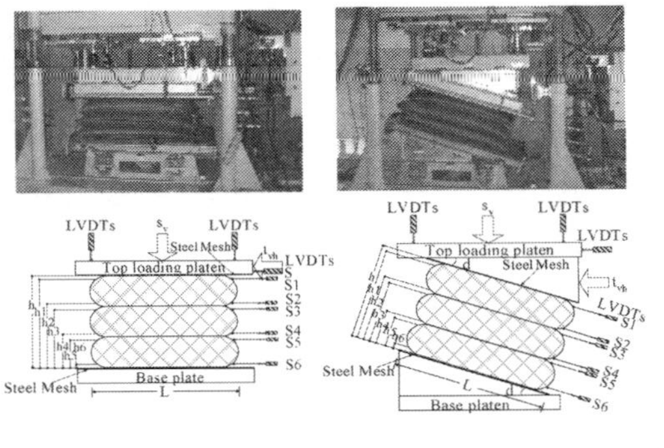

Figure 15. Specimen with bedding planes at an angle of 18 degrees relative to the horizontal set in a lateral shearing apparatus.

the global direction of compressive load is largely inclined from the vertical direction. In such a case, the strength and deformation characteristics, which are essential for a reliable seismic design, are not known. The strength under lateral shear is likely to be very different from that under vertical compression due to the highly anisotropic structure of stacked soil bags. In view of the above, a series of full-scale lateral shear tests on a stacked soil bag system was conducted, and effective measures to increase the strength when the system was subjected to lateral shear were devised (Matsushima et al., 2006).

To accurately evaluate the shear strength and stiffness of a soil bag system, a single stack of full-scale soil bags was used as shown in Figure 15. Also, to

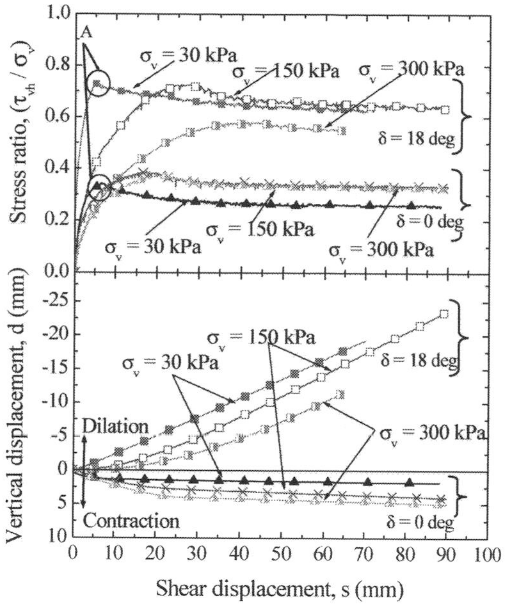

Figure 16. Relationship among stress ratio, shear displacement and vertical displacement for tests at different σ_v and $\delta = 0$ & 18 degrees (Matsushima 2006).

increase the shear strength of the soil bag system, soil bags were arranged with an inclined interface between vertically adjacent bags. The angle, δ, of the direction of the interface relative to the horizontal was set at 18 degrees as well as at 0 degrees for reference. To investigate the failure mechanism of the soil bag system, in particular to examine whether failure is caused by shear deformation of bags or by slippage at the interface between vertically adjacent soil bags, lateral displacements of soil bags were measured with a set of pulley-type LVDTs.

The respective full-scale soil bags were each filled with 120 kg of air-dried Toyoura sand. For shear tests at $\delta = 0$, the soil bags were placed on the horizontal base plate of the lateral shearing apparatus. To prepare the specimens for tests at $\delta = 18$ degrees, soil bags were placed on the base plate of the shear apparatus.

Figure 16 shows the relationship among the stress ratio, τ_{vh}/σ_v, the shear displacement, s, and the vertical displacement, d, both at the top of the specimen relative to the bottom, from six tests on a single stack of soil bags filled with Toyoura sand. Figures 17a & b show the distributions of lateral displacement along the specimen lateral face at different shear displacements from test cases 4 and 6 (at low and high σ_v). These figures reveal the following trends. Both the peak τ_{vh}/σ_v value and the pre-peak stiffness increased significantly upon changing the angle δ from 0 to 18 degrees. The volumetric change also changed significantly from a contractive one to a dilative one.

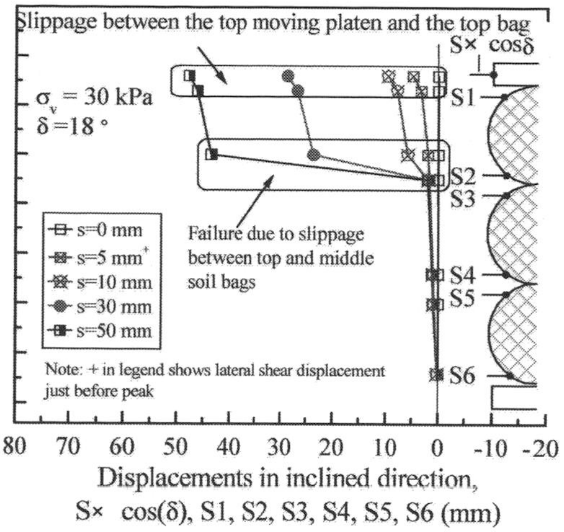

Figure 17a. Distribution of lateral deformation in soil bags along the height at low normal stress (Matsushima 2006).

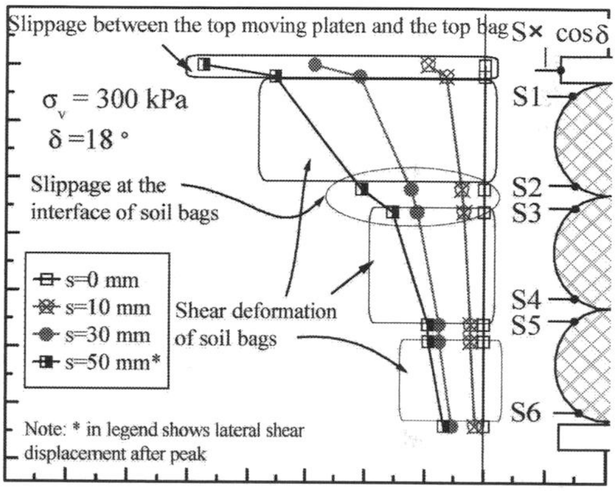

Figure 17b. Distribution of lateral deformation in soil bags along the height at high normal stress (Matsushima 2006).

At low σ_v (i.e., 30 kPa), the slippage between soil bags controlled the shear strength of the soil bag system. Slippage started at the interface between the top and middle soil bags (Figure 17a). As the slippage started at points A shown in Figure 16, the τ_{vh}/σ_v value suddenly started dropping. It is very likely that, if slippage had not taken place, the peak stress ratio would have become much higher than that observed.

At high σ_v (i.e., 300 kPa), the peak τ_{vh}/σ_v value was controlled by the shear failure in the sand inside the soil bags. Slippage started taking place only after the peak stress state (at s ≈ 50 mm, Figure 17b). At smaller shear displacements, the distribution of shear deformation of the soil bag system was uniform, as seen from the rather constant slope of the relation at s = 10 mm (Figure 17b). Slippage between the upper moving plate and the top bag was not significant at s = 10 mm.

These test results indicate that a soil bag system becomes much more stable by simply being placed inclined so that the direction of the bedding planes of the soil bags becomes normal to the principal direction of applied compressive load. Moreover, to maximize the shear strength it is important to prevent slippage at the interface between adjacent soil bags, particularly when the applied normal stress is low.

4 PHYSICAL BREACH TEST FOR SOIL BAG SYSTEM

The mechanics of embankment erosion during overflowing have been widely documented in the literature, such as for the numerous small earth-fill dam failures that occurred in the 1960s by overtopping outflow. Ralson (1987) provided a good description of the mechanics of embankment erosion. For cohesive soil embankments, breach takes place by headcutting. A first small headcut is formed near the toe of the dam in general and then progresses to the upstream side until the crest of the dam is breached, as shown in Figure 18. A series of stair step headcuts forms on the downstream face of the dam. If the embankment contains a cohesive core that is symmetrical about the axis of the dam, the core will be eroded in a manner similar to that for a cohesive embankment. If the core is sloped such that the downstream shell provides structural support for the core, the core will fail structurally as the downstream shell is eroded away, as shown in Figure 19.

Powledge et al. (1989b) described three hydraulic flow regimes and erosion zones for flow overtopping an embankment, as shown in Figure 20. In the critical flow region on the dam crest, energy slopes, velocities, and tractive stress are relatively low and erosion will occur only if the crest materials are highly erodible. A transition to supercritical flow occurs on the downstream portion of the crest. Energy slopes and traction stress are higher in this region, and erosion is sometimes observed at the knick point at the downstream edge of the crest. Another point of erosion is the downstream face of the dam, on which the flow accelerates at supercritical depths until reaching uniform flow conditions. Tractive stresses are very high, and changes in slope or surface discontinuities cause the stress to concentrate, thereby initiating erosion.

Erosion may initiate at any point on the slope, but the toe is the most common location. Once erosion has been initiated, a headcutting behavior is generally observed in which the scour hole moves upstream and widens.

Powledge summarized by noting six factors affecting embankment erosion:

(a) Embankment configuration, materials and densities of fill

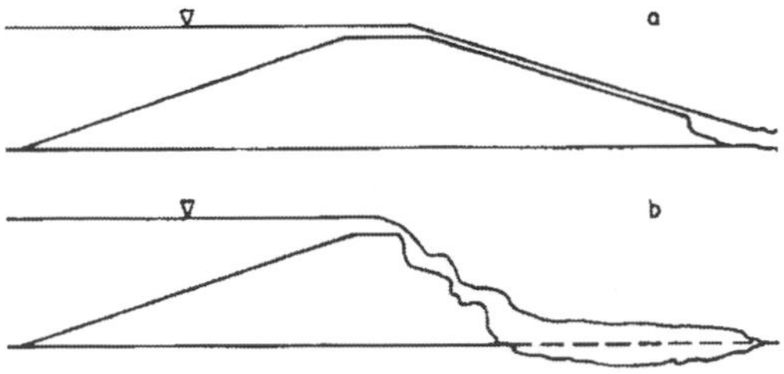

Figure 18. Progressive headcutting breach of cohessive soil embankment (Powledge et al., 1989).

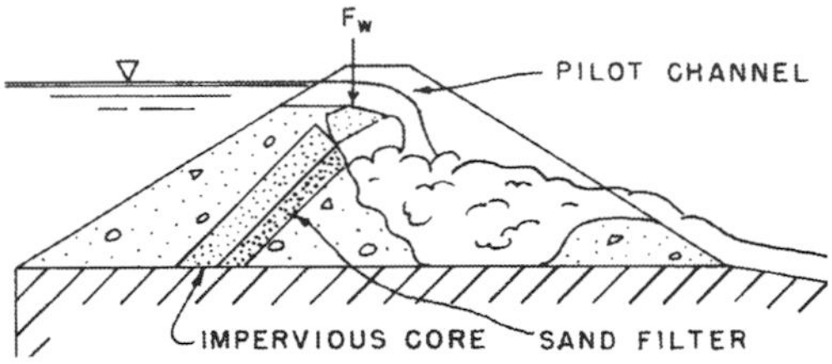

Figure 19. Schematic of the breaching process for the embankment with a sloped core section (Pugh, 1985).

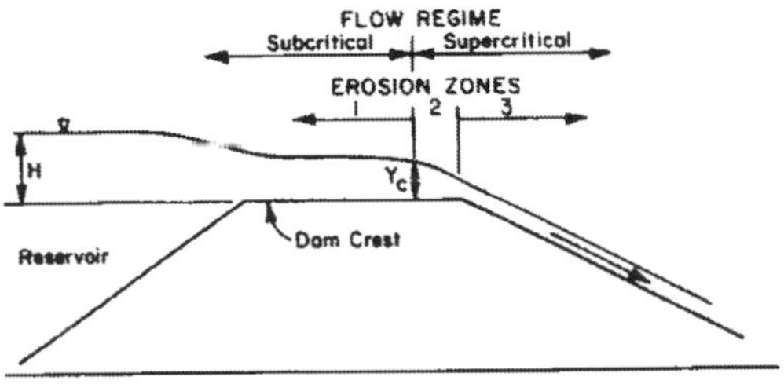

Figure 20. Flow and erosion regimes in embankment overtopping (Powledge et al., 1989).

(b) Maximum velocity attained by flow
(c) Discontinuities, cracks, or voids in the slope, and appurtenances or anomalies at the toe
(d) Presence and depth of tail-water on the downstream slope
(e) Flow concentration at low points along the embankment or at abutment groins
(f) Toe drain, blanket drains, or highly erodible materials in the abutments or foundation that will cause undercutting of cohesive fill materials and accelerate headcut advance.

Dodge (1988) reported on the results of early tests by Reclamation of overtopping flows over model embankment dams. The tests were designed to evaluate the effectiveness of various crest and embankment face protection schemes that would permit overtopping flow without causing dam breach. Reclamation has pursued the development of embankment protection techniques that would permit safe overtopping of embankment dams.

Table 3. Small scale breaching tests cases (Matsushima et al., 2005).

Case	Slope protection	Soil bag type and dimensions (thickness × width × length, cm)
1	non-woven geo-textile sheets	Sheet
2	soil bags of type A without over-lapping	A: 2 × 5 × 6
3	soil bags of type A with over-lapping	A: 2 × 5 × 6
4	soil bags of type B with over-lapping	B: 2.5 × 8 × 8
5	soil bags of type C with over-lapping	C: 2 × 8 × 12
6	soil bags of type D with over-lapping	D: 2 × 5 × 18

The observations of Ralston (1987) and Powledge et al. (1989a,b) indicate that headcut formation and advance are critical processes in embankment dam breach. The mechanics of headcut erosion causing breach of an earth spillway were discussed in detail by Temple (1989). Robinson and Hanson (1993, 1995) reported the results of large-scale tests of two-dimensional headcut erosion on two different soil types (sandy clay and silty sand) under constant hydraulic conditions (drop height, discharge, and tail-water level). Soil properties were further varied by placing materials at different moisture and density conditions. Headcut advance rates were heavily dependent on soil conditions, varying by a factor of more than 100. The advance rate declined as the unconfined compressive strength of soil increased.

4.1 *Small-scale breach tests for soil bag system*

An overlaid downstream slope covered by impermeable materials such as geosynthetics, gabions and concrete blocks should have a high resistance to overflowing during a flood. However, if the overflow passes into the dike under the surface layer, the downstream slope may suffer severe erosion as the advance rate dramatically increases. To validate the above and confirm the behavior of the soil bag facing system on the slope during overflowing, a series of model tests was performed. Six small models having different arrangements for facing, listed in Table 3, were prepared. The model consisted of a 20 cm-thick base ground, on which a 50 cm-high model earth-fill dam with a slope of 1V : 2H on a scale of 1/10 of an assumed prototype was constructed as shown in Figure 21a. The fill material was Hokota sand (specific gravity $G_s = 2.676$, mean diameter $D_{50} = 0.184$ cm, coefficient of uniformity

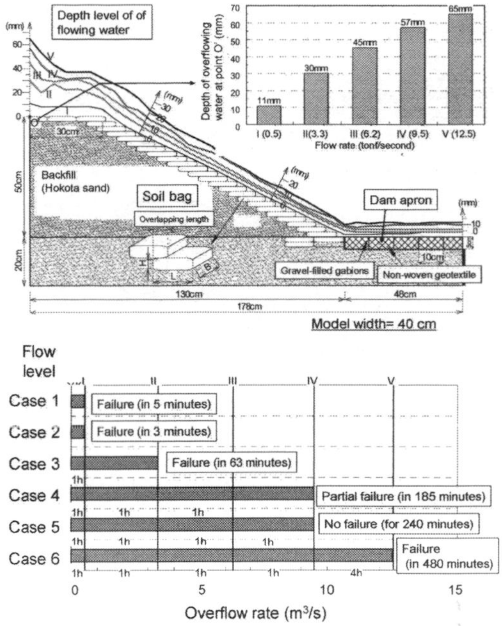

Figure 21. Overflow line and results for small scale breaching tests (Yamazaki et al., 2005).

$U_c = 5.82$, maximum dry density $\rho_{dmax} = 1.517$ g/cm^3 and optimum water content $w_{opt} = 14.3\%$), compacted at water content $w = 1.1.5\%$ to a relative density D_r equal to 85% (not very dense). The soil bags were made of polyester. With the respective model, the water flow was continued while increasing the flow rate every one hour as shown in Figure 21.

Figure 21 also illustrates the stream line on the slope at the constant flow rate for every case when the slope may suffer failure by washing out of the slope surface and the depth of overflowing water at the crest. The test results showed that the stability against overflow was increased by protecting the slope by using GSET with sufficient overlapping between vertically adjacent soil bags (Figure 21b). The first phase of failure of a slope embankment by overflowing was initial failure of the protective cover on the downstream face of the fill. Of course, a higher velocity and higher turbulent flow over the slope face may induce more rapid erosion of face materials.

Case 1: The slope soil was eroded gradually with time as water flowed below the non-woven geosynthetic sheets, finally resulting in a sudden shallow failure of the slope.

Case 2: The slope was protected by placing the soil bags on the slope surface with no overlapping between vertically adjacent soil bags. Sliding failure between the slope and the soil bags occurred by water flowing into the interface even when the flow

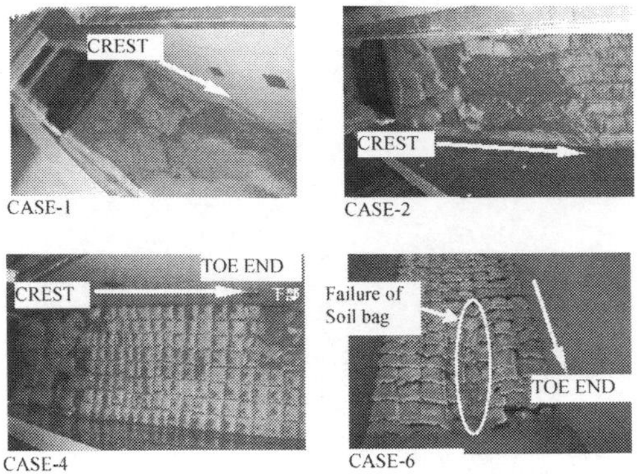

Figure 22. Surface erosion or failure patterns for each case after overflowing test.

rate was still low. This sliding failure resulted in a sudden sliding down of all the soil bags 3 minutes after the water started to flow.

Case 3: The slope was protected by placing the soil bags with overlapping vertically, with a lapping length of about 2 cm. This slope type was not eroded by flowing water level I, but after rising up to the second flowing water level, the cover layer with soil bags was washed out by the flow, because water entered the interface between the soil bags and slope.

Case 4: This type, which had a longer overlap of 3 cm compared with Case 2, suffered no damage at the third water level. Some soil bags around the center level were pulled out suddenly, and the slope toe failed by washing out of the soil bags.

Cases 5 and 6: By using longer soil bags with larger overlap (8 cm and 14 cm) between vertically adjacent bags, the slope became more stable with less erosion of slope soil. No rupture occurred in both cases at the fourth water level. In Case 6, the top surface of the soil bags suffered damaged by cutting of the soil bag materials. Of course, the infill materials were quickly eroded at the fourth water level, but the amount of this erosion gradually decreased because the infill material of damaged soil bags was caped the surface by the soil bag material, resulting into no failure at the fourth water level.

4.2 *Large-scale breaching tests for GSET-dam model*

Physical hydraulic model testing may include two-dimensional modeling of earth-fill dam sections of the entire dike, and will help clarify the fundamental mechanism of downstream slope breaching. In order to confirm the overall behavior of the dike and the formation of breaching, physical modeling should be performed on a large scale to overcome the problem

Table 4. Test steps for overflowing

Step	Quantity (t/s)	Depth (cm)	Time (hours)	
Step1	0.1	21.7	2.0	
Step2	0.2	26.4	2.0	–
Step3	0.4	34.1	2.0	–
Step4	0.8	40.5	3.5	–
Step5	1.1	48.4	2.0	–
Step6	0.8	40.5	2.0	:Failure of bag
Step7	1.5	58.5	3.0	–

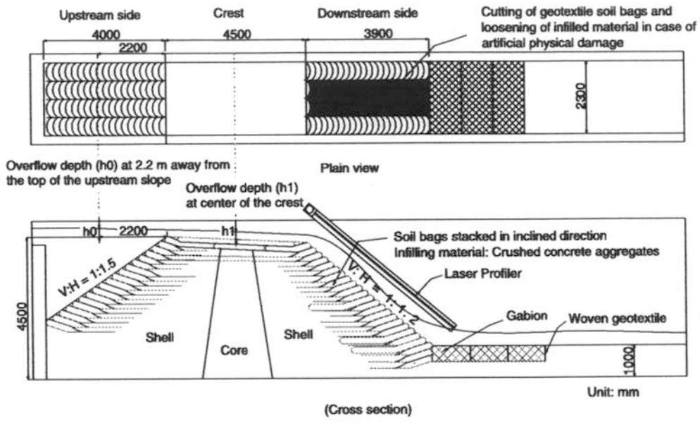

Figure 23. Test overview (Yamazaki 2007).

of simultaneous scaling of both hydraulic conditions and material properties. Small-scale model testing has already been performed, showing the mechanical behavior and discrepancies for some overlying types for downstream slope against overflowing. However, material properties do not scale uniformly and are difficult to reproduce at small scales, whereas large-scale modeling tests allow the use of near prototype size materials, making the results more reliable and easier to interpret the overall behavior.

Even for a small earth-fill dam, it will be difficult to conduct full-scale hydraulic model testing, so the scale should be sufficiently large to allow the use of prototype embankment materials, soil bags, and infill materials.

Half-scale model tests were conducted to develop the advanced soil bag system, providing full-scale soil bags in the flume of the earth fill dam.

We constructed a 4.5 m high embankment model with soil bags used for an actual earth-fill dam to conduct an overflow test. As shown in Table 4, the overflow depth was gradually increased to confirm the deformation of the embankment and the damage of soil bags at every step. This model has a downstream slope of 1V:1.2H and a depth of 2.3 m. Kasama sand ($\rho_s = 2.650$ g/cm^3, $\rho_{dmax} = 1.935$, $w_{opt} = 11.6\%$) and fine-grained fraction mixed materials ($\rho_s = 2.617$ g/cm^3, $\rho_{dmax} = 1.470$, $w_{opt} = 24.6\%$, Kanto loam 1: Kasama sand 1.5) were used for the embankment and for the core soil, respectively. As shown in

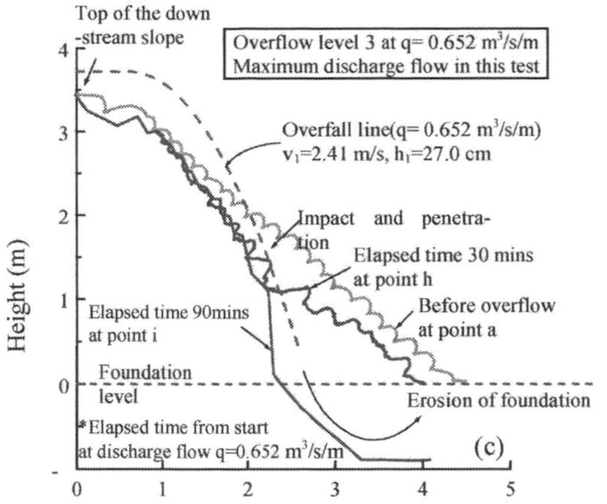

Figure 24. Lateral displacement of soil bag surface after overflowing tests (Matsushima 2007).

Figure 25. Eroded surface of Soil bags (Yamazaki 2007).

Figure 23, we stacked soil bags on the upstream and downstream faces of the embankment model to construct a surface that is resistant to overflow. These soil bags were stacked with a 15-degree angle to the foundation ground in order to increase the shear resistance of the stacked soil bags and enhance the safety of the model. Referring to the results obtained through a compression test, we used recycled crushed stone (RC-40) as a infill material for the soil bags to improve the bag's strength, and established an embankment model having enhanced overall safety.

PP sheet having tensile strength of 1.25 kN/m and tensile strain of 29.8% was used for the soil bag material. Unlike a normal large soil bag, this soil bag basically has a flat shape with a large aspect ratio (L/H = 5–8). As shown in Figure 9, it has a wing and a tail at its side and rear end, respectively. The wing is inserted into neighboring soil bags to prevent separation between soil bags, and also to help the whole soil bag system to serve as a wall similar to a reinforced earth retaining wall because the tail is buried in the embankment.

The flow rate was gradually increased from Step 1 to Step 7. Overflow water with a low flow rate went along the surface of the soil bag and was basically in the state of nappe flow, however, a small hydraulic jump occurred on each soil bag steps. In Steps 2 and 3, the overflow ran down the downstream slope of the embankment with the strong entrainment of air and entered a state of skimming flow.

As a result, the running water and soil bag surface started to separate from each soil bag surface, resulting in more sand being suctioned from the soil bags. From Step 4, the water's flow velocity increased, causing a hydraulic jump at the crest.

Figure 24 shows the lateral displacement of soil bag surface after overflowing tests.

In Steps 1, 2 and 3 with a low flow rate, these soil bags did not vary, although a very small amount of soil filled among the soil bags was drained.

In Steps 4 and 5, as shown in Figure 25, fill materials in the soil bags were released, so their surface sank. This phenomenon of washout occurred uniformly across the downstream slope of the embankment, but was centered on the area where overflow water jumped from the crest landed.

For Step 6, given that soil bag materials were damaged and deteriorated, we gave artificial damage (two soil bag column on the center from 4 to 23 steps: 2 × 20 = 40 soil bags) to the interface between the soil bags' surface and running water to continue this test.

As the damage expanded, some of the fill materials were released. Thereafter, however, the bag materials covered the upper fill materials again and adhered tightly to the whole surface of the embankment, so the overflow water did not erode the embankment's materials directly, resulting in no overall collapse of the dam. In Step 7 (1.5 m^3/s), where water with a higher flow rate flowed continuously, a soil bag where the overflow water landed directly broke, causing the successive destruction of lower soil bags. For this performance, the lower part of the embankment rose perpendicularly from the exact part where the overflow water landed but was not deformed further, so destruction of the whole dam did not occur.

Based on these findings, it was demonstrated that GSET-dam had great resistance to piping and suction and remained stable without complete collapse against large overflow. In addition, it is effective to place materials such as soil-cement and vegetation on the surface of the soil bags in order to ensure the durability of small earth-fill dams.

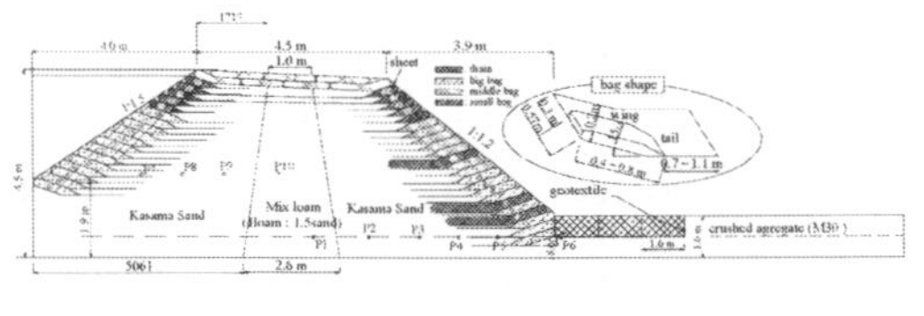

Figure 26. Cross section for in situ test earth-fill dam (Yamazaki 2007).

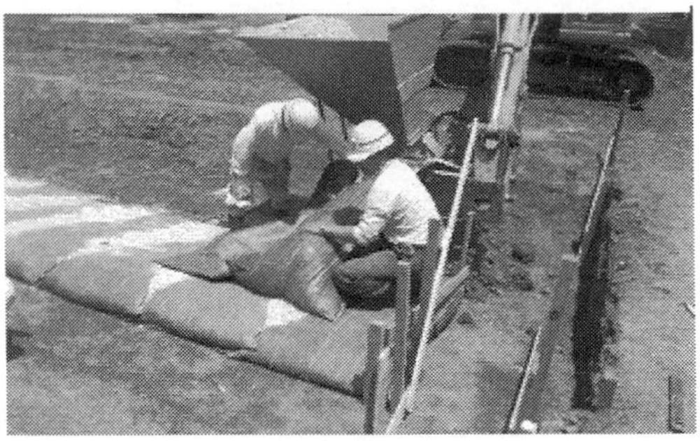

Figure 27. A typical screw extruder for soil bag (Matsushima 2007).

5 IN SITU TEST (IN NIRE)

NIRE constructed a full-scale earth-fill embankment using GSET as shown in Figure 26, examined its construction performance, observed its behavior during and after the construction and conducted a test to evaluate the long-term durability of the model (Matsushima et al., 2006; Yamazaki et al., 2006).

For this demonstration test, an embankment with a height of 3.2 m, an upstream slope of 1H:1.8H, a downstream slope of 1V:1H and a body width of 21 m (embankment made of soil bags containing recycled crushed rock was constructed: 10 m, embankment made of soil bags containing Kanto loam: 11 m). Kanto loam was used for embankment materials, but for two types of embankment structure, or for Type 1 and Type 2, recycled crushed rock (RC-40) and Kanto loam were used as infill materials for soil bags, respectively.

For the preparation of soil bags, we developed a unit that makes soil bags as shown in Figure 27. This machine enables us to prepare the bags directly on-site and substantially improves construction performance without infilling and transportation. In addition, a soil bag after being placed was fully compacted with a vibrator and was stacked at a 15-degree angle in order to increase the rigidity of the soil bag. After completing the embankment, we retained water to conduct long-term observations of the behavior of this earth-fill dam. Regarding the deformation volumes of Type 1 and Type 2 soil bags one year after completion, a maximum deformation of ±10 mm immediately after

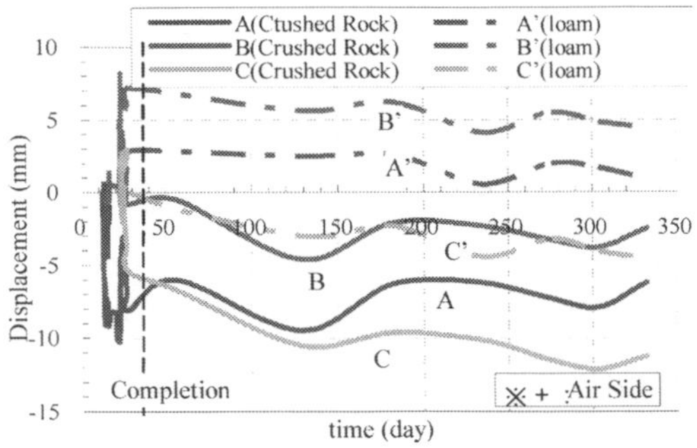

Figure 28. Lateral displacement of soil bag surface after construction (Yamazaki 2006).

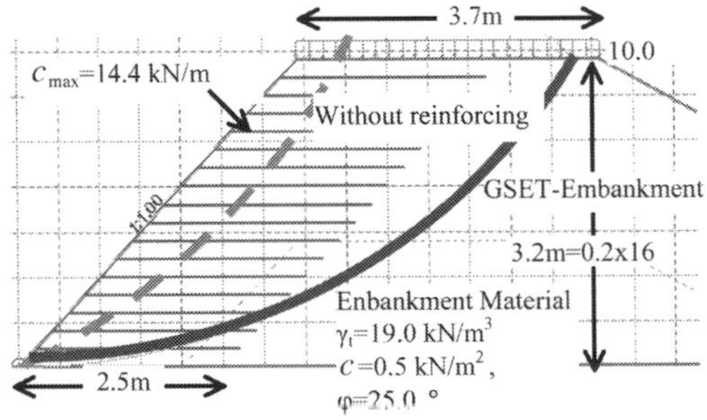

Figure 29. Limit state analysis for soil bag system (Yamazaki 2006).

Table 5. Limit state analysis for soil bag system.

Type	Static Fs	Seismic Fs
Normal embankment	0.578	0.461
GSET embankment	1.736	1.491

Fs: Safety factor.

construction was recorded for both, but both the bags then remained very stable with a variation of ±5 mm or less (Figure 28).

Figure 29 shows the results of stability calculations. For the embankment without soil bags, its safety factor was 0.578 at static condition. However, when the downstream slope was constructed with GSET to create the embankment, its safety factor was increased to 1.736. Even during an earthquake, the improved dam presented a safety factor of 1.491. In this demonstration test, the embankment was overlaid with soil-cement and vegetation to confirm the effectiveness of these methods of preventing UV deterioration of soil bags in the final stage.

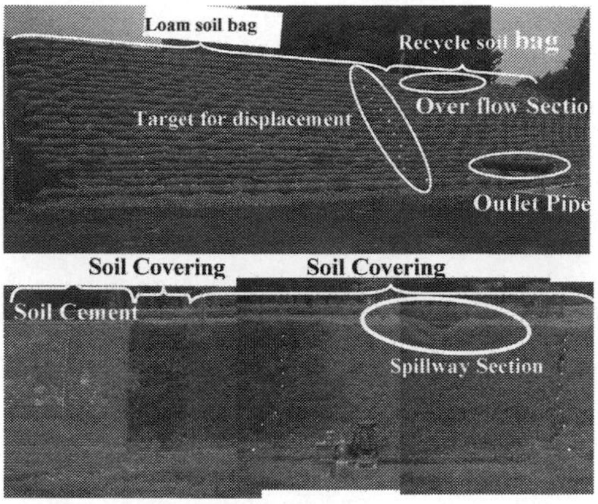

Figure 30. Surface view after piled up the soil bags and overlying the soilcement, vegitations.

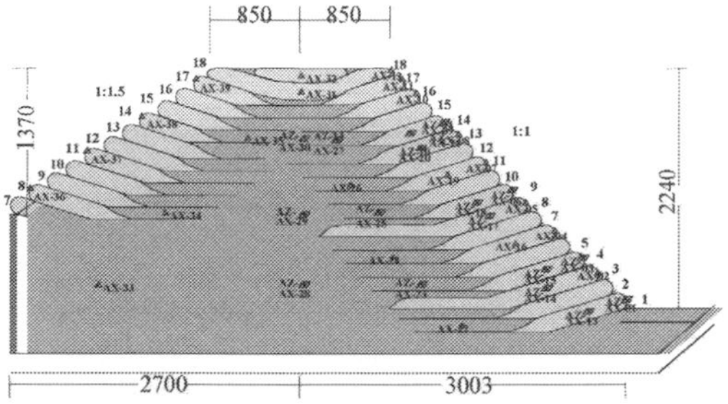

Figure 31. Cross section of the geosynthetics soil bag system reinforced earth dam.

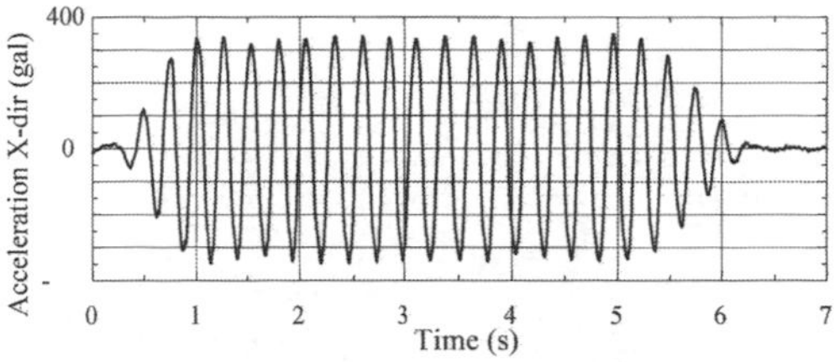

Figure 32. Input acceleration, sine shape $a_{max} = 348.718$ gals, $a_{min} = -348.718$ gals, frequency $f = 3.8$ Hz.

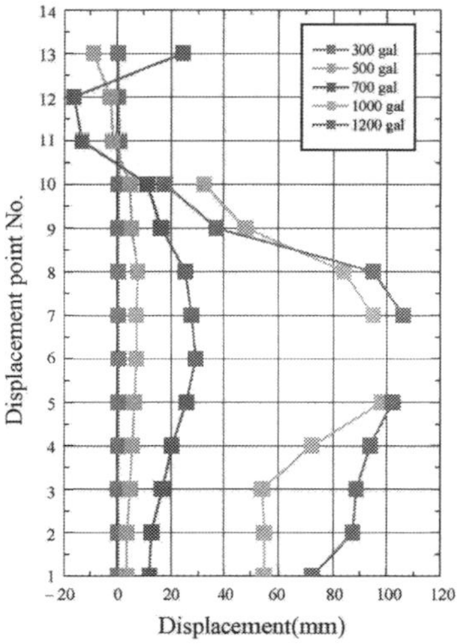

Figure 33. Relative lateral displacements of soil bag surface between two shakings (at the end of shaking).

6 SHAKING TEST FOR GSET-DAM AND ANALYSIS

In order to evaluate the applicability and design procedure of the newly proposed high-performance earth-fill dam with GSET (Mohri et al., 2005), NIRE conducted analyses and experiments (Goran, 2006; Matsushima et al.). The experiments consisted of shaking table tests of the model dam. The shaking table test data were compared with the analytical data in order to verify the analytical solution, and a detailed finite element analysis was performed.

6.1 *Experiments*

The model earth-fill dam had a height of 2.2 m, crest width of 1.7 m, downstream slope of 1V:1H and upstream slope of 1V:1.5 H, as shown in Figure 31. The body of the dam was constructed with silty sand with density of $Dr \approx 90\%$. The geosynthetic soil bags were filled with recycled concrete. It should be noted that the soil bags were embedded into the body of the dam with the tail extending from the soil bags.

The input acceleration had a sinusoidal shape with frequency $f = 3.8$ Hz and five levels of intensity, $a_{max} \approx 300$ gals; $a_{max} \approx 500$ gals; $a_{max} \approx 700$ gals; $a_{max} \approx 1000$ gals; $a_{max} \approx 1200$ gals (Figure 32).

The experimental results can be summarized as follows:

Lateral displacements of the soil bag surface were as shown in Figure 33. On the downstream slope when the intensity of input acceleration was low ($a_{max} = 300$ gals and 500 gals), displacement was limited to 0.85 mm and 7.85 mm, respectively. In the case of input acceleration $a_{max} = 700$ gals, the residual deformation was 29.06 mm. In the cases with stronger amplitudes of shaking $a_{max} = 1000$ gals and 1200 gals, the dam exhibited large horizontal deformation of 98.28 mm and 106.08 mm, respectively.

The crest of the dam subsided with large vertical displacement only during strong shaking.

After the experiment the dam was carefully sliced longitudinally in order to investigate the inner mode of deformation. Figure 34 shows that the failure mode for the downstream slope consisted of two parts: a circular sliding surface in the part with no geotextile, and

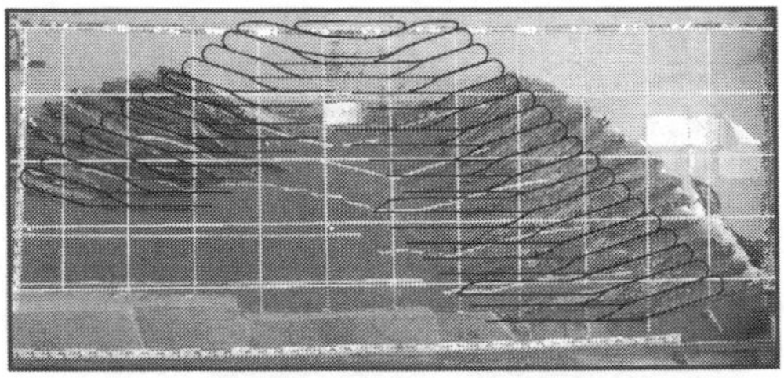

Figure 34. Sectional view and deformation of soil bag system after strong shaking.

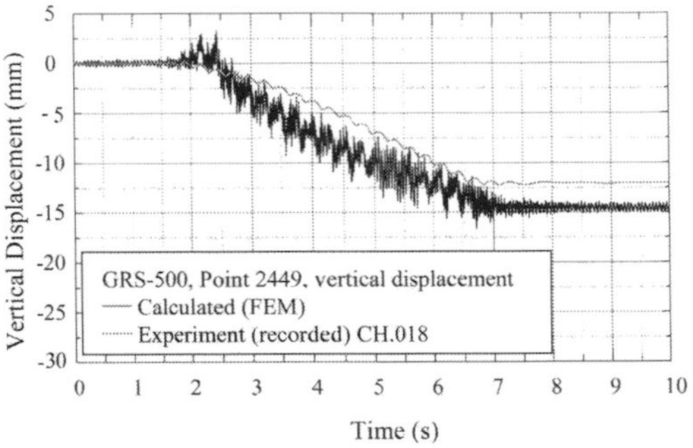

Figure 35. Comparison of FEM analysis and experimental data, vertical displacement at the middle of the top of the dam, case GRS-500.

horizontal sliding at the interface between the soil and geotextile.

The experiments revealed that GSET structure gives good seismic stability to the dam. The part of the dam with GSET is considerably strong and stiff, and so the sliding surface appeared at the dipper layer of the dam and significantly increased its stability.

6.2 *Analysis for GSET-dam*

The displacement obtained by the finite element analysis showed very close correlation with the data obtained experimentally from the shaking table test. This shows that the finite element analysis taking into account only the main soil parameter characteristics and using plastic modeling can successfully simulate the actual behavior of the dam. Figure 35 shows good predictions at the main points of interest such as the vertical settlement of the top of the dam as well as the displacement at the toe of the dam.

In general, in the case of lower intensity of shaking ($a_{max} = 300$, 500 and 700 gals) the data on the vertical settlement at the top of the dam showed closer correlation between experimental and analytical data. In the case of greater intensity of shaking ($a_{max} = 1000$ and 1200 gals) the data on the horizontal displacement at the toe of the downstream slope showed closer correlation between experimental and analytical data.

The finite element analysis thus successfully simulated the experimentally observed sliding surface.

7 CONCLUSIONS

A great number of small earth-fill dams for agricultural irrigation, which have an embankment height of 10 m or less, damaged and completely failed during recent several severe earthquakes and overflowing in events of flooding, which exceeds the drainage capacity of a spillway discharge system. In these case histories for earth-fill dams are described in this paper were built using empirical construction method.

This paper focused to develop new construction method for the small earth-fill dam, which can tolerate overflowing due to flooding and increase the earthquake resistance. Reinforced earth-fill dams, which were stacked by geosynthetic soil bags with extended tail(GSET), have good performance against both erosion by overflowing and earthquake resistance.

ACKNOWLEDGEMENTS

The authors wish to extended special thanks to Prof. Tatsuoka, F. (Tokyo University of science) for his advice and assistance in the series of research, and to Prof. Tanaka, T. (Tokyo University) for his assistance in the numerical analysis by FEM.

REFERENCES

Aqil, U., Matsushima, K., Lohani,T.N., Mohri,Y., Yamazaki, S., and Tatsuoka, F. 2005. Large scale shearing tests of stacked soil bags, Proc. of the 40th Japan National Conference on Geotechnical Engineering, Vol.40, 735–736.

Aqil, U., Matsushima, Mohri, Y., Yamazaki, S., and Tatsuoka, F. 2006. Lateral shearing tests on geosynthetic soil bags. 8th International Conference on Geosynthetics, 1703–1706.

Arangelovski, G., Mohri, Y., Matsushima, K., and Yamazaki, S. 2006. Comparison of analytical and experimental residual displacement of an earth dam improved with geotextile soil bag system. Proc. 41th Annual conference, JGS., 689–690.

Bathrust, R.L., Hatami, K. 1998. Seismic response analysis of geosynthetic reinforced soil retaining wall, Geosynthetics International, 5, 127–166.

Dodge, Russell, A. 1988. Overtopping Flow on Low Embankment Dams – Summary Report of Model Tests, REC-ERC-88-3, U.S.Bureau of Reclamation, Denver, Colorado, August 1988, 28p.

Fujii, H., Shimada, K., and Nishimura, S. 1991. Damage to small earth-fill dam in Okayama pref. at Typhoon 19th in 1990. Final report of Grants-in-Aid for Scientific

Research, Ministry of Education, Culture, Sports, Science and Technology, 101–130. (in Japanese).
Hori, T., Tagashira, H., Yasunaka, M., and Tani, S. 1998. Damage to earth dams in 1997. Proceedings Eighth International Congress IAEG, 241–248.
Lohani, T. N., Matsushima, K., Mohri, Y., and Tatsuoka, F. 2004a. Stiffness of soil bags filled with recycled concrete aggregate in compression. International Conference on Geosynthetics and Geoenvironmental Engineering, India, 106–112.
Lohani, T. N., Matsushima, K., Mohri, Y. and Yamazaki, S. 2004b. Deformation behavior of recycled concrete aggregate confined with geosynthetics soil bag. Proc. 39th Annual conference, JGS.
Lohani, T.N., Matsushima, K., Aqil, U., Mohri, Y., and Yamazaki, S. 2005. Applicability of geosynthetics soil bags in permanent civil engineering works., Proc. 40th Annual conference, JGS., 733–734.
Matsuoka, H., Liu, S., and Yamaguchi, K. 2001. Mechanical properties of soil bags and their applications to earth reinforcement. Landmarks in Earth Reinforcement, Ochiai et al. (eds), 587–592.
Matsushima, K., Yamazaki, S., Mohri, Y., Aqil, U. and Tatsuoka, F. 2005a. Overflow model test of small dam with soil bag, Proc. 40th Annual conference, JGS., 1995–1996. (in Japanese)
Matsushima, K., Lohani, T.N, Mohri, Y., Aqil, U., Tatsuoka, F., and Yamazaki, S. 2005b. Study on compression characteristic of geosynthetic soil bags., Geosynthetic Journal. Vol.20, JCIGS. (in Japanese)
Matsushima, K., Mohri, Y., Aqil, U., Aranglovski, G., Hironaka, J. and Yamazaki, S. 2006a. Shaking table model test on small earth dam with geotextile soil bags; Proc. 41th Annual conference, JGS., 685–686. (in Japanese)
Matsuhima, K., Yamazaki, S., Mohri, Y., Aqil, U. and Tatsuoka, F. 2006b. Structural features of earth dams allowing overtopping and a full-scale construction test. Annual conference, JSIDRE. (in Japanese)
Matsushima, K., Aqil, U., Mohri, Y., Tatsuoka, F., and Yamazaki, S. 2006c. Shear characteristics of geosynthetic soil bags stacked in tilted and horizontal directions., Geosynthetic Journal. Vol. 21, JCIGS. (in Japanese)
Matsushima, K., Yamazaki, S., Mohri, Y., Hori, T., Ariyoshi, M., Gotoh, M., and Tatsuoka, F. 2007. Overflow hydraulic test on full scale dam allowing overtopping, Proc. 42th Annual conference, JGS. (in Japanese)
Matsushima, K., Yamazaki, S., Mohri, Y., Hori, T., Ariyoshi, M., Gotoh, M., and Tatsuoka, F. 2007. Overflow hydraulic break test on full-scale earth dam allowing overtopping, Proc. 42th Annual conference, JGS. (in Japanese)
Mohri, Y., Matshushima, K., Hori, T., and Tani, S. 2005. Damage to small-size reservoirs and their reconstruction method, Special Issue on Lessons from the 2004. Niigataken Chu-Etsu Earthquake and Reconstruction, Foundation Engineering and Equipment (Kiso-ko), October, pp. 62–65. (in Japanese).
Powledge, George R., Ralston D.C., Miller, P., Chen, Y.H., Clopper, P.E. and Temple, D.M. 1989a. Mechanics of Overflow Erosion on Embankments. I: Research Activities, Journal of Hydraulic Engineering, vol.115, No.8, August 1989, 1040–1055.
Powledge, George, R., Ralston, D.C., Miler, P., Chen, Y.H., Clopper, P.E., and Temple, D.M. 1989b. Mechanics of Overflow Erosion on Embankments. II : Hydralic and Design Considerations., Journal of Hydraulic Engineering, vol.115, No.8 August 1989, 1056–1075
Pugh, Clifford, A., 1985. Hydraulic Model Studies of Fuse Plug Embankments, REC-ERC-85-7, U.S. Bureau of Reclamation, Denver, Colorado, December 1985, 33p.
Ralston, David, C., 1987. Mechnics of Embankment Erosion During Overflow. Hydraulic Engineering Proceedings of the 1987. ASCE National Confernce on Hydraulic Engineering, Williamsburg, Virginia, August 3–7, 1987, 733–738.
Robinson, K.M., and G.J.Hanson, 1993. Large-Scale Headcut Erosion Testing. Presented at the 1993 ASAE International Winter Meeting, Paper No. 93–2543, Chicago, Illinois, December 14–17.
Robinson, K.M., and G.J.Hanson, 1995. Large-Scale Headcut Erosion Testing. Transactions of the ASAE vol.38, no.2, 429–434.
Tani, S., 1996. Damage to earth dams, Special issue of Soil and Foundations, Japanese Geotechnical Society, 263–273.
Tatsuoka, F., Tateyama, M, Uchimura, T., and Koseki, J. 1997. "Geosynthetic-Reinforced Soil Retaining Walls as Important Permanent Structures", 1996–1997. Mercer Lecture, Geosynthetic International, Vol.4, No.2, 81–136.
Tatsuoka, F., Koseki, J., Tateyama, M., Munaf, Y., and Horii, N. 1998. Seismic stability against high seismic loads of geosynthetic-reinforced soil retaining structures, Proc. 6th Int. Conf. on Geosynthetics, 1, 103–142.
Temple, Darrel, M. 1989. Mechnics of an Earth Spillway Failure. Transactions of the ASAE, vol.32, No.6, 2015–2021.
Tournier, J.P. 2006. Safity of fill dams. ICOLD 2006, general report, Q86, 1327–1381.
Yamazaki, S., Matsushima, K., Mohri, Y., and Arangelovski, G. 2006. Prototype Test of Small Dam with Soil Bag, Proc. 41th Annual conference, JGS. (in Japanese)
Yamazaki, S., Matsushima, K., and Mohri, Y. 2007. Over flow test of small dam with soil bag, Proc. 42th Annual conference, JGS. (in Japanese)
Yasunaka, M., and Tagashira, H., 1994. Damage and rehabilitation for small earth-fill dam by heavyrain fall and earthquake, Material and construction committee of The Japanese Society of Irrigation, Drainage and Reclamation Engineering (JSIDRE), Vol.32, 33–44. (in Japanese).

New design philosophy of reinforced soil wall

New Horizons in Earth Reinforcement – Otani, Miyata & Mukunoki (eds)
© 2008 Taylor & Francis Group, London, ISBN 978-0-415-45775-0

Working stress design for geosynthetic reinforced soil walls

R.J. Bathurst
GeoEngineering Centre at Queen's-RMC, Department of Civil Engineering, Royal Military College of Canada, Kingston, Ontario, Canada

ABSTRACT: The paper briefly reviews current practice with respect to working stress design (WSD) for geosynthetic reinforced soil walls with a focus on North America and Japan. A number of issues are identified which have implications to current and future WSD for geosynthetic reinforced soil walls.

1 INTRODUCTION

This paper is focused on major features of working stress design (WSD) methods for reinforced soil walls constructed with extensible polymer reinforcement. A common feature of working stress design is the use of a factor of safety applied to prescribed failure modes. This represents a classical geotechnical approach to the design of earth structures and, at least in North America and Japan, represents the current state of the practice. Within the brief space available a number of issues are identified which have implications to current and future WSD for geosynthetic reinforced soil walls and future methodologies based on a limit states design format.

2 BRIEF HISTORY

The first geosynthetic reinforced soil walls were built in France in 1970 and 1971 (Leflaive 1988, Leclercq et al. 1990, Puig et al. 1977). A review of early French experience can be found in the paper by Allen et al. (2002).

Geosynthetic reinforced walls have been in use in the United States since 1974. Bell and Steward (1977) describe some of these early applications, which were primarily geotextile wrapped-face walls supporting logging roads in the northwestern United States. The history of geosynthetic wall design in North America has been summarized by Allen and Holtz (1991) and Berg et al. (1998).

3 GUIDANCE DOCUMENTS

Limit states design methods have been developed in the UK (BS8006 – BSI 1995), Hong Kong (Geoguide 1 – Geo 1993), and Australia (RTA 2003). In North America the most recent issue of the AASHTO (2007) highway bridge design code uses a limit state design approach for these structures (called LRFD – load and resistance factor design). However, these methods have at their core the deterministic equations found in WSD. Furthermore, limit state design methods have been calibrated by fitting to WSD rather than a formal calibration using a rigorous reliability-based framework and measured load and resistance data.

In the USA, the most recent fully WSD-based AASHTO guidance document for reinforced soil walls is AASHTO (2002). In Canada, the CFEM (2007) guidance document published by the Canadian Geotechnical Society is available. The design manual by the National Concrete Masonry Association (NCMA 1997) is focused on the design of geosynthetic reinforced segmental (modular block) walls. At present, both these guidance documents adopt a WSD approach.

Based on examination of both WSD and limit-states design methods, it can be argued that the underlying general approach for the key deterministic equations for reinforcement load prediction resistance have not changed significantly over the last 30 years.

4 CURRENT WSD PRACTICE

In North American working stress design practice, factors of safety are assigned to failure modes that are broadly classified as external, internal or facing stability modes of failure (Figure 1). In this section the general approach is reviewed with some examples. However, the reader is advised that there are many variations in the details of method implementation between guidance documents.

External modes of failure treat the composite facing and reinforced soil zone as a monolithic block. Factor

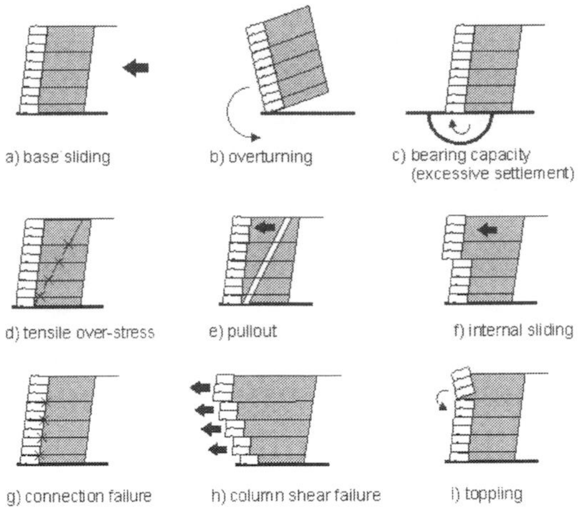

Figure 1. Modes of failure for geosynthetic reinforced soil walls: a), b), c) External; d), e), f) Internal; and g), h), i) Fac-ing (after CFEM 2007).

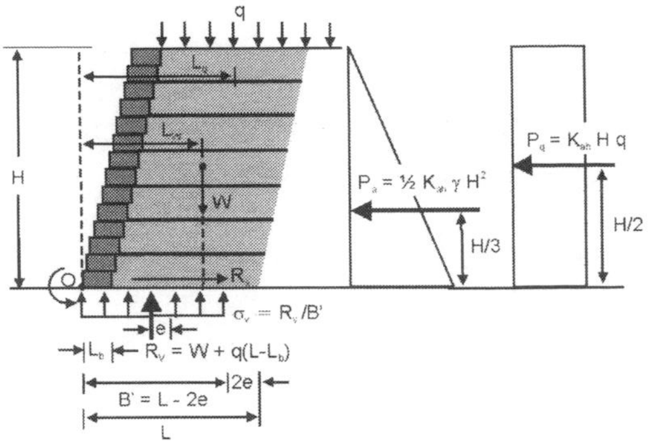

Figure 2. Free body diagram for external stability calculations (after CFEM 2007).

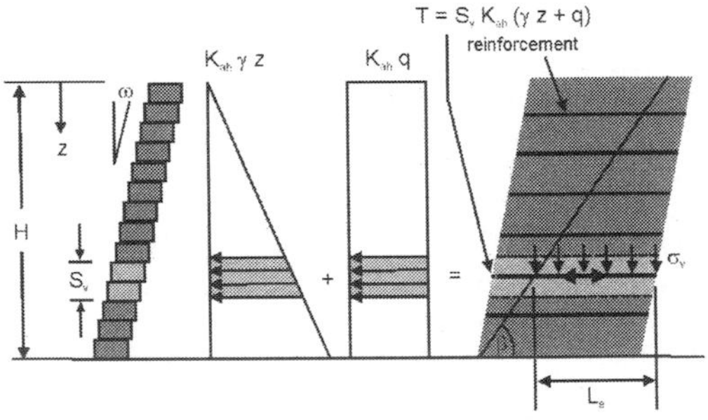

Figure 3. Free body diagram for internal stability calculations (after CFEM 2007).

of safety expressions are limit equilibrium-type and are the same as those used for conventional gravity wall structures. The total active earth force is calculated using the horizontal component of Rankine or Coulomb earth force. Coulomb theory has the advantage of explicitly including the influence of wall batter. Interface friction, δ, between the wall facing and soil is often taken as zero, and all soil volumes above the wall crest are treated as an equivalent uniform surcharge. The zero interface friction assumption is consistent with US design codes and practice. However, it is recognized that this assumption is likely conservative for the calculation of active earth forces used in internal stability design. The geometry and forces associated with external modes of failure are summarized in Figure 2. The minimum length of the reinforced soil zone taken from the face of the structure (L) is typically prescribed as $L/H = 0.6$ to 0.7 regardless of the magnitude of factors of safety against overturning or sliding.

In North America, the Simplified Method (also known s the Tie Back Wedge Method) is used to compute reinforcement loads and to establish minimum anchorage lengths in internal stability design and to compute loads acting on the facing. This is a limit equilibrium approach, and contains two key assumptions for the calculation of reinforcement load:

1. The magnitude of tensile load in each reinforcement layer is proportional to the soil overburden stress. Hence, reinforcement load will increase linearly with increasing depth of soil below the crest of the wall.
2. Tensile load in the reinforcement is a direct indicator of the state of stress in the soil since the reinforcement layer is assumed to carry the full lateral active earth pressure in the soil in the vicinity of the layer (i.e. contributory area approach).

The expression for the maximum reinforcement load is:

$$T_{max} = S_v K \sigma_v = S_v K(\gamma z + q) \tag{1}$$

where K is the lateral earth pressure coefficient computed using Rankine or Coulomb earth pressure theory, σ_V is the vertical pressure acting at the reinforcement layer located at depth z below the crest of the wall, S_V is the reinforcement spacing, γ is the unit weight of the soil, and q is a uniformly distributed surcharge pressure. It can be noted that BS8006 (1995) calculates the vertical stress σ_V using a Meyerhof approach. This increases the vertical stress at the level of each reinforcement layer and thus the tensile load assigned to the layer is larger compared to the AASHTO approach. In view of the comments made later in this paper, it may be argued that the Meyerhof approach adds to additional conservatism with respect to design against reinforcement overstressing.

The predicted maximum load level T_{max} in a reinforcement layer using Japanese practice PWRC (2000)

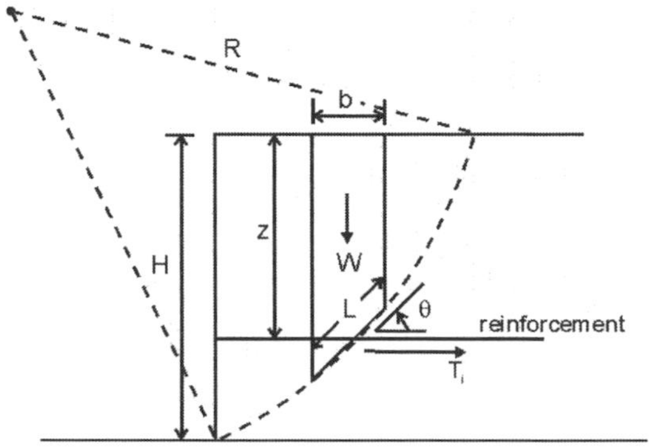

Figure 4. Circular slip analysis showing parameters used to compute required tensile load capacity of reinforcement layers.

is also calculated using Equation 1. However, K is computed as:

$$K = \frac{2 \ \Sigma T_{req}}{\gamma \ H^2} \qquad (2)$$

Here the required minimum sum of tensile load capacity for the reinforcement (ΣT_{req}) intersecting a circular failure surface is calculated as:

$$\Sigma T_{req} = \max \frac{F_s \ \Sigma(W \sin\theta) - \Sigma(cL + W\cos\theta \tan\phi)}{\Sigma\left\{\frac{2}{H^2} z\, b \tan\theta\, (\cos\theta + \sin\theta \tan\phi)\right\}} \qquad (3)$$

The other parameters used in the conventional circular slip analysis for an unreinforced soil mass (Bishop's Method of Slices) are shown in Figure 4. The recommended factor of safety is $F_s = 1.2$. Miyata and Bathurst (2007a) showed that setting $F_s = 1.0$, $q = 0$ and assuming a purely frictional soil gave the same the values for T_{max} as the AASHTO Simplified Method. Nevertheless, this factor should not be interpreted as being equivalent to the overall factor of safety FS used in Equation 4 (discussed below) in the context of the AASHTO method. In the Japanese approach, the factor of safety term F_s applies only to uncertainty related to the load side.

The allowable long-term tensile load of the reinforcement T_{allow} according to AASHTO (2002) and PWRC (2000) practice can be calculated as:

$$T_{allow} = \frac{T_{ult}}{FS \times RF_{design}} \qquad (4)$$

where FS is the overall factor of safety and the term $RF_{design} = RF_D \times RF_{CR} \times RF_{ID}$ is the product of partial factors to account for installation damage ($RF_D \geq 1.1$), creep ($RF_{CR} \geq 1.2$ – typically) and environmental degradation ($RF_{ID} \geq 1.1$), respectively. For the AASHTO method, FS = 1.5 in Equation 4.

For current Japanese practice, the denominator in Equation 4 is calculated using FS = 1.0 and $RF_{design} = 1.67$ or 3.33 depending on the reinforcement. It can be argued that the factor (F = 1.12 or 1.21) used to reduce the average ultimate strength value from tensile tests, is qualitatively similar to the overall factor of safety term FS in the AASHTO method. However, it must be pointed out that FS in the AASHTO method accounts for uncertainty that is related to both the load and resistance side of internal stability design equations while in the Japanese approach, F is related only to the resistance term (reinforcement capacity).

With the exception of wrapped-face walls, loads are transmitted from the reinforcement layers to the facing (Figure 1g). Where these facings are structural (e.g. concrete panel or modular block), the connections must be designed to have adequate design capacity computed as:

$$T_{conc} = \frac{T_{ult} \times CR_u}{FS \times RF_{CR} \times RF_D} \qquad (5)$$

where FS = 1.5 is the overall factor of safety. The other partial factors are for installation damage ($RF_D \geq 1.1$) and creep ($RF_{CR} \geq 1.1$). The reduction factor CR_u is determined from connection tests and is the ratio of connection strength to the index rupture strength of the intact reinforcement. For static loading cases using AASHTO design, the connection load is computed using the Simplified Method described earlier and without modification (i.e. connection loads are not increased or decreased from maximum internal tensile design loads). The value for RF_{CR} should be based on creep connection test (FHWA 2001). However, typically, this value is taken from creep-reduced strength values determined from in-isolation creep-rupture data.

5 SOME ISSUES WITH RESPECT TO DESIGN OF GEOSYNTHETIC REINFORCED SOIL WALLS

5.1 *Accuracy of load predictions*

A fundamental feature of current WSD approaches is the scaling of failure loads and resistance at limit-equilibrium to working stress conditions using one or more factors of safety or partial factors as described in the previous section. However, the stresses at incipient collapse cannot be simply scaled to working stress conditions in this manner.

Predicted versus measured T_{max} values are plotted with normalized depth below the crest of the wall in

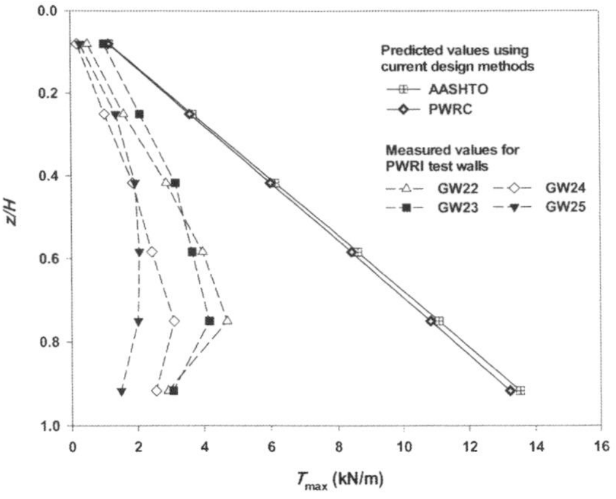

Figure 5. Predicted versus measured values of T_{max} using AASHTO (2002) and PWRC (2000) design methods and PWRI case histories (Miyata and Bathurst 2007a).

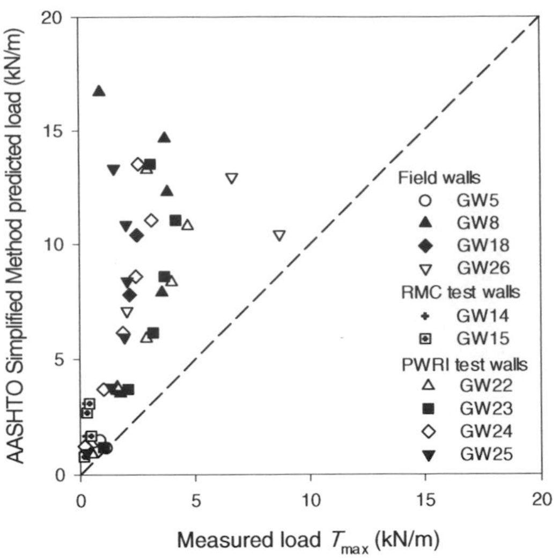

Figure 6. Predicted versus measured values of maximum reinforcement load using the AASHTO (2002) Simplified Method and structures with frictional backfill soils (Miyata and Bathurst 2007a).

Figure 5 for a set of 6 m-high PWRI walls in Japan that varied only with respect to facing type. The figure shows that for practical purposes both AASHTO (2002) and PWRC (2000) design methods give the same load predictions but are conservative with respect to measured loads.

Allen et al. (2002) and Miyata and Bathurst (2007a) carefully estimated the loads in reinforced layers from a large database of instrumented and monitored full-scale field and laboratory walls that used a frictional backfill. The loads were compared to predicted values using the Simplified Method. The comparisons were made using an estimate of the peak plane strain friction angle for the soil in each case. This required increasing the friction angle from triaxial or direct shear tests using published equations. The data are plotted in Figure 6. The measured loads are taken within the soil backfill away from the connections which may be locally higher as discussed later in the paper. The data show that predicted loads are consistently greater than the measured loads. Miyata and Bathurst showed that the ratio of measured to predicted loads was on average was about one third. This means that current WSD design methods are conservative with respect to predicting measured loads under operational conditions by at least a factor of three. This ratio is even greater if peak triaxial or direct shear strengths are used to compute predicted reinforcement loads since these values are typically less than plane strain values.

Bathurst et al. (2006) explored the accuracy of the Simplified Method with respect to influence of selection of soil friction angle to be used in reinforcement load computations and to identify those conditions when the general approach may be expected to give accurate estimates. They compared the reinforcement loads measured in 3.6-m high full-scale reinforced

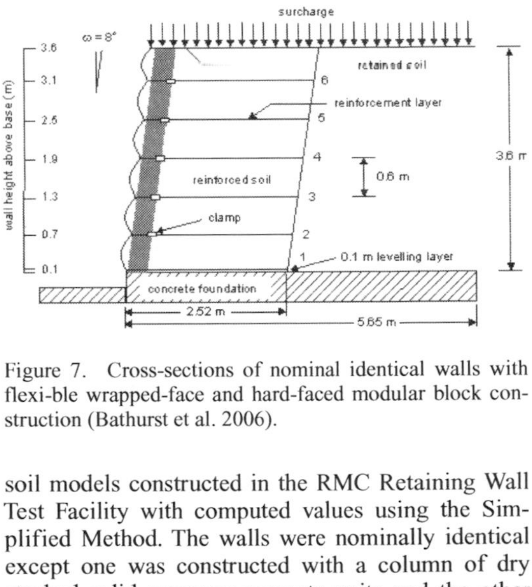

Figure 7. Cross-sections of nominal identical walls with flexi-ble wrapped-face and hard-faced modular block construction (Bathurst et al. 2006).

soil models constructed in the RMC Retaining Wall Test Facility with computed values using the Simplified Method. The walls were nominally identical except one was constructed with a column of dry stacked solid masonry concrete units and the other with a very flexible wrapped-face construction (Figure 7). They showed that for the most critical reinforcement layer in the wrapped-face wall, the measured and predicted reinforcement loads were in reasonable agreement provided that the peak plane strain angle was used. The discrepancy between predicted and measured loads increased in the order of peak direct shear friction angle and constant volume friction angle. For the companion wall with a hard (structural) facing, the Simplified Method grossly over-estimated the measured reinforcement loads even when the peak plane strain friction angle of the soil was used. The last observation is consistent with the database of results for a large number of field and laboratory walls that show that current limit-equilibrium methods are excessively conservative when reinforcement loads are computed for walls with structural facings. This is because the wall facing resists earth pressures, and load capacity is not due to just the mobilized shear strength of the soil and mobilized tensile capacity of the reinforcement. This is particularly true under operational conditions when the wall deformations are least and mobilized extensible reinforcement tensile capacity is low, as opposed to incipient (limit equilibrium) collapse conditions.

5.2 *Residual strength of geosynthetic reinforcement*

Conventional practice with regard to the selection of creep-limited strength values is to develop creep-rupture curves. Bernardi and Paulson (1997) and Greenwood (1997) summarized observations from the results of index tensile tests carried out on geosynthetic reinforcement materials after long-term creep loading. They concluded that the rupture strength reduction of PET and polyolefin reinforcement products does not vary linearly with logarithm of time. Rather, the residual index strength of polymeric reinforcement products is always greater than what is assumed based on conventional log-linear creep-rupture curves. Residual strength curves for materials with an index tensile strength, T_o, are illustrated in Figure 9. The residual strength curves are assumed to intersect the conventional creep-reduced strength curve at static and dynamic design strength values, T_{DS} and T_{DD}, respectively. In North American practice the design load under seismic loading can be increased by 33%. Hence, $T_{DD} > T_{DS}$ in this figure. Importantly, a reinforcement layer at a value of T_{DD} can be expected to have an available residual strength $T_{RDS} >> T_{DD}$. This additional strength is not considered in current limit-equilibrium methods of design and is a potential source of conservatism. An implication of observations reported in this section to seismic design is that the available strength and stiffness of geosynthetic reinforcement products under earthquake loading is not less than conventional estimates of available reinforcement strength in static load environments and may indeed be very much greater.

5.3 *Connections*

As noted earlier, current practice is to assume that connection loads at the facing of a wall are the same as those computed for internal stability design (i.e. for tensile over-stress and for pullout). However, there is evidence from monitored walls that connection loads are the highest loads in a layer of reinforcement. This

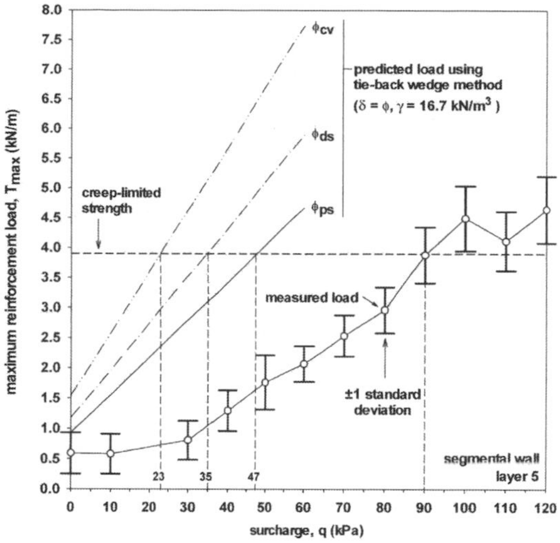

a) hard-faced wall

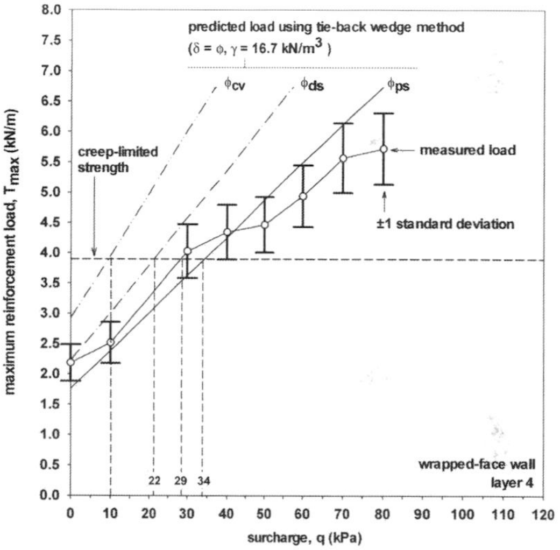

b) flexible wrapped-faced wall

Figure 8. Predicted and measured maximum reinforcement tensile loads at the end of construction and during surcharging for flexible wrapped-face wall and stiff-face segmental retaining wall (Bathurst et al. 2006).

can be attributed to relative downward movement of the soil backfill with respect to the relative vertically stiffer facing column or panel as the wall facing moves outward and as the soil is compacted and settles.

Figure 10 shows normalized peak strain values collected by the writer and colleagues from a total of 16 instrumented field walls. The range bars in the figure represent ± 1 standard deviation on the mean of data sets grouped according to distance intervals. The local peak at the free end of the reinforcement

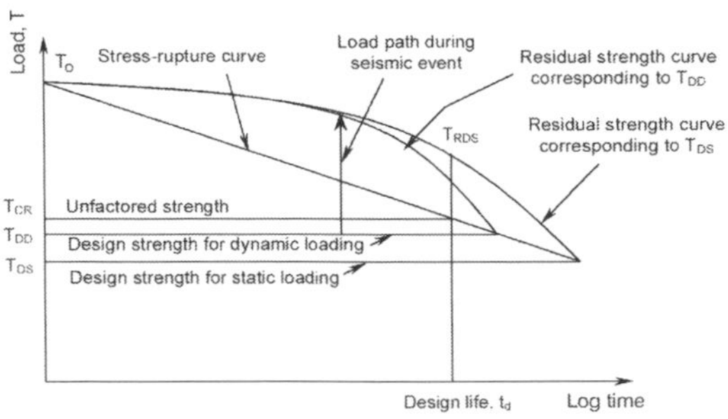

Figure 9. Concept of residual strength available to reinforce-ment layer under static and dynamic loading (Bathurst et al. 2002).

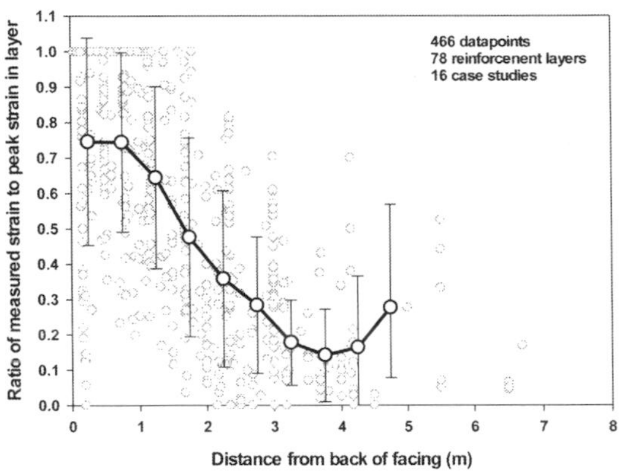

Figure 10. Normalized peak strain values for instrumented full-scale field walls constructed with a hard facing.

layers is likely an artifact of the limited data available at distances greater than about 4.5 m from the back of the facing column. The figure shows a clear visual trend in support of the hypothesis that connection loads are, on average, the highest loads in a reinforcement layer attached to a hard facing. Hence, if the estimate of maximum reinforcement is assumed to be correct, the same estimated load for the connection is non-conservative. Fortunately, internal tensile loads appear to be excessively over-designed and this may explain why connection failures are not systemic in these types of structures.

An example set of data for a creep connection test carried out in accordance with the recommendations found in FHWA (2001) is presented in Figure 11. The reinforcement material in this study was a biaxial coated woven polyester geogrid. It was connected to commercially available hollow masonry blocks with a discontinuous shear key. Hence the connection was a frictional-mechanical type. The connection-creep test protocol calls for a series of constant loads to be applied to the free end of the reinforcement and the

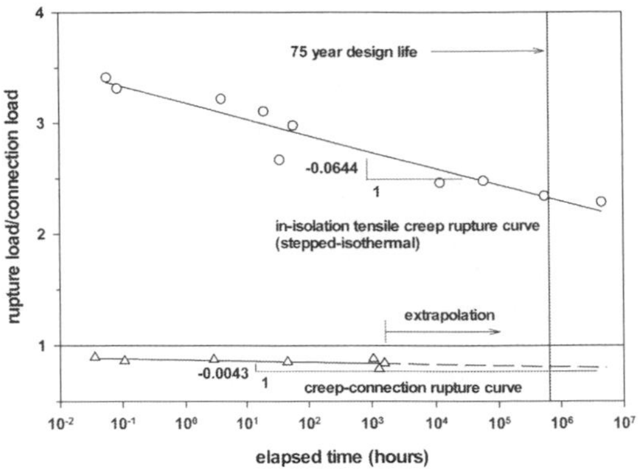

Figure 11. Normalized creep rupture curves for reference data and from creep-connection testing for woven polyester geogrid in combination with a typical hollow masonry facing unit.

time to rupture recorded for each test. The tensile loads are selected to generate a series of data points up to 1000 hours. The data presents as a straight line on a conventional semi-log plot. The curve can be extrapolated to a prescribed design life – in this example equal to 75 years. In this particular plot all tensile rupture loads have been normalized with respect to the index connection strength from a rapid constant rate of displacement test as described in the conventional ASTM D6638 standard. This has been done to focus on the qualitative features of the data rather than the actual block and geogrid materials used in the testing. Superimposed on the plot is the result of conventional creep-rupture data reported by the manufacturer for the same product. The important observations that can be made from this plot are that the index connection load is less than the short-term isolation strength. This is almost always the case for block-geogrid connections. However, the log-linear creep rate is much less for the connection test. This is, in the experience of this writer very typical. Hence, conventional practice which is to use the slope of the in-isolation creep data to reduce index connection test results for creep is conservative.

5.4 *Use of cohesive-frictional backfill soils*

In North America, the use of frictional well-draining backfill soils is preferred and recommended particle size distributions are presented in the guidance documents cited earlier. These granular materials are desirable because they have relatively high strength and stiffness, are easy to compact and are free-draining. However, in many cases and particularly on a worldwide basis, these good quality soils are not available or are cost-prohibitive. Nevertheless, many walls constructed with cohesive-frictional soils have performed well. Current practice is often to ignore the cohesive strength component of a backfill soil and to carry out load computations based only on the frictional component of soil strength. The argument offered is that the cohesive strength component may be attenuated due to moisture content increases and hence may not be available for the life of the structure. However, this may lead to conservative design for structures in which moisture content increases are not permitted to occur due to good drainage design and implementation.

Miyata and Bathurst (2007b) proposed a method to compute an equivalent secant friction angle from laboratory direct shear and triaxial c-ϕ Mohr-Coulomb shear strength data that can be used in Simplified Method calculations. Their approach removes a portion of the conservatism that occurs by simply setting $c = 0$ in load calculations. Nevertheless, similar to experience with walls constructed with purely frictional soil backfill, they showed that the Simplified Method still results in excessively conservative estimates of reinforcement loads based on measured load values from a database of monitored walls constructed with c-ϕ soils. For example, ratios of measured to predicted reinforcement loads were in the range of 1/3 to 1/5. Again, this may explain the good performance of many of these walls even for cases when they may have been poorly compacted and/or the soils have been wetted up due to poor soil surface drainage management.

6 CONCLUSIONS

This paper provides a brief review of some key aspects of current conventional working stress design (WSD) for geosynthetic reinforced soil retaining walls. This technology has been largely driven by economics since it has been shown that these systems can be constructed at up to 50% of the cost of conventional gravity wall systems (Koerner et al. 1998). Nevertheless, test methods and design methods have lagged. In particular, the distribution and magnitude of reinforcement loads predicted using conventional Tie Back Wedge Methods of analyses are likely excessively conservative. Fortunately, the writer and collaborators have collected over a number of years a database of wall performance data that can be used to formulate calibrated design methods that preserve features of conventional approaches but hold promise to better predict reinforcement loads. A companion paper by Bathurst et al. (2007) that appears in this conference proceedings describes recent developments in the K-stiffness Method which is a step in this direction.

ACKNOWLEDGMENTS

The work described in this paper is the result of efforts of many graduate students and colleagues over many years. The writer would like to recognize in particular the contributions of my long-term collaborators D. Walters, Y. Miyata, A. Nernheim and T.M. Allen for much of the data presented this paper. However, the opinions expressed in this paper are solely those of the writer.

REFERENCES

Allen, T.M., Bathurst, R.J. and Berg, R.R. 2002. Global Level of Safety and Performance of Geosynthetic Walls: An Historical Perspective, *Geosynthetics International*, Vol. 9, Nos. 5–6, pp. 395–450.

Allen, T.M. and Holtz, R.D. 1991. Design of Retaining Walls Reinforced with Geosynthetics, Geotechnical Engineering Congress 1991, McLean, F., Campbell, D.A. and Harris, D.W., Editors, *ASCE Geotechnical Special Publication* No. 27, Vol. 2, Proceedings of a congress held in Boulder, Colorado, USA, June 1991, pp. 970–987.

American Association of State Highway and Transportation Officials 2002. *Standard Specifications for Highway Bridges*, Seventeenth Edition. Washington, DC, USA.

American Association of State Highway and Transportation Officials 2007. *LRFD Bridge Design Specifications*. AASHTO, Fourth Edition, Washington, DC, USA.

ASTM D 6638. Standard Test Method for Determining Connection Strength Between Geosynthetic Reinforcement and Segmental Concrete Units (Modular Concrete Blocks), American Society for Testing and Materials, West Conshohocken, PA, USA.

Bathurst, R.J., Miyata, Y. and Allen, T.M. 2007. Recent Developments in the K-Stiffness Method for Geosynthetic Reinforced Soil Walls, *Proceedings of IS-Kyushu 2007*, Fukuoka, Japan, 6 p.

Bathurst, R.J., Vlachopoulos, N., Walters, D.L., Burgess, P.G. and Allen, T.M. 2006. The influence of facing rigidity on the performance of two geosynthetic reinforced soil retaining walls, *Canadian Geotechnical Journal*, Vol. 43, No. 12, pp. 1225–1237.

Bell, J.R. and Steward, J.E. 1977. Construction and Observations of Fabric Soil Walls, *Proceedings of the International Conference on use of Fabrics in Geotechnics*, Paris, France, Vol. 1, pp. 123–128.

Berg, R.R., Allen, T.M. and Bell, J.R. 1998. Design Procedures for Reinforced Soil Walls – A Historical Perspective, *Proceedings of the Sixth International Conference on Geosynthetics*, IFAI, Vol. 2, Atlanta, Georgia, USA, pp. 491–496.

Bernardi, M. and Paulson, J. 1997. Is creep a degradation phenomenon? In: Wu, J.T.H. (Ed.), *International Symposium on Mechanically Stabilised Backfill*, Denver, Colorado, USA, February 1997, Balkema, pp. 289–294.

BS8006. 1995. Code of Practice for strengthened/reinforced soil and other fills. *British Standards Institution*, London, UK.

Canadian Foundation Engineering Manual (CFEM), 2007. 4th Edition, *Canadian Geotechnical Society*.

FHWA. 2001. Mechanically Stabilized Earth Walls and Reinforced Soil Slopes Design and Construction Guidelines, FHWA NH1-00-043, National Highway Institute, *Federal Highway Administration U.S. Department of Transportation*, Washington, DC, USA.

GEO 1993. Guide to Retaining Wall Design, (Geoguide 1). *Geotechnical Engineering Office*, Hong Kong.

Greenwood, J.H. 1997. Designing to residual strength of geosynthetics instead of stress-rupture. *Geosynthetics International*, Vol. 4, No. 1, pp. 1–10.

Koerner, J., Soong, T-Y. and Koerner, R.M. 1998, Earth Retaining Costs in the USA, GRI Report #20 for FHWA, Geosynthetics Institute, Folsom, PA, USA.

Leclercq, B., Schaeffner, M., Delmas, Ph., Blivet, J.C. and Matichard, Y. 1990. Durability of Geotextiles: Pragmatic Approach Used in France, *Proceedings of the Fourth International Conference on Geotextiles, Geomembranes and Related Products*, Balkema, Vol. 2, The Hague, Netherlands, May 1990, pp. 679–684.

Leflaive, E. 1988. Durability of Geotextiles: The French Experience, *Geotextiles and Geomembranes*, Vol. 7, Nos. 1–2, pp. 23–31.

Miyata, Y. and Bathurst, R.J. 2007a. Evaluation of K-Stiffness method for vertical geosynthetic reinforced granular soil walls in Japan, *Soils and Foundations*, Vol. 47, No. 2, pp. 319–335.

Miyata, Y. and Bathurst, R.J. 2007b. Development of K-Stiffness method for geosynthetic reinforced soil walls constructed with c-ϕ soils. *Canadian Geotechnical Journal*, (in press)

NCMA. 1997. Design manual for segmental retaining walls (2nd edition, J. Collin Editor). *National Concrete Masonry Association*, Herdon, Virginia, USA, 289 p.

Puig, J., Blivet, J.-C. and Pasquet, P. 1977. Earth Fill Reinforced with Synthetic Fabric, *Proceedings of the International Conference on the use of Fabrics in Geotechnics*, Vol. 1, Paris, France, April 1977, pp. 85–90.

PWRC. 2000. Design and Construction Manual of Geosynthetics Reinforced Soil, (revised version), Public Works Research Center, Tsukuba, Ibaraki, Japan, 305 p. (in Japanese)

RTA 2003, Design of Reinforced Soil Walls, QA Specification R57, Roads and Traffic Authority of New South Wales, Australia.

New Horizons in Earth Reinforcement – Otani, Miyata & Mukunoki (eds)
© 2008 Taylor & Francis Group, London, ISBN 978-0-415-45775-0

Seismic design of geosynthetic reinforced soils for railway structures in Japan

J. Koseki
Institute of Industrial Science, the University of Tokyo, Japan

M. Tateyama & M. Shinoda
Railway Technical Research Institute, Japan

ABSTRACT: Focusing on seismic design of geosynthetic reinforced soils, several features of a newly-revised design standard for railway earth structures in Japan are reported. Three ranks of seismic performance against level 1 and 2 earthquakes are assigned. Prescriptive measures are admitted, where use of primary and/or secondary geosynthetic reinforcements is mandated for embankments. Recommendation of verification procedures is made, where Newmark sliding block analysis is adopted against the level 2 earthquake load. No creep reduction is considered in setting the tensile strength of geosynthetic reinforcements against earthquake loads.

1 INTRODUCTION

1.1 *Background*

Recently, Japan suffered from several large earthquakes which caused extensive damage to earth structures. In some of them, reinforced soils performed very well (e.g., Tatsuoka et al., 1997 among others), and thus they were considered more frequently for the replacement of conventional earth structures in reconstruction works (Koseki et al., 2006 and Shinoda et al., 2007 among others).

Since the 1995 Hyogoken-Nanbu (Kobe) earthquake, in particular, the level of the design seismic load has been raised significantly, while introducing the concept of so-called level 2 earthquake load, which is defined as the maximum possible level over the design life of civil engineering structures (JSCE 2006). Meanwhile, the principle of performance-based design has been introduced as well. Under such circumstances, several design guidelines for new construction works of civil engineering structures in Japan have been revised or are under revision process, and reinforced soils are not the exceptions.

1.2 *Scope of the report*

In view of the above, it is attempted in this report to describe briefly the features of the design standard for railway earth structures (RTRI 2007) that has just been revised, focusing on the seismic design of geosynthetic reinforced soils (hereafter called as GRS).

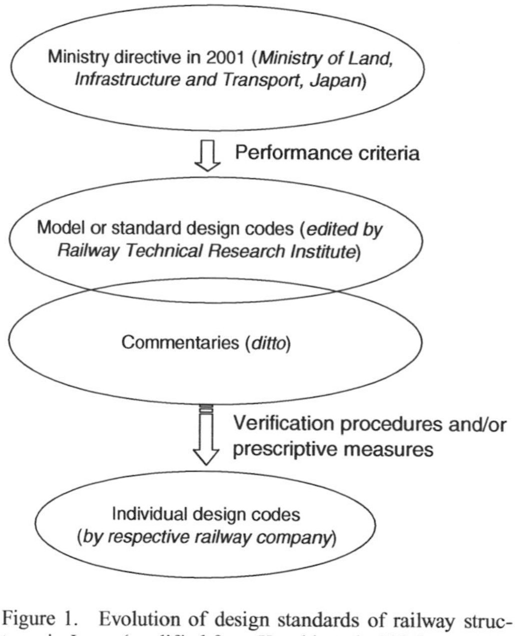

Figure 1. Evolution of design standards of railway structures in Japan (modified from Koseki et al., 2006).

As shown schematically in Figure 1 (Koseki et al., 2006), the revision was made following a directive by the Ministry of Land, Infrastructure and Transport in Japan issued in 2001. This directive mandated that structures should be designed based on performance criteria. It should be noted that the design standards for

Table 1. Composition of the design standard for railway earth structures and its commentaries (RTRI 2007).

Chapters	Pages
1. General	30
2. Design principles	32
3. Embankment	90
4. Cut and natural slopes	36
5. Base course	40
6. Subgrade	21
7. Reinforced soil (general)	28
8. Reinforced soil wall	17
9. Reinforced soil abutment	17
10. Reinforced cut slope	24
Appendix	365

foundations and retaining walls and for seismic design principles are currently under revision. Refer to Koseki et al. (2006) for the details of the relevant revision history of the whole design standards of railway structures in Japan.

As also shown in Figure 1, the design standard is accompanied by commentaries. Based on these, individual design codes will be implemented by railway companies in Japan.

2 DESIGN STANDARD OF RAILWAY EARTH STRUCTURES IN JAPAN

2.1 *Composition of the standard*

The table of contents of the newly revised design standard for railway earth structures and its commentaries (RTRI 2007) is listed in Table 1.

As one of the standard construction methods, soil reinforcing techniques including GRS are adopted in chapters 7 through 10, covering a subtotal volume of 86 pages that is about a quarter of the total volume excluding the appendix. The reinforced soil wall (chapter 8) and the reinforced soil abutment (chapter 9) deal with GRS retaining walls with a full-height rigid facing. The reinforced cut slope (chapter 10) deals with retaining walls to support cut slopes that are reinforced with nailing, micropiling or doweling.

In addition, in chapter 3 on embankment, use of secondary geosynthetic reinforcements is standardized. In this report, therefore, design of embankments will be also described in sections 2.2 and 2.4.

2.2 *Design flow of embankments*

Figure 2 illustrates the flow chart of the performance-based design of embankments that is specified in the design standard (RTRI 2007). After setting up the required performance in terms of safety, serviceability and repairability, decision is made whether prescriptive measures are adopted or not.

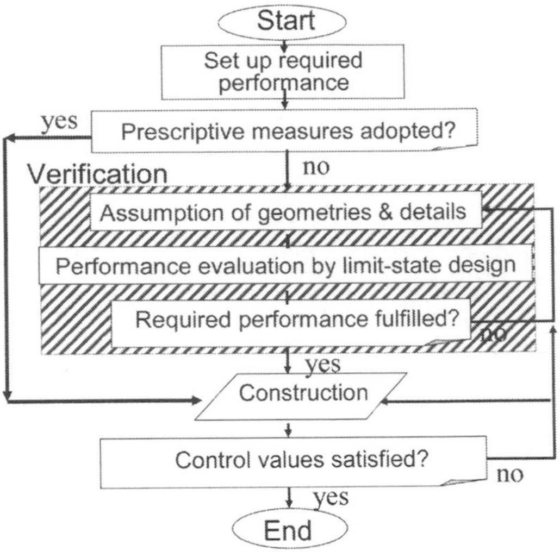

Figure 2. Flow of performance-based design of embankments.

The prescriptive measures have three different levels. Each of the prescriptive measures has been verified in advance to fulfill the corresponding performance rank, and thus no additional verification is required at the design stage.

On the other hand, if the prescriptive measures are not adopted, one needs to proceed to the verification process, including assumption of the geometries and structural details, performance evaluation based on limit-state design and confirmation of the required performance. In case the required performance is not fulfilled through the verification, one should modify the assumption and repeat the same procedures.

It should be noted that, in designing structures other than embankments, prescriptive measures are not available, and thus the verification process shall be implemented.

2.3 *Required seismic performance*

Table 2 summarizes the required performances of railway earth structures against two levels of design earthquake loads. In this table T_{des} is the design life of the structure. In general, the design life of a railway earth structure is assumed as 100 years.

For a level 1 earthquake load that is highly expected over the design life, it is required that all the earth structures will maintain their design functions without requiring repair work, i.e. will not exhibit excessive displacements (Performance rank III).

Against a level 2 earthquake load, which is defined previously in section 1.2, it is required that important earth structures can be restored to design function conditions with minimal repair (Performance rank II),

Table 2. Performance requirements for railway earth structures in Japan (modified from Koseki et al., 2006).

Action (design earthquake loads)	Level 1 (highly expected for T_{des})	Level 2 (maximum possible for T_{des})
Important structures	Performance I: Will maintain their expected functions without repair works (no excessive displacements)	Performance II: Can restore their functions with quick repair works
Others		Performance III: Will not undergo overall instability

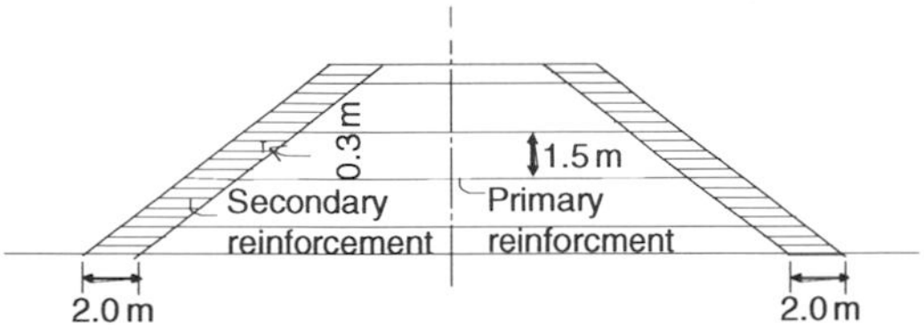

Figure 3. Cross-section of 6m-high embankment with performance rank I specified as prescriptive measure.

Table 3. Configurations of 6m-high embankment with different performance ranks specified as prescriptive measures.

	Rank I	Rank II	Rank III
Slope	1:1.8	1:1.5 (upper) 1:1.8 (lower)	1:1.5
Primary reinforcement	Yes	Basically no	No
Secondary reinforcement	Yes	Yes	Yes

while the other earth structures will not undergo overall instability (Performance rank I).

2.4 *Prescriptive measures for embankments*

When the prescriptive measures are adopted for embankments, different configurations of reinforcement arrangement as well as the slope geometry are employed depending on the performance rank.

For example, for a 6 m-high embankment, the primary reinforcements are used with rank I, as shown in Figure 3 and Table 3. They are placed for the full width of the embankment at a vertical spacing of 1.5 m. With ranks II and III, on the other hand, only the secondary reinforcements are used to protect the slope for the width of 2.0 m and to enhance the specified height (=0.3 m) of fill lift during construction. The design tensile strength of the primary and secondary reinforcements shall be 30 and 2 kN/m, respectively.

For a 6 m-high embankment, the slope shall have no berm, while its angle is varied depending on the performance rank as listed in Table 3. It should be noted that the type of the fill material and the required degree of compaction to be secured during construction are also varied depending on the performance rank.

2.5 *Design tensile strength of geosynthetics reinforcement*

The design strength of geosynthetics reinforcements are determined based on the following procedures.

The design strength T_d is assigned by applying a material factor γ_g to the characteristic strength T_a (Eq. 1), which is obtained as a product of the derived strength T_k and the modification factor ρ_m (Eq. 2).

$$T_d = \gamma_g \times T_a \tag{1}$$

$$T_a = \rho_m \times T_k \tag{2}$$

The derived strength T_k is evaluated by Eq.3 using the average value T_{AVE} and the standard deviation σ_x of the measured strengths, where the factor a is set equal to 0.67.

$$T_k = T_{AVE} - a \times \sigma_x \tag{3}$$

The material factor γ_g is given as a product of several reduction factors α_i (Eq. 4), while the modification factor ρ_m is set equal to unity against accidental actions such as the level 2 earthquake load and to 0.8 otherwise.

$$\gamma_g = \prod \alpha_i \tag{4}$$

Table 4 summarizes the combination of reduction factors α_i in evaluating the design tensile strength of reinforcements under five different situations. Reductions are made considering the combined effects of alkalis (α_1), damage during construction (α_2), creep load (α_3), impact load (α_4) and cyclic load (α_5). It should be noted that the design strengths against earthquake loads as well as train load can be assigned without considering the possible effects of creep reduction. This is consistent with recent findings on the load-strain-time behavior of geosynthetic reinforcement products as reported by Greenwood et al. (2001) and Tatsuoka (2003) amongst others.

2.6 *Verification against level 1 earthquake load*

As explained in section 2.2, if the prescriptive measures are not adopted, one needs to proceed to the verification process. In case of GRS retaining walls,

Table 4. Combination of reduction factors.

Design situations (action combinations)	Reduction factors				
	α_1	α_2	α_3	α_4	α_5
Permanent	Yes	Yes	Yes	No	No
Permanent + variable (train load)	Yes	Yes	No	No	Yes
Permanent + major variable (level 1 earthquake) + minor variable	Yes	Yes	No	Yes	No
Permanent + accidental (e.g., level 2 earthquake) + minor variable	Yes	Yes	No	Yes	No
Temporary state during construction	No	Yes	No	No	No

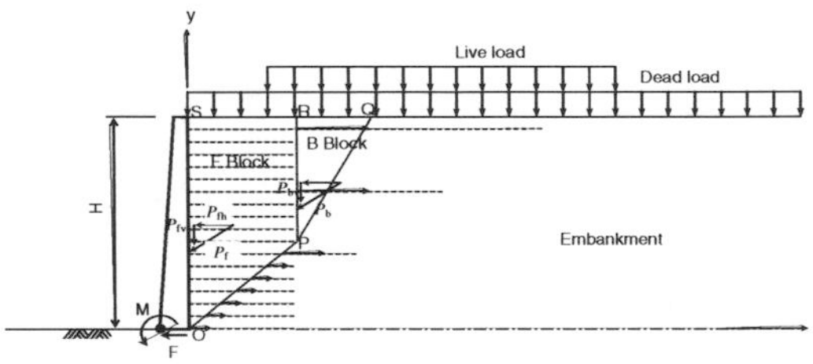

Figure 4. Modeling of GRS retaining wall for internal stability analysis.

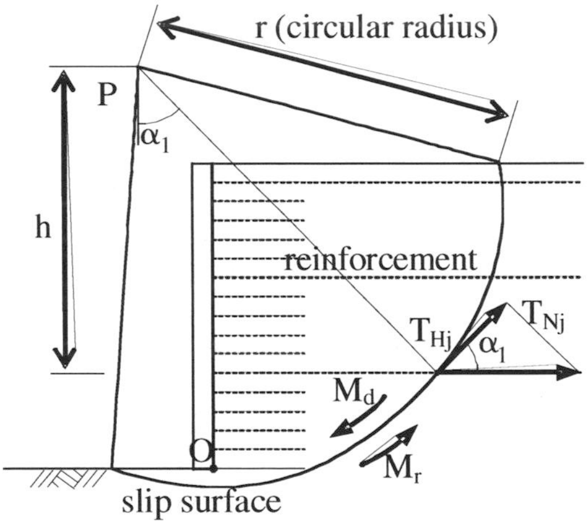

Figure 5. Modeling of GRS retaining wall for external stability analysis.

the performance requirement for level 1 earthquake load as explained in section 2.3 is verified through stability analyses using load and resistance factors against internal instability (Figure 4), external instability (Figure 5), and facing failure (Figure 6).

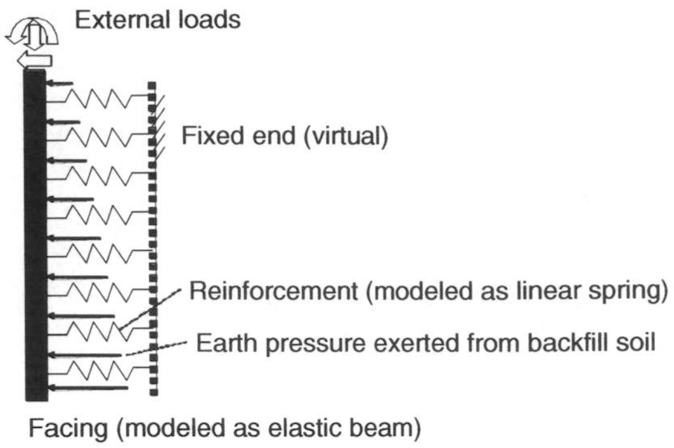

Figure 6. Modeling of facing and reinforcements.

Internal stability analysis is conducted with respect to base sliding and overturning. For example, the stability against base sliding is verified using the following equation:

$$\gamma_i \times \frac{H_{Rd}}{H_{Ld}} \leq 1.0 \tag{5}$$

where γ_i is the structure factor (set equal to unity in general); H_{Rd} is the response value of base sliding force; H_{Ld} is the limiting value of base sliding force.

The response value of base sliding force is evaluated as:

$$H_{Rd} = \gamma_H \times (P_{fh} + W_{EQ} + F_H) \tag{6}$$

where γ_H is the load factor (set equal to unity in general); P_{fh} is the horizontal component of the resultant force of earth pressure exerted from the backfill; W_{EQ} is the horizontal inertia force of facing; F_H is the external load applied to the top of the facing (e.g., due to the existence of noise barrier). As schematically shown in Figure 4, the resultant force of earth pressure is evaluated based on the two-wedge method.

The limiting value of base sliding force is evaluated as:

$$H_{Ld} = f_{ri} \times (\sum T_i + W_{BS} + W_{hp}) \tag{7}$$

where f_{ri} is the resistance factor (set equal to 0.80 against level 1 earthquake load); T_i is the design tensile resistance of reinforcement; W_{BS} is the design shear resistance mobilized at the bottom of facing; W_{hp} is the design horizontal resistance mobilized at the embedded part of the facing. The design tensile resistance of reinforcement is evaluated as the smaller value between the design tensile strength of reinforcement T_d that has been explained in section 2.5 and

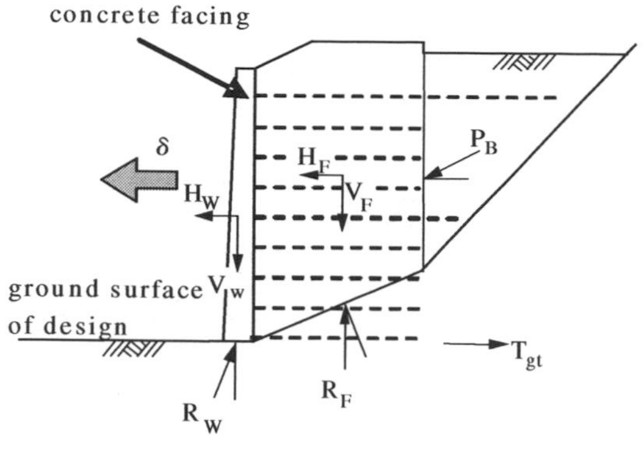

Figure 7. Modeling of GRS retaining wall for evaluation of base sliding.

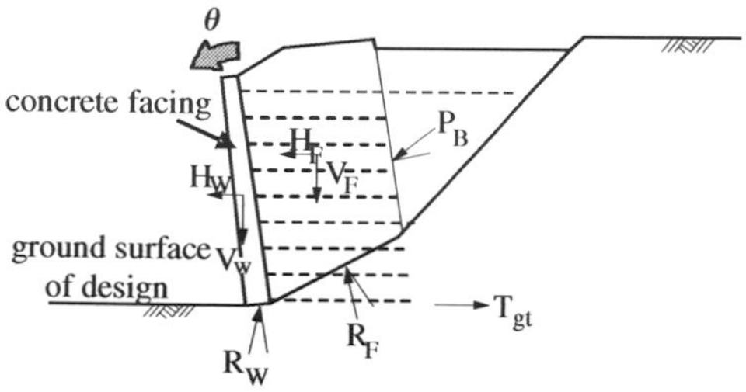

Figure 8. Modeling of GRS retaining wall for evaluation of overturning.

the pull-out resistance of the reinforcement T_p. that is evaluated as:

$$T_p = f_{rg} \times (\sigma_{vi} \times \tan\phi \times 2l_i + c \times 2l_i) \quad (8)$$

where f_{rg} is the resistance factor (set equal to 0.80 against level 1 earthquake load); σ_{vi} is the effective vertical stress acting on the ith reinforcement; l_i is the effective length of the ith reinforcement; ϕ and c are the internal friction angle and cohesion of the backfill soil, respectively.

2.7 *Verification against level 2 earthquake load*

For structures with performance ranks II and III, performance requirement for level 2 earthquake loads is verified in terms of their residual deformations using Newmark sliding block analyses and other numerical analyses. In case of GRS retaining walls, base sliding displacement of the retaining wall (Figure 7), overturning displacement (Figure 8), and shear deformation of the reinforced backfill (Figure 9) are evaluated.

It should be noted that, the residual shear deformation of the reinforced backfill has not been considered in many of the other existing design codes which adopt the assumption that the reinforced backfill behaves as a rigid body.

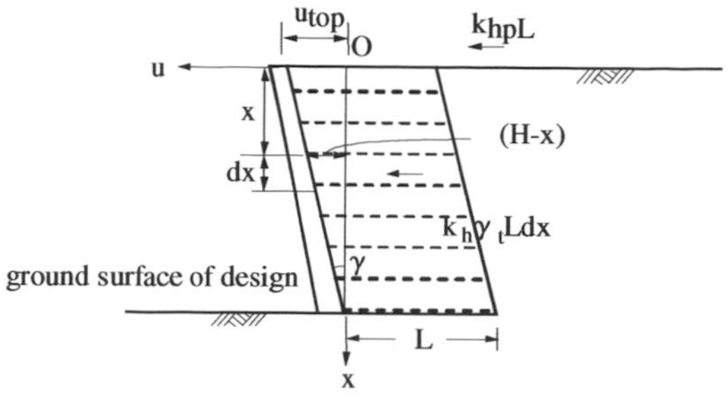

Figure 9. Modeling of GRS retaining wall for evaluation of shear deformation of reinforced backfill.

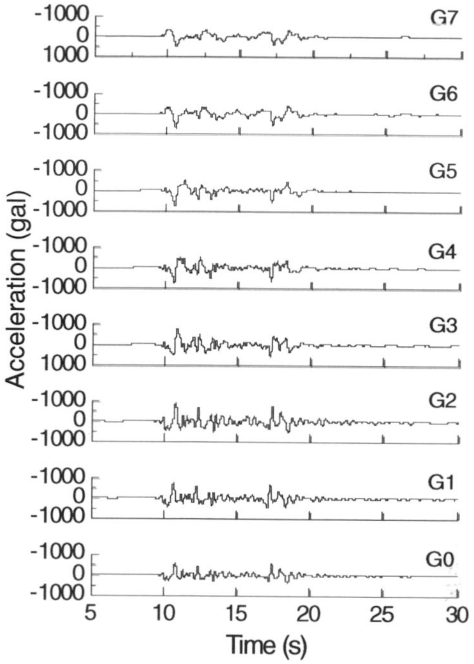

Figure 10. Time history of level 2 design earthquake motions (refer to Tables 5 and 6 for ground type classification and amplitude of maximum acceleration).

In conducting the Newmark sliding block analysis, one needs to specify the design earthquake motions. They are specified in the design standard as shown in Figure 10. They were obtained by applying a band-pass filter (0.3–4.0 Hz) to the design motions specified at the ground surface levels in the relevant design standard (RTRI, 1999). Depending on the natural period T_g of the ground, which is evaluated using Eq. 9 based on the profile of shear wave velocities, different wave forms and amplitudes are assigned as listed in Tables 5 and 6. The peak accelerations a_{max} are in the range

Table 5. Ground type classification based on natural period T_g (unit in seconds).

G0-G2	G3	G4	G5	G6	G7
Less than 0.25	0.25–0.5	0.5–0.75	0.75–1.0	1.0–1.5	More than 1.5

G0: Rock deposit; G1: firm base deposit; G2: Pleistocene deposit; G3: moderate; G4: moderate to soft; G5 and G6: soft; G7: very soft.

Table 6. Maximum acceleration of level 2 design earthquake motions (unit in gals).

G0	G1	G2	G3	G4	G5	G6	G7
578	732	924	779	−718	−741	−694	−501

Table 7. Results from Newmark sliding block analyses on 6 m-high embankments with different performance ranks.

	Rank I	Rank II		Rank III
Slope	1:1.8	1:1.5	1:1.8	1:1.5
Primary reinforcement	Yes	No	No	No
Secondary reinforcement	Yes	Yes	Yes	Yes
Residual displacement*	10.6 cm	36.0 cm	61.8 cm	96.5 cm

* Against level 2 design earthquake motion for G2 ground.

between 500 and 920 gals, and the largest value of a_{max} is assigned for the G2 ground consisting mainly of Pleistocene deposits.

$$T_g = 4 \times \sum_{i=1}^{N} \left(\frac{h_i}{V_{si}} \right) \tag{9}$$

where N is the total number of soil layers; h_i and V_{Si} are the thickness and the shear wave velocity of the ith layer, respectively.

For example, results from the Newmark sliding block analyses on the 6m-high embankment with different performance ranks specified as prescriptive measures (see section 2.4) are shown in Table 7. In setting the level 2 design earthquake motion, the severest ground condition (i.e., the condition of G2 ground) was assumed. It should be noted that the embankment with performance rank II was simplified into two kinds of configurations having a uniform slope angle.

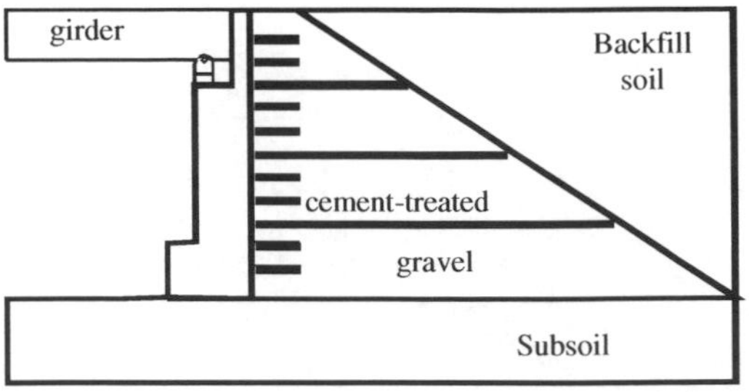

Figure 11. GRS abutment having cement-treated gravel for reinforced backfill (Aoki et al., 2005).

As a result, the embankment with performance rank I suffered from a residual displacement of about 10 cm, while those with performance rank II underwent residual displacements in the range of 40 to 60 cm. Further, the embankment with performance rank III suffered from a residual displacement of about one meter. Such different performances are expected when they are subjected to level 2 earthquake motion.

2.8 *Seismic design of reinforced soil abutment*

In chapter 9 of the design standard, design of GRS abutments having cement-treated gravel for reinforced backfill, as developed by Aoki et al. (2005) and schematically shown in Figure 11, is described.

In their seismic design, the abutment body and the reinforced backfill are verified with respect to safety against level 1 earthquake load and repairability against level 2 earthquake load.

For example, in verifying the repairability of the abutment body, a pseudo-static non-linear push-over analysis is conducted against the inertia force of the girder, while considering the tensile reaction of the reinforcements and assuming that the reinforced backfill does not exert any earth pressure to the body.

3 SUMMARY

The features of seismic design of GRS structures as specified in the newly-revised design standard for railway earth structures in Japan can be summarized as follows.

Three ranks of seismic performance against level 1 & 2 earthquakes are assigned considering the importance of the structure. The level 2 design earthquake motions have the maximum acceleration levels in the range of 500 to 920 gals.

Use of prescriptive measures is admitted in designing embankments. In setting their standard cross-section, use of primary and/or secondary geosynthetic reinforcements is mandated.

Recommendation of verification procedures is made. Against the level 2 earthquake, Newmark sliding block analysis is adopted as well as the evaluation of residual shear deformation of reinforced backfill in case of GRS retaining walls.

No creep reduction is considered in setting the tensile strength of geosynthetic reinforcements against earthquake loads.

Cement-treated gravel is used for reinforced backfill in case of GRS abutments to support bridge girders.

REFERENCES

Aoki, H., Yonezawa,T., Tateyama, M. Shinoda, M. and Watanabe, K. 2005. Development of aseismic abutment with geogrid-reinforced cement-treated backfills, *Proc. of 16th International Conf. on Soil Mechanics and Geotechnical Engineering*, 3, 1315–1318.

Greenwood, J.H., Jones, C.J.F.P. and Tatsuoka, F. 2001. Residual strength and its application to design of reinforced soil in seismic areas, *Landmarks in Earth Reinforcement*, Balkema, 1, 37–42.

JSCE 1996. *Proposal on earthquake resistance for civil engineering structures*, Special Task Committee of Earthquake Resistance of Civil Engineering Structures, Japan Society for Civil Engineers, 14p. http://www.jsce.or.jp/committee/earth/index-e.html

Koseki, J., Bathurst, R.J., Guler, E., Kuwano, J. and Maugeri, M. 2006. Seismic stability of reinforced soil walls, Keynote lecture, Proc. of 8th International Conference on Geosynthetics, Yokohama, 1, 51–77.

RTRI. 1999. *Seismic design standard for railway structures*, Railway Technical Research Institute (ed), Maruzen, 467p. (in Japanese).

RTRI. 2007. *Design standard for railway earth structures*, Railway Technical Research Institute (ed), Maruzen, 703p. (in Japanese).

Shinoda, M., Watanabe, K., Kojima, K., Tateyama, M. and Horii, K. 2007. Seismic stability of reinforced soil structure constructed after the mid Niigata prefecture earthquake, this symposium.

Tatsuoka, F., Koseki, J. and Tateyama, M. 1997. Performance of reinforced soil structures during the 1995 Hyogo-ken Nanbu Earthquake, *Earth Reinforcement*, Balkema, 2, 973–1008.

Tatsuoka, F. 2003. Technical report -wall structures session, *Landmarks in Earth Reinforcement*, Balkema, 2, 937–948.

New Horizons in Earth Reinforcement – Otani, Miyata & Mukunoki (eds)
© 2008 Taylor & Francis Group, London, *ISBN 978-0-415-45775-0*

Limit state design of reinforced soil

C.J.F.P. Jones
Newcastle University, United Kingdom

S.P. Corbet
Faber Maunsell, United Kingdom

ABSTRACT: The Limit State design method is established as a comprehensive and logical approach to the design of complex structures; the method is particularly suited to the design of reinforced soil. The philosophy of the method is to design against the occurrence of either ultimate or serviceability limit states. The limit states are conveniently identified as limit modes of failure. Over the years the method, originally identified in BS8006 and recently in Hong Kong Geoguide 6 has produced reliable safe structures using a wide range of materials and structural forms.

1 INTRODUCTION

The first limit state code of practice for the design of reinforced soil was published in 1995 as British Standard 8006. BS 8006 was written by a group of experts between 1985 and 1995, but in reality the document was finalised in 1990, and the next five years were spent in getting industry to agree the contents. BS 8006 is therefore a document which is now nearly 20 years old.

BS8006 is used not only in the UK, as the basis for the design for reinforced and strengthened soils, but is also used in some European countries, and a number of counties in the Far East or has been the inspiration for other national design codes such as Hong Kong Geoguide 6 (2002).

The art of reinforced soil design has moved on significantly since BS8006 was originally published and the document is currently being revised to reflect recent developments. In the absence of the new version of BS 8006, Geoguide 6 as the currently the most modern code and is used in this paper to illustrate the limit state concept.

The use of limit state design for reinforced soil is logical in that it is compatible with current civil and geotechnical engineering design documents such as Eurocode 7, EN 1997-1 Geotechnical Design, Hong Kong Geoguide 1 and BS 5400 for the design of bridges. The advantage of being able to use a common design philosophy can be illustrated by considering the design of a bridge. With a simply supported bridge it is possible to adopt different design philosophies for the substructure (abutments) and the superstructure (deck) but in the case of an Integral bridge, which is the favored form of construction in many situations, the use of different design philosophies for different parts of the structure can lead to conflict in design logic.

2 LIMIT STATES

The philosophy followed in Geoguide 6 and similar codes is to design against the occurrence of a limit state. For the purposes of reinforced soil design a limit state is deemed to be reached when one of the following occurs:

(a) Total or partial collapse;
(b) Deformation in excess of acceptable limits;
(c) Other forms of distress or minor damage which render the structure unsightly, require unforeseen maintenance, or shorten the expected life of the structure.

The performance of reinforced soil structures and slopes are considered in accordance with the ultimate limit state and the serviceability limit state criteria. The ultimate limit state and the serviceability limit state are defined as:

ultimate limit state – a state at which failure mechanisms can form in the ground or within the reinforced soil structure or slope, or when movement of the reinforced soil structure or slope leads to severe damage to its structural elements or in nearby structures or services.

serviceability limit state – a state at which movements of the reinforced soil structure or slope affect its appearance or its efficient use or nearby structures or services which rely upon it.

The condition defined in (a) above is the ultimate limit state and (b) and (c) are serviceability limit states. The use of the limit state methodology permits various limit states to be considered separately in the design and their occurrence is either eliminated or is shown to be sufficiently unlikely.

3 MODES OF INSTABILITY

The modes of instability (i.e. failure mechanisms) relating to the external and internal ultimate limit states are illustrated in Figures 1 and 2 respectively. The modes of failure relating to the compound ultimate limit states are illustrated in Figure 3. The limit states relating to serviceability are shown in Figure 4.

Other modes of failure may be appropriate in certain circumstances and have to be checked accordingly, for example:

(a) three-dimensional effects could influence the overall failure mechanism;
(b) modes of failure could be governed by seismic or cyclic loading; and
(c) complex modes of failure could be caused by excessive movement of the structure.

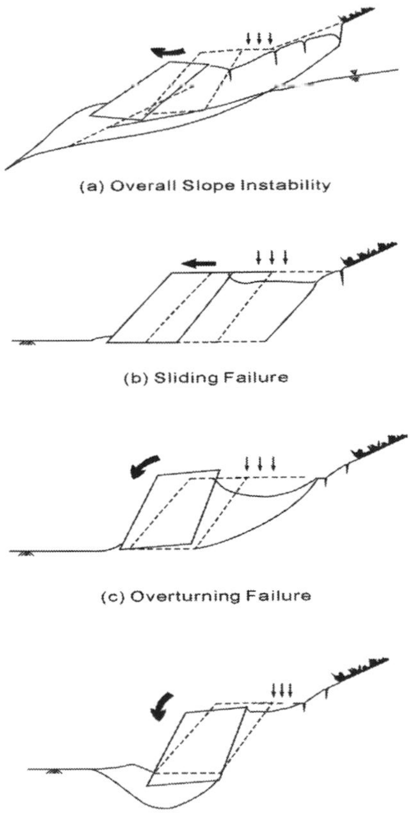

Figure 1. Ultimate limit states – external instability (from Geoguide 6, 2002).

4 FACTORS OF SAFETY

The reliability of reinforced soil design depends not only on the method of analysis, but also on the way in which factors of safety are defined, the reliability of the geotechnical model and the built quality required to be achieved. Limit state codes can accommodate any accepted analytical procedure and it is usual to use a tie back wedge analysis for extensible (polymeric) reinforcements and the empirical coherent gravity hypothesis for steel reinforcements.

4.1 *Overall stability*

When designing against overall slope instability, Figure 1, the global-safety-factor approach is usually adopted.

4.2 *Partial safety factors*

4.2.1 *General*

The current approach to applying factors of safety for reinforced soil structures and slopes uses partial consequence factors, material factors and load factors.

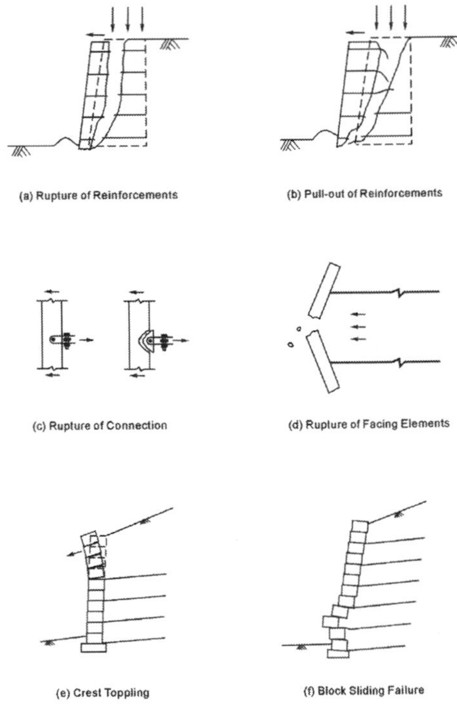

Figure 2. Ultimate limit states – internal instability (from Geoguide 6, 2002).

The partial factor format is appropriate to reinforced soil design where a range of materials may be used for structures of different design lives and where the consequence of failure depends upon the end use.

Design values of reinforcement parameters, geotechnical parameters and loading, as defined below, are used directly in the design calculations:

$$R_D = R/\gamma_n\gamma_m \quad (1)$$

$$G_D = G/\gamma_m \quad (2)$$

$$F_D = \gamma_f F \quad (3)$$

where,

R_D = design value of reinforcement parameters, R

G_D = design value of geotechnical parameters, G

F_D = design value of loading, F

$\gamma_n, \gamma_m, \gamma_f$ = partial consequence factor, material factor and load factor respectively.

The geotechnical parameters include shear strength of the fill and the permeability of drainage materials.

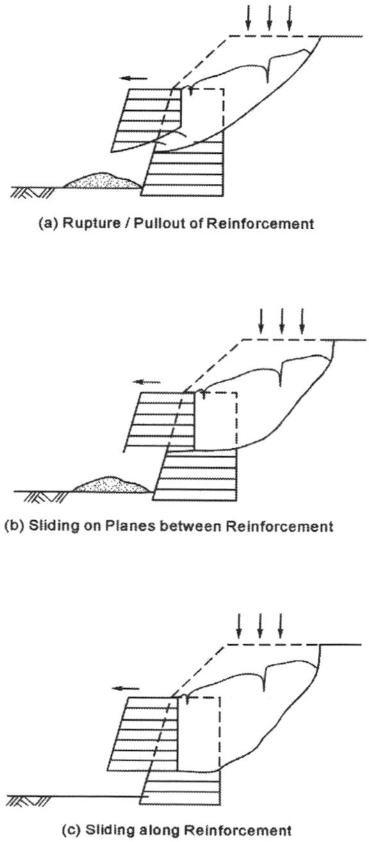

Figure 3. Ultimate limit states – compound instability (from Geoguide 6, 2002).

The reinforcement parameters include tensile strength of the reinforcing elements, fill-to-reinforcement interaction, facing unit-to-unit and facing unit-to-reinforcement interaction.

4.2.2 *Partial factors for consequence of internal failure*

In order to account for the consequence of internal failure, both BS 8006 and Geoguide 6 apply a partial consequence factor γ_n to the reinforcement parameters in accordance with Equation 1. As the application of increased (factored) external loads to a reinforced fill structure or slope is not always unfavourable due to the fact that increased stresses in a granular fill results in an enhanced shear strength, the application of γ_n to the reinforcement parameters is a more consistent approach to a margin of safety than if it was applied to the loads.

The values of γ_n for the different consequence categories are given in Table 1. Typical examples of failures relating to the consequence-to-life category and the economic consequence category are provided in Tables 2 and 3 respectively.

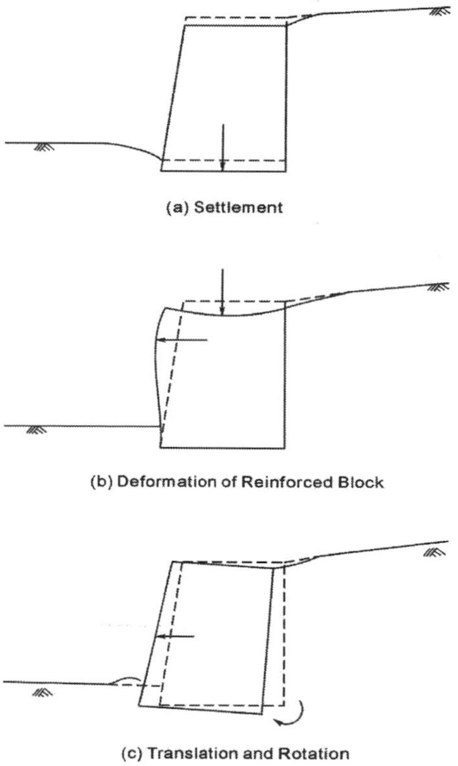

Figure 4. Serviceability limit states (from Geoguide 6, 2002).

Table 1. Recommended partial consequence factors for the design of reinforced soil structures and slopes (from Geoguide 6, 2002).

	Consequence-to-life Category		
Economic consequence	1	2	3
Category A	1.1	1.1	1.1
Category B	1.1	1.0	1.0
Category C	1.1	1.0	1.0

Table 2. Examples of failures in each consequence-to-life category (From geoguide 6, 2002).

	Consequence -to- Life		
Examples	Cat 1	Cat 2	Cat3
1. Failure affecting occupied buildings	√		
2. Failures affecting buildings storing dangerous goods	√		
3. Failures affecting recreational facilities		√	
4. Failures affecting heavily trafficked roads		√	
5. Failures affecting country parks			√
6. Failures affecting low trafficked roads			√

4.2.3 *Partial factors for reinforcements*

The required minimum value of the partial material factor γ_m for reinforcement should cover the effects of material variability, construction damage, environmental effects on material durability and other special factors including hydrolysis, creep and stress rupture that are related to polymeric reinforcements. Typical recommended minimum values for the partial material factor, γ_m, prescribed for reinforcements are given in Table 4.

For polymeric reinforcement, the partial material factor, γ_m is given by:

$$\gamma_m = \gamma_d \cdot \gamma_{cr} \cdot \gamma_{cd} \tag{4}$$

where,

γ_d = partial factor on reinforcement to allow for durability

γ_{cr} = partial factor on reinforcement to allow for creep

γ_{cd} = partial factor on reinforcement to allow for construction damage.

The partial factors γ_d and γ_{cr} on the reinforcement to allow for durability and creep respectively are not only governed by the design temperature but also the design life of the structure or slope.

Table 3. Examples of failures in each economic consequence category (from geoguide 6, 2002).

	Economic consequence		
Examples	Cat A	Cat B	Cat C
1. Failures affecting buildings causing excessive structural damage		√	
2. Failures affecting essential services	√		
3. Failures of strategic roads	√		
4. Failures affecting essential services for a short time		√	
5. Failures affecting country parks			√

Table 4. Partial material factors for the design of reinforced soil structures and slopes (after Geoguide 6, 2002).

	Partial material factor	
Material parameter	ULS	SLS
Fill: unit weight, γ	1.0	1.0
effective shear strength, tan ϕ^1	1.2	1.0
Ground: effective shear strength	1.2	1.0
base friction, tan δ_b	1.2	1.0
Granular fill and drainage material: Permeability, k	10.0	–
Structural elements:		
Reinforcement strength	1.5	–
Facing strength	as per relevant Structural code	–
Fill-to-reinforcement interaction:		
Sliding resistance	1.2	–
Pullout resistance	1.2	–
Facing units interaction:		
Unit-to-unit resistance	1.2	–
Unit-to-reinforcement Resistance	1.2	–

4.2.4 *Partial factors for fills*

The minimum value of partial material factor γ_m on the shear strength of fill should cater for possible difference between the shear strength of the fill measured in laboratory samples and that mobilised in the field.

4.2.5 *Partial factors for fill-to-reinforcement interaction*

Two possible fill-to-reinforcement interaction mechanisms are usually considered

(a) fill-to-reinforcement interaction where a potential failure surface crosses a layer of reinforcement. The fill-to-reinforcement interaction mechanism in this case is related to tensile and pull-out resistance.

Table 5. Partial load factors for the design of reinforced soil structures and slopes (from Geoguide 6, 2002).

	Partial factor, γ_f	
Loading	ULS	SLS
Dead load due to weight of fill	1.0	1.0
Dead load due to weight of facing	1.0	1.0
External dead load	1.5	1.0
External live load	1.5	1.0
Seismic load	1.0	1.0
Water pressure	1.0	1.0

1. γ_f should be zero for loads which produce a favourable effect
2. The external loads to which the partial load factors are associated should be characteristic values in the original unfactored state.
3. The worst credible water pressure loading should be considered.

Table 6. Partial load factors for load combinations for reinforced soil walls and slopes (from Geoguide 6, 2002).

	Load combination Partial load factors, γ_f		
Loading	A	B	C
Dead load due to weight of fill	1.0	1.0	1.0
Dead load due to weight of facing	1.0	1.0	1.0
External dead load on top of structure	1.5	1.0	1.0
External earth loading behind structure	1.0	1.0	1.0
External live loads:			
(i) on reinforced soil block	1.5	0	0
(ii) behind reinforced soil block	1.5	1.5	0
Temperature effects on external loads	1.5	1.5	–

(b) fill-to-reinforcement interaction where the potential failure surface coincides with a layer of reinforcement. The fill-to-reinforcement interaction mechanism in this case is related to sliding resistance.

Typical recommended minimum values for the partial material factor, γ_m, prescribed for pull-out resistance and for sliding resistance are given in Table 4.

4.2.6 *Partial factors for facing units interaction*

In the design of reinforced segmental block retaining walls, stability checks are required to ensure the column of block units remains intact, hence, the available shear capacity at any unit-to-unit or unit-to-reinforcement interface level will have to be assessed. Typical recommended minimum values for the partial material factor, γ_m, prescribed for unit-to-unit and unit-to-reinforcement interactions are given in Table 4.

4.2.7 *Partial load factors*

Typical minimum values of partial load factor γ_f recommended for use in the design of reinforced soil structures and slopes are listed in Table 5.

The most adverse loads and load combinations likely to be applied to reinforced soil structures and slopes should be considered in design. Different load combinations are identified with different scenarios. Typical partial factors to be applied to each component of the different load combinations for reinforced soil retaining walls and bridge abutments are recommended in Table 6.

In Table 6 Load combination A considers the maximum values of all loads and normally generates the maximum reinforcement tension and foundation bearing pressure; Load combination B considers the maximum overturning loads together with the minimum self weight and superimposed loading: Load combination C considers dead load only with unit partial factors. This combination is used to determine foundation settlements and reinforcement tension for checking the *serviceability limit state*.

5 DESIGN STRENGTHS

5.1 *Steel reinforcement*

For steel reinforcement the design tensile strength, T_D, per unit width of reinforcement is given by:

$$T_D = \quad ar\ \sigma_t/b\ \gamma_n\ \gamma_m \tag{5}$$

Where;

a_r = cross sectional area of the reinforcement min potential corrosion losses
σ_t = the ultimate tensile strength of steel
b = width of the reinforcement
γ_n = partial factor to account for consequence of internal failure (Table 1)
γ_m = partial factor on tensile strength of reinforcement (Table 4)

For design, the selected value for σ_t should be the minimum ultimate tensile strength guaranteed by the manufacturer.

5.2 *Polymeric reinforcement*

The design strength of polymeric reinforcements should be derived on the basis of the following principles:

(i) During the life of the structure, the reinforcement should not fail in tension.
(ii) At the end of the design life of the structure, the strain in the reinforcement should not exceed a prescribed value.

Thus, the design tensile strength, T_D, per unit width of reinforcement is given by:

$$T_D = T_{ult} / \gamma_n \gamma_m \quad (6)$$

where,
T_{ult} = ultimate tensile strength per unit width of the polymeric reinforcement.

5.3 *Fill materials*

For good quality fill materials which satisfy required grading and plasticity requirements it is generally sufficient to adopt a $c' = 0$ soil strength model for design purposes, (BS8006 accepts the use of cohesive – frictional fill where c' does not exceed 5 kPa). Such a model gives a conservative estimate of the shear strength of the backfill and is simple to apply in design. The design shear strength parameters, ϕ'_{des}, can be determined from:

$$\phi'_{des} = \tan^{-1}[\tan\phi'/\gamma_m] \quad (7)$$

where,
γ_m = the partial material factor on the shear strength of fill, Table 4.

5.4 *Fill-to-reinforcement interaction*

The design coefficients of interaction μ_{pD} and μ_{dsD} relating to pullout and direct sliding instabilities respectively are given by:

$$\mu_{pD} = \alpha_p \tan\phi' / \gamma_n \gamma_m \quad (8)$$

$$\mu_{ds\,D} = \alpha_{ds} \tan\phi' / \gamma_n \gamma_m \quad (9)$$

where,
μ_{pD} = design coefficient of interaction against pullout
$\mu_{ds\,D}$ = design coefficient of interaction against direct sliding
γ_n = partial factor to account for consequence of internal failure, Table 1
γ_m = partial factor for fill-reinforcement interaction, Table 4
α_{ds} = interaction coefficient relating to direct sliding
α_p = interaction coefficient relating to pullout.

In the case of the proprietary reinforced soil products the partial factors γ_m may be identified in certification procedures.

5.5 *Facing units interaction*

The stability of segmental block retaining wall facing depends on the shear capacity between block units and the resistance mobilised at the block-to-reinforcement interface. The ultimate shear capacity at any unit-to-unit or unit-to-reinforcement interface can be calculated using the following equation:

$$V_u = a_u + N_u \tan\lambda_u \quad (10)$$

where,
V_u = ultimate shear capacity per unit length of wall acting at the interface
N_u = normal load per unit length acting at the interface
a_u = ultimate adhesion at the unit-to-unit or unit-to-reinforcement interface
λ_u = peak friction angle at the unit-to-unit or unit-to-reinforcement interface.

The design coefficients a_{des} and λ_{des} for the design shear capacity at any unit-to-unit or unit-to-reinforcement interface can be determined from:

$$\lambda_{de} = \tan^{-1}[\tan\lambda_u / \gamma_n \gamma_m] \quad (11)$$

$$a_{des} = a_u / \gamma_n \gamma_m \quad (12)$$

where;
γ_n = partial factor for consequence of internal failure, Table 1
γ_m = partial factor for interaction between facing units, Table 4

6 CONCLUSIONS

The Limit State design method is established as a comprehensive logical approach to design of complex structures, the method is particularly suited to the design of reinforced soil. The approach is flexible in being able to accommodate different structural forms and analytical procedures and with different and new construction materials.

The first limit state design code BS8006 has been used widely throughout the world and has been shown to be a robust design document which produces safe durable structures, the current revisions will bring the code up to date. Other limit state codes such as Geoguide 6 have been developed or are in preparation.

REFERENCES

British Standards Institution (1990) BS 5400: Part 4: "Steel, concrete and composite bridges. Code of practice for design of concrete bridges". London.

British Standards Institution (1995) BS 8006: "Code of Practice for strengthened/reinforced soil and other fills". London.

GEO (1993) "Guide to Retaining Wall Design" (Geoguide 1). Geotechnical Engineering Office, Hong Kong.

New Horizons in Earth Reinforcement – Otani, Miyata & Mukunoki (eds)
© 2008 Taylor & Francis Group, London, ISBN 978-0-415-45775-0

Design philosophy for reinforced soil walls. Noteworthy aspects of European standards

P. Segrestin
Consulting Engineer, France

ABSTRACT: The paper first outlines the concerted process of the development of new norms in Europe, especially the ones about reinforced soil, of which a few aspects, thought of particular interest, are then selected. Attention is paid to the concern for sound conceptual design and careful quality control of work, deemed often more important for safety than accurate calculations. Topics related to computation are nevertheless discussed. Differences between ultimate and serviceability limit states, as understood in Europe, are clarified. The need for both internal and compound stability analyses is commented. Lastly, attention is drawn to the requirement for durability samples.

1 INTRODUCTION

Before attempting to highlight, in a second part of this paper, a few key features and outstanding points of some design standards recently formalized in Europe for reinforced soil structures, it is likely useful, at least for the non-European readers, to first outline how such standards are now established.

2 THE EUROPEAN STANDARDIZATION SYSTEM

2.1 *The CEN organization*

A considerable effort is made throughout the European Union in order to progressively develop, in all possible domains, a series of common norms. It is managed by CEN, i.e. the "Comité Européen de Normalisation" or, European Committee for Standardization. The main purpose of the European standards is to promote free trade within the Union, together with safety of workers and consumers, public procurement, environmental protection, as well as exploitation of "R and D" programmes and, interoperability of networks.

This effort presently involves the standards bodies of 30 countries, including 3 which are not (or not yet?) members of the European Union. The European norms, also known as EN, are developed through a consensus process. Participants in the technical committees and working groups represent (mainly through their national standards bodies) all interests concerned: industry, authorities and civil society.

Draft standards are made public for consultation. The final, formal vote is binding on all members. The EN Standards must be then transposed into national standards and conflicting standards withdrawn.

2.2 *The Eurocodes*

Of course, the European standardization effort also concerns the building and civil engineering industry.

As far as design is concerned, civil and structural engineering works will be based on the principles set by a series of 10 Eurocodes, also known as EC, which are now all published.

The Eurocodes are applicable to whole structures and to individual elements of structures and cater for the use of all the major construction materials. Some have a broad-spectrum subject, such as EC0 and EC1 which respectively deal with "Basis of design" and "Actions on structures". Others have a specific field: for example EC2 relates to "Design of concrete structures", while EC7, issued in 2005, is about "Geotechnical design", i.e. the geotechnical aspects of the design of buildings and civil engineering works.

2.3 *Execution of special geotechnical works*

The EC7 Eurocode regarding geotechnical design is intended to be used in combination with other standards that cover the construction, or execution, of special geotechnical works (as well as laboratory and field testing of soil).

Special geotechnical works for which EN execution standards have already been issued include, for example, sheet-pile walls, ground anchors and, since early 2007, reinforced fill structures, known as EN 14475.

3 NATIONAL NORMS COMPLYING WITH EN STANDARDS

3.1 *National annexes to Eurocodes*

Although they establish the fundamental principles and requirements, the Eurocodes still leave a few options open for a national endorsement or, adjustment (such as the values to be given to some design factors, so that every country may keep a chance of sticking more or less to the level of safety it was used to). So, every country is meant to issue its own National Annex to every Eurocode.

In France, for example, the National Annex to EC7 was published in 2006. It does not deviate from the preferred and recommended options.

3.2 *Application standards*

As said above, the Eurocodes essentially set the basic principles, define good engineering practices and highlight what should be cared about in design. But, they generally do not lay down precise design procedures for particular types of structures. They are mainly a basis for working out further specific European or national standards (in the same way as EC0 and EC1 were primarily a basis for elaborating the other Eurocodes).

This is why a lot of application standards now need to be developed, or updated, in true compliance with the Eurocodes.

For example, as far as the design of special geotechnical structures is concerned, France is currently in the process of elaborating several "national application norms", supplementary to EC7, which respectively deal with topics such as embedded walls, shallow foundations, piled foundations, or, of uppermost interest to us, reinforced soil. The latter, provisionally known as Pr NF P 94-270, is presently submitted to the traditional public enquiry. It will cover soil nailing as well as reinforced fill walls, using all types of reinforcement, geosynthetics, metals (sheets, grids, and strips) and meshes. If, one day, a will arises to work out a common European standard for the design of reinforced soil structures, P 94-270 might be one of the helpful reference manuals.

4 IDENTIFICATION OF KEY ISSUES FOR REINFORCED SOIL DESIGN

Now that things are likely clearer and we have a better understanding of the ramification of the norms about reinforced soil which closely fit in with the CEN framework and principles, let us point out in this group of standards a few aspects which may be found particularly significant, or even excellent, regarding the design of reinforced soil structures, as the theme of our special session suggests.

We will first focus on conceptual design and quality of construction, then give attention to computational design.

5 CONCEPTUAL DESIGN AND QUALITY OF CONSTRUCTION

5.1 *Relevance*

When a discussion is launched, or a panel is invited to debate about the topic of design, the "academics" generally instinctively think about design models, limit states or working stress methods, load and resistance design factors, computer programs and so on.

However, although there is now, throughout the world, a good number of records and case stories about failures of reinforced soil walls, of all kinds, the fact is that, so far, very few (really very few, if any) can be attributed to a deficient computational design of the reinforced soil body itself, properly speaking.

On the other hand, those who would rather introduce themselves as "practitioners" (the ones who are often the first ones called to the sites, in order to look at the damages and try to figure out the causes and the remedies...) would likely first put other concerns forward. Their experience is that real failures are often a consequence of such causes as defective drainage (resulting for instance in washing out of soil or, should water be polluted, accelerated corrosion). Or else, improper identification of the in-situ soil properties (resulting in excessive settlements or, worse, punching shear failure or, overall sliding). Or, lack of care regarding the compaction requirements and/or the fill specifications (resulting in large deformations or, over-stresses or, in other circumstances, untimely degradation). Or, unwise combinations of mismatched structures, systems or technologies (resulting in converging loads and, ruptures).

This list is certainly not exhaustive and it would be fair to further comment, illustrate and scrutinize every allusion. Imperfect though it is, it shows that the problems which are actually encountered mainly lie with the conceptual design of the whole work and the quality of its implementation. One good thing with the set of standards which we are looking at here is that it does acknowledge it and, emphasize it.

5.2 *EC7 and NF P 94-270*

From the very beginning indeed, EC7 strongly states the following in an introductory chapter (clause 2.4.1(2)) about the basis of geotechnical design:

"It should be considered that knowledge of the ground conditions depends on the extent and quality of the geotechnical investigations. Such knowledge and the control of workmanship are usually more significant to fulfilling the fundamental requirements than is precision in the calculation models and partial factors".

This is repeated, in the exact same terms, in the foreword of the French norm P 94-270 regarding the design of reinforced soil structures.

It is quite significant that norms which essentially deal with calculation models and procedures, partial safety factors and so on, modestly acknowledge that there are more important things to look at for insuring the resistance, stability, serviceability and durability of the structures.

5.3 *EN 14475*

The subject of EN 14475 is execution, not design. It applies to the construction of all types of reinforced fill structures, for practically all possible applications, using nearly all existing technologies. As one can easily imagine, working it out was quite a challenge because of such a wide and varied scope (not to mention the contrasted origins, experiences and motivations of the working group members ...).

So, it soon became obvious that the norm ought to be limited to common and essential requirements. It also turned out that, among them, one had to mention several stipulations or recommendations related to the proper selection of materials and products as well as to the suitable ways of combining them in a same structure, depending on its function and environment. This was initially somewhat disputed, on the grounds that such matters may rather come within design than within execution. But it was finally agreed, for two main reasons.

First, "conceptual design" is usually not addressed by the true design standards (maybe because it is viewed as a matter of engineering judgment, or expertise, which should not be codified?). So, if there are things which do need to be stated, the norm about execution is the only place available for that.

Second, the contractor in charge of building a reinforced fill wall often has some latitude for selecting the materials he will use. This is routinely the case for the fill, almost up to the last moment, but sometimes also for the reinforcement and facing, when the contract makes allowance for alternatives. It is therefore important that the norm about execution draws both the contractor's and owner's attentions to the necessary compatibility of the various materials, between them, as well as regarding the performance of the structure.

5.4 *EN 14475 and differential settlement*

As a matter of example, one issue which is particularly stressed in this respect in the EN 14475 standard and its informative annexes is the risk of differential settlement between the reinforced fill mass and the facing,

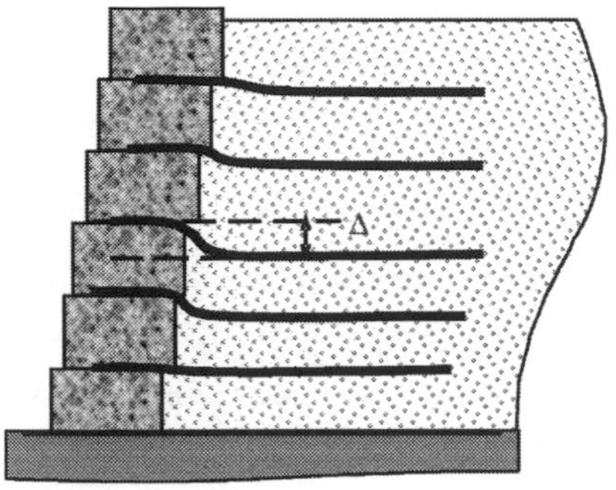

Figure 1. Potential differential settlement between reinforced fill mass and facing.

which may result from the compression of the fill during, and sometimes after, construction. What the norm states is, in short, as follows:

If the reinforcement is structurally connected to the facing units, without allowance for potential differential movement, then additional loads may be imposed on the fill reinforcement (fig.1). Such differential movement will mainly be affected by the quality of the selected fill, and the way it is compacted.

More stringent specifications should therefore apply to a fill material used with a less flexible facing system. Or, conversely, a facing system should be more flexible if the selected fill is prone to settle or not easy to compact.

For semi-flexible systems made of partial height facing panels, moderate differential movements are usually accommodated by the use of compressible bearing pads installed in the horizontal panel joints.

For rigid facing systems such as full height panels without moving connections, and segmental blocks packed without compressible filler, deformation in the region of the face connections may occur. Additional loads imposed on the connections and reinforcements should be mitigated, as far as possible, by proper selection, placement and compaction of the fill material. Otherwise, it is clear that such additional loads could not be reasonably estimated and, moreover, they might be incommensurable with the tensile loads computed according to the routine design procedures and standards. This is not all imagination: it did happen in several cases and resulted in the collapse of walls. It does confirm that conceptual design and quality control of execution may prevail over theoretical design models and partial factors.

6 COMPUTATIONAL DESIGN

6.1 *Preconditions*

Of course, the importance of conceptual design and quality control does not make it exempt from running any calculations. For them to be valid, EC0 and EC7 lay down a few major prerequisites:

- Data needed for design should be collected and interpreted, structures designed and execution carried out by qualified personnel having the appropriate skill and experience
- Adequate continuity and communication should exist between the personnel involved in data collection, design and construction
- Adequate supervision and quality control should be provided in plants and on site.

This being born in mind, let's now pick out from the future French P 94-270 application design standard a few noteworthy features.

6.2 *Ultimate and Serviceability Limit States*

As already mentioned, the NF P 94-270 norm is meant to apply to soil nailed as well as reinforced fill walls (including bridge abutments) made with polymeric as well as metal reinforcements, of any shape in use. Therefore, the core of the norm focuses on concepts and principles which do apply to all of them, while details pertaining to particular subjects, materials or products are to be found in annexes.

Coming at the head of the general rules for the justification of the works, is the differentiation between Ultimate Limit States (ULS) and Serviceability Limit States (SLS), which are linked with different sets of partial factors.

In strict compliance with EC0, the norm reminds that, by definition, ultimate limit states are associated with the conditions of collapses caused either by the loss of stability, the rupture, or excessive deformation of either some parts or, of the entirety of the structure, including those due to the effects of time.

On these grounds, there is no doubt that the tensile breakage of the reinforcements, especially when resulting from ageing or corrosion, relates to ULS (and so does the loss of adherence). But, it must be also acknowledged that large elongations of the reinforcements which could result in detrimental deformations must also be considered under the ULS heading. We are not contemplating here deflections which would only affect the aspect or, aesthetics of the wall. We are thinking of deformations which could, for example, possibly bring about some dislocation of the facing, hence result in progressive leaching and washing out of the soil, then eventually lead to collapse.

For that reason and, as a matter of example, as far as steel reinforcements are concerned, the NF P 94-270 norm concurrently takes into account for ULS design:

- The rupture stress, in cross-sections of the steel members where their tensile resistance might be locally most affected by corrosion (fig.2)
- The yield stress, supposing it prevails over a certain length where corrosion is assumed to be virtually uniform. Yield, if exceeded, might indeed entail elongations well in excess of 10%. It is understandable that the yield criterion controls as long as corrosion is small in comparison with the cross section.

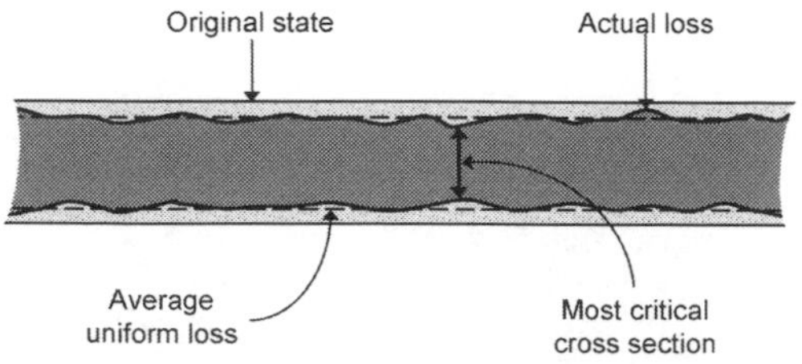

Figure 2. Schematic effect of corrosion along a reinforcing steel member.

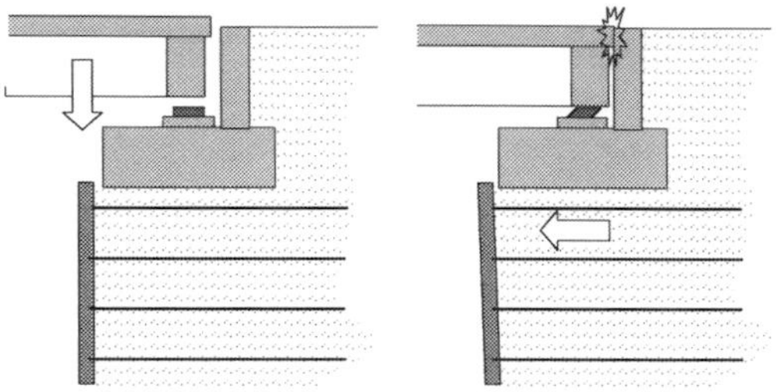

Figure 3. Example of a potential serviceability limit state at the top of a reinforced soil bridge abutment.

As to serviceability limit states, the norm reminds that they relate to situations which might be harmful to a proper utilisation of the structure (or the ones in its vicinity) in its habitual conditions of service. The relevant criteria essentially concern the deformations, movements or displacements of the reinforced soil body or its foundation (which may not be easily assessed, but with numerical models) in place of the loads and stresses.

An instructive example of serviceability limit state is described in the annex of the norm which deals with bridge abutments. It is aimed at the strain which may almost instantly affect the upper reinforcing layers (all the more so if extensible) when the bridge deck is put down on its bearings. It can be anticipated that the elongation of the top reinforcements could bring about a frontward displacement of the beam-seat, hence a distortion of the bearing pads or a closing of the bridge joint unacceptable for a good functioning of the bridge (fig. 3).

In short, the main point which we wanted to emphasize here is that, in the EN meaning, a serviceability limit state should not be confused with what may be

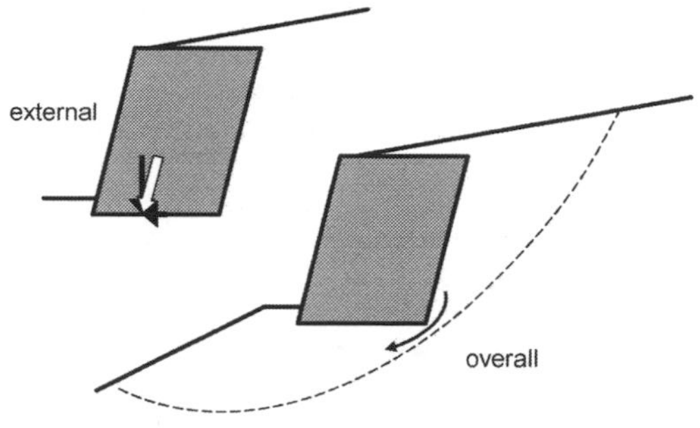

Figure 4. ULS design of external and overall stabilities of the reinforced soil body.

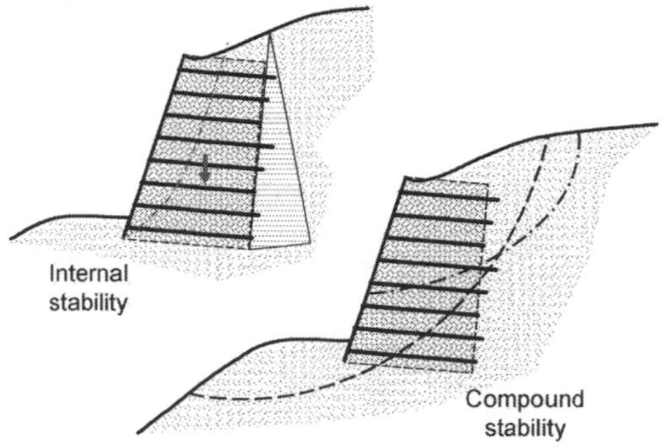

Figure 5. ULS design of internal and compound stabilities of the reinforced soil structure.

called in other codes either "in service" or "working stress" conditions.

6.3 *Four stages ULS design*

The French P 94-270 design standard states that the ultimate limit states justifications of a reinforced soil structure should be presented in four successive stages.

The first two deal with the shape of the reinforced soil body, considered as a block (fig. 4): "external stability" (sliding on the base, bearing capacity) and "overall stability" (potential failure surfaces outside the block). Up to a certain point, they could be considered as a kind of pre-sizing of the reinforced soil structure, which, at least, sets the lengths of the reinforcing layers.

The others relate to two different ultimate limit states but both deal with the arrangement of the reinforcements inside the block: types, spacing and numbers.

First, the "internal stability" is based on what is known, from experimental data, about the actual behaviour of similar structures, i.e. the likely distribution of stresses and forces once in service and, as a result of the construction procedure. Second, the "compound stability" analyses the potential risk of shear failure along lines which intersect some reinforcing layers (fig. 5).

6.4 *Internal and compound: both necessary*

The P 94-270 norm stresses that designing for internal stability and checking compound stability are both necessary but, none is sufficient.

On one hand, internal stability aims at placing what is needed where it is actually needed, in order to balance the maximum forces which are expected to build up in the reinforcing layers and keep more or less steady over the whole service life of the structure. The ULS is clearly linked in that case to the long-term ageing of the reinforcements. However, the usual procedures (such as the so-called "coherent gravity method" for reinforced fill walls) only take into account simplified effects of the soils retained by the structure. What is more, they can't consider at all the characteristics of the foundation and their possible impact on the response of the structure.

On the other hand, whereas the usual compound stability models can take into account the varied natures of the layers or zones of in-situ soils and imported fills which compose and surround the structure, they often only consider sorts of global equilibriums, without worrying about any particular layout of the reinforcements. Therefore, some models might possibly validate a distribution which is unsuitable regarding internal stability. Above all, the usual "at-failure" models are generally based on assumptions which are somewhat disputable or unrealistic. Some assume that both the maximum tensile capacity and pull-out capacity can be mobilised simultaneously, which is doubtful. Yet, they impose on the pull-out capacity of some layers to not exceed a given long-term tensile capacity, which is illogical or (so to speak ...) unfair. Others derive the calculated forces from assumed displacements, which do not and cannot take place, unless the structure is failing and is already beyond an ultimate limit state.

So, the P 94-270 norm, having warned against the limits and deficiencies of both the internal and compound stability analyses, advises to use both of them, provided the most is made of sound engineering judgment and previous successful experience. In unusual cases, resorting to numerical models is allowed.

Before moving to the closing subject, let's go back to at-failure models (without displacement) for a final remark. It could make sense to cope successively with the pull-out capacity of the reinforcements, recognized as a short-term issue (in a non geotechnical sense of the word) then with their tensile capacity, which definitely is a long-term issue. In the short term, once the structure is completed and subjected to its service loading, one can first and only see to it that there is no

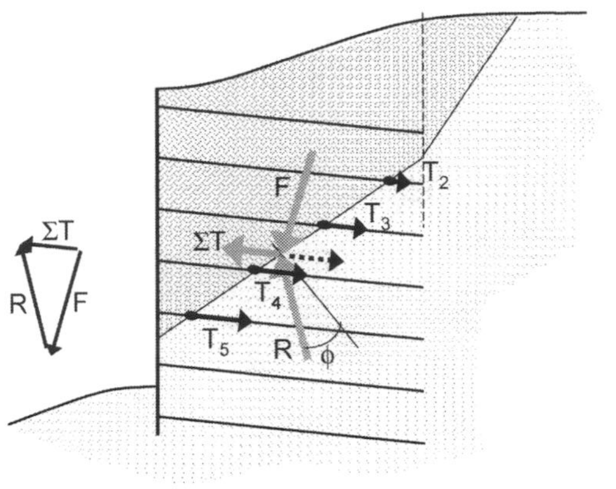

Figure 6. Equilibrium of forces and moments in a potential slip failure analysis.

risk of sliding. This involves the pull-out capacity of the intersected reinforcing layers and allows determining what their minimum width or perimeter should be, without yet worrying about their tensile capacity. One hypothesis has to be made, however, for sharing out the total required force between the intersected layers, depending on what the shortest resisting lengths merely can bear up, so that the equilibriums of forces and moments are satisfied (as suggested on figure 6 for a simple example with bilinear failures lines).

A series of potential failure lines can thus lead for each layer to the maximum tensile load it has to withstand up to the end of the service life and, in a second step, to its required minimum tensile capacity. Starting from the top can even allow optimising the reinforcing layers one after the other (and better understand how the pull-out capacities which are available here affect the tensile loads there).

7 DURABILITY SAMPLES

In closing, let's discuss something at the junction of design and construction. In its chapter about "Design and construction considerations" for retaining structures (Ch. 9.4), the Eurocode EC7 sets the following requirement: "As far as possible, retaining walls should be designed in such a way that there are visible signs of the approach of an ultimate limit state. The design should guard against the occurrence of brittle failure, e.g. sudden collapse without conspicuous preliminary deformations".

This is faithfully implemented in the French P 94-270 by requiring that durability samples are installed, during construction, in all types of permanent reinforced soil structures. They are meant to be extracted at scheduled intervals (after 10, 30, 50 years or so) and monitored in order to make sure that nothing abnormal is taking place. Should the case occur, the owner would have plenty of time for thinking about what can be done. The only exempt walls are those whose collapse would merely entail negligible consequences and the ones which would be no longer accessible once built.

8 CONCLUSION

It is obviously impossible to summarise, in just a few leaves, a set of three thorough norms which, in total, weigh round about four hundred densely made up pages. Claiming that their major points, as far as reinforced soil is concerned, have been all clearly spotted above would be by far quite pretentious. At least, it is hoped that those which have been selected and addressed here will contribute towards a stimulating panel debate and, hopefully and later on, towards some further progresses and international harmonization of the design principles and standards.

REFERENCES

EN 1990:2002. Eurocode (0). Basis of structural design

EN 1991-1:2002. Eurocode 1. Actions on structures (7 parts).

EN 1992. Eurocode 2: Design of concrete structures (3 parts).

EN 1997-1:2004. Eurocode 7. Geotechnical design. Part 1: General rules.

EN 1997-2:2007. Eurocode 7. Geotechnical design. Part 2: Ground investigation and testing.

EN 14475:2006. Execution of special geotechnical works. Reinforced fill.

Pr NF P 94-270. Geotechnical design. Retaining structures. Reinforced fill and soil nailed structures.

Recent case histories of earth reinforcement

New Horizons in Earth Reinforcement – Otani, Miyata & Mukunoki (eds)
© 2008 Taylor & Francis Group, London, ISBN 978-0-415-45775-0

The influence of backfill settlement or wall movement on the stability of reinforced soil structures

C.J.F.P. Jones & D. Gwede
Newcastle University, UK

ABSTRACT: In 1986 a number of reinforced soil walls constructed across steep ravines on a highway in eastern Tennessee, USA collapsed. It was observed that in most of the failed sections large deflections of the reinforcing strips had occurred. The subsequent investigation of the failures raised issues that were not adequately considered in design practice. Tests have been conducted to simulate differential settlement of the fill relative to the wall facing in reinforced soil structures using full scale reinforcements. The tests modeled the scale of the differential movements observed in Tennessee. The results of the tests showed that some forms of reinforcement slip when differential settlement occurs. This has serious implications, the structure is no longer coherent and the coherent gravity hypothesis used in some designs is no longer valid. It can be concluded that settlement should be included as part of the analysis of high reinforced soil structures carrying a heavy surcharge load.

1 INTRODUCTION

The construction of a scenic highway along the foothills of the Great Smokey Mountains in eastern Tennessee was started in 1984. Two sections of the highway are located in rugged terrain and are built in steep cuts and on high embankments. The design of these two sections incorporated 14 reinforced soil walls ranging in height up to 18 m which supported embankments up to 25 m high. In April 1986 one of the project walls collapsed while the embankment above it was being constructed. In May 1986, a second wall, which had been constructed eight months earlier, collapsed. Construction of the walls was suspended and a series of investigations was initiated, Lee *et al* (1994). The investigation revealed widespread deficiencies in a number of the other reinforced soil walls, two of which were in imminent danger of collapse. All the walls were constructed using reinforced concrete facing panels, steel strip reinforcement and frictional fill. The design of the walls was based upon the empirical coherent gravity method which is widely used for the design of reinforced soil structures using "inextensible" reinforcement.

A feature of the Great Smokey Mountains wall collapses was that the failure mechanism did not conform to the failure modes assumed in the design. Failure of the two walls which collapsed was initiated by part of the facing at ($1/3-1/2$) height exploding outward. Inspection of the collapsed walls immediately after the failures showed that the reinforcement had ruptured at the connections with the facing and that there was major differential settlement of the backfill with respect to the facing, which in both walls measured 490 mm below the reinforcement strip connection levels. Similar differential settlements of the reinforcement relative to the facing have been have been observed in collapsed walls in Japan, as with the Tennessee walls rupture of the connections also occurred, (Tatsuoka, 2006).

The failure mechanism of the walls was identified as being complex, involving forward translation of the structures, loss of adhesion of the reinforcement, mechanical instability of the facing leading to rupture of the reinforcement/facing connections. The investigation identified a number of answered questions including:

(i) The transfer of tension in the reinforcement strips which preceded collapse and in particular why the failures occurred in the reinforcement/facing connections when current theoretical models assume that maximum reinforcement tension occurs at a position remote from the facing.
(ii) The seriousness of stress induced by deflection of the reinforcement.

This paper considers the influence of backfill settlement or wall movement on the stability of reinforced soil structures and reports on laboratory studies undertaken using full scale reinforcements into the fill/reinforcement mechanisms which develop.

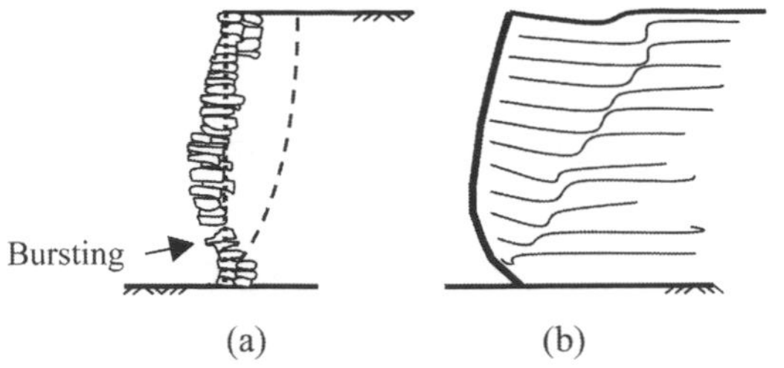

Figure 1. Failure mechanisms for (a) Victorian masonry walls (after Jones, 1979), (b) model reinforced soil walls (after John, 1983).

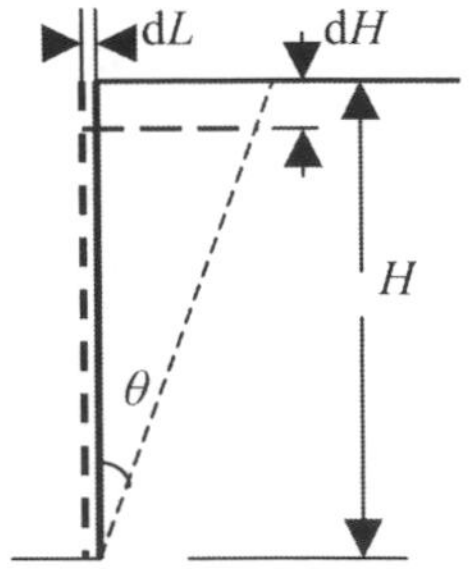

Figure 2. Wall-induced settlement (see equation (1)).

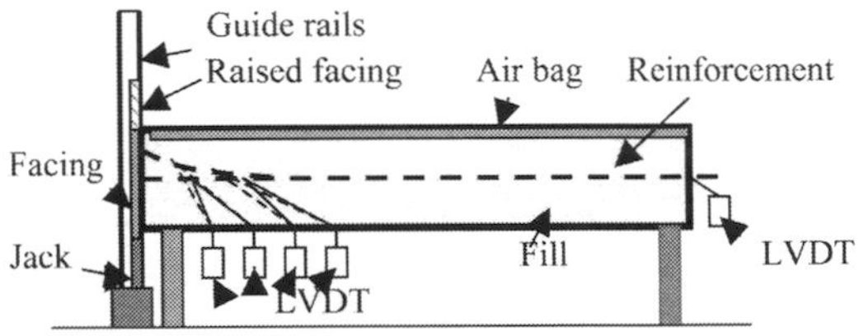

Figure 3. Settlement box.

2 FAILURE MODES

Current design methods for reinforced soil structures are based upon limit states which may be defined in terms of limit modes covering external, internal and combined failure mechanisms. Analytical procedures adopted for reinforced soil structures often consider the various limit modes separately, although this is explicitly warned against in some codes such as Hong Kong Geoguide 6, (GEO, 2002). Critically, the implication of the development of one limit mode on another may not always be considered or appreciated. Consideration of the Tennessee wall failures indicate that combined limit modes can lead to unexpected failure mechanisms. In particular, the development of the Limit Modes of Sliding, Bearing, Reinforcement Adhesion and Deformation could have a major influence on the Limit Modes associated with Reinforcement Rupture and Rupture of the Facing.

Field observations of reinforced soil structures indicate that all backfill materials settle but that the settlement is less than 1% of wall height, (Findlay, 1978; Jones *et al*, 1990; Jones and Hassan, 1992). The settlement of the backfill in the Tennessee structures exceeded 3 per cent of the height.

Settlement of the backfill can occur with outward movement of the wall face produced as a result of sliding, lack of reinforcement adhesion or deformation. Figure 1 shows the deformation of dry stone retaining walls which collapse in a manner similar to the Tennessee walls in that they burst outwards at ($\frac{1}{3}$–$\frac{1}{2}$) height. Figure 2 shows a reinforced soil wall of height H and reinforcement length L that has translated a distance dL forwards. If the total volume of the backfill remains constant, the translation would result in a backfill settlement of dH where:

$$dH = \frac{(dL + H\tan\theta - \sqrt{[(dL + H\tan\theta) - 2HdL\tan\theta]}}{\tan\theta} \qquad (1)$$

In Tennessee a translation of 300 mm would have resulted in the observed deflection of 490 mm of the reinforcement relative to the facing. In addition to translation, isolated bulging of the facing could also induce strip deflections. Bulging of the facing of the walls in Tennessee was observed.

3 LABORATORY STUDIES

In order to study the effects of differential settlement of the reinforced fill relative to the facing a special settlement box was developed. The settlement box was constructed as a conventional reinforcement pull out box having a front wall mounted between two vertical steel columns which allowed the face to move up or down in a vertical plane. Two hydraulic jacks were attached to the front wall to power the vertical movement, Figure 3. The overall dimensions of the box were (3.0 × 0.7 × 0.6 m) for the length, breadth and depth respectively permitting the use of full scale reinforcement.

Overburden pressure was simulated by the application of a uniform normal stress applied to the top of the fill by an airbag. The airbag was capable of developing a normal stress of 140 kN/m^2 equivalent to a fill height of approximately 7 m. Thus the stress conditions associated with the Tennessee failures could be replicated in the laboratory studies.

3.1 *Fill*

The fill used in the tests was Leighton Buzzard Sand. This was placed in the settlement box in layers and compacted using a hand-held vibrating plate compactor. An average density of 16.0 kN/m^3 was achieved. The specific gravity of the fill was 2.65 and the angle of internal friction was; ($\phi_{cv} = 32°$; $\phi_{peak} = 42°$).

3.2 *Reinforcement*

Two forms of reinforcement were used in the tests. The main study focused on the use of "high adherence" steel strip reinforcement as this was the reinforcement used in the Tennessee walls. This reinforcement develops greater resistance to pull out than plane strip, by having raised ribs 3 mm high cast into the surface. The reinforcement used was 50 mm wide, 6 mm thick and extended beyond the end of the test box. Tensile tests on the reinforcement showed the yield strength to be 490 MN/m^2 and the ultimate tensile strength to be 650 MN/m^2. Attachment of the reinforcement to the wall face was provided using a bolted connection similar to that used in field applications.

The second form of reinforcement which was studied for comparative purposes was a polymeric geogrid formed from high density polyethylene (HDPE). The reinforcement had a characteristic tensile strength of 80 kN/m. A strip of 0.5 m width was used in the tests. The polymeric grid reinforcement was connected to the moving wall face using two methods. In the first method a complete length of reinforcement was cast into the concrete facing panel. In the second a short starter piece of reinforcement was cast into the facing to which the main length of reinforcement was connected using a flat HDPE bodkin. Both of these connection systems are used in field applications.

3.3 *Displacement and force measurements*

Displacements of the steel strip reinforcement tested were measured at the wall face connection, the free end of the reinforcement and at two locations along the embedded length close to the facing. The displacements were measured using linear variable differential transformers (LVDT) attached to a computer. Based upon the measured displacements, the deflection profile of the reinforcement associated with any vertical location of the facing could be determined.

When extensible polymeric grid reinforcement was tested, the deflection profiles were determined using a water level formed from polymeric tubing attached to the reinforcement passing the full length of the cell and through the facing and rear of the pullout box. (This procedure could not be used with the steel strip reinforcement as it produced too much interference with the strip/fill adhesion characteristics.)

Tensile loads generated in the reinforcement during the tests were determined in two ways:

(a) By the use of strain gauges attached to the steel reinforcement.
(b) By the use of BISON gauges attached to the geogrid reinforcement.

Table 1. Settlement tests.

Test	Overburden pressure (kN/m^2)	Facing deflection (mm)	Reinforcement type	Connection method
1a	0	N/A	nil	–
2b	28	N/A	nil	–
3	21	200	geogrid	fixed
4	28	150	geogrid	fixed
5	28	150	geogrid	bodkin
6	28	200	geogrid	fixed
7	28	150	geogrid	fixed
8	85	180	steel strip	bolt
9	105	250	steel strip	bolt
10	116.5	160	steel strip	bolt
11	140	200	steel strip	bolt

Note: a. test with no backfill or reinforcement.
b. test with no reinforcement

3.4 *Test programme*

A total of 11 settlement tests were undertaken using overburden pressures ranging from (21–140 kN/m^2), Table 1. The lower overburden pressures were used with the polymeric reinforcement as this material had a very high soil/reinforcement adhesion which was fully developed at low pressures. The tests in the steel strip reinforcement used the maximum overburden capacity of the test box.

To simulate differential settlement of the fill relative to the facing the front face of the pullout box was displaced vertically. The rate of movement was 5 mm/minute and the maximum vertical displacement was 250 mm. Tests 1 and 2 were undertaken to determine the forced needed to move the wall facing alone and when fill but no reinforcement was present.

In addition to the settlement tests, five pull out tests were undertaken using the steel strip reinforcement. These were conducted at two overburden pressures of 60 and 100 kN/m^2, to determine the apparent friction coefficient (μ^*) of this form of reinforcement.

4 TEST RESULTS

The forces required to move the facing vertically under different conditions is shown in Figure 4. The deflection profiles of the steel reinforcement are shown in Figure 5 and those of the grid reinforcement in Figure 6.

Figure 7 shows the strain recorded in the geogrid reinforcement in Tests 3 to 7. Many of the strain gauges attached to the steel reinforcement failed during the test and little data was recorded, at no time did recorded strain exceed 0.08%.

The values of the apparent friction coefficient, μ^*, of the high adherence steel strip reinforcement

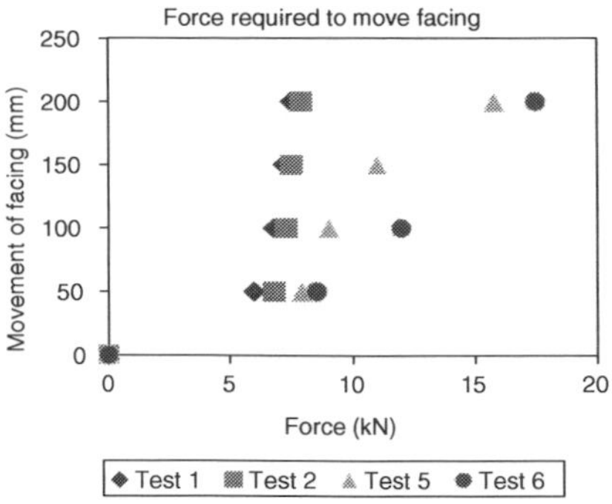

Figure 4. Force required to move facing.

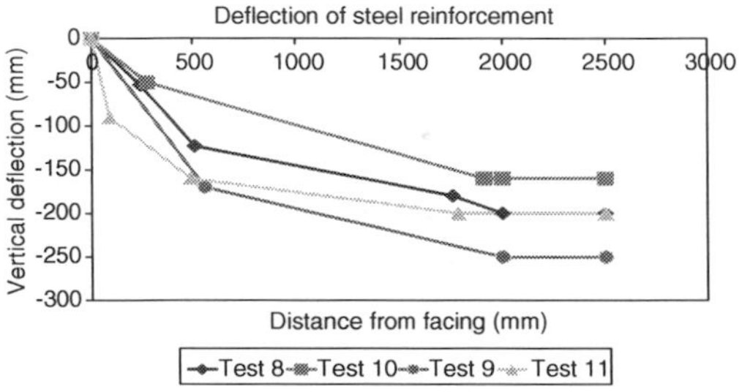

Figure 5. Deflection of the steel reinforcement with vertical movement of the facing.

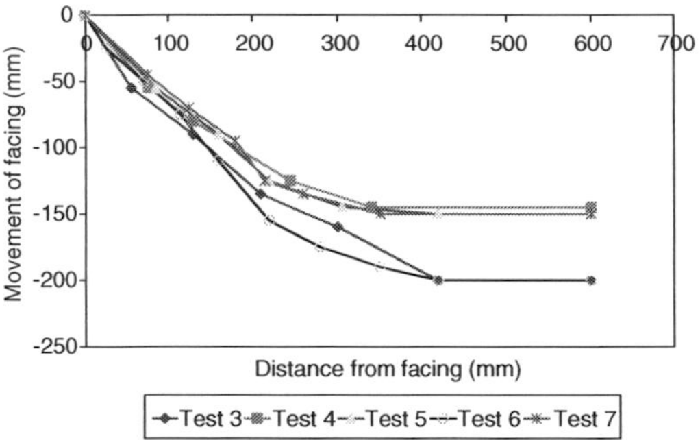

Figure 6. Deflection of the grid reinforcement with movement of the facing.

obtained from the pullout tests and calculated from strain gauge readings are shown in Tables 2 and 3.

5 DISCUSSION

5.1 *Force required to move facing*

The force required to move the facing, equivalent to the back wall friction of fill settling relative to the facing, ranged from 6–8 kN/m^2. This increased substantially when reinforcement was present, Figure 4.

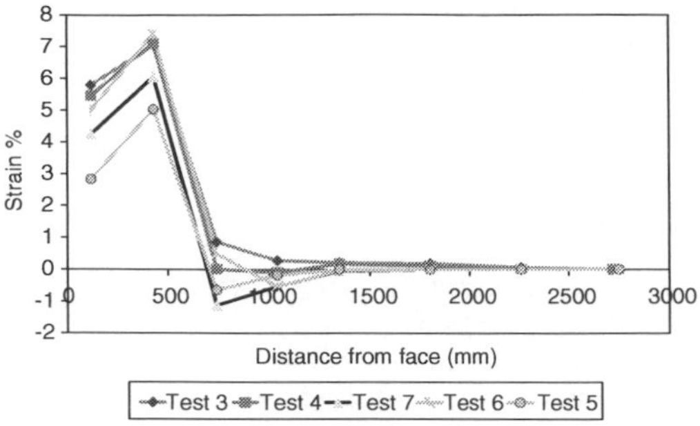

Figure 7. Strain in geogrid reinforcemt.

Table 2. Apparent friction coefficient, μ*, from pullout tests.

Overburden pressure (kN/m^2)	Displacement at peak load (mm)	Apparent friction coefficient, μ*
60	33	1.69
60	32	1.50
100	32	0.80
100	35	1.01
100	35	1.06

Table 3. Values of μ* calculated from strain gauge readings.

Test No.	Pressure (kN/m^2)	Max. Tension in reinforcement	Apparent friction coefficient, μ*
9	105	63.5–77.4	1.8–2.2
10	116.5	40–53.8	1.0–1.4
11	140	81.2–95.0	1.7–2.1

5.2 *Deflection profile of the reinforcement*

In the Tennessee wall failures a uniform deflection profile of the steel reinforcement was observed and this was replicated in the laboratory trials. In the failing walls, reinforcement frictional force was mobilized and the reinforcement slipped, becoming part of the failure mechanism. The steel reinforcement in the laboratory trials also slipped. It was observed that the steel reinforcement – facing connections had undergone significant strain with cracking of the galvanizing coating clearly visible. This implies that bending stresses were generated at the connections.

The deflection profile of the steel reinforcement in Tennessee prior to and following failure and in the laboratory trails can be expressed by the polynomial, (Lee *et al*, 1994):

$$S = A\,[1 - e\,(-K^2\,\chi^2)] \qquad (2)$$

Where, S = deflection of the reinforcement; A and K = constants; χ = horizontal distance measured from the facing.

Equation (2) can also be used to describe the deflection profile of the polymeric grid reinforcement but a better fit is obtained using a cubic polynomial:

$$S = H_{max} - 0.75\chi + 0.014\chi^2 / \sqrt{H_{max}} - 9x10^{-7}\chi^3 \qquad (3)$$

Where H_{max} = maximum movement of the face.

5.3 *Strain and slip in the reinforcement*

Observations in the laboratory trials showed that the steel reinforcement slipped in all the tests. Slip started to occur at vertical movements of the facing of 20 to 40 mm. This confirms the stiffness of the reinforcement, that there was little extension of the materials and that it would be best to identify the material as effectively being inextensible. Slipping was initiated at strain levels well below that required to bring the reinforcement to yield even allowing for a length of 5m. Assuming no elongation Lee *et al* (1994) deduced that slip, S, can be defined by:

$$S = \int_{x_1}^{x_2} [1 + f'(x^2)]dx \qquad (4)$$

Where $f'(x)$ is the equation describing the deflected profile of the reinforcement.

Applying Equation (4) the measured slip in the laboratory tests can be compared with the calculated slip, this is shown in Table 4.

The tests also produced evidence that differential settlement of the fill relative to the facing can introduce significant bending stresses in the reinforcement, particularly if the connections are rigid. A deflection of 175 mm at a distance of 1000 mm from the wall face is theoretically able to induce a stress, σ_y = 350 MPa. In reality some slackness will reduce this significantly.

The deflected profile of the polymeric reinforcement is a function of the total differential movement between the fill and the facing. The maximum strain following a 200 mm differential movement of the facing relative to the fill was 7 per cent which occurred at a distance of approximately 500 mm behind the facing, Figure 7. The reduction in strain at the facing could be a result of fixity of the reinforcement to the concrete facing. In none of the tests was geogrid reinforcement located 2000 mm behind the facing subjected to strain. This observation leads to the conclusion that grid reinforcement does not slip. When a bodkin joint was introduced the maximum strain in the geogrid reinforcement reduced to approximately 5 per cent, Test 5 Figure 7. The forces needed to move the facing also reduced, Figure 4.

Table 4. Measured slip compared with calculated slip.

Test No.	Measured slip (mm)	Calculated slip (mm)
8	55	33–68
9	40	30–50
10	27	22–32
11	30	25

5.4 *Apparent friction coefficient* μ^*

The value of the apparent friction, μ^*, in the pullout tests changed with the level of stress on the reinforcement, Table 2. The peak shear was mobilized at about 35 mm displacement. The values of μ^* calculated from the strain gauges were higher than those obtained from the pullout tests, Table 3. The results compare with values obtained by other researchers and from pullout tests. Schlosser and Elias (1978) give a range for μ^* of 0.33 to 2.5 for ribbed steel reinforcement in fine sand.

6 CONCLUSIONS

The reinforcement deflection profiles obtained in the laboratory study agreed closely with the observed deflections associated with the Tennessee failures. Once the fill settles > 120 mm the steel reinforcement will slip. Failure to slip would result in rupture of the reinforcement. As soon as the reinforcement slips the concept of a *coherent* structure is lost and the analytical model using this concept is invalid. When reinforcement slip occurs forward movement of the structure is likely, leading to additional differential settlement of the fill relative to the facing. Settlements of the fill in the range of 120 mm could occur at the base of high walls, particularly those supporting embankments and this should be a consideration in the design.

Differential settlement of the fill relative to the facing induces bending stresses in the connections which are not usually considered during design; this could be part of the reason why the observed failures involved rupture of the connections.

Polymeric geogrid reinforcement was able to accommodate the differential settlements imposed during the tests and no slipping occurred. However the reinforcement was subjected to strains greater than those usually accepted. The use of bodkin connections appeared to have a positive effect in reducing reinforcement strain.

REFERENCES

Findley, T.W., (1978) "Performance of reinforced earth structures at Granton", *Ground Engineering*, **11**, 42–44.

John, N.W.M., (1983) *"Behavior of fabric reinforced soil walls"* PhD thesis, Portsmouth University, UK

Jones, C.J.F.P., (1979) "Current practice in designing retaining structures", *Ground Engineering*, **12**, 40–45

Jones, C.J.F.P., Cripwell, J.B., and Bush, D.I., (1990) "Reinforced earth structure for Dewsbury ring road" *Institution of Civil Engineers, Proc. Part 1*, **88** (April), 321–345.

Jones, C.J.F.P., & Hassan, C.A., (1992) "Compression of a reinforced soil wall during construction", *International Conference, Geotrika '92*, Johor Bahru, Malaysia.

Jones, C.J.F.P., (2002) "Guide to reinforced fill structure and slope design" *Geoguide 6*, Geotechnical Engineering Office, Government of Hong Kong SAR, p. 236.

Lee, K., Jones, C.J.F.P., W.R. Sullivan, and W. Trollinger (1994) "Failure and deformation of four reinforced soil walls in eastern Tennessee". *Geotechnique*, **44**(3), 397–426.

Schlosser, F., & Elias, V. (1978) "Friction in reinforced earth". *Proceedings of symposium on earth reinforcement,* Pittsburg. 735–763. New York, American Society of Civil Engineers.

Tatsuoka, F., (2006) "Twenty years of geosynthetic-reinforced soil retaining walls with full-height rigid facing and related structures in Japan" *5th BGA/IGS Invitation Lecture*, London, October.

New Horizons in Earth Reinforcement – Otani, Miyata & Mukunoki (eds)
© 2008 Taylor & Francis Group, London, *ISBN 978-0-415-45775-0*

An innovative connection between a nailed slope and an MSE structure: Application at Sishen mine, RSA

N. Freitag
Terre Armée Internationale, Vélizy-Villacoublay, France

A.C.S. Smith & H.J.L. Maritz
Reinforced Earth (Pty) Ltd, Johannesburg, South Africa

ABSTRACT: The expansion project at Sishen Mine Northern Cape, RSA includes the construction of a primary crushing facility. The primary crusher building is founded at a depth of 35 metres below the natural ground level. MSE wing-walls are required to provide access for 400^T trucks to the tipping areas. The wing-walls are tiered and split into 14 m and 21 m high structures, which in effect are equivalent to higher structures because the dry density of the hematite backfill is 33 kN/m^3. Insufficient space was available for placing the reinforcing strips on one of the lower walls viz. Wall B. Additional excavation to accommodate the strips was not feasible and a solution was required to connect the narrow MSE structure to the existing cut slope. This has been undertaken by the installation of nails into the cut slope; attaching a high friction strip to the nail heads and then lapping shortened reinforcing strips attached to the cladding with the friction strip. The frictional interaction between the MSE reinforcing strips and the extended nails ensures the internal stability of the MSE mass. Monitoring of the load at the nail head has been attempted by means of strain gauges. This paper describes the MSE wing-walls and in particular the design and stability of Wall B.

1 INTRODUCTION

The Sishen Mine is one of the largest open pit mines in the world. It is undertaking an expansion project to increase its production from 29 to 42 million tons per year. This entailed the construction of a primary crusher building. The building is 35 metres high. Mechanically stabilized earth (MSE) tip walls and wing-walls to crushing plants have become standard practice in Africa (Smith 1999) and were required for the wing-walls for this particularly massive structure.

2 SISHEN MINE PROJECT

The new primary crusher is situated at the edge of the existing pit and in an area where the existing pit had been backfilled. Excavation through the natural inside edge of the pit as well as the backfilled material was undertaken in order to found the structure at the minus 35 m level. Figure 1.

2.1 *General*

The backfill material is Hematite and has a bulk density of 33 kN/m^3. This has the effect of increasing the stresses in the MSE of the order of 65% or increasing the effective total height of the 35 m high tiered structures to $35\,\text{m} \times 1.65 = 58\,\text{m}$. The reinforcing strips needed to be stiff to reduce deflection and bulging. Hot-rolled ribbed strips with 50×4 and 45×5 mm cross-section were used. The 45×5 mm strips are padded to prevent any loss of strength at the connection and these were used for all strips with length longer than 11 m. The maximum length of strip was

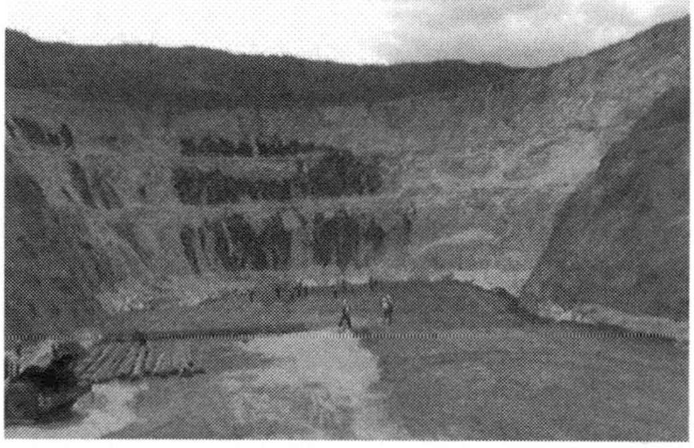

Figure 1. Excavation completed.

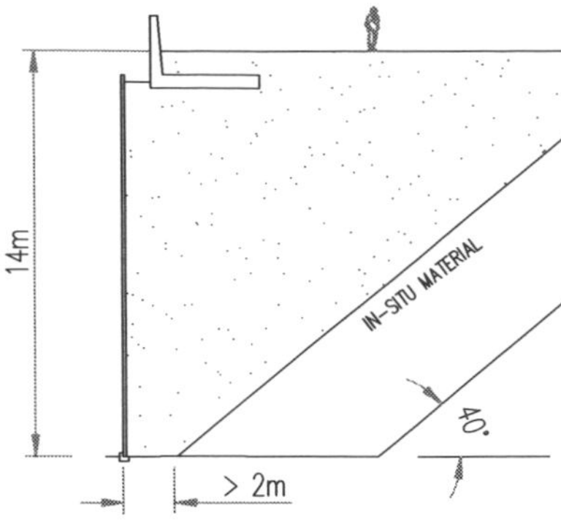

Figure 2. Wall B – Typical cross section.

23 m. The cladding needed to be strong yet flexible enough to accommodate differential settlements as well as the large number of reinforcing strips required. A weldmesh cladding with 8 mm, 10 mm and 12 mm diameter bars and a height of only 320 mm was used. Once excavation had been completed and construction of the crusher building was underway Wall B was moved several metres nearer to the excavated cut face. In order to improve the stability of the cut face soil nails were installed in particular locations.

2.2 *Wall B*

The existing slope was, in one location, as close as 2 meters from the projected position of the wall facing, with an inclination of about 40° about the horizontal. Further excavation of this excavated slope was avoided on account of overall stability concerns. The position of Wall B was relocated several metres nearer the excavation line and it became necessary to find a cost effective solution for a 14 m high vertical structure, with an available space at the bottom of about 2 m, i.e. 15% of the height (Figure 2). The standard practice for MSE retaining walls lies generally between 60% and 70%, with a minimum set in the major design codes to 40% at the base. The recent Shored Mecahnically Stabilized Earth (SMSE) Wall Systems Design Guidelines, edited in February 2006 by the US Central Federal Lands Highway Division, states that the width at the base should be at least 0.3 H and concentrates on vertical and near-vertical backslopes.

2.3 *Fill material: hematite backfill*

The backfill material commonly available at the site is an iron ore called hematite. Hematite (Fe_2O_3) has a particular high density, compared to common quarry materials. The grain density is 5.24 g/cm^3, and the design fill density is 3.4 t/m^3(33 kN/m^3). This material is insoluble. Its internal friction angle was taken as 40°.

Figure 3. Comparative corrosion testing between saturated hematite and de-mineralised water.

A test was done to compare the corrosion behaviour of a galvanised steel sample placed into saturated hematite, with a similar sample placed in tap water. The rates measured showed similar behaviour. The resistivity of the hematite ore saturated with de-mineralised water at 20°C is 11,600 Ω.cm, and the pH level is slightly alkaline at 8.3. About 1% of fine elements is enough to give to the water a strong red colour, similar to blood, which gives its name (in Greek) to the ore.

3 PROPOSED SOLUTION FOR WALL B

In order to ensure a satisfactory level of safety, the structure had to be tied to the existing slope. This is commonly done by anchoring a concrete beam or plinth to the slope with the help of tie-backs. The solution presented made it possible to save a large quantity of concrete, by using the locally available granular material.

The use of hollow self drilling nails allowed rapid installation of anchoring points.

The height of the structure and density of the backfill led to the choice of the use of inextensible steel reinforcing strips to link the existing slope with the new facing. The use of a direct link would have required the use of adjustable devices, such as turnbuckles, in order to cope with the naturally distorted alignment of the existing slope. This was considered to be impractical, expensive and difficult to install. Moreover a rigid link between the slope and the new face may have led to a high concentration of loads, since it would not provide sufficient flexibility.

3.1 *Friction connection*

The conception of this solution is based on the observation that the use of friction between the reinforcements

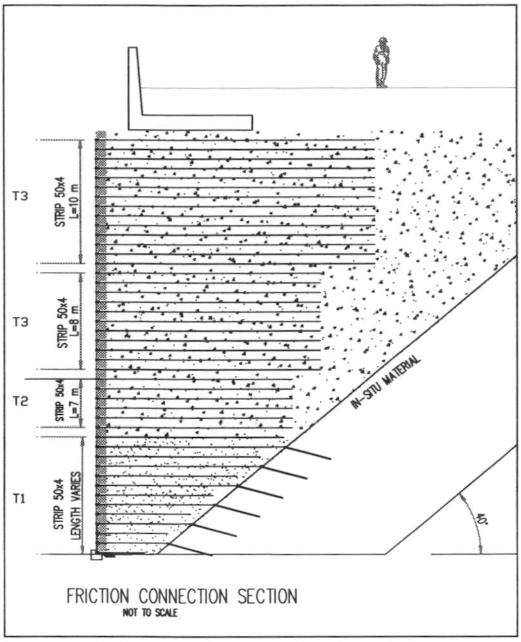

Figure 4. Sishen cross section.

and the backfill is at the basis of the performance and reliability of MSE structures.

There have been previous attempts, in the design of retaining walls to disconnect the main reinforcements from the cladding elements. Berg et al. (1986) presented an MSE retaining wall in Lithuania where the main reinforcements, made of HDPE geogrids, were not positively connected to the facing, but overlapped geogrids embedded in the cladding panels, with a minimum thickness of fill between them. Construction was reported to be difficult, which may be a consequence of the flexibility and extensibility of the geogrids. Freitag et al. (2004, 2005) presented a similar concept, but with the use of discrete reinforcements; strips and geostraps.

3.2 *Sishen concept*

A similar layout was proposed for the Sishen structure. Instead of trying to make a positive connection between soil nails and the new facing, leading to issues described above, it was decided to make use of the friction connection between reinforcing strips extending from the facing and specially designed highly frictional ladder reinforcements connected to the nails. The elements were designed so that an overlap of 2 metres would be enough to ensure stability. A patent on this concept was filed in January 2006 by Terre Armée Internationale.

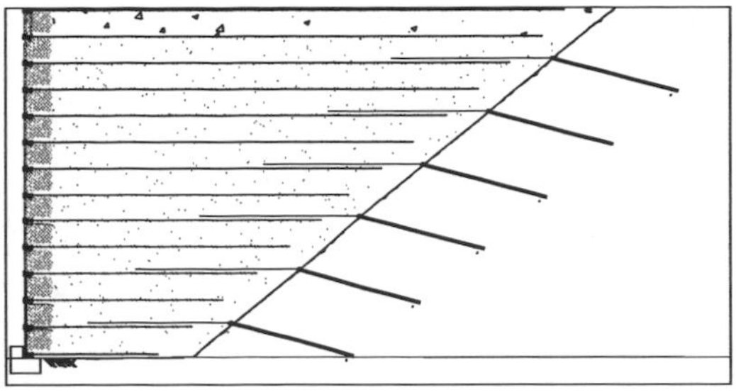

Figure 5. Friction connection.

3.3 *Technology*

The facing was made of galvanized wire-mesh, backed with stone. Due to its flexibility and engineered compressibility, this combination delivers a high quality in terms of constructability and appearance. The height of the structure in combination with a very dense backfill material led to the use of low height panels, increasing the number of reinforcement layers, in order to minimize the facing deflections due to bending or bulging.

The cross section is basically divided into two sections. The lower section on the RHS of the crusher building was too narrow for standard MSE technology, so the friction connection was used to link the reinforced soil mass to the slope. The upper section was constructed using standard MSE technology.

The facing was connected to standard high-adherence steel strips, which extend up to the slope in the lower section, and have common lengths in the upper sections.

Since the nails are independent of the reinforcing strips the positioning of the nails in the excavated slope does not have to comply with the position of the reinforcing strips. The nails were approximately twice as strong as the reinforcing strips and consequently their density was approximately half that of the strips. A friction strip was required to bond two reinforcing strips to one nail and led to a specific design of the friction ladder. Depending on the level, one or two such ladders were connected to one nail head. Figure 6 shows a typical top view with one ladder per nail.

3.4 *Construction*

The schedule for construction of wall B was tight. Firstly, all the nails were installed in early June 2006 (Figure 7), followed by the construction. Free draining crushed stone backfill was specified for the 2 m wide zone extending from the excavated face. This material facilitated compaction of the backfill around the protruding nail heads.

In places, the spacing between cladding and cut slope was less than 2 m. To ensure sufficient strength at these locations, concrete was used instead of backfill.

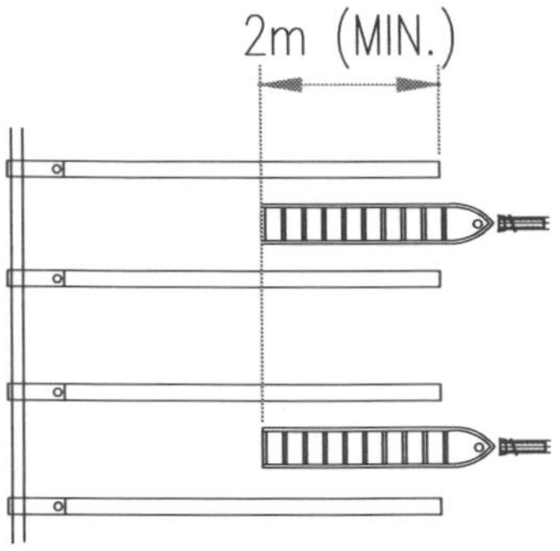

Figure 6. Typical top view of a layer comprising steel strips connected to the facing and ladders connected to the nail heads.

Figure 7. Nail installation.

A mass concrete foundation was constructed below the entire length of the nailed section of Wall B to ensure adequate bearing capacity of the foundation. After one or two lifts, when the spacing reached 2m, the hematite backfill was used. In early August 2006 construction of wall B reached the stage shown in the photograph (Figure 8).

4 DESIGN/JUSTIFICATION

4.1 *Internal stability*

The internal stability was checked using the standard "coherent gravity" method. It was applied to the long term strength of the high-adherence strips throughout the section and the adherence capacity of the strips in the upper section. This was considered to be conservative in this case, as the nailed back-slope bears a portion of the vertical stress due to the fill weight and it does not apply an active earth pressure at the base. In parts where the cut slope was not rock the active earth pressure is resisted by the friction and the soil nails.

Figure 8. Wall B close to completion.

As explained in §4.4, the internal stability was confirmed by numerical modeling.

4.2 *Block horizontal stability*

A conservative analysis of the horizontal stability was undertaken, considering that the foundation soil did not offer any sliding resistance. Consequently all the horizontal force due to the earth pressure in the upper section and to the sliding on the slope in the lower section was distributed among the soil nails and ladders. The total resisting force R_h is determined so that the horizontal stability is ensured (Figure 9).

4.3 *Wedge and circular stability analysis*

The Bishop method with slip circles and the Perturbation method with broken lines were used to check that the proposed solution satisfied these general stability criteria.

4.4 *Finite difference analysis*

All previous analysis only dealt with static force and stress equilibrium, with no consideration of the deformations of the structure during the construction. However the novelty of this layout made it necessary to make a step by step analysis of the stress/strain build-up in order to validate and refine the design.

This was done with the help of the geotechnical finite difference program, Flac 5.0, which incorporates a specific structural element to model discrete reinforcements as steel strips or ladders. The modeling was made with the "large displacements" option which allows better simulation of arching and membrane effects (so-called second order effects). Figure 10 shows the overall model.

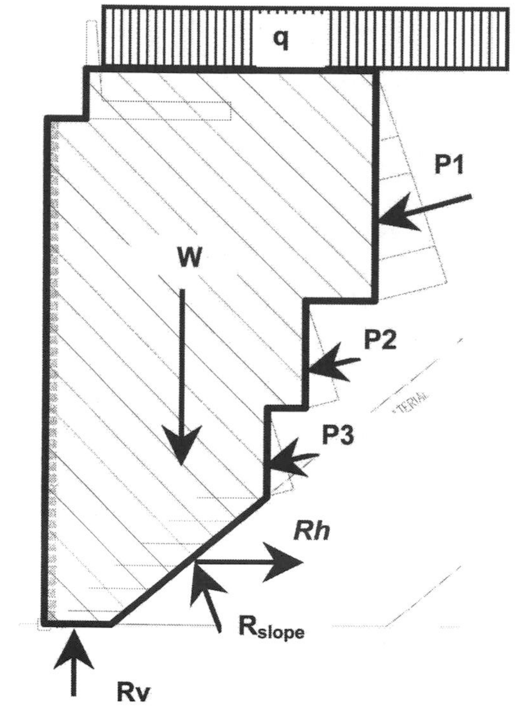

Figure 9. Block horizontal stability.

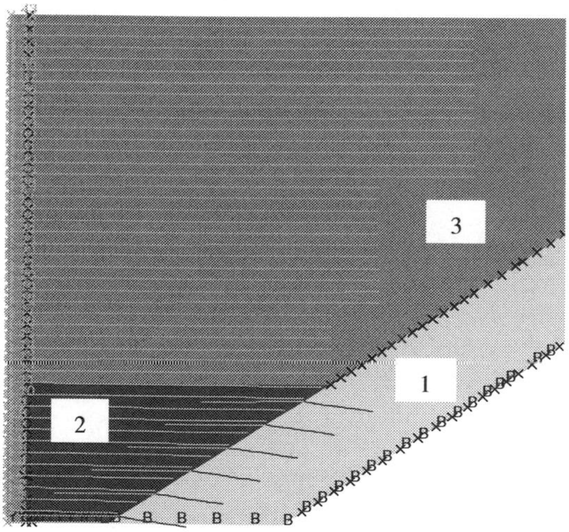

Figure 10. Flac model showing the nails embedded in the existing slope (1), the overlapping reinforcements in the bottom part (2) and the conventional layout in the upper part (3).

This modeling confirmed that overall stability is ensured, but the upper two of the six layers of ladders connected to nail heads are more stressed than the four lower ones, which required an increase in the

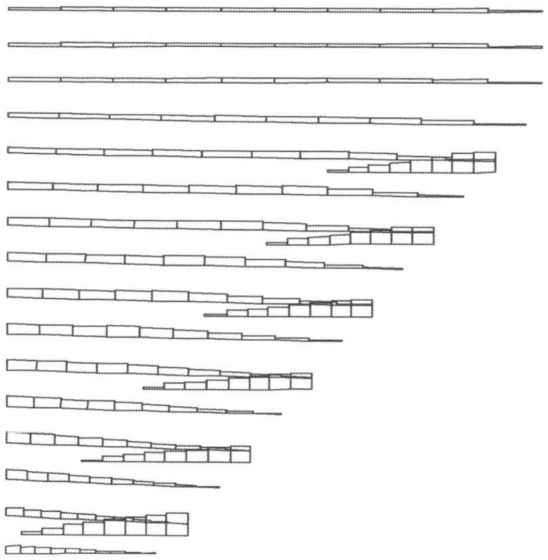

Figure 11. Relative tensile forces along the reinforcements of the lower section – not to scale.

density of ladders for these two layers. This is due to the fact that the existing slope is stiff, and settlement of the reinforced fill during construction, tends to overstress the upper connector levels. One could comment that once again the classical slope analysis programs do not allow us to represent any type of overstressing due to deformation or differential settlement. Figure 11 shows the estimated relative tensile forces in the reinforcements of the lower section. The higher tensile force in the lower ladder is artificial due to the assumption of zero friction capacity at the bottom of the fill.

5 INSTRUMENTATION

The connection between the nail and friction ladder is by means of an eye bolt and a shackle. A Wheatstone strain–gauge bridge was used to instrument six shackles. These were calibrated at the University of Pretoria and installed in pairs at three levels viz. 0.32 m, 1.28 m and 2.20 m above the foundation. The lowermost and middle sets showed almost zero tensile stress. One of the top shackles showed a tensile load of 85.8 kN while the other top shackle did not respond, probably due to failure of the electronics in a harsh environment. These results agree with the numerical analysis and confirm that the upper ladders are more stressed than the lower ones.

The construction site is in a remote place, and the construction needed to be fast, so the situation was as often frustrating for us engineers as far as instrumentation is concerned. It was not possible to monitor face displacements or settlements. However, from the excellent visual alignment of the face, we can conclude that the structure behaved as expected for Mechanically Stabilised Earth structures reinforced with steel elements.

6 CONCLUSION

The construction of MSE wing-walls at Sishen has been successfully tested in practice. It has been shown that tensile forces can be transferred in friction from one reinforcing strip to another. In addition it was verified that narrow MSE structures with base width less than 15% of the height can be constructed if they are bonded to reinforcements installed into in-situ material in excavated back slopes.

This type of solution has also applications in the construction of MSE structures in cut situations and also for the widening of fills without the need for bulk excavation.

REFERENCES

Berg R.R., Bonaparte R., Anderson R.P. and Chouery V.E. – Design, Construction and Performance of Two Geogrid Reinforced Soil Retaining Walls – 3rd Int. Conference on Geotextiles, 1986, Vienna, Austria.

Freitag N., Morizot J.-C., Segrestin P. and Fernandes K.S. – What connection to facing for polyester strips in MSE walls – Eurogeo 3; 2004, Munich, Germany.

Freitag N., Morizot J.-C., Berard G., Fernandes K.S. – Innovative Solution for Reinforced Earth walls: Friction Connection – 16 ICSMGE, 2005, Osaka, Japan.

Shored Mechanically Stabilized Earth (SMSE) Wall Systems Design Guidelines – FHWA-CFL/TD-06-001 – February 2006 – USA.

Smith A.C.S. – Reinforced Earth® structures for headwalls to crushing plants in Africa – 12th Regional Conference for Africa on Soil Mechanics and Geotechnical Engineering, October 1999, Durban, South Africa.

New Horizons in Earth Reinforcement – Otani, Miyata & Mukunoki (eds)
© 2008 Taylor & Francis Group, London, ISBN 978-0-415-45775-0

Walls over compressible soils and unstable slopes. Examples

A. Ramirez, F. Valero & C. Perez
Tierra Armada S.A., Madrid, CAM, Spain

ABSTRACT: It is well known that earth reinforcement structures are the best solution when a compressible soil with a poor quality is present at the foundation. Choosing an earth reinforcement wall instead of other solutions allows us to solve the geotechnical problem; this is one of the most common applications for this kind of structures. But, there are far more advanced applications, such as...

When there is a risk of overall slope stability failure, the use of an earth reinforcement structure instead of classical solutions allows for an increase of the stability safety factor at both short and long terms.

1 INTRODUCTION

Traditionally, the MSE walls have been the solution adopted for structures laying over foundation soils having a too low bearing capacity.

The characteristic flexibility of this kind of structures is what gives its proper behavior. But, what happens when we push up to its limit the bearing capacity of the foundation soil? How good is the behavior of this kind of walls for situations in which other kind of structural solutions are discarded?

In many cases MSE walls are showing adequate performance even over soils where other structures have been declared as both technically and economically useless. For these cases, the MSE structures show themselves as a clear alternative when dealing with inadequate, too soft and compressible subsoils.

This is a compilation of recent samples of MSE structures (with metallic reinforcement) designed and erected in Spain. After the analysis of these samples, we will be able to extract some conclusions quite generic about the factors that condition not only the design, but also the construction and service performance of these structures. Some of the walls samples herein exposed were designed and built knowing in advance the bad subsoil conditions, while for other samples the problem was exposed during the service life of the structure.

2 EXAMPLES

2.1 *Icod de los vinos (Tenerife)*

This is a peculiar case because the project was split in different phases. Initially, an 11 m height wall was designed at the bottom of an embankment; a highway

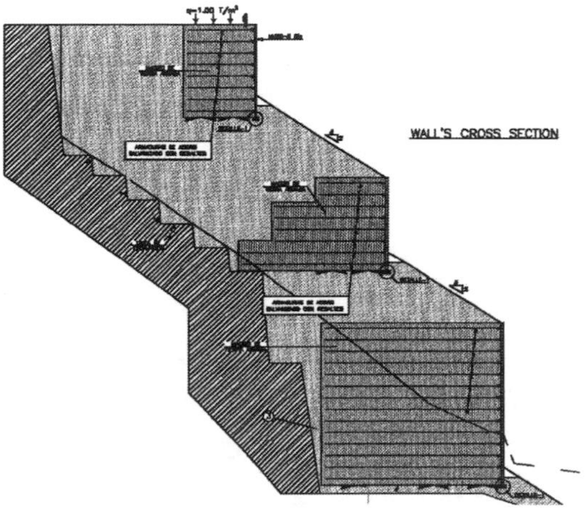

Figure 1. Caption of a typical cross section. *Icod de los Vinos* (Tenerife, Spain).

passing over the embankment needed to be widened. So, in order to achieve this goal, the above mentioned wall was built; allowance for a width increase would be possible due to the addition of backfill material over the embankment, being the backfill to be withstood by the wall at the base.

After having designed and erected the structure taking into consideration the imposed design requirements, it was found that there was a potential risk of lack of slope stability due to the self weight of the planned backfill at the top of the first (the lower one) MSE wall designed and the critical geometrical configuration of the original cross section .

Consequently, after the first wall was built, it was necessary to design (after conducting a global stability

Figure 2. Tiered walls at Icod de los Vinos (Tenerife, Spain).

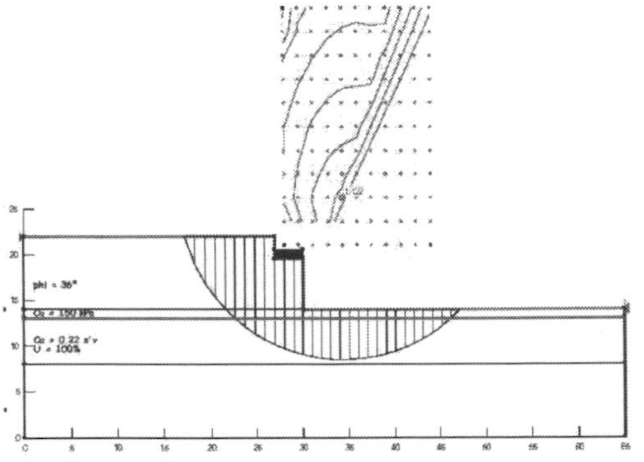

Figure 3. Monzon's detour abutment's long term circular slip stability analysis. (1.5 Security coefficient not reached).

analysis using TALREN, a Bishop Method's commercial software) a group of retaining walls inserted in the new slope in order to stabilize it.

It was decided to place two MSE walls which allowed for an increase of the global stability safety coefficients at the short and the long term for the analysed failure surfaces.

The contribution of the two walls to the increase of stability was not equal: it was the intermediate one who more effectively contributed to the slope stabilization. That wall was designed using a trapezoidal cross section, but the longer reinforcements were placed at the lower block instead of on the upper, as these walls are typically projected. (Figs. 1–2)

The length of the steel reinforcement strips (and also the reinforcement density) were increased section by section as demanded in order to 'interrupt' the most critical failure circles and achieve the imposed security coefficients.

The average security coefficients (from the preliminary client's studies) before designing the definitive solution (two tiered MSE walls placed at the top of the lower one) had a value of 1.2. At this point, it was compulsory to act section by section, up to reach a 1.5 security coefficient (at least) by increasing the length and the reinforcement quantities.

Such a configuration, composed of three tiered MSE walls allowed for the stabilization of a slope that initially was deemed as unstable; this was achieved without altering the design of the first wall, which was not the optimum within the scope of the more complex final solution.

The technical specifications (calculation cases, security coefficients, method, etc...) took into account during the design time was the *French Technical Specs for MSE Structures*: NF P 94-220-0.

2.2 *Monzon's detour (Zaragoza, Spain)*

A detailed analysis was accomplished during the construction of the Monzon's detour (Zaragoza, Spain); this analysis took into account the possibility of having some global stability problems (Fig. 3) in some of the backfills, slopes and even some of the MSE abutments due to the possibility of finding inadequate substratum soil. That is, it would be necessary to modify the design of some structures some of them already partially built, to ensure their stability.

More specifically, there was a problem with two MSE abutments partially erected which didn't have the necessary reinforcement dimensions to have a stable structure. After reaching this construction stage, the possibility of dismantling the structure and redesigning a new one was discarded. In order to solve the problem quickly, it was necessary to use the elements which had been already pre-cast and used to partially build the structure, (reinforcement strips and pre-cast panels). After analysing several possibilities, the following solution was adopted: additional reinforcement was placed as needed beginning from the highest level already built, in order to allow for an increase of the structure's global stability safety coefficient. Again, it was necessary to 'interrupt' the most critical failure surfaces.

As a result of the application of these design criteria, two additional reinforcement levels had to be placed, each one having a density of 3.5 strips per linear meter (metallic strips, 45 mm × 5 mm rectangular section) and 20 m length. It wasn't possible to link those additional reinforcements to the facing, which was already pre-cast; in addition, such a linkage was not necessary at all, given the projected strip lengths and the way the reinforcement would work (Fig. 4).

Such a configuration allowed for an adequate stabilization of the structure, increasing as already mentioned the global stability safety coefficient mentioned the global stability safety coefficient.

2.3 *Wall #19 (River Ballonti Area)*

This case corresponds to a wall laying over extremely soft and compressible soils. A badly planned and

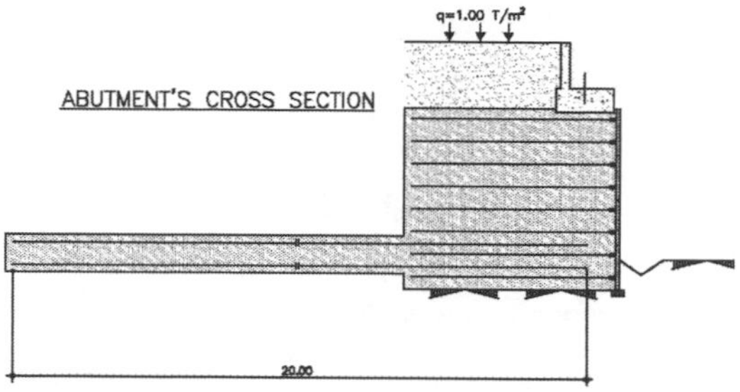

Figure 4. Caption of an abutment's cross section. Focus on two (20 m. long) metallic reinforcement layers added.

Figure 5. Huge settlements (0.5 m–0.75 m) suffered along the wall's facing with no cracked panels reported at all. See the horizontal joint's deflection.

inadequate site characterisation campaign prevented from knowing in advance the real characteristics of the substratum under the structure. Later, the real soils foundation characteristics were discovered.

The structure partially laid over the Ballonti's actual river bed, while its ends laid over anthropic soil. The average thickness for these anthropic deposits was around 3.0 m. Under that layer, there was a superficial stratum of alluvial-colluvial clayey sediments having a 0.70 m average thickness. Bellow it, there was a 7 m layer of alluvial material, mainly silts & mud, showing an allowable bearing pressure of 0.1 MPa approximately.

The results of the in-situ & laboratory tests gave a friction angle from 14° to 16° and cohesion values varying from 0.014 to 0.03 MPa.

Because of this, big settlements were observed; although some settlement was already expected, it was not in the same amount as it was observed. The wall suffered from a maximum settlement of nearly 0.75 m (Fig. 5). In spite of this settlement, the structure showed no disruption along its length. This behaviour was to be expected and, as the wall was built in several construction stages, some added measures were considered.

Figure 6. Repairing jobs. Elastic joint of expanded polystyrene placed at the footing.

There were some zones of the structure in which there was a possibility of interaction between the MSE wall and the footings of some piles which were cast in place close to the facing (0.5 m at most). In order to minimize the effect of the above mentioned footing over the MSE wall, some vertical joints were placed at adequate locations of the facing. But, due to a mistake in the construction of the piles' footing, the wall facing was allowed to lie over a corner of the footing, while the backfill had no restraint for its vertical movement.

Once the settlement of the foundation soil started, the backfill followed this movement, while the facing started cracking because of the vertical displacement restraint. As a consequence, some of the ties that linked the reinforcement to the facing panels broke. It was necessary to repair the facing and to make an EPS' elastic joint at the footing in order to reduce the remaining settlement (Fig. 6).

2.4 *Bilbao - behobia highway*

This is the case of two MSE abutments under an isostatic bridge (No intermediate piles, 25 m span, 2 decks 12 m width each one, and an average weight on beam-seats of 300 kN/m) built over extremely compressible soils in which the amount of maximum settlement to be expected was known in advance; in this job the different design and construction stages were carefully planned.

A detailed site exploration including lab and on site tests was conducted in order to know the strength parameters of the subsoil and the maximum expected settlements.

In this case, after a standard stability analysis, the global stability wasn't considered as a critical matter, so no special study should be necessary (Fig. 7).

The abutments were monitored in order to have real time data during the construction about the settlement evolution, allowing for the possibility of making modifications on the design of the structure if

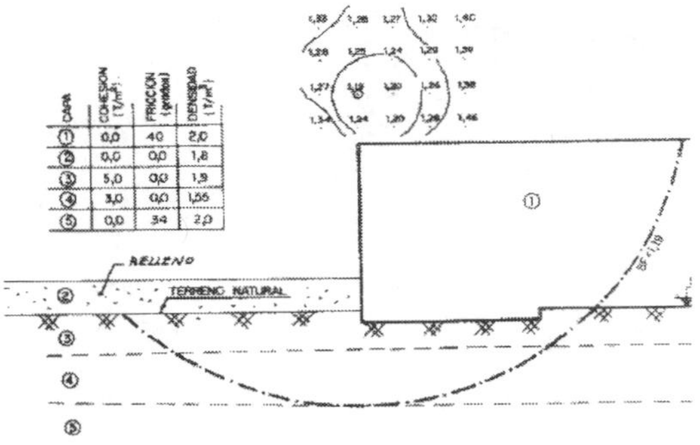

Figure 7. Bilbao – Behovia abutments. Standard circular slip stability analysis. No special study needed.

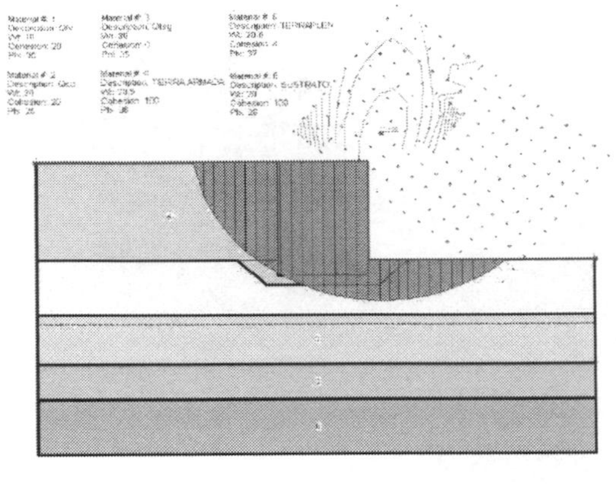

Figure 8. Abutment's global stability assured due to a foundation's soil replacement.

necessary, according to the so called Observational Method.

In order to limit the bridge settlements its construction was delayed up to two months alter finishing the underlying MSE abutments. The abutments' design height was increased in order to take into account the expected vertical settlement, around one meter.

In spite of keeping finally all this cautions, two months alter the construction of the bridge the mean value of the observed abutment settlements were approximately 0.2 m higher than expected; the bridge was not affected because it was an isostatic structure.

In spite of all the above mentioned, the observed differential settlement along the abutment facing was in the order of 1.8%. It should be mentioned that the behaviour of the MSE blocks has been excellent given this level of deformations; some panels showed just some cracking on the corners, which under no circumstances, compromised the proper functionality of the structure.

This structure was put on service without any problem and it is still open to traffic.

2.5 *Subsoil replacement under an abutment (Zaragoza)*

This example is quite representative of a fairly common solution within the MSE walls design area.

This case showed the typical problem of the presence of a substratum layer of very low bearing capacity. We should add to this the existence of a preliminary estimation of very high short and long term settlements (total and differential). As a consequence, the structure (a MSE abutment) would have suffered a series of vertical movements (non uniform) that were not compatible with the serviceability of the bridge itself. In addition, the global stability of the whole (abutment + foundation) was compromised, showing unacceptable values for the safety coefficients.

The adopted solution in this case was to replace the foundation material up to 2 m depth with a selected material which greatly increased the bearing capacity and showed out to conform a foundation soil for the MSE abutment. (Fig. 8)

3 CONCLUSIONS

The MSE structures are and will remain for a long time as the most adequate to be built over soft soil with low bearing capacity. The flexibility is the main quality of this kind of structures when comparing them to any other "rigid" solution that is discarded at first sight when these geotechnical problems are present, because technically and also economically there are no valid competitors against the MSE solutions.

The above mentioned consideration should not be used as a valid excuse to forget about the need of conducting a proper site exploration and testing, with a detailed analysis of the foundation soil characteristics.

When dealing with a global stability problem, the MSE solutions answer to the problem by themselves, without the need of resorting to external solutions, that is, the structure itself due to its basic conception (backfill + reinforcements) together with an adequate design is capable giving an adequate answer to the stability problem.

REFERENCES

Abraham, A. and Sankey, J. 1999. Design and Construction of Reinforced Earth Walls on Marginal Lands. *Geotechnics of High Water Content Materials, ASTM STP* 1374.

Manual para el proyecto y ejecución de estructuras de suelo reforzado 1989. MOPU.

Rodríguez-Miranda, M.A. and Villaroel, J.M. 1978. La Tierra Armada utilizada en la cimentacón de una estructura en suelos blandos. *Obras Públicas Magazine* 3163.

Valero, F. 1995. Muros de suelo reforzado con armaduras metálicas y paramento de hormigón; consideraciones, patología y corrección. *Retaining Structures Symposium* (1.08).

New Horizons in Earth Reinforcement – Otani, Miyata & Mukunoki (eds)
© 2008 Taylor & Francis Group, London, ISBN 978-0-415-45775-0

SeaTac third runway: Design and performance of MSE tall wall

J.E. Sankey
The Reinforced Earth Company, Vienna, Virginia, USA

M.J. Bailey & B. Chen
HartCrowser, Inc., Seattle, Washington, USA

ABSTRACT: Expansion of the SeaTac International Airport in the State of Washington (USA) included construction of a third runway, which required installation of a series of single face and multi-tiered Mechanically Stabilized Earth (MSE) wall structures. The tallest of the MSE walls consisted of a four-tier structure with a total exposed height of approximately 43 meters (45 meters with wall base embedment). This paper discusses the investigation, design, construction and performance of the West Wall, one of the tallest MSE walls built in the world.

1 INTRODUCTION AND BACKGROUND

1.1 *Location, project purpose and subsurface conditions*

In 1999, HartCrowser, Inc. started subsurface investigations for the addition of a third runway as part of expansion to the Seattle Tacoma (SeaTac) International Airport. The airport is a major international hub located in the northwest part of the United States between the cities of Seattle and Tacoma, Washington. Major embankment construction (13,000,000 m^3) was necessary to accommodate the added runway on the western part of the airport; however, the presence of a stream and wetlands confined the expansion to the steeply sloping topography that bounded the area.

As a result of the space limitations posed by the site, consideration was given to the construction of a series of single face and multi-tiered retaining walls along the runway right-of-way alignment. The topographic features required consideration of exposed wall heights approaching 43 meters. The tallest such wall location was designated as the West Wall.

The area of the West Wall was the subject of a detailed geotechnical investigation. The results of the investigation identified the presence of fill and recent alluvium overlying recessional outwash, glacial till and advance outwash. The basal layer addressed in the investigation consisted of very dense silty sand with gravel (glacial till) and very stiff to hard silt and clay. In summary, the physical features of the area were influenced by subsurface conditions containing peat, liquefiable sands and the potential for excess pore pressures in silt and clay soils. Subgrade improvement was considered to address these issues.

1.2 *Selection of MSE wall system*

The significant height of the retaining walls for the project left few viable options from engineering and economic standpoints. The design team made a preliminary evaluation of more than sixty retaining walls and slope geometric relationships before selecting steel-reinforced Mechanically Stabilized Earth (MSE) walls for the three main retaining walls. It is notable that a comparison of wall technologies provided by FHWA (1995) indicates that steel-reinforced MSE walls are a stand alone selection for heights exceeding 20 meters. On this basis, a qualifications-based selection process by the Port of Seattle (Owner) and HNTB (Architect/Engineer) found that Reinforced Earth ®technology had a reliable international track record with regard to retaining walls exceeding 30 meters in height. The technology for tall Reinforced Earth walls typically consists of cruciform facing panels connected to discrete steel reinforcing strips in a select granular fill matrix. The design of the panels and the material components that make up the Reinforced Earth volume needed to be based on a combination of internal, external and compound stability evaluations.

2 DESIGN BASIS FOR WEST WALL

2.1 *Wall geometry*

The West Wall was designed using four terraces to break up the sight lines of the structure. Space limitations for the wall limited offsets to a little more than 2 meters between terrace levels. The ratio of the minor offsets in the tiers compared to the overall

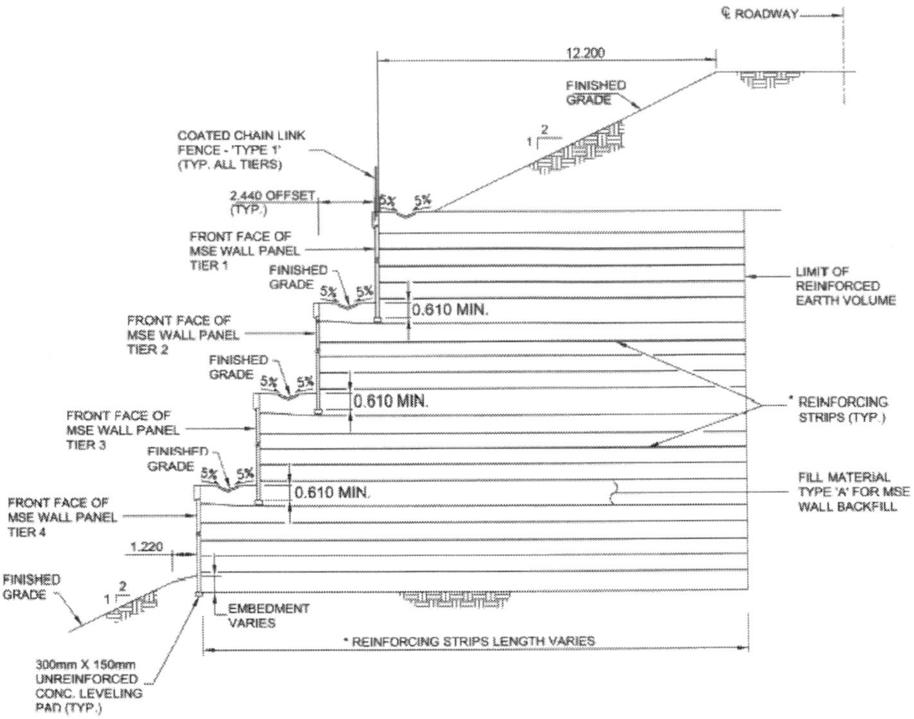

Figure 1. West wall typical section.

wall height dictated that the lines of maximum tension within the Reinforced Earth volume be evaluated as a single structure (FHWA 1997). In other words, the limited offsets provided no reduction in the influence of the upper tiers on the lower tiers of the wall. A typical section of the West Wall is shown on Figure 1.

2.2 *MSE wall stability*

Borrow source requirements for the select granular fill in the Reinforced Earth volume were identified early in the design process. Specific fill sources were selected and pre-qualified on the basis of shear strength and other tests during the construction bidprocess. The design parameters for the select fill were established at 37° for the friction angle and 22 kN/m^3 for the unit weight. The random fill behind the Reinforced Earth volume had design parameters selected at 34° for the friction angle and 21 kN/m^3 for the unit weight.

Lengths for the reinforcing strips were evaluated on a terrace-by-terrace basis and maintained at a minimum of at least 70% of the overlying wall height. The longest reinforcing strips of approximately 30 meters were therefore located at the tallest wall section with exposed height of 43 meters. Evaluations were made for pullout, tensile capacity, sliding and overturning in both static and seismic conditions. It is noted that the lower tier of the wall was embedded up to 4 meters below the finished grade. Upper terraces had the base panels embedded approximately 1 meter into the next lower terrace, which allowed reinforcing strips to be installed without interference between strip levels.

The significant loads imposed by the overall height of the West Wall required increasing the design reinforcing strip thickness, as well as the thickness of the lower precast panels themselves. In the case of the 50 mm wide reinforcing strips, the thickness was increased from the 4 mm standard normally used in the United States to a modified 6 mm thickness. While precast panels in the upper terraces could be maintained at the standard 140 mm thickness, the lower panels were increased to 178 mm thickness.

The nominal plan dimensions of the cruciform panels were maintained at 1.5 m by 1.5 m. However, even with the increased strip thickness considerations, it was necessary to use a large number of strips in the lower terrace panels. The number of reinforcing strips in the upper terrace was maintained at a nominal density of 4 per panel; while in the lower level the density varied from 12 to 21 reinforcing strips per standard-sized panel (some larger special panels at the base had up to 28 strips – Figure 2).

As a final design consideration, the number and thickness of elastomeric bearing pads separating the

Figure 2. Tie strips ready to receive reinforcing strips in lower tier panels.

precast panels at the horizontal joints needed to be adjusted by terrace level. Upper terraces could use two pads per panel with a nominal thickness of 19 mm each, while the lowest terrace needed four evenly spaced pads per panel at a nominal thickness of 25 mm each. The selection of the number and thickness of bearing pads was critical in maintaining the structural integrity of the panels and corresponding joints, as well as the overall wall appearance when architectural treatment was added to the fascia casting.

2.3 *Settlement and global stability*

The bearing pressures imposed by the West Wall varied with the height and number of wall terraces. The maximum bearing pressure was considered at 1.1 MPa. Settlements of as much as 500 mm were originally estimated without any ground or wall improvement measures due to the presence of compressible and liquefiable bearing soils. The Probabilistic Seismic Hazard Analysis for the wall gave a basis for peak ground horizontal acceleration of 0.36 g for a 475-year event and 0.47 g for a 975-year event.

Following a test of subgrade improvement using stone columns (Chen and Bailey 2004), design measures were made to remove and replace the subgrade soils of concern. In addition, slip joints were added to the wall to better accommodate differential settlements at critical elevation changes in the wall. The additional design measures resulted in a reduction in total settlement to approximately 150 mm, with a maximum differential settlement of about 1/100 to 1/200.

Global stability analyses were conducted using the computer program SLOPE/W (Geo-Slope 1998). Both external and composite failure planes were evaluated using the limit equilibrium based methods available in the program including Janbu, Bishop Spencer and Morgenstern-Price Methods. Other computer programs were used for additional evaluation (e.g., Newmark analysis) of the global stability under seismic conditions. Reinforcing strip lengths, thickness and/or depth of embedment were modified in some cases to meet stability requirements, based on analyses of different sections for the three main walls.

2.4 *Numerical analyses*

Critical sections of the West Wall were selected for numerical analysis using the FLAC computer program (Itasca 2000). The purpose of the FLAC analyses was to provide additional information to the design team on anticipated wall performance to supplement AASHTO design analyses (1996 thru 2000). Results of the FLAC analyses were not intended to replace design analyses accomplished in accordance with AASHTO code.

Information was developed for input to the FLAC analyses including wall geometry, soil analyses, concrete facing properties and steel reinforcing strip properties. A dynamic time history for seismic shaking was developed based on a site-specific design response analysis. The FLAC analyses verified that predicted stresses in the reinforcing strips would be maintained close to the performance criteria allowed by AASHTO (0.55 times yield). Furthermore, predicted wall settlements and horizontal displacements were deemed acceptable for both steady state (normal) and seismic conditions. Horizontal displacements at the tallest wall section under static loading were analyzed to be less than 90 mm; subsequent monitoring showed end-of-construction displacements ranged up to 150 mm.

3 CONSTRUCTION AND MONITORING

3.1 *Performance*

Construction of the West Wall started in October 2004 along with the four other Reinforced Earth walls designed for the runway extension. Installation was performed by TTI Constructors, a joint venture of Seattle, Washington firms Fiorito Construction, Scarsella Construction and Tri-State Construction. Work on the West Wall was completed in September 2005, though progress was split with the other walls and main embankment being constructed during the same period.

The select granular backfill considered in design development was tested and confirmed during its use throughout construction of the walls. The West Wall itself covered a face area of approximately 12,100 m^2, with a length along the top tier measuring approximately 450 m and a length along the bottom tier measuring approximately 190 m. Considerable demands were placed not only in the delivery of select backfill (approximately 1,500 trucks per day delivering 110 million kilograms of fill during peak construction), but

Figure 3. Reinforcing strip placement.

Figure 4. Architectural treatment.

also on the panel and reinforcement components delivered to the site. Reinforcing strips were manufactured and galvanized at longer lengths of approximately 12 meters to minimize splicing needs (Figure 3). Over 250 form liners were manufactured to accommodate the architectural appearance of the panels, many used only once per panel (Figure 4). Material delivery was expedited to meet project and weather-related deadlines.

3.2 *Monitoring*

An extensive instrumentation and monitoring system was provided during construction of the West Wall. Instrumentation included survey points on selected panel faces, strain gages attached along the length of selected reinforcing strips, piezometers and inclinometers. Monitoring of the instrumentation was conducted during wall construction to confirm tolerances

Figure 5. View of completed four tier Reinforced Earth wall.

established during design. In addition to the instrumentation, bearing pads at selected joints were also measured for compression. As a final measure, durability samples were installed in the West Wall select fill volume to monitor the integrity of the reinforcing strips over the 100 year design life of the structure.

Results of the monitoring during construction generally confirmed the design tolerances set for the West Wall. The survey points on the panel faces measured up to 150 mm of lateral deflection and a maximum of 178 mm of settlement. The inclinometers showed somewhat less deformation of the overall wall volume at a lateral movement ranging between 10 to 74 mm. The strain gages generally showed deformation and calculated stresses at less than predicted FLAC values. Piezometers showed normal seasonal fluctuations of up to 2.1 m in the shallow unconfined aquifer, with no discernable head gain due to the consolidation of the underlying very stiff to hard sediments. Finally, the bearing pad compression varied from 12% to 53%, which is well within tolerable horizontal joint maintenance.

4 CONCLUSION

The West Wall for the SeaTac Airport represents the tallest MSE structure built in the United States and one of the tallest walls in the world. Modifications made to the basic components of the Reinforced Earth system proved that ever increasing wall heights may be considered for MSE technology using steel reinforcements. With the use of increasing wall heights comes the need to incorporate both the standard codes used in typical MSE wall design along with numerical modeling tools for detailed evaluations. Instrumentation and monitoring during construction may be compared to the design evaluations for verification of stability. The excellent quality of the select fill and well planned delivery and placement of the wall components, in strict compliance with the project specifications, is necessary to achieve a reliable and aesthetically pleasing MSE wall.

REFERENCES

AASHTO. 1996–2000. Standard Specifications for Highway Bridges – Retaining Walls. 16th Edition, 1996, with periodic addenda through 2000. American Association of State Highway and Transportation Officials.

Chen, B.S. and Bailey, M.J. 2004. Lessons Learned From A Stone Column Test Program in Glacial Deposits. In *GeoSupport 2004*, edited by J.P. Turner and P.W. Mayne, Geotechnical Special Publication No. 124, published by the GeoInstitute, ASCE and ADSC, pp. 508–519.

FHWA Report No. SA-96-071. 1997. Mechanically Stabilized Earth Walls and Reinforced Soil Slopes, Design and Construction Guidelines.

FHWA HI-95-038. 1995. Cost Comparison for Retaining Walls.
Geo-Slope. 1998. Slope/W software for stability analysis, Version 4. Geo-Slope International Ltd. Calgary, Alberta.
Itasca. 2000. FLAC software for deformation analysis, Version 4.0. Itasca Consulting Group, Sudbury, Ontario.
Sankey, J.E. and Soliman, A. 2004. Tall Wall Mechanically Stabilized Earth Applications. *Proceedings of Geo-Trans 2004, ASCE, pp. 2149–2158.*

New Horizons in Earth Reinforcement – Otani, Miyata & Mukunoki (eds)
© 2008 Taylor & Francis Group, London, ISBN 978-0-415-45775-0

Case study of a MSE wall supporting a multi-story building

F.W. Fordham
The Reinforced Earth Company, Atlanta, Georgia, USA

M. Louis
The Reinforced Earth Company, Lake Forest California, USA

K.M. Truong
The Reinforced Earth Company, Vienna, Virginia, USA

ABSTRACT: The proposed construction of a multi-story building supported by a Mechanically Stabilized Earth (MSE) retaining wall provided many unique engineering challenges. Foremost was that the building's shallow strip and spread footings were founded directly on top of and behind the reinforced MSE retaining wall mass. These footings imparted maximum loads of 120kPa onto the reinforced volume. The design was complicated by the presence of a permanent lake adjacent to the MSE wall, which meant evaluating rapid draw-down and dam breach conditions as well as increased metal loss issues in submerged wall applications. Further design complexities included a required one horizontal to eight vertical (1 H: 8 V) front face batter and continuous stone fascia.

This paper presents the design methods and modeling used for evaluating internal stability (bond and strip rupture), external stability (overturning and sliding) and global stability. All designs were checked for normal (steady state) conditions, as well as rapid draw-downs. The engineering design also included global stability analysis for normal conditions as well as rapid draw-down and the unlikely event of a catastrophic dam failure. This discussion will also include the design modifications necessary to alter a standard vertical face MSE wall system in order to meet the required batter with continuous stone fascia. Special considerations given for drainage, select backfill and geotextile selection for joint cover in submerged applications will be presented.

1 INTRODUCTION

1.1 *Preliminary assumptions*

Bid plans for the site work for *The Tom Harkin Global Communications Center at the Centers for Disease Control and Prevention* in Atlanta, DeKalb County, Georgia indicated that a Mechanically Stabilized Earth (MSE) retaining wall was required adjacent to the building plaza. The wall was to provide a grade separation between the plaza and a permanent lake as well as providing a scenic overlook. Structural plans were not part of the site work and it was assumed by The Reinforced Earth Company® (RECo), when preparing their bid, that the building included a basement. The wall being in close proximity to the building meant that the discrete metallic strips used to reinforce the wall mass extended to the assumed basement in some areas. The benefit of this was that there is not a lateral load due to active earth pressure against the reinforced mass in that case.

1.2 *Final reality*

RECo was successful in their bid and final retaining wall plans were prepared from the site package. The wall construction plans were forwarded to the wall installer, MC Inc., and to the General Contractor, Turner Construction. These plans were in turn submitted to the Architect, Thompson, Ventulett, Stainback & Associates, and to U.S. Government reviewing agencies for review and comment. Plans were returned not approved with annotations concerning building loads on top of the wall. Building plans and loading were then provided and re-engineering of the retaining wall was initiated.

2 DESIGN

2.1 *Design requirements*

Normally RECo is responsible only for the internal stability of the MSE mass, including pullout resistance

Figure 1. The Tom Harkin global communications center.

and rupture of the soil reinforcement. The external stability of the structure; sliding, overturning and slope stability (i.e. compound global stability) is usually the domain of the Geotechnical Engineer. On this project, however, the responsibility was placed on Construction Manager and the Retaining Wall Engineer for the overall wall stability.

Although the required design life was seventy five (75) years, calculations for stresses and factors of safety were based on one hundred (100) years. This was due to critical nature of the project and a portion of the wall being in water. It should be pointed out that the steel design is based on 0.55 of the yield stress of steel (f_y) at the end of the design life. Design stresses and factors of safety were calculated for working stress in accordance with 1996 AASHTO Specifications for Highway Bridges.

2.2 *Loading conditions*

Initially the preliminary design of the wall only considered the case of a level backfill with 6kPa pressure from pedestrian surcharge with no regard to the building. Final design considered in addition to the active earth pressure loads, that the MSE wall was to support footing loads from a four-story building bearing directly on the reinforced mass. The loads imposed by the building were concentrated on shallow spread footings and strip footings. The vertical pressures imposed by the spread footings and the strip footings reached 135 kPa and 125 kN/m respectively. These values were converted to horizontal pressures using empirical Boussinesq's formulae.

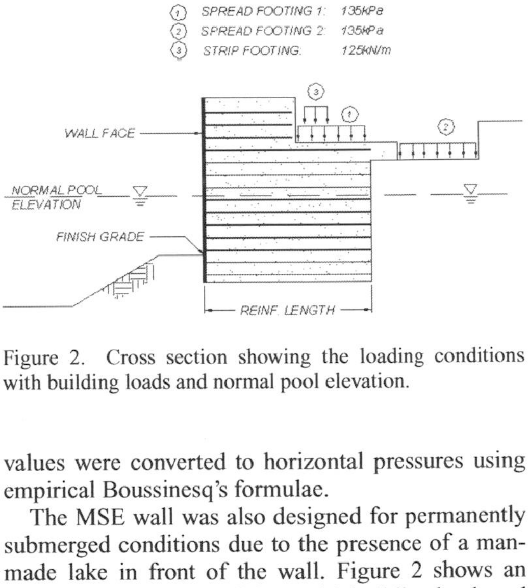

Figure 2. Cross section showing the loading conditions with building loads and normal pool elevation.

The MSE wall was also designed for permanently submerged conditions due to the presence of a man-made lake in front of the wall. Figure 2 shows an example of loading conditions with building loads and permanent water elevation (normal pool). Because of a potential of high water elevation, the wall was designed to withstand 0.90 m of rapid drawdown from the 100-year flood elevation. The 100-year flood elevation was approximately 3.20 m higher than the permanent water elevation. The wall was also modeled with hydrostatic loads corresponding to a rapid drawdown from the 100-year elevation to the toe elevation on the downstream side of the lake dam (up to 6.25 m of rapid drawdown).

Table 1. Design parameters.

	Structural backfill*	Retained backfill	Residual foundation
Internal friction angle (degree)	45	30	30
Cohesion	0	0	0
In-place unit weight (kN/m^3)	15.1	18.1	18.1
Saturated unit weight (kN/m^3)	17.3	18.9	18.9

* Structural backfill is open graded stone.

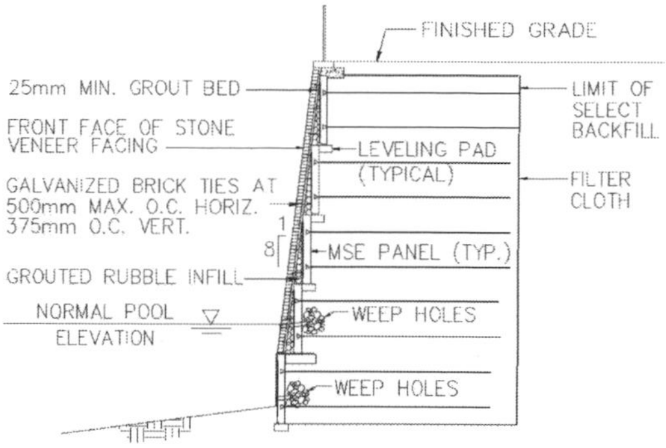

Figure 3. Cross section showing weep holes, filter fabric and stone fascia.

2.3 *Subsurface conditions*

Based on the Geotechnical Engineer's (MACTEC) recommendations, the walls were to be founded on residual soils or structural fills. In selected areas, between 1.00 m and 1.50 m of existing fills were excavated underneath the wall and replaced with structural fill (compacted open graded stones). This provided a solid foundation for the MSE wall and thus controlled the amount of differential settlement between the shallow spread footings of the four-story building.

Based on in-situ and laboratory tests, geotechnical parameters were determined to use for the design of the wall. These design parameters are presented in Table 1. Please note that the internal friction angle for the structural backfill tested at 49 degrees maximum, but 45 degrees was used based on a corresponding 12 mm of movement in the direct shear test.

2.4 *Design considerations*

Several design considerations were addressed to satisfy the unique conditions of this MSE wall. Special considerations were given to submerged conditions, the complexity of the wall and the importance of the building structure that it supports.

First, fluctuation of the water in the reinforced mass induces temporary hydrostatic pressures, which need to dissipate. To make this possible, the backfill consisted of a free draining open graded stone. On a typical MSE wall a 20 mm open joint between the panels would allow dissipation of the water. However, because of the specified stone masonry veneer to be installed in front of the wall, the MSE panel joints were sealed and waterproofed. Therefore, to mitigate potential hydrostatic pressure, weep holes (refer to Figure 3) were installed every 1.50 m just above the finish grade at the bottom of the wall and also provided at 1.50 m centers at 150 mm below normal pool.

Second, to prevent the retained backfill, which contains materials passing the #200 U.S. sieve, from migrating through the reinforced mass, it was recommended to install a non woven geotextile between the open graded stone backfill and the retained backfill.

Finally, compound global stability calculations were performed using the STABL program modified by RECo to model the metal reinforcing strips shearing resistance. Compound global stability calculations do not restrict the failure surface from crossing through the reinforced mass. These calculations were necessary because of the complexity of the wall and the critical nature of the building it supports. The slope at the toe of the wall affected the factor of safety for compound global stability as well. Therefore, it was prudent to provide more embedment at the bottom of the walls (between 1.00 m and 1.50 m).

2.5 *Factors of safety*

The coherent gravity method (Meyerhoff) was utilized to calculate the horizontal pressures against the wall for the internal stability of the wall. Each reinforcing strip tension was designed not to exceed 32kN for permanent conditions. The maximum allowed tension was increased by 25%, in the rare case of a high water elevation or a dam breach. Each reinforcing strip was designed to provide a minimum factor of safety of 1.50 for pullout. This factor of safety was reduced to 1.20 for the temporary condition of high water case and the unlikely case of dam breach. It was noted that the closer to the wall face the footing of the building were, the more strips per unit area were necessary in the upper most layers.

Factor of safety against sliding and overturning was computed. For sliding, a minimum of 1.50 was maintained for the permanent conditions and 1.20 for the high water case and the dam breach case. For overturning, a minimum of 2.00 was maintained for the permanent conditions and 1.50 for the high water case and the dam breach case. These conditions did not govern the design of the soil reinforcement length.

Table 2. Compound global stability results.

Case no.	D	Fore Slope	H	B	B/H	Factor of safety (a)	(b)	(c)
1	N/A	Flat	7.0	5.8	0.83	1.71	1.50	1.41
2	4.9	Flat	9.5	8.5	0.90	1.60	1.52	1.46
3	6.4	4:1	7.3	10.0	1.38	1.65	1.57	1.35
4	N/A	3:1	5.2	5.8	1.29	1.66	1.53	1.10
5	1.4	3:1	5.0	11.0	2.18	1.52	1.50	1.29

D = Distance from edge of building footing to face of wall (m)
H = Wall height from top of leveling pad to top of coping (m)
B = Reinforcement length (m)
(a) Permanent submerged conditions
(b) 0.90 m rapid drawdown from 100-year water elevation
(c) Dam breach at 100-year water elevation.

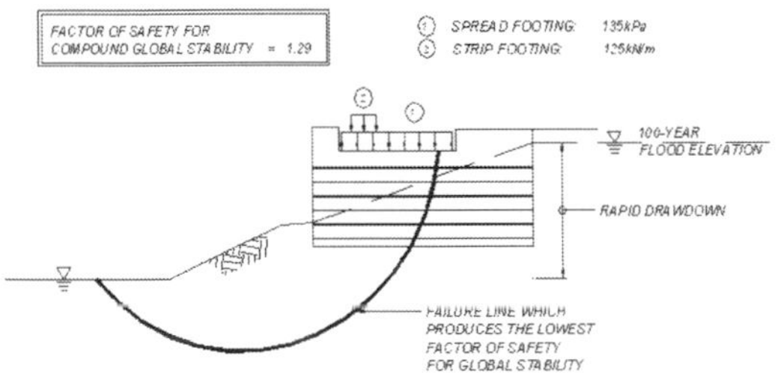

Figure 4. Example of calculated failure mode for compound global stability – Case 5, dam breach.

In the calculations of the compound global stability, the geotechnical parameters for the residual materials or the structural fill were conservative to allow for any unknowns in the subsurface. A minimum factor of safety of 1.50 was required for the permanently submerged conditions and also for the 100-year water elevation with 0.90 m of rapid drawdown. A minimum factor of safety of 1.1 was used in the highly unlikely event of a dam breach at the 100-year water elevation.

This 100-year water elevation with 0.90 m of rapid drawdown criterion governed the length of the soil reinforcement. It is noted that the reinforcement length over wall height ratio rapidly increased for a design case with a slope at the toe of the wall or when the building loads to the wall face.

Table 2 shows the compound global stability results at five locations with three hydrostatic conditions each.

Figure 4 illustrates an example of calculated failure mode for compound global stability in the event of a dam breach. In this example, it is noted that the failure line is predicted in the middle of the reinforced mass. This is due to the significant building loads being located very close from the wall face. The calculated factor of safety of 1.29 is conservative for a temporary and unlikely loading condition.

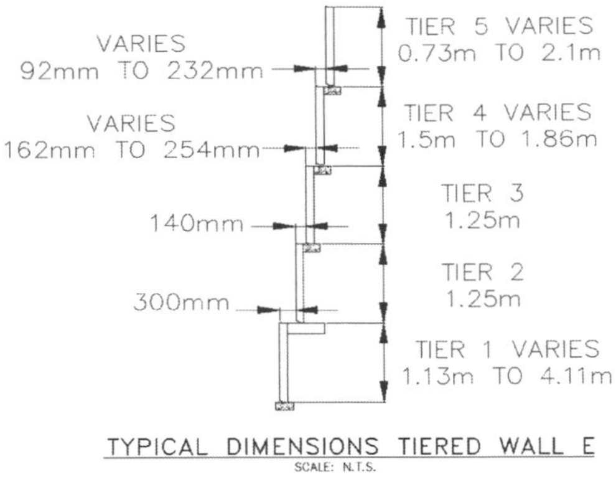

Figure 5. Typical MSE wall section.

3 AESTHETICS

3.1 *Requirements*

A crab orchard stone masonry veneer was specified to cover and embellish the precast MSE panels. The masonry veneer was specified to have a permanent batter of 1H : 8V. Tree wells were also specified close to the wall.

3.2 *Resolutions*

Attaching the veneer was a simple matter of casting dovetail insert slots into the front face of the MSE panels. Dovetail anchors were designed to attach into the slots and support the loads of the veneer.

Achieving the required batter was a more difficult task, due to MSE walls are designed to be plumb. The first proposed solution was to construct a plumb full-height MSE wall with a 0.15 m offset from the horizontal alignment of the final course of stone. This idea was dismissed because of the 1.20 m width of stone required at the bottom of a 9.50 m wall. This would mean excess veneer in excess of 1.00 m as the minimum fascia thickness was 0.15 m.

The resolution arrived at was to split the wall into tiers. The tiers were designed to minimize the stone fascia by reducing the width of stone required at the bottom of each tier. Now a 1.13 m wall height only required a 0.30 m width at the bottom (0.15 m + 1.13 m /8). Figure 5 shows the final typical section.

4 CONSTRUCTION

Construction of five almost parallel walls on reversing curves was a challenging task. It was made more difficult due to the upper two walls had a variable wall offset depending on wall height. Mr. Rod Kindoll of MC Inc. must be commended for his expert wall construction.

5 CONCLUSION

Though MSE walls are routinely used to support spread footing abutments with higher loads (192 kPa) than exerted by the building, the higher tolerances of the building come in to play. Of great concern to RECo was the potential for differential settlement of the footings on top of the reinforced mass and those of the footings behind the mass.

The use of a MSE wall supporting a multi-story building is a highly unusual application, but it is a viable one.

REFERENCE

Gathany, James (CDC/CCHIS/NCHM) Figure 1. Photograph of *The Tom Harkin Global Communications Center,* 2007.

New Horizons in Earth Reinforcement – Otani, Miyata & Mukunoki (eds)
© 2008 Taylor & Francis Group, London, *ISBN 978-0-415-45775-0*

Multiple applications of Reinforced Earth technologies for industrial mining structures – Georgia Pacific Mining design/build project

P. Proctor
Reinforced Earth Company Ltd., Truro, Nova Scotia, Canada

P. Wu
Reinforced Earth Company Ltd., Mississauga, Ontario, Canada

ABSTRACT: Georgia Pacific Mining Canada required expansion of their gypsum mining operation in Nova Scotia, Canada. Herein describes various design-build industrial mining structures using Reinforced Earth® technologies under dynamic and ever-changing surcharge loading conditions. Three separate but inter-connected structures were required to extract the raw materials for the new mining area, namely: 1. Truck Dump MSE Walls and Distribution Slab supporting Cat 773 haul trucks. 2. Surcharge Tunnel with dynamic and fluctuating surface loads applied and 3. Escape Tunnel using MSE walls and concrete roof slabs with heavily loaded spread footing forces applied. All three structures had the added components of conceptual design, final detail design, supply and construct under a "Design-Build" contract. Georgia Pacific Mining Canada required this turnkey project to be in operation for the transition period between the old mining site to the new facilities with guarantees of performance of all structures. The following case history illustrates the flexibility of MSE, TechSpan and the innovative applications of engineering solutions in meeting the needs of a challenging site environment.

1 INTRODUCTION

Georgia Pacific (GP) Mining Canada required expansion of their gypsum mining operation in Nova Scotia, eastern Canada due to the mine extraction limits nearing its end at the Sugar-Camp Mine. Geotechnical survey at a nearby location named Melford proved positive for the eventually expansion to the new mining site. The initial planning of infrastructure for the Melford site required entirely new facilities for every aspect in the mining sequence because the Sugar-Camp facility would remain in operation for un-interrupted supply of processed gypsum to their port location. Upon completion of the new mine site structures, an overlap period would commence to supplement the gypsum stock to the port, and then a de-commissioning at Sugar-Camp would follow, while the Melford Mine would increase their daily output.

Continuous supply of gypsum was paramount to be maintained throughout the planned mine relocation. The critical schedule demanded each structure to be built on-time without delay. Therefore, chosen contractors had to plan, schedule, construct and guarantee the performance of each structure so that the overall relocation schedule met the owners' requirements for the new Melford Mine.

2 DESIGN CONCEPTS – STRUCTURES

The project consultants CBCL Limited, through terms of reference from GP, were charged with the task of laying out the new Melford Mine facility in its entirety. Three main structures were required for the processing of the gypsum raw material. A conceptual design by CBCL and GP was initially available, but the detailed design, layout, scope of structures, and orientation relative to each other required developing through a design-build contractor. Reinforced Earth Company Ltd. Canada and Alva Construction formed a joint venture (RAJV) to design-build Mechanically Stabilized Earth (MSE) structures in conjunction with a pre-cast concrete arch tunnel to suit the needs of the GP Melford mine site. The design also included structural load distribution and roof slabs associated with the main MSE and arch structures.

3 DESIGN-BUILD CONSTRUCTION

Initially, the overburden material had to be removed. Drilling and blasting of the underlying gypsum bedrock to rough base grade elevation occurred next.

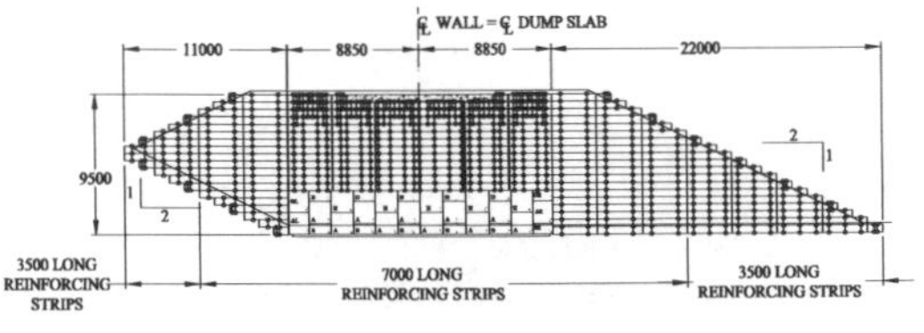

Figure 1. TDW MSE wall elevation.

The evolving industrial structure design was based on the site conditions post excavation. The geotechnical firm provided settlement, allowable bearing capacity and overall global stability assessments to RAJV. In a Gypsum Mine, quite often soft pocket of clay are trapped between bedrock layers. In some cases, the supporting foundation required improvement either by sub-excavation and rebuilding with engineered fill or by use of mud-slabs to have clean and relative firm surface to begin the construction.

The general layout of the structures in relation to each other began and scope of work was then refined to determine the finish grades at each location. The three industrial mining structures were 1) Truck Dump Wall (TDW) with a Distribution Slab, 2) Surcharge Tunnel and 3) Escape Tunnel (man-way). The TDW was needed to unload gypsum raw material from the mine haul trucks into a large hopper. Sufficient grade separation was required so that the GP crusher unit had enough head-room to accept, crush, and expel the processed gypsum by means of a conveyor system. The required design height for the TDW was 9.90 m, which had to resist the surcharge pressure produced by a Cat 777D mine haul trucks. Although the smaller Cat 773 trucks were used at the Melford mine site, future upgrade in truck size to 777D was required for design of all surcharges on MSE and tunnel structures. A pressure distribution slab and curb stop on top of the wall were installed to support the mine haul trucks. Impact loads, breaking loads, and vibration loads from the truckers and crusher were also taken into consideration for proper design of the TDW. The geometry of the wall was left to the wall designer with basic grades, slopes and turning radius of the Cat 777D provided in the terms of reference from CBCL.

Understanding the general requirements for the TDW, RAJV provided a wall layout design incorporating wire face MSE wall for the upper and wing wall area due to short design life required 30 years, and pre-cast face panels for the lower section near the crusher for ease of clean up. The spilled processed gypsum material was able to be removed in this area with a front end wheel loader without damaging the wall face due to the smooth pre-cast concrete finish. The wall was approximately 50 m in length and 9.9 m tall at the highest point underneath the distribution slab. (see Figure 1).

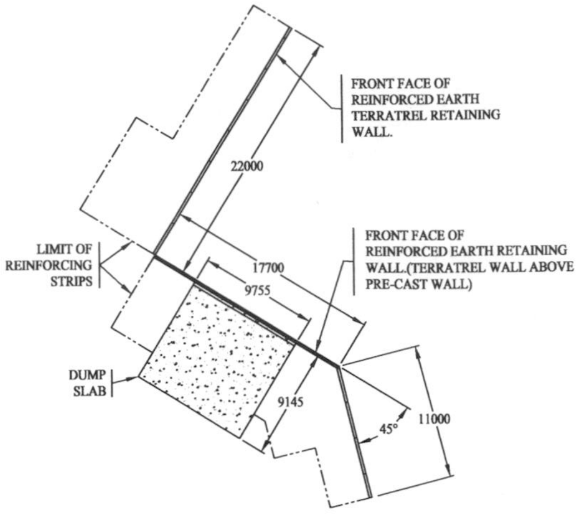

Figure 2. TDW and distribution slab in plan view.

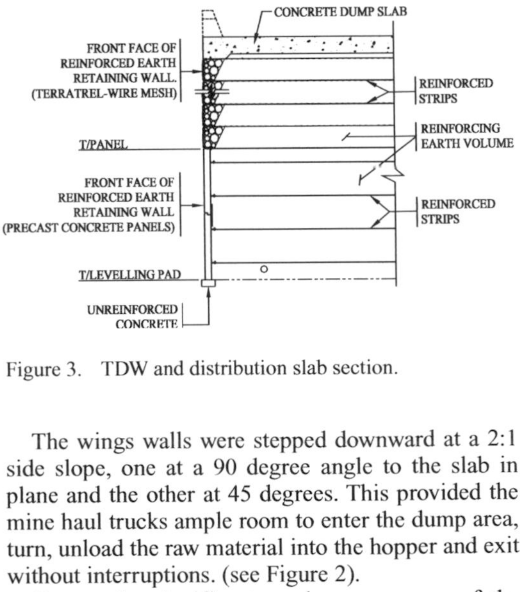

Figure 3. TDW and distribution slab section.

The wings walls were stepped downward at a 2:1 side slope, one at a 90 degree angle to the slab in plane and the other at 45 degrees. This provided the mine haul trucks ample room to enter the dump area, turn, unload the raw material into the hopper and exit without interruptions. (see Figure 2).

Due to the significant surcharge pressure of the Cat 777D mine haul truck, 100 kPa, the distribution slab was required to disperse the wheel, breaking and impact loads over a larger surface area. This slab also provided a wearable driving surface at the hopper dump area that could be maintained by GP. A curb-stop at the outer limit of the dump slab near the hoppers' upper edge was incorporated in the design. This item was quite important to the truck drivers for two reasons, 1) Safety, as to not let the truck fall over the edge of the wall into the hopper and 2) not dump the load too soon where excessive clean up would be required on the dump slab. (see Figure 3).

Once the TDW was constructed, the next mining structure required was the surge tunnel with

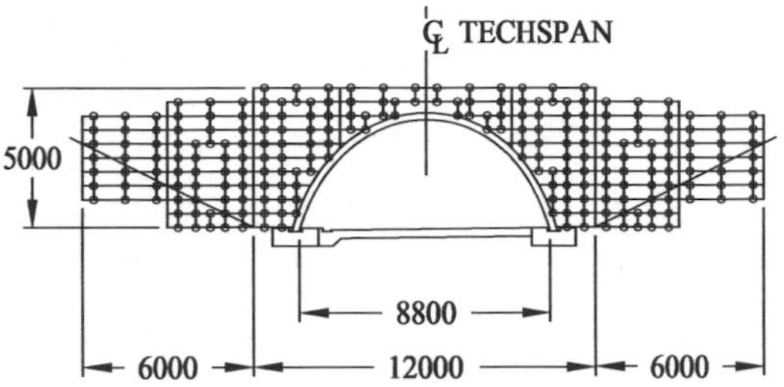

Figure 4. Load out entrance MSE wall elevation.

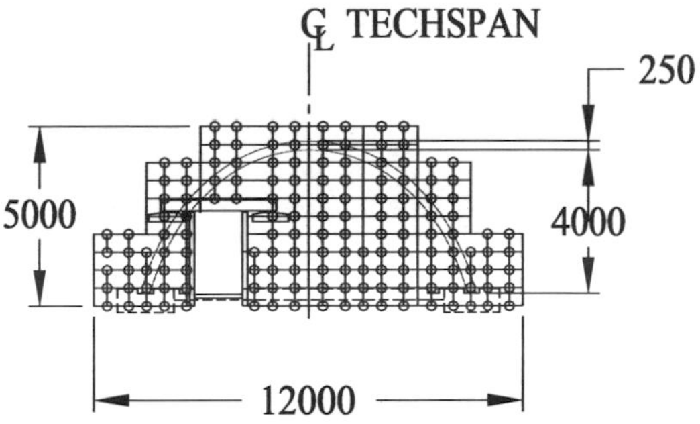

Figure 5. Capped end MSE wall elevation.

interior-exterior MSE walls at either end of the tunnel. TechSpan®, designed and supplied by Reinforced Earth Company Ltd., a three hinged segmental arch was chosen to be the surge tunnel structure (STS). At the load out end of the STS, a wire faced MSE wall was designed to withstand ever-changing top slope surcharge loading conditions. (see Figure 4).

On the capped end of the STS, the orientation of the MSE wire wall was reversed so that the face of the MSE wall was used to close off the STS and the soil reinforcement extended into the embankment parallel to the STS. A doorway was made in the end of the STS to provide access to the escape tunnel. (see Figure 5).

The STS was required to support the large processed gypsum surge pile above the crown of the arch while providing space for a working conveyor system to expel the live gypsum material to the load out building. A series of hoppers and feeders were suspended from the inside crown on the roof of the STS requiring several specially reinforced large openings in the STS design. (see Figure 6).

Finally, the escape tunnel (ET) was required in the design to provide a second safe passage out of the capped end of the STS. Two parallel pre-cast concrete facing MSE walls running longitudinal to the STS were designed. (see Figures 7 and 8).

Roof slabs supported by cast-in-place (CIP) footings were designed to support the surge pile and the foot-print of a large tower footing with bearing pressures in excess of 200 kPa for the conveyor system

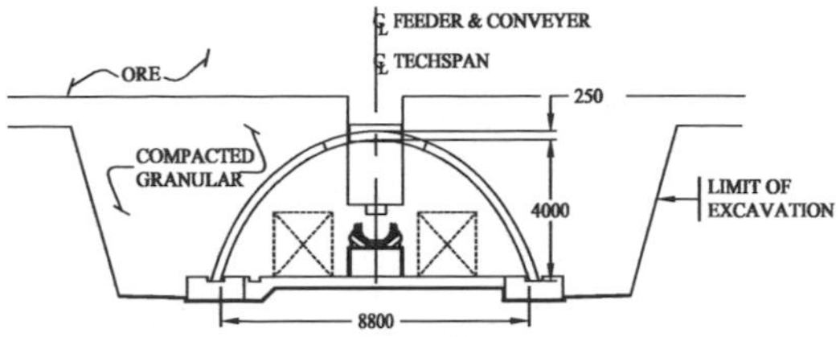

Figure 6. Hopper and feeder hanging from STS.

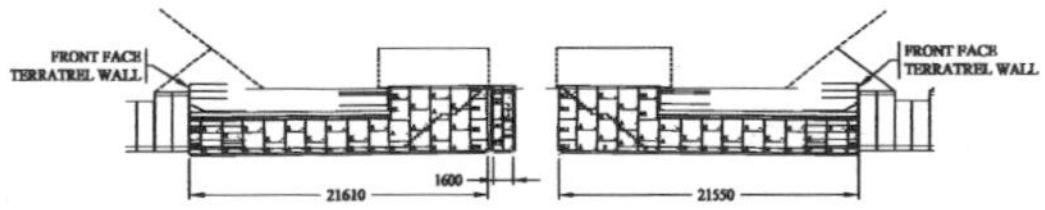

Figure 7. Escape tunnel MSE walls elevation.

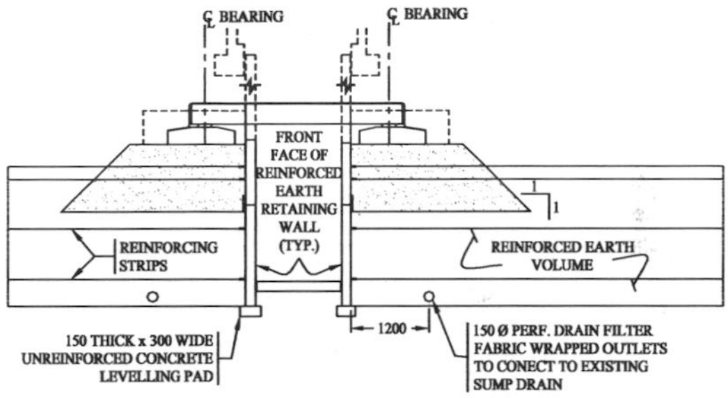

Figure 8. Escape tunnel – roof slabs.

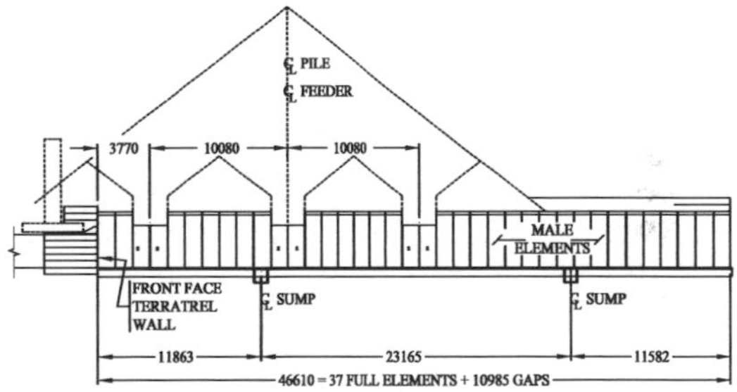

Figure 9. Surge pile and bent foot print.

above the surge pile. (see Figure 9). The CIP spread footings under the roof slabs were design for a bearing pressure of 300 kPa. The ET had a stair-well at its exit, accommodated with pre-cast MSE walls for its closure. The ET required MSE walls in all four planes and orientation to cap the end of the stair-well. (see Figures 7, 10, 11).

The ET was finished off with a concrete floor, stairs, lighting and entrance building. The concrete floor was sloped towards the STS to accommodate for drainage. (see Figure 12).

See annexed photos of truck dump, surge tunnel, and escape tunnel.

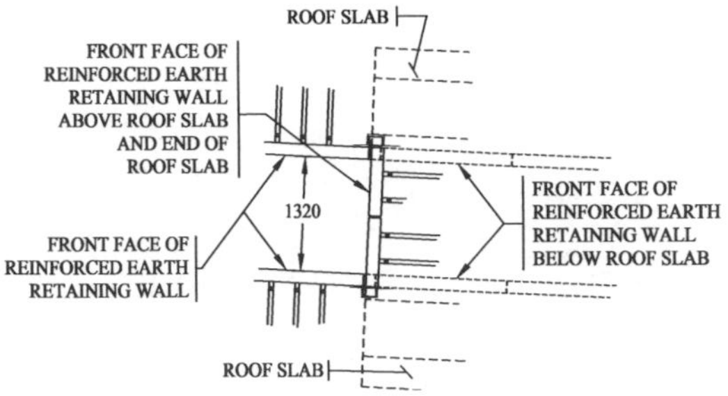

Figure 10. Escape tunnel stair-well far wall end.

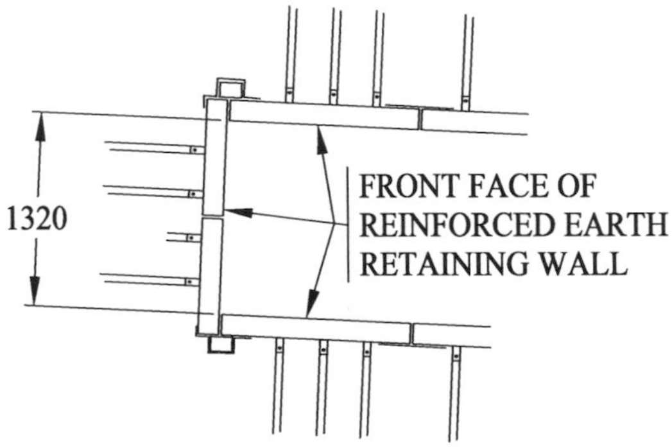

Figure 11. Escape tunnel stair-well near wall end.

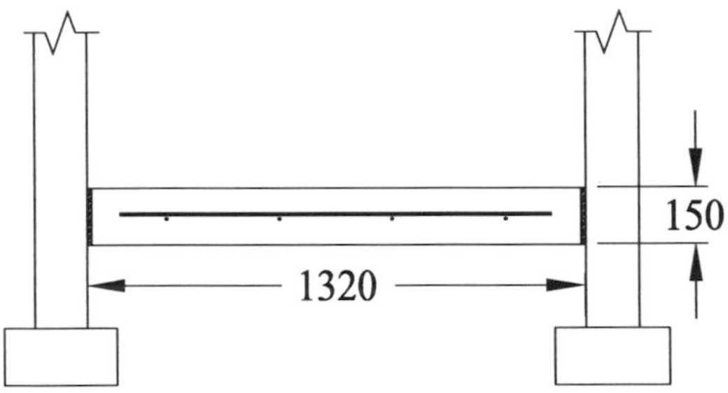

Figure 12. Escape tunnel sloping floor.

4 CONCLUSIONS

Georgia Pacific Mining Canada had a conceptual design of the mining structures required to open the new Melford Mine site. It pursued an experienced design-build contractor team to plan, layout and build the three mining structures, applying a variety of the team's MSE and arch technologies as the turnkey project developed.

This paper illustrates the many ways Reinforced Earth wall and TechSpan arch technologies can be adapted to the challenging local site conditions, geometry, and intended functions for industrial mining applications. The evolving structure designs and the inherent flexibility of the MSE system provided the RAJV to design-build the structures according to the requirements of Georgia Pacific Mining to the local site conditions as they were un-covered. The project was successfully completed on time and budget in 2002, and a second phase is currently being developed for construction in 2007.

ACKNOWLEDGMENT

Georgia Pacific Mining Canada
Alva Construction Limited
Jacques Whitford and Associates Limited
CBCL Limited
EARTH Tech Engineering Limited

REFERENCES

Wanschneider, H. and Wu, P. 1985. *Reinforced Earth Bridge Supporting Structures. Transportation Forum,* Ottawa:

Brockbank, W. J. and Segrestin, P. 1995. *Precast Arches as Innovative Alternate to Short Span Bridges. Fourth International Bridge Engineering Conference,* San Francisco:

Brockbank, W.J. Dunphy, R. and Yasinko, L.1994. *Innovative Underpass for Highway 63, International Road Federation Conference and Exposition Proceedings, Volume 9, Alberta:*

Annexed Photos
Georgia Pacific Mining

Man-way for Escape Tunnel

Escape Tunnel - Partially Complete

Surge Tunnel - Partially Constructed

Surge Tunnel Head walls

Truck Dump Wall Complete

Surge Tunnel Complete

New Horizons in Earth Reinforcement – Otani, Miyata & Mukunoki (eds)
© 2008 Taylor & Francis Group, London, ISBN 978-0-415-45775-0

Reinforced fill for temporary work in Hong Kong

G. Ng
Director, G and E Co. Ltd., Hong Kong.
Southeast Asia Area Manager, Tenax International B.V., Italy

ABSTRACT: This paper reviews the use of reinforced fill in three case studies (Belcher Garden at Pokfield Road, Po Tat Estate at Po Lam Road and Tseung Kwan O Subway South Station) as temporary structure with different applications and under different site conditions. The flexibility of using reinforced fill solution for difficult site conditions has appealed to engineers in temporary application in construction industry and perhaps leads to more innovative and ingenious usage of the technique.

1 INTRODUCTION

In the last decade, we have seen reinforced fill construction matured, design innovated, new aesthetic facings emerged and advance geogrid debut, so much so that a great many renowned projects have become to look like monuments. While the technique is well established, applications are largely adopted towards permanent works. Temporary works, which often do not require stringent submission, spectacular appearance and long term performance assurance, have received much less enthusiasm. It could have been that temporary features are put up and removed over a relatively short period rendering their existence unnoticeable. But this down to earth and practical solution to create convenience is made possible with reinforced fill construction. The merits and advantages are being reviewed in three projects, with different applications and under different site constrains.

2 BACKGROUND

Reinforced fill design and construction in Hong Kong is governed by Geoguide 6 – Guide to Reinforced Fill Structure and Slope Design prepared by the Geotechnical Engineering Office of the Hong Kong Government. The guide provides good practice in stability design and construction and a model specification which stipulates quality of material, standard of workmanship, testing method and acceptance criteria for reinforced fill construction. For temporary reinforced fill construction, some of administrative requirements have been streamlined, e.g. design submission and supervision report can be undertaken in the site office instead of going through the Central Government. Maintenance manual is also not necessary. For polymeric reinforcement elements, approval is governed by a material certification system and design and reinforced fill construction is therefore closely monitored. Prior to the inception of Geoguide 6 in 2002, Geospec 2 and GEO Report No. 34 were adopted as the design guideline.

The three case studies were designed and constructed based on these documents.

3 CASE STUDIES

3.1 *Belcher Garden, Pokfield Road*

A residential complex was developed over a hilly terrain and site formation was a challenge when bore pile crane was to be brought in an already heavily congested site. A temporary reinforced fill structure was proposed to extend and widen a working platform. The platform allowed equipment to be mobilized close to pile A5, A6, A7 and A8 which would otherwise be inaccessible from elsewhere. Refer to Figures 1 & 2.

A reinforced fill structure of 7 m × 12 m × 6.5 m height was put forward. This extended platform had two right angle facings and the facing inclination was 85°. Polymeric geogrid was designed for reinforcement and its length was 6.5 m. A wrapped back type of construction was chosen. The facing concrete blocks were placed with a set back between layers of reinforcement and were slightly tilted inwards. Refer to Figure 3.

Reinforced fill was thought of because:

- No other access is workable to mobilize the crane;
- There were abandon supply of fill material which, if not used, would have to find temporary storage space;

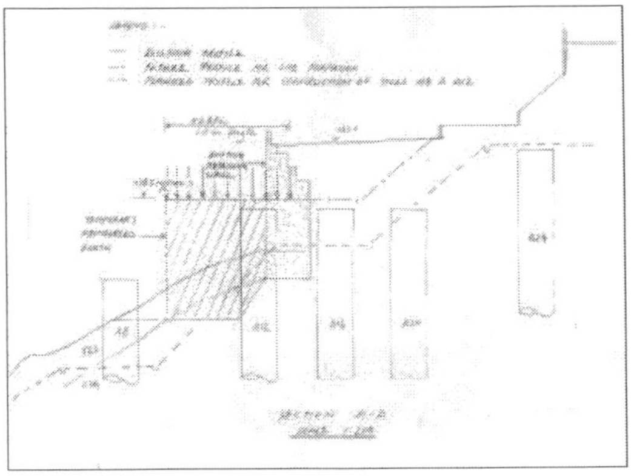

Figure 1. Cross section of temporary platform (Belcher Garden).

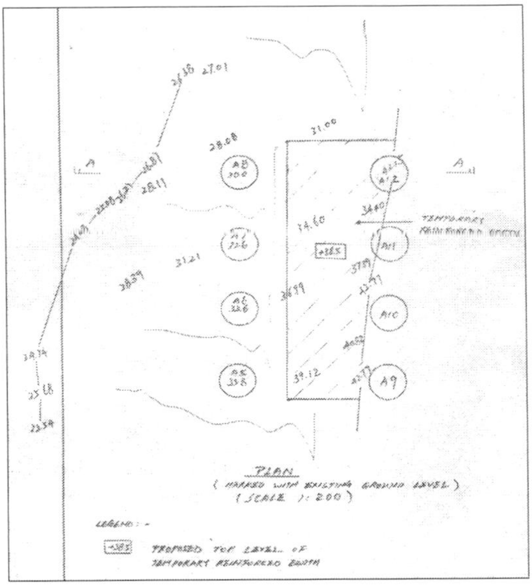

Figure 2. Plan of temporary platform (Belcher Garden).

- The availability of concrete block from completion of temporary surcharge, and;
- A similar platform made from steel structure or reinforced gabion structure would be more expensive and more time consuming to design, construct and dismantle.

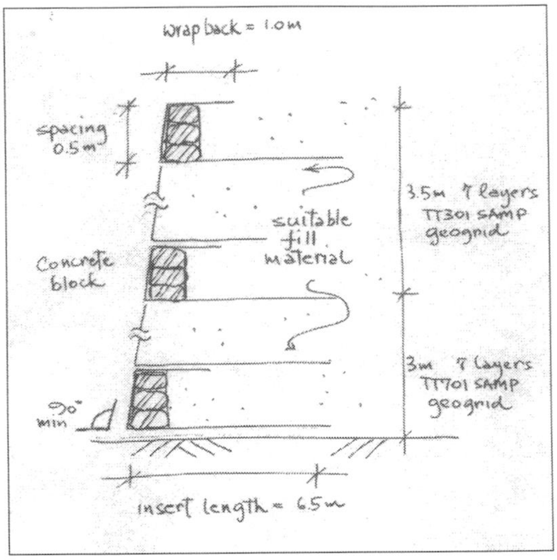

Figure 3. Reinforcement configuration (Belcher Garden).

Photo 1. Temporary work platform with access ramp behind (Belcher Garden).

Table 1. Project details of Belcher Garden.

Contractor	Sunley Engg. & Const. Co Ltd
Client	Sun Hung Kai Properties Ltd
Consultant	JMK Consultant Engineers
Geogrid	HDPE mono directional
Design/Approval period	6 months
Construction Period	8th to 17th October 1998
Period of Usage	About 6 months

3.2 *Po Tat Estate, Po Lam Road*

Public housing blocks were being developed on a terrace 15–20 m below the main road. The future site access was under construction and traffic logistic was insufficient to allow concurrent activities. A temporary haul road was necessary to increase construction traffic capacity before the permanent access road was put into operation.

The two lanes temporary haul road was a reinforced fill embankment built alongside and against the main road down side slope. It was about 80 m long and maximum height of 9 m. To occupy as little space as possible, the slope angle was designed

Photo 2. Construction of haul road (Po Tat Estate).

Photo 3. Estate road occupies early haul road (Po Tat Estate).

to 50° minimum balancing the demarcation limitation and the cost of steeper slope. A wrapped back geogrid method was adopted with 6.5 m wrapped back length. Sand bags were used as facing. The embankment was subsequently hydroseeded, an erosion protection measure.

Reinforced fill slope was taken because:

- Steel frame structure would have taken much too long to come into use. Reinforcement fill construction was quick;
- There was availability of fill material;
- Reinforced fill construction did not require additional heavy equipment thereby overloading already congested space;
- Reinforced fill construction was less expensive to implement and dismantle, and;
- A vegetation facing matched the contractor's environmental friendly motto.

3.3 *Tseung Kwan O Subway South Station*

The site was from reclaimed land on which the main access road of the future residential development lies.

Table 2. Project details of Po Tat Estate.

Contractor	Gammon Construction Limited
Client	Hong Kong Housing Authority
Consultant	Hsin Hieh Architects
Geogrid	HDPE mono directional
Design/Approval period	About 3 months
Construction Period	January to May 1999
Period of Usage	About 12 months

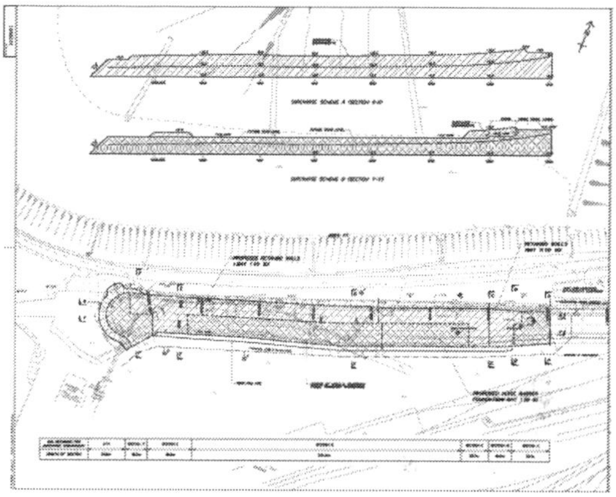

Figure 4. Layout of reinforced surcharge block (Tseung Kwan O).

A design with prefabricated vertical drain and temporary surcharge was sought to encourage ground consolidation. The surcharge required a 24,000 m^3 volume of fill. Reinforced fill construction was considered a good option to build this surcharge embankment.

The geometry of this rectangular embankment was 2.0 m to 8.5 m high, 20 m width and ran 150 m long with a facing angle of about 85 degree. Reinforcement was applied to all sides of the structure. Concrete blocks were placed at the base in certain locations for retaining surcharge fill, primarily because of its early availability to meet a tight program. In this area, no geogrid was applied (refers to Figure 4).

The reinforced fill embankment was constructed with wrap-around facing. Steel wire mesh and woven geotextile were used at the face to retain fill material. The vertical spacing of the primary geogrid was 500 mm and the reinforcement length was between 3.0 m to 7.5 m. A facing set back of 50 mm had been introduced to each layer of construction (refer to Figure 5).

Why reinforced fill was chosen:

- Ground consolidation was only required in a close defined proximity. Reinforced fill technique beings the rectangular surcharge block within this boundary. Typical fill construction would have to have toe line extended far beyond this boundary;

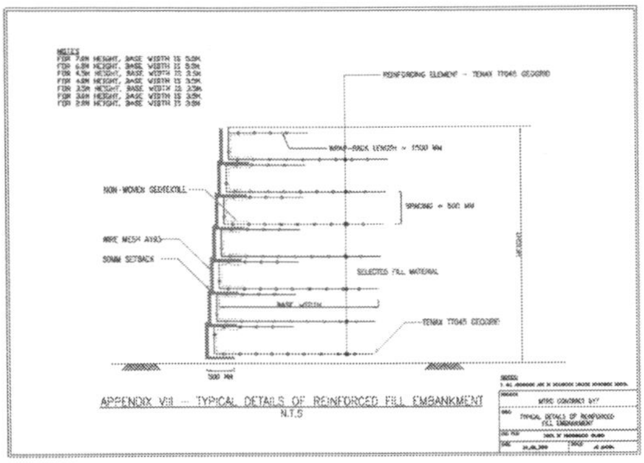

Figure 5. Cross section of reinforced surcharge block (Tseung Kwan O).

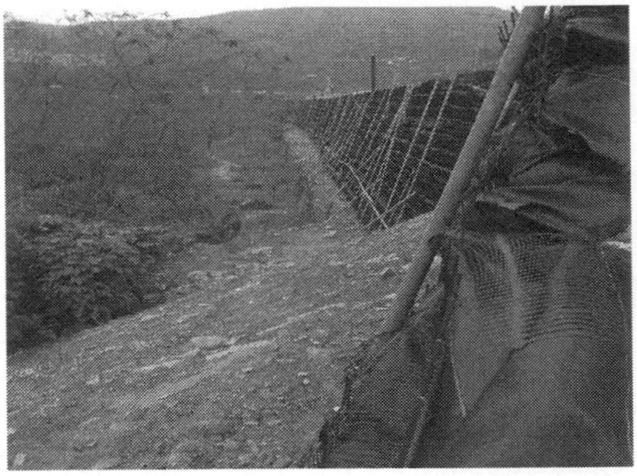

Photo 5. Vertical facing along site boundary (Tseung Kwan O).

Photo 4. End view of reinforced surcharge block (Tseung Kwan O).

Photo 6. Surcharge embankment at site boundary (Tseung Kwan O).

- Geogrid reinforcement was a simple and efficient method even to vertical facing;
- Unavailability of large quantity of concrete block (estimate 3,500 m^3), the handling, logistics and its disposal were unmanageable, and;
- Expensive option of gabion buttress, mechanical stabilized wall and reinforced concrete construction, in particular their disposal.

4 OVERVIEW

The three projects utilized reinforced fill as a "construction convenience" to provide temporary facilities, a working platform, a haul road and a surcharged containment. All took the advantages of construction simplicity (no particular skill and equipment), cost effectiveness (no expensive steel frame structure and concrete fabrication), overall time saving (as fast as any fill and compaction work) and less construction waste (many of the fill material were reusable for landscaping). Site conditions were also favorable such as space for laying reinforcement and the availability of fill material.

Permanent reinforced fill design methodologies are well established and design soft wares are abundant. The same is applicable for designing temporary works. However, the design, acceptance, testing and approval can be undertaken with less stringent submission procedures. Two examples are the exemption of pull out test and the necessity of full submission to the Authority. The Contractors are made responsible for their design and endorsed by an Independent Checking Engineer. Therefore design can be more aggressive, taking advantage to maximize the reinforcement spacing and width and optimize the full design strength

Photo 7. Surcharge embankment under construction (Tseung Kwan O).

Table 3. Project details of Tseung Kwan O Subway South Station.

Contractor	Maeda Corporation
Client	Mass Transit Railway Corp.
Consultant	Maunsell Consultants Asia Ltd
Geogrid	HDPE mono directional
Design/Approval period	About 2 months
Construction Period	November 2006 to March 2007
Period of Usage	About 9 months

without taking into the effects of temperature and creep, thereby minimize the cost of the geogrids, with the stability design and reinforcement configuration conforms fully the statutory practice.

One of the attractiveness with reinforced fill is the flexibility in change of design geometry during construction. An access ramp or a platform can be widened, steepened and turned, where structures are often too massive to change at ease. In Belcher Garden, the work platform can be extended larger simply by building it wider, using more fill material and geogrid. In Po Tat Estate, the haul road widening can be made possible with altering the crest alignment. And in Tseung Kwan O Station, the reinforced block can be built over layers of concrete blocks, catching a tight program. This type of temporary construction can be considered as a valuable tool in site management. It is, therefore, to the interest of contractors, to look into reinforced fill method whenever site constrains, access requirement or other complications are encountered.

Temporary application is viewed as a catalyst to bring forward the experience, to build up the confidence and to exercise the practicality of employing reinforced fill techniques. It is through more accustomed applications that engineering will excel.

REFERENCES

Geotechnical Control Office, Geospec 2 – Model Specification for Reinforced Fill Structure, Civil Engineering Services Department, Hong Kong, July 1989.

Geotechnical Engineering Office, GEO Report 34 – A Partial Factor Method for Reinforced Fill Slope Design, Civil Engineering Department, Hong Kong, July 1993.

Geotechnical Engineering Office, Geoguide 6 – *Guide to Reinforced Fill Structure and Slope Design*. Civil Engineering Department, Hong Kong, 236 p.

Geotechnical Engineering Office, Hong Kong, Reinforced Fill Product Certificate – RF1/97, 1997, RF3/98, 1998, RF1/00, 2000, RF3/01, 2001, RF3/03, 2003, RF1/05, 2005 and RF1/07, 2007.

Jones, C.J.F.P. 1996. *Earth Reinforcement and soil Structures*. Thomas Telford, 379 p.

Koerner, R.M. 1998. *Designing with Geosynthetics,* Prentice-Hall, Upper Saddle River, New Jersey 1998.

Li Lai Kuen. Review of Reinforced Fill Application in Hong Kong, The University of Hong Kong, 2003.

New Horizons in Earth Reinforcement – Otani, Miyata & Mukunoki (eds)
© 2008 Taylor & Francis Group, London, ISBN 978-0-415-45775-0

Design of roadside barrier systems for MSE retaining walls

P.L. Anderson
The Reinforced Earth Company, North Reading, MA., USA

R.A. Gladstone
Association for Metallically Stabilized Earth, McLean, VA., USA

K. Truong
The Reinforced Earth Company, Vienna, VA., USA

ABSTRACT: There is confusion among civil engineers in the United States regarding the applicable design method and the appropriate impact load for sizing the moment slab of roadside barrier systems atop MSE retaining walls. The design method and impact load discussed herein have been used successfully for more than 15 years to size the barrier moment slab and to determine the magnitude of loads applied to the supporting MSE wall. The source of confusion by civil engineers is explained and the current research to eliminate this confusion is described.

1 INTRODUCTION

Concrete safety barriers have been constructed on MSE walls in the United States since the early 1980s. Wall-mounted barriers were developed in France and crash tested by Service D'Études Techniques Des Routes et Autoroutes (SETRA) and Terre Armée Internationale (TAI) in 1982 (TAI, 1982). Hundreds of kilometers of both cast-in-place and precast barriers are in service and performing successfully throughout the United States and around the world.

Safety barriers and their supporting MSE walls are designed by a pseudo-static design method developed more than 20 years ago. The resulting moment slab dimensions (typically ±350 mm thick × 1250 mm wide by 6 m long minimum) are reasonable and barrier performance has been excellent, with no reports of failures despite numerous impacts by both passenger vehicles and trucks. Figure 1 shows a typical precast concrete barrier and moment slab designed by the pseudo-static design method. This or similar barrier designs have been constructed atop thousands of MSE retaining walls from 1985 to 2000 and performance has been excellent.

Since 1994, The American Association of State Highway and Transportation Officials (AASHTO) specifications for the design of Mechanically Stabilized Earth (MSE) walls have included the pseudo-static barrier design method (AASHTO, 1994).

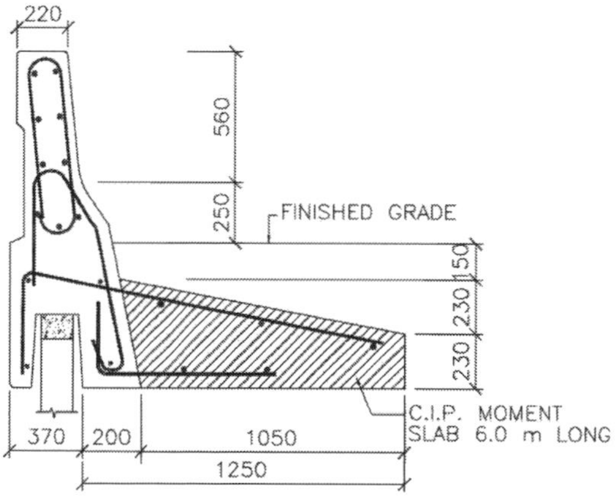

Figure 1. Typical barrier and moment slab 1985–2000.

Recently, AASHTO established new bridge railing and concrete barrier performance levels and higher dynamic impact loads, based on roadway type, speed, and percentage of truck traffic (AASHTO, 2002). Although the pseudo-static design method has not changed, engineers are attempting to design MSE barriers using the new dynamic loads. The result is unreasonable barrier designs having moment slabs with 2 to 3 times the mass required to withstand traditional pseudo-static design loading. Thus, the new AASHTO dynamic impact loads are significantly increasing barrier costs while providing no apparent benefit over the performance of long-proven designs.

2 DESIGN OF ROADSIDE BARRIER SYSTEMS

A roadside barrier system must be designed to contain and safely redirect a vehicle during an impact event. In addition, the barrier system must not transfer high impact forces to the precast concrete facing panels of the MSE wall below. Therefore, parapet shape, internal strength and overall mass stability must all be considered in barrier system design.

A variety of traffic barrier shapes with predictable deflection characteristics are in use throughout the United States. For overall mass stability against rotation and sliding, barrier systems atop MSE walls, whether cast-in-place or precast, are designed to resist the impact load by calculations using simple statics over a 6 m length of barrier and moment slab. Internal strength of the barrier is determined using appropriate reinforced concrete design procedures.

To preclude the transfer of high impact loads to the MSE wall panels below the barrier, a 20 mm gap is provided between the throat of the precast barrier and the back side of the facing panels. When casting a barrier in place, a 20 mm thick compressible foam material is placed on the back side of the facing panels prior to pouring the moment slab. Since there is no barrier-to-panel contact, due to the gap or the compressible material, the horizontal impact force is transferred to the reinforced soil by shear stresses that develop beneath the barrier slab. The influence depth of these shear forces is a function of the soil shear strength, the width of the barrier slab, and the stiffness of the reinforced soil structure. The stiffer the structure, the deeper the shear forces will distribute, thus reducing the concentration of these forces at the top of wall.

Due to the instantaneous nature of the impact loading, the apparent coefficient of friction between the soil and the reinforcements becomes virtually infinite as the load is applied. As seen from full scale crash testing, pullout of the reinforcements does not have time to occur before the impact loading ends. Therefore, only tensile stress in the MSE soil reinforcements needs to be checked, and pullout during impact may safely be ignored. Considering the minimum reinforcement density used in Reinforced Earth wall design (4 strips across a 3 m width of wall), the allowable tensile resistance of the top row is more than adequate to resist the AASHTO-specified pseudo-static 45 KN impact load. The excellent performance of hundreds of kilometers of both precast and cast-in-place traffic barrier atop Reinforced Earth structures having minimum reinforcing strip density (4), and in many cases minimum

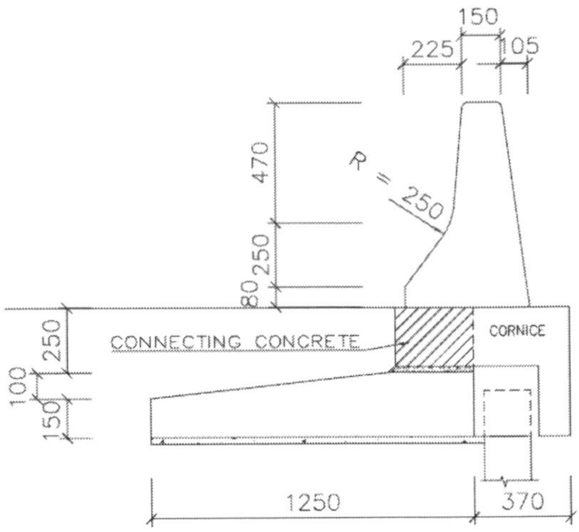

Figure 2. Crash tested barrier (TAI, 1982).

length (2.4 m), is testament to the appropriateness of this design method and pseudo static impact load.

3 FIELD TEST OF A ROADSIDE BARRIER ATOP A REINFORCED EARTH WALL

In 1982 SETRA and TAI jointly conducted crash tests on a roadside barrier system atop a Reinforced Earth wall.

The wall and barrier were constructed on the test site of the Organisme National de la Sécurité Routière in Bron, France. The tested barrier (Figure 2) was a so-called Jersey shape, 800 mm high from the roadway to the top of barrier, 150 mm thick at the top, and 480 mm thick at the roadway surface. Six 1500 mm long precast coping units (labeled "cornice" on Figure 2), connected by three 1250 mm wide junction slabs totaling 9000 mm in length, formed the base of the cast-in-place test barrier.

There was almost no concrete reinforcement in the Jersey barrier shape, with only 2 longitudinal 12 mm bars (Figure 3). The tension members connecting the 9000 mm long cast-in-place barrier sections to the junction slab would be considered extremely light by today's standards, consisting of two 12 mm longitudinal bars and 8 mm stirrups at 250 mm on center.

The SETRA/TAI crash test vehicle was a Berliet PHN 8 bus. It weighed 12 metric tonnes and impacted the barrier at a speed of 71.2 km/hr and a 20° angle. During the event there were two distinct impacts, the first from the front of the bus and the second as the rear of the bus slid into the barrier. Sensors on the front and rear axles recorded the deceleration due to impact.

Damage to the precast barrier system was limited to the parapet itself. A 2200 mm-long V-shape area was

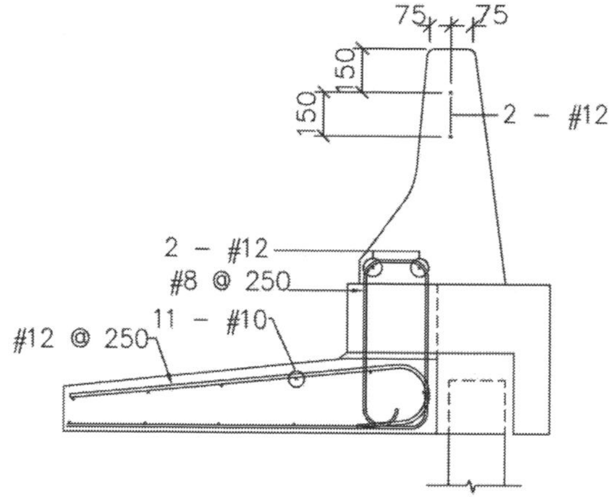

Figure 3. Concrete reinforcement in tested barrier.

ruptured, with the depth of rupture being 500 mm at the center of the V. Fragments of concrete from the rupture were contained on the roadway side of the barrier.

Dynamic displacement of the Reinforced Earth wall was limited to 4.9 mm during the event, with permanent deformation of 1.5 mm after rebound. There was no loss of adherence, and there was no failure of any of the 5 m long reinforcing strips in the top level, despite use of minimum strip density (four 40 × 5 mm strips per 3 m horizontally). The maximum force recorded on the most highly stressed reinforcing strip was 29 KN, less than the reinforcing strip long term allowable tension.

The SETRA/TAI crash test was instrumental in developing an understanding of the required dimensions of the roadside barrier system, including the width and length of the moment slab, and in development of a pseudo-static design method and appropriate impact load for roadside barriers mounted atop Reinforced Earth walls.

4 PSEUDO-STATIC DESIGN METHOD

The instantaneous nature and magnitude of the applied load cannot be modeled by static computations. Therefore, it is recommended to use the pseudo-static design method given in the 1994 AASHTO Interims. Using this method, the traffic barrier and junction slab system are designed for a (pseudo-static) 45 KN impact load applied at the top of barrier and distributed over a 6 m continuous junction slab length. The junction slab is joined to adjacent sections with either shear dowels or continuous reinforcement through the construction joints. Concrete design is by a strength design method, while overall stability of the barrier system is checked by calculations using simple statics.

The resulting barrier is proportioned and reinforced conservatively compared to the barrier that was crash tested by SETRA and TAI.

The minimum factors of safety for barrier/slab sliding and overturning should be 1.5 and 2.0, respectively, when using the pseudo-static 45 KN impact load applied to the top of the barrier. The full soil reinforcement length is considered effective in resisting pullout during the impact event. Since the impact load is distributed over a 6 m junction slab length, the full 6 m length of junction slab would need to move out as a unit for the barrier to move at all.

To check reinforcement tension, the 45 KN impact load is distributed over a 1.5 m length of wall. With the minimum reinforcing strip density, 4 strips per 3 m horizontally, the sum of the impact load plus the tensile load from soil retention results in a calculated total tensile load of 29 KN per strip. This total load must be less than the long-term allowable load for a reinforcing strip. Measurements of reinforcing strip tension during the TAI/SETRA crash tests were in excellent agreement with the pseudo-static design calculations and the top layer of reinforcing strips was loaded within allowable limits during the crash event. The calculated and measured 29 KN load is less than the 32 KN long term allowable tension for standard 50 × 4 mm reinforcing strips used in United States design practice.

5 DYNAMIC LOADS FOR YIELD LINE ANALYSIS OF RAILINGS

AASHTO recently established new bridge railing and concrete barrier performance levels, with associated dynamic impact loads, based on roadway type, speed, and percentage of truck traffic. The dynamic loads are presented in Table 1. These dynamic loads are for use in yield line analysis of metal bridge railings and for strength design of reinforced concrete parapets, but engineers have attempted to use them (notably the TL-4 loading condition) for dimensioning the moment slabs of barriers atop MSE walls. Considering the resulting confusion and unrealistic designs, it is instructive to compare the SETRA/TAI crash test to the TL-4 requirements.

AASHTO Test Level 4 is considered "...generally acceptable for the majority of applications on high-speed highways, freeways, expressways, and interstate highways with a mixture of trucks and heavy vehicles" (AASHTO, 2002). AASHTO defines a typical TL-4 test vehicle as a single unit van truck weighing 8.2 tonnes, traveling at 80 kph and impacting the barrier at 15°.

From Table 1, the expected transverse impact load is 240 KN. The SETRA/TAI crash test vehicle significantly exceeded those requirements, however. Multiplying the filtered rear axle deceleration (11.4 g) by one-half the weight of the vehicle (6 tonnes)

Table 1. AASHTO Table A13.2-1 Design Forces for Traffic Railings (AASHTO, 2002).

Parameter	Railing test levels					
Designations/ Design forces	TL-1	TL-2	TL-3	TL-4	TL-5	TL-6
F_t Transverse (KN)	60	120	240	240	550	780
F_l Longitudinal (KN)	20	40	80	80	183	260
F_v Vertical Down (KN)	20	20	20	80	355	355
L_t and L_L (mm)	1220	1220	1220	1070	2440	2440
L_v (mm)	5500	5500	5500	5500	12200	12200
H_e (min) (mm)	460	510	610	810	1070	1420
Rail Height (min) (mm)	685	685	685	810	1070	2290

the calculated dynamic force from the back of the bus impacting the barrier was 680 KN. This dynamic force was 2.83 times the recommended TL-4 value and even exceeded the TL-5 value by 23%. Yet the corresponding peak tensile force in the most highly stressed reinforcing strip indicated that the impact load reaching the soil reinforcements was only 45 KN over 1.5 m of wall, exactly as assumed in the pseudo-static design method. The 1250 mm wide moment slab and unreinforced parapet of the TAI-tested barrier proved adequate for the impact condition.

Civil engineers are attempting to use the dynamic loads specified in Table 1 in the pseudo-static design method. These loads were not intended for use in the pseudo-static design method for barriers atop MSE walls, however, and they were not added to the MSE section of the specifications. Indeed, the MSE specification is unchanged and continues to specify the pseudo-static impact load for barrier design. Since the pseudo-static design method has not been changed, MSE-mounted traffic barriers designed using TL-4 dynamic loads have unreasonable dimensions, such as moment slabs with 2 to 3 times the mass required by designs using the 45KN load. Figure 4 shows such a barrier; note the 2440 mm wide moment slab, fully 2.3 times the width of the in-service barrier in Figure 1 that was designed using the pseudo-static method and a 45 KN impact load.

The new AASHTO dynamic impact loads are significantly increasing barrier costs while providing no apparent benefit over the performance of proven designs.

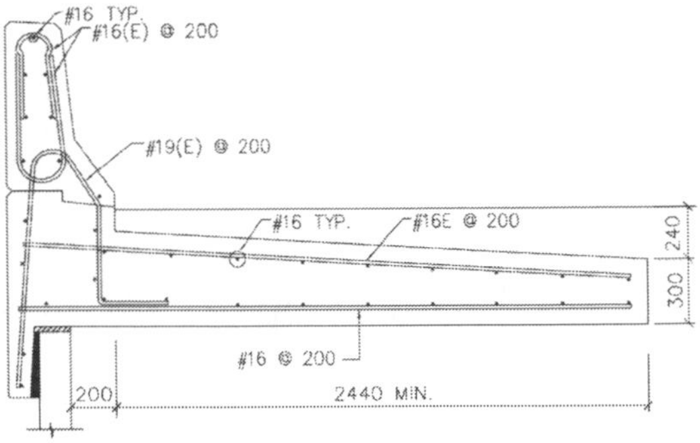

Figure 4. Moment slab sized using TL-4 dynamic loads.

6 RESEARCH CURRENTLY UNDER WAY

National Cooperative Highway Research Program (NCHRP) 22–20, *Design of Roadside Barrier Systems Placed on MSE Retaining Walls*, was begun in July 2004 to develop standardized procedures for economical design of roadside safety barrier systems placed on MSE retaining walls (NCHRP, 2004). Computer modeling and full scale crash testing are being used to develop these standardized design procedures. The results of this study, to be completed in 2008, should return barrier design to a more economical level, similar to that used successfully in the United States from 1985 to 2000.

REFERENCES

Terre Armée Internationale (TAI), Field Test of a GBA Safety Barrier erected on a Reinforced Earth wall, TAI Report No. R22, May 1982, not published.

American Association of State Highway and Transportation Officials (AASHTO), Standard Specifications for Highway Bridges; 15th Edition 1992, including 1994 Interim Specifications, Division I – Design, Section 5 Retaining Walls, Paragraph 5.8.9 Special Loading Conditions.

American Association of State Highway and Transportation Officials (AASHTO), LRFD Bridge Design Specifications; SI Units, Second Edition 1998, including 2002 Interim Specifications, Section 13 Railings.

National Cooperative Highway Research Program (NCHRP) Design of Roadside Barrier Systems placed on MSE Retaining Walls, July 2004, *www.trb.org/trbnet/project display.asp?projectid=693*

New Horizons in Earth Reinforcement – Otani, Miyata & Mukunoki (eds)
© 2008 Taylor & Francis Group, London, *ISBN 978-0-415-45775-0*

Design considerations of earth reinforced structures using inextensible reinforcements in heavy load surcharge support capacity

P. Wu & W.J. Brockbank
Reinforced Earth Company Ltd., Mississauga, Ontario, Canada

ABSTRACT: This paper presents the Reinforced Earth® design considerations carried out in using In-Extensible reinforcement in the application of Mechanical Stabilized Earth (MSE) structures in the support of heavily loaded surcharged and high retaining structures. Design methods and considerations, as well as critical performance criteria are demonstrated through actual project case histories in MSE applications in:

1. Mining crusher and dump structures supporting huge trucks in the oil sands operations in Alberta, Canada,
2. MSE Abutment walls loaded with footing pressures of 550 kPa, and
3. Heavily loaded rail structures Cooper E90 load in British Columbia and Ontario, Canada

The load-supporting capacity of in-extensible reinforcement MSE structures is illustrated and discussed against horizontal deflection and movement design criteria, settlement consideration, and post-construction performance and safety requirements.

1 INTRODUCTION

The main difference between MSE walls supporting heavy surcharges and conventional MSE walls is the high stress level to which the walls are exposed. Under this high stress condition there is potential for large deformations to occur unless the components of the wall are sufficiently stiff to resist the high loads. Since high deformations are generally not acceptable in these structures it is necessary to use high modulus materials to control the strains. Vertical consolidation of the fill is generally controlled by the selection of high modulus backfill, consisting of well graded granular material compacted to a high relative density. Horizontal deformations are controlled with the selection of high modulus or inextensible steel soil reinforcement.

Also of great importance in highly loaded MSE walls is the issue of strain compatibility. A clear example of this can be seen in comparing the compression of the MSE backfill with the compression of the facing. In order to not overstress the soil reinforcement's connection to the facing it is necessary to have a compressible facing. This is accomplished in two different ways. The first way is to use a flexible wire mesh facing, which can compress and bend as the backfill behind it consolidates. The second way is to introduce a compressible component into the facing. This is done in the case of precast faced structures by introducing compressible pads in the horizontal joints. By allowing these pads to compress as illustrated in Figure 1, the facing in effect, will consolidate at the same rate as the backfill behind.

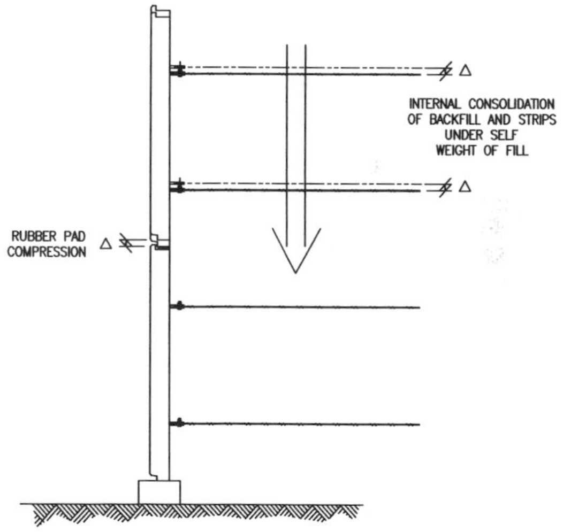

Figure 1. Internal settlement accommodated with compressible pads between precast panels.

2 HISTORY

An early example of an MSE wall supporting a high load is the industrial wall at Dunkirk constructed as early as 1970, where a Reinforced Earth wall supports a Gantry Crane with a wheel load of 1,200,000 kg one to three meters from the face.

Following this in 1988, a test wall was constructed in France by the Reinforced Earth international group to confirm the design theory and failure mode of an MSE load supporting wall. The wall was lightly reinforced with the intention of loading the wall to failure. MSE abutment walls are generally designed to support a surcharge footing pressure on top of the wall of 200 kPa. In the case of this test wall, the load was increased to 800 kPA before conclusion, or four times the normal design pressure. No failure was achieved. The test proved that MSE structures are capable of supporting loads higher than previously imagined. (Reference Bastick, M., et al, 1990)

3 ECONOMICAL LOAD SUPPORT

Reinforced Earth walls have been used for many years to support high loads. The reason is these walls can support high loads very efficiently and therefore economically, due to their basic nature. The vertical surcharge loads in an MSE wall are taken entirely by the soil underneath the load, and since soil is relatively inexpensive, compared to steel or concrete, MSE load supporting structures are economical. In the case of an MSE wall, loads are supported by the soil itself, and additional loads do not require any additional vertical structural elements. In contrast to this in the case of a pile supported structure, where the higher the load, more or bigger piles are required to support this load. This results in a direct increase in structure cost with additional vertical load. It is true that with higher loads more horizontal soil reinforcement is required; however, since the horizontal stress in a structure increases at a rate of about 25%, to the rate of the increase of the vertical stress, this means that the cost of structural elements in an MSE wall increases only at this similar rate.

4 DEFORMATIONS UNDER HEAVY LOADS

4.1 *Vertical deformations*

4.1.1 *Foundation settlement*

Compression of the foundation soils under an MSE wall is estimated using well known geotechnical principles by the projects geotechnical consultants and is not described in this paper.

It is noted that external foundation settlement estimate analysis should consider the entire mechanical stabilized embankment, including its bearing eccentricity and not just the facing.

4.1.2 *Internal settlement*

Vertical internal consolidation of the fill in an MSE wall depends on three aspects. First is the vertical stress, second is the volume of backfill influenced by the stress, and third is the property of the fill. It is important to make the point that vertical consolidation is not a function of the density of horizontal reinforcement.

In the case of non-cohesive fills the verticalconsolidation occurs as the load is placed on the structure, and the compression can be controlled with the selection of well graded, easier compactable sands and gravels. In the case of cohesive fills, the consolidation will occur over a period of time and the consolidation of cohesive fills tend to lead to much higher settlement values. The volume and the behavior of backfill influenced depend on the magnitude of pressure, the overall size of the footing being supported, and on the height of the MSE wall.

Although well-compacted sand and gravel is usually much preferred for highly loaded structures, the authors have been involved in the successful construction of highly loaded MSE walls with fine backfills. In the use of these fine backfills a flexible bar meshed facing has been used. By monitoring the walls, constructed with fine lean oil sand fill, it has been found that the internal consolidation of the fill is at least 3%. To accommodate this high level of compression, the bar mesh facing bends and bulges out in the horizontal direction. The bending capacity, durability, the mesh opening dimensions and the structural integrity of these bar mesh facing need careful evaluation in the design. Thin wires are generally not capable of sustaining these demands.

The percentage of fine material in the lean oil sand fill in this case ranges from approximately 40% to 60%. In these types of structures, not only is it necessary to have a strain compatible facing, but it is also necessary to monitor the excess pore pressures that occur in the backfill as the wall height and vertical stress increases. It should be pointed out that extreme caution must be exercised in controlling moisture content and compaction when using fine fill in MSE walls.

Numerous walls referred to as truck dumps have been designed with Reinforced Earth and consist of vertical walls in excess of 20 m in height, where large mining trucks back up to the edge and dump their load into an adjacent hoppers. (see Figures 2 and 3).

For MSE walls with precast facing exposed to high loads, it is necessary to have compressible pads in the horizontal joints between modular precast panels. The smaller the panel height, the more total number of horizontal joints is available to accommodate settlement. Since the precast panels themselves, particularly in rigid big size units, can obviously not compress, it is essential to have highly compressible pads in the

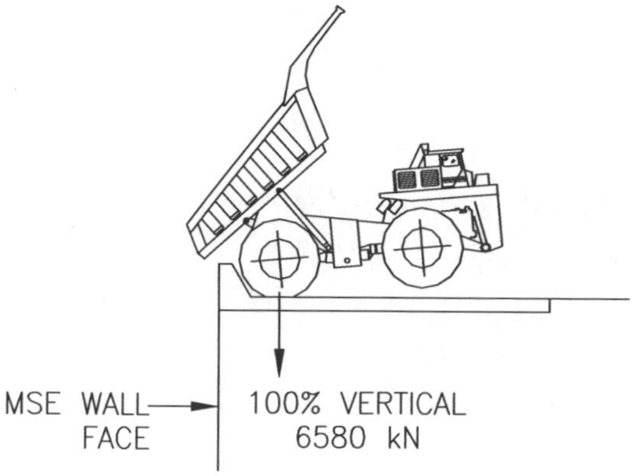

Figure 2. While dumping its ore, the heavy hauler trucks exert a high load near the face of the MSE dump wall.

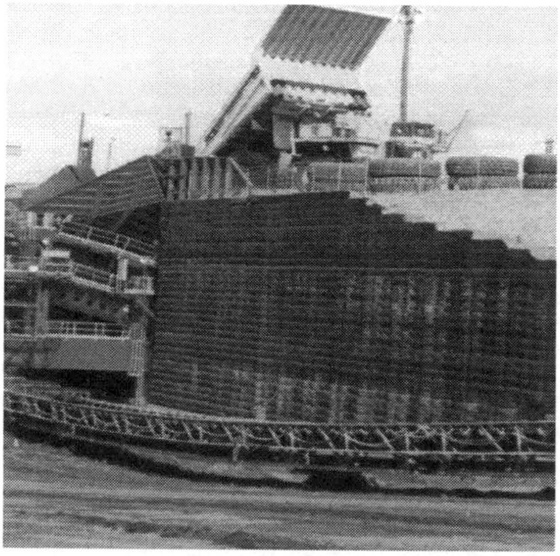

Figure 3. Heavy hauler mining truck dumps into hopper/crusher while supported by bar mesh temporary MSE wall.

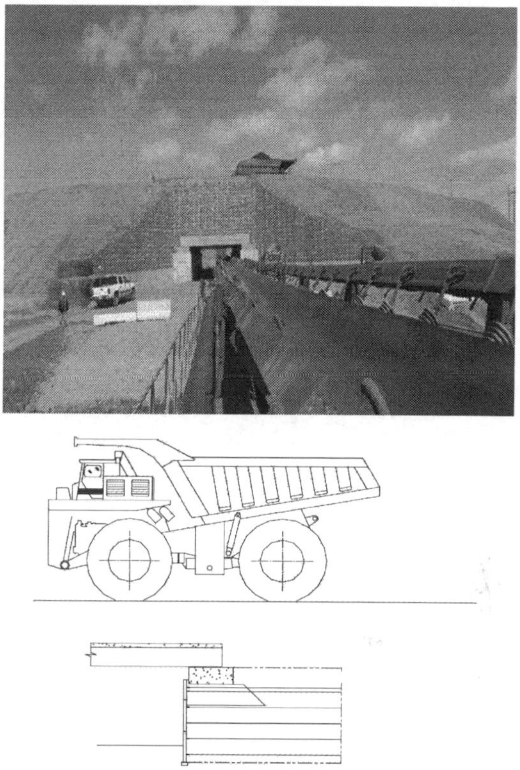

Figure 4. A 700 ton load of truck and ore crosses the bridge and MSE abutment wall over an uninterrupted conveyor. Figure is drawn to scale.

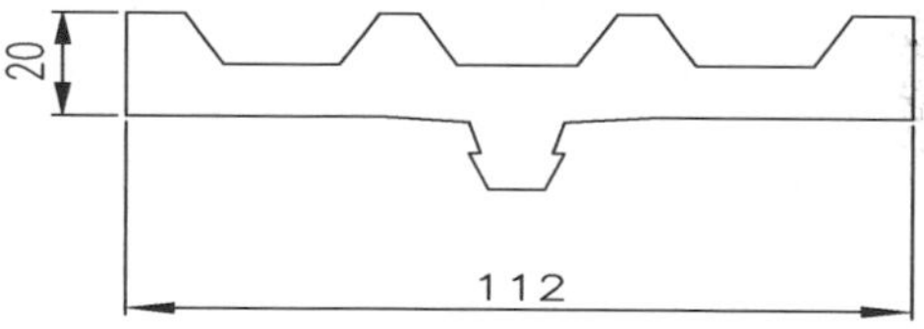

Figure 5. Geometry of a standard compressible joint pad.

horizontal joints so that the overall vertical consolidation of the facing can match that of the backfill behind the facing (strain compatibility). If this is not done and the backfill can settle to a greater degree than the facing, overstressing of the soil reinforcement connection can occur. Particularly when very compressible backfill is used, it will consolidate to such a large extent that the connections to the precast panels shear off and will allow the facing panels to separate from the MSE wall.

Non-standard compressible pads should be carefully designed and tested to accommodate higher than normal backfill compressions, usually accomplished by making the rubber pads thicker.

It has been confirmed that the compressible pads in the horizontal joints compress more under higher loads. This was shown in a survey that the Reinforced Earth Company in Canada performed in March 2007. The compression of rubber pads was measured at 35 different locations under various loading conditions simplified here as low, medium and high vertical load. The corresponding compression of the rubber ribs averaged out to be respectively 30%, 65% and 80%. The 80% rib compression was observed in a true MSE abutment wall where the dead and live loads exerted a 550 kPa footing pressure. (see Figure 4).

The geometry of the standard pads is shown in Figure 5. The top section of the bearing pads are in

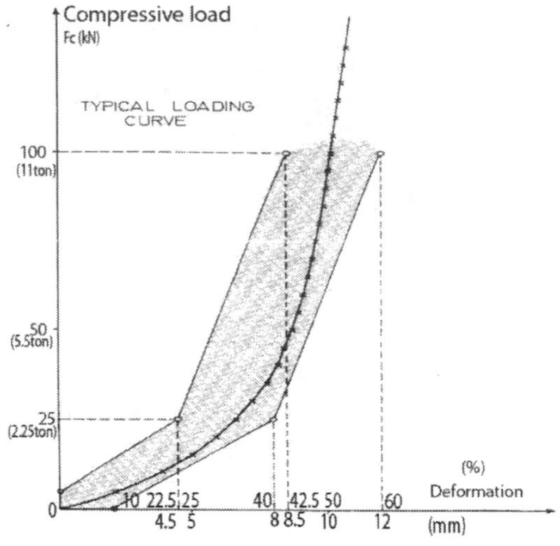

Figure 6. Laboratory results plotted as load versus deformation for compressible joint pads.

the form of 4 nibs to allow easier initial compression. A main solid portion of the rubber pad compresses less easily and prevents the pads from completely squashed, ensuring the panels will not contact each other. Figure 6 shows the compression characteristics of the rubber pads as tested in the laboratory, showing stiffness increasing as a function of increased deformation.

Another way in which internal consolidation can be accommodated for high walls is by breaking the wall into several tiers with small setback distances. This allows the facing to accommodate additional consolidation over and above that which is available through the compressible pads.

4.2 *Horizontal deformations*

Horizontal deformation of an MSE wall can be broken into two categories.

4.2.1 *External movement*

External movement is caused by the deformation of the foundation soils. Since in the case of highly loaded MSE walls the foundation soils are required to have good bearing capabilities, the horizontal deformation of the soils are generally not that large. This aspect can be ascertained by conventional geotechnical design and is also not described in this paper.

4.2.2 *Internal movement*

The internal deformation of an MSE wall is very much a concern of the MSE wall designers and occurs as a result of two different modes. The first is that which is caused by the slip of the soil reinforcement, or the amount of horizontal movement the soil reinforcement

Figure 7. MSE wall supports heavy rail load in Vancouver, Canada.

undergoes to mobilize the required fictional resistance. In the case of the types of the structures which this paper is addressing, the soil reinforcement is generally relatively long. When soil reinforcement is long, frictional pull-out and frictional capacity is generally not an issue. What is of much more interest in highly loaded walls is the elastic or plastic elongation of the soil reinforcement under high load. For walls designed for high security, high modulus steel soil reinforcement is selected. Since the design of steel in tension is done only in the elastic range there is no plastic deformation to consider and the elongation is very low. In fact, for a 20 m high wall assuming about 5 m of steel strip reaches approximately 50% of yield only 5 mm of elongation occurs. This predictable and controlled deformation is paramount to highly loaded structures since very strict tolerance is usually required.

5 SECURITY OF HIGHLY LOADED MSE WALLS

The consequences of failure of a load supporting MSE wall are much greater than a conventional MSE wall, especially for the examples given in this paper, which include walls supporting bridges, walls supporting heavy rail (see Figures 7 and 8) and walls supporting heavy mining trucks. In addition to the potential loss of life, there are in each of these cases potential for millions of dollars worth of damage in equipments and through the loss of income caused by the facility being out of service. In the case of a truck dump wall, the cost of the trucks, and of the hopper, which is situated approximately 300 mm from the face of the wall, is prohibitively high to repair or replace. The loss of income that would result in stoppage or delay of facility operation would be unthinkable. It is obvious then that the owners of these structures must have extreme confidence in the selection of the retaining wall structures, and their designers.

Figure 8. Heavy rail load on MSE wall in Toronto, Canada (Design load of Cooper E90).

Safety of a highly loaded MSE wall depends on the soil reinforcement's resistance against rupture or pullout. For highly loaded MSE walls the most critical internal safety issue is that of the tensile capacity. For the design of these walls, the material of choice has been structural steel. The safety against the rupture or breaking of soil reinforcement is ensured in several ways. The durability of the strip is addressed by disregarding the thickness of the steel which is anticipated to corrode over the life of the structure, which ensures that all of the required factors of safety are met even at the end of the structures design life. Secondly, the uncertainty in loads and material properties are accounted for with the use of limit state load factors and capacity reduction factors. In addition to these two standard design approaches, there is additional safety provided with the use of structural steel. The strength of the structural steel is governed by strict specifications which ensure that the minimum yield strength is guaranteed. This provides a level of security over some other materials which are designed using average tensile test values.

Recently the Finite Difference program FLAC has been used to further the understanding of the direction and magnitude of stresses and strains internal to an MSE wall under high loads. (see Figures 9 and 10). Figure 9 shows the predicted vertical settlements of an MSE wall under 100 kPa train loading. Maximum predicted settlement was 35 mm. Figure 10 shows that the direction of displacements is different between the reinforced and unreinforced zones verifying that the MSE wall behaves as a composite material.

6 CONCLUSION

With the careful selection of material properties, including high modulus fills, high modulus soil reinforcement and detailing of facing panels to account for consolidation, MSE walls can be successfully designed and constructed to support extremely high surcharge and pressures.

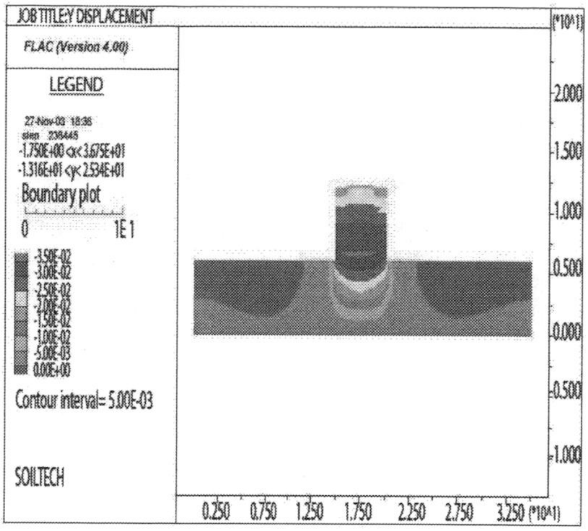

Figure 9. FLAC analysis predicted a maximum settlement of 35 mm under Cooper E90 rail load for the Vancouver MSE wall.

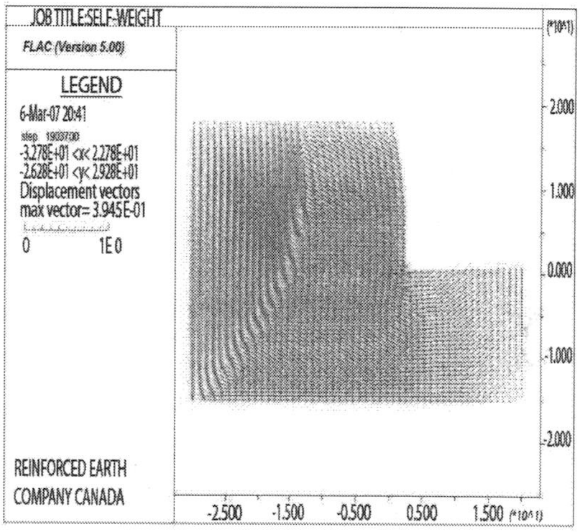

Figure 10. FLAC displacement vectors for a 20 m high truck dump MSE structure.

ACKNOWLEDGMENT

The authors acknowledge the many owners, structural and geotechnical engineers who have put their confidence over several decades in the Reinforced Earth designs of MSE walls to support very high loads.

REFERENCES

Bastick, M., Schlosser, F., Segrestin, P., Amar, S., Canepa, Y. 1990: *Experimental Reinforced Earth Structure of Burron Marlotte: Slender Wall and Abutment Test.* Report TAI.

Brockbank, W.J., Mimura, W. & Scherger, B. 2005. *Case Study: Highly Loaded MSE Bridge Supporting Structure, Syncrude NMAPS Conveyor Overpass, Proceeding, 58th CGS Conference*, Saskatoon:

Brockbank, W.J., Essery, D. & Wu, P. 2003. *Design of Reinforced Earth Wall for Contingent Loading and Longevity, 56th CGS Conference*, Winnipeg:

Brockbank, W.J., Ropret, M. & Wu, P. 2002. *Advanced loading aspects of Reinforced Earth Mechanically Stabilized Earth (MSE) Design. 55th CGS Conference*, Niagara Falls:

Brockbank, W.J., Kalynchuk, G. & Scherger, B. 1992. *Use of Oil Sand for Construction of a Reinforced Earth Dump Wall, Proceeding, 45th Canadian Geotechnical Society Conference*, Toronto:

Brockbank, W.J. & Weinreb, D. 1986. *Construction of Reinforced Earth Abutment Walls at the Harvey Creek Crossing, Proceeding, Second International Ottawa 1986 Conference on Short and Medium Span Bridges, CSCE*, Ottawa:

Segrestin, P. 1987. *Bearing pads in EPDM, Internal report by Terre Armee Internationale.*

Vidal, H. 1969. *The Principle of Reinforced Earth, Highway Research Record, No. 282.*

Wandschneider, H. & Wu, P. 1985, *Reinforced Earth Bridge Supporting Structures, Canadian Transportation Research Forum*, Ottawa:

New Horizons in Earth Reinforcement – Otani, Miyata & Mukunoki (eds)
© 2008 Taylor & Francis Group, London, ISBN 978-0-415-45775-0

Environmental friendly reinforced retaining wall by using traditional stone masonry

N. Fukuda & Y. Kameda
Fukken Co., Ltd, Consulting Engineers, Hiroshima, Japan

T. Yoshimura & K. Abe
Kikkouen Co., Ltd., Mine Yamaguchi, Japan

K. Watanabe & T. Hara
Tokuyama College of Technology, Shunan Yamaguchi, Japan

Y. Kochi
K's lab Co., Ltd, Yamaguchi, Japan

ABSTRACT: The ancient civil structures had been constructed by using natural resources such as soil, stone or timber. In modern structures, those construction materials have been replaced by artificial materials such as concrete, steel or plastic, and such change has contributed to rather speedier and larger scale construction. This paper introduces the newly developed reinforced retaining wall construction method which is expected to harmonize the surrounding ecology and also ensure the stability of structure itself with combination of conventional stone masonry and RC pre-cast reinforcement called Branch Reinforcing Method.

1 INTRODUCTION

We have inherited the ancient civil and architectural structures which had been constructed by natural resources such as soil, stone or timber several hundreds or thousands years ago. In modern structures, the major construction materials have been replaced by artificial materials such as concrete, steel or plastic. Such change has enabled rather speedier and larger scale construction and contributed to the economic development significantly.

Figure 1. Shape of RC Branch Block (L = 1.5 m).

However, recently, the durability of those modern structures has been put in question and the repairing/rehabilitation method to prolong the serviceability has become the one of social proposition.

This paper introduces the new retaining wall construction method by combining the natural and artificial resources i.e natural stones as major material and RC Branch Block (ref. Figure 1, hereafter called as block) as pre-cast reinforcing members.

Configuration of this method is shown as Figure 2. The concept of design and construction of the method is also discussed.

2 GENERAL

2.1 *Aims of proposed wall method*

Since the applicable construction height by dry masonry is limited to approx. 2 m, the wet masonry has commonly been adopted with using the concrete blocks instead of natural stones due to its advantage in productivity.

Also for the revetment works, the priority has been put into flood control, so that the "Three faces

Figure 2. Configuration of proposed block wall using traditional stone masonry for river revetment in February 2003.

revetment method made by concrete" has commonly been adopted and that has been said as one of major influence factor to the surrounding ecology.

According to the amendment of river law in 1997, "the maintenance and protection of river environment" became the one of the main purpose of public works in addition to the flood control and irrigation, henceforward many kinds of ecology/environment friendly blocks have been developed.

From the environmental point of view, the characteristics of this method are listed as below;

(1) Using natural stones available nearby site,
(2) Not using the concrete which will deteriorate the water quality nearby site,
(3) By applying the dry masonry method, the void between each stone provides the life space for small animals and fishes.
(4) Enabling the sodding and planting works between each stone which contribute the good landscaping and surrounding ecology, and,
(5) Enabling the quick dewatering of excess ground water due to its high permeability of facing.

2.2 *Construction sequences*

Construction sequences are shown as Figure 3. The advantages of this method from the viewpoint of construction sequences are listed as below;

(1) Since the weight of one block is relatively light with 240kg, the big construction equipment is not required,
(2) Utilizing the hexagon shaped by blocks as a guide, the skilled mason is not necessarily required,
(3) Enabling the works in curved alignment (minimum radius = 3 m),
(4) Flexible to the undulation of ground condition. The base ground is not necessarily horizontal.

(a) Piling up blocks for second layer

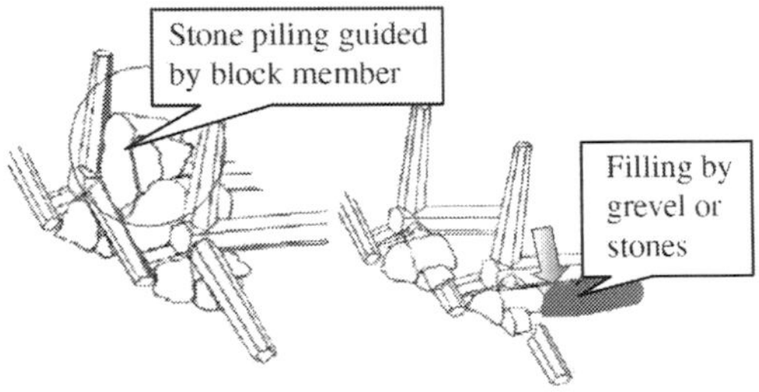

(b) Piling wall facing with stones and blocks

Figure 3. Construction details of proposed wall.

3 DESIGN CONCEPT

In order to establish the design concept of this method, each function of stones and blocks have to be clearly identified.

Figure 2 showed the photo of river revetment works implemented as a pilot project in Yamaguchi city. In this case study, assuming that only the block structure bears the whole earth pressure, it was calculated that this structure is no longer stable. Therefore, in order to incorporate the function for stones to share certain loading into design, the following model is assumed.

Figure 4 explains the mechanism of resistance against earth pressure as below:

(1) Firstly, the stone masonry (average width $b =$ approx.35 cm) resists against the earth pressure by backfilling material as a leaning type retaining wall. The stability of this phase is calculated by force diagram method.
(2) When the pressure exceeds the resistance capacity of stone masonry, the main member of block starts to resist against tensile force transferred through block member.

As long as the tensile force is within the allowable capacity of main member, this structure is internally stable and able to resist as a mass enveloped by dot line against the earth pressure as externally stable condition. The stability of this phase is calculated same as gravity retaining wall i.e. checking for sliding, overturning and bearing capacity.

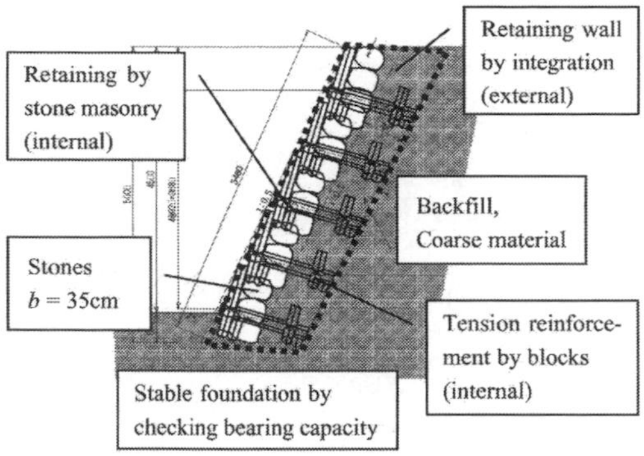

Figure 4. Model of load sharing mechanism.

Figure 5. Setup of pull out test of block by hydraulic jacks for 4m-high test wall (Nov. 2004).

Here, it is predicted that the bending moment will occur on two blanch members underneath the block which tend to deflect toward outside, as well as on main member. Therefore, it is important to fill the void by stones laid underneath the blocks during construction. Durability is estimated over 50 years for RC pre-cast reinforcement.

Above design concept has been proved as appropriate by implementing the several monitoring during the trial project with 4m height i.e. monitoring the stress of reinforcement bar, during construction and pullout test of block as shown in Figure 5.

4 INVESTIGATION AND EXPERIMENTATION ON EFFECTIVENESS AS REVETMENT

In order to confirm that this method is effective on ecology and capable for flood control as the revetment, several investigations and experimentations were carried out at two revetment project sites in Yamaguchi prefecture, i.e. investigation of surrounding ecology, investigation during flood and implementation of hydraulic test in laboratory.

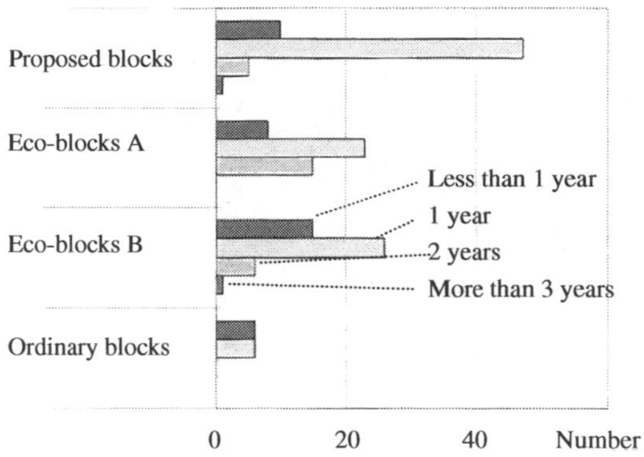

Figure 6. Comparison of number of "Kawamutsu" (*Zacco temminckii*) at each revetment block type.

4.1 *Investigation on ecology*

The investigation was implemented on ecology of fishes, aquatic animals and plants nearby site. Nearby the site in Mine-city, the existing revetment had been constructed by using two kinds of environmental friendly blocks (eco-block) and ordinal blocks so that the investigation on those locations was also implemented as a comparison.

Figure 6 shows the number of fish named Kawamutsu (*Zacco temminckii*) as a sample of investigations. It was observed that the number of Kawamutsu living nearby the proposed method site is larger than the other sites. Also, existence of Kawamutsu older than 3 years was observed at proposed block site. Thus, it is confirmed that the void in the proposed block provides the appropriate condition for "Kawamutsu" to grow up.

4.2 *Calculation of velocity of flow at flooding*

After the completion of revetment, the investigation was carried out during the flood as shown on Figure 7 (a) and the mean velocity of flow V was back-analyzed by equation (1) and (2) based on the section of river as shown on Figure 7 (b).

$$V = \frac{1}{n} R^{2/3} I^{1/2} \quad (1)$$

$$R = \frac{A}{S} \quad (2)$$

(a) Flood condition of revetment at Yamaguchi City

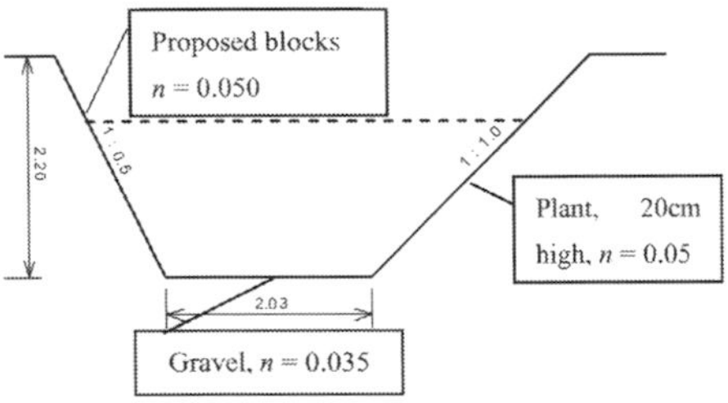

(b) Estimation of flow velocity during flood by observation

Figure 7. Calculation of velocity of flow during flood.

where n = coefficient of roughness; R = hydraulic radius; I = inclination of river bottom (0.0189); A = water area; and S = wetted perimeter.

As a result, the velocity of flow was calculated as 3.5 m/s when the flood was 2 m depth. In addition, the strength of this method against hydraulic force was checked by using "Sliding–Unit" model which is based on the Revetment Design Method and also incorporates the result of pull-out test for block. As a result, this structure is confirmed as stable until velocity of flow reaches 10 m/s.

4.3 *Laboratory test by using hydraulic model*

When applying this method as a revetment, it is known that the blanch members of the proposed blocks have the special roughness in repeated geometric pattern on its surface. Flow characteristics of this type revetment have been studying by using the hydraulic model in 1:40 scale. In order to measure the velocity of flow, PTV (Particle Tracking Velocimetry) is utilized. This method is to let the tracer particles with 100μm diameter flow from upstream, and to capture the movement of those particles by digital camera when passing the film of light shaped by radiating through the 2mm slit in horizontal and vertical direction as shown in Figure 8(a), (b). Figure 8(c) shows the example of experimentation results of turbulence in cross section in front of revetment (Watanabe, K. et al. 2006, 2007).

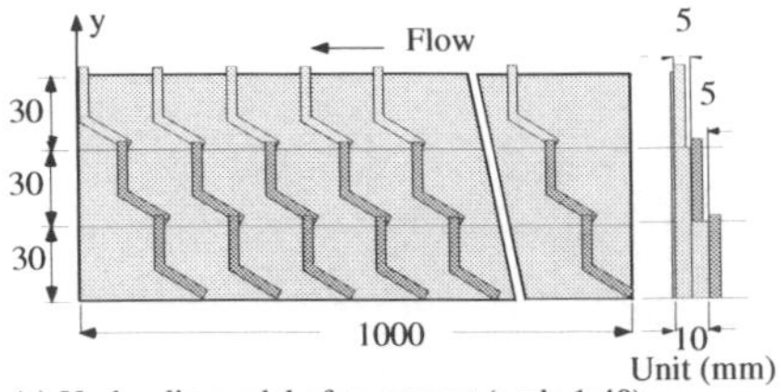

(a) Hydraulic model of revetment (scale 1:40)

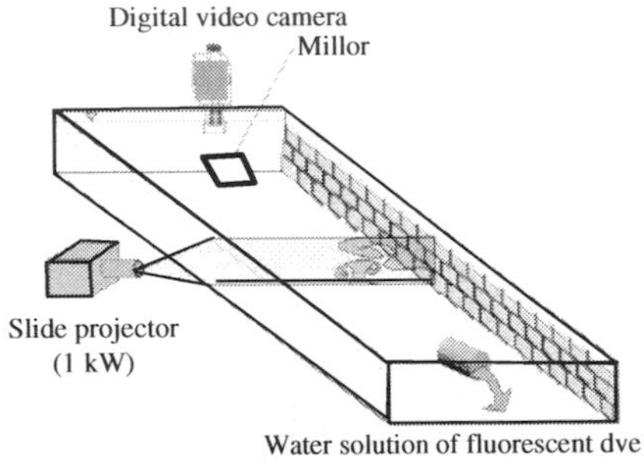

(b) Setup of hydraulic model test to measure turbulence

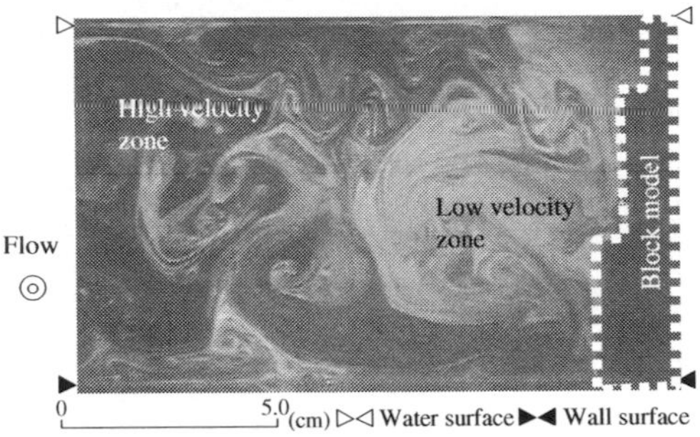

(c) Cross sectional turbulence in front of the proposed block revetment model

Figure 8. Laboratory test by using hydraulic model.

Through those experimentation results, it was confirmed that the application of this method is able to reduce the velocity of flow near the revetment surface and move the high-speed flow zone toward the centre of river. In other words, this method is able to improve the stability of revetment during the flood.

5 APPLICATION

The proposed wall method was applied to 5 river revetments and 7 embankment walls of reconstruction works after disaster induced by heavy rainfalls in Yamaguchi and Shimane prefectures. Figure 9 introduces the typical example of river revetment adjacent drop structure for reconstruction after flood disaster.

Figure 9. Reconstruction of river revetment after flood disaster.

6 CONCLUSIONS

This method has been developed by Branch Block method study group from June 2005. The proposed method has the following advantages;

- Unnecessary big construction machines.
- Flexible to the undulation of ground condition.
- Friendly for surrounding ecology due to require no curing of concrete, and provide life space for small animals and fishes between void of stones.
- In case the method is adopted as river revetment, it is expected to provide rather stable and durable function than ordinary environmental friendly blocks due to use of natural stones and effect of turbulence by unique roughness of the surface of the structure.

REFERENCES

Watanabe, K. Yoshimura, T. Hara, T. Fukuda, N. & Kochi, Y. 2006. Characteristics of mean velocity profiles near the Branch Block protection model. *Evaluation and Application of Flow-induced in Hydraulics, JSCE.* 4: 7–12. (in Japanese).

Watanabe, K. Yoshimura T. & Hara, T. 2007. Characteristics of flow structure near the Branch Block bank protection model. *Journal of Hydraulic Engineering, JSCE.* 51: 739–744. (in Japanese).

New Horizons in Earth Reinforcement – Otani, Miyata & Mukunoki (eds)
© 2008 Taylor & Francis Group, London, ISBN 978-0-415-45775-0

Steel nails for stabilizing forested slopes

N. Iwasa, M. Q. Nghiem & T. Ikeda
Nippon Steel & Sumikin Metal Products Co. Ltd.

ABSTRACT: In Japan, slope failure often occurs under influences of earthquake and rainfall. Annually, it causes serious loss of dead, affects economy and threatens the safety of cultural or landscape heritage sites. However, traditional landslide countermeasures are often damaged the landscapes by clearing off the vegetation. To stabilize forested slopes while preserving the vegetation, a new nailing method, named Non-frame, was proposed. It is structured from three main parts: steel nail, fixed plate at nail head and the wire net connecting nail heads. Non-frame stabilizes a forested slope by fixing the unstable soil layer into the bedrock. The reinforcement of Non-frame mainly depends on the axial force that was born by settlement of fixed plate in combing with skin friction force between steel nail and slope and shear reinforcement was born by nail deforming. The wire net ties steel nail heads and topsoil moving together as a block to reduce partial failures that occurred very often in natural slopes. This paper introduces reinforcement mechanism of Non-frame and its applications in fields.

1 INTRODUCTION

The stabilization of forested mountains avoiding natural disaster such as landslide and slope failure is one of most important work of forest landscape conservation. On these papers, we discuss about Non-frame, a new method that stabilizes mountain slopes while preserving forest landscape in the strict conditions of typhoon and earthquake. The new method fulfills all the basic requirements of a landslide countermeasure such as producing enough reinforcement for design factor of safety. That is easy and simple of construction with light and standardized parts for setting up in fields. It also has to fulfill extra requirements for protecting landscape and cultural heritage sites such as: will not cut off vegetation, perfect soundproof under construction time, suitable for narrow scaffold and steep slope (up to 80°), suitable for various weather such as heavy rainfall and frost heave. In other side, we knew that tree roots could only penetrate into soft topsoil. If a traditional stabilization method such as soil nail, anchor or retaining wall was chosen, the weak topsoil must be cleared off and thus trees on slope with root reinforcement must be cleared also.

Non-frame protects vegetation and topsoil of the forested slopes those easily displace when factor of safety is smaller than 1.0. The displacement of topsoil, however, causes steel nail deformed and vertical soil pressure occurred under settled fixed-plate those respectively born shear reinforcement and vertical resistant. These reinforcement are not only depend on skin friction as a traditional soil nail, but also depends on axial resistant of settled fixed-plate and shear reinforcement of deformed steel nail.

2 MECHANISMS OF NAILING METHOD IN JAPAN AND NON-FRAME FEATURES

The mechanism of soil nailing methods is to fix unstable part of slope to the stable bedrock. It can be classified into two main types: Concrete-frame and Non-frame. Concrete-frame includes a number of nails in combining with a concrete frame covering slope

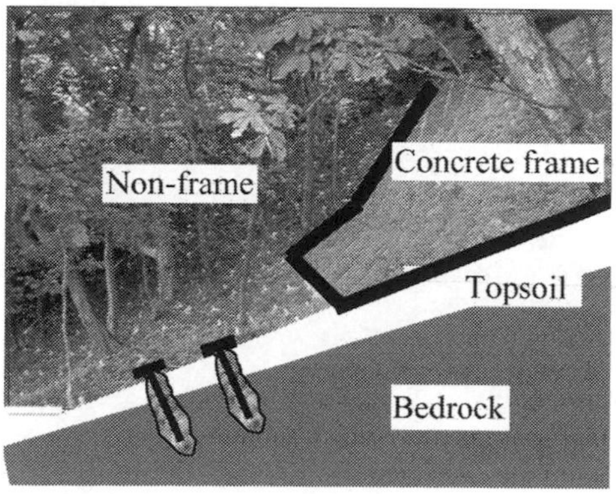

Figure 1. Non-frame stabilizing forested slope.

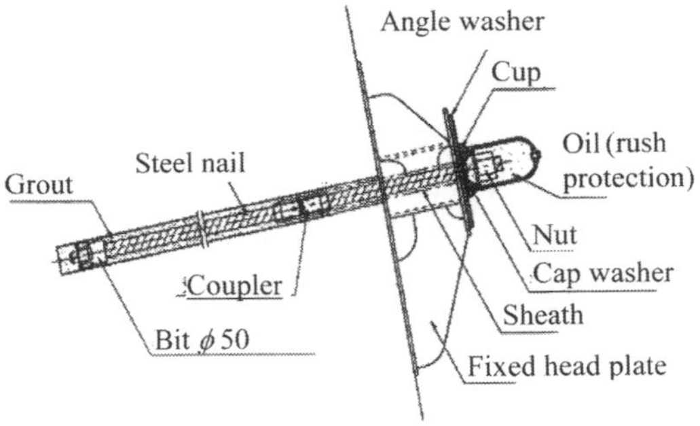

Figure 2. Structure of Non-frame.

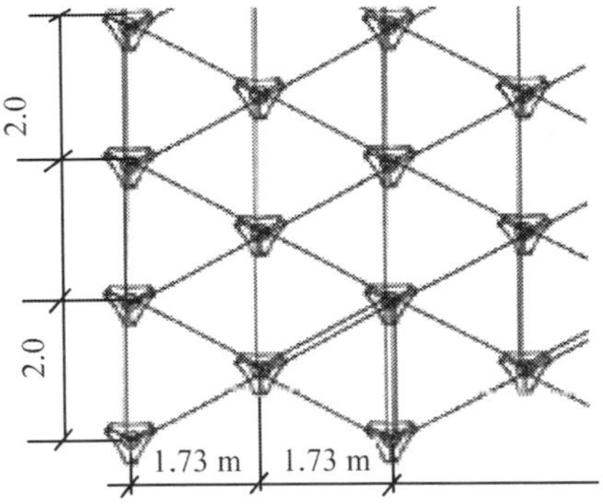

Figure 3. Distribution of Non-frame.

surface (see Figure 1). Soil nails fix topsoil tidily into the bedrock while concrete frame connects soil nails into each other as a frame. Hereby, concrete frame, soil nails and topsoil are locked as a rigid block. Concrete frame nailing method is suitable for stabilizing the slope of highway, port, infrastructures and other urban constructions where slope had good properties with large skin friction. In practice, concrete frame methods cannot be applied at cultural or landscape heritage sites where the natural landscape including vegetation must be conserved as well as possible. Furthermore, to build the construction by heavy machines, the environment around is disturbed. Non-frame, a new nailing method has the similar structure compare to the concrete frame method, but steel nails and unstable soil of slope are connected to each other by a flexible wire net system (not by a concrete frame) thus the trees on the slope surface can be totally remained as original situation.

The distribution of Non-frame on slope is showed in Figure 3 and the structure of Non-frame is showed in Figure 2. Steel nails stabilize a slope by flexibly reducing the movement of unstable slope. The vertical axial force is born by vertical settlement of fixed plate, which fixes unstable soil into bedrock. The triangular wire net connects steel nail heads to each other to avoid

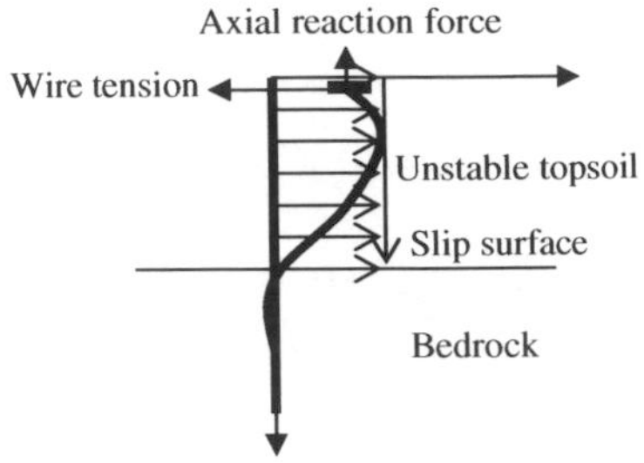

Figure 4a. Mechanism of Non-frame.

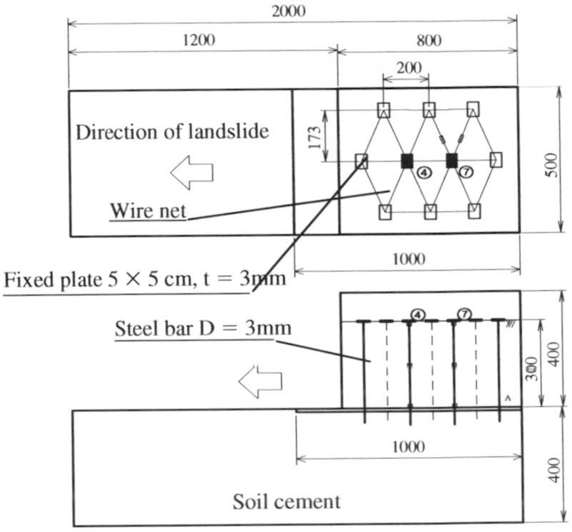

Figure 4b. Experiment of steel bar with fixed plate.

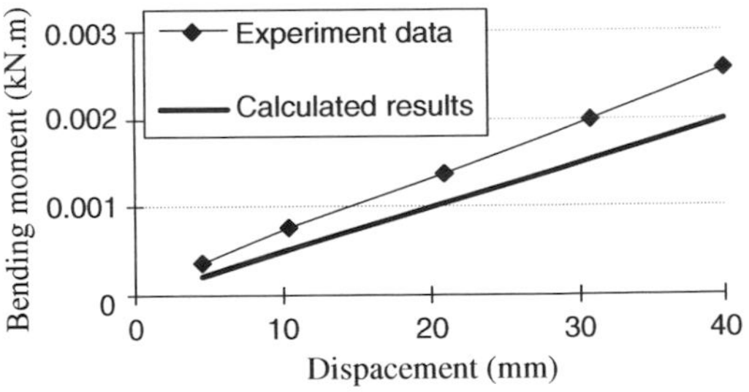

Figure 4c. Bending moment of steel bar at slip surface.

the partial failure of slope. Base on the equilibrium condition of a steel nail (see Figure 4), it reinforce slopes by transporting resisting force from bedrock to topsoil, equations (1) and (2) (Nghiem et al, 2004) are written:

$$EI \frac{d^4 y}{dx^4} + Es(y - p) = P_x \frac{d^2 y}{dx^2} \tag{1}$$

$$P_x = Kv \cdot S_p \cdot \Delta x \tag{2}$$

Table 1. Properties of soil in upper box.

Properties of soil	Value	Unit
Grain diameter	<4.75	mm
Unit weight of soil	2.647	g/cm3
Unit weight of dried soil	1.769	g/cm3
Saturation	13.9	%
Cohesion	0.9	kN/m2
Shear resistance angle	36.7	degree

where: E, I Young modulus and bending stiffness of steel nail, p: soil displacement, Px: axial force, Sp: area of fixed plate, (y, x): horizontal and vertical axes, Δx: vertical settlement of fixed plate, Kv: coefficient of vertical subgrade reaction, Es: Young's modulus of soil. Solutions of equations (1) are deflection, deflection angle; bending moment, shear force and axial force of steel nail. Solution of equation (2) leads to the axial force P. There are three main factors of reinforcement that influences the reinforcement of Non-frame: shear force of steel nail, axial reaction force of soil acting on fixed plate and wire tension at nail head. These three factors can be calculated by equations (1), (2) considering influences of wire tension. Reinforcement of steel nails on the slip surface is given by Equation (3):

$$R_c = S + S_w + P_x \cos(\phi) \quad (3)$$

where Rc: reinforcement of steel nail, S: shear force of steel nail on slip surface, Sw: wire tension, ϕ: inertial friction angle of soil. P_x is often calculated from skin friction between soil and grout around steel nails, it sometime is called pullout resistance. A reinforcement coefficient that shows the increase of resistance force by influences of shear force, fixed plate and wire tension compare to resistance force was born by only pullout resistance is previously calculated. Then the reinforcement is simplicity calculated by P_x and the reinforcement coefficient. A simple experiment was conducted to compare the results of the theoretical solution (Nakamura et al, 2005). Two steel boxes were used in the experiment. The upper box containing soil is the model of topsoil layer, the lower one containing soil-cement represents bedrock.

When the driving force became greater than the resisting force, the upper box started to slide down. The strain gauges glued upon steel bar recorded the stress distribution along the steel bar. Figure 4c shows a good agreement of comparison between the calculated data and experiment data of bending moment.

The factor of safety (F_s) of slope can be presented by the Equation (4):

$$F_s = \frac{Resisting\ force + Rc}{Driving\ force} \quad (4)$$

Figure 5. Shrine was buried by land-slide (Nagaoka, Niigata Prefecture).

Incase of forested slope, the factor of safety is usually chosen as 1.2 for common design and is usually chosen as 1.0 for seismic design.

3 APPLICATION OF NON-FRAME

Japanese consider that forest is not only a place just has many plants; but they also considered forest as a complex system of plants, animals, society and culture. The trees sometimes become a sacred symbol that the people respect and protect with their high spirit. In that context, stabilization of slope must consider about the protection of the trees. As mentioned above, Non-frame was born to prevent shallow landslide while protect the vegetation on slopes. It was applied in many fields concerning the natural environment and landscape conservation issues. However, some of the cultural sites were damaged by landslide since the managers did not find out a suitable method such as Non-frame for stabilizing slope. Figure 5 shows a shrine was buried by shallow landslide after Chuetsu earthquake in Niigata Pref. of Japan. The structures at center and at right side of picture were ruined by deposit of landslide. Hereby, typhoon 23rd in combination with earthquake caused terrible earth disasters occurred not only on the natural slope, but also on the cement slope. The rainfall weakened topsoil, reduced skin friction between slope and soil nails, and then the slope and concrete frame were probably shaken to move separately until slope was failed. Incase of Non-frame, the flexible steel nail can move together with soil as a block that can avoid partial movement between steel nail and soil. In only three days 20~22nd March. 2005 more than 10 times earthquake M $\geq$ 4 and three times of rainfalls about 10 mm occurred in Fukuoka city. The field was surveyed on Mar. 24th, 2005. Figures 6a and 6b were taken at that time show Kumano shrine, which was protected by Non-frame. The slope

Figure 6a. Slope failure by earthquake and rainfall.

Figure 7a. Concrete frame method used in Ma-tsuyama castle (CG).

Figure 6b. Non-frame protected Kumano shrine.

Figure 7b. Non-frame method for conserving Matsuyama castle (1 year).

near the shrine (on the left side of Figure 6b) was stabilized successfully while the slope on the left side of Figure 6a was fail by the effects of rainfall and the West off Fukuoka earthquake.

In addition, the Non-frame slope was greened while the slope without Non-frame (Figure 6a) was eroded seriously. Hereby, the wire net system played importance role in avoiding surface erosion and in protecting vegetation growing stable on the steep slope. Fixed plates also have similar role to reduce the surface erosion by stop topsoil movement.

Figures 7a, 7b show Non-frame protecting Matsuyama castle. It is the symbol of Matsuyama city, also is a very precious cultural heritage and landscape site in Japan. In 2001, the lower part of slope under the castle became unstable that threatened the safety of the castle at the top. The slope instability also affects to the safety of buildings at the foot of the slope. In this case, Non-frame was applied not only for preserving the castle, but also for the safety of the people living in the buildings. Figure 7a shows imagination of concreted slope, one of most common method in Japan. The sightseeing would become terrible with a part of slope covered in concrete without green.

To avoid such a problem, Non-frame was used at upper slope where people can see and concrete frame was only applied at lower places behind the building (where people cannot see the trees were cut off). Fig 7b was made one year after Non-frame was constructed that shows the sightseeing was successfully preserved while stabilizing the forested slope.

On Jun. 14th, 1998, about five events of 50 mm rainfalls caused area B (Figure 8) of Izuinatori slope failed (Fig 9b). There is a national rout at top of slope (Figure 8, area A) and Izu express railway in combining with other rout at foot of the slope. The slope must be stabilized carefully. In other side, the vegetation need to be preserved as well as possible because of Izuinatori is a famous landscape spot. Izu peninsula also was known as a center of Nankai earthquake. On Jul. 12nd, 2005 earthquake M4 and typhoon occurred at the same time that caused many shallow landslides close to the Izuinatori slopes occurred (Figure 10). Figures 10a, b show shallow landslide occurred at Imaihama area, about 1 km South of Izuinatori slope. Figure 10d shows

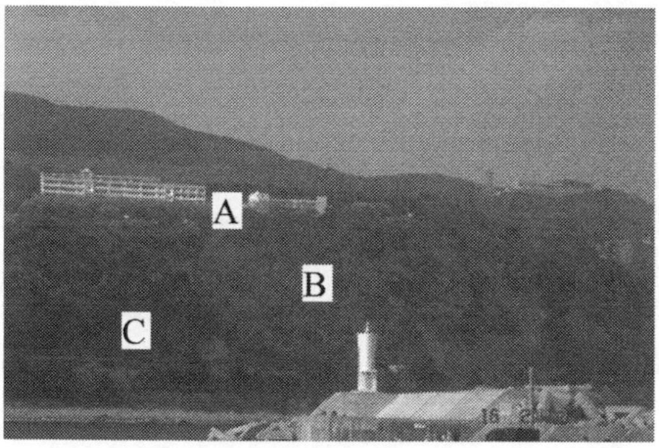

Figure 8. Non-frame stabilizing Izuinatori slope.

Figure 9. Izuinatori slope before and after rein-forcing by Non-frame.

slope failure at Kawazu tunnel very close to the Non-frame field and Figure 10c shows other landslide in upstream area of Izu peninsula. Such a good example of Non-frame application, Izuinatori slope is perfectly stable, Figure 9a.

Figure 10. Four shallow landslides occurred near the Izuinatori slope (12nd. Jul. 2005).

4 CONCLUSION

By innovative method of steel nail flexibly, it nails top-soil into the bedrock in combining with block efficient of wire net to avoid the partial failure and thus keep slope more stability, Non-frame can protect vegetation while stabilizing slope. It fulfills the complicated requirements of conservation works for a cultural or landscape heritages. Three fields in Japan show Non-frame preserving the forest slopes that is successfully under effects of typhoons and earthquakes.

REFERENCES

Nakamura, H., Inoue T., Iwasa N., 2005. Influence of fixed plate on reinforcement of a new nailing method, proceeding of annual conference, JSCE.

Nghiem, M. Q., Nakamura, H., and Shiraki, K., 2004. Slope stability of forested slopes considering effect of tree root and soil nail reinforcement. *Journal of the Japan Landslide Society*, 3: 262–272

New Horizons in Earth Reinforcement – Otani, Miyata & Mukunoki (eds)
© 2008 Taylor & Francis Group, London, *ISBN 978-0-415-45775-0*

Study of a 15 m vertical soil nailed wall at Capella@Sentosa

S.A. Tan (Harry) & A. Rumjeet
Civil Engineering Department, National University of Singapore

ABSTRACT: The Knolls at the Capella (a six star resort hotel) of Sentosa Island involved the construction of a vertical soil nail shotcrete retaining structure to support an open excavation of 15 m height, and to preserve the existing historic. The design follows the French code, Recommendation Clouterre 1990. During stage construction of soil nail walls, significant ground displacements were expected as soil nail is a passive soil reinforcement system. The ground displacements were predicted using FEM program Plaxis. Wall deflection and ground settlements were monitored and minimal damage were caused to Tanah Merah house throughout the process of soil nail wall construction. A parametric study has been carried out on important parameters inherent to the soil and soil nails. This study showed which parameters are significant in the design of such a retaining structure.

1 INTRODUCTION

Soil nailing is an in-situ ground reinforcing method for retaining excavations and stabilising slopes by passive inclusions. Soil nails are extremely effective in stabilising existing slopes or where slopes have to be steepened.

The six star hotel Capella@Sentosa is an example where a permanent vertical slope has been stabilised using soil nailing reinforcement. Various field instruments, including inclinometers, settlement markers and water stand pipes, have been placed on site as the near vertical slope is being constructed and measurements taken with progression of the excavation. The measurements included settlements of existing structures (Old Tanah Merah House) and lateral displacement of the cut slope from inclinometer readings, and ground water levels.

Predictions of the deformation behaviour of a soil nailed structure are required to ensure that displacement limits set by the authorities are not exceeded. The case at Sentosa provides opportunity to validate the use of a finite element analysis for a soil nailed problem. For closely spaced soil nails, equivalent 2D FEM models can give good results (Tan et al, 2005). Once the models are calibrated to fit the deformations obtained on site, a parametric study on was conducted to determine the sensitivity behavior of the soil nailed system. With the parametric study, engineers will be able to know which parameters are more critical in the design of soil nailed walls.

2 GEOLOGY AND SOIL PROFILE

The prevailing geological formations underlying the site are Rimau Facies and St. John Facies of Jurong Formation. The Geological Map is presented on Figure 1. Eight boreholes (Figure 2) revealed that the site is underlain by the Residual Soils of Jurong Formation. The subsurface strata of the site consist of weathered surficial fill, predominantly of yellowish brown silty sand. The density index of the fill is loose. This unit is approximately 1 m to 4 m thick, with an average thickness of 3 m. The fill is underlain by the Residual Soils of Jurong Formation which appeared in the form of silty sand. The relative densities of this unit were found to vary between loose to very dense and generally improving with depth. Table 1 showed the four idealized layers of soil according to SPT blow counts.

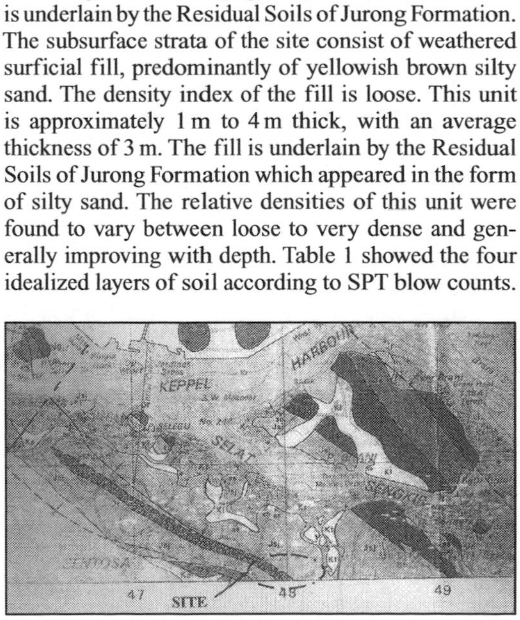

Figure 1. Geological map of site at Sentosa.

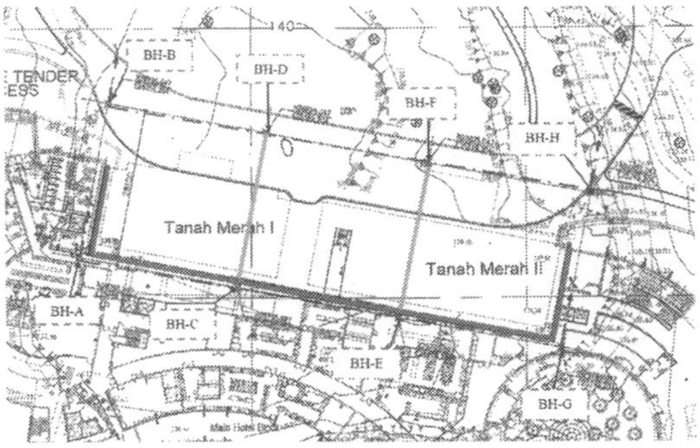

Figure 2. Borehole location.

Table 1. Soil layers found on site.

Layer	Identification	SPT N blows /30 cm
1	Loose silty sand [<10]	<10
2	Medium dense silty sand	10–40
3	Dense silty sand	40–100
4	Very dense silty sand	>100

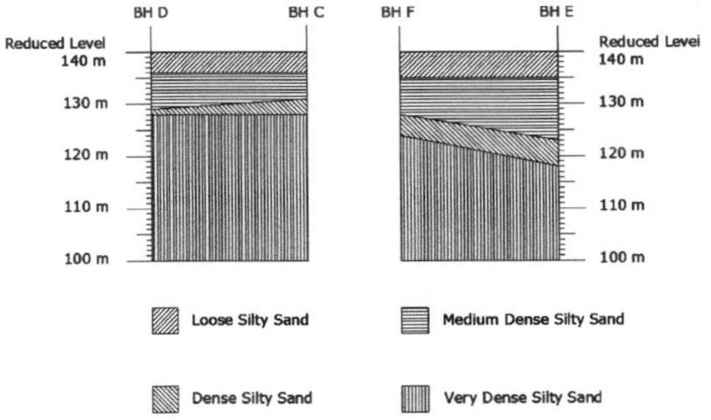

Figure 3. Soil profile.

Two sections have been chosen for the FEM study of the soil nailed wall behavior. These sections are CD and EF as shown in Figure 3.

3 INSTRUMENTATION RESULTS

Figure 4 showed the instruments installed on site.

3.1 *Inclinometer*

Unfortunately only one inclinometer is near the two analyzed sections, namely inclinometer I2 near borehole C and also the inclinometer was installed quite late, after some excavation had taken place. There were also other problems related to the readings. Figure 5 shows the readings retrieved as the excavation was being carried out. It shows the readings taken after the excavation reached a 5 m depth, a 10 m depth and finally a 15 m depth when the excavation was completed.

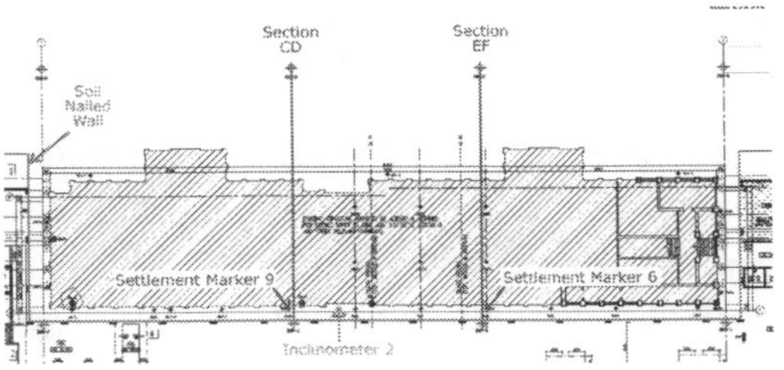

Figure 4. Instrument location.

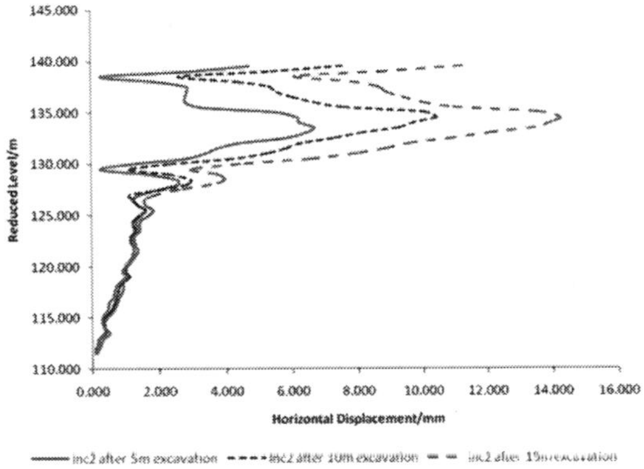

Figure 5. Inclinometer readings.

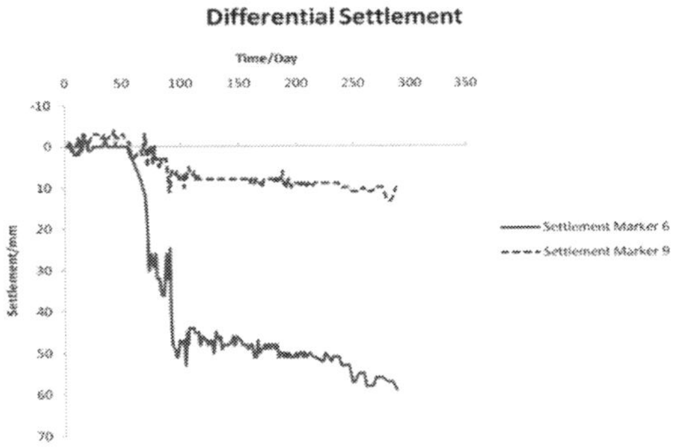

Figure 6. Settlement readings.

3.2 *Settlements*

Various settlement markers were placed around and inside the old Tanah Merah House. Settlement marker 6 was placed just behind borehole E and settlement marker 9 was placed behind borehole C. The readings from these two settlement markers are relevant to the 2 sections analyzed in this paper. Figure 6 showed the settlement readings and consequently the differential settlement induced between the two sections.

Table 2. Maximum settlement.

Settlement marker 9 [Behind BH-C] [mm]	Settlement marker 6 [Behind BH-E] [mm]
51	7

The maximum settlement registered by the settlement markers occurred after the full excavation for the wall was completed are summarized in Table 2.

3.3 *Water standpipe*

Figure 7 shows the water standpipe readings. The data showed that the ground water levels were below the base of the excavation (RL124 m) throughout construction. Being very stiff soils, a drained analysis is more appropriate for this site.

4 FEM MODELLING

A Mohr-Coulomb model was used for the stiff residual soils assuming drained behavior.

4.1 *Soil Mohr-Coulomb parameters*

From consolidated undrained triaxial compression test c' and ϕ' were determined for the residual layers. Based on other excavation experience in similar soils, it is estimated that a correlation of E = between 1 N and 2 N MPa would generally apply for these residual soils. All soil materials used have a Poisson's Ratio of 0.3 and a permeability of 0.01 m/day. A correlation of E = 1 N was used for the worst credible soil profile namely Section EF and a correlation of E = 2 N was used for Section CD to obtain reasonable values for the 2 models done in Plaxis. Both models were run in fully drained conditions. The parameters used are listed in Tables 3–5.

4.2 *Soil nail and shotcrete properties*

The soil nails and shotcrete were modeled using the equivalent thin plate theory. The EI and EA of the plate elements used in Plaxis are actually EI/S_h and EA/S_h, where S_h is the horizontal spacing of the nails. For the shotcrete, the equivalent EI and EA were calculated per metre run of wall. The equivalent of nail stiffness E is obtained from calibration to pullout tests results.

All the plate elements representing the nails and shotcrete were modeled as elastic materials with a Poisson's ratio of 0.1 and a weight of 1 kN/m^2. Listed below are the material properties for the plate elements used in the Plaxis model for both sections CD and EF.

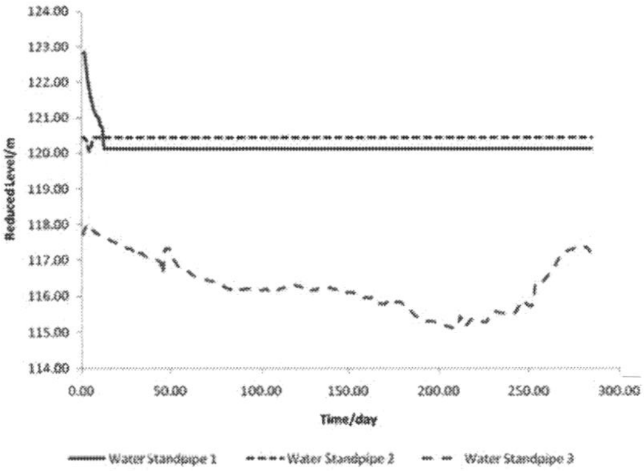

Figure 7. Water standpipe readings.

Table 3. Soil type legend.

Layer no.	Soil type
1	Loose silty sand
2	Medium sense silty sand
3	Dense silty sand
4	Very dense silty sand

Table 4. Soil properties section CD (E = 2N MPa).

No	γ_unsat kN/m^3	γ_sat kN/m^3	E_ref kN/m^2	c_ref kN/m^2	ϕ °	ψ °
1	18	20	10000	5	30	0
2	18	20	40000	10	32	2
3	18	20	120000	15	35	5
4	19	21	200000	20	40	10

Table 5. Soil Properties Section EF (E = 1N MPa).

No	γ_unsat kN/m^3	γ_sat kN/m^3	E_ref kN/m^2	c_ref kN/m^2	ϕ °	ψ °
1	18	20	5000	5	30	0
2	18	20	20000	10	32	2
3	18	20	60000	15	35	5
4	19	21	100000	20	40	10

5 FEM RESULTS

FEM study for sections CD and EF and parametric studies are presented below.

5.1 *Deflection of wall*

The results of wall deflection predictions at 5 m, 10 m and 15 m depths are shown in Figure 8.

Table 6. Plate element properties for soil nails and shotcrete.

Name	EA [kN/m]	EI [kNm^2/m]	M_p [kNm/m]	N_p [kN/m]
Shotcrete	5000000	16667	1.00E + 15	1.00E + 15
Soil nail type 1	169000	42200	1.00E + 15	1.00E + 15
Soil nail type 2	141000	35200	1.00E + 15	1.00E + 15
Soil nail type 3	141000	35200	1.00E + 15	1.00E + 15

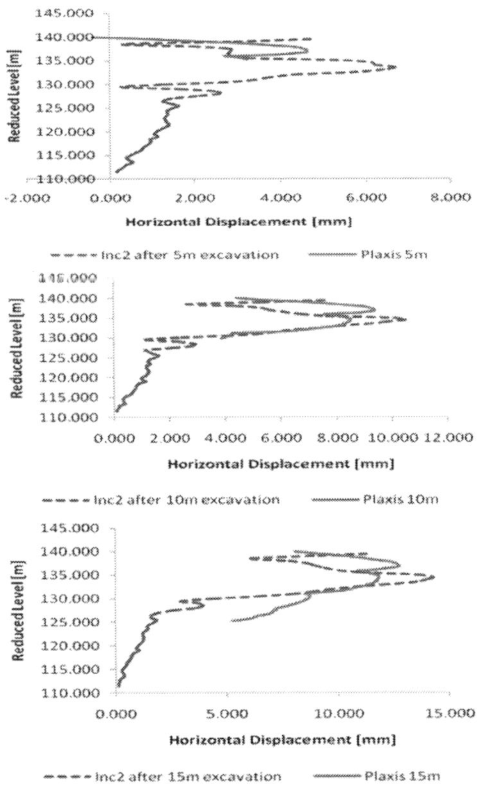

Figure 8. Wall deflections at 5 m, 10 m and 15 m depths.

The stiff soils combined with closely 1 m square grid spaced nails results in very small wall deflection of less than 0.1% of wall height, in this case.

5.2 *Ground settlements*

The predicted settlements under Tanah Merah house for sections CD and EF are shown in Figure 9.

These indicate that the maximum settlements of section CD would be about 15 mm, and the weaker section EF would be about 50 mm, consistent with the measurements in Figure 6.

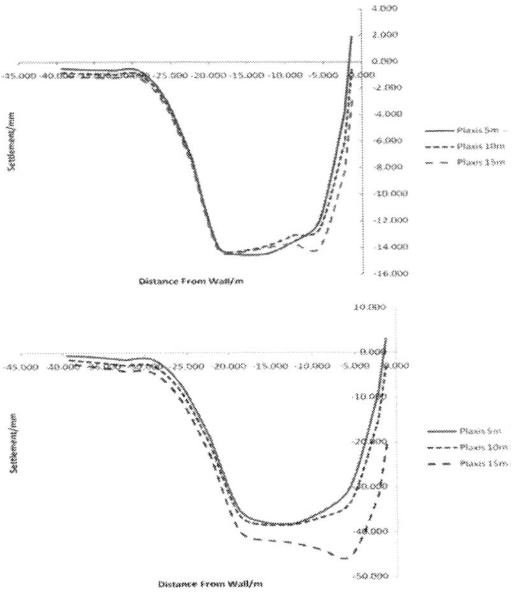

Figure 9. Predicted settlements of sections CD and EF.

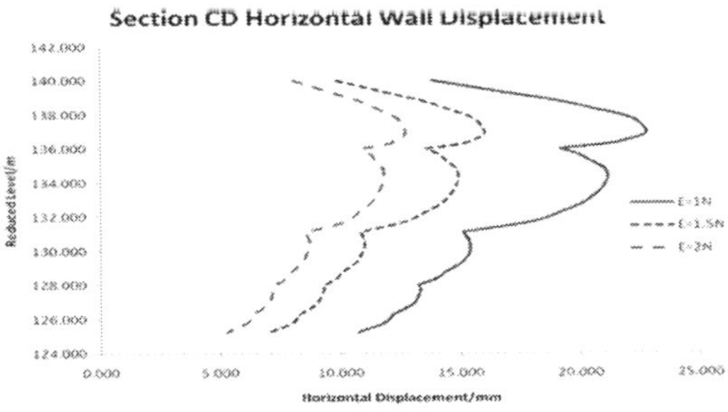

Figure 10. Influence of soil stiffness on deflections at CD.

5.3 *Parametric study*

The study was done to examine the influence of soil strength and stiffness, and nail axial and bending stiffness and lengths on the soil nailed wall.

Soil Stiffness was varied using different correlations between SPT N values and Young's Modulus of soil, E = 1 N, 1.5 N and 2 N MPa. Increased soil stiffness would reduce wall deflection near proportionately as these stiff soils remain essentially elastic when stiffen with closely spaced soil nails as in Figure 10.

Similarly, ground settlements are near proportionately reduced with increase of soil stiffness, as in Figure 11. However, the improved results from E = 1 N to 1.5 N is larger than from E = 1.5 N to 2 N.

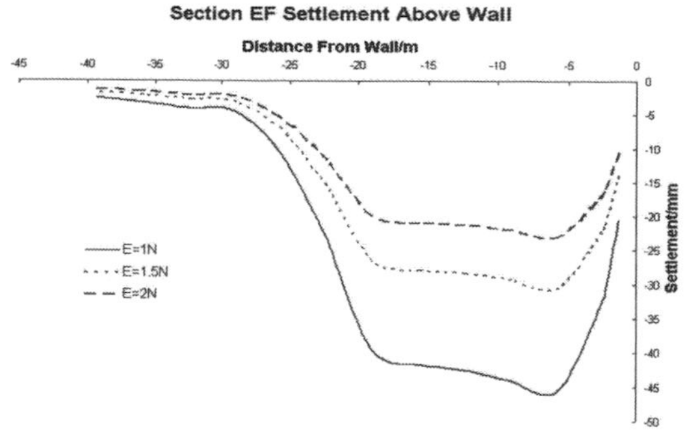

Figure 11. Influence of soil stiffness on settlements at EF.

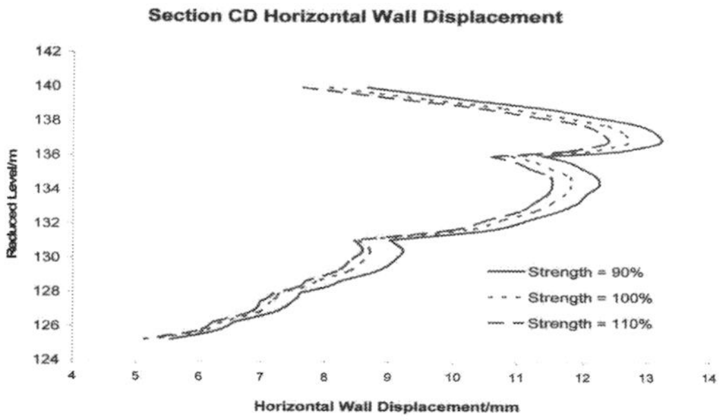

Figure 12. Influence of soil strength on deflections at CD.

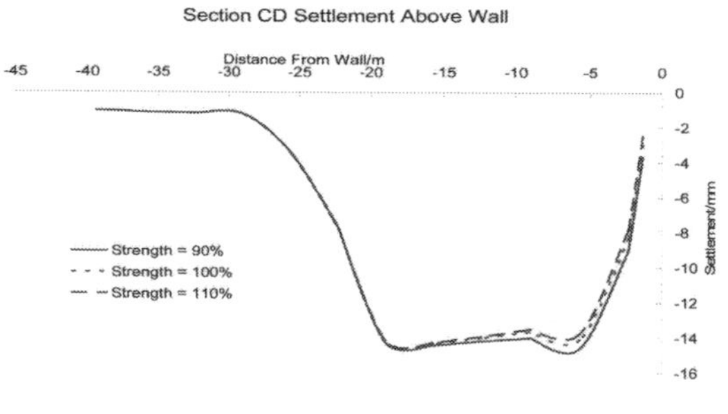

Figure 13. Influence of soil strength on settlements at EF.

Soil strength were varied by plus minus 10% from initial models. Three models were run with soil strengths of 90%, 100% and 110%. For the soil strength, c' and tan ϕ' were varied concurrently since the equation of shear strength is based on Mohr-Coulomb criteria as in Equation 1.

Figures 12 and 13 showed that soil strength is not sensitive as the soils response are essentially elastic with relatively small amount of soil yielding.

The nail bending stiffness of the soil nails were varied between 50% and 150% of the initial Plaxis models. Three models were run with nail stiffness of 50%, 100% and 150%. The results in Figures 14 and 15 showed that nail bending stiffness has little influence on wall deflections and ground settlements. This is

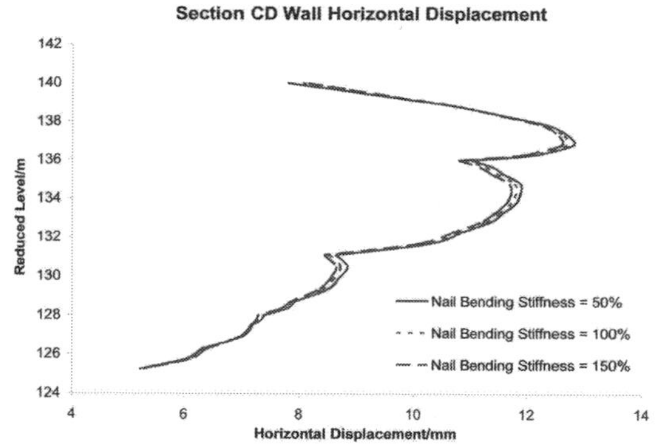

Figure 14. Influence of nail bending stiffness on deflections at CD.

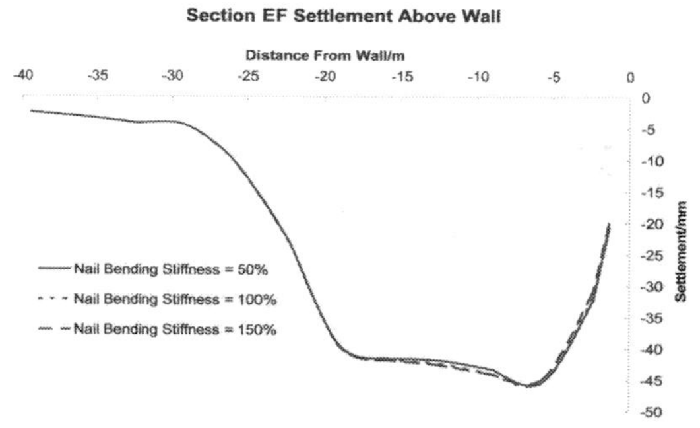

Figure 15. Influence of nail bending stiffness on settlements at EF.

consistent with Clouterre 91 that bending stiffness would contribute less than 15% to nail capacity.

The nail axial stiffness of the soil nails was varied between 50% and 150% of the initial Plaxis models. Three models were run with nail stiffness of 50%, 100% and 150%. The results in Figures 16 and 17 showed that nail axial stiffness has strong influence on wall deflections and little influence on ground settlements. This is consistent in that nail axial stiffness acts mainly in the horizontal direction restraining wall lateral movement but not ground vertical settlements.

The length of the soil nails was varied between 80% and 120% of the initial Plaxis models. Three models were run with nail stiffness of 80%, 100% and 120%. The results in Figures 18 and 19 showed that length of nails had greater influence on wall deflection and little effects on ground settlements.

6 CONCLUSIONS

It can be seen that the soil stiffness and the axial stiffness of the soil nail are quite important in the design of such earth retaining structures in stiff soils.

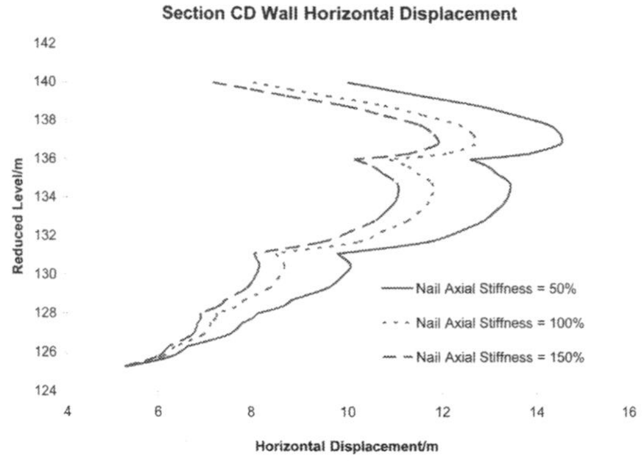

Figure 16. Influence of nail axial stiffness on deflections at CD.

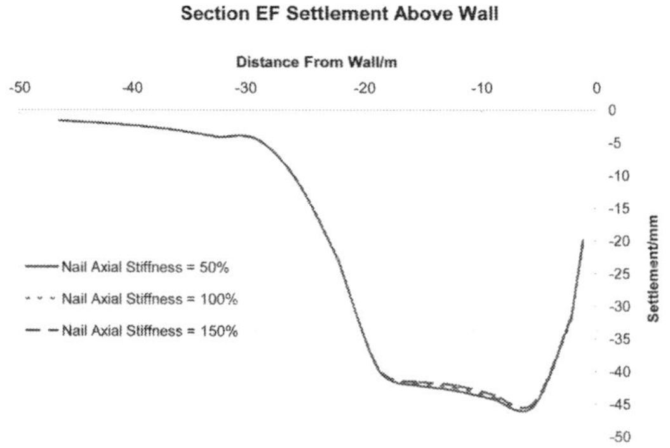

Figure 17. Influence of nail axial stiffness on settlements at EF.

Table 7. Influence of parameters on soil nails.

Parameter		Horizontal displacement	Settlement
Soil	Stiffness strength	Significant small	Significant minimal
Nail	Bending stiffness	Minimal	Minimal
	Axial stiffness	Significant	Minimal
	length	Significant	Minimal

This is so as there is little soil yielding, and the soil remains essentially elastic making soil stiffness more

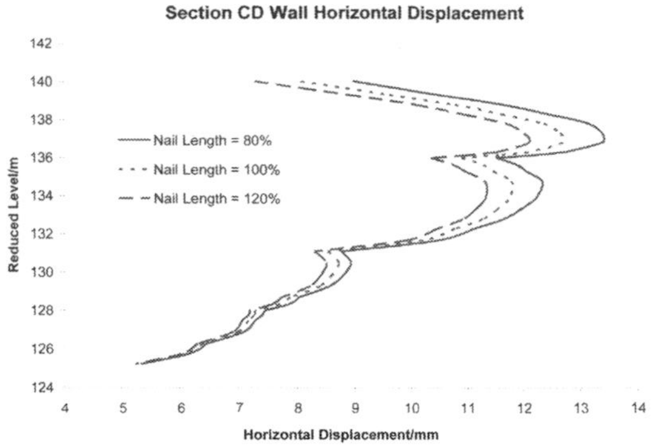

Figure 18. Influence of nail lengths on deflections at CD.

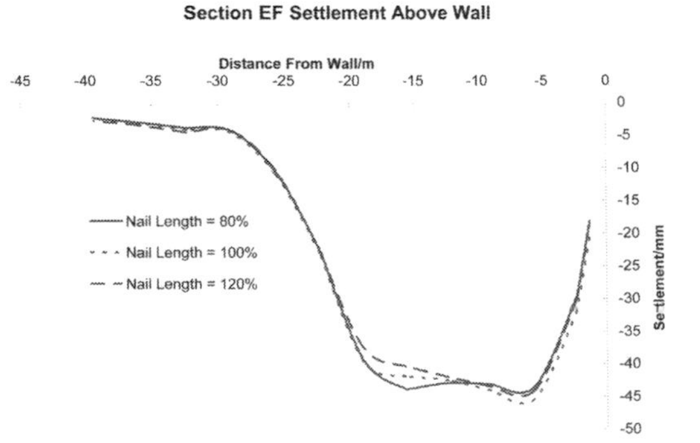

Figure 19. Influence of nail lengths on settlements at EF.

significant than strength in the nail responses. The bending stiffness of the soil nails has little influence on soil nail and ground deformations. Nail lengths has greater influence on wall deflection and little impact on ground settlements.

The findings of this paper are summarized in the following Table 7:

REFERENCES

Plumelle, F. Schlosser, 1990, "A French National Research Project on Soil Nailing: Clouterre, *Performance of Reinforced Soil Structures*", McGown, A. *et al.*, (eds), Thomas Telford, London.

Tan SA, GR Dasari and Lee CH, 2005, "Effects of 3D soil nail inclusion on pullout with implications for design", Ground Improvement, 9, No.3: 119–125.

New Horizons in Earth Reinforcement – Otani, Miyata & Mukunoki (eds)
© 2008 Taylor & Francis Group, London, ISBN 978-0-415-45775-0

Case study of Geotextile Method on extremely soft ground

K. Iwataki
Coastal Development Institute of Technology, Tokyo, Japan

K. Zen
Graduate School of Kyushu University, Fukuoka, Japan

K. Sakata
Kyushu Regional Development, Kitakyusyu, Japan

H. Yoshida
Kyushu Regional Development, Shimonoseki, Japan

N. Kitayama & T. Fujii
Fukken Co., Ltd., Hiroshima, Japan

ABSTRACT: The New Kitakyushu Airport is an off-shore airport constructed on the ground filled up by dredged soils from various port projects. Those dredged soils are classified as very soft clayed soil with over 300% water content immediately after discharged. For airport construction, the settlement acceleration work was necessary to be conducted with a view to secure trafficability and soil stabilization, and also to improve bearing strength of the ground. The Geotextile Method has been adopted for this purpose. As a large-scale rupture of geotextile occurred during construction, some countermeasures have been proposed to achieve safe construction based on those work experiences. Test soil stabilization works were carried out together with laboratory analyses to confirm the trafficability of the ground after sand spreading as a cover. The analyses confirmed that the geotextile rupture occurred due to differential strength distribution of ground, and non-uniform thickness of soil spreading accompanied by load concentration on geotextile.

1 INTRODUCTION

The New Kitakyushu Airport is an off-shore airport constructed about 3km off-shore from Kitakyushu City, Fukuoka Prefecture. The air port ground was filled up by dredged soils disposed from the adjacent construction works of Kanmon Navigation Channel and New Moji Navigation Channel. The disposal site of dredged soils was made up of four separate divisions as shown in Figure 1.

The dredged soil brought over to the disposal site were composed of supersoft clayey soil with water content of 2000% just after dredging and with 300% immediately after discharged into the disposal site. Furthermore, it is a clayey soil that undergoes consolidation settlement over a long period of time.

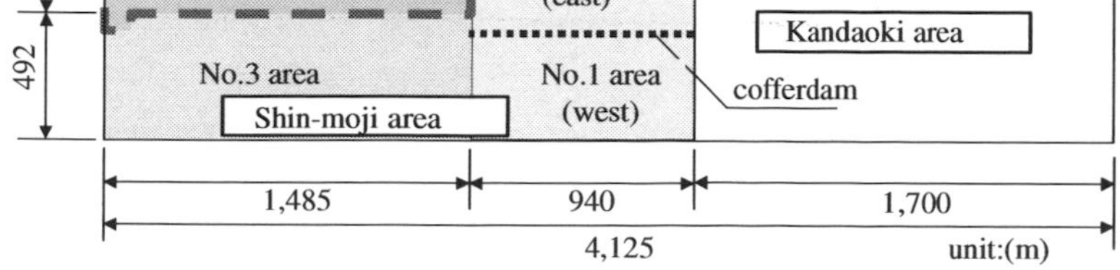

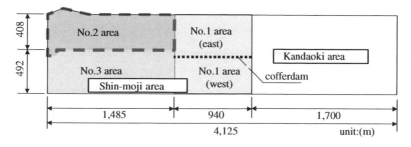

Figure 1. Plan view of New Kita-Kyushu airport construction site.

To construct an airport according to a fixed time schedule on such an extremely soft ground, it is necessary to improve the latter to a degree of hardness with the residual settlement brought down to an allowable limit. With the idea in mind, first the upper soil-layer treatment was conducted to secure trafficability and then the ground was improved to obtain sufficient bearing strength by means of consolidation settlement acceleration method and also to secure hardness gain by means of settlement acceleration method.

In this project geotextile method was adopted as a kind of surface treatment method for Shin-moji No.1 Work Section. However a large-scale geotextile rupture occurred during sand spreading. Hence, following that incident, studies were conducted to find causes of and countermeasures against geotextile rupture in order to safely conduct surface treatment method and settlement acceleration method. Furthermore, conducting loading tests by the use of ground improvement equipment, the trafficability of construction machines on the ground was checked. And then on the basis of the results, the new ground improvement and management system was formulated and applied to the project and the ground improvement works were successfully completed. In this paper, the outline of the case of geotextile rupture that occurred in Shin-moji No.1 Work Section shall be described together with the results from loading tests conducted by using ground improvement equipments.

2 THE OUTLINE OF GROUND AT SHIN-MOJI NO.1 WORK SECTION

Figure 2 shows the composition of ground at Shin-moji No.1 Work Section just after dredged soils were discharged (Egashira 2003). It is a typical presumed cross-sectional view of the ground. From this figure, it is learnt that the dredged clayey soil of thickness about 15 m were discharged onto the soft alluvial clay layer of thickness 3 m ~ 7 m.

Figure 3 shows the distribution with depth of physical characteristics and strengths of dredged clayey soil layer. From this, the shear strength of this ground is $\tau = 0.055 \sim 1.149\,kN/m^2$, water content $w_n = 120 \sim 210\%$ and wet density $\rho_t = 1.25 \sim 1.40\,g/cm^3$. These characteristics reveal that it is a super-soft clayey soil.

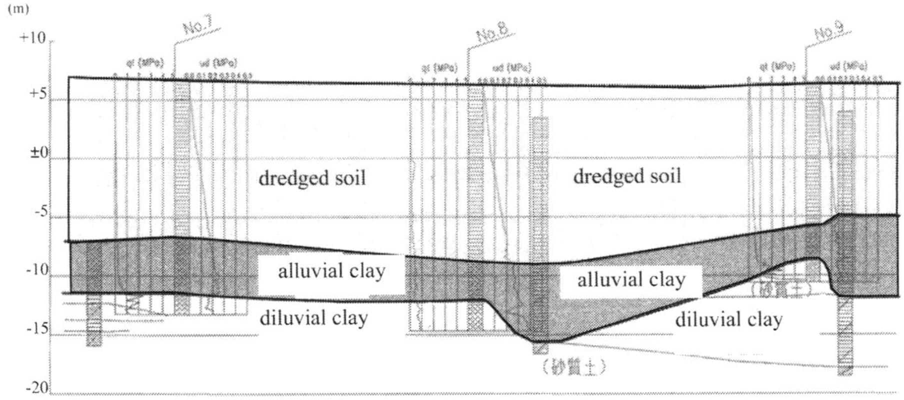

Figure 2. Presumed cross-sectional view of ground (Shin-moji NO.1 work section).

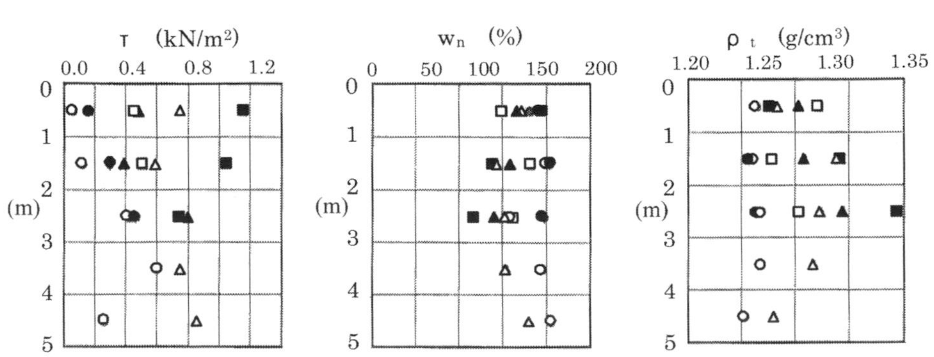

Figure 3. Strength and physical characteristics of dredged clayey soil layer (Shin-moji No.1 work section).

3 SUMMARY OF CONSTRUCTION WORKS AT SHIN-MOJI NO.1 WORK SECTION

As stated above, the dredged clayey soil has an extremelty low strength and in such a condition, it cannot be developed or improved to become a ground for construction works. Hence surface-layer treatment method was employed, in which sand spreading was done by hydraulic conveyor method after geotextiles were laid on the soil. The tensile strength of geotextile was determined by equation 1 and Figure 4, T was taken 80 kN/m

$$q_d = \alpha c N_c + T\left(\frac{2\sin\theta}{B} + \frac{N_q}{r^2}\right) + \gamma_t D_f N_q \qquad (1)$$

q_d = ultimate bearing strength
c = cohesion of clayey ground
N_c, N_q = Terzaghi's coefficients of Bearing Strength
γ_t = unit weight of soil
α = shape factor
T = Tensile strength of geotextile
D_f = Length of geotextile Settlement
r = The radius of ground when its displaced shape around loading domain is assumed appropriately as a circle
θ = Angle of inclination with geotextile

Geotextile were laid using a special reserved boat. After laying geotextiles, sand spreading was carried out. In sand spreading, first 0.9 m thick sand layer was spread using a hydraulic conveyor technique in which a micro-pump boat was employed. And then 0.6 m thick sand was spread by means of on-land conveying method in which on-land construction equipments such as bulldozer and dump truck were used. It is planned in on-land conveying method to make use of granulated slag instead of sand for weight reduction ($\gamma_t = 15\text{kN/m}^3$).

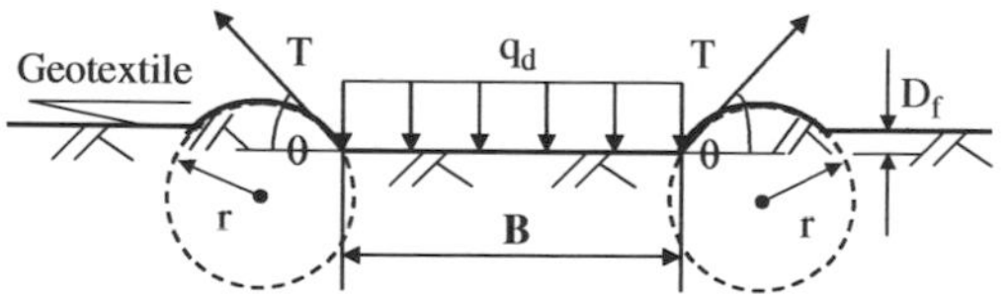

Figure 4. Transformation model concept chart of geotextile.

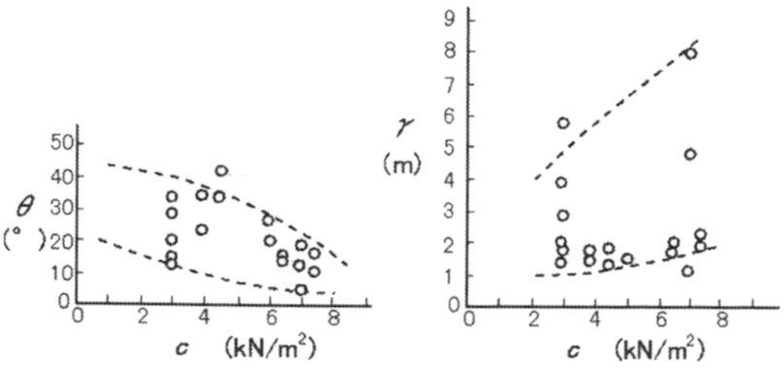

Figure 5. The relationship between c and r, θ (Nishibayashi 1980).

4 GEOTEXTILE RUPTURE

While around 50% of on-land conveying work that made use of granulated slag was in progress after the end of sand spreading by hydraulic conveyor technique, geotextile rupture suddenly occurred with the blow-out of dredged clayey soil extending to a wider area. First geotextile rupture occurred one point and got enlarged along revetment and inner separation embankment. In Figure 6, the scope of dredged soil blow-out and in Figure 7 the cross-sectional view is shown. From this it is found that the final dredged soil blow-out area is 440 m × 330 m; it is around 100,000 m^2 and the ground surface just before blow-out settled down more than 7 m with the pile-up of dredged soil on the upper layer.

5 ESTIMATION OF CAUSES OF GEOTEXTILE RUPTURE

The estimated causes of geotextile rupture are as follows.

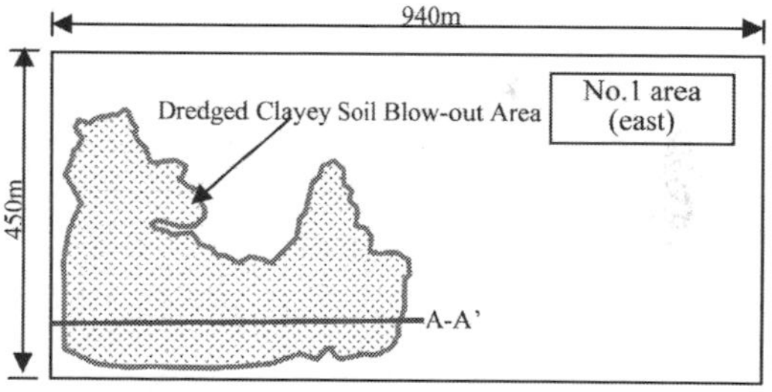

Figure 6. The scope of dredged clayey soil blow-out.

Photo 1. View of dredged clayey soil blow-out.

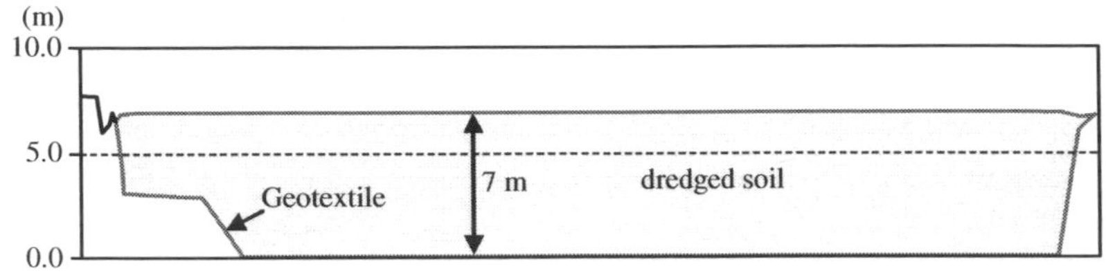

Figure 7. Cross sectional view (A-A′ Section).

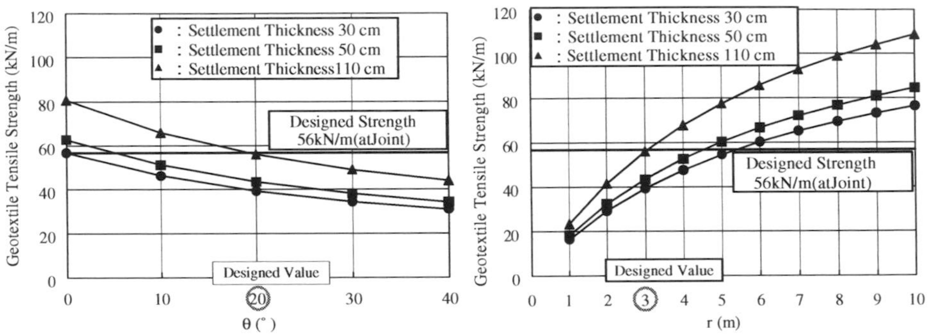

Figure 8 The relationship between r, θ and geotextile strength.

5.1 *The strength of dredged clayey soil is unevenly distributed*

When checking the procedure of land reclamation by dredged soil the site where geotextile was first ruptured n has been used as a spillway during filling of dredged soil. Generally the dredged soil was pumped in pointing toward the outer perimeter. Hence the coarse materials are prone to collect at the outer section and it is presumed that fine particles with high water content settled down at the spillway area. Thus this area had a rather low bearing capacity. And around this area some heavy equipments were passing when geotextile rupture occurred. Hence the repeated travel of heavy equipments possibly weakened the area leading to geotextile rupture.

5.2 *Sand spreading thickness is not uniform*

As stated in 5.1, that spillway area had a rather low bearing strength compared to others. Hence the settlement at this area was great leading to greater settlement by increased sand filling to compensate for the loss. Thus the non-uniform thickness of sand spreading was speculated.

5.3 *Tensile stress higher than the designed value was working on geotextile*

By equation 1, the required tensile strength of geotextile was determined assuming r and θ from ground's estimated bearing strengths.

Figure 8 shows the changes of geotextile's required tensile strengths with the changes of r and θ. From this, when r is greater than the designed value, it is possibly estimated that a great tensile stress higher than the designed value was exerting on the geotextile.

5.4 *High tensile stress was working on edge section of geotextile*

Since the ruptured geotextiles had their edges fixed to the inner separation embankment, there occurred sliding of geotextiles around that embankment as a result of ground settlement or ground bogged down and concentrated tensile stresses were exerting on a part of geotextile possibly leading to rupture.

As stated above, it is considered that various factors were involved in the rupture of geotextiles. The rupture occurred not for a single factor but for various reasons. A single rupture triggered other ruptures at other places covering a wider area.

6 COUNTERMEASURES AFTER RUPTURE

To rehabilitee the geotextile ruptures, the following countermeasures were made up on the basis of the above stated causes and improvement were made as described in Table 1.

- Increase the strength of geotextiles
- Enforce strict construction management to control sand pile-up as much as possible
- Manage Strictly the thickness of sand spreading

Table 1. Comparison of surface layer treatments before and after countermeasures.

Type of work		Before	After
Geotextile laying	Tensile strength	80 kN/m	100 kN/m
	Construction method	By boat	By boat
Sand spreading (sand)	Layer thickness	30 cm/layer × 3	15 cm/layer × 2 + 30 cm/layer × 2
	Construction method	Hydraulic conveyor	Hydraulic conveyor
	Confirmation method	From boat	By driver
Sand spreading (Granulated slug)	Layer thickness	30 cm/layer × 2	30 cm/layer × 2
	Construction method	On-Land conveyor	Dry hydraulic conveyor
	Dump load	15 kN/m²	4 kN/m²
Drains installation	Water level	Drain by pump up	High water level as much as possible

Table 2. Results from FEM analysis and physical parameters.

	Slag			Sand			Dredged soil				
Case of analysis	E (kN/m²)	γt (kN/m²)	ν	E (kN/m²)	γt (kN/m³)	ν	Surface Cu (kN/m²)	Surface E (kN/m³)	γt (kN/m³)	ν	Geotextile EA (kN/m)
Case 1	18660 ($\phi = 35°$)						0.15	$E = 210\,c_u$ $= 31.5$			
Case 2	E = 105 qu =21000	13.0	0.35	10500 ($\phi = 30°$)	18.0	0.35			13.5	0.4	49000
Case 3	qu = 2000 kN/m²						0.80	$E = 210\,c_u$ $= 168$			

- Decrease the surcharge; the transport of granulated slug should be changed to dry hydraulic conveyor method and keep water at high level as much as possible
- Conduct thorough crisis management such as dynamic observation of the ground behavior.

7 LOADING TESTS

In the present work section, there occurred rupture of geotextile during sand spreading. Hence Loading tests were conducted using ground improvement machine (PD driving machine) with a view to checking the stability of passage of construction machine before settlement acceleration works. The stability of passage of construction machine was confirmed by comparing the test results with those of FEM analysis. Furthermore, contemplating the construction of surcharge embankment after ground improvement, the stability was checked with regard to loading by dump truck.

7.1 *Summary of loading test and test results*

The ground improvement machines used for loading tests were of two types having dead weight 184 ~ 342 kN (loading 20 ~ 25 kN/m²). Tensile stresses on geotextile and settlements of geotextile were recorded while ground improvement machine was passing in the vicinity of geotextile. From the results of tests the geotextile settled down 170mm at maximum and it became approximately zero when the machine got out of the test area. And the tensile stress exerting on geotextile was around 35kN/m at maximum; that was far below the allowable tensile strength of geotextile : (80kN/m).

7.2 *Checking by finite elements model (FEM) analysis*

The results from loading tests were verified by FEM analysis. FEM method utilized in this project was two dimensional soil and ground water coupling method. Cases of analysis and physical parameters used in the analyses were shown in Table 2.

Furthermore, the results from FEM analyses were shown in Figure 9. From this it is judged that CASE 3 reproduced the results obtained from loading tests.

In this FEM model, it is assumed that soil foundation and the geotextile material are linear elastic bodies and the geotextile material has no strength against the compression stress. In terms of drainage condition of this model, sand layer spread on the dredged soil is assumed to be impervious by taking rapid loading into consideration and be pervious layer of granulated blast furnace slag is assumed as pervious.

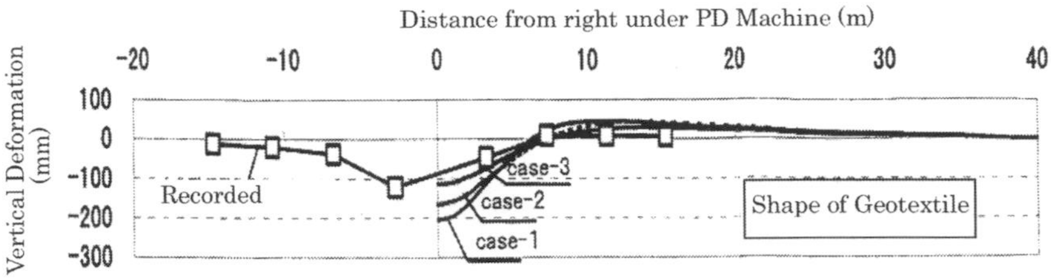

Figure 9. Results from FEM analysis.

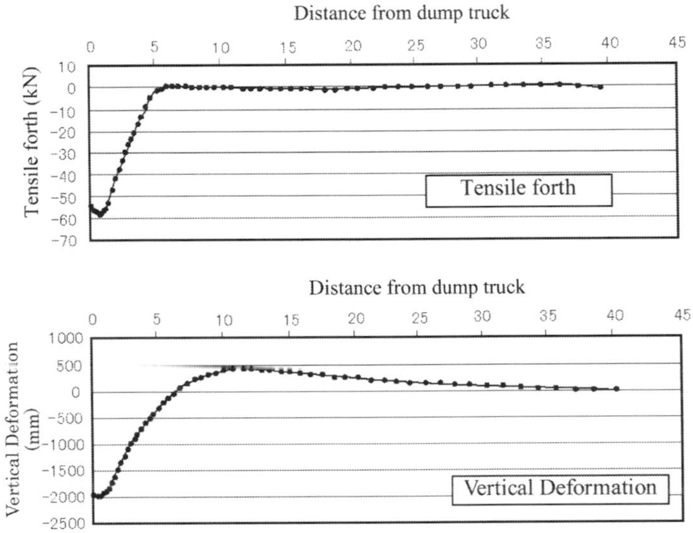

Figure 10. Results from FEM analysis.

7.3 *Analysis during loading by dump truck*

Analysis was conducted on the behavior of geotextile in case the dump truck that shall be used for construction of the after-improvement loading embankment was passing nearly. The FEM analysis (7.2) that could reproduce the loading test results was employed for this purpose.

The results from the analysis were shown in Figure 10. From this, the maximum tensile stress exerting on geotextile was 58 kN/m and geotextile settlement was 1.95 m.

Thus the maximum exerting tensile stress was far below the allowable tensile strength (80 kN/m) of the geotextile used in the project. Hence it was judged that the dump truck could safely be employed for the loading embankment construction.

8 CONCLUSION

With the objective of constructing an airport on the disposal site of dredged soils produced from port development projects, the site had been subjected to surface treatment (geotextile Laying + Sand Spreading). However a large scale geotextile rupture occurred during work execution. Countermeasures were envisaged to execute works safely estimating the causes of rupture and trafficability of construction equipment was checked by loading tests. As a result, sand spreading and settlement acceleration works were safely conducted leading to successful completion of the project.

The results were employed to neighboring disposal sites.

REFERENCES

Egashira, K. 2003. A study about the settlement and the stability used super soft dredging clay in the airport construction. *Doctoral dissertation, Kyusyu University.* (in Japanese).

Nishibayashi, K. 1980. The surface processing method (sheet), *Civil Engineering Journal, vol21, No13, 48–56.* (in Japanese).

New Horizons in Earth Reinforcement – Otani, Miyata & Mukunoki (eds)
© 2008 Taylor & Francis Group, London, ISBN 978-0-415-45775-0

Reactivation of a geogrid-bridged sinkhole: A real life solution approval

D. Alexiew
HUESKER Synthetic GmbH, Gescher, Germany

ABSTRACT: In 1993 a critical huge sinkhole funnel in a karstic area on the German Federal Highway B180 near Eisleben was bridged and secured for the first time in Germany using extremely high-strength low-strain geogrids. Philosophy, design and construction of the high-strength geogrid solution are shortly described. In October 2001 the sinkhole funnel re-opened. The geogrid system hold the road for over one hour (although the owner asked in 1993 for 15 minutes) over a funnel of more than 15 meters, which was enough to stop the traffic. The solution proved to be successful in preventing disasters of this type. It is the first case known when a geogrid sinkhole-bridging was tested by real life.

1 INTRODUCTION

For a long time the village of Neckendorf, south of the town of Eisleben in Germany, has been the repeated object of intense geological and geotechnical interest. This was prompted by a series of spectacular sinkholes and secondary ground failures near and on the German federal highway B180 (previously called the F180). In June 1987 a sinkhole, when it reached its fully developed state (the funnel was over 15 m diameter at the surface and 25 m deep), caused the complete destruction of the road and the closure of the unsafe section. The sinkhole was backfilled soon after the incident and a temporary diversion built. The increase in traffic after the German reunification and the generally unsatisfactory situation with regard to a temporary diversion prompted highways authority to commence planning the safe reopening to traffic. Bridging the sinkhole with a geosynthetic solution was put forward as the preferred option. The German Federal Highways Office (BASt) approved and confirmed this decision in 1992. This prepared the way for the first use in Germany of geosynthetic reinforcement for bridging a sinkhole.

2 PHILOSOPHY, CONCEPT AND DESIGN OF THE SINKHOLE BRIDGING STRUCTURE

The top layers of the affected zone consist of around 160 m of thickly-bedded soils; principally silts and clays, gypsum, anhydrite and limestone. Pronounced leaching effects are present particularly in the so called Zechstein layers, with cavernous gypsum sometimes with open voids, residue from leaching and seepage deposits. Numerous depressions and minor sinkholes with some major sinkholes are typical for this Karst region. The structure of this type of sinkhole can be simplified to a large cavern deep underground, a vertical chimney passing upwards from the cavern and a much wider sinkhole funnel on the surface.

2.1 *An engineering view of the problem*

The problem is the result of a natural process; the formation of a cavern deep underground – the chimney extends upwards – and a sinkhole funnel appears at the ground surface. After filling a chimney and funnel there is always the risk of secondary failures because the leaching processes in the caverns continue. A prognosis at any particular time cannot be given. The only engineering solution in such cases is to neutralize the consequences of the sinkhole for the road on the surface. The sinkhole funnel can form in a relatively short time. In 1987 the B180 road at Neckendorf near Eisleben was destroyed by such a major sinkhole, which occurred as a result of the above phenomenon. The road collapsed over its whole width. The layers in the lower section of Zone 1 collapsed first and later those in Zone 2 followed due to further subsidence (Fig. 1). In 1987 the crater was completely filled with loosely placed imported stone and sand. However, there was no information about the density and stability of the failed soils and the new fill material in the sinkhole, nor about the water flows and the processes continuing in the caverns deep underground. Therefore the important traffic route B180 remained closed on safety grounds until 1993. A temporary diversion had to be used.

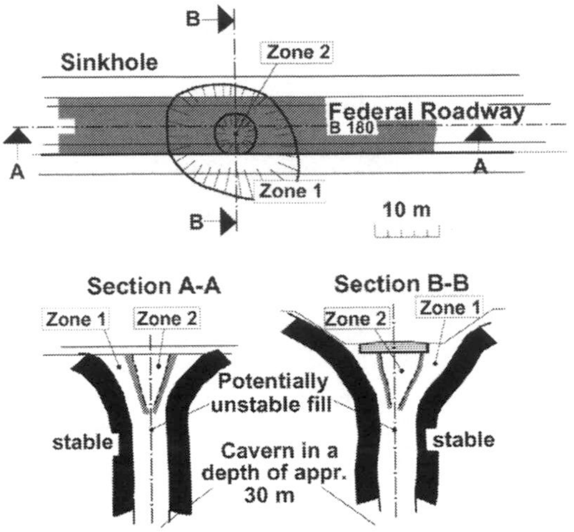

Figure 1. Geometry and potential sinkhole zones under the B180.

2.2 *Concept and philosophy of the reconstruction*

In 1992, the regional highway authority decided that the 1987 temporary diversion was no longer acceptable. Based on the history of the sinkhole and the measurements taken after filling, it did, however, seem plausible to predict that there was a higher probability of fresh failures (secondary subsidence) in the smaller Zone 2 and a lower probability of fresh failures in the larger Zone 1. The worst-case scenario would be a catastrophic failure of the whole chimney and with it Zone 1 (Fig. 1). There were principally two possible solutions under discussion: a bridging reinforced concrete slab ("hidden bridge") and, for the first time in Germany, a geosynthetic-reinforced soil solution. The reinforced concrete slab was discounted mainly due to one decisive reason: the brittle failure mode in the event of the ultimate load capacity being exceeded ("brittle failure without warning").

Therefore the concept of a innovative solution involving a heavily geosynthetic-reinforced gravel cushion was preferred. This system would retain its load-carrying capacity and remain fit for use up to a very large deformation. Approaching failure, it is ductile rather than brittle and thus would undergo a "failure with warning" after an adequate time period and in a suitably safe manner.

The final safety philosophy and concept included the following significant key characteristics and requirements:

- The primary consideration was the safety of the driver and the vehicle traveling at up to 100 km/h on the B180 right where a new large sinkhole opens (Zone 1, Fig. 1). The longitudinal and transverse deflections (or bending or settlement) of the carriageway had to be kept within acceptable limits and the carriageway should not crack or collapse locally over the underlying gaping "large" funnel (Zone 1, Fig. 1). No sharp edges / steps should form on the carriageway.
- The system would have to safeguard the traffic in this way for a short time only (10 minutes at the most). This limit arose partly for safety and engineering reasons (it was a new first-time application in Germany), but above all for economy.
- Within the 10 minute period, a detection and warning system should stop the traffic in both directions at a distance of several hundred meters by means of automatic stop signs.
- The area to be protected was located in a cutting with the effect that only a flat, thin, geosynthetic-reinforced cushion placed almost directly under the road construction could be considered. The solution would involve the minimum of excavation and fill.
- In the worst case, the solution would have to bridge over a funnel with a diameter of up to 15 m.
- In this worst case, the relative deflection of the carriageway (ratio of the settlement in the centre of the depression to the diameter) must not exceed 0.06–0.07.
- The project, the first such in Germany, represented a major and unique engineering challenge, as it would even today.

2.3 *Structural analysis, design and detailing*

The structural analysis, design and detailing was carried out in spring 1993 in the Engineering Department of HUESKER Synthetic GmbH (HUESKER, 1993). Three possible design methods were examined: (Giroud, 1982), (Giroud et al, 1990) and (BSI, 1995), which were all based on the use of the membrane theory for the geosynthetic reinforcement solution, the theory being considered safe and plausible at that time. BS 8006 was then only available in draft. The BS 8006-method (Fig. 2) was preferred due to different reasons (Alexiew, 1997 & Alexiew, 2007).

Later on for other sinkhole projects modified design procedures were developed (Alexiew et al, 2002b). One of the focal points is that the limitation of strain in reinforcement reduces the deflection (d/D) and finally the deflection on the surface(ds/Ds) (Fig. 2). Further details of the model and analysis can be found in (BSI, 1995). Generally, a geogrid was preferred as reinforcement (instead of e.g. a woven geotextile) mainly due to its higher coefficient of interaction (bond) to the soil.

The determination of the design strength of the reinforcement was carried out in accordance with the requirements of the German Guidance Note (FSGV, 1994), which was only available in draft form in 1993. The final design calculations were performed in a way to minimize deflection (ds/Ds, Fig. 2) to 0,02–0,03

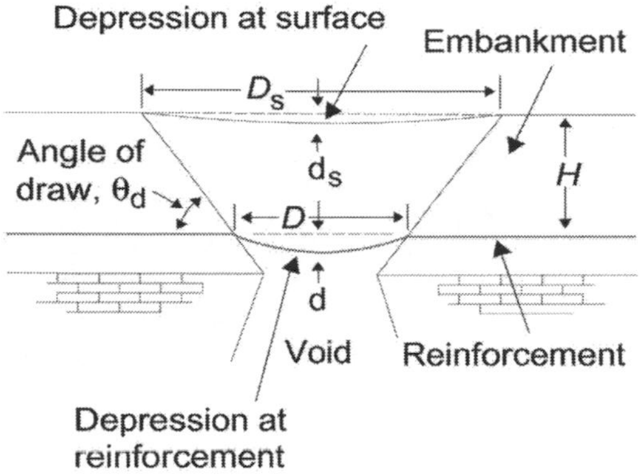

Figure 2. Analysis model in accordance with BS 8006.

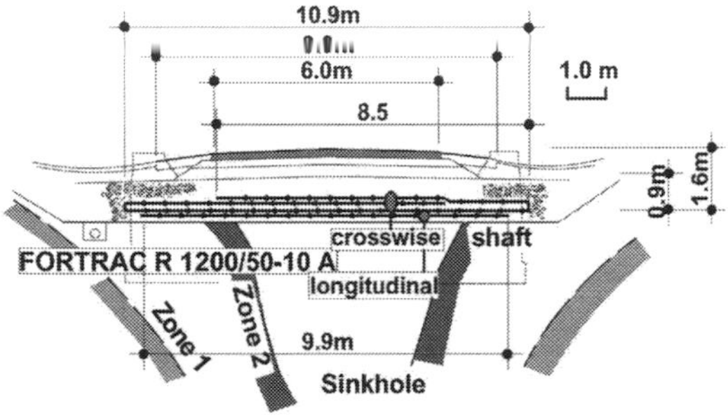

Figure 3. Schematic cross section of the sinkhole bridging system.

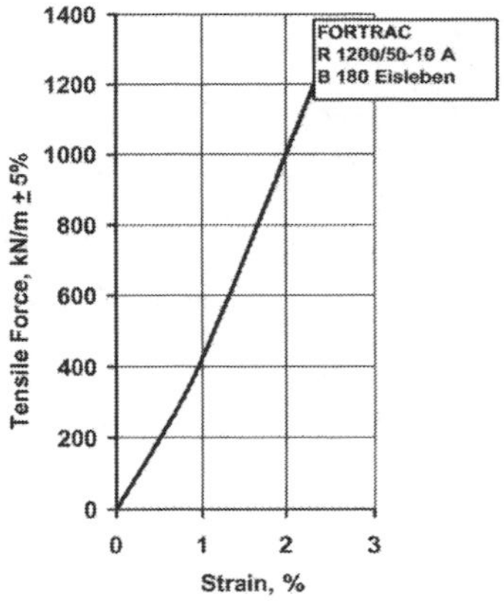

Figure 4. Tensile force/strain graph for the new geogrid used.

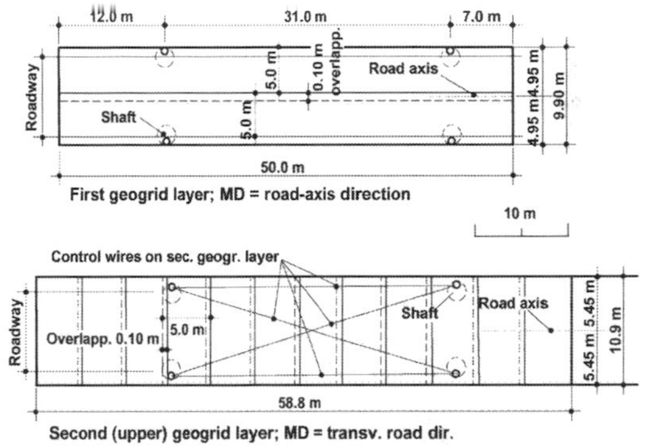

Figure 5. Simplified layout of the reinforcement and the warning system.

for bridging the smaller funnel (Zone 2) and to 0,06–0,07 for bridging the worst case (failure of Zone 1, big funnel) (Fig. 1).

The analysis, the polyvariant calculation with various load cases and variations of other properties showed that a uniaxial low-creep geogrid with a mobilisable tensile force in the rolled-out (or "machine direction" MD) direction of 1200 kN/m at ≤3.0% strain and 600 kN/m at ≤1.5% strain (short-term) would be required. More detailed explanation regarding the design assumptions and the engineering background for the final solution (Fig. 3) can be found in (Alexiew, 2007). Geosynthetic reinforcement with these properties was not yet available in 1993. Thus, a new geogrid had to be developed for this project with extreme strength, high tension modulus and low creep. The choice was a yarn made from Aramid. A five meter wide uniaxial geogrid was specially developed, manufactured and tested. A world first. The typical tensile force/strain graph (short-term) in the rolled-out direction (MD) is shown in Figure 4.

Figure 5 shows a plan view of the geogrid reinforcement (simplified execution drawings from (HUESKER, 1993)) and the warning system (see section 2.2). A high-quality clean well-graded gravel (soil classification GW) within the 0.1/56 fraction was specified for the material for the reinforced gravel layer. It had to be compacted to a relative Proctor density of Dpr ≥ 103% in order to ensure good mechanical properties and the composite action of the gravel-geogrid-system.

3 CONTROL AND WARNING SYSTEM

Due to the great importance of the B180 and the high traffic flows it carries, the complexity of the problem and the impossibility of an exact analysis of all the theoretically possible situations and the then uniqueness of the project (see "Philosophy" in Section 2.2), the client decided to install a control and warning system with steel wires and control chambers. The layout is shown in Fig. 5. The almost non-extendible wires were connected to electrical sensors (contacts) in the chambers.

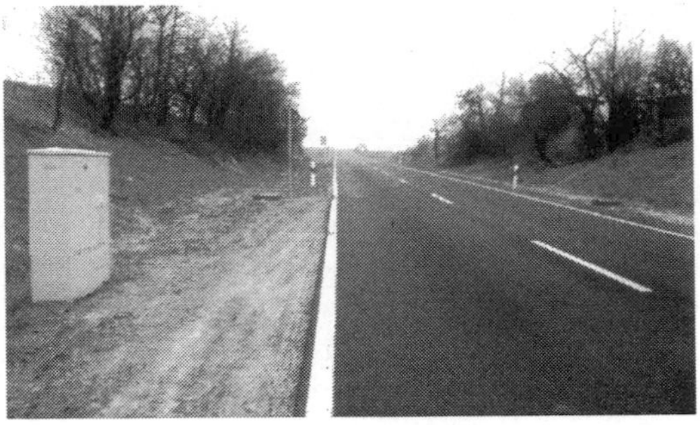

Figure 6. The B180 after completion: the carriageway, two of the control chambers and the warning system automatic switching cabinet.

Figure 7. The B180 operating again: a warning stop sign in the foreground, the area made safe in the background.

If the movement of the ends of the wires indicates a settlement of the reinforced soil structure equal to a strain of 1.5% (ca. half the ultimate, Fig. 4) in the geogrid, a warning system comes into operation and traffic in both directions is stopped at a safe distance on both sides of the critical area by electronic warning signs. The warning system was designed to be activated before the complete opening of the "large" Zone 1 (Fig. 3).

4 EXPERIENCE DURING THE CONSTRUCTION PHASE AND AFTERWARDS

The geogrids were supplied prefabricated in the placement lengths required by the design. The flexibility of the geogrid and its relatively low weight per unit area (high specific strength) meant that handling was easy and a small team was able to install the geogrids on site.

A cross beam was used to install the geogrid. It had already been realized that for bridging sinkholes, it was most important for the geogrid to be tightened, as even the best high-modulus geosynthetic reinforcement loses efficiency if it does not immediately react (activate) as a result of improper placement. For more details concerning construction technology for bridging sinkholes see (Alexiew et al, 2002a & Sobolewski, 2001).

The sandy gravel fill layers were compacted to $Dpr \geq 103\%$. The entire system over a length of about 60 m was constructed in a week in October 1993.

The reconstructed section of the B180 was back in operation in October 1993 (Figs. 6 and 7). The road has been monitored for deflection ever since using the measurement and inspection chambers.

A more detailed information regarding design and construction can be found in (Alexiew, 1997 & Alexiew, 2007).

5 RENEWED SINKHOLE ACTIVITY IN OCTOBER 2001

Between 1993 and 2001 any settlement of the carriageway was visually monitored at regular intervals and the sensors attached to the wires in the warning system inspection chambers were checked for any displacement. No deformation was detected in eight years of monitoring. The mechanics and electrics of the warning system were inspected and maintained.

Then on 17.10.2001 (eight years after the construction of the sinkhole bridging system) there was something to measure: Renewed sinkhole activity and reopening of the sinkhole funnel under the B180.

What follows is the chronological reconstruction of the events according to eye-witness statements of the occupiers of the allotments near the protected zone.

About 18:00: The first noises from the side slopes, which were starting to move (area in cutting, see above). A sinkhole funnel appeared in the slope to the east of the protected zone. The traffic on the B 180 continued to flow.

About 18:30: Settlement on the carriageway surface could now clearly be seen. At this point in time many vehicles were still passing over the site at undiminished speeds of 100 km/h. The warning system, which was intended to stop traffic in both directions at a safe distance away in the event of increased deflection (settlement) (see above), did not react.

About 18:45: The deformations continued to increase and affected a very large area. The local people managed to stop the traffic and informed the authorities. The warning system did not react.

About 19:00: One hour after the start of renewed sinkhole activity, the whole of the carriageway area was undermined and the cutting slope to the west side of the road collapsed. The sinkhole funnel was already bigger than the width of the reinforced system (Fig. 3) i.e. the pavement width plus the trenches on both sides

Figure 8. View in the morning after the renewed sinkhole activity.

Figure 9. Cracks in the asphalt in the geogrid-anchorage zone; the funnel is to the right, just out of the photograph; one of the control chambers in front.

(say 12 to 13 m); the "large" Zone 1 (Fig. 1) had collapsed. The carriageway was severely deformed but was still intact as a whole unit.

About 19:30: The sinkhole funnel had greatly increased in size in all directions. The carriageway is severely deformed but is still "standing" as a whole unit. At this point the soil structure had been bridging an irregularly-shaped sinkhole funnel with a diameter estimated to be between 12 and at least 15 m for more than half an hour. One and a half hours had elapsed since the renewal of sinkhole activity. Shortly after 19:30, the carriageway collapsed (by tearing) (i.e. including the geogrid) over the increasingly widening sinkhole and fell into the funnel, which had also increased in size. For understandable reasons it cannot be said with any certainty whether the diameter was 16 or 18 or 20 m. Investigations would not reveal whether the reinforcement tore exactly in the middle of the funnel. The system's anchorage zones under the road in front of and behind the funnel remained intact.

6 THE DAY AFTER

Next morning the irregularly-shaped funnel had a diameter of over 20 m (Fig. 8). It cannot be clearly established whether and by how much the hole in the carriageway (on the previous evening) had grown after the tear. Its position corresponded with the estimated position of Zone 1 in 1993 (see Fig. 1). It was obvious that the bridging system had more than met the requirements and expectations of the 1993 design and construction in terms of load carrying capacity, deformation and behaviour over time. This was all the more crucial as the warning system and the stop signs had not reacted.

The following three main points were now of greatest interest: checks of the geogrid behaviour (or its current condition) and the anchorage zone and the answer to the question of why the warning-stop signs did not react (warning system).

6.1 *Geogrid and anchorage*

Several square metres were cut out and recovered from the various geogrid layers before being tested. Visually the recovered geogrid appeared to be in very good condition, even near the tear site. The values obtained from the tests were compared with the records of the load-strain wide strip tests from 1993 of the geogrid as produced. The only recorded change was a slight increase in the tension modulus but no loss of strength, even after eight years in the reinforced gravel layer under heavy traffic loading on the B180, with the geogrid very close to the carriageway surface and following loading to failure. The two anchorage zones under the road in front and behind the funnel had remained intact. The geogrid had not pulled out of them despite loading to failure over the funnel.

Tensile forces of at least 1200 kN/m would have been carried (anchored) over a relatively short stretch. The wide cracks in the asphalt surfacing in the anchorage zone near the sinkhole funnel gave an indirect indication of the large loads on the area (Fig. 9).

6.2 *Warning system*

The question of why the warning system and the stop signs did not operate was investigated in detail. The investigation revealed that the failure to react was not due to the (simple and reliable) design concept nor to the construction of the warning system but rather that at the last inspection two weeks earlier someone had forgotten to switch the power back on to the electronics.

7 COMMENTS AND CONCLUSION

Upon the reopening of the sinkhole, the first sinkhole bridging system incorporating geosynthetic reinforcement to be designed and constructed in Germany functioned better than predicted: it bridged a sinkhole that was larger than the design requirements for longer than was required. The 1993 system had been correctly planned, designed and constructed. The events proved the philosophy of a "ductile failure with warning", upon which the project had been based. The bridging of an oval funnel with uniaxial reinforcement is possible and functions well with proper design and implementation. A flat soil-geogrid bridging system close to the road surface can be feasible and will function correctly with suitable choice of design and reinforcement. The first project in the world to incorporate an Aramid® geogrid (in this case with a short-term strength of 1200 kN/m) was proven in practice. The method of analysis and design of sinkhole bridging systems according to (BSI, 1995) is sufficiently correct at least for relatively thin bridging layers using non-cohesive fills; the same applies for the "elastic membrane theory". For warning systems it should be the aim to eliminate human involvement technically or logistically as much as possible. And: the project described shortly herein considers a very rare event – the occurrence of the "worst case design scenario" (somewhat like the 1 in 100-year earthquake). As far as is known, this is the first time a sinkhole bridging solution has been tested in real life.

8 EXPRESSIONS OF THANKS

Several institutions and people were involved in the concept, design, approval, construction etc. of the 1993 project and in the procedures following the renewal of sinkhole activity in 2001. All of them deserve thanks for their readiness to innovate and their courage in adopting new efficient engineering solutions and for engaging in the open and beneficial discussions.

REFERENCES

Alexiew D., 1997, *Bridging a sink-hole by high-strength high-modulus geogrids*, Proc. of Geosynthetics '97 Conference, March 1997, Long Beach, USA, pp. 13–24.

Alexiew D., Sobolewski J., Pohlmann H., 2000, *Projects and optimized engineering with geogrids from 'non-usual' polymers*. Proc. of 2nd European Conference EuroGeo 2, Bologna, 2000, pp. 239–244.

Alexiew D., Sobolewski J., Ast W., Elsing A., Hangen H., 2002a, *Erdfallüberbrückungssystem Eisenbahnknoten Gröbers: Zur Berechnung, Bemessung, Ausführungsplanung und Ausführung*. Proc. KGeo 2003, February 2003, Munich, in Special Issue "Geotechnik", DGGT, Essen, August 2003, pp. 235–241.

Alexiew D., Elsing A., Ast W., 2002b, *FEM-Analysis and dimensioning of a sinkhole overbridging system for high-speed train at Gröbers in Germany*, Proc. 7th ICG, Nice 2002, pp. 1167–1172.

Alexiew D., 2007, *Reaktivierte Erdfallsicherung mit hochfesten Geogittern*. Bautechnik, Sonderpublikation Bauen mit Geotextilien, Februar 2007, Ernst & Sohn, Berlin, pp. 17–24.

British Standards Institution – BSI, 1995, BS 8006 *Code of practice for strengthened/reinforced soils and other fills*. London, UK.

Forschungsgesellschaft für Straßen- und Verkehrswesen – FSGV, 1994, *Merkblatt für die Anwendung von Geotextilien und Geogittern im Erdbau des Straßenbaus*. Ausgabe 1994. Cologne. (in the meantime Issue 2005).

Giroud J.P., 1982, *Designing of geotextiles associated with geomembranes*. Proc. 2nd Int. Conf. on Geotextiles, Vol. 1, Las Vegas, USA. pp. 37–42.

Giroud J.P., Bonaparte R., Beech J.F., Gross B.A., 1990, *Design of soil layer -geosynthetic overlaying voids*. Geotextiles and Geomembranes, Vol. 9. pp. 11–50.

HUESKER Synthetic GmbH, 1993, D. Alexiew: *Project and design calculations for the re-opening of the Road B 180: Reinforcement with high-strength geogrids*. Gescher, March 1993 (unpublished).

Sobolewski J., 2001, *Erdfallsicherung mit einaxialer geosynthetischer Bewehrung: Ortsumfahrungen Zeitz-Theißen und Dingelstädt*. Proc. KGeo 2001, February 2001, Munich, in Special Issue "Geotechnik", DGGT, Essen, August 2001, pp. 233–239.

New Horizons in Earth Reinforcement – Otani, Miyata & Mukunoki (eds)
© 2008 Taylor & Francis Group, London, ISBN 978-0-415-45775-0

Two examples of recent innovation linked to optimization of soil reinforced structures

Ph. Delmas
Cnam

A. Nancey
Tencate Geosynthetics France

ABSTRACT: The paper presents 2 cases of innovative design. Both are linked to a previous research on reinforcing soil above soil subsidence area. The first one allowed reducing the vertical loads on large water pipes under a high embankment and the second allowed solving the problem of crossing a soil subsidence area by a motorway in a cut-off, by using a bimodulus reinforcing geosynthetic.

1 INTRODUCTION

The optimisation of the earth reinforcement structures is a permanent task of the civil engineer and the high innovation potential of the geosynthetics offers the engineers a great potential. Two structures built recently in France show the interest of such developments. Both of them are based on further developments of a previous research program "Rafael" (Delmas & al., 1999) on geosynthetic reinforcement in case of risks of soil subsidence.

2 REDUCTION OF VERTICAL LOAD ON A LARGE WATER PIPE BY USING GEOSYNTHETIC REINFORCEMENT

In Valenton, city close to Paris, the stresses applied by the high embankment (>10 m high) of a new motorway (A86) on large water pipes (2.76 m & 2.24 m diameter), which may be partly under water during certain periods of the year, could be reduced by using a specific reinforcement structures and allowed an important reduction of costs of the structure.

2.1 *Principle of the solution proposed*

The principle of the proposed solution consists in reducing the vertical stresses above the pipes by using a compressible material above the pipes and creating above this compressible area a reinforced mattress which transfers the vertical loads laterally on two reinforced structures.

Additionally a specific layer of reinforcement geosynthetic was placed above the pipes and anchored laterally counterbalancing the uplift forces due to the highest potential water table.

The principle of the proposed structure is presented in the figure 1, with the two designs which where realised depending on the number of pipes (1 or 2).

This design allowed using HDPE DN2000 and HDPE 2500 pipes (instead of metallic pipes) which reduces highly the cost of the structure.

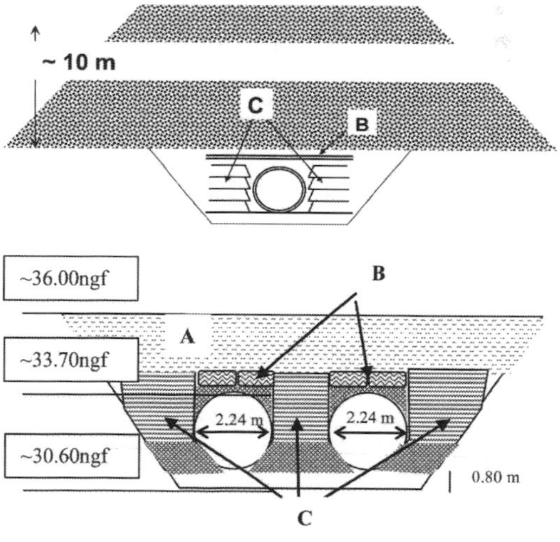

Figure 1. Principles of the structure – 1 pipe design & 2 pipes design: (A) reinforcing mattress transferring the vertical loads laterally, (B) compressive layer, (C) lateral reinforced structures supporting the vertical loads transferred from mattress (A).

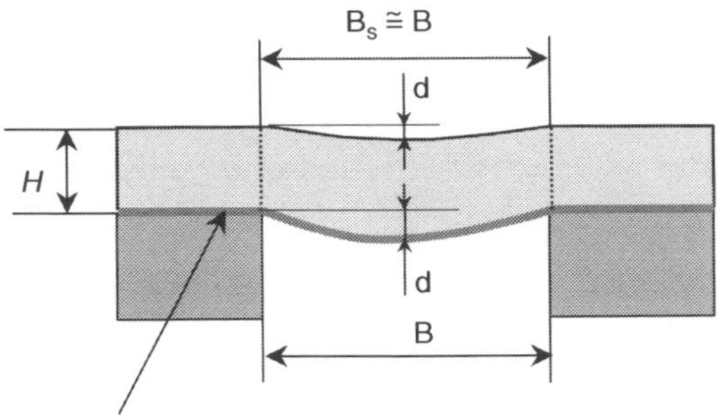

Figure 2. Membrane design principle.

2.2 *Design of the reinforcements*

The specification of the water pipes limits the vertical deformation to a maximum of 6% of the diameter.

A layer of old tires is placed just above the pipes to avoid the vertical load transfer. A reinforced mattress above the tires acts as membrane like in the case of soil subsidence (figure 2).

The reinforcements of the mattress have been designed using the method developed during the "Rafael" research program (Blivet et al., 2001). The width of the membrane effect is considered equal to the diameter of the pipes.

According to the results of this program, the influence of the dilatancy of the granular soil above the reinforcement layers has been taken into account to evaluate the load transfer on the geosynthetics.

Long term design strength has been taken into account as far as the load will last during the service life of the structure (100 years). This leads to a reinforcement by 4 layers of Rock PEC 200 (T_{max} = 230 kN/m) anchored laterally on a length of 3.3 m on the side of the side reinforced structures.

The side reinforced structures have been designed considering the added vertical surcharge due to the transfer of load by the reinforced mattress. Both external and internal long term design have been realised. This leads to reinforced structures of 1.5 m width using the same geosynthetic with a spacing of 0.4 m corresponding to an optimum thickness for the compaction of the soil used.

To control the potential uplift of the pipes linked to the changes of the water table, especially in the case of empty pipes, a reinforcement geosynthetic as been placed above the pipes and anchored laterally under the side reinforced structures.

2.3 *Construction of the structure*

The photo 1 shows the construction of the reinforcing system with the compressive layer. Several years after construction, the behaviour of the pipes is satisfactory, confirming the interest of the solution chosen.

Photo 1. View of the different construction phases.

3 USING A SPECIFIC BIMODULUS GEOSYNTHETIC FOR CROSSING A SOIL SUBSIDENCE AREA UNDER A TREATED SOIL STRUCTURE

The southwest Meaux bypass is partly situated in an area of old gypsum quarries. The detailed investigations as well as the preventive treatments of these quarries by fillings and grouting let remained a high risk of collapse.

The figure 3 shows the geological profile of the project. The east part of the project is concerned by gyps quarries excavated from the surface but also from the slope. The galleries and the rooms are situated between 25 m and 30 m under the surface; this means between 15 m and 30 m under the level of the finished project.

The "old" quarries (XIX century) situated under the slope are not regular and the evaluation of average destruction level is around 75%. The height is estimated to 1.8 m.

The photo n°2 confirms that risks are real. It shows a subsidence discovered during the realisation of the road close to an injection borehole. During the injections the cavity has not been discovered.

3.1 *Design of the solution*

In similar situation the geotechnical state of the art in France plans to realise injections with a square pattern of 5 m × 5 m.

For the "old" quarries, the solution proposed consists in reinforcing the base of the structure by a

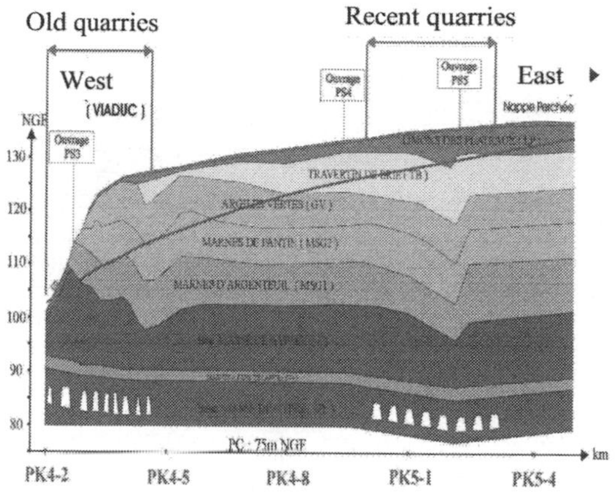

Figure 3. Geological profile of the project. The line corresponds to the final level of the road.

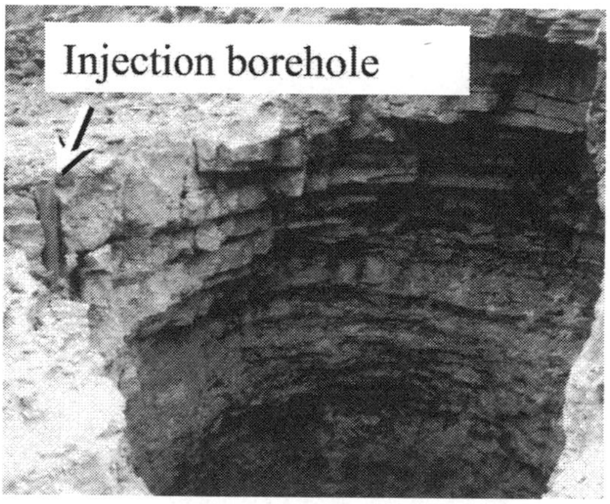

Photo 2. Cavity discovered during the construction close to an injection borehole.

geosynthetic which allow enlarging the pattern of injections to 10 × 10 m under the road and 15 × 15 m under the slopes. For the "geometrical" part, the spacing between the boreholes depends on the density of the columns in the quarries.

The road structure consists in 1.10 m with (1) 47,5 cm pavement $\gamma = 22\,\text{kN/m}^3$, (2) 35 cm base layer treated with 2% of lime and 6% cement ($c = 50$ kPa et $\varphi = 35°$, $\gamma = 20\,\text{kN/m}^3$), (3) 27,5 cm silt layer treated with 2% of lime ($c = 30$ kPa, $\varphi = 30°$ et $\gamma = 20\,\text{kN/m}^3$).

The design of the geosynthetic required to take into account:

(1) cavity of a diameter of 2 m,
(2) a maximum vertical displacement at the surface d_s under the structure own weight of 10 cm
(3) the following reduction factors (in accordance with the French regulations)

- $F_{inst} = 1,1$ (installation damage),
- $F_{env} = 1,05$ (coefficient linked to the environmental behavior, chemical degradation),
- $F_{géo} = 1,2$ (safety factor on the geosynthetic),
- $F_{flu} = 1,54$ (creep factor linked to the reinforcement cables used for a duration of the loads of 1 year maximum).

This means a global factor of:

$$F = 1,1 \times 1,05 \times 1,2 \times 1,54 = 2,13$$

The ultimate design strength is calculated under the load of the structure with a vertical load of the 1/2 of a 13 to axle.

The polymers used shall resist to the physicochemical conditions of the soil. In this case the use of treated soil with lime and cement induces the requirement to resist to a pH of 11.

As the choice of the polymer has been driven by the mechanical behaviour, it has been decided to separate the geosynthetic from the treated soil by 2 HDPE géomembranes on both side of the reinforcement. This means that the friction angle is 11°.

3.2 *Analytical method*

According the results of the real size experimentation "Rafael" (see §2.2) a specific design method was developed.

Considering the assumption of a geosynthetic stiffness of $J = 2900$ kN/m and $\varphi' = 30°$, $c' = 0$ the results of the design are the following:

- vertical displacement of the surface $d_s = 10,5$ cm with a dilatancy of 3% and $d_s = 16,5$ cm without dilatancy;
- maximum strain of the geosynthetic $\varepsilon_{max} = 1,8\%$;
- service tensile strength $T_{max} = 54$ kN/m under the structure own weight and $T_{max} = 87$ kN/m under the 1/2 axle load.

It shall be considered that the "Rafael" method has been developed for granular materials. This means that the cinematic of the displacements of the soils is conform to the figure 2.

In this case, the failure zone is a cylinder having the diameter of the cavity. The vertical load applied by the soil on the geosynthetic can be considered as uniform; the deformation of the geosynthetic is parabolic and the dilatancy of the soil reduces the vertical displacement at the surface compared to the one at the level of the geosynthetic.

Nevertheless the road fill material in Meaux is treated and in case of failure it will be most probably brittle and create a rigid bloc above the geosynthetic (Figure 4).

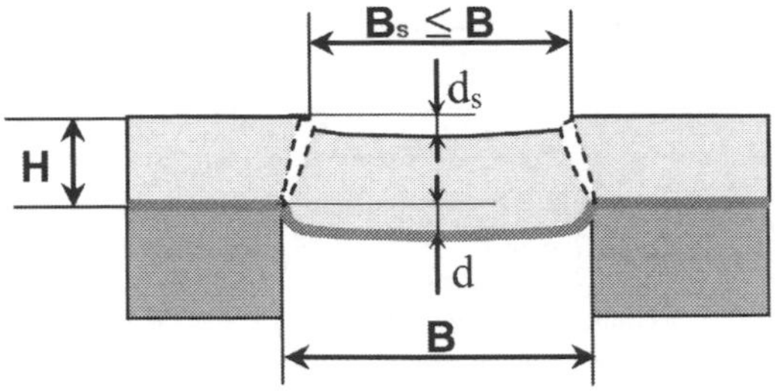

Figure 4. Cinematic principles in case of treated soils.

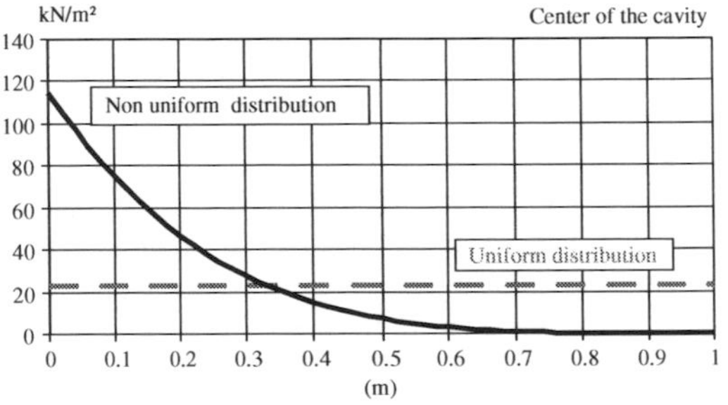

Figure 5. Distribution of vertical stresses: uniform and non-uniform assumptions.

In this case, the stress applied on the geosynthetic is non-uniform and most probably more important on the side of the cavity than in the center. To evaluate the influence of the rigidity of the soil a new design approach has been used.

3.3 *Model using a specific FEM code*

The failure mode of a cohesive soil is a difficult task and the FEM codes have difficulties to take into account the mechanisms including cracks and large deformations.

A tentative of simplified model has been realised in Lirigm to understand and analyse the effect of a non uniform distribution of stress on the membrane deformation of the geosynthetic.

When the failure occurs, it can be considered that the soil will be mainly supported by the periphery of the geosynthetic. The distribution of the vertical stress taken into account for the preliminary design is presented in the Figure 5.

The maximum stress of the soil is estimated to 110 kN/m^3 (compression resistance of the treated soil). The total load on the geosynthetic corresponds to the weight of the fill and the traffic load (q = 22.95 kN/m).

A Finite Element Code has been used (Villard et Giraud, 1998). Specific tools have been developed to allow modelling the fibre structure of the geosynthetic, the friction and the sliding between the geosynthetic

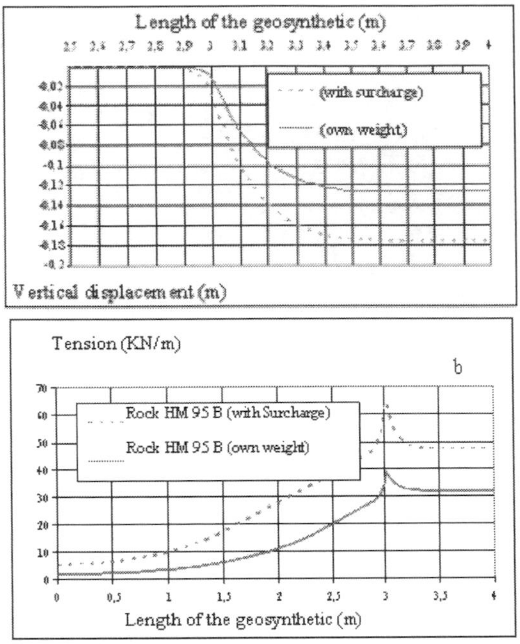

Figure 6. Distribution of the displacements (a) and of the tensions (b) in the geosynthetic reinforcement.

and the soil in anchorage area and reaching large displacements.

The action of the non-uniform vertical load on the geosynthetic has been realised by applying forces on the nods of the model (no model of the embankment fill).

The main difference with the "Rafael" model (§2.2) is the distribution of the vertical forces and the consideration of the possible slippage of the geosynthetics in the anchorage area. This changes the deformation from a parabolic shape to a more flat shape and creates larger displacements of the geosynthetic and changes in the tension.

Most of the results are presented in the figure 6. Under the own weight, the displacement of the geosynthetic is around 12 cm. If it is considered that the dilatancy of the soil is equal to 0, the displacement at the surface is the same (12 cm).

In this case, the needed anchorage length is 2.5 m if the friction angle between soil and geosynthetic is 25°. This length will be more than double in the case of a friction angle of 11° (friction between the geosynthetic and the geomembranes). The tensile load is 40 kN/m under the own weight and the horizontal displacement of the anchorage is 1 cm.

An other validation, using FLAC model, realised by Scétauroute confirms these results (Blivet et al., 2006) and the design of the structure.

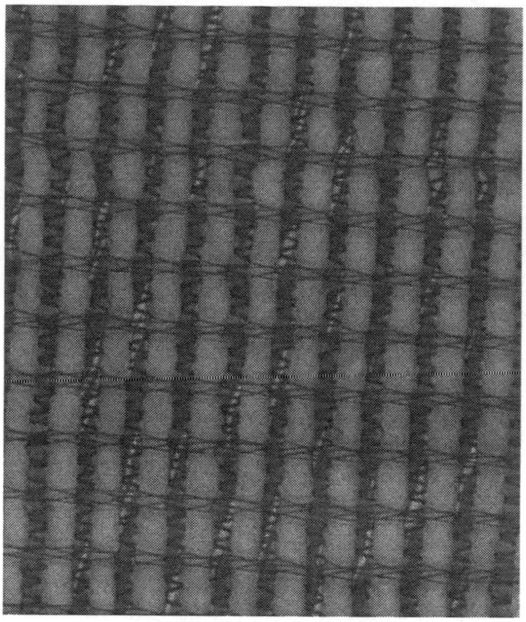

Photo 3. View of the composite reinforcement geosynthetic.

3.4 *Innovative bi-modulus reinforcement geosynthetic used for optimisation of the structure*

To optimise the answer to the requirements on both the maximum displacement at the surface and the long term safety against failure, an innovative solution has been designed and developed.

The reinforcement geosynthetic developed is a bi-modulus composite realised with combined aramid and polypropylene cables knitted on a non-woven support (Photo 3). The principle of this product has been patented in many countries of Europe, America and Asia.

The product stress–strain curve is characterised by 2 zones (Figure 7 – table 1). Between 0 and 4.5% the strength of the aramid cables is added to the one of the polypropylene yarns. This allows reaching a very high stiffness, with a linear curve. After 5% the geosynthetic behaves following the performance of polypropylene yarns until failure.

Considering the first part of the curve under the own weight of the structure, the factor of safety is greater than 2.6 and the displacement is in accordance with the above requirements. This allows limiting the elongation of the geosynthetic to 1.5%.

Considering the ultimate tensile strength of the geosynthetic, it allows to secure a factor of safety greater than 3.1. This value is considered in accordance with the expected service life (maximum of one year).

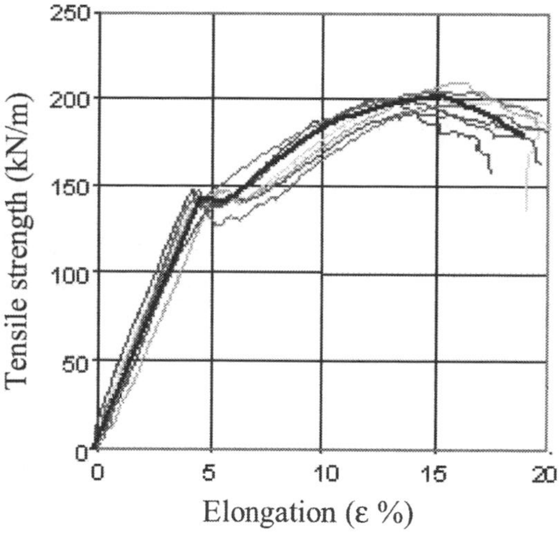

Figure 7. Stress-strain curve of the reinforcement geosynthetic.

Table 1. Main characteristics of the reinforcement geosynthetic.

Maximum tensile strength for the first part of the curve	145 kN/m
Maximum elongation for the first part of the curve	4.75%
Stiffness at 4.75% elongation	3 052 kN/m
Maximum tensile strength for the second part of the curve	200 kN/m
Maximum elongation for the second part of the curve	15%

The optimisation of the design of this product has been finalised by the Grenoble University (Lirigm) using the FEM. The tensile behaviour of product has been tested by a French accredited laboratory.

3.5 *The design of the structure*

The design is realised considering a brittle failure of the soil above the cavity.

This induces a non-uniform distribution of the vertical stresses on the geosynthetic. The main analysis has been realised using a Finite Elements Method developed by the University of Grenoble (Lirigm) which allows taking into account the friction behaviour and the eventual slippage at the anchorage level.

The photo 4 shows the installation of the geosynthetic under the road structure.

4 CONCLUSION

The interest of using geosynthetics for reinforcing platforms above soils with risk of subsidence has been clearly shown.

Photo 4. Installation of the reinforcement composite between 2 layers of geomembranes.

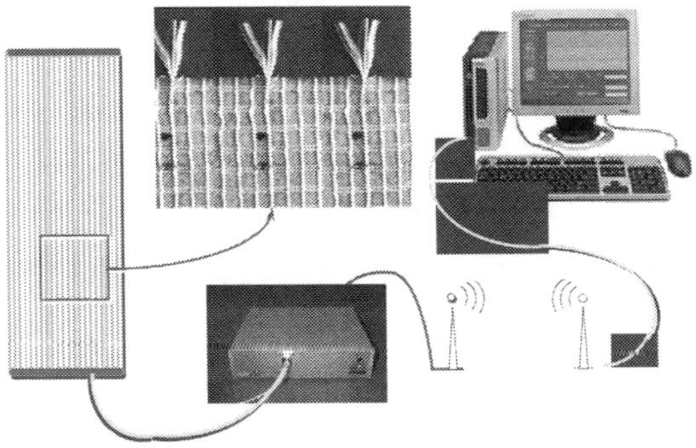

Figure 8. Principle of the Geodetect product, including optical fibbers allowing the strain measurements and the close monitoring of the site.

This solution has many technical advantages but has also economical interests, like it has been shown in the case of Meaux bypass. The bimodulus geosynthetic solution allowed to enlarge the injection borehole pattern from 5 m × 5 m to 10 m × 10 m, which represents for this project a total win for the owner of more than 8 millions €.

In addition, considering the recent development of innovative geosynthetic which offer the possibility of realisation of strain measurements and detection, it might allow an increased reduction of costs by avoiding totally the injections. This is the case of the recently developed Geodetect product (figure 8) which has been installed by the French railways under the railway in Arbois (France) above a geological fault area.

REFERENCES

J.C. Blivet, P. Garcin, A. Hirschauer, A. Nancey, P. Villard (2006): Reinforcement with geosynthetic over probable sinkholes: example of southwest Meaux bypass Rencontres Géosynthétiques Montpellier, pp. 281–288.

Blivet J.C., Khay M., Gourc J.P., Giraud H. (2001) "Design considerations of geosynthetic for reinforced embankments subjected to localized subsidence". Proc. Geosynthetics'2001. Conference, February 12–14, 2001, Portland, Oregon, USA, pp. 741–754.

Delmas Ph.., Gourc J.P., Villard P., Giraud H., Blivet J.C., Khay M., Imbert B., Morbois, A., (1999), "Sinkholes beneath a reinforced earthfill – A large scale motorway and railway experiment", Proc. Conf. Geosynthetics'99, Boston, Massachusetts, USA, April 1999, Vol. 2, pp. 833–846.

Villard P., Giraud H. (1998) "Three-Dimensional modelling of the behaviour of geotextile sheets as membrane". Textile Research Journal, Vol. 68, N° 11, November 1998, pp. 797–806.

New Horizons in Earth Reinforcement – Otani, Miyata & Mukunoki (eds)
© 2008 Taylor & Francis Group, London, ISBN 978-0-415-45775-0

Special process techniques with project specified geosynthetics for sludge lagoon covers

O. Syllwasschy, O. Detert, D. Brokemper & D. Alexiew
Huesker Synthetic GmbH, Gescher, Germany

ABSTRACT: The construction of a capping system on top of a sludge lagoon is a technically hard to please task. The weak sediments and their low shear strength afford higly sophisticated and plant optimized solutions. In the past the use of plant specified fabricated geosynthetics and a detailed planned construction work results in cost and time effective solutions. Based on two topical projects the main characteristics for choosing the reinforcement, the design and the work schedule will be presented.

1 INTRODUCTION

The design of the cover respectively the liner system normally depends on the requirements of the customer or the appropriate authority and the intended use after capping. It may consist of a simple reinfocing geotextile and fill material or a multi layer combination of geotextiles, high quality soils and a qualified liner system with gas and water drainage.

The geotextile reinforcement stabilizes the soft subsoil and enables workers and construction machinery to work carefully on the sludge lagoon. The sludge maintains its geotechnical characteristics and has e.g. not to be stabilised chemically or treated in any other way.

2 TECHNICAL DESIGN

Design basis is an intensive soil investigation. There may be a higher effort compared to normal sites. Often sludge lagoons can only be walked on directly just at extreme weather (frost or drought) or with boats and pontoons.

The following data are necessary for design:

- Geotechnical parameters of sludge/fill material
- Stratification of the subsoil
- Dimensions of the sludge lagoon
- Water level
- Live load of contsruction machinery

According to the deposition history, dewatering and weather conditions the hydraulic and mechanical characteristics of the sludge may be subject to larger differences in depth and across the area.

In large lagoons with inhomogenous sludge parameters in different areas you might use reinforcement with varying and adapted design strength. In smaller lagoons or lagoons where areas with different sludge characteristics are not well known it is advisable to use the most conservative design parameters for the whole lagoon.

2.1 *Dimensioning*

At present there exists no geotechnical design analysis that correctly describes the failure mode of a typical sludge lagoon cover with geosynthetics. Nevertheless in the past analytical designs with circular or polygonal slip surfaces showed good results with sufficient factors of safety (Fig. 1).

Sometimes analysis based on wedge or slice methods (e.g. Janbu) may be advantageous compared to circle methods (e.g. Bishop, Krey) because of the better considerations of the forces of the geotextile

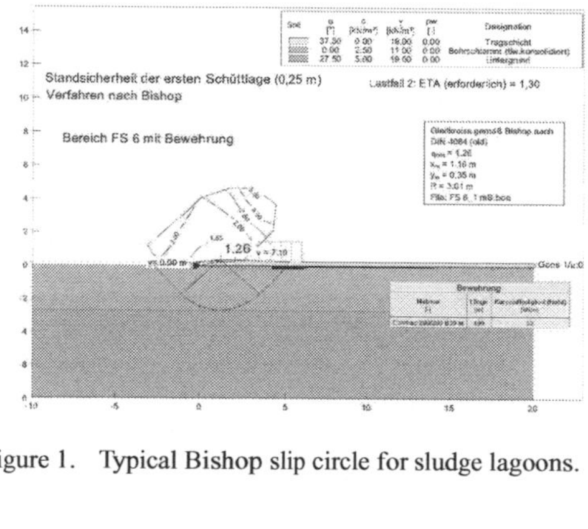

Figure 1. Typical Bishop slip circle for sludge lagoons.

reinforcement. Within the circle design methods the reinforcement is taken into account as a moment that is strongly influenced by the choice of the midpoint and so by the length of the lever arm. This can lead to inadequate consideration of the horizontal forces of the geotextile.

Regarding the stability of the system the stratification of the cover layers and their soil parameters have to be checked carefully in each stage of filling. Especially the first soil layers up to 0.6 to 1.0 m may become critical. At this stage the shear resistance of the weak sludge as well as the counter pressure activated by surcharge is negligible small to prevent ground failure. In this situation the use of light machinery, e.g a lightweight bulldozer like the so called pistenbully, with broad chains and very small live load starting from 4.2 kN/m^2 normally used in skiing areas, and additionally the filling of thin soil layers up to a maximum of 30 cm helps to solve the problem.

2.2 *Settlement*

Settlement due to consolidation may be very important in case of qualified sealing systems where the drainage of the cover soil is of major importance and specified inclinations are required. Soil investigation should contain consolidation tests to specify settlement during and after construction period.

If there are no data available a settlement assumption has to be done, based on experience in comparison with similar soils. Based on this assumption multiplied by a factor of safety the required gradient of the drainage layer should be constructed sufficiently. It may be useful to install measuring points on the surface of the sludge to allow control of settlement during the filling process and verify the settlement prediction thereby adjusting the gradient if necessary.

2.3 *Choice of raw-materials and type of reinforcement*

There may be used woven fabric or geogrid or combinations like woven/geogrid or geogrid/non-woven. The function of woven fabrics and geogrids is primary transferring tensile stresses, so in case of sludge lagoons the load distribution across a larger area into e.g. an anchor trench. Like non-wovens woven fabrics additionally may take the part of a filtration and separation layer.

Depending on the chemical characteristics of the sludge different raw materials have been established. Normally polypropylene (PP), polyesther (PET, PES), polyethylene (HDPE) and polyvinylalcohol (PVA) can be used in a normal pH-range from 4 to 9.5. In areas with pH 2–4 or 10–13 only PP and PVA may be used, if long term stability has to be considered in the design.

2.4 *Laying of geotextiles*

It may be differenciated between small and large sludge lagoons. Small lagoons have an area up to 20,000–30,000 m^2 or a maximum width of 150–200 m independent of the length. These areas can be covered by one or more large panels, consisting of several geotextiles sewed together.

An important point with panels is placement. First there is a staging area needed for preparation and sewing. Panels may be prefabricated in plant, but the size is limited by the allowable weight of the rolled panel during transport, i.e. loading, unloading and handling on site. A maximum roll weight of about 1.5 t allows good handling but should not be exceeded. So a larger panel may consist of several prefabricated smaller panels. These have to be sewed along the lagoon to form the complete panel, i.e. there has to be at least the width to place the roles and 1 m at the side where the sewing will take place. The unrolling of the smaller panels may be done by a cable winch. Using other machinery you have to add some place for driving the machinery, e.g. 3 m for an excavator. Driving directly on the geotextiles should be avoided.

The panel can be installed by winches or excavators. Cable winches need very little place and provide the possibiliy of a very continuous pulling in contrast to the stepwise pulling of an excavator. The connection between cable and panel can consist of loops and a traverse. Furthermore it is important to pull the panel as far as needed to even anchor it in the trenches at the side of the lagoon.

The materials used are woven fabrics or combinated products because they can easily be sewn and provide good seam strength. The higher price of a prefabricated panel will be more than compensated by the faster installation, e.g. a panel of about 10,000 m^2 can be installed in one day including adjusting. So the covering can start the next day.

Large lagoons with length of more than 200 or 300 m should be covered by single sheets of geotextiles because the seam strength may become too small to transfer the tensile forces between the geotextiles.

The ability to walk on the sludge is of major importance for installation. On some lagoons you might be able to walk on. In this case the geotextiles can be installed by men with the help of winches. On other lagoons you can walk on just after the installation of the first non-woven or geogrid. So here the installation has to be done in advance to the workers. The great benefit of a walkable lagoon is the good control of the installation, positioning and e.g. overlapping. Lagoons you can not walk on have to be covered just by machine, additional guidance by handheld ropes helps in positioning the geotextiles.

A special case is the installation under water where it is useful to install prefab panels from pontoons and cover them with soil in one step.

3 CASE EXAMPLES

3.1 *Open pit mine "Grube Hoffnung"*

The "pit hope" located in the southeast of Helmstedt in the former German Democratic Republik was used as a hazardous waste landfill after formerly being used as clay mine. Geology showed a naturally clay sealed pit that has been filled with ashes and sludges from the coal refining industry (Fig. 2).

Since closure in the 1990s the pit filled with rain water up to more than 2 m above sludge level. In 2002 the responsible authority decreed to cover the landfill to avoid direct contact to the environment. The cover fill surface level should have a maximum height of 5 m including a clay barrier of 30 cm.

The sludge showed a thixotropic behaviour, incrustations on the surface and a high pH-level due to high salinity. Consolidation took place during time only by self weight, so the undrained shear strength in the water covered area was very small, just about $c_u = 0.5\,kN/m^2$ or even less.

Before remediation could start nearly 65,000 m^3 of wastewater had to be pumped and treated on site before feeding the next receiving water course.

3.1.1 *Design*

The geotextile part of design and engineering, done by HPC Harress Pickel Consult, Merseburg, consisted of a 250 g/m^2 non-woven and a PP-geogrid of short term tensile strength of $P_{ult} = 60\,kN/m$ respectively 80 kN/m. In the northern stiffer part as single layer and in the southern weaker part as double crosswise geogrid reinforcement (Fig. 3). Additionally the southern part was divided by dams made from dumped concrete debris into four smaller lagoons (Fig. 8).

Figure 2. Open pit mine "Grube Hoffnung".

3.1.2 *Construction*

After profiling steep slopes down to an inclination of 1v:3h the installation of the non-woven and the geogrid started in the stiffer northern part, anchoring them in embankments at the side of the lagoon (Figs. 4, 5). In this part the non-woven could be placed by hand and the help of cross beams.

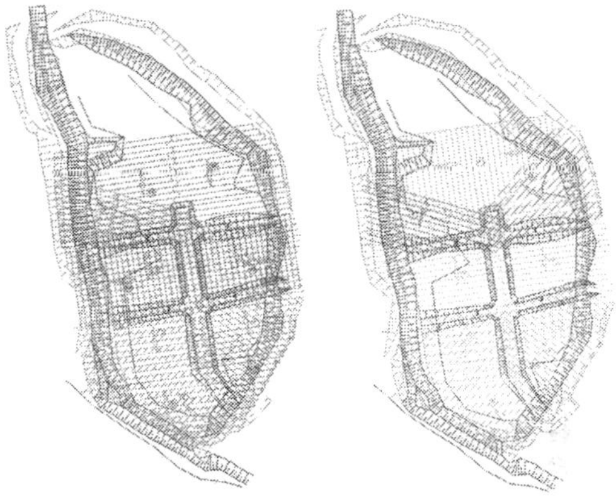

Figure 3. Laying pattern of crosswise geogrid reinforcement.

Figure 4. Profiling the slopes to 1v:3h.

Figure 5. Anchoring of geogrids in the embankments.

Figure 6. Ground failure at the border of the lagoon due to excavator works.

Figure 7. Pistenbully at work (Picture HPC).

The non-woven stabilized the sludge so far that it was possible to work on it safely although local vibrations could provoke liquefaction in the sludge.

Due to the round shaped geometry of the lagoon there had to be multiple overlaps installed to guarantee complete cover and anchorage in several areas. The work, done for the first time by Jaeger Umwelttechnik (Berneburg), required special measures especially after a ground failure due to the load of machinery (Fig. 6).

The cover using high-class sub-base gravel was done by a pistenbully (light-weight bulldozer). The material was pushed from the side to the middle with small height up to 30 cm (Fig. 7).

The southern part of the lagoon, that was covered by water for a long time, showed a substantial weaker consistency. The vane shear test showed not usable results with shear strength less than $c_u = 0{,}5\,\mathrm{kN/m^2}$.

After building seperation dams with concrete debris with size up to 1 m the installation of the geotextiles could start. During the cold winter period the frozen sludge surface enabled an easier installation. The sub-base gravel could only be moved in by a first thin layer of about 10 cm. 3 to 5 overruns of the pistenbully were

Figure 8. Construction of seperation dams through weak sludge (Picture HPC).

Figure 9. Profiling the final cover layer (Picture HPC).

enough for sludge liquifaction. After a second layer of 20 cm the following layers could be built in with normal height of 30 to 60 cm.

In late summer 2006 the pit was filled up to 5 m above original sludge level with sub-base course and fill material. Finally a 30 cm clay liner was built in, compacted dynamically by a 16 t sheepsfoot roller. At least about 200,000 m^3 of fill material were built in and 80,000 m^3 of water were pumped and treated (Fig. 9). The project was finished in 2007 after 1.5 years of construction time.

3.2 *Drilling mud lagoon Victorbur*

In the north western part of Germany close to the city of Aurich the drilling mud lagoon Victorbur has the size of approximately 10,000 m^2. At the base it is sealed by a HDPE flexible membrane layer (FML). The responsible authority ordered the surface cover with FML, gas drainage and recultivation soil layer.

The period for design and construction from summer to autumn 2006 was very short, so a time and cost saving solution had to be found.

The lagoon could be divided into two sections. The southeastern "dry surface" and the northern part, covered by water (Figs. 10, 11). The highly basic drilling mud showed a very soft to liquid consistency. In some parts the surface was covered by a crust men could

Figure 10. Southeastern Part (Photo EN-PRO-TEC).

Figure 11. Northern Part (Photo EN-PRO-TEC).

walk on, but at least 60 cm below surface the SPT showed values between N = 0.5 to 3, or even less.

The geotechnical data base of the mud was rather small and no time for additional investigation, so the design based on the SPT results and the review of the designer from EN-PRO-TEC, Nordhorn.

Due to the rather small size of the lagoon a panel design was chosen (Fig. 12). The required seam strength of the panel and the high pH-value pH = 10.6 required to use a biaxial PVA-geogrid combined with a 300 g/m² PP-non-woven. The use of a PP-geogrid was not taken into account due to the worse seam strength. The higher price of the PVA material in contrast to PP was more than compensated by the highly reduced amount of work as with single layer installation.

In accordance with the designing engineer the panel had to be anchored in- and outside of the lagoon. The inner anchot trench was designed to withstand pullout forces of the panel, whereas the outer anchor trench guarantees that no sludge can squeeze out between reinforcement and the basic FML at the edge. This solution was very well appreciated by the responsible authority that forbade any pollution of the surrounding moreland (Fig. 13).

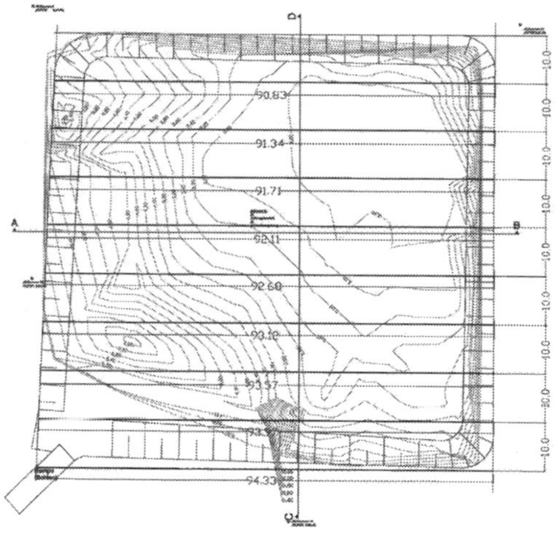

Figure 12. Plan view of lagoon and panel.

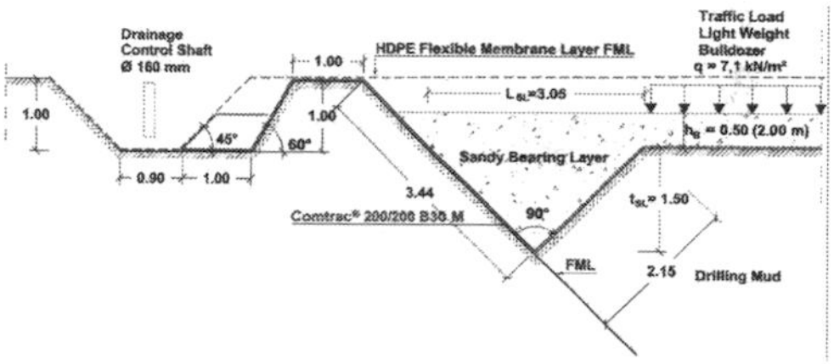

Figure 13. Detail of anchor trenches.

Figure 14. Preparation of the panel on site.

The panel was produced in three steps. At first the non-woven was sewed on the geogrid creating the combined product type Comtrac®.

In the second step two adjacent sheets were seamed together in plant in accordance to the installation plan. These sheets (width = 10 m, weight = 850 kg) were coiled in the right direction so they could be unwound correctly on site, the plant seam on the right hand side, the left hand side near to the lagoon ready for the seam to be done using hand sewing machines

Figure 15. Installation of the panel.

Figure 16. Filling the inner anchor trench.

(Fig. 14). The seam was tested in the laboratory with $F_{ult} = 57$ kN/m providing enough strength even for the first seam to carry the load of the whole panel during installation.

The total weight of the panel was about 10,200 kg. 11 prefab sheets were seamed together by hand with a double seam on site. The total seam length was about 2200 m done by two workers on two days. The sewing workers were supported by three helpers who held the sheets in position.

Before installation the lagoon surface was levelled and the anchor trenches inside the lagoon were prepared and partially supported by additional sand fillings. Normally these "sludge locks" are built by first installing the reinforcement on the even ground and secondly pressing soil and geotextile into the soft sludge. But in this case the building contractor Knoll (Haren-Ems) chose the inverted way to protect the sealing layer from additional surload or shear stress.

The anchor trenches were excavated by a long-bow excavator. The sustaining sand improved the drainage of the sludge filling the trench with leachate water. Even sand profiled areas with supposed stable conditions showed flowing conditions.

The installation of the panel was done within four hours in the morning using two movable cable winches and two excavators at the side of the lagoon. In the afternoon the panel was aligned laterally so the filling of the inner anchor trench could start the next morning. The outer trench was excavated and refilled after partially covering the panel inside the lagoon.

After filling the outer anchor trench the panel was fixed and could not be removed by wind. The filling was done by a long-bow excavator at the side and a lightweight bulldozer pushing and spreading the sand on top of the panel.

The sandy bearing layer with a maximum grain size of 1.5 mm was installed with a nominal gradient and an additional gradient to compensate settlement. The cover system had a height of 1.4 m above the equalizing bearing layer on top of the sludge surface. After installation of a HDPE-membrane layer, a geosynthetic drainage layer and the cover soil the construction could be finished in autumn 2006.

4 CONCLUSION

This paper provides recommendations for design and construction of sludge lagoon covers. Boundary conditions are declared for the influence of the choice of material and installation pattern. Anchorage of the reinforcement can be performed in- and/or outside the lagoon according to the place available and the danger of squeezing out of the sludge. Different raw materials allow the use of geotextiles in nearly every chemical environment. Geocompounds or wovens have a reinforcing and seperating function and are predestinated for sewing larger panels. Geogrids in combination with non-wovens or wovens can be used for all size of lagoons.

Small lagoons or parts of larger but narrow lagoons should be covered by large panels shortening the installation period essentially. Large lagoons should be covered by single but longer geotextiles.

A lot of projects worldwide show that each sludge lagoon is unique. Each project had its own specifics regarding geology, products and construction. Furthermore the experience is that most of the building contractors are doing the job for the first time and even though they are guided as good as possible during design and construction, there always is a process of learning by doing, changing and optimizing the original design.

REFERENCES

HPC AG, Geusaer Str. 1, D-06217 Merseburg (Design and Photographies).

EN-PRO-TEC Umwelt + Technik Beratungs- und Planungs-GmbH, Kotthook 8, D-48529 Nordhorn (Parts of Design and Photographies).

New Horizons in Earth Reinforcement – Otani, Miyata & Mukunoki (eds)
© 2008 Taylor & Francis Group, London, ISBN 978-0-415-45775-0

Construction of a large geogrid reinforced fill structure to increase landfill capacity

J.W. Cowland
GeoSystems Ltd, Hong Kong, China

ABSTRACT: Geogrid reinforced fill has recently been used to increase the capacity of a valley landfill in Hong Kong. A geogrid reinforced fill wall, 8 metres high, was constructed across the narrow opening of the valley to increase the landfill void space. Then the capacity of the landfill was substantially increased by the construction of a large reinforced fill embankment to fill a gap in the rim of the valley bowl. This reinforced fill embankment structure is 30 metres high and 300 metres long. This paper summarises the design of the embankment structure, the selection and quality control of the geogrids and fill material, and the facing details to prevent the outside face of the structure being damaged during a possible hillside grass fire.

1 INTRODUCTION

A solid waste landfill in Hong Kong, China, was designed in 1993 for a capacity of 40 million tonnes of waste. This landfill is now being operated and is about one third full. The landfill is lined to prevent the escape of leachate.

The landfill occupies a bowl shaped valley, which has a narrow opening at the downstream end leading onto a flat plain. The waste will eventually fill the valley, with a maximum depth of 140 metres. The rim of the bowl has a variable elevation, and in one location there is a 30 metre drop in the level of the top of the rim over a length of 300 metres.

1.1 *First use of geogrid reinforced fill to increase landfill capacity*

A geogrid reinforced fill wall, 8 metres high, was constructed across the narrow opening of the valley in 2003 and 2004 to increase the capacity of the landfill. This allows the placing of an extra 8 metres depth of waste across a large part of the 40 hectare landfill, which will increase the waste capacity by about 2 to 3 million tonnes.

This wall was formed with 80 kN/m and 120 kN/m uniaxial HDPE geogrids at a vertical spacing of 400 mm tied to concrete block facing units on the side of the wall facing away from the waste. The side of the wall facing the waste was formed with a geogrid wrap around arrangement, and was then lined with an HDPE geomembrane connected to the main landfill liner. The fill material was a silty sand obtained from within the landfill.

Placing of landfill waste against the wall commenced in 2007.

1.2 *Second use of geogrid reinforced fill*

The capacity of the landfill was again increased in 2006 by the construction of large geogrid reinforced fill embankment to fill the gap in the rim of the valley bowl. This structure is 30 metres high and 300 metres long, and again utilises excavated material from within the landfill. The geogrids have been placed in a wrap around arrangement.

The embankment was designed to withstand the lateral stresses from the waste, and it will be lined on the inside face to contain the landfill leachate. It is being monitored to determine settlement and lateral movement deformations.

The large reinforced fill embankment, which fills in the rim of the valley bowl, provides an interesting case history. This paper summarises the design, the selection and quality control of the geogrids and the fill material, and the construction of the structure. The outside face of this embankment structure has been designed to prevent the geogrids from being damaged during a possible hillside grass fire.

2 SHAPE OF THE EMBANKMENT

The embankment follows and fills the dip in the rim of the landfill. In plan it follows a variable curve. It is 30 metres high for over 200 metres of its length; the rest being a smaller transition towards the higher edges of the dip.

The sides of the embankment are 70° to the horizontal, with 2 metre wide benches at 7.5 metre vertical spacing. The top of the embankment is 5 metres wide, and the base about 40 metres wide on average.

3 DESIGN

The purpose of the embankment is to retain an extra 30 metres height of waste within the landfill.

Geogrid reinforced fill structures in Hong Kong are designed in accordance with the Guide to Reinforced Fill Structure and Slope Design (2002). With sides sloping at 70°, the embankment was designed in accordance with reinforced fill slope design.

The long term design strength of the geogrids was determined from tests for tensile strength, installation damage, durability and creep. Geogrid creep characteristics were determined at 30°C as Hong Kong is located within the tropics.

The waste loading on the side of the structure was accounted for by a slope stability analysis, whereby a series of potential slip surfaces from the top of the landfill through the embankment were analysed. The properties of Hong Kong waste have been reported by Cowland et al (1993).

It was decided that the lining of the side of the embankment facing the waste will not be carried out until the monitoring of the embankment shows that settlement has mostly ceased. It was also decided that the side of the embankment facing away from the waste should be designed to resist the occasional hillside grass fires that occur in Hong Kong.

3.1 *Fill material*

At the design stage, granular fill material was chosen to minimise the settlement of the embankment.

3.2 *Geogrids*

HDPE geogrids were chosen because of their better resistance to landfill leachate than geogrids made from other polymers.

The design resulted in uniaxial geogrids with three different ultimate tensile strengths being used. Geogrids with a strength of 90 kN/m were used in the bottom third of the embankment, 60 kN/m in the middle third and 45 kN/m in the top third. These geogrids were placed in a wrap around arrangement with a vertical spacing of 400 mm.

Secondary biaxial geogrids were placed between the main uniaxial geogrids to help maintain alignment of the embankment faces. These secondary geogrids were attached to the centre of each wrap around and had a penetration of 1 metre into the embankment.

A geotextile was placed inside the wrap around to keep the fill material in place.

Figure 1. Gravel facing for fire resistance.

Figure 2. Completed gravel facing.

3.3 *Lining*

The lining of the side of the embankment facing the waste, which will contain the leachate and gas in the landfill, will not be carried out until the monitoring of the embankment shows that settlement has mostly ceased. A temporary liner, 1 mm thick, has been placed on this face to protect it until the permanent liner is installed.

If some distortion of the face were to occur with the embankment settlement over this time, then it is thought that shotcrete could be used to smooth the face for the permanent lining.

3.4 *Resistance to grass fires*

In order to protect the side of the embankment facing away from the waste from the occasional hillside grass fires that occur in Hong Kong, a 100 mm thick layer of gravel was placed on this face. This was attached to the face using a steel mesh (see Figure 1).

The gravel is intended to keep the temporary heat of a moving grass fire away from the geogrids (see Figure 2). The thickness of the gravel was chosen based on the work of Austin (1997), and was confirmed with a rudimentary fire test.

It was recognised that the effect of fire on a reinforced fill embankment is unlikely to cause structural failure. The maximum tension in the geogrids is within the embankment, some distance from the faces, where the geogrids will be protected from fire.

Only the wrap around portion of the geogrid on the faces is likely to be affected by a fire. In this location, after settlement of the embankment, the geogrids are not in tension and their localised removal would not cause structural failure. However, the absence of this portion of the geogrid would cause a serviceability problem with time as the compacted fill material would start to escape from the face of the embankment.

Therefore, protection of the wrap around with a flexible gravel facing was thought to be prudent. It was thought that a shotcrete facing applied in the early life of the embankment would be too rigid and might crack as the embankment settled.

4 CONSTRUCTION

4.1 *Foundation*

Hong Kong soils are saprolitic in nature and relatively strong. During excavation for the base of the embankment, small areas of weaker soils were removed and replaced by a thin concrete foundation.

4.2 *Fill material*

The fill material was obtained from a nearby excavation within the landfill to increase the void space for waste.

The excavation was being carried out in a profile of saprolitic soils and igneous rocks. The saprolitic soils were encountered first in the excavation and some consideration was given to changing the design of the reinforcement to utilise these fine grained soils for the fill material. However, it was realised that the use of fine grained soils would increase the likelihood of unacceptable settlements of the embankment, especially if the fill became wet during placement.

It was then decided to use the igneous rock beneath. This was reduced in particle size as much as possible during the excavation blasting, and then the rock material was passed over a vibrating screen with a screen opening size of 40 mm. The material that passed through the screen was a remarkably consistent sandy gravel that was perfect for the construction of a reinforced fill structure.

4.3 *Geogrids*

In addition to the factory quality control of the geogrids, samples were taken on site for testing. These samples were tested for ultimate tensile strength.

Figure 3. Placing geogrids on a curved alignment.

Figure 4. Compaction and testing of fill material.

Interface friction tests were carried out for each of the three grades of uniaxial geogrids against the fill material in a large shear box.

Due to the variable curved shape of the alignment of the embankment, and the wrap around configuration of the geogrids at the faces, it proved to be an interesting exercise to place geogrids from one side of the embankment to the other (see Figure 3).

4.4 *Compaction*

The fill material was compacted using vibrating rollers. Heavier rollers were used in the centre of the embankment and lighter rollers near the edges.

Compaction control was achieved using a nuclear density gauge (see Figure 4). Due to the consistent nature of the granular fill, the required compacted density was consistently achieved.

In addition, this allowed a consistently good vertical alignment of the faces.

4.5 *Haul roads*

The construction of a 30 metre high reinforced fill embankment required considerable thought to be given

Figure 5. The embankment nearing completion.

Figure 6. Completed reinforced fill embankment on the side facing the waste.

to how the fill material was to be transported to the narrow top without constructing another large temporary embankment alongside.

In addition, considerable thought had to be given to how the compaction equipment would be removed from the completed top of the embankment (see Figure 5).

4.6 *Safety*

Temporary scaffolding was installed on the faces of the embankment during construction to prevent workers from falling down the steep and high faces.

Permanent handrails are being installed on the benches on the side of the embankment facing away from the waste for the same purpose.

Figure 7. Completed reinforced fill embankment on the side facing away from the waste.

4.7 *Monitoring*

Monitoring of embankment movement has been carried out by standard surveying techniques. Due to the use of high quality granular fill, and good compaction during placement, very little movement has occurred.

5 CONCLUSIONS

The use of geogrid reinforced fill has proved to be a feasible method to increase the capacity of waste landfills in Hong Kong.

This paper has presented some practical aspects of the design and construction of a 30 metre high and 300 metre long embankment for this purpose. Figures 6 and 7 show the completed structure.

REFERENCES

Austin R.A. 1997. The Effect of Installation Activities and Fire Exposure on Geogrid Performance, *Geotextiles and Geomembranes*: 15 (4–6) 367–376.

Cowland J.W., Tang K.Y. & Gabay J. 1993. Density and Strength Properties of Hong Kong Refuse *Fourth International Landfill Symposium*, Sardinia, Italy: 1433–1446.

Guide to Reinforced Fill Structure and Slope Design 2002. Geotechnical Engineering Office, Government of the Hong Kong Special Administrative Region, 236p.

New Horizons in Earth Reinforcement – Otani, Miyata & Mukunoki (eds)
© 2008 Taylor & Francis Group, London, *ISBN 978-0-415-45775-0*

Tire-chips for geotechnical applications

K. Yasuhara
Department of Urban and Civil Engineering, Ibaraki University, Hitachi1, Ibaraki, Japan

H. Hazarika
Department of Architectural Environmental Systems, Akita Prefectural University, Akita, Japan

Y. Mitarai
Technical Research Center, Toa Corporation, Tokyo, Japan

A.K. Karmokar
Technical Research Center, Bridgestone Corporation, Tokyo, Japan

ABSTRACT: This paper introduces recent Japanese experiences related to scrapped tires for geotechnical applications with a focus placed on tire-chips. Techniques for using tire shreds and chips are classifiable into two categories: tire chips mixed and not mixed with soil. The former techniques include cement-treated clay with tire chips, which possess high ductility and toughness, and the latter describes the cushion effects of tire chips' placement installed close to quay walls which are intended to reduce lateral earth pressures, residual deformations of backfills during earthquakes and ongoing projects of the latter category include tire-chip drains that replace gravel drains as a countermeasure against liquefaction in sand.

1 INTRODUCTION

Scrapped tires provide numerous advantages from the viewpoint of civil engineering practices. They have light weight, high vibration-absorption, high elastic compressibility, high hydraulic conductivity, and temperature-isolation potential. New techniques have emerged to utilize these advantageous characteristics for practical purposes. Mainly, two types of scrapped tire materials are used for civil engineering applications: with and without shredding or cutting into small pieces of several tens of centimeters' or several centimeters' diameter. The former, without shredding, is useful for infrastructural retaining walls and foundations. The material is sometimes reinforced with geosynthetics. On the other hand, in cases with shredding or cutting into pieces of similar size, as with gravel and sand (sometimes called tire-derived aggregate, TDA after Humphrey, 2007), they can be put into practical use for geotechnical application. This latter material type might be more costly than unshredded or uncut material.

Among these possible techniques using the used tires, this paper describes attempts to introduce two possibilities of tire-chips for geotechnical applications. Techniques for using tire-chips are classifiable as cases of tire chips mixed and not mixed with soil. The former techniques include cement-treated clay with tire chips, which possess high ductility and toughness, and non-cement-treated sand mixed with tire chips, which is intended to reduce lateral pressures, lateral displacements and liquefaction potential during earthquakes.

2 UTILIZATION OF USED-TIRE CHIPS WITHOUT ADDITIVES

The tire chips were made from used tires by crushing and subsequent removal of the constituent textiles and metal fibers. The chips therefore have a rough or serrated surface. The specific gravity of tire chips used in this study is 1.15 and the grain size is 2 mm on average. Tire chips are an elastically compressible material, of which Poisson's ratio is nearly 0.5 and the elastic modulus is approximately 4–6 MPa on average, corresponding to 0–15% strain.

2.1 *Application of tire-chips without soil mixing to damage mitigation during earthquakes (Hazarika et al. 2006a, b)*

2.1.1 *Sandwiched backfilling technique*

The environmentally friendly and the cost-effective disaster mitigation technique involve placing a

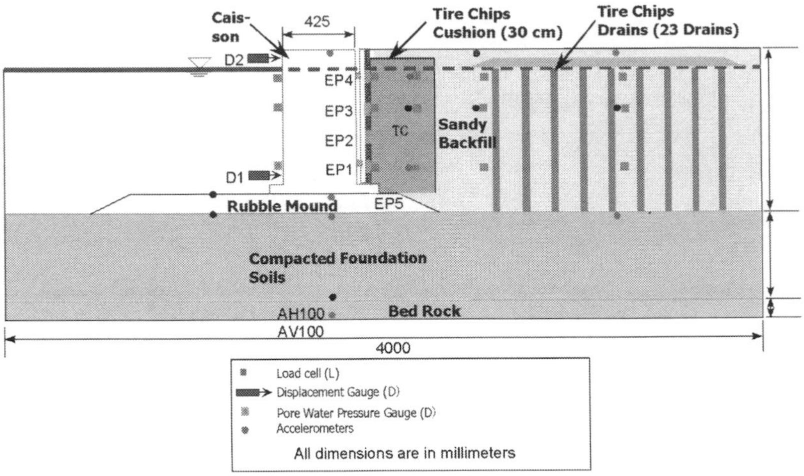

Figure 1. Cross section of the experimental.

cushioning layer of tire chips as a vibration absorber immediately behind the structure. The beneficial effects of such a sandwiched cushioning technique have been described in Hazarika et al. (2006a). In addition, vertical drains made out of tire chips can be installed in the backfill to prevent soil liquefaction. Yasuhara et al. (2004) used tire chips in vertical drains for reducing liquefaction-induced deformation. One function of the cushion is to reduce the load against the structure caused by the energy absorption capacity of the cushion material. Another function is to curtail permanent displacement of the structure attributable to compressible material.

2.1.2 *Underwater shaking-table testing*

Large three-dimensional underwater shaking table assemblies of the Port and Airport Research Institute (PARI) were used in the testing program. The shaking table is circular with 5.65 m diameter; it is installed on a 15 m long × 15 m wide × 2.0 m deep water pool. The details of the shaking table are available in Iai & Sugano (2000).

A caisson-type quay wall (model to prototype ratio of 1/10) was used for testing. Figure 1 shows a cross section of the soil box, the model caisson and the locations of the various measuring devices (load cells, earth pressure cells, pore water pressure cells, accelerometers and displacement gauges). The model caisson (425 mm in breadth) was made of steel plates filled with dry sand and sinker to bring its center of gravity to a stable position. The caisson consists of three parts: the central part (width 500 mm) and two dummy parts (width 350 mm each). All the monitoring devices were installed at the central caisson to eliminate the effect of sidewall friction on the measurements. The soil box was made of a steel container 4.0 m long, 1.25 m wide, and 1.5 m deep. The foundation rubble beneath the caisson was prepared using Grade 4 crushed stone with 13–20 mm particle size. The backfill and the seabed layer were prepared using Sohma sand (No. 5).

The dense foundation sand representing the seabed layer was prepared in two layers. After preparing each layer, the entire assembly was shaken with 300 Gal of vibration starting with a frequency of 5 Hz and increasing up to 50 Hz. Backfill was also prepared in stages using free falling technique; it was then compacted using a manually operated vibrator. After constructing the foundation and the backfill, and setting up of the devices, the pool was filled with water, which gradually elevated the water depth to 1.3 m to saturate the backfill. This submerged condition was maintained for two days so that the backfill can attain a complete saturation stage.

2.1.3 *Test cases*

As was shown in Figure 2, two test cases were examined. In one case (Case A), a caisson with a rubble backfill with conventional sandy backfill behind it was used. In another case (Case B), behind the caisson, a cushion layer of tire chips (average grain size 20 mm) was placed vertically; its thickness was 0.4 times the wall height. In actual practice, the design thickness will depend upon many other factors such as the height and rigidity of the structure, in addition to compressibility and stiffness of the cushion material. In compressible buffer applications, there seems to be an optimum value for the cushion thickness, beyond which increased thickness will not engender a proportionate decrease of the load. The effect of the cushion thickness shaking table test was

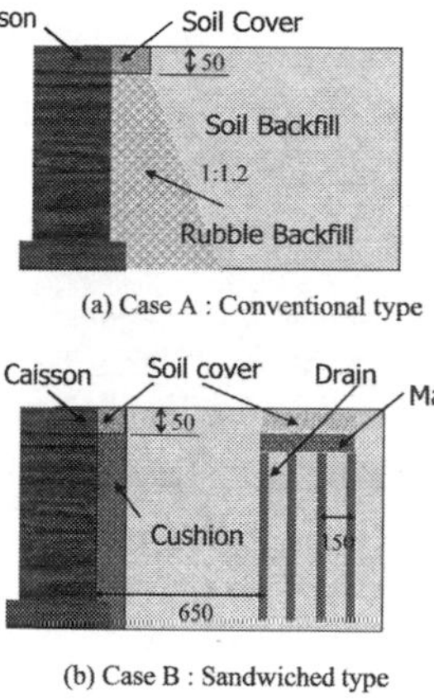

Figure 2. Test cases of backfill.

described using a small-scale model in Hazarika et al. (2006b).

2.1.4 *Test procedures*

The cushion layer was prepared by filling the tire chips (average grain size 10 mm) inside a geotextile bag. Then, the presence of geotextiles prevents flowing of sand particles into the chip structure. The average dry density of the tire chips achieved after filling and tamping was 0.675 t/m^3.The relative densities that were achieved after the backfill preparation were about 50% to 60%, implying that the backfill soil is partly liquefiable. The foundation soils were compacted with a mechanical vibrator to achieve a relative density of about 80%, implying a non-liquefiable foundation deposit. Vertical drains made out of tire chips (average grain size 7.0 mm) were installed in the backfill. They were then installed with a spacing of 150 mm in triangular pattern. The drain diameter was chosen as 50 mm. The tops of all drains were covered with a 50-mm-thick gravel layer underlying a 50-mm-thick soil cover. The purpose of such a cover layer is twofold: one is to allow the free drainage of water and other is to prevent the likely uplifting of the tire chips during shaking because of its lightweight nature. Earthquake loadings of different magnitudes were imparted to the soil-structure system during the tests. The input motions selected were: 1) the Port Island (PI) wave – the strong motion acceleration record at the Port Island, Kobe, Japan during the 1995 Hyogo-ken Nanbu earthquake (M 7.2); and 2) the Ohta Ward (OW) wave – a scenario synthetic earthquake motion assuming an earthquake that is presumed to occur in the southern Kanto region with its epicenter at Ohta ward, Tokyo, Japan. It is noteworthy that the 1995 Hyogo-ken Nanbu earthquake is an intra-plate earthquake, while the scenario earthquake (synthetic)

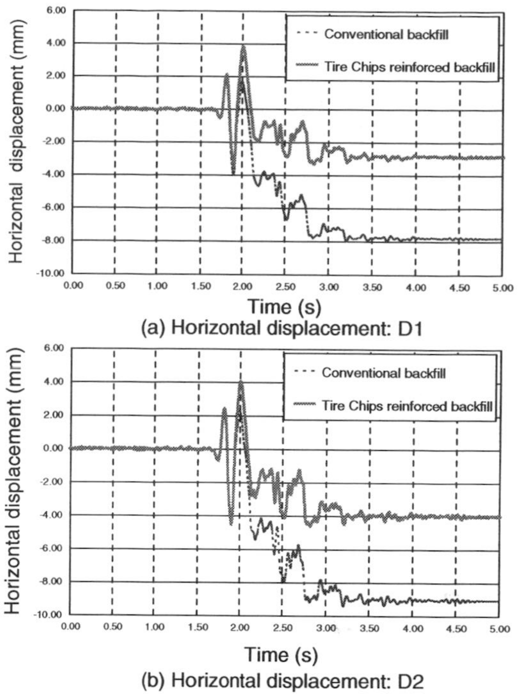

Figure 3. Time history of the caisson displacement.

was constructed assuming an inter-plate earthquake. The loading intensities were varied using the various maximum acceleration ratios (0.5, 1.0, 1.2, and 1.5) of the target acceleration to the actual acceleration. Durations of the shaking in the model testing were based on the time axes of these accelerograms.

2.2 *Test result and discussion*

Various types of earthquake motion with different magnitudes were adopted for this study. However, the discussion here will be mostly limited to series no. 3 (PI 1.0). The PI 1.0 data are the actual recorded data at Port Island, Kobe, with the time axes scaled to fit the model to prototype ratio of 1/10.

The time histories of the horizontal displacements (D1 and D2 in Figure 1) during the loading for the two test cases are compared in Figure 3. Comparisons reveal that the maximum displacement experienced by the quay wall with a tire-chip reinforced caisson (thick continuous line) is toward the backfill, in contrast to the quay wall without any reinforcement (shown in dotted line), for which case it is seaward. The compressibility of the tire chips renders flexibility to the soil-structure system, which allows the quay wall to bounce back under its inertia force; this tendency ultimately (at the end of the loading cycles) aids in preventing the excessive seaward deformation of the wall.

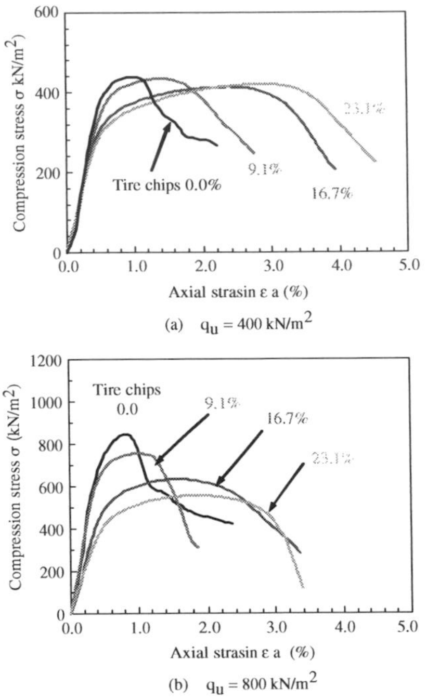

Figure 4. Stress vs. strain curves for unconfined-compression tests on CTCT specimens.

However, the wall with a conventional backfill experiences very high seaward displacements right from the beginning because of its inertia. As a consequence, the structure can not move back to the opposite side and ultimately suffers from a huge permanent seaward displacement.

3 CEMENT-TREATED MARINE CLAY MIXED WITH TIRE-CHIPS (MITARAI ET AL., 2006)

3.1 *Preparation of specimens*

Dredged clay was used to make specimens of the cement-treated clay with addition of tire chips (CTCT). The dredged clay was taken from Tokyo Bay ($\rho s = 2.72\,g/cm^3$, $wL = 100\%$ and $Ip = 70$). Its percentage of fine-grained fraction is about 90%. Slurry dredged clay with initial water contents of 250% was mixed with seawater to produce a $1.25\,g/cm^3$. The cement used was normal Portland cement. The tire chips were made from used tires, and these average grain sizes are 2 mm which particle density is $1.15\,g/cm^3$. The specimens were prepared to provide two kinds of strength of cement-treated clay, and four kinds of tire chips' contents. The targeted unconfined compressive strength of cement-treated clays were $qu = 400$ or $800\,kN/m^2$. The mixing conditions of cement-treated clay and the percentage of added tire chips were 0%, 9.1%, 16.7%, and 23.1% within a whole volume of specimen.

3.2 *Undrained behavior*

Figure 4 shows the stress-strain relation of unconfined compressive test (UC-test) on CTCT. These results imply the following; 1) The deformation property of cement-treated clay was changed from brittle into tough, merely by addition of tire chips, 2) The larger the percentage of tire chips in the mixture is, the larger the failure strain becomes. In the case of the cement-treated clay without adding tire chips, the stress-strain relation shows a marked strain-softening behavior. On the other hand, CTCT show strain-hardening behavior, 3) The larger the strength of cement-treated clays is, the smaller the tangent elastic modulus by adding tires chips becomes.

For comparison with the results from UC-tests, undrained triaxial compression tests (TC-test) were conducted under the following condition consolidation stress was 200 kPa (2 hr) and after consolidation, during undrained shear, the confining pressure was maintained constant with 300 kPa. The results from TC-tests were interpreted to investigate the effects of tire-chips contents and hardness of tire chips on toughness improvement. Figures 5 show the stress vs. strain curves for specimens with different contents of tire chips for the target undrained strengths of $400\,kN/m^2$ and $800\,kN/m^2$, respectively, although the strain at peak stress for specimens with $800\,kN/m^2$ is greater than that for specimens with $400\,kN/m^2$. Figure 5 illustrates a set of comparisons between results in which the target unconfined strength was $800\,kN/m^2$. Both tests were carried out on three classes of tire-chip contents. This test series was carried out for verifying the effect of confinement on toughness characteristics. It is apparent from comparison between both test results that: 1) the stress and strain curves in triaxial tests exhibit ductile behavior, whereas those in unconfined compression test show brittle behavior, independently of tire-chip content; 2) the ductile behavior in triaxial tests was improved according to the increase in the tire-chip content; 3) as in the ductile behavior improvement, less increase in triaxial undrained strength was observed, even with increasing the tire-chip content, while unconfined compressive strength rather decreases, even with increased tire-chip content.

3.3 *Improvement of hydraulic conductivity*

As shown above, the cement-treated clay shows brittle deformation. Cracks develop with progress of shear deformation; then it is expected that the hydraulic conductivity increases by increasing deformation. But, in

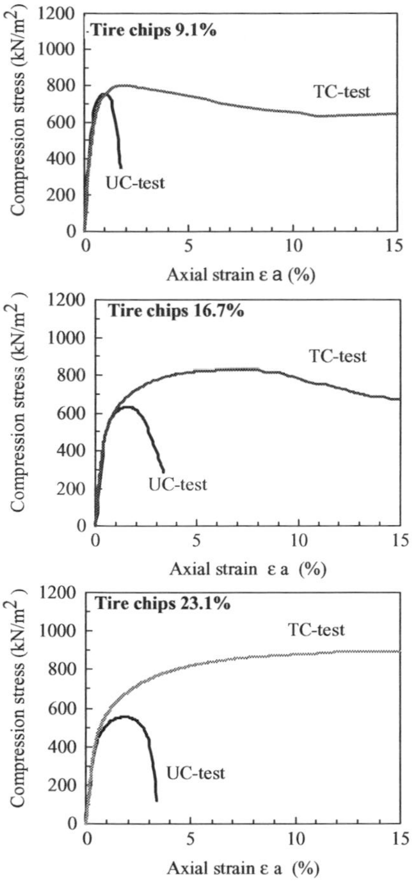

Figure 5. Effect of horizontal stress (Comparison UC-test with TC-test; Target undrained strength of cement treated clay part: qu = 800 kN/m^2).

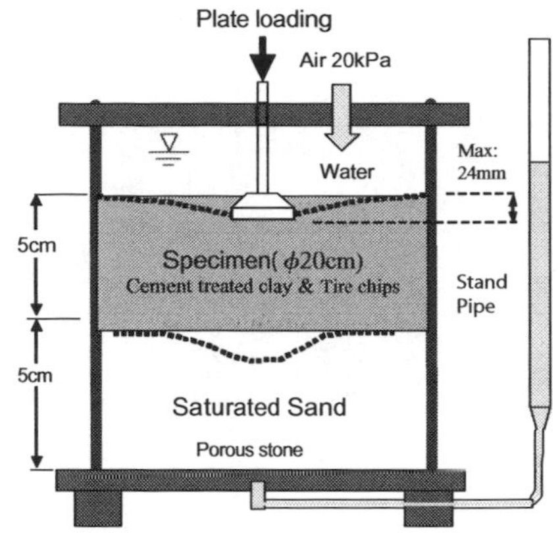

Figure 6. Plate loading tests with measurement of hydraulic conductivity.

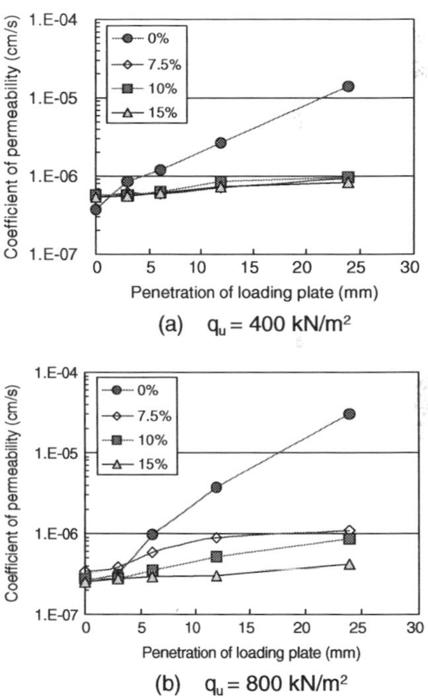

Figure 7. Plate loading tests with measurement of hydraulic conductivity.

the case of adding tire chips, it can be expected that the hydraulic conductivity changes during deformation because the toughness was improved by adding tire chips. To verify that fact, the hydraulic conductivity was measured during shear deformation in the plate loading tests on CTCT. To examine the hydraulic conductivity of a no deformed specimen, we conducted a falling head permeability test. The authors adopted an acrylic fiber cell of 20 cm diameter and 20 cm height, as shown in Figure 7. First, the saturated Toyoura-sand was poured at the bottom of the cell, and then the CTCT was poured on to the sand surface. The thickness of the sand tire and specimen was 5 cm. After the specimen was cured for 7 days under the high moisture condition of more than 95% and constant temperature of 20°C ± 2°C, the falling head permeability test was conducted under the water pressure of 20 kPa applied onto the specimen surface. Figure 7 shows the result of the permeability tests of the cement treated clay. To examine the variations of the hydraulic conductivity of CTCT with deformation in the plate loading tests as shown in Figure 7, we conducted the following examination: the specimen and devices were the same as above, and the specimen was loaded into the center using a steel loading plate, which had 4 cm diameter. The rate of loading was 15 mm/min.The loading to the specimen was applied gradually, and the permeability test was conducted after unloading. This process

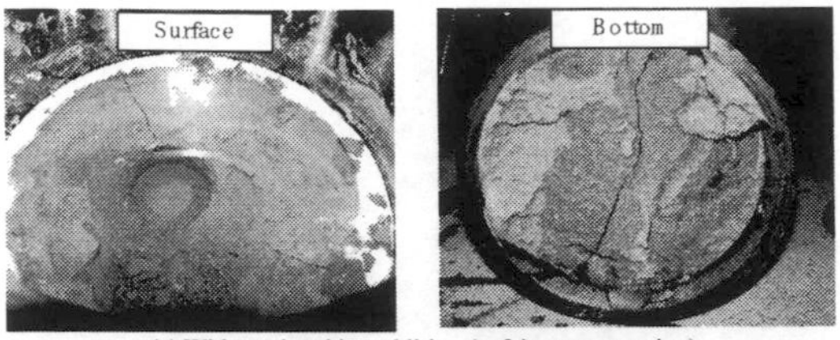

(a) With no tire chips addition (at 24 mm peneration)

(b) With 10% tire chips addition (at 24 mm peneration)

Figure 8. Observed appearance of specimens after penetration tests.

was repeated until cracks were readily apparent. In this examination, the loading plate was applied in the center of specimen at a rapid speed; then, the loading plate was penetrated into the specimen. Consequently, the bending deformation was developed in the specimen. Then, it was speculated that the coefficient of permeability might increase by developing some cracks, as compared to that before deformation. Figure 7 shows the relation between the penetration displacement and the coefficients of permeability of CTCT specimen. In this figure, the specimen deformation is shown in the value of penetration by loading plate. As shown in this figure, the change of the coefficient of permeability with deformation was very small in the case of added tire chips (CTCT). On the other hand, the hydraulic conductivity of the cement-treated clay without added tire chips (0% tire-chips) increases with deformation. Figure 8 shows a photograph of the specimen that was taken at 24 mm penetration of the loading plate. It is apparent that the specimen without added tire chips was cracked, but that the specimen of CTCT was not cracked. These observations indicate that the fewer cracks in CTCT are attributable to improving the ductility of cement-treated clay with the added tire chips.

4 CONCLUSIONS

This paper presented descriptions of some recent experiences of the utilization of used tire chips for geotechnical practices. Two techniques were proposed using tire chips that were processed by cutting them into smaller pieces of several centimeters' diameter on average. In one case, a series of model shaking table tests under a 1 *g* gravitational field was conducted to examine the performance of a newly developed sandwiched backfilling technique for earthquake disaster mitigation. In that technique, sandwiched cushions and vertical drains made out of an emerging and smart geomaterial known as tire chips were used as retaining materials behind massive rigid structures such as caisson quay walls. To overcome a salient disadvantage of cement treated clay (CTC), *brittleness*, an attempt was made to mix tire chips with CTC. This technique, abbreviated as CTCT, provided an additional characteristic, toughness or ductility, which was useful for resistance against occurrence of cracks in CTCT during development of shear displacement. This was already put into practice at a disposal yard where CTCT was used as a sealing material to protect leakage of contaminated materials.

ACKNOWLEDGEMENTS

The studies described in this paper have been financially supported by the Grants-in-Aid from the Ministry of Land, Transportation and Infrastructure and the Ministry of Education, Culture, Sports, Science and Education of the Japanese Government. The author is indebted to Drs. Sugano and Yoshiaki Kikuchi of Port and Airport Research Institute, Dr. Takao Kishida of Toa Corp., and Mr. Hideo Takeichi of Bridgestone Corp. for their strong support in carrying out the research project successfully.

REFERENCES

Hazarika, H., Kohama, E., Suzuki, H. & Sugano, T. 2006a. Enhancement of Earthquake Resistance of Structures using Tire chips as Compressible Inclusion. *Report of the Port and Airport Research Institute* 45(1): 1–28.

Hazarika, H., Sugano, T., Kikuchi, Y., Yasuhara, K., Murakami, S., Takeichi, H., Karmokar, A.K., Kishida, T. & Mitarai, Y. 2006b. Evaluation of Recycled Waste as Smart Geomaterial for Earthquake Resistant of Structures. *41st Annual Conference of Japanese Geotechnical Society, Kagoshima*, pp. 591–592.

Humphrey D.N. 1998. Civil Engineering Applications of Tire Shreds, *Manuscript Prepared for Asphalt Rubber Technology Service*, SC, USA.

Humphrey, D. N., Whetten, N., Weaver, J., Recker, K. & Cosgrove, T. A. 1998. Tire shreds as lightweight fill for embankments and retaining walls. *Recycled Materials in Geotechnical Applications, ASCE, Geotechnical Special Publication* 79: 51–65.

Humphrey, D. N. 2007. Tire derived aggregates as lightweight fill for embankments and retaining walls, Special Invited Lecture, *Proc. International Workshop on Scrap Tires Derived Geomaterials*, pp. 56–79, Yokosuka, Japan.

Iai, S. 1989. Similitude for Shaking Table Tests on Soilstructure- fluid Model in 1 *g* Gravitational Field. *Soils and Foundations* 29(1): 105–118.

Iai, S. & Sugano, T. 2000. Shake Table testing on Seismic Performance of Gravity Quay Walls, *12th World Conference on Earthquake Engineering, WCEE*, Paper No. 2680.

Mitarai, Y. et al. 2006. Application of the cement treated clay with tire-chips added to the sealing materials of coastal waste disposal site. *Proc. 6th International Congress on Environmental Geotechnology, Cardiff, UK*, Vol. 1.

Yasuhara, K., Komine, H., Murakami, S., Taoka, K., Ohtsuka, Y. & Masuda, T. 2005. Tire chips drain for mitigation of liquefaction and liquefaction-induced deformation in sand. *Proc. of Symposium on Technology of Using Artificial Geomaterials, Fukuoka, Japan*, pp. 115–118 (in Japanese).

Yasuhara, K., Karmokar, A.K., Kato, Y., Mogi, H. & Fukutake, K. 2006. "Effective utilization technique of used tires for foundations and earth structures." *Kisoko* 34(2): 58–63 (in Japanese).

Yasuhara, K. 2007. Recent Japanese experiences on scrapped tires for geotechnical applications, Keynote Lecture, *Proc. International Workshop on Scrap Tires Derived Geomaterials*, pp. 17–40, Yokosuka, Japan.

New Horizons in Earth Reinforcement – Otani, Miyata & Mukunoki (eds)
© 2008 Taylor & Francis Group, London, ISBN 978-0-415-45775-0

Jute geotextile and its application in civil engineering, agri-horticulture and forestry

P.K. Choudhury & A. Das
Indian jute industries' research association, Kolkata, India

T. Sanyal
Jute manufactures development council, Kolkata, India

ABSTRACT: Jute geotextile (JGT) & Agrotextile (JAT) made from fibres of jute plant have proved to be effective to address soil related problems in civil engineering and agriculture. Years of research coupled with concurrent field applications have made it possible to develop appropriate fabrics for JGT & JAT utilizing its distinctive features. In India use of JGT & JAT is gradually gaining ground. JGT has been highly successful in controlling riverbank erosion and stabilizing roads. JGT is also the most sought after fabric for surficial soil erosion control. JAT has been effective in weed suppression, afforestation in arid zones, soil conservation and as facilitator of plant/sapling growth. Depleting petroleum reserves and deteriorating environment should make these two natural textile fabrics attractive to the end users not only from economical and technical considerations, but also as a natural intervention to rein in engulfing environmental pollution. The paper highlights issues that make JGT & JAT distinctive.

1 INTRODUCTION

Jute, a natural, eco-friendly biodegradable and annually renewable bast fibre grows abundantly in India and Bangladesh in particular. In India annual production of jute is of the order of 1.6 million tons with jute sacks being the potent product. Jute industry in India is one of the oldest agro-industries in the world. In India alone about 0.7 million people are dependent on jute production, its manufacture and marketing for their livelihood.

Ingress of man-made polymers is posing threats to the jute industry which is why diversification of jute products has become an imperative necessity. Indian Jute Industries' Research Association (IJIRA) has developed a number of jute diversified products like Jute Geotextile (JGT) and Jute Agrotextile (JAT) through extensive R & D work utilizing the unique intrinsic properties of jute fibres like high initial tensile strength, low extensibility, high water absorbency, excellent drapability and spinnability. Varieties of JGT and JAT namely, woven, non-woven, open mesh woven, pre-fabricated vertical jute drain (PVJD), jute sleeve etc. have been developed by IJIRA with the support of Jute Manufactures Development Council (JMDC). Laboratory study followed by successful field applications has established the efficacy of these products. It is relevant to mention that all geotextiles act as change agents to soil to improve its engineering performance and its long term durability is not a technical necessity. Bio-degradability is therefore both a technical and environmental advantage. Man-made geotextiles are environmentally questionable despite their longer durability. The stress is now on adopting bio-engineering measures to address soil-related problems in civil engineering. Depletion of petroleum reserves and deteriorating environment in the planet should make JGT & JAT more attractive to the end-users.

This article indicates the salient properties of JGT & JAT along with references to a few case studies substantiating efficacy of these two products.

2 JUTE, THE GOLDEN FIBRE

Jute, known as a golden fibre for its gold-like texture, is a unique textile grade lingo-cellulosic fibre. Its main chemical constituents are alfa cellulose (62%) , hemicelluloses (24%) , lignin (12%) and others (2%). Its high tenacity (40 gm/tex), low elongation at break (1.4%) and moisture content (13%) at 65% RH hold an edge over other geotextiles.

3 JUTE GEOTEXTILE (JGT) AND JUTE AGROTEXTILE (JAT)

JGT is a permeable textile fabric available in woven or non-woven form used in or on soil to improve its engineering performance. JAT is woven, or non woven or fabricated product used for achieving higher agricultural yield by enhancing the agronomical characteristics of soil and suppressing unwanted growth of vegetation like weeds.

3.1 *Types of JGT and JAT available along with their specifications*

JGT and JAT are tailor-made products. Site-specific products can be manufactured depending upon end-use requirements. Stronger (40 kN/m) and wider (upto 5 m) fabric with finer porometry (100 micron) can easily be produced. However, specifications of some standard JGT & JAT are given as reference in Table 1 & Table 2.

4 POTENTIAL APPLICATIONS OF JGT

4.1 *Case studies with JGT in India*

A good number of field applications has been carried out with JGT. General findings of some of the significant field applications carried out in India are presented below.

Table 1. Specifications of widely used Jute Geotextile.

Properties type	Weight (g/m^2)	Thickness (mm)	Strength (MD × CD) (kN/m)
Woven	760	2	20 × 20
Nonwoven	500	4	4 × 5
Open mesh	500	4	10 × 7.5

Table 1. (Contd.)

Properties type	Elongation (MD × CD) (%)	Water holding capacity (%)	Pore size (O_{90}) Micron
Woven	10 × 10	400	200
Nonwoven	20 × 25	600	400
Open mesh	–	500	–

Table 2. Specifications of widely used Jute Agrotextile.

Properties type	Thickness (mm)	Strength (MD × CD) (kN/m)	Water holding capacity (%)
Jute sleeve	1	7.5 × 6.5	300
Open mesh	3	12 × 12	400

4.1.1 *Road construction*

In all the field applications in roads, it has been observed that sub-grades, despite being expansive, experienced increase in CBR in the range of 1.5 to 3.0 times the control value. As soil consolidation is a time-dependant process, with the passage of time CBR shows a sustained rise even after a period of 8/9 years. Void ratio and compression index of the sub-grade soil registered a downward trend while its dry density increased. In most of these road applications, woven JGT having tensile strength of 20 kN/m to 25 kN/m were used. In busy and heavy roads use of woven JGT of higher tensile strength (30 kN/m) and above may be called for.

4.1.2 *Slope stabilization*

Open weave JGT was used for stabilization of slopes in hills and embankments followed by creation of vegetative cover that protects the slopes from destabilization on bio-degradation of JGT. In almost all case studies, the yield of vegetation improved by around 5 times the usual yield (Kg per ha.) after 3 years of plantation. Moisture conservation, reduction of the velocity of surface run-off and quality improvement of soil were in evidence in all the applications. The extent of improvement of soil character depends on selection of the vegetation, soil type and climatic features. The choice of open weave JGT (500 gsm or above) depends on the amount and intensity of precipitation, and inclination of slope.

4.1.3 *River bank protection*

Woven JGT smeared with bitumen (90/15 grade) was used in all application related to erosion control in river and waterways. This was done with a view to protecting JGT yarns from direct exposure to water. Moreover, time of bank soil consolidation in river depend on the nature of flow (two way or one way), existence of vortices at the toe of the bank slope and other hydraulic factors. In the tidal reach of a river (e.g., the Hugli in West Bengal, India), mangrove implant was tried to add to the strength of the bank soil through its roots. In non–tidal reaches (one way flow), usual boulder rip rap was used. The schematic diagram of installation of JGT in road construction, slope stabilization and river bank protection are shown in Figs. 1, 2 and 3.

Use of JGT in river bank protection proved to be an effective alternative to conventional method of granular overlay on the eroded bank in respect of capital investment and recurring maintenance cost. Normally, 1 to 2 years is observed to be adequate for optimizing bank soil consolidation in unidirectional rivers and 2 to 3 years for tidal rivers.

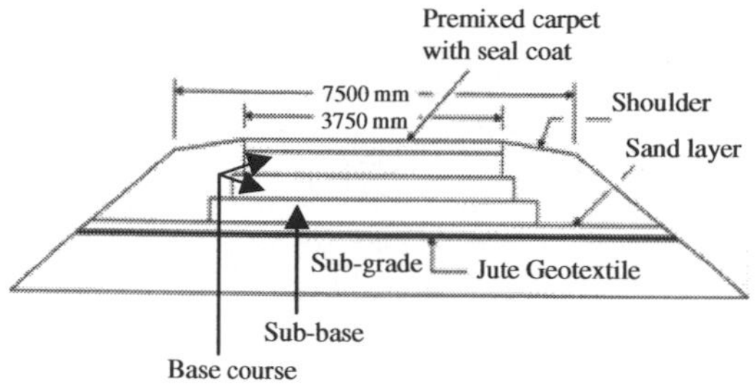

Figure 1. Laying of JGT over sub-grade of road.

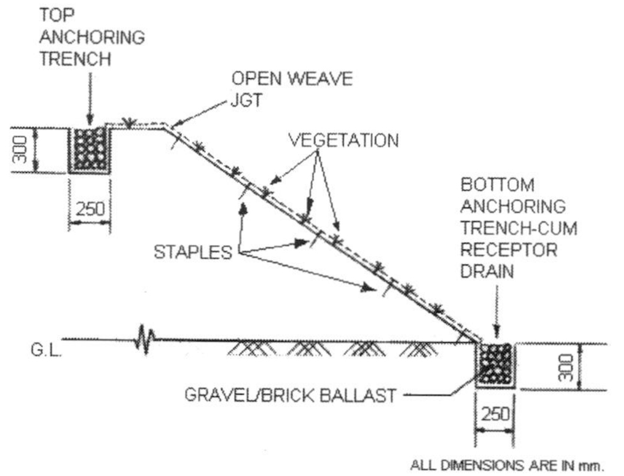

Figure 2. Installation of JGT on slope.

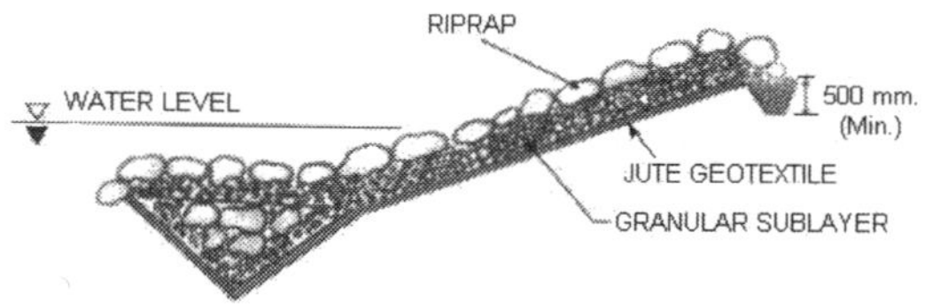

Figure 3. Installation of JGT over bank of river.

5 POTENTIAL APPLICATIONS OF JAT

5.1 *Case studies with JAT in India*

In absence of vegetative cover, the top soil is eroded by overland run-off that usually flows as sheet. The rate of top soil erosion increases, understandably, with the steepness of ground and the intensity and duration of precipitation. It takes years to re-vamp a denuded ground through deposition of new soil. In fact, denuded areas become vulnerable to erosion if they are not overlain by vegetation of the right type in time. Moreover, continued surficial soil erosion causes aggradation of beds of rivers, canals and drainage channels depleting their drainage capacity and enhances the probability of floods caused due to excessive rain fall. JAT aids in quick growth of vegetation and checks loss of nutrients in soil. The mulching effect of JAT on its biodegradation aids "greening" of the soil for a substantial period due to creation of a conducive micro-climate including retention of optional humidity of the top soil and preservation of the existing nutrients. JAT adds nutrients to the soil at micro level.

5.2 *Surficial soil erosion control*

Field trial was undertaken in a tea estate in Assam to observe the effect of JAT on control of soil erosion and nutrient loss. Data obtained after two years of monitoring indicated that JAT was effective, when used in combination with green crop, in reducing soil erosion by around 96% and controlling nutrient loss to the tune of 95%. Usually 500 gsm non-woven JAT is used.

5.3 *Weed suppression*

Another application of JAT is in the field of weed suppression. The objective of weed suppression is to suppress growth of weeds on the one hand and to foster growth of the desired vegetation on the other. Proficiency of non-woven JAT is established in two field trials conducted in two tea-gardens in the north-eastern region of India.

Field trial with non-woven JAT was given for weed suppression in Tea Garden in Assam. The fabric was laid in between two rows of plants over the existing weeds and fixed to the ground by wooden staples. Weedicides were also used in conjunction to study the comparative effects.

Results confirmed that under the given climatic soil and topographical conditions non-woven JAT could reduce weed emergence by 65%. Addition of weedicides could enhance the extent of suppression by a further 13% maximum. Usually dense non-woven JAT (500 gsm/1000 gms) are used for this purpose.

5.4 *Afforestation in semi-arid zone*

Study was conducted to find the effectiveness of JAT in fostering plant-growth in a semi-arid laterite soil zone in West Bengal. Fabric was laid in between the rows of plants and pegged to the ground. The height of plants was measured after 3 and 6 months. The growth of plants was found to be more pronounced with Jute Agrotextile and was almost double the growth under control conditions.

5.5 *Growth of sapling in jute sleeve*

Study was conducted to observe the efficacy of Jute sleeves in raising sapling as compared to conventional poly-sleeves at nurseries in high altitudes on the lap of the Himalayas (Sikkim in India).The trial was conducted at different altitudes ranging from 1350 metres to 2600 metres above M.S.L. Jute Sleeves (size 22.5 cm × 15 cm) with one end open and the other end closed were used to fill soil bulb with the root. The growth rate was observed for one year. The grown up plants were then transplanted directly into the soil. Removal of Jute Sleeves was not necessary as jute was supposed to coalesce with the soil and became an integral part of it ultimately. The soil bulb contained compost manure of cow dung and sand. It was observed that with jute sleeve there was faster & better growth of sapling and better aeration. Extremes of temperature could be evened out. No ice formation was noticed within the sleeves. High survival rate (90%), spread of roots through fine openings of the Jute Sleeves. No extraction at the transplantation stage was necessary, unlike in the case of poly sleeves. Jute Sleeves are found to be superior to poly-sleeves in performance, installation and after care. The cost of Jute Sleeves is, of course, higher than poly sleeves.

6 FURTHER R & D WORK

IJIRA&JMDC are jointly continuing extensive R&D activities to develop newer varieties of such products with improved physical, mechanical and hydraulic properties like strong, durable and cost- effective fabrics for various applications including paving fabric.

The case studies referred to are not exhaustive and there remains areas that have not been experimented. Efforts in this direction have taken off.

7 INFERENCE

In the case studies presented above, Jute Geotextile was found to have performed as expected, having executed the basic functions in three different applications in civil engineering. The case studies confirmed that Jute Geotextile principally acted as catalyst or as change agent in improving the engineering performance of soil. Improvement was achieved basically as a result of ridding of the soil of the entrapped water usually in plastic soils, on the one hand and retention of fines on the other (filtration function). JGT also acted as a drain within its own thickness. Its inherent capacity of water absorption helped in lateral dispersion of water. Design of porometry assumes significance in this respect. Retention of fines and permittivity act under contrasting porometric parameters. Experience plays a big role indeciding the right porometry on the basis of the appropriate established empirical relations. The installation stresses were absorbed by high initial strength of JGT.

Durability of JGT beyond two season cycles as earlier indicated, was found not a technical necessity, as soil on reaching the maximum dry density becomes self-reliant and independent of the functions of any Geotextile. In the case studies with Jute Agrotextile in which strength of the fabric is of lesser significance, its effectiveness as mulch, temperature–conditioner and humidity–regulator is established. Dense non-woven JAT acts as a barrier to air and sunlight and helps in suppressing growth of weeds. JAT also supplied nutrients to soil though at a micro level. In fine both JGT and JAT can, to a large extent, reduce concerns for polluting environment in their respective applications. Techno-economically the products have been found to be highly effective, the cost of the products are available in site-www.jmdcindia.com .These two jute diversified products-more aptly technical textiles-fit in with the emerging global trend in adopting eco-friendly as well as bio-engineering measures and deserve bigger support from sensible end-users.

ACKNOWLEDGEMENT

The authors gratefully acknowledge the support and guidance provided by Dr. J. Srinivasan, Director, IJIRA and Sri A. Bhattacharya-IAS, Secretary, JMDC in publishing the paper.

REFERENCES

Barooah, A. K. Goswami H. & Dutta, U. 1997. Biodegradable Jute Geotextile for Integrated Soil & Crop Management in Tea Estates – Soil Conservation (part – I) . Workshop on Jute Geotextile.

Choudhury, P. K. Chatterjee, P. K. and Dutta, U. 1999 A Low Tech. Approach for Forests. 1st Asia – Pasific Conference on Water & Bioengineering for erosion control & slope stabilization.

Juyal, G. P. and Dadhwal, K. S. 1996. Geojute for erosion control with special reference to mine-spoil rehabilitation. Indian Journal of Soil Conservation, Vol. 24,

Ramaswamy, S. D. and Aziz, M. A. 1989. Jute Geostextile for Roads. International Workshops on Geotextiles,

Rao, P. J. Bindumadhava & Venisiri, N. 1998 Construction of Highway Embankment on Soft Marine Soil using Jute Geotextile. 6th International Conference on Geosynthetics.

Sanyal, T. 1992 Control of Bank Erosion Naturally – A Pilot Project in Nayachara Island in the River Hugli. National Workshop on Role of Geosynthetics in Water Resources Projects.

Sanyal, T. & Choudhury, P. K. 2003. Prevention of Railway Track Subsidence with Jute Geotextile – A Case study under Eastern Railway. Workshop on Applications of Geosynthetics in Infrastructure Projects.

New Horizons in Earth Reinforcement – Otani, Miyata & Mukunoki (eds)
© 2008 Taylor & Francis Group, London, ISBN 978-0-415-45775-0

Geotechnical problems on reinforcement soil ground in Kazakhstan

A. Zhusupbekov & R. Lukpanov
Eurasian National University, Astana, Kazakhstan

ABSTRACT: The priority task of the development of modern construction is improvement the reliability and longevity of building materials along with economical effectiveness which satisfy mass high volume growth in the term of progressive intensification of constructions. Geosynthetic reliability and durability criterion under the interest of engineers, and reinforced soil model is one of the progressive solution of engineering.

1 INTRODUCTION

Reinforcement means to use special elements in soft soil constructions which allow increase the mechanical property of soil. Dealing with soil the reinforced elements redistribute load among construction parts, providing the transmission of stress from the overloading zone to the adjacent underloading zone. Nowadays there are a lot of different reinforcing materials in a world practice. The most part of them consists of geosynthetic – materials on base of synthetic polymer fiber which are made of polypropylene (PP) or polyester (PET). The geosynthetics subdivide into geogrid (PET material) and geotextile (PP material). Geotextile – material is produced from fabric by the method of needle punching and might be woven or not woven and geogrid for soil reinforcement might be as volumetrically or flat (biaxial or uniaxial too) according to their assignments. Although there are many cases when composite materials by combining geogrid with geotextile methods have been used. Geosynthetic material provides its high chemical inertness against acid and alkaline, stability against termooxidizely process. The material is fast against ultra-violet rays and it is although green product. Physical and mechanical properties of geosynthetics are shown in Table 1.

Under the highest possible loading, geosynthetic has till 45 percent elongation. It depends on the applicable thickness of material. In this way local damages do not lead to the destruction of materials. Due to the high index of elastic modules, the material can bear considerable load, implementing function of reinforcement at not great deformation (F. Tatsuoka etc.).

Choice of reinforced material does not depend on its characteristics of strength. The polymer which is produced from reinforced material has substantial degree. By way of illustration geosynthetic made from polypropylene is used in dynamic loading as polypropylene has high index of creep, that is it has ability to long the term extension under the dead load. Therefore the material is used in road building in pavement capacity. Geosynthetic, produced from polyester, with very low index of creep, usually is used in case of static load or exists probability uneven development of settlement in the result of heterogeneousness soil. As example, we can give retaining wall, strengthening of embankment, reinforcment the heterogeneousness soils which have very low index of bearing capacity.

Table 1. Physical and mechanical properties of geosynthetics.

Characteristics	Geotextile (PET)	Geotextile (PP)	Composite
Surface density, g/м^2	250	250	250
Tensile strength, κN/м^2	4,2	2,8	8,4
Thickness at the load 2 MPa, мм	3,2	3,2	3,2

2 REINFORCEMENT MODEL

2.1 *Construction of retaining wall within the reinforcement model*

Construction of retaining wall within the reinforcement model is usually used to strengthen slope covers of railways and highways in bridge abutments, foundations of different constructions (Figure 2)

As tests have shown the destructing load for these types of constructions exceeds the design load. This is explained as that geotextile possesses high index of tensile strength and follows for deformation of soil of the construction creating general state of

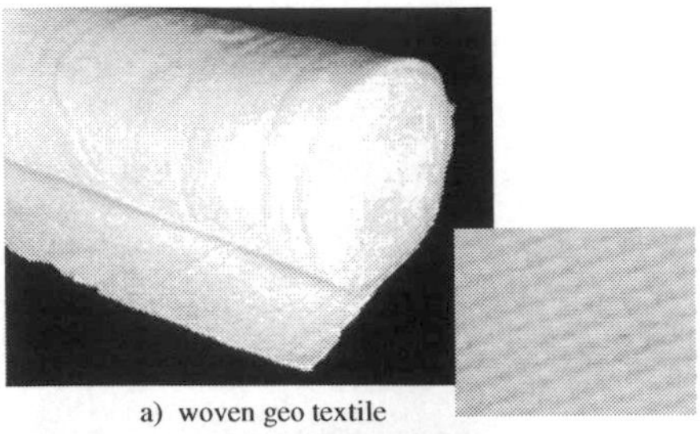

a) woven geo textile

b) volumetrically biaxial geogrid

c) flat biaxial geogrid

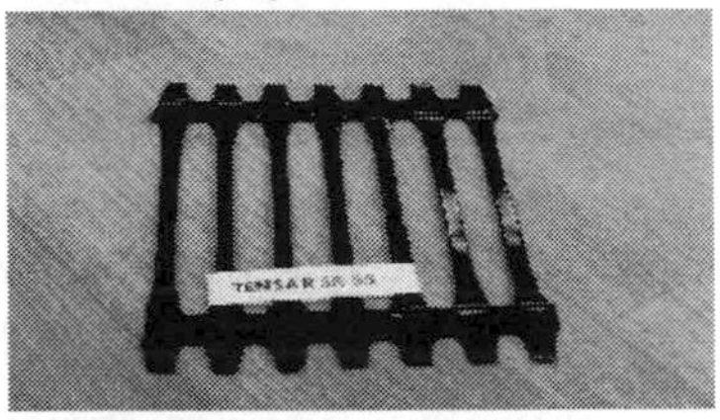

d) flat uniaxial geogrid

e) combined (woven geotextile with flat biaxial geogrid)

Figure 1. Types of geosynthetic reinforced materials.

Figure 2. Construction of retaining wall within the reinforcement model.

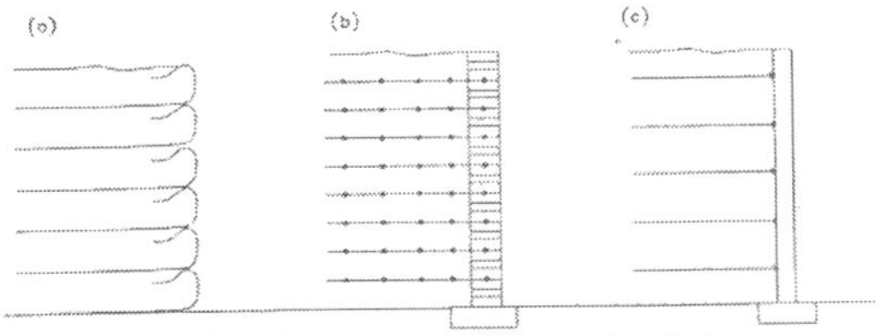

(a) with wrap-around geotextile facing,
(b) with segmented precast concrete or timber panels
(c) with full-height precast panels

Figure 3. Reinforced retaining wall system using geotextile.

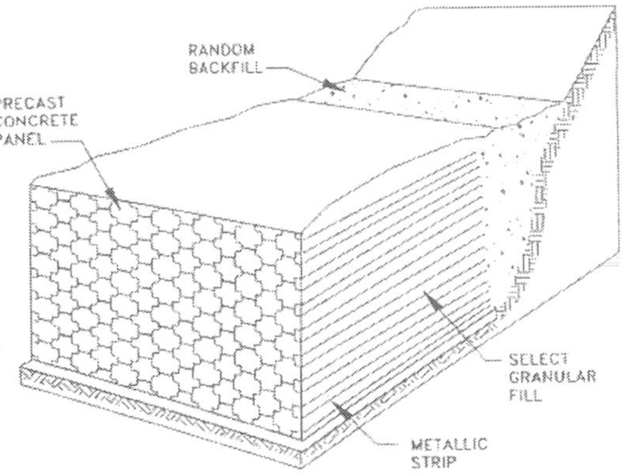

Figure 4. Component parts of a reinforced earth wall.

stress and increasing construction stability (E.C. Shin etc.).Model of geotextile retaining walls and consisting reinforced elements are given in Figures 3 & 4.

A number of approaches to geotextile and geogrid reinforced retaining wall design have been proposed, and these are summarized by Christopher and Holtz (1985), Mitchell and Villet (1987), Christopher, et al. (1989), and Claybourn and Wu (1993). The most commonly used method is classical Rankin earth pressure theory combined with tensile-resistat tie-backs, in which the reinforcement extends beyond an assumed Rankin failure plane. Figure 5 shows an system and the model typically analyzed. Because this design approach was first proposed by Steward, Williamson, and Mohney (1977) of the U.S. Forest Service, it is

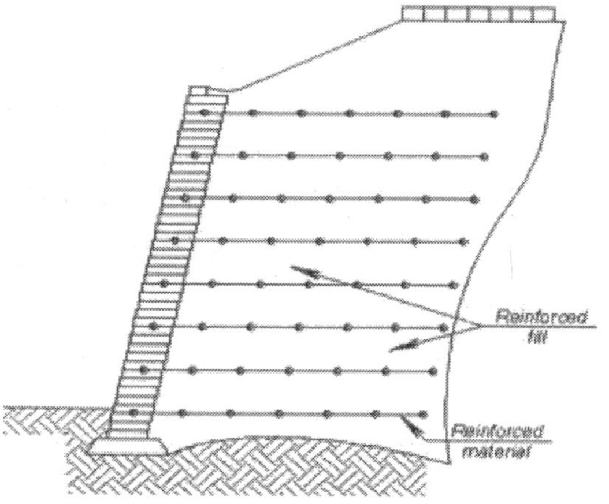

Figure 5. Actual geosynthetic reinforced soil wall in contrast to the design model.

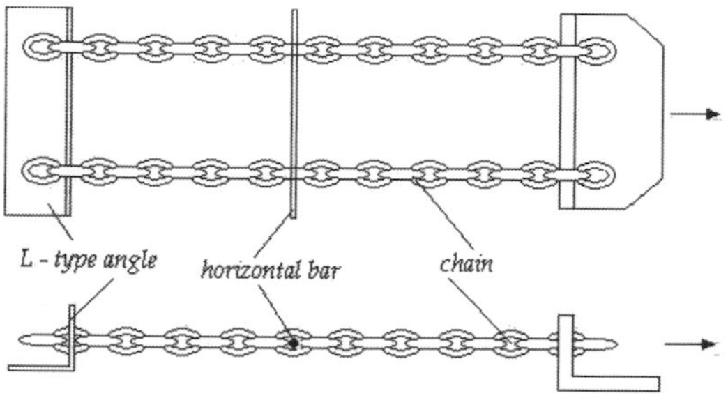

Figure 6. Type of chain reinforcement.

often referred to as the Forest Service or tie-back wedge method (E.C. Shin etc.).

Except geosynthetic material to other materials can be used make reinforcement. So during the pullout test chains from non-rusting steel have considerable figures of resistance (Figure 6, Table 2).

The Pullout resistance by chain reinforcement can be defined by the following equation:

$$F_{tc} = F_1 + F_2 + F_3 \quad (1)$$

where F_1 = the frictional force between chain and soil skeleton; F_2 = the shearing resistance with including the soil inside the chain; F_3 = the passive resistance in cross sectional area of chain.

The earth pressure resistance of horizontal bar is defined as $F_{ri} \cdot F_{bi}$ is the pullout force with a L type angle.

2.2 *Reinforcement of road building*

In road building reinforcement fulfills the function of layer separation. This permits to increase the index of bearing capacity largely due to of its stress redistribution. By way of illustration – The model of reinforcement installation of "The new western road" project (city of Atyrau, Kazakhstan, 2003). In the result of the research, which was held on road building "The new western road" project we want to say that is very difficult to compact natural soil to required coefficient of compaction because the natural soil (loamy soil) has very low index of bearing capacity. To increase of durability and deformation property of road basement the model of reinforcement with the following steps were decided to choose (A. Zhusupbekov etc.).

1. The natural loamy soil is compacted by road-roller to ultimate level according to required standards. Evening of surface (Filling the pits, pot-holes and another local damage where the water may stay for a long period)
2. Installation of reinforced material (Figure 7)
3. Filling of the soil (Figure 7), with height no more that 200 mm, and its compaction to required coefficient of compaction standard.

The benefit of reinforcement was determined by examine of surface during three years service. The economical efficiency diagram which has been determined by comparing appearance of pits, pot-holes represent on Figure 8.

Initially the reinforced pavement cost more but after a certain period of time the reinforced pavement is a lower total cost.

The next research represents that the effective work of reinforced materials depends on its shape of geosinthetic (Figure 9) besides its type (PET, PP).

Efficiency of geogrid application serviceability with comparing geotextile is represented in Figure 10. For the initial data the appearance of serviceability pits and pot-holes were considered.

A cost comparison for reinforced versus other types of retaining walls is present in Figure 11.

Geosynthetic is recommended for use in soft soil subgrade because is the less expensive. Application of reinforcement materials allow to decrease thickness (ellipse in Figure 12) of stone base simultaneously require to demands of reliability and durability.

However there exists several variations (Figure 13) of choosing reinforced materials for soft soil condition, the final selection is based on technical and economical comparing.

Consequently one of the traditional types of road construction – asphalt pavement has the best characteristics of serviceability but not perfect. Working in various temperature and considerable dynamic load influence lead to the appearance of cracks because of low index of asphalt tensile strength. Even the low level of tensile load leads to appearance of crack and decreases serviceability properties and durability of asphalt pavement. Therefore the most

Table 2. The results of pullout test with different chain lengths.

Length of chain	Vertical pressure (kgf/cm²)	Pullout force (kgf)						Sum (kgf)	
		F_1	F_2	F_3	F_{tc}	F_{ri}	F_{bi}	$F_{tc} + F_{ri}$	$F_{tc} + F_{bi}$
2.0 m	0.4	90.93	68.20	81.15	200.27	108.69	360	308.96	560.27
	0.8	101.85	136.39	162.30	400.53	184.45	720	584.98	1120.53
	1.2	152.78	204.58	243.44	600.80	260.21	1080	861.01	1680.80
2.5 m	0.4	63.49	85.02	101.17	249.68	108.69	360	358.37	609.68
	0.8	126.98	170.04	202.34	499.37	184.45	720	683.82	1219.37
	1.2	190.47	255.06	303.51	749.05	260.61	1080	1009.26	1829.05
3.0 m	0.4	76.06	101.85	121.19	299.10	108.69	360	407.79	659.10
	0.8	152.12	203.70	242.39	598.20	184.45	720	782.65	1318.20
	1.2	228.17	305.55	363.58	897.30	260.61	1080	1157.51	1977.30

Figure 7. Installation of reinforced soil of "The new western road" object (Atyrau city, Kazakhstan, 2003).

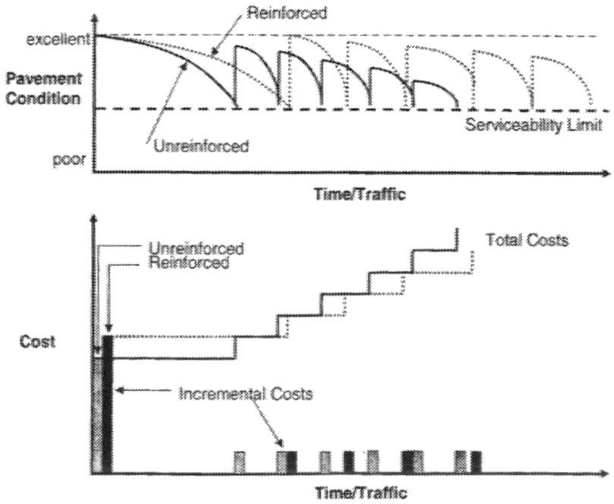

Figure 8. Economical efficiency of reinforcement model.

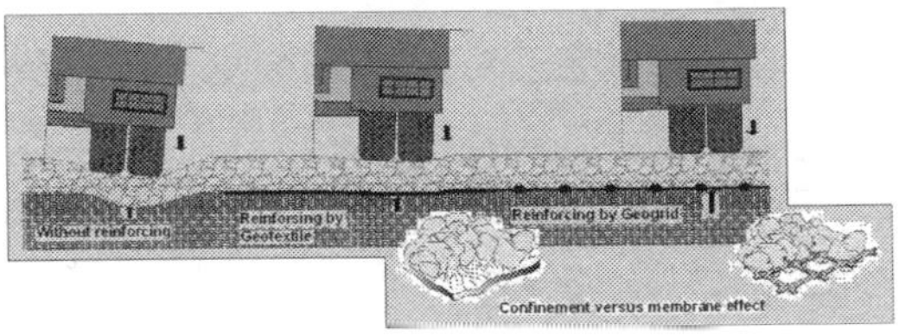

Figure 9. The work of reinforcement model.

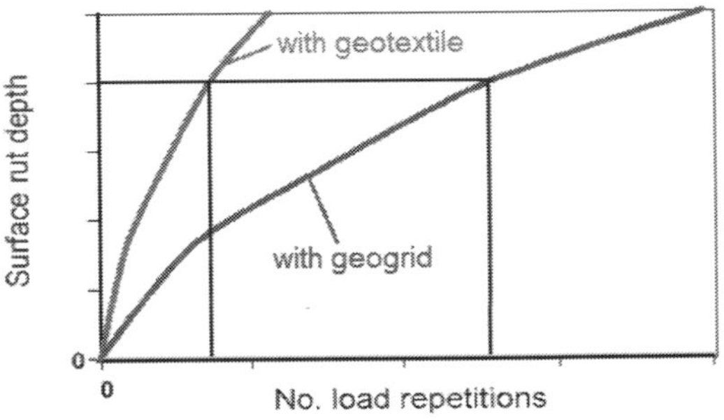

Figure 10. Efficiency of geogrid application serviceability.

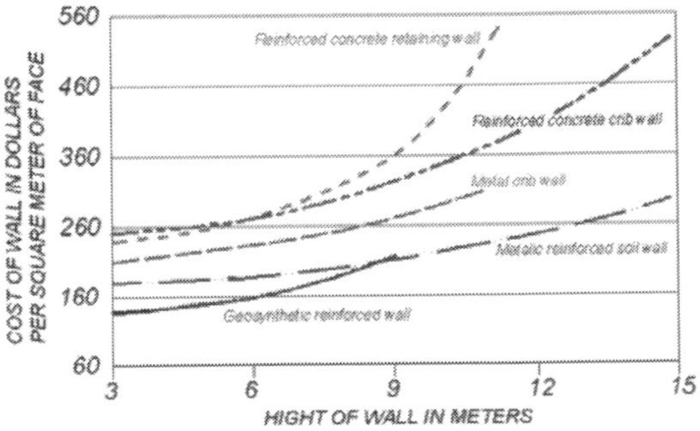

Figure 11. Cost comparison of reinforced system.

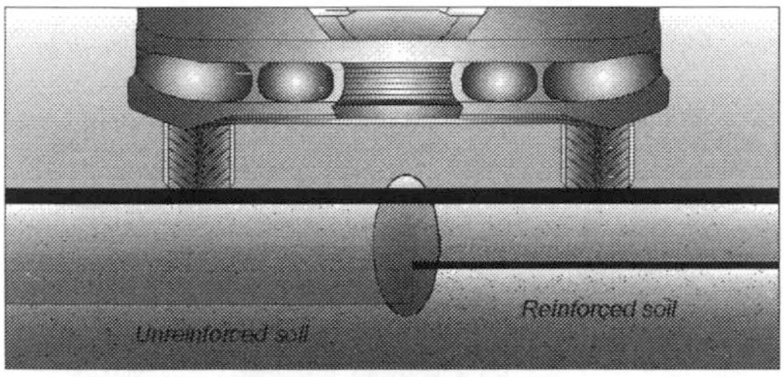

Figure 12. The difference in required thickness of stone base is then compared with the cost.

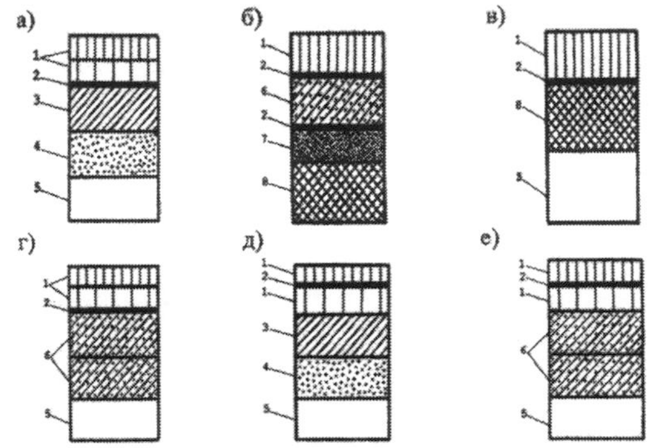

1 – asphalt, 2 – reinforced materials, 3 – gravel or crashed stone, 4 – sand, 5 – earth, 6 – gravel processed by bitumen, 7 – leveling coat (sand), 8 – existing road.

Figure 13. Type of road reinforcement.

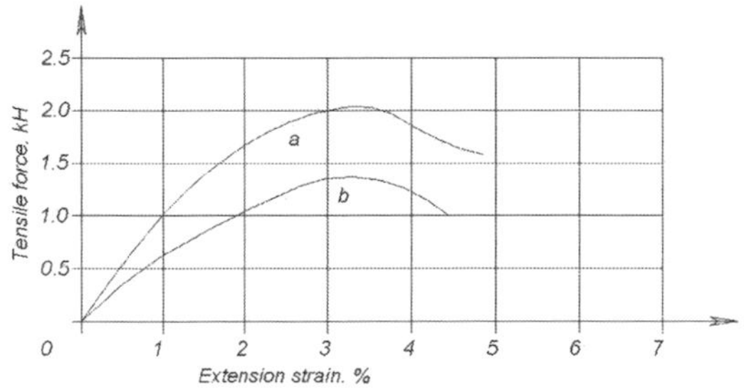

Figure 14. Dependence extension strain different asphalt pavement from tensile force.

progressive solution, based on durability and reliability of construction which excepts such problems, is reinforcement. Influence of reinforced geogrid of asphalt pavement samples are given in Figure 14 (A. Zhusupbekov etc.).

Usually in contrast in non-reinforced asphalt pavement samples where we can see big cracks appear than small distributed cracks will appear in reinforced sample.

3 CONCLUSIONS

Retaining walls with use geosynthetic are generally less expensive than conventional earth retaining system. Using geogrids or geotextiles as reinforcement has been found to be 30 to 50% less expensive than other reinforced soil construction with concrete facing panels. Due to their greater flexibility, this model offer significant technical and cost advantage over conventional gravity or reinforced concrete cantilever walls at site with poor foundations and slope conditions.

As the results of research work show that the application of reinforced construction will be proved from the economical point in case if that height of retained construction are higher than three meters. The cost of one meter reinforced wall with reinforcement is cheaper for 2 or 3 times than the price for one meter reinforced concrete.

From the point of economical and technical expediency the reinforcement application is conformed by its wide usage in developed countries of the world and the base of its successful application that will provide to increase its serviceability road period for 2 times.

ACKNOWLEDGEMENTS

The authors express deep thanks to Professor Eun Chul Shin (Incheon University, Korea) for his geotechnical consulting and advising of this research work.

REFERENCES

Shin, E.C. & Young, I. O. 2006. Case Histories of Geotextile Tube Construction Project in Korea. In E.S. Shin & J.G. Kang etc. (ed.), *New Developments in Geoenvironmental and Geotechnical Engineering; Proc. intern. symp., 9–11 November 2006, Incheon,* Korea: IETeC.

Tatsuoka, F., Tateyama, M. & Koseki, J. 1995. Performance of soil retaining walls for railway embankment. *A Special Issue of Soils and Foundation on Geotechnical Aspects, 17 January 1995.*

Zhusupbekov, A. & Lukpanov, R. 2007. The experience of long-term performance of reinforcement soil structure. *International Workshop on Scrap Tire Devided Geomaterials "Opportunities and Challenges", 23–24 March 2007, Yokosuka,* Japan: PARI..

New Horizons in Earth Reinforcement – Otani, Miyata & Mukunoki (eds)
© 2008 Taylor & Francis Group, London, ISBN 978-0-415-45775-0

Case study on deep excavation works by soil nailing on adjacent building

Y.S. Cho & H. Imanishi
Samsung Corporation, Korea

ABSTRACT: It was done the introduction that it was easy for an on-site example and the Korean ground where deep excavation was carried out in the place that was near to a building in Korea. And, it introduced the pre-soil nailing method executed to control the deformation of the retaining wall. Deformation control of the retaining wall of this pre-soil nailing method was able to be confirmed. The forecast of the wall deformation by the back-analysis method of using the measurement is executed by using FEM and while excavation. As a result, the prediction method was very effective.

1 INTRODUCTIONS

Soil nailing is a typical in-situ reinforcement method used for a wide variety of construction applications, such as, stabilization of cut slopes and excavation retaining walls(Juran & Elias 1991, Hanna, Juran, Levy & Benslimane 1998). By the way, in major big cities of the world, many kinds of infrastructures have been highly developed. Therefore, it is very important to predict the deformation behavior of ground due to the excavation work and to estimate the effect of ground deformation on the neighboring structures. Therefore, this paper tries to introduce the example of constructing about 35ṁ in underground excavation depth by using soil nailing and the ground anchor in Seoul Korea. Digging up 35 m in the underground is that a lot of the construction cases are rare in Seoul Korea by using soil nailing either. Various earth retaining wall methods were applied in this construction and there was a building that was very adjacent and, on the other hand, a variety of reinforcement methods were executed. The steel pipe grouting is executed from the center of the earth retaining wall by the industrial method to reinforce the lower side of an existing building and the ground has been improved. The pre-nailing method was executed, and moreover, the deformation of the earth retaining wall was controlled, and, as a result, the pre-nailing method that controlled the settlement of an existing building was executed. And, because the back analysis is executed when under construction and the forecast in the future was executed compared with the measurement result, it tries to introduce this. Finally, it tries to introduce the risk management method that executes it from the first stage of construction. This risk management method is a new risk management technology while using it making it in Samsung Corporation.

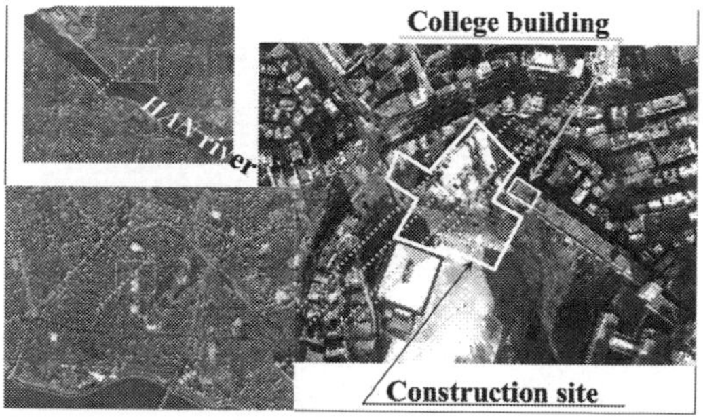

Photo 1. Aerial photograph of site.

2 OUTLINE OF CONSTRUCTION

2.1 *Outline of site*

This site is situated in South Korea Seoul, and an architectural scale is the 16th floor on the ground, and the sixth floor in the underground. And the excavation area is about 94,508 m^2, and the maximum underground excavation depth is about GL-34 m. A building and a gymnasium and a private house approach the excavation site circumference. Figure 1 shows a site circumference.

The constitution of the ground is 0 ~ 3 m fill, 3 ~ 12 m residual soil, 12 ~ 19 m weathered rock, 19 ~ 20 m weak rock, 20 ~ 35 m hard rock. The initial water level is under the ground in the weathered rock neighborhood.

In addition, a CIP+E/A method is applied to the method of the retaining wall. However, a strut method was applied for the section (college building south side section) that difficulty was expected for execution by interference when that executed the work in E/A. In addition, a S/N method of construction was applied for

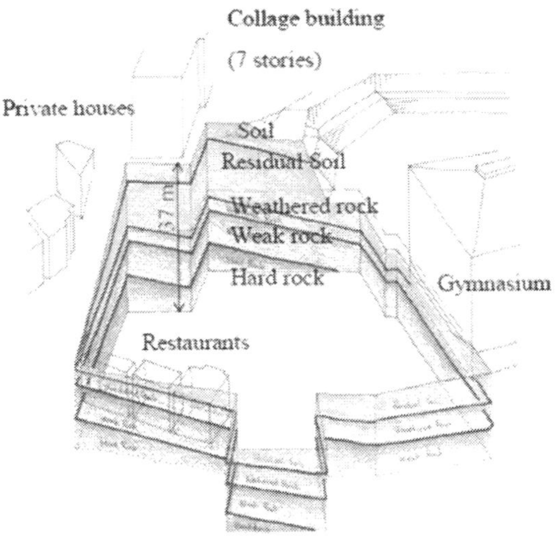

Figure 1. Excavation area and Geological profile.

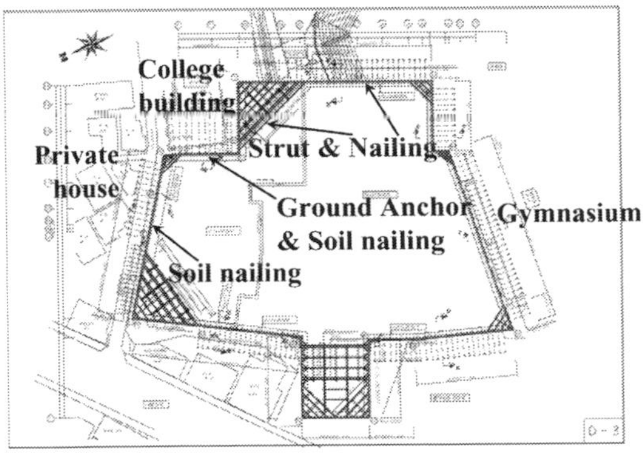

Figure 2. Support method of the excavation plane.

the part section (residential area part section) that was not able to obtain the outskirts's consent in E/A execution. CIP was carried out for weathered soil section. It was applied H-pile + shotcrete for the bedrock section. Figure 2 shows the support method of the excavation plane.

2.2 *Outline of measurement*

The setting position of the measure was shown in figure 3. In this site, a excavation section is very complicated. In addition, this site where excavation depth is very deep, and an adjacent building approaches it, and it is executed the work very hard. Therefore, it was judged deformation measurement of the back ground and that the deformation measurement of the adjacent structure was very important. Inclinometer, piezometer, tiltmeter, crack gage, settlement gage is installed around an adjacent building. In addition, multi-cell liquid settlement systems was installed for the

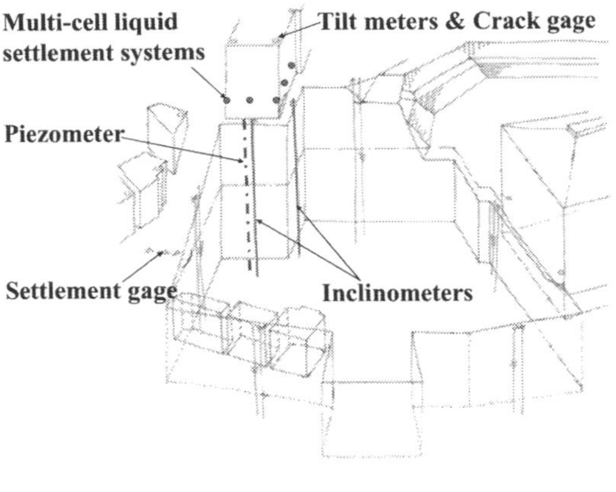

Figure 3. Setting position of the measure.

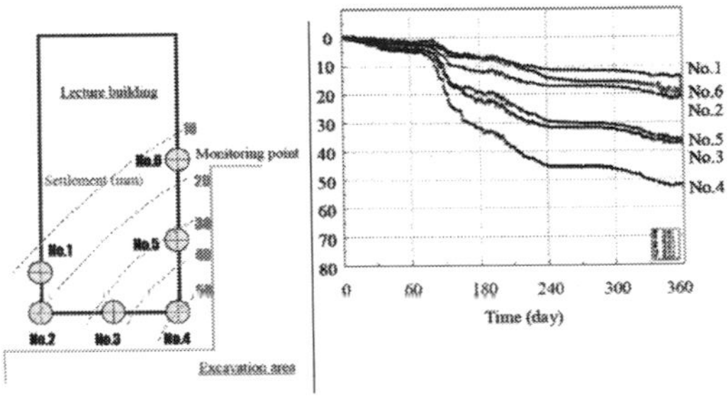

Figure 4. Measurement position and actual settlement.

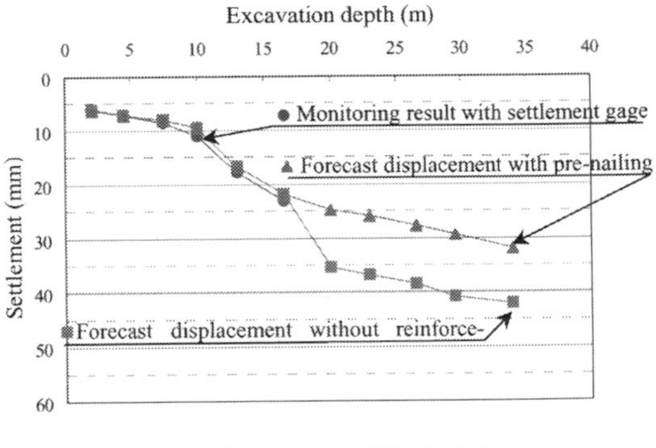

Figure 5. Expected settlement of the building.

college building and precision was high and measured the settlement of the building.

2.3 *Result of monitoring*

Figure 4 show the actual settlement at the collage building in Figure 1 during deep excavation.

Figure 5 showed relations between excavation depth and the settlement of the college building.

Figure 5 estimated quantity of settlement of the college building occurred by the deformation of the retaining wall by the back analysis method. An analysis

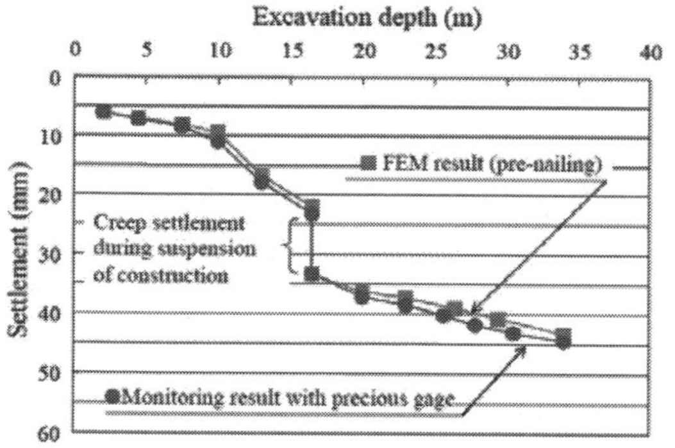

Figure 6. Calculated settlement and actual settlement.

result, it was estimated that quantity of settlement of a building exceeded about 40 mm to last excavation. Deformation when it applied pre-soil nailing to restrain settlement of the building was estimated together. As a result, when it applied a pre-soil nailing method of construction, quantity of expected settlement was estimated by it being occurred to about around 10 mm small. Therefore, it was applied a pre-soil nailing method of construction.

Figure 6 shows volume of deformation and the relations of the real volume of deformation predicted during excavation. Quantity of expected settlement accords with quantity of real occurred settlement well. Therefore, expectation by the back analysis mothod that used volume of deformation occurred during excavation is very effective.

3 CONSTRUCTION METHOD

3.1 *Pre-soil nailing method*

As a result of having expected the displacement of the wall, because the inequality settlement of the college building exceeded a creterion, it was judged that excavation was performed by an existing method, and a pre-soil nailing method was applied. The concept of pre-soil nailing showed in figures 7 and 8. A Pre-soil nailing method is a method to minimize liberation from stress by the excavation. It is a method to minimize the deformation of the wall body by it leave soil before excavation and restrain the deformation of the wall by the liberation from stress, and executing the work in nailing.

Photo 2 shows the construction of pre-soil nailing in site.

3.2 *Steel pipe reinforcement*

Photo 3 is a scene reinforcing the building retainer by a steel pipe grouting method of construction from the wall for the settlement restraint of the building.

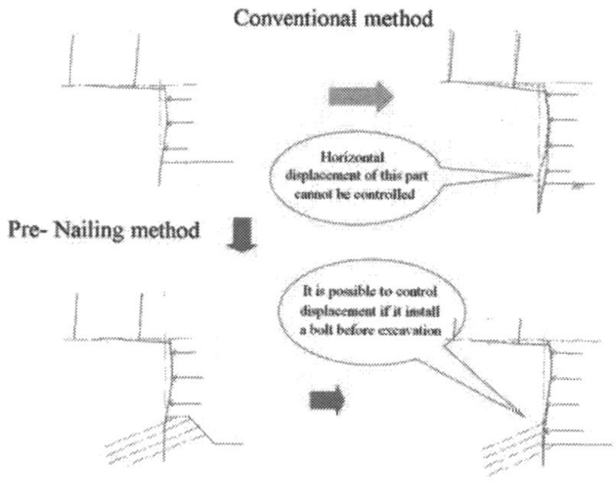

Figure 7. Conventional method and pre-soil nailing.

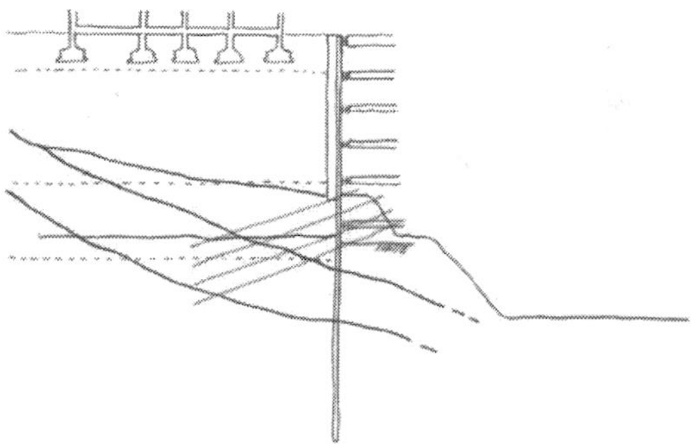

Figure 8. Concept of pre-soil nailing.

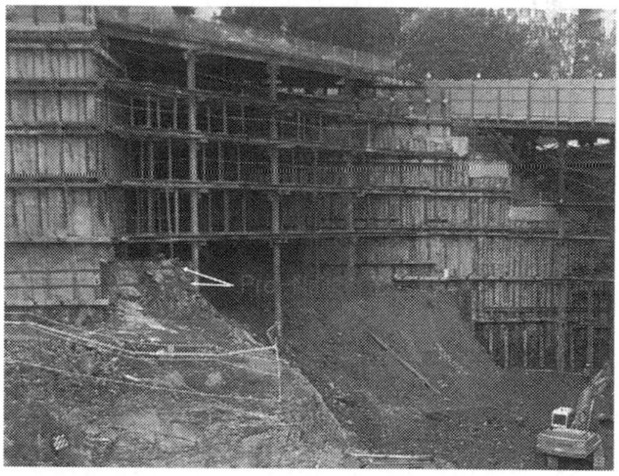

Photo 2. Construction of pre-soil nailing.

As for the grouting, the deformation of the wall body and the possibility that the deformation that it cannot expect occurred of the building is expected and reinforced it with minimum pressure and quantity of injection. Therefore, it managed 5 kg/cm^2 with

Photo 3. Reinforcement of building foundation.

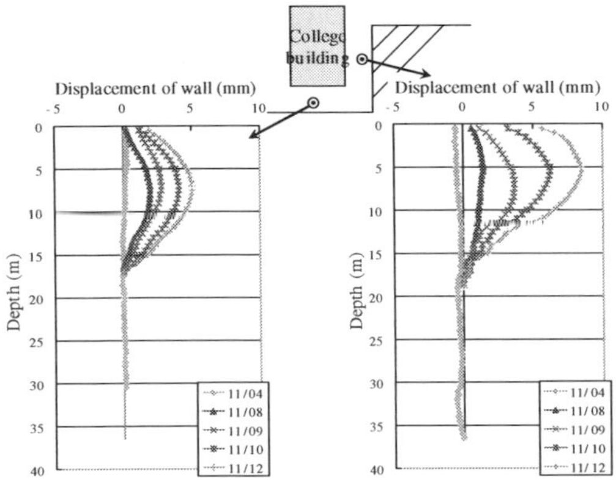

Figure 9. Wall displacement during steel pipe grouting.

injection pressure and volume of injection with 3% of ground.

Figure 9 shows the displacement of the retaining wall at steel pipe grouting. Wall body displacement of about 5 ∼ 8 mm occurred by pressure at the time of the grouting additionally around a grouting position. In addition, it was shown because the displacement that almost resembled the displacement that was expected before steel pipe grouting reinforced it according to upper figure 6 occurred. Therefore, as for steel pipe reinforcement to restrain the settlement of the existing building which there was on the back of the wall, the deformation of the wall body was increased with outbreak of the creep deformation by the perforation of the structure lower part, and effect of the reinforcement was confirmed in the thing that was not so big.

4 CONCLUSION

This article analyzed the excavation method that was appropriate in excavation through in-situ measurement and numerical analysis in an object in the site that did a building and very neighboring excavation, a reinforcement method. When it arranges the result, it seems to be next.

(1) It introduced the ground in South Korea with the case with the site where it touched the building in South Korea and a deep excavation had been executed.
(2) The FEM analysis that used a measurement result to predict the deformation of the retaining wall was carried out. The predicted value by the back analysis method that used measurement data was able to confirm that it was effective in reliability improvement.
(3) Steel pipe reinforcement was carried out to restrain the settlement of the back existing structure of the retaining wall. However, the deformation of the wall was increased with outbreak of the creep deformation by the perforation of the lower part of the structure. Therefore, the effect of the reinforcement was confirmed in the thing that was not so big.
(4) Pre-nailing method of construction was carried out to decrease the displacement of the retaining wall and displacement of the back ground by it in digging as much as possible. As a result, a pre-nailing method of construction was able to confirm that a displacement restraint effect was good.

REFERENCES

Juran, I. & Elias, V. 1991, Ground Anchors and soil Nails in retaining Structures, *Foundation Engineering Handbook*, Chapter 26, Second edition, Van Nostrand Reinhold, New York, pp. 868–905

Hanna, S., Juran, I., Levy O. & Benslimane, A. 1998, Recent Developments in Soil-Nailing – Design & Practice, *Soil Improvement for Big Digs*, Geotechnical Special Publication, ASCE, No. 81, pp. 259–284.

Cho, Y. S., Oda, K., Matsui, T., 2002, Analytical study on deformation characteristics of soil-nailed systems due to excavation, 4*th* International Conference on Ground Improvement Techniques, Malaysia, Vol. 1, pp. 277–282.

Materials and new testing

New Horizons in Earth Reinforcement – Otani, Miyata & Mukunoki (eds)
© 2008 Taylor & Francis Group, London, ISBN 978-0-415-45775-0

Durability evaluation of various geogrids by index and performance tests

H.Y. Jeon
Division of Nano-Systems Engineering, INHA University, Incheon, Korea (South)

M.S. Mok
SAGEOS, Saint-Hyacinthe, Quebec, Canada

ABSTRACT: This study is focused on the test method that used for evaluating the long-term design strength of the various type of geogrid and suggestion of improved test method. Estimated long-term creep deformation indicate that the 65% of T_{ult} loading level is the optimum value that satisfying the creep criteria (curve becomes asymptotic to a constant strian line, of 10% or less) in the case of woven geogrid and 60% in warp knitted geogrid and 30 ~ 35 % in membrane drawn geogrid. All the tested geogrids showed the good resistance under chemical, biological and UV circumstances. The total reduction factor was determined by creep deformation, installation damage, chemical and biological degradation. Creep data interpretation is performed by using performance limit strain. From this procedure we can obtain more accurate reduction factor by creep deformation at the aim design life.

1 INTRODUCTION

Rheological models were adapted in order to describe the creep of geosynthetics(Navarrete, 2001; Mano and Sousa, 2001). For the study related to the installation damaged geosynthetics, more site performance tests were accomplished compared to index test. Several studies have suggested that the level of damage induced by construction to a polymer geogrid is a function of the following primary factors (Giroud 2002, DeMerchant 2002, Lin and Shi 2001); geogrid thickness, compacting effort and lift thickness, type and weight of construction equipment used for fill spreading, type and weight of compaction of backfill, angularity of backfill, etc. In this study, to estimate the RF (reduction factor) values, short-term tensile test, creep test, installation damage test and durability test were performed and the test result compared among geogrids. To review the index test results, more site-specific and material specific test and data analysis methods were proposed especially through creep deformation, installation damage, and chemical and biological resistance test.

2 EXPERIMENTAL

2.1 *Preparation of geogrids*

Three types of geogrids were employed in this study. The textile type of geogrid is divided into woven geogrid and warp knitted geogrid again and made of polyester high tenacity yarn coated with PVC (polyvinyl chloride) resin. The membrane drawn type geogrid is made of melted high density polyethylene with uni-directional conformation. So, for better comparison of these two types of geogrids, geogrids having same nominal strength (e.g., 8 ton/m, 10 ton/m) are selected as references. And, all the tests were performed to only longitudinal directions because the uni-directional drawn geogrid samples were used in this study.

2.2 *Evaluation of engineering properties*

To evaluate more optimum tensile strength of geogrid, ASTM D4595, wide-width tensile test method was adopted. ASTM D5262 was used to determine the creep deformation behavior of geogrid under constant temperature and load condition. Installation damage test was done under consideration the real installation field conditions. Chemical resistance test were performed by modified EPA 9090 standard. For biological resistance test, samples were incubated in two types of conditioned box which was filled with weathered granted soil and sewage sludge. The tensile strength values before and after incubation were determined by GRI test method GG-1.

3 RESULTS AND DISCUSSION

3.1 *Tensile properties*

Tensile strength of each specimens are higher than its product strength about 4 ~ 13%. In the case of warp knitted type geogrid specimens, the extra tensile strengths are above 13% to the design strength, and about 4 ~ 12% higher in woven type geogrids. And the tensile strength of membrane drawn type geogrids are higher than the products strength about 8 ~ 13%. From these extra tensile strengths, we may say that the additional factor of safety has been connoted in the geogrids. Also, the tensile strain at the ultimate strength are about 11.0 ~ 14.0%, 9.0 ~ 12.0% and 8 ~ 12% about the each geogrid samples. All of the geogrids in this study showed the good elongation properties. However, in the case of membrane drawn type geogrid, the additional elongation possibility exists in its inner structure. Figures 1 ~ 3 show the results of tensile test of each geogrid samples.

3.2 *Creep deformation behaviors*

To obtain master curves for long-term creep deformation, time-temperature superposition principle was applied. Conventional ambient creep test results and accelerated test are shown for each geogrid samples in Figures 4 ~ 6. From these test results, we can estimate the strain values that the curve becomes asymptotic to a constant strain line, of 10 percent or less. In the case of textile geogrids, each woven and warp knitted geogrids have 60% of UTS (ultimate tensile strength). And, the 30% and 35% of UTS is asymptotic to a line of 10 percent in membrane drawn type geogrid. From the creep testing, it was observed that polyester geogrids resist creep strain better than HDPE (high density polyethylene) geogrids at similar temperatures and load levels, However, for both HDPE and polyester geogrid specimens the increase in temperature and load level have a strong effect on the creep strain behavior, relatively larger for HDPE specimens. The increase in load level also increase the amount of creep strain in the specimens, but the influence is not as large as that due to the temperature. However, higher the temperature, the larger is the influence of the increase in load level.

3.3 *Installation damage*

Table 1 showed the strength retention of geogrids by installation damage.

The weathered granite soil having less than 20 mm particle size was used in this test. From these results it was confirmed that there were some tensile strength decreases in each samples, in the case of textile geogrid, about 6 ~ 7% strength decreases were founded for each textile geogrids (woven and warp knitted type). And strength decreases about 1.4 ~ 2.2% were observed in membrane drawn geogrid samples.

3.4 *Chemical resistance*

In the case of membrane drawn geogrid which was made by HDPE, there was merely small amount of decease (+0.7% (increase) ~ −2.6 (decrease)) in both acidic and alkaline conditions. While in the cases of

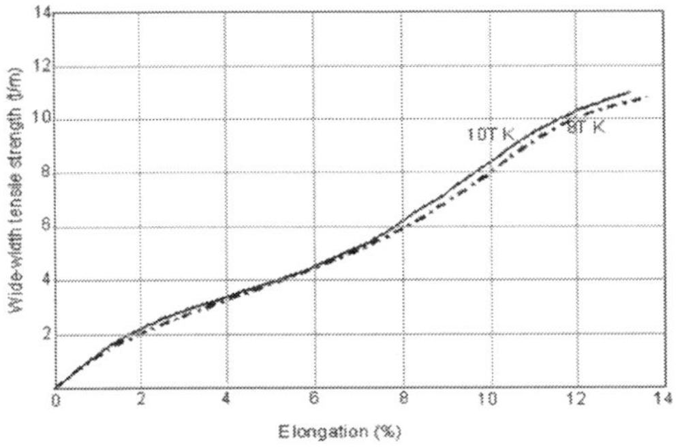

Figure 1. Wide-width tensile strength-elongation curves of warp knitted geogrids (longitudinal direction).

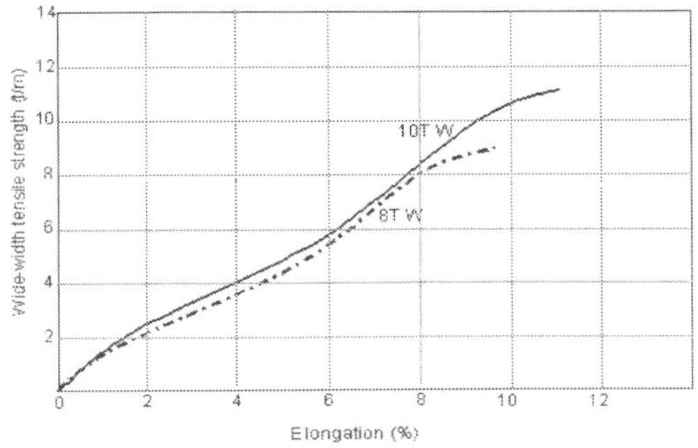

Figure 2. Wide-width tensile strength-elongation curves of woven geogrids (longitudinal direction).

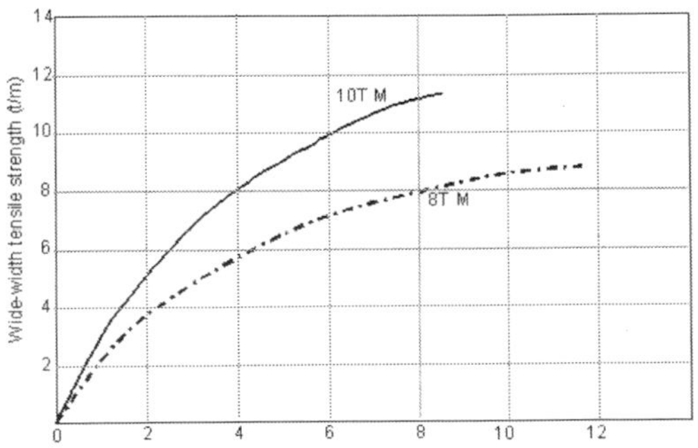

Figure 3. Wide-width tensile strength-elongation curves of membrane drawn geogrids (longitudinal direction).

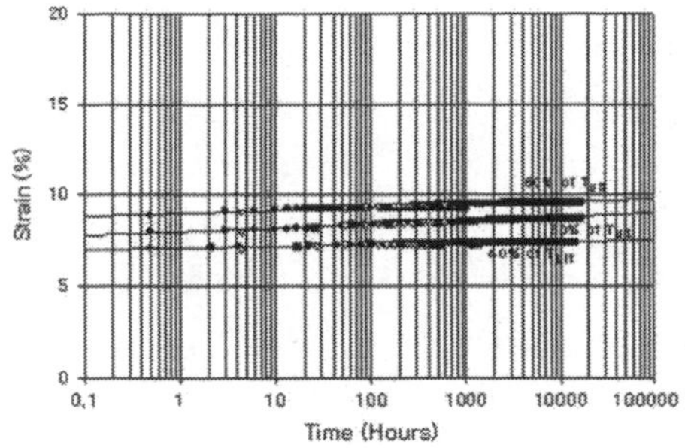

(a) 8TK

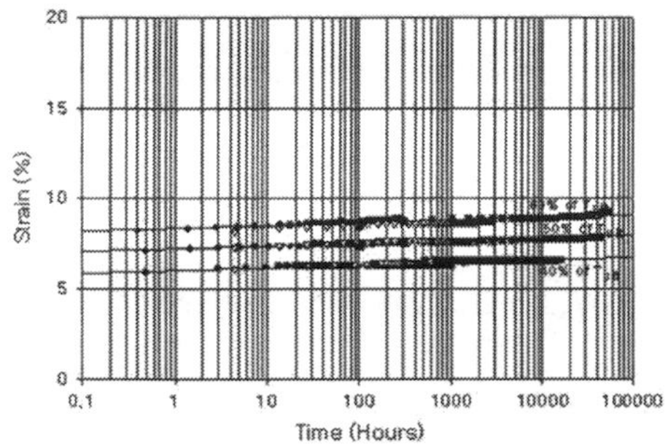

(b) 10TK

Figure 4. Creep deformation curves of warp knitted geogrids.

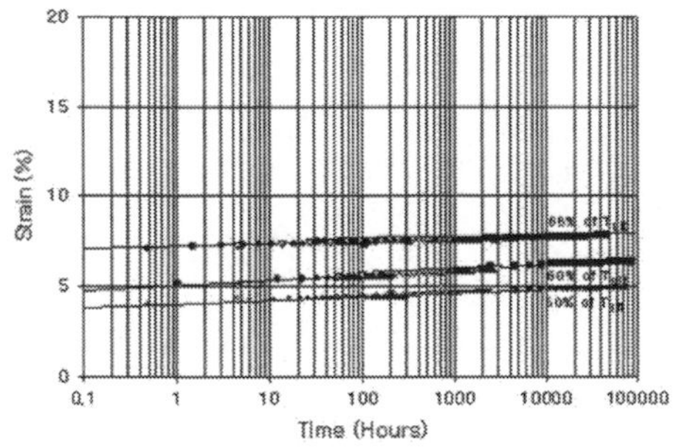

(a) 8TW

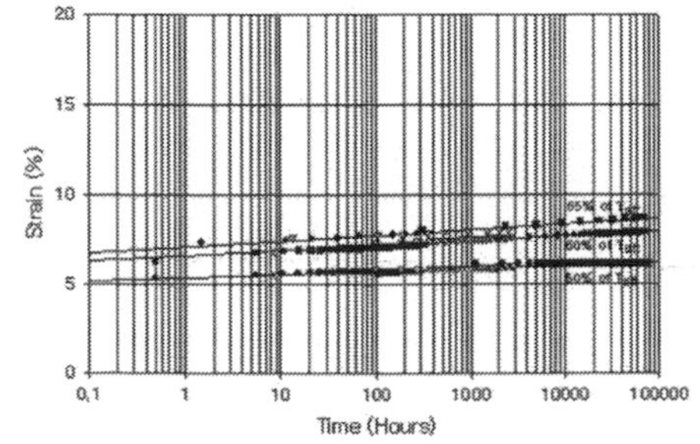

(b) 10TW

Figure 5. Creep deformation curves of woven geogrids.

the textile geogrids, it resulted in very similar tendency with the membrane drawn geogrid in acidic conditions but the tensile strength decreased about max. 45% in severe alkaline condition, NaOH, pH = 13, especially in knitted type geogrid. However, the strength decrease of woven type geogrid resulted in a reduction (%) as 17%, and this is smaller value compared with the results of knitted type geogrid. However, the actual site-specific condition (pH = 8.5 ~ 9.5) is considered, both type of geogrids can be used without problems and related durability safety factor for any soils having pH = 9.

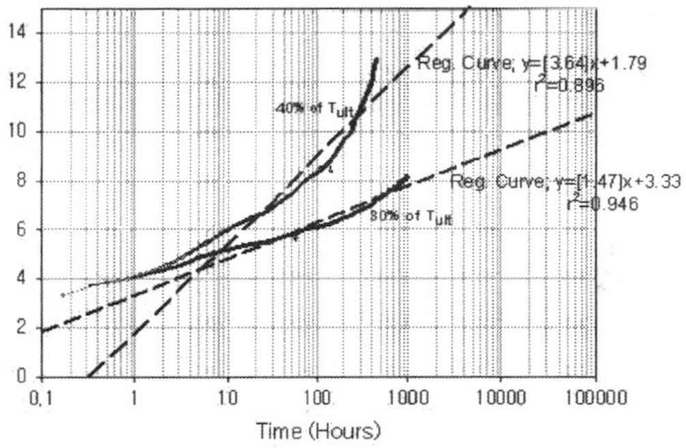

(a) 8TM

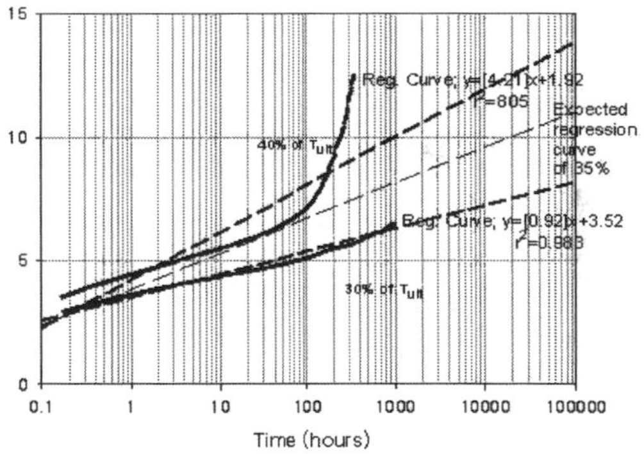

(b) 10TM

Figure 6. Creep regression curves of membrane drawn geogrid.

Table 1. Wide-width tensile properties of geogrids after installation damage test (longitudinal direction).

Geogrid	Strength reduction (%)
8TK	7.7
10TK	10.6
8TW	5.9
10TW	10.9
8TM	1.3
10TM	2.1

Table 2. Reduction factors of warp knitted geogrids.

Geogrid specimen reduction factor		8TK	10TK
RF_{ID}		1.09	1.12
RF_{CR}		1.67	1.67
	$RF_{CD}(pH \leq 9)$	1.05	1.05
RF_D	RF_{BD}	1.0	1.0
	Total $RF_D(pH \leq 9)$	1.05	1.05
Total RF		*2.0*	*2.0*

Table 3. Long-term design strength of warp knitted geogrids.

	Geogrid specimen	
Property	8TK	10TK
Nominal tensile strength (t/m)	8	10
Total RF	2.0	2.0
Long-term design strength (t/m)	*4.0*	*5.0*

3.5 *Biological Resistance*

There are some decrease of strength for the exposed sampled (<3.0%), but these values can be contained within specimen variation and test errors. From these results it can be concluded that all of these geogrid samples are not affected by any of biological affects.

3.6 *Total reduction factor and long-term design strength*

Tables 2 ~ 7 shows the total reduction factor and long-term design strength by creep deformation, installation damage, chemical and biological degradation. So, the long-term design strength of the geogrids will be reduced by this reduction factors. In designing with geogrid reinforcement, considering of these reduction factor and applying to designing process is very important for more safe structure within the aimed design life time. The GRI test method GG-4 that used world widely is applying to determine the reduction factors in this study.

Table 4. Reduction factors of woven geogrids.

Geogrid specimen reduction factor		8TW	10TW
RF_{ID}		1.12	1.12
RF_{CR}		1.54	1.54
	RF_{CD} $(pH \leq 9)$	1.05	1.05
RF_D	RF_{BD}	1.0	1.0
	Total $RF_D(pH \leq 9)$	1.05	1.05
Total RF		*1.90*	*1.90*

Table 5. Long-term design strength of woven geogrids.

	Geogrid specimen	
Property	8TW	10TW
Nominal tensile strength (t/m)	8	10
Total RF	1.90	1.90
Long-term design strength (t/m)	*4.21*	*5.26*

Table 6. Reduction factors of membrane drawn geogrids.

Geogrid specimen reduction factor		8TM	10TM
RF_{ID}		1.05	1.05
RF_{CR}		3.3	2.8
	$RF_{CD}(pH \leq 9)$	1.0	1.0
RF_D	RF_{BD}	1.0	1.0
	Total $RF_D(pH \leq 9)$	1.0	1.0
Total RF		*3.46*	*2.94*

Table 7. Long-term design strength of warp knitted geogrids.

	Geogrid specimen	
Property	8TM	10TM
Nominal tensile strength (t/m)	8	10
Total RF	3.46	2.94
Long-term design strength (t/m)	*2.3*	*3.4*

4 CONCLUSION

In this study, the engineering properties related with the total reduction factor were evaluated by creep deformation, installation damage, chemical and biological degradation. For warp-knitted type geogrids, total reduction factor is estimated as 2.00. Installation reduction factors under the grain condition of <19 mm, was estimated as 1.09 and 1.12 respectively (8TK, 10TK). Warp knitted geogrid has low resistance to the alkaline circumstance (pH > 12) and high temperature. The biological resistance of the warp knitted geogrid was estimate that it has very strong resistance to the biological environment. For woven type geogrids, the total reduction factor is estimated as

1.90. Estimated long-term creep deformation indicates that the 65% of T_{ult} loading level is the optimum value that satisfying the creep criteria. Installation reduction factors under the grain condition of < 19 mm, was estimated as 1.12. In the case of low alkaline conditions (= site-specific conditions), it has satisfied resistance to its circumstance. The biological resistance of the woven geogrid was estimate that it has very strong resistance to the biological environment. The total reduction factor of membrane drawn geogrids having different nominal strength (8TM, 10TM) is estimated as 3.46 and 2.94 respectively. Estimated long-term creep deformation from creep test results indicate that the 30% (8TM) and 35% (10TM) of T_{ult} loading level is the optimum value that satisfying the creep criteria. Installation reduction factors under the grain condition of <19 mm, was estimated as 1.0. Also, membrane drawn geogrid had high resistance to the critical alkaline and acidic conditions because of its polymeric property of HDPE. The biological resistance of the membrane drawn geogrid was estimate that it has very strong resistance to the biological environment.

REFERENCES

Navarrete, F. 2001, "Creep of Geogrid Reinforcement for Retaining Wall Backfills", Proc. of Geosynthetics Conference 2001, pp. 567–578.

Mano, J. F., and Sousa, R. A. 2001, "Viscoelastic Behaviour and Time-Temperature Correspondence of HDPE with Varying Levels of Process-Induced Orientation", Polymer, Vol. 42, pp. 6187–6198.

Giroud, J. P. 2002, "Lessons Learned from Successes and Failures Associated with Geosynthetics", Proc. of 2nd European Geosynthetics Conference, pp. 77–118.

DeMerchant, M. R. 2002, "Plate load tests on geogrid-reinforced expanded shale lightweight aggregate", Geotextiles and Geomembranes, Vol. 20, pp. 173–190.

ASTM D 4595, "Standard Test Method for Tensile Properties of Geotextiles by the Wide-Width Strip Method", ASTM International, W. Conshohocken, PA, USA, 2005.

ASTM D 5262, "Standard Test Method for Evaluating the Unconfined Tension Creep and Creep Rupture Behavior of Geosynthetics", ASTM International, W. Conshohocken, PA, USA, 2006.

EPA 9090, "Compatibility Test/ Wastes & Membrane Liners", Environmental Protection Agency, USA, 1992.

GRI Test Methods and Standards GG1, "Standard Test Method for Geogrid Rib Tensile Strength", 1998.

Lin, M. T., and Shi, J. L. 2001, "Microstructure and Creep Behaviour of an Y-α-β Sialon Composite", Journal of the European Ceramic Society, Vol. 21, pp. 833–840.

New Horizons in Earth Reinforcement – Otani, Miyata & Mukunoki (eds)
© 2008 Taylor & Francis Group, London, ISBN 978-0-415-45775-0

Lifetime prediction of PET geogrids under dynamic loading

H. Zanzinger
SKZ – TeConA GmbH, Würzburg, Germany

H. Hangen & D. Alexiew
HUESKER Synthetic GmbH, Gescher, Germany

ABSTRACT: The lines of 'damage begins' and specimen break for dynamic loading of a geogrid were determined in a series of laboratory testing. The load ratio was set to $R = 0.5$, loading frequency $f = 10$ Hz and $f = 3$ Hz. The test results show clearly that the chosen procedure for the determination and analysis of the beginning of damage and break is reproducible and allow for safe extrapolation for lower load levels. Furthermore the method chosen enables explicit decrease of the required testing time. The assumption of linear damage accumulation was examined in two-step-trials. The number of load cycles to break evaluated in 'one-step-tests' compared with those of 'two-step-tests' is practically the same. The existence of damage lines for the examined geogrid in a dynamic pulsating load of 10 Hz and 3 Hz and an R-value of 0.5 could be verified. Thus there are considerably high numbers of load cycles the geogrid can withstand without any damage caused by the dynamic load. This area is bounded by the damage line. Damage of the samples occurs only for load-cycles lying between the damage line and the 'Wöhler-curve' (fatigue curve). Higher testing frequencies present the critical case when it comes to designing against damage beginning or break.

1 INTRODUCTION

High strength geosynthetic reinforcement materials are nowadays frequently used for the construction of highly trafficked structures like motorways and railway embankments. Measurement data of highly loaded structures typically railroad structures has been reported e.g. in Auersch & Rücker (2005). For such applications where the stress-strain behaviour is important designers also like to know how dynamic loading affects the tensile properties of the reinforcement. To gain more detailed information about the dynamic behaviour of geosynthetic reinforcement products and to avoid over-conservative design approaches it was decided to carry out a detailed research program.

This paper will present the results of an extensive research program which has been carried out recently with single strip specimens of a polymeric coated high strength knitted geogrid manufactured from PET type Fortrac® R 560/115-15T (with an ultimate tensile strength of 560 kN/m).

2 TEST PROCEDURE, BASIC CONVENTIONS

2.1 *Loading*

Prior to the start of the endurance tests some basic arrangements had to be set. One is the type of dynamic loading. Based on field experience of earlier projects, it was decided that the dynamic service loads of a train could be represented well using a sine shaped loading arrangement. A loading scheme as shown in Fig. 1 was

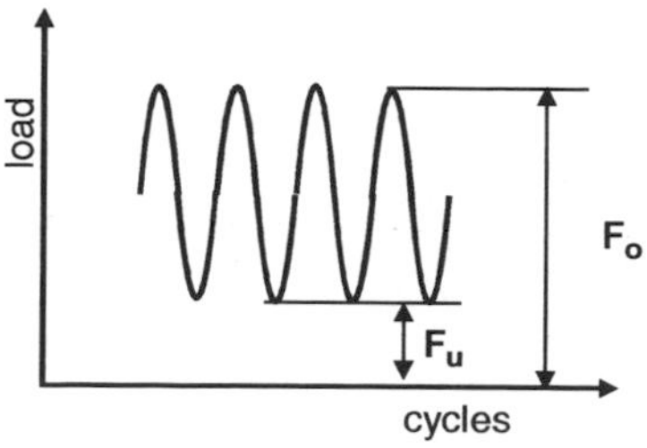

Figure 1. Nomenclature to specify the dynamic loading.

applied during all the tests. Load ratio R is defined as ratio between lower load F_u and maximum load F_0. With regard to the number of different variables of dynamic loading it was decided to carry out two test series with load frequencies of 3 Hz and second with 10 Hz, the load ratio R was always kept constant at R = 0.5. This value was chosen as it was seen the most critical condition for a particular railway project. Each series was subdivided into three different load levels where every load level was performed with 10 single specimens.

2.2 *Wöhler-curve*

The 'Wöhler-curve' named after August Wöhler, 1819 – 1914, is an appropriate method to present the relation between the number of load cycles to break and the loading amplitude.

Depending on the nature of the tested material different shapes of 'Wöhler-curves' or fatigue curves may occur: in a half or double logarithmic plot the classic 'Wöhler-curve' has the shape like curve 1 in Fig. 2. It is composed of a branch with load dependent limited life time followed by a branch parallel to the x-axis, the endurance limit, where the number of load cycles N sustained before the break/rupture is independent from the load. Many steels show similar behaviour. Curve 2 has two branches with different slopes but no endurance limit. This type of fatigue curve is typical for some aluminium alloys.

Thermoplastic materials behave more or less elastic-viscous therefore an endurance limit like shown in curve 1, Fig. 2 can be excluded. The 'Wöhler-curve' of these materials should be comparable to curve 3 or somewhat between curves 2 and 3. Similar behaviour can also be observed due to creep under constant static loading. In that regard it seems reasonable and conservative to assume that the fatigue behaviour of the geogrid examined here will be comparable to curve 3, Fig. 2.

It should be noted that Fig. 2 curve 3 is only a straight line when either axes or only the x-axis are in logarithmic scale.

2.3 *Begin of specimen damage*

Due to the magnitude of different influence parameters, it can be expected that the process of material deterioration is controlled by multiple, non-additive effects. In such cases the break events on every load level are often log-normal distributed. With this assumption the expected 'Wöhler-curve' for each series has the appearance of a straight line in a double logarithmic plot. The 'Wöhler-curve' can be interpreted as a line of 100% damage which is equal to the rupture of the specimen.

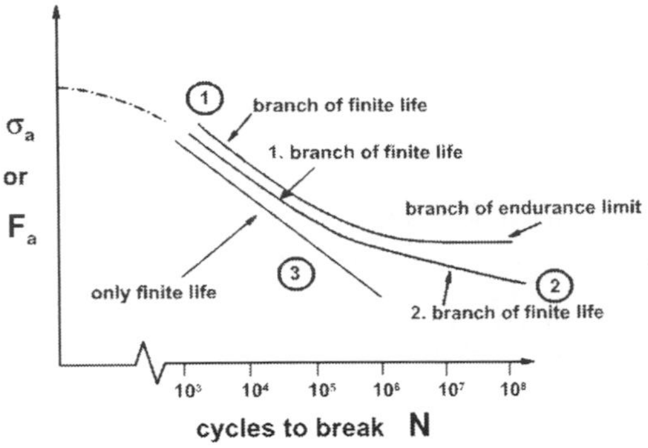

Figure 2. Different possible shapes of 'Wöhler-curves'.

During dynamic loading material damage can start with the first cycle or with any later one. In the latter case this means that there is a first life time cycle, of course depending on the load level, which the specimen can sustain free of any damage corresponding to 0% damage. This is followed by the second life time cycle when the specimen is continuously damaged, damage >0% which is accumulating until the rupture, damage = 100%, is reached.

This idea was formulated by Palmgren (1924) and Miner (1945). To quantify and to detect the limit for the start of the damage which is the end of the first life time cycle and the beginning of the second cycle many researchers, e.g. Renz et al. (1986), have used the hysteresis and its properties. This is appropriate because of its high sensitivity.

3 TESTING

3.1 *Test set-up*

The dynamic tests were performed in the load controlled mode using servo hydraulic machines (SKZ, 2005).

The set-up is depicted in Fig. 3 and shows that the specimen is located in the middle of each clamp, and then wound around the smaller and the larger half of each clamp. Additionally both halves were screwed together so that both ends of the specimen are fixed in its clamp. Although no influence of the cross machine direction was expected it is to be noted that the specimen is prepared and tested always including the node points.

3.2 *Data acquisition*

The strain was measured in the middle of each specimen using a clip-on-gage. Additionally the movement of the plunger was recorded using a 'built-in transducer', not visible in Fig. 3. The load cell is located at

Figure 3. Test set-up showing one thread (warp direction of the geogrid), clamping arrangement, extensometer (in the middle) and three thermocouples.

the top of the rig, ref. Fig. 3. For further analysis it was also required to record the specimen temperature, this was carried out using thermocouples at three locations of the specimen: close to the lower and the upper grip and in the middle. All data was recorded and used to compute online several derived functional values like the hysteresis, the loss and the stored work, the specimen stiffness and the amplitude of the extensometer stroke.

4 EVALUATION OF DATA AND RESULTS

4.1 *Specimen rupture*

Analysis of the test results showed sensible scattering of the data. To that reason it was decided to arrange the number of load cycles to break with increasing number of cycles. The 'break events' then were logarithmized. The calculated probabilities were then plotted into a probability paper. After this linearisation the corresponding length values were used to calculate a linear regression for each load level. The regression line for the three load levels showed high determination factors r^2. This means that the assumption of log-normal distributed break events is reasonable.

After determining the mean or median value for each level the data was ready to perform a regression to generate the 'Wöhler-curve', see Fig. 5.

4.2 *Definition of the damage*

The detection or definition of the start point of the damage process of a specimen is not necessarily easy as there is usually no evident sign to be registered. In this case useful feature of the evaluation and presentation software was used to identify this point: Fig. 4 shows four different plots for one typical test, where the x-axis, displaying the number of cycles, is identical in every plot. The y-axis in contrast is displaying one of the following four parameters: amplitude, loss work, stiffness and temperature. Loss work is the area within the hysteresis loop. It will be analysed online during a running test. When moving the cursor along a master curve, the program would display the corresponding parameter in the remaining plots at the same time. Using this feature specific points of the curve could be marked simultaneously.

Starting with the Fig. 4a, the screenshot shows the extensometer amplitude, Fig. 4b shows the loss work and Fig. 4c the temperature in the middle of the specimen. It is evident that the cursor marks local minima of amplitude and loss work which correspond to the local maximum of the stiffness plot at Fig. 4d. This determination method was applied for every data set. There were some gradual variations in the curve appearances; however, the main tendency was the same as stated in the example given in Fig. 4.

The raw data generated for the 'damage begins' was processed the same way as reported for the break events (section 3.1). The fitting of the regression lines is quite convincing and supports the assumption of the existence of a limit line for the 'damage begins'. Furthermore it can be followed that these events are reproducible and with a sufficient approximation log-normal distributed as well. The r^2-values (determining factors) for the regression lines of the 'damage begins' work out also to high r^2-values.

4.3 *Generation of the life-time diagram/ 'Wöhler-curve'*

The last step of the evaluations is to display all data points for the 'break events' and for the assumed 'damage begins' in one lifetime diagram.

This is shown in Fig. 5 for the results of 10 Hz loading frequency. The upper curve is the 'Wöhler-curve' for specimen break, the lower line represents the appertaining limit line for the 'damage begin'. Both regression functions and their scatter factor are given in the plot.

As stated before, tests have been carried out with 3 Hz frequency following the same procedure resulting in a similar plot shown in Fig. 6. The number of cycles to reach break or to reach the beginning of damage is higher with lower frequency. This corresponds to a shift of the 'Wöhler-curve' for 10 Hz frequency towards the 'safer side' (Fig. 6).

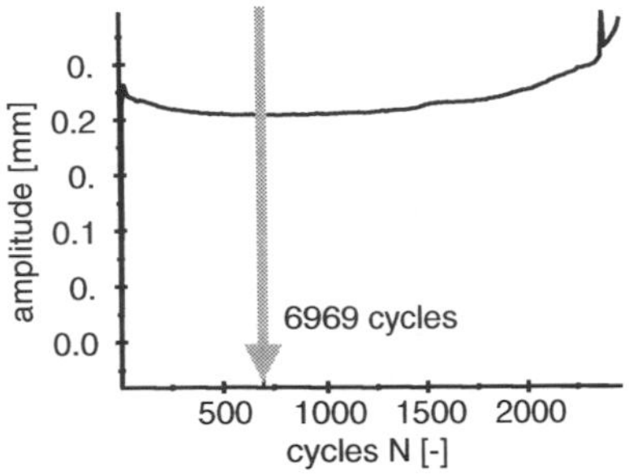

a) Screenshot showing extensometer amplitude

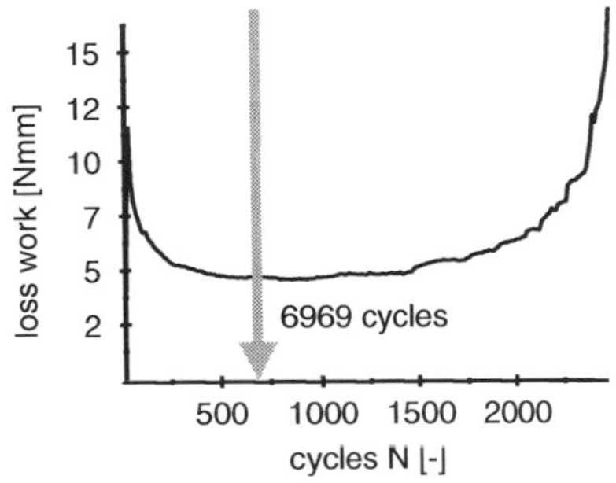

b) Screenshot showing loss work

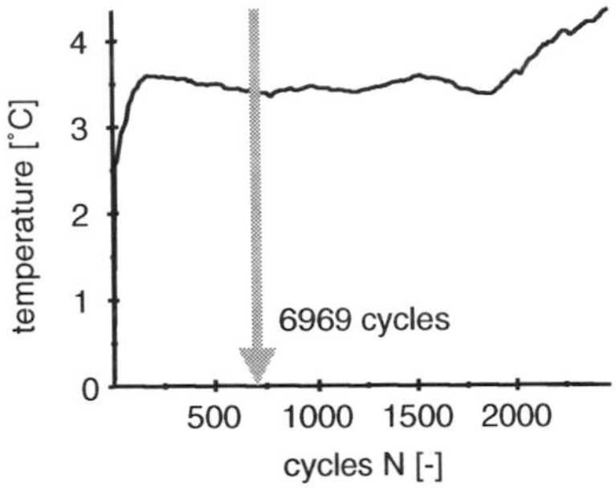

c) Screenshot showing specimen temperature

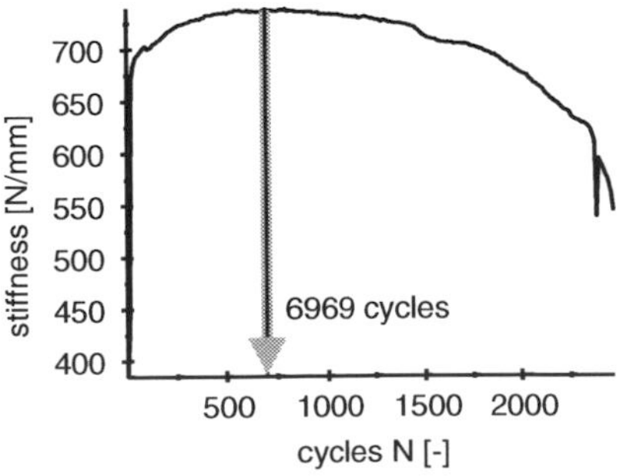

d) Screenshot showing stiffness

Figure 4. Typical screenshots of the computer program. In this example the position marks the elapsed number of cycles until a local min. or max. value is reached. Here the upper load is 7.5 kN, frequency is 10 Hz.

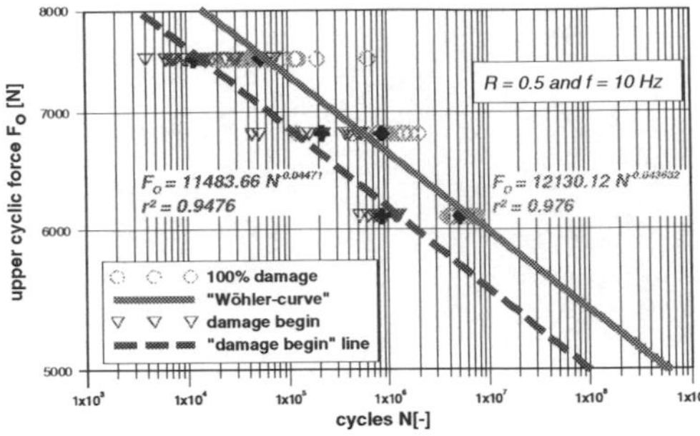

Figure 5. 'Wöhler-curve' and the 'damage begins' line for 10 Hz, both regressions were calculated with the median values. Both regressions have a probability of approximately 50%.

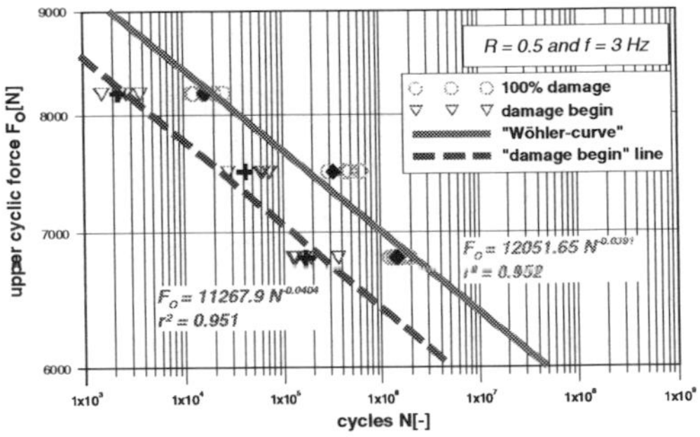

Figure 6. 'Wöhler-curve' and the 'damage begins' line for 3 Hz, both regressions were calculated with the median values. Both regressions have a probability of approximately 50%.

4.4 *Proof of the 'damage begins' line*

In the theoretical derivation of the damage accumulation Zanzinger et al. (2007) show simple ideal 'Wöhler-curves' and ideal 'damage begins' lines. In reality the records show considerable scatter, such that a statistical evaluation is needed, ref. section 4.1. For the proof of the existence of a 'damage begins' line this scattering would be difficult to handle and further measures would be necessary. But just to or proof, that the 'damage begins' line exists, its exact position in the lifetime-diagram is of secondary importance. To proof the existence is the essential point – this will be done using 'double-step-testing'.

4.5 *Double-step tests*

Zanzinger et al. (2007) explained the application of the linear Miner rule by means of ideal examples for double- or multiple-step tests.

Figure 7 shows the 'Wöhler-curve' and the associated 'damage begins' line for a loading frequency of 10 Hz and an R-value of 0.5 each for single-step tests

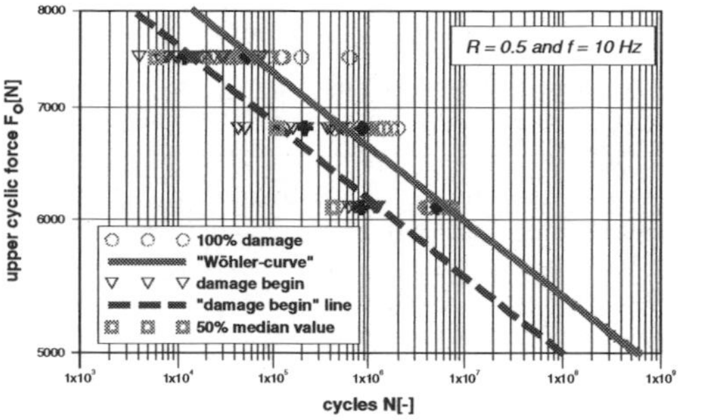

Figure 7. Commencement of the cycles of the first step, geogrid Fortrac®R 560/115-15T (here: 50% of median values 'damage begins').

together with the measurement points and the median values. It is apparent, that the events 'damage begins' as well have a significant scattering. Due to that reason the length of the first loading step of the 'double-step tests' was chosen such that only 50% of the hypothetical 'damage free' lifetime is covered where the median values of each load level is assumed as 'damage begins', Figs 7 and 8 show, that all first steps are largely outside the scattering of the data. This means that the specimen does not experiences any damage by the dynamic loading and that the test can be continued without interruption and relief in an other load level. This could be a higher or lower load level and the break of the specimen will occur as it will occur at the corresponding single-step tests for this load level.

The definition of the load for second step should be made such as there is sufficient distance between each level (about 10% of the maximum upper load) and the differences of the medium cycles to break shouldn't become too much.

If the assumption of a 'damage begins' line is not true, the specimen would get damaged in the first step and would therefore break earlier on the second step if compared with the corresponding 'single-step test' of this load level (SKZ, 2006).

4.6 *Results of double-step tests*

Figure 8 presents the number of load cycles to reach the rupture of the specimen recorded for both single- and double-step tests together with the 'Wöhler-curve' generated form results of single-step tests.

It is obvious that the first loading step of the double-step tests have caused practically no pre-damaging of the samples. Thus the assumed damage begins defined by the 'damage begins' line (≙ 0% damage), can be considered to be right.

Based on that one could consider the first step as 'not existent' and each second step can therefore be interpreted as a sovereign single-step test and can be analysed like was described under sections 4.1–4.3.

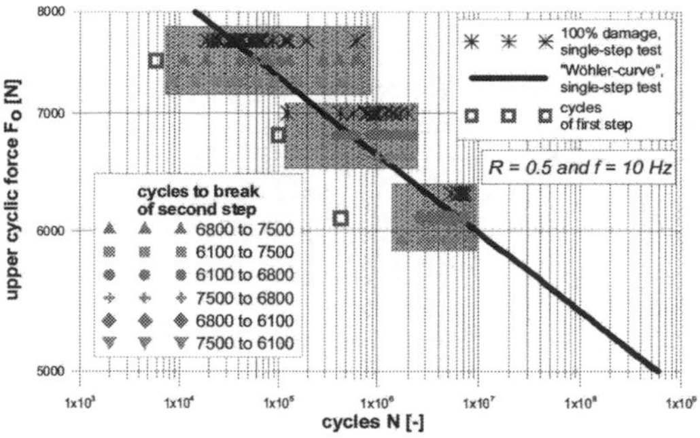

Figure 8. Evaluation of the break events of the second step of the double-step tests. Note: The data points have been spread around the horizontal centreline of the blue boxes for better visibility.

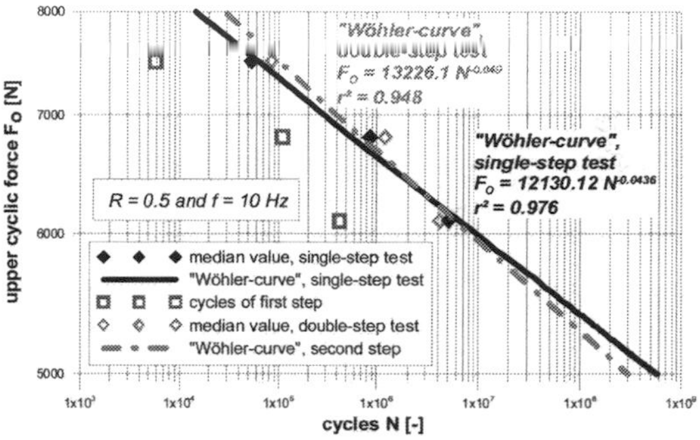

Figure 9. Comparison of results from double-step tests with those from single-step tests.

Fig. 9 shows the comparison of this data with the damage begins- and the specimen rupture-curves of the regular single-step tests. The dashed line is the regression curve for the double-step tests, the full line symbolizes the break curve for the single-step tests. The 'Wöhler-curve' for the double-step tests (calculated with regression analysis) fits very well in comparison with the 'Wöhler-curve' of the single-step tests; there is just a minor deviation in the inclination of the curves. This confirms clearly that the chosen procedure to analyse the available test data was right, but basically the existence of 'damage begins line' for the tested geogrid.

To conclude that testing with 3 Hz loading frequency would give comparable results is highly probable but as explained before, fatigue at 3 Hz is less critical as at 10 Hz frequency. A verification at 3 Hz frequency was not pursued for that reason.

5 CONCLUSIONS

Extensive dynamic testing allowed to establish 'damage begins lines' and 'Wöhler-curves' for the knitted PET geogrid Fortrac® R 560/115-15T.

In line with the statistical evaluation of the data the presented results show clearly, that the procedure used is reproducible for the generation of break and damage curves. The existence of a load depending, 'damage begins' offers additional safety for dimensioning of geogrids reinforced structures under dynamic loading.

The selection of the load levels for testing was made such as, potential changes of mechanical properties caused by elevated temperature did not influence the tests at the same allowing for a sufficiently high number of tests in relative short testing time. In consideration of elementary graphical methods the lifetime curves appear as straight lines – predictable with high statistical safety. The extrapolation to lower load levels is possible and on the safe side. In any case it is conservative. For an extrapolation of the regression curves to times above 50 years, which is about 10^8 cycles, the general aging of the plastics has also to be considered.

The evaluation of the tests provide clearly to the results, that specimens at 3 Hz frequency have a longer lifetime than specimens at 10 Hz – whereas the difference becomes higher with falling upper loads. The higher frequencies are the more critical case in a design against 'damage begins' or against break.

Through double-step tests the existence of a 'damage begins' line should be verified. The results of the double-step tests and the comparison with single-step tests can be summarised as follows:

- The online-acquisition of specimen temperature, loss work, stiffness and amplitude of the extensometer allows to establish a valid criteria for the definition of the 'damage begins'.
- The loads cycles for the break events registered for the double-step tests show excellent agreement with those of single-step tests, both for the distribution and for the range of the values.
- The results could be shown in plots visually, qualitatively and also, with statistical calculations, quantitatively.
- The assumption of a linear damage accumulation (linear Miner rule) was verified using 'double-step' tests. Good correlations were found, independent of whether the second load level was defined as an increased level or as a reduced level.
- It was statistically proven, that the number of load cycles to reach specimen rupture for single-step tests and for double-step tests – if the second step was counted – had a log-normal-distribution.
- It can be concluded, that there is existing a 'damage begins' line, nearly parallel to 'Wöhler-curve', which is defining the number of load cycles the tested geogrid can be loaded with at 10 Hz frequency and an R-value of 0.5 without damaging the material. The damaging of the sample only happens later for elevated numbers of load cycles which are located between the 'damage begins' line and the 'Wöhler-curve'.
- An enormous advantage of the described procedure is the reduced testing time required, thereby reducing considerable the testing cost. For example, an increase of around 1000 N of the upper cyclic load reduced the cycles to break with a factor of around 30.

REFERENCES

Auersch, L. & Rücker, W. 2004. Dynamic Loads of Railway Traffic, Proceedings of Geosynthetics, ECI-Conference, Pillnitz, Germany, Essen, Glückauf 2005.

Miner, M.A. 1945. Cumulative Damage in Fatigue, journal of appl. mech. trans., ASME 12, H.3, pp 154–164.

Palmgren, A. 1924. Die Lebensdauer von Kugellagern. VDI-Z. 58, pp 339–341, (in German).

Renz, R., Altstätt, V. & Ehrenstein, G.W. 1986. Hysteresis measurements for characterizing the dynamic fatigue of R-SMC, 41st Conference Reinforced Plastic/Composite Institute, SPS, Atlanta.

SKZ, 2005. Gutachten über die Durchführung von Dauerschwingversuchen bei Zugschwellbelastung mit 3 Hz und mit 10 Hz am Geogitter "Fortrac R560/115-15T", (in German).

SKZ, 2006. Gutachten zum Nachweis der Existenz einer Schadenslinie durch zweistufige Dauerschwingversuche bei Zugschwellbelastung am Geogitter "Fortrac R560/115-15T" unter Einbeziehung von einstufigen Dauerschwingversuchen bei Zugschwellbelastung am gleichen Geogitter, (in German).

Zanzinger, H., Hangen, H. & Alexiew, D. 2007. Ermüdungsverhalten von Geogittern unter dynamischer Belastung, 10. Informations- und Vortragstagung über "Kunststoffe in der Geotechnik". geotechnik Sonderheft 2007, DGGT, Essen, pp 23–31, (in German).

New Horizons in Earth Reinforcement – Otani, Miyata & Mukunoki (eds)
© 2008 Taylor & Francis Group, London, ISBN 978-0-415-45775-0

Load-deformation behaviour of virgin and damaged non-woven geotextiles under confinement

M.J.A. Mendes & E.M. Palmeira
University of Brasilia, Department of Civil and Environmental Engineering, FT, Brasilia, DF, Brazil

ABSTRACT: This paper examines the influence of confinement on the mechanical behaviour of virgin and mechanically damaged nonwoven geotextiles. In-soil tensile tests were performed on different types of geotextiles and confining materials under normal stresses between 25 kPa and 150 kPa. Three types of sands and geotextiles were used in the tests. The influence of different types and dimensions of damages were also investigated under in isolation and in-soil conditions. The effects of confinement and the presence of the damages were quantified. It was observed that confinement reduces the detrimental effects of the damages.

1 INTRODUCTION

Reinforced soil structures are arrangements of two elements with different properties and complementary functions: the soil, which can be chosen to present good compression strength and the reinforcement, which can present good tensile strength. The combination of these two materials results in a stronger and less deformable structure than the soil alone. The behaviour of a reinforced soil structure depends on the soil strength and on the mechanical properties of the reinforcement. Stiffer reinforcements require less deformation to mobilize significant tensile forces in the reinforcement, which will yield to a less deformable reinforced mass. Extensible reinforcements must not be used in situations where deformations of the reinforced structure are limited by stability or serviceability constraints.

Despite nonwoven geotextiles being considered extensible reinforcements, several examples of old structures reinforced with these materials have presented little deformations due to the geotextile stiffness increase caused by the confinement by the surrounding soil (McGown et al. 1982). In this context, the study of the mechanical behaviour of these materials under confinement, using in-soil tensile tests, is important to improve design parameters and to increase the use of non-woven geotextiles in reinforced soil structures, particularly those of low to moderate heights.

Because of the relevance of confinement to the tensile stiffness of nonwoven geotextiles, this work examines the mechanical behaviour of these materials in tension confined by different soils, including the influence of confinement on the tensile properties of mechanically damaged geotextiles. Some aspects relevant to the in-soil behaviour of nonwoven geotextiles are investigated and discussed in the following sections.

2 TEST APPARATUS AND MATERIALS

2.1 *Test Apparatus*

A test apparatus developed at the University of Brasília was used for the in-soil tensile tests (Palmeira 1996). The main characteristics of the apparatus are presented in Figure 1.

The in-soil tensile cell is a metallic box 20 cm wide, 22 cm long and 6 cm high, laterally open. The geotextile specimen (200 mm × 100 mm) is installed at the centre of the box and clamps allow for the application of the tensile force. A pressurized rubber bag provides a uniformly distributed confining stress on the top layer of the confining soil material.

The movable pair of clamps that hold the geotextile specimen is connected to a hydraulic cylinder,

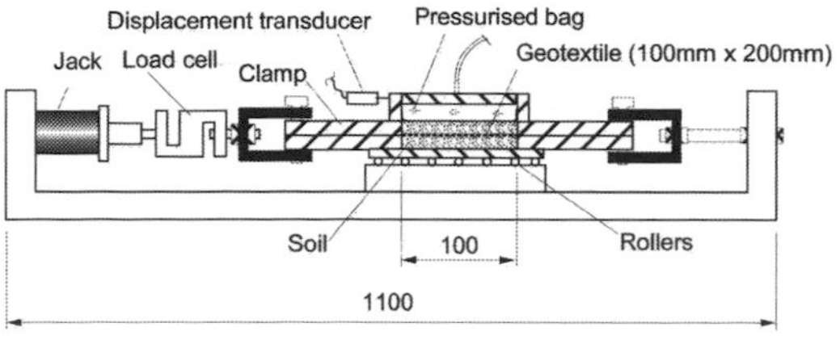

Figure 1. In-soil tensile test apparatus.

Table 1. Some characteristics of the geotextiles used.

Geotextile	$\mu^{(1)}$ (g/m^2)	t_{GT} (mm)	J (kN/m)$^{(2)}$
GA	200	2.9	21
GB	300	2.6	31
GC	400	3.8	42

Notes: (1) μ = mass per unit area, t_{GT} = geotextile thickness; (2) J = tensile stiffness from wide strip tensile tests in isolation.

which applied the tensile load at a constant rate of strain of 2%/min. The arrangement of the test is such that negligible friction is developed along the soil-geotextile interface during the test as both materials deform horizontally by the same amount, in contrast to what occurred in previous similar apparatus (McGown et al. 1982, Leshchinsky & Field 1987, Siel et al. 1987, Kokkalis & Papacharisis 1989, Palmeira et al. 1996, for instance). Tensile loads and displacements of the geotextile ends are measured by a load cell and four displacement transducers, respectively.

2.2 *Materials*

The geotextiles used in the tests were nonwoven needle punched geotextiles formed by polyester monofilaments. To minimise the scatter of test results due to the variation of geotextile mass per unit area, the geotextiles specimens were weighted one by one and those with mass per unit area varying more than 5% from the average weight were discarded. The relevant characteristics of the geotextiles tested are summarised in Table 1.

The materials used to confine the geotextiles were a coarse and uniform sand (Leighton Buzzard sand 14/25 – code LBS), with particle diameters varying between 0.6 mm and 1.2 mm, a fine sand from Corumbá river, Brazil (code CRS), with particle diameters between 0.06 mm and 0.42 mm, a uniform sand (code SSB) slightly coarser than soil LBS and wooden plates (WP) with plan, even and lubricated surfaces to minimise the friction between the plates and the geotextile specimen. The reduction of the friction between plates and geotextile was achieved by using double layers of a thin plastic film and grease on the surfaces of the plates. The tests with wooden plates allowed the study of the influence of confinement only, without the effect of geotextile impregnation by soil particles. The main characteristics of the three sands used in the tests are presented in Table 2.

Damaged geotextile specimens were also tested for the evaluation of the influence of confinement on their tensile responses. Different types of damages were investigated including circular holes, horizontal and vertical cuts, inclined cuts and "Y" shaped cuts.

Table 2. Confining materials characteristics.

Property	Soil LBS	Soil CRS	Soil SSB
G (g/cm^3)	2.66	2.68	2.58
D_{10} (mm)	0.60	0.61	0.64
D_{50} (mm)	0.80	0.20	1.14
D_{85} (mm)	1.05	0.38	1.68
C_u	1.3	4.1	1.4

Notes: G = soil particle density, D_{10}, D_{50} and D_{85} = diameters for which 10%, 50% and 85% of the particles in weight are smaller than those diameter, respectively; C_u = coefficient of uniformity (= D_{60}/D_{10}).

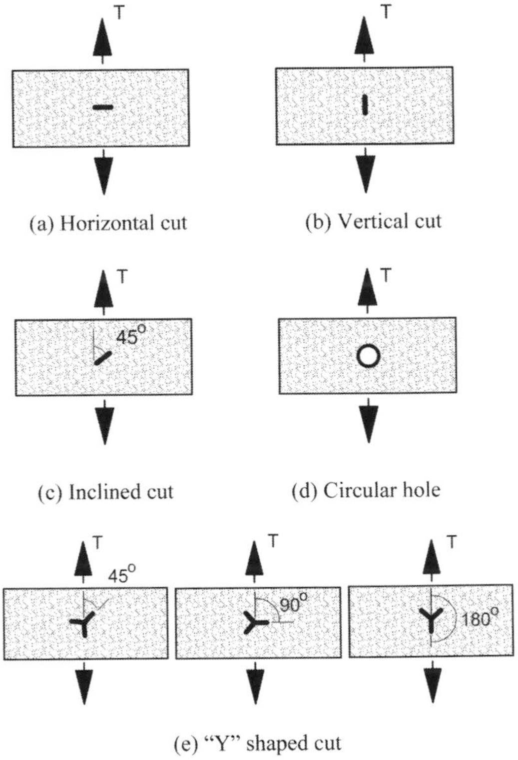

Figure 2. Types of damages investigated.

Figures 2(a) to (e) schematically presents the types of damages investigated.

Additional information on test equipment and methodology can be found in Mendes (2005) and Mendes and Palmeira (2006).

3 RESULTS OBTAINED

Tensile tests on nonwoven geotextiles were performed varying the confining material, geotextile confining

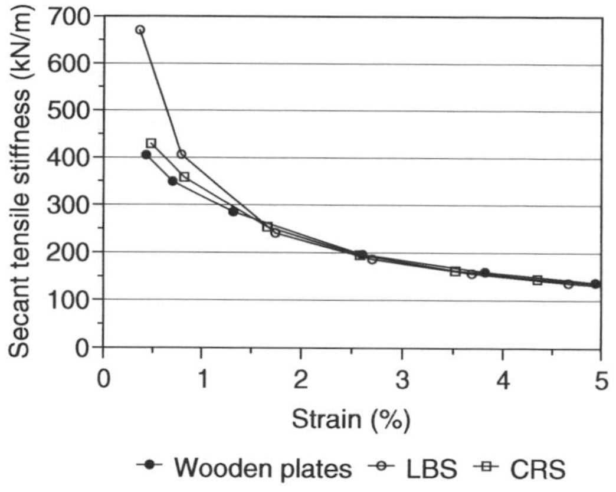

Figure 3. Secant stiffness against tensile strain for different confining materials – confining stress of 50 kPa.

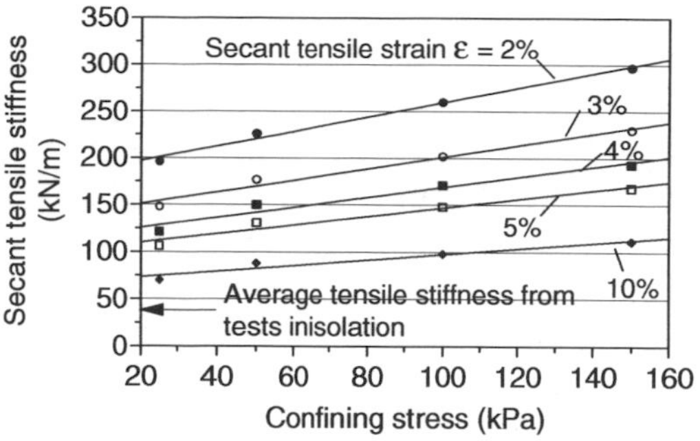

Figure 4. Influence of the confining stress on geotextile tensile properties – geotextile GC, soil LBS.

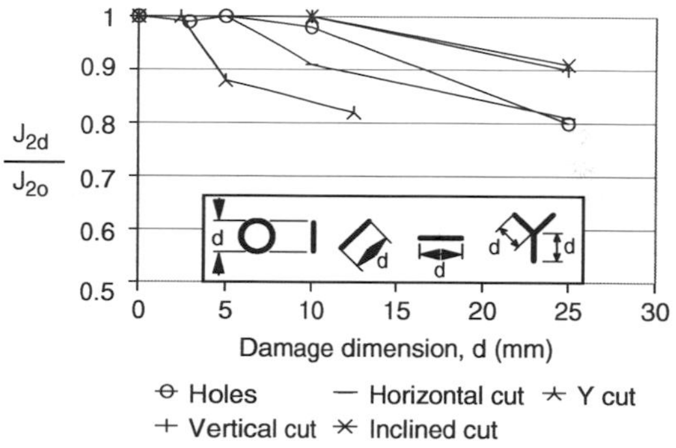

Figure 5. Tensile tests on damaged geotextile specimens of geotextile GB in isolation.

stress and type of mechanical damage to study the influence of these variables on the geotextile tensile stiffness. The confining stresses used varied between 25 kPa and 150 kPa.

3.1 *Tests on virgin geotextile specimens*

Figure 3 presents the variation of secant tensile stiffness of geotextile GA versus tensile strain for test under 50 kPa normal stress and different confining materials. The results show no significant influence of the confining material for the type of apparatus used, except for the test with sand LBS and for strains below 1%. This greater influence for sand LBS can be attributed to the angular shape of the particles of this soil which are likely to interlock more efficiently with the geotextile fibres.

The influence of the confining stress on the secant stiffness of geotextile GC obtained for different strains in tests with soil LBS is shown in Figure 4. The result obtained in wide strip tests in isolation (in air) is also presented for comparison. The results show a significant increase on the secant tensile stiffness due to confinement. It can also be noticed a rather linear relationship between secant stiffness and confining pressure.

3.2 *Tests on damaged geotextiles*

Figure 5 presents the ratio between secant tensile stiffness values at 2% tensile strain for virgin (J_{2o}) and damaged (J_{2d}) geotextiles in tests in isolation. It can be observed that for the dimensions of the damages the circular hole, the horizontal cut and the "Y" shaped cut were the most detrimental for the stiffness of the geotextile. The damages caused reductions of tensile stiffness up to 20%.

The shapes of some of the damages on geotextile GB at the beginning of the test and at a tensile strain of 30% are presented in Figures 6(a) to (c). Both the initial circular hole (Fig. 6a) and the horizontal cut (Fig. 6b) tend to degenerate to elliptical shapes during the test, but with the open area of the hole being significantly greater than that of the cut. The "Y" cut evolves similarly to a heart like shape with increasing strains (Fig. 6c).

Figures 7(a) and (b) show results of in-soil tensile tests on damaged specimens of geotextile GA confined in soil SSB under a confining stress of 100 kPa. Reductions of secant tensile stiffness were observed for cut lengths (d) equal or greater than 25mm, particularly for strains below 1% (Fig. 7a). For d = 50 mm the reduction of secant tensile stiffness of the damage geotextile was of the order of 20% for tensile strains between 0.5 and 2%. The results in Figure 7(a) obtained for horizontal cuts up to d = 25 mm were close to those obtained for the virgin specimens, which shows that confinement tends to reduce the detrimental effects of this type of mechanical damage. The "Y" shaped cut caused secant tensile stiffness reductions between 22% and 34% for tensile strains between 0.5%

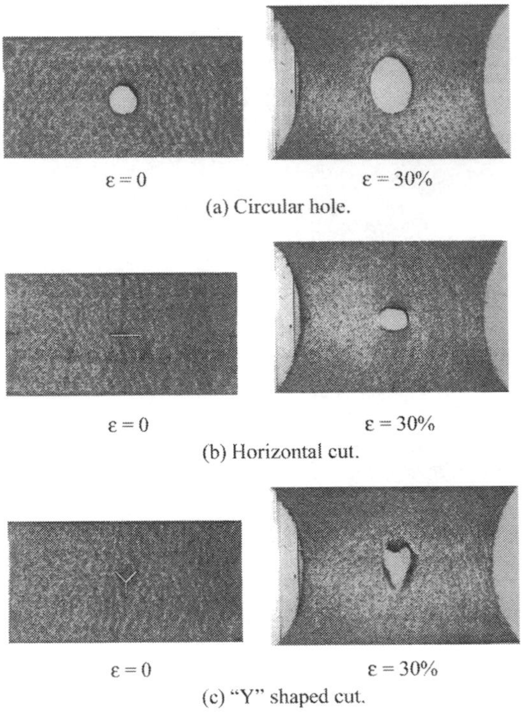

Figure 6. Shapes of damages at different strain levels – geotextile GB.

and 2% (Fig. 7b). This type of damage had a more important effect on the secant tensile stiffness than the horizontal cut.

4 CONCLUSIONS

This paper presented a study on the influence of confinement on the mechanical properties of virgin and damaged geotextiles. The results obtained showed significant increases of geotextile tensile stiffness due to confinement. It is important to point out that for the type of testing equipment used these increases were smaller than those obtained in tests with previous in-soil tensile test equipment, where friction develops between soil and geotextile. In the latter type of apparatus the shear stresses on the geotextile surface provide an additional constraint to geotextile fiber stretching, yielding to stiffer responses of the geotextile. For tests in the equipment described in this paper the behaviour of the geotextile was rather independent on the type of confining material used for tensile strains above 2%.

The circular hole, horizontal cut and Y shaped cut were the types of damages that caused the most detrimental effects on the geotextile mechanical properties in tests on virgin geotextiles in isolation. It was also observed that the confinement of the geotextile by the soil reduced the detrimental effects of the damages.

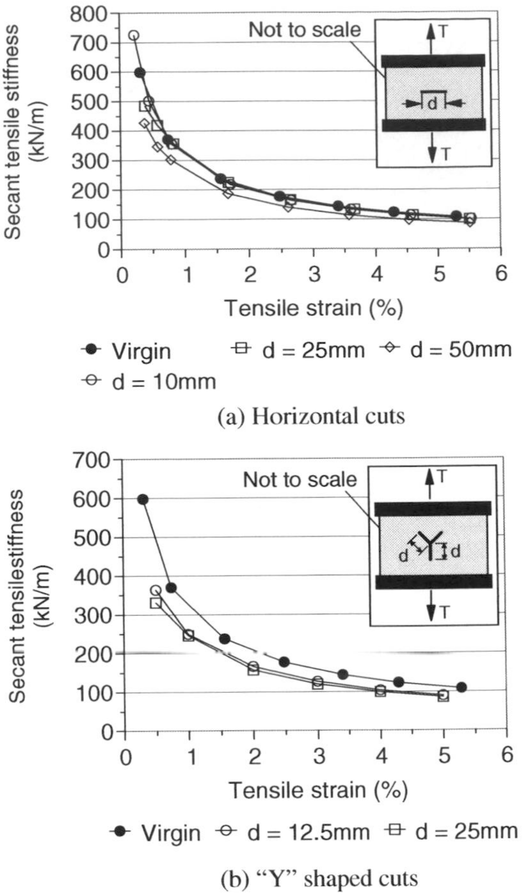

Figure 7. Tensile behaviour of confined damaged geotextile GA.

ACKNOWLEGEMENTS

The authors are indebted to the following institutions that supported the research activities described in this paper: University of Brasilia, FINATEC-UnB, CAPES-Brazilian Ministry of Education and CNPq-Brazilian National Council for Scientific and Technological Development.

REFERENCES

Kokkalis, A. and Papacharisis, N. 1989. A simple laboratory method to estimate the in soil behaviour of geotextiles. *Geotextiles and Geomembranes*, 8(2): 147–157.

Leshchinsky, D. & Field, D.A., 1987. In-soil load, elongation, tensile strength and interface friction of nonwoven geotextiles. *Geosynthetics '87*, New Orleans, USA, Vol. 1, pp. 238–249.

McGown, A., Andrawes, K.Z. & Kabir, M.H. 1982. Load-extension testing of geotextiles confined in soil. 2nd

International Conference on Geotextiles, Las Vegas, USA, Vol. 3, pp. 793–798.
Mendes, M.J.A. (2006). Load-deformation behaviour of geotextiles under confinement. *MSc. Thesis*, University of Brasília, Brasilia, Brazil, 151 p. (in Portuguese).
Mendes, M.J.A., Palmeira, E.M. & Matheus, E. (2007). Some factors affecting the in-soil load-strain behaviour of virgin and damaged nonwoven geotextiles. *Geosynthetics International*, 14(1): 39–50.
Palmeira, E.M. 1996. An apparatus for in-soil tensile tests on geotextiles-version 2. *Research Report*, University of Brasília, Brasília, Brazil, 37 p. (in Portuguese).
Palmeira, E.M., Tupa, N. & Gomes, R.C. 1996. In-soil tensile behaviour of geotextiles confined by fine soils. *III International Symposium on Earth Reinforcement – IS Kyushu 1996*, Fukuoka, Kyushu, Japan, Vol. 1, pp. 129–132.
Siel, B.D., Tzong, W.H. & Chou, N.N.S. 1987. In-soil stress-strain behavior of geotextile. *Geosynthetics'87*, New Orleans, USA, Vol. 1, pp. 260–265.

New Horizons in Earth Reinforcement – Otani, Miyata & Mukunoki (eds)
© 2008 Taylor & Francis Group, London, ISBN 978-0-415-45775-0

Evaluating in-plane hydraulic conductivity of non-woven geotextile and plastic drain by laboratory test

K. Hara
Technical Research Center, Taiyokogyo Corporation (SUN), Osaka, Japan

J. Mitsui, K. Kawai & S. Shibuya
Kobe University, Japan

T. Hongoh & T.N. Lohani
Geo-research Institute, Japan

ABSTRACT: In this study, the hydraulic conductivity associated with in-plane flow of geosynthetics was investigated in the laboratory. Three kinds of testing methods employed were; an in-plane hydraulic conductivity test for geosynthetics based on JGS T-932 (plan), a similar test on in-soil geosynthetics, and an in-plane flow test of geosynthetics in a triaxial cell. The examination focused on not only the effects of in-soil flow conductivity of geosynthetics, but also the effects of boundary conditions around the geosynthetics in the laboratory tests. It was successfully demonstrated the in-plane hydraulic conductivity was more stable against pressure for the plastic board drain than non-woven geotextiles, noting that the hydraulic conductivity of non-woven geotextiles was greatly reduced with the sustained pressure. It was also found that the triaxial apparatus was more suitable for measuring the hydraulic conductivity of plastic board drain that exhibited a high discharge rate.

1 INTRODUCTION

Geosynthetics made of synthetic resin such as planar non-woven geotextiles and strip plastic drains are often employed in earth fills and foundations so as to facilitate seepage water flow in the earth fill and also consolidation of the foundation soil.

It is expected that the drainage capacity of these geosynthctics in the horizontal direction, i.e., in-plane hydraulic conductivity, deteriorates to some extent owing to reduction in cross sectional area of the material when pressurized. In addition, such reduction in the hydraulic conductivity varies with the soil grading.

In practical design, it is thus important to evaluate quantitatively such deterioration of the material's conductivity. In so doing, it is urgently needed to establish rational testing method for assessing in-soil conductivity of geosynthetics. At present, in-plane hydraulic transmissivity of geosynthetics is usually evaluated by using an in-plane hydraulic conductivity testing system after the standard plan described by the Japanese Geotechnical Society, JGS T-932 (Plan). On the other hand, a similar testing method using a triaxial apparatus is presumably superior in a respect that it is capable of applying confining pressure uniformly to the specimen of geosynthetics.

Accordingly, the in-plane hydraulic transmissivity of non-woven geotextiles as well as a plastic board drain was measured by using two sets of apparatus; i.e., a device designed after the standard plan from the Japanese Geotechnical Society, JGS T-932 (Plan), and a modified triaxial apparatus. Discussion was made on the effects of the suspained pressure with and without the soil.

2 EXPERIMENTAL

In this paper, three types of test employed are tentatively called "***normal test***", "***in-soil test***", and "***triaxial-cell test***", respectively (Table 1).

"Normal test" employs an in-plane hydraulic conductivity testing device of geosynthetics after standard plan described by the Japanese Geotechnical Society, JGS T-932 (plan). As seen in Fig.1, the normal compressive stress was applied to the bare geosynthetic specimen by using an air-bag.

"In-soil test" was carried out in the same testing device in which the geosynthetic specimen was sandwiched in Toyoura sand. As seen in Photos 1 and 2, the geosynthetic specimen was covered with a Poly Vinyl den Chloride (PVDC) film so that the

Table 1. Testing method.

Test name	Description
"Normal test" (See Fig.1)	In-plane hydraulic conductivity test using bare geosynthetic after JGS T-932 (plan)
"In-soil test" (See Photo 1)	In-plane hydraulic conductivity test using in-soil geosynthetic after JGS T-932 (plan)
"Triaxial-cell test" (See Fig.2)	In-plane hydraulic conductivity test using bare geosynthetic in a triaxial cell

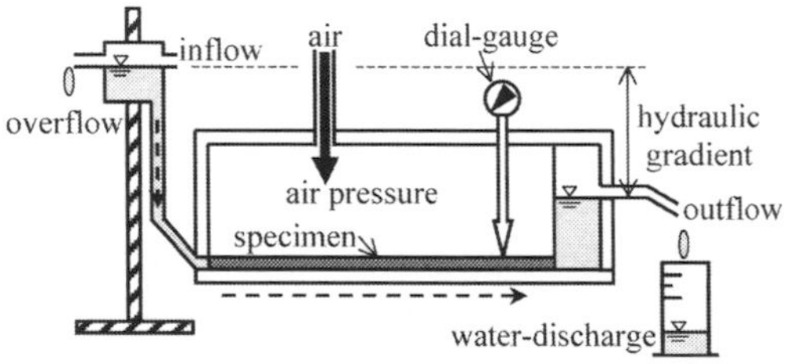

Figure 1. Configuration of *"Normal test"*.

Photo 1. Snap for the *"In-soil test"*.

water flow in the soil can be separated from that in the geosynthetics.

"Triaxial-cell test" refers to a similar test performed in a modified triaxial apparatus. As shown in Fig.2, the bare geosynthetic specimen can be subjected to a uniform confining pressure through a flexible rubber membrane.

Table 2 shows three kinds of geosynthetics tested. Figure 3 shows the cross-sections of *"Non-woven geotextile"* and *"Non-woven geotextile reinforced fiber"*. Photo 3 shows *"Plastic board drain"*.

In all the tests, the geosynthetic specimen was subjected to incremental pressures of 20, 40, 100, and 200 kPa. In tests using *"NW"* and *"NW-RF"*, the conductivity was examined at each stage by using the

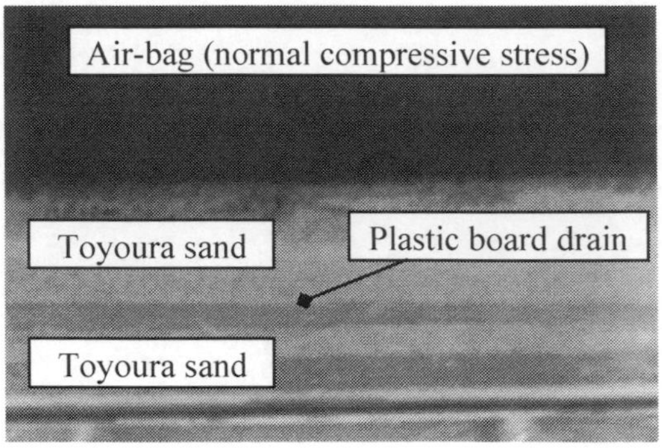

Photo 2. Geosynthetic specimen in the *"In-soil test"*.

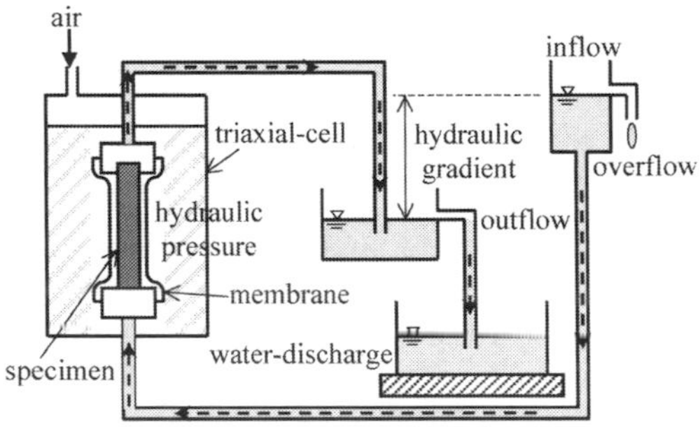

Figure 2. Configuration of *"Triaxial-cell test"*.

Table 2. Geosynthetics tested.

Name	Type of drainage materials
"NW"	non-woven geotextile (no reinforcement)
"NW-RF"	non-woven geotextile with reinforced fiber
"PD"	plastic board drain

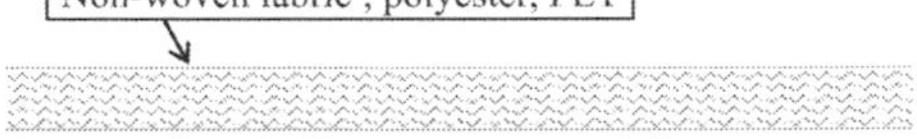

(a) Cross-section of *"Non-woven geotextile"*.

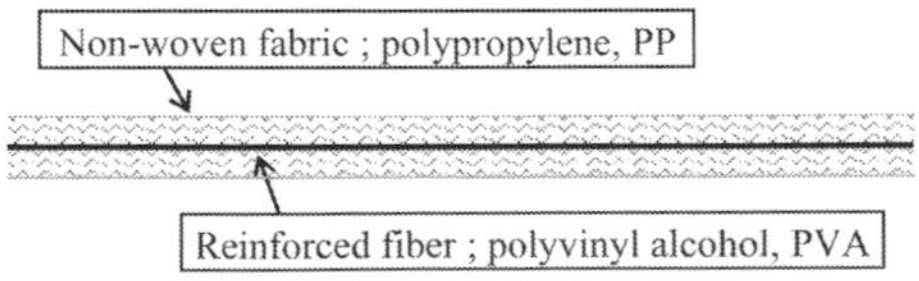

(b) Cross-section of *"Non-woven geotextile reinforced fiber"*.

Figure 3. Cross-sections of *"Non-woven geotextile"* and *"Non-woven geotextile reinforced fiber"*.

hydraulic gradients of 0.25, 0.5, and 1.0. On the other hand, the test using *"PD"* employed the values of hydraulic gradient of 0.025, 0.05, 0.1, 0.2, 0.25, 0.5 and 1.0 in each step.

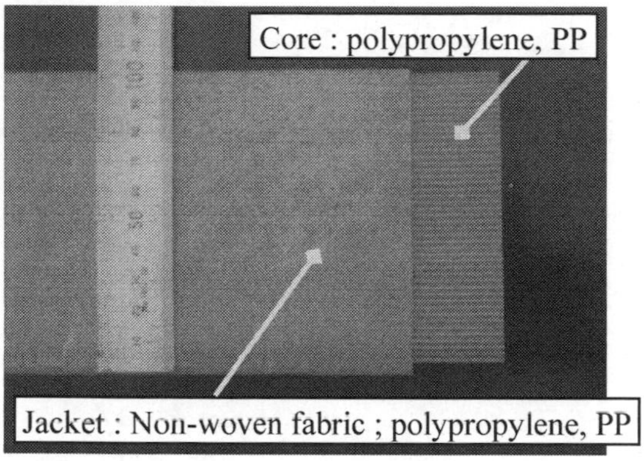

Photo 3. A snap for *"Plastic board drain"*.

3 RESULTS AND DISCUSSION

3.1 *Definitions for hydraulic conductivity*

The hydraulic conductivity in the plane of geosynthetics is characterized in terms of the coefficient of in-plane hydraulic transmissivity θ_h or in-plane permeability k_h.

$$\theta_h = \frac{Q}{W(\Delta h/L)} \quad \text{and} \quad k_h = \frac{Q}{W(\Delta h/L)H_g} = \frac{\theta_h}{H_g}$$

where Q is the rate of discharge, W and L are the width and length of the specimen in the flow direction, Δh is the total head loss in the geosynthetics, and $I(= \Delta h/L)$ is the hydraulic gradient in the geosynthetics, and H_g is the current thickness of the specimen. It should be mentioned that the H_g value was measured in a separate test in which the dead load was applied to the specimen.

3.2 *Comparison between "normal test" and "in-soil test"*

Figures 4, 5 and 6 show the relationship between the in-plane hydraulic transmissivity and the normal compressive stress P in *"in-soil test"* and *"normal test"* by using three kinds of geosynthetics of *"NW"*, *"NW-RF"* and *"PD"*, respectively. Figures 7, 8 and 9 show similar results in terms of the relationship between the in-plane hydraulic transmissivity and the hydraulic gradient in *"in-soil test"* and *"normal test"* for *"NW"*, *"NW-RF"* and *"PD"*, respectively.

In tests using *"NW"*, the transmissivity decreased considerably as the P increased, implying that the transmissivity in *"in-soil test"* was almost zero at $P = 200$ kPa (see Figs.4 and 7). Note also that the transmissivity in *"in-soil test"* was higher than *"normal test"* when P was less than say 40 kPa. However, the trend was reversed for $P > 100$ kPa. These observations may be attributed to an influence that the thickness of the geotextile decreased locally at around $P = 100$ kPa involved with intrusion of soil grains into the geotextile.

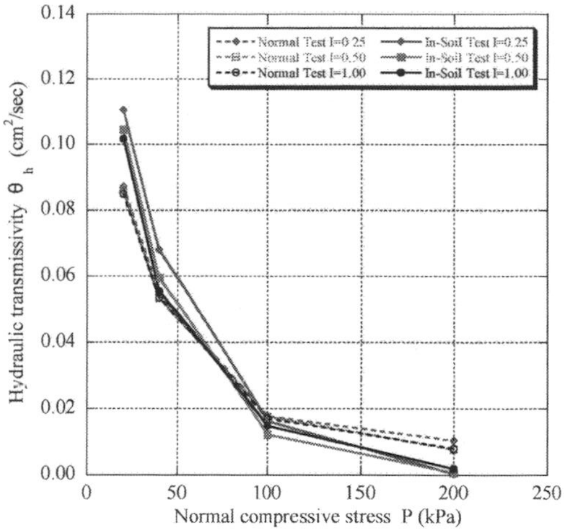

Figure 4. Relationship between hydraulic transmissivity and normal compressive stress for *"Non-woven geotextile"*.

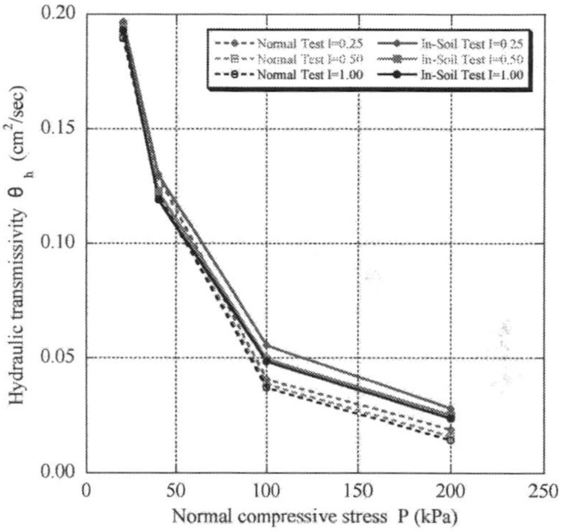

Figure 5. Relationship between hydraulic transmissivity and normal compressive stress for *"Non-woven geotextile reinforced fiber"*.

In comparative tests using *"NW-RF"*, the transmissivity in both of *"in-soil test"* and *"normal test"* steadily decreased when P increased in value. When P ranged between 20 kPa and 40 kPa, the transmissivity in *"in-soil test"* was approximately equal to *"normal test"*. However, when P was more than 100 kPa, the transmissivity in *"in-soil test"* was more than that of *"normal test"* (see Figs.5 and 8). The effect of reinforcement to the non-woven geotextile was obvious in that the hydraulic conductivity was improved at higher

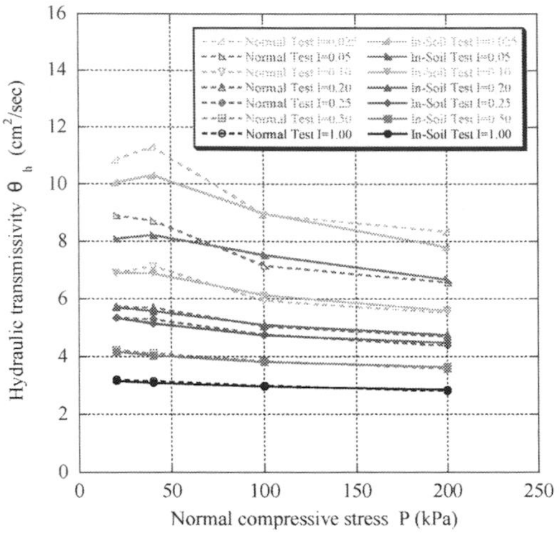

Figure 6. Relationship between hydraulic transmissivity and normal compressive stress for *"Plastic board drain"*.

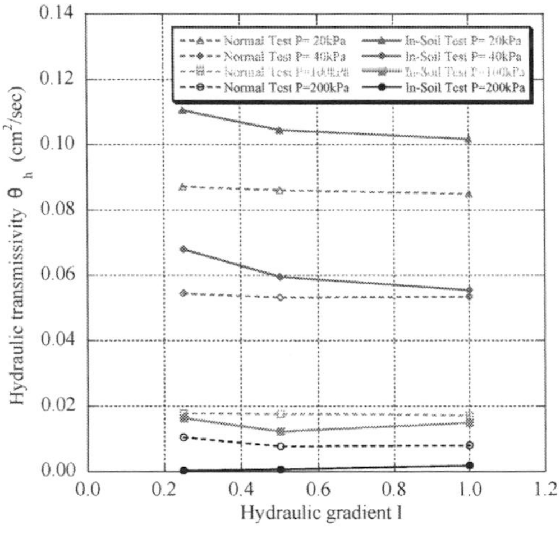

Figure 7. Relationship between hydraulic transmissivity and hydraulic gradient for *"Non-woven geotextile"*.

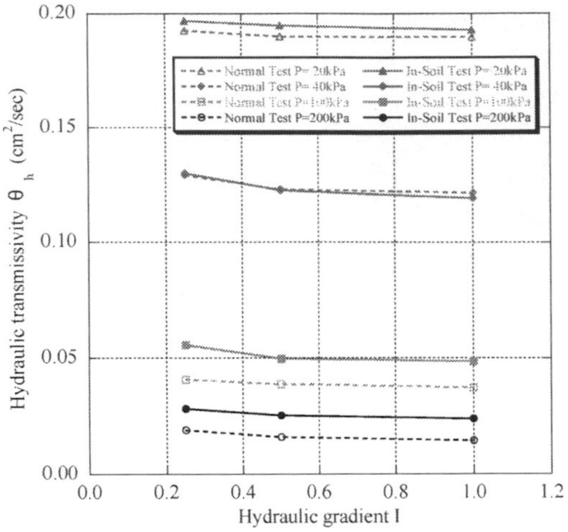

Figure 8. Relationship between hydraulic transmissivity and hydraulic gradient for *"Non-woven geotextile reinforced fiber"*.

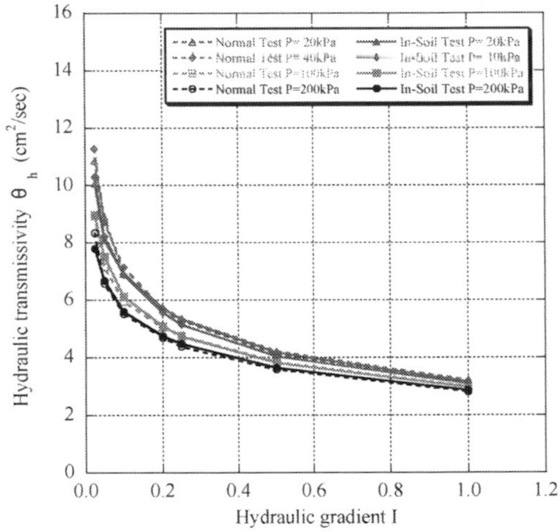

Figure 9. Relationship between hydraulic transmissivity and hydraulic gradient for *"Plastic board drain"*.

pressures since *"NW-RF"* prevented soil grains from penetrating into the material.

As seen in Figs.4 and 5 (or Figs.7 and 8), the result of *"NW"* and *"NW-RF"* for $P > 100$ kPa is opposite to each other between *"in-soil test"* and *"normal test"*. The difference may be attributed to the effects of reinforced fiber in *"NW-RF"*.

As seen in Figs.6 and 9, the transmissivity of *"PD"* was substantially larger by approximately ten-fold. Besides, it was virtually the same between *"in-soil test"* and *"normal test"*. Moreover, the transmissivity was unaffected by the normal compressive stress for the range of *P* examined. The rigidity of *"PD"* is much higher compared to non-woven geotextiles, which in turn may result in independence of transmissivity against *P*.

Effects of the hydraulic gradient in tests using non-woven geotextiles showing lower transmissivity were insignificant. Conversely, the transmissivity of the *"PD"* was significantly influenced by the hydraulic gradient in a manner that the transmissivity apparently decreased when the hydraulic gradient increased. This may be attributed to energy loss possibly due to turbulent flow in the testing system. Accordingly, the hydraulic gradient should be low enough to prevent any occurrence of turbulent flow in test using a high transmissivity geosynthetic like *"PD"*.

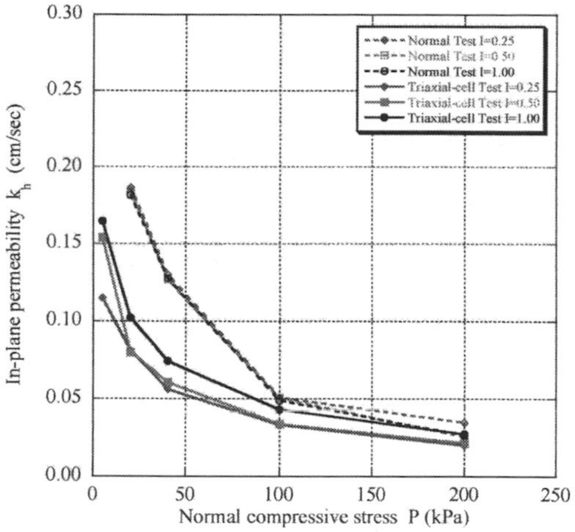

Figure 10. Relationship between in-plane permeability and *P* for *"non-woven geotextile"*.

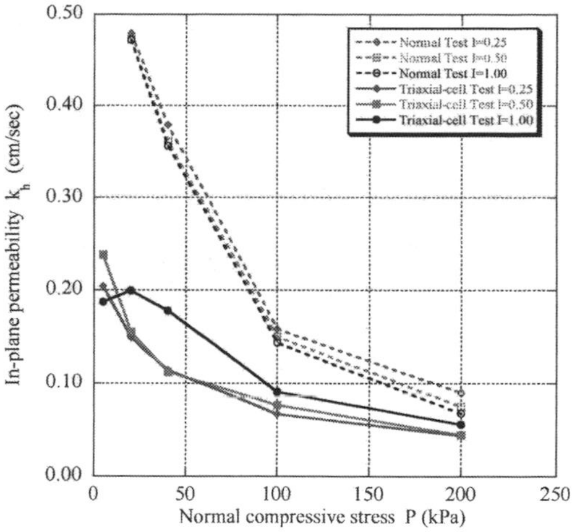

Figure 11. Relationship between in-plane permeability and *P* for *"non-woven geotextile with reinforced fiber"*.

In summary, the *"PD"* seems more efficient as a drain material when used in earth fill, since the transmissivity is higher, and unaffected by in-soil pressure as compared to non-woven geotextiles.

3.3 *In-plane permeability in two testing devices*

Figures 10, 11, and 12 show the relationship between in-plane permeability and *P* for *"NW"*, *"NW-RF"*, and *"PD"* as examined using *"triaxial-cell test"* and *"normal test"*.

As for *"NW"* and *"NW-RF"*, a trend was clear in both the tests that the permeability decreased as *P* increased. It may be attributed to the fact that the density of fiber in *"NW"* and *"NW-RF"* increased with *P* involved with decrease in the thickness of the geotextile. On the other hand, the permeability was unaffected by the hydraulic gradient applied, since no turbulent flow took place in these tests with relatively low rate of discharge. Note also that the permeability in *"normal test"* was noticeably higher compared to that in *"triaxial-cell test"* over a range of *P* from 20 kPa to 100 kPa. In *"normal test"*, the normal pressure on the specimen may have been much less than the pressure applied in the air-bag possibly due to the effects of friction between the airbag and the sidewall. It is a potential drawback involved in *"normal test"* with the rigid boundary.

Conversely, in tests on *"PD"*, the permeability did not depend on the *P* very much, since the *"PD"* was stiff enough against the applied pressures over a range examined. However, the permeability of *"PD"* varied with the hydraulic gradient in a manner that the permeability decreased with the hydraulic gradient. The tendency was more significant in *"normal test"*.

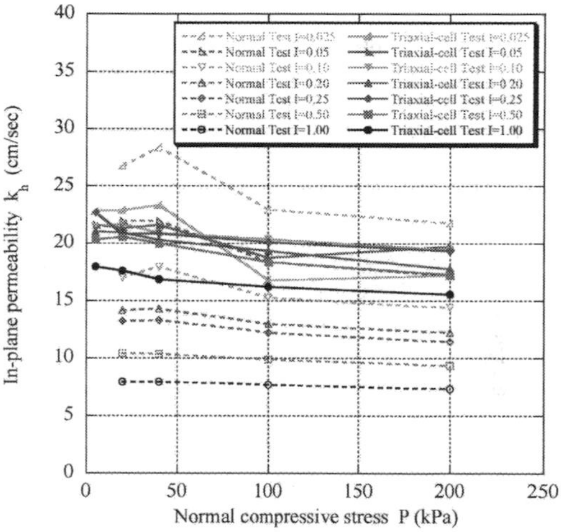

Figure 12. Relationship between in-plane permeability and *P* for *"plastic board drain"*.

Figure 13 shows the relationship between the in-plane hydraulic transmissivity of *"PD"* and *P* in *"triaxial-cell test"* and *"normal test"*. The transmissivity of *"PD"* is examined against the hydraulic gradient in Fig.14. As stated earlier, the transmissivity hardly varied with *P*, and varied significantly with the hydraulic gradient. As for the dispersion of the transmissivity against the hydraulic gradient, *"normal test"* was more significant compared to the results in *"triaxial-cell test"*. The transmissivity in *"triaxial-cell test"* hardly depended on the hydraulic gradient. On the other hand, the transmissivity in *"normal test"* varied greatly with the hydraulic gradient. Moreover, when the hydraulic gradient was relatively low, the

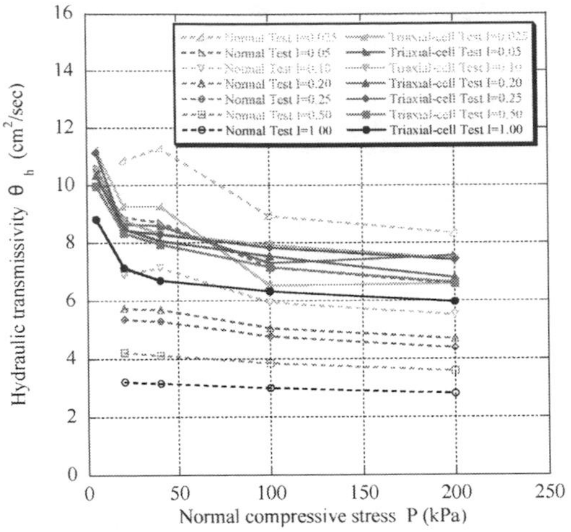

Figure 13. Relationship between hydraulic transmissivity and *P* for *"Plastic board drain"*.

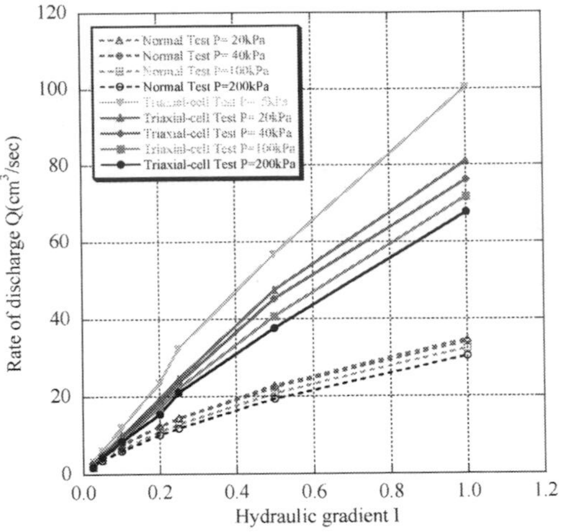

Figure 15. Relationship between the rate of discharge and hydraulic gradient for *"Plastic board drain"*.

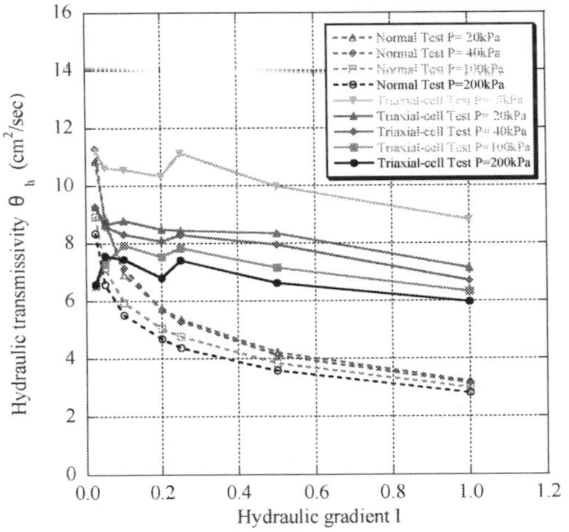

Figure 14. Relationship between hydraulic transmissivity and hydraulic gradient for *"Plastic board drain"*.

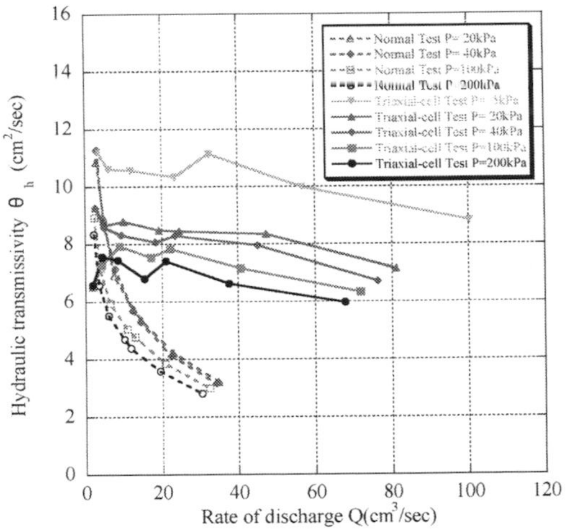

Figure 16. Relationship between the rate of discharge and hydraulic transmissivity for *"Plastic board drain"*.

transmissivity in *"normal test"* was approximately equal to that in *"triaxial-cell test"*.

Figure 15 shows the relationship between the rate of discharge and the hydraulic gradient for *"PD"*. The relationship between the in-plane hydraulic transmissivity and the rate of discharge in *"triaxial-cell test"* and *"Normal test"* is shown in Fig.16. As seen in these figures, the rate of discharge of *"PD"* was much lower in *"normal test"* than in *"triaxial-cell test"*, implying that the transmissivity was relatively low in *"normal test"*. Provided that the results of *"triaxial-cell test"* are correct, it may be surmised that some loss of discharge took place in *"normal test"* associated with the characteristic configuration of the testing device. The results suggest that the *"normal test"* is not suitable for measuring correctly the in-plane transmissivity of geosynthetics when the hydraulic gradient is relatively high (i.e. when the rate of discharge is relatively high).

4 CONCLUSIONS

i) The in-plane hydraulic transmissivity of in-soil non-woven geotextile decreased with the sustained normal compressive stress, reaching approximately to zero at 200 kPa. This may be attributed to the increase in fiber density.

ii) The behavior of "non-woven geotextile reinforced fiber" was slightly improved by showing slower decrease against *P*.

iii) The in-plane hydraulic transmissivity of "plastic board drain" was higher by ten-fold compared to those of the non-woven geotextiles. Moreover, due to a high stiffness of the "PD", it hardly depended on *P* as examined up to 200 kPa.

iv) iv) Therefore, the "plastic board drain" seems potentially more effective geosynthetic to promote drainage in the earth fill.

v) In tests using "non-woven geotextile" and "non-woven geotextile reinforced fiber", the in-plane permeability in *"normal test"* was higher than *"triaxial-cell test"* only when the normal compressive stress was relatively low, say less than 100 kPa,

vi) On the other hand, the in-plane conductivity of "plastic board drain" was lower in *"normal test"* than in *"triaxial-cell test"*, and it increased with the increase in the hydraulic gradient, i.e. the increase in the rate of discharge.

vii) When the rate of discharge is relatively high, *"triaxial-cell test"* seems superior to *"normal test"* for the purpose of measuring the in-plane conductivity of geotextiles correctly.

REFERENCES

Ling, H.I., Jonathan, T.H.WU, Nishimura, J., & Tatsuoka, F. 1989. In-plane Hydraulic Conductivity of Geotextiles, 4th Geosynthetics symposium. *Japan Chapter of International Geosynthetics Society, 56–63.*

Kitajima, T., Suwa, Y., Pradhan, T., & Kamon, M. 1990. Factors Affecting the Discharge Capacity of a Prefabricated Band Shaped Drain. *25th Geotechnical Research Meeting, The Japanese Geotechnical Society, 1717–1718.*

Park, Y.M, Kim, Kim, S.S. & Jeon, H.Y. 2004. An Evalutation on Discharge Capacity of Perfabricated Vertical Drains using Large scale test. *Peoc.of the Asian Regional Conference on Geosynthetics, Korean Geosynthetics Society and International Geosynthetics Society, 330–337.*

Mitsui, J., Hara, K., & Shibuya, S. 2007. Evaluating in-plane permeability of geosysnthetics in-soil. *42th Geotechnical Research Meeting, The Japanese Geotechnical Society, 1581–1582.*

New Horizons in Earth Reinforcement – Otani, Miyata & Mukunoki (eds)
© 2008 Taylor & Francis Group, London, ISBN 978-0-415-45775-0

A theoretical method to predict the pullout behaviour of extruded geogrids embedded in granular soils

N. Moraci, G. Cardile & D. Gioffrè
Department MECMAT, Faculty of Engineering, "Mediterranea" University of Reggio Calabria, Italy

ABSTRACT: Pullout tests are necessary in order to study the interaction behaviour between soil and geosynthetics in the anchorage zone; hence the test results have direct implications in the design of reinforced soil structures. A new theoretical method to determine the pullout behaviour of extruded geogrids embedded in a compacted granular soil was developed. In particular, in the theoretical method: the frictional and the bearing component of the pullout reistance of geogrids are evaluated by simple equations (considering the scale effects negligible), the mobilisation of both frictional and bearing components of pullout resistance were determinated using the load transfer functions approach and the geometry of the elements on which the bearing resistance mobilizes, the soil dilatancy effects and the geogrid extensibility were taken into account. The comparison between theoretical and experimental results showed a good agreement, thus confirming the reliability of the proposed approach.

1 INTRODUCTION

Pullout tests are necessary in order to study the interaction behaviour between soil and geosynthetics in the anchorage zone; hence the resulting properties have direct implications on the design of reinforced soil structures.

In order to analyse the internal stability of reinforced earth structures, it is necessary to evaluate the pullout resistance of reinforcement, mobilized in the anchorage zone.

The pullout resistance in a pullout test can be described by the following equations:

$$P_R = 2L\sigma'_v f_b \tan\phi' \quad (1)$$

$$P_R = 2L\sigma'_v \mu_{s/GSY} \quad (2)$$

$$P_R = 2L\sigma'_v \alpha F^* \quad (3)$$

where P_R = pullout resistance (per unit width); L = reinforcement length in the anchorage zone; σ'_v = effective vertical stress; ϕ' = soil shear strength angle; f_b = soil–geosynthetic pullout interaction coefficient; $\mu_{S/GSY}$ = soil–geosynthetic interface apparent coefficient of friction; F^* = pullout resistance factor; and α = scale effect correction factor account for a non-linear stress reduction over the embedded length of highly extensible reinforcements (FHWA, 2001).

The soil–geosynthetic pullout interaction coefficient f_b may be determined by means of theoretical expressions (Jewell et al., 1985), whose limits have been investigated by different researchers (Moraci and Montanelli, 2000; Palmeira and Milligan, 1989), or by back-calculation from pullout test results. In this case previous experimental studies (Moraci and Montanelli, 2000; Palmeira and Milligan, 1989) have shown that the values of f_b are largely influenced by the choice of the value of the soil shear strength angle.

According to FHWA (2001) the scale effect correction factor can be obtained from pullout tests on reinforcements with different lengths or derived using analytical or numerical load transfer models which have been "calibrated" through numerical test simulations. In the absence of test data, $\alpha = 0.8$ for geogrids and $\alpha = 0.6$ for geotextiles.

Nevertheless, it is important to define the role of all the design (and test) parameters on the mobilisation of the interaction mechanisms (frictional and passive) in pullout condition, including geosynthetic length, tensile stiffness, geometry and shape, vertical effective stress (acting at the geosynthetic interface) and soil shear strength. In the present paper, on the basis of the test results obtained by the authors (Moraci and Recalcati, 2006; Moraci and Gioffrè, 2006), a new theoretical method to determine the pullout response of extruded geogrids embedded in a compacted granular soil was developed.

The proposed method is able to evaluate both the passive and the frictional components of pullout resistance taking into account the reinforcement extensibility and geometry, as well as, the non linearity of the failure envelope of the backfill soil. This paper deals with some results of the theoretical research carried out in order to evaluate the pullout behaviour of extruded geogrids.

1.1 *The pullout resistance*

The main interaction mechanisms affecting the pullout resistance of extruded geogrids are the skin friction (Fig. 1a), between soil and reinforcement solid surface, and the bearing resistance, that develops against transversal elements (Fig. 1b).

The pullout resistance of a geogrid, assuming that the different interaction mechanisms act at the same time with maximum value and that they are independent of each–other, may be evaluated using the following equation:

$$P_R = P_{RS} + P_{RB} \tag{4}$$

where P_{RS} is the skin friction component of pullout resistance and P_{RB} the bearing component of pullout resistance.

According to the experimental results obtained by Moraci and Recalcati (2006), which may explain the different behaviour of the three geogrids used in the research and according to the previous analytical work carried out by the authors (Moraci and Gioffrè, 2006), the determination of the pullout resistance of geogrids embedded in granular soil for which the scale effects are negligible (i.e. S/B larger than 40 and S/d_{50} larger than 1000) may be evalutated using the equation:

$$P_R = 2C_{\alpha S}\alpha_s L_R \tau + n_t n_{tb} A_b \sigma'_b \tag{5}$$

where $C_{\alpha S}$ is the reduction coefficient of geogrid area where skin friction develops; $n_t = (L_R/S)$ is the number of geogrid bearing members; n_{tb} = number of nodes in a transversal element; A_b is the area of each rib element where the bearing resistance can be mobilized; σ'_b is the bearing stress.

In order to separate the two components of the pullout resistance it is possible to perform pullout tests on geogrid specimens where a portion of transverse reinforcing elements are removed (Alagiyawanna et al., 2001; Alfaro et al., 1995; Matsui et al., 1996; Palmeira and Milligan, 1989). Therefore experimental research was carried out to evaluate and calibrate the load transfer curves essentials to validate the proposed method.

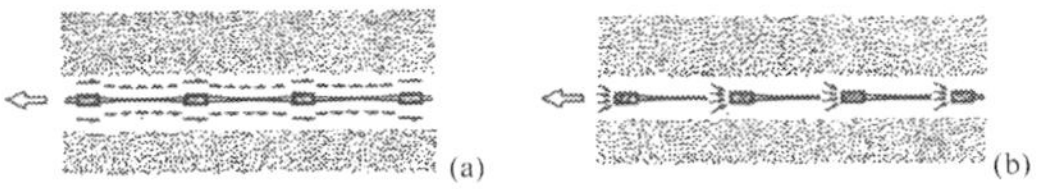

Figure 1. The two mechanisms for bond between reinforcement and soil (Jewell et al., 1985).

2 LOAD TRANSFER CURVES: EXPERIMENTAL RESEARCH

2.1 *Test apparatus and materials*

The load transfer curves evaluation was carried out performing pullout tests on a HDPE geomembrane (Fig. 2a) and on a HDPE extruded mono-oriented geogrid with isolated transversal bar (Fig. 2b) by varying the applied vertical effective pressures (10, 25 and 100 kN/m^2). The displacement rate was 1.0 mm/min in all tests.

The test apparatus used is composed by a pullout box (1700 × 600 × 680 mm), a vertical load application system, a horizontal force actuator device, a special clamp, and all the required instrumentation (Moraci and Recalcati, 2006). For the geogrid, a more detailed analysis of the transversal bar geometry has also shown a non-uniform shape with greater thickness at the rib intersection. The passive interaction mechanisms develop both at the node embossments and at the transversal bars. Therefore, the node embossment and the transversal bar geometry have been carefully determined to calculate the effective passive resistance surfaces.

A granular soil was used in these tests. The soil was classified as a uniform medium sand with uniformity coefficient $U = d_{60}/d_{10} = 1.5$ and average grain size $d_{50} = 0.22$ mm. Standard Proctor compaction tests gave a maximum dry unit weight $\gamma_{dmax} = 16.24$ kN/m^3 at an optimum water content $w_{opt} = 13.5\%$.

The results of pullout tests are reported in Figure 3, that shows the frictional stress τ versus the displacement d, measured at the edge attached to the clamp,

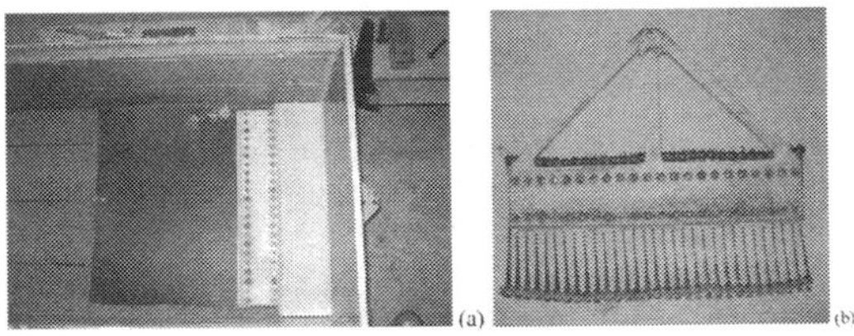

Figure 2. Pullout tests on a HDPE geomembrane (a) and on a HDPE extruded mono-oriented geogrid with isolated transversal bar (b).

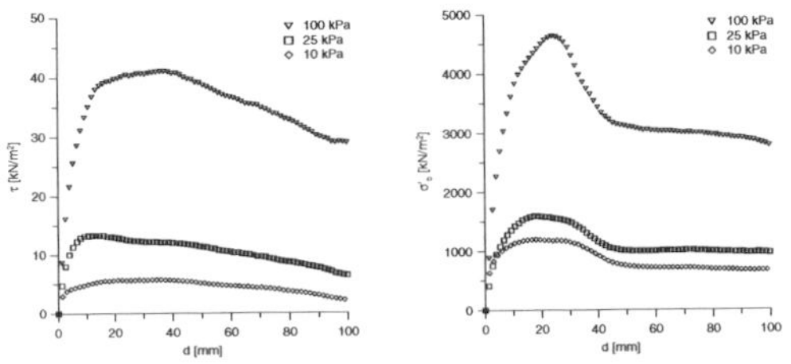

Figure 3. Load transfer curves: experimental results.

for the HDPE geomembrane and the bearing stress σ'_b versus the displacement d for the geogrid with isolated transversal bar. The different curves on the graphs are referring to the different applied confining pressures.

Tests performed both on HDPE geomembrane and on geogrid with isolated transversal bar show a strain softening behaviour, with a progressive decrease of pullout resistance after the peak.

3 THE METHOD

The load-transfer method is probably the most widely used technique to study the settlement of single axially loaded piles, and is particularly useful when the soil behavior is clearly nonlinear.

The proposed simple method involves modeling the geogrid as a series of elements supported by discrete nonlinear springs, which represent the resistance of the soil in skin friction (τ-d springs), and a nonlinear spring at the transversal bar representing the bearing (σ'_b-d) spring.

The springs are nonlinear representations of the shear stress τ and of the bearing stress σ'_b mobilization versus displacement (d) as shown schematically in Figure 4.

The method may be summarized in the following way:

- The geogrid can be divided in n elements within transversal bars (Fig. 5);
- Considering the element n, it assumes a displacement d_n;
- Evaluate the passive stress σ'_b using the load transfer curve σ'_b-d and then the passive component P^n_{RB} using the following expression:

$$P^n_{RB} = n_{tb} A_b \sigma'_b \tag{6}$$

- Assuming a displacement d_{gn} in the middle of the element n (in first analysis $d_{gn} = d_n$), it can possible evaluate the frictional stress τ using load transfer curve τ-d;

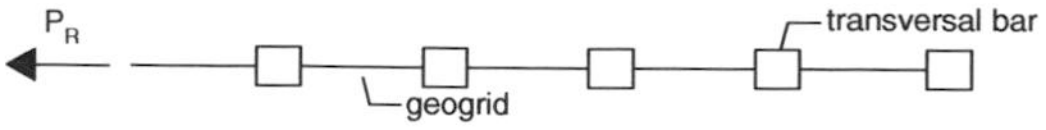

Figure 4. Idealized model used in load transfer analyses.

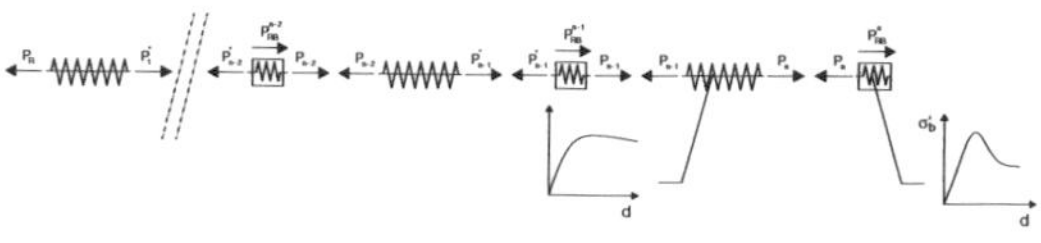

Figure 5. Method scheme used in load transfer analyses.

- The pullout resistance at top of the block n can be evaluated using the following expression:

$$P_{n-1} = 2C_{\alpha S}\alpha_s L_R \tau + P^n_{RB} \tag{7}$$

- Block deformation can be evaluated, assuming a linear variation of load:

$$\varepsilon_n = \frac{P_m W_R}{EA} \tag{8}$$

where

$$P_m = \frac{P_n + P_{n-1}}{2} \tag{9}$$

- A new displacement in the middle of the block n can be evaluated:

$$d^*_{gn} = d_n + \varepsilon_n \frac{L_n}{2} \tag{10}$$

- When the difference between $|d_{gn} - d_{gn}^*|$ is less than a tolerance, it is possible to pass to the next block n-1 evaluating the pullout resistance P_{n-1}

$$P^*_{n-1} = P_{n-1} + P^{n-1}_{RB} \tag{11}$$

- It is proceeded until obtaining the values P_R and d at the top of geogrid.
- Repeating the procedure for various values of d_b it is possible to obtain the pullout curve (P_R, d).

According to experimental results on a HDPE extruded mono-oriented geogrid with isolated transversal bar and on a HDPE geomembrane (see Table 1 and Table 2), the validation of the method was carried out using simplify transfer load curves (Fig. 6). The load transfer curves for friction stress component of pullout resistance are schematized as a bilinear model. The load transfer curves for bearing stress are schematized with an non linear model taking into account the

Table 1. Load transfer curve: Frictional stress component.

σ'_v [kN/m^2]	τ_{max} [kN/m^2]	d(τ_{max}) [mm]	τ_{min} [kN/m^2]	d(τ_{min}) [mm]
10	5	3.00	2.5	100.00
25	15	3.00	7	100.00
100	45	3.00	28.5	100.00

Table 2. Load transfer curve: Bearing stress component.

σ'_v [kN/m^2]	T_i [–]	d($\sigma'_{b\,max}$) [mm]	$\sigma'_{b\,max}$ [kN/m^2]	$\sigma'_{b\,min}$ [kN/m^2]
10	2000	20	1200	670
25	2000	20	1600	980
100	2000	20	5200	2800

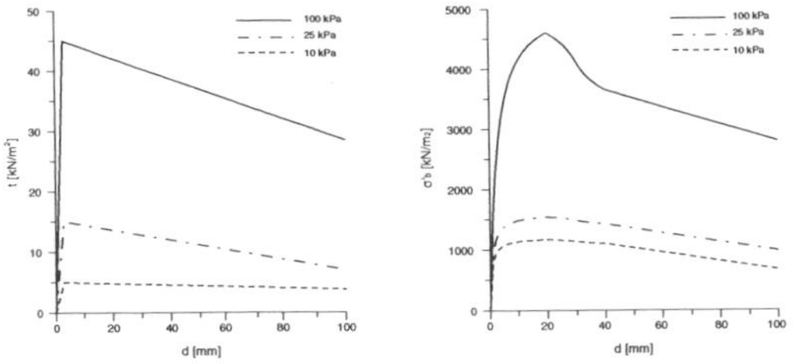

Figure 6. Load transfer curves used: a) friction stress component; b) bearing stress component.

decrease of bearing stress after the peak value. Using the following expression:

– Friction stress component

$$\text{if } d \le d(\tau_{max})$$

$$\tau = \frac{\tau_{max}}{d(\tau_{max})} \cdot d$$

$$\text{if } d > d(\tau_{max}) \tag{12}$$

$$\tau = \tau_{max} - \frac{\tau_{max} - \tau_{min}}{d(\tau_{min}) - d(\tau_{max})} \cdot (d - d(\tau_{max}))$$

– Bearing stress component

$$\text{if } d \le d(\sigma'_{bmax})$$

$$\sigma'_b = \frac{d}{\frac{1}{T_i} + \frac{d}{\sigma'_{bmax}}}$$

$$\text{if } d(\sigma'_{bmax}) < d \le 1.5d(\sigma'_{bmax})$$

$$\sigma'_b = \frac{d^*}{\frac{1}{T_i} + \frac{d^*}{\sigma'_{bmax}}}$$

$$\text{with } d^* = 2d(\sigma'_{bmax}) - d \tag{13}$$

$$\text{if } 1.5d(\sigma'_{bmax}) < d \le 2d(\sigma'_{bmax})$$

$$\sigma'_b = 2\sigma'^{**}_b - \frac{d^{**}}{\frac{1}{T_i} + \frac{d^{**}}{\sigma'_{bmax}}}$$

$$\text{with } d^{**} = 3d(\sigma'_{bmax}) - d; \qquad \sigma'^{**}_b = \frac{0.5d(\sigma'_{bmax})}{\frac{1}{T_i} + \frac{0.5d(\sigma'_{bmax})}{\sigma'_{bmax}}}$$

$$\text{if } d > 2d(\sigma'_{bmax})$$

$$\sigma'_b = \sigma'^{***}_b - \frac{\sigma'^{***}_b - \sigma'_{bmin}}{d(\sigma'_{bmin}) - 2d(\sigma'_{bmax})} d^{***}$$

$$\text{with } d^{***} = d - 2d(\sigma'_{bmax}); \qquad \sigma'^{***}_b = \frac{d(\sigma'_{bmax})}{\frac{1}{T_i} + \frac{d(\sigma'_{bmax})}{\sigma'_{bmax}}}$$

Table 3. Structural characteristics of the different geogrids.

	W_r (mm)	W_t (mm)	B_r (mm)	B_t (mm)	A_b (mm^2)
GG1	11.26	6.6	3.80	3.57	66.35
GG2	11.86	6.0	4.65	4.48	82.03
GG3	12.36	5.5	5.16	4.85	90.45

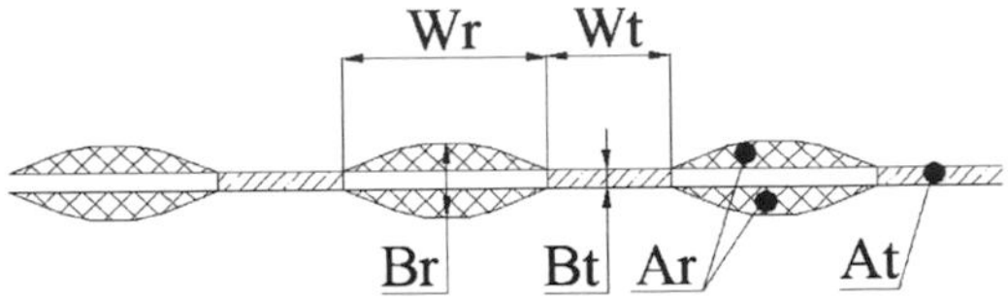

Figure 7. Schematic cross section of the geogrid bar.

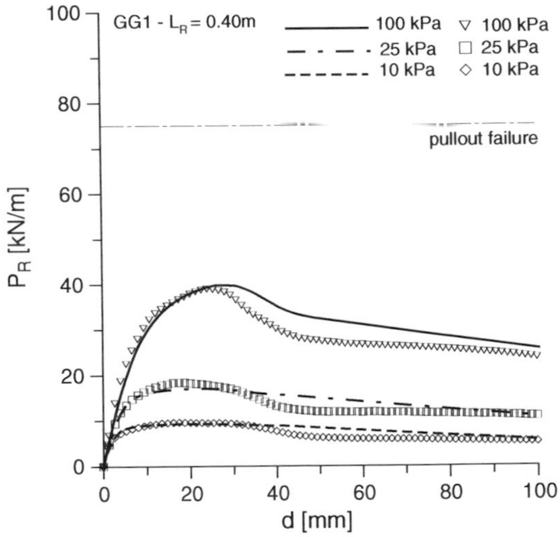

Figure 8. Geogrid GG1: Comparison of pullout curves for $L_R = 0.40$ m.

4 EXPERIMENTAL VALIDATION OF PROPOSED METHOD

In order to verify the stability of the algorithm and the reliability of the numerical results, the developed method is applied and compared with experimental results obtained by Moraci and Recalcati (2006) using the simplify experimental load transfer curves obtained for the three applied vertical effective pressures (10, 25 and 100 kN/m^2) investigated.

Reader may obtain the test details on geotechnical observation, test procedure and instrumentation etc. in Moraci and Recalcati (2006).

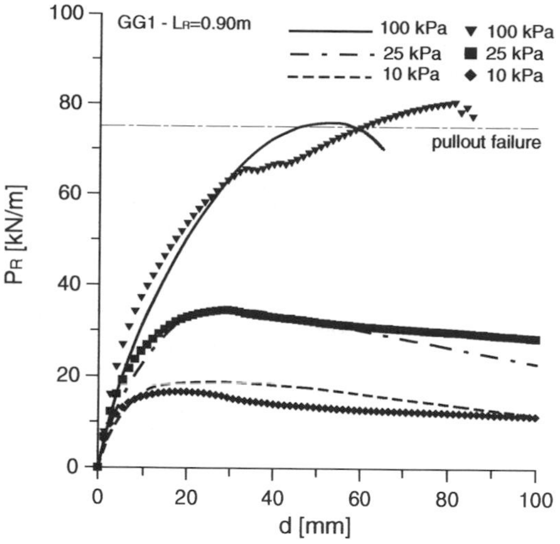

Figure 9. Geogrid GG1: Comparison of pullout curves for $L_R = 0.90$ m.

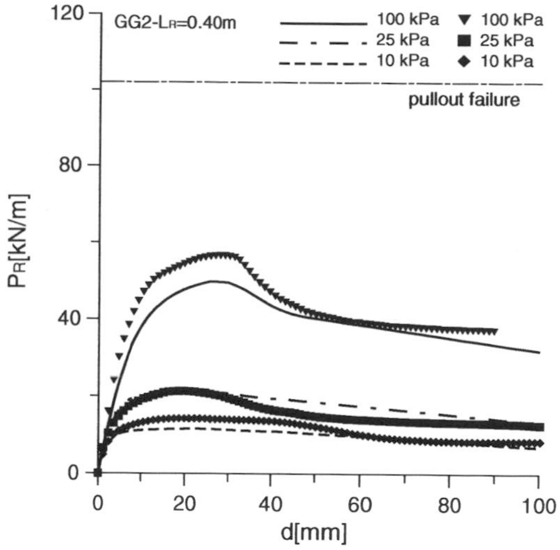

Figure 11. Geogrid GG2: Comparison of pullout curves for $L_R = 0.40$ m.

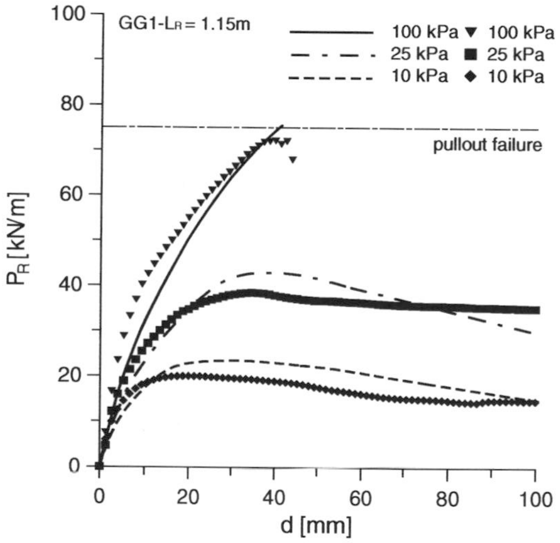

Figure 10. Geogrid GG1: Comparison of pullout curves for $L_R = 1.15$ m.

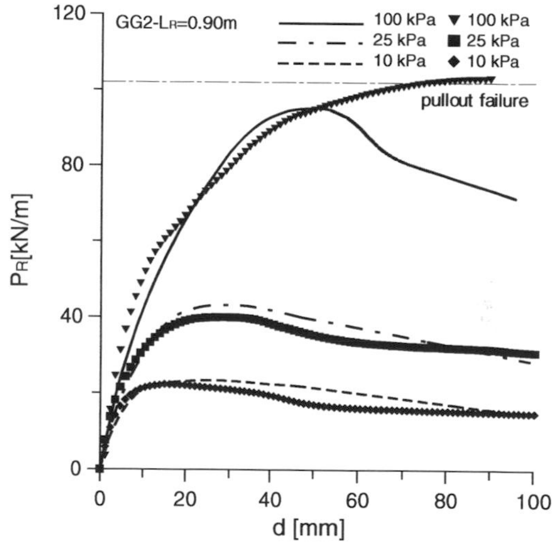

Figure 12. Geogrid GG2: Comparison of pullout curves for $L_R = 0.90$ m.

A more detailed analysis of the transversal bar geometry has also shown a non-uniform shape with greater thickness at the rib intersection. The passive interaction mechanisms develop both at the node embossments and at the transversal bars. Therefore, the node embossment and the transversal bar geometry have been carefully determined to calculate the effective passive resistance surfaces.

The results of this analysis are reported in Table 3, where W_r and B_r are, respectively the node width and thickness; W_t and B_t are, respectively the width and thickness of the bar portion between two nodes (Fig. 7), and A_b is the effective area of each rib element (composed of the node embossment and of the bar portion between two nodes $A_t + A_r$) where the passive resistance can be mobilized.

The direct comparison between experimental data (symbols) and numerical results (continuous and broken lines) show in Figures 8–16.

The presented comparisons clearly illustrate the application of the above formulation. The proposed

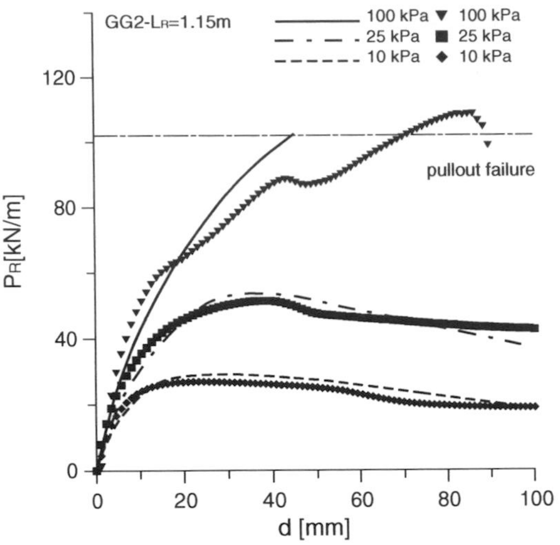

Figure 13. Geogrid GG2: Comparison of pullout curves for $L_R = 1.15$ m.

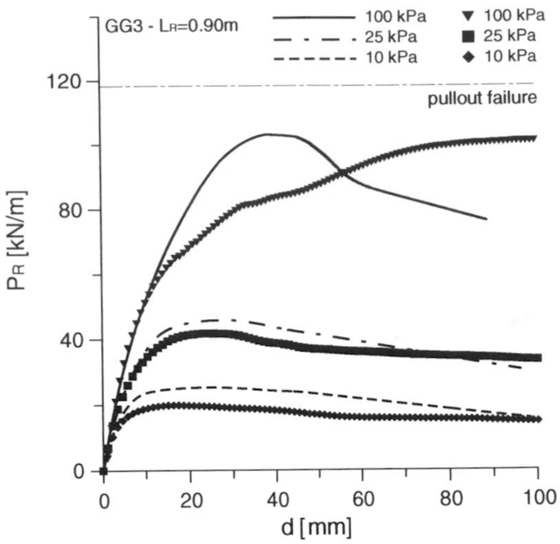

Figure 15. Geogrid GG3: Comparison of pullout curves for $L_R = 0.90$ m.

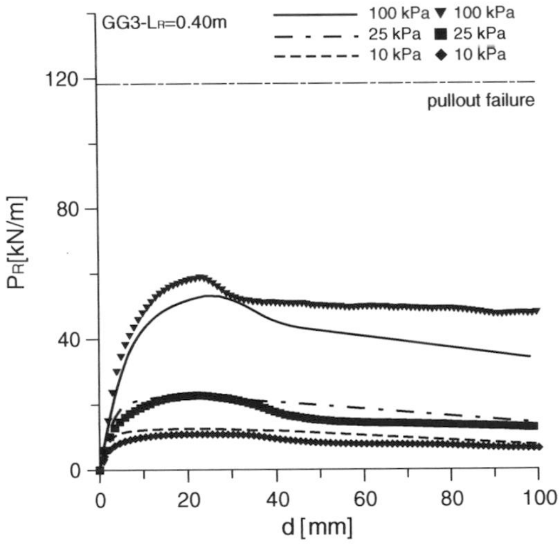

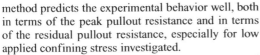

Figure 14. Geogrid GG3: Comparison of pullout curves for $L_R = 0.40$ m.

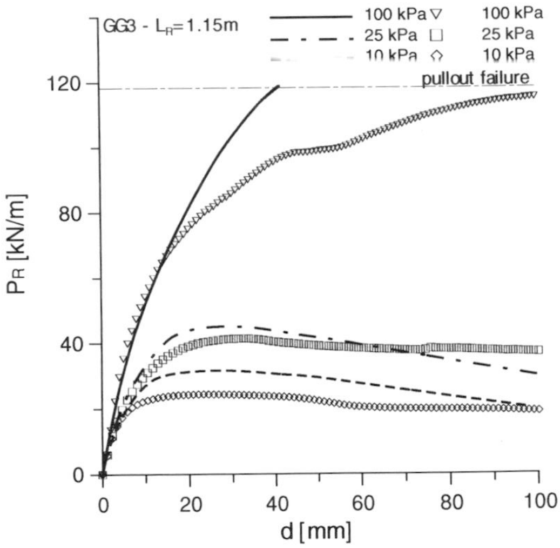

Figure 16. Geogrid GG3: Comparison of pullout curves for $L_R = 1.15$ m.

method predicts the experimental behavior well, both in terms of the peak pullout resistance and in terms of the residual pullout resistance, especially for low applied confining stress investigated.

Vice versa, the comparison between experimental data and numerical results for high applied confining stress are mainly influenced by the method assumption of a displacement at the bottom of geogrid therefore in this simple theoretical method is difficult to evaluate the geogrid rupture achieved for long reinforcement at high applied confining stress.

5 CONCLUSIONS

Comparison between the theoretical and experimental results permits the following conclusions to be drawn:

- the proposed method predicts the pullout curves well, especially for low applied confining stress;
- using this method it is possible to predict the pullout response for different reinforcement lengths and stiffness and different confining stresses on the base of simple pullout tests performed in order to evaluate the load transfer curves.

REFERENCES

Alagiyawanna, A.M.N., Sugimoto, M., Sato, S. and Toyota, 11, 2001. Influence of longitudinal and transverse members on geogrid pullout behaviour during deformation. *Geotextiles and Geomembranes, Vol. 19*, 483–507, Elsevier.

Alfaro, M.C., Miura, N. and Bergado, D.T. 1995. Soil-Geogrid Reinforcement Interaction by Pullout and Direct Shear Tests. *Geotechnical Testing Journal, Vol. 18*, 157–167, ASTM.

Jewell, R.A., Milligan, G.W.E., Sarsby, R.W. and Dubois, D.D. 1985. Interactions Between Soil and Geogrids. *Proceeding from the Symposium on Polymer Grid Reinforcement in Civil Engineering*, 18–30. Thomas Telford, London.

Matsui, T., San, K.C., Nabesahirna, Y. and Arnii, U.N. 1996. Bearing mechanism of steel reinforcement in pull-out test. *Proceedings of the International Symposium: Earth Reinforcement*, 101–105. Fukuoka, Kyushu, Japan. Balkema Publishers.

Moraci, N. and Montanelli, F. 2000. Analisi di prove di sfilamento di geogriglie estruse installate in terreno granulare compattato. *Rivista Italiana di Geotecnica N. 4/2000*, 5–21. Patron Editore.

Moraci, N. and Gioffrè D. 2006. A simple method to evaluate the pullout resistance of extruded geogrids embedded in a compacted granular soil. *Geotextiles and Geomembranes 24*, 116–128.

Moraci, N. and Recalcati, P.G. 2006. Factors affecting the pullout behaviour of extruded geogrids embedded in a compacted granular soil. *Geotextiles and Geomembranes 24*, 220–242.

Palmeira, E.M. and Milligan, G.W.E. 1989. Scale and Other Factors Affecting the esults of Pull-out Tests of Grid Buried in Sand. *Géotechinique Vol. 11, N. 3*, 511–524. Thomas Teldford, London.

New Horizons in Earth Reinforcement – Otani, Miyata & Mukunoki (eds)
© 2008 Taylor & Francis Group, London, ISBN 978-0-415-45775-0

Ultimate pullout forces of orthogonally horizontal-vertical geosynthetic reinforcement

M.X. Zhang, S.L. Zhang & J. Huang
Department of Civil Engineering, Shanghai University, Shanghai, PR China

A.A. Javadi
Department of Engineering, School of Engineering and Computer Science, University of Exeter, Exeter, United Kingdom

ABSTRACT: The geosynthetic reinforced soil, as a new reinforcement technique, has come to play a rapidly increasing role in a variety of civil and geotechnical engineering applications. In conventional reinforced soils, the reinforcements are often laid horizontally in the soil. A new concept of soil reinforced with orthogonally horizontal-vertical (H-V) geosynthetics was proposed. In the proposed H-V reinforced soil, besides conventional horizontal reinforcements, some vertical reinforcing elements are also placed upon the horizontal ones. The remarkable function is that the vertical elements can not only restrict the lateral deformation of soil, but also form strengthened zones and provide passive resistances to soil enclosed within the H-V reinforcing elements. Moreover, it can change the stress distribution and deformation of reinforced soil effectively, that will increase the strength and stability of soil. The interface behaviour would be significant to reinforcing mechanism, bearing capability and stability of the soil retaining structure reinforced with orthogonal H-V inclusions. In this paper, a series of pullout tests of orthogonal H-V geosynthetics were carried out to study the interface behaviour between sand and orthogonal H-V inclusions in terms of load-displacement relationship and pullout resistances. Comparison was made between load-displacement relationship and pullout resistances of the soil reinforced with horizontal inclusions and with orthogonal H-V ones. The influences of the height, horizontal space of vertical reinforcing elements, and kind of reinforcement material on the interface behaviour between sand and orthogonal H-V inclusions were discussed. From the test results, the coefficient of apparent pullout friction was evaluated. The interaction mechanism between sand and orthogonal H-V inclusions was analyzed and a new theoretical model was proposed to determine the pullout resistance. The comparison between theoretical values and experimental results was in good agreement.

1 INTRODUCTION

The behavior of interface between soil and reinforcements are the main influential factors in the safety and stability of reinforced structure. Due to its importance, many investigations have been carried out to study the pullout mechanism experimentally and theoretically. Jewell et al. (1984) & Rowe et al. (1985) investigated the pullout mechanism. Irsyam & Hryciw (1991) analyzed the friction and passive resistance in soil reinforced by plane ribbed inclusions. Raju & Fannin (1997) studied pull-out resistance of geogrids under monotonic and cyclic load. Racana et al. (2003) studied the pull-out response of corrugated geotextile strips. Hong et al. (2003) analyzed the pullout resistance of single and double nails. The concept of three-dimension inclusions was studied (Zhang et al. 2006), and a series of triaxial tests were carried out to investigate the behaviour and strength of the soil reinforced with three-dimension inclusions (Zhang & Min 2006). The contributions related to new reinforcing styles have played an active role in development of reinforced soil technology.

A new concept of soil reinforced with H-V geosynthetics was proposed. In H-V reinforced soil, besides conventional horizontal elements, some vertical reinforcements were also placed upon the horizontal ones. In this paper, the interaction mechanism between sand and H-V inclusions was analyzed and a new theoretical model was proposed to determine the pullout resistance. A series of pullout tests of H-V geosynthetics were carried out to prove the model.

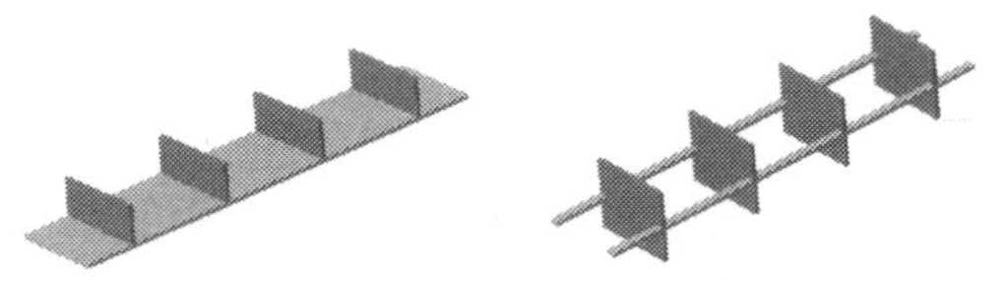

Figure 1. The typical H-V reinforcing elements: (a) horizontal-vertical reinforcing elements with the different width; (b) horizontal-vertical reinforcing elements with the different width.

2 TYPE OF SOIL REINFORCED WITH DENTI-STRIP REINFORCEMENTS

In 3D reinforcements, some kinds of reinforcing structure schemes have been established. The H-V reinforcement is one specific example of 3D reinforcements. A typical H-V reinforcing element with the same width is shown in Fig. 1(a), while H-V reinforcement with the different width is shown in Fig. 1(b). For the former, the horizontal reinforcements provide a friction force to the soil, and the vertical inclusions also provide a resistance force, but, for the latter the soil restricted mainly by the vertical reinforcements.

3 PULLOUT RESISTANCE MODEL FOR THE H-V REINFORCEMENT

3.1 *Mechanism analysis*

In conventional reinforced soil, the reinforcements are laid horizontally. The soil is restricted only by the frictional stress between soil and the reinforcement. In H-V reinforced soil, the vertical inclusions block the soil to a whole system. Besides the τ_{h1} and τ_{h2}, the vertical reinforcements block a part of soil, and provide stress $(\sigma_P - \sigma_a)$ to restrict the soil. The top of the vertical reinforcements also provide a frictional stress τ_v to the soil (see Fig. 2).

3.2 *Analysis of pullout resistance model*

A conventional reinforcement (such as strip inclusions) of width B and length L is embedded in soil. A pullout force T is applied at the end of the strip. According to Mohr-Coulomb theory, at limiting equilibrium, the ultimate pullout resistance can be calculated as follows:

$$T_0 = 2A_0(c_0 + \sigma_0 f) \tag{1}$$

where, T_0 is pullout resistance (kN), A_0 is contact area between soil and reinforcement (m^2), c_0 is cohesion, σ_0 is normal stress (kPa), f is the coefficient of friction and $f = \tan\delta$, where δ is friction angle between soil and inclusion (°).

According to the above analysis, the pullout resistance is equal to the sum of passive resistance component of vertical reinforcements, frictional components of horizontal and vertical elements. So the following relationship can be given:

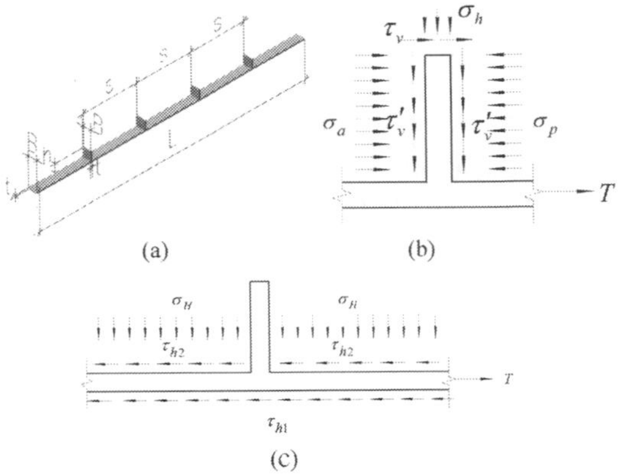

Figure 2. Mechanism analysis of H-V reinforced soil.

$$T = T_h + T_v + E_v \tag{2}$$

where T is ultimate pullout resistance; T_h is ultimate frictional resistance of horizontal reinforcements; T_v is ultimate frictional resistance of vertical reinforcements; E_v is passive resistance of vertical reinforcements.

(1) The friction resistance of horizontal reinforcements

If f_1^* is the coefficient of interface friction determined by horizontal reinforcements and sand, A_{h1} is the top contact area of horizontal reinforcements and $A_{h1} = B(L - nt)$, A_{h2} is the bottom contact area of horizontal reinforcements and $A_{h2} = BL$, σ_H is the normal stress of the interface of horizontal inclusions; γ is unit weight of sand (kN/m^3); H is the distance from the top of sand to the top of horizontal reinforcements (m). Then,

$$T_h = A_h \sigma_H f_1^* = (A_{h1} + A_{h2})\sigma_H f_1^* \tag{3}$$

(2) The friction resistance of single vertical element

It can be assumed that the two sides of vertical reinforcements arrive at active limiting equilibrium and passive limiting equilibrium at the same time. In comparison with the thickness of sand laid on the horizontal reinforcements, the height of vertical reinforcements is small. The following relations are given.

$$E_v = E_p - E_a \tag{4}$$

$$E_p = \left(\sigma_H K_p + 2c\sqrt{K_p}\right) A_1 \tag{5}$$

$$E_a = \left(\sigma_H K_a - 2c\sqrt{K_a}\right) A_1 \tag{6}$$

$$E_v = \left(\sigma_H\left(K_p - K_a\right) + 2c\left(\sqrt{K_p} + \sqrt{K_a}\right)\right)A_1 \quad (7)$$

where, A_1 is the profile area of vertical reinforcements and $A_1 = Bh$.

The friction resistance of the vertical reinforcements developed by top contact can be expressed as

$$T_v = A_v \sigma_h f_2^* \quad (8)$$

where, A_v is the top area of vertical reinforcements and $A_v = Bt$; f_2^* is the coefficient of interface friction determined by vertical reinforcements and sand.

Finally, integrating equations (3), (7), (8) and (1), gives the theoretical pullout model, *i.e.*

$$T = (2BL - nBt)\sigma_H f_1^* + nBt\sigma_h f_2^*$$

$$+nBh\left(\sigma_H\left(K_p - K_a\right) + 2c\left(\sqrt{K_p} + \sqrt{K_a}\right)\right) \quad (9)$$

If the distribution of earth pressure is assumed as the area of rectangle, i.e.

$$\sigma_H = q + \gamma H \quad (10)$$

$$\sigma_h = q + \gamma(H - h) \quad (11)$$

where, q is surcharge (kN/m^2).

Similarly, if the distribution of earth pressure is assumed as the area of trapeziform, i.e.

$$\sigma_H = q + \frac{1}{2}\gamma H \quad (12)$$

$$\sigma_h = q + \frac{1}{2}\gamma(H - h) \quad (13)$$

4 EXPERIMENTAL RESULTS AND COMPARISON

Twenty-four series of pullout tests (6 horizontal reinforcements and 18 H-V reinforcements) were performed to investigate the effects of test parameters on the behavior of sand reinforced with H-V inclusions. Uniform, clean, beach sand was used. The physical properties of the sand are presented in Table 1. The reinforcements used in the tests were geosynthetics (e.g., plexiglass) with a thickness of 3 mm shown in Fig. 3(a). The configurations of vertical elements included 5, 10, 15 mm. The thickness and width of the vertical inclusion were 3 and 15 mm.The width and length of the horizontal elements were 15 and 550 mm. The distance from the top of sand to the top of horizontal reinforcements was 150 mm. The dimension of pullout box was 650 mm (length) × 800 mm (width) × 1100 mm (height), as shown in Fig. 3(b), various reinforced inclusions were installed in the central location.

Table 1. Physical properties of sand.

Unit weight γ (kN/m^3)	Moisture content w (%)	Specific gravity Gs	Void ratio e
15.99	0.15	2.643	0.5855

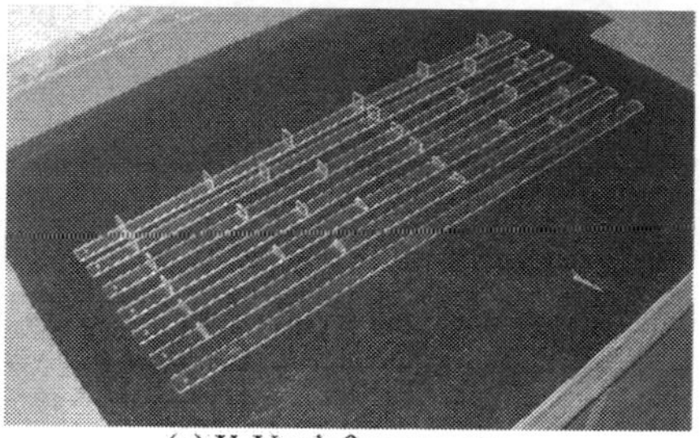

(a) H-V reinforcements

(b) pullout box

Figure 3. Layout of pullout test.

A layout of pullout test was used for testing specimens of sand reinforced with H-V reinforcements. The data collected in these tests include displacement of reinforcement and pullout force.

The aim of these tests was to verify the interface behavior theory of reinforced sand with different configuration of H-V reinforcements. The H-V reinforcements used in this study were composed of vertical reinforcing elements with different height and space. The typical pullout load-displacement curves of H-V reinforced sand are presented in Fig. 4.

Figure 4 indicated that the reinforced sand with H-V elements increases the pullout resistance considerably, compared with horizontally reinforced soil. Compared with sand reinforced with shorter vertical inclusions, the sand reinforced with higher vertical inclusions provides greater ultimate pullout resistance.

The results calculated from equation (9) were compared with the force corresponding to mutational displacement of the H-V reinforcement during tests, as shown in Table 2 and Fig. 5. It can be found that theoretical values are in good agreement with the experimental ones. The percentage error was mostly smaller than 10%.

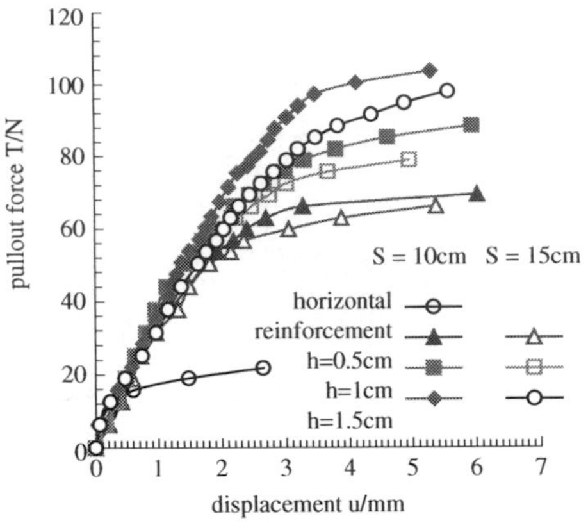

Figure 4. The curves of pullout force versus displacement under different height of vertical elements. Note: S is the spacing of the adjacent vertical elements.

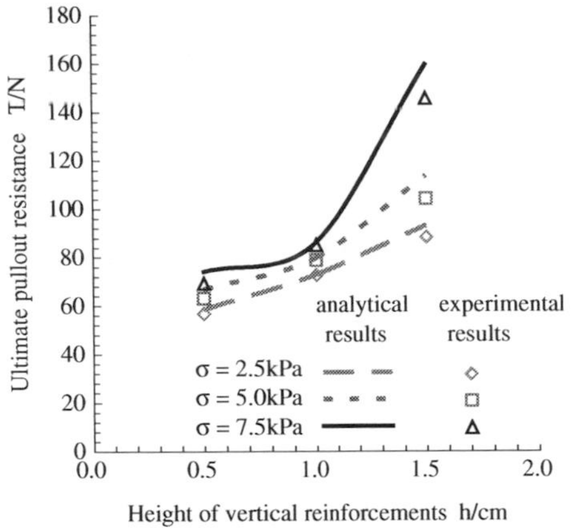

Figure 5. Comparison of the experimental results and analytical ones.

5 CONCLUSIONS

In this paper, a new concept of soil reinforced with H-V reinforcements was proposed, to change conventional reinforcing types. The interaction mechanism between sand and H-V inclusions was analyzed and a new theoretical model was proposed to determine the pullout resistance. In order to study the interface behavior of the reinforced sand under different configurations of H-V reinforcements, a series of pullout tests on dry sand reinforced with H-V reinforcements were carried out. The following conclusions can be drawn from the results:

(1) The ultimate pullout resistance of sand reinforced with H-V reinforcements increases with the increment of height of the vertical reinforcement.

Table 2. Comparison of experimental results and analytical ones.

σ_H (kPa)	S (cm)	h (cm)	T proposed model (N)	T test results (N)	*Error* (%)
2.5	10	0.5	66.86	63.12	4.16
		1.0	84.68	79.02	6.68
		1.5	105.6	97.38	7.78
	15	0.5	58.57	56.76	3.09
		1.0	72.93	72.66	0.37
		1.5	93.21	88.56	4.98
5.0	10	0.5	71.48	66.3	7.25
		1.0	95.93	85.26	11.1
		1.5	123.7	107.52	13.1
	15	0.5	66.81	63.12	5.52
		1.0	79.86	78.9	1.20
		1.5	113.8	104.34	8.31
7.5	10	0.5	87.42	78.90	9.75
		1.0	114.2	97.86	14.3
		1.5	201.6	177.36	12.0
	15	0.5	74.23	69.30	6.64
		1.0	86.38	85.14	1.44
		1.5	159.6	145.56	8.79

*σ_H is normal stressing on interface of horizontal inclusions.

(2) The comparison between theoretical values and experimental results was in good match.

REFERENCES

Hong, Y.S., Wu, C. & Yang, S.H. 2003. Pullout resistance of single and double nails in a model sand box. *Canada Geotechnique Journal* 40: 1039–1047.

Irsyam, M. & Hryciw, R.D. 1991. Friction and passive resistance in soil reinforced by plane ribbed inclusions. *Geotechnique* 41 (4): 485–498.

Jewell, R.A., Milligan, G.W.E. & Sarsb,Y.W. 1984. Interaction between soil and geogrids. *On Polymer Grid Reinforcement in civil engineering*. London: Tomas Telford.

Racana, N., Grediac, M. & Gourves, R. 2003. Pull-out response of corrugated geotextile strips. *Geotextile and Geomembrances* 21: 265–288.

Raju, D.M. & Fannin, R.J. 1997. Monotonic and cyclic pillout resistance of geogrids.*Geotechnique* 47(2): 331–337.

Rowe, R.K., Ho, S.K. & Fisher, D.G. 1985. Determination of soil geotextiles interface strength Properties. *2nd Canada Symposia On Geotexiles*, 25–34.

Zhang, M.X., Javadi, A.A. & Min, X. 2006. Triaxial tests of sand reinforced with 3D inclusions. *Geotextiles and Geomembranes* 24(8): 201–209.

Zhang, M.X. & Min, X. 2006. Behavior of sand reinforced with one-layer 3D reinforcement by triaxial tests. *Chinese Journal of Geotechnical Engineering* 28 (8): 931–936.

New Horizons in Earth Reinforcement – Otani, Miyata & Mukunoki (eds)
© 2008 Taylor & Francis Group, London, ISBN 978-0-415-45775-0

Pullout response study for cellular reinforcement

M.S. Khedkar & J.N. Mandal
Department of Civil Engineering, Indian Institute of Technology Bombay, Powai, Mumbai, India.

ABSTRACT: Various forms of reinforcing elements have been used for construction of reinforced soil retaining wall i.e., sheets, grids, meshes, strips, bars, rods etc. In ultimate limit state design, considering the internal stability analysis, the reinforced soil wall may fail in tension/rupture or pullout type of failure. Pullout tests results are commonly used to predict actual field pullout performance of reinforcements. Various authors have studied different kinds of shapes and sizes of reinforcements for pullout test, but much attention has not been given to conduct pullout test with cellular reinforcement like geocell. Though Soil reinforcement using of cellular reinforcement (geocell) has been utilized successfully in many other areas of geotechnical engineering, there is still need to study the probable use of cellular reinforcement in reinforced soil retaining wall. In the present paper, cellular type geometry of reinforcement is proposed for reinforced soil applications. Pullout analysis of such reinforcement is performed. Also, the pullout test conditions are simulated in finite element method, with the help of readymade software, Plaxis-V8 and the outputs are visualized. The results are found supportive to the assumptions made in the analysis.

1 INTRODUCTION

In conventional reinforcement soil structures two dimensional reinforcing elements are seen till date. In this paper, concept of cellular type geometry of reinforcement is introduced.

1.1 *Reinforcement geometry*

Internal stability of reinforced soil structures rely very much upon reinforcing elements. Three types of reinforcement geometry can be considered according to FHWA (2001), i.e., (1) linear unidirectional, (2) composite unidirectional and (3) planar bidirectional. Yang and Wang (1999) proposed a new reinforced soil structure composed of horizontal reinforced concrete grids. Xie (2003) presented a new type of reinforcement, reinforcing ring, whose mechanical function is to turn lateral earth pressure to stress within the reinforcing ring. Zhang et al. (2006) have conducted several triaxial tests on sand reinforced with a single layer of three dimensional (3D) reinforcing elements, and demonstrated the acceptability and better performance of 3D reinforcement over horizontal two dimensional (2D) reinforcement.

1.2 *Pullout resistance of reinforcement*

Jewell et al. (1985) and Palmeira & Milligan (1989) have demonstrated that, in case of grid reinforcement bearing mechanism governs, which changes to more frictional in nature as the spacing between bearing members is reduced.

Pullout test analysis of grid reinforcement has been conducted by several researchers in literature (Sobhi and Wu, 1996; Bakeer et al., 1998; Gurung and Iwao, 1999; Perkins and Cuelho, 1999; Gurung, 2000; Sugimoto et al., 2001 and Palmeira 2004). In particular, Jewell et al. (1985) have given the range to calculate the bearing stress ratio (σ'_b/σ'_n) depending upon the angle of friction of soil and has set upper and lower limit as general shear and punching shear failure; where, σ'_b = bearing stress and σ'_n = normal stress. Matsui et al. (1996) proposed an equation to calculate (σ'_b/σ'_n) which lies between the general shear and punching shear limits given by Jewell et al. (1985).

Bergado (1987) found that bamboo grids have a higher pullout resistance than Tensar SS2 geogrids provided that each has the same plan area. The reason may be attributed to the thicker transverse member for the bamboo than the Tensar geogrid. Various researchers like, Palmeira and Milligan (1989); and Ghionna et al. (2001) have shown that values of pullout bearing resistance are largely influenced by the reinforcement geometry, extensibility and soil dilatancy.

From literature (e.g., Madhav et al., 1998 and Moraci & Gioffre, 2006) it is very clear that, pullout study is essential and has direct implications on design of reinforced soil structures. However, no work

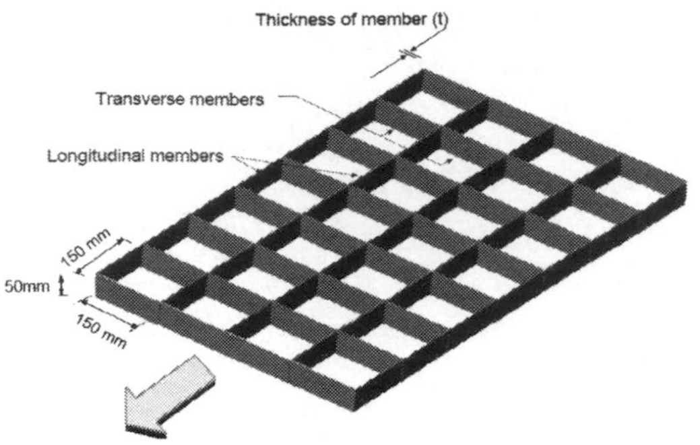

Figure 1. Typical biaxial cellular reinforcement.

is available in literature demonstrating the pullout analysis of 3D reinforcement.

The present paper, concentrates mainly on bearing type of mechanism of cellular reinforcement. pullout analysis for cellular reinforcement is performed and results are visualized with finite element software, Plaxsis-V8, by simulating the pullout test of cellular reinforcement.

2 CELLULAR REINFORCEMENT

Cellular reinforcement is a type of reinforcement in which in addition to the length and breadth as like of the conventional two dimensional 2D reinforcement; the third dimension in the form of depth of reinforcement is added. Materials like steel or geosynthetics may be used to manufacture cellular reinforcements. Such reinforcement is proposed as an alternative to conventional horizontally placed 2D reinforcement in soil reinforcement techniques like reinforced soil walls.

Figure 1 shows the longitudinal and transverse reinforcing elements, connected at right angle to each other forming a three dimensional, honeycombed, cellular like structure, called as 'cellular reinforcement'. The beneficial effects of the reinforcement pullout are derived from the passive resistance provided by transverse elements, along with the frictional resistance of cellular structure. A better reinforcing element thus formed is expected to perform well in tension behavior as well as pull-out behavior, than the conventional two dimensional reinforcements.

3 PULLOUT RESISTANCE OF CELLULAR REINFORCEMENT

Figure 2 shows points of tangency for grid and cellular reinforcement. In case of grid reinforcement as shown in Figure 2(a); if height of reinforcement (H) is increased, for a constant spacing (S) then point of tangency appears closer and the reinforcement practically would become a sheet as shown in Figure 2(b), decreasing the bearing resistance to a negligible value. But, in case of cellular reinforcement as shown in Figure 2(c), if height of reinforcement (H) is increased, for a constant spacing (S) then point of tangency does not seem to be closer and the reinforcement practically would not be seen as sheet, as shown in Figure 2(d), increasing the bearing resistance. However, in this case, the bearing resistance may influence due to interference of bearing members in bearing mechanism.

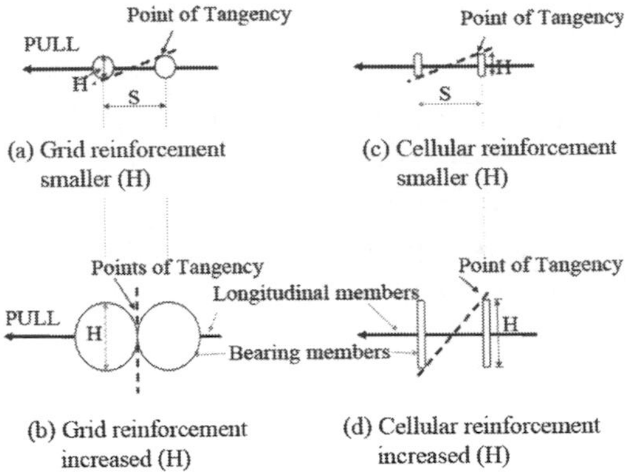

Figure 2. Points of tangency for grid and cellular reinforcement.

3.1 *Pullout analysis of cellular reinforcement*

The analysis of cellular reinforcement assumes the bearing members of reinforcement as a strip footing in deep soil. Pullout analysis of cellular reinforcement can be performed on some what similar lines as like bond for geogrid.

In case of cellular reinforcement, assuming the bearing member as a strip footing (Fig. 3a, b), the zone of influence is assumed as 3 times the height of bearing members. Therefore, the interference between bearing members in bearing resistance development can be taken in account after the S/H ratio decrease below 3.

In analysis of bond for cellular reinforcement (i) It is assumed that the two components of bond i.e., frictional resistance and bearing resistance are independent and can be calculated separately and added, (ii) Frictional resistance can be calculated from the top and the bottom part reinforcement area which is solid, and can be obtained from pullout test conditions as,

$$(Pp)_f = 2\,\alpha_S\, L_r\, \sigma'_n\, \tan\delta \tag{1}$$

and (iii) Bearing resistance can be calculated as given by Jewell et al. (1985).

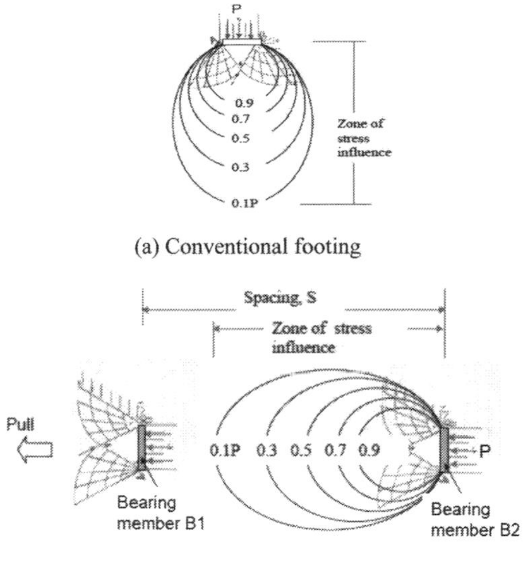

(a) Conventional footing

(b) Cellular reinforcement (plan view).

Figure 3. Zone of interference between bearing members.

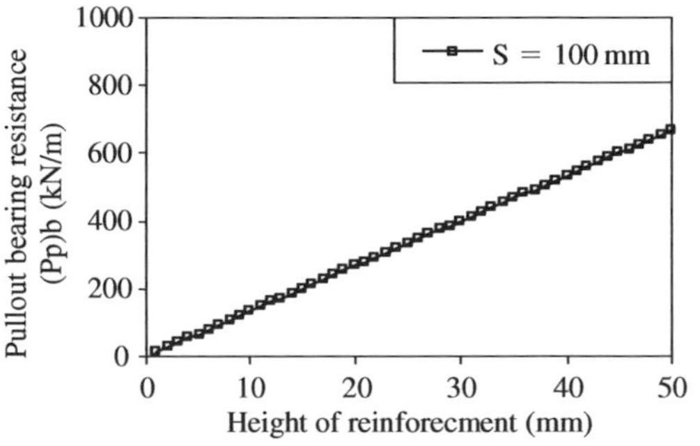

Figure 4. Relation between bearing resistance and height of reinforcement.

$$(Pp)_b = \left(\frac{L_r}{S}\right) \alpha_b \; H \; \sigma_b' \tag{2}$$

Here, σ_b' can be calculated using general shear failure mechanism,

$$\frac{\sigma_b'}{\sigma_n'} = e^{(\pi \tan\phi')} \tan^2\left(45 + \frac{\phi'}{2}\right) \tag{3}$$

where, $(Pp)_f$ is the frictional contribution in pullout of reinforcement; α_s = fraction of reinforcement plan area which is solid; L_r = reinforcement length; σ_n' = normal stress; δ = skin friction angle between soil and reinforcement; $(Pp)_b$ is the bearing contribution in pullout of reinforcement, S = spacing between bearing members; α_b = fraction of total area available for bearing; H = height of reinforcement; σ_b' = bearing stress and Φ = angle of soil friction. For a particular set of data with a unit width of reinforcement ($\alpha_b = 1$, $\sigma_n' = 100$ kN/m, Lr/S = 4 and S = 100 mm), the relation between pullout bearing resistance $(Pp)_b$ and reinforcement height (H) is determined with the help of computer program with Matlab-7, and plotted as shown Figure 4. Here, as H is increasing, $(Pp)_b$ is also increasing.

Table 1. Data set taken for FEM analysis.

Reinforcement	Spacing (S)	100 mm
	Height (H)	3 mm to 50 mm
	Length (Lr)	400 mm
Plate element	EA	8400 kN/m
	Depth (d)	2.1 mm
Soil	C	10 kPa
	Φ	35°
	γ	18 kN/m^3
Interface	R	0.100
Normal load	σ_n	100 kN/m
Pullout load	Pp	1 to 100 kN/m

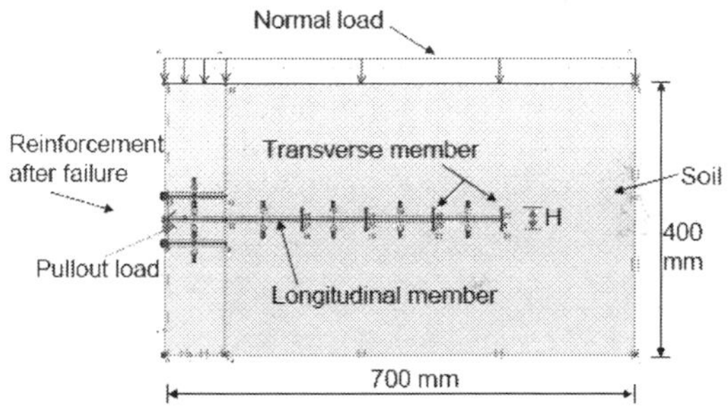

Figure 5. Geometry showing FEM simulation of pullout test of cellular reinforcement.

3.2 *Finite element analysis of cellular reinforcement*

The assumptions made in the theoretical analysis are checked by visualizing outputs from finite element method (FEM) with software, Plaxsis-V8. A numerical set of data considering a unit width of reinforcement, as shown in Table 1 is assumed for finite simulation purpose. Figure 5 shows the geometry of Pullout test simulation.

4 RESULTS AND DISCUSSION

4.1 *Load verses displacement for cellular reinforcement*

Pullout failure is simulated in FEM by increasing pullout load and the displacement is observed for different

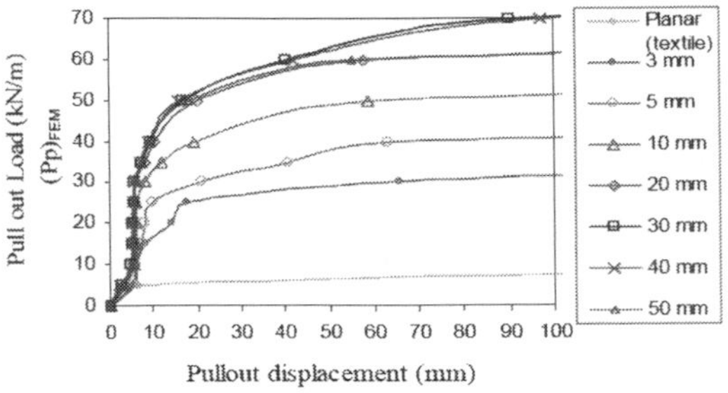

Figure 6. Relationship between pullout load and displacement for cellular reinforcement.

Table 2. Pullout load for various height of cellular reinforcement at a displacement of 20 mm.

Height of cellular reinforcement (mm)	Pullout load (kN/m)
3	25.5
5	29.5
10	40.0
20	49.5
30	52.5
40	52.5
50	49.0

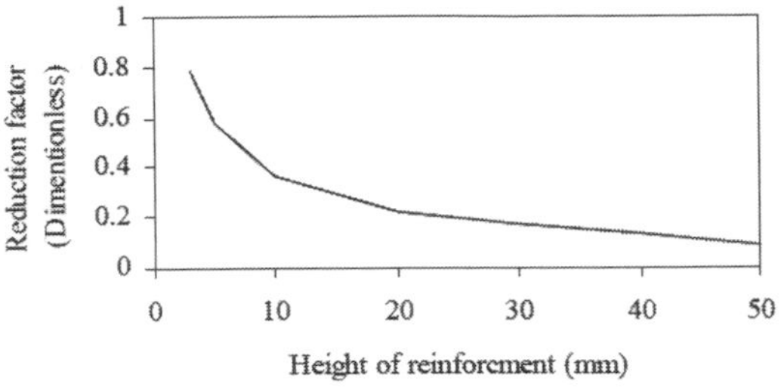

Figure 7. Relation between reduction factor and height of cellular reinforcement, for normal load of 100 kN/m.

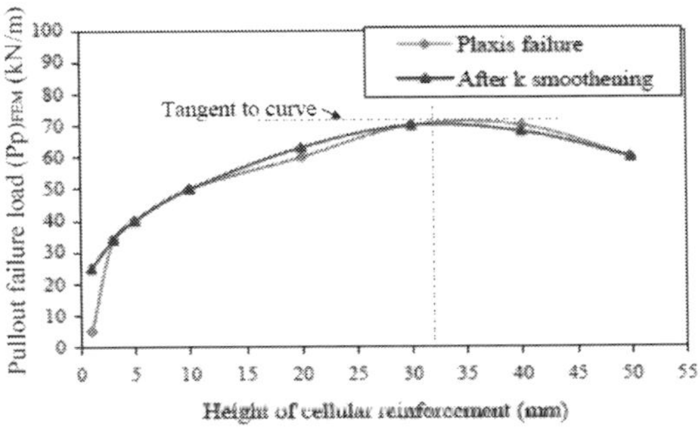

Figure 8. Relationship between pullout failure load and height of cellular reinforcement.

heights of reinforcement. Figure 6 shows pullout load and displacement curve for various heights of cellular reinforcement (3 mm–50 mm, including 2D planar textile); for a spacing of 100 mm and normal load equal to 100 kN/m. Here, for a displacement of 20 mm, pullout load is observed increasing with increasing in height of cellular reinforcement up to a height of 40 mm but for 50 mm height of cellular reinforcement pullout load is observed decreased, as shown in Table 2. However, the rate of increase in pullout resistance is observed decreasing after the pullout displacement of 20 mm.

The bearing resistance obtained from FEM analysis for pullout test is compared with the bearing resistance obtained in Figure 4. It is found that in case of FEM analysis, the rate of increase in pullout resistance is reduced than that of Figure 4. Hence a reduction factor k is empirically adopted. Reduction factor k can be defined as,

$$k = \left(\frac{(Pp)_b}{(Pp)_{FEM}} \right) \tag{4}$$

where, pullout bearing resistance, $(Pp)_b$ can be obtained from Eq^n 2 and pullout bearing resistance from FEM analysis, $(Pp)_{FEM}$ can be obtained from Plaxis V-8 simulation. For each height of cellular reinforcement reduction factor k is obtained for a normal load of 100 kN/m and is plotted as shown in Figure 7. The reduction factor is observed decreasing with increase in height of cellular reinforcement.

4.2 *Pullout failure load verses height of cellular reinforcement.*

Figure 8 shows relationship between pullout failure load and height of cellular reinforcement. Ultimate pullout resistance is observed increasing with increase in height of cellular reinforcement up to a height of 32 mm i.e., (S/H) ratio of 3.1 (approximately). Further increase in reinforcement height corresponds to decrease in pullout resistance. This shows that there is interference between bearing members after (S/H) ratio decreases below 3.1 (approximately) which supports the assumptions made in analysis.

4.3 *Pullout displacement verses height of cellular reinforcement.*

Figure 9 shows the relation between pullout displacement and height of cellular reinforcement, taken from Figure 6; measured at a particular pullout load of 50 kN/m. It is observed that the pullout displacement is reducing with increase in height of cellular reinforcement. It is observed that the pullout displacement for 30 mm height of cellular reinforcement is equal to

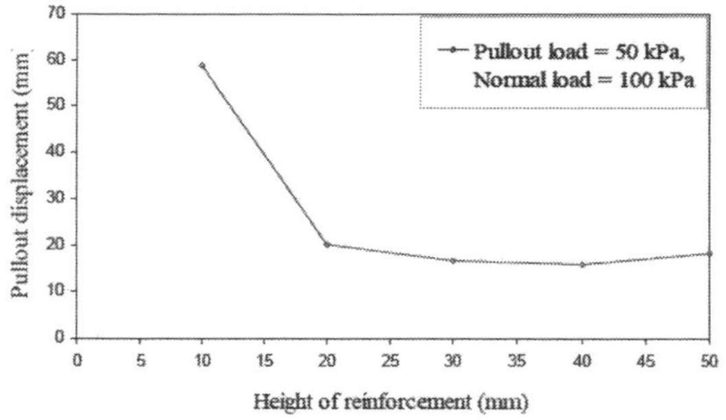

Figure 9. Relationship between pullout displacement and height of cellular reinforcement for the pullout load of 50 kN/m.

Table 3. Displacement at pullout load of 50 kN/m for various height of cellular reinforcement.

Height of cellular reinforcement (mm)	Displacement (mm)
10	58.91
20	20.24
30	16.62
40	16.04
50	18.2

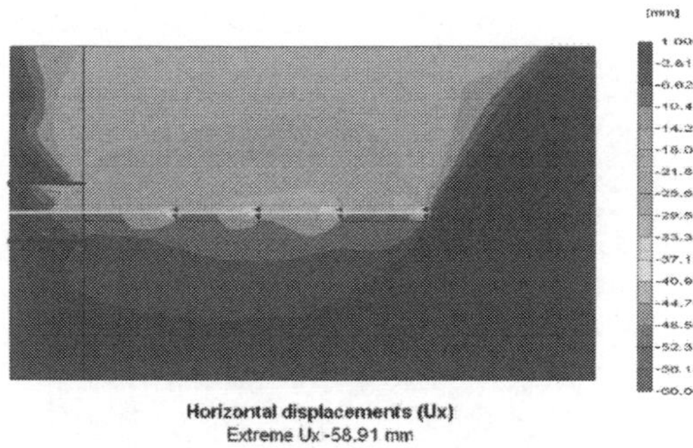

(a) Height of cellular reinforcement = 10 mm

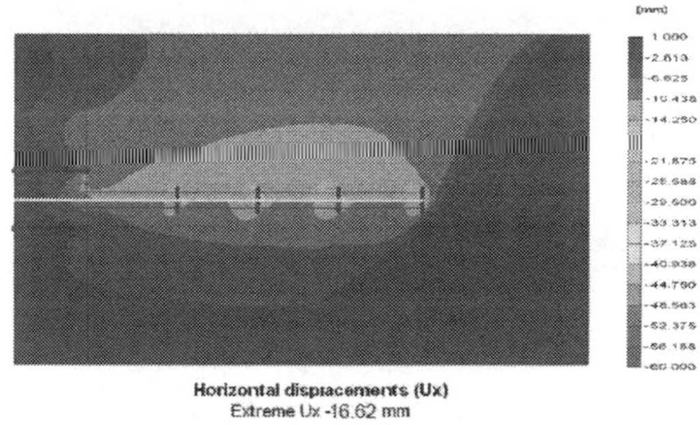

(b) Height of reinforcement = 30 mm

Figure 10. FEM visualization of horizontal displacement for different heights of cellular reinforcement.

16.62 mm as shown in Table 3. The reduction in displacement is negligible after 30 mm height of cellular reinforcement i.e., below the (S/H) ratio of 3, which increases for a reinforcement height of 50 mm.

4.4 *FEM visualization of horizontal displacement for different heights of cellular reinforcement*

Figure 10 shows FEM visualization of horizontal displacement, from Plaxis V-8, for different heights of cellular reinforcement with the pullout load of 50 kN/m. It is observed that in case of 10 mm height of cellular reinforcement, the extreme horizontal displacement, in front of bearing member is 58.91 mm, as shown in Figure 10(a). While, in case of 30 mm height of cellular reinforcement, the extreme horizontal displacement, in front of bearing member is observed reduced to a value of 16.62 mm, as shown in Figure 10(b). This shows that there is significant amount of reduction in horizontal displacement with increase in height of cellular reinforcement.

4.5 *FEM visualization of horizontal stresses for different heights of cellular reinforcement*

Figure 11 shows FEM visualization of horizontal stresses, from Plaxis V-8, for different heights of cellular reinforcement with the pullout load of 50 kN/m. It is observed that for 10 mm height of cellular reinforcement, the extreme horizontal stress is 1.47×10^{-3} kN/mm^2, as shown in Figure 11(a) and for 30 mm height of cellular reinforcement, extreme horizontal stress is 0.933×10^{-3} kN/mm^2, as shown in Figure 11(b), which shows that there is reduction in horizontal stress with increase in cellular reinforcement height.

5 CONCLUSION

Cellular type geometry of reinforcement is found to be better performance in pullout analysis than that of 2D type reinforcement. Pullout capacity is observed increasing with increase in height of cellular reinforcement up to spacing to height ratio (S/H) of 3.1. If the (S/H) ratio is less than 3.1, the pullout capacity is decreasing with increase in height of cellular reinforcement. Pullout displacement is found decreasing significantly with increase in height of cellular reinforcement, up to 30 mm height of reinforcement, after which the decrease in displacement is negligible. Finite element results for pullout load and horizontal displacement are found supportive to the assumptions

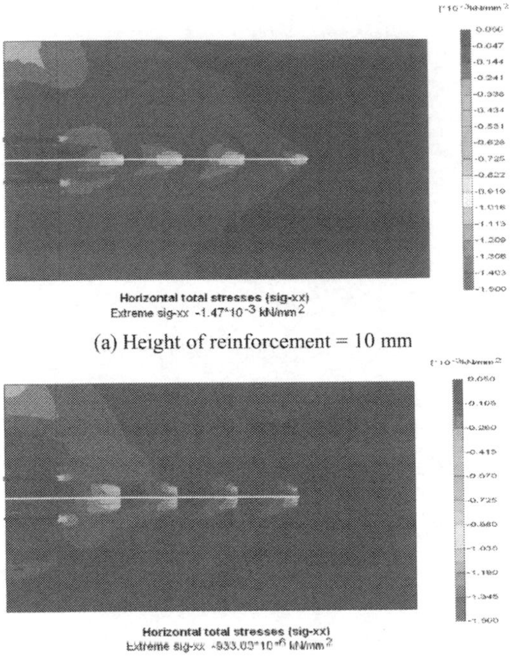

(a) Height of reinforcement = 10 mm

(b) Height of reinforcement = 30 mm

Figure 11. FEM visualization of horizontal stresses for different heights of cellular reinforcement.

made in the theoretical analysis of cellular reinforcement. Also, it is visualized in Plaxis V-8, that there is significant reduction of horizontal stress with increase height of cellular reinforcement.

REFERENCES

Bakeer, R. M., Abdel-Raheman, A. H. & Napolitano P. J. 1998. Geotextile friction mobilization during field pullout test, *Geotextiles and Geomembranes,* 16, pp. 73–85.

Bergado, D. T., Bukkanasuta, A. & Balasubramaniam, A. S. 1987. Laboratory pull-out tests using bamboo and polymer geogrids including a case study, *Geotextiles and Geomembranes,* 5, pp. 153–189.

Federal Highway Administration (FHWA), Elias, V., Christopher, B. R. and Berg, R. R. (2001). Mechanically stabilized earth walls and reinforced soil slopes design & construction guidelines. Technical Report, FHWA-NHI-00-043, 394 pp.

Ghionna, V. N., Moraci, N. & Rimoldi, P. 2001. Experimental evaluation of the factors affecting pullout test results on geogrids, *In: Pro. of the Int. Symp.: Earth Reinforcement, Fukuoka, Japan, 14–16 November, 2001, IS Kyushu 2001, "Landmarks in Earth Reinforcement",* vol. 1. Balkema Publisher, pp. 31–36.

Gurung, N. 2000. A theoretical model for anchored geosynthetics in pull- out tests, *Geosynthetics International journal,* 7 (3), pp. 269–284.

Gurung, N. & Iwao, Y. 1999. Comparative model study of geosynthetic pull-out response, *Geosynthetics International journal,* 6 (1), pp. 53–68.

Jewell, R. A. 1990. Reinforcement bond capacity, *Geotechnique,* 40 (3), pp. 513–518.

Jewell, R. A., Milligan, G., Sarsby, R. W., & Dubois, D. 1985. Interaction between soil and geogrids, *In: Pro., Symp. on Polymer Grid Reinforcement, Civil Engineering,* London, pp. 18–29.

Madhav, M. R., Gurung, N. & Iwao, Y. 1998. A Theoretical Model for the Pull-Out Response of Geosynthetic Reinforcement. *Geosynthetics International Journal,* 5 (4), pp. 399–424.

Matsui, T., San, K. C., Nabesahirna, Y. & Arnii, U.N. 1996. Bearing mechanism of steel reinforcement in pull-out test. *In: Pro. of the Int.Symp.: Earth Reinforcement,* Fukuoka, Kyushu, Japan. Balkema Publisher, pp. 101–105.

Moraci, N. and Gioffre, D. (2006). A simple method to evaluate the pullout resistance of extruded geogrids embedded in a compacted granular soil. *Geotextiles and Geomembranes,* 24, pp. 116–128.

Palmeira, E. M., & Milligan, G. W. E. 1989. Scale and Other Factors Affecting the Results of Pullout Tests of Grids Buried in Sand, *Geotechnique,* 39, (3), pp. 511–524.

Palmeira, E., M. 2004. Bearing force mobilization in pull-out tests on geogrids. *Geotextiles and Geomembranes,* 9, pp. 481–509.

Perkins, S.W. & Cuelho, E.V. 1999. Soil–geosynthetic interface strength and stiffness relationships from pull-out tests, *Geosynthetics International Journal,* 6, (5), pp. 321–346.

Sobhi, S. & Wu, J. T. H. 1996. An interface pullout formula for extensible sheet reinforcement, *Geosynthetics International Journal,* 3 (5) pp. 565–582.

Sugimoto, M., Alagiyawanna, A. M. N. & Kadoguchi, K. 2001. Influence of rigid and flexible face on geogrid pullout tests, *Geotextiles and Geomembranes,* 19, pp. 257–277.

Xie, W. 2003. Consideration for modifying reinforced retaining wall Nonferrous Mines, 32, (3), pp. 46–48.

Yang, G. & Wang, Y. 1999 Strength property of RC net reinforced earth retaining structure and its experimental study, *Chinese Journal of Geotechnical Engineering,* 21 (5), pp. 534–539.

Zhang, M. X., Javad, A. A.& Min, X. 2006. Triaxial tests of sand reinforced with 3D inclusions, *Geotextiles and Geomembranes,* 24, pp. 201–209.

New Horizons in Earth Reinforcement – Otani, Miyata & Mukunoki (eds)
© 2008 Taylor & Francis Group, London, ISBN 978-0-415-45775-0

Resistance of steel chain in pullout tests with and without sliding box

M. Fukuda & T. Hongo
Geo-Research Institute, Japan

A. Kitamura & Y. Mochizuki
Showa Kikai Shoji, Japan

S. Inoue & E. Fujimura
Kinki Polytechnic College, Japan

M. Kimura
Kyoto University, Japan

ABSTRACT: Notable features of a steel chain are its flexibility in surrounding a deformed soil and a high pullout resistance larger than those expected from steel bars and plates. However, little is known about its mechanism of the pullout resistance. Therefore, a new experimental apparatus is developed to support the experienced hypothesis on the pullout resistance. This new apparatus also makes it possible to measure the pullout resistance under a high axial tensile force at a low confining pressure, which is commonly observed in the back fill of a retaining wall. In this study, two test procedures are conducted separately. One measures the resistance using a simple pullout test and the other one with a sliding box. This paper shows the parametric test results examined by combining chain shape, strip, plate and round bar with representative soils, relative density and confining pressure. Also, similar behaviour in resistance displacement curve is shown by comparing both test results.

1 INTRODUCTION

Design method for steel chain as a reinforcement material developed to stabilize fill slope has not been yet established even though the chain has an effective rigidity and shape with respect to pullout resistance. Therefore, to account for the pullout resistance characteristic of the steel chain, based on experimental fact data, a chain pullout testing device was developed. Various expected test conditions in practice were examined with this apparatus using chains of different shape and length, and comparative study done with steel plates with smooth surface with small projections, and round steel bars for different soils ranging from coarse to fine.

Moreover, this testing apparatus is added to have an advanced procedure to slide the surrounding soil in a container box along the steel chain under high axial tension and low confining pressure. This operation is aimed at studying the pullout resistance characteristic observed within the back fill of a reinforced retaining wall. Herein, two types of test are defined, a standard type test and a sliding box test. The complicated behaviour in the region close to the retaining wall is impossible to reproduce by the standard test in general. In this paper, basic equations that govern the pullout resistance of the steel chain are derived from the standard test by considering the effect of internal friction angle and dilatancy. Furthermore, confining pressure dependency and its correction method on the pullout resistance is introduced. Finally, the sliding box test results are indicated to follow the governing equations obtained from the standard test.

2 DEVELOPEMENT OF PULLOUT RESISTANCE UNDER HIGH AXIAL TENSION SITUATION

Axial tensile force reacting to geosynthetics materials in the back fill of a retaining wall is not of concerned in this research but only the pullout resistance generally localized in the vicinity of sliding surface in the reinforced fill far away from the retaining wall is focused on in this research.

Development of full resistance stretched over a total length of reinforcement is easily recognized to raise its function more efficiency in ideal. However, the lower overburden pressure acting on reinforcement materials in the back fill of retaining wall and the shrinkage of the reinforcing materials lead to the neglecting of the resistance in the backfill. Therefore, to elaborate its reinforcement function, it is necessary to evaluate the sliding resistance in the back fill of the retaining wall.

3 CHAIN PULLOUT TEST APPARATUS AND ADDITIONAL DEVICES OF OUTER BOX SLIDING

A cubic container box for soils of dimension 50 cm × 50 cm × 50 cm with an inner volume of 0.125 m^3 that was about 10 times larger than the outer width of chain and the maximum diameter of the compacted soil particles used to conduct the test was designed as shown in Figure 2.

Five kinds of sensors are set up as load sensors for measuring vertical external pressure, there are also sensors for measuring the displacement of chains in the box, and earth pressure acting on chain in the soils and the side wall of the box.

The sliding box test is designed to measure pullout resistance of chain under high tensile force. Values of 12 kN and 15 kN of tensile force are chosen to act during the test since these values are considered to be a little bit lower than ultimate tensile strength of about 17 kN as recorded in Figure 3 for M624 (chain of 21 mm outer width). The testing materials were pulled out on three different confining pressures of 30, 90 and 150 kN/m^2 applied on the cover plate.

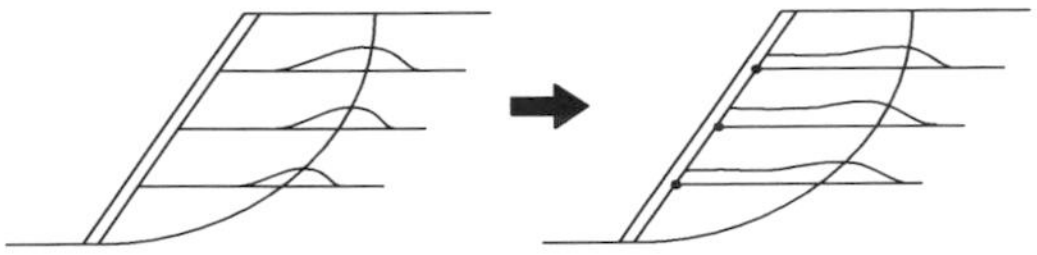

(a) General distribution (b) Efficiency distribution of strain

Figure 1. Wall confining effect and strain distribution pattern.

4 PULLOUT RESISTENCE AND CYLINDER MODEL

Equations (1) to (3) were proposed to predict the pull-out resistance of steel chain that assumes a cylindrical shape of soil block that envelop the inside space of the chain and outer surrounding soil. Figure 4 shows

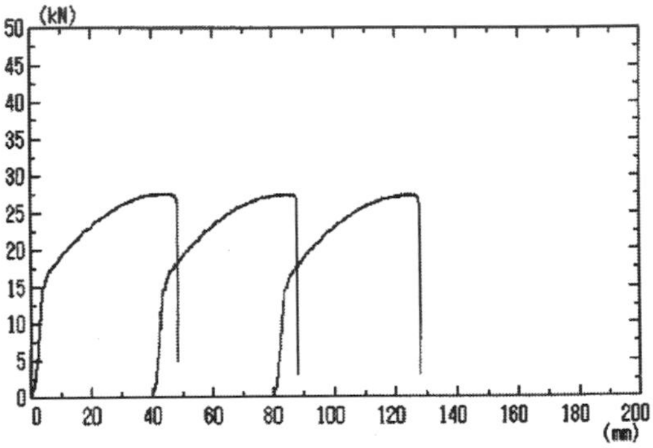

Figure 3. Tensile loading test result for chain.

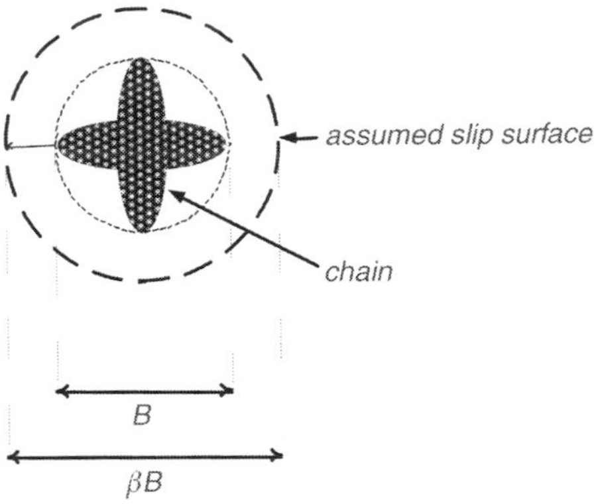

Figure 4. Cylinder model of sliding block attached to chain.

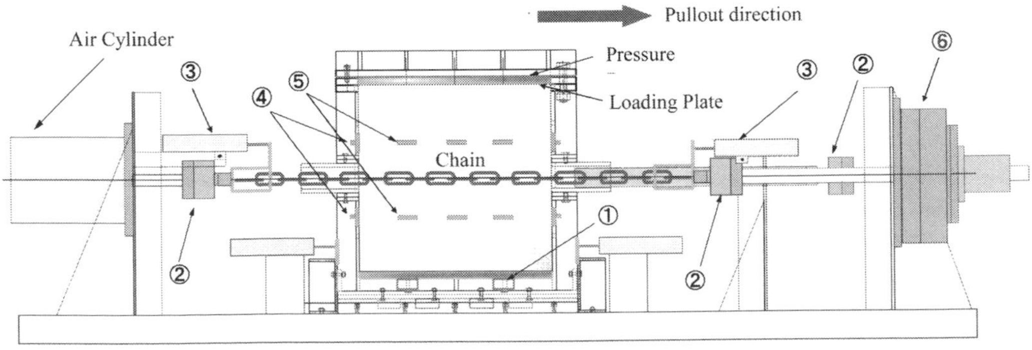

①,② Load Cell ③Displacement Indicator ④,⑤Earthpressure Cell

Figure 2. Schematic diagram of pullout testing apparatus.

the schematic diagram of sliding model used in this research.

$$F_f = \pi B \times \beta \times L \times \tan\phi' \times (1+K_0)/2 \times \sigma_v' \quad (1)$$

$$\alpha = \beta(1+K_0)/2 \quad (2)$$

$$F_f = \alpha \times \pi B \times L \times \sigma_v' \times \tan\phi' \quad (3)$$

Where F_f: pullout resistance, B: the outside width of the chain, ϕ: internal friction angle, K_0: lateral earth pressure coefficient, σ_v: effective vertical earth stress on the surface of chain, L: the chain length. Since F_f, B, L, and σ_v are directly obtained from the test results. Equation (3) is transformed to the equation (4) so as to obtain a factor α, the outer surface adjustment coefficient of the cylindrical model which is the targeted to of this research

$$\alpha = F_f/(\pi B \times L \times \sigma_v \times \tan\phi') \quad (4)$$

5 PULLOUT TEST CONDITION

5.1 *Filling material*

Figure 5 and Table 1 show physical properties of materials used for the tests. Toyoura sand compacted to 85% relative density, mixture filled with crushed stone and adjusted to diameters of 1 mm and 5 mm of 50% passing finer in weight compacted to 90% and 95% relative density and DL-clay and silica sand to 95% were used in this study.

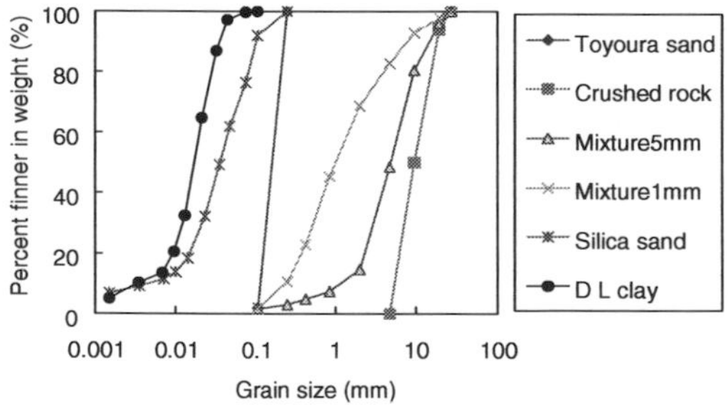

Figure 5. Grain size distribution curves of soils used for tests.

Table 1. Soil properties.

Material	Density of particle g/cm^3	Maximum dry density g/cm^3	Opitcal moisture %	Relative density %	Dry density g/cm^3	c′ kN/m^2	ϕ °
Crushed rock	2.661	1.696	5.6	95	1.61	26	39.2
Toyoura sand	2.65			85	1.58	4	37.9
Mixture 5 mm	2.656	1.927	8.3	95	1.83	13	41.4
				90	1.73	8	37.9
Mixture 1 mm	2.653	1.951	11.1	95	1.85	7	37.1
				90	1.76	5	35.9
Silica	2.634	1.365	10.3	90	1.23	4	33.8
DL clay	2.641	1.498	21.7	90	1.35	1	33.2

5.2 *Reinforcement materials used for test*

Since the purpose of the study was to investigate shape effect of chain on frictional resistance, chain of different shape and sizes were examined. Table 2 shows the dimensions of testing materials used in the study. Typical shape of the chain used in the study is as in shown in Figure 6. The outer width of the chains ranged from 1.5 cm to 3.1 cm.

A stripe steel plate of 3 mm high, steel plate of smooth surface 5 cm wide and a round steel bar of 2.2 cm in diameter used for comparative study are as shown in Photo 1.

6 BASIC PHENOMENA OF PULLOUT RESISTENCE

Figure 7 shows the pullout resistance obtained for different materials using the Toyoura sand filled in

Table 2. Chain specifications used for pull out tests.

No.	Name	Diameter D(mm)	Inner pitch p(mm)	Outer diameter b(mm)	links/ 50 cm
1	M6–Normal	6	24	21	21
2	M6–Long	6	37	21	14
3	M6–Short	6	18	21	28
4	M6–Wide	6	24	22.8	21
5	M6–Small	6	24	19.2	21
6	M6–Bar	6	24	21	21
7	M6–Knob	6	24	21	21
8	S6–24	6	24	21	21
9	L6–24	6	24	21	21
10	304–624S	6	24	21	21
11	M5–Square	4 × 5	19	15	18
12	B6–Cross	6	25.5	22	20
13	M8–38	8	38	31	14
14	L8–32	8	32	28	16

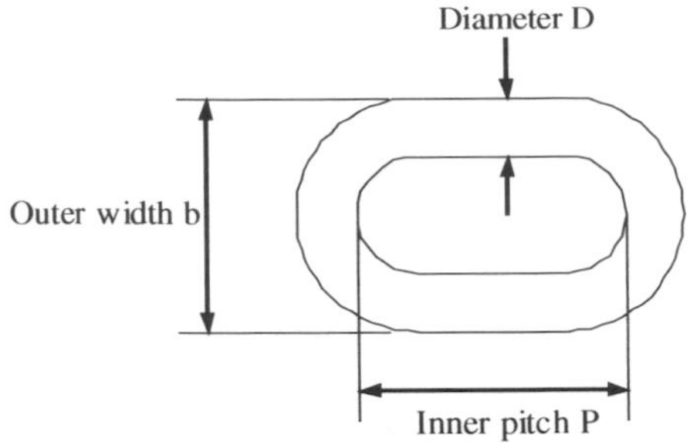

Figure 6. Shape definition of each chain.

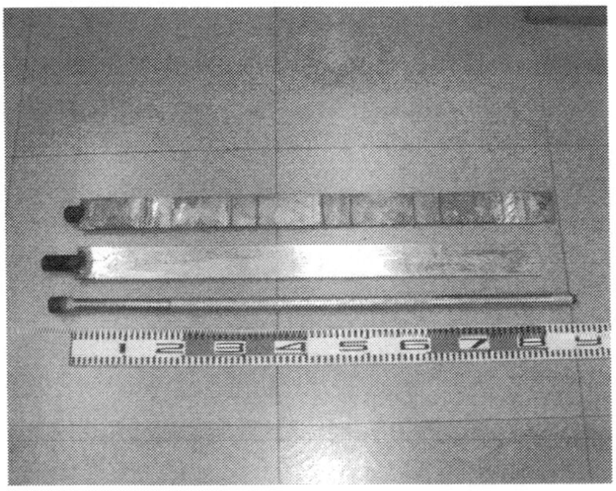

Photo 1. Sub-materials used for comparative study.

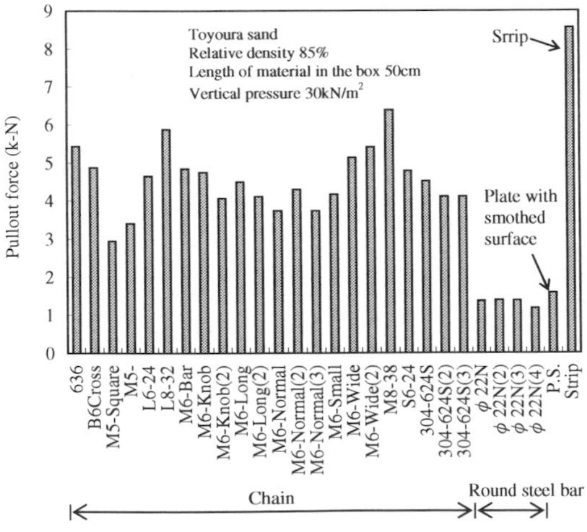

Figure 7. Pull out resistance obtained by pull out tests.

the testing box and compacted to 85% of the relative density under a vertical testing load of 30 kN/m^2.

Test results shows that the resistances of the chain ranged from 3 kN to 6 kN, on the other hand, the resistance of round steel bar was about 1 kN, while that

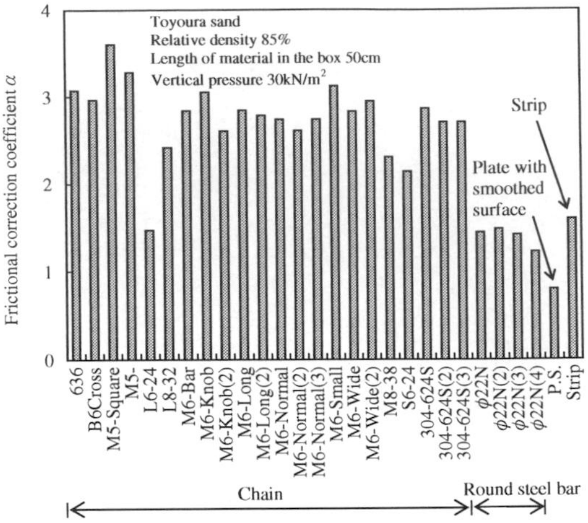

Figure 8. Variation of correcting factor with pullout force.

of steel plate with smooth surface was 1.5 kN and for the stripe with projection ranged from 8 kN to 9 kN. Although the resistances were plotted in large variety, the largest of the resistance of a particular material is taken compared with that of the strip with projection because of width effect. The resistance of plate with smooth surface is less than that of chain. This difference relating to both surface and type of plate describes the effect of shape on strength.

There can also be a different inspection for resistance, when focusing on the mechanism of resistance generation. Figure 8 indicates transformed values of the pullout force based on equations (3) and (4). The vertical earth pressure σ_v, measured in the fill is used for the calculation and not the load intensity applied on the cover plate of the box.

The vertical earth pressure measured close to the area surrounding the chain in the fill is found to be higher than the load intensity applied to the cover plate. This shows that the fill tends to swell due to dilatancy subject by pulling out the chain, however the swelling is restricted by the side wall effect.

As mentioned above, although the largest pullout resistance is obtained in the strip with projection, however, larger value of corrected friction factor is obtained in the chain. This means that the chain is more superior in generating resistance than the strip with projection. The variation of the friction correction factor for the various reinforcing materials used for the test is as shown in figure 8.

The frictional correction factor obtained for the various shapes of the chain ranged between 2.0 to 3.5 while that obtained for strip with projection was 1.6, for steel bar, the range was from 1.2 to 1.5, and for smooth steel plate it was 0.8.

Although the diameters of chains and round steel bars used for the test were the same (about 22 mm),

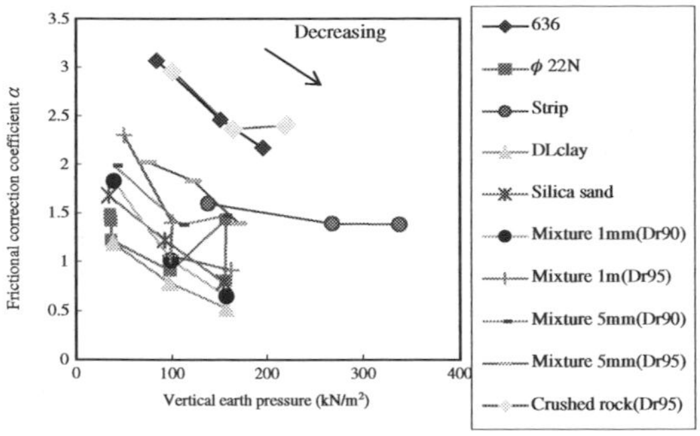

Figure 9. Decrease of resistance as confining pressure increases.

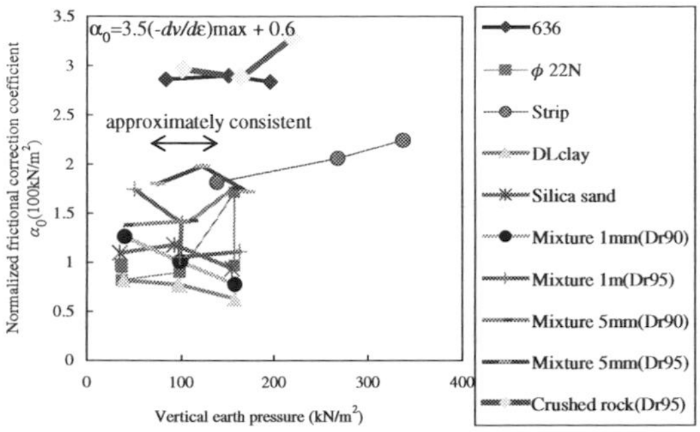

Figure 10. Normalized frictional correction factor by confining pressure.

the friction correction factors obtained for chains were larger than that of round steel bar. This is clear evidence that the reinforcement effect of the chain is more efficient.

Figure 9 shows the variation of the frictional correction factor with the measured vertical earth pressure. As the vertical earth pressure increases, the friction correcting factor tends to decrease although the surface and shape of different material are of concern. This decreasing behaviour of the frictional correcting factor can be corrected using equation (5) to give similar value almost independent of the confining pressure.

$$\alpha_0 = \alpha \times (\sigma_v / 100)^{0.4} \quad (5)$$

In this paper, the frictional correction factor adjusted at the stress of 100 kN/m^2 is called a normalized frictional correction factor α_0.

The adjustment is as shown in Figure 10 when normalization is done at the vertical earth pressure level of 100 kN/m^2 and raise a power factor 0.4 for the confining earth pressure as given in equation(5).

In this study, the normalized frictional correction factor is a parameter necessary for design of the steel

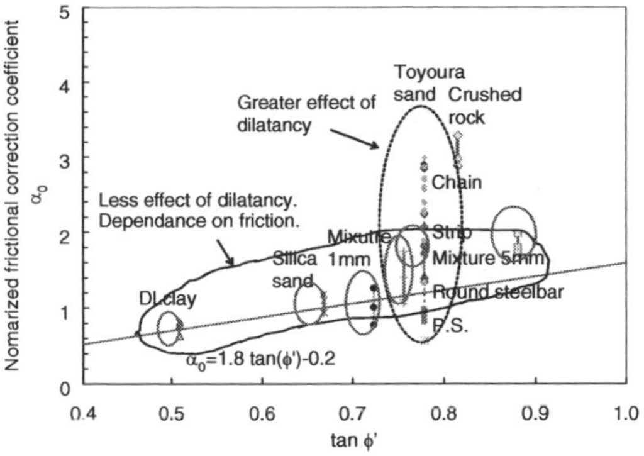

Figure 11. Design method considering dilatancy effect.

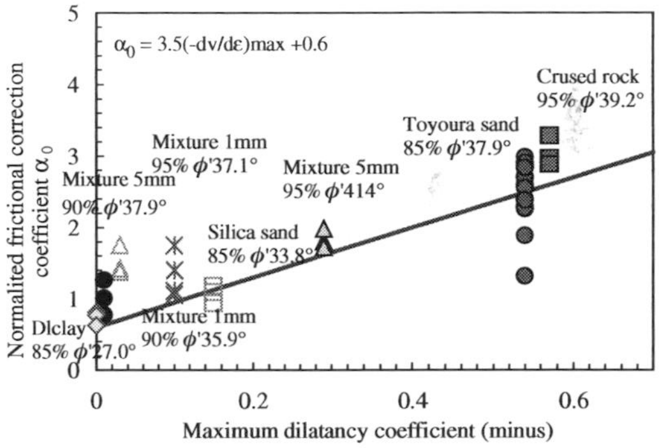

Figure 12. Dilatancy coefficient vs. correcting factor.

chain reinforcement. Its property is divided into two, one in which the effect of the dilatancy is expected and the other which is strongly affected by internal friction angle but dilatancy effect is not of concern.

Figure 11 shows relationship between the internal Friction angle and correction factor. A slope can be drawn through average data, if the vertical variation of data group is neglected at a friction coefficient of 0.8 that corresponding to Toyoura sand, crushed rocks and other soil types, while data for material type of smooth surface plate and round steel are plotted well along the estimation line.

Figure 12 shows the relationship between the dilatancy and the friction correction factor defined at the pressure 100 kN/m^2. The normalized friction correction factor with respect to chain tends to increase as an absolute value of the dilatancy coefficient increases. This relation for the increase in absolute value can be approximated by equation (7).

$$\alpha_0 = 1.8 \tan\phi' - 0.2 \quad (6)$$

$$\alpha_0 = 3.5\left(-\frac{dv}{d\varepsilon}\right) + 0.6 \quad (7)$$

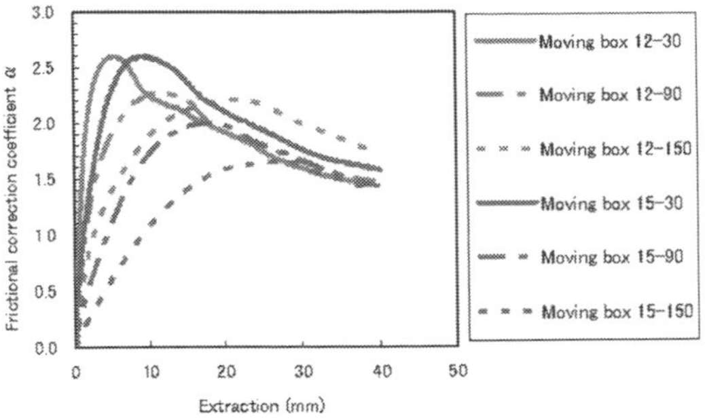

Figure 13. Curves figured by sliding box test results.

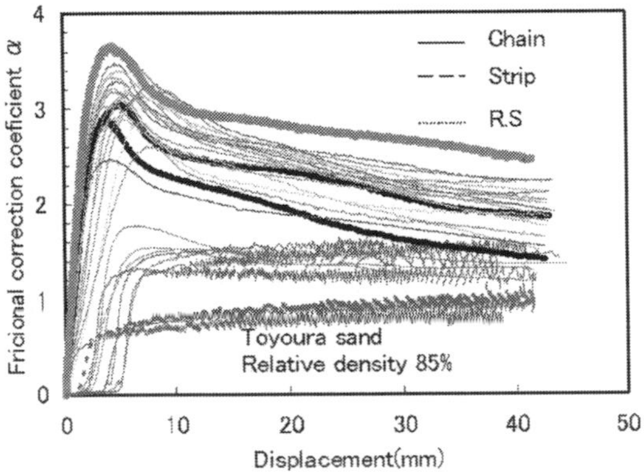

Figure 14. Curve modes subjected to standard pull out testing.

7 SLIDING BOX TEST RESULT

Figure 13 shows the relation between the frictional correction factor and the amount of extraction displacement of chain subject to the sliding box test. A set of shape of curve is divided into two groups; the first group consist of curves which have a hardening and sequent softening that well suits the behaviour of over consolidated soils, and the other group consists of curves with only hardening behaviour similar to the ones of normally consolidated soil. This phenomenon is similar to the standard test result shown in Figure 14.

Figure 13 are the result of test subjected to a vertical applied load of 30 kN/m^2 while figure 14 show a typical curve relating to standard test results. Comparing both figures, it is evident that the same curve patterns are obtained.

Figure 16 shows the relationship between normalized frictional correction factor and the ratio of confining earth pressure to the load intensity applied at the cover plate. The pressure ratio reveals a degree of dilatancy effect, because when the dilatancy effect is large, the ratio becomes larger. This figure shows the same trend for those obtained with the standard test results.

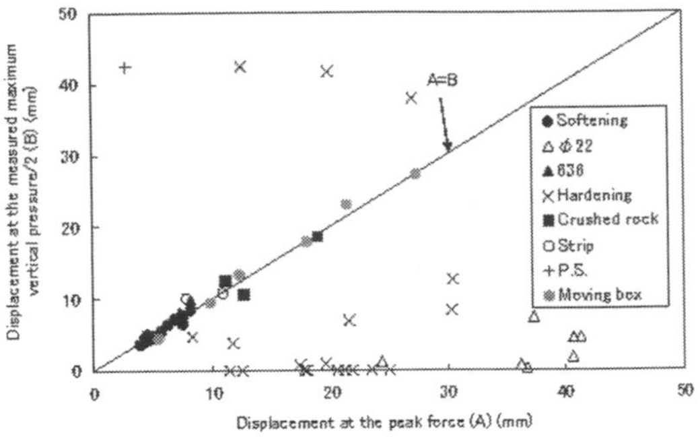

Figure 15. Maximum frictional correction coefficient and measured vertical earth pressure.

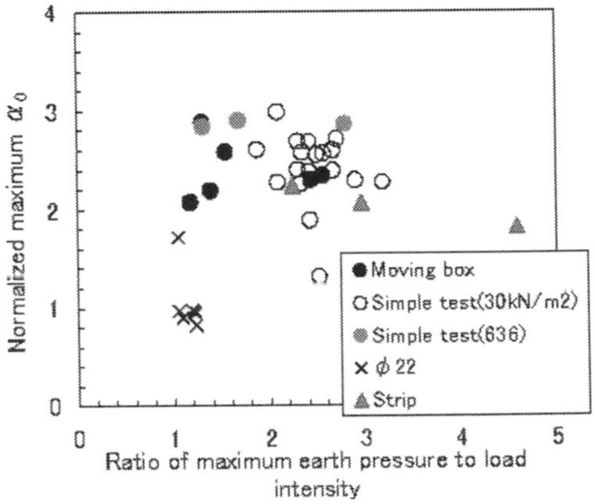

Figure 16. Frictional coefficient vs. earth pressure acting to chain.

Photo 2. Arrangement of chain and its attachments.

8 SITE CONSTRUCTION EXAMINATION

Photo 2 shows the chain arrangement work in the field for the installation of chain, anchor and the wall of steel frame set at the 50 cm spacing.

Photo 3. Completion of test embankment.

Photo 4. Affluent surrounded by green grass.

Photo 3 shows the figure of completed slope, and Photo 4 shows the growth of grass around the wall.

9 CONCLUSIONS

The sliding box test apparatus was made for trial purposes to examine the chain pullout resistance under standard condition and to compare the sliding box test properties with the standard test results. As a result, similar and compatible results were obtained from the both tests. From the result of both tests, the following points can be notes.

1) The governing operation for predicting resistance is summarized into the set of equations presented in this paper.
2) It is proved that the frictional correction factor can unify the degree of resistance among various kinds of reinforcements.
3) It was shown that the chain is a good reinforcement material that demonstrates the effect of dilatancy of the soil.
4) Similar characteristics in pullout test result generated in the standard test and sliding box test with regards to chain used are recognized.

REFERENCES

Fukuda M., Hongo T., Kitamura A., Mochizuki Y., Fujimura E., and Kimura M. (2006). "Dilatancy effect on pull out frictional force of chain", Proceedings of 7th National Symposium on Ground Improvement, The Society of Material Science, Japan, pp. 65–70 (In Japanese).

Fukuda M., Hongo T., Kitamura A., Mochizuki Y., and Kimura M. (2006). "General formuras of geosynthetic mechanics", Japan Chapter of International Geosynthetics Society, 21th Annual symposium (In Japanese).

New Horizons in Earth Reinforcement – Otani, Miyata & Mukunoki (eds)
© 2008 Taylor & Francis Group, London, ISBN 978-0-415-45775-0

Pullout load tests of the anchor plates in compacted sand used for typical backfill embankment in Thailand

J. Sunitsakul & A. Sawatparnich
Bureau of Road Research and Development, Department of Highways, Ministry of Transport, Thailand

ABSTRACT: In Thailand, soil-reinforced walls have been constructed increasingly every year upon the needs of road widening and highway reconstruction. It is necessary to characterize the relationship of pullout load and displacement for backfill materials used. Therefore, the full-scale reinforced wall tests with anchors plates were performed. Three anchor plates were installed at the depth of 0.1, 0.5, 0.9, 1.3, and 1.7 meters from the top of compacted sands. Pullout load tests of tie rod with and without anchor plates in compacted sand at various depths were carried on. Regarding to test results, the force coefficient correlates with the embedment ratio. It is shown that the force coefficient increases exponentially when the embedment ratio increases. The theoretical method proposed by Merifield and Sloan (2006) could be used for a good estimation of the pullout load of anchor plates in compacted sand.

1 INTRODUCTION

Department of Highways (DOH), Thailand, is the main agency to construct and maintain highways in Thailand. Recently, Department of highways operates over sixty thousand kilometers long of highways and over fifteen thousand highway bridges throughout the country. In hilly terrain especially in the northern part of Thailand, numbers of landslides occur along the highways. DOH spends hundred million baht of annual road maintenance budget to maintain highway backfill slopes with mechanical stabilized earth and soil-nail wall systems.

With limiting budget on constructing highway bridges, DOH adopts backfill embankment as a bridge approach. Moreover, Department of highways applies concrete facing with metal strips as a reinforcement in the area of limited right of way. Therefore, numbers of constructions of reinforced walls increases every year.

In the design and analysis of soil reinforced wall, guidelines and manuals are available in literature (e.g., FHWA/RD-82-047, 1982); however, the understanding of the relationship of pullout load and displacement for backfill material specifically in Thailand is very scare. In 1995, highway engineers at the Center of Highway Research and Development constructed the full-scale reinforced wall with steel anchor plates on compacted crushed rock to investigate the two considerations: a) the pullout load and displacement relationship and b) the performance of the reinforced wall. Bearing capacity and stability failures of the reinforced wall are not concerned on this study since the walls are constructed on the compacted crushed rock. Three anchor plate dimensions are installed at the depth of 0.1, 0.5, 0.9, 1.3, and 1.7 meters from the top in compacted sand. Pullout load tests of tie rod with and without anchor plates in compacted sand were carried out. Test results are studied and compared with available case histories and standard practices.

2 CONSTRUCTION OF TRIAL REINFORCED WALL

2.1 *Backfill material*

Backfill material used in this construction is cohesionless and classified as SP-SM following the unified soil classification system (USCS). In addition, gradation of this material is shown in Figure 1. Maximum dry

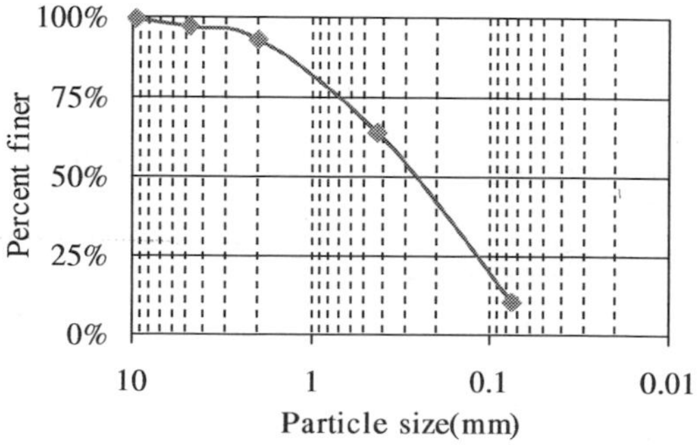

Figure 1. Gradation plot of the backfill used in the study (After Leerakomson and Charoenpon, 1996).

Table 1. Square plate dimensions (After Leerakomson and Charoenpon, 1996).

No.	Width (m)	Thickness (m)
1	0.10	0.010
2	0.15	0.010
3	0.2	0.015

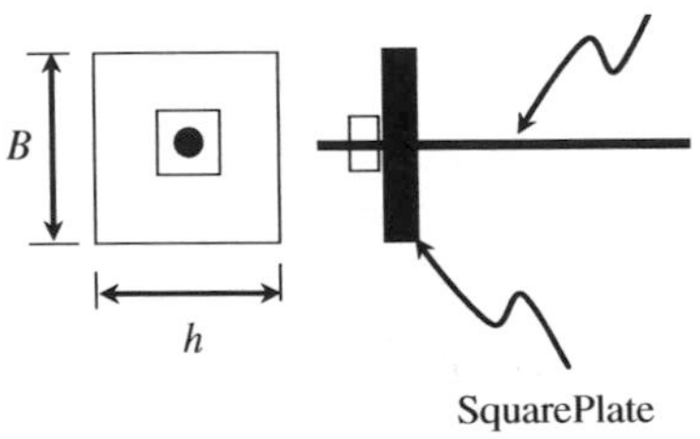

Figure 2. Square plate and tie rod cross-section used in the study (After Leerakomson and Charoenpon, 1996).

density and optimum moisture content of the standard proctor are 1750 kg/m^3 and 14.4 percent, respectively. California Bearing Ratio (CBR) at 95 percent compaction of the standard proctor is 17 percent. Internal friction angle from direct shear tests is 36 degree.

2.2 *Reinforced concrete block, tie rod and steel anchor plate*

Facing of the constructed wall is made from reinforced concrete block; a reinforced rebar is a deformed rebar with a diameter of 12 millimeters and the yield strength is 3,000 ksc (SD30). Concrete ultimate compressive strength is 135 kg/cm^3. The dimension of the reinforced concrete block is 0.2, 0.4 and 0.15 meter in width, length, and thickness, respectively.

Tie rod is a deformed bar with a diameter of 25 millimeters and the yield strength of 3,000 ksc (SD30). Yield strength from tensile tests in laboratory indicates average yield and ultimate strengths are 3,900 and 5,700 ksc, respectively. The square anchor plate is made from SR24 steel type.

The dimension of the square plates is shown in Table 1. Tie rod and anchor plate connection is shown in Figure 2.

2.3 *Reinforced wall construction*

The reinforced wall is constructed on a 40 centimeters compacted crushed rock. The construction of the reinforced wall is by compacting sand with a vibratory

Table 2. Field density tests of the compacted sand backfill (After Leerakomson and Charoenpon, 1996).

Number	Depth (m)	Average moist density (kg/m^3)
1	0.41	1,950
2	0.81	2,028
3	1.04	2,037
4	1.39	2,020
5	1.81	2,016

Table 3. Characteristics of the anchor plates and tie backs (After Leerakomson and Charoenpon, 1996).

Type	Depth (h_e:m)	Diameter of tie rod (mm)	Anchor plate dimension (m)
1	0.1, 0.5, 0.9, 1.3	25	–
2	0.1, 0.5, 0.9, 1.3	25	0.10 × 0.10 × 0.01
3	0.9, 1.3	25	0.15 × 0.15 × 0.10
4	0.5, 1.3, and 1.7	25	0.20 × 0.20 × 0.15

compactor in a layer of 20 centimeters at the water content approximately around the optimum moisture content till the total wall height of 1.8 meters. Field density tests, listed in Table 2, are performed at depth in which tie backs and anchors are installed. Average moist density of the backfill sand is 2016 kg/m^3. In addition, reinforced concrete block and tie rod connection is similar to that of the anchor plate and tie rod connection (see Figure 2).

2.4 *Pullout load tests*

Pullout load tests are performed on both tie rods only and tie rods with anchor plates. All pullout load tests are performed after finishing constructing the reinforced wall. An application of pullout load is by a hydraulic jack, used in prestressed concrete construction. Dial gauges are installed at a fix steel column to measure horizontal displacements during pullout load tests.

3 LOAD TEST RESULTS AND EVALUATIONS

3.1 *Pullout load test results*

All key parameters used in this study are listed in Figure 3. Since the yield load of tie rod in this study is approximately four times the maximum pullout load in this study, an elongation of the tie rod during the pullout load test is not taken into account.

Pullout load and displacement relationship at the depth of 0.9 meter are shown in Figure 4. The rest of the pullout load test results are provided in Leerakomson and Charoenpon (1996). From the series of testing

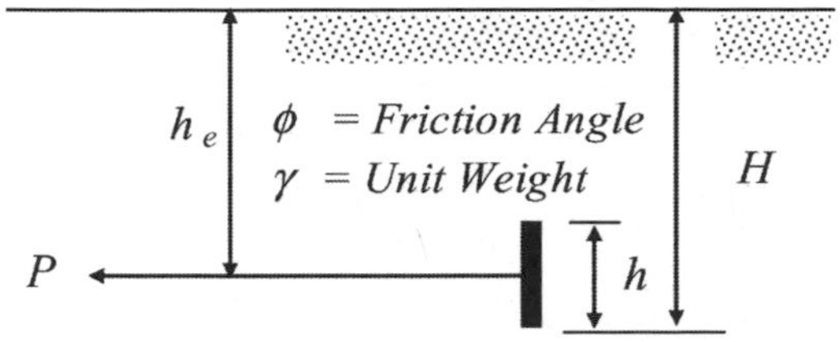

Figure 3. Parameters used in this study.

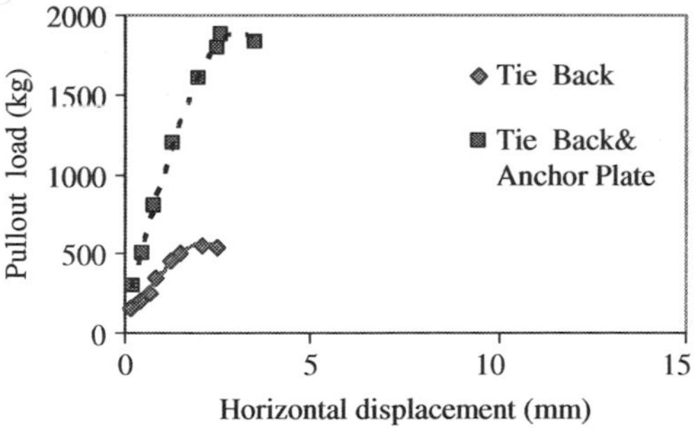

Figure 4. Pullout load test of tie back and anchor plate at the depth of 0.9 meter (After Leerakomson and Charoenpon, 1996).

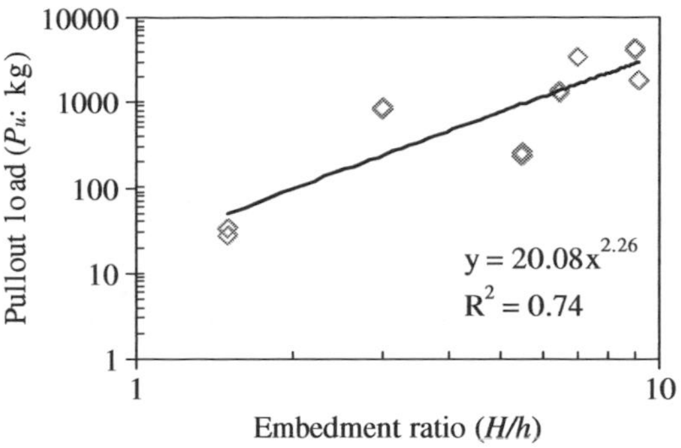

Figure 5. Maximum pullout loads, subtracting load resistances from tie rod, and the embedment ratio (H/h) relationship.

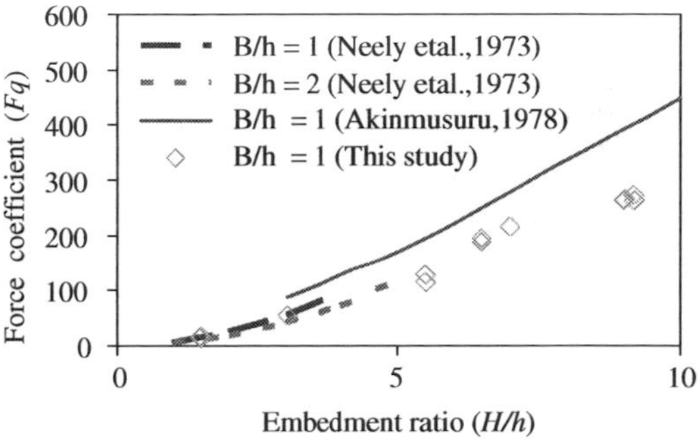

Figure 6. Variation of the force coefficient with embedment ratio.

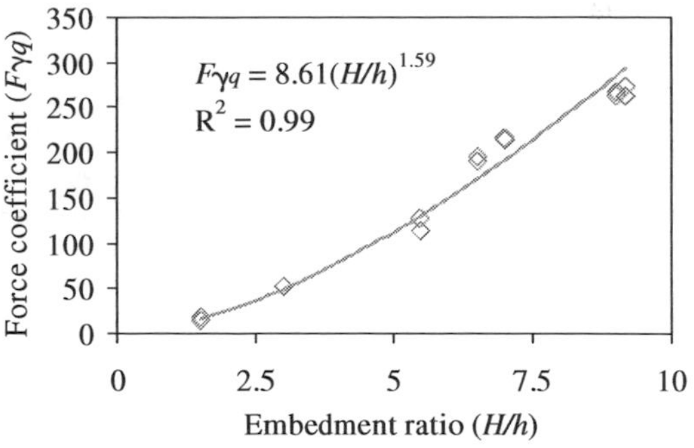

Figure 7. Correlation of the force coefficient with embedment ratio.

results, it is found that tie back with anchor plate enhances their pullout capacities about three to fours times that of the tie back rod only.

In addition, from the full scale load tests, the relationship between pullout load and embedment ratio (H/h) can then be characterized as shown in Figure 5.

3.2 *Pullout load test evaluations*

All Pullout load test data are analyzed and compared with case histories. Neely et al. (1973) and Akinmusuru (1978) introduced the normalized pullout load, called force coefficient ($F_{\gamma q}$), as shown in Equation 1.

$$F_{\gamma q} = \frac{P}{Bh^2\gamma} \tag{1}$$

Where P = the ultimate pullout load; B = the width of the plate; h = height of the plate and γ = the unit weight of the backfill material. Pullout loads and embedment ratio are scatter (Figure 5), however, force coefficient and embedment ratio are much less scatter (Figures 6 and 7). For embedment ratio not less than three, field data from this study, Figure 6, are matched with data presented by Neely et al. (1973) with the embedment ratio equal to 1. Whereas the embedment ratio is over four, field data from this study are outside the range proposed by Neely et al. (1973), see Figure 6. However, they lower than the range proposed by Akinmusuru (1978). In addition, the force coefficient and embedment ratio as shown Figure 7 indicates that this relationship is highly correlated.

Glaly (1997) collected several case histories and proposed the pullout capacity as shown in Figure 8. Since the backfill sand is compacted beyond the density in the direct shear test, the internal friction angle of the backfill sand should be over thirty six degree. Following NAVFAC (1982), the friction angle of the

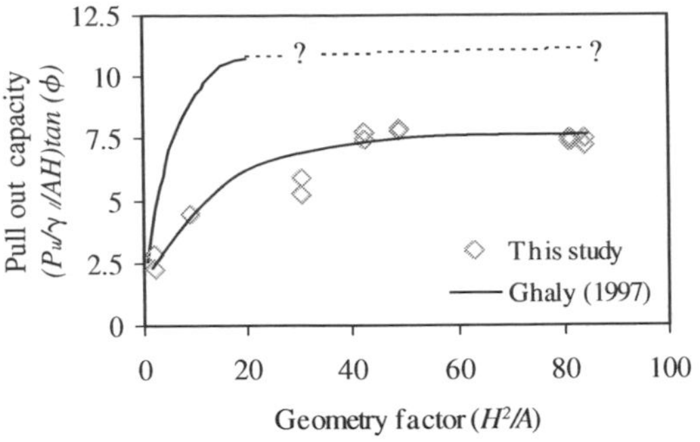

Figure 8. Relationship between the pullout capacity and geometry factor (After Ghaly, 1997).

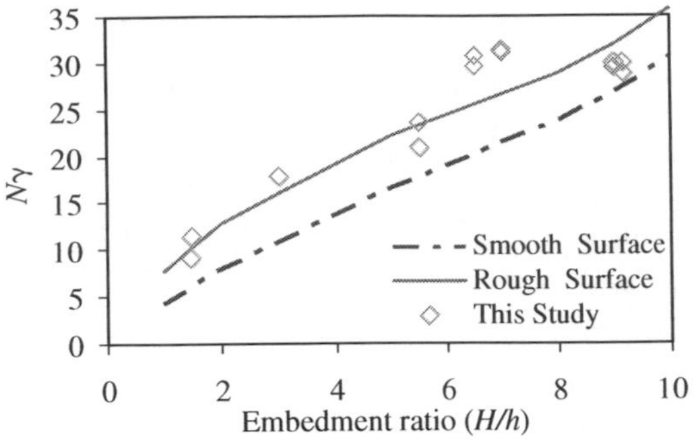

Figure 9. Theoretical solutions by Merifield and Sloan for backfill material with friction angle of 40 degree (After Merifield and Sloan, 2006).

backfill material used in this study is around forty degrees. However, pullout capacity of this study is lower than that proposed by Glaly (1997).

3.3 *Theoretical solutions for pullout load estimation*

Merifield and Sloan (2006) proposed a theoretical method to predict the maximum pullout load resistance of the anchor plate by the application of limit analysis of the upper and lower bound theorem, see Merifield and Sloan (2006) for more information. The ultimate pullout load capacity is presented following Equation 2, in which, $N\gamma$ is the anchor break-out factor. The anchor breakout factor by Merifield and Sloan (2006) with pull out load test data is presented in Figure 9.

$$P_u = \gamma H N_\gamma \tag{2}$$

In addition, pullout load test data in this study are plotted in Figure 9 to verify the effectiveness of Merifield and Sloan method. Regarding to pullout load test data in Figure 9, the method by Merifield and Sloan tends to provide a good estimation of the pullout resistance of the anchor plate for backfill sand in this study.

4 CONCLUSIONS

The full-scale reinforced wall tests with anchors plates were successfully performed. Pullout load tests of tie rod with and without anchor plates in compacted sand at the depth of 0.1, 0.5, 0.9, 1.3, and 1.7 meters from the surface were carried on. Regarding to test results, the force coefficient correlates with the embedment ratio. In addition, the theoretical method by Merifield and Sloan (2006) provides a good estimation of the pullout load of the compacted sand.

ACKNOWLEDGEMENTS

The authors would like to express the profound gratitude to Dr. Pichit Jamnongpipatkul, the director of Bureau of Road Research and Development, Department of Highways for his guidance and support throughout the study, Mr. Chayan Chareonpon and his associates for diligent works in the field.

REFERENCES

Akinmusuru, J.O. 1978. Horizontal loaded vertical plate anchors in sand. *Journal of the soil mechanics and foundations division* 104(2): 283–286.

Glaly, A.G. 1997. Load displacement prediction for horizontal loaded vertical plates. *Journal of geotechnical and geoenvironmental engineering* 123(1): 74–76.

Leerakomson, S. and Charoenpon, C. 1996. Vertical anchor plates in compacted sand. 83 p. (in Thai).

Merifield, R.S. and Sloan, S.W. 2006. The ultimate pullout capacity of anchors in frictional soils. *Canadian geotechnical journal* 44: 852–868.

Neely, W.J., Stuart, J.G. & Graham, J. 1973. Failure loads of vertical anchor plates in sand. *Journal of the soil mechanics and foundations division* 99(9): 669–685.

US Department of Navy, 1982. NAVFAC DM-7.1: Soil mechanics. Naval facilities engineering command, VA: 348p.

Weatherby, D.E. 1982. Tiebacks. FHWA-RD-82-047. 249p.

New Horizons in Earth Reinforcement – Otani, Miyata & Mukunoki (eds)

© 2008 Taylor & Francis Group, London,
ISBN 978-0-415-45775-0

Pullout resistance of reinforcement bar due to bearing capacity of expanded toe

T. Hayashi & T. Konami
Okasanlivic Co., Ltd. Tokyo, Japan

H. Ito & T. Saito
Dai Nippon Construction Co., Ltd. Tokyo, Japan

ABSTRACT: Effective measures are awaited for preventing the embankment or natural slope from the failure caused during large-scale earthquake in Japan. The reinforcing bar described in this paper has expanded toe and the capacity of bearing resistance enough to apply for the soft ground and can stabilize the slope. This paper describes the results of pullout tests using the reinforcing bar with expanded toe. It first shows the differences of pullout resistance between vertically installed reinforcing bars with expanded toe which have three types of diameter respectively. Secondary, it is verified that the value of pullout resistance of the reinforcing bar with expanded toe can be described as sum of the value of skin resistance and bearing resistance, and proposes the design method of ultimate pullout resistance through the pullout tests horizontally installed in the soil. Finally, safety factor of pullout resistance is proposed.

1 INTRODUCTION

Effective measures are awaited for preventing embankment slopes of highways and housing lots from the failure during large-scale earthquake. Conventional reinforcing methods, such as nailing, stabilize slopes by increasing the skin frictional resistance between the soil and grout, and require a relatively large number of long bars to be installed to stabilize the slope of embankments. This paper proposes a new method for stabilizing slopes of even small skin frictional resistance, which involves use of reinforcing bars with expanding toes. Several series of pullout tests using new type reinforcing bars with expanded toes were carried out to investigated 1) the differences in pullout resistance by differences in expanded toe diameter when the bars were vertically installed, 2) the validity of the method for assessing the pullout resistance of the reinforcing bars with expanding toes when installed horizontally, and 3) the safety factor and its adequacy. This paper describes the results of the study.

2 TESTING METHODS

2.1 *Vertical pullout test*

An overview of a reinforcing bar with expanded toe is shown in Fig. 1, whose toe consists of ten steel rods. The steel rods are compressed and spread radially when the inner steel pipe (rod section) is pulled out

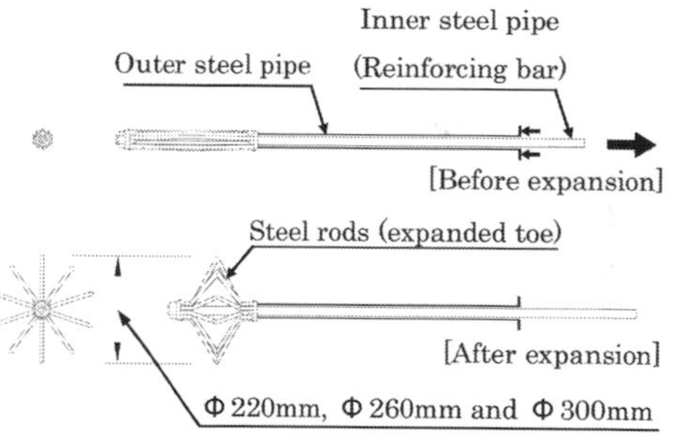

Figure 1. An overview of a reinforcing bar with expanded toe.

from the outer steel pipe, which serves as the reaction force. Grout can be poured through the inner pipe to the toe and to the bore hole.

Ground consisting of fine sand was built by heaping up 0.2 m spreading depth soil layers compacted by 700 kg weight vibration roller respectively in an indoor soil tank (width: 5 m, depth: 4 m), and vertical bore holes of a diameter of 65 mm and a length of 1400 mm were drilled. Reinforcing bars of a rod length of 1.0 m with expanded toes were inserted into the holes, and the toes were expanded using a hydraulic jack. Cement milk (W/C = 50%) was grouted and cured for seven days. A pullout test was conducted by controlling the load and using a center-hole type hydraulic jack. Loads to be applied in steps were determined by preliminarily

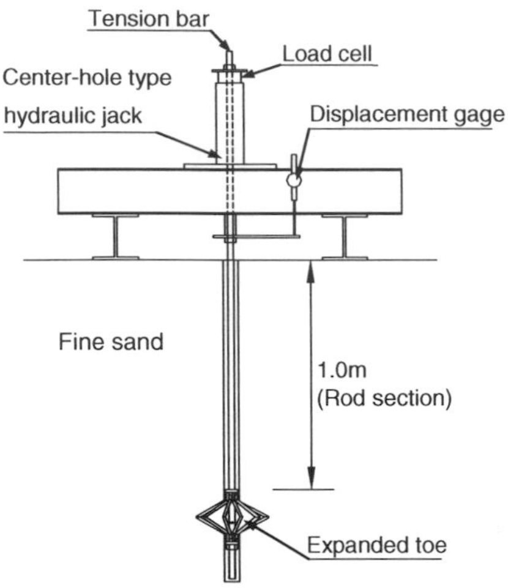

Figure 2. An overview of the vertical pullout test.

Table 1. Soil properties of fine sand (vertical pullout test).

Adopted parameters	Value
Cohesion, c	12.8 kN/m^2
Maximum friction angle, ϕ	34.2°
Wet unit weight, γ	17.1 kN/m^3
Water content	13.9 %
Maximum diameter of soil particle	4.25 mm
Fine content	11.1 %

calculating the loads that reach the ultimate pullout resistance in ten steps. The load was retained for five minutes at each step. During the test, the pullout resistance P and pullout displacement δ were measured at the heads of the reinforcing bars, and the test was stopped when δ reached 50 mm. An overview of the vertical pullout test is shown in Fig. 2, and the soil properties of the ground tested are shown in Table 1.

The toes were expanded to diameters of Φ220 mm, Φ260 mm, and Φ300 mm. Two bars were tested for each diameter, one of which was entirely grouted and the other was grouted only at the expanded toe section. All bars had a rod length of 1.0 m. To understand the ultimate skin friction force of the prepared ground, a pullout test was similarly conducted using reinforcing bar without expanded toes (Fig. 3). Test cases are shown in Table 2.

2.2 *Horizontal pullout test*

A slope was prepared in the indoor soil tank so as to have a slope length of 1.0 m and an inclination of 1:0.3. Bore holes of a diameter of 65 mm and a length of 2.4 m were drilled perpendicular to the slope surface, and the reinforcing bars were installed so that the

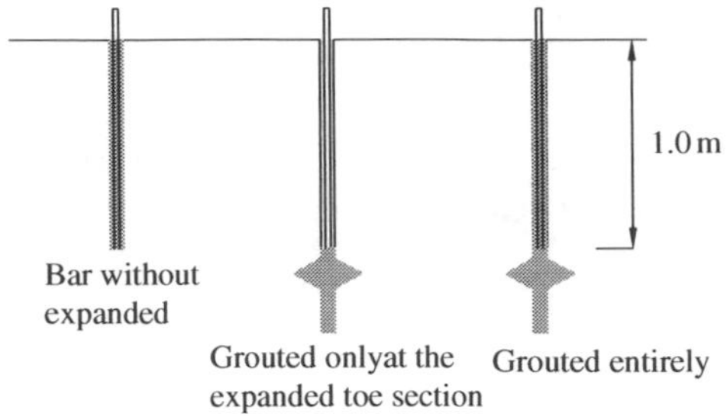

Figure 3. Type of reinforcing bars in vertical pullout test.

Table 2. Type of vertical pullout test.

Name	Type of toe	Grout filling
A	No expanded	Entirely
B-1	Φ 220 mm expanded	Entirely
B-2	Φ 260 mm expanded	Entirely
B-3	Φ 300 mm expanded	Entirely
C-1	Φ 220 mm expanded	Only expanded section
C-2	Φ 260 mm expanded	Only expanded section
C-3	Φ 300 mm expanded	Only expanded section

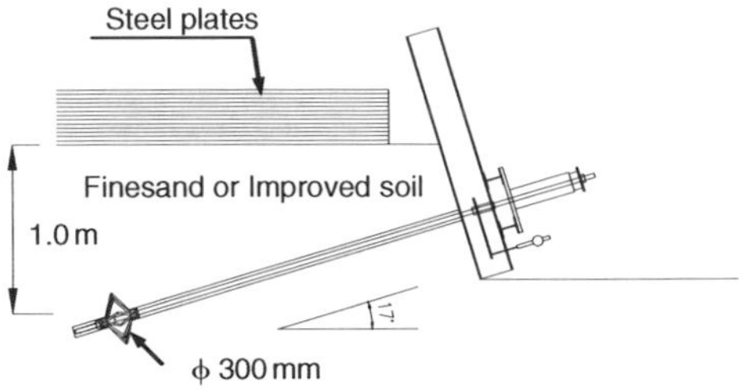

Figure 4. An overview of the horizontal pullout test.

expanded toes were 1.0 m from the ground surface. The diameter of the expanded toes was Φ300 mm, and the rod length was 2.0 m. An overview of the test is shown in Fig. 4. The pullout test was performed as in the vertical pullout test.

Two types of ground were tested: fine sand which was the same soil shown in Table 1 and improved soil prepared by mixing 1% of cement to the fine sand. Placing and spreading process was equivalent to vertical pullout test. The soil properties of the improved soil are shown in Table 3.

To understand the effects of the vertical stress σ_v acting on the expanded toe on the pullout resisting force P, the steel plates were loaded on the upper part of the slope to adjust the surcharge. Test cases are shown in Table 4. The ultimate skin friction force between the soil and grout was determined by conducting a pullout test of reinforcing bars without expanded toes on the same grounds.

Table 3. Soil properties of improved soil (horizontal pullout test).

Adopted parameters	Value
Cohesion, c	19.1 kN/m^2
Maximum friction angle, ϕ	30.0°
Wet unit weight, γ	18.1 kN/m^3
Water content	14.3%
Maximum diameter of soil particle	4.25 mm
Fine content	11.1%

Table 4. Type of horizontal pullout test.

Name	Surcharge	Type of soil
D-1	0 kN/m^2	Fine sand
D-2	26.1 kN/m^2	Fine sand
D-3	52.2 kN/m^2	Fine sand
E-1	0 kN/m^2	Improved soil
E-2	26.1 kN/m^2	Improved soil
E-3	52.2 kN/m^2	Improved soil

3 TEST RESULTS

3.1 *Vertical pullout test*

The relationship between the pullout resistance and displacement is shown in Fig. 5 for the reinforcing bars with and without expanded toes for each step. The reinforcing bars without expanded toes showed a peak of pullout resistance at a small displacement δ. On the other hand, the pullout resistance of the bars with expanded toes showed no clear peak but kept increasing. The P-δ curves of expanded toes can be fitted by hyperbolic curves as shown in Fig. 5 and the asymptotic line values were used as the ultimate pullout resistance P_u. The figure shows that the larger the diameter of the expanded toe, the larger the pullout load. This suggests that the differences in the projected area of the expanded toe toward the pulling direction affected the pullout resistance of the bar.

3.2 *Horizontal pullout test*

The relationship between pullout resistance P and pullout displacement δ in the horizontal pullout test with fine sand is shown in Fig. 6. As in the vertical pullout test, the values were subjected to hyperbolic approximation, and the results are shown in the figure with lines. The pullout resistance of the bars with expanded toes was larger than that of the reinforcing bars without expanded toes. The pullout resistance increased as the surcharge increased.

Calculated pullout strength P_d and the measured ultimate pullout resistance P_u of the reinforcing bars

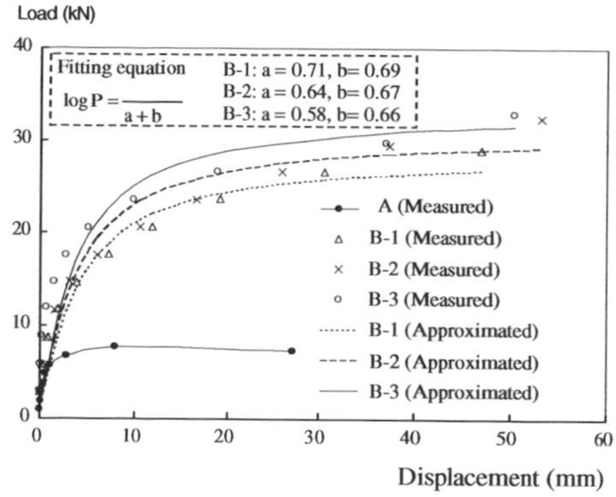

Figure 5. The relationship between the pullout resistance and displacement in vertical pullout tests.

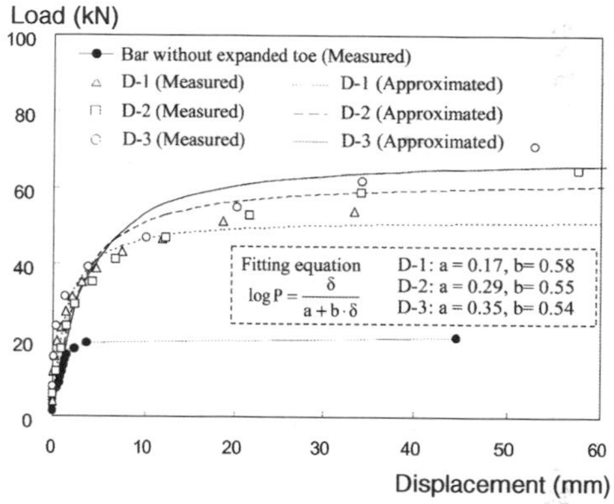

Figure 6. The relationship between the pullout resistance and displacement in horizontal pullout tests (fine sand).

with expanded toes are shown in Fig. 7. Here, the pullout strength P_d was calculated as the sum of the bearing resistance of the expanded toe and the skin frictional force between the soil and grout at the rod section, based on the results of the vertical pullout test, using the following equation:

$$P_d = \tau \cdot \pi \cdot D \cdot L + (c \cdot N_c + N_q \cdot q_p) \cdot A_p \quad (1)$$

where, τ is the ultimate skin frictional force between the soil and grout, D is the diameter of the grout, L is the length of the reinforcing bar, c is the cohesion of the soil, q_p is the horizontal confining pressure at the expanded toe, N_c and N_q are the bearing capacity factors, and A_p is the projected area of the expanded toe (toward the pullout direction). The calculated values are shown in Table 5 for each case.

The ultimate pullout resistance P_u measured in both soil conditions of fine sand and improved soil was larger than the calculated ultimate resistance P_d, which was the sum of the bearing resistance and the skin frictional force. The figure shows that the measured pullout resistance P_u increased as the surcharge was increased, showing a trend similar to that in which design pullout strength P_d increased as the horizontal confining pressure at the expanded toe increased. Then, the increment in pullout displacement during each steps (Δt) was put to be $\Delta\delta$ and the changes in logarithmic value during Δt was put to be $\Delta \log t$ to draw a curve that shows the relationship between $\Delta\delta/\Delta \log t$ and P by assuming that the pullout resistance at the break point was the measured pullout resistance at yield P_y. The relationship is a method used to judge the yielding load of piles.

Figures 8 and 9 show the relationships between the vertical stress at the toe section σ_v and the pullout resistance at yield P_y, the ultimate pullout resistance P_u, which were both measured in the fine sand and improved soil grounds, and the calculated pullout resistance P_d. The pullout resistance at yield P_y was 0.60 to 0.85 times of the ultimate pullout resistance P_u. Therefore, the pullout displacement at a load of about 1/2 of the calculated pullout force P_d is likely to be controlled small in practice, enabling a safety factor of 2.0 to be proposed for the design pullout force P_d.

4 CONCLUSIONS

A pullout test of reinforcing bars with expanded toes which were installed vertically to the ground showed

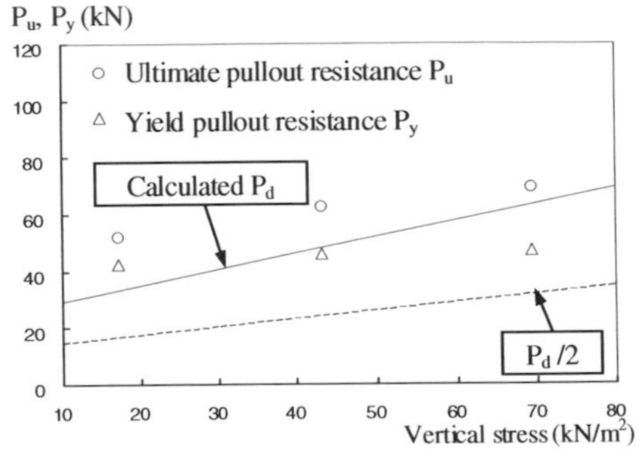

Figure 8. The relationships between the vertical stress at the toe section σ_v and the pullout resistance at yield P_y, the ultimate pullout resistance P_u (fine sand).

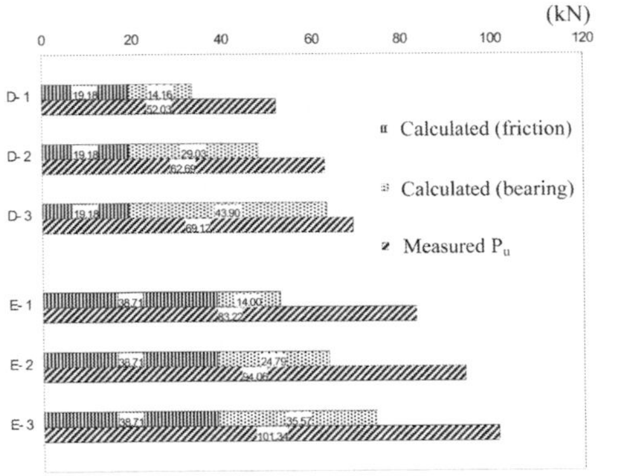

Figure 7. Calculated pullout strength P_d and the measured ultimate pullout resistance P_u.

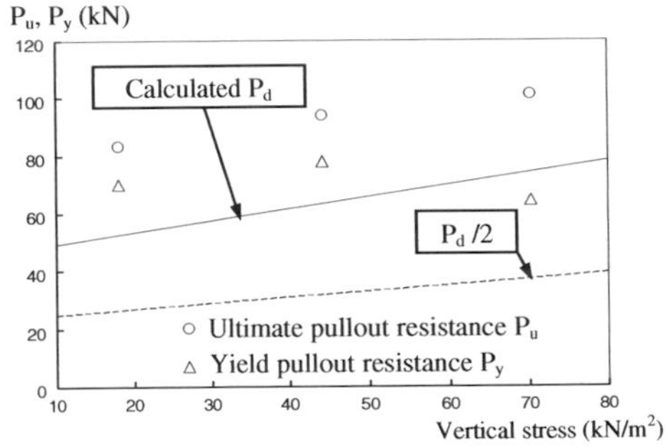

Figure 9. The relationships between the vertical stress at the toe section σ_v and the pullout resistance at yield P_y, the ultimate pullout resistance P_u (improved soil).

Table 5. Calculated values of ultimate skin frictional forces and ultimate bearing resistances.

Case	τ [kN/m^2]	$\tau \cdot \pi \cdot D \cdot L$ [kN]	K_a	N_c	N_q	q_p [kN/m^2]	$(c \cdot N_c + q_p \cdot N_q) \cdot A_p$ [kN]
D-1	49.96	19.18	0.28	5.1	30.2	4.80	14.16
D-2	49.96	19.18	0.28	5.1	30.2	12.12	29.03
D-3	49.96	19.18	0.28	5.1	30.2	19.14	43.90
E-1	94.78	38.71	0.33	5.1	18.4	6.02	14.00
E-2	94.78	38.71	0.33	5.1	18.4	14.72	24.79
E-3	94.78	38.71	0.33	5.1	18.4	23.42	35.57

*K_a: Coefficient of active earth pressure $[=(1-\sin\varphi)/(1+\sin\varphi)]$ *L = 2.0 m, D = 0.065 mm, z = 1.0 m, $A_p = 0.067\ m^2$.

that the ultimate pullout resistance increased as the projected area of the expanded toe increased.

A pullout test of reinforcing bars with expanded toes that were installed horizontally into two different kinds of ground resulted in the ultimate pullout resistance larger than the design pullout resistance, which was the sum of the bearing resistance of the expanded toe and the friction resistance at the rod section, showing that the method for assessing the pullout resistance was valid. The pullout resistance at yield was confirmed to be 0.60 to 0.85 times of the ultimate pullout resistance, and a safety factor of 2.0 was proposed for pullout resistance. Field pullout tests will be conducted to obtain precise data and assess the stability of reinforced soil structures.

REFERENCES

Hayashi T., Konami T., Ito H and Saito T. 2006. Pullout resistance of reinforcing bar with expanded head installed in the soil horizontally, Proc. of the 41st annual meeting of JGS, pp.1841–1842.

Ito H., Saito T., Konami T., Yamashita T and Hayashi T. 2005. Pullout resistances of reinforcing bars with expanded heads which have different diameters, Proc. of the 60th annual meeting of JSCE, 989–990.

New Horizons in Earth Reinforcement – Otani, Miyata & Mukunoki (eds)
© 2008 Taylor & Francis Group, London, *ISBN 978-0-415-45775-0*

Soil/reinforcement interface characterization using three-dimensional physical modeling

A. Abdelouhab, D. Dias & Y. Bourdeau
INSA de Lyon, LGCIE, France

N. Freitag
Terre Armée Internationale/c/o Ménard-Soltraitement, France

ABSTRACT: Several extraction tests on geosynthetic strips and ribbed steel strips, developed by the *Terre Armée Internationale*, were carried out at the LGCIE laboratory of the INSA Lyon. These tests permit to follow the imposed tension as well as the displacements of several points along the strip (geosynthetic and metal) and to highlight their stiffness influence. The analysis of the tension/displacement curves makes it possible to define the interaction law between the soil mass and the reinforcement by an analytical modeling. From the analysis of these results, the influence of the reinforcements extensibility is highlighted.

1 INTRODUCTION

The soil mass reinforcement by horizontal strips constitutes a complex interaction soil/structure problem. For a fine understanding of this interaction, a specific knowledge of the behaviour of the constitutive materials (ground, strips) as well as interface properties between these elements is necessary. The Reinforced Earth method uses strips embedded in layers to grip soil strongly enough to enable the construction of high vertical banking, large embankment or abutments and is included in this category. Safety for short and long-term must be ensured. It requires stability checking and control of the strain level reached during the life cycle of such works.

The reinforcements used in Reinforced Earth structures are most commonly made of ribbed steel strips, however in aggressive environments smoother geosynthetic strips based on high-tenacity polyester are used, which exhibit some relative elongation. The design methods created for the structures reinforced by metallic reinforcements and thus inextensible were brought to be extrapolated to extensible materials. The difference in behaviour of these two types of reinforcement induces the definition of elongation limits beyond which the behaviour of the structure may be different. In order to adapt and to improve these methods, a better knowledge of the interaction between the soil mass and the reinforcement strips seems necessary in order to develop the comprehension of strip-reinforced structures.

Previous studies Schlosser (1981), Segrestin and Bastick (1996) studied the behavior of the geosynthetic straps and established soil/reinforcement interaction laws. However, the studies concerning the dilatancy influence in the grounds, reinforced by extensible reinforcements, are very few Lo (2003). The realization of the tests at various levels of surcharge permits to deduce this influence.

This paper presents the geosynthetic straps behaviour used in the soils reinforcement. It consists of pull out tests of steel and geosynthetic strips in a three-dimensional physical modelling. This type of modeling is carried out in a three-dimensional metal tank. It permits to reproduce the influence of various parameters such as the soil dilatancy which plays an important part in this type of reinforcement. Each of these test have been realized in controlled and instrumented conditions. From this experimental study, the Soil/Reinforcement interface is deduced by analytical methods and permits to improve the current knowledge state.

2 MATERIAL AND METHODS

2.1 *Reinforcements*

The tests were carried out on two types of reinforcement (figure 1), the extensible geosynthetic strips and ribbed steel strips. Extensible inclusions are presented in the form of geosynthetic strips containing

Figure 1. Ribbed steel strip and OMEGA geostrap.

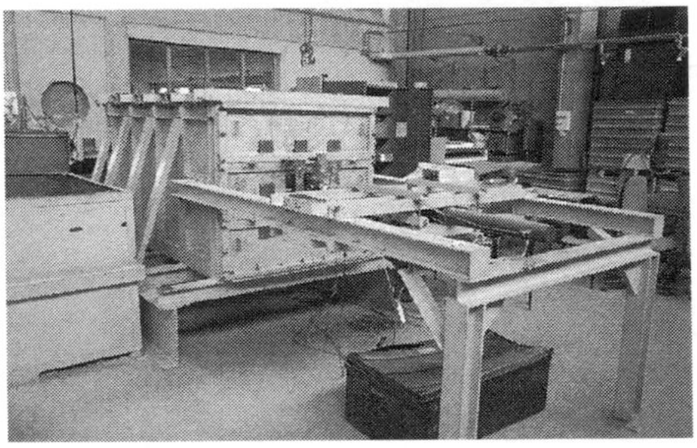

Figure 2. Test tank.

polyester fibers at high tenacity protected by a low density polyethylene sheath. The dimensions of these strips are: 50 mm width and 2 mm thickness. The *Terre Armée Internationale* makes use of these straps for the fully synthetic Omega® system. The advantage of this system lies in the fact that it removes any metal intermediary (thus corrodible) between the concrete facing and the reinforcement strips. The steel strip is made of galvanized and ribbed steel and known under the name of High Adherence strip (HA 50x4). In this case, the dimensions are: 50 mm width and 4 mm thickness.

2.2 *The test tank*

The test tank has large inner dimensions: 1.10 m width, 1.10 m height and 2.0 m length. An air cushion makes it possible to apply an overload. It is placed between the sand and the tank closure. This cushion is inflated under pressure which is controlled by a pressure gauge. Externally, an extraction jack permits to extract the strips and measure the tension load. All the measurements sensors are connected to a computer.

2.3 *The material (sand of Hostun RF)*

The material used in the tests is fine sand known under the name of Hostun RF sand. Various authors were interested in this sand (Gay 2000, Gaudin 2002). Main physical and granulometric characteristics are deferred on table 1.

Table 1. Characteristics of the Hostun RF sand.

Characteristics	Value
Granulometry (mm)	0.16–0.63
D50 (mm)	0.35
Maximum index of vacuums	1.041
Minimal index of vacuums	0.648
Unit weight of the grains (kN/m^3)	26.5
Maximum volumic weight (kN/m^3)	15.99
Minimal volumic weight (kN/m^3)	13.24
Angle of friction	38°

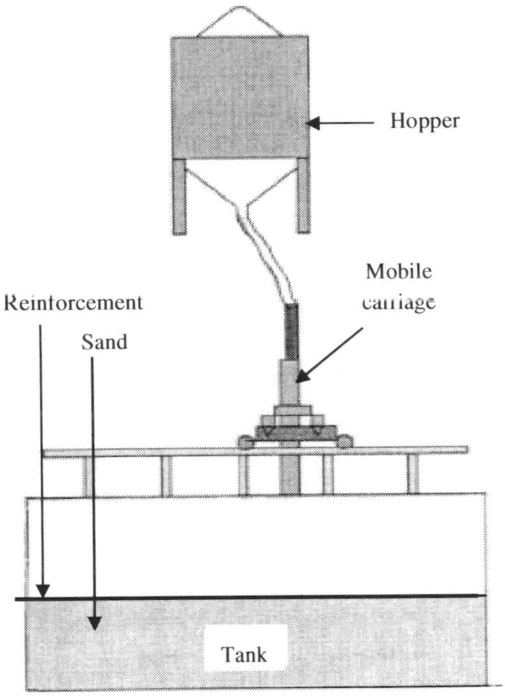

Figure 3. Pluviation system.

2.4 *Sensors*

Two types of incremental position sensors were used. A wire sensor placed in the front of the tank, allowing measuring displacements of the strip and LVDT sensors, placed on supports at the tank rear, allowing measuring displacements along the reinforcement. In order to avoid any friction effect with sand during the extraction test, these cables are threaded in Teflon sheaths. To measure the tensile force, an annular load sensor is placed at the end of the extraction jack.

2.5 *Pluviation system*

The pluviation method is defined as a technique of granular sample reconstitution by material discharge.

Among other techniques, it allows the control of the density of the sand set up and to simulate the reconstitution of a sandy ground formed by sedimentation. An automatic system, double axis, allowing that the whole tank surface was set up. It is remoted by a computer and moves at constant speed in the two directions of the tank. A hopper is placed above the tank and connected by a flexible pipe to a mobile carriage being on the automatic axis system. Sand runs out of the hopper towards the carriage by the means of a ring of diameter equal to 20 mm. This system makes it possible to control the flow of sand. An adjustment, making it possible to obtain a density of approximately 1.55, was selected by L. Martinez & B. Novelli (2006).

3 TESTS

3.1 *Procedure*

The operation starts with the filling of the tank with the pluviation system. Then, the reinforcement strips, equipped with the sensors, are installed before the total filling and the closing of the tank. In order to apply an overload similar, we inflate the air cushion to a pressure equivalent to a height of fill.

The tests were carried out at an extraction speed of 1 mm per minute.

3.2 *Test routine*

Nine tests were carried out on the two types of reinforcements at various levels of surcharges. Five tests on the geosynthetic strips and four tests on the steel strips. Concerning the geosynthetic straps, the tests were doubled for each load level to obtain better results.

4 ANALYSIS AND EXPLOITATION

The tests made it possible to determine the tension load, the maximum friction parameter mobilized as well as displacements in several points of the strips. These results (tables 2 and 3), show that the friction parameter decreases with the increase on loads on the two types of reinforcement. This behavior is related to the dilatancy decrease of the ground with the increase of the vertical pressure. For the geosynthetic straps, with a pressure of 50 kPa, the friction parameter is equal to 0.81 (slightly higher than tan(ϕ) for $\phi = 38°$) and for a pressure of 100 kPa, it is equal to 0.74 (lower than tan (ϕ)). These values are higher than the results obtained from smooth steel reinforcements pull-out tests carried out by Alimi (1977) on small-scale model. However, better results were obtained for higher densities ground. The grading limits of soil presents also an influence on the dilatancy soil and consequently on the friction parameter. This last increases with the increase on grading limits of soil. It is also necessary to note that in the tank, the geosynthetic strip is placed on a relatively plane surface whereas in a actual work it is placed on an irregular surface of the ground, which increases its adherence.

The analysis of the behavior of the two types of reinforcement (figure 4) shows that the steel strip starts to move over all their length, as soon as a load is applied at the head. Displacements in the medium and the rear of the strip are close to displacement at the head and friction is mobilized uniformly on the reinforcement. However, in the case of the polyester strips, the tension

Table 2. Summary of the test results obtained on the synthetic strips.

Test	1	2	3	4	5
Vertical stress (kPa)	50	50	75	100	100
Maximum tension (kN)	16.44	15.46	22.5	28.17	28.27
Tensile stress (kPa)	43.26	40.68	59.21	74.13	74.39
friction parameter	0.87	0.81	0.79	0.74	0.74
friction Angle(°)	40.9	39.1	38.3	36.6	36.6

Table 3. Summary of the test results obtained on the steel strip.

Test	6	7	8	9
Vertical stress (kPa)	50	75	100	120
Maximum tension (kN)	14.09	20.9	24.03	27.7
Tensile stress (kPa)	74.16	110	126.47	145.8
friction parameter	1.48	1.47	1.26	1.21
friction Angle(°)	56.0	55.7	51.7	50.5

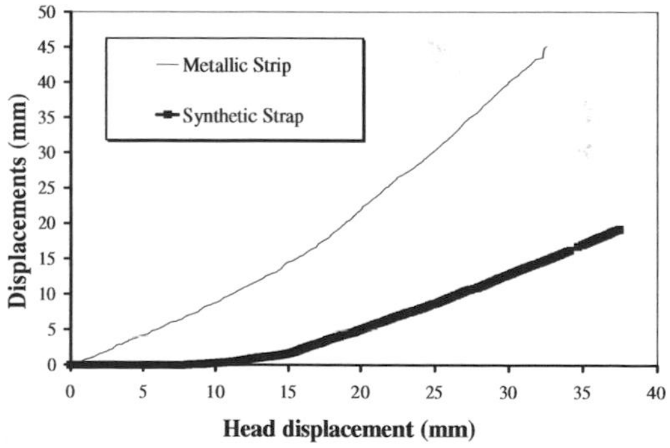

Figure 4. Displacement of the points located at the center of the steel strip and of the polyester strap (confining stress equal to 50 kPa).

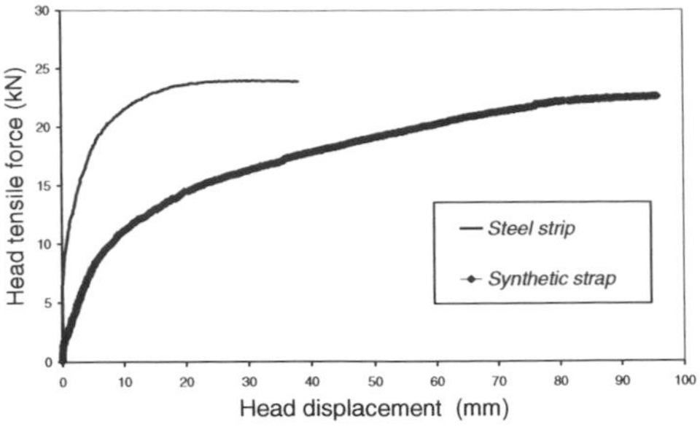

Figure 5. Tensile force at the head of steel and synthetic strips (loading constraint equal to 100 kPa).

Table 4. Effort mobilized for a displacement at the head of 1% inclusion length.

Test	Uh	T′ (kN)	T (kN)	T′/T
5	19(mm)	17.32	28.30	0.61
8	19(mm)	23.45	24.01	0.98

T′. Tensile force for 19 mm of the head displacement. T. Tensile force at saturation of friction strength. Uh : Displacement at the head of the strip.

in the strip is gradually mobilized with the increase of the tension at the head of the strip. Friction is thus mobilized gradually along the band and displacement at the head is requested for low tensile stresses (figure 5). This behaviour is anticipated by Segrestin & Bastick (1996).

Table 4 highlights the difference in tension/displacement curves between the two types of reinforcement for the same displacement at the head of the strip.

The steel strip makes it possible to work at loads levels of order of magnitude as those taken again by a surface of Omega band twice higher. The ratio between the 50 mm wide steel strip friction parameter and the one for two parallel 50 mm polyester strips varies between 1.7 and 1.8.

5 CONCLUSIONS

The results obtained from the pull-out tests highlight that the steel strip friction is slightly higher than the Omega strip friction. These results highlight two differences on these two types of strips: in the load curve (tension versus displacement) as well as the displacements delay in the geosynthetic strip. In fact, the steel strip starts to move over all their length, as soon as a load is applied at the head and friction is mobilized uniformly on the reinforcement. While, in the case of the synthetic strips, the tension and the friction are mobilized gradually along the strip. The behaviour difference of the two types of reinforcement, confirms that it is necessary to develop new design methods more adapted for the structures reinforced by extensible reinforcements. The analytical modeling of displacement curves versus tension, using the test results, will make it possible to define the interaction law between the ground and geosynthetics and to develop the knowledge for a new design method.

In fact, this article is an introduction to an ambitious research work of the reinforced earth structures behavior and which is the subject of collaboration between the LGCIE laboratory of the INSA Lyon and the *Terre Armée Internationale*. Within the framework of this collaboration, a doctorate thesis is currently in progress. The subject of this thesis is around three principal axes. An experimental axis aims to perform pull out laboratory tests in a test tank and to instrument in situ reinforced earth structures in order to understand the operation mechanism of this soil mass and to validate numerical calculations. A numerical part of the research will permit us to better understand the influence of various parameters and to use a new structure safety approach. Finally, the seismic impact on structure will be studied. Dynamic approach can be completed by a comprehensive study of the structures under dynamic loading of TGV type.

REFERENCES

Alimi, Bacot, Lareal, Long, Schlosser, 1977. "Etude de l'adhérence sol-armature", Proc. of the 9th Int. Conf. on Soil Mechanics and Found. Eng. 1977 Tokyo.

Gaudin, C. 2002, "Modélisation physique et numérique d'un écran de soutènement autostable, application a l'étude de l'interaction écran-fondation", *thèse de doctorat en géotechnique, LGCIE, INSA Lyon*, 2002.

Gay, O. 2000, "Modélisation physique et numérique de l'action d'un glissement lent sur des fondations d'ouvrages d'art", *thèse de doctorat en Mécanique, Laboratoire 3S, Grenoble 1*, 2000.

Lo, S.R. 2003, "The influence of constrained dilatancy on pullout resistance of strap reinforcement", *Geosynthetics international* 10, No.2: 47–55.

Martinez, L. Novelli, B. 2006, "Renforcement des sols par inclusions souples, approche expérimentale de laboratoire", *Projet d'initiation à la recherche et développement, LGCIE, INSA Lyon.*

Schlosser, F. Guilloux, A. 1981, "Le frottement dans le renforcement des sols", *Revue française de géotechnique*, no 16: 65–77.

Segrestin, P., Bastick, M., 1996, "Comparative Study and Measurement of the Pull-Out Capacity of Extensible and Inextensible Reinforcements", *Earth Reinforcement, Ochiai*: 81–86.

New Horizons in Earth Reinforcement – Otani, Miyata & Mukunoki (eds)
© 2008 Taylor & Francis Group, London, ISBN 978-0-415-45775-0

Installation method and overburden pressure on soil nail pullout test

K.C. Yeo
Maunsell Geotechnical Services Limited, Hong Kong, China

S.R. Lo
School of Aerospace, Civil & Mechanical Engineering, University of New South Wales at ADFA, Australia

J.H. Yin
Department of Civil and Structural Engineering, Hong Kong Polytechnic University, Hong Kong, China

ABSTRACT: Laboratory soil nail pullout tests have been recently carried out in Hong Kong to investigate into the effect of installation influence and surcharge pressure on the pull-out resistance of soil nails. Based on the findings of a preliminary numerical analysis, published laboratory results, together with data obtained from field tests, this paper discusses the possible factors which may influence the pull-out resistance of soil nail, in particularly, the installation procedure and overburden pressure.

1 INTRODUCTION

There are very few publications available on the pull-out resistance of soil nail pull-out tests carried out on site. Most of the design methods assumed a threshold value of soil nail skin friction (i.e. 120 kPa) or taking a calculated soil nail skin friction based on the overburden pressure, as discussed by Yeo & Leung (2001).

Heymann & Rohde (1992) cited results of 40 field pull-out tests of soil nails, 100 mm in diameter with a bond length of 1.0 m to 1.5 m, in residual Andesite and Granite in South Africa to demonstrate the importance of soil dilatancy in the prediction of soil nail pull-out resistance. It was observed from the results that the ultimate shear stresses of the test nails are independent of depths of the nails where the nails were embedded, which varied from 2 m to 6 m below ground.

Franzen & Jendeby (2000 and 2001) conducted full-scale field tests on pull-out capacity of different types of soil nails including a grouted-nail type. The results indicated that at a low overburden pressure (about 25 kPa) the pull-out capacity is dependent of a local stress field rather than the overburden pressure. The results further indicated that the local stress field around the nail would mainly depend on the installation method and the volume of sand that the nail would displace during the installation and loading.

Li & Lo (2007) also pointed out that the behaviour of soil nail is highly dependent on the method of construction. In Hong Kong, soil nails are commonly installed in existing slopes or newly formed slopes. The construction method involves forming a drillhole, inserting a deformed bar and grouting under gravity. When a stable drillhole is formed, the radial stress in the vicinity of the soil face will be close to zero, therefore completely different from the test conditions as carried out in the laboratory pull-out tests. The variation of local stress field of soil around the soil nail may be resulted from arching effect of soil around the drillhole during installation and from soil dilatancy during loading of the soil nail. Soil arching and constrained dilatancy should therefore be addressed in studying the behaviour of soil nail in laboratory in order to evaluate the ultimate capacity of pull-out resistance of the soil nail in field.

2 LABORATORY WORKS ON PULL-OUT TESTS

Many laboratory studies of soil nail behaviour have been attempted in recent years. Davies and Le Masurier (1997) made use of a steel shear box and an air bag on top to apply confining pressure for soil nails (2.8 m long and 25 mm diameter steel or aluminium bars without grout) in medium dense sand under confining pressures of 100 kPa and 200 kPa. Strain gauges and soil pressure cells were installed on each test nail and soil around it respectively for measurements of the soil and nail responses. It was observed in the study that the maximum shear stress increased with an increase in confining pressure, but not proportionally. The authors pointed out that it was because at lower confining pressures, the sand displayed relatively higher dilatancy and higher apparent cohesion.

Lee et al (2001) & Pradhan et al (2003) investigated the pull-out resistance of nails in loose fill as a function of vertical pressure and relative compaction using a pull-out box. They obtained similar findings on the effect of overburden pressure on the pull-out resistance of soil nails when carrying out pull-out test of soil nails for three different vertical pressures, i.e., 25, 75 and 125 kPa. Similar findings about the effect of overburden pressure on the pull-out resistance of soil nails were also obtained by Chu & Yin (2005) when carrying out soil nail pull-out tests in shear box and pull-out box under the influence of vertical pressure.

In addition, Chu & Yin (2003) carried out a series of laboratory pull-out tests on grouted soil nails in soil of two different degrees of saturation of 74% and 78% for testing under a normal pressure of 300 kPa. The results indicated that the shear strength decreased by 11.43% for an increase in degree of saturation of only 4%. However, test results were very limited to indicate a clear trend of the effect of degree of saturation of soil on the pull-out resistance of soil nail. The installation procedure was not simulated in this study.

There seems to be inconsistent results between the findings of laboratory pull-out tests under different overburden pressures and the results of field pull-out tests on grouted nails. This may be attributed to the fact that in the laboratory studies, the overburden pressure has been applied after the nail was installed. Therefore the effect of installation method on behaviour of soil nail has warranted further investigation.

3 INFLUENCE OF INSTALLATION METHOD AND OVERBURDEN STRESS

The investigation of the influence of installation method can be looked into from three different perspectives, namely analytical, experimental, field performance.

3.1 *Preliminary numerical study*

A segment of the soil nail was idealized as a unit cell as illustrated in Figure 1. This simplifies the problem from 3D to axi-symmetric but allows an investigation into the local stress near the nail-soil interface and the effect of constrained dilatancy in the vicinity of the interface (Lo 2003).

3.1.1 *Numerical model*

The radial distance to the far boundary is set at an artificially large distance of 5 m so that fixity at this boundary will only have a small effect on the local stress generated by constrained dilatancy around the nail.

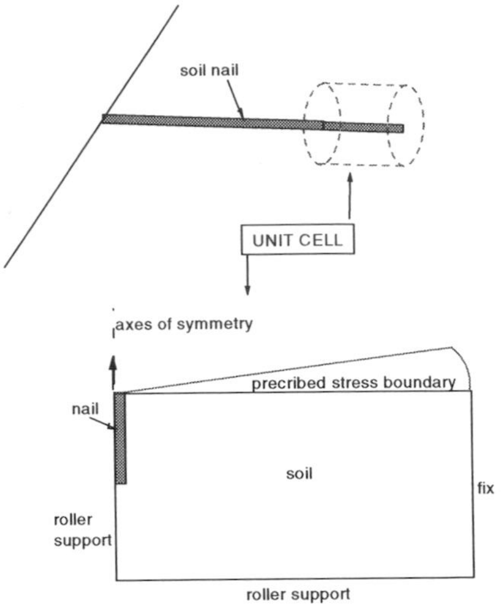

Figure 1. Unit cell for modeling a soil nail.

The other dimensions are: height of unit cell = 2 m, length of nail segment = 1 m, nail radius = 0.067 m. As the soil nail is of a grout-in installation, it included both steel bar and grout, the latter gave a perfectly rough interface with the surrounding soil. The soil was modeled as a Mohr Coulomb elastic plastic material. The soil parameters are:

G = elastic shear modulus = 10 MPa,
ϕ = friction angle = 36°,
ν = dilatancy angle = 0 to $\phi/2$, c = cohesion = 20 kPa during nail installation and reduced to 0 prior to application of nail pullout.

The drop in cohesion implies that the cohesion is an apparent value due to matric suction. Therefore, the critical design condition is one after prolong wetting matric suction being reduced to a negligible value. Although the above are assumed parameters they are considered as reasonable and probably conservative values.

The stress state was first initialized by applying the following boundary stresses:

- σ_r, radial stress, of 100 kPa representing the vertical and lateral stresses in the soil.
- σ_z, stress along axes of symmetry, of 50 kPa representing the stress in the soil along the direction of nail.

The drilling of the nail hole was modeled by the removal of soil elements at the nail location. This led to a change in local stress around the nail hole. The insertion of the nail and grouting was then modeled by re-activating the elements at nail location back on as elastic elements. At this stage the radial stress at the nail-soil interface was reduced to zero. The

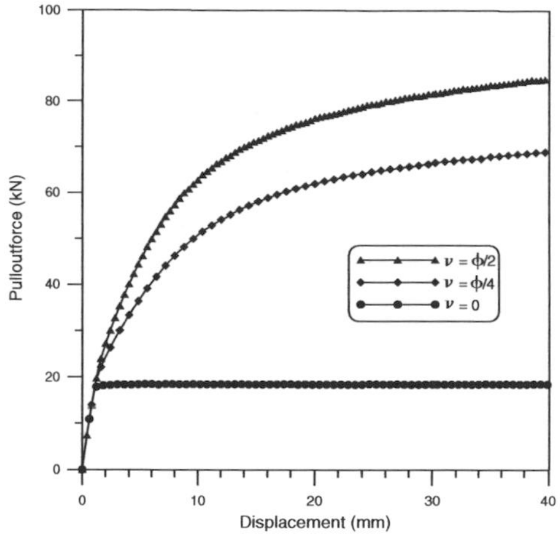

Figure 2. Pullout force versus displacement.

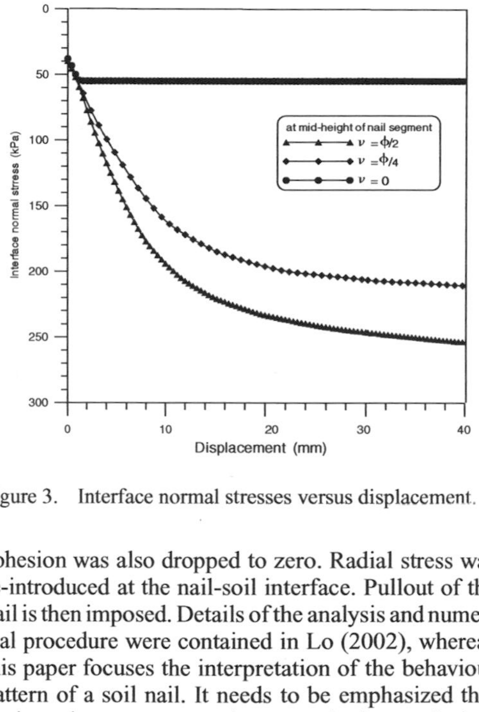

Figure 3. Interface normal stresses versus displacement.

cohesion was also dropped to zero. Radial stress was re-introduced at the nail-soil interface. Pullout of the nail is then imposed. Details of the analysis and numerical procedure were contained in Lo (2002), whereas this paper focuses the interpretation of the behaviour pattern of a soil nail. It needs to be emphasized that the intention was not to make a quantitative prediction. Rather, this is a qualitative investigation.

3.1.2 *Results of analysis*

The computed pullout displacement relationships for 3 different values of dilatancy angle, ν, are presented in Figure 2. It can be seem that the computed pullout response of the nail segment is highly dependent on the dilatancy angle of the soil.

Figure 3 examines the radial stress along the nail-soil interface. This is the local normal stress that

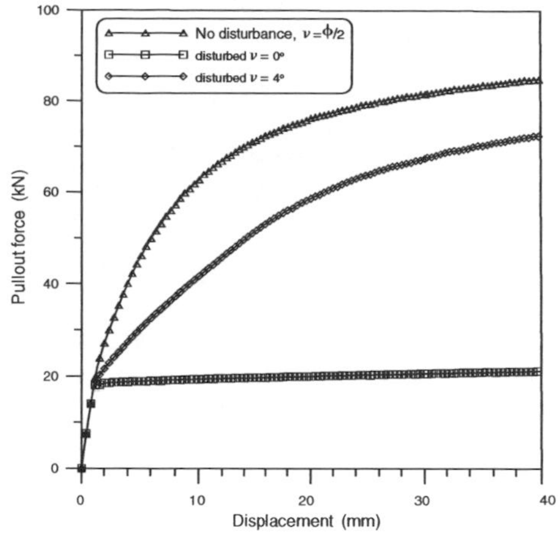

Figure 4. Influence of disturbed zone.

determines the failure shear stress along the interface. This local radial stress was initially at a small value due to the nail installation procedure. However, it increases with pullout displacement and thus enhancing the pullout force that can be mobilized. This local stress generated by the pullout displacement is due to the constrained dilatancy mechanism (Lo 2003) and thus strongly dependent on ν, as evident in Fig. 3. For the cases of $\nu \geq \phi/4$, the local interface normal stress generated during pullout is several times that at the commencement of pullout. This indirectly infers that the in-situ overburden stress is of a minor effect.

The influence of hole disturbance was investigated by setting the dilatancy angle in a 40 mm zone around the nail a reduced value of either 0 or 4°. The "undisturbed" dilatancy angle was taken to be $\phi/2$. As evident from Fig. 4, this can have a considerable adverse effect on the pullout response of the soil nail. For $\nu = 4°$ in the disturbed zone, the pullout response is similar to that of an undisturbed dilatancy angle of $\phi/4$. If the dilatancy of this small zone around the nail is completely lost, then the pullout resistance dropped to ~25% of the undisturbed value.

3.2 *Laboratory studies*

Yin & Su (2006) and Su et al (2006) carried out laboratory investigation using a fully instrumented pull-out box to study the influence of overburden pressure, degree of saturation of compacted completely decomposed granite and influence of installation procedure on pull-out resistance of grouted nail. The general layout of the box is as shown in Figure 5. The internal dimensions of the pull-out box are 1000 mm in length, 600 mm in width and 830 mm in height. The soil chamber has equipped with set of instrumentation to monitor the soil nail performance and soil

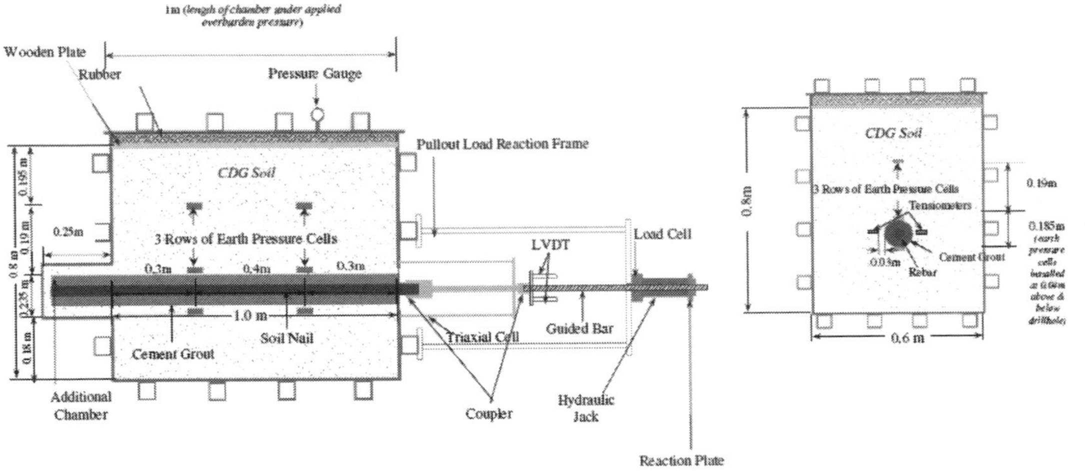

Figure 5. Layout of instrumented pull-out box.

mass response during the course of the laboratory testing. In addition, the soil chamber has provision of an access hole for installation of the test soil nail. The soil is compacted and overburden stress applied prior to installation of soil nail as described below.

An electric drilling machine was used to drill a 100 mm diameter drillhole in the soil through the access hole. Soil nail was then installed in place. Grout was subsequently pumped into the drillhole by using soil sample extruder driven by a motor. After curing for about 5 days when the cement grout had reached a strength of at least 21 MPa, soil nail was pulled out by a hydraulic jack. The soil nail was initially pulled out by a number of load increments (held for about 1 hour). As the peak load reached, the pull-out test continued with a constant rate of 1 mm/min. The procedure adopted is similar to the field test procedure as recommended in Design Technical Guideline No. 11 (GEO, 2004).

Earth pressure cells are installed at different locations within the soil chamber to record the soil responses before and after the formation of the drillhole during installation of test nail. It was reported that uncased drillhole remained stable under a high overburden pressure of 200 kPa.

Su et al (2006) reported a number of pull-out tests carried out under different overburden pressures of range from 40–300 kPa and with different degrees of saturation of 38% and 75%. A typical response of changes of average total earth pressure at different stages of testing for tests with soil at 38% degree of saturation is presented in Figure 6. The authors concluded from the test results that the pull-out resistance of soil nail is dependent on the local stress state of soil around the drillhole at the time of pull-out. The stress in the soil around the drillhole was largely released after drilling and the recovered stress was very small in comparison with the applied overburden pressure. The authors also reported that during the constant rate pull-out test, the average earth pressure measured increased with the development of pull-out resistance and then decreased with subsequent displacements after the peak value had attained. It is apparent that this increase in normal stress was due to the effect of constrained dilatancy caused by shearing of the dense granular soils around grouted nail. Therefore in design of soil nailing system, the normal pressure on the soil nail surface should not be taken as the weight of the soil above the soil nail as a matter of course. This is consistent with the observations of field pull-out tests by Cartier and Gigan (1983) and Clouterre (1991).

3.3 *Field testing*

Cheung et al (2005) reported the behaviour of two soil nails instrumented with strain gauges under pull-out tests. Drillholes were sunk in close proximity to the testing locations to obtain SPT 'N' values and pressuremeter tests were performed at the bond section levels of the test nails.

A grouted length of approximately 2 m was formed for the bond section. Packers were used to ensure the integrity of the grouted bond length of the soil nail. Pull-out tests were carried out using standard set up and test procedures as recommended by Design Technical Guideline No. 11 (GEO, 2004). The general set up of the soil nail pull-out test are as shown in Figure 7.

The authors reported that the overburden pressure of the test nails was about 8 m, with average SPT 'N' values of 75 and 40 at the level near the bond section of the test nails. The average bond strength of the two

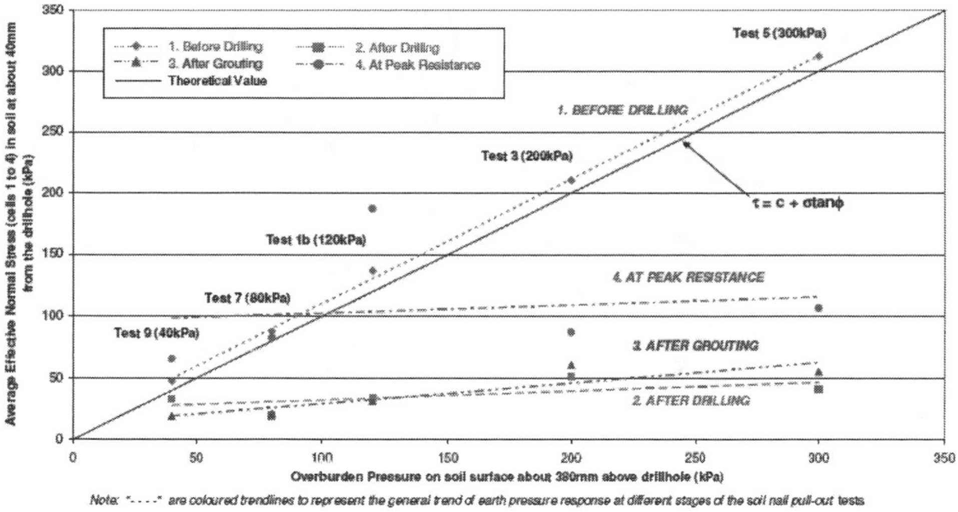

Figure 6. (Typical) average effective normal stress at various stages of soil nail installation vs overburden pressure (for Sr = 38%).

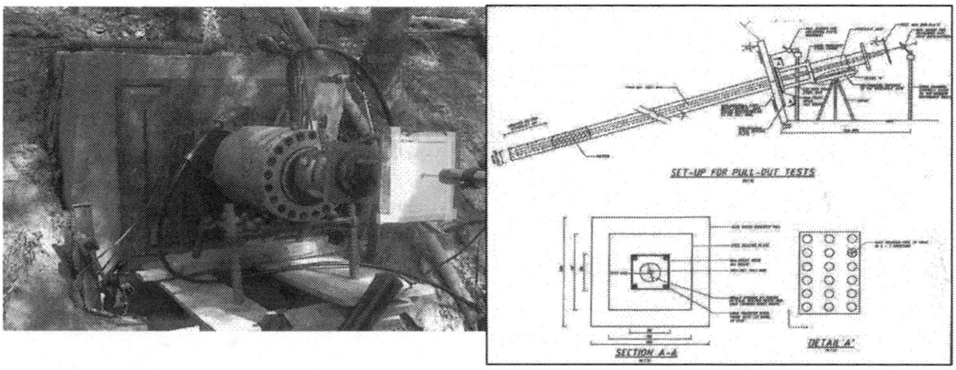

Figure 7. Testing seup of the field pull-out test.

test nails, namely A1 & A2, at the maximum test load was 193 kN/m^{2}and 183 kN/m^2.

Leung & Fu (2005) reported the failure mechanisms of three soil nails using packers under pull-out test, adopting the set up and test procedures as recommended in Design Technical Guideline No. 11 (GEO, 2004). The test nails were installed at about 3 m below ground into colluvium of hard stratum. It was reported that the test nails, namely B1, B2 & B3 were terminated at maximum test loads corresponding to bond strength of 426 kPa, 348 kPa & 451 kPa respectively. The pull-out resistance of soil nail from the field data can be summarized as in Table 1.

It can be seen that the pull-out resistance observed from the field tests do not increase with the increase in overburden height, but may likely directly affected by the SPT 'N' value and by the local ground conditions.

Table 1. Summary of Max. Pull-out Resistance vs SPT 'N' Value, Overburden Height and Local Ground Conditions.

Site/Nail	Resistance (kPa)	SPT 'N'	Overburden height (m)	local ground Conditions
A1	193	75	8	CDV IV/V
A2	183	40	8	CDV V
B1	426	NA	2.5	Hard Stratum
B2	348	NA	2.5	Hard Stratum
B3	451	NA	2.5	Hard Stratum

4 DISCUSSION & CONCLUSION

Finding of the preliminary numerical analysis is in line with field test results reported by other researchers. In

Table 2. Recommended values of bond strength at soil/grout interface.

Countries	Soil types	Recommended soil/grout bond strength in kPa	References
China	Dense Sandy Soil	160 to 200	(CECS 1997)
Japan	Sandy soil of N-Value from 30 to 50	180 to 240	(JH 1998)
United States	Very dense silty sand and gravel	120 to 240	(FHWA 1994 and 1998)
France	Sand of limit pressure at pressuremeter test ranging from 0.5 to 3 MPa	50 to 125	Clouterre (1991)

essence, the installation of soil nail changes the local stress around the soil and subsequent pullout response is strongly influenced by the dilatancy characteristic of the soil.

Laboratory tests on pull-out resistance of grouted soil nails under a constant displacement rate were carried out in compacted completely decomposed granite under different overburden pressures and different degrees of saturation.

The test results indicated that the soil nail pull-out resistance was dependent on the local stress state of soil around the drillhole at the time of pull-out. The stress in the soil around the drillhole was largely released after drilling and the recovered stress was very small in comparison with the applied overburden pressure. Therefore the pullout resistance is largely dependent on the local interface normal stress generated by constrained-dilatancy. The pull-out resistance may decrease as the degree of saturation increases.

It has also observed from the field tests carried out that the effect of overburden pressure on the pull-out resistance was not obvious. In addition, the results indicated that the pull-out resistance observed on site is affected by the local ground (stress) condition (may be also the ground water condition) during installation. The local soil condition probably affects the constrained dilatancy. The ground water condition is probably related to the extent of disturbance during nail installation. It should be noted it is important to ensure that field testing are adopting a similar set up and testing procedure which enables inconsistency in testing method to be minimized and allows pull-out test results for different site conditions to be compared meaningfully. The adoption of using consistent grouting technique (i.e. packers) enables inconsistency in grouting bond length to be ascertained. The pullout resistance from the field tests indicated a bonding stress of 200 kPa to 400 kPa may develop depending on the local ground conditions which may be reflected using SPT 'N' value. The ultimate bond strength of under 200 kPa at the soil/grout interfaces may be comparable with those recommended in the codes of practice for soil nails in various countries as shown in Table 2.

REFERENCES

Cartier, G. & Gigan, J.P. (1983). Experiments and observations on soil nailed structures. Proc. 8th European Conf. Soil Mechanics and Foundation Engineering, Helsinki, Vol. 2, 473–476.

CECS 96:97 (1997). Technical Specification for Excavations Supported by Soil Nails.

Clouterre (1991). Recommendation Clouterre. English Translation Published by FHWA (1993). Publication No. FHWA-SA-93-026.

Cheung C.T., Chan, C.F., Chiu, S.L. & Tang, K.L. (2005) Behaviour of two soil nails under pull-out test. The HKIE Geotechnical Division 25th Annual Seminar, Safe and Green Slopes, Proceedings, pp. 222–229.

FHWA (1998). Manual For Design & Construction Monitoring Of Soil Nail Walls. Federal Highway Adminstration 1998. Publication No. FHWA-SA-96-069R.

FHWA (1994). Soil Nailing Field Inspectors Manual. Federal Highway Administration 1994. Publication No. FHWA-SA-93-068.

Chu, Lok-Man & Yin, Jian-Hua (2003). Shear strength at the interface between unsaturated soil and concrete grout. The 2nd Asian Conference on Unsaturated Soils, 15–17 April 2003, Osaka, Japan.

Chu, Lok-Man & Yin, Jian-Hua (2005). Comparison of Interface Shear Strength of Soil Nails Measured by Both Direct Shear Box Tests and Pullout Tests. Journal of Geotechnical and Geoenvironmental Engineering, ASCE, September, Vol.131, No.9, pp 1097–1107.

Davies, M. C. R., & Le Masurier, J. W. (1997). Soil/nail interaction mechanisms from large direct shear tests. Proceedings of the 3rd International Conference on Ground Improvement Geosystems, London, pp 493–499.

Franzen, G. & Jendeby, L. (2000). Pullout capacity of soil nails. Proceedings of the 4th International Conference on Ground Improvement Geosystems – Grouting, Soil Improvement and Geosystems including Reinforcement, Finnish Geotechnical Society, pp 347–354.

Franzen, G. & Jendeby, L. (2001). Prediction of pullout capacity of soil nails. Proceedings to 15th International Conference on Soil Mechanics and Geotechnical Engineering, Vol 3, pp 1743–1747.

GEO (2004). Pull-out Test of Soil Nails in Hong Kong. Design Division Technical Guideline No. 11. Geotechnical Engineering Office, Hong Kong.

Heymann, G., Rohde, A. W., Schwartz, K. & Friedlaender, E. (1992). Soil nail pull out resistance in residual soils. Proceedings International Symposium on Earth Reinforcement Practice, Fukuoka, Japan, pp 487–492.

JH (1998). Design & Works Outlines on the Soil Cutting Reinforcement Soil Works. Japan Highway Public Corporation.

Lee, C. F., Law, K. T., Yue, Z. Q., Tham, L G. & Junaideen, S. M. (2001). Design of a large soil box for studying soil-nail interaction in loose fill. Proc. of the 3rd International Conference on Soft Soil Engineering, Hong Kong, pp 413–418.

Leung, K.Y. & Fu, K.F. (2005) Failure mechanisms of three soil nails under pull-out test. Paper submitted to HKIE for competition.

Li, K.S. & Lo, S.R. (2007). Discussion of "Comparison of Interface Shear Strength of Soil Nails Measured by Both Direct Shear Box Tests and Pullout Tests" by Chu & Yin. Journal of Geotechnical and Geoenvironmental Engineering, ASCE, March, pp 344–346.

Lo S.R. (2002). Soil nail pull-out : influence of dilatancy angle. Workshop on Numerical Modelling in Geotechnical Engineering, CRPD, HK. Dec .

Pradhan, B., Yue, Z, Q., Tham, L. G., & Lee, C. F. (2003). Laboratory study of soil nail pullout strength in loosely compacted silty and gravelly sand fills. 12th Pan American Conference on Soil Mech & Geot Engr., MIT, 2139–2146.

Su, L.J., Chan, C.F. Shiu, Y.K., Chiu, S.L. & Yin, J.H. (2006). Laboratory testing study on the influence of overburden pressure on soil nail pullout resistance in compacted completely decomposed granite fill. Paper submitted to ASCE Geotechnical & Geoenvironmental Engineering, January, in-press.

Yeo, K.C. & Leung, S K (2001). Soil nail design with respect to Hong Kong conditions. Proc. Symposium on Reinforced Soil in Kyushu, Japan, 14–16 November 2001.

Yin, Jian-Hua & Su, L.J. (2006). An innovative laboratory box for testing nail pull-out resistance in soil. Geotechnical Testing journal, Vol. 29, No. 6, ASCE International, Paper ID GTJ100216.

New Horizons in Earth Reinforcement – Otani, Miyata & Mukunoki (eds)
© 2008 Taylor & Francis Group, London, ISBN 978-0-415-45775-0

Shear tests on fibre reinforced sand

A. Diambra, E. Ibraim & D.M. Wood
Department of Civil Engineering, University of Bristol, Bristol, United Kingdom

A. Russell
School of Civil and Environmental Engineering, University of New South Wales, Sydney, Australia
(formerly Department of Civil Engineering, University of Bristol)

ABSTRACT: The concept of reinforcing soils by introducing tension-resisting elements such as fibres is becoming widely accepted by geotechnical engineers. This follows research which shows that mixing sands with random discrete flexible fibres causes a significant increase in strength and a reduced post-peak strength loss. In this research Hostun RF sand reinforced with flexible discrete polypropylene fibres has been tested using direct shear tests and conventional triaxial tests. The moist tamping technique was used for specimen preparation. The results of the direct shear tests indicate that inclusion of fibres increases the peak shear strength and the strain required to reach the peak. A continuous increase of the deviatoric stress on the stress-strain response up to 30–40% of axial strain was recorded for conventional drained triaxial tests. A more dilative volumetric behaviour was systematically observed for the reinforced specimens.

1 INTRODUCTION

Reinforcement with randomly distributed discrete flexible fibres is an effective and reliable technique for increasing the strength and stability of granular soils in a variety of applications ranging from retaining structures and embankments to subgrade stabilization beneath footings, pavements and sport pitches.

The influence and the contribution of fibre reinforcement to shear strength of sand have been examined by various investigators (Gray & Ohashi 1983; Michałowski & Čermák 2003 and Heineck et al. 2005 among others). Several parameters such as confining stress, fibre type, volume fraction, density, length, aspect ratio, modulus of elasticity, orientation and soil characteristics including particle size, shape, gradation have been studied using monotonically loaded direct shear tests, consolidated drained triaxial tests or unconfined compression tests.

Presented in this paper are results of direct shear tests and drained triaxial compression tests for a fine sand, Hostun RF sand, reinforced with randomly distributed discrete flexible polypropylene fibres. The objective of this research is to examine the influence of fibre addition on stress-strain and volumetric behaviour of the fibre-reinforced composites.

2 EXPERIMENTAL CONDITIONS

2.1 *Materials*

The sand tested in this study was Hostun RF (S28) sand. The material characteristics of the sand were: mean grain size $D_{50} = 0.32$, coefficient of uniformity $C_u = 1.70$, coefficient of gradation $C_g = 1.1$, specific gravity $G_s = 2.65$ and minimum and maximum void ratio $e_{min} = 0.62$ and $e_{max} = 1.00$ respectively (Ibraim 1998).

Loksand™ discrete polypropylene crimped fibres (Fig. 1) of length $l_f = 35$ mm and of circular cross section with diameter $d_f = 0.1$ mm have been used. Other physical properties of Loksand™ fibres are reported in Table 1.

2.2 *Specimen preparation*

The specimen preparation involved two stages: mixing and compaction. In order to avoid the floating of fibres and segregation, the sand was mixed with some amount of water. The results of Modified Proctor compaction tests on fibre reinforced specimens indicated the existence of an optimum moisture content of 10%, independent of the amount of fibres (Ibraim & Fourmont 2006). Once the water is added to the sand, the fibres are progressively added and mixed through

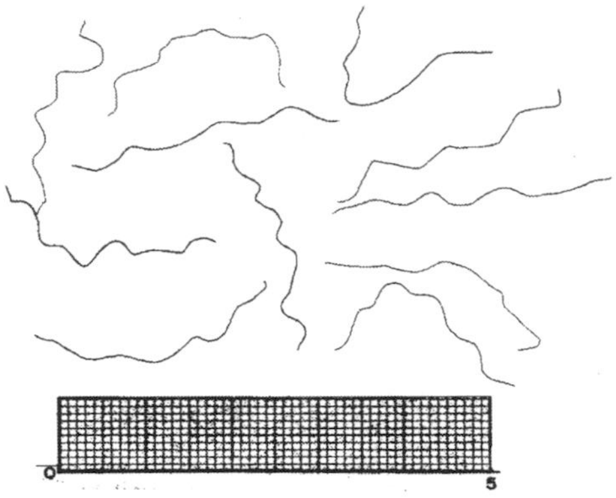

Figure 1. Specimen of individual crimped polypropylene fibres (scale in mm).

Table 1. Characteristics of Loksand™ fibres.

Weight (Denier)	Tensile strength (MPa)	Specific gravity	Elongation at break	Moisture regain
50	225	0.91	160%	<0.1%

using a small spoon. Mixing was stopped when, by visual examination, it was considered that the fibres were evenly distributed throughout.

Rectangular specimens of 100 × 100 mm and 45 mm height were used for the direct shear tests, whereas specimens with diameter 70 mm and height 70 mm were prepared for the triaxial tests. In both cases, three layers of soil were used. The amount of mixture required to form the first layer was delicately deposited into the specimen's mould to ensure a zero drop height and minimal disturbance to the fibre distribution. The mixture was then compacted by tamping with a light rectangular or circular hammer up to the desired layer height. It was observed that the already compacted layers did not under-compact during the formation of the subsequent layers.

Previous research (Diambra et al. 2007) has shown that fibres in reinforced specimens prepared with the moist tamping technique have a near horizontal orientation. 97% of fibres are orientated within 45° of the horizontal.

The average concentration of fibres added in the composite is defined as a fraction of dry mass of sand:

$$w_f = \frac{W_f}{W_s} \times 100\ (\%) \tag{1}$$

where W_f is the weight of fibres and W_s is the weight of the dry sand.

The maximum amount of fibres that can be mixed with a given amount of sand, placed into a given volume and compacted using moist tamping has been determined by Ibraim & Fourmont (2006). The results are presented in the Figure 2 where the minimum and maximum void ratios of the Hostun sand are also plotted. A value of 2% fibre content seems to be a limit beyond which sand reinforced specimens cannot be prepared.

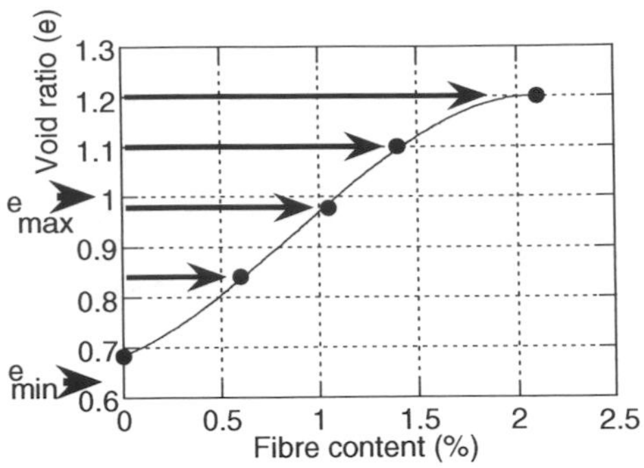

Figure 2. Maximum amount of fibres that can be mixed with a fixed amount of sand without leading to a change in specimen volume using moist tamping.

3 DIRECT SHEAR TEST RESULTS

A recently improved Direct Shear Apparatus (Dietz 2000; Lings & Dietz 2004) was first used for fibre reinforced specimen testing. Three different nominal fabrication void ratios (e) of 0.8, 0.9 and 1.0 (which respectively corresponds to relative densities, D_r, 0%, 26% and 53%) have been chosen for the experimental testing programme. Details of the apparatus, measurement conditions and specimen reinforcement details are given elsewhere (Ibraim & Fourmont 2006). For consistency, similar densities and fibre contents have been used for the experiments conducted using the triaxial apparatus. In all cases, void ratio refers only to the sand matrix since fibres are considered as a part of voids.

Some typical direct shear responses for Hostun RF sand (e = 1.0) reinforced with randomly distributed polypropylene fibres are presented in Figure 3. The figure presents the variation of the shear stress and vertical displacement (v_y) with the horizontal displacement (v_x). The fibre contents are specified on the figure. As can be observed, the inclusion of fibres increases the shear strength of very loose specimens. Also, as observed by Palmeira & Milligan (1989), Kaniraj & Havanagi (2001), Jewell & Wroth (1987) and Shewbridge & Sitar (1989), the amount of vertical dilation increases with the amount of fibres.

Overall, as presented in Figure 4, the increase of the peak shear strength is very close to being a linear function of fibre content. For specimens of lower density at relatively low effective normal stresses the rate of the

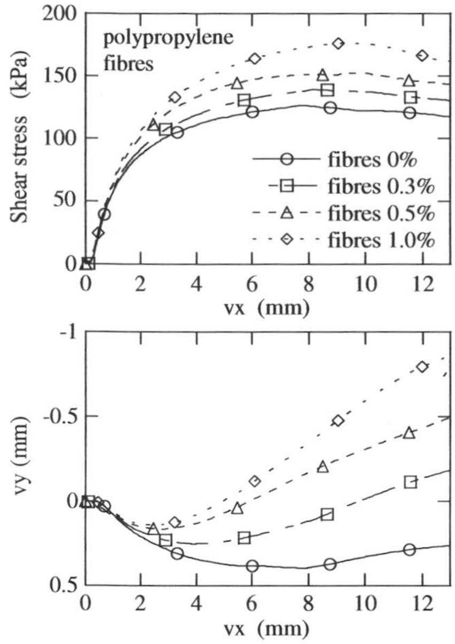

Figure 3. Typical direct shear box test results for Hostun RF sand (approximately 0% relative density) reinforced with polypropylene fibres.

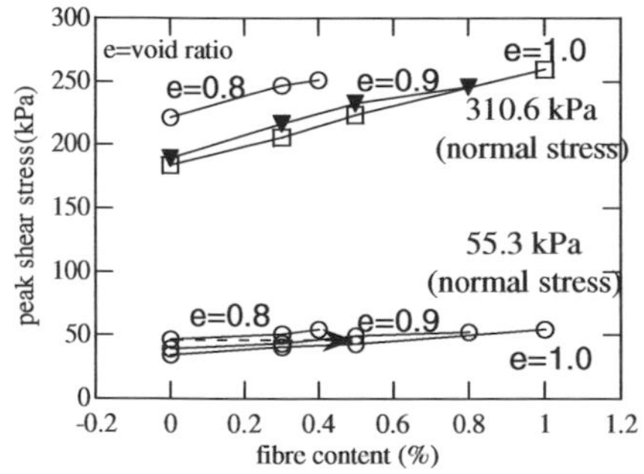

Figure 4. Evolution of the peak shear stress with fibre content, density and effective normal stress.

increase is less than for higher densities and confining stresses. Nevertheless, above some limiting fibre content, the peak stress increase approached an asymptotic upper limit, more pronounced for denser specimens and higher normal stresses. A similar trend was observed by Gray & Al-Refeai (1986), Ranjan et al. (1996) and Murray et al. (2000) but for lower confining stress levels. The loosest specimens at the highest fibre content (1.0%) revealed 40 to 60% gain in strength.

4 TRIAXIAL TEST RESULTS

Triaxial tests were conducted on fully saturated specimens using effective confining pressure ranging from 100 kPa to 300 kPa. Saturation has been achieved by firstly flushing CO_2 trough the specimen and then flushing water slowly trough the same. Satisfactory saturation was monitored in each test, ensuring a B value of at least 0.95 for drained tests and 0.97 for the undrained tests. A back pressure of 300 kPa was used in all the tests.

Axial strain was monitored with the use of an LVDT placed outside the cell and following the movement of the loading ram. Volumetric strains were measured by a double chamber volume change gauge which uses a LVDT to monitor the relative position of the two chambers. Pore pressure was measured by a pore pressure transducer with a capacity up to 800 kPa and a resolution of ± 0.05 kPa. The measurement system was completed by an internal 5 kN load cell.

Lubricated ends were used at the top and bottom of the specimen. The lubrication consisted on silicone grease and two or three latex rubber disks at the bottom and the top of the specimen, respectively. By visual inspection, the homogeneous shape of the specimen was well preserved at least up to 20% of axial strain (ε_a). Lubricated ends induced bedding error for the strain response, but no any correction has been applied to the results presented here. No membrane penetration correction has been considered either.

Table 2 reports a list of all the triaxial tests performed in this study. The test name gives an

Table 2. List of the triaxial test performed ($p' = (\sigma'_1 + 2\sigma'_3)/3$, $q = \sigma'_1 - \sigma'_3$ and ψ_{max} is the maximum angle of dilation).

Test	Void ratio before starting the test	q/p' (at 20% strain)	ψ_{max} (degrees)
L100-00	0.991	1.374	0
L100-03	0.983	1.736	0.78
L100-06	0.979	1.984	2.51
L100-09	0.987	2.260	2.42
L200-00	0.980	1.326	0
L200-03	0.975	1.434	1.46
L200-06	0.969	1.808	1.90
L200-09	0.962	2.020	2.12
M100-00	0.914	1.352	2.78
M100-03	0.911	1.778	5.77
M100-03/2	0.936	1.704	4.30
M100-06	0.919	2.051	7.77
M100-06/2	0.931	1.994	7.24
M200-00	0.915	1.328	3.42
M200-03	0.918	1.615	4.62
M200-06	0.907	1.882	5.20
M300-00	0.928	1.314	2.31
M300-03	0.916	1.528	2.89
M300-06	0.928	1.773	4.59
D100-00	0.833	1.473	16.66
D100-03	0.832	1.947	18.52

*$q/p'q/p'$ at 15% of strain

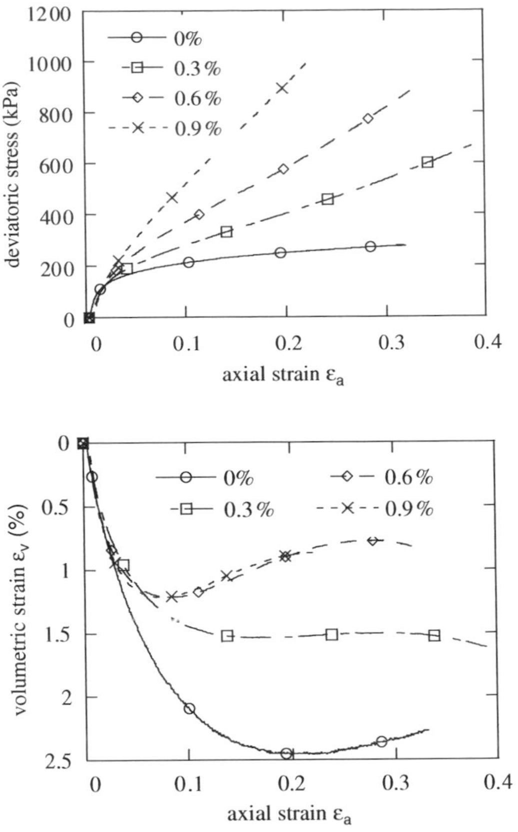

Figure 5. Stress-strain and volumetric behaviour for drained tests on L100 series specimens (legend gives the fibre content).

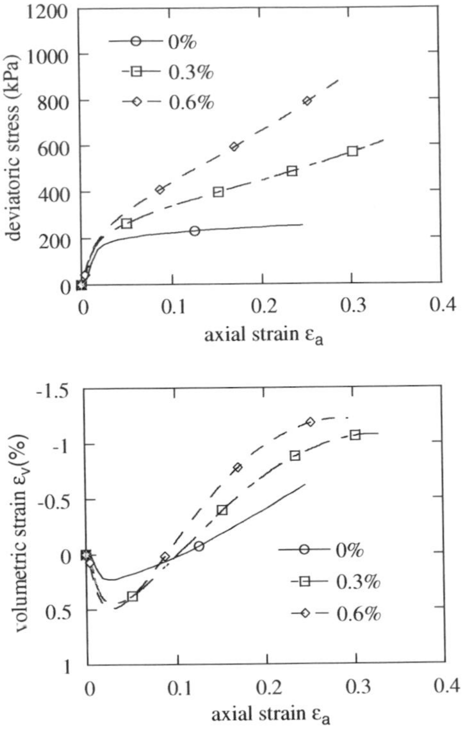

Figure 6. Stress-strain and volumetric behaviour for drained tests on M100 series specimens (legend gives the fibre content).

explanation for the test conditions: the first letter indicates the nominal density (L stands for loose, M for medium and D for dense), the following three figures indicate the testing cell pressure in kPa, and finally the used percentage of fibres (w_f) is mentioned. For example L200-06 means a test for a specimen with a void ratio close to 1, at a constant cell pressure of 200 kPa, reinforced with 0.6% of fibres. For two tests, the length of the fibres has been reduced by half (symbol '/2').

Typical results of drained triaxial tests are presented in Figures 5–7 where the variations of the deviatoric stress (q) and the volumetric strain (ε_v) are presented with the axial displacement (ε_a). The results for the unreinforced sand are in accord with the already published results on identical sand specimens (Ibraim 1998) and the maximum angles of friction mobilised at 20% of axial strain varies between 32°–34° for the test series L, M and 36° for the tests D.

The stress responses for unreinforced and reinforced sand seem to be similar at low displacements suggesting that the initial stiffness of the composite is not influenced by the presence of fibres (Heineck et al. 2005). With increasing axial strain, the contribution of fibres to the strength of the composite increases and the deviatoric stress for reinforced sand specimen is greater than for an unreinforced one. An increase of the mobilised angle of friction of up to 60% was recorded at 20% of axial strain for the loosest specimen at a cell pressure of 100 kPa. Despite the fact that some tests were carried out until the axial strain reached 30 to 40%, the reinforced specimens show a linear stress-strain relationship that does not seem to flatten. This pattern seems to suggest that the interaction mechanism between the fibres and the sand matrix is not weakened by the deformation process, the fibres are not pulled out and they do not break either. Although the use of a scanning electron microscope for the inspection of polypropylene fibres would be more recommended for a possible indication of the fibre deformation mode, at this stage of the research only visual inspection of the fibres was carried out. Exhumed specimens revealed no appreciable plastic deformation of the fibres and no breakages.

Figure 8 shows near linear trends of deviatoric stress increase with the fibre content added to the specimen.

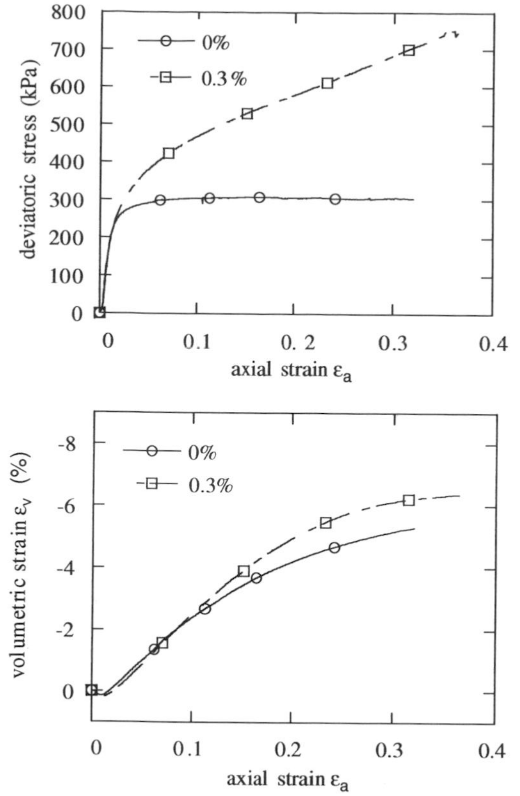

Figure 7. Stress-strain and volumetric behaviour for drained tests on D100 series specimens (legend gives the fibre content).

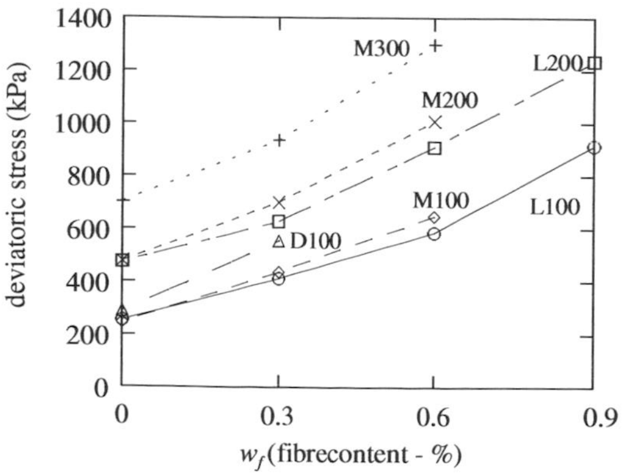

Figure 8. Trend of the deviatoric stress at 20% strain versus the fibre content for the tests performed.

As mentioned earlier, the specimens prepared with a moist tamping technique have a marked horizontal orientation of fibres. As in a triaxial test the direction of the tensile strain is horizontal, the contribution of

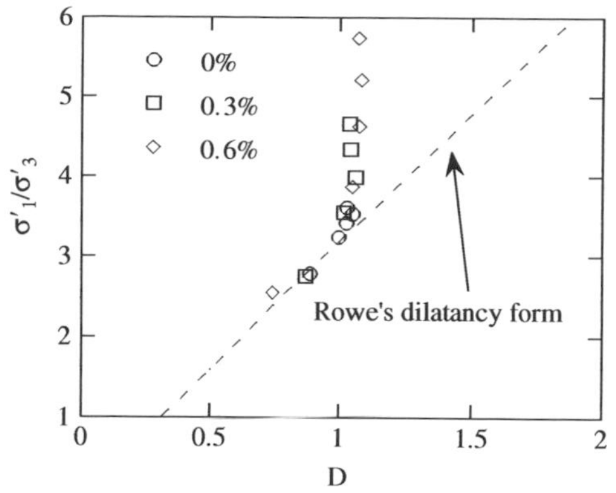

Figure 9. Stress dilatancy ratio for the test series M200 (legend express fibre content).

the fibres to the composite performance is enhanced by this particular orientation distribution.

For all the densities considered in this study, the shape of the volumetric curve response for reinforced sand, as for the unreinforced sand, showed an initial contraction followed by dilation. The addition of fibres to the composite resulted firstly in a decrease of the compression of the specimen followed then by a more dilative response, as also observed in the direct shear tests. The maximum recorded dilation angle, ψ_{max}, is given in the Table 2 and ψ is defined as:

$$\tan\psi = -\frac{d\varepsilon_v^p}{d\varepsilon_a^p} \qquad (2)$$

where the superscript p stands for plastic.

For some technical reasons, the volumetric change for the medium unreinforced sand at 100 kPa cell pressure (test: M100-00) did not follow the expected trend and this test will be performed again.

Figure 9 reports the trend of the dilatancy ratio evolution (defined as $D = 1 - dV/V/d\varepsilon_a$) with the principal stress ratio. Comparison with Rowe's relationship (Rowe 1962) for interparticle friction $\varphi_\mu = 31.5°$ is also shown. For a reinforced specimen, when the volumetric behaviour is constant, there is still an increase on the supported deviatoric stress leading to a vertical trend on the stress-dilatancy plane and the data diverge from the Rowe's relationship.

Few experiments were also conducted to investigate the influence of fibre length on the behaviour of the composite. Fibres having half the normal length were used in these tests, but for comparison the average concentration of fibres was kept constant. Although it was not possible to cut the fibres exactly in their mid point, the qualitative influence of fibre length on the stress-strain behaviour of the composite could still be assessed. Results demonstrated that the

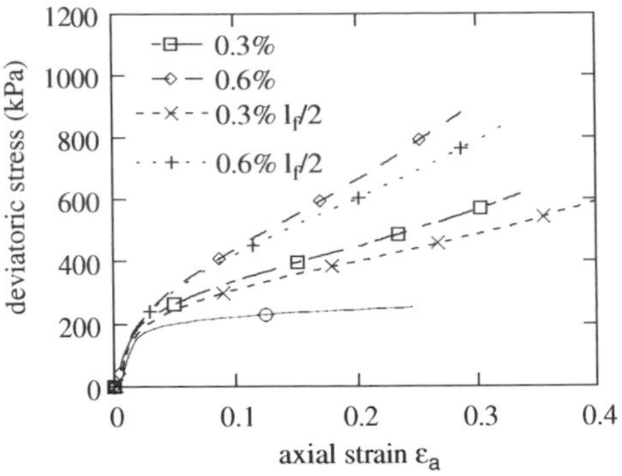

Figure 10. Influence of fibre length on the stress strain relationship on M100 specimens (legend express fibre content and fibre length).

performance of the composite was affected by decreasing the length of the fibres as shown in Figure 10. A drop of about 2° has been recorded for the mobilised friction angle at 20% of axial strain. Increasing the length of fibre reinforcements increases the shear strength of the fibre-sand composite (Al Refeai 1991). However, this increase was effective only up to a point beyond which any further increase in fibre length had no effect on shear strength.

5 CONCLUSIONS

An experimental program was undertaken to investigate the individual effect of randomly distributed crimped polypropylene fibres on the mechanical behaviour of the Hostun RF sand. Direct shear tests and drained triaxial compression tests were performed on unreinforced and reinforced sand.

In both tests the addition of polypropylene fibres increases the shear strength response of the specimen and increased its dilative behaviour. For the highest fibre concentration used (1.0%), a gain on the peak strength of 60% was recorded in the direct shear test. In the triaxial tests an increase of the deviatoric stress apparently without limit was observed and an increase of the mobilised angle of friction (at 20% of axial strain) up to 60% was also observed.

Dilatancy ratio evolution with the principal stress ratio for reinforced specimens shows a different relationship from the one proposed by Rowe (1962).

The strength of the reinforced specimens was found to be dependent on the length of the fibres. A small reduction of the mobilised friction angle was recorded when shorter fibres were used.

These experimental results highlights the potential use of flexible discrete random fibres for the reinforcement of fine granular materials with applications including more resistant earthfills, foundations for buildings, heavy trafficked pavements and slope stabilization.

ACKNOWLEDGEMENTS

The authors would like to thank Drake Extrusions Ltd for support on providing the Loksand fibres.

REFERENCES

Al-Refeai, T.O. 1991. Behaviour of granular soils reinforced with discrete randomly oriented inclusions. *Geotexiles and Geomembranes*, 10:319–331.

Diambra, A. Russell, A.R. Ibraim, E. & Muir Wood, D. 2007. Determination of fibre orientation distribution in reinforced sand, *Geotechnique (in press)*.

Dietz, M.S. 2000. Developing an holistic understanding ofinterface friction using sand within the direct shear apparatus. *PhDthesis*, University of Bristol.

Lings, M.L. & Dietz, M.S. 2004. An improved direct shear apparatus for sand. *Géotechnique*, 54(4):245–256.

Gray, D.H. & Ohashi, H. 1983. Mechanics of fiber reinforcement in sands. *Journal of Geotechnical Engineering, ASCE* 109(3): 335–353.

Gray, D.H. & Al-Refeai, T. O. 1986. Behaviour of fabric – versus fiber-reinforced sand. *Journal of Geotechnical Engineering.* 112(8): 804–820.

Heineck, C.S., Consoli, N.C. & Coop, M.R. 2005. Effect of microreinforcement of soils from very small to large shear strains. *Journal of Geotechnical & Geoenvironmental Engineering*, 131(8):1024–1033.

Ibraim, E. 1998. Différents aspects du comportement des sables à partir d'essais triaxiaux: des petites déformations à la liquéfaction statique. *PhD thesis*, ENTPE Lyon

Ibraim, E. & Fourmont, S. 2006. Behaviour of sand reinforced with fibres. In Ling, H. and Callisto (ed.), *Soil stress-strain behaviour: Measurement, Modelling and Analysis, Geotechnical symp., Roma, 16–17 March 2006.*

Jewell, R.A. & Wroth, C.P. 1987. Direct shear tests on reinforced sand. *Géotechnique* 37(1): 53–68.

Kaniraj, S.R. & Havanagi, V.G. 2001. Behavior of cement-stabilized fiber-reinforced fly ash-soil mixtures. *Journal. of Geotechnical and Geoenvironmental Engineering*, 127(7): 574–584.

Michałowski, R.L. & Čermák, J. 2003. Triaxial compression of sand reinforced with fibers. *Journal of Geotechnical & Geoenvironmental Engineering*, 129(2):125–135.

Murray, J.J. Frost, J.D. & Wang, Y. 2000. Behaviour of sandy silt reinforced with discontinuous recycled fiber inclusions. *Transportation Research Record*, 1714: 9–17.

Palmeira, E.M. & Milligan, G.W.E. 1989. Large scale direct shear tests on reinforced soil. *Soils andfoundations*, 29(1): 18–30.

Ranjan, G. Vasan, R.M. & Charan, H.D. 1996. Probabilistic analysis of randomly distributed fiber-reinforced soil, *J. of Geotech. Eng.* 122(6): 419–426.

Rowe, P.W. 1962. The stress dilatancy relation for static equilibrium of an assembly of particles in contact. *Proc. R. Soc. London*, Ser. A, 269:500–527.

Shewbridge, S.E. & Sitar, N. 1989. deformation characteristics of reinforced sand in direct shear. *Journal of Geotechnical Engineering.* 115(8): 1134–1147.

New Horizons in Earth Reinforcement – Otani, Miyata & Mukunoki (eds)
© 2008 Taylor & Francis Group, London, *ISBN 978-0-415-45775-0*

Effect of plasticity index and reinforcement on the CBR value of soft clay

S.A. Naeini & M.R. Yousefzadeh
Department of Civil Engineering, Imam Khomeini International University, Qazvin, Iran

ABSTRACT: In recent years, soil reinforcement is considered of great importance in many different civil projects. One of the most significant applications of soil reinforcement is in road construction. Sub grade soil and its properties are very important in the design of road pavement structure. Its main function is to give adequate support to the pavement from beneath. Therefore, it should have a sufficient load carrying capacity. One of the most appropriate methods for increasing this parameter of the soil is to reinforce it by means of geogrid which is one kind of geosynthetic materials.Geogrid reinforcement of sub grade soil is achieved through the increase of frictional interaction between the soil and the reinforcement. Further, if the weak sub grade is stabilized and reinforced, the crust thickness required will be less that would be more cost saving. Thus, in this paper the effects of plasticity index and also reinforcing soft clay on CBR values are studied. Three samples of clay with different PI values are selected and tested without reinforcement. Then by placing one and two layer of geogrid at certain depth within sample height, standard CBR tests with ASTN D1883 method are carried out. The result of these tests shows that increase in PI of the clay will decrease the CBR value and reinforcing clay with geogrid will increase the CBR value.

1 INTRODUCTION

Concrete or asphalt pavement can not be constructed on weak soil, because in this case the pavement will be easily cracked. As sub grade soil function is to transfer applied loads from pavement to the layer beneath, it should have a sufficient load carrying capacity. One of the methods for improving the weak sub grade soil strength and stability is to reinforce it with geogrid.

In general, geogrids are sheets made of polymer material whose main characteristic is its invulnerability against corrosive elements in soil. Hence, from this point of view, geogrids have many different applications in geotechnical engineering and improving soil properties.

The presence of high friction not only prevents the sliding between soil and the reinforcement element, but also helps the process of transferring stress from soil into the reinforcement element. Lack of integrity of geogrid in different levels causes some types of interlocking with soil particles. In addition, it's the little stiffness of geogrids that makes it possible to refer the increase in strength properties of soil to tensile strain created in geogrids. Using reinforcements in sub grades can increase safety coefficient of embankment stability and also decrease displacements. Furthermore, if the weak sub grade is stabilized or reinforced, the crust thickness required will be less, which results in less repairs and overall economy.

As it is known, in road construction, one of the most significant parameters for designing road sub grades is CBR value. In some projects, because of soft clay soils, CBR value is low, thus different methods such as reinforcing with geogrids are used to improve soil behavioral characteristics. The purpose of this research is to study and measure the effects of PI and also reinforcing soft clay on CBR values.

2 LITERATURE SURVEY

Rao et. al. (1989), Shetty (1989), Rao and Raju(1990), Gopal Ranjan and Charan(1998) presented the results of series of laboratory CBR tests (soaked and unsoaked) on silty sand (SM) reinforced with randomly distributed polypropylene fibers. The test result showed that CBR value of the soil increase significantly with increase in fiber content. The increase in CBR was observed to be 175% and 125% under soaked and unsoaked conditions, respectively with addition of 3% fibers (by weight).

Cancelli et. al. (1996) Montanelli et. al.(1997), Perkins and Ismeik (1997) analyzed the results of a full scale pavement test conducted on several reinforced sections by use of geogrids with saturated silty clay soil having the in-situ CBR value of about 1% to 8%. The test result showed that multi layer geogrids provide the best base reinforcement results for sub base soil having

CBRs equal to 3% or lower. No major differences were found between different single layer integral geogrids. The higher tensile modulus geogrids have shown better contribution at CBRs 3% or lower. The percent reduction of rutting, between reinforced and unreinforced sections, increases with reducing the sub grade CBR, for all geosynthetics. The Traffic Improvement Factor for road service life increases for deep allowed ruts, lower CBR values and lower pavement structural number.

Gosavi, et. al.(2004), Mittal and Shukla (2001) investigated the strength behavior of locally available black cotton soil reinforced with randomly mixed geogrid woven fabric and fiberglass. CBR value of black cotton soil is 4.9% without geogrid. Soaked California Bearing Ratio test results show considerable increase in the CBR value for black cotton soil when reinforced. CBR value of black cotton soil increases 42% to 55% when 1% woven fabrics and fiberglass, respectively are added randomly. The rate of increase in CBR value with 2% addition of fibers is less and the absolute value of CBR still decrease with more addition of fibers. Increase in % CBR is more for higher aspect ratio of fibers. This may be because of higher tensile strength of the woven fabrics. For addition of 3% woven fabrics the rate of increase in CBR value decreases. Increase in the CBR value is due to the compaction characteristics of the fiber reinforced soil. Higher compaction in their study achieved by addition of fiber with higher aspect ratio up to certain limit. It was concluded that for flexible pavement design, higher value of CBR (percentage) for sub grade soil gives lesser pavement thickness and which proves to be economical solution in the pavement construction.

3 TESTED MATERIAL

Geogrid used as reinforcement was cut in to circular pieces with the same diameter of CBR mould (15.2 cm). It was used as an artificial reinforcement. Properties of the reinforcement are given in Table 1.

Soft clay soil was collected locally from Khatoon Abad (located in Semnan road) and was used for experimental work. Bentonite was used as a material for changing PI of the clay soil samples. Properties of the soil with different percentage of bentonite are given in Table 2.

Table 1. Properties of the reinforcing material (geogrid).

Name	Material	Std.weight g/m2	Mesh Aperture, mm	Mesh Thickness, mm
GS 50	LDPE	300	2	1

4 EXPERIMENTAL WORK

7 kegs of unsoaked soft clay soil (CL) passed through No.40 sieve are mixed with optimum water content of 11.4% (by weight) which is obtained from modified compaction test. The mixture achieved is then hammered in 5 layers within CBR mould and become ready for performing unsoaked CBR test. However, for soaked CBR test, the mould should be placed under water, until it is completely saturated. In the next stage the soil is reinforced at layer 2, in the way that geogrid is put between layer 2 and 3 (from the top). The soaked and unsoaked CBR test is carried out. CBR test is also performed in 2 layers, that in this case geogrid is put at the first and the third layer. Thus, soaked and unsoaked CBR of soil is achieved. By performing activities explained above, required information related to the CBR of soil with PI 16 without any geogrid, with 1 layer of geogrid and 2 layers of geogrid under soaked and unsoaked condition is achieved.

For preparing the next sample, soil is mixed with 10% (by weight) of bentonite, in the way that it becomes 7kgs totally. PI of soil is 16. Optimum water content of 12.2% is added to the soil and is completely blended until the mixture obtained becomes homogeneous and then compacted in CBR mould. CBR test is carried out under soaked and unsoaked and also reinforced and unreinforced condition.

The last sample is prepared by mixing soil with 20% (by weight) of bentonite with total weight of 7 kgs. The soil PI is 23 and 13.9% of optimum water content is added to it. CBR test is carried out under both soaked and unsoaked condition. In addition, the soil CBR is obtained by putting geogrid at layer 2 and also at layer 1 and 3.

5 RESULTS OF TESTS

The results of performing soaked and unsoaked CBR tests and reinforced and unreinforced situation for various PIs are shown in Table 3 and also Figures 1 to 6.

Table 2. Properties of soft clay soil and soils tested.

Soil Type	Color	Maximum Dry density (MDD) KN/m3	Gs	LL %	PL %	PI %
CL	Brown	19.4	2.62	25.5	15.5	10
CL + 10% Bentonite	Light Brown	18.8	2.60	34.9	18.9	16
CL + 20% Bentonite	Gray	18.2	2.57	46	23	23

From Figure 1, it is observed that in unsoaked CBR test with increase in PI, the CBR value decrease, because when soil PI become more, OMC (optimum moisture content) of soil rises and so the water content increases and water particles replace the soil particles and cause the soil to be more ductile. Thus the soil strength decreases and its CBR value declines.

With comparison between soaked and unsoaked CBR test, it is inferred that in saturated condition, the CBR value is remarkably lower compared with unsoaked state. For example, for PI 10, soaked CBR 92% decreases than unsoaked CBR. This trend continues for the rest of PIs. However, soaked CBR decreases with increase in PI. This is shown in Figure 2.

Figure 3 and 4 shows respectively reinforced soil at layer 2 and at layers 1 and 3 in both states soaked and unsoaked condition. As it is expected, in both figures a clear difference is between soaked and unsoaked CBR. For example, for PI of 16 soaked CBR as compared with unsoaked one decreases by 97.6% and 96.9% respectively. This is because of the effect of water in the test.

Now only unsoaked CBR value in different reinforced condition is considered. As it is observed in Figure 5, CBR value for PI = 10 is 55.3 that in case of adding geogrid in the second layer it increased by 39.42%. If geogrid is placed in layer 1 and 3 CBR value becomes 23.69%. Similarly, for plasticity indexes of 16 and 23, CBR value without geogrid is 50.6 and 44.6 which in case of being reinforced in layer 2, it increased by 39.92% and 38.79% respectively. If geogrid is placed in layer 1 and 3, CBR value increases by 29.45% and 40.58% in comparison with the initial state. By putting geogrid at layer 1 and 3 CBR values grow considerably as compared with the state of putting no geogrid. However this growth in comparison with putting geogrid at layer 2 is less, because of by placing geogrid at layer 2 more braced forces is produced in geogrid and interlocking between soil

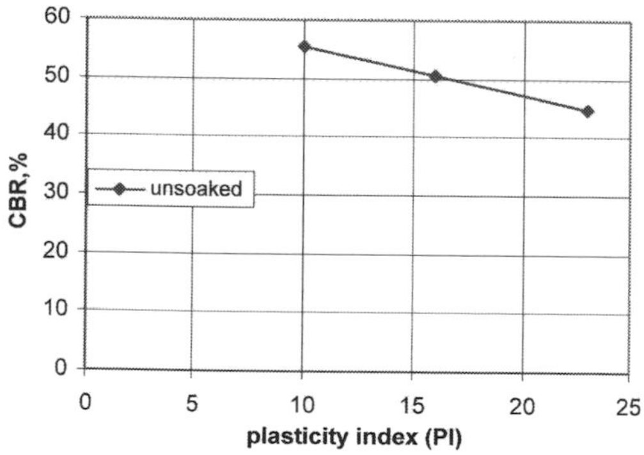

Figure 1. Dipicts unsoaked CBR for various PIs.

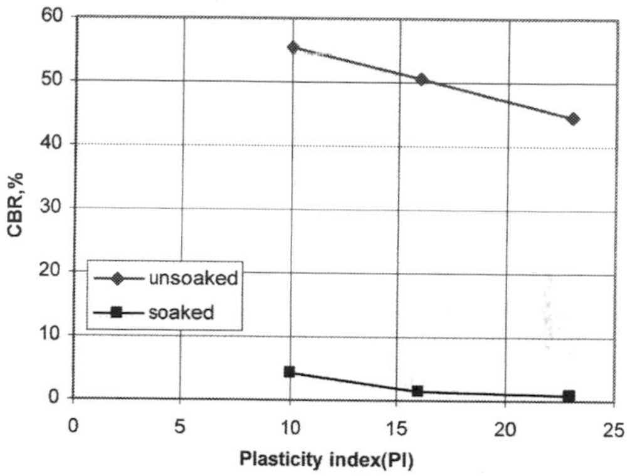

Figure 2. Compare soaked and unsoaked tests without geogrid.

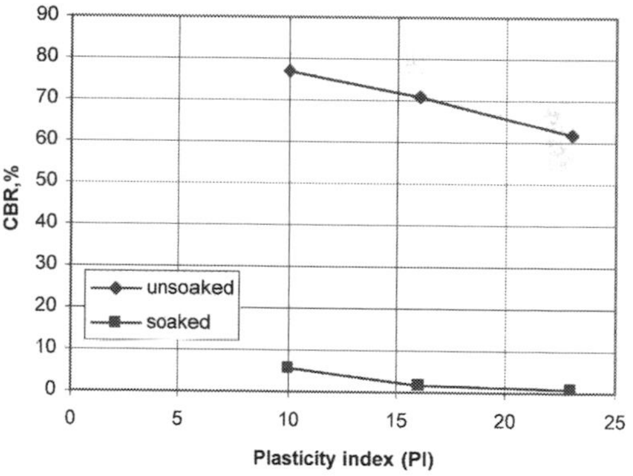

Figure 3. Compare soaked and unsoaked tests with geogrid in layer 2.

Table 3. Summery of the results of the CBR tests performed.

PI	Gs	Maximum Dry density (MDD) KN/m3	Optimum Moisture content (OMC), %	CBR, % (unsoaked) No Geogrid	Geogrid in layer 2	Geogrid in layer 1 & 3	CBR, % (soaked) No Geogrid	Geogrid in layer 2	Geogrid in layer 1 & 3
10	2.62	19.4	11.4	55.3	77.1	68.4	4.4	5.61	6.00
16	2.60	18.8	12.2	50.6	70.8	65.5	1.5	1.73	2.03
23	2.57	18.2	13.9	44.6	61.9	62.7	0.9	0.92	1.20

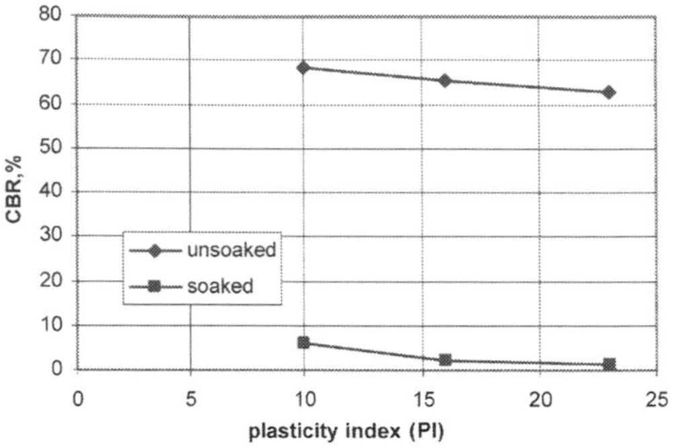

Figure 4. Compare soaked and unsoaked tests with geogrid in layers 1,3.

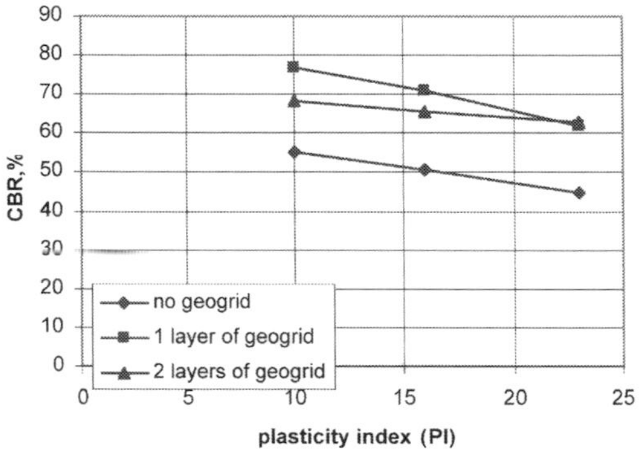

Figure 5. Compare (1), (2) and no layers of geogrid in unsoaked tests.

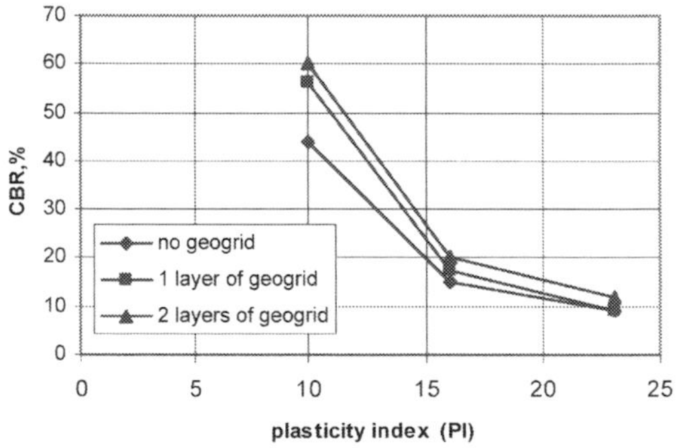

Figure 6. Compare (1), (2) and no layers of geogrid in soaked tests.

and geogrid increases. Thus placing geogrid is more appropriate. See Figure 5.

In Figure 6 only soaked CBR is depicted with different types of reinforcement. It is interesting that with reinforcing at layer 1 and 3 the maximum CBR for various PIs is obtained. This result is opposite to what we had in former test (in which CBR was more at layer 2). It can be inferred that in saturated state because of accumulation of more water in soil the effect of geogrid at layers nearer to the soil layer of themselves are great. As shear strength and braced force of geogrid with soil decreases, only tensile strength is determining that this strength in upper levels is more effective.

6 OMC AND MDD

From Table 3 it is observed that the OMC (Optimum Moisture Content) and MDD (Maximum Dry Density) of clay soil with PI 10 are 11.4% and 19.4 respectively. With increase of the soil plasticity index to number 16, OMC increased by 7% and MDD decreased by 3 %. In the same way, with increase of the soil plasticity index to 23 again OMC increased by 22%.and MDD decreased by 6% compared with the primary state.

It is found that the rate of increase in OMC and decrease in MDD rises by increase in the soil PI. The reason might be that with increase in PI (by adding bentonite to clay soil) the number of fine particles in the mixture become more in comparison with the primary state and require more water to reach to OMC state that this causes increase in soil MDD, because water replace the soil particles.

7 CONCLUSIONS

In this study we had three types of soil with various PIs of 10, 16 and 23 that it is achieved by adding different percents of bentonite to clay soil. The samples were initially tested without geogrid in soaked and unsoaked conditions. Then by placing a single layer of geogrid at the second layer of the sample CBR tests were performed on the reinforced soil. Consequently, geogrid was placed at the first and the third layer and CBR tests were repeated. The results obtained from tests done above are as follows:

1. With increase of PI in all kinds of reinforcements the CBR value of soaked and unsoaked decreases due to the presentation of water through the soil and reduction of its strength. It is obvious that when soil becomes more plastic the penetration piston requires less pressure.
2. From Figure 2, 3 and 4 it is observed that soaked CBR values are lower than unsoaked CBR values. It can be explained that in soaked situation because of more water content soil particles slides easily as water causes less interlock of soil particles with each other. The less the interlocking of the soil (more ductility), the less would be the CBR value such as Figure 2.
3. Using single layer of geogrid at layer 2 causes a considerable increase in CBR value compared with unreinforced soil in both soaked and unsoaked conditions.

4. Using two layers of geogrid at layer 1 and 3 causes an increase in unsoaked CBR value compared with unreinforced soil but this increase is less than the case in which geogrid is placed at layer 2. However the soaked CBR value is more than both single and no layer of geogrid.

REFERENCES

A. Cancelli, F. Montanelli, P. Rimoldi & A. Zhao (1996) 'Full Scale Laboratory Testing on Geosynthetics Reinforced Paved Roads'. Proceedings of the International Symposium of Earth Reinforcement, pp. 573–578.

ASTM D1883:87 (1987) 'Standard Test Method for CBR (California Bearing Ratio) of Laboratory-Compacted Soils.' Annual Book of ASTM Standards.

F. Motanelli, A. Zhao & P. Rimoldi (1997) 'Geosynthetics-Reinforced pavement system: testing and design'. Proceeding of Geosynthetics '97, pp.549–604,

G. Gosavi, K. A. Patil & S. Saran (2004) 'Improvement of Properties of Black Cotton Soil Sub grade through Synthetic Reinforcement'. Department of Civil Engineering, IIT Roorkee.

Gopal Ranjan & H. D. Charan (1995) 'Soil Improvement through Randomly Mixed Fibers.' Indian Geotechnical Society, Indore.

Gopal Ranjan & H. D. Charan (1998) 'Randomly Distributed Fiber Reinforced Soil—the State of the Art.' Journal of the Institution of Engineers (India), Vol 79, pp. 91–100.

G. V. Rao, K. K. Gupta & P. B. Singh. (1989) 'Laboratory Studies on Geotextiles as Reinforcement in Road Pavement.' Proceedings of the International Workshop on Geotextile, Bangalore, (1989) Vol 1, pp. 137–143.

G. V. Rao & G. V. S. Raju (1990) 'Pavements.' Tata Mc Graw Hill, New Delhi, pp. 283–306.

K. R. Shetty & P. P. Shetty (1989) 'Reinforced Soil layers in Pavement Construction.' Proceedings of the International Workshop of Geotextile, Bangalore, Vol 1, pp. 177–183.

S. Mittal & J. P. Shukla. (2001) 'Soil Testing for Engineers.' Khanna Publishers, New Delhi.

S.W. Perkins & M. Ismeik (1997) 'A Synthesis and evaluation of geosynthetic-reinforced base layers in flexible pavements: part I', Geosynthetics International, pp.549–604.

New Horizons in Earth Reinforcement – Otani, Miyata & Mukunoki (eds)
© 2008 Taylor & Francis Group, London, ISBN 978-0-415-45775-0

Effects of palm fibers on CBR strength of fine sand

H. Ghiassian & H. Sarbaz
College of Civil Engineering, Iran University of Science and Technology Tehran, Iran

ABSTRACT: Soil often lacks the tensile strength but this defect may be resolved with the incorporation of reinforcing elements with proper tensile strength. Elements made of metal, synthetic or natural materials have been used for this purpose. Many research studies have been reported in the literature about the soils reinforced with natural fibers such as coir, jute, sisal, flax, reed and wood fiber. However, less attention has been given to palm fibers and their influences on the soil strength behavior. This paper discusses the influence of palm fibers on CBR strength of fine sand. Samples were prepared with the fiber dry weight ratios of 0.5% and 1.0%, with the lengths of 20 mm and 40 mm. CBR tests were conducted under dry and submerged conditions. The results show that the addition of palm fibers increases the CBR strength of the sand specimens considerably.

1 INTRODUCTION

According to history findings in ancient times, natural fibers such as hey, wood, and bamboo were used for the improvement of construction materials [5]. The use of appropriate elements in soil improves its engineering properties such as strength, hardness and deformability. Materials used for the reinforcement are usually made of metal, geosynthetics or natural materials like plant roots and stems.

Nowadays, natural fibers as Kenaf, Coir, Banana, Jute, Flax, Sisal, Palm, Reed, Bamboo and Wood Fibers are used for soil reinforcement and stabilization [1, 2, 6, 8-11]. The most advantages of using natural materials are due to environmental and economical considerations.

Many research studies have been reported on soil reinforced with natural fibers [1, 2] and [3-8]. Ghavami et al., (1999) observed that the addition of 4% coconut and sisal fibers to soil causes its deformability to increase significantly. Besides, the crack creation in dry seasons was highly lessened. A study by Prabakar and Sridhar (2002) on soil specimens reinforced with sisal fibers showed that both fiber content and aspect ratio have important influences in shear strength parameters (c, ϕ). They observed that an optimum value for the fiber content exists such that the shear strength decreases with increasing the fiber content over this value. Mesbah et al., (2004) performed tensile tests on soil specimens reinforced with sisal fibers and concluded that the fiber length and the tensile strength of fibers are the most important factors affecting the tensile strength of the soil composite. Bouhicha et al., (2005) working on reinforced soil composites made of barely fibers observed that the presence of the fibers causes shrinkage and curing time to decrease. In addition, they reported that the shear, bending, and compression strengths of specimens increase for specific fiber contents.

2 EXPERIMENTAL STUDY

2.1 *Materials*

The utilized soil is fine sand (SP) supplied from Kerman city. Palm fibers were obtained from Bam city in Kerman province. The properties of these materials are given in Tables 1 and 2 respectively.

2.2 *Preparation of samples*

CBR specimens were prepared and tested according to ASTM D-698B procedure. Compression tests were performed under both moist and submerged conditions.

Table 1. Properties of the sand.

Property	Value
Sand (%)	95.0
Silt and Clay (%)	5.0
Unified Classification	SP
AASHTO Classification	A-3
G_s	2.66
C_u	2.85
C_c	1.2
Plasticity	NP
w_{op} (%)	16.0
γ_d (kN/m^3)	15.86
ϕ	43°

Table 2. Physical and mechanical properties of palm fibers.

	Palm fiber		
Property	Lower	Upper	Mean
Diameter (mm)	0.11	1.4	0.42
Length (mm)	115	900	295
Density (kN/m^3)	8.4	9.76	9.06
Natural moisture content (%)	3.8	7.8	5.9
Water absorption upon saturation (%)	139	159	149
Tensile strength (MPa)	77.15	151.39	123.23
Modulus of elasticity (GPa)	1.75	3.26	2.47
Strain at failure (%)	3.7	6.3	5.1

To evaluate the effects of the fiber length, two sizes of 20 mm and 40 mm, for fiber percentage of 0.5, 1, and 2 were examined. Optimum water content was obtained from the standard Proctor test as about 16% for plain and reinforced specimens. The required water was added in two stages in order to prepare more homogenous specimens [3]. In the first stage, the half of the water was added to the mixture of the soil and fibers, and followed by 15 min continuously mixing with hand. Then, the second portion of the water was added, followed by 5 min hand mixing. Submerged specimens were placed in water for 48 hours, then taken out and allowed to drain before being loaded.

3 RESULTS AND DISCUSSION

3.1 *Moist specimens*

Figures 1 and 2 presents the results of CBR tests for moist specimens reinforced with 20 mm or 40 mm fibers, and fiber contents of 0.5, 1 or 2%. It is seen that adding 0.5 to 1% fibers enhances the CBR strength significantly up to 56% increase in the strength with respect to plain specimens. However, this effect gradually diminishes at higher fiber contents such that the strength at 2% fiber content decreases slightly. Obviously, the presence of fibers in the soil, more than what required for optimum reinforcement, can substitute soil particles with weaker materials; therefore, reducing the bearing strength of the soil. In addition, it appears that longer fibers contribute further to the strength. This can likely be attributed to the more mobilized frictional resistance around the fibers, and consequently, higher tensile stresses developed in the fibers. Another words, the failure mode of fibers seems to be more pullout rather than breakage, as observed also during the experiments. This trend is expected to lessen for longer fibers such that an optimum length is obtained for any fiber content. Previous studies show that the optimum fiber length becomes larger as the fiber content decreases (Prabakar and Sridhar, 2002).

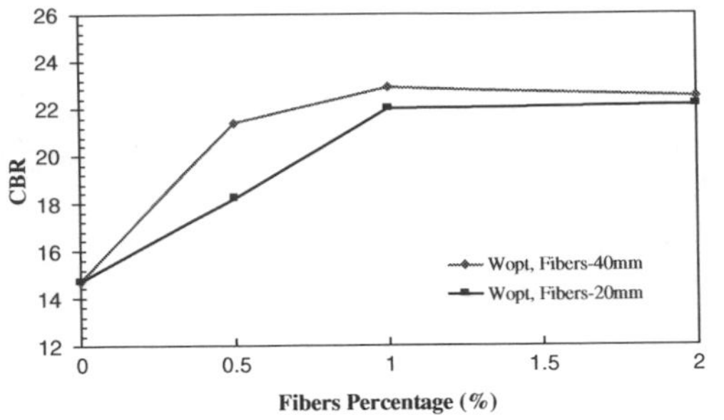

Figure 1. Effect of Palm fiber content on CBR strength (moist condition).

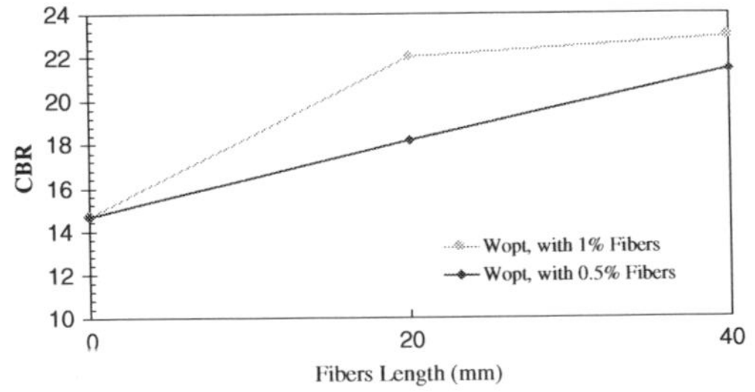

Figure 2. Effect of Palm fiber length on CBR strength (moist condition).

3.2 *Submerged specimens*

Figures 3 and 4 present the CBR results for the fiber-reinforced specimens under submerged condition. Similar to the results for moist specimens, the CBR values show some improvement due to the reinforcement with the maximum of 41% increase for the specimen with 1% of 40 mm fibers. Longer fibers also have resulted in higher CBR values. However, these increases are not as pronounced as those for the moist specimens where 56% increase was observed for the specimen with 1% of 40 mm fibers (Fig. 1 or 2). Besides, the effect of fiber length has evidently diminished for submerged specimens particularly for those with 0.5% fiber content. Saturation has obviously important influences on the soil behavior that can be explained in view of three aspects. First, the strength and modulus of soil itself decrease because of the water interaction with fine, cohesive particles. Second, the loss of capillarity because of saturation reduces the effective stress, and consequently, the soil bearing capacity. Third, the frictional resistance between fibers and soil particles reduces as water lubricates the surfaces of soil particles and fibers, and thus reduces the pullout capacity of the fibers. This aspect is more important if it is realized that the CBR strength of the specimens are greatly controlled by the pullout, rather than the breakage, behavior of the fibers as explained for the moist specimens. A comparison between the

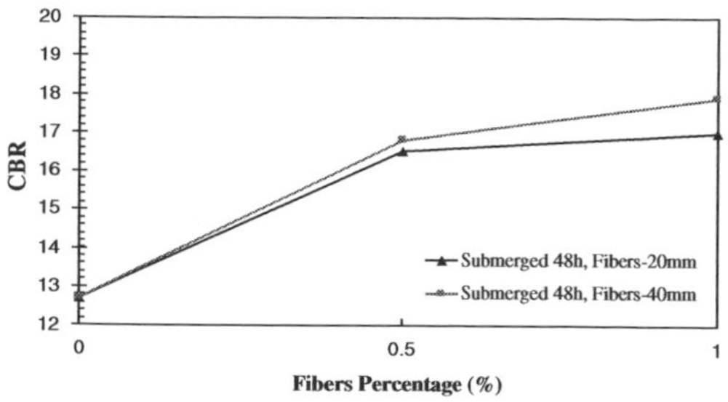

Figure 3. Effect of Palm fiber content on CBR strength (submerged condition).

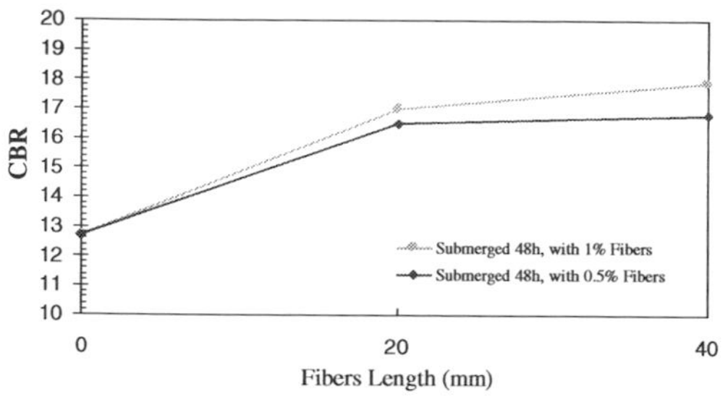

Figure 4. Effect of Palm fiber length on CBR strength (submerged condition).

results of moist and submerged specimens can be seen in Fig. 5 in which the major roles of water in the reduction of CBR values due to both soil and fiber interactions with water are apparent.

Similar to the results in figures 1 and 2, it can also be seen in figures 3 and 4 that there appear to be some optimum fiber contents or fiber lengths for which maximum bearing strength values for the specimens can be achieved.

4 CONCLUSIONS

1. Fibrous palm waste can be converted into a value-added product for soil reinforcement.
2. Sand specimens reinforced with palm fibers show some increase in CBR strength.
3. Submergence of plain and reinforced specimens causes the CBR bearing strength to decrease considerably. This can be attributed to the water interaction with soil particles, and the reduced frictional resistance of fibers caused by water.
4. There appear to be some optimum fiber contents or fiber lengths for which maximum bearing strength values can be obtained for a given soil condition.
5. The failure mode of fibers appears to be more pullout than breakage.
6. Further research is under way to evaluate the durability of these fibers when mixed with soil.

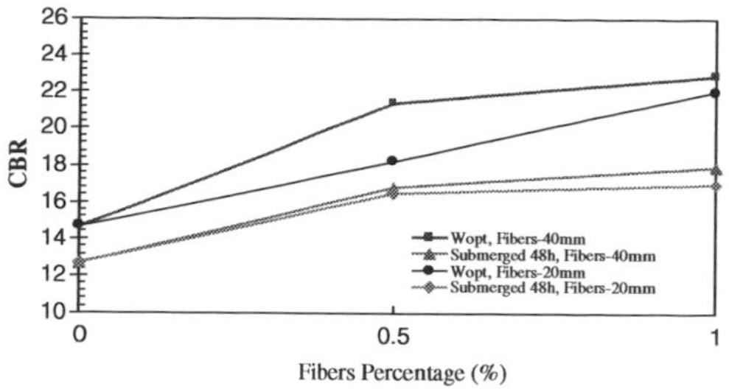

Figure 5. Comparison between the results of moist and submerged specimens.

REFERENCES

Bouhicha M., Aouissi F., Kenai S., (2005). "Performance of composite soil reinforced with barley straw." Cement & Concrete Composites, Vol. 27, 617–621.

Ghavami Kh., Romildo D., Toledo Filho, Normando P. Barbosa (1999). "Behavior of composite soil reinforced with natural fibres." Cement and Concrete Composites, Vol. 21, 39–48.

Ghiassian, H., Poorebrahim, G., Gray, D., 2004, "soil reinforcement with recycled carpet wastes". Waste management & research 22–2, pp. 108–104.

John V.M., Cincotto M.A., Sjostrom C., Agopyan V., Oliveira C.T.A. (2005). "Durability of slag mortar reinforced with coconut fibre." Cement & Concrete Composites, Vol. 27, 565–574.

Khedari J, Suttisonk B, Pratinthong N, Hirunlabh J., (2001) "New lightweight composite construction materials with low thermal conductivity." Cement Concr Composites, Vol. 23, 65–70.

Mesbah A., Morel J.C., Walker P., and Ghavami Kh., (2004) "Development of a Direct Tensile Test for Compacted Earth Blocks Reinforced with Natural Fibers." Journal of Materials in Civil Engineering, Vol. 16, No. 1, February, 95–98.

Murali Mohan Rao K., Mohan Rao K. (2005). "Extraction and tensile properties of natural fibers: Vakka, date and bamboo." Composite Structure, Available Online at www.sciencedirect.com.

Prabakar J., Sridhar R.S., (2002) "Effect of random inclusion of sisal fibre on strength behaviour of soil." Construction and Building Materials, Vol. 16, 123–131.

Ramakrishna G. and Sundararajan T., (2005) "Studies on the durability of natural fibres and the effect of corroded fibres on the strength of mortar." Cement and concrete composites, vol.27, Issue 5, May 2005, Pages 575–582.

Romildo D. Toledo Filho, Khosrow Ghavami, George L. England, Karen Scrivener, (2003) "Development of vegetable fibre–mortar composites of improved durability." Cement & Concrete Composites, Vol. 25, pp. 185–196.

Santoni, R.L., Tingle, J.S., Webster, S.L., 2001. Engineering properties of sand–fiber mixtures for road construction. Journal of geotechnical and Geoenvironmental Engineering, ASCE 127 (3), 258–268.

New Horizons in Earth Reinforcement – Otani, Miyata & Mukunoki (eds)
© 2008 Taylor & Francis Group, London, ISBN 978-0-415-45775-0

Shear behavior of waster tire chip-sand mixtures using direct shear tests

M. Ghazavi
Civil Engineering Department, K. N. Toosi University of Technology, Tehran, Iran

F. Alimohammadi
Civil Engineering Department, Faculty of Engineering, Azad University, Tehran-Center, Tehran, Iran

ABSTRACT: In this paper, waste tire chips in various contents have been added to sand to investigate changes in shear strength parameters of the sand. The chip-sand mixtures are composed of two types of relatively uniform sand and tire chips in two grain size distribution. The sand is granular shaped. Four chip contents of 15, 30, 50, and 100% by volume have been chosen and mixed with the sand to obtain uniformly distributed mixtures. Two moisture contents and two compaction states have been considered. The results show that the influencing parameters on shear strength characteristics of sand-chip mixtures are normal stress, sand matrix unit weight, chip content, and moisture content. Moreover, the initial friction angle is mostly affected by compaction and moisture content. From environmental viewpoint and based on findings in this paper, it appears that if mixtures of tire chips and sand are mixed and properly confined, they can be used as lightweight materials for backfilling in highway applications.

1 INSTRUCTION

Typically waste tires are disposed in huge open piles, causing the environmental problems. Thus, finding appropriate applications for them are necessary. Nowadays, waste tires are used for reinforcing soft soil in road construction (Bosscher et al., 1997), to control ground erosion (Poh and Broms, 1995), for stabilizing slopes (O'Shaughnessy and Garga, 2000a), as lightweight material for backfilling in retaining structures (Bosscher et al., 1997; Tatlisoz et al., 1997; Allman and Simundic, 1998; O'Shaughnessy and Garga, 2000a), as aggregates in leach beds of landfills (Hall, 1991; Ahmed and Lovell, 1993), as an additive material to asphalt (Foose et al., 1996; Heimdhal and Druscher, 1999), as sound barriers (Hall, 1991), as limiting for freezing depth (Humphrey et al., 1997), as a source for creating heat (Lee et al., 1999), as a fuel supplement in coal-fired boilers (Ahmed and Lovell, 1992), for vibration isolation (Eldin and Senouci, 1993), as cushioning foams (Ahmed and Lovell, 1992), for low strength but ductile concrete (Eldin and Senouci, 1003), for varying shear strength parameters of soils (Foose et al, 1996; Lee et al., 1999; Ghazavi, 2004, Ghazavi and Amel Sakhi, 2005a), and for CBR improvement (Ghazavi and Amel Sakhi, 2005b).

In the present study, large direct shear tests were carried out on rubber particles mixed with sands. The main goal of the tests was to investigate the influence of the waste tire particles on shear strength parameters of tire chip-sand mixtures as lightweight materials.

2 MATERIALS

Two types of relatively, uniform and rounded sand (S_1, S_2) and two types of waste tire chips (R_1, R_2) were chosen for present experiments. Various contents of waste tire chips have been mixed with sand at two loose and slightly dense states in two dry and saturated conditions. The sand alone was tested in direct shear tests. The grain size distribution was obtained based on ASTM D854-63 (1995). Figure 1 shows the grain size distribution of sands and tire chips.

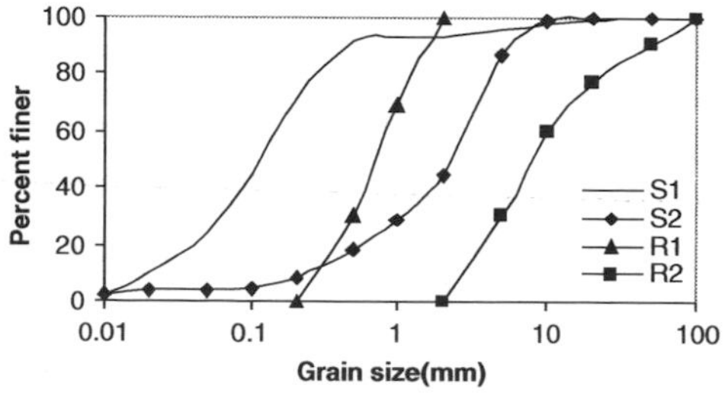

Figure 1. Particle size distribution of sands and chips.

Table 1. Properties of two types of sand.

Description	Value		
Sand type		S_1	S_2
Effective size, D_{10}		0.02	0.35
D_{30}		0.08	1.1
Mean size, D_{50}		0.11	2.71
D_{60}		0.15	3.2
Uniformity coefficient		7.5	9.14
Curvature coefficient		2.13	1.1
Unit weight at lower compaction (kN/m^3)		14.8	15.9
Unit weight at higher compaction (kN/m^3)		19	2.88
Friction angle at lower compaction (degree)	Dry	40	41
	Saturated	18	35
Friction angle at higher compaction (degree)	Dry	42	51
	Saturated	26	43
Specific gravity		2.64	2.67
Sand classification	Sandy silt		SW

A value of 2.64 and 2.67 were measured for the specific gravity of S_1, and S_2 according to ASTM D92-854 (1995). The minimum and maximum values of unit weights of S_1 were 14.8 kN/m^3 and 19 kN/m^3, and for S_2 were 15.9 kN/m^3 and 20.2 kN/m^3, respectively using the procedure described in loose and dense case by ASTM D4254, and ASTM D4253 (2003). Other details of the sands are presented in Table 1.

A special simple machine was developed by local industries to produce chip from waste tires. The produced chip grains were granular and angular. An average value for the specific gravity of the chip grains was found to be about 1.1 as tested several times based on ASTM D854-92 (1992).

3 TESTING PROCEDURE

Small direct shear test apparatus with square mold having a width of 63 mm was used to perform shear tests on sand-chip samples. The thickness of samples depended on tire chip content. The tests were carried out based on the procedure described by ASTM D 3080-99. In all tests, three normal stresses of 49.05, 98.1, and 147.15 kN/m^2 were used. Samples consisting of sands alone, tire chip alone, and various mixtures of sand-chip were tested. A matrix unit weight for the sand is used, as defined and used by others (Foose et al., 1996; Ghazavi and Amel Sakhi, 2005a; Ghazavi and Amel Sakhi, 2005b).

Numerical calculations were made to determine the amount of the sand and the chip grains for each mixture and after that poured in a container. The materials were then carefully mixed with a blade. The mixed materials were poured steadily into the shear box with a circular motion using the blade. At each stage, the materials in the container were mixed by means of the blade and then poured in the shear box. Segregation did not occur during sample preparation and transferring into the shear box. This was controlled carefully by a continuous mixing the materials in the container and with careful observation.

Table 2. Unit weight of various sand-chip mixtures.

Mixture unit weight (kN/m^3)				
	Sand			
	S_1		S_2	
Chip content (%)	Loose	Dense	Loose	Dense
0	14.8	19	15.6	20.2
15	13.9	17.9	15.4	19.1
30	13.93	17.2	14.5	17.8
50	13.43	16	14.5	17.2
100	5.2	6.1	5.2	6.1

To obtain loose mixtures, the mixtures were poured in the shear box from a very low height. For preparing slightly more compacted samples, a square steel plate was located at the top of the sample. Then a weight with a mass of 0.5 kg was dropped from a height of 15 cm on the plate five times. Table 2 shows unit weights of loose and slightly compacted samples. For saturated case, the procedure was repeated to a situation when the shear box became full of water for 8-12 hours each time, depending on the grain size of tire chip and sand.

When a certain chip-sand mixture was prepared at a prescribed sand matrix unit weight, the normal stress was applied and then the sample was sheared. The procedure was identically repeated for another normal stress. All shear tests were conducted using a controlled-displacement procedure. The shear rate in standard direct shear test instrument is normally controlled by an electric motor and a multi-speed drive unit, typically providing 240 speeds ranging from 5 mm/min to about 0.0003 mm/min (Head, 1982). In the present experiments, the shear rate was 0.5 mm/min and this was kept constant for all tests. This speed was suitable for shearing mixtures containing sand S_1. According to ASTM D3080 (1999), all tests were continued until the shear stress becomes essentially constant or until a maximum shear deformation of 7 mm has been reached. The maximum shear stress, in almost all samples, was achieved at deformation less than 7 mm (Figures 6 and 7). The inclusion of all figures in this paper makes the paper lengthy. Only a limited number of them are presented here.

Figures 2 and 3 illustrate the variation of shear stress versus normal stress for S_1 mixed with R_1 in two moisture contents for various chip contents of the loose

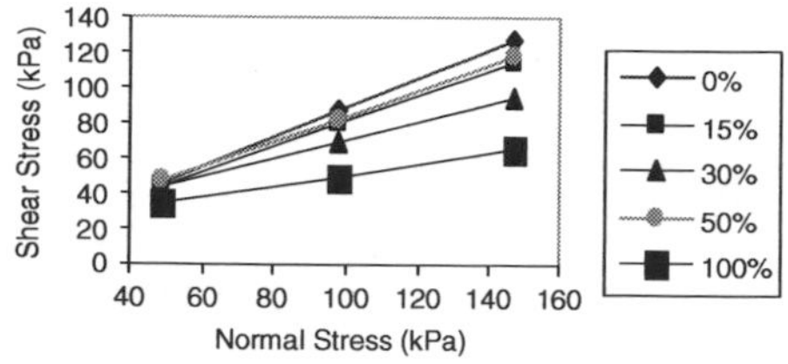

Figure 2. Variation of shear stress with normal stress for loose dry samples containing $S_1 + R_1$.

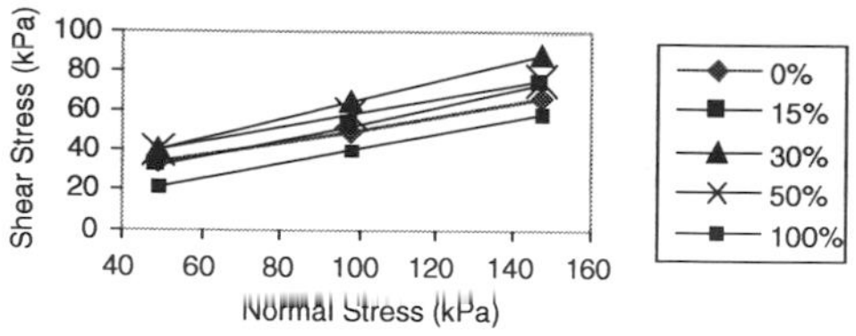

Figure 3. Variation of shear stress with normal stress for loose saturated samples containing $S_1 + R_1$.

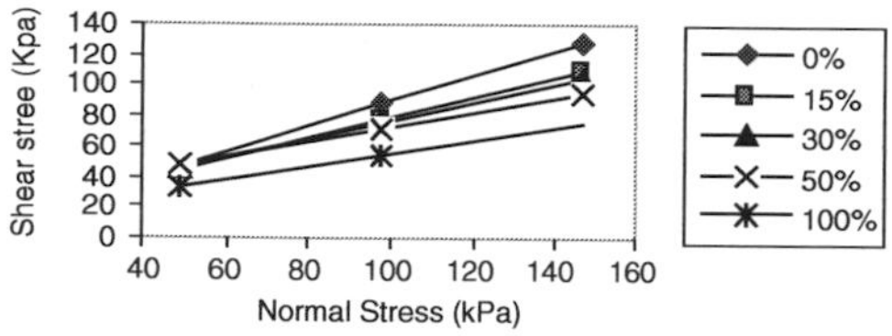

Figure 4. Variation of shear stress with normal stress for loose dry samples containing $S_1 + R_2$.

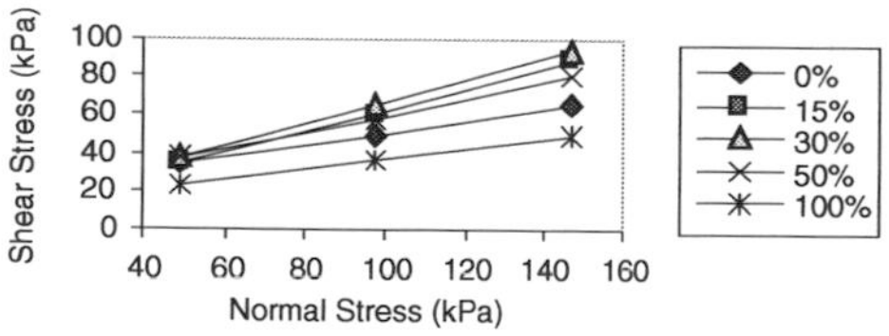

Figure 5. Variation of shear stress with normal stress for loose saturated samples containing $S_1 + R_2$.

mixtures. Figures 4 and 5 illustrate the variation of shear stress versus normal stress for S_1 mixed with R_2 in two moisture contents for various chip content of the loose mixtures. Similar trends were observed for mixture in slightly compacted and sand S_2 at the same condition (Alimohammadi, 2005).

4 RESULTS FOR SAND-CHIP TIRE LOOSE MIXTURES

Figures 6 and 7 show the variation of shear stress versus horizontal displacement for various chip grain

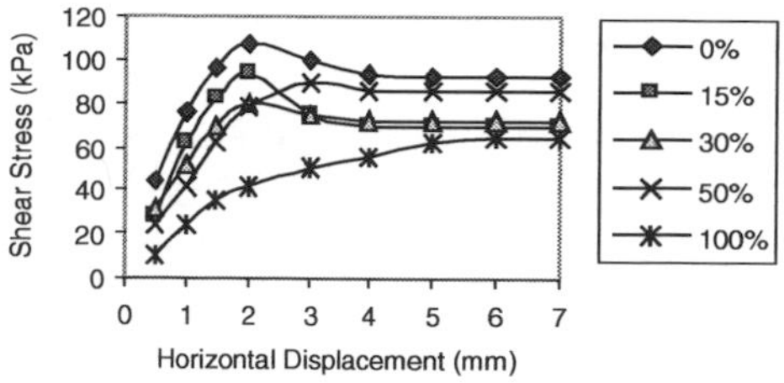

Figure 6. Variation of shear stress with horizontal displacement for slightly compacted $S_1 + R_1$ mixtures and at normal stress of 98.1 kPa in dry state.

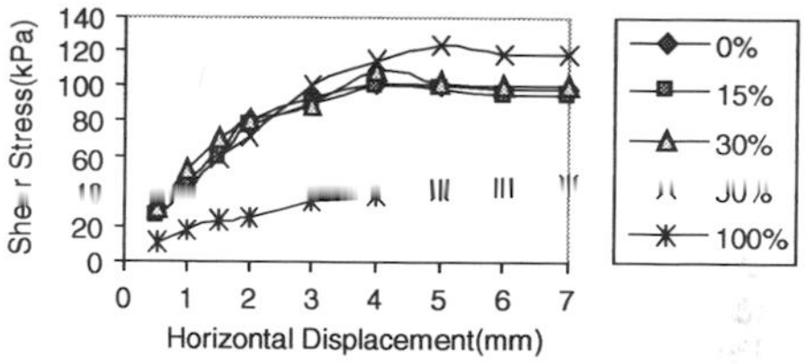

Figure 7. Variation of shear stress with horizontal displacement for loose with $S_2 + R_2$ mixtures and at normal stress of 147 kPa in saturated state.

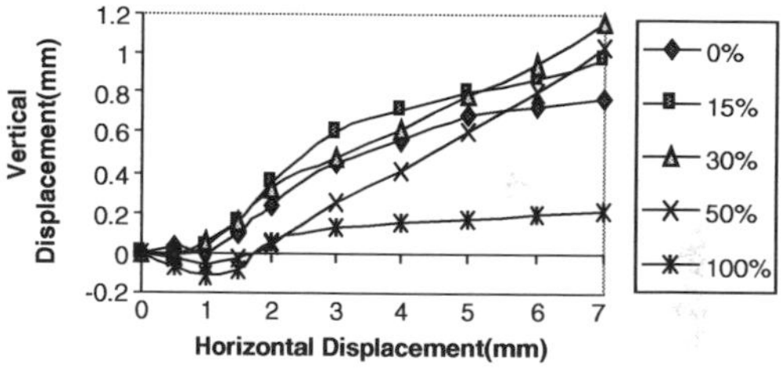

Figure 8. Variation of vertical displacement whit horizontal displacement for slightly compacted whit $S_1 + R_1$ mixtures under 98.1 kPa in dry states.

content of the loose mixtures having S_1 with R_1 in dry and saturated conditions, respectively. The normal stress was 98.1 kPa. As seen, for all chip grain contents in the mixtures, except 100%, a peak shear stress is observed, explaining shear strength of the mixtures.

Figures 8 and 9 show the variation of vertical displacement versus horizontal displacement for slightly compacted $S_1 + R_1$ mixtures under 98.1 kPa in dry and saturated conditions, respectively.

Figure 10 illustrates the variation of initial friction angle versus chip content for $S_1 + R_1$ mixtures in dry and saturated state with loose and slightly compacted samples.

Some remarks may be extracted from Figures 2–11 as follow:

1) For a given normal stress applied on specimens, the shear resistance of the sand-chip mixtures is

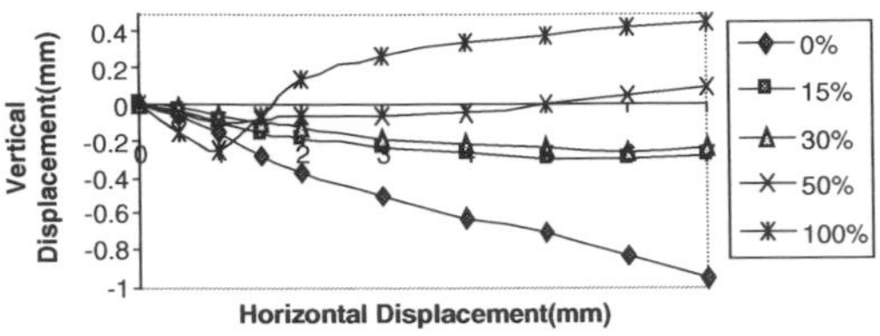

Figure 9. Variation of vertical displacement whit horizontal displacement for slightly compacted whit $S_1 + R_1$ mixtures under 98.1 kPa in saturated states.

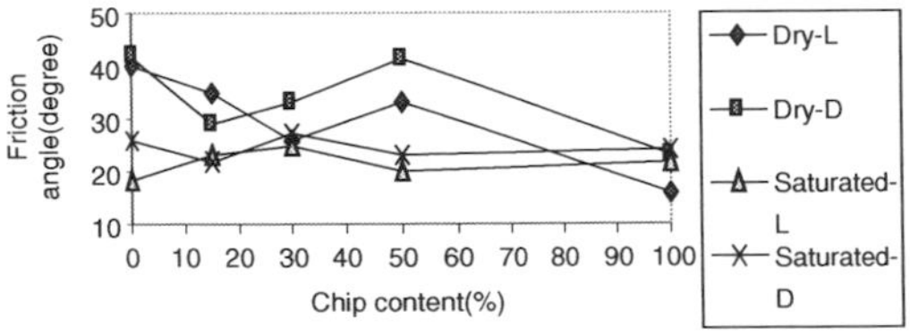

Figure 10. Variation of initial friction angle versus chip content for $S_1 + R_1$ mixtures in dry and saturated state with loose and slightly compacted samples (L = Loose ; D = Dense).

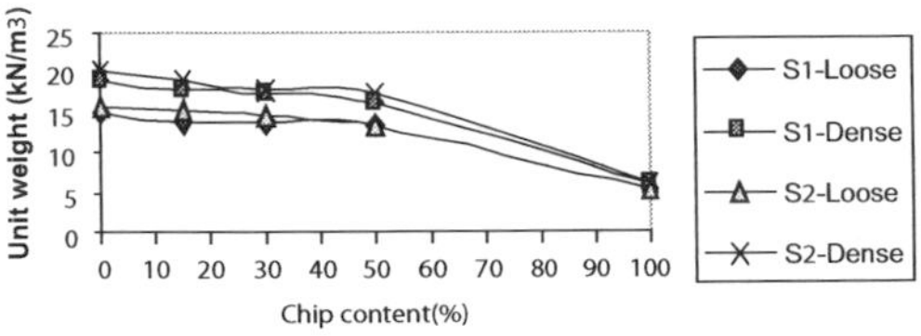

Figure 11. Variation of mass density of different mixtures versus chip contents at two compaction states for $S_1 + R_2$ mixtures.

greater than that of the sand alone at the same compaction state, especially in saturated samples containing 30% chips. However, this phenomenon for dry condition is revered and shear resistance in mixtures is smaller than that of the sand alone.

2) A relatively clear peak can be observed in shear resistance of almost all mixed samples, regardless of compaction level and chip contents, except for pure chips for which the shear stress still increases slightly with increasing the horizontal deformation in dry state. For saturated samples, the shear stress peak is not reached and it still increases slightly with increasing the horizontal deformation (Figures 6 and 7).
3) The inclusion of chip grains decreases the shear resistance of the mixtures in dry state whereas it tends to increase in saturated cases.
4) The shear resistance of the mixtures does not increase in a regular manner with increasing chip.
5) An apparent cohesion is obtained in samples containing chips.
6) The Mohr-Coulomb envelope is almost linear. The linear envelope obtained in the present paper is similar to that found by Tatlisoz et al. (1997) who mixed sandy silt with tire chips with dimensions of 11–33 mm and tested in large direct shear test apparatus. In the present experiments, the same trend was observed for saturated mixtures. It should be noted that the use of larger reinforcing tire shreds mixed with sands may result in nonlinear envelopes (Foose et al., 1996; Ghazavi and Amel-Sakhi, 2006). Nonlinear envelopes were also reported by Gray and Ohashi (1983) for fiber-reinforced dense sand. It seems that small grain sizes of chip mixed with sand and distributed randomly in sand-chip mixtures make a continuum material and have a granular behavior. Thus, it is likely to have a linear behavior in Mohr-Coulomb envelope similar to pure sand (Ghazavi, 2004). In contrast, randomly distributed discontinuous long inclusions such as shreds, strips, and fibers in the mixtures can cause nonlinearity in Mohr-Coulomb envelopes.
7) When a normal stress of 98.1 kPa is used, slightly compacted mixtures containing chip grains in all content of chip in mixture, at first until a little displacement condensed and then dilated under shear stress. The dilation of mixtures increases with increasing chip volume. It is significant in dense state with high content of chip. The sand alone dilates after a small compressing. These findings are I agreement with those reported by Lee et al. (1999), who found that rubber-sand mixtures tend to contract initially and then begin to expand. This is a typical behavior of sands, but the range of strains for which there is contraction is wider than that for sands, and dilation much less in saturated condition and in slightly compacted state for S_1+R_1 mixtures. All mixtures present smooth variations in vertical displacements upon shear displacements, except mixtures containing 100% chips. In such mixtures, the samples show clearly dilation characteristics.
8) The initial friction angle (ϕ) of dry mixtures containing up to 15% chips by volume initially decreases. By the use of chip contents in excess of 15%, ϕ increases and then decreases. For saturated loose mixtures, ϕ first increases slightly with increasing chip contents up to 30% and then decreases smoothly toward the friction angle of the sand alone. In compacted mixtures, there is no significant change in the friction angle values in terms of chip content variation in the mixtures.

The values of initial friction angle for different sand-chip mixtures at two compaction conditions and at two moisture contents are shown in Figure 10. The results for sand and tire chips alone are also shown at the same compacted effort and saturated conditions to demonstrate the effectiveness of chip inclusion. Figure 10 clearly shows that the addition of chip to sands cannot increase the friction angle effectively.

5 CONCLUSIONS

In the present research, shear strength behavior of sand-tire chip grain mixtures have been investigated in order to assess their use as a lightweight backfill material. Small direct shear apparatus has been used to determine the influence on shear response of waste tire chips mixed with two types of sand. Tire chips alone, sand alone, and sand-tire chip mixtures having 15%, 30,% and 50% waste tire chip by volume have been used. Two different compaction degrees, two size of gradation for chips and sand and two dry and saturated conditions have been considered. An apparent cohesion is obtained in samples containing chip grains. Moreover, adding tire chips to sand decreases the friction angle in both cases of moisture contents. Within the materials used and regardless of moisture contents and compaction degrees, it has been generally shown that an addition of tire chips to sand the friction angle does not vary significantly, but light mixtures are obtained which can be used as lightweight material for back-filling. Also the mixtures of tire chips and sand can be used as a free draining material in construction of subgrade where water table level is close to the ground. Dilation characteristics have been observed in sand-chip mixtures, especially in samples having greater tire chip content and more compaction. For dry and saturated mixtures, dilation has delay versus dry state.

ACKNOWLEDGMENTS

The writers wish to thank the Technical and Engineering Laboratory of Ministry of Road and Transportation for the laboratory facilities provided for this research.

REFERENCES

Ahmed, I. and Lovell, C. 1993 Use of rubber tires in highway construction. *In:Utilization of waste materials in civil engineering construction, ASCE*, New York, N.Y., pp. 166–181.

Alimohammadi, F. 2005. Investigation of sand-chip tire mixtures behavior using direct shear & CBR tests. *Msc thesis, Department of Civil Engineering,* Islamic Azad university-Central Unit of Tehran, Iran.

Allman, M.A., and Simundic, G. 1998. Testing of a retaining wall constructed of waste tires. Proc., *3rd Int. Congress on Environmental Geotechnics, Vol. 2,* Lisbon, Portugal, 655–660.

Amel-Sakhi, M. 2001. Influence of optimized tire shreds on strength of sand–reinforced with tire shreds. *MSc thesis, Department of Civil Engineering, Isfahan University of Technology,* Isfahan, Iran.

ASTM D422-630. 1989. Standard method for particle-size analysis of soils. *Annual Book of Standard, Vol .04.08,* West Conshohoken, pp. 86–92.

ASTEM D698-78. 1989. Standard test method for moisture-density relation of soil and soil-aggregate mixtures using 5.50 lb (2.49 kg) ram and 12 in (305 mm) drop, Annul Book of Standards, Vol. 04.08, West Conshohoken, pp. 155–159.

ASTM D854-83. 1989. Standard test method for specific gravity of soil. *Annual Book of Standards, Vol. 04.08,* West Conshohoken, pp. 162–164.

ASTM D3080-72. 1989. Standard method for direct shear test of soil under consolidated drained conditions. *Annual Book of Standards, Vol. 04.08,* West Conshohoken, pp. 376–378.

Bader, C. 1992. Where will all the tires go?. *Municipal Solid Waste Management*, 2(7), pp. 26–30.

Basheer, I.A. and Najjar, Y.M. 1996. Rubber tires and geotextiles. *Journal of Performance of Constructed Facilities*, 10 (1), pp. 40–44.

Black, B.A. and Shakoor, A.A. 1994. Geotechnical investigation of soil-tire mixtures for engineering applications. *Proceedings of the First International Congress on Envoronmental Geotechnics*.

Bosscher, P.J., Edill, T.B., and Eldin, N.N. 1992. Construction and performance of a shredded waste tire test embankment. *Transportation Research Record 1345, Transportation Research Board*, Washington, DC.

Council of Ministers for the Environment (CCME). 1994. Harmonized economic instruments for used tires. *Final Rep. Prepared for the Canadian Council of Ministers for the Environment, Apogee Ref. 363 CCME, No. CCME-SPC-EITG-92E.*, Canada.

Edil, T.B., and Bosscher, P.J. (1992). Development of engineering criteria for shredded waste tires in highway applications. *Final report." Res. Rep. No. GT-92-9, Wisconsin Dept. of Transportation*, Madison, Wis.

Edil, T.B., and Bosscher, P.J. 1994. Engineering properties of tire chips and soil mixtures. *Geotech. Test. J.*, 17(4), pp. 453–464.

Eldin, N.N., and Senouci, A.B. 1993. Rubber-tire particles as concrete aggregate. *J. Mater. Civ. Eng.*, 5(4), pp. 478–496.

Foose, J., Benson, H., and Bosscher, J. 1996. Sand reinforced with shredded waste tires. *J. Geotech. Eng., ASCE*, 122(9), pp. 760–767.

Garga, V. K., and O'Shaughnessy, V. 2000. Tire-reinforced earthfill. Part 1: Construction of a test fill, performance, and retaining wall design. *Can. Geotech. J.*, 37, pp. 75–96.

Ghazavi, M. 2004. Shear strength characteristics of sand-mixed with granular rubber. *Geotechnical and Geological Engineering, An International Journal*, Vol. 22, No. 3, pp. 401–416

Ghazavi, M., and Amel-Sakhi, M. 2005a. Influence of optimized tire shreds on shear strength parameters of sand. *International Journal of Geomechanics, ASCE*, Vol. 5, No. 1, March, pp. 58–65.

Ghazavi, M., and Amel-Sakhi, M. 2005b. Optimization of aspect ratio of waste tire shreds in sand-shred mixtures using CBR tests. *Geotechnical Testing Journal (ASTM)*, Vol. 28, No. 6, pp. 564–569.

Gray, D. H., and Ohashi, H. 1983. Mechanics of fiber reinforced in sands. *J. Geotech. Eng., ASCE*, 109 (3), pp. 335–353.

Hall, T. 1991. Reuse of shredded tire material for leachate collection system. *Proc., 14th. Annual Conf. Dept. of Engineers. Prof. Devel*, Univ. of Wisconsin, Madison, Wis., pp. 367–376.

Humphrey, D.N., Katz, L.E., and Blumenthal, M. 1997. Water quality effects of tire chip fill placed above the groundwater table. *ASTM STP 1275, M. A. Wasemiller and K. B. Hoddinott, eds*., Philadelphia, pp. 299–313.

Humphrey, D.N., and Nickels, W.L. 1997. Effect of tire chips as lightweight fill on pavement performance. *Proc., 14th Int. Conf. on Soil Mechanics and Foundation Engineering*, New Delhi, India, pp. 1617–1620.

O'Shaughnessy, V., and Garga, V.K. 2000a. Tire-reinforced earthfill. Part 2: Pull-out behavior and reinforced slope design. Can. Geotech. J., 37, pp. 97–116.

O'Shaughnessy, V., and Garga, V.K. 2000b. Tire-reinforced earthfill. Part 3: Environmental assessment. *Can. Geotech. J.*, 37, pp. 117–131.

Poh, P. S.H., and Broms, B.B. 1995. Slope stabilization using old rubber tires and geotextiles. *J. Perform. Constr. Facil.*, 9(1), pp. 76–80.

Tatlisoz, N., Benson, C., and Edil, T. 1997. Effect of fines on mechanical properties of soil-tire chip mixtures. *ASTM STP 1275*, M. A. Wasemiller and K. B. Hoddinott, eds., Philadelphia, 93108–93108.

New Horizons in Earth Reinforcement – Otani, Miyata & Mukunoki (eds)
© 2008 Taylor & Francis Group, London, ISBN 978-0-415-45775-0

Mechanical properties of lightweight treated soil under water pressure

Y.X. Tang
Kanmon Kowan Construction, Shimonoseki, Japan

T. Tsuchida
Hiroshima University, Higashihiroshima, Japan

ABSTRACT: An artificial lightweight geomaterial (SGM) has been developed as a backfill to reduce the lateral earth pressure behind waterfront structures. The unit weight is reduced by mixing lightening ingredient, either air foam or EPS beads, with slurry of dredged soft clay, while certain cement is used as stabilizer to warrant compressive strength. This experimental investigation is conducted to characterize the strength and deformation properties of the lightweight treated soil. Considering expectable compressibility under pressured casting, curing or loading, samples were cured under various pressures, and subjected to undrained shear tests on triaxial apparatus modified able to detect volumetric change.

1 INTRODUCTION

1.1 *Application of SGM in port structures*

SGM is a lightweight technique developed to control unit weight and shear strength as desired, and to make beneficial reuse of surplus soil at the same time (Tsuchida, *et al.* 1996). Generally, unit weight of dredged soil is reduced by use of either air foam or expended polystrol (EPS) beads, while shear strength is obtained with cement.

Figure 1 presents two applications of the lightweight soil used in port structures. Probably, most of port structures are planned on soft grounds. So it is believed a ration option to decrease unit weight of backfill so as to increase the structure stability and to decrease the consolidation settlement.

1.2 *Underwater deployment*

As shown in Figure 1, the majority of lightweight soil needs to be placed underwater as deep as 10 m. Some peculiar techniques of underwater placement are required in order to prevent soil separate. There are working vessels available for producing lightweight soil and casting it underwater. Photo 1 shows one of them, when it was conducting backfill of lightweight soil (Yamamoto *at al.* 1997).

In addition to construction techniques, we need to comprehend mechanical properties of this developed material when deployed under large water pressure, because lightening ingredient (air foam or EPS beads) is significantly compressible. This paper presents a series of experimental study of the lightweight soil with triaxial apparatus.

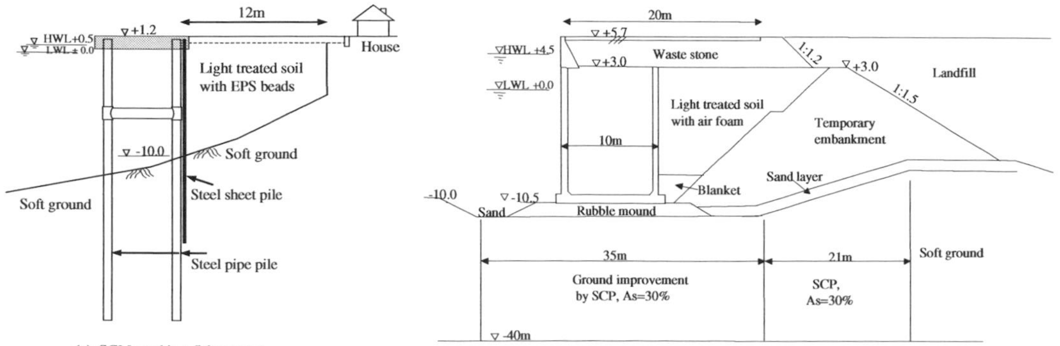

(a) SGM used in a fishery port

(b) SGM designed in a wharf section (after T.Sato, 2002)

Figure 1. Applications of lightweight soil in port structures.

Photo 1. A working vessel placing lightweight soil underwater behind port structure.

2 PROGRAM OF EXPERIMENT

2.1 *Modified triaxial apparatus*

Lightweight treated soil consists of compressible lightening material, either air foam or EPS beads. So it is expected to behave like unsaturated soil, even though the samples are prepared and sheared under saturated condition. To study deformation behavior, it is necessary to perform triaxial tests with volumetric change measured. For this purpose, a conventional triaxial apparatus was modified as shown in Figure 2, where an acryl cylinder peripheral to the specimen was erected inside the triaxial cell. Volumetric change of specimen was measured by monitoring water level of the inner cell by a differential pressure sensor (Tang, *et al.* 1996).

2.2 *Preparation and pressured curing of samples*

The original soft clay was dredged from Kawasaki Port. Table 1 shows the physical properties of Kawasaki clay.

Table 2 describes the designed proportions among clay slurry, lightener and cement. With target unit weight at $12\,kN/m^3$, the original clay was diluted to slurry states with water content about $2.5w_L$ for air foam mixing or $1.8w_L$ for EPS beads mixing. After examination of slurry density, certain quantity of cement based on target strength (e.g. $q_u = 200$ or $400\,kN/m^2$) was added in and agitated for 3 minutes. Then, pre-calculated quantity of lightener was mixed. The mixture was churned further for 3 minutes in EPS beads cases, or 30 seconds in air foam cases. It was identified that the unit weight of the created sample hereby met with $12 \pm 0.3\,kN/m^3$ adequately. Otherwise, further adjustment was conducted by slight addition of lightener when it was greater than $12.3\,kN/m^3$ or by charging a little more cement mixed slurry when it was less than $11.7\,kN/m^3$.

The prepared lightweight slurry was poured into curing molds ($\phi 50\,mm \times H100\,mm$) for specimen making. The molds were set into different containers, which were teemed with seawater, then sealed and applied by different curing pressure, $p_{cure} = 50$, 100, 200 and $300\,kN/m^3$, respectively. Underwater pressured curing was kept for 28 day in each case.

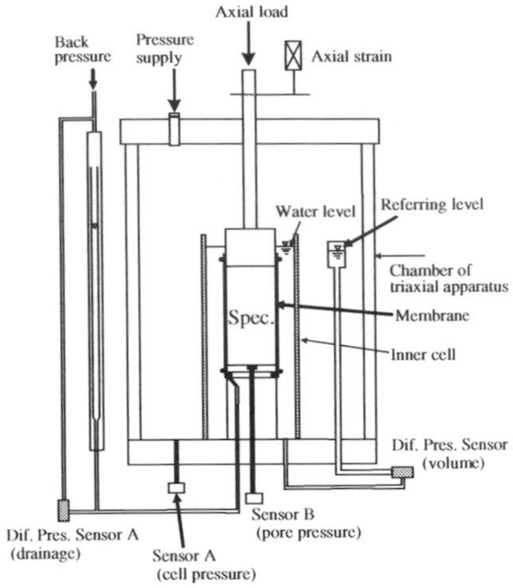

Figure 2. Triaxial apparatus modified to measure water drain-age and volumetric change.

Table 1. Physical properties of Kawasaki clay

Physical index	Grain density 25.8 kN/m	Liquid limit 76.1%	Plastic limit 45.8%	Plasticity index 31.3
Grain gradation	Coarse grain 0.3%	Sand 6.6%	Silt 77.1%	Clay 16.0%

Table 2. Target strengths, curing conditions, and designed proportions among slurry, lightener and cement.

Lightener type (kN/m^2)	Target strength (kN/m^2)	Curing pressure (kN/m^3)	Slurry (kN/m^3)	Lightener (kN/m^3)	Cement
Air foam	200	50,100,200,300	11.30	0.04	0.66
	400	50,100,200,300	11.20	0.04	0.76
EPS beads	200	50,100,200,300	11.33	0.03	0.64
	400	50,100,200,300	11.24	0.03	0.73

*Unit weight was targeted at $12\,kN/m^3$ for all cases.

2.3 *Procedures of triaxial experiment*

After pressured curing, the pressure in the sealed container was released. With the top-end of molded sample trimmed, the specimen was mounted onto triaxial apparatus. At first, cell pressure was resumed to the curing pressure (p_{cure}), while excess water or air

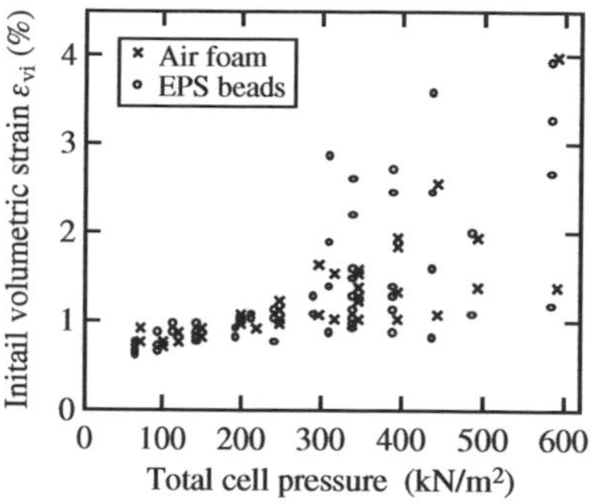

Figure 3. Initial volume change under cell pressure.

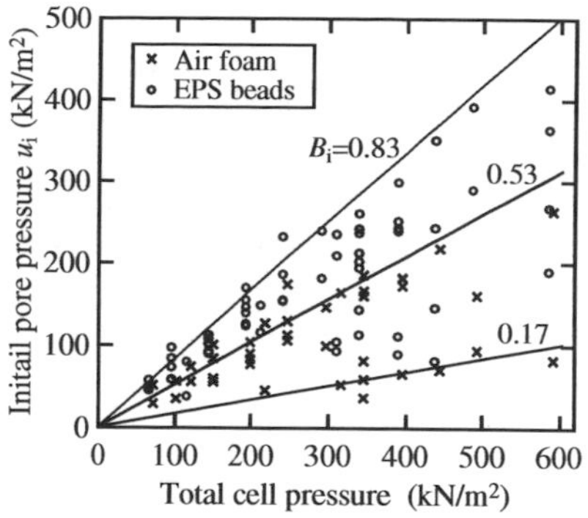

Figure 4. Pore water pressure response under cell pressure.

was squeezed out of the specimen. 10 minutes later, the specimen was switched to undrained state, and additional isotropic cell pressure was applied on respective specimen by $\Delta\sigma_c = 20$, 50, 100, 150 and 300 kN/m². The second procedure was also kept for 10 minutes. Soon, undrained shear was carried out on specimens at an axial strain rate of 0.2%/min with confining cell pressure constant.

3 MECHANICAL PROPERTIES

3.1 *Initial responses*

As the lightweight soil samples were resumed to the curing pressure and then applied by additional cell pressure (total cell pressure $= p_{cure} + \Delta\sigma_c$), they exhibited volume reduction and pore water pressure increase simultaneously. Figure 3 shows initial volumetric strain related to total cell pressure before the specimens were subjected to undrained shear. It can be seen that volume reduction of specimen increases with total cell pressure. The scatter looks more significant as total cell pressure exceeds 300 kN/m². Figure 4 shows the pore water pressure response. Even though efforts were made to produce saturated samples, the pore water pressure response were considerably retarded, with pore water coefficient B_i ranging from 0.17 to 0.83.

3.2 *Undrained shear behaviors*

During the undrained shear process, both pore water pressure and volume change of the specimen were measured. Figures 5 and 6 show some examples of experimental results for air foam mixed and EPS beads mixed samples.

From these figures, compressive strengths q_{max} are obtained by the maximum values of deviator stress, which are equivalent to unconfined compression strength q_u excepting that various cell pressures exerted on the specimens. It is found that air foam mixed samples shows significantly greater

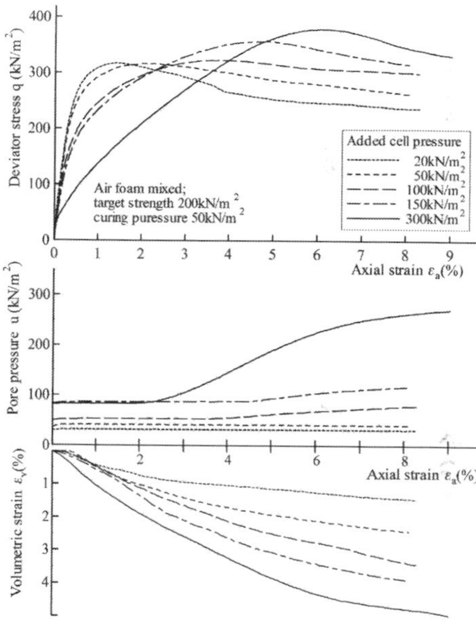

Figure 5a. Undrained shear curves (air foam, qu = 200 kN/m²).

compressive strength q_{max} than the designed target strength q_u. For EPS bead mixed samples, on the other hand, q_{max} seems nearly equal to or slightly less than the target strength q_u.

During the undrained shear process, specimens' volume kept shrinking. Beside the volume reduction depended mainly on total cell pressure. Pore water pressure generally exhibited plus response, which also depended on the condition of cell pressure.

3.3 *Factors affecting strength*

It is usually recognized that increment of unit weight contributes to strength gain for soil samples. Figure 7 shows the change of unit weight with increasing curing

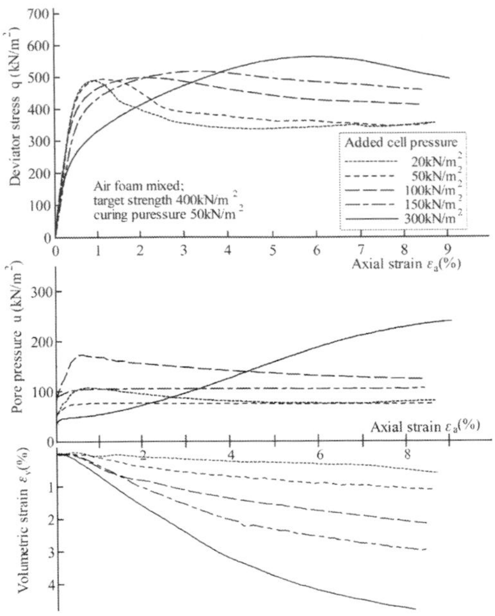

Figure 5b. Undrained shear curves (air foam, qu = 400 kN/m²).

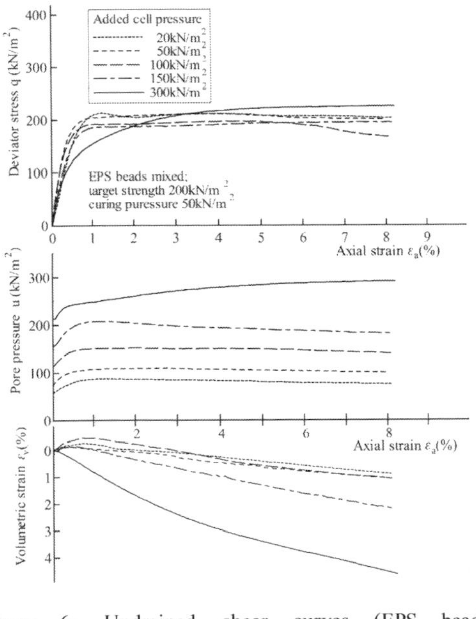

Figure 6a. Undrained shear curves (EPS beads, qu = 200 kN/m²).

pressure. Both air foam mixed and EPS beads mixed samples exhibit the similar tendency of unit weight increase induced by pressured curing. Yet, the unit weight increase is small and acceptable.

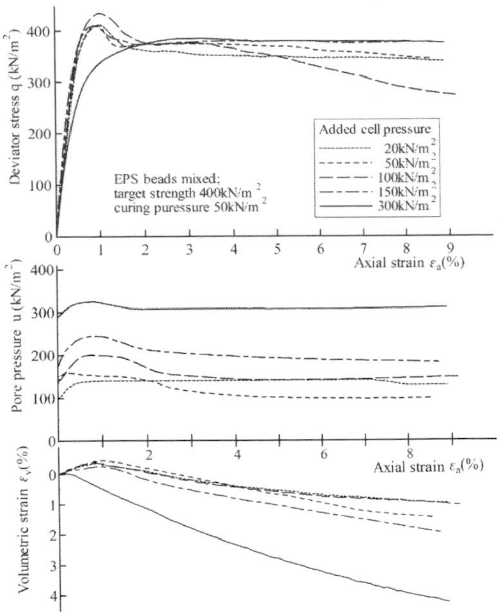

Figure 6b. Undrained shear curves (EPS beads, qu = 400 kN/m²).

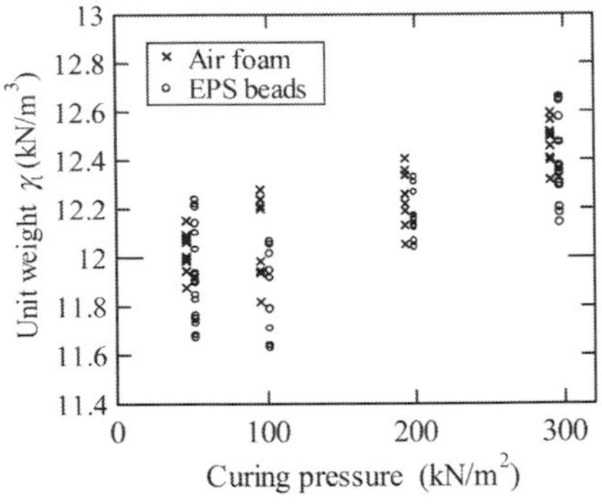

Figure 7. Unit weight changing with curing pressure.

Figure 8 shows the relation between compressive strength q_{max} and unit weight γ_t. Certainly, the compressive strength increases with unit weight, but the tendency is more prominent for air foam mixed samples than that for EPS beads mixed ones. Although cement contents to ensure strength were designed nearly equal for each mixed cases as shown in Table 2, there arose a noticeable split in strength gain. It is suggested that such a discrepancy attributes to the pressured curing process.

In Figure 9, the difference of compressibility between air foam and EPS beads is illustrated. For air foam mixed slurry, the enclosed gas within individual trapped bubbles was compressed under curing pressure instantaneously. If assuming that volume

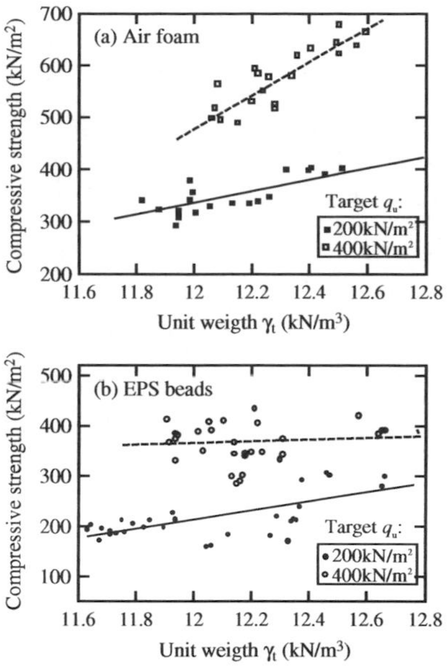

Figure 8. Correlation between compressive strength and unit weight.

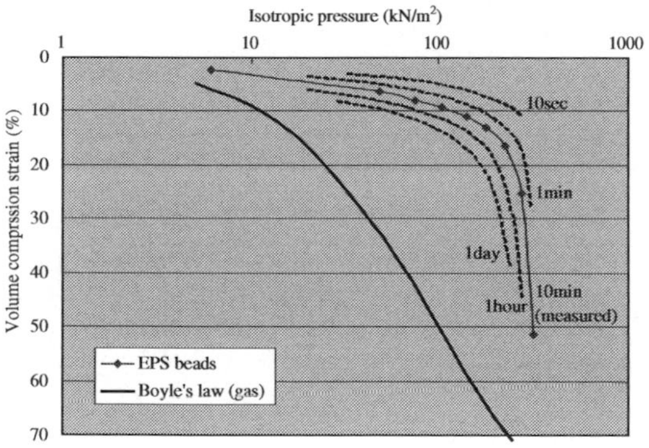

Figure 9. Comparison of compressibility between enclosed gas and EPS beads.

compression takes place in accordance with boyle's low as shown in Figure 9, the enclosed gas could be regarded as elastic material.

In contrast, EPS beads is a polymeric material compounded of abundant gas. It is highly compressible with distinct viscosity. The compression curves by the dotted line group in Figure 9 illustrate time-dependent property of EPS beads. During the pressured curing process, hereby, the bound structure by cement hardening was presumably destroyed to a certain degree, because of the continual compression of EPS beads under water pressure.

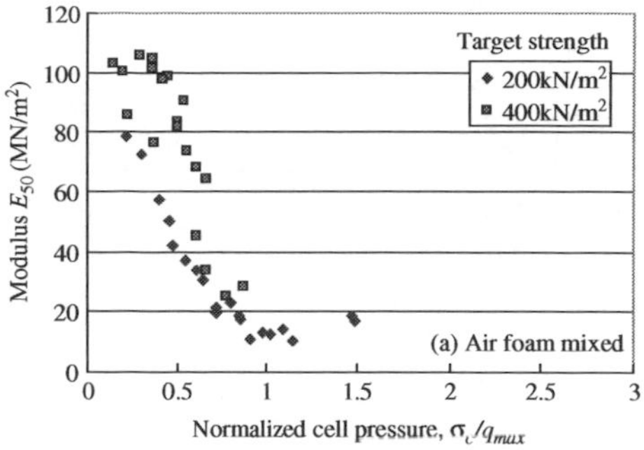

Figure 10a. Secant modulus E50 descending with normalized cell pressure (air foam mixed case).

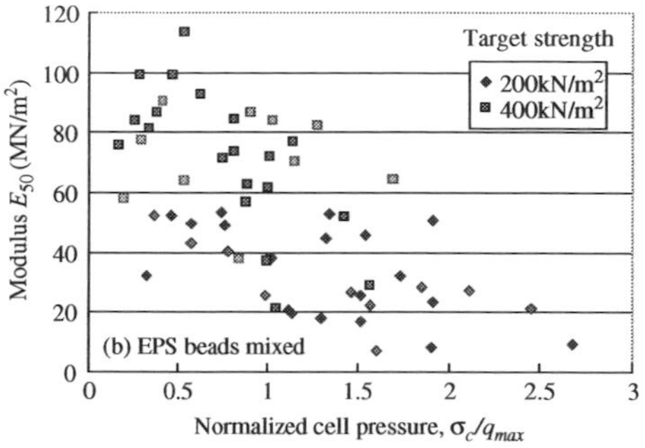

Figure 10b. Secant modulus E50 descending with normalized cell pressure (EPS beads mixed case).

3.4 *Deformation characteristics*

The secant modulus E_{50} is an important parameter to assess deformation characteristics. Here, E_{50} is defined as the gradient when a secant passing through the origin and the point at half maximum of deviator stress. That is, $E_{50} = q_{max}/2/\varepsilon_{50}$. Here, ε_{50} is axial strain ε_a at $q_{max}/2$.

Figure 10 shows the secant modulus with relation to total cell pressure, which is normalized with compressive strength q_{max}. By normalizing with q_{max}, we can evaluate the relative degree of confining pressure σ_c. It can be seen that E_{50} varies to such an extent that it decrease from over 100 MN/m^2 to less than 10 MN/m^2. For air foam mixed cases in Figure 10(a), there is an apparent correlation between E_{50} and σ_c/q_{max}. As total cell pressure σ_c surpassed compressive strength q_{max}, the bound structure due to cement hardening was destroyed radically, thus E_{50} descended to an ultimate low value.

For the EPS beads mixed samples as shown in Figure 10(b), there exhibits the same descent tendency as that of air foam mixed ones. However, the correlation seems much more obscure. This result suggests that

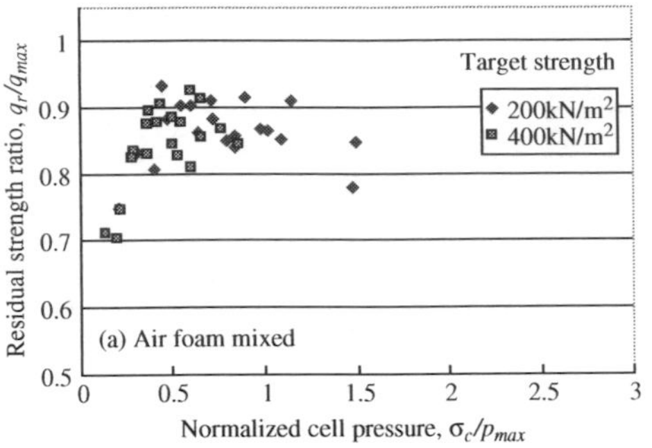

Figure 11a. Relation between residual strength and normal-ized cell pressure (air foam mixed case).

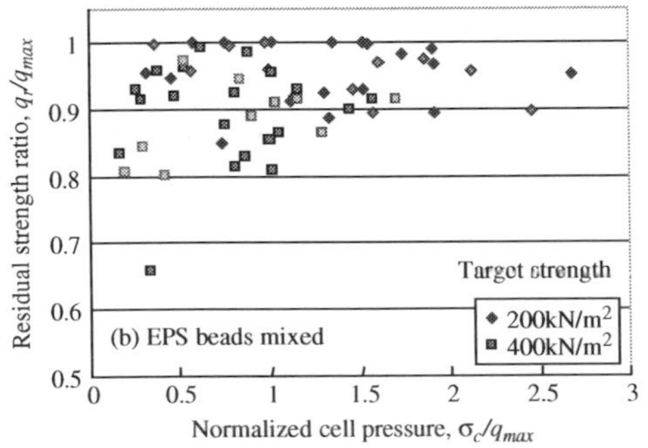

Figure 11b. Relation between residual strength and normal-ized cell pressure (EPS beads mixed case).

the role EPS beads plays in lightweight soil is more intricate than air foam does.

3.5 *Residual strength*

Figure 11 shows the residual behavior of compressive strength q_r. Here, axial strain ε_a when undrained shear experiment was terminated is regarded as ultimate state. Usually, the experiments were terminated around $\varepsilon_a = 8\%$. It was found that residual strength distributes within a range of $0.8{\sim}0.9q_{max}$ for air mixed samples, comparing with $0.8{\sim}1.0q_{max}$ for EPS beads mixed ones. As the lightweight samples underwent ultimate axial strain, bound structure due to cement hardening radically destroyed. So the deviator stress was born by soil particles under confined stress for air foam mixed samples, but EPS beads also shared a small part in the EPS beads cases.

Even though there is subtle discrepancy between different lightening ingredients, it is confirmed that residual strength is fairly larger than what is observed in unconfined compression tests. This result implies that lightweight soil treated with cement is not a vulnerable material as long as certain confining stresses exert on it.

4 CONCLUSIONS

Lightweight treated soil has found various applications in port structures. The new technology is aimed to reduce the earth pressure and the consolidation settlement, to enhance the stability of structure, and to make beneficial reuse of surplus dredged soft soil at the same time. Underwater deployment of lightweight soil requires us to make sure that the artificial geomaterial be adequately placed without soil separate. It is also important to confirm whether or not water pressure should inhibit the lightweight soil's mechanical properties as we designed. For this purpose, a series of undrained triaxial experiments were carried out on lightweight samples cured under various water pressures. The result may be concluded as followings.

1) Pressured curing resulted in an increase of unit weight, which remained in an acceptable range. It was observed that increase of unit weight induced strength gain, consequently.
2) Though prepared in saturated state, the lightweight samples exhibited unsaturated-like behaviors, such as volumetric compressible, retarded pore water response when subjected to undrained compression. During undrained shear, the specimens kept shrinking, showing plus pore water pressure.
3) Compressive strengths q_{max}, given by maximum of deviator stress, were significantly greater than the designed target strength for air foam mixed case, but nearly to or slight smaller than the target value for EPS beads mixed case. The reason for this result is unclear, but the strong viscosity of polymeric EPS beads is suspected as most important culprit.
4) Secant modulus E_{50} descends with increasing confining pressure. Residual strength q_r remains greater than $0.8q_{max}$ as long as certain confining stresses exist under ground environment.

REFERENCES

Tsuchida, T., Takeuchi, D., Okumura, T. and Kishida, T. 1996. Development of light-weight fill from dredgings. *Proc. of 2nd Int. Congress on Environmental Geotechnics*, 415–420.

Sato, T. 2002. Practical studies of lightweight treated soil by use of dredged clay. *Doctoral Dissertation of Kyushu University*. (in Japanese).

Yamamoto, T., Takayama, M., Yoshida, S. and Kitamura, R. 1997. Reconstruction of earthquake damaged quaywall by the use of PMC method for lateral earth pressure reduction. *Tsuchi-To-Kiso JGS* No.478, Vol.45, N.11, 21–23. (in Japanese).

Tang, Y.X., Tsuchida, T., Takeuchi, D., Kagamida, M. and Nishida, N. 1996. Mechanical properties of lightweight cement treated soil using triaxial apparatus. *Technical note of PHRI*, No.845, 29. (in Japanese).

New Horizons in Earth Reinforcement – Otani, Miyata & Mukunoki (eds)
© 2008 Taylor & Francis Group, London, ISBN 978-0-415-45775-0

Analysis of geofiber reinforced soils

A.T. Sway (T. Swe)
HSA Engineers and Scientists, Tampa, FL, USA

S. Bang
South Dakota School of Mines and Technology, Rapid City, SD, USA

ABSTRACT: Various types, shapes, and amounts of geofibers mixed with sand and clay were tested and analyzed in this study. Sixty-six consolidated undrained triaxial compression tests were run and multivariate regression analyses were carried out to obtain the stress-strain relationships. The confining pressure was found to be the most influential parameter. As for the fiber properties, the fiber dosage had greater influence than the fiber aspect ratio. In addition, a finite element analysis program was developed to analyze the behaviors of the geofiber-reinforced soil as a sub-base in a highway system. Analyses indicate that the fibrillated geofiber reinforcement can improve the soil strength better than the tape geofiber reinforcement, but the tape geofiber reinforcement provides more consistent behaviors than the fibrillated geofiber reinforcement.

1 INTRODUCTION

The use of geofibers for reinforcing earth masses has been implemented with great success in recent years. Geofibers consist of relatively small fiber inclusions, such as the polypropylene, distributed randomly as an additive throughout the soil mass (Figure 1). Therefore, the geofiber may be classified as a micro-reinforcement. Unlike traditional macro-reinforcements, such as the Reinforced Earth, the soil nail walls, geofabrics, and geogrids, the geofibers have no preferred orientation, i.e., the material orthotropy does not exist. Therefore, the geofiber-reinforced soil mass can be treated as an isotropic continuum with material properties influenced by the addition of geofibers (Swe et al. 2000). Furthermore, one of the main advantages of using geofibers is the maintenance of the strength isotropy and the absence of potential planes of weakness that can develop parallel to the oriented reinforcements (Gray & Maher 1989, Maher 1988).

The mechanism of the soil strength increase associated with the geofiber-reinforced soils includes: (1) the pullout resistance due to the friction between the individual fiber and the surrounding soil; (2) the adhesion between the individual geofiber and the surrounding soil (in cohesive soils); (3) the micro-bearing capacity of the soil mobilized by the pullout resistance of individual fibers looped across the shear plane; and (4) the increased localized normal stress in the soil across the shear surface resulting from the pullout resistance of the geofibers during shearing of the soil (Gregory & Chill 1998).

Figure 1. Field-mixed geofiber-reinforced soil.

Numerous studies have been conducted in the past to investigate the effects of natural and synthetic fibers on the improvement of the soil shear strength, on the constitutive relationship, and on others through CBR tests on cohesive soils (Hoare 1977, Setty & Rao 1987, Setty & Murthy 1990), direct shear tests on cohesive soils (Gregory & Chill 1998) and cohesionless soils (Gray & Ohashi 1983, Shewbridge & Sitar 1989), triaxial tests on cohesive soils (Gregory & Chill 1998) and cohesionless soils (Gray & Al-Refeai 1986, Setty & Rao 1987, Setty & Murthy 1990), unconfined compression tests on cohesionless soils (Santoni et al. 2001), laboratory model earth walls (Arenicz & Chowdhury 1988), and strength tests on fly-ash soil (Karniraj & Havanagi 2001). They indicate various degrees of effects of the fiber length, fiber diameter, fiber dosage, fiber aspect ratio, and confining pressure. In addition, Shewbridge & Sitar (1990) developed a closed form solution for determining the development of tension in soil reinforcements from their direct shear tests. Their model accounted for the plastic work to deform the soil and the elastic work to deform the reinforcements in tension and bending. They concluded that the mobilization of tension in the reinforcements was a function of the reinforcement properties and the deformation characteristics of the reinforced soil. They also stated that the relationship between the strength increase of the reinforced soil and the percentage reinforcement was non-linear.

The main objectives of this study are to determine the constitutive relationship of the geofiber-reinforced soils and to investigate the effects of various geofiber properties through an application of geofiber-reinforced soils as a sub-base material in highway construction.

2 MATERIALS AND LABORATORY TESTS

The geofibers used in this study are the fibrillated polypropylene (FIBERGRIDS, 0.034 KN/m^3) and the tape polypropylene. Both are manufactured by the Synthetic Industries, Chattanooga, Tennessee, and they are 2.54 to 5.08 cm in nominal length. These geofibers can be classified as ASTM- D 4101-group1/class1/grade 2. They have a specific gravity of 0.91. The geofibers have a carbon content (ASTM-D1603) of 0.6%, tensile strength (ASTM- D 2256) of 275.8 MPa minimum, tensile elongation (ASTM-D2256) of 25% maximum, and Young's modulus (ASTM-D2101) of 4.19 GPa minimum. They have a coefficient of friction of 0.6 to 0.8 of the tangent friction angle of the soil. The amounts of fibers added into the soil (fiber dosage) were 0.2% and 0.4% by the soil dry unit weight.

Consolidated Undrained (CU) triaxial compression tests with the pore water pressure measurement were conducted for this study. Entire tests were performed at the A.G.T laboratory of the Synthetic Industries. Specimens with a diameter of 7.29 cm and a length of 14.73 cm were tested. The details of the sample preparation methods are well described in reference (Gregory 1996).

A total of 66 triaxial tests were run, including 19 tests conducted with lean clay (CL) samples and 47 tests conducted with poorly graded sand (SP) samples. The lean clay samples have a liquid limit of 26, plastic limit of 14, plasticity index of 12, effective friction angle of 13°, effective cohesion of 21.64 KPa, and dry density of 17.67 KN/m^3. The samples also have a moisture content of 12.9% and the percentage of particles passing number 200 sieve is 55%. The sand properties include an effective friction angle of 33°, moisture content of 10%, and dry density of 15.66 KN/m^3. The specific gravities for clay and sand samples were assumed as 2.7 and 2.65, respectively.

Confining pressures of 68.95 KPa, 137.89 KPa, and 275.79 KPa were applied to the clay samples. Reduced confining pressures of 34.47 KPa, 68.95 KPa, and 137.89 KPa were, however, used for the sand samples.

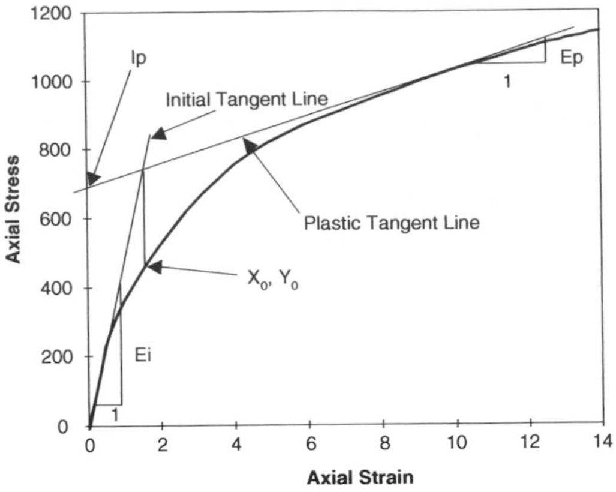

Figure 2. Bilinear parameters.

3 BILINEAR PARAMETERS

From the observation of the triaxial test data for the geofiber-reinforced soils, it was concluded that the stress-strain relationship follows a bilinear pattern, i.e., a steep initial elastic behavior followed by a relatively flat plastic response. Figure 2 shows the bilinear curve used to simulate the triaxial test results. As described in the figure, five bilinear parameters; the initial tangent modulus Ei, the plastic tangent modulus Ep, the intersection of the axial stress axis and the plastic tangent line Ip, the strain value at the intersection of initial and plastic tangent lines X_0 and the corresponding stress value Y_0, can completely describe the bilinear stress-strain relationship. The value of X_0 can however be determined analytically from Ei, Ep and Ip.

Bilinear parameters were graphically obtained from the CU triaxial test results. For the clay samples, three tests were conducted with unreinforced samples and 16 tests were conducted with fibrillated and tape geofiber-reinforced samples. For the sand samples, three tests were conducted with unreinforced and 20 and 24 tests were conducted with the fibrillated and tape geofiber-reinforced samples, respectively. Three geofiber dosages, 0%, 0.2%, and 0.4%, were applied to both clay and sand samples. Three aspect ratios (length to width ratio) of 0, 8, and 43 were used for the clay samples and five aspect ratios of 0, 8, 15, 21, and 43 were used for the sand samples. Please note that the zero dosage and zero aspect ratio indicate that the soil samples are unreinforced.

4 LABORATORY TEST RESULTS

Complete laboratory triaxial test results on all 66 samples with different fiber types, fiber widths, fiber dosages, fiber aspect ratios, and confining pressures can be found in reference (Swe 2002).

From the careful observation of the stress-strain curves of the geofiber-reinforced soils, it was observed that they experienced the strain-hardening behavior.

Variations of the bilinear parameters were obtained with respect to three parameters, i.e., the confining pressure, the fiber dosage, and the fiber aspect ratio. It was observed that the values of all bilinear parameters for all samples increased with respect to the increase in these parameters. Among these three parameters, the confining pressure was recognized by far most influential, which was followed by the fiber dosage.

The geofibers tend to slip during deformation at confining pressures below the threshold value or the "critical confining pressure." Above the critical confining pressure, they either yield or break. To reach the critical confining pressure with a typical fiber aspect ratio and fiber dosage, an extremely high overburden pressure would be required. However, the results suggest that the confining pressures used in this study are below the threshold value. Therefore, it can be concluded that the failure mechanism of the geofiber-reinforced soils, under virtually all practical conditions, will be the pullout of the geofibers (Gregory 1999).

5 CONSTITUTIVE RELATIONSHIP

To establish relationships among the bilinear parameters and the fiber dosage, the fiber aspect ratio, and the confining pressure; multivariate regression analyses were carried out on the bilinear parameters obtained from the triaxial test results. The following relationships were assumed based on their observed variations:

$$
\begin{aligned}
E_i &= a_1 e^{a_2\beta + a_3\alpha + a_4\sigma_3}\\
E_p &= b_1 e^{b_2\beta + b_3\alpha + b_4\sigma_3}\\
I_p &= c_1 e^{c_2\beta + c_3\alpha + c_4\sigma_3}\\
Y_o &= d_1 e^{d_2\beta + d_3\alpha + d_4\sigma_3}
\end{aligned}
\qquad (1)
$$

where a_1 through d_4 = coefficients; β = fiber dosage in percentage by the soil dry unit weight; α = fiber aspect ratio (length to width) and σ_3 = confining pressure in KPa. The values of E_i, E_p, I_p, and Y_o are in KPa.

Table 1 shows the optimized values of the coefficients a_1 through d_4.

The bilinear relationship describes the transition of the stress-strain variation, i.e., the initial tangent modulus, the plastic modulus as the asymptote, and the intercept of the asymptote with the ordinate, with the following boundary conditions.

1. At $\varepsilon = 0$, the slope, i.e., the tangent modulus, $\left(\frac{\partial\sigma}{\partial\varepsilon}\right)$ should be equal to the initial modulus;
2. At $\varepsilon = \infty$, the slope should be equal to the plastic modulus and
3. At $\varepsilon = X_0$, the calculated stress should be equal to Y_0.

The following equation describes the bilinear relationship with a smooth transition between elastic and plastic portions and also satisfies the required boundary conditions (Cho 1992, Preber et al. 1995):

$$
\sigma = (\mathrm{Ip} + \mathrm{Ep}\varepsilon)\left[1 - \exp\left(-C\varepsilon^2 \frac{\mathrm{Ei}\varepsilon}{\mathrm{Ip}}\right)\right]
$$

where (2)

$$
C = -\frac{\mathrm{Ei}}{\mathrm{Ip}\,X_0} - \frac{1}{X_0^2}\ln\left[1 - \frac{Y_0}{\mathrm{Ip} + \mathrm{Ep}X_0}\right]
$$

ε = axial strain
σ = axial stress

Figure 3 shows a comparison of the stress-strain relationships between the measured and the predicted from Equation 2 for three soil samples reinforced with tape fibers with the fiber dosage of 0.2 %, the aspect ratio of 42, and the confining pressures of 68.95 KPa, 137.89 KPa, and 275.79 KPa.

Table 1. Values of coefficients defined in Equation 1.

	Clay, fibrillated	Clay, taped	Sand, fibrillated	Sand, taped
a_1	6649.881	7858.875	4501.105	2721.757
a_2	2.248218	1.724888	1.279834	2.466681
a_3	0.002291	0.000533	0.002393	0.004825
a_4	0.0046372	0.004407	0.010610	0.012907
b_1	80.72060	80.22342	130.9817	69.68024
b_2	0.597407	0.566442	3.210653	3.750844
b_3	0.005283	0.006299	0.008667	0.099270
b_4	0.005890	0.005806	0.008960	0.010756
c_1	39.81462	40.39581	93.27565	68.49521
c_2	0.328688	0.604583	1.174416	1.125804
c_3	0.000516	1.801e-11	0.003862	0.005238
c_4	0.004699	0.004469	0.008375	0.010631
d_1	29.50490	31.93911	95.23840	65.33820
d_2	0.002854	0.290047	0.774288	0.877247
d_3	0.001247	7.951e-12	0.003688	0.004097
d_4	0.004616	0.004449	0.006850	0.008835

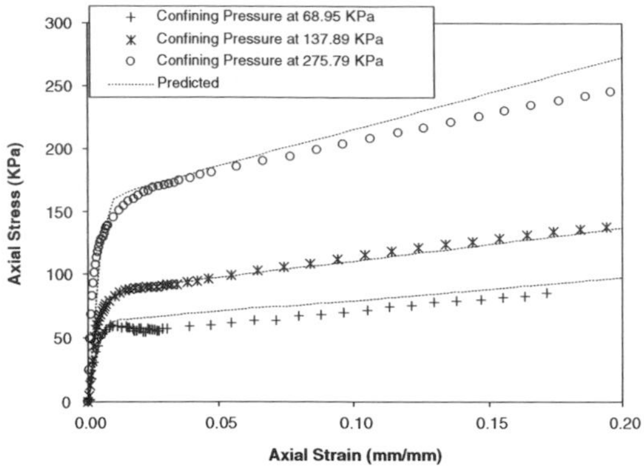

Figure 3. Raw and predicted stress-strain curves for the clay reinforced with tape geofiber.

The tangent modulus of a geofiber-reinforced soil can be obtained by differentiating σ in Equation 2 with respect to ε.

6 FINITE ELEMENT ANALYSIS

A two-dimensional plane strain finite element analysis software for the geofiber-reinforced soils was written to investigate a typical highway cross-section. The program can analyzes the behavior of the soil by three different models: the linear elastic model, the non-linear hyperbolic model (Duncan et al. 1980) and the bilinear model as described above.

The highway system cross-section considered in this study is a typical interstate freeway system that was provided by the South Dakota Department of Transportation. The cross-section consists of a rigid pavement layer having 7.31 m in width and 0.20 m in thickness of continuously reinforced Portland cement concrete. It has a 3.05 m wide shoulder next to the pavement. The highway system also includes base and sub-base courses with variable thicknesses.

The finite element mesh used for the analysis includes the widths of the pavement of 7.31 m, the shoulder of 3.05 m, the base course of 17.07 m, the sub-base of 31.70 m and the subgrade of 34.75 m. The thickness of the pavement layer was 0.20 m, whereas those of the base and the sub-base were 0.41 m and 1.22 m, respectively. In this study, two different thicknesses of 0.46 m and 0.61 m of the sub-base were used.

The design point load was calculated based on the American Association of State Highway and Transportation Officials' Load and Resistance Factor Design (LRFD) specifications, proposed by Nowak et al. (1993) and Nowak (1999), which increases the design truck loading by combining the standard truck HS20-44 load with a lane load of 472 KN/m.

Three cases were studied: Case 1, fibrillated and tape geofiber-reinforced clay as the sub-base with a clay subgrade; Case 2, fibrillated geofiber-reinforced sand as the sub-base with a sand subgrade; and Case 3, tape geofiber-reinforced sand as the sub-base with a sand subgrade.

Linear elastic constitutive model was used for the pavement and the hyperbolic model was used for the base and the subgrade elements. Bilinear model was applied to the geofiber-reinforced sub-base elements. For fiber reinforced sub-base, the following variables were used: (1) soil: clay and sand, (2) fiber dosage: 0.2% and 0.4%, (3) fiber aspect ratio: 8, 15, 21, and 43, and (4) sub-base thickness: 0.46 and 0.61 m. Following soil material input parameters were used for the analysis:

Pavement: linear, $E = 27.58$ GPa, $\gamma = 23.56$ KN/m^3, $\upsilon = 0.3$
Base: hyperbolic, $k = 600$, $n = 0.4$, $R_f = 0.7$, $C = 0$, $\phi_0 = 42°$, $\Delta\phi = 9°$, $k_b = 175$, $m = 0.2$

Sub-base: Bilinear
Clay subgrade: hyperbolic, $k = 90$, $n = 0.45$, $R_f = 0.7$, $C = 9.57$ KPa, $\phi_0 = 30°$, $\Delta\phi = 0°$, $k_b = 80$, $m = 0.2$
Sand subgrade: hyperbolic, $k = 200$, $n = 0.4$, $R_f = 0.7$, $C = 0$, $\phi_0 = 33°$, $\Delta\phi = 3°$, $k_b = 50$, $m = 0.2$

where γ = unit weight, υ = Poisson's ratio, E = Young's modulus, k = loading modulus, n=loading modulus exponent, R_f = failure ratio, C = cohesion, ϕ_0 = initial friction angle, $\Delta\phi$ = increment in friction angle over 10 fold increase in confining pressure, k_b = bulk modulus, m = bulk modulus exponent.

Twelve runs were completed for Case 1. Fourteen and 16 analysis runs were completed for Cases 2 and 3, respectively. The results of the surface deflection, the vertical displacement, and the maximum compression were investigated and discussed below.

6.1 *Effect of fiber dosage*

Both the fibrillated and tape fibers exhibit similar patterns for the maximum vertical displacement distribution. However, a detailed inspection suggests that the tape geofiber-reinforced sub-base is generally stronger than the fibrillated one.

For surface deflections, in all cases, the lesser values are associated with the thicker sub-base. For all cases, the smallest surface deflection is observed when the sub-base is reinforced with the fiber dosage of 0.4 % and the aspect ratio of 43, which corresponds to the largest amounts of the fiber dosage in this study. With the clay sub-base, increasing the sub-base thickness yields significantly smaller values of the surface deflection. The effects of the fiber dosage and the aspect ratio are not distinctly observed with the clay sub-base, although they exhibit some visible differences. Maximum surface deflection of 1.26 cm is noted in Case 1. However, the effect of the fiber dosage can be clearly seen with the sand sub-base. This effect is more pronounced with the fibrillated geofiber-reinforced sub-base. For Case 3, for both sub-base thicknesses, only negligible differences are observed with the fiber dosage of 0.2%. Maximum surface deflections of 0.97 cm and 0.89 cm are observed in Cases 2 and 3, respectively. This suggests that the use of the tape geofiber- reinforced soil as the sub-base can slightly reduce the surface deflections.

For the maximum compressive stress developed within the pavement, all cases indicate that the increase in fiber dosage decreases the maximum compression. Maximum compression changes with respect to the sub-base thickness are more pronounced in Cases 1 and 2 than in Case 3. This suggests that the soil beneath the pavement becomes stronger when the fiber dosage and the sub-base thickness increase.

For maximum compressive stress developed within the sub-base, all cases show that the thickness of the sub-base has little effect on the maximum compression. All cases reveal that the maximum compression increases as the fiber dosage increases.

6.2 *Effect of fiber aspect ratio*

In all cases, as the fiber aspect ratio increases, the maximum compression slowly decreases. The decrease in the sub-base thickness, however, increases the maximum compression significantly. In Case 2, almost identical values of the maximum compression in the sub-base are noted with the thickness of 0.46 m and the fiber dosage of 0.4% when compared to that with the thickness of 0.61 m and the fiber dosage of 0.2%.

For the maximum compressive stress in the sub-base, all cases show that the thickness of the sub-base attributes only a small decrease in maximum compression. As for the effect of the fiber aspect ratio, the maximum compression slightly increases as the aspect ratio increases.

For the maximum compressive stress developed within the subgrade, in all cases, the maximum compression increases as the fiber aspect ratio increases. However, the maximum compression becomes relatively smaller when a 0.61 m sub-base configuration is used.

7 CONCLUSIONS

From the measured stress-strain curves of the geofiber-reinforced soils, it was observed that they revealed a strain-hardening behavior. The stress-strain curves for the samples reinforced with tape geofibers exhibited more consistent behavior in improving the soil strength than those reinforced with fibrillated geofibers. In general, the effect of the fiber dosage is found to be more influential than the effect of the fiber aspect ratio.

It was observed that the values of all bilinear parameters for all samples increased as the confining pressure, the fiber dosage, and the fiber aspect ratio increased. Among these three variables, the confining pressure was recognized as the most influential one, followed by the fiber dosage. Although, it was evidenced that the soil strength improved with respect to the increase in the fiber aspect ratio, the effect was not as well defined as those of the confining pressure and the fiber dosage.

It was clear that the soil strength increased with respect to the increase in the fiber dosage. This was more distinct for the tape geofiber-reinforced clay.

It was observed that tape geofibers increased the clay strength more than fibrillated geofibers. However, for the sand, fibrillated geofibers demonstrated slightly stronger soil strength than tape geofibers.

It was also observed that the clay sub-base reinforced with fibrillated geofibers yielded lesser vertical displacements than the tape geofiber-reinforced sub-base. Even though they experienced more or less the same maximum vertical displacements, contour plots suggested that the tape geofiber-reinforced sub-base exhibited stronger soil strength characteristics than the fibrillated one.

Observation on the maximum compressions indicated that the increase in the sub-base thickness had relatively minor effects, as compared to the increase in the fiber dosage and the aspect ratio.

This study was entirely based on experimental laboratory tests conducted on two types of soils (CL and SP). In the future, studies should consider other types of soils and also use a wider variation of the confining pressure.

The study indicates that the fiber dosage has more significant effects on the soil strength improvement than the fiber aspect ratio. Therefore, tests with more diversified fiber dosages are recommended to be conducted in the future.

Based on the limited number of tests, the fibrillated geofiber reinforcement provides slightly better improvements in soil strength, whereas the tape geofiber reinforcement yields more consistent behaviors. Again, additional tests, including the full-scale field tests with detailed instrumentation, need to be conducted to further generalize the results of this study.

ACKNOWLEDGEMENTS

The authors are grateful for the financial and technical supports provided by Garry Gregory of Gregory Geotechnical and the Synthetic Industries, Chattanooga, Tennessee.

REFERENCES

Arenicz, R.M.& Chowdhury, R. 1988. Mechanics of fiber reinforcement in sand. *Journal of Geotechnical Engineering,* ASCE, 109(3), 335–353.

Cho, Y. 1992. Behavior of Retaining Wall with EPS Blocks as Backfill. MS thesis submitted to the South Dakota School of Mines and Technology.

Duncan, J.M., Byrne, P., Wong, K.S. & Mabry, P. 1980. Strength, stress-strain and bulk modulus parameters for finite element analyses of stresses and movements in soil masses. Report No. UCB/GT/80-01, Department of Civil Engineering, University of California, Berkeley.

Gray, D.H. & Al-Refeai, T. 1986. Behavior of fabric versus fiber-reinforced Sand. *Journal of Geotechnical Engineering.*, ASCE, 122(8), 804–820.

Gray, D.H. & Maher, M.H. 1989. Admixture stabilization of sands with discrete, randomly distributed fibers. *Proceedings of XIIth International Conference On Soil Mechanics And Foundation. Engineering*, Rio de Janeiro, Brazil, 1363–1366.

Gray, D.H. & Ohashi, H. 1983. Mechanics of fiber reinforcement in sand. *Journal of Geotechnical Engineering.*, ASCE, 109(3), 335–353.

Gregory, G.H. 1996. Design guide for fiber-reinforced soil slopes, version 1.0. A Design Guide submitted to Synthetic Industries.

Gregory, G.H. 1999. Theoretical shear-strength model of fiber-soil composite. *Proceedings of ASCE Texas Section Spring Meeting*, Longview, Texas.

Gregory, G.H. & Chill, D.S. 1998. Stabilization of earth slopes with fiber reinforcement. *Sixth International Conference on Geosynthetics*, Atlanta, Georgia.

Hoare, D.J. 1977. Laboratory study of granular soils reinforced with randomly oriented discrete fibers. *Proceedings of International Conference. On Use of Fabrics in Geotech.*, Paris, France, 1, 47–52.

Karniraj, S R. & Havanagi, V.G. 2001. Behavior of cement-stabilized fiber-reinforced fly ash-soil mixtures. *Journal of Geotechnical Engineering*, ASCE, 127(7), 574–584.

Maher, M.H. 1988. Static and dynamic response of sands reinforced with discrete, randomly distributed fibers," Ph.D. thesis submitted to the University of Michigan.

Nowak, A.S. 1999. Calibration of LRFD Bridge Design Code, *NCHRP Report 368, Transportation Research Board*, Washington, D.C.

Nowak, A.S., Nassif, H. & DeFrain, L. 1993. Effect of Truck Loads on Bridges, *Journal of Transportation Engineering*, ASCE, 119 (6), 853–867

Preber, T., Bang, S., Chung, Y. & Cho, Y. 1995. Behavior of Expanded Polystyrene Blocks, *Transportation Research Record*, No. 1462, pp. 36–46.

Santoni, R.L., Tingie, J.S. & Webster, S. L. 2001. Engineering properties of sand-fiber mixtures for road construction. *Journal of Geotechnical Engineering*, ASCE, 127(3), 258-268.

Setty, K.R.N.S. & Murthy, A.T.A. 1990. Characteristics of fiber reinforced lateritic soil. *IGC* (87) Bangalore, India, 329–333.

Setty, K.R.N.S. & Rao, S.V.G. 1987. Evaluation of geotextile-soil friction. *Indian Geotechnical Journal*, 18(1), 77–105.

Shewbridge, S.E. & Sitar, N. 1989. Deformation characteristics of reinforced sand in direct shear. *Journal of geotechnical Engineering*, ASCE, 115(8), 1134–1147.

Shewbirdge, S.E. & Sitar, N. 1990. Deformation-based model for the reinforced sand. *Journal of geotechnical Engineering*, ASCE, 116(7), 1153–1170.

Swe, T. 2002. Analysis of geofiber reinforced soils with bilinear constitutive relationship, Ph.D. thesis submitted to the South Dakota School of Mines and Technology.

Swe, T, Gregory, G.H. & Bang, S. 2000. Constitutive relationship of geofiber reinforced soils. "*Proceedings of the 35th Symposium on Engineering Geology and Geotechnical Engineering*, Pocatello, Idaho, 250–259.

New Horizons in Earth Reinforcement – Otani, Miyata & Mukunoki (eds)
© 2008 Taylor & Francis Group, London, ISBN 978-0-415-45775-0

Particle and shear characteristics of granulated coal ash as geomaterial

N. Yoshimoto, M. Hyodo, Y. Nakata & R.P. Orense
Yamaguchi University, Ube, Japan

T. Hongo & A. Ohnaka
Ube Industries, Ltd., Ube, Japan

ABSTRACT: In this study, single particle crushing tests were carried out on various kinds of granulated coal ash to evaluate the crushing characteristic of each individual grain. Then, drained monotonic triaxial compression tests were carried out under different confining pressures on specimens prepared by granulation. The effects of the confining pressure on the shear characteristics, as well as its relation to shear strength, were examined taking into account the crushing strength. As a result, it became clear that contrary to that observed in natural sands, the particle crushing strength of granulated coal ash did not depend on the particle size. Moreover, the crushing strength affected the shear characteristics, i.e., as the crushing strength increased, the shear stiffness became higher and the shrinkage of the volumetric strain became smaller. Therefore, crushing strength can be a useful parameter in evaluating the shear characteristic of the granulated coal ash.

1 INTRODUCTION

In recent years, the challenge concerning the depletion of good geomaterials and the effective use of renewable resources have become a social responsibility. Therefore, the applicability of various by-product materials as geomaterial should be examined. Coal ashes are inevitably produced by burning coal. In Japan, based on the law to "promote usage of renewable resource" (resources recycling law) which was enforced in 1991, the coal ashes generated from thermal power plants are specified as by-product of electric industry and there is a duty to promote their effective usage. Moreover, in the fiscal year 1999, 7.6 million tons of coal ashes were generated, and this quantity tends to increase every year. At the present, about 80% of the coal ashes are used effectively and the majority is used as blended material in cement, such as in civil engineering field. Since a large quantity can be utilized at once in this field of application, it is expected that civil engineering usage can be an acceptable alternative for the predicted increase in coal ash utilization.

The works on the application of coal ashes to geotechnical engineering have been performed by many researchers, such as Horiuchi (1996) and Sawa et al. (2002). The authors have been studying the mechanical properties of granulated coal ash formed by milling process with small amount of cement added and whose particle size is almost equivalent to that of sand or fine gravel. The use of granulated coal ash has many advantages, such as the suppression of leaching of heavy metals and the possibility of outdoor curing. In addition, since granulated coal ashes are produced artificially, the particle strength can be grasped easily. Furthermore, another advantage is that it is possible to control the particle strength, something which cannot be carried out in natural sands. The present research was carried out in order to investigate the possibility of the positive utilization of such advantage.

In this study, single particle crushing tests were carried out on various kinds of granulated coal ash to evaluate the crushing characteristic of each individual grain. Then, drained monotonic triaxial compression tests were carried out under different confining pressures on specimens prepared by granulation. The effects of the confining pressure on the shear characteristics, as well as its relation to shear strength, were examined taking account into the crushing strength.

2 SAMPLE PROPERTIES

In this research, four kinds of sand-size granulated coal ashes were used. The compositions of coal ashes and cement additive differ in Type A and Type B. The composition of Type C was the same as that of Type A, but manufactured using different method. Finally, Type C and Type D were similar except for the kind of coal ash used. The composition, machine type and curing condition of granulated coal ashes are shown in Table 1. Table 2 shows the physical properties of the samples used. The specific gravities of the particles of granulated coal ashes are very low because of the presence of air vesicles in individual grains. In particular, the

Table 1. Composition, machine type and curing condition.

Sample	Composition (%)			Type of machine	Curing condition
	Coal ash	Cement	Addition		
A	85	5	10	eirich	natural
B	80	10	10	mixer	dry
C	85	5	10	mortar	w=
D	85	5	10	mixer	40 ~ 50%

Table 2. Physical properties of each granulated coal ash.

Sample	G_s	e_{max}	e_{min}	d_{50}(mm)
A	2.35	2.544	1,916	0.385
B	2.36	2.679	1968	0.467
C	2.41	2.222	1.522	0.561
D	2.28	2.280	1.512	0.368
P.I. Masado	2.62	0.967	0.491	0.546

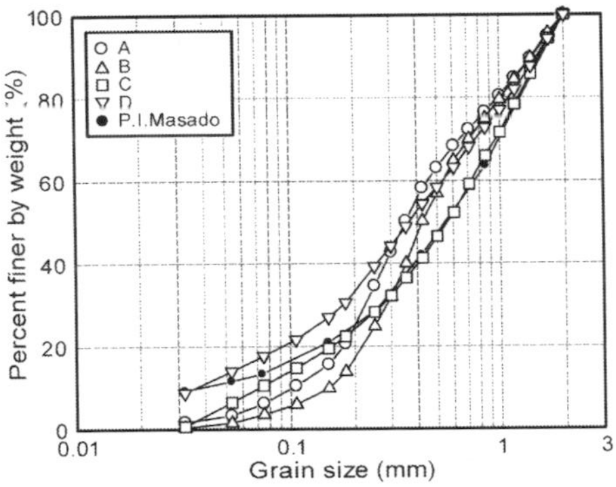

Figure 1. Grain size distribution curves.

maximum and minimum void ratios, which were determined using the methods specified in the Japanese Geotechnical Society Standards, were found to be very large for granulated coal ash. The particle size distribution curves depicted in Figure 1 show practically similar curves for granulated coal ashes and P.I. Masado. These materials contain about 10 ~ 20% fines and are well-graded with high coefficient of uniformity.

3 SINGLE PARTICLE CRUSHING TEST

The test was carried out by placing a particle on the lower plate in the most stable direction and then moving the upper plate at a constant rate of displacement to crush the particle (Nakata et al., 2001). Since the plates are flat, the loads during test are applied at two points in the vertical direction. Force and displacement were measured during test. The load measuring capacity was 4.91×10^2 N with a resolution of 9.81×10^{-3} N. On the other hand, the displacement was measured by un-contacted type with

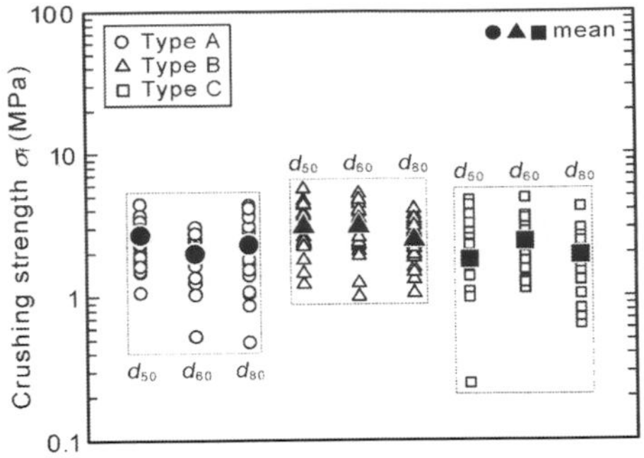

Figure 2. Crushing strength for each grain size of granulated coal ash.

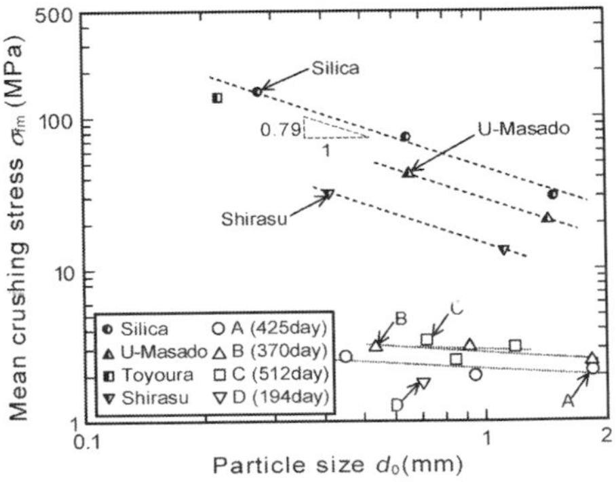

Figure 3. Effect of particle size on mean crushing strength.

measuring capacity of 2.00 mm with a resolution of 1.00×10^{-3} mm. Displacement rate of 0.1 mm/minute was applied. Single particle crushing test was carried out on d_{50}, d_{60}, d_{80} size particle of each granulated coal ash. Crushing strength σ_f is defined as in Eq. (1). The average value of each crushing strength σ_f in the same condition is defined as mean crushing strength σ_{fm} (Eq. (2)).

$$\sigma_f = F_f / d_0^2 \quad (1)$$

$$\sigma_{fm} = \sum_{j=1}^{n} \sigma_f / n \quad (2)$$

where F_f is the maximum load, d_0 is the initial height of particle, n is total number of particles tested.

Figure 2 shows all the values of crushing strength for each grain size (d_{50}, d_{60}, d_{80}) of granulated coal ashes (Types A–C). The variation of crushing strength σ_f and the value of mean crushing strength σ_{fm} for each grain size of granulated coal ashes are similar. Figure 3 shows the mean crushing strength against particle size. The figure shows data for the granulated coal ashes (Types A–D) and, for comparison purposes, include data for natural sands. It is already known that for natural sands under the same conditions, a smaller particle has a higher crushing strength, and a

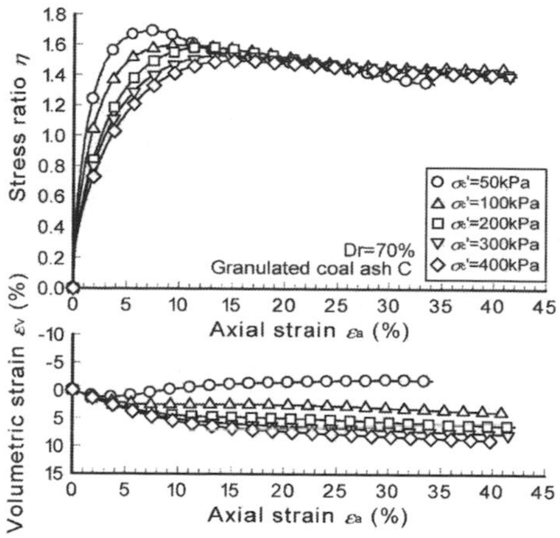

Figure 4. Stress ratio $\eta(=q/p')$ and volumetric strain ε_v plots against axial strain ε_a.

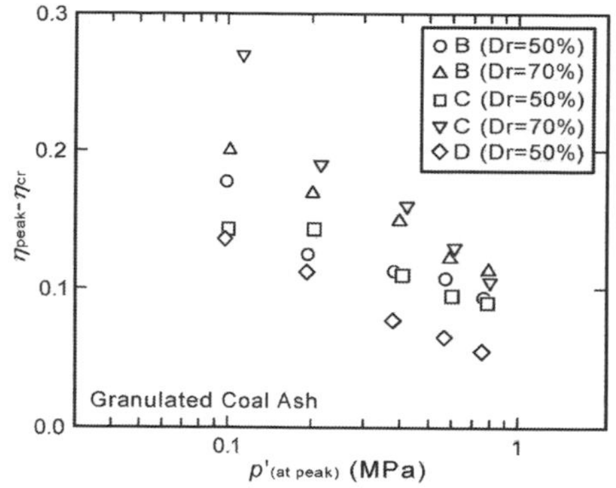

Figure 5. Relationship between η_{peak}-η_{cr} and effective mean principal stresses p' at peak stress ratio.

regression line drawn through this data on a log-log scale has a slope of −0.79 (Nakata et al., 1999). In the comparison among mean crushing strengths for granulated coal ashes, it was found that Type B with the most cement content has higher strengths than Type A and Type C. In granulated coal ashes, the grain size dependency of strength as observed in natural sands was not seen. This would be because the calcium silicate hydrate is present uniformly and the internal structure is homogeneous.

4 TRIAXIAL COMPRESSION TEST

Triaxial compression tests were carried out under consolidated and drained condition for granulated coal ashes. To prepare the test specimens, granulated coal ash was water-pluviated into a mould which was then gently tapped until a relative density of 50% and 70% were achieved. Specimens were isotropically consolidated to 50, 100, 200, 300, and 400 kPa. Subsequently, the specimen was sheared under a drained condition at an axial strain rate of 0.1%/min until the critical state. Because consecutive voids from the surface to the particle interior side are present, it is not easy to completely saturate granulated coal ash. The saturated specimens were made using de-aired granulated coal ashes. The de-aired granulated coal ashes were soaked in de-aired water and were placed in a vacuum tank for three or four days.

Figure 4 shows the stress ratio and volumetric strain plots against axial strain for granulated coal ashes. The granulated coal ash had a much greater range of peak stress ratio η_{peak} depending on the confining stress, with a marked decrease occurring in the peak stress ratio η_{peak} as σ'_c increased. It was dilative at 50 kPa and only became compressive at $\sigma'_c \geq 100$ kPa with large compressive strains occurring at $\sigma'_c = 400$ kPa. The residual stress ratio shows a constant value, and is not influenced by the confining pressure. It is thought that this behaviour is caused by the particle crushing and by the fact that particle strength is low. This result is consistent with the research results obtained previously on particle crushing of the sand.

Several experimental results have shown that the peak friction angle ϕ_p of sands increases with increasing relative density Dr and decreases with increasing confining stress p (e.g., Vesic and Clough (1968), Bolton (1986)). Using experimental results for several sands, Bolton (1986) showed that the combined effects of relative density and confining stress on the peak friction angle can be represented by a mathematical expression. Gutierrez (2003) revised this expression by using the peak stress ratio instead of the peak friction angle. The expression relating the peak stress ratio to the confining pressure is formulated as:

$$\eta_{peak} = \eta_{cr} + CDr\ln\left(p_{cr}'/p'\right) \tag{3}$$

where C is a material parameter, Dr is relative density, and, η_{cr} is stress ratio at the critical state. The critical state was defined as the state at which a soil mass is continuously deforming at constant volume, and constant effective stresses. The p'_{cr} was defined as the effective mean principal stress when the peak is equal to the stress ratio at the critical state.

The relationship between η_{peak}-η_{cr} and effective mean principal stresses p' at peak stress ratio is presented in Figure 5. As the effective mean principal stresses p' increases, η_{peak}-η_{res} approaches 0. It is thought that the drained shear strength demonstrated by the dilatancy of granulated coal ash decreases due to particle crushing associated with stress increase. The value when η_{peak}-η_{res} becomes 0 is p'_{cr}. The p'_{cr} was determined based on each individual experiment result. The findings are summarized in Table 3. From the table, the p'_{cr} does not depend on the relative density and indicates almost the same value. Figure 6 shows the relation between $(\eta_{peak}$-$\eta_{cr})/Dr$ and $p'_{cr}/p'_{(at\ peak)}$.

Table 3. p'_{cr} from each experiment results.

		Sample		
		B	C	D
p'_{cr}	$Dr = 50\%$	7.788	8.524	3.941
	$Dr = 70\%$	7.871	7.389	–

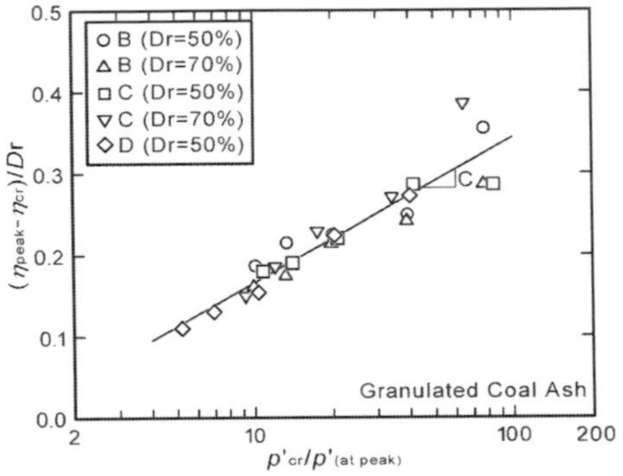

Figure 6. Relationship between η_{peak}-η_{cr}/Dr and $p'_{cr}/p'_{(atpeak)}$.

Although the data points are scattered, the relation can be represented by a straight line.

5 EVALUATION OF SHEAR STRENGTH OF GRANULATED COAL ASH BY PARTICLE STRENGTH

The relationship between p'_{cr} and σ_{fm} is presented in Figure 7. Although the data points are few, a unique relation seems to exist between the two parameters. The relation between $(\eta_{peak}$-$\eta_{res})/Dr$ and σ_{fm}/p' is shown by using this unique relation in Figure 8, which indicates a good correlation. From this relation, the following expression is obtained when solving for η_{peak}.

$$\eta_{peak} = \eta_{cr} + C' Dr \ln(\sigma_{fm} / p') \tag{4}$$

where C' is a material parameter. From Eq. (4), crushing strength can be a useful parameter in evaluating the shear characteristic of the granulated coal ash.

6 CONCLUSIONS

The effects of the confining pressure on the shear characteristics, as well as its relation to shear strength, were examined taking account into the crushing strength. As a result, it became clear that contrary to that observed in natural sands, the particle crushing strength of granulated coal ash did not depend on the particle size. Moreover, the crushing strength affected the shear characteristics, i.e., as the crushing strength increased, the shear stiffness became higher and the shrinkage

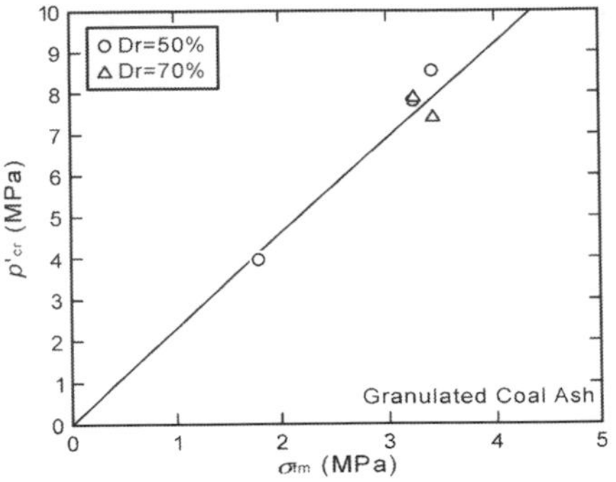

Figure 7. Relationship between p'_{cr} and σ_{fm}.

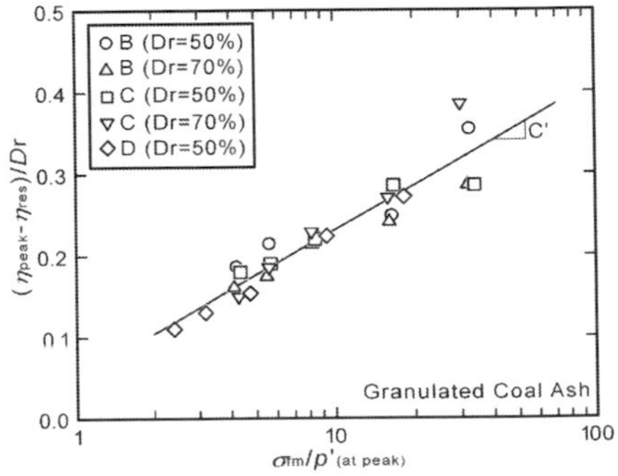

Figure 8. Relationship between $(\eta_{peak}$-$\eta_{cr})/Dr$ and $\sigma_{fm}/p'_{(at\ peak)}$.

of the volumetric strain became smaller. Therefore, crushing strength can be a useful parameter in evaluating the shear characteristic of the granulated coal ash.

REFERENCES

Bolton, M. D. 1986. The Strength and Dilatancy of sands. *Géotechnique*, Vol. 36, No. 1, pp. 65–78.

Gutierrez, M. 2003. Modeling of the Steady-State Response of Granular Soils. *Soils and Foundations*, Vol. 43, No. 5, pp. 93–105.

Horiuchi, S., Tamaoki, K. and Yasuhara, K. 1995. Coal ash slurry for effective underwater disposal. *Soils and Foundations*, 35(1), 1–10.

Nakata, Y., Hyde, A.F.L., Hyodo, M. and Murata, H. 1999. A probabilistic approach to sand particle crushing in the triaxial test. *Géotechnique*, Vol. 49, No. 5, pp. 567–583.

Nakata, Y., Hyodo, M., Hyde, A.F.L., Kato, Y. and Murata, H. 2001a. Microscopic particle crushing of sand subjected to high pressure one-dimensional compression. *Soils and Foundations*, Vol. 41, No. 1, pp. 69–82.

Sawa, K., Tomohisa, S., Maruyama, S. and Ogawa, A. 2002. Strength characteristics of cement-treated sludge mixed with coal fly ash. *Journal of the Society of Materials Science*, 51(1), 30–35 (in Japanese).

Vesic, A. and Clough, G. W. 1968. Behavior of granular materials under high stresses. *J. Soil Mech. Fnd. Div. ASCE*, Vol. 94(SM3), pp. 661–688.

New Horizons in Earth Reinforcement – Otani, Miyata & Mukunoki (eds)
© 2008 Taylor & Francis Group, London, ISBN 978-0-415-45775-0

Improvement of geotechnical characteristics of silts deriving from washing quarry gravel

R. Meriggi, M.D. Fabbro & E. Blasone
Dipartimento di Georisorse e Territorio, University of Udine, Udine, Italy

ABSTRACT: Quarrying activities worldwide produce large amounts of silt from washing gravel now considered as waste but, if treated, it might be utilized in civil and environmental works. The aim of the study is to quantify the strength and permeability variations of silts mixed with cement, bentonite, or both in percentages of between 2.5% and 5.0%. Laboratory compaction, unconfined compression and permeability tests have been performed on natural material and soil mixtures. Independently of the additive used, the values of maximum compression strength and dry unit weight rise with increased water content, reach a maximum at the optimum and then decrease for higher moisture. Nevertheless the efficacy of cement treatment on unconfined compression strength depends on water content and cement percentage. Whatever the water content, the stress-strain behaviour of the silt-cement mixture remains brittle, while for silt-bentonite mixtures it depends on moisture content. Natural silts permeability may be reduced by adding small quantities of bentonite to construct hydraulic barriers.

1 INTRODUCTION

Every year quarrying activity produces large amounts of silts, as a residual product of gravel washing, which are stockpiled in large basins near the washing plant. This material is not generally used because of its poor geotechnical characteristics and so is disposed of as waste.

This paper reports the preliminary results of an experimental study, as yet not completed, to determine how geotechnical properties of silts may be improved so that they can become available as a low-cost resource.

With this aim, compaction tests, constant head permeability in triaxial cell tests and unconfined compression tests have been performed on both natural material and mixtures with low percentages (2.5% and 5%) of bentonite and cement.

Cement is usually added to high plasticity natural soils in order to improve their strength and stiffness (Chew et al. 2004; Horpibulsuk et al. 2004; Bellezza et al. 1995), and to modify permeability properties of soils with different grain size distributions and plasticity characteristics (Bellezza & Fratalocchi 2006).

The use of bentonite as an additive to reduce soil permeability is widespread and well-documented (Daniel 1991), but there is little information (Bellezza et al. 1999) on the effectiveness of adding cement and bentonite to silty soils deriving from industrial processes.

2 EXPERIMENTAL PROGRAMME

The investigated material came from the clarification of water used to wash the gravel extracted from a quarry on the fluvial-glacial plain of Friuli Venezia Giulia (north-eastern Italy).

After extracting the material, the fine fraction has been separated from coarse fraction by washing, then deposited in special basins and lastly dewatered by a filterpress.

The first stage of the research investigated the main geotechnical properties of silt without additives; two samples were prepared for this by means of quartering material collected from different deposit zones.

The values of index properties (Tab. 1) measured for the two samples resulted as almost the same, so the material was classified as CL in the USCS system. The very low values of plasticity and activity indexes depend on the mineralogical composition of these silts, with a prevalence of carbonates and just traces of clayey minerals with low activity, such as illites and mica.

The tests included Standard Proctor compaction at different moistures, to determine the optimum water content and maximum unit dry weight. Oedometric and drained and undrained triaxial tests were only performed on samples at the optimum Proctor (Tab. 2).

The compressibility values are typical of low compressible and low swelling soils; the values of the coefficient of consolidation are instead dependent

Table 1. Main soil characteristics.

	Grain size distribution					Plasticity				
Sample n.	Sand (%)	Silt (%)	Clay (%)	D_{60} (mm)	D_{10} (mm)	LL (%)	PI (%)	Activity AI	Gs	Classif. USCS
C1	6.35	79.92	13.73	0.0178	0.0015	24.5	5.2	0.38	2.73	CL
C2	7.5	79.19	13.31	0.0182	0.0010	25.7	7.0	0.52	2.73	CL

Table 2. Unconfined compression strength and deformability characteristics of compacted silt.

Unconfined compression test			Oedometer test		Triaxial test	
q_u kPa	E_{25} MPa	ε_r %	C_c	C_s	c' kPa	ϕ' °
180	15.2	2.2	0.110	0.017	12	16

E_{25} = Young's modulus at 25% of unconfined shear strength q_u.

Table 3. Main index properties of silt–cement–bentonite mixtures.

Silts deriving from washing gravel with:	LL (%)	PL (%)	PI (%)	Gs
No additives	25.7	19.3	7.0	2.73
2.5% bentonite	35.3	18.0	17.3	2.77
5.0% bentonite	40.0	17.8	22.2	2.79
2.5% cement	36.3	22.2	14.1	2.76
5.0% cement	39.1	23.4	15.7	2.76
2.5% bentonite + 2.5% cement	43.1	22.9	20.2	2.79
5.0% bentonite + 5.0% cement	49.7	25.7	24.0	2.78

on the stress level and range between 8.76×10^{-3} and 1.37×10^{-2} cm^2/s. In undrained conditions compacted silt shows brittle behaviour characterized by high values of both stress-strain modulus and undrained shear strength, although the latter is reached for very low vertical strain and then quickly decreases. The very low effective shear strength value is typical of compacted silty soils.

To study the possibility of improving the geotechnical properties of washing silts, six series of samples were prepared mixing the untreated material with 2.5% and 5% of bentonite, cement and both respectively. Table 3 shows the main index properties and specific gravity of every blended soil investigated. An increase in liquid and plastic limits for low plasticity soils treated with low percentages of cement has also been reported by Chew et al. (2004) and Locat et al. (1990) explain this behaviour with the aggregation and cementation of particles into larger size clusters caused by the hydraulic binder.

Another possible reason is the water trapped within intra-aggregate pores which increases the apparent water content without really affecting interaction between aggregates (Locat et al. 1996). The increased plasticity in the specimen of silts mixed with bentonite is due to the high activity of clayey minerals in this additive.

All specimens were prepared mixing the additive (Portland 32.5 cement and/or sodium bentonite with a high content of montmorillonite) to the dry silt (ASTM 1997) and then wetting to a predetermined moisture content. The blended material was compacted with a Standard Proctor effort and two specimens then collected. At present, constant head permeability tests into a triaxial cell have only been completed for silts mixed with 2.5% bentonite.

The unconfined compression tests were performed after 28 days or 7 days curing for silt-cement mixtures and silt-bentonite mixtures respectively. Table 4 shows the values of W_{opt}, γ_{dmax}, q_u and stress-strain modulus (E_{25}) for the samples of Optimum Proctor compacted wet, and the ratio of the variation of these parameters from those of the untreated soil. All the blended samples show a reduction of compaction characteristics, with a decrease in dry density and an increase in optimum moisture, counterbalanced by a higher unconfined compression strength, but not always coupled with higher values of E_{25}.

3 INFLUENCE OF CEMENT CONTENT

The compaction curves for the mixture with cement are lower and to the right of the curve of the untreated silt (Fig. 1a): the reduction in dry density, of between 1.7% and 5.5%, is coupled with an increase in optimum wet of between 1.7% and 2.8%, in order to hydrate the added cement.

For both mixtures (2.5% and 5.0% of cement) the unconfined compression strength increases with water content until the optimum moisture and then rapidly decreases (Fig. 1b). The values of stress-strain modulus E_{25} for both series of specimens lie along an almost symmetric curve with respect to W_{opt} (Fig. 1c). Moreover, with the lower water contents the unconfined compression strength seems to be independent of the cement percentage.

Table 4. Compaction and unconfined compression strength characteristics of silt – cement – bentonite mixtures compacted at Optimum Proctor. (*7 days curing) (**28 days curing).

Silts with	W_{opt} (%)	γ_{dmax} (kN/m^3)	q_u (kPa)	E_{25} (MPa)	ΔW_{opt} (%)	$\Delta\gamma_{dmax}$ (%)	Δq_u (%)	ΔE_{25} (%)
No additive	14.1	18.1	180	15.2	–	–	–	–
2.5% bentonite*	17.9	17.3	204	3.2	3.8	−4.4	13	−78.9
5.0% bentonite*	17.2	17.6	232	5.4	3.1	−2.8	29	−64.5
2.5% cement**	15.8	17.8	1004	181.9	1.7	−1.7	458	+1097
5.0% cement**	16.9	17.1	1376	197.6	2.8	−5.5	664	+1200
2.5% bentonite + 2.5% cement**	17.8	17.4	1240	153.6	3.7	−3.9	589	+910
5.0% bentonite + 5.0% cement**	18.0	17.3	1503	114.0	3.9	−4.4	735	+650

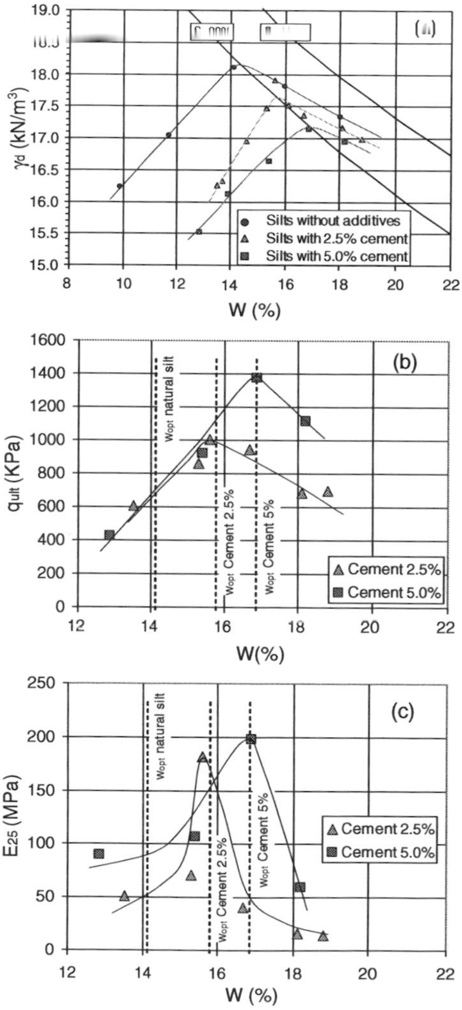

Figure 1. Silts mixed with cement: compaction curves (a); unconfined compression strength (b); stress-strain modulus (c).

The stress-strain modulus quickly increases with water content, reaches maximum for the optimum wet and then rapidly decreases to values three-four times lower than the maximum.

This behaviour may be explained considering that there must be enough water in the mixture to complete the hydration reactions, so that higher percentages of cement could develop further cementation bonds among grains.

The best water content to generate this process seems to be the optimum wet, instead for higher moistures the increase of the water-cement ratio in the mixture causes an increase in voids percentage not filled by hydraulic binder, with a consequent loss of strength and stiffness (Helal & Krizek 1992, Krizek & Helal 1992).

4 INFLUENCE OF BENTONITE CONTENT

Compaction curves for specimens with bentonite (Fig. 2a) also show reduced maximum dry density of between 2.8% and 4.4% compared to the untreated silts because of the larger quantity of fine fraction in the mixture.

In the dry side of the compaction curve the unconfined compression strength depends on bentonite percentage (Fig. 2b); for wet higher than optimum, q_u are instead almost the same, independently of the additive percentage. The trend of E_{25} always decreases with increasing water content (Fig. 2c) and appears to be independent of the percentage of bentonite in the mixture.

5 PERMEABILITY TESTS

Only two preliminary constant head permeability tests in triaxial cell have so far been done to evaluate the reduction in permeability related to the addition of bentonite.

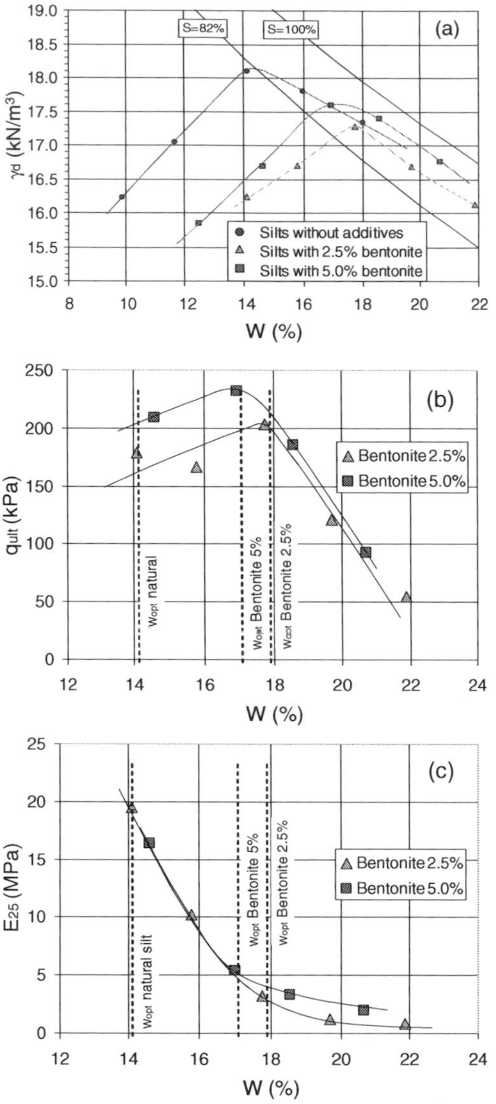

Figure 2. Silts mixed with bentonite: compaction curves (a); unconfined compression strength (b); stress-strain modulus (c).

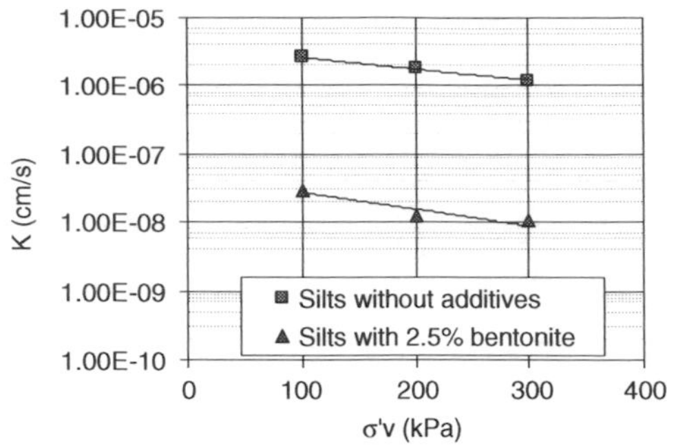

Figure 3. Results of preliminary constant head permeability tests.

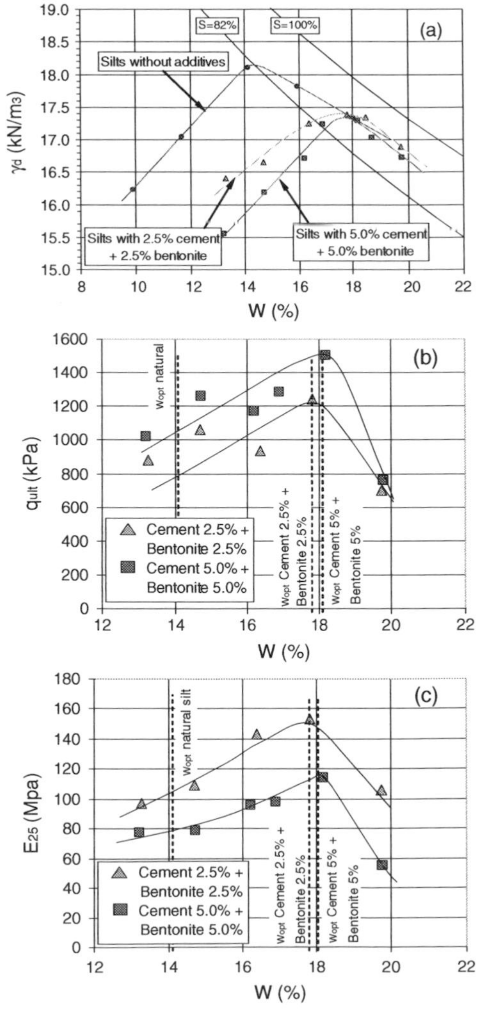

Figure 4. Silts mixed with cement and bentonite: compaction curves (a); unconfined compression strength (b); stress-strain modulus (c).

The first was performed on the silts without additives, the second on the mixture with 2.5% bentonite; both tests were performed on specimens compacted to Proctor optimum wet and subsequently saturated and consolidated.

As is well known, bentonite has marked characteristics of imperviousness and an appreciable reduction in permeability can be obtained even with small contents (2.5%) (Fig. 3).

In the interval of consolidation pressures imposed (100–300 kPa) and with a hydraulic gradient of

30 the decrease reaches two orders of magnitude compared to the silts without additives, from $k = 2.6 - 1.2 \times 10^{-6}$ cm/s to $k = 2.9 - 1.0 \times 10^{-8}$ cm/s. The latter values meet the international standards for hydraulic barriers, which recommend a permeability value lower than 1×10^{-7} cm/s.

6 INFLUENCE OF CEMENT AND BENTONITE CONTENT

Compaction characteristics for both cement and bentonite mixtures show comparable behaviour with similar values of W_{opt} and γ_{dmax} (Fig. 4a).

For $W < W_{opt}$ the unconfined compression strength depends on the percentage of additives (Fig. 4b); it reaches a maximum when $W = W_{opt}$ and then rapidly decreases to converge at the same value for both percentages of additive in the samples.

Instead, the stiffness of compacted mixtures depends on the water content, increasing for $W < W_{opt}$ and decreasing for $W > W_{opt}$ (Fig. 4c). The higher values of E_{25} for the samples prepared with 2.5% of cement and 2.5% of bentonite can be observed; this may be explained by the lower fine contents (especially less bentonite/montmorillonite), which reduce deformability.

The greater stiffness due to the higher percentage of cement (5%) is on the other hand partially counterbalanced by the increased deformability due to the 5% of bentonite.

7 EFFECTIVENESS OF ADDITIVES

In order to evaluate the efficacy of mixing silty soils with additives and the behaviour of the mixtures, the values of q_u and E_{25} have been normalized against the respective values at optimum wet obtained for the untreated soil $q_u = 180$ kPa and $E_{25} = 15.2$ MPa.

For mixtures with cement (Tab. 5), the maximum values of efficacy have been reached for moisture values very close to optimum wet (1%–2% near optimum); the mixture with 5.0% cement offers greater efficacy only for moisture higher than the optimum wet of the mixture with 2.5% cement.

As expected, the addition of bentonite to the silt doesn't improve the mechanical behaviour, and for values of moisture higher than optimum it sensitively reduces the unconfined compression strength (Tab. 6), while for moisture lower than optimum there are negligible improvements. The only effect of the mixing with bentonite is in reducing stiffness of the silt and changing its behaviour from brittle to ductile, so that it can undergo large deformations.

Table 5. Efficiency of the cement addition on the properties of washing silt.

2.5% cement			5.0% cement		
W (%)	$E_{25}/E_{25\,nat}$	$q_u/q_{u,nat}$	W (%)	$E_{25}/E_{25,nat}$	$q_u/q_{u,nat}$
13.5	3.3	3.3	12.8	5.9	2.3
15.3	4.7	4.8	15.4	7.1	5.1
15.6 (w_{opt})	12.0	5.6	16.9 (w_{opt})	13.0	7.6
16.7	2.7	5.3	18.2	3.9	6.2

Table 6. Efficiency of the bentonite addition on the properties of washing silt.

2.5% bentonite			5.0% bentonite		
W (%)	$E_{25}/E_{25\,nat}$	$q_u/q_{u,nat}$	W (%)	$E_{25}/E_{25,nat}$	$q_u/q_{u,nat}$
14.1	1.3	1.0	14.6	1.1	1.2
15.8	0.7	0.9	17.2 (w_{opt})	0.4	1.3
17.8 (w_{opt})	0.2	1.1	18.6	0.2	1.0
19.7	0.1	0.7	20.7	0.1	0.5

Table 7. Efficiency of the cement-bentonite addition on the properties of washing silt.

2.5% cement + 2.5% bentonite			5.0% cement + 5.0% bentonite		
W (%)	$E_{25}/E_{25\,nat}$	$q_u/q_{u,nat}$	W (%)	$E_{25}/E_{25,nat}$	$q_u/q_{u,nat}$
13.3	6.4	4.9	13.2	5.1	5.7
14.7	7.2	5.9	14.7	5.2	7.0
16.4	9.4	5.2	16.2	6.3	6.5
17.8 (w_{opt})	10.1	6.9	16.9	6.4	7.1
18.7	7.4	2.9	18.2 (w_{opt})	7.5	8.4
19.8	7.0	3.9	18.7	4.5	4.6

The most important benefits in terms of effectiveness are obtained for mixtures with cement and bentonite at the same percentages (Tab. 7).

The increase of both additives at the same percentage improves, to the same level, the effectiveness of the unconfined compression strength ($q_u/q_{u,natural}$), which rises until maximum values of about 7–8 times the soil values without additives are reached; this is true for the optimum wet, whereas the trend quickly decreases for small increments of water content.

As far as the strain behaviour is concerned, the effectiveness of mixing still depends on the water content, while larger percentages of bentonite reduce stiffness.

8 CONCLUSIONS

Compaction, unconfined compression and permeability tests performed in the laboratory show that it is possible to improve the geotechnical characteristics of silts obtained from washing quarry gravel by mixing them with different additives.

- The use of cement as additive results as the most effective method to improve shear strength characteristics. Nevertheless, the efficacy of treatment, for the cement contents investigated, mainly depends on water content: there must be enough in the mixtures to allow the completion of hydration reactions, so that larger percentages of binder can develop further cementation bonds between grains. The highest shear strength and stress-strain modulus values are reached for moistures $W = W_{opt} \pm 1\%$, while for higher values of w they quickly decrease. On the other hand the addition of cement confers greater brittleness on the silts compared to the natural material, and this limits the application to a narrow range of stress states.
- The use of bentonite as additive produces a mixture with similar behaviour to compacted clayey soils: shear strength considerably reduces for $W > W_{opt}$, and at the same time the stress-strain behaviour becomes progressively ductile. The significant decrease in hydraulic conductivity obtained by mixing the silts with small percentages of bentonite suggests hydraulic barriers as a possible field of application.
- In the mixtures obtained blending silts with the same percentages of both additives, the cement is responsible for the shear strength increase but, as previously observed for the mixture with cement alone, its efficacy depends not only on the added percentage but also on the water-cement ratio. The strain behaviour instead primarily depends on the percentage of bentonite, and it makes even the mixture with the higher percentage of cement ductile. Permeability tests on these mixture types have not yet been completed, so no conclusion can be made about the effectiveness deriving from the concurrent addition of two types of additives.

The laboratory investigations performed on silts deriving from washing quarry gravel mixed with small amounts of different additives show that it is possible to improve the mechanical and hydraulic behaviour of the silt and transform it from a waste product into useful resource.

REFERENCES

ASTM (1997). Test method for Mixture-Density Relations for Soil-Cement Mixtures. ASTM International, West Conshohocken, PA, D558.

Bellezza I., Fratalocchi E. (2006). Effectiveness of cement on hydraulic conductivity of compacted soil – cement mixtures. *Ground Improvement*, 10, No. 2: 77–90.

Bellezza I., Fratalocchi E., Pasqualini E. (1995). Permeabilità e resistenza al taglio di terreni compattati additivati con cemento. *XIX Convegno Nazionale di Geotecnica. Pavia 19–21 Settembre 1995.*

Bellezza I., Pasqualini E., Stella M. (1999). Reuse of an industrial waste in road constructions; *Proc. of the 12 European Conf. of Soil Mechanics and Geotechnical Engineering, Vol.1: 83–88.* Amsterdam.

Chew S.H., Kamruzzaman A.H.M., Lee F.H. (2004). Physicochemical and Engineering Behavior of Cement Treated Clays. *Journal of Geotechnical and Geoenvironmental Engineering*, 130, No. 7: 696–706.

Daniel D. E. (1991). Compacted clay and geosynthetic clay linings; *Atti delle Conferenze di Geotecnica di Torino XV ciclo: La ingegneria geotecnica nella salvaguardia e recupero del territorio. Torino, 20-22 novembre.*

Horpibulsuk S., Bergado D.T., Lorenzo G.A. (2004). Compressibility of cement-admixed clays at high water content. *Géotecnique* 54, No. 2: 151–154.

Helal M., Krizek R.J., (1992). Preferred orientation of pore structure in cement-grouted sand. *Grouting Soil Improvement and Geosynthetics*, Vol. 1: 526–540.

Krizek R.J., Helal M (1992). Anisotropic behaviour of cement grouted sand. *Grouting Soil Improvement and Geosynthetics*, Vol. 1: 541–550.

Locat J., Berube M.A., Choquette M.(1990). Laboratory investigations of the lime stabilization of sensitive clays: shear strength development. *Canadian Geotechnical Journal*, 27: 294–304.

Locat J, Tremblay H., Leroueil S. (1996). Mechanical and hydraulic behaviour of a soft inorganic clay treated with lime. *Canadian Geotechnical Journal*, 33: 654–669.

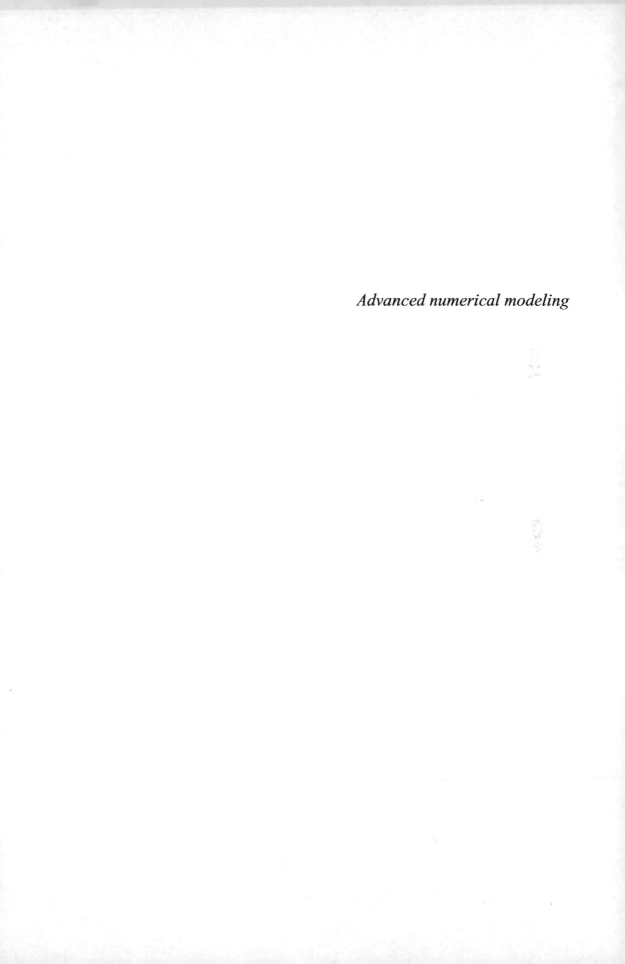

Advanced numerical modeling

New Horizons in Earth Reinforcement – Otani, Miyata & Mukunoki (eds)
© 2008 Taylor & Francis Group, London, ISBN 978-0-415-45775-0

Anisotropy of fiber-reinforced soil and numerical implementation

R.L. Michalowski
University of Michigan, Ann Arbor, U.S.A

ABSTRACT: While reinforcement in form of geosynthetic sheets or metal strips is used predominantly to improve structural integrity of the retaining walls, highway embankments, etc., short fiber reinforcement can be used to improve properties of the soil used as fill in geotechnical structures. A development of an anisotropic yield condition for fiber-reinforced sand is presented. As the composite's properties are anisotropic, the traditional kinematic approach of limit analysis needs to be modified to account for anisotropy. An example is shown where fiber-reinforced sand is used as backfill behind a retaining wall.

1 INTRODUCTION

Estimates of contribution of short fibers to the soil strength were made earlier for isotropic reinforced soils where the orientation of fibers was distributed evenly in all directions (Michalowski & Zhao 1996, Michalowski & Čermák 2003). Because of the technology of deposition of fiber-reinforced soils (mixing and compacting), the orientation of fibers has clearly a preferred bedding plane, giving rise to anisotropy of mechanical properties of the mixture. The research presented in this paper is concentrated on anisotropy of the strength properties, and on application of limit analysis to anisotropic fiber-reinforced sand.

The interest in fiber reinforcement of soils began in the 1970's with an attempt to estimate the influence of the plant/tree roots on stability of earth slopes (Walderon 1977). Subsequent models were based on analysis of a single fiber intersecting a band of soil with localized shear strain (Gray & Ohashi 1983, Maher & Gray 1990. While these efforts revealed the complexity of the soil-fiber interaction, they did not produce a yield condition in terms of the components of the stress tensor. Consequently, the early models were not implemented in the numerical tools, such as the finite element method, or in the numerical optimization schemes used in kinematic approach of limit analysis.

A yield condition for fiber-reinforced soil with ellipsoidal distribution of the fiber orientation is presented here. The plastic state and the kinematic discontinuities in the deformation fields will be discussed. The internal friction angle of an anisotropic composite is dependent on the orientation (direction), and a method to extract this angle from the anisotropic yield condition will be presented. Finally, an application example will be shown.

2 DEVELOPMENT OF THE YIELD CONDITION

The distribution of fiber orientation is considered in a form of an ellipsoid (Fig. 1), with a preferred bedding plane xOz.

This distribution is axi-symmetric and is a function of angle θ

$$\rho(\theta) = \frac{ab}{\sqrt{a^2 \sin^2 \theta + b^2 \cos^2 \theta}} \quad (1)$$

where a and b are the half-axes of the distribution. For convenience, the distribution aspect ratio is introduced, defined as

$$\xi = \frac{b}{a} \quad (2)$$

The total content of fibers is defined by their average concentration $\bar{\rho}$

$$\bar{\rho} = \frac{V_r}{V} \quad (3)$$

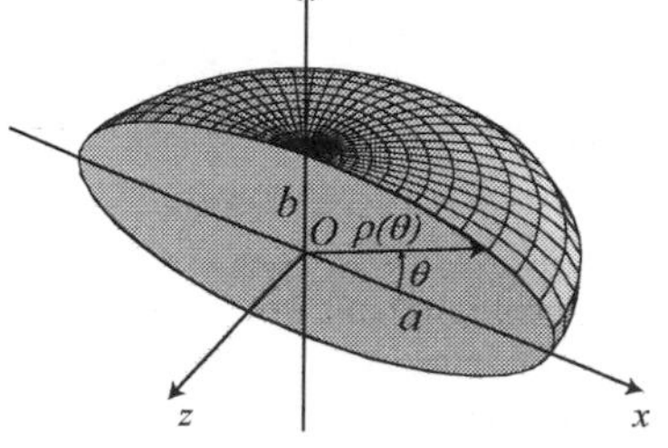

Figure 1. Ellipsoidal distribution of fiber orientation.

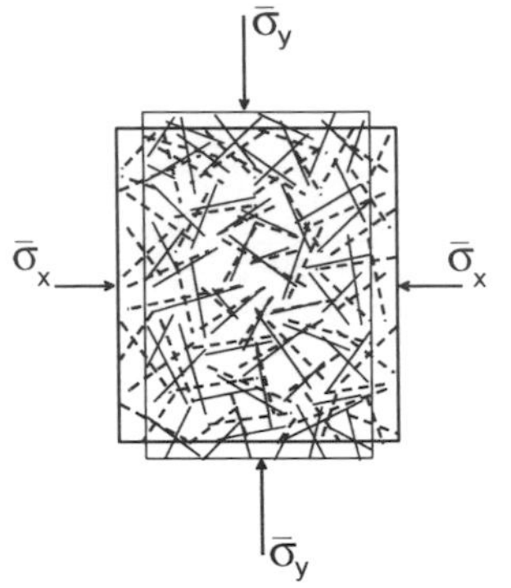

Figure 2. Deformation of a representative element.

where V_r and V are the volume of fibers and the volume of the representative element, respectively. The aspect ratio η of the cylindrical fibers of length l and radius r is defined as

$$\eta = \frac{l}{2r} \tag{4}$$

The yield condition is developed using a homogenization method where an incipient plastic deformation process is considered, as in Figure 2.

It is required that the internal work $D(\dot{\varepsilon}_{ij})$ in the representative element be balanced by the work of the average (macroscopic) stress $\bar{\sigma}_{ij}$ on the boundaries of the element

$$\bar{\sigma}_{ij}\dot{\bar{\varepsilon}}_{ij} = \frac{1}{V}\int_V D(\dot{\varepsilon}_{ij})\,dV \tag{5}$$

The internal work is dissipated in the irreversible process, and it is due to the frictional slip of the fibers in the matrix material (sand). Because the composite is anisotropic, the principal directions of the stress and strain rates, in general, do not coincide.

Calculations of the internal work rate in eq. (5) are tedious, and the details are omitted here (relevant earlier research can be found in di Prisco & Nova 1993, Michalowski 1997, Michalowski and Čermák 2002). The yield condition is sought in the following form

$$f = R - F(p, \psi) = 0 \tag{6}$$

where R is the maximum shear stress

$$R = \frac{1}{2}\left[(\bar{\sigma}_x - \bar{\sigma}_y)^2 + 4\bar{\tau}_{xy}^2\right]^{\frac{1}{2}} = \left(q^2 + \bar{\tau}_{xy}^2\right)^{\frac{1}{2}} \tag{7}$$

and p and ψ are the in-plane mean stress p ($p = (\bar{\sigma}_1 + \bar{\sigma}_3)/2$) and the angle that the major principal stress makes with axis x, respectively. Some numerical results are shown in Figure 3.

The contribution of fibers to the strength can be characterized as a function of total fiber content $\bar{\rho}$

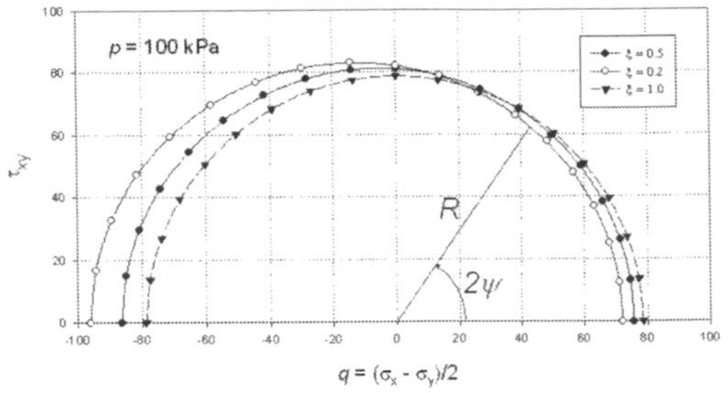

Figure 3. Calculated yield condition for fiber-reinforced sand.

(eq. (3)), distribution aspect ratio ξ (eq. (2)), fiber aspect ratio η (eq. (4)), angle of the sand-fiber interface friction ϕ_w, and also the internal friction angle of the sand (matrix material), ϕ. The analysis revealed that the number of independent parameters can be reduced to three: ϕ, product $\bar{\rho}\eta\tan\phi_w$, and ξ. The calculated yield surfaces in Figure 3 are for $p = 100$ kPa, $\phi = 36°$, $\bar{\rho}\eta\tan\phi_w = 0.89$, and 3 distribution ratios ξ: 0.2, 0.5, and 1.0 (isotropy).

It is interesting to notice that the trace of the yield surface at $p =$ const is approximately circular, independent of the distribution aspect ratio ξ. It is convenient then to approximate the yield function with a circle whose radius and the shift of the center along q-axis (Fig. 3) depend on internal friction angle of the sand ϕ, product $\bar{\rho}\eta\tan\phi_w$ characterizing the fibers, and ratio ξ that carries information about the anisotropic fiber orientation distribution.

3 LIMIT ANALYSIS WITH ANISOTROPIC SOIL

3.1 *Anisotropic yield condition*

Kinematic approach of limit analysis has been used often in geotechnical engineering, but the applications have been limited predominantly to isotropic soils. However, the distribution of the fibers in Figure 1 leads clearly to an anisotropic composite. For constant mean stress p, the trace of the yield condition for the fiber-reinforced soil can be conveniently represented as a circle; this is confirmed by the numerical homogenization calculations that led to the yield contours in Figure 3. For an isotropic distribution of fibers, $\xi = 1.0$, the yield condition can be graphically represented as a circle with its center at the origin of coordinate system q, τ_{xy} (Fig. 3).

The calculations indicated that for anisotropic distribution of the fibers the yield condition can still be represented by a circle, but this circle is now shifted along axis q by distance Q, as indicated in Figure 4.

The maximum shear stress R is now different from the radius of the circle R_0, and it is dependent on inclination angle ψ of the major principal stress, which

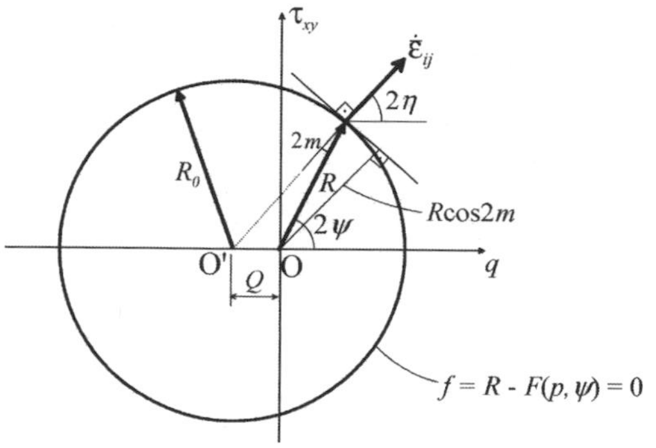

Figure 4. Anisotropic yield condition for fiber-reinforced sand.

Table 1. R_0/p and Q/p for anisotropic fiber-reinforced sand.

ϕ	$\bar{\rho}\eta \tan\phi_w$	ξ	R_0/p	Q/p
30°	0.20	1.0	0.5406	0
		0.5	0.5456	0.0110
		0.2	0.5524	0.0251
	0.40	1.0	0.5812	0
		0.5	0.5912	0.0220
		0.2	0.6048	0.0502
35°	0.20	1.0	0.6175	0
		0.5	0.6228	0.0116
		0.2	0.6298	0.0265
	0.40	1.0	0.6614	0
		0.5	0.6721	0.0233
		0.2	0.6861	0.0530

is characteristic of anisotropy. It is also clear from Figure 4 that the principal directions of the macroscopic stress and strain rate tensors do not coincide: $2\eta \neq 2\psi$ (normality rule is used to describe the deformation rate).

For practical purposes it is convenient to represent the radius R_0 and the shift Q as functions of angle ϕ, coefficient $\bar{\rho}\eta \tan\phi_w$, and distribution ratio ξ. A sample of the calculation results is given in Table 1. The results for $\xi = 1.0$ represent the yield condition of isotropic fiber-reinforced soil, with shift $Q = 0$. Since function $F(p, \psi)$ in eq. (6) is a linear and homogeneous function of p, the results in Table 1 are given as ratio R_0/p and Q/p.

3.2 *Evaluation of yield surface parameters*

This paper relates only to the theoretical aspects of the model and analysis. However, the parameters of the yield condition can be validated experimentally. Two plane-strain compression tests are needed to estimate R_0 and Q, one with the major principal stress perpendicular to the plane of fiber preferred orientation, and the second one with the major principal stress coinciding with the trace of the preferred bedding plane. Consequently, maximum and minimum of q on the yield surface in Fig. 4 can be determined, and R_0 and Q (or R_0/p, Q/p) evaluated, since the trace of the yield condition was approximated as a circle. In practice, more than two tests are needed, to assure repeatability of the results.

3.3 *Mechanisms with velocity discontinuities*

Kinematic approach of limit analysis requires consideration of velocity discontinuities, since they are part of admissible failure mechanisms. It can be demonstrated that the differential equations describing the plastic stress field and the plastic deformation are of hyperbolic type (Booker & Davis 1972). Similar to the isotropic case, the characteristics of the two sets of equations coincide (normality flow rule is used), but their analytical representation differs slightly for anisotropy

$$\frac{dy}{dx} = \tan(\psi - m \pm \nu) = \tan(\eta \pm \nu), \qquad \alpha, \beta \tag{8}$$

where angles m, η and ψ are illustrated in Figure 4, angle ν for the soil with linear dependence on p becomes

$$\nu = \frac{\pi}{4} - \frac{\phi_a}{2} \tag{9}$$

It follows from kinematics considerations that velocity discontinuities must coincide with velocity characteristics, and the dilatancy along these discontinuities is governed by internal friction angle ϕ_a. However, angle ϕ_a is dependent on orientation of the characteristic line. In kinematic approach of limit analysis the geometry of velocity discontinuities is not given *a priori*, but their inclination is subject to variation in search for the best bound to the limit load. Therefore, angle ϕ_a also will vary in the numerical procedure. Calculations of angle ϕ_a as function of orientation (direction) are illustrated in the next subsection.

3.4 *Internal friction angle in anisotropic soil*

For isotropic material the internal friction angle ϕ is independent of ψ

$$\sin\phi = \frac{R_0}{p} \tag{10}$$

Internal friction angle ϕ_a for anisotropic composite is not a constant, and it varies with the change in the physical orientation (direction). The relation among ϕ_a, R and p now becomes

$$\sin\phi_a = \frac{R\cos(2\psi - 2\eta)}{p} = \frac{R\cos 2m}{p} \tag{11}$$

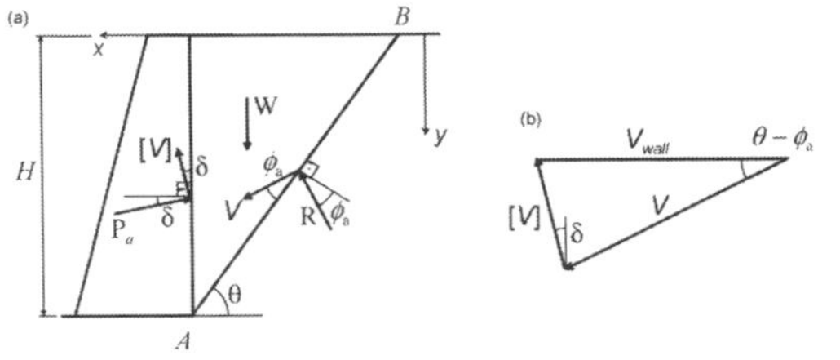

Figure 5. (a) Schematic of a retaining wall, (b) hodograph.

Table 2. Wall load coefficient K_a.

ϕ	δ	$\bar{\rho}\eta \tan\phi_w$	ξ	K_a
30°	15°	0	–	0.301
		0.20	1.0	0.271
			0.5	0.260
			0.2	0.245
		0.40	1.0	0.242
			0.5	0.221
			0.2	0.193

where angle $2m$ and the graphical interpretation of $R\cos 2m$ are shown in Fig. 4. Based on eq. (8), the characteristics (and therefore the velocity discontinuities) are inclined to axis x at angles $\eta \pm (\pi/4 - \phi_a/2)$, but angles η and ϕ_a are related through eq. (11). Hence the calculations of angle ϕ_a need to be iterative. A simple "classroom" example is presented in the next section.

4 EXAMPLE

This example relates to the problem of active backfill load on a retaining wall. It is expected that fiber reinforcement in a backfill will lead to reduction of the load. Of particular interest is how the anisotropy of fiber distribution affects the loads.

We consider a vertical but rough wall, Figure 5(a). The kinematic approach of limit analysis is used, and the hodograph for one-block (Coulomb-type) mechanism is shown in Figure 5(b). The roughness of the wall is included in calculations by assuming relative wall/soil motion vector [v] being inclined at interface friction angle δ to the wall.

The velocity discontinuity (or failure surface) AB is a characteristic of β-family, therefore its inclination angle θ can be related to the direction of the major principal strain rate η through eq. (8) as

$$\theta = \eta - \frac{\pi}{4} + \frac{\phi_a}{2} \qquad (12)$$

In the process of finding the best solution to the wall loading, angle θ is varied until the maximum of pressure coefficient K_a is found (the kinematic approach yields the lower bound here, because force P_a is a reaction rather than an active force). The numerical results are presented in Table 2.

As the fibers reinforce the backfill, the load from the backfill on the wall is reduced, and this reduction increases with an increase in $\bar{\rho}\eta \tan\phi_w$. Significant reduction in the load on the wall for given amount of fibers is achieved by making the fiber distribution "more anisotropic" (reduction in the fiber distribution ratio ξ).

5 FINAL REMARKS

A model for fiber-reinforced soil with anisotropic distribution of orientation of fibers was developed, and application of the kinematic approach of limit analysis to anisotropic soils was presented. The method becomes more elaborate, compared to its application to isotropic soils, because the internal friction angle depends now on orientation. The method is effective in solving problems with anisotropic materials. A simple example of retaining wall with fiber-reinforced backfill was presented.

ACKNOWLEDGEMENT

The work presented in this paper was carried out while the author was supported by the Army Research Office, grant No. DAAD19-03-1-0063. This support is greatly appreciated.

REFERENCES

Booker, J.R. & Davis, E.H. 1972. A general treatment of plastic anisotropy under conditions of plane strain. *J. Mech. Phys. Solids* 20: 239–250.

Gray, D.H. & Ohashi, H. 1983. Mechanics of fiber reinforcement in sand. *J. Geot. Eng.*, 109: 335–353.

Maher, M.H. & Gray, D.H. 1990. Static response of sands reinforced with randomly distributed fibers. *J. Geot. Eng.*, Vol. 116(11): 1661–1677.

Michalowski, R.L. 1997. Limit stress for granular composites reinforced with continuous filaments. ASCE *Journal of Engrg. Mechanics*, 123(8): 852–859.

Michalowski, R.L. & Čermák, J. 2003. Triaxial compression of sand reinforced with fibers. *Journal of Geotechnical and Geoenvironmental Engineering*, 129(2): 125–136.

Michalowski, R.L. & Čermák, J. 2002. Strength anisotropy of fiber-reinforced sand. *Computers and Geotechnics*, 29(4): 279–299.

Michalowski, R.L. & Zhao, A. 1996. Failure of fiber-reinforced granular soils. *J. Geot. Geoenv. Engrg.*, ASCE, 122(3): 226–234.

di Prisco, C. & Nova, R. 1993. A constitutive model for soil reinforced by continuous threads. *Geotextiles and Geomembranes*, 12: 161–178.

Waldron, L.J. 1977. The shear resistance of root-permeated homogeneous and stratified soil. *Soil Sci. Soc. Am.*, 41: 843–849.

New Horizons in Earth Reinforcement – Otani, Miyata & Mukunoki (eds)
© 2008 Taylor & Francis Group, London, ISBN 978-0-415-45775-0

Modelling fibre reinforced sand

A. Diambra, E. Ibraim & D.M. Wood
Department of Civil Engineering, University of Bristol, Bristol, United Kingdom

A. Russell
School of Civil and Environmental Engineering, University of New South Wales, Sydney, Australia (formerly Department of Civil Engineering, University of Bristol)

ABSTRACT: Results of triaxial compression tests for a reinforced sand show that the effectiveness of fibres as a reinforcing agent depends largely on their concentration as well as their orientation with respect to tensile strains which develop. In this paper a modelling approach for implementing the effects of fibre reinforcement into a constitutive model is presented. Application of the approach is demonstrated using an example where it is assumed that the soil stress:strain behaviour may be described by the elastic-perfectly plastic Mohr-Coulomb model. The modelling approach is based on the rule of mixtures in which the stresses in the fibres and the stresses in the sand matrix are superposed according to their volumetric fraction. A fibre stiffness matrix is defined in which it is assumed that fibres are working in their elastic domain. It accounts for fibre concentration and fibre orientation distribution. Drained triaxial compression test results for reinforced sand are presented as are model outputs. The model simulates the major characteristic features such as a monotonically increasing shear stress with increasing shear strain.

1 INTRODUCTION

Mixing sands with random discrete flexible fibres increases their strength and alters their deformation characteristics. Reinforced sands of this type are gaining acceptance by engineers as a cost effective and strong geomaterial.

Most research in this area has focused on the experimental stress:strain behaviour of reinforced sands (Michałowski & Čermák 2003 and Heineck et al. 2005 among others). There have been only a few attempts to model the constitutive behaviour of reinforced sands. Most were limited to the determination of the shear strength (Zornberg 2002 and Michałowski & Čermák 2002, among others). However, there are a few examples of modelling the stress:strain behaviour. The first (Villard et al. 1990) used the principle of superposition, where the stress:strain behaviour of soil and fibres were treated separately. Ding & Hargrove (2006) proposed a nonlinear stress:strain relationship for reinforced soil by using a volumetric homogenization technique. However, none of these models are able to account for non-uniform fibre orientation distributions.

In this research the fibre reinforced soil is treated as a composite with a stress:strain behaviour that obeys the rule of mixtures. The stiffnesses of the fibres and sand are defined separately and then superposed according to their volumetric fraction. Emphasis is placed on accounting for the concentration and orientation distribution of the fibres in a systematic way, and how the corresponding fibre stiffness matrix may be combined with any sand stiffness matrix. Application of the modelling approach is demonstrated using the simple Mohr-Coulomb elastic-perfectly plastic model for the sand, although more complex models for sand could be used in the same way.

A range of stress–strain behaviours of the composite are presented and compared to those observed from triaxial tests. The modelling approach enables simulation of the major characteristic features such as a monotonically increasing shear stress with increasing shear strain. It is also shown that the model outputs are very sensitive to the fibre orientation distribution.

2 A MODELLING FRAMEWORK

2.1 *Rule of mixtures*

The rule of mixtures may be used when studying the stress:strain behaviour of a composite material. Each component of the composite satisfies its own constitutive law. When each component is homogeneously distributed throughout the composite, their individual contributions to the overall composite behaviour are linked to their volumetric fractions. It follows that, for a reinforced sand, the general definitions for the stress and strain of a composite, expressed in terms of stresses and strains of the sand matrix and fibres, are:

$$\sigma = \sigma_m v_m + \sigma_f v_f \quad (1)$$

and

$$\varepsilon = \varepsilon_m v_m + \varepsilon_f v_f \quad (2)$$

where $\sigma, \sigma_m, \sigma_f, \varepsilon, \varepsilon_m$ and ε_f are the stresses and strains in the composite, matrix and fibres, respectively. The concentration factors v_m and v_f (for the matrix and the fibres, respectively) are defined as:

$$v_m = \frac{V_m}{V} = \frac{V - V_f}{V}, \quad v_f = \frac{V_f}{V} \quad (3)$$

where V, V_m and V_f are the volumes of the composite, matrix and fibres, respectively. Notice that concentration factors are related by:

$$v_m + v_f = 1 \quad (4)$$

These definitions for the concentration factors are universally accepted when each component of the composite behaves elastically. Even when different components behave plastically it is common for researchers using the rule of mixtures to adopt the same definitions as for the elastic range. Examples include Dvorak & Bahei-El-Din (1987) and Voyaiadjis & Thiagarajan (1995) for reinforced metal-matrix composites, Car et al. (2000) for anisotropic elastoplastic behaviour for fibre reinforced material at large strains, Luccioni (2006) for fibre-reinforced laminates, Ortiz & Popov (1982) when model the behaviour of concrete as a composite of mortar and aggregates, and Villard et al. (1990) and di Prisco & Nova (1993) when modelling fibre reinforced soils. However, there are a few exceptions. For a reinforced metal-matrix composite Dvorak & Bahei-El-Din. (1982) determined instantaneous concentration factors to evaluate the instantaneous stresses in the matrix and in the fibres in the plastic region.

One of two main hypotheses must be assumed when using the rule of mixtures: Voigt's hypothesis or Reuss's hypothesis. The former generally assumes that the strain fields in the composite coincide with those of its components:

$$\varepsilon = \varepsilon_m = \varepsilon_f \quad (5)$$

In the later an equality of stresses is assumed:

$$\sigma = \sigma_m = \sigma_f \quad (6)$$

To maintain simplicity in the formulations that follow it is supposed that sliding between sand grains and fibres is absent, fibres only act in tension and elastically and Voigt's hypothesis applies.

A very minor modification to the classical rule of mixtures was also made for simplicity. More specifically, the concentration factors are defined as the (slightly approximate) expressions:

$$v_m = 1 \text{ and } v_f = \frac{V_f}{V} \quad (7)$$

which is reasonable since the volume of the fibres is very small compared to the volume of the composite, i.e. $V_f << V$. The final form of the stress:strain relationship for the reinforced soil, when expressed incrementally, is:

$$\dot{\sigma} = \dot{\sigma}_m + v_f \dot{\sigma}_f = [M_s]\dot{\varepsilon} + v_f [M_f]\dot{\varepsilon} \quad (8)$$

where M_s is the stiffness matrix for the sand and M_f is the stiffness matrix for the fibres.

2.2 *Modelling fibres behaviour*

2.2.1 *Single fibre*

In reinforced composites fibres are embedded in the sand matrix and according to the Voigt's hypothesis the deformation in the fibres coincides with the deformation in the sand. Thus the entity of deformation in a single fibre depends on its orientation. Also, assuming an elastic behaviour for the fibres, even the entity of the stress in the fibre depends on fibre orientation as will now be shown. Considering conventional triaxial conditions, the following relationship between the strains at any angle θ from the horizontal (ε_θ) and the principal strains (ε_1 in vertical direction and ε_3 in horizontal direction) can be easily obtained:

$$\varepsilon_\theta = \varepsilon_1 \sin^2(\theta) + \varepsilon_3 \cos^2(\theta) \quad (9)$$

The stress carried by an elastic fibre oriented by an angle θ from the horizontal is then:

$$\sigma_\theta = E_f \varepsilon_\theta \quad (10)$$

and it is possible to derive the contribution of a single fibre to the principal stresses, $\sigma_{1f}(\theta)$ and $\sigma_{3f}(\theta)$, by decomposing σ_θ to give:

$$\sigma_{1f}(\theta) = \sigma_\theta \sin^2(\theta) \quad (11)$$

$$\sigma_{3f}(\theta) = \sigma_\theta \cos^2(\theta)/2 \quad (12)$$

In expanded form these can be rewritten as:

$$\sigma_{1f}(\theta) = E_f\left(\varepsilon_1 \sin^4(\theta) + \varepsilon_3 \cos^2(\theta)\sin^2(\theta)\right) \quad (13)$$

$$\sigma_{3f}(\theta) = E_f \frac{\left(\varepsilon_1 \sin^2(\theta)\cos^2(\theta) + \varepsilon_3 \cos^4(\theta)\right)}{2} \quad (14)$$

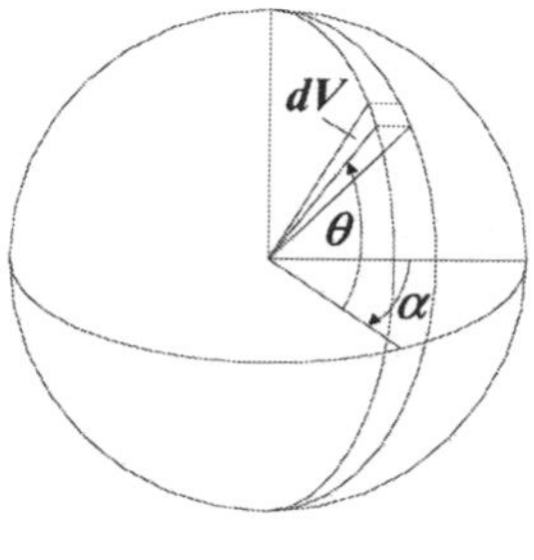

Figure 1. Spherical coordinates used on the definition of the fibre orientation distribution.

2.2.2 *Agglomerate of fibres*

A procedure similar to the one proposed by Zhu et al. (1994) for the determination of the strength of short-fibre reinforced metal-matrix has been used here. Fibres have a non-uniform orientation distribution, and using the spherical coordinates shown in Figure 1, a general fibre orientation distribution function $\rho(\theta)$ is now introduced. $\rho(\theta)$ represents the volumetric concentration of fibres in an infinitesimal volume $\mathrm{d}V$ (Fig. 1) having an orientation of angle θ above the horizontal and its main property has to be:

$$\bar{\rho} = \frac{1}{V}\int_V \rho(\theta)\mathrm{d}V \tag{15}$$

where V is the volume of the reference sphere (made of the composite material) shown in Figure 1 and $\bar{\rho}$ is the average fibre concentration defined by:

$$\bar{\rho} = V_f / V \tag{16}$$

According to its definition, $\rho(\theta)$ is the concentration factor v_f representing the contribution of fibres within the composite which have an orientation of θ above the horizontal plane. Expressions for the overall contribution of fibres within the composite in the directions of the principal stresses can then be obtained by integration using the two expressions:

$$\sigma_{f1} = \frac{1}{V}\int_V \rho(\theta)\sigma_{1f}(\theta)\mathrm{d}V \tag{17}$$

$$\sigma_{f3} = \frac{1}{V}\int_V \rho(\theta)\sigma_{3f}(\theta)\mathrm{d}V \tag{18}$$

If the orientation distribution is symmetrical with respect to the horizontal plane, (17) and (18) can be expanded to give:

$$\sigma_{f1} = \frac{E_f}{2}\left(\varepsilon_1 \int \rho(\theta)\sin^4(\theta)\cos(\theta)\mathrm{d}\theta + \varepsilon_3 \int \rho(\theta)\cos^3(\theta)\sin^2(\theta)\mathrm{d}\theta\right) \tag{19}$$

$$\sigma_{f3} = \frac{E_f}{4}\left(\varepsilon_1 \int \rho(\theta)\sin^2(\theta)\cos^3(\theta)\mathrm{d}\theta + \varepsilon_3 \int \rho(\theta)\cos^5(\theta)\mathrm{d}\theta\right) \tag{20}$$

It must now be considered that fibres can only contribute to the stresses acting on the composite by acting in tension. According to Mohr's circle for deformation, the direction of zero strain is:

$$\theta_0 = \arctan\sqrt{-\frac{\varepsilon_3}{\varepsilon_1}} \tag{21}$$

and the integrations in (19) and (20) should only be performed within the limits $0 \le \theta \le \theta_0$. It follows that the concentration factor multiplied by the stiffness matrix for the fibres, in terms of principle stresses, is:

$$v_f[M_f] = E_f \begin{bmatrix} \int_0^{\theta_0} \rho(\theta)\cos(\theta)\sin^4(\theta)\mathrm{d}\theta & \int_0^{\theta_0} \rho(\theta)\cos^3(\theta)\sin^2(\theta)\mathrm{d}\theta \\ \frac{1}{2}\int_0^{\theta_0} \rho(\theta)\cos^3(\theta)\sin^2(\theta)\mathrm{d}\theta & \frac{1}{2}\int_0^{\theta_0} \rho(\theta)\cos^5(\theta)\mathrm{d}\theta \end{bmatrix} \tag{22}$$

2.3 *Modelling sand*

A simple elastic-perfectly plastic Mohr-Coulomb model has been employed for sand. Considering a formulation in the principal directions, in the elastic region the increments of the principal effective stresses are related to the increments of strains through Young's modulus E and the Poisson's ratio μ:

$$\begin{bmatrix} \dot{\sigma}'_1 \\ \dot{\sigma}'_3 \end{bmatrix} = \frac{E}{(1+\mu)(1-2\mu)} \begin{bmatrix} 1-\mu & 2\mu \\ \mu & 1 \end{bmatrix} \begin{bmatrix} \dot{\varepsilon}'_1 \\ \dot{\varepsilon}'_3 \end{bmatrix} \tag{23}$$

Yielding occurs when the following relation is satisfied:

$$\dot{\sigma}'_1 = \frac{1+\sin(\varphi)}{1-\sin(\varphi)}\dot{\sigma}'_3 \tag{24}$$

where φ is the friction angle of sand and beyond first yield the relationship between principal strains is:

$$\dot{\varepsilon}_3^p = -\frac{1+\sin(\psi)}{2(1-\sin(\psi))}\dot{\varepsilon}_1^p \tag{25}$$

where ψ is the dilation angle of the sand.

3 APPLICATION

Application of the model will now be demonstrated and outputs will be compared to results of conventional drained triaxial tests performed on Hostun RF(S28) sand reinforced with Loksand type polypropylene fibres. Specimens 70 mm high and 70 mm diameter were prepared with the moist tamping technique in three different layers and a detailed explanation of the formation procedure is given in Diambra et al. (2007). Hostun RF(S28) sand is a fine sand with

Table 1. Details of Loksand™ fibres.

Weight (Denier)	Tensile strength (MPa)	Specific gravity	Elongation at break	Moisture regain
50	225	0.91	160%	<0.1%

mean grain size $D_{50} = 0.32$ mm, coefficient of uniformity $C_u = 1.70$, coefficient of gradation $C_g = 1.1$, specific gravity $G_s = 2.65$ and minimum and maximum void ratios $e_{min} = 0.62$ and $e_{max} = 1.00$, respectively (Ibraim 1998). Loksand discrete polypropylene crimped fibres have a length $l_f = 35$ mm and are of circular cross section with diameter $d_f = 0.1$ mm. Other physical properties of Loksand^{TM} fibres are reported in Table 1. Results of triaxial tests with two different initial void ratios (e) of 0.9 and 1.0 (which respectively correspond to relative densities of $D_r = 0\%$ and 26%) and differrent fibre contents ($\bar{\rho} = 0.43\%$, 0.87% and 1.30%) have been selected here for comparisons with the model outputs. The σ_3' of the tests was 100 kPa.

3.1 *Parameters for fibres*

Modelling fibre behaviour requires definition of the elastic modulus (E_f) and the orientation distribution of the fibres. E_f was determined experimentally by performing a number of tension tests on the fibres. Using a machine with low precision, the results revealed an approximate value of $E_f = 600$ MPa. As will be illustrated later this leads to a reasonable fit between model outputs and experimental data. Fibre orientation distribution for specimens used in the experimental work was determined through an experimental procedure reported in Diambra et al. (2007) and it was found that the fibres mostly had a near horizontal orientation. The form of fibre orientation distribution proposed by Michałowski & Čermák (2002) was used:

$$\rho(\theta) = \bar{\rho}\left(A + B\left|\cos^n \theta\right|\right) \tag{26}$$

and the parameters A, B and n defining the orientation distribution were calibrated to be $A = 0$, $n = 5$ and $B = 2.04$. Difficulties were encountered in the integration of the fibre stiffness matrix when (26) was implemented so a slightly modified orientation distribution function $\rho(\theta)$ was used in this research. The new function for $\rho(\theta)$ is:

$$\rho(\theta) = \bar{\rho}\frac{2ab^2\left|\cos(\theta)\right|}{\cos(\theta)^2(b^2 - a^2) + a^2} \tag{27}$$

where $a = 1.02$ and $b = 0.46$ were determined by forcing equality in (26) and (27) at $\theta = 0$ and by satisfying (15). The similarity between the two orientation distribution functions is illustrated in Fig. 2.

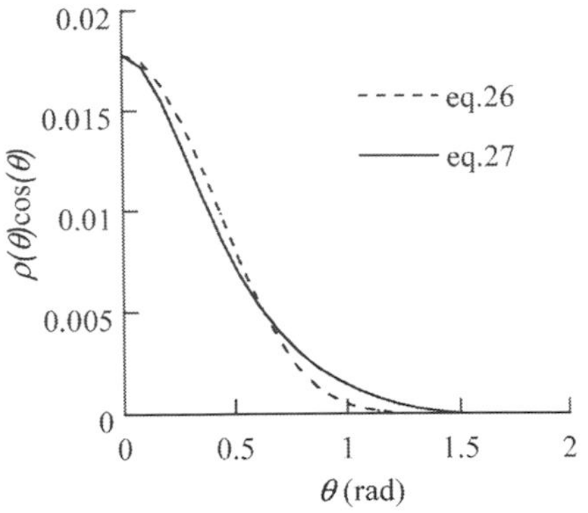

Figure 2. Comparison between the assumed orientation distribution function in eq. 26 and eq. 27.

3.2 *Parameters for sand*

Only a few parameters are needed for the definition of the simple Mohr-Coulomb model. Parameters which give as reasonable fit between model outputs and experimental data include $E = 3600$ kPa, $\mu = 0.3$ and $\phi = 36°$ and were found by trial and error. A value of $\psi = 0°$ was assumed and the appropriateness of this is discussed later.

3.3 *Results and discussion*

Figures 3 and 4 show the drained triaxial test results and model simulations in the $q \sim \varepsilon_q$ plane, where $q = \sigma_1 - \sigma_3$ and $\varepsilon_q = 2(\varepsilon_1 - \varepsilon_3)/3$ are the usual triaxial shear stress and shear strain, respectively. In qualitative terms the model adopted here reproduces the stress:strain response for the reinforced sand quite well. The difference between experimental results and model outputs is mainly due to the simple elastic-perfectly plastic model's inability to account for the non-linearity of unreinforced sand behaviour, especially at small strains. The use of more complex models which capture this non-linearity would result in smoother simulated curves and a better match between experimental data and simulation.

More specifically, the experimental stress:strain responses include an initial non-linear part followed by a linear part that does not flatten with increasing shear strain. The initial non-linear part is largely unaffected by the presence of fibres and the initial sections of the model outputs exhibit this characteristic also. In these initial sections the fibre contribution to the composite behaviour is quite small, which is not surprising as ε_3 is also small, preventing the fibres from elongating and mobilising large tensile forces. At larger shear

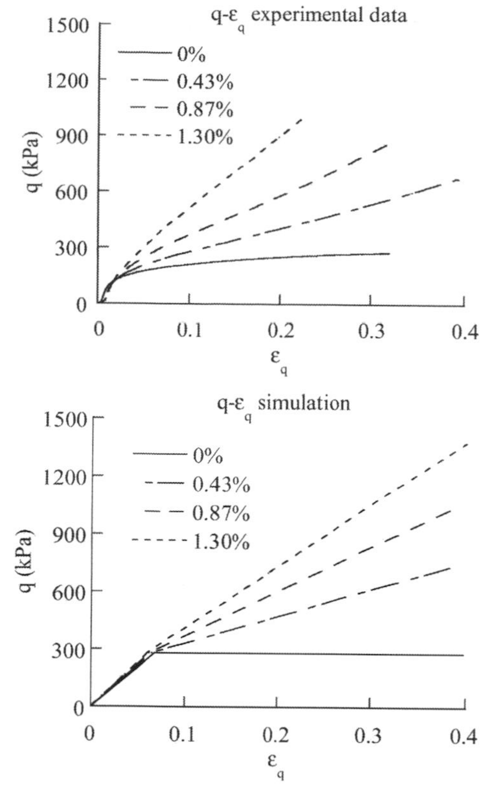

Figure 3. Experimental drained conventional triaxial tests and model simulation for void ratio $e = 1.0$ and cell pressure $\sigma_3' = 100$ kPa. (legend expresses initial fibre content).

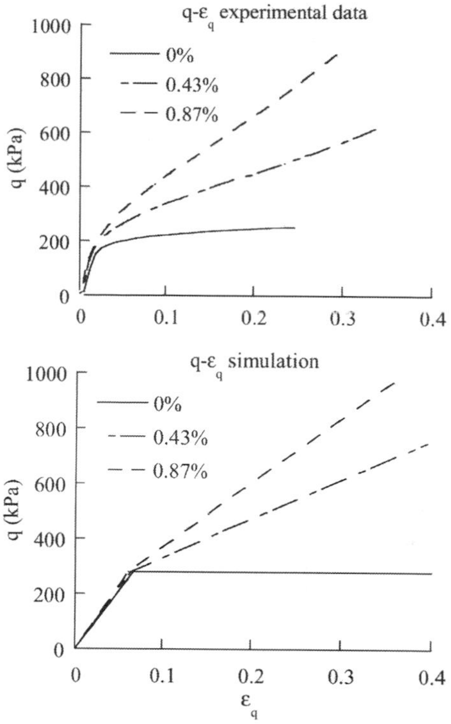

Figure 4. Experimental drained conventional triaxial tests and model simulation for void ratio $e = 0.9$ and cell pressure $\sigma_3' = 100$ mkPa. (legend expresses initial fibre content).

strains ε_3 goes into tension and the shear stress supported by the composite increases, apparently without limit. Fibres go into more and more tension as shear strain is increased. A plausible explanation for this is that, within the range of strains considered, tension is not released due to the fibres breaking, slipping or 'pulling out' of the composite.

The model captures this general behaviour, and the assumptions that fibres are working within their elastic range without slipping seem reasonable. In fact, careful examination of the fibres at the end of the triaxial tests showed no signs of fibre breakage or plastic deformation. In short, for the conventional triaxial test load path, fibres increase the confinement on the sand matrix as shear strain increases, allowing the sand to support greater deviatoric stresses. However, at larger shear strains, the tension in the fibres may be released, for example by breakage, and suppress the increase in deviatoric stress.

The fibre orientation distribution has a major influence on the behaviour of the composite. For the samples prepared here 97% of fibres were orientated within 45° of the horizontal plane (Diambra et al. 2007). This anisotropic fibre orientation distribution results in an anisotropic behaviour of the composite.

Any tendency for ε_3 to decrease (go into tension) is counteracted by a large amount of resistive tension developing in the fibres. If the orientation distribution was isotropic then the effectiveness of the fibres to increase the deviatoric stress supported by the composite will be greatly reduced for the conventional triaxial test load path, as shown by the results of the simulations in Figure 5. Results on samples reinforced with fibres uniformly orientated were not available for the analysed tests, but results published in literature (e.g. Michałowski & Čermák 2002) demonstrate the same influence of fibre orientation as predicted by the proposed model.

The modelling approach outlined above is quite versatile since any fibre orientation distribution can be implemented. Also, the evolution of the effectiveness of the fibres as the sample changes shape, linked to the variation of θ_0 in the integration, is readily accommodated. Furthermore, through minor changes of the framework presented here, different assumptions can be made on the constitutive behaviour of the fibres, such as including a non-linear elasticity or a pull out mechanism between fibres and the sand matrix.

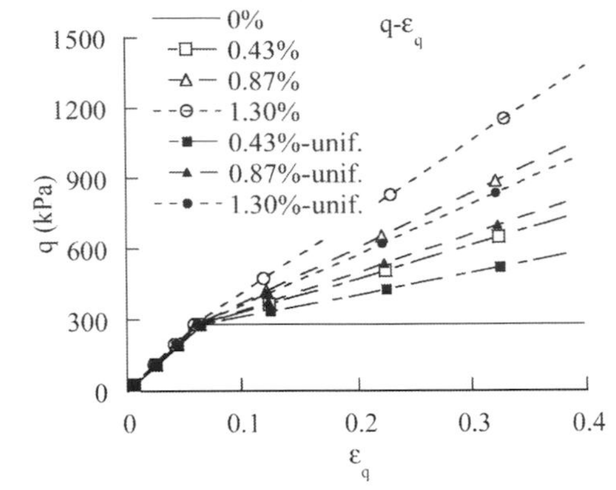

Figure 5. Model simulations for the determined fibre orientation distribution and for an assumed uniform orientation distribution for void ratio $e = 1.0$ and cell pressure $\sigma_3' = 100$ kPa (legend expresses initial fibre content and fibre orientation).

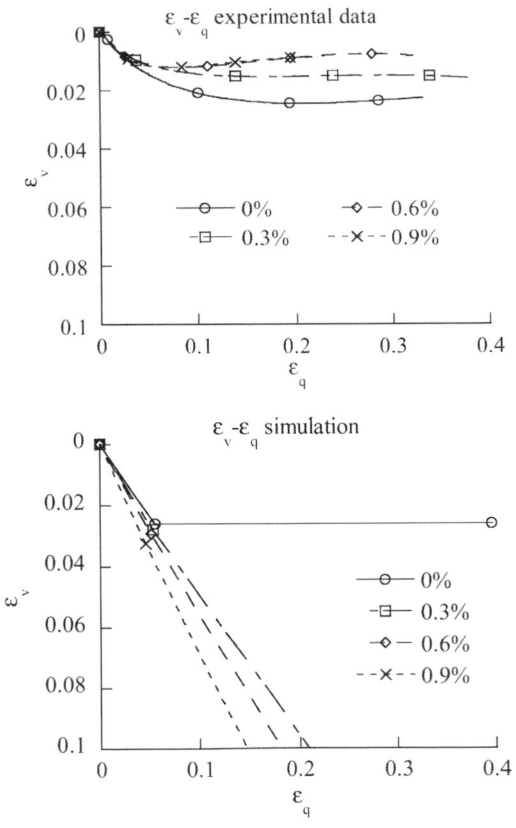

Figure 6. Experimental drained conventional triaxial tests and model simulation for void ratio $e = 1.0$ and cell pressure $\sigma_3' = 100$ kPa. (legend expresses initial fibre content).

Only poor simulations of the volumetric behaviour were achieved for the assumed $\psi = 0°$ (Fig.6). Experimental results showed that the reinforced sand exhibits a more dilative behaviour than the unreinforced sand and the model is unable to capture this behaviour when using constant dilation angle, irrespective of its value. Further research is in progress to achieve a better understanding of this phenomenon and a better simulation of the experimental data.

4 CONCLUSIONS

A simple modelling approach for fibre reinforced sand has been presented, based on the rule of mixtures. Its application was demonstrated using a simple sand model, and model outputs for the composite exhibit key features that were also observed in triaxial test results. Discrepancies between simulation and experimental results can be attributed to the simplicity of basic sand model and the assumed dilatency relationship. A better simulation would be obtained by adopting a more complex model that captures the non-linear behaviour.

The modelling approach permits the use of any fibre orientation distribution function. It also accounts for the evolution of fibre orientation as the composite deforms.

In any case, the main objective of this work – to outline a very simple modelling tool for fibre reinforced sand – has been met. Furthermore, the very simple model outlined can easily be implemented into a numerical analysis.

REFERENCES

Car, E., Oller, S. & Oñate, E. 2000. An anisotropic elastoplastic constitutive model for large strain analysis of fiber reinforced composite materials. *Computer methods in applied mechanics engineering* 185: 245–277

di Prisco, C. & Nova, R. 1993 A constittuitve model for soil reinforced by continuous *threads. Geotexiles and Geomembranes* 12:161–178

Diambra, A., Russell, A.R., Ibraim, E. & Muir Wood, D. 2007, Determination of fibre orientation distribution in reinforced sand. *Geotechnique (in press)*

Ding, D. & Hargrove, S.K. 2006. Nonlinear stress-strain relationship of soil reinforced with flexible geofibers. *Journal of Geotechnical & Geoenvironmental Engineering* 132(6): 791–794

Dvorak, G.K. & Bahei-El-Din, Y.A. 1982. Plasticity analysis of fibrous composites. *Journal of applied mechanics* 49: 327–335

Dvorak, G.K. & Bahei-El-Din, Y.A. 1987. A bimodal plasticity theory of fibrous composite materials. *Acta Mechanica* 69: 219–241

Heineck, C.S., Consoli, N.C. & Coop, M.R. 2005. Effect of microreinforcement of soils from very small to large shear strains. *Journal of Geotechnical & Geoenvironmental Engineering* 131(8):1024–1033

Ibraim, E. 1998. Différents aspects du comportement des sables à partir d'essais triaxiaux: des petites déformations à la liquéfaction statique. *PhD thesis*, ENTPE Lyon

Luccioni, B.M. 2006. Constitutive model for fiber-reinforced composite laminates. *Journal of applied mechanics* 73: 901–910

Marakova, I.S. & Saraev, L.A. 1990. Theory of elastic-plastic deformation of randomly reinforced composite materials. *Journal of applied mechanics and technical physics* 32(5): 768–772

Michałowski, R.L. & Čermák, J. 2002. Strength anisotropy of fiber-reinforced sand. *Computers and Geotechnics* 29:279–299

Michałowski, R.L. & Čermák, J. 2003. Triaxial compression of sand reinforced with fibers. *Journal of Geotechnical & Geoenvironmental Engineering* 129(2):125–135

Ortiz, M. & Popov, E.P. 1982. Plain concrete as a composite material. *Mechanics of Materials* 1:139–150

Villard, P., Jouve, P. & Riou, Y. 1990. Modélisation du compertment mécanique du Texsol. *Bulletin liaison Labo. P et Ch.* 168: 15–27

Voyadijs, G.Z. & Thiagarajan, G. 1995. An anisotropic yield surface model for directionally reinforced metal-matrix composites. *International journal plasticity* 11(8): 867–894

Zhu, Y.T., Zong, G. Manthiram, A. & Eliezer, Z. 1994 Strength analysis of random short-fibre-reinforced metal matrix composite materials. *Journal of material science* 29:6281–6286

Zornberg, J.G. 2002 Discrete framework for equilibrium analysis of fibre-reinforced soil. *Géotechnique* 52(8): 593–604

New Horizons in Earth Reinforcement – Otani, Miyata & Mukunoki (eds)
© 2008 Taylor & Francis Group, London, ISBN 978-0-415-45775-0

Numerical analysis of fibre-reinforced granular soils

E. Ibraim
Department of Civil Engineering, University of Bristol, UK

K. Maeda
Department of Civil Engineering and Environmental, Nagoya Institute of Technology, Japan

ABSTRACT: Laboratory experimental tests of sand reinforced with discrete random flexible fibres show that fibre reinforcement could be an effective technique for improving their strength and deformation characteristics. The behaviour of fibre reinforced sand is influenced by many factors such as density and stress level, fibre type, content and orientation. A two dimensional DEM (Distinct Element Method) biaxial compression simulation of mixtures of sand and fibres is applied to understand how randomly distributed flexible fibres generate a bond within the soil and affect the kinematics of the granular matrix. Sand particles are modelled by rigid disks with conventional contact elements in DEM. The fibre element is modelled by connecting small circular particles with a bond contact algorithm, where the strength of the bond is fairly high. The specimens have been reinforced with different fibre fractions. The effect of fibre reinforcement is discussed.

1 INTRODUCTION

The concept of reinforcing soils by tension-resisting elements included in soil is widely accepted in geotechnical engineering practice. Traditional methods of earth reinforcement for embankments, earth-retaining structures and foundations use a large variety of continuous planar synthetic inclusions such as strips, fabrics or geotextiles (Koerner and Welsh, 1980). The reinforcement inclusions are normally oriented in a preferred direction, dependent on the geometry of the structure, the nature of the applied loads and the expected principal tensile strains in the soil. There has been much research on this topic over the last few decades. However, far less information is available on reinforcements with non-traditional inclusions, such as natural or synthetic short and flexible fibres randomly distributed.

Laboratory experiments show that the engineering behaviour of randomly distributed fibre reinforced soils is controlled by several parameters: stress level; fibre type, volume fraction, aspect ratio, modulus of elasticity, strength, orientation; and soil density, particle size, shape, gradation. The benefits of fibre reinforcement come from fibre-soil particle interaction. Modelling of fibre reinforcement effects has concentrated on prediction of the contribution of fibres to strength increase. For example, an attempt to describe the shear strength increase through consideration of soil/fibre interaction in a localised shear band by using a simple force-equilibrium model was made by Gray and Ohashi (1983), Jewell and Wroth (1987) – for oriented fibres – and Maher and Gray (1990) – for random distributed fibres. Using an energy-based homogenisation approach, a failure criterion was derived by Michalowski and Zhao (1996) while, more recently, Zornberg (2002) proposed a framework to predict failure of different soil types mixed with different fibre contents and fibre length/diameter ratios based on the superposition of the sand-fibre effects. Diambra et al. (2007) have shown that moist tamping technique produces specimens with 97% of fibres orientated within ±45° to the horizontal.

In this paper a micromechanical approach is used in order to reveal the reinforcement mechanism of how randomly distributed flexible fibres with kinematics of the granular matrix. A two dimensional DEM (Distinct Element Method) (Cundall and Strack 1979) simulation of the mixture of sand and fibre is developed and numerical results under isotropic compression and shearing processes are presented and discussed in association with the observed experimental patterns.

2 EXPERIMENTAL FINDINGS

Published laboratory experimental test results, and pilot tests performed at Bristol (Ibraim and Fourmont 2006), show that reinforcement of sands with discrete random flexible fibres represents an effective

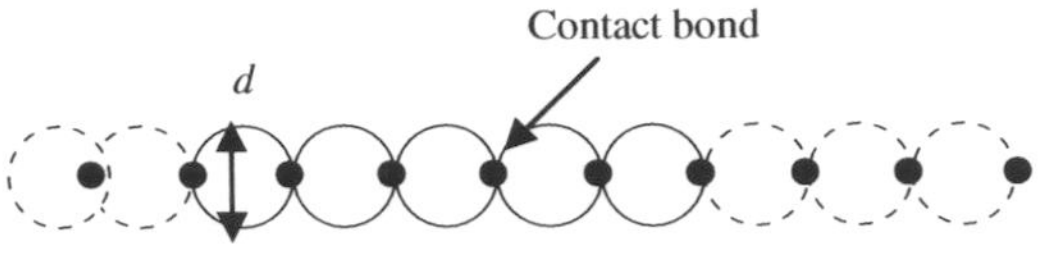

Figure 1. Typical model for fibres in DEM.

technique for improving their strength, ductility and reducing post-peak strength loss. Strength increase of the reinforced sand seems to be linear with the amount of fibres. Nevertheless, above some limiting fibre content, this increase seems to approach an asymptotic upper limit (Gray and Al-Refeai 1986, Murray et al. 2000). The amount of vertical dilation increases with the amount of fibres (Palmeira and Milligan 1989, Kaniraj and Havanagi 2001, Jewell and Wroth 1987, Shewbridge and Sitar 1989, Ibraim and Fourmont 2006).

Typically, sand reinforced with randomly distributed discrete fibres exhibits either curved-linear (for uniform, rounded sand) or bilinear (for well-graded or angular sands) failure envelopes (Gray and Ohashi 1983, Maher and Gray 1990). Above a threshold confining stress (critical confining stress, σ_{crit}) failure envelopes for the reinforced sand are parallel to the unreinforced sand envelope. Below the σ_{crit}, it is considered that the reinforcing mechanism is not fully mobilized and with the shearing process the fibres tend to slip or pull out.

Recent experimental results presented by Heineck et al. (2004) show that the fibre reinforcement seems to have no influence in the small strain domain.

3 DEM PROCEDURE

3.1 *Modelling of fibres and granular matrix*

The fibre was modelled by particles of diameter d connected by contact bond with tension and shear strengths and without rotation constraints like a hinge connection, as shown in Figure 1 (Ibraim et al., 2006). Since both strengths were considered to be fairly high, in this paper, fibre elements were flexible but were not allowed to break. The basic parameters for fibres are presented in Table 1.

The granular matrix is composed of rigid disks. The interaction between elements was modelled by contact elements (springs, dash-pots, sliders and non-extensional elements); these coefficients were introduced elsewhere by Maeda et al. (2003). Table 2 presents the parameters for the granular matrix.

3.2 *Sample preparation*

Figure 2 shows the fibre reinforced granular soil specimen model for the DEM. Gray and black balls represent granular matrix (approximately 1000 disks) and fibres, respectively. Bi-axial stresses are controlled by the movement of four boundary walls. After isotropic compression, the sample is sheared axially under constant strain-rate and lateral pressure.

In order to study the effect of the change in soil fabric due only to the imposed stress states, the specimens have been constructed under zero gravity condition. Initially fibre elements and matrix particles (void ratio of matrix phase: 0.24; $e_{max} = 0.27$, $e_{min} = 0.21$) were randomly generated. Then, all radii were multiplied

Table 1. Basic parameters for DEM simulations: fibres.

Parameter	Unit	Value
ρ_f	(Mg/m^3)	9.1
Diameter: d	(mm)	$1 (d/D_{max} = 0.1)$
Length: l	(mm)	$150 (l/D_{max} = 15)$
Aspect ratio: $\lambda = l/d$	(mm)	150

*d = diameter, l = length, λ = aspect ratio ($\lambda = l/d$).

Table 2. Basic parameters for DEM simulations: granular matrix.

Parameter	Unit	Value
ρ_s	(Mg/m^3)	2.65
Shape of grain	–	circle
Grain size distribution		Gaussian distribution in weight
D_{max}	(mm)	10
D_{min}	(mm)	5
D_{50}	(mm)	7.1
C_u		1.3
C_g		1.1

*D_{max} = maximum grain size, D_{min} = minimum grain size, D_{50} = mean grain size, C_u = coefficient of uniformity (D_{60}/D_{10}), C_g = coefficient of gradation ($D_{30}/D_{60}{}^*D_{10}$), ρ_s = density.

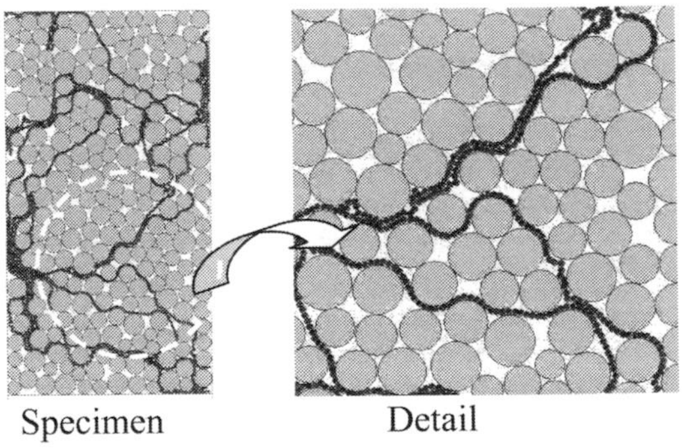

Figure 2. Example of fibre-reinforced constructed specimen for DEM simulation.

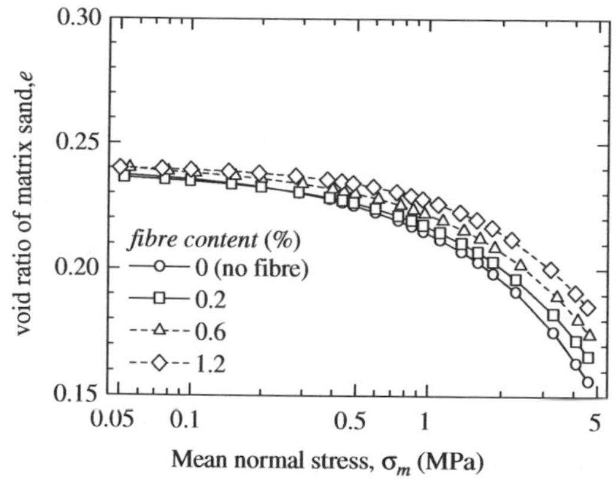

Figure 3. Typical results of isotropic compression tests analyzed by DEM on fibre-reinforced mixture: void ratio is obtained only from granular matrix phase.

gradually to the prescribed values in order to obtain an isotropic specimen. The detailed techniques for particle generation and density control of specimen are given by Maeda et al. (2003).

4 MACRO BEHAVIOURS ANALYZED

Figure 3 shows deformation behaviour of fibre-reinforced granular specimens under isotropic compression tests in a wide range of lateral pressure from 0.05 to 5 MPa. For pressures lesser than 0.5 MPa, the deformation responses for all specimens are almost identical. However, starting with 0.5 MPa, it is clearly that the specimen reinforced with higher fibre content shows a lower compressibility. For each fibre content, distinct normal compression lines are obtained and they tend to become parallel, as experimentally observed by Consoli et al. (2005).

Figure 4 shows deformation-failure behaviour of fibre-reinforced granular specimens analyzed. Different percentages by weight of matrix of fibre contents are used. All tests were conducted with constant lateral pressure σ_c of 0.1 MPa in x-axis direction and constant strain rate in y-axis direction.

The presence of fibres enhances the response of the reinforced mixture. The peak stress ratio, τ_m/σ_m, where τ_m is the maximum shear stress and σ_m the mean effective stress, increases with the amount of fibres. After an initial contractant phase, the simulations show a dilative volumetric response of the specimens. However, the dilation is not inhibited by the presence of fibres, but increases with the fibre content.

Figure 5 shows the dilatancy ratio evolution with the principal stress ratio for all the test simulations. The figure shows also Rowe's relationship (Rowe 1962)

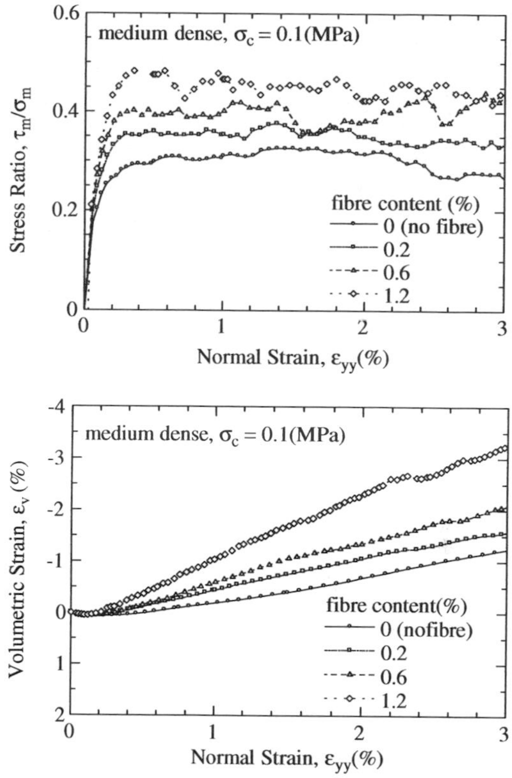

Figure 4. Typical results of biaxial compression tests given by DEM on fibre-reinforced mixture.

for an inter-particle friction angle, ϕ_μ, of 25°. There is no difference among the numerical results under contractant phase. Otherwise, the dilation phase shows different inclinations of the curves with the highest slope corresponding to the highest fibre content. The data diverge from Rowe's flow rule. These results may suggest that for fibre-reinforced granular materials there is no longer a unique flow rule which links the strength to the volumetric response.

5 MICRO OBSERVATIONS

5.1 *Fibre interactions during isotropic compression*

The averaged tensile stress computed for all the bond contacts on the fibres developed with the macro deformation under isotropic compression is plotted in Figures 6. A slight increase of the averaged tensile stress was recorded for σ_m values below 0.5 MPa. However, as can be observed in Figure 6, above 0.5 MPa of medium confining pressure, the averaged tensile stress increases remarkably. These results show that even if the specimens are undergoing isotropic

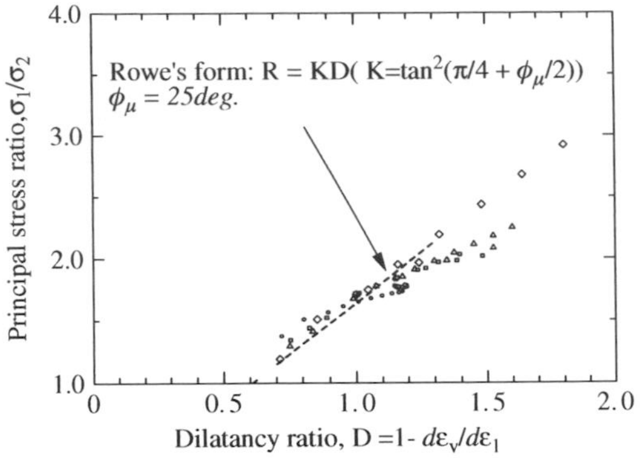

Figure 5. Stress – dilatancy ratio relations for the performed simulations (legend as given in Figure 4).

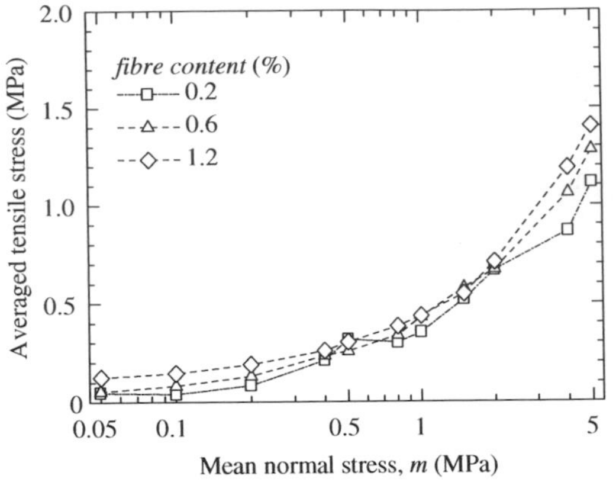

Figure 6. Averaged tensile stress in fibres developed with macro deformation under isotropic compression.

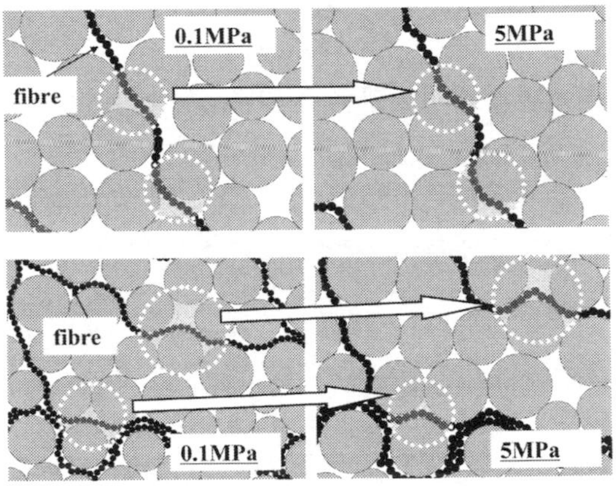

Figure 7. Change in configuration of fibre during isotropic compression from 0.1 MPa to 5 MPa.

compression, important tensile stresses are developing for the fibres. As the averaged tensile stress increases with the applied stress, for certain fibre types with relatively low individual tensile stress resistances, the breaking point can be reached. Extended and broken fibres have been found on isotropically compressed fibre reinforced specimens by Consoli et al. (2005). The variation of the fibre content does not seem to have a big influence on the computed averaged tensile stress.

It is interesting to note the change in configuration of fibres during isotropic compression, as we can observe changes in curvature of fibres at some local areas (Figure 7). For lower stress levels, the interaction mechanism between fibre and granular matrix in is not fully mobilized; otherwise it is induced by rearrangement of particles by applied higher stresses.

Figure 8 shows the spatial distribution of the tensile stress along the fibres located in the central area of the specimen (0.6% of fibre content) at the normal strain $\sigma_m = 5$ MPa. The distribution of the tensile stress appears to be random but highly non-homogeneous.

5.2 *Fibre interactions during shear*

The averaged tensile stress in the fibres developed with macro shear deformation is plotted in Figure 9. In the small and medium strain domains, ε_{yy} less than 0.2%, the initial tensile stress induced locally by the particle/fibre interaction during the isotropic compression phase remains unchanged; the interaction mechanism between the fibres and matrix is not mobilized and the behaviour of the composite appears to be dominated by the matrix properties. These results are in agreement with experimental observations presented by Palmeira and Milligan 1989, Jewell and Wroth (1987), and Heineck et al. (2004).

For axial strains higher than 0.2%, the averaged tensile stress in the reinforcement begins to increase. This trend appears to coincide also with the onset of the dilation phase of the mixed material. The tensile stress continues to increase well beyond the corresponding peak stress of the mixed material. As shown in Figure 9, higher fibre tensile stresses are mobilized in the specimens with higher fibre contents.

Figure 10 shows the spatial distribution of the tensile stress along the fibres located in the central area of the specimen (0.6% fibre content) at the normal strain $\varepsilon_{yy} = 2\%$. The distribution of the tensile stress appears to be highly non-homogeneous. The highest tensile stresses seem to be mobilized on fibres oriented towards the minor principal strain direction (tensile strain direction). The intensity of the tensile stress can be up to two times the averaged tensile stress. The beneficial effect of fibre reinforcement is higher when the fibres are oriented in the direction of the tensile strain, as experimentally observed by Gray and

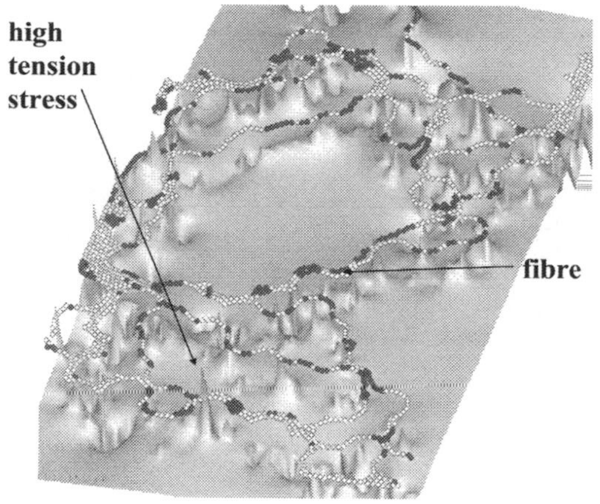

Figure 8. Spatial distribution of tensile stress in fibres in a zone cut in specimen: fibre content = 0.6% and $\sigma_m = 5$ MPa.

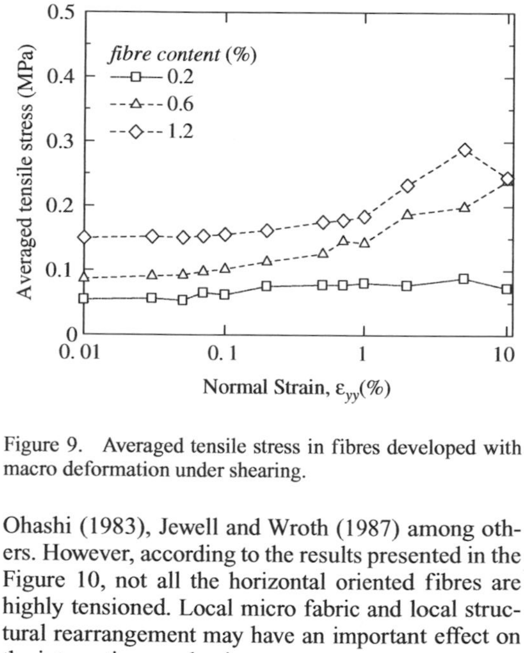

Figure 9. Averaged tensile stress in fibres developed with macro deformation under shearing.

Ohashi (1983), Jewell and Wroth (1987) among others. However, according to the results presented in the Figure 10, not all the horizontal oriented fibres are highly tensioned. Local micro fabric and local structural rearrangement may have an important effect on the interaction mechanism.

In order to assess the fabric change of the granular matrix during deformation, two fabric indexes N_cF_1 and N_cF_2 are examined. The numerical results are plotted in Figure 11. The geometry index N_c represents the coordination number, and F_1 and F_2 are principal values of the fabric tensor corresponding to the major principal direction (1) and minor principal direction (2), respectively. Here, major principal direction corresponds to y-axis. The coordination number represents the averaged number of contacts per particle and indicates the overall stability of the fabric. In general, N_c has been found to be mainly affected by the normal

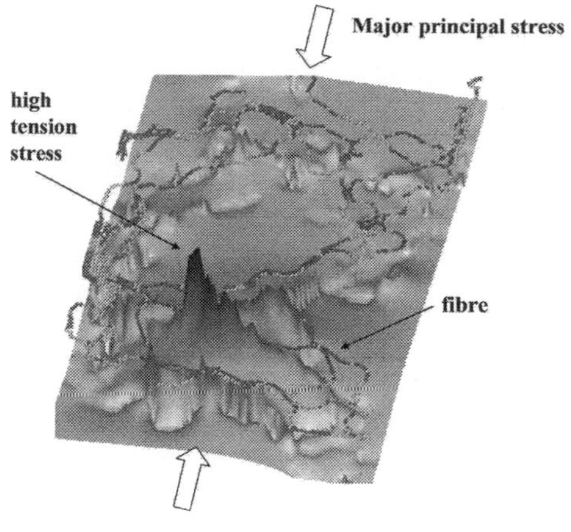

Figure 10. Spatial distribution of tensile stress in fibres in a zone cut in specimen: fibre content = 0.6% and at normal strain $\varepsilon_{yy} = 2.0\%$.

stress or stress ratio (Maeda et al. 2003; Maeda and Hirabashi, 2006). The major, F_1, and minor, F_2, principal values of fabric tensor are proposed by Satake (1982). Hence, F_1 and F_2 indicate the concentration intensity of normal direction of contact planes; the ratio F_1/F_2 means intensity of anisotropy. Maeda et al. (2003) pointed out that the directions of F_1 and F_2 agree with major and minor principal stress directions and the ratio F_1/F_2 is controlled by stress ratio and is independent of material properties and test conditions. Therefore, N_cF_1 and N_cF_2 could be considered as representative indices for fabric intensity over the principal directions.

As presented in Figure 11, for the case without fibres, the fabric intensities, N_cF_1 does not change but N_cF_2 decreases with macro deformation; this indicates a loss of contacts in lateral direction. For the cases with fibre, N_cF_1 increases for medium and large strains. This may explain, at macro-scale, the strength increase and the post-peak ductile behaviour experimentally observed for real fibre reinforced soils. In addition, the reduction of N_cF_2 reinforced is significantly smaller compared with the unreinforced case. These results suggest that the presence of fibres can prevent the loss of fabric in the direction of minor principal stress with beneficial consequences for mixed material.

The spatial distributions of the mean normal stress and deviator stress at the normal strain $\varepsilon_{yy} = 2\%$ for granular matrix phase located in the central area are shown in Figure 12 for an un-reinforced specimen and in Figure 13 for a reinforced specimen (0.6% of fibres content). In both cases, the stresses are distributed highly non-homogeneously. However, in the case of fibre reinforcement (Fig. 13), the vertical contact chains are much closer compared with the unreinforced specimen (Fig. 12). Higher mean normal

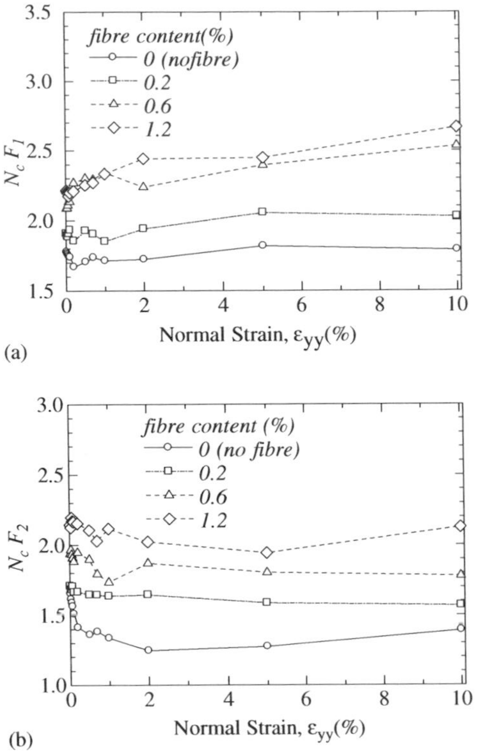

Figure 11. Contact fabric for granular matrix phase; (a) and (b) shows fabric in the direction of major and minor principal stress (y-axis and x-axis), respectively.

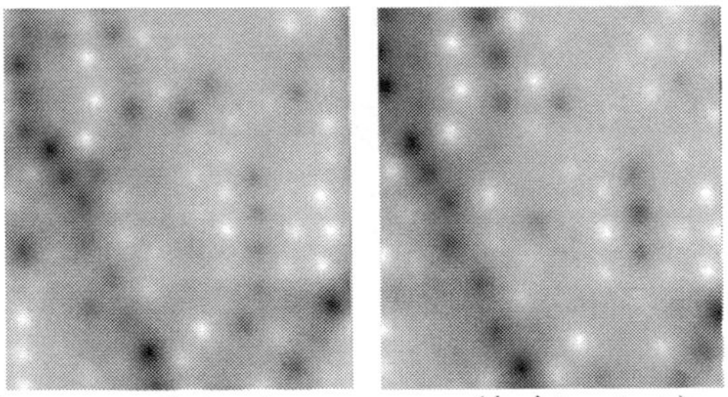

(mean normal stress) (deviator stress)

Figure 12. Stress distribution in micro zone for granular matrix phase in a zone cut in unreinforced specimen at normal strain $\varepsilon_{yy} = 2.0\%$ (stress scale in Pascals given on Figure 13).

and deviator stresses are localised on particles adjacent to the fibres, including horizontal directions.

6 CONCLUSIONS

A two dimensional DEM has been developed for the micromechanical analysis of mixtures of granular

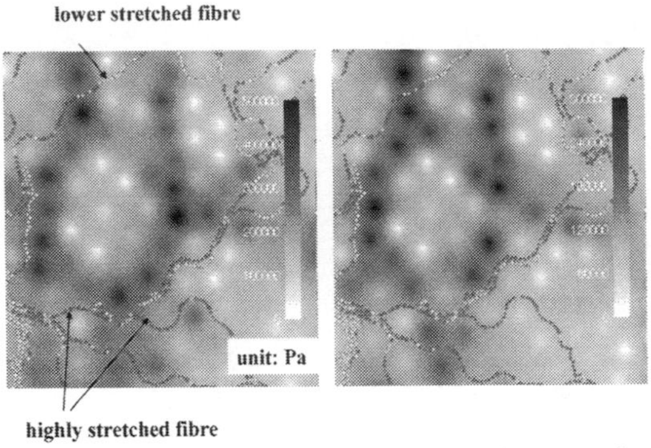

(mean normal stress) (deviator stress)

Figure 13. Stress distribution in micro zone for granular matrix phase in a zone cut in reinforced specimen at normal strain $\varepsilon_{yy} = 2.0\%$.

materials and flexible, discrete, randomly distributed fibres under isotropic compression and biaxial compression conditions. Evidently a two dimensional analysis implies that the fibres are completely separating the sand particles whereas the real material will have a continuity of stress transmission through the granular material provided by the third dimension.

Nevertheless, the DEM analysis describes well the experimentally observed strength increase induced by the presence of fibres. The dilatancy is equally affected by the amount of inclusions. The interaction mechanism does not seem to mobilise over the small and medium strain domains. Then, the fibre tensile stresses increase with the strain level and higher fibre tensile stresses are mobilized in specimens with higher fibre contents. It is also clearly shown that the highest tensile stresses are mobilized on fibres oriented towards the minor principal strain direction. However, local micro fabric and local structural rearrangement can affect the interaction mechanism and this was clearly shown for the isotropic compresses specimens. The presence of fibres also creates additional micro-confinement for the granular matrix.

REFERENCES

Consoli, N.C., Dal Toe Casagrande, & M. Coop, M.R. 2005. Effect of fibre reinforcement on the isotropic compression behaviour of a sand. *J. of Geotech. and Geoenv. Eng.* 131 (11): 1434–1436.

Cundall, P.A. 1971. A Computer Model for Simulation Progressive, Large Scale Movement in Blocky rock system. *Symp. ISRM*, Vol.2: 129–136.

Diambra, A., Russell, A.R., Ibraim, E. & Muir Wood, D. 2007, Determination of fibre orientation distribution in reinforced sand, *Geotechnique (accepted for publication)*.

Gray, D.H. & Ohashi, H. 1983. Mechanics of fibre reinforcement in sand, *J. of Geotech. Eng.* 109 (3): 335–353.

Gray, D.H. & Al-Refeai, T.O. 1986. Behaviour of fabric – versus fiber-reinforced sand, *J. of Geotech. Eng.* 112 (8): 804–820.

Heineck, K.S., Coop, M.R. & Consoli, N.C. 2005. Effect of microreinforcement of soils from very small to large shear strains. *J. of Geotech. and Geoenv. Eng.* 131 (8): 1024–1033.

Ibraim, E. & Fourmont, S. 2006. Behaviour of sand reinforced with fibres. *Int. Geotech. Symposium*, March, Roma. 10 p.

Ibraim, E., Wood, D.M., Maeda, K. & Hirabashi, H. 2006. Fibre-reinforced granular soils behaviour, *International Symposium on Geotechnics of Particulate Media*: 443–448.

Jewell, R.A. & Wroth, C.P. 1987. Direct shear tests on reinforced sand. *Géotechnique* 37 (1): 53–68.

Kaniraj, S.R. & Havanagi, V.G. 2001. *J. of Geotech. and Geoenv. Eng.* 127 (7): 574–584.

Koerner, R.M. & Welsh, J.P. 1980. *Constructional and geotechnical engineering using synthetic fabrics.* New York; Chichester: Wiley.

Maeda, K., Hara Y. & Ohno, R. 2003. Interaction between piles with different skin roughness and granular ground by DEM. *VIIth International Conference on Computational Plasticity (COMPLAS)*, Barcelona, CD-ROM.

Maeda, K. & Hirabayashi, H. 2006. Influence of grain properties on macro mechanical behaviors of granular media by DEM, *Journal of Applied Mechanics, JSCE*: 623–630.

Maher, M.H. & Gray, D.H. 1990. Static response of sands reinforced with randomly distributed fibers. *J. of Geotech. Eng.* 116 (11): 1661–1677.

Michalowski, R.L. & Zhao, A. 1996. Failure of fiber-reinforced granular soils. *J. of Geotech. Eng.* 122 (3): 226–234.

Murray, J.J., Frost, J.D. & Wang, Y. 2000. Behaviour of sandy silt reinforced with discontinuous recycled fiber inclusions. *Transportation Research Record*, 1714: 9–17.

Palmeira, E.M. & Milligan, G.W.E. 1989. Large scale direct shear tests on reinforced soil. *Soils and foundations*, 29(1): 18–30.

Rowe, P.W. 1962. The stress Dilatancy relation for static equilibrium of an assembly of particles in contact. *Proc. R. Soc.* London, Ser. A., 269, 500–527.

Satake, M. 1982. Fabric tensor in granular materials. *IUTAM-Conference on Deformation and Failure of Granular Materials*: 63–68.

Shewbridge, S.E. & Sitar, N. 1989. deformation characteristics of reinforced sand in direct shear. *J. of Geotech. Eng.* 115 (8): 1134–1147.

Zornberg, J.G. 2002. Discrete framework for limit equilibrium analysis of fibre-reinforced soil. *Géotechnique*, 52 (8): 593–604.

New Horizons in Earth Reinforcement – Otani, Miyata & Mukunoki (eds)
© 2008 Taylor & Francis Group, London, ISBN 978-0-415-45775-0

Development of multiphase model of reinforced soils considering non-linear behavior of the matrix

E.S. Hosseinina & O. Farzaneh
Faculty of Civil Engineering, University of Tehran, Tehran, Iran

ABSTRACT: The multiphase model is developed to include non-linear behavior of the soil (matrix). In this macroscopic model, the mass of soil and the reinforcements are considered as continuously distributed in the entire medium. The soil behavior is considered as nonlinear elastic, while the reinforcement phase acts as a linear elastic-perfectly plastic material. Equations of equilibrium and constitutive laws of the method are derived and presented. The relations are used for simulation of an undrained triaxial compression test. The results are compared with some tests and it is shown that there is good agreement between the results of modeling and those of laboratory tests.

1 INTRODUCTION

Soils reinforced by linear inclusions can be considered as homogenous but anisotropic materials. The bolts in the tunnel wall, the vertical piles under the foundation and the geo-synthetics layers back the retaining walls are the examples of these materials where a number of inclusions exist directionally. A so called "Multiphase model" has been proposed (de Buhan & Sudret, 2000; Bennis & de Buhan, 2003) and developed (Hassen & de Buhan, 2005) which provides a mechanically consistent framework to set up appropriate design methods for theses structures. The main advantage of this kind of modelisation is the dramatically reduced computational effort related to that where each inclusion and the soil media are simulated individually.

2 TWO-PHASE MODEL FOR REINFORCED SOILS WITH LINEAR INCLUSIONS

2.1 *Basic relationships for the two-phase medium*

The equivalent material is modeled as a homogenous continuum. It is assumed that the reinforcements, which can only resist the tension-compression forces, are distributed every where at each particle of the soil medium as shown in Figure 1. The constitutive equations of the medium can be obtained by applying *Virtual Work principle*. This principle consists of two statements; the first one states that the internal forces of a medium will be zero in a rigid movement and the latter denotes the equality of internal and external works in a free movement. Using these two conditions for the homogenized medium and the individual parts separately, it is possible to reach the equilibrium equations in each phase.

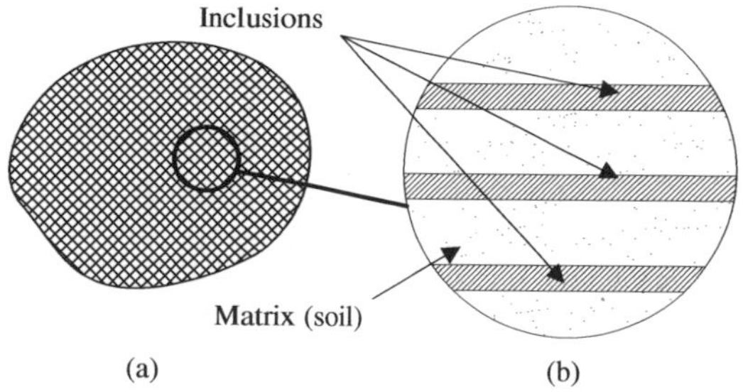

Figure 1. Definition of two-phase medium; (a) homogenized medium (macroscopic scale), (b) microscopic view of soil and inclusions.

The equilibrium equation of each phase is as follows:

– *Soil phase*

$$div\underline{\underline{\sigma}}^m + \rho\underline{F}^m + \underline{I} = 0 \quad (1)$$

– *Reinforcement phase*

$$div(\sigma^r \underline{e}_r \otimes \underline{e}_r) + \rho^r \underline{F}^r - \underline{I} = 0 \quad (2)$$

where σ = stress tensor, ρF = external body force vector, I = volume density of interaction force vector and $\otimes$ denotes the tensor product. The superscripts m and r denote matrix (soil) and reinforcement respectively. Since the interaction forces are the same for two phases with opposite sign, combining these two equations can give us a global equilibrium equation as follows:

$$div\underline{\underline{\Sigma}} + \rho^i \underline{F}^i = \underline{0} \tag{3}$$

where $\underline{\underline{\Sigma}} = \underline{\underline{\sigma}}^m + \sigma^r \underline{e}_r \otimes \underline{e}_r$ and $i = m, r$. The $\underline{\underline{\sigma}}^m$ and σ^r are partial stresses of the phases. Thus, it is noted that regarding Equations 1 and 2, it is possible to consider the behavior of each phase separately while Equation 3 holds for the whole body.

2.2 *Constitutive equation for soil (matrix)*

In the previous works and modeling, the behavior of matrix phase was considered as linear elastic-perfectly plastic. It is clear that this hypothesis is not accepted for all soils especially the cohesionless soils such as sands. Also, there are different factors such as confining pressure which define the relation of stress-strain in the soil. Regarding the non-linear behavior of soil, several non-linear models are proposed based on hyperelasticity and hypoelasticity. The premier type is consistent with the thermodynamic laws in reverse to the latter, but the soil behavior is not path dependent. However, it is possible to consider the path dependency of soil behavior in hypoelasticity. The most famous soil model of this type is the Duncan-Chang model, also known as the hyperbolic model (Duncan & Chang, 1970). This model captures the soil behavior in a very tractable manner on the basis of only the initial engineering parameters of soil as friction angle (ϕ), cohesion (c) and the initial soil stiffness (E_i). Also, there are a few parameters to relate the soil stiffness to the confining pressure.

The stress-strain relationship of the hyperbolic model for the triaxial mode is as follows:

$$(\sigma_1 - \sigma_3) = \frac{\varepsilon_1}{a + b\varepsilon_1} \tag{4}$$

where σ_1 = major principal stress, σ_3 = minor principal stress, ε_1 = axial strain, a = the reciprocal of E_i, $b = R_f / (\sigma_1 - \sigma_3)_f$. R_f is about 0.75–1.0 (Kondner et al, 1963). The failure strength of the soil can be expressed as follows:

$$(\sigma_1 - \sigma_3)_f = \frac{2c\,Cos\phi + 2\sigma_3 Sin\phi}{1 - Sin\phi} \tag{5}$$

2.3 *Constitutive equation for reinforcement*

Since the reinforcement acts as linear elastic-perfectly plastic, the stress-strain equation is:

$$\sigma^r = E^r \varepsilon^r, \ \sigma^r \le \sigma_0^r \tag{6}$$

where E^r = Young modulus, σ_0^r = ultimate yield stress. To distribute the reinforcement, Young modulus (E^{inc}) and the ultimate yield stress of inclusions (σ_0^{inc}) have been already multiplied by the volumic ratio (η) of the reinforcement (V_{renf}) to the soil volume (V_{soil}).

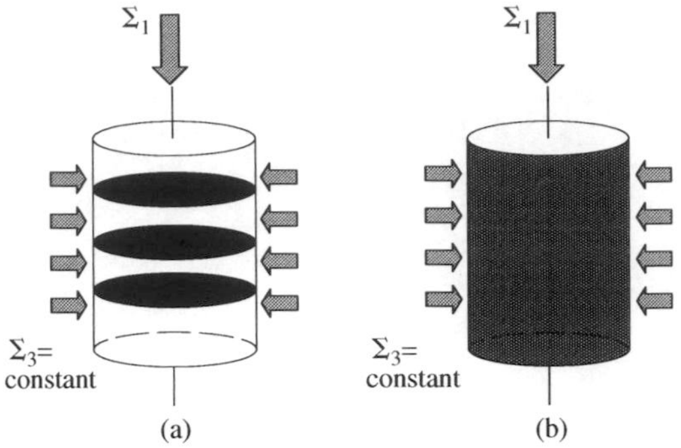

Figure 2. Presentation of stress field on the reinforced soil sample; (a) real medium, (b) homogenized model.

$$\eta = \frac{V_{renf}}{V_{soil}}, \ E^r = \eta E^{inc}, \ \sigma_0^r = \eta \sigma_0^{inc} \tag{7}$$

3 MODELING A TRIAXIAL COMPRESSION TEST

Figure 2 shows a triaxial homogenized soil sample in which the reinforcement plates are located in horizontal layers among the sand medium. The test is modeled under the consolidated undrained (CU) condition. It means that $\varepsilon_v = 0$ and thus, $\varepsilon_1 + 2\varepsilon_3 = 0$. Also, it is supposed that there is a perfect bonding between reinforcements and the soil body unless the soil goes to the failure strength. The sample is loaded by a constant confining pressure Σ_3 and the major principal stress Σ_1. The compression is taken as positive and the tension as negative. Since the reinforcements are oriented horizontally, it can be written as:

$$\Sigma_1 = \sigma_1^m, \Sigma_3 = \sigma^r + \sigma_3^m \Rightarrow \sigma_3^m = \Sigma_3 - \sigma^r \tag{8}$$

Substituting Equation 8 into Equation 4 gives the following:

$$(\Sigma_1 - \Sigma_3) = \frac{\varepsilon_1}{a + b\varepsilon_1} - \sigma^r \tag{9}$$

It is possible to find the initial Young modulus of the composite material (E_i^H) by derivation of the above equation related to ε_1 and using the strain compatibility. We obtain:

$$E_i^H = \left.\frac{d(\Sigma_1 - \Sigma_3)}{d\varepsilon_1}\right|_{\varepsilon_1 \to 0} = \left.\frac{a}{(a + b\varepsilon_1)^2}\right|_{\varepsilon_1 \to 0} + 0.5E^r$$
$$E_i^H = E_i^m + 0.5E^r \tag{10}$$

Since the soil is cohesionless ($c = 0$), the relation between the stresses at the failure based on

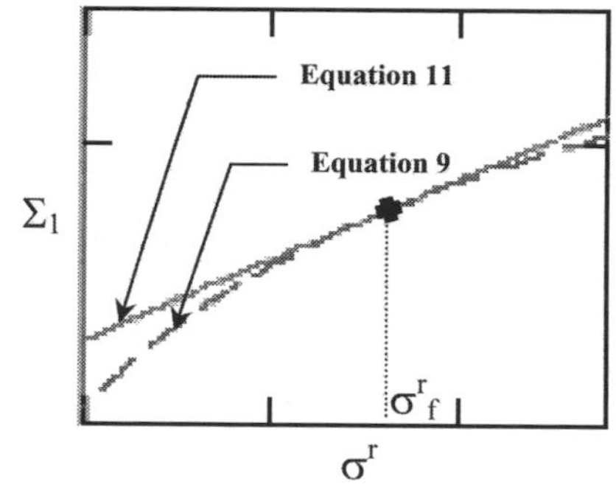

Figure 3. Presentation of variation of Equation 9 & 11 with σ^r.

Mohr-Coulomb criterion (Equation 5) is as follows (using $R_f = 1$):

$$\Sigma_1 - (\Sigma_3 - \sigma^r) = Sin\phi(\Sigma_1 + \Sigma_3 - \sigma^r) \tag{11}$$

Writing Equation 9 in terms of Σ_1, Σ_3 and σ^r and combining it with Equation 11, the quadratic equation related to σ^r is obtained.

$$\left(\frac{2Sin\phi}{1-Sin\phi}\right)\Sigma_3 + \left(\frac{2}{1-Sin\phi}\right)\sigma^r + \frac{2\sigma^r}{\frac{E^r}{E^{im}} - 2b\sigma^r} = 0 \tag{12}$$

The unity of solution implies that Equation 12 should have double root, or say the discriminant equals to zero. It follows to the other quadratic equation based on b. The responses of the equation are as follows, where the smallest one satisfies the true solution (the smaller b gives the strength of sample even less than the soil sample and thus it is not true):

$$b = \frac{2}{\Sigma_3}\left[\frac{1}{Sin\phi} + \frac{E^r}{E_i^m} - 1 \pm 2\left(\frac{E^r}{E_i^m}(\frac{1}{Sin\phi} - 1)\right)^{\frac{1}{2}}\right] \tag{13}$$

Figure 3 shows the variation of Equations 9 & 11 with σ^r. As can be seen, the two curves will intersect only in one point (σ_f^r) where the soil reaches the ultimate strength.

The value of the σ_f^r is assessed as follows:

$$\sigma_f^r = \frac{1}{4b}\left(2b\Sigma_3 + \frac{E^r}{E_i^m} + 1 - \frac{1}{Sin\phi}\right) \le \sigma_0^r \tag{14}$$

In other words, the maximum stress in the reinforcement can only reach the value of σ_f^r which is smaller than the ultimate yield stress (σ_0^r). It should be reminded that the aforementioned relations are based on the assumption that the composite material should be failed while the soil reaches the ultimate state. The ultimate strength of the reinforced soil can be calculated from Equation 9 while the ε_1 goes to infinity as shown below:

$$(\Sigma_1 - \Sigma_3)_{ult} = \left(\frac{\varepsilon_1}{a + b\varepsilon_1}\right)_{\varepsilon_1 \to +\infty} - \sigma_f^r = \frac{1}{b} + |\sigma_f^r| \tag{15}$$

It is reminded that the reinforcement tolerates the tension stress and thus it is negative.

Table 1. Properties of geosynthetics used in tests.

Material type	Thickness (mm)	Ultimate tensile strength (kN/m)	Secant modulus at 5% strain (E) (kN/m)
Woven geotextile	1	51	120
Geogrid	0.275	3.75	62
Polyester film	0.1	Not failed in test limits	100

Table 2. Parameters of geosynthetics used in the simulations.

Material type	Volumic ratio η (%)	Ultimate tensile strength (kN/m^2)	Radial modulus* (kN/m^2)	Poisson's ratio ν –
Geotextile	11.0	6000	14000	0.05
Geogrid	3.0	407	9400	0.20
Film	1.1	–	15300	0.20

*Radial modulus = E/[(1+ν)·(1−2ν)].

4 PREDICTIONS OF THE MODEL

4.1 *Parameters of the model*

A series of triaxial compression tests on river sand reinforced by three types of geosynthetics (woven geotextile, geogrid and polyester film) are performed by Latha & Murthy (2007) in the consolidated undrained (CU) condition. The aim of these tests was to understand the strength improvement in sand due to reinforcement in different forms. Among them, the tests performed with three types in the planar form are selected herein to predict the behavior of the composite material by the proposed model. The tests were conducted at three confining pressures; 100, 150 and 200 kPa. The woven geotextile is made of Polypropylene and geogrid is from Polyethylene. The properties of the geosynthetics are presented in Table 1. Applying Equation 7, the modified parameters of the geosynthetics are as shown in Table 2.

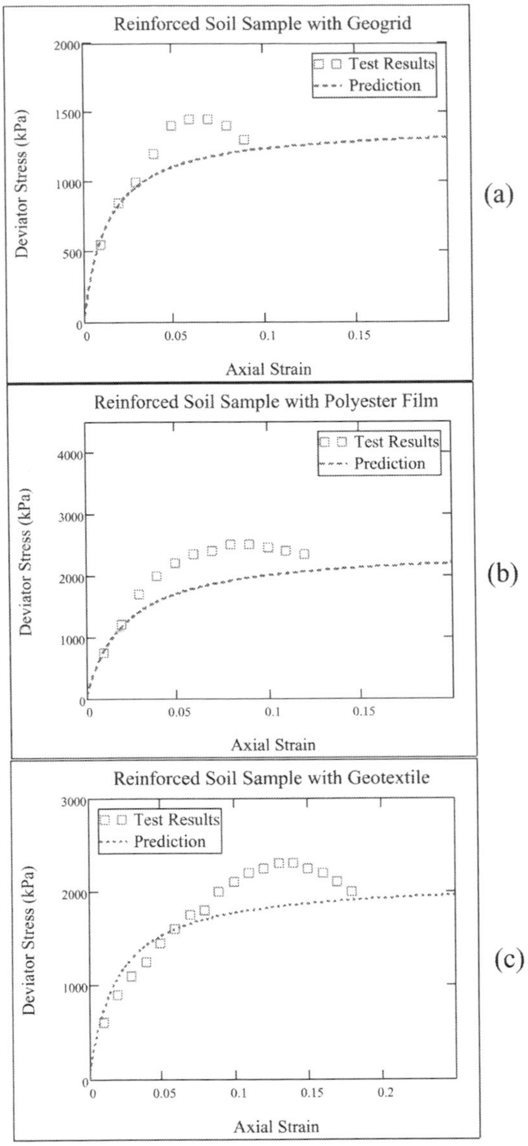

Figure 4. Presentation of stress-strain curve in confining pressure of 100 kPa for (a) geogrid; (b) Polyester film; (c) geotextile.

The shear strength parameters for the unreinforced sand at 70% relative density are obtained as $c = 0$ and $\phi = 42^o$. Also, the initial Young modulus of the soil is measured form the stress-strain curve as $E_i = 100$ MPa at confining pressure of 100 kPa.

4.2 *Results and discussion*

Figure 4 shows the test results and the predictions of stress-strain curve of the reinforced soils in confining pressure of 100 kPa for different types of geosynthetics. In the reference paper, only the stress-strain curves corresponding to confining pressure of 100 kPa are presented.

As can be figured out, the predictions for different types of layers are acceptable regarding the initial slope and the ultimate shear strength. In these curves, except for the geotextile type, there is no good coincidence between the test results and the predictions from the strain of 4% to then. In this range, the curves tolerate a hardening of the strength showing a peak and then the strength reduces slightly reaching a stable limit. It is because the hyperbolic relation goes directly to the ultimate strength in large strains and it can not predict the softening behavior, predicting the initial slope and ultimate shear strength very well, though. About the sample reinforced with geotextile, the stress-strain prediction is not so good as the others from small strain like 1%. The reason might be because of the high volume ratio of the reinforcement (11%). In the future study, the reason will be investigated more.

To show the high ability of the proposed model to predict the ultimate shear strength of the composite soil, Equation 15 is applied for all geosynthetics types and the results are compared with the data published in the reference paper. The comparison of the results is shown in Figure 5, which indicates a very good agreement between real and predicted values.

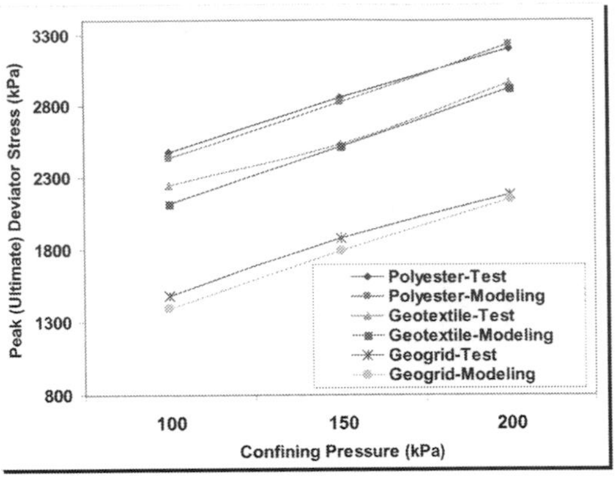

Figure 5. Comparison of peak and ultimate shear strength in tests and modeling in different confining pressures.

5 CONCLUSION

A closed analytical form is presented based on some assumptions for prediction of the stress-strain curve of the triaxial compression tests in CU conditions. The model predicts the initial modulus and the ultimate shear strength of the composite material very well, while there is some inconsistency in the stress-strain curve due to the existence of the peak value in the curves.

REFERENCES

Bennis, P. & de Buhan, P. 2003. A multiphase constitutive model of reinforced soils accounting for soil-inclusion intera behavior. *Math. Comp. Model.*, 37: 469–475.

de Buhan, P. & Sudret, B. 2000. Micropolar multiphase model for materials reinforced by linear inclusions. *Eur. J. Mech. A /Solids*, 19: 669–687.

Duncan, J.M. & Chang, C.Y. 1970. Nonlinear Analysis of Stress and Strain in Soils. *Journal of Soil Mechanics and Foundation Division, SM5*, 96: 1629–1653.

Hassen, G. & de Buhan, P. 2005. A two-phase model and related numerical tool for the design of soil structures reinforced by stiff linear inclusions. *Eur. J. Mech. A /Solids*, 24: 987–1001.

Kondner, R.L. & Zelasko, J.S. 1963. A Hyperbolic Stress-Strain Formulation for Sands. *Proceeding Second Pan-American Conference on Soil Mechanics and Foundation Engineering*, Brazil. 1: 289–324.

Latha, G.M. & Murthy, V.S. 2007. Effects of reinforcement form on the behavior of geosynthetics reinforced sand. Geotextiles and Geomembranes 25: 23–32.

New Horizons in Earth Reinforcement – Otani, Miyata & Mukunoki (eds)
© 2008 Taylor & Francis Group, London, ISBN 978-0-415-45775-0

Numerical analysis of stability of slope reinforced with piles subjected to combined load

T.K. Nian, M.T. Luan & Q.Yang
School of Civil and Hydraulic Engineering and State Key Laboratory of Coastal and Offshore Engineering, Dalian University of Technology, China

G.Q. Chen
Department of Civil Engineering, Kyushu University, Fukuoka, Japan

ABSTRACT: By using elasto-plastic finite element method based on the technique of shear strength reduction, stability of slope reinforced with piles and performance of load-bearing piles in slopes subjected to combined load are numerically analyzed. The iteration non-convergence criterion conventionally used for assessing the instability state of slopes is employed to evaluate the limit-equilibrium state of pile-soil-slope system. Moreover, considering the complexity of pile-soil interaction, the criterion based on the uncontrolled displacement at a certain characteristic nodes on the slope surface is used for evaluate the limit state as an assistant criterion in addition to the iteration non-convergence criterion of solution in order to get more reasonable and reliable solution from numerical results. Finally, the effect of the combinations of vertical load and horizontal load on the stability of pile-soil-slope system is investigated through numerical analyses.

1 INTRODUCTION

The stabilizing piles are widely used in the reinforcement engineering of slopes and mitigation and prevention of natural geological disasters induced by landslides. However, because of its rather sophistication, the working mechanism of stabilizing piles and stability of piles in the slopes are not clarified especially when the piles are subjected to the combined actions of both components of horizontal load and vertical load imposed by the superstructures. Under such a circumstance, the piles in slope will play two functions, one is as reinforcement to induce the instability and another is to bear the loads transferred by structures. In fact, the pile-soil-slope will constitute an interaction system. Therefore, it will be theoretically important and practically significance to examine the mechanism of piles in both stabilizing the slope and carrying the loads and to discover the interaction effect of piles-soil-slope. In this paper, the stability of slopes reinforced with piles against potential sliding and bearing capacity behavior of piles in the slopes subjected to combined loads as well as interaction mechanism of pile-soil-slope are numerically investigated.

2 SHEAR STRENGTH REDUCTION ELASTO-PLASTIC FEM CONSIDERING CONTACT BEHAVIOR

2.1 *Interaction between surfaces*

The interaction of contacting surface between pile and soils consists of two components: one normal to the surfaces and one tangential to the surfaces. The tangential component consists of the relative motion (sliding) of the surfaces and, possibly, frictional shear stresses. Each contact interaction can refer to a contact property that specifies a model for the interaction between the contacting surfaces. There are several contact interaction models available in ABAQUS soft, the friction behavior of interface is adopted here.

2.2 *Contact algorithm*

ABAQUS examines the state of all contact interactions at the start of each increment to establish whether slave nodes are open or closed. In Figure 1, p denotes the contact pressure at a slave node and h represents the penetration of a slave node into the master surface. If a

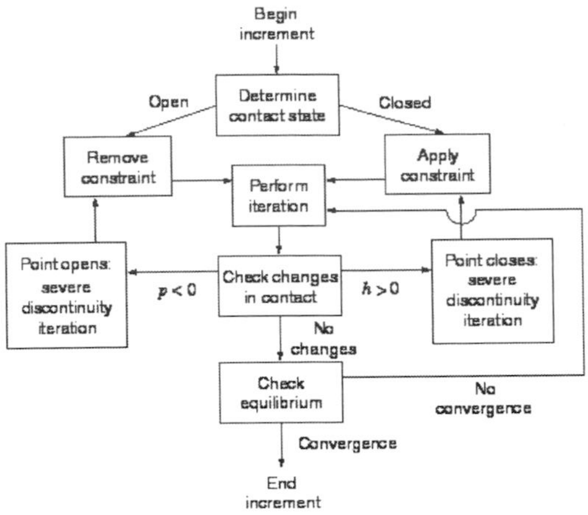

Figure 1. Contact algorithm(from ABAQUS6.3).

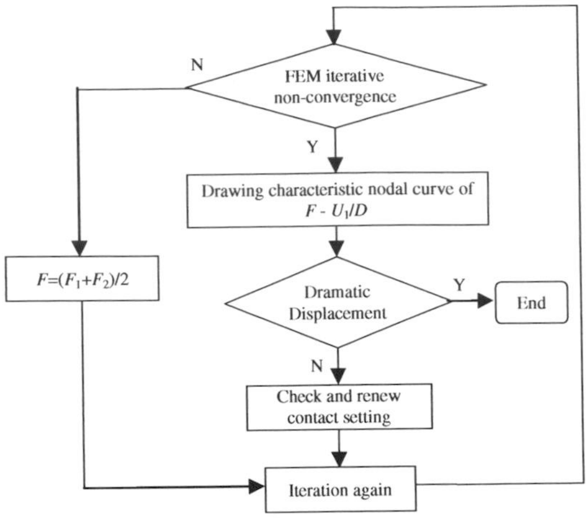

Figure 2. SSR-E-P-FEM procedure considering contact behavior.

node is closed, ABAQUS determines whether it is sliding or sticking. ABAQUS applies a constraint for each closed node and removes constraints from any node where the contact state changes from closed to open. ABAQUS then carries out an iteration and updates the configuration of the model using the calculated corrections. By default, ABAQUS abandons any increment where it needs more than twelve severe discontinuity iterations and tries the increment again with a smaller increment size. If there are no severe discontinuity iterations, the contact state is not changing from increment to increment (see Figure 1).

2.3 *Shear strength reduction elasto-plastic FEM considering contact behavior*

Finite element analysis of slope stability does not provide an explicit factor of safety (FS) but utilizes the so-called shear strength reduction technique (Zienkicwicz,1975; Ugai,1989; Matsui & San,1992). The FS of a slope is defined as the number by which the shear strength parameters must be factored down to bring the slope to failure. Using this technique, the mobilized shear strength parameters, c_m and $\tan\phi_m$, are obtained by dividing c and $\tan\phi$ by the strength reduction factor (SRF) as follows:

$$c_m = \frac{c}{F}, \quad \tan\phi_m = \frac{\tan\phi}{F} \tag{1}$$

where F is defined as shear strength reduction factor.

According to the above ideas, an elasto-plastic FE computation programming using Fortran90 is developed based on ABAQUS software. Python language is used to write a judgement program and evaluate the numerical convergence (see Figure 2). The non-convergence criterion combined with dimensionless displacement increment at some characteristic nodes on slope surface is employed to analyze the stability of slopes with stabilizing piles and load-bearing piles under combined load mode.

Table 1. Material parameters for FE analysis.

Property	Soil	Pile
Young's modulus, E(MPa)	50.0	26000.0
Poisson's ratio, ν	0.3	0.2
Unit weight, γ(kN/m^3)	1800.0	2400.0
Internal friction angle, ϕ(°)	35.0	–
Effective cohesion, c(kPa)	10.0	–
Dilation angle, ψ(°)	0.0	–

3 NUMERICAL ANALYSIS OF THE SLOPES CONTAINING LOAD-BEARING PILES UNDER COMBINED LOAD

3.1 *Numerical analysis procedure and model parameters*

An example of pile-soil-slope system in an idealized elastic, perfectly plastic soil are investigated using finite element program ABAQUS (Hibbitt, Karlsson & Sorensen, INC. 1978). A single pile is considered in the FEM Model. Materials parameters for finite element analysis are listed in Table 1. Figure 3(a) and (b) show the plan view and FE computational model, respectively. Seen from Figure 3(a), a 15 m high cut slope with an average slope angle of 32° is modeled. The concrete piles are assumed to be 2 m in diameter, 30 m in length, and with 22.5 m penetration. The cut slope is formed by a series of excavation processes

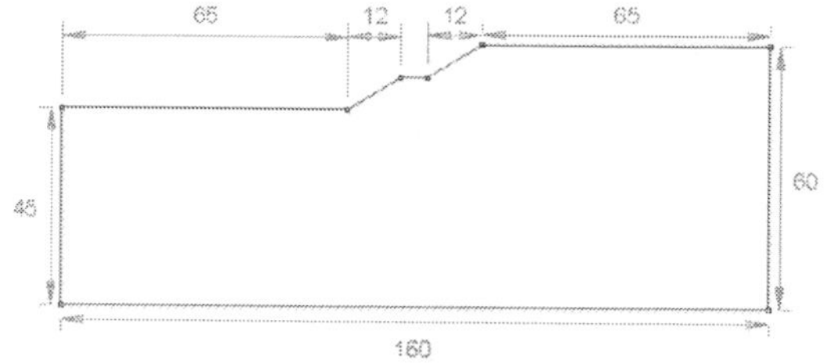

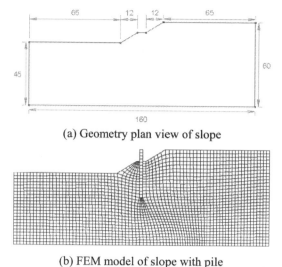

(a) Geometry plan view of slope

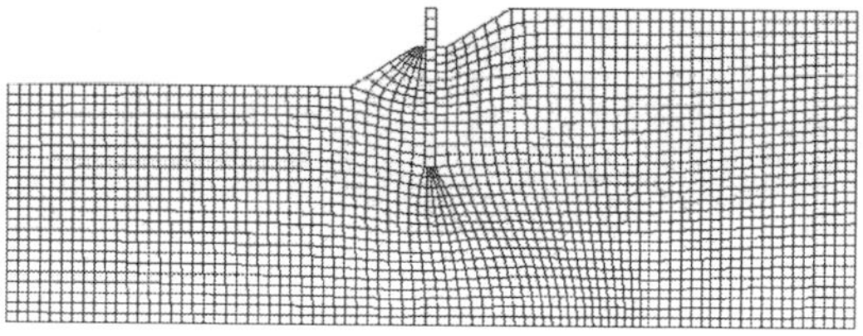

(b) FEM model of slope with pile

Figure 3. Slope geometry and FEM computational model.

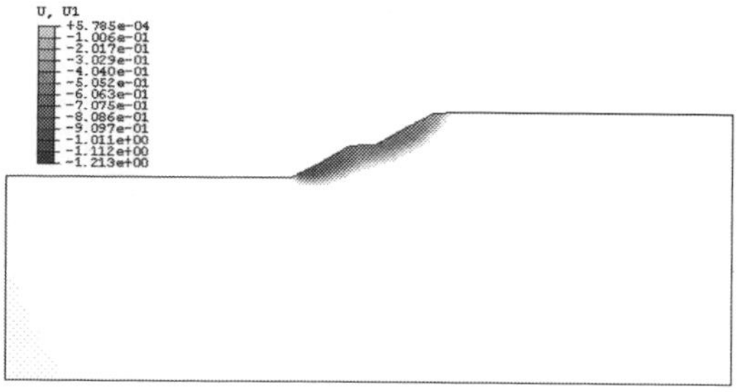

(a) Horizontal displacement distribution

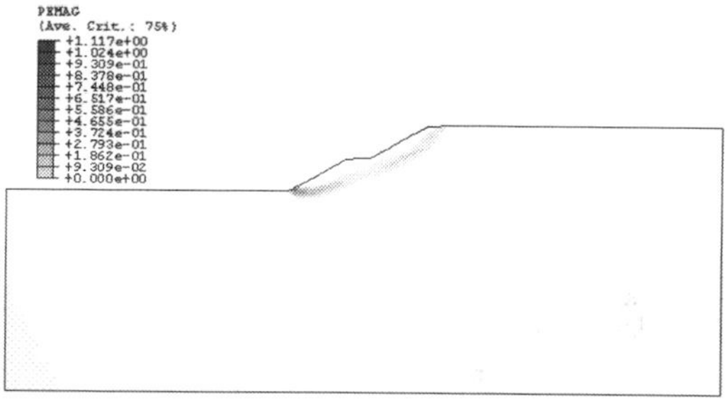

(b) Equivalent plastic strain distribution

Figure 4. Slope deformation and plastic strain distribution at $F = 2.13$ (global failure).

simulated numerically. Each step excavates 2.5 m of soil and the excavation continues until a slope height of 15 m has been reached. The piles are formed by concrete in the middle of the cut slope. For simplicity, the groundwater table is taken to be far below the pile tip in the analysis (Charles et al., 2001).

In the numerical analysis, soils adopt idealized ealsto-plastic constitute model based on the Mohr-Coulomb failure criterion utilizing eight-node quadrilateral elements with reduced integration and non-associated flow laws, pile is regarded as elastic body, the interaction between pile and soils abides by the afore-mentioned contact algorithm. The non-convergence option is taken as being a suitable indicator of failure, and the relationships between SRF and the displacements at some characteristic nodes on slope surface are also analyzed to evaluate the global failure of the slope. Slope failure and numerical non-convergence occur simultaneously, and are accompanied by a dramatic increase in the nodal displacements within the mesh.

3.2 *Numerical analysis of slope stability*

According to the technique of shear strength reduction and the non-convergence criterion of FEA, the distributions on horizontal displacement and equivalent plastic strain representing the global failure of slope are given in Figure 4, the corresponding safety factor of slope is $F_s = 2.13$.

3.3 *Stability of slope with stabilizing piles*

In order to check the effect of reinforcing pile on the stability of slope, a row of stabilizing piles are inserted on the middle platform of multi-stage slope. FEA non-convergence criterion is adopted to evaluate the stability of pile-slope systems, the distribution of horizontal displacement and equivalent plastic strain at the time of non-convergence with $F = 2.20$ are showed in Figure 5. Compared with the figures, the internal displacement in slope remarkably increases at the time of non-convergence than that of before non-convergence.

Figure 6 shows the relationships between the dimensionless characteristic nodal displacement on slope surface and strength reduction factor SRF. In the Figure, T_1 indicates adjacent node behind the pile, T_2 indicates adjacent node in front of the pile, B is the node at the toe of the lower slope, M is the node situated on the first-stage slope surface, which displacement is largest. Through analyzing the curves in Figure 6, the four characteristic nodes on the slope surface all show uncontrolled displacement increase before the non-convergence.

3.4 *Stability of slope with load-bearing piles under combined load mode*

In order to investigate the mechanical behavior of load-bearing pile and the stability of pile-slope system, the above-mentioned stabilizing pile is again studied here. A vertical load as $P_v = 1000$ kN, 2000 kN and 4000 kN is respectively applied on the top of the stabilizing pile. The instability criterion of numerical computation

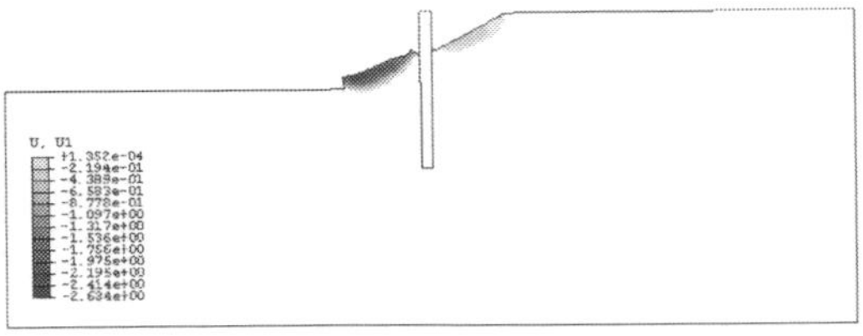

(a) Horizontal displacement distribution

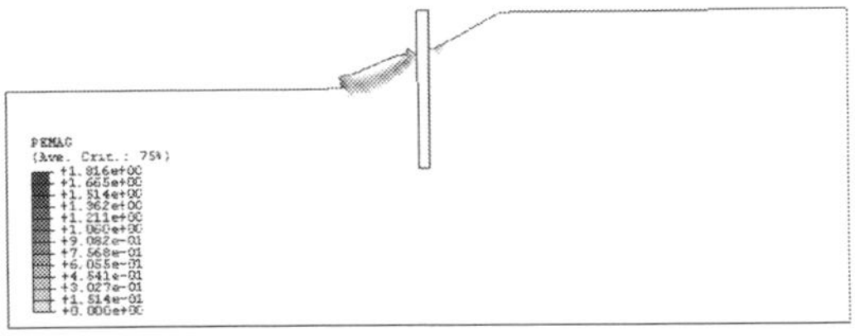

(b) Equivalent plastic strain distribution

Figure 5. Deformation and plastic strain of slope stabilized by piles at $F = 2.20$(at the time of non-convergence).

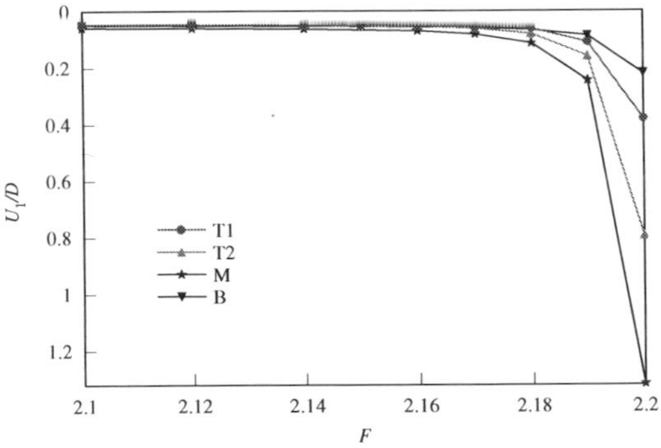

Figure 6. Relationship between SRF and dimensionless nodal displacements.

non-convergence combined with key nodal dramatic displacement on the slope surface is employed to evaluate the global stability. According to the numerical results, the global stability factor is 2.20 as same as that condition of only stabilizing pile. But the maximum horizontal displacement at the characteristic nodes on the slope surface rather increases with increasing vertical load. The above analysis shows that load-bearing pile has no effect on the stability of slope with high safety factor when the vertical load of load-bearing pile is not beyond the admissible value of the standard.

On the other hand, the global stability of pile-slope system is also computed under four different horizontal load mode as $P_h = 500$ kN, 1000 kN, 1500 kN and 2000 kN. According to the numerical results, the corresponding safety factor of pile-slope system is respectively 2.13, 1.65, 1.33 and 1.11 as shown in

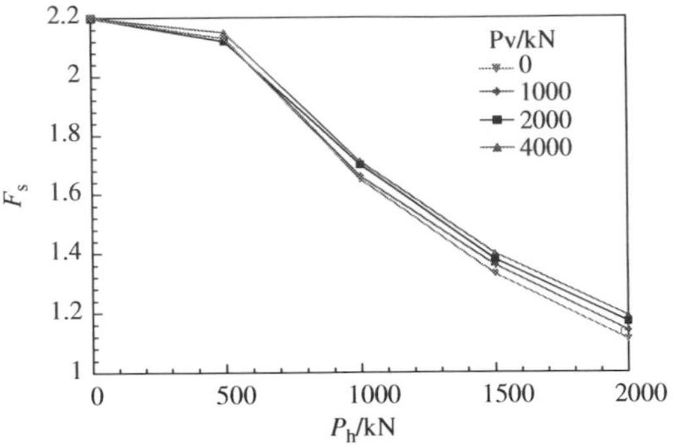

Figure 7. The relationship between safety factor of slope with stabilizing and load-bearing piles and combined loads.

Figure 7, which are far lower than that of the slope only containing stabilizing pile or load-bearing pile, the maximum decreasing degree of the former is more two times than the latter. Thus, the effect of horizontal load on the global stability of pile-slope system can not be ignored.

In addition, the combined load mode including horizontal and vertical load is computed to investigate the global stability of pile-slope system. Under the combined load modes, the horizontal load acting on the top of pile is respectively $P_h = 500$ kN, 1000 kN, 1500 kN and 2000 kN, the vertical load is respectively $P_v = 1000$ kN, 2000 kN and 4000 kN, which are the same as the afore-mentioned. Figure 7 plots the variations of the global safety factor F_s with applied horizontal load P_h on the top of pile under different vertical load P_v. Seen from Figure 7, the global safety factor of pile-slope system obviously decreases with increasing the horizontal load under a given vertical load on the top of pile. A nonlinear relation between the two sides can be drawn by the analysis of curves. Moreover, the global safety factor of pile-slope system will slightly increases with increasing the vertical load under a given horizontal load. In contrast, the key nodal maximum horizontal displacement on the slope surface yet takes on some increase within 10%. It indicates that load-bearing and stabilizing piles cause a vertical load transfer, reducing the horizontal and vertical stresses in the shallow layers of the slope but increasing the stresses in the deeper layers. Due to the reduced stress in the shallow depths in front of pile, the local stability of the shallow layers of the slope will be improved. In addition, due to a very higher global safety factor of the slope without containing stabilizing piles, the impacts of the load-bearing and stabilizing piles on the global stability of the slope are not expected to be significant, but the horizontal loads become a controlled factor on the global stability of slope, and will not be ignored.

4 CONCLUSIONS

(1) By using elasto-plastic finite element method based on shear strength reduction, stability of slope with stabilizing piles and load-bearing piles under combined load are numerically analyzed.
(2) The iteration non-convergence criterion combined with the uncontrolled displacement at a certain characteristic node on the slope surface is reasonable and reliable to evaluate the stability of slope containing piles.
(3) Numerical results show that the impacts of the load-bearing and stabilizing piles on the global stability of the slope are not expected to be significant for a slope with higher safety factor, but the horizontal loads become a controlled factor on the global stability of slope.

ACKNOWLEDGEMNETS

The project was supported by Training Foundation for Distinguished Young Teachers of DUT (893211) and State Key Laboratory Foundation of Coastal and Offshore Engineering, PRC (LP0608)

REFERENCES

Zienkiewicz, O. C., Humpheson, C. and Lewis, R. W. Associated and non-associated visco-plasticity and plasticity in soil mechanics [J]. Geotechnique, 1975, 25(4): 671–689.
Ugai, K. A method of calculation of total factor of safety of slopes by elasto-plastic FEM [J]. Soils and Foundations, 1989, 29(2): 190–195(In Japanese).
Matsui, T. and San, K. C. Finite Element Slope Stability Analysis by Shear Strength Reduction Technique [J]. Soils and Foundations, 1992, 32(1): 59–70.
Charles, W. W. Ng, Zhang, L. M. Three-dimensional analysis of performance of laterally loaded sleeved in sloping ground [J]. Journal of Geotechnical and Geo-environmental Engineering, ASCE, 2001, 127(6): 499–509.
Hibbit, Karlsson and Sorensen, Inc. ABAQUS/Scripting Manual (Version 6.3)[M]. Providence: HKS Inc., 2002.

New Horizons in Earth Reinforcement – Otani, Miyata & Mukunoki (eds)
© 2008 Taylor & Francis Group, London, ISBN 978-0-415-45775-0

Effect of restraint deformation on stability of cut slope with soil nailing

T. Nishigata
Kansai University, Osaka, Japan

S. Araki
Dia Consultants Co., Ltd., Osaka, Japan

Y. Nakayama
Kansai Geo-Environment Research Center, Osaka, Japan

ABSTRACT: In this study, the reinforcement mechanism on restraint effect is firstly investigated by distinct element method, and secondly influence of the interval of reinforcements on the development of the restraint effect is examined by finite element method. It concludes that the restrain effect is capable of designing a reinforced slope with small displacement in high priority road system. If the restraint effect of deformation is satisfactorily guaranteed in the construct condition, only the analysis of external stability of reinforced zone like a overturning of retaining wall is required.

1 INTRODUCTION

In recent years, reinforcing methods for natural slopes have been developed for a wide variety of situations. Current design methods for reinforcing natural slopes are based on the conventional limit equilibrium method and consider the effects of the components of the tensile forces oriented parallel and orthogonal to the slip surface (internal stability), however, as the tensile force developed in the reinforcing material is closely related to deformation of the slope soil, the number of reinforcement must be determined with due consideration of the allowable deformation.

There is an approach to the design of such reinforcement: external stability, which considers the entire reinforced region as a pseudo-retaining wall. Fundamentally, internal and external stability cannot coexist, since the former allows deformation and the second is based on the presumption that deformation is prevented (restrained). The design methods of external stability are becoming more important as specifications increasingly require permanent reinforced structures that minimize soil deformation. So, it is necessary to identify the mechanisms responsible for the formation of a pseudo-wall structure (restraint of soil deformation) in the reinforced region.

The present study is an investigation of the mechanisms of slope deformation is restrained by a reinforcement, based on distinct element and finite element methods and observations of the influence of the internal stresses and friction angle of the soil on the restraint of deformation. The study also involves an investigation of the number of reinforcing materials necessary to obtain restraint of deformation and observations of the relationship between the spacing of reinforcements and slope deformation.

2 DEM ANALYSIS OF REINFORCED SLOPE

Most existing methods of soil reinforcement were developed to stabilize slopes made up of sandy soils, and their effectiveness is generally attributed to the particulate nature of the soil. For this reason, we employed a DEM(PFC) as a method of observing the behavior of particulate masses and predicting the mechanisms of slope reinforcement.

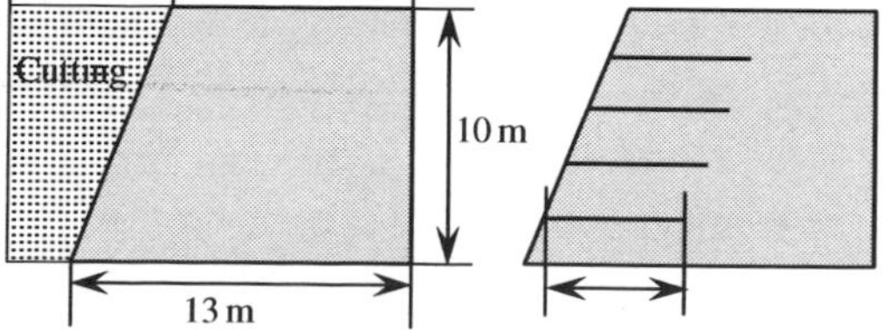

(a) Without reinforcement (b) With reinforcements

Figure 1. Slope models for distinct element method.

Table 1. Physical constant for DEM analysis.

	Normal and shear stiffness (kN/m)	Friction coefficient	Bonding parameter (kN)
Soil Particles	1×10^4	0.5, 0.7	2.3
Reinforcements	1×10^{10}	0.5	–

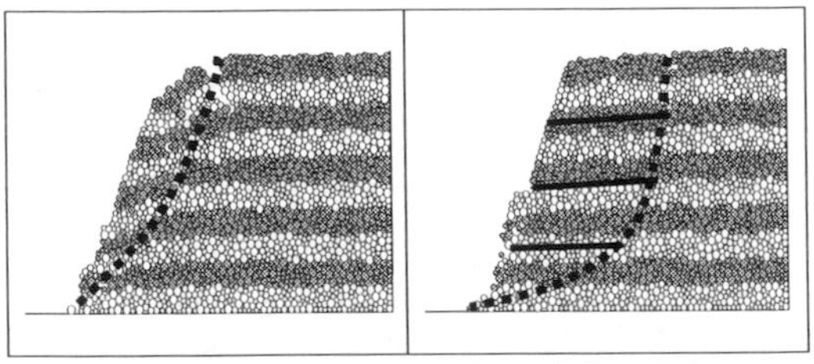

(a) Without reinforcement (b) Reinforced slope

Figure 2. Slope models for distinct element method.

Figure 1 shows the shape of the slope used for analysis. The distinct element method was performed as follows. We generated circular discrete elements with 10–30 mm in diameter and a density of 2.6 g/cm^3 in the rectangular region of 16×10 m. A initial stress condition accounting for the weight of the discrete elements was established, and the cutting portion at the front, as shown in Fig. 1, was removed to perform the analysis of the resulting deformation after the stress release. The reinforcements were handled in this analysis as rigid bodies and constructed of rigid elements of 10 cm in diameter and 5 m in length. Next, reinforcing was installed to the model in horizontal positions in an evenly spaced configuration. Table 1 shows the physical properties employed in the analysis. To study the influence of the internal friction angle ϕ of the slope soil on the reinforcing effect, the soil particles were given the two friction coefficient values shown in Table 1.

Figure 2 shows how the deformation occurred after the cutting as due to stress release. When there was no reinforcing (Fig. 2a), a slip surface developed at a shallow location in the slope. With three nails inserted (Fig. 2b), a little bulging is observed at the toe of the slope, but overall the slip surface has moved behind the reinforcing: the reinforcing has restrained deformation of reinforced region.

Figures 3 and 4 show the distribution of horizontal displacement along a vertical plane in the reinforced region with varying frictional coefficients (0.5 and 0.7). The arrows in the figures show the locations of the reinforcements. If we assume that the coefficient of friction of the discrete elements has the same physical quantity as ϕ in ordinary soil, the two figures clearly

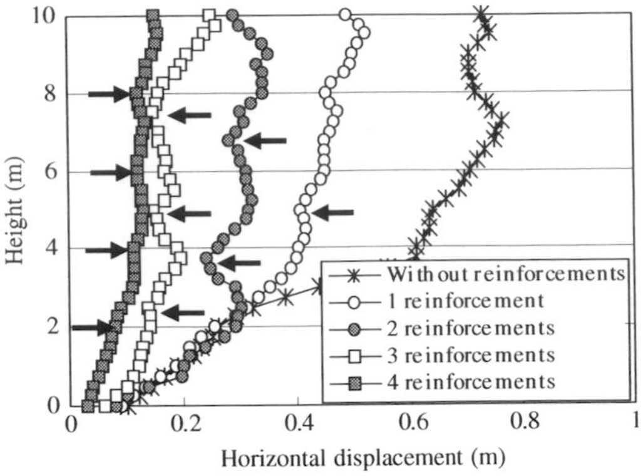

Figure 3. Horizontal displacement in reinforced area for the case of friction coefficient = 0.5.

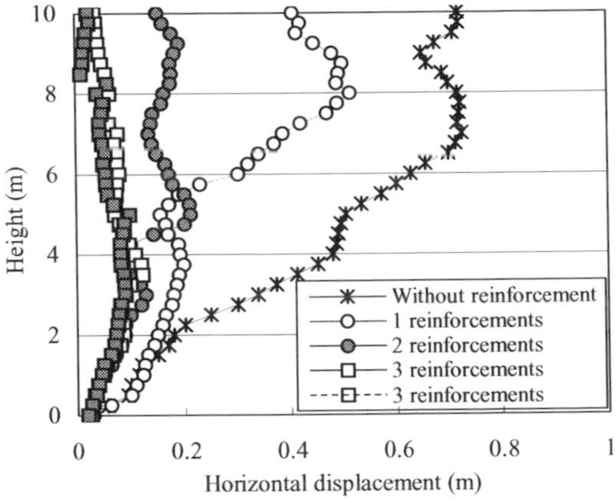

Figure 4. Horizontal displacement in reinforced area for the case of friction coefficient = 0.7.

indicate that the greater ϕ gives the lower the magnitude of deformation. In addition, both cases show that for the case of small numbers of reinforcements the local horizontal deformations are large in the spaces between the bars. Thus, the distribution of deformation has a wavy shape. For the case of greater numbers of bars, the restraint zones near the reinforcements begin to interact, broadening the extent of restraint to the entire slope, and the wavy profile disappears. This phenomenon corresponds to the complete restraint of the reinforced region by the reinforcements. Figure 5 shows a conceptual picture of progressing of restraining effect throughout the reinforced region. It has been said that reinforcement had a extent to restrain a soil particle around it and when these extents contact each other, the reinforcement effect comes up to maximum. The analytical results of Figs. 3 and 4 indicate the progress of the restrain effect shown in Fig. 5.

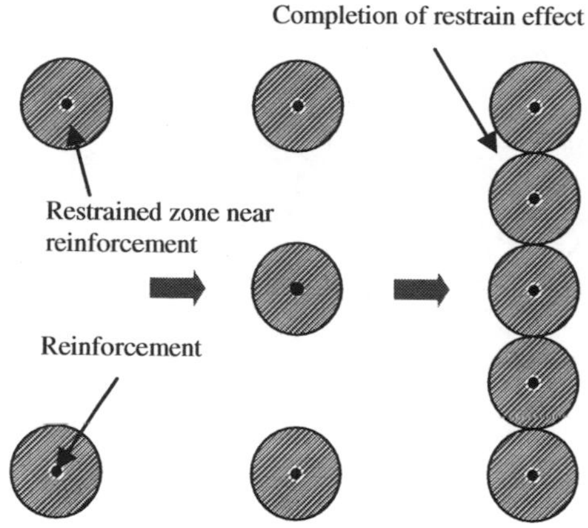

Figure 5. Progress and completion of restrain effect.

3 INFLUENCE OF SOIL PROPERTIES ON RESTRAINT EFFECT

It is necessary to quantitatively identify the size of the restrained zone influenced by a single reinforcement in order to find the optimal spacing of the bar. This restrained zone is also affected by the internal friction angle ϕ of the slope soil and its stress condition (overburden pressure). The authors continued this investigation using FEM.

Figure 6 shows the model of the slope used in the FEM(PLAXIS); it had the same dimensions as the model used in the DEM, being 10 m high, with an 80° slope and 5 m reinforcements. The bars were installed horizontally in evenly spaced positions, adjusted to the number of bars in each case. The initial stress condition due to the weight of the soil was applied throughout the analytical region, then cuts were taken and reinforcements were installed. The soil was treated as an elasto-plastic material and the reinforcing material was treated as elastic. Joint elements were inserted between the soil and the reinforcements, and the friction angle in the joint elements was set at two-thirds the value of ϕ of the soil. Table 2 shows the other physical characteristics employed in this analysis. Fore values were used for the internal friction value of the soil in order to observe its influence.

Figure 7 shows the distribution of horizontal displacement along the vertical plane in the reinforced area under the condition of $\phi = 30°$. This FEM result shows the wavy distribution of deformation as can be seen in the DEM results. The locations of the local small deformation correspond to the locations of the bars. In other words, the horizontal displacement of the soil on the slope was locally restrained in the vicinity of the reinforcements, while in the spaces between the bars, the displacement was large because it was unrestrained. In the case of seven reinforcements, the wavy

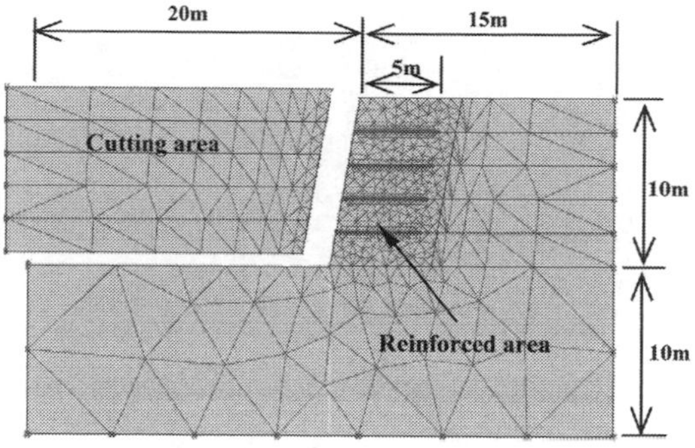

Figure 6. Reinforced slope model for FEM Analysis.

Table 2. Physical constants for FEM analysis.

	Slope soil	Reinforcement
Unit weight (kN/m^3)	18	–
Elastic modulus (kN/m^2)	33000	1.6×10^5
Poisson's ratio	0.35	0.3
Internal friction angle (°)	25, 30, 35, 40	–
Cohesion (kN/m^2)	10	–
Bending stiffness (kN · m^2)	–	40

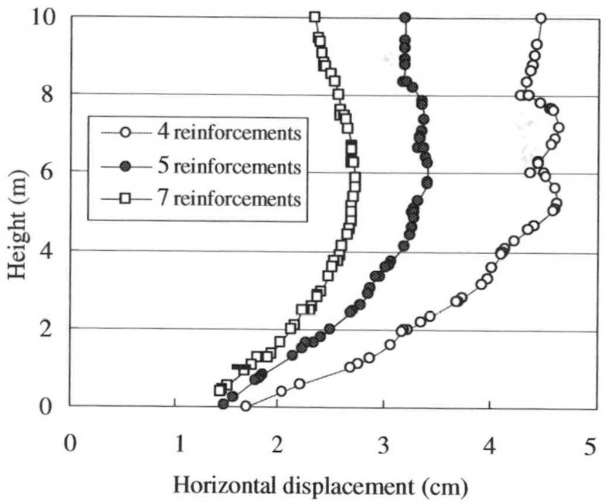

Figure 7. Horizontal displacement in reinforced area by FEM analysis for internal friction $\phi = 30°$.

shape of the displacement curve almost disappears. This indicates that the soil displacement was restrained between the bars. These results quantitatively agree with those found in the DEM. Figures 8 and 9 similarly show the results for $\phi = 35°$ and 40°, respectively. These figures indicate that the zone of influence of the bars is a function of the angle of internal friction. From Figs. 7, 8 and 9, the number of reinforcement which

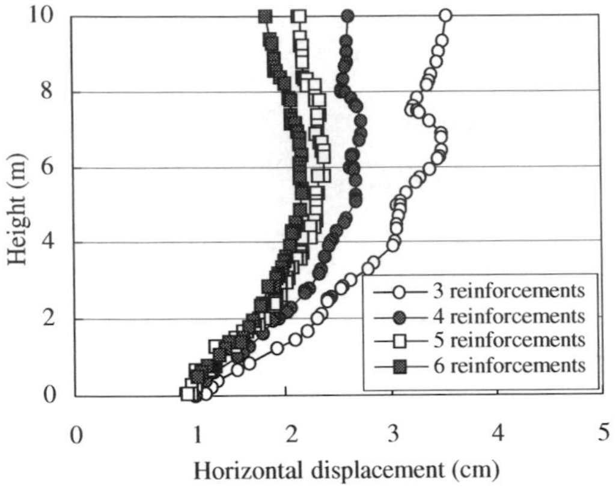

Figure 8. Horizontal displacement in reinforced area by FEM analysis for internal friction $\phi = 35°$.

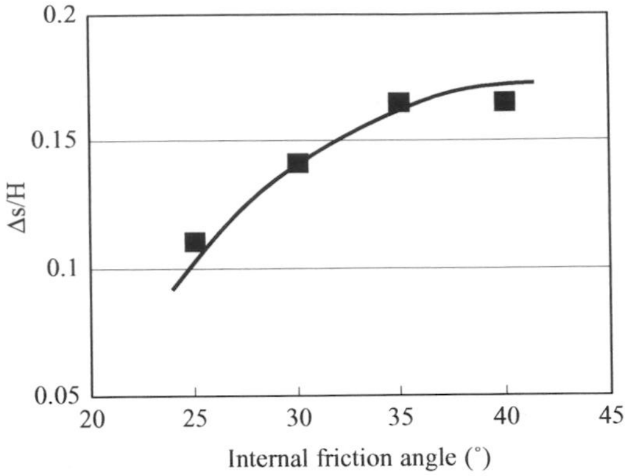

Figure 10. Relationship between friction angle and space of reinforcement on condition that the restraint effect becomes effective.

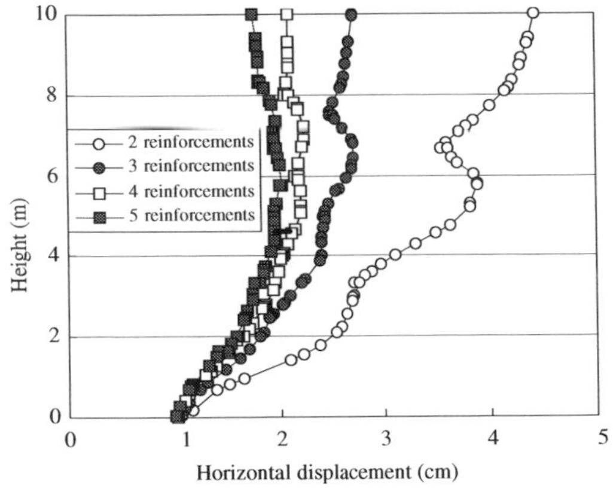

Figure 9. Horizontal displacement in reinforced area by FEM analysis for internal friction $\phi = 40°$.

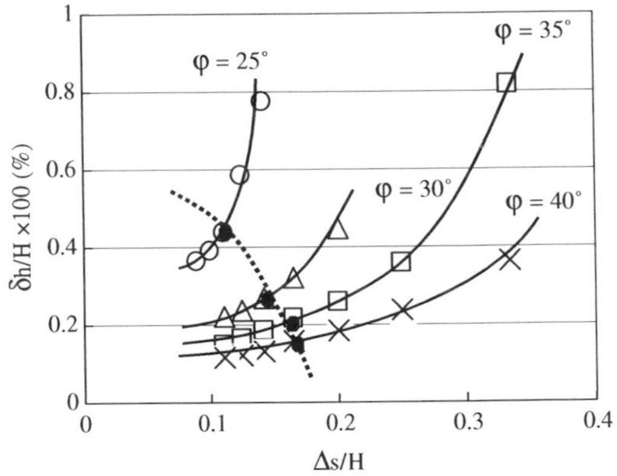

Figure 11. Relationship between deformation at top of slope and space of reinforcement.

the wavy shape of the displacement curve disappears is decreasing with increasing the internal friction ϕ.

Figure 10 shows relationship between the number of reinforcement which the wavy shape of the displacement curve disappears and ϕ. The vertical axis of the figure represents the spacing between reinforcements (Δs) divided by the height of the slope (H = 10 m).This figure can be considered to show the relation between the maximum spacing that the restraint effect works and the angle of internal friction of the slope. Since the object of study in this research was 10-m slopes, Figure 11 is applicable to reinforced cut slopes of approximately 10 m in height. This figure also clearly indicates that the spacing of reinforcements in ordinary slopes of sandy soil (with $\phi = 30$–$40°$) is about $\Delta s/H = 0.15$–0.17 (actual spacing of $\Delta s = 1.5$–1.7 m). As most actual worksites use spacings of 1–2 m between reinforcements, the above results imply that the restraint effect works sufficiently under the actual design.

4 DEFORMATION OF REINFORCED SLOPE AND RESTRAINT EFFECT

Usually, execution management in a reinforcement works is carried out while observing the extent of displacement that occurs during the project. Fig. 11 shows the results obtained from the FEM analysis for predicting the relationship between the spacing of reinforcements ($\Delta s/H$) and the horizontal displacement ($\delta h/H \times 100\%$) of the top of the slope. The spacing of reinforcements shown in Fig. 10 that a complete restraint effect of deformation works are plotted using solid circles on the lines of constant ϕ, and a dashed line connects the each solid circles. This Figure indicates that the restrain effect works sufficiently, if the

Table 3. Provided horizontal deformation for management of safety by Japan Highway Public Corporation.

	Safety level (%)	Caution level (%)	Unsafe level (%)
Soil	$\delta h/H \leqq 0.20$	$0.20 \leqq \delta h/H \leqq 0.40$	$0.40 < \delta h/H$
Soft rock	$\delta h/H \leqq 0.15$	$0.15 \leqq \delta h/H \leqq 0.30$	$0.30 < \delta h/H$
Hard rock	$\delta h/H \leqq 0.10$	$0.10 \leqq \delta h/H \leqq 0.20$	$0.20 < \delta h/H$

spacing between reinforcements that is used at the appropriate value of soil ϕ (Fig. 11) is less than the values indicated by the dashed line in the figure.

Table 3 shows the levels of deformation provided in the stability management of cutting procedures mandated by Japan Highway Public Corporation for work on slopes. Comparing the values shown in the table for soil with the present results shown in Fig. 11, Fig. 11 shows that at a low internal friction angle of 25° the reinforcement restrains deformation; however, the predicted deformation ($\delta h/H$) exceeds 0.4%, an unsafe level according to the regulations. In contrast, to meet the safe level of ($\delta h/H$) $\leq 0.2\%$ for soils with $\phi = 30°$, the normalized spacing of reinforcements ($\Delta s/H$) must be less than 0.1; in soils with $\phi = 35°$, $\Delta s/H \leq 0.17$ is acceptable. In other words, for soils with internal friction angles in the range of 30–40°, the values given in Fig. 11 will satisfy the required safety levels given in Table.

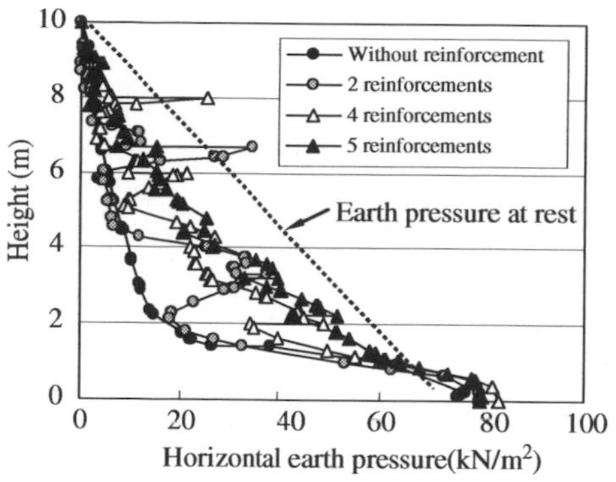

Figure 12. Distribution of horizontal earth pressure in reinforced area.

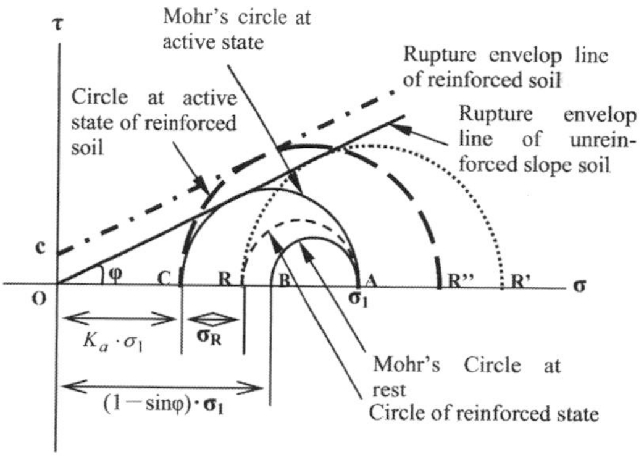

Figure 13. Expression of Mohr's stress circles for reinforced slope.

5 STRESS CONDITIONS IN THE REINFORCED AREA

Next, we observe the restraint effect of deformation from the viewpoint of the stress condition within the reinforced slope. Figure 12 shows the horizontal stress distribution on a plane that passes through the top of the reinforced slope with internal friction $\phi = 40°$. As seen in the cases with 2 or 4 reinforcements, high magnitudes of horizontal soil stress occur locally in the vicinity of the bars, whereas no such high stresses are present in the spaces between members. These local minima in the horizontal stress are of exactly the same significance as the minima seen in the results in Figs. 7–9. For large numbers of reinforcements (5), no local high earth pressure points were observed and the pressure was uniformly high. The figure simultaneously shows the distribution of earth pressure at rest (coefficient of static earth pressure, $K_0 = 1 - \sin\phi$). The stress condition of the slope without the reinforcement reaches to the active state, and the stress conditions with reinforcement remain the intermediate position between that in the active and rest conditions. Since deformation is more highly restrained with increasing numbers of reinforcements, the reinforcements are clearly holding high soil pressures. Thus, when soil deformation is restrained by reinforcing material, the soil zones become interlocked, and it considers that variations in the stress condition depend on the characteristic of dilatancy of slope soil.

Mohr diagrams were constructed to describe the stress conditions and the mechanics of the deformation restrained effect in the reinforced area (Fig. 13). The cohesion of the slope soil was neglected for simplicity. Since the stress condition must be constant in the static condition regardless of whether reinforcing is present, it is represented by a circle whose diameter is AB (point A at x = maximum principal stress = vertical stress σ_1'; point B at x = minimum principal stress = $(1 - \sin\phi) \cdot \sigma_1$). When the slope is deformed by making a cutting, the horizontal soil pressure is reduced and deformation occurs. The horizontal soil pressure in the case of without reinforcement drops during deformation, and the stress condition is represented by a circle whose diameter is AC that the soil is in the active state. When a reinforcement is installed, the stress condition can be represented by a circle whose diameter is AR, where R is a point on the x-axis somewhere between B and C. Thus, the horizontal stress condition is held at a higher level than the active pressure by the amount σ_R (see Fig. 13). From the viewpoint of soil strength, this σ_R plays a role that is equivalent to suction in unsaturated soil. When the minimum principal stress is at point R, the Mohr circle for the soil elements at rupture has a diameter RR′. As with the suction effect, CR″ becomes the circle that represents the apparent strength, and this becomes the Mohr stress circle for the reinforced soil. This means that the increase in strength in the reinforced soil can be estimated as the increase in *c* in the slope soil.

6 SUMMARY

This study presented investigations of the mechanisms of restraint effect of deformation in soil slopes with the use of reinforcements. The following results were obtained.

(1) A distinct element method was carried out to examine the restrained zone of soil that is influenced by the reinforcements. When the reinforcements have sufficiently closely spaced, their effects are mutually additive, the restrained effect of the soil over a wider volume.

(2) Use of the finite element method also revealed the restrained zone of influence of the reinforcements. The relationship between internal friction angle and the spacing of the reinforcement was stated quantitatively.

(3) The spacing of the reinforcement which slopes first began to affect the restraint effect was investigated. In soils with internal friction angles of 30–40°, restraint of deformation was observed at spacings of $\Delta s/H = 0.15$–0.17 (spacings of 1.5–1.7 m for a slope of height $H = 10$ m).

(4) Slope deformations maintain the stable levels that are designated by Japan Highway Public Corporation when the reinforcements are installed at the spacing found to provide restraint effect of deformation in the present research.

(5) Observations of the stress condition in the reinforced area of the slopes revealed uniformly high horizontal soil stresses when deformation was restrained.

REFERENCES

Nishigata, T., Nishida, K. & Kuramochi, K. 2004. Study on Reinforcement Mechanism and Design Method of Soil Nailing in Cut Slope, Jour. of the Society of Materials Science, Vol. 53, No. 1, pp. 1–4.

Okuzono, S., Nagao, T., Okamura, M., Innan, S. & Yamauchi, H. 1984. Experiment on Reinforcement by Soil Nailing, Proc. of 27th Japan National Conf. on Soil Mechanics and Foundation Engineering, pp. 1167–1168.

Okabayashi, K., Kawamura, M., Okamura, M., Minami, M. & Kitayama, K. 1998. Relation Between the Wall Displacement and Optimum Amount of Reinforcements on the Reinforced Retaining Wall, Proc. of 33rd Japan National Conf. on Soil Mechanics and Foundation Engineering, pp. 2411–2412.

Ninomiya, Y., Ochiai, H., Yasufuku, N. & Omine, K. 2002. Confining Effect of Geogrid-reinforced Soil and Its Application to Design Method, Proc. of 37th Japan National Conf. on Soil Mechanics and Foundation Engineering, pp. 1711–1712.

Tayama, S., Ogata, K., Nagayoshi, T. & Takeuchi, T. 2000. Stability Management Based on Deformation of Slope Excavated with Soil Nailing, Jour. of Construction Management and Engineering, JSCE, No. 644, VI-46, pp. 113–122.

New Horizons in Earth Reinforcement – Otani, Miyata & Mukunoki (eds)
© 2008 Taylor & Francis Group, London, ISBN 978-0-415-45775-0

An in-depth numerical analysis of 25 m tall Reinforced Earth wing walls, built back-to-back and supporting a bridge approach

K.M. Truong & S. Aziz
The Reinforced Earth Company, USA

N. Freitag
Terre Armée Internationale, France

ABSTRACT: The project on State Route 288 in Richmond, Virginia, USA includes several Reinforced Earth walls (generically called Mechanically Stabilized Embankment or MSE walls) (Figure 1). The 25 m high walls supporting a bridge approach at Interstate 64 are the tallest MSE walls in the state. They also have a slender back-to-back configuration, with a 15 m separation. Due to the unusual height and slenderness of the structure, the Virginia Department of Transportation required a numerical analysis to verify the walls stability prior to construction. Field instrumentation and monitoring were also implemented. Based on his numerical analysis and the collected field data, the state's design consultant published a paper that raised questions about the validity of the current design practice for back-to-back walls. In response, The Reinforced Earth Company conducted a new numerical analysis study of a more representative model of the MSE structure. The results, reported herein, validate the current design methods and improve our understanding of back-to-back MSE wall behavior.

1 DESCRIPTION OF THE PROJECT

Figure 2 illustrates the slenderness of the structure, Wall 1, the tallest at 25 m, and Wall 5 (hidden behind wall 4 and parallel to Wall 1) are the subject of this study. The wall facing consists of cruciform shape concrete panels 1.5 m × 1.5 m × 0.14 m thick. The soil reinforcement is made of ribbed steel strips 50 × 4 mm. Figure 3 shows the elevation and plan lay-out of the bridge and approach walls. Figure 4 shows the walls unfolded elevations.

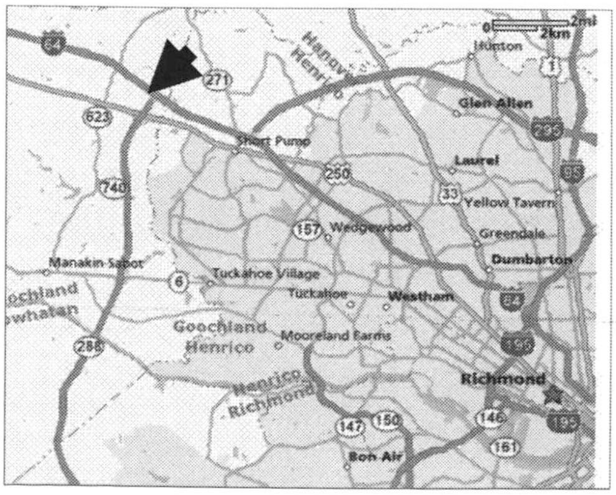

Figure 1. Project Location, West Side of Richmond.

Figure 5 shows the upper 15 m of wall 1 is backed by wall 5. The soil reinforcements from both walls overlap over 14 m within this zone. The wall 5 design procedure followed the guideline established by the

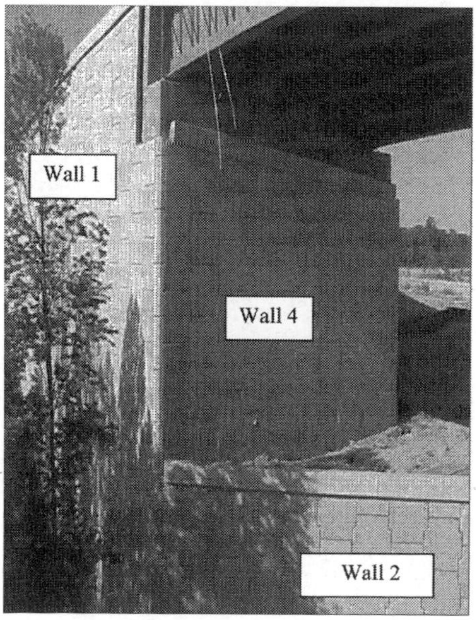

Figure 2. Completed project.

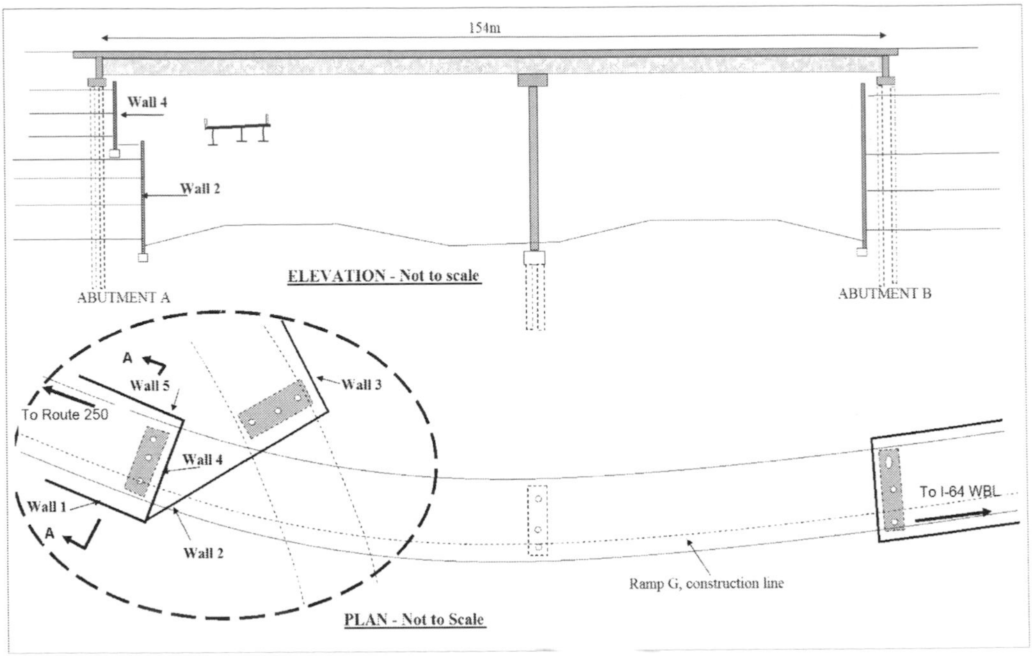

Figure 3. Walls lay-out – Elevation and plan.

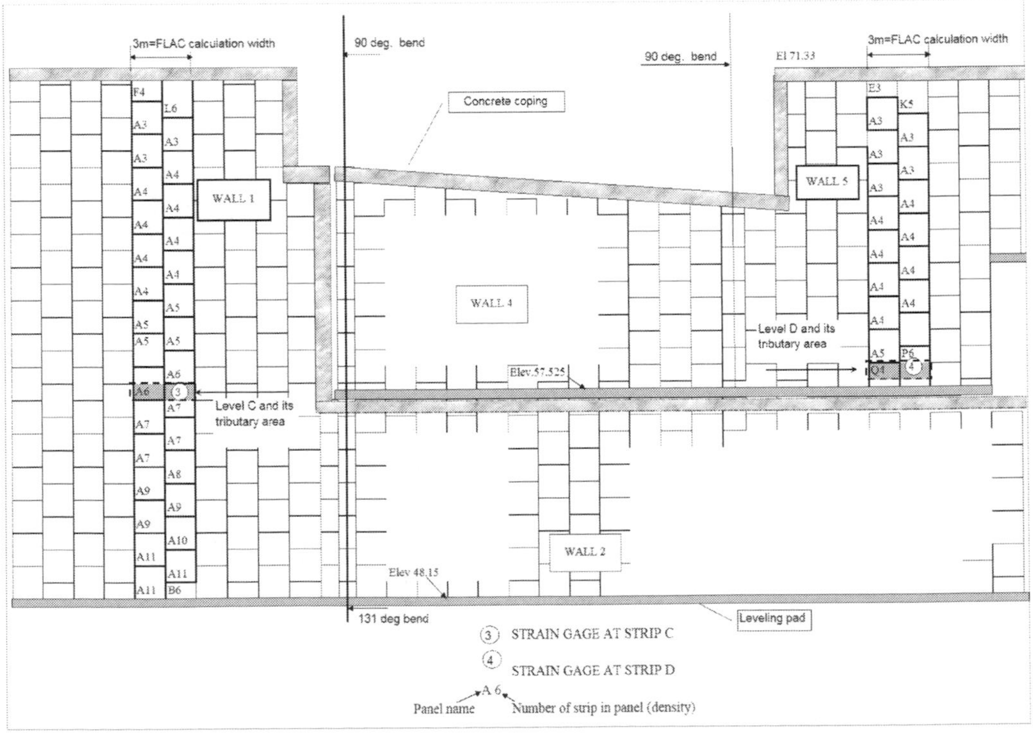

Figure 4. Walls unfolded elevation.

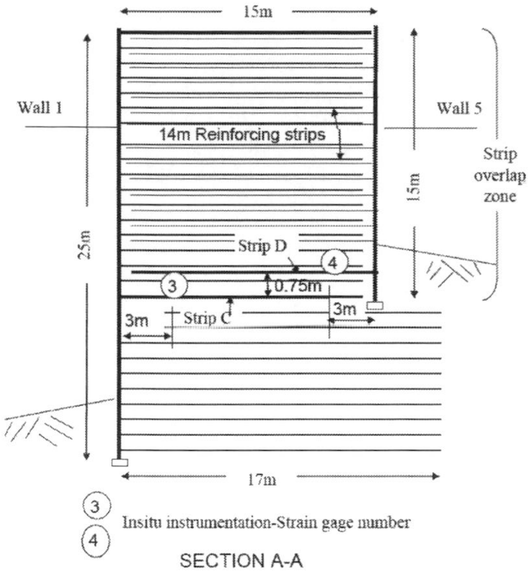

Figure 5. Section A-A through walls 1 & 5 with location of strain gages 3 & 4 shown.

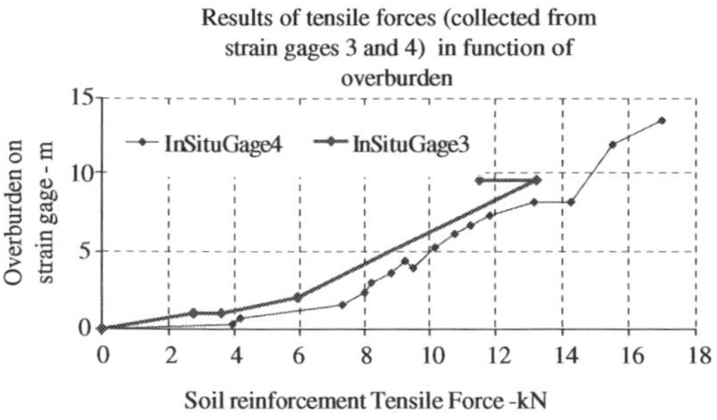

Figure 6. Results collected from strain gages 3 and 4.

US Federal Highway Administration which allows a reduction of the lateral earth pressures based on the overlap length of the soil reinforcement. However, wall 1 was conservatively designed as a stand–alone wall (non back-to-back) over its entire height. Strip C is tied to wall 1 and carries the strain gage 3. Strip D is tied to wall 5 and carries the strain gage 4. It is 0.75 m above strip C. Both strain gages are approximately 3 m away from their respective panel facings. These locations were selected based on the assumption they would record the maximum tension in their respective reinforcing strips C and D.

Figure 6 shows that at an approximate depth of 10 m, strain gage 4 (wall 5) recorded a 30% higher tensile force than strain gage 3 (wall 1). This recording of forces prompted the questions by the state's consultant in his paper. Does it mean that the lateral

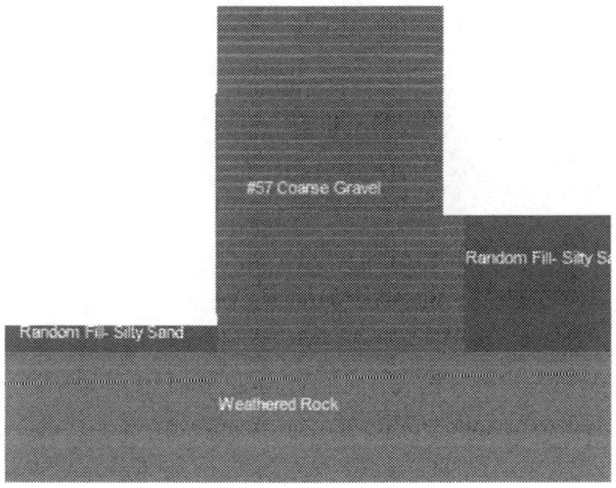

Figure 7. FLAC model with 3 soil types.

Table 1. Soil Type Properties.

	1-Foundation	2-Select fill	3-Random fill
Type	Mohr-Coulomb	Mohr-Coulomb	Mohr-Coulomb
Young's modulus	1.14 GPa	156 MPa	104 MPa
Poisson's ratio	0.3	0.3	0.3
Density	2450 kg/m^3	1650 kg/m^3	2000 kg/m^3
Cohesion	0	0	1.0 kPa
Friction angle	35°	40°	33°
Dilation/Tension	0	0	0

pressure on wall 5 is larger than that on wall 1? If so, is the standard practice in the USA for the design of these back-to-back walls still adequate? To answer to these questions, we created an accurate FLAC model (Figure 7) representing the section A-A shown on Figure 5.

2 THE FINITE DIFFERENCE MODEL

RECO used FLAC 5.0 to create the model shown in Figures 5 and 7. FLAC (Fast Lagrangian Analysis of Continua) is a two-dimensional explicit finite difference program for engineering mechanics computation. The model was based on the values in Tables 1, 2, 3 and 4.

The select fill used is coarse aggregate. The model was built from the leveling pad up, in lifts of 0.375 m, re-creating the same construction sequences as in situ. The levels of soil reinforcements spaced at 0.75 m vertically were inserted every other lift. The strips are allowed to slip to mobilize their resistance in friction. Each layer of fill was computed with the application of a uniform live load of 12 kPa to simulate the compaction operation then recomputed after the removal

Table 2. Physical Properties of the High Adherence Steel Reinforcing Strips.

Modulus of elasticity	2.1×10^5 MPa
Strip width/thickness	50 mm × 4 mm
Calculation width	3 m
Number of strips per calc. width	12 max- 3 min*
Max. apparent coefficient of friction f*	2.0
Minimum f*	$\text{Tan}(\phi) = 0.84$
Tensile strain	0.12
Shear stiffness	1.0×10^6 kPa
Transition Confining pressure	120 kPa
Tensile/Compressive force	24 kN

*See Figure 4 for the densities in both walls

Table 3. Physical Properties of Concrete Panels.

Modulus of elasticity (concrete with creep)	10 GPa
Panel thickness	0.14 m
Moment of inertia	2.3×10^{-4} m^4

Table 4. Physical Properties of Interface between Concrete Panels and Select Fill.

Normal stiffness	1×10^3 MPa
Shear stiffness	1×10^3 MPa
Friction angle	26°

of it. The interface between the panels and the fill was modeled with an angle of friction of 26 degrees.

3 THE RESULTS FROM THE FLAC MODEL

3.1 *Tensile forces in strips*

Figure 8 show the FLAC generated total tensile forces on the tributary areas at levels C and D. At a depth of 10 m, the lateral force on level C is 30% higher than that on level D because the tributary of Level C (half of an "A" panel = 1.5 m tall) is 30% larger than that of Level D (half of a "Q" panel = 1.125 m tall). This finding reconfirms the current principle of calculating $T = \sigma A$, where T = total tensile force, σ = lateral stress and A = panel tributary area at the strip level.

Although the total lateral force at level C (wall 1) is larger by 30% than that at level D wall 5, level C contains 50% more strips than level D (Figure 9). Thus the tensile force in each strip at level C is smaller than that in level D (Figure 10).

Figures 8, 9, 10 and table 5 show that when the wall construction is completed, the strips in wall 5 at level D get a tensile force 28% higher than those in wall 1 at level C. This finding concurs with the instrumentation results at strain gages 3 and 4.

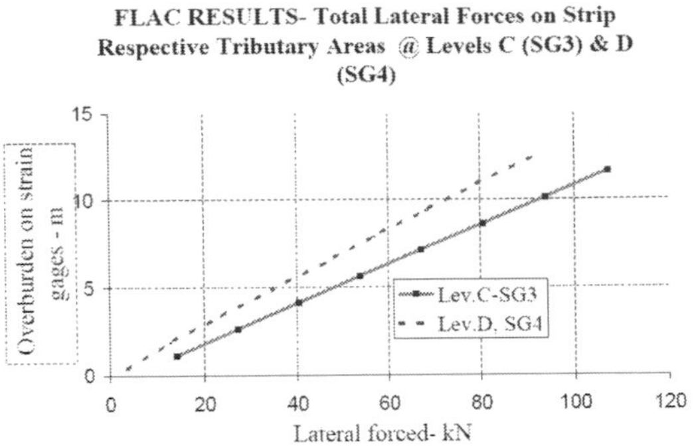

Figure 8. Total lateral forces on tribute areas of strips at levels C and D.

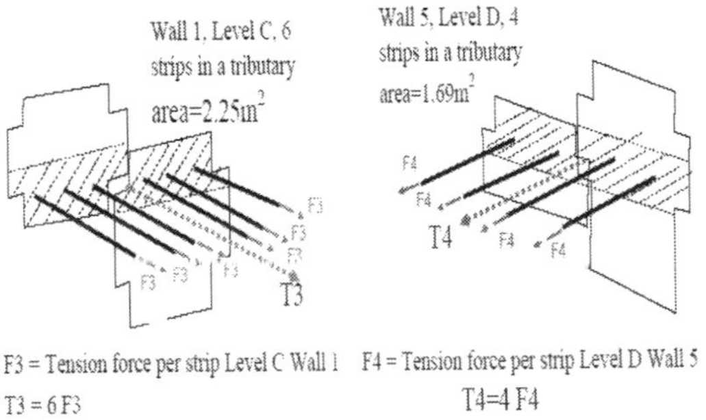

Figure 9. Existing densities within the respective tributary areas: 6 strips at level C wall 1, 4 strips at level D wall 5.

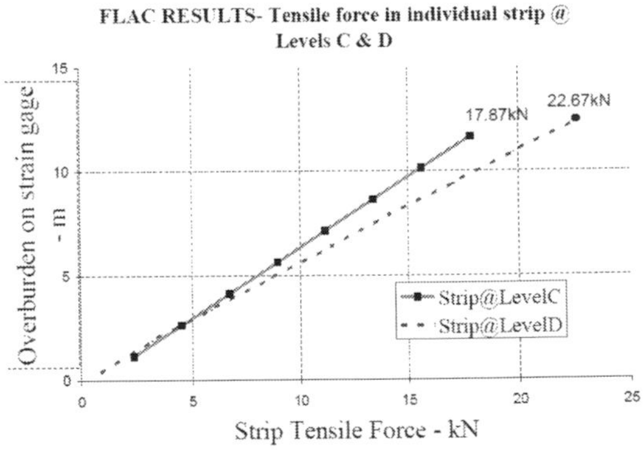

Figure 10. Tensile forces in one strip at levels C and D at consecutive backfilling steps.

Table 5. Summary of FLAC final results (walls completely built).

	Level C, SG3		Level D, SG4
Tributary area	2.25 m^2		1.69 m^2
Number of strips	6	>	4
Total lateral forces	T3 = 107.19 kN	>	T4 = 90.66 kN
Tensile force/strip	17.87 kN = F3	<	22.67 kN = F4

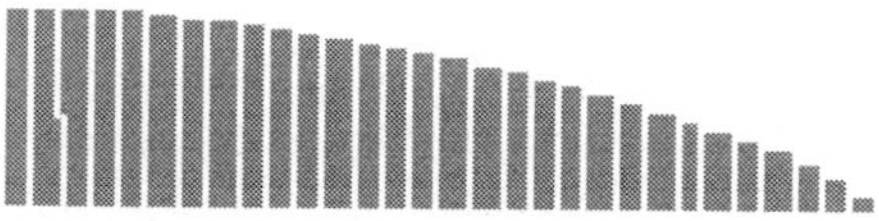

Figure 11. FLAC results-Typical tensile force diagram in a reinforcing strip in the back-to-back zone.

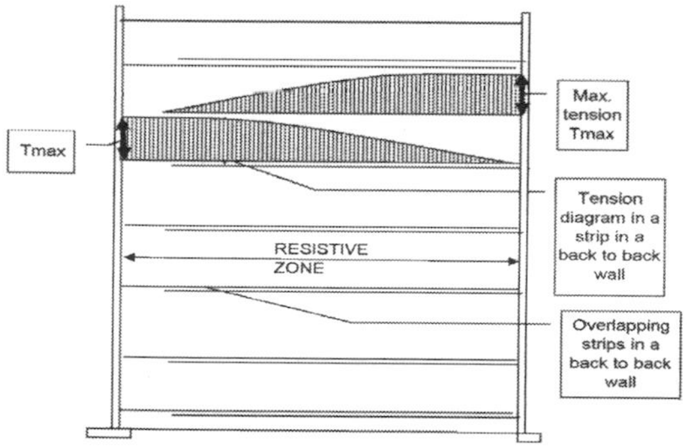

Figure 12. No active zone in a back to back wall, line of maximum tension coincides with the vertical panel facings.

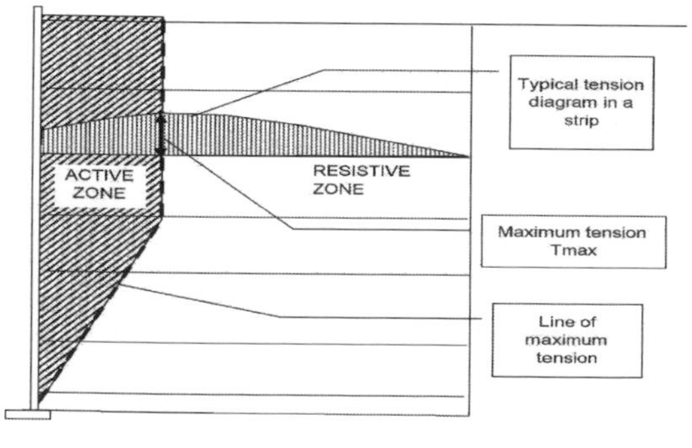

Figure 13. Active zone and line of maximum tension in a standard, stand-alone wall.

3.2 *Typical tensile force diagram in a reinforcing strip in a back to back wall*

Figure 11 shows the tensile force in the strip reaches a maximum at the face of the panel, where there is no more overlapping effect. This demonstrates that in a back-to-back wall, the overlapping effect did not allow the formation of the active zone (Figure 12), as in the case of a standard stand-alone (non back-to-back) wall, as per the long accepted theory of MSE wall behavior (Figure 13)

The free end of strip is allowed to slip to mobilize the necessary resistance in bond with the surrounding soil. Therefore the tension at the free end is always

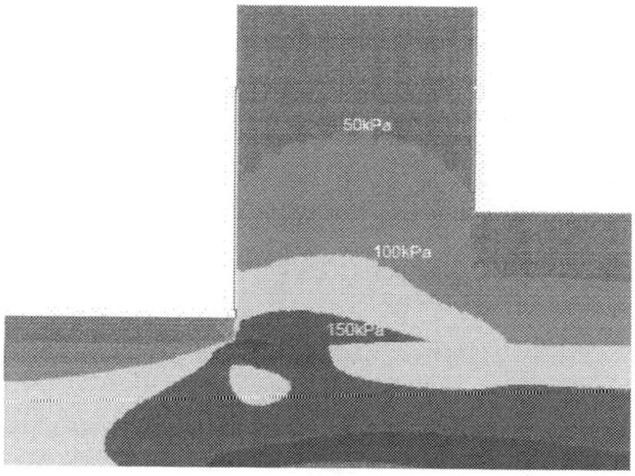

Figure 14. Contours of lateral pressures on partially back-to-back walls (our study case).

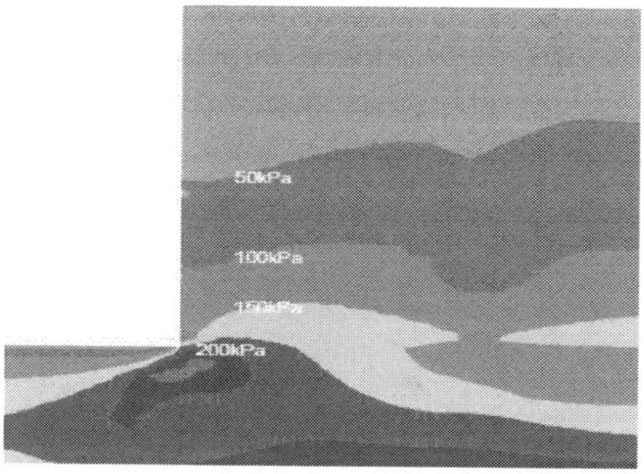

Figure 15. Contours of lateral pressures in a stand-alone wall.

equal to zero. At 3 m away from the panel facings, the strain gages 3 and 4 recorded about 90% of their respective maximum tensile forces.

3.3 *Comparison of lateral stresses on back-to-back walls versus stand-alone wall*

Figure 14 shows the contours of lateral earth pressures in walls 1 and 5. To compare the effect of back-to-back walls on the lateral pressures, a second model for a stand-alone wall was created (Figure 15). The models show that the lateral pressures and tensile forces at 15 m below the top of the stand-alone wall are about 5% higher than those at the same level in our study case (60% back-to-back). At the bottommost reinforcement layer of the walls, the pressures in the stand-alone wall are 30% higher than those in our study case. It reconfirms that the lateral pressures and reinforcement tensile forces in partially or entirely back-to-back walls are lower than those in stand-alone wall (Figure 16).

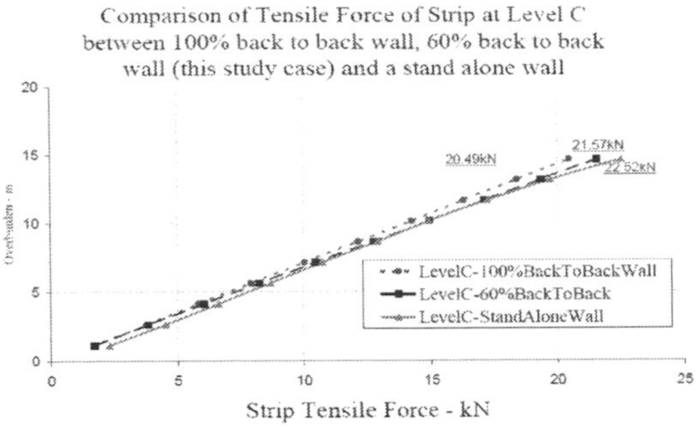

Figure 16. At 15 m below top of wall, the tensile force in each strip in the stand alone wall is 10% higher than that in a 100% back to back wall, and 5% higher than that in a 60% back to back (our study case).

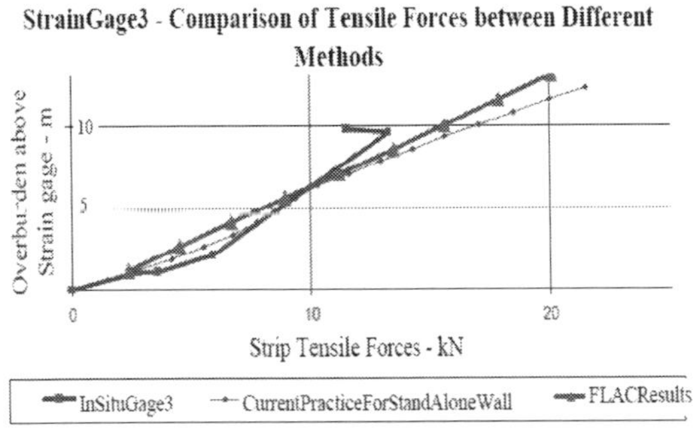

Figure 17. Graphs of tensile forces at strain gage 3 obtained by the 3 methods mentioned above.

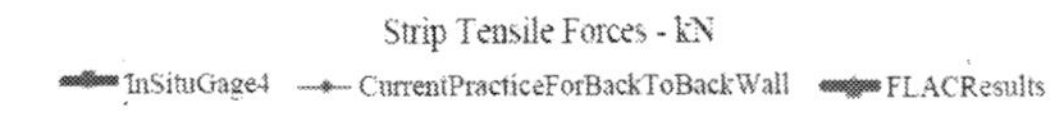

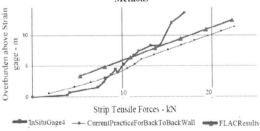

Figure 18. Graphs of tensile forces at strain gage 4 obtained by the 3 methods mentioned above.

3.4 *Comparison of the Strain gage results with outputs from FLAC and the current pratice in the USA*

Let's compare the tensile forces resulting from 3 methods: a) strain gages, b) FLAC analysis and c) the current practice in the USA.

For overburden less than 6 m, Figures 17 and 18 show that the in-situ strain gages 3 and 4 recorded higher tensile forces than those from the FLAC model. This is caused by the point loads generated by the wheels of the heavy compaction equipment, whereas in the FLAC model, the compaction equipment was implemented as a uniform pressure of 12 kPa over the entire Reinforced Earth area. Nevertheless, the recorded tensile forces (of about 11 kN) are only 1/3 of the maximum allowed tensile force of 32 kN. For overburden greater than 6 m (strain gage 3 overburden = 14.875 m, strain gage 4 overburden = 14.125 m), fig. 17 and 18 show that the standard practice and FLAC calculates higher tensile forces in the soil reinforcement than those collected in situ.

4 CONCLUSIONS AND FINDINGS

4.1 *Conclusions*

We conclude the current standard practice for the design of back-to-back walls is validated through the field instrumentation and our FLAC numerical analysis. The following principles remain valid:

1. The total tensile force "T" of all strips in a tributary area "A" is equal to the lateral stress σ at the strip level multiplied by this tributary area. $T = \sigma A$
2. The lateral stresses in back-to-back walls are lower than those in stand alone wall.

4.2 *Findings*

In back-to-back walls where soil reinforcements overlap more than 90% of their lengths, the maximum tension in each reinforcement level occurs at the panel facing. The overlapping effect does not allow the formation of the active zone (Fig. 12). Therefore the entire length of the soil reinforcement can be used to calculate its frictional resistance against pull-out. The overlapping effect also reduces the lateral pressures in partially back-to-back walls, by as much as 30% less than those in stand-alone wall. In light of this finding, we plan on an in-depth parametric study, which will lead to significant savings in the soil reinforcement quantities in future walls of similar condition.

REFERENCES

Performance of MSE walls supporting bridge foundations. (Farouz et al. 2002), GeoTrans Conference, California, USA.

Federal Highway Administration. Reinforced Soil Structures. Design and construction guidelines.

New Horizons in Earth Reinforcement – Otani, Miyata & Mukunoki (eds)
© 2008 Taylor & Francis Group, London, *ISBN 978-0-415-45775-0*

Parametric analysis of a 9-m high reinforced soil wall with different reinforcement materials and soil backfill

B. Huang
GeoEngineering Centre at Queen's-RMC, Department of Civil Engineering, Queen's University, Kingston, Ontario, Canada

K. Hatami
School of Civil Engineering and Environmental Science, University of Oklahoma, Norman, Oklahoma, USA

R.J. Bathurst
GeoEngineering Centre at Queen's-RMC, Department of Civil Engineering, Royal Military College of Canada, Kingston, Ontario, Canada

ABSTRACT: A verified FLAC model is used to investigate the influence of three different soil materials in combination with three different reinforcement materials on the behavior of otherwise identical modular block walls 9 m in height. The soils are a high quality sand backfill and two lower-quality c-ϕ materials. The reinforcement properties correspond to a uniaxial HDPE geogrid, a woven polyester geogrid and a welded wire mesh reinforcement product. The numerical results demonstrate the combined influence of reinforcement stiffness and soil mechanical properties on wall response. The numerical results show that for the same reinforcement type the largest deformations occurred when soil parameters corresponding to a CL backfill soil were assumed. However, the quantitative behavior of walls with ML soil properties and the nominal identical walls built with well-graded sand were similar with respect to wall deformations and reinforcement loads.

1 INTRODUCTION

A series of 11 full-scale instrumented reinforced soil walls has recently been completed at the Royal Military College of Canada (RMC). The test walls were 3.6 m in height. Most were constructed with solid modular block (segmental) facing units placed at a target batter of 8 degrees from the vertical. The walls were 3.3 m wide and the backfill extended about 6 m from the wall toe. The soil in each test was a high quality washed sand. The reinforcement materials were a uniaxial punched and drawn high density polyethylene (HDPE) geogrid, a woven polyester (PET) geogrid and a welded wire mesh (WWM) material. Some walls were constructed with different numbers of reinforcement layers, different facing batters and a wrapped-face configuration. Examples of two recent walls have been reported by Bathurst et al. (2006).

One objective of the physical test program was to generate a comprehensive set of physical test data that can be used to verify numerical codes. These codes can be used in turn to extend the physical test program to reinforced soil wall structures with a wider range of reinforcement materials, reinforcement spacing and length, different soils, facing types and batter angles. The measured results of four RMC walls were used to verify a numerical model using the program FLAC (Hatami & Bathurst 2005, 2006). A unique feature of the verification exercise was that a wide range of measured wall performance was compared to predicted values. These included wall deformations, footing loads, foundation pressures, reinforcement strains and connection loads.

In this paper an updated version of the FLAC code is now used to investigate the influence of three different soil materials on the behavior of nominally identical walls to the 3.6-m high RMC walls with a modular block facing, but extended to a wall height of 9 m. The soils were taken as a high quality sand backfill and two lower-quality c-ϕ soils. Three different reinforcement materials are used in the simulations.

2 NUMERICAL MODELS

2.1 *General*

The numerical simulations were carried out using the computer program FLAC (Itasca 2005). Sequential bottom-up construction of each segmental wall model and compaction of the soil was numerically simulated

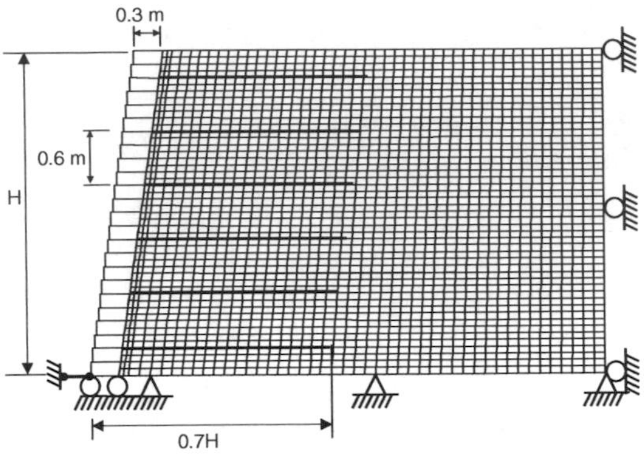

Figure 1. Typical FLAC numerical mesh.

following the procedures described by Hatami & Bathurst (2005). Computations were carried out in large-strain mode to ensure sufficient accuracy in the event of large wall deformations or reinforcement strains and to accommodate the moving local datum as each row of facing units and soil layer was placed during construction simulation. The same reinforcement length to height ratio of 0.7 was used in the numerical models. Similarly, the height of wall to length of soil mass in the cross-plane strain direction was kept the same as the original RMC physical wall models (Figure 1).

2.2 *Materials*

The compacted backfill soil was assumed as a homogenous, isotropic, nonlinear elastic material using the Duncan-Chang hyperbolic model. The elastic tangent modulus is expressed as:

$$E_t = \left[1 - \frac{R_f(1-\sin\phi)(\sigma_1-\sigma_3)}{2c\cos\phi + 2\sigma_3\sin\phi}\right]^2 K_e p_a \left(\frac{\sigma_3}{p_a}\right)^n \quad (1)$$

where: σ_3 = minor principle stress, p_a = atmospheric pressure and other parameters are defined in Table 1. The original Duncan-Chang model was developed for axi-symmetric (triaxial) loading conditions. However, the boundary conditions for the RMC experimental walls, and for most walls in the field are closer to plane strain conditions. Hatami & Bathurst (2005) showed that the Duncan-Chang parameters back-fitted from triaxial tests on the RMC sand under-estimated the stiffness and strength of the same soil when tested in a plane strain test apparatus. This discrepancy is believed to be due to under-estimation of the average confining pressure of the soil specimens using the original Duncan-Chang formulation for bulk modulus, which is a function only of σ_3, specifically:

$$B = K_b p_a \left(\frac{\sigma_3}{p_a}\right)^m \quad (2)$$

Table 1. Soil properties.

Property	Soil Type SW	ML	CL
K_e (elastic modulus number)	950	440	120
K_{ur} (unloading-reloading modulus number)[(1)]	1140	528	144
n (elastic modulus exponent)	0.60	0.40	0.45
R_f (failure ratio)	0.70	0.95	1.00
ν_t (tangent Poisson's ratio)	0–0.49	0–0.49	0–0.49
ϕ (friction angle) (degrees)[(2)]	48	37	17
c (cohesion) (kPa)	2	28	62
B_i/p_a (initial bulk modulus number)	74.8	48.3	21.2
ε_u (asymptotic volumetric strain value)	0.02	0.06	0.13
ρ (kg/m^3) (density)	2250	2030	1900

[1] $K_{ur} = 1.1 \times K_e$; [2] increased from peak triaxial values increased by 10% to adjust to peak plane strain values.

Hatami and Bathurst increased the value of parameter K_e by a factor of two in order to achieve satisfactory agreement with the plane strain test results. In the current study, the bulk modulus formulation proposed by Boscardin et al. (1990) was shown to give accurate predictions of plane strain test results for the RMC sand without using a multiplier applied to the elastic modulus number. The bulk modulus is expressed as:

$$B_t = B_i\left[1 + \frac{\sigma_m}{B_i\varepsilon_u}\right]^2 \quad (3)$$

where: σ_m = mean pressure = $(\sigma_1+\sigma_2+\sigma_3)/3$; B_i and ε_u are material properties that are determined as the intercept and the inverse of slope from a plot of $\sigma_m/\varepsilon_{vol}$ versus σ_m in an isotropic compression test. An additional correction to triaxial test results was to increase the peak friction angle by 10% to reflect plane strain conditions.

Soil properties are summarized in Table 1. These parameters have been taken from Boscardin et al. (1990) with some adjustments. They represent soils with a wide range of mechanical properties. The highest quality soil designated as SW using the Unified Soil Classification System is a sand material. The lowest quality material is the CL soil. However, it is important to note that we are focused on the influence of mechanical strength and stiffness properties. Clearly, the CL soil is a less desirable backfill material from the point of view of ease of compaction, creep and potential loss of strength due to increases in moisture content.

Example stress-strain plots for the backfill soils used in this study are presented in Figure 2. Typical of non-linear elastic hyperbolic models of the type

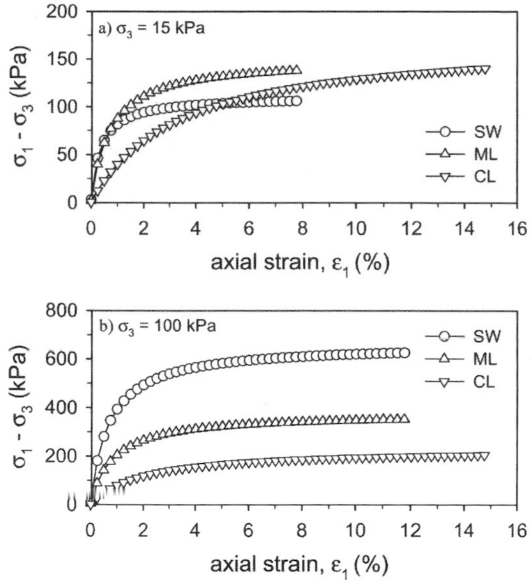

Figure 2. Computed example triaxial compression test results using soil parameters in Table 1.

discussed above, there is no strain-softening behavior. It should be noted that the focus of this paper is on the prediction of wall performance under operational conditions (i.e. working stress conditions). Hence, accurate modeling of the post-peak shear behavior of the soil backfill that can be expected to occur at incipient wall collapse is not a concern. Provided that strains in the reinforcement and soil remain low, the simple constitutive soil model used here is useful. This has been demonstrated in the verification studies reported by Hatami and Bathurst who successfully modeled the RMC walls that used a sand backfill up to reinforcement strain levels as great as 3%. Allen et al. (2003) reviewed a large number of monitored reinforced soil walls with granular backfills and concluded that contiguous failure zones and other obvious signs of poor wall performance did not occur if polymeric reinforcement strain levels were kept to less than about 3%. For cohesive soils, strain hardening can be expected to occur up to much larger strains. Miyata and Bathurst (2007) investigated the performance of monitored reinforced soil walls with c-ϕ backfill soils and noted that good performance was observed when reinforcement strains remained less than 4%. These reinforcement strain levels are benchmark values that can be used as indicators of the onset of contiguous zones of plasticity in numerical models. Hence, they can also be used to establish limits on the confidence that can be placed on numerical results generated using the constitutive soil model employed here.

Three different reinforcement materials were used to represent a range of materials with different

Table 2. Reinforcement properties.

Reinforcement type	Equation 4 and t = 1000 hours: J_o (t) (kN/m)	η(t)	T_f (t) (kN/m)	Ultimate (index) strength $T_y^{(1)}$ (kN/m)
PET	285	0	NA	80
HDPE	1650	0.89	32.5	72
WWM	9300	0	NA	42

Notes: [1] Based on peak strength measured during 10% strain/minute constant-rate-of-strain (CRS) test; NA = not applicable for PET and WWM case with $\eta(t) = 0$.

load-strain-time characteristics and overall stiffness. The stiffness and strength properties for the PET and WWM have been scaled up from the material properties used in the original RMC physical tests and simulations reported by Hatami & Bathurst (2006).

A generalized time-dependent reinforcement tangent stiffness function $J_t(\varepsilon, t)$ proposed by Hatami & Bathurst (2006) was used to characterize the load-strain-time properties of the reinforcement materials:

$$J_t(\varepsilon,t) = \frac{1}{J_o(t)\left(\dfrac{1}{J_o(t)} + \dfrac{\eta(t)}{T_f(t)}\varepsilon\right)^2} \tag{4}$$

where: $J_o(t)$ is the initial tangent stiffness, $\eta(t)$ is a scaling function, $T_f(t)$ is the stress-rupture function for the reinforcement and, t is time. The values assumed in this study are given in Table 2 and correspond to a duration of loading of 1000 hours which is a reasonable elapsed time for a typical reinforced soil wall to come to equilibrium from start of construction (Allen et al. 2003). The relative stiffness of the reinforcement materials increases in the order of PET, HDPE and WWM in this investigation and varies by a factor of 30.

2.3 *Boundary conditions*

The interfaces between dissimilar materials were modelled as linear spring-slider systems with interface shear strength defined by the Mohr-Coulomb failure criterion. Direct shear tests were carried out on the solid masonry blocks used in the reference RMC experimental walls. The value of interface stiffness between modular blocks was selected to match the direct shear test results.

A fixed boundary condition in the horizontal direction was assumed at the numerical grid points on the backfill far-end boundary, representing the bulkheads that were used to contain the soil at the back of the RMC test facility. A fixed boundary condition in both horizontal and vertical directions was used at the foundation level matching the test facility concrete strong

Table 3. Interface properties.

Interface	Value
Soil-Block	
δ_{sb} (friction angle) (degrees)	48, 37, 17*
c_{sb} (cohesion) (kPa)	2, 28, 62*
ψ_{sb} (dilation angle) (degrees)	6, 2, 0
K_{nsb} (normal stiffness) (MN/m/m)	100
K_{ssb} (shear stiffness) (MN/m/m)	1
Block-Block	
δ_{bb} (friction angle) (degrees)	57
c_{bb} (cohesion) (kPa)	46
K_{nbb} (normal stiffness) (MN/m/m)	1000
K_{sbb} (shear stiffness) (MN/m/m)	40
Backfill-Reinforcement	
ϕ_b (friction angle) (degrees)	48, 37, 17*
s_b (adhesive strength) (kPa)	1000
K_b (shear stiffness) (kN/m/m)	1000

*Assumed equal to the friction angle of the backfill soil

floor. The toe of the facing column was restrained horizontally by a very stiff spring element with properties matching those measured at this boundary in the RMC physical tests. Interface properties are summarized in Table 3. The reinforcement-backfill interface properties were selected to prevent slip. There is no information available at present to select suitable quantitative values for any combination of reinforcement and soil. However, experience with the high quality sand used at RMC and measured reinforcement displacements suggests that this is a reasonable assumption for these conditions. In order to keep the interpretation of results as simple as possible, the same no-slip interface was considered for the c-ϕ soil cases in this numerical study. The reader is directed to the paper by Hatami & Bathurst (2005) for details of how the remaining interface material properties were selected.

3 EXAMPLE RESULTS

Figure 3 shows the out-of-alignment wall profiles at the end of construction. The horizontal datum at each elevation is the location of the wall if the facing blocks could be placed at the target 8-degree batter from the vertical without any movement. The data in Figures 3a, 3b and 3c show that the relative qualitative trends in the three profiles in each figure are similar. However, average quantitative deformations are less in the order of PET, HDPE and WWM (i.e. average deformations decrease with increasing reinforcement stiffness). However, at the top of the wall the maximum deformations are independent of reinforcement type. The explanation for this is that the soil controls wall deformation under working stress conditions (i.e. low reinforcement strain levels). The profiles shown here should not be confused with the relative displacement

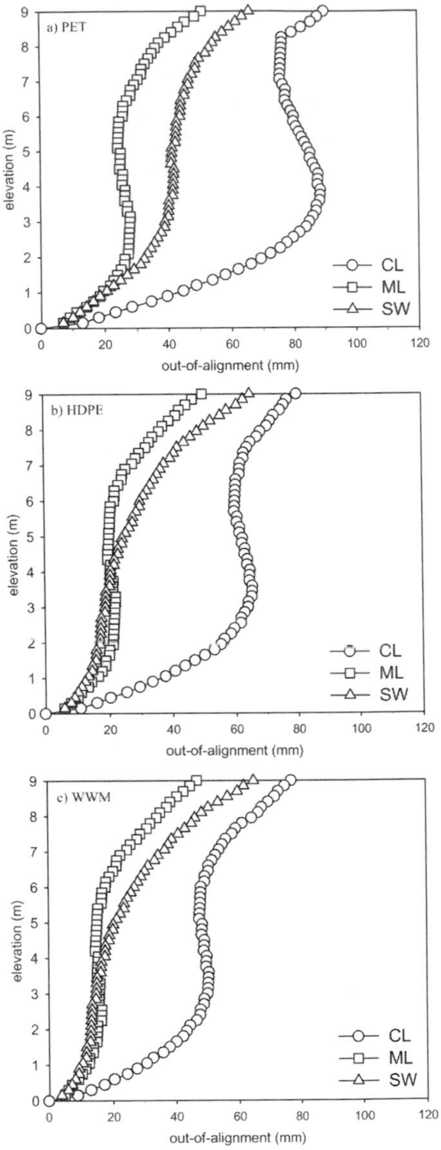

Figure 3. Out-of-alignment wall profiles at end of construction.

of the wall, which is the wall profile that is created when the relative movement of the block is measured from the time of installation. One example is illustrated in Figure 4. The shape of these plots was the same for all three reinforcement materials with most of the relative movement occurred over the bottom third of the wall height. Maximum relative displacements are summarized in Table 4b.

Figure 5 shows plots of connection and horizontal toe loads recorded for each wall at the end of

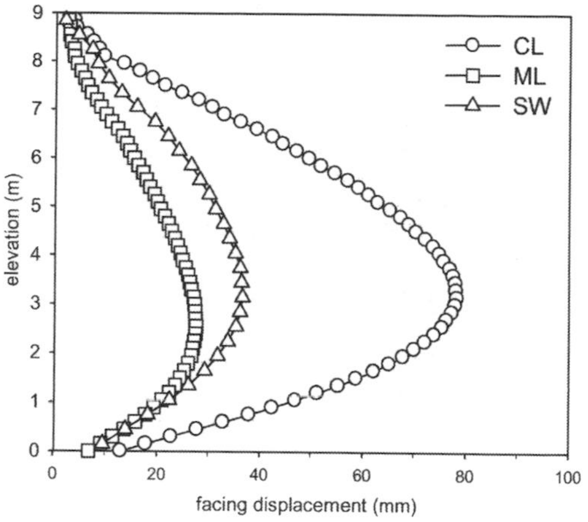

Figure 4. Relative facing displacement (PET reinforcement).

construction. For the polymeric reinforcement cases there are small but detectable lower reinforcement loads for the lower stiffness PET geogrid case compared to the HDPE case. The stiffer WWM wall generated the highest connection loads. The polymeric reinforcement cases show a more uniform distribution of loads while the stiffer metallic reinforcement cases show a trend of increasing load with depth below the top of the wall. Maximum connection loads and horizontal toe loads are summarized in Tables 4c and 4d, respectively. It can be seen in both Figure 5 and the tabulated results that there are significant toe loads generated at the base of the modular block facing as a result of the near-rigid horizontal boundary at this location. For the relatively extensible polymeric reinforcement cases the connection loads are about 30 to 60% of the total horizontal load summed over all connections and the restrained toe (Table 4e). For the stiffer WWM cases the connections loads range from 70 to 80% of the sum of total horizontal loads.

Example relative distributions of reinforcement loads (or strains) from numerical simulations are shown in Figure 6 as bar graphs along each reinforcement layer. For the case of a polymeric reinforcement material, Figure 6a shows that in general, the largest loads in the reinforcement occur at the connections. This is attributed to the effect of relative vertical settlement of the backfill soil with respect to the facing column, which becomes more pronounced with height above the toe. Figure 6b shows the same data but for a WWM reinforced soil wall case. It can be noted that the reinforcement loads are slightly more uniform and propagate deeper into the reinforced soil zone at the lower elevations. This can be attributed to the greater stiffer of the metallic reinforcement. The maximum strains in the reinforcement layers are summarized in Table 4f. The maximum strain levels for

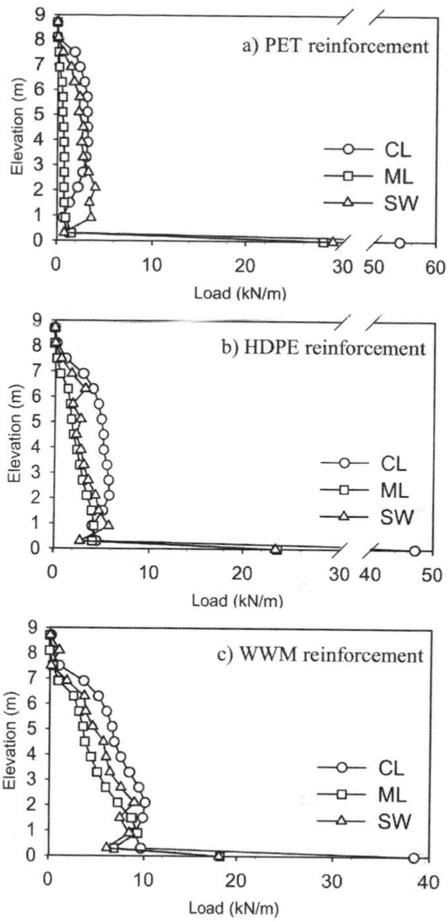

Figure 5. Connection and horizontal toe loads.

the polymeric reinforcement cases are consistent with working stress conditions as discussed in Section 2.2. The maximum strain in the metallic reinforcement is well within the yield strain limit of this reinforcement material (Hatami & Bathurst 2006).

4 CONCLUSIONS

A numerical parametric study is reported for three different soil types and three different reinforcement materials used in the construction of otherwise identical 9-m high modular block retaining walls. The numerical results show that for the same reinforcement type the largest deformations occurred when soil parameters corresponding to a CL backfill soil were assumed. However, the quantitative behavior of walls with ML soil and well-graded sand were similar with respect to wall deformations and reinforcement loads. These results show that a c-ϕ soil with the equivalent

Table 4. Results of numerical simulations.

	Soil type		
Reinforcement	CL	ML	SW
a) maximum out-of-alignment (mm)			
PET	90	51	66
HDPE	80	50	65
WWM	77	47	65
b) maximum relative facing movement (mm)			
PET	78	29	37
HDPE	58	23	16
WWM	46	17	13
c) maximum connection loads (kN/m)			
PET	3.3	1.6	4.1
HDPE	5.9	4.2	5.8
WWM	10.1	9.3	9.0
d) horizontal toe loads (kN/m)			
PET	85.6	37.6	61.2
HDPE	106.9	55.6	63.9
WWM	132.6	80.5	90.5
e) ratio of sum of connection loads to sum of connection loads plus horizontal toe load (%)			
PET	37	26	53
HDPE	56	58	63
WWM	71	77	80
f) maximum reinforcement strains (%)			
PET	1.14	0.56	1.54
HDPE	0.50	0.30	0.36
WWM	0.15	0.11	0.11

mechanical properties of the ML soil in this study, can result in the same quantitatively good performance as a wall built with a purely frictional sand soil. Walls with relatively inextensible (metallic) reinforcement resulted in good performance in all cases but the magnitude and distribution of wall deformations and reinforcement loads were different from the structures with polymeric reinforcement.

Finally, while the performance of the wall models with the weakest and most compressible soil used in this investigation can be judged to have given satisfactory performance, it is important that walls with c-ϕ soils be carefully constructed and compacted, protected from surface water accumulation, and internally well drained so that the mechanical properties of the backfill are not allowed to degrade.

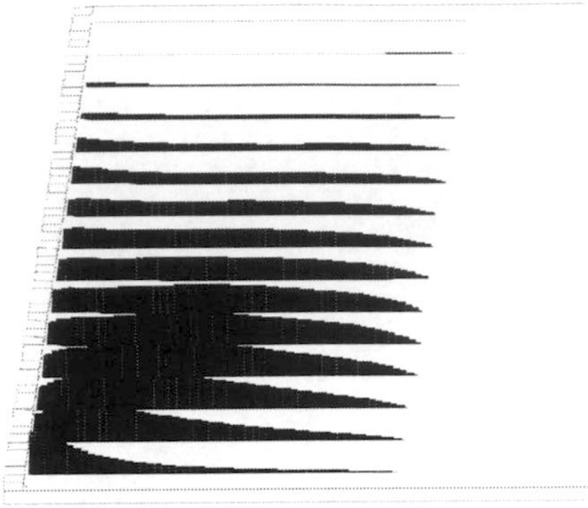

a) PET reinforcement with ML backfill soil (maximum strain = 0.56%).

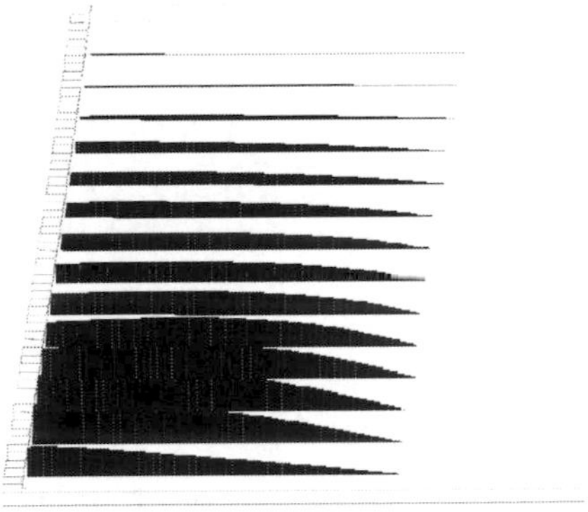

b) WWM reinforcement with ML backfill soil (maximum strain = 0.11 %)

Figure 6. Example distributions of reinforcement load.

REFERENCES

Allen, T.M., Bathurst, R.J., Holtz, R.D., Walters, D.L. & Lee, W.F. 2003. A new working stress method for prediction of reinforcement loads in geosynthetic walls. *Canadian Geotechnical Journal*, Vol. 40, pp. 976–994.

Boscardin, M.D., Selig, E.T., Lin, R.S. & Yang, G.R. 1990. Hyperbolic parameters for compacted soils. ASCE *Journal of Geotechnical Engineering*, Vol. 116, No. 1, pp. 343–375.

Bathurst, R.J., Vlachopoulos, N., Walters, D.L., Burgess, P.G. & Allen, T.M. 2006. The influence of facing rigidity on the performance of two geosynthetic reinforced soil retaining walls. *Canadian Geotechnical Journal*, Vol. 43, No. 12, pp. 1225–1137.

Hatami, K. & Bathurst, R.J. 2005. Development and verification of a numerical model for the analysis of geosynthetic reinforced soil segmental walls under working stress conditions. *Canadian Geotechnical Journal*, Vol. 42, No. 4, pp. 1066–1085

Hatami, K. & Bathurst, R.J. 2006. A numerical model for reinforced soil segmental walls under surcharge loading. ASCE *Journal of Geotechnical and Geoenvironmental Engineering*, Vol. 132, No. 6, pp. 673–684.

Itasca Consulting Group, 2005. FLAC: Fast Lagrangian Analysis of Continua, version 5.0. Itasca Consulting Group, Inc., Minneapolis, Minnesota, USA.

Miyata, Y. & Bathurst, R.J. 2007. Development of K-Stiffness method for geosynthetic reinforced soil walls constructed with c-ϕ soils. *Canadian Geotechnical Journal*, (in press).

New Horizons in Earth Reinforcement – Otani, Miyata & Mukunoki (eds)
© 2008 Taylor & Francis Group, London, ISBN 978-0-415-45775-0

Parametric study of geosynthetic reinforced soil retaining structures

S.J. Chao
National Ilan University, Ilan, Taiwan

ABSTRACT: Geosynthetic reinforced soil retaining structures (GRSRS) are composed of backfill materials and reinforcements, which are relatively complicated considering the soil-structure interaction. The complex soil-reinforcement system of GRSRS can be best analyzed by the finite element method (FEM). Finite element method is used in this study to analyze the geosynthetic reinforced soil retaining structures for more understanding. A parametric study is performed using the finite element model to comprehend the mechanical behavior of geosynthetic reinforced soil retaining structures. The factors affecting the wall performance, including backfill material, wall height, wall inclination, and offset for two-tiered construction technique are investigated. Last of all, design recommendations for GRSRS are proposed.

1 INTRODUCTION

Geosynthetic reinforced soil retaining structures (GRSRS) are used commonly in geotechnical engineering practices in Taiwan recently (Chou, 1992) as well as in the whole world (AASHTO, 1996; FHWA, 1997; CERF, 1998; GEO, 2000). Sand and gravel are preferred to be the backfill materials for constructing geosynthetic reinforced soil retaining structures. However, the soil deposits in the construction site can be any kind of materials. Following the principle of balancing the total amount of cutting and filling to avoid construction pollution, accepting cohesive soils as the backfill materials for the purposes of economical and ecological considerations is unavoidable today in Taiwan. Therefore, the range of the acceptable backfill materials covers between GW and CL nowadays.

On the other hand, GRSRS are composed of backfill materials and reinforcements, which are relatively complicated in considering of the soil-structure interaction. Fortunately, the complex soil-reinforcement behavior of GRSRS can be best analyzed by the finite element method (FEM). In this study, a commercial finite element analysis program PLAXIS is used as a numerical tool to capture the mechanism of GRSRS. PLAXIS is specifically intended for the analysis of deformation and stability in geotechnical engineering projects.

A parametric study is performed using the PLAXIS finite element program to understand the mechanical behavior of reinforced soil retaining structures. The factors affecting the wall performance, including backfill material, wall height, wall inclination, and offset for two-tiered construction technique are investigated. Finally, design recommendations for geosynthetic reinforced soil retaining structures are proposed.

2 PROBLEM DESCRIPTIONS

2.1 *FEM model*

In the finite element numerical model, the GRSRS are assumed to be plain strain condition. The backfill materials are simulated using the Mohr-Coulomb model while the reinforcements simply using the elastic tensile model. The boundary conditions are chosen to be fixed on the bottom for both directions and on the backside for horizontal direction.

A special option termed as ϕ-c reduction is available in PLAXIS to compute safety factors. In the ϕ-c reduction approach, the soil shear strength parameters $\tan\phi$ and c of the soil are successively reduced until failure of the reinforced soil retaining structure occurs. The strength of interfaces, if used, would be reduced in the same way.

The factor of safety (FS) of the GRSRS is used to define the value of the soil strength parameters at a given stage in the analysis:

$$FS = \frac{\tan\varphi_{input}}{\tan\varphi_{reduced}} = \frac{c_{input}}{c_{reduced}} \qquad (1)$$

where the strength parameters with the subscript *input* refer to the properties entered in the material sets and parameters with the subscript *reduced* refer to the reduced values used in the analysis. The strength parameters are successively reduced repeatedly until failure of the structure occurs. At this point the factor of safety is given by:

$$FS = \frac{\text{available strength}}{\text{strength at failure}} \quad (2)$$

This approach resembles the method of calculation of safety factors conventionally adopted in slip-circle analyses. When using ϕ-c reduction in combination with advanced soil models, these models will actually act as a standard Mohr-Coulomb model, since stress-dependent stiffness behavior and hardening effects are excluded. The stress-dependent stiffness modulus at the end of the previous step is used as a constant stiffness modulus during the ϕ-c reduction calculation.

2.2 *Material properties*

The typical properties of the backfill materials used in the simulation are chosen as follows (unless mentioned elsewhere in this paper): the unit weight of the sand = 19.5 kN/m^3, the Elastic modulus E = 18900 kN/m^2, the Poisson ratio $\nu = 0.3$, the friction angle $\phi = 27 \sim 48°$; on the other hand, the unit weight of the clay = 17 kN/m^3, the Elastic modulus E = 9800 kN/m^2, the Poisson ratio $\nu = 0.35$, the friction angle $\phi = 0°$, while the unconfined compression strength $c_u = 25 \sim 50$ kN/m^2. The typical backfill material properties of the sand and the clay used in the PLAXIS program are listed in Table 1.

The geosynthetic reinforcements are slender objects with a normal stiffness for tension but with no bending stiffness. That is to say, reinforcements can only sustain tensile forces and no compression. Finite element methods have been used extensively to study this type of elements. In PLAXIS program, the geosynthetic reinforcements are modeled as Geotextile elements. The only material property of the Geotextile element is elastic axial stiffness EA entered in units of force per unit width. Geotextile element cannot sustain compressive forces. The material property of geosynthetic reinforcement used in this study, based on the test conducting in the laboratory, EA = 6000 kN/m.

Table 1. Backfill material properties of the sand and the clay.

Parameter	Name	Sand	Clay	Unit
Material model	Model	Mohr-Coulomb	Mohr-Coulomb	–
Soil unit weight	γ	19.5	17	kN/m^3
Young's modulus	E	18900	9800	kN/m^2
Poisson's ratio	ν	0.3	0.35	–
Cohesion	c	0	25 ~ 50	kN/m^2
Friction angle	ϕ	27 ~ 48	0	°

3 PARAMETRIC STUDY

3.1 *Backfill material*

Taking into consideration the application of GRSRS, it would be the most benefit for using the site soil as the backfill materials. Therefore, a variety of backfill materials constructing the GRSRS are simulated by FEM to evaluate the influence on FS. A typical profile of the geosynthetic reinforced soil retaining wall is shown in Figure 1.

The influence on the factor of safety due to different values of cohesion of the backfill material is investigated first. According to Hunt (1985), the cohesion of clay ranges form 25 kN/m^2 to 50 kN/m^2. Therefore, 10 different cohesions with an increment of 2.5 kN/m^2 are used to perform the finite element analysis to predict the FS of the GRSRS.

The predicted results for different clayey backfill materials are shown in Figure 2. From Fig. 2, it can be seen that the FS for the reinforced clay wall is in the range of 2 ~ 4. The values of FS of the GRSRS increase with increasing the values of cohesions.

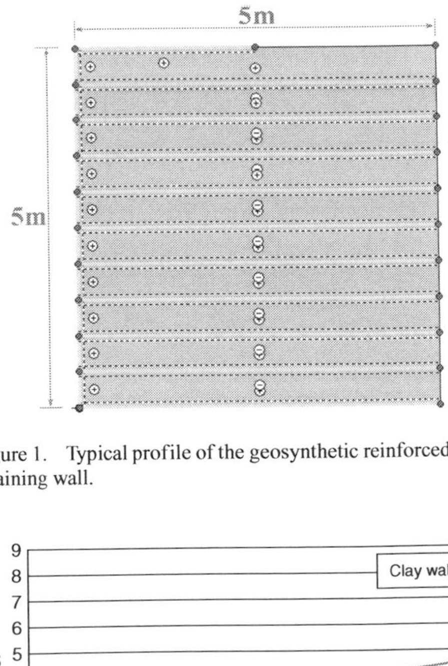

Figure 1. Typical profile of the geosynthetic reinforced soil retaining wall.

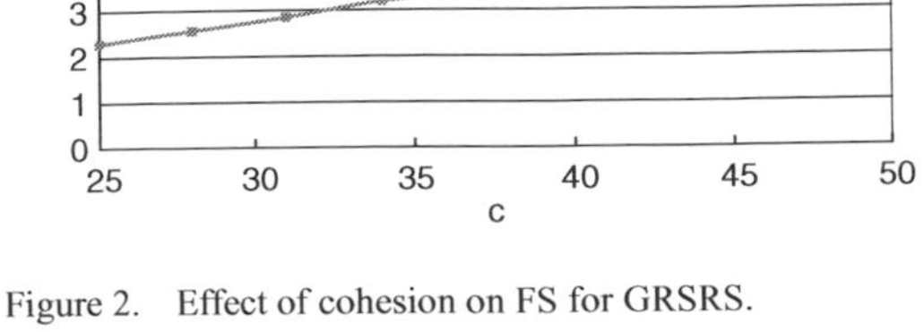

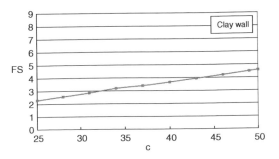

Figure 2. Effect of cohesion on FS for GRSRS.

The influence on the factor of safety due to different values of soil friction angles of the backfill material is also studied. According to Das (1994), the soil friction angle typically ranges between 27° ~ 48°. Similarly, 10 different friction angles with an increment of 2° are selected to perform the analysis to predict the FS of the GRSRS.

The predicted results for different granular backfill materials are illustrated in Figure 3. From Fig. 3, we can find that the values of FS of the GRSRS increase with increasing the values of soil friction angles quite linearly.

3.2 *Wall height*

The designed wall height of the GRSRS in the geotechnical practice in Taiwan are challenging worldwide all the time. Thus, the influence on FS due to wall height is necessarily to be considered.

The limitation of the wall height is assumed to be 10 meter in this study. According to the general regulation, the length of the reinforcement has to be equal to 70% of the wall height at the least. Therefore, 10 different walls in height ranging from 10 m to 1 m with a constant width of 7 m are analyzed.

In order to study the effect of the wall height on the values of FS for the GRSRS, the value of soil friction angle is set to be 45° for sandy backfill material, while the value of cohesion is set to be 30 kN/m^2 for clayey backfill material at this point. The predicted results for the granular and the clayey retaining walls are both shown in Figure 4. From Fig. 4, it can be clearly seen that higher wall dimension obtains lower FS for the GRSRS.

In addition, the stability of sandy retaining wall is generally safer than that of clayey retaining wall by examining Fig. 4. It is noted that the value of FS of sandy retaining wall is lower than that of clayey retaining wall under the extreme condition with very low wall dimension. The reason for this result is that under such condition, the overburden pressure is not big enough to provide adequate frictional resistance for sandy reinforced wall comparing to the contribution of cohesion for clayey reinforced wall. Anyhow, the present regulation for the limitation of wall height for each individual tier (5 meter) is rather conservative according to the predicted results for both the sand walls and the clay walls.

3.3 *Wall inclination*

The effect of wall inclination of the GRSRS is also an interesting issue and worthy to do some research. Usually, we can separate the GRSRS from slope and wall by a boundary wall inclination angle as 70°. In this section, 10 different values of wall inclination angles between 70° and 90° are used. The predicted results with different wall inclination angles for both the sand and the clay retaining walls are shown in Figure 5. From Fig. 5, it can be seen that the steeper wall provides lower values of FS for sand wall. However, the values of FS for clay wall are almost remained unchanged with different wall inclination angle.

3.4 *Offset distance*

Due to the fact that tensile stresses in the reinforcements increase rapidly with height, current design requires multi-tiered system for high GRSRS (Leshchinsky and Han, 2004). Therefore, the effect of

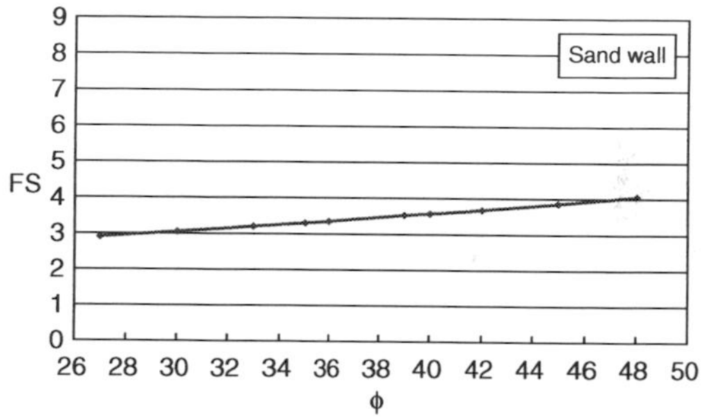

Figure 3. Effect of soil friction angle on FS for GRSRS.

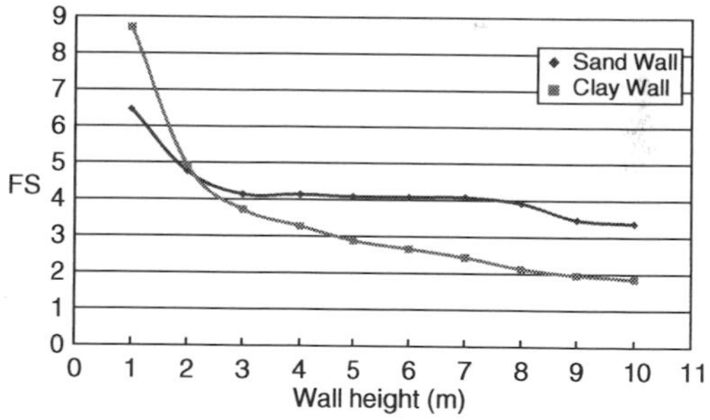

Figure 4. Effect of wall height on FS for GRSRS.

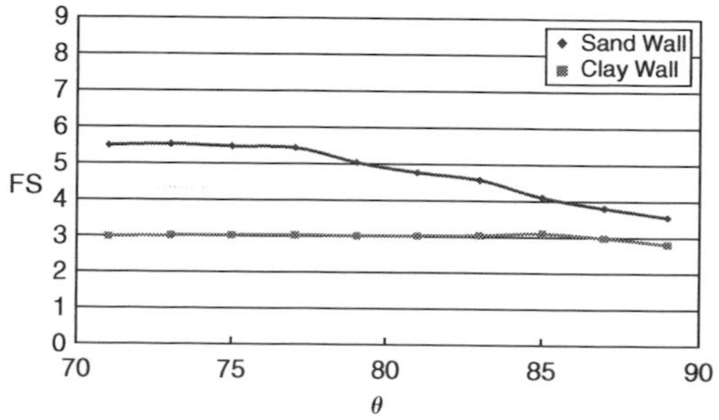

Figure 5. Effect of wall inclination on FS for GRSRS.

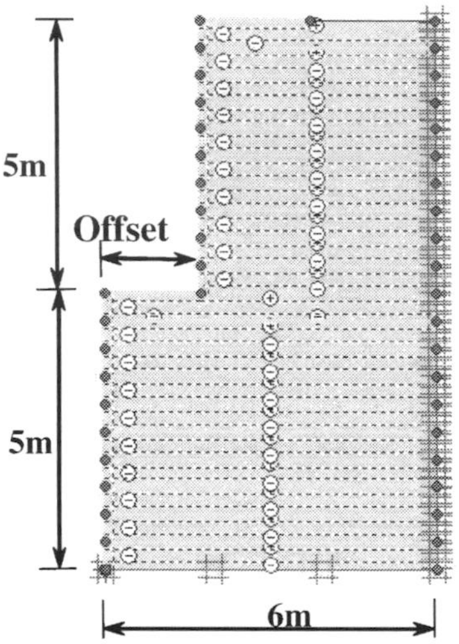

Figure 6. Profile of a two-tiered system for GRSRS.

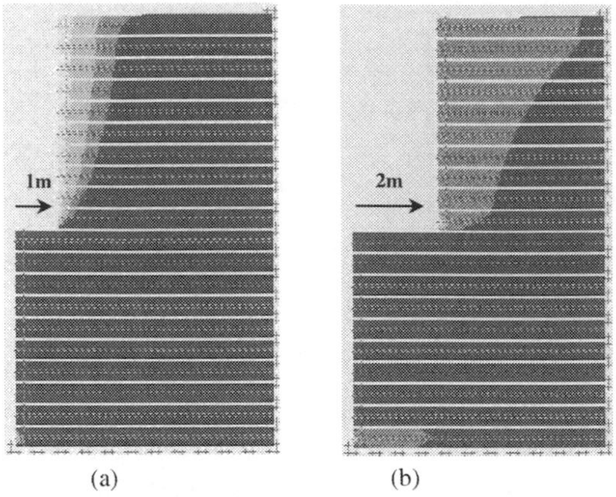

Figure 7. Typical upper tier failure conditions of (a) sand wall and (b) clay wall.

offset distance between adjacent tiers within a GRSRS system is worthy examined in detail. For the purpose of simplification, a two-tiered system is used in this section for both the cases of the sandy reinforced walls and the clayey reinforced walls. Each tier height in this model is set to be a constant value of 5 m as shown in Figure 6.

The effect of offset distance is investigated by performing the finite element analysis to predict the values of FS for both the sand walls and the clay walls with various offsets distances. The offset distances are chosen from 0 m to 2 m with an increment of 0.25 m for each simulation process.

For the sand wall, it can be found that the failure surface always pass through the two-tiered system all the way down to the base of the wall for those cases with offset distance smaller than 0.75 m. when the offset distance increases to 1 m and further, the failure surface can only be found in the domain of the upper tier region as shown in Figure 7(a). On the other hand, for the clay wall, the offset distance needs to be as large as 2 m to reach the condition of purely upper tier failure as shown in Figure 7(b). The offset distance for a clay wall system is thus suggested to be larger comparing that for a sand wall system.

4 CONCLUSIONS

FEM can be utilized to simulate the complicated behaviors of geosysthetic reinforced soil retaining structures. In this study, a commercial finite element analysis program PLAXIS is used as a numerical tool to capture the mechanism of GRSRS and thus provides useful information in detail.

A comparative parametric study for GRSRS, including backfill material, wall height, wall inclination, and offset distance, is carried out and described in detail. Finally, design recommendations for geosysthetic reinforced soil retaining structures are proposed as follow.

1. The predicted results for different clayey backfill materials indicate the values of FS of the GRSRS increase with increasing the values of cohesions.
2. The predicted results for different granular backfill materials prove the values of FS of the GRSRS increase with increasing the values of soil friction angles linearly.
3. The predicted results for the granular and the clayey retaining walls both demonstrate that the higher wall dimension provides lower FS for the GRSRS.
4. Steeper wall provides lower values of FS for sand wall, but not the case for clay wall.
5. The offset distances for two-tiered GRSRS system are suggested larger than 1 m for sand wall, while 2 m for the clay wall. Adopting the suggesting values of the offset distance can guarantee the GRSRS to develop the purely upper tier failure condition and thus reduce the tensile stresses in the reinforcements within the lower tier.

REFERENCES

AASHTO, 1996. Standard Specifications for Highway Bridges, With 197 Interims, American Association of State Highway and Transportation officials, Fifteenth Edition, Washington, D.C, USA.

Chou, N.N.S. 1992. Performance of Geosynthetic Reinforced Soil walls, Ph.D. Dissertation, University of Colorado.

Das, B.M. 1994. Principles of Geotechnical Engineering, PWS Publishing Company.

FHWA, 1997. Mechanically Stabilized Earth Walls and Reinforced Soil Slopes Design and Construction Guidelines, Pub. No. FHWA-SA-96-071.
Geotechnical Engineering Office, The Government of the Hong Kong Special Administrative Region, 2000. Technical Guidelines on Landscape Treatment and Bio-engineering for Man-made Slopes and Retaining Walls, GEO Publication No. 1/2000.
Highway Innovative Technology Evaluation Center, a service center of Civil Engineering Research Foundation, 1998. Guidelines for Evaluating Earth Retaining Systems, CERF Report No. 40334, March 1998.
Hunt, R.E. 1985. Geotechnical Engineering Techniques and Practices, McGraw-Hill Book Company.
Leshchinsky, D. and Han, J. 2004. Geosynthetic Reinforced Multitiered Wall, Journal of Geotechnical and Geoenvironmental Engineering, Vol. 130, No. 12, pp. 1225–1235.

New Horizons in Earth Reinforcement – Otani, Miyata & Mukunoki (eds)
© 2008 Taylor & Francis Group, London, ISBN 978-0-415-45775-0

Influence of interference on bearing capacity of strip footing on reinforced sand

M. Ghazavi & A.A. Lavasan
Civil Engineering Department, K.N. Toosi University of Technology, Tehran, Iran

ABSTRACT: Numerical evaluation of bearing capacity of interfered strip footing on unreinforced and reinforced sandy soils has been performed in this paper using finite difference method based on commercially available code, FLAC^{3D} (Fast Lagrangian Analysis of Continua). The failure criterion for the soil has been assumed to be based on Mohr-coulomb with non-associative flow rule by considering $0 \leq \psi < \varphi$. To ensure the accuracy of the constructed numerical models, the results obtained have been compared with available experimental and theoretical data. This comparison has validated the numerical modeling. Parametric studies have been carried out to determine the best locations for reinforcing layers in the forms of normalized ratios, for example width ratio, depth ratio, geogrid layer distance ratio, etc.). This facilitates to achieve the greatest values for bearing capacity of closely spaced strip footings. The results show that at low footing spacing, the bearing capacity increases sometimes up to three times compared with the case where no reinforcement is used. It has also found out that there is a certain spacing beyond which the bearing capacity decreases with increasing the distance between footings. With further increase in footing distance, the interference effect vanishes.

1 INTROUDUCTION

Due to heavy loads exerted from superstructures to closely constructed shallow foundations on the ground surface. There is interference between footings. The interference between closely spaced footings may have effects on bearing capacity, settlement, and rotation of closely spaced footings.

Each of aforementioned conditions can affect design factors qualitatively. Some studies have been performed to investigate the bearing capacity of interfered footings on unreinforced soil (Stuart, 1962; Das & Larbi-Cherif, 1983 a, b; Graham et al., 1984; Kumar & Saran, 2003; Wang & Jao, 2002). With growing technology, significant promotion has been achieved in soil reinforcement, Thus, the subject of bearing capacity improvement has been of concern significantly. Different types of reinforcement have been used to reinforce soil beneath footings, for instance, metal strips (Binquet & Lee, 1975; Fragaszy & Lawton, 1984; Huang & Tatsuka, 1988), metal bars (Huang & Tatsuka, 1990), rope fibers (Akinmusuru & Akinboladeh, 1981), geotextiles (Guido et al., 1986), and geogrids (Guido et al., 1986; Yetimoglu et al., 1994; Omar et al., 1993a,b; Adams & Collin, 1997; Das & Shin, 1999). These studies have shown more encouragement in the use of geogrid to improve behaviors of spread footing mainly because of the stiffness of geogrid. Further research work has led to developing non-dimensional bearing capacity ratio for soil reinforcement effects showing benefits of reinforcement. This non-dimensional ratio, BCR, is defined as

$$BCR = \frac{q_{u(R)}}{q_u} \tag{1}$$

where $q_{u(R)}$ is the ultimate bearing capacity with soil reinforcement and q_u is the ultimate bearing capacity without reinforcement.

Figure 1 shows two interfered shallow strip foundations of width B supported by a soil reinforced with layers of geogrid. Character N represents the number of geogrid layers. The width of each geogrid layer is denoted by b. Parameter u depicts the depth of the closest geogrid layer from footing. The vertical distance between consecutive layers of geogrid is shown by h. Center to center spacing between two interfered footings is illustrated with Δ.

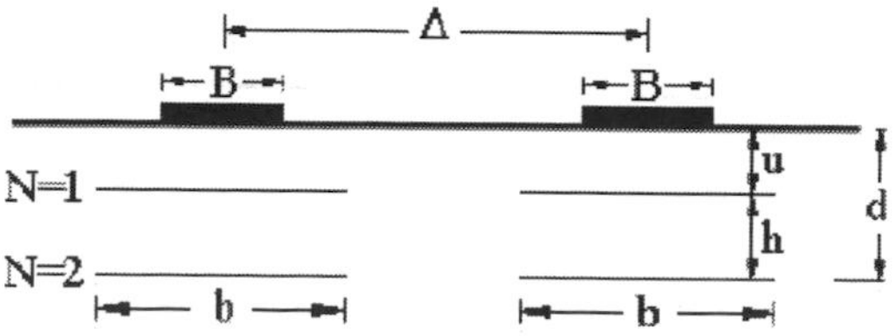

Figure 1. Geometry of two interfered strip footings supported by geogrid-reinforced soils.

To evaluate the bearing capacity of interfered footings on reinforced soil, the interference factor, I_f, may be defined as:

$$I_f = \frac{q_{uN(\text{interfered})}}{q_{u(\text{single})}} \qquad (2)$$

where $q_{uN(interfered)}$ is the ultimate bearing capacity of interfered footing with N layers of reinforcement and $q_{u(single)}$ is the ultimate bearing capacity of same single footing with no reinforcement.

In recent years, some attempts have been devoted to the failure mechanism of reinforced soil (Huang & Tatsuka, 1988, 1990; Yamamoto & Otani, 2002; Michalowski & Shi, 2003). Two different mechanisms are offered. "Deep footing effect" which occurs in soil with short reinforcement and "width slab effect" associated with reinforcement extending considerably beyond the influenced zone by the footing.

The failure mechanism of interfered footings on both unreinforced and reinforced soil has not been considered comprehensively. Thus, this paper focuses on determining the bearing capacity of interfered strip footings on reinforced and unreinforced sand and also investigating the failure mechanism in different condition.

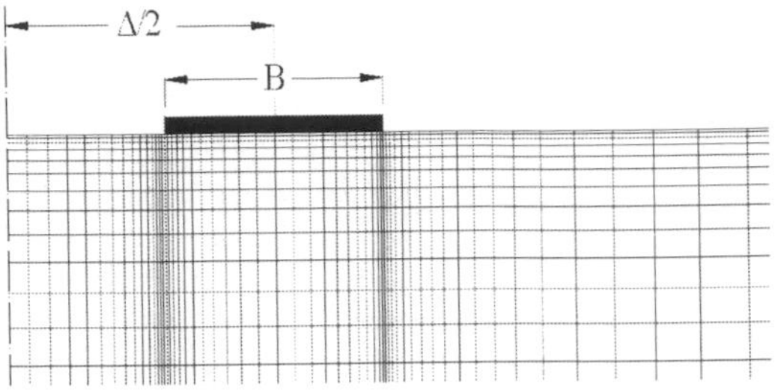

Figure 2. Typical mesh shape used in numerical FLAC model of interfered strip footing.

Table 1. Mechanical properties of soil and reinforcement.

Soil parameters		geogrid parameters	
Bulk modulus	2×10^4 kPa	Elasticity modulus	5.0×10^6 kPa
Shear modulus	1×10^4 kPa	Poisson ratio	0.3
		interface parameters	
cohesion	0.5 kPa		
friction	35°	Stiffness per unit area	2.39×10^6 kN/m³
Dilation	20°	Cohesion	0
		friction	28°

2 NUMERICAL ANALYSIS PROCEDURE

In the present numerical study, finite difference program FLAC^{3D} (Itasca Group, 2002) was used to model strip interfered footings constructed on unreinforced and reinforced sand. It uses an explicit, time marching method to solve the governing field equations. The Mohr-Coulomb failure criterion was used for prediction of soil behavior. Due to the symmetry of the soil-footing system and decrease the analysis time, only half part of the system was simulated. Rigid rough-base footings were assumed in parametric studies. It is assumed that the strip footing has a width of 1 m. To ensure the independency between bearing capacity and both boundary conditions and model dimension, the width and depth of soil-footing system was assumed to be 10B in both lateral and vertical directions, where B is the footing width. A maximum settlement of s = 10%B was applied to all models with a constant velocity of 5×10^{-7} m/step. Typical mesh of interfered model is shown in Figure 2.

2.1 *Soil properties*

Mechanical parameters of the soil, which were used in numerical modeling are presented in Table 1. The difference between ϕ and ψ represents a non associated plastic flow rule which means the plastic potential surface is not identical to the yield surface. Yin et al. (2001), Erickson & Drescher (2002), Frydman & Burd (1997) and De Borst & Vermeer (1984) found that the dilation angle has a significant influence on the numerical estimation of the footing bearing capacity. This dependence is more significant for higher values of the friction angle. According to previous study on determination of bearing capacity with FLAC, more accurate results can be obtained by considering dilation angle of soil about 2/3 friction angle. By using less dilation angle, local shear failure appears and by increasing dilation angle it tends to change to general shear failure. The difference between φ and ψ dictates the use of non-associated flow rule. Mechanical properties of soil are shown in Table 1.

2.2 *Reinforcement properties*

In FLAC^{3D}, the geogrid behaves as an isotropic linear elastic material with no failure limit. A shear directed (in the tangent plane to the geogrid surface) frictional interaction occurs between the geogrid and the soil grids, and the geogrid is slaved to the grid motion in the normal direction.

Because the settlement ratios were also small at failure for both unreinforced and reinforced sand in the analysis (i.e. s/B < 5% at failure), the strains developed in the geogrid reinforcement were likely to be very small, too. Therefore, a constant modulus of elasticity of E = 5.0×10^6 kPa was used for numerical

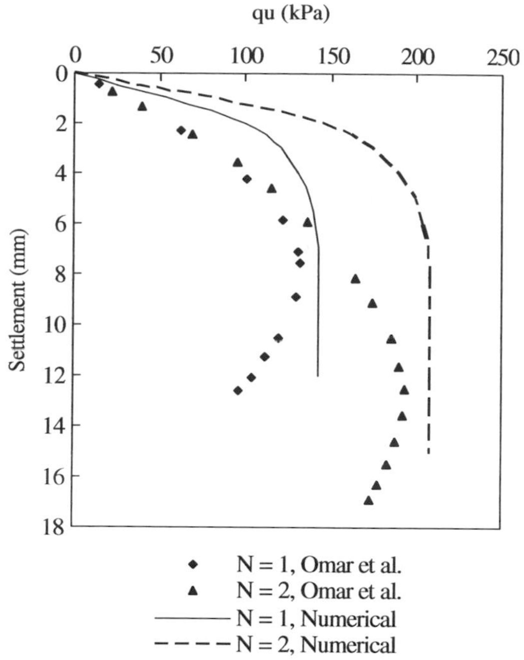

Figure 3. Results comparison of numerical and experimental methods.

analyses. To permit sliding between soil and geogrid, an interface element was used on both sides of reinforcement layers. The shear behavior of the geogrid-soil interface is cohesive and frictional in nature and is controlled by the coupling spring properties of: (1) stiffness per unit area; (2) cohesive strength; and (3) friction angle.

3 VERIFICATION OF NUMERICAL MODELING

To ensure the accuracy and capability of the numerical modeling, laboratory test of Omar et al. (1993 a, b) on bearing capacity of strip footing on reinforced silica sand with geogrid was simulated numerically and the results were compared. They used a $1.1 \times 0.914 \times 0.304$ m tank for their tests. The strip footing was 76.2×304 mm in plan.

The sand had an average dry unit weight of $\gamma d = 17.14\,\text{kN/m}^3$ and a relative density of 70%. These were the same in all tests. The peak friction angle of the sand was 41°. A biaxial polypropylene polymer geogrid with nominal thickness of 1 mm was used for reinforcement. A comparison of numerical and experimental results is presented in Figure 3. As seen, a good agreement between numerical and experimental results exists and this indicates the capability of numerical modeling to predict the behavior of reinforced soil.

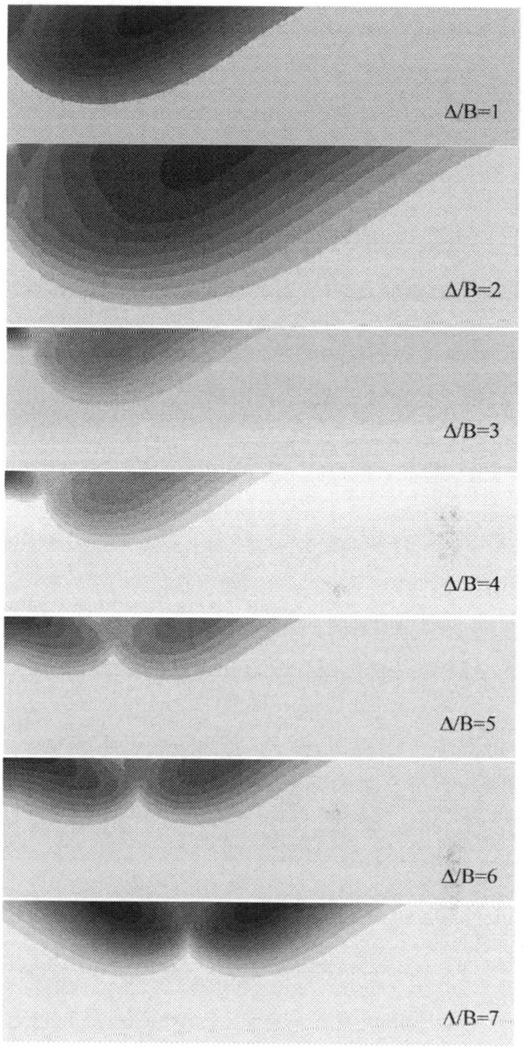

Figure 4. Displacement mechanism at different spaces.

4 ANALYSIS RESULTS AND DISSCUSIONS

4.1 *Interfered strip footing on unreinforced sand*

The first group of numerical analysis was conducted on unreinforced sand. Variations of displacement mechanism at different spacing for half of model were shown in Figure 4. As seen in Figure 4a, when $\Delta/B = 1.0$ (no distance exists between footings) system acts like a single foundation with a width equal to $2B$. The mechanism in Figure 4a coincides to that proposed by Prantdl (1920). At $\Delta/B = 2.0$ (Figure 4b), the "blocking" occurs and both footings act as a single foundation with width more than 2B. an increase in the shape of spiral confirms this postulate. By increasing the distance between two neighboring

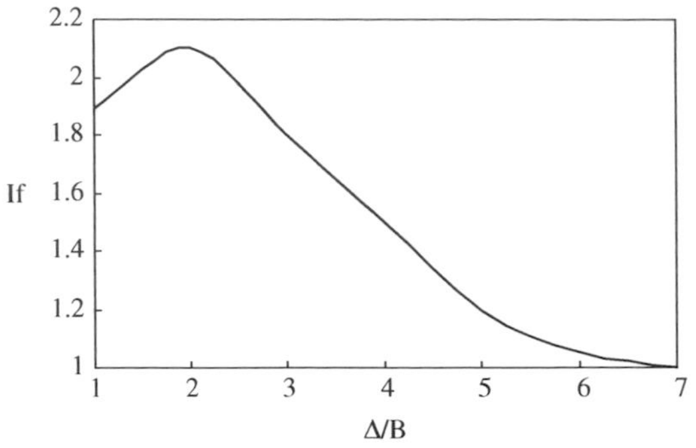

Figure 5. Variation of interference factor at different spacing on unreinforced sand.

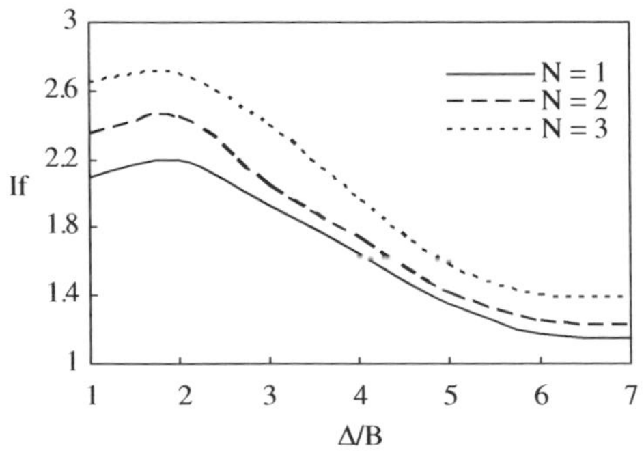

Figure 6. Variation of interference factor at different spacing on short width reinforced sand ($b/B = 1.5$; $u/B = h/B = 0.3$).

foundations, the influence of interference on footing behavior decreases.

The variation of the failure mechanism of unreinforced soil is in accordance with that proposed by Stuart (1962). A fluctuation of interference factor, I_f, at different spacings for interfered strip on unreinforced sand is exhibited in Figure 5. It is obvious from Figure 5 that by increasing Δ/B from 1 to 2, the bearing capacity increases. This is due to the blocking effect on failure mechanism as mentioned before. For Δ/B greater than 2, I_f value decreases gradually and interference effect on the bearing capacity disappears at $\Delta/B > 6$.

4.2 *Interfered strip footing on reinforced sand with short layers of geogrid*

The second numerical analysis group was conducted on reinforced sand with maximum 3 layer of short width geogrid ($b/B = 1.5$). These analyses were performed to evaluate the variation of bearing capacity with different number of reinforcement layers. For these tests, the parameters were $u/B = h/B = 0.3$; $b/B = 1.5$ and $N = 1, 2$ and 3.

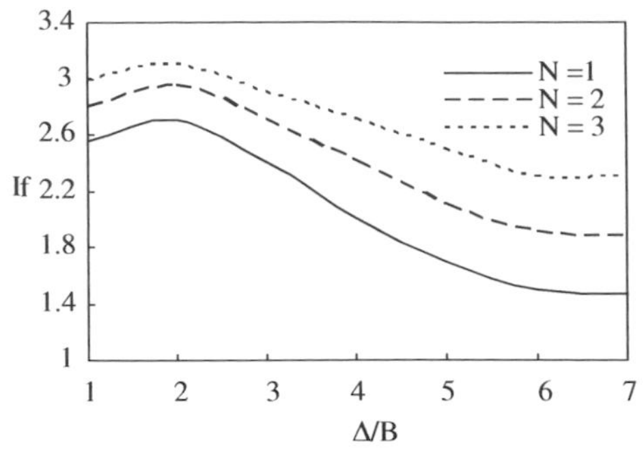

Figure 7. Variation of interference factor at different spacing on short width reinforced sand ($b/B = 5$; $u/B = h/B = 0.3$).

As shown in Figure 6, an increase in the number of reinforcement layers causes the interference factor to increase.

4.3 *Interfered strip footing on reinforced sand with wide layers of geogrid*

The last series of analysis was conducted to evaluate the effect of using wide layers of reinforcement layers on interference factor. These tests were performed on geometric parameters $u/B = h/B = 0.3$; $b/B = 1.5$ and $N = 1, 2$ and 3. The variation of I_f with space between closely spaced strip footings s shown in Figure 7. The difference between vertex quantity of curves and other parts decreases by using wide reinforcement layers.

5 DESIGN CHARTS

Based on preceding discussions, the variation of I_f at different spacings for three combination of soil reinforcing $N = 0$, 1, 2 and 3, design charts of interfered strip footings on unreinforced and reinforced sand are presented. For practical design purposes, the distance between closely spaced footings is normally determined by architectural restriction. In the subsequent section, the first step is to determine the number of reinforcing geogrid for the expected bearing capacity. Therefore, in this charts, variations of interference factor at each spacing ratio for any number of reinforcement between 0 to 3 are illustrated. By the use of these charts, the width and number of reinforcement layers could be determined to receive required bearing capacity (Figures 8–9).

6 CONCLUSIONS

The results of a number of numerical analyses on surface rough strip foundation on unreinforced and

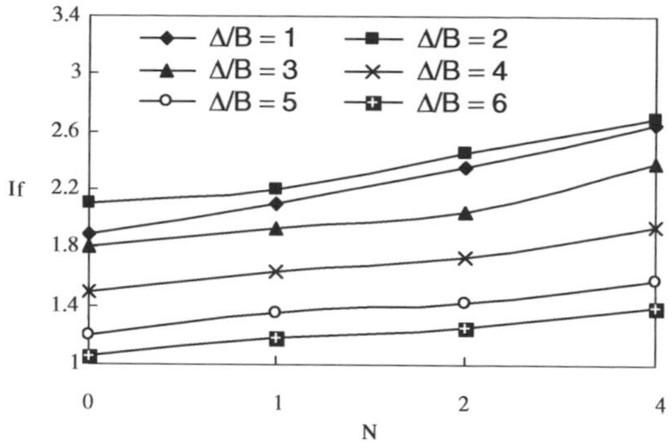

Figure 8. Variation of interference factor with respect to number of reinforcement at different footing spacings.

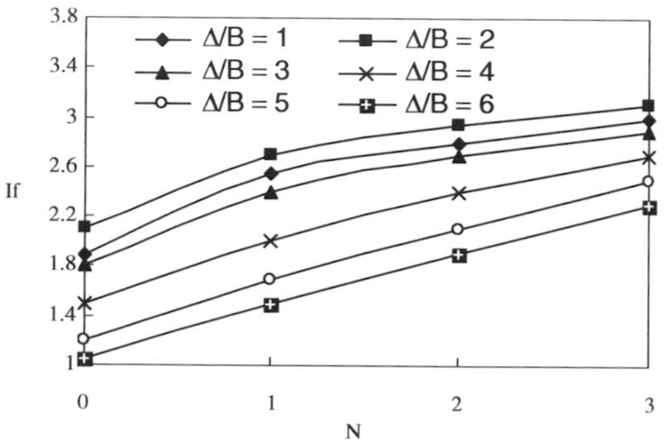

Figure 9. Variation of interference factor at different spacing.

reinforced sand were presented. Based on these analyses, following general conclusions can be pointed out:

1. When two neighboring footings are constructed besides (no distance exists between 2 footings), both footings act like a single footing with 2B width in both unreinforced and reinforced sand.
2. The bearing capacity of interfered strip footing is maximized at $\Delta/B = 2$ for reinforced and unreinforced sand.
3. The influence of interference disappears at footing spacing more than about 6B.
4. Using wide layers of reinforcement leads to greater bearing capacity.

REFERENCES

Adams, M.T. & Collin, J.G. 1997. Large Model Spread Footing Load Tests on 0Geosynthetic Reinforced Soil Foundations. *Journal of Geotechnical and Geoenvironmental Engineering,* ASCE, Vol. 123, No. 1, pp. 66–72.

Akinmusuru, J.O. & Akinboladeh, J.A. 1981. Stability of Loaded Footings on Reinforced Soil. *Journal of Geotechnical and Geoenvironmental Engineering,* ASCE, Vol. 107, No. 6, pp. 819–827.

Binquet, J. & Lee, K.L. 1975. Bearing Capacity Tests on Reinforced Earth Slabs. *Journal of Geotechnical and Geoenvironmental Engineering*, ASCE, Vol. 101, No. 12, pp. 1241–1255.

Das, B.M. & Larbi-Cherif, S. 1983a. Bearing Capacity of Two Closely Spaced Shallow Foundations on Sand. *Soils and Foundations*, Vol. 23, No. 1, pp. 1–7.

Das, B.M. & Larbi-Cherif, S. 1983b. Ultimate Bearing Capacity of Closely Spaced Strip Foundations. *TRB, Transportation Research Record*, Vol. 945, pp. 37–39.

Das, B.M. & Shin, E.C. 1999. Bearing Capacity of Strip Footing on Geogrid-Reinforced Sand. *11th Asian Regional Conference on Soil Mechanics and Geotechnical Engineering*, Hong, Rotterdam, pp. 189–192.

De Borst, R. & Vermeer, P.A. 1984. Possibilities and Limitations of Finite Elements for Limit Analysis. *Geotechnique*, Vol 34, No. 2, pp. 199–210.

Erickson, H.L & Drescher, A. 2002. Bearing Capacity of Circular Footings. *Journal of Geotechnical and Geoenvironmental Engineering*, ASCE, Vol. 128, No. 1, pp. 38–43.

Fragaszy, R.J. & Lawton, E.C. 1984. Bearing Capacity of Reinforced Sand Subgrades. *Journal of Geotechnical and Geoenvironmental Engineering*, ASCE, Vol. 110, No. 10, pp. 1500–1507.

FLAC-Fast Lagrangian Analysis of Continua, 2002, Version 2.1., *ITASCA Consulting Group*, Inc., Minneapolis.

Graham, J. & Raymond, G.P. & Suppiah, A. 1984. Bearing Capacity of Two Closely Spaced Footings on Sand. *Geotechnique*, Vol 34, No. 2, pp. 173–182.

Guido, V.A. & Chang, D.K. & Sweeney, M.A. 1986. Comparison of Geogrid and Geotextile Reinforced Earth Slabs. *Canadian Geotechnical Journal*, Vol. 23, pp. 435–440.

Huang, C.C. & Tatsuoka, F. 1988. Prediction of Bearing Capacity in Level Sandy Ground Reinforced With Strip Reinforcement, *Proceedings of International Geotechnical Symposium on Theory and Practice of Earth Reinforcement,* Fukuoka, Kyushu, Japan, pp. 191–196.

Huang, C.C. & Tatsuoka, F., 1990. Bearing Capacity of Reinforced Horizontal Sandy Ground, *Geotextiles and Geomembrans*, Vol. 9, No. 1, pp. 51–82.

Kumar, A. & Saran, S. 2003. Closely Spaced Footings on Geogrid-Reinforced Sand. *Journal of Geotechnical and Geoenvironmental Engineering*, ASCE, Vol. 129, No. 7, pp. 660–664.

Michalowski, R.L. & Shi, L. 2003. Deformation Patterns of Reinforced Foundation Sand at Failure. *Journal of Geotechnical and Geoenvironmental Engineering,* ASCE, Vol. 129, No. 6, pp. 439–449.

Omar, M.T., Das, B.M., Yen, S.C., Puri, V.K. & Cook, E.E. 1993a. Ultimate Bearing Capacity of Rectangular Foundations on Geogrid-Reinforced Sand. *Geotechnical Testing Journal*, ASTM, Vol. 16, No. 2, pp. 246–252.

Omar, M.T., Das, B.M., Puri, V.K., Yen, S.C. 1993b. Ultimate Bearing Capacity of Shallow Foundations on Sand with Geogrid Reinforcement. *Canadian Geotechnical Journal*, Vol. 30, pp. 545–549.

Prandtl, L. 1920. Uber die Härte plastischer Körper. *Nachr. Königl. Ges. Wissensch., Göttingen; Mathematisch physikalische Klasse,* pp. 74–85.

Stuart, J.G. 1962. Interference Between Foundations with Special Reference to Surface Footings in Sand. *Geotechnique*, Vol 12, No. 1, pp. 15–23.

Wang, M.C. & Jao, M. 2002. Behavior of Interacting Parallel Strip Footing. *Electronic Journal of Geotechnical Engineering*, Vol 7, part A.

Yetimoglu, T., Wu, J.T.H. & Saglamer, A. 1994. Bearing Capacity of Rectangular Footings on Geogrid Reinforced Sand. *Journal of Geotechnical and Geoenvironmental Engineering*, ASCE, Vol. 120, No. 12, pp. 2083–2099.

Yamamoto, K. & Otani, J. 2002. Bearing Capacity and Failure Mechanism of Reinforced Foundations Based on Rigid-Plastic Finite Element Formulation. *Geotextiles and Geomembrans*, Vol. 20, No. 1, pp. 367–393.

Yin, J.H. & Wang, Y.J. & Selvadurai, A.P.S. 2001. Influence of Nonassociativity on the Bearing Capacity of a Strip Footing. *Journal of Geotechnical and Geoenvironmental Engineering*, ASCE, Vol. 127, No. 11, pp. 985–989.

New Horizons in Earth Reinforcement – Otani, Miyata & Mukunoki (eds)
© 2008 Taylor & Francis Group, London, ISBN 978-0-415-45775-0

The counteracting effects of rate of construction on reinforced embankments on rate-sensitive clay

R.K. Rowe & C. Taechakumthorn
GeoEngineering Centre at Queen's-RMC, Department of Civil Engineering, Queen's University, Canada

ABSTRACT: Previous research has shown that for conventional soils, a slower construction rate leads to higher embankment stability, while for rate-sensitive soils faster construction mobilizes higher short-term strength as a result of soil viscosity. Thus for rate-sensitive soils, the critical period with respect to the stability of the embankment is after the end of construction. This paper examines the effects of construction rate and PVDs on the short-term failure height and the role pore pressure dissipation during the construction can have on stability. The interaction between pore pressure dissipation and geosynthetic reinforcement is investigated. The implications with respect to (a) the construction rate and design of PVDs, as well as (b) the development of reinforcement strains and the selection of reinforcement are discussed. Practical implications are highlighted.

1 INTRODUCTION

The behaviour of reinforced embankments constructed on typical soft soils has been extensively studied. However, the effect of the viscous behaviour of rate-sensitive foundations on the short-term and long-term performance of reinforced embankments has only received limited attention.

A study by Rowe et al. (1996), on the behaviours of the Sackville test embankment, showed that in order to accurately predict the responses of embankment on the rate-sensitive soil, a constitutive model considering the viscous behaviours of the soil is essential. Rowe & Hinchberger (1998) proposed and demonstrated that an elasto-viscoplastic constitutive model could adequately describe the behaviour of the Sackville test embankment. Rowe & Li (2002) showed that the long-term stability of the reinforced embankment on the rate-sensitive soil decreases after the end of construction due to delayed build up of excess pore water pressures as a result of soil viscosity. Installation of prefabricated vertical drains (PVDs) has potential to reduce the effect of delayed excess pore pressures. However, the effect of PVDs on the performance of reinforced embankments on the rate-sensitive soil has not been studied.

The objective of this paper is to perform a parametric study of the combined effects of PVDs and geosynthetic reinforcements on the behaviour of embankments on soft rate-sensitive soils. The short-term stability of the reinforced embankment will be compared with the result from the conventional elasto-plastic model. The influence of factors such as the stiffness of reinforcement, rate of construction and spacing of PVDs will be examined with respect to the time-dependent responses of excess pore pressure and reinforcement strains.

2 FINITE ELEMENT MODELING

The finite element program AFENA (Carter & Balaam, 1990), previously modified by Rowe & Hinchberger (1998) to incorporate an elasto-viscoplastic constitutive model, was adopted in this study. Drainage elements (Russell, 1990) implemented by Li & Rowe (2001) were utilized for studying the effects of PVDs. The results presented here were obtained for embankments with 2H:1V side slopes overlaying 15 m of soft rate-sensitive clay above the rigid and permeable sand layer. A typical mesh is shown in Figure 1.

The finite element mesh included a total of 1815 six-noded linear strain triangular elements, with 4003 nodes to discretise the embankment and foundation soils. Two-noded bar elements were used for modeling the reinforcement and two-noded interface joint elements (Rowe & Soderman, 1985) were used for modeling the interfaces. For PVDs modeling, two-noded drainage elements (Li, 2000) were utilized.

The centerline of the embankment and far field boundary, located 100 m away from centerline, were taken to be smooth-rigid boundaries. The bottom boundary of the finite element mesh was assumed to be free draining and rough-rigid. The embankment construction was simulated by gradually turning on

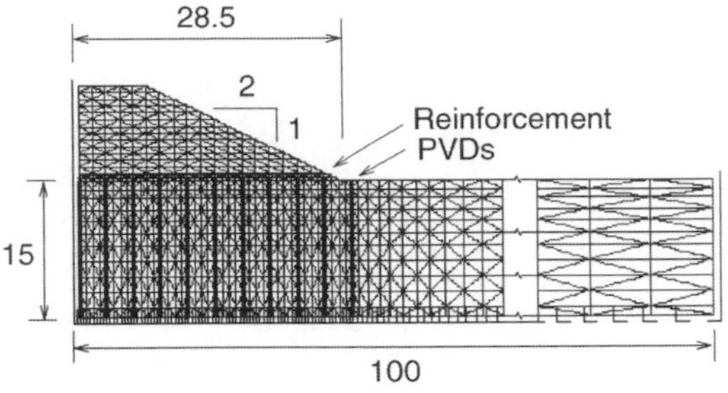

Figure 1. Finite element mesh discretisation.

the gravity of the embankment in 0.75 m thick lifts at a rate corresponding to the construction rate of the embankment. The PVDs fully penetrated the 15m thick clay layer and were arranged in a square pattern with three different spacing; S = 1 and 3 m. Zero excess pore water pressure was assumed along drainage elements. The details of the elasto-viscoplastic constitutive model and drainage element are presented in the previous papers by Rowe & Hinchberger (1998) and Li & Rowe (2001), respectively.

3 CONSTITUTIVE PARAMETERS

3.1 *Foundation soil properties*

The soft rate-sensitive soil examined is denoted here as soil CR1. Constitutive parameters used for soil CR1 are similar to the estimated soil foundation properties at the Sackville test embankment (Rowe & Hinchberger, 1998). The various parameters for CR1 are listed in Table 1.

The hydraulic conductivity of soft rate-sensitive clay was taken to be a function of void ratio as detailed in Rowe & Hinchberger (1998).

3.2 *Backfill properties and construction rate*

The purely frictional granular soil is used to model the embankment fill. The assumed properties are friction angle $\phi' = 37°$, dilation angle $\psi = 6°$ and unit weight $\gamma = 20\,kN/m^3$. The non-linear elastic behaviour of the fill was modeled using Janbu's (1963) equation:

$$\frac{E}{P_a} = K\left(\frac{\sigma_3}{P_a}\right)^m \quad (1)$$

where E is the Young's modulus; P_a is the atmospheric pressure; σ_3 is the minor principle stress and K and m are material constants selected to be 300 and 0.5, respectively.

The construction rates for the two cases examined in this study were 2 m/month and 6 m/month.

Table 1. Details of foundation soil properties.

Soil parameter	Soil CR1
Failure envelope, $M_{N/C}(\phi')$	0.96 (29°)
Cohesion intercept, c_k (kPa)	0
Failure envelope, $M_{O/C}$	0.75
Aspect ratio, R	1.25
Compression index, λ	0.16
Recompression index, κ	0.034
Coefficient of at rest earth pressure, K_o'	0.75
Poisson's ratio, υ	0.3
Reference hydraulic conductivity, k_{vo} (m/s)	2×10^{-9}
Hydraulic conductivity ratio, k_h/k_v	4
Unit weight, γ (kN/m^3)	17
Initial void ratio, e_o	1.50
Viscoplastic fluidity, γ^{vp} (1/hour)	2.0×10^{-5}
Strain rate exponent, n	20

3.3 *Interface parameters and reinforcements*

Rigid-plastic joint elements (Rowe & Soderman, 1985) were used to model the fill/reinforcement and fill/foundation interface. The fill/reinforcement interface was assumed to frictional with $\phi' = 37°$. The fill/foundation interfaces had the same shear strength as the foundation soil at ground surface.

Geosynthetic reinforcements were modeled as an elastic material with tensile stiffness, J, of 0 (no reinforcement), 500 and 1000 kN/m.

4 RESULTS AND DISCUSSIONS

4.1 *Effects of reinforcement and construction rate on the short-term stability of embankment*

The stability of the embankment can be assessed in term of the fill thickness at which the net embankment height above the original ground surface reaches a maximum value as illustrated in Figure 2.

For the four cases examined in Figure 2, the response is initially linear and this is then followed by a non-linear response and eventually failure. The fill thicknesses giving rise to short-term embankment failure are 3.5, 5.8, 6.0 and 7.2 m for case I, II, III and IV respectively. In case I, the unreinforced embankment was constructed on the rate-insensitive foundation having the same properties as those of other cases except for the viscoplastic characteristics to illustrate the effect of rate-sensitivity on short-term behavior. The results from case II and III demonstrate the effects of construction rate on the short-term failure height of the embankment. The faster construction rate resulted in higher short-term failure height of the embankment due to soil viscosity.

The short-term failure height of embankment was significantly improved by using geosynthetic reinforcement. The results from case III and case IV

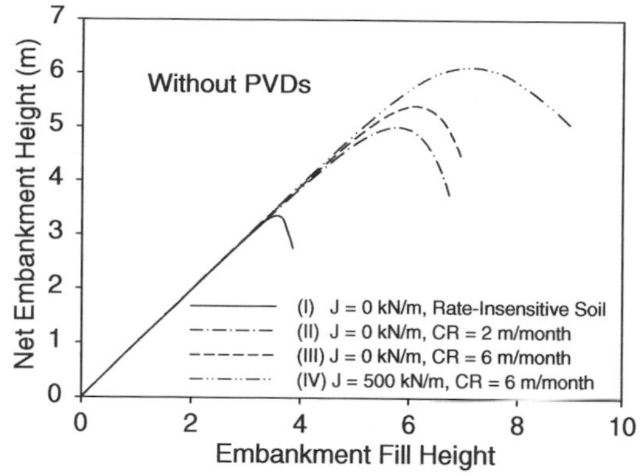

Figure 2. The effects of reinforcement and construction rate on the short-term stability of embankment.

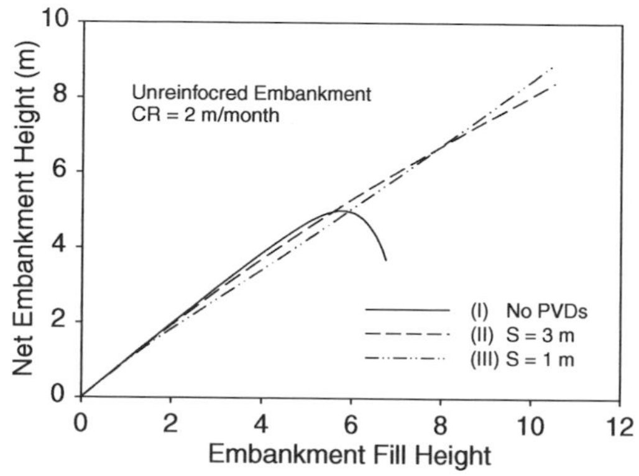

Figure 3. The effect of PVDs on the short-term stability of the embankment.

show that the failure height of reinforced embankment was improved 20% compared to the unreinforced embankment.

4.2 *Effects of PVDs on the short-term stability of embankment*

The main function of PVDs is to increase rate of excess pore water pressure dissipation by reducing the length of the drainage path. The consequence is an increase degree of partial consolidation as well as shear strength of the foundation soils. As shown in Figure 3, the short-term stability of embankment was improved drastically using PVDs and no failure occurred for fill thicknesses up to 10.5 m.

During the initial stage of construction, the smaller PVDs spacing (case III) resulted in larger settlement due to the higher degree of partial consolidation. This higher degree of partial consolidation also resulted in smaller overstress developed in the foundation and

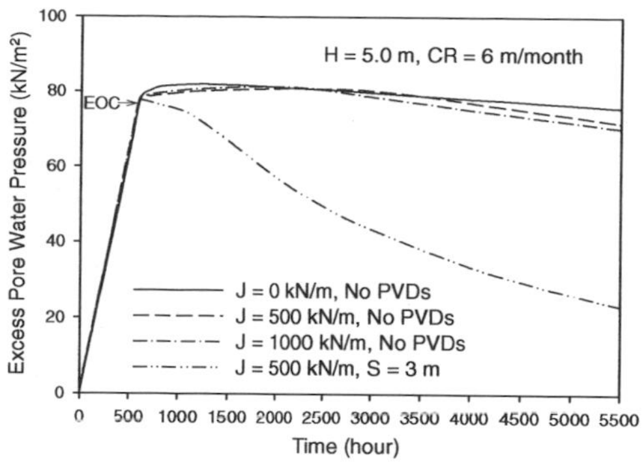

Figure 4. The effects on excess pore water dissipation.

accordingly less viscoplastic deformation was generated. Thus, as the fill thickness approached 10 m, the net embankment in case III was higher than for case II.

4.3 *Effects on the excess pore water pressure dissipation*

In order to investigate the long-term behaviour of the embankments, a number of embankments were numerically constructed to 5 m on a rate-sensitive soil. The calculated excess pore pressures at 5 m below the original ground surface under the embankment crest are given in Figure 4. These results show that the excess pore pressures kept increasing even after the end of construction, when the external fill load was constant. This phenomenon is similar to that observed at the Sackville test embankment (Rowe & Hinchberger, 1998).

The inclusion of reinforcement slightly reduced the maximum excess pore pressure and resulted in greater apparent dissipation. This is due to less generation of pore pressure resulting from less overstress and hence less creep induced pore pressure. The effect increased with increasing reinforcement stiffness, although the overall effect was not large in this case.

The use of PVDs resulted the peak pore pressure occurring at the end of construction and this was followed by relatively rapid dissipation of the excess pore pressures.

4.4 *Effects on the reinforcement strain*

The constructions of three 5 m height reinforced embankments were simulated in order to study the effects of PVDs and construction rate on the reinforcement strain as shown in Figure 5.

Results from Cases I and II show the effect of construction rate on the mobilized reinforcement strains. Case I (slower construction rate) resulted in larger reinforcement strain at the end of construction because

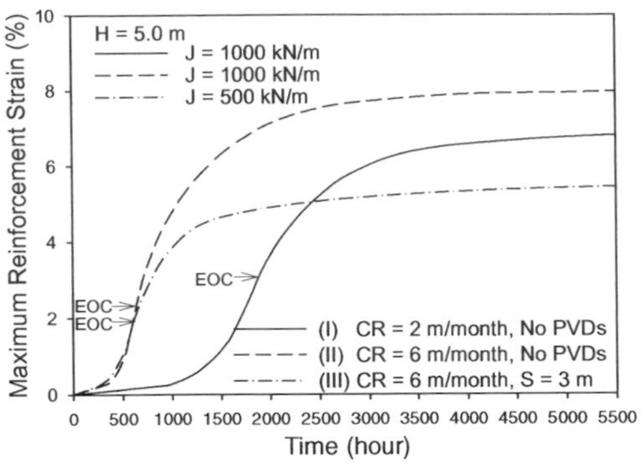

Figure 5. The effects on mobilized reinforcement strain.

soil tended to transfer more load to the reinforcement since it had lower strength at the lower strain rate. However, the slower construction rate allowed greater degree of partial consolidation and this reduced the amount of overstress in the soil and consequently reduced the long-term viscoplastic deformations in the soil. Accordingly, smaller long-term reinforcement strains were mobilized. Commonly designers aim to limit reinforcement strains to 5%–6%. The results for Cases I and II correspond to long-term reinforcement strains of 6.8% and 8.0%, respectively.

The effect of PVDs on the reinforcement strain is also presented in Figure 5. The use of PVDs allowed the use of lower stiffness reinforcement and also limited the long-term mobilized reinforcement strains, due to the fact that PVDs significantly increased the rate of excess pore water pressure dissipation which, in turn, reduced the amount of overstress in the system as well as the long-term reinforcement strains.

5 CONCLUSIONS AND DISCUSSIONS

For the rate-sensitive soil, a faster rate of construction resulted in higher short-term stability of the embankment. However a larger amount of overstress was generated in the soil and this resulted in large viscoplastic deformations. The excess pore water pressures continued to increase and reached its maximum value after the end of construction. Thus the critical period regarding the stability for these embankments may occur after the end of construction. The use of reinforcement resulted in less overstress in the soil for a given embankment fill thickness and this resulted in less viscoplastic deformation of the soil. The use of PVDs significantly increased the rate of excess pore water pressure dissipation; minimizing the effects of overstress and the long-term reinforcement strain.

This study demonstrates that the behaviour of rate-sensitive soil can have significant effects on the engineering performances of the reinforced embankment, especially after the end of construction. Therefore, the viscosity and viscoplastic characteristic of the soil should be considered in the design and construction of earth structures on the rate-sensitive soil.

ACKNOWLEDGEMENTS

The research reported in this paper was supported by the Natural Sciences and Engineering Research Council of Canada (NSERC).

REFERENCES

Carter, J.P., & Balaam, N.P. 1990. *AFENA – A general finite element algorithm: Users manual.* N.S.W: School of Civil Engineering and Mining Engineering, University of Sydney, Australia.

Janbu, N. 1963. Soil compressibility as determined by oedometer and triaxial tests. *Proc. of the European Conf. on Soil Mechanics and Foundation Engineering*, Wiesbaden, Germany. 1:19–25.

Li, A.L. 2000. *Time dependent behaviour of reinforced embankments on soft foundations.* Ph.D. thesis, London: The University of Western Ontario, Canada.

Li, A.L., & Rowe, R.K. 2001. Combined effects of reinforcement and prefabricated vertical drains on embankment performance. *Canadian Geotechnical Journal* 38: 1266–1282.

Rowe, R.K., & Hinchberger, S.D. 1998. The significance of rate effects in modelling the Sackville test embankment. *Canadian Geotechnical Journal* 33: 500–516.

Rowe, R.K., & Li, A.L. 2002. Behaviour of reinforced embankments on soft rate sensitive soils. *Geotechnique* 52(1): 29–40.

Rowe, R.K., & Soderman, K.L. 1985. An approximate method for estimating the stability for geotextiles reinforced embankments. *Canadian Geotechnical Journal* 22(3): 392–398.

Rowe, R.K., & Taechakumthorn, C. 2007. Behaviour of reinforced embankments on soft and rate-sensitive foundation. *Proc. of Geosynthetics Conf. 2007*, Washington D.C., USA: 86–98.

Rowe, R.K., Gnanendran, C.T., Landva, A.O., & Valsangkar, A.J. 1996. Calculated and observed behaviour of a reinforced embankment over soft compressible soil. *Canadian Geotechnical Journal* 33: 324–338.

Russell, D. 1990. An element to model thin, highly permeable materials in two dimensional finite element consolidation analyses. *Proc. of the 2nd European Specialty Conf. on Numerical Method in Geotechnical Engineering*, Santander, Spain: 303–310.

New Horizons in Earth Reinforcement – Otani, Miyata & Mukunoki (eds)
© 2008 Taylor & Francis Group, London, ISBN 978-0-415-45775-0

Numerical simulation of stone column installation using advanced elastoplastic model for soft soil

Z. Guetif, M. Bouassida & F. Tounekti
Unité de recherche "Ingénierie géotechnique", Tunisia

ABSTRACT: A numerical procedure is proposed to simulate the stone column installation in a normally consolidated soft clay layer. This simulation is performed by the use of Plaxis software version 8.1. Numerical computations are conducted by implementing a hardening soil model of normally consolidated soft clay which take account of the stress-stiffness dependency. From the obtained results it has been predicted both the increase of soft clay Young modulus and the radius of the area where this improvement is occurred.

1 INTRODUCTION

Stone columns are widely used to improve weak cohesive soft soils. The columns technique is an economical solution as foundation for structures having large loading area, e.g. embankments, storage tanks (Priebe 1995)... Stone columns serve three advantages namely: increase of bearing capacity, settlement reduction, and acceleration of consolidation by providing shorter drainage paths and higher permeability of columns material.

It should be noted that the reinforcement by stone columns has been largely reported in looking mainly for the predictions of bearing capacity and settlement, while the settlement acceleration was not always the prime interest.

Focusing on predictions of settlement, a variety of either analytical or numerical methods was proposed, (Balaam & Booker 1981, Van Impe & De Beer 1983, Bouassida et al. 2003, Guetif 2004). In these contributions the settlement reduction is the consequence of a partial substitution of the native soil by a stiffer material and equivalent stiffness characteristics of the reinforced soil have been established. However it is shown from experimental data (Vautrain 1980, Sanglerat 2002, Dhouib & Blondeau 2005, Alamgir & Zaher 2001) that the stiffness of native soil is also increased after the stone column installation.

The column installation is generally modeled by pouring stone (or gravel) into pre-formed holes with prescribed radius. It should be noticed that such a type of modeling is not realistic because the installation is accompanied by a lateral expansion of column in the native soft clay (Guetif et al, 2007).

Most of the design methods (Priebe, 1995, etc.) for stone columns assume unchanged Young modules of the native soft soil E_s and of the column material E_c. However, in soft clay deposits, the stiffness is variable with depth, or equivalently, with the degree of confinement (Biarez et al. 1998, Schanz et al. 1999).

The assumption of unchanged stiffness of the improved soil shall lead to an underestimation of settlement reduction.

After field data of soil improvement projects, during the consolidation, the increase of mechanical characteristics of soft clay reinforced by stone columns was well observed (Vautrain 1980, Sanglerat 2002, Dhouib & Blondeau 2005, Alamgir & Zaher 2001). The increase of native soil stiffness appeared to be an obvious consequence of the increase of effective mean stress during the consolidation period, (Biarez et al. 1998).

This paper presents a simulation of vibrocompacted column installation performed to investigate its effect on the stiffness of native soil. For this purpose, a recent contribution revealed a significant increase of native soil stiffness by adopting Mohr Coulomb model for soft clay reinforced by vibrocompacted stone column (Guetif et al, 2007). As continuation, in this paper, a hardening soil model (HSM) which shall give a more realistic characterization of the soft clay behavior is adopted (Brinkgreve and Vermeer 1998, Schanz et al. 1999). By using such a model the increase of soil stiffness with stress level can be taken into account. But, as disadvantage, the needed time for numerical computation will be greater.

The aim of this work is, then, to predict the increase of Young modulus of a native normally consolidated soft clay reinforced by vibrocompacted stone columns, and the radius of improved zone.

First, the constitutive model is presented for normally consolidated soft clay.

Second, the simulation of column installation is presented from which the prediction of stiffness improvement is derived.

The paper ends with some conclusions and remarks related to the obtained results. An outlook on further developments is proposed.

2 BEHAVIOUR OF NORMALLY CONSOLIDATED SOFT CLAYS

This section is dedicated for analyzing the behavior of normally consolidated soft clays. In some investigations the hardening soil model (HSM) has been considered to characterize the behavior of soft clays (Schanz et al. 1999, Guetif et al. 2006). Also, the HSM was largely adopted to predict the behavior of sands.

After stress-strain curves recorded from drained triaxial tests carried out on soft clays, it is shown the consolidation pressure significantly affects the behavior in the range of small strains as well as large strains. As examples the results obtained from experiments conducted by Biarez & Hicher (1994) and Tounekti et al, 2007) illustrated a strong non-linear stiffness which depends on the stress level. This fact can not be reproduced if Mohr Coulomb behavior model is envisaged for soft clays. Meanwhile the use of HSM permits to predict the increase in stiffness as a result from effective stress consolidation, Schanz et al (1999), Brinkgreve and Vermeer (1998). Compared to Mohr Coulomb model, the implementation of an elastoplastic HSM provides the advantage in controlling the stress-stiffness dependency for a given loading path.

The stress-strain behavior for primary loading is characterized by the stiffness modulus E_{50}, which is considered instead of initial tangent modulus E_i for small strain. For this latter the experimental determination appears more difficult. The modulus E_{50} is determined from deviatoric stress-axial strain curves recorded after drained triaxial tests. E_{50}, which corresponds to the half of the maximum shear strength, is calculated from the stress-stiffness relation (Schanz et al, 1999):

$$E_{50} = E_{50}^{ref}\left(\frac{\sigma_3 + c' \cot g\varphi'}{\sigma_3^{ref} + c' \cot g\varphi'}\right)^m \quad (1)$$

c' and φ' are drained strength characteristics.
m is the degree of stress dependency, for normally consolidated soft clays it is recommended $m = 1$.

The HSM does not involve a unique relationship between the drained triaxial modulus E_{50} and the oedometric modulus E_{oed} which is determined independently from:

$$E_{oed} = E_{oed}^{ref}\left(\frac{\sigma_1 + c' \cot g\varphi'}{\sigma_1^{ref} + c' \cot g\varphi'}\right)^m \quad (2)$$

Table 1. Hardening soil model parameters for soft clay.

γ kN/m^3	E_{50}^{ref} kPa	E_{ur}^{ref} kPa	E_{oed}^{ref} kPa	m	c' kPa	$\varphi'°$	$\psi°$	k_0	k_x m/day
18/17	2500	20000	1300	1	2	25	0	0.7	2.10^{-4}

γ = unit weight saturated/unsaturated; ψ = dilatancy angle; k_x = horizontal permeability.

For unloading and reloading stress paths, the recognized modulus is:

$$E_{ur} = E_{ur}^{ref}\left(\frac{\sigma_3 + c' \cot g\varphi'}{\sigma_3^{ref} + c' \cot g\varphi'}\right)^m \quad (3)$$

E_{oed}^{ref} is determined for a referenced major principal stress $\sigma_1 = 100$ kPa.

E_{50}^{ref}, E_{ur}^{ref} are determined for a referenced minor principal stress $\sigma_3 = 100$ kPa.

More details related to the HSM constitutive law can be found in Brinkgreve and Vermeer (1998) and Schanz et al. (1999).

In order to study the influence of stone column installation on soft clay behavior parameters of the HSM are adopted in the numerical simulation (Biarez & Hicher 1994). These parameters are calibrated with experimental data from oedometric and drained triaxial tests carried out on normally consolidated clay.

Using the HSM parameters proposed in table 1 for soft clay, the loading paths of oedometer and drained triaxial tests are well reproduced (Biarez et Hicher 1994).

It was also reported that the use of HSM makes possible in reproducing other loading paths such as the pressumeter test (expansion of cylindrical cavity) (Biarez et al. (1998), Brinkgreve and Vermeer (1998), Schanz et al. (1999).

It is, then, agreed a more wide use of the HSM, for normally consolidated soft clays, permits to simulate various loading paths as those occurring in several soil mechanics in situ tests. For this reason, the implementation of HSM is intended to analyze, first, the soft clay behavior that results during and after stone column expansion and, second, the expected improvement in term of Young modulus.

3 SIMULATION OF COLUMN INSTALLATION

3.1 *Reinforced soil characteristics*

The saturated subsoil comprises a sand layer up to 5 m thickness followed by normally consolidated soft clay up to 10 m thickness overlying a rigid and impervious stratum (Fig. 1). The soft clay is reinforced by vibrocompacted stone columns with 1 m of final diameter and 2.7 m of triangular grid spacing. The characteristics of soft clay are given in table 2.

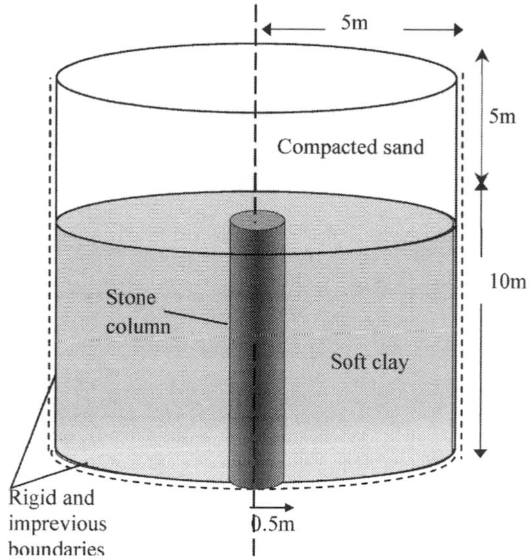

Figure 1. The composite cell model.

Table 2. Characteristics of the reinforced soil.

	γ kN/m^3	E′ kPa	c′ kPa	ψ°	φ'°	k m/day
Column material	20	32000	1	8	38	100
Initial sand	18	25000	1	5	35	10
Compacted sand	20	50000	1	8	38	10

Column material and sand layer are modeled as Mohr Coulomb materials which characteristics are given in Table 2.

To simulate the stone column installation a composite cell model is considered with external radius of 5 m (Fig. 1). The influence of boundary condition has been discussed in previous contributions (Debats et al. 2003, Bouassida et al. 2003b). From the latter it has been indicated that the soft clay is not influenced by the column installation beyond a radius of 5 m that is adopted for the composite cell model. It should be reminded that the installation of vibrocompacted stone column is due to the soft clay expansion from initial diameter of 0.25 m to a final diameter of 0.5 m (Guetif, 2004).

The numerical analysis is carried out by implementing an axisymmetric study with 15-noded triangular elements. The initial stresses are generated by assuming a coefficient of lateral earth stress $k_0 = 1 - \sin\varphi' = 0.69$. In this analysis the up-dated mesh option is used to take into account the large prescribed displacement especially around the interface soil-column.

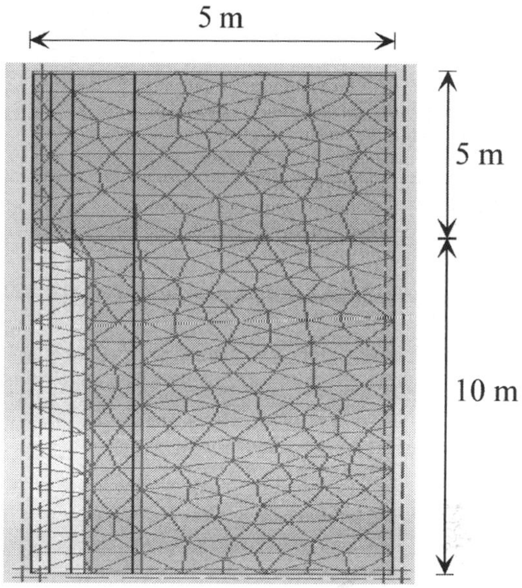

Figure 2. Deformed mesh of the composite cell model.

The finite element mesh of composite cell is shown in Figure 2 after the installation of vibrocompacted column.

3.2 *Simulation of column installation*

The numerical procedure adopted to simulate the installation of vibro-compacted column in soft clay comprises four stages which are:

- The vibro-probe is generated by the creation of a hole of initial diameter of 0.5 m.
- The withdrawal of the vibro-probe generates a hole in which the ballast will be incorporated.
- The ballast is expanded laterally: this stage is simulated at the interface soft clay-column where a prescribed lateral displacement is applied up to a final diameter of 1 m.
- In the upper sand layer, the soil improvement is pursued by vibrocompaction until a complete withdrawal of the vibro-probe.

After the column installation, the purpose is to study the soft clay improvement resulting after a given period of consolidation (eleven months) which is assumed as taking place horizontally around the column. The deformed mesh of the composite cell after column installation is presented in Figure 2.

4 PREDICTION OF SOFT CLAY IMPROVEMENT

The expansion of vibrocompacted stone column represents the loading of the surrounding soft clay where

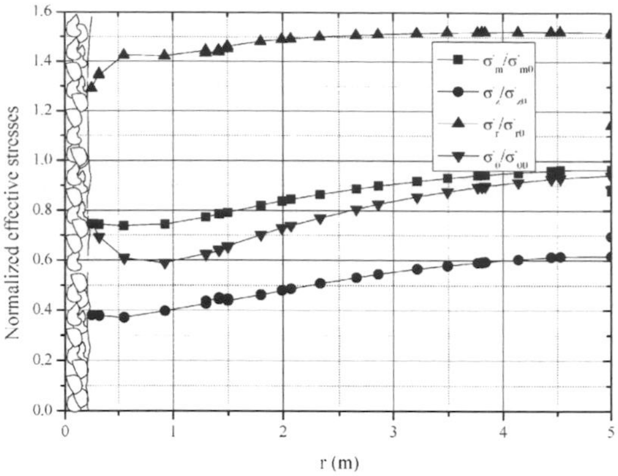

Figure 3. Normalized effective stresses distribution at mid-thickness of soft clay, before consolidation.

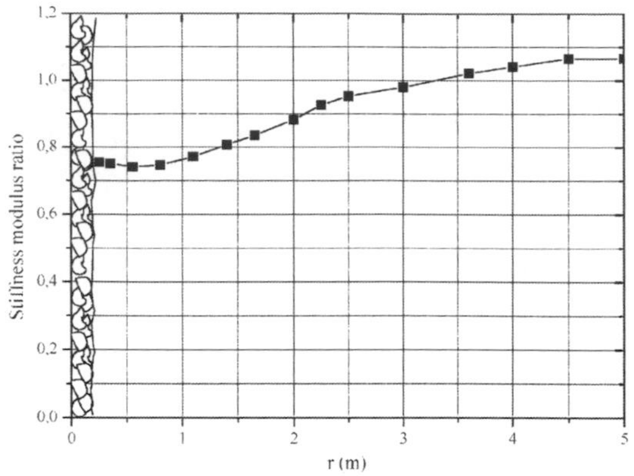

Figure 4. Stiffness modulus ratios at mid-thickness of soft clay, before consolidation.

the consolidation process takes place and the constitutive ballast plays the rule of vertical drain due to its high permeability (Table 2). Therefore excess pore pressures generated in the soft clay especially at the column's vicinity will be dissipated by radial drainage towards the column. As result, the effective state of stress in soft clay is modified from which the improvement of Young modulus will be determined from Eq (2).

4.1 *Results of numerical simulation*

The evolution of effective stress is examined at mid-thickness of the soft clay layer at the end of two phases. The first phase corresponds to the end of pouring column material which is accompanied by vibroprobe withdrawal. At this stage predicted results of numerical simulation showed up a significant excess pore pressure generated at the vicinity of the column. Contrarily, the effective mean stress distribution is slightly modified regarding that occurring before column installation (Fig. 3).

In addition the column expansion induces large plastic strains, close to soft clay-column interface, which vanish progressively when radial distance increases. Consequently, a lateral confinement is exerted by the expanded soft clay on the column.

In Figure 4, it is predicted immediately after the column installation a minor decrease of soft clay Young modulus. However, this fact might be counterbalanced since the radial drainage will occur during the consolidation of soft clay for a period of eleven months.

The second phase corresponds to the end of radial consolidation in soft clay that was estimated roughly of eleven (11) months. Therefore excess pore pressures generated in the soft clay, especially at column vicinity, are dissipated by radial drainage towards the column. Figure 5 shows the degree of consolidation of soft clay as a function of horizontal distance.

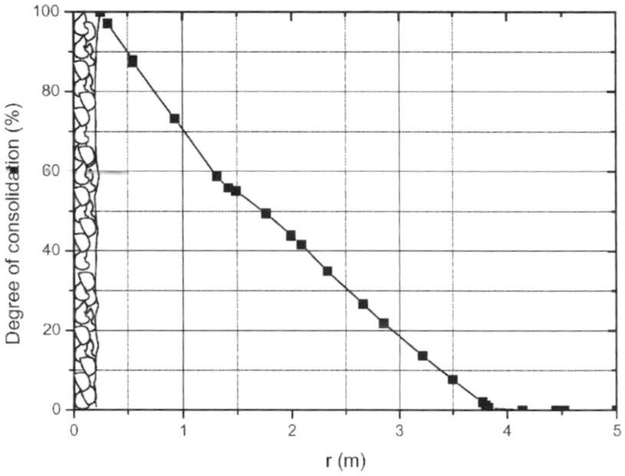

Figure 5. Degree of consolidation at mid-thickness of soft clay, after consolidation.

The corresponding stress distribution in terms of normalized ratios, with respect to initial values, is presented in Figure 6.

The vertical effective stress is slightly modified after the consolidation of soft clay. But the significant increase of horizontal effective stress is quasi-identical to that of effective mean stress.

By using the hardening soil model the expected increase of secant Young modulus in soft clay is determined (Figure 7). Such a stiffness improvement is due to the increase of effective horizontal stress. From which it is estimated the soft clay improvement stops at 2 m distance from the columns axis. The mean value of stiffness modulus increase is of about 40% which is not negligible.

5 CONCLUSIONS AND PERSPECTIVES

In this study the aim was to predict the improvement of soft clay Young modulus resulting from stone column

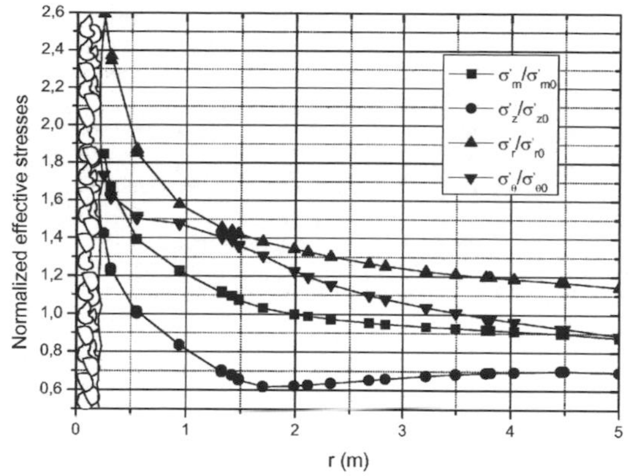

Figure 6. Normalized effective stresses distribution mid-thickness of soft clay, after consolidation.

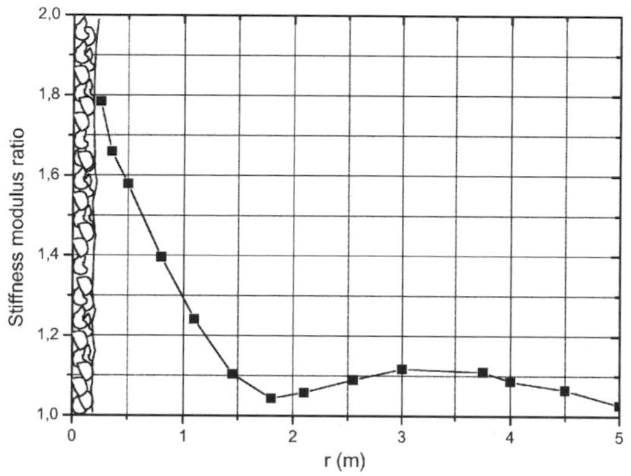

Figure 7. Stiffness modulus ratios at mid-thickness of soft clay, after consolidation.

installation. For this purpose, a numerical simulation was performed using the finite element code Plaxis 8.1. This procedure involves the use of a hardening soil model which takes into account the stiffness-stress dependency for soft clay. The calibration of the constitutive model parameters is done by a simulation of laboratory tests and the comparison between experimental results and predicted ones.

Then, the increase of stiffness is evaluated at the middle of soft clay layer after the consolidation generated by the column installation.

The numerical implementation of hardening soil model, showed up the significant influence of column installation on the increase of soil stiffness after the period of consolidation. The predicted increase of soft clay Young modulus is in average of about 40% the initial value. This improvement occurs in distance equals three times the column radius.

The advantage in determining the influence zone radius is to optimise the spacing between columns along which the improvement of stiffness modulus is completely effective. The regardless of this improvement may also induce the current design methods to overestimate the required quantities of incorporated column material.

The consolidation period preceding the loading has an important role on this improvement and it should be considered for settlement prediction.

This contribution only focused on the effect of column installation and consolidation preceding the loading of the reinforced soil. The influence of final loading on the behavior of reinforced soil shall be investigated in a further work.

REFERENCES

Alamgir, M. & Zaher, S. M. 2001. Field investigation on a soft ground of Bangladesh reinforced by granular piles. *Landmarks in earth Reinforcement, Ochiai and al Editors*: 517–522.

Balaam, N.P. & Booker, J. R. 1981. Analysis of rigid rafts supported by granular piles. *International Journal for Numerical and analytical Methods in Geomechanics, Vol 5, N 4*: 379–403.

Biarez J. & PY Hicher. 1994. Elementary Mechanics of soil behaviour saturated remoulded soils, Balkema, Rotterdam.

Biarez, J., Gambin, M., Gomes-Corriea, A., Falvigny, E. and Branque, D. 1998. Using pressumeter to obtain parameters to elastoplastic models for sands. *Proof the first international conference on site Characterization, ISC'98, 19–22 April*: 747–752. ATLANTA GEORGIA, USA.

Bouassida, M., Guetif, Z de Buhan, P. and Dormieux L. 2003a. Estimation par une approche variationnelle du tassement d'une fondation rigide sur sol renforcé par colonnes ballastées. *Revue Française de Géotechnique, N° 102*: 21–29.

Bouassida, M., Guetif, Z and Debats, J. M. 2003b. Parametric study of the improvement due to vibrocompacted columns installation in soft soils. *Int. Workshop on Ground Improvement, TC 17, August 29th*. Prague.

Brinkgreve, R. B. T. and Vermeer, P. A. 1998. Plaxis- Finite Element Code for soil and Rocks Analysis. *Version 7 and 8, A. A.* Balkema-Rotterdam-Brookfield.

Debats, J. M., Guetif, Z. and Bouassida, M. (2003) Soft soil improvement due to vibro-compacted columns installation. *Workshop on Geotechnics of Soft Soils- Theory and Practice*: 551–557. Noordwijkerhout, The Netherlands.

Dhouib A. et Blondeau F. 2005. Colonnes ballastées. *Edition des presses de l'école nationale des ponts et Chaussées* : 33–193, Paris.

Guetif Z. 2004. Sur l'estimation du tassement d'un sol renforcé par colonnes'. *Thèse de doctorat de l'Ecole Nationale d'Ingénieurs de Tunis.*

Guetif, Z., Bouassida, M. and Tounekti, F. 2006. Numerical simulation of vibro-compacted stone column installed in Tunis soft clay. Proc. 1st Euro Mediterranean in Advances on Geomat. and Struct. May 3rd-5th, Tunisia, 233–238.

Guetif, Z., Bouassida, M. and Debats, J.M. 2007. Improved Soft Clay Characteristics Due to Stone Column Installation. *Computers & Geotechnics*. 34 (2007), 104–111.

Priebe, H. J. 1995. The design of vibro replacement. *Ground Engineering*: 31–37.

Sanglerat, G. 2002. Contrôle des colonnes ballastées à l'aide du pénétromètre statique AMAP'sols'. *Jubilé Jimenz Salas,* Madrid.

Schanz, T., Vermeer P. A. and Bonnier, P. G. 1999. The hardening soil model: Formulation and verification. *Beyond 2000 in Computational Geotechnics_ 10 years of Plaxis,*: 1–16. Balkema, Rotterdam.

Soyez, B., Besançon G. and Iorio J. P. 1984. Analyse des paramètres de calcul intervenant dans le dimensionnement des colonnes ballastées. *C. R. Colloque international: Renforcement des sols et des roches*: 119–126. Paris.

Tounekti F., Bouassida, M, Marzougi, K.and Klay, M. 2007. Etude expérimentale en vue d'un modèle de comportement pour la vase de Tunis. Accepted in *Revue Française de Géotechnique.*

Van Impe, W. and De Beer E. 1983. Improvement of settlement behaviour of soft layers by means of stone columns', *C. R. 8th European Conference on SMFE, Helsinki, May, Vol. 1*: 309–312.

Vautrain, J. 1980. Comportement et dimensionnement des colonnes ballastées. *Revue Française de Géotechnique 11* : 59–73.

New Horizons in Earth Reinforcement – Otani, Miyata & Mukunoki (eds)
© 2008 Taylor & Francis Group, London, ISBN 978-0-415-45775-0

Rigid plasticity based stability analysis of reinforced slope

S. Ohtsuka
Nagaoka University of Technology, Nagaoka, Japan

Y. Inoue
JR East Company, Tokyo, Japan

T. Tanaka
Nagaoka Univeristy of Technology, Nagaoka, Japan

ABSTRACT: Rigid plastic finite element method is applied to stability assessment for reinforced slope with anchors or piles. Penalty method is introduced into rigid plastic constitutive equation to express an indeterminate stress component. Anchors and piles are modeled into beam elements with rigid plastic constitutive equations. Friction model is also considered between soil and countermeasures. Applicability of proposed method is discussed through simple case studies.

1 INTRODUCTION

It is very important to take into account an interaction force between soil and countermeasures in stability assessment of reinforced slope with anchors and piles. However, the interaction force depends on the failure mechanism of slope and it is difficult to determine prior to analysis. Slope stability has been estimated by the limit equilibrium method. It is simple and useful, however, adopts a simplified model on interaction phenomenon to apply a modeled interaction force on slip line.

In this study, rigid plastic constitutive equation is developed to express the behavior of countermeasures as anchors and piles at limit state. It is derived after Tamura (1991). Interface element is taken into account between soil and countermeasures, too. It is modeled into rigid plastic behavior with friction model. Applicability of proposed constitutive equation to stability assessment of reinforced slope is discussed through case studies.

2 SLOPE STABILITY ANALYSIS

2.1 *Constitutive equation of soil and rigid plastic finite element method*

Rigid plastic constitutive equation is derived after Tamura (1991) in this study. Drucker-Prager type yield function is employed for soil as follows:

$$f = \frac{\alpha}{Fs} I_1 + \sqrt{J_2} - \frac{k}{Fs} = \hat{\alpha} I_1 + \sqrt{J_2} - \hat{k} = 0 \qquad (1)$$

In slope stability assessment, factor of safety, Fs has been defined by a strength reduction coefficient at limit state. α and k are material constants of soils. The kinematic condition of soil is derived from Equation 1 with the associated flow rule.

$$h = \dot{\varepsilon}_v - \frac{3\hat{\alpha}}{\sqrt{3\hat{\alpha}^2 + 1/2}} \dot{e} = \dot{\varepsilon}_v - \frac{3\hat{\alpha}}{\hat{\lambda}} \dot{e} = \dot{\varepsilon}_v - \hat{\beta}\dot{e} \qquad (2)$$

$\dot{\varepsilon}_v$ and $\dot{e}$ express the volumetric strain rate and the norm of strain rate, respectively. After Tamura (1991), the indeterminate stress in associated flow rule is derived by introducing the kinematic condition of Equation 2 with penalty method.

$$\boldsymbol{\sigma} = \frac{\hat{k}}{\hat{\lambda}} \frac{\dot{\boldsymbol{\varepsilon}}}{\dot{e}} + \kappa\left(\dot{\varepsilon}_v - \hat{\beta}\dot{e}\right)\left\{\boldsymbol{I} - \hat{\beta}\frac{\dot{\boldsymbol{\varepsilon}}}{\dot{e}}\right\} \qquad (3)$$

κ is a penalty coefficient (arbitrary large number).

In rigid plastic constitutive equation, the norm of strain rate is basically indeterminate. It is necessary to fix the norm of strain rate in rigid plastic finite element method. In rigid plastic finite element method, the relative distribution of strain rate inside slope is important to assess the stability. The following equation is employed as the constraint condition in finite element discretized form.

$$\int_V \boldsymbol{b}^T \dot{\boldsymbol{u}}\, dv - 1 = 0 \qquad (4)$$

$\boldsymbol{b}$ is the body force vector. Discretized equilibrium equation is expressed with a load factor ρ for body force to achieve the limit state in the followings.

$$\int \boldsymbol{B}^T \boldsymbol{\sigma}\, dv = \rho \int \boldsymbol{b}^T \dot{\boldsymbol{u}}\, dv \tag{5}$$

In the above equation, σ reveals the dicretized stress vector. Load factor is revealed by introducing the constraint condition of Equation 4 with the use of penalty method,

$$\rho = \mu\left(\int \boldsymbol{b}^T \dot{\boldsymbol{u}}\, dv - 1\right) \tag{6}$$

where μ is a penalty coefficient (arbitrary large number). Equation 5 is solved by using Equations 3 and 6. When the obtained load factor is $\rho \geq 1$, the slope is still safe for reduced strength by Equation 1 with Fs employed. In order to obtain the factor of safety at limit state, an iterative computation is necessary. Fs is updated by the obtained load factor such that

$$Fs = Fs \times \rho\,. \tag{7}$$

When the load factor converges to $\rho = 1$, Fs is finally obtained.

2.2 *Constitutive equation of countermeasure*

Countermeasure as anchor and pile is modeled into a beam in this study. In beam, stress vector is composed of axial force, N and bending moment, M. The yield function of beam is assumed based on the strength reduction concept as follows:

$$f = (\varpi N)^2 + (\xi M)^2 = \left(\frac{\sigma_o}{Fs}\right)^2 = \hat{\sigma}_o^2\,. \tag{8}$$

Strain rate vector is composed of axial strain rate, $\dot{\varepsilon}_n$ and rate of bending angle, $\dot{\varepsilon}_\vartheta$ in beam. Constitutive equation for countermeasure is easily derived in the same way with soil in the followings.

$$\begin{pmatrix} N \\ M \end{pmatrix} = \frac{\hat{\sigma}_o}{\hat{e}} \begin{bmatrix} \dfrac{1}{\varpi^2} & 0 \\ 0 & \dfrac{1}{\xi^2} \end{bmatrix} \begin{pmatrix} \dot{\varepsilon}_n \\ \dot{\varepsilon}_\vartheta \end{pmatrix} \tag{9}$$

In the above equation, the norm of strain rate vector is defined as follows:

$$\hat{e} = \sqrt{\left(\frac{\dot{\varepsilon}_n}{\varpi}\right)^2 + \left(\frac{\dot{\varepsilon}_\vartheta}{\xi}\right)^2}\,. \tag{10}$$

It is noted that the stress vector of beam is uniquely determined for the strain rate vector as Equation 9.

2.3 *Constitutive equation of interface element between soil and countermeasures*

Interface element is introduced into between soil and countermeasures. Constitutive equation of interface element is derived by friction model in the three dimensional condition. Traction vector at interface is composed of t_s and t_t in tangential direction and t_n in normal direction. Yield function is afforded by the Mohr-Coulomb criteria with a strength reduction coefficient such that

$$\begin{aligned} g &= \sqrt{t_s^2 + t_t^2} - \frac{c_s}{Fs} + t_n \frac{\tan\phi_s}{Fs} \\ &= \sqrt{t_s^2 + t_t^2} - \hat{c}_s + t_n \tan\hat{\phi}_s = 0. \end{aligned} \tag{11}$$

Traction vector is correlated with differential velocity $\Delta\dot{\boldsymbol{u}}$ at interface. Dilation property at interface is introduced into the constitutive equation to determine the indeterminate stress in associated flow rule of Equation 11.

$$\begin{aligned} \boldsymbol{t} = \begin{pmatrix} t_s \\ t_t \\ t_n \end{pmatrix} &= \frac{\hat{c}_s}{1+\tan^2\hat{\phi}_s}\frac{\Delta\dot{\boldsymbol{u}}}{|\Delta\dot{\boldsymbol{u}}|} \\ &+ \varsigma_s\left(\sqrt{\Delta\dot{u}_s + \Delta\dot{u}_t}\tan\hat{\phi}_s - \Delta\dot{u}_n\right)\frac{1}{|\Delta\dot{\boldsymbol{u}}|}\begin{pmatrix} \Delta\dot{u}_s \tan\hat{\phi}_s \\ \Delta\dot{u}_t \tan\hat{\phi}_s \\ |\Delta\dot{\boldsymbol{u}}| \end{pmatrix} \end{aligned} \tag{12}$$

Penalty method is employed in the same way where ςs is a possible large number.

3 CASE STUDIES

3.1 *Applicability of interface element*

Slope stability assessment is conducted for Figure 1. It includes various discontinuous lines the inclination angles of which are different (D1, D2 and D3). Effect of discontinuous line on both failure mode and factor of safety is investigated. Soil constants of slope are set such that $c = 16.5$ kPa (cohesion), $\phi = 23.9°$ (angle of shear resistance) and $\gamma_t = 18.0$ kN/m^3 (density). Discontinuous line of D1 to D3 expresses a seam layer inside slope for example. Figure 2 indicates the obtained failure mode in case of no discontinuous line inside sloe. The failure mode is apparently the circular arc slip line. Factor of safety is obtained as $Fs = 1.53$.

Figure 3 reveals the failure mode of slope in case of discontinuous line D1. Shear strength of discontinuous line is set as $c_s = 10.0$ kPa and $\phi_s = 0°$. Due to low shear strength of D1, slope fails along D1. Factor of safety is computed as $Fs = 1.14$. It is apparent that shear zone develops in slope and slope fails as a wedge block sliding mode. Figure 4 expresses the failure mode of slope in case of D2. Shear strength

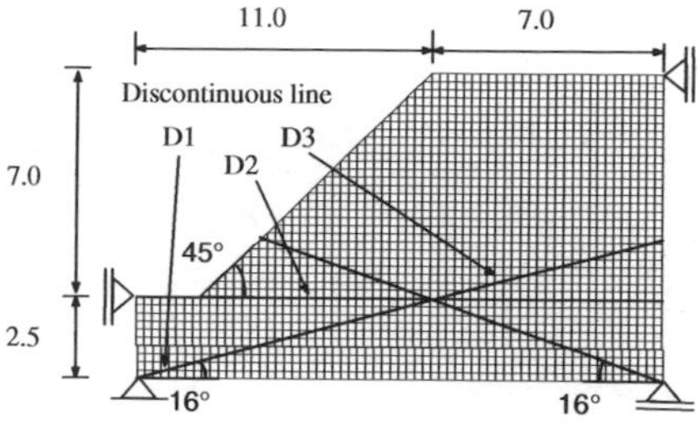

Figure 1. Slope with discontinuous lines inside (unit:m).

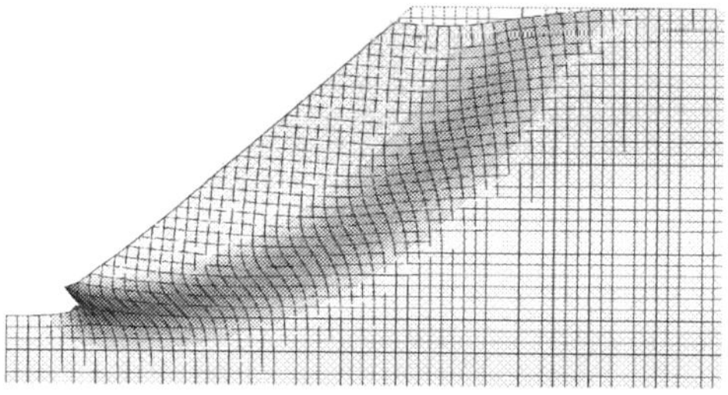

Figure 2. Failure mode in case of no discontinuous line (*Fs* = *1.53*).

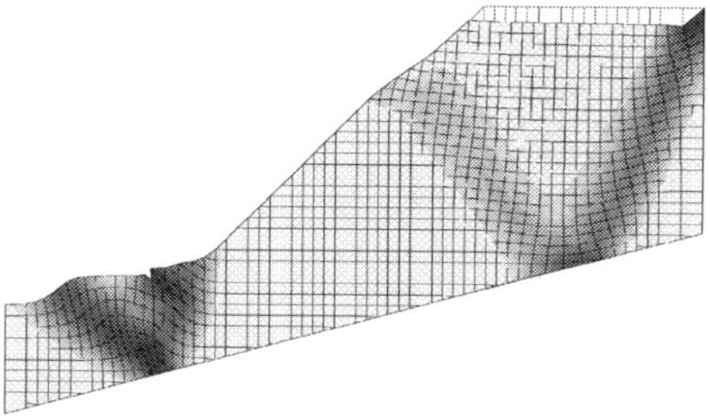

Figure 3. Failure mode in case of discontinuous line D1 (*Fs* = *1.14*).

of discontinuous line is set as $c_s = 1.56$ kPa and $\phi_s = 12.1°$. Factor of safety is obtained as $Fs = 1.23$. It is also seen the localized shear zone the width of which is finite. Velocity field reflects the dilation property at discontinuous line. Figure 5 reveals the failure mode of slope in case of D3. Shear strength of discontinuous line is set as $c_s = 1.56$ kPa and $\phi_s = 12.1°$. Failure mode is obtained almost same with that of Figure 2. It is because D3 inclines rightward and slope is difficult to fail along the discontinuous line D3 even though the shear strength parameters of D3 are same with D2. Factor of safety is obtained as $Fs = 1.56$ and almost coincides with the case of slope with no discontinuous line.

3.2 *Application to anchor (extension part)*

Stability of slope with anchor is assessed. Figure 6 shows an anchor and interface elements. Interface

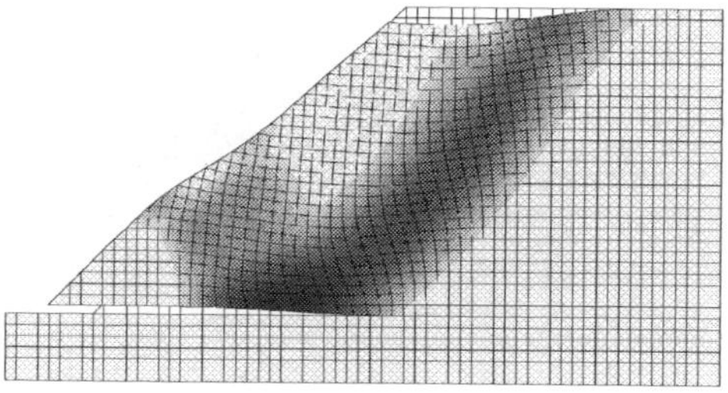

Figure 4. Failure mode in case of discontinuous line D2 (Fs = *1.23*).

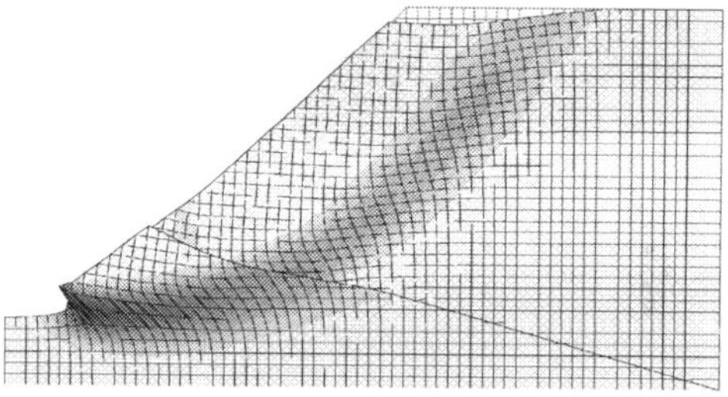

Figure 5. Failure mode in case of discontinuous line D3 (*Fs* = *1.52*).

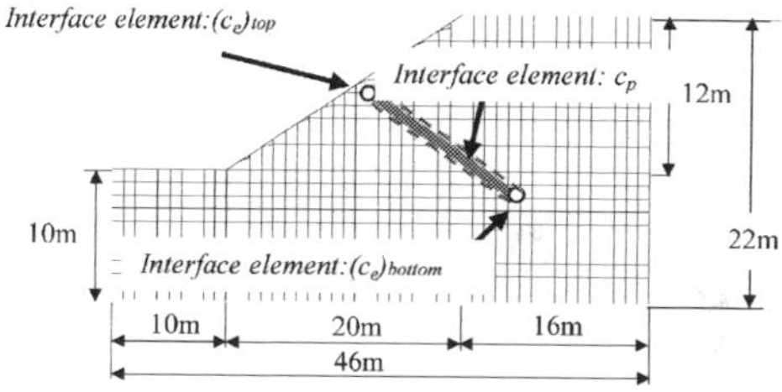

Figure 6. Boundary condition for case study (unit:m).

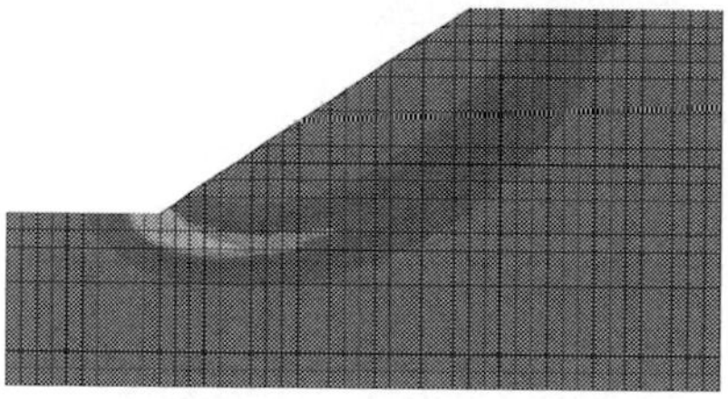

Figure 7. Failure mode in case of no anchor (*Fs* = *1.064*).

elements are set at not only contact plane between soil and anchor, but also at both ends of anchor. To simplify the problem, a cohesion model is focused where c_p and c_e denote cohesions at contact plane and both ends of anchor. Figure 6 also reveals the boundary condition of slope. Soil constants are $c_s = 19.6$ kPa, $\phi_s = 10°$ and $\gamma_t = 19.6$ kN/m^3. Figure 7 expresses the failure mode of slope obtained for the case without anchors (extension parts). Factor of safety is $Fs = 1.064$.

In order to discuss the effect of anchor on slope stabilization, two simple cases are considered. In case

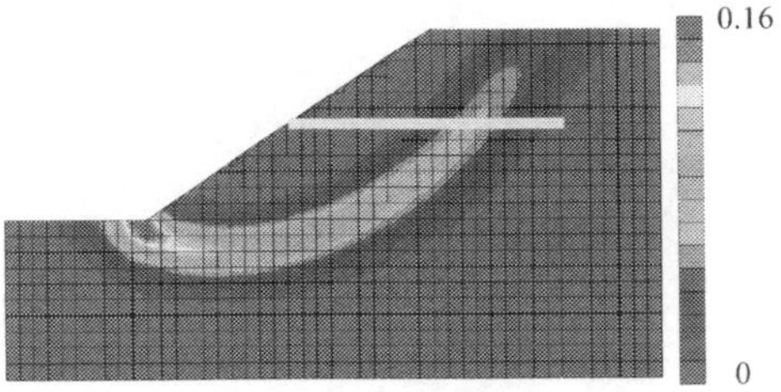

Figure 8. Failure mode of Case A ($Fs = 1.073$).

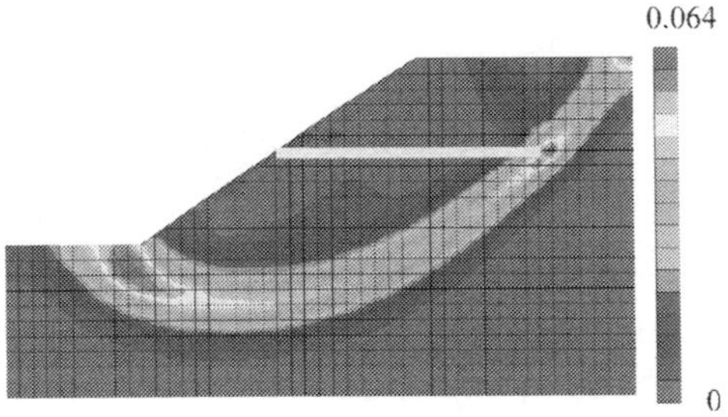

Figure 9. Failure mode of Case B ($Fs = 1.148$).

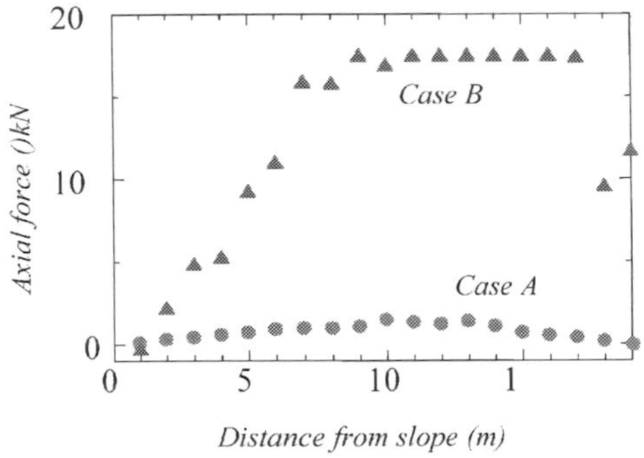

Figure 10. Axial force distribution of anchor.

A, the effect of anchor is set low as $c_p = 0.01$ kPa and $c_c = 1.0$ kPa, where the anchor is easily pull out. In case B, The effect of anchor head is taken into account as $(c_e)_{top} = 10000$ kPa; $(c_e)_{bottom} = 0.01$ kPa and $(c_p) = 10000$ kPa. Anchor is set to yield at 200 kN in axial force and 0.01 kNm in bending moment. The diameter is 0.2 m. Figures 8 & 9 reveal the obtained results of failure modes. In Case A, it is clear that the failure mode and the factor of safety are almost coincident with those in case of no anchors. On the contrary, Case B reveals wider failure mode including the anchor and the factor of safety is obtained higher. Axial force distribution of anchor is exhibited in Figure 10. In Case A, axial force is recorded maximum at slip line and the anchor is found to slip along interface element. On the other hand, it is seen that the axial force increases with distance from the top and attain to the yield limit at the middle point of anchor in Case B.

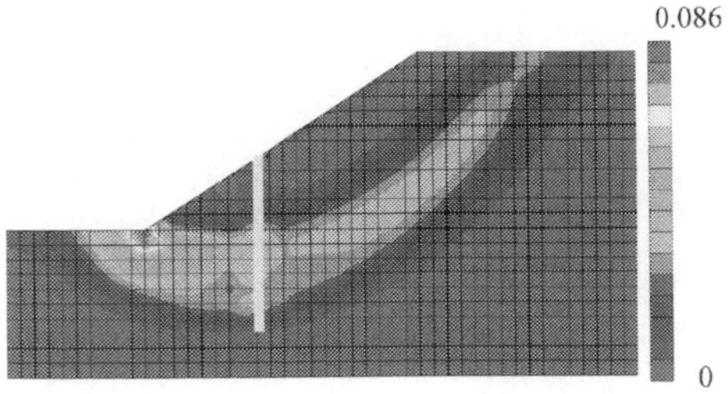

Figure 11. Failure mode of slope in case of pile ($Fs = 1.105$).

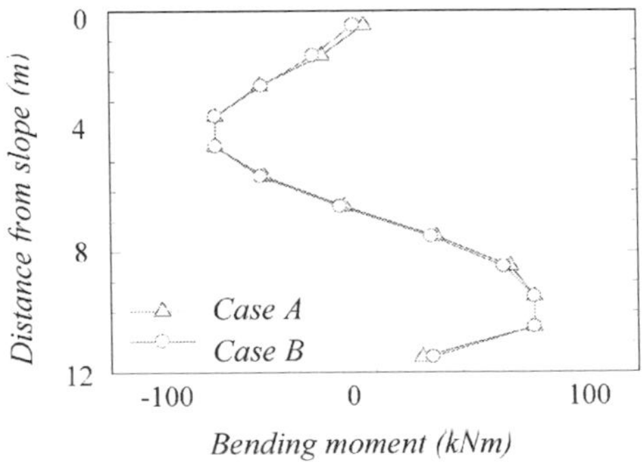

Figure 12. Bending moment distribution of pile.

3.3 *Application to pile*

Pile is modeled to yield at 100 kN in axial force and 100 kNm in bending moment. The diameter is set as 0.5 m. Condition of interface element is free at both ends of pile as $(c_p) = 0.01$ kPa and the cohesion at contact plane, c_e is varied whether $c_c = 1.0$ kPa or $c_c = 10000$ kPa. Figure 11 reveals the failure mode expands in comparison with Figure 7. Factor of safety is obtained as $Fs = 1.105$ and higher than that of no pile. Bending moment of pile is obtained rationally as shown in Figure 12.

4 CONCLUSION

Rigid plastic constitutive equation of countermeasures for slope stabilization was developed. Interface element was also introduced to simulate the sliding between soil and countermeasures. Applicability was examined through simple case studies on anchor and pile problems. It was clear to simulate the interaction force between soil and countermeasures.

REFERENCES

Tamura, T. 1991. Rigid-plastic finite method in geotechnical engineering. *Computational Plasticity, Current Japanese Material Research*, 7: 135–164. Elsevier.

Ohtsuka, S., Miyata, Y., Ikemoto, H. & Iwabe, T. 2001. Slope stability analysis with rigid plastic finite element method. *J. of Japan Landslide Society*. 38, 3: 75–83.

New Horizons in Earth Reinforcement – Otani, Miyata & Mukunoki (eds)
© 2008 Taylor & Francis Group, London, ISBN 978-0-415-45775-0

Static analysis of slopes reinforced with stone columns

M. Ghazavi
Civil Engineering Department, K. N. Toosi University of Technology, Tehran, Iran

A. Shahmandi
Civil Engineering Department, Faculty of Engineering, The University of Yazd, Yazd, Iran

ABSTRACT: The stabilization of slopes has been of great concern to geotechnical engineers. Various methods may be used to increase the safety factor of slopes prone to failure. These include retaining walls, piles, and geosynthetics, etc. An alternative solution is the use of stone columns. Such columns have been used since 1950 normally for cohesive soil improvement. A potential application of stone columns may be to stabilize slopes against instability. In this paper, a numerical approach in conjunction with an analytical approach is used to investigate the stability of slopes reinforced with stone columns. The slope soil is assumed to be soft, cohesive, and undrained. The present solution has been verified using the finite element method (FEM) as coded into GEO-OFFICE software. The results obtained from the developed method have shown that the factor of safety of slope-reinforced with stone columns increases. Moreover, it has been found that to achieve the greatest safety factor for slopes, the best location of the column is at the top the slope. Parametric studies have been performed to determine the best location of stone columns in the slope to achieve the maximum factor of safety. Further parametric studies have been performed to determine the influencing factors such as stone column diameter, friction angle of stone column material, distance between stone columns.

1 INTRODUCTION

Many parameters are effective on slope stability. Among these, the most important ones are soil unit weight, slope geometry, tectonic, earthquake, vibration, heterogeneousness of soil, strength parameters of soil, and pore water pressures.

Engineering stabilizing is generally referred to stop or return the instability process. Preventing the movement of a slope or increasing the safety factor (SF) is possible by using structural or geotechnical methods. Among techniques which increase resisting forces and basically act externally on the soils or rocks sliding are geometrical methods, structural barriers such as rigid walls and piles, permeable or impermeable coverage at surface, hydraulic improvement, physical improvement, chemical improvement, mechanical improvement, reinforcing with geosynthetics, an soil nailing, etc. (Jorge & Zornberg, 2002; Komak and Panah, 1994; Ausilio et al., 2001; Hassiotis et al., 1997; Jorge & Zornberg, 2002).

Stone column is another method for slope stabilization. It is a hole with circular section which is filled by gravel, rubble and etc and is an effective method to increase the shear strength on the slip surface of clayey slopes. The most important cases for utilizing stone columns (School of Civil Engineering Georgia Institute of Technology Atlanta, 1983) are:

1. Slopes stabilization
2. Stabilizing the retaining walls
3. Decreasing the liquefaction potential of sandy soils
4. Increasing the bearing capacity of shallow foundations situated on soft soils.

The performance of stone columns for reinforced and improved soil is easier and cheaper in comparison to other methods such as geotextile, grouting, compaction, etc. In some cases, it offers better results than other methods. Usually the diameter of stone column varies between 0.3 to 1.2 m and their intervals between 1.5 to 3 m. Stone columns are often performed in multiple rows (depending on soil condition).

In this paper, two rows of stone columns have been located within the slope. A two-dimensional finite element software (Geo-Slope software version 5.04) has been used for slope stability analysis, therefore, 3-D stone columns must be changed to 2-D. To do this, an equivalent column is replaced with one row successively and with distance s between two neighboring columns. Thus their centers are replaced by a continuous stone strip with equivalent width W. The volumes of stone column materials are identical in both two and three dimensional conditions. On the basis of equality of volume, equivalent strip width for each row of the stone columns is obtained from (Cheung, 1998):

$$W = \frac{\pi R^2}{s} \tag{1}$$

where R = radius of 3-D stone columns and s = distance between centers of 3-D stone columns in each row.

2 ANALYSIS METHOD

The SF of slope reinforced with stone columns varies with changing the column location within the slope. Therefore, it is necessary to change the column location along the slope to determine the greatest SF. To achieve this, first, by the use of Taylor method, the SF and the critical slip surface of the slope is found and then this normal slope is simulated using Geo-Slope software. The value of nH is then found, where H is slope height and n is a factor representing the location of toe of slip surface in Taylor method. By the use of same nH in Geo-Slope, the same slip surface with the same SF is obtained. When this critical surface is found, the stone column is displaced along the slope and the variation of SF is determined.

3 SLOPE STABILITY ANALYSIS REINFORCED BY A ROW OF STONE COLUMN

Figure 1 shows the geometry of a slope reinforced with a row of stone columns. Tables 1 and 2 show the geotechnical properties of clayey soil and stone column materials, respectively. The critical slip surface is first determined with displacing the row of stone columns in the horizontal direction along the slope. The variation of SF is captured.

In Figure 1, x = horizontal distance of stone column from the slope crest and h = slope height.

Using parameters shown in Tables 1 and 2 in conjunction with three values of 27, 38 and 45° for slope angles, and three values of .50, 0.65, and 0.80 m for the column diameter, the variation of SF was determined with respect to the varying location of the column along the slope. Figure 2 shows that when the stone column is located at the upper part of the slope, the greatest SF is achieved. With moving the column toward down the slope, the column effect on the slope stability decreases.

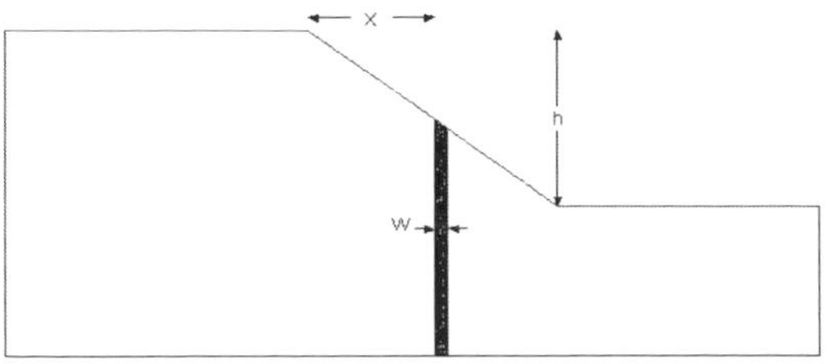

Figure 1. Geometry of slope reinforced with stone column.

4 SLOPE STABILITY ANALYSIS REINFORCED BY TOW ROWS OF STONE COLUMNS

It is customary to use several rows of stone columns for slope reinforcement. In this section, the influence of other contributing parameters such as number of rows of columns, location of columns, and the distance between two subsequent rows are investigated.

Table 1. Geotechnical properties of clayey soil.

Elastic modulus	4000 kN/m^2 4500 kN/m^2 5000 kN/m^2
Undrained cohesion	25 kN/m^2 40 kN/m^2
Unit weight	17 kN/m^3 18 kN/m^3
Friction angle	0.0 degree
Poisson's ratio	0.48

Table 2. Geotechnical properties of stone columns materials.

Elastic modulus	40000 kN/m^2 50000 kN/m^2 55000 kN/m^2
Cohesion	0.0
Unit weight	22 kN/m^3
Friction angle	35 degree 40 degree 45 degree
Poisson's ratio	0.25 0.30

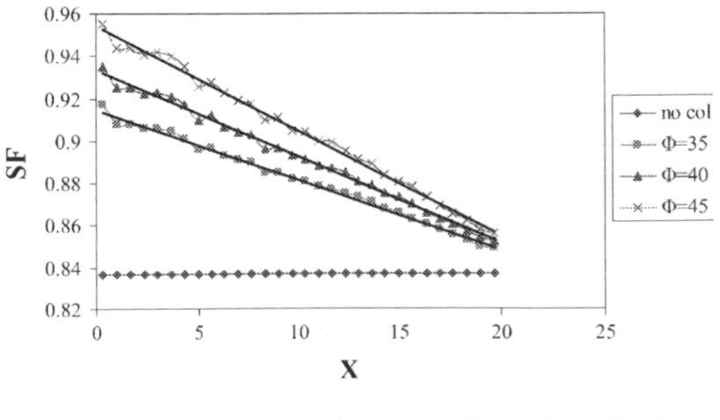

Figure 2. Variation of SF in terms of location of column from the sloe crest (Φ = friction angle of stone column materials).

For brevity, only some analyses are presented herein. The results of these limited analyses have shown that generally when the stones in rows are located close to the slope crest, the greatest SF is achieved. It was found that the location of two rows of columns is very effective on SF and when two rows of columns are placed in upper half slope face, the greater SF. With moving rows of columns toward the middle of the slope face and toward down the slope, the value of SF and the stability decreases.

It was further investigated that by the use of both one and two rows of columns, the stability increases markedly. Furthermore, it was observed that with increase equivalent width of stone column and friction angle of column materials, SF increases. In addition, with increasing the undrained cohesion of the slope soil, the effect of column on increasing the SF decreases.

5 STONE COLUMN EFFECT ON SLOPE STABILITY

Stone columns have two impacts on increasing slope stability:

1. Reduction of pore water pressure by dissipation
2. Increasing the shear strength on the slip surface due to high friction angle of stone column materials.

Stone columns can drain well and reduce pore water pressure with time elapse. However, in short time, columns are unable to perform this mechanism. The presence of the column causes pore pressures dissipate considerably on the slip surface-column intersection. The pore water pressure at each level in column is approximately equal to $u = \gamma_w h$ where $\gamma_w =$ water unit weight and h=water depth from the slope surface at the column location.

The main reason for SF increase is high column material friction angle, which is offered at the failure surface. The presence of stone column causes the SF value suddenly increases for the slip surface. This surface passes the column material.

With accuracy to this mechanism a new analytical equation has been found.

6 CLOSED-FORM SOLUTION FOR STONE COLUMN-REINFORCED SLOPE

6.1 *Normal slope*

Taylor (1937, according to Das, 1941) presented an equation to determine SF of homogenous undrained ($\Phi = 0$) clayey slopes (Figure 3):

From Figure 3, it is seen that the average shear strength of the soil is $\tau_f = c_u$ where $c_u =$ undrained shear strength of clay, $\tau_f =$ shear strength.

The mobilized shear strength on the slip surface is $\tau_d = c_d$. Therefore, the sliding moment is given by:

$$M_d = w_1 x_1 - w_2 x_2 \tag{2}$$

The resisting moment is expresses as:

$$M_R = c_d R^2 \theta \tag{3}$$

Using equations (2) and (3) gives SF as:

$$SF = \frac{M_R}{M_d} = \frac{c_d R^2 \theta}{w_1 x_1 - w_2 x_2} \tag{4}$$

6.2 *Reinforced slope by a row of column*

With the presence of column on the slope face, the shear strength on the slip surface is mobilized. Thus from Figure 4, following equations are expressed:

$$DE = L_1;\ EF = L_2;\ FA = L_3;\ \text{and } DEFA = L.$$

Figure 5 shows exerted forces on the equivalent strip stone column between surface of slope and slip surface.

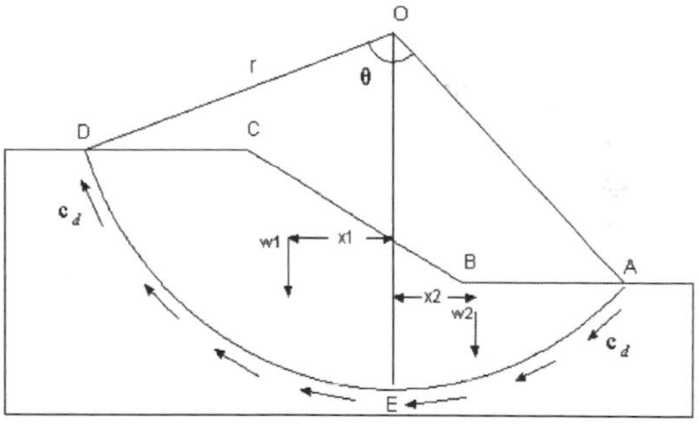

Figure 3. Slope stability analysis of homogenous saturated clay ($\Phi = 0$).

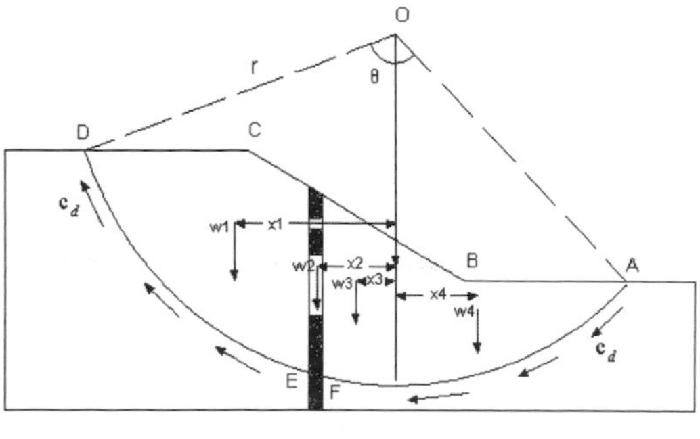

Figure 4. Slope stability analysis of homogenous saturated clay ($\Phi = 0$) reinforced by a row of column.

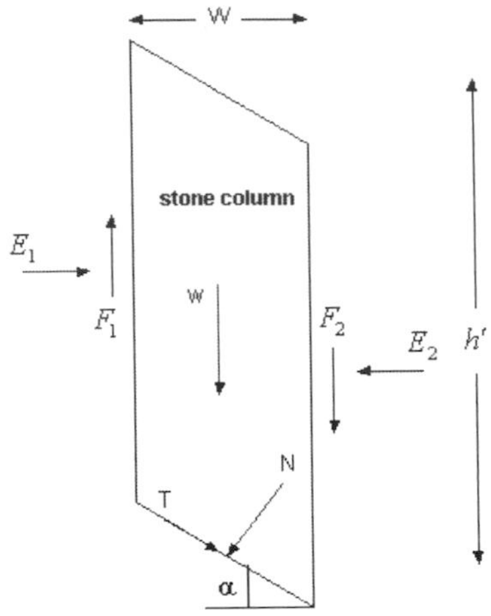

Figure 5. Exerted forces on strip stone column.

With respect to Figures 4 and 5 and assuming that $E_1 = E_2$ and $F_1 = F_2$:

$$w = \gamma_{sat} W h' \tag{5}$$

where w = weight of strip stone column; γ_{sat} = saturated unit weight of strip stone column; and h' = height of strip stone column between surface of slope and slip surface.

$$N = w\cos\alpha = \gamma_{sat} W h' \cos\alpha \tag{6}$$

$$T = w\sin\alpha = \gamma_{sat} W h' \sin\alpha \tag{7}$$

$$L_2 = W/\cos\alpha$$

The total normal stress on the base of stone column is:

$$\delta = N/(W/\cos\alpha) = \gamma_{sat} h' \cos^2\alpha \tag{8}$$

The shear strength on the base of the stone column is given as:

$$\tau = T/(W/\cos\alpha) = \gamma_{sat} h' \cos\alpha \sin\alpha \tag{9}$$

The shear strength on the slip surface of slope-column system is determined from:

$$\tau_f = c_u + \Delta\tau_f = c_u + \delta' \tan\phi = c_u + (\delta - u)\tan\phi$$

$$= c_u + \gamma' h' \cos^2\alpha \tan\phi \tag{10}$$

The shear strength mobilized on the slip surface with the presence of the column is expresses as:

$$\tau_d = c_d + \gamma' h' \cos^2\alpha \tan\phi_d \tag{11}$$

The sliding moment is computed from:

$$M_d = w_1 x_1 + w_2 x_2 + w_3 x_3 - w_4 x_4 \tag{12}$$

The resisting moment is given by:

$$M_R = c_d L_1 r + \gamma' h' \cos^2\alpha \tan\phi_d L_2 r + c_d L_3 r$$

$$= c_d r(L_1 + L_3) + r\gamma' h' \cos\alpha W \tan\phi_d \tag{13}$$

The SF value for unreinforced slope can be determined using:

$$(SF)_{col} = \frac{c_u r(L - L_2) + r\gamma' h' W \cos\alpha \tan\phi}{w_1 x_1 + w_2 x_2 + w_3 x_3 - w_4 x_4} \tag{14}$$

Similarly, SF of reinforced slope by a row of stone column is obtained from:

$$(SF)_{col} = \frac{c_u r(r\theta - \frac{W}{\cos\alpha}) + r\gamma' h' W \cos\alpha \tan\phi}{w_1 x_1 + w_2 x_2 + w_3 x_3 - w_4 x_4} \tag{15}$$

where α = angle of stone column failure with horizontal direction on the slip surface and γ' = buoyant unit weight of stone column materials.

With use geometrical method and with obtained equation we can determine SF of slope.

By splitting Equation (15), the ratio of SF for reinforced slope to SF for un-reinforced slope, SF_{ratio}, is determined from:

$$SF_{ratio} = \frac{(SF)_{col}}{(SF)_{no-col}} = 1 - \frac{W}{r\theta\cos\alpha} + \frac{\gamma' h' W \cos\alpha \tan\phi}{c_u r\theta} \tag{16}$$

Equation (16) may be converted to:

$$SF_{ratio} = \frac{(SF)_{col}}{(SF)_{no-col}} = 1 - \frac{\pi R^2}{sr\theta\cos\alpha} + \frac{\gamma' h' \pi R^2 \cos\alpha \tan\phi}{sc_u r\theta} \tag{17}$$

Similarly, SF_{ratio} can be computed for reinforced slopes with several rows of columns.

7 VERIFICATION OF DEVELOPED CLOSED-FORM SOLUTION

Parametric studies were performed using Geo-Slope software to verify the new developed closed-form solution. The results obtained from closed-form solution are comparable with those obtained from Equation (16), as shown in Figures 6 and 7. Figure 6 illustrates SF_{ratio} values with respect to column location.

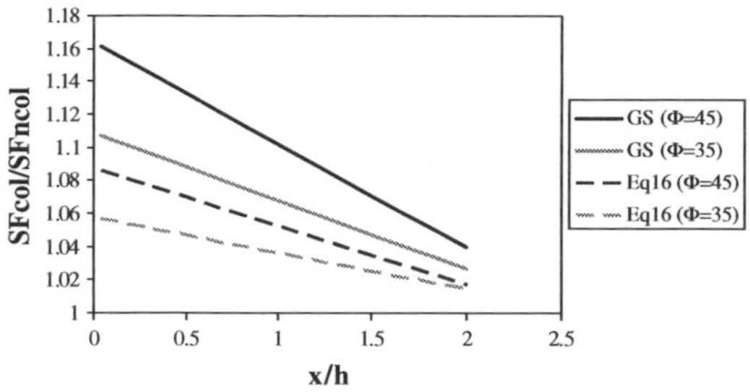

Figure 6. Comparison between relative safety factors given by Equation (16) and those given by Geo-Slope (GS stands for Geo-Slope).

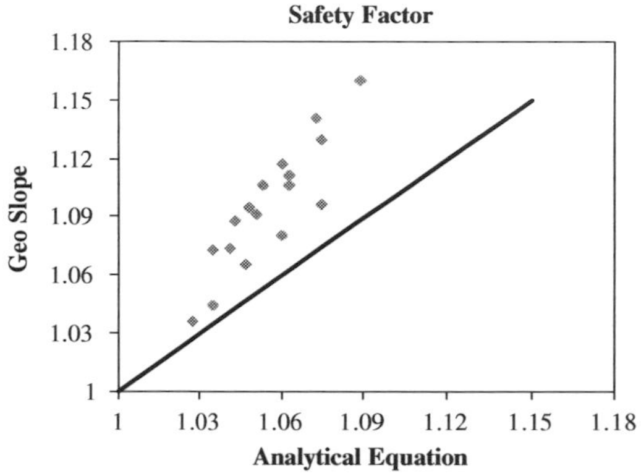

Figure 7. Comparison of SF values from closed-form solution and Geo-Slope.

Figures 6 and 7 show that the maximum difference between the results of two methods is less than 10%. The closed-form solution gives lower SF_{ratio} than Geo-Slope software, and this is on the safe side. This difference may be partly attributed to ignoring total external forces exerted on the strip stone column.

8 CONCLUSIONS

The results of analyses performed in this paper indicate that:

1. The SF values of stone column-reinforced slopes are influenced by various parameters including geometrical specifications of slope, slip surface, geotechnical properties of soil and stone column materials, center to center of columns, location of columns , number of column rows.
2. If the slope is reinforced by a row of column, the maximum SF is achieved when the column is located in the upper slope head.
3. The SF decreases with moving the column from the slope crest toward the slope toe.
4. With increasing sliding active force, for example due to low undrained shear strength of slope soil, increasing the slope height, or the slope angle, the influence of column on SF values increases.
5. With increasing equivalent width of stone columns and friction angle of column material, SF values increase remarkably.
6. An analytical equation verified by Geo-Slope was developed to determine the SF accurately for practical purposes.
7. The SF values for slopes reinforced with two rows of columns are higher for cases when columns are located in the upper slope part and column rows are very close.

REFERENCES

Ausilio, E., Conte, E. & Dente, G. 2001. Stability analysis of slope reinforced with piles. *Computers and Geotechnics*. 28: 591–611.

Bowles, J. E. 1984. *Physical and Geotechnical Properties of Soils*. Singapore: Mc graw hill Inc.

Cheung, K. 1998. Geogrid reinforced light weight embankment on stone columns. *Roading Geotechnics*. 273–278.

Das, Braja. M. 1941. *Principles of Geotechnical Engineering*. Boston: PWS publishing company.

Desai, C. S. & Abel, J. F. 1972. A numerical method for engineering analysis. *Introduction to the Finite Element Method*. Melbourne: Van nostrand reinhold company.

Fredlund, D. G. & Scoular, R. E. G. 1999. Using limit equilibrium concepts in finite element slope stability analysis. *Slope Stability Engineering*. Saskatoon: University of Saskatchewan.

Geo Slope International Ltd. 2002. *Users Guide Geo Slope Office*. Alberta.

Griffiths, D. V. & Lane, P. A. 1999. Slope stability analysis by finite elements. *Geotechnique* 49(3): 387–403.

Hassiotis, S., Chameau, J. L. & Gunarante, M. 1997. Design method for stabilization of slopes with piles. *Journal of Geotechnical and Geoenvironmental Engineering* 123(4): 314–323.

Jorge, G. & Zornberg, M. 2002. Geosynthetic reinforcement in landfill design. *American Society of Civil Engineering.*

Komak Panah, A. 1994. *Slope Remediation*. Tehran: International institute of earthquake engineering and seismology.

Rocscience Inc. 2004. *Application of Finite Element Method to Slope Stability. Toronto.*

School of Civil Engineering Georgia Institute of Technology Atlanta. 1983. Design and construction of stone columns. *Federal Highway Administration Office of Engineering and Highway Operations Research and Development.* Volume I and II.

New Horizons in Earth Reinforcement – Otani, Miyata & Mukunoki (eds)
© 2008 Taylor & Francis Group, London, ISBN 978-0-415-45775-0

Bearing capacity of reinforced foundation subjected to pull-out loading: 3D model tests and numerical simulation

T. Nakai, F. Zhang, M. Hinokio, H.M. Shahin, M. Kikumoto & S. Yonaha
Nagoya Institute of Technology, Nagoya, Japan

A. Nishio
Mitsubishi Heavy Industry, Mihara, Japan

ABSTRACT: The pile foundation with reinforced bars is put into practical use for increasing uplift bearing capacity of transmission towers and others For investigating the mechanism of such type of reinforced foundation under various loading conditions in real 3D conditions, 3D model tests and the corresponding numerical analyses were performed, in which the insertion direction of reinforcements and the position of reinforcement are different. The test results show that the reinforcements protruded diagonally downward is the most effective under vertically uplift loading in the same way as those in 2D condition. On the other hand, when the direction of the uplift load is inclined, the reinforcing effect decreases with increasing inclined angle. The numerical results in which mechanical behavior of the soil and the reinforcement and frictional behavior between the soil and the reinforcement are taken into account properly describe well the experimental results.

1 INTRODUCTION

Foundations with reinforcements protruded diagonally downward or horizontally from the side of the foundation were developed and put into practice to increase the uplift bearing capacity of electric transmission tower and others (Matsuo and Ueno, 1989; Tokyo Electric Power Company and Dai Nippon Construction, 1990). 2D numerical simulation and model tests have been carried out to investigate the mechanism of reinforcement (Nakai et al., 2001; Hinokio et al., 2007). The numerical and experimental results show that the reinforcements protruded diagonally downward from the side of the foundation are the most effective, when the foundation is uplifted vertically. However, the reinforcements set up at the side of the foundation are not effective against inclined uplift loading.

In the present study, 3D model tests of the foundation with the flexible reinforcements, which have only tensile stiffness, have been carried out under not only vertical uplift loading but also inclined uplift loading to investigate the reinforced mechanism and to develop the most effective foundation under various kinds of loading conditions. We pay attention particularly to the direction of the reinforcements and the position of the reinforcements. In the finite element analyses, subloading t_{ij} model (Nakai & Hinokio 2004) is used as an elastoplastic constitutive model. This model can describe typical stress deformation and strength characteristics of soils such as the influence of intermediate principal stress, the influence of stress path dependency of plastic flow and the influence of density and/or confining pressure.

In addition to the comparison between tests results and computed results in 3D conditions, these results are compared with those in 2D conditions described in another paper (Hinokio et al, 2007).

2 DESCRIPTION OF MODEL TEST AND ANALYSIS

Figure 1 shows the schematic diagram of the model test apparatus. The size of the model ground is 100 cm in width, 80 cm in length and 50 cm in height. Alumina balls, having diameters of 2.0 mm and 3.0 mm and mixed with the ratio of 1:1 in weight, are used as the model ground (unit weight of the mass is 21.5 kN/m^3). The foundation with the length of 23 cm and 6 cm in diameter is set up in model ground where the penetration depth of the foundation is 18 cm. The arrangement patterns of the reinforcements are illustrated in Figure 2. In the case that the reinforcements are set up at the depth of 15 cm of the side of foundation, the length of the reinforcements (5 cm) should be less than the diameter of the foundation (6 cm). This is due to the construction condition – in the actual case, the reinforcements are protruded from the inside of the foundation after construction of the caisson type foundation. Three kinds of the protruded directions

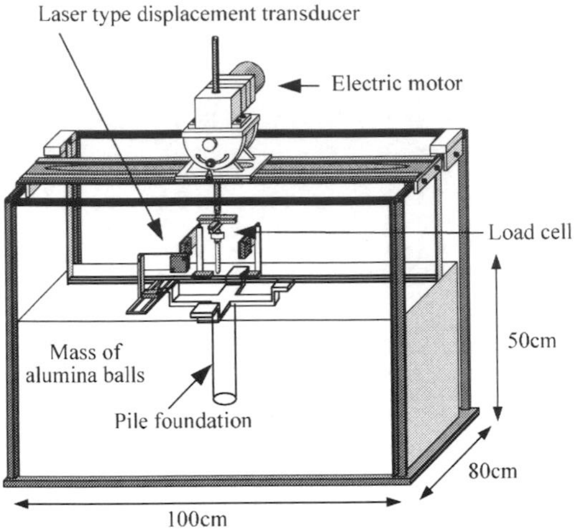

Figure 1. Schematic diagram of the model tests apparatus.

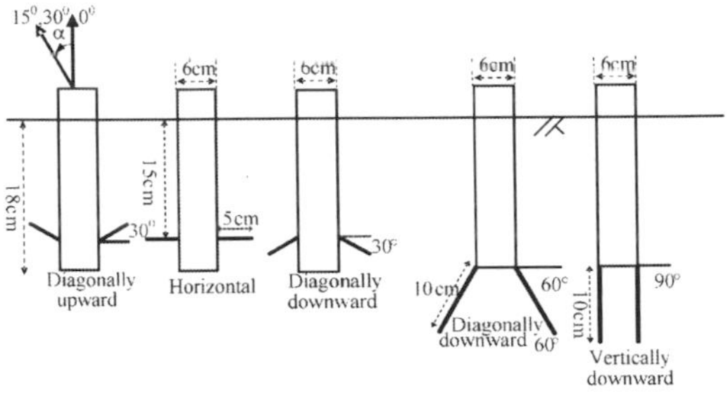

Figure 2. Arrangement of reinforcements.

of reinforcements (diagonally upward, horizontal and diagonallydownward) are employed. In addition to these cases, model tests in which the reinforcements are set up vertically and diagonally downward at the bottom of the foundation are carried out. For these cases, longer reinforcement (10 cm) is protruded from the inside of the foundation.

Aluminum plates with the thickness 0.1 mm and the width of 1 cm are used as the reinforcements for every case. Reinforcement with the thickness of 0.1 mm has enough stiffness against tension but has no bending stiffness. Aluminum plates on which aluminum rods of 1.6 mm in diameter are glued at an equal spacing of 1 cm are used. The friction angle between the reinforcement and the model ground is about 14.5°. Vertical and inclined (angle α is 0°, 15° and 30°) upward displacements are imposed continuously to the foundation. The direction of the uplift load and the plan view of the arrangement of the reinforcements are illustrated in Fig. 3. The uplift load of the foundation is measured by the load cell, and the displacements and rotations of the foundation in three-dimensional space

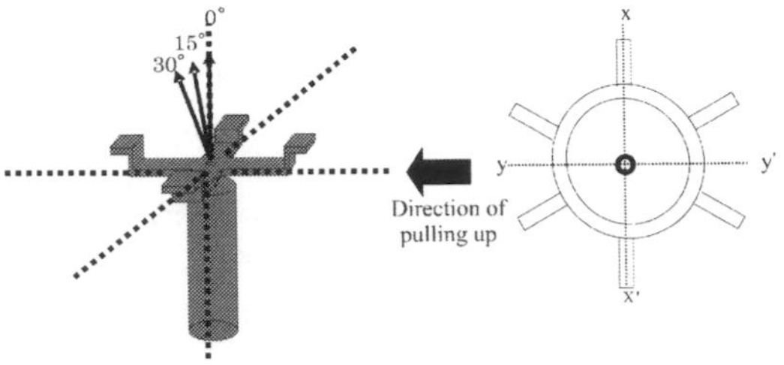

Figure 3. Direction of uplift load and plan view of the arrangement of reinforcements.

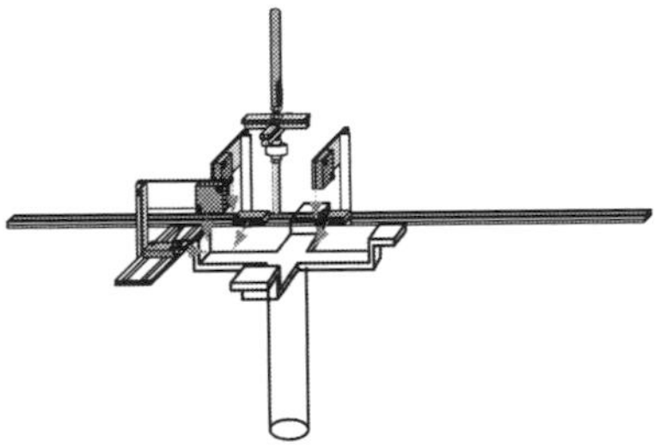

Figure 4. Arrangement of laser type displacement transducer.

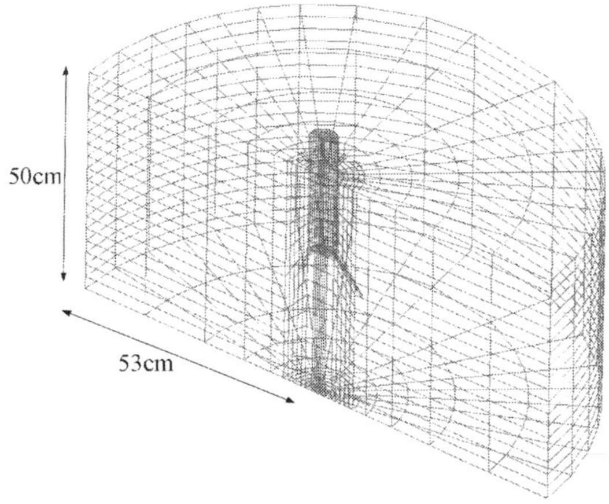

Figure 5. Finite element mesh with diagonally downward reinforcements.

can be grasped by six laser type displace transducers which is arranged around the foundation as shown in Fig. 4. Axial force and bending moment of the reinforcements are measured by the strain gauges that are glued at both sides of the reinforcements.

3D finite element analyses under drained condition are carried out in the same scale as the model tests. Finite element meshes for the cases that the reinforcements are protruded diagonally downward from the bottom is shown in Fig. 5. Elastoplastic constitutive model for soils named subloading t_{ij} model (Nakai and

Table 1. Parameters of mass of alumina balls.

Parameters	Value
λ	0.024
κ	0.014
N (e_{NC} *at* $p = 98\,kPa$ & $q = 0\,kPa$)	0.78
$R_{CS} = (\sigma_1/\sigma_3)_{CS(comp.)}$	2.0
β	2.0
ν_e	0.2
a	150

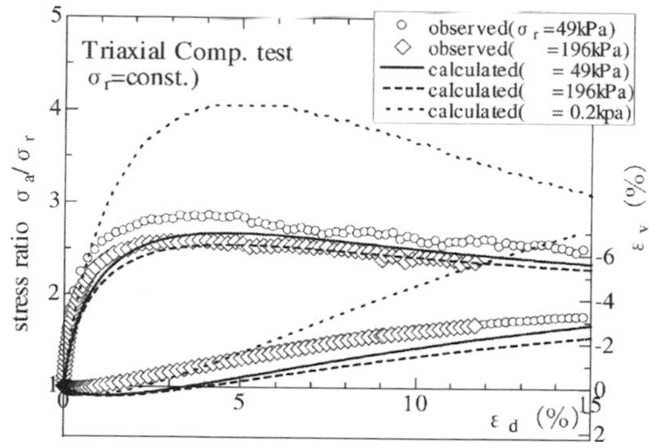

Figure 6. Relationships of stress-strain of alumina balls mass.

Hinokio, 2004) is used. This model can describe properly the following typical characteristics of soils, in spite of its small numbers of parameters: (1) Influence of intermediate principal stress on the deformation and strength of soils. (2) Influence of stress path on the direction of plastic flow, and (3) Influence of density and/or confining pressure. The values of soil parameters for the alumina balls mass are listed in Table 1. Where, λ and κ are the slope of loading and unloading curve of *e-lnp* graphs at the loosest state. N is the void ratio at mean principal stresses (p) 98 kPa in the above mentioned loading curve. β is the model parameter, ν_e is the Poisson's ratio and 'a' represents the influence of density and/or confining pressure. The dots and the solid curves in Fig. 6 are the calculated results corresponding to the observed stress-strain relations (dots) in the triaxial compression tests of the mass of alumina balls, and the dotted curves are the calculated results in which the initial confining pressure is assumed to be two orders smaller in magnitude. This is because the initial confining pressure in model tests is much smaller than that in the triaxial tests. The initial state of the model ground is created by simulating the one-dimensional self-consolidation. The foundation is assumed to be an elastic material with enough stiffness. The reinforcements are simulated by shell elements. Axial stiffness and bending stiffness of each reinforcement is EA $= 7.03 \times 10$ kN and

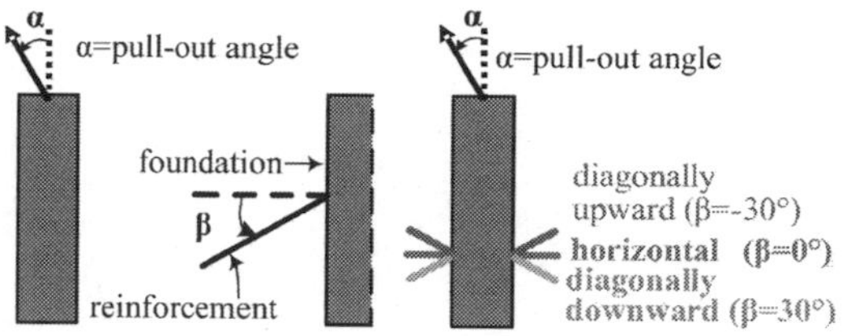

Figure 7. Description of the factors α and β.

EI $= 5.86 \times 10^{-8}$ kN-m^2. In order to model the frictional behavior between the foundation and the ground and between the reinforcements and the ground, an elastoplastic joint element is inserted between them (Nakai, 1985). The friction angle used in the analysis between the foundation and the ground is 8°, and those between the reinforcements and the ground is14.5°.

3 RESULTS AND DISCUSSIONS

3.1 *Reinforcements protruded from the side of foundation*

Figures 8–10 show the observed and computed variations of uplift load and rotation angle of the foundations against the displacements of the foundation in which the reinforcements are set up at different directions from the side of the foundation. In the figures the upper part of the vertical axis from x = 0 represents load in Newton, and the lower part denotes rotation angle (θ) of the foundation. In these figures, curves without marks show the results of the foundation without reinforcements. The factors α and β of the legends are described in Figure 7. Here, α is the angle of uplift loading which is measured from the vertical direction, while β denotes the placement angle of reinforcement which is measured from the horizontal direction. It can be seen from Figure 8(a) that the reinforcements protruded diagonally downward from the side of the foundation is the most effective against vertical uplift loading. The results of the diagonally upward and the horizontal reinforcements are almost same, but smaller than that of the diagonally downward reinforcement. The observed and the computed results are similar to those obtained in the previous works (Nakai et al., 2001). The results are also very much close to the results of the two-dimensional analysis (Fig.15). The results of two-dimensional analysis (Hinokio et al, 2007) are illustrated in the APPENDIX. This research is conducted using the flexible reinforcements which act as tensile reinforcements alone.

From Figures 9 and 10, it is revealed that for the inclined uplift loading the reinforcements set up at the side of the foundation are not as effective as observed for the vertical uplift loading regardless the placement angles of the reinforcement. Although for the

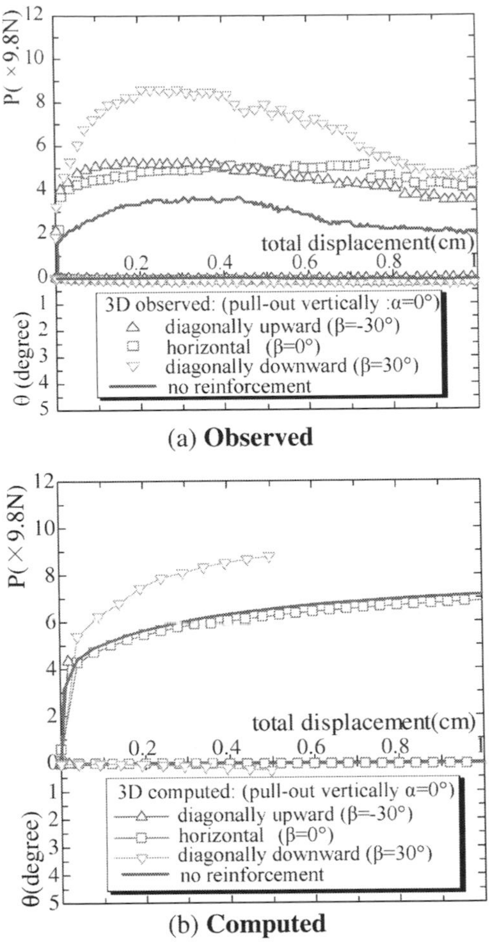

Figure 8. Load and rotation vs. displacement under vertical uplift loading ($\alpha = 0°$): reinforcements protruded from side.

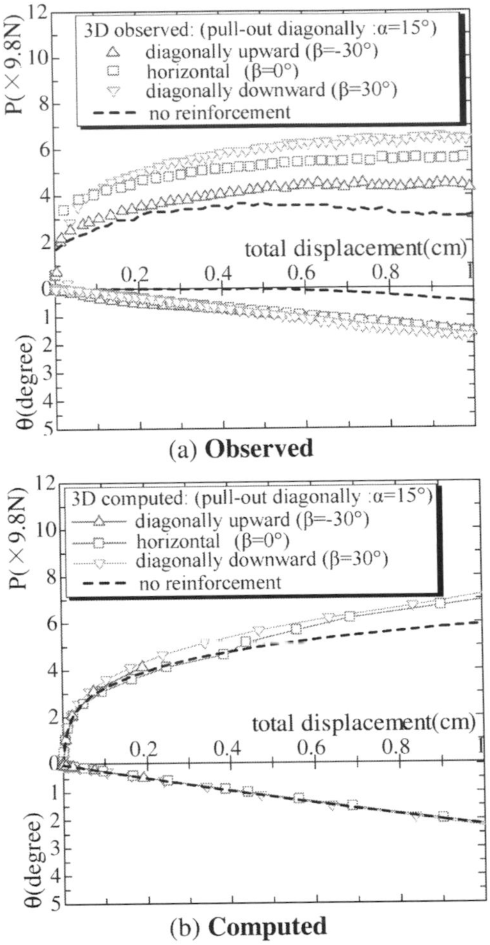

Figure 9. Load and rotation vs. displacement under inclined uplift loading ($\alpha = 15°$): reinforcements protruded from side.

3D condition the movement inside the ground is not possible to visualize by taking the photographs of the ground unlike the 2D condition, we can guess that for the diagonally downward reinforcements the deformed zone of the ground spreads wider region from the position of the reinforcements compare to the other positions of the reinforcements. For this reason, the diagonally downward reinforcements have still advantage over the horizontal and diagonally upward reinforcements. In the case of the inclined uplift loading, the frictional force between the ground and the reinforcements in the loading side works in the upward direction and acts as a negative resistance which diminishes the positive resistance of the reinforcements of the other side. Hence, the effectiveness of the reinforcements decreases with the increase of the inclination of the uplift loading. The computed results capture well the observed behavior of the foundations qualitatively and quantitatively. The observed and computed results appear to be in agreement with the results of the two-dimensional observation (Figs.16 and 17).

3.2 *Reinforcements protruded from the bottom of foundation*

Figures 12 to 14 show the observed and computed results of uplift load and rotation angle of the foundation, in which the reinforcements are protruded from the bottom of the foundation. Figure 11 denotes the description of angles α and β, where $\beta = 90°$ represents the reinforcement placed in the vertically downward direction. It is seen from these figures that these types of reinforcements increase the uplift bearing capacity significantly not only against vertical uplift load but also against inclined uplift load. Though the

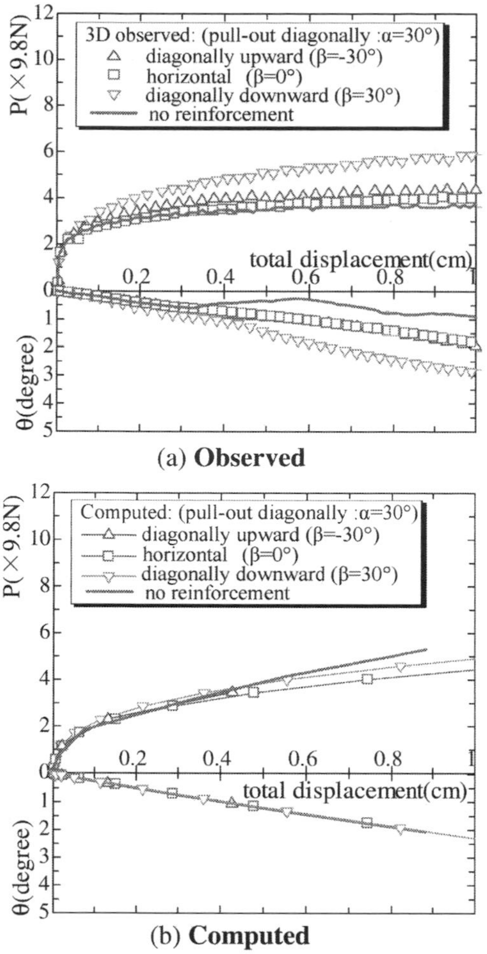

Figure 10. Load and rotation vs. displacement under inclined uplift loading ($\alpha = 30°$): reinforcements protruded from side.

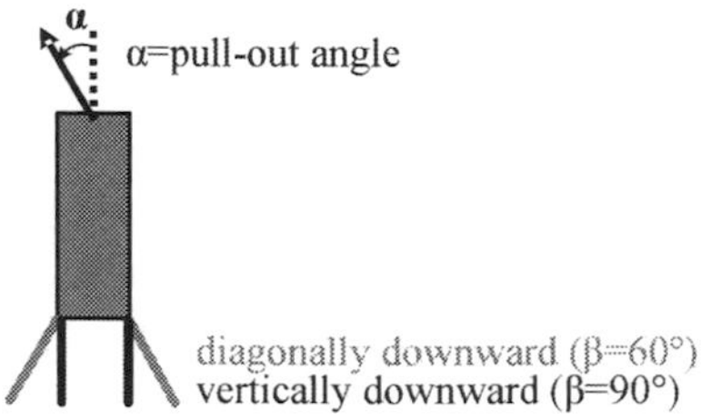

Figure 11. Description of reinforcement protruded from bottom.

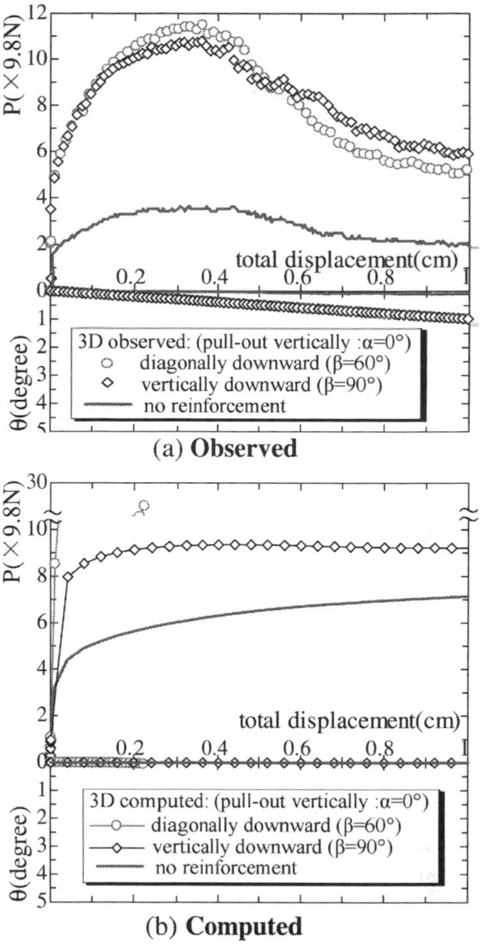

Figure 12. Load and rotation vs. displacement under vertical uplift loading ($\alpha = 0°$): reinforcements protruded from bottom.

uplift bearing capacity for diagonally downward and vertically downward reinforcements is almost same in the model tests, in the numerical analyses it is slightly larger for the diagonally downward reinforcement than that for the vertically downward reinforcement. However, there are good agreement between the results of the model tests and the numerical simulations. There are also quite similarities of the results of the three-dimensional model tests and numerical analyses with the two-dimensional ones (Figs.18 to 20). Therefore, it can be said that the 2D model tests can represent well the behavior of 3D model tests in general three-dimensional stress conditions. Similar to the model tests 2D plane strain analyses well simulate the behavior of general three-dimensional stress condition in predicting the uplift bearing capacity. Comparing the results of the reinforcements protruded from the side with the reinforcements set up at the bottom, it is found that the uplift bearing capacity for the later case is much larger than that for the former one for both loading conditions.

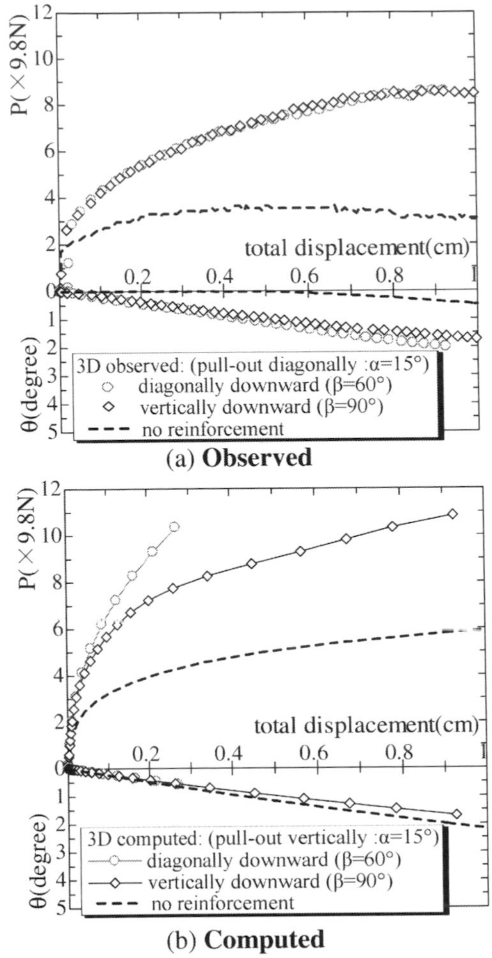

Figure 13. Load and rotation vs. displacement under inclined uplift loading ($\alpha = 15°$): reinforcements protruded from bottom.

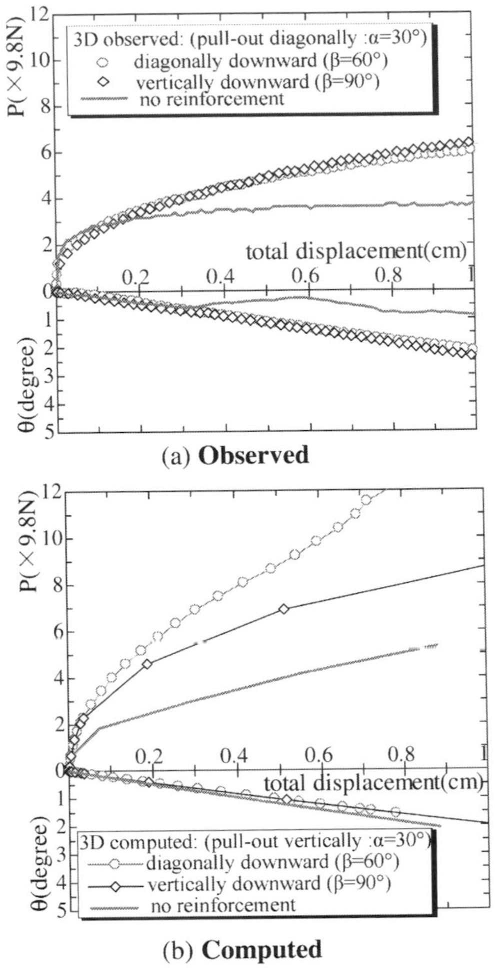

Figure 14. Load and rotation vs. displacement under inclined uplift loading ($\alpha = 30°$): reinforcements protruded from bottom.

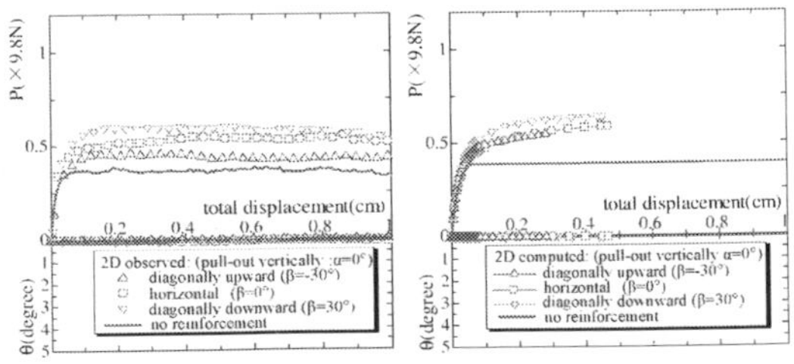

Figure 15. Load and rotation vs. displacement under vertical uplift loading ($\alpha = 0°$): reinforcements protruded from side.

4 CONCLUSIONS

The following conclusions are obtained from the model tests and numerical simulations of the reinforced foundation under uplift loadings:

(1) When reinforcement bars are set up at the side of the foundation, the reinforcements protruded diagonally downward is the most effective against vertically uplift loading. However, the reinforcement set up at the side of the foundation is not effective against inclined uplift loading (inclination angles of the uplift loads are 15° and 30° in the present study).

(2) The reinforcements which are protruded downward from the bottom of the foundation are effective not only under vertical uplift loading but also inclined uplift loading.

(3) 2D model tests and numerical analyses can represent well the behavior of 3D model tests and

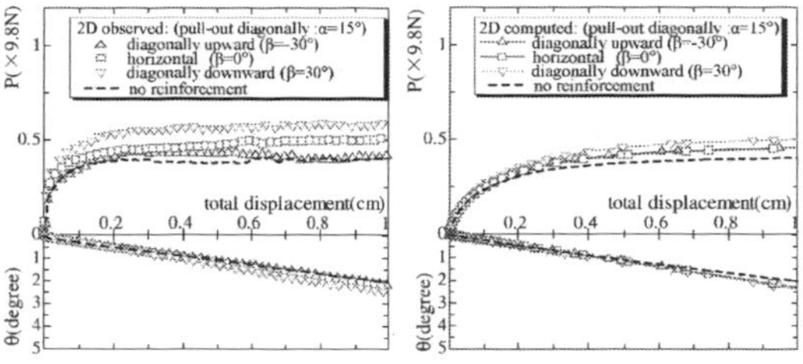

Figure 16. Load and rotation vs. displacement under inclined uplift loading ($\alpha = 15°$): reinforcements protruded from side.

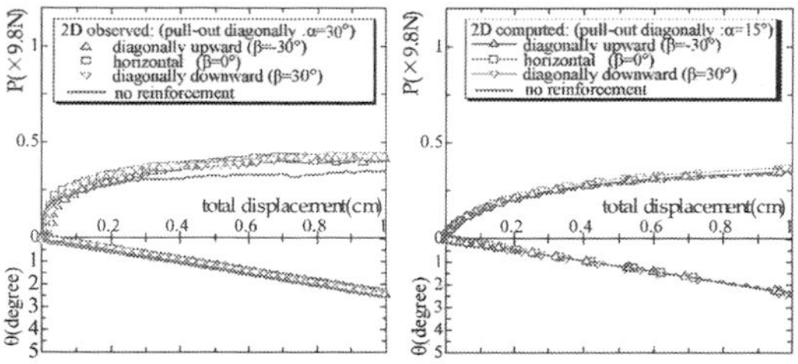

Figure 17. Load and rotation vs. displacement under inclined uplift loading ($\alpha = 30°$): reinforcements protruded from side.

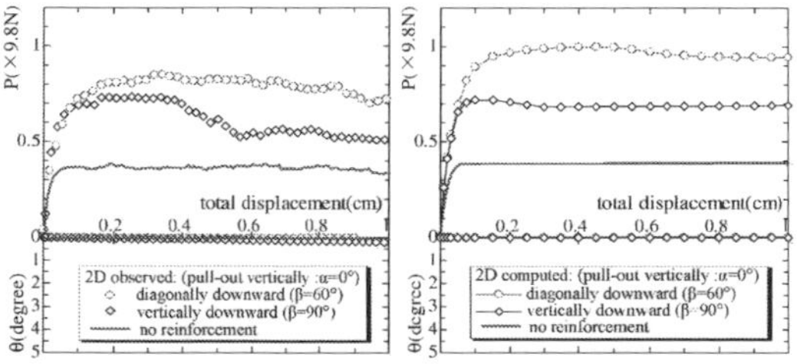

Figure 18. Load and rotation vs. displacement under vertical uplift loading ($\alpha = 0°$): reinforcements protruded from bottom.

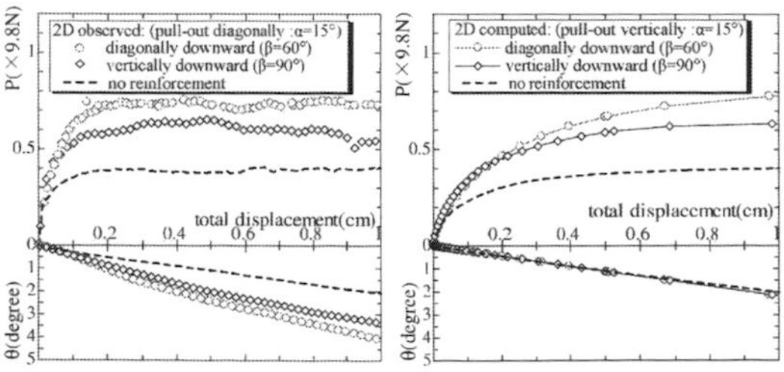

Figure 19. Load and rotation vs. displacement under inclined uplift loading ($\alpha = 15°$): reinforcements protruded from bottom.

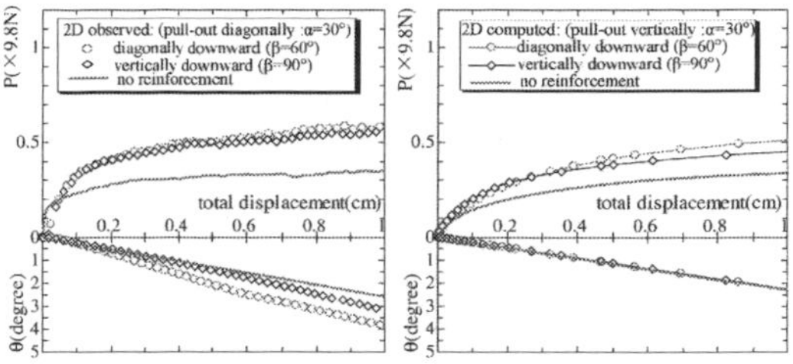

Figure 20. Load and rotation vs. displacement under inclined uplift loading ($\alpha = 30°$): reinforcements protruded from bottom.

numerical analyses in general three-dimensional stress conditions.

The analysis in which typical mechanical behavior of soils is appropriately taken into account can predict well the behavior of the reinforced foundation under uplift loading.

REFERENCES

Hinokio, M., Nakai, T., Yohona, S. and Nishio, A. (2007): Bearing capacity of reinforced foundation subjected to pull-out loading and its reinforced mechanism, Proc. of 13th Asian Regional Conference on Soil Mechanics and Geotechnical Engineering, India, December (in press).

Matsuo, M. and Ueno, M.(1989) – Development of ground reinforcing type of foundation, Proc. of 12th ICSMFE, 2, pp1205–1208.

Nakai, T. (1989) – Finite element computations for active and passive earth pressure of retaining wall, Soils and Foundatons, 25(3), 98–112.

Nakai, T., and Hinokio, M. (2004) – A simple elastoplastic model for normally and over consolidated soils with unified material parameters. Soils and Foundations. 44(2): 53–70.

Nakai, T., Teranishi, T., Hinokio, M. and Adachi, K. (2001) – Behavior of reinforced foundation under uplift and push-in loading: model tests and analyses, Proc. of IS-Kyushyu 2001, 593–598.

Tokyo Electric Power Company and Dai Nippon Construction (1990) – Report on bearing capacity of caisson type foundation with reinforcing bars, 4, (in Japanese).

APPENDIX: RESULTS OF 2D CONDITION

The results of the two-dimensional model tests and numerical analyses conducted by the authors (Hinokio et al, 2007) are illustrated in this section for the comparison with the results of the three-dimensional condition.

New Horizons in Earth Reinforcement – Otani, Miyata & Mukunoki (eds)
© 2008 Taylor & Francis Group, London, *ISBN 978-0-415-45775-0*

Numerical simulation on bearing capacity of soilbag-reinforced ground considering finite deformation

D. Muramatsu, F. Zhang & H.M. Shahin
Nagoya Institute of Technology, Japan

ABSTRACT: In this research, we conduct a numerical analysis of the capacity problem of soilbag-reinforced ground with elastoplastic finite element method (FEM) to simulate a series of two-dimensional model tests a ground made of aluminum rods. The FEM conducted here is based on subloading t_{ij} model (Nakai and Hinokio, 2004). Finite deformation scheme is adopted in the FEM due to a large deformation happened in the model tests. The numerical results coincide with the test results not only qualitatively but also quantitatively to some extent.

1 INSTRUCTIONS

Soilbag reinforcement is one of the most ancient and efficient methods used by human beings to improve a ground we live. It is a very simple, cheap method and the soilbag can be easily manufactured and constructed in many construction sites of civil engineering. It does not need any heavy construction machine in its use. Recently, the soilbag reinforcement has been wildly used in ground improvement and its mechanism has been studied, especially with laboratory tests and field tests. Continuum-theory based numerical analyses on soilbag, however, were mainly focused on simplified sheet-shaped reinforcement consisted of a large number of soilbags, instead of individual soilbag. There are some calculations using DEM to simulate the mechanical behavior of soilbag, in which the complicated stress-strain relation of the combination of soil and soilbag is still found to be difficult to determine. Finite element method (FEM) seems to be a very useful tool in describing the bearing capacity problems of soilbag-reinforced ground if the soil is properly modeled with a suitable constitutive model. In this paper, finite element analysis based on subloading t_{ij} model (Nakai & Hinokio, 2004) is conducted to simulate a series of model tests on the capacity problem of a soilbag-reinforced model ground made of Aluminum bars (Matsuoka and Liu, 1999; Iwai & Okuda, 1993). Due to large deformation of the ground happened in the tests, a finite deformation scheme using Jaumann rate of stress tensor is adopted in the analysis.

2 MODEL TESTS

In the two-dimensional (2D) models conducted by Matsuoka and Liu (1999), a ground made of aluminum rods was used. The aluminum rods have two types, with the diameters of 1.6 mm and 3.0 mm and a length of 50 mm. The weight ration of the two rods is 3:2 and the unit weight of the ground is 20.4 kN/m^3. The size of the model ground is 120 cm in width, 45 cm in height. In the model test, a rigid footing made of brass, with a width of 20 cm and a depth of 2 cm was laid on the surface of the ground and a vertical load was applied on the footing, as shown in Figure 1. Soilbags were just laid beneath the footing. During the loading, Load cell and displacement transducer were installed to measure the vertical load and the settlement of the footing. The material of soilbag is a thin paper made of Polypropylene.

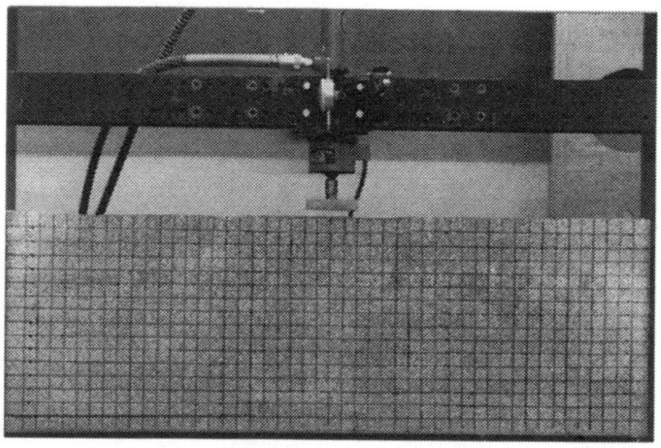

Figure 1. Model test device for bearing capacity (Iwai & Okuda, 1993).

In the model tests, Matsuoka & Liu (1999) conducted four types of grounds, CASE A, without soilbag; CASE B, one soilbag with a size of 15 × 15 cm, CASE C, six soilbags with a size of 15 × 2.5 cm and CASE D, twelve soilbags with a size of 4 × 1.5 cm, as shown in Figure 2.

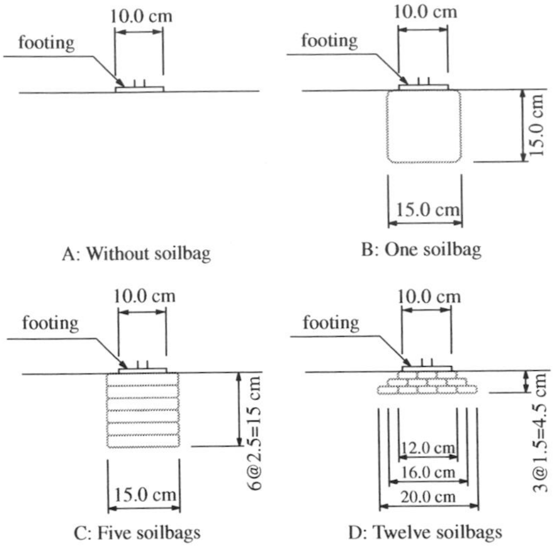

Figure 2. Patterns of reinforcement with soilbag in model tests.

3 NUMERICAL ANALYSIS WITH FEM IN FINITE DEFORMATION SCHEME

In this paper, a 2D finite element analysis using finite deformation scheme was conducted based on a simple constitutive model for clay and sand named subloading t_{ij} model, proposed by Nakai & Hinokio (2004), in which the following important aspects of soft soils can be properly described:

a) Influence of intermediate principal stress on deformation and strength of soils
b) Stress-path dependency on the direction of incremental plastic strain tensor
c) Positive dilatancy in strain hardening region
d) Influence of density and/or confining pressure on deformation and strength of soils.

The constitutive model has seven parameters among which the first five parameters listed in Table 1 are the same as those of Cam-clay model. Other two parameters are those parameters that determine the shape of yield function and consider the influence of density and confining stress. Detailed description about the model and the method to determine the values of these parameters can be referred to corresponding reference (Nakai & Hinokio, 2004). Figure 3 shows the test results of Aluminum bar with two-axial compression test and its theoretical prediction. From the figure, it is know that the model can predict well the behavior of model ground made of Aluminum bars at element level. In the element test, the vertical stress (σ_1) was kept constant while the lateral stress (σ_2) was reduced up till failure.

In finite element analysis, the modeling of soilbag is also very important. It is modeled with a truss element that has a nonlinear stress-strain relation in which, the stiffness of the truss in compressive state is very small so that the soilbag only resist tensile force. Figure 4 shows the stress-strain relation for soilbag.

Table 1. Material parameters of model ground.

λ	0.008	Parameters which are the same as Cam-clay model
κ	0.004	
N (e_{NC} at $p = 98$ kPa & $q = 0$ kPa)	0.3	
R_{cs} ($(\sigma_1/\sigma_3)_{cs}$)	1.8	
ν_e	0.2	
β	1.2	Parameter which determines the shape of yield function, $\beta = 1$ means the same as Cam-clay model
a	1300	Parameter which considers the influence of density and confining stress

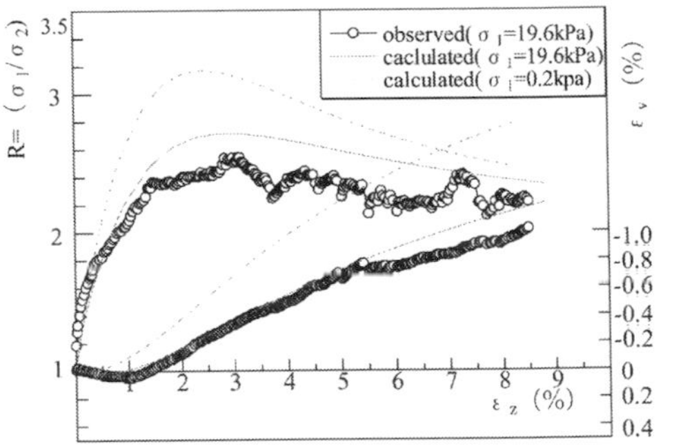

Figure 3. Test results of Aluminum bar with two-axial compression test and its theoretical prediction.

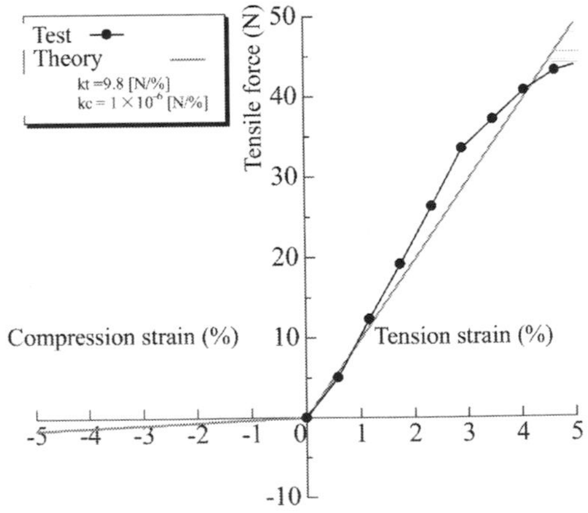

Figure 4. Nonlinear model for truss element that simulates soilbag material in FEM.

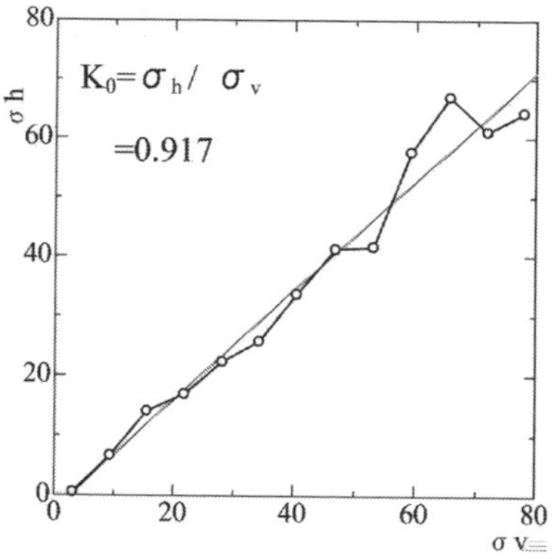

Figure 5. Tested distributive k_0 values of model ground.

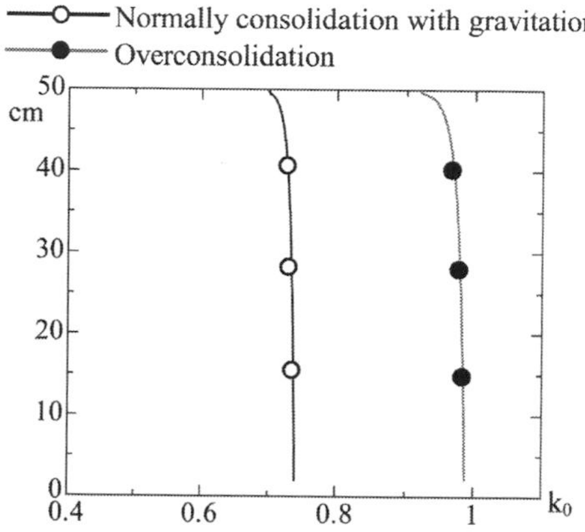

Figure 6. Calculated distributive K_0 values of model ground.

In the numerical analysis, accurate evaluation of initial stress field is very important. Figure 5 shows the test result of the distribution of k_0 value in initial stress field of the model ground made of Aluminum bars. The ground is relatively dense and the test value of k_0 is about 0.91. In finite element analysis, if the initial stress field is calculated with gravitational field under the condition of normally consolidated condition, then the calculated k_0 value will be much smaller than the observed one. Figure 6 shows the calculated distribution of k_0 value in initial stress field of the model ground with consolidation processes. In the figure, normally consolidation with gravitation means the result is calculated under the condition that initial stress analysis is simply evaluated with 1G gravitation; while overconsolidation here means the result is calculated under the condition that initial stress analysis is firstly evaluated with 1.5G gravitation and then unloaded with 0.5G gravitation to reproduce the overconsolidated state. Compared the results from the test shown in Figure 5, it is understood that the initial

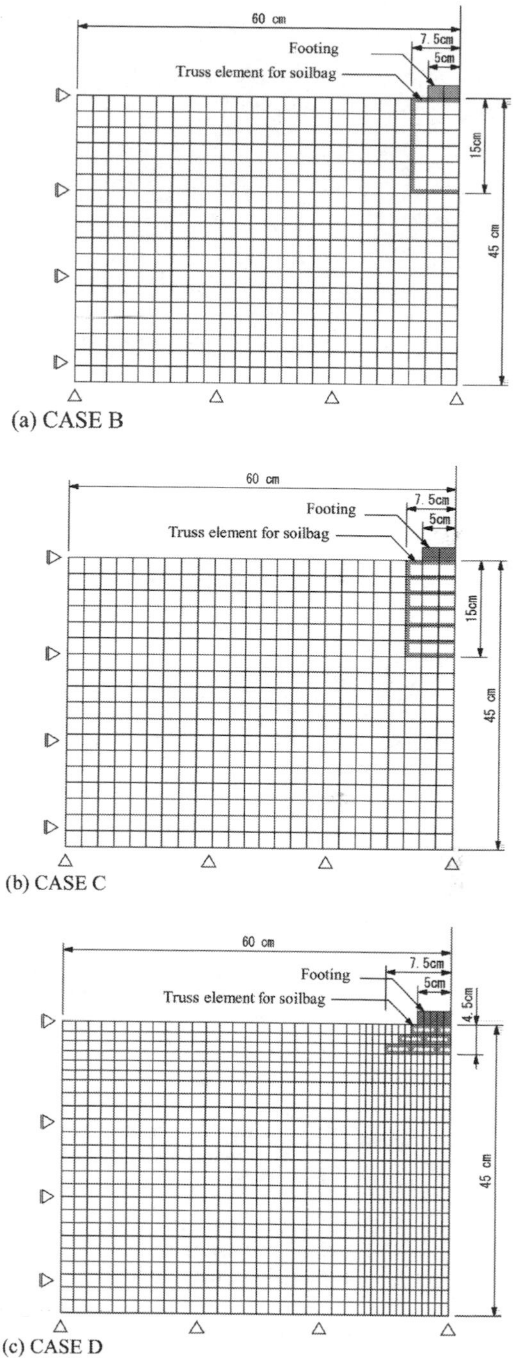

Figure 7. Finite element meshes.

stress state can be accurately evaluated with a suitable overconsolidation simulation.

Figure 7 shows the finite element meshes used in the finite element analyses. The mesh used in CASE

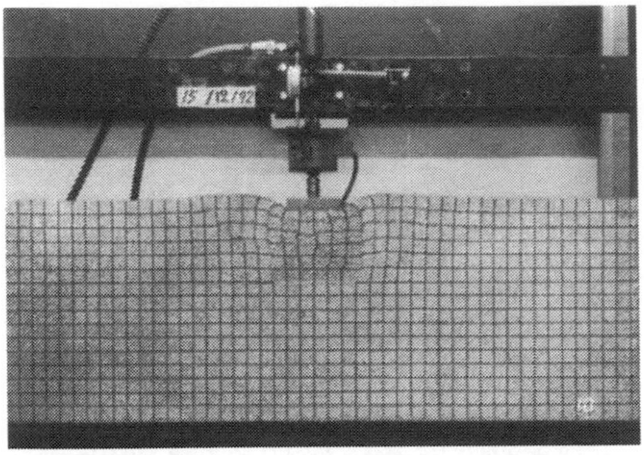

(a) v/B = 0.15~0.20

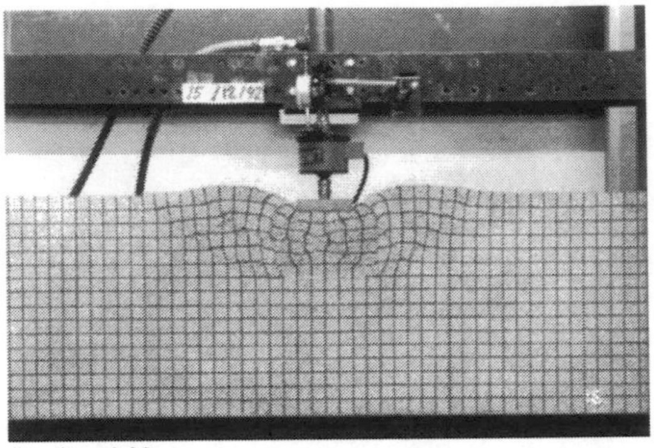

(b) v/B = 0.30

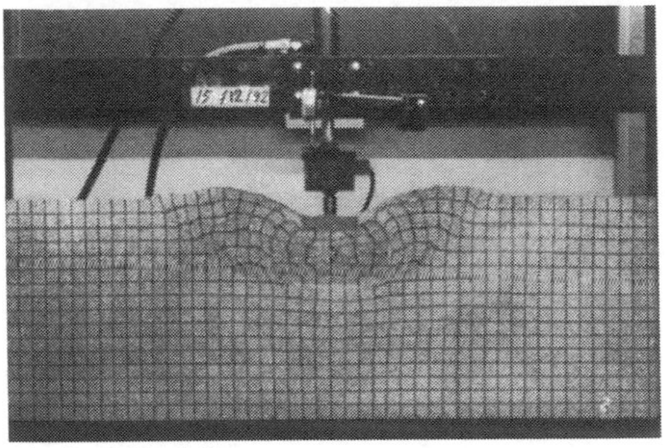

(c) v/B = 0.50~0.55

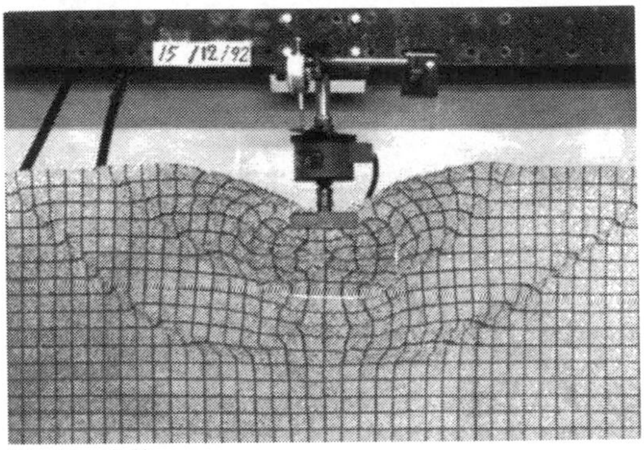

(d) v/B = 0.70

Figure 8. Observed deformation patterns of ground and soilbag at different loading stages (CASE B, Iwai & Okuda, 1993).

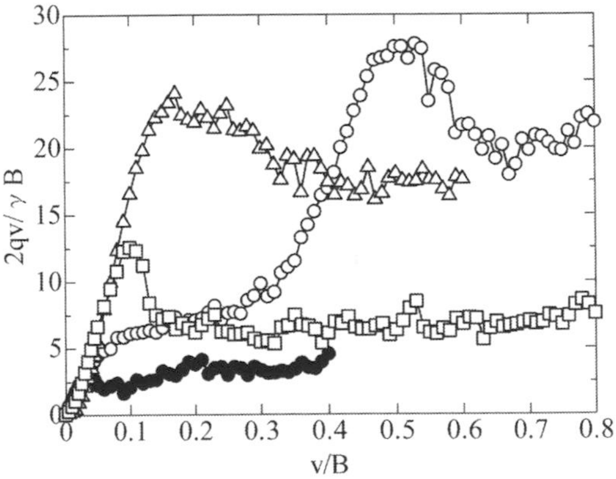

(a) Test results (Matsuoka & Liu, 1999)

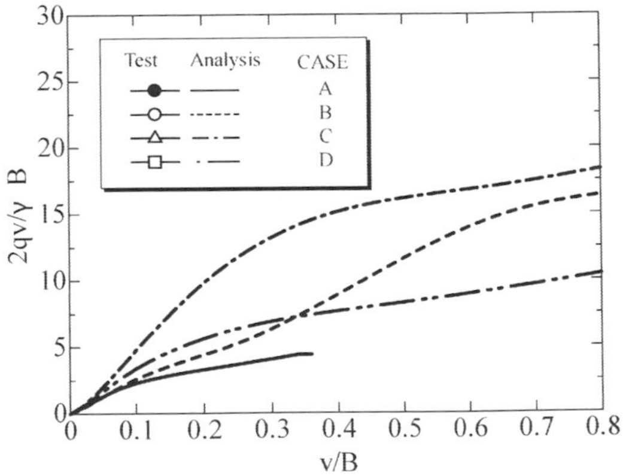

(b) Calculated results

Figure 9. Comparison of load-displacement relations between test and calculation.

A is the same as the one in CASE B but without truss elements. Due to the symmetrical condition, only half of the domain was considered in the calculation. Boundary condition of the domain is as follow: the bottom is fixed and the side is fixed in horizontal direction and free in vertical direction.

Figure 8 shows the observed deformation patterns of ground and soilbag at different loading stages of CASE B. Compared with the bearing capacity problem of green field, soilbag-reinforced ground shows a particular deformation pattern. When the settlement of footing (10 cm in width) reaches a very large value

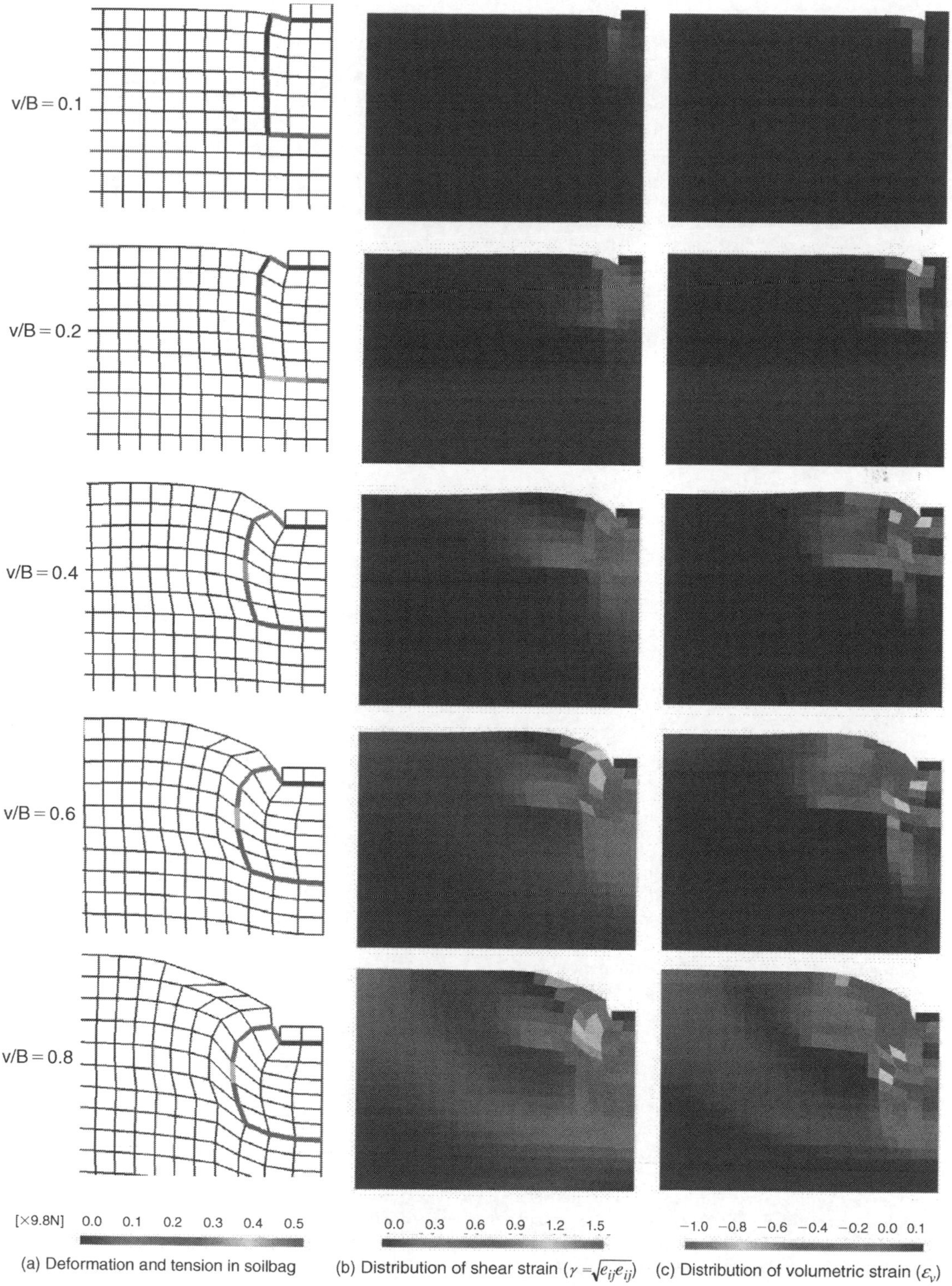

Figure 10. Calculated deformation and strain distribution of model ground (CASE B).

(7 cm), two sliding surfaces were formed within the ground, which is different from green field in which only one sliding surface could be seen. At first, the soilbag deformed largely until the settlement reached a value of 5.0 cm to 5.5 cm (v/B = 0.50 ~ 0.55) and formed the first sliding face. Then the deformed soilbag settled again keeping the shape of the deformed soilbag unchanged to form the second sliding face.

Due to finite deformation happened in the test, a finite deformation scheme using Jaumann rate of stress tensor was used in the numerical analysis. In the scheme, updated-Lagrangian method, as described in Eq. 1, is used for describing current coordinates.

$$x_i^{(t+\Delta t)} = x_i^{(t-\Delta t)} + \dot{u}_i^{(t-\Delta t)} \Delta t \qquad (1)$$

Strain rate tensor (Stretching) is defined as:

$$\dot{\varepsilon}_{ij} = \frac{1}{2}\left(\frac{\partial \dot{u}_i}{\partial x_j} + \frac{\partial \dot{u}_j}{\partial x_i}\right) \qquad (2)$$

The Jaumann rate of stress tensor is defined as

$$\overset{\nabla}{\sigma}_{ij} = \dot{\sigma}_{ij} + (\omega_{ik}\sigma_{kj} - \sigma_{ik}\omega_{kj}) \qquad (3)$$

where, σ_{ij} is Cauchy stress tensor and spin tensor ω_{ij} is defined as:

$$\omega_{ij} = \frac{1}{2}\left(\frac{\partial \dot{u}_i}{\partial x_j} - \frac{\partial \dot{u}_j}{\partial x_i}\right) \qquad (4)$$

Figure 9 shows the comparisons between observed and calculated load-displacement relations for different soilbag-reinforced grounds. It is known from the figure that for green field, the bearing capacity of the ground is much lower than those of reinforced grounds. For CASE B, with one soilbag, the load dropped down firstly when the settlement reached a certain value and then grown up again until a second sliding face formed within the ground. Calculated result also showed the same tendency. Among three different soilbag-reinforced grounds, CASE C, with six long and thin soilbags seems to be the most effective in improving the bearing capacity of the ground. Calculation also shows the same tendency as the test.

Figure 10 shows the calculated deformation and strain distribution of soilbag-reinforced ground of CASE B. Compared with the observed results of deformation shown in Figure 8, it is clear that calculated deformation of soilbag and ground coincide well with the observed ones. Especially the first and second sliding surfaces observed in test can also be well simulated in the numerical calculation in the same loading stages.

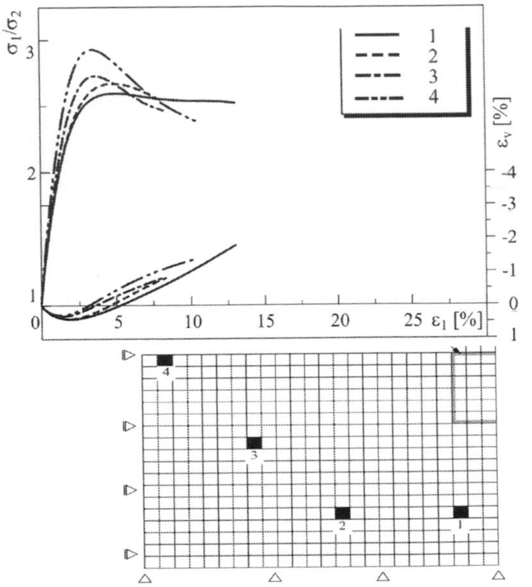

(a) CASE B

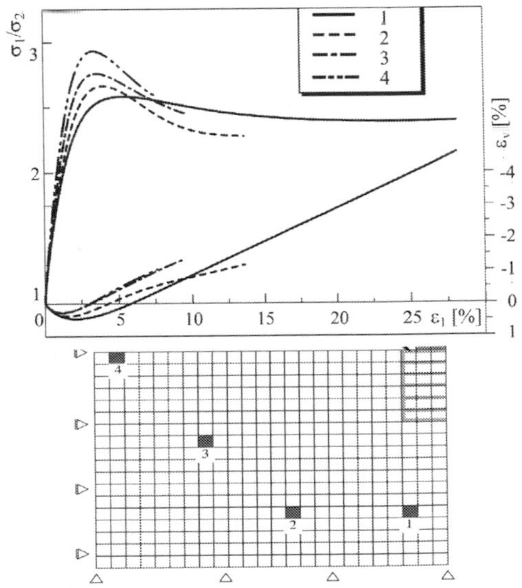

(b) CASE C

Figure 11. Calculated stress-strain-dilatancy relations of the elements along the sliding faces obtained from numerical calculations.

It is known that the maximum tension in soilbag happened at the bottom of the soilbag. The maximum shear strain in all cases reached a value of 100%, a typical finite deformation region, which clearly shows the necessity to use finite deformation scheme in the finite element analyses in this paper.

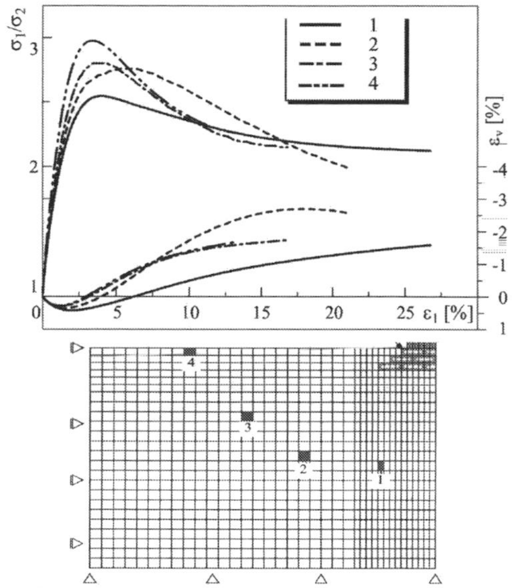

(c) CASE D

Figure 11. (Continued)

Figure 11 shows the calculated stress-strain-dilatancy relations of the elements along the sliding faces obtained from both the model tests and the corresponding numerical simulations. It is known from these figures that the stress-strain-dilatancy relations of all the elements along the second sliding surface of the soilbag-reinforced ground show positive dilatant and strain softening behaviors at residual state, similar to the element behavior of the ground material shown in Figure 3.

4 CONCLUSIONS

In this paper, a 2D FEM based on subloading t_{ij} model and finite deformation scheme, was used to simulate the capacity problem of soilbag-reinforced ground. The following conclusions can be given.

The material parameters of the model ground made of Aluminum bars are determined with biaxial compression tests. The initial stress state of model ground, which is under overconsolidated state in tests, is also carefully calculated with suitable consolidated loading in gravitational field. By correct evaluation of mechanical behavior of model ground at element level and initial stress condition, it is possible to conduct a relative accurate calculation. As the result, some typical mechanical behaviors of the model ground observed in the tests, such as the load-displacement relation, the deformation pattern of model ground, the formation of the first and the second sliding surfaces occurred within the soilbag-reinforced grounds can be properly simulated with the numerical analyses. Some other physical quantities that cannot be directly measured in the tests, such as the tensile force of soilbags and the dilation of largely deformed ground, can also be evaluated quantitatively, which could be useful in designing reinforcement of ground with soilbags in field. Further study should be down in its application to field in the future.

REFERENCES

Nakai, T. and Hinokio, M., 2004. "A simple elastoplastic model for normally and over consolidated soils with unified material parameters", Soils and Foundations, Vol.44, No.2, 53–70.

Matsuoka, H. and S. H. Liu, 1999. "Bearing capacity improvement by wrapping a part of foundation", JSCE, No617/III-46, 235–249 (in Japanese).

Iwai, S. and Okuda, N., 1993. "Research on the bearing capacity of reinforced ground", Bachelor Thesis of Nagoya Institute of Technology, (in Japanese).

New Horizons in Earth Reinforcement – Otani, Miyata & Mukunoki (eds)
© 2008 Taylor & Francis Group, London, ISBN 978-0-415-45775-0

3D soil reinforcement modeling by means of embedded pile

E.G. Septanika & P.G. Bonnier
Plaxis BV, Delft, The Netherlands

K.J. Bakker & R.B.J. Brinkgreve
Plaxis BV, Delft, Netherlands, Delft University of Technology, Delft, The Netherlands

ABSTRACT: Currently the embedded pile element – consisting of beam elements crossing the soil elements interior, embedded interface to model skin interaction and embedded nonlinear spring to model end capacity – is successfully implemented in the Plaxis 3D Foundation Beta Program. The present embedded pile approach, in which the beam elements can cross the soil element independent of the global mesh structure, makes it also very efficient for modeling large number of piles in e.g. earth reinforcement problem. The interfaces represent both the stiffness and the strength of such pile-soil interaction system. Rigid/flexible connection or inelastic interaction can be modeled by choosing appropriate characteristic of interfaces. For illustration purposes, the 3D modeling capability of the present embedded pile approach is demonstrated by considering slope reinforcement.

1 INTRODUCTION

In the Plaxis 3D Foundation Beta, a so-called embedded pile approach has recently been implemented. Within this approach the pile is assumed as line elements (slender beam elements) instead of volume elements. The slender beam element may have arbitrary inclination and can cross the soil elements at any arbitrary position. The connection between the beam and the soil is established by means of special-purposed interface elements representing the pile-soil contact at the skin and special-purposed non-linear spring representing the pile-soil contact at the base (Septanika, 2005a). In addition to the approach of Sadek and Shahrour (2004), the present embedded pile approach also considers: (i) different types of skin traction/slippage model (constant/linear traction, multi-linear diagram & layer-dependent), and (ii) foot interaction model. Maximum foot resistance is represented by a user-defined maximum value. This value corresponds to the maximum force that can be sustained by the non-linear spring at the pile foot during compression. In case of soil reinforcements, foot interaction is of minor importance.

This paper presents a short description of finite element formulation of the embedded pile approach and it shows the 3D modelling capability of embedded pile in case of slope reinforcement.

2 BACKGROUND

2.1 *Schematization*

The proposed model considers the pile as line elements (i.e. slender beam elements). The beam can cross the bulk soil elements at any arbitrary position and with an arbitrary inclination. Along the axis and at the intersection points between the beam and the soil elements, extra nodes are generated representing the pile nodes.

The displacement of the soil $\mathbf{u}_s$ and of the pile $\mathbf{u}_p$ at any soil-pile contact point $\xi = (\xi, \eta, \zeta)$ follow from the shape functions matrices of the soil element & the pile element and the corresponding nodal displacement vectors of the soil nodes & the pile nodes.

2.2 *Skin interface model*

First the soil-pile contact at the skin/mantle will be described. The soil-pile contact can be represented by a so-called skin traction $\mathbf{t}$ (traction in kN/m = force in kN per circumference in meter). For this purpose, a special-purposed interface element has been developed for connecting the soil element and the pile element. The traction $\mathbf{t}$ at the skin interface is assumed to obey the following constitutive relation

$$\Delta \mathbf{t}^{skin} = \mathbf{T}^{skin} \Delta \mathbf{u}_{rel} \quad (1)$$

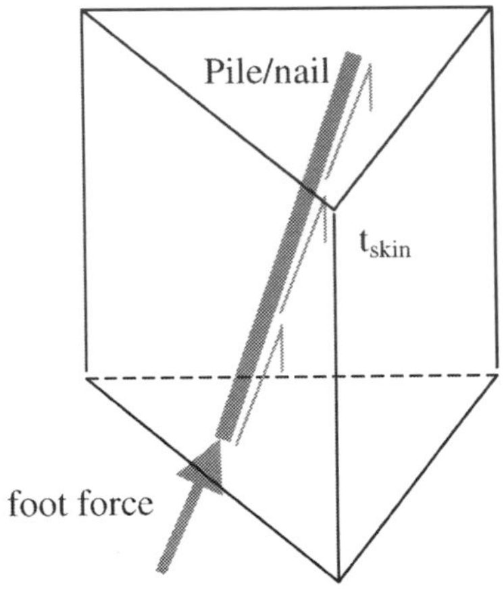

Figure 1. Schematization of a single embedded pile in one soil element.

$$\Delta \mathbf{u}_{rel} = (\Delta \mathbf{u}_p - \Delta \mathbf{u}_s) \tag{2}$$

where $\Delta \mathbf{t}$ is the traction increments at the contact points, $\mathbf{T}^{skin}$ is the material stiffness matrix of the pseudo skin interface and $\Delta \mathbf{u}_{rel} = (\Delta \mathbf{u}_p - \Delta \mathbf{u}_s)$ represents the relative displacement vector between the soil and the pile. The element stiffness matrix $\mathbf{K}_{skin}$ representing the pile-soil interaction at the mantle can be derived based on the following internal virtual work consideration ($\Delta \mathbf{a}$ represents the displacement increments at the corresponding nodes adjacent to the pile skin)

$$\delta(\Delta W^{skin}) = \delta \mathbf{a}^T \mathbf{K}_{skin} \Delta \mathbf{a} \tag{3}$$

where $\mathbf{K}_{skin}$ represents the stiffness contribution of the newly defined pile nodes, the contribution from the soil elements around the pile, and the mixed-terms. Note that since only small displacement differences are desired in the elastic regime, the stiffness of the matrix $\mathbf{T}^{skin}$ should be sufficiently "large" with respect to the bulk soil material. Next, to include slippage at the pile-soil contact, one may limit the shear-traction components. In the current version, the following traction/slippage models are available:

(a) Constant/Linear model relates the allowable traction t_s and the depth y.
(b) Multi-linear diagram by means of a set values of allowable traction values and the corresponding depth (with respect to the pile head); This option can be used to model non-linear skin forces profile which may obtain from pile tests.
(c) Layer-dependent relates the allowable traction with the adjacent soil layer.

Note a rigid connection between pile/reinforcement and the soil can be modeled by assuming very large values of the stiffness components in $\mathbf{K}_{skin}$.

2.3 *Foot interface model*

Next, for the completeness the foot interaction will be also described. The interaction at the foot is modeled by a special-purposed spring element to represent the foot stiffness against the relative movements at the foot. For this purpose, a so-called foot force $\mathbf{F}^{foot}$ vector has been considered obeying the following force-displacement relation

$$\Delta \mathbf{F}^{foot} = \mathbf{D}^{foot} \Delta \mathbf{u}_{rel}^{foot} \tag{4}$$

$$\Delta \mathbf{u}_{rel}^{foot} = (\Delta \mathbf{u}_p - \Delta \mathbf{u}_s) \tag{5}$$

where $\Delta \mathbf{F}^{foot}$ is the force increment at the foot points, $\mathbf{D}^{foot}$ represents the material stiffness matrix of the spring element at the foot, $\Delta \mathbf{u}_{rel}^{foot} = (\Delta \mathbf{u}_p - \Delta \mathbf{u}_s)$ represents the relative displacement vector between the soil and the pile at the foot. By using a similar internal virtual work approach as for the skin interaction the following relations for the foot stiffness matrix $\mathbf{K}_{foot}$ can be obtained ($\Delta \mathbf{a}$ represents the displacement increments at the corresponding nodes adjacent to the pile foot)

$$\delta(\Delta W^{foot}) = \delta \mathbf{a}^T \mathbf{K}_{foot} \Delta \mathbf{a} \tag{6}$$

where $\mathbf{K}_{foot}$ contains the stiffness contribution of the newly defined foot nodes, the contribution from the soil element at the foot, and the mixed-terms. For the maximum foot resistance representing the failure (due to penetration or pulled-out) at the pile foot, the following simplified criterion has been utilized

$$F_{axial}^{foot} | \leq F_{max}^{foot} \quad \text{(for compression)} \tag{7a}$$

$$F_{axial}^{foot} = 0 \quad \text{(for tension)} \tag{7b}$$

where F_{axial}^{foot} is the axial component of the force at the pile foot. In case of reinforcement Eq.(7b) will be valid for both compression and tension.

2.4 *Elastic zone approach*

Embedding the sub-pile/reinforcement to only one adjacent soil element appears to be insufficient and may lead to mesh-dependent behaviour, i.e. the smaller the element, the stronger the local effect. To reduce/eliminate this effect a so-called elastic zone approach has been employed. In this approach all soil points, which fall inside the pile radius (or its equivalent radius), will assume to remain "elastic". This approach appears to be quite robust and sufficient for reducing/eliminating the undesirable mesh-dependent effects (Engin, Septanika and Brinkgreve 2007).

3 NUMERICAL EXAMPLE

The validation of the embedded pile approach in the Plaxis 3D Foundation Beta Program has been presented in previous studies (Septanika 2005a, 2005b; Engin, Septanika and Brinkgreve *et al.* 2007; Septanika, Bonnier, Brinkgreve and Bakker 2007).

3.1 *Previous study on single pile*

3.1.1 *Mesh dependency issue*

Mesh-dependency is highly unpleasant since the total pile capacity may strongly depend on the mesh size. Based on the previous study (Engin, Septanika and Brinkgreve 2007), the application of the elastic zone approach appears sufficient for producing the total pile capacity which is independent of the mesh size. By excluding the elastic zone, the soil elements inside the pile zone will undergo undesirable high inelastic deformation (lowering the capacity). By using the elastic zone mesh independent results are obtained.

3.1.2 *Compression pile test*

To validate the embedded pile for simulating the real case, the Alzey Bridge pile load test (carried out in Frankfurt) has been analyzed. Load cells were I installed at the pile base to measure the loads carried directly by pile base. The layout of the pile load test arrangement is given (El- Mossallamy *et al.* 1997 & 1999). The upper subsoil consist of silt (loam) followed by tertiary sediments down to great depths. These tertiary sediments are stiff plastic clay similar to the so-called Frankfurt clay, with a varying degree of over-consolidation. It is located completely in the over-consolidated clay. Skin friction curves are obtained by subtracting the base resistance from total load–displacement curve. It was shown that embedded pile model is quite in agreement with the pile load test results (Figure 2). For more details one may refer to Engin, Septanika and Brinkgreve (2007).

3.1.3 *Tension pile test*

The tension tests on bored piles in cemented desert sands (which were carried out in Kuwait) have also been analyzed using the Plaxis 3D Foundation Program. The details of the geometry and soil parameters are given in Ismael *et al.* (1994). The load transfer of bored piles in medium dense cemented sands was investigated by field tests at two sites. The first site (South Surra) has a profile of medium-dense and very-dense weekly cemented calcareous sand. Two short bored piles were tested in axial tension to failure. It was also shown that the total pile capacity according to embedded pile is in reasonably agreement with the results of the tension tests (Figure 3). For more details one may refer to Septanika, Bonnier, Brinkgreve and Bakker (2007).

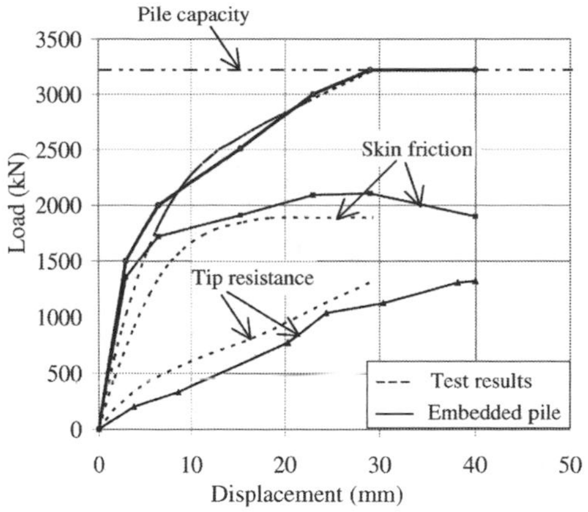

Figure 2. Load-displacement curve of the Alzey bridge pile load test together with Embedded pile results.

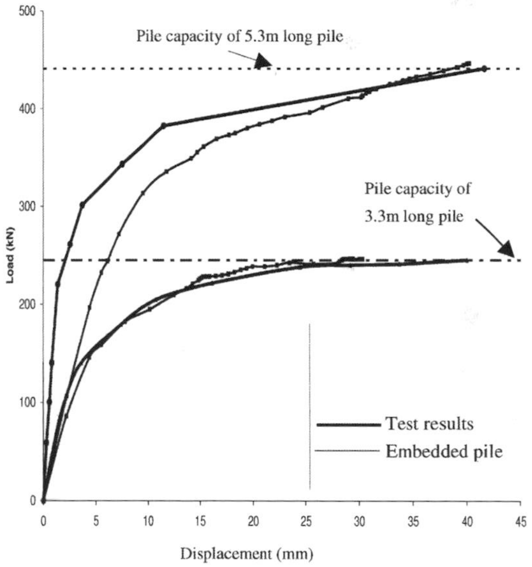

Figure 3. Load-displacement curve of the South Surra pile load test together with Embedded pile results.

3.2 *Pile group application*

For pile grouping application, it is possible to generate a 3D finite element model in which the piles can be inclined with respect to the vertical axes. Without mentioning further details, the Figure 4 below shows a raft foundation of a building supported by inclined piles. It can be seen that the mesh structure of the soil is not affected by existence of the inclined piles. The interaction between the pile and soil is realized, firstly by taking into account for the pile stiffness (also including the elastic zone of each pile) that reinforces the adjacent soil, and secondly a proper modeling of skin

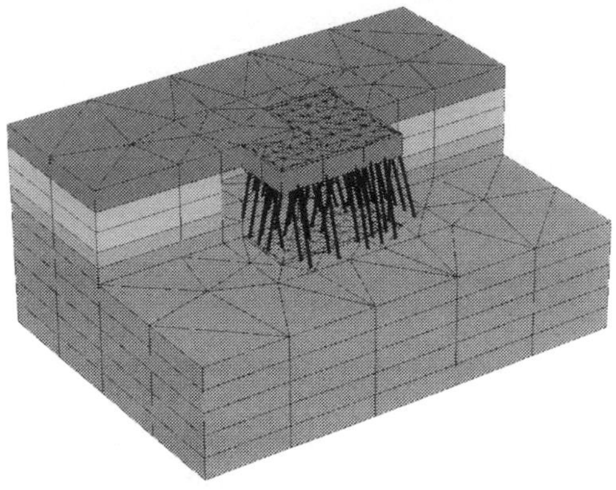

Figure 4. 3D-model of pile raft foundation using embedded piles.

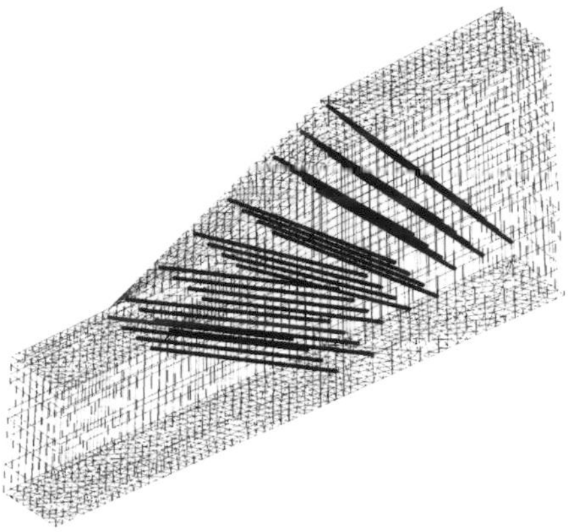

Figure 5. 3D finite element modeling of reinforced slope using embedded piles.

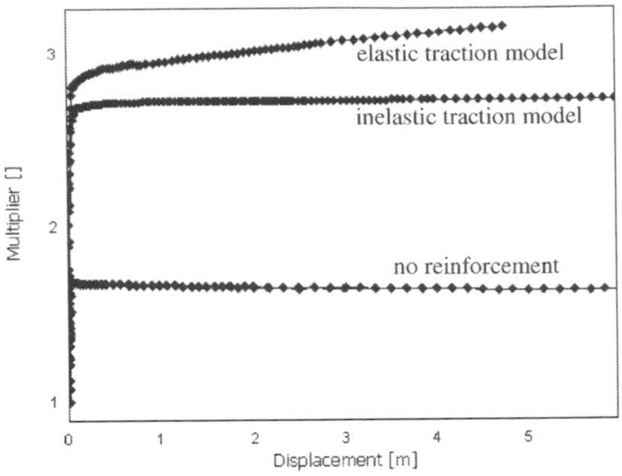

Figure 6. Safety analysis of the slope stability problem using elastic & inelastic skin traction model at embedded pile.

and foot resistances defines the strength of the pile-soil connection. Note that when using volume piles, each pile is modeled by means of a number of volume elements that leads to huge small soil element around the piles and a complex 3D mesh structure due to pile inclination and different soil layers.

3.3 *Reinforcement of slope*

For illustration purposes the application of embedded pile as reinforcements in a reinforced slope problem has been considered (Figure 5). This numerical example shows the 3D modeling capability of embedded pile approach in the Plaxis 3D Foundation Program. Due to the existence of nails – modeled by means of inclined embedded (micro) pile – the slope increases its stiffness. In contrast to the classical reinforcement approach in which the reinforcement is rigidly connected to the soil elements, the strength of the present type of nailing element can be limited by a user-defined maximum value of shearing traction/force along the pile skin. Rigid connection is simply modeled by using very high stiffness and very high maximum value of shearing traction/force, while a more realistic traction limit value can be obtained from e.g. pull-out tests.

For the present analysis a slope of 10 m high has been considered (with inclination of around 45° with respect to the horizontal axis). The soil behaviour is according to the Mohr-Coulomb model with the following parameters: Young's modulus $E = 2.10^4$ kPa, Poisson's ratio $\nu = 0.3$, cohesion $c = 10$ kPa, friction angle $\varphi = 30°$. The result of safety analyses of the reinforced slope using embedded pile model in Plaxis 3D Foundation Beta program is shown in Figure 6. It is show that the slope with no reinforcement has a lower factor of safety FS as compared to the reinforced soil (i.e. employing the phi-c method). Two cases have been considered: (i) based on inelastic skin traction model by presuming a certain skin traction limit, (ii) based on the "elastic" model using a very large value of skin traction limit. Note that the elastic model is comparable to reinforced soil, excluding the relative movements between soil and the reinforcements. It can be seen that the elastic traction model leads to an (unrealistic) overestimated behaviour. The results also show the usefulness of the present inelastic skin traction option in modelling the failure behaviour of reinforced slope.

Further, the total deformation mechanism in case of no reinforcements and reinforced slope are shown in Figure 7 and Figure 8. The axial force distribution and the skin traction distribution for a typical reinforcement are shown in Figure 9(a)–(b). It has to be note that the present results are based on a simplified inelastic traction model and are purposed for demonstration

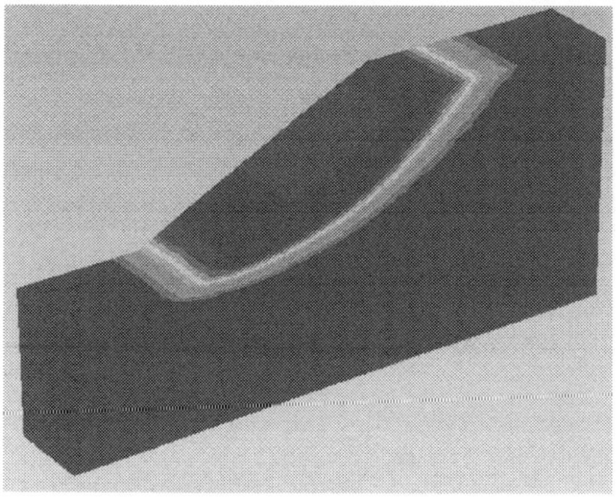

Figure 7. Deformation mechanism in case of no reinforcements.

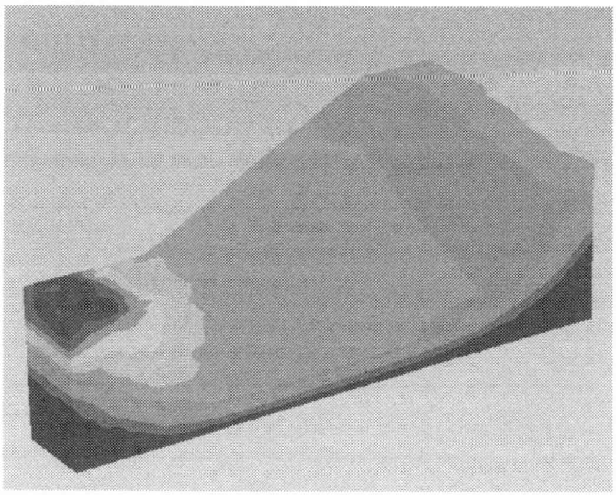

Figure 8. Deformation mechanism in case of reinforced slope.

only. More realistic skin traction distribution can be estimated by using advanced inelastic skin traction models, based on more accurate soil data and pull-out test data.

4 CONCLUDING REMARKS

This paper shortly describes the embedded pile approach, followed by a review of validation tests that have been done previously. The accuracy of single pile model has been validated, by considering the pile compression tests in Frankfurt and the pile tension tests in Kuwait. For both cases, the results are reasonably in agreement with the field test results. It is important to mention that an accurate modelling of the soil part also plays an important role for simulating the behaviour

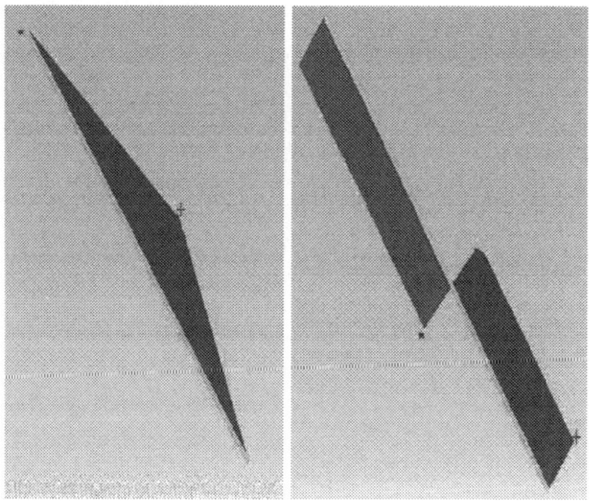

(a) (b)

Figure 9. (a) Axial force distribution at the reinforcement, (b) Skin traction distribution at the reinforcement.

of the real tests. With the embedded pile approach it has been shown that 3D modelling of multi piles application (with or without inclination) can be relatively easy generated. The resulting 3D mesh structure is unaffected by the existence of the pile. Finally, a reinforced slope problem has been considered to illustrate the modelling capability of the present embedded pile in the Plaxis 3D Foundation Program. It was shown that the factor of safety FS in the reinforced slope is much higher than without the reinforcement. In the near future more study will be performed concerning to further evaluate the capability of embedded pile in pile group applications and soil reinforcement problems.

REFERENCES

El-Mossallamy, Y. and Franke, E. (1997). Numerical Modelling to Simulate the Behaviour of Piled Raft Foundations, August 1997, Darmstadt, Germany.

El-Mossallamy, Y. (1999). Load-settlement behaviour of large diameter bored piles in over-consolidated clay. *Proc. of the 7th intern. symp. on numerical models in geotechnical engineering* – NUMOG VII, Graz, 1–3 September 1999, 443–450. Rotterdam: Balkema.

Engin, H.K., Septanika, E.G. and Brinkgreve, R.B.J. (2007). Improved embedded beam elements for the modelling of piles. *International Symposium on Numerical Models in Geomechanics* – NUMOG X, Rhodes, Greece. April 2007.

Ismael, N.F., Al-Sanad, H.A. and Al-Otaibi, F. (1994). Tension tests on bored piles in cemented desert sands. *Canadian Geotechnical Journal*, Vol. 31 (4), 597–603.

Sadek, M. and Shahrour, I. (2004). A three dimensional embedded beam element for reinforced geomaterials. *International journal for numerical and analytical methods in geomechanics* 28:931–946.

Septanika, E. G. (2005a). A finite element description of the embedded pile model. Plaxis internal report.
Septanika, E. G. (2005b). Validation testing embedded pile in Plaxis 3D Foundation. Plaxis internal report.
Septanika, E.G., Bonnier, P.G., Brinkgreve, R.B.J. and Bakker, K.J. (2007). An efficient 3D modelling of (multi) pile-soil interaction. *Proceeding of Third International Geomechanics Conference*, 11–15 June 2007, Nessebar, Bulgaria.
Skempton, A.W. (1951). The Bearing Capacity of Clays. *Proceeding of Building Research Congress*, Vol. 1, 180–189.

Design and measurement on full-scale behavior of reinforced structure

New Horizons in Earth Reinforcement – Otani, Miyata & Mukunoki (eds)
© 2008 Taylor & Francis Group, London, ISBN 978-0-415-45775-0

Full-scale model test and numerical analysis of reinforced soil retaining wall

K. Arai
University of Fukui, Fukui, Japan

K. Yoshida, S. Tsuji & Y. Yokota
Maeda Kosen Co., Ltd., Fukui, Japan

ABSTRACT: This paper reports the results of field observation and numerical analysis of full-scale model test for a new reinforced soil retaining wall system, having a vertical layer which absorbs the deformation between facing concrete brocks and reinforced backfill. The field observations and FE-analysis show that little earth pressure acts on the facing blocks and that the structure system is stable. A dynamic elastic-plastic FE-analysis which investigates the interaction behavior between the facing blocks and reinforced backfill, shows that the structure system is stable on the condition of earthquake.

1 INTRODUCTION

Many types of reinforced soil retaining wall have been proposed and have been built worldwide. However, because earth pressure acts directly on facing material, there is the possibility of deformation of facing and lack of soil compaction near the facing. We developed a new reinforced soil retaining wall system, having a vertical layer which absorbs the deformation between facing concrete brocks and reinforced backfill. The layer prevents the earth pressure from exerting directly on the facing brocks. We call the system "double wall structure". Subjected to an actual structure of this system, we performed field observations in order to evaluate the performance of system. In the observations, we monitored the horizontal earth pressure against facing blocks, deformations of facing blocks, strains on reinforcement geogrids, vertical earth pressure at the base of backfill, and so on. At first, we carry out an elastic-plastic FE-analysis which employs Mohr-Coulomb yield criterion, a simple non-associated flow rule and the initial stress method, and which represents failure mode more realistically. The results of field observations and FE-analysis show that little earth pressure acts on the facing blocks and that the structure system is stable. Secondly we perform a dynamic elastic-plastic FE-analysis in order to investigate the interaction behavior between the facing blocks and reinforced backfill. The result shows that the structure system is stable also on the condition of earthquake.

2 STRUCTURE

The structure of the reinforced soil retaining wall is shown in Figure 1 (Yoshida, K. et al., 2006). It has a vertical layer which absorbs the deformation between facing concrete brocks and reinforced backfill. The reinforcement geogrid in which aramid fibers are inserted in polyethelene net, is shown in Figure 2. The wall system has two facing walls such as an outer wall and an inner wall. The facing block exists as the outer wall, and the inner wall consisting a non-woven fabric and L-form wire net exists in the inside. Reinforced backfill is stable state by the inner wall and geogrid. In order to reduce the earth pressure acting from backfill to facing concrete block, facing concrete block and backfill separate perfectly. Between the facing concrete block and backfill, an absorption layer for deformation consisting single sized crushed stone is set. Facing concrete block is not connected with geogrid laying in backfill. In addition, the facing concrete block and reinforced backfill are connected by fiber belt made in non-corrosive polyester (below called the connection belt). Characteristics of this system are as follows: 1) The facing concrete block and reinforced backfill separate perfectly so that it can compact backfill near the inner wall surface sufficiently. 2) It can reduce the earth pressure acting on the facing concrete block, with reducing the external stress with the deformation of compression of backfill after construction by the absorption layer for deformation. 3) Since material of the connection belt is

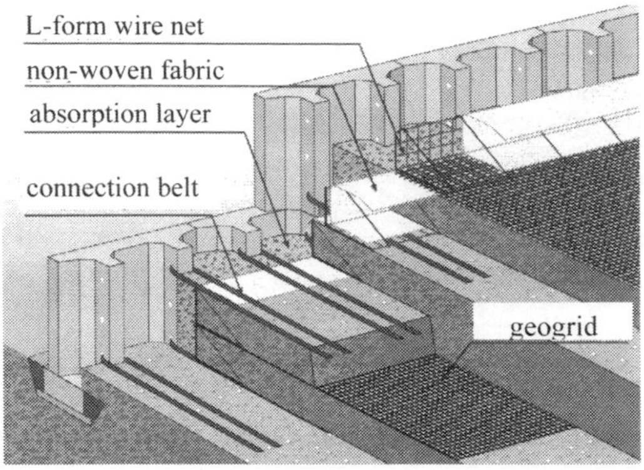

Figure 1. Structure of reinforced soil retaining wall.

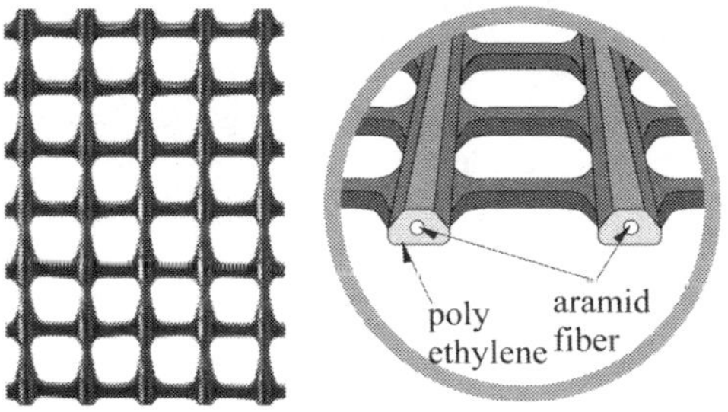

Figure 2. Geogrid reinforcement.

flexible, it can respond to the deformation due to the consolidation of reinforced backfill and the stress concentration in the connection part of facing concrete block is prevented.

3 FIELD OBSERVATIONS

The reinforced soil retaining wall used in the full-scale test is shown in Figure 3. The wall was constructed to have a vertical height of 9 m by piling up 0.9 m high facing concrete blocks in ten rows. The construction procedures are follows: 1) grading the foundation ground to enable the facing concrete blocks to be horizontally installed, spreading crushed stones for the foundation, and building concrete foundation, 2) installing the facing concrete blocks on the foundation concrete, 3) installing the connection belt, installing a form for crushed stones on the inner wall, 4) installing the geogrid reinforcement, backfilling and compacting soil, and 5) filling the layer for absorbing deformation with single sized crushed stones. A front view and a cross section of the reinforced soil retaining wall are shown in Figure 4. The earth pressure acting on the back of the facing concrete blocks was measured by using an earth pressure meter. For monitoring the displacement of the wall, several targets were installed on the concrete blocks. The displacement of the wall during and after the construction was monitored using an electro-optical distance meter. An earth pressure meter was also installed on the bottom of the reinforced soil retaining wall to monitor the ground reaction at the bottom of the wall. Strain gauges were attached to geogrid reinforcement to monitor the strain of the geogrid.

Figure 3. Overall view.

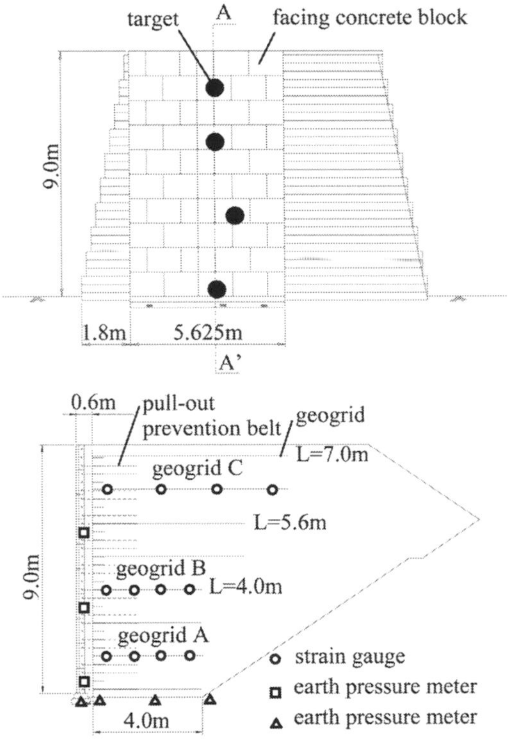

Figure 4. Front view and cross section.

4 NUMERICAL ANALYSIS

The ground is expressed as a plane strain element, and the shift between the ground and the wall blocks

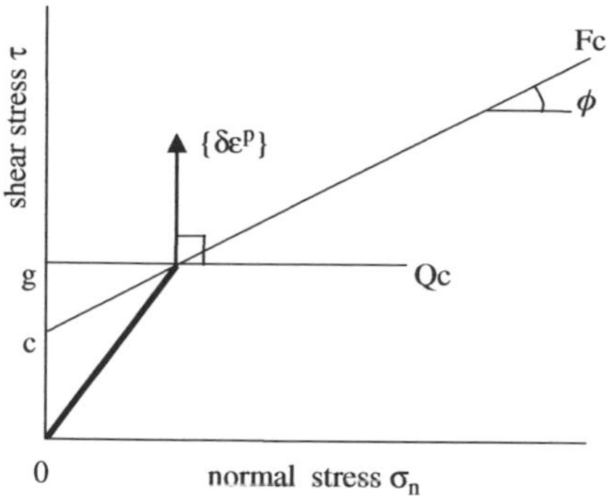

Figure 5. Yield surface and flow rule (Coulomb).

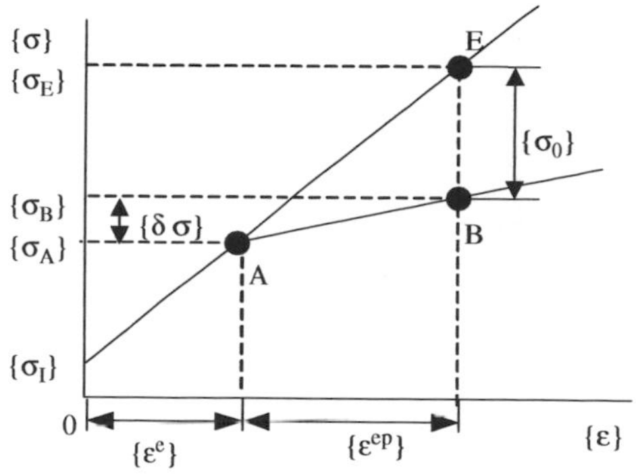

Figure 6. Initial stress method.

is expressed in an interface element. Mohr-Coulomb failure criterion is applied to the ground, and Coulomb failure criterion is applied to the interface element (Desai, C. S. et al. 1984).

Mohr-Coulomb:

$$F_M = \left\{(\sigma_x - \sigma_y)^2 + 4\tau_{xy}^{\ 2}\right\}^{1/2} - \left\{(\sigma_x + \sigma_y)\sin\phi + 2c\cos\phi\right\} = 0 \quad (1)$$

$$\text{Coulomb}: \ F_C = |\tau| - c - \sigma_n \tan\phi = 0 \quad (2)$$

Where, $\sigma_x, \sigma_y, \tau_{xy}$: the stress components in the entire coordinates, $\sigma_{n,4}\ \tau$: the stress components on the slip surface and c, ϕ: Mohr-Coulomb strength parameters.

When the confining pressure σ_3 or the vertical stress σ_n of the shear surface continues increasing by the load of banking, the curve is likely to move on the yield line after yielding. As shown in Figure 5, the increment of plastic strain, when the curve moves on the yield line, was assumed to follow the non-associate flow rule with a dilatancy angle of 0. Plastic potential is Mohr-Coulomb:

$$Q_M = \left\{(\sigma_x - \sigma_y)^2 + 4\tau_{xy}^{\ 2}\right\}^{1/2} - 2g = 0 \quad (3)$$

$$\text{Coulomb}: \ Q_C = |\tau| - g = 0 \quad (4)$$

where, g: unnecessary parameter since plastic potential is used in a differential form. As shown in Figure 6, the entire load is applied at a single loading stage. In Figure 6, Point A shows the yield point, and Point B is the final equilibrium point. $\{\varepsilon^e\}$: elastic strain, $\{\varepsilon^{ep}\}$: elastic-plastic strain. $\{\sigma_I\}$: actual initial stress and $\{\sigma_0\}$: initial stress in the initial stress method which is determined by an iteration loop (Arai, K. et al., 1996).

The FE meshing is shown in Figure 7. The facing concrete block is represented by beam elements, the geogrid reinforcement is expressed using truss elements, and the backfill and the layer for absorbing deformation are represented by plane strain elements. Interface elements are put between the facing concrete blocks and the deformation absorbing layer and

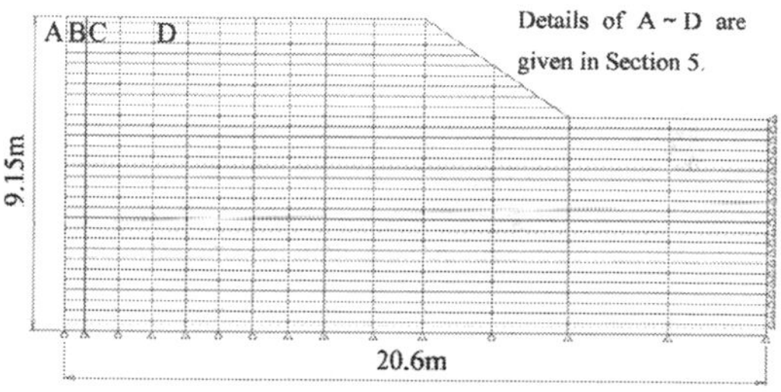

Figure 7. FE meshing.

Table 1. Soil parameters.

	E(kN/m^2)	ν	c(kN/m^2)	ϕ (degree)
Embankment	1.0×10^4	0.3	0.0	36.0
Absorption layer	1.0×10^4	0.3	0.0	30.0
Interface	1.0×10^4 (G = 100kN/m^2)	0.3	0.0	20.0

	E(kN/m^2)	A(m^2)	I(m^4)
Facing concrete block	1.4×10^6	0.14	2.3×10^{-4}
Geogrid	4.1×10^6	5.0×10^{-4}	–
Connection belt	2.0×10^5	4.8×10^{-4}	–

between the deformation absorbing layer and reinforced backfill. The geogrid is assumed to not bear the compression stress. Soil parameters used for numerical analysis are shown in Table 1, in which E: elastic modulus, ν: Poisson's ratio, G: shear modulus at the interface, γ: unit weight, A: cross sectional area, and I: moment of inertia. The joints between two vertically adjacent facing concrete blocks were assumed to have a moment of inertia of 1/100 of that of facing blocks to reproduce the drops in bending stiffness. Note that the elastic modulus of backfill gives little effect to earth pressure and wall displacement.

The calculated and measured horizontal earth pressures acting on the facing concrete blocks are

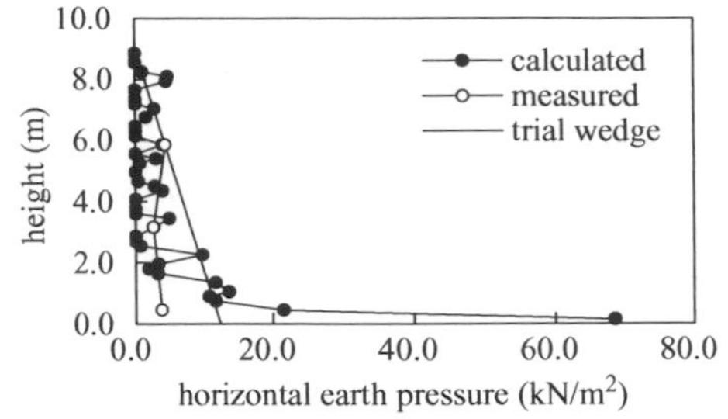

Figure 8. Horizontal earth pressure.

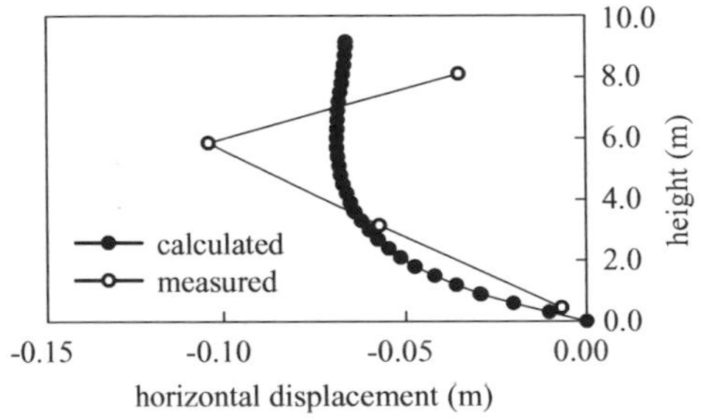

Figure 9. Horizontal displacement.

comparatively shown in Figure 8. The earth pressure values calculated by the trial wedge analysis are also shown in the figure, which were determined by assuming that the inner wall was stable natural ground. Both the measured and calculated values were much smaller than the values determined by the trial wedge analysis, showing that the earth pressure acting on the facing concrete blocks was small. The calculated values well reproduced the measured values although the calculated values were larger than the measurements at the lower part of the wall, because of restricting displacements at the boundary. The calculated and measured horizontal displacements of the facing concrete blocks are shown in Figure 9. The measured horizontal displacement at the completion of the reinforced soil retaining wall was about 10cm the maximum at a height of 6 m and was bulging frontward. The calculations reproduced the measured displacements almost completely. A comparison between the calculated and measured tensile force of the geogrids after the completion of the reinforced soil retaining wall is shown in Figure 10. Large strains were observed near the wall on Geogrid A and at the centers of Geogrids B and C.

5 BEHAVIOR DURING EARTHQUAKE

To simulate the behavior of the reinforced soil retaining wall during an earthquake, a non-linear dynamic analysis is performed (see Owen et al., 1980). The

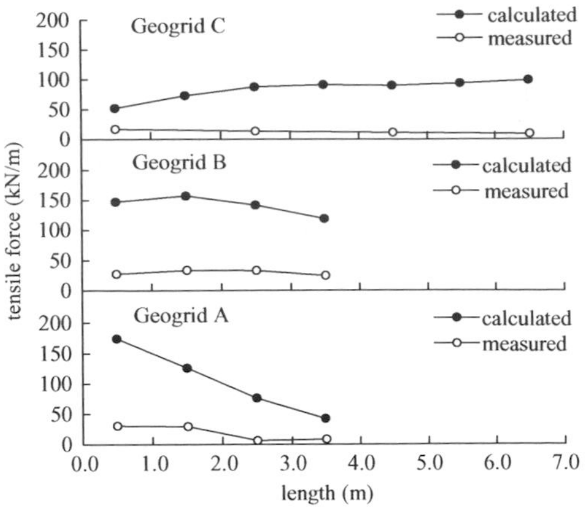

Figure 10. Tensile force of geogrid.

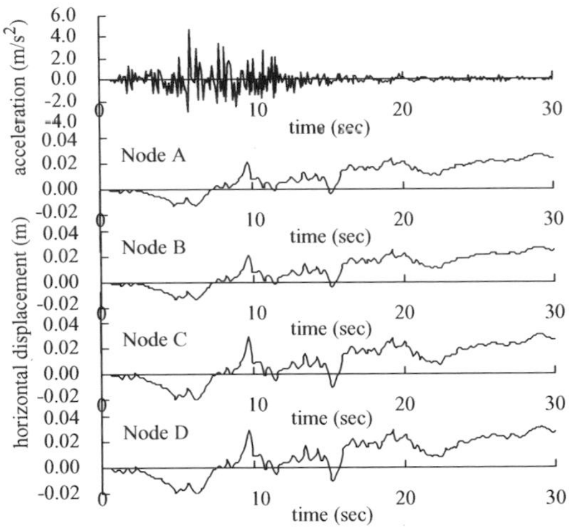

Figure 11. Result of dynamic FE-analysis.

elastic-plastic stress and strain relationship is used for the ground, which is the same as the one employed in the static analysis. The attenuation matrix is derived by the modified Rayleigh attenuation so that the attenuation properties are independent of frequency. For time integration, Newmark's β method is applied. The earthquakes along the north-south direction monitored at 17 km from the epicenter during the Niigata Chuetsu Earthquake in 2004 are given to the FE meshing shown in Figure 7. The attenuation parameters α is 36.7 and β is 0.0021. Displacements are calculated at Nodes A to D shown in Figure 7. Node A is on the facing concrete blocks, Node B is on the front of the deformation absorbing layer, Node C is on the front of the inner wall, and Node D is within the reinforced soil. The calculated horizontal displacement is shown in Figure 11 for the given acceleration waveforms. When an earthquake motion is given, a displacement difference of about 5 mm is observed between the facing concrete blocks and the reinforced soil at the crown of the wall, but there are no phase differences among the facing concrete blocks, the deformation absorbing layer and the reinforced soil, and all behaved as a united structure.

6 CONCLUSIONS

A field observation and a numerical analysis were carried out on a reinforced soil retaining wall that was installed with a layer for absorbing deformation between facing concrete blocks and reinforced soil to mitigate the earth pressure acting on the facing concrete blocks by the deformation of the soil during and after the construction of the wall. The results are: 1) The field observation and numerical analysis showed that the horizontal earth pressure acting on the facing concrete blocks was very small and 2) The dynamic finite element analysis showed that there were no phase differences among the facing concrete blocks, the deformation absorbing layer and the reinforced soil, and all behaved as a united structure.

REFERENCES

Yoshida, K., Yokota, Y., Tatta, N., Tsuji, S. & Arai, K. 2006. Field observation of reinforced soil retaining wall of "double wall structure, *Pro. of the 8th International Conference on Geosynthetics*, Vol. 3: 1133–1136.

Desai, C. S., Zaman, M. M., Lightner, J. G. & Siriwardane. H. J. 1984. Thin-layer element for interfaccs and joints, *Int. J. Numer. Anal. Methods Geomech.*, Vol. 8: 19–43.

Arai, K. & Kasahara, K. 1996. Limit design of earth reinforcement methods considering displacement field, *Proc, Int. Symposium on Earth Reinforcement, Fukuoka/JAPAN*: 191–196.

Nayak, G. C. & Zienkiewicz, O. C. 1972. Elasto-plastic stress analysis, A generalization for various constitutive relations including strain softening, *Int. J. Numer. Methods Eng.*, Vol. 5: 113–135.

Owen, D. R. J. & Hinton, E. 1980. Finite elements in plasticity: Theory and practice, Pineridge Press.

New Horizons in Earth Reinforcement – Otani, Miyata & Mukunoki (eds)
© 2008 Taylor & Francis Group, London, ISBN 978-0-415-45775-0

Stability analysis of back-to-back MSE walls

J. Han
Department of Civil, Environmental and Architectural Engineering, the University of Kansas, Lawrence KS, USA

D. Leshchinsky
Department of Civil and Environmental Engineering, the University of Delaware, Newark, DE, USA

ABSTRACT: Back-to-back MSE walls are commonly used for embankments approaching bridges. However, available design guidelines are limited. The distance between the two opposing walls is a major parameter used for determining the analysis methods in FHWA/AASHTO Guidelines. Two extreme cases are identified: (1) reinforcements from both sides overlap, and (2) the walls are far apart, independent of each other. However, existing design methodologies do not provide a clear answer how the required tensile strength of reinforcement changes with respect to the distance of the back-to-back walls. The focus of this paper is to investigate the effect of back-to-back distance on stability of MSE walls under static conditions. Finite difference method incorporated in FLAC software and limit equilibrium method (i.e., the modified Bishop method) in ReSSA software were used for this analysis. Parametric studies were carried out by varying two important elements, the wall back-to-back distance and the quality of backfill material, to investigate their effects on the critical failure surface and the required tensile strength of reinforcement. The results of the parametric studies imply that the back-to-back distance of MSE walls influences the required reinforcement tensile strength when the walls are relatively close.

1 INTRODUCTION

Design of back-to-back MSE walls is considered as a special situation having a complex geometry in FHWA Demonstration Project 82 (Elias and Christopher, 1997). In this FHWA design guideline, two cases are considered based on the distance of two opposing walls, D, as illustrated in Figure 1. When D is greater than Htan $(45° - \phi/2)$, full active thrust can be mobilized. For this case, the typical design method for MSE walls can be used. When D is equal to 0, two walls are still designed independently for internal stability but no active thrust is assumed from the backfill. The guideline indicates that when D is less than H tan $(45° - \phi/2)$, active thrust cannot be fully mobilized so that the active thrust is reduced. However, the guidelines do not provide any method how to consider the reduction of the active thrust, thus, no method is provided to calculate the required tensile strength for reinforcement.

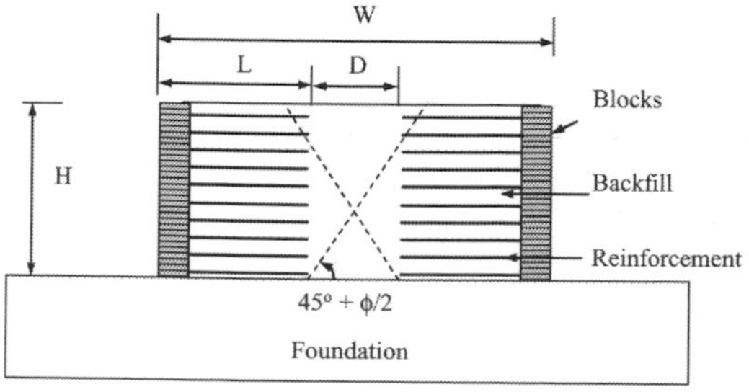

Figure 1. Back-to-back MSE wall and definitions.

Limit equilibrium and numerical methods have been successfully used to evaluate the stability of MSE walls (for example, Leshchinsky and Han, 2004; Han and Leshchinsky, 2006; Han and Leshchinsky, 2007) and yield close results in terms of factors of safety and critical failure surfaces. In this study, these two methods were also adopted to investigate the effect of the wall back-to-back distance and the quality of backfill material on the required tensile strength of reinforcement.

2 METHODS OF ANALYSES

2.1 *Limit equilibrium method*

Bishop's simplified method, utilizing a circular arc slip surface, is probably the most popular limit equilibrium method. Although Bishop's method is not rigorous in a sense that it does not satisfy horizontal force limit equilibrium, it is simple to apply and, in many practical problems, it yields results close to rigorous limit equilibrium methods. In this study, Bishop's simplified method was modified to include reinforcement as a horizontal force intersecting the slip circle, which is incorporated in ReSSA(2.0) software, developed by ADAMA Engineering (2002). This modified formulation is consistent with the original formulation by Bishop (1955). The mobilized reinforcement strength at its intersection with the slip circle depends on its long-term strength, its rear-end pullout capacity (or connection strength), and the soil strength. The analysis assumes that when the soil strength is reduced by a factor, a limit equilibrium state is achieved (i.e., the system is at the verge of failure). The slip circle for which the lowest factor (i.e., the largest mobilized soil strength) exists is the critical slip surface for which the factor of safety is rendered. Under this state, when the factor of safety is a unit, the soil and reinforcement mobilize their respective strengths simultaneously.

2.2 *Numerical method*

The finite difference program (FLAC 2D Version 5.0, developed by the Itasca Consulting Group, Inc.) was adopted in this study. A shear strength reduction technique was adopted in this program to solve for a factor of safety of stability. In this technique, a series of trial factors of safety are used to adjust the cohesion, c and the friction angle, ϕ, of soil. Adjusted cohesion and friction angle of soil layers are re-inputted in the model for limit equilibrium analysis. The factor of safety is sought when the specific adjusted cohesion and friction angle make the slope become instability from a verge stable condition (i.e., limit equilibrium). The critical slip surface often can be identified based on the contours of the maximum shear strain rate.

3 MODELING

3.1 *Baseline case*

The geometry and material properties of the baseline model used in this study are shown in Figure 2. Since the factor of safety is determined based on a state of yield, or verge of failure, it is insensitive to the selected elastic parameters: Young's modulus (E) and Poisson's ratio (ν) when using FLAC. If the system contains soils with largely different elastic parameters, it will take longer time to solve for the factor of safety; however, the effects on this factor would be small since it depends mainly on Mohr-Coulomb strength parameters. Hence, constant values of $E = 100\text{MPa}$ and $\nu = 0.3$ were used in FLAC. The effect of wall facing cohesion on the required tensile strength of reinforcement in the numerical analysis will be discussed in the next section. Mohr-Coulomb failure criteria were used for strength between stacked blocks, the

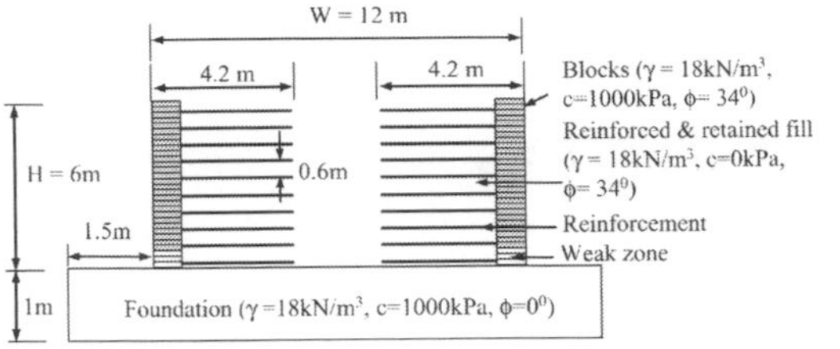

Figure 2. Dimensions and parameters of the baseline case.

reinforced and retained fill, and the foundation soil. Reinforcement is modeled as a cable with grouted interface properties between cable and soil. The bond strength between reinforcement and reinforced fill was assumed equal to 80% the fill strength, same as in the limit equilibrium analysis when considering pullout resistance. A weak zone at the toe of the MSE wall with a dimension of 0.3m wide and 0.4m high having cohesion equal to 0 but the same friction angle as the fill was assumed to ensure the critical failure surface of passing through the toe of the MSE wall.

In this baseline case, the back-to-back wall width (W) /height (H) ratio is equal to 2.0 and the distance at back of two walls, D is equal to 3.6m, which is slightly greater than $H \tan(45° - \phi/2) = 3.2\text{m}$. Based on the FHWA design guideline, a typical design method for a single wall can be adopted. The reinforcement length, $L = 4.2\text{m}$, was selected based on the typical reinforcement length/wall height ratio of 0.7 recommended by the FHWA design guideline.

Two important parameters, the back-to-back wall width and the quality of backfill material, were selected in this study to investigate their influence on the critical failure surface and the required tensile strength of reinforcement. In addition to $W/H = 2.0$ for the baseline case, two other W/H ratios (1.4 and 3.0) were used. One parameter in the baseline was changed at a time while all others were unchanged. The same models were used in numerical and limit equilibrium analyses. The required tensile strength of reinforcement was determined to ensure the factor of safety of the MSE wall equal to 1.0.

3.2 *Effect of wall facing cohesion*

The effect of wall facing cohesion was examined in this study. As shown in Figure 3, the factor of safety of the back-to-back MSE wall increases with an increase of the wall facing cohesion. However, it becomes constant after the cohesion is greater than 100 kPa. In this case, the potential failure of the MSE wall would only pass through the toe of the MSE wall. In all analyses discussed below, the cohesion of the wall facing was assumed to be 100kPa except a weak zone close to the toe. Figure 3 also shows that the effect of the wall facing cohesion for the case with low-quality

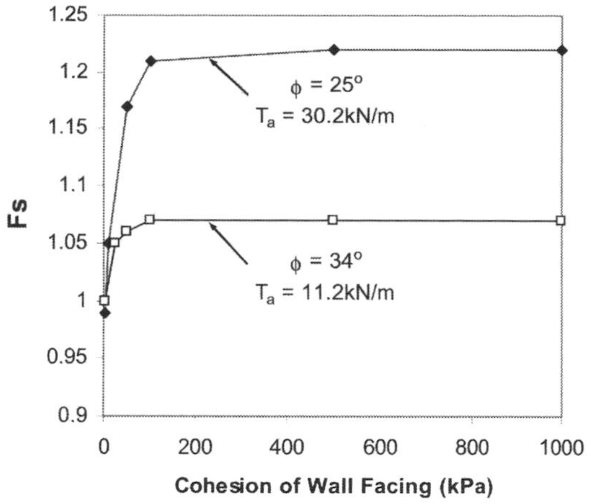

Figure 3. Effect of wall facing cohesion.

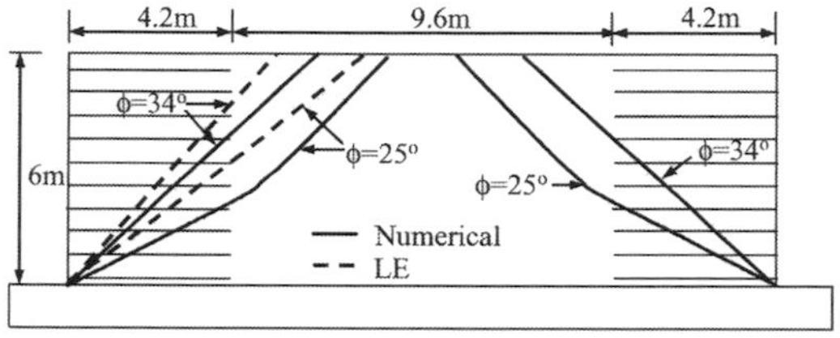

Figure 4. Critical failure surfaces within walls at W/H = 3.

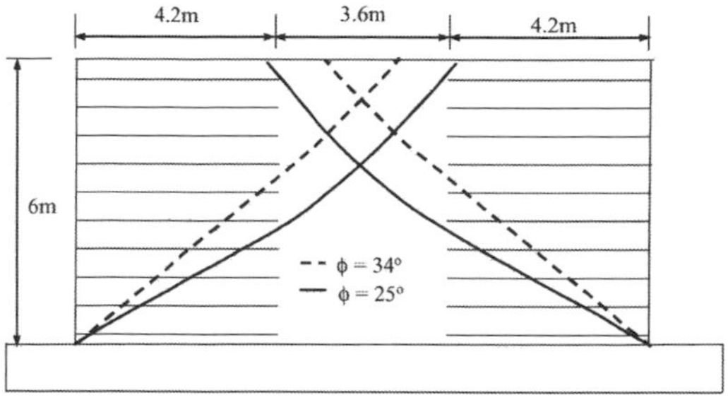

Figure 5. Critical failure surfaces within walls at W/H = 2.0.

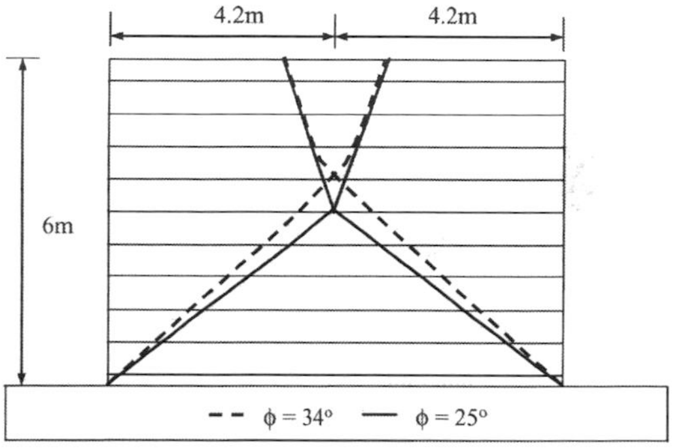

Figure 6. Critical failure surfaces within walls at W/H = 1.4.

backfill ($\phi = 25°$) is more significant than that with high-quality backfill ($\phi = 34°$).

4 RESULTS

4.1 *Critical failure surfaces*

The locations and shapes of critical failure surfaces of the back-to-back walls at different wall width/height ratios (W/H) were determined based on the contours of shear strain rate in the numerical analysis and presented in Figures 4, 5, and 6. Figure 4 shows that the critical failure surfaces in two opposing walls do not intercept each other, therefore, they behave independently. The critical failure surfaces by the LE method are also shown in Figure 4 and have slightly steeper angles than those by the numerical method.

Figure 5 shows the critical failure surfaces within back-to-back walls at W/H = 2, which intercept each other from two sides. More interactions occur for the case with a low-quality backfill. (i.e., $\phi = 25°$). For both cases, the critical failure surfaces do not enter the reinforced zone on the opposing side. In other words, the potential failure surface is constrained by the reinforced zone on the opposing side. Based on the FHWA formula (D > 3.2m) using $\phi = 34°$, there should be no interaction between these two walls. Apparently, this assumption is not supported by the numerical result. However, the FHWA assumption leads to more conservative results.

Figure 6 shows critical failure surfaces developed within the back-to-back walls when there is no retained fill between these two walls (i.e., D = 0m). In both cases, reinforcement layers are not connected at the back of two walls. The numerical results show the interactions of critical failure surfaces in two opposing walls. In both cases, the failure surfaces enter the reinforced zone from another side.

The comparisons of locations and shapes of critical failure surfaces at different W/H ratios but the same quality of fill are presented in Figure 7. Figure 7 shows that the locations and shapes of the critical failures are almost same for W/H = 3 and 2. This result can be explained as the failure surfaces not entering the reinforced zone on the opposing side. For W/H = 1.4, however, the locations and shapes of the critical failure surfaces deviate from others as the failure surfaces enter the reinforced zone on the opposing side.

4.2 *Required tensile strength*

The required maximum tensile strengths of reinforcement for all the cases discussed above are presented

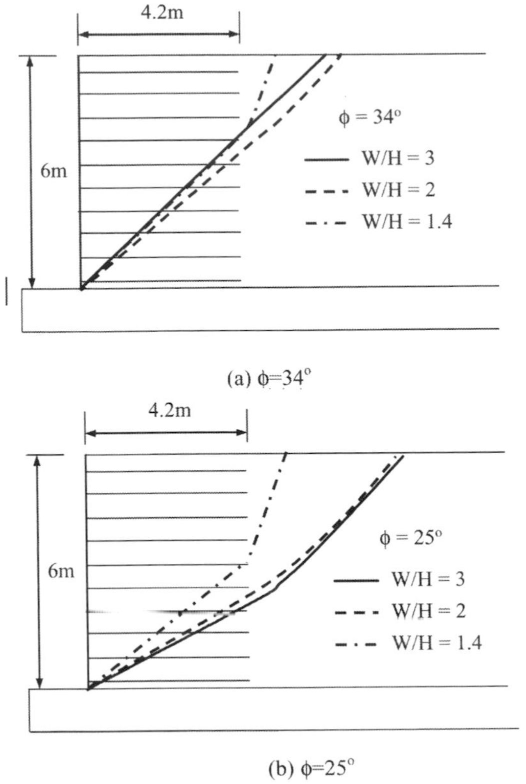

Figure 7. Critical failure surfaces at different W/H ratios.

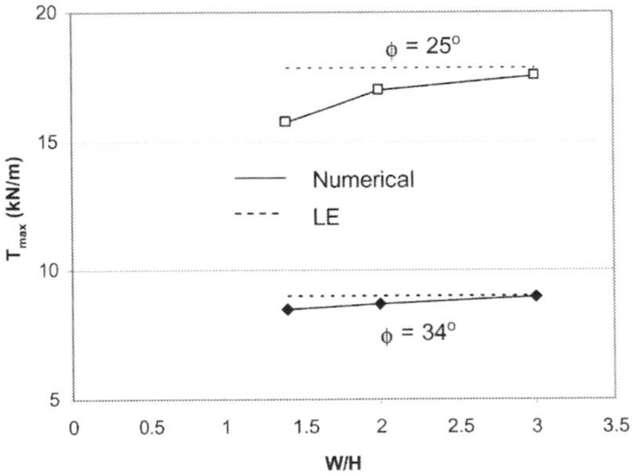

Figure 8. Required maximum tensile strength of reinforcement.

in Figure 8. The results from the LE method were based on the analyses of one side wall, therefore, no interaction of two opposing walls was considered. In other words, the required tensile strengths do not change with the W/H ratios. Figure 8 clearly shows that a decrease of W/H ratio from 3 to 1.4 reduce the required maximum tensile strength of reinforcement. The LE method without considering the interaction of the opposing walls would provide conservative design of back-to-back MSE walls. The difference in the maximum tensile strength of reinforcement with and without considering the interaction is within 12% based on the cases investigated in this study. The required maximum tensile strengths can be used for the selection of geosynthetics in the back-to-back MSE walls.

5 CONCLUSIONS

The study using the numerical and limit equilibrium methods shows that two back-to-back walls interact when they are close. This interaction will change the location and shape of critical failure surface. When the distance of the walls gets closer, the required maximum tensile strength decreases.

ACKNOWLEDGEMENT AND DISCLAIMER

This paper is based upon the work supported by the National Science Foundation under Grant No. 0442159. Any opinions, findings, and conclusions or recommendations expressed in this paper are those of the authors and do not necessarily reflect the views of the National Science Foundation.

REFERENCES

ADAMA Engineering, Inc. 2002. *ReSSA Version 2.0*. Newark, Delaware, USA.

Bishop, A.W. 1955. The use of the slip circle in the stability analysis of slopes, *Geotechnique*, 5: 7–17.

Elias, V. and Christopher, B.R. 1997. Mechanically Stabilized Earth Walls and Reinforced Soil Slopes Design and Construction Guidelines. Publication No. FHWA-SA-96-071.

Han, J. and Leshchinsky, D. 2006a. General analytical framework for design of flexible reinforced earth structures. ASCE *Journal of Geotechnical and Geoenvironmental Engineering,* 132 (11): 1427–1435.

Han, J. and Leshchinsky, D. 2006b. Stability analyses of geosynthetic-reinforced earth structures using limit equilibrium and numerical methods. Proc. of the 8th Int. Geosynthetics Conf., 18–22 Sept., Yokohama, Japan, 1347–1350.

Itasca Consulting Group, Inc. (2006). *FLAC5.0 user's guide*, Minneapolis.

Leshchinsky, D. and Han, J. (2004). "Geosynthetic reinforced multitiered walls", ASCE *Journal of Geotechnical and Geoenvironmental Engineering*, 130 (12), 1225–1235.

New Horizons in Earth Reinforcement – Otani, Miyata & Mukunoki (eds)
© 2008 Taylor & Francis Group, London, ISBN 978-0-415-45775-0

Analysis of RE wall using oblique pull for linear subgrade response: Coherent gravity approach

P.V.S.N. Pavan kumar
V.N.R. Vignana Jyothi Institute of Engg. and Tech., Hyderabad, India

M.R. Madhav
J.N.T.U College of Engg., Hyderabad, India

ABSTRACT: In the design of RE wall most of the available methods assume that reinforcement is subjected to only axial pull. In reality, the reinforcement is subjected to oblique pull due to oblique sliding of failure wedge. In the present work, the effect of oblique pull on the stability of RE wall is studied considering a coherent gravity failure mechanism. The factor of safety modified to incorporate the effect of obliquity and is evaluated and compared with the conventional one. Parametric study quantifies the significance of length of reinforcement, number of reinforcement layers, angle of shearing resistance of backfill, interface bond resistance, global subgrade stiffness factor and magnitude of displacement on the modified factor of safety.

1 INTRODUCTION

The available methods of design of RE wall consider only axial pullout of reinforcement. But in practice, the reinforcement is subjected to transverse/oblique pull (Fig. 1). The equilibrium of RE wall is affected since the additional normal stresses acting on the reinforcement in the resistant zone increase thereby increasing the pullout resistance.

The obliquity of failure surface with respect to the orientation of the reinforcement was considered by Gray & Ohashi (1983), Leschinsky & Reinschmidt (1985), Degenkamp & Dutta (1988), Shewbridge & Sitar (1989), Leschinsky & Boedeker (1990), Athanasopoulos (1993), Neubecker & Randolph (1994), Burd (1995) and Bergado et al. (2000).

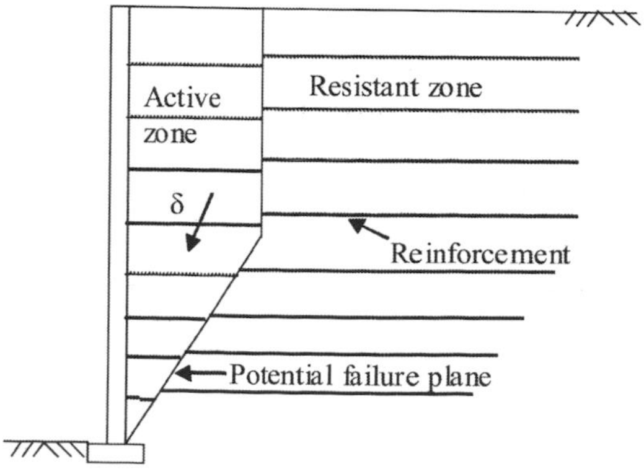

Figure 1. Oblique pullout of reinforcement, bilinear failure mechanism.

But the problem of reinforcement subjected to transverse force at end was identified (Fig. 2) and solved by Madhav & Umashankar (2003). The analysis is carried out assuming the reinforcement to be inextensible, transverse displacement at the free end to be small (<1% length of reinforcement), Winkler type response for ground with linear stress – displacement response for subgrade soil and full mobilization of interface bond resistance. A relation is developed between transverse force and free end displacement. A comprehensive parametric study illustrates the significance of depth of embedment, length of reinforcement, interface characteristics and stiffness of ground on the overall response of the reinforcement. This formulation is extended for large transverse displacements (displacement > 1% of reinforcement length) by Madhav & Manoj (2004).

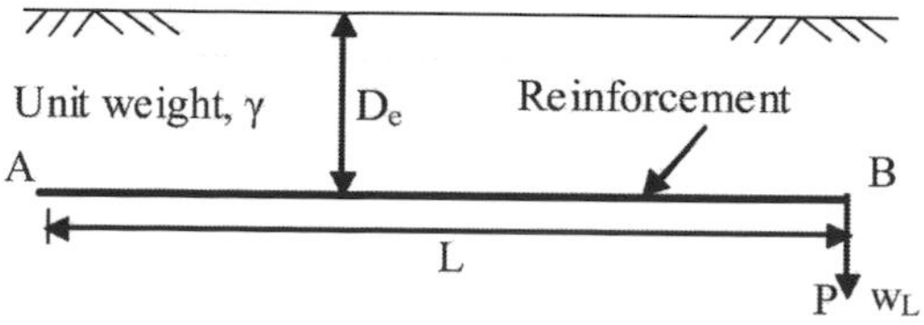

Figure 2. Reinforcement subjected to transverse force at end (Madhav & Umashankar 2003).

2 PROBLEM DEFINITION AND ANALYSIS

A reinforced earth wall (Fig. 3) of height, H, to retain a granular backfill of friction angle, φ and unit weight, γ, is considered. Inextensible reinforcement sheets (n layers) of length, L, and interface friction angle, ϕ_r are laid inside the backfill. The reinforcement sheets having a uniform spacing of $S_v = H/n$ were arranged in the backfill, with spacing of $S_v/2$ at the top and bottom of backfill. The RE wall is designed to satisfy external stability of reinforced earth structure as a unit, including sliding, rotation, bearing failure. The internal stability is essentially associated with bond and tension failure mechanisms.

2.1 *Characteristics of coherent gravity analysis*

The tensile strains developed in (steel) reinforcement (strips, grids/anchors) under working stress conditions are generally less than 1% which is insufficient to generate the active (k_a) stress state. In such conditions the coherent gravity analysis described below is adopted.

- The reinforced mass is divided into two zones, active and resisting zones, separated by the line of maximum tension in the reinforcement (Fig. 3).
- The state of stress within the reinforced mass varies from at rest state i.e. $k_{des} = k_0$ at ground level to active state i.e. $k_{des} = k_a$ at mid height of the wall of the structure and is entirely in active state below the mid depth (Fig. 3).
- Meyerhof type pressure distribution is assumed to exist beneath and within the reinforced fill.

In Fig. 3 a typical arrangement of reinforcement is presented with coherent gravity failure mechanism. L_{ei} is the effective length of ith layer of reinforcement located at a depth of z_i from the top of the wall.

$$z_i = \left(i - \frac{1}{2}\right)\frac{H}{n} \quad (1)$$

$$\text{If } z_i \le \frac{H}{2}, \; L_{ei} = L - (0.3 \times H) \quad (2)$$

$$\text{and for } z_i > \frac{H}{2}, \; L_{ei} = L - \{\tan(\frac{\pi}{4} - \frac{\varphi}{2})(H - z_i)\} \quad (3)$$

$$\text{If } z_i \le H/2, \; k_{des} = k_0\left(1 - \frac{z_i}{6}\right) + k_a\frac{z_i}{6} \quad (4)$$

$$\text{and for, } z_i > H/2, \; k_{des} = k_a \quad (5)$$

where k_0 and k_a are coefficients of earth pressures at-rest and active conditions respectively. Tension in each layer is obtained from the following equation

$$P_{ai} = \sigma_{vi}\, k_{des}\, S_{vi} \quad (6)$$

where σ_{vi} is modified vertical stress and S_{vi} is the spacing of reinforcement.

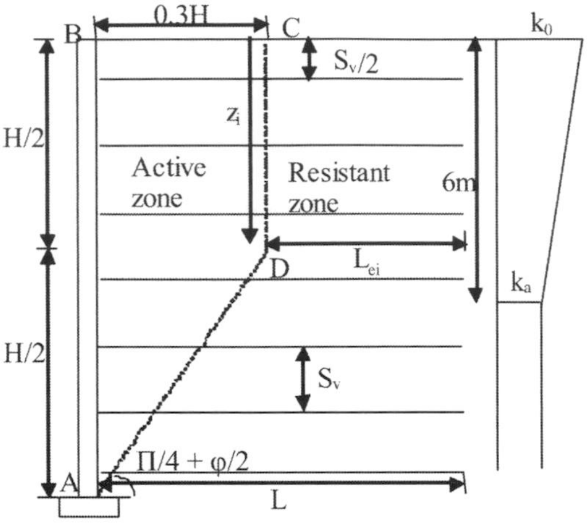

Figure 3. Coherent gravity analysis of RE wall.

The axial pullout resistance of the reinforcement sheet is obtained as follows

$$T_i = 2\gamma z_i L_{ei} \tan\varphi_r \quad (7)$$

Conventional factor of safety (FS_{conv}) is the ratio of total pullout resistance mobilized in all the reinforcement layers to the total tension or active force to be resisted, as

$$FS_{conv} = \frac{\sum_{i=1}^{n} T_i}{\sum_{i=1}^{n} P_{ai}} \quad (8)$$

2.2 *Analysis considering oblique pull*

The failure wedge ABCD undergoes an oblique displacement, δ, thus subjecting each reinforcement layer to transverse/oblique displacement along the surface ADC. Along DC reinforcement is subjected to transverse pull, δ, and along AD the displacement is oblique to the alignment of reinforcement, hence can be resolved into vertical and horizontal components, $\delta\cos\theta$ and $\delta\sin\theta$ respectively (Fig. 4).

A transverse force, P_i, is mobilized on either side of failure plane due to transverse displacement (Fig. 4). The force, P_i, is the resultant of the normal stresses mobilized along the reinforcement – backfill interface. Additional shear resistance is mobilized along the soil – reinforcement interface due to increased normal stresses leading to an increased pullout resistance. The procedure for evaluation of transverse force, P_i, and pullout resistance along each reinforcement layer is explained below.

Madhav and Umashankar (2003) quantified the transverse force mobilized due to transverse pull (w_L)

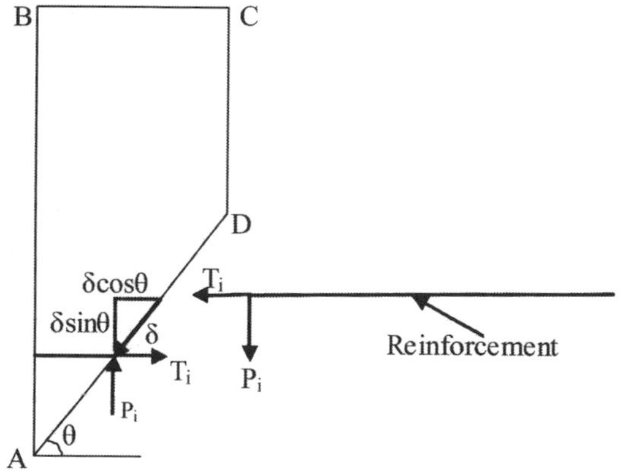

Figure 4. Equilibrium of forces for oblique pullout.

at the free end of reinforcement for the problem identified in Fig. 2. This analysis however does not predict the redistribution of stresses above the reinforcement because of the transverse pull. The normalized transverse force is obtained as

$$P^* = \frac{w_L}{L}\,\mu \int_0^1 W\,dX \tag{9}$$

where μ is relative subgrade stiffness factor, W is normalized transverse displacement and X is the normalized horizontal dimension.

The depth of reinforcement, z_i, and the effective length, L_{ei}, of each layer of reinforcement in the passive zone of RE wall is utilized to evaluate the normalized transverse displacement and relative subgrade stiffness factor as follows. Normalized transverse displacement of ith layer:

$$\text{If } z_i \le \frac{H}{2}, \quad \frac{w_L}{L_{ei}} = \frac{\delta}{L_{ei}} \tag{10}$$

$$\text{and for } z_i > \frac{H}{2}, \quad \frac{w_L}{L_{ei}} = \frac{\delta \sin\theta}{L} \tag{11}$$

Relative subgrade stiffness factor of ith layer:

$$\mu_i = \frac{\mu_{global}\, L_{ei}\, H}{L\, z_i} \tag{12}$$

$$\text{where } \mu_{global} = \frac{k_s L}{\gamma H} \tag{13}$$

which is the same as μ defined by Madhav and Umashankar (2003).

Substituting the above values of normalized transverse displacement and relative subgrade stiffness factor in Eq. 9, the normalized transverse force (P^*_i) for each reinforcement layer in passive zone is obtained and the corresponding transverse force is evaluated from the following equation

$$P_i = P^*_i \times \gamma \times z_i \times L_{ei} \tag{14}$$

Due to the transverse displacement, each reinforcement layer in RE wall is subjected to transverse force P_i obtained above and an equal force is applied on reinforcement in active zone as shown in Fig. 4. In the present work only the effect of transverse force mobilized in reinforcement of resistant zone is considered in terms of the improvement in pullout resistance as follows.

$$T_{iT} = 2\gamma\, z_i\, L_{ei} \tan\varphi_r + P_i \tan\varphi_r \tag{15}$$

The ratio of total pullout resistance to the total tension in all layers is defined as modified factor of safety, F_T, as

$$F_T = \frac{\sum_{i=1}^{n} T_{iT}}{\sum_{i=1}^{n} P_{ai}} \tag{16}$$

$$\text{Improvement ratio: } R_T = \frac{F_T}{FS_{conv}} \tag{17}$$

3 RESULTS AND DISCUSSION

To elucidate the effect of oblique pullout in stability of RE wall, the variation of modified factor of safety and improvement ratio for a wide range of following parameters is presented. Length of reinforcement L = 0.5 H–0.8 H, angle of shearing resistance of backfill $\varphi = 30°$–$35°$, interface friction angle $\varphi_r = (2/3)\varphi$ to φ, number of reinforcement layers n = 3 to 6, global subgrade stiffness factor, $\mu_{global} = 10$ to 1000 and oblique displacement, $\delta = 0.001$ L – 0.1 L.

The conventional factor of safety increases linearly with increase in length of reinforcement, (Fig. 5) since the effective length of reinforcement in passive/resisting zone increases thereby increasing the pullout resistance of reinforcement. For $\varphi = 30°$, FS_{conv} increased from 2.76 to 6.45 with increase in length of reinforcement from 0.5 H to 0.8 H.

The increase in angle of shearing resistance of soil increases the conventional factor of safety due to increase in pullout resistance (T_i) of reinforcement, with simultaneous reduction of active pressure force (P_{ai}). It can be observed that with increase in φ from 30° to 35°, FS_{conv} increased from 4.0 to 5.94 for length of reinforcement L = 0.6 H.

The variation of modified factor of safety with length of reinforcement is presented in Fig. 6. Due

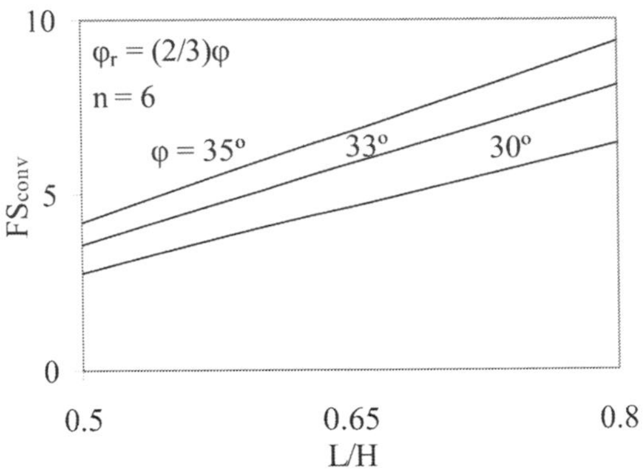

Figure 5. Variation of FS_{conv} with L/H – Effect of φ.

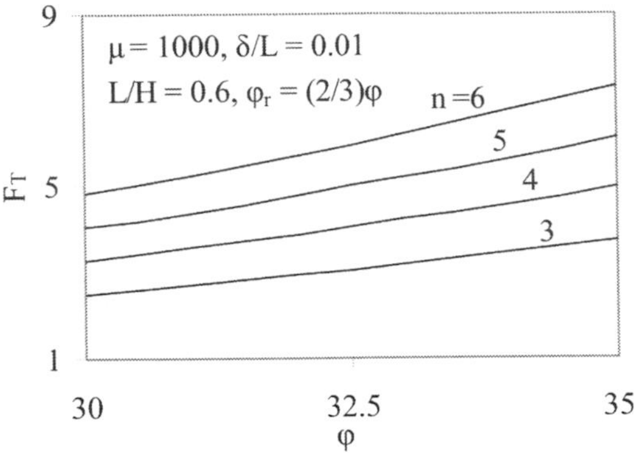

Figure 7. Variation of F_T with φ – Effect of n.

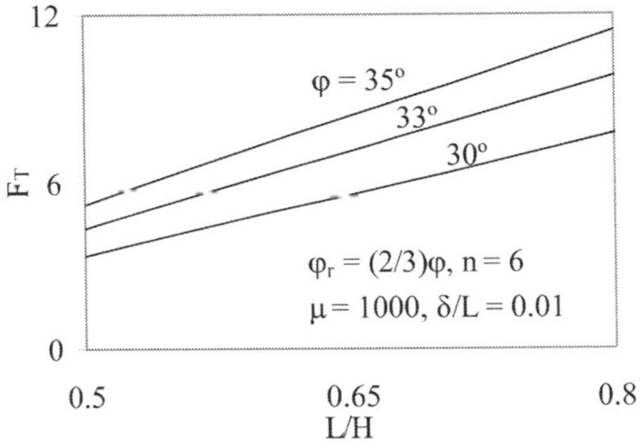

Figure 6. Variation of F_T with L/H – Effect of φ.

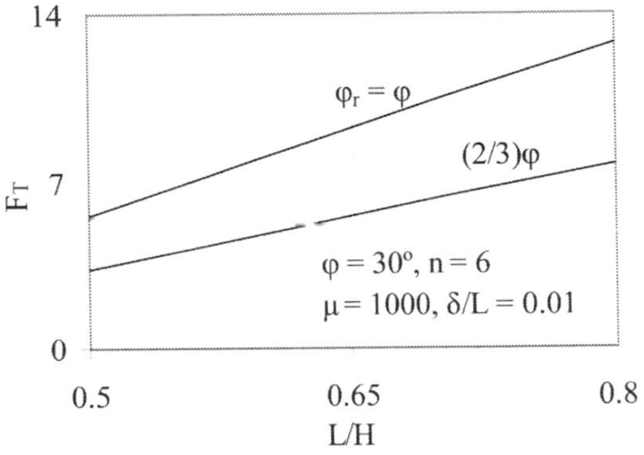

Figure 8. Variation of F_T with L/H – Effect of φ_r.

to increase in length of reinforcement, the extent of soil affected above the reinforcement increases thus inducing larger normal stresses on reinforcement. The additional normal stress increases the pullout resistance, hence F_T increases from 3.33 to 7.7 with increase in length of reinforcement from 0.5 H to 0.8 H for $\varphi = 30°$. The rate of improvement in F_T and FS_{conv} are uniform with increase in length of reinforcement.

The effect of friction angle of soil and number of reinforcement layers is presented in Fig. 7. For six layers of reinforcement, F_T increases from 4.82 to 7.3 with increase in φ from 30° to 35°. As mentioned earlier the active earth pressure force decreases with simultaneous increase in pullout resistance due to increase in friction angle of soil. Due to consideration of oblique pull the rate of improvement in F_T is more compared with the corresponding increase in FS_{conv}. The increase in number of reinforcement layers increases modified factor of safety, since the tension is distributed in all layers and also the total pullout resistance of RE wall increases. F_T increased from 2.46 to 4.82 with increase in number of layers from 3 to 6 for $\varphi = 30°$. The rates of improvement of F_T and FS_{conv} are almost similar with increase in number of reinforcement layers.

The modified factor of safety increased from 4.82 to 8.0 with increase in interface friction angle from $(2/3)\varphi$ to φ for length of reinforcement L = 0.6 H due to increase in pullout resistance of reinforcement (Fig. 8). The rate of improvement of F_T with increase in interface friction angle is greater compared with the increase in FS_{conv}.

The increase in modified factor significantly depends on two factors – global subgrade stiffness factor and oblique displacement. The influence of global subgrade stiffness factor on modified factor of safety is depicted in Fig. 9. With increase in stiffness of subgrade the transverse force required to mobilize an oblique displacement increases, hence F_T increased from 4.08 to 4.82 with increase in μ_{global} from 10 to 1000 for length of reinforcement L = 0.6 H. The increase is linear and marginal for μ_{global} less than 200 and beyond 200 the modified factor of safety increased substantially.

The variation of modified factor of safety with oblique displacement is presented in Fig. 10. The increase of oblique displacement of reinforcement increases the normal stresses acting on reinforcement thus increasing the total pullout resistance of RE wall.

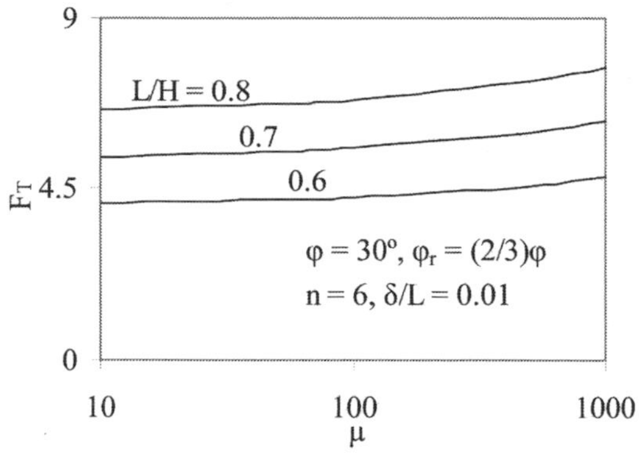

Figure 9. Variation of F_T with μ – Effect of L/H.

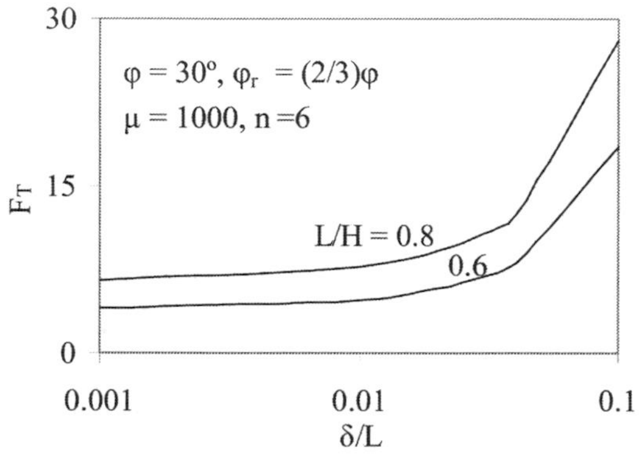

Figure 10. Variation of F_T with δ/L – Effect of L/H.

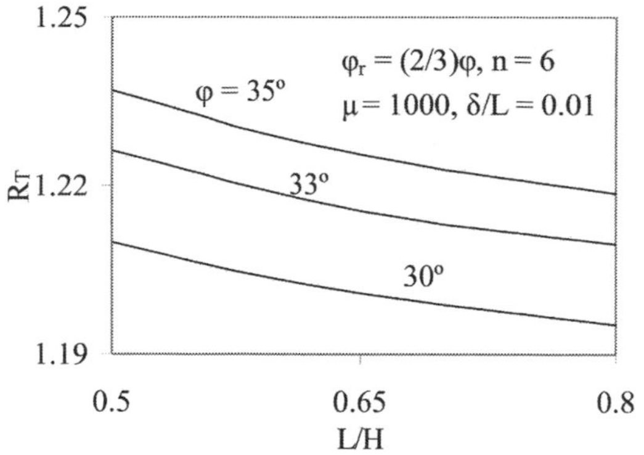

Figure 11. Variation of R_T with L/H – Effect of φ.

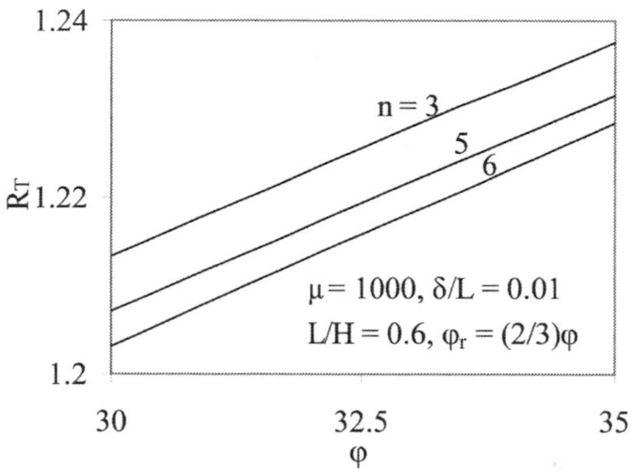

Figure 12. Variation of R_T with φ – Effect of n.

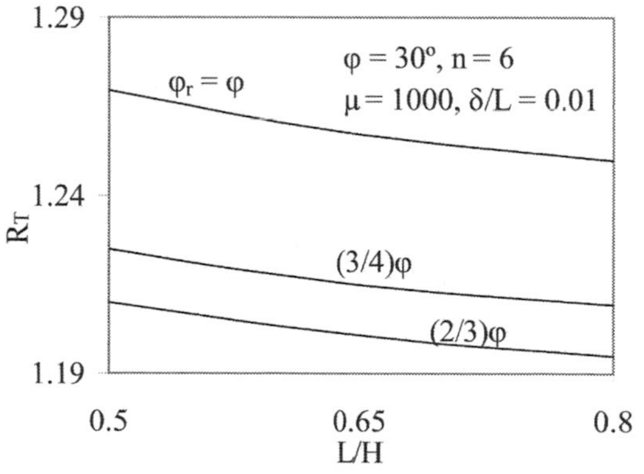

Figure 13. Variation of R_T with L/H – Effect of φ_r.

F_T increases from 4.1 to 18.4 with increase of oblique displacement δ from 0.001 L to 0.1 L for length of reinforcement L = 0.6H. The conventional factor of safety will not depend on variations of global subgrade stiffness factor and oblique displacement of reinforcement.

As mentioned earlier, both FS_{conv} and F_T increase with length of reinforcement but the rate of increase of F_T is smaller compared with FS_{conv}, hence the improvement ratio decreases slightly with increase in length of reinforcement (Fig. 11). The improvement ratio varies around 1.21 with increase in length of reinforcement L = 0.5 H to 0.8 H for friction angle $\varphi = 30°$.

The variation of improvement ratio for different friction angles of soil and number of reinforcement layers is shown in Fig. 12. The improvement ratio increased from 1.2 to 1.23, i.e. by 3% with increase in φ from 30° to 35° for six layers of reinforcement. But the effect of number of layers of reinforcement on improvement ratio is similar to the length of reinforcement mentioned earlier, the improvement ratio varies around 1.22 with increase in number of layers from 3 to 6 for $\varphi = 30°$.

The improvement ratio R_T increased from 1.2 to 1.26 i.e. by 6% with an increase in interface friction angle from (2/3)φ to φ for length of reinforcement L = 0.6 H (Fig. 13).

The influence of global subgrade stiffness factor and oblique displacement on improvement ratio is

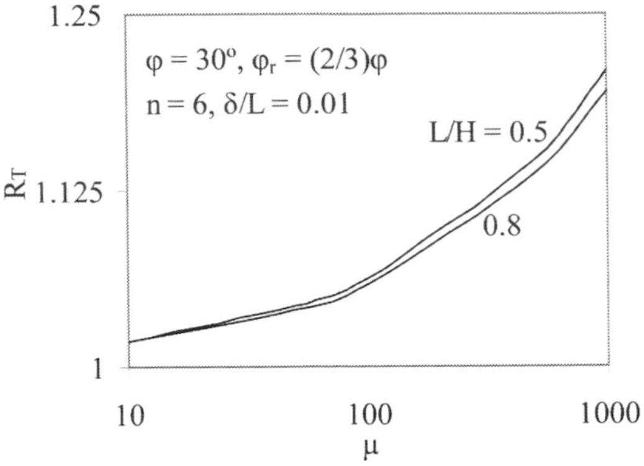

Figure 14. Variation of R_T with μ – Effect of L/H.

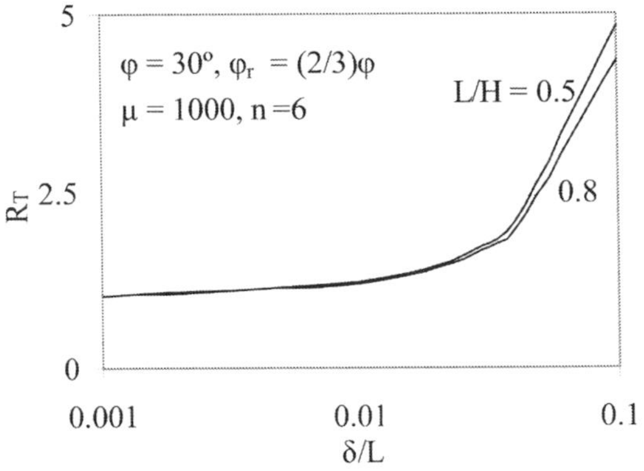

Figure 15. Variation of R_T with δ/L – Effect of L/H.

depicted in Fig. 14 & Fig. 15. The improvement ratio increased substantially from 1.02 to 1.20 with increase in μ_{global} from 10 to 1000 for length of reinforcement L = 0.6 H (Fig. 14). R_T increases significantly from 1.02 to 4.6 with increase in oblique displacement, δ, from 0.001L to 0.1 L for length of reinforcement, L = 0.6 H (Fig. 15). In both the cases curves merge for different lengths of reinforcement. This confirms that the influence of global subgrade stiffness factor and oblique displacement of reinforcement on improvement ratio is predominant compared with length of reinforcement.

4 CONCLUSIONS

The oblique pullout of reinforcement and its influence on the stability of RE wall is investigated for coherent gravity failure mechanism. A linear stress – displacement response of the backfill is assumed with full shear mobilization along the reinforcement soil interface. The oblique displacement causes mobilization of additional normal stresses along the reinforcement in the passive zone leading to additional shear resistance to counteract active forces. A formulation is presented to evaluate the transverse force in each reinforcement layer and a modified factor of safety incorporating the additional resistance is defined.

The variations of modified factor of safety with length of reinforcement, friction angle of soil, interface friction angle and number of reinforcement layers are presented and compared with conventional one to illustrate the significance of oblique pull vis a vis the axial pull in the stability of RE wall. The improvement ratio varies from 1.19 to 1.27 due to the influence of above parameters.

The improvement ratio varied from 1 to 1.38 with global subgrade stiffness factor and 1 to 4 for oblique displacement of reinforcement. Hence the global subgrade stiffness factor and oblique displacement of reinforcement have relatively greater significance than the other parameters.

REFERENCES

Athanasapoulos, G.A. 1993. Effect of particle size on the mechanical behavior of sand – geotextile composites. *Geotextiles and Geomembranes* 12: 252–273.

Bergado, D.T., Teerawattanasuk, C. & Long, P.V. 2000. Localized mobilization of reinforcement force and its direction at the vicinity of failure surface. *Geotextiles and Geomembranes* 18: 311–331.

Burd, H.J. 1995. Analysis of membrane action in reinforced unpaved roads. *Canadian Geotechnical Journal* 32: 946–956.

Degenkamp, G. & Dutta, A. 1989. Soil resistance to embedded anchor chain in soft clays. *Journal of Geotechnical Engineering* 115 (10): 1420–1437.

Grey, D.H. & Ohashi, H. 1989. Mechanics of fiber reinforcement in sand. *Journal of Geotechnical Engineering* 109 (3): 335–353.

Leschinsky, D. & Boedekaer, R.H. 1989. Geosynthetic reinforced soil structures. *Journal of Geotechnical Engineering* 115 (10): 1459–1478.

Leschinsky, D. & Reinschmidt, A.J. 1985. Stability of membrane reinforced slopes. *Journal of Geotechnical Engineering* 111 (11): 1285–1300.

Madhav, M.R. & Umashankar, B. 2003. Analysis of inextensible sheet reinforcement subject to transverse displacement/force: Linear subgrade response. *Geotextiles and Geomembranes* 21: 69–84.

Madhav, M.R. & Manoj, T.P. 2004. Response of geosynthetic reinforcement to transverse force/displacement with linear subgrade response, *Proc., Int. Conf. on Geotechnical and Geoenvironmental Engg., Mumbai.*

Neubecker, S.R. & Randolph, M.F. 1995. Profile and frictional capacity of embedded anchor chains, *Journal of Geotechnical Engineering* 121 (11): 797–803.

Shewbridge, S.E. & Sitar, N. 1989. Deformation characteristics of reinforced sand in direct shear. *Journal of Geotechnical Engineering* 115 (8): 1134–1147.

New Horizons in Earth Reinforcement – Otani, Miyata & Mukunoki (eds)
© 2008 Taylor & Francis Group, London, ISBN 978-0-415-45775-0

Recent developments in the K-stiffness Method for geosynthetic reinforced soil walls

R.J. Bathurst
GeoEngineering Centre at Queen's-RMC, Department of Civil Engineering, Royal Military College of Canada, Kingston, Ontario, Canada

Y. Miyata
National Defense Academy, Yokosuka, Japan

T.M. Allen
Washington DOT, State Materials Laboratory, Olympia, United States

ABSTRACT: The K-stiffness Method is an empirically-developed working stress method used to compute reinforcement loads for the internal stability design of geosynthetic reinforced soil walls under operational conditions. Recently, the writers have collected data from Japanese case studies of monitored full-scale geosynthetic reinforced soil walls constructed with different facing batters, facing types and a range of c-ϕ soils. These data have been used to adjust the method calibration and to extend the utility of the K-stiffness Method to a wider range of wall configurations and soil types. The paper describes the essential features of the new method. The improvement of the modified K-stiffness Method over the AASHTO Simplified Method is demonstrated quantitatively by statistical analysis of the ratios (bias) of average measured to predicted reinforcement load values. The new K-stiffness Method holds promise as a more accurate design tool for internal stability of reinforced soil walls in North America, Japan and worldwide.

1 INTRODUCTION

The current approach for the internal stability design of geosynthetic reinforced soil walls in North America is the AASHTO (2002) Simplified Method. The approach is based on limit-equilibrium of a "tied-back wedge" and its origins can be traced back to the early 1970's (Allen et al. 2002).

Allen et al. (2002, 2003) and Bathurst et al. (2005) investigated quantitatively the accuracy of the Simplified Method using careful interpretation of a database of 11 well-documented and monitored full-scale field walls. They concluded that the Simplified Method tends to be excessively conservative with respect to the selection of the reinforcement required for good wall performance under typical operational conditions corresponding to end of construction. Furthermore, the distribution of reinforcement loads in the instrumented walls was seen to be generally trapezoidal in shape rather than linear with depth as is assumed in the Simplified Method for walls with uniform reinforcement spacing (Allen & Bathurst 2002).

Miyata & Bathurst (2007a) showed that the same level of conservatism was true for vertical walls with granular backfill soils using the current PWRC (2000) guidelines in Japan. In the Japanese approach, the earth loads to be carried by the horizontal reinforcement layers are computed using a circular slip method. However, for vertical geosynthetic reinforced soil walls with a granular backfill, the predicted loads using the PWRC and AASHTO methods were sensibly identical and hence equally excessively conservative.

Miyata & Bathurst (2007b) and Bathurst et al. (2007) extended the original database of walls used to originally investigate the accuracy of the Simplified Method to include a total of 18 vertical wall sections and 11 battered walls. Most of the new data were for instrumented and monitored Japanese walls reported in documents published in Japanese. A total of 15 walls in the new database were constructed with frictional soils ($c = 0$, $\phi > 0$) and 14 walls with soils characterized as having both frictional and cohesive strength components ($c > 0$, $\phi > 0$).

The conservatism in load predictions using the current AASHTO Simplified Method can be appreciated from the data plotted in Figure 1. In almost all cases the predicted loads (T_{max}) fall above the 1:1 correspondence line. For c-ϕ soils, an additional

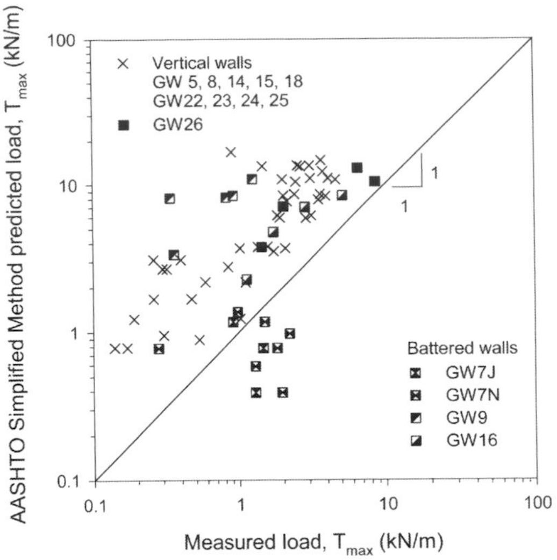

a) granular soil backfill

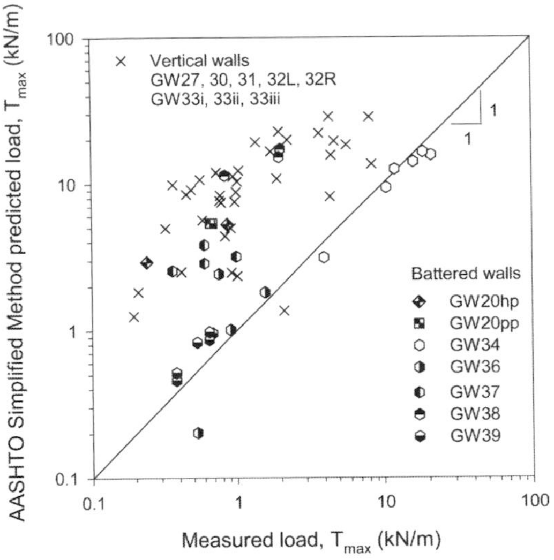

b) c-ϕ soil backfill

Figure 1. Predicted versus measured reinforcement loads (T_{max}) using AASHTO Simplified Method.

source of conservatism is due to the AASHTO recommendation to set $c = 0$ for cohesive-frictional soil in load calculations. The calculation of the magnitude of conservatism is discussed later in this paper.

The focus of this paper is on a description of the K-stiffness Method as it has evolved over the last few years, and illustration of the quantitative improvement in the prediction of reinforcement loads in geosynthetic reinforced soil walls using this new method compared to the current AASHTO Simplified Method.

2 K-STIFFNESS METHOD

2.1 *General*

The K-stiffness Method as originally proposed by Allen et al. (2003) was developed to overcome the deficiencies in the current Simplified Method noted above. This empirically developed working stress method was limited to geosynthetic reinforced soil walls constructed with granular backfills. For these conditions, the method was demonstrated to give more accurate predictions of reinforcement loads based on the statistics for the mean and coefficient of variation of the ratio (bias) of measured to predicted loads.

The general expression that is the core of the current K-stiffness Method (Bathurst et al. 2007) is:

$$T_{max} = \tfrac{1}{2} K \gamma (H + S) S_v D_{tmax} \Phi_g \Phi_{local} \Phi_{fs} \Phi_{fb} \Phi_c \quad (1)$$

Here, K = lateral earth pressure coefficient; γ = unit weight of the soil; H = height of the wall; S = equivalent height of uniform surcharge pressure q (i.e. $S = q/\gamma$); S_v = tributary area (equivalent to the vertical spacing of the reinforcement in the vicinity of each layer when analyses are carried out per unit length of wall); D_{tmax} = load distribution factor that modifies the reinforcement load based on layer location. The remaining terms, Φ_g, Φ_{local}, Φ_{fs}, Φ_{fb} and Φ_c are influence factors that account for the effects of global and local reinforcement stiffness, facing stiffness, face batter, and soil cohesion, respectively. The coefficient of lateral earth pressure is calculated as $K = 1 - \sin\phi$ with $\phi = \phi_{ps}$ = peak plane strain friction angle of the soil. However, it should be noted that parameter K is used as an index value and does not imply that at-rest soil conditions exist in the reinforced soil backfill according to classical earth pressure theory.

In the original K-stiffness Method equation proposed by Allen et al. (2003) for walls with granular backfill soils, the influence factor for soil cohesion does not appear (or $\Phi_c = 1$ for soils without a cohesive soil strength component). The restriction to granular soils was imposed largely because of the lack of high-quality data for walls constructed with c-ϕ soils. Furthermore, the original database used to calibrate the method was limited with regard to the range of facing batter angle available.

The proposed model as expressed by Equation 1 captures all qualitative effects due to reinforcement stiffness, soil strength, facing stiffness and reinforcement arrangement expected by experienced wall design engineers. Furthermore, the general structure of Equation 1 may be familiar to geotechnical engineers using classical earth pressure theory in combination with a tributary area approach for the distribution of earth pressures to the internal reinforcement layers. For example, the load carried by a reinforcement layer will decrease as soil friction angle increases (i.e. because the magnitude of coefficient of earth pressure K decreases). The reinforcement load will increase as soil unit weight (γ) and reinforcement spacing (S_v) increases.

Recently, the writers have used the additional data from Japanese case studies mentioned in the previous section to recalibrate the method and to extend the utility of the K-stiffness Method to a wider range of wall configurations and soil types (Miyata & Bathurst 2007a,b, and Bathurst et al. 2007). The additional monitored full-scale geosynthetic reinforced soil walls were constructed with different facing batters, facing types and a range of c-ϕ soils. For brevity in the following text, references to the K-stiffness Method refer to the modified model proposed by Bathurst et al. (2007).

2.2 *Calculation of K-stiffness Method parameters*

The general expression that is the core of the K-stiffness Method has been given in Equation 1.

Parameter Φ_g is a global stiffness factor that accounts for the influence of the stiffness and spacing of the reinforcement layers over the entire wall height and is calculated as follows:

$$\Phi_g = \alpha\left(\frac{S_{global}}{p_a}\right)^{\beta} \tag{2}$$

Here, S_{global} is the global reinforcement stiffness and α and β are constant coefficients equal to 0.25. The non-dimensionality of the expression is preserved by dividing the global reinforcement stiffness by $p_a = 101$ kPa (atmospheric pressure). The global reinforcement stiffness value for a wall is calculated as:

$$S_{global} = \frac{J_{ave}}{(H/n)} = \frac{\sum_{i=1}^{n} J_i}{H} \tag{3}$$

Here, J_{ave} is the average tensile stiffness of all n reinforcement layers over the wall height and, J_i is the tensile stiffness, at the end of wall construction, of an individual reinforcement layer expressed in units of force per unit length of wall. The practical result of the calculation for the global stiffness factor is that reinforcement loads will increase as average reinforcement stiffness increases and the spacing between reinforcement increases. The method has been calibrated against measured reinforcement loads deduced from in-isolation isochronous stiffness values corresponding to reinforcement strain at end of construction. To implement the method for design, a default time of 1000 hours is reasonable since most walls are constructed within 1000 hours and the end of construction stiffness value is taken at 2% strain. Hence, $J_i = J_{2\%}$ in Equation 3. Results of in-isolation constant load (creep) and constant-rate-of-strain tests on the polyolefin reinforcement products used in the case studies have shown that the $J_{2\%}$ secant stiffness is a constant value for practical purposes at or beyond 1000 hours (Walters et al. 2002).

Parameter Φ_{local} is a local stiffness factor that accounts for the relative stiffness of the reinforcement layer with respect to the average stiffness of all reinforcement layers and is expressed as:

$$\Phi_{local} = \left(\frac{S_{local}}{S_{global}}\right)^{a} \tag{4}$$

where a is a constant coefficient and S_{local} is the local reinforcement stiffness for reinforcement layer i calculated as:

$$S_{local} = \left(\frac{J}{S_v}\right)_i \tag{5}$$

Back fitting of measured versus predicted reinforcement loads by Allen et al. (2003) gave a = 1 for geosynthetic reinforced soil walls. Local deviations from overall trends in reinforcement load can be expected when the reinforcement stiffness and/or spacing of the reinforcement change from average values over the height of the wall (i.e. $S_{local}/S_{global} \neq 1$; Hatami et al. 2001). This effect is captured by the local stiffness factor Φ_{local} expressed by Equation 4.

Parameter Φ_{fb} in Equation 1 accounts for the influence of the facing batter and is computed as:

$$\Phi_{fb} = \left(\frac{K_{abh}}{K_{avh}}\right)^{d} \tag{6}$$

where, K_{abh} is the horizontal component of active earth pressure coefficient accounting for wall face batter, K_{avh} is the horizontal component of active earth pressure coefficient (assuming the wall is vertical), and d is a constant coefficient. The form of the equation shows that as the wall face batter angle $\omega \rightarrow 0$ (i.e. wall facing batter approaches the vertical) the facing batter factor $\Phi_{fb} \rightarrow 1$. The value of the coefficient term d is taken as 0.25. However, for vertical walls the value of this coefficient is not consequential (i.e. $K_{abh}/K_{avh} = 1$) and its value is insignificant for near-vertical walls.

The influence factor for facing stiffness (rigidity) Φ_{fs} is computed as:

$$\Phi_{fs} = \eta(F_f)^{\kappa} \tag{7}$$

where

$$F_f = \frac{1.5H^4 p_a}{ELb^3 (h_{eff}/H)} \quad (8)$$

is the facing column stiffness parameter. Here, b = thickness of the facing column, L = unit length of the facing (e.g., L = 1 m), H = height of the facing column, and E = elastic modulus of the "equivalent elastic beam" representing the wall face. The two expressions used to compute the facing stiffness factor show that as the wall becomes higher (H) and less stiff (ELb^3), its rigidity becomes less and hence more load is carried by the reinforcement layers (i.e. Φ_{fs} is larger in Equation 1). This effect has been quantitatively demonstrated by careful tests on two full-scale reinforced soil walls tests reported by Bathurst et al. (2006). The 3.6-m high structures were nominally identical except one was built with a relatively stiff modular block facing and the other with a wrapped-face. The loads in the most heavily loaded reinforcement layers were 3.5 times greater in the flexible-face wall than those in the stiffer modular block wall.

The term h_{eff} is the equivalent height of an un-jointed facing column that is 100% efficient in transmitting moment through the height of the facing column. The ratio h_{eff}/H is used to estimate the efficiency of a jointed facing system to transmit moment along the facing column. The non-dimensionality of Equation 8 is preserved by the use of $p_a = 101$ kPa. Further discussion regarding the selection of parameters in Equation 8 can be found in the earlier papers by Allen et al. (2003) and Miyata & Bathurst (2007a, b) and the design guidance document (WSDOT 2005). Based on back-analyses performed by Miyata & Bathurst (2007b) the coefficient terms η and κ were determined to be 0.55 and 0.14, respectively.

The effect of soil cohesion is captured by the cohesion (influence) factor ϕ_c computed as:

$$\Phi_c = 1 - \lambda \frac{c}{\gamma H} \quad (9)$$

where the cohesion coefficient $\lambda = 6.5$. Examination of Equation 9 with $\lambda = 6.5$ reveals that the practical limit $0 \geq \Phi_c \geq 1$ requires $c/\gamma H \leq 0.153$. It is possible that a combination of a short wall height and high cohesive soil strength could lead to $\Phi_c = 0$. In practical terms this means that no reinforcement is required.

The load distribution factor $D_{tmax} = T_{max}/T_{mxmx}$ is used to modify the reinforcement load T_{max} based on layer location. Parameter T_{mxmx} is the maximum reinforcement load from all reinforcement layers. The distribution of D_{tmax} plotted against normalized height of wall is trapezoidal in shape as originally proposed by Allen et al. (2003) and illustrated in Figure 2. Note that the same distributions apply to battered walls (Bathurst et al. 2007). The value of T_{mxmx} can be calculated by setting $D_{tmax} = 1$ in Equation 1.

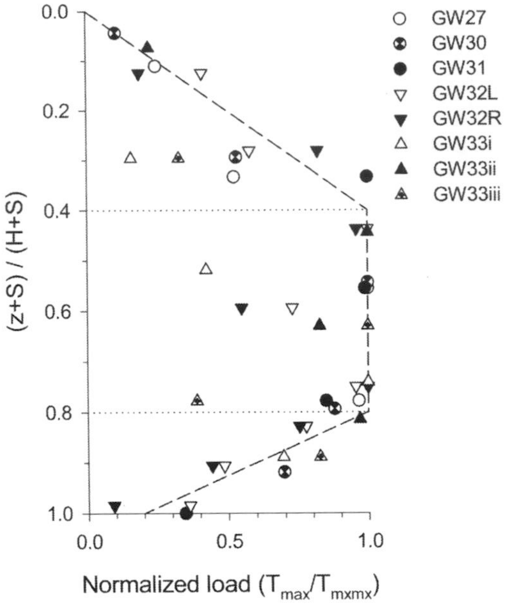

a) granular soil backfill

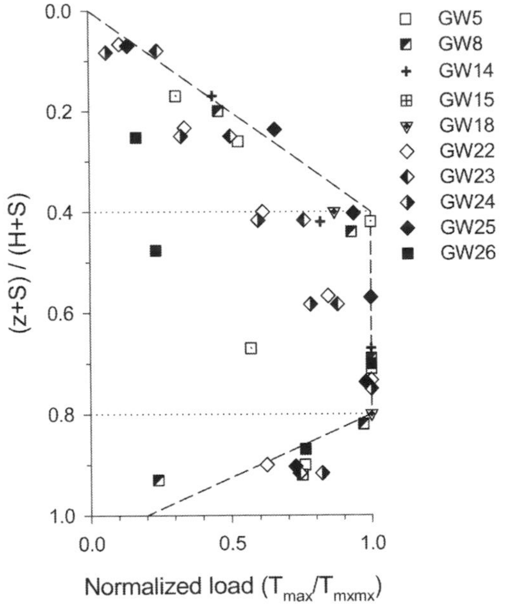

b) c-ϕ backfill soil

Figure 2. Measured values and distribution of $D_{tmax} = T_{max}/T_{mxmx}$.

2.3 *Accuracy of K-stiffness Method*

The improvement in the prediction of reinforcement loads using the K-stiffness Method is visually apparent when Figure 3 is compared to Figure 1. The improvement in the accuracy of the K-stiffness Method compared to the current AASHTO Simplified Method can be quantified using the statistics for the ratio (bias) of measured to predicted reinforcement loads T_{max} and T_{mxmx}. In the limit of a perfect deterministic model, the mean of the bias values is one and coefficient of variation (COV) is zero. For practical design, a mean value for the bias statistics equal to one or slightly less than one is desirable together with a low coefficient of variation (COV) value, although the choice of an acceptable COV value is subjective.

The bias statistics are summarized in Table 1. The number of data points for T_{max} and T_{mxmx} refer to the number of reinforcement layers and the number of wall sections in the database, respectively. Table 1a shows that for walls with granular backfill soils, the AASHTO Simplified Method over-estimates the reinforcement loads (T_{max}) by a factor of about two, *on average* and the COV of the ratio of measured to predicted loads is about 130%. The K-stiffness Method gives a value of one and 50% for these two values, which is a significant improvement. It should be noted that in practice many design engineers use lower values of peak friction angle based on triaxial and direct shear tests to compute reinforcement loads using the Simplified Method. The Simplified Method statistics shown in Table 1 have been computed using less conservative plane-strain estimates of peak friction angle in order to focus the comparison on the design methodology rather than the choice of strength parameters. The reader is directed to the earlier papers by the writers for a discussion on selection and computation of shear strength parameters using the K-stiffness Method for both fictional and c-ϕ backfill soils. The data in Table 1b shows that the measured reinforcement loads for c-ϕ backfill soils are about one third of predicted values, *on average*, using the Simplified Method and the COV of the ratio of measured to predicted loads is about 140%. The bias statistics for the K-stiffness Method are much better.

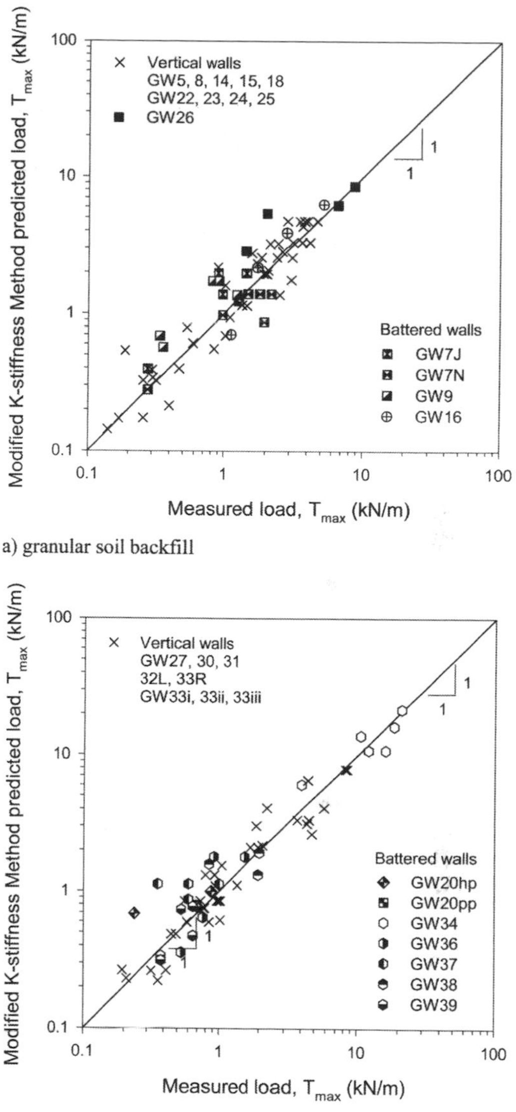

a) granular soil backfill

b) c-ϕ soil backfill

Figure 3. Predicted versus measured reinforcement loads (T_{max}) using K-stiffness Method.

3 CONCLUSIONS

This paper has briefly reviewed recent progress in the refinement of the K-stiffness Method to include c-ϕ backfill soils. In addition, the method has been recalibrated against a larger database of case studies including walls with a wider range of facing batter angles than was available at the time the method was originally developed. Measured loads are compared to predicted loads using the current AASHTO Simplified Method and the new modified version of the K-stiffness Method. The AASHTO Simplified Method is shown to be excessively conservative (on average) and to be inconsistent with respect to accurate prediction of the distribution of reinforcement loads when compared to measured values. The improvement of the modified K-stiffness Method over the AASHTO Simplified Method is demonstrated quantitatively by statistical analysis of the ratios (bias) of average measured to predicted reinforcement load values. The new K-stiffness Method holds promise as a more accurate

Table 1. Summary of statistics for ratio (bias) of measured to predicted reinforcement loads.

Parameter	Number of data points	Simplified Method AASHTO (2002)	K-stiffness Method (Bathurst et al. 2007)
a) *walls with granular soil backfill (c = 0)*			
Mean T_{max}	67	0.59	1.00
COV T_{max} (%)	67	133	50
Mean T_{mxmx}	14	0.60	1.10
COV T_{mxmx} (%)	14	105	35
Parameter	Number of data points	Simplified Method AASHTO (2002)	K-stiffness Method (Bathurst et al. 2007)
b) *walls with cohesive soil backfill (c > 0)*			
Mean T_{max}	63	0.30	0.96
COV T_{max} (%)	63	138	40
Mean T_{mxmx}	14	0.31	1.03
COV T_{mxmx} (%)	14	82	32

design tool for internal stability of reinforced soil walls in North America, Japan and worldwide.

ACKNOWLEDGMENTS

The work reported in this paper was supported by grants to the first author from the Natural Sciences and Engineering Research Council (NSERC) of Canada, the Ministry of Transportation of Ontario, the Department of National Defence (Canada) and the following state departments of transportation in the USA: Alaska, Arizona, California, Colorado, Idaho, Minnesota, New York, North Dakota, Oregon, Utah, Washington and Wyoming. The second author is grateful to the Japanese Defense Agency and National Defense Academy of Japan, which allowed him sabbatical leave as a visiting scholar at the GeoEngineering Centre at Queen's-RMC at RMC where the work described in this paper was carried out.

REFERENCES

AASHTO. 2002. Standard Specifications for Highway Bridges. *American Association of State Highway and Transportation Officials* (AASHTO), 17th ed., Washington, D.C.

Allen, T.M., Bathurst, R.J. & Berg, R.R. 2002. Global level of safety and performance of geosynthetic walls: An historical perspective. *Geosynthetics International*, Vol. 9, Nos. 5–6, pp. 395–450.

Allen, T.M. & Bathurst, R.J. 2002. Soil reinforcement loads in geosynthetic reinforced walls at working stress conditions. *Geosynthetics International*, Vol. 9, Nos. 5–6, pp. 525–566.

Allen, T.M., Bathurst, R.J., Holtz, R.D., Walters, D.L. & Lee, W.F. 2003. A new working stress method for prediction of reinforcement loads in geosynthetic walls. *Canadian Geotechnical Journal*, Vol. 40, pp. 976–994.

Bathurst, R.J., Allen, T.M. & Walters, D.L. 2005. Reinforcement loads in geosynthetic walls and the case for a new working stress design method. *Geotextiles and Geomembranes*, Vol. 23, pp. 287–322.

Bathurst, R.J., Miyata, Y., Nernheim, A. & Allen, T.M. 2007. Refinement of K-stiffness Method for geosynthetic reinforced soil walls constructed with a facing batter. *Geosynthetics International* (in review).

Bathurst, R.J., Vlachopoulos, N., Walters, D.L., Burgess, P.G. & Allen, T.M. 2006. The influence of facing rigidity on the performance of two geosynthetic reinforced soil retaining walls. *Canadian Geotechnical Journal*, Vol. 43, No. 12, pp. 1225–1137.

Hatami, K., Bathurst, R.J. & Di Pietro, P. 2001. Static response of reinforced soil retaining walls with non-uniform reinforcement. ASCE *International Journal of Geomechanics*, Vol. 1, No. 4, 2001, pp. 477–506.

Miyata, Y. & Bathurst, R.J. 2007a. Evaluation of K-Stiffness method for vertical geosynthetic reinforced granular soil walls in Japan. *Soils and Foundations*, Vol. 47, No. 2. pp. 319–335.

Miyata, Y. & Bathurst, R.J. 2007b. Development of K-stiffness Method for geosynthetic reinforced soil walls constructed with c-ϕ soils. *Canadian Geotechnical Journal*, (in press).

PWRC 2000. Design and Construction Manual of Geosynthetics Reinforced Soil (revised version). *Public Works Research Center*, Tsukuba, Ibaraki, Japan, 305 p. (in Japanese).

Walters, D.L., Allen, T.M. & Bathurst, R.J. 2002. Conversion of geosynthetic strain to load using reinforcement stiffness. *Geosynthetics International*, Vol. 9, Nos. 5–6, pp. 483–523.

WSDOT. 2005. Geotechnical Design Manual M46-03. *Washington State Department of Transportation*, Washington, USA.

New Horizons in Earth Reinforcement – Otani, Miyata & Mukunoki (eds)
© 2008 Taylor & Francis Group, London, ISBN 978-0-415-45775-0

High capacity geostrap reinforcement for MSE structures

M.J. Grien
The Reinforced Earth Company, Dallas, Texas, USA

J.E. Sankey
The Reinforced Earth Company, Vienna, Virginia, USA

ABSTRACT: High capacity geostraps, combined with new methods for connecting reinforcing elements to facing panels, have offered durable design solutions for mechanically stabilized earth (MSE) structures in sea water and aggressive environments. Analysis of the internal stability of the MSE structure is required to verify that the maximum tension applied to the reinforcing element does not exceed the long-term allowable tension and that the reinforcing length is sufficient to prevent pullout of the strip. Although physical considerations (i.e. material type, failure surface, coefficient of friction and lateral earth pressure coefficients) vary between reinforcement types, the internal design approach for MSE walls using high capacity geostraps has been found to be similar in many ways to discrete steel strips. The focus of this paper is to evaluate internal design considerations for a high strength polyester geostrap based upon results of a field test study, laboratory tests and numerical analysis.

1 INTRODUCTION

Recent developments in high strength polyester-based geostraps for use as reinforcing elements in MSE walls offer a cost-efficient design solution in marine and aggressive environments. The synthetic geostraps are noted for their high resistance to both chemical and biological degradation. When attached with a fully-synthetic connection to a reinforced concrete facing panel, as shown in Figure 1, reliable information on the geostrap's long-term design strength can be assessed considering design life and ambient temperature of the MSE wall.

Both laboratory and field test studies were conducted to select the parameters necessary for internal design of a MSE wall using the geostraps. From these studies, results were developed using geostraps for the line of maximum tension, lateral earth pressure coefficients and apparent coefficient of friction in MSE wall applications with width to height ratio of 0.7 and no heavily loaded structures. A design application was then developed using AASHTO Specifications (2002) for MSE walls, which were compared in turn to numerical analysis results for accuracy.

Figure 1. Construction of an MSE wall with geostraps.

1.1 *Tensile strength design*

The required allowable tension to prevent rupture of the geostrap may be checked against the maximum applied tension based on the lateral earth pressure from externally imposed loads. Studies of lateral earth pressure at different geostrap levels were used to back calculate the ratio for the coefficient of lateral earth pressure, K vs. K_a (active earth pressure coefficient). These results were then compared to values given in AASHTO Specifications for MSE walls, as shown in Figure 2, were referenced to determine a conservative design approach using geostrap reinforcement.

1.2 *Pullout resistance design*

For pullout prevention, the frictional properties and delineated line of maximum tension for the geostraps contribute to internal stability verification. The selection of design considerations for a MSE wall depend on the extensibility of the reinforcing element. Figures 3 and 4 from AASHTO Specifications define the

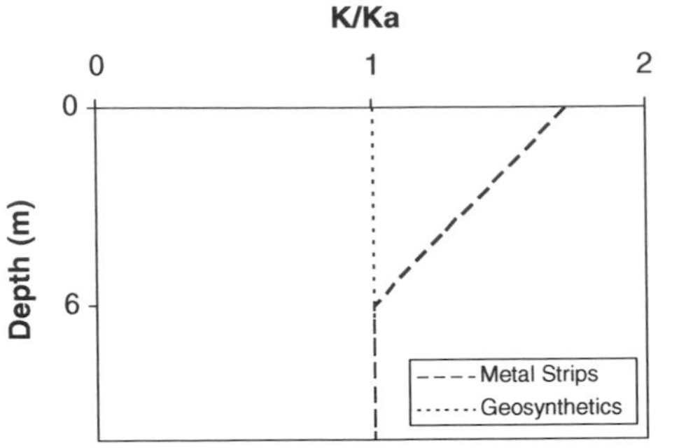

Figure 2. Ratio of lateral earth pressure coefficient, K/K_a.

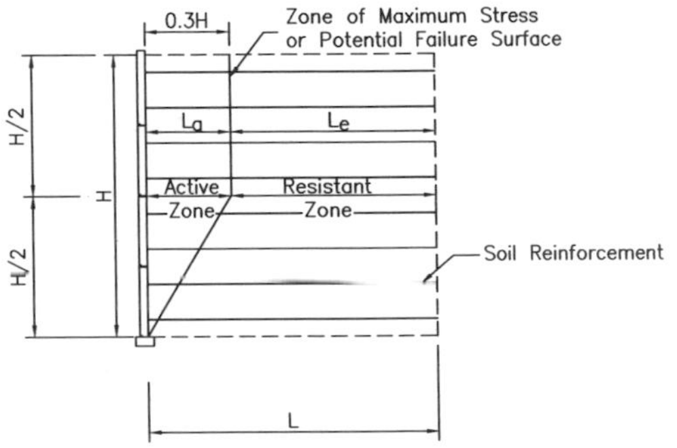

Figure 3. Inextensible failure surface.

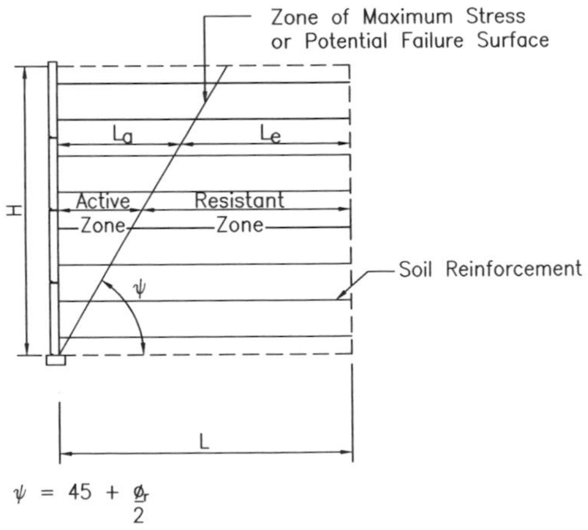

Figure 4. Extensible failure surface.

failure surface (line of maximum tension) based on reinforcing elements being either inextensible (steel) or extensible (geotextiles and geogrids). For the purpose of this paper, the actual values from the field study and numerical analysis were compared to AASHTO Specifications.

Default values from AASHTO for coefficient of friction (f*) between the reinforcing strip and soil

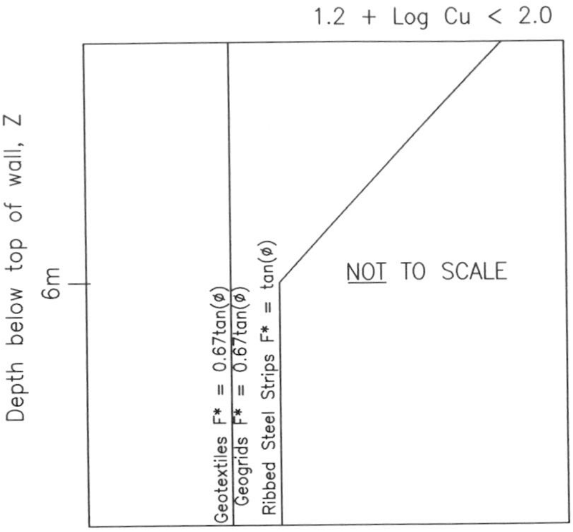

Figure 5. Default values of f*.

are shown in Figure 5. The actual values found from laboratory testing are compared to values in Figure 5.

2 DESCRIPTION OF GEOSTRAP AND CASE STUDIES

2.1 *Description of geostrap*

The geostrap reinforcing strip used in the study consisted of high tenacity polyester fibers (HTPET) encased in a polyethylene sheath. The high tenacity polyester is the load bearing element, while the sheath protects the yarns from installation damage and degradation. The durability of the geostrap has been increased by the polymerization process and is only for use in soil environments characterized by $3 < pH < 9$, with no detrimental affect on the strip due to low resistivity backfill or from backfills with high chloride or sulfate content. Ambient temperatures of the retaining wall site and design life are considered in the determination of the long-term allowable reinforcement tension.

Material Properties:
Tension = 22 kN with Strain; $\varepsilon = 4\%$
Strip Width; $b = 2 \times 50$ mm.

2.2 *Field study of MSE wall with geostrap*

A 6.4 m high MSE wall, located in St. Remy Les Chevreuse, France, was instrumented with 560 strain gauges on the geostraps and 176 strain gauges on the connections to the precast facing panels (Hoteit, Price and Schlosser, November 1993).

Properties of Backfill Material:
Average Dry Density; $\gamma = 15.3$ kN/m^3

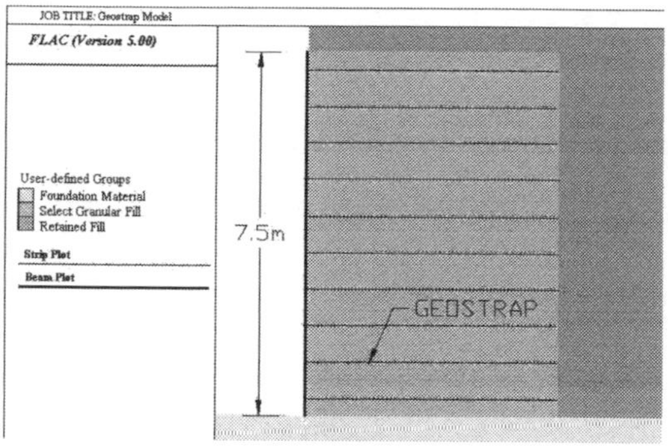

Figure 6. FLAC model.

Internal angle of friction; Ø = 37°
Cohesion; c = 5 kPa.

2.3 *Numerical analysis of MSE wall with geostrap*

The numerical analysis study was carried out using a finite difference program, Fast Lagrangian Analysis of Continua (FLAC) version 5.0, a 2D geotechnical program developed by the consulting company Itasca in the USA. The model was built using a sequential wall construction: first placing a 1.5 m high facing panel, followed by placement of level geostraps (5.5 m long) spaced uniformly at 750 mm, followed by reinforcing backfill and repeated up to a maximum design height of 7.5 m, as shown in Figure 6. After construction of the wall, a live load surcharge of 10 kPa was added. Maximum tensile forces (per linear meter) over the entire length of each strip were used to locate the line of maximum tension and to calculate actual lateral earth pressure coefficients.

Properties of Select Granular Fill Material:
Mohr-Coulomb Model
Density; $\gamma = 1900\,kg/m^3$
Internal angle of friction; Ø = 34°
Cohesion; c = 0 kPa
Young's Modulus; E = 80,000 kPa
Poisson's ratio; $\upsilon = 0.33$

Properties of Foundation Material:
Mohr-Coulomb Model
Density; $\gamma = 1900\,kg/m^3$
Internal angle of friction; Ø = 30°
Cohesion; c = 0 kPa
Young's Modulus; E = 40,000 kPa
Poisson's ratio; $\upsilon = 0.33$.

3 RESULTS

3.1 *Line of maximum tension*

Numerical analysis results show the maximum tension for the geostrap (per linear meter) close to the back of

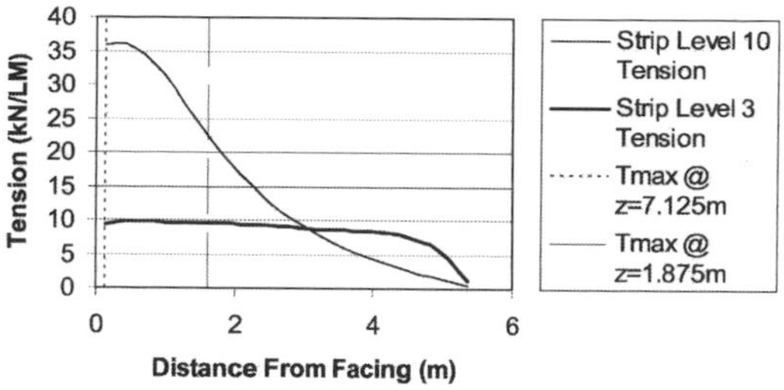

Figure 7. Numerical analysis results of tension throughout two geostrap levels.

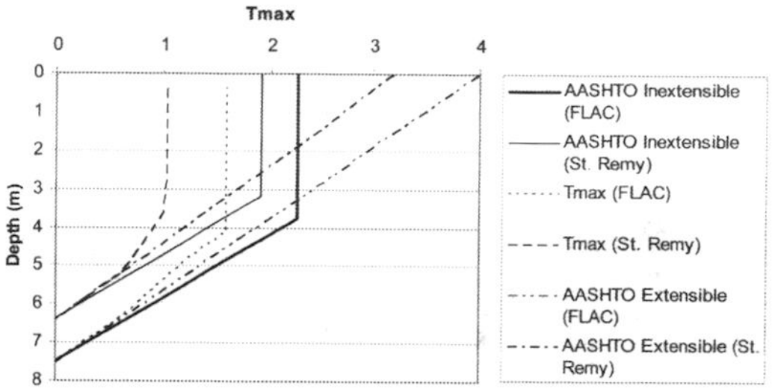

Figure 8. Actual T_{max} vs. Theoretical T_{max}.

the panel facing then becoming unloaded on all strip levels beyond 4.5 m of reinforcement length. Geostrap tensions from the numerical analysis in levels 3 and 10, at depths of 1.875 m and 7.125 m, respectively, are shown in Figure 7. For the upper reinforcing levels, it is noted that the length of tensile forces over the entire length of the strip is greater than the lower strip levels.

Field study and numerical analysis results for lines of maximum tension (T_{max}) in the geostrap (per linear meter), shown in Figure 8, are both within 0.3H (wall design height), 0.16H and 0.2H, respectively. These lines for the geostrap were more closely represented in Figure 3 rather than Figure 4. This has a significant affect on the effective length (L_e) of the reinforcing strip, especially at the top of the structure. For example, the top level (z = 0.375 m) would have an effective length for resistance to pullout of the geostrap of 3.25 m, using the resistant zone shown in Figure 3, and approximately 1.51 m, using Figure 4.

3.2 *Lateral earth coefficients*

Actual values for maximum geostrap tension were compared in Figures 9 and 10 to calculate maximum tension using known values in Equation 4 for lateral earth coefficients, K_o (Equation 2) and K_a (Equation 1). T_{max} over the top 5 m of the field study wall was greater than calculated values, using K_a. Below 5 m, T_{max} was less than values obtained using K_a.

$$K_a = \tan^2(45 - \tfrac{\phi}{2}) \tag{1}$$

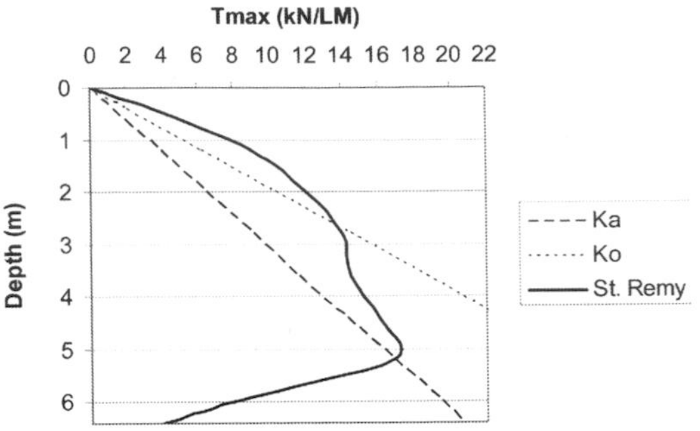

Figure 9. Field study values of T_{max} vs. K_a and K_o.

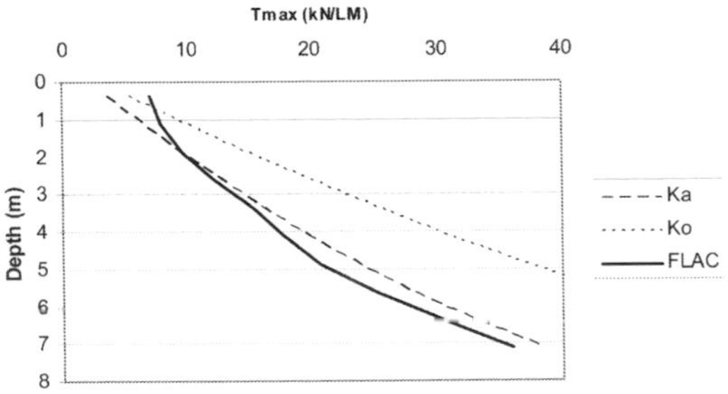

Figure 10. Finite difference analysis values of T_{max} vs. K_a and K_o.

$$K_o = 1 - \sin(\phi)^{\;2} \tag{2}$$

$$Tributary\ Area;\ A = Vertical\ Strip\ Spacing \times Horizontal\ Strip\ Spacing \tag{3}$$

$$T_{max} = K * \sigma_v * A \tag{4}$$

Similar to the field study, the finite analysis results of T_{max} over the top 2 m of the wall was greater than calculated values using K_a. Below 2 m, T_{max} was less than values obtained using K_a.

In a comparison between AASHTO and both studies, more than 20% percent of the top portion of the overall wall had a K/K_a ratio greater than what is required for geotextiles. Below 5.5 m, this ratio was less than 1.0 for both structures. Therefore in determining the applied tensile forces to the geostrap, simply using the line for a geotextile would not be sufficient.

3.3 *Apparent coefficient of friction*

Coefficient of Friction based on laboratory testing; f*

$$f^* = T_{max} / 2bL\sigma_v \tag{5}$$

where:
T_{max} = maximum pullout load
L = embedment length of geostrap
σ_v = total vertical stress.

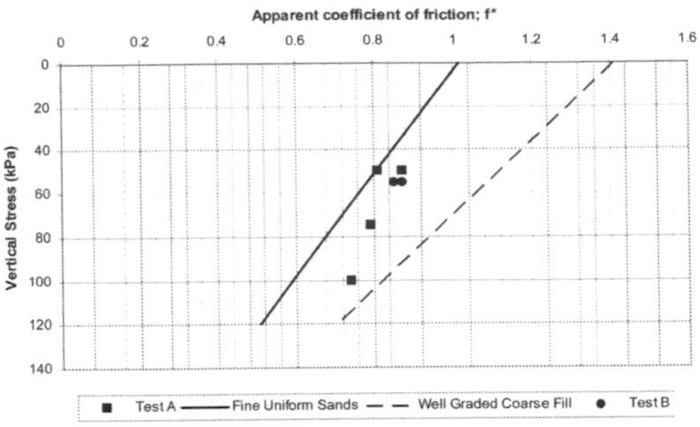

Figure 11. Coefficient of friction; f*.

The lab results show the coefficient of friction between the geostrap and soil was between the values for a ribbed steel strip and a geotextile, as shown in Figure 11. Internal stability analysis for pullout of the geostrap using the results for the coefficient of friction and effective length of the geostrap were more similar to a ribbed steel strip, as shown previously in Figures 3 and 5, respectively.

4 CONCLUSIONS

Field testing, laboratory testing and numerical analysis results were used in this paper to evaluate a MSE wall design using geostrap reinforcing elements. The evaluation showed some differences between the AASHTO design standards and actual test results. A possible reason for the differences could be that the geostrap used in the evaluation were of higher strength and less extensible than standard geotextiles originally considered by AASHTO. Additional design development is needed in AASHTO to account for higher strength polymeric reinforcing elements with closer behavior to steel reinforced structures.

Overall, high strength geostraps exhibit greater tensile capacity and lesser extensibility than other geosynthetic reinforcing elements used in MSE wall applications. At the same time, the frictional interaction is more akin to discrete reinforcement elements than the planar elements typical of geogrids and geotextiles. Therefore, where aggressive backfill conditions limit consideration for steel reinforcements, a cost-effective and more durable alternative is a high strength geostrap MSE wall application.

REFERENCES

AASHTO, "Standard Specifications for Highway Bridges," seventeenth edition, 2002.

Hoteit, N, Price, D. I., Schlosser, F., "Instrumented Full Scale Freyssisol-Websol Reinforced Soil Wall," technical article presented at the Conference for Soil Reinforcement – Full Scale Experiments of the 80's, November 1993.

New Horizons in Earth Reinforcement – Otani, Miyata & Mukunoki (eds)
© 2008 Taylor & Francis Group, London, ISBN 978-0-415-45775-0

Ultimate bearing capacity tests on an experimental geogrid-reinforced vertical bridge abutment without stiffening facing

D. Alexiew
Engineering Department, HUESKER Synthetic GmbH, Gescher, Germany

ABSTRACT: The paper deals with geogrid reinforced soil as solution for bridge abutments. Most important results are presented of a real scale test of a 4.5 m high geogrid reinforced vertical soil block loaded directly on top near the edge by a sill beam, high-lightening the low settlements and horizontal displacements measured. In one test, the reinforced embankment was nearly lead to failure, what occurred with a load in the order of 3 times the usual one for this kind of structures.

1 INTRODUCTION

Steep slopes and walls from geosynthetic reinforced soil (GRS) became very popular and established practice due to their advantages: cost-effectiveness, blending in well with the landscape, fine-tuning for optimum functionality etc. The broad range of available geosynthetic reinforcement allows optimisation and eliminates any limits to height and load capacity. The next step to be expected was the use of GRS in an "exclusive" area like bridge abutments, which are heavily loaded and have to fulfil stringent requirements with regard to load capacity and any kind of deformation. They are an intersection of traditional construction types (i.e. reinforced concrete (RC)), soil mechanics and foundation engineering with what today is loosely called "geosynthetic engineering".

2 BRIEF OVERVIEW AND BACKGROUND

The first steps began with the use of GRS to form the front face or part of the wing walls of bridge abutments. The experience with "conventional" GRS slopes and walls (see e.g. Herold & Alexiew (2001), Herold (2002) and Alexiew (2005)) was used.

Here the GRS is not loaded directly by the sill beam: The sill is supported on separate conventional (i.e. RC) load-bearing systems. The GRS wall is built around this conventional system. The bearing and serviceability requirements do not generally differ much from other vertical GRS walls. A more recent example of this type of bridge abutment is presented in Sobolewski & Alexiew (2005).

It is also possible to relieve abutment walls and/or wing-walls of earth pressure by placing a GRS block at the back of them. The GRS only carries loads from the earth pressure of the backfill and traffic load (Jossifowa & Alexiew 2002).

However, for what we should call "real" GRS abutment the sill beam is seated directly on the GRS block. Typical are the significant contact pressures under the sill of 150 to 250 kN/m^2 over a limited area (the width of the sill is usually < 2.0 m) positioned very close to the top edge of the GRS, frequently within 1.0 to 1.5 m. The allowable vertical (most importantly) and horizontal deformations are very much more limited, depending on the bridge system, than for a "normal" GRS wall.

One of the first non-experimental structures of this type was built at the beginning of the 1990s near Ullerslev in Denmark for a road bridge over a railway using high-tenacity PET geogrids Fortrac 110 (Kirschner & Hermansen 1994). Design, detailing and construction were completed without problems and the structure has performed well in service. At the time of construction it was considered a pioneering project.

Despite the Ullerslev project and other activities along the same lines (e.g. Uchimura et al. 1998, Zornberg et al. 2001), it was a long time before a "real" GRS bridge abutment in the German highway network was constructed at Ilsenburg on the German National Road K 1355 (Herold 2002). The knowledge gained is of great importance, including the "psychological" aspects on the part of the client.

Recently a "jointless bridge abutments" research programme was introduced in Germany the experimental part being carried out in cooperation with the author. In a test pit a 5 m high GRS block with vertical wraparound "soft" facing was installed at the back of a moving inwards and outwards RC-wall simulating the movements due to temperature changes in a jointless bridge. Well graded crushed sandy gravel and Fortrac[R] 80/30–35 M geogrids from PVA (Alexiew et al 2000) at 0.5 m spacing were used. The GRS was loaded only horizontally in these tests. The system is recommended for practical implementation (Pötzl & Naumann 2005a, 2005b). For a more detailed information see Alexiew (2007).

After completion of this test series the GRS test wall remained in the test pit for a year.

3 TEST WALL AS A "REAL" BRIDGE ABUTMENT

3.1 *Test set-up and comments*

Then it was suggested that the structure could serve for testing a "real" bridge abutment. Vertical loads could be applied directly from a reinforced concrete beam acting as a sill beam.

The focal point was not the "internal life" in the sense of e.g. internal stresses and strains, but the overall behaviour of the structure:

- What is the contact pressure under the sill that would drive the GRS to failure; this is the only way of estimation of load capacity resources and safety margins (ultimate limit state – ULS).
- How large are the settlements of the sill in the usual pressure range of 150 to 250 kN/m^2 (very important) and how large are the displacements of the facing (serviceability limit state – SLS).

A risky tendency to exclusively concentrate on the serviceability (SLS) in GRS structures and to a greater or lesser extent "neglect" the ULS has been noticed recently in a few publications. In doing this one loses sight of the fact that in these cases the SLS only is not in any respect relevant to safety.

For the new testing programme a rigid slab was fixed in position at a certain distance in front of the GRS (Fig. 1). This meant the structure remained a vertical GRS wall with a "soft" facing of the type "wraparound" without any stiffening elements. Two layers of a smooth membrane between the fill and the side walls of the test pit were installed to minimise friction. A 1.00 m × 2.70 m RC-block was placed as a sill beam 1.0 metres behind the front edge loaded vertically by means of hydraulic jacks (Fig. 1 & 2). Twelve displacement transducers were attached to the facing to measure its horizontal displacement (Fig. 2). Reflectors were attached over the whole surface and to the RC-beam to measure the settlements and any tilting by a precision level (Fig. 1). A data acquisition system recorded permanently all load and deformation data. Figure 3 shows the test set up.

3.2 *Constraints and boundary conditions*

The equipment available and the history of the project imposed certain constraints and boundary conditions

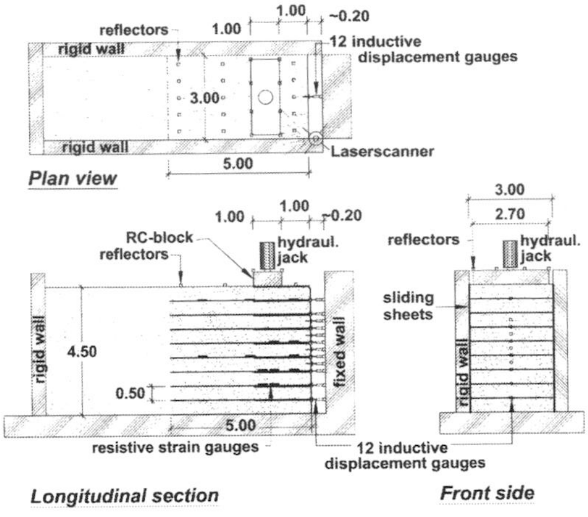

Figure 1. Schematic of the test.

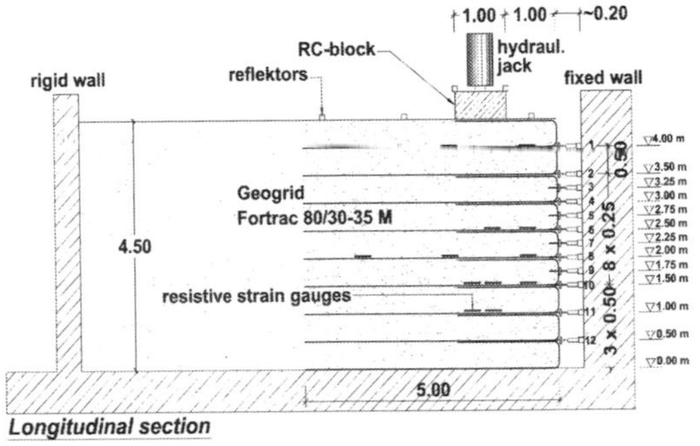

Figure 2. Horizontal displacement transducers.

Figure 3. Overview of the test set-up.

to the tests. The height had to be reduced to 4.50 m. The compaction of the crushed sandy gravel fill was found to be only ca. $D_{pr} = 95\%$ in the upper (critical) zone. It was also not clear whether the extreme outer area of the fill at the facing had experienced some loss of density as a result of the earlier tests "jointless bridges".

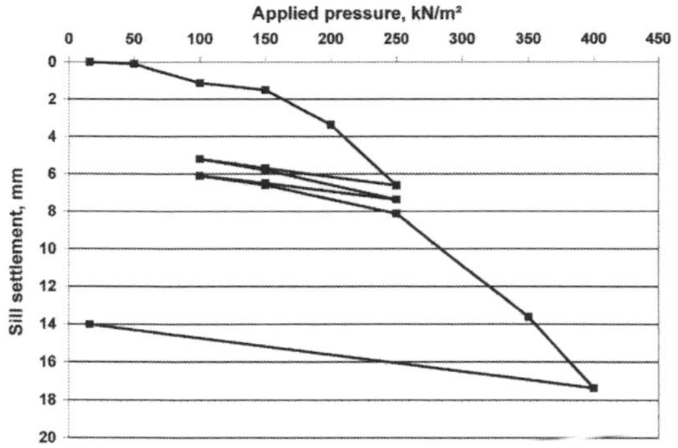

Figure 4. Settlement of the sill beam in Test 1.

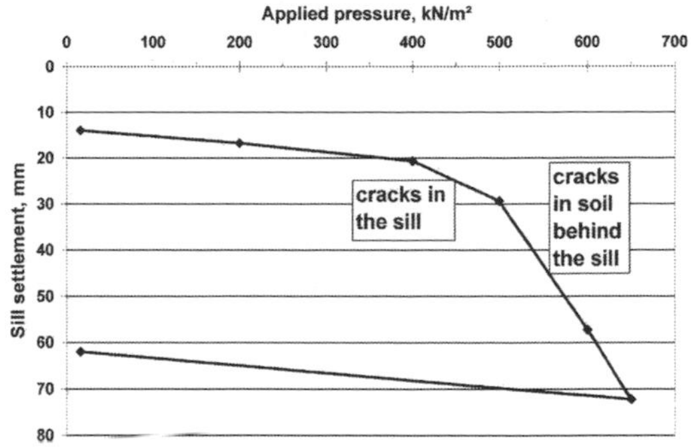

Figure 5. Settlement of the sill beam in Test 2.

The pressure under the loading RC-block (sill) could be taken up to a maximum of ca. 650 kN/m^2.

3.3 *Test procedure*

Two separate tests were carried out. In Test 1 the maximum load was 400 kN/m^2, i.e. twice the contact stress normally experienced under a sill beam (Fig. 4). After each increment a pause took place until there was a reduction in the change of settlement meeting the requirements of the plate bearing tests in accordance with DIN 18134.

The aim of Test 2 was to take the GRS block to failure using the full capacity of the jacks of 650 kN/m^2 (Fig. 5). Following values were recorded (cf. Section 3.1): contact pressure, average settlement of the loading RC-block, the horizontal displacement of the facing and the settlements of the reflectors on the top of the GRS system and on the RC-block.

3.4 *Important test results and comments*

Only the most important results have been included herein. Figure 4 shows the relationship between load and sill beam settlement in Test 1 unsmoothed. The shape of the graph suggests that a certain amount of further compaction may have taken place between 150–250 kN/m^2. It should be born in mind that the top zone of fill had only $D_{pr} = 95\%$ (Section 3.2), and that some loosening of the front part near the beam may have occurred as a result of the horizontal loading of the front area in the earlier "jointless bridges" tests (Section 2).

Let us make an analogy to the well known loading plate test. The increase in settlement in the first loading cycle in the range 150–250 kN/m^2 is ca. 5.9 mm and in the second loading cycle ca 1.3 mm, what indicates an increase in compressive stiffness of $5.9/1.3 = 4.5$. This is an unusually high value and indicates an additional compaction; a value of about 2.0 would have been expected here for a well compacted fill. Obviously, in Test 1 a recompaction of the fill directly under the block and perhaps in the possibly loosened front area takes place together with a higher mobilisation of the reinforcing geogrid in combination with this recompaction. With an a priori good compaction in the critical area the sill beam settlement would be even less than 5–6 mm in the relevant loading range, although even 5–6 mm would satisfy common requirements.

It should be noted that on Figure 4 the "unloading" line from 250 to 100 kN/m^2 is flatter than the "loading" one and that the hysteresis between 100–250–100 kN/m^{2}is parallel to the unloading part of the graph for unloading from 400 kN/m^2 down to 0 kN/m^2. This also indicates an increase in stiffness of the system and a tendency towards identical elastic behaviour at higher loads, or after recompaction and full mobilisation of the system.

Figure 5 shows the sill settlement for Test 2. The graph starts at the residual settlement of 14 mm after unloading in Test 1.

The results for Tests 1 and 2 can be converted into a modulus of subgrade reaction M , MPa/m, according to Winkler: the approximate equivalents are 57 MPa/m for Test 2 and 31 MPa/m and 16 MPa/m for Test 1. The system behaviour in Test 2 is clearly stiffer. Interestingly the common, generally accepted values of modulus of subgrade reaction for a gravel-sand mixture (as here) with good compaction are approximately 50 to 60 MPa/m (similar to 57 MPa/m here in Test 2) and with poor compaction 25 to 35 MPa/m (similar to 16 to 31 MPa/m here in Test 1) (Alexiew et al. 1989). This is one indication more for an insufficient compaction of the fill at the beginning of Test 1. Note: The common values of M apply to loading on a laterally infinite plane, in our case there is a 4.5-high, vertical slope only 1.0 meter away. That a similar M to that on a plane is achieved on the top of a GRS vertical block with a strip load applied close to its edge (whether compacted or not) shows that the geogrids used are acting very efficiently.

In Test 2 at approximately 450 kN/m^2 several fine vertical cracks were visible on the bottom edge of the

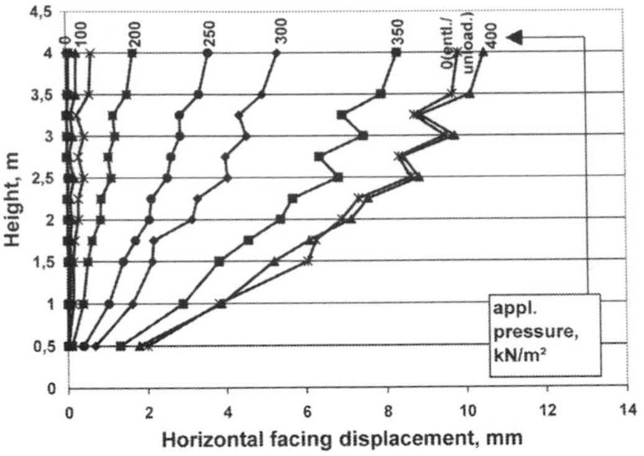

Figure 6. Horizontal displacements of the measured points on the facing for Test 1.

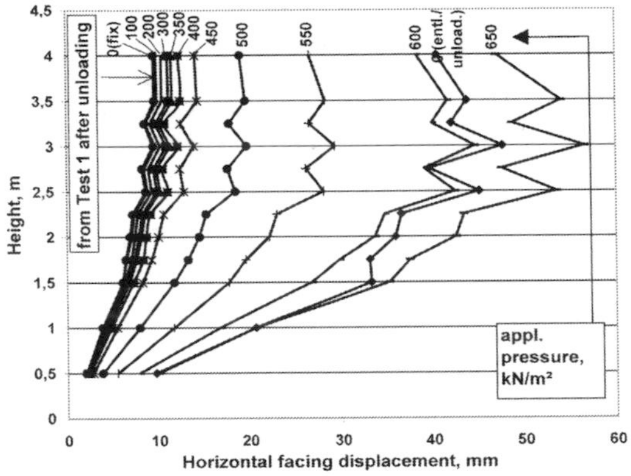

Figure 7. Horizontal displacements of the measured points on the facing for Test 2.

heavily reinforced RC sill beam, whilst in the GRS wall itself there was still nothing significantly amiss. Up to 600 kN/m^2 there were no noticeable symptoms of failure to be seen. Between 600 and 650 kN/m^2 a small irregular crack finally appeared in the fill surface behind the sill beam and extended towards the rear along the test pit walls. At 650 kN/m^2, the full capacity of the jacks was reached and increasingly accompanied by the above mentioned initial signs of failure. A clear failure, such as a failure body slipping forwards and downwards as might be expected, never occurred. It is a question of interpretation as to whether the ultimate limit state was reached or not, but the situation could be used as an "On-the-safer-side-ULS-benchmark".

Now to refer briefly to the horizontal displacement of the "soft" facing. Figures 6 & 7 show the results of Tests 1 and 2 (polygonal, not smoothed!).

Note, that the GRS wall sat directly on the concrete bed of the test pit and was surrounded by smooth membranes only at the sides. As expected, a bowing out at the middle of the front soft "single pillow" was noticed as compared to the points directly at the planes of the reinforcement (Fig. 2). However, the "local" bowing out was very small, probably because of the tensile stiffness of the flexible Fortrac 80/30–35 M geogrids used. Figure 7 (Test 2) starts at the residual displacements after unloading of Test 1 (no reset).

The maximum displacements occur for up to 400 kN/m^2 near the top of facing and in both tests amounts to a maximum of ca. 10 mm (measured in the worst position at a "bellying out" point). Analogically to the "vertical" behaviour of the system, Figures 6 and 7 also indicate an increasing stiffness in the horizontal direction after "recompaction".

A more detailed analysis of this and the relationships between the "vertical" and "horizontal" behaviour will be published separately.

From around 500 kN/m^2 (i.e. in Test 2) the character of the distribution of the deformation changed – the maximum values were no longer near the top. A "global bellying out" was increasingly noticeable between approximately 2.0–2.5 m and the 3.5 m level, together with an equally noticeable increasing curvature to this "bellying out". The position and height of this zone corresponds fairly accurately to the projection of the 1 m strip load (Fig.1) under ca. 45° down to the right at the facing. All this appears very plausible and corresponds well with common earth pressure theories. The maximum displacement of the soft facing amounted to 56 mm at 650 kN/m^2 (between two geogrid planes at H = 3.0 m) – a fairly large value, but under an extreme load. However, the "local" bowing of the geogrid at the front between the geogrid planes was still only ca. 9 mm. In relation to the layer spacing of 0.5 m it represents a low value and indicates that the wraparound geogrid is highly efficient. From a load of approximately 500 kN/m^2 there is an increase in the rate of deformation. The (relatively) large displacement from ca. 550 kN/m^2 could be taken as a trend in the direction of failure, however up to the end of the test at 650 kN/m^2 there was no visible breakthrough movement of any failure body at the facing. The results may well speak for themselves as to the remarkable reserve capacity of the GRS block; on removal of the load from 650 kN/m^2 to zero and despite the initial indications of failure the facing moved back ca. 10 mm. In plan view the front remained straight with no bowing in the middle; evidently the slip membrane layers at the pit walls were effective and the system can be idealised as 2-dimensional. This is important for the proposed further analyses and comparison with calculations, which will be published separately.

4 FINAL REMARKS

The tests presented herein on a geogrid-reinforced soil block simulating a real bridge abutment under a sill beam are in no way intended to be a comprehensive scientific analysis. The exercise is much more about testing the behaviour of a system and its reserves (!) in a situation directly related to practice. The use of an already constructed and used for other purposes test object after modification was advantageous in terms of time and costs, but it also brought its own restrictions and deficiencies, e.g. that we would have to live with the known insufficient compaction in the upper zone and the possible looser fill zones near the facing resulting from previous tests. The tests described herein are still fairly recent; and so the following remarks are a first, rather incomplete overview, but the most important points are readily recognisable and can be translated into practice.

The tested arrangement should be seen as a "worst case" scenario:

- The sill beam was only 1.0 m wide and placed only 1.0 m away from the edge
- The front face was vertical
- The facing had no special stiffening elements, being only a geogrid-wrapped-back wall
- The density of the fill in the most sensitive upper zone was only $D_{pr} = 95\%$, with probably loosened zones in the front area near the loading beam, some probably as a result of the previous tests.

The following remarks can be made:

- A contact pressure under the sill beam of up to 650 kN/m^2(approx. 3 times the pressure normally experienced) led to no obvious component or system failure. However, because there were signs of serious effects taking place, the situation could be used as a benchmark for the ultimate limit state.
- A contact pressure of up to 400 kN/m^2(approximately twice the usual value) resulted only in completely acceptable deformations.
- The tested system exhibited technically advantageous, ductile behaviour with no discontinuities and seems to have a substantial reserve capacity.
- The overall performance can be considered very good despite the previously found soil density deficiencies.
- The facing consisting of flexible geogrids had no bending stiffness but showed only small local and global deformations (marginal in the relevant load range).
- The settlement behaviour of the sill (indirectly assessed by the modulus of subgrade reaction) was as if it had been sitting on an infinite horizontal plane and not near a vertical slope; the only plausible explanation is the apparently high effectiveness of the incorporated geogrids.

The author would have no reservation using the structure as built and tested (and ideally with better fill compaction) in practice.

5 ACKNOWLEDGEMENTS

The author would like to thank Professor Pötzl, Coburg; without the tests for the "jointless bridges" the tests presented briefly herein would probably not have taken place. Thanks are also due to Mr. Straußberger, LGA Nuremberg for his encouragement and Mr. Homburg and colleagues, LGA Nuremberg for their great commitment to the execution of this project.

REFERENCES

Alexiew A. et al. 1989 *Manual of Soil Mechanics and Foundations*. Vol. 2. Technika, Sofia (in Bulgarian), p. 46.

Alexiew D., Sobolewski J., Pohlmann H. 2000 *Projects and optimized engineering with geogrids from "non- usual" polymers*. Proc. 2nd European Geosynthetics Conference EuroGeo 2000, Bologna, 2000, pp. 239–244.

Alexiew D. 2005 *Design and construction of geosynthetic-reinforced "slopes" and "walls": commentary and selected project examples*. Proc. 12th Darmstadt Geotechnical Conference. Darmstadt Geotechnics No. 13, TU Darmstadt, Institute and Laboratory of Geotechnics, Darmstadt, March 2005, pp. 167–186.

Alexiew D. 2007 *Belastungsversuche an einem 1:1 Modell eines geogitterbewehrten Brückenwiderlagers*. Proc. KGeo 2007, Munich (to be published).

DIN 18134, September 2001 *Soil, Testing procedures and testing equipment – Plate load test*

Herold, A. 2002 *The First Permanent Road-Bridge Abutment in Germany Built of Geosynthetic-Reinforced Earth*. Proc. 7th ICG, Nice, France, A. A. Balkema Publ., pp. 403–409.

Herold, A., Alexiew, D. 2001 *Bauweise KBE (Kunststoffbewehrte Erde) – Eine wirtschaftliche Alternative?* Proc. 3rd Austrian Geotechnical Conference, Vienna, pp. 273–288.

Jossifowa, S., Alexiew, D. 2002 *Geogitterbewehrte Stützbauwerke und Böschungen an Autobahnen und Nationalstraßen in Bulgarien*. Geotechnik 25 (2002) No. 1, Essen, pp. 31–36.

Kirschner, R., Hermansen, E., 1994 *Abutments in Reinforced Soil for a Road Bridge*. Proc. 5th IGS Conf., Singapore, pp. 259–260.

Pötzl, M., Naumann, F. 2005a *Fugenlose Betonbrücken. Final report to research project No. 1700402*. German Federal Ministry of Education and Research, Berlin, 2005.

Pötzl, M., Naumann, F. 2005b *Fugenlose Betonbrücken mit flexiblen Widerlagern*. Beton- und Stahlbetonbau 100 (2005) Issue 8, Ernst & Sohn, Berlin, p. 675 ff.

Sobolewski J., Alexiew D. 2005 *Erdbewehrte Blockwände – System Terrae®– an Widerlagern einer Brücke auf der "Via Baltica" bei Riga*. Proc. 9th German Geosynthetic Conference "Kunststoffe in der Geotechnik", Munich, February 2005, Special Issue Geotechnik 2005, Essen. pp. 271–274.

Uchimura, T., Tatsuoka, F., Tateyama, M. and Koga, T. 1998 *Preloaded-Prestressed Geogrid-Reinforced Soil Bridge Pier*, Proc. 6th ICG, Atlanta, IFAI, pp. 565–572.

Zornberg, J. G., Abu-Hejleh, N. and Wang, T. 2001 *Geosynthetic Reinforced Soil Bridge Abutments*. GFR Magazine, Vol. 19, No. 2, March, pp. 52–55.

New Horizons in Earth Reinforcement – Otani, Miyata & Mukunoki (eds)
© 2008 Taylor & Francis Group, London, ISBN 978-0-415-45775-0

Full-scale behavior of a surface loaded geosynthetic reinforced tiered segmental retaining wall

C. Yoo, S.B. Kim & Y.H. Kim
Sungkyunkwan University, Suwon, Korea

ABSTRACT: This paper presents the results of a load test for a full-scale geosynthetic reinforced segmental retaining wall (GR-SRW) in a tiered arrangement. A four year old, 5 m high tiered SRW, originally constructed to investigate short and long term behavior, was load tested using a large precast concrete box culvert filled with ready mix concrete, simulating a surcharge loading condition of a GR-SRW in bridge abutment application. Measured items included horizontal displacement at the wall face and strains in the reinforcement. The measured results revealed that the GR-SRW's response was well within the serviceability limits and within the range of those predicted based on the current design guideline. Design implications and the findings from this study are discussed.

1 INTRODUCTION

The use of geosynthetic reinforced segmental retaining wall (GR-SRW) in both private and public sectors is increasing worldwide. Although the currently available limit equilibrium-based design approaches, such as NCMA (Collin, 1997) and FHWA (Elias and Christopher, 1997) design guidelines, are considered to be conservative on account of several assumptions regarding the wall behavior, much still needs to be investigated to bridge the gap between the theory and the practice. In addition, despite the fact that many geosynthetic reinforced soil walls have been safely constructed and are performing well to date, there are many areas that need in-depth studies to develop a more generalized design approach that will help safely construct GR-SRW systems under more aggressive and complex boundary conditions.

Recently GR-SRWs are frequently used in bridge construction, as the form of geosynthetic-reinforced soil (GRS) bridge-supporting structure (Lee and Wu 2004). The GRS bridge-supporting structure can be constructed using either rigid or flexible facings. A "rigid" facing is either precast or cast-in-place type while a "flexible" facing takes the form of wrapped geosynthetic sheets, segmental blocks, or gabions (Lee & Wu 2004). Lee & Wu (2004) synthesized measured data of four in-service GRS bridge abutments and six full-scale field experiments, and concluded that GRS bridge abutments with flexible facing are indeed a viable alternative to conventional bridge abutments.

A number of studies concerning GRS bridge-supporting structures with flexible facings are available, i.e., Mannsbart and Kropik (1996), Won et al. (1996), Wu et al. (2001), and Abu-Hejleh et al. (2000). Although these studies provided valuable information as to the performance of GRS-bridge-supporting structures with flexible facings, much still need to be investigated to better understand the response of GRS abutment to bridge loading.

In the present investigation, a four year old, 5 m high, two tier GR-SRW was load tested at Geotechnical Experimentation Site (GES) in Sungkyunkwan University, located in Suwon, Korea. A primary objective of the test was to evaluate the performance of a GR-SRW under a surcharge load, simulating a loading condition when used as a bridge abutment. This paper describes the test wall, the load test program, details of the observed performance, and finally, design implications.

2 DESIGN CONSIDERATION OF GR-SRW ABUTMENT

According to the FHWA design guideline, a GR-SRW bridge abutment can be designed as being a wall with surcharge load at the top of the wall. The internal stability calculations are performed by taking account of both vertical and horizontal components of the surcharge load. The reinforcement force T_i at ith level is computed based on the lateral pressure $\sigma_{H,i}$ and the tributary area $A_{t,i}$ as given in Eq. (1).

$$\sigma_{H,i} = K(\gamma z_i + \Delta\sigma_{V,i}) + \Delta\sigma_{h,i} \quad (1)$$

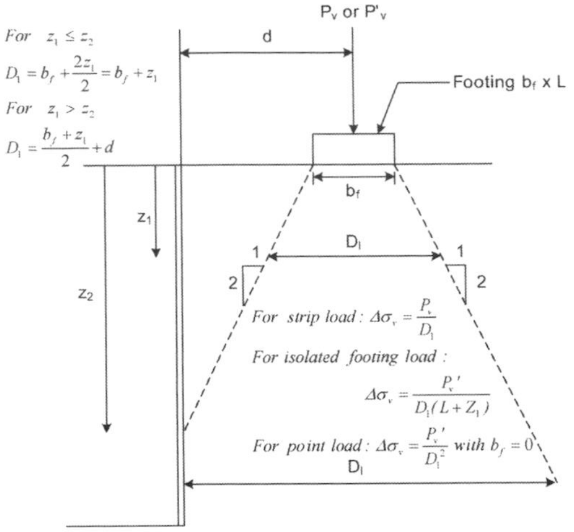

Figure 1. 2V:1H pyramid distribution.

where K is the lateral earth pressure coefficient, $\Delta\sigma_{V,i}$ is the increment of vertical stress due to the concentrated vertical surcharge assuming a 2V:1H pyramid distribution (Figure 1), $\Delta\sigma_{h,i}$ is the incremental horizontal stress due to the horizontal loads, and γz_i is the vertical stress at ith level.

3 WALL DESCRIPTION

3.1 *Site condition*

The ground under which the test wall was situated consists of approximately 3.0 m thick miscellaneous fill material including sand and gravel. Underlying the fill layer is a 3.0 to 4.0 m thick alluvial sandy clay deposit followed by a 6.0 to 8.0 m thick weathered granite residual soil overlying a slightly weathered granite rock stratum (Figure 2). Details of the site condition are given in Yoo & Jung (2004).

3.2 *Wall design and construction*

The wall was originally constructed in 2002 in order to investigate the short and long term performance of a two tier GR-SRW. The measured performance during construction based on an extensive field instrumentation has been reported in Yoo & Jung (2004). A brief discussion of the wall design and construction is given in this section.

The test wall had an exposed height of 5 m and consisted of two tiers, i.e., a 3.4 m high lower tier and a 2.2 m high upper tier as illustrated in Figure 3. The upper and lower tiers had no pre-batter angle and the face of the upper tier was 1.0 m away from the lower tier face, thus giving an offset distance of 1.0 m. As seen in Figure 4, eleven layers of PET reinforcement, having a

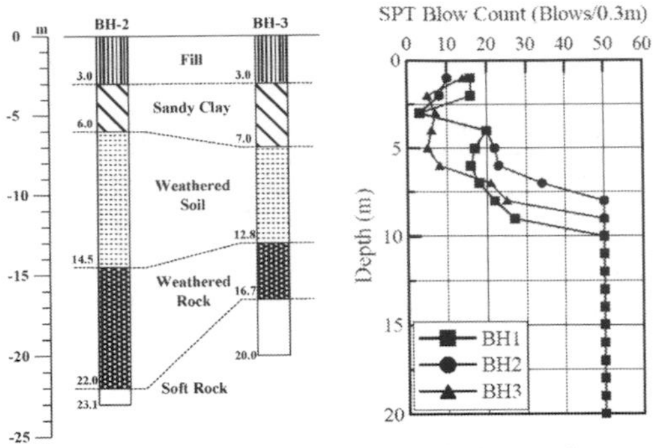

(a) soil profile (b) SPT results

Figure 2. Foundation soil profile & SPT blow counts for GES.

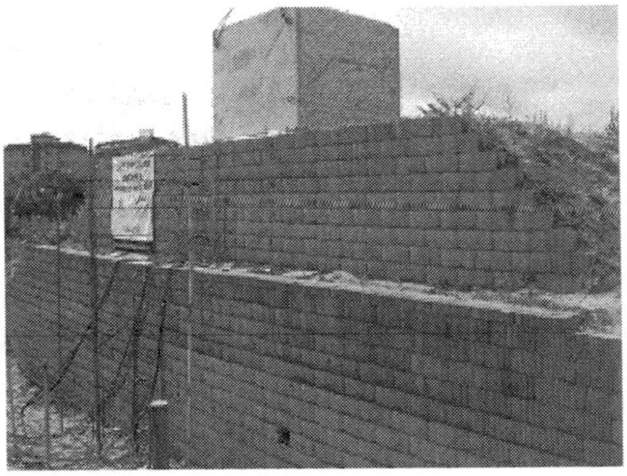

Figure 3. Photo of load test.

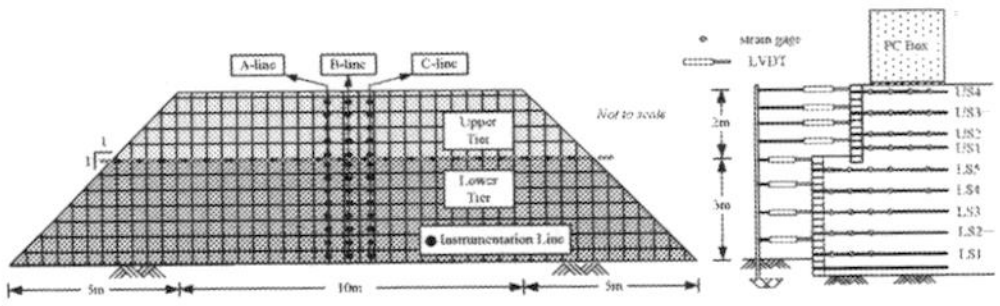

Figure 4. Schematic view of load test setup.

tensile strength of 55 kN/m at strain of 12.5% with an average axial stiffness of J = 500 kN/m, were placed at a maximum vertical spacing of 0.6 m. For each tier, the reinforcement length ratio with respect to the respective tier height was kept constant at 1.0. Note that the facing blocks are 450 × 330 mm in plan × 200 mm in height, having a compressive strength of 21 MPa with a maximum water absorption of 6∼8% for standard weight aggregates. Shear transfer between the blocks is developed primarily through shear keys formed on each block. Note that no provision was made for any future surcharge loading in terms of the internal and external stability at the time of wall design and construction.

Table 1. Results of internal stability calculations.

		Internal stability			
		FS_{to}		FS_{po}	
Layer	Elev. (m)	NCMA	FHWA	NCMA	FHWA
B1	0.2	1.59	1.32	33.21	28.07
LS1	0.6	1.38	1.15	24.28	21.30
LS2	1.2	1.31	1.09	17.37	15.84
LS3	1.8	1.53	1.27	14.50	14.12
LS4	2.4	1.84	1.52	11.64	12.40
LS5	3.0	2.30	1.66	8.78	9.35
US1	3.6	3.68	3.8	17.82	13.44
US2	4.0	4.78	4.05	15.53	9.72
US3	4.6	6.90	5.70	10.18	6.09
US4	5.2	33.12	21.89	8.55	4.19

A weathered granite soil, classified as SW-SM according to Unified Soil Classification System (USCS) was used as select fill and compacted to 95% of its maximum unit weight to create both the reinforced and retained zones. The estimated internal friction angle using a large scale direct shear test at a density corresponding to the as-compacted state was approximately 35° with a cohesion of 10 kPa. The results of the internal and external calculations for as-built design according to the NCMA and FHWA design guidelines are given Table 1.

4 LOAD TEST

4.1 *Test setup*

The load test was carried out in August of 2006, four years after the wall construction. The load was applied using a precast concrete box frame (2.4 m × 2.4 m in plan) for sewage drainage together with ready mixed concrete and a steel frame, totaling approximately 348 kN (Figure 3). As seen in Figure 4 showing the test setup, the concrete box was placed 0.5 m away from the upper tier facing units. The applied load exerted approximately 62 kPa of vertical pressure at the top of the wall. Such a pressure is within the typical design pressure for single span bridge deck when a GR-SRW is used to as a bridge abutment, and thus can be considered as a working load. During the test the load was applied in five increments by controlling the amount of remicon put into the concrete box as summarized in Table 2. During the test, the test load was applied in 20 kN increments by pouring in the remicon of 3~4 m^3 in the concrete box. For each load increment a sufficient amount of time was allowed for the wall displacement to stabilize before a next load increment. A total of five and a half hours were required to complete the test.

Table 2. Summary of load application process.

Step	Incremental load (kN)	Cumulative load (kN)	% of total load	Loading description
1	78	78	22.4	placement of concrete box
2	69	147	42.2	1st Remicon of 3 m^3
3	69	216	62.1	2nd Remicon of 3 m^3
4	92	308	88.5	3rd Remicon of 4 m^3
5	40	348	100.0	Steel frame
6	−40	308	88.5	Unload steel frame

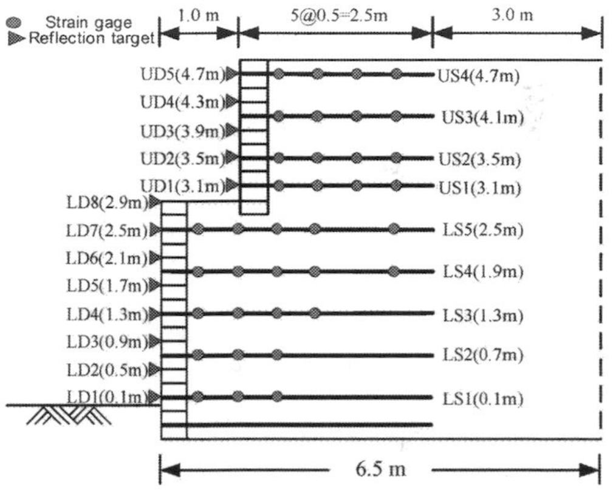

Figure 5. Instrumentation layout.

4.2 *Instrumentation*

The response of the test wall to the surface loading was evaluated in terms of the wall facing displacement and the reinforcement strains. Figure 5 shows the schematic layout of instrumentation. The wall facing displacement was measured by leveling using a 3D total station (MONMOS Model NEA2A) together with the reflection targets installed on the wall face. For redundancy of the wall facing displacement measurements eight potentiometers were additionally placed on a vertical row as shown in Figure 4.

Reinforcement strains were measured using strain gauges, high elongation bonded resistance strain gages, manufactured by Tokyo Sokki Kenyujo Company (Model YFLA-5-1L), that had survived at the time of the load test after their installation. Note that of the approximately 70% of xx strain gauges installed during construction had survived. The three instrument arrays for the leveling targets and the reinforcement strain measurements are shown in Figure 5. Table 3 summarizes the details of the instrumentation.

Table 3. Details on instruments.

Array/Instrumentation	Location
Array A, B, C	−1.0, 0, +1.0 from wall center line
Optical survey target on wall facing column	0.1, 1.5, 0.9, 1.3, 1.7, 2.1, 2.5, 2.9 3.1, 3.5, 3.9, 4.3, 4.7 m above wall base
Strain gage	0.5, 1.0, 1.5 m behind wall facing (LS1–LS3)
	0.5, 1.0, 1.5, 2.5 m behind wall facing (LS4–LS5)
	0.5, 1.0, 1.5, 2.5, 3.0 m behind wall facing (US1–US4)

5 RESULTS

5.1 *Lateral wall displacement*

Figure 6 shows the progressive development of wall displacements at the monitoring point. As seen in this figure, stepwise increases in the lateral wall displacements are evident due to the stepwise increase in the surcharge load. It is seen in Figure 6(a) only minimal displacements are measured in the lower tier at the final stage of the loading, showing a maximum of 0.5 mm. As expected, larger lateral displacements are measured in the upper tier with a maximum of 1.7 mm at the upper most measuring point, as shown in Figure 6(b). Such results are well reflected in the lateral wall displacement profiles at the final loading stage in Figure 7, in which a cantilever type movement prevails.

5.2 *Reinforcement strains*

The progressive development of strains in the reinforcement layers due to the surface loading is shown in Figure 8. No appreciable strains in layers LS1~LS4 in the lower tier were recorded and therefore are not given here. The influence depth for the surface load can thus be inferred as being slightly larger than the upper tier height of 2 m. As seen for layer LS5 in Figure 8, a maximum strain of 0.05% was developed at the location close to the end of the layer with essentially negligible strains elsewhere. A decrease in strain of approximately 0.1% is noticed at the mid location of the layer (LS5), although the cause for such a trend is not immediately clear.

In the upper layers, as seen in Figure 9(a)~9(c), stepwise increases in strains caused by the stepwise increase in the load are evident, showing a maximum strain of 0.1% occurring at the top layer US4. As one can expect, the absolute maximum increase in each layer increases with increasing the proximity to the load. Of interest trends in this figure are twofold. First, for each load increment a sharp increase in strain is noticed immediately after the load increment, followed

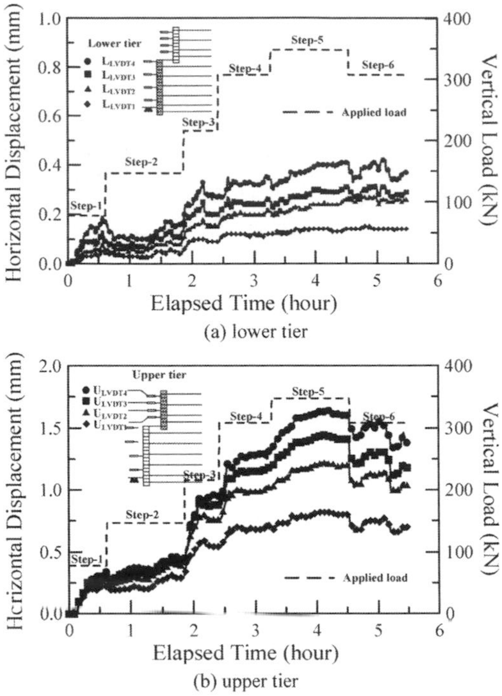

Figure 6. Progressive development of wall displacements at monitoring points.

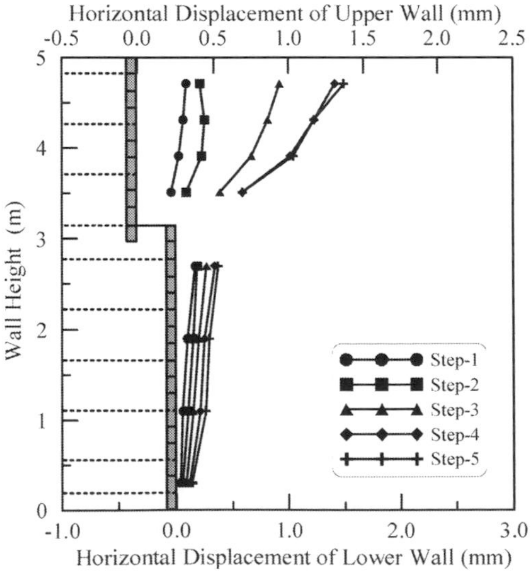

Figure 7. Wall displacement profiles at various loading stages.

by gradual convergence to a certain value. Note that such a trend is in fact similar to the observation in a sustained loading test on a reduced geosynthetic reinforced wall (Yoo et al., 2006). Another of interest

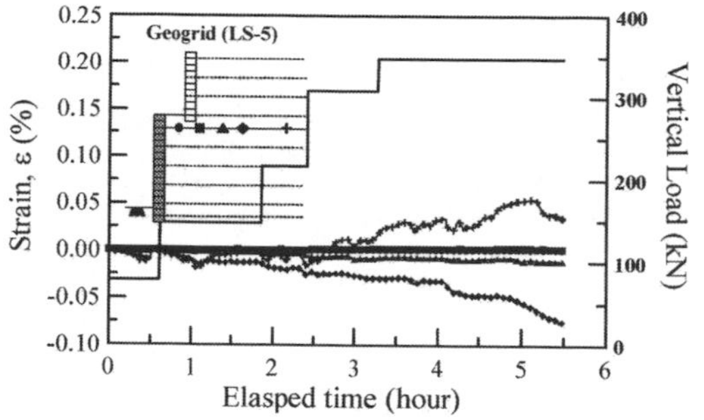

Figure 8. Progressive development of wall displacements at monitoring points (lower tier).

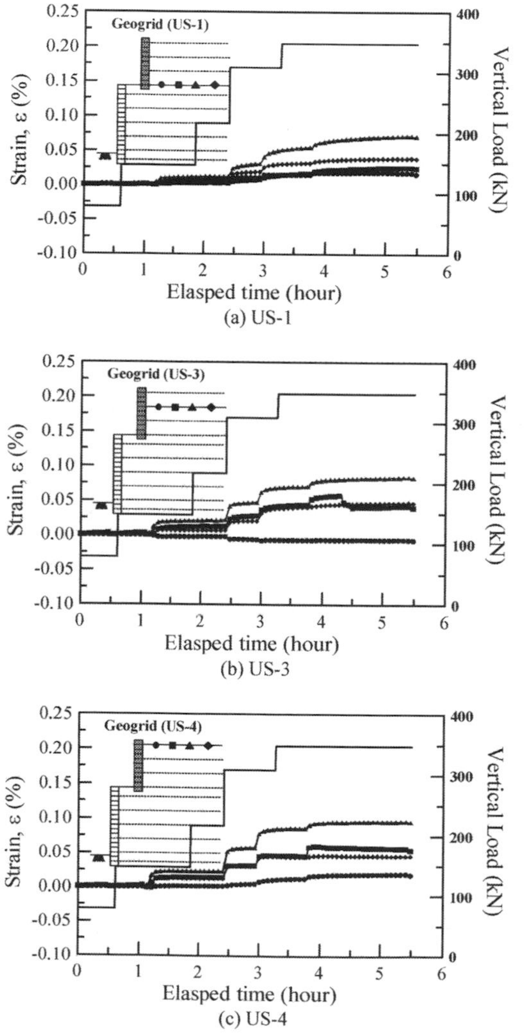

Figure 9. Progressive development of wall displacements at monitoring points (upper tier).

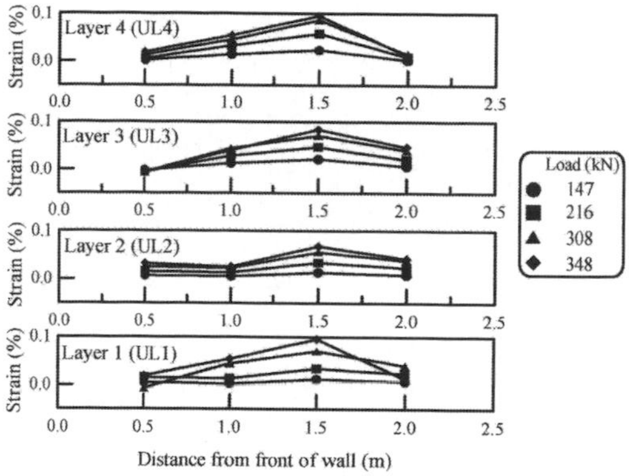

Figure 10. Reinforcement strain distributions for reinforcements in upper tier at final loading stage.

Table 4. ΔT according to FHWA design guideline.

Layer	h (m)	$\Delta\sigma_V$ (kPa)	$\Delta\sigma_H$ (kPa)	A_t (m^2)	ΔT (kN/m)
US4	4.7	48.01	14.4	0.6	8.6
US3	4.1	32.14	9.6	0.6	5.8
US2	3.5	23.01	6.9	0.5	3.5
US1	3.1	18.93	5.7	0.4	2.3
LS5	2.5	14.58	4.4	0.7	3.1
LS4	1.9	11.57	3.5	0.6	2.1
LS3	1.3	9.41	2.8	0.6	1.7
LS2	0.7	7.80	2.3	0.6	1.4
LS1	0.1	6.57	2.0	0.5	1.0

trend is that location of the largest strain increase for a given layer. As seen in these figures, a largest increase occurs approximately at the location 0.7 L from the wall facing for a given layer. Such a trend is well reflected in Figure 10 in which the reinforcement strain distribution for a given layer follows a concave-up shape showing a maximum strain at approximately 0.7 L from the wall facing, not directly under the loading center. The lateral wall yielding (displacement) may be attributed such a trend and therefore, a non-uniform strain increase along a layer should therefore be expected for a loading case similar to the one considered in this study.

Table 4 summarizes increases in the reinforcement forces ΔT computed according to the FHWA design guideline. The computed values in fact more than twice the measured values. This will be studied further in a future study.

In short, the surface loading of 60 kPa on a 1.5 m × 1.5 m loaded area induced reinforcement strains in the upper tier less than 0.1%, while negligible reinforcement strains were developed in the lower tier. These strain values are is well within the serviceability limits of the reinforcement. The calculation results based on the FHWA guideline yielded are however 50% larger than measured ones, suggesting some degree of conservatism in the current design approach. Further study is necessary in this area to further refine the calculation model adopted in the FHWA design guideline.

6 CONCLUSIONS

This paper presents the results of a load test of a full-scale geosynthetic reinforced segmental retaining wall (GR-SRW) in a tiered arrangement. A four year old, 5 m high tiered SRW, constructed to investigate short and long term behavior, was load tested using a large box culvert filled with ready mix concrete, simulating a loading condition of a SRW in bridge abutment application. Measured items included horizontal deformation at the wall face and strains in the reinforcement.

The measured results revealed that the SRW's response was well within the serviceability limits and within the range of those predicted based on the current design guideline. Also shown is that the calculation results based on the FHWA guideline yielded however 50% larger than measured ones, suggesting a conservatism in the current design guidelines. In short, the surcharge load of 348 kN did not significantly increase the wall performance in terms of the wall displacement and the reinforcement strains, although the surcharge load was not accounted for during design. Such a result demonstrates that a GR-SRW can be effectively used in surcharge loading situations.

ACKNOWLEDGEMENT

This work was supported by Grant No. R01-2004-000-10953-0 from the Basic Research Program of the Korea Science & Engineering Foundation. The financial support is gratefully acknowledged.

REFERENCES

Abu-Hejleh, N., Wang, T., Zornberg, J.G. 2000. Performance of geosynthetics-reinforced walls supporting bridge and approaching roadway structures. *ASCE Geotechnical Special Publication No. 103, Advances in Transportation and Geoenvironmental Systems using Geosynthetics*: 218–243.

Collin, J. 1997. Design Manual for Segmental Retaining Walls, 2nd Ed. NCMA, Virginia.

Elias, V. & Christopher, B.R. 1997. Mechanically Stabilized Earth Walls and Reinforced Soil Slopes, Design and Construction Guidelines. *FHWA Demonstration Project 82*, FHWA, Washington, DC, FHWA-SA-96-071.

Jappelli, R. & Marconi, N. 1997. Recommendations and prejudices in the realm of foundation engineering in Italy: A historical review. In Carlo Viggiani (ed.), *Geotechnical engineering for the preservation of monuments and historical sites*; *Proc. intern. symp., Napoli, 3–4 October 1996*. Rotterdam: Balkema.

Lee, K.Z.Z. & Wu, J.T.H. 2004. A synthesis of case histories on GRS bridge-supporting structures with flexible facing. *Geotextiles and Geomembranes*: Vol. 22, No. 4: 181–204.

Mannsbart, G. & Kropik, C. 1996. Nonwoven geotextile used for temporary reinforcement of a retaining structure under a railroad grack. In DeGroot, M.B., Hoedt, G., Termaat, R.J.(Eds.), *Geosynthetics: Applications, Design and Construction*: 121–124. Rotterdam: Balkema.

Won, G.W., Hull, T., De Ambrosis, L. 1996. Performance of a geosynthetics segmental block wall structure to support bridge abutments. In Ochiai, H., Yasufuku, N., Omine, K. (Eds.), *Earth Reinforcement*: Vol. 1. 543–548. Rotterdam: Balkema.

Wu, J.T.H., Ketchart, K., Adams, M. 2001. GRS bridge piers and abutments. Report FHWA-RD-00-038. FHWA, US Department of Transportion, 136pp.

Yoo, C. & Jung, H.S. 2004. Measured behavior of a geosynthetic-reinforced segmental retaining wall in a tiered configuration. *Geotextiles and Geomembranes*: Vol. 22, No. 5: 359–376.

Yoo, C., Kim, S.B., Kim, Y.H. 2006. Behavior of geosynthetic reinforced wall under sustained and cyclic loads – Reduced-scale model test. *Proc. Fall KGS Conference*: 24–25.

New Horizons in Earth Reinforcement – Otani, Miyata & Mukunoki (eds)
© 2008 Taylor & Francis Group, London, ISBN 978-0-415-45775-0

Subgrade reaction of Reinforced Earth Wall underneath the facing panel

T. Kumada & K. Watanabe
Hirose & Co.,Ltd , Tokyo, Japan

ABSTRACT: Typical facing elements of the Reinforced Earth Wall include the use of steel strip reinforcements, steel plate materials, and concrete or steel rod grid panels. The erection of facing components made of steel plate materials or steel rod grid panels may not require concrete foundation. However, the concrete facing panels require a concrete foundation pad to facilitate proper placement of concrete panels and distribute the vertical loads from facing to subgrade soil. Flexible joint materials are installed to allow movements of concrete facing to follow settlement of backfill soil. Some measurement results show that subgrade reaction in the foundation is larger than facing panel weight. Subgrade reaction is used for assessing external stability such as bearing or tilt failure mechanisms. This paper presents a statistical study based on full-scale model tests and some measurements in actual structures. A relationship between subgrade reaction and tensile force in reinforcements has been obtained.

1 INTRODUCTION

Reinforced Earth Walls consist of granular soil, soil reinforcing elements and facing materials. Typical facing elements are made of reinforced concrete panels. Some type of facing elements made of steel plate materials or steel rod grid panels. In these cases, the erection of facing components may not require concrete foundation. The facing elements cannot support vertical loads given that steel plate and a steel grid system behaves as flexible components.

The other hand, the concrete facing panels require a concrete foundation pad to facilitate proper placement of concrete panels and distribute the vertical loads from facing to subgrade soil. A construction procedure using concrete panels is following steps. First, concrete foundation pad is placed. Usual size of foundation is 0.4 m widths and 0.2 m heights.

Concrete facing panels are set on a foundation pad, then backfill is spread and compacted in lifts up to level of strip connecting layer. And reinforcing strips are connected to facing panels with the bolt. It is repeated a cycle of filling and compacting of backfill, connecting strips, setting panels until design height is reached.

Wall surface is divided structure, and flexible joint materials are installed to prevent concrete-to-concrete contact between the concrete facing elements. Therefore, these constitutions allow movements of concrete facing to follow settlement of backfill soil.

It was known that some measuring results of the vertical loads from facing are bigger than self-weight of facing panels. It is important to estimate subgrade reaction of a foundation pad for assessment of external stability such as bearing or tilt failure mechanisms.

Design method of Reinforced Earth wall consists of rupture analysis of soil reinforcement and pull out resistance from embankment of soil reinforcement. And, it is calculated against estimated tensile force of soil reinforcement.

However, this design method is not necessary to estimate reaction underneath facing panels or concrete foundation pad. In conventional design method of Gravity Retaining Wall, subgrade reaction of wall calculates from vertical earth pressure along wall surface, obtain by multiplying horizontal earth pressure by angle of wall friction.

In case of Reinforced Earth Wall, we assume that the horizontal earth pressure may calculate from tensile force of soil reinforcement near the facing panel. Some measurement data of subgrade reaction and of soil reinforcement tensile force, available on 4 structures, which are an actual wall or experimental walls. It is shown below as relation between subgrade reaction and tensile force of reinforcement.

2 GENERAL MEASURING RESULT

The profile of 4 structures with the measurement results of subgrade reaction and tensile force in reinforcing strips is shown. The characteristic of backfill material, which is used for each structure, is shown in Table 1 and the overview of facing panel is shown in Table 2.

Table 1. Characteristic of backfill material.

Name of Fill Material		Field Test(1)	Field Test(2)	Field Test(3)	Load Test	Shaking test	Moving test
Soil particles	Gs	2.737	NA	2.706	3.648	2.721	NA
Grain Size	Gravel (%)	66.8	67.7	55.9	0	0	0
Distribution*	Sand (%)	25.8	20.8	26.3	98.3	81.0	91.6
	Silt (%) } Clay (%)	7.4	11.5	17.8	1.7	15.0	5.4
	Clay (%)					4.0	3.0
Shear Strength	Cohesion (kN/m^2)	NA	NA	NA	Cd=0	C'=0	Cd=2.0
	Angle of shear resistance (deg)	NA	NA	NA	φd=36.6	φ'=34.4	φd=37.8
Construction	Unit weight γt (kN/m^3)		20.2 (Average)		17.5 (Average)	13.4	15.5 (Average)
	Water content w (%)		NA		3.6	4.3	3.4

* Gravel (2–75 mm), Sand (75 μm–2 mm), Silt (5–75 μm) (%), Clay (<5 μm) (%)

Table 2. Type of facing panel.

Type	Nominal dimension (m)	Weight (kg)
Thick type	1.5 × 1.5 × 0.18	950
Thin type	1.5 × 1.5 × 0.1	650
	1.5 × 1.5 × 0.13	750
Small type	1.0 × 1.0 × 0.11	270

2.1 *Field measurement*

This field measurement in an actual wall (Height is 12.75 m) is executed to survey the behavior of the structure, which used thin panels (thickness 0.1 m/0.13 m) or thick panels (thickness 0.18 m). And it compared the difference of the effect for the behavior between using thin panels and using thick panels.

The cross-section is as shown in Figure 1. As for the backfill material, "Field test(1)" is used for the range from the base to 5.25 m high, "Field test(2)" is used for the range from 5.25 m high to 9.75 m high, and "Field test(3)" is used for the range from 9.75 m high to 12.75 m high. Load cell is installed underneath the concrete facing panel at the concrete foundation pad for measuring subgrade reaction. And tensile force in reinforcement strips were measured at behind a wall of all facing panels, above a load cell.

Figure 2 shows the relation between the height of the fill and the subgrade reaction during fill process. The measurement results up to 12.15 m fill height is shown. The subgrade reaction is assumed in which the weight of the facing panel is subtracted from the measurement result with the load call. And it shows as a load for 1 m along wall.

As for the gradient of an increase between subgrade reaction and fill height, thin type panels and thick type panels are the same until 8 m high. But in the thick type, increase of subgrade reaction is less than thin type panels from 8 m high. On the other hand, subgrade reaction of thin type increases in linear with fill height.

Relation with the total tensile force of reinforcing strips and subgrade reaction is shown in Figure 3. As for tensile force of reinforcing strips at the same subgrade reaction, both thick type panels and thin type panels are approximately the same. For maximum value of subgrade reaction, in case of thick type panels, subgrade reaction is 52kN/m and tensile force of reinforcing strips is 138 kN/m. In case of thin type, subgrade reaction is 138 kN/m and tensile force of reinforcing strips is 95 kN/m.

2.2 *Loading test*

The loading test is conducted to understand a behavior of Reinforced Earth wall when the high load surcharge at the wall top of the experimental wall.

The cross-section is as shown in Figure 4. In CASE1 and CASE3, vertical distance between all reinforcements layers is 0.5m. On the other hand, in CASE2, reinforcing strips install 0.25 m in the vertical interval from the top of wall to 1 m depth. A small type is used for the facing panels.

As backfill procedure in CASE2 and CASE3, after completion of load test CASE1, once backfill material is removed from the top of wall to a depth of 2 m, and then reconstruct a experimental wall for next case. As surcharge procedure, the load beam is set on the top of the wall, then the load beam is loaded using a hydraulic jack. In CASE1 and CASE2, a maximum surcharge load is 500 kN/m2, although it is loaded until 1200 kN/m2 in CASE3.

For the filling process, relation with fill height and subgrade reaction is shown in Figure 5. The calculation of subgrade reaction is the same as section "2.1 Field measurement". The subgrade reaction with an increase of fill height is increasing in a linear relation at any

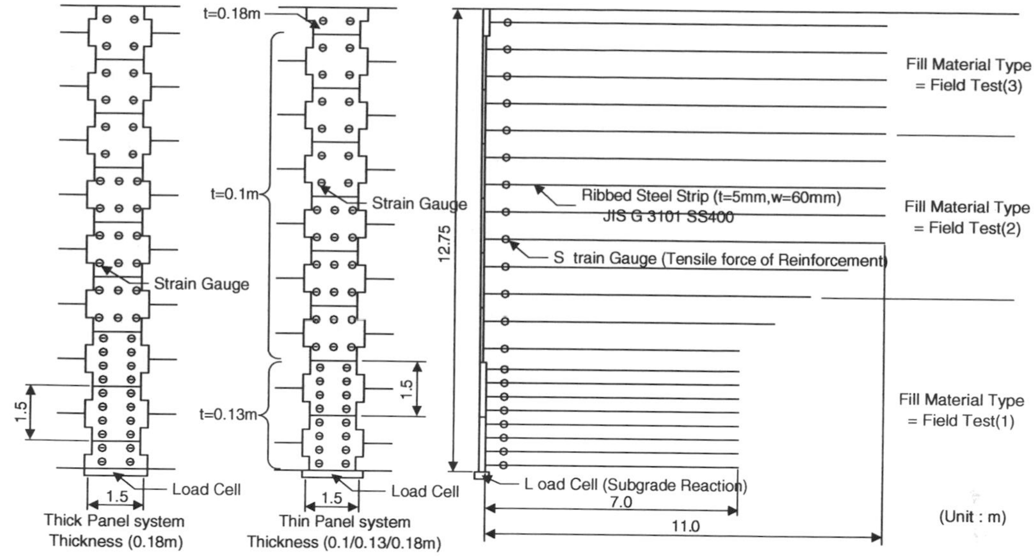

Figure 1. Cross section of actual wall for field measurement.

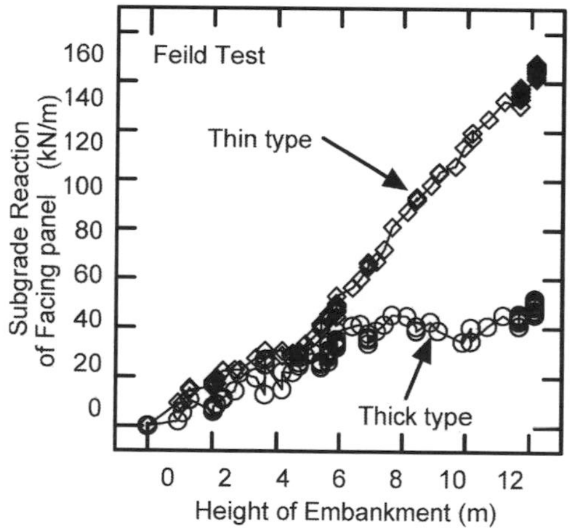

Figure 2. Subgrade reaction versus fill height.

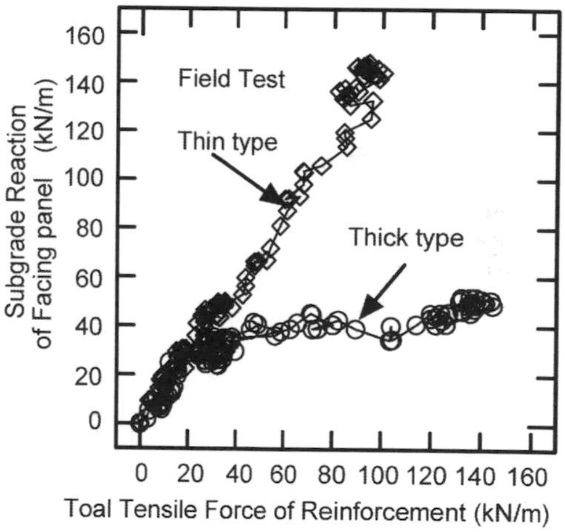

Figure 3. Subgrade reaction versus tensile force.

cases. Relation with the total tensile force of soil reinforcement and subgrade reaction is shown in Figure 6. It shows a direct proportional relationship. For maximum value of subgrade reaction, in CASE1, subgrade reaction is 11 kN/m and tensile force of reinforcing strips is 37 kN/m. In CASE2, subgrade reaction is 12 kN/m and tensile force of reinforcing strips is 39 kN/m. In CASE3, subgrade reaction is 13 kN/m and tensile force of reinforcing strips is 36 kN/m.

As the result of the Loading process, the relationship between a surcharge load and subgrade reaction is shown in Figure 7, and the relation between subgrade reaction and tensile force of reinforcing strips is shown in Figure 8. Data of subgrade reaction and tensile force of strips shows as incremental load during the loading process.

In CASE1, when a surcharge load is 507 kN/m^2, subgrade reaction is 70 kN/m and tensile force of reinforcing strips is 193 kN/m. In CASE2, when a surcharge load is 499 kN/m^2, subgrade reaction is 82 kN/m and tensile force of reinforcing strips is 200 kN/m. The tensile force of reinforcement is measured up to roughly 1000 kN/m^2 in CASE3. When a surcharge load is 450 kN/m^2, subgrade reaction is

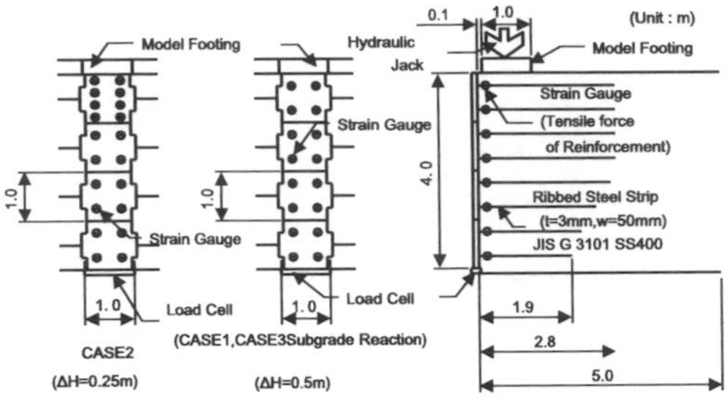

Figure 4. Cross section of experimental wall for load test.

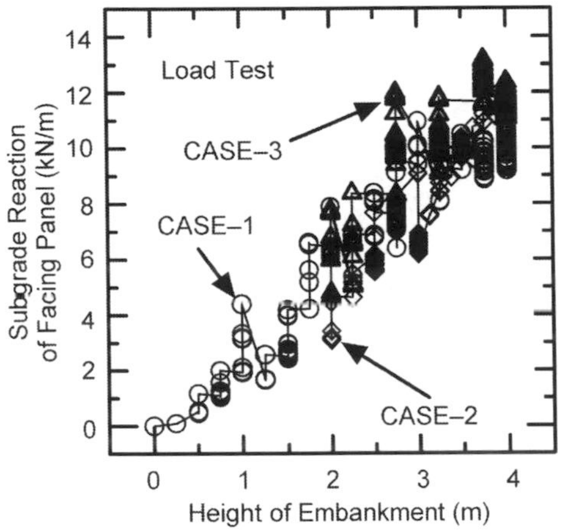

Figure 5. Subgrade reaction versus fill height.

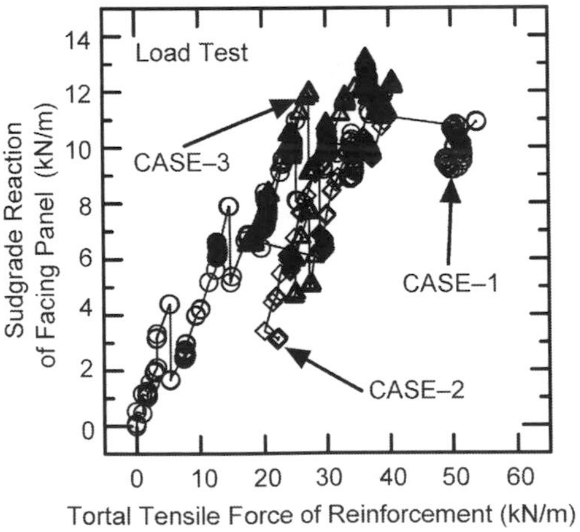

Figure 6. Subgrade reaction versus tensile force.

77 kN/m and tensile force of reinforcing strips is 182 kN/m. and when a surcharge load is 1045 kN/m^2, subgrade reaction is 260 kN/m and tensile force of reinforcing strips is 567 kN/m.

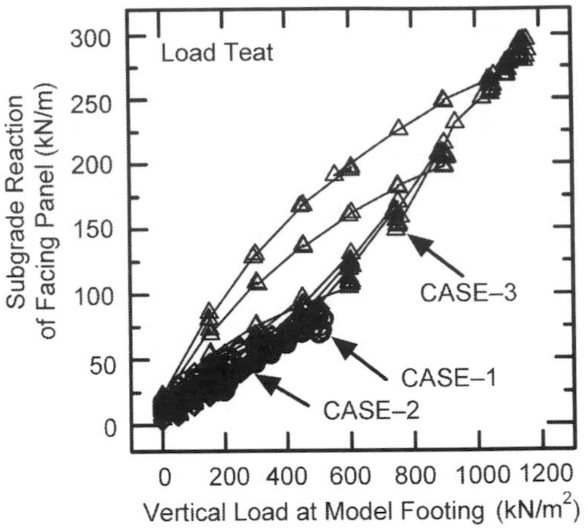

Figure 7. Subgrade reaction versus surcharge load.

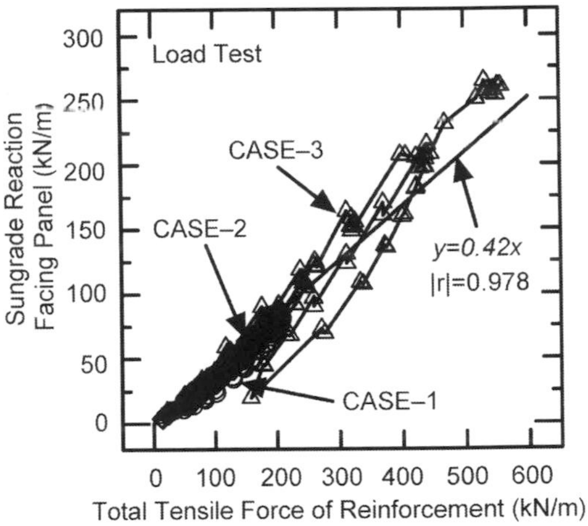

Figure 8. Subgrade reaction versus tensile force.

In Figure 7, which shows relation of a surcharge load and subgrade reaction, the inclination of subgrade reaction is changing in roughly 600 kN/m^2 in a surcharge load.

The other hand, the relation between subgrade reaction and tensile force of reinforcement are an approximately direct proportional relationship in Figure 8. And, a regression equation is estimated "y = 0.42x" in linear expression.

2.3 *Shaking test*

This experiment[1] is executed to understand a behavior of Reinforced Earth wall during an earthquake. But in this paper, only the result of the filling process is shown. The cross-section is shown in Figure 9. The facing panel is used thick type panels. Relation with

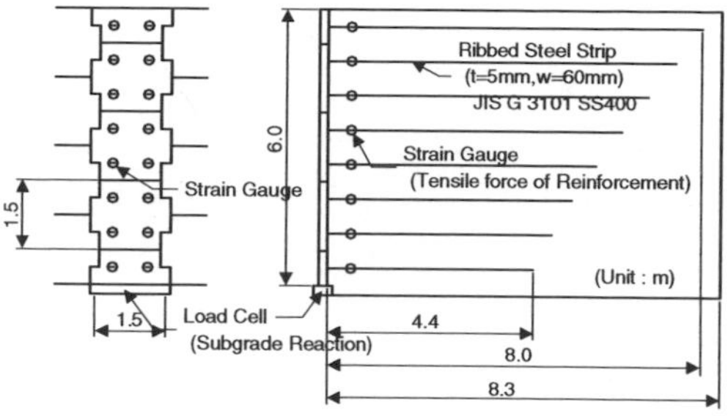

Figure 9. Cross section of experimental wall for shaking test.

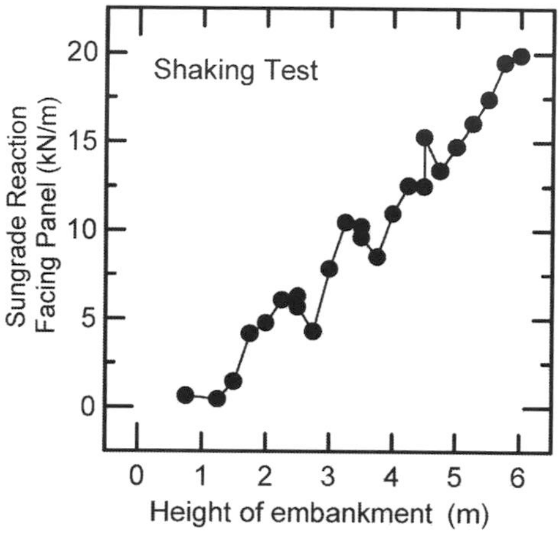

Figure 10. Subgrade reaction versus fill height.

fill height and the subgrade reaction is shown in Figure 10. The subgrade reaction is calculated also the same as case of "2.1 Field measurement". The subgrade reaction is increasing almost linearly with the increase in fill height. Relation with the total tensile force of reinforcing strips and subgrade reaction is shown in Figure 11. It shows also a direct proportional relationship. When subgrade reaction indicates maximum value, subgrade reaction is 20 kN/m and tensile force of reinforcing strips is 65 kN/m.

2.4 *Moving test*

This Moving test is conducted to make out of behavior of Reinforced Earth wall with footing foundation, when footing moved laterally due to its active earth pressure. It is assumed that failure mechanism is sliding mode. The procedure of the footing movement is to gradually loosen the support member, which set at a front of a footing, then a footing foundation move to front of a wall by itself. The cross-section is shown in Figure 12.

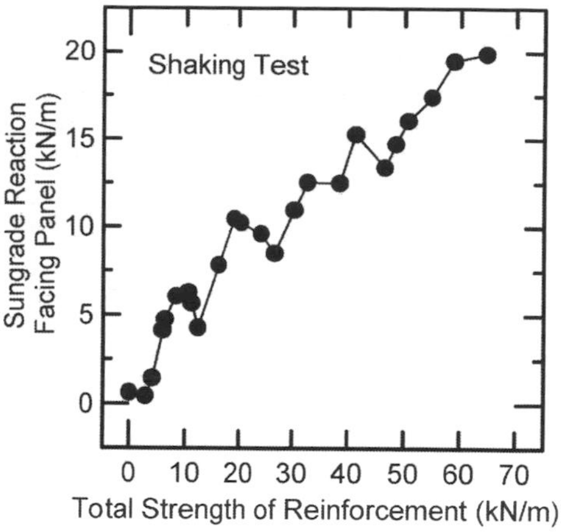

Figure 11. Subgrade reaction versus tensile force.

In CASE1, reinforcement material length is arranged 4 m at all layers to cross an assumed active failure surface, which occurs from the bottom of footing foundation. As for CASE2, the length of reinforcing strips is arranged to be in inside of an assumed failure surface. The load cell measured the vertical load underneath facing panels in 3m of wall-developed length. The measuring position of tensile force is indicated mark "o" in Figure 12. We assume that measured data is representative tensile force in the area with 3.0 m width and 1.0 m heights. Therefore, total tensile force of reinforcing strips is calculated by measured data, which multiplies 12 in this area.

Relation with fill height and the subgrade reaction is shown in Figure 13. The subgrade reaction is estimated by the same way as "2.1 Field measurement". The subgrade reaction increases direct proportional relationship with an increase of fill height. Relation with the total tensile force reinforcing strips and subgrade reaction is shown in Figure 14. It shows also a direct proportional relationship. Tensile force of reinforcing strips in CASE2 is smaller than CASE1 in same subgrade reaction. When the peak value of subgrade reaction is indicated 19 kN/m, tensile force of reinforcing strips is 39 kN/m in CASE1.

In CASE2, when the maximum value of subgrade reaction is indicated 22 kN/m, tensile force of reinforcing strips is 26 kN/m.

Relation with the displacement of the footing foundation and subgrade reaction is shown in Figure 16. Subgrade reaction is almost same up to 20 mm in CASE1 and CASE2. The footing does not move in CASE1 by 18 mm or more, and subgrade reaction is 4.5 kN/m at this point. In CASE2, when the displacement of footing is 15 mm, subgrade reaction indicates 5.9 kN/m. and when the final displacement is 50 mm, subgrade reaction become 11.5 kN/m. Figure 16 shows

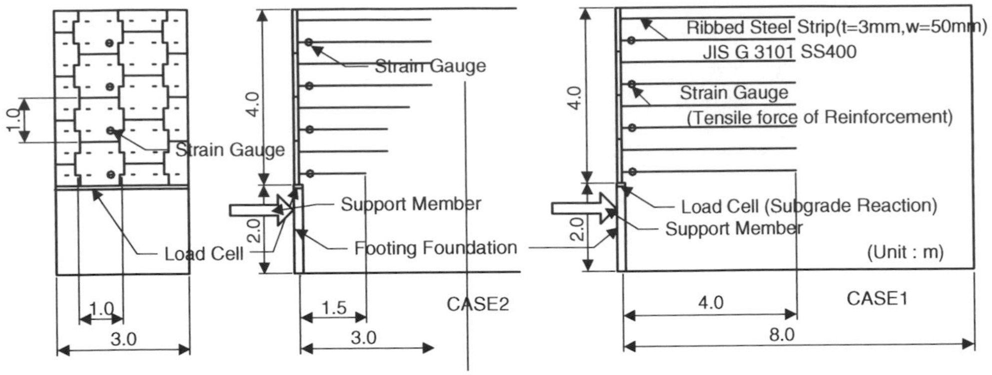

Figure 12. Cross section of experimental wall for moving test.

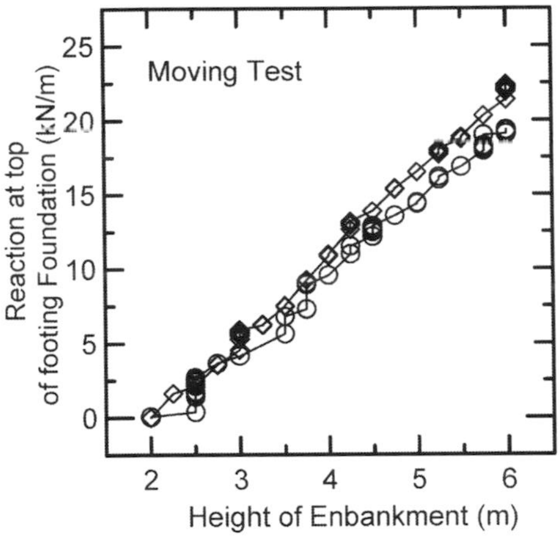

Figure 13. Subgrade reaction versus fill height.

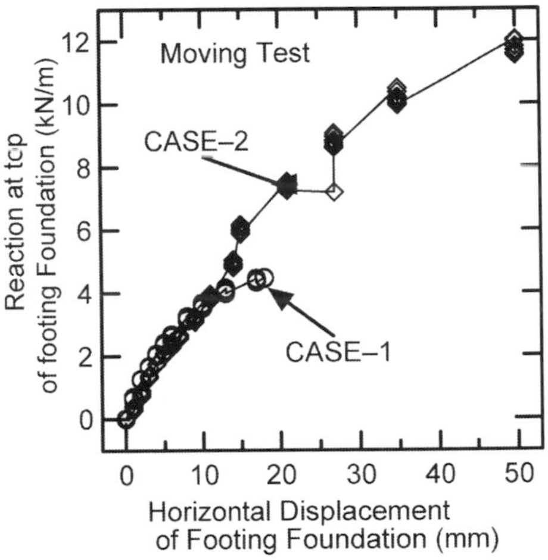

Figure 15. Subgrade reaction versus footing displacement.

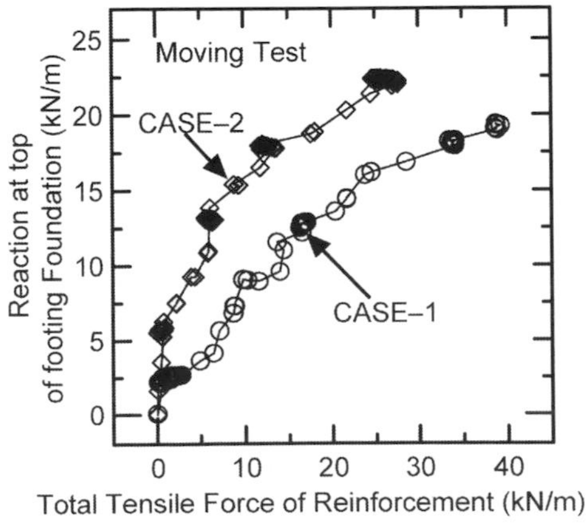

Figure 14. Subgrade Reaction versus Tensile Force.

the relation between the subgrade reaction and the reinforcement tension. In CASE1, the subgrade reaction increases linearly with the tension of reinforcement.

And, a regression equation is estimated "$y = 0.11x$" in linear expression in Figure 16. But, in Case 2, subgrade reaction increases though tensile force of reinforcing hardly increases.

3 RELATION BETWEEN SUBGRADE REACTION AND TENSILE FORCE

The relation between the reinforcement tension and the subgrade reaction is shown in Figure 17 with all data of fill process. When there is measurement data of same fill height, the first measured data is adopted in this graph. The subgrade reaction is proportional relation with a sum of tension of reinforcement. But, it varies

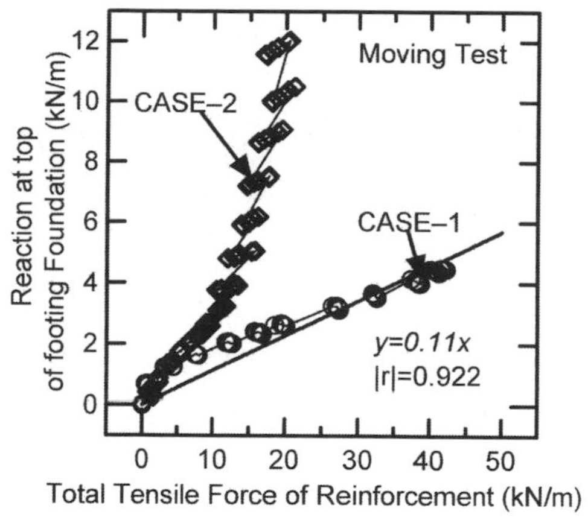

Figure 16. Subgrade reaction versus tensile force.

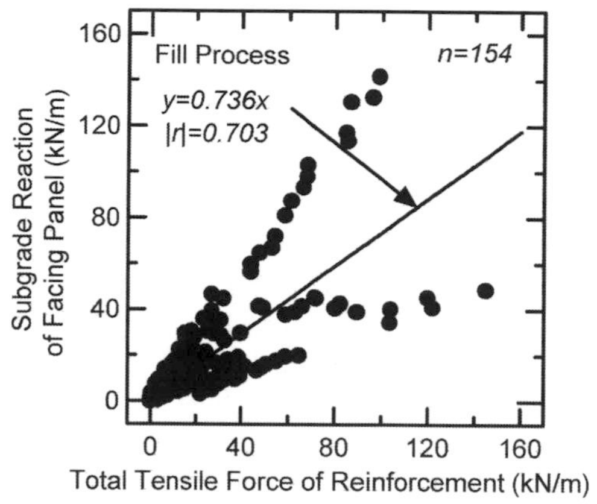

Figure 17. Subgrade reaction versus tensile force.

widely. A regression equation is estimated "y = 0.62x" in linear expression with the use of all data. Therefore, it is calculated that the angle of wall friction is $\delta = 32$(deg), using relation between tensile force of reinforcement and reaction. However, some date indicate that subgrade reaction becomes bigger than the tensile force of the reinforcement. Inclination of regression line is in range from 0.3 to 1.4. At this point, it calculated lateral earth pressure using the formula of Coulomb's active earth pressure with the assumption that the angle of wall friction is $\delta = 0$(deg). Comparison of earth pressure and tensile force is shown in

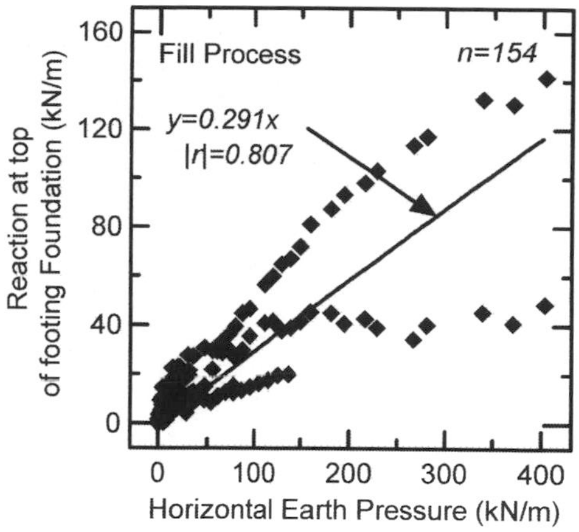

Figure 18. Subgrade reaction versus lateral earth pressure.

Figure 18. At this time, data about the angle of shear resistance is not obtained in the field measurement therefore we assume $\varphi = 35$(deg). With the average, it becomes 0.27 times of the horizontal earth pressure.

4 CONCLUSIONS

1 Subgrade reaction becomes by 0.62 times of the total tensile force of reinforcing strips in a filling process.
2 Subgrade reaction becomes by 0.27 times and the total tensile force of reinforcing strips is 0.44 times the design earth pressure to use Coulomb's earth pressure.
3 The total tensile force of reinforcing strips and the vertical load on a foundation pad has a relation. And tensile force of reinforcing strips and the lateral earth pressure has a relationship too. Therefore it is possible to estimate subgrade reaction from the design earth pressure.

REFERENCES

Futaki, M., Ogawa, N., Sato, M., Kumada, T., and Natsume, S. (1996). "Experiments about Seismic performance of Reinforced Earth Retaining Wall." *Proceedings of Eleventh World Conference on Earthquake Engineering*, ElsevierScience, Ltd. (CD-ROM)

New Horizons in Earth Reinforcement – Otani, Miyata & Mukunoki (eds)
© 2008 Taylor & Francis Group, London, ISBN 978-0-415-45775-0

Performance of auxiliary bearing plates in active zone for multi-anchored reinforced soil retaining wall

T. Konami & Y. Kudo
Okasan Livic Co., Ltd., Tokyo, Japan

K. Miura
Toyohashi University of Technology, Aichi, Japan

T. Tatsui
Ttechno-Sole Co., Ltd., Tokyo, Japan

S. Morimasa
Toyohashi University of Technology, Aichi, Japan

ABSTRACT: The aim of this study is to clarify the reinforcement mechanism of reinforced soil retaining walls and to reduce the construction cost of the structure. The retaining wall whose backfill is reinforced with multi-anchors is investigated to demonstrate the performance of the auxiliary bearing plates and propose new type of reinforced soil retaining wall. Three types of real scaled model tests were conducted to investigate the efficiency of the auxiliary bearing plates in active zone. It is confirmed that the bearing plates in active zone behind wall facing and passive zone both are effective not only on the reinforcement of backfill soil, but also on the reduction of the earth pressure on the wall facing. Based on the results of the model test, a design method for the multi-anchored type reinforced soil retaining wall with the auxiliary bearing plates is introduced.

1 INTRODUCTION

Multi-Anchored Reinforced Soil Retaining Wall shown in Fig.1 is a type of the widely used reinforced soil retaining walls including Terre Armee wall and Geosynthetics-reinforced retaining wall, where the backfill soil is reinforced with multiple anchors. As shown in Fig.1 the anchors are connected with the panels of wall facing through tie bars, and the pull-out resistance of the anchors acts as reaction to the earth pressure on the back of wall facing. The internal stability of the reinforced soil wall is examined based on the equilibrium condition between the earth pressure and the pull-out resistance in the practical design method (Public Works Research Center; 2002). As the backfill soil is confined between the wall facing and the group of anchor plates and its mechanical properties are improved with the confining effect, the confined region is regarded as an integrated mass body (or rigid body) in the external stability calculations (Higashihara et al.;1996).

The introduction of auxiliary anchor plates on tie bars near wall facing is investigated in this study. The efficiency of the auxiliary bearing plates on the confinement of backfill soil and the reduction of the earth pressure on the back of wall facing are examined through model loading tests. If the backfill soil is reinforced more effectively with the auxiliary bearing plates and its mechanical properties such as stiffness and strength are improved, the more stable performance of the reinforced retaining wall will be expected during earthquake as well as in ordinary condition. And if the earth pressure on the back of wall facing is reduced due to the existence of auxiliary bearing plates, wall facing panels can be require less stiffness and strength of themselves. Thus the thickness of the wall facing plates can be reduced, and it becomes possible that the weight of wall facing and the vertical load on the foundation beneath it are reduced. As a result it will be expected that the introduction of auxiliary bearing plates leads to the reduction of construction cost as well as the improvement of the structural stability.

The performance of the auxiliary bearing plates was first examined in a series of aluminum-rods-stack model tests (Konami et al.; 2005). In the tests the model with different types of anchors with and without

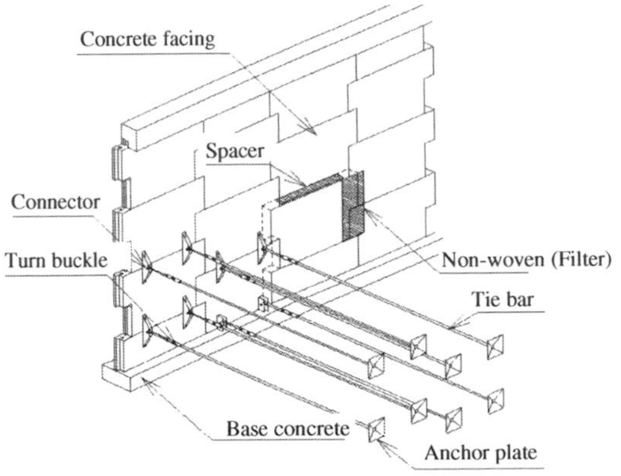

Figure 1. Composition of Multi-Anchored Reinforced Retaining Wall.

auxiliary bearing plates are tested under various types of loading conditions. The performance of the models were observed and examined comparatively. As a result, the efficiency of the auxiliary bearing plates on the improvement of the stability and the reduction of the earth pressure was recognized qualitatively. And the real-scaled models of the reinforced soil retaining wall were planed to confirm the effect quantitatively in real scale; a series of loading tests were conducted. In this paper the test results of the model loading tests are presented and examined, and finally the efficiency of the auxiliary bearing plates are discussed for the practical design method.

2 REAL-SCALED MODEL LOADING TEST

2.1 *Model retaining walls*

Three real-scaled model reinforced soil retaining walls which were constructed for this study are shown in Fig. 2. The height and wall facing inclination were common in all the models, and however different types of reinforcement were employed in each of the models. The wall was 4.5 m in height and 2.5 m in width, and the inclination ratio was 1:0.3. The wall facing consisted of five steps of the blocks made of reinforced concrete. In all the models the backfill soil material are common, and the backfill material was compacted by means of the same vibro-compaction method with the density of 90% of its maximum dry density; the mechanical properties of the compacted backfill material are listed in Table 1.

In the model Case-1 no reinforcement was employed, which was constructed to evaluate the effect of reinforcement comparatively. In the models Case 2 and Case 3, auxiliary bearing plates as well as an anchor plates are set on each of tie bars, both of which are square in shape and 15 cm in size. The tie bars

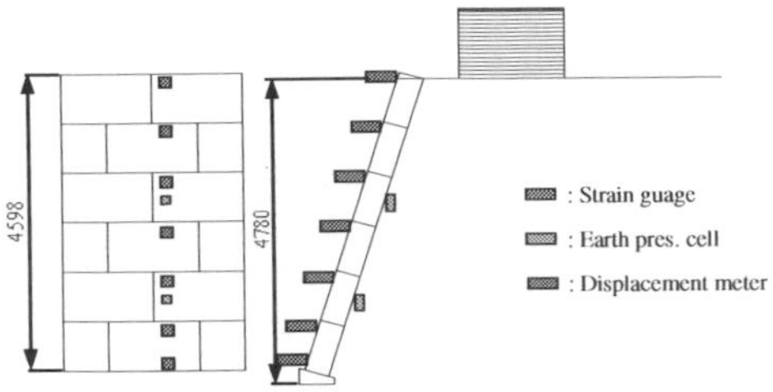

(a) Case 1 without reinforcement

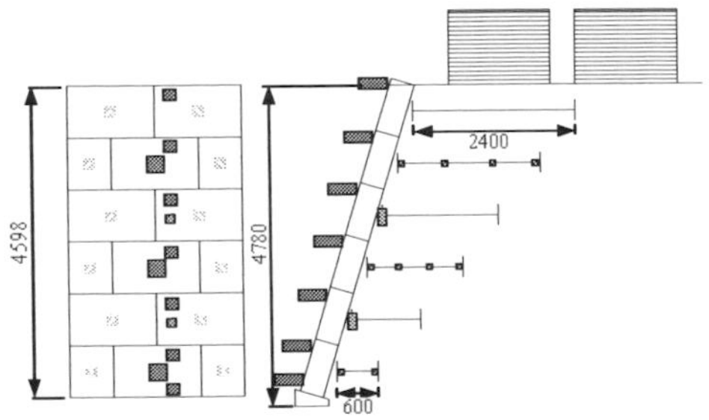

(b) Case 2 with reinforcement but without connetion

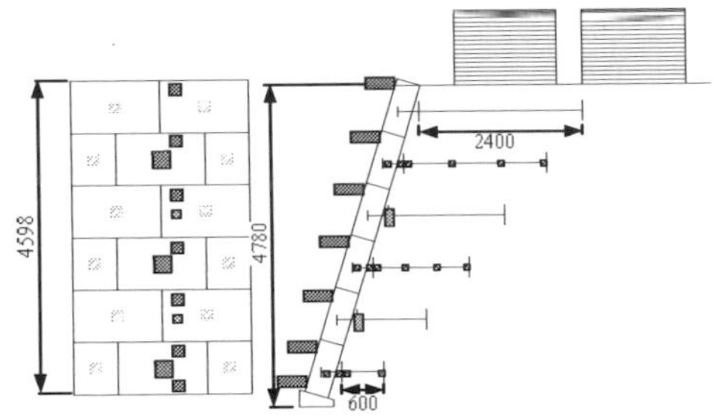

(c) Case 3 with reinforcement and connetion

Figure 2. Real-scale test model.

Table 1. Mechanical properties of backfill soil material.

Soil type	Fine sand with gravels
Wet density (g/cm^3)	1.635
Dry density (g/cm^3)	1.587
Water content (%)	3.0
Cohesion c_d(kN/m^2)	0.0
Internal friction angle ϕ	37.0 deg.

were not connected to the wall facing in Case 2, but connected to the wall facing in Case 3. Sections of the models are show in Fig. 2; the anchor plates were located 1.2 m behind the active soil wedge which was determined by Coulomb's earth pressure theory with cohesion c of 0 and ϕ of 30 deg.

The arrangement of the sensors for the displacement of the wall facing and the tension along anchors are shown in Fig. 2. Horizontal displacements of each of the wall facing blocks were directly measured with linier displacement gauges. The earth pressure on the wall facing was measured by using earth pressure cells. The tensions along the anchors were measured by using strain gauges glued on the surface of tie bars. The earth pressure on the fall facing was measured by using earth pressure cells placed behind the fall facing, or calculated from the tensions in the anchors near the wall facing.

2.2 *Construction and Loading on the model retaining walls*

After the construction of the three model retaining walls, the model retaining walls were loaded by using stacks of steel plates. The stack of steel plates was statistically placed in some steps on the active soil wedge. The maximum intensity of the overburden pressure was 42 kN/m^2 in Case-1, and 84 kN/m^2 in Case-2 and Case-3. Next the stack of steel plates was placed behind the active soil wedge in Case-2 and Case-3. The deformation behavior of the model retaining walls and anchor were observed during the construction and the loading.

3 TEST RESULTS

3.1 *Displacement of wall facing*

Shown in Fig. 3 is the horizontal displacement of the wall facing during the loading. At the overburden pressure of 42 kN/m^2 the displacement of the top of retaining wall was rather large and reached nearly 80 mm in Case-1 in which the model was not reinforced. On the other hand the displacement was less than 10 mm under the same intensity of overburden pressure of 84 kN/m^2, in Case-2 and Case-3. At the higher over burden pressure of 84 kN/m^2 the displacement was increased; the maximum horizontal displacement was about 30 mm and 40 mm in Case-2 and Case-3, respectively.

3.2 *Tension in anchors*

The distribution of tension in anchors observed during the loading is shown in Fig. 4 for Case-3, where the anchors were connected to the wall facing. The distribution of tension was almost uniform between the anchor plate and the auxiliary bearing plate; this tendency was observed also in Case-2. The tension is, however, much smaller behind the wall facing in front of the auxiliary bearing plate. The behavior observed suggests that major part of the confinement was induced between the anchor plate and the auxiliary bearing plate.

In the conventional type of multi-anchored reinforced soil retaining wall without the auxiliary bearing plates, backfill soil are confined between the wall facing and anchor plates, and the certain degree of strength is required for the wall facing. On the other hand, the confining effect is induced behind the auxiliary bearing plates, and as a result the earth pressure on the wall facing is fairly reduced. The effect of the auxiliary bearing plate surely allows the wall facing panels to become thin and/or light.

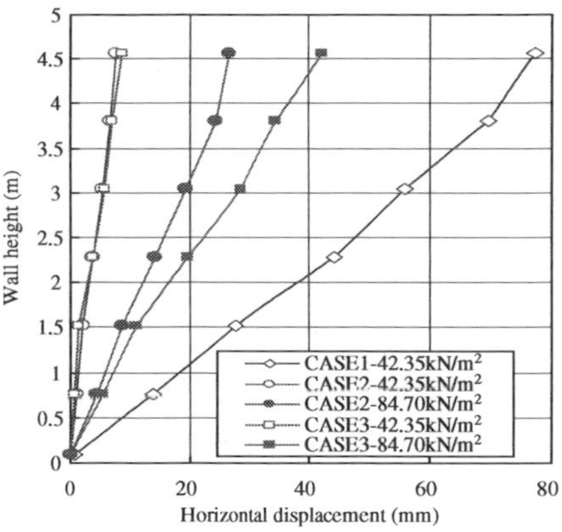

Figure 3. Displacement of wall facing during loading.

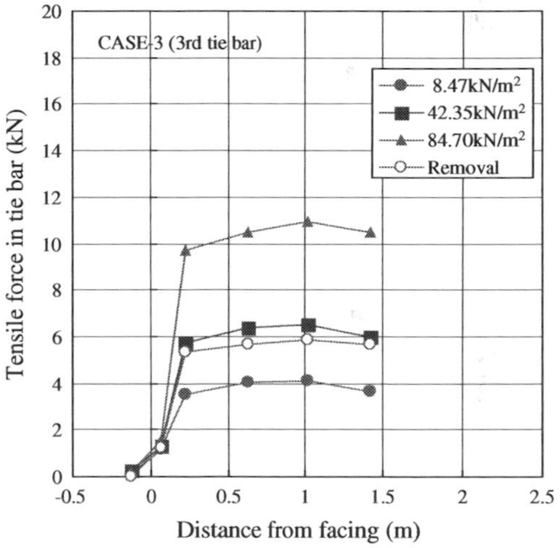

Figure 4. Distribution of tension in tie bar during loading.

4 EVALUATION OF CONFINING EFFECT WITH AUXILIARY BEARING PLATE

4.1 *Bearing resistance of auxiliary bearing plate*

For the simplicity of analytical condition the wall facing is assumed vertical. In Fig. 5 the vertical section

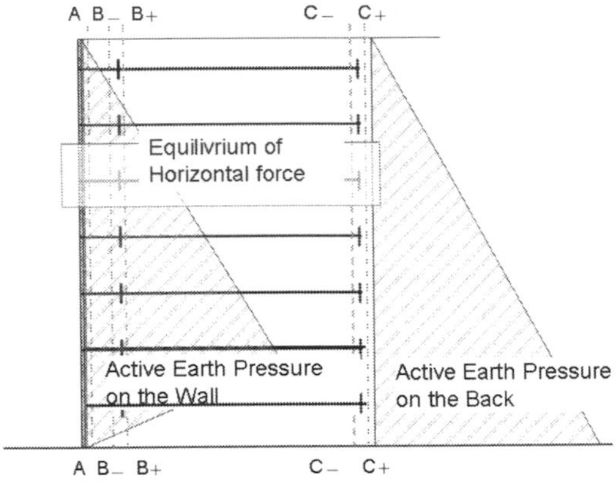

Figure 5. Vertical section of reinforced soil retaining wall.

of the retaining wall and some vertical lines are drown for the equilibrium of horizontal forces; section A-A is located just behind the wall facing, sections B_--B_- and B_+-B_+ are before and behind the auxiliary bearing plates, and sections C_--C_- and C_+-C_+ are before and behind the anchor plates. Active earth pressure acts on C_+-C_+ line, and the active earth pressure is assumed to act also on B_+-B_+ according to the assumption employed for internal stability. The earth pressure on the wall facing is p', and averaged earth pressure on B_+-B_+ is p_a'.

Load bearing behavior of the auxiliary bearing plate is analyzed in the equilibrium of horizontal load along the tie bar as shown in Fig. 6. The tension in the tie bar is T' between the wall facing and the auxiliary bearing plate, and T between the auxiliary bearing plate and the anchor plate. As in Fig. 7, bearing load capacity is assumed to be mobilized due to the plastic local failure.

From the equilibrium condition on section B_+-B_+ the following equation can be derived.

$$p_a'A = p'(A - A_p) + q'A_p = T \tag{1}$$

where A is the area of wall facing corresponding to a single tie bar, A_p is common area for an anchor plate and auxiliary bearing plate, and q' is bearing resistance on the auxiliary bearing plate.

According to local plastic failure mechanism the following formula can be used for an ultimate state (Miura et al.;1994)

$$q' = cN_c + p'N_q \tag{2}$$

where N_c and N_q are bearing capacity coefficients used generally in bearing capacity formula for foundation.

Also the tension in tie bar behind the wall facing T' can be given by the equilibrium on section A-A.

$$T' = Ap' \tag{3}$$

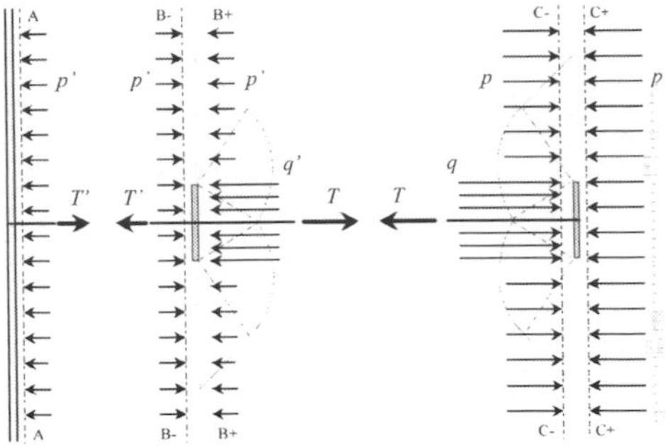

Figure 6. Equilibrium of horizontal forces along tie bar.

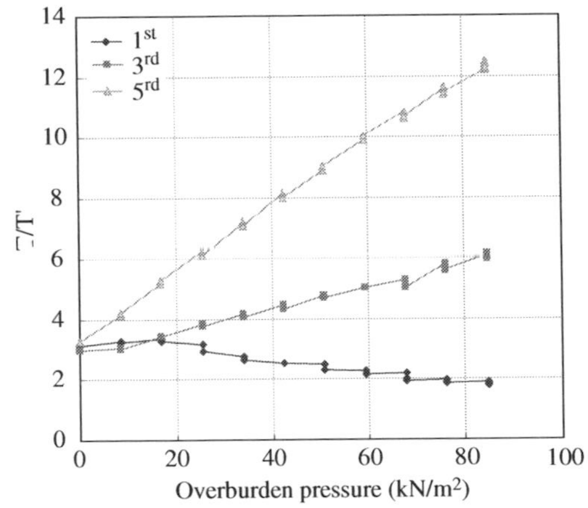

Figure 7. The ratio of tensions in tie bar.

4.2 *Reduction of earth pressure on the wall facing*

We can know from Eqs. (1) and (3) that the reduction ratio of earth pressure p'/pa' is equal to the ratio of tensions in the tie bar T'/T. The variation of T'/T is plotted against the overburden pressure in Fig. 7. At the end of construction, i.e. at the beginning of the loading, the ratio T'/T was about 3 in all the tie bars; however, it increased or decreased during the loading depending of the depth of the tie bar. It can be said that the ratio becomes larger with a decrease of the depth. On the test condition employed in this study, the earth pressure can be reduced to the half at least on the wall facing.

From Eqs. (1) through (3) the ratio T'/T can be expected to be constant irrespective of the depth,

$$\frac{T}{T'} = 1 + \left(N_q - 1\right)\frac{A_p}{A} \tag{4}$$

where the cohesion c is neglected as shown in Table 1. Equation (4) tells that the ratio can reach about 10 at imaginary ultimate plastic state. It should be noted that the bearing resistance of the auxiliary bearing plate which can be evaluated with Eq. (2) is nor full mobilized in the model loading tests conducted in this study. The inclination of the wall facing of the models may have some amount of effects on the test results.

5 CONCLUDING REMARKS

A series of loading tests were conducted on real-scaled model multi-anchored reinforced soil retaining walls with and without reinforcement. And the efficiency of the auxiliary bearing plates was examined based on the test results. As a result of this study, The auxiliary bearing plates are effective on;

- the stability of the retaining walls.
- the reduction in the earth pressure on the wall facing.

The evaluation method for the reduction of earth pressure was derived by means of load bearing mechanism of plasticity; however, the degree of mobilization of the bearing resistance was not constant depending on the depth during loading. Further experimental and theoretical examination will be necessary for the practical design of the reinforced soil retaining wall of this type.

REFERENCES

Public Works Research Center 2002. Design and construction manual for Multi-Anchored Reinforced Soil Wall (in Japanese)

Higashihira, N., Miura, K., Nagano, T., Dobasghi, K. and Misawa, K. 1996. Confining effect of Anchors in Multi-Anchored Reinforced Soil Retaining Wall, Proc. of the 31st annual meeting of JGS, 2421-2422. (in Japanese)

Konami, T., Miura, K., Misawa, K., Tatsui, Y. and Morimasa, M. 2005. Aluminum rods stack model tests on the confining effect in Multi-Anchored Reinforced Soil Retaining Wall, Journal of Geotechnical Engineering, vol. 50, pp. 279–286. (in Japanese)

Miura, K., Nomiyama, T., Kusakabe, O. and Sakai, T. 1994. Model test of Muti-Anchored Reinforced Soil Retaining Wall and Numerical Analysis of Pull-out Resistance, Proc. of the 29th annual meeting of JGS, pp. 2445–2448. (in Japanese)

New Horizons in Earth Reinforcement – Otani, Miyata & Mukunoki (eds)
© 2008 Taylor & Francis Group, London, ISBN 978-0-415-45775-0

Stability assessment of geogrid reinforced soil wall by using optical fiber sensor

S. Tsuji, K. Yoshida & Y. Yokota
Maeda Kosen Co.,Ltd., Fukui, Japan

A. Yashima
Gifu University, Gifu, Japan

ABSTRACT: This paper reports that we can assess a total stability of a geogrid reinforced soil wall during and after construction by using an optical fiber sensor. We developed an "optical fiber sensor geogrid" in which an optical fiber sensor was installed in geogrid. We applied the optical fiber sensor geogrid to measure the strain distributions of an actual reinforced soil wall. The stability of the reinforced soil wall can be evaluated during and after construction continuously.

1 INTRODUCTION

To assess the total stability of geogrid reinforced soil wall during and after construction, field observation is carried out. However, it is difficult to assess the total stability from the result of field observation. We try to get a clear picture of strain distribution of geogrid during and after the construction of a geogrid reinforced soil wall to understand the total stability of the structure. A strain gauge has been widely used for strain measurement of geogrid inside the soil. However the strain gauge lacks durability, continuity of measuring point and ease of installing. In order to resolve these practical problems within a strain gauge, we have newly developed an "optical fiber sensor geogrid" in which optical fiber sensor was installed in geogrid. We applied the optical fiber sensor geogrid to measure the strain distributions of an actual reinforced soil wall. The reinforced soil wall consists of facing material, geogrid reinforced backfill and intermediate vertical layer which absorbs earth pressure from backfill to wall surface. To verify the reliability of optical fiber sensor, the strain gauges were also equipped at the same time. Results of strain measurement reveals as follows: 1) optical fiber sensor is more sensitive than the strain gauge, 2) it can measure a continuous strain distribution of geogrid and detect a local abnormality, and 3) we can provide the stability level of measured strain to assess the stability of reinforced soil wall, so that the stability of the reinforced soil wall can be evaluated during and after construction continuously. This paper reports the measuring system of an optical fiber sensor, manufacturing method of senor geogrid and real measuring results of geogrid reinforced soil wall.

2 OPTICAL FIBER SENSOR GEOGRID

The structure of the optical fiber sensor geogrid is shown in Figure 1. A single mode optical fiber of a diameter of 0.9 mm is fixed by adhesives in a protecting stainless tube of an inner diameter of 1.6 mm. Since the stiffness of stainless tube is lower than the stiffness of aramid fiber, the stainless tube does not influence on strain measuring. The optical fiber monitors strain using the BOTDR method, which is irradiating light pulses through the fiber and monitoring changes in the frequency of Berlian scattered light, which changes its frequency proportionally to the strain of the optical fiber (Tatta, N. et al., 2005). Optical fibers have

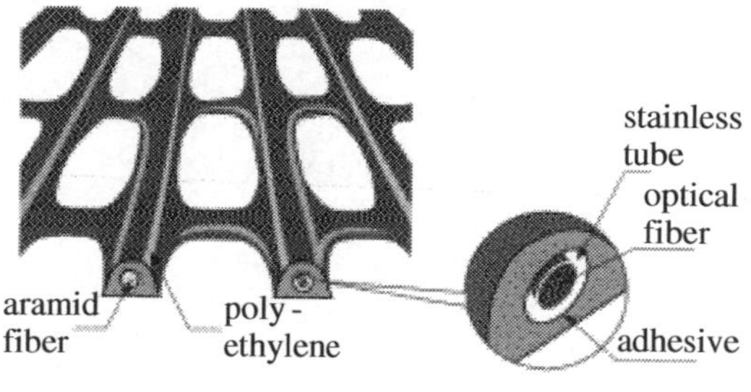

Figure 1. Optical fiber sensor geogrid.

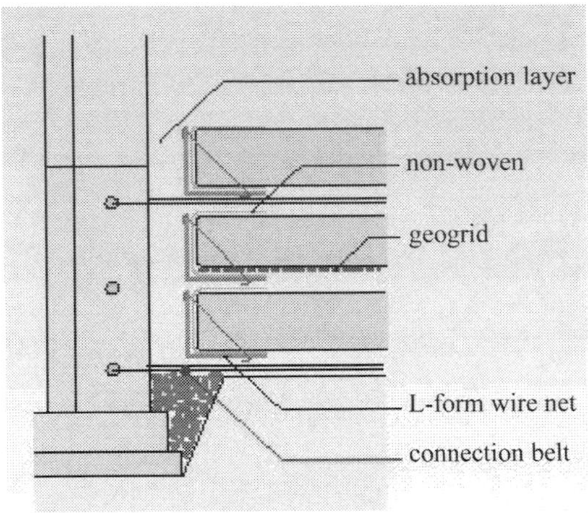

Figure 2. Structure of geogrid reinforced soil wall.

various advantages in monitoring strain distribution, such as 1) optical fiber is durable and it can be used to measure strains over a long period of time compared to strain gauges, and 2) the fibers can measure the strain in the entire geogrid.

The structure of reinforced soil wall is shown in Figure 2 (Yoshida, K. et al., 2006). The wall has a double wall system with a vertical layer (called absorption layer) between a wall of facing concrete blocks and a reinforced backfill. The absorption layer consists of single sized crushed stones. The facing concrete blocks and the reinforced backfill are connected with many belts of polyester fiber (called connection belt). The procedure of construction is follows: 1) grading the foundation ground to enable the facing concrete blocks to be horizontally installed, spreading crushed stones for the foundation, and building concrete foundation, 2) installing the facing concrete blocks on the foundation concrete, installing the connection belts and geogrids, backfilling and compacting soil, and 3) filling the absorption layer with single sized crushed stones. When the foundation ground is soft, the wall can be constructed after banking the reinforced backfill. The facing concrete blocks are installed after consolidation of ground. The installation of the optical sensor geogrid is shown in Figure 3. Considering that strain would be about 3% for the design tensile strength of the geogrid, the stability assessment index is established for assessing the stability of the measured strain values as shown in Table 1 (Yoshida, K. et al., 2006).

3 EXAMPLES OF STRAIN MONITORING

3.1 *Example 1*

A cross section of the reinforced soil wall and the placement of geogrids are shown in Figure 4, where *L* is the length of geogrid. The wall was 15 m high and soil produced by excavating a tunnel was used for the backfill. The strain of geogrids during the construction of the wall was monitored to assess the stability of the wall during construction. The strains were monitored using optical fibers on Geogrids A, B, C and D shown in Figure 4. To verify the reliability of strain monitoring using optical fibers, the strain was also measured on Geogrids A and B using strain gauges.

The strain distribution monitored using optical fiber is shown in Figure 5, and the strain distribution

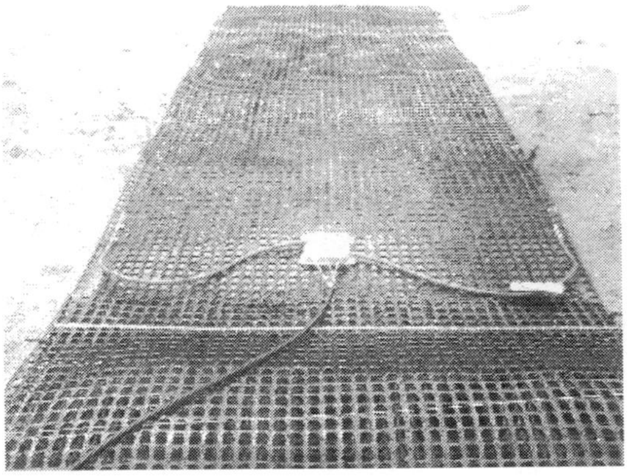

Figure 3. Installation of optical fiber sensor geogrid.

Table 1. Stability assessment index.

State	Maximum value of measured strain (%)
Stable	0.0~3.0
Warning	3.0~4.0
Unstable	Greater than 4.0

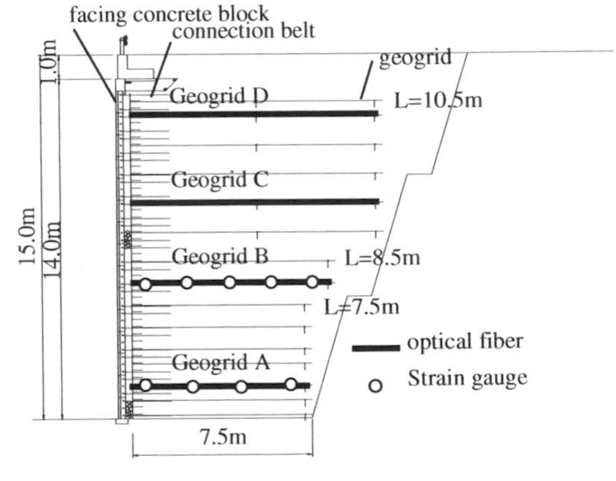

Figure 4. Cross-section.

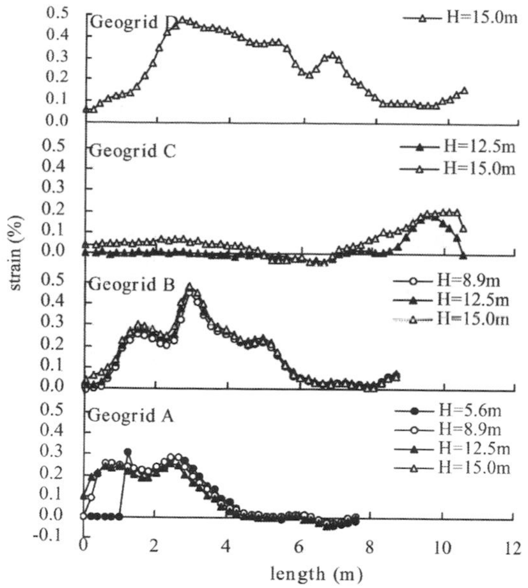

Figure 5. Strain distribution measured by optical fiber.

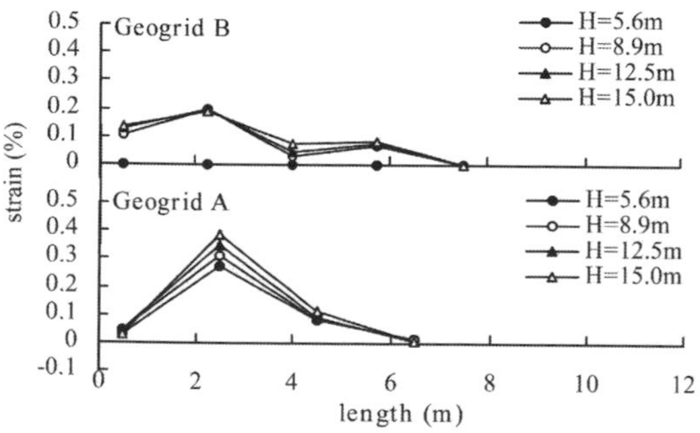

Figure 6. Strain distribution measured by strain gauge.

monitored using strain gauge is shown in Figure 6. It shows the increment in strain from the initial values monitored immediately after installing the geogrids, and *H* is the construction height of the wall. The optical fibers monitored the strain continuously throughout the geogrids. As banking progressed, the strain increased. The strain is largest near the wall on Geogrid A, which is installed low, and in the middle to the rear on Geogrids B and C, which are installed at intermediate heights. Stains of Geogrid C measured by optical fiber are small compared with other geogrid, because of error of measuring initial value of strain. The values monitored by the optical fibers and strain gauges are slightly different from each other but showed similar strain distributions. The maximum value of strain is about 0.5%, which shows that the wall is stable state.

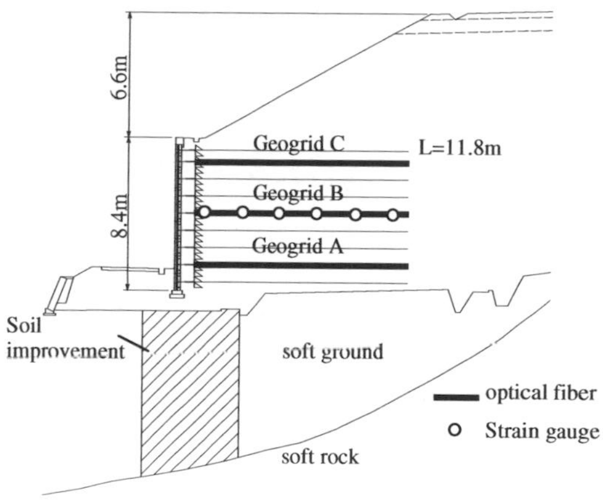

Figure 7. Cross-section.

3.2 *Example 2*

A cross section of the reinforced soil wall and the placement of geogrids are shown in Figure 7. The foundation ground consisted of an alluvial clayey soil layer (SPT N-value is about 1) of a thickness of about 5 m and a layer of sandy soil (SPT N-value is about 6) of a thickness of 5 m. Under the soft alluvial layer, the bed rock exists. Since the foundation ground was soft, there were risks of sliding failure of the foundation ground, liquefaction of the saturated sandy soil layers during an earthquake and immediate settlement of the foundation ground. To overcome these risks, the soil was stabilized using the deep mixing method at the site on which the facing concrete block wall was to be constructed. After the compression and consolidation of the non-stabilized ground section, which were caused by the construction of the geogrid reinforced backfill, ceased, the facing concrete blocks were piled up. Strain was monitored using optical fibers on Geogrids A, B, and C in Figure 7, and using strain gauges on Geogrid B.

The strain distribution monitored using optical fiber is shown in Figure 8, and the strain distribution monitored using strain gauge is shown in Figure 9. As banking progressed, the strain increased. Local large strain is observed on Geogrid A at about 3.5 m. This is likely attributable to the difference in settlement between the backfill built on stabilized and non-stabilized ground sections. The strain gauges cannot detect this behavior. Local increases in strain are observed in the reinforced backfill on non-stabilized ground. The maximum value of strain is about 1.8%, which shows that the wall is stable state.

3.3 *Example 3*

A cross section of the reinforced soil wall and the placement of geogrids are shown in Figure 10. After

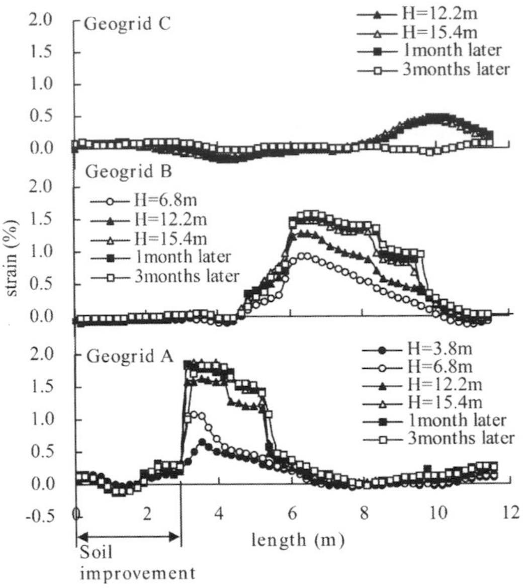

Figure 8. Strain distribution measured by optical fiber.

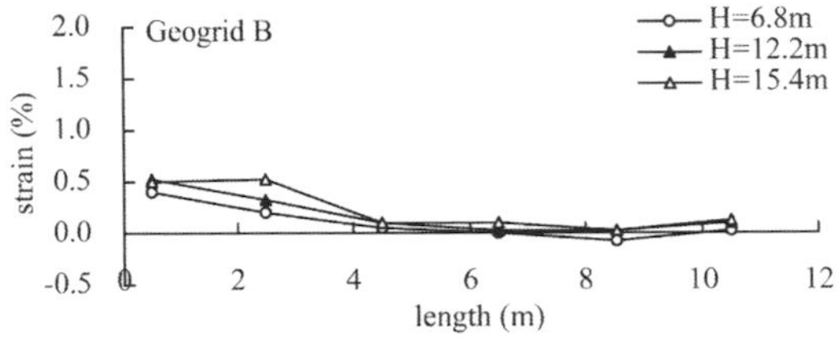

Figure 9. Strain distribution measured by strain gauge.

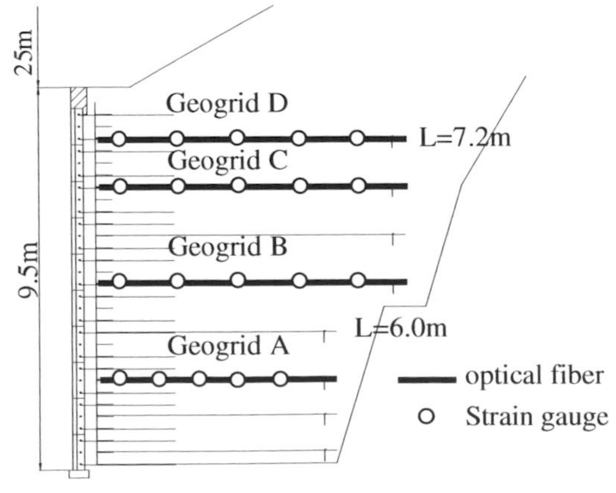

Figure 10. Cross-section view.

constructing the reinforced soil wall of 9.5 m high, the embankment slope of 25 m high was constructed. Strain was monitored using optical fibers and strain gauges on Geogrids A, B, C, and D in Figure 10.

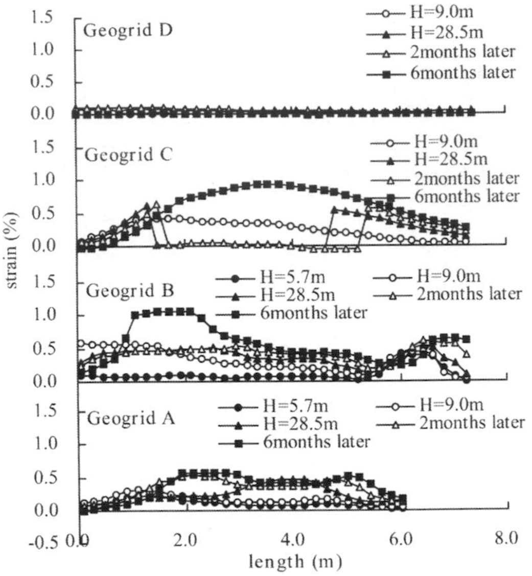

Figure 11. Strain distribution measured by optical fiber.

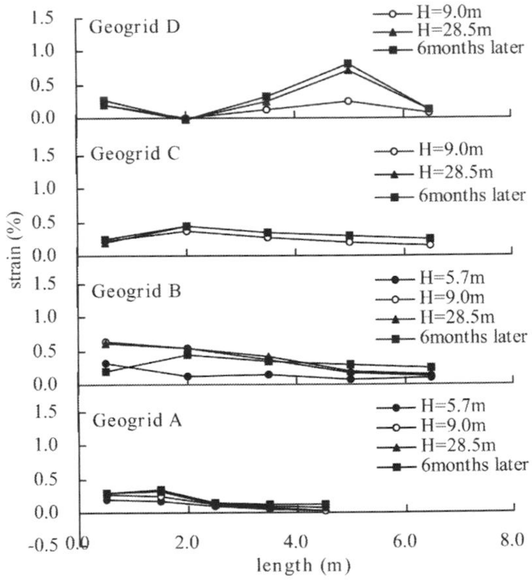

Figure 12. Strain distribution measured by strain gauge.

The strain distribution monitored using optical fiber is shown in Figure 11, and the strain distribution monitored using strain gauge is shown in Figure 12. The strain value equal to 0 on Geogrid C is attributable to mistaken setting of the frequency range to monitor. The strain is the largest near the wall on Geogrids A and B and in the middle on Geogrids C and D. Stains of Geogrid D measured by optical fiber are small compared with other geogrid, because of break at the part of fusion splice of optical fiber. The monitoring using optical fibers shows a trend similar to that of values monitored using strain gauges. The maximum value of strain is about 1.0%, which shows that the wall is stable state.

4 CONCLUSIONS

To assess the stability of geogrid reinforced soil walls during and after construction, strain of geogrid is monitored using optical fiber. The results are: 1) Optical fiber sensor geogrid can monitor the strain of geogrid during and after construction. 2) Optical fiber sensor geogrid is effective in monitoring strain on geogrid more continuously than strain gauge and identifying points of local deformation. 3) The stability of the geogrid reinforced soil walls can be assessed from monitored strain values using the stability assessment index of geogrid. 4) Strain monitoring using optical fiber sensor geogrid is effective in precisely detecting deformations inside of reinforced backfill. 5) The geogrid reinforced soil walls subjected in this paper are stable state without abnormality, such as deformation of wall.

Thus, the monitoring method using optical fiber sensor geogrids is shown to be an effective method for assessing the stability and deformation of reinforced soil walls during and after construction and after long service. This method will be used in a number of projects and long-term monitoring will be conducted. The reliability of optical fibers will be thoroughly examined by also installing strain gauges since some data monitored using optical fibers in this study did not agree with the strain monitored by using strain gauges.

REFERENCES

Cao, J., Yokota, Y., Tatta, N. & Ito, S. 2003. The performance confirmation test of geogrid with sensor function, *Geosynthetics Engineering Journal*, Vol. 17: 207–210.

Tatta, N., Yoshida, K., Yokota, Y. & Yashima, A. 2005. Field test of the geogrid with optical fiber, *Geosynthetics Engineering Journal*, Vol. 20: 305–308.

Yoshida, K., Yokota, Y., Tatta, N., Tsuji, S. & Arai, K. 2006. Field observation of reinforced soil retaining wall of "double wall structure, *Proceedings of the 8th International Conference on Geosynthetics*, Vol. 3: 1133–1136.

Yoshida, K., Atarashi, M., Tsuji, S., Yoshida, Y. & Yashima, A. 2006. Stability evaluation of geogrid reinforced soil wall by using optical fiber sensor geogrid, *Geosynthetics Engineering Journal*, Vol. 21: 73–76.

New Horizons in Earth Reinforcement – Otani, Miyata & Mukunoki (eds)
© 2008 Taylor & Francis Group, London, *ISBN 978-0-415-45775-0*

Deformation measurements of test embankments reinforced by geocell

H. Omori
Fujiko, Co. Ltd., Japan

T. Ajiki, M. Okuyama, K. Yazawa, K. Kaneko & K. Kumagai
Hachinohe Institute of Technology, Japan

M. Horie
Tokyo Printing Ink, Co. Ltd., Japan

ABSTRACT: We constructed two full-size test embankments reinforced by a geocell. One of them was made by only loam soil which is recognized as problematic soils, while other was made by loam soil and macadam. We performed failure tests of the embankments after measuring the horizontal displacements and the heights of them for about two years. We can understand from the measurements and failure tests that the embankments made only by loam soils is stable and safe as well as other one. We can confirm the possibility of using surplus soils for the filling materials.

1 INTRODUCTION

Recently, soil slopes reinforced by geocells (Sitharam et al. 2005; Krishnaswamy et al. 2000), which are one of the three-dimensional geosynthetics materials, have been constructed gradually. If we can use local surplus soil as fillings in the geocell, the geocell is very useful for reinforcement of soil. The geocell-reinforce method with surplus soils has many advantages for natural environments and construction costs. However, in the present circumstances, because mechanical behavior of geocell-reinforced soil structure is complex and not clarified, we use macadam, which is considered comparatively good quality. Moreover, a reasonable design method of the geocell-reinforced soil structures made with the surplus soil has not been established.

We constructed two full-size test embankments reinforced by the geocell in Hachinohe which is very cold region (omori et al. 2006). The shape of these embankments is the frustum of cone. We used the macadam and the loam clay for one of them, while we used only loam clay for other one. We measured the deformation of them for about 700 days. Measured items for embankments are the horizontal displacements and the height. In addition, we made failure tests of them in order to investigate the strength and the failure mechanism. we also examined the condition of inner soils such as the moisture content. Our main purpose in this study is investigation of the possibility of valid utilization of local surplus soils as inside materials of geocells.

2 OUTLINE OF TEST EMBANKMENTS

Figure 1 shows a photograph of one of the test embankments. We illustrates a outline of the test embankments in Fig. 2. We made the shape of the test embankments

Figure 1. Test embankment in 700 days since construction.

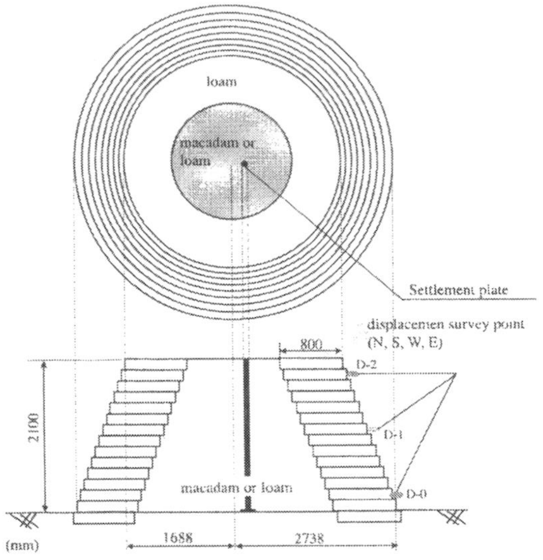

Figure 2. Outline of test embankments

the frustum of cone, in order to be isotropic state. (Radius of bottom 2738 mm, radius in top 1688 mm and height 2100 mm.) The geocell (omori et al. 2006) used in this study is made by high density polyethylene (HDPE) and is honeycomb-like continuous cells.

We constructed Embankment A by using only Hachinohe loam soil for the filling and backfill materials, while we constructed Embankment B by using macadam C-40 for the backfill. Our main purpose in this study is to examine the possibility of the employment of the surplus soil in the geocell reinforced soil wall construction method. Therefore, the loam soil that is one of the most difficult geo-materials was used for the filling materials.

3 LONG-TERM SURVEY OF TEST EMBANKMENTS

3.1 *Height of embankment*

Figure 3 shows the variation of the height of Embankment A and Embankment B for about 700 days. Variations of height both of Embankment A and Embankment B hardly change 700 days after construction and were under 4 mm. The variation of Embankment A is fewer than that of Embankment B. Moreover, we can understand from this figure that they are stable perfectly after about 300 days. We can say that even if we use the loam clay as the filling in the geocell, the height of the geocell-reinforced embankment is almost invariable.

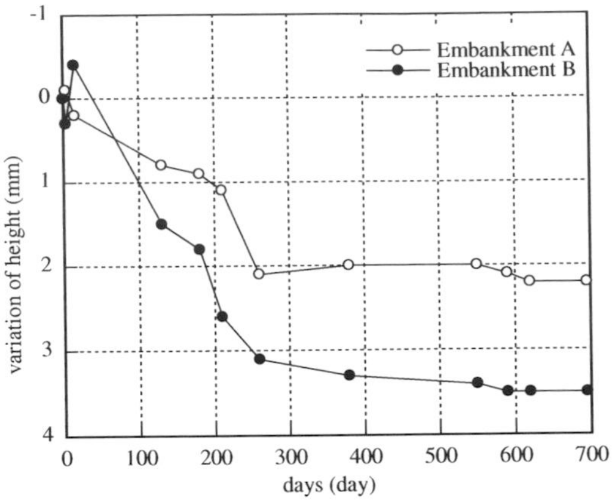

Figure 3. Variation of height of embankments

3.2 *Horizontal displacement*

Figure 4 shows the horizontal displacement results, which were measured by the electro-optical distance apparatus in measurement point D-0, D-1 and D-2 shown in Fig. 2. Figure 4 shows the relation between the lapsed days and the displacement, which was measured in four azimuths of north, south, east, and west.

The displacement both of Embankment B and Embankment A shows almost same tendency. The displacement of Embankment A is small rather than Embankment B. In about 200 days, the variation of the displacement is rapidly increase. We can consider that its reason is the first rainy season. After the first rainy season, they are very stable, although we can observe a little variation in the second rainy season that is about 550 days. In Embankment A, the horizontal displacement increases suddenly on about 200 days. On the other hand, in Embankment B, the displacement increases gradually for 200 days.

4 FAILURE TESTS OF EMBANKMENTS

4.1 *Outline of failure test*

Outline of horizontal loading for failure is shown in Fig. 5. We cut down the upper part from the half of the height of the embankments, though 1/5 part was left for failure tests. We pulled out to the horizontal direction the left part in the upper part of Embankment A by using a large-weight backhoe through the wire. The load cell of 20 tons in capacity was installed in the wire. If we assumed that unit weight of the loam soil is 2.0 g/cm^3, own weight of the left upper part of Embankment A is about 6.8 tons. When the horizontal load reached prescribed value, we measured a

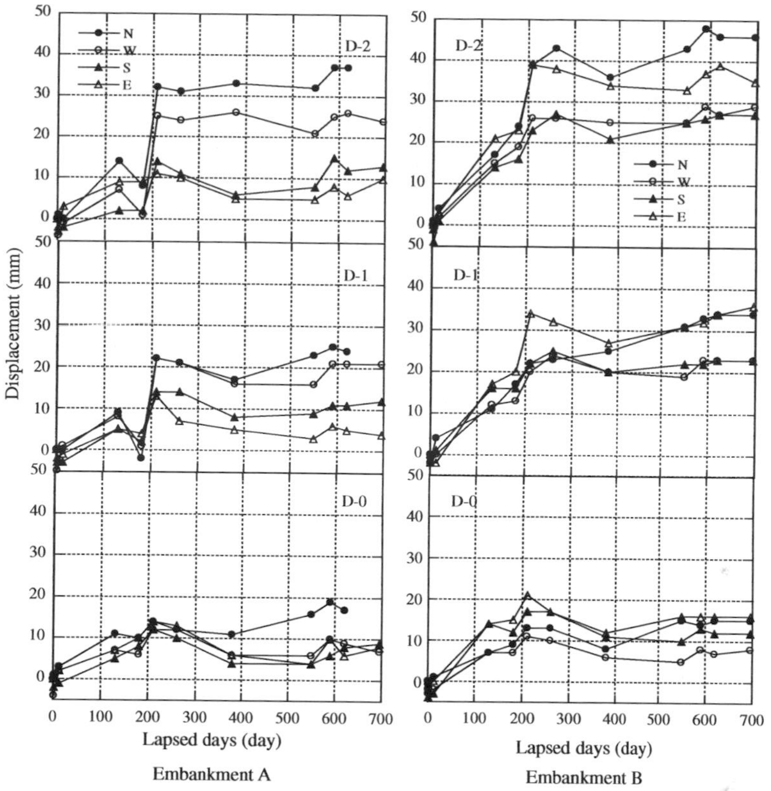

Figure 4. Results of displacement measurement

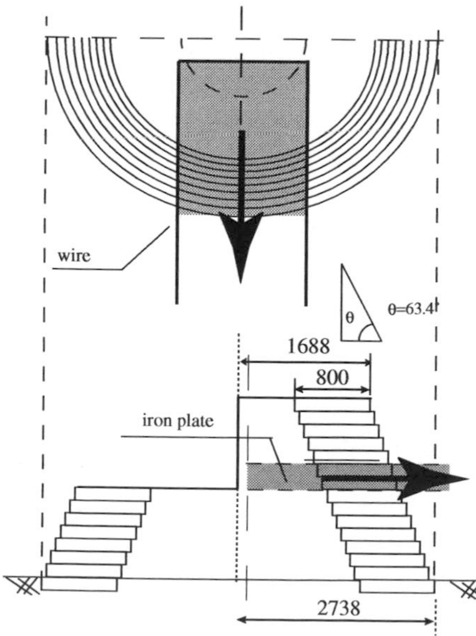

Figure 5. Outline of horizontal loading

horizontal displacement and observed the condition of Embankment A.

We also measured the moisture content in both Embankment A and Embankment B. The soil samples were gathered at the failure tests carefully. We put the samples in airtight containers, and carried them to our laboratory. We measured the moisture content by following JGS 0122.

4.2 *Results*

The situations of the failure test are shown in Fig. 6. Figure 6(a) shows the initial situation of the test for Embankment A. Figure 6(b) shows the state at the failure, and Figure 6(c) shows the situation after the failure. We can observe that Embankment A failed by a sliding at a boundary between lower and upper part.

We shows the relationship between the horizontal force and the horizontal displacement in Fig. 7. The horizontal displacement is very small until the horizontal force reaches 3 tons. The displacement begins to increase gradually when the force exceeds 3 tons, and the displacement increases rapidly when 5 tons are exceeded. Embankment A failed at the horizontal

(a) Preparation of failure test

(b) Situation at the failure

(c) Situation after the failure

Figure 6. Situation of failure test (Embankment A).

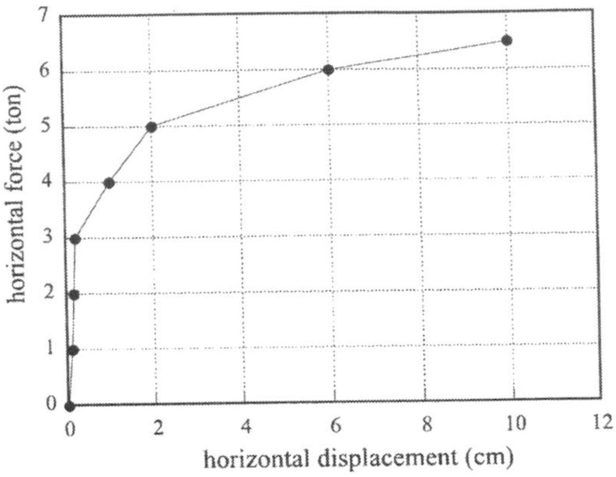

Figure 7. Relationship between the horizontal force and displacement.

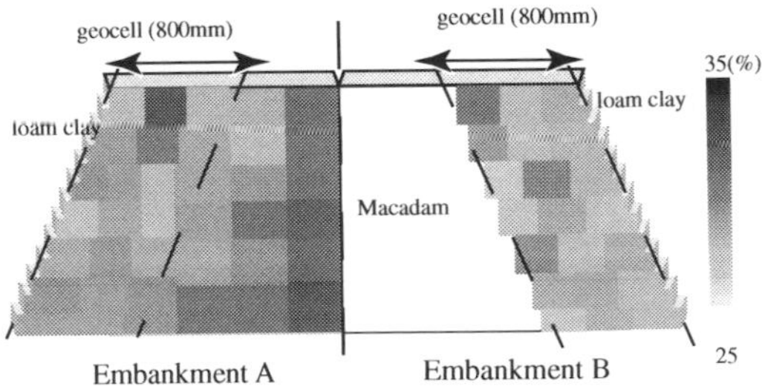

Figure 8. Distribution of moisture content in embankments

force 6.5 tons, which is a little smaller than the self weight.

We show the distribution of moisture contents in the Embankment A and B in Fig. 8. The moisture contents of Embankment B, which is made by using macadam for the backfill material, is somewhat smaller than that of Embankment A. In Embankment A, the moisture contents of the center part are larger, and maximum value is almost 34%. We confirmed that the drainage was better in using macadam for the backfill. However, we can say that the moisture content doesn't increase perilously even when we made the embankment by using only the loam clay, which is known as one of the difficult soil.

5 SUMMARY

In this paper, we constructed two full-size test embankments reinforced by the geocell. The test embankments were made by using loam clay and macadam for filling and backfill materials. We performed the failure tests of them after long term deformation survey. We can understand from these experimental results that the embankments made only by loam soils is stable and safe as well as other one. In addition, we can confirm the possibility of using surplus soils for the filling materials.

REFERENCES

Krishnaswamy, N. R., K. Rajagopal, and L. G. Madhavi (2000). Model studies on geocell supported embankments constructed over soft clay foundations. *ASTM Geotechnical Testing Journal Vol. 23*, 45–54.

Omori, H., M. Shimada, K. Kaneko, M. Horie, and K. Kumagai (2006). Field observation and deformation measurements of geo-cell reinforced retaining walls. *Geosynthetics Engineering Journal 21*, 23–30, in Japanese.

Sitharam, T. G., S. Sireesh, and D. S. K. (2005). Model studies of a circular footing supported on geocell-reinforced clay. *Canadian Geotechnical Journal Vol. 42, No. 2*, 693–703.

New Horizons in Earth Reinforcement – Otani, Miyata & Mukunoki (eds)
© 2008 Taylor & Francis Group, London, ISBN 978-0-415-45775-0

Full-scale experiments on bend of pressure pipeline using geogrid

Y. Sawada, T. Kawabata & K. Uchida
Faculty of Agriculture, Kobe University

A. Totsugi
Development Section, Taisei Kiko Co., LTD

J. Hironaka
Research and Technology Development Division, Mitsui Chemicals Industrial Products, LTD

ABSTRACT: In a bend of pressure pipeline, thrust force is generated. Commonly concrete block is set up on the bend in order to resist the thrust force. However, such heavy concrete block becomes a weak point during earthquake. Therefore a lightweight thrust restraint using geogrids and an anchor plate was suggested in previous study. In the present study, full-scale experiments were conducted using a pipeline (ϕ300) to verify the effect for the proposed method in actual size. As the results, the lateral movement of bend in case of the proposed method was reduced in comparison with a bend without the restraint. In addition it was clarified that the effect was depended on the stiffness and the length of geosynthetics.

1 INTRODUCTION

In a bend of pressure pipeline, thrust force is generated depending on the pressure level and the angle of the bend. Commonly a concrete block is placed on the pipe bend in order to resist the thrust force (M.A.F.E., 1988). However it was reported that thrust block on the bend caused damage of pipeline during earthquake since it moved largely due to inertia (Mohri et al., 1995). In addition, it can be expected that such heavy concrete block induces a differential settlement in pipeline on soft ground.

For these issues, new lightweight thrust restraint with geogrids and anchor plate was suggested by Kawabata et al. (2004). In the new method, geogrids and anchor plate were connected with the pipe bend as shown in Fig. 1. In addition, Kawabata et al. (2004) conducted the lateral loading tests using the model pipe (ϕ90) in order to clear the effect of the proposed thrust restraint. As the results, it was clarified that the lateral resistance in case of the proposed method increased by approximately 60% comparing with a model pipe without the restraint.

In this study, large-scale experiments were conducted using a pipeline (ϕ300) in order to clear the effect of the new method in actual size. In addition, the influences of the stiffness and the length of geogrid on the lateral resistance were discussed.

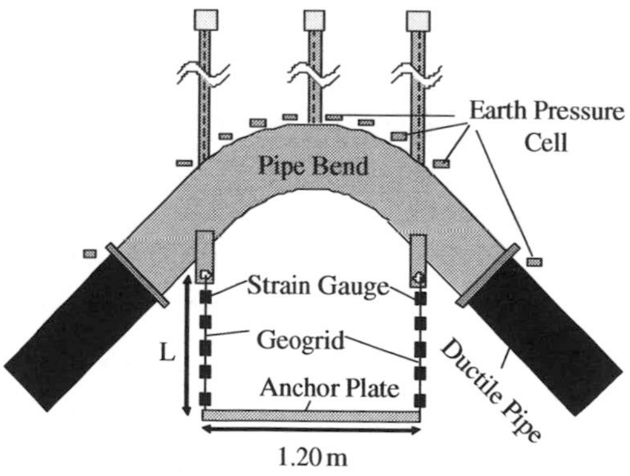

Figure 1. Test Setup.

2 EXPERIMENTAL PROGRAM

2.1 *Test procedure and materials*

Fig. 1 shows a test setup. A series of experiments described in this paper were carried out in a pit having the width of 5.4 m and the length of 8.4 m. In the pit, a bed having the thickness of 1 m was prepared and a pipeline having a diameter of 300 mm was set on the bed. The pipeline was consists of a bend (90°) and

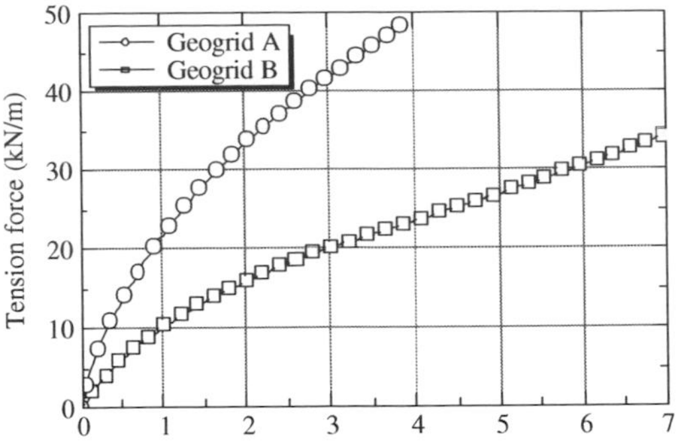

Figure 2. Results of tensile tests of geogrids.

4 short ductile iron pipes. After setting the pipeline, geogrids and an anchor plate were connected with the bend and were backfilled up to 0.6 m. Two types of geogrids (Geogrid A and Geogrid B) were used and results of tensile tests are indicated in Fig. 2. As shown Fig. 2, the stiffness of Geogrid A is larger than that of Geogrid B. In addition, a rigid steel plate (having the width of 1200 mm and the height of 300 mm) was used as the anchor plate. Ground was compacted by a vibration compactor every layer of the thickness of 0.15 m and the average dry density of ground was 1.75 g/cm^3. As backfill materials, screenings were used and its average particle size was 1.2 mm. The internal friction angle and the cohesion obtained from direct shear tests were about 38 degrees and 0 kPa respectively. After backfilling, internal water pressure was loaded using hydrostatic pump.

2.2 *Measurements*

In order to measure the lateral displacement of the bend, pulley type displacement transducers were used in front of the bend as shown in Fig. 1. In addition, loaded internal pressure was measured using a water pressure cell. Furthermore, in order to investigate the horizontal earth pressure acting on the bend, ten earth pressure cells were installed on the center level of the bend as shown in Fig. 1. In cases of using geogrid, strain gauges were used to measure tensile strain in geogrids in extensional and transverse direction.

2.3 *Cases of experiments*

Cases of experiments are indicated in Fig. 3. In Case-A, the pipe bend was buried without the thrust restraint and in Case-B, Case-C, Case-D and Case-E the pipe bend was connected to geogrids and an anchor plate. In Case-B and Case-C, high stiffness geogrids (Geogrid A in Fig. 2) were used and in Case-C and Case-E, low

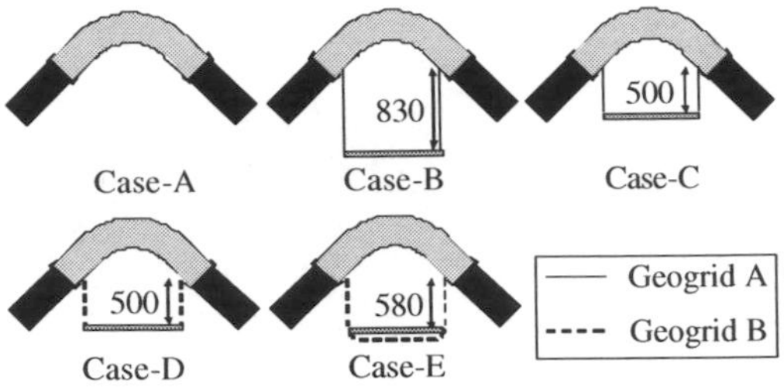

Figure 3. Cases of experiments.

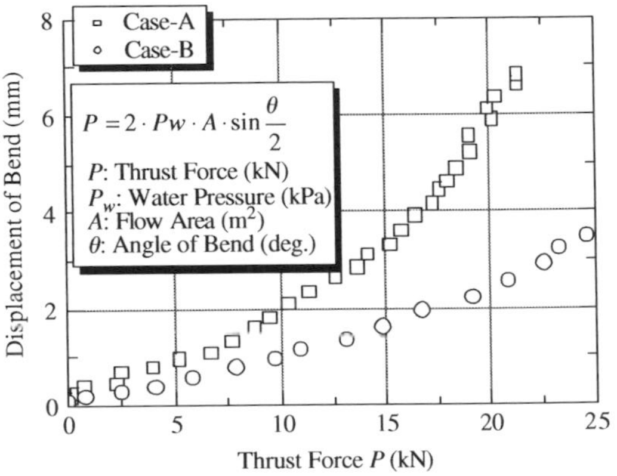

Figure 4. Relationships between thrust force and displacement of bend.

stiffness geogrids (Geogrid B) were used. In addition, geogrids were fixed on the side of the anchor plate in Case-B, Case-C and Case-D. On the other hand, geogrid was set as wrapping the ground and the anchor plate in Case-E as shown in Fig. 3.

3 TEST RESULTS AND DISCUSSION

3.1 *Lateral displacement of pipe bend subjected to thrust force*

Fig. 4 shows relationships between the thrust force and the lateral displacement in Case-A and Case-B. The thrust force P was calculated from internal water pressure, bending angle and cross-sectional area of bend.

From Fig. 4, it is found that the displacement of the bend increases with the thrust force in both cases. In addition, the displacement is about 6 mm for the thrust force of 20 kN in Case-A. On the other hand, in Case-B, the displacement is only 2.5 mm for thrust force of 20 kN. Therefore it is clarified that the lateral movement of the bend is reduced by proposed method.

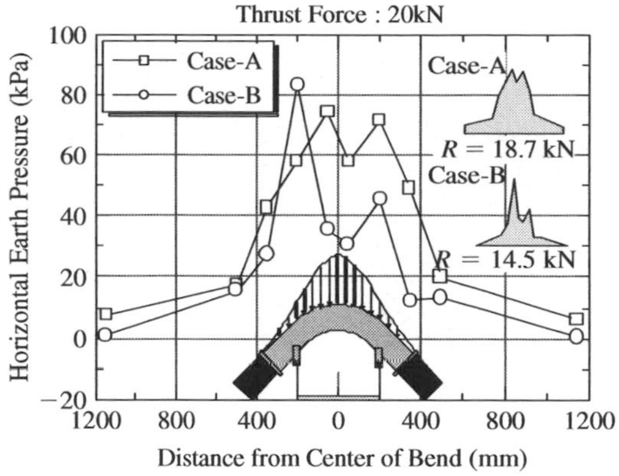

Figure 5. Distributions of earth pressure acting on bend.

3.2 *Earth pressure acting on pipe bend*

A passive earth pressure and an active earth pressure act on a bend when the bend is laterally moved due to thrust force. Commonly the active pressure can be neglected since it is extremely small. In this study, ten earth pressure cells were installed as shown in Fig. 1 in order to measure passive earth pressure acting on the bend. Note that eight earth pressure cells were placed in front of the bend and two pressure cells were placed in front of the straight short pipes.

The earth pressure distributions for thrust force of 20 kN are shown in Fig. 5. It is found that the passive earth pressure around the center of the bend is relatively large and the distribution is not uniform. This distribution is close to the results in 3 D finite element analyses presented by Fujita et al. (1994). However, in the current design (M.A.F.E, 1988), the earth pressure acting on the bend is assumed as a uniform distribution. Thus, it is thought that the further examinations are required to improve the design method.

In addition, from in Fig. 5, it is found that the resistance force calculated using the earth pressure distributions is 18.7 kN in Case-A and the value is close to the thrust force (20 kN). On the other hand, in Case-B, the resistance force is 14.5 kN and the value is smaller than that in Case-A. Therefore it can be considered that other factors (i.e. geogrid and anchor plate) contribute to the lateral resistance in Case-B.

3.3 *Strain distribution in geogrids*

Tensile strains are generated in geogrids when the bend subjected to thrust force is moved laterally. In Case-B, Case-C, Case-D and Case-E, strain gauges were equipped on geogrids in extensional and transverse direction respectively.

Fig. 6 shows tensile strain distributions in geogrids in extensional direction at the displacement of 7 mm.

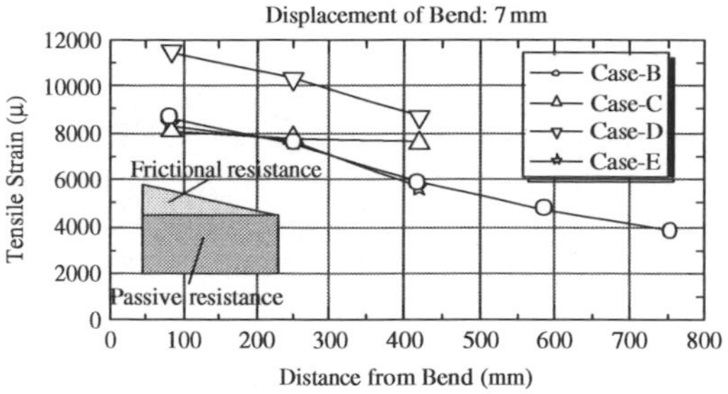

Figure 6. Tensile strain distributions in extensional direction.

These results are averaged in both side geogrids. From Fig. 6, it can be seen that the maximum strain is generated at the front position of geogrids and the minimum strain is generated at the end of geogrids. This behavior is similar to results of pull-out tests of geogrids in dry sand. However in general pull-out tests, tensile strain is not generated in the end of geogrids. These strain distributions are close to strain distributions in pull-out test under the condition, in which the end of geogrid was fixed (Nakamura, T. et al., 2003). It is thought that the end was fixed by the anchor plate in the proposed method. Therefore it can be understood that the tensile strain at the end of geogrid is generated due to the passive resistance acting on the anchor plate. Judging from these results, it can be assumed that tensile strain in geogrid is generated due to two components of pull-out resistance and passive resistance. The tensile strain from pull-out resistance is indicated with a triangle and the strain from passive resistance is constant and indicated with a rectangle.

Furthermore, from Fig. 6, it can be seen that the inclination of the strain distribution in Case-C is smaller than that in other cases. It can be considered that the pull-out resistance was small since the density of ground around geogrid was small. In addition, comparing Case-D and Case-E, the tensile strain in Case-E is smaller than that in Case-D. This result is considered to be due to difference of the installation of geogrids.

Fig. 7 shows tensile strains in geogrid in the transverse direction. Generally it is thought that strain at deep position is larger than that at shallow position since the horizontal earth pressure acting on geogrids increases with the depth. However, this tendency can not be seen in Fig. 7. For this reason, it can be considered that the height of geogrid was short (300 mm) and the variation of earth pressure in direction of depth was relatively small. Thus it is thought that variation of strain in vertical direction can be neglected in case of short geogrid in practical design.

Fig. 8 shows tensile strains in geogrid behind the plate in Case-E. From Fig. 8 it is found that tensile strains are approximately 2000 μ. It can be understood

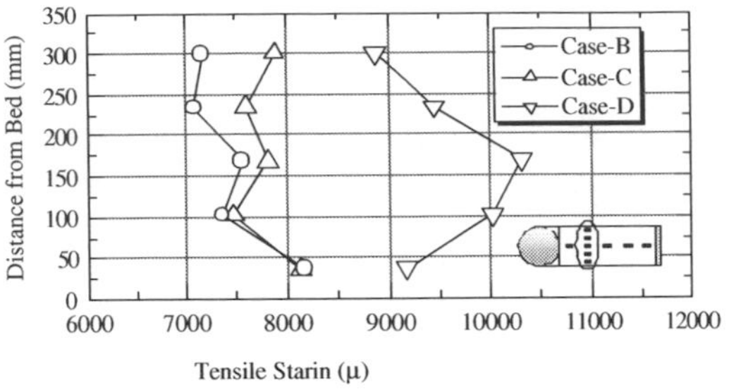

Figure 7. Tensile strain distributions in direction of depth.

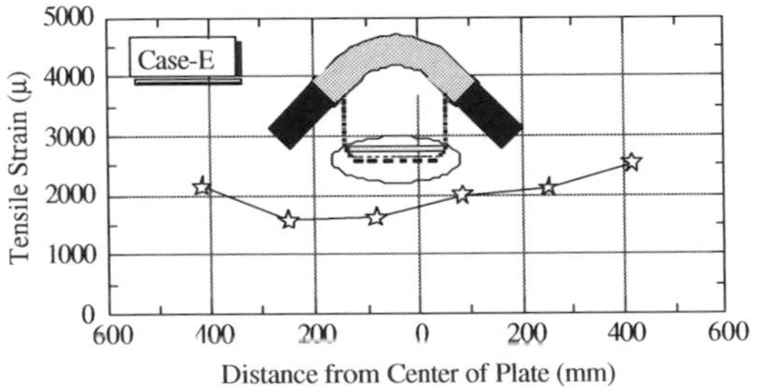

Figure 8. Tensile strain distribution behind anchor plate.

that, as shown in Fig. 6, strains in side geogrids is reduced due to the tensile strain behind the plate.

3.4 *Incremental resistance*

In case of the proposed method, it was clarified that geogrid and the anchor plate contributed to the lateral resistance as shown in Fig. 6. If the total resistance provided by the proposed method is defined as incremental resistance, it is equivalent to the tensile force in the front of geogrid. Thus, the incremental resistance can be estimated by multiplying the strains by the stiffness of geogrids per unit length. The stiffness of geogrids can be obtained in Fig. 2.

Fig. 9 shows variations of the incremental resistance with the horizontal displacement of the bend in Case-B, Case-C, Case-D and Case-E.

From Fig. 9, it is found that the increment resistance increases with the horizontal displacement. In addition, comparing 4 cases, it can be seen that the increment resistance in Case-D and Case-B is much larger than that in Case-C and Case-E. From this result, it can be consider that the most important factor for the incremental resistance is the stiffness of geogrid.

In addition, the incremental resistance in Case-B is as large as that in Case-C although these cases are different in the length of geogrids. As shown in Fig. 6, pull-out resistance in Case-B is larger than that in Case-D since the contact area between geogrid and ground is large. On the contrary, with respect to the passive resistance acting on the anchor plate, Case-B is smaller than Case-C since geogrid is long and it is easy to be deformed. Therefore it can be thought that there is an optimum length of geogrid for the maximum resistance.

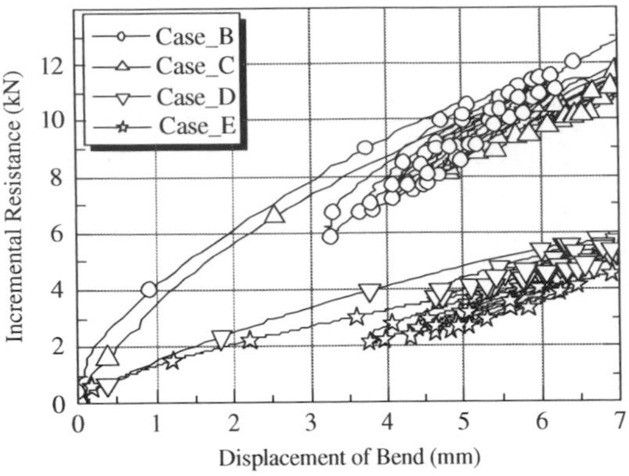

Figure 9. Relationships between displacement of bend and incremental resistance.

Further, it is found that the incremental resistance in Case-E is the smallest of all cases. The result indicates that passive resistance was slight since geogrid behind the anchor plate was elongated as shown in Fig. 8.

From the above discussion, important proposals on the stiffness and the length of and the installation of geogrids are found. I) The large incremental resistance can be expected in case of high stiffness geogrids since the large passive resistance is provided by the anchor plate. II) The large pull-out resistance can be expected in case of long geogrids. However the passive resistance acting on the anchor plate can be reduced in some degree. Therefore it is important to determine the optimum length of geogrids. III) In case of setting of geogrid as Case-E, the incremental resistance can be reduced due to the elongation of geogrid behind the anchor plate.

4 CONCLUSIONS

In this paper, large-scale tests for a buried pipe bend under internal water pressure were conducted in order to clear an effect of a proposed thrust restraint using geogrids and an anchor plate. Results in these tests are summarized as described below.

1. Horizontal displacement of the pipe bend was reduced in case of proposed method. Therefore the effect of proposed method was verified at actual size.
2. Tensile strain distributions along geogrids in extensional and transverse direction were discussed. As the results, it was found that the strain distribution

was the trapezium shape. In addition, the triangle part and the rectangle part were corresponding to the tensile strain from the pull-out resistance along the geogrids and the passive resistance acting on the anchor plate respectively. Furthermore it was cleared that the variation of strain distribution in vertical direction was slight and in case of small diameter bends, this variation can be ignored.

3. Incremental resistance due to geogrids and an anchor plate was calculated from results of tensile strains. As the results, it was found that the most effective factor for the lateral resistance was the stiffness of geogrid. In addition, it was indicated that there was the optimum length of geogrid for maximum lateral resistance. Therefore it is important for detail design to determine the optimum length of geogrid. Furthermore it was clarified that the incremental resistance was reduced due to the elongation of geogrid behind the plate in case of setting geogrid as wrapping the ground with the anchor plate.

REFERENCES

Fujita, N., Kawabata, T. and Mohri, Y. 1994. Behavior of the Bend Corner in Buried Pipeline under Internal Pressure. *Proceedings of the 29th Japan National Conference on Soil Mechanics and Foundation Engineering*: 2007–2008.

Kawabata, T., Uchida, K., Ling, H. I., Nakase, H., Sawada, Y., Hirai, T. and Saito, K. 2004. Lateral Loading Tests for Buried Pipe with Geosynthetics. *Proceedings of Geo-Trans 2004*(1): 609–616.

Ministry of Agriculture, Forestry and Fisheries. 1988. *Design Standard for Pipeline.*

Mohri, Y., Yasunaka, M. and Tani, S. 1995. Damage to Buried Pipeline Due to Liquefaction Induced Performance at the Ground by the Hokkaido-Nansei-Oki Earthquake in 1993, *Proceedings of First International Conference on earthquake Geotechnical Engineering*, IS-Tokyo: 31–36.

Nakamura, T., Mitachi, T. and [illegible], I. 2003. Estimating Method for The In-Soil Deformation Behavior of Geogrid Based on The Results of Direct Shear Test. *Soils and Foundations* 43(1): 47–57.

New Horizons in Earth Reinforcement – Otani, Miyata & Mukunoki (eds)
© 2008 Taylor & Francis Group, London, ISBN 978-0-415-45775-0

Simplified design method for reinforced slopes considering progressive failure

J.-C. Jiang & T. Yamagami
Department of Civil and Environmental Engineering, The University of Tokushima, Japan

S. Yamabe
Geotechnical Division, CTI Engineering Co., Ltd., Japan

ABSTRACT: A limit equilibrium-based design method for reinforced slopes considering progressive failure is developed. A local safety factor is first defined at the base of each slice so as to describe local failure along a slip surface. An incremental approach is constructed to solve the equations and an empirical interslice force function is introduced to enhance the efficiency and robustness of the solution procedure. Then, a new design scheme for reinforced slopes is proposed. In this scheme, an optimization technique is contrived to determine tensile resistances of reinforcement elements that will be required to provide an adequate factor of safety of reinforced zones. The proposed procedure can be used to determine suitable layout of reinforcements and required tensile strength of reinforcing material. Finally, results obtained from an example are presented and discussed to provide a guideline for practical design of reinforced slopes using the proposed method.

1 INTRODUCTION

In general, slip surface development in an actual slope is a progressive phenomenon. This is especially true for reinforced slopes that contain foreign materials (e.g., Huang et al. 1994). This behavior of slope failure cannot be simulated using traditional limit equilibrium methods as they are based on a single value factor of safety analysis (Yamagami et al., 2001).

A limit equilibrium-based method has been developed by the authors (Yamagami et al., 2001, 2002) to analyze the stability of unreinforced slopes considering progressive failure. In this paper, an empirical interslice force function is introduced to enhance the efficiency and robustness of the solution procedure. Then, the method by Yamagami et al. (2001) is extended so as to present a design scheme for reinforced slopes. In this scheme, an optimization technique is contrived to determine tensile resistances of reinforcement elements that would be required to provide an adequate factor of safety of reinforced zones. The proposed procedure can be used to determine suitable layout of reinforcement and required tensile strength of reinforcing material.

2 THEORY OF STABILITY ANALYSIS

2.1 *Equations*

Figure 1 shows forces acting on an infinitesimal slice. The symbols in this figure are taken from Yamagami et al. (2001). A local factor of safety was defined at the base of each slice (Eq. (1)) and a relationship between interslice normal and shear forces (Eq.(2)) was assumed.

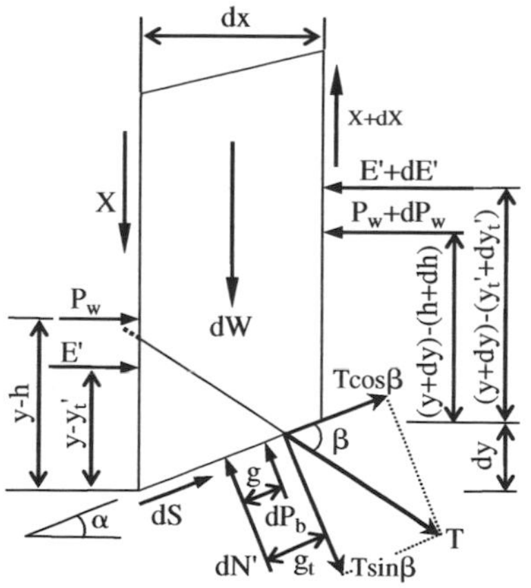

Figure 1. Forces acting on a typical slice.

$$F = \frac{1}{dS}\left[(c'dx\sec\alpha + dN'\tan\phi) + T(\cos\beta + \sin\beta\tan\phi')\right] \quad (1)$$

$$X = \lambda f(x)E \quad (2)$$

where T is tensile force due to reinforcement, λ is an unknown constant, and $f(x)$ is an insterslice force function. For a sliding mass divided into n slices, the basic equations can be obtained using a similar derivation to the Morgenstern-Price's procedure. They are two recurrence relations, as shown bellow.

$$E_i = \frac{1}{L_i + K_i b_i}\left[E_{i-1}L_i + \frac{1}{2}N_i b_i^{\,2} + (P_i + R_i)b_i\right] \qquad (3)$$

$$\begin{aligned} M_i &= M_{i-1} + \int_{x_{i-1}}^{x_i} E[\lambda f(x) - A]dx - T_i \sin\beta \cdot g_t \\ &= M_{i-1} + \int_0^{b_i} \frac{[\lambda f(x) - A_i]\left[E_{i-1}L_i + P_i x + \frac{1}{2}N_i x^2\right]}{L_i + K_i x}dx \\ &\qquad - T_i \sin\beta \cdot g_t \end{aligned} \qquad (4)$$

where $b_i = x_i x_{i-1}$ ($i = 0, 1, 2, \cdots, n$), x_i, x_{i-1} are horizontal coordinate of the left and right side of slice i, respectively, $M_i[= E_i(y_{ti} - y_i)]$ is a moment of E_i about the rightmost point of the base of slice i. Note that tensile force T_i in Equation (4) is equal to zero for slices where reinforcement is not included.

2.2 *Solution procedure*

Suppose the solution process has reached the (i-1)th slice, starting from the first one, and hence F_{i-1}, E_{i-1}, etc. have become known. By assuming a value for the local factor of safety F_i of slice i, E_i can be calculated from Equation (3). Substituting E_i obtained in such a way will not usually satisfy the equality, because the F_i value used is assumed. Equation (4) is then regarded as a function only of F_i and is solved for it iteratively by, for example, the Newton-Raphson method. Note that during this iteration E_i is kept at the value obtained immediately before.

Returning to Equation (3) with F_i obtained above, we calculate a new value for E_i, and check whether or not Equation (4) is satisfied using the new value of E_i. This process is repeated until the equality of Equation (4) becomes valid within a prescribed tolerance. Then, we can proceed to the next slice (i + 1) and carry out the same process as above. The complete solution must satisfy the boundary condition: $E = \overline{E}_n$, in which $\overline{E}_n$ is a prescribed value at the end point of the slip surface; usually this is zero.

2.3 *Introduction of an interslice force function*

Use of an interslice force function was advocated by Morgenstern and Price (1965). It was suggested that integration of the interslice shear and normal forces acting along vertical planes through the soil mass could provide the forces necessary for an appropriate interslice function. Fan et al. (1986) used an elastic theory approach (i.e. finite element method) to compute the normal and shear stresses along vertical planes through a sliding mass. The stresses were then integrated vertically and the ratio of the shear to normal interslice forces was computed. The interslice force functions for simple slope geometries were found to be bell-shaped. Based on the examination of several hundred interslice slice functions obtained from the linear finite element stress analysis, Fan et al. (1986) proposed a generalized equation for expressing the interslice force function:

$$f(x) = Ke^{(-d^m \omega^m)/2} \qquad (5)$$

where K is the magnitude of interslice force function at mid-slope (i.e., maximum value), d is a variable defining the inflection points near the crest and toe of a simple slope, m is a variable specifying the flatness or sharpness of curvature of function, and ω is the dimensionless position relative to the midpoint of each slope. Fan et al. (1986) presented the charts which can be used to determine values of the above-mentioned parameters for a given slope.

2.4 *Optimization of y_t*

An analysis using the Morgenstern-Price method (1965) and the $f(x)$ defined by Equation (5) yields a set of values of λ and y_t. The λ, $f(x)$ and y_t values so obtained are used to solve Equations (3) and (4). It has been shown that for simple slope geometries solutions which meet the boundary condition $E = \overline{E}_n$ can usually be reached. For cases where the solution procedure does not converge, Yamagami et al.(2002) suggested an optimization problem in which E_n is considered to be a function of λ and y_t:

$$\left|E_n - \overline{E}_n\right|^2 = \mathrm{Fun}[\lambda, y_{t1}, y_{t2}, \cdots, y_{tn-1}] \rightarrow \mathrm{Minimize}(=0) \qquad (6)$$

Solving Equation (6) will lead to a set of appropriate values of λ and y_t (Yamagami et al., 2002).

2.5 *Load incremental procedure (LIP)*

A load incremental procedure proposed by Yamagami et al. (2002) is an effective approach for solving Equations (3) and (4). In this procedure, the self-weight and subsequent surface load is subdivided into several increments. The solution for each step of loading is obtained, and the incremental process is repeated until the total load has been reached. During the solution process, if a local failure takes place, the local factor of safety for that region will be kept at unity in subsequent steps. Note that the definition of load increments in the proposed method differs slightly from that for the usual finite element stress deformation analysis as seen in the following.

The self-weight W and external load P is divide into N and M increments, respectively:

$$W = \sum_{k=1}^{N} \Delta W_k ; \qquad P = \sum_{k=1}^{M} \Delta P_k \tag{7}$$

Then, the following load increments are defined:

$$W_i = \sum_{k=1}^{i} \Delta W_k \qquad \text{where } i = 1, 2, \ldots, N \tag{8a}$$

$$P_j = W + \sum_{k=1}^{j} \Delta P_k \qquad \text{where } j = 1, 2, \ldots, M \tag{8b}$$

Note that $W_N = W$ and $P_M = W + P$. The analysis procedure using the LIP is shown as follows (Yamagami et al., 2002):

1) The solution procedure in Section 2.2 is carried out for each incremental load, starting from W_1.
2) Suppose that a local safety factor less than unity has appeared for the first time at the base of a slice when the load at ith loading step, W_i, is employed.
3) An iterative calculation is done so that the factor of safety for the locally failed slice becomes equal to unity under the load W_i. As a result of this calculation, if a slice other than the slice mentioned above has had a factor of safety less than unity, then the calculation must be repeated until the factor of safety of each slice is not less than unity.
4) The procedure is continued with next load W_{i+1}; however, in the subsequent process the factors of safety are known (=unity) for all the slices which have already failed in the preceding steps.
5) Hereafter, the process described above can be repeated using an incremental load one after another. The required solution is provided by the results obtained at the final load step, i.e. using load W_N, or P_M if an external load is applied. At the final step of loading, the factor of safety of each slice in local failure zone must be equal to unity.

In this way the LIP procedure continues with gradually increasing loads, and once a failed zone occurs the local factor of safety for the region will be kept at unity in the subsequent process.

It should be pointed out that softening of soil can be easily considered and handled in a LIP, and an overall factor of safety can be computed from the ratio between the sum of the mobilized shear forces and the sum of the available shear strengths along the slip surface (Yamagami et al. 2002).

3 DESIGN METHOD FOR REINFORCED SLOPES

3.1 *Basic idea*

With predetermined tension forces of reinforcements the procedures described in the foregoing section can be used to yield the local factors of safety along a given slip surface in reinforced slopes. This also means, however, that the reinforced slope problems cannot be solved without knowing the tensile forces of the reinforcement elements. This is a weak point associated with the limit equilibrium methods including the proposed approach when solving the stability problem of reinforced slopes. Unlike pretension type anchor works, it is virtually impossible to know the mobilized tensile forces of passive reinforcements such as nails or steel bars in a reinforced slope, without resorting to some numerical means like the finite element method. In practice, therefore, designers often simply assume values for the reinforcement forces in advance when using a limit equilibrium approach. With regard to this, a novel method was presented by Yamagami et al. (2002) that could be used for design of reinforced slopes in practice. For the completeness, this approach is presented by using a hypothetical situation in Figure 2.

Suppose that the slope shown in Figure 2 (a) is potentially unstable, having a local failure zone along the slip surface. Suppose also that reinforcement elements (nails) are inserted passing through the bases of all slices in the failure zone, as shown in Figure 2(b). Note that nails may also be further installed in the other portions of the slip surface out of the local failure zone if necessary. And conversely they do not necessarily have to cover the whole of the local failure zone.

Here we introduce the following two very important premises:

I) Even after nails are installed, failure of the slope, if it occurs, takes place initially from the reinforced zone, never from some part of the unreinforced zones.
II) It is possible to calculate mobilized tensile forces of the nails at an inception of the failure of the reinforced zone.

Then, the slope never fails along the slip surface if the nails can actually resist the mobilized tensile forces mentioned in the second premise II) provided that the first premise I) is ensured.

The idea briefly above described enables us to design reinforced slopes rationally; in the following its details will be discussed focusing on an optimization scheme that will be contrived to realize the two premises.

First, the solution procedure described in Section 2.2 is performed to compute the local factors of safety with the constraints that the factor of safety for each of the reinforced slices must be equal to unity. How to solve this problem is described later. Then, it is necessary to make sure that in the obtained result local safety factors have become all greater than unity for the remaining unreinforced zones. It should be noted that only if this condition is satisfied, premise I) can be realized. And if not, i.e. somewhere in the unreinforced region if there is at least one slice whose safety factor is less than 1.0, the analysis has to be performed again by changing the arrangement of nails. A detailed explanation for this is also given later. Next, take note of the tensile forces mobilized in the nails. These tensile forces, when actually applied to the reinforcements in Figure 2(b), have functions to render the reinforced region to be in a limit equilibrium state having the factors of safety of just unity, and to render the unreinforced region to be in a stable state with the factors of safety greater than unity. Therefore, if the designer employs reinforcement materials whose strengths are sufficient to resist the tensile forces, then the slope would never fail.

As seen from the above discussion, the current problem can eventually be condensed into how to establish an approach which substantiates premises I) and II) to identify the tensile forces corresponding to the factors of safety of unity for the reinforced zone. To this end, Yamagami et al. (2002) have contrived an optimization approach which is explained below.

3.2 *Layout and required strength of reinforcements*

The proposed approach by Yamagami et al. (2002) consists of two stages.

First stage

1) Stability analysis of a (potentially unstable) slope without reinforcement is performed in terms of the procedure described in Section 2.2, and thus local failure zones where local factors of safety are lower than unity are found.
2) A reinforcing element is installed at each of the slices within the failure zones (and in other places along the slip surface if necessary). Tensile forces T_i ($i=1, 2, \cdots, M$), where M is the number of reinforcements installed, are evaluated by solving the following optimization problem (Yamagami et al., 2002):

$$U = U(T_i) = \sum_{i=1}^{M}(F_0 - F_i)^2 \quad \text{minimize } U(\rightarrow 0) \qquad (9)$$

where F_0 is a target value of local factors of safety taken to be unity for all slices with reinforcements along the slip surface; F_i ($i=1, 2, \cdots, M$) is the calculated local factor of safety.

Values for T_i ($i=1, 2, \cdots, M$) obtained by solving the optimization problem shown in Equation (9) make the local factors of safety of all nailed slices equal to the target value (i.e., unity) In addition, T_i values should be those which yield local safety factors greater than unity for the unreinforced zones in order to realize premise I) requiring the failure to start at the reinforced zones.

3) If the values of local factors of safety of all unreinforced slices are greater than unity in the solution of the optimization problem, the result obtained is regarded as the required solution that satisfies premises I) and II) in practice. The condition that the factors of safety must be greater than unity for unreinforced regions is usually satisfied because reinforce-ments are inserted covering the local failure zones for the unreinforced state. Nevertheless this condition may not be satisfied though it is rare. A situation in which the condition is not satisfied implies that in the unreinforced regions there exist some slices whose factors of safety are smaller than or equal to unity. Thus, if this is the case, there is no other choice but to install additional reinforcements for the failed slices and repeat steps 2) to 3) until the required solution is obtained.

Second stage

A factor of safety is defined for each of the reinforcements as follows:

$$F_{sr,i} = T_{f,i} / T_i \ (i = 1, 2, \cdots, M) \qquad (1)$$

where T_i is the computed tensile force, $T_{f,i}$ is the available tensile strength of each reinforcement material, and $F_{sr,i}$ = the factor of safety for each of the reinforcements. The stability of the slope (along the slip surface) is ensured if the design is conducted using sufficient values for $F_{sr,i}$.

While no discussion on the overall factor of safety has been made so far, actually it may be necessary to investigate from a viewpoint of the overall factor of safety. That is, we must occasionally investigate further when the overall factor of safety that has been obtained at the end of the first stage is not enough in its magnitude. Even in such a case, theoretically failure is not presumed to occur as long as sufficient values for $F_{sr,i}$ are assigned. However, an attempt to realize the values for $F_{sr,i}$ may require impractically high strength for the reinforcement material, indicating a failure in design. For this situation, the problem can be easily resolved by adding reinforcement to the slice with the minimum factor of safety in the unreinforced regions and by carrying out steps 2) to 3) again. More detailed explanation regarding this matter will be given in the following section based on an example problem.

4 EXAMPLE PROBLEM

A 7 m tall fill slope (embankment) with an inclination of 1:1.2 is designed using the proposed method. The

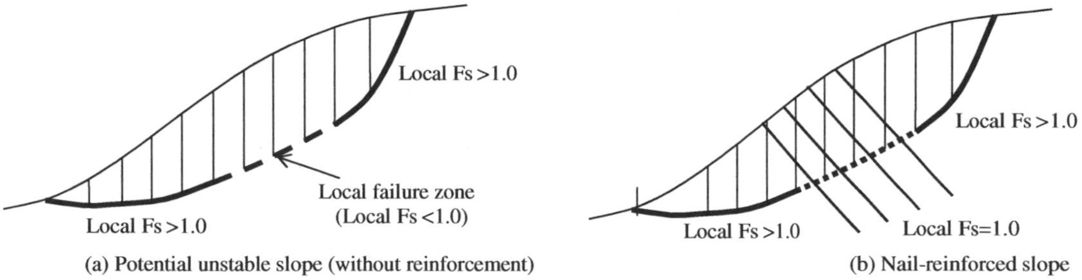

Figure 2. A schematic slope without and with reinforcement elements.

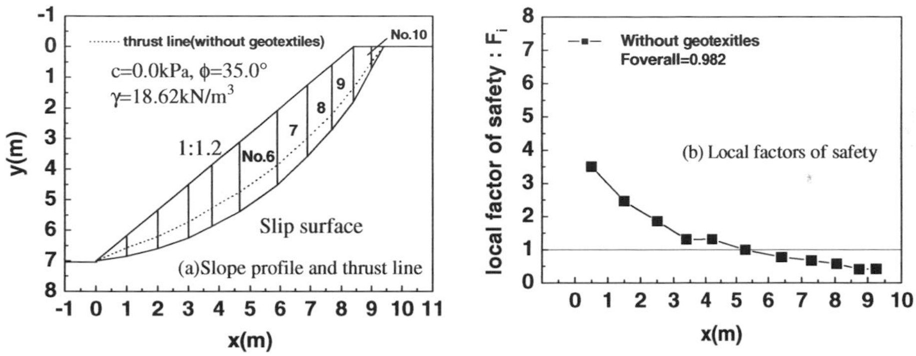

Figure 3. Example fill slope.

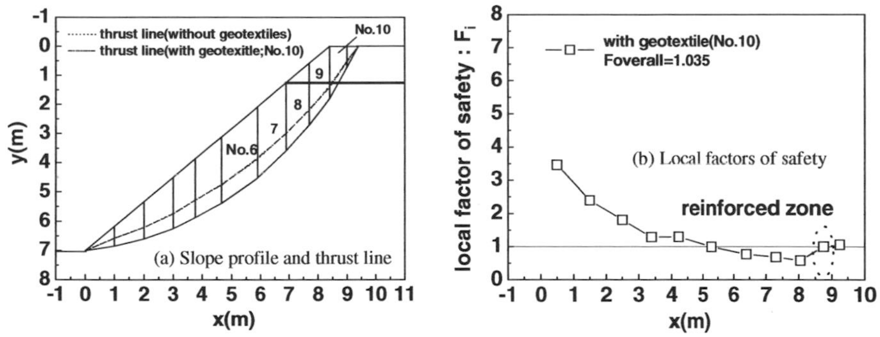

Figure 4. Results for the slope with one layer of geotextile.

slope geometry, soil parameters, and the critical slip surface obtained from the Morgenstern-Price method are shown in Figure 3 (a). Distribution of local factors of safety in terms of the solution procedure in Section 2.2 is shown in Figure 3 (b) by solid squares. It is seen that the slope has a local failure zone that covers the bases of five slices from No.7 to No.11. The overall factor of safety is 0.982.

Geotextile is used to enhance the overall factor of safety. The analysis was made by increasing the number of layers of geotextile one by one from one to four. Results are shown in Figures 4~7. All these are obtained under the condition for a target factor of safety F_0 to be unity for the reinforced slices.

For one layer reinforcement case the safety factor for slice No.10 has definitely become unity as targeted (Figure 4). However, the factors of safety of slices No.7~No.9 are still below unity. Similarly, when two reinforcements were installed at slices No.9 and No.10, the condition of local factors of safety to be

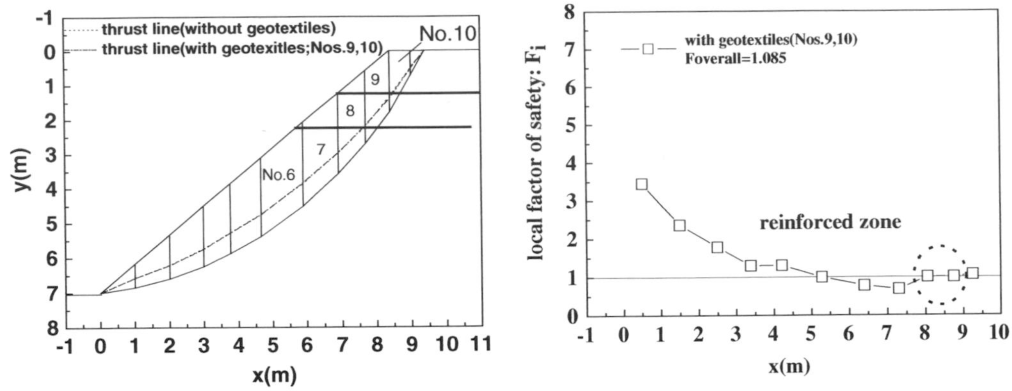

Figure 5. Results for the slope with two layers of geotextile.

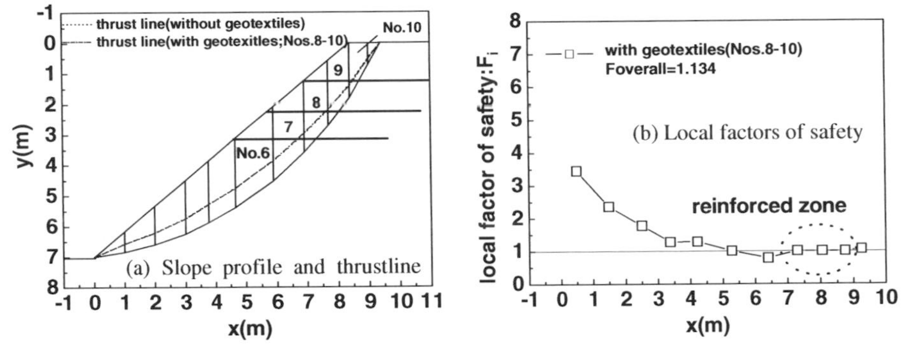

Figure 6. Results for the slope with three layers of geotextile.

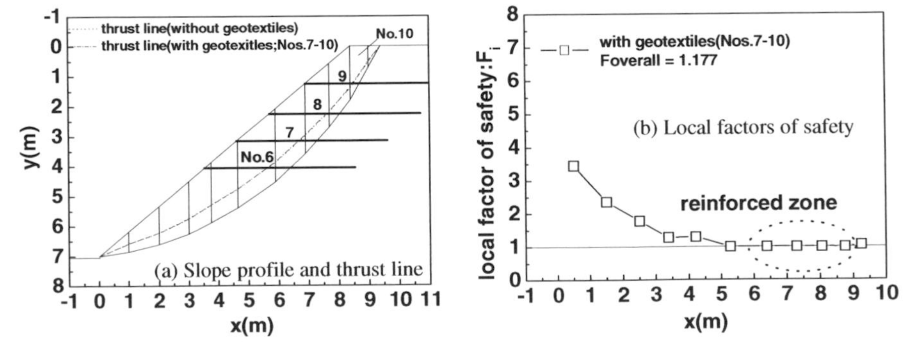

Figure 7. Results for the slope with four layers of geotextile.

unity has been satisfied, but slices No.7 and No.8 still have a factor of safety lower than unity. When the slope is reinforced with three layers of geotextile at slices No.8~No.10, all the failed slices except for No.7 have a safety factor of unity, and the overall factor of safety is about 1.13.

When four layers of geotextile are installed for slices No.7 to No.10, local factors of safety are all equal to unity for the reinforced region and are greater than unity for the remaining unreinforced regions, as shown in Figure 7. The overall safety factor value was found approximately to be 1.18.

Table 1. Tensile forces (kN) in geotextiles for the example.

Number of geotextiles	Slice number No.7	No.8	No.9	No.10	Overall safety factor
1 one layer	–	–	–	3.75	1.035
2 two layers	–	–	3.46	3.71	1.085
3 three layers	–	3.17	3.44	3.71	1.134
4 four layers	2.75	3.12	3.44	3.71	1.177

Table 1 shows calculated tensile forces in reinforcements mobilized at an incipient failure of the reinforced zone. It can be said that failure never occurs provided that reinforcement elements which are able to sustain the tensile forces with a good safety margin are actually used in practice.

5 CONCLUSIONS

A limit equilibrium-based design method for reinforced slopes considering progressive failure was developed. An empirical interslice force function was introduced to enhance the efficiency and robustness of the solution procedure. In the proposed design scheme for reinforced slopes, an optimization technique was contrived to determine tensile resistances of reinforcement elements that would be required to provide an adequate factor of safety of reinforced zones. The proposed procedure can be used to determine reasonable layout of reinforcements and required tensile strength of reinforcing material. The effectiveness of the approach has been demonstrated by the results of a geotextle-reinforced slope. Further research is needed to apply the proposed procedure to practical design of reinforced slopes.

REFERENCES

Fan, K, Fredlund, D. G. & Wilson, G. W. 1986. An interslice force function for limit equilibrium slope stability analysis, *Can. Geotech. J.*, 23, 287–296.

Huang, C. C., Tatsuoka, F. & Sato, Y. 1994. Failure mechanisms of reinforced sand slopes loaded with a footing, *Soils and Foundations* 34 (2): 27–40.

Morgenstern, N. R. & Price, V. E. 1965. The analysis of the stability of general slip surfaces. *Geotechnique* 15: 79-93.

Yamagami, T., Jiang, J.-C. Yamabe, S., & Taki M. 2001. Stability analysis of reinforced slopes considering progressive failure. *In Proceedings of the Int. Symp. on Earth Reinforcement*, Vol. 1, pp. 749–754, Rotterdam: Balkema.

Yamagami, T., Jiang, J.-C. & Yamabe, S. 2002. LEM-based progressive failure analysis and its application to nail-reinforced slopes. *In Proceedings of the 3rd Int. Conf. on Landslides, Slope Stability and The Safety of Infra-Structures*, pp. 83–96.

Physical modeling

New Horizons in Earth Reinforcement – Otani, Miyata & Mukunoki (eds)
© 2008 Taylor & Francis Group, London, *ISBN 978-0-415-45775-0*

Steep slope reinforcement with geogrids – deformation behaviour under static & cyclic loading

G. Heerten Naue
GmbH & Co. KG, Espelkamp-Fiestel, Germany

J. Klompmaker
BBG Bauberatung Geokunststoffe GmbH & Co. KG, Espelkamp-Fiestel, Germany

ABSTRACT: The construction of geosynthetic reinforced slopes and retaining walls is continuously increasing and has already developed to a state-of the art technology. Whereas the load-carrying capacity of geosynthetic reinforced soil structures has sufficiently been examined and proven, details for the proof of the deformation behaviour of these structures, especially under cyclic loading rarely exist. This paper will present results from laboratory and large–scale model tests, which have investigated the interaction behaviour between soil and different types of geosynthetic reinforcing elements in combination with the resulting deformation behaviour of the reinforced soil structures under constant and cyclic loading. The paper ends with a case history about the construction of several geogrid reinforced soil structures, which were necessary to establish 10,000 m^2 of building land in the mountainous area of Marbella (Andalusia) in Spain.

1 IMPACTS ON GEOSYNTHETIC REINFORCED SOIL STRUCTURES

Vertical static impacts on reinforced soil structures consist of the dead load of the soil and vertical stresses resulting from permanent structures. The static loading on the geosynthetic reinforcement can approximately be determined by the resulting earth pressure on the facing system and its distribution over the connected reinforcing elements. Dynamic impacts can result from:

- Dynamic impacts on the foundation resulting from traffic areas, construction operation as well as due to dynamic stresses of structures
- Dynamic impacts on structural parts due to crushes resulting from collision
- Earthquakes

2 DEFORMATION BEHAVIOUR OF GEOSYNTHETIC REINFORCED SOIL STRUCTURES

Several investigations concerning the load-carrying capacity of geosynthetic reinforced soil structures have come to the conclusion, that the installation of horizontal reinforcing elements in soil can generate an extremely complex composite material, which offers a behaviour that is different from the summation of the individual material parameters (geosynthetic and soil).

From a mechanical point of view, the properties of compound materials are mainly based on the following aspects:

Geosynthetic:	– Stress-strain behaviour (modulus)
	– Polymer/manufacturing technique
	– Aperture size
Soil:	– shear strength parameters
	– Grain size
Compound:	– Geosynthetic layer spacing
	– Type and magnitude of impact load
	– Stiffness of facing system
	– Allowable deformations

The load carrying capacity of geosynthetic soil structures has been proven in large scale model tests (Bathurst et al. 2003), with quasi monolithic sliding wedges (Figure 1).

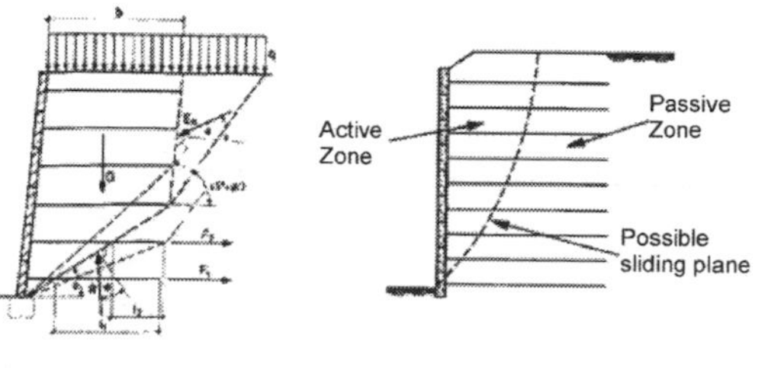

Figure 1. Determination of tensile forces in the geosynthetic reinforcing element.

The sliding soil wedge is moving against the resisting soil. Due to sufficiently high tensile forces and a sufficiently high embedment length of the reinforcing element into the passive zone, the stability of the sliding wedge is assured. The proof of the serviceability is implied as being evidenced, when the load- carrying capacity can be warranted. Especially for geosynthetic soil structures which are susceptible to settlements (e.g. bridge abutments or retaining walls in railway applications), it is desirable for the warranty of the durability and the serviceability, to be able to forecast the deformations resulting from the effective loads and to compare those with the allowable deformations.

Several carried out and monitored projects confirm that the in-situ measured tensile forces of the geosynthetic reinforcing elements partly show lower values as they have been predicted by the design.

Due to the rheological properties of the geosynthetics only a small increase of plastic and thus non-reversible deformations develop at low load levels after the construction phase. With the low activated tensile forces under service load, the available tensile force reserves of the reinforcing element are not utilised in comparison to the calculated long-term design strength.

High geosynthetic tensile forces can only be activated after the loss of serviceability and lead to a high safety potential of the structure under service loads.

The safety potential of geosynthetic reinforced soil structures constitutes the necessity to adjust the load carrying capacity within the scope of the serviceability as well as the bearing capacity to the real conditions. This can only be achieved if the real load distribution inside the compound structure is determined. With the knowledge of the progress of stresses inside the reinforced soil structure, the effective forces can be estimated more precisely and the deformations under service load can be forecasted.

For the investigation of the deformations of geosynthetic reinforced soil structures under service load, the mechanical behaviour of the compound material, the influence of the individual material properties (soil and geosynthetic) and the interaction behaviour between both components has to be known.

2.1 *Laboratory Model test – Deformation under static loads*

The main aim of the model test (Bussert 2006) which is described in the following is the determination of the influencing factors for the interaction between geosynthetic and soil and the determination of the load distribution behaviour under static service loads.

For this, a particular section of a geosynthetic retaining wall is examined in laboratory tests. With the variation of possible soil and geosynthetic properties, the influence on the interaction and compound

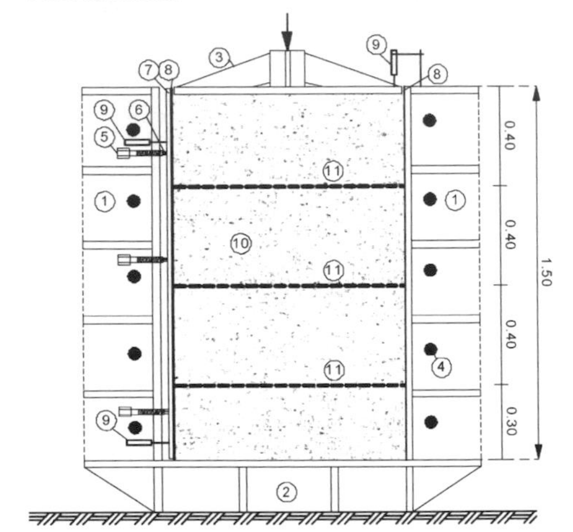

Figure 2. Cross Section of Biaxial test apparatus.

behaviour between soil and geosynthetic and the load carrying capacity of the resulting compound structure is determined.

With the distance-dependent registration of the activated stresses and strains inside the geosynthetic reinforced soil structure, the load carrying capacity-reserves can be determined as a function of the deformations in the facing area.

For the determination of the boundary conditions of the laboratory test a continuous long geosynthetic reinforced soil structure was assumed. Based on the cinematic degree of freedom, only a horizontal deformation, transverse to the embankment axis can occur. Due to shear deformations inside the compound structure, horizontal deformations lead to vertical deformations at the top surface. The deformation is a two-dimensional state, where deformations only appear transverse but changes in the state of stress also appear parallel to the embankment axis. Deformations in the subgrade or the backfill are neglected.

For the investigation, a 9 m high retaining wall is assumed. From deformation measurements it is known that the biggest deformations and horizontal forces appear approx. at 1/3 or ½ of the wall height away from the toe of the structure (Bathurst et al. 2004, Rankilor 2004), which results in a decisive vertical load of 120 kPa on top of a 9 m high wall. The test apparatus is shown in Figure 2.

It consists of a base plate and four rigid side elements. The deformation is forced by a translative movable rigid steel plate, which is hanging in front of six plug gauges at one longitudinal side. There is no further contact of the movable steel plate to the four side elements besides at the plug gauges.

The stress which is acting on the steel plate during movement can thus be determined without any loss.

The horizontal deformation of the movable plate is measured by three spacers that are mounted to the plate. On top of the spacers displacement transducers are attached in the centre of the plate as well as in the area of the bolts.

The vertical load on top of the test setup is applied by a hydraulic cylinder and measured by a pressure cell. The settlements of the load plate are monitored by inductive displacement transducers at the edges and in the centre of the plate. The stresses transverse to the direction of movement are measured by strain gauges which are applied to the plug gauges.

As fill material 3 different soil types were used:

- silica sand (0–2 mm)
- silica gravel (2–12 mm)
- gravel sand (0–36 mm)

To achieve constant test conditions, all soils were installed in dry condition. To guarantee a constant density of the installed soil layers, the sand or gravel was rippled through a punched plate. The density was controlled by the drop height and the hole diameter.

Several geosynthetics of different manufacturing types (woven, laid, stretched and knitted), different polymers (PET, PP, HDPE) and different stress-strain characteristics (modulus) were used in the tests.

The effectiveness of different geosynthetic reinforcing elements can explicitly be derived from the effective horizontal forces at the front face of the biaxial test apparatus. Due to the external loading and the steel plate movement at the front face, grain rearrangement and strains in the geosynthetic reinforcement lead to shear and tensile force activation in the compound material, which change the effective horizontal stresses.

The test results show that the interaction of soil and geogrid mainly depends on the geosynthetic layer spacing, soil grain size, geosynthetic aperture size as well as strength of shape and extensional stiffness of the geogrid product. Opposite to the presently used design methods no correlation between geosynthetic tensile strength and serviceability of the geosynthetic reinforced soil structure can be accomplished. The stress reduction at the front wall, by moving the front wall in x-direction (Fig. 2), caused by different geogrid products and with no reinforcement are shown in Fig. 3. With "stiff", welded and extruded geogrids a reduction of the stress level is already given before any front wall movement is initiated. A reinforcing effect occurs immediately without any deformation of the front wall whereas the woven geogrids need an initial deformation to activate a reinforcing effect. When fully activated, the soil/geogrid composite material is

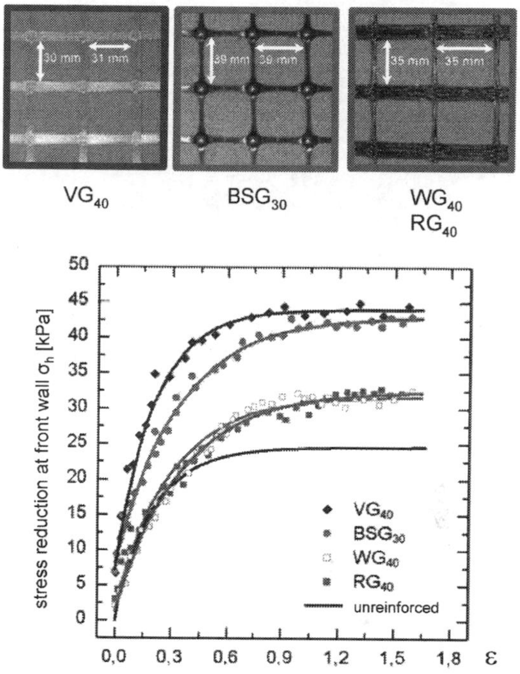

Figure 3. Earth pressure/stress reduction at movable front wall with different/no reinforcement of soil body.

characterised by significant smaller effective horizontal stress than the unreinforced soil.

Depending on the manufacturing technique (laid, stretched, woven & knitted) of the used geogrid a different effectiveness in relation to the maximum stress absorption can be determined.

In general it can be noticed that the laid geogrid shows the highest effectiveness in gravel as well as in sand. Due to a much higher flexibility of the woven geogrid in comparison to the stretched and laid geogrids, no effective fixed support of the soil is given, which allows a higher deformation capability of the compound material. A higher stiffness of the geogrid also leads to a higher stiffness of the compound material which encourages the stress absorption and low deformations.

Further tests within the investigation also document that woven and knitted products need an initial deformation before a reinforcing effect can be measured compared to an un-reinforced system.

Due to the production related pre-stressing of laid and stretched geogrids, immediate stress absorption without primary deformations of the compound material takes place. Dimensionally stable and high-modulus geogrids assure an immediate frictional connection with the surrounding fill and increase already the stress absorption capacity of the compound material prior to the movement of the facing.

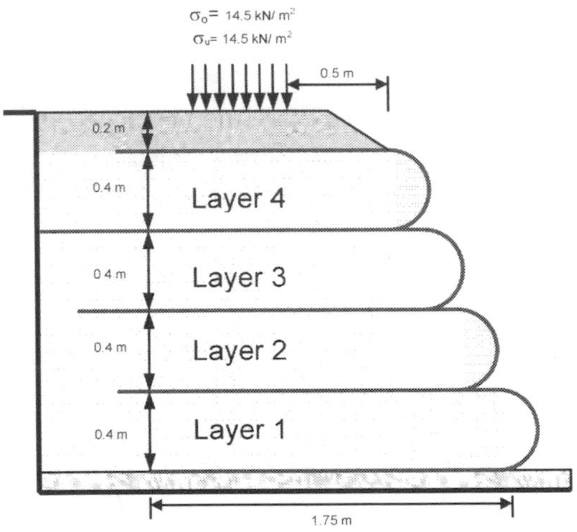

Figure 4. Cross section of test setup.

The achievable maximum stress absorption of woven and knitted geogrids is much lower due to a missing stabilising effect in the beginning and the resultant soil movement within the compound material.

Whereas the overall bearing effect of compound materials reinforced with stretched and laid geogrids consists of a stabilising and reinforcing part, only a reinforcing bearing effect can be achieved with woven or knitted products.

The activation of the reinforcing effect is further a result of the stress-strain behaviour of the geosynthetic product. Due to the fact that a higher modulus of the used geogrid normally results in an increase of the dimensional stability, it is incidental that an influence on the bearing strength of the compound material is also given.

2.2 *Laboratory Model test – Deformation under dynamic loads*

There is still a need for clarification concerning the behaviour of reinforced soil structures under constant dynamic loads, as they are typical e.g. in the safety-relevant field of railway traffic loads.

To monitor the behaviour of geosynthetic reinforced soil structures under such dynamic loads, extensive large scale tests have been carried out at the University for Engineering and Economy Dresden (FH) in a 1:1 scale (Göbel, Großmann; 2006).

The reinforced soil structure consisted of 4 geogrid layers, which were installed with a vertical spacing of 0.4 m. The embedment length of the installed geogrids was 1.75 m and the width of the reinforced slope was 3 m. The slope was built with an inclination of 70° (Figure 4). As fill material sandy gravel (0/32 mm) has been used.

Figure 5. Laid & welded geogrid.

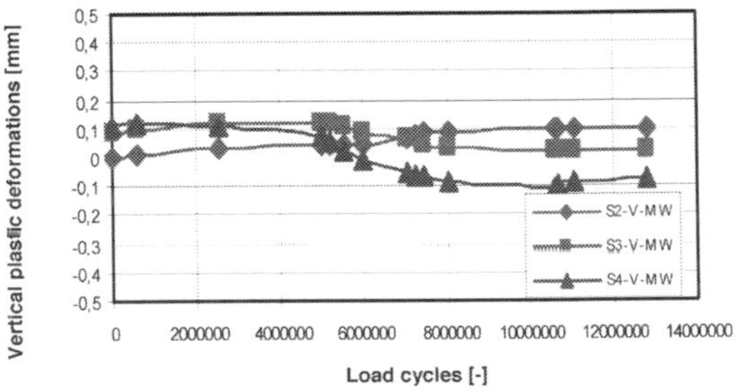

Figure 6. Measured vertical plastic deformations.

As reinforcing element a laid and welded geogrid with a tensile strength of 60 kN/m (md & cmd) was used (Figure 5).

To be able to monitor the dynamic stability of the reinforced soil structure under realistic railway traffic loads, extensive devices for the measurement of the deformation as well as for the oscillating rates have been installed inside the structure.

As dynamic load an alternating load with a lower load level of $\sigma_u = 14.5\,kN/m^2$ and an upper load level of $\sigma_o = 94.5\,kN/m^2$ at a frequency of 7 Hz was applied to the structure with a hydraulic loading device. These loads typically represent a train with a speed of 200 km/h. In addition to those typical loads, also extreme dynamic loads (frequency of 28 Hz & 40 Hz), which rarely occur in practice, have been applied to the reinforced soil structure.

As a result of 12.8 million applied load cycles, maximum vertical deformations of 0.1 mm were measured (Figure 6), whereas the measured horizontal deformations were within the accuracy of measurement.

The applied vibration rates are very quickly absorbed over the depth as shown in Figure 7.

At applied vibration rates of 15 mm/s, remaining values of <5 mm/s have been measured already at a depth of 0.50 m underneath the load plate. The magnitude of dynamic impacts on the geogrid directly depends on the depth. It is thus of great importance,

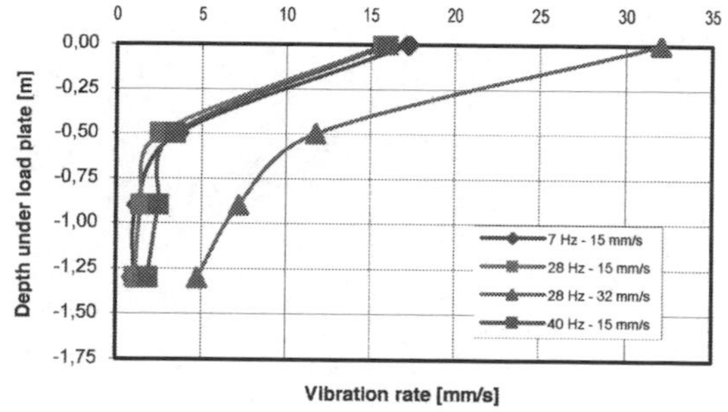

Figure 7. Measured vibration rates.

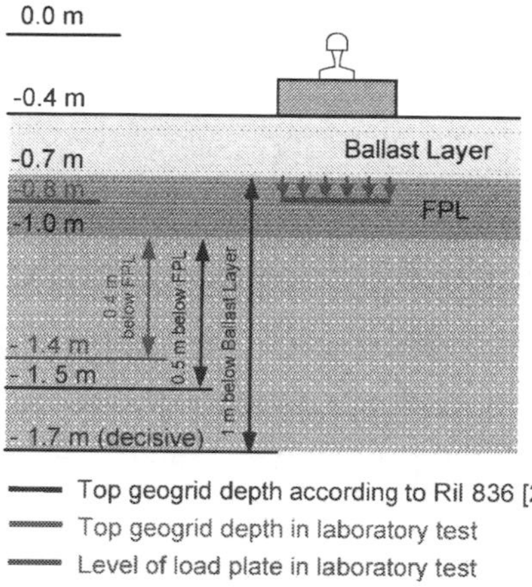

Figure 8. Top geogrid level in laboratory test compared to Ril 836.

at which depth underneath the dynamic load level the first geogrid level is installed.

With regard to this respect it is defined in Ril 836 that a minimum vertical distance from the top geogrid layer

- to the foundation level of 0.5 m
- to the formation level of 1.0 m &
- to the upper level of the rail track of 1.7 m

is required (Figure 8). These regulations have been made on the basis of the present standard of knowledge about the dying out of dynamic loads over depth and can be judged as very conservative.

Due to this fact a very small distance between load impact level and top geogrid layer of 0.6 m was chosen in the large-scale test. Even under these extreme test conditions, the reinforced soil structure has proven sufficient load carrying capacity and serviceability.

After the test the top geogrid layer has been excavated. The results of the visual inspection can be summarized as follows:

- No damages of the geogrid could be recognized
- Between geogrid and fill soil a good interlocking has been detected
- Abrasion of the geogrid due to high dynamic loads could not be registered

To investigate the mechanical properties of the top geogrid layer after applying 12.8 Mio dynamic load cycles, samples have been tested to measure the remaining tensile strength of the geogrid. Based on the carried out tensile test according to DIN EN ISO 10319 a 3 % lower tensile strength of the installed geogrid was measured, in comparison to a sample from the same production lot that was tested during quality control.

Based on this results, a safety factor for installation damage and dynamic loads of $SF_{Inst+Dyn} = 1.03$ can be derived. According to Hubal (2000), a safety factor SF_{Dyn} for the determination of the long-term design strength of the reinforcing element, considering dynamic effects resulting from railway traffic, have to be considered. Recommended values of SF_{Dyn} are defined according to the depth of the geogrid layer underneath load level:

- $\leq$1.5 m below load level : $SF_{Dyn} = 1.5$
- $\geq$4.0 m below load level : $SF_{Dyn} = 1.0$

Interim values can be interpolated.

Based on the gained test results from the large scale laboratory test it can be concluded that the current standards underestimate the resistance of the tested geogrid against dynamic loads, resp. overestimate the propagation of vibrations over depth and thus have to be reconsidered to allow even more economic geosynthetic alternative solutions in comparison to conventional construction methods.

3 REINFORCED SLOPE IN MARBELLA, SPAIN

In the hilly landscape of the Andalusia coastal city Marbella, located about 50 km west of Malaga, land is extremely expensive as well as difficult to develop resulting from the natural topography, which is characterised by a terrain inclination of about 45°.

For the establishment of real estate, the Spanish private owner planned to fill up the hilly site to create an area of 10,000 m^2 of building land. As an attractive landscaping with natural sea view, the owner separated the total area into 3 main plateaux, which were stabilised by geogrid reinforced retaining walls and steep slopes. To meet an attractive landscaping, different facing systems have been chosen for the different wall and slope sections.

The final task was to construct three independently located houses including complexes of recreation facilities, access roads and gardens separated into individual sections with an area of 5,400 m^2, 3,000 m^2 and 3,000 m^2. The original situation of the particular area is shown in Figure 9.

The aim was to provide reinforced earth structures to reach the maximum terrain level of 150 m ASL

Figure 9. Original site situations in the hills of Marbella/Spain.

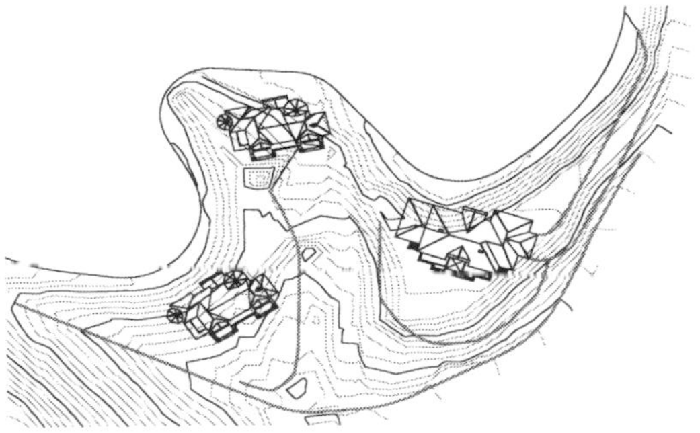

Figure 10. Site plan view with 5 wall gradients separating the area.

Figure 11. Construction of the base wall (10 m high).

Figure 12. Constructed villa on reinforced soil structure (May 2006).

(nominal level above sea level) starting from the lowest level of 130 m ASL by limiting the required maximum height with characteristics of terraces.

In total 5 main geogrid reinforced retaining walls and slopes have been constructed to create the above mentioned necessary building land.

Single geogrid reinforced earth structures with lengths varying between 60 m and 200 m were required due to the existing topography and design requirements (Figure 10).

Two different types of a laid and welded geogrid, manufactured by NAUE GmbH & Co. KG, have been installed for these civil works:

- Secugrid® 80/20 R6 (primary reinforcement, which is installed with wrap-around-method)
- Secugrid® 200/40 R6 (installed at half of normal layer spacing as secondary reinforcement)

Secugrid® is a laid geogrid made of stretched, monolithic polyester (PET) flat bars with welded junctions. The subsoil characteristics are based on in-situ probes.

The mountainous region of Marbella is characterised by rocky slopes with bedrock. It is assumed that the subsoil consists of undisturbed rock with sufficient bearing capacity for the planned structures.

Inclinations of 90° (with small intermediate berms) and 70° (continuous slope) have been realised.

Figure 11 shows the first and largest (10 m high) geogrid reinforced earth structure during construction.

Due to the extremely dry summers in the south of Spain and the steep inclination of the bottom wall, an artificial irrigation system, consisting of slotted pipes, has been installed to the facing system.

The geogrid reinforced soil structures as flexible alternative to concrete retaining walls have allowed a cost effective, attractive development of real estate in a difficult topography.

In total 5 different geogrid reinforced soil structures have been realised in Marbella to provide terraced plateaux, filled up with 40,000 m^3 of soil. The presented solution provides significant advantages concerning the flexibility in geometry and cost-effectiveness in relation to the regionally expensive land prices and conventional construction methods.

4 CONCLUSION

Based on the presented test results from the investigated reinforcing elements in large scale model tests it can be concluded that current design standards for reinforced soil structures clearly underestimate the effectiveness of geosynthetic reinforcing elements with regard to:

- Resistance to dynamic loads
- Interaction behaviour between soil & geosynthetic
- Load transfer mechanism inside the compound material (soil & geogrid)

The results of the presented investigations also show that product properties of different reinforcing materials, like e.g. modulus and dimension stability (stiffness) have a decisive influence on the stress absorption of the compound material and its deformation behaviour. Current design standards do not completely allow for the consideration of these influencing factors.

Geosynthetic reinforced soil structures often offer ecological and economical advantages compared to conventional construction methods like e.g. gravity or angular retaining walls as shown in the documented case history of geogrid reinforced soil structures in Marbella, Spain.

Soong and Koerner (1999) have shown that the costs for geosynthetic reinforced retaining walls are only half, compared to those of gravity walls.

The consideration of the new cognitions in combination with further investigations will help to design safe and even more economical steep reinforced soil structures.

REFERENCES

Bathurst, R.J., Blatz, J.A., Burger, M.H., 2003. Performance of instrumented large-scale unreinforced and reinforced embankments loaded by a strip footing to failure, Canadian Geotechnical Journal

Bathurst, R.J., Allen, T., Walters, D., 2004. Reinforcement loads in geosynthetic walls and the case for new working stress design method, Proc. Of the 3rd European Geosynthetics Conference, Munich, Germany

Bussert, F., 2006. Verformungsverhalten geokunststoffbewehrter Erdstützkörper–Einflussgrößen zur Ermittlung der Gebrauchstauglichkeit, Schriftenreihe Geotechnik und Markscheidewesen, TU Clausthal, Heft 13/2006, Clausthal, Germany

Göbel, C., Großmann, S., 2006. Geokunststoffbewehrte Erde unter dynamischer Belastung- Weiterentwicklung des Systems, 7. Sächsisches Bautextilien-Symposium Bautex 2006, Chemnitz, Germany

Hubal, H., 2000. Geotextilien im Eisenbahnbau-Einsatzbereiche, Erfahrungen und Entwicklungen, Eisenbahningenieur (51) 4/2000

Rankilor, P., 2004. Soil Reinforcement Theory-We may have been wrong for Forty Years, Int. Conference on the Use of Geosynthteics in Soil Reinforcement and Dynamics, 5.-8. September 2004, Schloss Pilnitz, Dresden, Germany

Ril 836, 1999. Erdbauwerke planen, bauen und instand halten, Deutsche Bahn AG, Munich, Germany

Saathoff, F., Werth, K., Klompmaker, J., Wittemöller, J., Vollmert, L., 2004. Stabilisation of new Real Estate in Marbella / Spain with geogrid reinforced soil structures, Proc. Of the 3rd European Geosynthetics Conference, Munich, Germany

Soon, T.-Y, Koerner, R.M., 1999. Geosynthetic reinforced and geocomposite drained retaining walls utilising low permeability backfill soils, GRI Report #24, Geosynthetic Research Institute, Drexel University, USA

New Horizons in Earth Reinforcement – Otani, Miyata & Mukunoki (eds)
© 2008 Taylor & Francis Group, London, ISBN 978-0-415-45775-0

Behavior of reinforced sand: Effect of triaxial compression testing factors

I.N. Markou & A.I. Droudakis
Department of Civil Engineering, Democritus University of Thrace, Xanthi, Greece

ABSTRACT: Triaxial compression tests were conducted in order to investigate the effect of specimen preparation parameters and testing parameters on the mechanical behavior of Ottawa 20–30 sand reinforced with woven and non-woven geotextiles. Strength and failure deformation of reinforced sand are always higher than the ones of unreinforced sand. Strength of reinforced sand increases with increasing number of geotextile layers and cell pressure and is not affected by the rate of axial displacement used in the tests. Axial strain at failure of reinforced sand increases as specimen size and number of geotextile layers increase. An empirical equation is proposed for the computation of the equivalent confining stress increase due to geotextile reinforcement in connection with specimen size, number of geotextile layers and cell pressure. The results obtained by this equation are in good agreement with the experimental results obtained in this investigation.

1 INTRODUCTION

The design of reinforced soil structures requires the knowledge of the mechanical behavior of composite material. The mechanical behavior of sand – geotextile composites has been extensively investigated in the past and several research efforts were based on the results of triaxial compression tests (e.g. Gray et al. 1982, Gray & Al-Refeai 1986, Baykal et al. 1992, Ashmawy & Bourdeau 1998, Haeri et al. 2000). An appropriate selection of parameters related to triaxial compression test is needed, before conducting this test for the study of mechanical behavior of geotextile reinforced sand. Although the effect of specimen preparation and testing parameters, such as specimen size, number and position of reinforcement layers and strain rate, on mechanical behavior of reinforced sand has been evaluated in some research efforts (e.g. Yang & Singh 1974, Gray & Al-Refeai 1986, Moroto 1992, Haeri et al. 2000), it is of merit to make an all-embracing assessment of the effect of all these parameters on strength and deformation characteristics of reinforced sand. The research effort reported herein, aims at the evaluation of the effect of specimen size, number of reinforcement layers, cell pressure and strain rate as well as the combined effect of the first three parameters on the results of triaxial compression tests conducted on sand reinforced with geotextiles. Based on the results of this investigation, the behavior of geotextile reinforced sand could be predicted by selecting the values of the above mentioned parameters.

2 EXPERIMENTAL PROCEDURES

Conventional laboratory triaxial compression equipment was used to conduct tests on geotextile reinforced sand. All tests were conducted using dry and dense Ottawa 20–30 sand. This sand has maximum and minimum void ratios of 0.77 and 0.46, respectively, and angle of internal friction, φ, equal to 36° at an average relative density of 84%. This value of angle of internal friction was found to be unaffected by the rate of axial displacement values used for conducting triaxial compression tests in this investigation. Two thermally bonded (TYPAR SF 56 and TYPAR SF 77), one needle-punched with thermally treated surfaces (FIBERTEX F 400) and one needle-punched (POLYFELT TS 65) non-woven polypropylene geotextiles, as well as one standard grade woven polypropylene geotextile (BONAR SG 80/80), were tested. These geotextiles are designated as TB1, TB2, TTS, NP and WSG, respectively. Properties according to the manufacturers of the geotextiles are presented in Table 1.

Tests were conducted using sand specimens reinforced with a number of geotextile discs, N, equal to 3, 5 or 7 placed as shown in Figure 1. The discs had diameter equal to that of specimens. The specimens had diameter, d, equal to 50 mm and 70 mm and overall height, H, of 101 mm and 141 mm, respectively. Specimen configurations same as that of Figure 1b have been used previously (Atmatzidis & Athanasopoulos 1994, Markou & Droudakis 2005) in laboratory investigations using triaxial compression tests. The sand was compacted using a special hand operated tamper. All tests were conducted at a relative density of sand between 77% and 89%. Triaxial compression tests were conducted with cell pressures, σ_3, equal to 50, 100, 200, 400 and 600 kPa and at a testing rate of axial displacement ranging from 0.1%/min to 5.94%/min.

3 RESULTS AND DISCUSSION

The effect of specimen size on strength and axial strain at failure of reinforced sand is shown in the diagrams of Figure 2. In these diagrams, the maximum deviator stress, $(\sigma_1 - \sigma_3)_{max}$, values and axial strain at failure, ε_f, values obtained from reinforced sand specimens with diameter of 50 mm are directly compared to the values obtained from specimens with diameter of 70 mm. The results were obtained from sand specimens reinforced with 3 and 5 layers of TB2, TTS and NP non-woven geotextiles. It can be seen (Fig. 2a) that, although a significant number of points is very close to the diagonal line, the majority of points is above this line, indicating that specimens of 70 mm in diameter give generally higher strength than the ones of 50 mm in diameter. The observed variations in maximum deviator stress can reach the value of 20% for tests conducted with low cell pressures and are limited to the value of 10% for tests conducted with higher cell pressures (Fig. 2a). It is clearly seen in Figure 2b that specimens with diameter of 70 mm generally present higher values of axial strain at failure than specimens having diameter of 50 mm. Therefore, it is concluded that the effect of specimen size on maximum deviator stress is not very strong, while the effect of it on axial strain at failure is significant.

The strength ratio, S_R, defined as the ratio of maximum deviator stress of reinforced sand to the maximum deviator stress of unreinforced sand for the same cell pressure, is used for the quantification of the strength increase due to reinforcement of sand. Strength ratio values obtained from specimens reinforced with 3, 5 and 7 layers of TTS geotextile are presented in Figure 3. It is observed that strength ratio increases with increasing the number of geotextile layers used for sand reinforcement and with decreasing cell pressure. The S_R values obtained from

Table 1. Geotextile properties.

Geotextile	Thickness mm	Mass per unit area g/m^2	Tensile test results: Max tensile load, kN/m	Tensile test results: Extension at max load, %
TB1	0.54	190	12.8	65
TB2	0.65	260	20.0	70
TTS	1.80	275	16.5/17.5*	52/55*
NP	1.10	285	21.5	80/40*
WSG	1.35	360	82.0/86.0*	20/11*

* Machine direction / Cross machine direction

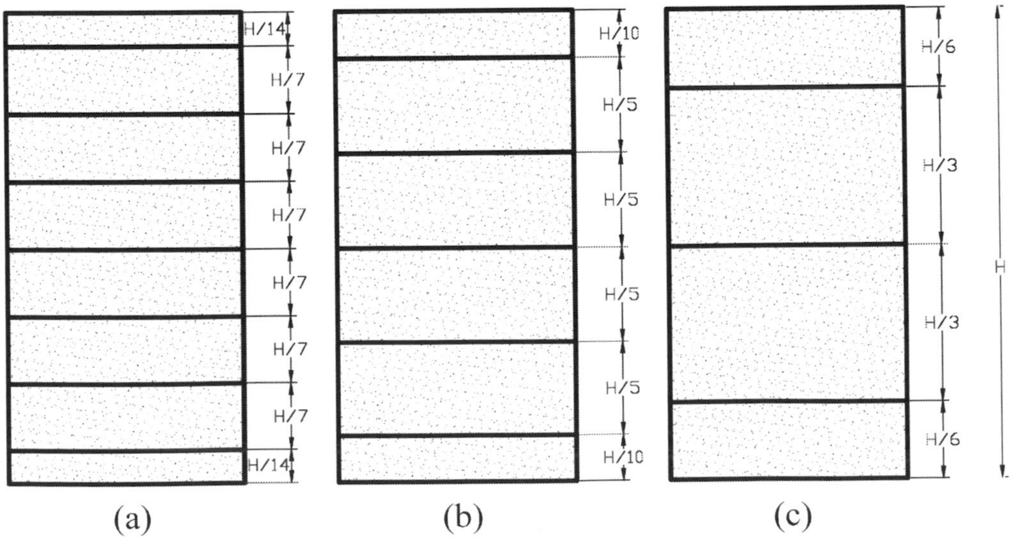

Figure 1. Sand specimens reinforced with (a) 7, (b) 5 and (c) 3 geotextile layers.

all specimen configurations used in this investigation, range between 1.34 (specimen with d = 50 mm, TB2 geotextile, N = 3, σ_3 = 400 kPa) and 5.97 (specimen with d = 70 mm, NP geotextile, N = 7, σ_3 = 50 kPa) showing that the use of geotextiles as reinforcement leads to a significant increase in the strength of sand.

Similarly, axial strain at failure values obtained from specimens reinforced with 3, 5 and 7 layers of NP geotextile are presented in Figure 4. Axial strain at failure increases as the number of geotextile layers increases. Furthermore, the axial strain at failure of reinforced sand is always higher than the one of unreinforced sand. Also, it seems that an increase in cell pressure leads to an increase in axial strain at failure. This observation is not always confirmed when a larger number of woven and non-woven geotextiles are considered (Markou et al. 2006a, b).

Yang (1972) presented a semi-empirical equation that relates the equivalent confining stress increase, $\Delta\sigma_3$, to the reinforcement spacing ratio, ΔH/d, in a triaxial compression test. The equation has the form:

$$\frac{\sigma_3 + \Delta\sigma_3}{\sigma_3} = \frac{1}{1 - CK_p(\frac{\Delta H}{2d})^m} \quad (1)$$

where ΔH = spacing between reinforcement layers; d = triaxial specimen diameter; $K_p = \tan^2(45° + \varphi/2)$; and C, m = empirical constants.

It was found that the experimental results and theoretical curve predicted by Equation 1 compare fairly well at large spacing ratios, but diverge at ratios less than 0.5 and that reinforcements placed at ΔH/d ratios more than unity have little effect. (Gray & Al-Refeai 1986).

In the present investigation, ΔH/A ratio (A = specimen cross-sectional area) is preferred than ΔH/d ratio, since it expresses the combined effect of number

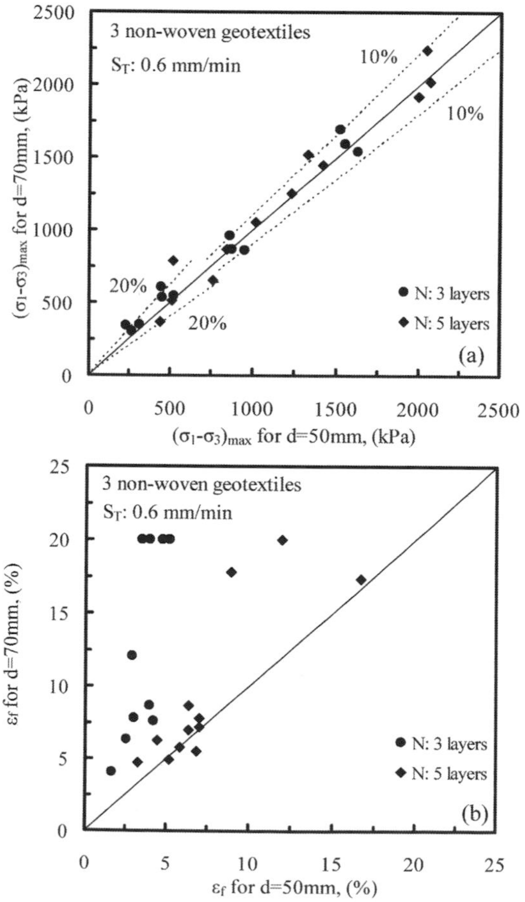

Figure 2. Effect of specimen size on (a) maximum deviator stress and (b) axial strain at failure of reinforced sand.

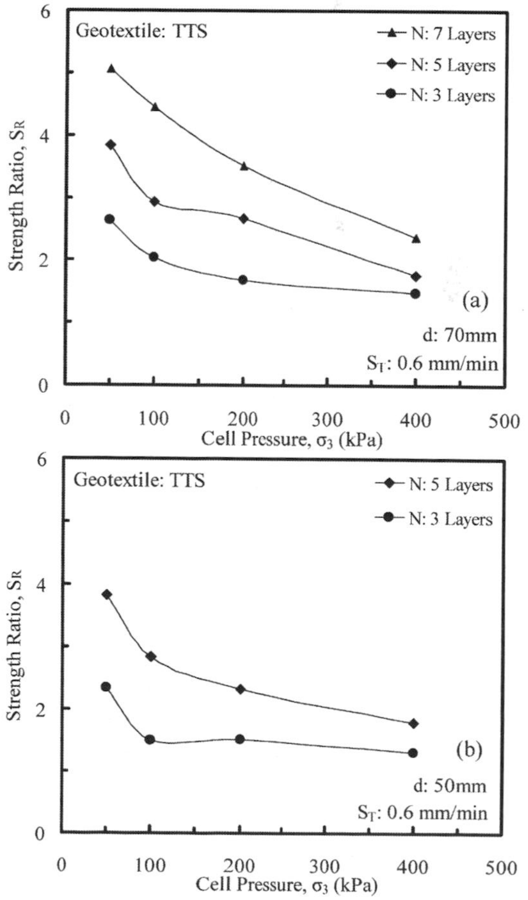

Figure 3. Effect of number of geotextile layers on strength increase of reinforced sand.

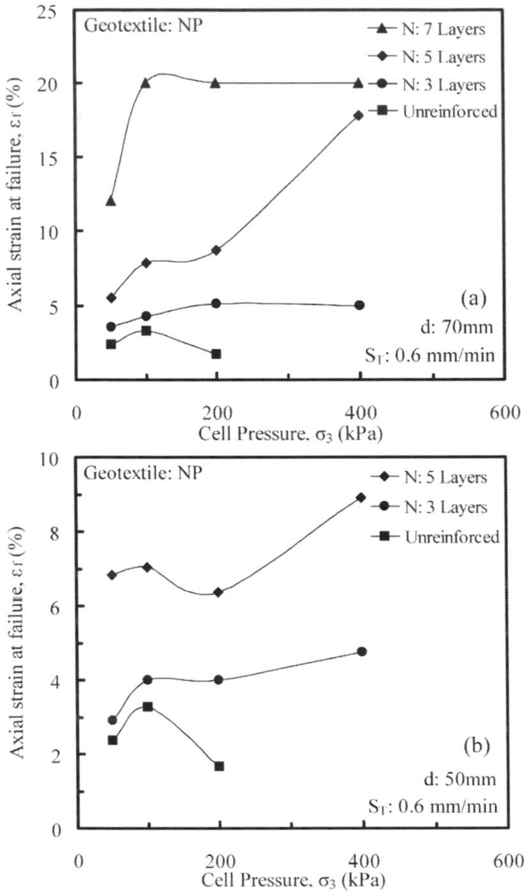

Figure 4. Effect of number of geotextile layers on axial strain at failure of reinforced sand.

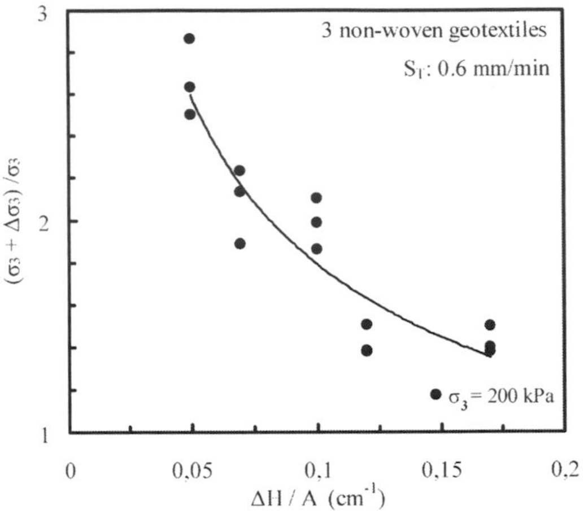

Figure 5. Combined effect of specimen size and number of geotextile layers on $(\sigma_3 + \Delta\sigma_3)/\sigma_3$ ratio.

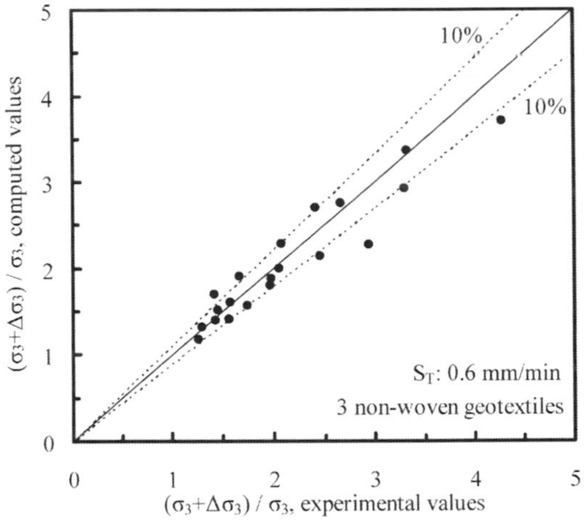

Figure 6. Comparison between computed and experimental values.of $(\sigma_3 + \Delta\sigma_3)/\sigma_3$ ratio.

of geotextile layers and of specimen size on triaxial compression test results. As it is typically shown in Figure 5, $(\sigma_3 + \Delta\sigma_3)/\sigma_3$ values decrease with increasing $\Delta H/A$ ratio. This behavior is mostly attributed to the number of geotextile layers since it was shown earlier that specimen size does not present a strong effect on reinforced sand strength (Fig. 2a). Based on this observation and on the experimental results obtained in this investigation, an alternative empirical equation is proposed for the quantification of the effect of specimen size, number of geotextile layers and cell pressure on the equivalent confining stress increase:

$$\frac{\sigma_3 + \Delta\sigma_3}{\sigma_3} = (0.0005\sigma_3 + 0.4247)\left(\frac{\Delta H}{A}\right)^{0.001\sigma_3 - 0.754} \quad (2)$$

The comparison between values of $(\sigma_3 + \Delta\sigma_3)/\ \sigma_3$ ratio computed with Equation 2 and values obtained experimentally is shown in Figure 6. As it can be seen, the results of Equation 2 compare fairly well with the experimental results, since the variation between them is lower than or equal to 10% in most cases.

Shown in Figure 7, is the effect of rate of axial displacement on maximum deviator stress of sand reinforced with a woven (WSG) and a non-woven (TB1) geotextile. For both geotextiles tested and for the range of axial displacement rates used, it is evident that maximum deviator stress is not affected by the axial displacement rate.

Failure envelopes obtained by triaxial compression testing of the unreinforced and reinforced sand, are

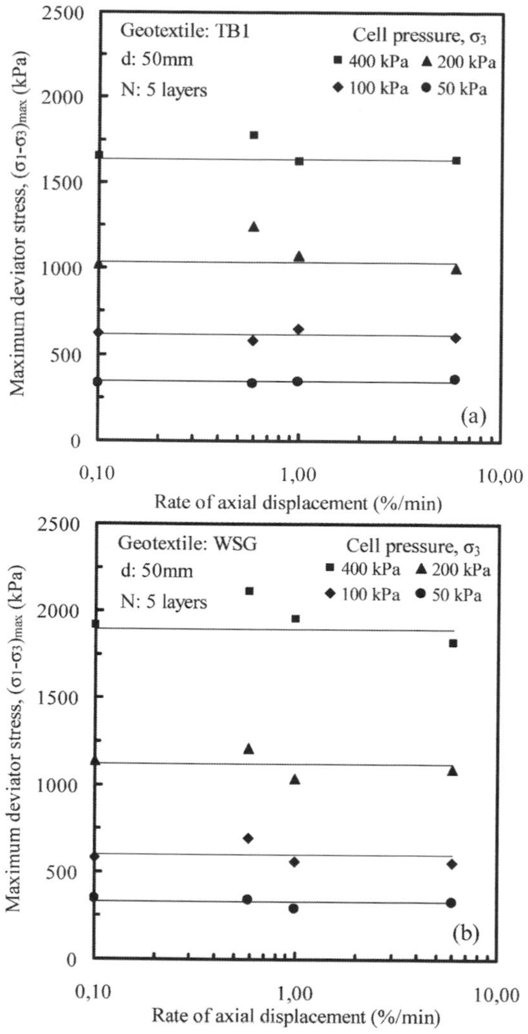

Figure 7. Effect of rate of axial displacement on maximum deviator stress of reinforced sand.

presented in Figure 8. The marks in Figures 8a–d represent the points at which failure envelopes are tangent to Mohr circles resulted from the triaxial compression tests. Such marks were not plotted in Figures 8e, f for clarity reasons. It can be observed that the triaxial compression tests yielded bilinear envelopes for the composite material in good agreement with the observations of other investigators (e.g. Gray et al. 1982, Gray & Al-Refeai 1986). In every case, reinforced sand presents higher shear strength than unreinforced sand (Figs 8a–d). Furthermore, it appears that the specimen sizes (diameter of 50 mm and 70 mm) used, do not affect significantly the shear strength of reinforced sand (Figs 8a, b). It is also observed, that shear strength of reinforced sand increases with increasing number of geotextile layers (Figs 8c, d). Failure envelopes from triaxial compression tests conducted at different testing speeds, S_T, in sand reinforced with a non-woven and a woven geotextile, are presented in Figures 8e, f, respectively. It is seen, that the range of testing speed (rate of axial displacement) used in this investigation has no effect on shear strength of reinforced sand. Consequently, all the above mentioned observations on the effect of triaxial compression testing parameters on the strength of geotextile reinforced sand arc also confirmed by the failure envelopes.

4 CONCLUSIONS

Based on the results of this investigation and within the limitations posed by the number of tests conducted and the materials used, the following conclusions may be advanced:

- Strength of reinforced sand increases with increasing number of geotextile layers, specimen size and cell pressure and is unaffected by the rate of axial displacement. However, the effect of specimen size is not considered as strong.
- Axial strain at failure of reinforced sand increases as the number of geotextile layers, specimen size and cell pressure increase. However, the increase of axial strain at failure with increasing cell pressure was not always observed when a larger number of woven and non-woven geotextiles were tested in previous research efforts.
- Strength and axial strain at failure of reinforced sand are always higher than those of unreinforced sand. The increase in maximum deviator stress of sand, due to geotextile reinforcement, ranges from 1.30 to 6.00 times.
- An empirical equation is proposed for the quantification of the effect of specimen size, number of geotextile layers and cell pressure on the equivalent confining stress increase due to geotextile re-inforcement. The results obtained by this equation are in good agreement with the experimental results obtained in this investigation.

ACKNOWLEDGEMENTS

The Research Committee of Democritus University of Thrace provided financing for the research effort reported herein. This financial support is gratefully acknowledged. Part of the triaxial compression tests described in this paper, were conducted by Mr. G. Sirkelis and Mr. Ch. Ioannou whose careful work is acknowledged.

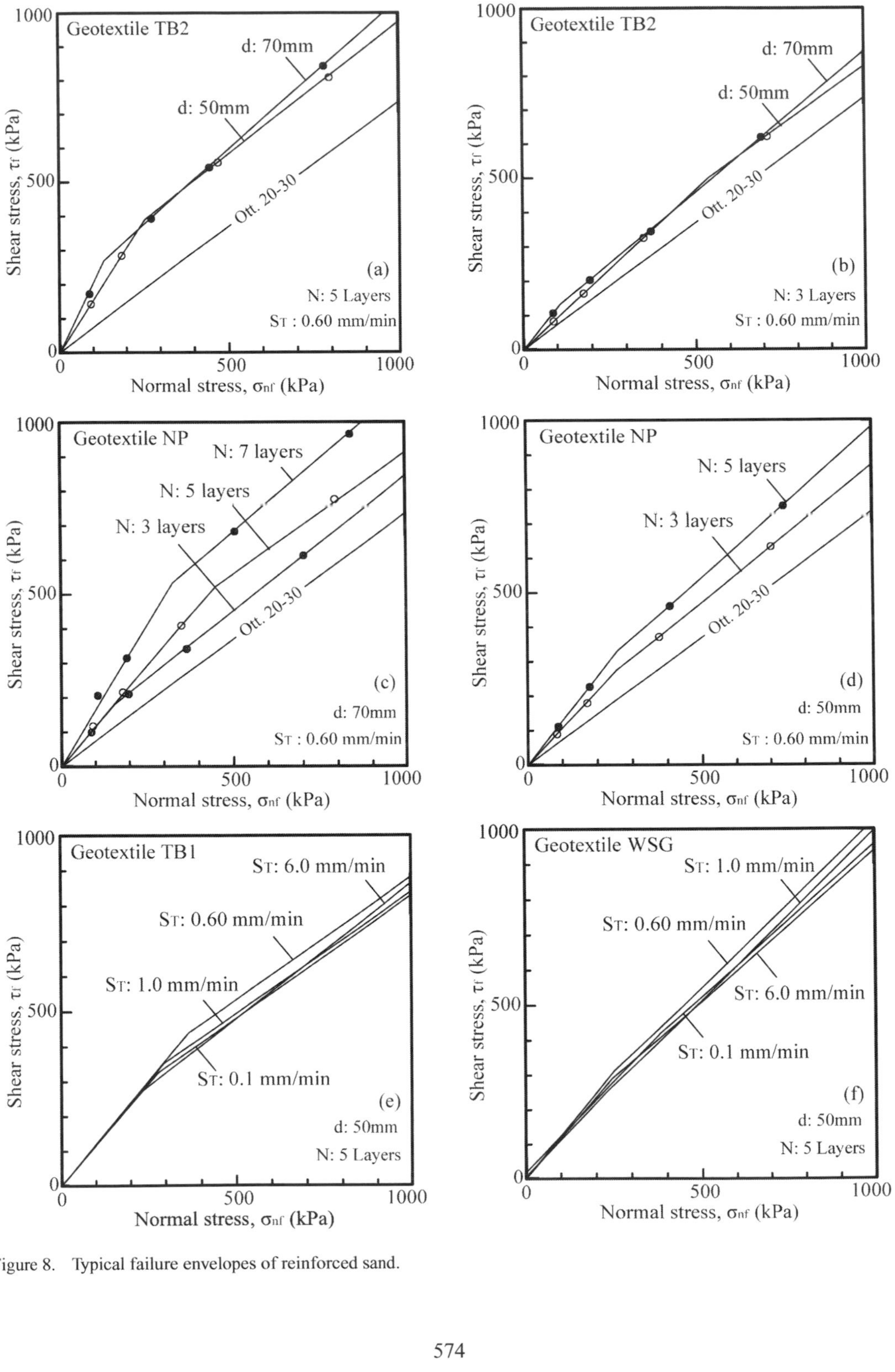

Figure 8. Typical failure envelopes of reinforced sand.

REFERENCES

Ashmawy, A.K. & Bourdeau P.L. 1998. Effect of geotextile reinforcement on the stress-strain and volumetric behavior of sand. *Proc., 6th Intern. conf. on geosynthetics, Atlanta, USA*, 2: 1079–1082.

Atmatzidis, D.K. & Athanasopoulos, G.A. 1994. Sand – geotextile friction angle by conventional shear testing. *Proc., 13th Intern. conf. on soil mechanics and foundation engineering, New Delhi, India*, 3: 1273–1278.

Baykal, G. Guler, E. & Akkol, O. 1992. Comparison of woven and nonwoven geotextile reinforcement using stress path tests. *Proc., Intern. symp. on Earth reinforcement practice, Fukuoka, Japan*, 1: 23–28.

Gray, D.H. & Al-Refeai, T. 1986. Behavior of fabric- vs. fiber-reinforced sand. *Journal of Geotechnical Engineering* 112: 804–820.

Gray, D.H., Athanasopoulos, G.A. & Ohashi, H. 1982. Internal/external fabric – reinforcement of sand. *Proc., 2nd Intern. conf. on geotextiles, Las Vegas, USA.*, 3: 611–616.

Haeri, S.M., Noorzad, R. & Oskoorouchi, A.M. 2000. Effect of geotextile reinforcement on the mechanical behavior of sand. *Geotextiles and Geomembranes* 18: 385–402.

Markou, I.N. & Droudakis, A.I. 2005. Evaluation of sand – geotextile interaction by triaxial compression and direct shear testing. *Proc., 11th Intern. conf. of IACMAG, Turin, Italy*, 2: 341–348.

Markou, I.N., Droudakis, A.I. & Sirkelis G. 2006a. Sand – non woven geotextile interface friction angle by triaxial compression tests. *Proc., 5th Hellenic conf. on geotechnical & geoenvironmental engineering, Xanthi, Greece*, 1: 399–406. (in Greek).

Markou, I.N., Droudakis, A.I. & Sirkelis G. 2006b. Sand – woven geotextile interaction by triaxial compression tests. *Proc., 5th Hellenic conf. on geotechnical & geoenvironmental engineering, Xanthi, Greece*, 1: 383–390. (in Greek).

Moroto, N. 1992. Triaxial compression test for reinforced sand with a flexible tension member. *Proc., Intern. symp. on Earth reinforcement practice, Fukuoka, Japan*, 1: 131–134.

Yang, Z. 1972. *Strength and deformation characteristics of reinforced sand.* Ph.D. dissertation, University of California at Los Angeles.

Yang, Z. & Singh, A. 1974. Strength and deformation characteristics of reinforced sand. *Preprint No. 2189; Intern. meeting on water resources engineering, Los Angeles, USA*.

New Horizons in Earth Reinforcement – Otani, Miyata & Mukunoki (eds)
© 2008 Taylor & Francis Group, London, ISBN 978-0-415-45775-0

Centrifuge model tests of static and dynamic behaviors of multi-anchored sea revetment

Y. Kikuchi & M. Kitazume
Port & Airport Research Institute, Yokosuka, Japan

ABSTRACT: In sea revetment construction, composite-type revetments in which huge concrete caissons are placed on a gravel mound to sustain the earth pressure induced by sea reclamation, have frequently been used in Japan. However, in several cases, serious disasters resulted from large displacement of the caisson in a giant earthquake. Therefore, research to find a new type of sea revetment with better static and dynamic performances was necessary. A type of tieback caisson, in which a concrete caisson of relatively small width is reinforced by multiple anchors, is one concept for satisfying this requirement. A series of centrifuge model tests of this new type of caisson in sea revetment construction was conducted to investigate its static and dynamic behaviors. This paper describes the test results and applicability of sea revetments with multi-anchors.

1 INTRODUCTION

Reinforced earth structures have been widely used in retaining structures such as Terre armee or anchored plates. On the other hand, few case records have been reported in connection with the construction of port facilities. In order to introduce this method in the construction of port facilities such as sea revetments, as shown schematically in Figure 1, the applicability of reinforced earth methods to sea revetment construction was examined in centrifuge model tests in a series of static and dynamic centrifuge model tests of this new type of caisson.

In the static tests, the effect of the stiffness of the foundation ground on the behavior of this kind of structure was examined. In the dynamic tests, the model ground was subjected to several earthquake motions in a 50 g centrifugal acceleration field until the ground failed. Based on these test series, this paper discusses the applicability of sea revetments with multiple anchors.

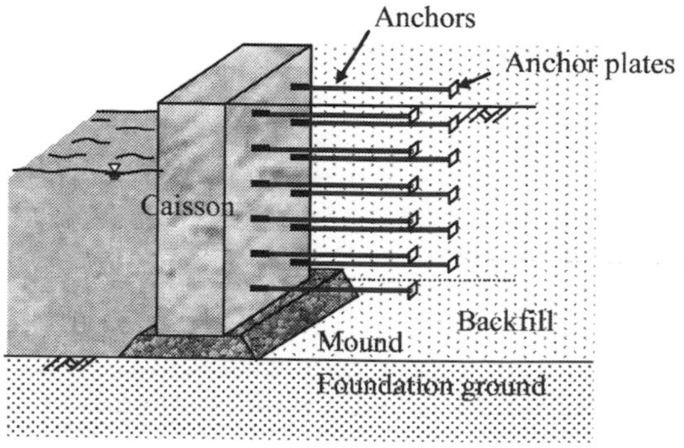

Figure 1. Schematic view of new type caisson.

Many types of retaining walls reinforced with anchors have been applied to various types of construction such as road and railroad embankments. Much research has investigated the static and dynamic behaviors of such walls, and a design manual was established for road and railroad embankments in Japan (Public Works Research Center, 1997). However, there has been little research on the behavior of sea revetments with multiple anchors for port facilities such as sea revetments. The authors therefore began to study the applicability of this new type of caisson reinforced by multiple anchors to sea revetment construction by conducting a series of centrifuge tests to investigate its static and dynamic behaviors. The model tests were conducted by changing the caisson width and number and length of anchors. This paper describes the model ground preparation, test results, and calculated results in detail.

2 CENTRIFUGE MODEL TESTS

2.1 *PARI centrifuge facilities*

The centrifuge used in this study is the PARI (Port & Airport Research Institute) Mark II geotechnical centrifuge. This centrifuge has a maximum acceleration of 113 g, maximum effective radius of 3.8 m, and maximum payload of 2,710 kg. For safe operation, the main part of the centrifuge is housed in an underground reinforced concrete pit. Two swinging platforms are hinged to a rotating arm via torsion bar systems to safely deliver the radial force at high acceleration to end plates at both ends of the arm. As the centrifuge drive unit, a 450 kVA DC motor is mounted on the underground floor. The centrifuge and its peripheral equipment were described in detail by Kitazume and Miyajima (1996).

2.2 *Static loading tests*

To investigate the effect of the stiffness of the foundation on the behavior of a reinforced revetment, two extreme ground conditions were simulated, as shown schematically in Figures 2 and 3. In Figure 2, a thin clay layer underlaid by the revetment was modeled. The caisson was estimated to fail by sliding failure with negligible settlement (series S). In Figure 3, a relatively thick, low strength clay layer was modeled, and the caisson was estimated to fail with relatively large vertical and horizontal displacements (series B).

A rigid specimen box having inside dimensions of 50 cm in length, 35 cm in height, and 10 cm in width was used in both cases. The front side of the box has an acrylic window to allow direct observation of the model behavior during the test.

The sand used was Toyoura sand. The clay was a mixture of two types of Kaolin clay (ASP100:5M = 1:1, $w_L = 64\%$, $I_P = 37$). The dense sand layer for the base was compacted to Dr = 70% in a wet condition. Clay having an initial water content

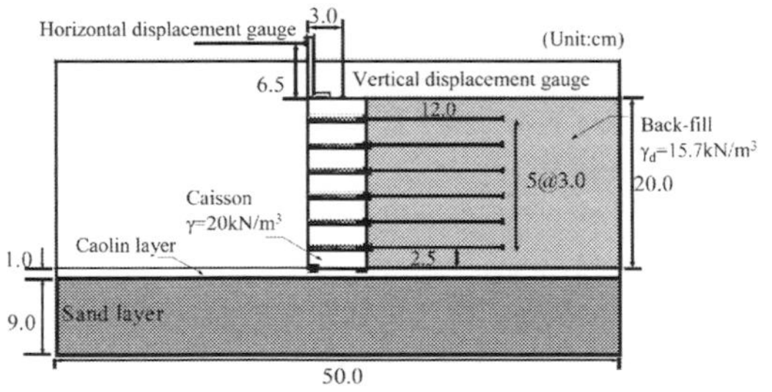

Figure 2. Schematic view of model (Series S).

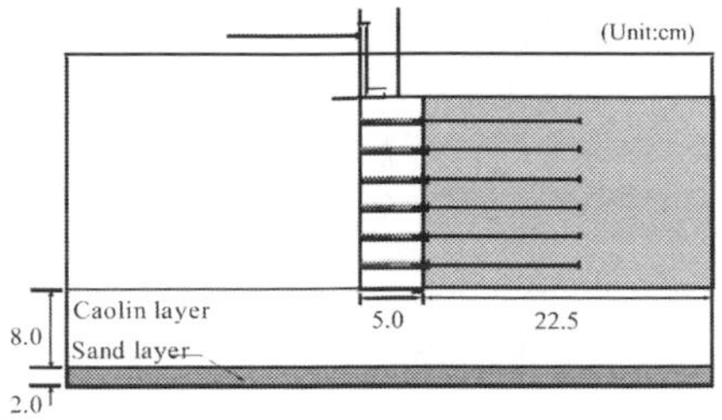

Figure 3. Schematic view of model (Series B).

of 120% was consolidated in the laboratory to the specified pressure. The conditions of the base ground are summarized in Table 1. Dry Toyoura sand, whose soil particle density and maximum and minimum void ratios were 2.652 g/cm^3, 0.992, and 0.624, respectively, was used for backfilling. Its unit weight was controlled to 15.4 kN/m^3 ($D_r = 80\%$, $\phi' = 40°$).

In the test series, two types of caisson were used with both ground conditions. One was a conventional type caisson (called C), the dimensions of which were 14.8 cm in width, 20 cm in height, and 9.6 cm in length. The other was a multi-anchored caisson (called R) with comparatively small width, in which several sets of anchor rods were installed on the rear side. The dimensions of the R caisson were 5 cm in width, 20 cm in height, and 9.6 cm in length. Each R type caisson was reinforced with either 0, 4, 8, or 12 steel anchor rods, which were inserted in 0, 2, 4, 6 levels in two rows from the bottom, respectively. Each anchor rod had a diameter of 0.45 mm and a length of 12 cm. Copper anchor plates, which were 1 cm square, were attached at the end of each rod. A strain gauge was installed on the copper plates, connecting each anchor rod at the caisson, to measure the tensile force induced in the test. The earth pressure acting on the caisson was measured by earth pressure gauges on the rear side of the caisson at five depths along its center.

The unit weight of both types of caisson was controlled to 20 kN/m^3 by filling the caissons with lead shot. Sandpaper was attached to the caisson bottom to simulate a rough condition.

In the static test series, centrifuge acceleration was increased continuously until the model ground failed. The height of the model caisson corresponded to 20 m in a prototype scale at a centrifugal acceleration of 100 g, which is within the previous case histories for larger sea revetments. Displacement of the caisson, earth pressure acting on the caisson, and the tensile force of each anchor rod were measured at 1 g increments. The ground behavior and displacement of the caisson during the test were also videotaped.

2.3 *Dynamic loading tests*

The shaking table used was specially designed for the centrifuge. Its main specifications are as follows; Maximum centrifugal acceleration is 50 g, Maximum mass to be shaken is 200 kg, Maximum frequency

Table 1. Base ground condition.

	Mode of failure	
Layer	Sliding (Series S)	Bearing (Series B)
Clay	1 cm (p = 196 kPa)	8 cm (p = 59 kPa)
Sand	9 cm (Dr = 70%)	2 cm (Dr = 70%)

is 250 Hz, Maximum acceleration is 18 g, Maximum stroke is 6 mm.

The specimen box was a 2-dimensional rigid box whose inside dimensions were 60 cm in length, 41 cm in depth, and 10 cm in width. In order to focus on the interaction of the ground and model caisson and avoid the complicated influence of seawater and liquefaction which might occur in the backfill under seismic loading, the model ground studied in this test series consisted of the caisson, base layer, and backfill with dry dense sand, as shown in Figure 4. Toyoura sand was used as the ground material.

The caissons were the same as those used in the static tests. Several anchors were installed in two rows on the rear plate of the caisson. The anchors were the same type as those used in the static tests. Several accelerometers were installed in the model ground and on the model caisson to measure their dynamic responses. Laser displacement transducers were installed to measure the displacement of the caisson.

The model ground was accelerated to a 50 g centrifugal field in order to simulate the prototype stress condition, and was then subjected to several seismic loadings of 50 sinusoid waves until the ground failed. During seismic loading, the accelerations in the model ground, earth pressure, tensile force along the anchors, and vertical and horizontal displacements of the caisson were measured. A total of five model tests were carried out using various caisson widths

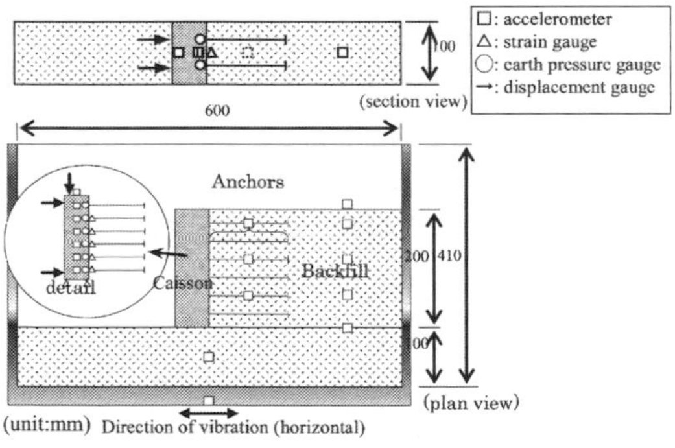

Figure 4. Schematic view of model ground for dynamic loading tests.

Table 2. Test cases for dynamic loading tests.

Test no.	Width of caisson	Number of anchors	Length of anchors
COD-1	14.8 cm	0	–
COD-2	5.0 cm	0	–
U3D-1	5.0 cm	12 (6 rows)	12 cm
UT1_18D	5.0 cm	4 (2 rows)	18 cm
UT1_6D	5.0 cm	4 (2 rows)	6 cm

and anchor conditions, as summarized in Table 2, and included non-anchored cases (COD-1 and COD-2).

3 TEST RESULT AND DISCUSSION

3.1 *Static loading tests*

3.1.1 *Acceleration field at failure*

Figure 5 shows the relationship between the number of anchor rods and centrifugal acceleration at failure for test series S. The upward arrows in the figure indicate that the model ground did not fail at the maximum centrifugal acceleration of 100 g. In the type R caisson, the acceleration at failure increased almost linearly as the number of rods increased. The measured acceleration in the conventional type caisson is also plotted as a full square in the figure. In this case, the model caisson was stable with very small displacement even at 70 g, at which centrifuge loading had to be terminated due to trouble in the displacement transducer. However, it can be concluded that the failure acceleration was higher than 70 g.

Figure 6 shows the relationship between the number of anchor rods and centrifugal acceleration at failure for series B. The failure acceleration increased with the number of rods, but became basically constant at

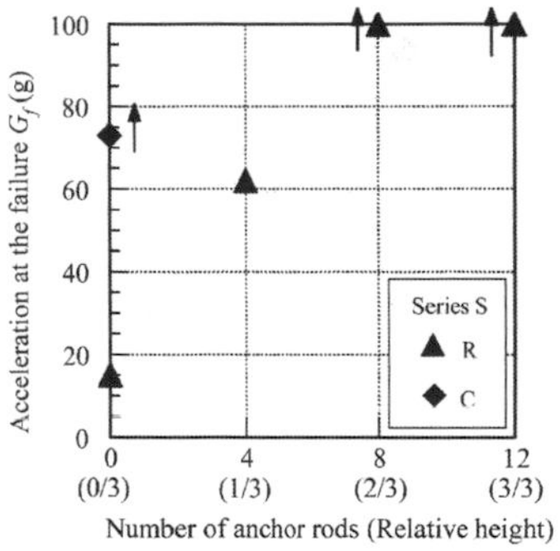

Figure 5. Acceleration at failure (Series S).

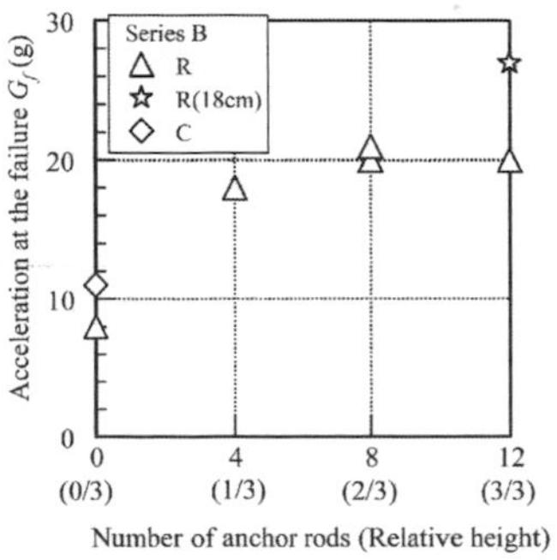

Figure 6. Acceleration at failure (Series B).

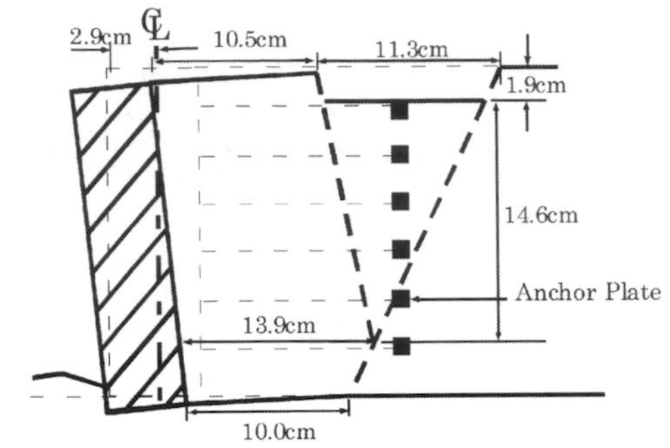

Figure 7. Slip in backfill (B-R-3/3).

8 rods. An additional experiment was carried out to investigate the effect of the anchor length, in which the length and number of anchor rods were 18 and 12 cm, respectively. The test result was plotted as a star mark in Figure 6. This result showed that the longer anchor rods gave higher resistance of this extent. The failure acceleration obtained in the conventional caisson test is also plotted as an open square in the figure. Even with 4 anchor rods, the failure acceleration of the R type caisson was higher than that of the C type.

Both figures show that the reinforced caisson has higher resistance than conventional caissons.

3.1.2 *Displacement of caisson and deformation of ground*

The displacement of the caisson and deformation of the ground were evaluated based on the video record. From this observation, only small deformation took place in the ground in series S until failure. In this series, the final failure mode of the caisson seemed to be rotational, and the slip deformation in the backfill was similar to the zone assumed in Coulomb's active earth pressure theory.

In series B, slip failure was clearly observed in the backfill layer. Figure 7 shows an example of the typical slip failure at the maximum centrifugal acceleration of 20 g. This kind of failure mode was observed only when the number of rods was 8 or 12. From this kind of deformation, the imaginary retaining wall width can be considered in the zone of anchor rods existing if the reinforced height is more than 2/3 of the caisson height.

3.1.3 *Change of earth pressure acting on caisson*

Figures 8 and 9 show the changes in the earth pressure acting on the caisson measured during increasing acceleration in series S and B, respectively. Several thin lines with various values of $K = \sigma_h/\sigma_v$ are also plotted in the figure, in which σ_h and σ_v are the horizontal and vertical earth pressures, respectively. It was found that the values of K obtained from the measured σ_h and estimated σ_v increased with the number of rods.

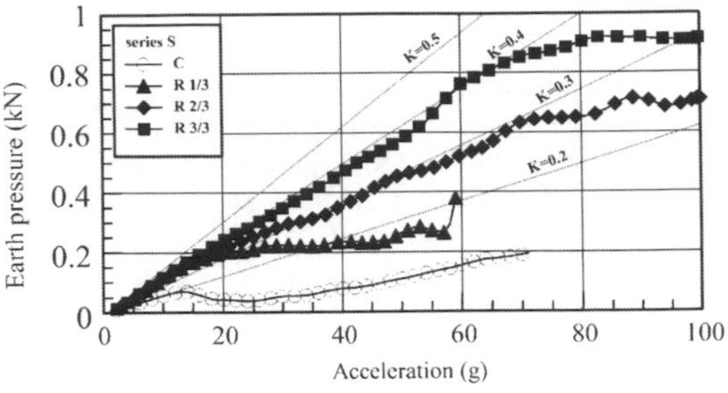

Figure 8. Relationship between acceleration and earth pressure (Series S).

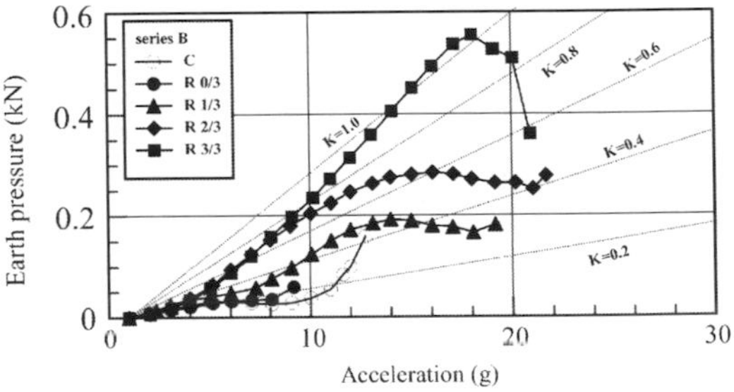

Figure 9. Relationship between acceleration and earth pressure (Series B).

It is considered that the sum of the measured tensile forces along the rods increased with increasing centrifugal acceleration in both series. A large value of K means that the backfill in the anchored zone was strengthened by the anchors. This phenomenon indicated that the group of anchor rods could also function to confine the backfill layer.

In series S, the values of K were substantially constant as long as acceleration remained low but decreased when acceleration increased. The decrease in K was dominant when the relative height of the anchored zone was small. In series B, the change in K was different from that in series S. The K estimated in the test was almost constant as long as the centrifugal acceleration remained relatively small but increased rapidly with increased acceleration for further acceleration.

In the non-anchored cases in both series, K values decreased temporarily and then increased or were constant as centrifugal acceleration increased. This kind of change in K was affected by the mode of caisson movement.

3.2 *Dynamic loading tests*

3.2.1 *Displacement of caisson*

Figure 10 shows the relationship between the base acceleration and horizontal displacements of the caisson at its bottom and top. The displacements in the figure were measured at the end of each seismic

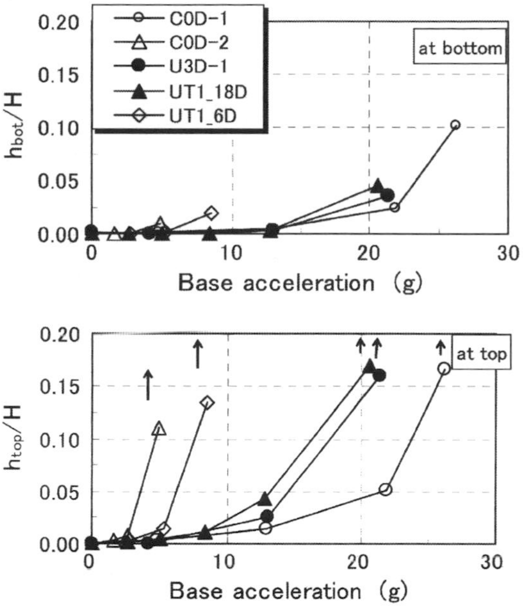

Figure 10. Displacement of caisson.

loading and normalized with respect to the caisson height. It was found that the displacements in all the test cases increased with base acceleration. The non-anchored caisson with a small caisson width, COD-2, showed a relatively large increase in displacement with increasing base acceleration and failed at about 5 g acceleration. In the anchored caissons, on the other hand, the displacements of the caisson in U3D-1 and UT1_18D increased gradually and the ground failed with large displacement at about 20 g acceleration. However, in UT1_6D, with short anchors, relatively large displacement took place and failure occurred at smaller acceleration. It was found that the anchors function to increase the base acceleration at failure, provided that the number and/or length of the anchors are sufficient.

In the non-anchored caisson, COD-1, the displacement at the top of the caisson is on virtually same order as that at the bottom of the caisson, which means that a caisson with a relatively large width moves almost horizontally under seismic loading. In the anchored caisson, on the other hand, the displacement at the top of the caisson is larger than that at the bottom, which means overturning displacement is dominant, rather than horizontal displacement. These phenomena indicate that the failure mode of the caisson becomes overturning failure rather than sliding failure when the caisson width is small.

3.2.2 *Interaction of caisson and backfill*

Several acceleration gauges were installed on the caisson and in the backfill in addition to the earth

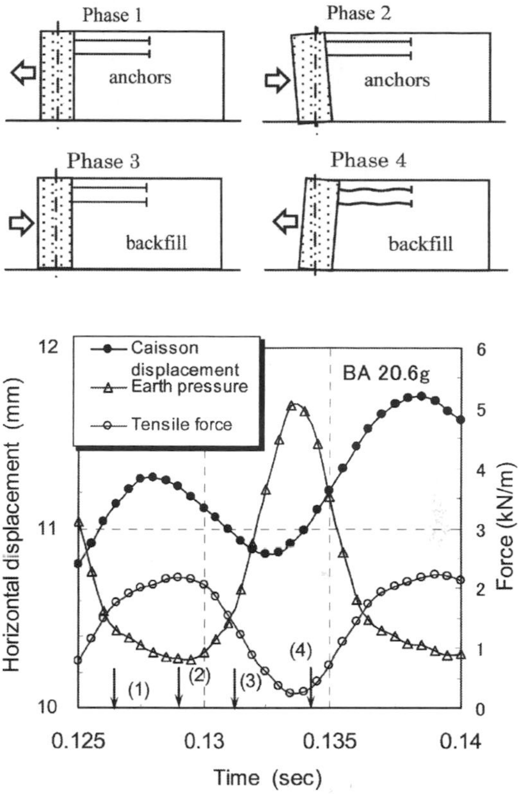

Figure 11. Interaction of caisson and backfill (UT1_18D; base acceleration is 20.6 g).

pressure gauges and tensile force gauges in order to investigate the interaction of the caisson and backfill. Figure 11 shows typical records measured in the test of UT1_18D, which include earth pressure, acceleration of the caisson, and the tensile force of the anchor. The figure also shows the horizontal displacement of the caisson at its center. Horizontal displacement of the caisson gradually increased with several rises and falls over the time duration. For ease in discussing the interaction of the caisson and backfill, it is convenient to divide the record into four phases.

In phase 1, in which the caisson and backfill move toward the sea side, the caisson moves faster than the backfill. The earth pressure acting on the rear side of the caisson decreases rapidly, whereas tensile force increases.

In phase 2, in which the caisson moves backward to the center, the earth pressure increases rapidly and the tensile force of the anchors decreases.

In phase 3, the caisson moves backward from the center, and the caisson movement is faster than the backfill movement, which causes an increase in earth pressure. Tensile force decreases continuously.

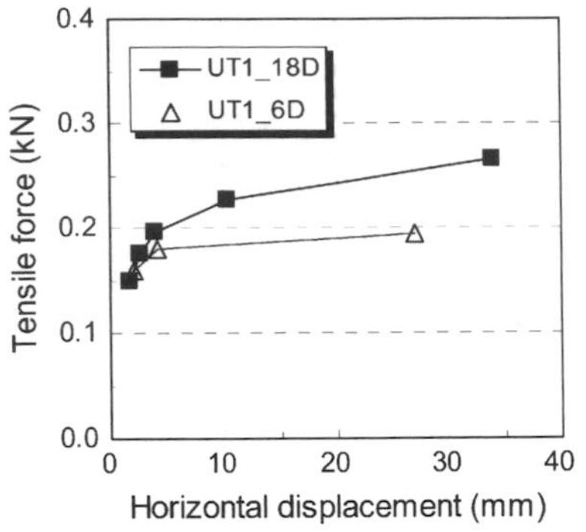

Figure 12. Tensile force.

In phase 4, the caisson moves forward, and the earth pressure decreases because the caisson moves forward faster than the backfill. Tensile force also increases.

These phenomena show that the earth pressure changes due to the interaction of the caisson and backfill, and the maximum earth pressure occurs when the caisson moves toward the backfill side.

3.2.3 *Dynamic earth pressure*

In the case of a non-anchored caisson, the maximum earth pressure increases very rapidly as base acceleration increases. The minimum pressure at each shaking is almost zero throughout loading. This is because the caisson moves faster toward the sea side than the backfill, as described in the previous section.

In the case of the anchored caisson, the maximum earth pressure is smaller compared with the non-anchored case and also shows a small increase as base acceleration increases. The maximum earth pressure occurs when the caisson moves toward the backfill side. This indicates that the magnitude of the maximum earth pressure is influenced not by the dynamic backfill pressure but by the inertial force of the caisson.

3.2.4 *Tensile force of anchors*

Typical measured tensile forces induced along the anchor rods are plotted in Figure 12 against the horizontal displacement at the top of the caisson for UT1_18D and UT1_6D, plotting the sum of the tensile forces. The tensile force in UT_18D was slightly larger than that in UT_6D even when the number of anchors was same. The anchors in UT_18D extended beyond the slip surface and were large enough to mobilize large tensile resistance, while the anchors in UT_6D were short and did not extend beyond the slip surface. Therefore, it can be concluded that the anchor length should be large and the anchors should extend beyond the slip surface in order to mobilize tensile force.

In the current Japanese design procedure for earth reinforcements, the maximum tensile resistance of the anchor is derived in a similar manner to the bearing capacity formula and can be formulated in the following equations (Public Works Research Center, 1997). The equations mean that the ultimate capacity of the anchor increases linearly as horizontal stress increases.

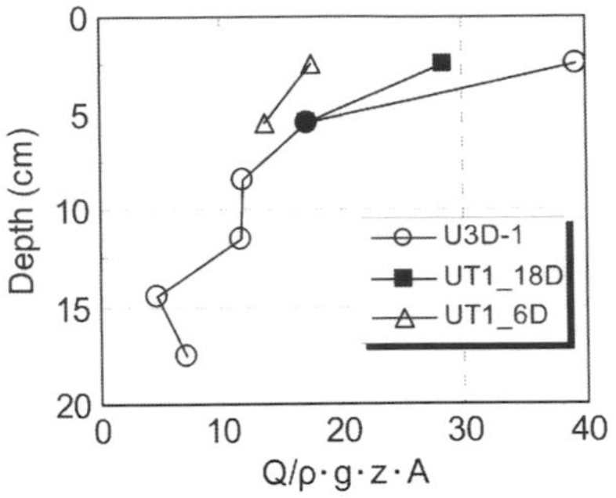

Figure 13. Tensile force distribution along the depth.

$$Q_{pu} = q_p \cdot N_q \cdot A \tag{1}$$

$$q_p = \rho \cdot g \cdot z \cdot K_a \tag{2}$$

where, A: sectional area of anchor plate, g: centrifugal acceleration, K_a: coefficient of active earth pressure, Q_{pu}: ultimate tensile resistance of anchor, N_q: bearing capacity factor, z: depth of each anchor, q_p: horizontal stress, ρ: density of backfill.

The measured tensile forces of each anchor are replotted in Figure 13 along the depth. Tensile force is normalized with respect to the calculated values. This figure shows that the mobilization ratio, $Q/(\rho \cdot g \cdot z \cdot A)$ at shallow depths is relatively large and then decreases rapidly in the depth direction. This is probably because the anchored caisson shows horizontal displacement in the deeper portion is not large enough to mobilize full resistance. This phenomenon indicates that only some anchors at relatively shallow depths mobilize their full resistance, and not all anchors achieve full capacity.

4 CONCLUSIONS

A series of static and dynamic centrifuge tests was performed to investigate the applicability of the anchored caisson to waterfront structures. The major conclusions derived from this study are summarized as follows:

1) A caisson reinforced with multiple anchor rods has higher resistance than conventional-type caissons.
2) From static loading tests, it was found that the earth pressure acting on the wall increases as the number of anchor rods increases. This increment is due to the increment of confining pressure in the

reinforced zone. An imaginary caisson width considering the safety of the structure can be estimated with a sufficient number of anchors.

3) The failure mode and resistance increment of the reinforced caisson are affected by the base ground condition.
4) Under dynamic loading conditions, the failure mode of an anchored caisson of relatively small width is overturning failure, whereas the failure mode of conventional-type caissons with large widths is sliding failure.
5) The tensile force on the anchors is influenced by the anchor depth and length at failure. A relatively large anchor length is required to mobilize large tensile force. The mobilization ratio of tensile resistance in the current design procedure decreases with increasing depth.

The authors wish to note that this research does not give sufficient consideration to actual construction problems. The conditions affecting construction of port facilities, such as the difficulty of backfill compaction in underwater construction and the effect of residual settlement of the base ground and backfill on anchor rods, should also be considered. Examination of problems of this type is important in the introduction of multi-anchored structures as sea revetments.

REFERENCES

Kikuchi, Y., Kitazume, M. and Kawada, Y. 1999. Applicability of Reinforced Earth Method to Sea Revetment. *Technical Note of the Port & Harbour Research Institute* (946): 36. (in Japanese).

Kitazume, M. and Miyajima, S. 1996. Development of PHRI Mark II Geotechnical Centrifuge. *Technical Note of the Port & Harbour Research Institute* (817): 33.

Public Works Research Center. 1997. *Design and execution manual for reinforced wall with multi-anchors*: 190. (in Japanese).

New Horizons in Earth Reinforcement – Otani, Miyata & Mukunoki (eds)
© 2008 Taylor & Francis Group, London, ISBN 978-0-415-45775-0

Surface holding conditions of reinforced slope and slope stability

Y. Nabeshima
Akashi National College of Technology, Hyogo, Japan

S. Kigoshi
Soken Engineering Co, Ltd., Osaka, Japan

ABSTRACT: A series of the direct sheer tests and model slope failure tests performed to investigate the surface holding effect of reinforced slope on the slope stability. As the results, it was confirmed that the shear force increased gradually as improving surface holding condition, and the bending and tensile strains of rock bolts increased as increasing the holding effect of slope surface. The shallow slope failure can be restrained by increasing holding effect of slope surface. The proposed reinforced earth method with the holding effect of slope surface was effective to stabilize the slope.

1 INTRODUCTION

In recent years, various reinforced earth methods for natural and cut slopes are proposed (e.g. Kusumi et al., 2002). The authors also proposed a new reinforced earth method which was composed of five items such as rock bolts, cup plates, pressure plates, steel wing plates and tie rods as shown in Figure 1. The rock bolts are used to stabilize a slope, and cup plates are used to fix rock bolts on the slope. Four steel wing plates are connected with a cup plate and put on the surface by a pressure plate. Tie rods connected the top of rock bolts. By combining these items, this reinforced earth method has a holding effect of slope surface, which is an advantage of the proposed reinforced earth method. However, the surface holding effect on the slope stability has not been fully clarified.

In this paper, a series of the direct sheer tests and model slope failure tests are performed to investigate the surface holding effect of reinforced slope on the slope stability. Direct shear tests are performed to investigate the reinforcing mechanism of the proposed reinforced earth method. Model slope failure tests are performed to investigate controllability of the shallow slope failure by the proposed method. The holding effect of slope surface on slope stability is discussed through the both test results.

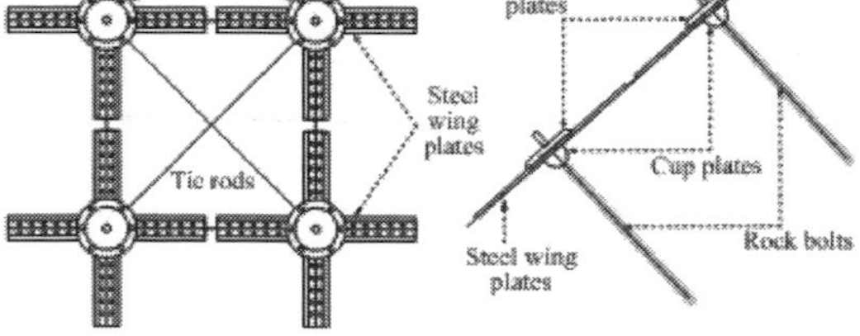

Figure 1. Schematic illustration of the proposed reinforced earth method.

2 DIRECT SHEAR TEST

2.1 *Test procedures of direct shear test*

The authors carried out a series of direct shear test to investigate the role of every items used in the proposed method. Figure 2 shows shear test apparatus used in this study. The soil box is formed by 20 steel frames. The inner size of soil box is 300 mm in length, 300 mm in width, and 410 mm in depth. Tooura sand was used as test sample to make model ground, and the model ground was prepared for the relative density of about 75%. In the direct shear tests, middle part of the soil box is sheared and the shear area is 100 mm between 145 mm and 245 mm from the surface as shown in Figure 2, and pushed out with 1.0 mm/min. Main items of the proposed method were truthfully modeled in the shear test. Rock bolts were modeled by brass bars, the diameter is 3.0 mm and the length is 320 mm, steel wing plates by acrylic boards which have 100 mm × 100 mm in length, 20 mm in width and thickness of 2 mm, and tie rods by brass bars, the diameter is 2.0 mm. And the model rock bolts have the screwed heads. Edges of a tie rod are equipped with

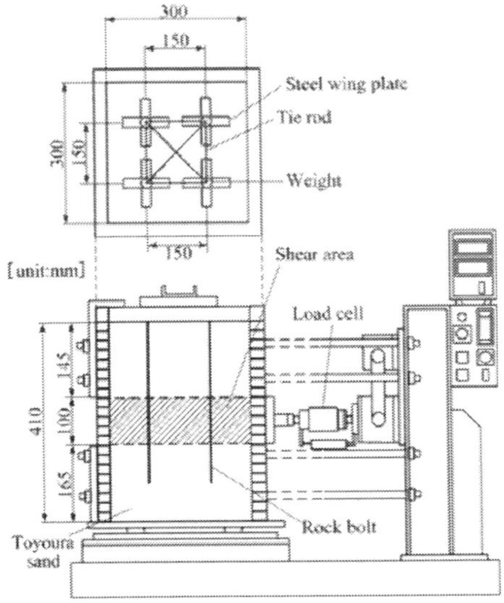

Figure 2. Direct shear test apparatus.

rings by which the heads of rock bolts can be connected. Details of the direct shear test are referable in the references (Kawajiri et al., 2005; Kawajiri et al., 2006).

2.2 *Test cases and test conditions*

Five cases of shear tests are performed, CASE1 is performed without any reinforcements, CASE2 is with rock bolts which are located in the white circles in Figure 3. CASE3 is with rock bolts and tie rods, CASE4 is with rock bolts and four steel plates, CASE5 is with rock bolts, tie rods, and four steel plates. Figure 5 shows appearances CASE2 to CASE5 in the model ground. Figure 4 shows a cross-sectional view of the model ground in the soil box. In CASE4 and CASE5, the weight of 200 g was put on every model steel plates to press down the ground surface. In addition, six strain gages are installed on rock bolts to measure tensile and bending strain. The direct shear test in this study was selected because the difference of surface condition on the shear test results was clearer than conventional direct shear test (Kawajiri et al., 2005).

2.3 *Test results*

Figure 5 shows relationship between shear force and displacement in all cases. As shown in this figure, the shear force in CASE2 to CASE5 is much bigger than that in CASE1. It is confirmed that the shear force increased by combining every items. In particular, the shear force in CASE5 is the biggest in all cases, and twice as big as that in CASE1 at the displacement of 20 mm. Thus, holding effect by combining every items contributes to an increase in shear force.

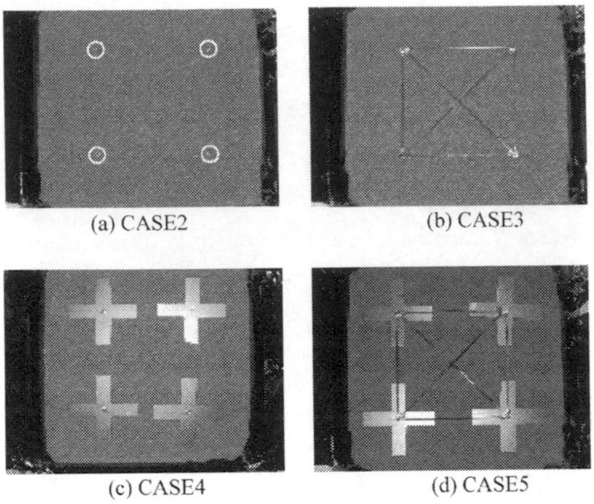

Figure 3. Appearance of direct shear tests in CASE2 to CASE5.

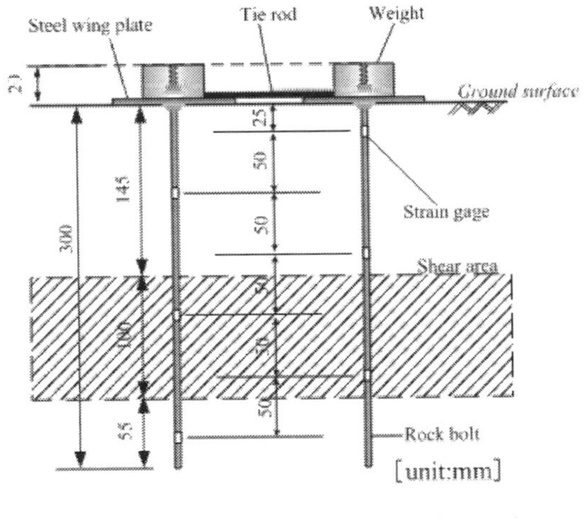

Figure 4. Cross-sectional view of the model ground.

Figure 6 shows the distribution of the tensile strains in CASE2 to CASE5 at the displacement of 20 mm. As shown in this figure, the tensile strains increase in the shear area. The tensile strain at 225 mm from ground surface in CASE4 obviously increase, it means that rock bolts with steel plates effectively work as reinforcements. While the tensile strain at the same position in CASE5 is smaller than that in CASE4, it attributes that tie rods distribute and average the load to all rock bolts. In addition, the tensile strain in CASE2 did not increase around surface area, although those in CASE3 to CASE5 increase. This is because holding effect by combining all items contributes to the increase in surface area. Figure 7 shows the distribution of the bending strains in CASE2 to CASE5 at the displacement of 20 mm. As shown in this figure, the bending strains increase in the shear area. And

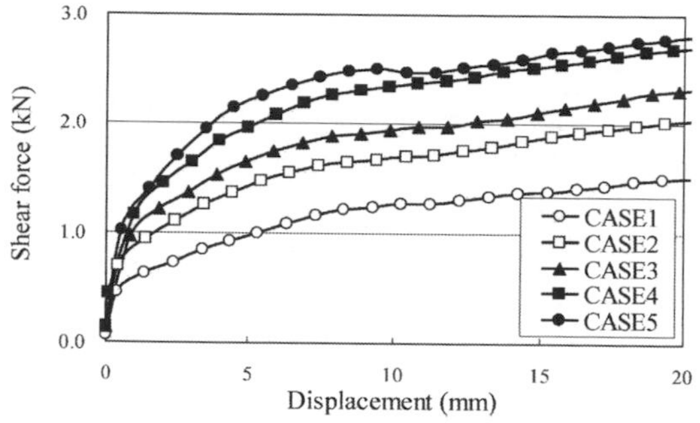

Figure 5. Relationship between shear force and displacement.

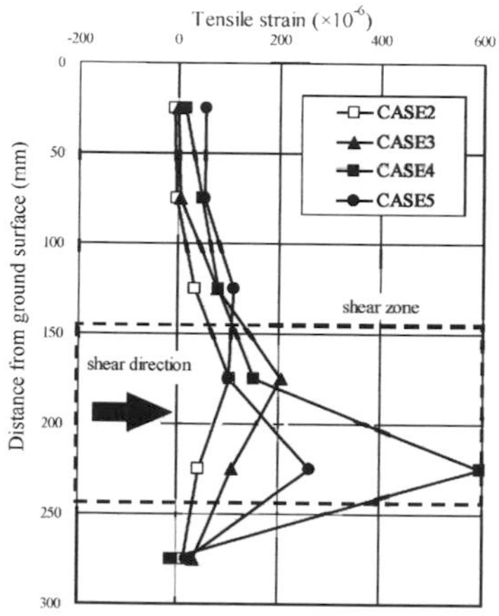

Figure 6. Tensile strains at displacement of 20 mm in CASE2 to CASE5.

in the shear area, the bending strains in CASE3 to CASE5 almost equal to that in CASE2, but in surface area, those in CASE3 to CASE5 are markedly smaller than that in CASE2. It seemed to be the holding effect of surface area by tie rods and steel plates. Thus, the holding effect of slope surface contributes to stabilize the slope judging from the behavior of bending strain.

3 SLOPE FAILURE TESTS

3.1 *Test procedures of slope failure test*

The surface holding condition was recognized as important factor to stabilize slope failure. The authors are carried out to investigate failure pattern of reinforced slope with and/or without surface holding effect.

In this study, the decomposed granite soil was used as test sample. The model slope with 1202 mm high, 1600 mm wide and 1700 mm long was constructed in

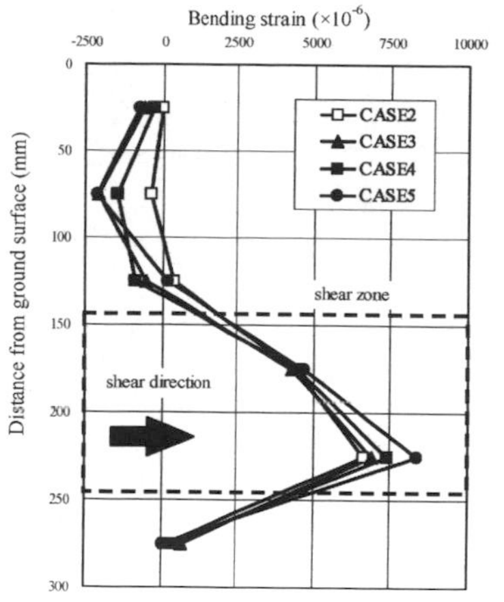

Figure 7. Bending strains at displacement of 20 mm in CASE2 to CASE5.

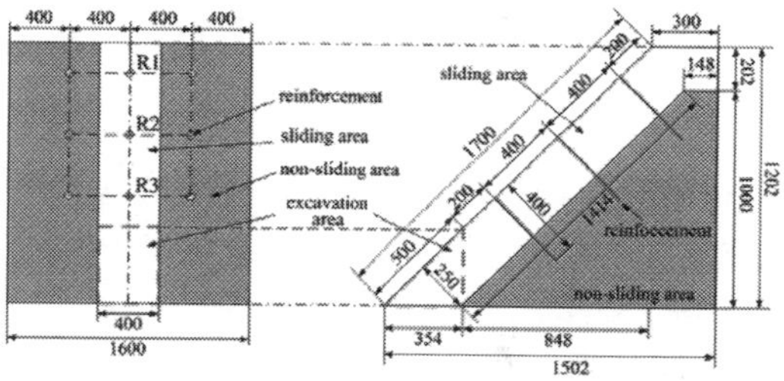

Figure 8. Appearance and schematic diagram of the model slope failure test.

the wooden box. Chloroethylene sheets were stuck on the wooden box to prevent absorption of water and thin rubber sheets and silicon grease were used as the lubrication. Figure 8 show the appearance and schematic diagram of the model slope, respectively. The model slope was divided into two areas, which were sliding area and non-sliding area. Thin Chloroethylene sheet was laid between two areas and the soil mass in the sliding area slid on the sheet. Decomposed granite soil in the non-sliding area was compacted to about 1.80 g/cm^3 of wet density with water content of about 14% which was the optimum water content. While decomposed granite soil in the sliding area was compacted to around 1.35 g/cm^3 of wet density with water content of about 7.5%.

3.2 *Test cases and test conditions*

In the model slope failure tests, a 1/5 model slope was made by a step construction method. Shallow slope failure was demonstrated by the excavation of the slope toe in the sliding area. The excavation area in the sliding area was shown in Figure 8 and every 50 mm

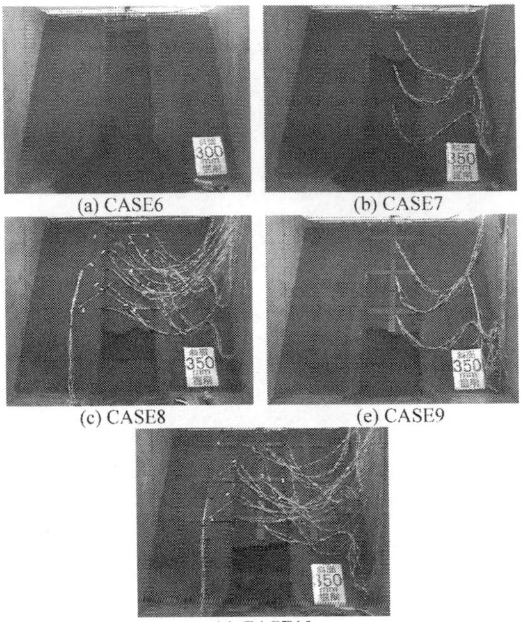

Figure 9. Slope failure pattern in CASE6 to CASE10.

of slope was excavated until 500 mm. Main items in the proposed method were modeled as faithfully as possible. Rock bolts were modeled by brass bars, which diameter is 3.0 mm and the length is 400 mm, steel plates by iron cross-shaped boards which have 380 mm × 380 mm in length, 50 mm in width and thickness of 2 mm, and tie rods by brass bars which diameter were 2.0 mm.

A series of slope failure test was carried out in conjunction with direct shear tests. Normal shallow slope failure without any reinforcements was demonstrated in CASE6, slope failure with rock bolts in CASE7, slope failure with rock bolts and tie rods in CASE8, slope failure with rock bolts and steel plates in CASE9 and slope failure with proposed method in CASE10.

Schematic arrangement of rock bolts in the slope failure test was also shown in Figure 8. Rock bolts were installed into non-sliding area. The settlement at the slope top and deformation of slope were measured at every excavation stages.

3.3 *Test results*

Figure 9 shows slope failure pattern of each case at the final excavation stages in which final excavation length were 300 mm in CASE6 and 350 mm in other cases. Whole slope failure was occurred in case B-1 and not occurred in other cases. It was clear that whole slope failure can be inhibited by reinforcements by comparing CASE6 and CASE7, because slope failure was limited at the middle and low part of slope in CASE7. In addition, surface slope can be hold by members of tie rods, steel plates and both combination as shown in CASE8, CASE9 and CASE10.

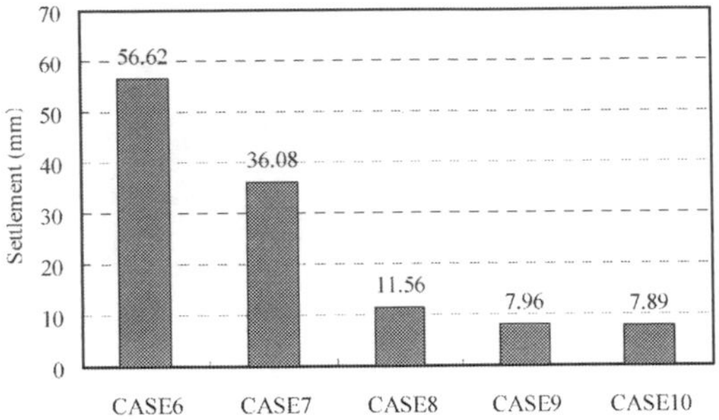

Figure 10. Final settlement at top of slope at final excavation stages.

Figure 10 shows the final settlement at top of the slope at the final excavation stages. As mention above, final excavation length was 300 mm in CASE6 and 350 mm in other cases. The final settlement in CASE7 was much smaller than that in CASE6, which was due to confliction effect of reinforcements. In addition, the settlement in CASE7 was much bigger than that in CASE8, CASE9 and CASE10. It was due to the surface holding effect of members such as tie rods, steel plates and their combination. From view point of surface holding effect, the slope stability increases as increasing the surface holding effect.

4 SURFACE HOLDING CONDITIONS AND SLOPE STABILITY

It was confirmed through the direct shear tests and model slope failure tests that surface holding conditions have a clear relationship between slope stability.

It was observed that shear force, tensile and bending strains increased as increasing surface holding effect in the direct shear tests and shallow slope failure could be controlled by changing the surface holding conditions in the model slope failure tests. Judging from both test results, shallow slope failure, which often becomes a trigger of whole slope failure, can be restrained by increasing holding effect of slope surface. Therefore, the proposed reinforced earth method with the holding effect of slope surface was effective to stabilize the slope.

5 CONCLUSIONS

In this paper, we proposed a new reinforced earth method, and performed a series of direct shear tests and model slope failure tests to investigate the holding effect of slope surface and the role of each part.

As the results, it was confirmed that the shear force increased gradually as combining each item. The tensile and bending strains of rock bolts increased as increasing the holding effect of slope surface. The shallow slope failure can be restrained by increasing holding effect of slope surface. Judging from both test results, the shear force and the holding effect of slope surface were increased by combining all items. The proposed reinforced earth method with the holding effect of slope surface was effective to stabilize the slope.

REFERENCES

Kusumi, H., Iwai, S., Fukumasa, T. and Kitamura, Y. (2002): Fundamental research on slope stability method of the natural slope considered the landscape and tree, Proceeding of 11th Japan Rock Mechanics Symposium, I-08.

Kawajiri, Y., Nabeshima, Y. and Kigoshi, S. (2005): Experimental study on a reinforced earth method with holding effect of slop surface, Proceedings of the 4th Korea/Japan Joint Seminar on Geotechnical Engineering, 153–158.

Kawajiri, Y., Nabeshima, Y., Tokida, K. and Kigoshi, S. (2006): Variation of reinforcing efficiency due to changing the surface holding conditions, Proceeding of 41st Annual Meeting of Japanese Society, 677–678.

New Horizons in Earth Reinforcement – Otani, Miyata & Mukunoki (eds)
© 2008 Taylor & Francis Group, London, ISBN 978-0-415-45775-0

Stability analyses of nailed sand slope with facing

C.C. Huang & W.C. Lin
Department of Civil Engineering, National Cheng Kung University, Taiwan

N. Mikami & K. Okazaki
Nittoc Construction Co. Ltd., Japan

D. Hirakawa & F. Tatsuoka
Department of Civil Engineering, Tokyo University of Science, Japan

ABSTRACT: Modified Bishop's and Fellenius' slice methods were used in the stability analysis of the behaviour of a set of unreinfored and nailed model sand slopes in loading tests in the laboratory. One of the Bishop's methods (i.e., method 4) successfully simulates the ultimate footing load and the location of slip surfaces for the unreinforced slope and most of the reinforced slopes reinforced slope with various types of facing. This method takes into account the soil strength increase induced by the reinforcement force. An exceptional case in which the experimental and the analytical results are inconsistent as observed in Test No. 1 on a reinforced slope is explained by the so-called 'wide-slab' effect that was not taken into account in this analysis.

1 INTRODUCTION

A series of loading tests on model nailed sand slopes with various types of facing (Fig. 1 & Table 1) was performed by Mikami et al. (2007). They reported that the use of continous facings (i.e., Test Nos. 1 and 3 listed in Table 1) substantially increased the ultimate bearing capacity and ductility of the nailed slope (Fig. 2). They also reported that higher values of T_o/T_{max} (T_o: reinforcement force activated immediately in back of the facing; and T_{max}: maximum reinforcement force in the reinforcement layer in the respective tests) were observed with the nailed slopes with higher local or global bending stiffness (Test Nos. 2 and 3 in Table 1). A series limit equilibrium analyses was performed on these test results listed in Table 1 to investigate the applicability of various slice methods in predicting the ultimate bearing capacity and the location of failure surface of nailed slope with facing.

The simplified Bishop's method (Bishop, 1955) is based on limit equilibrium satisfying: 1) force equilibrium in the vertical directions for all vertical slices; and 2) moment equilibrium of the circular failure mass around the center of rotation. The Fellenius' method (or conventional method, Fellenius, 1936) is also based on limit equilibrium satisfying: 1) force equilibrium in the direction normal to the slice base; and 2) moment equilibrium for the circular failure mass around the center of rotation. The simplified Bishop's method implicitly assumes that the resultant inter-slice force is activated

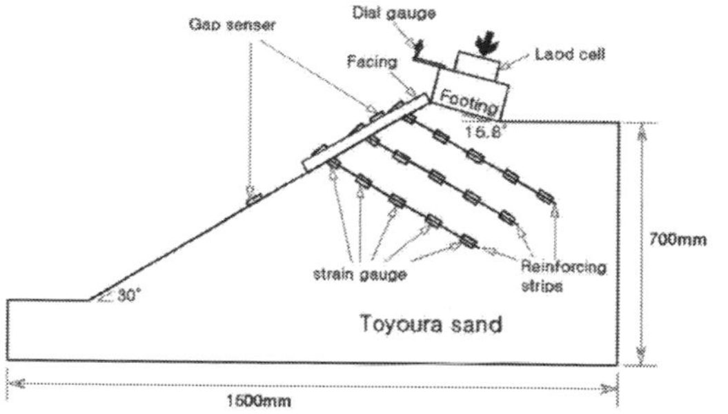

Figure 1. Configuration and geometry of loading test.

horizontal and equal magnitudes of the forces at both sides of the respective slices. The Fellenius' method implicitly assumes all the resultant inter-slice forces are activated in the directions parallel to the respective slice bases. These implicit assumptions have some influence on the calculated safety factor (F_s) of a given slope. In general, Fellenius' method provides underestimated values of F_s compared with those provided by the simplified Bishop's and other rigorous methods, such as Spencer's and Mogenstern and Price's methods.

2 STABILITY ANALYSES

The following four definitions of safety factor against circular sliding failure based on the modified Bishop's

Table 1. Summary of test conditions and results.

Test No.	Reinforcement	Facing type	ϕ_t (°)	ϕ_p (°)	Dry unit weight (kN/m^3)	q at 5 mm (kN/m^2)	q at 10 mm (kN/m^2)	q at 20 mm (kN/m^2)	Yielding point (kN/m^2)
1	YES	Agar + Cotton yarns (0.4 m × 0.3 m)	39.6	44.2	14.75	54.9	53.6	77.3	57.5 at 5.65 mm
2	YES	Bearing plate (15 mm × 15 mm)	39.3	43.8	14.70	32.7	37.0	47.0	33.8 at 7.34 mm
3	YES	Cement Bentonite + Polyester yarns (0.4 m × 0.3 m)	39.8	44.4	14.79	58.1	71.0	86.8	70.5 at 9.04 mm
4	YES	Non-woven filter (0.4m × 0.3m)	39.0	43.5	14.66	37.5	42.2	56.7	37.2 at 4.51 mm
5	YES	NO	40.9	45.8	14.98	30.9	33.3	42.8	32.7 at 9.29 mm
6	NO	NO	42.6	47.7	15.28	34.0	31.7	39.6	33.0 at 11.22 mm

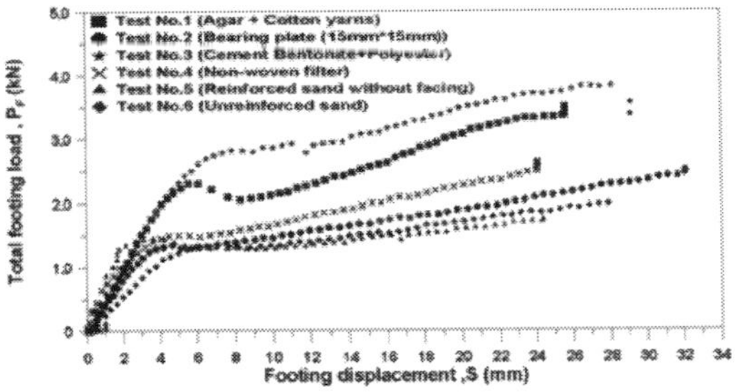

Figure 2. Total footing load (P_F) vs. footing displacement (S) relationships for reinforced and unreinforced slopes.

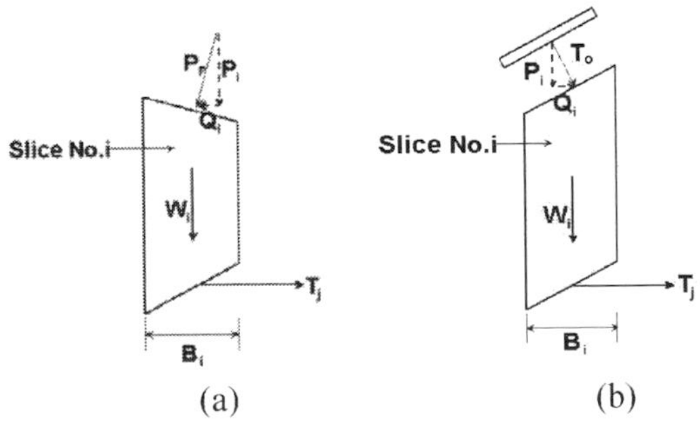

Figure 3. (a) Schematic figure of a slice subjected to the footing load and reinforcement force, (b) Schematic figure of a slice subjected to the reinforcement force and facing confinement.

and Fellenius' methods (e.g., Huang and Tatsuoka, 1994) are used:

Method 1 (Fellenius 1): incorporating the resisting moment induced by the reinforcement force and the confining force applied by the facing into the conventional Fellenius' method, defining the safety factor (F_{s_1}) as follow:

$$F_{s_1} = \frac{\sum\{[(W_i + P_i)\cdot\cos\alpha_i - Q_i\cdot\sin\alpha_i]\cdot\tan\phi_i\cdot R\} + \sum(T_j\cdot Y_{t_j})}{\sum[(W_i + P_i)\cdot R\cdot\sin\alpha_i + Q_i\cdot Y_{q_i}]} \quad (1)$$

in which (see Figs. 3a and 3b),

W_i : self-weight of slice No. i

P_i : vertical force applied at the top of slice No. i (positive downward)

Q_i: horizontal force applied at the top of slice No. i (positive outward)

α_i: inclination angle of slice base

ϕ : internal friction angle of the soil

R : radius of the circular failure surface

T_j : reinforcement force of layer No. j mobilized at the reinforcement-slice base intersection

Y_{t_j} : arm of rotation for reinforcing layer No. j

Y_{qi} : arm of rotation for Q_i

Method 2 (Fellenius 2): incorporating the shear strength increase of the soil induced by the reinforcement force and the confining force applied by the facing, into the conventional Fellenius' method:

$$F_{s_2} = \frac{\sum\{[(W_i + P_i)\cdot\cos\alpha_i - Q_i\cdot\sin\alpha_i]\cdot\tan\phi\cdot R\} + \sum(T_j\cdot\sin\alpha_j\cdot\tan\phi\cdot R)}{\sum[(W_i + P_i)\cdot R\cdot\sin\alpha_i + Q_i\cdot Y_{q_i}] - \sum(T_j\cdot\cos\alpha_j)} \quad (2)$$

Method 3 (Bishop 1): incorporating the resisting moment induced by the reinforcement force and the confining force applied by the facing into the conventional simplified Bishop's method:

$$F_{s_3} = \frac{\sum\left\{\left[(W_i + P_i)\cdot \sec\alpha_i \cdot \tan\phi\right]/\left(1 + \tan\alpha_i \cdot \tan\phi/F_{s_3}\right)\right\} + \sum\left(T_j \cdot Y_{t_j}\right)}{\sum\left[(W_i + P_i)\cdot R \cdot \sin\alpha_i + Q_i \cdot Y_{q_i}\right]} \tag{3}$$

Method 4 (Bishop 2): incorporating the shear strength increase of the soil induced by the reinforcement force and the confining force applied by the facing into the conventional simplified Bishop's method :

$$F_{s_4} = \frac{\sum\left\{\left[(W_i + P_i)\cdot \sec\alpha_i \cdot \tan\phi\right]/\left(1 + \tan\alpha_i \cdot \tan\phi/F_{s_3}\right)\right\}}{\sum\left[(W_i + P_i)\cdot R \cdot \sin\alpha_i + Q_i \cdot Y_{q_i}\right] - \sum\left(T_j \cdot \cos\alpha_j\right)} \tag{4}$$

In addition to the possible error induced by the basic assumptions used in the Felleniu's and simplified Bishop's methods as discussed previously, different approaches for taking into account the contribution of reinforcement force to the slope stability may yield some inherent errors in terms of the value of F_s. In general, approaches (1) and (3) generate under-estimated results due to the ignorance of soil strength increase induced by the normal stress increase along the failure surface. This drawback can be improved by taking into account the interaction between the reinforcement force and the soil shear stress along the failure surface, as described in Eqs. (2) and (4).

3 ANALYTICAL RESULTS AND DISCUSSIONS

The stabilities in all the test cases summarized in Table 1 were analyzed by the four methods represented by Eqs. (1)–(4) using the total footing load measured at the top of the footing, the measured reinforcement forces and the facing reaction forces that were inferred from the measured reinforcement forces immediately in back of facing. In each analysis, a trial-and-error grid consisting of 400 rotation centers with 20 mm horizontal and vertical spacings was used. For each rotation center, trial-and-error circular arc with various arms of rotation were used. The arm of a trial-and-error circle was increased at progressively at a small increment of 20 mm until it touches the boundary of the model slope. A minimum value of F_s for the slope was found from all the values of F_s calculated for all trial-and-error surfaces. Fig. 4 compares the calculated minimum values of F_s for all the test cases obtained based on Eqs. (1)–(4) taking into account the reinforcing effects and the confining effects of facing. Two soil internal friction angles, namely, the peak value from drained plane compression tests, ϕ_p, and the one from triaxial compression tests, ϕ_t were used in the all cases. The upper and lower bound values of F_s in the respective analysis cases shown in Fig. 4 are the values of F_s

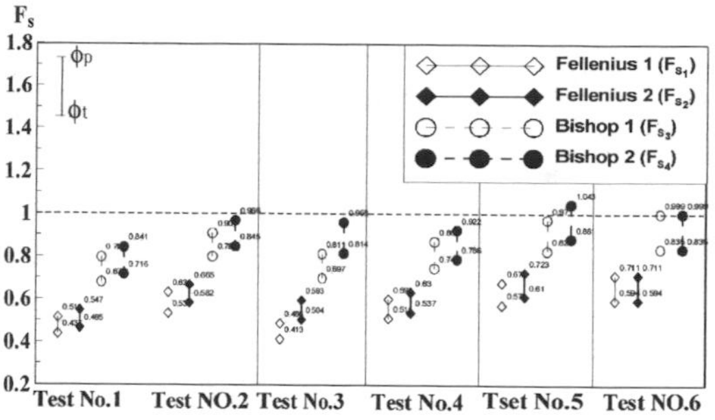

Figure 4. Calculated values of safety factors using various methods and ϕ's for the footing load measured at yielding point(facing's confining stress considered).

obtained by using the ϕ_p and ϕ_t values. The following can be seen from Fig. 4:

(1) With the base line tests (Nos. 5 and 6), $F_s = 0.99$ (unreinforced, Test No. 6) and 1.04 (reinforced without facing, Test No. 5) are obtained by Bishop 1 and Bishop 2 methods using $\phi = \phi_p$, which are in good agreement with the failure in the experiments (i.e., $F_s = 1.0$).

(2) With all the reinforced slopes with facing, the values of F_s using 'Bishop 2' method using $\phi = \phi_p$ generate $F_s = 0.92 - 0.97$, except $F_s = 0.84$ in Test No. 1. The inconsistency found for Test No. 1 results from relatively small values of T_{max} and T_o/T_{max} measured in this test, which may be attributable to the fact that the so-called 'wide-slab' effect was not taken into account in this slice method. An experimental observation of the possible 'wide-slab' effect for Test No. 1 is reported in Mikami et al. (2007).

(3) In all the cases of reinforced slopes, Tests Nos. 1–5, the F_s values by methods 1 and 3 are noticeably lower than the respective corresponding values by methods 2 and 4. This result indicates that methods 2 and 4, in which soil shear strength induced by the reinforcement force is neglected, are conservative, sometime overly. The same conclusion has been obtained by Huang and Tatsuoka (1994).

(4) In all the cases investigated (i.e., Tests Nos. 1–6), the values of F_s obtained by Fellenius' approach (i.e., methods 1 and 2) are significantly smaller than those obtained by Bishop's approach (i.e., methods 3 and 4). This intrinsic conservatism comes from the less realistic assumptions regarding the limit equilibrium formulations.

It is considered that the benefits of using a facing for slope stabilizing associated soil nailing is two fold:

(1) An increase in the tensile reinforcement force as represented by T_{max} via an increase in the effective lateral confinement of soils by facing; and

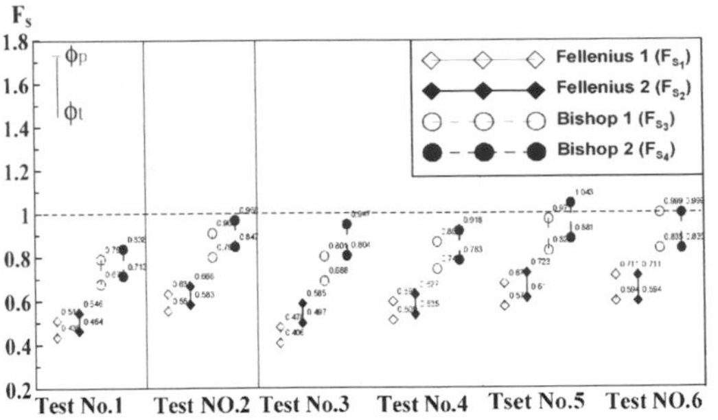

Figure 5. Calculated values of safety factors using various methods and ϕ's for the footing load measured at yielding point(facing's confining stress not considered).

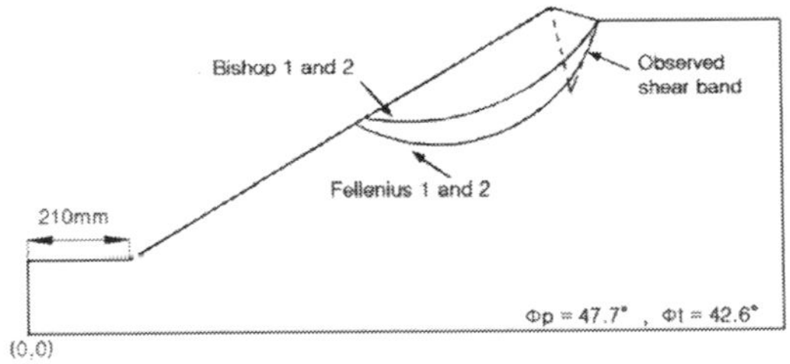

Figure 6. (a) Failure surfaces for the unreinforced slope (Test No.6) obtained using various methods under the yield-point footing load.

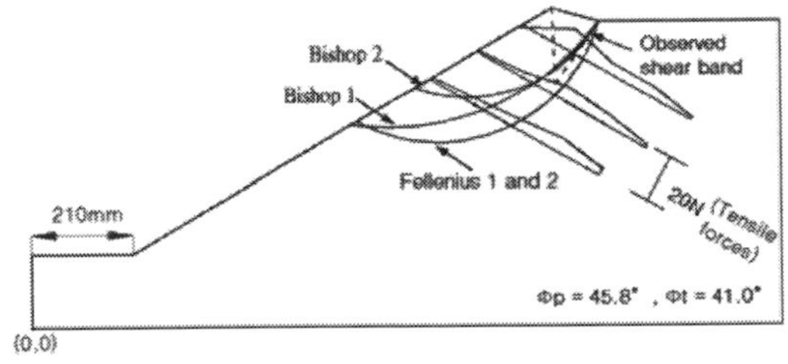

Figure 6. (b) Failure surfaces for the reinforced slope without facing (Test No.5) obtained using various methods under the yield-point footing load.

(2) An increase in the external force T_o (or P_i and Q_i) on the slope surface.

Figure 5 shows results from similar analysis as Fig. 4 but not taking into account the confining pressure activated by the facing in the analysis. Note that the confining effects of facing are reflected in the measured reinforcement forces and these are used in the analysis. The difference between the F_s values presented in Figs. 4 and 5 in the respective cases is generally less than 1%. This result indicates that the effects of factor 1) above are more pronounced than factor 2).

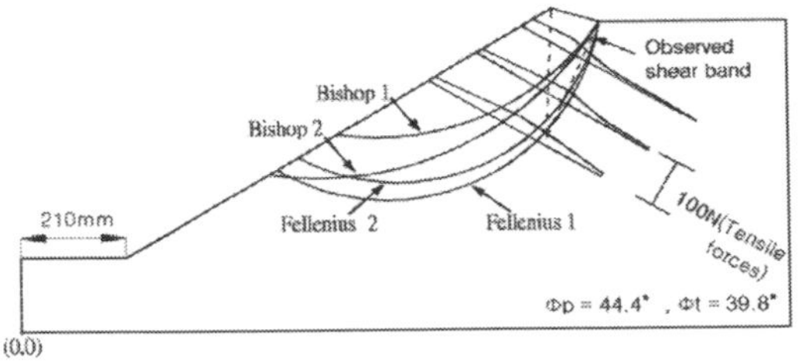

Figure 6. (c) Failure surfaces for the reinforced slope with Cement Bentonite + Polyester yarns facing (Test No.3) obtained using various methods under the yield-point footing load.

In the present analysis, the effects of strength anisotropy and progressive failure on the soil shear strength mobilized at the moment of slope failure are not taken into account. Then, the F_s values calculated using the ϕ_p should be larger than 1.0 if no inherent errors are included in the analysis. As these F_s values obtained by using the ϕ_p values are generally close to 1.0, or lower than 1.0 even based method 4, it seems that method 4 still include some inherent errors.

Figures 6(a)–6(c) show the critical failure surfaces obtained by the four slice methods (i.e., Eqs. 1-4) for Test Nos. 6, 5 and 3, respectively, which are typical of those obtained by the present analysis. The shear bands observed at large footing settlements in the experiments are also plotted for comparison. It was found that the location and shape of potential failure surface are not influenced by the used ϕ values, ϕ_p or ϕ_t. Furthermore, deepest failure surfaces are obtained by analysis for Test No. 3, in which the greatest bearing capacity was obtained among the tests performed. This consistency reveals the relevance of the stability analysis method described in this paper.

4 CONCLUSIONS

The stability of nailed model slopes with various types of facing were analysed using modified Bishop's and Fellenius' methods. The following conclusions were obtained:

(1) The modified Bishop's methods (Bishop 1 and Bishop 2) successfully simulate the ultimate footing load and slip surface of the unreinforced slope and the reinforced slope without facing when ϕ_p is used.
(2) The modified Bishop's method (Bishop 2) also successfully simulate the ultimate footing load and the failure surface for three reinforced slopes with various types of facing (Test Nos. 2, 3 and 4). The stability (or ultimate footing load) for the reinforced slope with facing type 1 (Test No. 1)

was underestimated, which could be due to the so-called 'wide-slab' effect, which was not taken into account in the analysis.

(3) The benefits of using facing is twofold: the increase in the tensile reinforcement force of via the effect of soil confinement via facing-confining; and direct effects of the facing-confining pressure on the slope surface. The results of the stability analysis showed that the effect of the former factor is much more significant in the stability analysis.

REFERENCES

Bishop, A.W. (1955) "The use of the slip circle in the stability analysis of slopes" Geotechnique, Vol. 5, No. 1, pp. 7–17.

Fellenius, W. (1936) "Calculation of the stability of earth dam" Trans. 2nd Cong. Large Dams, Vol. 4, pp. 445.

Huang, C.C. and Tatsuoka, F. (1994) "Stability analysis for footings on reinforced sand slopes" Soils and Foundations, Vol. 34, No. 3, pp. 21–37.

Mikami, N., Okazaki, K., Hirakawa, D., Tatsuoka, F. and Huang, C.C. (2007) "Effect of facing rigidity on the stability of nailed sand slope in model tests" Proc. 5th Int. Sympo. on Earth Reinforcement, IS Kyushu '07- New Horizon in Earth Reinforcement, Nov. 14–16, 2007 at Kyushu University, Fukuoka, Japan.

New Horizons in Earth Reinforcement – Otani, Miyata & Mukunoki (eds)
© 2008 Taylor & Francis Group, London, ISBN 978-0-415-45775-0

Effect of facing rigidity on the stability of nailed sand slope in model tests

N. Mikami & K. Okazaki
Nittoc Construction Co., LTD., Japan

D. Hirakawa & F. Tatsuoka
Department of Civil Engineering, Tokyo University of Science, Japan

C.C. Huang
Department of Civil Engineering, National Cheng Kung University, Taiwan

ABSTRACT: A series of loading tests on nailed model sand slope covered with various types of facing was performed to investigate the effect of facing rigidity on the load-settlement behavior of nailed slopes loaded on its crest. The test results showed that the ultimate bearing capacity (i.e., the stability) of the nailed slope was effectively increased with an increase in the facing rigidity by using continuous stiff and/or flexible panel facings. It was also shown that the post-yielding strain hardening behavior of the nailed slope becomes significantly ductile by using effectively facing having a relevant rigidity.

1 INTRODUCTION

Facings are conventionally used as one of the essential structural components for reinforced or nailed slopes. However, the contribution of facings to the stability of reinforced slopes is rather poorly understood and is yet to be addressed in the current design guidelines. Gutierrez and Tatsuoka (1988) pioneered an experimental study focusing on the mechanical contribution of facing to the stability of reinforced slopes by performing loading tests on a footing placed on the crest of reinforced sandy slope. They found that the bearing capacity of reinforced slope increases by using a rigid facing of which connected to the nails. However, reports of similar experiments using facings having different degrees of rigidity cannot be found in the literature.

The present study investigated the effects of facing rigidity on the stability of nailed slope by performing loading tests on a footing placed on the crest of nailed model sandy slopes with a wide variety of facing rigidities. Numerical stability analysis on these tests is reported in the companion paper (Huang et al., 2007).

2 LOADING TESTS ON MODEL SLOPES

Figure 1 shows the test set-up in the present study. The model sandy slope was constructed using

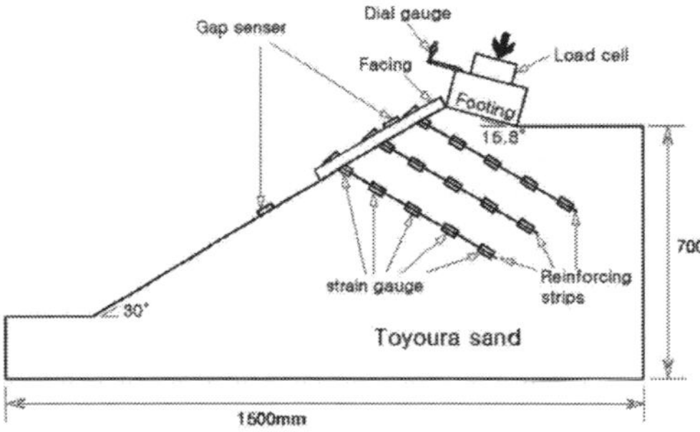

Figure 1. Configuration and geometry of loading test.

Toyoura sand, which is a uniform subangular fine sand ($e_{max} = 0.933$, $e_{min} = 0.624$, $G_s = 2.650$, $D_{50} = 0.179$ mm). The model was prepared by pluviating air-dried particles of Toyoura sand through air using multiple sieves. Table 1 shows the average dry unit weight (γ_d) of the model slopes. The surface of a sand mound was trimmed to form a slope with an angle of inclination from the horizontal, $\alpha = 30°$, and the top of slope was trimmed to form an inclined loading plane (15.8° inclined-backward) for placing a 100 mm-wide rigid footing as shown in Fig. 1.

Four types of facing having largely different rigidity (or strength) were prepared by using different natural or artificial materials listed in Table 2. Phosphor bronze strips, 3 mm-wide, 0.5 mm-thick and

Table 1. Summary of test conditions and results.

Test name	Reinforcement	Facing type	Φ_t	Φ_p	Dry density (kN/m^3)	q at 5 mm (kN/m^2)	q at 10 mm (kN/m^2)	q at 20 mm (kN/m^2)	Yielding point (kN/m^2)
1	Yes	Agar + cotton yarns	39.6	44.2	14.75	54.9	53.6	77.3	57.5 at 5.65 mm
2	Yes	Bearing plate (15 mm* 15 mm)	39.3	43.8	14.7	32.7	37.0	47.0	33.8 at 7.34 mm
3	Yes	Cement bentonite + Plyester yarns	39.8	44.4	14.79	58.1	71.0	86.8	70.5 at 9.04 mm
4	Yes	Non-woven filter	39	43.5	14.66	37.5	42.2	56.8	37.2 at 4.51 mm
5	Yes	No	40.9	45.8	14.98	30.9	33.3	42.8	32.7 at 9.29 mm
6	No	No	42.6	47.7	15.28	34.0	31.7	39.6	33.0 at 11.22 mm

Table 2. Facing types and material properties.

Facing type	Materials	Size	Tensile strength (100 mm-wide)
1	Agar + Cotton yarns	0.4 m*0.3 m	25 N
2	Bearing Plate	0.15 m*0.15 m	>300 N
3	Cement bentoine + Polyester yarns	0.4 m*0.3 m	>100 N
4	Non-woven filter	0.4 m*0.3 m	84 N

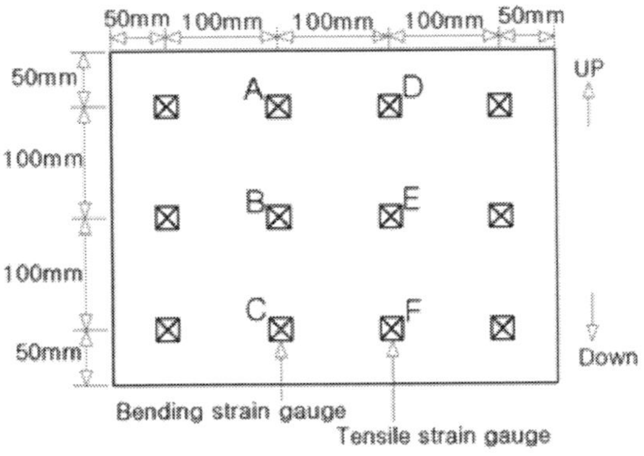

Figure 2. Front view of facing plate and locations of reinforcing bars.

250 mm-long, were used as model tensile reinforcement members simulating prototype soil nails. These strips were placed with a vertical and horizontal center-to-center spacing of 100 mm and the top ends were fixed to the back face of facing in some of the tests. An unreinforced slope without facing (Test No. 6) and a reinforced slope without facing (Test No. 5) were performed as baseline tests. The front view of the facing and the reinforcement configurations are show in Fig. 2. Electric-resistant strain gages were attached to the reinforcing strips at the positions shown in Fig. 3. The surface of strips and strain gages were coated using epoxy and Toyoura sand particles to simulate a rough soil-reinforcement interface condition.

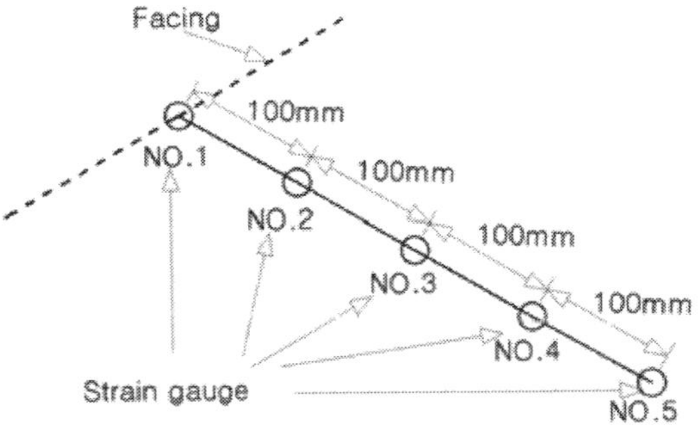

Figure 3. Schematic view of reinforcing strip and locations of strain gauges.

A 100 mm-wide rigid footing was loaded at a constant displacement rate in the direction normal to its base. Five two-component load cells, which measure normal and shear load simultaneously with negligible coupling effects, were mounted on the central third of the base of the 400 mm-long footing to eliminate the influence of the boundary friction from the sidewalls of the sand box. Displacement of the footing was measured using a displacement transducer and the deformation of facing and slope surface were measured using proximity transducers (i.e., gap sensors) as shown in Fig. 1.

3 TEST RESULTS AND DISCUSSIONS

Figure 4 shows the relationships between the total normal footing load (P_F) and the footing settlement (S) in the direction normal to the footing base. As this load was measured using a load cell arranged between the footing and the loading piston, it includes the boundary friction and it is greater by about 10% than the value measured using the load cells mounted at the central third of the footing. The following trends can be seen from Fig. 4:

1) The behaviors of the unreinforced slope and the reinforced slope without facing (Test Nos. 5 and

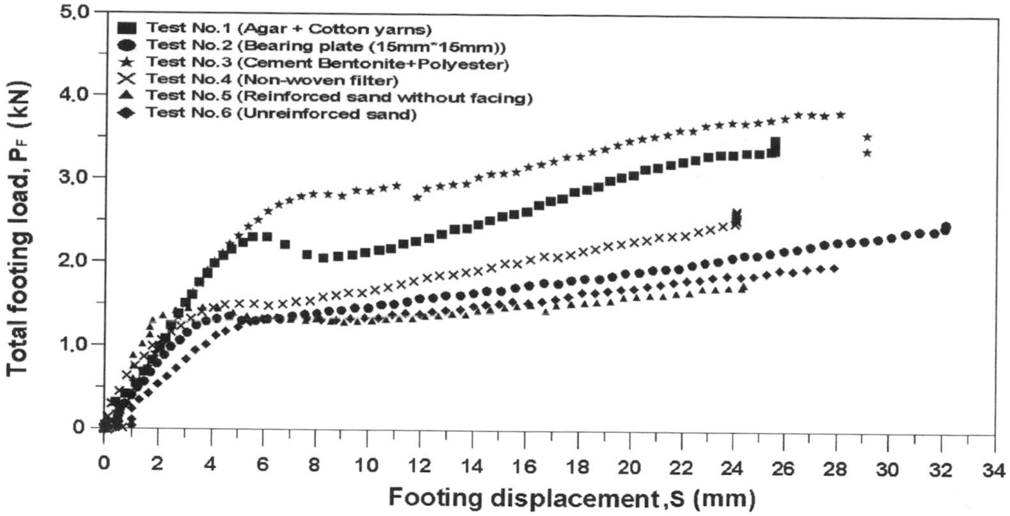

Figure 4. Total footing load (PF) vs. footing displacement (S) relations for reinforced and unreinforced slopes.

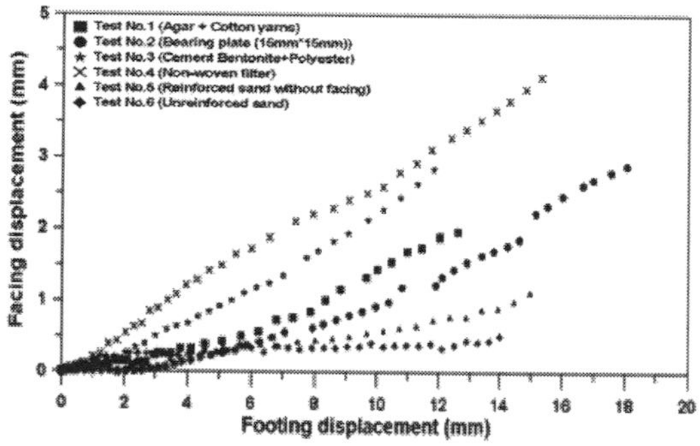

Figure 5a. Normal displacement of facing at 100 mm from the slope crest.

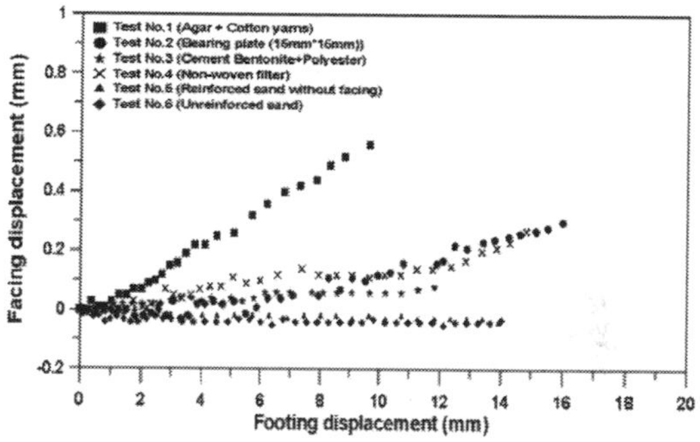

Figure 5b. Normal displacement of slope face at 400 mm from the slope crest.

6) are rather similar. However, when taking into account the fact that the unreinforced sand slope was denser than the reinforced slope without facing and comparing the behaviours for the same sand density, the unreinforced slope should become noticeably weaker than the reinforced slope without facing.

2) Two reinforced slopes with type 1 and type 3 facings (Test Nos. 1 and 3) are much stronger than these two slopes described above. Again, when taking into account that the slopes in Tests Nos. 1–4 are generally looser than those in Tests Nos. 5 & 6 and comparing the behaviours for the same sand density, the effects of facing rigidity should have become larger than those seen from Fig. 4.

Fig. 5(a) shows the heaving of the facing in the direction normal to the slope face measured at a 100 mm distance from the slope crest (see Fig. 1). It can be seen the deformations of the reinforced slopes are generally larger than the unreinforced slope and the reinforced slope without facing. It is like that these two slopes failed under the 'punching' mode due to a low strength of the sand zone below the footing because of no reinforcing (Test No. 6) or small reinforcing effects resulting from free top conditions of reinforcement (Test No. 5). On the other hand, it is like that the failure of the reinforced slope with facing was more general associated with load re-distribution and soil confinement within the more stabilized reinforced zone. Fig. 5(b) shows the heaving of slope surface measured at a 400 mm distance from the slope crest. It can be seen that only the test using type 1 facing (agar + cotton yarns) has a distinctly large deformation compared with the others. This trend may be attributed to the load re-distribution

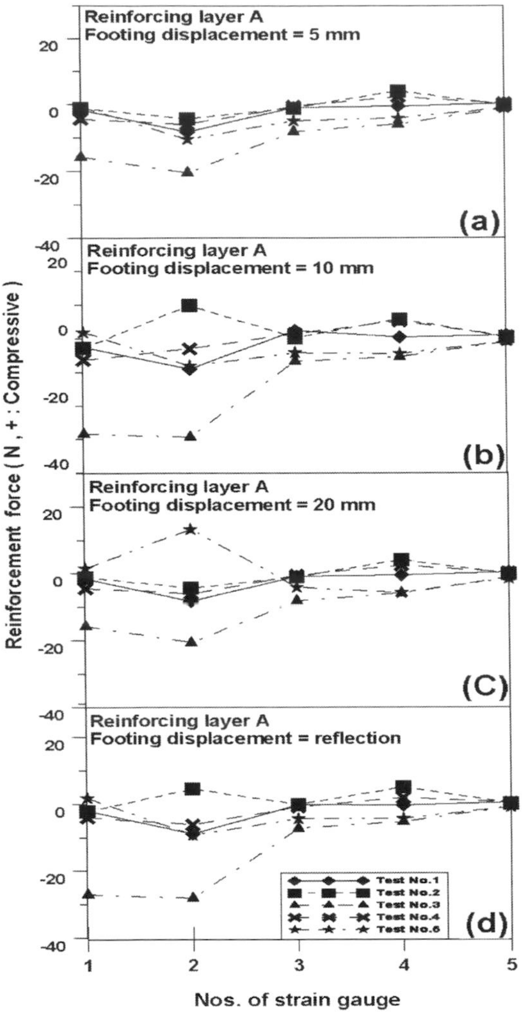

Figure 6. Reinforcement force activated in the top reinforcement layer measured at various footing settlements: a) 5 mm, b) 10 mm and c) 20 mm, and d) at yielding point.

mechanism induced by the 'wide-slab' effect formed in the reinforced zone as reported by Huang and Tatsuoka (1994). This wide-slab mechanism may account for the large value of q (=57.5 kN/m^2 at yield point which is second largest one among the six tests performed).

Figs. 6a)–d) show the tensile force in the uppermost layer of reinforcement 'A' at various strain gage locations (shown in Fig. 1) measured when the footing settlement S was equal to 5 mm, 10 mm and 20 mm and when the load-settlement relation exhibited a yielding point. The tensile force is always highest in Test No. 3 among all the tests regardless of the footing settlements. This is consistent with the largest footing pressure q in this test, as summarized in Table 1. It

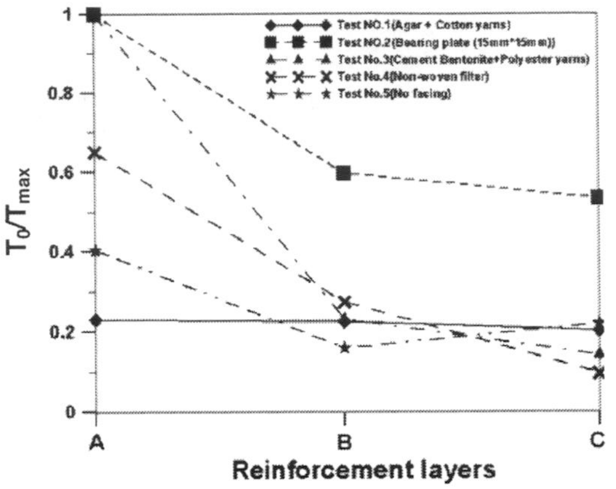

Figure 7. Measured values of T0/Tmax in the top reinforcement layer measured at the yield point.

is also important to note that the test results indicate high connection force at the back of the facing, which is essential to activate high reinforcing effects.

Fig. 7 compares the values of T_o/T_{max}, where T_o is the reinforcement force developed adjacent to the slope surface, and T_{max} is the maximum tensile force among the three reinforcement layers A, B and C in all the tests, measured at the yielding point of the footing load-settlement (P_F – S) curve. It can be seen that Test Nos. 2 and 3 using facings that had locally or globally highest bending stiffness exhibit highest values of T_o/T_{max}. The values of T_o/T_{max} in the other tests fall between 0.2 and 1.0. The second highest value of q measured in Test No. 1 is inconsistent with the fact that the measured reinforced force is relatively low. This trend of behavior may be related to the 'wide-slab' effect and a future study into this point is necessary. Despite that the above, the general trend is an increase in the footing pressure (i.e., an increase in the slope stability) with an increase in the T_o/T_{max} value associated with an increase in the facing rigidity.

4 CONCLUSIONS

From the test results obtained in a series of loading test on a set of model nailed sand slopes using various types of facing, the following conclusions can be derived:

1) Ultimate bearing capacity and ductility of nailed slopes increases effectively by using rigid facing panels to which the head of nails are connected.
2) The use of panel facings with a local or global bending stiffness (facing types 2 and 3) effectively increase the tensile reinforcement force, in particular adjacent to the back of the facing.
3) Relatively high tensile reinforcement force including a high value in back of the facing measured in

Test No.3 are consistent with the highest value of footing pressure (i.e., the highest stability) in this test among all the tests investigated.

REFERENCES

Huang, C.C., Mikami, N., Okazaki, K. Hirakawa, D. and Tatsuoka, F. (2007) "Stability analysis of nailed sand slope with facing" Proc. 5th Int. Sympo. on Earth Reinforcement, IS Kyushu '07- New Horizon in Earth Reinforcement, Nov. 14–16, 2007 at Kyushu University, Fukuoka, Japan.

Huang, C.C., Tatsuoka, F. and Sato, Y. (1994) " Failure mechanisms of reinforced sand slopes loaded with a footing" Soils and Foundations, Vol. 34, No. 2, pp. 27–40.

Gutierrez, V. and Tatsuoka, F. (1988) "Role of facing in reinforcing cohesionless soil slopes by means of metal strips" Proc. Int. Geotech. Sympo. On Theory and Practice of Earth Reinforcement, Fukuoka, Japan. Yamanouchi, T., Miura, N. and Ochiai, H (Eds.), Balkema, pp. 553–558.

New Horizons in Earth Reinforcement – Otani, Miyata & Mukunoki (eds)
© 2008 Taylor & Francis Group, London, ISBN 978-0-415-45775-0

The effect of inclination of reinforcement on the horizontal bearing capacity of the ground reinforcing type foundation

J. Izawa & H. Kusaka
Department of Civil Engineering, Tokyo Institute of Technology, Japan

M. Ueno & N. Nakanani
Engineering division, Dai Nippon Construction, Japan

H. Sato
Tokyo Electric Company, Japan

J. Kuwano
Geosphere Research Institute, Saitama University, Japan

ABSTRACT: The ground reinforcing type foundation, which is a caisson type pile foundation with steel reinforced bars around the pile, was developed with the aim of increasing compression and uplift bearing capacity of the foundation. Compared to current popular foundation, the ground reinforcing type foundation can dramatically reduce the construction cost because of the smaller in size of the new foundation. Form the past practical experiences, it was found that this method can improve not only uplift bearing capacity but also horizontal capacity. It is inferred that inclination of reinforcements is effective to the horizontal bearing capacity. Therefore, this research is focusing on the effect of inclination of reinforcements on the horizontal capacity of the caisson type pile foundation by using of steel reinforced bars. For that purpose, the horizontal loading tests in the centrifugal acceleration were then conducted. Testing results showed that the horizontal capacity of foundation that placing the reinforcement in the diagonal direction is higher than that of the foundation placing the horizontal reinforcement because normal force can act on the reinforcements effectively.

1 GENERAL

An auxiliary geo-reinforcing type foundation (GRF) around a caisson type foundation increases compression and uplift bearing capacity of the main foundation. The mechanism of reinforcement and the evaluation method of compression as well as uplift bearing capacity were clarified by Matsuo et al. (1989) and Nakai et al. (1996) respectively. This paper is focused on the optimal inclination of reinforcement against horizontal loading during earthquake load or wind load. When a caisson foundation with GRF receives horizontal loads, it acts like a rigid foundation and displacement by rotation becomes dominant as indicated in Figure 1. In consideration of such behaviour, it is thought effective to arrange reinforcements in a slant direction for following reasons.

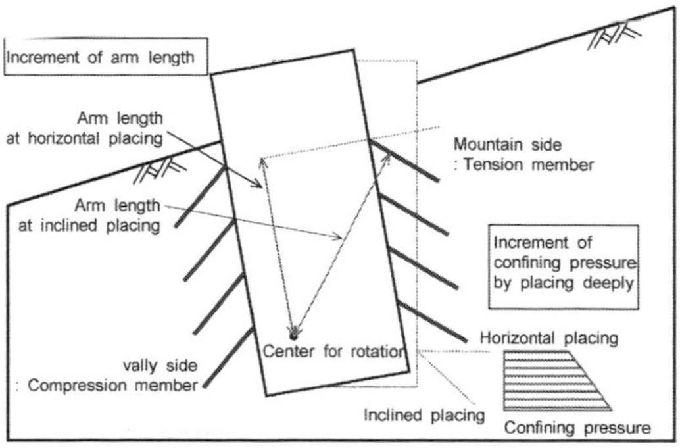

Figure 1. Schematic view of horizontally loaded caisson foundation with inclined reinforcement.

- Resistance moment increases as arm length between the rotation centre and each reinforcements of GRF foundation becomes large.
- Since reinforcements can be placed at various depths with large confining pressure, frictional force between reinforcement and ground increases.

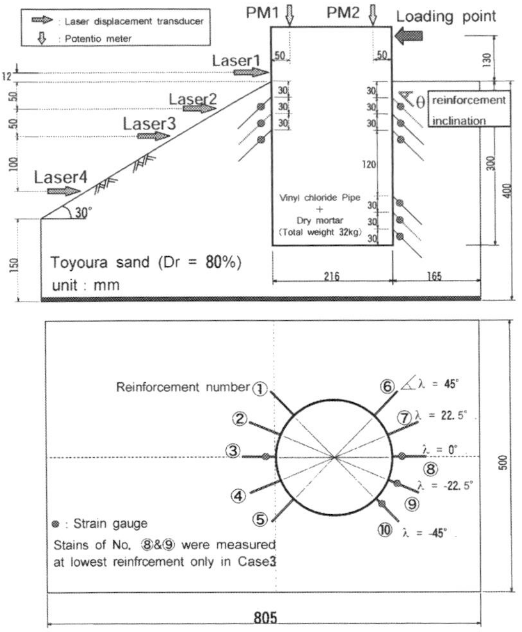

Figure 2. Centrifuge model of caisson foundation with GRF.

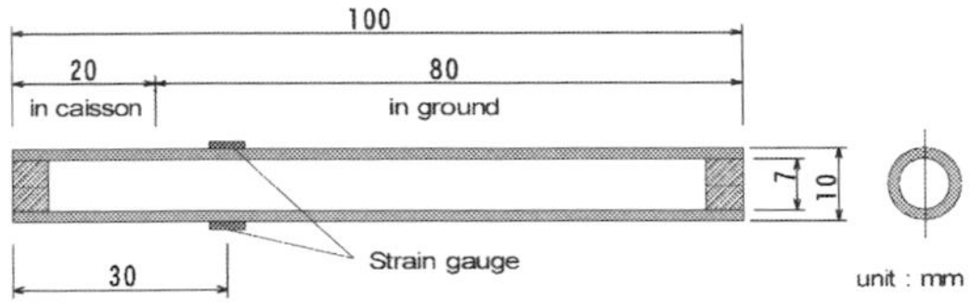

Figure 3. Model reinforcement.

In order to evaluate these influences, this research is focused on horizontal loading tests and centrifuge model tests were performed.

2 THE OUTLINE OF THE TEST

Figure 2 shows a configuration of a model caisson foundation with reinforcement. The foundation was installed on the foundation with 30 inclinations supposing the case where it applies to an alpine area. Model was created with dry Toyoura sand with relative density of 80% ($D_{50} = 0.19$ mm, $U_c = 1.56$, $\phi = 41$ deg.) in a rigid container of 805 mm (length) × 500 mm (width) × 400 mm (height). A vinyl chloride pipe with outside diameter of 216 mm and thickness of 8 mm was used for model caisson. The pipe was filled up with cement mortar with compression strength of 24 MPa. The total weight of model caisson was 32 kg. The hollow bakelite stick was used for the model reinforcement as shown in Figure 3. Tensile rigidity (EA) of the model was coincided to a prototype model. In order to measure sectional force acting on reinforcement, strain gauges were pasted on upper and lower surface of reinforcement. Toyoura sand was pasted on the surface of model reinforcement in order to achieve sufficient friction between soil and reinforcement. Friction characteristics obtained from pullout test were indicated in Figure 4. The Reinforcements were arranged in all the cases as shown in Figure 2 and Table 1. The model was loaded horizontally by using electric jack with displacement velocity of 1.0 mm/min and jack displacement up to 20 mm at loading height in centrifugal acceleration of 50 G. Loading point was at 13 0mm from ground surface in model scale.

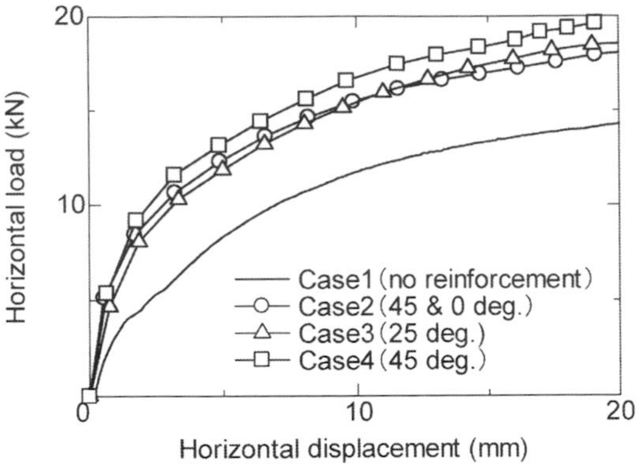

Figure 4. Horizontal load vs Horizontal displacement.

Table 1. Test case and inclination of reinforcement.

	Vally side	Mt. side upper	Mt. side bottom
Case 1	No reinforcement		
Case 2	45°	0°	45°
Case 3	25°	25°	25°
Case 4	45°	45°	45°

3 TEST RESULTS

Figure 4 shows relationships between horizontal load and horizontal displacement of caisson. Case-4 with placing angle of 45 degree shows larger horizontal bearing capacity in compare to Case-3 with reinforcement inclination of 25 degree. In Case-2 which showed lowest horizontal bearing capacity among the reinforced cases, an increment of horizontal bearing capacity could be seen by about 25% as compared with Case-1 with no reinforcement.

Figure 5 shows schematic views of the movement of the caisson foundation, which were calculated from three displacement measurements: Laser1, PM1 and

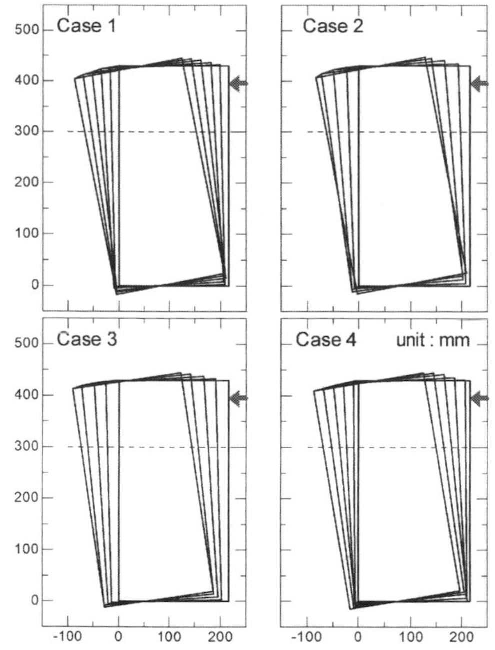

Figure 5. Behavior of model caisson.

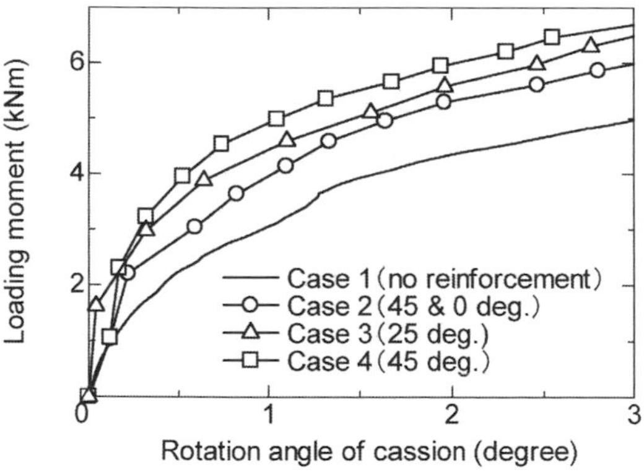

Figure 6. Loading moment vs Rotation angle of caisson.

PM2, before rotation angle of the caisson goes to 10 degree. From these figures, it can be confirmed that most horizontal displacements were depended on rotation and sliding was very small. Therefore, relationships between rotational moment at loading point and rotation angle of the caisson were calculated and are shown in Figure 6. Here, center of rotation was assumed to be center of bottom of the caisson. If the moment was the same as in Figure 6, rotation of the caisson could be greatly restricted by the reinforcements. The effect of these reinforcements was the largest in Case-4 with placing angles of 45 degrees and increment of moment capacity was about 15%

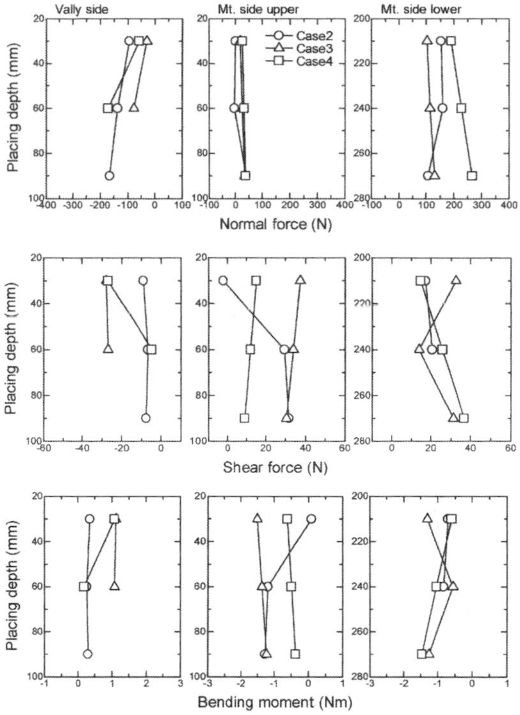

Figure 7. Sectional force acting on reinforcement (Rotation angle = 10 deg.).

in compare with that of Case-2 with smallest bending capacity. Case-3 with 25 degrees placing angles shows a little larger value than Case-2. From these results, highest reinforcement effect can be obtained by arranging them at 45 degree angle.

Figure 7 shows distributions of normal force, shear force and bending moment acting on reinforcement when rotation angle of the caisson was 1.0 degree. Here, sectional forces during a centrifugal acceleration rise were neglected as they were much smaller than those in horizontal loading phase. Since the strain gauges were pasted on a position of 10 mm from junction of reinforcement and caisson, sectional forces were converted into values at the junction. In next chapter, reinforced mechanism and effect of inclination of reinforcement are mentioned by using sectional force results.

4 THE GENERAL EXPECTED MECHANISM

4.1 *Reinforced mechanism*

A caisson type foundation with no reinforcement resists horizontal load by bearing capacity of ground and friction between caisson and ground. In a case of Geo-reinforcing type foundation, resistance against

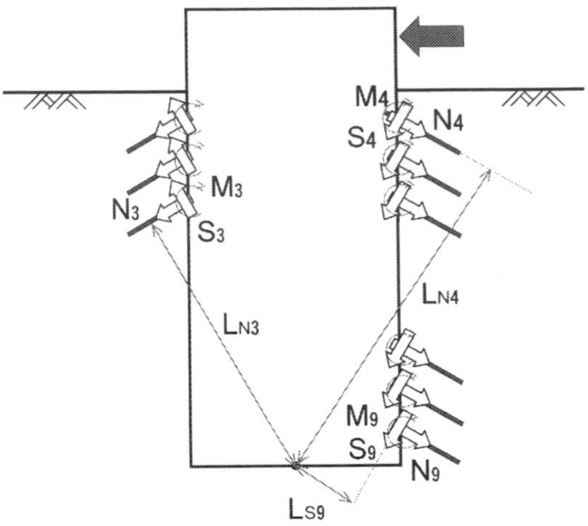

Figure 8. Structure effect.

horizontal loading can be increased by "structural effect" and the "reinforced soil effect" of reinforcement. Therefore, the increment of resistance moment is expressed by following.

$$\Delta M_R = \Delta M_{RS} + \Delta M_{RR} \quad (1)$$

ΔM_{RS}: Increment by structural effect.
ΔM_{RR}: Increment by reinforced soil effect

In the structural effect, reinforcement shares a part of load acting on the caisson foundation as shown in Figure 8. As a result, sectional force occurs in reinforcement. Accordingly, increment of moment resistance by structural effect can be calculated by Equation (2).

$$\Delta M_{RS} = \Delta M_{SN} + \Delta M_{SS} + \Delta M_{SM} \quad (2)$$

$\Delta M_{SN} = \sum N_i \cdot L_{Ni}$: Increment by normal force
$\Delta M_{SS} = \sum S_i \cdot L_{Si}$: Increment by shear force
$\Delta M_{SM} = \sum M_i$: Increment by bending moment
N_i: Normal force acting of reinforcement i
S_i: Shear force acting of reinforcement i
M_i: Bending moment force acting of reinforcement i
L_{Ni}: Arm length to reinforcement I (Normal force)
L_{Si}: Arm length to reinforcement I (Shear force)

In the reinforced soil effect, confining pressure of the ground is increased by sectional force of reinforcement and frictional force between reinforcement and ground increase as shown in Figure 9. The moment resistance increment by the reinforced soil effect can be calculated by Equation (3).

$$\Delta M_{RR} = \sum\left(N_i \cos\theta - S_i \sin\theta\right) \cdot \tan\phi \cdot L_{Ri} \quad (3)$$

L_{Ri}: Arm length to reinforcement (reinforced soil effect)

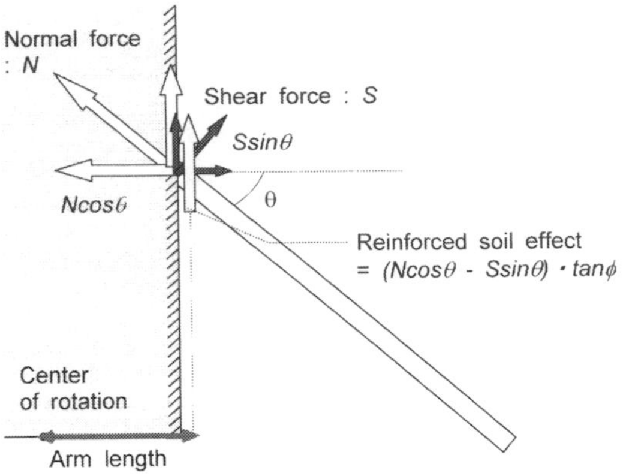

Figure 9. Reinforced soil effect.

Validity of reinforcement mechanism as assumed above is verified by using Equations (1) with sectional force obtained from the centrifuge tests.

Loading moment acting on model caisson can be expressed by Equation (4).

$$M_D = F_D \times L_D \quad (4)$$

M_D: Loading moment
F_D: Horizontal Load
L_D: Arm length from rotation center to loading point

Loading moment in Case-1 "$MD_{\_Case1}$" was in agreement with that of caisson with no reinforcement. Therefore, the sum of resistance moment increment obtained from Equation (1) and $MD_{\_Case1}$ are resistance moment M_R of Geo-Reinforcing Type foundation. That is, it can be expressed as following equation.

$$M_R = M_{D\_Case1} + \Delta M_{RS} + \Delta M_{RR} \quad (5)$$

In order to validate the assumed mechanism, resistance moments were calculated by using Equation (5) and sectional force obtained from tests. Then, calculated results are compared to loading moment measured in the centrifuge test as shown in the figure.

4.2 *Validation of assumed mechanism*

Figure 10 shows resistance moment calculated by Equation (1) and loading moment measured in the tests. In calculation, all sectional force of the reinforcement radially arranged at the same height was assumed to be equal. As shown in this figure, since calculated values show good agreement with measured loading moment, it can be said that the reinforce mechanism currently assumed is appropriate.

The amount of resistance moment increment contributed by normal force, shear force and bending

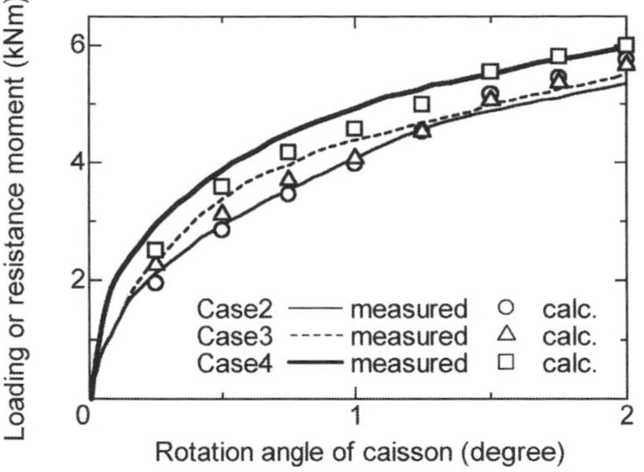

Figure 10. Comparison between measured results and calculation results.

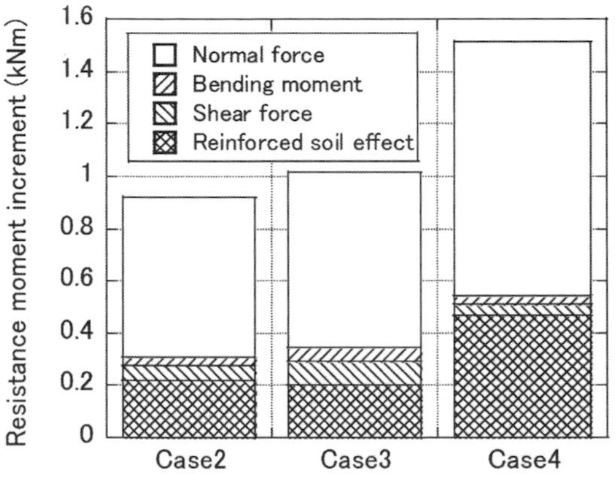

Figure 11. Items of amount contributed in resistance moment increment.

moment were shown in Figure 11. The contribution of structural effect by shear force and bending moment were very small and were about 5 ~ 10% of total resistance moment increment. Reinforcement could not resist horizontal loading sufficiently as well as shear and bending because bending rigidity of reinforcement was lower than stiffness of ground. On the other hand, contribution of normal force was very large in all cases and it accounts for about 90% of total resistance moment increment. Moreover, structural effect by normal force is larger than reinforced soil effect.

4.3 *Effect of reinforcement inclination*

From these results, it turned out that arrangement of reinforcement to which normal force can act on reinforcement effectively is optimal method. Moreover, 45-degree placing was the most effective in the test series. In this chapter, effect of the inclination of reinforcement is evaluated.

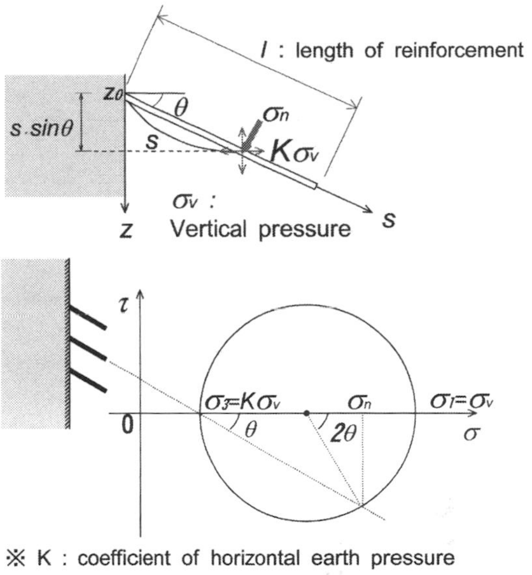

Figure 12. Confining pressure of reinforcement.

It can be thought that normal force acting on reinforcement is caused by pullout resistance between reinforcement and ground. Pullout resistance can be calculated by using following equation obtained from pullout test:

$$T_P = A \cdot c_p + P_c \cdot \tan\delta_p \tag{6}$$

T_P: Pullout resistance
A: Surface area of reinforcement
c_p: apparent cohesion of pullout
δ_p: friction angle between soil and reinforcement
P_c: confining force acting on reinforcement

Since stress state around the reinforcement is considered as shown in Figure 12, confining force P_c can be calculated by following equation

$$\begin{aligned} P_c &= \int_0^l \pi r(1+K)\sigma_n ds \\ &= \pi r\rho(1+K)\left(\frac{1+K}{2}+\frac{1-K}{2}\cos 2\theta\right)\left(z_0 l+\frac{1}{2}l^2\sin\theta\right) \end{aligned} \tag{7}$$

ρ: density of soil
K: horizontal earth pressure coefficient

Tateyama et. al. (1993) suggested that a correction coefficient in consideration of the direction of minor principal strain of the ground should be multiplied for reinforced effect. The correction coefficient is given by following equation.

$$f(\theta) = \frac{2\cos^2\left\{\theta-\left(45^\circ-\nu/2\right)\right\}-\left(1-\sin\nu\right)}{1+\sin\nu} \tag{8}$$

ν: dilatancy angle of soil

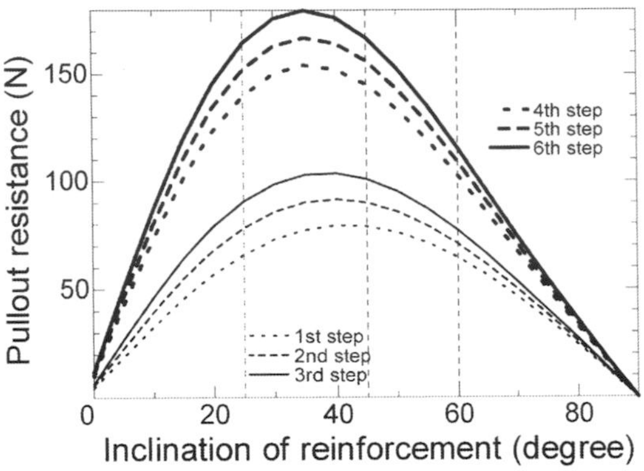

Figure 13. Pullout resistance vs reinforcement inclination.

As mentioned before, since normal force acting on reinforcement is caused by pullout resistance, it can be evaluated by calculating pullout resistance obtained from following equation.

$$T_{PC} = \left(2\pi r \cdot l \cdot c_p + P_c \cdot \tan\delta_p\right) \cdot f(\theta) \tag{9}$$

Figure 13 shows relationships between calculated pullout resistance and inclination of reinforcement. Here, coefficient of Rankin's passive earth pressure, which is expressed as $K_A = \tan^2(45 - \phi/2)$, was used in Equation (7). Smaller dilatancy angle of 5 degrees was selected because displacement between soil and reinforcement was much smaller until rotation angle of caisson is about 3 degrees. As shown in Figure 13, the largest pullout resistance can be obtained near 40 degrees. Therefore, in the centrifuge tests also, large normal force was acquired at reinforcement inclination of 45 degrees, and it is thought that rotation of caisson could be controlled efficiently.

5 CONCLUSIONS

By introducing Geo-Reinforcing Type foundation for a caisson foundation, horizontal bearing capacity could be increased. Reinforced mechanism of it is divided in to "structural effect" and "reinforced soil effect". Such reinforced mechanism was validated by result of centrifuge horizontal loading tests.

Additionally, since structural effect by normal force acting on reinforcement showed about 90 percent of total resistance moment increment, inclination of reinforcement on which normal force acts effectively is an optimal. Such optimal inclination can be evaluated in consideration of pullout resistance between reinforcement and ground, confining pressure of reinforcement, and a correction coefficient relevant to the direction of minor principal strain of ground.

REFERENCES

M. Matsuo and M. Ueno, Development of ground reinforcing type foundation, Proceedings of 12th ICSFE, pp. 1205 ~ 1208, 1989

T. Nakai and M. Ueno, Numerical study on uplift bearing capacity of caisson type pile with reinforcing bars, Proceedings of the International Symposium on Earth Reinforcement, pp. 629–634, 1996

M. Tateyama, F. Tatsuoka, H. Kishida, T. Urakawa and Y. Tamura, Consideration on reinforcing effect in Bar Like Reinforcing members, Proceedings of 28th annual conference of Japan Geotechnical Society, pp. 2787–2790, 1993 (in Japanese)

New Horizons in Earth Reinforcement – Otani, Miyata & Mukunoki (eds)
© 2008 Taylor & Francis Group, London, ISBN 978-0-415-45775-0

Visualization of failure pattern of reinforced soil with face bolts on direct shear tests

D. Takano, J. Otani & T. Mukunoki
Kumamoto University, Kumamoto, Japan

N. Lenoir
Laboratoire 3S-R, University Joseph Fourier, Grenoble, France

ABSTRACT: The objective of this paper is to investigate precise soil-structure interaction behavior on the problem of face bolting method for tunneling using X-ray CT with direct shear test on reinforced sand. First of all, a direct shear apparatus of soil-bolts interaction under X-ray CT is newly developed. After conducting a series of shear test with and without bolts, the precise interaction behavior and failure pattern of reinforced soil are discussed in three dimensions. Here, not only visualization of the behavior but also its quantitative discussion is conducted. Finally, it is confirmed that the proposed test apparatus is useful for clarifying the behavior of soil-structure interaction.

1 INTRODUCTION

In recent years, mountain tunneling method such as NATM (New Austrian Tunneling Method) has been used for tunnel construction even in the urban area. In Japan where land area is small, it is required that tunnel is constructed into unconsolidated ground or low overburden ground. When excavating a tunnel, tunnel face becomes very unstable due to stress release caused by excavation. It is a main issue to maintain the stability in tunnel face during construction. To create the condition for safe construction, it is necessary that tunnel face is reinforced by auxiliary method. Kasama and Mashimo (2003) clarified the effect of typical auxiliary methods such as face bolting, forepoling and vertical pre-reinforcement bolting on the face stability using centrifugal model and distinct element method. It has been concluded from their studies that the length and arrangement of bolts are important factors for stabilization of tunnel face. These results were discussed with the observation of failure mechanism using experimental models of two or quasi-three-dimensional planes. However, failure zone due to tunneling is a behavior in three dimensions and the mechanism of face failure in three dimensions has not been well known yet.

In order to evaluate the mechanism of tunnel face failure, authors developed a system of tunnel pull-out model test system that could be carried out in the system of X-ray CT scanner (Otani et al, 2005) and using this system, failure pattern due to tunneling without reinforcement and with face bolting method were visualized using X-ray CT and as shown in Figure 1, it is observed that face bolting method prevents development of failure zone due to tunneling (Takano et al, 2006).

The objective of this study is to investigate precise soil-structure interaction behavior on the problem of face bolts for tunneling using X-ray CT with direct shear test on reinforced sand. Here, a new direct shear test apparatus is developed with the use of X-ray CT scanner and a series of shear test with and without bolts are examined. The specimen is then investigated using X-ray computed tomography (CT) to visualize shear zone in three dimensions. And the influence of reinforcement orientation is visually investigated in three dimensions. Finally, it is concluded that the interaction behavior between soil and bolts can be investigated both qualitatively and quantitatively in three dimensions by the proposed test apparatus.

2 OUTLINE OF TEST

2.1 *Test apparatus*

In this study, a new direct shear apparatus is developed for the use with system of industrial X-ray CT scanner for the purpose of observation on the close interaction behavior in three dimensions. The arrangement of this apparatus is shown in Photo. 1 and its features are described in details with Figure 2. The specifications of this apparatus are summarized as follows:

1) All the operations for this apparatus are conducted remotely using a personal computer from outside the shield room in X-ray CT and the whole loading process was recorded electronically.
2) In order to scan the specimen horizontally during direct shear test, there should be no obstacles (e.g., steel bars) around the compression cell because such materials would interfere with X-ray CT imaging.
3) Loading is stopped during scanning operation, which takes about 2.5 min. for one cross sectional image.

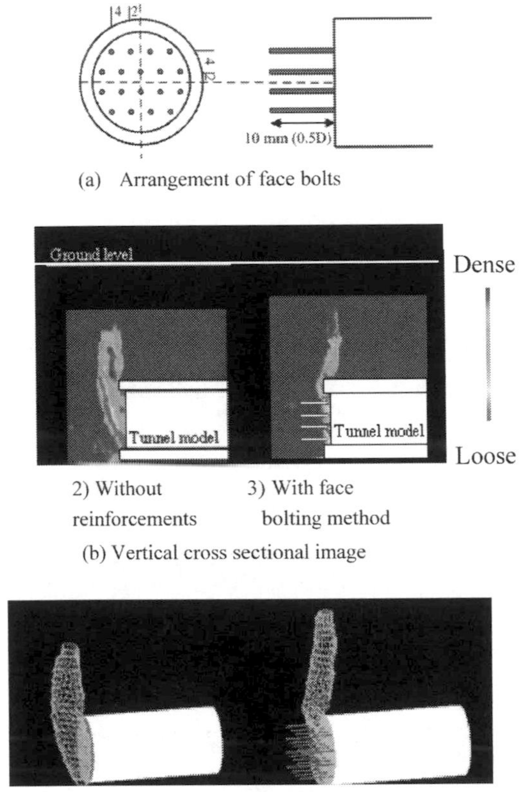

(a) Arrangement of face bolts

2) Without reinforcements 3) With face bolting method

(b) Vertical cross sectional image

(c) Three-dimensional wire frame image of failure zone

Figure 1. Failure pattern due to tunneling without reinforcement and with face bolting method.

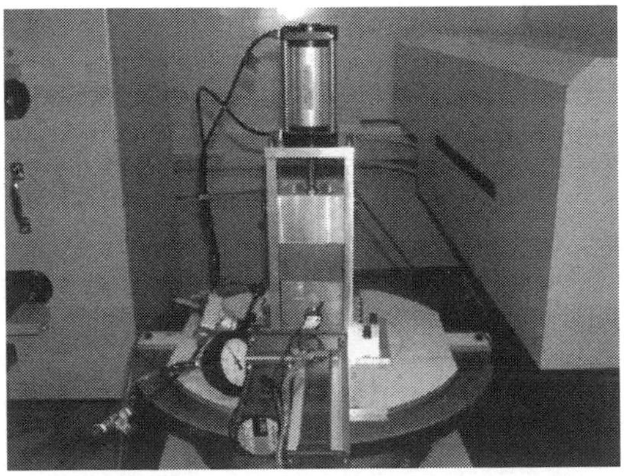

Photo 1. General picture of the sirect shear apparatus.

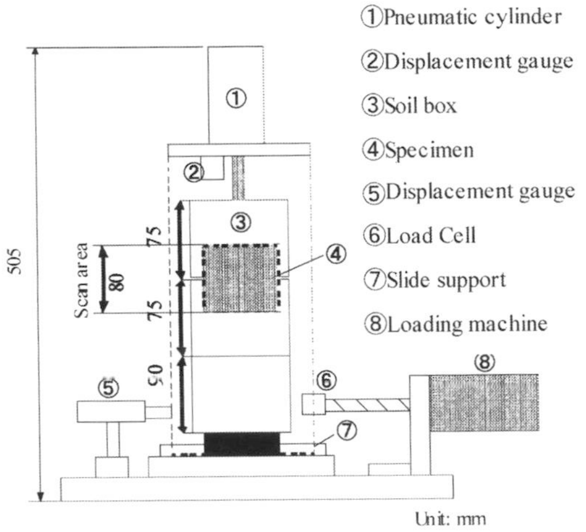

Figure 2. Schematic of the direct shear apparatus.

Table 1. Property of Toyoura sand.

Maximum dry density	1.66 t/m^3
Minimum dry density	1.34 t/m^3
Mean grain size	0.2 mm
Rerative density	85%

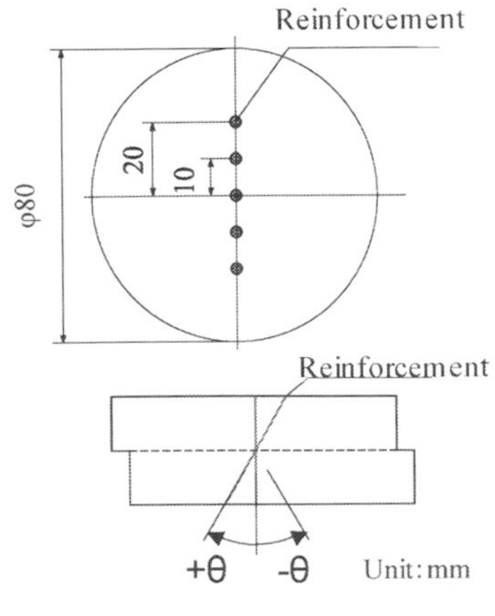

Figure 3. Arrangement of reinforcement.

2.2 *Test condition*

The size of the specimen was 80 mm in diameter and 30 mm in height. The space between upper and lower box was 0.2 mm. In this study, air-dried Toyoura sand was used and a shearing speed of 1.0 mm/sec was applied. Dense specimens with a relative density of Dr = 85% were prepared under vertical pressure of $\sigma = 100$ kPa (constant). Four different specimens (without reinforcement and reinforced by bolts) were examined. Figure 3 shows the arrangement of bolts in the specimen and Table 2 shows orientation of reinforcement. In reinforced specimens, three different reinforcement orientation θ was examined. Face bolts 2 mm diameter and 30 mm long were used as reinforcement. Its surface was made practically rough.

All the specimens were scanned at four different shear displacements which were 1 mm, 2 mm, 5 mm and 8 mm and three dimensional images were reconstructed.

3 TEST RESULT

Figure 4 shows relationship between shear displacement and shear stress. An effect of reinforcement on strength is noticeable. Peak strength and shear displacement at peak increace in order of Case 3, Case 2, and Case 4. The influence of reinforcement orientation θ was analyzed by Jewell (1897), and it was concluded that the reinforcement orientation parallel to the principal tensile strain would give rise to the maximum rate of strength increase. The comparable result is obtained from this test. On the other hand, vertical displacement is the longest in Case 1.

Table 2. Orientation of reinforcement.

	Case 1	Case 2	Case 3	Case 4
Orientation of reinforcement	Without reinforcement	0°	30°	−30°

Vertical cross sectional image of all test cases at four different shear displacements which are 1mm (before peak), 2 mm (peak), 5 mm (after peak) and 8 mm (the end of shear test) are shown in Figure 5. CT image is constructed by the degree of x-ray absorption in the materials. CT images are presented with shaded grey or black color for low density and light gray or white color for high density. Total number of levels on these colors is 256. It is noted that more precise contents about X-ray CT scanner can be obtained by the reference of Otani et al. (2000). It is realized that the low density area can be observed around edge of shear box on images of 1 mm displacement. This area can be considered to be an area of strain localization or shear band after CT scanning. The shear bands develop to center of specimen at 2 mm displacement. At 5 mm of shear displacement, shear strength reach to residual strength and the shear band extends to the whole of the test piece. Finally, at 8 mm of shear displacement, though shear band extends in a perpendicular direction in Case 1 (without reinforcement), it is not remarkable for other cases. It is realized that newly developed shear band can be observed after shear strength reaches to the residual strength and progressive failure occurs in direct shear. When paying attention to area placed between two shear bands in Case 1, density decrease cannot be observed. From this result, it is realized that volumetric strain does not occur in this area. Table 3 shows the amount of maximum vertical displacement and thickness of shear bands. In each test case, maximum volumetric strain and thickness of shear bands is almost the same among all the cases. It is also seen that volumetric strain occurs only in shear band.

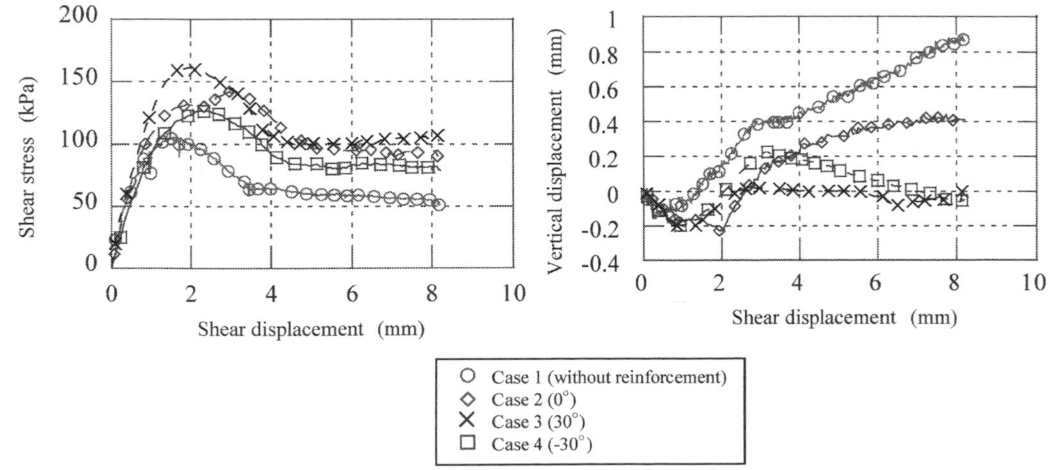

Figure 4. Relationship between shear displacement and shear stress.

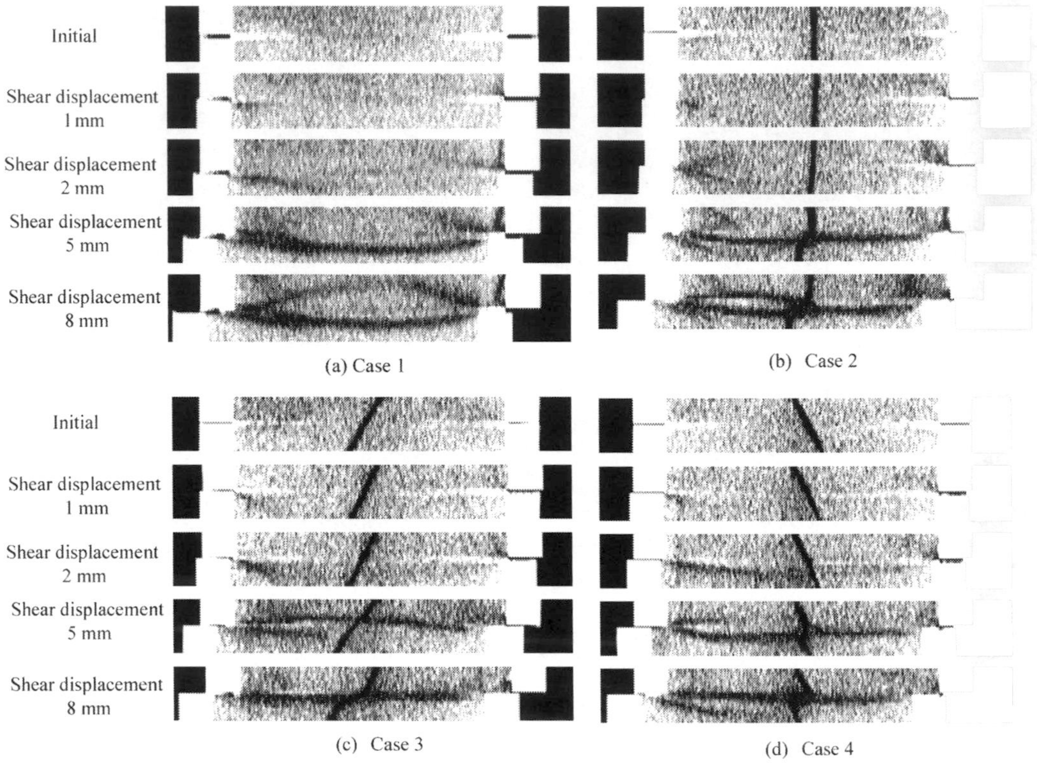

Figure 5. Vertical cross sectional images.

Table 3. Amount of maximum vertical displacement and thickness of shear band.

	Maximum vertical displacement (mm)	Thickness of shear band (mm)
Case 1	0.888	1.24
Case 2	0.441	0.7
Case 3	0.023	0.43
Case 4	0.226	0.39

4 CONCLUSIONS

A new direct shear test apparatus was developed for the use in the system of industrial X-ray CT scanner. Using this apparatus, a series of direct shear tests and CT scanning were conducted. Following conclusions are drawn from this study;

1) The shear band due to direct shearing can be visualized in three dimensions using X-ray CT scanner.
2) The peak shear stress and the shear displacement to reach the peak stress increased for the case of reinforced sand.
3) The progressive failure can be observed from CT image.
4) The maximum volumetric strain and thickness of shear zone is almost the same amount among all the cases.

A newly developed direct shear test apparatus made it possible to evaluate interaction behavior between soil and bolts using an industrial X-ray CT scanner. More quantitative discussion should be required using this apparatus and this should be done in the future study.

REFERENCES

Kasama, H. and Mashimo, H., 2003. Centrifuge model test of tunnel face reinforcement by bolting, *Tunneling and Underground Space Technology*, 18: 205–212.

J. Otani, D. Takano, and H. Nagatani, 2005. Evaluation of passive failure at tunnel face using X-ray CT, *Proceedings of 16th International Conference on Soil Mechanics and Geotechnical Engineering*, pp. 1639–1642.

D. Takano, H. Nagatani, J. Otani, and T. Mukunoki, 2006. Evaluation of auxiliary method in tunnel construction using X-ray CT, *The sixth international conference on physical modeling in geotechnics*, pp. 1189–1194.

R.A. Jewell, and C.P. Wroth, 1987. Direct shear test on reinforced sand, *Geotechnique 37*, No. 1, 53–68.

New Horizons in Earth Reinforcement – Otani, Miyata & Mukunoki (eds)
© 2008 Taylor & Francis Group, London, ISBN 978-0-415-45775-0

Improvement in bearing capacity of shallow improvement ground by mixing short fibers

H. Matsui, H. Ochiai, K. Omine, N. Yasufuku, T. Kobayashi & R. Ishikura
Department of Civil Engineering, Kyushu University, Japan

ABSTRACT: Ground improvement technology such as a cement stabilization method has been applied to very soft ground in Japan. As for this method, cement-treated soil has relatively low tensile strength and shows very brittle behavior. It is therefore expected to use a fiber reinforcing material for improving mechanical property of the cement-treated soil. In this study, short fibers produced from waste paper were used as a reinforcing material in shallow improvement ground for increase of the tensile strength. Furthermore, model test of the improved ground was performed using loading device for applying uniform vertical stress corresponding embankment load. Improvement in bearing capacity of the shallow improvement ground with deep mixing columns was investigated under the different test conditions. The effectiveness of short fibers for improvements of deformation property and bearing capacity of the improved ground was confirmed from the test results.

1 INTRODUCTION

In the view point of geotechnical engineering, ground improvement technology such as a cement stabilization method has been applied to very soft ground in Japan. Cement-improved soil has some defects such as low tensile strength and brittle behavior, so that the defects may become a problem for application of shallow improvement method. It is also expected to use waste material as a fiber reinforcing material for improving mechanical property of cement-treated soil. Improvement in brittle behavior regarding deformation-strength property was discussed based on triaxial compression test of cement stabilized ground with various types of short fibers[1),2)].

In this paper, at first, in order to clarify improvement effect of reinforcing material with short fibers, simple bending test and tensile splitting test are performed.

Secondly, in order to evaluate bending stress of shallow stabilized ground with floating type deep mixing columns under uniform vertical stress conditions, bending stress model is proposed based on the idea of stress distribution ratio *b*.

Furthermore, model test of the improved ground is performed using loading device for applying uniform vertical stress corresponding embankment load. Short fibers produced from waste paper are used as a reinforcing material in shallow improvement ground. Improvement in bearing capacity of the shallow improvement ground with deep mixing columns is investigated under the different test conditions. The effectiveness of short fibers for improvements of tensile strength and bearing capacity of the improved ground and the validity of bending model are discussed based on the test results.

2 DEFORMATION-STRENGTH PROPERTY OF CEMENT-TREATED SOILS MIXED WITH SHORT FIBERS

2.1 *Soil sample and testing method*

First, short fibers were made by agitating waste paper with water in food processor and breaking into flocculate. Photo 1 shows the shapes of waste short fiber. Kaolin clay ($w_L = 50.6\%$, $I_p = 19.6$ and $\rho_s = 2.70\ g/cm^3$) is used as soil sample. After adding Portland cement to sample with water content of 100%, short fibers are mixed with it. The cement content C is 200 and 300 kg/m^3. After curing those specimens during 7 days in the thermostatic chamber of 20°C, each test is performed. Waste paper fiber content is 2.5 kg/m^3. Optimum fiber content changes in shape,

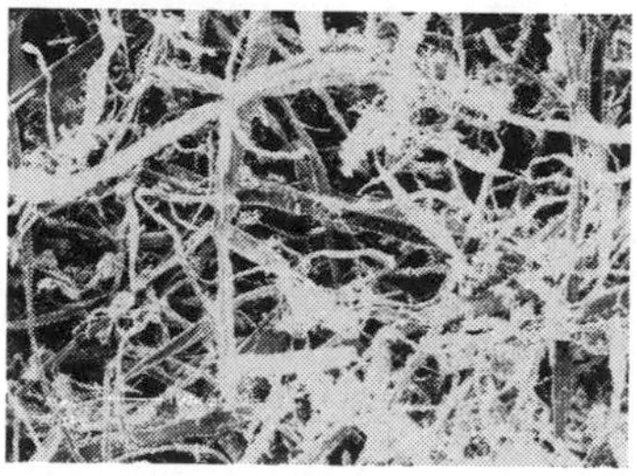

Photo 1. Shape of short fibers made from waste paper.

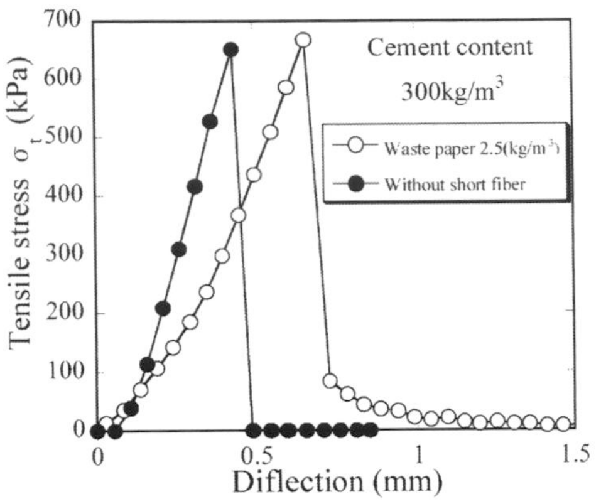

Figure 1. Relationship between tensile stress and deflection obtained from the bending test.

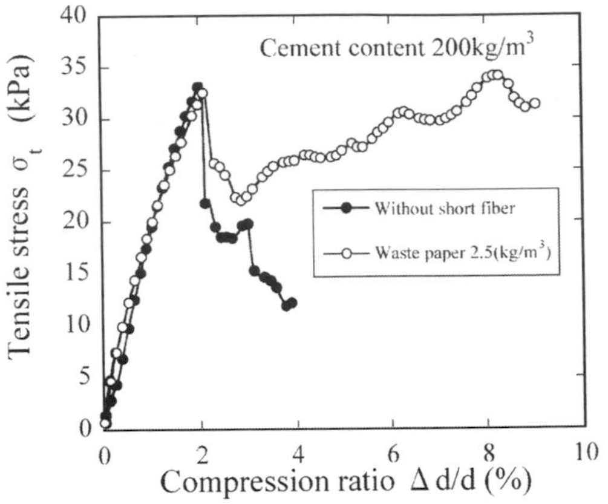

Figure 2. Relationship between tensile stress and compression ratio obtained from tensile splitting test.

length and kind of short fiber, and therefore the fiber content was decided in consideration of preliminary experiment. The specimen size of bending test is in the height of 40 mm and the width of 40 mm and the length of 160 mm. The specimen is supported by two fulcrums just below the specimen, and one is roller. Line load is applied on the center of upper surface by displacement control condition. The specimen size of the tensile splitting test is in the diameter of 150 mm and the height of 75 mm.

2.2 *Test result and discussions*

Figure 1 shows the relationship between tensile stress σ_t and deflection obtained from the bending test. Deflection is defined as the value of displacement center of specimen. As shown in this figure, tensile stress of the cement treated soil without waste paper becomes zero after reaching to the peak strength and shows a brittle failure. On the other hand, such brittle behavior is improved by mixing short fibers.

Figure 2 shows the relationship between tensile stress and compression ratio obtained from the tensile splitting test. Compression ratio is defined as the ratio between compressive displacement Δd and diameter of specimen d. As shown in this figure, the peak strengths of these two specimens are almost same. However, cement-treated soil without waste paper has a brittle failure after reaching to the peak strength. On the other hand, cement-treated soil with short fibers has a large residual strength after reaching to the peak strength.

In these tests, the effectiveness of short fibers for tensile strength of cement-treated soil was confirmed. In next chapter, in order to evaluate bending stress of shallow stabilized ground with floating type deep mixing columns under uniform vertical stress conditions, bending stress model is proposed.

3 ESTIMATION OF BENDING STRESS OF SHALLOW STABILIZED GROUND WITH FLOATING TYPE DEEP MIXING COLUMNS

In shallow stabilized ground with floating type deep mixing columns, vertical stress distribution has occurred between improved columns and soft clay just below the shallow stabilized ground. It is considered that larger bending stress of the shallow ground with the columns will occur than that of only shallow stabilized ground. So it is important to evaluate bending strength of shallow stabilized ground in these type improved conditions.

In this chapter, in order to evaluate tensile strength of shallow stabilized ground in uniform loading condition, an estimation method of bending stress of the shallow stabilized ground is proposed.

Figure 3 shows the concept of shallow stabilized ground with floating type deep mixing columns. As shown in this figure, bending stress of the shallow stabilized ground is discussed based on 1 unit of these type improved ground. Figure 4 shows the flow of estimation of bending stress σ_m of the shallow stabilized ground. In order to estimate bending stress of the shallow stabilized ground, it is important to determine stress distribution ratio b.

3.1 *Evaluation of stress distribution ratio with consideration of improved column rigidity and skin friction*

Figurer 5 shows the concept for the stress distribution ratio just below the shallow stabilized ground. The parameters of m_{vs} and m_v^* denote the coefficients of the volume compressibility of improved column and soft clay, respectively. When uniform vertical load p_1 is applied to the shallow stabilized ground, stress distribution just below the shallow stabilized

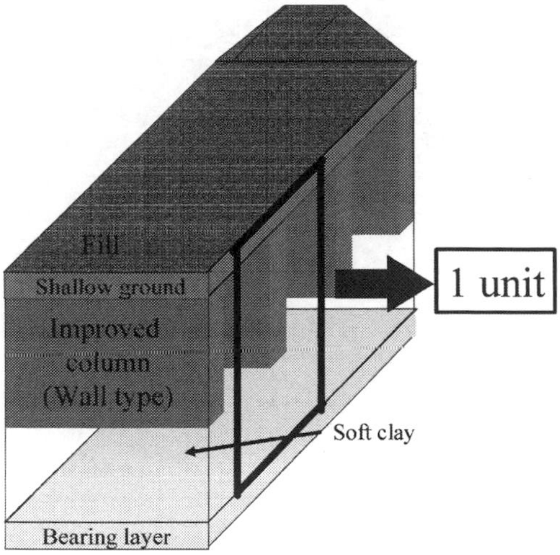

Figure 3. Concept of 1 unit of improved ground.

ground is represented using average stresses of soft clay and improved column, namely $\overline{p_2}$ and $\overline{p_3}$. This type improved ground remains soft clay just below the improved columns, so it is considered that skin friction around surface of the improved columns has occurred. The authors proposed stress distribution model with consideration of column rigidity and skin friction[3]. Stress distribution ratio b just below the shallow stabilized ground, which is defined by the ratio of $\overline{p_2}$ and $\overline{p_3}$, is expressed by Eq.(1).

$$b = \frac{\overline{p_3}}{\overline{p_2}} = \frac{2R(D+1)p_1 + \left[\frac{(1-f_s)}{f_s}D + \frac{(f_s D+2)}{f_s(1-f_s)}R\right]\frac{\overline{\tau} A_s}{A_0}}{2(D+R)p_1 - \left[D + \frac{(f_s D+2)}{f_s(1-f_s)}R\right]\frac{\overline{\tau} A_s}{A_0}} \quad (1)$$

where, R is the rigidity ratio ($= m_v^*/m_{vs}$), D is the depth ratio($= H_1/H_2$), A_0 is the area of improved ground, A_s is the sum of area around surface of improved column, f_s is ratio of the improved column in the improved ground, $\overline{\tau}$ is average skin friction around surface of improved column. In this paper, it is assumed that $\overline{\tau}$ is equal to the shear strength of soft clay c_u. The b value is calculated by the improved ground properties and improvement condition.

3.2 *Estimation of bending stress of shallow stabilized ground based on stress distribution ratio*

By the determination of stress distribution ratio b, bending stress σ_m is calculated as follows[4]. Figure 6 shows the concept of bending moment M of shallow stabilized ground. L_1 means distance between columns and L_2 means width of improved column. M is expressed in Eq.(2).

$$M = M_1 + M_2 \quad (2)$$

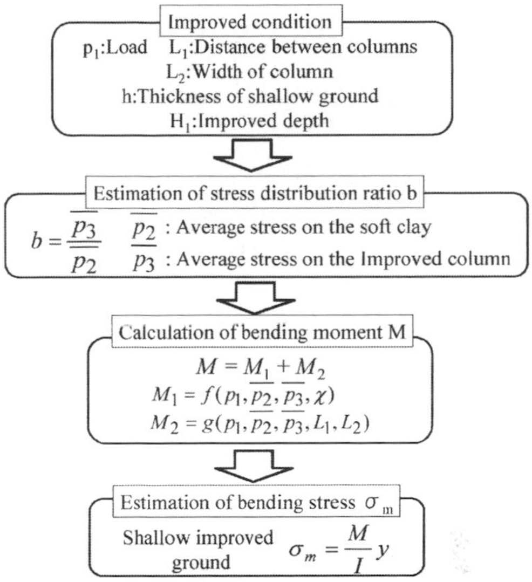

Figure 4. Flow of estimation of bending stress σ_m of shallow stabilized ground.

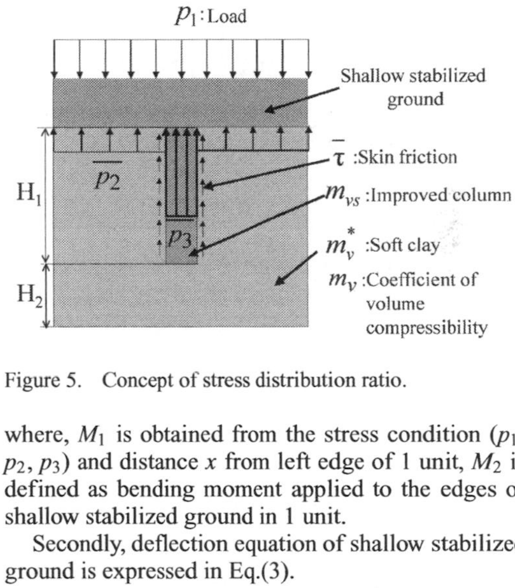

Figure 5. Concept of stress distribution ratio.

where, M_1 is obtained from the stress condition (p_1, p_2, p_3) and distance x from left edge of 1 unit, M_2 is defined as bending moment applied to the edges of shallow stabilized ground in 1 unit.

Secondly, deflection equation of shallow stabilized ground is expressed in Eq.(3).

$$\frac{d^2 y}{dx^2} = -\frac{M}{EI} \quad (3)$$

where, E is deformation modulus of the cement-treated soil, I is geometrical moment of inertia of the shallow stabilized ground.

When deflection angle θ is equal to zero on the edges of the shallow stabilized ground in 1 unit, by substituting Eq.(2) to Eq.(3), M_2 is obtained as the function of p_1, $\overline{p_2}$, $\overline{p_3}$, L_1 and L_2, and consequently M value is obtained as the sum of M_1 and M_2.

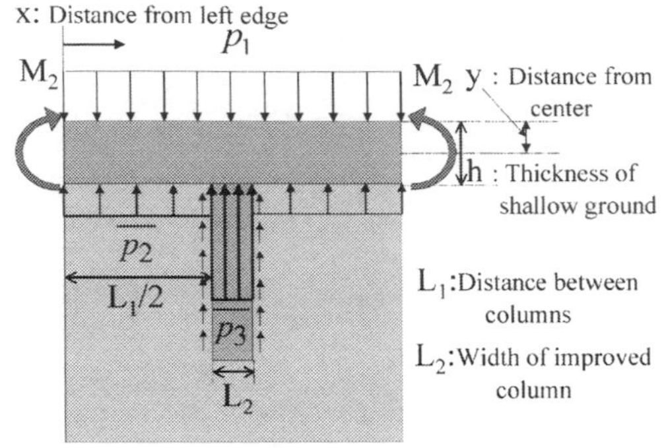

Figure 6. Concept of bending moment of shallow stabilized ground.

Finally bending stress of the shallow stabilized ground σ_m is obtained by the Eq.(4).

$$\sigma_m = \frac{M}{I} y \tag{4}$$

where, y is the distance from the center of thickness of shallow stabilized ground.

By substituting Eq.(2) to Eq.(4), bending stress σ_m of the shallow stabilized ground under uniform vertical stress condition is obtained. In next chapter, in order to confirm the validity of this model, loading model tests of the shallow stabilized ground with deep mixing columns under uniform vertical stress condition are conducted.

4 LOADING MODEL TEST OF THE SHALLOW STABILIZED GROUND WITH DEEP MIXING COLUMNS UNDER UNIFORM VERTICAL STRESS CONDITION

4.1 *Apparatus used in model test*

Figure 7 shows a schematic illustration of model test device. This device consists of a container box, loading device and 13 split loading plates. Displacements are measured on the loading plates of A and B in Figure 7. This loading model tests for the improved model ground were conducted under uniformed vertical stress-controlled condition using bellofram cylinders.

4.2 *Preparation of model ground and test procedure*

The model ground was prepared using soft clay and model columns. Kaolin clay was used as the soft clay. The Kaolin clay was mixed with a water content of about 80% and the slurry was poured into the container box up to a depth of approximately 260 mm. Soft clay was consolidated under pre-consolidation

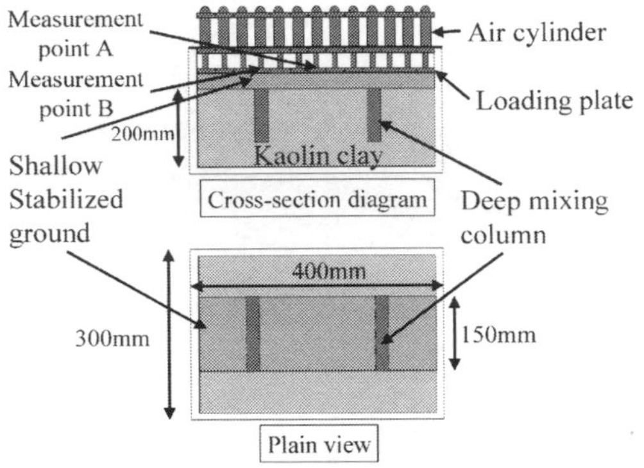

Figure 7. Schematic illustration of model test device.

Table 1. Test conditions of loading model test.

	Shallow ground thickness h (mm)	Depth deep mixing column H_1 (mm)	Cement content C kg/m^3	Waste paper content kg/m^3	Pre-consolidation stress p_0 kPa
Case-1	30	150	300	2.5	10
Case-2	30	150	200	2.5	10
Case-3	30	150	300	0	10
Case-4	30	150	300	0	20

pressures of 10 and 20 kPa using a bellofram cylinder for around 2days. After pre-consolidation, a model ground with a height of 200 mm was obtained and the column was installed into the ground. A model column was used instead of the actual deep mixing columns in order to homogenize the rigidity of the columns. These columns are made from urethane and have a width of 20 mm and a height of 150 mm. By the unconfined compression test, the deformation modulus E of this column was approximately 50 MPa. Shallow stabilized ground was made by cement-treated soil for curing of 7 days.

4.3 *Test conditions*

After completing model ground, the loading model tests were carried out under an undrained condition.

Loading stress is applied at a whole area of shallow stabilized ground using 13 split loading plate of 30 mm width. In order to apply the load uniformly, model ground is covered with sand of approximately 20 mm height. Displacements are measured in two points , namely (A) center between model columns and (B) upper side of the model column. Table 1 shows test conditions of loading model test. Improvement ratio f_s is 10% in each case. Shallow stabilized ground thickness is 30 mm and depth of deep mixing column is 150 mm in all test conditions. In order to confirm the effectiveness of short fiber for improvement

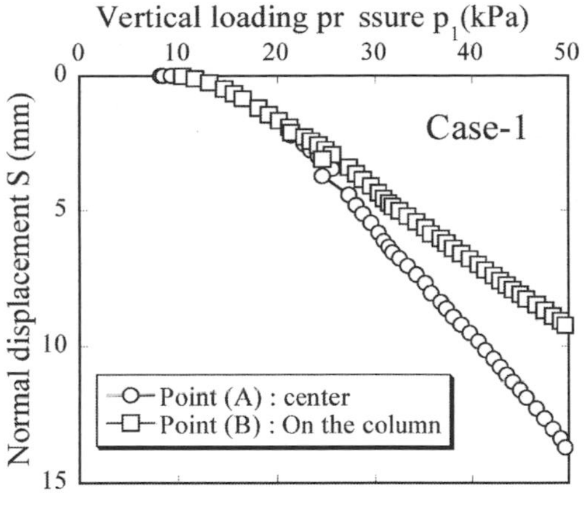

Figure 8. Relationship between vertical loading pressure p_1 and normal displacement S (Case-1).

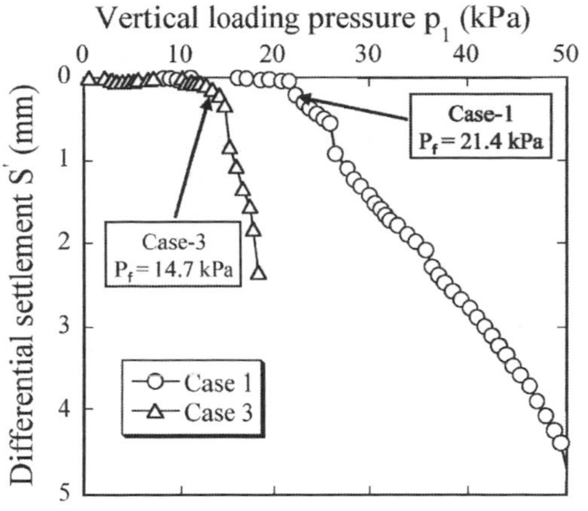

Figure 9. Relationship between vertical loading pressure p_1 and differential settlement S'.

of tensile strength, cement-stabilized ground in Case-1 and 2 contains waste paper. For making clear the influence of soil property, test condition of different pre-consolidation pressure p_0 was used in Case-3 and 4 of non-mixture.

4.4 *Test result and discussions*

Figure 8 shows the relationship between vertical loading pressure p_1 and normal settlement S of measurement point (A) and measurement point (B) in Case-1. As shown in this figure, the improved ground does not show clear peak strength and brittle failure such as observed in the bending tests. So it is difficult to recognize tension failure of the shallow stabilized ground under uniform vertical loading condition. On the other hand, difference of settlement between measurement points (A) and (B) becomes large with an increase in the vertical loading pressure p_1.

Figure 9 shows the relationship between differential settlement S' and vertical loading pressure p_1 in Case-1 and 3. Differential settlement S' means difference of settlements between measurement points (A) and (B). Each differential settlement S' is very small in low vertical loading pressure level. However, differential settlement S' increases with increase in the vertical loading pressure p_1 in all test cases. In this model test, yielding pressure p_f of the shallow stabilized ground is defined as vertical loading pressure at the rapid increase of differential settlement in each case. As shown in this figure, yielding pressure p_f of shallow stabilized ground in Case-1 is larger than that of Case-3, that is to say, the effectiveness of short fibers for tensile strength of cement treated soil was confirmed under the uniform vertical stress condition.

Improvement in bearing capacity of the shallow improvement ground by mixing short fibers is also confirmed by vertical loading pressure-differential displacement curve in comparison with Case-1 with short fiber and Case-3 without short fiber.

Table 2. The parameter used in the estimation of bending stress σ_m of shallow stabilized ground.

	Case-1	Case-2	Case-3	Case-4
Tensil strength σ_t (kPa)	618	294	296	541
Calculated value of stress distribution ratio b	5.01	7.37	7.71	8.34
Soft clay of coefficient of volume compressibility m_v^* (m^2/MN)		6.007		1.628
Improved column of coefficient of volume compressibility m_{vs}(m^2/MN)		0.02		

4.5 *Comparison between experimental and calculated values*

In this section, in order to confirm the validity of bending stress model based on stress distribution ratio in chapter 3, experimental values of vertical loading pressure p_f are compared with calculated values in occurring tension failure of the shallow stabilized ground.

Table 2 shows the parameters used in the estimation of bending stress σ_m of shallow stabilized ground. Tensile strength σ_t means the average value for 3 specimens of bending test in each case. In order to evaluate tensile strength of shallow stabilized ground using in the model test size, it is considered that scale effect on strength of cement-treated soils is one of influence factors. Omine et al. suggested an estimation method for predicting the scale effect[5]. Tensile strength of shallow stabilized ground for the model test size can be

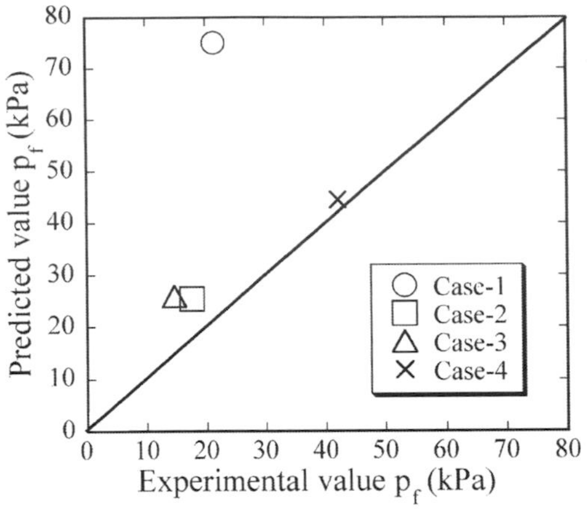

Figure 10. Comparison between calculated and experimental value p_f of uniform vertical stress.

estimated from that of standard size specimen by this model. Tensile strength of shallow stabilized ground for the model test size σ_s is expressed by tensile strength reduction coefficient α and σ_t. α is obtained from average tensile strength value and standard deviation of bending test in each case. Relationship between σ_s and σ_t is $\sigma_s = \alpha\sigma_t$. In calculating this model test, the value of α value becomes 0.66.

Stress distribution ratio b just below the shallow stabilized ground is estimated by Eq. (1). Parameters within Eq. (1) are determined by the initial ground and boundary conditions and vertical loading pressure p_1, that is to say, b value is depended on vertical loading pressure p_1. Bending moment M of shallow stabilized ground is determined by $\overline{p_2}$ and $\overline{p_3}$ obtained from relationship between b and p_1. By substituting bending moment M into Eq.(4), bending stress σ_m is obtained. Namely, when p_1 value is determined, bending stress σ_m is also obtained.

Calculated value p_f at tension failure of shallow stabilized ground is defined as the value obtained from relationship between p_1 and σ_m when σ_m is equal to σ_s of tensile strength of shallow stabilized ground with consideration of scale effect.

Table 2 also shows the calculated value of stress distribution ratio b when σ_m is equal to σ_s.

Experimental value p_f is defined as the p_1 value at the rapid increase of differential settlement shown in Figure 9.

Figure 10 shows the comparison between the calculated and experimental value p_f. As shown in this figure, calculated values in Case-2 and 3 approximately correspond with the experimental results. Calculated value in Case-1 is larger than that of experimental. It is considered that tensile strength σ_t in this Case become larger than that of other cases in spite of the same curing day and cement content. In order to investigate the influence of soil property, pre-consolidation pressure p_0 in Case-4 is different from other cases. As shown Table 2, b value in this Case becomes larger than other cases. It is considered that the undrained shear strength of soft clay c_u increases with the increase in pre-consolidation pressure p_0. Calculated value in Case-4 also approximately corresponds with the experimental result in different condition of soil property.

5 CONCLUSIONS

The main conclusions obtained from this study are as follows;

1) Brittle behavior of cement-treated soil in bending test and tensile splitting test are improved by mixing short fibers.
2) Yielding pressure of shallow stabilized ground mixed with short fibers is increased under the condition of uniform vertical pressure and bearing capacity characteristic is also improved.
3) Bending stress of shallow stabilized ground with deep mixing columns under the condition of uniform vertical pressure can be estimated by the bending stress model based on stress distribution ratio.

REFERENCES

1) Takayama, E., Ochiai, H., Yasufuku, N., Omine, K. & Kobayashi, T. 2005. Improvement effect on mechanical properties of cement-stabilized soils by mixing various types of short fibers, *JGS, No. 134, Proc., the symposium on cement treated soil*, pp. 335–338 (in Japanese).
2) Omine, K., Ochiai, H., Yasufuku, N., Kobayashi, T. & Takayama, E. 2006. Strength properties of air-formed lightweight soils with various types of short fibers, *Proc., of the 8th International Conference on Geosynthetics (8ICG)* vol.4, pp. 1659–1662.
3) Ishikura, R., Ochiai, H., Omine, K., Yasufuku, N. & Kobayashi, T. 2006. Estimation of the settlement of improved ground with floating-type cement-treated columns, *Proc., of Fourth International Conference on Soft Soil Engineering*, pp. 625–635.
4) Matsui, H., Ochiai, H., Omine, K., Yasufuku, N., Kobayashi, T. & Ishikura, R. 2007. Evaluation of bending stress on improved ground by combination of shallow soil stabilization and floating type of columns, *Proc., of 42th annual meeting, JSSMFE* (in Japanese).
5) Omine, K., Ochiai, H. & Yasufuku, N. 2005. Evaluation of scale effect on strength of cement-treated soil based on a probabilistic failure, *Soils and Foundations*, vol.45, No.3, pp. 125–134.

New Horizons in Earth Reinforcement – Otani, Miyata & Mukunoki (eds)
© 2008 Taylor & Francis Group, London, ISBN 978-0-415-45775-0

Fundamental mechanical properties of geocell reinforced sands

K. Yazawa, T. Ajiki, H. Ohmori, K. Kaneko & K. Kumagai
Hachinohe Institute of Technology, Japan

ABSTRACT: In the geocell reinforced soil structures, we can use the various materials as the material into the geocells. Mechanical properties of geocell reinforced soils change by the kind of filling materials. Therefore, if we want to predict the deformation and strength of them, it is necessary to grasp the relation between the characteristics of the filling materials and the mechanical properties of the geocell reinforced soil structures. In this study, at first, we perform the tri-axial compression tests to examine the mechanical properties of the sandy filling materials. In addition, fundamental experiments on the compressibility and the frictional property for the cell structures are performed. Finally, we analyze the relationship between results of these experiments.

1 INTRODUCTION

The geocells (Sitharam et al 2005) are one of the three-dimensional geosynthetics materials to reinforce earth structures (Fig. 1). They are made by high density polyethylene, and they are excellent at the chemical stability, the rigidity, impact strength, and the low temperature property. Yang's modulus is about about 2.2×10^5 KN/m^2, Tensile strength of seam is about 14.2 KN/m. The geocell reinforced earth structure is reinforced 3-dimensionally by geocells and filling materials, which are put in the cells. Various materials can be used as a filling (Yazawa et al 2006.) Moreover, there are a lot of usages in the geocell reinforcement such as slope stability and protection, foundation and river bulkhead (omori et al 2006). The application to disaster recovery can also be expected, because the geocell is very small and light and easy construction.

Figure 1. Geocell (GeoWeb® made by ALCOA).

In geocell reinforce method, if we can use the local generation soils as the filling materials effectively, we can get large advantages for environments and costs. However, since geocell reinforced earth structures are combined structure made by polyethylene and arbitrary soils, their mechanical properties are very complex and difficult. In the present circumstances, macadam, which is considered a comparatively good quality material as a filling, is bought and used in consideration of the safety aspect.

In this study, aiming at the establishment of the construction method that effectively uses the local generation soil as a filling in geocells, we examine the correlation of the mechanical properties of the filling and the mechanical properties of geocell structure. First of all, we performed the triaxial compression tests for some sandy materials to know their fundamental mechanical properties. Next, we also performed fundamental experiments concerning the compressibility and the shear resistance characteristics of the geocell structure when those materials are used as a filling. Finally, we discuss the relationship between the mechanical properties of filling materials and geocell-structures.

2 TRI-AXIAL COMPRESSION TESTS FOR FILLING MATERIALS

In this study, our main purpose is to examine the relation between the mechanical property of the filling material and the mechanical property of the geocell structure that uses it in the cell. In this section, we perform the triaxial compression test (CD) for the sample used as a filling material in geocells to examine

Table 1. Fundamental properties of sample.

Sample no.	Collection point (material name)	Uniformity coefficient	Coefficient of curvature	Soil particle density (g/cm^3)	Minimum density (g/cm^3)	Maximum density (g/cm^3)
A	Gonohe, Aomori	5.14	1.38	2.711	1.254	1.522
B	Hachinohe, Aomori	2.42	1.23	2.671	1.635	2.007
C	Aomori, Aomori	3.36	0.95	2.705	1.423	1.727
D	(Silica sand No.5)	2.29	1.21	2.603	1.321	1.642
E	Kamo, Niigata	2.69	1.03	2.567	0.837	1.156
F	Inzai, Chiba	2.20	1.26	2.713	1.283	1.588
G	(Ferronickel slaggy)	28.9	0.75	3.232	1.543	2.246

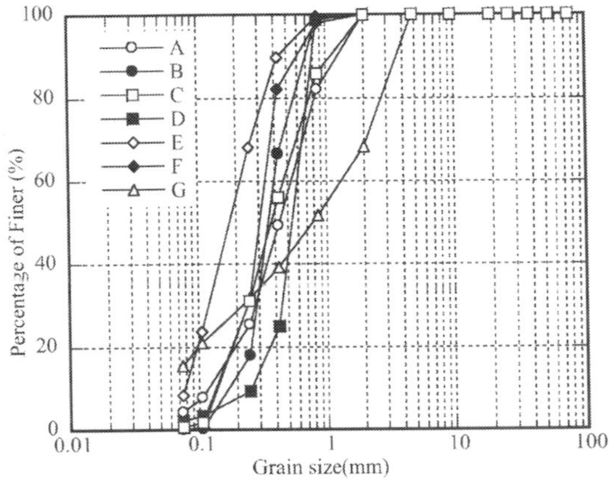

Figure 2. Grain size distribution curve.

internal frictional angle, cohesion, and deformation modulus. The triaxial compression test is one of the general shear tests. Triaxial compression test (CD) is on the condition without excess pore water pressure. Because the pore water is drained by the compression process, it is applied to the soil with a good permeability like sand.

2.1 *Outline of tests*

In this study, we use seven kinds of sandy materials, and Table 1 shows fundamental properties of these samples. Figure 2 shows grain size distribution curves of these samples. The distribution of the grain diameter is greatly different only sample G. Sample G is ferronickel slaggy, which is waste generated when the nickel alloy is refined. Moreover, the particle density of ferronickel slaggy is larger than general sand. While the one with a large grain diameter is chiefly used as aggregate for concrete effectively, it is hoped that ferronickel slaggy of the grain diameter at the sand level is used as geomaterials.

We perform the consolidated and drained triaxial compression tests (JGS 0524) for above-mentioned seven samples. Test conditions are shown in Table 2.

Table 2. Test conditions.

Sample	Pressure (kPa)	ρ_t (g/m^3)	ρ_d (g/m^3)	D_r(%)
A	100	1.510	1.455	85.0
	200	1.494	1.450	
	300	1.501	1.453	
B	100	1.902	1.879	85.0
	200	1.905	1.881	
	300	1.905	1.881	
C	100	1.679	1.671	85.0
	200	1.678	1.670	
	300	1.688	1.673	
D	100	1.597	1.584	85.0
	200	1.596	1.583	
	300	1.594	1.582	
E	100	1.144	1.090	85.0
	200	1.161	1.093	
	300	1.141	1.086	
F	100	1.567	1.533	85.0
	200	1.549	1.528	
	300	1.556	1.530	
G	100	2.104	2.088	85.0
	200	2.135	2.102	
	300	2.115	2.090	

The sample size is 5 cm in the diameter and 10 cm in height, and the relative density is about 85%.

2.2 *Results*

From the results, we calculate the internal friction angle and the cohesion of each sample. Moreover, we define the deformation modulus E_{di} (MPa) by the following expression as an index that shows the deformation property.

$$E_{di} = \sigma_d/\varepsilon_d \tag{1}$$

where the σ_d is half an axial stress of maximum axial stress, ε_d is the axial strain at ε_d and i shows the lateral pressure. Because deformation characteristics depends on the lateral pressure, we calculate the deformation moduli in each case of pressures.

Table 3. Mechanical properties of filling materials.

	Cohesion c_d (kPa)	Internal friction angle ϕ_d (°)	E_{d100} (MPa)	E_{d200} (MPa)	E_{d300} (MPa)
A	16.1	39.0	7.8	10.7	12.6
B	6.8	36.3	16.2	25.9	36.1
C	3.5	40.3	10.8	15.4	48.4
D	12.8	37.5	14.1	20.9	25.5
E	2.7	31.1	2.8	4.8	7.1
F	12.0	32.3	5.7	8.8	11.6
G	9.9	43.4	12.1	24.4	26.9

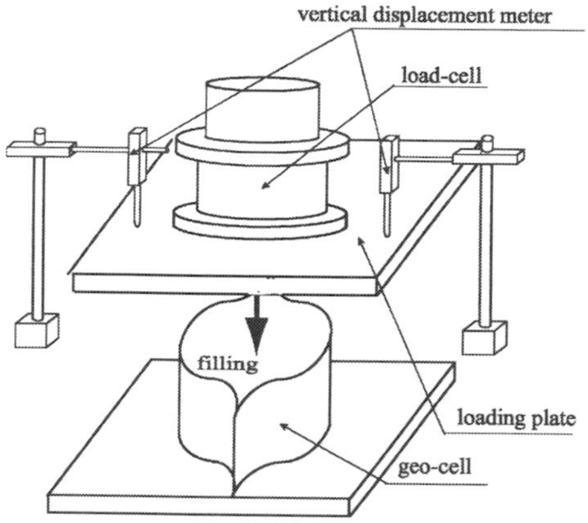

Figure 3. Compression test device.

The strength and deformation characteristics of all samples are shown in Table 3. Cohesion is very small value compared with the maximum principal stress difference, and we can consider almost 0.

3 COMPRESSION CHARACTERISTICS OF GEOCELL REINFORCED SAND

To use various materials as the filling in geocell, it is necessary to understand mechanical correlations between the filling and the geocell structural body. In this section, we examine the correlation between the deformation characteristics of fillings and the compressibility of geocell structure. We use seven kinds of above-mentioned sandy materials as the filling.

3.1 *Outline of compression test*

Figure 3 shows an outline of the compression test device. The filling materials are filled separately for the 3 layers in the cell. Each layer is compacted 25 times respectively so that the relative density might become constant. We use the unit geocell structure for the compression tests. The compression tests are performed by controlling the vertical

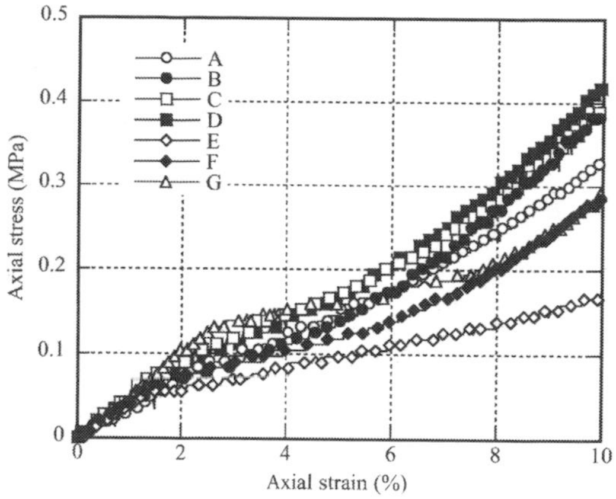

Figure 4. Stress-strain curves.

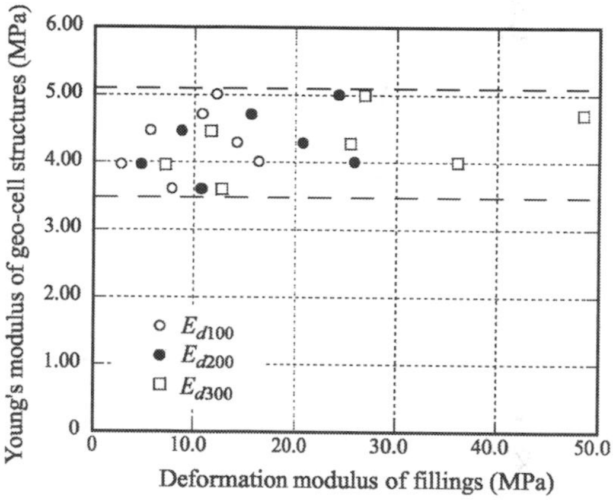

Figure 5. Relation between Young's modulus of geocell structure and deformation modulus of filling.

force (0.25 ton/minute). The vertical displacements are measured. The axial stress and the axial strain were calculated by the method similar to the compression test for the soil.

3.2 *Results and discussions*

Figure 4 shows the relation between the axial strain and the axial stress of the geocell structure obtained by the compression tests. Here, the results up to the axial strain 10% was shown. Even when the examination was continued up to about 30%, complete failure of the geocell structure was not seen. We can consider that it is not completely destroyed under a realistic loading condition. Therefore, it is important that we evaluate the mechanical properties of the geocell structure according to the deformation characteristics.

Figure 5 shows relation between deformation modulus of fillings and Young's modulus of geocell

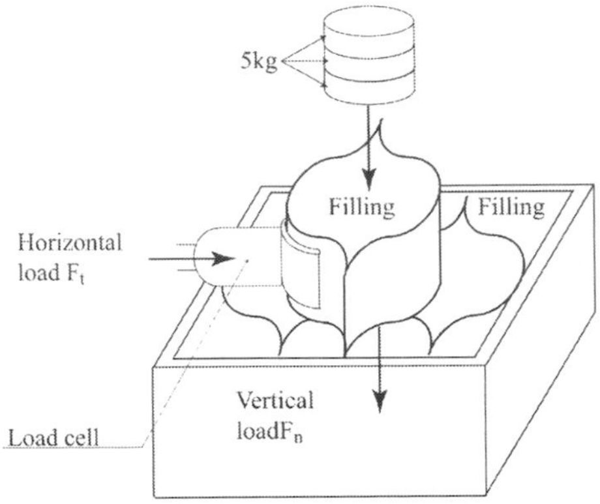

Figure 6. Shear resistance test device.

reinforced sand. In this study, we consider that the Young's modulus of geocell structure can be defined by the ratio of the axial stress and the axial strain in initial state of loading. As for Young's modulus of the geocell reinforced soil, it enters between 5MPa from 3MPa in all cases. The clear correlation between deformation modulus of the filling and Young's moduli of geocell strucutres can not be seen from Fig. 5. A similar deformation property is shown, when the sandy materials is used as the filling of the geocell structure. In the geocell method, it is guessed that there is hardly influence by the few difference of deformation property of the filling.

4 SHEAR RESISTANCE OF GEOCELL REINFORCED SAND

As for the geocell reinforced soil, there are a lot of cases vertically using the cell repeatedly. In general of these cases, because upper and lower cells are not connected and each step is independent, shear resistance between upper and lower cells is important. In this section, the correlation between internal frictional angle ϕ_d of the filling and the shear resistance between geocell structures is examined.

4.1 *Outline of shear resistance test*

Figure 6 illustrates an outline of the shear resistance examination between cells. Lower two geocells with filling are set in the box. We put an upper geocell with filling on the lower cells. The weight is put on the upper cell to change the vertical force F_n. The one side center of the upper geocell is horizontally pushed at 2 mm/min by the hydraulic lifter. Horizontal reaction forces and horizontal displacements are measured.

4.2 *Results and discussions*

First of all, the results of the case of sample A is shown in Fig. 7 as an example. We can say from this figure that in the first stage of the horizontal loading, the horizontal reaction forces increase linearly. When the horizontal reaction forces reach a maximum value, slippage occurs. We assume the maximum value as shear resistance force F_s. The shear resistance coefficient μ, which is so called static friction coefficient, is calculated as follows.

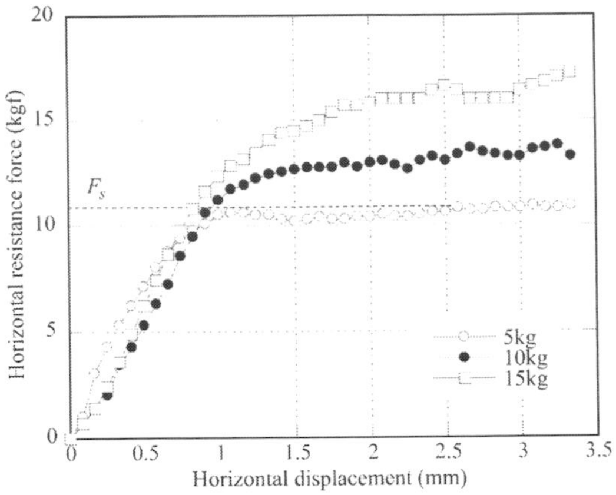

Figure 7. Relation between horizontal displacement and horizontal load (sample A).

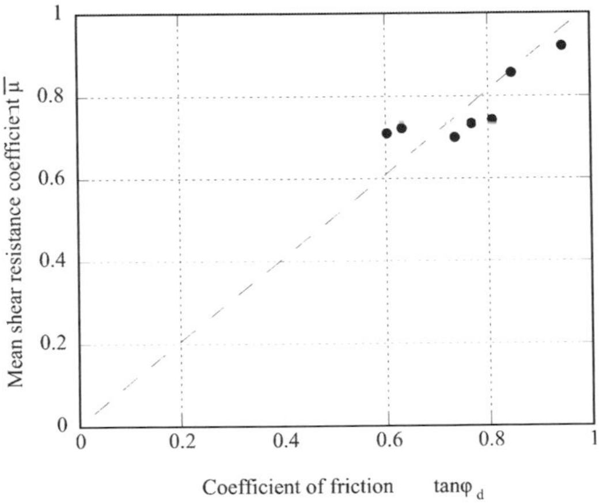

Figure 8. Relation between internal friction coefficient and mean shear resistance coefficient.

$$\mu = F_s/F_n. \tag{2}$$

Moreover, we calculate the mean shear resistance coefficient $\bar{\mu}$ by averaging the results of the three cases of tests, which change vertical force, in each filling material. Fig. 8 shows the relationship between the internal friction coefficient $\tan\phi_d$ of filling material and the mean shear resistance coefficient between geocells $\bar{\mu}$. As can be seen from this figure, $\bar{\mu}$ grows by ϕ_d large and it is almost linear relationship.

5 SUMMARY

In this study, to develop the geocell construction method by using the local generation soil effectively, we performed some experiments and we discussed the relationships between the mechanical properties of filling materials and the mechanical property of the geocell structures. The results of our experiments are shown as follows, When we use the sandy materials as the filling in the geocells, Young's modulus of the unit geocell structures is from 3MPa to 5MPa. In the geocell method, the few difference of deformation characteristics of the filling hardly influence the compressibility of the geocell structure. When we use sandy materials as filling, the mean shear resistance coefficient between cells grows by the internal friction angle of the filling large and it is almost linear relationship. As the next task, it will be necessary to make the same investigation when the clay soil is used as a filling.

REFERENCES

Omori, H., M. Shimada, K. Kaneko, M. Horie, and K. Kumagai (2006). Field observation and deformation measurements of geo-cell reinforced retaining walls. *Geosynthetics Engineering Journal 21*, 23–30, in Japanese.

Sitharam, T.G., S. Sireesh, and D.S.K. (2005). Model studies of a circular footing supported on geocell-reinforced clay. *Canadian Geotechnical Journal Vol. 42, No. 2*, 693–703.

Yazawa, K., H. Omori, K. Kaneko, M. Horie, and K. Kumagai (2006). Influence of filling materials to mechanical properties of geo-cell reinforced soil. *Geosynthetics Engineering Journal 21*, 31–36, in Japanese.

New Horizons in Earth Reinforcement – Otani, Miyata & Mukunoki (eds)
© 2008 Taylor & Francis Group, London, ISBN 978-0-415-45775-0

Experimental study on bearing capacity of geocell-reinforced soil

T. Ajiki, H. Ohmori, K. Yazawa, K. Kaneko & K. Kumagai
Hachinohe Institute of Technology, Japan

ABSTRACT: This paper reports the results from a series of laboratory model tests and in-situ tests carried out to investigate the performance of the geocell reinforcement for bearing capacity of soils. At first, we carried out the middle-size model tests for sands to examine the tendency of supporting performance and supporting mechanisms. The bearing capacity, the vertical displacement under footing and the distribution of the local deformation are measured. Next, we carry out the in-situ simple bearing capacity tests to investigate the effect of geocell reinforcement for loam clay.

1 INTRODUCTION

Recently, geocell reinforced method for soils is developed. The geocell (Horie et al. 2006; Omori et al. 2006), which is made by high density polyethylene (HDPE), is one of the three-dimensional geo-synthetics materials, and is honeycomb-like continuous cells. Geocells are conventionally filled with free-drainage material such as sand and gravel. However, if we can use surplus soil as filling materials in the cells suitably, it is very useful and has many advantages for natural environments and construction costs. It is an important problem to utilize the surplus soils usefully for sustainable society. A series of our study on geocell aims to use the surplus soil as the filling materials.

In this paper, we performed a series of laboratory model tests and in-situ tests carried out to investigate the performance of the geocell reinforcement for bearing capacity of soils. At first, we carried out the laboratory bearing capacity tests for sands to examine the tendency of supporting performance and supporting mechanisms. We carried out some cases of the tests by changing patterns of the arrangements of geocells. The bearing capacity and the vertical displacement of footing were measured. In addition, to examine the reinforced mechanisms, we analyzed the distribution of the local deformation by using the digital images, which are taken continuously at the test. Next, we carried out the in-situ simple bearing capacity tests to investigate the effect of geocell reinforcement for loam clay. We used the loam clay as the filling materials that assumed the surplus soils. Moreover, we conducted same experiment to other soil-reinforcement techniques, such as geogrid-reinforced methods and gravel replacing, for the comparison.

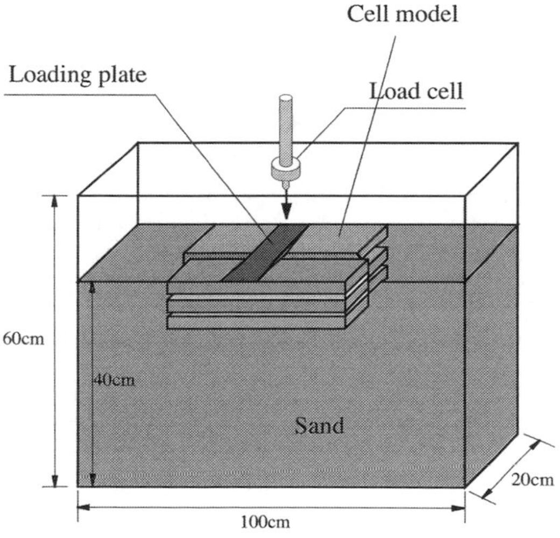

Figure 1. Model of soil tank.

2 LABORATORY TESTS FOR BEARING CAPACITY

2.1 *Outline*

In this study, we use a medium soil tank for laboratory bearing capacity tests. Figure 1 shows the experimental apparatus. The size of the soil tank frame is 60 cm × 100 cm × 20 cm. The front side of the soil tank is made from acrylic to take pictures for the PIV (Particle Image Velocimetry) analysis, which is one of the digital image analysis method to examine the distribution of velocity.

We put samples into the soil tank up to 40 cm in depth. The loading plate, whose size is

Table 1. Fundamental properties of a sample.

Coefficient of uniformity		2.29
Coefficient of curvature		1.21
Density of particles	(g/cm^3)	2.603
Minimum density	(g/cm^3)	1.434
Maximum density	(g/cm^3)	1.639
Maximum dry density	(g/cm^3)	1.644
Optimum moisture content	(%)	15.02

Table 2. Cases of bearing capacity tests.

Case no.	Situation of sand	Density (g/cm^3)
Case 1	Loose sand	1.33
Case 2	Dense sand	1.52
Case 3	One layer of geocell	1.37
Case 4	Two layers of geocell	1.37
Case 5	Three layers of geocell	1.37

1 cm × 15 cm × 20 cm, is put at the center on the sample. We give a vertical displacement to the center of the loading plate compulsory. Loading speed is 1.6 mm/minute.

We measure the reaction forces by the load cell, which is set to the place where displacement is given. When the reaction force is not changing for three minutes, we end the examination. We define the reaction force at this time as the limit bearing capacity.

We model geocell using celluloid sheets. In this experiment, a quarts sand is used as a sample. Table 1 shows fundamental properties of it.

We carry out five cases of bearing capacity tests and the test conditions of each case are shown in Table 2. Case 1 and 2 are loose and dense state, respectively. Case 3, 4, 5 have 1, 2, 3 layers of geocell with loose state of sand, respectively.

2.2 *Results and discussions*

Relationships between reaction forces and vertical displacements, which are obtained by the bearing capacity tests, are shown in Fig. 2. We can understand from this figure that the limit bearing capacity rises by geocell reinforcement effectively. We calculate limit bearing capacities of each case from these results. The limit bearing capacities are as follows:

Case 1, 146 kgf, Case 2, 159 kgf,
Case 3, 153 kgf, Case 4, 181 kgf,
Case 5, 191 kgf.

The limit bearing capacities of cases reinforced by geocell, which are Case 3–5, are larger than that of Case 1 that is non-reinforced state. Moreover, the limit

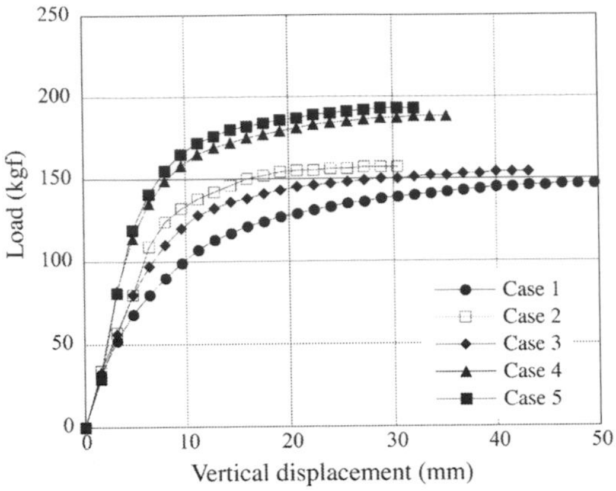

Figure 2. Load-displacement curves.

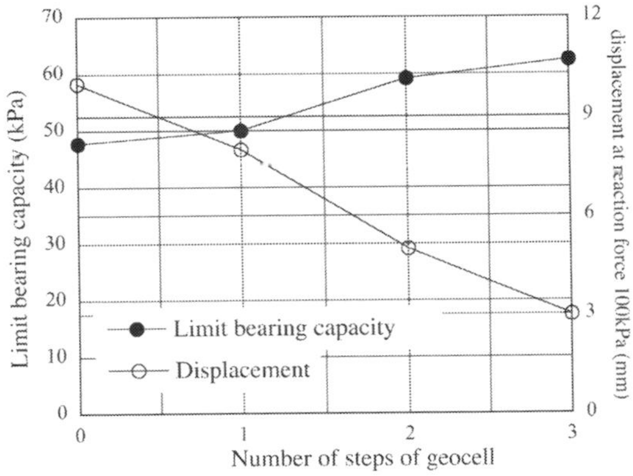

Figure 3. Effects of geocell reinforcement.

bearing capacities of Case 4 and 5 are larger than that of Case 2 whose density is dense. In these experiments, the bearing capacity when three layers of the geocell is set up is the largest. Relationship between the number of geocell layers and the limit bearing capacity is shown in Fig. 3. Relationship between the number of geocell layers and the displacement when the reaction force reaches 100 kgf is also shown in Fig. 3. From this figure, we can understand for limit bearing capacity to rise when the number of geocell layers is increased and to be able to restrain vertical displacement. In addition, when the number of geocell layers is adjusted to two or more, limit bearing capacity is up by 25%. At this time, the vertical displacement became about 1/2.

Distributions of local velocities of the sand under loading, which are obtained from PIV image analysis, are shown in Fig. 4. The PIV analysis is performed by using the pictures that are taken at the tests. Figure 4 shows the distributions of velocity at when the vertical displacement is during the 19.2 mm from 14.4 mm. The length of vector means speed of motion of the

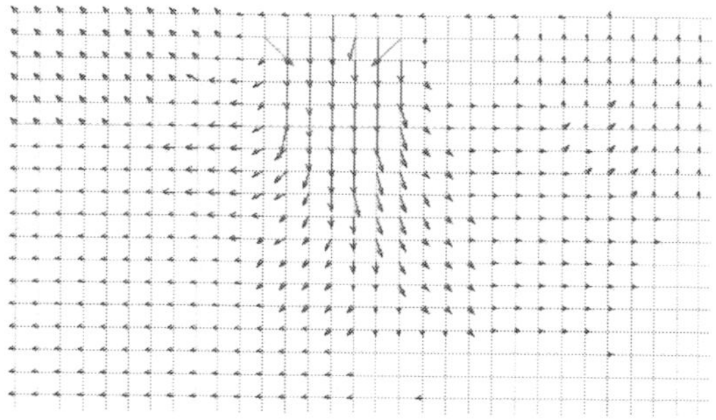

Case 1

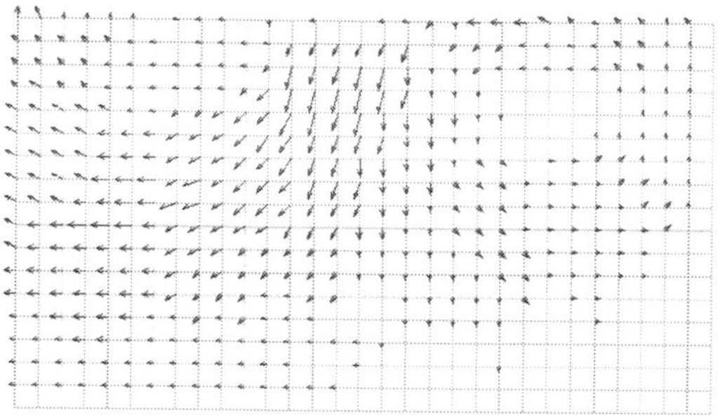

Case 4

Figure 4. Distributions of velocities of sands (vertical displacement: 14.4–19.2 mm).

sand. The velocity vector distribution in Case 1 shows that the sands mostly move from loading plate in a under direction. The speed of motion in this direction is relatively very fast. However, the velocity vector distribution of Case 4 shows that the sands under the loading plate move to the horizontal direction. In the Case 4, the range of moving area became large though the speed of each vector became small by the geocell. From these results, we can say that the displacement is distributed by geocell and vertical displacement under loading plate is decreased. As a result, the limit bearing capacity rises.

3 IN-SITU IMPACT TESTS FOR BEARING CAPACITY

3.1 *Outline*

We use a simple bearing capacity instrument (Fig. 5), which is called CASPOL, in order to measure in-situ bearing capacity of geocell-reinforced soil. The rammer of 50 mm in the diameter and 4.5 kg in weight is freely dropped from the height of 45 cm and an impact value is measured. A coefficient of the ground reaction is calculated by a following equation.

$$K_{30} = -37.58 + 8.554 I_a \qquad (1)$$

Figure 5. Simple bearing capacity instrument.

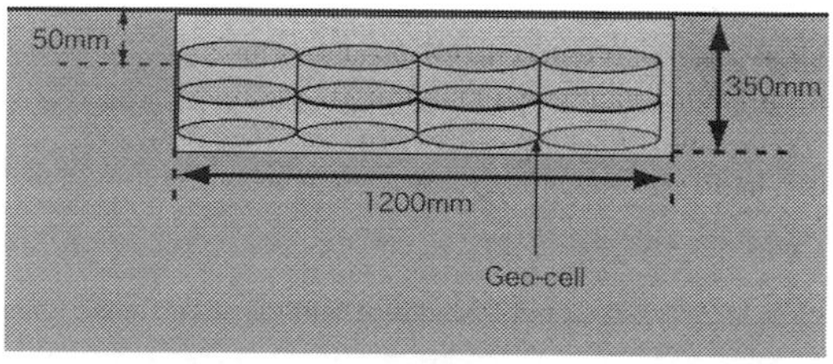

Figure 6. Geocell reinforcement soil.

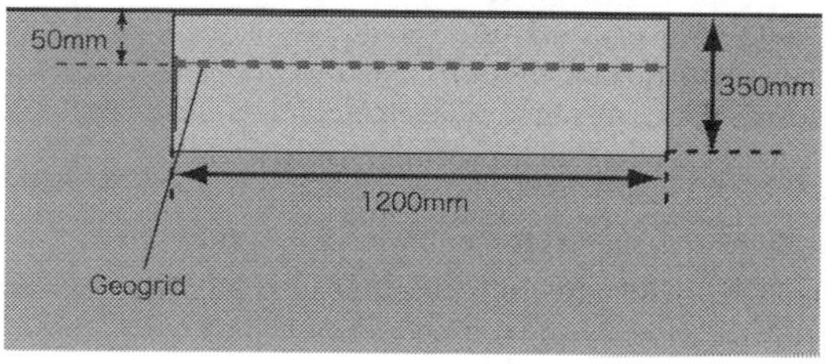

Figure 7. Geogrid reinforcement soil.

where K_{30} is the coefficient of the ground reaction and I_a is an impact value measured by tests. By using the simple bearing capacity instrument, we can perform a prompt measurement because carrying is easy. In this study, we dig a hole on the site ground and construct the reinforce soil structures. The hole size is 800 mm in length, 1200 mm in width and 350 mm in depth.

We perform following five cases of tests:

1. Geocell-reinforced soil
2. Geogrid-reinforced soil
3. Geocell and geogrid reinforced soil
4. Re-filling surplus soil
5. Macadam replacement.

Figure 6, 7, 8, 9 and 10 show the outline of these cases of in-situ tests, respectively.

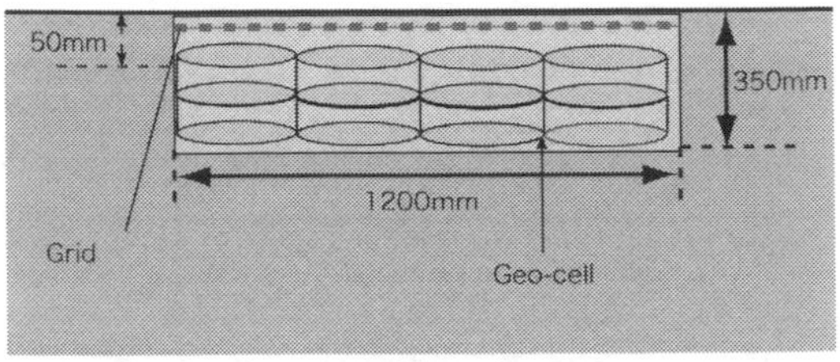

Figure 8. Geocell and geogrid reinforcement soil.

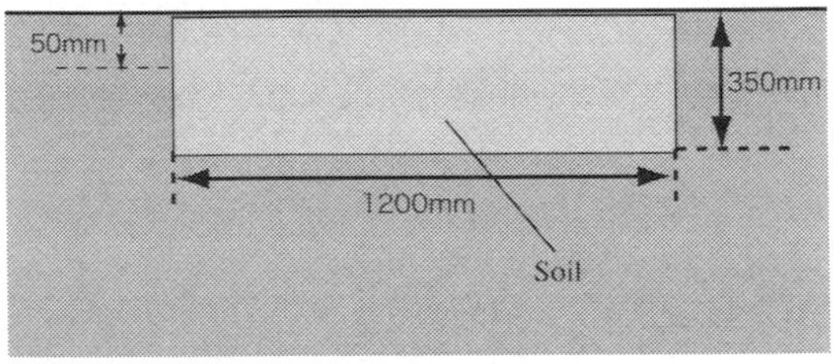

Figure 9. Re-filling surplus soil.

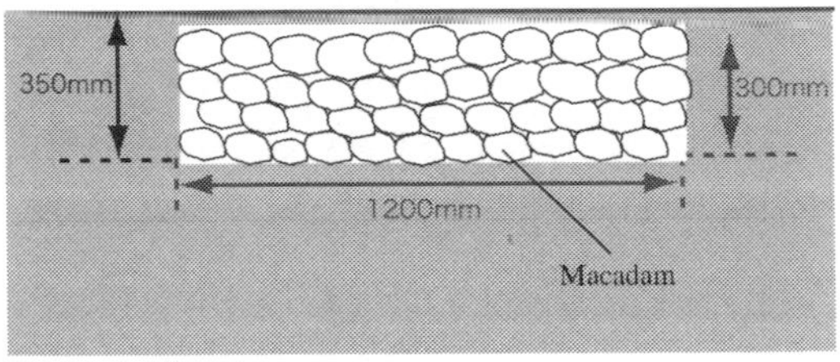

Figure 10. Macadam reinforcement.

3.2 *Results and discussions*

Table 3 shows the ground reaction coefficients K_{30} obtained by the simple in-situ bearing capacity tests. We carried out the tests three times in each cases and calculated the average value. From these results, we can say that the value of the ground reaction increase by the reinforcement. The most effective reinforcement in these cases is the method of using the geocell with geogrid. The ground reinforced by the geocell and geogrid indicates twice the reaction coefficient compared with the method of re-filling the surplus soils. K_{30} of the ground reinforced by only the geocell is nearly equal to that of the replacement to the macadam. K_{30} of the ground reinforced by only the geogrid is nearly equal to that of the method of re-filling the surplus soils.

Table 3. Ground reaction coefficient.

Test condition	K_{30}	K_{30}	K_{30}	Average
Geocell	23.3	26.3	24.2	24.6
Geogrid	18.3	17.8	16.1	17.4
Geocell and Geogrid	36.8	35.3	31.4	34.5
Re-filling surplus soil	14.5	15.3	18.3	16.0
Macadam	23.7	20.7	32.2	25.5

4 CONCLUSIONS

In this study, we carried out the laboratory and in-situ tests to examine the effect of the geocell reinforcement for bearing capacity. We confirmed the effect and mechanism of geo-cell reinforcement for the bearing capacity from the laboratory tests. In addition, we performed the simple in-situ tests to demonstrate the effect of reinforcement by the geocell. Moreover, we compared the effect of reinforcement with other reinforce method by in-situ test. The results of our experiments are shown as follows:

1. The limit bearing capacity rises by the geocell-reinforcement effectively.
2. As the displacement is distributed by geocell, the limit bearing capacity increases and the vertical displacement under loading decreases.
3. The most effective reinforcement in this study is the method of using the geocell with geogrid. The ground reinforced by the geocell and geogrid has twice reaction coefficient than re-filling surplus soils.
4. K_{30} of the geocell-reinforced ground is nearly equal to that of the replacement to the macadam.

REFERENCES

Horie, M., H. Omori, T. Hirose, K. Kaneko, E. Adachi, and K. Kumagai (2006). Fundamental study on variable slope concrete gutter reinforced by geocell. *Geosynthetics Engineering Journal 21*, 17–22, in Japanese.

Omori, H., M. Shimada, K. Kaneko, M. Horie, and K. Kumagai (2006). Field observation and deformation measurements of geo-cell reinforced retaining walls. *Geosynthetics Engineering Journal 21*, 23–30, in Japanese.

New Horizons in Earth Reinforcement – Otani, Miyata & Mukunoki (eds)
© 2008 Taylor & Francis Group, London, ISBN 978-0-415-45775-0

Effect of soil dilation on performance of geocell reinforced sand beds

S.K. Dash
Department of Civil Engineering, Indian Institute of Technology Guwahati, India

ABSTRACT: This paper presents results of a series of laboratory model tests, designed to bring out the influence of dilation of the fill soil on the performance improvement due to the geocell reinforcement, in sand beds. The benefit due to geocell reinforcement in foundation soil is found to increase with increase in dilation of infill soil. With geocell reinforcement offering three dimensional confinement the dilation induced benefit is substantially high. Therefore, for effective utilisation of geocell reinforcement, the infill soil should be compacted well.

1 INTRODUCTION

Soil reinforcement in the form of geocell has been utilised successfully in many areas of geotechnical engineering (e.g. Bathurst and Jarrett 1989, Bush et al. 1990, Dash et al. 2001). The geocell reinforcement arrests the lateral spreading of the infill soil and creates a stiffened mat to support the foundation, thereby, giving rise to higher load carrying capacity. This paper presents results of a series of laboratory model tests designed to bring out the influence of placement density of the fill soil, on the performance of geocell reinforced sand beds.

2 TEST DETAIL

The model tests were conducted in a steel tank having a length of 1200 mm, width of 332 mm and a height of 700 mm. The model footing used was made of steel and measured 330 mm length × 100 mm width × 25 mm thickness. The bottom surface of this footing was made rough by cementing a thin layer of sand with epoxy glue. The footing was centered in the tank with the length of the footing parallel to the width of the tank. As the length of the footing is almost equal to the width of the tank, plane strain conditions prevailed in the tests. The soil used in this investigation is a poorly graded river sand. The maximum and minimum dry density of the soil are found to be 17.41 kN/m^3 and 14.30 kN/m^3 respectively.

The model tests were performed at relative densities of 30%, 40%, 50%, 60% and 70%. The friction angle of the sand at three relative densities, 30%, 50% and 70%, as determined from standard triaxial compression tests, are 39.2°, 41° and 42.2° respectively. The dilation angles of the sand were found to be 0°, 6°, 20°; at these relative densities respectively. The geocell layers were formed (Bush et al. 1990), using a biaxial geogrid of aperture opening size of 35 × 35 mm and ultimate tensile strength of 20 kN/m. The joints of the geocells were formed using 6 mm wide and 3 mm thick plastic strips cut from commercially available bodkins made of low-density polypropylene (Fig 1a). All the tests were performed with single layer of geocell reinforcement. To achieve uniform density of the fill soil in the test tank sand raining technique was used. The height of fall to achieve a desired relative density was determined through calibration tests a priori.

The load was applied through a hydraulic jack in small increments, in stress controlled fashion. The actual load transmitted to the footing was measured through a pre-calibrated proving ring placed between the hydraulic jack and the footing. The settlements (s) of the footing were measured using two dial gauges, placed diagonally on opposite sides of the footing centerline. The deformations on fill surface (settlement/heave, δ) were recorded through dial gauges placed at a distance of 2.5 B on either side of the footing (shown through ▼ in Fig. 1). The strains developed in the geocell were measured through electrical resistance type strain gauges fixed in horizontal direction on the geocell wall along its width at mid height. In many cases the strain gauges were damaged prematurely. However meaningful observations can be made from the available data. The load transferred to the footing, settlements of the footing, deformations on fill surface and strain in geocell wall, were recorded after stabilisation of settlement under each load increment. The geometry of the problem is shown in Fig. 1. In order to understand the influence of dilation of soil on the overall performance, in all the tests the geometry of the geocell layer was kept constant as, height, $h/B = 1.6$; width, $b/B = 8$; depth of placement, $u/B = 0.1$. The pocket size (d) that is considered as

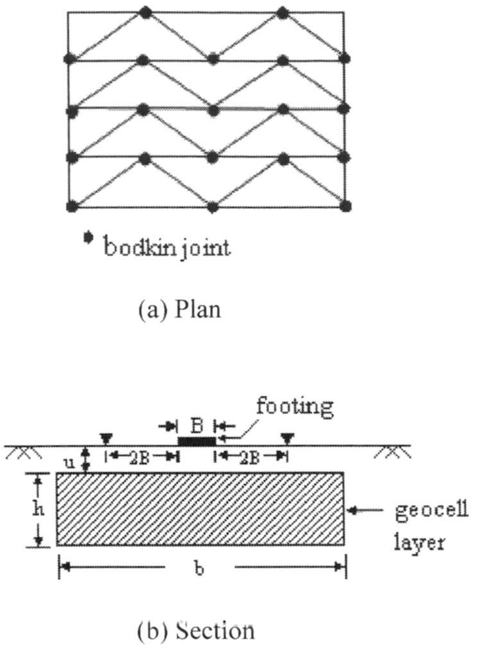

Figure 1. Geometry of geocell reinforced sand bed.

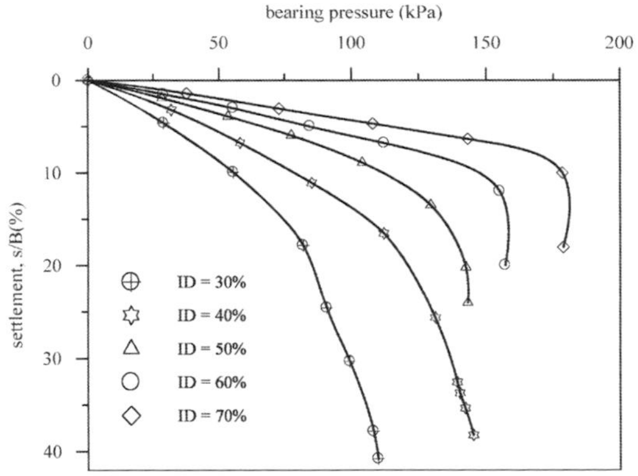

Figure 2. Pressure-settlement plots for unreinforced soil.

the equivalent circular diameter of the geocell pocket opening was kept equal to 1.2 B in all the tests.

3 RESULTS AND DISCUSSION

Bearing pressure versus settlement responses for unreinforced soil are shown in Fig. 2 and that for geocell-reinforced soil are shown in Fig. 3.

It could be observed that, while, the responses for unreinforced soil have a clear break point with reduction in slope indicating shear failure in soil, the geocell

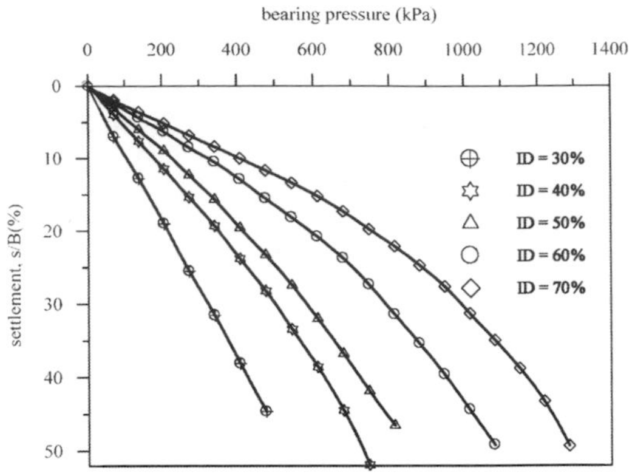

Figure 3. Pressure-settlement plots for geocell reinforced soil.

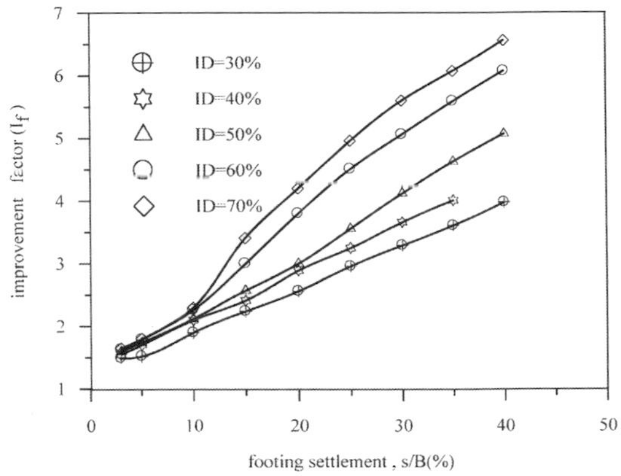

Figure 4. Improvement factor-settlement plots for different relative densities of soil.

reinforced soil beds without showing any such failure continue to sustain increased footing loading till a settlement as high as 50% of the footing width.

The improvement due to the provision of geocell reinforcement is represented using a non-dimensional improvement factor (I_f) which is defined as the ratio of footing pressure (q) with geocell at a given settlement to the pressure on unreinforced soil (q_o) at the same settlement. If the footing has reached its ultimate capacity at a certain settlement, the bearing pressure q_o is assumed to remain constant at its ultimate value for higher settlements. The variation of improvement factors with footing settlement for different relative densities of soil are presented in Fig. 4.

From Fig. 4 it could be observed that the performance improvement due to the geocell reinforcement increases with increase in density of infill soil and footing settlement. The soil with higher relative density dilates more that induces higher frictional resistance at the geocell soil interface thereby increasing the

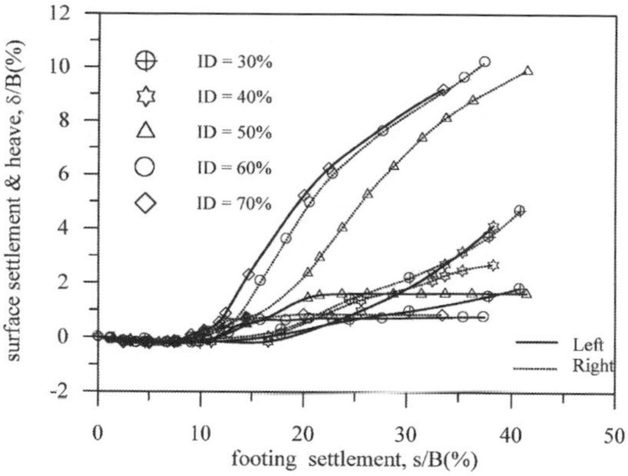

Figure 5. Surface deformation-footing settlement for unreinforced soil.

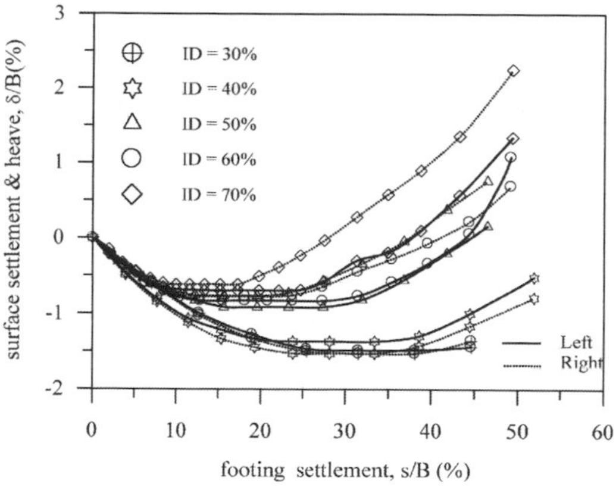

Figure 6. Surface deformation-footing settlement for geocell reinforced soil.

resistance to downward penetration of sand and hence a higher improvement in load carrying capacity. It is also believed that loose soil contracts under deformation therefore more strain is required before stress transfer to the geocell occurs. Whereas, soil with higher relative density dilates. This leads to a compact structure that mobilises higher strength in the geocell reinforcement. Therefore, the geocell reinforcement gives enhanced performance with higher relative density of fill soil.

In the case of soils with 30% and 40% relative density, the rate of increase of I_f at large settlements is the same as that at small settlements. On the other hand, the soils with higher relative density have exhibited higher rate of increase of I_f at higher settlements as compared to that at smaller settlements.

In Fig. 5 and Fig. 6, that depict the pattern of deformation on fill surface at left and right side of the footing, heave is shown through (+) sign and settlement through (−) sign. It could be observed that the unreinfiorced sand bed has undergone heaving equal to about 10% of footing width (Fig. 5), while with geocell reinforcement it is less than 2.5% (Fig. 6). Besides, the soils with higher relative density have undergone higher surface heaving, because of volumetric expansion. From Fig. 5 it can be observed that for dense soil (ID = 70% and 60%) heaving starts at a settlement equal to about 10% to 15% of footing width and it is at about 20% to 25% of footing width for loose soil (ID = 40 % and 30%). While in case with geocell reinforcement (Fig. 6) there occurs a zone where there is no surface deformation, indicated through a segment of horizontal line. In case of dense soil the no-surface-deformation zone starts at a footing settlement equal to about 10% to 15% of footing width. It may be the stage where the geocell by virtue of its confinement suppresses dilation in the soil, thereby, higher strength of geocell material is mobilised. This could be the reason for rapid increase in I_f values beyond settlement (s)

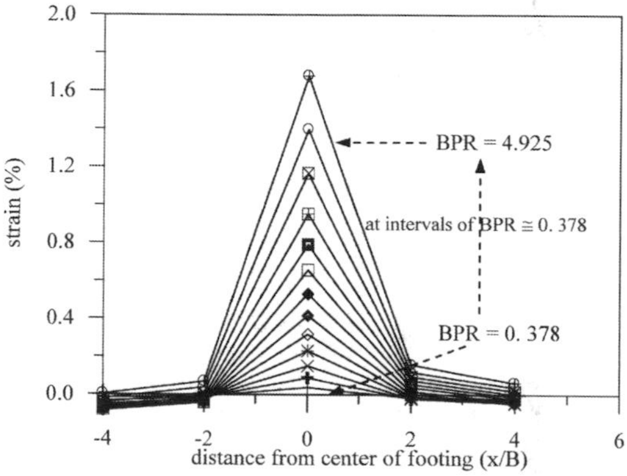

Figure 7. Strain in geocell wall for soil at 70% relative density.

of about 0.1B as compared to that at lower settlements (Fig. 4). However, in case of loose soil the improvement factor increases almost linearly with increase in footing settlement. This is because of non-dilative nature of the loose soil, which is reflected in the observation that in case of loose soil the surface deformation is completely in the settlement range (Fig. 6).

Besides, the surface settlements are generally found to be lower for higher density of soil, which may be due to the increase in subgrade strength and end anchorage from soil that resists the downward deflection of the geocell mattress, giving rise to higher performance improvement. In the tests with loose soil fill, the two free ends of the geocell mattress were visible at the soil surface at large footing settlements. This result clearly shows that there was not enough anchorage at the two ends of the geocell mattress in the case of loose soil.

Fig. 7 and Fig. 8 depict the pattern of strain variation in geocell wall for dense soil (relative density 70%) and loose soil (relative density 30%) respectively.

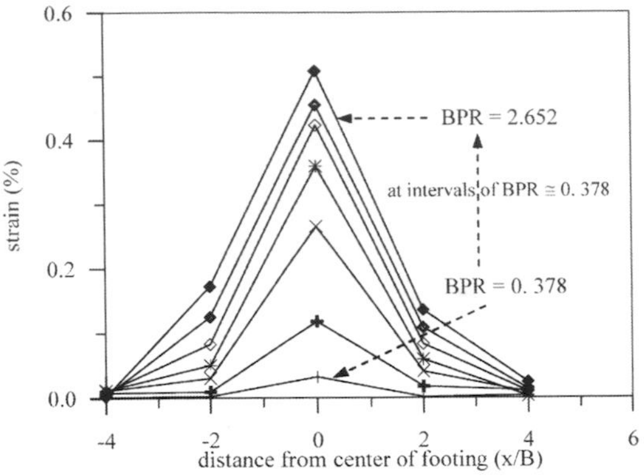

Figure 8. Strain in geocell wall for soil at 30% relative density.

The strain measurements are reported at various normalised footing load levels (BPR). The Bearing Pressure Ratio (BPR) is defined as the ratio between the footing pressure with geocell (q) and the ultimate footing pressure (q_{ult}) in tests on unreinforced soil. For uniformity in comparison of data from different tests, the q_{ult} is taken as the ultimate pressure on unreinforced soil at 70% relative density uniformly for all the tests. The compressive strains are shown with negative sign and the tensile strains with positive sign.

From Fig. 7 it could be observed that, in case of dense soil (ID = 70%) compressive strains are induced at (or near) two free ends of the geocell mattress. This compressive strain is found to be maximum at mid-height of the geocell. Due to dilation of soil in the regions of loading, there develops a volume expansion of sand, through aperture opening of geogrid walls. This expansion of soil is mostly in the transverse direction, in the vertical direction there is infinite zone of soil. This localised transverse expansion is restrained by the sand in the adjacent stable region. Such a restraint produces compression in the soil mass thereby giving rise to compression in the geocell wall. Fig. 8 shows that in the case of loose sand (ID = 30%), compressive strains have not developed anywhere in the mattress. This is because of the absence of dilation induced volume expansions in the loose soil. This observation once again reinforces the earlier conclusions regarding the infill soil dilation induced behaviour of geocell mattress.

The present observations in case with geocell reinforcement is in contrary to the findings of Fragaszy and Lawton (1984) in case of planar reinforcement that at 4% settlement (s = 0.04B) the values of BCR (which is same as I_f in the present study) for ID = 51%, 61%, 70%, 80%, 90% are 1.2, 1.2, 1.4, 1.5 and 1.5 respectively while at settlement close to failure (i.e. s = 10% of B) it is 1.6, 1.7, 1.7, 1.7 and 1.6 respectively. These results indicate that at relatively lower settlement range the performance improvement increases with increase in relative density of soil whereas at higher settlement range close to failure, the performance improvement is independent of soil density. In case of planar reinforcement system, the reinforcing effect is mostly due to interfacial frictional resistance mobilised through deformation and at settlement close to failure the soil shears away with the large strength of reinforcement remaining unmobilised. Whereas, the geocell system being an all-round confining system, restrains soil flow, thereby the encapsulated soil does not shear away. Hence, with increased density the dilation induced benefit is more at higher settlement.

4 CONCLUSIONS

From this study it could be said that the benefit due to geocell reinforcement in foundation soil increases with increase in dilation of soil. With geocell reinforcement offering three dimensional confinement the dilation induced benefit is substantially high compared to the case with planar reinforcement where the soil shears away easily. Therefore, for effective utilisation of geocell reinforcement, the infill soil should be compacted to higher density that it achieves higher dilation angle.

REFERENCES

Bathurst, R.J. and Jarrett, P.M. (1989) Large-scale tests of geo-composite mattresses over peat subgrades. *Transportation Research Record* 1188, pp. 28–36.

Bush, D.I., Jenner, C.G. and Basset, R.H. (1990) The design and construction of geocell foundation mattress supporting embankments over soft ground. *Geotextiles and Geomembranes*. Vol: 9, pp. 83–98.

Dash, S.K., Krishnaswamy, N.R. and Rajagopal, K. (2001) Bearing capacity of strip footings supported on geocell-reinforced sand. *Geotextiles and Geomembranes*, Vol:19, pp. 235–256.

Fragaszy, R.J. and Lawton, E. (1984) Bearing capacity of reinforced sand subgrades. *Journal of Geotechnical Engineering Division*, ASCE, Vol. 110, pp. 1500–1507.

New Horizons in Earth Reinforcement – Otani, Miyata & Mukunoki (eds)
© 2008 Taylor & Francis Group, London, ISBN 978-0-415-45775-0

Laboratory investigation into effectiveness of thixotropic gel compaction method

A.M. El-Kelesh & K. Tokida
Osaka University, Osaka, Japan

T. Oyama & S. Shimada
Kyokado Engineering, Tokyo, Japan

ABSTRACT: In the Thixotropic Gel Compaction (TGC) method a grout material that is basically a fly-ash mortar is injected into the soil. During injection, this grout material increases in size and displaces the surrounding soil without mixing with or penetration into the soil. This mechanism of grout expansion renders the method effective in improving the in-situ loose soils and in compensating the ground settlement associated with underground excavation and bored tunneling. In this paper, the method is examined through physical modeling in the laboratory using a large-scale double-wall calibration chamber. The calibration chamber and testing procedure are described herein. Also typical results of injection into sandy soil are presented and discussed to illustrate the grouting performance and the variation of soil improvement and compensation effectiveness with injection. The presented discussions reveal significant observations that can be used to effectively plan the TGC grouting works.

1 INTRODUCTION

In the Thixotropic Gel Compaction (TGC) method a grout material that is basically a fly-ash mortar is injected into the soil. The composition and characteristics of the TGC grout are described by Shimada et al. (2005). During injection, the grout material increases in size and remains in a homogeneous state resulting in displacement of the surrounding soil through a distinct grout-soil interface.

The mechanism of treatment by displacement provides for controlled treatment, as the required amount of grout can be injected where needed. This has the advantage of the possibility of varying the grout volume to account for the variation of soil properties, so that given target or uniform results can be obtained. The method is applicable to all soils that are compactable. Therefore, it can be used as a ground improvement technique to increase the bearing capacity or to minimize the liquefaction potential of the in-situ loose soils. The treatment mechanism renders the method also capable of inducing upheave of the soil, and as a result can be used to compensate the ground settlement associated with underground excavation or bored tunneling.

The grout material can be injected into the soil through injection pipes to form discrete grout bulbs or grout piles. An economically feasible procedure is to inject the grout material in stages while going upward. In such a procedure, after drilling to the bottom boundary of treatment zone, the lowermost bulb is injected. Then, the pipe is pulled upward for a given depth interval and the next bulb is injected. This procedure is followed till reaching the upper boundary of treatment zone. An alternative procedure is to inject the grout material through a sleeved tube known as a *tube à manchette* (TAM). The TAM is a plastic or steel pipe with pairs of holes drilled at intervals along the length of the pipe. Such holes are covered by tight rubber sleeves that act as non-return valves. The use of TAM provides more flexibility in performing the grouting works, where the grout material can be injected at any pre-determined order in both of the vertical and lateral directions throughout the treatment zone. Such a control of the injection sequence is useful where the injections can be planned, so that the previous injections provide confining actions for the subsequent ones in both of the vertical and lateral directions. The use of TAM along with the TGC grout that is thixotropic has an additional advantage of allowing re-injection of grout from the same port, if needed. The use of the appropriate procedure depends on the site conditions and economic considerations.

In this paper, the effectiveness of the TGC method is examined for both of the ground improvement and settlement compensation works. For this purpose, the method has been physically modeled in the laboratory using a large-scale calibration chamber.

This chamber is of the flexible double-wall type and has been developed to allow for preparing, consolidating and injecting soil samples under conditions approximating the actual in-situ ones. The paper starts with a brief description of the calibration chamber and the testing procedure. Then, typical results of injection into sandy soil are presented. These results are further discussed to illustrate the grouting performance in terms of the corresponding soil stresses and deformations and the variation of soil improvement and compensation effectiveness with injection.

2 EXPERIMENTAL SETUP AND PROCEDURE

Figure 1 shows the setup of the calibration chamber and the setting of injection pipe. The chamber is of the flexible double-wall type that well simulates the in-situ conditions (Holden 1971) and consists of seven major components: chamber base, loading piston, side membrane, double-wall barrel, retaining cylinder, lid and assembly rods. The chamber details, testing procedure and differences with the other double-wall chambers in use, as well as the required corrections for the effects of chamber size and boundary conditions, are reported by El-Kelesh and Matsui (2006). The principal features of the chamber are summarized in the following (see Fig. 1):

- The sample, 1.40 m in diameter and 1.45 m in height, is enclosed inside a cylindrical rubber membrane.
- The vertical stress is applied at the top of sample via the piston by pressurized air, while the lateral stress is applied by water filling the annular space between the side membrane and barrel.
- It utilizes a double-wall barrel to control the K_0-condition and impose different boundary conditions. The annular space between sample and inner wall (inner cell) as well as that between the inner and outer walls (outer cell) are filled with de-aired water. On increasing the vertical stress, the pressure of inner cell increases due to the tendency of sample to deform laterally. An electro-pneumatic (E/P) regulator is used to control the pressure of outer cell and keep it equal to the developed pressure of inner cell. This control assures no deflection of the inner wall, and therefore an average zero lateral deformation of the sample.
- It provides for upheave of the sample surface and can independently control the sample vertical stresses and lateral stresses and strains.
- To simulate the free-field conditions, the required boundary conditions of the sample are expected to be between the limits of constant stress and zero average strain on both the vertical and horizontal boundaries. The chamber panel of controls can impose these boundary conditions.

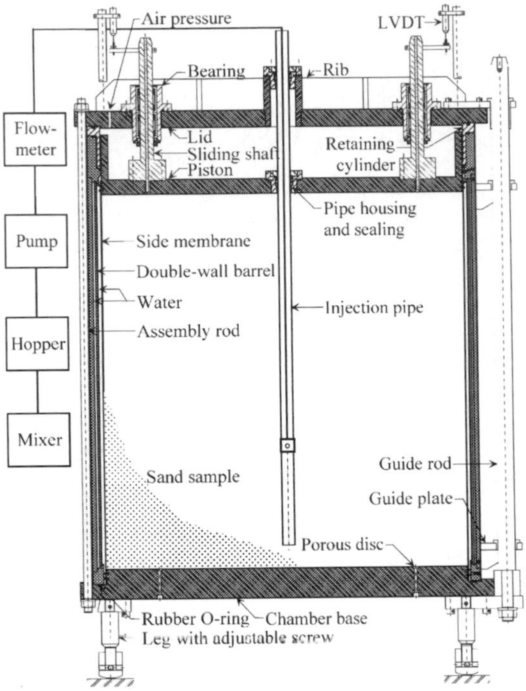

Figure 1. Setup of calibration chamber and injection system and positioning of injection pipe.

- It utilizes a carefully designed vertical sliding system consisting of four shafts fixed to the piston and slide through linear bearings bolted to the lid. This system provides for sample vertical compression of 125 mm and upheave of 25 mm from the initially prepared sample surface. It also allows for externally measuring the displacement of sample surface through the four sliding shafts and four Linear Variable Displacement Transducers (LVDTs) attached to the lid.
- The piston slides through a rubber O-ring recessed in an annular shoulder machined in the inside of the barrel.

The injection system consists of mixer, hopper, squeeze pump, flow meter and double packer. The flow meter is used to measure the injected volume of grout. The injection pipe passes through central holes in the lid and piston. It is perforated at 1.0 m from the sample surface. The pipe perforations are covered with a cylindrical rubber sleeve that acts as a non-return valve for the grout material.

The test is conducted in five stages: sample preparation, chamber assembly, sample consolidation, injection and chamber disassembly. A funnel system consisting of hopper, flexible tube and rigid PVC pipe is used for sample preparation. The flexible tube connects the hopper to the PVC pipe. The hopper filled

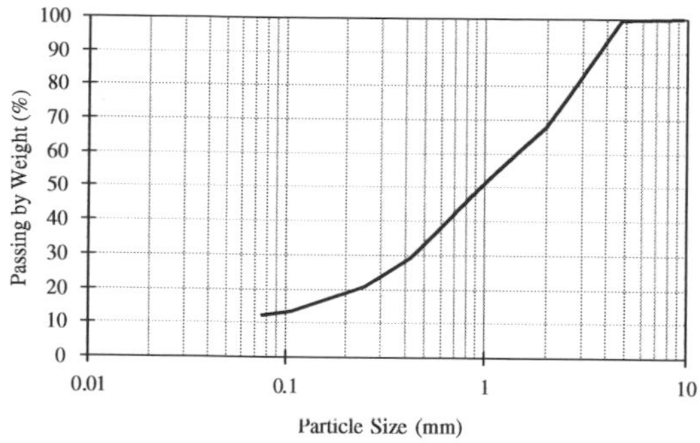

Figure 2. Gradation of test sand.

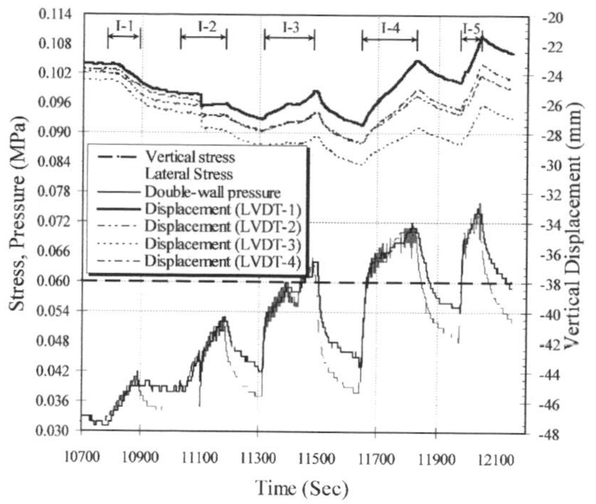

Figure 3. Vertical and lateral stresses and vertical displacement of soil during injection and suspension.

with sand is hung by an overhead crane and the sand is dispensed by opening the hopper outlet valve. The sand is deposited inside the rubber membrane (that is stretched inside a split sample former), after positioning the injection pipe guided by a centralizing bar, in horizontal thin layers by horizontally moving the PVC pipe forward and backward in parallel lines. The density is controlled by varying the falling height of sand that is the distance between the lower end of the PVC pipe and the surface of deposited sand. After filling the split former, the sand surface is leveled, the piston is gently lowered on the top of sand and the side membrane is sealed against the piston by means of metallic O-rings. Then, vacuum is applied to the sample and the split former is disconnected leaving a vacuum-stabilized sample. The chamber is then assembled around the sample, connected to the panel of controls and filled with de-aired water. After releasing the vacuum, the sample is consolidated under the K_0-condition by increasing the vertical stress in increments till reaching the consolidation stress. Then, the double packer is lowered into the injection pipe and inflated, the pump is set to the desired injection rate and injection is proceeded. After hardening of the injected grout material, vacuum is applied to the sample, the panel of controls is disconnected, the water is emptied and the chamber is disassembled.

The test soil is an air-dried natural sand, the gradation of which is shown in Figure 2. The specific gravity, maximum void ratio and minimum void ratio of the test sand are 2.650, 0.923 and 0.562, respectively. The sand was deposited at a relative density of 50% and consolidated by increasing the vertical stress in increments until reaching a consolidation stress of 0.06 MPa. The injection process was volume controlled and performed under the boundary condition of constant vertical stress and zero average lateral strain (BC3). A grout material of 46 L in volume was injected in five stages (I-1 to I-5); about 10 L/stage for I-1 to I-4 and 6 L for I-5. The injection pump was set to an injection rate of 5 L/min. and a time lag of 2–3 min. between every successive injection stage was considered, resulting in five suspensions of injection (S-1 to S-5).

3 GROUTING PERFORMANCE

The grouting performance may be understood by examining the variation of soil stresses and deformations with injection. Figure 3 shows the soil vertical and lateral stresses, the double-wall (outer cell) pressure and the vertical displacement of soil surface (as measured by the four LVDTs) measured during the injection and suspension stages. The initial values shown in the figure are those attained at the end of consolidation and before starting the injection. It is seen that the E/P regulator was satisfactorily responding to the changes in the lateral stress. The lateral stress and vertical displacement are re-plotted in Fig. 4 to show their variation with the injected grout volume (as measured by the flow meter). The vertical displacement in Fig. 4 represents the average of the four displacements shown in Fig. 3.

The results in Figs. 3 and 4 reveal that, as a result of the expanding grout, the soil experienced gradually increasing lateral stress with injection. However, in terms of the vertical displacement, the soil experienced gradually increasing settlement during I-1, followed by practically no change during I-2, and then continuously increasing upheave during I-3, I-4 and I-5.

During suspension, the soil experienced gradually decreasing lateral stress and increasing settlement with time. The boundary condition BC3 was imposed on the sample during injection and suspension. This along with the fact that the rubber sleeve of the injection pipe acts as a non-return valve indicates that during suspension the soil did not experience a displacement at the

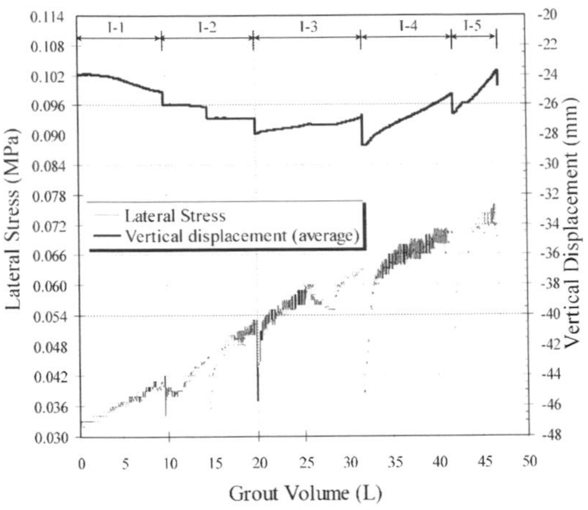

Figure 4. Variation of soil lateral stress and vertical displacement with the injected grout volume.

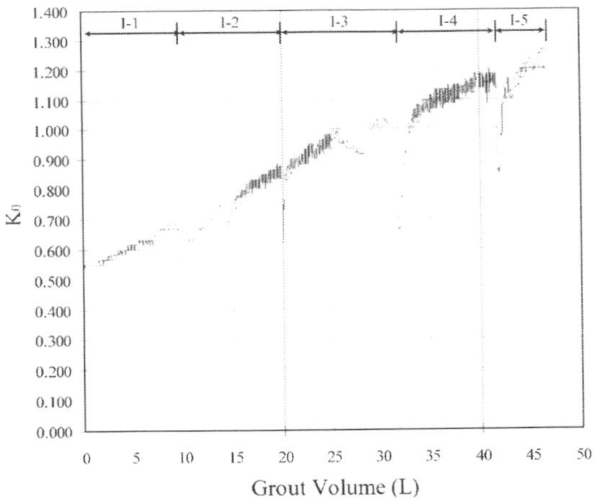

Figure 5. Variation of coefficient of earth pressure at rest (K_0) with injected grout volume.

lateral boundary of the soil or at the gout-soil interface. Nonetheless, it experienced gradually decreasing lateral stress and increasing surface settlement. This implies that immediately after the termination of injection and during the suspension, the soil exhibited both of lateral stress relaxation and creep deformations, as a result of re-arrangement of the soil particles till reaching equilibrium.

By carefully examining the results in Fig. 4, it is seen that immediately after resuming the injection (after a given suspension) the lateral stress increased sharply. Then, with continued injection it followed almost the same rate of increase as that of the previous injection stage. It may therefore be reasonable to conclude that the increase of lateral stress follows a practically linear relationship with the injected volume of grout. The results however reveal that the suspension process has a significant effect on the performance in terms of the vertical displacement. It is seen that the variation of displacement during a given injection stage is significantly different from that during the preceding one. Such a difference is attributed to the creep deformation experienced during the suspension stages. These observations reveal the significance of the soil rheological properties in terms of stress relaxation and creep deformation during suspension and indicate that during post-suspension injection the developed lateral stress is not significantly influenced by the suspension process, while the developed soil deformation is significantly influenced.

4 TREATMENT EFFECTIVENESS FOR GROUND IMPROVEMENT WORKS

The soil improvement due to treatment by the TGC grout may be evaluated in terms of the coefficient of earth pressure at rest (K_0). K_0 is an important parameter and is usually considered as a target criterion in most of the ground improvement problems. It is defined as the ratio of the soil lateral stress to the vertical one at the condition of zero lateral strain. The boundary condition BC3 (constant vertical stress and zero average lateral strain) was imposed during injection and suspension. Therefore, the measured lateral stress may be considered as a direct measure of K_0. Figure 5 shows the variation of K_0 with the injected grout volume. It is seen that K_0 increases continuously with injection and has a practically linear relationship with the injected grout volume.

The soil improvement may also be evaluated by examining the variation with injection of the soil volume change that is the change of volume of voids. The soil volume change may be estimated from the injected volume of grout and the corresponding vertical displacement of sample surface, according to the following expression:

$$\Delta V = V_g + V_d \quad (1)$$

where ΔV = soil volume change; V_g = injected grout volume; and V_d = cross-sectional area of sample multiplied by the vertical displacement of soil surface (positive for settlement, negative for upheave).

Figure 6 shows the variation of the soil volume change as estimated by Eq. 1 and the net vertical displacement of soil surface (due to injection) with the injected grout volume. It should be mentioned that the volume of grout actually expanding into the soil is not strictly the same as that measured by the flow meter. The grout essentially loses an amount of water during injection. This amount depends on the bleeding characteristics of grout. The grout volume considered in Eq. 1 and represented in Fig. 6 is that measured by the

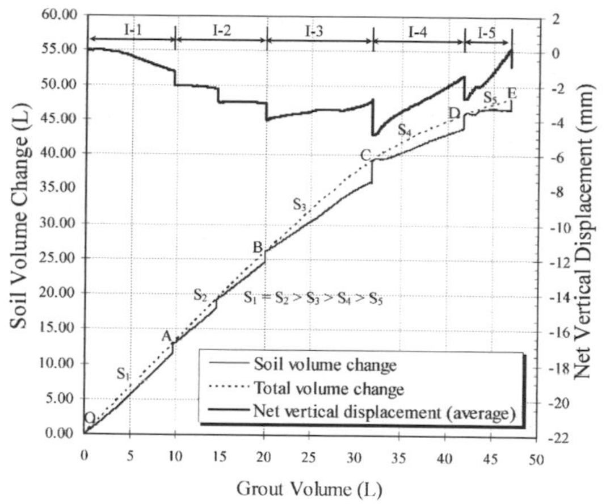

Figure 6. Variation of soil volume change and net vertical displacement with the injected grout volume.

contribution of injection to upheave. In other words, it can be said that, before the upheave onset the injection is totally contributing to improvement of the treated soil. However, after the upheave onset, the injection is contributing to both of improvement and upheave. Moreover, the contribution to upheave is increasing with continued injection after the upheave onset.

The above observations on the increase of K_0 and soil volume change with injection are important in planning the TGC grouting for ground improvement works, especially in deciding the grout point spacing and the grout volume to be injected. They imply that the most effective treatment may be attained by planning the grouting works, so that upheave of the soil surface is avoided. However, should larger grout volumes be injected because of economic considerations, the variation of improvement with grout volume should be taken into account.

flow meter without being corrected for the water loss. The following observations however can be made (see Fig. 6):

- The volume change increased continuously with injection.
- For a given injection stage, the rate of volume change is practically constant.
- During all the suspension stages, the volume change (compressive) is gradually increasing with time as a result of creep deformation.
- The rate of volume change is decreasing with stage injection. This is explained by the gradual transition of vertical displacement from settlement during I-1 through constant (no change) during I-2 to upheave during I-3, I-4 and I-5.
- It is interesting to note that the onset of upheave (at the beginning of I-3) characterizes two distinct stages of improvement. Before the upheave onset, the lines representing the total volume change due to injection and suspension (OA for I-1 and S-1, AB for I-2 and S-2) have the same slope. However, for the post-upheave onset stages, the slope of the line representing the total volume change (BC for I-3 and S-3, CD for I-4 and S-4, DE for I-5 and S-5) is decreasing with stage injection. The slopes of the lines representing the total volume changes due to the five stages of injection and suspension may be expressed as:

$$S_1 = S_2 > S_3 > S_4 > S_5$$

where S_i = slope of line representing the total volume change due to both of I-i and S-i (i = 1, 2, 3, 4 and 5).
- The variation of improvement before and after the occurrence of upheave is explained by the relative

5 TREATMENT EFFECTIVENESS FOR SETTLEMENT COMPENSATION WORKS

Because of its treatment mechanism (by pure displacement), the TGC method is capable of inducing upheave of the treated soils. This renders the method effective for compensating the ground settlement caused by underground excavation and bored tunneling. The effectiveness of compensation is usually evaluated in terms of the ratio of the upheave volume to the injected grout volume.

Ideally, if injection is made quickly in saturated clayey soil so that soil deformation occurs in undrained condition (incompressible behavior), upheave will occur immediately after the start of injection and its amount will be equal to the volume of injected grout. However, this does not occur in practice, since the volume of grout that expands into the soil is generally smaller than the injected volume (owing to the bleeding of grout water). For sandy soils, in addition to the bleeding of water, the soil does not experience incompressible behavior, but will experience some compression before the occurrence of upheave. With continued injection the attained amount of upheave will be dependent on the soil compressibility characteristics and the corresponding water bleeding. After termination of injection, the upheave attained at the end of injection will tend to decrease with time, because of two reasons: (1) soil compression till reaching equilibrium, and (2) tendency of the excess water trapped in the grout to bleed into the soil as the grout itself shrinks. The occurrence of upheave and its variation with injection and with time therefore depend on the soil compressibility characteristics, the grout rheological characteristics and the injection procedure (especially the injection rate). The remaining part of this section discusses for the test soil and procedure considered herein, as well as for the standard TGC grout material, the occurrence of soil surface upheave and its variation during injection and suspension.

The variation of upheave with injection may be understood by examining the results of the last three stages that were upheave-inducing. For these stages, the following observations can be made (see Fig. 6):

- Upheave did not occur immediately after starting the injection into the sandy soil. However, it occurred after reaching a certain degree of improvement.
- For the procedure of alternate injection and suspension, as considered herein, the upheave increases at a practically constant rate during the injection of a given stage.
- The rate of upheave increase during injection increases with stage injection.
- Once the injection is suspended, the attained upheave decreases gradually with time as a result of creep deformation until reaching equilibrium.
- During suspension the soil particles get in a more compact and stiffer structure (than that of the particles at just before the suspension) as may be represented by the creep deformation. A post-suspension resumed injection into a more compacted soil results in a larger contribution of the injection to the upheave. This is corroborated by the decreasing volume change of soil during both of injection and suspension and the increasing rate of upheave with stage injection. It is interesting to note that there was almost no volume change of the soil during I-5, which implies that the injection was almost entirely contributing to the upheave.

6 CONCLUSIONS

In this paper, the TGC method was examined through physical modeling in the laboratory using a large-scale double-wall calibration chamber and a commercially available injection system. The used systems were described in the paper and typical results of injection into sandy soil were presented. These results were then discussed to illustrate the effectiveness of the method in improving the soils and in compensating the ground settlement. Based on the presented results and discussions, the following conclusions can be made on the performance and effectiveness of the TGC method:

- During injection, the soil exhibits significant deformations and increases in the lateral stress. For the soil and stress levels considered herein, the test soil experienced gradual transition from settlement to upheave during injection. After termination or during suspension of injection, the soil experiences creep deformation and relaxation of the lateral stress till reaching equilibrium. The suspension of injection does not significantly influence the increase of lateral stress with resumed injection. However, the soil deformation due to injection of a given grout volume depends on the deformations experienced during both of injection and suspension.
- The coefficient of earth pressure at rest, K_0, increases continuously with injection and has a practically linear relationship with the injected volume of grout (for the soil, procedure and grout volume considered in this paper).
- The soil improvement due to treatment by the TGC method in terms of soil volume change (that is the change of volume of voids) increases continuously with injection. The improvement increases practically linearly with the injected grout volume till the occurrence of soil surface upheave. Then, with continued injection (upheave-inducing injection) the improvement increases at a decreasing (attenuating) rate.
- For the settlement compensation works in sandy soils, the occurrence of upheave should be expected after reaching a certain degree of improvement. After the onset of upheave, continued injection results in a gradually increasing rate of soil upheave. Once the injection is suspended, the attained upheave decreases gradually with time as a result of creep deformation until reaching equilibrium.

REFERENCES

El-Kelesh, A.M. & Matsui, T. 2006. Development of calibration chamber for compaction grouting. *Submitted to Geotechnical Testing Journal*, ASTM.

Holden, J.C. 1971. Laboratory research on static cone penetrometers. Report No. CE-SM-71-1, Department of Civil Engineering, University of Florida, Gainesville, Fla.

Shimada, S., Oyama, T., Sasaki, T., Indutsu, T., Kurisaki, K. & Kajima, T. 2005. Development of thixotropical gel compaction pile. *Proc., 60th Japan National Conference on Civil Engineering*, JSCE (In Japanese).

New Horizons in Earth Reinforcement – Otani, Miyata & Mukunoki (eds)
© 2008 Taylor & Francis Group, London, ISBN 978-0-415-45775-0

Deformation behaviour of clay cap liners of landfills from centrifuge and full-scale tests – influence of reinforcement inclusion

J.P. Gourc & S. Camp
Lirigm-LTHE, University Joseph Fourier-Grenoble, France

B.V.S. Viswanadham & S. Rajesh
Department of Civil Engineering, Indian Institute of Technology (IIT) Bombay, Mumbai, India

ABSTRACT: The behaviour of the cover barrier of a site for storing nuclear waste of a very low activity is studied. The risk of a bending of the clay layer in case of differential settlements within underlying waste is particularly studied. Field bending tests ("bursting tests") are performed in France. Initiation and propagation of cracks are studied. The limit value of the extension strain of the clay layer without cracking is characterized. In the same time, centrifuge tests are carried out in a large beam centrifuge facility available in India to simulate bursting tests executed in the field and to test the reinforcement effect of a geosynthetic sheet included in the cover barrier.

1 INTRODUCTION

Landfills have a top barrier including, in particular, a cap cover of compacted clay. In spite of the different stresses and solicitations, the capping cover must retain its physical, mechanical and hydraulic characteristics during the life of the landfill (operation and monitoring). Imperviousness of clay liners is essential, especially in landfills in order to safeguard the environment from contamination (leakage of biogas in the atmosphere, with greenhouse effect). However, this barrier meets many problems, in particular those related to its implementation and to the mechanical solicitations after closing the cell.

In case of differential settlements, cracks can occur within the clay that can modify permeability of the clay barrier (Cheng et al., 1994). In the framework of an important research programme, the implementation conditions, that is to say water content and compaction energy, are studied in order to optimize the characteristics of the clay layer and more precisely, its flexibility. The influence of a geosynthetic inclusion to prevent or limit cracking of the clay layer is also considered.

1:1 full-scale bursting type tests were conducted in the field at two different moulding water contents to simulate the initiation of cracking of cap liner at the onset of differential settlements. Similar configuration was tested in a large beam centrifuge at 12.5 g. This paper intends to bring-out the susceptibility of cracking of cap liners (with nominal overburden) to differential settlements.

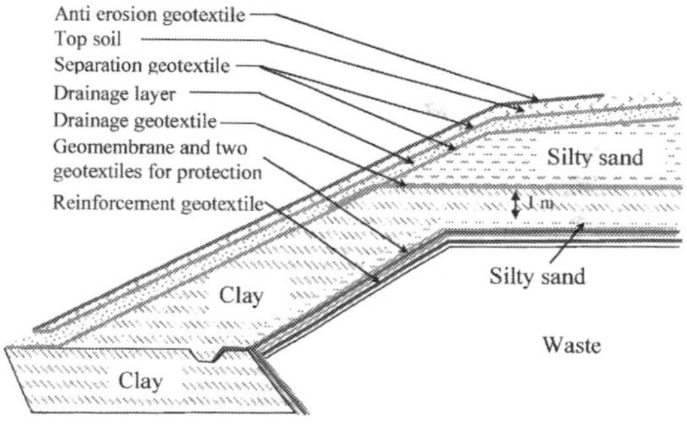

Figure 1. Structure of the cover of the storage cells.

2 BACKGROUND OF THE SITE

Since 2003, the French radioactive waste agency ANDRA is responsible for the site of storage of very low level nuclear waste, located in the Aube, in France. To ensure the radioactive waste containment, the confinement of the storage cells by a cap cover comprises of compacted clay liner (CCL) and a geomembrane (Camp et al., 2005). Figure 1 depicts schematic cross-section of the cap cover. Characteristics of the clay are summarised in Table 1.

Waste is stored in blocks of variable shape, in the form of big bags, tanks and barrels and spaces between blocks are filled with sand. Due to this type of storage and prevalence of voids which, settlements within waste are likely to occur. Hence, there is a risk

Table 1. Characteristics of the clay on site.

Composition	50% kaolin	40% illite	10% chlorite
Particle size distribution	<2 μm 45%	<80μm 93%	
Atterberg limits	Liquid Index 22	Plastic limit 22%	Liquid limit 44%
Compaction characteristics	γd_{max} 17.7 kN/m^3	W_{opn} 17%	
Strength parameters	ϕ_u 20°	c_u 90 kPa	

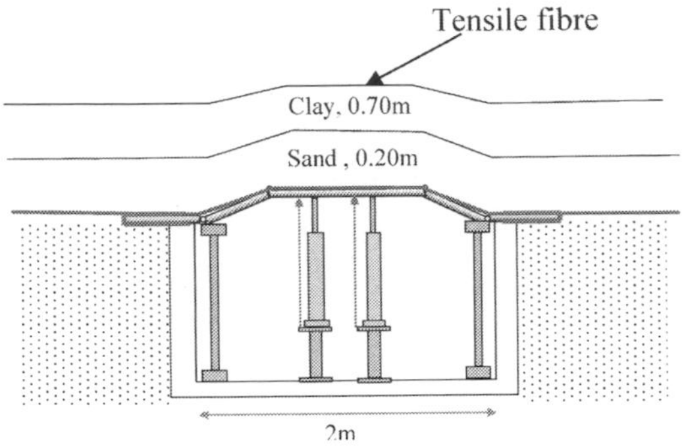

Figure 2. Bursting test.

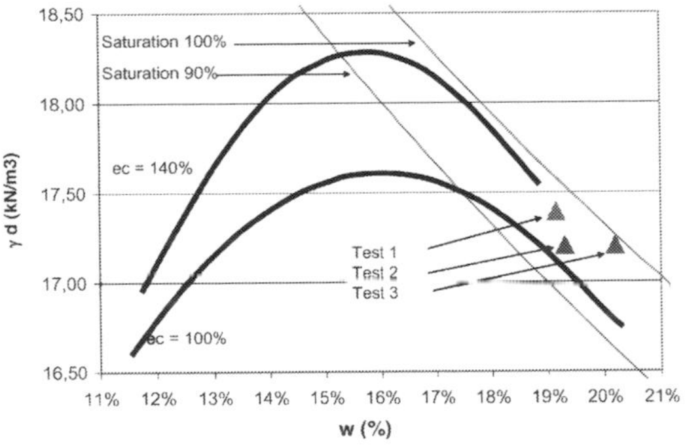

Figure 3. Characteristics of the clay for the three field tests.

of deformation of the clay layer due to differential settlements within underlying waste.

3 FIELD BURSTING TESTS

3.1 *Field test procedure*

The influence of differential settlements within the waste mass on the cap cover can be modelled by submitting a clay layer to bending stresses (Jessberger and Stone, 1991; Viswanadham and Mahesh, 2002). In the framework of the present paper only bursting tests which induce the most critical situation with regard to the risk of cracking are presented (Figure 2). Indeed the stretched zone in this test is located at the top of the layer and consequently not confined. These tests can be considered as inverted bending tests. They don't represent the actual situation in case of differential settlements but these tests present the additional benefit of allowing observation of initiation of cracking along the most stretched fibre of the liner.

In a second phase of this on-going research programme, it is proposed to induce differential settlements to the overall structure of the cap cover, including the geomembrane and the reinforcement geotextile layer.

For the implementation of these tests, a rigid pit made of reinforced concrete (breadth: 2 m) was built, and an articulated steel plate (2 m × 2 m) placed over the pit. Plane strain state is considered as existing for a central profile. A system of four vertical hydraulic jacks fitted in the pit allow the plate to induce vertical movement of the central plate. These jacks are synchronized such a way that they impose identical displacement at all four corners. Performance of the jacks was monitored with the help of displacement transducers and load cells. A maximum vertical movement of 250 mm is possible with a rate of 7 mm/min. A grid of markers is installed on the surface of the clay layer and displacements of markers with reference to markers could be obtained optically.

The clay used for these tests is identical to the one that is used for the actual cap cover (passive barrier) of the site for storing low level nuclear waste (see Table 1). Characterization of this clay in detail discussed elsewhere earlier by Camp et al., 2005). Firstly, a 0.2 m thick loosely tamped sand layer was placed above the articulated plate. The purpose of this layer is to prevent stress non-uniformities to the clay layer and also to facilitate uniform compaction of clay layer. A total thickness of 0.7 m was achieved in two lifts, with each compacted layer thickness being 0.3 m and 0.4 m and the kneaded clay is compacted with a tamping compactor uniformly.

For tests T1 and T2, a moulding water content equal to 19% (which is $w_{opn} + 2\%$) and a moulding water content of 20.5% was maintained for test T3 (which is $w_{opn} + 3.5\%$). The clay in tests T1 and T2 possess identical water content but the average dry unit weight in test T1 is 17.4 kN/m^3 and test T2 is 17.2 kN/m^3 respectively. This corresponds to the difficulties in ensuring adequate compaction energies in the field. Figure 3 presents dry unit weight-moisture content relationship for the clay used along with achieved moulding water contents and dry unit weights for tests T1, T2, and T3 (with e_c: compaction energy/Proctor standard compaction).

3.2 *Results of bursting tests in the field*

Importantly, field tests have confirmed that clay is very sensitive to flexural tensile stresses. In all tests, cracks were observed to appear along the surface of the clay symmetrically for vertical movements as small as 30 mm. Based on the analysis of photos (see Figure 4) allow to determine the strain at initiation of crack of the clay along the outer fiber of the clay layer. Where L_0 is the initial distance between the two farthest markers and L_i is the distance between the two same markers at the onset of cracking of clay. Strain at crack initiation ε_i can be calculated using $\varepsilon_i = (L_i - L_0)/L_0$, which is an average strain along the outer fiber of the clay layer due to integral displacement of markers. The value of ε_i is 0.6% for the test T3 (w_{opn} + 3.5%) and 0.3% for tests T1 and T2 (w_{opn} + 2%). A slight delay in observing strain at crack initiation can be noted for the clay layer compacted towards wet side of its optimum water content. These results confirm the observations made in the laboratory through bending tests on small beams made out of same soil where strain at initiation of cracks is found to be in the range of 0.04%–0.4%, depending upon the moulding water content. Hence, strain at initiation of crack appears to be well adressed satisfactorily through laboratory tests. Earlier, Ajaz and Parry (1975) have reported that strain at crack initiation is in the range of 0.1–0.7% and Edelmann et al. (1996) in the range of 0.2 to 1.3%.

For a vertical movement of a central plate of 250 mm, cracks were observed to completely develop through all the thickness of the clay layer and found to have widths in the range of 40–70 mm. (see Figure 5). A clear and distinct difference in behaviour was observed among tests T1 and T2 and to that of T3. For tests T1 and T2, two main cracks were observed along the width and depth of the clay layer. Whereas for the clay (w_{opn} + 3.5%), a single crack of 410 mm deep was observed to take place centrally. The bottom portion of the clay layer was found to be intact. This conformed to the influence of moulding water content on the behaviour of the clay layer at the onset of flexure. The clay layer was found to experience cracking almost its entire depth in tests T1 and T2. Formation of a gap was observed between the sand layer and the clay layer or within the clay layer. This is attributed to the rigidity of the clay layer and non-uniform compaction along the interface of both sand and clay layers.

Undisturbed samples were sampled from the clay layer out of the damaged area. Triaxial compression (Unconsolidated and Undrained) and unconfined compression tests were performed. When the water content (higher than w_{opn}) increases and for a same compaction energy, the dry unit weight, the cohesion were observed to decrease: for w_{opn} + 2%, $\gamma_d = 17.39\,\text{kN/m}^3$ and undrained $C_u = 81$ kPa and for w_{opn} + 3.5%, $\gamma_d = 17.23\,\text{kN/m}^3$ and undrained $C_u = 71$ kPa.

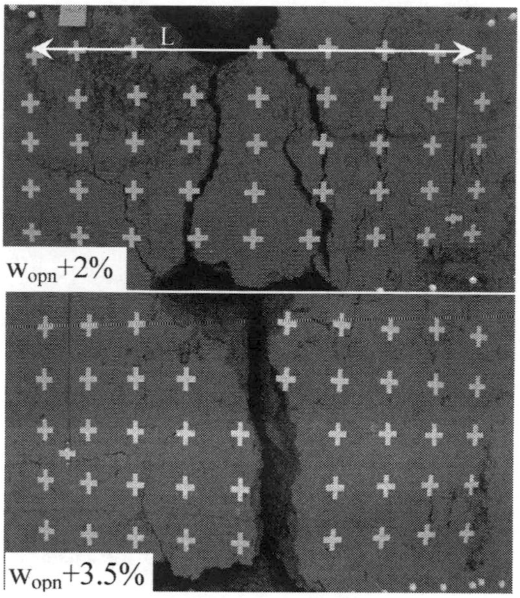

Figure 4. Final top view of the field test (vertical displacement of the plate a = 0.25 m).

Figure 5. Final cut of the field tests (vertical displacement of the plate a = 0.25 m).

4 SIMULATION OF BURSTING TESTS IN A LARGE BEAM CENTRIFUGE

4.1 *Centrifuge modeling technique*

A geotechnical centrifuge can be used to perform tests on models that represent full-scale prototypes under normal field conditions. A 1/N scale model tested at a centrifugal acceleration N times the earth's gravity (g) experiences stress conditions identical to those in the prototype. The above technique of modelling in geotechnical engineering can be extended

Table 2. Main features of the tests at the real scale and in the centrifuge.

Scale	Test legend	Acceleration level	Thickness of sand layer	Geomembrane inclusion	Thickness of clay liner	Moulding water content
1:1	T1 , T2	1 g	200 mm	No	700 mm	$w_{opn} + 2\%$
1:1	T3	1 g	200 mm	No	700 mm	$w_{opn} + 3.5\%$
1:12.5	C1	12.5 g	*24 mm	No	64 mm	$w_{opn} + 2\%$
1:12.5	C2	12.5 g	*40 mm + 40 mm	^0.25 mm thick	64 mm	$w_{opn} + 2\%$

* In addition, 30 mm thick sacrificing coarse-sand layer was placed below this layer
^ A thin geomembrane sandwiched between two woven polypropylene geotextiles (75 gsm) was placed

to model cap clay liners with an aim to study their response to differential settlements created artificially in a centrifuge. The 4.5 m radius large beam centrifuge at Indian Institute of Technology Bombay (IIT Bombay), India was used in the present study. The centrifuge capacity is 250 g-ton with a maximum payload of 2.5 t at 100*g* and at higher acceleration of 200*g* the allowable payload is 0.625 t (Viswanadham and Muthukumaran, 2007).

4.2 *Developed test-package and test procedure*

With an aim to simulate the response of cap clay liners for low-level radioactive waste depositories to differential settlements, a first series of centrifuge model tests were carried-out. Schematic cross-section for testing bursting mode of failure of a cap liner is shown in Fig. 2 and it was simulated by developing a custom designed set-up for inducing bursting type of failure to the liner during centrifuge test at 12.5 g. Table 2 gives summary of centrifuge tests and field tests executed. In order to simulate the dimensions in the full-scale field trial, a g-level of 12.5 was selected. In this paper, the first results of centrifuge tests C1 and C2 are presented. Centrifuge test simulates field tests T1 and T2 and whereas centrifuge model test C2 is a preliminary test conducted in order of verifying effect of reinforcement inclusion in the form of a geomembrane layer cushioned between two woven geotextile layers on the integrity of the clay liner. Configuration adopted in the centrifuge model test C2 simulates a reinforcement of the sand layer in charge of the bursting displacement spreading. Until now, this test is not performed at a full scale. Figure 6 presents a cross-sectional view of a test package used in the present study for performing centrifuge experiments. The test package consists of a hydraulic cylinder at the centre. The experiment consists in creating an upward movement to the central plate during centrifuge test at 12.5 g. The plate is initially horizontal during the implementation of the soil layers. Thereafter, air pressure was applied to an oil reservoir tank placed behind the container gradually to initiate the movement of the plate at 0.90 mm/min. With this arrangement a vertical displacement of 28 mm is possible, which corresponds to a vertical movement of a = 0.35 m at scale 1:1 at the centre of the plate at 12.5 g.

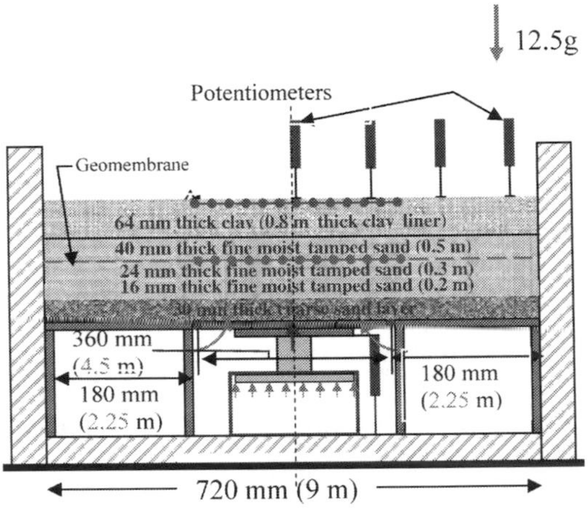

Figure 6. Cross-sectional view of test package [Model: C2].

In case of model test C1, a 24 mm thick sand layer was placed above 30 mm sacrificing layer and thereafter a moist-compacted clay liner of 64 mm thick was placed. In centrifuge model tests, compacted clay liner was constructed by using a blend of 80% kaolinite and 20% silica sand (referred herein as model soil) by dry weight. The model soil was found to have liquid limit equal to 38% and 16% plasticity index. The maximum dry unit weight of the model soil is 15.9 kN/m^3 and its optimum moisture content is 22% (standard Proctor compaction). To some extent, the model soil adopted in the present study resembles the properties of soil used in full-scale tests (table 1, figure 3). Figure 7 shows measured variation of deformed profiles at the surface during centrifuge test for model C1 (based on potentiometer data). It was noted that a clear visual crack was noticed after attaining a central burst displacement a = 0.125 m. However, crack initiation would have taken place at a = 0.05 m. Figure 8 presents observed deformation profiles for model test C2.

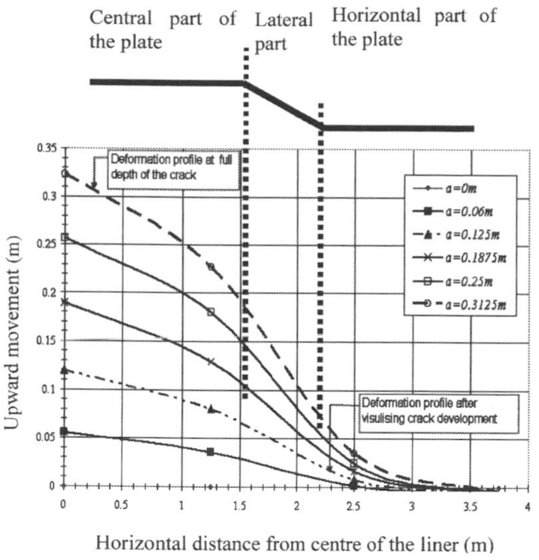

Figure 7. Measured surface deformation profiles [Model: C1].

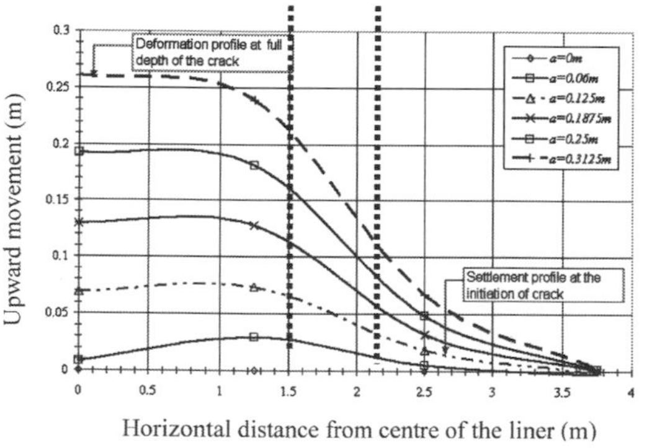

Figure 8. Measured surface deformation profiles [Model: C2].

4.3 *Results of bursting tests in a centrifuge*

Figure 9 shows front elevation of the model during centrifuge test at 12.5 g after subjecting to a vertical movement equal to 0.25 m. A clear distinct formation of cracks is observed to develop in both the centrifuge models tested. This observation is consistent with the field observations (Figure 5 up) with especially in the two cases, two symmetrical cracks/plate axis. Consequently centrifuge modelling seems clearly as a viable alternative in a first step to the full scale experimentation.

Figure 10 depicts status of the surface of the tested cap liner after subjecting to a vertical movement equivalent to a = 0.3125 m at the end of centrifuge test.

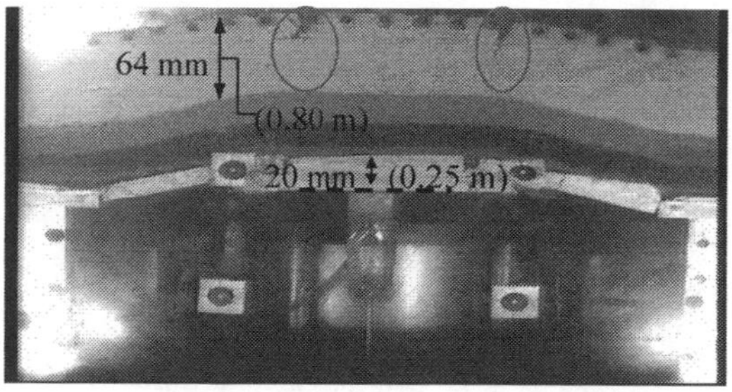

Figure 9. Front elevation of model during centrifuge test for $a = 0.25$ m [Model: C1].

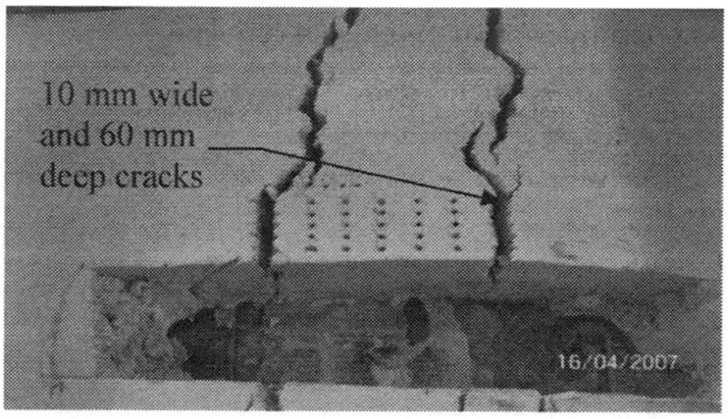

Figure 10. View of model clay liner surface at the end of the centrifuge test [Model: C1].

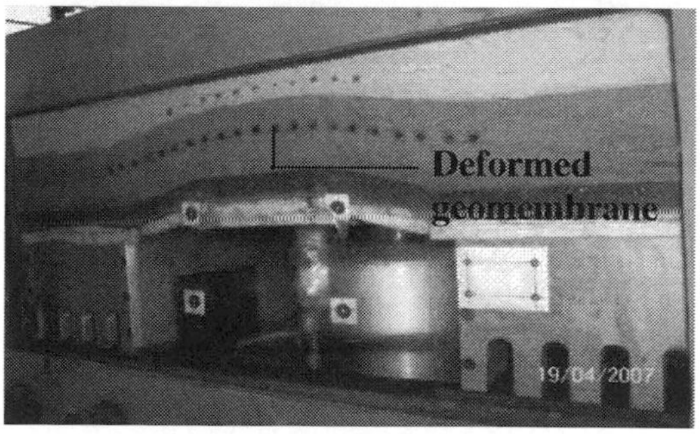

Figure 11. View of model clay liner at the end of the centrifuge test [Reinforced Model: C2].

Figure 11 presents view of the model C2 at the end of the centrifuge test. Stretched view of the geomembrane installed within the sand layer can be noted. The influence of the reinforced sand layer in limiting deformations of the clay layer is demonstrated, comparing figures 7 and 8.

Based on the preliminary analysis and interpretation of both field and centrifuge test results, a reasonable agreement between the observed deformation behaviour of cap clay liners was observed. The average strain at the onset of cracking in centrifuge tests is estimated as 0.7% for a 0.8 m thick clay liner. This is found to be in good agreement with the one in the field test.

5 CONCLUSIONS

Field bursting tests have been performed to characterize the behaviour of a clay layer submitted to differential settlements and to study particularly the initiation of cracks. However it's worth noting that cracks would probably appear later in cover system. The average strain of the stretched fibre at the initialization of cracks on site is less than 0.6%. Centrifuge tests are in progress in collaboration with the Indian Institute of Technology Bombay (Viswanadham and Muthukumaran, 2007) to reproduce the field tests. The centrifuge tests will permit to test the influence of several parameters like thickness of the clay liner, compaction energy and settlement rate, and influence of reinforcement inclusions.

ACKNOWLEDGEMENTS

The work in France is performed in the framework of a research agreement LTHE/Scetauroute and is supported by ANDRA and Benedetti and the authors would like to acknowledge the support extended by the staff at National Geotechnical Centrifuge Facility of Indian Institute of Technology Bombay, Mumbai, India for their assistance throughout the centrifuge study.

REFERENCES

Ajaz, A., and Parry, R.H.G. (1975). Stress-strain behaviour of two compacted clays in tension and compression, *Géotechnique 25*, No. 3, pp. 495–512.

Camp, S., Rey, D., Kaelin, J.L. (2005). Presentation of a new French site for storing very low level radioactive waste, *International Workshop Hydro-Physico-Mechanics of Landfills "HPM1", LIRIGM, Grenoble 1 University, France*, 21–22 March 2005.

Camp, S., Gourc, J.P., ; Plé, O.; Villard, P.; Rey, D. (2005) Landfill cap cover issue : improvement of the capability to sustain differential settlement, *Tenth International Waste Management and Landfill Symposium (SARDINIA 2005), S. Margherita di Pula, Sardinia Italy,* 3–7 October 2005.

Cheng S.C., Larralde J.L., Martin J.P. (1994) Hydraulic conductivity of compacted clayey soils under distortion or elongation conditions. *In David E. Daniel and Stephen J. Trautwein, Eds., Hydraulic conductivity and waste contaminant transport in soil, Philadelphia, ASTM,* 1994, pp. 266–283, (ASTM STP, 1142).

Edelmann L., Katzenbach R., Amann P., Weiss J. (1996) Large-scale deformation tests on soil layers for landfills, Proc. Of the 2nd int. *Congress on Environmental Geotechnics, Osaka, Japan.*, pp 205–209.

Jessberger, H.L., and Stone, KJL (1991) Subsidence effects on clay barriers, *Geotechnique*. London., Vol. 41, no. 2, pp. 185–194.

Viswanadham, B.V.S., and Mahesh, K.V. (2002) Modeling deformation behaviour of clay liners in a small centrifuge. *Can. Geotech. J.* 39, pp. 1406–1418.

Viswandham, B.V.S., and Sengupta, S.S. (2005). Deformation behaviour of compacted clay liners in a geocentrifuge" *Proc. Tenth International Waste Management and Landfill Symposium (SARDINIA 2005), S. Margherita di Pula, Sardinia Italy,* 3–7 October 2005.

Viswanadham, B.V.S., and Muthukumaran, A.E. (2007). Influence of geogrid layer on the integrity of compacted clay liners of landfills. *Soils and Foundations* (In press).

New Horizons in Earth Reinforcement – Otani, Miyata & Mukunoki (eds)
© 2008 Taylor & Francis Group, London, ISBN 978-0-415-45775-0

Geosynthetic liners on landfill cover slope: Possible reinforcement of the stability of veneer soil layer

J.P. Gourc
LTHE-Lirigm, Grenoble University, Grenoble, France

H.N. Pitanga
EESC, São Paulo University, São Carlos, Brazil, PhD Student LTHE-Lirigm, France

ABSTRACT: Stability of Geosynthetic Lining Systems on cap cover slope of landfill or reservoir slope is a difficult matter. Several specific geosynthetics were recently proposed by the manufacturers in order to mobilize a higher friction at the interface with the soil veneer. A sophisticated Inclined Plane Test device is used to assess the efficiency of this new geosynthetic in comparison with standard materials.

1 INTRODUCTION

Stability of Geosynthetic Lining Systems (GLS) on landfill slope (Fig.1) is, from a geotechnical standpoint, a complex matter. The GLS design is based on the separation of the different functions. The main components are from the bottom (in interface with waste) to the top a geomembrane for sealing, a geospacer for runoff water drainage, a geotextile for filtration to avoid clogging of the geospacer, and a veneer cover soil for protection (thickness between 0.20 and 0.50 m).

Unfortunatly several superficial failures were observed due to the soil sliding down the smooth geosynthetic interface (Fig.2). Consequences are often severes because in the same time the global GLS is very often pull out, consequence of the upper anchorage failure and should be completely replaced.

It was demonstrated previously by Gourc and Reyes-Ramirez (2004) that Inclined Plane Test is appropriated for analysing this phenomenon.

For reinforcing the stability of the veneer soil, many geosynthetic manufacturers propose a new kind of geotextile (geotextile reinforced by a geomat, "GT mat") in place of the conventional filter with both functions, filtration (for the geospacer) and reinforcement (for the soil veneer). A comprehensive study of the interface behaviour was carried out, testing a lot of different "GT mat" structures in contact with soil, using a sophisticated Inclined Plane device.

Initial sliding conditions and residual friction corresponding to large displacements are assessed. It is demonstrated that a significative effect could be obtained for a stability point of view. However this effect is not miraculous since only the local stability in a limited layer of soil in the vicinity of the geotextile (GT) is concerned. Design of the GLS should be modified, taking into account these new conditions.

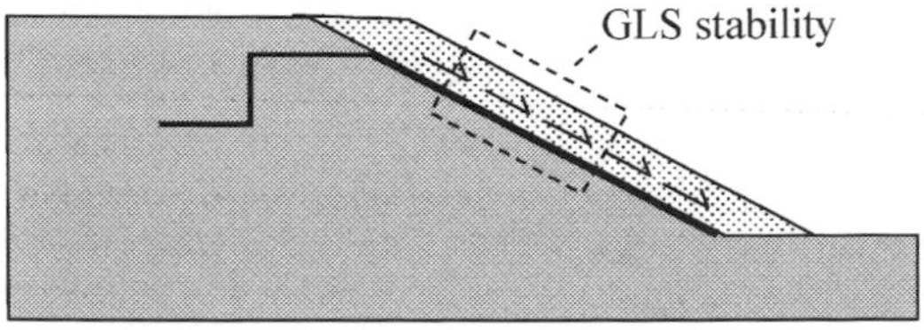

Figure 1. Landfill slope barriers: stability problems considered.

2 SPECIFIC APPROACH OF THE INTERFACE FRICTION TEST AT THE INCLINED PLANE

The Inclined Plane Test is used to determine either soil-geosynthetic or geosynthetic-geosynthetic interface

Figure 2. Example of sliding of the veneer layer along the geosynthetic interface.

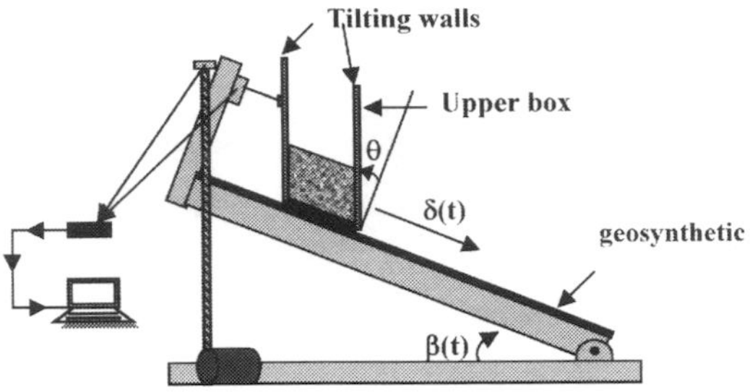

Figure 3. Main features of the inclined plane device.

properties, especially in cases where the stress normal to the interface is small (σ' less than approximately 30 kPa). The interface friction test, when performed in accordance with the current European test standard, only provides a value of the interface friction angle. The parameter usually deduced from the Inclined Plane Test is the interface friction angle ϕ (Izgin and Wasti, 1998; Lala Rakotoson et al., 1999; Palmeira *et al.*, 2002). This friction angle is calculated for a conventional displacement ($\delta = 50$ mm) in accordance with the standard Pr EN ISO 12957-2 (2001) denoted here as angle ϕ_{50}^{stat}.

In this report, the potential for drawing considerably greater information from the test is demonstrated. It has been shown in two previous paper (Reyes-Ramirez and Gourc, 2003) that the in-depth study of the diagram of the tangential displacement along the interface (δ), as a function versus the inclination (β), prior to "non-stabilized sliding" (obtained for an inclination $\beta = \beta_s$), enabled distinguishing the behavior of interfaces which display an identical value of the standard friction angle (ϕ_{50}^{stat}).

This procedure requires the interpretation of an entire dynamics phase, in particular the phase during which the upper box is engaged in uniformly-accelerated movement. Two new parameters are also defined, called angle (ϕ_o) of initial friction (static conditions) and angle (ϕ^{res}) of residual friction (dynamic conditions).

2.1 *Study of the dynamic sliding phase of the inclined plane*

Under standard test conditions, a geosynthetic layer is installed bonded to the base plane. This plane is inclined at a constant speed ($d\beta/dt = 3°/min$) and the upper box, whose displacement (δ) is measured, has been filled with a soil used for measuring interface friction with respect to the geosynthetic (Fig. 3).

In the classical case, the behavior may be separated into three phases, as follows:

- Phase 1 (static phase): upper box practically immobile ($\delta = 0$) over the inclined plane until reaching an angle $\beta = \beta_0$;
- Phase 2 (transitory phase): for an increasing value of inclination ($\beta > \beta_0$), upper box moving gradually downwards;
- Phase 3 (non-stabilized sliding phase): upper box undergoes non-stabilized sliding at an increasing speed ($d\delta/dt$), even if plane inclination is held constant ($\beta = \beta_s$).

As indicated both in the previous paper (Reyes-Ramirez and Gourc, 2003), one can distinguish a mechanism of "sudden sliding" where $\beta_0 = \beta_s$ and a mechanism of "gradual sliding" where $\beta_0 < \beta_s$ which corresponds to the majority of tests presented here. The non-stabilized sliding (dynamic, Phase 3) arises very often for plane displacement values of less than the value $\delta = 50$ mm conventionally considered when measuring the standard friction angle (ϕ_{50}^{stat}).

From the inclination value $\beta = \beta_s$, the sliding rate of the upper box becomes significant and the mechanical analysis must definitively be conducted using a dynamic approach (taking into account the displacement acceleration, γ) and not using a static approach as is typical practice. A constant dynamic friction angle (ϕ^{res}) is found, characterizing the interface friction during the phase 3 so long as the acceleration γ is taken into account:

$$\tan\phi^{res} = \frac{(m_b + m_s)\cdot g\sin\beta - T_{guide}^{dyn} - (m_b + m_s)\cdot\gamma}{(1-\alpha)\cdot m_s\cdot g\cdot\cos\beta} \quad (1)$$

where $m_b \cdot g$ = weight of the upper box; $m_s \cdot g$ = weight of soil in the upper box; and T_{guide}^{dyn} = friction of the box guides.

In the general case of a correctly-built device, the guides of the box absorb not only the normal component of the box weight, but ultimately a portion (**α**) of the normal component of the weight of soil contained in the box (by friction along the box walls), as therefore:

$$N_{guide} = (m_b g \cos\beta) + \alpha\ (m_s g \cos\beta) \quad (2)$$

with: $0 \leq \alpha << 1$.

Equation 1 naturally applies to the special "static" case ($\gamma = 0$), which strictly accurate for the end of Phase 1 ($\beta = \beta_0$, $\delta = 0$). The new equation is given below:

$$\tan\phi_o^{stat} = \frac{(m_b + m_s)\cdot g\sin\beta_0 - T_{guide}^{stat}}{(1-\alpha)\cdot m_s\cdot g\cdot\cos\beta_0} \quad (3)$$

ϕ_o^{stat} is considered as a characteristic value for the friction interface. The static limit equilibrium is reached for $\beta = \beta_0$.

Now consider the case of the present experiment. Friction due to guidance system is independent of the movement conditions ($T_{guide}^{stat} = T_{guide}^{dyn} = T_{guide}$, Fig. 4).

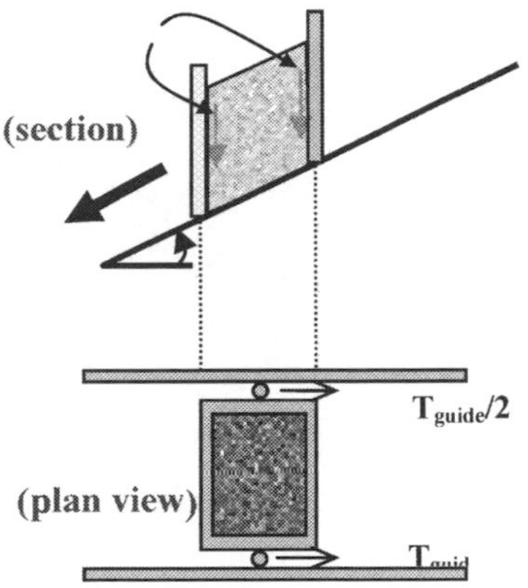

Figure 4. Conditions of sliding of the inclined box.

In addition it was assumed that the sides of the box are smooth and that no load was transferred to the box walls (i.e. $\alpha = 0$) (Reyes-Ramirez and Gourc, 2003):

$$\tan\phi_0^{stat} = \frac{(m_b + m_s)\cdot g\sin\beta_0 - T_{guide}}{m_s \cdot g \cdot \cos\beta_0} \quad (4)$$

ϕ^{res} is variable with the acceleration (γ). It will be shown, from tests results, that a generally uniformly accelerated movement (constant acceleration $\gamma = \gamma_c$) is reached at $\beta = \beta_s$ after an intermediate period necessary to go from $\gamma = 0$ to $\gamma = \gamma_c$. Therefore a second characteristic for the interfaces, the residual friction angle is obtained corresponding to the uniformly accelerated movement:

$$\tan\phi^{res} = \frac{(m_b + m_s)\cdot g\sin\beta - T_{guide} - (m_b + m_s)\cdot \gamma_C}{m_s \cdot g \cdot \cos\beta} \quad (5)$$

It's worth noting that, following the standard (Pr EN ISO 12957-2, 2001), the friction angle is conventionally determined for an inclination β_{50} corresponding to a sliding displacement $\delta = 50$ mm, with the assumption of a static equilibrium (Eq. 4 with β_{50} in place of β_0) which is a rough approximation.

2.2 *Experimental adaptation of the inclined plane device*

The displacement length available for the upper box of standard devices along the plane is generally insufficient for dynamic tests. The box was thereby modified in order to enable a trajectory greater than 500 mm. The box length in the direction of the slope, initially 1000 mm (Lala Rakotoson *et al.*, 1999), was subsequently shortened to L=180 mm. The width, measured transversally, was maintained at B=700 mm (Reyes-Ramirez and Gourc, 2003). The standard test does not generally provide an accurate enough measurement of the sliding displacement (δ) vs. time (t) to enable measurement of the speed (υ) and acceleration (γ_c) of either the box or the geosynthetic support plate. A wire sensor at the top of the inclined plane was installed to allow continuous displacement measurements to be taken with a recording rate of once every 0.05 seconds.

The following parameters were assessed and calculated during testing:

- β_0, plane inclination corresponding to the initialization of the upper box movement;
- β_s, plane inclination corresponding to the non-stabilized sliding;
- ϕ_0^{stat}, static (or initial) friction angle (arbitrary defined for $\delta = 1$ mm representative of a small relative displacement);
- ϕ^{res}, residual friction angle for $\gamma = \gamma_c$.

The initial normal stress ($\beta = 0$) is equal to σ_0' and for a plane inclination β:

$$\sigma' = \sigma'_0 \cdot \cos\beta \quad (6)$$

3 PERFORMANCE OF SMOOTH GEOSYNTHETICS IN INTERFACE WITH SOIL (REFERENCES TESTS)

Before to present the diagrammes corresponding to the rough geosynthetics with a mat dedicated to stabilization of soil veneer layer on slope, it was relevant to carry out tests on common geosynthetics which are assumed to exhibit less friction. These results will be used as a reference for the rough geosynthetics.

Four different materials were selected:

- High Density Poliethilene (HDPE) geomembrana ("GM hdpe") considered as the smoothest interface;
- Polipropilene woven geotextile ("GT woven");
- Heatbonded non-woven geotextile ("GT heatbonded");
- Needlepunched non-woven geotextile ("GT needlepunched").

The tests were repeated on two or three different samples for each value of the normal stress σ' and the soil was a sandy sand at $\gamma_t = 14.2$ kN/m^3 and a water content w = 6.5%, commonly used on cap covers of landfill slopes.

$\sigma' = \sigma_o$' initial normal stress (for horizontal plane: 2.8 kPa, 5.9 kPa and 10.4 kPa). The typical behaviour for these different geosynthetics is presented on Figure 5 for $\sigma_o' = 5.9$ kPa.

The values of residual friction obtained for different standard geosynthetics versus normal stress are presented on Figure 6.

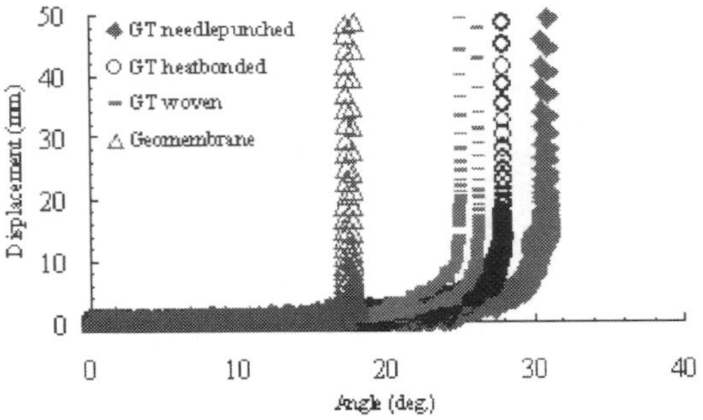

Figure 5. Sliding displacement versus the inclination β for $\sigma_o' = 5.9$ kPa for standard geosynthetics.

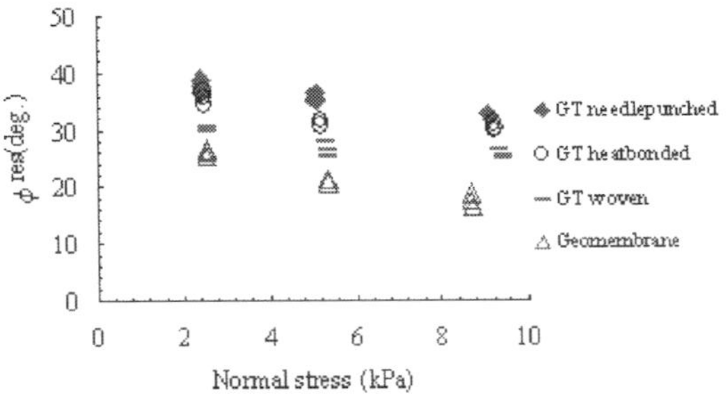

Figure 6. Summary of the values of residual friction obtained for different standard geosynthetics versus normal stress.

The main conclusions are:

- ϕ values are decreasing significantly when the increasing compression σ' (thickness of soil veneer);
- The interface soil-geomembrane exhibits a friction which could be considered as the "bottom value": ϕ(GM hdpe) < ϕ(GT woven) < ϕ(GT heatbonded) < ϕ(GT needlepunched);
- $\phi_0 < \phi^{res}$ for almost all the tests, in agreement with the observed behaviour "gradual sliding".

So the inclined plane is a relevant test to distinguish the friction performance of different geosynthetics under low values of normal stress.

4 PERFORMANCE OF GEOTEXTILE OF REINFORCEMENT + MAT IN INTERFACE WITH SOIL

To assess the efficiency of the geotextile associated to a mat for stabilizing a soil veneer, this product (see Fig. 7) is compared to the geotextile which exhibits the maximum friction (GT needlepunched).

The comparative results obtained at the inclined plane test for $\sigma'_0 = 5.9$ kPa are presented on the Figure 8. A surprising result is obtained since the same limit value of the inclination for complete sliding (β_s) is got for the mat and the non-woven needle punched. The corresponding friction characteristics are calculated on the Table 1.

Figure 7. Macro view of a geotextile of reinforcement + mat ("GT reinf Mat").

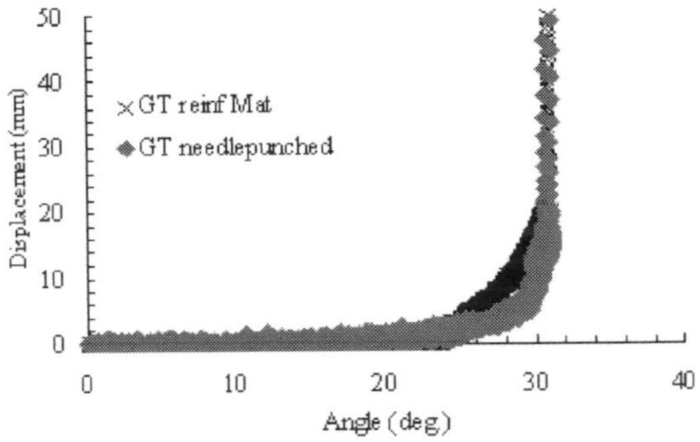

Figure 8. Comparison of the performance of a "GT reinf Mat" with a "GT needlepunched" for $\sigma'_o = 5.9$ kPa.

Table 1. Friction values of the characteristics angles.

Geosynthetic	σ'_0 (kPa)	ϕ_0 (deg.)	ϕ_{res} (deg.)	ϕ_{50}^{stat} (deg.)
GM		16	26	28
GT np	2.8	23	37	46
GT reinf Mat		46	37	47
Soil/soil			40	
GM		16	21	23
GT np	5.9	21	35	40
GT reinf Mat		30	36	40
Soil/soil			36	
GM		13	17	18
GT np	10.4	17	32	34
GT reinf Mat		24	31	34
Soil/soil			31	

The residual friction ϕ^{res} values are equivalent for the two geosynthetics despite the difference of interface structure. In addition the conventional friction for a displacement $\delta = 50$ mm and using a (wrong) static calculation are also quite identical. However the initial friction ϕ_0 corresponding to the inclination of sliding starting are different and allow an identification of the difference of shape of the two diagrams (Fig. 8).

To explain the identical ϕ^{res} value , complementary tests with soil in place of geosynthetic on the lower support were carried out (soil/soil tests), and it is demonstrated (Tab. 1) that ϕ_{res} obtained with soil /geosynthetic reachs the limit ϕ^{res} for soil /soil tests, limit which is logically impossible to pass beyond.

A remaining question is pending: what the actual meaning of the initial friction ϕ_0? One is authorize to interpretate this angle in term of angle corresponding to the initialization of the layer sliding. In this condition the "GT reinf Mat" exhibits a higher efficiency than the "GT needlepunched".

5 CONCLUSION

Design of GLS on slopes is a difficult matter. A large programme of Inclined Plane tests with a sophisticated device was carried out, specifically in order to assess the relative efficiency of geotextile of reinforcement with a mat interface which are presently proposed by several manufacturers .Comparing with a simple nonwoven needlepunched, the result is not at all obvious, in these specific conditions of very low normal stresses.

REFERENCES

EN-ISO 12957-2 Standard, 2001. Geosynthetic – Determination of friction characteristics, Part2, Inclined Plane Test. Brussels, European Committee for Standardization.

Gourc,J.P. , 2004. Geosynthetics in landfill applications. In 3rd Asian Regional Conference on Geosynthetics (GeoAsia 2004) , Keynote Lecture, Seoul, Korea, June 2004.

Gourc,J.P., Pitanga,H.N., Reyes-Ramirez,R. and Jarousseau,C., 2006. Questions raised regarding the interpretation of inclined plane results for geosynthetics interfaces. In 8th International Conference on Geosynthetics,Yokohama,Japan.

Gourc,J.P. and Reyes-Ramirez,R., 2004. Dynamics-based interpretation of the interface friction test at the inclined plane. In Geosynthetics International Journal, Thomas Telford, Vol.11, No 6, pp 439–454.

Izgin, M. and Wasti, Y., 1998. Geomembrane-sand interface frictional properties as determined by inclined board and shear box tests. In Geotextiles and Geomembranes, Vol. 16, No. 3, pp. 207–219.

Lala Rakotoson, S.J., Villard, P. and Gourc, J.P., 1999. Shear Strength Characterization of Geosynthetic Interfaces on Inclined Planes. In Geotechnical Testing Journal, Vol. 22, No. 4, pp. 284–291.

Palmeira, E.M., Lima, Jr. N.R. and Mello L.G.R., 2002. Interaction Between Soils and Geosynthetic Layers in Large-Scale Ramp Tests. In Geosynthetics International, Vol. 9, No. 2, pp. 149–187.

Reyes-Ramirez, R. and Gourc, J.P., 2003. Use of the inclined plane test in measuring geosynthetic interface friction relationship. In Geosynthetics International Journal, Thomas Telford , Vol.10,N°5, pp 165–175.

Combined technologies

New Horizons in Earth Reinforcement – Otani, Miyata & Mukunoki (eds)
© 2008 Taylor & Francis Group, London, ISBN 978-0-415-45775-0

Mechanism of reinforcement using soil nails, rope nets for slope stability

H. Kimura
Graduate School of Science and Technology, Kobe University, Kobe, Japan

T. Okimura
Research Center for Urban Safety and Security, Kobe University, Kobe, Japan

ABSTRACT: Shake table tests for 1/10-scaled slopes built on a platform with 45-degree inclination were conducted to verify the effectiveness of the reinforcement with soil nails and rope nets. Shear deformation was observed in the homogeneous single-layered slopes using Masado. Sliding deformation was observed in the two-layered slopes using a cohesive soil. Collapse was observed in both slopes without reinforcement at the ultimate stage. Yet, it was found that the combination of soil nails and rope nets was effective for resisting a mass movement and reducing deformations. The author suggested two calculation models for slope deformations; a shear deformation model based on the elasticity and a sliding deformation model using an exponential function derived from the shear tests. The idea of the shear resistance by rhizomes was utilized for explaining how the reinforcement was effective. So, the mechanism of the reinforcement can be expressed in the proposed models.

1 INTRODUCTION

Seismic slope stability is a difficult subject since earthquake records and ground displacements in mountainous slopes have been rarely taken. Dynamic behavior of natural slopes has not been understood so much. The Hyogo-ken nanbu earthquake brought heavy damages to natural slopes in Kobe. This event awaked administrators and researchers to pay attentions to the importance of prevention works for natural slopes during an earthquake. A research project on a new reinforcement method using soil nails and rope nets for seismic slope stability was initiated in such a movement.

Photo 1 shows the soil nails and rope nets used for stabilizing a natural slope in JAPAN. The soil nails, reinforcement steel bars with grouting, are embedded in the ground. The rope nets, steel wires, are directly connected to the head of soil nails with a head plate, a steel round plate, on the ground surface.

This method has some features. It has a flexible structure to stabilize a slope while rope nets and soil nails follow the ground deformations during an earthquake. In another aspect, there is no need to remove the vegetation and to scrape the ground surface, which makes it very efficient and economical in practice.

For this study, the deformation behavior of a slope and the mechanism of reinforcement were examined in shake table tests. Fukumasa et al (2001) reported that single-layered slopes using Masado with and without reinforcement were tested on a shaking table. In the same way, Nishihara et al (2006) reported that two-layered slopes using a cohesive soil with and without reinforcement were tested. The shake table test results showed the effectiveness of the combination of soil nails and rope nets. From the observation throughout the tests, it was found that the failure patterns could be classified according to the slope displacement or shear strain.

Photo 1. Soil nails and rope nets constructed on a natural slope.

Hence, we conceived an idea that a displacement could be an index to evaluate slope stability (Murakami et al, 2002). Then we formulated equations to simply estimate the amount of a displacement. Methodology

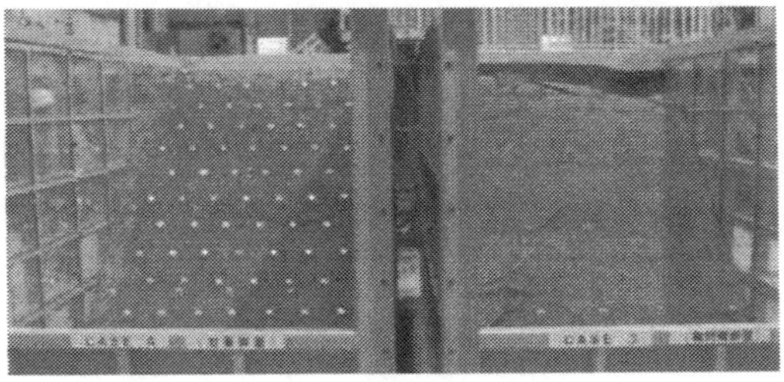

Photo 2. Front view of a shaking table test.

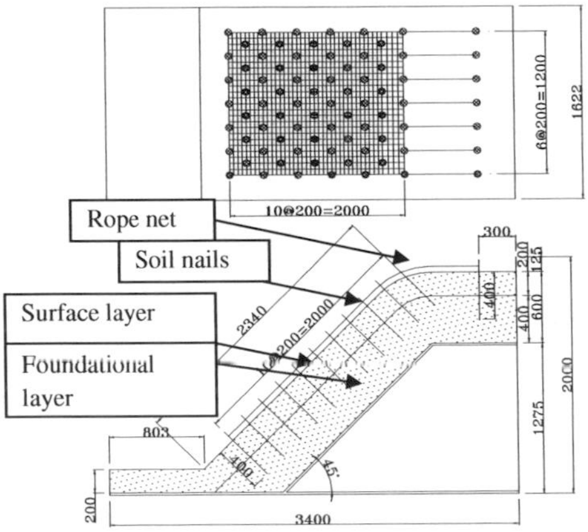

Figure 1. Plan and section of slope.

to describe the mechanism of reinforcement for seismic slope stability was discussed in this paper.

2 SHAKE TABLE TESTS

Shake table tests for slopes with and without reinforcement were performed at the same time (Photo 2). The scale was 1/10. The single layer of Masado had 0.6 m thickness. The size of the model was about 3.4 m long and 1.5 m high. The slope angle was 45-degree. The cohesive soil slope had mostly same dimensions except two layers; 0.2 m thickness for the surface layer and 0.4 m for the foundational layer. The total thickness of soil above the platform was 0.6 m. Fig.1 shows the plan and section of a slope with two layers.

The wet density of the Masado homogenously had 1.6 g/cm^3. The natural water content was about 10 percent. The cohesion C and internal friction angle ϕ were respectively 8 kN/m^2 and 30 degree from the laboratory tests. The grain size distribution of both Masado and cohesive soil is shown in fig. 2.

The two layers of the cohesive soil had different wet densities; 1.3 g/cm^3 and 1.65 g/cm^3, with the same natural water content 21%. The strength properties

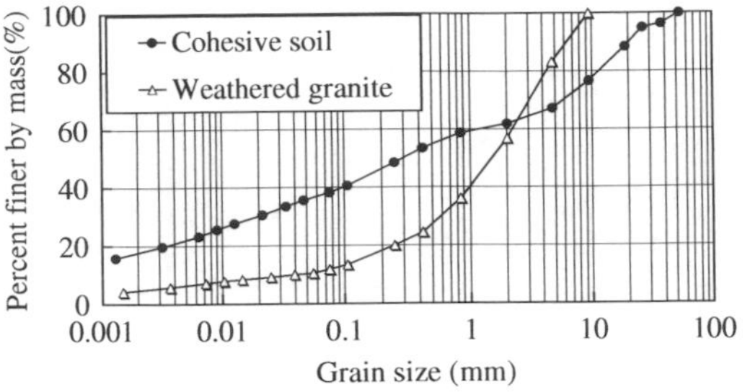

Figure 2. Grain size distribution of Masado and cohesive soil.

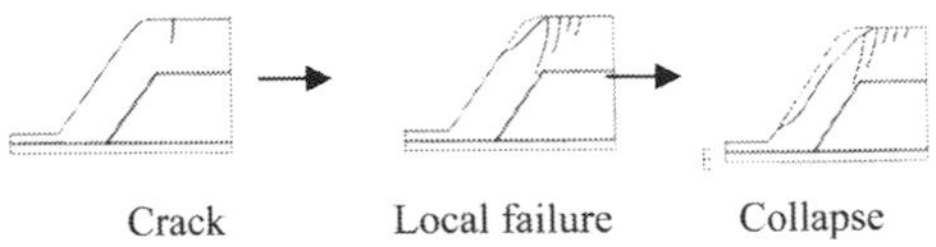

Figure 3. Illustration of failure phases.

were $C = 5$ kN/m^2, $\phi = 28$ degree, and $C = 21$ kN/m^2, $\phi = 28$ degree from the laboratory tests respectively.

The input motions were sine waves with 5 Hz frequency. Each test consists of multiple loading steps. In the beginning, the test started from the maximum acceleration of 100 Gal (100 Gal = 1.0 m/s^2). Then the input acceleration gradually stepped up by 50 Gal until sliding failure occurred in the model.

In the failure process of the slopes, three major phases were categorized (fig. 3), although there were some differences dependant on the conditions, the kind of a soil and the existence of reinforcement.

In the first phase, some local hair cracks were observed at the crest of the slopes but the displacement of the slopes remained small. There was no local failure in the slopes. Then, some cracks developed in the tension zone in the upper flat of the models.

As the input acceleration increased, the second phase resulted in visible deformations along with the widen cracks at the crest. Permanent deformations, such as settlements and heaving at the toe, accompanied with local failures. Soil particles and small broken blocks fell down from the crest.

The last phase was sliding failure in which the surface layer collapsed rapidly with large deformations throughout the most part of the slopes from the top to the toe unless the reinforcement prevented.

Displacement curves for the Masado slope without and with reinforcement were drawn in fig. 4, and for the cohesive soil slopes (fig. 5). The displacement curves for both soil slopes without reinforcement showed larger increments than those for the reinforced ones as the level of input acceleration became high.

On the other hand, the reinforcement changed the deformation behavior. The deformations of the

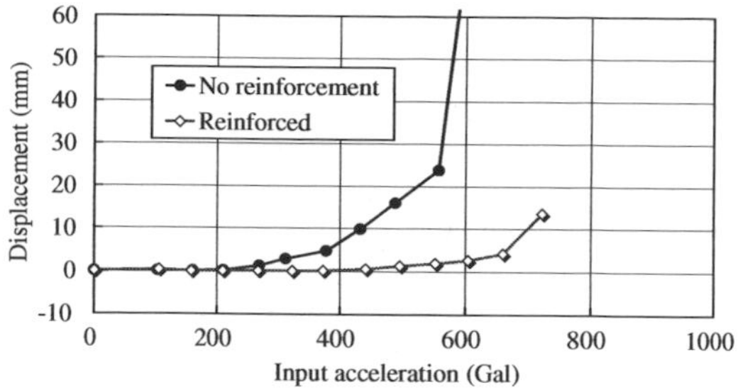

Figure 4. Displacement curves for Masado slope.

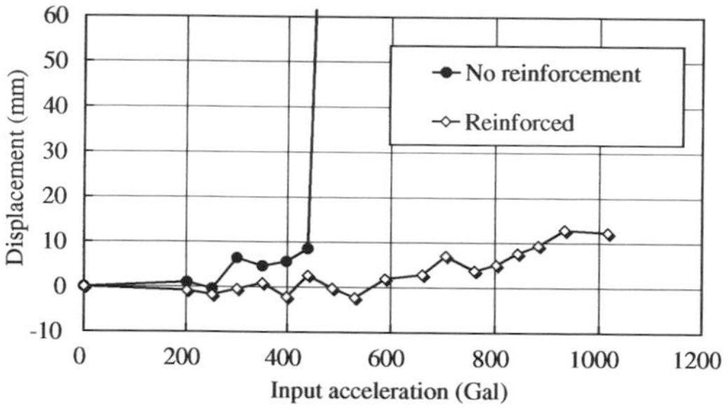

Figure 5. Displacement curves for cohesive soil slope.

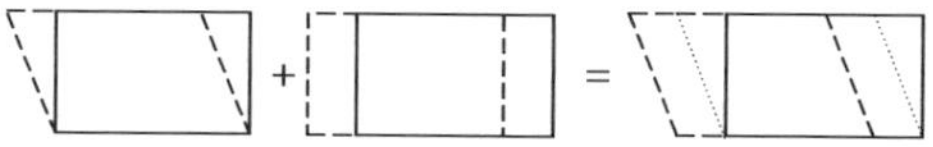

Figure 6. Shear and sliding deformation.

reinforced slopes were less than those of the slopes without reinforcement in the same level of the input motion. Collapse did not occur even in the much higher loading level, nearly twice, than the loading level in which the cohesive soil slope without reinforcement collapsed. A mass movement was not clearly seen. Soil particles and partial broken blocks were retained under the rope nets coverage.

The effect of the reinforcement did not appear in the low level of the input motion. Each model shows differences in deformations as the level of input acceleration became higher.

3 CALCULATION MODELS

It is assumed that the deformation of a soil element could be divided into two components; shear and sliding (Fig. 6). These two types of deformation were mostly dependant on soil strata. The Masado slope had a single layer, so the deformation was mostly shear deformation. The cohesive soil slope had two layers. The large portion of the upper layer formed a soil mass and slid over the surface of the foundational layer.

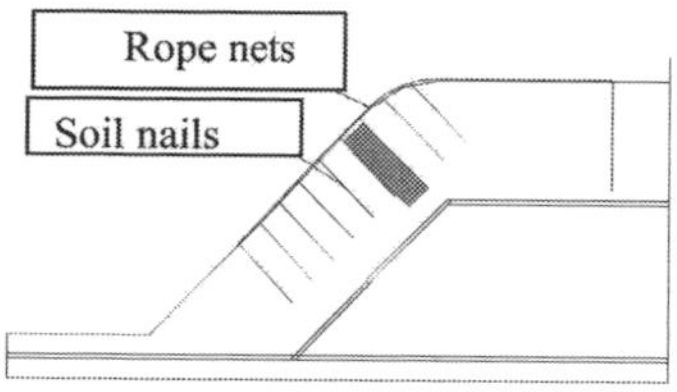

Figure 7. Soil block model.

Table 1. Parameters.

Unit weight	γ_e	16.0 kN/m^3
Height of a soil block	h	0.4 m
Pitch of soil nails	B	0.2 m
Thickness of a soil block	L	0.2 m
Dilatancy angle	v	12.0 Degree
Young's moduli of a soil nail	E	7.00×10^7 kN/m^2
Cross sectional area of soil nails	A_{RB}	2.83×10^{-5} m^2
Initial shear modulus	Go	14,556 kN/m^3

Therefore, it is considered that two calculation models represent each deformation mode.

3.1 *Shear deformation model*

Referring to the elastic shear deformation of a solid, the authors created a shear deformation model for a single layered slope. A soil block surrounded by 4 soil nails at each corner deforms in a way that a rectangular look from the side tilts and becomes a parallelogram in a seismic force (Fig. 7). In this method, it is assumed that a soil block keeps the balance of internal and external work following the principle of virtual work. A seismic force coupled with the gravity, acts as an external force so that a soil block deforms in a shear mode. Strain energy equivalent to the external work is produced in the soil block at the same time.

To predict accurate a displacement, the shear modulus G of a soil is an important parameter. The stress-strain relationship of a soil can be considered linear only at a very low strain level. A laboratory test gives the strain dependant shear modulus in the strain range from 10^{-6} to 10^{-2}. The shear modulus of the Masado obtained from a cyclic tri-axial test is shown in Fig. 10. The vertical axis of the graph is shear modulus G normalized by the initial shear modulus Go. The horizontal axis is shear strain γ. Table 1 shows the parameters for the calculation.

In this method, Go is given as an input parameter and the related shear strain is calculated. Cumulatively, new shear modulus is estimated from the experimental data (Fig. 8) and calculation repeats. This iterative process continues until the difference becomes small.

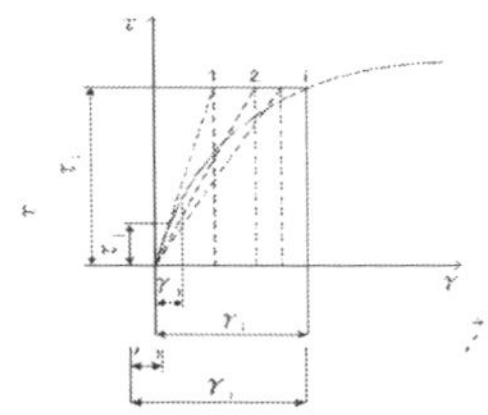

Figure 8. Iterative scheme for stress-strain relationship of soil.

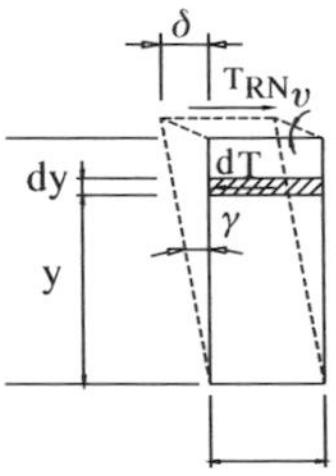

Figure 9. Soil block and shear deformation.

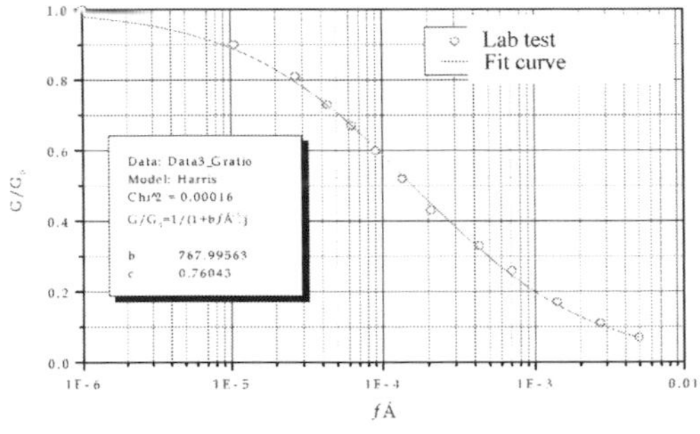

Figure 10. Shear strain and shear modulus.

A soil block deforms under the gravity and a seismic force (fig. 9). The self-weight of a soil block multiplied by the coefficient of seismicity, a ratio of input acceleration to the gravity acceleration in this paper, gives a seismic force. Tensile forces of soil nails and rope nets satisfy the force equilibrium of a reinforced slope. The equations were formulated as follows.

The external virtual work of the system *WE* can be written in the combination of the work done by a seismic force and the tensile force of rope nets T_{RN}.

$$W_E = \int_0^h \gamma . y . dT - T_{RN} . \gamma . h \tag{1}$$

The seismic force can be written as

$$dT = k_e \gamma_e BL.dy \tag{2}$$

where, k_e is the coefficient of seismicity in slope direction, γ_e is the unit weight of a soil and B, L, h are the dimensions of a soil block.

Substituting eq.2 into eq.1 gives

$$W_E = \frac{1}{2} k_e \gamma_e BLh^2 \gamma - T_{RN} . \gamma . h \tag{3}$$

The internal work of the system consists of the work done by the shear deformation of a soil block and the work done by the axial deformation in soil nails.

$$W_I = \frac{1}{2} G \gamma^2 BLh + \frac{1}{2} E \varepsilon^2 A_{RB} h \tag{4}$$

where, E, A_{RB} are the Young's modulus and the cross sectional area of 4 soil nails. ε is a normal strain of a soil nail which can be calculated directly form the shear deformation of a soil block accompanying by the volumetric deformation with a dilatancy angle υ, since it is assumed that soil nails and a soil block deform together. Hence,

$$\varepsilon = \frac{\sqrt{(h + h.\gamma.\tan \upsilon)^2 + (h.\gamma)^2}}{h} - 1 \tag{5}$$
$$= \sqrt{1 + 2\tan \upsilon.\gamma + (1 + \tan^2 \upsilon).\gamma^2} - 1$$

Expanding the function and neglecting higher orders of γ leads to

$$\sqrt{1 + 2\tan \upsilon.\gamma + (1 + \tan^2 \upsilon).\gamma^2} = 1 + \tan \upsilon.\gamma + \frac{1}{2}\gamma^2 + \ldots$$

Substituting in to eq.5 gives

$$\varepsilon \approx \tan \upsilon.\gamma + \frac{1}{2}\gamma^2 \tag{6}$$

The internal work of the system can be re-written as eq.7 by substituting eq.6 into eq.4.

$$W_I = \frac{1}{2} G\gamma^2 BLh + \frac{1}{2} EA_{RB} h.(\frac{1}{4}\gamma^4 + \tan\upsilon\gamma^3 + \tan^2\upsilon\gamma^2) \tag{7}$$

Equalizing the internal and external work of the system given in eq.3 and 7 and solving for γ yields to the basic shear strain of a reinforced soil slope as,

$$\gamma^3 + 4\tan\nu \cdot \gamma^2 + \left[\frac{4GBL}{EA_{RB}} + 4\tan^2 \nu\right] \cdot \gamma - \frac{(4k_e W - 8T_{RN})}{EA_{RB}} = 0 \tag{8}$$

In this equation: γ = shear strain: W = self-weight of a soil block (kN) ; L = thickness of a soil block (= length of soil nail) (m) ; G = shear modulus of soil material (kN/m^2) ; ke = coefficient of seismicity in slope direction; T_{RN} = axial force of rope nets (kN) ; E = Young's modulus of a soil nail (kN/m^2); A_{RB} = cross sectional area of 4 soil nails (m^2) ; ν = dilatancy angle.

Fig. 11 compares the calculated displacement of a reinforced slope to the measured of the Masado slope

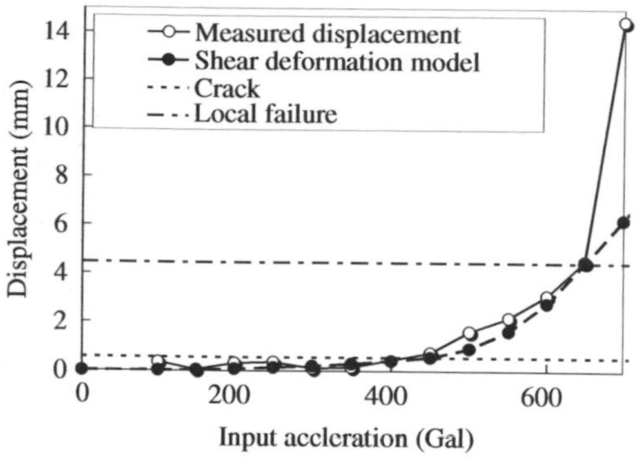

Figure 11. Measured and shear deformation model displacements of a single layered slope with the reinforcement.

with a single layer. The calculated displacement at the low level was in good agreement with the measured. At the high level, the measured displacement was larger than the calculated because of failures.

The axial forces of rope nets and soil nails for both methods were associated with the slope displacement in the shake table tests. Since those member forces were proportional to the increase of the slope displacement, once a slope displacement is gained in other slopes, member forces can be presumed using the empirical relationship.

3.2 *Sliding deformation model*

Observing the shake table tests for two-layered slopes, we assumed that a sliding deformation was proportional to the factor of safety along a failure surface. Fujii (2001) reported a similar prediction method for a displacement of an abutment during an earthquake. Although only an invisible deformation was observed in the low level of input acceleration in the shaking table tests, the residual deformation became large and be accumulated as the level high level of input acceleration was repeatedly loaded with an increment. Non-linearity in a sliding deformation behavior was needed to be described.

A fitting curve written as eq.9 for the shear stress vs. shear displacement of the shear test (fig. 12) is utilized as an exponential function to predict a sliding displacement of a surface layer in a slope without reinforcement.

$$\tau = 12 \cdot \delta_f^{\ 0.33} \tag{9}$$

where, τ = shear stress (kN/m2); δ_f = shear displacement (mm).

Transforming the function gives the following eq.10. X, a ratio of a sliding displacement to the length of a failure surface, is named a sliding strain in this paper. Fs is factor of safety along a failure surface. Y is a reciprocal of Fs. Fig. 13 shows the relationship.

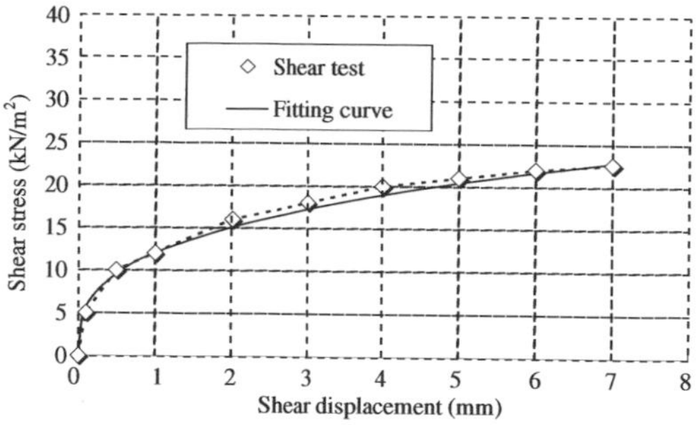

Figure 12. Shear stress and displacement of a shear test.

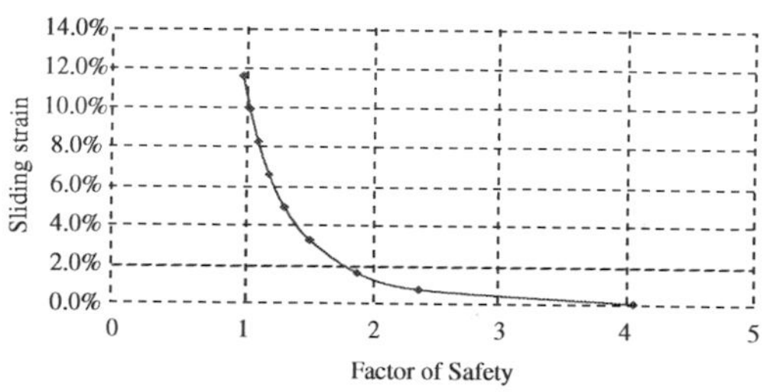

Figure 13. Factor of safety and sliding displacement.

$$X = \frac{1}{h} \cdot \left(\frac{\tau_f \cdot Y}{12} \right)^{3.030303} \tag{10}$$

where $X = \delta/h$ and $Y = 1/F_s$.

Since there is no residual displacement in a static state, an initial sliding strain in a static state should be deducted as eq.11.

$$\delta_e = h \cdot (X_e - X_i) \tag{11}$$

Where δ_e = sliding displacement due to a seismic force; h = length of a failure surface; X_e = sliding strain under static and seismic forces; X_i = sliding strain in a static state

Let a soil block on a slope (Fig. 14) is under the gravity and a seismic forces. Factor of safety during an earthquake FSe is calculated as in eq.12 to 14,

$$FS_e = \frac{Sre}{Se} \tag{12}$$

$$S_{re} = (W \cdot \cos\alpha - kH \cdot W \cdot \sin\alpha) \cdot \tan\phi + c \cdot \ell \cdot B \tag{13}$$

$$S_e = W \cdot \sin\alpha + kH \cdot W \cdot \cos\alpha \tag{14}$$

where Sre = shear strength during an earthquake; Se = seismic force; kH = horizontal coefficient of

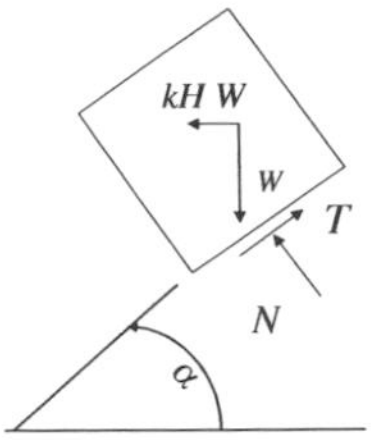

Figure 14. Soil block.

Table 2. Parameters.

Unit weight of a soil	γt	13.0 kN/m^3
Height of a soil block	H	0.2 m
Span of soil nails	B	0.2 m
Length of a soil block	L	0.2 m
Internal friction angle	ϕ	28.0 degree
Cohesion	C	3 kN/m^2

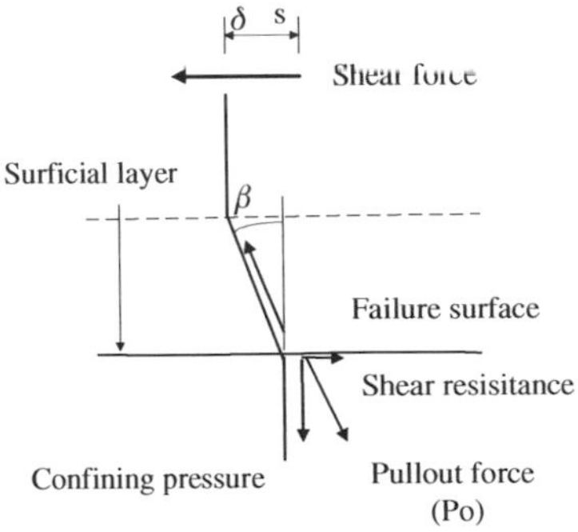

Figure 15. Resistance force of a soil nail on a slip surface and pullout displacement of a soil nail.

seismicity; W = self-weight of a soil block (N) ; α = slope angle(degree); ϕ = internal friction angle of a soil (degree); c = cohesion of a soil (kN/m^2); ℓ : = length of a soil block (m); B = width of a soil block (m). Parameters are shown in Table 2.

In case of a reinforced slope, soil nails resist a soil block sliding. The increment of the shear strength on a failure surface due to a soil nail is calculated by the method derived for the effect of tree roots (Abe, 2004). He suggested that a pullout force of a tree root was gained from a field test. We propose that a shear force causes pullout of a soil nail when a surface layer moves down along a failure surface (Fig. 15). The tensile force during the pullout is divided into two components. One is a shear resistance force against sliding and another is a confining pressure normal to the failure surface. Both forces work for adding shear resistance against a slope failure. The geometric relationship between a sliding displacement, a soil nail's inclination and its slippage from a surface layer (Fig. 16) are expressed in eq.15 and 16.

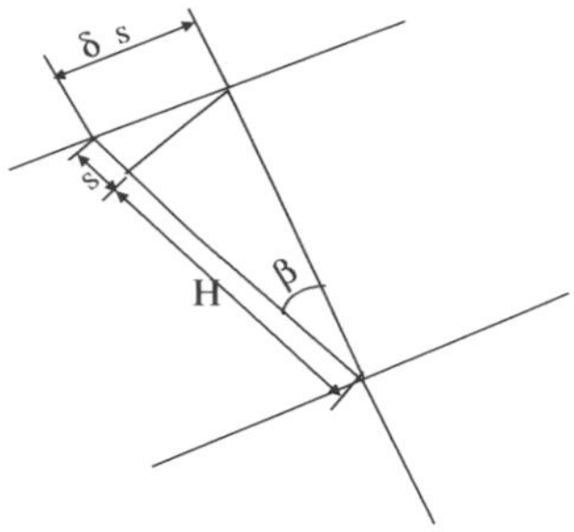

Figure 16. Deformed rock bolt.

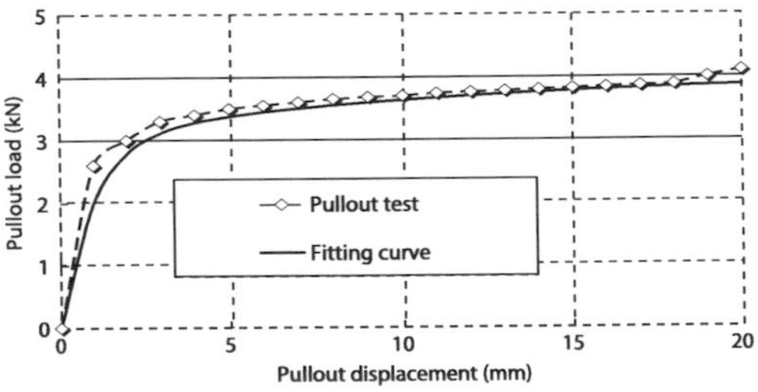

Figure 17. Load vs. displacement of a soil nail in the pullout test.

$$s = \delta_s / \sin\beta - H \tag{15}$$

$$\beta = \tan^{-1}(\delta_s / H) \tag{16}$$

where β = soil nail's inclination; δ_s = sliding displacement of a surface layer; H = length of a soil nail in a surface layer; s = slippage between a soil nail and a surface layer.

Laboratory pullout tests were performed for soil nails and rope nets with head plates. The relationship between the slippage and the pullout force (Fig. 17) was formulated in the fitting function, eq.17. Shear resistance increment can be calculated from eq.18 as the same equation for tree roots. Then, factor of safety is re-calculated using this shear resistance. Factor of safety affects a sliding displacement again. This process is recursive. Once again, the sliding displacement causes change in the pullout force of a soil nail in calculation. This calculation repeats until conversion.

$$P = 3/4 \cdot \{1-\exp(-s)\} + \{1-\exp(-0.1 \cdot s)\} \tag{17}$$

$$\Delta S = P \cdot (\cos\beta \cdot \sin\phi + \sin\beta) \tag{18}$$

where, s = slippage between a soil nail and a surface layer (mm); P = pullout force of a soil nail (kN); β = soil nail's inclination.

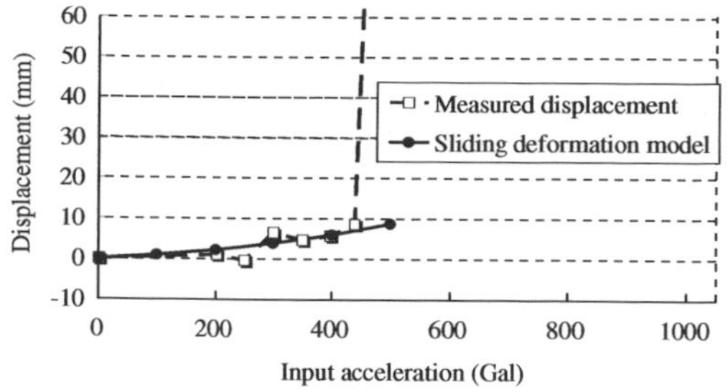

Figure 18. Measured and calculated displacement of the two layered slope without reinforcement.

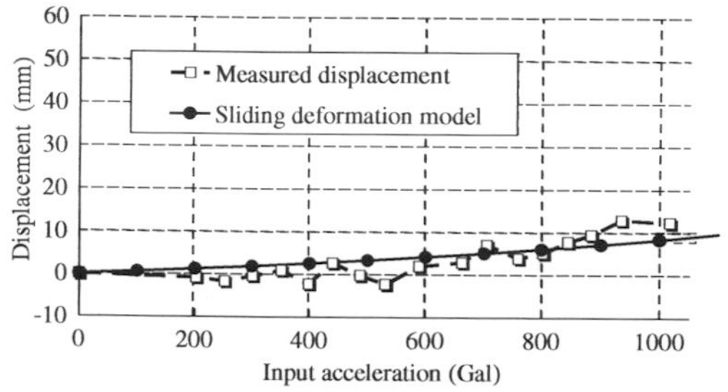

Figure 19. Measured and calculated displacement of the two-layered slope with the reinforcement.

The results of the calculation for two-layered slopes using a cohesive soil with and without reinforcement are compared with the measured displacements in fig. 18 and 19. Although an abrupt increase in the measured displacement in fig. 18 shows collapse, both displacement curves agree.

4 SUMMARY

Based on the shake table tests, two calculation models; shear and sliding, were invented to predict displacements so as to describe the dynamic behavior of a slope and the mechanism of the reinforcement with soil nails and rope nets.

The shear deformation model is a model for calculating a shear deformation of a soil block between soil nails with rope nets under the gravity and a seismic force. It rotationally deforms downward according to the loading condition. Based on the principle of virtual works, the equations were formulated. The strain-dependant characteristic of a shear modulus is considered in the iterative calculation. The calculation result was in good agreement with the measured displacements in the one-layered slopes using the Masado.

The calculation for a sliding deformation has two major steps. Firstly, sliding deformation of a soil block without reinforcement is proportional to the factor of safety. Then, the reinforcement is considered in a way that shear resistance increases by the induced pullout force in a soil nail during sliding deformations. Displacement curves for slopes with and without reinforcement are both in good agreement with the measured displacements in the two-layered slopes using the cohesive soil.

There are some limitations. Further study will lead us to establish a new design criterion based on displacements for the reinforcement.

REFERENCES

Abe et.(2004):Method for evaluating thinning influences on a forest's ability to prevent shallow landslides, Landslides, Journal of the Japan Landslides Society, Vol.41 No.3, pp.9–19

Fukumasa T, et al., (2001): Experiment and Analysis on Reinforcement Method for Seismic Slope Stability (Part 1), 36th Conference on Geotechnical Engineering, JAPAN, pp.2313–2314

Fujii et al, (2003): An Aseismic Assessment for Embankment Considering the Deformation, Japanese Geotechnical Society, Soil and Foundation, 50-1, pp.10–12

Murakami H. et al, (2002): A new displacement-based design method for a new remedy of rope nets and rock bolts to stabilize natural slopes, 3rd International Conference on landslides, slope stability & the safety of infra-structures, 11–12 July 2002, Singapore

Nishihara R., et al, (2006): Shake table tests for reinforcement method about natural slope with cohesive soil, 41st Conference on Geotechnical Engineering, JAPAN, pp.2095–2096

New Horizons in Earth Reinforcement – Otani, Miyata & Mukunoki (eds)
© 2008 Taylor & Francis Group, London, ISBN 978-0-415-45775-0

Development of rational design method for the geogrid reinforced soil wall combined with soil cement and its application

H. Ito, T. Saito & M. Ueno
Engineering division, Dai Nippon Construction, Japan

J. Izawa
Department of Civil Engineering, Tokyo Institute of Technology, Japan

J. Kuwano
Geosphere Research Institute, Saitama University, Japan

ABSTRACT: A geogrid reinforced soil wall (GSW) combined with soil cement was recently developed and the application of this method has been used increasingly. From the results of the past centrifuge shaking table tests, it is clearly shown the effectiveness of the GSW to increase seismic stability. This current research was then continued to study the effect of the arrangement of geogrid and soil cement wall in order to develop a new rational design method of GSW for practical use in the real construction site. To achieve this target, a series of centrifuge shaking table teats were then carried out. Results show that even if the size of the width of the cement wall was reduced remaining only 2/3 of the full width but the seismic stability of the GSW is still the same and effect of the length of geogrid laid at the upper part of the wall plays the higher important role to the seismic stability than that of in the lower part. Based on these two results, the new design concept was developed. It was shown in details in this paper about the new design method and an example of case study in Japan using this new design concept to reduce the length of geogrid and the width of soil cement.

1 GENERAL INSTRUCTIONS

Geogrid reinforced soil wall (GSW) method was developed in recent years. It is considered that the demand for a steep slope is getting higher for the reduction of construction cost by the effective use of site or the reduction of purchase fee and so on. In order to establish the economical and reasonable construction method of the reinforced soil wall, Ito et al., 2001 developed a new type GSW, whose wall was made of soil cement as shown in Figure 1. Saito et al. studied seismic behavior of the wall and indicated that it showed higher seismic stability than GSW with only divided panel type wall. In this paper, it is focused on arrangement of soil cement wall and geogrid for more rational design. For that purpose, centrifuge shaking table tests were conducted in order to investigate both static and seismic stability of this method with shorter width of soil cement wall or length of geogrid because arrangement of soil cement wall and gaogrid are determined by evaluation of seismic stability. Finally, case histories, which were designed by new rational method, are reported.

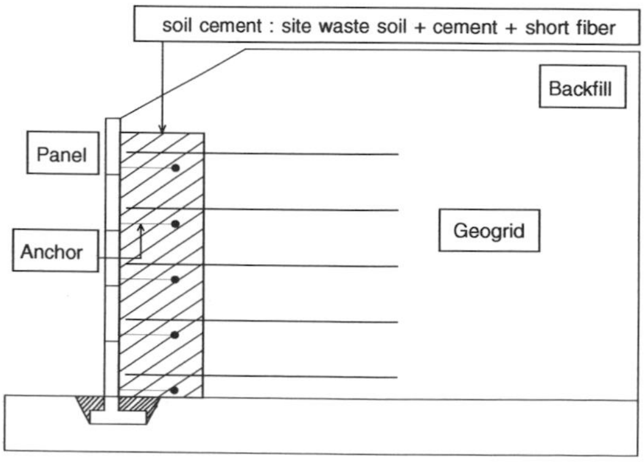

Figure 1. Schematic view of the reiforced soil wall combined with soil cement.

2 OUTLINE OF THE TEST

2.1 *Model GSW*

A typical model soil wall is shown in Figure 2 and Photograph 1. The soil cement wall had a height of 200 mm, 10 m in prototype scale when the test is

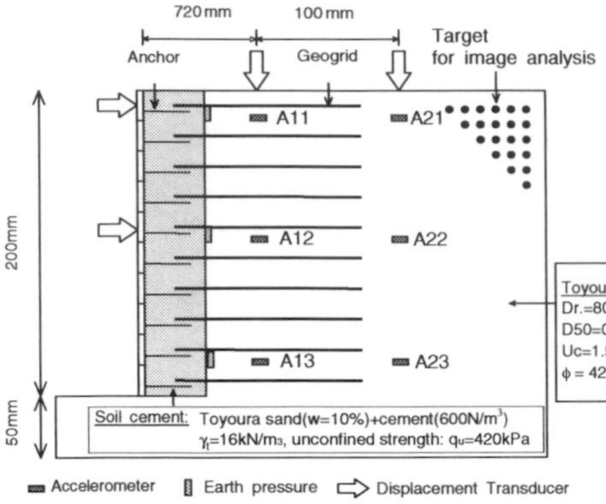

Figure 2. Schematic view of model setup.

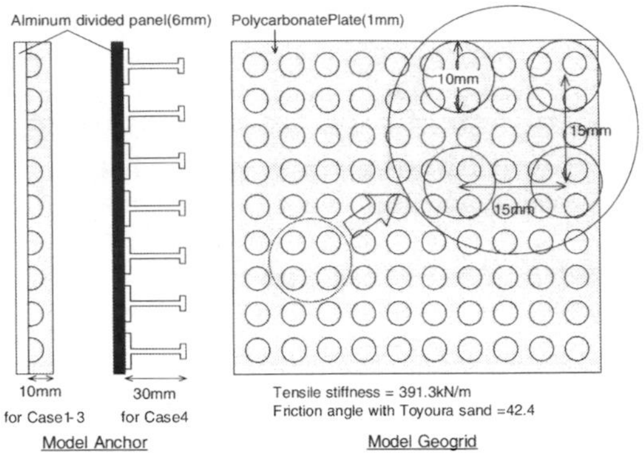

Figure 3. Details of model geogrid and anchor.

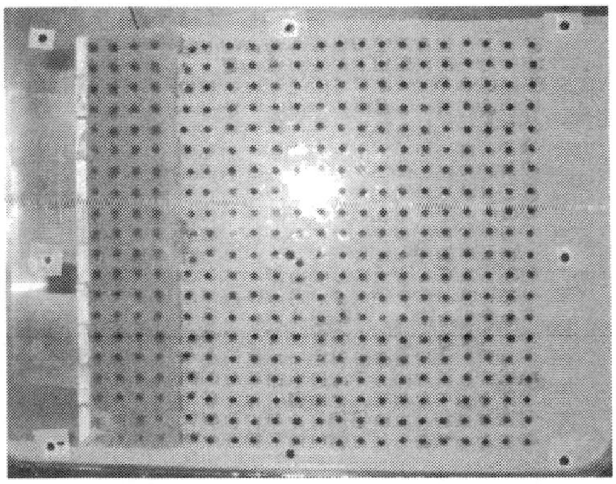

Photo 1. Model of the reinforced soil wall.

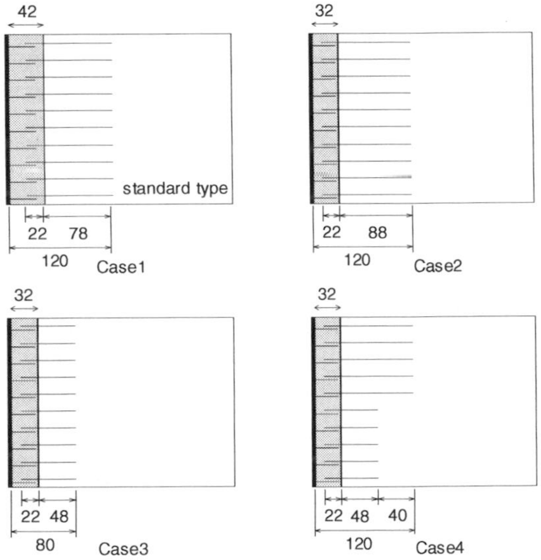

Figure 4. Test cases.

performed at 50 g. Vertical spacing of the reinforcement was also at 20 mm. The backfill used was air dry Toyoura sand with relative density of 80%. The properties of Toyoura sand are shown in Figure 2.

The soil cement was composed of Toyoura-sand and high early strength Portland cement of 600 N/m^3. The wet density and water content of Toyoura-sand were 16 kN/m^3 and 10% respectively. The curing time was 7 days in order to obtain the unconfined shear strength of $q_u = 420$ kPa. Vinylon short fibers with length of about 10 mm and diameter of 43∞m which was mixed in the model soil cement wall was the same as used in situ. Ductility of soil cement against the seismic loading can be improved by mixing short fiber.

Model geogrid was made of polycarbonate with 1 mm in thickness. The schematic view of model geogrid is showed in Figure 3. The holes with a diameter of 10 mm were made at 15 mm interval. Tensile stiffness of geogrid and friction angle between soil and geogrid, which were investigated by tensile test and pullout test respectively, were summarized in Figure 3. They were almost the same as those of the geogrid used in situ. The aluminum panel with 5 mm thickness was used as wall panel.

The test cases were shown in Figure 4. Case1 is the standard case and Model of it was designed by using previous design procedure. Model of Case2 had shorter width of soil cement wall as compared with that of Case1. But End of geogrid was the same with Case1. In Case3, length of geogrids was also shorter. According to the past study, not so large strains were not observed in bottom geogrids. Therefore, it seems that length of bottom geogrids may be shorter and length of upper geogrid is more important for stability of GSW. Consequently, model of Case 4 had shorter length of bottom geogrids and usual length of upper ones.

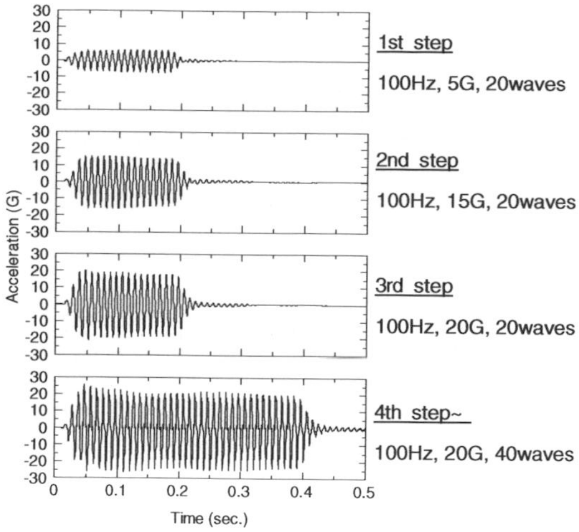

Figure 5. Time histories of input seismic wave.

2.2 *Test procedure*

The centrifuge shaking table tests were performed at the centrifugal acceleration of 50G. Some sinusoidal input seismic waves shown were applied to the model reinforced soil wall with gradually increasing amplitude of acceleration. Properties of seismic wave are shown in Figure 5. During the test, displacements, earth pressures and acceleration responses were measured by some transducer and accelerometers which were shown in Figure 2. Deformation of the model wall was monitored by a CCD camera through the Perspex window of the container. Image analyses were done by using the digital image captured from CCD camera.

3 TEST RESULTS

3.1 *Stability at centrifugal acceleration of 50G*

Figure 6 shows distributions of horizontal earth pressure acting on soil cement wall and horizontal displacement of wall at centrifugal acceleration of 50G in order to evaluate the stability in ordinary condition. Broken line as shown in Figure 6(b) indicates horizontal earth pressure distribution calculated by using Coulomb's equation. Smallest horizontal earth pressure was observed in Case1, which was designed by past design code and its values were much smaller than that obtained from Coulomb's equation. That is to say, horizontal earth pressure acting on the wall could be reduced by laying geogrids in back fill. In Case 2, whose width of soil cement wall was 10 mm shorter than that of Case1 but end of geogrid was the same as the end of Case1, about 4 times as large horizontal earth pressure as Case1 was measured. It turns out that horizontal earth pressure which acts on the wall

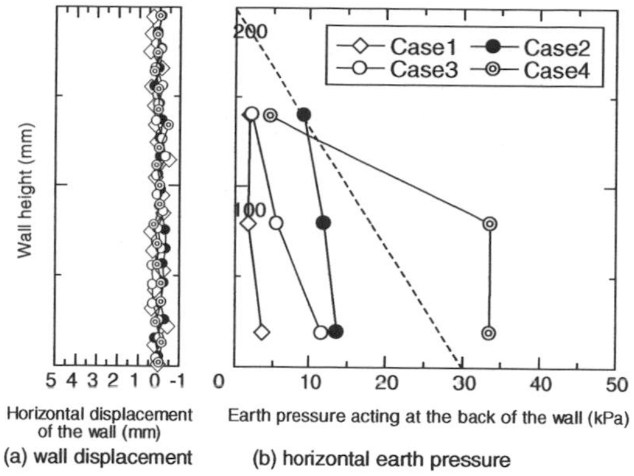

Figure 6. Wall displacement and horizontal earth pressure acting on soil cement at centrifugal.

becomes large if reinforced area becomes large. On the other hand, in Case3, whose width of soil cement wall and length of geogrids were shortest of all cases, horizontal earth pressure was smaller than that of Case2. The reason is why the soil cement wall in Case3 moved horizontally and stress state in reinforced backfill was active state as compared with in Case1 and 2. But it was not confirmed because difference of horizontal displacement among all cases was not observed. In Case4, whose length of upper and lower geogrids was the same as Case2 and Case3 respectively, movement of the wall was restricted by upper geogrids. But large earth pressure acted on the lower part of the wall due to shortage of geogrids.

As mentioned above, sufficient stability in ordinary state could be observed in all cases. But earth pressure distribution which acts on a soil cement wall differs greatly by influence of width of the wall or length of geogrids.

3.2 *Seismic stability*

Figure 7 shows displacement vectors after finishing 4th seismic step of all cases. Although the soil cement wall overturned during shaking, progressive failure was prevented by tensile force of geogrids in all cases.

It has been recognized that displacement of reinforced soil wall during earthquake accumulates with shaking in the past studies (Izawa et al. 2002). Therefore, acceleration power is used as index to indicate seismic scale. Acceleration power can consider not only seismic intensity but also duration of shaking and it is calculated by Equation (1).

$$I_E = \int_0^T a^2(t)dt \qquad (1)$$

a : *Input acceleration*
T : *Shaking time*

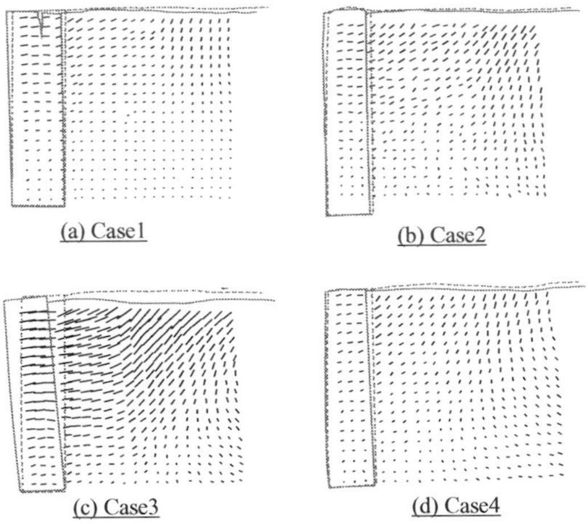

Figure 7. Displacement vectors after step 4.

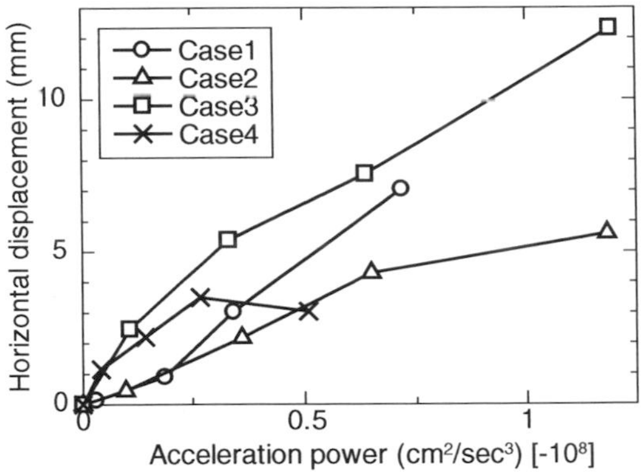

Figure 8. Relationships between horizontal displacement at top of the wall and cumulated acceleration power.

Figure 8 shows relationships between horizontal displacement at top of the wall and cumulated acceleration power. Almost the same horizontal displacement were observed in Case1 and Case2 until the acceleration power reached about $0.25 \times 10^8 (cm^2/sec^3)$. After that, increment of displacement in Case1 was larger than that in Case2 because the soil cement wall cracked at the top of the wall. As shown in Figure 9, tension force occurred in the soil cement wall due to inertia force of the wall and tension force of geogrid to prevent from overturning of the wall during shaking. As a result, the soil cement wall collapsed. On the other hand, such tension crack was not observed in Case2. If the soil cement wall is made small, inertia force of the wall decrease with reduction in weight. Accordingly, large tension force did not occur in the soil cement wall in Case2 and generation of crack could be prevented.

In Case3, the soil cement wall overturned and almost collapsed as shown in Figure 8 due to shortage

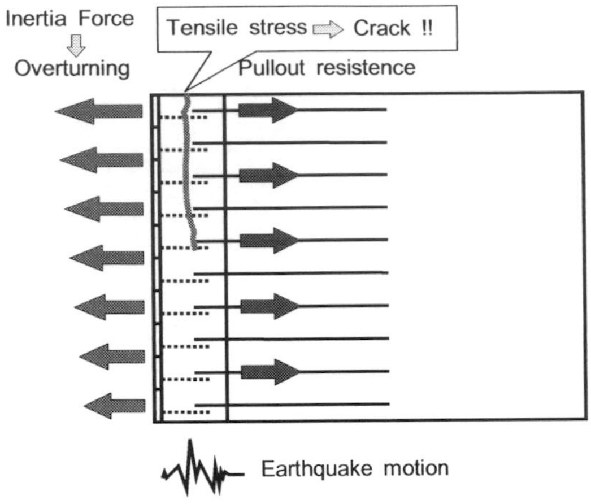

Figure 9. Illustration of generating crack in the soil cement wall.

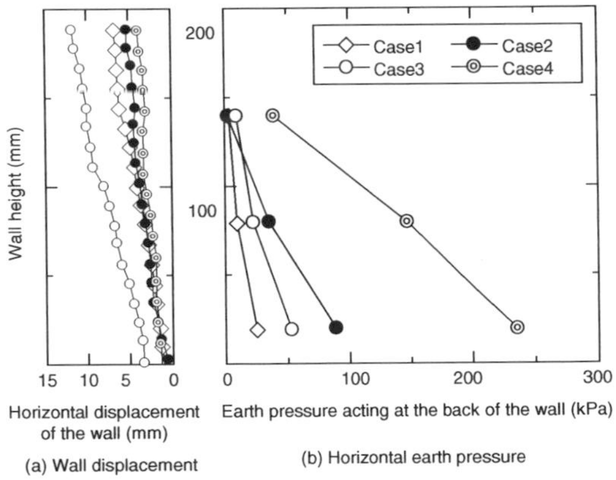

Figure 10. Wall displacement and horizontal earth pressure acting on soil cement after 4th shaking step.

of geogrid. Such large deformation was not observed in Case4 because length of upper part geogrid was longer and it acted effectively in the prevention from overturning of the wall. But large horizontal earth pressure acted on the wall as shown in Figure 10.

3.3 *Summary*

If the width of soil cement wall is reduced about 20%, the wall has sufficient static and seismic stability. Moreover, generating of a crack in the soil cement wall can be prevented due to reduction of inertia force acting on the wall. On the other hand, length of geogrid is very important for seismic stability of the soil cement wall. Although collapse could be prevented if the length of upper part geogrid was excelled, large horizontal displacement and horizontal earth pressure acting on soil cement wall were observed.

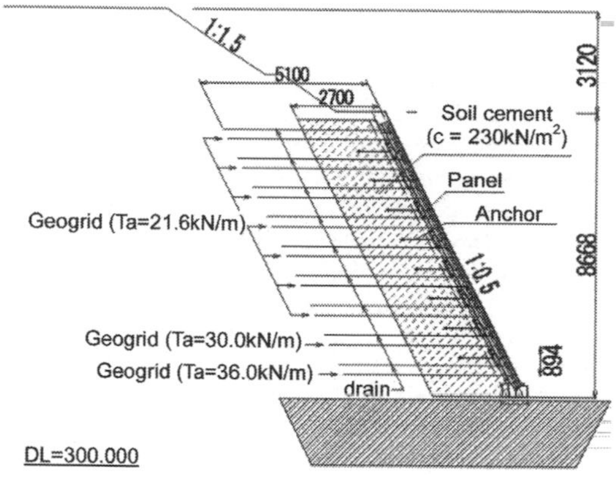

Figure 11. Configuration of Application site 1.

Photo 2. Application site 1.

New design code was developed for the purpose of cost reduction of construction in consideration of test results. In the code, width of soil cement wall can be reduced as compared with past design code because almost the same seismic stability with usual one was shown in the centrifuge model test. In next chapter, two applications, which designed by using new design code, were reported.

4 EXAMPLE OF CASE STUDY

4.1 *Application for reclaimed site*

New rational design method was applied to reclaimed area in Kumamoto prefecture, Japan. Volcanic cohesive soil, which is the peculiar volcanic soil around Mt. Aso area in Kumamoto, was used for soil cement wall. Backfill material is the other sandy soil in order to achieve sufficient fiction between soil and geogrid. Soil properties are shown in Table 1. Height and slope of GSW is 8.7 m and 1:0.5 respectively as shown in Figure 11 and Photograph 2. Moreover, there is embankment with 3.2 m in height above the GSW.

Table 1. Soil properties in application site 1.

	Back fill	Soil cement wall
Specific gravity	2.595	2.632
Natural water content w_n (%)	43.6	58.2
Gradation		
Gravel (%)	38.7	0.2
Sand (%)	43.7	63.6
Silt (%)	14.3	30.9
Clay (%)	3.3	5.3
Maximum dry density ρ_{dmax} (kN/m³)	11.85	10.70
Optimal water content w_{opt} (%)	39.6	39.5
c_u (kN/m²)*	27.5	–
ϕ_u (deg.)*	27.2	–

*$\rho = 10.67$ (kN/m³) at water content $= w_n$

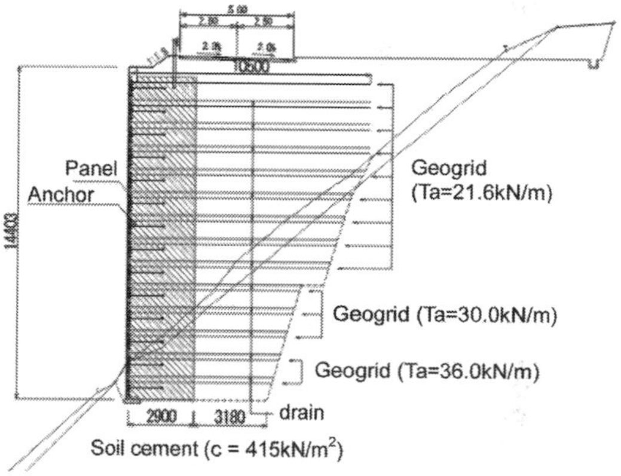

Figure 12. Configuration of Application site 2.

Mixture was determined by unconfined compression test as shown in Table 1. It was conformed that soil cement wall in-situ had the necessary strength. In this site, construction cost could be reduced about 6% because soil cement wall width could decrease from 3.2 m to 2.7 m by using new design code.

4.2 *Application for road embankment*

Second application is at road embankment site also at Kumamoto Prefecture. Maximum height of the wall is over 14m as shown in Figure 12 and Photograph 2. Backfill material is gravelly soil, which was made with crashed weathered rock at the construction site. Properties of backfill soil are shown in Table 2. The same gravelly soil used as Backfill material was also used for soil cement wall. Also at this site, width of the soil cement wall width could make less from 3.4 m

Photo 3. Application site 2.

Table 2. Soil properties in application site 2

	Back fill
Specific gravity	2.814
Natural water content w_n (%)	6.8
Gradation	
Gravel (%)	92.2
Sand (%)	5.6
Silt (%)	2.2
Clay (%)	
Maximum dry density ρ_{dmax} (kN/m^3)	21.3
Optimal water content w_{opt} (%)	6.2
c_u (kN/m^2)*	60.7
ϕ_u (deg.)*	32.6

*$\rho = 19.17$ (kN/m^3) at water content $= w_n$

to 2.9 m by using new design code. As a result, construction cost could be reduced about 5%. This wall is showing very high stability in spite of a very high perpendicular wall.

5 CONCLUSIONS

Results of centrifuge model tests indicated that the reinforced soil wall combined with soil cement has sufficient stability both in ordinary and seismic state although its soil cement wall width is reduced about 20% as compared with the model designed by past design code. On the basis of results of centrifuge model tests, new design method, which can reduce width of soil cement wall, was made. New design procedure was applied to two construction site at Kumamoto prefecture in Japan. Construction cost could be reduced about 5% in both site by using new design method. Additionally, they show high stability in spite of severe conditions.

REFERENCES

Ito, H., Saito, T., Izawa, J. & Kuwano, J., In situ construction of the Geogrid reinforced soil wall combined with soil cement, Proceedings of 8th International Conference of Geosynthetics, pp. 1181–1184, 2006

Saito, T., Ito, H., Izawa, J. & Kuwano, J., Seismic stability of the geogrid reinforced soil wall combined with soil cement, Proceedings of 8th International Conference of Geosynthetics, pp. 1511–1514, 2006

Izawa, J., Kuwano, J., and Takahashi, A., "Behavior of steep geogrid-reinforced embankments in centrifuge tilting tests", Proceedings of Physical modeling in geotechnics, pp.993–998, 2002

New Horizons in Earth Reinforcement – Otani, Miyata & Mukunoki (eds)
© 2008 Taylor & Francis Group, London, ISBN 978-0-415-45775-0

Effect evaluation for the geocomposite reinforced embankment of cohesive soil

Y. Tanabashi, Y. Jiang & S. Sugimoto
Department of Civil Engineering, Engineering Faculty, Nagasaki University, Nagasaki, Japan

R. Katoh
Kyushu Regional Development Bureau, Ministry of Land, Infrastructure and Transport, Fukuoka, Japan

K. Tsuji
Graduate School of Science and Technology, Nagasaki University, Nagasaki, Japan

ABSTRACT: In recently years, the low quality soft clay from construction site increases rapidly, due to the development of urban and underground space. Combining the geosynthetics reinforcement with this surplus soil is one of the effective recycling methods. Considering a steep slope banking with high water content and low quality soil, the reinforcement effect of geocomposite is evaluated by using the finite difference analysis. In this study, it is clarified that the reinforcement mechanics of geocomposite for the improvement on the tension and drainage capability of embankment, and also some suggestions for the design of geocomposite reinforced embankment are proposed.

1 INTRODUCTION

Recently, the surplus soils, especially the low quality soft clay from construction sites, rapidly increase due to the development of urban area and underground space. In the metropolitan area, about 120 million m^3 surplus soils are generated per year. 70% among them are high water content and low quality soil such as silt, Kanto Loam and other clays, which are difficult to be disposed properly and economically. On the other hand, environment protection has attracted more and more society concern and become an indispensable issue for the construction industry. Therefore, it raises an important problem to reuse such low quality surplus soils properly.

Combining with geosynthetics reinforcement, the surplus soils can be reused as an embankment material. The reinforcement material, geocomposite can greatly improve the tension strength and drainage capability of the embankment, which makes a steep slope embankment feasible and more economic. In *Design and Construction Manual for Geotextile Reinforced Embankment* (Civil Engineering Research Center, 1993), the reinforcement evaluations for drainage and tension are introduced, however, there is no methodology that can take both the drainage and tension-reinforcement effects into account.

Saving the issues mentioned above, a steep slope embankment with high water content and low quality soil is applied for evaluating the reinforcement effect of geocomposite by using the finite difference analysis. Some suggestions for the design of geocomposite-reinforced embankment are also proposed in this study.

2 STABILITY ANALYSIS OF VOLCANIC ASH CLAY BANKING

2.1 *The outline of analyses model and method*

Using the finite difference analysis (simulation codes: $FLAC^{3D}$), several cases with different reinforcement layers, embankment heights and consolidation periods, as list in Table 1, are studied. The constitutive law employed for the banking soil, i.e. the Kisarazu Kanto loam with 98% water content, is the Mohr-Coulomb model. The geocomposite TRF-31 (abbreviated as GC hereafter) is selected as the reinforcement material, and the drainage and tension strength are focused in these case studies. The embankment is supported by a drainage foundation with enough loading capability.

The properties of the banking material and the reinforcement material that employed in these studies are listed in Table 2, which are partly referenced

Table 1. The study cases.

Height of embankment (m)	8	12	16
Interval of laying reinforced material (cm)	No reinforcement(N), 45(GC45 cm) 90(GC90 cm)		
Speed of work (hour)	240, 480, 720	480, 720	720, 960

Table 2. The properties of banking soil and GC.

Value of banking material (kanto lome)
Volumetric elastic coefficient bulk modulus: K(kPa) = 500
Cohesion: c (kPa) = 19.6
Density: $\rho(g/cm^3) = 1.366$
Angle of internal friction: ϕ_{cu}(deg) = Formula (1)
Dilatancy angle (deg) = 0
Limit of tensile stress: σ'(kPa) = 19.6
Value of geocomposite
Cohesion of interface: c_{cus}(kPa) = 4.41
Angle of internal friction of interface: ϕ_{cus}(deg) = Formula (2)
Rotation of elastic modulus: (kPa) = 28000

to the former researcher's studies (Y.Tanabashi & H.Nagashima, 2002). The strength characteristic in each layer differs from each other, since they are banked step by step at different periods. Through a series of constant volume simple shear tests, an empirical equation (1) is proposed to generalize the relation between the friction angle of each banking layer and its consolidation period. For the same method, the friction angle of GC-soil interface is formulated as empirical equation (2). As for the cohesion, it can be regarded as a constant during all consolidation periods.

$$\phi_{cu}(t_c) = 20.2 - \frac{1}{\exp\left(-3.00 + 0.941\sqrt{t_c}\right)} \tag{1}$$

$$\phi_{cus} = 22.0 - \frac{1}{\exp\left(-3.09 + 0.944\sqrt{t_c}\right)} \tag{2}$$

Here, t_c is the consolidation period (unit is hour), ϕ_{cu} and ϕ_{cus} are the friction angles of the banking material and the GC-soil interface, respectively.

The membrane element is selected to simulate the GC, since they cannot subject to bending moment. And the tension capability of GC is set to 40% of its designation capability due to the embankment's creep effect during its lifetime. The consolidation period for each layer is assumed to be same, as illustrated in Fig. 1. After banking, a footing loading is loaded at every 5–10 kPa step to the crown surface of the embankment, as illustrated in Fig. 2, and the embankment's deformation behavior is evaluated in these studies.

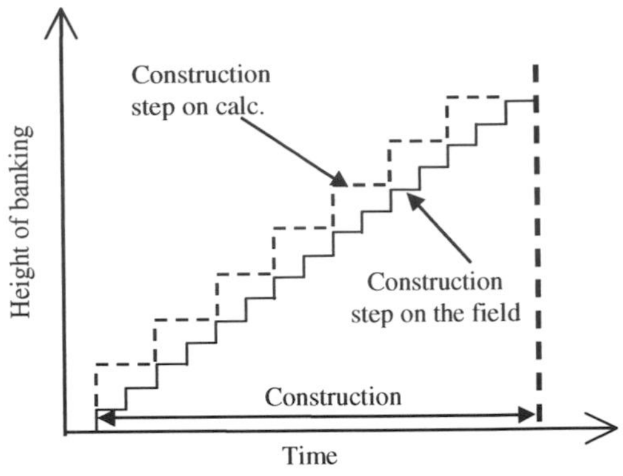

Figure 1. The relationship between construction process and construction period of the banking.

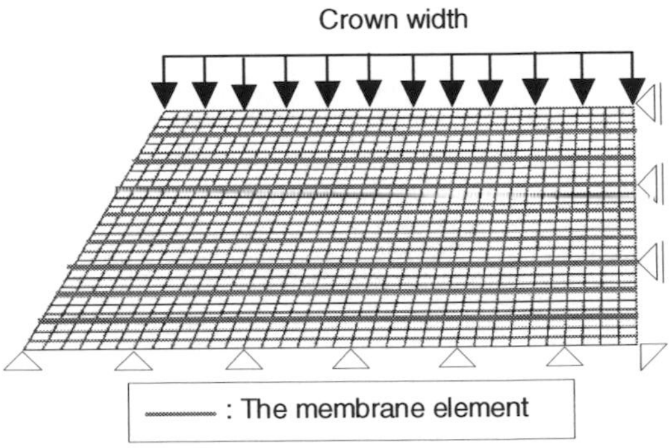

Figure 2. Analysis model.

2.2 *Conversion for the banking consolidation period*

The conversion consolidation time was introduced in order to handle the condition of the consolidation in the banking as a result of different layers. Corresponding to different consolidation coefficient and consolidation period at every 90 cm/layer, the frictional angle of Kanto loam, as well as the friction angle of GC-soil interface, was evaluated. From layer 1, the subsequent layers were heaped step by step, and actual consolidation period for every layer was calculated. In other words, various stress history is converted into a unique stress history, then the properties employed in each layer can be calculated according to its consolidation period. Some constants used for consolidation period conversion could be obtained from simple shear tests.

$$k = 0.0042(m/\min)$$
$$c_v = 8.25\times10^{-6} + 2.72\times10^{-6}\ln(\sigma_c/9.8)(m^2/\min)$$

where k is coefficient permeability, c_v is coefficient consolidation and σ_c is the consolidation pressure.

In the GC45 cm case, a unique c_v is used for every two soil layers. And the conversion of the time factor to

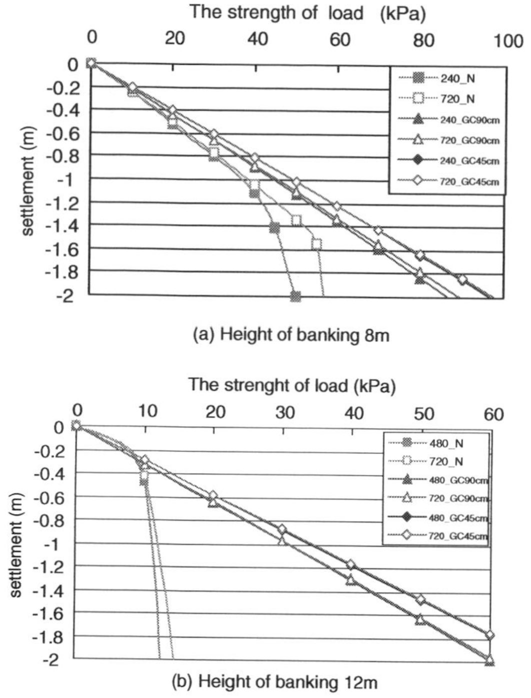

Figure 3. Loading – settlement curves.

real time is required, since strength characteristics of soil and the frictional property between GC (made of no woven fabric commonly) and soil are experimental parameter. They are related to the consolidation degree of soil. Then in a real ground or a test specimen, a unique consolidation degree corresponds to a unique consolidation period. Therefore, they can be converted interchangeably by using the Terzaghi's square law, and the conversion is carried out in the real time.

3 ANALYSIS RESULTS AND DISSCUSIONS

3.1 *Loading-settlement curve*

Loading-settlement curves of banking height of 8 m and 12 m are shown in Fig. 3, respectively. In the cases without geocomposite reinforcement, there is the depression effect of the settlement by lengthening the construction period, as shown in Fig. 3a. It is proven that the loading capability of embankment increases with the consolidation degree. In the 720_N case, the settlement developed rapidly after it was subjected to a surface loading of 50 kPa. It can be inferred that there is a limit in the strengthening effect, even if the construction period is lengthened in the non-reinforced banking.

In the cases with GC reinforcement, the loading-settlement curves are almost the same with those

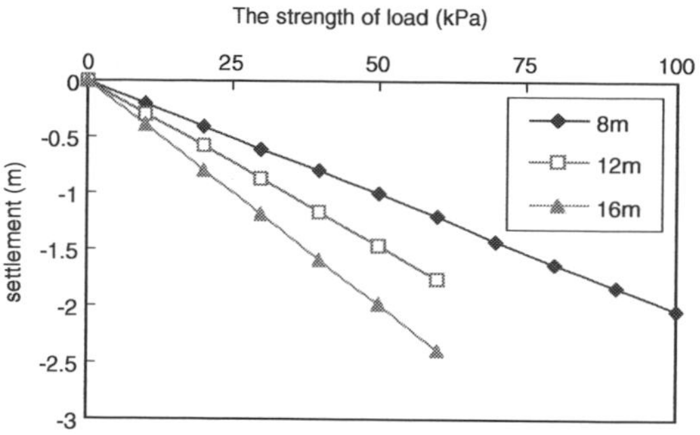

Figure 4. Crown settlements on each embankment heights.

without reinforcement before a loading threshold of 30 kPa. And the slope of loading-settlement curve remained constant even the embankment was subjected to a load of 50 kPa.

It can be inferred from Fig. 3b that the embankment tends to fail after the 10 kPa loading on the cases with reinforcement. From this result, near 12 m seems to be the limitative banking height.

In cases of GC90 cm and GC45 cm, the difference between the reinforcement effects of the settlement is remarkably observed in comparison with those cases with a banking height of 8 m. It can be inferred that there is the strengthening effect from the case in which the load strength is smaller, as the banking rises. The effect of consolidation period is not remarkable, as illustrated in both Fig. 3a and Fig. 3b.

The crown settlements of different height cases with GC45 cm reinforcement are illustrated in Fig. 4. At a same loading, the settlement of the higher embankment is larger.

3.2 *Load strength-deformation of slope*

The curves of loading versus slope deformation for the case with an embankment height of 8 m and 16 m are shown in Fig. 5. In the cases without reinforcement, a large deformation occurred under a surface load of 50 kPa. In the cases with GC90 cm or GC45 cm reinforcement, the horizontal displacement of the slope is suppressed about 50~65%, the crown settlement is also suppressed about 50~65% comparing to those cases without reinforcement. Moreover, there is no large deformation over the foot of the embankment, and it can be inferred that the whole embankment is stable. In the cases without reinforcement, large displacement occurred at the foot of the embankment, while in the cases with reinforcement, the displacement at the middle slope is the largest, which is quite differed from the former ones.

The analyses results of embankment with the height of 12 m are shown in Fig. 5b. In cases without reinforcement, the embankment failed under the surface

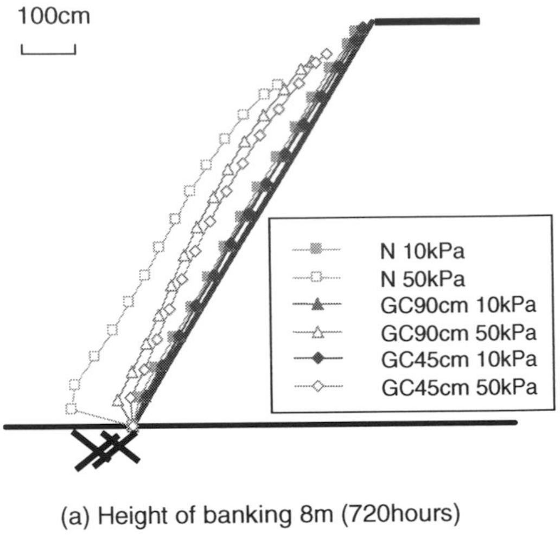

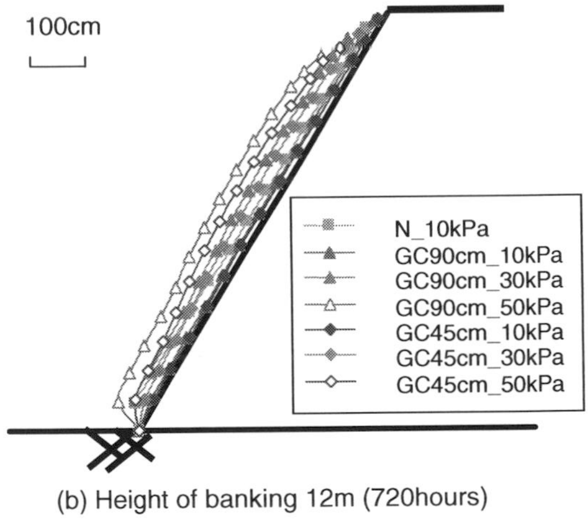

Figure 5. Load strength-deformations of slope.

load of 15~20 kPa. Therefore, there were no more displacement records in following the deformation of the slope. In the cases with reinforcement, it is effective to restrain the deformation by the difference between the laying intervals. In case GC45 cm, the deformation is small, and the deformation at the foot of the embankment is restrained apparently comparing with the case of laying interval 90 cm. The tendency that the stress is concentrated at the foot of slope is observed for the embankment. It seems that the stress concentration at the foot of embankment could be reduced by increasing the laying interval dense. Without a rapid collapse the 12 m high embankments with GC reinforcement can be regarded as stable under a maximum surface loading of 50 kPa.

3.3 *Shear failure region and displacement vector*

The displacement vector and the shear failure region distribution of the 8 m high embankments after a construction period 720 hours are shown in Fig. 6, where the difference can be confirmed from the displacement vector figures. In the case without reinforcement, excessive deformation and plasticity occurred near the slope at the load strength of 50 kPa. Then the embankment failed along this sliding surface. The shear fracture region figure is also depicted in this figure. In comparison with cases of GC90 cm and 45 cm, it is shown that the displacement of the whole embankment is suppressed in the GC45 cm case. In both cases with reinforcement, the displacement at the middle and the foot of embankment is restrained, due to the effect of GC reinforcement.

There is a small failure region at the foot of slope on both cases. The stress concentrates in this part subject to the effect of the load, and the destruction seems to have locally been generated. However, there is no progress of the breakdown region to banking upper part. It is considered that the reinforcement near the foot of slope must be sufficiently considered from this result, when the banking is constructed in the field.

3.4 *Tensile stress the reinforcement material*

The resultant tensile stress in the reinforcement after construction period 720 hours is shown in Fig. 7. The legend is the distance from the embankment bottom. It is proven that the stress is also concentrating each GC near 4~6 m from the slope, which indicates the position of the sliding surface. Especially, a larger tensile stress occurred near 270~540 cm from the embankment bottom, which agrees with the behavior of the displacement vector figure mentioned above. Moreover, the tensile stress locally increases at the foot of embankment right after the slope. This seems to be the effect by the stress concentration to the top of slope, as shown in the distribution diagram of shearing stress. In comparison with both cases, the tensile stress in GC45 cm case is smaller than that in the GC90 cm cases. It tends to be similar even in the case of high fill (12 m, 16 m), and load diffusion effect of each GC seems to stabilize the whole embankment.

3.5 *Limit of embankment height without reinforcement*

The smallest safety factor for the no reinforcement case after a construction period of 720 hours can be calculated and depicted in Fig. 8, which is derived from exponent extrapolation. For example, a height of 11.3 m corresponds to a design safety factor of 1.2. The Limit of embankment height (corresponding to the

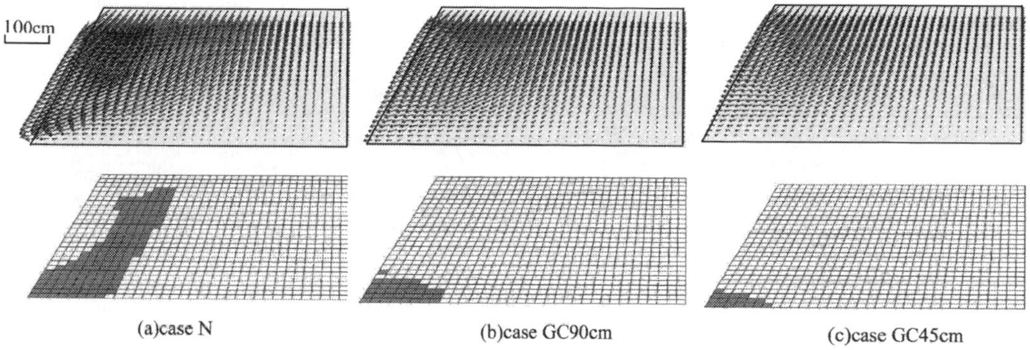

Figure 6. Displacement vector and shear failure region.

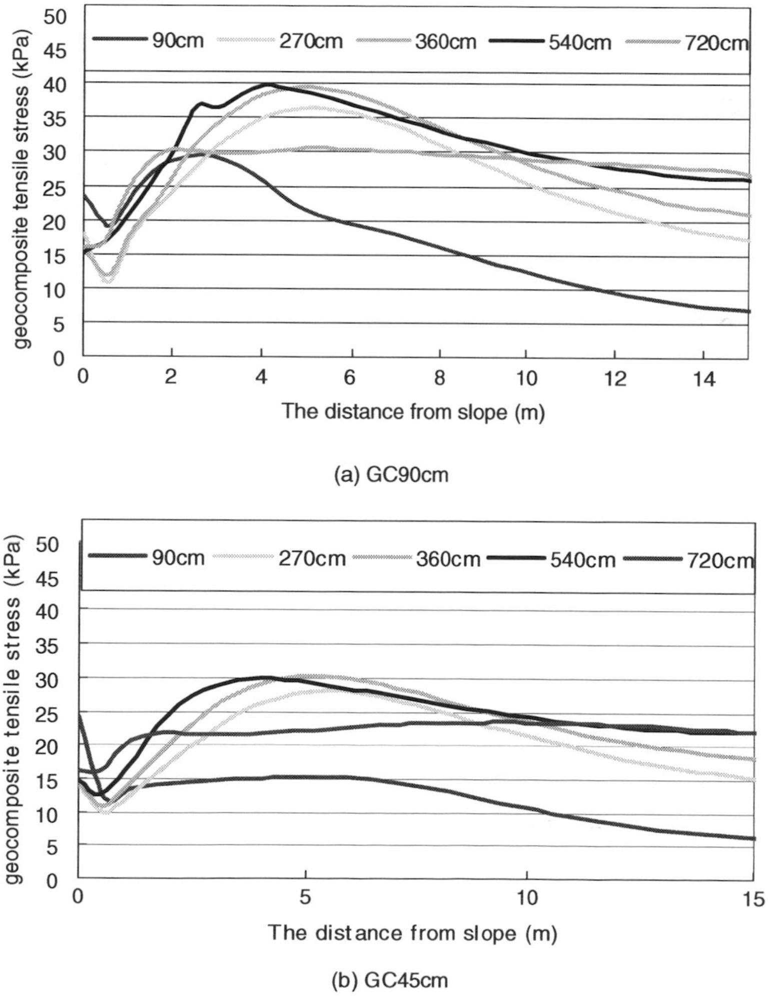

Figure 7. Tensile stresses in the reinforcement material.

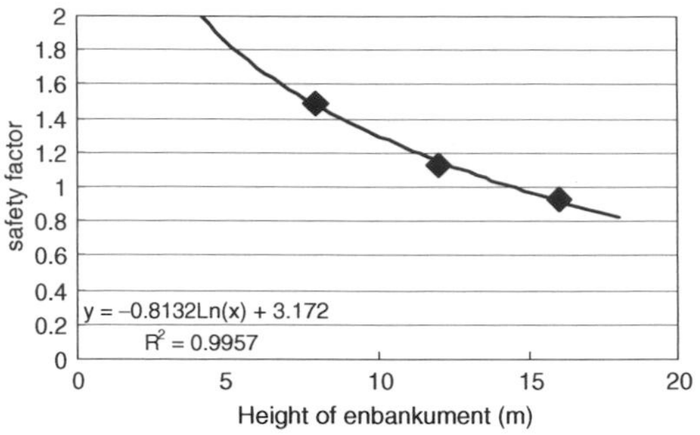

Figure 8. Safety factor against embankment height with out reinforcement material.

safety factor of 1.0) is 14.4 m, which approximately agrees with the results from 12 m high cases.

4 CONCLUSIONS

Considering a steep slope banking with high water content low quality soil, the reinforcement effect of geocomposite is evaluated by using finite difference analysis. The settlement at the crown, the displacement at the slope and the resultant stress in geocomposite are studied in this paper. Some suggestions for the design of geocomposite-reinforced embankment are also proposed. It has been clarified as follows.

This analysis method was able to simulate the increase of strength that related to the consolidation of embankment itself and the consolidation construction period; therefore, it can approximate a real embankment by the analysis.

In case of the reinforced embankment, the effect of restraining the displacement of the slope is much greater than the effect of restraining the settlement of the crown.

In the cases without reinforcement, large displacement occurred at the foot of the embankment, while in the cases with reinforcement, the displacement at the middle slope is the largest, which is quite differed from the former one.

The characteristics of resultant tensile stress in the reinforcement and the stress concentration in the embankment are analyzed, which can help the design of geocomposite-reinforced embankment.

In the future, this analysis method can be verified and generalized by comparing the analyses results with construction case in the field.

REFERENCES

Civil Engineering Research Center (1993): Design and Construction Manual for Geotextile Reinforced Embankment.

Y. Tanabashi and H. Nagashima (2002): Geocomposite design method tentative plan. *Journal of Geotechnical Engineering* JSCE, III-58, pp. 145–153.

Y. Kitamoto and Y. Abe (1999): Deformation of the reinforced embankment which originates for the banking material and countermeasure in the construction. *Journal of Geosynthetics*, Vol. 14, pp. 142–147.

H.G. Poulos and E.H. Daivis (1974): Elastic Solutions for Soil and Rock Mechanics.

New Horizons in Earth Reinforcement – Otani, Miyata & Mukunoki (eds)
© 2008 Taylor & Francis Group, London, ISBN 978-0-415-45775-0

Toughness improvement of hybrid sandwiched foundations and embankment reinforced with geosynthetics

S. Yamazaki, K. Yasuhara, S. Murakami & H. Komine
Department of Urban and Civil Engineering, Ibaraki University, Hitachi, Ibaraki, Japan

ABSTRACT: We have developed a new construction technique called Hybrid Sandwiched Reinforcement (HBS) method in which thin sand layers are placed above and beneath the geosynthetic fabric to increase the mechanical potential of cohesive soil embankment and foundations. This reinforcement method offers advantages of reinforcement improvement and maintenance of hydraulic conductivity. Successive to the authors' previous works, this paper describes small-scale model tests on embankments with and without reinforcement and a sand layer.

Results from model tests are interpreted with emphasis on improved toughness of HBS-reinforced embankments. Results clarified that HBS not only controls embankment deformation; it also improves the toughness using placement of geosynthetics in an embankment comprising cohesive soils such as Kanto loam of volcanic-ash origin. As an important finding related to toughness improvement of HBS earth structures, the model test results show that the improved toughness of foundations and embankments are independent of the sand layer thickness, which implies that the sand thickness in sandwich-type earth structures is sufficient for maintaining the HBS structures' hydraulic conductivity and avoiding clogging of the geosynthetics.

1 INTRODUCTION

Great demand has arisen for effective utilization of high-water-content viscosity soil because of a lack of good soil and difficulty securing construction sites. For those reasons, we have developed a new construction technique called Hybrid Sandwiched Reinforcement (HBS) method, by which thin sand layers are placed over and beneath the geosynthetic fabric (GS) to increase the mechanical potential of cohesive soil embankments and foundations. The HBS method is used together with sand layer not only for protection of non-woven clogging but also for a new function: toughness improvement. This reinforcement method is advantageous for reinforcement improvement and retention of hydraulic conductivity. Both are greater than in cases of reinforcement without the sand layer. Successive to the authors' previous works, for this study, we performed small-scale model tests on embankments with and without reinforcement, and with and without the sand layer. This report specifically addresses toughness improvement. We executed model tests to assess reinforced conditions with sand and GS.

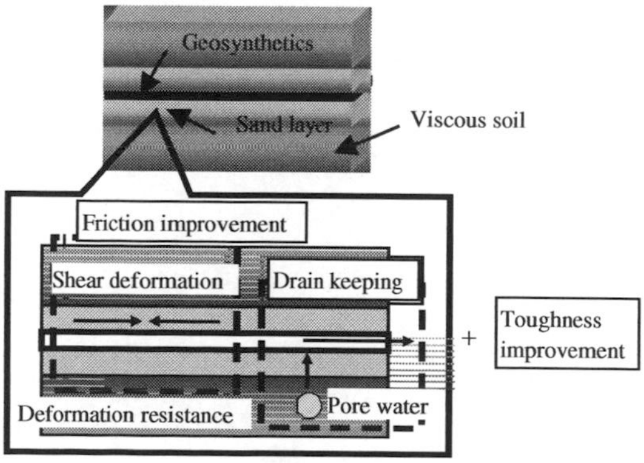

Figure 1.1. Outline of HBS method.

2 TOUGHNESS AND TOUGHNESS VALUE CALCULATION

We define toughness as that which is not prone to incidence of sliding and movement. Figure 2.1 shows the toughness value calculation method. We estimated the

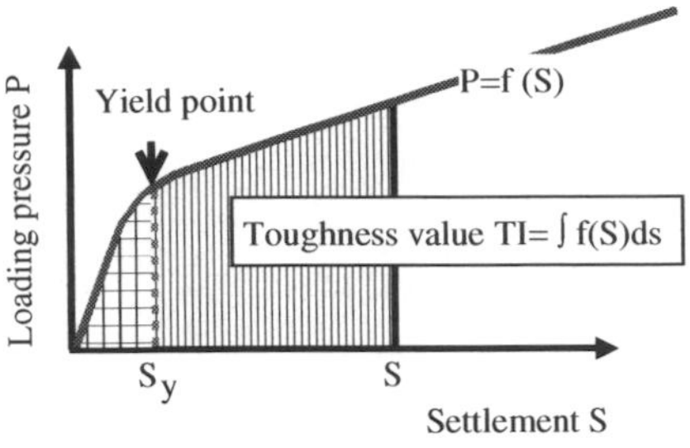

Toughness value to yield point: $TI_y = \int_0^{S_y} f(S)ds$
Toughness value to settlement: $TI_s = \int_0^{s} f(S)ds$
Increase rate of toughness: $E_s = TI_s / TI_y$

Figure 2.1. Toughness value calculation method.

HBS method toughness value (TI) from the loading pressure-settlement curve to involve shear deformation. We defined the area from the original to the yield point as TI_y and from the original to each certain settlement point as TI_s. We calculated the reinforcement toughness values as TI_{ysr}, TI_{ssr} and no-reinforcement toughness values as TI_{ysn}, TI_{ssn}. Concretely, we estimated the rate of toughness improvement (E_{sr}/E_{sn}) using (1).

$$E_{sr} \Big/ E_{sn} = \frac{TI_{ssr}/TI_{ysr}}{TI_{ssn}/TI_{ysn}} = \frac{\int_0^s f(S)_{sr}ds \Big/ \int_0^{s_y} f(S)_{sr}ds}{\int_0^s f(S)_{sn}ds \Big/ \int_0^{s_y} f(S)_{sn}ds} \quad (1)$$

E_{sr}: increase rate of toughness in reinforcement
E_{sn}: increase rate of toughness in non-reinforcement
TI_{ssr}, TI_{ysr}, TI_{ssn}, TI_{ysn}: toughness value in reinforcement and non-reinforcement

3 OUTLINE OF MODEL TEST

3.1 *Test system and measurement method*

This 50-cm-high and 19.5-cm-deep model had a slope gradient of 1 : 0.6. Kanto loam (64.5% initial water content, 0.83 g/cm^3 dry density, 2.77 void ratio, 77.4% degree of saturation) soil was used. The soil was spread, then compacted using a hand vibrator; the slope was reset with a pallet. In addition, Toyoura sand was used in HBS method case as the sand layer with geosynthetics. The loading plate width was 10 cm and the loading speed was 0.2 mm/min. Figure 3.1 shows an outline of a model test. Loading pressure, settlement and slope displacement were measured. Furthermore, the slope displacement was recorded using a video camera. Measured positions were at 20 cm, 25 cm, 30 cm, 35 cm, and 40 cm.

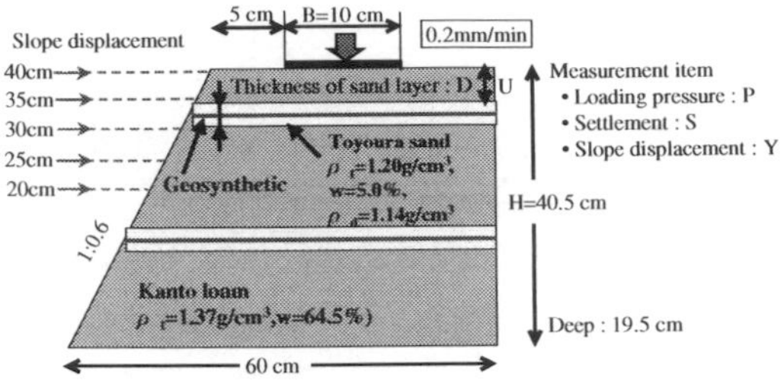

Figure 3.1. Outline of model test.

Table 3.1. Material data of Toyoura sand and Kanto loam.

Toyoura sand		Kanto loam	
Soil particle density	2.64 g/cm^3	Soil particle density	2.72 g/cm^3
Maximum void ratio	0.977	Natural water content	68.7%
Minimum void ratio	0.605	Liquid limit	84.5%
Fine-grained soil content	0%	Plastic limit	61.1%
		Plasticity index	23.4%

Photo 3.1. Non-woven

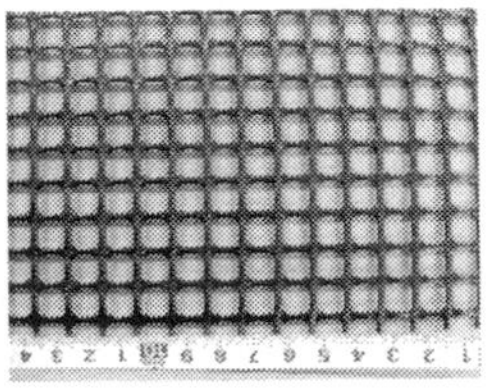

Photo 3.2 Geonet

3.2 *Material of soil and geosynthetics*

3.2.1 *Soil*

We used volcanic Kanto loam as the embankment material and Toyoura sand as the sand layer, for which data are shown in Table 3.1.

3.2.2 *Geosynthetics*

We used two types of GS. One is non-woven (Photo 3.1), with tensile strength of 5.8 kN/m (strain is 110%). Another is Geonet (Photo 3.2), with tensile strength of 3.6 kN/m (strain is 10.6%). Table 3.2 shows GS characteristics.

3.3 *Model test case*

We tested reinforcement effects and toughness improvement effects of the HBS method. Table 3.3

Table 3.2. Characteristics of GS materials.

	Tensile strength	Strain	Coefficient of permeability		Space	Material
			Vertical	Horizontal		
Non-woven	5.8 kN/m	110%	5×10^{-1} cm/s	4×10^{0} cm/s	–	PP
Geonet	3.6 kN/m	10.6%	–	–	10 mm	PE

Table 3.3. Model test case.

Case	Reinforced method	Geosynthetics	Sand thickness	layer	Single layer depth
1	Non-rein forcement	–	–	–	–
2	Non-woven + sand	Non-woven	1 cm	1 (top)	2 cm
3	Non-woven + sand	Non-woven	1 cm	1 (top)	4.5 cm
4	Non-woven + sand	Non-woven	1 cm	1 (top)	7 cm
5	Sand	–	1 cm	1 (top)	4.5 cm
6	Non-woven	Non-woven	–	1 (top)	4.5 cm
7	Sand	–	3 cm	1 (top)	4.5 cm
8	Non-woven + sand	Non-woven	3 cm	1 (top)	4.5 cm
9	Geonets	Geonets	–	1 (top)	4.5 cm
10	Geonets + sand	Geonets	3 cm	1 (top)	4.5 cm
11	Non-woven + sand	Non-woven	1 cm	2	4.5 cm
12	Non-woven + sand	Non-woven	1 cm	3	4.5 cm

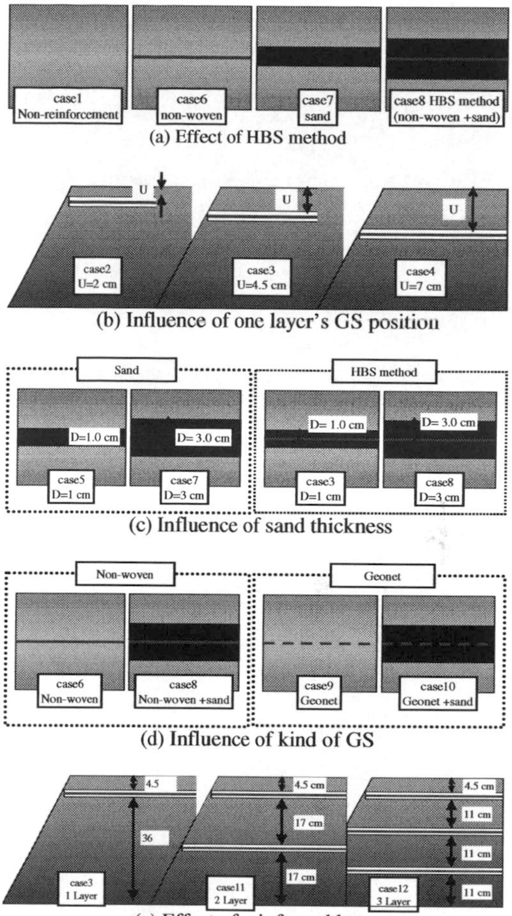

Figure 3.2. Test Case.

shows the test case. As Fig 3.2 shows, we devoted particular attention to effects of: (a) HBS method, (b) one layer's GS position, (c) sand thickness, (d) the GS type, and (e) a reinforced layer.

4 RESULTS AND DISCUSSION

4.1 *Effect of HBS method*

We confirmed the HBS method effects. Fig 4.1 shows settlement of the embankment with loading pressure in cases 1, 6, 7, and 8. Fig 4.2 shows changes in rate of toughness improvement with each settlement. These results show that case 8 of HBS method (non-woven + sand) offers the most rigidity and most rate of toughness improvement. Especially, the rate of toughness improvement in case 8 was 1.8, which contrasts to case 1, with 1.0, in 40 mm settlement. The rate of toughness improvement was about 2 times by using HBS method. These results verify that HBS method prevents clogging of the GS and increases toughness. Fig 4.3 shows slope displacement that occurs with loading pressure in cases 1, 6, 7, and 8. In addition, we calculated the experimental displacement two times to show the difference of displacement. The top of the slope showed large deformation in case 1 (non-reinforced): maximum displacement was about 3 cm. Moreover, the maximum displacement was about 1 cm in case 8 (non-woven + sand). Results confirmed that conditions in case 8 improved rigidity and deformation, in addition to toughness, using HBS method.

4.2 *Influential factors on HBS method toughness improvement*

4.2.1 *Influence of one layer's GS position*

We confirmed the influence of one layer's GS position on toughness improvement. Fig 4.4 shows

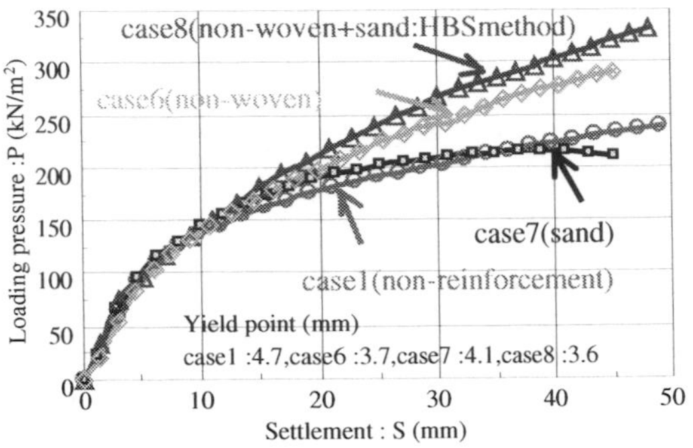

Figure 4.1. Settlement of loading position and loading pressure.

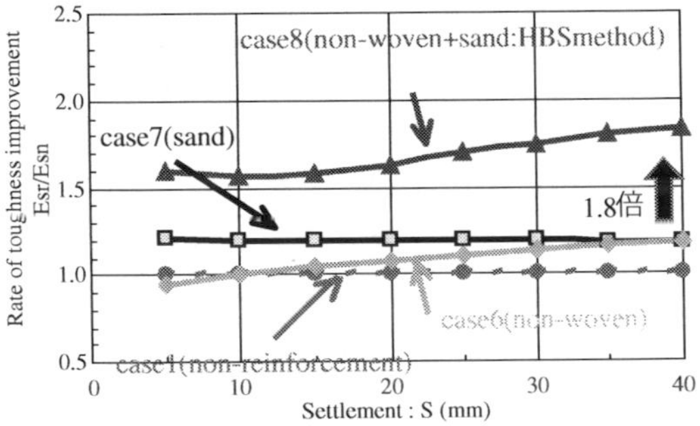

Figure 4.2. Rate of toughness improvement (effect of HBS method).

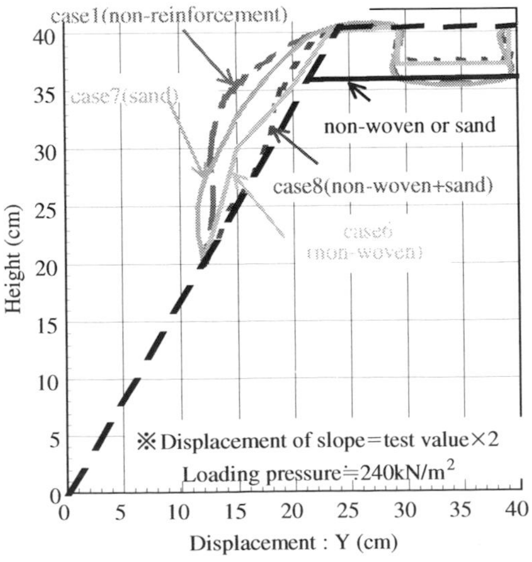

Figure 4.3. Slope displacement (effect of HBS method).

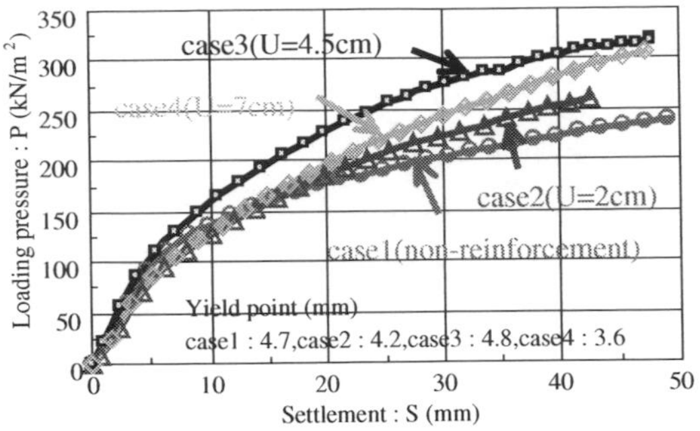

Figure 4.4. Settlement of loading position and loading pressure.

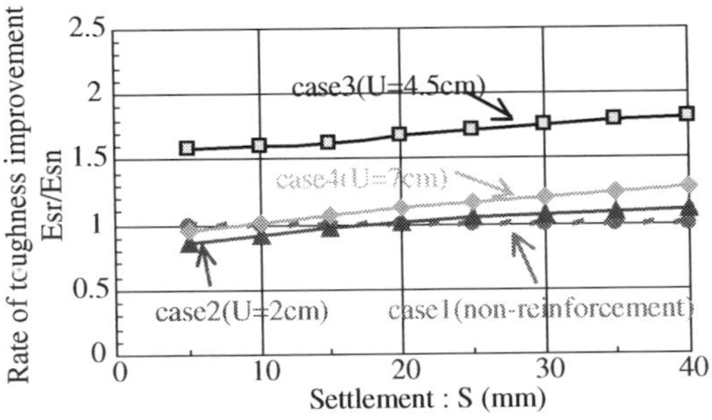

Figure 4.5. Rate of toughness improvement (influence of one layer's GS position).

embankment settlement with loading pressure in cases 2, 3, and 4. Fig 4.5 shows changes in rate of toughness improvement with each settlement. These results show case 3 (4.5 cm reinforcement layer position) as the most rigid and largest effect of toughness improvement. The GS pressure was confirmed as weak in case 2 (2.0 cm reinforcement layer position) because the GS tensile force was not sufficiently secured. We considered that the embankment underwent slope failure above the reinforcement layer position in case 4 (7.0 cm reinforcement layer position).

Fig 4.6 shows slope displacement with loading pressure in cases 1–4. From that result, in case 2 ($U = 2.0$ cm), the maximum displacement was 2.0 cm. The top of the slope shows large deformation that is comparable to those of cases 3 and 4; the GS of case 2 did not control slope deformation. Furthermore, we confirmed 1 cm displacement above the reinforcement layer position in case 4 ($U = 7.0$ cm). However, displacement of case 4 was less than the displacement of case 1 (non-reinforcement). There were reinforcement effects of GS and HBS method in case 3 ($U = 4.5$ cm) where GS is lain in a deep position. Fig. 4.4 demonstrates the effectiveness of HBS method because the rigidity of case 4 actually increased with settlement.

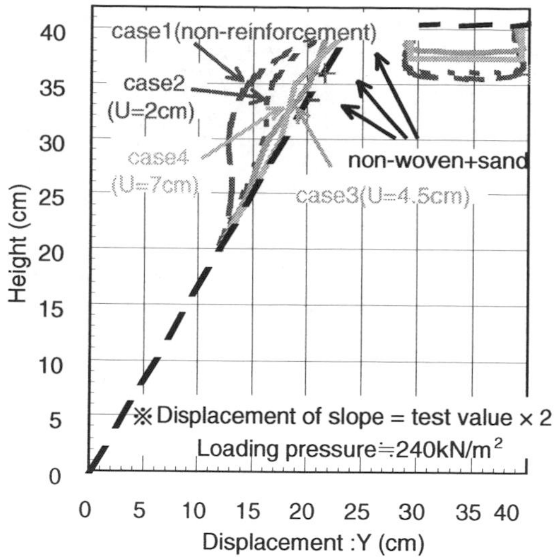

Figure 4.6. Slope displacement (influence of one layer's GS position).

4.2.2 *Influence of sand thickness*

We confirmed the influence of sand thickness on toughness improvement. Fig 4.7 shows embankment settlement with loading pressure in cases 3, 5, 7, and 8. Fig 4.8 shows changes in toughness improvement with each settlement. Cases 3 and 8 use HBS method (non-woven + sand), cases 5 and 7 use only sand. In addition, sand thickness of cases 3 and 5 are 1 cm, sand thickness of cases 8 and 7 is 3 cm. Comparison of sand thickness of 1 cm (cases 3 and 5) to sand thickness of 3 cm (cases 7 and 8) shows that rigidity and toughness improvement rate were equivalent. Therefore, sand thickness showed no difference: if the lay area of GS and sand are equal, the friction characteristic is independent of the sand thickness. However, to maintain drainage characteristics of non-woven GS (clogging prevention), we must distinguish the sand layer thickness from the drainage characteristics.

Fig 4.9 shows the slope displacement with loading pressure in cases 3, 7, and 8. Displacement for case 5 could not be measured because of a measurement error. From this result, the maximum displacement was 0.5 cm in case 3 (sand thickness $D = 3.0$ cm) and 1.0 cm in case 8 (sand thickness $D = 3.0$ cm), implying that displacement was determined by sand thickness. Especially, case 8 showed large displacement at the nearby reinforced layer. This test differs from actual behavior because of the low confining pressure model test. For a thick sand layer ($D = 3.0$ cm), Kanto loam and Toyoura sand were not identical; embankment slip failure might have occurred because of sand layer weakness.

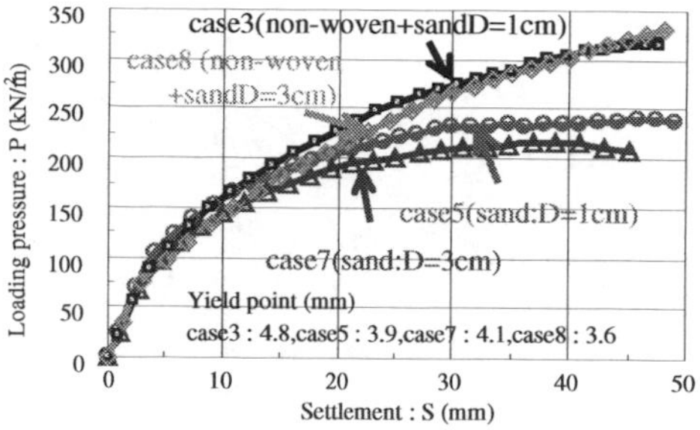

Figure 4.7. Settlement of loading position and loading pressure.

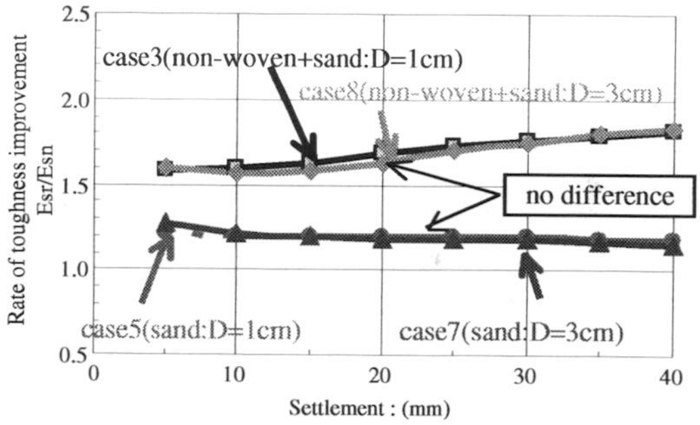

Figure 4.8. Rate of toughness improvement (influence of sand thickness).

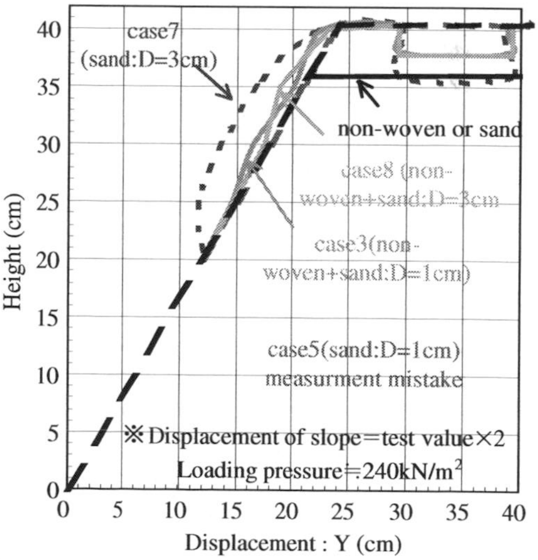

Figure 4.9. Slope displacement (influence of sand thickness).

4.2.3 *Influence of kind of GS*

We confirmed that the GS type influences toughness improvement. Fig 4.10 shows settlement of the

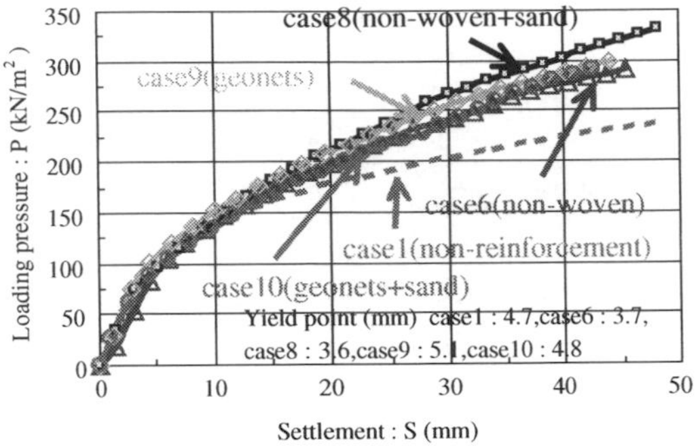

Figure 4.10. Settlement of loading position and loading pressure.

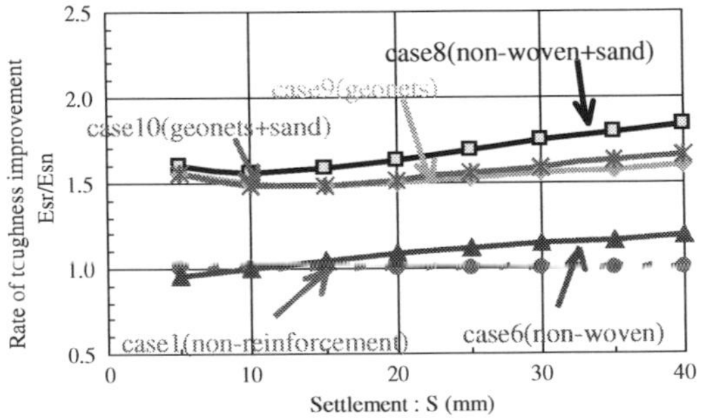

Figure 4.11. Rate of toughness improvement (influence of kind of GS).

embankment with loading pressure in cases 6 and 8–10. Fig 4.11 displays changes of toughness improvement for each settlement. These results show no clear difference of non-woven (cases 6 and 8) and Geonet (cases 9 and 10) for rigidity. However, cases 6, 8–10 have high rigidity compared to case 1 (non-reinforced). Furthermore, toughness improvement of case 8 (non-woven + sand) in 40 mm settlement was 1.8, in contrast to that of case 10 (Geonet + sand) in 40 mm settlement, which was 1.6. Case 8 (non-woven + sand) exhibits a large effect of toughness improvement.

Fig 4.12 shows slope displacement with loading pressure in cases 6, 8–10. The maximum displacement was 2.0 cm in case 10 (Geonet), and the maximum displacement was 3.5 cm in case 1 (non-reinforced). Consequently, embankment displacement was controlled using Geonet. However, because the maximum displacement was 1.0 cm in case 8 (non-woven + sand), we confirmed that using non-woven is most effective for displacement control. This test showed that the interaction effect of sand and GS was superior to that of Geonet, but the possibility exists that Geonet can have a larger carrying capacity using a different kind of sand. This test used effective high-content viscosity soil to advance consolidation and increase strength.

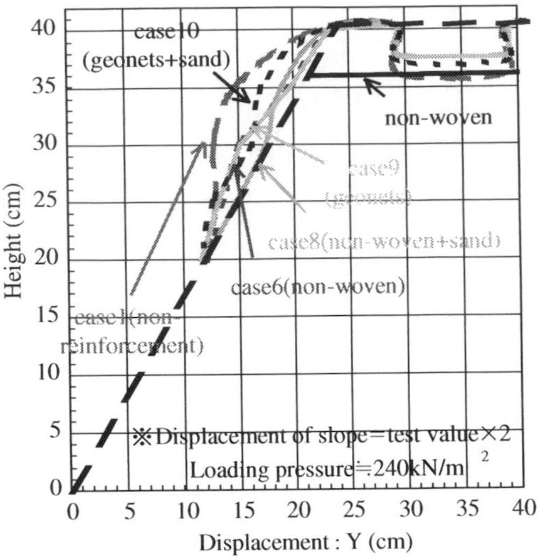

Figure 4.12. Slope displacement (influence of kind of GS).

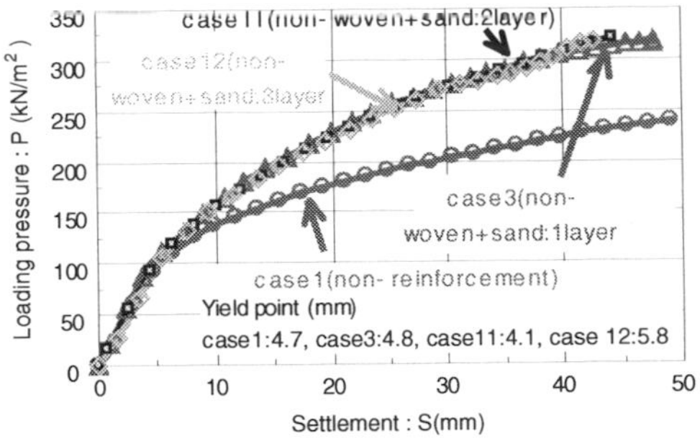

Figure 4.13. Settlement of loading position and loading pressure.

Therefore, we considered that using non-woven and Geonet GS is suitable.

4.2.4 *Effect of reinforced layer*

Results of 4.1 effect of HBS method confirmed the effect of HBS method. Next, we studied the number of reinforced layers' influence on toughness improvement. Fig 4.13 shows embankment settlement of with loading pressure in cases 1, 3, 11, and 12. Fig 4.14 shows changes of toughness improvement with each settlement. Rigidity is nearly equal and toughness improvement is nearly equal in cases 3, 11, and 12. There was no increase of rigidity and toughness value (carrying capacity) by changing the reinforced layer. Fig 4.15 shows slope displacement with loading pressure in cases 1, 3, 11, and 12 because slope displacement of cases 3, 11, and 12 was smaller than that of case 1 (non-reinforced).

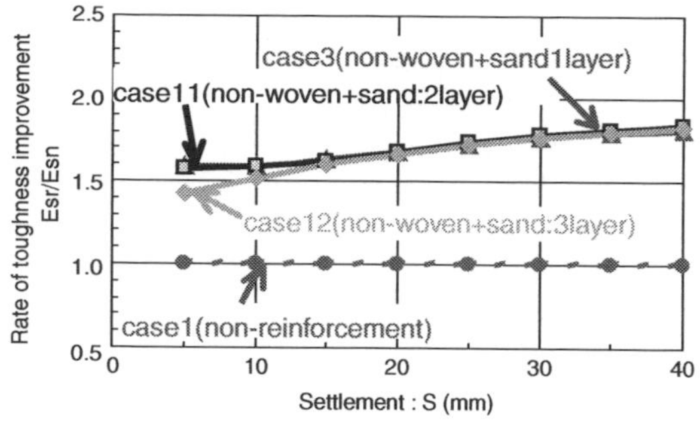

Figure 4.14. Rate of toughness improvement (effect of reinforced layer).

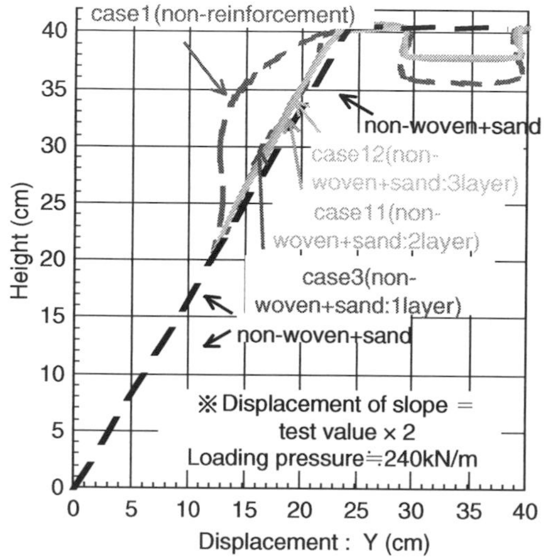

Figure 4.15. Slope displacement (effect of reinforced layer).

Results show that deformation was controlled, but no difference was apparent from changing the reinforced layer. From Fig 4.16, we considered the position of the sliding surface was less than that of the second layer. Therefore, we must explore numerous reinforcement layers and positions of reinforcement layer in reinforcement function (carrying capacity) and drain function in future studies.

5 CONCLUSION

From the results described in this paper, the following are concluded.

(1) Results clarified that HBS improves toughness more markedly than conventionally adopted placement GS in clay embankments.

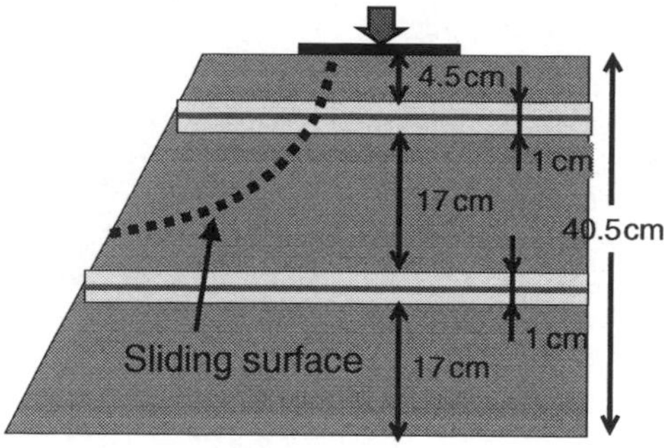

Figure 4.16. Sliding surface (case 11).

(2) Effects of toughness improvement did not depend on sand thickness in these model tests. It is an important subject to decide sand thickness not to decrease drain characteristics of non-woven materials.
(3) Embankment material and friction of sand and GS exert a large influence.
(4) These results are inferred to be applicable to conditions at many sites.

REFERENCES

K. Yasuhara., Chandan Ghosh., T. Sakakibara., S. Murakami., H. Komine., (2004); "Advantage Aspects of Hybrid-Sandwich type Reinforced Earth", Proc. of Geosynthetics Engineering Journal, Vol. 19, pp. 139–146 (in Japanese).

T. Sakakibara., K. Yasuhara., S. Murakami., H. Komine., (2005); " Toughness Improvement Using Hybrid-sandwich Earth Structures with Geosynthetics", Proc. of Geosynthetics Engineering Journal, Vol. 20, pp. 81–88 (in Japanese).

K. Yasuhara., S. Murakami., H. Komine., T. Sakakibara., (2006); "Hybrid-sandwiched foundations reinforced with geosynthetics", Proc. of Geosynthetics 8th ICG, pp. 1005–1010.

S. Yamazaki., K. Yasuhara., T. Satou., (2006); "Effect of Improvement Toughness by Hybrid-Sandwich Method", Proc. of Japanese Society of Civil Engineers 61th, pp. 79–80 (in Japanese).

S. Yamazaki., K. Yasuhara., (2006); "Influencing Factors on Toughness Improvement of Hybrid Sandwich Reinforced Embankment", Proc. of Geosynthetics Engineering Journal, Vol. 21, pp. 81–88 (in Japanese).

New Horizons in Earth Reinforcement – Otani, Miyata & Mukunoki (eds)
© 2008 Taylor & Francis Group, London, ISBN 978-0-415-45775-0

Electrokinetic soil nailing for the strengthening or repair of failures of clay slopes and cuttings

J. Lamont-Black, D. Huntley & C.J.F.P. Jones
Electrokinetic Ltd., United Kingdom

S. Glendinning & J. Hall
Newcastle University, United Kingdom

ABSTRACT: Traditional methods for the repair of cuttings have included slackening the slope by the provision of dwarf walls at the toe or the acquisition of additional land, although the latter is seldom possible. Other methods of repair can include soil nailing and/or the provision of reinforced soil berms to increase stability. As the failures are predominantly caused by the development of residual shear strength conditions or uncontrolled pore water pressures, an obvious remedial method would be to effect a reduction in pore pressures and an increase in the shear strength of the soil. This can be achieved by electroosmosis as part of an electrokinetic treatment. Electroosmosis can be used, either to aid construction of remedial works or as a means to effect permanent improvement. A major advantage of the process is that it can be installed without excavation of the cutting and with the use of limited equipment. Treatment is rapid and the strengthening can be permanent.

1 INTRODUCTION

There are a growing number of shallow slip failures in existing railway cuttings and embankments involving fine-grained soils and an expectation that this problem will increase with climate change. Failures in cuttings pose particular problems, not the least of which are the need to maintain rail traffic and the difficulty of access for large/heavy plant. Traditional methods for the repair of cuttings have included slackening the slope by thc provision of dwarf walls at the toe or the acquisition of additional land, although the latter is seldom possible. Other methods of repair can include soil nailing and/or the provision of reinforced soil berms to increase stability. The paper considers the use of electrokinetic geosynthetic (EKG) drains and electrokinetic soil nails to strengthen failing cuttings and slopes and provides details of the design and analytical procedures that can be adopted together with details of a case history.

1.1 *Use of electrokinetics in slope stabilisation*

Cassagrande (1983), explained the benefits of using electroosmosis in civil engineeing when he wrote: "Electroosmosis when applied to fine grained soils is an effective method for increasing strength. This increase is principally the result of tension produced in the pore water as the water content is reduced and menisci are formed, and is also partially due to bonding and cementation of the soil particles.

Electroosmosis is also effective for control of seepage forces, which is highly beneficial for stabilizing slopes in fine grained cohesionless soils.

For cohesive soils, electroosmosis usually results in an increase in the plasticity index, especially near the anode. In combination with a reduction in the water content, the change in plasticity more quickly reduces the liquidity index, and thus more effectively increases the strength of the soil.... The coefficient of electroosmostic permeability (k_e) may be 2 or 3 orders of magnitude greater than the hydraulic coefficient of permeability (k_h), which explains the relatively rapid benefits of electroosmosis".

The difference in the coefficients of electroosmotic and hydraulic permeabilities referred to by Casagrande and which lends such and advantage to electroosmosis when applied in fine grained soils is illustrated in Figure 1. Despite the apparent benefits of electroosmosis and the success of some historical applications, the limitations of conventional electrodes have meant that the technique has received relatively little attention in geotechnical engineering. These limitations related to problems associated with the electrodes including, corrosion, build-up of surface electrical resistance, excessive power consumption, poor drainage of water, undesired entrapment of gas, and the inability to effect polarity reversal.

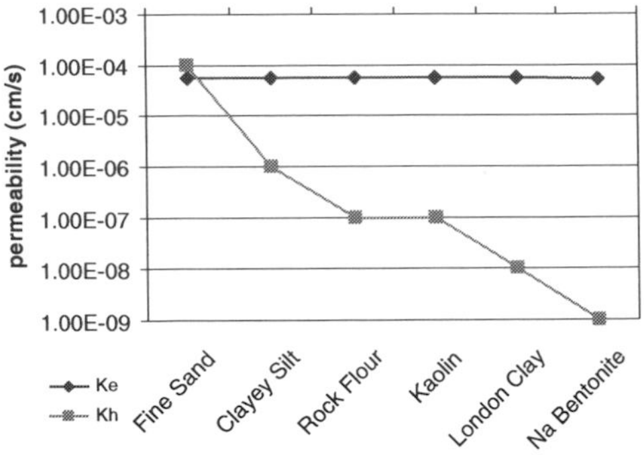

Figure 1. The coefficients of electroosmotic permeability k_e and hydraulic permeability k_h for a range of soils. Note that whilst k_h decreases exponentially with decreasing grain size k_e remains effectively constant across all grain sizes.

EKG technology has been developed to overcome the practical problems associated with the application of electroosmosis (Jones*et al.*, 2005).

2 EKG TO STABILISE CLAY SLOPES

The primary objective in the maintenance of slopes is to identify potential failing slopes and to return them to full stability before failure occurs. In the case of railway cuttings this can be difficult, as although the identification of potential failures is possible, treatment is difficult. The use of heavy equipment at the top of the slope is not desirable as it could initiate failure and treatment of the cutting from track level results in disruption of trains and the incurrence of fines.

As the failures are predominantly caused by the development of residual shear strength conditions or uncontrolled pore water pressures, an obvious remedial method would be to effect a reduction in pore pressures and an increase in the shear strength of the soil. This can be achieved by electroosmosis. Electroosmosis can be used, either to aid construction of remedial works or as a means to effect permanent improvement. A major advantage of the electroosmosis process is that it can be installed without excavation of the cutting and with the use of limited equipment. Treatment is rapid and the strengthening can be permanent.

In using EKG to stabilise clay slopes, EKG treament offers immediate effects and long-term benefits. The concept of EKG strengthening of a slope is shown in Figure 2. The orientation of the electrodes is selected to intercept any potential failure plane. Electoosmotic dewatering of the slope results in a reduction in pore water pressure and an increase in the shear strength of the soil, reducing the risk of a slip plane developing.

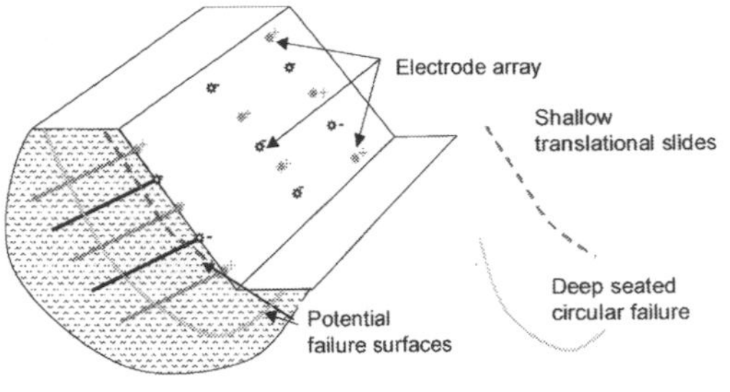

Figure 2. Electrokinetic strengthening of slopes.

2.1 *Immediate effects*

The immediate effect of the application of an electrokinetic force is a reduction of pore water pressure and a reduction of water content. Reductions in pore water pressure and electroosmotic flow are given by, (Mitchell 1993);

$$u = - k_e/k_h . g_w . V \quad (1)$$

$$Q = k_e . V/L . A \quad (2)$$

where: u = porewater pressure, k_e = coefficient of electroosmotic permeability, k_h = coefficient of hydraulic permeability, g_w = density of water, Q = electroosmotic water flow, V = voltage applied between electrodes, L = spacing between electrodes (anode – cathode), A = area.

During soil treatment, electroosmotic flow is independent of hydraulic permeability and the degree of negative pore water pressure or suction that builds up is proportional to the ratio of the coefficients of electroosmotic and hydraulic permeabilities. Therefore, electroosmosis is most effective in fine-grained soils such as clays and silts. By adjusting the parameters of electrode spacing and voltage control, different factors can take priority. For example, if treatment time is critical then the use of close electrode spacing is appropriate. On the other hand if cost is the main driver a wider spacing of electrodes can be used to reduce the number of electrodes and spread the treatment out over a longer duration, Figure 3.

2.2 *Long-term benefits*

By choosing the appropriate combination of anode and cathodes and a specific electrode array or style of installation, the treatment will provide the following long-term benefits, Figure 3.

2.2.1 *Reinforcement*

The anodes remain in-situ as stiff soil nails. The effects of the soil nail array on slope stabilization factors of safety can be analyzed using standard slope engineering software.

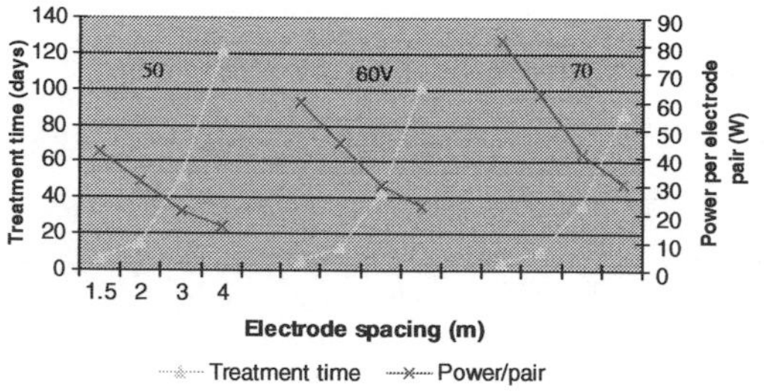

Figure 3. Relationship of electrode spacing applied voltage and treatment time for a slope stabilization project.

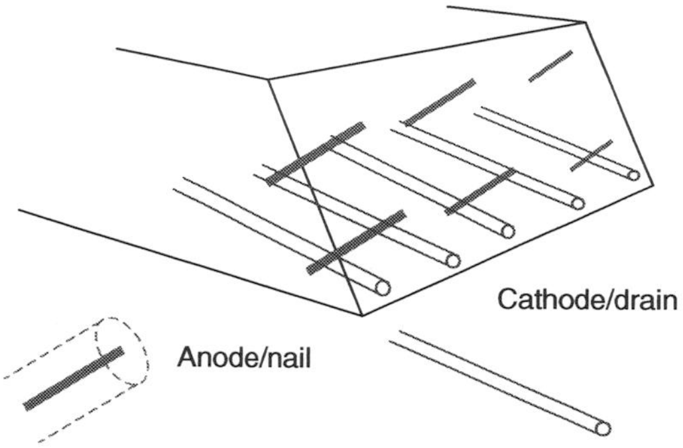

Figure 4. Arrangements of EKG electrodes to stabilize slope and then act as long term soil nails and drains.

2.2.2 *Cementation*

The anodes can be formed to have a partially sacrificial function which will produce several distinct effects:

- ions precipitated from solution cement the clay around the anode thus stiffening the clay and forming a 'mini pile'; effects such as these have been noted (Milligan, 1994).
- create a very strong bond with the anode/nail.
- the formation of a larger effective surface area for the nail thus increasing its pullout capacity.

2.2.3 *Drainage*

Cathodes can be inserted into the ground and oriented so as to act as permanent drains after active electrokinetic treatment has ceased, Figure 4. This sets up an irregularly shaped electric field and attention has to be given when designing the array to focus the strongest parts of the electric field on the weakest zones of soil in the slope.

2.2.4 *Analysis*

The following analytical procedure can be assumed base on (HA 68/94: 1994).

1. Determine slope geometry and soil properties including the ground water flow conditions.
2. Identify potential failure planes.
3. Identify the relationship between shear strength and water content of the soil in the cutting.
4. Select target shear strength to produce a stable cutting.
5. Compute the reduction in water content to provide shear strength in (4).
6. Select electrical layout.
7. Determine treatment time and current design.
8. Following electro kinetic treatment reanalyse with the anode electrodes operating as soil nails (base bond strength nail/soil on an estimation of the residual negative pore water pressure).

2.2.5 *Depth of treatment*

Slope stabilization with EKG can be designed to treat either shallow translational slides or deep circular failures. With shallow failures the top 2 m of soil can be treated. With deep circular slips, field data and the results of global stability analysis can be used to identify a target depth requiring treatment. The electrode array can then be installed with electrical insulation around the upper parts of the electrodes to ensure their targeted action.

3 CASE STUDY

A number of cuttings on Brunel's London – Bristol main railway line constructed circa 1850 are cut at approximately 35° into clay materials. Recently these slopes have shown repeated movement, especially after heavy rain. Factors involved in the instability are thought to be a combination of: high pore pressure in the clay; high water content in the clay; and the ineffectiveness of counterfort drains (which were installed as a measure to improve slope drainage). A number of slope remediation techniques have been implemented on the site, including weighting of the toe of the slope, and excavation and trimming the slope to a shallower angle. These methods have been generally successful but they have limitations in that they involve:

- Significant time to mobilise and complete.
- Do not deal with the core problem i.e. the high weakness of the clay itself and its high pore water pressures.
- Require heavy plant and machinery, which may pose safety hazards when situated at the top of a moving slope.
- Taking possession of the track thus incurring cost penalties.

3.1 *EKG treatment*

Stabilization of the cutting in the area of the counterfort drains is being undertaken using a single treatment which combines electroosmotic stiffening of the soil, electroosmotically enhanced soil nails and permanent

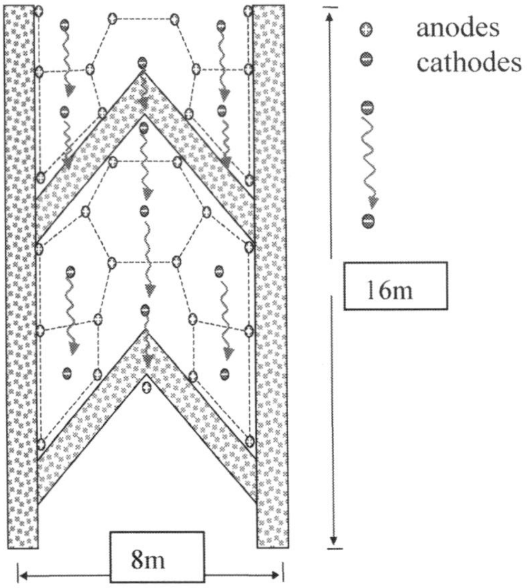

Figure 5. EKG treatment array, based on a hexagonal array which has been adapted to fit within the layout of a slope panel.

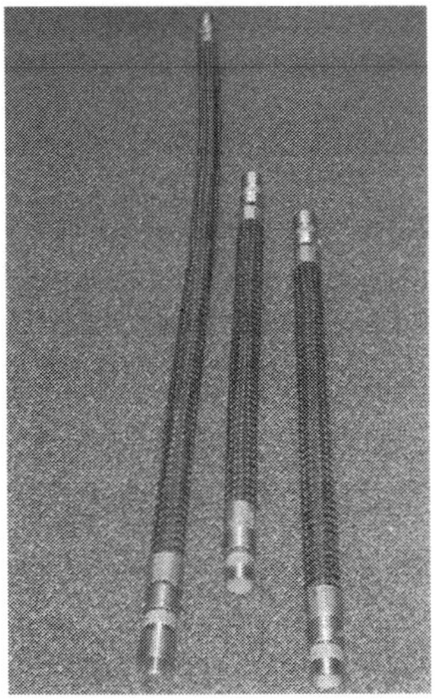

Figure 6. EKG cathode and drain.

drainage. The objectives of the remedial treatment are both short and long term. The short term objectives are to achieve stability and reduce the pore water pressures. The long term objectives are to increase stability of the cutting, improve drainage and improve the strength of the Lias clay forming the cutting.

Calculations using assumptions of the site characteristics show that treatment using an electrode spacing of 2 m and an applied potential of 60 V would produce a theoretical negative pore water pressure of – 300 kPa and a 2% reduction in water content in 6–10 days. Figure 5 shows the layout of the anode and cathode electrodes relative to the existing counter-fort drains into which the cathodes discharge. Figure 6 shows the form of an EKG electrode/drain.

REFERENCES

Casagrande, L. (1983) "Stabilisation of soils by means of electro-osmosis, state-of-the-art." *Journal of the Boston Society of Civil Engineers Section,* **69**(2), 255–302, ASCE.

Design Manual for Roads and Bridges (1994) "Design Methods for the reinforcement of highway slopes by reinforcing soil and soil nailing techniques", HA 68/94, *Geotechnics and Drainage, Part 4.*

Glendinning, S., Jones, C. J. F. P. & Pugh, R.C., (2005) Reinforced soil using cohesive fill and electrokinetic geosynthetics (EKG), *International Journal of Geomechanics,* **5** (2), 138–146.

Milligan, V. (1994) "First application Of Electro-Osmosis To improve Friction Pile Capacity-Three Decades Later", *13th Int. Conf. on Soil Mech. and Found. Eng.,* **5**, 1–5, New Delhi, India, Pub. Balkema.

Mitchell, J.K. (1993). *Fundamentals of Soil Behaviour*. 2nd Edn., Pub. John Wiley & Sons Inc, New York, USA.

New Horizons in Earth Reinforcement – Otani, Miyata & Mukunoki (eds)
© 2008 Taylor & Francis Group, London, ISBN 978-0-415-45775-0

Stability analysis of a new type of reinforced earth slope

Y. Yokota
Maeda Kosen Co., Ltd., Fukui, Japan

K. Arai
University of Fukui, Fukui, Japan

S. Tsuji
Maeda Kosen Co., Ltd., Fukui, Japan

H. Ohta
Tokyo Institute of Technology, Tokyo, Japan

ABSTRACT: In 1999, a high and steep earth slope was constructed which have a maximum height of about 28 m and a slope gradient of 1:1.0. During the project, field observation was performed since the project involved widening of an old embankment that consisted of unknown soil (muck) in a steep topography with an elevation difference exceeding 50 m. To ensure the long-term stability of the embankment and protect the environment, TOGA wall method was applied, which reinforces the slopes by the compression force of geogrid reinforcements. TOGA is the name of place where the method was applied for the first time. The slope inclination and the arrangements of the reinforcements were determined using a standard manual of geogrid reinforced soil structures (Public Works Research Center, 2000). A FE analysis was carried out for predicting failure types by considering the deformation of the ground, and the analysis well duplicated the behaviors of the embankment monitored during the field observation. The results of FE analysis and the limit equilibrium method using circular slip surface method are compared to examine the validity of the soil parameter used for the embankment, and a design method to be used in the future is proposed.

1 INTRODUCTION

In 1999, a geogrid reinforced earth slope in which was 28 m high and 1:1.0 gradient of slope was designed and constructed as the widening of old embankment consisted of unknown soil (muck) in a steep topography. The usual slope gradient (1:1.5 to 1:2.0) would result in a bank extending long and spreading very wide at the toe of the slope, and would require a huge quantity of soil to bring in and adversely affect the natural environment. The bank was decided to be as tall as about 30 m, have a steep slope, and be constructed without excavating the natural ground. Since the geological and topographical conditions were complex and the bank was large, various methods were investigated to ensure the long-term stability. TOGA wall method was decided, which reinforces the bank slopes by the compression force of geogrid reinforcements. At the time, geogrid reinforcements had little been used as permanent structures of tall banks. Loosening of the soil, deformation of the slope, and surface failure during an earthquake and/or by seepage of rainwater were concerned for since the bank slopes could not be thoroughly compressed and the pressure of the bank was small. Thus, vertical compression force was applied to the slope sections of the terraced bank slope (each step had a height of 1 m, a width of 0.8 m, and a gradient of 1:0.2), to increase the soil density, control the expansion of soil volume during shear, and improve the shear resistance and stiffness.

The arrangements of geogrid (installation intervals and length) were determined using the standard manual of geogrid reinforced soil structures (called the standard manual), in which circular slip surface method is used. Cohesion was considered as a design parameter of the banking materials to reflect the reinforcement effects of TOGA wall method and the effects of rolling compaction. Compared with the result of a design made on a simple calculation of soil tests, TOGA wall design was very economic but was on the dangerous side. Thus, a field observation was conducted to examine the long-term stability of the bank.

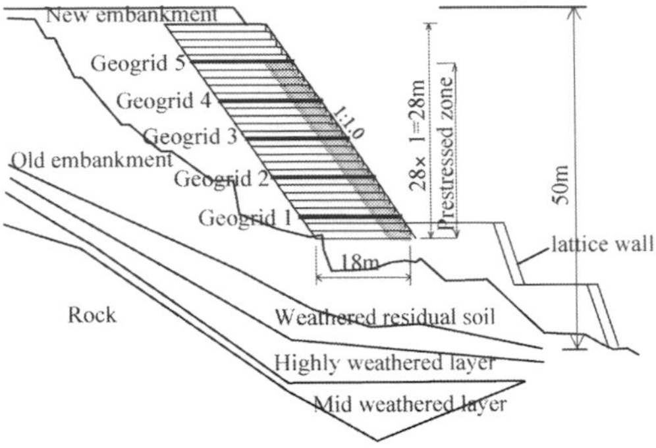

Figure 1. Cross section of TOGA wall.

Figure 2. Overall view of TOGA wall.

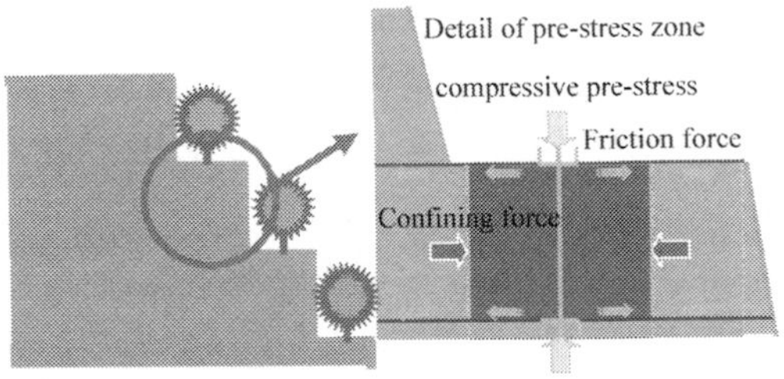

Figure 3. Mechanism of reinforcement.

This paper describes the results of a FE analysis carried out to estimate the failure types by considering the deformation of the ground and a comparison of the results with the data monitored in the field. The long-term stability of the bank was assessed also using data monitored in this study. The failure types determined by the FE analysis were compared with the results of the design method described in geogrid design manual using circular slip surface method. A design method to be used in the future is also proposed.

2 OUTLINE OF CONSTRUCTION

2.1 *Geogrid reinforced earth slope*

This project was executed as a part of a project for constructing a 317 m long highway. The project involved construction of TOGA wall of a height of 25 to 30 m, a lattice wall of a height of 15 m to 20 m, and cutting soil of about 50,000 m^3. A cross section of TOGA wall is shown Figure 1. The foundation ground was broadly classified into talus deposit, the old embankment and the natural ground. The old embankment was difficult to excavate since there were surplus earth and earth filling (SPT-N value is about 5 to 10) of the previous project extending over the embankment. The overall sliding through the old embankment or talus deposit was decided to be controlled by counterweight fill using lattice walls. The inclination of the bank slope of TOGA wall method was 1:1.0 to effectively use the earth discharged during the work and to ensure the stability as a permanent structure of high seismic performance and durability. The shape of the slope was decided to be terraced and consist of steps of a height of 1 m, a width of 0.8 m, and an inclination of 1:0.2 to structurally control compression force and protect the environment by planting vegetation as shown in Figure 2 (Ohta, H. et al., 2006).

TOGA wall has two features. One is the step form in which slope face consists of steel panel of 1 m high and flat area of 0.8 m wide. Another feature is, by adding compressive pre-stress vertically to the about 2 m area sandwiched by geogrids in the flat area using steel plate and steel bar, to suppress volume expansion (positive dilatancy) generated at the occurrence of shearing deformation of the soil (Figure 3). The slope face helps maintenance as the space that is tense again when the pre-stress being relaxed. The embankment of the area sandwiched by geogrids and received pre-stress remarkably increase the shear resistance, and its ductility is drastically improved. A zone of this slope face acts as pseudo-wall of soil and adds shear resistance against slipping and retaining effect by the weight of wall surface body. As a result, from the generation of the apparent cohesive strength with the increase of slip protection force and the increase of lateral pressure, the decrease of necessary tensile strength of geogrid and the shortening of laying longitude are expected.

2.2 *Soil parameters of embankment and foundation ground*

Soil parameters used in the stability analysis are shown in Table 1. Where γ: unit weight, c: cohesion and ϕ: angle of shear resistance. For the embankment materials, direct shear test (constant volume) was carried out for the actual gravel soil used for banking. Considering long-term stability, the shear strength in drainage condition was used in the stability analysis. We determined the strength parameters of banking material considering effect of pseudo-wall of TOGA wall. Although cohesion of banking material $c = 0$ kN/m^2 determined by the direct

Table 1. Soil parameters.

	γ (kN/m^3)	c (kN/m^2)	ϕ (degree)
New embankment	19.0	0(30)	30.0
Old embankment	19.0	0	38.0
Weathered residual rock	18.5	50	23.3
Highly weathered layer	20.0	60	23.3
Mid weathered layer	20.0	70	23.3
Rock	23.0	0	42.5

Table 2. Comparison of design parameters.

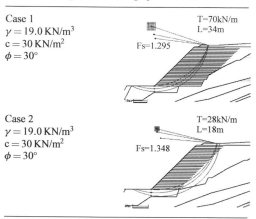

Case 1 $\gamma = 19.0$ KN/m^3 c = 30 KN/m^2 $\phi = 30°$	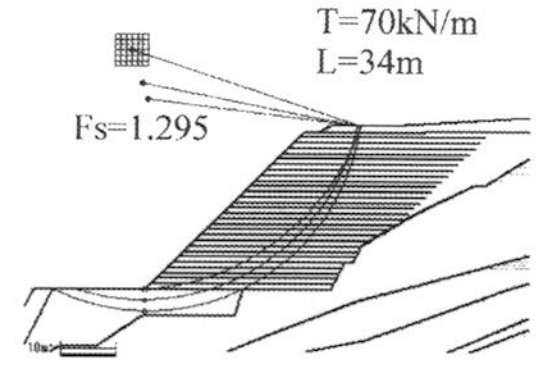
Case 2 $\gamma = 19.0$ KN/m^3 c = 30 KN/m^2 $\phi = 30°$	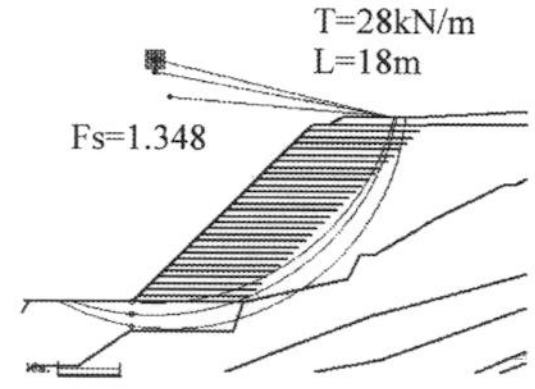

shear test, we assume to $c = 30$ kN/m^2 by considering confining effect and soil compaction effect of geogrid reinforced zone. We designed TOGA wall using strength parameters $c = 30$ kN/m^2 and $\phi = 30°$. The designs using $c = 0$ kN/m^2 and $c = 30$ kN/m^2 are compared in section 2.3. Validity of value of c is verified in section 4.3 by applying FE analysis and field observation. For the foundation ground, soil parameters were determined by laboratory test using undisturbed specimen and on standard penetration test.

2.3 *Geogrid arrangement*

As geogrid reinforcement, we used the aramid fiber reinforced geogrid. It has high strength, low ductility (4 to 6%), small deformation, and the design tensile strength of 27 kN/m. To determine the arrangement of the geogrids, a slope stability analysis was carried out which involved examining the inner stability against fracture and pull-out using the method described in the standard manual, and the overall stability was examined by investigating all slip surfaces including the slips in and outside the reinforced zone. Since the bank to built was as tall as $H = 28$ m, value of cohesion was changed from $c = 0$ kN/m^2 to $c = 30$ kN/m^2 in designing the earth slope since the former gave a very long geogrid length of $L = 34$ m, which was difficult to excavate. The results of the slope stability analysis are shown in Table 2 for each soil parameter.

Table 3. Results of field observation (Ohta, H. et al., 2006).

Tensile force of geogrid	Strain gauge	Force increases 9 kN/m during construction and reaches 32 kN/m after introducing prestress.
Force of steel Bar	Strain gauge	Force acting on steel bar is reduced to 1/3 after introducing prestress.
Earth pressure	Earth pressure meter	Horizontal earth pressure in the prestressed zone is smaller than active earth pressure calculated by earth pressure theory.
Displacement of foundation ground	Clinometer	Foundation ground deforms 50 mm horizontally.
Displacement of foundation ground	Optical distance measuring	Vertical displacement is 63 cm and horizontal displacement is 37 cm in middle part of slope.

2.4 *Field observation*

Field observation was carried out for 5 data items such as tensile force of geogrid, stress of steel bar for introduction of compressive force, vertical and horizontal earth pressures for each pre-stressed zone, displacement of the foundation ground and the toe of slope, and displacement of the slope face. The results of field observation at the completion of embankment are shown in Table 3.

3 NUMERICAL ANALYSIS

The soil is expressed as a plane strain element, and the shift between the soil and the wall blocks is expressed in an interface element. Mohr-Coulomb failure criterion is applied to the soil, and Coulomb failure criterion is applied to the interface element (Desai, C. S. et al. 1984).

Mohr-Coulomb:

$$F_M = \left\{(\sigma_x - \sigma_y)^2 + 4\tau_{xy}{}^2\right\}^{1/2} - \left\{(\sigma_x + \sigma_y)\sin\phi + 2c\cos\phi\right\} = 0 \quad (1)$$

$$\text{Coulomb}: \; F_C = |\tau| - c - \sigma_n \tan\phi = 0 \quad (2)$$

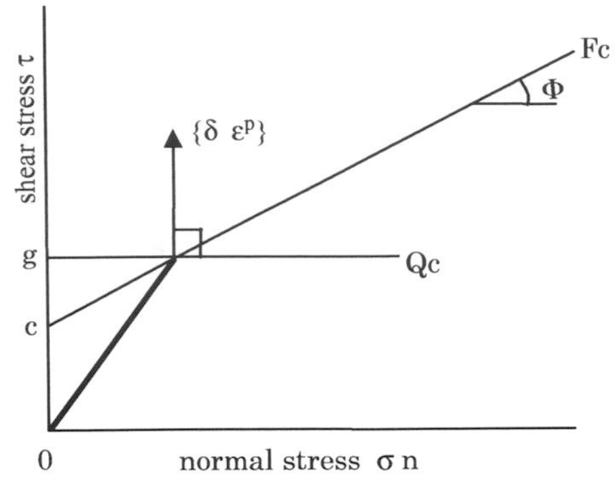

Figure 4. Yield surface and flow rule.

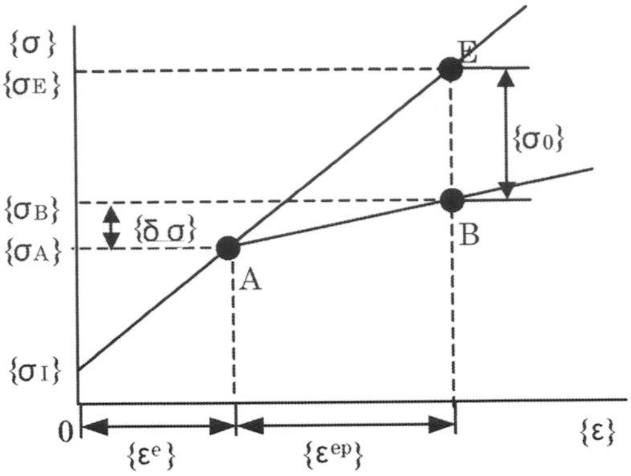

Figure 5. Initial stress method.

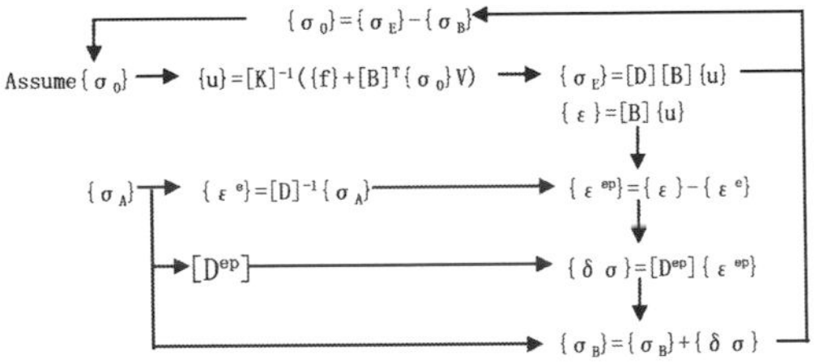

Figure 6. Flow of initial stress method.

where, $\sigma_x, \sigma_y, \tau_{xy}$: stress components in the global coordinates and σ_n, τ: normal and shear stresses in interface element.

When confining pressure σ_3 or σ_n continues increasing by the load of banking, the stress state is assumed to move on the yield surface after yielding. As shown in Figure 4, the increment of plastic strain $\{\delta\varepsilon^p\}$, when the curve moves on the yield line, was assumed to follow the non-associate flow rule with a dilatancy angle of 0. In Figure 4, Q_C is plastic potential for Coulomb interface.

Mohr-Coulomb :

$$Q_M = \{(\sigma_x - \sigma_y)^2 + 4\tau_{xy}^{\ 2}\}^{1/2} - 2g = 0 \tag{3}$$

$$\text{Coulomb}: \ Q_C = |\tau| - g = 0 \tag{4}$$

where, g: unnecessary parameter since plastic potential is used in a differential form. As shown in Figure 5, the entire load is applied at a single loading stage. In Figure 5, Point A shows the yield point, and Point B is the final equilibrium point. Up to Point A, the material is regarded as a linear elastic body and where, $\{\varepsilon^e\}$: elastic strain and $\{\varepsilon^{ep}\}$: elastic-plastic strain. $\{\sigma_I\}$ is the actual initial stress, $\{\sigma_A\}$, $\{\sigma_B\}$ and $\{\sigma_E\}$ are the stresses at each point, and $\{\sigma_0\}$ is the initial stress in the initial stress method, which is determined by an iteration loop. The flow of calculating $\{\sigma_0\}$ is shown in Figure 6 and where, $\{u\}$: nodal displacement vector, $[K]$: stiffness matrix, $\{f\}$: total load vector, $[B]$: matrix for calculating strain from $\{u\}$, $[D]$: stress-strain matrix, $[D^{ep}]$: elastic-plastic stress-strain matrix and V: volume of element (Arai., K. 1993). Since this numerical procedure treats only the final equilibrium state after the completion of construction, the procedure does not simulate at the construction on steps of the wall.

4 LONG-TERM STABILITY OF REINFORCED EARTH SLOPE

4.1 *Numerical analysis and field observation*

The soil parameters of the banking materials and the foundation ground used in the numerical analysis and those of the geogrid are shown in Table 4. In the table, E: elastic modulus, μ: Poisson's ratio, A: sectional area of geogrid, and T: tensile strength of the geogrid. The soil parameters of the banking materials were determined by laboratory tests. Geogrid is expressed as truss materials that bear no compressive stress. To find the final stress state of TOGA wall, the self-weight of embankment is loaded at one loading step in the numerical analysis. The calculated and monitored vertical and horizontal displacements of the ground surface and the tensile forces generated on the geogrid are compared. A comparison of the horizontal displacement of the slope surface is shown in Figure 7. Vertical displacements are compared in Figure 8. The monitored horizontal displacement at the time of completing banking is about 0.38 m the maximum at a height of about 15 m, showing that the slope bulged forward. The calculated vertical displacement values are similar to the monitored values. Since the compressive prestress was applied after banking, the calculated displacement does not agree with measured

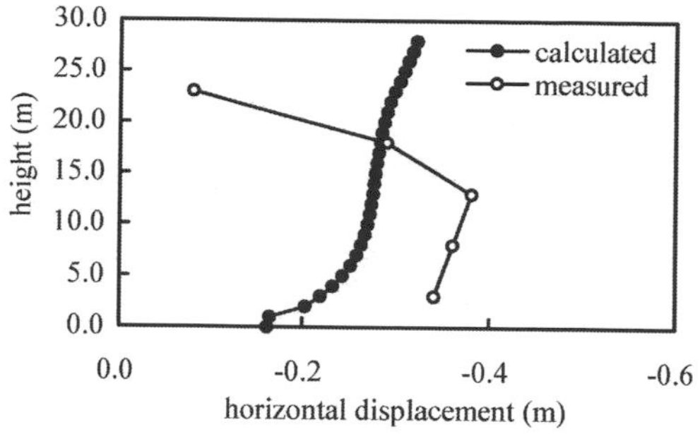

Figure 7. Horizontal displacement.

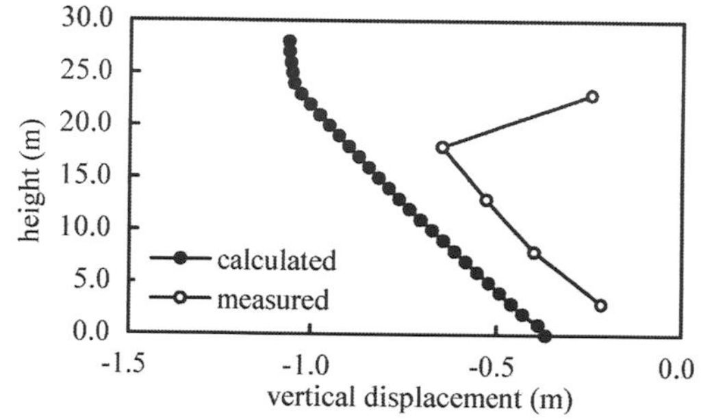

Figure 8. Vertical displacement.

Table 4. Parameters using in numerical analysis.

	E(kN/m²)	μ	c (kN/m²)	ϕ (degree)
New embankment	1.0×10^4	0.3	30	30.0
Old embankment	1.0×10^4	0.3	0	38.0
Weathered residual rock	3.0×104	0.3	50	23.3
Highly weathered layer	3.0×10^4	0.3	60	23.3
Mid weathered layer	3.0×10^4	0.3	70	23.3
Rock (elastic body)	3.0×10^4	0.3	–	–
	E(kN/m²)		A(m²)	T(kN/m)
geogrid	2.6×106		5.2×10^{-4}	28.0

displacement completely. The monitored and calculated tensile forces on the geogrid are compared in Figure 9. The calculated forces are slightly different from the monitored values. Possible causes include incorrect assessment of the modulus of deformation during banking and the difficulty of monitoring the tensile forces on geogrids.

Although the results of the numerical analysis are slightly different from the monitored values in slope

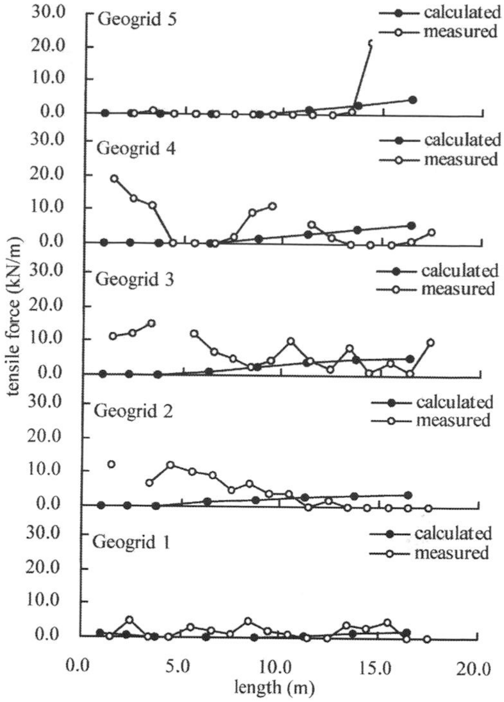

Figure 9. Tensile force of geogrid.

surface displacement and tensile force on the geogrid, the values are sufficiently close with each other in relation to a large bank height of 28 m and shows similar trends. Thus, the numerical analysis is judged to have correctly predicted the behavior of the embankment.

4.2 *Long-term stability*

To verify of long-term stability of TOGA wall, we have carried out the field observation for displacement of slope face and tensile force of geogrid after construction of TOGA wall. When the natural disasters occur such as earthquake and heavy rainfall, we are checking the settlement of road surface. We conform that the settlement of road surface and the tensile force of geogrid converge, so that we can evaluate that TOGA wall is stable state.

The stability against the failure of TOGA wall (safety factor) is predicted by using the numerical analysis. The safety factor of the earth slope is defined as follows. Generally, to represent a phenomenon of failure in which the deformation of the earth slope becomes infinitively at a certain loading step is difficult. In this paper, the failure state of the earth slope is defined that the yield finite elements connect in the earth slope like a slip surface as shown in Figure 10.

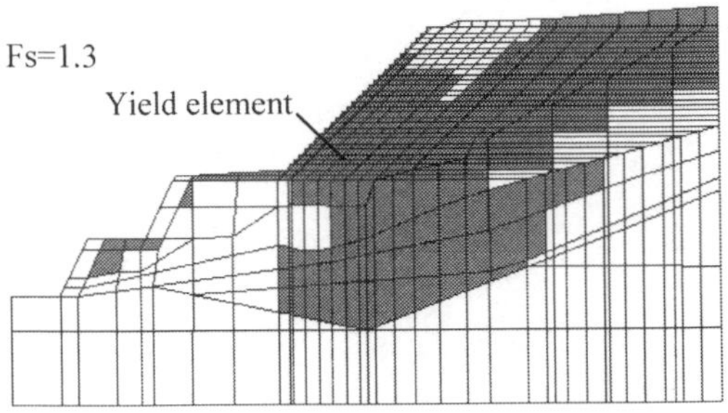

Figure 10. Distribution of yeild element.

Table 5. Effect of design parameters.

Case	Limit equilibrium method	Numerical analysis
$\gamma = 19.0\,\text{KN/m}^3$ $c = 30\,\text{KN/m}^2$ $\varphi = 30°$	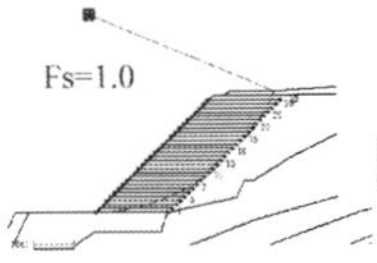	
Case 2 $\gamma = 19.0\,\text{KN/m}^3$ $c = 30\,\text{KN/m}^2$ $\varphi = 30°$	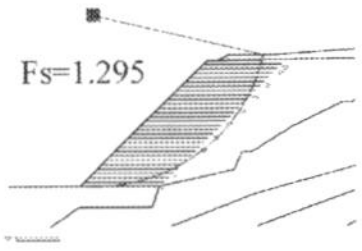	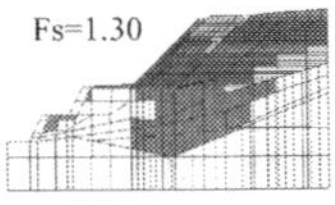

To find the failure state, hypothetical strength parameters $\underline{c}$ and are mobilized using the actual strength parameters c and ϕ and safety factor F_S as:

$$\underline{c} = c/F_S\,, \quad \tan\underline{\phi} = \tan\phi/F_S \tag{5}$$

The safety factor ofTOGA wall is $F_S = 1.3$ as shown in Figure 10, so thatTOGA wall has high safety factor. We can verify that TOGA wall is stable state sufficiently based on field observation and numerical analysis over long term.

5 DETERMINATION OF SOIL PARAMETERS FOR STABILITY ANALYSIS

5.1 *Numerical analysis and limit equilibrium method*

The results of stability analysis of TOGA wall calculated using the limit equilibrium method for each design parameter and the results of the numerical analysis are shown in Table 5. The conventional design method, which uses the limit equilibrium method, assesses the tensile strength of geogrids. On the other hand, the proposed numerical procedure evaluates the stiffness of the geogrid. Thus, the values of safety factor of the two methods cannot be directly compared, but the safety factors shown in Table 5 are mutually similar. Thus, the proposed procedure should be effective for assessing the stability of actual banks.

5.2 *Evaluation of soil parameters*

When cohesion c was assumed to be zero, the proposed procedure could not calculate safety factor since the iteration procedure diverged. This occurred because the displacement and stress of the bank increased too much disabling a stable final state to be found and the iteration procedure to converge. The results of a numerical analysis conducted by assuming a safety factor of $F_S = 1.0$, which involved 50 iteration procedures, are shown in Table 5. The yield region spread throughout the bank not satisfying any of the presently arranged geogrids. Since the earth slope is confirmed to be stable based on field observation and the results of numerical analysis, designing a slope by disregarding cohesion would result in a very uneconomical design.

6 CONCLUSIONS

The long-term stability of a reinforced earth slope, which was completed about seven years ago, is examined by conducting the numerical analysis, by comparing the calculated and monitored result and by performing visual inspection. Since the results of field observation can be represented by the numerical analysis, the degree of allowance to failure is examined by calculating the safety factor, which is found to be as stable as $F_S = 1.3$. The slope is confirmed to be stable over a long period of time judging from the states of slope and highway surfaces, which are visually inspected. Safety factors calculated by the proposed procedure are compared with circular slip surface method. The comparison shows that the proposed procedure can reproduce the safety factor determined by the limit equilibrium method. The study also confirmed that design parameters should be set by considering cohesion and not by just following existing standard manual.

The slope reinforcement project was executed by carefully and thoroughly draining underground water from the foundation and the bank since the project involved large-scale banking. The reinforced earth method is likely to be highly stable and earthquake resistance in principle, but thorough drainage measures should be taken while designing and constructing reinforced earth structures to deal with intense storms observed these years. The design manual used today as a standard for designing reinforced earth structures considers cohesion very little since cohesion is difficult to assess and banking materials and soil qualities may be non-uniform, and limits the cohesion of banking materials to be $c = 10\,\text{kN/m}^2$ or less.

Since most civil engineering works are public works, which must ensure safety with least expenses, economical and appropriate designs will be increasingly demanded. The authors propose a method that combines the conventional design method and FE analysis described in this paper. The new method can evaluate the safety of circular slip surfaces and safety factors as in the conventional method but only using simple soil parameters (elastic constants and c, ϕ). The method is also effective in understanding the deformation and stress states inside embankments, enabling economic and highly reliable embankment to be constructed.

REFERENCES

Ohta, H., Iwata, K., Arai, K., Kawamura, K., Nishimoto, Y., Yokota, Y. & Tsuji, S. 2006. Design and performance of a new type of reinforced earth slope, *Proceedings of the 8th International Conference on Geosynthetics*, Vol. 3: 1201–1204.

Public Works Research Center. 2000. The standard manual of geogrid reinforced soil structures.

Desai, C. S., Zaman, M. M., Lightner, J. G. & Siriwardane, H.J. 1984. Thin-layer element for interfaces and joints, *Int. J. Numer. Anal. Methods Geomech.*, Vol. 8: 19–43.

Zienkiewicz., O. C., Valliappan., S. & King., I. P. : 1969. Elastoplastic solutions of engineering problems 'initial stress', finite element approach, *Int. J. Numer. Methods Eng.*, Vol. 1: 75–100.

Zienkiewicz., O. C., Valliappan., S. & King., I. P.: 1968. Stress analysis of rock as a 'no tension' material, *Geotechnique*, Vol. 18: 56–66.

Arai., K. 1993. Active earth pressure founded on displacement field, *Soils and Foundations*, Vol. 33, No. 3: 54–67.

Yokota., Y, Arai., K., Haguro., T. & Tsuji., S. 2006. Stability analysis of reinforced soil slope considering deformation and stiffness, *Journal of Applied Mechanics JSCE* Vol. 9: 445–454.

Kubo., T. & Yokota., Y. 1999. About steel wall materials in a geogrid reinforcement soil wall construction, *Geosynthetics Engineering Journal*, Vol. 14: 72–81.

Kubo., T., Yokota., Y., Ohta., H. & Yamagami., N. 1999. Confirmation of an effect of a compression power prestress in face of slope of embankment reinforcement construction, *Geosynthetics Engineering Journal*, Vol. 14: 82–91.

Hirata., M., Iizuka., A., Ohta., H., Yamagami., N., Yokota., Y. & Omori., K. 1999 : Finite element analysis of geosynthesitics reinforcement embankment, considering dilatancy, *Proc. of Japan Society of Civil Engineering*, No.631/III-48: 179–192.

Ito., M., Yokota., Y., Kubo., T. & Arai., K. 2000. In-situ loading experiment to the protective embankment retaining wall reinforced by geosynthetics, *Geosynthetics Engineering Journal*, Vol. 15: 340–349.

New Horizons in Earth Reinforcement – Otani, Miyata & Mukunoki (eds)
© 2008 Taylor & Francis Group, London, ISBN 978-0-415-45775-0

Pullout resistance of strip embedded in cement-treated soil layer for reinforced soil walls

M. Suzuki
Graduate school of Science and Engineering, Yamaguchi University, Ube, Japan

Y. Tasaka & O. Yoneda
Ube Mitsubishi Cement Research Institute Corporation, Ube, Japan

A. Kubota
Hazama Corporation, Tokyo, Japan

T. Yamamoto
Graduate school of Science and Engineering, Yamaguchi University, Ube, Japan

ABSTRACT: This paper describes pullout resistance of a metallic strip embedded in a cement treated backfill soil. In this study, a test apparatus was newly developed to pullout a flat strip from a reinforced soil wall model such as the Terre Armee wall. A series of tests was performed on soil sample cured under different conditions of vertical stress, cement quantity and curing time. As a result, the pullout resistance of non-standard soil whose fine fraction content is over 25% increased remarkably by cement treatment technique.

1 INTRODUCTION

Earth reinforcement techniques are a useful and economical option for solving the problem of engineering on narrow tracts of land. With regard to reinforced soil walls, the Terre Armee method, which was invented by Vidal in 1963, has developed remarkably. A Terre Armee wall consists of facing (a concrete skin), reinforcement (metallic strips) and backfill soils. As shown in Figure 1, an earth retaining structure can be stabilized by the equilibrium between active earth pressure acting on the skin and the pullout resistance of strips embedded in the soil layer. Since frictional resistance is expected to be well mobilized on contact between the soil and strip, the fine fraction content and maximum grain size of backfill soil is regulated according to GRSW, RRR, Terre Armee, and MARW manuals (Miyata et al. 2001). The applicable backfill soil is sandy soil. However, the supply of sandy soil has recently been exhausted. Even if construction generated soil is fine-particle soil such as clay, it has been occasionally used by executing a chemical stabilization technique (JGS, 2006). The strength characteristics of cement-stabilized soil have not been sufficiently taken into account in the design and execution procedures for reinforced soil walls.

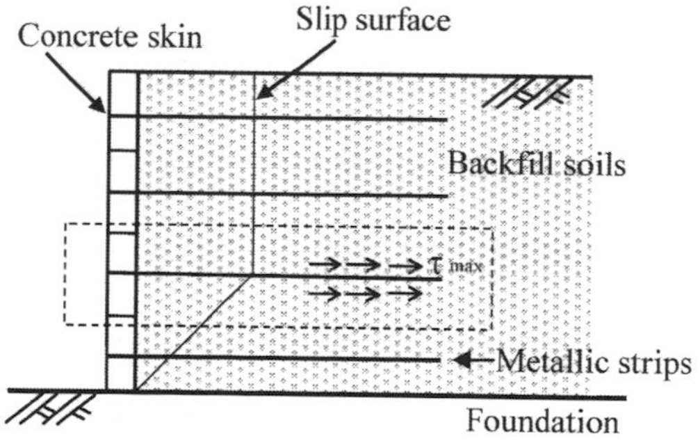

Figure 1. Concept of reinforcement mechanism in reinforced soil wall.

The aim of this study is to clarify the pullout resistance characteristic of reinforcement installed in a cement-treated soil layer. In this study, a test apparatus was newly developed to pull out a strip from a reinforced soil wall model. The validity of this apparatus was demonstrated through a comparison with previous in-situ test results. A series of tests was conducted with different conditions of vertical stress, cement quantity and curing time. This paper describes mechanism of pullout resistance mobilized on the contact between a strip and cement-treated soil based on the test results.

Table 1. Physical properties of soil samples.

Soil sample		Jiseiji clay	Kawakami silt	Nakayama sand	Shimonoseki sand
Natural water content	(%)	48.1	22.2	17.5	23.3
Density of soil particle	(g/cm^3)	2.834	2.638	2.739	2.624
Gravel fraction content	(%)	0.7	18.5	21.3	20.2
Sand fraction content	(%)	18.0	34.6	43.7	56.5
Silt fraction content	(%)	30.6	34.9	1.4	11.7
Clay fraction content	(%)	50.7	12.0	33.6	11.6
Fine fraction content	(%)	81.3	46.9	35.0	23.3
Liquid limit	(%)	76.2	43.5	37.6	–
Plasticity index		41.8	18.4	13.0	–
Soil classification		CH	SFG	SFG	SFG

2 BACKFILL SOILS IN DESIGN PROCEDURE OF REINFORCED SOIL WALL

The design procedure of the Terre Armee method is summarized as follows. Design conditions such as place, use, etc. are checked in detail. Earth pressures acting on the skin and strip are evaluated respectively. The spacing and length of strips is determined based on pullout resistance and the allowable tensile strength of the strip. Internal stability is examined by assuming a fixed slip line method (bi-linear). The overall stability and settlement of the foundation are checked. Backfill soils with high shear resistance and low compressibility are required. Soils having fine fraction content, $F_c < 25\%$ and maximum grain size, $D_{max} < 300$ mm are recommended and soils having $25\% < F_c < 35\%$ and $D_{max} < 75$ mm are accepted in the Terre Armee manual. If the fine fraction content of backfill soils is over 25% or their maximum grain size is over 300 mm, their physical properties can be improved by various chemical stabilization techniques. However, the increase of apparent cohesion in cement-stabilized soil is not sufficiently considered in the design procedure.

2.1 *Soil sample and cement stabilizer*

Soil samples used in the experiment were "*Jiseiji clay*", "*Kawakami silt*", "*Nakayama sand*" and "*Shimonoseki sand*". These soils were sampled at several construction sites in Yamaguchi Prefecture, Japan. The physical properties of the soil samples are listed in Table 1. The grading carves of soil samples are shown in Figure 2. The fine fraction contents of all samples except for Shimonoseki sand were higher than 25%. Since such samples could not be utilized as backfill soils, they were improved by application of cement stabilization. The used stabilizer was an ordinary general-purpose cement stabilizer.

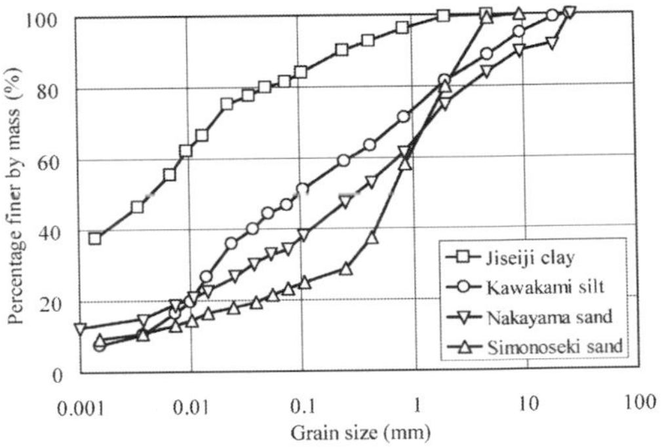

Figure 2. Grading curves of soil samples.

2.2 *Test apparatus*

A test apparatus was newly produced to simulate stress and deformation conditions of the soil around a strip as shown in Fig. 1. Actually the pullout resistance of a strip laid under a ground has been examined by an in-situ pullout test. However, an in-situ test could not be carried out on the completed state of the structure. The stress condition of the soil around the strip was not simulated in the execution process. Our apparatus was capable of measuring the horizontal pullout force of a strip in a soil layer with high accuracy. Figure 3 illustrates the essential features of the apparatus (see Photo 1). The apparatus is composed of a soil tank, a strip, a retaining wall, a loading plate for vertical pressure, two dial gauges for vertical and horizontal displacements, a load cell for horizontal force, a motor and gear box, and data recorder. The dimensions of the tank were 70 cm long $\times$ 20 cm wide $\times$ 30 cm high. The tank has double drainage layers on top and bottom. The wall was fixed through a series of tests. The strip used in the experiment was a flat type 6 cm wide and 0.5 cm thick. Except for Shimonoseki sand,

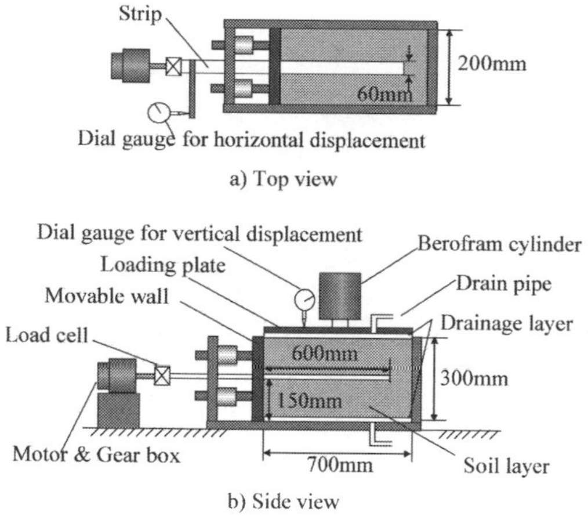

Figure 3. Schematic diagrams of test apparatus.

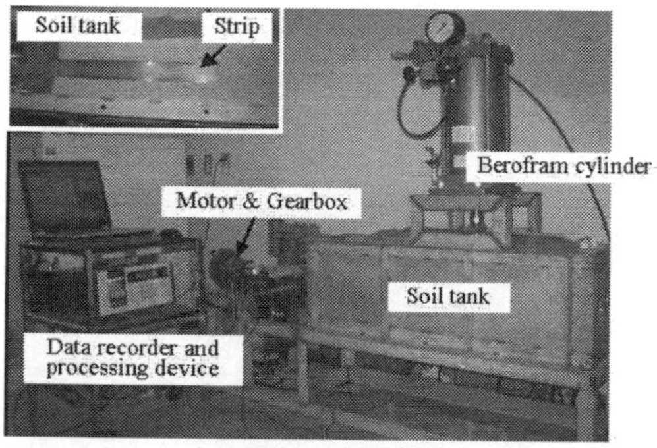

Photo 1. Overview of test apparatus.

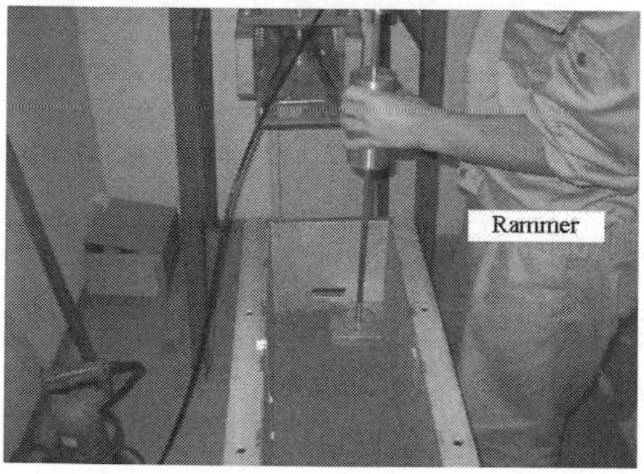

Photo 2. Soil compaction by rammer.

the initial water content of the soil sample was prepared to be at its liquid limit. After the soil sample was compacted using a vibrator or a rammer in the tank, the top surface of the soil layer was leveled at a height 15 cm from the bottom (see Photo 2). A 60-cm strip was placed on the smoothed surface of the soil

Photo 3. Condition of strip around clay after pullout test (Jiseiji clay).

layer. The strip was covered by the soil sample and consolidated by applying a constant vertical stress, σ_v, in the range of 50 kPa to 150 kPa. In the case of cement-treated soil, immediately after the soil sample was cured under a constant vertical stress during a curing period in a room with controlled temperature and humidity, the pulling out test was carried out. During the test, the horizontal force and displacement were measured. According to the preliminary test, it was shown that the rate effect was negligible in the range of 0.12 to 1.20 mm/min. The rate of the pullout test was uniformly set at 1.0 mm/min. The pullout test was finished when horizontal displacement reached about 10 mm. Photo 3 shows the condition of strip around Jiseiji clay after pullout test.

3 RESULTS AND DISCUSSIONS

3.1 *Typical behavior of untreated sample*

The test cases and results are listed in Table 2. Figure 4 shows the relationship between horizontal pullout stress, τ, and the horizontal displacement, δ, of the untreated sample (Shimonoseki sand). Here τ is derived from an equation dividing horizontal force by the surface area of the strip. As δ increased, τ increased monotonously. The $\tau \sim \delta$ curve for $\sigma_v = 100$ kPa becomes higher than that of $\sigma_v = 50$ kPa. This behavior was similar to that of other untreated samples. The maximum value of τ, τ_{max}, was determined based on the relationship between τ and δ. Figure 5 shows the relationship between τ_{max} and σ_v for test results including the in-situ test results quoted from a previous study (Ogawa et al. 1995). Figure 6 shows the relationship between F_c and τ_{max} of all data as mentioned above. As can be seen from Fig. 5, τ_{max} for Shimonoseki sand is the highest among those for our tested samples, but was low compared with previous data. It can be seen

Table 2. Test cases and results.

Test No.	Soil sample	Initial water content w_o (%)	Quantity of cement Q_c (kg/m^3)	Curing time T_c (day)	Compaction energy E_c(kJ/m^3)	Vertical stress σ_v (kPa)	Maximum pullout stress τ_{max} (kPa)	Unconfined compressive strength q_u (kPa)
1-1	Jiseiji clay	70.0	0	–	0	50	10.6	–
1-2						100	19.8	–
1-3			60	3		0	41.8	231.5
1-4			80			0	70.6	337.3
1-5			100	1		0	69.4	530.1
1-6				3		0	83.4	1009
2-1	Kawakami silt	40.0	0	–		50	6.8	–
2-2						75	11.9	–
2-3						100	16.4	–
2-4			60	1		0	19.6	314.4
2-5				3		0	37.1	402.7
2-6						50	65.3	657.3
2-7						100	78.8	744.0
2-8				7		0	42.2	455.2
2-9			80	3		0	56.3	776.0
2-10			100			0	67.1	1209
3-1	Nakayama sand	34.0	0	–		50	9.4	–
3-2						100	17.4	–
3-3			50	1		0	29.2	301.3
3-4						50	54.4	477.5
3-5			70			0	43.2	601.7
3-6			90			0	84.0	772.8
3-7		20.0	50		104.5	0	39.0	305.0
3-8					156.8	0	44.1	343.0
3-9					209.0	0	53.6	425.0
4-1	Shimonoseki sand	13.5	0	–	–	50	16.9	–
4-2					–	100	35.0	–

–: None in particular.

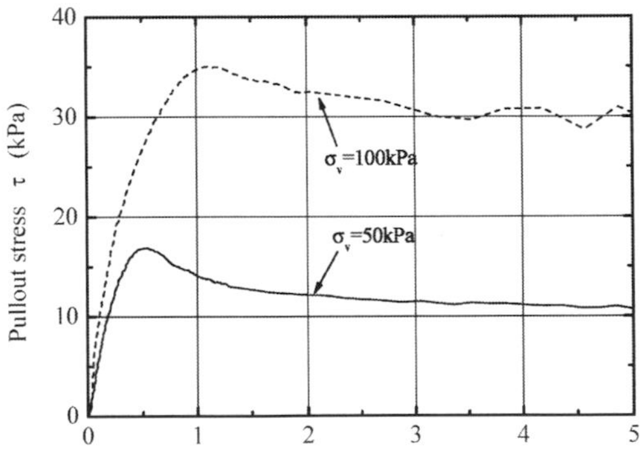

Figure 4. Pullout behavior of Shimonoseki sand without cement treatment.

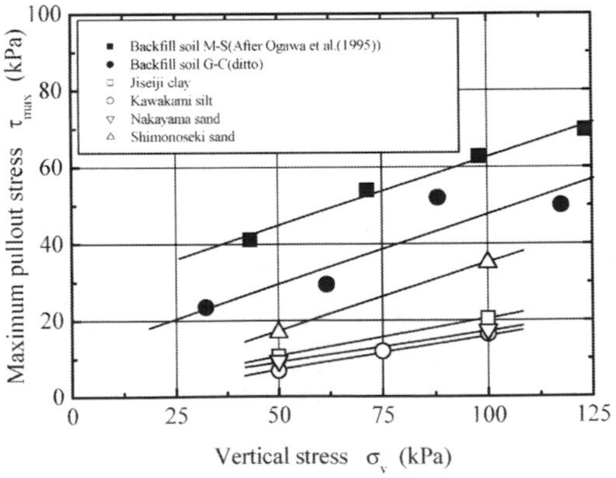

Figure 5. Comparison between laboratory and in-situ pull-out tests.

from Fig. 6 that τ_{max} decreased as the fine fraction content increased. The results obtained by this apparatus are in good agreement with those by the in-situ tests.

3.2 *Typical behavior of treated sample*

Figure 7 shows $\tau \sim \delta$ curves of untreated and treated soil samples (Kawakami silt). The τ_{max} of the treated sample was much higher than that of untreated sample.

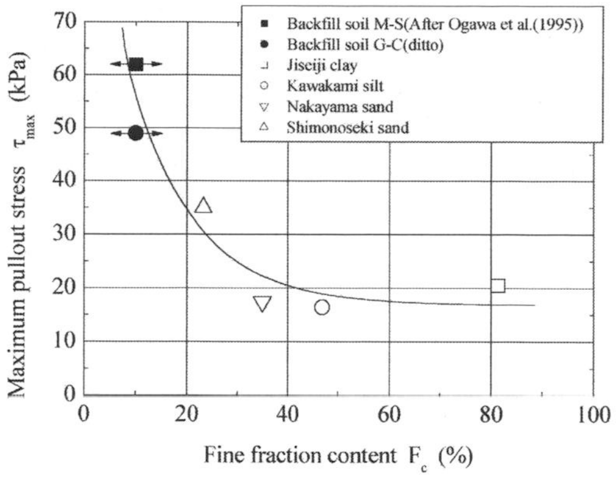

Figure 6. Relationship between F_c and τ_{max}.

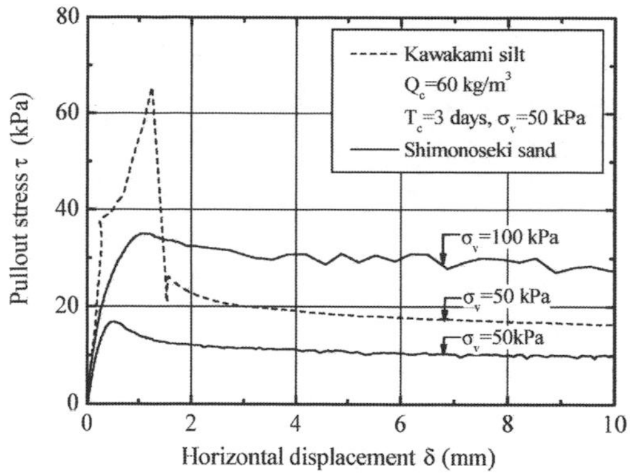

Figure 8. Pullout behavior of cement treated soil and recommended backfill soil.

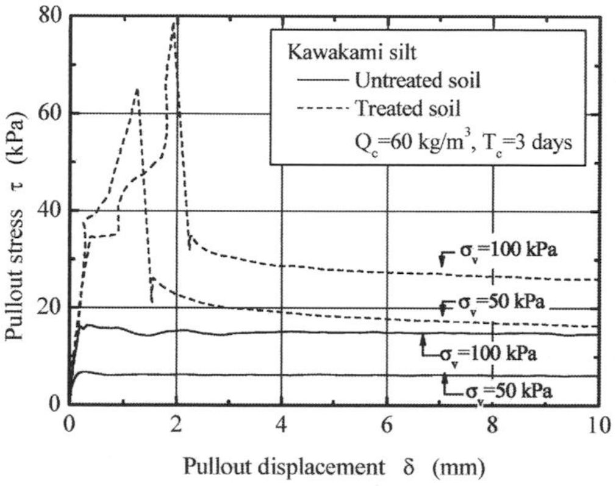

Figure 7. Pullout behavior of Kawakami silt with and without cement treatment.

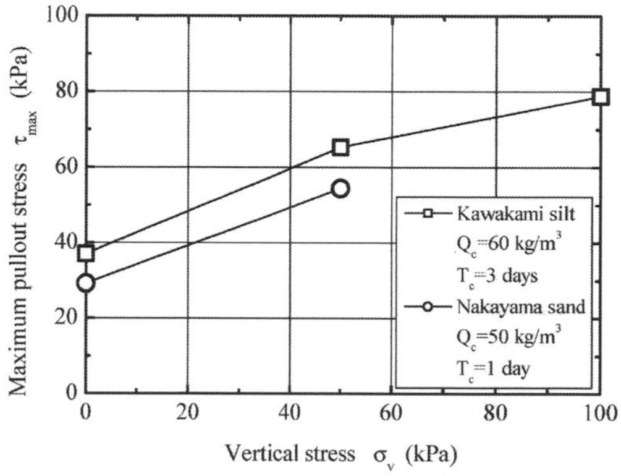

Figure 9. Relationship between σ_v and τ_{max}.

In the case of $\sigma_v = 50$ kPa, the τ_{max} of the treated sample was 10 times bigger than that of the untreated sample. In the case of $\sigma_v = 100$ kPa, the τ_{max} of the treated sample was 5 times bigger than that of the untreated sample. This tendency may be due to an increase in the cohesion component of the pullout resistance. Immediately after exhibiting a peak value, the treated sample shows remarkable reduction of pullout resistance. This brittleness behavior may be attributable to exfoliation of the cemented part formed between the soil and the strip. After τ reached a peak value, the $\tau \sim \delta$ curve showed unstable behavior. It is suggested that this behavior may be due to step-by-step exfoliation. Figure 8 shows test results obtained from a non-standard sample (treated Kawakami silt) and a standard sample (untreated Shimonoseki sand). Shimonoseki is classified into the recommended backfill soils, because of its low fine fraction content ($F_c = 23.3\%$). In the case of $\sigma_v = 50$ kPa, τ_{max} and the residual value, τ_{res}, of treated Kawakami silt become higher than those of untreated Shimonoseki sand. Therefore, the non-standard sample improved by cement stabilization could achieve as high a pullout resistance as a standard sample.

3.3 *Effects of vertical stress, cement quantity and curing time*

Figures 9 to 11 show the relationships of τ_{max} to σ_v, cement quantity, Q_c, and curing time, T_c, respectively. As shown in Fig. 9, the τ_{max} of cement-treated samples increases linearly with an increase in σ_v. The resistance increase may be due to the increasing contact area between the soil and the strip. As can be seen from Figs. 10 and 11, τ_{max} increased with the increase in either Q_c or T_c. The resistance increase was due to the development of cementation. These results suggested that adhesion was newly generated in the soil around the strip.

Figure 12 shows the distributions of unconfined compressive strength, q_u^*, and the water content, w, in the soil layer. The strip is situated in the center part of the soil tank. The unconfined compressive strength

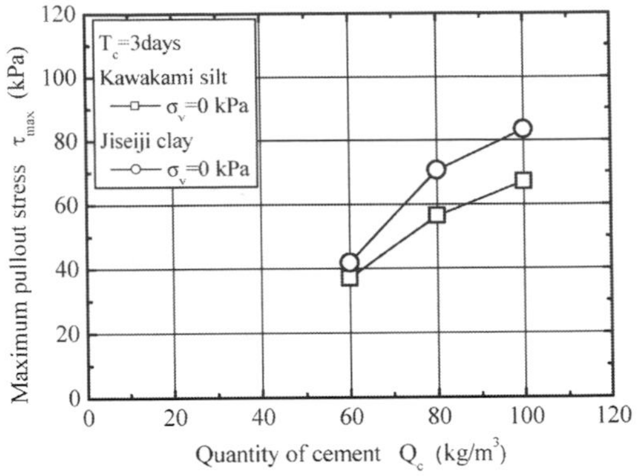

Figure 10. Relationship between Q_c and $\tau_{\max}$.

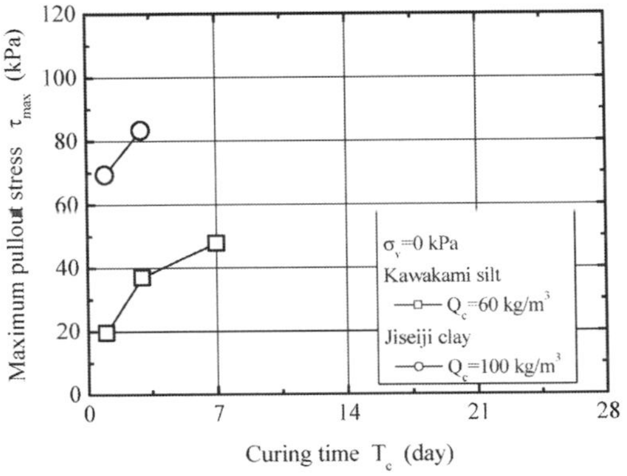

Figure 11. Relationship between T_c and $\tau_{\max}$.

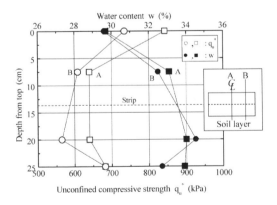

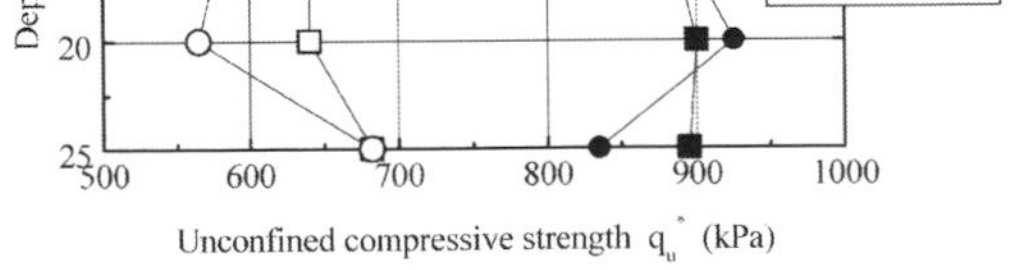

Figure 12. Distribution of q_u^* and w in soil layer.

was estimated from the index obtained by a soil hardness tester. The water content of the top and bottom becomes lower than that of the center part. On the other hand, the q_u^* of the top and bottom becomes higher than that of the center part. In fact, the measured unconfined compressive strength of the center part corresponds to

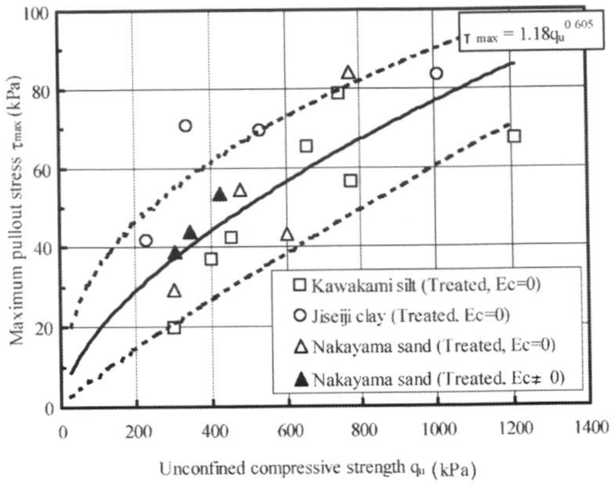

Figure 13. Correlation between q_u and $\tau_{\max}$ for various cement treated soil.

640 kPa. This tendency may be due to the process of consolidation near the drainage layer. Furthermore, the cementation in between soil particles was caused by gradual cement hydration, so that the pore water was constrained and entrapped.

3.4 *Correlation between of pullout resistance and unconfined compressive strength*

Figure 13 shows the correlation between $\tau_{\max}$ and q_u for treated samples of Jiseiji clay, Kawakami silt and Nakayama sand. The unconfined compressive strength was determined under the same conditions as the pullout test. The plotted data are obtained under different conditions of soil type, cement stabilizer content, curing time and applied vertical stress. Although the data are more or less scattered, $\tau_{\max}$ tends to increase with an increase in q_u. There seems to be a clear correlation between $\tau_{\max}$ and q_u. Considering the spacing and length of strips in the design stage, the pullout resistance of cement-stabilized soil may be accurately estimated from the results of an unconfined compression test on a soil sample.

4 CONCLUSIONS

The main conclusions of this study can be summarized as follows.

1) The pullout resistance measured by our apparatus was almost consistent with that by the in-situ test. The developed apparatus is suitable for evaluating the pullout resistance characteristics of strips in a soil layer.
2) The pullout resistance of non-standard soil whose fine fraction content is over 25% increased remarkably by cement stabilization. It was demonstrated that soil with a high fine fraction content can be utilized as backfill soil in reinforced soil walls.

3) The pullout resistance of cement-treated soil increased with increasing vertical stress. The increase of pullout resistance increased with increasing cement content and curing time. Stiff behavior soil may result from cementation developing in contact between the soil and the strip.
4) There seems to be a correlation between the pullout resistance and the unconfined compressive strength for cement-treated soil. Therefore, the pullout resistance can be estimated by conducting a conventional unconfined compression test.

ACKNOWLEDGEMENT

The authors express special thanks to Mr. T. Kaneshiro and Mr. K. Yamada for useful advice, and to Mr. N. Takazane for experimental assistance.

REFERENCES

Miyata, T., Fukuda, N., Kojima, K., Konami, T. and Otani, Y. 2001. Design of reinforced soil wall – Overview of design manuals in Japan. *Landmarks in Earth Reinforcement, Proc. of the International Symposium on Earth Reinforcement*, Vol. 2: 1107–1114.

Ogawa, N., Ohta, H., Fuseya, H. and Sakai, S. 1995. Shear strength developing in strip of Terre Armee method. *Proceedings of the thirtieth Japan national conference on soil mechanics and foundation engineering*, pp. 2367–2368 (*in Japanese*).

The Japanese Geotechnical Society. 2006. New Application of Earth Reinforcement-Combined Geotechnical Technology (*in Japanese*).

New Horizons in Earth Reinforcement – Otani, Miyata & Mukunoki (eds)
© 2008 Taylor & Francis Group, London, *ISBN 978-0-415-45775-0*

Design and construction of a composite nailed and mechanically stabilized embankment structure across a talus slope

T. Bergmann
HHO Africa, Cape Town, South Africa

A.C.S. Smith
Reinforced Earth (Pty) Ltd, South Africa

ABSTRACT: A new section of road on a greenfields alignment, over grasslands and indigenous and exotic forest is under construction from the Ncembu Plateau to the Langeni Sawmill, Eastern Cape Province, South Africa. Retaining wall number 1 starts as the road leaves the escarpment and runs beneath a dolerite cliff face. The wall is founded on steeply sloping boulder and clay talus. A mechanically stabilized earth (MSE) solution was chosen for construction of the retaining wall. The contractor's hard rock source was located at the top of the escarpment and the haul road providing access to this rock source fell within the road prism of retaining wall 1. A solution became necessary to construct this wall, while keeping the haul road open to traffic. This paper outlines the solution which comprises a composite structure, combining the use of self drilling, self grouting hollow soil nails, and an MSE structure using synthetic reinforcing strips. Although this type of solution is not new, the particular circumstances on this project are unique and called for a particular solution.

1 INTRODUCTION

The construction of a road, between the town of Ugie and the Langeni Sawmill in the Eastern Cape Province of South Africa, is presently nearing completion. This is a 110 million dollar government project to link extensive timber plantations in the Ugie-Maclear district to a large sawmill at Langeni. The final 17 kilometres of the road crosses the Ncembu Plateau, before descending the 500 metre high Langeni escarpment to Langeni Sawmill. The escarpment is dominated by sheer cliffs of dolerite, extending for many kilometers in each direction. The key to the route down the escarpment is a break in the cliff line which takes the form of a steeply sloping "nose". This is illustrated in Figure 1.

By cutting deeply into the dolerite cliffs of the escarpment edge, it was possible to locate a route down and around this "nose". The road gradient is 11 percent and a 680 metre long MSE has been used to carry the road across the slope. The route down the escarpment presented many challenges, leading to innovative design and construction. This paper describes specifically a 150 metre length, in two sections, of the MSE where soil nails were used to form a composite structure, satisfying both the space constraints and stability requirements.

Figure 1. Road and temporary haul road crossing talus slope.

2 GEOLOGY

The dolerite cliffs are the remnants of a massive sill intrusive into Beaufort Group mudrocks, siltstones and sandstones. The reason for what appears to be a talus slope located in the break between the line of the dolerite cliffs is not clear. The core drilling investigations carried out found that a layer of transported unweathered dolerite boulders in a doleritic soil matrix overlies undisturbed, open jointed dolerite bedrock. The size of the boulders varies from small to several metres across.

3 INVESTIGATIONS CARRIED OUT

24 core boreholes were drilled along the 680 metre length of the embankment. It proved difficult to identify the base of the transported layer with any certainty. Generally the thickness of this layer varied between about 5 metres and 12 metres in depth. The open jointed and blocky dolerite extended to depths of between about 10 metres and 20 metres below ground level. A typical geological section is shown in Figure 2. Generally the water table levels were found to lie deep within the slope, often near where the undisturbed dolerite level was estimated to be.

Six undisturbed block samples of the doleritic matrix soil were taken from depths of about 3 metres below ground. 9 shear box and 3 triaxial tests were carried out on these samples. No successful tube samples were obtained from the boreholes. The matrix soil is a clayey silt, with a high natural moisture content. The average effective stress parameters obtained from the triaxial tests were cohesion = 12 kPa and internal friction = 35 degrees. Drained shear box tests gave average values of cohesion = 29 kPa and internal friction = 31 degrees.

An air photograph study was carried out using both recent and 30 year old photographs. Although pine trees masked much of the area on the recent photographs, the older photographs showed the talus to have a consistent slope with none of the typical signs which may indicate previous or incipient instability. This was verified by careful field reconnaissance, which yielded no sign of any instability of the slope.

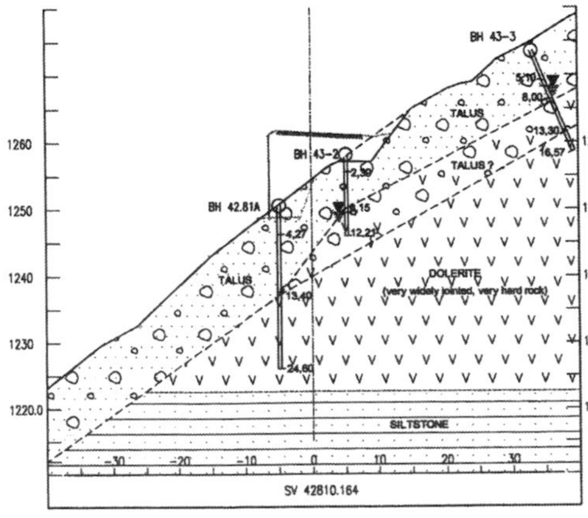

Figure 2. Geological section.

Figure 3. Cutting in talus slope.

4 SHEAR STRENGTH AND STABILITY OF THE TALUS SLOPE

The maximum slope of the talus surface is 40 degrees. A stability analysis of the natural slope using the results of the shear box and triaxial testing on the talus matrix gave factors of safety ranging between 0.84 and 1.07. This implies that the slope should be showing signs of incipient instability, which was clearly not the case. The soil/boulder mix therefore has a higher shear strength than the soil matrix alone. In order to develop a solution and analyse its effect on the stability of the slope, an estimate had to be made of the strength of the boulder and soil mix. Due to the large variation in size and spatial distribution of the boulders, in-situ testing was not considered to be a realistic option.

Based on the borehole core logs, the initial assessment of the proportion of boulders in the talus was 60 percent. Subsequent monitoring during soil nail installation proved that this figure was about 45 percent. Boulder size varied from small to several metres across. Figure 3 shows a photograph of a cutting through the talus and illustrates the boulder distribution.

Since no signs of significant soil creep or hummocky ground due to past slumping were evident, the assumption of a natural slope factor of safety of 1.2 was considered reasonable. A back analysis gave effective stress parameters of cohesion = 25 kPa and internal friction = 38 degrees. The increase in factor of safety using these parameters over that obtained using the shear strength of the matrix only was about 25 percent. An internet search produced very limited information on the shear strength of boulder-soil mixes, but did suggest that an increase in factor of safety of 20 percent to 25 percent for boulder-soil mixes with a boulder content of 50 percent was realistic. Some modelling of failure through the boulder-soil mix within the talus was carried out using Geo-slope's program SLOP/W with various boulder distributions. An exhaustive analysis of this type would be time consuming. The limited analyses which were carried out confirmed that the tortuosity induced in the failure surface by the boulders resulted in a substantial increase in the factor of safety.

Initial analyses and cost comparisons during the preliminary design stage indicated that the use of a flexible mechanically stabilized embankment to cross the talus slope was likely to be more cost effective than stabilizing a deep cut slope in this material.

5 FORMULATION OF SOLUTION

5.1 *Original MSE design*

Wall 1 was originally designed as a standard MSE structure with cross section as shown in Figure 4.

The structure is a flexible one and is able to accommodate settlements and differential settlement. The width of the structure, equal to the length of the reinforcing strips is dependent on overall stability and economical considerations. In this case due to space limitations the narrowest trapezoidal shaped structure possible, satisfying overall sliding and overturning criteria as well as internal stability requirements, was designed. In order to further limit the bulk excavation required for preparation of the foundation the structure was designed without embedment. The boulder and clay matrix talus foundation was deemed sufficiently strong and the risk of erosion sufficiently small to eliminate the need for such embedment.

Constituent materials: Cladding: The road grade is steep. In order to facilitate construction and survey of the wall and also to ensure a smooth continuous top level of the cladding the structure is designed on grade. The cladding is a 'weldmesh' with 100 mm × 100 mm apertures and 8, 10 and 12 mm diameter bars. Each cladding unit is 3 metres long and 720 mm high. Tie strips envelop the lower and upper waler bars in order to connect the cladding by way of a single bolt in double shear to the reinforcing strips. The cladding is backed with rock and the rock in turn with geo-fabric to provide a durable and aesthetically pleasing appearance with free draining properties while preventing loss of fines through the cladding.

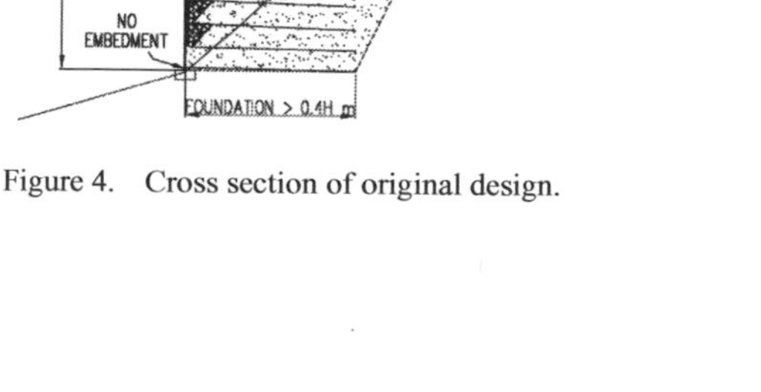

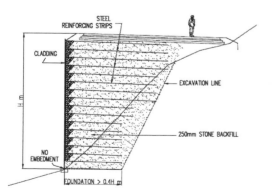

Figure 4. Cross section of original design.

Figure 5. Original design construction.

Reinforcing strips: The reinforcing strips are made of medium tensile steel with 50 × 4 mm cross-sectional area. The length of the reinforcing strips varies from 6 metres to 10 metres. The strips are of medium tensile steel and are hot dip galvanised. They are ribbed to improve frictional properties with the backfill.

Backfill: Rockfill, with maximum size 250 mm was specified for wall 1. This material eliminates the need for drainage and is easily placed in all weather conditions. The MSE backfill is defined as the volume contained by the face area of the cladding and the length of the reinforcing strips. Figure 5 shows the section of wall, unaffected by the haul road, constructed according to the original design.

5.2 *Redesign of section of wall 1 to accommodate haul road*

At an early stage in the construction process it was necessary to construct a temporary haul road across the talus slope to haul crushed aggregate from the crusher, located on top of the escarpment, down to the site of a 280 metre long reinforced concrete viaduct located on the lower slopes. Environmental controls were such that the haul road had to be constructed within the earthworks prism. This limited the width of excavation which could be carried out below the embankment to about 3 metres.

Over two sections of the wall a solution was required to enable the MSE structure to be constructed around and over the temporary haul road. In addition the construction of the 12 metre high embankment across the talus slope would result in a reduction in the assumed overall factor of safety of the natural slope of

1.20. Some form of slope stabilization was therefore required which would raise the factor of safety above this value.

6 DESIGN OF COMPOSITE MSE AND SOIL NAIL STRUCTURE

The use of a composite structure satisfied all the design criteria: 15 metre long soil nails intercepted deep seated failure surfaces and increased the overall factor of safety of the embankment and the hillside to in excess of 1.40; the soil nails provided stability for the steep cut face below the haul road; and the use of a polyester strap as earth reinforcement, connected to the soil nail heads with a yoke, allowed the construction of an MSE structure in a confined space.

For the stability analysis the water table depth was assumed to be slightly less than that measured during the investigations. Very high water loss was experienced during the core drilling investigations and the excavation for the temporary haul road exposed several voids in the talus. It was therefore considered that the slope was too permeable for a general rise in the water table so near the surface. This assumption was borne out by the very high grout takes during grouting of the soil nails.

6.1 *Design philosophy*

Although sufficient space was available to place the upper reinforcing strips, insufficient space was available to place the lowermost reinforcing strips. In order to ensure internal stability of the abbreviated MSE mass the lowermost strips could be connected to nails driven into the talus. In order to ensure and enhance overall stability these nails could be extended into the talus to ensure that slip surfaces did not develop behind and beneath the structure. The bearing capacity of the foundation should be such that it would be able to support the now narrow based and tied back MSE structure.

6.2 *Internal stability*

An MSE structure should behave as a coherent gravity mass dependent on the strength and frictional interaction of the reinforcing strips and the backfill. At each level of reinforcing strips the horizontal stresses in the strips are determined by calculation of the vertical stresses on the strips and applying to it a coefficient of earth pressure. The strips should be strong enough to resist the horizontal stresses and should be long enough to ensure that they are able to mobilise these stresses in friction and do not pull out of the backfill. In this case the nails are designed to be strong enough, both in strength and pull out, to resist the loads imposed on them by the reinforcing strips and by overall stability requirements.

Figure 6. Placing of synthetic link reinforcement.

6.3 *The connection of the reinforcing strips to the nails*

A system of connecting synthetic reinforcing straps had been previously developed and was considered practical for use on this project. All steel strips in the standard cross section which were too long to be placed were replaced with synthetic straps consisting of discrete channels of closely packed, high tenacity, polyester fibres encased in a polyethylene sheath. The strap is supplied in a 100 metre long coil and is laced between tie points and hooks on the cladding and the yokes placed over the head of the nails. A nut was used to hold the yoke onto the nail head and to tension the link reinforcing straps between the yoke and the cladding. The density and position of the nails was determined by the density and position of the reinforcement straps which were to be attached to them. Since each nail head is attached to two tie points on the cladding, the vertical and horizontal spacing of the nails was 720 mm and 1450 mm respectively. The nails were also positioned on grade to match the position of the reinforcing strips.

6.4 *FLAC analysis confirmed that extensible and inextensible strips cannot be mixed in section*

Inextensible steel reinforcing strips could not be used above the extensible polyester straps. A "FLAC" numerical analysis showed that the extensible strips accept load until such time as the first layer of inextensible strips is placed. Thereafter the inextensible strips collect all future stresses. The design was finalized with synthetic strips placed from bottom to top of the structure; connected to the nails at the lower part of the structure and simply placed in the backfill where space permitted it above the level of the nails.

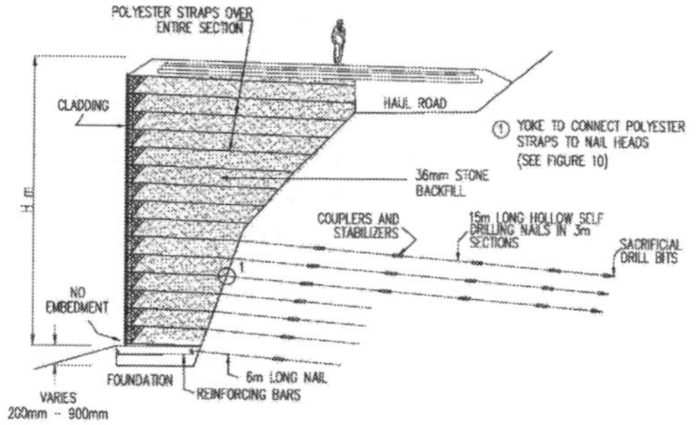

Figure 7. Typical section as redesigned to accommodate haul road and enhance the overall stability.

Figure 8. Construction of combined MSE / nail solution.

6.5 *Foundation of composite structure*

On account of the fact that there was no embedment and that the MSE structure was narrow at its base and differential settlement between the nailed slope and the MSE backfill would impose large loads on the foundation a reinforced concrete base with minimum thickness 250 mm was specified. The lowermost reinforcing strap was omitted and the lowermost row of nails was cast into the foundation. The bottom of the cladding was also cast into the foundation.

The main constraints and components of the solution are illustrated in Figure 7 and construction is shown in Figures 8 and 9.

7 BACKFILL

The coarse rockfill used for the bulk of wall 1, although suitable for steel reinforcing strips, would probably have damaged the synthetic strap reinforcement. A material that would not damage the straps was required; ie fill the irregularities of the boulder slope around the nails without the need for compaction and also provide adequate drainage of the cut slope face. An evenly graded 36 mm stone backfill

Figure 9. Installation of the soil nails.

met all these requirements and was specified for the entire sections where synthetic reinforcing straps were used.

8 THE NAILS

The nails: Pre-drilling and later installation of nails into grouted holes was deemed to be too onerous. Hollow self-drilling, self-grouting nails were required for installation into the massive boulder and clay talus. These nails are supplied in 3 metre lengths. A sacrificial drill bit with diameter 72 mm is attached to the first 3 metre length. Additional 3 metre lengths are joined together during the drilling operation by means of couplers. A centraliser is placed at each coupler. The larger diameter drill bit and centralisers position the nail in the centre of the hole. During the drilling operation grout is pumped down the central hollow core of the nail, flows through the drill bit and back up the drill hole created by the larger diameter drill bit. Figure 9 illustrates the installation of the soil nails.

9 DURABILITY

The structures are designed for a service life of 70 years. The hollow nails are not stressed and are covered with a thickness of at least 20 mm of grout. The nail is also approximately twice as strong as required and consequently has considerable sacrificial thickness. The ground waters are not particularly aggressive. The nailed talus slope structure is an indeterminate one and should failure of a nail occur the loads would be redistributed to the other nails. For these reasons it was only felt necessary to galvanise the last 3 metre nail piece which protrudes into the backfill without grout covering and onto which the hot dip galvanised yoke and nut are attached. The MSE part of the composite structure is designed for durability according to standard MSE design codes (see references).

Figure 10. Connection of synthetic reinforcing strip to nail head.

10 CONSTRUCTION

The solution adopted was only developed once the construction had been underway for several months. All parties on site reacted positively to the proposed solution and this attitude facilitated its successful construction. In particular the installation of the nails to precisely surveyed positions on an unstable massive boulder and clay talus slope in a high and persistent rainfall area was a considerable challenge readily accepted by the contractor. In order to install the nails a drill rig was required and a 3 metre wide temporary shelf had to be created in front of the toe of the cut face. The average rate of installation of the nails varied from about 40 metres to 100 metres per day. A total length of 9000 metres of nail was required. Many days were lost to rain. There was also a substantial learning curve period during which a number of difficulties in the installation process had to be overcome. The nail installation period took approximately 10 months using on average 2 drill rigs. The placement of backfill, cladding and link reinforcement proceeded smoothly, the pacing item being the placement of the backfill. Figure 10 illustrates the connection between synthetic reinforcing strip and the nail head.

11 COSTS

Over the cladding area affected the cost of the re-designed solution, excluding bulk earthworks, was approximately 10 times the cost per square metre of the originally specified solution. No other solution was considered practical or economical.

12 CONCLUSIONS

The solution to the problem proved to adequately meet the technical, economical and environmental requirements. The environmental requirements were to minimize the visual impact of the road on the side of the mountain as well as the area of indigenous forest which had to be cleared. The solution allowed the width of the earthworks prism to be substantially reduced and so achieved these requirements. The appearance of the structure is unchanged from the specified solution. The combination of in-situ reinforcement and MSE structures opens up many possibilities. This solution may be used to found and widen roads on steep slopes without disruption to traffic and at the same time enhance the overall stability of slopes on which the road is founded.

REFERENCES

Design Code: *Soil Nailing Recommendations*, Clouterre 1991.

Design Code: *Terre Armee Internationale design guidelines*.

Price, GV, Smith, ACS and Muhajer, A (2001). *Innovative solution for a Kei Cutting problem*. Proceedings of International Symposium on Earth Reinforcement, IS Kyushu 2001.

Ziai, F, Paris, A. *Review and design approach of a retaining structure built against and connected to a nailed embankment*, EuroGeo 3 – Paper A03700 – Munich – 2004.

Federal Highway Administration (Feb 2006), *Shored mechanically stabilized earth (SMSE) wall systems design guidelines*, Publication No. HWA – CFL/TD-06-001.

New Horizons in Earth Reinforcement – Otani, Miyata & Mukunoki (eds)
© 2008 Taylor & Francis Group, London, ISBN 978-0-415-45775-0

Shaking table model tests on retaining walls reinforced with soil nailings

S. Nakajima
University of Tokyo, Tokyo, Japan

J. Koseki
Institute of Industrial Science, Tokyo, Japan

K. Watanabe & M. Tateyama
Railway Technical Research Institute, Tokyo, Japan

ABSTRACT: Based on results from shaking table model tests on retaining walls reinforced with soil nailings, effects of the soil nailings as an aseismic countermeasure for existing retaining wall is discussed. It is also attempted to apply Newmark's sliding block method with the introduction of the effect of the nailings so as to develop a procedure to predict the residual displacement of the walls with the nailings. Computed sliding displacements are in good agreement with the measured ones, while computed tilting displacements are much larger than the measured ones due possibly to overestimation of overturning moment used in the analysis.

1 INTRODUCTION

In Japan, use of geosynthethic-reinforced soil retaining wall with full height rigid facing for new permanent structures is continuously increasing because of their higher seismic performance than conventional type retaining walls (e.g. gravity, leaning and cantilever type retaining wall), which suffered severe damages in recent earthquakes (Koseki et. al. 2006). However, there exist large numbers of the conventional walls which support road or railway embankments. Installations of aseismic reinforcements to these walls are required especially in case such walls support important embankments.

Currently, large diameter soil nailing, which is called as LDN hereafter, is used for railway structures in Japan as an aseismic countermeasures for the existing retaining walls. It consists of a column of improved soil and a tension bar that is located at the center of the column. Based on shaking table model tests results (Kato et. al. 2002), it was reported that seismic performance of the retaining wall on sloped subsoil could be effectively improved even though such walls without nailings were severely damaged during the Chi-Chi earthquake in Taiwan (Huang et. al. 2005). On the other hand, it is essential to evaluate the effects of the aseismic counetermeasures in terms of the amount of reduced residual displacements in performance based design.

In view of the above background, in this study, the results from a series of shaking table model tests on retaining walls reinforced with LDNs (Nakajima et. al. 2007) are analyzed, and it is also attempted to develop a procedure to predict the residual displacements of the walls after the earthquake.

2 MODEL TEST PROCEDURES

After a wooden leaning type retaining wall model having a height of about 500 mm was placed on a sloped subsoil layer, a backfill layer was prepared. Both the backfill and the sloped subsoil layers were made of Toyoura sand having a relative density of about 90%, which were prepared by air pluviation using a sand hopper.

Cement-treated sand and phosphor bronze bar having a thickness of 0.8 mm and a width of 5 mm were used to model the improved column and the tension bar of the LDN, respectively. Strain gages were pasted on the surface of the column to measure mobilized tensile resistance. Sand particles were also glued on the surface of the column to achieve enough frictional resistance. As shown in Fig.1, the LDN models were installed at the bottom of the footing and wall facing. It should be noted that former ones did not restrict tilting of the wall because they were fixed with the footing by hinges. The schematic view of the model without the LDNs, which would be refereed to in this study, is also shown in Fig.1. The subsoil thickness in this study was increased as compared with the previous study by Kato et.al. (2002) so as not to restrict

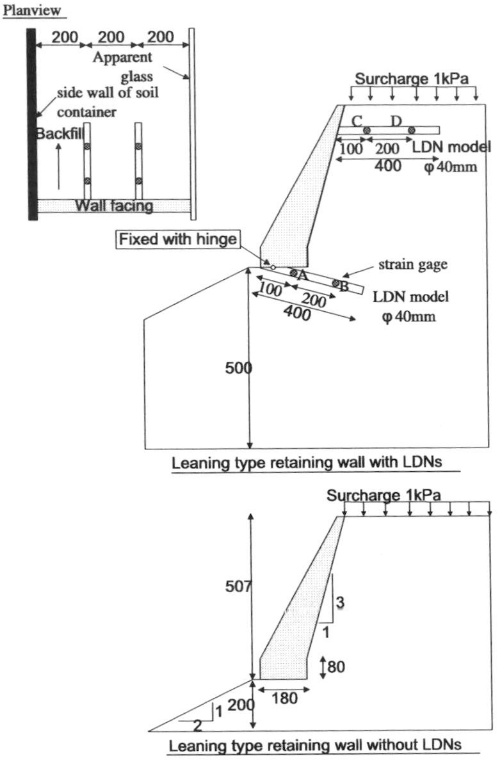

Figure 1. Schematic diagram of models (unit in mm).

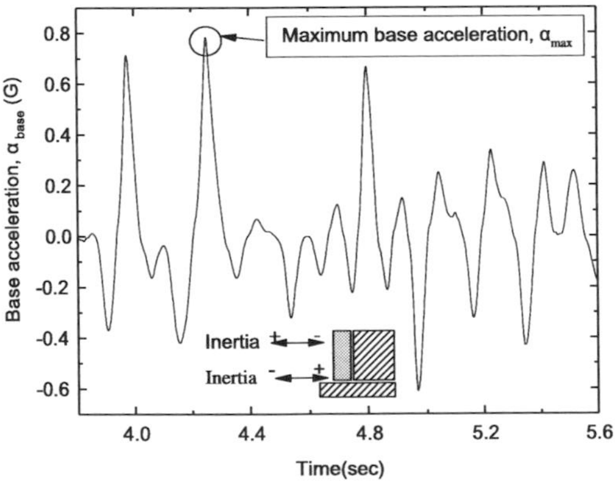

Figure 2. Typical time history of irregular excitation.

formation of failure plane passing through both the backfill and subsoil layers, which indicates the global instability of this model.

Seismic load was applied by shaking the soil container horizontally by using irregular excitations as typically shown in Fig. 2, while the maximum acceleration was gradually increased at an increment of about 100 gals until the wall displaced largely.

3 MODEL TEST RESULTS

3.1 *Failure process and residual displacements*

Figs.3a to 3d show the failure processes of the models. The sums of the residual displacements at each shaking step are plotted versus the maximum base acceleration α_{max} in Fig.4. Sliding displacement d_s, and tilting angle θ of the wall, and settlement of the backfill d_v are concerned in this study. The values of d_v were measured at a surface of the backfill having a horizontal distance of 200 mm from the wall facing as shown in Fig.1.

In case without the LDNs, increment of the residual displacements was accumulated largely after the formation of the failure plane in the backfill layers during the shaking at the α_{max} of 578 gals (Fig.3a). Residual wall displacements increased drastically after appearance of failure plane in the subsoil layer which indicated the bearing capacity failure during the shaking at the α_{max} of 636 gals (Fig.3b). In Fig.4, the value of permissible differential settlement of the backfill in the Level 2 earthquake for bridge abutment (damage level 3, which requires a minor retrofit work, in RTRI 1999) is also indicated. It was evaluated by reducing the actual value in prototype scale to one-tenth (i.e $200\,mm \times 1/10 = 20\,mm$), considering the difference in the model and prototype scales. As shown in Fig.4, the value of d_v in the case without the LDNs exceeded the permissible one during shaking at the α_{max} of 578 gals.

On the other hand, as also clearly shown in Fig.4, residual displacements decreased effectively by adding the LDNs. In case with the LDNs, failure plane in the backfill layers wasn't observed even after the shaking at the α_{max} of 640 gals (Fig.3c). The value of d_v was also reduced to about 5.7 mm, which was much smaller than the permissible value as evaluated above (i.e. 20 mm). With the increase of the α_{max}, residual displacements increased gradually, while no drastic increase of the wall displacements was observed during the whole shaking steps. However, after the formation of failure plane in the backfill and subsoil layers, wall displacements accumulated largely during the shaking step at the α_{max} of 1038 gals (Fig.3d). As also indicated in Fig.3d, the failure plane in the backfill layer was formed at a position just outside of the LDNs at the wall facing.

3.2 *Mobilized resistance by LDNs*

Mobilized tensile resistances by the LDNs at the wall facing (T_{top}) and bottom of the footing (T_{bottom}) at the timing of the α_{max} (i.e. the maximum inertia force was applied to the wall) are plotted versus the α_{max} in Fig.5. It should be noted that the values of T_{top} and T_{bottom} are converted to those per unit width considering the horizontal spacing of the LDNs as shown in Fig.1.

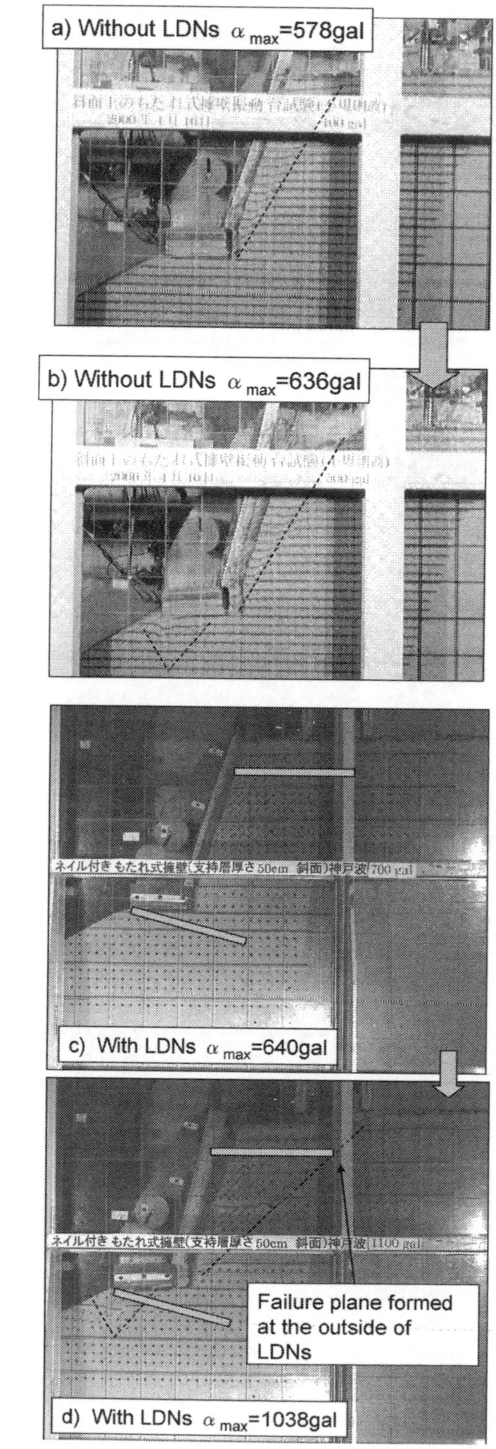

Figure 3. Failure processes of models.

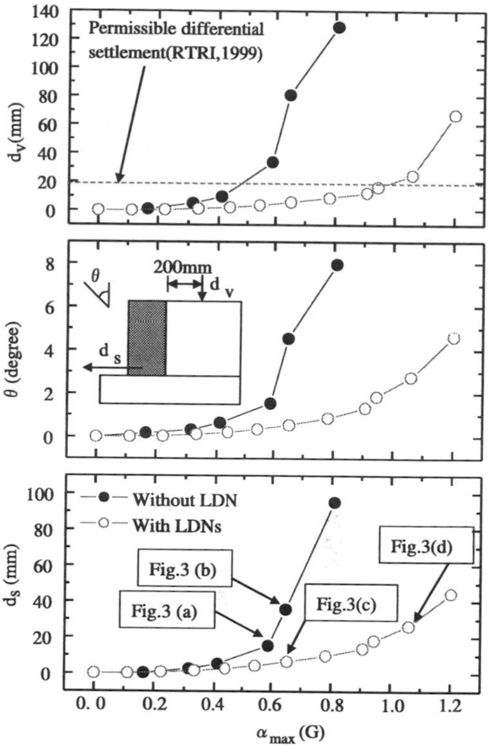

Figure 4. Comparison of residual displacements.

The values of T_{top} by gage C and gage D were almost equal to each other, while the T_{bottom} value by gage A, which is mobilized just beneath the footing, was much larger than the one by gage B which was pasted at a far position from the footing. This behaveior implies that the LDNs at the bottom could work effectively to restrict formation of failure plane in the subsoil layer that is associated with bearing capacity failure. As indicated in Fig.5, it was also observed that the T_{top} and T_{bottom} values were accumulated largely after the formation of failure plane in the backfill and subsoil layers during the shaking step at the α_{max} of 1038 gals. The levels of computed peak and residual tensile resistances as discussed later, are also indicated in Fig.5.

4 DISPLACEMENT ANALYSIS

4.1 *Newmark's sliding block method*

In order to evaluate the sliding displacement of gravity type retaining walls during earthquakes, Richards and Elms (1979) proposed to employ the Newmark's sliding block method. Relative displacement of the wall can be computed using the concept of threshold acceleration and the double integration proposed by Newmark (1965).

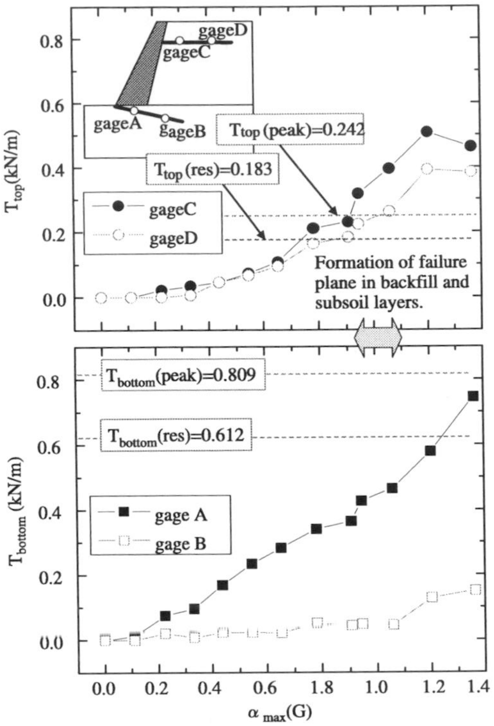

Figure 5. Mobilized tensile resistances by the LDNs.

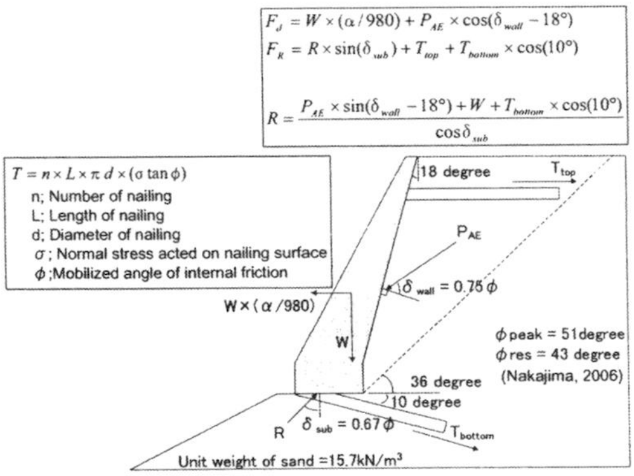

Figure 6. Pseudo static analyses against sliding.

As summarized in Fig.6, the threshold acceleration in this study was determined from a pseudo static analysis using the soil strength obtained from the relevant plane strain compression tests on dense Toyoura sand and the values of peak and residual resistances by the LDNs. The peak and residual mobilized angle of internal friction were set equal to 51 and 43 degrees in this analysis base on a result from relevant plane strain compression tests on dense Toyoura sand (Nakajima, 2006).

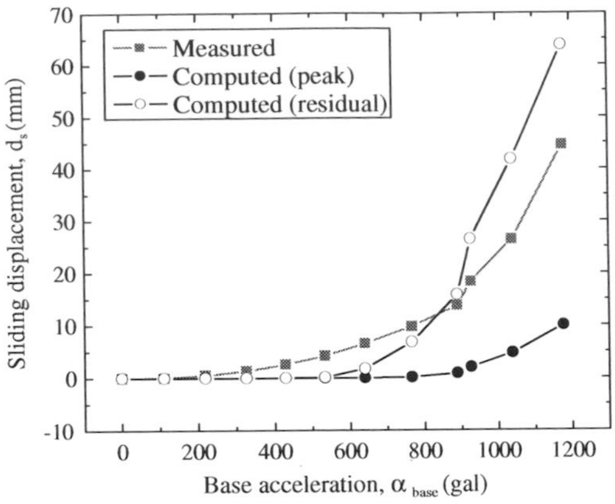

Figure 7. Comparison of sliding displacements.

It was assumed in the analysis that a driving force, which would induce the sliding of the wall, was the sum of the horizontal component of the earth pressure acted on the wall facing and the inertia force of the wall, while a resistant force against the sliding was assumed to be the sum of the horizontal component of the resistances by the LDNs and the frictional resistance at the bottom of footing.

The dynamic earth pressure was computed using a pseudo static approach proposed by Koseki et. al. (1998), while the angle of the failure plane in the backfill was set equal to 36 degrees based on the observation of shaking table model test as indicated in Fig.6. The resistances by LDNs were evaluated using the equation indicated in Fig.6. It should be noted that constant resistance was assumed to be mobilized in the relevant design guideline (RTRI, 1999) irrespective of the induced displacement. However, the measured resistance showed displacement-dependant behavior as shown in Fig.5.

Based on the results from the pseudo static analysis, the threshold acceleration was determined as the acceleration when the factor of safety against the sliding became unity. In this analysis, it was 711 gals when the peak soil strength was employed and 472 gals when the residual soil strength was employed. Computed sliding displacements using the above threshold accelerations and the Newmark's method are plotted versus the base acceleration in Fig.7. The computed displacement with threshold acceleration using the peak soil strength was smaller than the measured one, while the computed displacement using the residual soil strength corresponded well with the measured one. It should be noted that no displacement was computed until the base acceleration exceeded threshold acceleration, although a certain extent of the displacement was observed in the corresponding model test results.

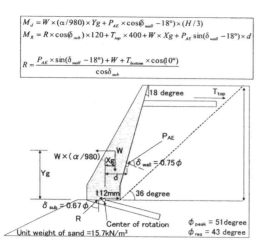

$$M_d = W \times (\alpha / 980) \times Yg + P_{AE} \times \cos(\delta_{wall} - 18°) \times (H/3)$$

$$M_R = R \times \cos(\delta_{sub}) \times 120 + T_{top} \times 400 + W \times Xg + P_{AE} \sin(\delta_{wall} - 18°) \times d$$

$$R = \frac{P_{AE} \times \sin(\delta_{wall} - 18°) + W + T_{bottom} \times \cos(10°)}{\cos\delta_{sub}}$$

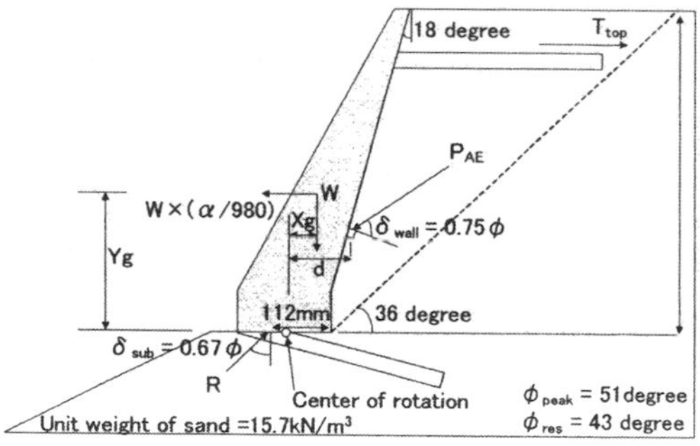

Figure 8. Pseudo static analyses against tilting.

One of the reasons causing such difference may be overestimation of the resistance by LDNs especially at lower acceleration levels. As shown in Fig.5, resistances by LDNs were not fully mobilized from the beginning of shaking. They increased gradually with the increase of the wall displacement. This behavior implies that displacement-resistance relationship of the LDNs as well as the maximum resistance, which is only focused in the current design procedure, should be taken into account in conducting displacement analysis like the Newmark's method.

In addition, the sliding displacement increment due to shear deformation of the subsoil (Koseki et. al. 2004) which was observed in the model test was also neglected in the above analysis. Therefore, further investigation on these issues is required.

4.2 *Newmark's rotating block approach*

Zeng and Steedman (2000) proposed to apply the Newmark's method to evaluate the overturning displacement of the walls. Following the same concept, evaluation of the threshold overturning moment is made in this study to compute overturning displacement of the wall.

Based on the results from the pseudo static approach as summarized in Fig.8, the threshold moment was evaluated. Horizontal component of the earth pressure and inertia force of the wall were taken into account to evaluate the driving moment, while the subsoil reaction force, the weight of the wall, vertical component of the earth pressure and the mobilized resistance by the top nailings were considered to evaluate the resisting moment against overturning.

It was assumed in the analysis that the center of the rotation of the wall was fixed at the center of the footing, and sum of the earth pressure would act at the middle third of the wall height. Based on the analysis of the model test in case without nailing, the resultant subsoil reaction was assumed to act at the point having the horizontal distance of 112 mm from the heel of the footing. In this analysis, measured value of the maximum mobilized resistance by top nailings, which was indicated in Fig.5, was also used because the computed ones using the equation in Fig.6 were different from the measured ones. Results and conditions of the pseudo static analysis are summarized in Table.1.

The computed tilting displacements of the wall are compared with the measured one in Fig.9. All the computed displacements were much larger than the measured one. This difference was possibly caused by overestimation of the driving moment. In this analysis, dynamic response of the retaining wall and backfill layer was not taken into account (i.e. the wall and backfill acceleration were set equal to the base acceleration). However, actual responses of the wall and the backfill were decelerated with the wall movement. In this point of view, further improvement on the evaluation of the driving moment is required.

Table 1. Summary of threshold moment.

Case name	Soil strength	Mobilized resistance by top nailing	Threshold moment (N/m)
peak1	ϕ_{peak}	0.242 kN/m (Evaluated)	331.8
res1	ϕ_{res}	0.183 kN/m (Evaluated)	162.5
peak2	ϕ_{peak}	0.400 kN/m (Measured)	277.55
res2	ϕ_{res}	0.400 kN/m (Measured)	263.55

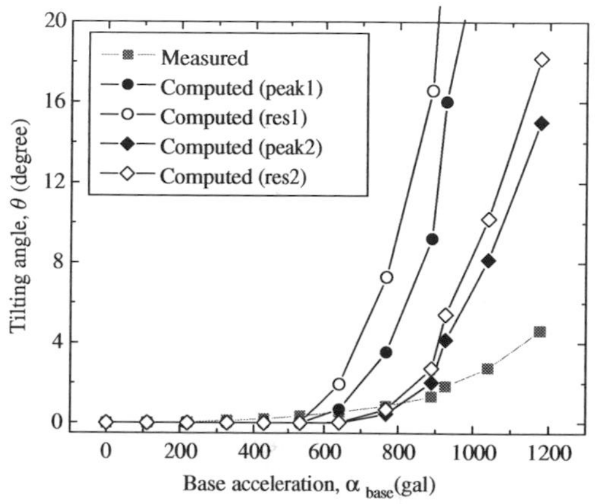

Figure 9. Comparison of tilting displacements.

5 CONCLUSIONS

Based on the results and analysis of the model test on retaining wall reinforced with large diameter soil nailings(LDNs), following conclusions were achieved;

1) Sliding, tilting displacement and settlement of the backfill in case with LDNs were effectively reduced compared with the case without LDNs.
2) Based on the measurement of the mobilized resistances by LDNs, it was found that the top nailings worked effectively to restrict the formation of the failure plane in the backfill layers, whilethe bottom ones resisted against the bearing capacity failure in the subsoil.
3) Newmark's method with the introduction of the effect of the LDNs was adopted to simulate the model test result. Computed sliding displacements corresponded well with the measured ones, while computed tilting displacements were much larger than the measured ones possibly because of the overestimation of the overturning moment.

REFERENCES

Huang, C.C., 2005, Seismic displacements of soil retaining walls situated on slope, *Journal of Geotechnical and Geoenvironmental Engineering*, ASCE, Vol.131, No.9, pp.1108–1117

Kato, N., Huang, C.C., Tateyama, M., Watanabe, K., Tatsuoka, F., and Koseki, J. 2002, Seismic stability of several types of retaining walls on sand slopes, *Proc. of 7th International Conference on Geosynthetics*, Vol.1, pp.237–240

Koseki J., Tatsuoka F., Munuf Y., Tateyama M., and Kojima K.,1998, A modified procedure to evaluate active earth-pressure at high seismic load, *Special issue of Soils and Foundations*, Japanese Geotechnical Society, pp.209–216

Koseki, J., Kato, N., Watanabe, K., and Tateyama, M., 2004, Effects of subsoil and backfill conditions on seismic displacement of gravity type retaining walls, *Cyclic Behavior of Soils and Liquefaction Phenomena (Triantafyllidis, ed.),Belkema*, pp.665–671

Koseki J., Barthurst, R.J., Guiler, E., Kuwano, J. and Maugeri, M., 2006, Seismic stability of reinforced soil walls, *Proc. of 8th International Conference on Geosynthetics*, Vol.1, pp.51–77

Nakajima, S., Koseki, J., Watanabe, K., and Tateyama, M., 2006, Evaluation of allowable displacements of retaining walls by shaking table model tests, International Conference on Physical Modeling in Geotechnics, Hong-kong, Vol.2, pp.1101–1106

Nakajima, S., Koseki, J., Watanabe, K., and Tateyama, M., 2007, Shaking table model tests on retaining walls with aseismic counter measures, *13th Asian Regional Conference on Soil Mechanics and Geotechnical Engineering*, Kolkata, (accepted)

Newmark, N.M., 1965, Effects of earthquake on dams and embankments, *Geotechnique*, Vol.15, No.2, pp.139–160

Railway Technical Research Institute, 1999, *Design guideline of railway structures Aseismic design*, Maruzen, pp.326 (in Japanese)

Richards, R. Jr., and Elms, D. G., 1979, Seismic behavior of gravity type retaining walls, *Journal of Geotechnical and Geoenvironmental Engineering*, ASCE, Vol.105, No.4, pp.449–469

Zeng, X., and Steedman., R.S., 2000, Rotating block method for seismic displacement of gravity walls, *Journal of Geotechnical and Geoenvironmental Engineering*, ASCE, Vol.126, No.8, pp.709–717

New Horizons in Earth Reinforcement – Otani, Miyata & Mukunoki (eds)
© 2008 Taylor & Francis Group, London, ISBN 978-0-415-45775-0

Geogrid reinforcement for cement stabilized soil

Y. Miyata, S. Shigehisa & K. Okuda
National Defense Academy, Yokosuka, Japan

ABSTRACT: This paper examined geogrid reinforcing effects for stabilized soil based on laboratory model test. In a series of the tests, influence of consolidation stress, drainage condition during loading, and loading speed on the strength properties of reinforced-stabilized soil are investigated. Main conclusions are as follows. 1) Geogrid prevents the propagation of cracks in stabilized soil. 2) Limit load resistance value increases by geogrid reinforcement. 3) Reinforcing effects can be expected as regardless of the consolidation pressure. 4) Reinforcing effects becomes larger according to the loading speed.

1 INTRODUCTION

In the construction of marine structure, utilization of dredged soil is important to reduce environmental impact due to construction. A popular utilizing method is to mix dredged soil with stabilizing material such as cement. However, the stabilized soil is generally brittle material. Improving method of ductility of the stabilized soil has been needed.

Authors and members in same project team in Port and Airport Research Institute, Japan have proposed a construction method of marine structure with dredged soil. This is combining method of geogrid reinforcement with cement stabilization. In this paper, influences of consolidation stress, drainage condition during loading, and loading speed on the strength properties of reinforced-stabilized soil are discussed.

2 CONCEPT OF GEOGRID REINFORCEMENT FOR CEMENT STABILIZED SOIL AND RESEARCH PROJECT

The invented method; SG-Wall method is combing technology of cement stabilization (S) of dredged soil and geogrid (G) reinforcement for quay wall (Wall). Typical cross section of the SG-wall is shown in Figure 1. The wall consists of cement stabilized soil, geogrid and sheet pile facing. Geogrid is used to reduce the tensile strain of stabilized soil and to prevent the propagation of cracks in the stabilized soil. Facing of the SG-wall is sheet pile. It is connected to geogrid directly. Sheet pile has high axial, shear and bending rigidity. The sheet pile is installed into base ground to fix the toe of facing. Tatsuoka (1992) noted that rule of facing rigidity on the stability of geosynthetics reinforced soil. El-Emam & Bathurst (2005) reported the effect of facing toe condition on response of geogrid reinforced soil wall. This facing system will make a rule in increasing of stability. In the first stage of this project, its feasibility was studied by performing 1/24-scale model shaking table tests (Miyata et al., 2006). In this test, high seismic performance of SG-wall was observed. In order to put this technology in practical use, the further examination has been required about design and construction method. Our project is now in second stage to investigate reinforcing mechanism in physical model test for the reduced-scale model.

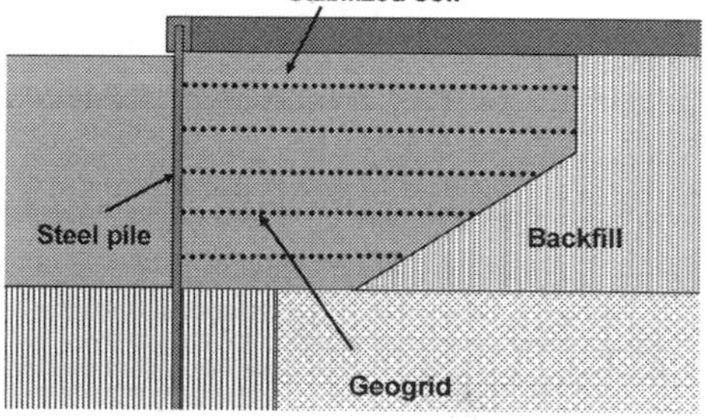

Figure 1. Outline of the SG-wall method.

3 LABORATORY TEST

In the case of cement stabilization of soft soil having high water content, overburden pressure will affect on the strength properties of the stabilized soil. The other

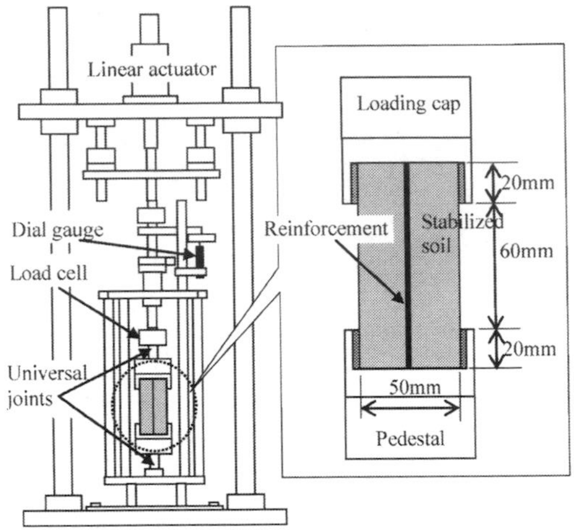

Figure 2. Laboratory test setup.

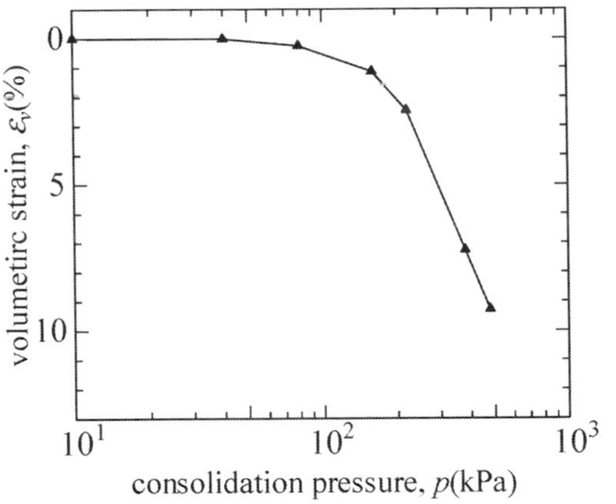

Figure 3. Triaxial test result of the stabilized soil used.

side, strength of geogrid depends on strain rate. This paper considers the rule of consolidation pressure and loading speed on the effect of geogrid reinforcement.

Performed test was loading test for reduced model with triaxial testing apparatus. The laboratory test setup is shown in Figure 2. The shape of the specimen resembled that of a square pillar (50 mm × 50 mm × 100 mm). Geogrid was placed at center of specimen vertically. After curing of sample, the sample was fixed to the loading cap with gypsum. The loading was performed by making the cap move at fixed speed. A series of isotropic consolidation and triaxial tensile loading were performed by referencing the JGS-0523. Consolidation time was determined by the 3t method.

The stabilized soil sample was prepare by mixing cement with Kibushi clay ($w_L = 92\%$, PI = 59,

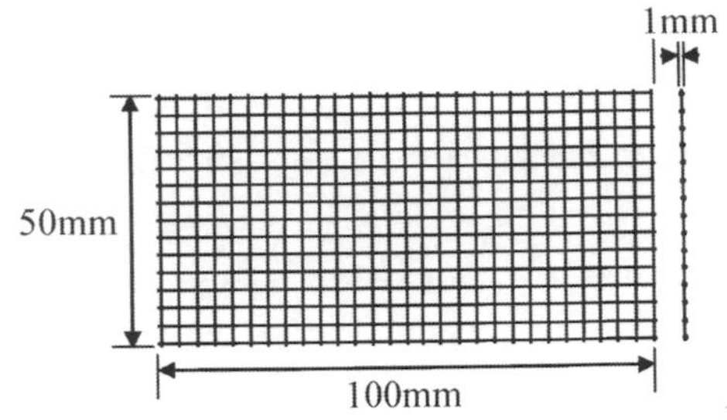

Figure 4. Outline of the geogrid used.

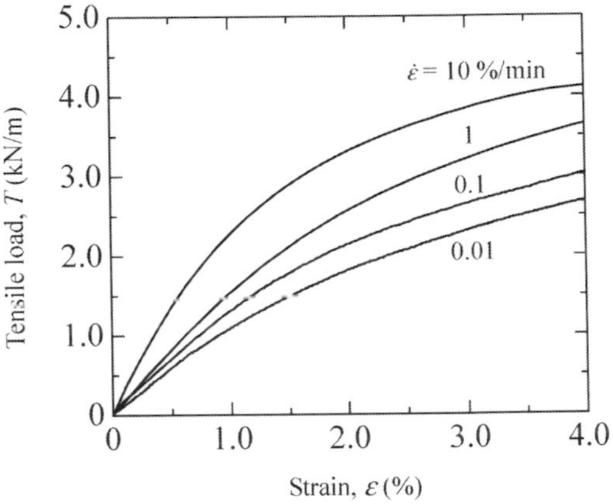

Figure 5. Tensile strength test results of the geogrid used.

$D_{50} = 0.0025$ mm) under the condition of high water content ($w = 135\%$). Cement content, which is the weight ration of cement to dried soil, a_w, was 17.3%. The result of triaxial consolidation test of this stabilized soil is shown in Figure 3. The consolidation yield stress, p_y can be estimated as $p_y = 150$ kPa. In a series of test, reduced-scale model of geogrid was used as shown in Figure 4. Tensile force per unit width – strain relations are shown in Figure 5. Curing of the sample was conducted in the mold. All of the test specimens were cured for 7 days in a humid room under atmospheric pressure at a temperature of 20 ± 3 °C.

In a series of laboratory test, the influence of confining pressure was investigated in the CD and the CU test. The range of investigated pressure was 0 kPa to a maximum of 196 kPa. The strain rate of loading was set to 0.1 %/min to the CU test, and 0.01%/min to the CD test, respectively. The influence of loading speed was investigated by the CU test. The CU tests were conducted at two kinds of consolidation stress as 0 kPa and 147 kPa. Range of investigated strain rates was 10^{-2} %/min to 10^{1} %/min.

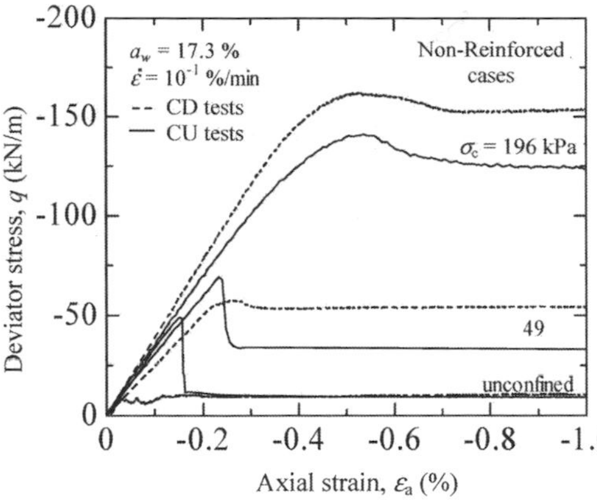

Figure 6. Stress – strain relations. (Non-reinforced cases).

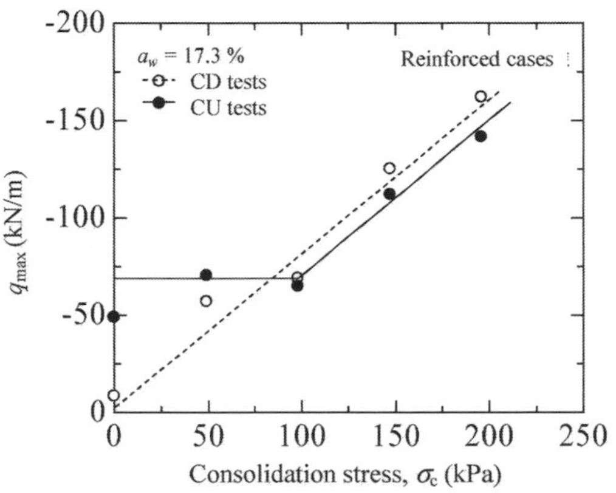

Figure 7. Peak deviator stress – consolidation pressure relations. (Non-reinforced cases).

4 TEST RESULTS AND CONSIDERATIONS

4.1 *Strength property of stabilized soil without reinforcement*

Figure 6 shows the relation between deviator stress, $q = \sigma_a - \sigma_r$, and averaged axial strain, ε_a. Where σ_a is axial stress, σ_r is lateral pressure respectively. In a series of tests, local deformation such as propagation of cracks was observed. However both the stress and strain were calculated by using typical length or area of samples. In the case of that consolidation pressure, σ_c is relatively smaller, stress-strain relations in the CU tests showed the peak value in each test. In the CD tests, such behavior was not observed. In the case of that σ_c is relatively larger, the difference between peak and residual strength was lower than one at that σ_c is relatively smaller. This does not depend on drainage condition during loading. The relations between peak deviator stress, q_{max} and σ_c are shown in Figure 7. The

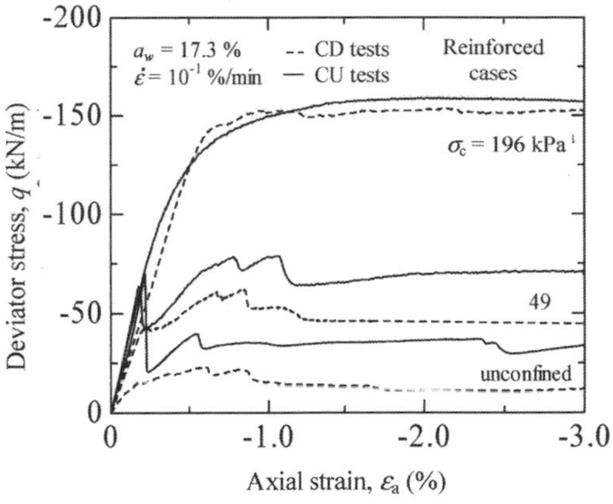

Figure 8. Stress – strain relations. (Reinforced cases).

$q_{max} - \sigma_c$ relations at the CD tests can be evaluated as follows.

$$q_{\max} = a \cdot \sigma_c \tag{1}$$

The $q_{max} - \sigma_c$ relations at the CU test can be evaluated as follows.

$$q_{\max} = q_0,\ \sigma_c < \sigma_y \tag{2}$$

$$q_{\max} = a' \cdot \sigma_c,\ \sigma_c < \sigma_y \tag{3}$$

Where σ_y is a certain stress that strength properties changes.

4.2 *Strength property of reinforced-stabilized soil*

Figure 8 shows the relation between q and ε_a. The q_{max} and residual values, q_r of reinforced cases are smaller than one of non-reinforced case. By geogrid reinforcement, ductility of stabilized soil was improved. The q_{max} of reinforced cases is almost same as one of non-reinforced case. However q_r of reinforced case is higher than one of non-reinforced case. The relations between residual strength, q_r and consolidation stress, σ_c are shown in Figure 9. The relation of CU test and one of CD test are almost same. From this result, it can be concluded that drainage condition does not affect on residual strength of reinforced stabilized soil.

The effect of loading speed on the peak strength of non-reinforced can be summarized in Figure 10. Influence of loading speed on the strength of stabilized soil can be almost omitted. The effect of loading speed on the residual strength of reinforced sample can be summarized in Figure 11. In the case that $\sigma_c = 0$, the effect of loading speed is same as non-reinforced case. However, In the case that $\sigma_c = 147$ kPa, strain rate effect is more remarkable when $\dot{\varepsilon} > 10^{-1}$(%). As shown in Figure 5, strength properties of the geogrid

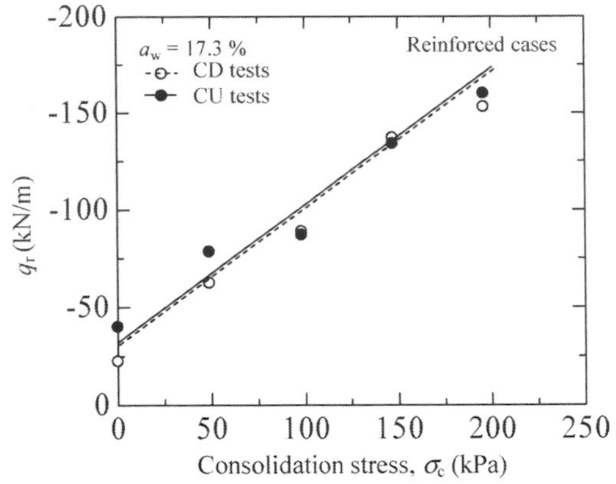

Figure 9. Peak deviator stress – consolidation pressure relations. (Reinforced cases).

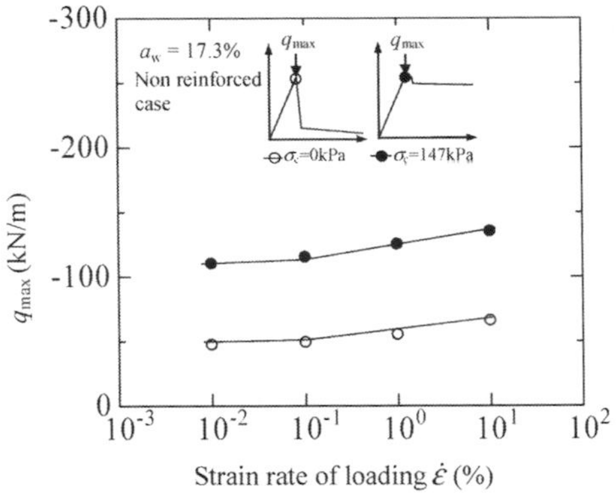

Figure 10. The effect of strain rate on the peak strength of non-reinforced sample.

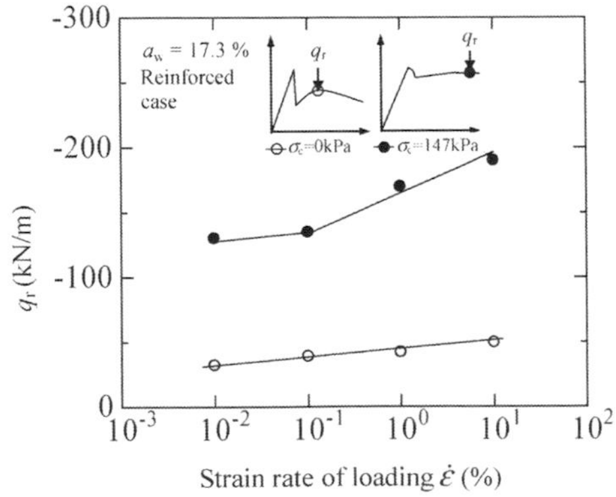

Figure 11. The effect of strain rate on the residual strength of reinforced sample.

depends on strain rate. This mean that geogrid resist to the applied load in reinforced cases.

The states at the end of testing for non-reinforced or reinforced sample are shown in Figure 12 (a), (b)

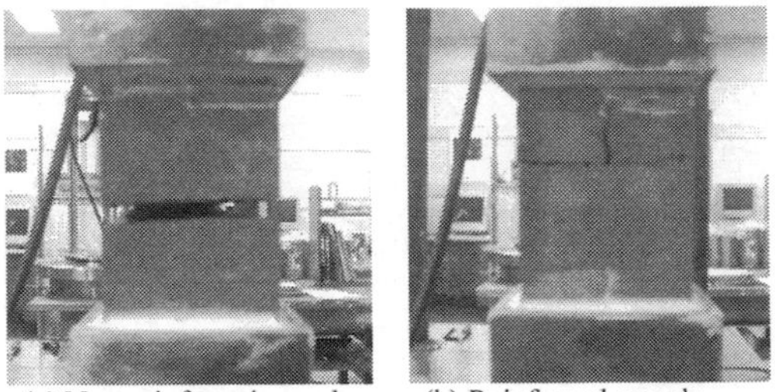

(a) Non-reinforced sample. (b) Reinforced sample.

Figure 12. The state at the end of test of no reinforced or reinforced sample.

respectively. Those results were obtained from unconfined test. In the case of un-reinforced case, failure plane was observed at the middle at of specimen. The other side, in the case of reinforced case, complicated failure mode was observed. Geogrid prevented the progressing of crack in stabilized soil.

5 CONCLUSION

Main conclusions are as follows. 1) Geogrid prevents the progressing of crack in stabilized soil. 2) Limit load resistance value of reinforced-stabilized soil increases. 3) Effect of the geogrid reinforcement can be expected as regardless of the consolidation pressure. 4) Effect of the geogrid reinforcement becomes larger according to the loading speed.

ACKNOWLEDGEMENTS

The authors are grateful to Prof. J. Koseki and Mr. Sato (IIS, the Univ. of Tokyo), for their kind suggestion in the developing of the testing apparatus. Additionally, member of project team concerning SG-wall method (Port and Airport Research Institute, Japan) is highly appreciated.

REFERENCES

Miyata, Y., Ichii, K., Suzuki, Y., Hirai, T., Takaba, Y., Fukuda, F. (2006) "Geosynthetics reinforcement for marine structure with soil stabilization", Proc. of 8th International Symposium on Geosynthetics, Yokohama, Japan, pp.883–888, 2006

El-Emam, M. M. & Bathurst, R. J. (2005). "Facing contribution to seismic response of reduced-scale reinforced soil walls", Geosynthetics International, 12, No. 3, 215–238.

Tatsuoka F. (1992). "Role of facing rigidity in soil reinforcing", Keynote lecture, Proc of 2nd International Symposium on Earth Reinforcement, Vol.2, Fukuoka, Japan, pp.831–870, 1992.

New Horizons in Earth Reinforcement – Otani, Miyata & Mukunoki (eds)
© 2008 Taylor & Francis Group, London, ISBN 978-0-415-45775-0

Bending tests on a beam of grid-reinforced and cement-mixed well-graded gravel

T. Uchimura
University of Tokyo

Y. Kuramochi & T.-T. Bach
Graduate School of Engineering, University of Tokyo

ABSTRACT: Concrete engineers have developed RCC (roller compacted concrete) and CSG (cemented sand and gravel) as low cost concrete for construction of dams, which are mixture materials of aggregate, cement, and water with lower cement contents than usual concrete. Recently, geotechnical engineers are developing cement-mixed well-graded gravels as high quality geo-materials for backfilling and embankments. Such materials are similar to each other in the quality of well-graded gravel (or aggregate), mixture ratio of cement and water, and compaction quality control methods. Thus, these materials could be dealt with unified concepts of evaluation, designing, and construction methods. In this view point, the authors have compared these materials on the basic properties of compaction and strength characteristics. In this paper, as an application of such materials, bending behaviours of scaled model beams consisting of cement-mixed well-graded gravel with reinforcement are discussed in comparison with steel-reinforced concrete beams.

1 INTRODUCTION

1.1 *Concrete and cement-mixed gravel*

Concrete engineers have developed RCD and CSG methods, which use mixture materials of aggregate, cement and water with lower cement contents than usual concrete (Nagayama and Jikan, 2003). Such materials can be filled and mechanically compacted as like geo-materials to construct massive structures of dams. On the other hand, geo-technical engineers are developing cemented well-graded gravels as a new backfill material, in order to construct more rigid and stable soil structures. An example of railway bridge abutment with such materials with geogrid reinforcement is reported by Kongsukprasert, et al. (2005).

These materials are quite similar to each other in the quality of aggregate (or well-graded gravel), mixture ratio of cement and water, and compaction quality control methods (Uchimura et. al., 2006). So, they could be dealt with unique concepts, intermediate between two disciplines of concrete engineering and geotechnical engineering, on their material design, structure design, quality control, and construction methods. Figure 1 shows typical ranges of mixture ratios of several materials mixed with cement. With this chart, we can compare these materials in a unified way. The lateral axis (w/g = weight of water per weight of aggregate) and the vertical axis (c/g = weight of cement per weight of aggregate) are in geotechnical terms. Then, each straight line from the origin corresponds to a water-cement ratio (w/c), which is the most important index in concrete engineering. Uchimura et. al. (2006) compared these materials on their compaction and strength characteristics with some experimental results, showing Figures 2a and 2b as conclusions. In these figures, the compaction density is the ratio of the weight of well-graded gravel versus the total volume, excluding the weight of cement and water. The compaction dry density of the gravel (excluding the weight of cement) with a standard energy of 4.56 MN/m^3, and the triaxial peak strength of the specimens with a confining pressure of 20 kPa are measured with various mixture ratios, and shown with their contours respectively. For the triaxial tests, the specimen density was chosen as the value obtained from the compaction test at each mixture ratio, so that the strength was evaluated under the same compaction energy. It was found in Figure 2a that the higher cement content results in higher optimum water contents and lower compaction density if tested under constant compaction energy.

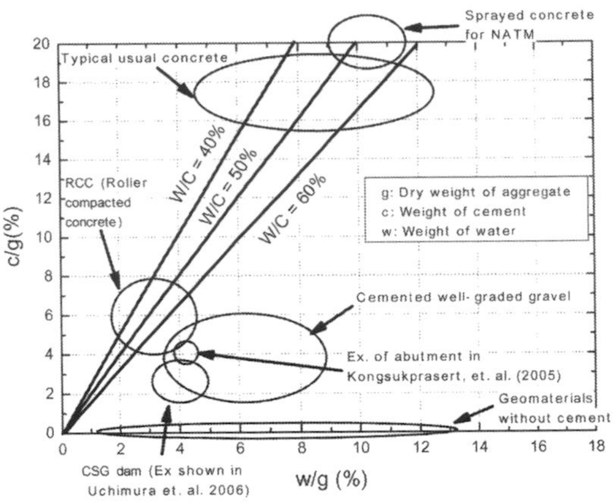

Figure 1. Mixture ratios of various materials with cement.

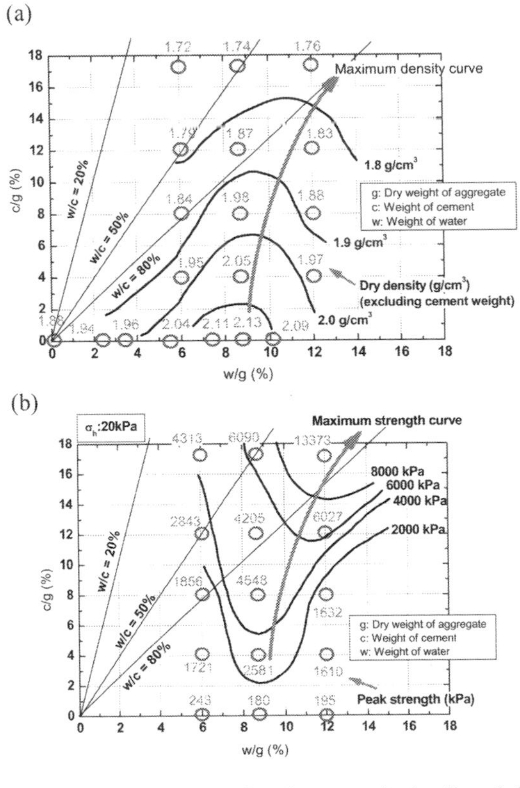

Figure 2. Material properties of cement-mixed well-graded gravel: a) compaction density, and b) triaxial strength.

Figure 2b shows that the strength of the specimens after curing was strongly dependent on the compaction density, rather than the effect of the cement-water ratio which is a dominant parameter in concrete engineering, if compared with the same cement contents.

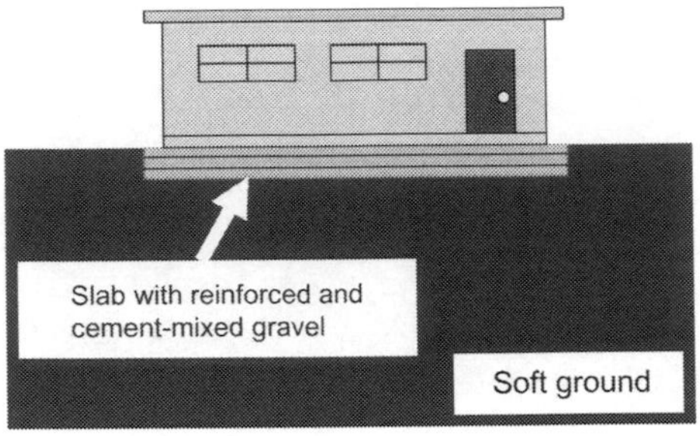

Figure 3. Example of use of reinforced and cement-mixed well-graded gravel for a foundation on a soft ground.

1.2 *Concepts of composite structure with reinforced and cement-mixed well-graded gravel*

In this paper, a combination technique of reinforcement with cement-mixed well-graded gravel is discussed. One of possible applications of this technique is schematically shown in Figure 3. When a small building is to be constructed on a soft ground, a slab of cement-mixed well-graded gravel is constructed with a high quality compaction, with layers of geogrid sandwiched. Such foundation structure could be low-cost and effective to prevent harmful differential settlement, as well as to achieve high seismic stability. High strength and rigidity are required to the slab, but it is hopefully ensured by the cementation and the function of geogrid layers. As the material components and structure are similar to those of reinforced concrete, the characteristics of grid-reinforced and cement-mixed well-graded gravel are possibly discussed in analogy with steel-reinforced concrete.

2 MODEL TESTS ON MODEL BEAMS

2.1 *Test method*

Bending tests were performed on short beams made of cement-mixed well-graded gravel, with one or two layers of geogrid or metal grid, and their deflection and failure sequences were observed.

The beam models were compacted in a steel mould into a rectangular shape (Length: 500 mm, Height: 200 mm, Width: 150 mm) as shown in Figure 4. The backfill material was well-graded gravel ($Gs = 2.71$, $D_{max} = 10$ mm, $D_{50} = 2.03$ mm, $Uc = 15.8$, fine contents = 4.3%), which is the same as the material used in the element tests shown in Figure 2. After standard portland cement and water were mixed with prescribed ratios, it was compacted to be the density shown in Figure 2a according to each mixture ratio assuming the same compaction energy. The mixture ratios and compaction density of the material were evaluated in the similar way

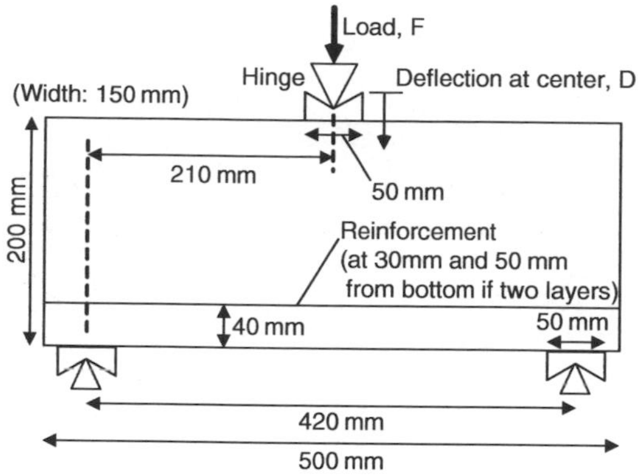

Figure 4. Beam models for bending loading tests.

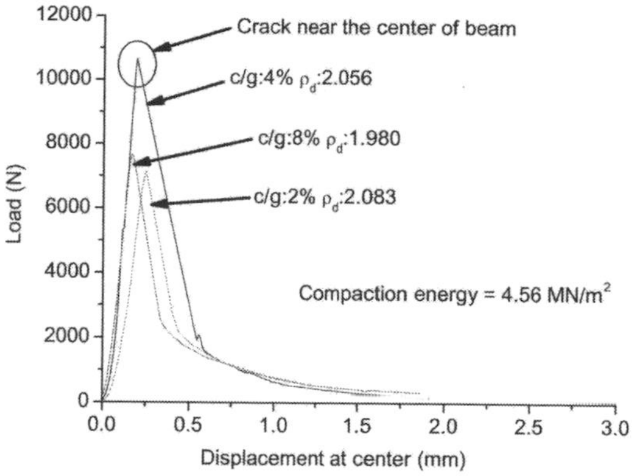

Figure 5. Behaviours of unreinforced beams with various cement contents.

as Figures 1 and 2. A geogrid (nominal rapture strength: 49.4 kN/m, aperture: 20 mm) and a metal grid (alminum, thickness: 1 mm, aperture: 10 mm, rib width: 3 mm) were used as reinforcement. They were laid at the height of 40 mm from the bottom of the beam for the models with one layer, and at 30 mm and 50 mm from the bottom for the models with two layers.

The beams were supported with hinges at both side of bottom with a beam length of 420 mm, and vertically loaded with a constant displacement rate at the center of the top surface with a hinge to make bending deformation. Steel plates with a length of 50 mm, which were attached to the surface of beam with gypsum, were placed at the hinges to prevent local failure due to stress concentration.

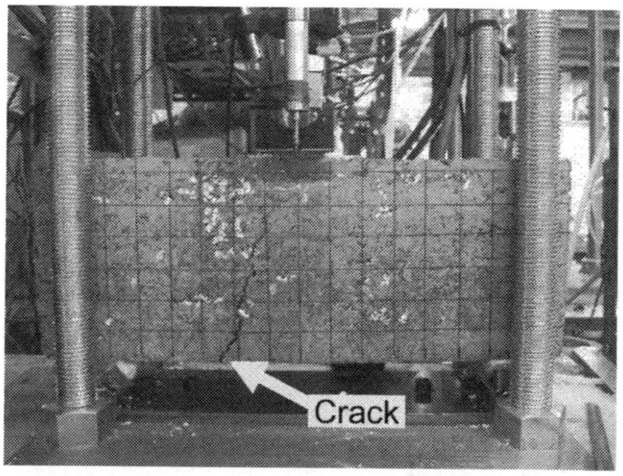

Photo 1. Failure of unreinforced beam model (c/g = 4%).

2.2 *Behaviours of unreinforced beams*

Figure 5 shows the relationships between the deflection and the load at the center of the beam without reinforcement. The cement mixture ratios for the backfill were and c/g = 2%, 4%, and 8%. The water contents was w/g = 8.75% for all the models, which is nearly the optimum water content for these cement ratios as shown in Figure 2a.

The curves show sharp peak strength at very small deflection of sub-millimeter order. A nearly vertical crack starting from the loading point was found at the peak stage of each beam as shown in Photo 1. The peak strength of the model with c/g = 4% was higher than that with a lower cement contents of c/g = 2%, to be easily understood. However, the peak strength of the model with higher cement contents of c/g = 8% was lower than that with c/g = 4%, probably due to the compaction density with c/g = 8% lower than that with c/g = 4% under the same compaction energy. Thus, the compaction density of cement-mixed well-graded gravel is important, as well as the cement contents.

2.3 *Behaviours of geogrid-reinforced beams*

Figure 6 shows the relationships between the deflection and the load at the center of the beam with one layer of geogrid, with c/g = 0% (uncemented), 2%, and 4%, and w/g = 8.75%.

As seen in Figure 6b, the beams with cement showed sharp peak strength at the beginning of loading, with a crack found at the center of the beam. The load and deflection at the peak were similar to those of the unreinforced models with corresponding cement contents shown in Figure 5. These results show that the reinforcement was not working at this stage with such small deformations.

After the peak, the load dropped similarly to the unreinforced models. However, some load was retained due to the effect of reinforcement, and the load started to increase again. At several stages, the load dropped with some extent, due to separation of the backfill from the surface of geogrid (Photo 2). The tension in geogrid was still functioning to resist the bending force even after the separation, and the load

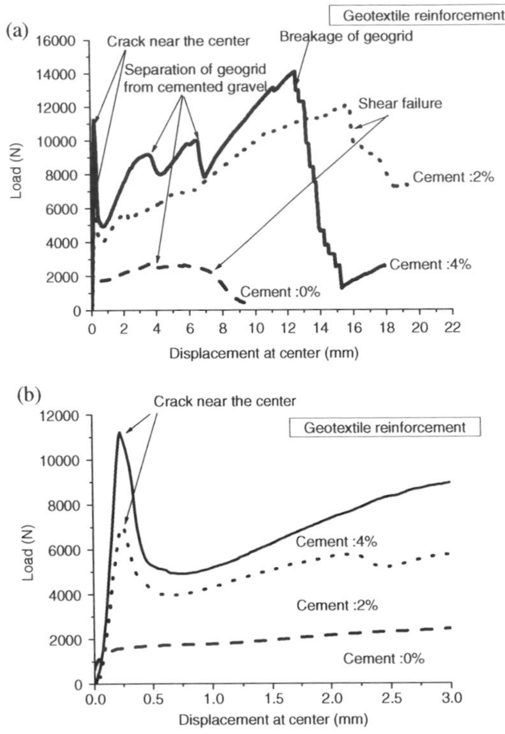

Figure 6. (a) Behaviours of geogrid-reinforced beams, (b) Detailed plot of (a) for the beginning of loading.

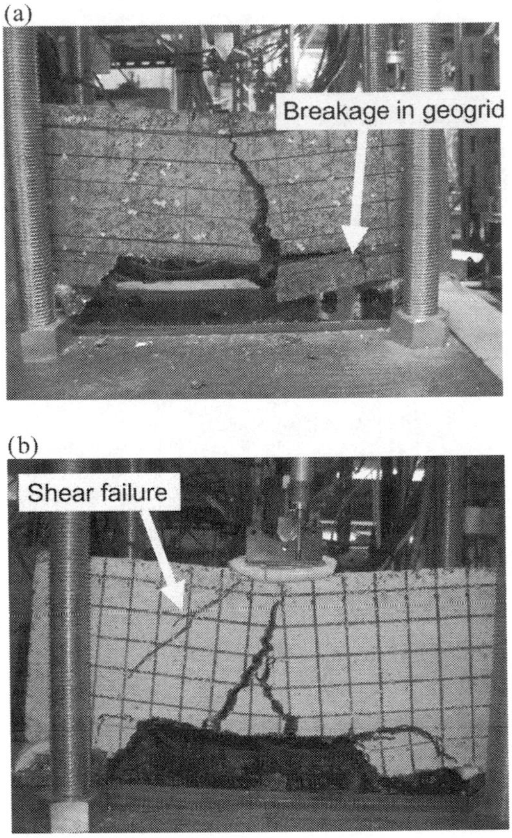

Photo 3. Ultimate failure of reinforced beam: a) breakage in geogrid (c/g = 4%); and b) shear failure in backfill (c/g = 2%).

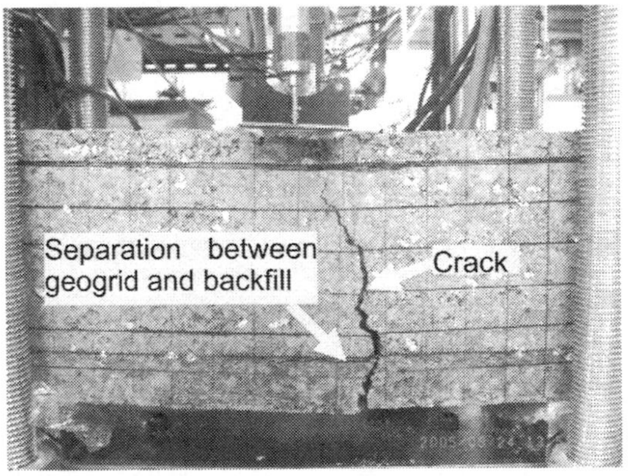

Photo 2. Crack at the center and separation along geogrid observed in reinforced beam (c/g = 4%).

continued to increase. Finally, the ultimate strength, which was higher than the first peak strength, was observed at breakage of failure in the geogrid or shear failure in the backfill (Photo 3a,b). The breakage in geogrid was observed at one end of the area of separation, not at the center of the beam.

The ultimate failure was observed at deflection of 12 to 16 mm, which is much larger than the deflection at the peak strength observed in the unreinforced beams. This means that the reinforced cement-mixed soil structures have significantly higher ductility than to unreinforced cases.

The tension in the reinforcement at the vertical crack near the center of the beam was estimated with a simplified assumption that the moment due to the vertical loading is fully supported by the tension in the reinforcement as shown in Figure 7. Figure 8 plots the obtained tension in the reinforcement when the first separation was observed against cement contents of the beams. The estimated tension at separation shows a good correlation with the cement contents. This correlation is probably corresponding to the difference in cohesion between the backfill material and reinforcement surfaces. This means that higher cement content contributes to the ductility of reinforced and cement-mixed well-graded gravel structures.

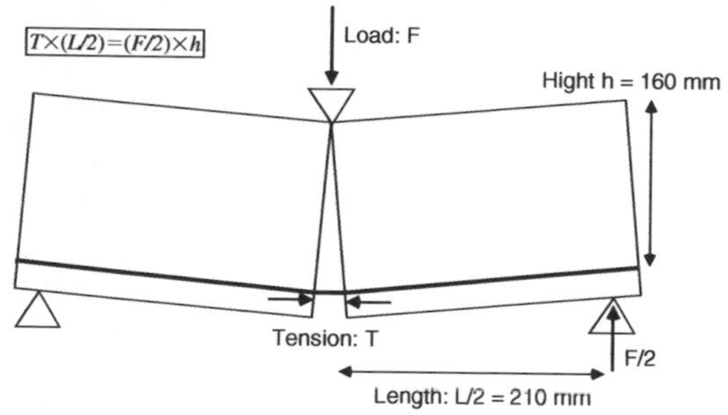

Figure 7. Schematic diagram of force equilibrium in a beam with reinforcement.

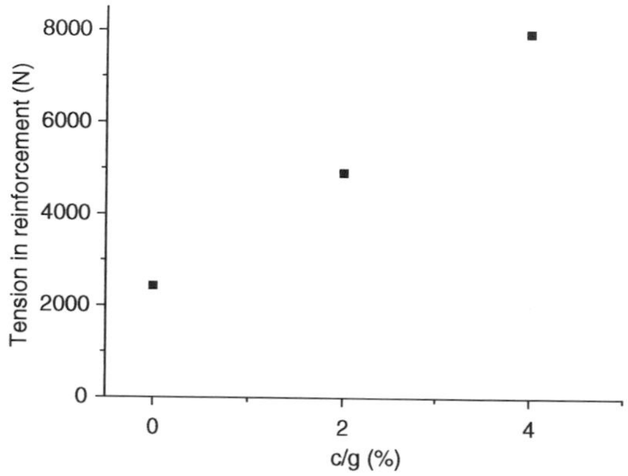

Figure 8. Relationships between load at separation along geogrid and estimated tension in geogrid.

2.4 *Behaviours of beams with stiffer reinforcement*

Figure 9 shows the relationships between the deflection and the load at the center of the beam with one layer of metal grid, with c/g = 2% and 4%, and w/g = 8.75%. Figure 10 compares its behaviour with those of the unreinforced beam and geogrid-reinforced beam, which are already shown in Figures 5 and 6, under the same cement content of c/g = 4%.

The metal-grid-reinforced beam again showed a peak at the beginning of loading, which is larger than the peak strength of the unreinfoeced and geogrid-reinforced beam. However, the load did not drop drastically, and then gradually increased to the ultimate failure. The ultimate failure mode was rupture in the metal grid or shear failure in the backfill depending on the cement contents. These behaviours are similar to what are typically observed in bending tests on steel reinforced concrete beams.

Figure 11 shows the relationships between the deflection and the load at the center of the beam with single and double reinforcement layers of geogrid,

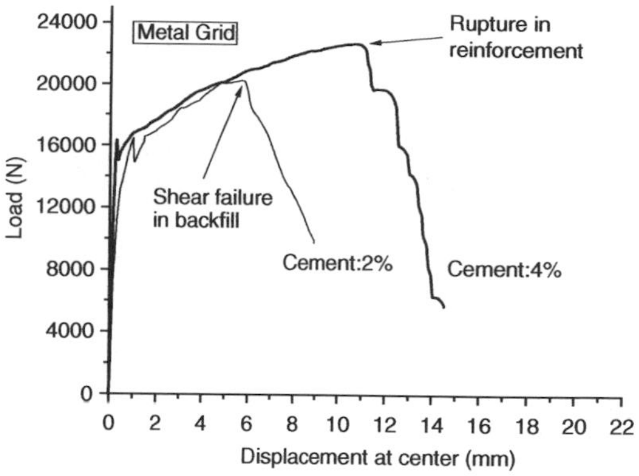

Figure 9. Behaviours of beams with metal reinforcement.

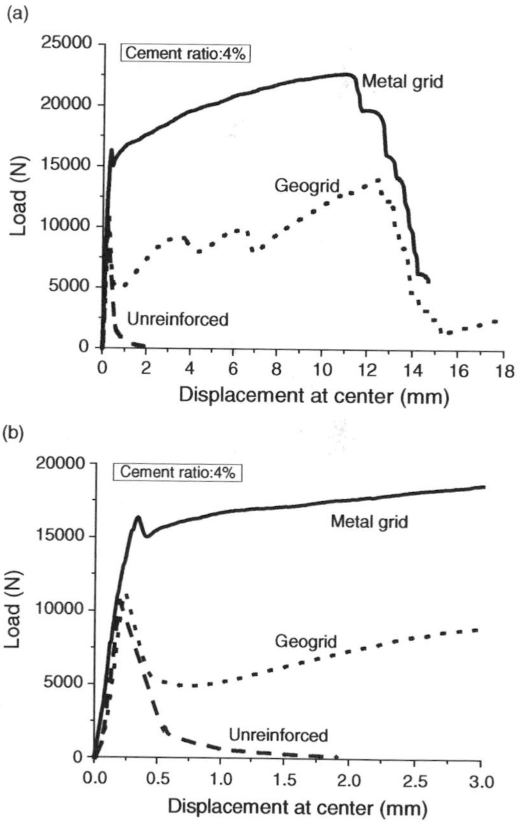

Figure 10. (a) Comparison of behaviours unreinforced, geogrid-reinforced, and metal-grid-reinforced beams; (b) Detailed plot of (a) for the beginning of loading.

with c/g = 4% and w/g = 8.75%. The behaviour of the beam with double reinforcement layers is similar to that with single layer. However, the amount of drop in the load after the first peak was smaller, and the

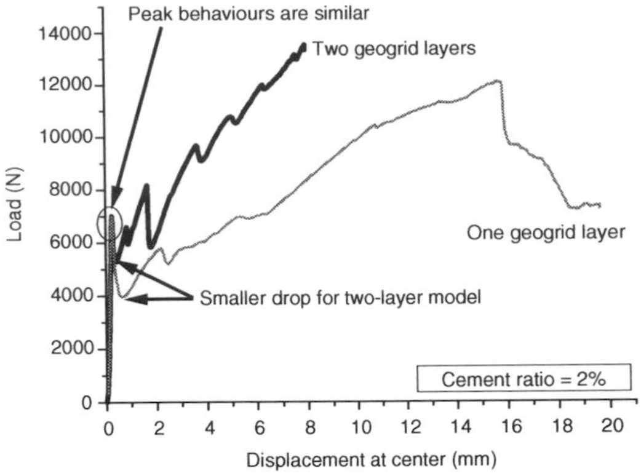

Figure 11. Behaviours of beams with single/double reinforcement.

ultimate strength was higher than the beam with single layer. It could be reasonable to understand that the beam became in an equilibrium state at smaller deformation of the double layer of reinforcement, and that resulted in lower drop in the load after cracking.

Thus, we can conclude that the behaviours of reinforced and cement-mixed well-graded gravel structures are largely affected by the stiffness of reinforcement, and it behaves like a reinforced concrete when the reinforcement is stiff enough.

3 CONCLUSIONS

Model beams made of reinforced and cement-mixed well-graded gravel were tested with bending load, and their characteristics were discussed with view points of concrete engineering and geotechnical engineering.

As like steel reinforced concrete beams, the reinforcement layers worked as a tensile member to resist to the bending force. The beams with reinforcement showed much more ductile behaviours compared to those of unreinforced beams. And their failure sequences are strongly affected by the stiffness of the reinforcement. When metal reinforcement was used, the beams behaved like reinforced concrete beam.

Besides, the beams showed some characteristics which are more familiar to geotechnical engineers. Not only the cement contents and water-cement ratio, but also the compaction density of the backfill material was an important factor for the strength of beams. The grid reinforcement did not contribute to the behaviour before a crack is observed at the center of beams, and after that, the load was dropped, and became supported by the tension in the reinforcement.

The progressive separation between the backfill and the reinforcement surfaces is also a specific characteristic of such structures.

In future, the reinforced and cement-mixed well-graded gravel could be useful as a new material who has higher quality than usual soils, as well as lower cost and better workability compared to concrete.

REFERENCES

Nagayama, I. and Jikan, S. : 30 Years History of Roller-compacted Concrete Dam in Japan, 4th International symposium on Roller Compacted Concrete (RCC) Dam, 2003.11.

Technical Material for Trapezoidal CSG Dam (2003). (in Japanese)

Technical Material for Cemented Reinforced Soil Abutment (2004). (in Japanese)

Kongsukprasert, L., Tatsuoka, F. and Tateyama, M. (2005): "Several factors affecting the strength and deformation characteristics of cement-mixed gravel", Soils and Foundations, Vol. 45, No. 3, pp.107–124.

Standard Specification for Concrete Structures -2002, Test Methods and Specifications, Japanese Society of Civil Engineering, 2002 (in Japanese)

Uchimura,T., Kuramochi, Y., and Bach, T.-T. (2006): Material properies of intermediate materials between concrete and gravelly soil, Proc. of Geotechnical Symposium in Roma. (to appear)

Uchimura, T. (2005): Intermediate Materials between Concrete and Geomaterials, Concret Journal, Vol. 43, No. 10, pp. 3–8, 2005. (in Japanese)

Kuramochi, Y., Kongsukprasert, L., Uchimura, T., and Tatsuoka, F. (2004): Effects of cement content and compaction degree on the strength and deformation characteristics of cement-mixed gravel., Proc. of Domestic Conference of Japanese Geotechnical Society, Vol. 1, pp. 793–794. (in Japanese)

Kuramochi, Y., Bach, T.-T., and Uchimura T. (2005): Mix proportion and compaction of intermediate materials between soil and concrete, Proc. of Domestic Conference of Japanese Geotechnical Society, Vol. 1, pp. 815–816. (in Japanese)

Uchimura, T. and Bach, T.-T. (2006): Bending loading tests on beams of cement-mixed well-graded gravel, Proc. of Domestic Conference of Japanese Geotechnical Society, Vol. 2, pp. 1835–1836. (in Japanese)

New Horizons in Earth Reinforcement – Otani, Miyata & Mukunoki (eds)
© 2008 Taylor & Francis Group, London, ISBN 978-0-415-45775-0

Support of MSE walls and reinforced embankments using ground improvement

F. Masse & S. Pearlman
DGI-Menard, Inc., Bridgeville, Pennsylvania, USA

R.A. Bloomfield
The Reinforced Earth Company, Vienna, Virginia, USA

ABSTRACT: Surcharge in combination with wick drains in highly compressible soils have traditionally been used to economically reduce post-construction settlements and construction time. Surcharging a wall structure or steep reinforced slope is more complicated given deep seated global stability concerns. Support of MSE structures with a ground improvement solution is economical for both cost and shortened time of construction. The use of Controlled Modulus Columns™ (CMC) is an ideal solution for the support of MSE Walls, steepened slopes and conventional embankments. CMC are pressure grouted auger displacement columns installed with a specially designed tool at the working end of a high torque, high down pressure drilling machine. To address sliding forces from retaining walls, high tensile geogrids are incorporated into a distribution layer and additional analyses are performed to check the lateral bending in the CMC elements. This paper summarizes the design approach and highlights two corresponding case histories.

1 CONTROLLED MODULUS COLUMNS

1.1 *Overview*

Controlled Modulus Columns (CMC) were developed first in Europe to meet technical, financial and quality requirements of a constantly more demanding ground improvement market. They are vertical semi-rigid inclusions designed to obtain a composite material (soil + inclusions) with controlled stiffness and they represent one of the best ground improvement technologies to date in terms of speed of construction, quality-control, reliability, range of applications and cost.

CMCs belong to the same class of ground improvement systems as the stone columns or the more recent vibro-concrete columns in the sense that they improve at the macroscopic level the overall stiffness of the foundation soils. More precisely, CMCs are filling the gap between the so-called rigid deep foundations (RDF such as piles, caissons and drilled shafts) and the more deformable foundation systems (DFS such as stone columns, rammed aggregate piers and dynamic replacement pillars). For RDF, the load of the structure is completely transmitted to the elements through a direct connection with the structure (pile caps or thick mats). For DFS, the modulus of deformation of these elements is compatible with the surrounding soils, creating a load sharing combination that results in a more deformable system with the structure supported on a load transfer platform usually made of densely compacted granular material. The CMC technology somewhat reconciles these two approaches by bringing together the advantages of both technologies into one hybrid solution, which offers better stiffness and better settlement reduction than the DFS without the difficulties and cost of a structural connection with the structure normally associated with RDF.

1.2 *Means and methods for the installation of CMCs*

The CMC technology has the following characteristics:

- A displacement hollow-stem auger is used to drill the inclusions. The auger has three main components: the bottom part penetrates into the ground and evacuates the cuttings upward – the middle part displaces the ground laterally by pushing the cutting to the sides – the upper part, with its flights in the reverse direction from the bottom part prevents any spoil or grout to reach the surface.
- In order to penetrate most ground, a high torque – high pull down drill rig is necessary.

In constructing CMC, the auger is first introduced into the ground and is advanced using the high torque and pull down available on the rig. No grout is inserted

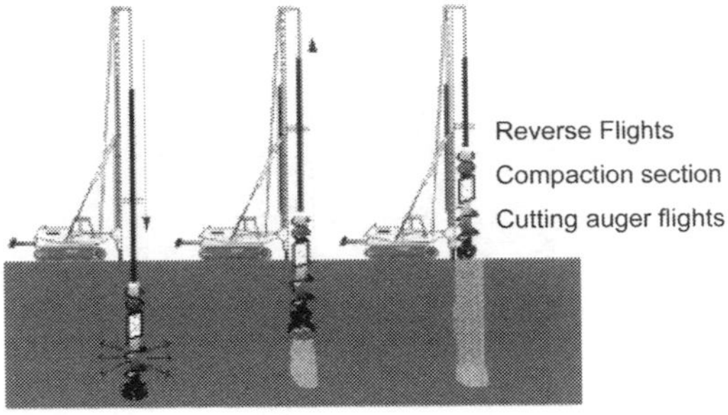

Figure 1. Installation of a controlled modulus column (CMC).

at this stage. When the required depth is reached, the grout is pumped through the hollow stem of the auger with sufficient pressure to overcome the gravity and lateral pressures at the tip of the auger. The auger is then extracted while turning in the same direction as during the drilling phase in order to avoid loss of grout and spoil migration along the shaft of the hole and along the kelly bar thanks to the reverse flights of the upper part of the displacement auger. (figure 1)

This results in virtually no spoil at the surface and no vibration during the whole process. The use of CMC is highly recommended for sites with constraints such as vibration limitations or for projects located on contaminated grounds as it eliminates the need for disposal of spoils.

Each inclusion is monitored by an on-board instrumentation device that can record the following parameters:

- Drilling phase: speed of penetration, torque, pull-down, depth, speed of rotation of the auger
- Grouting phase: pressure of grout, volume of grout, speed of rotation and extraction

The integration of these parameters by the on-board computer allows the visualization in real-time of the actual profile of each column. All the parameters can be recorded for later reporting.

2 USE OF CMC FOR SUPPORT OF MSE WALLS AND REINFORCED SLOPE EMBANKMENTS

When designing a CMC solution for support of an embankment (MSE Wall or reinforced slope), several factors have to be taken into account:

- Bearing capacity requirements
- Final elevation of the road as compared to the initial existing grade (loads)
- Width of influence of the embankment (i.e. depth of influence of new stresses)
- Risks of slope failure (i.e. global stability)
- Analysis and design of the load transfer mechanism
- Lateral spreading and lateral displacement of the elements.

CMCs are designed using numerical modeling techniques that include the effects of load sharing from the wall or the slope to the distribution layer, the columns and the surrounding improved soils. In particular, it is critical to understand the behavior at the interface MSE Wall/Load Transfer Platform (LTP)/CMC in order to accurately model the load transfer mechanism and to avoid large differential settlement of the CMC into this layer.

Design calculations are usually performed using PLAXIS or an equivalent software package and leads to the selection of the spacing of the CMCs. Depending upon the amount of tolerable construction settlement, i.e., sufficient strain to engage the tensile strength of a geogrid, the necessity of reinforcement by geogrid in the transfer layer can be selected. The evaluation also gives the stresses in the ground and in the column resulting from the stress distribution model. It is thus possible to refine the design parameters (diameter of the columns, grid of installation, thickness of the transfer layer and compression strength of the grout) to optimize the total cost of the solution. Once the design parameters have been chosen at the discrete level of a single column, a global elasto-plastic calculation can be performed using the same numerical modeling program to take into account specific boundary conditions such as:

- variable height of fill along the same section or non-symmetric loading conditions
- horizontal loads due to train braking friction on tracks for a railway embankment
- rapidly varying thickness of compressible ground along a given section
- variable CMC grid of installation

This second calculation usually allows the confirmation of compliance with the deformation criteria for the structure and allowable stresses inside the columns. Because 2D elements are usually used for the calculations, it is necessary to replace the layer of CMC + surrounding soil(s) by a global uniform layer defined by equivalent characteristics.

3 CASE STUDY #1: (1H:4V) MSE WALL IN KINGSTON, JAMAICA

The project consisted of the construction of a new section of a Tollway located on the coastal shores between Portmore and Kingston, Jamaica. The 7 km section goes through a mangrove swamp underlain by an organic peat layer and very soft clay layers up to 22 m deep. The road was generally set about 2 to 3 m above the existing ground elevation and wick drains

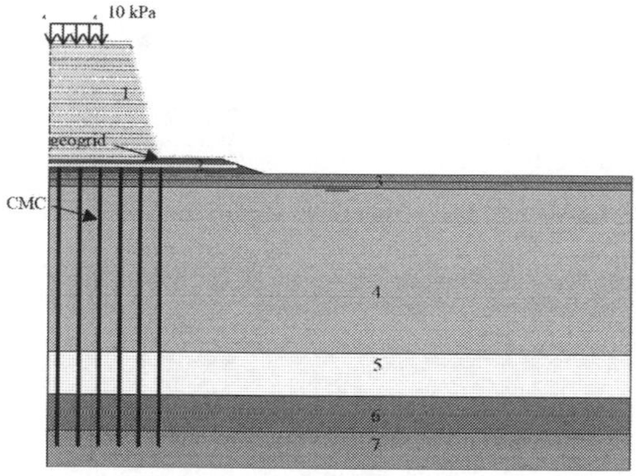

Figure 2. Typical numerical model – MSE embankment (1H: 4V) and CMCs – not scaled.

Figure 3. Site works in Jamaica – View of the bridge abutment area.

and surcharge were designed to accelerate the consolidation of the compressible layers. Nevertheless, three major interchanges, overpasses and a toll plaza were also to be built and the schedule did not permit the use of a classical solution for the consolidation of these compressible layers. The approach embankment to these overpass bridges reached up to 10 m on some portions of the highway. A CMC-supported embankment solution was designed for these sections of the job. A standard unreinforced embankment using 2H:1V or 3H:1V slopes was not feasible without creating additional costly and time-consuming requirements for land purchase. A solution using a steep (1H:4V) geosynthetic-reinforced embankment with a wire-faced MSE approach supported by CMC was designed for the project (Figure 2).

The project overall specification was to limit the residual settlement after the opening of the road to 200 mm over the next 35 years and to ensure a long term factor of safety against slope failure above 1.3. To deal with long term secondary consolidation of the soft compressible layers and to accommodate the differential settlement between the bridge on piles and

Figure 4. Wire-faced MSE Wall – 1H:4V Wall.

the approach abutments, two resurfacing programs were priced into the overall cost of the maintenance of the highway by the contractor.

The construction sequence was selected based on the results of the evaluation. As shown below, the stress ratio used for the soft layer was around 98%, which showed limited long term settlement due to the load of the MSE wall. It was therefore decided that no surcharge or construction of the wall in steps was necessary to meet the long term settlement requirements of the project.

The results of the 2D numerical analysis are summarized below:

Case	MSE wall
Embankment height	9 m
CMC length	19.5 m
Geogrid reinforcement	1 layer
Stress in the geogrid layers	79.1 kN/m
Geogrid deformation	3.9%
Settlement under static load from embankment	184.5 mm
Settlement under live loading from traffic	20 mm
CMC Load	400 kN
Stress distribution ratio	98%

4 CASE STUDY #2: REINFORCED EARTH EMBANKMENT – MIAMI, FLORIDA

4.1 *Description of the project*

This project involved the creation of an auxiliary ramp along the existing SR836 operated by the Miami-Dade Expressway Authority in Miami, Florida. Due to extremely constrained site conditions, the contract documents called for a design-built solution using a column-supported MSE wall for the abutments to the bridge. The wall height ranged from 4 m to a maximum of 10 m at the bridge connection with bearing pressures including overturning of up to 0.25 MPa. Part of the road extension was located under the existing slopes of the current road, while the rest was located beyond the existing slope, creating a risk of differential settlement as well as increased risk of slope failure. Also, part of the existing slope was "demucked" prior to construction to replace soft organic layers with crushed limestone material and sand, while the outer part of the wall was seated on highly compressible organic peat.

4.2 *Technical requirements/specifications*

Because of the presence of existing buildings along the alignment of the new lane, the settlement criteria were extremely strict for this project. The design-build specifications allowed "a maximum 1.3 cm deflection of the reinforced platform at the middle distance between two" inclusions. They also stipulate that the ground improvement system should be designed so as to "produce negligible settlements to adjacent structures and to avoid punch-through of the Load Transfer Platform". The total length of the project was 140 m for Wall 1A and 300 m for wall 3A, with reinforcing strip lengths varying from 3 to 8 m.

4.3 *Subsurface conditions/typical soil profile*

The available geotechnical information for the project was somewhat limited with less than 10 standard penetration test values obtained along the alignment of the wall outside the actual footprint of the wall. These borings showed heterogeneous conditions along the axis of the proposed lanes. The boring that indicated the poorest subsurface conditions was located at the bridge abutment where the wall was at its maximum height. The typical cross section was as follows:

- Upper 1 m, dense crust consisting of cemented sands, limerock fill with silts
- Below this crust, a 3 to 4.5 m thick layer of silt and organic silts (ML to CL-ML) with varying sand and clay contents, very soft to stiff depending on the location
- The bearing layer was comprised of a silty to clean sand (SP to SM) underlain by limestone.

4.4 *Design using numerical analysis*

In order to fully account for the pressure due to the MSE wall, it was necessary to incorporate in the calculations the effect of the moment created by the active forces acting at the back of the wall volume. The direct effect of this overturning moment was to create an eccentricity "e" in the line of application of the gravity load of the wall and the traffic load. As a result, the bearing pressure increased at the toe and decreased at the heel of the wall. The Meyerhof method was used to compute these pressures, whereby the weight of the

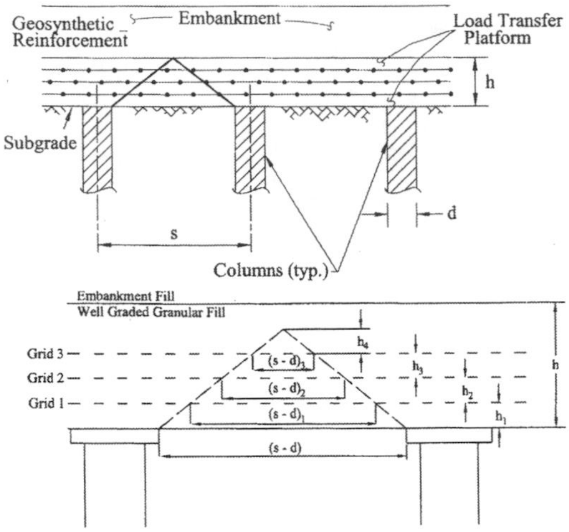

Figure 5. Load transfer platform design concept.

wall and the traffic load were assumed to be applied on a reduced width of B-2e (B is the strip length and e is the eccentricity). The overturning effect increased the bearing pressure under the wall by as much as 0.85 ksf at the maximum height of the wall.

4.4.1 *Design of the Load Transfer Platform (LTP)*

Based on the project requirements, an extensive design of the LTP was performed across the project. The design was based on the Collin method (figure 5). This method is based on the premises that the reinforcement (minimum of three layers of geogrid) creates a stiffened beam of reinforced soil that will distribute the load of the embankment or MSE wall above the LTP to the inclusions below the LTP. The primary function of the reinforcement in that case is to provide lateral confinement of the fill in the LTP to facilitate arching within the thickness of the LTP. The reinforcement supported the wedge of platform below the arch in order to avoid settlement in-between the inclusions.

The vertical load carried by each layer of reinforcement is a function of the column spacing and the vertical spacing of the reinforcement. Each layer of geogrid is designed to carry the load of the LTP within the soil wedge below the arch. As a result, the vertical load on any layer (n) of reinforcement (Wtn) can be determined from the equation below:

Wtn = [area of geogrid layer n + area of geogrid layer n + 1]/2 × layer thickness × LTP density/area of geogrid layer n.

The tensile load in the geogrid is then determined based on tension membrane theory and is a function of the strain in the reinforcement.

4.4.2 *Numerical analysis*

A series of Plaxis calculations were performed for different loading cases as well as different soil profiles.

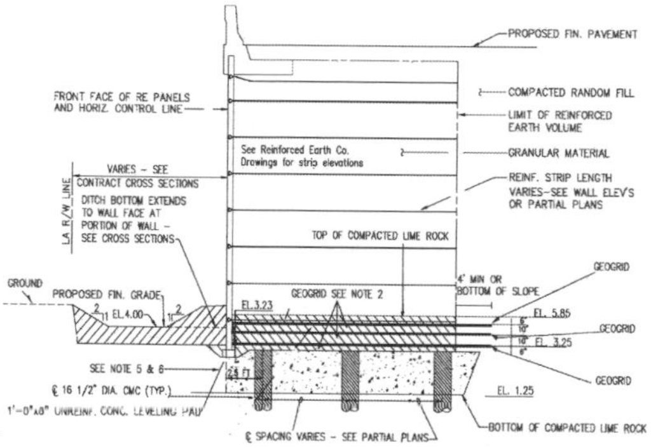

Figure 6. Typical cross-section of final design.

On wall 1A, an analysis was performed for a 6 m, 8 m and a 10 m high wall section with grids of installation of respectively 2.05 m, 1.83 m and 1.52 m (center-to-center on a square pattern). On Wall 3A (figure 6), an 8 m high wall and a 10 m high wall were analyzed. Long term settlement of the system ranged between 0.4 to 0.6 cm; well within the tolerance of the specifications.

A total of more than 950 CMCs were installed to support the two MSE walls.

4.5 *Site work pictures*

Figures 7 and 8. View of CMC rig and MSE Wall during construction.

5 CONCLUDING REMARKS

The Controlled Modulus Columns (CMC) foundation is one technique of ground improvement for support of industrial or residential structures as well as embankments (unreinforced and reinforced slopes) and MSE walls. The design methodology and case histories presented herein demonstrate the effectiveness of the CMC foundation system in terms of settlement performance and speed of construction. When used with a properly designed Load Transfer Platform, it provides suitable and economical support for MSE walls over compressible subsurface conditions.

CMC technology does not generate spoils and does not bring contaminated soils to the surface. In a challenging world where development of marginal sites is a necessity for the survival of future generations, this technology offers a competitive sustainable alternative to classical deep foundations.

REFERENCES

Collin, J.G. & al (2004) – FHWA – NHI Ground improvement manual – Technical summary #10: Columns supported embankment – FHWA – 2004

Dumas, C. & al (2003) Innovative technology for accelerated construction of bridge and embankment foundations in Europe – FHWA-PL-03-014, FHWA 2003

Masse, F. & al. (2004) CMC: potential application to Canadian soils with a new trend in ground improvement – CGS 2004, Winnipeg, Canada

Plomteux, C. & al (2003) – "Reinforcement of Soft Soils by Means of Controlled Modulus Columns" – Soil and Rock America 2003, pp 1687–1694

Plomteux, C. & al (2003) – "Controlled Modulus Columns (CMC): Foundation system for Embankment support: a case history" – Geosupport 2004, Orlando, USA, pp 980–992

Porbaha, A. & al (2007) – "Design and monitoring of an embankment on controlled modulus columns" – TRB paper #06-1743 – Transportation Research Board, 2007

Plaxis finite element code for soil and rock analysis – user's manual – Plaxis V8 – 2007

New Horizons in Earth Reinforcement – Otani, Miyata & Mukunoki (eds)
© 2008 Taylor & Francis Group, London, ISBN 978-0-415-45775-0

Two geogrid-reinforced steep slopes as combined structures on columns and piles: Case histories

D. Alexiew
Engineering Department, HUESKER Synthetic GmbH, Gescher, Germany

S. Jossifowa
GeoKraft, Sofia, Bulgaria

H. Hangen
Engineering Department, HUESKER Synthetic GmbH, Gescher, Germany

ABSTRACT: Geosynthetic-reinforced steep slopes and walls became very popular worldwide during the last 15 years because of their financial, technical, ecological and landscape-related advantages. Meantime, a wide range of solutions is possible based on the use of modern reinforcing geosynthetics. It is possible to combine them with other elements or systems e.g. piles or columns especially to ensuring external stability. Two specific geogrid-reinforced steep slopes set on columns or piles are described: The first one in Bulgaria on a steep slope in a seismic region, the second one in Germany on an instable slope of softer soils.

1 INTRODUCTION

Steep slopes and walls from geosynthetic-reinforced soil (GRS) became very popular because of their cost-effectiveness, blending in well with the landscape, fine-tuning for optimum functionality, ductile behaviour etc. (GRI 1998, Alexiew 2005). They become an increasingly adopted and well-established solution. The broad range of available geosynthetic reinforcement allows optimisation and eliminates any limits to height and load capacity.

Nevertheless, problems may arise like compound stability (Berg & Meyers 1997) as well as (quite often) external stability when the GRS-slope is positioned on steep or unstable natural slopes, sometimes in combination with seismic impact. In the latter cases "dowelling" columns or piles below the GRS-wall or slope can be used to increase the external stability to the level required.

Two typical projects of this type are shortly described.

2 PROJECT YUGOVSKO HANCHE, BULGARIA

A short stretch of a road in the Rhodopa Mountains in Bulgaria had to be stabilized and widened due to increasing traffic and indications of slope instability. During construction the traffic had to be kept running at least on the right lane of the road (Fig. 1), thus the existing old stone masonry retaining wall had to remain in place during widening and stabilization works.

Note, that both the terrain and the rock bed become even steeper to the right in Figure 1, which is not shown in the figure itself due to the lack of place. Additionally to the requirement of keeping the right half (on Fig. 1) of the existing road under traffic, the Road Authority asked for a solution which should be safe, easy to built, cost efficient and landscape friendly.

Critical sliding surfaces were identified below the road and the old wall through the local slope soil. It consists of silts and gravels (with some infiltrating water from the right on Figure 1) on a weathered rock substratum (Fig. 1).

Generally, the road and the masonry wall were built in the early 30ies. The stability analyses performed now were based on actual geotechnical data and resulted – especially with the planned widening to the left (Fig. 1) – in global factors of safety of 1.0 to 1.1 and even <1.0 under seismic conditions, which was alarming.

The stability analyses included both circular (Bishop) and polygonal (Janbu) failure surfaces according to DIN 4084.

A combined solution was found to be the optimal one (Fig. 1): a new GRS-wall was added in front of the

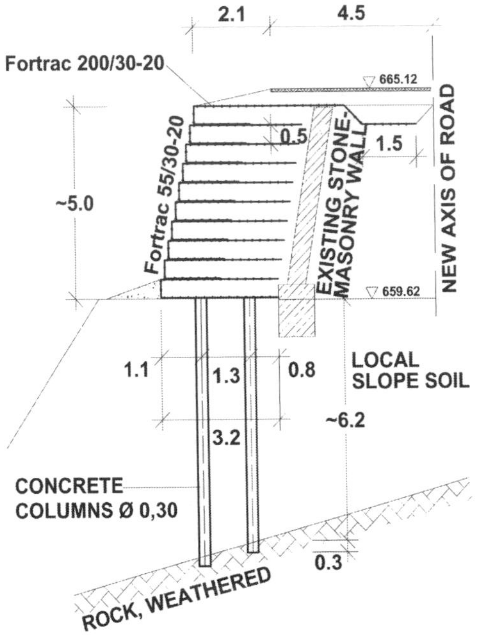

Figure 1. Yugovsko Hanche: typical geotechnical situation and solution with a combined system.

Figure 2. Yugovsko Hanche: 2/3 of the GRS-wall ready; old masonry wall to the left; geogrids not connected to it.

old stone masonry using crushed well-graded gravel and geogrids Fortrac®55/30-20; the GRS-package was founded on dowelling concrete columns. To make construction easier and to save time due to the critical situation (see above) the geogrids were not connected to the old retaining wall (Fig. 2). The high coefficients of interaction in both shear and pullout modi of the flexible geogrids used and the fill, evaluated in corresponding tests, made this efficient solution possible: no inner interface sliding could occur, and the design resulted in short anchorage lengths as well. Note: a

Figure 3. Yugovsko Hanche: anchoring of the upper final geogrid behind the old masonry wall (see Figure 1).

single stronger Fortrac®200/30-20 was applied on top and anchored behind the old wall and below the outer lane of the road to ensure integrity of new and old structure in both horizontal and vertical direction even in the case of an earthquake (Figs. 1 and 3).

Small diameter (0.3 m) concrete columns in two rows were installed below the GRS-wall to allow the use of light drilling equipment in the narrow space in front of the old wall; their position and spacing were optimised during design.

The combined system presented was cheaper and easier to build than e.g. a second new concrete or RC-wall. A new type of protective "green" facing was finally added to the wraparound geogrid wall (the so called "Muralex®") to meet the environmental (landscape) requirements. The entire system was successfully built up in less than two months by a local contractor without any problems despite its lack of experience with geogrid reinforcement, confirming the easiness of construction. An "inner formwork" of well sand-filled bags was applied during compaction in the front area. The technique is very easy and efficient in combination with the flexible geogrids used (Figs. 2 and 4).

The new widening and retaining system is since some years under traffic without any indications of movement, differential settlements etc. The solution seems to be successful.

3 RETAINING STRUCTURE NEAR OBERSTDORF, GERMANY

A new road had to be built in the mountains across a slope near Oberstdorf. A GRS structure turned out to be the optimum solution from the points of view of value for money and natural appearance. Above all, the project had to be completed inside two months: As the project was in a scenic mountainous area, ecology and economy made it necessary to use local cohesive soils

Figure 4. Yugovsko Hanche: completing GRS-wall and road widening (before installation of the final Muralex® facing).

Figure 5. Oberstdorf: failure due to insufficient external stability on the soft slope.

from the natural slope (clayey silts and silty clays) to be used as fill in the GRS.

The possible problem of external stability had not been really identified because of some lack of detailed geotechnical investigation and soil testing and overestimation of the shear strength of the soft natural slope especially when wet. Thus, the first GRS-structure failed quite quickly at beginning of construction because of insufficient external stability (Fig. 5). The part of the GRS-wall moved and rotated downslopes. It is a situation like in a learning book for soil mechanics.

Consequently, the consultants decided to put the GRS-wall on slim RC-piles with beams on top, bridging the 3 m distance between the beams also with geosynthetic reinforcement (Figs. 6 and 7). The idea was to use the typically ductile behaviour of GRS-systems being able to adapt some small slope movements (despite the dowelling piles) without problems. Thus, the system was different from the solution for

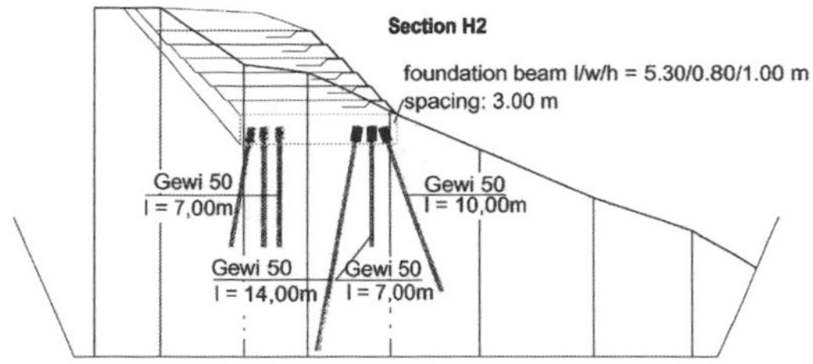

Figure 6. Oberstdorf: typical cross section of the retaining structure set on piles.

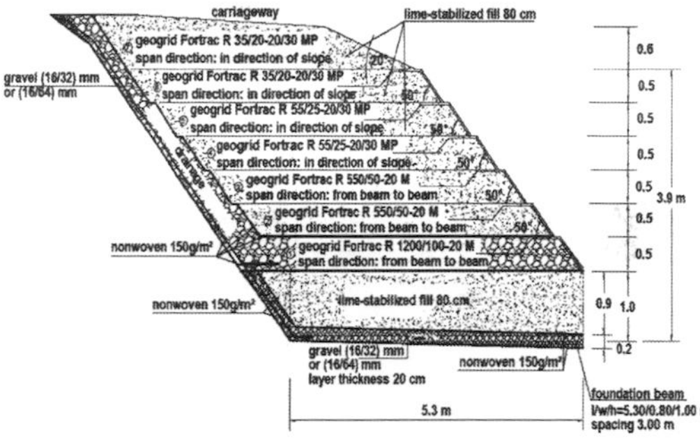

Figure 7. Oberstdorf: geogrid reinforcement (the bottom geogrids spanning over the beams are installed with their strong MD parallel to the road axis).

Yugovsko Hanche, but the same principle of external stabilisation with piles was applied. The design was based on existing methods of analysis and on experience dealing with pile-supported fill bodies (Alexiew & Vogel 2001, Alexiew 2006), applying uniaxially spanning design procedures in this case. The use of the cohesive local soils as fill could only be considered after lime-stabilisation, after which it would be strongly alkaline (pH > 12). Under such conditions, the use of Polyester is forbidden (FGSV 1994).

Polyvinylalcohol (PVA) as raw material is very appropriate in such cases. The geogrid family. Fortrac® M & Fortrac® MP from PVA Kuralon® was chosen. The reinforcement is durable in high pH environments and combines high tensile strength at low strains, very low creep (Alexiew et al 2000) and high coefficients of interaction to stabilised cohesive soils (Aydogmus et al 2006). Positive experience with many projects over recent years was already available. A special feature of this project is the twin load-carrying action of the lower geogrid layers, Fortrac®1200/100-20 M and Fortrac®550/50-20 M, which are installed with their strong unrolling machine direction (MD) of 1200 and 550 kN/m respectively parallel to the road axis and span over and between the beams. At the same time they act as "conventional" slope reinforcement transversely to the longitudinal axis of the road with their

Figure 8. Oberstdorf: installation of the first (bottom) geogrid layer parallel to the road axis.

Figure 9. Oberstdorf: front view of the GRS-body on the piled beams, deflection of the bottom geogrid layer between the beams (span 3 m).

weaker cross-machine directions (100 and 50 kN/m respectively), see Figure 7.

Figure 8 shows the installation of the first geogrid layer, and Figure 9 the deflection between the beams (at 3 m distance) after installation and compaction of the first fill layer. This deflection is typical for pile bridging systems and is evened out as more of the system is constructed. High strength low-creep reinforcement gives rise to no significant deformation in service (Alexiew 2006).

At Oberstdorf this was confirmed once more.

The GRS-structure is since some years in operation without any problems. Both the very specifically reinforced GRS-body and the piling system seem to be designed and constructed appropriately.

4 CONCLUDING REMARKS

Retaining walls and slopes from geosynthetic-reinforced soil (GRS) are very flexible and adaptive in shape and geometry and demonstrate an advantageous ductile behaviour in the sense of soil mechanics. The wide range of modern geosynthetic reinforcements (especially geogrids) eliminates practically any limitation to heights and loads in terms of internal and compound stability. Today it is possible to find an optimal (from the point of view of geometry, loads, ultimate and serviceability limit states and specific environment) appropriate reinforcement for any GRS-wall or slope. Possible problems with the external stability can be solved successfully by combining GRS-walls with other techniques like columns and piles. The two projects presented shortly herein from two European countries – designed and constructed for different purposes and subject to different restraints and requirements – are believed to demonstrate the above mentioned.

REFERENCES

Alexiew, D., Sobolewski, J., Pohlmann, H. 2000: *Projects and optimised engineering with geogrids from "non-usual" polymers.* Proc. 2nd European Geosynthetics Conference EuroGeo 2000, Bologna, pp. 239–244.

Alexiew, D., Vogel, W. 2001: *Reinforced embankments on piles for railroads: German experience.* Proc. XVth International Conference on Soil Mechanics and Geotechnical Engineering, Istanbul, 2001. A. A. Balkema, pp. 2035–2040.

Alexiew, D. 2005: *Design and construction of geosynthetic-reinforced "slopes" and "walls": commentary and selected project examples.* Proc. 12th Darmstadt Geotechnical Conference. Darmstadt Geotechnics No. 13, TU Darmstadt, Institute and Laboratory of Geotechnics, Darmstadt, March 2005, pp. 167–186.

Alexiew, D. 2006: *Piled embankments: Overview of methods and significant case studies.* XIII. Danube-European Conference on Geotechnical Engineering, Ljubljana, May 2006, Vol. II, pp. 185–192.

Aydogmus, T., Alexiew, D., Klapperich, H. 2006: *Evaluation of interaction properties of PVA geogrids in stabilized cohesive soils: Shear and pullout tests.* Proc. 8th International Conference on Geosynthetics, Yokohama, September 2006, pp. 1427–1430.

Berg, R.R., Meyers, M.S. 1997: *Analysis of the collapse of a 6,7 m high geosynthetic-reinforced wall structure.* Proc. Geosynthetics '97. Long Beach, Vol. 1, pp. 85–114.

FGSV 1994: *Merkblatt für die Anwendung von Geotextilien und Geogittern im Erdbau des Straßenbaus*, FGSV, Cologne. (*Recommendations for the use of geotextiles and geogrids in eartworks for roads*)(new Issue 2005)

Geosynthetic Research Institute GRI 1998: *GRI Report No. 20 from June 18.* GRI, Drexel University.

Jossifowa, S., Alexiew, D. 2002: *Geogitterbewehrte Stützbauwerke und Böschungen an Autobahnen und Nationalstraßen in Bulgarien.* Geotechnik 25 2002, Nr. 1, Essen, pp. 31–36.

New Horizons in Earth Reinforcement – Otani, Miyata & Mukunoki (eds)
© 2008 Taylor & Francis Group, London, ISBN 978-0-415-45775-0

Reinforced soil wall and approach embankment for Cliff Street overpass constructed on stabilized foundations

N. Fok
VicRoads Geopave, Melbourne, Victoria

G. Power
The Reinforced Earth Company, Sydney, Australia

P. Vincent
Austress Menard Pty Limited, Sydney, Australia

ABSTRACT: The Portland Transport Strategy called for a new bridge to be constructed over the Henty Highway and Portland/Hamilton railway line at Cliff Street, Portland in Victoria. The three span trough beam bridge was to be supported on piles inside Reinforced Soil Structures (RSS) abutments. The foundation for the Northern approach to the new bridge consisted of unconsolidated fill underlain by alluvium consisting of soft to firm clayey/sandy silt to a depth of 10 m below the existing surface. Groundwater table was measured to fluctuate between 1 m and 3 m. The northern RSS abutment was supported on a 400 mm deep layer of cement stabilized sand raft, reinforced with geogrids and geotextile that was constructed on top of an arrangement of capped reinforced concrete (RC) piles. The balance of the approach embankment was supported on an arrangement of shallow stone columns constructed using the Dynamic Replacement (DR) technique.

This paper will detail the development of the design from initial investigation through to construction with particular emphasis on the design of the foundation stabilisation works and the interaction with the RSS abutment and verification of the ground improved by DR during construction. The performance of the approach embankment and the RSS abutment is being monitored and the actual overall settlement will be compared to the prediction of 50 mm maximum after two years on completion of the construction. Cliff Street overpass opened for traffic in January 2007. There has been no report of cracking of the road pavement constructed on the approach embankment fill thus far.

1 INTRODUCTION

The township of Portland was established in 1834 and since then it has become one of the major seaports in Australia. The Port of Portland is a deep-water bulk port strategically located between the capital city ports of Melbourne and Adelaide. It has facilities capable of handling the berthing of all types of bulk and general cargo vessels. The port is well served by a road and rail network. Typical daily truck movements to and from the port are forecast to grow by 225% by 2030 (VicRoads Report 2004). The Cliff Street overpass project consists of a 3-span bridge with prestressed concrete beams as the superstructure. The main span is approximately 35 m long with two end spans each approximately 20 m long. The Project was approved in 2005 with an estimated cost of $15 million. The construction of the Cliff Street Overpass project was awarded to Akron Construction Pty Ltd.

This paper presents the findings of the geotechnical investigations at this site. It also provides detailed discussion on the design, construction, and testing of piles, ground improvement and a reinforced soil structure.

2 GEOLOGY

The south side of the site comprises Quaternary age igneous rock consisting of Iddingsite basalt which comprises the volcanic flows of the Portland area (Geological Survey of Victoria). Basaltic materials were not however encountered beneath the actual bridge site. The northern side of the site comprises Quaternary alluvium consisting of flood plain and river terrace deposits. At depth, the entire site is underlain by a sequence of Quaternary age weathered calcareous sands (aeolinites). From historical maps of the site, it appears that a pre-existing drainage channel near to the northern area of the site has been in filled and replaced with a man-made canal.

3 PERFORMANCE CRITERIA FOR CLIFF STREET OVERPASS

The following performance criteria were specified:

- Bridge foundation – maximum differential settlement 10 mm
- Reinforced soil wall facing panels – maximum
- differential movement 10 mm, and
- Bridge approach fill embankment – maximum differential settlement 50 mm.

4 GEOTECHNICAL INVESTIGATION

A geotechnical investigation was undertaken at this site by VicRoads GeoPave with reference to Australian Standard 1726 (AS1726). Because of the expected difficult ground conditions, the investigation sites were carefully selected to ensure that adequate geotechnical information would be available for design and construction purposes. Details of the field-work are as follows:

- Thirteen (13) investigation boreholes, four of which have incorporated a standpipe for subsequent groundwater monitoring. Four of the boreholes were extended to a depth of 60 m in order to ascertain the strength of the underlying weathered limestone (calcareous sand),
- Eight (8) test pits for the proposed road improvement work,
- Five (5) Cone Penetration Tests (CPT), and
- Four (4) test holes to determine the existing pavement composition at tie-in locations between the proposed road improvement work and the existing road pavement.

The boreholes were advanced by auguring and wash boring. NW casing was required to prevent caving of the boreholes when drilling encountered very loose silty sands and/or soft clays. Field sampling and testing of the soils consisted of Standard Penetration Tests (SPTs) in sandy soils and undisturbed tube samples in predominantly clayey soils. Sampling was undertaken at approximately 1.5 m intervals. Drilling was extended below the soft material until a suitable founding medium was found, typically 60 m at the piers and north abutment location. Following the finding of the weak layers from the drillings, it was necessary to determine the extent and frequency of these layers. CPT was selected to supplement the drilling investigation. The CPTs were performed using penetrometer test vehicle. CPT testing was extended to effective refusal at depths ranging from 17.6 m to 45.6 m.

Figure 1. Typical test pit showing the condition of the uncontrolled fill.

Table 1. Generalized sub-surface conditions.

Thickness	Description
0–4.0 m	Uncontrolled FILL.
0–5.6 m	Extremely weathered BASALT (south side only)
2.0 m–15 m	Firm to stiff silty CLAY,
8.5 m–40 m	Loose to very dense calcareous SAND.
26 m–36 m	Very soft clay (North side location only)
36 m–60 m	Medium dense to very dense calcareous SAND.

Test pits performed in the uncontrolled fill, Figure 1, confirmed the presence of weak material and shallow ground water depth. All of the test pit walls collapsed drilling excavation.

Table 1 summarizes the subsurface conditions at this site and the strength parameters adopted in the pile design. The overlapping of layer thickness indicates the variability of stratification across the site.

Groundwater monitoring indicated that groundwater was at depths between 1.8 m (North approach embankment) and 1.35 m (Road cutting along the Link Road) below the existing surface. The presence of groundwater at shallow depth needed to be addressed carefully, particularly if remediation measures were required to treat the uncontrolled fill. This will be discussed later in this paper.

5 DESIGN CONSIDERATIONS

Based on the investigation results, the following issues were addressed to ensure that satisfactory performance of the structure would be maintained over its entire design life. These issues were:

- Bridge structure foundation,
- Reinforced soil structure foundation,

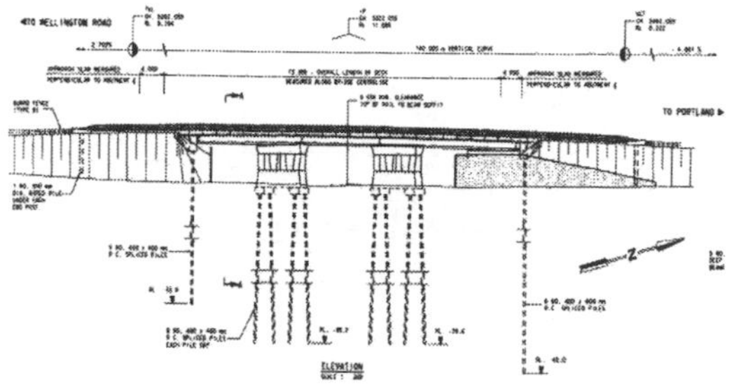

Figure 2. Elevation – General foundation layout.

– Ground improvement work for construction of embankment fill on the North approach.

Figure 3. View of completed reinforced earth wall.

5.1 *Bridge structure foundation*

Various foundation types, including steel shell piles were considered. Based on costs and availability considerations, reinforced concrete driven piles were considered the most suitable foundation type at this location. In accordance with AS 5100.3 Foundation, a material strength reduction factor of 0.5 was adopted in the geotechnical strength design. The design computations were based on "Pile Design and Construction Practice", 4th edition, M.J. Tomlinson. In order to satisfy the performance criteria for the bridge foundation, the piles were designed to be driven into the weathered limestone to founding depths between 35 m and 47 m below the existing surface. The general layout of the overpass is depicted in Figure 2.

5.2 *The RSS and foundation*

The Reinforced Earth Company designed and supplied the RSS and VicRoads GeoPave designed the foundation treatment.

The Reinforced Earth wall design combines galvanized, medium tensile steel reinforcing strips with granular backfill that was compacted against precast concrete facing panels in a single stage operation to create a strong yet flexible retaining wall. Chemical and electrochemical testing confirmed the backfill to be suitable for steel reinforcing strips and a design service life in excess of 100 years. The steel strips are a hot-rolled, ribbed flat bar with minimum tensile and yield strength of 520 MPa and 355 MPa respectively at 22% elongation. The strips are rolled with a localized thickening at the site of the bolted connection to ensure that the capacity of the 45 mm × 5 mm strip is not controlled by the connection strength. With the bridge loads directly supported on piles the peak bearing pressure at the base of walls was 230 kPA.

The project architects specified a complex finish to the precast concrete facing panels and a tapered coping to the top of the wall (Figure 3). The panels were 2 m × 2 m × 140 mm embossed with 30 mm and 50 mm deep relief in three different patterns. 50 mm × 5 mm galvanized steel tie points were cast into the panels for connecting the ribbed steel strips. The arrangement of the patterned panels formed a random appearance in the finished wall.

Based on the investigation results, it was estimated that long-term settlement in the subsurface stratifications beneath the 8 m high RSS would be in the order of 300 mm at the North abutment location. The differential settlement was expected to be about 100 mm. Therefore, the risk of rotational and vertical movement of the RSS would be very high. As a consequence, the bridge abutment piles would likely be subjected to excessive lateral and vertical (down drag) loading induced by movements of the RSS (Stewart 1999).

To reduce the potential for movement of the RSS, the RSS was designed to be supported on piles. Based on ease of construction, 350 × 350 mm RC driving piles were selected as the foundation for the RSS. The entire footprint of the RSS was designed to be supported on piles installed at a 2 m square grid. The founding depth of the piles was 15 m below the existing surface (i.e. R.L. −13 m). The design pile capacity was 800 kN (Ultimate Limit State) per pile. To ensure that the RSS loading was distributed as evenly as possible to the piled foundation, a raft consisting of 2 layers of bi-oriented geogrid reinforcing and cement-treated sand was constructed over the piled area. The geogrid was of polypropylene type with a design ultimate tensile strength (max strain 10%) of 40 kN/m in each direction. In addition, a pile cap, 600 mm × 600 mm, was provided for each of the piles for transfer of loads from the raft. The pile layout at the north abutment, geogrid reinforce raft and arrangement at the pile top is shown below (Figures 4 and 5).

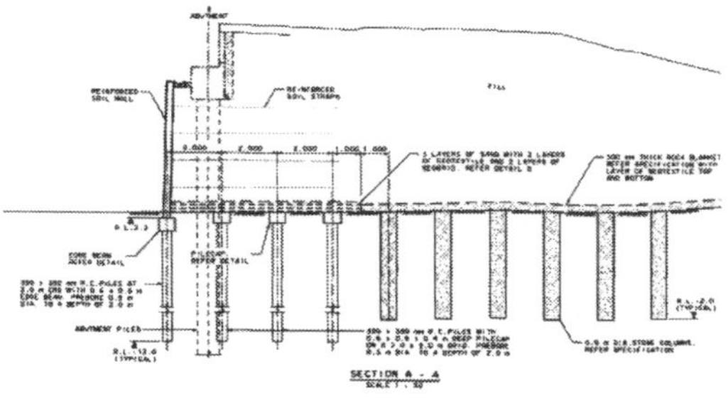

Figure 4. Layout of piled raft foundation for placement of bridge approach embankment fill at North abutment.

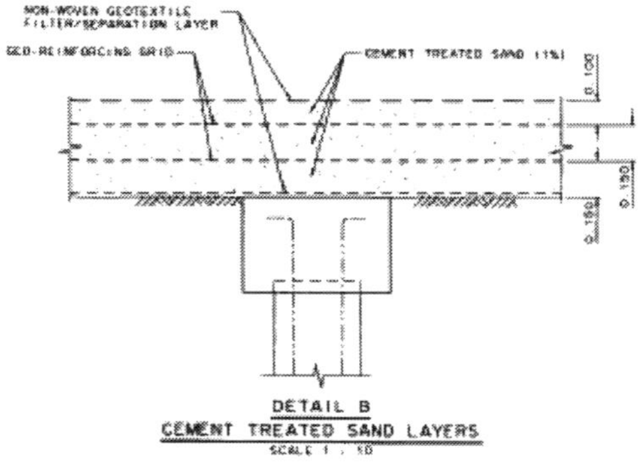

Figure 5. Geogrid reinforced raft and top of pile.

The specification for the geogrid reinforced raft was as follows.

1. Ultimate tensile strength (max strain 10%): 40 kN/m in both longitudinal and transverse directions
2. Loaded at 2% strain: 14 kN/m in both longitudinal and transverse directions
3. Loaded at 5% strain: 28 kN/m in both longitudinal and transverse directions
4. Junction Strength: 95% of (i) to (iii) above.
5. The geogrid reinforcing was of Polypropylene type
6. Shall have properties to inhibit attack by UV light
7. Shall be unaffected by all chemicals, including acids, alkalis and salts, and shall not be affected by micro-organisms in the soil.

5.3 *Embankment fill foundation*

5.3.1 *Ground treatment options*

The design criteria required that the total settlement and differential settlement must not exceed 50 mm and 50 mm respectively, over the design life of the structure, nominally 100 years. Based on investigation results, it was considered difficult to satisfy these design criteria as it would be difficult to predict the behavior of the 4 m thick uncontrolled fill

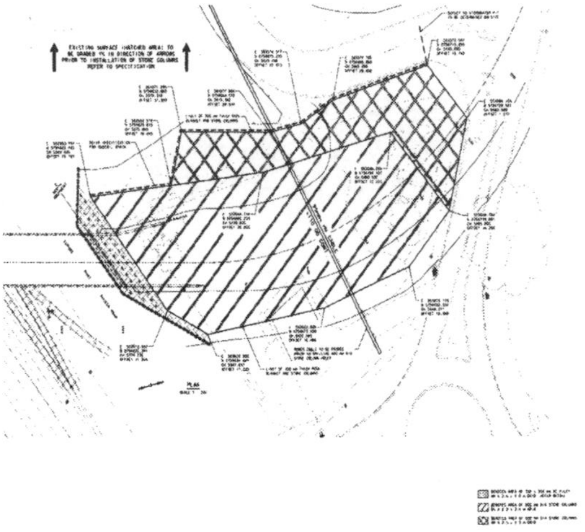

Figure 6. Area required DR ground improvement.

when it is subjected to the weight of the bridge approach embankment fill. It was decided that ground improvement would be required in order to satisfy this design requirement. The total area to be improved was 3,500 m^2 approximately.

Ground improvement using the conventional surcharge method was not considered due to the tight construction time frame. Several ground treatment options including lightweight fill were considered. The lightweight fill option was not favored as it required construction of containment structures. Other options such as dynamic compaction, dynamic replacement, grout injection and stone columns were also considered (Arulrajah & Abdullah 2002a, b, Arulrajah et al. 2004). The stone column option was selected since it was considered to have the least risk of causing excessive ground vibration and noise during construction. However, DR was accepted as an alternative solution after the contractor demonstrated that ground vibration and noise could be managed and minimized with appropriate construction controls.

The specification for DR proposed by the Contractor is as follows:

1. Targeted depth of improvement is 4 m minimum
2. DR columns shall not be spaced greater than 3 m on a square grid.
3. Impact hammer weight 8 to 25 tons, dropped in free fall from 15 to 25 m
4. Noise level: less than 75 db at a distance 50 m from the source of impact
5. Vibration level: less than 3 mm ppv at a distance 70 m from an impact source of 168 tonne-metres.
6. Material to be granular fill with D (max) < 100 mm and percentage (by mass) of fines passing the 75 micron sieve to be less than 10%.

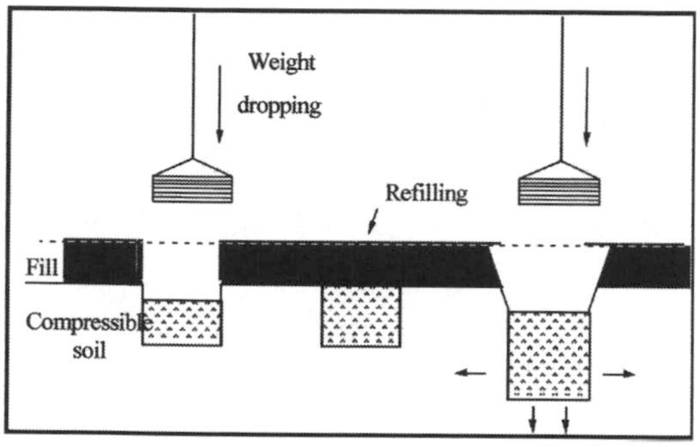

Figure 7. Schematic principle of method.

Ground improvement was undertaken by Austress Menard Pty Ltd using the DR technique.

5.3.2 *Dynamic replacement description and history*
DR is a method in which columns of large diameters are formed with granular material based on the techniques developed for Dynamic Consolidation (DC) in highly compressible and weak soils. This technique is similar to DC however the pounding is used to form large diameter granular pillars through the material to be improved. The columns of granular material formed are called "pillars". This method combines the advantages of DC with those of Stone Columns whilst providing an economical edge since excavation of the weak soil is avoided. Also, high internal shearing resistance is provided within the pillars. These pillars also act as large vertical drains and induce a reduction in the consolidation period. Schematic principle of the DR method is illustrated in Figure 7.

The equipment used for DR is similar to the DC equipment i.e. heavy rigs and pounders. However, usually pounders with smaller areas are used to facilitate the penetration capacity (Menard Soltraitement 2006). Heavy Dynamic Replacement (HDR) columns are made with boulders and cobles using energies exceeding 400 tm per blow. The relationship between the effective depth of attained improvement, the pounder weight and the height of the drop is expressed as reminded in equation (1):

$$D = \alpha\sqrt{W.H} \quad (1)$$

where:

D = maximum depth of improvement in metres;
α = damping factor (varies 0.3 to 0.7)
W = falling weight in metric tons;
H = height of drop in meters (Mitchell & Gallagher 1998).

On the Cliff street project, pre-excavation was performed down to 2 meters in order to penetrate the hard top layer and to allow for the installation of deeper columns especially in the clayey materials encountered.

Construction control methods of DR operation on site are similar to those of DC and include heave penetration tests, measurement of volume of stone used, number of drops per print and overall platform settlement. Once the DR has occurred conventional soil investigation can be performed such as CPT, SPT and pressure meter tests (PMT) (Robertson and Campanella 1988).

5.3.3 *Model*
In order to estimate the settlement of the improved layer, a finite element analysis was performed using the software Plaxis.

The concept is to perform a settlement calculation over an axi-symmetrical model representing one column and its surrounding soil over one cell: (ie. column and soil over a 5 m × 5 m grid).

The equivalent radius of the model is 2.82 meters and the DR pillar was found to be 1.5 meters in diameter after in-situ measurements. The pillars were assumed to extend 4.5 meters deep and a young modulus parameter of 40 MPa was retained.

Finally a service load of 10 kPa was taken in consideration.

5.3.4 *Soil improvement results*
A total of 81 CPT tests were performed after completion of the DR works between the DR prints. The compilation of these soil parameters highlights the consistent improvement throughout the treated layer. An average q_c (measured cone resistance) value of 4 MPa was found between pillars and after improvement this represents an improvement of q_c values of 50 to 100% (refer figure 8). CPT tests were performed 2–3 weeks on average after the installation of the DR pillars and the improvement measured is likely to account for some consolidation due to the increase in horizontal permeability as well as the improvement obtained by means of compaction. Figure 8 shows the CPT results before and after improvement.

Improvement can be found up to depths of 6 to 7 meters. Considering the weight and height used of 12 tonnes and 17 meters respectively we find an $\alpha = 0.5$ approximately (after Mitchell & Gallagher).

Thirteen SPT tests were also performed within the DR prints and they showed consistent improvement down to depths of 4 meters with N (SPT) results between 25 and 35 illustrating the good quality of the DR prints obtained.

6 CONSTRUCTION

6.1 *Piling installation.*

Wagstaff Piling Pty Ltd was the piling contractor for installation of the piles. A six tonne hydraulic hammer

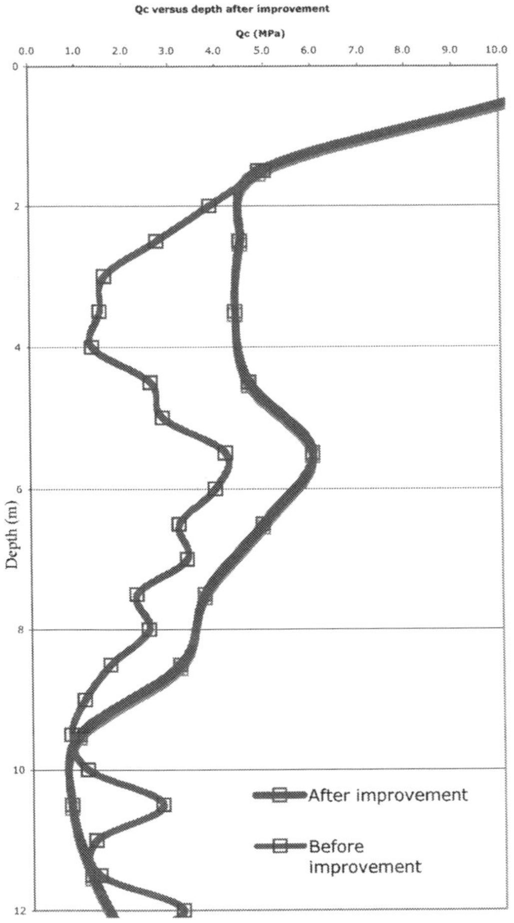

Figure 8. Estimated q_c before and after DR works.

was employed to drive the piles. Hammer drop heights for the bridge foundation piles ranged from 600 mm to 800 mm and typically 600 mm, whereas the drop height was 150 mm for all of the RSS foundation piles. No major construction issue was reported during construction. In general, piles were driven to depths as expected although relatively significant variation in pile toe levels were experienced between bridge support locations. The difference in pile penetration depths within the same pile group was judged likely due to a reduction of the thickness of the weak layers at that location.

6.2 *Ground improvement*

Ground vibration monitoring results indicated that the vibration level was within the contract specification. Noise level was within the acceptable level with no complaint having been received from the local community during construction. The required ground improvement was generally achieved. At the DR columns, the required density was confirmed by the drilling with SPT values generally above N20 (N15 was required) through to a depth of 5 m (4 m was required).

6.3 *Reinforced soil structure wall*

The RSS wall was constructed by Akron Construction Pty Ltd with assistance from Austress Freyssinet Pty Ltd as a specialist subcontractor. During construction of the wall there was no evidence of panel misalignment or distress due to settlement of the piled foundation support.

7 CONCLUSION

Various geotechnical designs were provided to suit specific foundation requirements for the overpass structure such as the bridge supports, the earth retaining abutments and the high embankment fill. The geotechnical designs utilized reinforced concrete piles driven to a depth in excess of 50 m for the bridge foundations. A piled raft foundation was provided for the RSS wall and the DR ground improvement technique was adopted to remedy the existing uncontrolled fill prior to placement of the bridge approach fill embankment. The ability to provide appropriate geotechnical design is attributed to a good understanding of the subsurface conditions at this site. Therefore, it may be concluded that an appropriate geotechnical site investigation is vital in ensuring that geotechnical issues can be addressed adequately during the detailed design stage.

Cliff Street overpass opened for traffic in January 2007. There has been no report of cracking of the road pavement constructed on the approach embankment fill thus far.

ACKNOWLEDGEMENT

The authors would like to acknowledge the support of VicRoads for allowing the use of sections of the construction drawings for the Cliff Street Overpass project in the preparation of this Paper.

REFERENCES

Arulrajah, A. & Affendi Abdullah Vibro Replacement Design of High-Speed Railway Embankments, , 2nd World Engineer Congress, Sarawak, Malaysia. 22–25 July 2002.

Arulrajah, A. Abdullah & H. Nikraz, Ground Improvement Techniques for the Treatment of High-Speed Railway Embankments, Australian Geomechanics Vol 39 No 2 June 2004.

AS 5100.3 Foundation.

AS 1726 – Geotechnical Site Investigation.

Geological Survey of Victoria 1:63,360 series, Part 928 Zone 6 (Portland)
Menard Soltraitement, 2006. Final Report for Soil Improvement works by DR for Cliff Street Overpass.
Mitchell, J. K. & P.M. Gallagher, 1998. Engineering guidelines on ground improvement for civil works structures and facilities, *U.S. Army Corp of Engineering Division Directorate of Civil Works*, Washington, D.C.
Robertson, P. K. & R.G. Campanella, 1988. Guidelines for Geotechnical Design Using CPT and CPTU Data, *Technical Report*, Civil Engineering Department, University of British Columbia, Vancouver, B.C., Canada.
Stewart, DP 1999 Analysis of piles subjected to embankment induced lateral soil movements., ASCE, 05/1999.
Tomlinson, M.J. "Pile Design and Construction Practice", 4th edition, M.J. Tomlinson.
VicRoads Report 2004, Cliff Street Overpass Project – Portland, PRC Report, VicRoads – South West Region.

New Horizons in Earth Reinforcement – Otani, Miyata & Mukunoki (eds)
© 2008 Taylor & Francis Group, London, ISBN 978-0-415-45775-0

Load transfer mechanism in reinforced embankment on pile elements

J. Hironaka & T. Hirai
Mitsui Chemicals Industrial Products, LTD, Saitama, Japan

J. Otani & Y. Watanabe
GeoX CT Center, Graduate School of Science and Technology, Kumamoto University, Kumamoto, Japan

ABSTRACT: When the embankment is constructed on soft ground, any kinds of geotechnical solutions, for example, pile foundation and deep mixing method, have been used for the purpose of reducing differential settlement. This paper deals with a combined method of pile elements with earth reinforcement technology using geogrids. This type of the combined method offers decrease of load concentration on pile elements due to the reinforcing effect by geogrids. And it is considered that a search for the load transfer mechanism in three dimensions is important for the performance-based design of this method. Recently, an industrial X-ray CT (Computed Tomography) scanner which is one of the nondestructive testing method has been developed and the inside behavior of material could investigate without any destructions in three dimensions. The objective of this paper is to visualize the load transfer mechanism over the pile element heads using industrial X-ray CT scanner. A series of model test were conducted. Then, the behavior in the soil was scanned during the settlement of the ground using X-ray CT scanner. Based on these results, the reinforcing effect by the geogrids and soil arching effect over the pile element heads were discussed precisely. And finally, the evaluation of load transfer mechanism of this system was examined, quantitatively.

1 GENERAL INSTRUCTION

Construction of embankment on soft ground often causes the differential settlement. Pile elements with earth reinforcement technology are used in order to reduce this settlement as shown in Fig-1. This combined method offers the reduction of the loading at the pile element head by using geogrid, because the embankment load is transferred with arching effect in the reinforced soil above the pile element heads and membrane effect of geogrid. And it is considered that a search on the load transfer mechanism under this system is important for the performance-based design of this method.

Recently, an industrial X-ray CT (Computed Tomography) scanner which is one of the nondestructive testing method has been used in geotechnical engineering field and the inside behavior of soils can be investigated without any destructions. Authors have conducted a series of studies on the application of industrial X-ray CT scanner to geotechnical engineering such as characterization of soil failure (Otani et al., 2000) and visualization of the failure in mixed soil with air foams (Otani et al., 2002) and others (Otani and Obara (2003) and Otani et al.(2005)).

The objective of this paper is to visualize the load transfer mechanism over the pile element heads using industrial X-ray CT scanner. A series of model test are conducted. Then, the behavior in the soil is scanned after the settlement of the ground using X-ray CT scanner. Based on these results, the reinforcing effect by the geogrids and soil arching effect over the pile element heads are discussed precisely. And finally, the evaluation of load transfer mechanism of this system is examined, quantitatively.

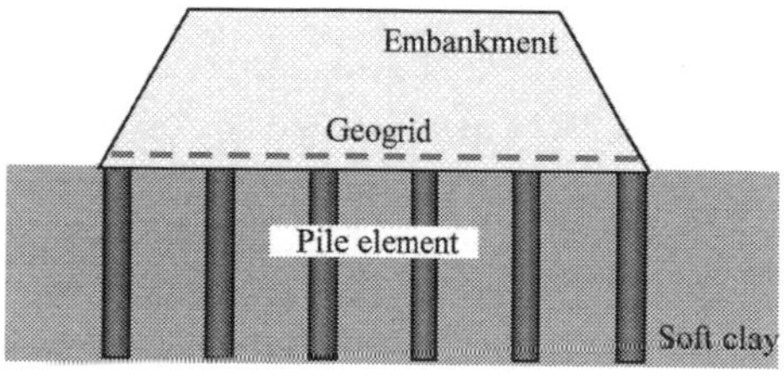

Figure 1. The outline of combined method.

2 X-RAY CT

The detected data are assembled and the cross sectional images are reconstructed using an image data processing device by means of the filtered back-projection method. By using all these cross sectional images around the circumference of the specimen, three

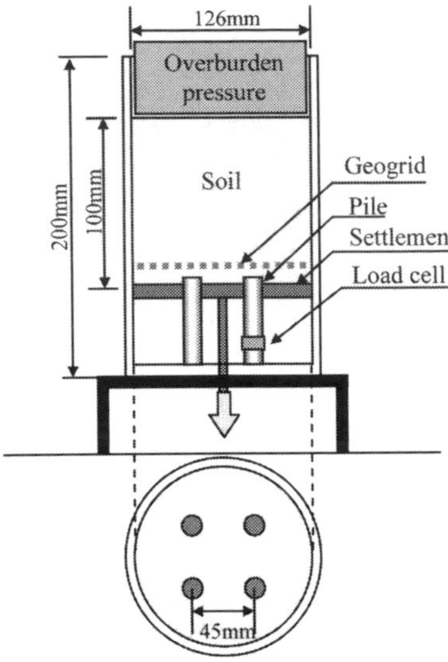

Figure 2. Settlement test apparatus.

dimensional (3-D) image can also be reconstructed. CT images are constructed by the spatial distribution of so called "*CT-value*" and this is defined as:

$$CT - value = (\mu_t - \mu_w)\kappa / \mu_w \quad (1)$$

where μ_t :coefficient of absorption at scanning point; μ_w: coefficient of absorption for water; and κ: constant (Hounsfield value). Here, it is noted that this constant is fixed to a value of 1000. Thus, the *CT-value* of air should be −1000 because the coefficient of absorption for air is zero. Likewise, *CT-value* for water is 0 from the definition of Eq. (1). CT images are presented with shaded gray or black color for low *CT-value* and light gray or white color for high *CT-value* in all the subsequent black and white colors. The total number of levels on these colors is 256. It is well known that this *CT-value* is linearly related to the material density. It is noted that the precise contents of X-ray CT method can be obtained in the reference by Otani et al. (2000).

3 TEST PROCEDURE

A series of model tests for different types of geogrids were conducted using newly developed test apparatus as shown in Fig-2. The case without reinforcement was also conducced in order to discuss the soil arching effect with pile elements. The soil box in the apparatus made by an acrylic mold, which is the size of 200 mm with the height of 126 mm diameter, was set in the CT room. A model pile, which was the size of 15 mm diameter, was set on the bottom of the soil box. And, total of four piles were installed at intervals of 45 mm between every adjacent two piles. The settlement plate, which

Table 1. Material properties.

	Effective particle size D_{50} (mm)	Uniformity coefficient U_C	Internal frictional angle φ (deg)
Toyoura sand	0.19	1.56	39.4
Silica sand No.7	0.15	1.63	36.0
Silica sand No.8	0.12	1.86	33.9
Dry clay powder	0.0026	10	25.9

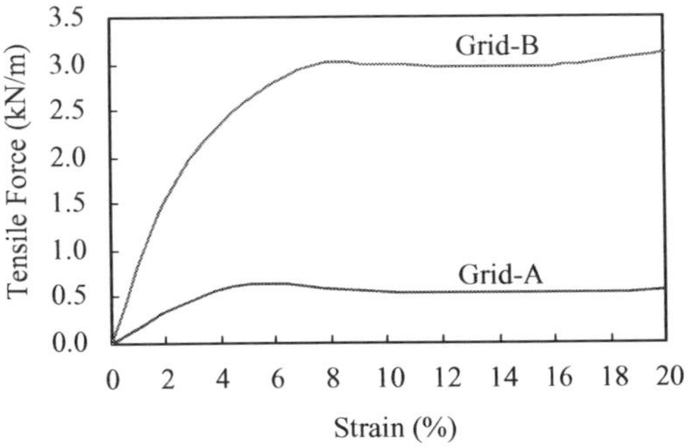

Figure 3. Tensile Force – Strain relationship.

can penetrate through the piles using automatic settlement plate apparatus, was set at the bottom of the ground. The method of pulling down this settlement plate at constant speed was assumed to be the consolidation settlement of the ground due to embankment load. In order to discuss the effect of different soils on the load transfer mechanism, Toyoura sand, Silica sands (No.7 and No.8) and Dry clay powder were used. The material properties of all the soils are shown in Table. 1. In this test, the density was fixed to be the relative density of 80% for all the soils and the overburden pressure of 3.2 kPa was applied by dead-load in order to apply relatively large confining pressure. Fig-3 shows tensile force – strain relationship of geogrids as the reinforcing materials used in this test, in which Grid-A is a geogrid with its spacing of 2 mm while Grid-B is a geogrid with that of 9 mm. In order to discuss the effect of different geogrids on the load transfer mechanism, Grid-A and Grid-B were used. These geogrids were installed at 5 mm height above the pile head in the soil. The settlement plate was pulled down with the loading speed of 1 mm/min under displacement control and the loading was stopped at the settlement of 5 mm. And the pile load that acted on the pile elements was measured using the load cell. For the CT scanning, the model grounds at initial and after 5 mm settlement were scanned with 1mm thickness until the height of 40 mm above the settlement plate. The test cases are listed in Table. 2, in which CASE1, CASE2, CASE3, and CASE4 are the cases with different types

Table 2. Test cases.

	Ground materials	Reinforcement
CASE1	Toyoura sand	Without
CASE2	Silica sand No.7	Without
CASE3	Silica sand No.8	Without
CASE4	Dry clay powder	Without
CASE5	Toyoura sand	Grid-A
CASE6	Toyoura sand	Grid-B
CASE7	Silica sand No.7	Grid-B
CASE8	Silica sand No.8	Grid-B
CASE9	Dry clay powder	Grid-B

of soils in order to discuss the soil arching effect without reinforcement, CASE5 and CASE6 are the cases with different types of geogrids in order to discuss the reinforcing effect, and CASE6, CASE7, CASE8, and CASE9 are the cases with different types of soils using a geogrid in order to discuss the load transfer effect of reinforcement. The load transfer effect is examined with CASE1 to CASE4 and CASE6 to CASE9.

4 RESULTS AND DISCUSSION

4.1 *Effect of soil arching*

Figure 4 shows the results of CT scanning which is the vertical cross sectional images for CASE1, CASE2, CASE3 and CASE4 at the end of the test. It is observed that there is the area of high density with cone shape at the pile head. And as easily realized, the density around the circumference of the pile head is decreased due to the settlement, which is the appearance of the area of ring shape in CASE1. These low density areas are interrupted in each other for the adjacent two piles at the area of within 10 mm height above the settlement plate. The angle, θ between these interrupted areas which is shown with dotted line in the image was about 38 degree for the CASE1. Although the behaviour of the density changes for CASE2 and CASE3 were almost the same as CASE1, the angle, θ was 48 degree and 55 degree, respectively. For CASE4, it is totally different from all other results and this angle was almost 90 degree. Based on these results, it is concluded that the smaller the angle, θ is, the wider the load transfer is and this angle becomes larger when the soil strength is smaller. Fig-5 shows 3-D extraction images of the interrupted areas for CASE1. This 3-D area is considered to be the area where some part of embankment load is applied except that applied on the pile heads. Fig-6 shows the model of the embankment load that acts on the ground between the pile elements at existing design method (Public Work Research Center (2000)). In this design, it is conducted the settlement calculation of the embankment using this proposed model. As easily realized, Fig-5 and Fig-6 are similar very well.

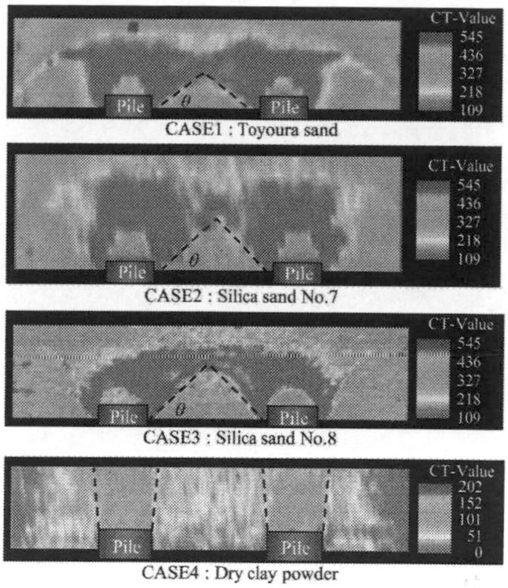

Figure 4. Vertical cross sectional images without reinforcement.

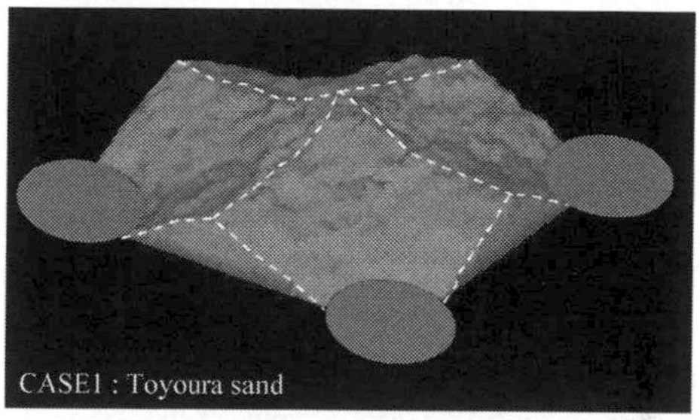

Figure 5. 3-D extraction image.

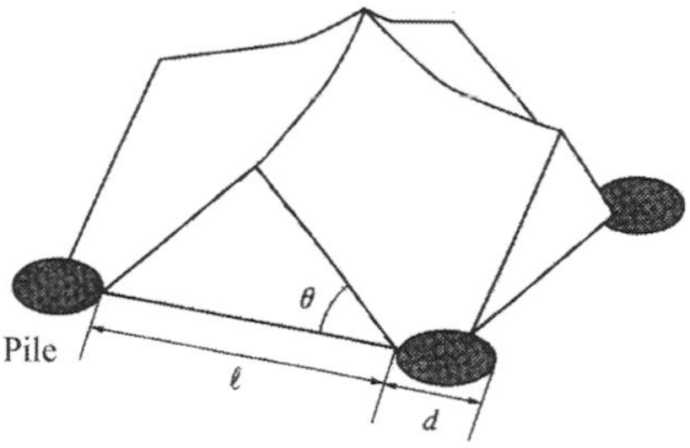

Figure 6. The proposed model of the embankment load that acts on the ground between the piles at existing design method (PWRC(2000)).

4.2 *Effect of reinforcement*

Figure 7 shows the vertical reconstruction images for the cases with geogrids. These images were reconstructed by the density change using different colours. For CASE5, the area of density change extends to horizontally due to the existence of geogrid. Thus, the angle of density change for CASE5 is

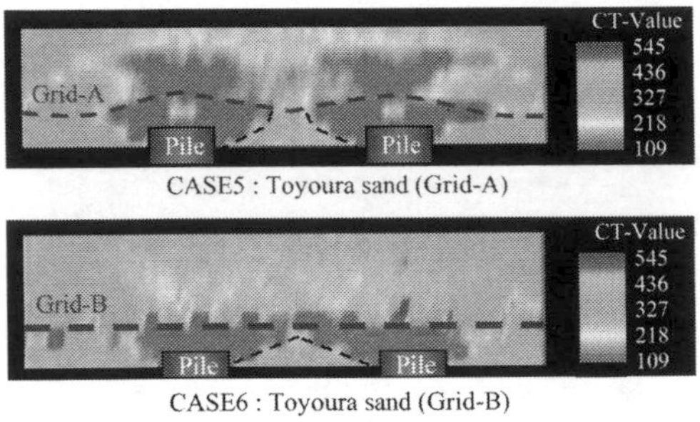

Figure 7. Vertical cross sectional images with reinforcement.

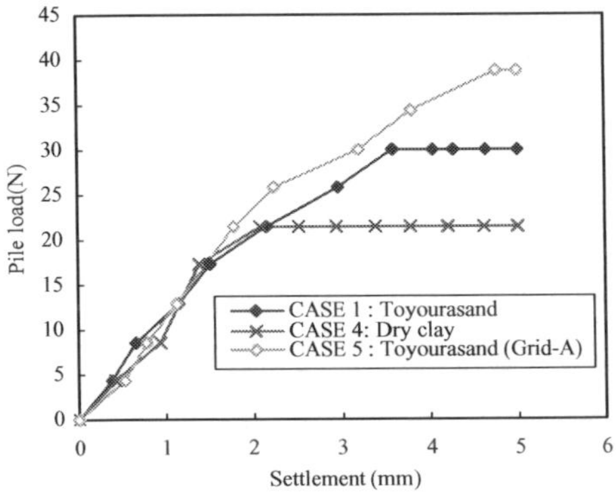

Figure 8. Pile load – settlement relationship.

decreased due to the effect of Grid-A. For CASE6, the area of changing density was not as large as that of CASE5 and it seems that the stress concentration is rather smooth. And the area of density change was observed between the piles and the geogrid. Here, if it is assumed that the transmission of the overburden pressure due to the settlement influences the density change over the pile head, it can be considered that the smaller the angle of this density change is, the wider the load distribution in the ground is. And this can be also considered to be the effect of soil arching and at the same time, the membrane effect due to existing of geogrid.

4.3 *Effect of load transfer*

Figure 8 shows some of the examples of pile load – settlement relationship of four piles measured by the load cell for CASE1, CASE4 and CASE5. For CASE4, the pile load is smaller than CASE1, because soil strength is small. For CASE5, the pile load is larger than CASE1, because of existing of geogrid. Fig-9 shows the relationship of the angle: θ of density change area between the piles by X-ray CT image – the internal friction angle: φ of soils by box shear test. For the cases without reinforcement, the relation between

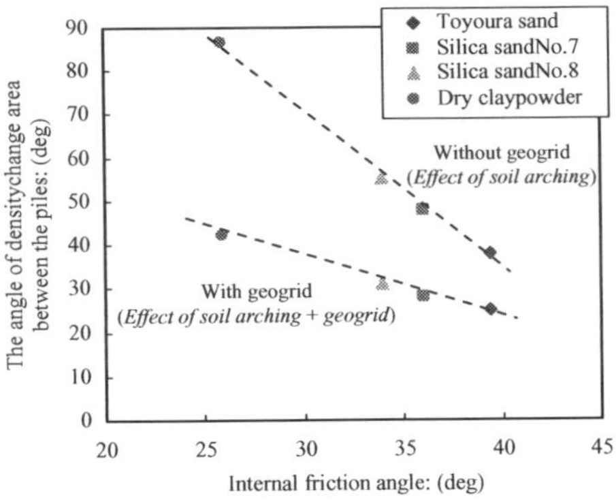

Figure 9. The angle: θ–Internal friction angle: φ relationship.

θ and φ is shown in a negative correlation. Thus, this relation is the effect of soil arching. And for the cases with reinforcement, the relation between θ and φ is shown in a negative correlation. Thus, this relation is the effect of soil arching effect and reinforcement due to existing geogrid.

5 CONCLUSIONS

The following conclusions are drawn from this study:

1) The load transfer mechanism in embankment beyond pile elements due to the settlement was observed using X-ray CT scanner; and
2) It may be said that earth reinforcement is effective with the use of pile elements for the purpose of stress re-distribution in the soil.

Finally, it is evident from all the discussion here that the industrial X-ray CT scanner promises to be a powerful tool even for the geotechnical engineering field.

REFERENCES

Otani, J., Mukunoki, T. and Obara, Y. (2000). "Application of X-ray CT method for characterization of failure in soils", Soils and Foundations, Vol.40, No.2, pp. 111–118.

Otani, J., Mukunoki, T. and Obara, Y. (2002). "Visualization for engineering property of in-situ light weight soils with air foams", Soils and Foundations, Vol.42, No.3, pp. 93–105.

Otani, J. (2003).State of the art report on geotechnical X ray CT research at Kumamoto University", *X-ray CT for Geomaterilas*, Balkema, Netherlands, pp.43–77.

Otani, J., Mukunoki, T. and Sugawara K. (2005). "Evaluation of particle crushing in soils using X-ray CT data.", Soils and Foundations, Vol.45, No.1, pp. 99–108.

Public Work Research Center. (2000). "Manual on design and execution of reinforced soil method with use of geotextiles", Second Edition, Public Work Research Center, pp. 248–256.

New Horizons in Earth Reinforcement – Otani, Miyata & Mukunoki (eds)
© 2008 Taylor & Francis Group, London, ISBN 978-0-415-45775-0

Laboratory test on the performance of geogrid-reinforced and pile-supported embankment

S.L. Shen
Department of Civil Engineering, Shanghai Jiao Tong University, China

Y.J. Du
Institute of Geotechnical Engineering, Southeast University, China

S. Hayashi
Institute of Lowland Technology, Saga University, Japan

ABSTRACT: This paper investigates the behavior of geosynthetic-reinforced and pile-supported (GRPS) embankments through laboratory model tests. The model ground was made using reconstituted soft Ariake clay with thin sand layers. Model piles were made of timber. Geogrids were placed at the top of pile and under the embankment. Model embankment was sand and additional load was exerted on the embankment top. The test results show that the increase in the length of both pile and reinforcement or decrease in the piles improvement ratio creates an economical solution for GRPS embankment system.

1 INTRODUCTION

Geosynthetic-reinforced and pile-supported (GRPS) embankments have been emerged as an effective alternative successfully adopted worldwide to solve many geotechnical problems. In the GRPS embankment system, the geosynsthetic reinforcement carries the lateral thrust from the embankment, creates a stiffened fill platform to enhance the load transfer from the soil to the piles, and reduce the differential settlement between pile caps. As a result, the GRPS system does not require inclined piles, large pile caps, and close pile spacing. Therefore, the GRPS system creates a more cost-effective alternative. In the GRPS-supported embankment system, the piles carry most of the loads from the embankment and the soil is only subjected to small loads.

The GRPS embankment systems have been used for a number of applications worldwide, which include: bridge approaching embankments; retaining walls; roadway widening; storage tanks; low height embankment; and buildings, etc. There are a few methods available to design the GRPS embankment system. British Standard BS8006 (1995) proposed a relatively comprehensive design method. However, Li et al. (2002) concluded that current design methods could not well predict the performance of constructed GRPS systems. Therefore, there is a need for developing more rational design methods for this emerging technology. For this reason, the objective of this study is to reveal the load transfer mechanism through laboratory model test.

2 EXPERIMENTAL PROGRAM

2.1 *Setup of test apparatus*

The set-up of the model test is illustrated in Fig. 1. A model box made by transparent acrylic has an inner dimension of 1.5 m in length, 0.6 m in width, and 0.8 m in height. An acrylic plate was fixed at the middle of the box along length direction to form two separated sub-model chambers with a width of 0.3 m. Two layers of geotextiles were placed at the bottom and two end vertical boundaries as drainage layers.

2.2 *Model ground*

The model ground was formed by four clay layers sandwiched three thin sand layers. The thickness of each clay layer was about 122.5 mm and about 20 mm for each sand layer. The clay used was remolded Ariake clay, which was sampled from a ground depth of 1 to 3 m from Saga Airport site. The initial physical properties of the clay sample are: specific gravity $Gs = 2.62$, natural water content $w_n = 120 \sim 130\%$ (higher than its liquid limit with the value of about 105 to 110%). The plastic limit is about 40 to 50% and plasticity

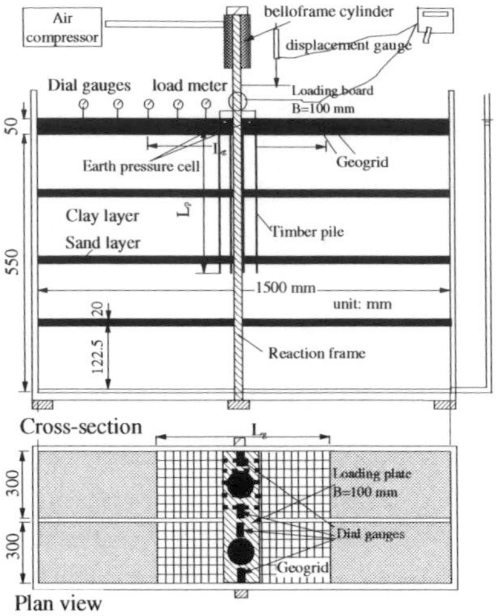

Figure 1. Set-up of the model test apparatus.

Table 1. Test cases with details of pile and geogrid.

Case label	Pile Length	Rows	Geogrid Length
0H0B	0	0	0
0H0N6B	0	0	600 mm (6B)*
3H2N5B	165 mm (0.3H)**	2	500 mm (5B)
3H4N2B	165 mm (0.3H)	4	200 mm (2B)
3H4N3B	165 mm (0.3H)	4	300 mm (3B)
5H2N3B	275 mm (0.5H)	2	300 mm (3B)
5H4N0B	275 mm (0.5H)	4	0
5H4N2B	275 mm (0.5H)	4	200 mm (2B)
5H4N3B	275 mm (0.5H)	4	300 mm (3B)
7H2N3B	385 mm (0.7H)	2	300 mm (3B)
7H2N5B	385 mm (0.7H)	2	500 mm (5B)

*B = width of loading plate (0.1 m); **H = thickness of model ground (0.55m).

index $I_p = 60$–70. The grain diameter of the sand is greater than 420 μm and less than 5 mm with specific gravity of 2.62, maximum density of 16.1 kN/m^3 and minimum density of 12.8 kN/m^3.

Before making the model ground, clay samples were completely remolded to the paste state using a hand controlling electric mixer by adding water to the water content of twice its liquid limit. The clay milk was put into the container in four layers with the thickness of about 175 mm. Among the clay layers sand was pulverized to a thickness of about 20 mm. After that a polywood board was placed over the soil. Then, a 10 kPa vertical consolidation pressure was applied to the soil for about 2 months. After the primary consolidation was finished, the consolidation load was removed. The thickness of the model ground was 550 mm and water content after consolidation was reduced to 77 ~ 80%.

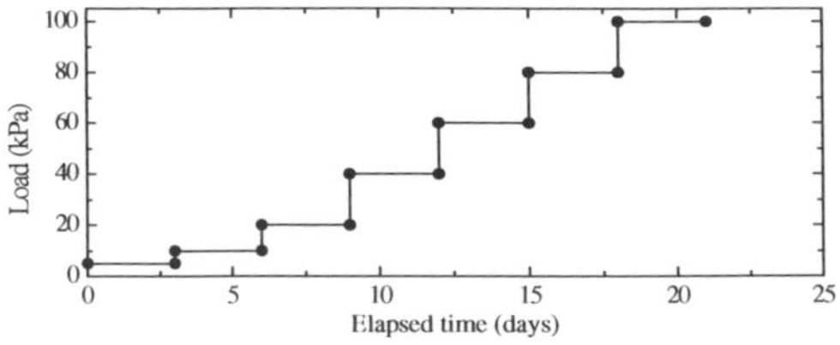

Figure 2. Embankment loading applied in the study.

The soil was sampled for strength and oedometer test. The test results indicate that the model ground had a compression index (C_c) of about 0.8, void ratio (e) of about 3.0, and undrained shear strength (S_u) of 4.5 kPa to 6.5 kPa (laboratory vane shear test).

2.3 *Test procedure*

Pile was made of timber with a diameter of 10 mm. Piles were inserted into model ground using a specific machine with the penetration rate of 10 mm/min. Model embankments (sand mat) with a height of 50 mm were compacted in three layers with the thickness of 15 mm, 20 mm, and 15 mm, respectively. At the bottom and top of the middle sand layer, two layers of geogrid in the sand mat were placed. The geogrid used was made of polyester and has a grid size of 6 mm by 6 mm, tensile strength of 5.2 kN/m (strain rate 1%/min). The stiffness is about 300 kN/m for less than 1% tensile strain condition. Due to the size of the model is about 1/20 to 1/30 of the prototype, the reinforcement was very strong and it can be regarded as "fully reinforced" (Jewell 1988). The second geogrid layer was connected with the top of pile. On top of the model embankment, a 100 mm wide loading plate was placed at the center. The load was applied stepwise by air pressure though a bello-frame cylinder with an pressure of 5, 10, 20, 40, 60, 80, and 100 kPa, respectively and the loading duration for each stage was about three days, as illustrated in Fig. 2. The same loading condition was maintained for both sub-models.

3 RESULTS AND ANALYSIS

3.1 *Effect of pile and geogrid*

The effect of pile and geogrid on the settlement and bearing capacity were investigated. Figure 3 show the results of three test cases: i) the unimproved case no pile and geogrid were applied; ii) subsoil was improved by four rows of piles with length of about 275 mm (0.5H); iii) in the third case, sand mat was reinforced by two layers of geogrid with the length of 600 mm (6B). The bearing capacity of unimproved subsoil is very

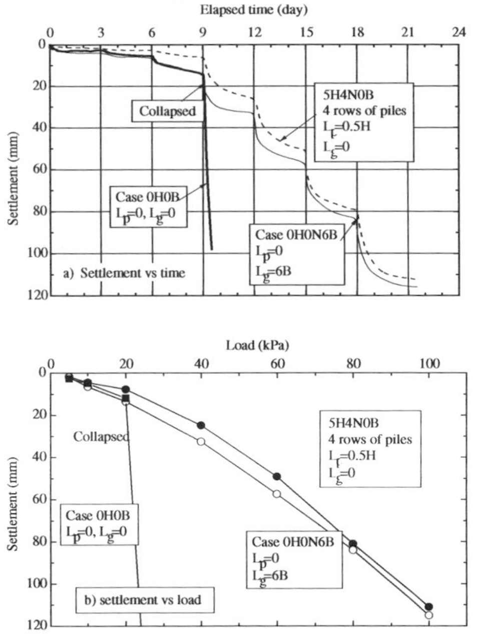

Figure 3. Effect of pile and geogrids on the settlement: a) settlement vs time, b) settlement vs load.

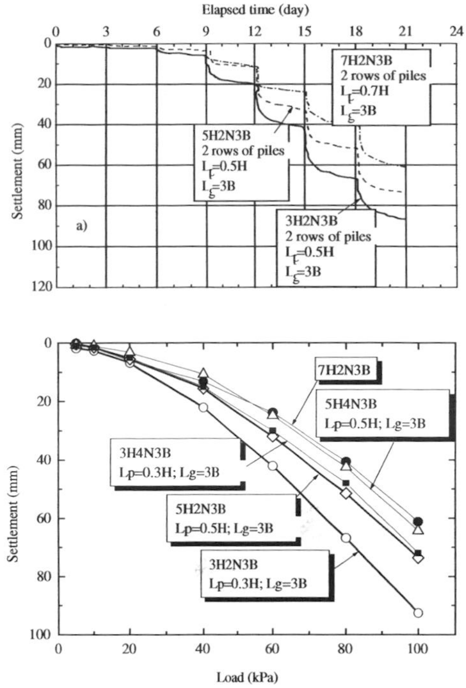

Figure 4. Effect of length and rows of pile on the settlement.

low with the collapsed load of only about 20 kPa. With piles or geogrids reinforcement, the bearing capacity of the subsoil was increased greatly. For the two improved cases with pile or geogrid, no collapse was found. It is clearly shown that pile and geogrid significantly reduced the settlement and increased the bearing capacity of the soft subsoil. In this test study, it is found that the reinforcement effect of two layers of geotextiles with length of 600 mm is similar to that of 4 rows of piles with the length of 275 mm. However, in the initial lower load (till to 20 kPa), geogrid reinforced case behaved in the same way as the unimproved case (see Fig. 3a). Moreover, when the load is less than 80 kPa, the settlement of geogrid improved case is greater than the pile improved case and after 80 kPa, difference of settlement between two case 0H0N6B and case 5H4N0B became smaller. The reason may be owing to the increased inward friction force at the interface between the geogrid and the soil with the increase of settlement, which might have played controlling role in reducing the settlement.

3.2 *Effect of pile length and rows*

Figure 4 shows the effect of length and rows of pile on the behavior of settlement. Increase in the length and rows of pile is very effectively in reducing settlement. If the length of pile increases from 0.3H (165 mm) to 0.7H (365 mm), at the load of 100 kPa, the settlement decreased up to about 30%. If the number of pile rows increased from 2 to 4, at the load of 100 kPa, the settlement decreased to about 20% for the case of 0.3H length pile and 15% for the case of 0.5H length of pile. Thus, the increase of pile length is much effective than the increase of pile row in settlement control. Two rows of 0.7H (365 mm) pile behaved in the same manner with 4 rows of 0.5H (275 mm) pile. Analogically, two rows of 0.5H (275 mm) pile behaved in the same manner with 4 rows of 0.3H (165 mm) pile.

3.3 *Effect of geogrid length*

Figure 5 shows the effect of length of geogrid on the behavior of settlement. Increase in the length of geogrid has the significantly effect on the reduction of settlement and the increase of bearing capacity. For the case of 0.3H pile length, if the length of geogrid increases from 2B (200 mm) to 3B (300 mm), at the load of 100 kPa, the settlement decreased to about 65%. Similarly, for the case of 0.7H pile length, if the length of geogrid increases from 3B (300 mm) to 5B (500 mm), at the load of 100 kPa, the settlement decreased to about 50%. Thus, the increase of georid length is much more effective for longer pile than that for the shorter pile in settlement control. Geogrid with 200 mm length over 4 rows of 0.5H pile behaves in the

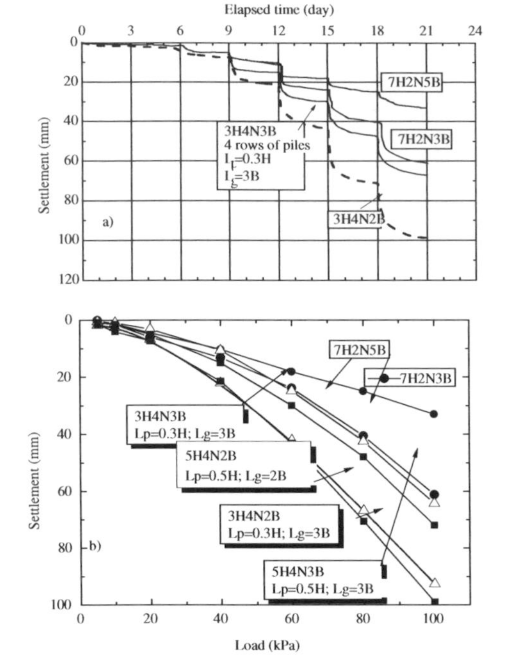

Figure 5. Effect of length and rows of pile on the settlement.

same manner with the 300 mm geogrid over 4 rows of 0.3H (165 mm) pile.

3.4 *Soil arching effect*

The degree of soil arching was defined as follows (as proposed by McNulty 1956):

$$\rho = \frac{p_b}{\gamma H + q_0} \tag{1}$$

where ρ = soil arching ratio; $\rho = 0$ represents the complete soil arching while $\rho = 1$ represents no soil arching; p_b = applied pressure on the top of the trapdoor in Terzaghi or McNulty's studies (geosynthetic for this study); γ = unit weight of the embankment fill; H = height of embankment; and q_0 = uniform surcharge on the embankment.

Figure 6 depicts the variation of soil arching ratio with the applied load. The test result shows that in the initial two load stages, the soil arching ratio decreases with an increase in the applied embankment load, which is in agreement with the experimental findings by McNulty (1965) and the numerical analysis results (Han and Gabr, 2002). However, in the later loading stages, the soil arching ratio increases with the increase of the applied load, which is contradict with the findings by other researchers. Moreover, the value of the soil arching ratio is lower than the results

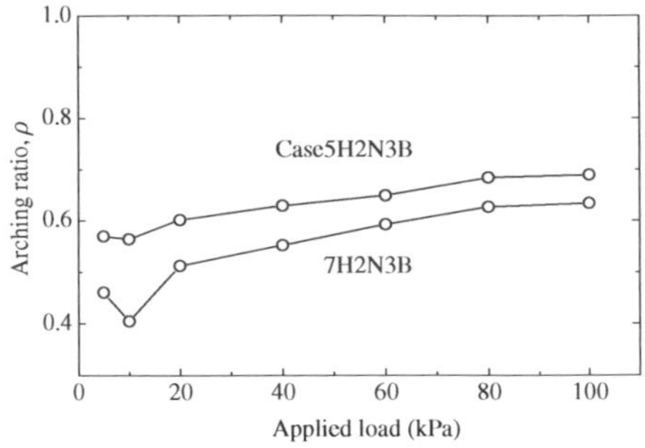

Figure 6. Variation of arching ratio with load.

by Han and Gabr (2002), which means that there is a very strong soil arching effect in the present study. The reasons of these discrepancies are: i) stiffness of the loading plate is much higher than the sand soil; ii) the sand mat covered the earth pressure meter is too thin. The strong arching effect is due to the loading plate. The sand mat is too thin to form the arching. Most of the load is directly transferred to adjoining supporting piles from loading plate (TTN: WM3 1989; Schmertmann (1999)). Arching effect also influenced by the pile length. From Fig. 6, the longer the pile, the stronger the arching effect. This is because for the shorter pile case, under the same load, the relative movement between the soil and pile is smaller so that the vertical pressure on the reinforcement is larger.

4 CONCLUSION

From the test studies, the following conclusions can be drawn:

1) Two layers of geogrids with length of 600 mm is equivalent to four rows of piles with length of 175 mm in terms of settlement behavior.
2) In settlement control, increase in the length of geogrid is more effective than the increase in the length of piles.
3) Increase in pile length is more effective than the increase in pile rows.
4) Further, numerical analysis should be conducted to investigate the optimum values of pile lengths, pile rows, the length and layers of geogrid, and the combination of pile and geogrid.

ACKNOWLEDGEMENTS

Part of the work was done during the first author's sabbatical stay in the Institute of Lowland Technology, Saga University, Japan as the Guest Professor from Shanghai Jiao Tong University, China. Laboratory tests were helped by Mr. T. Nanoka, the former graduate of Saga University.

REFERENCES

British Standard BS 8006, 1995. *Code of Practice for Strengthened/Reinforced Soils and Other Fills*, British Standard Institution, London, 162p.

Han, J. and Gabr, M.A. 2002. A numerical study of load transfer mechanisms in geosynthetic reinforced and pile supported embankments over soft soil. *Journal of Geotechnical and Geoenvironmental Engineering, ASCE*, 128(1): 44–53.

Jewell, R. A. 1988. The mechanics of reinforced embankments on soft soils. *Geotextiles and Geomembrane*. **7**: 237–273.

Li, Y., Aubeny, C., and Briaud, J.L. 2002. *Geosynthetic Reinforced Pile Supported (GRPS) Embankments (Draft)*. Texas A&M University, College Station, Texas, USA, 222p.

McNulty, J.W. 1965. An experimental study of arching in sand. *Report No. I-674, U.S. Army Engineer Waterways Experiment Station*, Corps of Engineers, Vicksburg, Miss., 170.

Schmertmann, J.H. 1999. Soil arching and spanning of voids. *Proc., 1999 ASCE/PaDot Geotechnical Seminar, Central Pennsylvania Section, ASCE and Pennsylvania Dept. of Transportation*, Hershey, Pa., 15.

Terzaghi, K. 1943. *Theoretical Soil Mechanics*, John Wiley & Sons, New York, 66–75.

New Horizons in Earth Reinforcement – Otani, Miyata & Mukunoki (eds)
© 2008 Taylor & Francis Group, London, ISBN 978-0-415-45775-0

Geosynthetic encased stone columns in soft clay

S.R. Lo
University of New South Wales, ADFA Campus, Canberra Australia

J. Mak
Roads and Traffic Authority, New South Wales, Sydney, Australia

R. Zhang
University of New South Wales, ADFA Campus, Canberra Australia

ABSTRACT: A preliminary numerical study on the reinforcing effects of stone columns in a very soft clay deposit was undertaken. The reinforcing effect of stone columns is dependent on the mobilization of confining stress by the surrounding soil. For very soft or weak soil, this confining stress may be augmented by the use of geosynthetic encasement. This hypothesis was examined by a preliminary numerical study. The findings of this preliminary study highlighted the potential benefit of reinforcing the stone columns by geosynthetic encasement.

1 INTRODUCTION

Stone columns have been used extensively to improve the bearing capacity of weak soils. They can be used in a small cluster as vertical load bearing members similar to piles. Alternatively, a large number of stone columns can be used to strengthen a weak soil stratum for supporting a fill structure such as a road embankment. This paper is for the latter application. The bearing capacity of a stone column can be assessed based on design guidelines such as FHWA (1983). Stone columns can also be used to reduce settlement. Such an application is useful for supporting a road embankment section that leads to a piled abutment. Oh et al (2007) reported the settlement performance of a 4 m high trial embankment constructed on soft estuarine clay improved by stone columns. The observed settlement at natural ground level (over a period of 457 days) of the stone columns treated section was only slightly less than that of the untreated section. The clay of this site, which is located in south-east Queensland, Australia, is very compressible, with a compressibility Index, C_C, exceeding 1.5. It is noted that there is no stiff crust overlying the soft clay layer, and water table is at a depth of 0.5 m. It was hypothesized that the stone columns bulged and compressed excessively because of lack of confinement. It is pertinent to note that such ground condition is not uncommon for estuarine deposits along the coast between northern New South Wales to south-east Queensland.

The observed settlement performance of the above trial embankment sections founded on a very soft clay strengthen with stone columns raises the question about the effectiveness of stone columns in reducing settlement of very weak deposits. The paper presents a preliminary numerical study on the performance of stone columns in very soft soil. The potential benefits of reinforcing stone column with a geosynthetic encasement are highlighted by this preliminary study.

2 THEORETICAL CONSIDERATIONS

2.1 *Stone columns as reinforcing elements*

Stone columns can be viewed as compressive reinforcements in a matrix of weak soils. They are constructed from stones which behave like a granular geo-material. Therefore, the strength and stiffness of a stone column is dependent on effective confining stress provided by the surrounding soil. A high effective confining stress can normally be induced by the installation process, with the stones being expanded against the surrounding soil. For very soil clay, and maybe stone columns installed at wider spacing, this may not be achieved effectively. The mobilization of additional confining stress on the stones, and thus the generation of higher bearing capacity, can still be realized during or after placement of fill as axial straining of the stones is always accompanied by lateral expansion against the surrounding clay. However, the stone columns may not be adequately effective in reducing settlement of the fill structure.

As one of the key issues is the generation of adequate confining stress to the stones prior to or

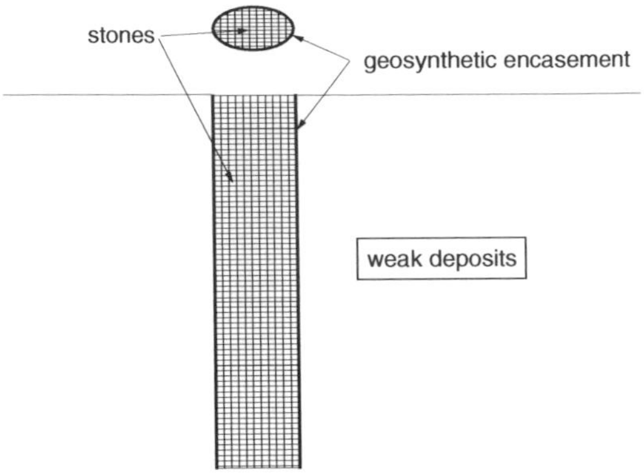

Figure 1. Geosynthetic encased stone column.

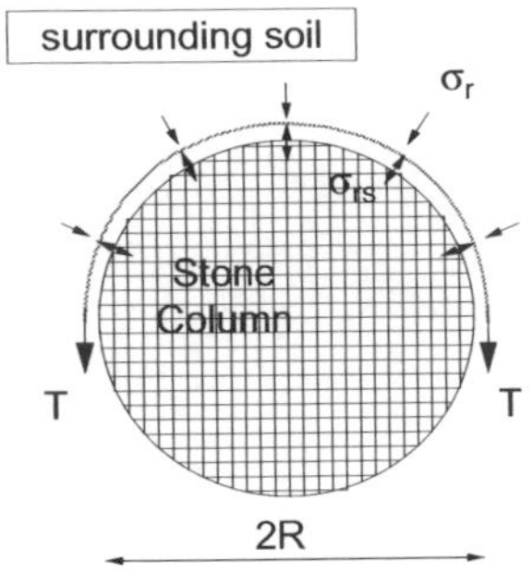

Figure 2. Hoop tension and radial stress in stone column.

during imposition of axial loading from the fill, it was decided to examine the option of encasing/wrapping the stone with geosynthetic reinforcement as illustrated in Figure 1. This geosynthetic encasement can be provided by a geogrid-geotextile composite. The geogrid functions as the reinforcement whereas the geotextile prevents loss of stones into the surrounding soft clay.

2.2 *Contribution of geosynthetic encasement*

Unless stated to the contrary, all stresses on geomaterials are effective stress. The radial stress in acting on the stone column, σ_{rs}, can be expressed in terms of the radial stress of the surrounding clay, σ_r, and the hoop tension, T, in the geogrid encasement as illustrated Figure 2.

$$\sigma_{rs} = \sigma_r + T/R \tag{1}$$

where R is the radius of the stone column. The second term can be view as the additional effective radial stress due to the geogrid encasement. Both T and σ_r can be decomposed into two parts, the initial value (ie after stone column installation) and the increase due to placement of fill and time dependent deformation. Therefore, Eqn (1) can be re-written as

$$\sigma_{rs} = \sigma_r(i) + T(i)/R + \Delta\sigma_r + \Delta T/R \tag{2}$$

where (i) denotes the initial (as-installed) state, and "Δ" denotes increase due to loading.

To enable the stones to develop adequate strength and stiffness, σ_{rs} has to be of adequate magnitude. If an adequately high $\sigma_r(i)$ can be generated, then both T(i) and ΔT are not needed. Indeed the value of $\Delta\sigma_r$ will also be low as the axial strain, and thus the radial expansion, of the stone column is small. However, one can compensate for a low $\sigma_r(i)$ value, say due to the surrounding clay being very soft, by the introduction of a significant T(i) value. The presence of a significant T(i) value automatically implies additional contribution to effective radial stress by the ΔT term. An initial tension in the geogrid encasement, T(i), can be induced by installing a stone column using the casing method, plus slightly "undersizing" the prefabricated geogrid encasement and controlling the compaction of the stone. This is more an in-principle statement on the possibility of installing a geogrid-encased stone column. The aim of this paper is to explore the potential beneficial effect of geogrid encasement.

Immediately after the imposition of fill loading imposed, the stone columns perform only a small reinforcing role as most of the total stress is taken by pore water pressure in the clay. It is only with dissipation of pore water pressure with time that the clay will settle and the weight of the fill will "arch over" to the stone column. During this process, the stone column will strain both axially and radially, the latter leading to both ΔT and $\Delta\sigma_r$. Some of the fill loading will still be transferred to the clay as effective stress and this also leads to $\Delta\sigma_r$. Therefore, the mechanism involves the interaction of the stone columns and dissipation of pore water pressure of the surrounding clay. The latter is a coupled process between mechanical behaviour (as governed by effective stress principle) and flow of pore water (as governed by Darcy law). In order to take into account all these complicated interaction, a fully coupled finite element analysis of a unit cell using elasto-plastic soil models will be conducted as explained in a subsequent section.

3 UNIT CELL ANALYSIS

We examined the condition where the fill area was large relative to the thickness of the soft clay and that a large number of stone columns were installed. With the exception of the stone columns near the edges, we can idealise the problem by a unit cell as illustrated in

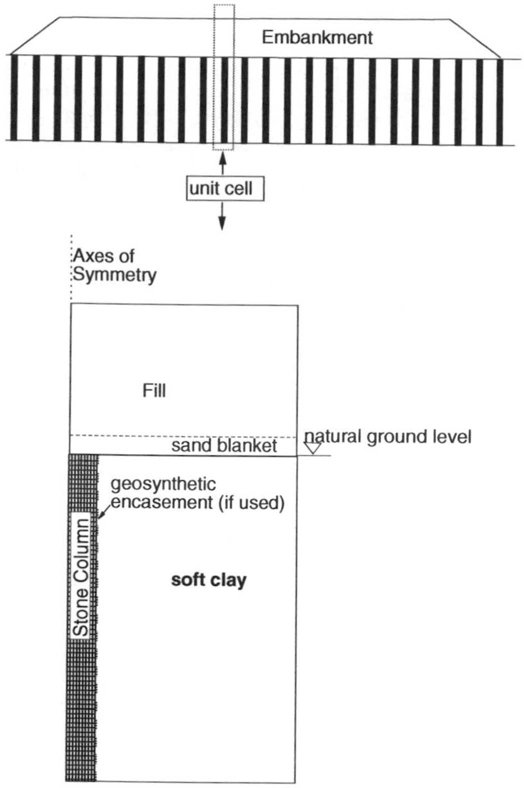

Figure 3. Unit cell.

Table 1. Parameters for unit cell analysis.

Item	Dimension
Embankment elevation	4.0 m
Sand blanket thickness	1.0 m
Diameter of stone column	0.6 m
Unit cell radius	2.0 m
Depth of ground water table	0.0 m
Thickness of soft clay	10.0 m

Figure 3. The top of the clay layer was modelled as a free draining boundary, whereas the edge of the unit cell was modeled as impermeable.

Placement of the embankment fill in layers was simulated in the analysis. Stone column construction commenced after the construction of a sand blanket. This was modeled by activating the hoop tension of the geogrid after placement of the sand blanket.

The assumed dimensions for the analysis were listed in Table 1. It was recognized that this was a rather extreme condition, but it served to highlight the beneficial effect, if any, of geogrid encasement.

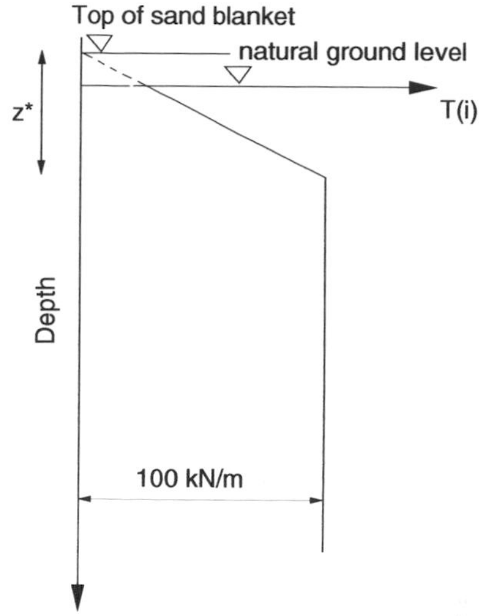

Figure 4. Initial hoop tension.

Two stone column configurations were analysed, one without geogrid and the other with geogrid encasement. For both configurations, the dissipation of pore water pressure and thus development of settlement after the completion of embankment was tracked numerically for 10 years.

3.1 *Modelling of geogrid encasement*

The geogrid, if used, was modeled as an anisotropic elastic material with a hoop stiffness of 2000 kN/m. The stiffness in the vertical direction was assigned a low value so that it will not provide any significant vertical support to the fill loading. The initial hoop tension, T(i), was also assigned a value of 100 kN/m except with the top zone where the value of T(i) was limited by triaxial extension failure of the stones.

From Eqn (2), the initial radial stress acting in the stone, $\sigma_{rs}(i)$, is given by:

$$\sigma_{rs}(i) = \sigma_r(i) + 3.33T(i) \qquad (3)$$

The condition of failure in triaxial extension means $\sigma_{rs}(i)$ is given by the following equation:

$$\sigma_{rs}(i) = K_p \sigma_{zs}(i) \qquad (4)$$

where $\sigma_{zs}(i)$ is the in-situ vertical stress (due to self weight of stones), $K_p = (1 + \sin\phi)/(1 - \sin\phi)$, and ϕ is the friction angle of the stones. Therefore, the distribution of T(i) is given in Figure 4, where T(i) attained the allowable value of 100 kN/m at a depth z* measured from top of sand blanket.

Assuming water table is at natural ground level, z* given by:

$$z^* = \frac{333/K_p - \gamma t_s}{\gamma'\left(1 - \rho K_0/K_p\right)} + t_s \qquad (5)$$

where γ' = effective unit weight of stones, γ = bulk unit weight of stones, ρ = ratio of effective unit weight of clay to that of stones = 0.65, t_s = thickness of sand blanket = 1 m, K_o = at-rest earth pressure coefficient of the soft clay = 0.535.

3.2 *Modelling of soft clay*

The soft clay was modeled by the modified Cam-Clay model. The relevant Cam-Clay soil parameters are: $\lambda = 0.65$, $\kappa/\lambda = 0.1$, M = 1.1, $e_{cs} = 4.0$.

In-situ stress was assigned based on an effective unit weight of 6 kN/m^3 and $K_o = 0.535$. As the analysis modeled the coupled process of time dependent dissipation of pore water pressure, permeability parameters were also needed. The horizontal permeability of the soft clay was 2.3×10^{-10} m/s, and with a horizontal to vertical permeability ratio of two. The above parameters were typical of very soft estuarine clay deposits of a road construction site along the coast of northern New South Wales, Australia. A typical undrained shear strength profile was also assumed. From this undrained strength profile, it was inferred, following Potts and Ganendra (1991), that the top 3 m was overconsolidated ever though the soil was soft. For this top 3 m, the inferred value of p'_c, the stress at the apex of the Cam-Clay ellipse, was 69 kPa at natural ground level reducing to 40 kPa at 3 m depth.

3.3 *Modelling of fill*

The fill was modeled as a Mohr Coulomb elastic-plastic material. The parameters adopted for the analysis were: unit weight = 20 kN/m^3, $\phi = 30°$, c = 20 kPa, E = 30 MPa and $\psi = 5°$, where E = Young's modulus and ψ = dilatancy angle.

3.4 *Modelling of stone columns*

The stones column was modeled as a free draining material with $\phi = 45°$, c = 5 kPa. A small cohesion of 5 kPa is used to suppress potential numerical problems and to take into account, approximately, that the failure surface of stones is curved.

Two different approaches, depending on whether geogrid encasement was used or not, were needed to complete the modeling of the mechanical behaviour of a stone column. For both approaches, the influence of confining stress on stiffness needs to be captured.

3.4.1 *The case of no-geogrid encasement*

It was assumed that high effective radial stress would not be generated and thus $\sigma_{rs}(i)$ is less than $\sigma_{zs}(i)$, the in-situ vertical stress (due to self weight of stones). Therefore, the major principal stress is initially vertical and will remain closer to vertical during and after placement of fill. The stones were modeled by the Duncan-Chang non-linear elastic equation for tangential Young's modulus, E.

$$E = K\left(\frac{\sigma_3}{p_a}\right)^n p_a\left(1 - r_f S\right)^2 \qquad (6)$$

Where K, n, and r_f are non-dimensional parameters for the Duncan-Chang model, σ_3 is the minor principal stress, p_a is the standard atmospheric pressure in consistent unit, and S is a function of the stress state, ϕ and c. S reflects the mobilization of the shear strength of soil and takes a value in the range of 0 to 1. The parameters for the stones were conservatively taken as: K = 1000, n = 0.6, $r_f = 0.7$.

3.4.2 *For the case with geogrid encasement*

The initial hoop tension, T(i) is related to an initial radial stress stones, $\sigma_{rs}(i)$ by Eqn (3). Therefore, together with the distribution of T(i) shown in Figure 4 (as established from earlier Section 3.2) and still using $\phi = 45°$ for the stones, the distribution of initial radial stress in the stones, $\sigma_{rs}(i)$, is given in Figure 5.

It is evident that $\sigma_{rs}(i) > \sigma_{zs}(i)$. Thus the major principal stress is initially horizontal. The imposition of loading on the stone column leads to an increase in σ_{zs}. During the early phase of loading, the stress states in the stones still satisfied the condition of $\sigma_{rs} > \sigma_{zs}$. Thus the stress ratio decreased with imposition of column load and the behaviour of the stones corresponded to that of "unloading"., and could be approximated by the initial Young's modulus, E_o given by:

$$E_0 = K\left(\frac{\sigma_{rs}}{p_a}\right)^n p_a \qquad (7)$$

During the later phase of loading when $\sigma_{zs} > \sigma_{rs}$, imposition of additional load on the stone column led to an increase in stress ratio. Furthermore, the major principal stress direction was closer to vertical. The Duncan-Chang equation (Eqn (6)) may apply. The tangential Young's modulus will be less than E_o, and is a function of the stress state of the stones. It has a minimum value of $E_0*(1 - r_f)^2$. As a first approximation, we use an "average" Young's modulus taken as 50% of E_o. Therefore:

$$\bar{E} = 0.5K\left(\frac{\sigma_{rs}}{p_a}\right)^n p_a \qquad (8)$$

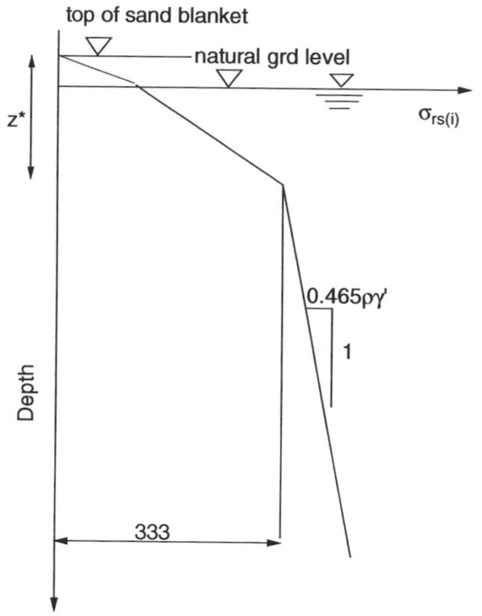

Figure 5. Initial radial stress in stone column.

Noting that $\sigma_{rs} > \sigma_{rs}(i)$, an additional conservative simplification was made by replacing σ_{rs} with $\sigma_{rs}(i)$. The average Young's modulus is given by:

$$\bar{E} = 0.5K\left(\frac{\sigma_{rs}(i)}{p_a}\right)^n p_a \tag{8a}$$

Note that $\bar{E}$ is a function of depth and initial hoop tension in the geogrid. It was used in conjunction with the Mohr Coulomb elastic-plastic model with an non-associative flow rule. The dilatancy angle, ψ, is taken as 0.33ϕ.

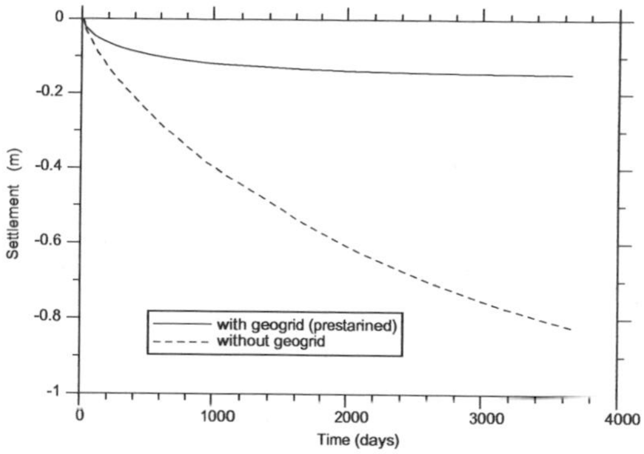

Figure 6. Development of settlement with time.

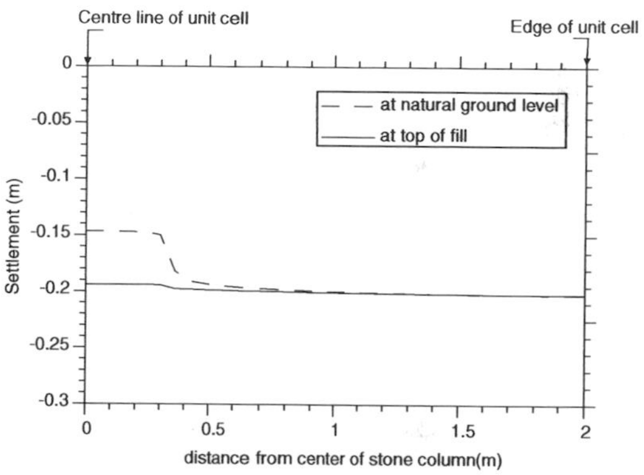

Figure 7. Settlement profile.

4 RESULTS OF FINITE ELEMENT ANALYSIS

4.1 *Settlement*

The developments of settlements with time for the two configurations are compared in Fig. 6 at top of stone column. It is evident that the settlement increases with time because the arching-over of the imposed load to the stone column is related to the settlement of the soft clay.

For the condition of nil strengthening by stone columns, the final settlement calculated based on conventional 1D consolidation was 1.55 m. Therefore, both stone column configurations reduced settlement by a significant amount. The without-geogrid stone column configuration had a computed settlement in excess of 800 mm in 10 years. The stone column with geogrid encasement had a considerably lower settlement, just slightly in excess of 150 mm in 10 years. For the configuration without geogrid, settlement is still increasing at a considerable rate after 10 yr. However, for the with-geogrid configuration, the 10-yr settlement of approaches closely to an asymptotic value. This aspect will be examined at a later sub-section with reference axial force in the stone column.

The settlement profiles at natural ground level and at top of fill are compared in Fig. 7. It was recognised that the natural ground surface already settled by a small amount at completion of fill placement, ie when settlement of top of fill just commenced. Therefore, the settlement profile at top of fill presented in Fig. 7 was shifted so that can be compared to that at natural ground level. The profile at natural ground level manifested a bump of 50mm over a distance of 0.2 m. This is believed to be due to the stiffening effect of the geogrid encased stone column. However, at top of fill, the settlement profile was smooth.

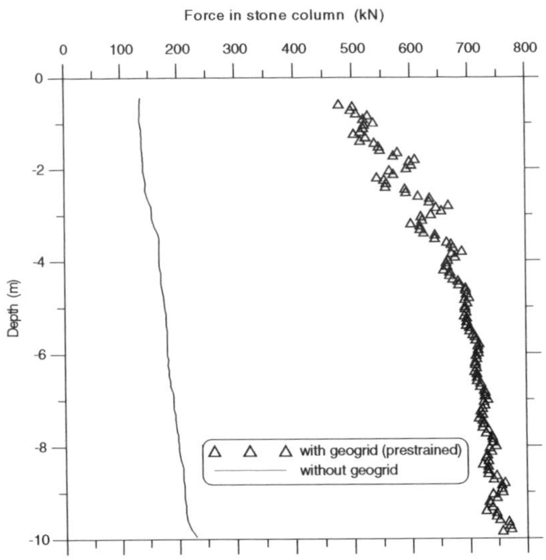

Figure 8. Force in stone column.

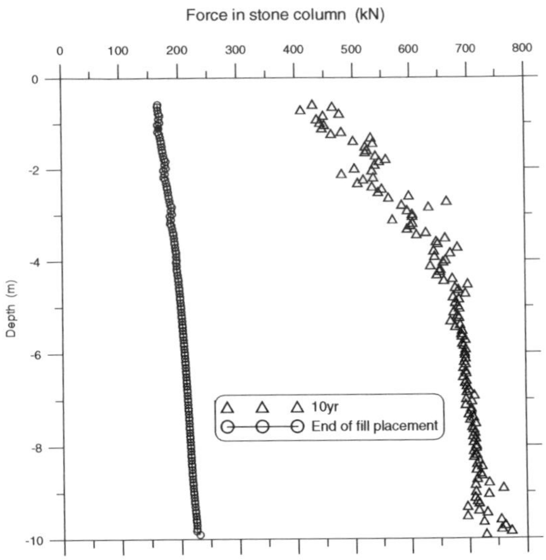

Figure 9. Development of column force with time (with geogrid encasement).

4.2 *Axial force in stone column*

The load transfer to the stone column was illustrated in Fig. 8 which showed the variation of axial load on stone column with depth. The stone column with geogrid encasement had a significantly higher load compared to that of without-geogrid configuration. For both configurations, the axial force in the stone column increased with depth. This was due to the drag-down force induced by the settlement of the surrounding clay relative to the stone column. The stiffer geogrid encased stone column attracted higher drag-down force and thus manifested a greater increase of axial force with depth, as evident from Figure 8.

The distribution of column load, for the geogrid encased configuration, at 10 year is compared to that at end of fill placement in Fig. 9. The stone column attracted an increasing amount of load with time. This means that the effective stress in the soft clay will attain an asymptotic value via two mechanism, dissipation of pore water pressure and increase in transfer of load to the stone column. The latter mechanisms will lead to the settlement reaching an asymptotic value faster for the geogrid encased configuration as illustrated in Fig. 6.

5 CONCLUSION

A preliminary numerical study was undertaken to examine the reinforcing role of stone columns in soft clay. For very soft clay, the results of the analysis indicated that the surrounding clay may not provide adequate confining stress to the stone columns. Thus the stone columns may not be effective in reducing settlement. This simplified numerical analysis suggested that if the hoop tension in the geosynthetic encasement can be mobilized upon stone column installation, then it will induced significant confining stress onto the stones despite the surrounding soil being very soft, which in turn enhance the reinforcing role of the stone columns.

ACKNOWLEDGEMENT

The last author acknowledged the financial supported provided by the the Cheung Kong Endeavour Award and the University College Postgraduate Research Award Scholarship, The University of New South Wales. The opinions expressed in this paper are solely those of the authors.

REFERENCES

FHWA (1983) Design and construction of stone columns. FHWA Report No. RD-83/026, 194p.

Oh E.Y.N., Balasubramaniam A.S., Bolton M., Chai G.W.K., and Huang M. (2007) Behaviour of a highway embankement on stone columns improved estuarine clay. 567–572.

Potts D.M. and Ganendra D. 1991. Discussion on "Finite element analysis of the collapse of reinforced embankment on soft ground by Hird C.C., Pyrah I.C., Russel D.". *Geotechnique*, **41[4]:**627–630.

New Horizons in Earth Reinforcement – Otani, Miyata & Mukunoki (eds)
© 2008 Taylor & Francis Group, London, ISBN 978-0-415-45775-0

FEM analysis of the effect of the prestress induced in micropiles

K. Miura & S. Morimasa
Toyohashi University of Technology, Aichi, Japan

Y. Otani
Hirose, Co. Ltd., Osaka, Japan

Y. Tsukada
National Institute for Land and Infrastructure Management

ABSTRACT: We have conducted model loading tests and field loading tests on the foundation reinforced with micropiles to reveal the mechanism of load bearing capacity of the foundation. The significance of the confining effects of a group of micropiles on the bearing capacity was clarified through the examination of the test results. And the prestress induced in the micropiles was certified to be effective on the enhancement of the confining effect. We also carried out FEM simulations to examine the confining effect, and discussed the applicability of the FEM analysis in the simulation of the behavior of micropile foundation considering the interactions between soil ground, micropiles and footing.

1 INTRODUCTION

Micropiles, which was pioneered by Lizzi (1971, 1978) in Italy, are now widely used both as structural supports in foundations as well as for in-situ earth reinforcement. Micropiles are considered as promising foundation elements in improving the bearing capacity of existing foundations which are deteriorating for one reason or another with minimum disturbance to structures, subsoil and the environment.

To clarify the mechanism of the bearing capacity and develop new rational method to improve the performance of micropile foundations, a study on model micropile has been conducted by authors continuously. In previous study, the method of model loading tests has been reported by Tsukada et al. (1999, 2006), the test results on three series of model tests (footing test, micropile test, and micropile foundation test) have been reported by Miura et al. (2000), and the effect of prestress on the improvement of bearing capacity has also been investigated in model micropile foundation study by Miura et al. (2001). From these studies, it is found that the network effect of a group of micropiles on the bearing capacity is mobilized positively with the appropriate confinement of the ground material beneath the footing. Large-scaled field loading tests on the footings reinforced with micropiles were also conducted on natural uniform loam ground, and the findings obtained in the series of laboratory model loading tests were observed.

The micropiles not only provide load bearing capacity directory through their skin friction, but also raise the base pressure on the footing with the confinement by the interaction between the footing and a group of micropiles. Also it was demonstrated that the bearing capacity is improved more efficiently with the prestress, which induces the confinement on the ground material beneath the footing at early stage of the loading process.

In this study, FEM simulations were carried out to examine the confining effect quantitatively, and discussed the potential of the FEM analysis in the simulation of the behavior of micropile foundation considering the interactions between soil ground, micropiles and footing. In the following, the results of the field loading tests are explained briefly, and the results of the FEM simulations are discussed.

2 FIELD LOADING TEST

2.1 *Test method*

The plan layout of the investigation is shown in Fig. 1. Four types of foundations were investigated as follows;

- FT; a footing without micropiles,
- S-MP; a single micropile,

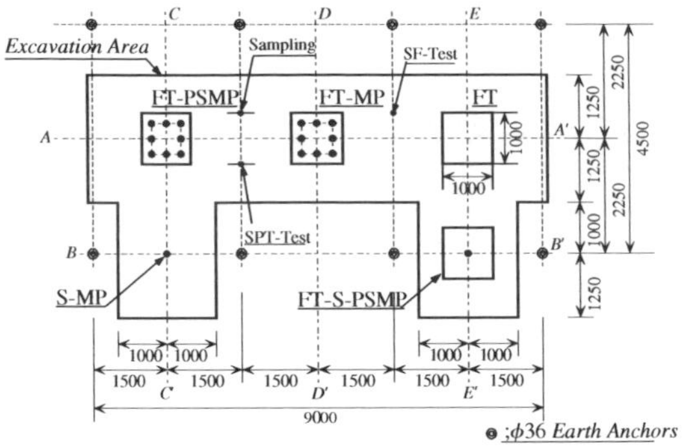

Figure 1. Layout of the field loading test.

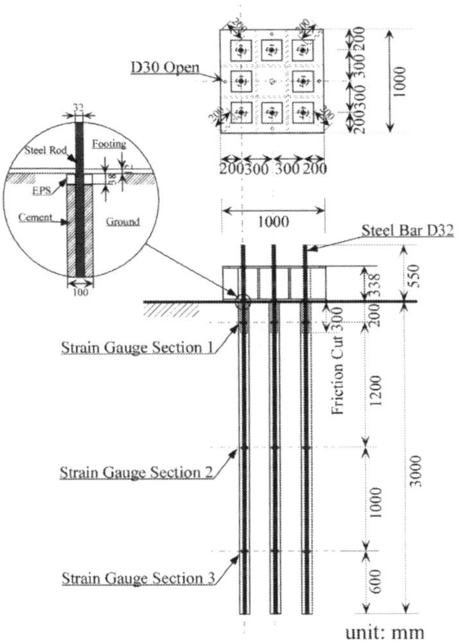

Figure 2. Micropile, footing and their connection (FT-MP and FT-PSMP).

- FT-MP; a footing reinforced with eight micropiles,
- FT-PSMP; a footing reinforced with eight prestressed micropiles.

Figure 2 shows the details of the micropile and the footing. The micropile (MP in short) used was 100 mm in diameter, 3.0 m in length and reinforced with D32 steel core bar. The MPs were instrumented with strain gauges arranged for the bending in two directions at three sections of −0.2 m, −1.4 m, and −2.4 m. The footing (FT in short) used is made of steel and 1000 × 1000 mm square in plane shape. The amount of prestress was 70% of the pullout resistance of the micropiles.

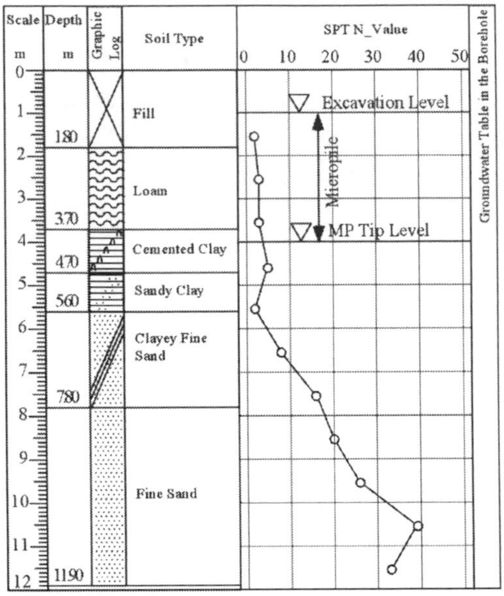

Figure 3. Geotechnical boring log of the test site.

Table 1. Soil properties of the investigation site.

Soils	Loam	Cemented clay	Fine sand
Depth (m)	2.0∼2.8 m	3.7∼4.5 m	8.0∼8.8 m
Wet density ρ_t (kg/m^3)	1.631×10^3	1.721×10^3	1.905×10^3
Dry density ρ_d (kg/m^3)	1.005×10^3	1.144×10^3	1.442×10^3
Water content ω (%)	62.7	50.6	32.2
Total stress c (kN/m^2)	23.54	46.09	42.17
Total stress ϕ (°)	17.6	16.8	43.7
Effective stress c' (kN/m^2)	10.79	7.85	3.92
Effective stress ϕ' (°)	28.4	33.7	37.0
B value	0.92∼0.96	0.94∼0.96	0.92∼0.96

The soil profile and the SPT N-value of the upper 10 m are shown in Fig. 3. The subsoil consisted of fill, loam, cemented clay, sandy clay, and fine sand, in order from ground surface to the 10 m depth. The fill, loam and clay are soft; the N values obtained are less than 5. The soil properties are tabulated in Table 1.

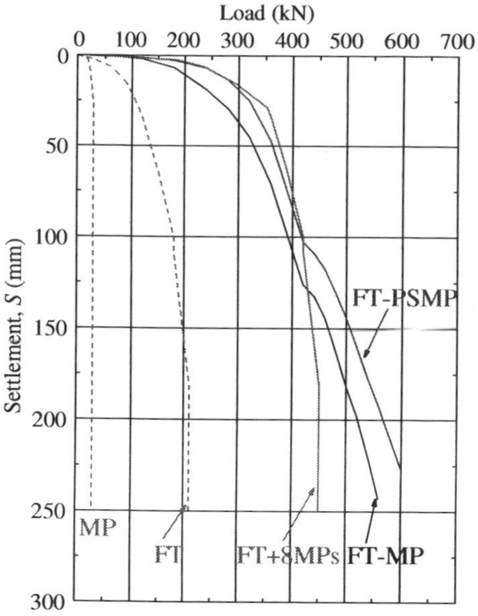

Figure 4. Load-settlement relations of FT, FT-MP, FT-PSMP and FT+8 MPs.

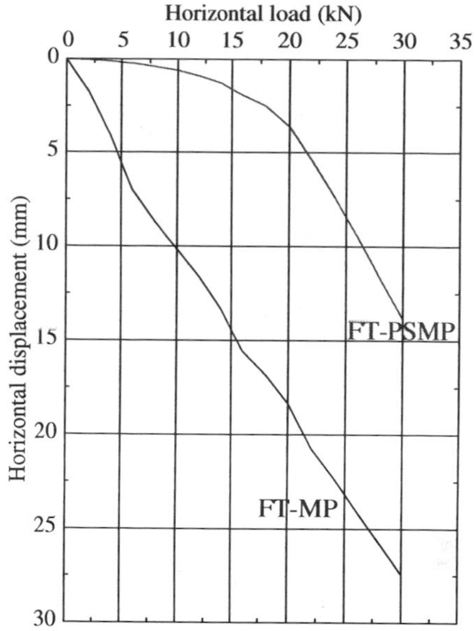

Figure 5. Movement of FT-PSMP and FT-MP under horizontal load.

2.2 *Test results*

2.2.1 *Vertical loading tests*

Figure 4 shows the load-settlement relationships of FT-PSMP and FT-MP. To investigate the MP group effect, the load carried by FT and the load taken by eight MPs are summated (FT+8 MPs). From this figure, MP group effect is not clear at the early loading stage when the settlement is small; the curve of FT-MP is even lower than that of FT+8 MPs, which may be due to loose contact between FT and the ground surface at the early stage. But with the increase of the load, FT-MP and FT-PSMP become higher than FT+8 MPs. From this figure, the curve of the FT-PSMP is always higher than that of the FT-MP. In other words, the coefficients of subgrade reaction were 1.86×10^4 kN/m^3 and 3.97×10^4 kN/m^3 in the tests on FT-PSMP and FT-MP, respectively. This comparison means the settlement was suppressed at the early loading stage and became almost half due to the effect of prestress on MPs.

2.2.2 *Horizontal loading tests*

Figure 5 shows the load-movement relationship of FT-PSMP and FT-MP. The effect of prestress on horizontal movement control is remarkable compared with on settlement control. The movement in the FT-MP seems linearly increased with load. The coefficient of subgrade reaction was 1.01×10^3 kN/m^3 and 17.1×10^3 kN/m^3 in the tests on FT-PSMP and FT-MP, respectively. This clearly showed the remarkable effect of prestress also in horizontal loading.

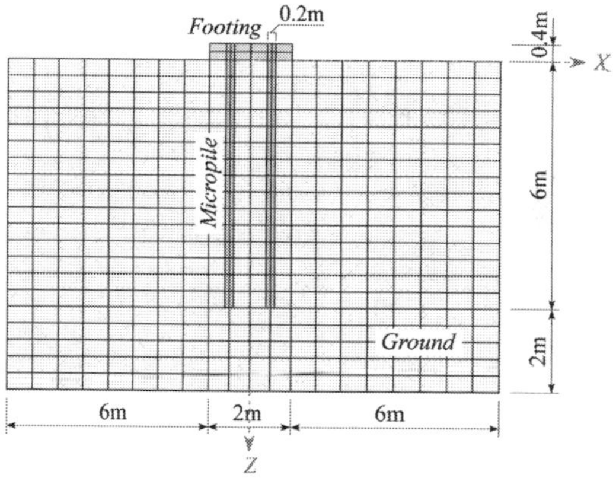

Figure 6. FEM mesh.

3 FEM SIMULATION

3.1 *Outline of FEM simulation*

Figure 6 shows the FEM mesh prepared for the simulation; the mesh consists of footing, micropiles and ground. The footing was reinforced with the micropiles and placed on the horizontal ground surface. The footing is detached with the ground surface in the simulation cases for the bearing capacity without micropiles. On the other hand, the footing contacts with the ground surface in the simulation cases for the bearing capacity with micropiles. Footing was modeled as rigid material where all the nodes on the

Table 2. Input parameters.

	Dense sand	Loose sand	Micropile
ϕ (°)	35	30	
c(kN/m^2)	0	0	
Dilatancy angle Ψ. (°)	10	−10	
Wet density ρ (kg/m^3)	2.00×10^3	1.90×10^3	2.40×10^3
Young's modulus E (kN/m^2)	1.04×10^5	0.35×10^5	70.0×10^5
Shear modulus G (kN/m^2)	1.20×10^5	0.40×10^5	2.69×10^6
Poisson's ratio ν	0.3	0.3	0.3

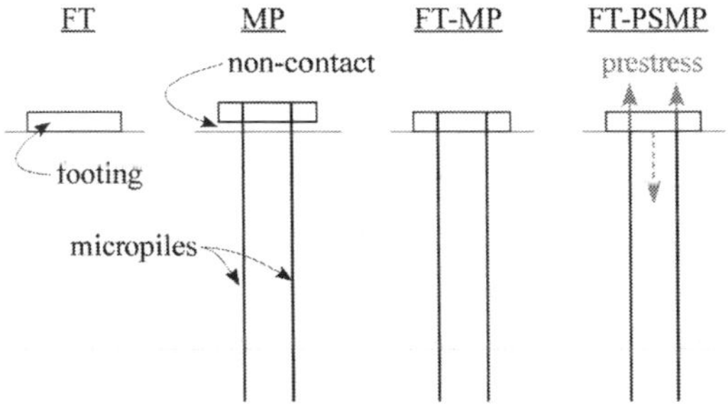

Figure 7. The types of foundations.

footing have the same displacement vector, and the micropiles were modeled as elastic bending material with second-order FEM elements. The ground was modeled with Drucker-Prager Type elasto-plastic model combining the Mohr-Coulomb Criteria for frictional failure and dilatancy. The interface between micropiles and ground is modeled with bi-linear slider elements with initial stress on the interface and internal friction angle of the ground. The mechanical properties, or input parameters for the models, are listed in Table 2. Two types of ground materials, i.e. dense and loose sands, were employed to clarify the effect of density and dilatancy on the bearing capacity. Not only the internal friction angle ϕ but also the dilatancy angle ψ is different with density.

Four types of footings were considered as follows (Fig. 7);

- FT Foundation; a spread footing without micropiles,
- MP Foundation; micropies where the footing was detached with the ground,
- FT-MP Foundation; a footing with micropiles, where the footing was attached with the ground,
- FT-PSMP Foundation; A footing with prestressed micropiles.

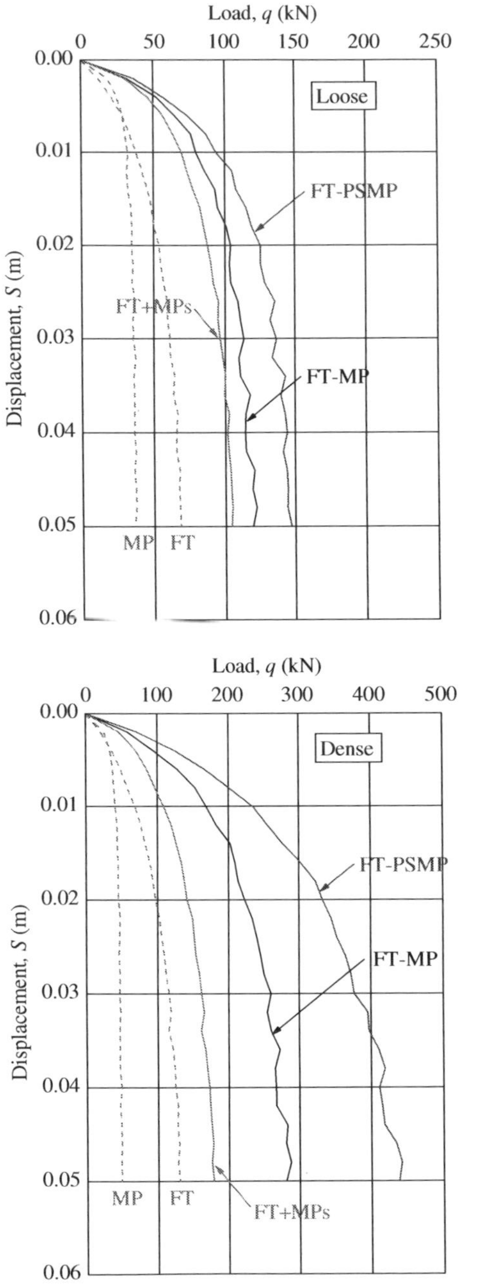

Figure 8. Load-settlement relations under vertical loading condition (loose and dense sand ground).

The prestess was applied before the loading; tension was induced in the micropiles, and the footing was pressed onto the ground at the same time as a reaction of the prestress. The amount of prestress was selected so as to be approximately 30% of the

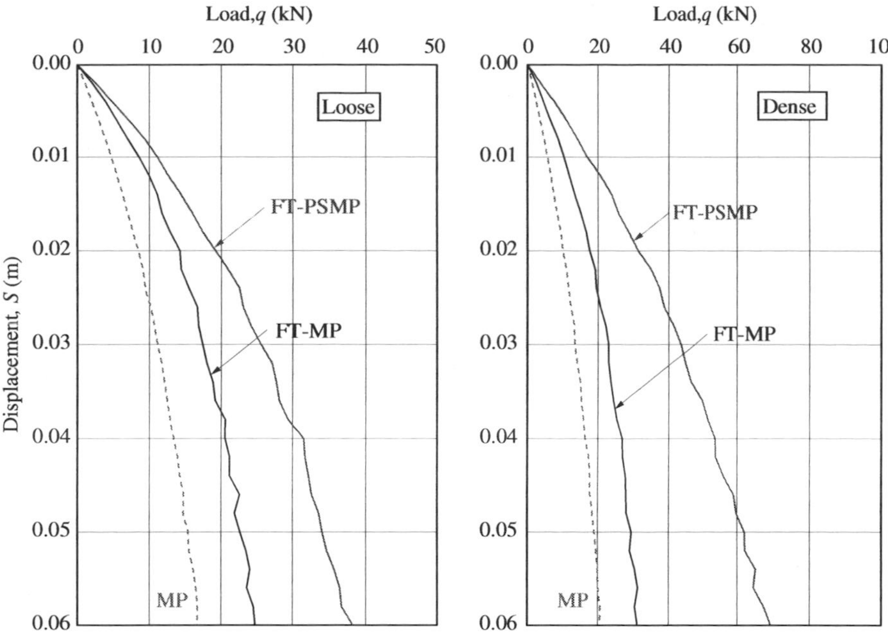

Figure 9. Load-movement relations under horizontal loading condition (loose and dense sand ground).

Table 3. The critical loads borne by the foundations without prestress and the ratio R.

	FT (kN)	MP (kN)	FT+MPs (kN)	FT−MP (kN)	R
Dense sand	124	45	169	307	1.82
Loose sand	64	35	99	123	1.24

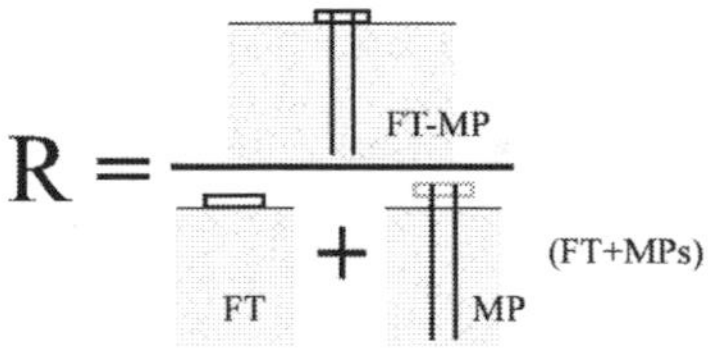

Figure 10. Confining effect ratio R.

bearing capacity of the micropiles. The loadings in two directions were conducted with incremental loading schemes with non-linearity of soil behavior and the interaction between soil and micropiles.

3.2 *Calculation results and discussion*

The calculated behaviors of the foundations under vertical loads are shown in Fig. 8, and those under horizontal loads in Fig. 9. The symbol FT+MPs is for the summation of the loads borne by FT Foundation and by MP Foundation. Table 3 lists the critical loads borne by the foundations without prestress. The ratio R designated for the critical bearing load of FT-MP Foundation to that of FT+MPs, shown in Fig. 10, was calculated and listed in the table. The value of the parameter R which indicates the effectiveness of the confinement of ground material with the footing and micropiles, was 1.82 in the dense sand ground, and 1.24 in the loose sand ground. The difference between the dense ground and loose ground is due to the dilatancy behavior of ground materials; the dense soil shows more dilative behavior compared with the loose sand. This significant effect of the dilatancy in the load bearing capacity was recognized also in the model loading tests (Tsukada, et al., 1999, 2006).

Table 4 shows the comparisons in critical bearing capacity and initial coefficient of subgrade reaction between the foundations with and without prestress: FT-MP Foundation and FT-PSMP Foundation. Remarkable effects of the prestress on both vertical and horizontal load bearing capacity can be recognized in the critical bearing capacity. On the other hand, the effects of the prestress cannot be seen clearly in the initial modulus of subgrade reaction. In the FEM simulation employed in this study, shear failure strength was properly increased with an increase in confining stress through Mohr-Coulomb Criteria. The modification of shear stiffness under confining stress was not sufficiently considered; i.e. the shear modulus of ground material was assumed constant

Table 4. Effect of prestress.

		Vertical loading		Horizontal loading	
		FT-MP	FT-PSMP	FT-MP	FT-PSMP
Dense sand	Critical bearing capacity (kN/m)	307	436	31	67
	Initial coefficient of subgrade reaction (kN/m^3/m)	31000	31000	760	820
Loose sand	Critical bearing capacity (kN/m)	123	156	23	38
	Initial coefficient of subgrade reaction (kN/m^3/m)	19000	20000	630	650

irrespective of the confining stress. The increase in critical load bearing capacity was clearly seen in the FEM simulation both in the vertical loading and horizontal loading. For example, in the case of horizontal loading on the dense ground, the critical load bearing capacity became large by the factor of two due to the prestress on micropiles. This effect of the prestress on the improvement of bearing capacity were recognized common in the model loading tests, field loading tests and FEM simulation.

4 CONCLUSIONS

To clarify the load bearing mechanism of the footing reinforced with micropiles, we conducted loading tests in laboratory and field, and FEM simulations. And the positive effect of prestress on the improvement of the bearing capacity was examined in the loading tests and FEM simulations. The following points were found in this study.

- The confining effects on the ground by the micropiles were clearly observed both in the loading tests and FEM simulations. The effect was remarkable in the case of dense ground which shows dilative deformation behavior,
- The effect of the prestress which induced the confinement on the subsoil by the interaction with footing and micropiles was recognized not only in the loading tests but also in the FEM simulations. The effect of the prestress was clearly seen both in vertical loading and horizontal loading.

The FEM simulation must be modified in order to take sufficiently account of the yielding behavior of ground under the horizontal loading on piles, and the increase in shear modulus of ground due to the confinement, and etc. Also the FEM simulation should be carried out in three-dimensional condition to simulate the network of micropiles and its confining effect on the bearing capacity.

REFERENCES

Lizzi, F. 1971. Special Patented Systems of Underpinning and more Generally, Subsoil Strengthening by Means Of Pali Radice (Root Piles) with Special Reference to Problems Arising from the Construction of Subways in Built-up Area, *Special Lecture given at university of Illinois at Urbana-Champaign*, etc.

Lizzi, F. 1978. "Reticulated Root Piles to Correct Land Slides," *Proceedings of ASCE Conference,* Chicago, Illinois, October 16–20. 1–25.

Miura, K., Tsukada, Y., You, G.L., Ishito, M., Otani, Y. & Tsubokawa, Y. 2000 International Geotechnical Conference. Model Investigation on the Bearing Mechanism of Footing Regarding the Interaction Between the Footing and a Group of Micropiles, *Proceedings of the 3rd international conference on ground improvement techniques*, September 2000, Singapore. 255–262.

Miura, K., Tsukada, Y., Otani, Y., Ishito, M. & You, G.L. 2001. Model loading tests on the footing reinforced with prestressed micropiles, *International Symposium of Earth Reinforcement*, Fukuoka, Japan.

Tsukada, Y. 1997. State-of-the-Art: Application of Micropiles in Japan, *Proceedings of 1st International Workshop on Micropiles*, Deep Foundations Institute, 265–279.

Tsukada, Y., Miura, K. & Tsubokawa, Y. 1999. Model Loading Tests on Micropile Foundation on Sand Ground, Japanese Geotechnical Society, *Tsuchi-to-kiso*, 47(1), 35–38.

Tsukada, Y., Miura, K., Tsubokawa, Y., Otani, Y., & You, G.L. 2006. Mechanism of Bearing Capacity of Spread Footings Reinforced With Micropiles, *Soils and Foundations,* 46, 3, 367–376.

Geo-hazards and mitigation

New Horizons in Earth Reinforcement – Otani, Miyata & Mukunoki (eds)
© 2008 Taylor & Francis Group, London, *ISBN 978-0-415-45775-0*

Earthquake performance of reinforced earth embankment subjected to strong shaking and ground deformations

C.G. Olgun & J.R. Martin II
Virginia Tech, Blacksburg VA, USA

H.T. Durgunoğlu
Boğaziçi University, Istanbul, Turkey

T. Karadayılar
Zetaş Earth Technology Corporation, Istanbul, Turkey

ABSTRACT: The 1999 Kocaeli Earthquake (M7.4) in northwestern Turkey provided the opportunity to study the field performance of a double-walled reinforced-earth (RE) embankment designed according to procedures essentially same as current guidelines used in the US. The embankment was located just a few meters from the ruptured fault and subjected to strong ground shaking and significant permanent ground deformations. Unanticipated liquefaction-related settlements also occurred beneath the embankment. Ground shaking at the site was much higher than what was accounted for using the k_h design value of 0.1. Estimated peak ground acceleration was about 0.25 g, corresponding to an estimated equivalent k_h of about 0.3. Despite being subjected to ground motions that significantly exceeded the design levels, the RE system performed well, suffering only minor damages related mainly to the ground deformations.

1 INTRODUCTION

Recent decades have seen an increased usage of reinforced soil structures, and designs are becoming more aggressive with taller walls and a wider variety of reinforcing elements and facing materials. The August 17, 1999 Kocaeli Earthquake ($M_w = 7.4$) that struck northwestern Turkey provided an important opportunity to study the field performance of a number of mechanically-stabilized and soil-nailed walls located in the affected region. The earthquake setting is shown in Figure 1. Of particular significance was the performance of a Reinforced Earth® approach embankment at the Arifiye Bridge Overpass. The embankment was immediately adjacent to the ruptured fault and underlain by soft and liquefiable soils. Peak ground accelerations during the earthquake are estimated at about 0.25 g for this site (Olgun 2003). The wall was designed using specifications similar to FHWA, but was designed for shaking levels much lower than those estimated to have occurred. Following the earthquake, a detailed field reconnaissance was made that included measurements of wall displacements, ground settlements, and fault-related ground movements. Despite being subjected to ground shaking levels above the design values, the RE system performed well, suffering only minor damage. This paper provides a description of the RE embankment, the seismic design of the structure, observed performance, and the numerical studies performed to better understand the seismic behavior.

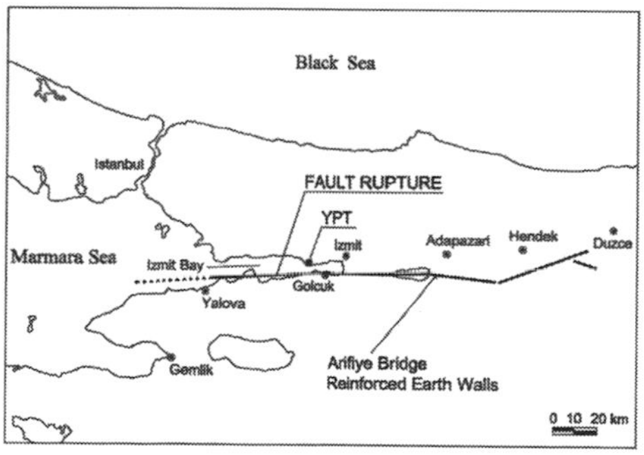

Figure 1. Setting of the August 17, 1999 Kocaeli Earthquake.

2 ARIFIYE BRIDGE OVERPASS AND RE EMBANKMENT

The Arifiye Bridge Overpass, which was constructed in 1988 and destroyed in the 1999 earthquake, consisted of four simply-supported spans resting on approach abutments and three mid-span pier supports. The two wing walls of the northern approach embankment were constructed using Reinforced Earth® (RE) technology, whereas the southern approach was a conventional earth embankment with sloping sides. The site is located along the Trans European Motorway adjacent to the ruptured fault as shown in Figure 2.

The northern approach ramp was 145 m long and 12.5 m wide with one traffic lane in each direction. The maximum height of the RE walls was 10 m. The bridge deck rested on a reinforced concrete abutment supported by a pile foundation. The RE walls that formed the embankment were of conventional design, consisting of cruciform and/or square, interlocking reinforced-concrete facing panels and ribbed, galvanized steel reinforcing strips. The facing panels were 150 cm × 150 cm in frontal area, and the reinforcing strips had a cross section of 40 mm × 5 mm. At the 10 meter high section, five strips per panel were used for the two lower panels, and four strips were used for the upper panels. The reinforcement length was 7 m along the section of the wall where the height ranged between 8–10 m. Reinforcement length was decreased progressively at shorter sections of the approach embankment. The reinforcing elements from each side of the wall overlapped 1.5 m at the center and were not connected.

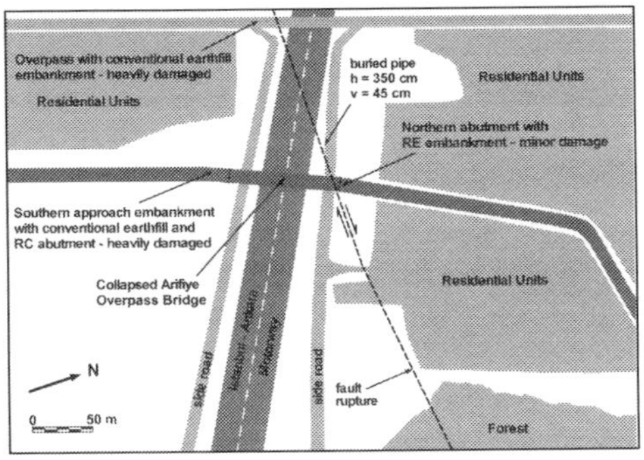

Figure 2. Plan view of Arifiye Overpass and the fault rupture.

The backfill soil was of high quality, consisting of sand and gravel that was compacted in lifts during wall construction.

As shown in Figure 3, a reinforced-concrete culvert passed beneath the wall. The culvert is located in a creek channel that runs beneath the site. Suspected liquefaction in the creek-bed soils beneath this culvert led to significant earthquake-induced settlements in this area of the wall. Also, slip joints were used along the height of the wall on both sides of the concrete culvert, as well as between the RE wall and the reinforced concrete bridge abutment. These special joints were used to mitigate anticipated static differential settlements, but apparently played an important role in limiting damage from earthquake-induced differential settlements.

The Arifiye Overpass site is situated within a deposit of Quaternary alluvial sediments consisting of alternating layers of medium clay/silt and loose sand with a shallow water table. The sounding from one of the Cone Penetration Tests near the culvert is shown in Figure 4. The upper 5 m of the profile consists of 2 m of silty/clayey sand fill underlain by a 3 m-thick medium clay layer with Q_c values of about 1.5 MPa. The clay is underlain by a 1 m-thick stratum of clean sand with an $I_c < 1.5$ and an estimated fines content <5%. Average value of normalized clean-sand-equivalent tip resistance ($q_{c1,CS}$) for the sand is 120. This layer is liquefiable under moderate levels of ground shaking. A mixed stratum of clay with lenses of interbedded sand is encountered below the sand between 5–8 m depth. Normalized clean-sand-equivalent tip resistance ($q_{c1,CS}$) for this stratum averages at 80. A medium-to-stiff clay stratum with an average Q_c of 2 MPa extends from a depth of 8 m down to 22 m where the CPT was terminated. The water table was encountered at a depth of about 2 m.

The sandy levels are potentially liquefiable under moderate shaking as evidenced by their low penetration resistance (Olgun 2003). Furthermore it was not possible to sample and test the clayey levels for a more detailed assessment of their cyclic vulnerability.

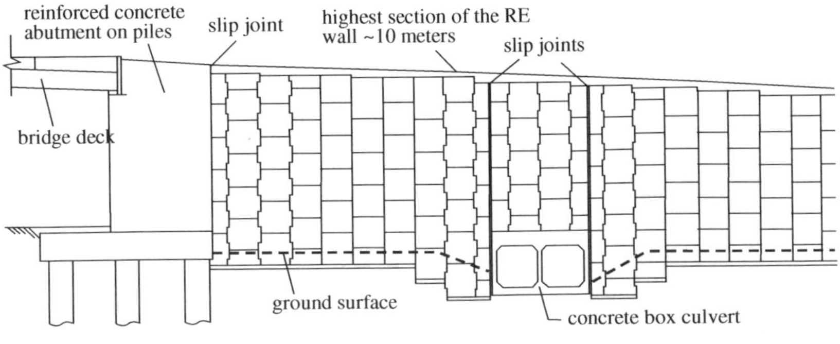

Figure 3. Side view of the Arifiye Overpass RE wall (northern approach embankment).

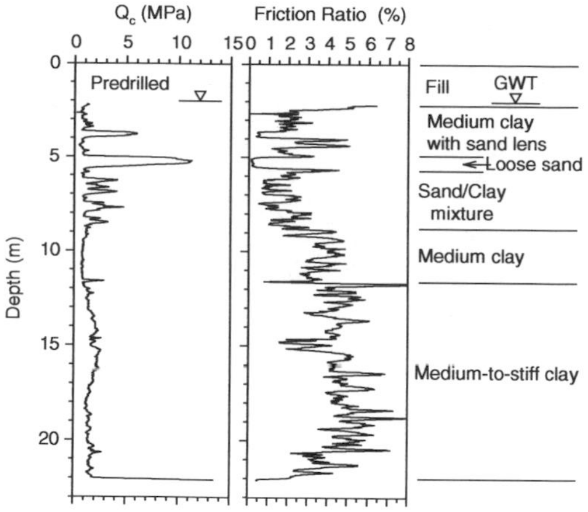

Figure 4. CPT sounding between the culvert and the reinforced concrete abutment.

However the observed ground failure pattern and the surface cracks near the base of the embankment strongly suggests such a potential vulnerability with these soils.

3 OBSERVED FIELD PERFORMANCE

Field reconnaissance at the Arifiye Overpass site was conducted following the August 1999 M7.4 earthquake (Mitchell et al. 2000). Peak ground acceleration at the site was in the range of 0.25 g, close to that recorded at other near-fault sites underlain by similar soil conditions such as the YPT station (Safak & Erdik 2000). In addition to significant ground shaking, ground displacements within a few meters of the RE walls were large, as the surficial fault rupture passed between the northern abutment and the center pier. The maximum horizontal and vertical ground displacements near the northern abutment were estimated at 3.5 m and 0.45 m, respectively, as inferred from the measured offset of a nearby ruptured pipe. Four spans of the bridge collapsed in a "saw-tooth" manner due to the resultant relative displacements between the piers and abutments along with beam-seat widths that were insufficient to accommodate the movements.

In addition to fault-related ground deformations, foundation settlements of up to 25 cm were observed. The resulting differential wall settlement caused the facing panels to become separated and misaligned, which allowed spillage of some backfill material. The maximum out-of-plane wall panel displacement was about 10 cm and occurred at 3 m or one-third of the wall height from the base. It appears that the heavy embankment and the relatively strong fill underneath the wall (2-m thick dry crust) punched through the soft foundation soil forming a localized "cone of depression" with the maximum settlement of 25 cm concentrated at the wall section overlying the culvert and creek bed. Numerous ground cracks running perpendicular to the wall were observed at ground surface near the embankment. The location and orientation of the cracks suggests they were associated with differential settlements along the base of the wall. In particular, the culvert appears to have settled during the earthquake, probably due to the presence of the soft and/or liquefiable creek bed sediments. As discussed later, it is likely that liquefaction occurred in the underlying 1-m sand layer located at a depth of 5 m. No sand boils were observed at the site or neighboring sites. However, the presence of a 2-meter thick dry crust would have likely prevented any surficial manifestation of liquefaction at depth. It is possible too that that the fined-grained silty and clay strata suffered strength loss and softening and allowed earthquake-induced undrained shear distortions and settlement under the weight of the structure. Significant earthquake-induced settlements of similar fine-grained soils under loaded areas were reported at other sites in the region during the earthquake (Sancio et al. 2002, Martin et al. 2004).

The slip joints used along each side of the concrete culvert and other sections of the wall apparently added to the flexibility and the structure's ability to tolerate differential settlements without being overstressed. The facing panels were intact and no signs of distress on the panels were noted. The wall was demolished several weeks following our site reconnaissance and no signs of tensile failures in connections or reinforcing strips were found.

Overall, the most notable observation was the relative lack of significant damage to the RE embankment despite being subjected to ground shaking levels higher than the design values and unanticipated large ground displacements. In stark contrast to this behavior, the conventional embankment at the southern approach suffered heavy damage and had been demolished when the reconnaissance team arrived. Also, a conventionally-constructed 10 m high approach embankment located about 250 m from the RE wall (see Fig. 2) suffered heavy damages during the earthquake, experiencing settlements of more than 1 m. The good performance of the RE embankment is thought to be particularly meaningful in demonstrating the seismic stability of conventionally-constructed walls of this type.

4 SIMPLIFIED SEISMIC ANALYSIS AND DYNAMIC DEFORMATION ANALYSIS

The Arifiye Overpass RE double-wall, which was constructed in 1988, was designed with a pseudo-static approach using a seismic coefficient $k_h = 0.1$ (Sankey & Segrestin 2001). To better interpret the performance of the RE walls and assess the adequacy of the seismic

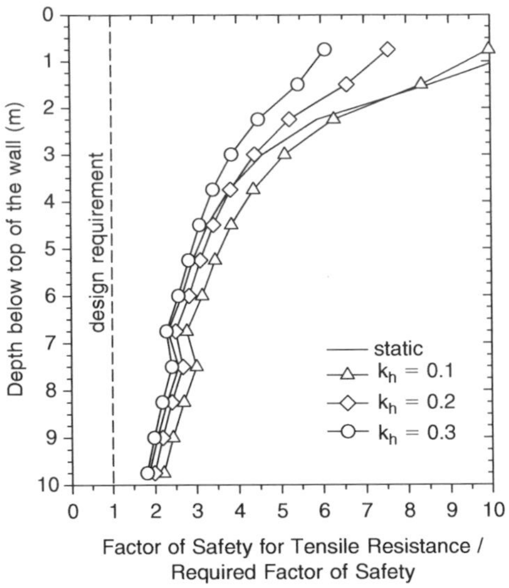

Figure 5. Ratio of the allowable tensile resistance (capacity) to that required for static and seismic design (demand).

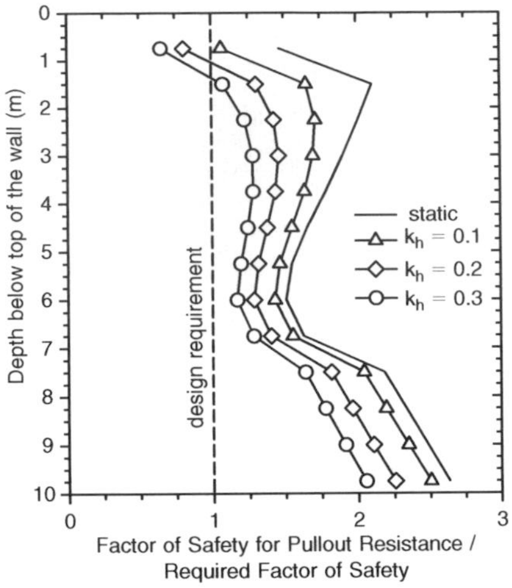

Figure 6. Ratio of the allowable pullout resistance (capacity) to that required for static and seismic design (demand).

design used, the authors performed a simplified seismic analysis as well as a numerical analysis to estimate the earthquake loading on the wall. As mentioned earlier, peak ground accelerations at the site were estimated at about 0.25 g during the M7.4 event.

For the simplified analysis, seismic loads on the RE walls, and the corresponding factors of safety for internal stability, were estimated using a pseudo-static approach where the earthquake-induced forces were represented via a horizontal seismic coefficient (k_h) as per the current FHWA design guidelines (Elias et al. 2001, AASHTO 2006). This approach provides a reasonable perspective for assessing the observed embankment performance, especially in that this approach is most often used in practice.

Simplified analyses were performed using three different seismic coefficients: $k_h = 0.1$, 0.2, and 0.3. The analyses were performed for $k_h = 0.1$ because this was the value used for design of the wall at the time. The value of 0.3 is the calculated "equivalent k_h" value that corresponds to the peak ground acceleration of 0.25g estimated to have occurred.

The calculated design forces for each case are compared with the allowable tensile resistance of the steel strips and the available pullout resistance at each level. Allowable tensile resistance is estimated as 55% of the tensile yield strength of steel ($\sigma_{yield} = 450$ MPa and $\sigma_{allowable} = 0.55 \cdot \sigma_{yield} = 247.5$ MPa). The design guidelines allow the use of 75% of the static factor of safety for seismic design (AASHTO 2006). In essence, this results in the allowable stress for seismic design as $0.73 \cdot \sigma_{yield,steel}$ (330 MPa). Allowable steel strength for static and seismic cases and the available reinforcement cross-section area per unit wall width were calculated at each elevation. Estimated design forces from the simplified approach and the allowable tensile resistance were calculated and comparison of these values are presented in Figure 5. It can be seen that the allowable capacity/demand ratio is well above 1.0, and is most critical at the lower reinforcement levels, ranging between 1.8–2.2 at the bottom reinforcement levels. The results suggest that the reinforcements should not have been stressed beyond their allowable strengths during shaking, but they were stressed much higher than the design values based on $k_h = 0.1$.

Design guidelines recommend the use of a minimum 1.5 for the factor of safety against pullout at each reinforcement level. Similar to the tensile capacity calculations, 75% of the static design requirement is allowed for the seismic design against pullout. The pullout resistance is calculated using the friction coefficient and the surface area of the reinforcement anchored to the passive zone defined by design guidelines. The friction coefficient used for seismic design is 80% of the value used for static design. The extra allowance by the reduced factor of safety is more-or-less offset by the use of a reduced frictional resistance for seismic design. Comparisons of the allowable pullout resistances required by design (pullout capacity/required factor of safety) to the estimated forces at each elevation are shown in Figure 6. Although not shown above, the actual pullout factor of safety ranges between 2.2 and 4.0 along the height of the wall for the static case where 1.5 is the required minimum for design. Therefore, the allowable-pullout-capacity to demand ratio varies between 1.5–2.7 for the static case as shown in the figure. The values of allowable

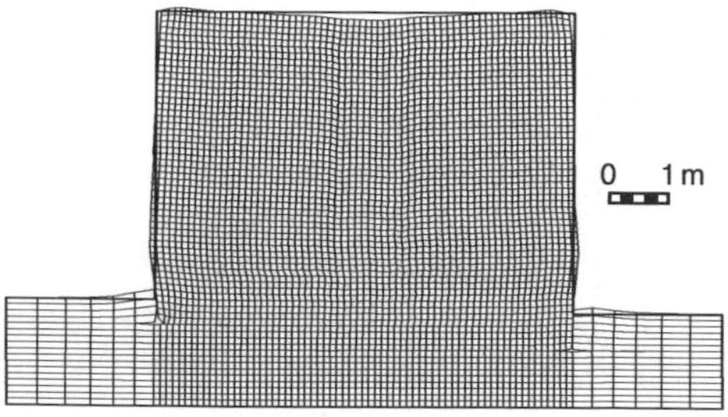

Figure 7. Deformed shape of the model wall (deformations at 1x, no exaggeration).

capacity-to-demand ratio drop slightly to 1.1–2.5 for a seismic coefficient of 0.1 which was used for design. Thus, seismic considerations probably did not govern the design of the wall and caused little or no increase to the reinforcement amount in comparison to the static case, as noted by Sankey and Segrestin (2001).

Values of allowable pullout capacity/demand ratio drop to a range of 0.8–2.3 for seismic coefficient of 0.2, and 0.7–2.1 for a seismic coefficient of 0.3. The values of allowable capacity to demand ratio for the reinforcements at the top 6 meters are only slightly above 1.0. The results suggests possible problems with pullout resistance at $k_h = 0.3$, the level suspected to have been induced during the earthquake.

In addition to the simplified seismic analysis, numerical analyses were also performed to better understand the seismic behavior of the structure (Olgun 2003). Using FLAC (Itasca 2000) with a two-dimensional analysis, we were successful in closely predicting the observed earthquake-induced deformation pattern and displacement magnitudes of the embankment. The deformed mesh shape from the seismic analyses is provided in Figure 7. For brevity purposes, the numerical analysis details are not provided, but the main findings are summarized:

1. A maximum earthquake-induced lateral wall displacement of 12–14 cm was predicted, compared to an observed value of 10-15 cm, and the predicted backfill settlement was 27 cm, consistent with the observed value of 25–30 cm.
2. The observed lateral wall bulging was due mainly to seismic shaking, whereas the backfill settlement and panel misalignment were due mainly to the differential ground distortions from liquefaction- and fault-related movements.
3. The earthquake-induced forces in the strips were on average 1.5–2 times larger than the design values. These higher than design values indicate that the reinforcements were most likely overstressed in the field during the earthquake.
4. Intolerable serviceability problems due to significant distortion of the structure, such as panel misalignment and significant backfill settlement, were likely to occur long before internal failure.
5. Consistent with the observations, the wall should have maintained its overall structural integrity despite the severe loading conditions it experienced. The numerical analysis added confidence to our interpretation of the field behavior and post-earthquake investigation of the structure.

5 SUMMARY AND CONCLUSIONS

The 1999 Kocaeli Earthquake (M7. 4) in northwestern Turkey provided the opportunity to study the field performance of a double-walled RE embankment designed according to procedures essentially same as those used for current FHWA guidelines and subjected to significant ground shaking and permanent ground deformations. The RE embankment formed the northern approach embankment for the Arifiye Bridge Overpass. The site was located immediately adjacent to the ruptured fault causing large horizontal and vertical ground movements within a few meters of the embankment. Unanticipated liquefaction-related settlements also occurred in the foundation, and the ground shaking levels that occurred were higher than those accounted for in design. The site was carefully documented and following the earthquake. Key observations and findings from the study are summarized as follows:

1. The structure was designed using guidelines similar to current FHWA Standards. Minimum FS values recommended for static and seismic conditions were maintained. Based on procedures in Turkey at the time, a seismic coefficient k_h of 0.1 was used for design of the RE walls and embankment. Liquefaction and/or fault-related movements in the foundation soils were not anticipated or designed for.
2. Ground shaking at the site was much higher than what was accounted for using the k_h design value of 0.1. Estimated peak ground acceleration was about 0.25 g, corresponding to an estimated "equivalent k_h" of about 0.3. Simplified and numerical analyses show reinforcement forces during shaking and pullout resistance greatly exceeded the design values.
3. In addition to strong ground shaking, the embankment was subjected to significant settlements and horizontal movements. Liquefaction and/or cyclic soil failure were responsible for differential foundation settlements, especially near the culvert that ran beneath the embankment- settlements of up to 25 cm occurred in this section. Also, the site was located immediately adjacent to the ruptured fault, as the fault rupture passed through the northern abutment and adjacent pier causing a lateral offset of more than 3.5 meters and a vertical offset of nearly 0.5 m.

4. Simplified and numerical analysis predicted that the steel reinforcements were probably not stressed beyond their yield strengths during shaking, but the induced forces in the strips were on average 1.5 times larger than the design values. Also, the pullout capacity may have been reached during the earthquake for the upper levels of the embankment.
5. Despite the high levels of shaking and unanticipated ground movements, the RE structure maintained overall integrity, exceeding the design provisions. (Although a peak acceleration of 0.25 g is estimated, it is also possible that the peak ground accelerations at the site were larger than 0.29 g which is the upper boundary of seismic loading where FHWA/AASHTO simplified design guidelines are applicable). The facing panels and reinforcements were undamaged, and if not for the ground movements, we suspect the RE embankment would probably not have suffered significant damage.
6. Numerical analyses predicted the observed lateral wall bulging was due mainly to seismic shaking, whereas the backfill settlement and panel misalignment were due mainly to the differential ground distortions from liquefaction- and fault-related movements. Also, intolerable serviceability problems due to significant distortion of the structure, such as panel misalignment and significant backfill settlement, were likely to occur long before internal failure. And, consistent with the observations, the embankment should have maintained its overall structural integrity despite the severe loading conditions experienced during the Kocaeli Earthquake.
7. Special slip joints were used to mitigate anticipated static differential settlements between different sections of the approach embankment (i.e., between the pile-supported abutment and RE embankment). The joints played important role in limiting damage associated with liquefaction-related foundation movements, allowing the wall to sustain significant differential foundation deformations without being overstressed.
8. In stark contrast to this behavior, two conventionally-constructed approach embankments located near the RE suffered heavy damages during the earthquake, experiencing settlements of more than 1 m. The good performance of the RE walls is thought to be particularly meaningful in demonstrating the seismic stability of conventionally-constructed walls of this type.
9. The study implies that well-designed conventional RE embankments constructed according to current design guidelines have an inherently high resistance to earthquake shaking and differential foundation movements. This performance is thought to be especially meaningful for illustrating the seismic resilience of these types of earth structures.

ACKNOWLEDGEMENTS

This research was performed under the auspices of the National Science Foundation awards CMS-0085281. and CMS-0201508. A major debt of gratitude is due to engineers and technicians at Zetas Earth Technology Corporation in Istanbul, Turkey. In particular, we are indebted to Dr. Canan Emrem who was instrumental in the data collection for this study and provided invaluable input. We also thank Mr. Murat Ozbatir, general manager of Reinforced Earth Corp.-Turkey (RECO) for kindly providing the design and construction details of the structures.

REFERENCES

AASHTO. 2006. LRFD Bridge Design Specifications. American Association of State Highway and Transportation Officials, Washington, D.C., 3rd Edition with 2005 and 2006 Interims.

Elias, V., Christopher, B.R. and Berg, R.R. 2001. Mechanically stabilized earth walls and reinforced soil slopes, design and construction guidelines. FHWA-NHI-00-043, Washington, D.C., 394 p.

Itasca Consulting Group. 2000. Fast Lagrangian Analysis of Continua, User Manual - Version 4.0. Itasca Consulting Group, Minneapolis, MN.

Martin, J.R., Olgun, C.G., Mitchell, J.K., Durgunoglu, H.T. 2004. High modulus columns for liquefaction mitigation. Journal of Geotechnical and Geoenvironmental Engineering, ASCE, vol. 130, no. 6, pp. 561–571.

Mitchell, J.K., Martin, J.R., Olgun, C.G., Emrem, C., Durgunoglu, H.T., Cetin, K.O. and Karadayilar, T. 2000. Chapter 9 - Performance of improved ground and earth structures. Earthquake Spectra, vol. 16, Supplement A to Volume 16, pp. 191–225.

Olgun, C.G. 2003. Performance of Improved Ground and Reinforced Soil Structures during Earthquakes - Case Studies and Numerical Analyses. PhD. Dissertation, Virginia Tech, Department of Civil and Environmental Engineering.

Sancio, R.B., Bray, J.D., Stewart, J.P., Youd, T.L., Durgunoglu, H.T., Onalp, A., Seed, R.B., Christensen, C., Baturay, M.B. and Karadayılar T. 2002. Correlation between ground failure and soil conditions in Adapazari, Turkey. Soil Dynamics and Earthquake Engineering, Volume 22, Issues 9–12, pp 1093–1102.

Safak, E. and Erdik, M. 2000. Chapter 5 – Recorded main shock and aftershock motions. Earthquake Spectra, vol. 16, Supplement A to Volume 16, pp. 97–112.

Sankey, J.E. and Segrestin, P. 2001. Evaluation of seismic performance in Mechanically Stabilized Earth structures. International Symposium on Earth Reinforcement Practice, IS Kyushu 2001, A.A. Balkema, K. Omine ed., November 14–16, 2001, Fukuoka, Kyushu, Japan, vol. 1, pp. 449–452.

New Horizons in Earth Reinforcement – Otani, Miyata & Mukunoki (eds)
© 2008 Taylor & Francis Group, London, *ISBN 978-0-415-45775-0*

Study on the performance of a reinforced earth wall during earthquake based on Tottori-ken Seibu earthquake event

K. Watanabe
Hirose & Co., Ltd, Osaka, Japan

T. Kumada
Hirose & Co., Ltd, Tokyo, Japan

ABSTRACT: The Tottori-ken Seibu earthquake of Mjj = 7.3 occurred on October 6, 2000. A lot of reinforced earth walls were built in a range of a radius of 10 km from epicenter. We investigated it after an earthquake disaster. The damage was not confirmed to most reinforced earth walls. However, the damage occurred at two reinforced earth walls that are near the epicenter comparatively. We carried out execution of repair and reinforcement for them. From a main shock 1 year and 6 month later, the aftershock of Mjj = 4.5 occurred on March 6, 2002. It is a point in time that passed for 1 year since we reinforce it. Therefore, we carried out investigation again. This paper is described about the investigation result that we performed after a main shock, the establishment of the technique to judge the damage of a retaining wall that is based on the investigation result, the implementation of a repair and reinforcement by the judgment technique.

1 INTRODUCTION

The Tottori-ken Seibu earthquake of Mjj = 7.3 around west of Tottori prefecture occurred on October 6, 2000. A lot of reinforced earth walls were built in a range of a radius of 10 km from epicenter. We investigated it after an earthquake disaster. By the those days, the technique to judge the damage of a retaining wall to be it was not established. Therefore, we inspected in the visual inspection and the hanging check. For all reinforced earth walls that we checked, we ranked a soundness evaluation by degree of the damage. The damage was not confirmed to most reinforced earth walls. However, the damage occurred at two reinforced earth walls, which are near the epicenter comparatively. (Site 1 in Figure 1) We examined repair and reinforcement for them. The two reinforced earth walls exceeded a serviceability limit. We carried out execution of repair and reinforcement for them.

From a main shock 1 year and 6 month later, in March 6, 2002, at the point near the main shock, the aftershock of Mjj = 4.5 occurred. It is a point in time that passed for 1 year since we reinforce it. Therefore, for the reinforced earth wall that we repaired and reinforced, we carried out investigation again.

This paper is described about the investigation result that we performed after a main shock, the establishment of the technique to judge the damage of a retaining wall that is based on the investigation result,

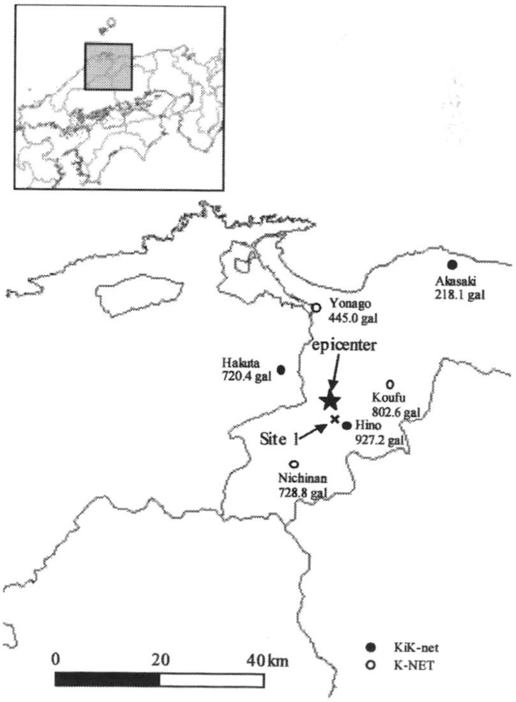

Figure 1. Location of the investigated site.

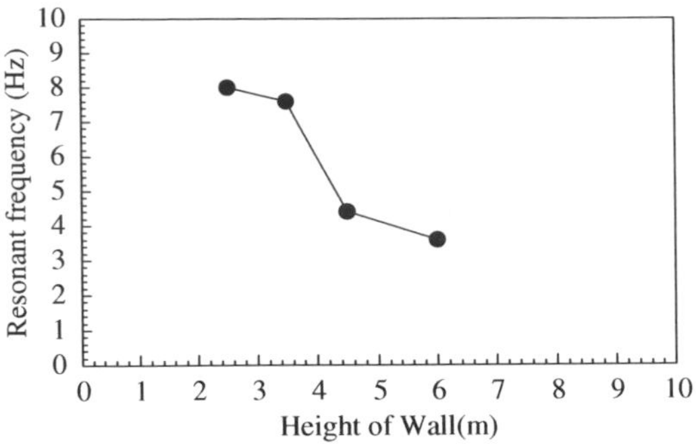

Figure 2. Resonant frequency for the wall height.

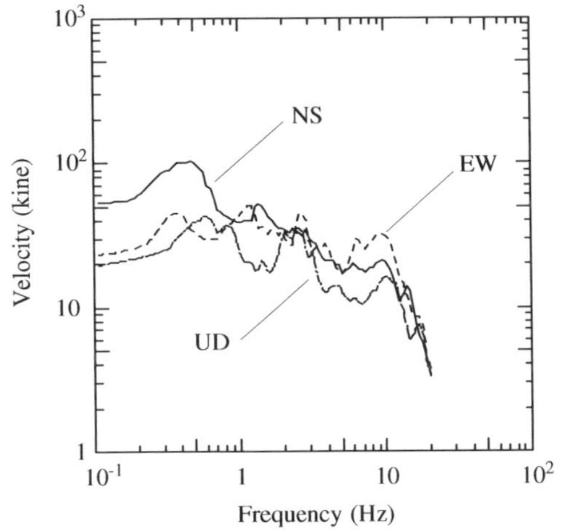

Figure 3. The spectrum of the main shock.

the implementation of a repair and reinforcement by the judgment technique.

It tried whether or not it was possible to use the full-scale experiments on the shaking table results for the sorting-out of an investigation object this time.

Relation between the wall height and the resonant frequency in the full-scale experiments on the shaking table results, which was implemented in the retaining wall with the same structure is shown in Figure 2. Futaki et al. said the amplification ratio is approximately 7.

Using the corrugated dater in which is provided from National Research Institute for Earth Science and Disaster Prevention (KiK-Net), a replying spectrum near the investigation place is shown in the Figure 3. The damped-ratio in this place used 5%. It doesn't lead by comparing with the shaking table results but this place shows that 0.6 Hz excel. This dater is one in the basis. It is possible to estimate that the wave that is similar approximately, because the object part is on the rock and doesn't leave with the KiK-net point occurred.

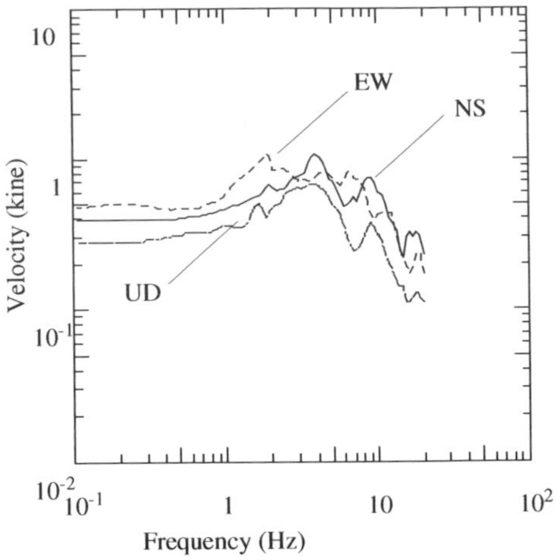

Figure 4. The spectrum of the aftershock.

It thought that it was the one where the wall is high in the earthquake this time from these results and that the damage occurred.

Also, a spectrum at the time of the aftershock is shown in the figure 4.

2 INVESTIGATION

We investigated it after an earthquake disaster. By the those days, the technique to judge the damage of a retaining wall to be it was not established. Therefore, we inspected in the visual inspection and the hanging check. For all reinforced earth walls that we checked, we ranked a soundness evaluation by degree of the damage. The damage was not confirmed to most reinforced earth walls. However, the damage occurred at two reinforced earth walls, which are near the epicenter comparatively.

As for one reinforced earth wall, because to have been built by the steep slope part, looseness on the slope in the front was seen. Also, opening in the vertical joint part was seen at the part, which is close to the other wall. As for another reinforced earth wall, damage to the wall was hardly seen but a partial hanging was seen. Therefore, the two reinforced earth walls were judged to have to be investigated in detail.

In this report, investigated Result of which the hanging in the seen one and it is on is described. In the amount of displacement at the wall, hanging a plumb bob from the wall upper end, it measured a distance from the plumb bob to the wall with the detailed investigation. It implemented measurement in the position of the upper and lower end of the panel near the vertical joint at the wall.

It defines the gradient of the wall as the percentage that removed the relative difference of the coping concrete upper side and the amount of displacement

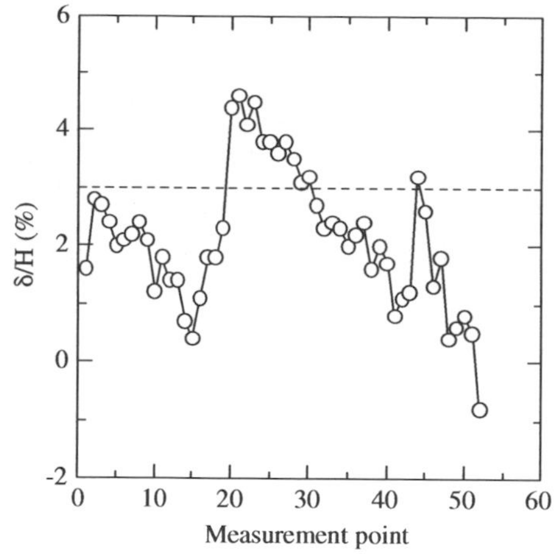

Figure 5. The measuring result after the main shock.

of the horizontal direction in the back filling surface position in wall height of the ground level. As for the measuring result, it is shown in the Figure 5. The meaning of Measurement point shows the part that was measured in the vertical joint from the edge in the wall to the direction of the extension. In the interval, it is 1.5 m. The vertical axis shows a gradient. The gradient is the one to have divided a horizontal direction amount of displacement by the height. Here, it does the central part finding of the thing that the hanging is big nearby.

3 REINFORCEMENT MEASURE

The part that exceeds 3% of the construction management standard value had the gradient of the wall facing and the wall decided to plan reinforcement. In case of selection of the reinforcement, it considered the following two. It decided to select from the inside of the reinforced earth method, which adopts the same mechanism and it adopted soil nailing of construction which is one of the reinforced earth methods. Because it becomes the structure which can resist the earth pressure of reinforced soil mass, being made from the reinforced concrete with 14 cm thickness, not to have a bad influence on the wall material by the reinforcement vs. Also, because it didn't range over the whole reinforced earth wall, the implementation section of the reinforcement decided to secure the system of the outward appearance. As for the design of the reinforcing rod insertion, the design earth pressure to use for the effect of the reinforcement without considering a stiffening effect by the strip in the reinforced earth wall uses the design earth pressure of the reinforced earth wall method of construction.

It calculates the resistance, which occurs to the reinforcing rod using the sought earth pressure. It bores the reinforcing rod insertion at the 15° angle in the

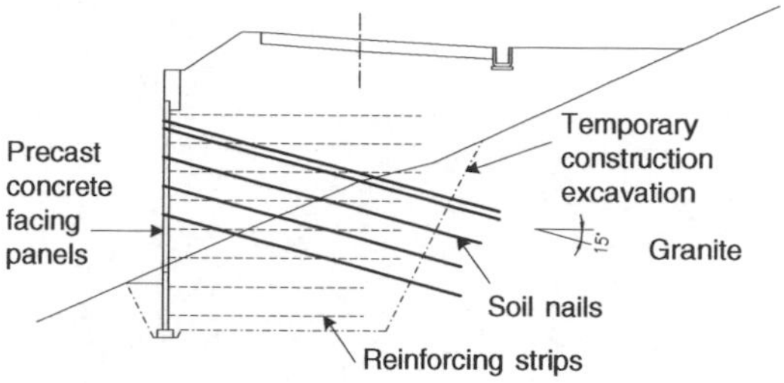

Figure 6. A cross section of the reinforcement range.

Figure 7. Boring status.

direction of the level more and the bottom. Here, it calculates resistance by the following equation.

$$R = T / \cos\alpha \tag{1}$$

where, R is the resistance which occurs to the reinforcing rod, T is the tension which was calculated from the earth pressure, α is the boring angle.

A necessary anchorage length is found by the following equation.

$$L = \frac{R \cdot f}{l \cdot \tau} \tag{2}$$

where, L is the necessary anchorage length, f is the safety factor, l is reinforcing rod circumference, τ is the allowable bond stress.

The crossing section of the reinforcement range is shown in the Figure 6.

Boring status is shown in the Figure 7. Reinforcement completion is shown in the Figure 8.

4 REINVESTIGATION

4.1 *Reinvestigation result*

After the aftershock occurs, it implemented reinvestigation. The time is time of the elapse in 1 year after reinforcement. The investigation method measured the gradient which used a visual examination and a plumb

Figure 8. Reinforcement completion.

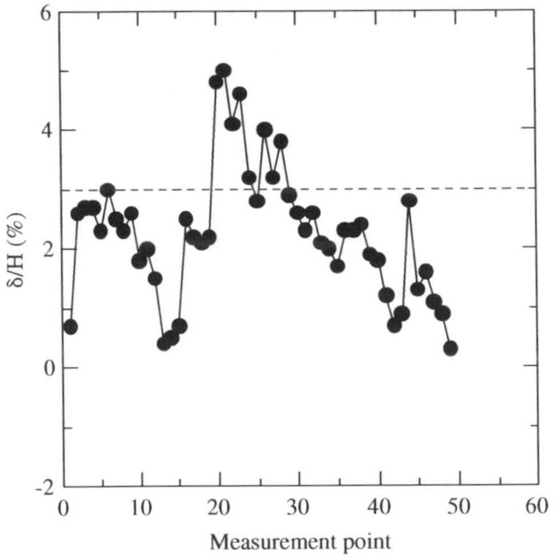

Figure 9. The measuring result after the aftershock.

bob like the after the main shock it. At the visual examination, specifically, a change was observed with after the main shock when there was not it. Here, this time, a measuring result is shown in the figure.

4.2 *Reinforced effect*

It gives the variation of the gradient in after the main shock and after the aftershock to be being reinforced by the following equation.

$$V = \frac{\Delta 2 - \Delta 1}{\Delta 1} \tag{3}$$

where, V is variation, $\Delta 1$ is The vertical degree after the main shock, $\Delta 2$ is The vertical degree after the aftershock.

The change of the vertical degree in the direction of the wall extension is shown in the figure. When the change quantity is positive, it shows that a wall is inclined on the side of the front. Oppositely, when the change quantity is a negative, it shows that a wall is inclined on the side of the mound.

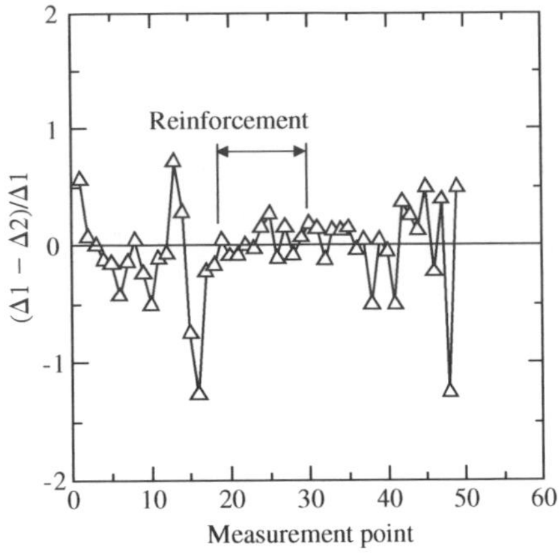

Figure 10. Variation of the gradient.

As for the section that was reinforced by the soil nailing range, there is not a change in the gradient before and after the aftershock. However, that the change has occurred to the vertical degree in the section that isn't reinforcing is confirmed.

5 CONCLUSIONS

That the reinforced earth method is earthquake-resistant could be confirmed from Research Result. With the investigation after the aftershock, the thing where the reinforcement, which uses the soil nailing is effective for the earthquake could be confirmed. As for the reinforcement, which uses the soil nailing, the system of the outward appearance can be secured. It thinks that the research method and the judgment method, which was used this case can be applied about the similar case, too. It thinks that the way of reinforcing this time in addition to the one which is occurs with the earthquake can be applied. To investigate the outward appearance of the whole wall immediately beforehand after building, too, is important.

ACKNOWLEDGEMENT

The corrugated dater used National Research Institute for Earth Science and Disaster Prevention (KiK-net). The spectrum analysis used the software that Prof. Kamada in Fukuyama University is providing.

REFERENCE

Futaki, M., Ogawa, N., Sato, M., Kumada, T., and Natsume, S. (1996). "Experiments about Seismic performance of Reinforced Earth Retaining Wall." *Proceedings of Eleventh World Conference on Earthquake Engineering*, Elsevier Science, Ltd.

New Horizons in Earth Reinforcement – Otani, Miyata & Mukunoki (eds)
© 2008 Taylor & Francis Group, London, *ISBN 978-0-415-45775-0*

Rainfall seepage analysis and dynamic response analysis for the railway embankments seriously damaged in the 2004 Niigata-ken Chuetsu earthquake

T. Matsumaru, M. Tateyama, K. Kojima, K. Watanabe & M. Shinoda
Railway Technical Research Institute, Tokyo, Japan

M. Ishizuka
Integrated Geotechnology Institute Limited, Tokyo, Japan

ABSTRACT: Due to the 2004 Nigata-ken Chuetsu earthquake, a lot of fill structures and retaining walls collapsed. It seems that strength of fill structures was loosed because of rainfall induced by typhoon before this earthquake. So, using a set of numerical analyses, we evaluated correlation between rainfall and earthquake in the damages of a fill structure. In this paper, we introduce the outline of the collapsed fill, methods of numerical analyses, and the results of seepage analysis and dynamic response analysis. We chose the fill in the vicinity Jouetsu Line 221 km000 m. This fill was damaged severely and reconstructed with reinforced soil. From the seepage analysis, it revealed that moisture-content state of the collapsed embankment increased before the earthquake and rainwater drainage of the reconstructed embankment improved. Moreover, from the dynamic response analysis, it was clearly shown that the seismic stability of the restored embankments was significantly improved as compared to that of the damaged embankment by evaluating the response accelerations and deformation of the damaged and restored embankment.

1 INTRODUCTION

The 2004 Nigata-ken Chuetsu earthquake caused wide-spread area damages of fill structures and retaining walls. In railway field, 86 soil structures collapsed at the Jouetsu Line, the Shinetsu Line, the Iiyama Line and the Tadami Line (Tateyama and Kato, 2005). A lot of damages of soil structures occurred on the river terrace and overlap the places where sediment disasters induced by rainfall have occurred. Nigata Prefecture was affected by the rainfall of typhoon No.23 just before this earthquake, so it seems that the main reason of these damages are the decrease of strength with the increase of the degree of saturation in embankments. But the precise estimation has not been conducted.

In this paper, a set of numerical simulations aimed at the embankment in the vicinity Jouetsu Line 221 km000 m are conducted to evaluate the main reason of damage and the seismic resistance of the reconstructed embankment. First, an outline is given for the damaged embankment and the reconstructed embankment. Then, methods of seepage analysis, dynamic response analysis and deformation analysis by Newmark's method are introduced briefly. Finally, we show the results of seepage analysis and dynamic response analysis. The results of deformation analysis by Newmark's method are shown in Shinoda et al. (2007).

2 OUTLINE OF DAMAGED EMBANKMENT AND RECONSTRUTED EMBANKMENT

2.1 *Damaged embankment*

In the vicinity Jouetsu Line 221 km000 m, the embankment on the river terrace by the Shinano River collapsed. The damaged area was about 65 m long and 4 to 12 m high. The amount of collapsed soil was estimated to about 13,000 m^3. Figure 1 shows the illustration of damaged embankment and Figure 2 shows the photograph. The location of this embankment is valley eroded by the Shinano River. In this area, Route 17 runs alongside the Jouetsu Line, so the embankment of this route was also collapsed. The bedrock is sandy rock with middle particle and the silty rock accumulates on the bedrock. The material which consists of the damaged-embankment on the silty rock is sand with gravel.

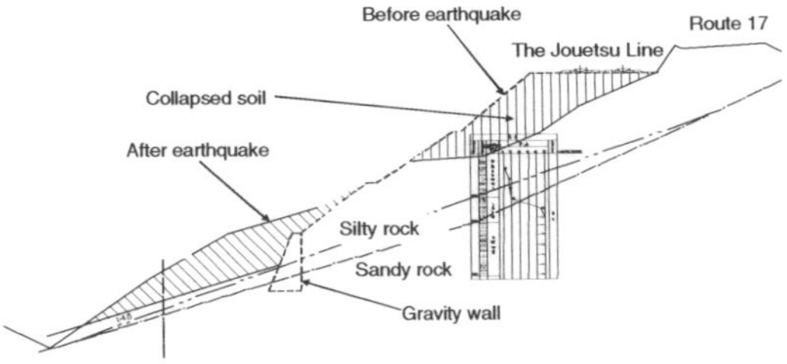

Figure 1. Illustration of damaged embankment.

Figure 2. Photograph of damaged embankment.

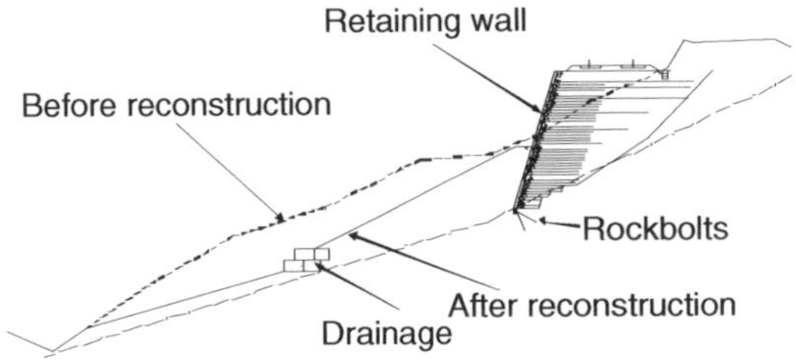

Figure 3. Illustration of reconstructed embankment.

2.2 *Reconstructed embankment*

The concept of reconstruction of the embankment was to reduce the amount of soil and days and to improve seismic resistance. Considering these conditions, geosynthetic-reinforced soil retaining wall was adopted. Figure 3 shows the illustration of reconstructed embankment and Figure 4 shows the photograph. Firstly, the collapsed soil was excavated until the silty rock. Before constructing the foundation of retaining wall, the rock bolts were installed in order to improve the stability of embankment. For the design of the reconstruction, load acting the road was also considered. The embankment was established until the height of 13 m with the geogrid every 30 cm. Furthermore, this embankment is located in drainage area and the boiling was observed after the earthquake.

Figure 4. Photograph of reconstructed embankment.

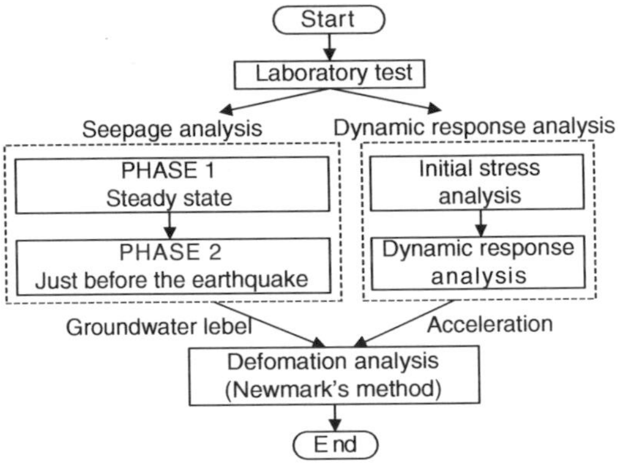

Figure 5. Flow of numerical simulations.

So, the improvement of drainage was also important. The amount of crushed stone used for reconstructing embankment was about 4,600 m^3 and concrete of RC wall was about 300 m^3.

3 OUTLINE OF NUMERICAL SIMULATIONS

In this section, we will show the outline of numerical simulations to evaluate the main reason of damage of embankment and the seismic improvement of reconstructed embankment. Figure 5 shows the flow of numerical simulations.

3.1 *Seepage analysis*

The degree of saturation in the damaged embankment may have increased because of the rainfall induced by typhoon No.23, so the drainage was important in the reconstructed embankment. We set the purpose of seepage analysis to evaluate the degree of saturation in the damaged embankment just before the earthquake and the improvement of drainage in the reconstructed embankment.

In this paper, the seepage analysis is conducted in two phases. In PHASE1, the seepage in the steady state is analyzed by using the rainfall for a long term equivalent to the annual rainfall. In PHASE2, the seepage just before the earthquake is analyzed by using the data observed at the site.

3.2 *Dynamic response analysis*

The purpose of the dynamic response analysis is to evaluate the response in the embankment subjected to the earthquake. In this analysis, after the initial stress is evaluated by the static analysis, the dynamic response is analyzed when the earthquake occurs in this stress state.

3.3 *Deformation analysis using Newmark's method*

The deformation of the embankment is analyzed by Newmark's method. In order to evaluate precisely the behavior in the earthquake, the response acceleration of the embankment acquired in the dynamic response analysis was considered. Furthermore, we take into consideration the strength parameters and the groundwater according to the degree of saturation seepage analysis. This part is introduced in the paper written by Shinoda et al. (2007).

4 SEEPAGE ANALYSIS

4.1 *Model of seepage characteristics*

In the seepage analysis, modeling the seepage characteristics of unsaturated soils is important. In this paper, we used the Van Genuchten model (1980) for the soil water characteristic curve. This model is often used in seepage analysis and generally described as

$$\Theta = \left[\frac{1}{1+(\alpha h)^n}\right]^m \tag{1}$$

where, Θ is the relative water content, h is the suction, and α, n and m are parameters determined by laboratory tests. Using the moisture water content θ, the relative water content is written as

$$\Theta = \frac{\theta - \theta_r}{\theta_s - \theta_r} \tag{2}$$

where, θ_s is the saturated moisture water content and θ_r is the lower limit of the moisture water content. From equation (1), the unsaturated moisture water content is described as

$$K_r(\Theta) = \Theta^{1/2}\left\{1 - \left(1 - \Theta^{1/m}\right)^m\right\}^2 \tag{3}$$

Table 1. Parameters of sandy loam.

θr	θs	α(1/cm)	m	n
0.065	0.41	0.075	0.471	0.5

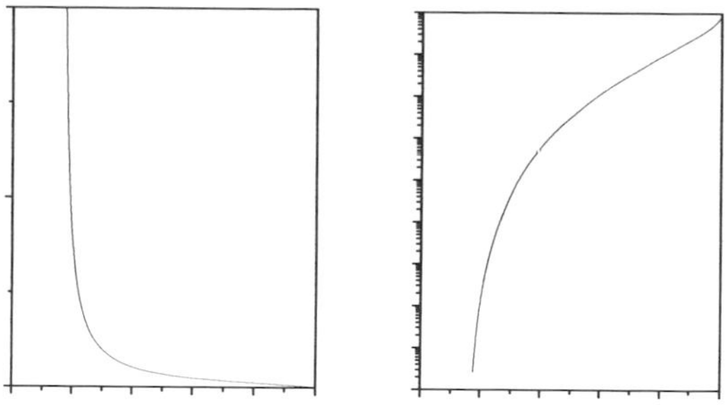

Figure 6. Relationship of the degree of saturation to the relative hydraulic conductivity and the suction.

where $K_r(\Theta)$ is the relative hydraulic conductivity, the ratio of the unsaturated hydraulic conductivity to the saturated conductivity.

As a usual, parameters in equation (1) are determined in the trial and error so that this equation describes the soil water characteristic curve obtained from laboratory tests. But, the laboratory tests of the collapsed soils were not conducted, so the parameters were determined according to the classification of soil characteristics in the standard of USDA (Carsel and Parrish, 1988). From this classification, all of the materials used in this analysis are classified as the sandy loam. Table 1 shows parameters of sandy loam, and Figure 6 shows the relationship of the degree of saturation to the relative hydraulic conductivity and the suction.

4.2 *Data of rainfall*

In this paper, seepage analysis is conducted in two phases. In PHASE1, the seepage in the embankment becomes in the steady state by using the annual rainfall, and in PHASE2 the degree of saturation just before the earthquake is analyzed by using the observed data of rainfall. Figure 7 shows the set of data of rainfall used in PHASE1 and PHASE2. The rainfall pattern in PHASE1 is 20.5 mm once every three days and 0 mm in the other days based on the annual average rainfall, 2500 mm in Nagaoka city. Using this rainfall pattern for 3,000 days, we have ensured that the degree of saturation in the embankment changes little. In PHASE2, the observed data of rainfall from September 1 until October 22 is used as the continual rainfall just before this earthquake.

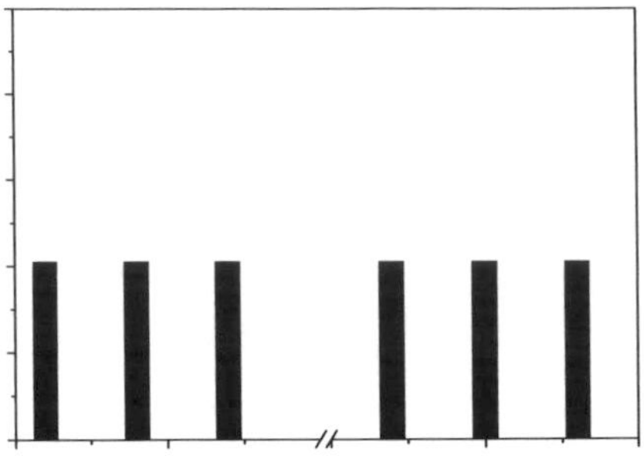

(a) Pattern of rainfall used in PHASE1

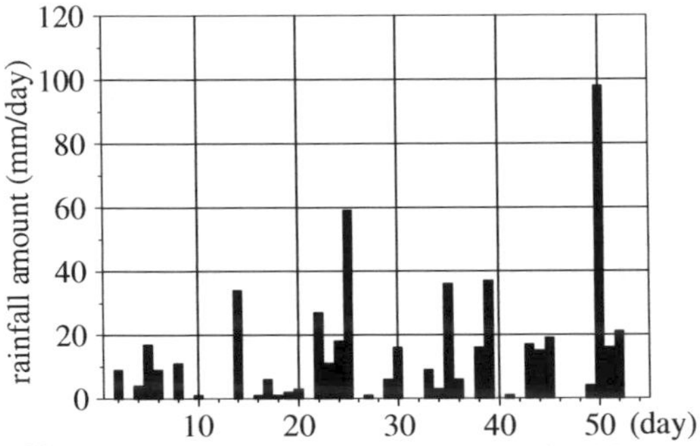

(b) From September 2004 until October 22

Figure 7. Set of data used in seepage analysis.

4.3 *Analytical conditions*

Figure 8 shows the analytical models. Both models consist of triangular elements, 31 m high by 90 m long. In these models, three materials; concrete, embankment material and weathering silt were considered. However, the concrete wall was not modeled as elements, but as the undrained boundary. From the situation of the damage, the rainfall just before the earthquake may have exceeded the capacity of the drainage, so the right edge was set under submerged condition. The hydraulic conductivity of the embankment 1.06×10^{-5}cm/s was determined from the laboratory test and conductivity of silty rock 1.00×10^{-3}cm/s was supposed from the comparison with the embankment material. Furthermore, the reconstructed embankment by crushed stone was also analyzed under the same rainfall. On the reconstruction, the drainage was improved, so the submerged condition was not considered. The hydraulic conductivity of the crushed stone 1.00×10^{-2}cm/s was supposed from the diameter.

4.4 *Results of analyses*

Figure 9 shows the distributions of degree of saturation in the embankment after the simulation in PHASE2. In the damaged embankment, the degree of saturation increases and the groundwater level appears along the boundary between the embankment and the silty rock. Though the capacity of the drainage in the embankment is not clear, the rainfall may exceed the capacity. So, this rainfall may have influenced the behavior of the embankment when the earthquake occurred. On the other hand, the degree of saturation in the reconstructed embankment was smaller than one in the damaged embankment. So, this indicates that the drainage in the reconstructed embankment was improved.

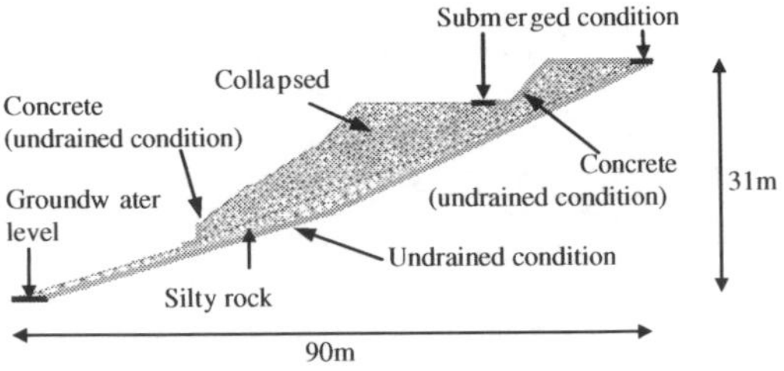

(a)Damaged embankment

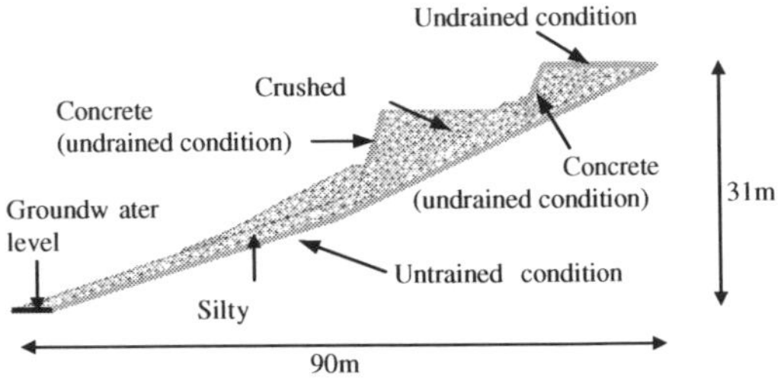

(b)Reconstructed embankment

Figure 8. Finite element models for seepage analysis.

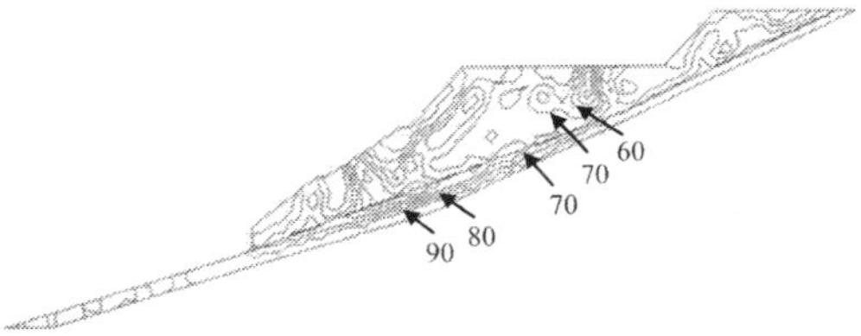

(a)Damaged embankment

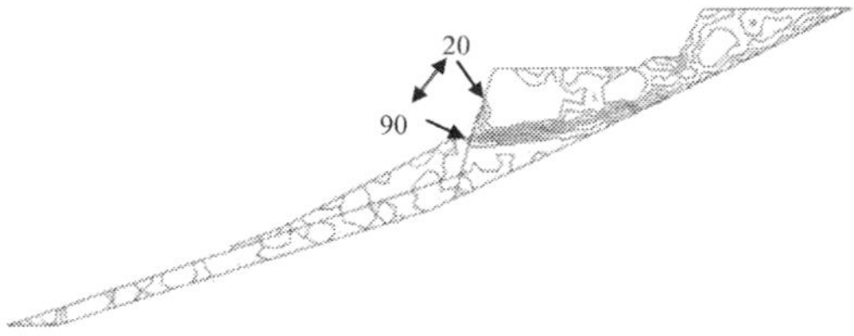

(b)Reconstructed embankment

Figure 9. Distributions of the degree of saturation (unit:%).

5 DYNAMIC RESPONSE ANALYSIS

5.1 *Modified GHE model*

In dynamic response analysis, modeling nonlinear characteristics of soil is important. R-O (Ramberg-Osgood) model or H-D (Hardin-Drnevich) model is often used because these models need not to determine a lot of parameters. But these models have difficulty of describing the relationship of shear strain to shear modulus and damping, from small-strain region to large-strain region. In this paper, modified GHE (General Hyperbolic Equation) model (Nishimura and Murono, 1999) as described in equation (4) is adopted. In this model, the hyperbolic model is applied to skeleton curve and Masing's rule is adopted to hysteresis curve. But as shown in equation (5) and (6), parameters C_1 and C_2 are modified to trace experimental results rather well.

$$\frac{\tau}{\tau_f} = \frac{\dfrac{\gamma}{\gamma_r}}{\dfrac{1}{C_1} + \dfrac{1}{C_2}\left(\dfrac{\gamma}{\gamma_r}\right)} \tag{4}$$

$$C_1(x) = \frac{C_1(0) + C_1(\infty)}{2} + \frac{C_1(0) - C_1(\infty)}{2} \cdot \cos\left\{\frac{\pi}{\alpha / x + 1}\right\} \tag{5}$$

$$C_2(x) = \frac{C_2(0) + C_2(\infty)}{2} + \frac{C_2(0) - C_2(\infty)}{2} \cdot \cos\left\{\frac{\pi}{\beta / x + 1}\right\} \tag{6}$$

where τ/τ_f is normalized shear stress, γ/γ_r is normalized shear strain, γ_r is standardized shear strain, defined by shear strength τ_f divided by initial shear modulus G_{max}, and $C_1(0), C_2(0), C_1(\infty), C_2(\infty), \alpha$ and β are parameters determined from laboratory tests. The equation which makes a correlation between damping factor h and shear strain γ is expressed by

$$h = h_{\max}\left(1 - \left|\frac{\tau_a}{\gamma_a}\right| / G_0\right)^{\beta_{P6}} \tag{7}$$

where γ_a and τ_a are shear strain and shear stress at the turn-round point and β_{P6} is parameter.

5.2 *Analytical conditions*

Figure 10 shows the analytical models. In dynamic response analysis, we also compare the response of damaged embankment and reconstructed embankment. The damaged embankment consists of 1828 elements and 1848 nodal points, and the re-constructed embankment consists of 1705 elements and 1749 nodal points. For boundary conditions of deformation, all nodes at the bottom are fixed and the nodes at the right edge are fixed only in lateral direction. In dynamic response analysis, we modeled the sandy rock, silty rock, damaged-embankment material, reconstructed-embankment material and concrete. The modified GHE model was adopted for sandy rock, silty rock, embankment materials, and elastic model was adopted for concrete. Table 2 shows the input parameters and Figure 11 shows the dynamic deformation properties of sandy stone and reconstructed-embankment material using the parameters in Table 1. The modified GHE model describes

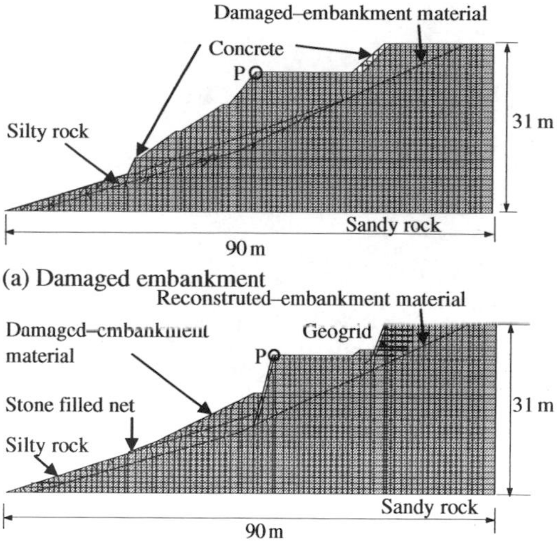

(a) Damaged embankment

(b) Reconstruted embankment

Figure 10. Analytical models for dynamic response analysis.

Table 2. Input parameters for ground, embankment and structure.

	Sandy rock	Silty rock and damaged-embankment material	Stone filled net and reconstructed embankment	Concrete
Shear modulus G_0(kPa)	830,000	30,275	338,679	1.04×10^7
Poisson's ratio ν	0.20	0.30	0.25	0.20
Cohesion c(kPa)	5.0	0.67	79.0	100.0
Internal friction angle ϕ (°)	49.0	33.63	47.8	0.0
Density ρ(t/m^3)	1.9	1.8	2.0	2.5
C1(0)	1.0	1.0	1.0	–
C2(0)	0.15	0.15	0.2	–
C1(∞)	0.2	0.1	0.2	–
C2(∞)	1.0	1.0	1.0	–
α	0.299	0.455	0.723	–
β	0.681	1.255	0.723	–
h_{max}	0.25	0.2	0.3	–
β_{p6}	1.1	1.5	0.08	–

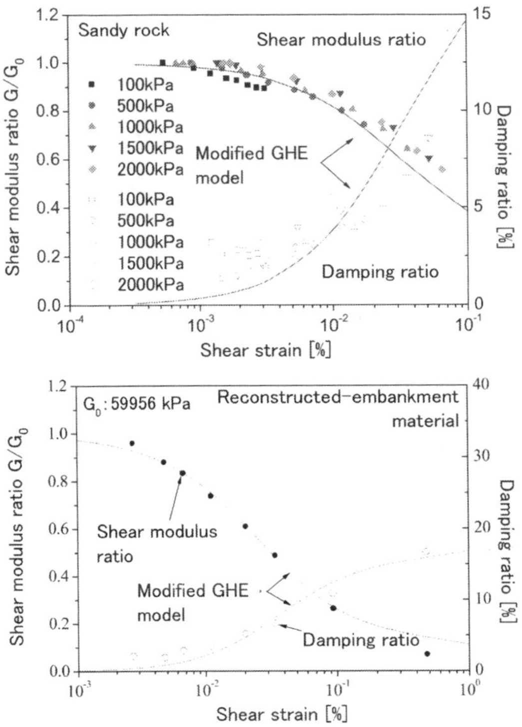

Figure 11. Dynamic deformation characteristics of sandy stone and reconstructed-embankment material.

Table 3. Input parameters for rock bolts and geogrid.

	Rock bolts	Geogrid
Young's modulus (kPa)	200,000,000	1,880,000
Moment of inertial (m^4)	1.17×10^{-8}	0.0002
Area (m^2)	9.60×10^{-5}	0.002

well dynamic characteristics obtained from laboratory tests. For reconstructed embankment, modeling the reinforcement materials is important. In this analysis, rock bolts and geogrid were modeled by beam elements and parameters of these materials were shown in Table 3.

The input wave is shown in Figure 12. This wave is inputted in horizontal direction.

5.3 *Results of analyses*

Figure 13 shows the distributions of maximum accelerations in the damaged embankment and the reconstructed embankment. In the damaged embankment, the maximum acceleration occurred at the top of slope. On the other hand, in the reconstructed embankment large acceleration occurred at the wide area. Figure 14

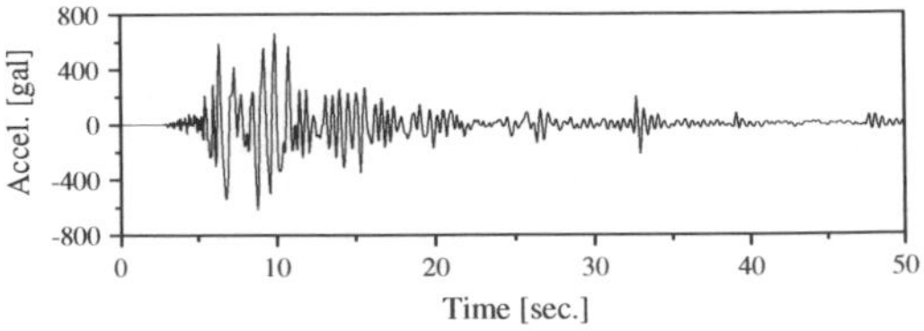

Figure 12. Input acceleration.

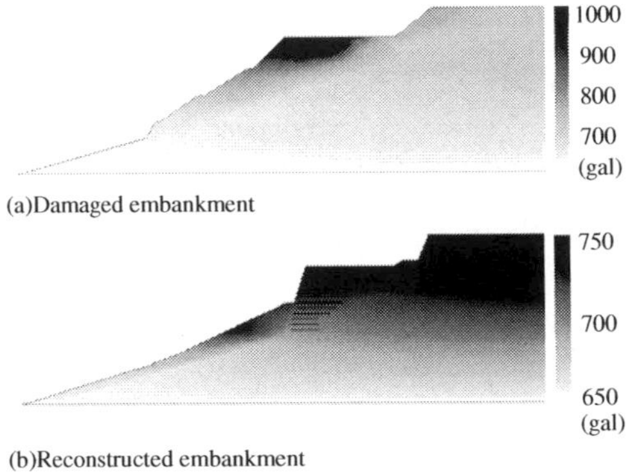

Figure 13. Distributions of maximum acceleration.

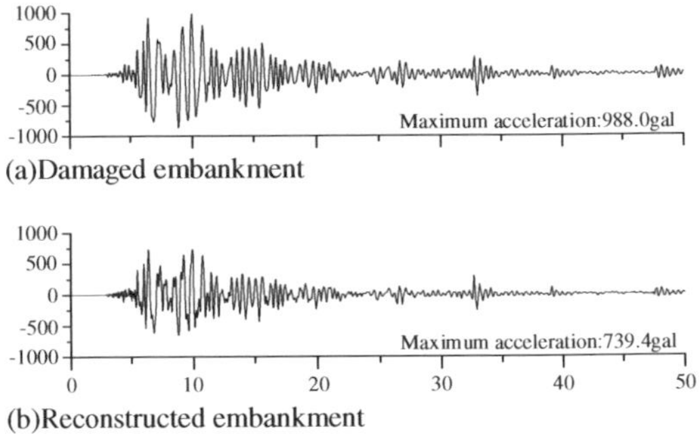

Figure 14. Time histories of acceleration at the top of embankment.

shows the time history of the accelerations at node P. The maximum acceleration in the reconstructed embankment was smaller than one in the damaged embankment. Figure 15 shows the deformation after this earthquake. The maximum lateral displacement of damaged embankment was about 10 cm, but that of the reconstructed embankment was very small. But the displacement obtained in the simulation of damaged embankment was considerably smaller than the observed displacement. From figure 15 (a), it is clear that a lot of finite elements at the top of embankment

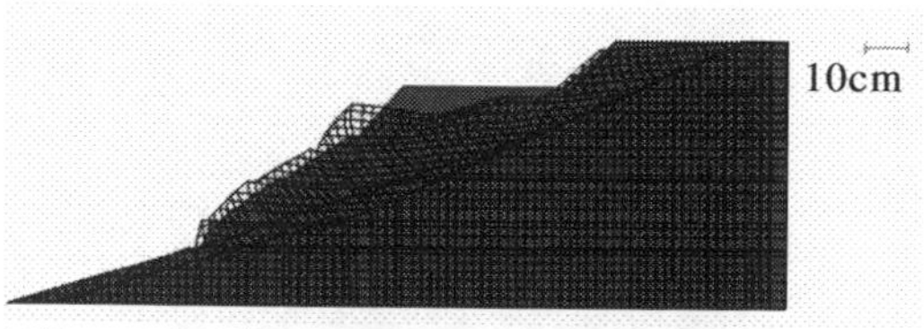

(a)Damaged embankment

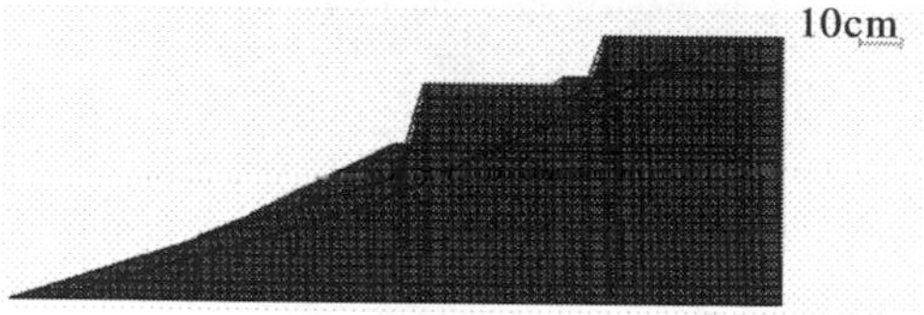

(b)Reconstructed embankment

Figure 15. Deformation after the earthquake.

were distorted. So, this indicates that this method has limitations for evaluating deformation.

6 CONCLUSIONS

In this paper, a set of numerical simulations were conducted in order to evaluate the main reason of collapse and the seismic resistance, aimed to the embankment in the vicinity Jouetsu Line 221 km000 m.

The seepage analysis provides the follow conclusions.

(1) In the damaged embankment, the degree of saturation increased and the groundwater level appeared along the boundary between the embankment and the silty rock.
(2) In the reconstructed embankment, the degree of saturation was smaller than one in the damaged embankment.

The dynamic response analysis provides the follow results.

(1) The maximum acceleration in the damaged embankment was larger than one in the reconstructed embankment. So, the seismic resistance of the reconstructed embankment was improved.
(2) The deformation obtained from the dynamic response analysis shows the same tendency of the maximum acceleration, but it revealed that the dynamic response analysis based on the FEM has the limitations to calculate the deformation of embankment.

According to the results obtained from the seepage analysis and dynamic response analysis, the deformation analysis using Newmark's method was conducted. The details are shown in Shinoda et al. (2007).

REFERENCES

Carsel, R. F. and Parrish, R. S. 1988. Developing joint probability distributions of soil water retention characteristics, Water Resources Research, No. 24, pp. 755–769

Nishimura, A. and Murono, Y. 1999. Proposal and experimental demonstration of nonlinear model of soil using the GHE model and the simple hysteretic rules, Proceedings of the JSCE Earthquake Engineering symposium, No. 29, pp. 309–312 (in Japanese)

Shinoda, M., Horii, K., Watanabe, K., Kojima, K. and Tateyama, M. 2007. Seismic stability of reinforced soil structure constructed after the mid Niigata prefecture earthquake, Proceedings of IS-Kyushu (submitted)

Tateyama, M. and Kato, S. 2005. Damages and lessons of railway structures in the 2004 Nigata-ken Chuetsu earthquake, the Foundation Engineering and Equipment, Vol. 33, No. 10, pp. 43–47 (in Japanese)

Van Genuchten, M. Th. 1980. A closed-form equation for predicting the hydraulic conductivity of unsaturated soils. Soil Sci. Soc. Am. J., No. 44, pp. 892–898

New Horizons in Earth Reinforcement – Otani, Miyata & Mukunoki (eds)
© 2008 Taylor & Francis Group, London, ISBN 978-0-415-45775-0

Seismic stability of reinforced soil structure constructed after the mid Niigata prefecture earthquake

M. Shinoda, K. Watanabe, K. Kojima & M. Tateyama
Railway Technical Research Institute, Tokyo, Japan

K. Horii
Integrated Geotechnology Institute, Tokyo, Japan

ABSTRACT: The mid Niigata prefecture earthquake in 2004 damaged a number of structures including embankments, earth slopes and retaining walls. This paper describes numerical techniques to evaluate the performance of structures constructed before and after the mid Niigata prefecture earthquake. Deformation analysis was conducted by the Newmark's sliding block analysis that implements the ground water level estimated by the seepage analysis and the response acceleration of the structure subjected to the earthquake by the finite element analysis. From the results of the current numerical analysis, the instillation of the reinforcement can increase the stiffness of the reinforced soil, resulting into the decrease of the deformation at the crest of the structure. The numerical analysis in this study revealed that the performance of the reinforced structure constructed after the earthquake significantly improved as compared to that of the structure constructed before the earthquake.

1 INTRODUCTION

1.1 *The mid Niigata prefecture earthquake*

The mid Niigata prefecture earthquake of magnitude 6.8 on Richter scale occurred Japan on October 23, 2004. The epicenter was lat. 37° 17′N. and 138°52′ E. About five thousand people were injured in this earthquake. The damage caused by this earthquake amounts to three trillion yen. The feature of this earthquake is strong aftershock in which the maximum magnitude exhibited 6.5. A number of structures including embankments, earth slopes and retaining walls were severely damaged. Among a large number of damaged earth slopes, this paper is focused one collapsed earth slope for railway. Moreover the reconstruction of the collapsed earth slope for railway was presented. The numerical analyses were conducted to evaluate the seismic stability of the collapsed earth slope and reconstructed retaining wall.

1.2 *Collapsed earth slope*

An earth slope supporting both up and down railway tracks collapsed on a length of 65 m and a height of 4–12 m as shown in Figures 1 and 2 (Tateyama & Kato 2005, Morishima, Saruya & Aizawa 2005). The total amount of collapsed soil volume was estimated at 13,000 m^3. The collapsed earth slope was located at the upper part of Shinano river terrace. The collapsed earth slope was located at the eroded valley walls by the

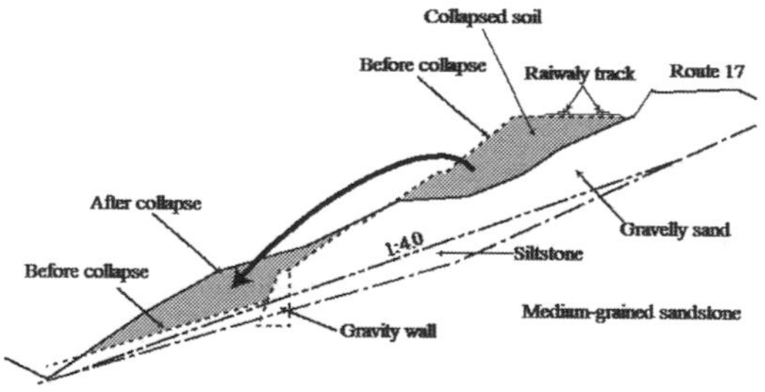

Figure 1. Collapsed earth slope at the Jouetsu-line.

Figure 2. Collapsed earth slope at the Jouetsu-line.

Shinano River with the tip of pond fed by the Ishida River following into the Shinano River. The geological feature is that a foundation is medium-grained sandstone; a sedimentary layer on the foundation is

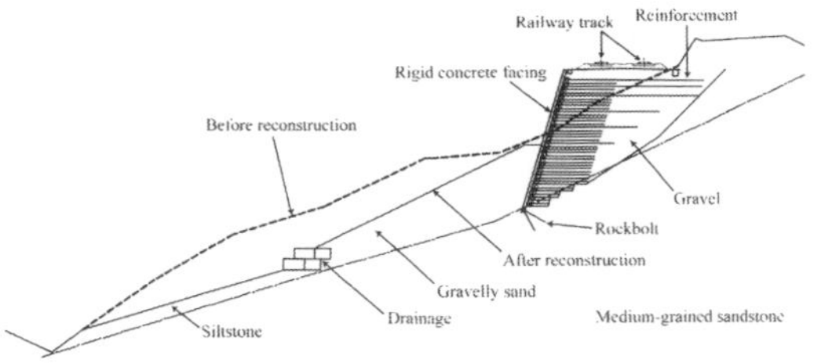

Figure 3. Schematic figure of reconstructed geosynthetic-reinforced soil retaining wall (GRS-RW) with a full height rigid facing.

siltstone and the backfill of collapsed earth slope is gravelly sand.

1.3 *Reconstructed retaining wall*

In the planning of the reconstruction of collapsed earth slope, the reconstruction using the same amount of collapsed backfill with the stable inclination of earth slope was thought to be practically difficult based on the current design code. It was important for the reconstruction that the small amount of backfill soil should be used, the permanent structure was preferable and the seismic stability should be higher than that of the collapsed earth slope. As a result, the geosynthetic-reinforced soil retaining wall (GRS-RW) with a full-height reinforced concrete facing was adopted as an alternative structure of the collapsed earth slope. For the GRS-RW with full-height rigid facing, it is very effective to use a rigid facing and to connect the reinforcement layers to the back of the facing to increase the seismic stability of reinforced soil RWs, as validated by high seismic performance of a number of reinforced soil RWs of this type during recent severe earthquakes, including the 1995 Hyogoken Nambu Earthquake (Tatsuoka et al. 1996). The GRS-RW with full-height rigid facing will be hereafter referred to as GRS-RW.

Figure 3 shows the schematic figure of reconstructed GRS-RW. For the reconstruction, the collapsed soil was excavated up to the surface of the siltstone. After the excavation, the rock bolts were installed into the siltstone to reinforce the foundation of the rigid facing of the GRS-RW. The backfill soil of the GRS-RW was constructed until the height of 13 m with the reinforcement installing at the vertical spacing of 30 cm. The amounts of used backfill soil and concrete for the facing of the GRS-RW for the reconstruction were respectively 4600 m^3 and 300 m^3. Figure 4 shows the reconstructed GRS-RW.

Figure 4. Reconstructed GRS-RW with full height rigid facing.

2 NEWMARK'S SLIDING BLOCK ANALYSIS

In this study, Newmark's sliding block analysis (Newmark 1965) was adopted for the seismic deformation analysis. It is a simplified procedure employed in the design code of railway structures in Japan (RTRI 2007), in which the seismic deformation of earth slopes or GRS retaining walls subjected to a strong ground motion can be calculated by integrating the equation of rotational motion of a soil mass contained within the critical circular slip surface by assuming the failure mass as a rotational rigid block. The equation of rotational motion is solved for the rotation caused by the difference between the driving and resisting moments. The critical slip surface is determined by the conventional modified Fellenius method (Fellenius 1927) using a specific acceleration or seismic coefficient to yield a safety factor of 1.0. Hereafter, this acceleration and seismic coefficient will be referred to as the yield acceleration and yield seismic coefficient, respectively. Newmark's sliding block analysis will be hereafter referred to as Newmark analysis. Refer to Shinoda et al. (2006) for an application of the Newmark analysis of earth and GRS slopes.

The seismic stability analysis is conducted with the conventional modified Fellenius method to determine the center and radius of the critical circular slip surface and yield acceleration. The safety factor in the above seismic stability analysis can be obtained from the following equation:

$$FS = \frac{M_r}{M_d} = \frac{M_{rw} + M_{rc} + M_{rt} - k_h M_{rk}}{M_{dw} + k_h M_{dk}} \quad (1)$$

where FS is the safety factor; k_h, seismic coefficient; M_r, overall resisting moment; M_d, overall driving moment; M_{rw}, resisting moment due to the self-weight of soil; M_{rc}, resisting moment due to soil cohesion; M_{rt}, resisting moment due to the design strength of reinforcement; M_{rk}, decrease in the resisting moment per unit seismic coefficient due to the self-weight of soil subjected to a seismic inertia force; M_{dw}, driving moment due to the self-weight of soil; and M_{dk}, driving moment per unit seismic coefficient due to the seismic inertia force. By substituting $FS = 1.0$ and arranging Equation 1, the yield seismic coefficient is obtained as follows:

$$k_y = \frac{M_{rw} + M_{rc} + M_{rt} - M_{dw}}{M_{dk} + M_{rk}} \tag{2}$$

Subsequently, after selecting the design ground motion, the seismic stability analysis is conducted by using the above-determined center and radius of the critical slip surface. The seismic coefficient is updated as follows:

$$k_h(t) = \frac{A(t)}{g} \tag{3}$$

where $A(t)$ is the acceleration time history, and g is the gravitational acceleration. The input ground motion was directly used for the standard Newmark analysis without considering the response of a structure as the above acceleration time history.

The above seismic stability analysis is performed up to the end of the acceleration time history. During the seismic stability analysis, the difference between the overall driving and resisting moments is calculated, and the equation of rotational motion is obtained as follows:

$$J\ddot{\theta}(t) = M_d(t) - M_r(t) = M_{dw} + k_h M_{dk} - M_{rw} + k_h M_{rk} - M_{rc} - M_{rt} \tag{4}$$

where θ is the rotational angle of the soil mass and J is the moment of inertia expressed as follows:

$$J = \Sigma\left(J_{g,i} + \frac{1}{g} \cdot R_{g,i}{}^2 \cdot W_i \right) \tag{5}$$

where $J_{g,i}$ is the polar moment of inertia of the i-th slice and $R_{g,i}$ is the distance between the center of the slice and that of the critical circular slip surface of the i-th slice. The angular acceleration, angular velocity, and rotation of the soil mass are obtained as follows:

$$\ddot{\theta}_{t+\Delta t} = \frac{1}{J} \Delta M_{t+\Delta t} \tag{6}$$

$$\dot{\theta}_{t+\Delta t} = \dot{\theta}_t + \frac{1}{2} \cdot \left(\ddot{\theta}_t + \ddot{\theta}_{t+\Delta t} \right) \cdot \Delta t \tag{7}$$

$$\theta_{t+\Delta t} = \theta_t + \dot{\theta}_t \cdot \Delta t + \frac{1}{6} \cdot \left(2 \cdot \ddot{\theta}_t + \ddot{\theta}_{t+\Delta t} \right) \cdot \Delta t^2 \tag{8}$$

The accumulated rotation of the soil mass is computed using Equation 8 only when the angular velocity is positive. Finally, the seismic deformation is obtained as follows:

$$d_t = R \cdot \theta_t \tag{9}$$

In this paper, the seismic deformation is defined as a rotational displacement along the critical slip surface of the failure mass according to the RTRI design code.

For the above standard Newmark analysis, the input ground motion was directly used without considering the response of a structure. However, when the height or stiffness of a structure affects the response, the numerical result by the Newmark analysis using the input ground motion may cause an impermissible error. To evaluate the seismic stability of the collapsed earth slope and reconstructed GRS-RW accurately, the response acceleration obtained from the dynamic analysis was used for the current Newmark analysis. Additionally, the ground water level was obtained from the nonstationary seepage analysis. Refer to Matsumaru et al. (2007) for detailed explanation of seepage and dynamic analyses.

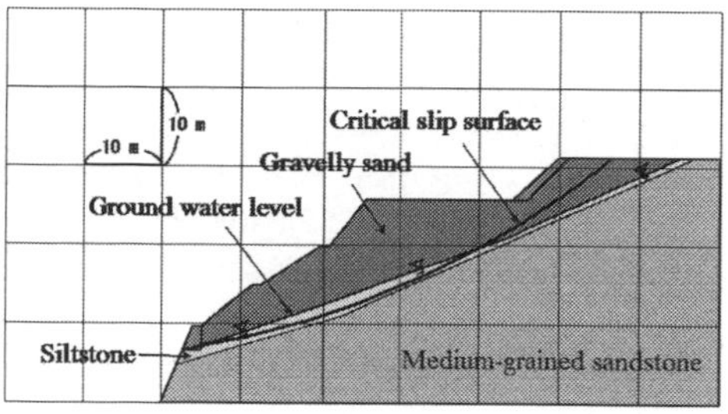

Figure 5. Analytical model for the current Newmark analysis.

3 NUMERICAL RESULT

Figures 5 and 6 show the analytical model of collapsed earth slope and reconstructed GRS-RW used for the current Newmark analysis. From the site observation after earthquake, a critical slip surface of the collapsed earth slope was assumed as shown in Figure 5. For comparison, the same critical slip surface was adopted in the calculation of the reconstructed GRS-RW. As mentioned above, the ground water lever was set over the layer of the siltstone from the result of the seepage analysis. Table 1 shows the soil properties used for the current Newmark analysis, which were obtained from laboratory tests. Two type of cohesion of the gravel at the peak and residual were used for the current analysis.

Tables 2 and 3 show the safety factors, yield accelerations and displacements obtained from the stability

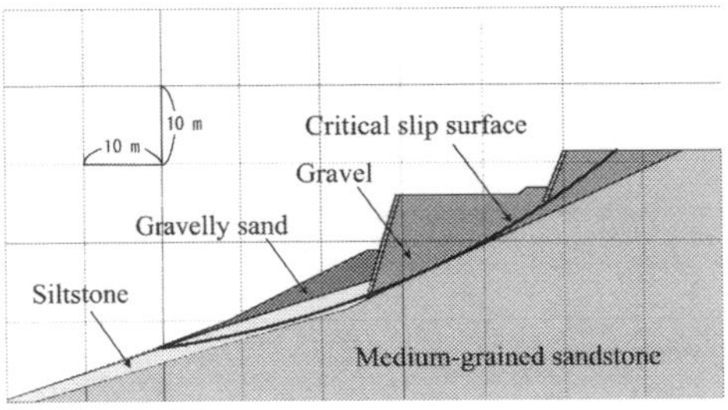

Figure 6. Analytical model for the current Newmark analysis.

Table 1. Soil properties used for the Newmark analysis.

	Unit weight (kN/m^3)	Cohesion (kPa)	Friction angle (degrees)
Siltstone	17.9	0.67	33.6
Gravelly sand	17.9	3.67	33.6
Gravel	19.8	81.7 (0.0)	47.1

Table 2. Result of collapsed earth slope.

Safety factor	Yield acceleration (gal)	Displacement (m)
1.48	0.163	1.42

Table 3. Result of reconstructed GRS-RW.

Safety factor	Yield acceleration (gal)	Displacement (m)
3.69	0.865	0.0

and Newmark analysis for the collapsed earth slope and reconstructed GRS-RW. The safety factor and yield acceleration of the reconstructed GRS-RW were higher than those of the collapsed earth slope. This is possibly due to the use of gravel as the backfill and geosynthetics as the reinforcement. The displacement of the collapsed earth slope obtained from the Newmark analysis was 1.42 m, while the displacement of the reconstructed GRS-RW exhibited zero, resulting into high seismic stability. Figure 7 shows the time history of displacement, angle velocity, angle acceleration and acceleration of the collapsed earth slope in the Newmark analysis. From the start of the earthquake, a large displacement was exhibited. For the reconstructed GRS-RW, because of the zero displacement, such time history could not be obtained.

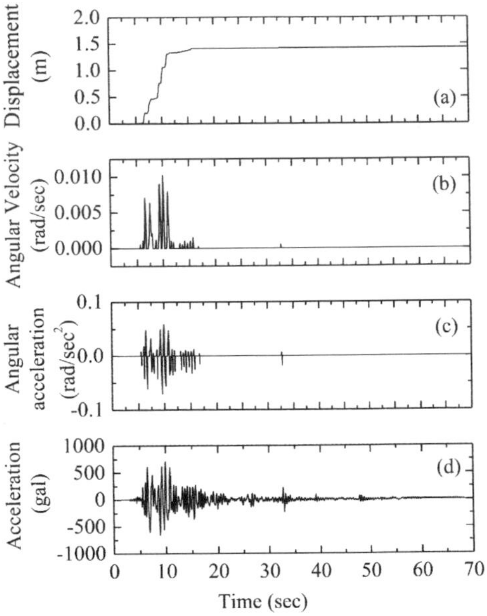

Figure 7. Time history of the collapsed earth slope in the Newmark analysis; (a) displacement, (b) angle velocity, (c) angle acceleration and (d) acceleration.

4 CONCLUSIONS

The mid Niigata prefecture earthquake in 2004 damaged a number of structures including embankments, earth slopes and retaining walls. The paper reports an evaluation of seismic stability of collapsed earth slope and reconstructed geosynthetic-reinforced soil retaining wall (GRS-RW) with full height rigid facing. A deformation analysis method adopted in this paper is Newmark's sliding block analysis based on the results of dynamic and seepage analysis, which can consider the effect of response of the structure.

The numerical results in this paper successfully show a discrepancy between the seismic stability of collapsed earth slope and reconstructed GRS-RW. The seismic stability of the collapsed earth slope was lower than that of the reconstructed GRS-RW. This is due to the use of geosynthetics as reinforcement and gravel as backfill soil for the reconstructed GRS-RW. The GRS-RW with full height rigid facing can be constructed as an important structure having a high seismic stability.

REFERENCES

Tateyama, M. & Kato, S. 2005. Damage and precept of railway structures subjected to the mid Niigata prefecture earthquake. *The Foundation Engineering & Equipment*. 33(10): 43–47.

Morishima, H., Saruya, K., Aizawa, F. 2005. Damage and reconstruction of the old railway at the section of sturuc-tures using soil. *The Foundation Engineering & Equipment*. 33(10): 78–83.
Tatsuoka, F., Tateyama, M. & Koseki, J. 1996. Performance of soil retaining walls for railway embankments. *Soils and Foundations*, Special Issue for the 1995 Hyogoken-Nambu Earthquake: 311–324.
Newmark, N. M. 1965 Effects of earthquakes on dams and embankment. *Geotechnique*, 15(2): 139–160.
Railway Technical Research Institute. 2007 Design standard for railway earth structures. Railway Technical Research Institute. Maruzen (in Japanese).
Fellenius, W. 1927: Erdstatische Berechnungen mit Reibung ind Kohaesion (Adhasesion) und unter Annahme kreizilin-drischer Gleiflaechen. W. Ernst und Sohn Verlag. Berlin.
Shinoda, M., Horii, K., Yonezawa, T., Tateyama, M. & Koseki, J. 2006. Reliability-based seismic deformation analysis of reinforced soil slopes, *Soils and Foundations*, 46(4): 477–490.
Matsumaru, T., Ishizuka, M., Tateyama, M., Kojima, K., Watanabe, K. and Shinoda, M. 2007. Rainfall seepage analysis and dynamic response analysis for the railway embankments seriously damaged in the 2004 Niigata-ken Chuetsu earthquake, *Proceedings of the 5th International Symposium on Earth Reinforcement*.

New Horizons in Earth Reinforcement – Otani, Miyata & Mukunoki (eds)
© 2008 Taylor & Francis Group, London, ISBN 978-0-415-45775-0

Damage to Terre Armée structures from the Mid-Niigata Earthquake and measures and actions taken to date

H. Nagakura & H. Oota
JFE SHOJI TRADE CORPORATION (Terre Armée Civil Engineering Dept.)

G. Berard
Terre Armée International

INTRODUCTION: In recent years, strong earthquakes have occurred frequently in various parts of Japan, subjecting many of the country's more than 20,000 Terre Armée(hereinafter called Reinforced Earth) Retaining Wall structures to significant seismic shakes. However, Reinforced Earth has resisted these large tremors without collapsing and, ever since the Kobe Earthquake, has been highly appraised for their considerable ability to withstand earthquakes. This report deals with the damages caused to the Reinforced Earth by the Mid-Niigata Earthquake, which extensively damaged many other embankments and with the measures and actions taken to date. The report treats of overall result of the inspections of the Reinforced Earth scrutinising focal points of the inspections and characters of the deformation and damage sustained, while classifying the earthquake's impact on the Reinforced Earth by the level of damages sustained. As two years have passed since the earthquake, during which time restoration and revitalisation of the stricken area have been undertaken. This report also introduces certain unusual cases of the structures of which serving environments have been drastically changed after the earthquake and a case of a structure reconstructed.

1 PROFILE OF EARTHQUAKE

At 17:56 on 23 October 2004, an earthquake with a Magnitude (M) of 6.8 in Richter Scale occurred at a depth of 13 km in the Chuetsu (i.e. mid-Niigata) Area of Niigata Prefecture. A seismic intensity of 7 on the Japanese Seismic Intensity Scale, the highest gradation in the scale and in this earthquake as well, was observed in the town of Kawaguchi. Table 1 contains data from the locations at which the Intensity of 5 or higher on the Japanese Scale was observed during the main shake. The intensity 7 denotes that ground is considerably distorted by large cracks and fissures and occasionally even highly aseismic buildings of reinforced concrete are severely damaged.

1.1 *Distinctive environment for earth structures*

The Mid-Niigata Earthquake had the following characteristics and it can be assumed that the ground, embankments and structures located there would have been subjected to intensively severe conditions.

- There was extensive aftershock activity following the main shake, including 10 aftershocks of seismic intensity 5 and above occurring on 23 October alone and a total of 17 aftershocks experienced in the two weeks following the earthquake.
- The Chuetsu Area is located at an area where soft plains stretch along rivers, such as Shinano River, while geological features of the slopes there are of soil deposits on the bedrock, which therefore tend to slide casily.
- Rainfall was particularly heavy in 2004, and the intense rainfalls accompanied by the Typhoon Nos.21 and 23 that struck the area immediately prior to the earthquake would have considerably saturated the ground and embankments.

2 SUMMARY OF RESULT OF INSPECTIONS OF REINFORCED EARTH

JFE Shoji Trade Corporation has constructed Reinforced Earth at a total of 408 sites in Niigata Prefecture, out of which 51 sites (72 walls altogether) are situated within about 35 km radius of the epicentre of the main shake and were inspected this time. 80% of these inspected were flat top structures without overburdens (Tables 3 & 4). The inspections were conducted twice immediately after the earthquake and before affected

Table 1. Data from the locations experienced an Intensity of 5 or above during the main shake.

Observation point	Longitude (deg.)	Latitude (deg.)	Maximum acceleration (gal)	SI value (kine)	Seismic intensity	
Ojiya	138.7930	37.3027	1501.87	166.01	6.73	7
Tokamachi	138.7500	37.1250	1750.17	75.96	6.20	6^+
Nagaoka Br.	138.8894	37.4231	920.88	90.69	6.11	6^+
Koide	138.9652	37.2302	639.49	47.77	5.55	6^-
Nagaoka	138.8463	37.4386	543.90	42.69	5.51	6^-
Naoetsu	138.2266	37.1577	220.20	24.48	5.22	5^+
Siozawa	138.8494	27.0333	363.63	27.00	5.12	5^+
Tsunan	138.6561	37.0116	427.92	40.32	5.03	5^+
Yasuzuka	138.4472	37.238	274.60	16.78	4.96	5^-
Kanose	139.4805	37.6833	315.32	18.93	4.93	5^-
Kasiwazaki	138.5611	37.3694	148.99	31.79	4.93	5^-
Sanjou	138.9591	37.6380	122.24	20.11	4.83	5^-
Numata	139.0816	36.6547	376.20	11.62	4.72	5^-
Tadami	193.3177	37.3461	188.90	14.44	4.72	5^-
Nishiaizu	139.6500	37.5972	164.97	11.55	4.55	5^-
Maki	138.8866	37.7608	135.73	15.24	4.54	5^-

Table 2. The time, intensity and epicentre where the highest intensity was observed in each of the 18 shakes occurred in a 2 weeks period.

					Epicenter		
No	Date	Time	Seismic intensity		Longitude	Latitude	Depth (km)
1	23/10/04	17:56	6.8	7	138.52	37.17	13
2		17:59	5.3	5^+	138.52	37.19	16
3		18:03	6.3	5^+	138.59	37.22	9
4		18:07	5.7	5^+	138.52	37.21	15
5		18:11	6.0	6^+	138.50	37.15	12
6		18:34	6.5	6^+	138.56	37.18	14
7		18:36	5.1	5^-	138.57	37.15	7
8		18:57	5.3	5^+	138.52	37.12	8
9		19:36	5.3	5^-	138.50	37.13	11
10		19:45	5.7	6^-	138.53	37.18	12
11		19:48	4.4	5^-	138.50	37.18	14
12	24/10/04	14:21	5.0	5^+	138.50	37.15	11
13	25/10/04	0:28	5.3	5^-	138.52	37.12	10
14		6:04	5.8	5^+	138.57	37.20	15
15	27/10/04	10:40	6.1	6^-	139.02	37.17	12
16	4/11/04	8:57	5.2	5^+	138.55	37.26	18
17	8/11/04	11:15	5.9	5^+	139.02	37.24	0
18	10/11/04	3:43	5.3	5^-	139.00	37.22	5

by snow load, in mid November and early December, as well as in June and October of the following year after snow had melted away.

The inspections were conducted primarily by the means of visual inspection taking drawings of the structures and inspection logs to record the site conditions in detail in them. The focal points of the surveys

Table 3. Type and cross section of the walls inspected.

Type of wall and facing		Number of Walls	Cross-section: Flat top	Gradient top
Vertical walls	Metal Panel	2	1	1
	Concrete Panel (180 mm thick)	25	16	9
	Concrete Panel (140 mm thick)	32	31	1
	Mini Terre Armée	8	7	1
	Collar Wall	2	2	0
Slope Walls	Terratrel & Terravale	3	0	3
Total		72	57	15

are indicated in Table 5. JFE Shoji Trade Corporation has been conducting site inspections whenever a strong earthquake occurred and has recorded the state and conditions of each structure studied at each site. And the data obtained have been used to evaluate structural soundness and to gauge the necessity for emergency measures. The surveyor completes a table including damage class based on the level of the damage sustained, by providing an objective assessment on the conditions rather than subjective view of the surveyor. And the table is used at such an occasion as of explaining the state of the structure to road administrator, etc.

3 INSPECTION RESULT

The result of the inspections is summarised in Table 6 and Fig. 1. In no case was there a collapse of a Reinforced Earth which had undergone displacement or

Table 4. Profile of structures Inspected.

Distance from Epicenter			
~10	10~20	20~30	30~40
21	38	9	4
Structural Characteristics of the Embankment			
Single	Multi	Cradle	Island
47	4	2	18
Water	Long Slope	Abutment	
0	0	1	
Wall Height			
$0 \leq H < 5$ m	$5 \leq H < 10$ m	$10 \leq H < 15$ m	15 m $\leq H$
25	39	6	2
Foundation Ground			
Ground	Replacement	Shallow Improvement	Deep Improvement
63	9	0	0

deformation due to seismic motion, and about 90% of the structures sustained either "light damage only" or "no deformation at all". The reinforcing effect of the embankment is considered to have mitigated deformation or damage to the road above, and it has been also confirmed through pull-out tests that even a Reinforced Earth subjected to seismic motion restores its resistance and, consequentially, its stability soon after seismic motions have subsided.

Common postures of the damages and deformation observed in the course of the inspections are introduced below.

Table 6. Summary of inspection result.

Damage class	State of structure	Number of walls
A^{++}	Loss of ultimate stability as Reinforced Earth (collapse of embankment)	0
A	Heavy and extensive deformation of embankment and damage to reinforcing elements. Partial removal of the reinforcing elements and backfill and reconstruction of the embankment are required. (Long term stability of the entire structure as Reinforced Earth embankment is considered difficult to guarantee.)	2
B	Although deformation of the embankment or damage to the reinforcing elements is localised, partial repair of the embankment or facing panels, etc. is required.	5
C	Despite a little deformation in the embankment (incl. only in the peripheral to the Reinforced Earth) or damage to the reinforcing elements, long-term structural stability is considered guaranteed. No repair work to Reinforced Earth is required.	60
D	No deformation to the Reinforced Earth embankments and to peripheral structures such as roads	5

3.1 *Deformation of reinforced earth and damage to its facing panels*

The inspections confirmed that the deformation in the cross-sectional direction mainly took a form of

Table 5. Inspection items and damage classification.

Location of deformation	Overall/Local/None	Damage class A^{++}/A/B/C/D
Wall Face	Transversal Direction	Bulging; Bending Forward; Buckling; Crevice; Distortion; Sliding; Backfill Runoff
	Longitudinal Direction	Crevice; Distortion; Sliding
Crown Type	Coping & others	Joint Opening; Crevice; Distortion; Sliding; Faulting; Cracking
	L-Shape, Gravity Type	Overturning; Slipping; Crevice; Distortion; Sliding; Faulting; Cracking
End of Wall	Structure Tie-in Points	Joint Opening; Crevice; Distortion; Sliding; Faulting; Cracking
	Slope Wrapping	Settlement at Wrapping Zone; Detachment
Facing Material	Cracks	Cracking at Corner; Cracking at Sleeve; Slits
	Damages	Buckling; Breakage from Load; Falling-down
Periphery of Reinforced Earth	On Top of Reinforced Earth	Settlement; Cracking; Faulting; Skidding
	In Front of Reinforced Earth	Settlement; Faulting; Damage in Peripheral Structures
Roads & Surface	Passable, Not passable	Cracking; Faulting; Opening between Subsidiary Structures
Necessity for Repair	Yes/No	Elements Repair; Local Reconstruction; Total Demolition

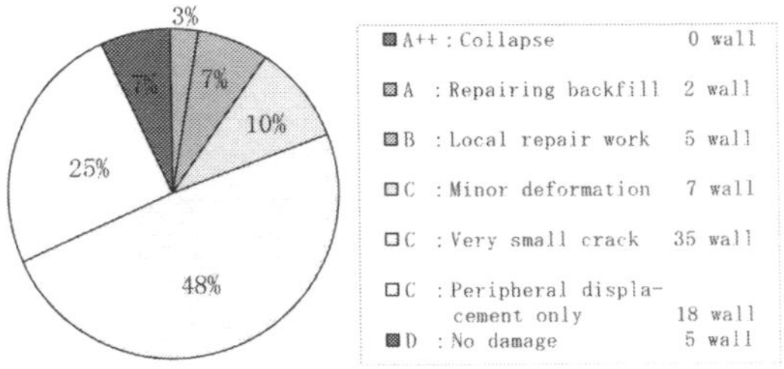

Figure 1. Aggregate of inspection result.

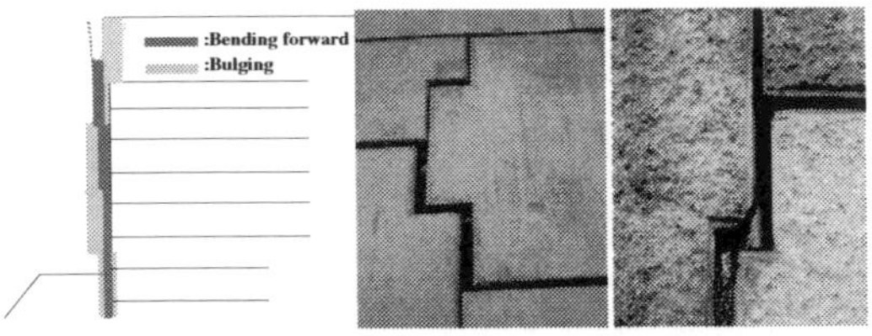

Figure 2. Form of lateral deformation (Bending forward; Bulging), Photo 1. Sliding and damage to wall facing panels.

Figure 3. Location of damage to road surface, Photo 2. State of damage to road paving.

bulging and/or bending forward of the entire facing panels (Fig. 2). Viewed from the front face of the wall, crevice, distortion or sliding of adjacent facing panels, or cracking or breaking at the corners of the facing panels were observed. (Photo 1). These states correspond to the types of the deformation and damages observed in the previous strong earthquakes. Moreover, as in the previous earthquakes, such critical damages, as collapse of entire embankment or falling-off of the facing panels, were not observed.

3.2 *Deformation and damage peripheral to reinforced earth and subsidiary structures*

The majority of the Reinforced Earth inspected this time were flat top type embankments. Extensive settlement, cracking and faulting of roads (paving) constructed on top of the embankments were observed. Most of the damages in the longitudinal direction occurred at the boundaries of the cut-and-fill sections or in the vicinity of the uppermost reinforcing steel strips, while they were seen at the tie-in points between structures in the transverse direction (Fig. 3, Photo 2).

Figure 4. Deformation of safety barrier base (Crushed stone), Photo 3. State of safety barrier base.

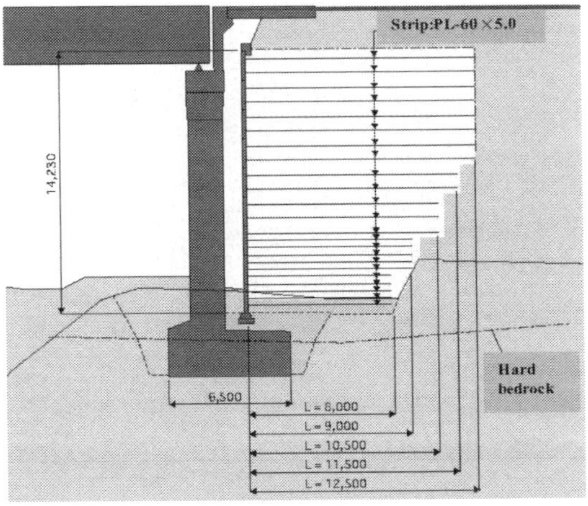

Figure 5. Cross-sectional plan as built.

In addition, there were some cases in which L-shaped bases for safety barrier installed on the crown of the Reinforced Earth had toppled inwards towards the road (Fig. 4, Photo 3). These L-shaped bases are rigid constructions structurally independent of the Reinforced Earth and the deformation is assumed to be due principally to the differences in the behaviour of the two structures.

4 INTRODUCING UNUSUAL CASES

4.1 *Structure A: Wall height: 14.98 m; Surface area: 768 m^2; 8 km from epicentre*

This case is extreme environment yet exertion of Reinforced Earth. This structure was constructed in 1995 of which deck was supported by a Reinforced Earth mixed-abutment. This is a case that the abutment was almost completely submerged to its wall crown. The flow of Imo River about 10 m below the foundation of the abutment turned to a natural dam created by a landslide the earthquake caused. The photographs show the flooded site immediately following the earthquake and the states of the site both in winter of the year and at

Photo 4. Immediately after completion of the abutment, Photo 5. 3 days after the main shake.

Photo 6. 3 months after the earthquake.

Photo 7. 6 months after the earthquake.

the time of an inspection conducted in the following year.

This abutment had been subjected to extremely severe conditions, not only because it was situated close to the epicentre and sustained frequent aftershocks therefore, but also because it was not designed for the use at waterfront, however, had been submerged due to a sudden change in water level. A new bridge was built and this Reinforced Earth played a new role as a construction road during the bridge construction period. Although detailed inspection of the Reinforced Earth was no longer possible because it was under

Photo 8. A new bridge is being constructed, June 2006.

Photo 9. Inauguration of the new bridge, September 2006.

the water, the structure supported the construction road safely as its surface was being measured and maintained on a daily basis.

4.2 *Structure B: Wall height: 7.48 m; Surface area: 273.5 m^2; 8.3 km from epicentre*

This structure constructed, as a retaining wall, for a local road facing to a junior high school is a case introduced. There were numerous kinds of structures in addition to a Reinforced Earth in this area where suffered a number of disasters including landslides, washouts, etc. when the earthquake occurred. This particular spot where the school stands was not an exception. The natural ground above the Reinforced Earth slid and many structures sustained displacements and deformation, which consequentially required extensive restoration works.

4.2.1 *States of the Reinforced Earth immediately after the earthquake (settlement of crown, bulging, displacement; sliding of L-shaped retaining wall)*

This case conducted a strip pull-out test for confirmed stability of Reinforced Earth. In this structure,

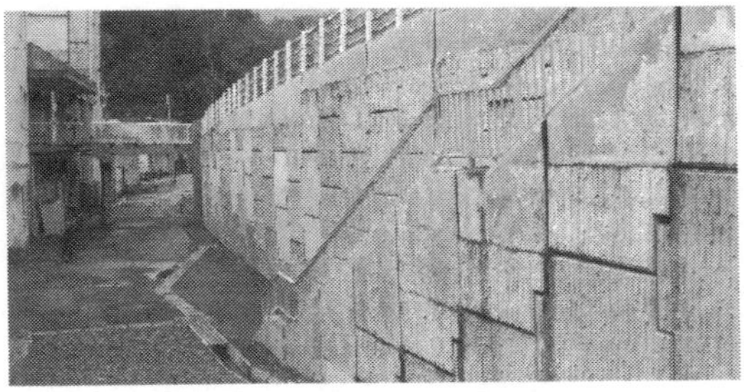

Photo 10. Panoramic view.

Photo 11. L-shaped retaining wall and tie-in point.

which consists of a precast L-shaped retaining wall, block masonry and a box culvert in addition to the Reinforced Earth, displacement was observed to have occurred at the tie-in points of the various structures due to differences in their behaviour. The rigid L-shaped retaining wall was displaced toward front side either by sliding or overturning, while flexible Reinforced Earth suffered displacement in its face due to bulging. The amount of maximum displacement of each structure was in the neighbourhood of 300 mm horizontally, while there was more than 50 mm of faulting at their tie-in point.

Distortion, crevice, sliding and cracking of the facing panels were particularly conspicuous at the junction with box culvert, while settlement, cracking and faulting in the road surface induced by the displacement were prominent at the top of Reinforced Earth. Although this deformation was not progressive, reinforcing strip pull-out tests were conducted to obtain one of the means to receive an idea for measures or actions to be taken.

4.2.2 *Pull-out test*

The reinforcing strip pull-out test is one of the methods to confirm that the reinforcing strips within a Reinforced Earth exhibit sufficient level of resistance. It was determined to verify the stability of this particular Reinforced Earth wall by way of confirming their resistance exhibited exceeds the design value.

Photo 12. Tie-in point with box culvert, Photo 13. State of the road.

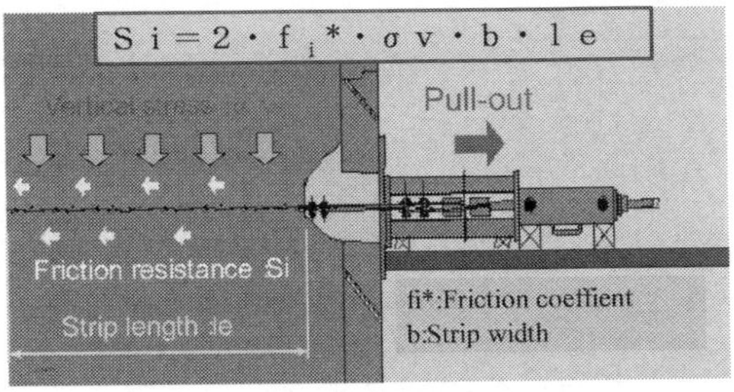

Figure 6. Diagram of pull-out test (Frictional resistance, Tester, Pulling).

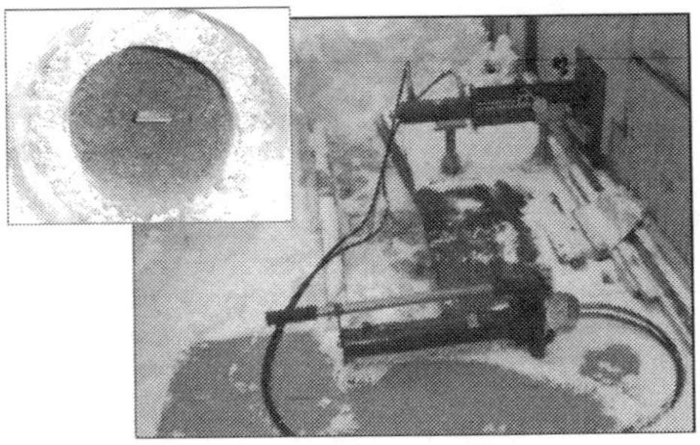

Photo 14. Conducting strip pull-out test.

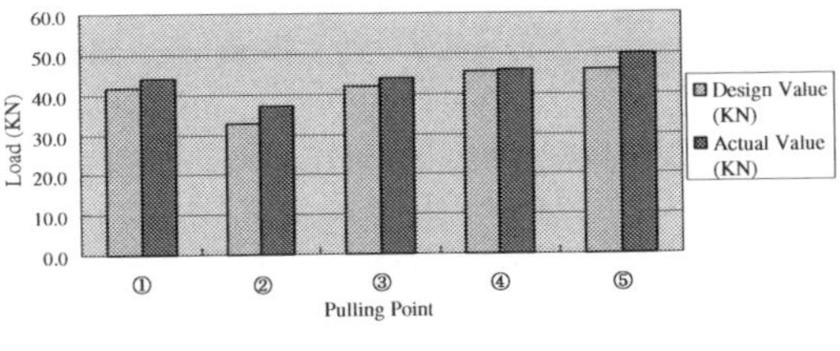

Figure 7. Result of pull-out tests.

The pull-out tests in this instance confirmed that the pulling resistance exceeded the structure's designed frictional resistance. The tie-strips connecting the concrete facing panels and the reinforcing strips were also examined when the holes were drilled for the pull-out tests and the result revealed no damage deformation was liable for.

Table 7. Summary of the pull-out tests.

Pulling point	Design value (KN)	Actual value (KN)	Pulling distance (mm)	Discovery
①	41.3	44.0	11.0	OK
②	32.5	37.0	12.0	OK
③	41.9	44.0	17.0	OK
④	45.7	46.0	18.0	OK
⑤	46.1	50.0	19.0	OK

Photo 15. Demolishing reinforced earth.

Since the pull-out tests were conducted on the strips of the structure in service, care was taken by controlling pulling force to slightly stronger than the design value in order not to affect the stability of the Reinforced Earth. It can be surmised that the frictional resistance would have been in excess of the values obtained this time should the limit pulling force was applied to the strips until they were physically pulled out.

As a result of these tests, it has been confirmed at the time of test that there has been no major reduction in the frictional resistance in the Reinforced Earth that was deformed by the earthquakc and the structure has maintained prescribed resistance. And no differences have been observed in relation to the magnitude of deformation.

4.2.3 *Restoration works*

The Reinforced Earth in this case faced a building of a junior high school and a large number of schoolchildren passed along the road in front of the wall. Even though the Reinforced Earth itself remained stable, the structure was considered as "under defective condition taking into account of its importance and appearance, in spite of its structural stability however", since it was felt that its bulged appearance would worry the local residents and the children. And, because the adjoining L-shaped retaining wall was in a state of requiring demolition and reconstruction, this Reinforced Earth too was eventually determined to demolish and rebuild.

Photo 16. Restoration works completed.

5 CONCLUSION

There were many conventional embankments collapsed and a number of roads interrupted by this earthquake. Meantime, Reinforced Earth sustained only a little damage compared to those of conventional earth structures and was confirmed through pull-out tests that Reinforced Earth subjected to seismic motion restores its resistance and stability soon after seismic motions subsided. These facts verify again that Reinforced Earth has very high aseismic characteristics and that Reinforced Earth is an effective and precautionary civil engineering method in order to prevent injuries and not to interrupt traffic caused by collapse of structures.

Various parties have anticipated that ways and means for inspections of reinforced earth structures sustained damages by seismic motions so as to ascertain their structural soundness and criteria for judgement of the inspection result are established. The assessment to the damage sustained and fundamental plans to the restoration works, etc. of the reinforced earth structures were discussed last year by the parties concerned. And relative associations et al. of the relevant methods presented a ground of basic concept of the criteria for determining extent of the damage as well as of a restoration work manual, which would be accomplished in the future as necessary.

A way of thinking, taking into consideration of the characteristics of Reinforced Earth, for inspection particulars and assessment for their result would be proposed henceforth so that a restoration work manual could be drawn up.

REFERENCES

Japan Meteorological Agency Press release 2004: No.28, December.

National Research Institute for Earth Science and Disaster Prevention: K-net data of quick report, November.

JFE Shoji Trade Corporation.2005,Report of Research on Terre Armée Structure Exposed to the Mid-Niigata Earthquake, July.

New Horizons in Earth Reinforcement – Otani, Miyata & Mukunoki (eds)
© 2008 Taylor & Francis Group, London, *ISBN 978-0-415-45775-0*

Behaviour of reinforced earth structures founded on soft silt deposit in seismically active hilly terrains

P. Mahajan
J&K Cell, RITES Ltd. New Delhi, India

S. Biswas & A. Adhikari
Reinforced Earth India Pvt. Ltd., New Delhi, India

ABSTRACT: The reinforced soil walls for Quazigund to Baramulla project, while under various stages of construction, experienced earthquake of magnitude about 7.0 on Richter scale with epicenter about 150 km away from the site, in Pakistan. The equivalent magnitude of tremor on Richter scale at site was 5.4. The character of the titled paper is to bring forth a unique case for reinforced earth wall construction on soft clayey silt deposit with settlements of large magnitude and share the experience of performance under actual seismic activity vis-à-vis other structures in the vicinity of the project, as constructed in difficult terrains and climatic conditions.

1 INTRODUCTION

Reinforced earth solutions were adopted in the design and construction of approach embankments to rail over bridges, being a part of the railway expansion project from Quazigund to Baramulla in the north bound state of Jammu & Kashmir, India.

The project involves the construction of eight rail over bridge approaches using reinforced soil walls (total about 50,000 square meters of face area) with ground treatment using high tenacity polyester Geogrids as transition course and to improve the safety factors against slip circle failure. Reinforced soil walls using discrete cruciform panels and high adherence steel strip reinforcement were adopted for the construction of approaches to rail over bridges. The static design of reinforced earth wall was done as per BS 8006, 1995 and the seismic design as per AFNOR NF P 94-220, July 1992. The backfill used was selected well-graded riverbed material, mined and screened specifically to meet the technical requirement of mechanical, physical and hydraulic properties, meeting the electrochemical criteria.

This paper narrates the investigation, design and analysis aspects of retaining walls, ground treatment, and the behavior of the structures under seismic load and post construction settlements. Special detailing adaptable to typical high altitude terrains, flash flooding incidences, high water table considerations and cold weather freeze and thaw cycles are discussed in the paper.

2 FOUNDATION SOIL INVESTIGATION AND GENERAL DESCRIPTION

Initially extensive soil investigation has been conducted by the main client for this project to evaluate the foundation soil behavior. Dynamic cone penetration tests (DCPT) and standard penetration tests (SPT) upto 15 m depth have been carried out in each km at an interval of 250 meters for whole 118 km stretch. The typical N-values are shown in Figure 1.

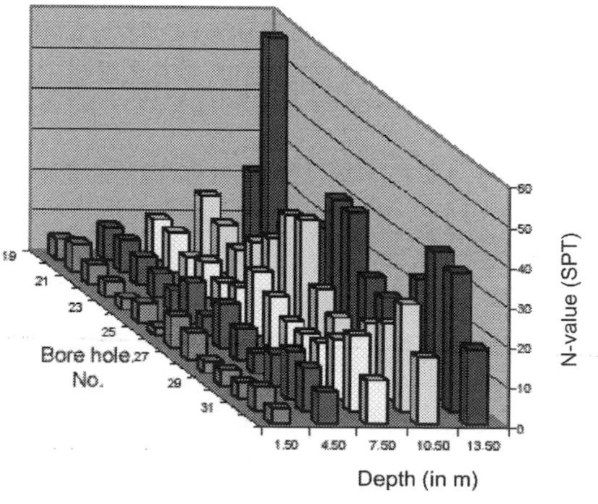

Figure 1. Typical SPT results from 22 km to 45 km.

The general description of foundation soil for most of the location is similar and consists of filled-up soil of 1–2 m and predominantly layers of fine grained clayey silt of low/medium plasticity upto 15–20 m depth. In some location sand has been encountered at 16 m depth. However, in general the top 15 m soil consists of clayey silt having 3–5% of sand, 80–95% of silt and 2–7% of clay. The top 6–10 m soil in all locations except for structure at 5B (for details refer the list of structure given in table no.1) consists of very loose soil having N-value (SPT) varying from 1 to 9.

3 FOUNDATION IMPROVEMENT

The maximum height of wall varied from 9.0 m to 12.0 m for the structures. The foundation soil did not have adequate bearing capacity to support high structures. The wall was therefore, proposed to be built in three stages. The gain in shear strength of subsoil due to consolidation under each stage of loading would allow construction of the next stage without any foundation failure. In order to accelerate the consolidation under each stage of loading the subsoil was initially proposed for treatment with prefabricated vertical drains (PVDs). However, after detailed soil investigation and recommendation from the design consultant the provision of PVD has been removed since the foundation soil is of low plasticity in nature and is self-draining. The reinforced soil structure walls are catered for total settlement of 400–650 mm for maximum wall height of 10 m depending on the location.

Layers of high strength polyester geogrids have been adopted to improve the stability of the structure against slip circle failure. Depending upon height, loading and foundation characteristics of sub-soil, and one or two layers of high tenacity polyester geogrids have been provided half meter below the reinforced earth wall panel. The analysis for slip circle failure is done using Talren 97 software for three stage of construction with FOS of 1.2 in seismic and typical out-put is shown in Figures 2, 3 and 4.

3.1 *Typical proposed foundation improvement programme*

For reinforced soil walls upto a height of 4m, no ground treatment was proposed. Walls exceeding 4m height, the ground was proposed to be treated with one or two layers of high strength PET geogrid, which, were extended 3 m on both sides beyond the structure width depending on detailed analysis.

Typical ground treatment proposed for 10 m high walls for various structures depending on type of foundation soil is as follows:

Two or three-stage construction was proposed for wall exceeding height of 4 m. The details of stage construction were as follows:

1. Stage I: Reinforced earth wall upto 4m high, were built first. Thereafter, 30–40 days pause period was provided for allowing consolidation of the subsoil.

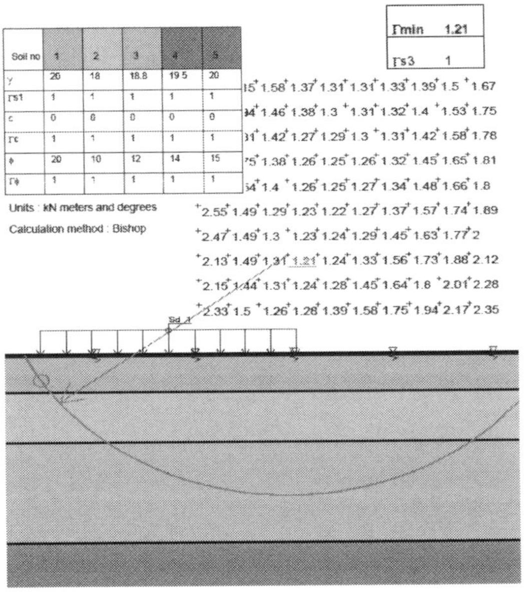

Figure 2. Slip circle analysis for Stage-I construction.

Table 1. Shear strength and consolidation parameters.

Site locations	Cohesion* 'c' (kg/cm^2)	Angle of internal fiction (ϕ^0)*	Coefficient of consolidation (c_c)	Initial void ration (e_0)
Bridge no.-127	0.03 to 0.20	10 to 20	0.18 to 0.27	0.70 to 0.90
Bridge no.-131	0.05 to 0.15	7 to 15	0.19 to 0.21	0.77 to 0.86
Bridge no.-139	0.10 to 0.20	10 to 25	0.18 to 0.22	0.64 to 0.82
Bridge no.-161	0.04 to 0.25	8 to 25	0.19 to 0.23	0.77 to 0.82
Bridge no.-165	0 to 0.15	7 to 20	0.18 to 0.21	0.79 to 0.87

* Test conducted both by direct shear and triaxial.

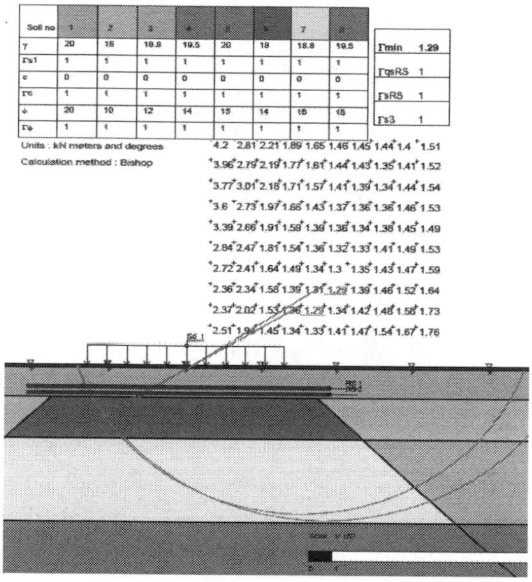

Figure 3. Slip circle analysis for Stage-II construction.

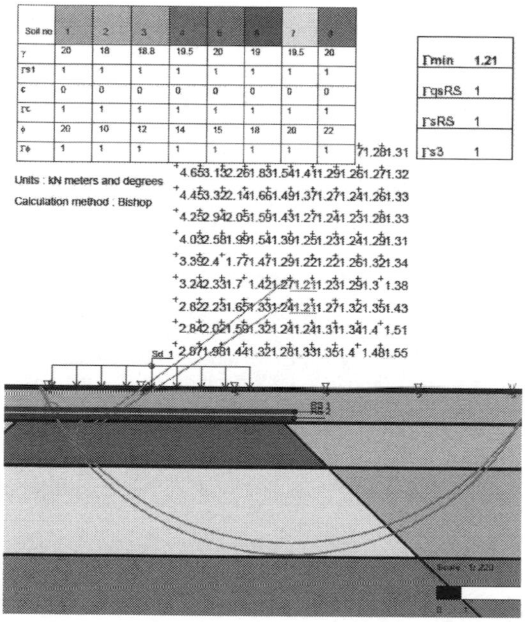

Figure 4. Slip circle analysis for Stage-III construction.

2. Stage II: In next stage another 2 m-compacted soil was filled up, increasing the wall height to 7 m. Thereafter, another 30–40 days pause period was provided for allowing for consolidation of the subsoil.
3. Stage III: Wall upto 10 m high was built after the pause period, including granular sub-base (GSB), wherever applicable. Thereafter, another pause period of 40 days was provided.
4. Then the pavement structure was built.

Table 2. Settlement recorded in trial embankment.

Site 1 (near bridge no. 127)		Site 2 (near bridge no. 139)	
Height of filling (in m)	Settlement (in mm)	Height of filling (in m)	Settlement (in mm)
1	13	2	16
5	28	5	32
9	155	8	160
Site 3 (near bridge no. 165)		**Site 4 (near bridge no. 161)**	
Height of filling (in m)	Settlement (in mm)	Height of filling (in m)	Settlement (in mm)
2	19	1	14
5	38	3	37
7	239	8	305

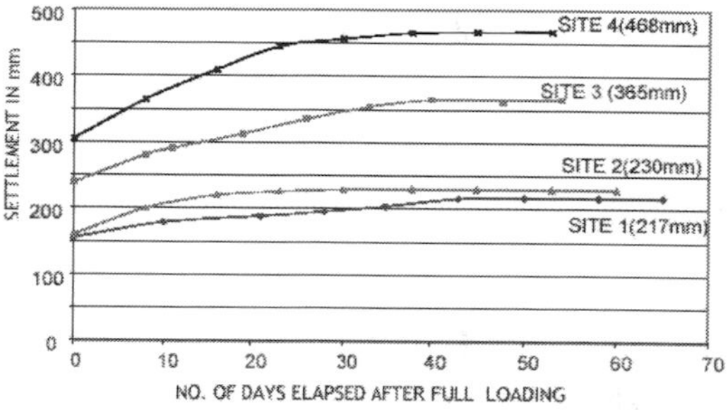

Figure 5. Recorded settlement at trial embankment constructed using sand bag.

4 FIELD TRIALS FOR SETTLEMENT

Test embankments were constructed at four places to determine the magnitude and rate of settlement in the embankment. The observed settlements are shown in Table 2 and in Figure 5. It is very clear from Figure 5 that the settlement is not increasing with time and the curve has become asymptotic after about 45 days after full loading. Some features observed during actual testing are:-

a) 65 to 70 % of total settlement took place during construction stage of embankment.
b) Post construction settlement became stable beyond 40–45 days.
c) No heaving was observed on either side of the test embankment.

Table 3. Actual panel settlement recorded till Jan, 2007.

Structure	Max. panel height of wall (in m)	Maximum recorded panel settlement (in mm)
Bridge no.-127	9.920	181
Bridge no.-131	10.670	555
Bridge no.-139	10.105	Not Available
Bridge no.-5B	9.170	10
Bridge no.-25	10.855	200
Bridge no.-161	11.605	400
Bridge no.-165	10.855	350
Bridge no.-178	8.980	75

d) Wet patches are observed near embankment at two sites where soil below is more slushy which may be due to dissipation/consolidation of base soil.

5 RECORDED PANEL SETTLEMENT

The panel settlement was recorded by measuring the panel top levels at each stage of construction. The total panel settlement recorded till date is as follows:

The settlement in test bank (Figure 5) is higher than the measured settlement in actual construction inspite of the fact that the earth is well compacted in actual banks whereas the test bank has been constructed by heaping sand bags without any compaction. This is due to the fact that the panel with steel strip reinforcement settles less than the fill and also use of basal geogrid reinforcement in the foundation has distributed the load over larger area and hence less settlement.

6 EARTHQUAKE EXPERIENCED DURING CONSTRUCTION

The Kashmir earthquake (also known as the South Asia earthquake or the Great Pakistan earthquake) of 2005, was a major earhquake whose epicenter was the Pakistan administered Kashmir. The earthquake occurred at 08:50:38 hr. Pakistan standard time (03:50:38 UTC) on 8*th* Oct. 2005. It registered 7.6 on the richer scale making it a major earthquake similar in intensity to the 1935 Quetta earthquake, the 2001 Gujarat earthquake, and the 1906 San Francisco earthquake. As of 8 November, the Pakistani government's official death toll was 73,276, while officials say nearly 1,400 people died in the Indian-administered Kashmir which is very close to construction site (Figure 6) and fourteen people in Afghanistan. The equivalent magnitude of tremor on Richter scale at site was 5.4.

All the reinforced earth structures on Baramula – Quazigund section has experienced the impact of this earthquake without any damage in the wall. The vertical alignment, individual panel joints, vertical and horizontal gap between the panels were found to be intact. No bulging, differential movement between the panels, sapling or any damage in the panels was observed after the earthquake (Photo 1). The constructed height of the wall was 6 m during this period. However, many residential structures in the vicinity were collapsed or damaged due to earthquake (see photos 2 & 3).

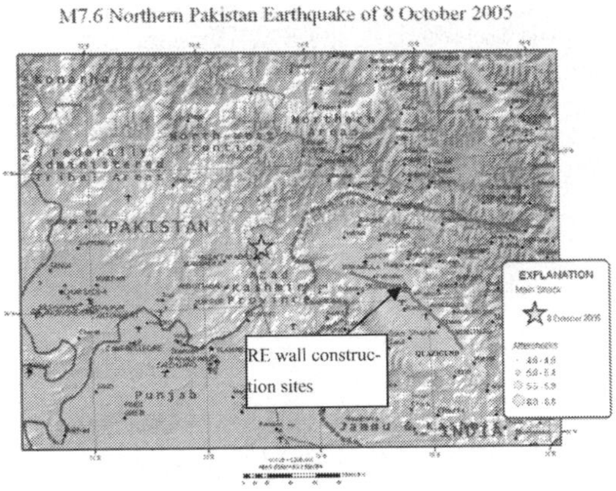

Figure 6. Map showing the epicenter of earthquake and construction site.

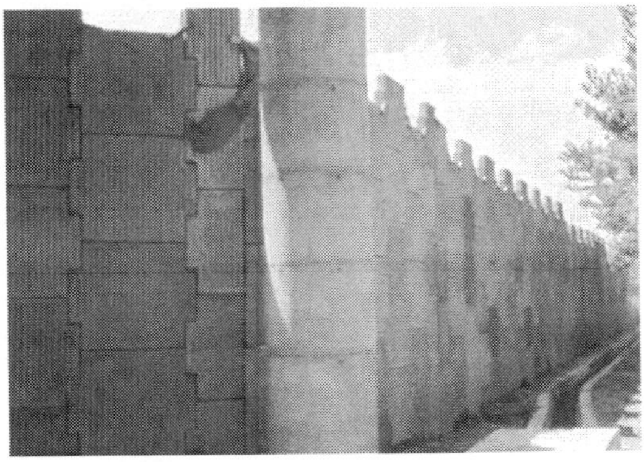

Photo 1. Reinforced earth wall for bridge no. 127.

7 SPECIAL DETAILING

The special detailing adopted for flooding incidences, high water table considerations and cold weather freeze and thaw cycles are discussed below.

7.1 *Typical detailing adopted for flooding*

One structures was adjacent to the existing canal running parallel to the approach retaining wall. The toe of the structure was 2 m away from the canal (see Figure

Photo 2. Damaged building at Uri about 30 km from the construction site.

Photo 3. Damaged building near the construction site.

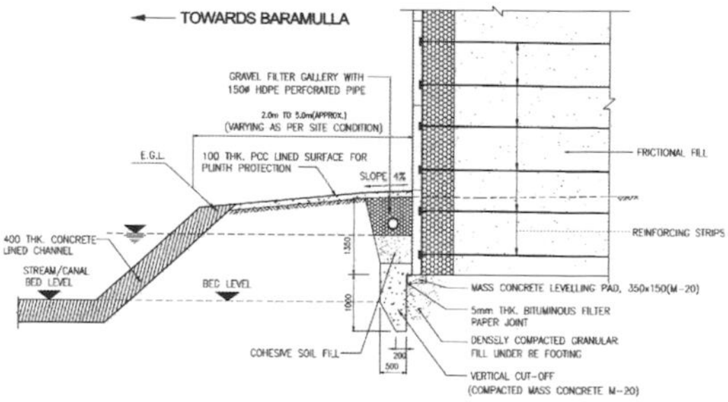

Figure 7. Typical toe detailing adopted for reinforced earth wall near canal.

7 and photo 4) and the top of the pad was above the base level of canal. To avoid seepage and any chance of toe erosion for long-term performance, a concrete cut-off wall was provided.

7.2 *Typical detailing adopted for snow fall area*

This part of India is subjected to extreme cold weather conditions and hence significant use of deicing salts

Photo 4. Reinforced soil wall running parallel to the canal.

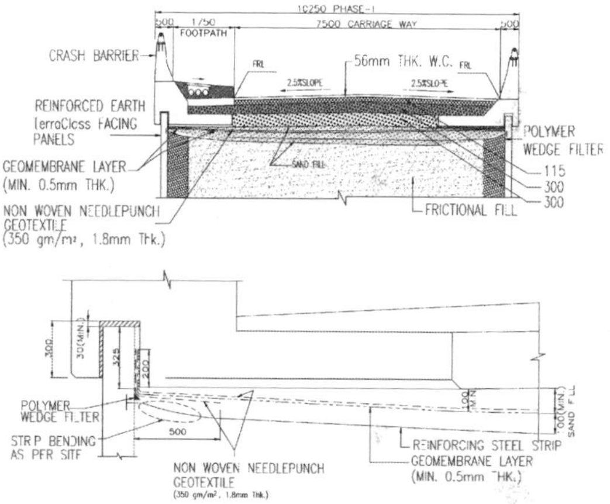

Figure 8. Geomembrane detailing at top to prevent salt percolation and corrosion steel strip.

is anticipated. To prevent the steel reinforcement from corrosion, an impervious barrier (one layer of geomembrane) beneath the pavement structure and just above the reinforced fill zone was provided. Sand drainage layer was adopted above the membrane layer to drain-out the seepage water from sub-surface layers (see Figure 8).

8 CONCLUSIONS

This paper presents a unique case of construction of reinforced soil structures on a foundation susceptible to high settlement and has experienced earthquake without causing any damage to the structures. Although the structure was not complete and not at its most critical state to make any conclusion on seismic behavior, but it was noted that there was no serviceability damage observed in any structures. Performance of reinforced soil retaining structures

Photo 5. Photograph showing installation of geomembrane.

Photo 6. Photograph of completed structure of Bridge No. 165.

constructed over soft foundation soil is also found to be satisfactory under seismic loading.

Reinforced soil wall being a flexible structure can be constructed over very soft soil where the expected settlement is very large and in high seismic prone area. However, special arrangement shall be provided to cater large differential settlement like provision of slip joints.

Reinforced soil structures can be built over loose soil having N-value (SPT) less than 10 without any major deep foundation treatment, subjected the soil is less plastic (Plasticity index less than 15) and clay content less than 10%. However, in such cases special attention shall be given against possibility of any slip circle failure from construction factors. High strength geogrid/geotextile shall be used in foundation as a stress transition layer to prevent any slip circle failure.

REFERENCES

AFNOR NF P 94–220, July 1992. " Soil Reinforcement, Backfilling structures with inextensible and flexible reinforcing strips or sheets." French Standard.

BS 8006: 1995. "Code of Practice for Strengthened/reinforced soils and other fills." British Standard.

Singh Kanwarjit, Bhanu Prakash & Chopra Rakesh. 2003. Construction of Railway abutment and bridges in the Kashmir valley for Qazigund – Srinagar – Baramulla project, IABSE symposium.

New Horizons in Earth Reinforcement – Otani, Miyata & Mukunoki (eds)
© 2008 Taylor & Francis Group, London, ISBN 978-0-415-45775-0

A new type integral bridge comprising of geosynthetic-reinforced soil walls

F. Tatsuoka, D. Hirakawa, M. Nojiri & H. Aizawa
Tokyo University of Science, Japan

M. Tateyama & K. Watanabe
Railway Technical Research Institute, Japan

ABSTRACT: A new type bridge combining an integral bridge and a pair of geosynthetic-reinforced soil (GRS) retaining walls having full-height rigid (FHR) facings, called the GRS integral bridge, is proposed. The geosynthetic reinforcement layers are connected to the facings (i.e., RC parapets) that are integrated with a girder. The GRS integral bridge is much more cost-effective in construction and long-term maintenance while having a higher seismic stability than conventional-type bridges having a girder via movable and fixed supports on either gravity-type abutments or GRS retaining walls and also than the conventional integral bridges. The GRS integral bridge alleviates several problems with the other types, mostly resulting from that the backfill is not reinforced and girder-supports are used, while taking advantage of their superior features.

1 INTRODUCTION

Despite its wide use until today over the world, the conventional gravity-type bridge abutment (Fig. 1) has a number of drawbacks, due mostly to that the backfill is not reinforced and therefore the abutment is a cantilever structure while a girder is placed on movable and fixed supports, as listed below:

1) As the abutment is a cantilever structure, piles are usually necessary to resist against stresses concentrated near the toe of the abutment base.
2) The RC abutment is not allowed to displace once constructed. After constructed, however, it is subjected to earth pressure and effects of the settlement and lateral flow in the subsoil via its effects on the piles associated with the backfill construction. To prevent such displacements of the abutment, it may become necessary to increase the number and size of piles.
3) The construction and long-term maintenance of girder-supports are generally costly.
4) The seismic stability of the unreinforced backfill as well as the abutment supporting the girder via a fixed-support is relatively low, as observed in many previous major earthquakes. Watanabe et al. (2002) and Aizawa et al. (2006) confirmed this point by model shaking table tests.
5) A bump is formed behind the abutment by residual deformation of the backfill due to its self-weight and traffic and seismic loads.

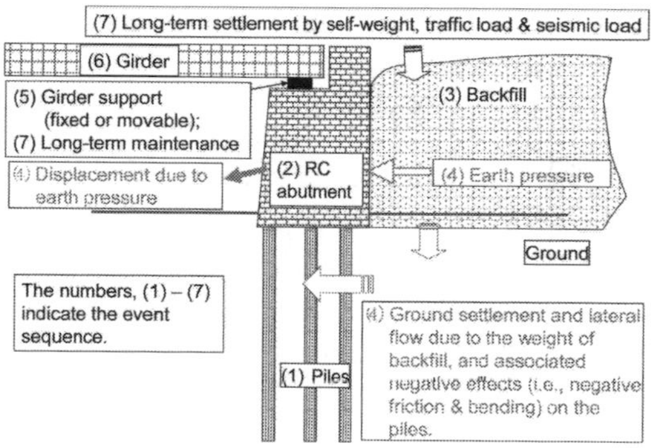

Figure 1. Conventional type bridge abutment (gravity type).

To develop bridge systems that are more cost-effective than the conventional bridge type while alleviating these problems described above, several solutions have been proposed as described in the next section.

2 PREVIOUS ATTEMPTS

2.1 *Improving the backfill*

One of the oldest attempts employed by the Japanese railways engineers is to construct a trapezoidal zone of well-compacted well-graded gravelly soil immediately behind the abutment (type *a*1 in Fig. 2). However,

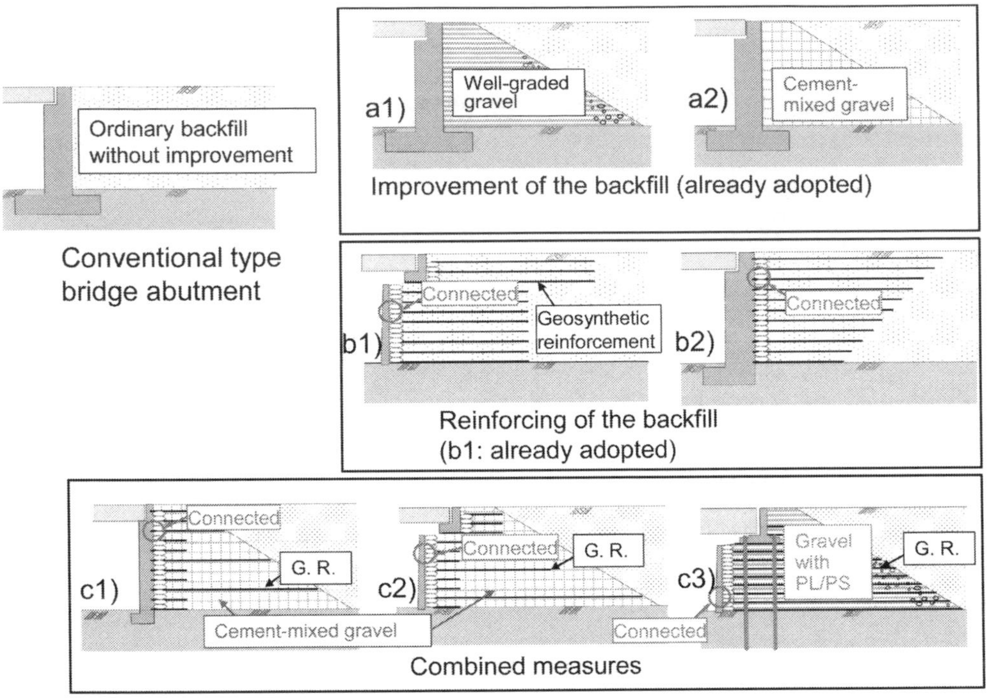

Figure 2. Conventional type versus proposed solutions of bridge (Tatsuoka, 2004; Tatsuoka et al., 2005).

the performance of this type of bridge during several previous earthquakes in Japan was not satisfactory (e.g., Tateyama et al., 2002). Watanabe et al. (2002) and Tatsuoka et al. (2005) confirmed the above by performing model shaking table tests.

They also showed that the seismic stability of another similar type constructing a trapezoidal zone of cement-mixed gravel (type *a*2, Fig. 2) is not sufficiently high either.

2.2 *Reinforcing the backfill*

Fig. 3 illustrates the staged construction of geosynthetic-reinforced soil (GRS) retaining wall (RW) with full-height rigid (FHR) facing. This is now one of the standardized RW construction technologies for railways in Japan while becoming popular also in other fields (such as highways). The main features of this technology are as follows:

1. The backfill is constructed with a help of gravel gabions placed at the shoulder of each soil layer.
2. Geosynthetic reinforcement layers are arranged with a vertical spacing of 30 cm. This small lift can facilitate a high compaction of the backfill.
3. After a geosynthetic RW is completed while sufficient compression of the backfill and settlement of the supporting ground has taken place, a FHR facing is cast-in-place directly on the wrapped-around wall face ensuring a strong connection to the reinforced backfill so that:

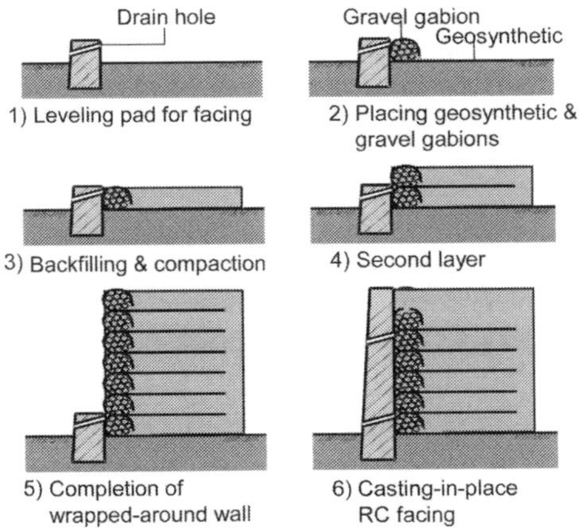

Figure 3. GRS RW with FHR facing (Tatsuoka et al., 1997).

a) negative interaction between the FHR facing and the compression of the backfill during

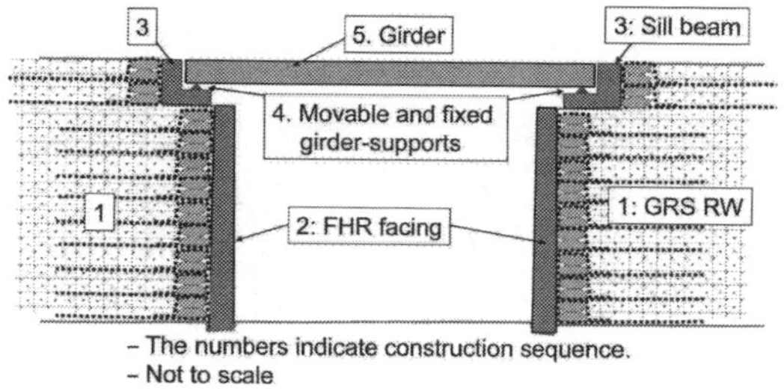

Figure 4. GRS RW bridge (type *b*1 in Fig. 2).

filling-up and compaction works can be avoided;

b) large compression of the supporting ground by the backfill construction can be accommodated ensuring the stability of wall;
c) the backfill immediately back of the wall face can be compacted dense with better mobilization of reinforcement tensile force; and
d) the alignment of completed wall face is easy.

Taking advantages of these features, a number of bridges comprising of a pair of GRS RWs with FHR facings that support a girder (type *b*1 in Fig. 2; Tatsuoka et al., 1997, 2005), which is herein called the GRS-RW bridge (Fig. 4), were constructed. Although this bridge type is structurally simpler and more cost-effective than the conventional type, it has the following limitations:

1. The girder cannot be very long due to low stiffness and potential large residual deformation of the backfill supporting the girder.
2. The construction and long-term maintenance of movable and fixed girder-supports is costly. This is the common problem with all of the bridge types presented in Fig. 2.
3. Despite that the dynamic stability of GRS RW with FHR facing is very high (e.g., Tatsuoka et al., 1998; Koseki et al., 2003), the dynamic stability of the sill beam on which a fixed girder-support is placed is not so (Aizawa et al., 2006; Hirakawa et al., 2007a). This is because the mass of the sill beam is much smaller than the inertia force of the girder while the anchorage capacity of the reinforcement layers connected to its back is small due to their shallow depths.

Type *b*2 (Fig. 2), placing a girder on the crest of the FHR facing, is dynamically more stable than type *b*1 (Watanabe et al., 2002; Tatsuoka et al., 2005). However, they also showed that the reinforced backfill behind the facing supporting the girder via a fixed-support would exhibit too large deformation when subjected to L2 design seismic load.

2.3 *Combining multiple-measures*

To substantially decrease long-term residual deformation of the backfill, it is very effective to vertically preload the reinforced backfill and then maintain some vertical prestress, which is typically about a half of the prestress, in the backfill during long-term service (i.e., the PL & PS technology). The above was validated by laboratory model tests (Shinoda et al., 2003a & b) and long-term performance of a prototype railway bridge pier (Uchimura et al., 2003). Moreover, Uchimura et al. (2003) and Tatsuoka et al. (2005) showed that the seismic stability of PL-PS reinforced bridge pier and abutment is very high. It is in particular the case if high prestress is maintained during dynamic loading and this can be ensured by using a ratchet mechanism as shown by model shaking table tests (Shinoda et al., 2003a & b). Type *c*3 in Fig. 2 consists of a PL-PS GRS RW with a ratchet system supporting a girder via a fixed-support. Its high seismic stability was validated by laboratory shaking table tests (i.e., Nakarai et al., 2002). Despite the above, any prototype bridge of this type has not been constructed, because possible long-term maintenance works of the ratchet system were not preferred by practicing engineers.

Types *c*1 and *c*2 were then proposed, which are combining types *b*2 and *b*1 with type *a*2. Type *c*1 was adopted by railway engineers and the first prototype was constructed for a new bullet train line in Kyushu (Tatsuoka, 2004; Tatsuoka et al., 2005). The conventional RC abutment (Fig. 1) supports the backfill with the earth pressure activated on its back. In comparison, with type *c*1 as well as type *b*2, the reinforced backfill zone laterally supports the RC parapet (i.e., facing) that is supporting a girder while without dynamic earth pressure activated on its back. Type *c*1 abutments are constructed by the staged procedure presented in Fig. 3.

3 CONVENTIONAL INTEGRAL BRIDGE

3.1 *State-of-the-art*

This type is very popular in the UK and the USA due mainly to high cost/performance by low construction and maintenance cost resulting from no use of girder-supports (Fig. 5). However, the backfill may exhibit large residual settlement by self-weight as well as traffic and seismic loads, while the seismic stability of both girder-parapet system and backfill is relatively low, as shown by Aizawa et al. (2006), Nojiri et al. (2006) and Hirakawa et al. (2007a) and also below. Moreover, as the girder is integrated with the parapets, seasonal thermal expansion and contraction of the girder results into cyclic lateral displacements at the top of the facings, which results in a gradual increase

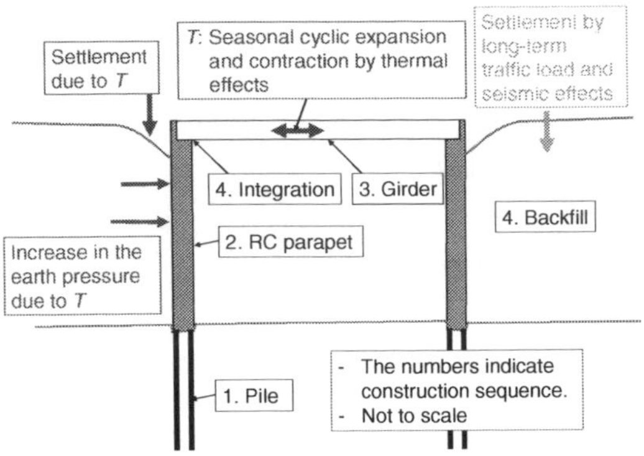

Figure 5. Integral bride and its inherent problems.

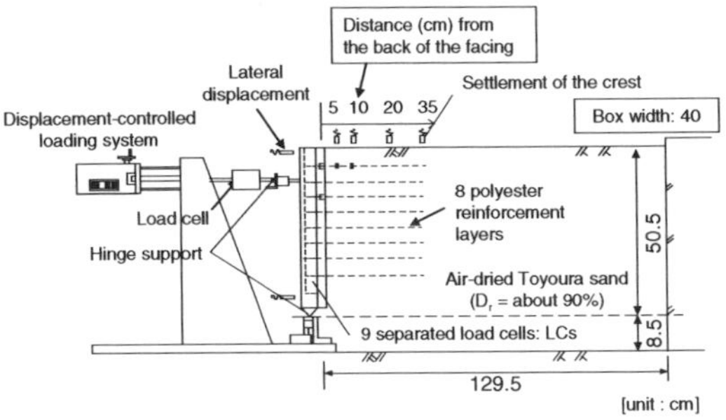

Figure 6. Static cyclic lateral loading test of the facing of RW model (Hirakawa et al., 2006, 2007b).

in the earth pressure and residual settlement in the backfill, as shown below.

3.2 *Effects of cyclic displacements of the facing*

Small-scaled model tests were performed in the laboratory (Fig. 6) to evaluate the detrimental effects of thermal cyclic displacements of the facing described above and also to examine whether this problem can be alleviated by reinforcing the backfill. The backfill was air-dried Toyoura sand produced by air-pluviation for the unreinforced backfill while by hand-tamping for the reinforced backfill. The reinforcement was a Polyester grid (strand diameter = 1 mm; spacing between the adjacent strands = 18 mm; covering ratio = 9.5%; and rupture tensile strength at an axial strain rate of 1.0%/min. = 19.6 kN/m). The FHR facing was cyclically displaced about the bottom hinge at a rotational displacement rate of 0.00053 degree/min.

Figure 7 summarizes the peak earth-pressure coefficients in the respective cycles, $K_{\text{peak}} = 2Q_{\text{peak}}/H^2\gamma$, where Q_{peak} is the peak total earth pressure per width in each cycle; H is the wall height (50.5 cm); and γ is the dry unit weight of the backfill (1.60 gf/cm^3), plotted against the ratio of the double amplitude of cyclic

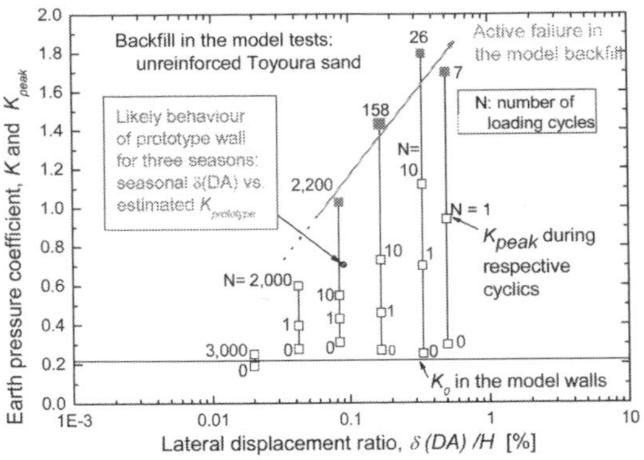

Figure 7. Peak earth pressure coefficients in the model tests and a field full-scale case (Hirakawa et al., 2007b).

displacement at the facing top to the facing height, $\delta(DA)/H$, at selected numbers of loading cycle, N. The facing top was allowed to move about 0.2 mm ($\delta/H = 0.04$ %) at the maximum toward the active direction associated with an increase in the earth pressure. The solid squares represent the cycles when the active failure plane appeared in the backfill. The earth pressure increases with an increase in $\delta(DA)/H$ and N. These test results are consistent with previous laboratory model tests (Ng et al., 1998; England et al., 2000) as well as the full-scale field behaviour for three seasons (i.e., $N = 3$). This earth pressure increase may result in structural damage to the facing and may push out the bottom of the facing. By reinforcing the backfill, this earth pressure increase does not reduce, but the facing is not structurally damaged and not pushed out at the bottom, as the FHR facing becomes a continuous beam supported by a number of reinforcement layers at a small spacing.

The other detrimental effect of cyclic displacement of the facing with unreinforced backfill is gradual but eventually large settlements in the backfill associated with the development of an active failure plane in the backfill (case NR in Fig. 8). The backfill settlement increases with an increase in the cyclic facing displacement, $\delta(DA)/H$. On the other hand, the backfill settlement becomes nearly null when the backfill is reinforced with reinforcement layers that are connected to the back of the facing (case R & C). Even slight heaving at the backfill crest takes place by dilatation of the backfill due to repeated passive movement of the facing. The benefits of reinforcing the backfill with reinforcement layers connected to the facing are as follow. Firstly, for the same thermal thrust from the girder, the displacements of the facing become smaller due to higher stiffness of the reinforced backfill. Secondly, for the same cyclic facing displacement, the residual settlement in the backfill decreases due to

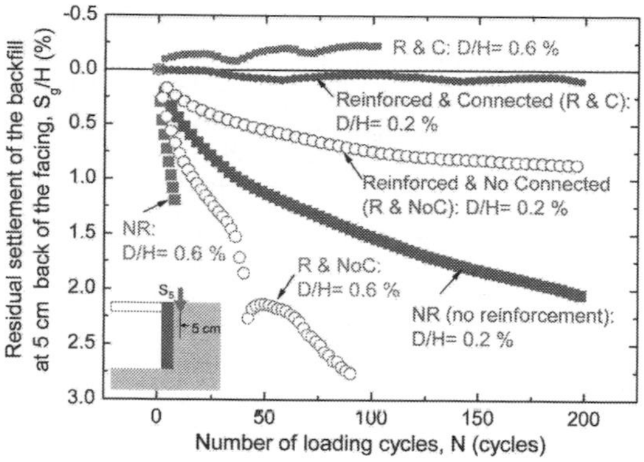

Figure 8. Residual settlement of backfill (when $\delta = 0$) by cyclic displacement of the facing and effects of reinforcing the backfill (modified from Hirakawa et al., 2006).

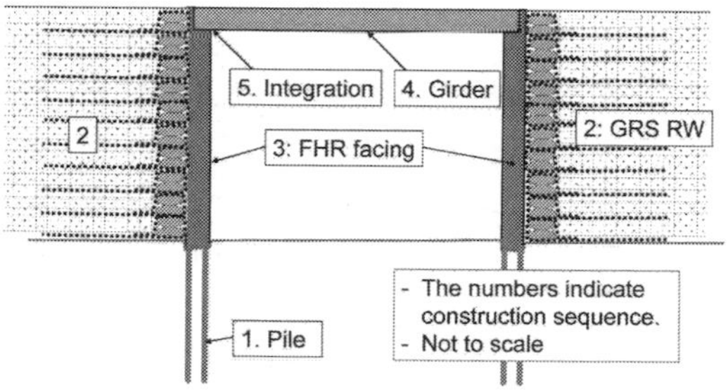

Figure 9. GRS integral bridge.

higher confining pressure in the backfill and membrane effects of reinforcement layers connected to the facing. It may also be seen from Fig. 8 that these positive effects of reinforcing the backfill become very small when the reinforcement layers are not connected to the facing (case R & NoC). This is because the deformation of the active zone cannot be effectively restrained by the reinforcement layers.

4 GRS INTEGRAL BRIDGE

4.1 *Features of IGS integral bridge*

A new bridge type (called the GRS integral bridge; Fig. 9), which is more cost-effective and more dynamically stable than the others described in this paper, is proposed. This type combines the GRS RW bridge (type b1, Fig. 2) and the integral bridge (Fig. 6), taking their advantages: i.e., stabilization of the backfill and non-soil structures by reinforcing the backfill (the GRS RW bridge) and simple and cost-effective non-soil structure without using girder-supports (the integral bridge), while alleviating their inherent problems. A GRS integral bridge may need a pile foundation to

Bridge type	Cost & period of construction	Maintenance cost	Seismic stability	Total
 Conventional (gravity)	1 A, B	1 C, D	1 252 gal*	3
 GRS RW	3	2 C	2 589 gal*	7
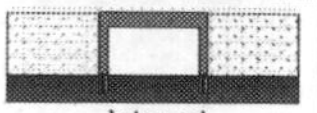 Integral	3	1 D, E	2 641 gal*	6
 GRS Integral	3	3	3 1,048 gal*	9

(* Acceleration at failure in model shaking table test)

A = heavy abutment structure as a cantilever structure
B = piles are usually necessary
C = high cost for construction and long-term maintenance of girder-supports
D = bump due to settlement of backfill by self-weight, traffic load and seismic load
E = settlement of the backfill and structural damage to the facing by cyclic lateral displacements of facing due to seasonal thermal expansion and contraction of the girder

Figure 10. Features of four different bridge types.

support the girder, but a lighter one than the integral bridge may be sufficient, as needs for a pile foundation are usually much lower with GRS RWs. As seen from Fig. 8, the residual settlement of the backfill reinforced with reinforcement layers connected to the facing is very small. Moreover, a high seismic stability with small deformation and displacements can be expected because of integrated performance of the whole bridge system, as shown below.

Figure 10 compares the advantages and disadvantages in the three items listed in the top line of the four bridge types: i.e., conventional gravity type, GRS RW, integral and GRS integral. At the accelerations shown in the second column from the right, the respective bridge models collapsed in the shaking table tests, as described below. Letters *A* through *E* denote negative factors with the respective bridge types discussed in the above. Full points equal to three are assigned to each item, which are reduced one by one when these negative factors are relevant. So, the total full points are equal to nine, which are assigned only to the GRS integral bridge.

4.2 *Model tests*

Shaking table tests of the four bridge types listed in Fig. 10 were performed to validate a high-seismic stability of the GRS integral bridge (Aizawa et al., 2006; Hirakawa et al., 2007a). Fig. 11 shows the IGS integral bridge model. Assuming a length similitude ratio equal

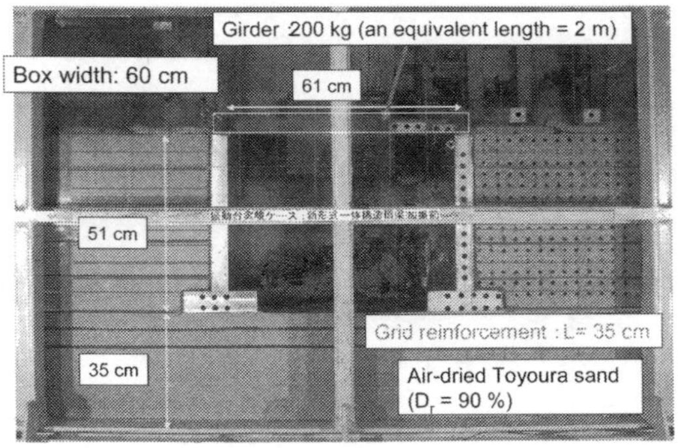

Figure 11. GRS integral bridge model (Aizawa et al., 2006).

to 1/10, the facings were 51 cm-high and the girder was 61 cm-long. By adding a mass of 200 kg at the center of the girder, the equivalent length became 2 m (i.e., 20 m in the assumed prototype). 20 sinusoidal waves with a frequency of 5 Hz was applied at the shaking table step by step while increasing the maximum acceleration $\alpha_{\max}$ with an increment of 100 gal per step.

Figure 12a shows the backfill settlements at 5 cm back of the sill beam supporting the girder via a fixed support with the GRS RW type (Fig. 4) and back of the facing with the other three types. Fig. 12b shows the lateral displacements at the top and bottom of the facing. In Fig. 12b, d_t is the displacement of the sill beam with the GRS RW type. In Figs. 12a and b, with the gravity and GRS RW types, the displacements of the abutment or facing on the side supporting the girder via a fixed-support are presented. It can be readily seen that the GRS integral bridge, together with the backfill, is much more stable than the other types, while the conventional gravity type and its backfill is least stable. It may also be seen that the pushing out of the facing bottom is the major failure mode with the integral and GRS integral bridges.

5 CONCLUSIONS

A new type bridge structure, the GRS integral bridge, is proposed, which comprises of geosynthetic-reinforced backfill and an integral bridge. Its advantageous features, which are due mostly to no use of girder supports and reinforcing of the backfill, are: 1) a high cost-effectiveness in construction and long-term maintenance; 2) essentially zero settlement in the backfill and no structural damage to the facing by an increase in the earth pressure caused by thermal cyclic expansion and contraction of the girder; and 3) a very high seismic stability of both backfill and non-soil structural component (i.e., a pair of parapets and a girder) due to integrated dynamic performance of both components.

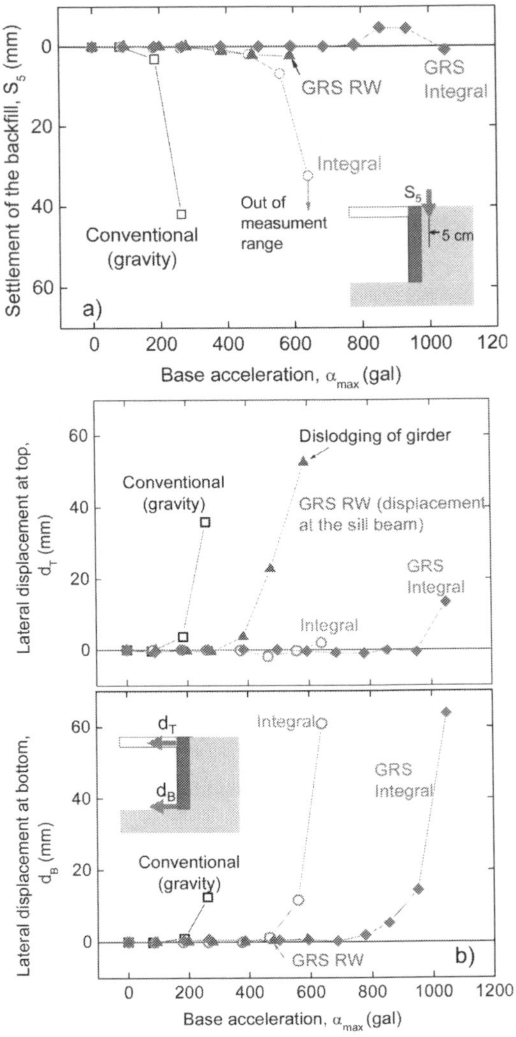

Figure 12. a) Backfill settlement; and b) outward lateral displacements of the facing in laboratory shaking table tests (Aizawa et al., 2006; Hirakawa et al., 2007a).

REFERENCES

Aizawa, H., Nojiri, M., Hirakawa, D., Nishikiori, H., Tatsuoka, F., Tateyama, M. and Watanabe, K. 2006, Comparison of cost-performance among conventional and new types of bridge system consisting of unreiforced or reinforced soil, *Geosynthetics Engineering Journal, IGS Japanese Chapter*, Vol. 21, pp. 175–182 (in Japanese).

England, G, L., Neil, C, M. and Bush, D, I. 2000, Integral Bridges, A fundamental approach to the time-temperature loading problem, *Thomas Telford*.

Hirakawa, D., Nojiri, M., Aizawa, H., Tatsuoka, F., Sumiyoshi, T. and Uchimura, T. 2006, Behaviour of geosynthetic-reinforced soil retaining wall subjected to forced cyclic horizontal displacement at wall face, *Proc.*

8th Int. Conf. on Geosynthetics, Yokohama, Vol. 3, pp. 1075–1078.

Hirakawa, D., Nojiri, M., Aizawa, H., Nishikiori, H., Tatsuoka, F., Tateyama, M. and Watanabe, K. 2007a, Effects of the tensile resistance of reinforcment embedded in the backfill on the seismic stability of GRS integral bridge, *Proc. of 5th International Symposium on Earth reinforcement (IS Kyushu* 2007).

Hirakawa, D., Nojiri, M., Aizawa, H., Tatsuoka, F., Sumiyoshi, T. and Uchimua, T. 2007b, Residual earth pressure on a retaining wall with sand backfill subjected to forced cyclic lateral displacements, *Soil Stress-Strain Behavior: Measurement, Modeling and Analysis, Geotech. Symposium in Roma* 2006 (Ling et al., eds.).

Koseki, J., Tatsuoka, F., Watanabe, K., Tateyama. M., Kojima, K. and Munaf, Y. 2003, Model tests of seismic stability of several types of retaining walls, *Reinforced soil engineering, Advances in Research and Practice*, Marcel Dekker, Inc. (Ling et al. eds.), pp. 317–358.

Nakarai, K., Uchimura, T., Tatsuoka, F., Shinoda, M., Watanabe, K. and Tateyama, M. 2002, Seismic stability of geosynthetic-reinforced soil bridge abutment, Proc. of 7th *Int. Conf. on Geosynthetics,* Nice, Vol. 1, pp. 249–252.

Ng, C., Springman, S. and Norrish, A. 1998, Soil-structure interaction of spread-base integral bridge abutments, *Soils and Foundations*, 38–1, pp. 145–162.

Nojiri, M., Aizawa, H., Hirakawa, D., Nishikiori, H., Tatsuoka, F., Tateyama, M. and Watanabe, K. 2006, Evaluation of dynamic failure modes of various bridge types by shaking table tests, *Geosynthetics Engineering Journal, IGS Japanese Chapter*, Vol. 21, pp. 159–166 (in Japanese).

Shinoda, M., Uchimura, T. and Tatsuoka, F. 2003a, Increasing the stiffness of mechanically reinforced backfill by preloading and prestressing, *Soils and Foundations,* 43–1, pp. 75–92

Shinoda, M., Uchimura, T and Tatsuoka, F. 2003b, Improving the dynamic performance of preloaded and prestressed mechanically reinforced backfill by using a ratchet connection, *Soils and Foundations,* 43–2, pp. 33–54.

Tateyama, M., Aoki, H., Yonezawa, T. 2002, Development of bridge abutment comprising of geosynthetic-reinforced cement-mixed backfill, Doboku Gijutsu, 57–2, pp. 54–61 (in Japanese).

Tatsuoka, F., Tateyama, M, Uchimura, T. and Koseki, J. 1997, Geosynthetic-Reinforced Soil Retaining Walls as Important Permanent Structures, 1996-1997 Mercer Lecture, *Geosynthetic International,* 4–2, pp. 81–136.

Tatsuoka, F., Koseki, J., Tateyama, M., Munaf, Y. and Horii, N. 1998, Seismic stability against high seismic loads of geosynthetic-reinforced soil retaining structures, Keynote Lecture, *Proc. 6th Int. Conf. on Geosynthetics,* Atlanta, Vol. 1, pp. 103–142.

Tatsuoka,F. 2004, Cement-mixed soil for Trans-Tokyo Bay Highway and railway bridge abutments, Geotechnical Engineering for Transportation Projects, *Proc. of Geo-Trans 04, GI, Los Angels, ASCE GSP No. 126 (Yegian & Kavazanjian eds.),* pp. 18–76.

Tatsuoka, F., Tateyama, M., Aoki, H. and Watanabe, K. 2005, Bridge abutment made of cement-mixed gravel backfill, *Ground Improvement, Case Histories, Elesevier Geo-Engineering Book Series, Vol. 3* (Indradratna & Chu eds.), pp. 829–873.

Uchimura, T., Tateyama, M., Koga, T. and Tatsuoka, F. 2003, Performance of a preloaded-prestressed geogrid-reinforced soil pier for a railway bridge, *Soils and Foundations*, 43–6, pp. 33–50.

Watanabe, K., Tateyama, M., Yonezawa, T., Aoki, H., Tatsuoka, F. and Koseki, J. 2002, Shaking table tests on a new type bridge abutment with geogrid-reinforced cement treated backfill, Proc. of *7th Int. Conf. on Geosynthetics*, Nice, Vol.1, pp. 119–122.

New Horizons in Earth Reinforcement – Otani, Miyata & Mukunoki (eds)
© 2008 Taylor & Francis Group, London, ISBN 978-0-415-45775-0

Effects of the tensile resistance of reinforcement in the backfill on the seismic stability of GRS integral bridge

D. Hirakawa, M. Nojiri, H. Aizawa, H. Nishikiori & F. Tatsuoka
Department of Civil Engineering, Tokyo University of Science, Japan

K. Watanabe & M. Tateyama
Research Engineers, Railway Technical Research Institute, Japan

ABSTRACT: A new bridge system comprising of a pair of geosynthetic-reinforced soil (GRS) retaining walls having full-height rigid facings unified with a girder is proposed. Shaking table tests were performed on five GRS integral bridge models to evaluate the effect of the tensile resistance of reinforcement layers on the seismic stability of the bridge. The dynamic stability of integral bridge increases by reinforcing the backfill and by increasing the tensile resistance of reinforcement layers, which increases with an increase in the number of reinforcement layer as well as an increase in the connection strength between the facing and the reinforcement layer for give pull-out strength and tensile rupture strength of reinforcement.

1 INTRODUCTION

A great number of conventional-type bridges with a girder supported by gravity or cantilever-type reinforced concrete (RC) abutments via fixed and moveable supports were seriously damaged or totally collapsed during previous major earthquakes, including the 1995 Hyogoken-Nambu (Kobe) earthquake and the 2003 Niigataken-Chuetu earthquake in Japan. The structural drawbacks of the conventional-type bridges include not only a relatively low seismic stability, but also a relatively high construction cost due to the use of girder-supports (and piles in many cases) but also long-term maintenance works for girder-supports. A bump between the backfill and the abutment that may be formed due to residual settlement of the backfill by long-term dead and live loads as well as seismic loads is another serious potential problem. Therefore, the development of a new cost-effective bridge structure having a high seismic stability against so-called Level 2 design seismic loads with a low cost for construction and maintenance has been required.

The integral bridge system (with unreinforced backfill) is now becoming popular in the UK and the USA due mainly to a low construction cost due to its simple structure. However, it has one inherent serious structural drawback even under static loading conditions. That is, by cyclic lateral displacements at the top of the abutment due to seasonal cyclic thermal expansion and contraction of the girder, the backfill gradually exhibits significant residual

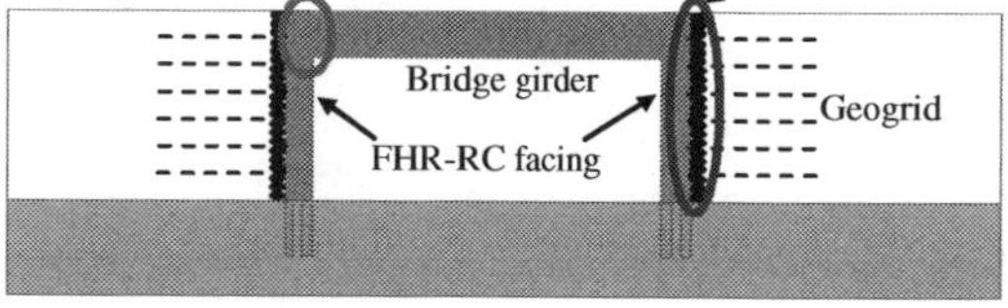

Figure 1. GRS integral bridge.

settlement and the earth pressure activated on the back of the abutment over years increases significantly (e.g., England, 2000). Hirakawa et al. (2006) showed that the above-mentioned detrimental effects are caused by "a ratcheting phenomenon" in the earth pressure and deformation behaviour of the backfill. They also showed that this problem can be effectively alleviated by reinforcing the backfill with geosynthetic layers with the ends connected to the back of the facing.

Based on the experiences described above, a new type bridge system consisting of a pair of geosynthetic-reinforced soil (GRS) retaining walls having full-height rigid (FHR) facings (i.e., GRS integral bridge; Fig. 1) was proposed (Tatsuoka et al., 2007; Aizawa et al., 2007). The proposed bridge system does not use girder supports but the girder is integrated with the FHR facings. Aizawa et al. (2007) performed a series of model shaking table tests and showed that the dynamic stability of the GRS integral bridge with a sufficiently high connection strength between the

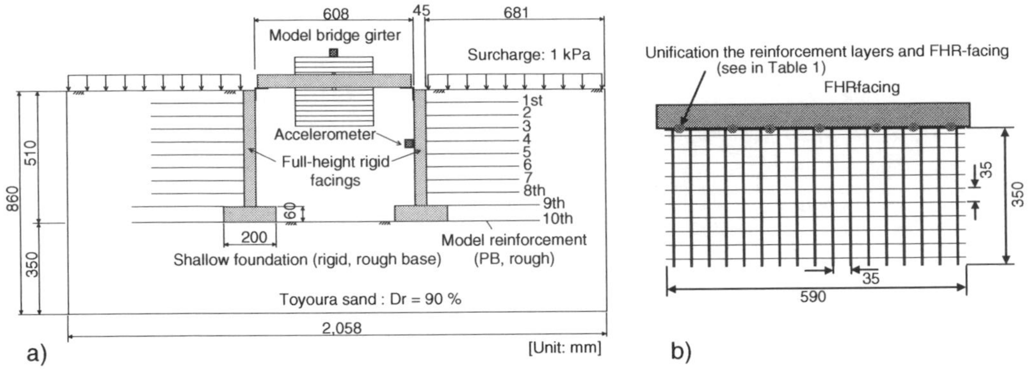

Figure 2. a) Cross-section of GRS-integral bridge model (Test 3 & 4), and b) plan of model reinforcement layer.

reinforcement layers and the facing was much more higher than conventional bridge types consisting of gravity type and GRS abutments as well as the integral bridge (with unreinforced backfill). They also showed that the center of the gravity of the non-soil structural part of the GRS integral bridge is located much higher than that of conventional-type bridge abutments, because:

1) the facings are integrated with a girder; and
2) the facings are much thinner and lighter than the conventional type abutment structures, as the facings act as a continuous beam with a number of supports (i.e., the geosynthetic reinforcement layers) with a small span between vertically adjacent supports (typically 30 cm).

They also showed that the collapse of the integral bride was associated with a large rotation of the facing with the bottom being pushed out and therefore large tensile force was activated in the reinforcement via the connection at the back of the facing in the lower part of the structure. That is, sufficiently high connection strength and pull-out strength of the reinforcement layers as well as high rupture strength of reinforcement are essential for high seismic stability of integral bridge.

In the present study, the effects of the tensile resistance of the reinforcement layers, in particular the number of reinforcement layer and the connection strength between the reinforcement and the facing, on the seismic stability of the proposed GRS integral bridge were investigated by performing a series of shaking table tests.

2 MODEL AND TESTING PROCEDURES

The GRS integral bridge model is described in Fig. 2. The model girder was unified with the model FHR facings that were placed on a subsoil layer while supporting the backfill. The subsoil and backfill were dense dry Toyoura sand at a relative density equal to 90% produced by pluviating air-dried sand particles through air via multiple sieves. Shallow foundations were arranged at the bottom of the FHR facings (Fig. 2a). No pile foundation was employed to observe the basic failure mode of integral bridge.

The model size was assumed to be 1/10 of a typical prototype bridge. A dead weight of 180 kg was attached to the center of the girder to make the girder length equivalent to 2 m (i.e., 20 m with the assumed prototype). The girder was integrated with the FHR facings via L-shape metal fixtures (3 mm-thick, 50 mm-wide and 200 mm-long). The fixtures were designed to start yielding when the facing rotates with the bottom being pushed out 6.5 mm, before the ultimate collapse of the model bridge system. It was actually the case in the model tests as shown by Aizawa et al. (2007) and later in this paper. It was considered that this collapse mode be likely with the prototype. The back face of the FHR facings and the bottom face of their foundations were made rough by gluing sandpaper #150 so that high shear stresses can be mobilized on these faces.

The reinforcement was a grid consisting of strands made of phosphor-bronze (PB) strips (0.3 mm-thick, 3 mm-wide and 350 mm-long) and ribs for transversal members made of PB (0.5 mm in diameter) (Figure 2b). Sand particles were glued on their surface. The reinforcement layers were arranged at a vertical spacing of 50 mm in the backfill.

Five model shaking tests were performed on integral bridge models (Fig. 3). With one model, the backfill was not reinforced, while, with four models, the backfill was reinforced changing the tensile resistance of reinforcement layers to evaluate its effects on the dynamic stability of GRS integral bridge. To this end, the number of reinforcement layers and the connection strength between the reinforcement and the facing were changed in these model tests, as listed in Table 1. In tests 3 and 4, ten reinforcement layers

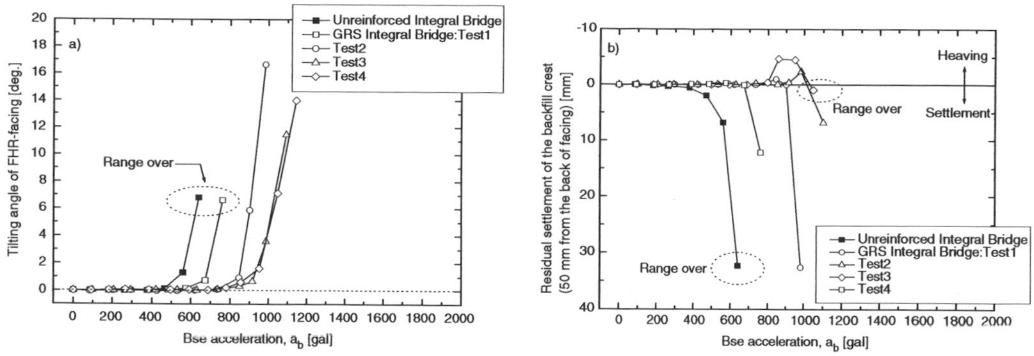

Figure 3. Relationships between base acceleration and a) residual tilting angle of the facing, and b) residual settlement of the backfill crest (50 mm from the back of facing).

Table 1. Summary of reinforcement properties.

Test name	Num. of reinforcement layers	Conection condition Conection strength [N/layer]	connection	Strand Rupture strength [N]	Num. of strand	Covering ratio, CR [%]	Friction angle at CR = 100% [deg.]
Test1	8	400	Melting, 4 points	207	17	10.1	35.0
Test2	9	520	Bolt(M3), 4 points	207	17	10.1	35.0
Test3	10	520	Bolt(M3), 4 points	207	17	10.1	35.0
Test4	10	1,070	Bolt(M3), 6 points	207	17	10.1	35.0

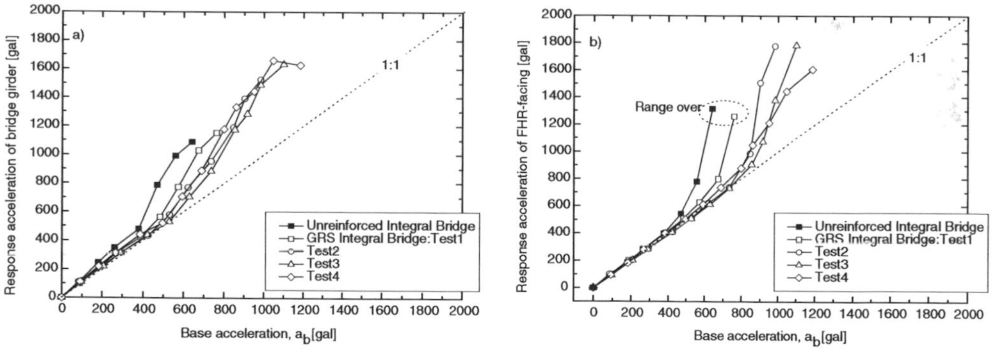

Figure 4. The relationships between base acceleration and a) response acceleration of the bridge girder, and b) response acceleration of the facing.

were arranged with 9th and 10th layers connected to the foundation of the respective facings (as shown in Fig. 2a). On the other hand, in tests 1 and 2, respectively eight and nine layers were arranged. As seen from Table 1, the connection strengths were different by a factor up to about 2.5 times among tests 1–4.

The input motion at the shaking table was 20 cycles of horizontal sinusoidal wave at a frequency of 5 Hz at each loading stage. The amplitude of acceleration at the table (α_b) was increased step by step from 100 gal with an increment of 100 gal.

3 TEST RESULTS AND DISCUSSIONS

Effects of backfill-reinforcing and connection strength: Fig. 3a presents the relationships between the residual tilting angle of the facing and the amplitude of base acceleration, a_b, while Fig. 3b shows the relationship between the residual settlement at the backfill crest at 50 mm back from the back face of the facing and a_b. Fig. 4 presents the relationships between the response acceleration of the bridge girder and FHR facing and a_b. The locations of the accelerometers are shown in

Fig. 2. The results from a shaking table test on the integral bridge with unreinforced backfill are also shown in Figs. 3 and 4. Fig. 5 compares the failure modes of these five models. The locations where the connection between the reinforcement and the facing failed are also indicated in Fig. 5. The following trends of behaviour may be seen from these figures:

1) In all the tests, the collapse of the integral bridge system was associated with a large rotation of the facing relatively to the girder about its top with the lower part being pushed out (Fig. 5).
2) The dynamic stability of the GRS integral bridge models (tests 1–4) was generally higher than the conventional integral bridge model with unreinforced backfill. The stability increasesd with an increase in the tensile resistance of reinforcement layers, which was achieved by increasing the number of reinforcement layer and the connection strength between the reinforcement and the facing (Figs. 3a & b). In particular, the tensile resistance of the reinforcement layers arranged at the lower level of the facing effectively resisted against the outward displacement of the facing, thereby increased the stability of the GRS integral bridge.
3) In all the tests, the residual deformation of the integral bridges (i.e., the tilting of the facing and the settlement of the backfill; Fig. 3) started increasing after the response ratio of acceleration at the girder and FHR facing to the input at the table (α_b) started increasing (Fig. 4). The increase in the response at the girder was caused by the passive yielding at the upper and lower levels in the backfill as well as the yielding of the fixtures between the girder and the facings.
4) The differences in the increasing rate of the residual deformation among the different tests (Fig. 3) were noticeably larger than those of the response ratio of acceleration (Fig. 4). This result indicates that the restraining effects of the reinforcing of the backfill and the increase in the tensile resistance of the reinforcement layers on the residual deformation of the integral bridge were more significant than those on the dynamic stiffness of the integral bridge.
5) The number of the connections between the reinforcement and the facing that failed decreased with an increase in the number of reinforcement layer or/and the connection strength among tests 1–4, which resulted in an increase in the stability of the bridge system. The failed connections were located more densely at lower places in accordance with the failure mode of the bridge systems (Fig. 5).
6) Despite that the connection strength in test 4 was significantly higher, by a factor of nearly two, than the one in test 3 for the same and largest number of reinforcement layer, the base acceleration,

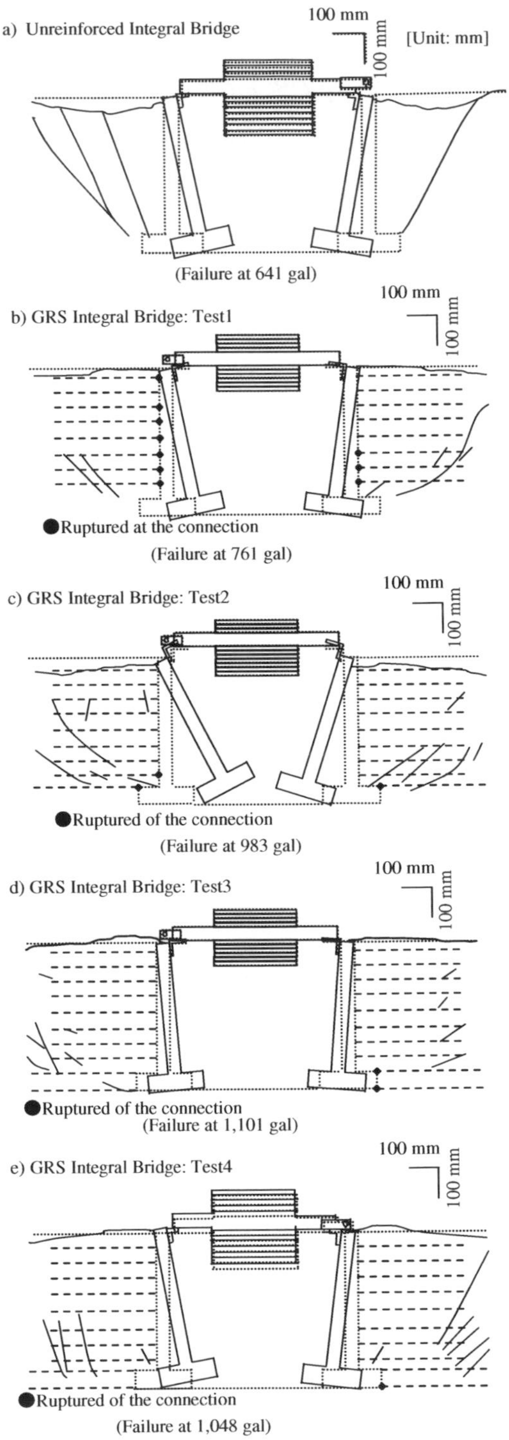

Figure 5. Failure modes of a) unreinforced integral bridge and b)-e) GRS integral bridges.

a_b, when the bridge collapsed was similar. Furthermore, the connection failure took place only at limited locations at the foundations of the facings. These results indicate that the collapse of the bridge system in tests 3 and 4 was associated with pull-out failure of the reinforcement layers at the lower level of the facing.

These results indicate that the seismic stability of GRS integral bridge is controlled in particular by the tensile resistance of the reinforcement layers at the lower level of the facing.

Failure mode: The following structural features control the dynamic failure mode of GRS integral bridge. The advantageous features are: 1) a girder and facings are integrated and the girder functions as a strut against the earth pressure acting on the facings; 2) the facing is laterally supported by a number of tensile reinforcement layers; and 3) the backfill is reinforced. An independent single bridge abutment of GRS retaining wall supporting a girder via a fixed support lacks the features 1) & 2), which results in a lower seismic stability with a less ductile failure compared with the GRS integral bridge (i.e., Koseki et al., 2006). On the other hand, the disadvantageous feature is that 4) the girder/facing system is relatively top-heavy while large lateral inertia force of the girder is activated to the top of the facing.

To develop a relevant seismic design procedure of GRS integral bridge taking into the above-mentioned features, the deformation and failure modes of the bridge observed in the five tests are analyzed more in detail below. Figs. 6a, 6b and 6c show the basic deformation modes as the input motion increased based on load and resistance components described in Fig. 7 that control the major failure mode of the bridge system.

Load components:

L1: Inertia force of the girder and the facings: In particular, large inertia force of the girder is activated on the top of relatively light facings, which results in large over-turning moment activated on the facings. The overturning moment increases with an increase in the weight of the girder (usually by an increase in the bridge length).

L2: Active earth pressure on the upper level of the facing on the right: This earth pressure is relatively small due to shallow concerned depths.

L3: Active earth pressure on the lower level of the facing on the left: This becomes more important when the forced rotation of the facing becomes larger after the passive yielding in the backfill has become large.

Resistance components:

R1: Passive earth pressure on the upper level of the facing on the left: This component cannot become large because of shallow concerned depths. The tensile reinforcement cannot help in increasing the passive earth pressure.

R2: Passive earth pressure on the lower level of the facing on the right: Full mobilization of this component needs relatively large pushing-in displacements of the facing into the backfill. Therefore, it is not relevant to expect full mobilization of this component in the design.

R3: Tensile force of the reinforcement layers at the upper level of the facing on the right: This cannot become very large because of low pull-out strength due to low confining pressure.

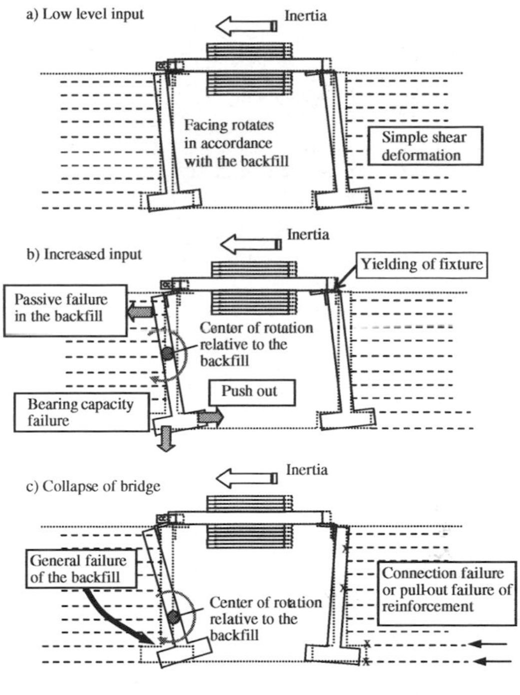

Figure 6. Failure mechanisms at different load levels.

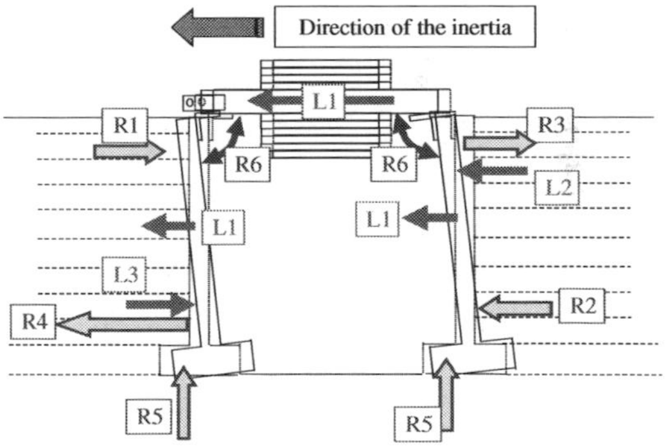

Figure 7. Load and resistance components when the inertia force of the structure is acting toward the left.

R4: Tensile force of the reinforcement layers at the lower level of the facing on the left: This is the most important resistance component, and analyzed in detail below.
R5: Bearing capacity of the subsoil at the bottom of foundations of the facings: This component becomes smaller at a fast rate as the eccentricity and inclination in the applied load becomes larger by an increase in the rotation of the facing.
R6: Bending strength of the fixtures between the girder and the facings: This component is activated by a small rotation of the facing and can effectively restrain the facing rotation when the input motion is relatively low. After the fixtures start yielding as the input motion increases, its importance becomes relatively small.

When the input load level is low and the dynamic behaviour of the bridge system is stable (Fig. 6a), the major deformation mode of the bridge system is in accordance with the basic deformation mode (i.e., simple shear) of the backfill. So the response ratio of acceleration at the girder is similarly very low (Fig. 5a) whether the backfill is reinforced and whether the tensile resistance of the reinforcement layers (i.e., R4) is different. At this stage, R6 is the most important resistance component while the other resistance components are not very active.

As the input load increases, the fixtures start yielding and the passive failure starts taking place in the upper level of the backfill on the left and in the lower level in the backfill on the right (Fig. 6b). Then, the response ratio of acceleration at the girder starts increasing associated with an increase in the tilting of the facing. As the passive failure is more significant in the upper part of the backfill on the left, the center of rotation of the facing on the left is then shifted downward, which increases the rotation moment acting on the facing for given load L1. Larger rotation of the facing relative to the backfill activates resistance components R1–R5. Then, if the connection strength is not sufficient, the connection failure may start taking place at all levels of the facing, as typically seen from Fig. 3b.

Fig. 6c illustrates the stage when the collapse becomes imminent. The push-out displacement at the lower part of the facing on the left becomes very large by a large rotation of the facing relative to the backfill. This is associated with a general rotational failure taking place in the backfill. Then, R4 becomes essential to prevent the ultimate collapse of the bridge, which may result into connection failure and/or pull-out failure of selected reinforcement layers at the lower part of the facing.

The tensile resistance of the reinforcement layers increases with an increase in: 1) the number of reinforcement layer, in particular at the lower level of the facing; and 2) the tensile resistance of the respective reinforcement layers. Factor 2) is equal to the minimum of: a) the strength at the connection between the reinforcement layer and the rigid facing; b) the pull-out strength of reinforcement; and c) the rupture strength of reinforcement (Tamura, 2006). The material properties (strength, stiffness, surface roughness ….), shape (strip or grid or sheet), length, arrangement in the backfill (vertical spacing ….) and so on of reinforcement layers should be determined by taking into account the above factors in the design of GRS integral bridge. In particular, if either the connection strength or the pull-out resistance is lower than the tensile rupture strength, the full capability of reinforcement cannot be activated in achieving high seismic performance of GRS integral bridge. Large connection force corresponds to large earth pressure activated on the back of the facing. This does not result into any serious structural damage of facing, because the facing is supported with the reinforcement layers at a small span (i.e., a small vertical spacing). A high connection strength also result in an increase in the tensile force that can be activated along the whole length of the reinforcement, which results in better reinforcing effects for the backfill (i.e., more stable behaviour of the backfill). To achieve high pull-out strength of the reinforcement, it is important to select an appropriate reinforcement type (e.g., a grid having stiff and strong enough not only longitudinal but also transversal members; a high friction angle between the surface of the reinforcement and the backfill material, a sufficient length and son on). Strip type reinforcement without any relevant anchorage system is not relevant.

4 SUMMARY

The following conclusions can be derived from the results from model shaking tests presented above:

1) The seismic stability of integral bridge increases substantially by reinforcing the backfill and connecting the reinforcement to the back of full-height rigid facings that are integrated with a girder. This type of integral bridge can also alleviate problems by cyclic displacements at the top of the facing due to seasonal thermal expansion and contraction of the girder as well as those by residual settlement of the backfill back of the facing by dead and live loads.
2) The stability increases with an increase in the tensile resistance of reinforcement layers, which increases with an increase in the number of reinforcement layer as well as an increase in the connection strength for given pull-out strength and rupture strength of reinforcement. The pull-out failure may take place when the connection strength became sufficiently large.

ACKNOWLEDGEMENT

This study is supported by the Japan Society for the Promotion of Society through the grant: "Development of a new type bridge structure by using geosynthetic-reinforced soil technologies".

REFERENCES

Aizawa, H., Nojiri, M., Hirakawa, D., Nishikiori, H., Tatsuoka, F., Tateyama, M. and Watababe, K. 2007, Validation of high seismic stability of A new type integral bridge consisting of geosynthetic-reinforced soil walls, *Proceeding of 5th International Symposium on Earth reinforcement* (submitted).

England, G.L., Tsang, N.C.M. and Bush, D.L., 2000, Integral Bridge, A fundamental approach to the time-temperature loading problem, Thomas Telford.

Hirakawa, D., Nojiri, M., Aizawa, H., Tatsuoka, F., Sumiyoshi, T. and Uchimura, T. 2006, Behaviour of geosynthetic-reinforced soil retaining wall subjected to forced cyclic horizontal displacement at wall face, *8th International Conference on Geosynthetics* Vol. 2, pp. 1075–1078.

Koseki, J., Bathurst, R.J., Güler, E., Kuwano, J. and Maugeri, M. 2006, Seismic stability of reinforced soil walls, 8th *International Conference on Geosynthetics* Vol. 1, pp. 51–78.

Tamura, Y. 2006, Lessons from construction of geosynthetic-reinforced soil retaining walls having full-height rigid facing for the last 10 years, *Proceedings of 8th International Conference on Geosynthetics*, Vol. 3, pp. 941–944.

Tatsuoka, F., Hirakawa, D., Nojiri, M., Aizawa, H., Tateyama, M. and Watababe, K. 2007, A new type integral bridge consisting of geosynthetic-reinforced soil walls, *Proceeding of 5th International Symposium on Earth reinforcement* (submitted).

Tatsuoka, F., Tateyama, M., Uchimura, T. and Koseki, J. 1997, Geosynthetic-reinforced soil retaining wall as important permanent structures, *1996–1997 Mercer Lecture, Geosynthetics International*, Vol. 4, No. 2. pp. 81–136.

New Horizons in Earth Reinforcement – Otani, Miyata & Mukunoki (eds)
© 2008 Taylor & Francis Group, London, ISBN 978-0-415-45775-0

Validation of high seismic stability of a new type integral bridge consisting of geosynthetic-reinforced soil walls

H. Aizawa, M. Nojiri, D. Hirakawa, H. Nishikiori & F. Tatsuoka
Tokyo University of Science, Japan

M. Tateyama & K. Watanabe
Railway Technical Research Institute, Japan

ABSTRACT: Shaking table tests were performed on scaled models of four different bridge types: 1) the conventional-type, comprising of a pair of gravity-type abutments retaining unreinforced backfill; 2) the GRS RW bridge, comprising of a pair of geosynthetic-reinforced soil (GRS) retaining walls with full-height rigid (FHR) facings directly supporting a girder on the crest of the reinforced backfill; 3) the conventional integral bridge with unreinforced backfill, unifying a pair of FHR facings with a girder; and 4) a new type integral bridge comprising of a pair of GRS retaining walls having FHR facings (called the GRS integral bridge). It is shown that the seismic stability of the GRS integral bridge is highest among the four types examined because of several inherent structural advantages resulting in a monolithic behaviour of the whole bridge system.

1 INTRODUCTION

In many previous major earthquakes, including the 1995 Hyogoken Nambu and 2004 Niigata Chuetsu earthquakes, a number of conventional type bridges, typically those comprising of a pair of gravity-type abutments supporting a girder on the top and the unreinforced backfill on the back, totally collapsed and it took long time to be reconstructed. As a line structure, such as a railway and a highway, may lose its function for a long period even by collapse of a single bridge, the development of cost-effective bridge system, while having a high stability against level 2 design seismic load, has been required. In view of the above, we proposed a new type bridge system, called the GRS integral bridge (Fig. 1), combining the geosynthetic-reinforced soil (GRS) retaining wall (RW) technology and the conventional integral bridge system (with unreinforced backfill) (Nojiri et al., 2006; Aizawa et al., 2006; Hirakawa et al., 2006, 2007; Tatsuoka et al., 2007).

In this study, a series of model shaking table tests was performed to evaluate the seismic stability of the GRS integral bridge in comparison with those of the following three conventional bridge types:

1. *Gravity-type abutment bridge*: a pair of gravity-type abutment supports a girder on their top via movable and fixed supports while unreinforced backfill on their back.

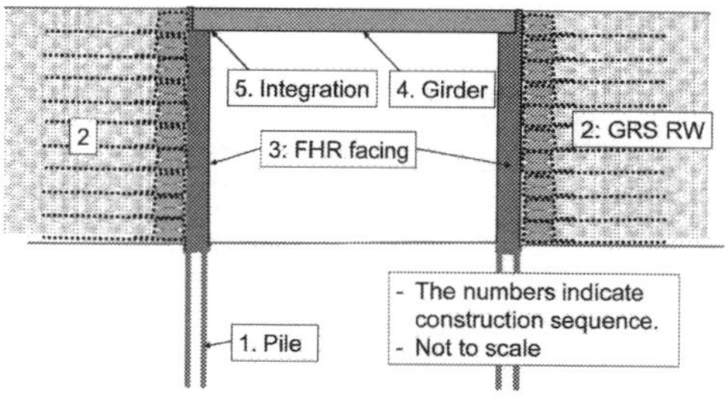

Figure 1. GRS integral bridge.

2. *GRS RW bridge*: a pair of geosynthetic-reinforced soil retaining walls (GRS RWs) with full-height rigid (FHR) facings directly support a girder via movable and fixed supports on sill beams placed on the reinforced backfill.
3. *Conventional integral bridge*: a girder is integrated with a pair of FHR facings while the backfill is unreinforced.

Tatsuoka et al. (2007) compare the structural features of the GRS integral bridge with those of these three conventional bridge types to highlight its advantages. Hirakawa et al. (2007) reports the effects of tensile resistance of reinforcement layers on the seismic stability on the GRS integral bridge.

2 TEST METHOD

2.1 *Apparatus*

A shaking table at Railway Technical Research Institute was used, which has a maximum excitation acceleration of about 1,500 gals with a maximum amplitude of horizontal displacement of ±5 cm when the load is 60 kN. The sand box, 205.8 cm-long ×60 cm-wide ×140 cm-high, was fixed on the shaking table. The front side wall of the box consists of a transparent tempered glass sheet, through which displacements and deformation of the model were observed. The other side wall consists of a steel plate with an inside face covered with a 0.2 mm-thick Teflon sheet to minimize side wall friction.

2.2 *Model subsoil and backfill and model bridges*

Fig. 2 presents the four bridge models. A 35 cm-high subsoil layer and 51 cm-high backfill were prepared by pluviating air-dried Toyoura sand throughout air using multiple sieves to a target initial relative density equal to 90%. Thin horizontal layers of black-dyed Toyoura sand particles were arranged in the subsoil layer and backfill to observe their deformation in the tests. The length similitude ratio of the bridge models was assumed to be 1/10. The model abutments and facings were made of basically duralumin. Their back face was made rough by gluing sandpaper #150. The model girder was made of steel having a mass of 25 kg and a length of 60.8 cm (i.e., the largest possible length to be accommodated in the sand box). By fixing a weight of 180 kg to the center of the girder, the equivalent girder length was made 6 m (i.e., 20 m in the prototype).

1. *Gravity-type abutment bridge (Fig. 2a):* A girder is supported by a pair of movable and fixed supports. The fixed support was allowed to rotate about a pin, while the movable support was allowed to horizontally slide along a linear rail.
2. *GRS RW bridge (Fig. 2b)*: A girder is supported in the same way as above. The bottom of the full-height rigid (FHR) facing was embedded 4 cm in the subsoil. The model reinforcement was a grid constituting of 0.2 mm-thick and 3 mm-wide phosphor bronze strips as the longitudinal members and 0.5 mm-diameter wire as the transversal members (Fig. 3). Sand particles were glued on their surface. Nine grid layers were placed at a vertical spacing of 5 cm in the backfill. Seven reinforcement layers were connected to the back of the FHR facings and two layers to the back of the sill beams.
3. *Integral bridge (Fig. 2c)*: A girder and FHR facings were connected to each other with a pair of L-shaped metal fixtures (20 cm-long, 5 cm-wide and 3 mm-thick). Fig. 4 shows the relationship between the moment, M, and the flexural angle, ϕ, from bending tests. It is likely that the initial low-stiffness

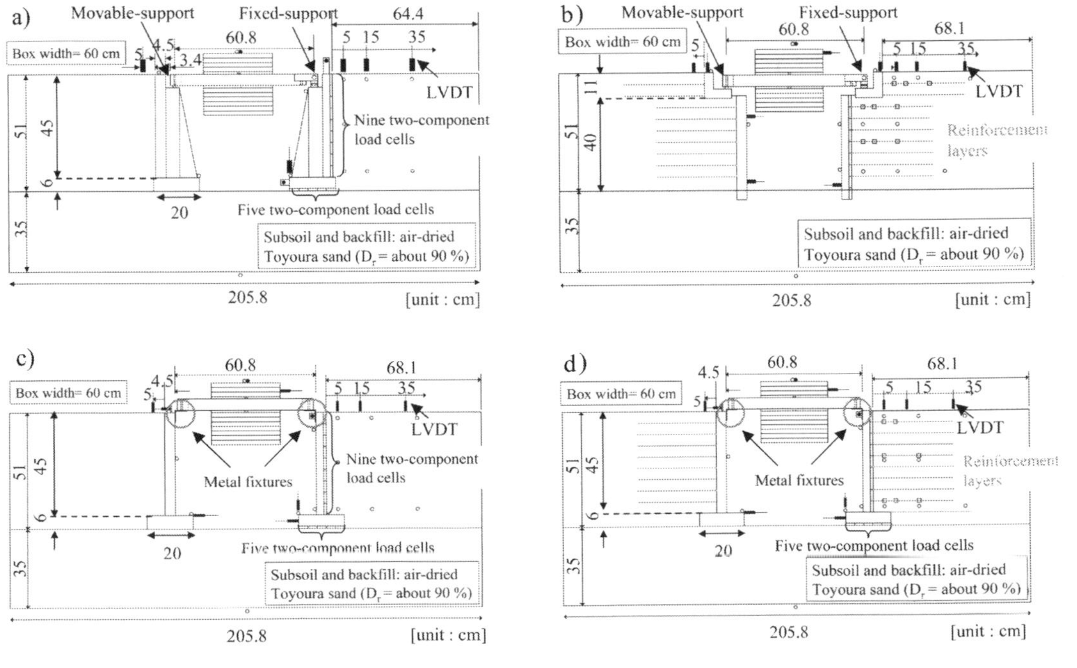

Figure 2. Four bridge models. a) gravity-type abutment bridge; b) GRS RW bridge; c) integral bridge with unreinforced backfill; and d) GRS integral bridge.

part of the measured M-ϕ relation is due to loose setting of the test specimen, so not reliable. It is assumed that the correct origin is located at point a. It was estimated that a flexural angle of about 0.9 degrees (from the point a) takes place by the weight of the bridge girder. So the metal fixture starts yielding at a flexural angle increment of about 2.6 degrees and exhibits the peak moment at an increment of 5.7 degrees. These are equivalent to shear strains (γ) in the backfill equal to about 4.5% and 9.9%. As the dense Toyoura sand exhibits the peak strength at $\gamma \sim 8$–12% (Tatsuoka et al., 1986), the fixtures have already started yielding and may start strain-softening when the backfill exhibits the passive failure. That is, the fixtures were designed not to become the major resisting structural component when the collapse of the bridge becomes imminent. It was considered that it is the case with proto-types.

4. *GRS integral bridge (Fig. 2d)*: A girder and a pair of FHR facings were integrated in the same way as the integral bridge model described above. In total ten reinforcement layers were arranged in the backfill, eight layers connected to the back of the facing at a vertical spacing of 5 cm and two layers to the back of the facing foundation at the vertical spacing of 6 cm. As the connection strength was sufficiently large, eventually pull-out failure took place as described by Hirakawa et al. (2007).

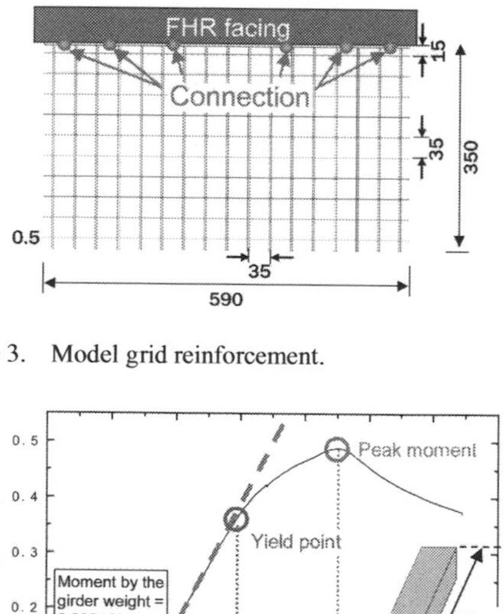

Figure 3. Model grid reinforcement.

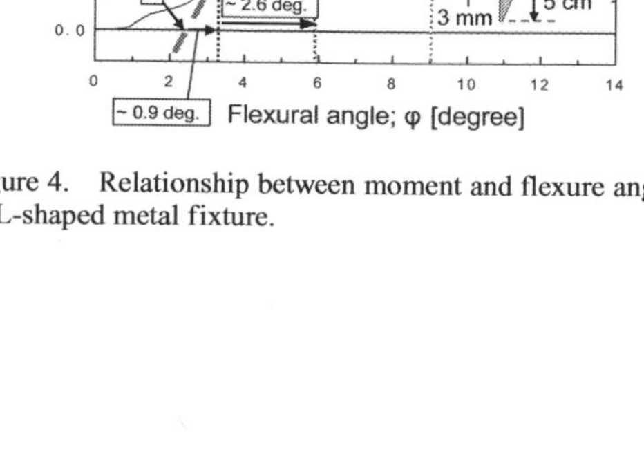

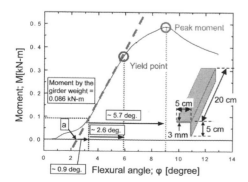

Figure 4. Relationship between moment and flexure angle of L-shaped metal fixture.

2.3 *Loading conditions*

A surcharge of 1 kPa was placed on the crest of the backfill simulating the railway conditions. The models were subjected to horizontal sinusoidal acceleration consisting of 20 waves at a frequency of 5 Hz at each loading stage. The input acceleration was increased incrementally at a step of 100 gals from 100 gals until the respective models ultimately collapsed.

3 TEST RESULTS

3.1 *Cumulative residual displacements*

Fig. 5a shows the cumulative residual rotational angle of the facing (θ) plotted against the amplitude of base acceleration (α_b). With the gravity-type abutment and GRS RW bridges, the displacements of the facing on the side supporting the girder via a fixed support is presented in Fig. 5a (and also in Fig. 5b). With the GRS RW bridge, the rotational angle of the sill beam is also presented. The angle θ is positive when the top of the facing overturns outward relative to the bottom (i.e., the active direction). Fig. 5b presents the cumulative residual outward lateral displacements (d_b) at the bottom of abutment or facing against α_b. With the GRS RW bridge, the displacements at the sill-beam are also presented. Figs. 6a and 6b present the cumulative residual settlements of the backfill at distances of 5 cm and 35 cm back of the abutment or facing on the side supporting the girder by a fixed support. The following trends of behaviour may be seen:

1. *Gravity-type abutment bridge*: When α_b became 180 gals, by large inertia of the girder at its top via a fixed-support, the abutment started overturning outward at the top and sliding outward at the bottom. At the same time, the backfill started settling at a distance of 5 cm back of the facing. The settlement at a distance of 35 cm was not noticeable until the end of the test. When α_b became 255 gals, the over-turning at the top and sliding at the bottom became significant and a general shear band was formed in the backfill (Fig. 7). No double, this type of bridge was weakest among the four bridge types.
2. *GRS RW bridge*: When α_b becomes 383 gals, the sill beam started largely overturning in the active direction associated with failure in the backfill immediately below the toe of the sill beam due to large moment caused by the inertia of the girder. On the other hand, the settlement in the backfill back of the sill beam still remained very small (Fig. 6b), showing that the GRS RW itself was still stable. As α_b increased to 472 gals, the rotation and outward lateral displacement of the sill beam became much larger, while the settlement increased and a shear band was formed in the backfill back of the sill

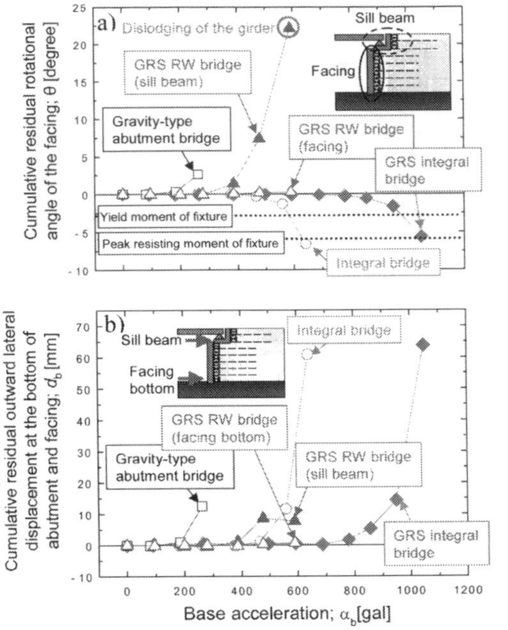

Figure 5. Relationships between a) facing rotation angle and b) outward displacement at the facing bottom and base acceleration.

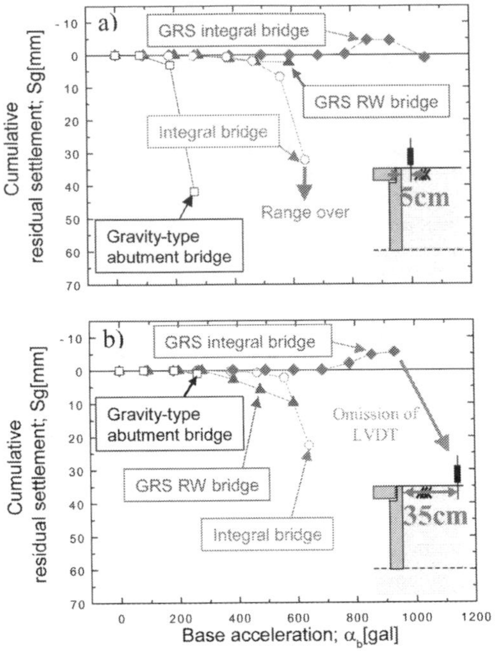

Figure 6. Relationships between cumulative residual settlement of the back-fill at positions of a) 5 cm and b) 35 cm from the top of wall and base acceleration.

beam (denoted as 1 in Fig. 8). At $\alpha_b = 589$ gals, the girder contacted and pushed inward the other sill beam having a movable support, which triggered the formation of a general shear band in the back-fill (2 in Fig. 8). The mass of the sill beam was too small to resist against the large inertia of the girder, while the tensile reinforcement layers attached at the back of the sill beam did not resist against this force. On the other hand, the deformation of the GRS walls remained still small (Fig. 5). It is clear that a low seismic stability of this type of bridge is due to a very low of dynamic stability of the sill beams.

3. *Integral bridge*: When α_b became 560 gals, the L-shaped metal fixtures, connecting the facings and girder, started yielding and the facings started noticeably rotating forward about the top, as the girder functioned as a strut against the earth pressure activated on the facings. The backfill started settling noticeably, while shear bands were formed in the backfill on the right (1 in Fig. 9). At $\alpha_b = 641$ gals, the facings started rotating significantly forward, and circular shear bands were formed (2 in Fig. 9), and the backfill settled down at distances of not only 5 cm but also 35 cm in back of the facing.
4. *GRS integral bridge*: It was only when α_b became as high as 799 gals that the facings started rotating noticeably like the integral bridge, as described

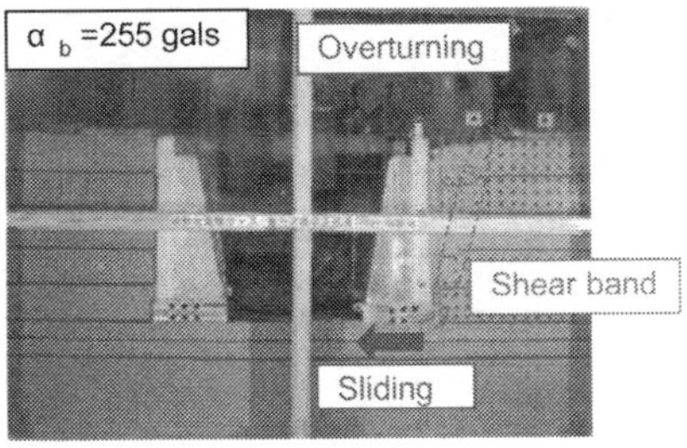

Figure 7. Gravity type abutment bridge after shaking at $\alpha_b = 255$ gals.

above. The backfill heaved slightly by passive movements at the top of the facing by the inertia of the girder. As α_b increased to 950 gals, the rotation of the facing increased but much more gradually than the other types of bridge. Shear bands were formed in the unreinforced backfill zone immediately back of the reinforced zones (Fig. 10). When α_b became 1,048 gals, the L-shaped metal fixtures yielded significantly and the bottom of the facings was largely pushed out associated with large rotational movements of the facing and reinforced backfill zone about the top of the facing. This failure mode can also be seen from significant settlement (about 2 cm by the photogrametric method)

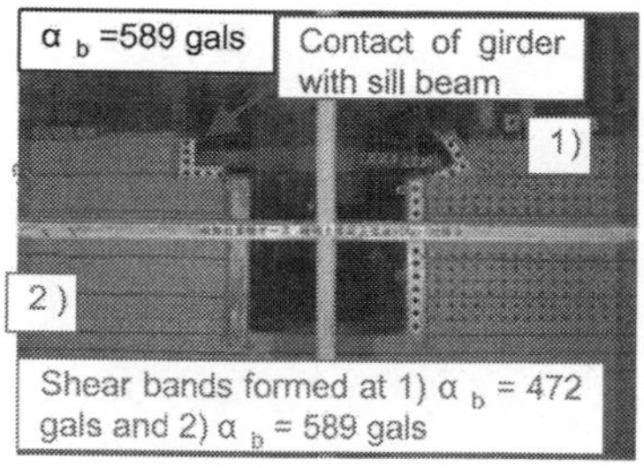

Figure 8. GRS RW bridge after shaking at $\alpha_b = 589$ gals.

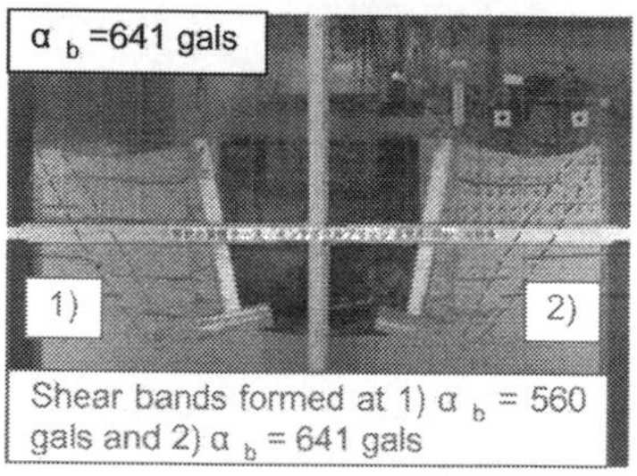

Figure 9. Integral bridge after shaking at $\alpha_b = 641$ gals.

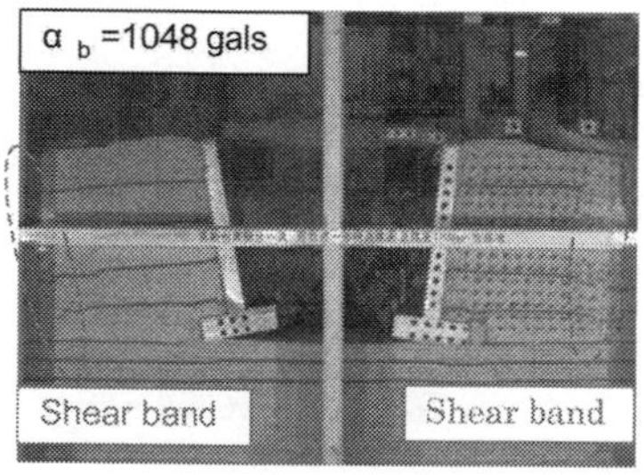

Figure 10. GRS integral bridge after shaking $\alpha_b = 1048$ gals.

of the backfill at a distance of 35 cm from the back of the facing, compared with small settlement at a distance of 5 cm (Fig. 10). This monolithic behaviour of the facing and backfill is due to firm connections between the full-height rigid facing and the reinforcement layers arranged in the backfill, which greatly improved the seismic stability of the integral bridge.

3.2 *Dynamic responses*

Fig. 11 shows the time histories of base acceleration at the shaking table, response acceleration at the girder and sill beam, average lateral earth pressure, p, on the back of the abutment supporting the girder via a fixed-support or the facing of integral bridge and average normal pressure, σ, at the bottom of the abutment or the facing footing. The earth pressure, p and σ, were obtained by averaging the loads measured with respective sets of local load cells (Fig. 2). With the GRS RW bridge, the horizontal acceleration at the sill beam supporting the girder via a fixed-support is presented instead of σ in the other cases. Acceleration is defined positive when the displacement of the facing on the side with the fixed girder-support is at the passive state (i.e., when the inertia of the girder acting to this facing is in the passive direction).

1. *Gravity type abutment bridge (Fig. 11a)*: During shaking at $\alpha_b = 255$ gals, the σ value at the abutment bottom became the minimum when the abutment top was at the passive state, which it was not very small initially. As α_b increased, the response of the girder increased associated with the bearing capacity failure in the subsoil below the abutment base, which resulted in a substantial decrease in the minimum of σ. Subsequently, the abutment base started moving toward the active direction while the abutment top was moving toward the passive direction. As a result, the minimum of σ started increasing associated with a decrease in the passive earth pressure of p.
2. *GRS RW bridge (Fig. 11b)*: The earth pressure, p, was much smaller than those with the other bridge types, because the facing and reinforced backfill behaved like a monolith in a rather stable manner. On the other hand, as α_b increased, the ratios of the response acceleration of the bridge girder (α_g) and sill beam (α_{sb}) to the base acceleration (α_b) became larger. During shaking at $\alpha_b = 589$ gals, both α_g and α_{sb} increased suddenly, which was due to sudden unstable over-turning of the sill beam supporting the girder via a fixed-support.
3. *Integral bridge (Fig. 11c)*: When α_b was 378 gals, both of pressures, p and σ, exhibited steady values. During shaking at $\alpha_b = 470$ gals, the response acceleration of the girder (α_g) was increasing at a noticeable rate, associated with a large increase in the displacements at the top of the facings. The σ value at the footing base became the maximum when the facing top was at the active state and the minimum when the facing top was at the passive state. The both decreased at a large rate with time, which was due to an increase in unstable large rotation of the facing: i.e., when the facing top was at the passive state, the toe area of the footing base separated from the subsoil while the subsoil below the heel area collapsed. A similar phenomenon took place when the facing top was at the active state.
4. *GRS integral bridge (Fig. 11d)*: At $\alpha_b = 799$ gals, the bridge was still rather stable, the girder and the facings together with the reinforced backfill behaving as an integrated structure. When α_b became 860 gals, both the maximum and minimum of σ on the facing footing base started decreasing by

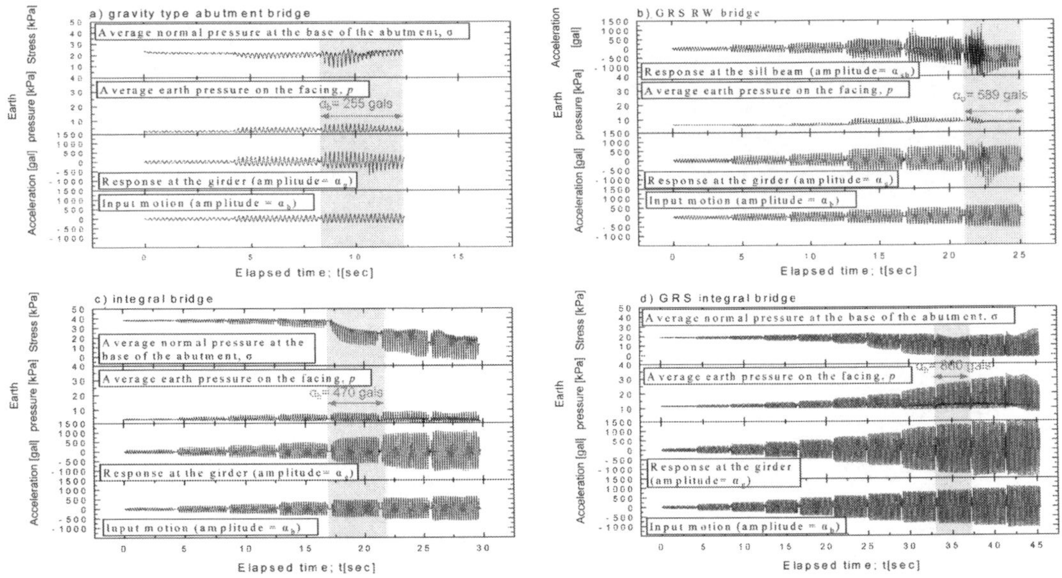

Figure 11. Time histories of measured physical quantities: a) gravity type abutment bridge; b) GRS RW bridge; c) integral bridge; and d) GRS integral bridge.

the mechanisms explained above with the integral bridge. When α_b became 1,048 gals, the response at the girder, α_g, increased significantly whereas the earth pressures, p, on the back of the facing started decreased. This was due to the monolithic rotational displacements about the facing top of the facing and reinforced backfill that became significant. The maximum of the earth pressure, p, with the GRS integral bridge was generally much larger than with the other types. This trend of behavior means that large tensile force was activated in the reinforcement, showing that the reinforcement contributed effectively to the stabilization of the bridge. This large increase in the earth pressure does not damage the facing, as the facing is a continuous rigid beam having many supports (i.e., the reinforcement layers) with a short span.

3.3 *Failure mode of GRS integral bridge*

Fig. 12 shows the distributions with depth of the earth pressure at 10*th* cycle at the respective shaking stages on the back of the facing of the GRS integral bridge when the top of the facing was at the largest passive and active displacements. Fig. 13 shows the relationships between the maximum tensile force in the reinforcement at selected points immediately back of the facing. The following trends of behaviour may be seen from Figs. 12 and 13:

1. When the facing top was at the passive state, the earth pressure was large (i.e., passive earth pressure) in the upper part of backfill while it was small (i.e., active earth pressure) in the bottom part of backfill. The trend was opposite when the facing top was at the active stage. Relatively low passive earth pressure near the facing bottom was due to the friction between the backfill and the subsoil. These trends of earth pressure are due to rotational displacements of the facing relative to the backfill.
2. The rotation displacements of the facing was resisted effectively only by the lower reinforcement layers (Fig. 13).
3. The passive earth pressure in the upper part of the backfill (Fig. 12a) was largest when $\alpha_b = 799$ gals and decreased as α_b increased subsequently. This means that the backfill exhibited passive failure, which resulted in an increase in the rotation of the facing relative to the backfill. Then, the displacement therefore the inertia of the girder increased and the center of the facing rotation relative to the backfill was shifted download, which all accelerated the passive yielding of the backfill, increasing the facing rotation associated with sliding at the facing foundation, and increased the tensile force in the reinforcement. Then, excessive tensile force in the lower reinforcement layers resulted in the connection failure or/and pull-out failure of the reinforcement.

These results indicate that the use of a sufficient number of tensile reinforcement layers having sufficiently high rupture strength, connection strength and

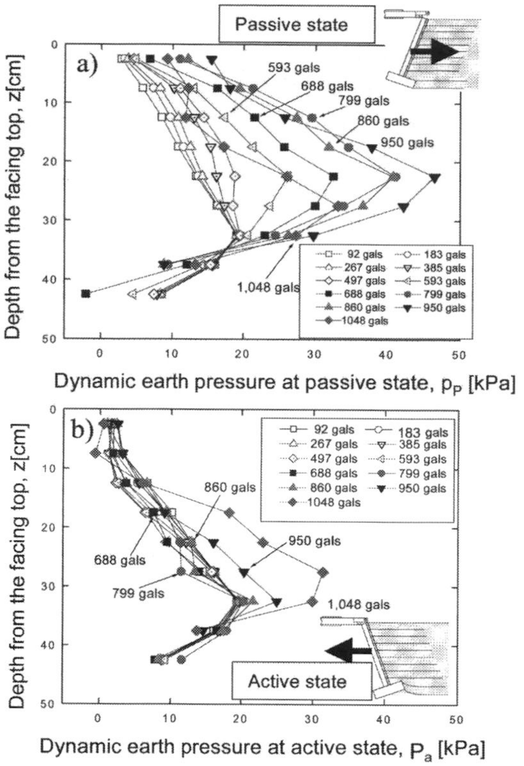

Figure 12. Earth pressure distribution with depth on the facing at 10th cycle at each stage, GRS integral bridge.

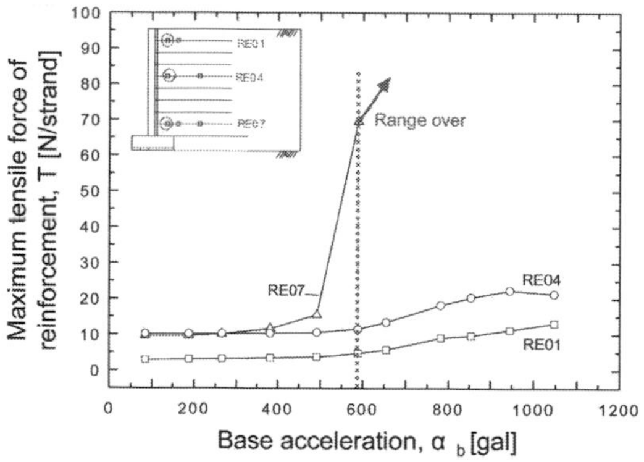

Figure 13. Relationship between tensile force of rein-forcement and base acceleration, GRS integral bridge.

pull-out strength, in particular at the lower part of the facing, is essential for a high seismic stability of GRS integral bridge.

4 CONCLUSIONS

The following conclusions can be obtained from the results from 1-g shaking table tests on four different bridge models presented in this paper:

1. The GRS integral bridge has the highest seismic stability among the different bridge types examined in the present study.
2. The collapse of the GRS integral bridge was associated with the rotation of the facing relative to the backfill with the bottom end being pushed out. The use of a sufficient number of reinforcement layers having a sufficient tensile resistance is essential for a high seismic stability of GRS integral bridge.

REFERENCES

Tatsuoka et al., (2007): A new type integral bridge consisting of geosynthetic-reinforced soil walls, *Proc. of IS Kyushu*, 2007 (this volume)

Hirakawa et al., (2007): Effects of the tensile resistance of reinforcement embedded in the backfill on the seismic stability of GRS integral bridge, *Proc. of IS Kyushu*, 2007 (this volume)

Aizawa et al., (2006): Comparison of cost-performance effectiveness of conventional and new types of bridge systems consisting of unreiforced or reinforced soil, *Geosynthetics Engineering Journal, IGS Japanese Chapter*, Vol. 21, pp. 175–182 (in japanese)

Nojiri et al., (2006): Evaluation of dynamic failure modes of various bridge types by shaking table tests, *Geosynthetics Engineering Journal, IGS Japanese Chapter*, Vol. 21, pp. 159–166.

Tatsuoka et al., (1986): Strength and deformation characteristics of sand in plane strain compression at extremely low pressures, *Soil and Foundations*, Vol. 26, No. 1, pp. 65 ~ 84, 1986

New Horizons in Earth Reinforcement – Otani, Miyata & Mukunoki (eds)
© 2008 Taylor & Francis Group, London, ISBN 978-0-415-45775-0

Improvement of earthquake resistance by reinforcing toe of embankments

K. Oda, K. Tokida, Y. Egawa & K. Tanimura
Graduate School of Engineering, Osaka University, Osaka, Japan

ABSTRACT: Recently, the development of economical and effective method for improving the seismic resistance of infrastructures is required strongly. In this paper, the availability of reinforcing the toe of embankment to improvement of seismic resistance of road embankment is discussed through the experiments and analyses. Both centrifuge model tests in the experiments and elasto-pastic finite element analyses in the analyses are carried out. Consequently, the effect of reinforcing the toe of embankment on both seismic resistance of embankment and sliding zone is elucidated.

1 INTRODUCTION

The 2004 Mid Niigata Prefecture Earthquake which was one of the typical near-field earthquakes caused the serious damage to many kinds of infrastructures in the central area of Niigata Prefecture. Especially, a large number of road embankments, which are main road structures in this area, suffered remarkable damage. The damage of road caused the disruption of the road network in this area, so that it was difficult to deal with an emergency and rehabilitation. About 70% of Japanese country is mountainous area. There are a large number of cities, towns and villages in mountainous areas in Japan. Therefore, the aseismic countermeasures for improving the earthquake resistance of road embankments are required to effectively preserve the road network even at an extreme earthquake.

In this paper, the availability of reinforcing the toe of embankments is investigated experimentally and analytically in order to develop efficient and economical techniques for improving the earthquake resistance of embankments. Firstly, the centrifuge model tests are carried out to experimentally demonstrate the improvement of earthquake resistance of embankments. Then, a numerical simulation, in which an elasto-plastic finite element method is applied, is carried out to discuss the applicability of numerical simulation to the sliding failure of embankment. Finally, the availability of reinforcing the toe of embankments to the improvement of earthquake resistance is confirmed. Furthermore, the residual deformation characteristics of embankments, especially, location and volume of sliding surface in the embankments are discussed.

2 TOE REINFORCEMENT

Figure 1 shows schematic diagrams of embankment both without reinforcement and with reinforcement at its toe. The deformation of embankment will be restricted with reinforcing the toe of embankment, so that its seismic resistance will be improved. The advantages of reinforcing the toe of embankments are as follows;

1) The construction of reinforcements can be carried out without suspending the traffic.
2) The sliding zone can decrease, because the substantial height of embankments decreases.

In this paper, the embankments without and with reinforcements are called as "Non-reinforce embankment" and "Reinforce embankment", respectively.

3 CENTRIFUGE MODEL TEST

Figures 2 & 3 show the outlines of model embankment used in centrifuge model test. The centrifuge model tests were carried out in Shimizu Institute of

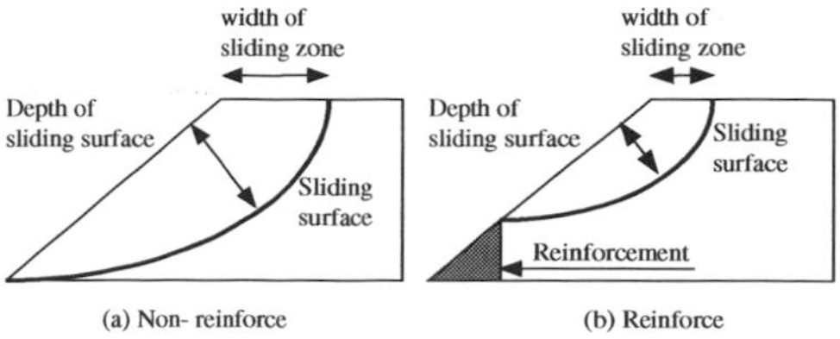

Figure 1. Schematic diagrams of both Non-reinforce and Reinforce embankment.

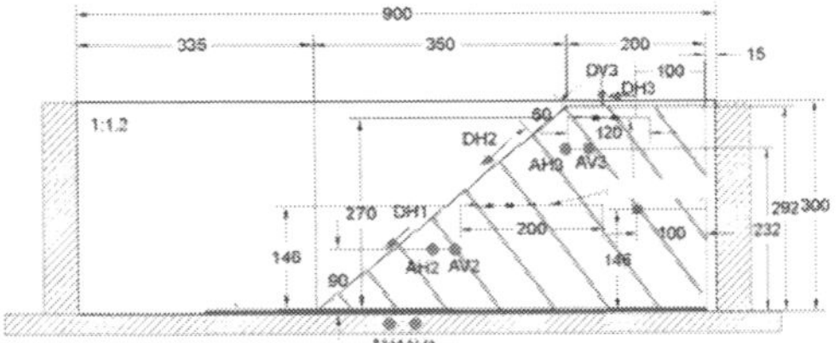

Figure 2. Outline of Non-reinforce embankment model in centrifuge model test.

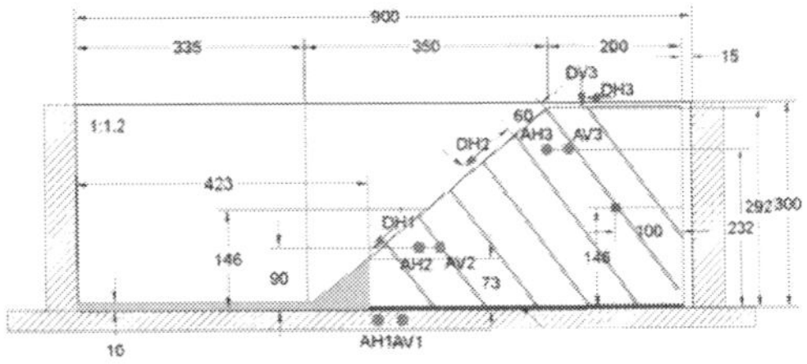

Figure 3. Outline of Reinforce embankment model in centri-fuge test.

Technology. The rigid container used in the centrifuge model test has 900 mm in length, 280 mm in width, and 300 mm in height. The model embankments with a single slope used in the test has 1:1.2 gradients. The centrifuge model tests were carried out under acceleration of 30 g. The model embankments were made of DL clays which were mixed with silicon oil at 5 % of oil content instead of water in order to match viscosity in the model tests to that in the prototype. The average density of DL clay was 1.51 g/cm^3. The shear wave velocity of model embankment under acceleration of 30 g was 175 m/s. An expanded polystyrene board with 15 mm thickness was set between wall and model embankment to absorb the vibration shock from a side wall of the container to model embankment. Silicon greases and rubber membranes were used to decrease the friction between side wall and model embankment. The reinforcement structure at the toe of embankment was made by mixing sand and cement with at the ratio of eight and two in the centrifuge model test. The shear wave velocity of this structure in 1 g field measured by embedded bender elements was 595 m/s. It was considered that the reinforcement structure is stiffer than the embankment body. The reinforcement structure was unified with 10 mm plate which was embedded to the left end of container. Also, the height of reinforcement structure is about quarter of the height of embankment. Therefore, the deformation in the bottom quarter part of Reinforce embankment will be restricted.

The sinusoidal wave as dynamic load was applied to the base of container under acceleration of 30 g. The sinusoidal waves applied were with 60 Hz and 30 cycles. Two types of amplitude of sinusoidal waves, 300 gal and 500 gal, were used in the centrifuge model tests. The accelerations in and to the model embankments were measured through acceleration sensors. Six acceleration sensors were installed as shown in Figures 2 & 3 with AH and AV marks. Also four laser displacement gages as shown in Figures 2 & 3 with DH and DV marks were installed to measure the deformation of the model embankments. Furthermore, the residual deformation in the model embankments were investigated by colored sand as shown in Figures 2 & 3.

Table 1 shows summary of test results. For the Non-reinforce embankments, they failed in the case of attitude of sinusoidal wave of not only 500 gal but also 300 gal. Therefore, the seismic resistance of Non-reinforce embankment might be less than 300 gal. For the Reinforce embankment, the model embankment did not failed in the case of attitude of sinusoidal waves of 300 gal. The seismic resistance of Reinforce embankment might be greater than 300 gal and less than 500 gal. Consequently, the seismic resistance of model embankment could be improved with reinforcing the toe of embankment.

Table 1. Summary of test results.

	Amplitude	Result
Non-reinforce embankment	300 gal	Fail
Non-reinforce embankment	500 gal	Fail
Reinforce embankment	300 gal	Not fail
Reinforce embankment	500 gal	Fail

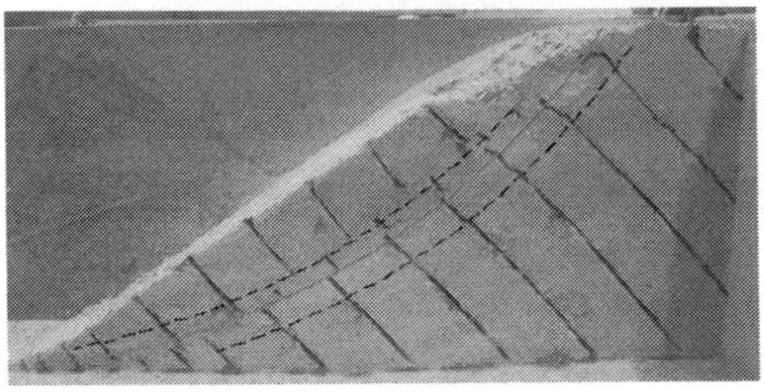

Photograph 1. Residual deformation of Non-reinforce embankment (500 gal).

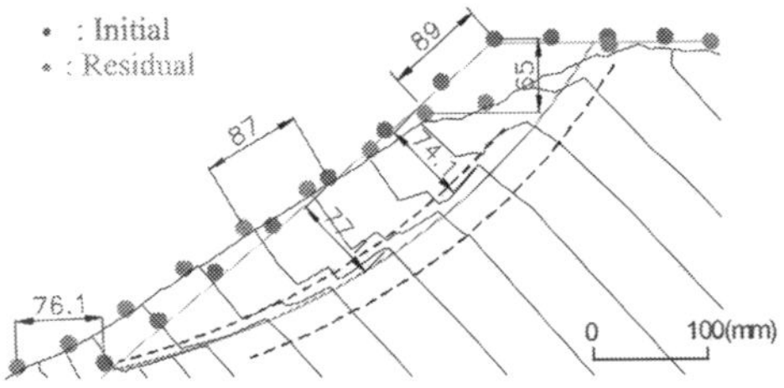

Figure 4. Sketch of residual deformation of Non-reinforce embankment (500 gal).

Photograph 1 shows the residual deformation of Non-reinforce embankment in the amplitude of

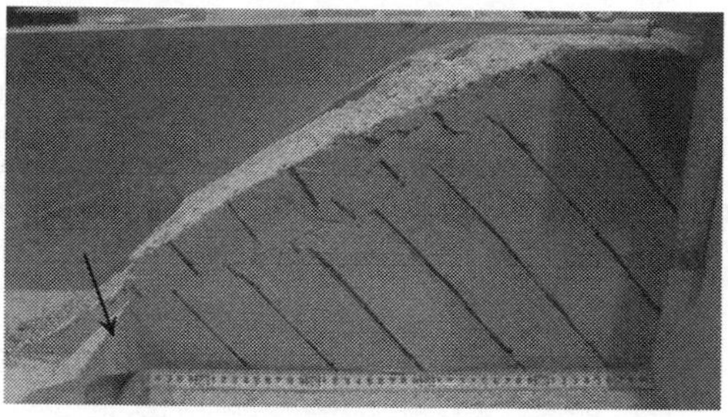

Photograph 2. Residual deformation of Reinforce embankment (500 gal).

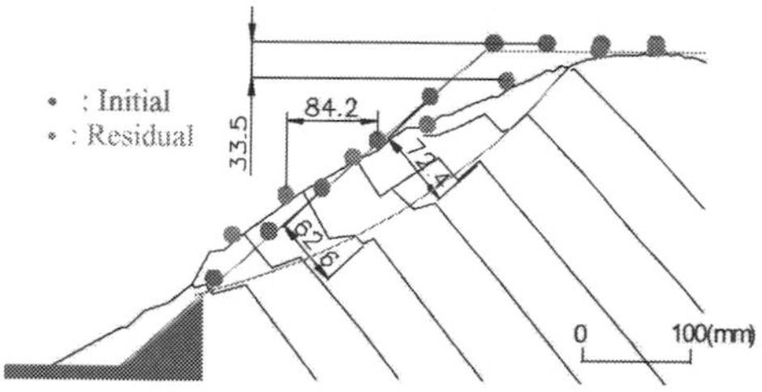

Figure 5. Sketch of residual deformation of Reinforce embankment (500 gal).

sinusoidal wave of 500 gal. Also, Figure 4 illustrates its sketch. Three sliding surfaces could be observed. The middle sliding surface would dominate the sliding failure of this model embankment, based on the displacement of colored sand. The depth of middle sliding surface was about from 70 mm to 80 mm. Also, the width of sliding zone, which is depicted in Figure 1, was about 100 mm. Furthermore, the residual displacement of sliding mass was about from 80 mm to 90 mm.

Photograph 2 shows the residual deformation of Reinforce embankment in the amplitude of sinusoidal wave of 500 gal. Also, Figure 5 illustrates its sketch. A single sliding surface could be observed. The depth of sliding surface was about from 60 mm to 70 mm. Also, the width of sliding zone was about 80 mm. Furthermore, the residual displacement of sliding mass was about from 80 mm to 90 mm. The volume of sliding mass in the Reinforce embankment is less than that in the Non-reinforce embankment. Also, the width of sliding zone in the Reinforce embankment is less than that in the Non-reinforce embankment. That is, the effect of sliding failure of slope on crest of embankment could be reduced. However, the residual displacement of sliding mass could not be shortened.

4 NUMERICAL SIMULATION

The numerical simulation, in which an elasto-plastic finite element method was applied, was carried out to

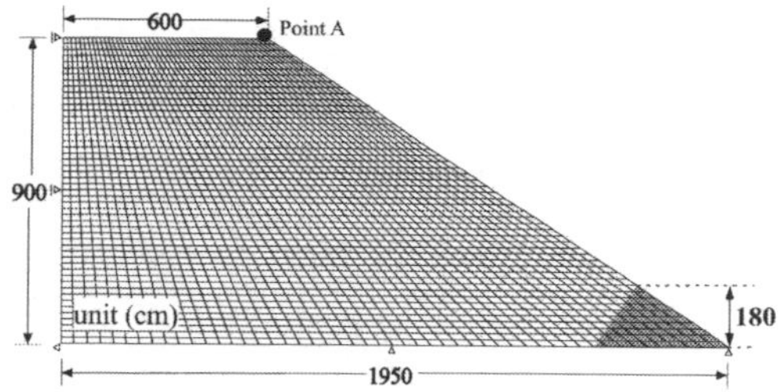

Figure 6. Analytical model used in numerical simulation.

Table 2. Analytical parameters.

a) Embankment body	
Elastic modulus	$2.45 \times 10^4 kN/m^2$
Poisson's ratio	0.25
Cohesion	$7.3 kN/m^2$
Internal friction angle	36.3°
b) Reinforcement	
Elastic modulus	$2.45 \times 10^8 kN/m^2$
Poisson's ratio	0.25

reproduce the deformation characteristics, especially the seismic intensity and the residual deformation of embankments (Oda et al, 2006). In this analysis, Drucker-Prager criteria is applied as both yield function and plastic potential. In the elasto-plastic calculation, the plastic strain increment is calculated with a return mapping algorithm (Oritz & Simo, 1986) based on non-associated flow role. Also, the spherical arc length procedure proposed by Crisfied (1991), one of the representative displacement control technique is applied, so that the stable numerical calculation can be carried out. In the analysis, firstly, self-weight analysis was carried out to determine the initial stress condition. In this analytical stage, the only vertical body force estimated from unit-weight, is applied in each finite element. Secondly, the horizontal body force increases gradually, so that the bearing resistance of embankments to the horizontal body force reaches to the ultimate state. Finally, the residual deformation in the embankment can be computed.

Figure 6 shows the analytical model used in the numerical simulation. This analytical model was based on the prototype in another series of centrifuge model test (Yoshino, 2006). Table 2 shows analytical parameters used in the numerical simulation. In the numerical simulation, the embankment body and reinforcement bulb were modeled as perfect elasto-plastic and elastic materials, respectively.

Figure 7 shows relationship between horizontal seismic intensity and vertical displacement at the edge of crest of embankment (Point A in Figure 6). The horizontal seismic intensities of Reinforce and Non-reinforce embankment at which the yielding

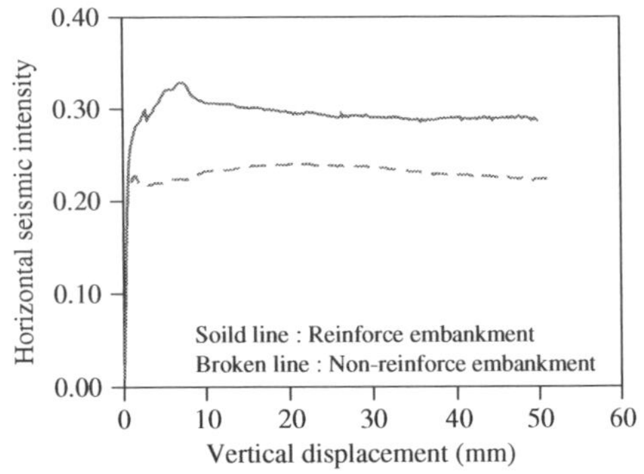

Figure 7. Relationship between horizontal seismic intensity and vertical displacement.

Figure 8. Distribution of shear strain at the final stage in Non-reinforce embankment.

occurs were 0.26 and 0.22, respectively. After the yielding occurred, the horizontal seismic intensity of Non-reinforce embankment hardly varied. On the other hand, that of Reinforce embankment gradually increases with the vertical displacement increasing. The horizontal seismic intensity was at the maximum, 0.33, at the vertical displacement of about 12 mm. Then, it decreased to about 0.30. After that, it hardly varied with the vertical displacement increasing. The horizontal seismic intensity at residual state in Reinforce embankment was about 1.3 times of that in Non-reinforce embankment. As was observed in the centrifuge model test, it was found that the seismic resistance of Reinforce embankment is greater than that of Non-reinforce embankment from the numerical simulation, too.

Figures 8 & 9 show the distribution of shear strain at the final stage of numerical simulation in Non-reinforce embankment and Reinforce embankment, respectively. In each case, a single narrow strap in which a remarkable shear strain occurs was formed. These straps must correspond to the sliding surface. The narrow strap in the Non-reinforce embankment reaches from the toe of embankment to the center of crest. Also, that in the Reinforce embankment reaches from the top of reinforcement bulb to the center of

Figure 9. Distribution of shear strain at the final stagte in Reinforce embankment.

crest, because the reinforcement bulb restricts deformation around toe of embankment. Therefore, the volume of sliding mass in the Reinforce embankment was less than that in the Non-reinforce embankment. However, there is little difference of width of sliding zone, which is depicted in Figure 1, between Reinforce and Non-reinforce embankment. The results of numerical simulation did not agree with those of centrifuge model tests. The further research would be required to resolve this difference.

5 CONCLUSIONS

In this paper, the improvement of seismic resistance of embankments by reinforcing their toe was discussed through experiments and numerical analyses. The main conclusions are summarized as follows.

1) The seismic resistance of embankments increases with reinforcing their toe.
2) The volume of sliding mass decreases with reinforcing the toe of embankments.
3) The width of sliding zone in Reinforce embankment is less than that in Non-reinforce embankment in the centrifuge model test.
4) There are little deference of width of sliding zone between Reinforce and Non-reinforce embankment.

REFERENCES

Crisfield, M.A. 1981. A fast incremental/iterative solution proceedure that handles snap through. Com. & Struc. 13, 55–62.

Oritz, M. & Simo, J.C. 1986. An analysis of a new class of integration algorithms for elastoplastic constitutive relations. I.J. Nu. Meth. Eng. 23, 353–366.

Tanimura, K., Oda, K., Tokida, K. & Egawa, Y. 2006. Applicability of elasto-plastic ultimate analysis to performance of road embankment at earthquake. Proc. of the 41th Annual meetings of JGS. 1279–1280.

Yoshino, T., Tokida, K., Nabeshima, Y. Nakahira, A. & Otsuki, A. 2006. Seismic centrifuge model test about slope failure of road embankment. Proc. of the 41th Annual meetings of JGS. 2087–2090.

New Horizons in Earth Reinforcement – Otani, Miyata & Mukunoki (eds)
© 2008 Taylor & Francis Group, London, ISBN 978-0-415-45775-0

Shaking table tests on the mechanism to stabilize slopes by steel nails during earthquakes

S.Yasuda, C. Higuchi & C. Ishii
Tokyo Denki University, Saitama, Japan

N. Iwasa
Nippon Steel & Sumikin Metal Products, Tokyo, Japan

ABSTRACT: Several shaking table tests were carried out to study the mechanism to stabilize slopes by steel nails. Weathered granite soil was filled in a soil container. Then the soil was strengthened by iron rods, plates and wire ropes. Five types of model slopes were selected: i) without iron rods, ii) with non-anchored iron rods, iii) with anchored iron rods, iv) with anchored iron rods, and connected by wires, and v) with iron rods, but the plates are small. Test results showed that this method is effective to prevent slope failure. And it was clarified that combination of iron rods, plates and wire ropes is important.

1 INTRODUCTION

One slope in Kashiwazaki City was protected by Non-frame method, which is a steel nailing method, in spring of 2004 in Japan. Heavy rain and Typhoon hit the slope in July and October. No deformation occurred due to two triggers. Moreover, the 2004 Niigataken-chuetsu earthquake hit this area on October 23. Liquefaction-induced damage to houses occurred at several sites in Kashiwazaki City. However, no deformation was induced at the treated slope. In 2005, Fukuokaken-seiho-oki earthquake occurred and caused slope failure at several sites. However, one slope protected by the method at Sawara-ku in Fukuoka City, shown in Photo 1, was also stable. These examples imply the effectiveness of this method during earthquakes. Seismic intensities at the treated slopes during two earthquakes were about 5 in JMA scale.

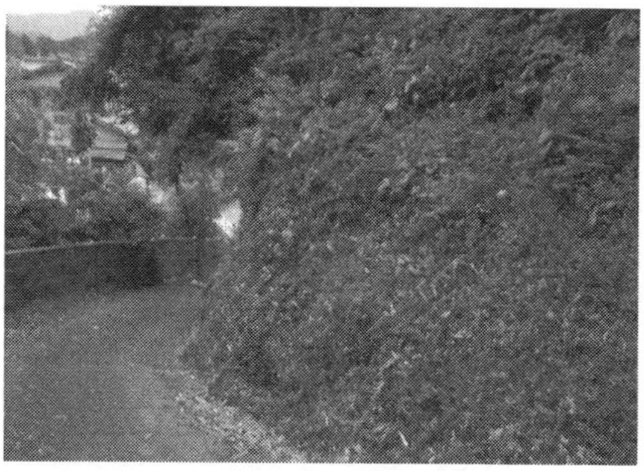

Photo 1. Treated slope by Non-frame method in Fukuoka.

Non-frame method is one of the nailing methods to stabilize natural slopes against heavy rains and earthquakes. In this method bore holes are drilled with triangle alignment as shown in Figure 1. Iron rods are inserted in the bore holes. Tips of the iron rods are inserted in stable rock or soil for 1 m. The bored holes are grout-filled with mortar. Then, place plates and caps at the top of the rods. The plates are connected by wire ropes. Thus unstable layers of a slope can be stabilized by the combination of the iron rods, plates and wire ropes. This method can be applied without cutting existing trees. Therefore, stabilizing effect by the roots of existing trees is not diminished. Moreover, treated slopes are beautiful because the slopes are covered with natural grasses and trees. As drilling machines are small and light, the method can be applied to steep slopes.

The method must prevent failure of slopes due to the following three effects:

(1) prevent the slide of soil mass by bending resistance of iron rods,
(2) prevent the deformation of slope surface by the pressure between plates and slope surface
(3) prevent the deformation of slope surface by the pressure between wire ropes and slope surface

However, these effects during earthquakes are not clear. Then the authors conducted shaking table tests to demonstrate the mechanism of stabilization and effect of each factor.

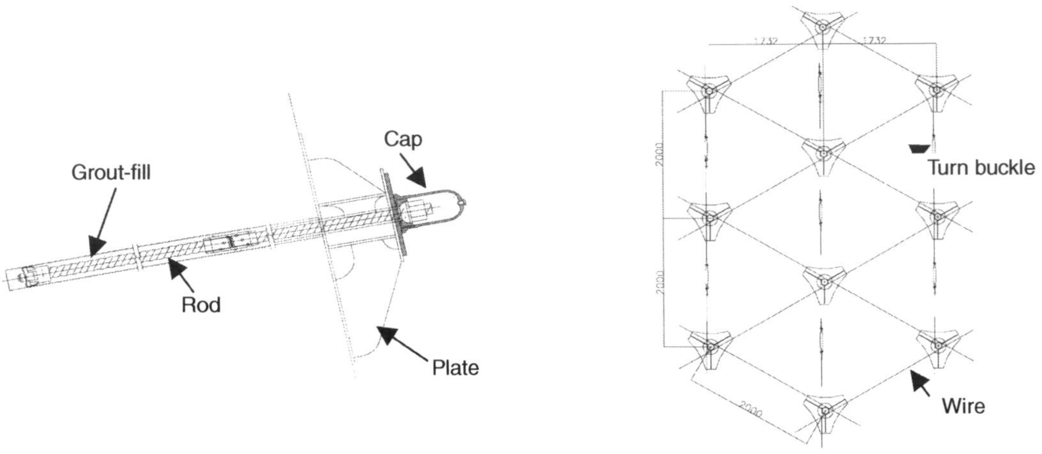

Figure 1. Non-frame method.

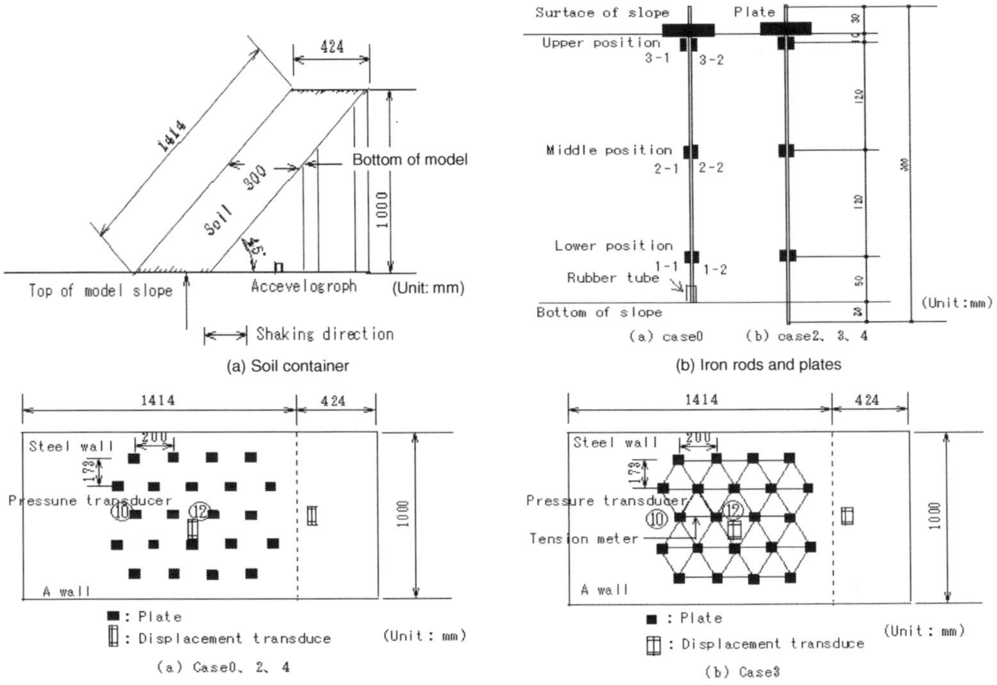

Figure 2. Soil container, iron rods and plates.

2 METHOD OF SHAKING TABLE TESTS

Several 1G shaking table tests were carried out to study the mechanism of the Non-frame method. In these tests, a soil container with a length of about 1,400 mm and a depth of about 400 mm was used as shown in Figure 2. Model iron rods were stood on the base plate of the soil container. Then weather granite soil was filled with a thickness of 100 mm. Fines content of the weathered granite is 14.2%. The soil was compacted in two densities, (1) loose: $\rho_t = 1.5\,\mathrm{g/cm^3}$, Dc = 70.8%, and (2) dense: $\rho_t = 1.6\,\mathrm{g/cm^3}$, Dc = 75.5%. After the compaction of the first layer, the following other layers were filled and compacted in the same way. Plates and wires were connected at the top of the iron rods. Then the soil container was tilted to make a slope.

Table 1 shows test conditions. Five types of model slopes were selected for loose soil: i) without iron rods,

Table 1. Test conditions and test results.

Test No.	Anchore of nails to bottom plate	ρ_t (g/cm^3) W(%) Dc(%)	Plate Shape	Plate Width (mm)	Plate Thickness (mm)	Tension of wire	Critical Acceleraiton (gal)
Case 0	Non-anchored	$\rho_t = 1.5$ W = 11.5	Square	50	5	–	650
Case 1	–	Dc = 70.8		–	–	–	295
Case 2	Anchored		Square	50	5	–	911
Case 3	Anchored		Square	50	5	3N	947
Case 4	Anchored		Square	30	5	–	770
Case D1	–	$\rho_t = 1.6$ W = 5.0	–	–	–	–	410
Case D2	Anchored	Dc = 75.5	Square	50	5	–	899

W: Water content, DC: Degree of compaction.

(a) Case 1 (b) Case 0 (c) Case 3

Photo 2. Failed slopes at critical acceleration.

ii) with non-anchored iron rods,, iii) with anchored iron rods,, iv) with anchored iron rods, and connected by wires, and v) with iron rods, but the plates are small. And, two types of model slopes were selected for dense soil.

Fifty cycles of sine waves at 5 cycles/s were applied to these model slopes. Acceleration of the shaking increased from 50 Gals up to the acceleration to cause slope failure, with an increment of 50 Gals. Occurrence of the slope failure was judged when some zone failed as shown in Photo 2. The input acceleration to cause the slope failure is called as the critical acceleration hereafter. During the shaking, accelerations on the ground surface, displacements of the ground surface, stresses of iron rod, stresses of wires and pressures of plates were measured, as shown in Figure 2.

3 TEST RESULTS

3.1 *Critical acceleration to cause failure*

Critical accelerations are shown in Table 1. In Case 1, settlement of top of the slope increased with input acceleration. Then entire slope failure occurred when input acceleration reached to 295 Gals, as show in Photo 2(a). After the test, the ground was excavated carefully. A clear slip surface was observed at the depth of about 10 cm. In Case 0, settlement of top of slope increased with input acceleration also. When the input acceleration reached to 650 Gals, non-treated zone failed as shown in Photo 2(b). As the critical acceleration was higher than that in Case 0, it can be said that the inserted iron rods resisted against sliding, even the iron rods were not anchored to base steel plate.

In Cases 2 and 3, no obvious settlement or deformation occurred in treated zone even input acceleration exceeded 900 Gals. Partial sliding occurred in non-treated zone when input acceleration exceeded 900 Gals, as shown in Photos 2(c). Anchored iron rods were very effective to resist against sliding. By comparing Case 2 and Case 3, deformation of the slid soil in Case 3 was smaller than that in Case 2. Therefore, it seemed that wire ropes prevented the deformation of slope.

In Case 4, slide occurred at 770 Gals. The critical acceleration to cause failure was smaller than that in Case 2. Size of plates must be important.

Critical acceleration in Case D1 was high as 410 Gals. It was about 1.5 times compared with the critical acceleration in Case 1. Soil density fairly affected the critical acceleration.

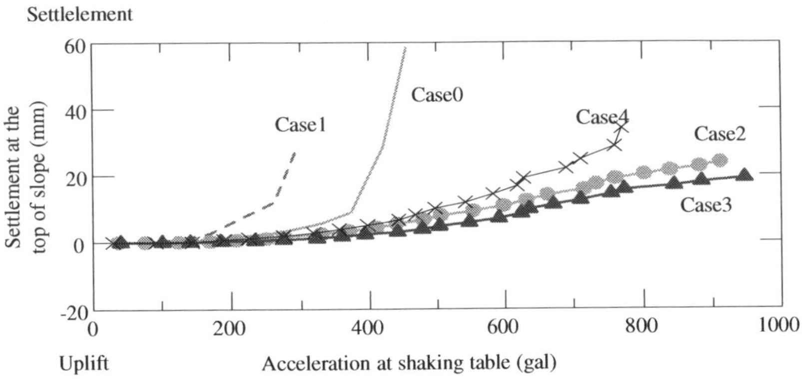

Figure 3. Relationships between input acceleration and settlement at the top of slope (loose soil).

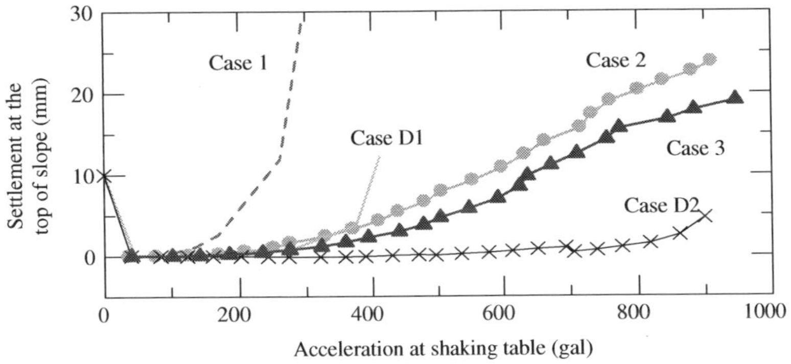

Figure 4. Effect of soil density on the settlement at the top of slope.

3.2 *Comparison of settlement of the top of slope*

Figure 3 shows relationships between input acceleration and settlement of top of slope. Settlement increased with input acceleration in all case. However, amount of settlement was different in each case. The settlement in Case 3 was smallest. And the settlement increased in the order of Case 2, Case 4, Case 0 and Case 1. The order was same as the order of critical acceleration to cause failure, as mentioned before.

At 600 Gals, the settlement in Case 2 was 3 to 6 mm less than that in Case 4. This difference must be attributed to the difference of size of plates. The settlement in Case 3 was about 6 mm less than that in Case 2. This implies the importance of wire ropes.

Figure 4 shows effects of soil densities on the settlement. Settlement increased gradually with increase of acceleration in loose soils. On the contrary, in dense soils, settlements occurred suddenly if acceleration exceeded the critical accelerations.

3.3 *Bending strain and axial strain of iron rods*

Figure 5 shows relationships between input acceleration and bending strain of iron rods in Case 0 and Case 2. In case 2, bending strain of the middle iron rods increased with input acceleration. At lower iron rods, bending strain decreased with input acceleration. In Case 0, on the contrary, bending strain did not increase up to the acceleration of 400 Gals. This means that iron rods were pulled out from base plate when input acceleration reached 400 Gals.

Figure 6 shows effects of soil density on the bending strain. In Case D2, bending strain did not increase with acceleration. Therefore, it seems that bending resistance can not be displayed in dense soils.

Relationships between input acceleration and axial strain are shown in Figure 7 for Case 0 and Case 2. In Case 2, axial strain decreased with input acceleration, because soil was compacted due to shaking. In contrast, axial strain in Case 0 increased with input acceleration. Iron rods must be stretched due to sliding of soil.

Figure 8 compares axial strains in Cases 2 and D2. Axial strain in Case 2 decreased with acceleration, mentioned above. On the contrary, axial strain in Case D2 increased with acceleration.

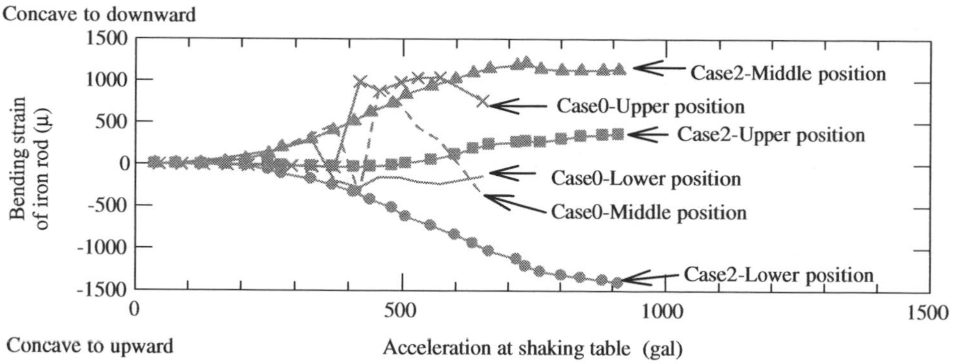

Figure 5. Relationships between input acceleration and bending moment of iron rod (Case 0, Case 2).

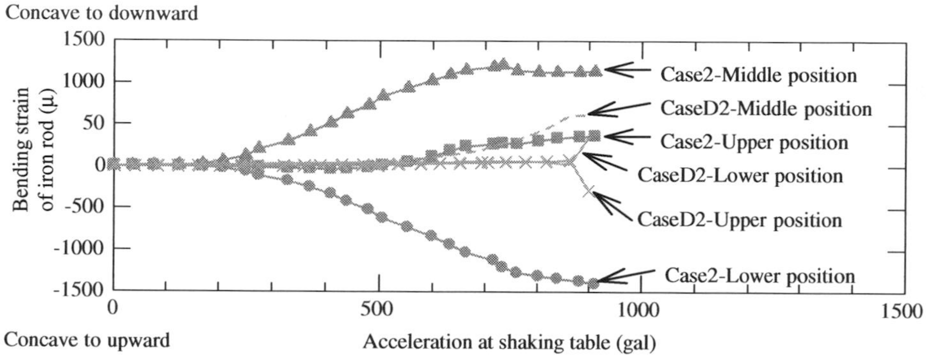

Figure 6. Effect of soil density on the bending strain of iron rod.

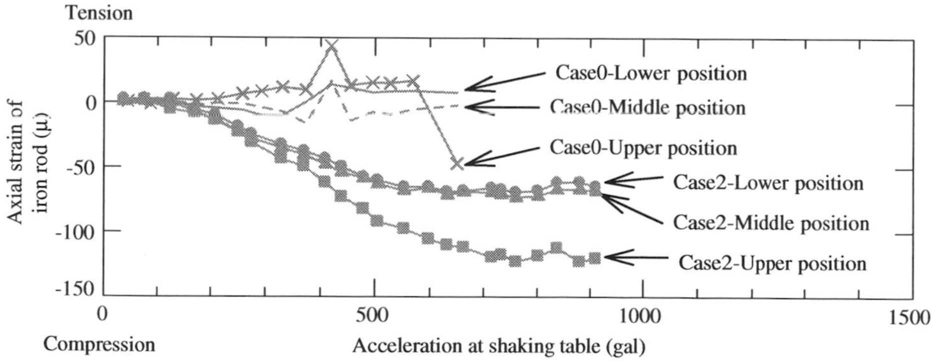

Figure 7. Relationships between input acceleration and axial strain of iron rod (Case 0, Case 2).

3.4 *Settlement of slope surface*

Figure 9 shows relationships between input acceleration and settlement measured at the middle of slope surface. In Case 2, settlement at the center of the slope surface increased with input acceleration. In contrast, slope surface heaved in Case 1 and Case 4. The heaving occurred when input acceleration reached to about 400 Gals and 600 Gals for Case 1 and Case 4, respectively. In Case 3, ground surface did neither settle nor heave. Wire ropes must prevent the settlement or heaving.

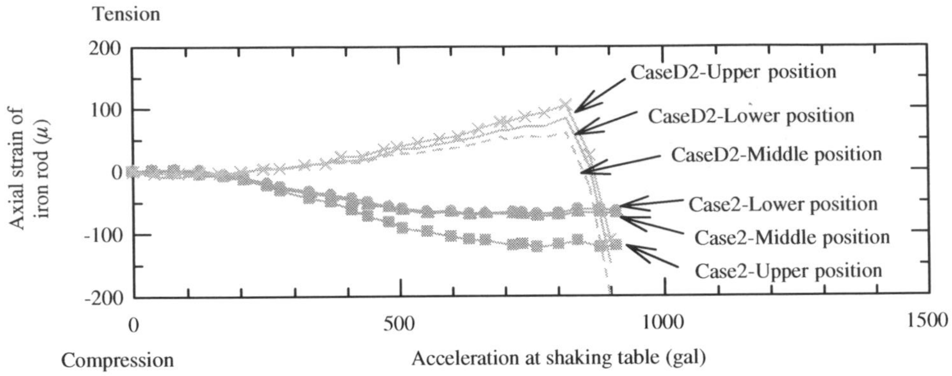

Figure 8. Effect of soil density on the axial strain of iron rod.

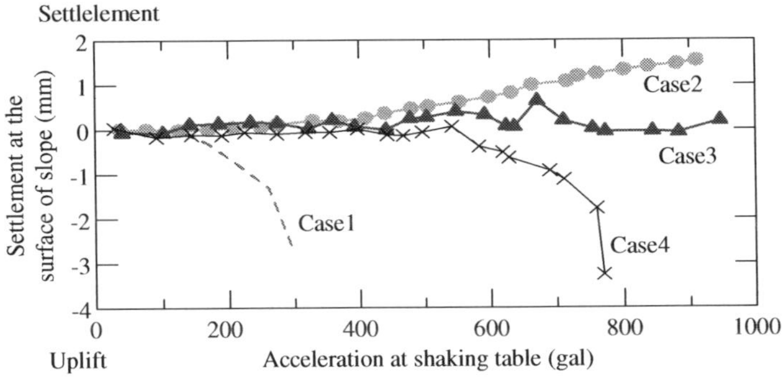

Figure 9. Relationships between input acceleration and settlement at the surface of slope.

4 CONCLUSIONS

Several 1G shaking tests were conducted to study the mechanism of the Non-frame method during earthquakes. The following conclusions were derived through the tests:

(1) Non-frame method is effective to prevent slope failure during earthquake.
(2) Iron rods must be fixed to stable bottom layer.
(3) Large plates are effective than small plates.
(4) Wire ropes can increase the resistance against failure.
(5) Bending and axial stresses acted on iron rods are affected by soil density.

ACKNOWLEDGEMENT

The authors would like to express their thanks to Miss A. Yamada, former students at Tokyo Denki University, for her cooperation in carrying shaking table tests.

REFERENCES

Yasuda S, Ishii C, Yamada A, Iwasa N, Higuchi C, 2006. Model test about the effect of slope works and influence of the anchor for natural slope at the time of the earthquake, *Proc. of the 41st Japan National Conference on Geotechnical Engineering*, pp. 2093–2094. (in Japanese)

Yasuda S, Ishii C, Yamada A, Iwasa N, Higuchi C, 2006. Model test to demonstrate the effectiveness of connection of nailing heads on the stability of natural slopes during earthquakes, *Proc. of the 61st Japan Society of Civil Engineers*, pp. 403–404. (in Japanese)

New Horizons in Earth Reinforcement – Otani, Miyata & Mukunoki (eds)
© 2008 Taylor & Francis Group, London, ISBN 978-0-415-45775-0

Shaking table test for lightweight spillway with geogrid

T. Kawabata, K. Uchida, T. Kitano & K. Watanabe
Faculty of Agriculture, Kobe University, Kobe, Hyogo, Japan

Y. Mohri
Soil Mechanics Laboratory, National Institute for Rural Engineering, Tsukuba, Japan

ABSTRACT: During 1995 Hyogoken-Nanbu earthquake, heavy concrete spillways on the embankment of small reservoir were significantly damaged. Especially, it was pointed out that some spillways were detached from the embankment because of the inertia force, and the incidents could be the cause of secondary disaster in downstream area in floods. However, the research on spillways on such embankment is lacking.

In this paper, firstly, a series of shaking tests at National Institute for Rural Engineering in Japan on heavy rigid spillway model was discussed. As the result, the detachment of the heavy spillway model from the embankment was confirmed. Moreover, newly-proposed countermeasures to fix the lightweight spillway by geogrid were discussed, and the superiority of the proposed method were confirmed.

1 INTRODUCTION

Japan has over 20 thousand small earth dams for irrigation. There is a spillway to discharge overflowed water safely to downstream during in high water level. In general, spillways on the embankment are made by reinforced concrete. On the other hand, the embankments themselves are soil structure. Therefore, in earthquakes, it was pointed out that the spillways detachment from the embankments or failure in the joint of the bottom plate and the sidewall might be occurred due to inertia force. Picture 1 shows the damage by the detachment of the concrete type spillway from the embankment during Hyogoken-Nanbu earthquake in 1995. In order to prevent the detachment, the researches on the safety of the structure are required.

As the countermeasure against earthquake, examinations on the behavior of geogrid-reinforced construction are discussed. For example, Watanabe et al. (2002) proposed a new seismic type bridge abutment with geogrid-reinforced cement treated backfill, Koseki et al. (2002) performed comparison of model shaking test results on reinforced-soil and gravity type retaining wall and Bathurst et al. (2002) conducted to assess the seismic performance of reinforced soil walls. However, most of the research on the geogrid-reinforced techniques has discussed the single wall or plate with the reinforced soil, there is lacking in discussion that could be applied to spillways in terms of the rectangular cross-section.

In this paper, a series of shaking table tests on the heavy spillway model, the lightweight PVC spillway models, and the newly proposed lightweight spillway models with geogrid were discussed.

Picture 1. The damage of small earth dam spillway at Hyogoken-nanbu earthquake (Hyogo pref.).

2 SHAKING TABLE TEST

2.1 *Equipments setup*

The shaking table for the tests had plane dimensions of 6.0 × 4.0 m, with the maximum loading capacity of 50 tf and the maximum displacement capacity of 150 mm. The test pit (2400 × 1900 × 1000 mm) was placed on the shaking table and the experimental models were placed. The embankment size was 2400 × 1900 × 800 mm with 45 degree slope and 300 mm width of the crest as shown in Picture 2. The spillway models were positioned in the center

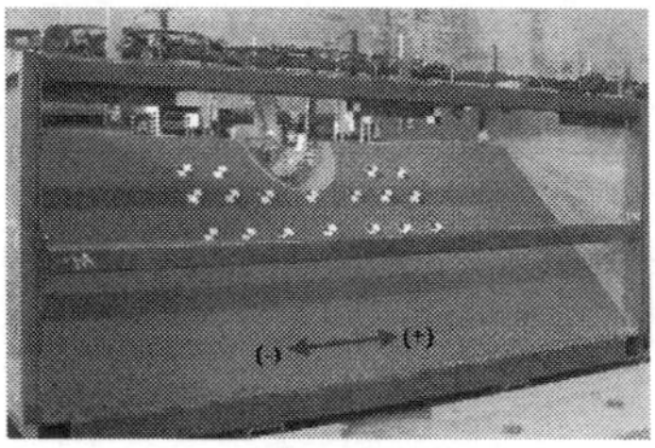

Picture 2. Embankment model (Case C).

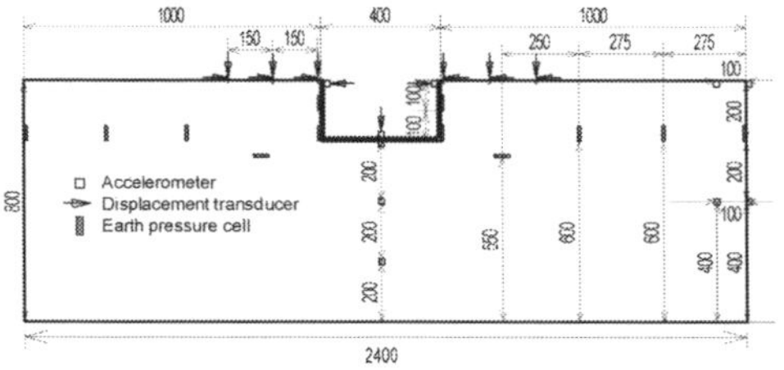

Figure 1. Elevation view of the test pit and positions of test equipments (Case A).

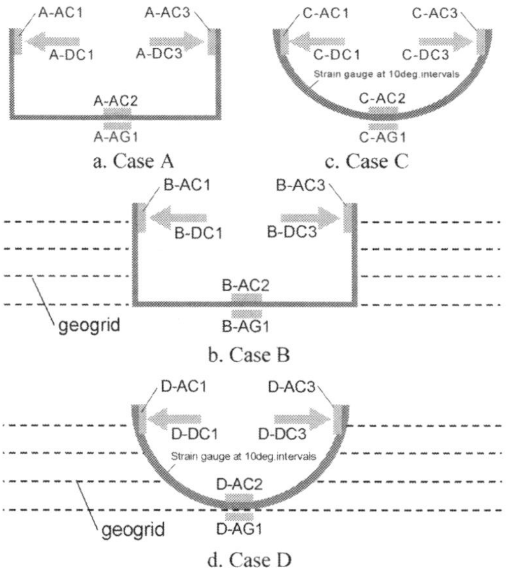

Figure 2. Discussed measure points in this paper (Initial letter is correspondent to the Cases).

of the crests. Figure 1 shows schematic elevation view of the test models. The models of embankments and spillways were instrumented extensively with instruments: accelerometers measured horizontal acceleration response of spillway models and grounds, displacement transducers measured displacement of the spillway models and ground level, and earth pressure cells measured earth pressure around soil of the models. As for the acceleration and displacement sign,

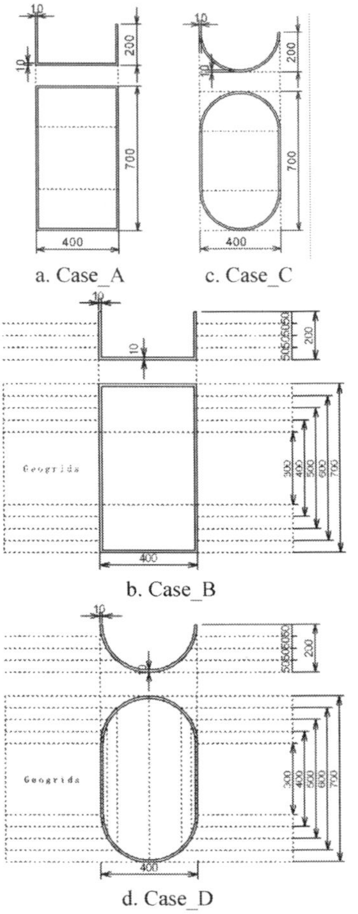

Figure 3. The spillway models.

right on the observer was positive in Figure 1. Note that the positions of the equipments were the same in every case. In addition, strain gauges were attached on inner surface of the spillway models and both surfaces along centerline of geogrid at appropriate intervals.

In this paper, the accelerometers of spillway models and the ground (AC1, AC2, AC3, and AG1 in Figure 2), displacement transducer of the spillway models (DC1, DC3) and strain gauges in the Case_C and D were mainly discussed.

2.2 *Materials and experimental program*

The soil material for the embankment was Kasumigaura sand with the mean particle diameter (D_{50}) of 0.40 mm and the uniformity coefficient (U_C) of 3.16. The relative density (D_r) was about 65.5%. The geogrid used in the tests was HDPE and the tensile strength was 3.5 kN/m.

Figure 3 shows the spillway models. The model in Case A was made of rigid steel plate; other models

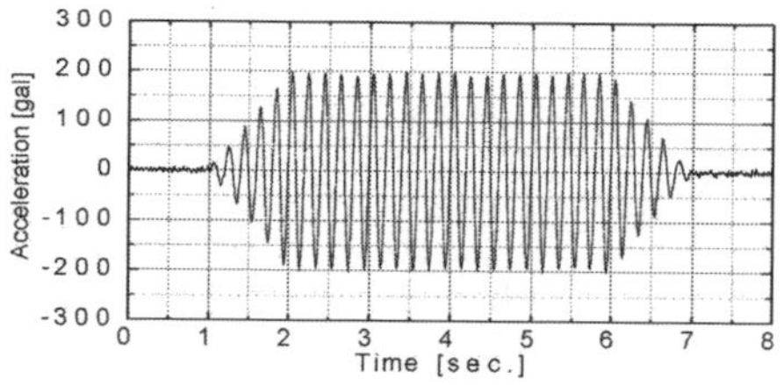

Figure 4. Input wave (200 gal).

were made of polyvinylchloride. The models in Case B and D were reinforced by geogrids.

In the experiments, 5 Hz horizontal sine wave was applied along the embankment axis. Figure 4 shows input wave. The shaking was applied at 200 gal, 400 gal, 600 gal, and 800 gal.

3 TEST RESULTS

3.1 *Response acceleration*

Figure 5 shows the response accelerations of spillway model and the ground beneath the model in 600 gal shaking.

In Figure 5a, it was observed that A-AC2, which was positioned the bottom of the spillway model showed larger response acceleration amplitude than A-AG1. From this fact, it was confirmed that there was difference in inertia force between the heavy spillway model and the embankment. Moreover, it was visually observed that the spillway model in Case A showed slight detachment from the ground after 600 gal shaking. From these facts, it was confirmed that heavy spillway was detached when the shaking was applied in the horizontal direction to the embankment axis. The response acceleration of heavy spill way model also showed the rapid decline after the peak response acceleration. The response might be given the stiffness of the spillway model. On the other hand, the response acceleration of B-AC2 and B-AG1 showed almost same amplitude. That implied the spillway model with geogrid and surrounding soil was moved simultaneously, and it was just conceivable that the model with geogrid was fixed to the embankment well.

In Figure 5b, it was observed that the geogrid-fixed spillway model in Case D showed the same acceleration level to the ground accelerations. From this result, it was safe to say that the cylindrical spillway model with geogrid was also fixed to the embankment during the shaking.

From the discussion of the response acceleration, it was also implied the geogrid-fixed spillway models were moved with the embankment under the models simultaneously although the reinforcement was equipped on the sidewalls. This facts suggested that

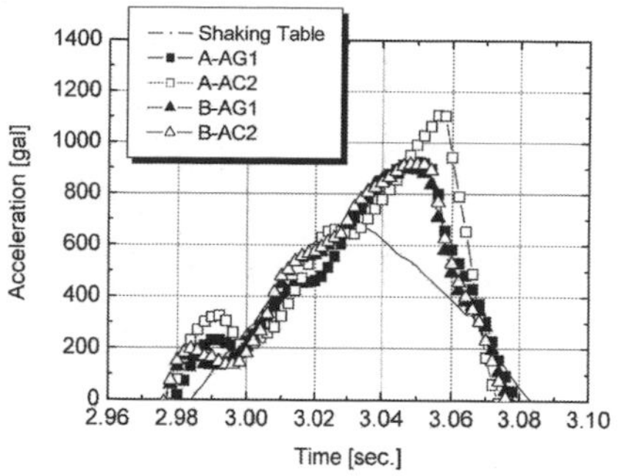

a. Case_A and B

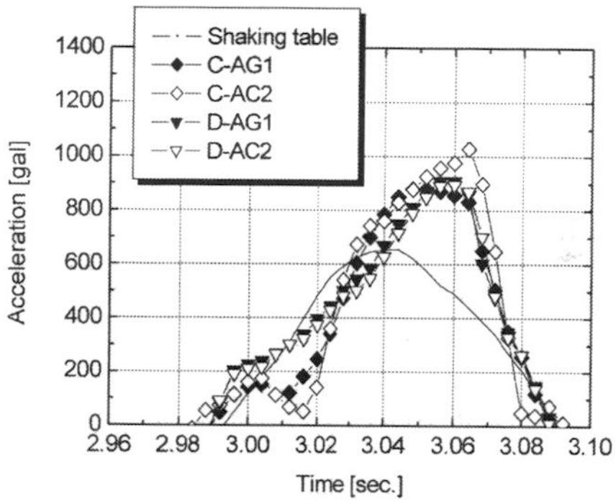

b. Case_C and D

Figure 5. Response accelerations of the spillway models in 600 gal shaking.

surrounding was moved by side positioned geogrid as soil mass.

3.2 *Response displacements*

Figure 6 shows response displacement in 800 gal shaking. In Case_A in Figure 6a, the response displacement of the spillway model was increased largely from 3.5 sec. and the maximum displacement amplitude approached to 15 mm. The spillway itself displaced about 10 mm as the residual displacement to the left from the initial position. On the other hand, in Case_B in Figure 6b, the response displacement spillway was increased from 5.0 sec., the maximum displacement amplitude was less than 5 mm. and the residual displacement was almost zero. From these response and residual displacements of Case A and Case B, the behaviors of the Case_B model showed the toughness as observed in reinforced soil compared with Case_A (Uchimura et al. 2002). Moreover, during the experiments, it was clearly and visually confirmed that the heavy spillway was detached from embankment in the 800 gal shaking. In addition, in reinforced rectangular spillway model in Case_B, it was observed that the both sidewalls were stretched to the backfills about

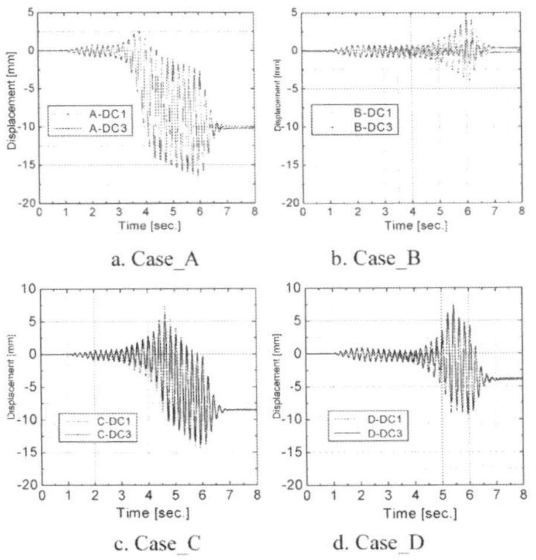

Figure 6. Response displacement in 800 gal shaking.

1.0 mm. The mechanism was not clarified, however, it was considered that the results was resulted by the mutual effect of geogrid and the embankment behind the both sidewalls. On this point, further discussion is needed to clarify the mechanism.

Figure 6c shows the response displacement of the model in Case_C increased rapidly from 3.0 sec. and residual displacement of the spillway model in Case_C was about 8.5 mm to left. On the other hand, Figure 6d shows the response displacement in Case_D increased from 4.0 sec. and the residual displacement of the spillway model in Case_D was about 4.0 mm to left. The reason was considered that Case_D model also showed the toughness as observed in reinforced soil, or Case B.

3.3 *Strain distribution*

Figure 7 shows transition of the deformation of the spillway models in Case_C and D while the shaking table was moving from the left end to the right end through the stroke.

Figure 7a–c shows the deformation transition in Case_C. In Figure 6a, the spillway model was concaved in 45deg. neighborhood. The spillway model was deformed by the active earth pressure at the left end of the shaking table. In addition, the spillway model was convexed in 135deg. neighborhood. In Case_C without geogrid reinforcement, the spillway model was deformed to the left when the shaking table was positioned at the left end and was deformed to the right when the shaking table was positioned at the right end.

Figure 7d–f shows the deformation transition of the spillway model equipped with geogrids. In Figure 7d, the spillway was convexed in the 45deg. neighborhood although the spillway model of Case C was concaved

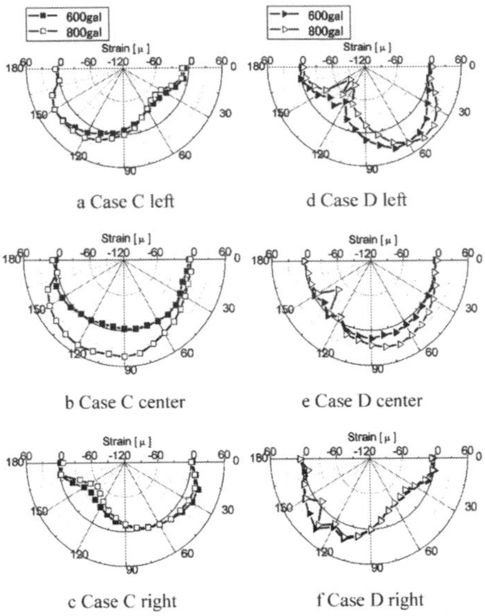

Figure 7. Strain distributions of Case C and D, during a stroke of shaking table.

in that area. In the transition, the shape of Case C was reversed to Case_D with geogrids although the time to measure the deformation was same. This tendency might be resulted from the mutual effect of geogrid, the spillway rigidity of the models, or earth pressure. However, more experiments, discussions and analysis are needed to clarify the mechanism.

4 CONCLUSION

In this paper, it was reported the results of shaking table test for the heavy spillway model assumed existing concrete-made spillway and the newly proposed lightweight countermeasures.

In the experiments, the heavy spillway model showed the detachment or large displacement by the shaking applied in the horizontal direction to the embankment axis, which was pointed out incidents in Hyogo-Nanbu earthquake.

As for the results of the proposed method, it was confirmed that the reinforced spillway models in Case_B and D were fixed to the embankment and showed the toughness during the shaking in terms of response displacement and showed less residual displacement than that of models without geogrids. Moreover, in Case_B, the sidewalls of the spillway model showed stretched to the backfill, or in Case_D, the deformation transition was opposite to the transition in Case_C. It was considered that these mechanisms were strongly tied to the rigidity of spillway model shapes and geogrid-reinforced earth, however, more experiment or discussion on these points are necessary to clarify the mechanism of newly proposed method.

REFERENCES

Bathurst, R.J., El-emam, M.M. & Mashhour, M.M. (2002) "Shaking table model study on the dynamic response of reinforced soil walls", *Geosynthetics-7th ICG-*, vol. 1, pp. 99–102.

Koseki, J., Watanabe, K., Tateyama, M. & Kojima, K. (2002) "Comparison of model shaking test results on reinforcd-soil and gravity type retaining walls", *Geosynthetics-7th ICG-*, vol. 1, pp. 111–114.

Uchimura, T., Tatsuoka, F., Shinoda, M., Sugimura, Y. & Kikuchi, T. (2002) "Roles of tie rods for seismic stability of preloaded and prestressed reinforced soil structures", *Geosynthetics-7th ICG-*, vol. 1, pp. 115–118.

Watanabe, K., Tateyama, M., Yonezawa, T., Aoki, H., Tatsuoka, F. & Koseki, J. (2002) "Shaking table tsets on a new type bridge abutment with geogrid-reinforced cement treated backfill", *Geosynthetics-7th ICG-*, vol. 1, pp. 119–122.

New Horizons in Earth Reinforcement – Otani, Miyata & Mukunoki (eds)
© 2008 Taylor & Francis Group, London, ISBN 978-0-415-45775-0

Shaking table tests on seismic behavior of sand slopes reinforced by carpet strips

H. Shahnazari, A. Fooladi & B. Ghosairi
College of civil engineering, Iran university of science and technology, Tehran, Iran

ABSTRACT: In order to investigate seismic behavior of slopes, made of reinforced sand, a series of shaking table tests were performed. Model slopes were prepared in a box with length of 1.8 m, width of 0.5 m and height of 0.7 m. Sand slopes were reinforced with randomly distributed carpet waste stripes. During application of cyclic loading on unsaturated soil, failure surface and acceleration were monitored in models. Result of this study shows the effects of soil reinforcement during cyclic loading. These results are very useful for study the behavior of different soil structure in which soil is reinforced by mentioned type of reinforcements.

1 INTRODUCTION

Engineering practitioners and researchers have shown a growing interest in laboratory tests for assessing the seismic stability of slopes and embankments. Element and model tests are two different useful methods to understand the behavior of soil structures subjected to different types of loading. Shaking table test is a model test, which has been used in many geotechnical researches. The recent trend in finding solution to geotechnical engineering works has resulted in various construction techniques that are based on developments of new materials and concepts. One of them is the use of reinforcement material such as waste carpet stripes that distributed in mixture with soil randomly.

Clough and Pirtz (1956) performed the first well-documented shaking table study on seismic slope stability. Later, Goodman and Seed (1965) performed shaking table tests on an inclined layer of sand. They focused on the yield acceleration (i.e., the acceleration required to bring the slope to a condition of marginal instability) and they found that calculated yield accelerations is comparable with measured test results. Arango and Seed (1974) used a modified version of the shaking table described above to investigate the seismic stability of slopes. They found that strong shaking resulted in development of a distinct "yield acceleration" that marked initiation of permanent deformation in their slopes. Lin and Wang (2006) studied the earthquake resistance of slopes. In their research large-scale shaking table tests were conducted to study the slope behavior under earthquake loading. They found that the failure surface appeared to be fairly shallow and confined to the slope surface, which was consistent with the field observations of earthquake induced landslides.

Many researches have also studied the use of reinforced soil as a material for geotechnical structures such as dikes, slopes and walls.

Rechardsons and Lee (1975) performed a well-documented shaking table study of seismic behavior of slopes, made of reinforced sand with strips of aluminum and Mylar. They investigated two mechanisms including yielding of strips and their pull out force. Perez (1999) used soil, reinforced with geotextile to study the seismic behavior of slopes during cyclic loading in shaking table. He investigated vertical and horizontal displacement in models. Wang et al (2000) used sand, which was reinforced by carpet strips in triaxial tests. They showed that these elements increase the strength and flexibility of soil.

In this research model experiments were performed on a 1*g* shaking table with the objectives of investigating the mechanisms of seismically induced permanent deformations and amplification of strong ground motion.

2 EXPERIMENTAL STUDY

The current study was performed by using a 500 mm wide, 1800 mm long and 700 mm height single-degree-of-freedom shaking table. Data acquisition software was used to acquire signals from different data channels.

The soil used in this study was uniform medium sand (Firouz-Kouh No. 161) which its particle-size distribution curve is shown in Fig. 1. Specimen were

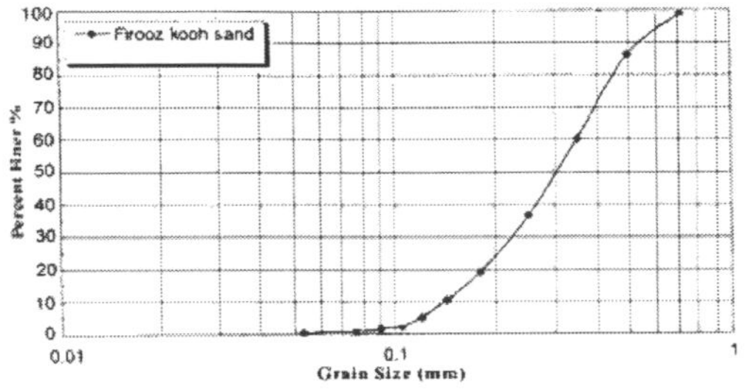

Figure 1. Particle-size distribution curve of the sand (Firouz *kouh* No. 161).

Table 1. Mechanical properties of materials.

Parameter	Reinforced sand
D_{50}	0.3
c' (kpa)	0.05
ϕ'	32
γ_d (kN/m^3)	14.92
e_{max}	0.85
e_{min}	0.608

prepared using volume controlled method of compaction with a water content of 5%.

Reinforced element which used in this research were discrete strips of carpets with 5 mm wide and aspect ratio (ratio of length to width) of 5.

These carpet strips were randomly distributed in sand. Direct shear tests were also performed on the specimens obtained from the slope models. Table 1 shows the results of tests.

3 MODEL PREPARATION

Specimens were compacted into the test box using the volume-controlled method to reach to a unit weight of 14.92 kN/m^3. In this method, specific weight of soil was put in the box first, and then compaction was being continued to get the target height based on density. Figure 2 shows the compaction of model.

After installing the accelerometer in the base of models, other soil layers were compacted similar to the first layer. Finally, another accelerometer was embedded on the crest. Figure 3 reveals the location of accelerometers.

4 LOADING AND TEST RESULTS

Experiments were performed on four slope models as following. The size of the model slope in first test was 0.35 m high, 0.5 m length of crest, and with both side slope of 1:2.5 (horizontal : vertical). Maximum amplitude of applied acceleration in the base of model was as 0.153 g. Figure 4 illustrates the time history of applied input acceleration in the base of the model in test No. 1. The maximum amplitude of acceleration in the crest was recorded as 0.188 g. Figure 5 shows the time history of acceleration on crest of model.

Model No. 2 had 0.35 m high, 0.5 m length of crest, and side slopes of 1 : 3 (horizontal : vertical). Similar to model No. 1, Model No. 2 was subjected to a low amplitude motion, and experienced relatively small

Figure 2. Compaction of model.

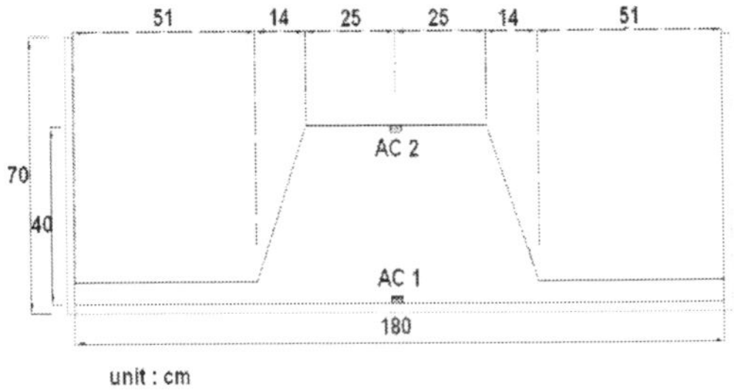

Figure 3. Position of accelerometers.

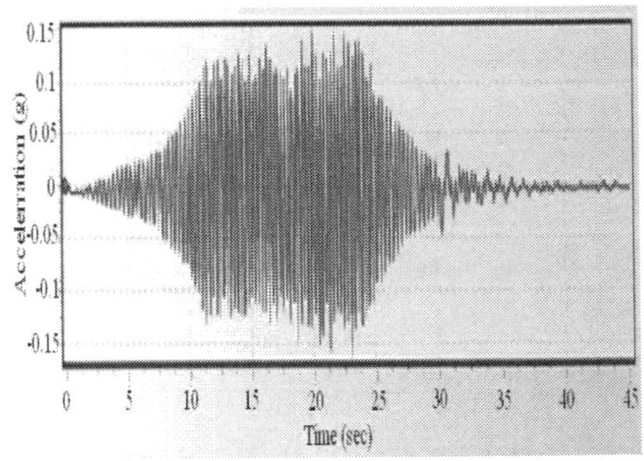

Figure 4. Time history of applied acceleration in base of test No. 1.

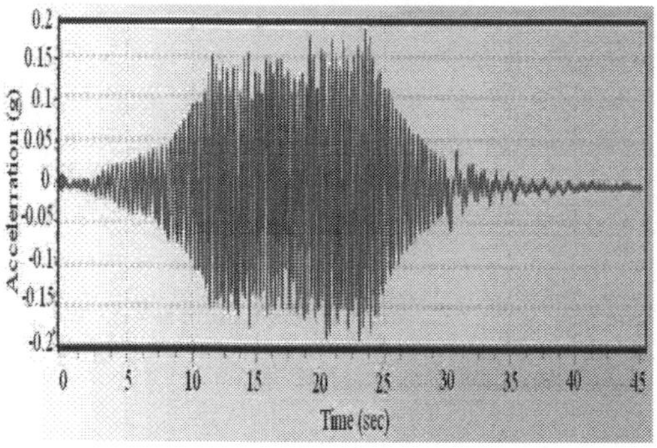

Figure 5. Time history of measured acceleration in crest of test No. 1.

Table 2. Geometry of models, recorded acceleration and amplification.

Model	Side slope H/V	Base Acc. (g)	Crest Acc. (g)	Amplification n
No. 1	1:2.5	0.153	0.188	1.229
No. 2	1:3	0.142	0.181	1.275
No. 3	1:3	0.164	0.205	1.250
No. 4	1:3	0.213	0.234	1.100

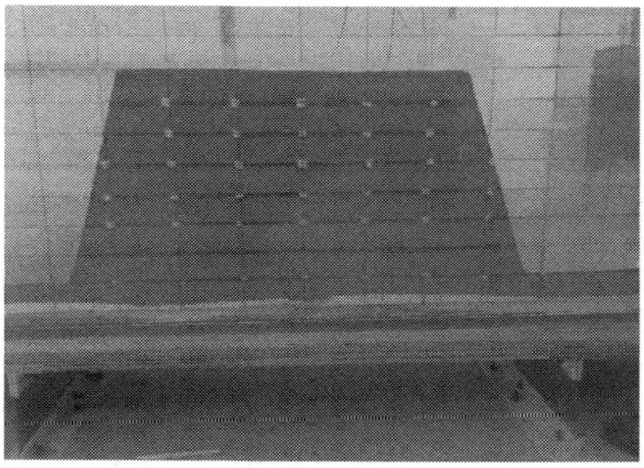

Figure 6. The model No. 4 before shaking.

deformations with development of a localized shear surface.

Models No. 3 and No. 4 were subjected to higher amplitude of acceleration (see Table 2). After application of dynamic loads to these models, model No. 3 experienced small deformation in the crest. However, in Model No. 4 failure surface happened and an unstable block moved downward. Figures 6 and 7 show the model No. 4 before and after shaking. Displacement of the moving block was more than 11 cm in horizontal and 24 cm in vertical direction. The observed failure surface appeared to be relatively deep in the face of the slope. Figure 7 also reveals that the failure surface was near to circular shape.

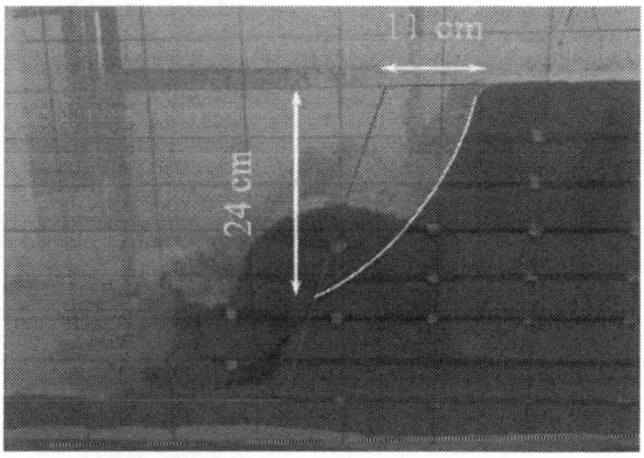

Figure 7. The model No. 4 after shaking.

Results of these tests can also be discussed as viewpoint of soil amplification. Table 2 shows the maximum-recorded acceleration of base, crest and the ratio of crest to base acceleration, which is defined as amplification factor. As it can be seen from this table in all samples amplification of acceleration happened. The range of amplification ratio is from 1.1 to 1.28. It is well known that the amplification of strong ground motion is depends on different parameters such as amplitude and frequency of input motion and geometry and density of soil structure.

Results of these tests and other similar tests can be used for study of dynamic behavior of reinforced sand when they combine with element tests results and numerical analysis.

5 CONCLUSION

In order to investigate the dynamic behavior of reinforced soil with carpet strips a series shaking table tests were performed. Results of these tests shows that:

1. Input acceleration is amplified in all tests. Range of amplification factor is from 1.1 to 1.28. It is known that this amplification depend on different parameters such as amplitude and frequency of loading, geometry and density of soil structure. More tests should be run to clear these effects.
2. Measured shape and geometry of failure surface in model No. 4 and displacement vector of downward moved block is very useful for study the behavior of real scale structures when theses results combine with results of different element test and numerical analysis

REFERENCES

Arango, I., Seed, H. B. (1974). "Seismic stability and deformation of clay slopes", Journal of Geotechnical

Engineering, American Society of Civil Engineering., 100(2), 139–156.
Goodman, R. E., Seed, H. B. (1965). "Displacements of slopes in cohesionless materials during earthquakes.", Rep. No. H21., Inst. Of Trans. and Traffic Engineering, Univ. of Calif., Berkeley, Calif.
Lin, M. L., Wang, K. L., (2006). "Seismic slope behavior in a large scale shaking table model test", Journal of Engineering Geology 86, 118–133.
Perez, A. (1999). "Seismic response of geosynthetic reinforced steep slopes" , M.S. Thesis, University of Washington.
Richardson, G. N. and Lee, K. L. (1975). "Seismic design of reinforced earth walls", Journal of the Geotechnical Engineering, Volume 101, No. GT-2, ASCE, pp. 167-188.
Wang, Y., Frost, J. D. Murray. J. Jones, A. (1999). "Utilization of carpet, textile and apparel waste for soil reinforcement.", ARC 99, SPE Annual Recycling.

New Horizons in Earth Reinforcement – Otani, Miyata & Mukunoki (eds)
© 2008 Taylor & Francis Group, London, ISBN 978-0-415-45775-0

Seismic design of mechanically stabilized wall structures

J.H. Wood
John Wood Consulting, Lower Hutt, New Zealand

D.E. Asbey-Palmer
Reinforced Earth Ltd, Auckland, New Zealand

C.W. Lawson
The Reinforced Earth Company, Hornsby, NSW, Australia

ABSTRACT: New Zealand limiting equilibrium (NZLE), Terre Armee Internationale, Australian Code and Federal Highway Administration earthquake design methods for mechanically stabilized walls are compared by applying them to typical wall geometries. These design procedures gave significantly different results for sliding stability and soil reinforcement density. It was concluded that the NZLE method produced designs that would not undergo significant permanent displacements in strong ground shaking, or alternatively, the method could be used in conjunction with Newmark sliding block theory to predict displacements. Walls designed in accordance with the other methods will move outwards an amount that cannot be readily estimated.

1 INTRODUCTION

Many retaining walls mechanically stabilized with steel strips (Reinforced Earth) have been subjected to strong ground shaking in major earthquakes in Japan, Turkey and USA (Kobayashi et al, 1996, Freyssinet, 2000, TAI, 1994). There have been reports of Reinforced Earth (RE) walls moving outwards or settling but there have been no reports of wall failures. Over 100 RE walls were subjected to strong shaking in the 1995 Hyogoken-Nanbu earthquake. Approximately 21 of these were subjected to peak horizontal ground accelerations (PGA's) greater than 0.4 g but none sustained significant damage. The facings of three of the walls moved outwards between 20 to 110 mm at the top of the walls (Wood and Asbey-Palmer, 1999).

Experimental research (Fairless, 1989) and full scale testing (Richardson and Lee, 1975) has shown that providing RE walls are designed to have ductile failure modes with either material yield in the strips or pull-out of the strips, they can undergo appreciable outward movement without loss of integrity. This characteristic is of considerable advantage and has been a factor leading to good performance of RE walls in shaking with peak ground accelerations considerably greater than design acceleration levels.

In New Zealand, and in other countries in regions of high seismicity, RE walls are specifically designed for earthquake loads. Two different design procedures have evolved. Terre Armee Internationale (TAI, 1989) was first to develop a method based on the active wedge theory used for static loads. A limitation of this method is that it is not possible to determine the critical acceleration at which the Reinforced Block (RB) commences to move outwards; or the extent of outward movement if the critical acceleration is exceeded. The New Zealand limiting equilibrium method (NZLE) was developed (Fairless, 1989, Wood and Elms, 1990) to allow the earthquake design of RE walls to be based on limiting the outward movement in strong ground shaking. Because testing and theory used as the basis of the design methods was unable to clearly identify the influence of soil amplification and the degree of coherence in accelerations over the long length of soil influencing the pressures on the facings, there remains uncertainty about the magnitude and point of application of the inertia forces acting on the RB and the retained soil behind the block.

In this paper, the two different design methods are compared by summarising results obtained for typical design parameters and wall configurations using four different design guideline or code documents. The NZLE design procedure specified in the Transit New Zealand Bridge Manual, 2003, and three different versions of the TAI active wedge procedure; TAI Design Guide, 1991, US Department of Transportation, Federal Highway Administration Guidelines (FHWA-NHI-00-043, 2001) and the Standards

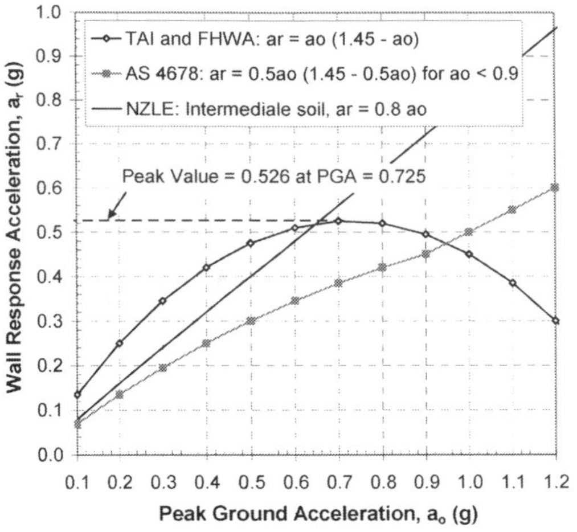

Figure 1. Response acceleration calculation.

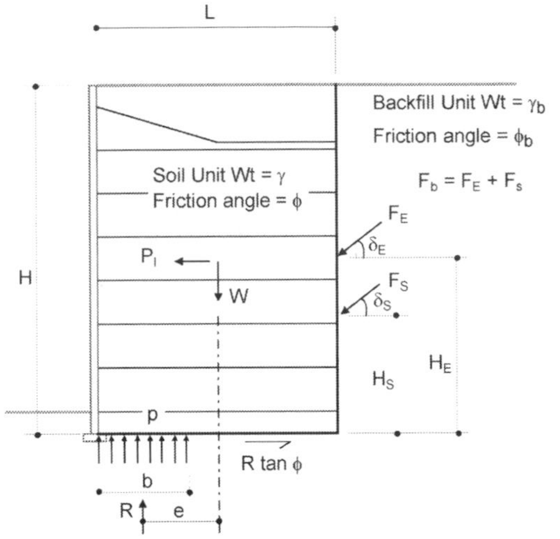

Figure 2. External stability analysis.

Australian Code for earth-retaining structures (AS 4678 – 2002) were investigated. In order to simplify the comparison, the walls were assumed to have rectangular reinforced blocks, horizontal ground surface behind the wall, zero live loading and to be free standing without other structures (such as bridge abutments and building foundations) loading them. The walls were assumed to be acted on by only horizontal earthquake accelerations without simultaneous vertical motions.

2 COMPARISON OF DESIGN METHODS

2.1 *Response acceleration*

The response acceleration acting on the RB and retained soil masses is calculated by modifying (reducing or increasing) the PGA. A comparison of the response acceleration calculation methods is shown in Figure 1. In the NZLE procedure the response acceleration is obtained by reducing the PGA by a constant factor related to the stiffness of the foundation soils. In the other design methods the response acceleration is assumed to vary as a function of the PGA. It both the TAI and FHWA methods the response acceleration is greater than the PGA up to a PGA of 0.45 g and less than the PGA at higher PGA's. In the AS 4678 method the response acceleration is less than the PGA for all values of PGA.

2.2 *External stability*

External stability is investigated by a similar procedure in both the NZLE and TAI active wedge methods although the assumptions regarding the magnitude and application of the earthquake forces on the RB, strength reduction factors and factors of safety differ considerably. External stability design involves checking the resistance to sliding on a plane through the base of the wall using horizontal equilibrium equations, and checking the base pressures (assumed to be uniform) using moment equilibrium equations. Gravity and earthquake forces acting on the RB are assumed to be those shown in Figure 2. The magnitude of the earthquake forces and their point of application are listed in Table 1.

2.3 *Internal stability*

In the NZLE method, a bilinear failure surface is assumed to develop at the toe of the wall and propagate up through the RB and the retained soil behind the RB. An upper-bound failure criterion is applied to find the critical failure surface inclination angles and the acceleration at which sliding develops. The disturbing forces acting on the sliding block are the soil block weight, inertia force, and the Mononobe-Okabe (M-O) pressure (Wood and Elms, 1990) on the back of the soil block. These are resisted by soil friction and cohesion (usually zero) on the failure plane and the forces in the reinforcing strips that cross the failure surface. Forces acting on the failure wedge are shown in Figure 3.

In the TAI active wedge method empirical rules are used to define the active wedges shown in Figure 4. The FHWA wedge is 18% larger in area than the TAI wedge. (Wedge details are not specified in the AS 4678 code.) The active wedge surface running through the block defines the line of maximum tension in the reinforcement strips as determined by experimental results. Strip lengths in the resistive zone behind the active zone are designed to resist the sum of the static

Table 1. External stability comparison.

Item	Symbol	NZLE (Transit NZ BM)	TAI	FHWA-NHI-00-043	AS 4678
RE block inertia force	P_I	$P_I = a_r L H \gamma$ (a_r = response accn)	$P_I = 0.5 a_r L H \gamma$	$P_I = 0.5 a_r H2 \gamma$	$P_I = 0.5 a_r L H \gamma$
Back face G + E horizontal force component	F_b	$F_b = 1.35 \times 0.5 \Delta K_{AE} \gamma_b H^2 \cos\delta + F_s$ $F_s = 1.35 \times 0.5 K_A \gamma_b H^2 \cos\delta$ K_A = Coulomb active coefficient 1.35 is load factor on earth pressures for G + E load	$F_b = 0.5 \Delta K_{AE} \gamma_b H^2 + F_s$ $F_s = 0.5 K_A \gamma_b H^2 \cos\delta$ K_A = Coulomb active coefficient $\Delta K_{AE} = K_{AE} - K_{AO}$ K_{AE} based on $0.5a_r$(a_r = resp accn)	$F_b = 0.25 \Delta K_{AE} \gamma_b H^2 + F_s$ $F_s = 0.5 K_A \gamma_b H^2$	$F_b = 1.25 \times 0.5 K_{AE} \gamma_b H^2 \cos\delta$ 1.25 is load factor on driving forces (see below).
Height of back face force	H_b	$H_b = 0.333\,H$ For total active force.	$H_s = 0.333H$ $H_E = 0.60\,H$	$H_s = 0.333H$ $H_E = 0.60\,H$	$H_b = 0.4H$ to $0.55H$ For total active force.
Back face friction angle	δ	Not specified but usually taken as: $\delta = 0.67\phi_b$ for both G & E comps.	$\delta = 0.8(1 - 0.7\,L/H)\,\phi_b$ Static component only.	Taken as 0 for both static and earthquake components.	Not specified but code indicates it is taken as non-zero.
Strength reduction		None specified.	None specified.	None specified.	0.95 or 0.9 factor on $\tan\phi$.
FOS for sliding under E, or load Factors	S_s	$S_s = 1.0$	$S_s = 1.2$	$S_s = 1.5 \times 0.75 = 1.125$	Driving forces from gravity are increased by 1.25 and resisting forces reduced by 0.8.

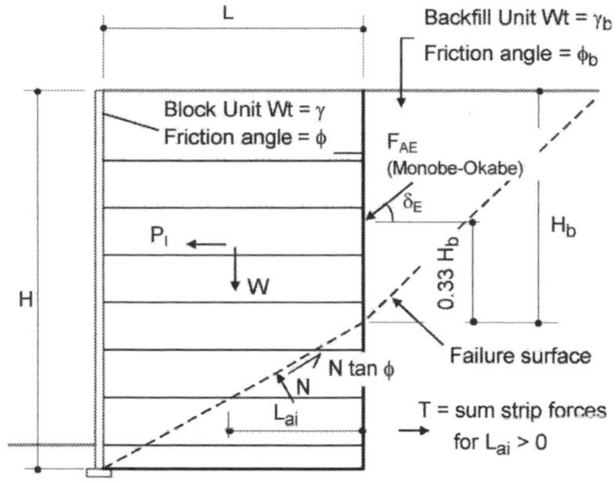

Figure 3. NZLE analysis.

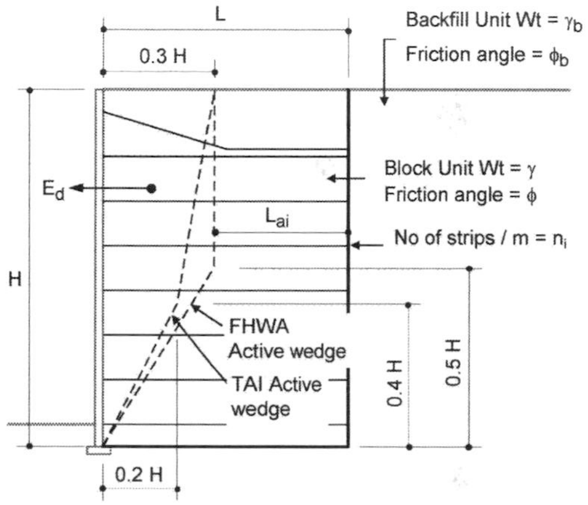

Figure 4. TAI and FHWA internal stability analyses.

and earthquake pressures generated on the wall facing. The total earthquake pressure on the facing is taken as the inertia force acting on the active wedge and is distributed to the strips in proportion to the product of their section area per unit length of wall and their resistive length.

The applied forces, pressure distribution assumptions, strength reduction factors, and factors of safety used in the internal stability design methods are summarised in Table 2.

3 ANALYSIS RESULTS

3.1 *External stability – sliding on base*

The PGA to produce the limiting factors of safety against base sliding are plotted against length/height (L/H) ratio of the RB in Figure 5. Similar results for response acceleration versus L/H are shown in Figure 6. To make these comparisons it was necessary to

Table 2. Internal stability comparison.

Item	Symbol	NZLE (Transit NZ BM)	TAI	FHWA-NHI-00-043
Earthquake load in soil reinforcement	T_{m2}	Inertia load assumed uniformly distributed over height of wedge with loads in strips determined by pull-out resistance or rupture strength.	Earthquake load from wedge, E_d, is distributed to strips using: $T_{m2} = E_d \frac{L_a}{\sum_{i=1}^{i=N} n_i L_{ai}}$ L_a = resisting length of strip and n_i = number of strips with uniform surface area per unit length of wall in layer *i*.	Earthquake load from wedge, E_d, is distributed to strips using: $T_{m2} = E_d \frac{L_a}{\sum_{i=1}^{i=N} L_{ai}}$ L_a = resisting length of strip. Text states that E_d is distributed in proportion to "resistant area" (equation incorrect).
Friction factor for strips	f^*	For ribbed type strips f^* varies from $f^*_o = 1.5$ at surface to $f^* = \tan \phi_b$ at depth of 6 m.	For ribbed type strips f^* varies linearly from $f^*_o = 1.2 + \log C_u$ at surface to $\tan \phi_b$ at depth of 6 m. C_u is soil uniformity and ϕ_b frict angle. Default, $f^*_o = 1.5$	For ribbed type strips f^* varies linearly from $f^*_o = 1.2 + \log C_u$ at surface to $\tan \phi_b$ at depth of 6 m. C_u is soil uniformity and ϕ_b frict angle. Default, $f^*_o = 1.8$
Backface force	F_z	Mononobe-Okabe pressure on the backface of sliding wedge section.	Gravity pressure force from backfill soil behind the block is used to calculate the vertical pressures at the wall face.	No backface soil pressure is used in the internal stability assessment.
Vertical pressure at facing	$\sigma_v(z)$	Not required in this analysis.	Calculated from weight of block and pressure on backface.	Calculated from weight of block.
Horizontal pressure coefficient	K	Not required in this analysis.	Horizontal pressure on facing calculated from vertical pressures using: $K = K_o$ at surface & K_a at 6 m depth. K_o is at rest coefficient and K_a is active coefficient.	Horizontal pressure on facing calculated from vertical pressures using: $K = 1.7\,K_a$ at surface & $1.2\,K_a$ at 6 m depth. K_a is the active pressure coefficient.
Strength reduction factors		0.8 for strip pull-out resistance. 0.9 on strip yield strength.	0.8 for strip pull-out resistance under E loads.	0.8 for strip pull-out resistance under E loads.
Factors of safety		1.0 to calculate critical acceleration for sliding failure.	For strip rupture = 1.5 on UTS generally or 1.65 on UTS for important walls. For strip pull-out = 1.0.	Strip rupture = 1.36 on strip yield stress (1/ 0.55 × 0.75). Strip pull-out = 1.125 (0.75 × 1.5).

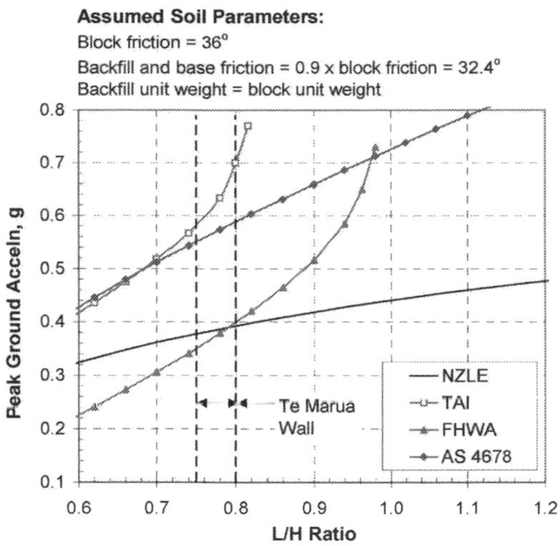

Figure 5. Sliding stability in terms of PGA.

define the ratios between the RB, backfill and base soil friction angles, and the ratio of the soil unit weights in the RB and backfill. For the comparisons shown in Figures 5 and 6 these ratios were taken as 1.0: 0.9: 0.9 and 1.0: 1.0 respectively. A PGA reduction factor of 0.8 was used in the NZLE method.

Typical L/H ratios for walls designed for seismic resistance vary from 0.75 to 1.0. Over this range the four methods investigated show a considerable variation in PGA or response acceleration to produce the specified limiting factors of safety with the NZLE method generally giving the most conservative PGA and response acceleration values.

3.2 *External stability – base pressures*

Base pressures calculated for a PGA of 0.3 g are compared in Figure 7. (Pressures were divided by the RB soil unit weight γ and the wall height H to give a dimensionless parameter.) The ratios of the soil friction angles and units weights were the same as used for the sliding stability comparison.

With the exception of the FHWA results, there is reasonable agreement between the base pressures. FHWA pressures are significantly higher than the others, especially for low L/H ratios.

3.3 *Internal stability*

To compare the strip distributions required to satisfy the internal stability requirements of the NZLE, TAI Guidelines and the FHWA methods, typical 10 m and 12 m high wall sections designed to support a highway at Te Marua, Wellington, New Zealand were used. (AS 4678 was not considered because it does not define an active wedge.) The geometry of the wall sections and

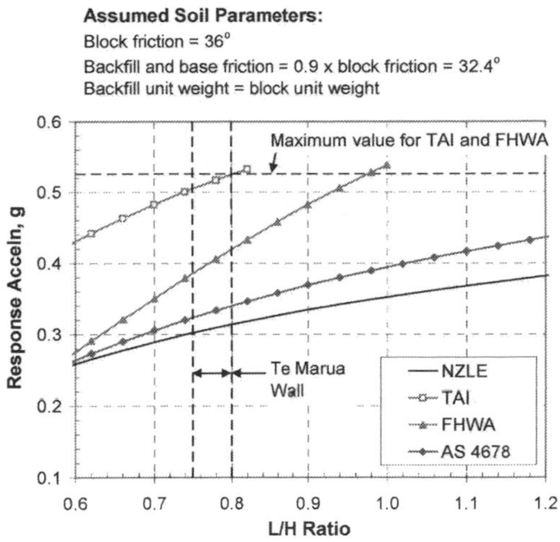

Figure 6. Sliding stability in terms of response acceleration.

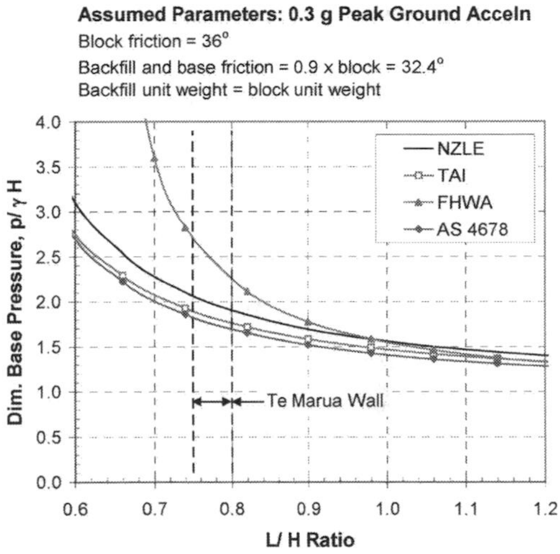

Figure 7. Dimensionless base pressures.

the assumed soil properties are given in Figures 8 and 9. The wall was designed using the NZLE method for a PGA of 0.4 g and a response acceleration of 0.32 g.

The TAI and FHWA active wedge geometries are compared with the NZLE sliding wedge in Figures 8 and 9. Also shown for comparison with the NZLE failure surface are the slip circles obtained by analysis of the wall sections with the STARES slope stability software (Balaam, 1999) which is based on the Bishop Simplified Method. The STARES input parameters for the soil and reinforcing strips were the same as used in the NZLE method. The circles shown in Figures 8 and 9 are the most critical circles passing through the wall facing footings. Slightly more critical circles passing through points on the wall facing were obtained but

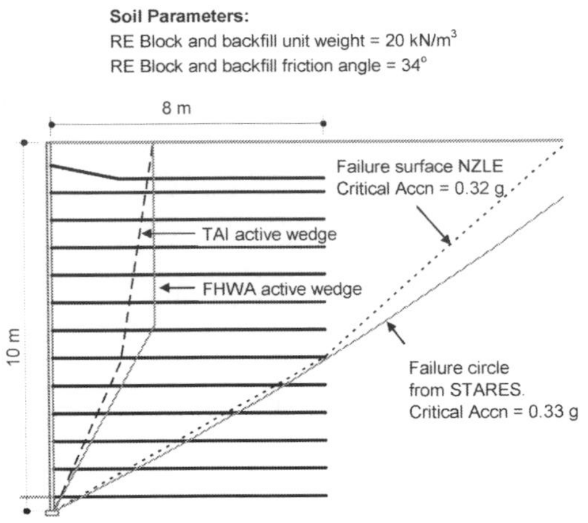

Figure 8. Te Marua 10 m high wall details.

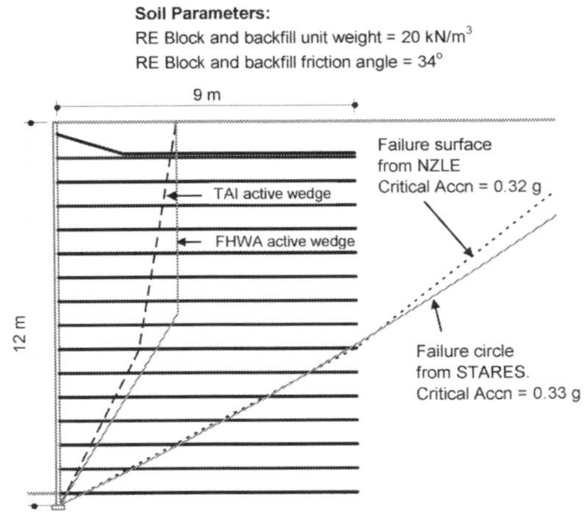

Figure 9. Te Marua 12 m high wall details.

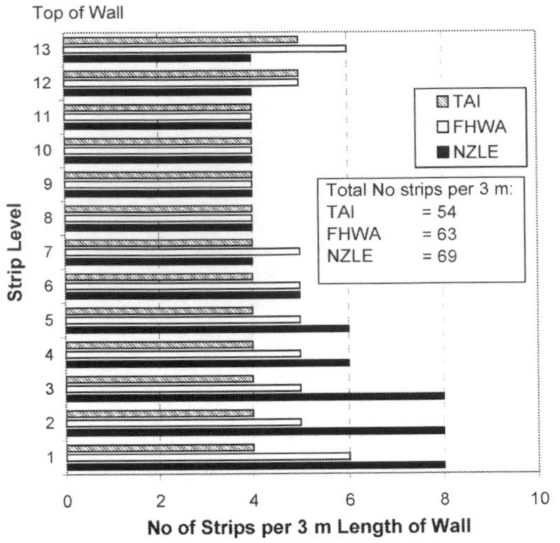

Figure 10. Strip density and distributions on 10 m high section to satisfy internal stability requirements.

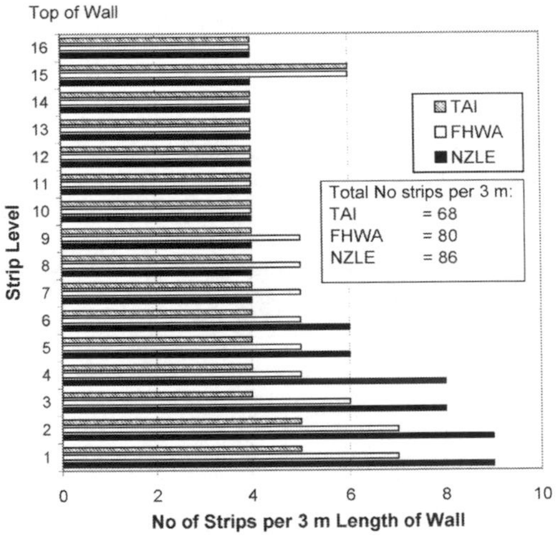

Figure 11. Strip density and distributions on 12 m high section to satisfy internal stability requirements.

the strength of the facing prevents these from developing at lower accelerations than the circle through the wall footing. The STARES critical circles have similar locations in the RB's to the NZLE failure surfaces and the critical acceleration of 0.33 g from STARES for both wall sections is in good agreement with the 0.32 g from the NZLE method.

Strip densities and the distribution over the wall height calculated to satisfy the internal stability requirements of the three design methods are shown in Figures 10 and 11 for the 10 and 12 m high sections respectively. For the 10 m high section the total number of strips per unit length of wall required by the FHWA and NZLE methods are 17 and 28% higher respectively than required by the TAI Guidelines. The TAI and FHWA strip densities were also analysed using the NZLE method which gave critical accelerations for the 10 m high section of 0.17 and 0.24 g respectively. Newmark sliding block theory (Newmark, 1965) indicates that sections with the TAI and FHWA strip densities would displace outwards about 80 and 20 mm respectively in the design level earthquake (0.32 g response acceleration by NZLE).

4 CONCLUSIONS

The currently used seismic design procedures for RE walls give significantly different results for sliding stability and soil reinforcement density.

The NZLE method is based on comprehensive theoretical and experimental research. Analyses for this paper, and back analyses of structures which exhibited permanent displacement as a result of the 1995 Hyogoken-Nanbu earthquake (Wood and Asbey-Palmer 1999) shows that this method gives failure predictions similar to those obtained using conventional slope stability analyses and also back analysis of actual structures subjected to strong ground shaking. Walls designed by the NZLE method are expected to remain elastic under the expected loading without significant permanent displacement. Alternatively, if permanent displacements are acceptable, structures can be designed to reduced acceleration levels and a specified outward displacement using the NZLE method in conjunction with the Newmark sliding block theory.

The active wedge method, in which internal stability is assessed with the inertia load on an active wedge resisted by the soil reinforcement in proportion to the product of the resistive lengths behind the wedge and section area per unit length of wall, and the external stability assessed using an assumed lack of coherency in soil mass accelerations is likely to result in significant permanent outward displacements in strong ground shaking.

REFERENCES

AS 4678, 2002 *Earth-retaining structures*, Standards Australia.

Balaam, N.P. 1999. *Stability analysis of reinforced soil: users manual for program STARES for Windows*. Centre for Geotechnical Research, University of Sydney, NSW, Australia.

Fairless, G.J. 1989. *Seismic performance of reinforced earth walls.* Research Report, 89–9, Dept of Civil Engineering, University of Canterbury, Christchurch.

FHWA, 2001. *Design of mechanically stabilized earth walls and reinforced soil slopes; design & construction guideline.* FHWA–NHI–00–043. Federal Highways Administration, US Department of Transportation.

Freyssinet, 2000. *Monograph: Performance of reinforced earth retaining walls near the epicentre of the Izmit earthquake (Turkey, August 17, 1999).* TA 2000 – A901. Technical Department, Soiltech.

Kobayashi K, Tabata H. and Boyd M. 1996. The Performance of reinforced earth structures in the vicinity of Kobe during the Great Hanshin Earthquake. *Proceedings First International Conference on Earthquake Geotechnical Engineering,* Tokyo, November, 49–54.

Newmark N.M. 1965. Effects of earthquakes on dams and embankments. *Geotechnique,* Vol XV, No 2, June.

Richardson G.N. and Lee K.L. 1975. Seismic Design of Reinforced Earth Walls. Proc ASCE, 101, GT2, February.

TAI, 1989. *Design for reinforced earth structures subject to seismic forces, documentation, synthesis and research results.* Research Report No 48 Terre Armee Internationale, Paris.

TAI, 1991. *Design guide: seismic design of reinforced earth retaining walls.* A-15/16. Terre Armee Internationale, Paris.

TAI, 1994. *Monography: performance of the reinforced earth structures near the epicentre of the Los Angeles Earthquake (January 17, 1994).* M12. Reinforced Earth Co, Vienna, USA.

Transit NZ, 2003. *Bridge Manual, Second Edition 2003,* (Amendments to 2004). Document SP/M/022, Transit New Zealand, Wellington.

Wood J.H. and Elms D.G. 1990. Seismic design of bridge abutments and retaining walls. *Road Research Unit Bulletin* 84, Vol 2, Transit New Zealand.

Wood J.H. and Asbey-Palmer D.E. 1999. Performance of reinforced earth retaining walls in the 1995 Hyogoken-Nanbu (Kobe) Earthquake. *Proceedings NZSEE Technical Conference*, Rotorua.

New Horizons in Earth Reinforcement – Otani, Miyata & Mukunoki (eds)
© 2008 Taylor & Francis Group, London, ISBN 978-0-415-45775-0

Physical and numerical modeling of EPS geofoam buffers for seismic load reduction on rigid walls

S. Zarnani
GeoEngineering Centre at Queen's-RMC, Department of Civil Engineering, Queen's University, Kingston, Ontario, Canada

R.J. Bathurst
GeoEngineering Centre at Queen's-RMC, Department of Civil Engineering, Royal Military College of Canada, Kingston, Ontario, Canada

ABSTRACT: The paper describes a series of experimental shaking table tests carried out at RMC to demonstrate that expanded polystyrene (EPS) geofoam materials can be used as seismic buffers to attenuate earthquake-induced dynamic forces developed against rigid retaining wall structures. Two different numerical modeling approaches are described that were used to predict the results of the physical experiments. The methods are based on: 1) a simple displacement block wedge approach, and; 2) a FLAC code. Both numerical approaches are shown to capture the peak dynamic-force time response of the seismic buffer wall models. The paper demonstrates both proof of concept and successful modeling approaches that can be used to predict load attenuation during earthquake using suitably selected EPS seismic buffers.

1 INTRODUCTION

Previous research has shown that the magnitude of static earth pressures against rigid wall structures can be reduced by placing a compressible vertical inclusion between the rigid wall and the retained soil.

One of the first reported field applications was described by Partos & Kazaniwsky (1987). In their study a prefabricated expanded polystyrene beaded drainage board 250 mm thick was placed between a 10-m high non-yielding basement wall and a granular backfill. Horizontally compressible platens were used by McGown et al. (1988) to construct 1-m high laboratory wall models in order to measure the effect of wall material compressibility on the magnitude of earth pressures and wall deformations. Physical test results reported by McGown & Andrawes (1987) and McGown et al. (1988) were used by Karpurapu & Bathurst (1992) to verify a non-linear finite element model (FEM) numerical approach. The numerical model was then used to develop a series of design charts for prototype-scale walls. The design charts can be used to select the thickness and elastic modulus of the compressible inclusion to minimize end-of-construction earth pressures against non-yielding retaining walls constructed to different heights and with a range of granular backfill materials compacted to different densities. A suitably selected vertical inclusion will allow sufficient lateral expansion of soil (controlled yielding) such that the retained soil is at or close to active failure and hence the earth pressures against the rigid structure are (according to classical earth pressure theory) at a minimum value.

Block-molded low-density expanded polystyrene (EPS) is the product of choice for the vertical compressible inclusion material today. EPS is classified as a geofoam material in modern geosynthetics terminology (Horvath 1995). A logical extension of the application of these systems to static load environments is the use of a geofoam inclusion for attenuation of seismic-induced dynamic earth loads. Specifically, potentially larger earth forces that develop during earthquakes can be reduced by a properly selected EPS geofoam material. In this paper, we refer to the compressible material as a seismic buffer.

Inglis et al. (1996) reported the first application of this technology in North America for seismic design. Panels of EPS from 450 to 610 mm thick were placed against rigid basement walls up to 9 m in height at a site in Vancouver, British Columbia. Their numerical analyses using program FLAC (Itasca 1996) showed that a 50% reduction in lateral loads could be expected during a seismic event compared to a rigid wall solution.

This paper first briefly describes an experimental test program that was carried out at the Royal Military College of Canada (RMC) to demonstrate proof of concept using the results of reduced-scale shaking table tests on rigid walls constructed with and without EPS geofoam seismic buffers.

Two different numerical model approaches were developed and verified against selected test results. The first is a simple displacement model and the second is a dynamic finite difference approach using the program FLAC. The paper shows that the results of both numerical models are in generally good agreement with experimental measurements.

2 EXPERIMENTAL INVESTIGATION

2.1 *Reduced-scale shaking table tests*

A bulkhead (rigid wall) with height of 1 m and width of 1.4 m was constructed at the front of a strongbox mounted on a shaking table at RMC. Granular backfill soil was placed in the strongbox to a distance 2 m beyond the model wall. A total of seven different tests were carried out. Figure 1 illustrates an example experimental test set up showing the arrangement of the non-yielding wall, geofoam seismic buffer, retained soil and instrumentation. No geofoam seismic buffer was installed in Wall 1 (control). In all other walls the geofoam buffer had a constant thickness of 0.15 m.

One of the key material properties that define the compressive behavior of geofoam is its density. Many relationships are proposed in the literature that correlate the density of the geofoam to its initial elastic tangent Young's modulus. A summary of these correlations for non-elasticized geofoam has been reported by Zarnani & Bathurst (2007). Elasticized EPS is manufactured by applying a load-unload cycle after manufacture. This process makes the EPS geofoam linear elastic up to about 10% strain during compression (i.e. strains are recoverable) and linear (proportional) up to about 40% strain. Non-elasticized EPS is linear elastic up to about 1% strain. However, at the same density the elasticized EPS geofoam has a lower elastic modulus compared to the non-elasticized material. Properties of the EPS geofoam material used in the shaking table experimental tests are summarized in Table 1.

Due to page limitations only the experimental results for Walls 1, 2, 3 and 7 are presented in this paper. Walls 2 and 3 were constructed with commercially available EPS geofoam with nominal densities of 16 and 12 kg/m^3, respectively. For Wall 7, the density and hence the stiffness of the seismic buffer was artificially reduced by mechanically removing material from the EPS panels. Additional details and experimental results can be found in the papers by Bathurst et al. (2007a) and Zarnani & Bathurst (2007).

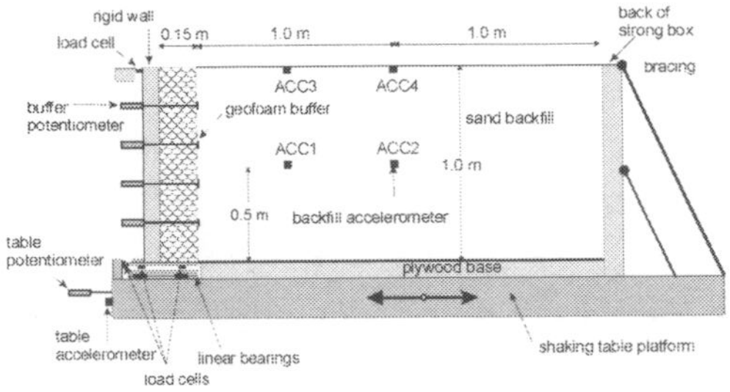

Figure 1. Example shaking table test configuration and instrumentation.

The backfill soil that was used for the experimental program was an artificial sintered synthetic olivine material (JetMag 30–60) which is silica-free and thus does not pose a health danger due to silica dust during material handling in an enclosed laboratory environment. Backfill soil properties are summarized in Table 2. The sand backfill was placed in 200-mm thick lifts and gently vibro-compacted using the shaking table. The same volume and placement technique was used in all of the experimental tests in order to have a consistent retained soil mass.

The rigid wall was made out of a 6-mm thick aluminum plate with aluminum stiffeners in all tests. Four load cells were used to rigidly brace the aluminium bulkhead to the shaking table platform. The entire wall system (bulkhead and seismic buffer) was seated on an instrumented footing supported by three frictionless linear bearings that were seated on five load cells. This arrangement made it possible to decouple the vertical and horizontal loads at the wall boundaries. The boundary conditions for the rigid wall prevented wall rotation and vertical and horizontal displacements. The lateral deformations at the geofoam-soil interface were measured by four potentiometers. The acceleration-time excitation history of the shaking table platform was measured by an accelerometer that was attached directly to the shaking table. In addition, four other accelerometers were embedded in the backfill soil at the locations shown in Figure 1.

The target base input excitation that was applied to all experimental tests was a stepped-amplitude sinusoidal record with a frequency of 5 Hz (Figure 2). This frequency (i.e. 0.2 s period) at 1/6-model scale corresponds to 2 Hz (i.e. 0.5 s period) at prototype scale according to the scaling laws proposed by Iai (1989). Frequencies of 2 to 3 Hz are representative of typical predominant frequencies of medium to high frequency earthquakes (Bathurst & Hatami 1998), and fall within the expected earthquake parameters for North American seismic design (AASHTO 2002). The displacement amplitude (i.e. actuator stroke) was increased at about 5-second intervals up to peak base acceleration amplitude in excess of 0.8 g and the test

Table 1. EPS geofoam buffer properties.

Wall #	Bulk density (kg/m³)	Initial tangent Young's modulus (MPa)	Thickness (m)	Type (ASTM C 578)
1		Control structure (rigid wall with no seismic buffer)		
2	16	4.8# (5.08 ± 1.89)‡	0.15	I
3	12	3.2# (3.31 ± 1.48) ‡	0.15	XI
4	14	1.3#	0.15	Elasticized
5	6† (50% removed by cutting strips)	0.53#	0.15	XI
6	6† (50% removed by coring)	0.6#	0.15	XI
7	1.32† (89% removed by coring)	0.38#	0.15	XI

Notes: † density of intact EPS geofoam = 12 kg/m³; # average back-calculated values from cyclic stress-strain measurements during experiments; ‡ average modulus and standard deviation using published correlations with density (Bathurst et al. 2007a).

Table 2. Backfill soil properties.

Property	Value
Density	15.5 Mg/m³
Peak angle of friction	51°
Residual friction angle	46°
Cohesion	0
Relative density	86%
Dilation angle	15°

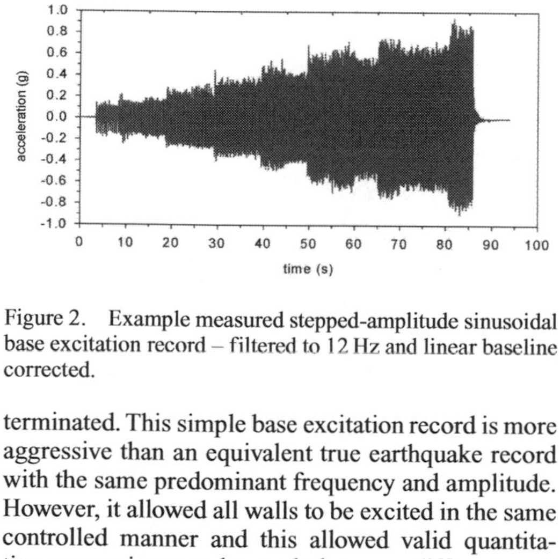

Figure 2. Example measured stepped-amplitude sinusoidal base excitation record – filtered to 12 Hz and linear baseline corrected.

terminated. This simple base excitation record is more aggressive than an equivalent true earthquake record with the same predominant frequency and amplitude. However, it allowed all walls to be excited in the same controlled manner and this allowed valid quantitative comparisons to be made between different wall configurations. Finally, it should be noted that the models were only excited in the horizontal cross-plane direction to be consistent with the critical orientation typically assumed for seismic design of earth retaining walls (AASHTO 2002).

2.2 *Experimental results*

The most important parameter to illustrate the relative influence of the seismic buffer on system response

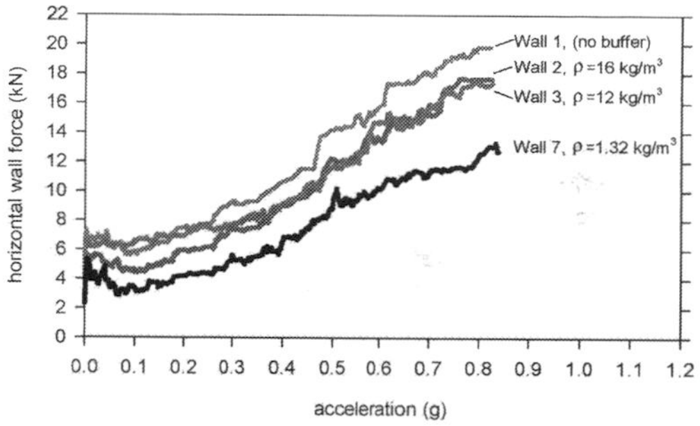

Figure 3. Horizontal wall forces recorded at peak base acceleration amplitude levels for Walls 1, 2, 3 and 7.

was the maximum lateral wall force-base acceleration history recorded at end of construction (initial static loading condition) and during subsequent excitation (Figure 3). The results illustrate that the control wall (Wall 1 with no seismic buffer) had the highest earth force and the model with the lowest bulk density geofoam buffer (Wall 7) had the lowest earth force. In order not to clutter the figure only the maximum horizontal wall force-peak base acceleration values at each stepped amplitude level are plotted in Figure 3. By installing an EPS seismic buffer at the back of the rigid wall in Walls 2, 3 and 7 the total lateral earth load was reduced by 11%, 15% and 40% of the value for the control wall, respectively, at peak acceleration of 0.75 g.

3 NUMERICAL MODELING

3.1 *Displacement model*

A simple one-block model was proposed by Bathurst et al. (2007b) to predict the dynamic response of the seismic buffer retaining walls described in the previous sections (Figure 4). This model is used here to simulate the experimental shaking table tests reported earlier.

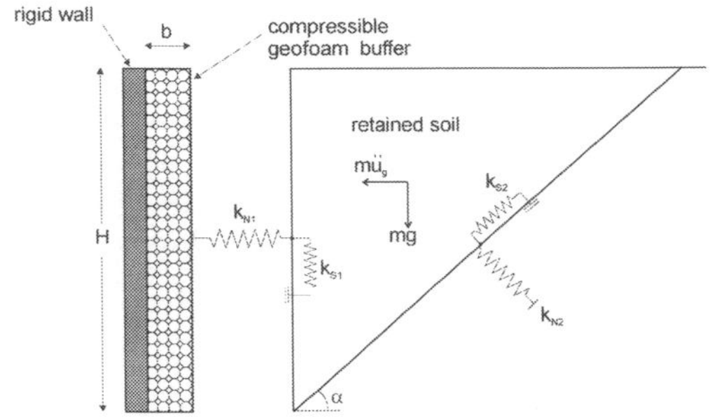

Figure 4. Single block displacement model.

In this model the soil wedge is considered as a rigid block under plane strain conditions. The displacement of the soil wedge is defined by horizontal and vertical displacements computed at the center of gravity of the mass. The small vertical deformations are ignored. The seismic buffer is located between the rigid retaining wall and soil. The failure plane in the backfill soil is assumed to be linear and to propagate from the heel of the buffer. A closed-form solution based on the classical Mononabe-Okabe wedge method was used to compute the wedge angle α (Bathurst et al. 2007b). This angle becomes smaller as the magnitude of horizontal acceleration increases. Linear spring models are used to compute the forces at the wedge boundaries as shown in Figure 4. A single linear compression-only spring is located at the geofoam-soil boundary.

During simulation runs this spring developed compression-only forces. The linear normal spring acting at the soil-soil wedge boundary permits tension and compression but was observed to develop only compressive forces during computation cycles. In order to allow plastic sliding at the block boundaries, the shear springs are modeled as stress-dependent linear-slip elements. An explicit time-marching finite difference approach that is commonly used for the solution of discrete element problems was used to solve the equations of motion for the sliding block. At each time step, the numerical scheme involves the solution of the equations of motion for the block followed by calculation of the forces.

Computed force-time responses for three test configurations with a compressible geofoam inclusion are presented in Figures 5a, 5b and 5c for Wall 2, 3 and 4, respectively. For clarity, only the peak values from the load-time records for each numerical simulation are plotted in the figures. The forces shown in these figures are the result of dynamic loading only. In other words, the datum for the plots is the end of construction.

The peak measured wall forces shown in Figure 5 are deduced from the sum of the readings from the horizontal load cells mounted against the back of the walls. There is generally good agreement between the physical and numerical models up to peak base input

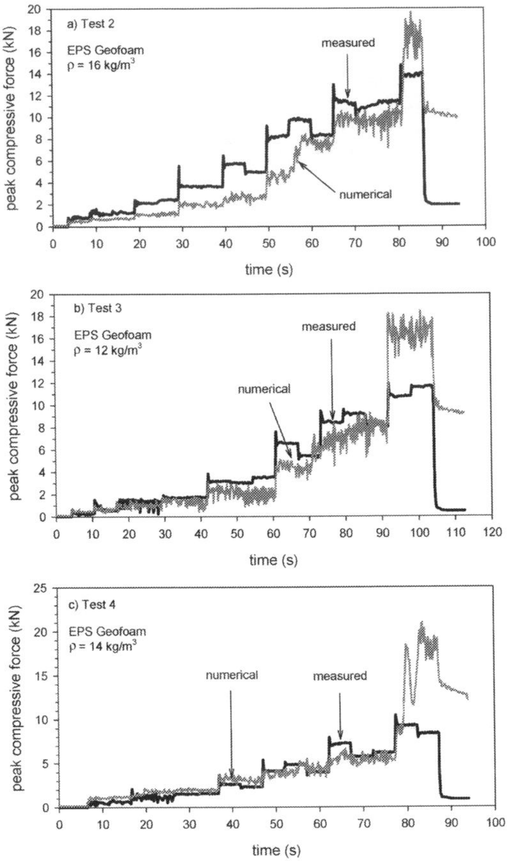

Figure 5. Wall force-time histories from physical tests and simple displacement numerical model.

acceleration of about 0.7 g. At higher accelerations there are likely more complex system responses that cannot be captured by the simple displacement model employed. For example, there are likely higher wall deformation modes at higher levels of base excitation. Nevertheless, the trends in the measured buffer force data for the three walls are generally captured by the numerical model up to about 0.7 g, and in many instances there is good quantitative agreement.

3.2 *FLAC modeling*

Program FLAC (Itasca 2005) was also used to simulate the RMC reduced-scale model walls. Figure 6 illustrates the FLAC numerical grid and boundary conditions used for the simulation of the geofoam buffer tests. The height and width of the numerical grid and thickness of the geofoam were kept the same as the physical tests.

The backfill soil was modeled as a purely frictional, elastic-plastic material with Mohr-Coulomb

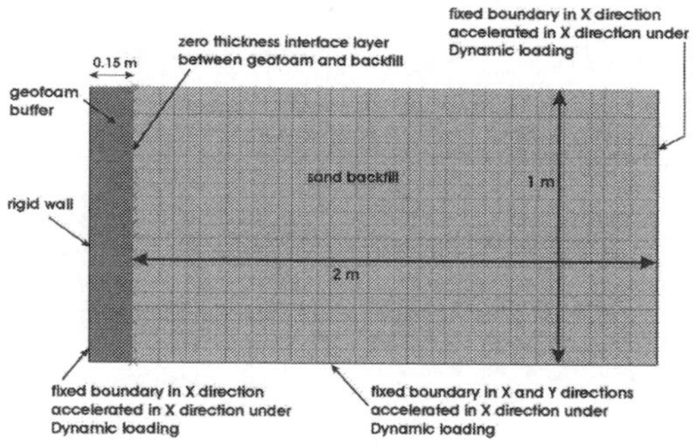

Figure 6. Numerical grid showing geofoam buffer, sand backfill and boundary conditions.

failure criterion. This model allows elastic behavior up to yield (Mohr-Coulomb yield point defined by the friction angle), and plastic flow at post-yield under constant stress. The geofoam buffer material was modeled as a linear elastic, purely cohesive material. While a more advanced non-linear strain hardening model could have been implemented in the FLAC code, the simple constitutive model adopted here was judged to be sufficient since the measured compressive strains in the physical models were less than the elastic strain limit of 1% determined from rapid uniaxial compression tests reported by the manufacturer. A constant Rayleigh damping ratio of 3% was used for the whole system at a frequency equal to the predominant frequency of the model (20 Hz).

The rough bottom boundary in the physical tests (i.e. a layer of sand was epoxied to the bottom of the strong box container) was simulated with a no-slip boundary at the bottom of the sand backfill. The interface between the buffer and the soil was a slip and separation boundary with zero thickness. The measured base excitation record during experiments for each test was applied to the base and the two vertical boundaries of the numerical models. The horizontal input acceleration was applied in the form of an equivalent velocity record (i.e. integrated acceleration record) with baseline correction to ensure zero displacement at the base at the end of shaking.

The peak magnitudes of horizontal wall force predicted at the end of construction and during base excitation of numerical models were investigated. The variation of maximum wall force with time for three physical tests and numerical simulations are presented in Figure 7. The vertical axis in the plots corresponds to the total horizontal earth force acting against the rigid wall per unit width of wall. The figures show that there is reasonably good agreement between measured and predicted results. There is a noticeable discrepancy between results at the beginning of the test for Wall 3. This is believed to be due to locked-in initial horizontal stresses that may have developed as a result of the

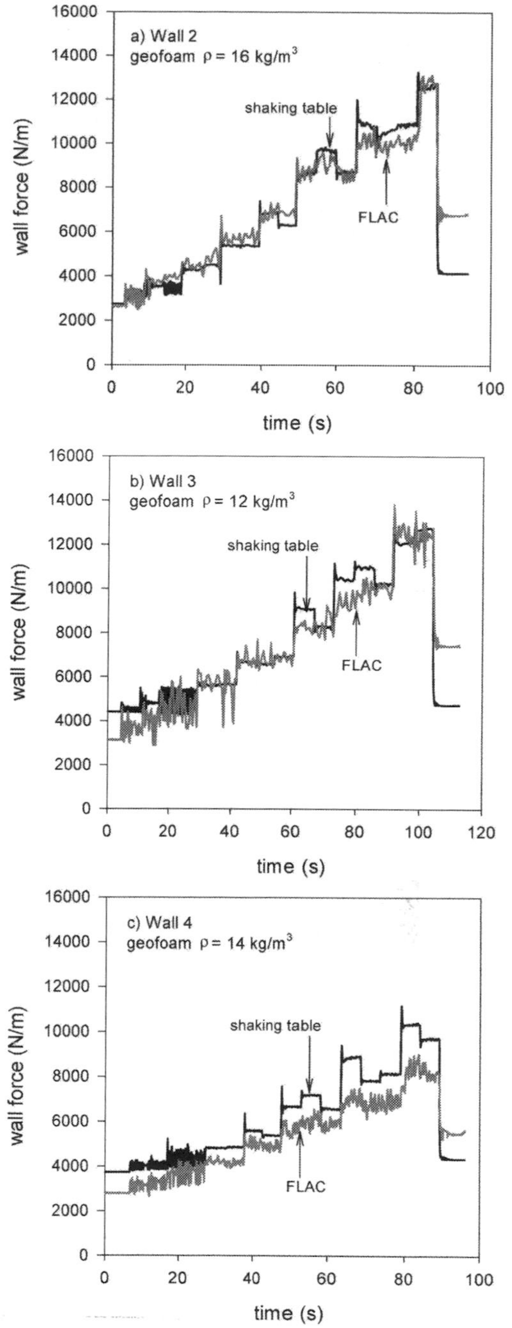

Figure 7. Wall force-time histories from physical tests and FLAC numerical simulations.

initial vibro-compaction technique that was used to densify the soil during placement of the sand layers in the strong box. The numerical model results are consistently lower than the physical test results for Wall

4 but the overall qualitative trends are in good agreement. The discrepancy is believed to be related to the selection of dynamic elastic modulus value used in the numerical model which was back-calculated from measured displacements and wall forces (Zarnani & Bathurst 2007).

4 CONCLUSIONS

A series of shaking table experimental tests are described that were carried out on 1-m high model rigid walls with and without a geofoam seismic buffer. The physical experiments demonstrate that a properly selected vertical compressible inclusion of EPS geofoam placed against a rigid retaining wall can be used to attenuate dynamic earth forces due to simulated earthquake. In this experimental program, modified EPS materials reduced dynamic loads by up to 40%.

Two different numerical approaches are described that can be used to simulate geofoam seismic buffer performance. One model is based on a sliding block approach and the other uses a dynamic finite difference approach implemented within a FLAC code. Both approaches captured the trend in measured data and in most cases there was good agreement between numerical and experimental results.

ACKNOWLEDGMENTS

The work described in this paper was supported by a research grant from the Natural Sciences and Engineering Research Council of Canada held by the second author.

REFERENCES

AASHTO, 2002. Standard Specifications for Highway Bridges. *American Association of State Highway and Transportation Officials*, Seventeenth Edition. Washington, DC, USA.

ASTM C 578-06, 2006. Standard Specification for Rigid Cellular Polystyrene Thermal Insulation. American Society for Testing and Materials, West Conshohocken, Pennsylvania, USA.

Bathurst, R.J. & Hatami, K. 1998. Seismic response analysis of a geosynthetic reinforced soil retaining wall. *Geosynthetics International* 5 (1&2): 127–166.

Bathurst, R.J., Hatami, K. & Alfaro, M.C. 2002. Geosynthetic-reinforced soil walls and slopes—seismic aspects. In: Shukla SK, editor, Geosynthetics and their applications, Thomas Telford, [chapter 14].

Bathurst, R.J., Zarnani, S. & Gaskin, A. 2007a. Shaking table testing of geofoam seismic buffers. *Soil Dynamics and Earthquake Engineering* 27 (4): 324–332.

Bathurst, R.J., Keshavarz, A., Zarnani, S. & Take, A. 2007b. A simple displacement model for response analysis of EPS geofoam seismic buffers. *Soil Dynamics and Earthquake Engineering* 27 (4): 344–353.

Horvath, J.S. 1995. *Geofoam Geosynthetic*, Horvath Engineering, P.C., Scarsdale, NY, 217 p.

Iai, S. 1989. Similitude for shaking table tests on soil-structure-fluid model in 1g gravitational field. *Soils and Foundations* 29: 105–118.

Inglis, D., Macleod, G., Naesgaard, E. & Zergoun, M. 1996. Basement wall with seismic earth pressures and novel expanded polystyrene foam buffer layer. *Tenth Annual Symposium of the Vancouver Geotechnical Society*, Vancouver, BC, Canada. 18 p.

Itasca Consulting Group, 1996, 2005. FLAC: Fast Lagrangian Analysis of Continua, version 3.3 and 5.0. Itasca Consulting Group, Inc., Minneapolis, Minnesota, USA.

Karpurapu, R. & Bathurst, R.J. 1992. Numerical investigation of controlled yielding of soil-retaining wall structures. *Geotextiles and Geomembranes* 11: 115–131.

McGown, A. & Andrawes, K.Z. 1987. Influence of wall yielding on lateral stresses in unreinforced and reinforced fills. *Research Report 113*, Transportation and Road Research Laboratory, Crowthrone, Berkshire, UK.

McGown, A., Andrawes, K.Z. & Murray, R.T. 1988. Controlled yielding of the lateral boundaries of soil retaining structures. In Proceedings of the *ASCE Symposium on Geosynthetics for Soil Improvement*, ed. R. D. Holtz, Nashville, TN, USA. pp. 193–211.

Partos, A.M. & Kazaniwsky, P.M. 1987. Geoboard reduces lateral earth pressures. In Proceedings of *Geosynthetics'87*, Industrial Fabrics Association International. New Orleans, USA. pp. 628-639.

Zarnani, S. & Bathurst, R.J. 2007. Experimental investigation of EPS geofoam seismic buffers using shaking table tests. *Geosynthetics International* 14 (3): 165–177.

New Horizons in Earth Reinforcement – Otani, Miyata & Mukunoki (eds)
© 2008 Taylor & Francis Group, London, ISBN 978-0-415-45775-0

Numerical assessment of the performance of protecting wall against rockfall

E. Sung, A. Yashima, D. Aminata & K. Sugimori
Dept. of Civil Engr., Gihu University, Gihu, Japan

K. Sawada
River Basin Research Center, Gifu University, Gihu, Japan

S. Inoue & Y. Nishida
Protech Engineering Co. Ltd., Niigata, Japan

ABSTRACT: We have to reduce the disaster caused by rockfall as well as conserve the surrounding environment. In order to satisfy these requirements, we have proposed new construction methods, which use on-site ground materials. Full-scale field tests have been performed by some of the authors. Three different types of protecting walls were constructed on the test site, type 1: the dyke-type wall composed of geo-grid reinforcement and soil, type 2: vertical wall composed of cast iron-panel and boulders, and type 3: vertical wall composed of wooden-panel and soil. It has been confirmed based on the full-scale tests that newly proposed protecting walls have a certain degree of potential with respect to the energy absorption against rockfall impact. However, the stability of the protecting walls against larger impacts has not been investigated yet. Therefore, in this study, a series of dynamic finite element analyses (LS-DYNA) were carried out in order to understand the details of stability of different types of protecting walls against larger impacts by falling rock. Firstly, the energy absorption during the full-scale tests was reproduced by employing an appropriate constitutive model for wall structures. Then, the stress, deformation and failure inside the wall by the larger impact forces than the observed in the field tests were discussed based on the numerical results. The efficiency of three types of protecting walls against rockfall is found to be quantitatively.

1 INTRODUCTION

A rockfall is a high frequency unexpected disaster among the slope disasters and has an effect on the road, railway and building. The protective countermeasures against this rockfall are very expensive and difficult. Therefore, a interest of protecting wall using a ground material like soil and boulders is increasing due to a energy absorption. However, The effect of rockfall on the ground has not been fully understood.

Recently, several studies on this interaction problem have been published. Among them Prisco, & Vecchiotti (2004) published the study about a rheological model for the description of rockfall on homogeneous ground and discussed impact load against vertical falling rockfall. Wu & Thomson (2007) also presents the interaction between a guardrail post and soil during quasi-static and dynamic loading by field tests and numerical analyses using LS-DYNA.

The interaction problem between rockfall and protecting wall using ground material should consider about constitutive model of ground and shape of protecting wall. Therefore, this paper presents full-scale model tests and numerical analyses to investigate the mechanical behaviors of protecting wall using ground material during rockfall. Acceleration, penetration and impact load due to rockfall are investigated with the newly developed three kinds of protecting walls and the influences of larger impact energy and impact point in the protecting wall are also investigated by numerical analyses.

2 DESCRIPTION OF FULL-SCALE FIELD TESTS

Full-scale field tests were carried out to investigate the effects of rockfall to three kinds of protecting walls. Figure 1 is a description of these full-scale field tests and an imitation rockfall for this purpose. The imitation rockfall, a 75 cm diameter and weight of 4.9 kN, was made by concrete and freely fallen by two cranes to the protecting wall from the height of 11 m as shown Figure 1. The impact energy in this full-scale field tests

Figure 1. Description of full-scale field tests.

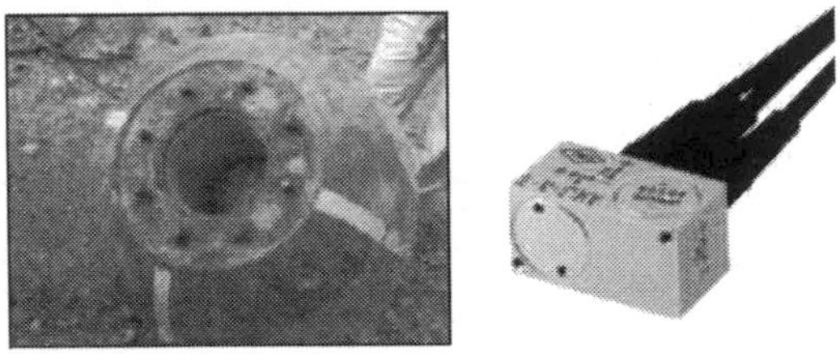

Figure 2. Concrete ball for rockfall and three-dimensional accelerometer.

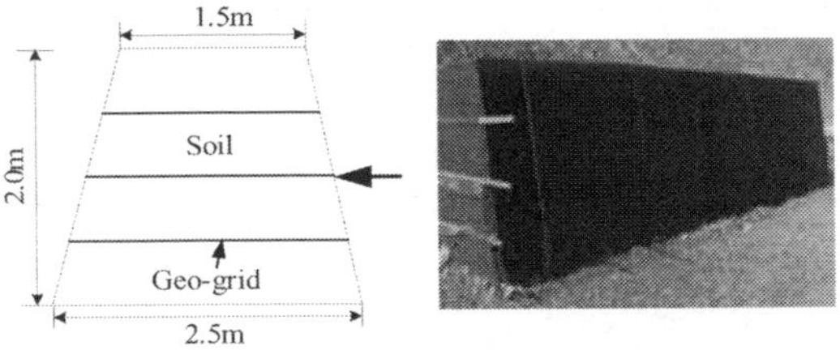

(a) Type 1

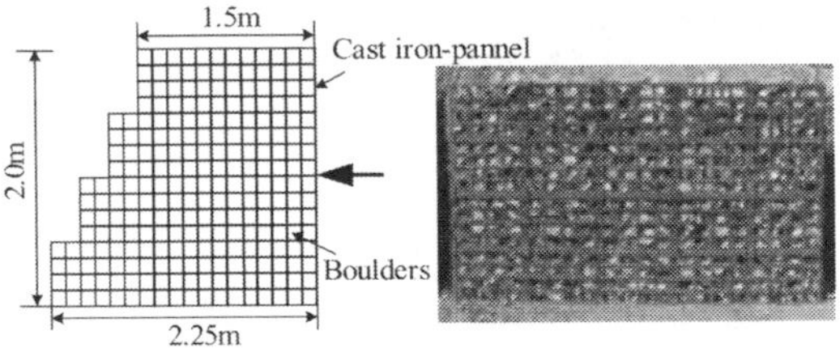

(b) Type 2

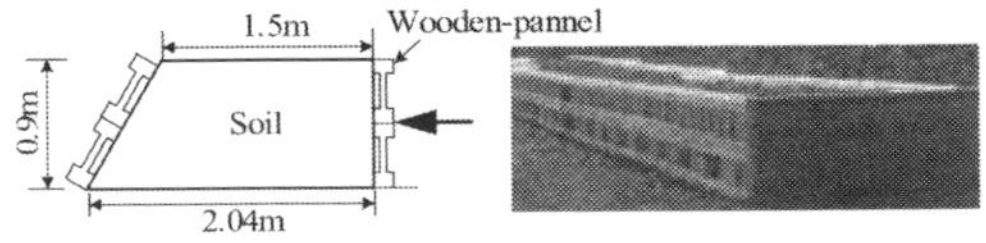

(c) Type 3

Figure 3. Three kinds of protecting walls in field tests.

is assumed to be 55.6 kJ. Acceleration of the rockfall at the moment of impact was measured continually by a three-directional accelerometer installed inside of rockfall as shown Figure 2. The measured acceleration (a, m/sec^2) was convert to the velocity v (m/sec^2), displacement u (m), load f (kN) and absorption energy Ea (kJ) by simple integration method.

Figure 3 represents three kinds of protecting walls used in the full-scale field tests. Type 1 is a dyke-type wall composed by reinforced soil with geo-grid. Type 2 is a vertical wall composed of cast iron-panel and boulders, and type 3 is a vertical wall composed of wooden-panel and soil. The heights of these protecting walls are 2 m for type 1 and 2 and 0.9 m for type 3. The length is 5 m. The rockfall is impacted at middle point in all types of protecting walls.

3 DESCRIPTION OF NUMERICAL ANALYSES

Numerical analyses by LS-DYNA (Hallquist, 2003) to investigate a deformation and stability of protecting wall is conducted in the same scale as the full-scale field tests. Figure 4 shows the 3D meshes for all types of protecting walls. Bottom faces of the 3D mesh are fixed and the other nodes are free in three directions. The geogrid, iron-cased panel and wooden panel are not considered in these analyses. All protecting walls are modeled by single material to investigate the effect between a ground material and an impact of rockfall. In order to simulate the impact of rockfall, initial velocity toward the protecting wall is applied. The initial velocity of free falling rockfall at 10 m height is about 14.76 m/s.

A concrete ball is assumed to be a rigid, and a visco-elastic (Mat_5) or elasto-perfect plastic model is used for protecting wall. The visco-elastic model can express the change on the stiffness of material according to time as following equation.

$$G = G_f + (G_i - G_f)\exp(-\beta t) \qquad (1)$$

Here, G is a shear modulus at the present. Gi and G_f are initial and final shear modulus respectively. β is a change rate of shear modulus with respect to the time. Figure 5 shows a simulation of triaxial test for visco-elastic model.

The Drucker-prager model is used to investigate the plastic and residual deformation of protecting wall. This model does not consider the strain hardening, softening and strain rate effect. On the other hand, this model is easy to get the material parameter (internal friction angle and cohesion). Figure 6 represents a stress-strain relation and dilatancy for triaxial test for this model. Expansions of material begin when stress reach at failure state as shown in Figure 6b and the material with larger internal friction angle shows larger expansion than those with smaller friction angles. Model parameters for these constitutive models are fitted based on the experiment results.

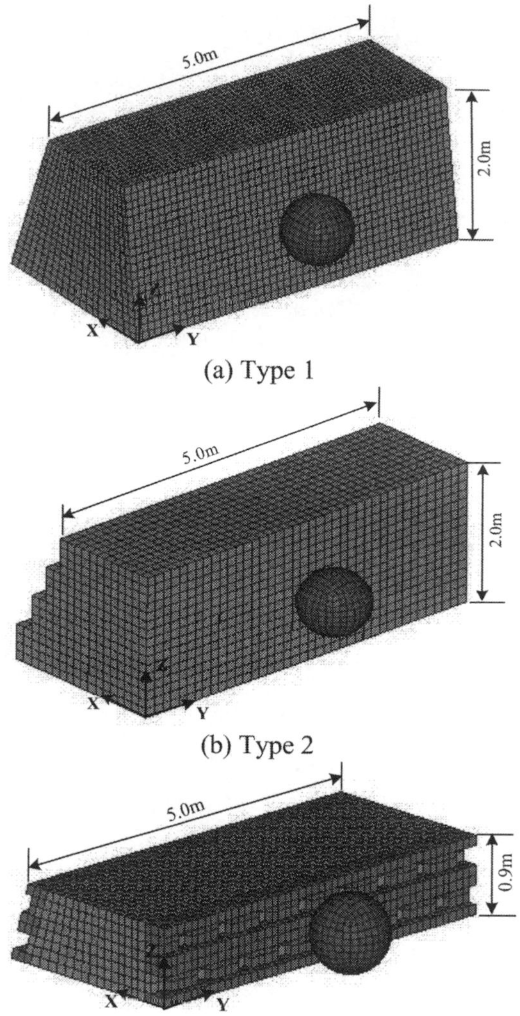

Figure 4. FEM meshes used in numerical analyses.

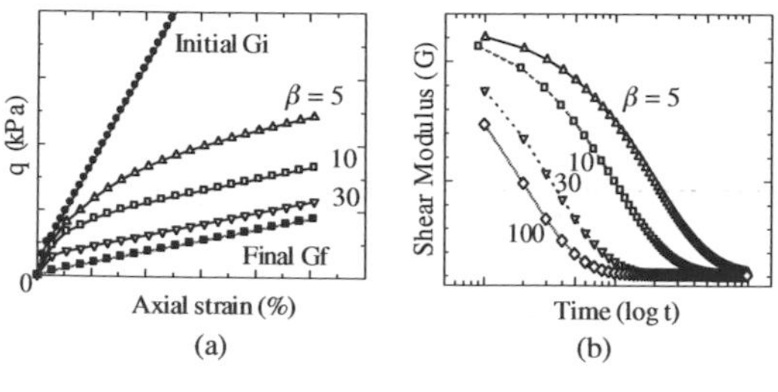

Figure 5. Results of numerical simulation for triaxial tests used in visco-elastic model (a) deviatric stress-axial strain relation (b) change of elastic modulus by time.

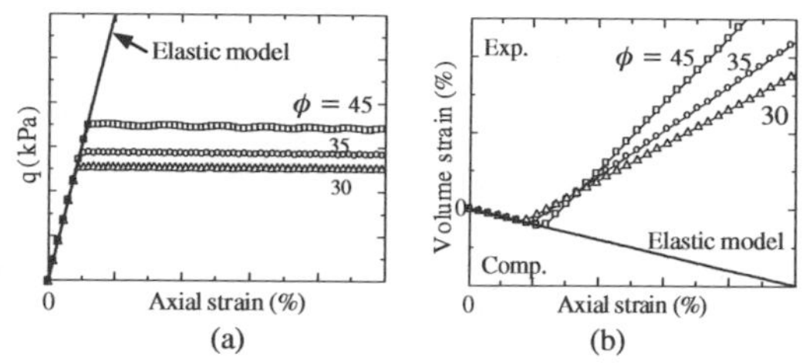

Figure 6. Results of numerical simulation for triaxial tests used in elasto-plastic model (a) deviatric stress-axial strain relation (b) dilatancy relation

Before the impact load is applied, the initial stress condition of a protecting wall are calculated by body forces to all elements under gravitational condition. The contact condition between rockfall and protecting wall is simulated by penalty method in LS-DYNA. The static and dynamic friction angle at the interface between rockfall and a protecting wall is assumed to be 35°.

4 RESULTS AND DISCUSSION

4.1 *Results of full-scale field tests*

Figure 7 shows the results of full-scale field tests for three kinds protecting walls. Type 1 composed by reinforced soil with geogrid represents the largest value of acceleration and impact load of rockfall among these three kinds of walls. The penetration amount of rockfall toward protecting wall for type 1 is smaller than other walls. This means that the stiffness of type 1 is the largest in three walls. The penetration amount of rockfall toward protecting wall for type 1 is smaller than other walls. This means that the stiffness of type 1 is the largest in three walls. On the other hands, the acceleration and impact load of rockfall of type 2 and 3 show lower values than type 1 because the deformations of wall in type 2 and 3 is larger than type 1 for the same impact energy. However, the repair of these protecting walls with soil and boulders is easy compared with concrete wall or steel frame type walls after rockfall.

4.2 *Comparisons between field tests and numerical analyses*

Firstly, the elastic analyses using various elastic moduli are carried out to get the initial elastic parameter of protecting wall by the comparison with test results. Figure 8 shows the acceleration time histories for type 1 according to various Young's moduli. All results of elastic analyses are different from test result. On the other hand, in the case of E = 3.0E + 07 Pa, the initial slope of acceleration curve is fit to the experiment.

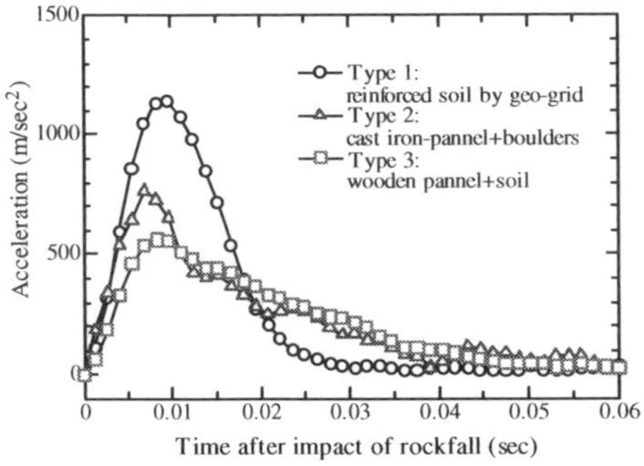

(a) Acceleration

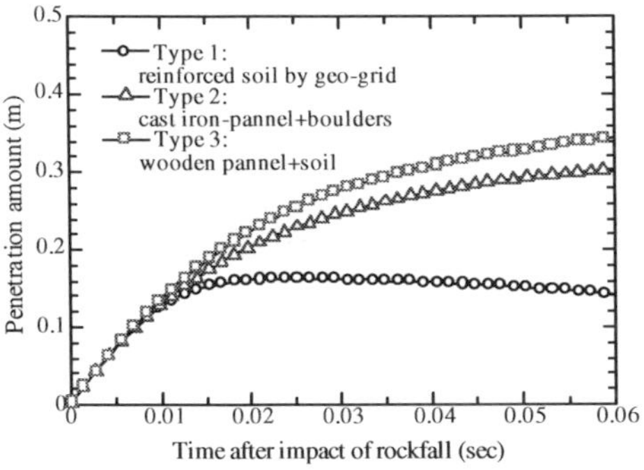

(b) Penetration

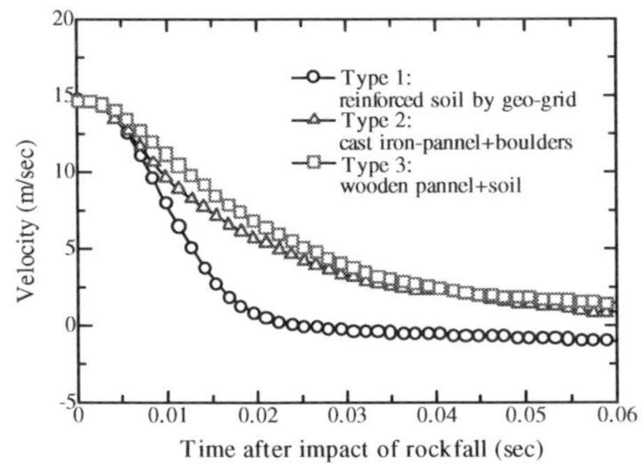

(c) Velocity

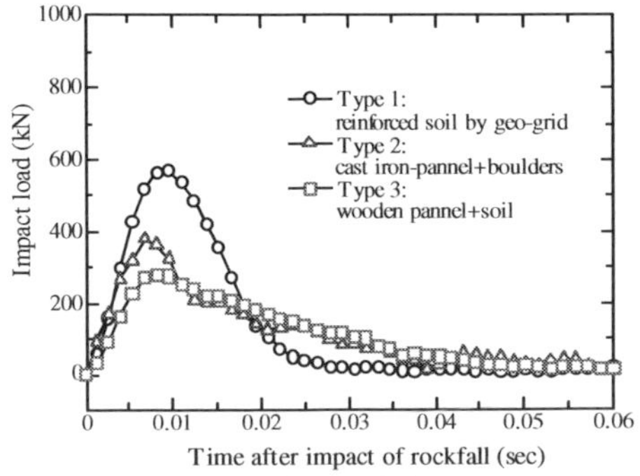

(d) Impact load

Figure 7. Results of full-scale field tests.

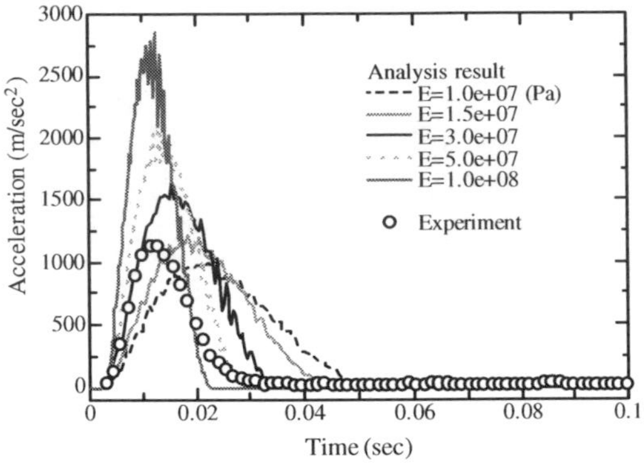

Figure 8. Elastic analyses of type 1 by various elastic constants.

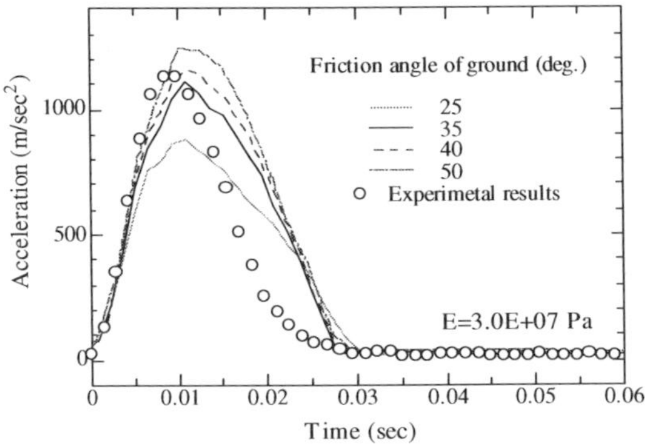

Figure 9. Elasto-plastic analyses for type 1 by various friction angles with E = 3.0E + 07 Pa.

Therefore, This Young's modulus is selected to all analyses for type 1. Those of type 2 and 3 are determined through this method.

Figure 9 shows the acceleration for type 1 by various internal friction angles. The internal friction angle, 35°, is selected due to the best fit to the test result. Parameters for another types are also decided by this way. The selected Young's modulus and friction angle are much smaller than those of usual soil and boulders. However, these parameters have been used to know the impact load and displacement of rockfall by numerical analyses.

Figure 10 represents the impact load with respect to time from numerical analyses and experiments for three kinds of protecting walls.

In these figures, three constitutive models (elastic, visco-elastic and elasto-perfectly plastic model) are indicated to compare with experiment results.

As shown in these figures, the impact load with visco-elastic and elasto-plastic model for protecting wall is well fit to those of experiments. The difference at the parts of descending curve after the maximum

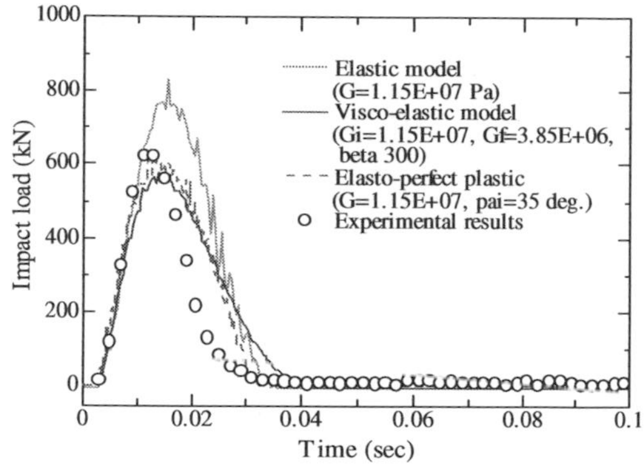

(a) Type 1

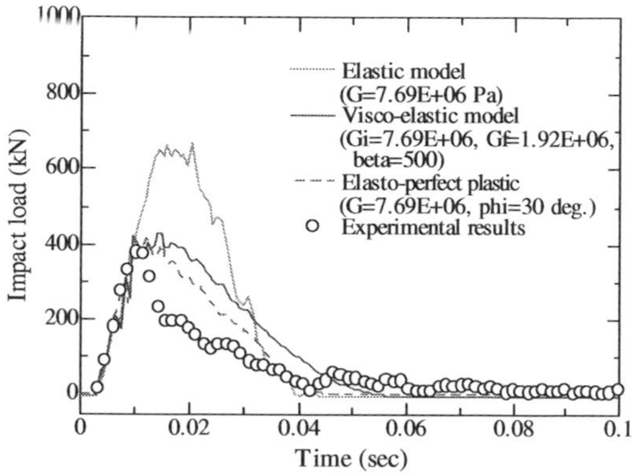

(b) Type 2

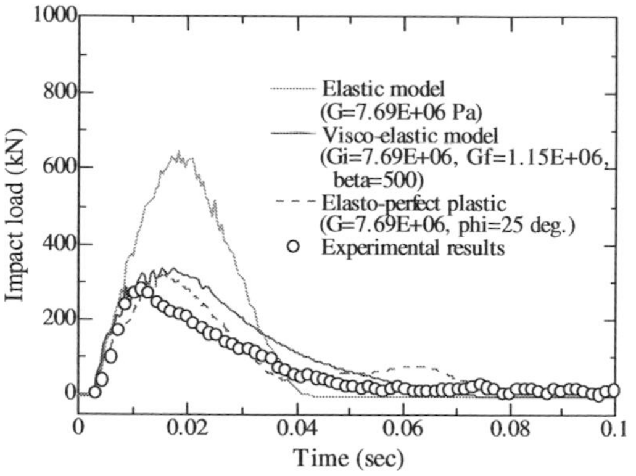

(c) Type 3

Figure 10. Comparisons with impact load of rockfall between numerical analyses and tests.

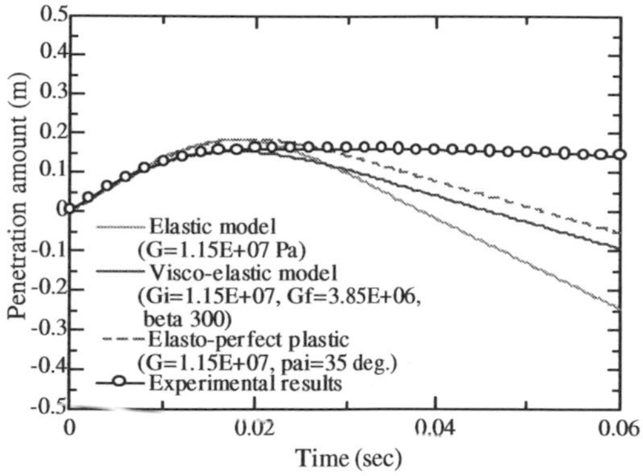

(a) Type 1

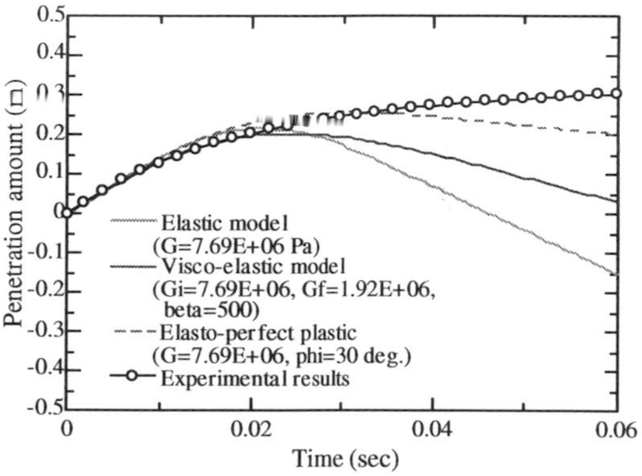

(b) Type 2

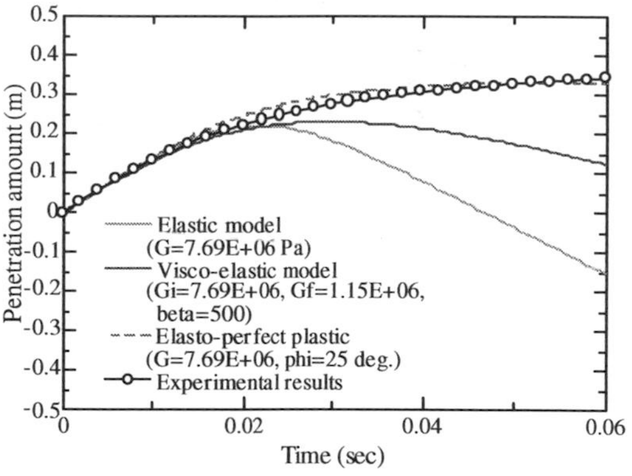

(c) Type 3

Figure 11. Comparisons with penetration amount between numerical analyses and tests.

impact load is due to the leading cable that connected to the rockfall and crane car.

The penetration amount of rockfall for three types by numerical analyses and experiment is represented in Figure 11. Although the maximum interpenetration amount of rockfall into the protecting wall is same to those of experiment results in the case of type 1 in all-material models, the difference between numerical analyses and test after peak penetration become larger with time.

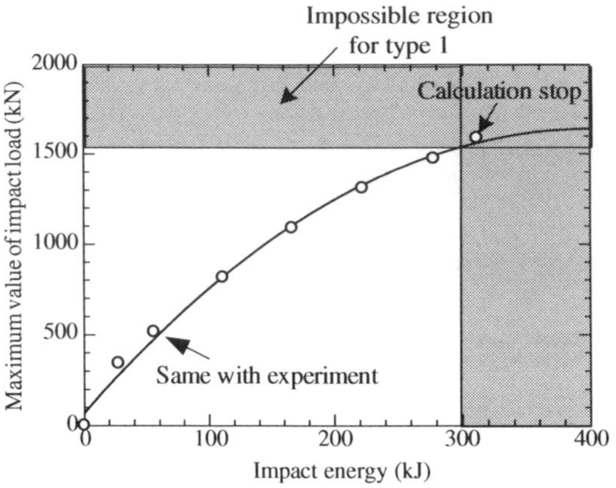

Figure 12. Calculated the maximum impact load of rockfall according to given impact energy.

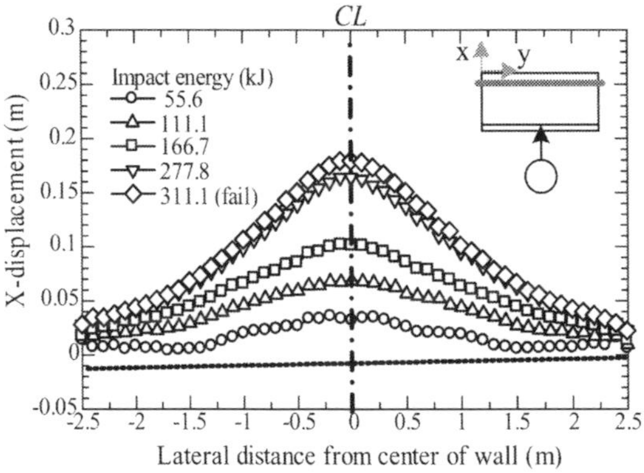

Figure 13. The maximum displacement of protecting wall according to given impact energy.

On the other hand, the results using elasto-perfect plastic model for protecting wall are more fit with those of experiment than the other material models in the case of type 2 and 3. This means that elastic model can't express large deformation of protecting wall such as type 2 and 3.

4.3 *The effect of impact energy and impact point*

Numerical analyses with elasto-perfectly plastic model are carried out to investigate the effect of impact energy at same impact point (a middle of protecting wall) and weight of rockfall (4.9 kN) for type 1.

Figure 12 represents the maximum impact load according to given impact energy. Calculation is impossible at about 300 kJ of impact energy and 1500 kN of impact load. These load and energy mean limit value in this protecting wall against rockfall.

Figure 13 shows the displacement along to the horizontal direction on top face of protecting wall

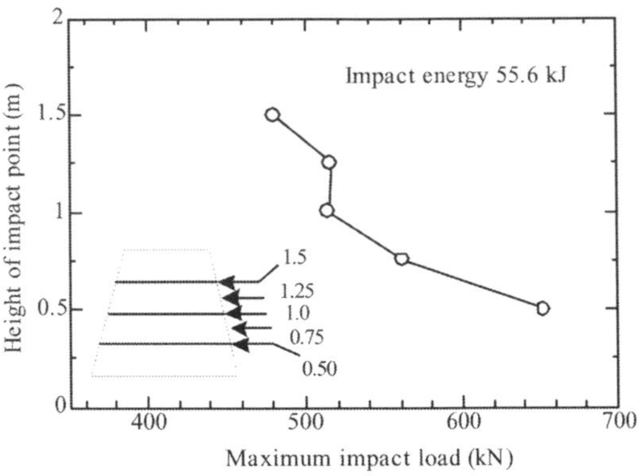

Figure 14. The maximum load according to given impact point in protecting wall (impact energy 55.6 kJ).

according to impact energy. The effect region and maximum displacement increase by impact energy. The effect distance of displacement is about 3 m in the case of impact energy 55.6 kJ. On the other hand, all parts on top face of protecting wall are influenced and amount of displacement become also larger according to the increase of the impact energy.

Figure 14 shows the effect of impact point in the protecting wall with impact energy 55.6 kJ. Although same impact energy is applied to each impact point, the value of impact load of rockfall is different.

In the case of impact point 0.5 m, the load is the largest than those of other impact point due to the fixed boundary condition and weight of protecting wall above impact point. On the other hand, in the case of high impact point such as 1.5 m, the impact load of rockfall become small due to the large deformation of wall.

5 CONCLUSION

Full-scale field tests and numerical analyses were carried out to investigate the interaction problem between a rockfall and a protecting wall. The influences of impact load and impact point on the stability of protecting wall were also investigated. From the results of the full-scale field tests and numerical analyses, the following conclusions are obtained:

(1) The efficiency of protecting wall against rockfall is very different according to the material used.
(2) Although the protecting wall using ground material indicates larger deformation than concrete wall, it can be used for low rockfall energy by its economic advantage and convenient repair.
(3) Numerical analyses with optimized material parameters reproduced experimental results quantitatively.

(4) A limit capacity of a protecting wall against rockfall can be calculated by numerical analyses.
(5) The calculated impact load is different according to the different impact point in the wall.

REFERENCES

Prisco, C. DI. & Vecchiotti, M. 2006. A rheological model for the description of boulder impacts on granular strata. *Geotechnique* 56, No. 7: 469–482

Hallquist, J. O. 2003. *LS-DYNA Theoretical Manual V.970.* Livermore, CA: Livermore Software Technology Corporation

Wu, W. & Thomson, R. 2007. A study of the interaction between a guardrail post and soil during quasi-static and dynamic loading. *Int. Jour. Impact engineering* 34: 883–898

New Horizons in Earth Reinforcement – Otani, Miyata & Mukunoki (eds)
© 2008 Taylor & Francis Group, London, ISBN 978-0-415-45775-0

Full-scale tests on a new type of debris flow trapping fence

H. Ohta
Tokyo Institute of Technology, Tokyo, Japan

K. Kumagai
East Nippon Expressway Company, Miyagi, Japan

H. Takahashi
Expressway Technology Center, Tokyo, Japan

H. Motoe & S. Hirano
Quest Engineer Co., Ltd, Ishikawa, Japan

Y. Yokota & S. Tsuji
Maeda Kosen Co., Ltd, Fukui, Japan

ABSTRACT: Performance of a new type of debris flow trapping fence was examined through a series of full scale tests in which debris flow the trapping fence after traveling a certain distance along a slope of about 5 m high. The fence is essentially a pair of geonet fence designed to prevent the possible traffic accidents by effectively trapping debris flows at the side of motorways. The debris flow trapping fence consists of two fences (0.7 m high × 1.5 m wide) one of which is standing vertically and the other of which is lying on the ground. The standing fence is connected to the lying fence by two hinges at a right angle at the bottom end of a standing fence and is also connected to the lying fence by two bars at the other ends. The debris flow hits the standing fence forcing the standing fence to fall down by rotating around the hinges resulting in the lying fence to stand up vertically. Although the fence is only 0.7 m high and 1.5 m wide, the trapping fence was found to be effective enough to trap 3.0 m^3 of debris flow of wet soil. Liquefied mud of 1.2 m^3 containing more water was also effectively stopped by the trapping fence with splashes of muddy water sprayed 3.5 m beyond the fence while the mud flow reached a distance of 7 m in case that there was no trapping fence at the toe of the slope. The test results ensured that the debris flow trapping fence is an economic measure to prevent the small scale debris flows and to protect the traffic on the motorway.

1 INTRODUCTION

In order to ensure the safety and smooth traffic flow under rainfall condition, the traffic control has been specified for each expressway section. Consequently, any disasters can be prevented when roadways are closed. East Nippon Expressway Company considers the criterion of traffic regulation, for example, taking road strengthening measures such as slope stability and analyzing accumulated disaster data. In low cut slope sections, simple control works are needed to prevent to hold up passing vehicles due to local slope failures.

We developed a debris flow trapping fence to prevent flowing debris from reaching the roadway. The characteristic of the debris flow trapping fence is to dissipate the impact energy of flowing debris by rotating itself and to prevent debris flow reaching roadway. Even when the trapping fence does not rotatc, the weight of the trapped debris acts as a counterweight, so that there is no need for a large foundation. This paper reports the characteristics of the debris flow trapping fence and results of full-scale field test.

2 DEBRIS FLOW TRAPPING FENCE

2.1 *Shape of the debris flow trapping fence*

We developed the debris flow trapping fence which consists of an L-shaped frame built with equal-leg angles and polyester netting, and is designed to rotate around an anchor bolt. The bottom is provided with a deformed bar (D35) anchor bolt which anchors the fence to the ground. The fence and bottom are connected together by high-tension bolts (HTB-M24)

Figure 1. Debris flow trapping fence.

Table 1. Specifications of debris flow trapping fence.

Dimensions	700 × 700 × 1500 mm
Frame	Equal leg angle
Net	Polyester raschel net Tensile strength : 30 kN/m
Anchor	Deformed bar, D35 × 500 mm
Connector	High-tension bolt M24
Design impact force	12.02 kN/m

and nuts. Figure 1 shows the structure of debris flow trapping fence, and Table 1 shows its specifications.

2.2 *Design of the debris flow trapping fence*

The impact force of debris flow considered for design of the debris flow trapping fence is calculated (Ministry of Land, Infrastructure and Transport, 2001). As a soil condition, we assume that soil density is $\rho_m = 1.4\,\mathrm{t/m^3}$ (loose cohesive soil), and that the design condition is "debris flow from a height of 0.5 m from the top of a two-stage 45 degree slope" as shown in Figure 2. The impact force of the debris flow is $F = 12.0\,\mathrm{kN/m}$ calculated as follows.

$$F_{sm} = \rho_m g h_{sm}\left[\left\{\frac{b_u}{a}\left(1-e^{\frac{-2aH}{h_{sm}\sin\theta_u}}\right)\cos^2(\theta_u-\theta_d)\right\}e^{\frac{-2ax}{h_{sm}}} + \frac{b_d}{a}\left(1-e^{\frac{-2ax}{h_{sm}}}\right)\right] = 48.08\mathrm{kN/m^2} \quad (1)$$

$$F = \alpha \cdot F_{sm} \cdot h_{sm} = 12.02\mathrm{kN/m^2} \quad (2)$$

where, F_{sm} : moving force of flow debris, ρ_m : density of flow debris, g : gravitational acceleration, h_{sm} : height of debris before flowing, θ_u : gradient of slope, θ_d : gradient of ground, H : height of slope, x : horizontal distance from toe of slope to fence, F : impact

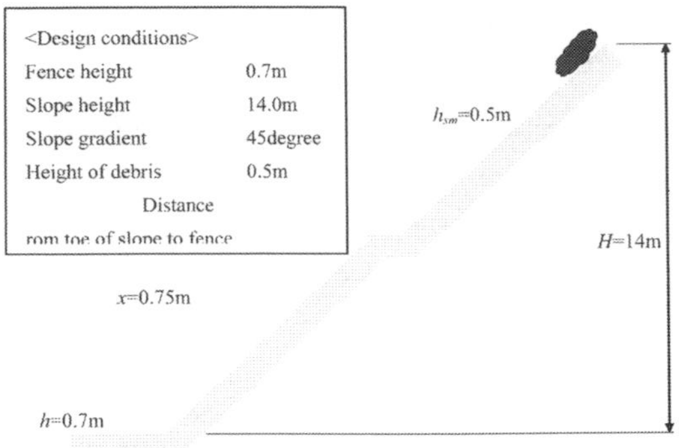

Figure 2. Design conditions and results.

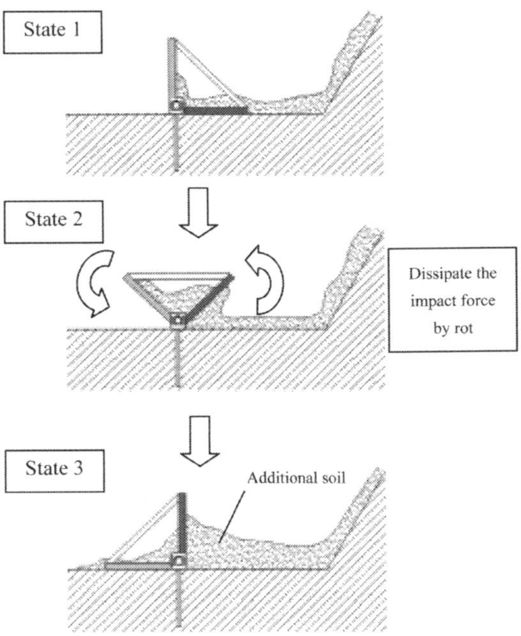

Figure 3. Mechanism of debris flow trapping fence.

force, α : relaxation coefficient of impact force and a, b_u, b_d : calculated by θ_u, θ_d, specific gravity, density, fluid resistance and angle of internal friction of debris.

2.3 *Rotation mechanism*

Figure 3 shows rotation mechanism of debris flow trapping fence. First, the flow debris hits the upright net of the L-shaped fence (State 1), the fence begins to rotate as it dissipates the impact force of the flow debris (State 2). The fence rotates around the high-tension bolt. After the rotation, the bottom face of the L-shaped fence acts as an upright fence surface as to

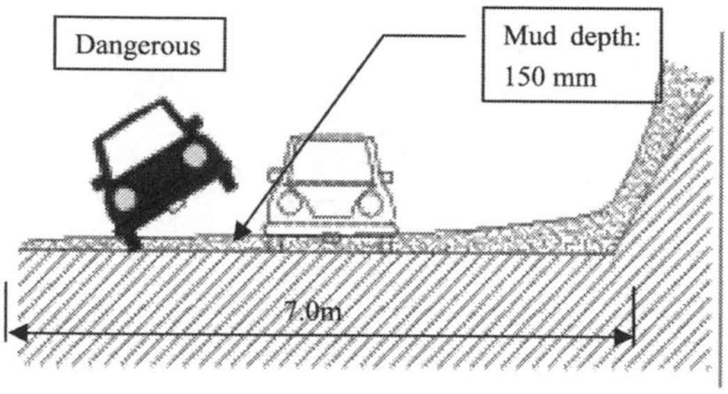

Figure 4. Failure of slope without trapping fence (mud flow directly hitting a passing vehicle).

Figure 5. Free flow of mud along slope without trapping.

continue to hold back the flow debris (State 3). Characteristic of the debris flow trapping fence is that the fence dissipates the impact force of debris by rotating itself and deflects the direction of energy flow from the horizontal direction to a downward direction, so that it reduces the amount of debris flow toward the road. A second characteristic is that even when it does not rotate, the debris deposited over the bottom net of the fence acts as a counterweight to retain a large volume of debris.

2.4 *Other characteristics*

The debris flow trapping fence has many advantages as follows : 1) After rotating, the fence can retain the flow debris. 2) Inexpensive. 3) Simple structure enables easy installation and removal. 4) Parts of the fence are easy to replace. 5) There is no need for a large foundation, and the amount of construction by-products is small.

3 RESULTS OF FULL-SCALE TESTS

3.1 *Dissipation of impact force by rotation*

In the field test, a $1.2\,m^3$ of mud mass ($0.8\,m^3/m$) was slid down freely along the slope in the test as shown

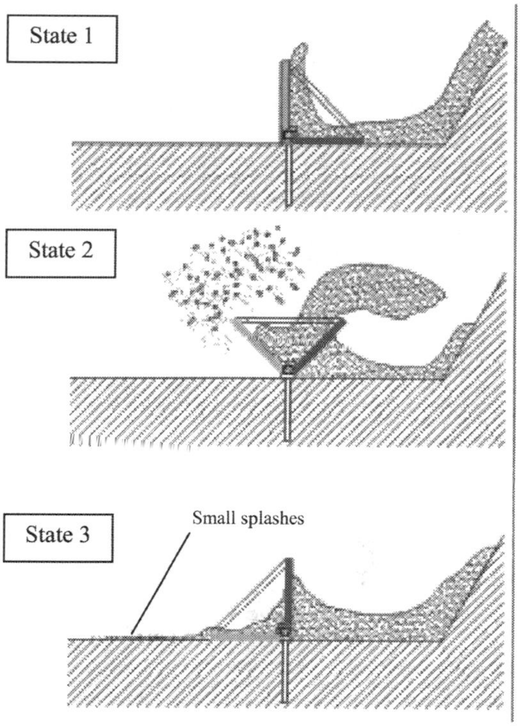

Figure 6. Behavior of mud with trapping fence.

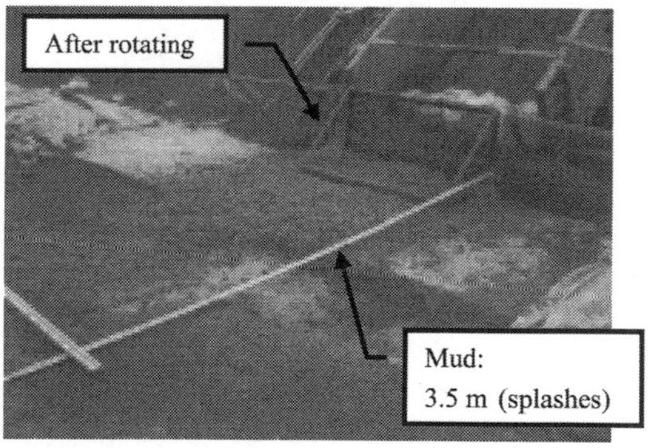

Figure 7. Flow of mud along slope with trapping fence.

in Figures 4 and 5. Figures 6 and 7 show that the mud mass was trapped by the trapping fence. In the case that the trapping fence was not used, the mud mass reached the roadway, and the front end of the mud mass was 7.0 m from the toe of the slope and the mud depth was 0.15 m. In the case that the trapping fence was installed and the mud mass was slid down the slope under the same conditions as the fenceless case, when the sliding mass reached the upright net of the

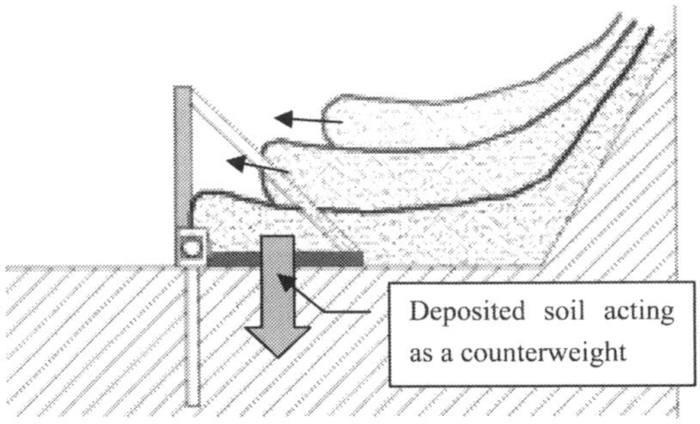

Figure 8. Trapping a flow of field product soil.

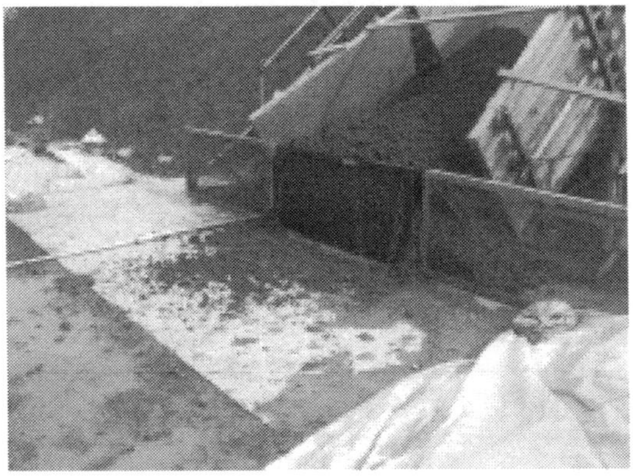

Figure 9. Trapping a flow of field product soil.

fence, much of the mud was thrown up (State 1). Part of the mud was thrown backward and the mud flow was prevented from reaching the roadway (State 2). Part of the mud was splashed (State 2) and reached a point 3.5 m from the toe of the slope. Since, however, these splashes were small, their influence on the roadway was considered to be small even if they reached the roadway (State 3). Concurrently, when the trapping fence began to rotate, it dissipates the impact force of the mud (State 2).

3.2 *Retention of debris by self-weight*

In the second field test, a 3.0 m^3 soil mass (2.0 m^3/m) of field product soil (sandy gravel containing cohesive soil) was slid down a single-stage cut slope toward the fence. Figures 8 and 9 show that the soil mass was trapped by the fence. The soil mass was retained by the upright net of the trapping fence. The debris deposited on the bottom net, and more debris was deposited on the material deposited earlier. In the test, the fence did not rotate, and the flow debris was successfully retained by the self-weight of the debris.

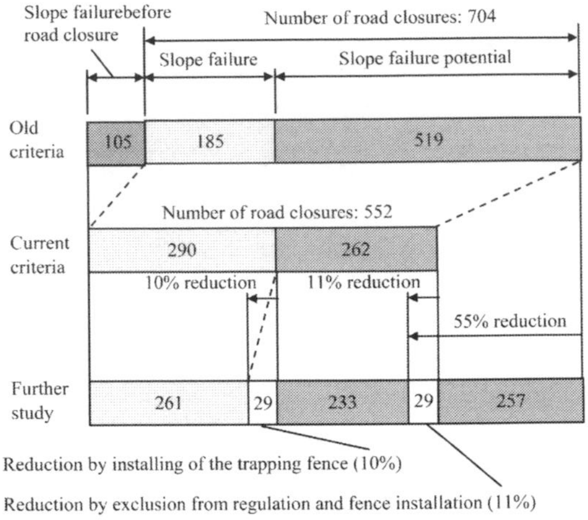

Figure 10. Road-closure-reducing effect.

4 VERIFICATION OF THE EFFECT OF DEBRIS FLOW TRAPPING FENCE

We verify that the closed road due to slope failure can be reduced by installing the debris flow trapping fence as shown in Figure 10. The volume of flowing debris retained by the debris flow trapping fence is assumed to 2.2 m^3. Of all cut slope failures in the past 10 years, single-stage cut slope failures account for about 35%. Of the single-stage cut slope failures that necessitated road closure, 28% are small-scale failures that involved 2.2 m^3 or less of collapsed soil. It will be possible to prevent road closure at 29 slope failure sites ($0.35 \times 0.28 \times 100 = 10\%$).

When we considered possible reductions in the number of road closures due to slope failure potential, 27% took place in low cut or fill slope sections. About 40% in low cut or fill slope sections can be prevented by installing the debris flow trapping fence. This indicates that it is possible to prevent 29 ($0.27 \times 0.4 \times 100 = 11\%$) road closures in the 262 road closures made because of slope failure potential. If the number of road closures (519) made because of slope failure potential in accordance with the old standards is taken into consideration, about 55% ($=(257 + 29)/519$) can be prevented.

When we consider the number of road closures that can be prevented, we find that 58 road closures made because of slope failure or slope failure potential can be prevented by installing the debris flow trapping fence. Therefore, it can be concluded that the debris flow trapping fences will contribute to better road management and reductions in the number of road closures.

5 CONCLUSIONS

The results of this study can be concluded as follows : 1) The debris flow trapping fence has been developed as a simple means of damage mitigation with the aim of preventing collapsing material from reaching the roadway in the event of a small-scale failure of a low cut slope. 2) The debris flow trapping fence has a number of advantages including its ability to dissipate the impact force of debris by rotation. 3) The advantages of the debris flow trapping fence have been verified through the field test. 4) We predict that better road management can be achieved, and that road closures can be prevented by installing the debris flow trapping fence.

The debris flow trapping fence is a means of damage mitigation based on a completely new concept that have been developed as simple control works following the reconsideration of criteria for road closure due to rainfall. By applying the debris flow trapping fence to actual road, it can be expected to contribute to better road management and reductions in the number of road closures due to small-scale slope failure. We believe that the debris flow trapping fences will contribute to safer and smoother traffic flows expected of expressways. We will continue to work for further improvement under the cold and snowy conditions.

REFERENCES

Ministry of Land, Infrastructure and Transport. 2001. Public Notice No. 332.

Expressway Technology Center. 2003. Study on Traffic Control Due to Rainfall.

New Horizons in Earth Reinforcement – Otani, Miyata & Mukunoki (eds)
© 2008 Taylor & Francis Group, London, ISBN 978-0-415-45775-0

Displacement and failure characteristics of model geogrid-reinforced structure subjected to impact load

N. Yasufuku, H.Ochiai, K. Omine & T. Kobayashi
Kyushu University, Fukuoka, Japan

K. Shomura
Graduate School of Kyushu University, Fukuoka, Japan

ABSTRACT: In order to clarify the failure mechanism of geogrid-reinforced structure affected by earthquake, the failure mode of reinforced structures are carefully investigated when it was subjected to the impact load, based on the pictures shot by the high speed camera. It is observed that as the acceleration increases, the sliding surface of the geogrid reinforced structure first appears from the back of the sliding surface predicted by Coulomb's active earth pressure. The fixing length of reinforcement is important to improve the deformation characteristics of the reinforced structure during the impact load.

1 INTRODUCTION

On March 20, 2005, an earthquake of magnitude 7.0, which was named as the 2005 Fukuoka-ken Seiho-oki Earthquake, occurred in the northwest part of Fukuoka City. A large number of retaining walls and houses in Genkai island located at 9km far from the epicenter had a huge damage under the predicted seismic intensity scale of 6 upper due to the earthquake (see Kobayashi et al., 2006). Two old geogrid reinforcement steep embankments existed which were constructed in 1980's survived with a little damage such as the clacks of top part of the embankments and the small forward displacement with the cutting of the front part of the geogrids. However, the degree of the damage inside of the reinforced embankment was not so clear. Thus, in order to properly judge what level of repairs are needed, it is important to clarify the effects of the damage mentioned above on the subsequent durability and resistance against the earthquake. It is desired to indicate the basic idea for judging the most suitable repair's method of the damaged reinforcement structures. In this study, as a first step of this object, the characteristics of the failure and deformation patters of the earth reinforcement embankment subjected to an impact load were experimentally investigated through the monitoring of the behaviours by using a high speed camera. It is noted that the discussion below is limited to be a qualitative one.

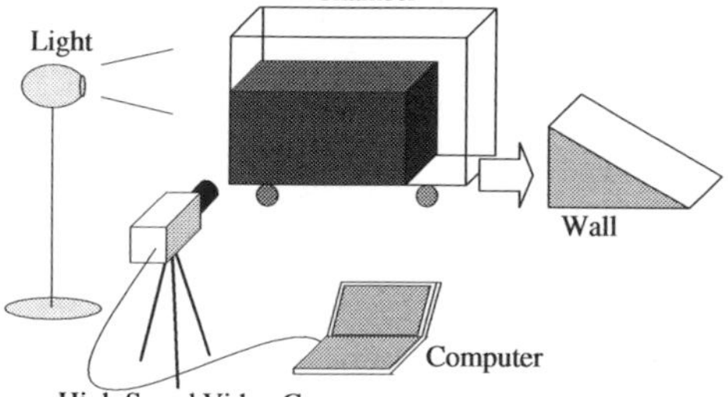

Figure 1. Outline of experimental system.

2 OUTLINE OF EXPERIMENTS

2.1 *Model apparatus used*

A model chamber with casters is used, where an impact load is applied by manually dashing the chamber against a wall. The model ground consists of the length with 41.5 cm, the height with 20 cm and the width with 10 cm, in which its side is made of an acryl plate to observe the movement of the ground during the test using a high speed digital video camera as schematically shown in Figure 1. This camera (Phantom V4.2; Vision research co. ltd.) can take 2024 scenes for 2 seconds with 512*512 pixel. A typical model reinforced ground is shown in Figure 2, together with the photo.

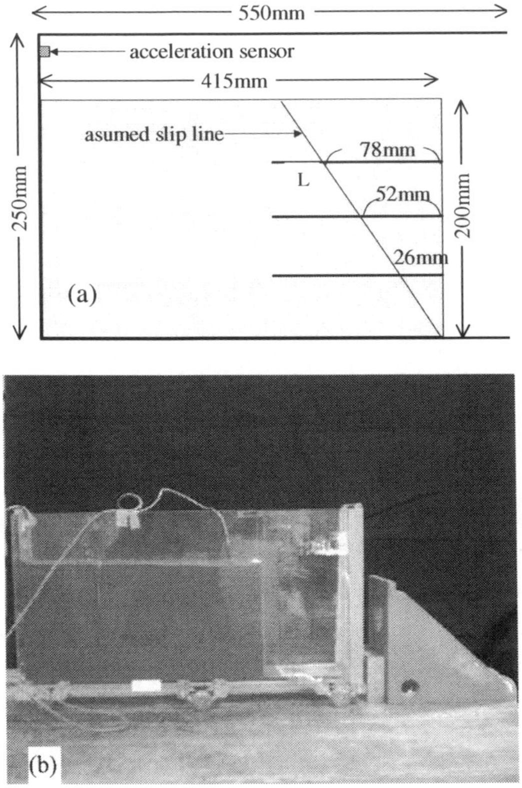

Figure 2. Earth reinforcement mode ground.

The magnitude of the impact load was evaluated by the acceleration sensor mounted in the top part of the chamber. As shown in this figure, each reinforcement material was horizontally laid in the ground with an equal vertical space. Length of the reinforcement material was determined based on the expected slip surface from the Coulomb's active earth pressure theory. In this study, three different embodiment lengths of the earth reinforcement materials were prepared, whose lengths were 67 mm and 33.5 mm longer, and 10 mm shorter than the length from the front of the ground to the slip line shown in Figure 2, where ΔL is the embodiment length defined.

2.2 *Earth reinforcement and geomaterial used*

A graph paper was selected as a model reinforced material, whose merits are to be easy to install in the ground and to adjust its length. A decomposed granite sandy soil called as "Masado" with $D_{50} = 0.5$ mm and $U_c = 30$ was used for making a model ground. The maximum size of the particle diameter was controlled to be less than 2 mm. The model ground were made by the compaction of Masado with the optimum water contents, which consists of four layers of Masado. Each layer is compacted to be 1.44 g/cm^3. A series of drained direct shear box tests of Masado with same sample conditions of the model ground were conducted under the constant normal stresses to obtain the strength parameters of this materials. As a result, the internal friction angle and cohesion are determined as around 39degrees and 4 kPa, respectively, which seems to reflect an well compacted model ground.

Table 1. Test condition.

Case1	Earth reinforcement with embodiment length ΔL of 67 mm
Case2	Earth reinforcement with embodiment length, ΔL of 33.5 mm
Case3	Earth reinforcement with embodiment length, ΔL of −10 mm
Case4	No earth reinforcement

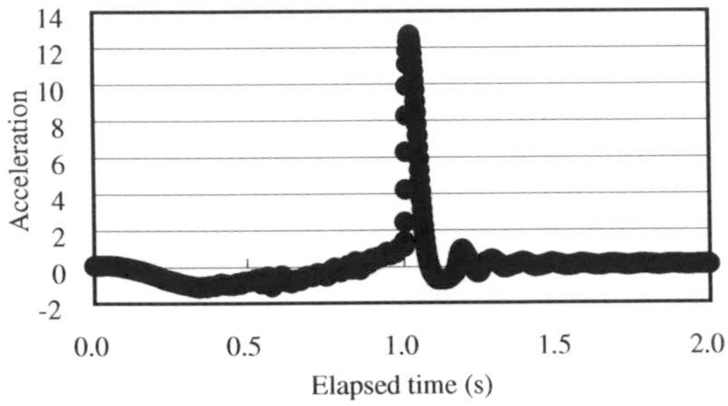

Figure 3. Typical acceleration time history.

2.3 *Experimental procedure and conditions*

A impact load is applied to the model ground by using an inertia force mobilized when the chamber strikes against a fixed wall (see Figure 1). The deformation and failure pattern of the ground are carefully observed by high speed video camera and then analyzed by PIV method through the test. The magnitude of the impact loads are relatively evaluated by the values of an acceleration sensor. The measured acceleration varies widely, and thus two or three tests were conducted under the same initial test conditions. Test cases conducted are summarized in Table 1, where the earth reinforcement materials with three layers are installed in the Cases from 1 to 3, whose differences are the length of the materials. On the other hand, for comparison, Case 4 is prepared, which treats the ground without any earth reinforcement materials. Figure 3 indicates a typical acceleration time history obtained.

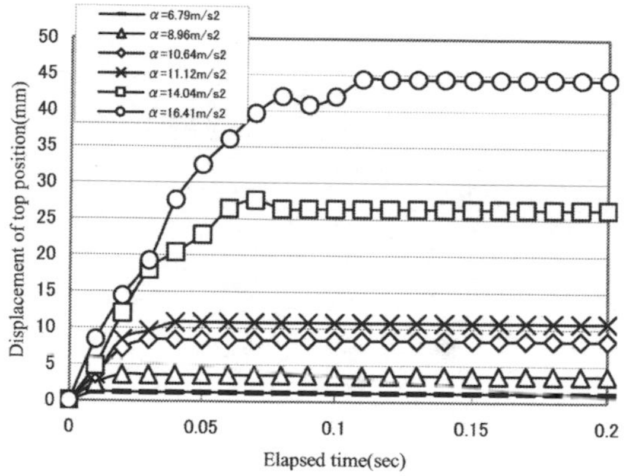

Figure 4. Displacement of earth reinforcement at top position against elapsed time.

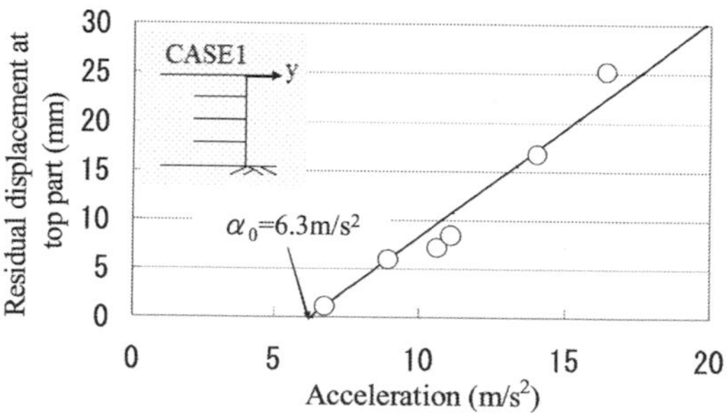

Figure 5. Residual displacement at top part of the ground against accelerations.

3 RESULTS AND DISCUSSIONS

3.1 *Displacement and failure properties of reinforced ground*

Figure 4 shows the displacements of the top of the ground in Case 1 against the elapsed times under the various acceleration conditions. The displacement begins to occur at the moment once an impact load is applied and proportionally increases with the increasing elapsed time. After that, the displacement for each case converges to a certain value being dependent on the magnitude of acceleration. Figure 5 indicates the relationship between the residual displacements at the top of the ground and the acceleration values in Case 1. It is found that the residual displacement begins to occur when the acceleration becomes greater than around 6.3 m/s^2 in this model earth reinforcement ground and then increases with the increasing acceleration. When the acceleration exceeds a certain value, it is observed that even if the inside of the earth reinforcement area has no damage, there is a case the slip layer can be formed behind the reinforcement area and the ground often reaches the failure. Therefore, the embodiment length of the earth reinforcement material seems to be important to increase the potential safety of the reinforcement ground. Figure 6 shows the typical failure patterns from Case 1 to Case 4 under the acceleration around 14.5 m/s^2 when the elapsed time reaches to 0.1 seconds. From the high speed video observation, the following behaviours can be seen. In Case 4 without any earth reinforcement and Case 3 in which there is no sufficient embodiment length for all the earth reinforcement materials, a clear clack such as a slip line first appears behind a Coulomb's slip surface. Many clacks inside of soil mass within the clear slip line are then presented and the length of clacks becomes longer and longer with the elapsed times. Finally the embankment in both cases reaches to a catastrophic failure. On the other hand, in Case 1, which has a sufficient embodiment length for all the earth reinforcement materials, a clear crack first appears which seems to be similar to the Case 1. After that, although the soil mass ahead of the clack moves forward, a catastrophic failure does not occur due to the earth reinforcement effects. The number in Case 1 (see Figure 6(a)) means the order of which the clack appears during the tests. It is found that the first clack appears fairly behind the theoretical Coulomb's slip line and each reinforcement material prevents the progress of the clack.

3.2 *Effect of embodiment length on the displacement behaviours*

Figure 7 shows the horizontal displacement at the top of the ground normalized by the wall length against the elapsed time with every 0.15 seconds after the impact load is applied under the acceleration of around 14.5 m/s^2. The typical results in Case 1, Case 3 and Case 4 are compared here. It is clear that the results in Case 3 and Case 4 have a same trend. It means that little reinforcement effect can be found in the case that the embodiment length is not sufficient. Figure 8 shows the horizontal displacements of the ground at depth of 50 mm, 100 mm and 150 mm from the bottom in the Case 1 and Case 2, paying attention to the accelerations normalized by α_0 defined in Fig. 5. It is noted that the embodiment length in Case 2 is half comparing with that in Case 1. When α/α_0 is relatively small, there is little difference of the displacement in each case. And then, when the normalized acceleration becomes larger, the displacement in Case 2 gradually becomes greater, comparing with that in Case 1. Such tendency tends to be remarkable for increasing acceleration. It is conformed that the embodiment length works well to reduce the horizontal displacement of the earth reinforcement structure when subjected to relatively large impact loads.

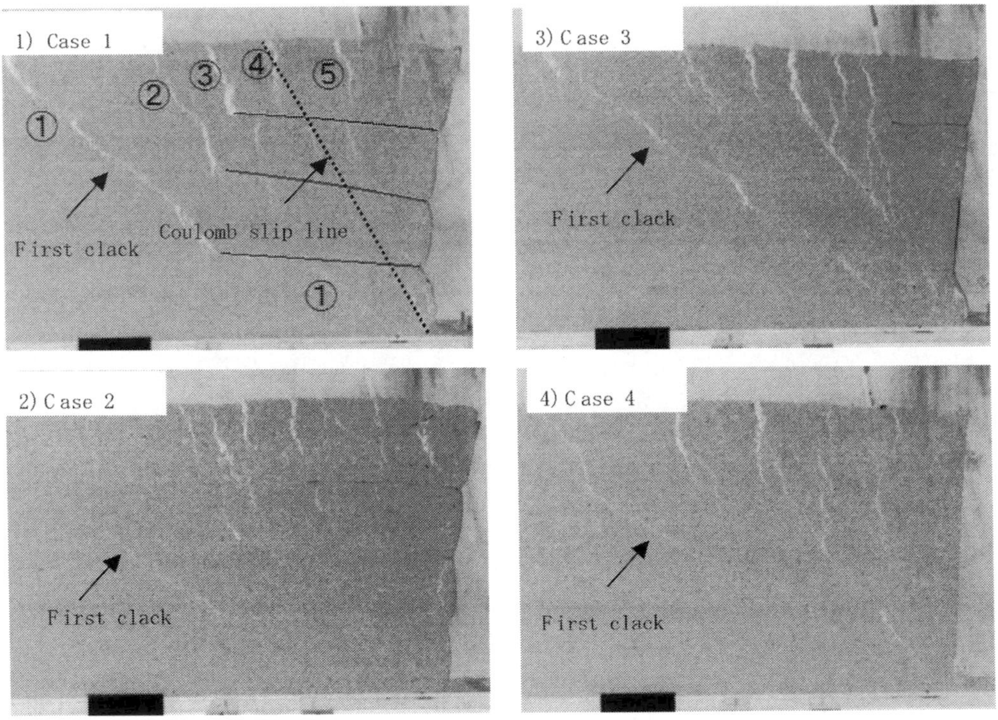

Figure 6. Typical failure patterns of each case at 0.1 seconds after subjected to an impact load.

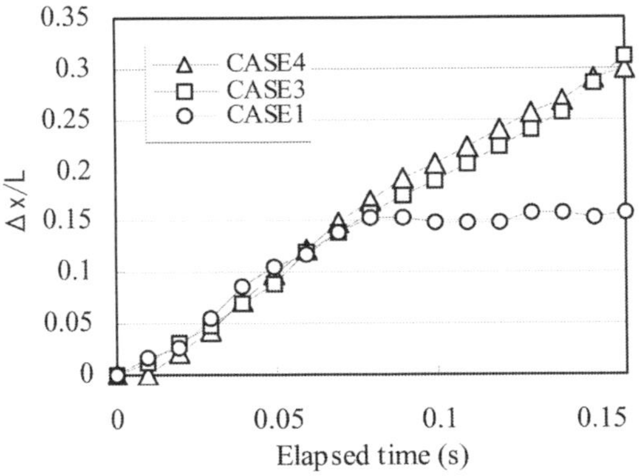

Figure 7. Normalized horizontal displacement at top of the ground against elapsed time for three cases with different embodiment length.

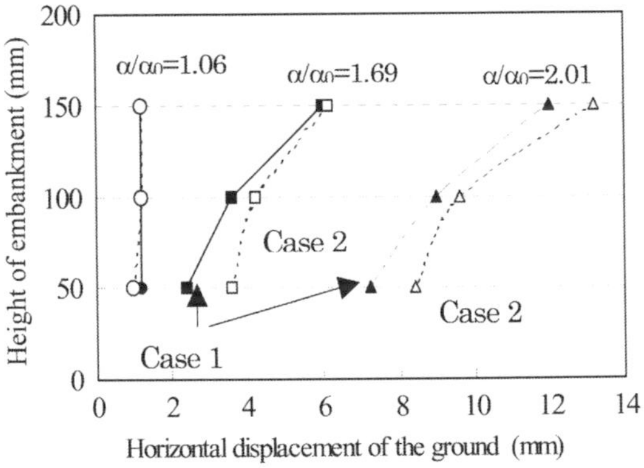

Figure 8. Comparison of horizontal displacement in Case 1 with those in Case 2 paying attention to the magnitude of the acceleration.

3.3 *Visualization of the movements of earth reinforcement ground*

Figure 9 indicates the displacement vectors of the earth reinforcement ground in Case 1 in the period of 0.006 seconds from 0.05 second just after subjecting the impact load. The displacement vectors tend to increase with approaching to the upper part of the ground and also to the earth reinforcement wall. A slip surface, which is first observed, is also depicted in this figure. It is confirmed that the soil mass inside of the slip surface mainly moves forward.

4 CONCLUSIONS

Failure and deformation patters of the earth reinforcement embankment subjected to an impact load were

Case1: α=14.5m/s²

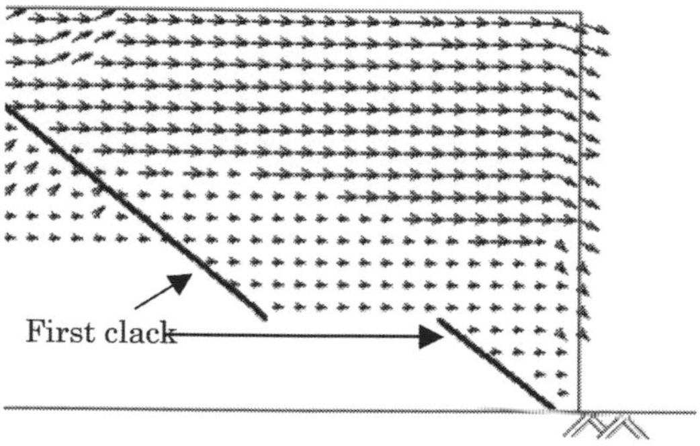

Figure 9. Typical displacement vectors of the ground in Case 1.

experimentally investigated through the monitoring by using a high speed camera. The following main conclusions are obtained in this study.

1) When the acceleration exceeds a certain value, even if the inside of the earth reinforcement area has no damage, there was a case that the slip layer can be found behind the reinforcement area.

2) Irrespective of the reinforcement condition, every time the first clack due to an impact load appeared fairly behind the active area by the Coulomb's theory. After the first clack appeared, many clacks were produced in the active area inside of the first clack within a very short elapsed time.

REFERENCES

Kobayashi, T., Zen, K., Yasufuku, N., Nagase, H., Chen, G., Kasama, K., Hirooka, A., Wada, H., Onoyama, Y. & Uchida, H. 2006. Damage to Residential Retaining Walls at the Genkai-Jima Island induced by the 2005 Fukuoka-ken Seiho-oki Earthquake, *Soils and Foundations*, Vol.46, No.6, pp.793–804.

New Horizons in Earth Reinforcement – Otani, Miyata & Mukunoki (eds)
© 2008 Taylor & Francis Group, London, ISBN 978-0-415-45775-0

Large-scale overflow failure tests on embankments using soil bags anchored with geosynthetic reinforcements

K. Matsushima
National Institute for Rural Engineering, Ibaraki, Japan

S. Yamazaki
Mitsui Chemicals Industrial Products, Ltd., Tokyo, Japan

Y. Mohri, T. Hori & M. Ariyoshi
National Institute for Rural Engineering, Ibaraki, Japan

F. Tatsuoka
Tokyo University of Science, Chiba, Japan

ABSTRACT: Every year, a great number of dams for agricultural irrigation are seriously damaged or totally destroyed by flood overflow exceeding the drainage capacity of a spillway. To increase the stability of the downstream slope of such small earth fill dams against overflow, it is proposed to protect the downstream slope by using inclined soil bags anchored with geosynthetic reinforcement. Two tests on hydraulic overflow-induced collapse of the downstream slope were performed on full-scale models, 3.5 m high and 2.3 m wide with a downstream slope of 1V: 1.2H. The first test was conducted on an unreinforced soil slope subjected to a fixed level of overflow and the other was on the geosynthetic soil bags with extended tails (GSET) in which soil bags were placed on the downstream slope and subjected to stepwise increased levels of overflow. Results showed that the GSET model protected by using the soil bag system was stable enough against temporary flooding at overflow levels required in the field. In particular, the rate of progressive erosion of the downstream slope when subjected to high overflow levels was significantly reduced by reinforcing the slope.

1 INTRODUCTION

1.1 *Background*

Across Japan, there are approximately 210,000 reservoirs with earth dams constructed for agricultural irrigation that are lower than 15 m in height. It is reported that, among them, approximately 20,000 earth dams have deteriorated and need urgent but cost-effective repair. Every year, a great number of dams are seriously damaged or even totally destroyed by flood overflow exceeding the drainage capacity of the spillway and by earthquakes, as typically seen in Photo 1. This can cause a serious disaster in the downstream area. To substantially increase the stability of the slopes of such small earth dams against overflow and seismic loads, Mohri et al. (2005) proposed to protect the downstream slope of a small earth-fill dam by using large-scale soil bags with additional sheet for anchoring into the embankment as shown Fig. 1. Furthermore, Matsushima et al. (2005) modified this soil bag system to use inclined soil bags anchored with geosynthetic reinforcements as shown in Fig. 2. This geosynthetic soil bags with extended tails (GSET) spillway is designed to function in emergency cases of temporary flooding.

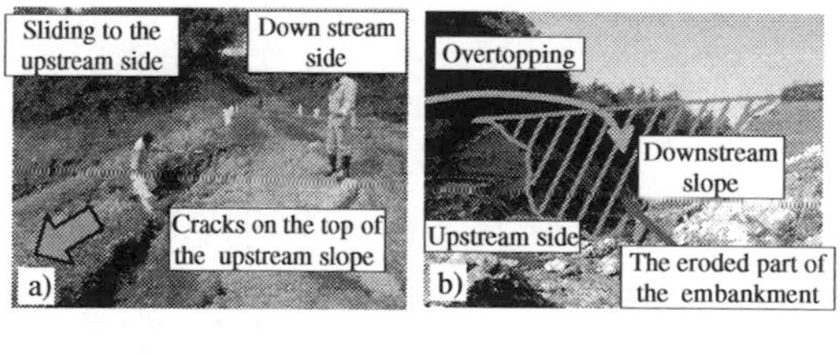

Photo 1. Failures of small earth dams: a) by the Niigata-chuetsu earthquake in 2004 in Kawaguchi town; and b) totally collapsed by heavy rainfall during Tokage typhoon No.200423 in Awaji Island.

The compressive strength of cohesionless soil that is not reinforced while located at the surface of a soil mass is essentially zero. On the other hand, soil located

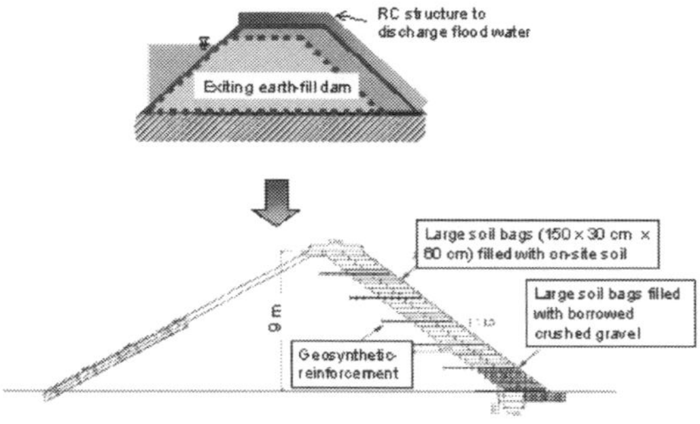

Figure 1. A new technology to rehabilitate existing old earth-fill dams to have a high flood discharge capacity and a high seismic stability (when applied to a 9 m-high typical earth-fill dam; Mohri et al., 2005).

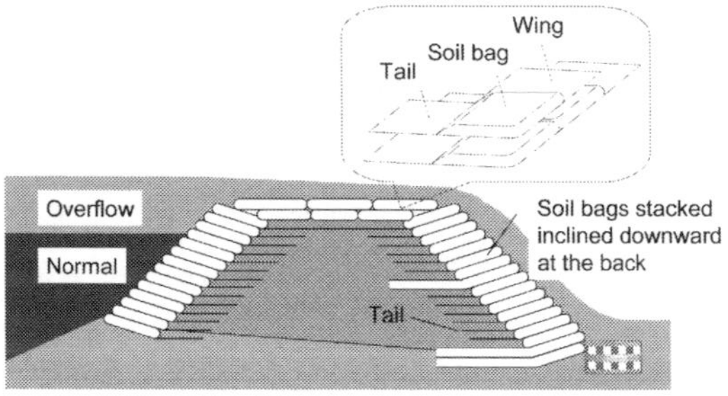

Figure 2. Basic components of geosynthetic soil bags with ex-tended tails (GSET) spillway for temporary flooding (Matsu-shima et al., 2005a&b).

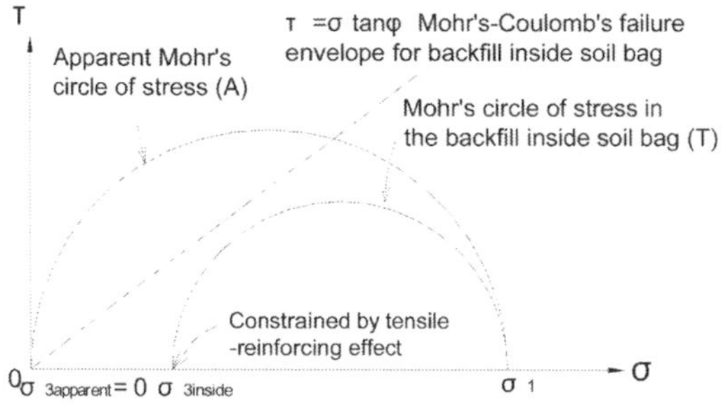

Figure 3. Stress states of soil bag located at shallow layer of slope represented by Mohr's circle of stress.

in a soil bag is subjected to additional confining pressure due to tensile reinforcing effects of the soil bag even when there exists no external confining pressure. Therefore, soil in a soil bag can exhibit compressive strength as represented by a Mohr's circle of internal stress denoted by T in Fig. 3 (Matsushima et al., 2006). The Mohr's circle of apparent stress denoted by A indicates the apparent stress condition of the soil bag. Due to the high compressive strength of the soil bag by this confining effect of the bag, not only slope stability but also washout resistance against overtopping can be increased, effectively protecting the slope face.

1.2 *Stream regimes of stepped spillways*

A GSET-spillway with stacked soil bags (Fig. 2) can be considered as one specific type of stepped spillway. Due to its high energy dissipating effect, such a stepped configuration has been widely used for concrete dams. A number of researchers have already studied the overflow characteristics of such dams (Hubert, 1994). The overflow characteristics can mainly be classified into the following three types (Fig. 4) depending on the step height, the discharge and the slope of the spillway:

1) nappe flow, characterized by the formations of a nappe and an air pocket at each step (Fig. 4a);
2) skimming flow, characterized by the formation of an eddy at each step (Fig. 4b); and
3) formation of free fall at the top of the slope (Fig. 4c).

However, no study on the influence of such regimes on the stability of GSET-spillway against overtopping has been performed. Figure 5 shows a conceptual illustration of damage mechanisms, as a function of overflow level, that may affect the stability of the GSET-spillway subjected to overtopping. The main damage mechanisms that are to be taken into account in the design of GSET-spillway against overtopping may include:

a) suction of backfill material by negative pressure; i.e., washout of soil particles from the slope through spaces between the soil bags;
b) attrition of geosynthetics by tractive force; and
c) breakage of geosynthetics by penetration force caused by free water fall.

To evaluate the stability of GSET-spillway against overtopping affected by these three damage mechanisms, a series of large-scale model hydraulic overflow failure tests was conducted. Based on observations of the tests, the damage patterns on the GSET-spillway were identified and categorized as a function of different stream regimes that depend on overflow levels.

2 OVERFLOW-INDUCED COLLAPSE TESTS

2.1 *Experiment models and materials*

Figure 6a shows a large-scale GSET-spillway model with a total of 24 soil bag steps placed on the down stream slope. The model is 3.5 m high and 2.3 m wide with a downstream slope of 1V: 1.2H. Figure 6b is a view of the completed downstream slope. Figure 6c shows a soil bag, integrated with a tail and a wing while approximately 0.2 m high, 0.6 m wide and 0.6–1.0 m long with a weight of 200 kg. The tail is

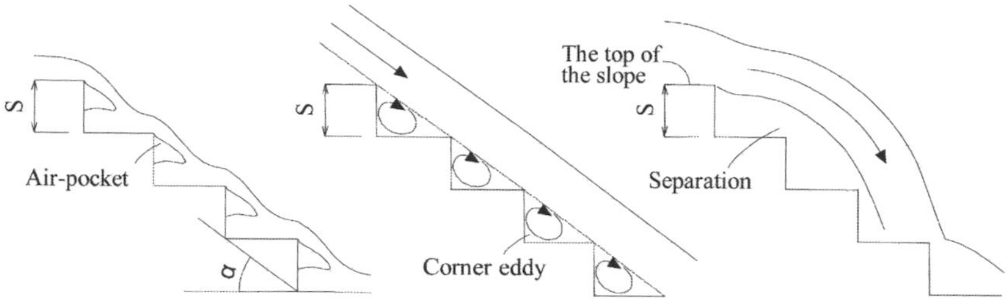

(a) Nappe flow (b) Skimming flow (c) Free fall flow

Figure 4. Stream regimes of stepped spillways (after Hubert, 1994).

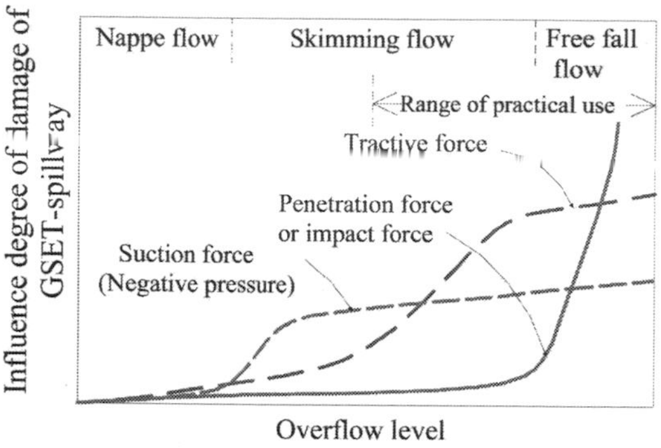

Figure 5. Damage mechanism of GSET-spillway as a function of overflow level.

designed to anchor a soil bag inside the slope while the wing connects horizontally adjacent soil bags. The bag is made of a woven polypropylene (PP) sheet. This PP geotextile is relatively cheap and widely used for agricultural-related purposes. Figure 7 shows its tensile load–strain properties. The infill material of the soil bags was a crushed concrete aggregate, having a slight inter-particle cementation due to residues of cement that had not been fully hydrated. The backfill materials in the shell and core zones of the slope, behind the soil bag, were Kasama sand and a mixture of Kasama sand and Kanto loam (proportion 1: 1 in weight). Figure 8 and Table 1 present the grading curves and physical properties of these soil bag infill materials as well as the slope backfill materials. The shell and core zones were compacted by manual tamping to degrees of compaction higher than 95% (D-value). The overflow depth (h_0 and h_1) on the upstream side and the center of the crest were measured by using water gauges. To evaluate the displacement distrbution of the downstream slope surface, a laser profiler, having a servo motor control system, was set in parallel with the downstream slope. For reference, a hydraulic test on a large-scale unreinforced embankment made using Kasama sand, 3.5 m high and 2.3 m wide having a downstream slope of 1V: 1.8H, was conducted at a discharge unit quantity flow $q = 0.050\ m^3/s/m$.

2.2 *Test conditions for GSET-spillway model*

Figure 9 shows the time history of discharge unit quantity flow in the overflow-induced collapse test on the GSET-spillway model. In this figure, the ranges of flow level in which respective damage levels were observed during the test (explained in the next section) are also presented. The test consisted of the following two stages:

Stage 1: Without artificial physical damage to the soil bags; and

Stage 2: With artificial physical damage to the soil bags by cutting the surface and loosening the infill material to simulate damage by floodwood, chemical or ultraviolet degradation.

As the downstream slope was stable even when the discharge unit quantity flow became as high as $0.48\ m^3/s/m$, it was decided to add the second stage. At the second stage, where the surface deformation by overtopping became noticeable, at every step increase of overflow level at selected stages denoted by a, b, c, d, e, f, g, h and i in Fig. 9, surface surveys of the downstream slope were conducted by laser profiler.

2.3 *Overflow level 1 (no or little damage)*

When the overflow depth, h_0, was less than 23.8 cm (i.e., S/d_c became 1.248–10.67, where S is step height and d_c is critical depth), a relatively steady flow in a staircase pattern was observed due to a high-energy dissipation caused by the soil bag steps. White water, which was actually rich-aerated flow, was formed while a small hydraulic jump impacted each soil bag step (Photo 2a and Fig. 10a). This stream regime was categorized into nappe or transition flow (Fig. 9). This observation is consistent with the lower limit of S/d_c at which a nappe flow is formed in stepped spillways: i.e., $S/d_c = 1.623$ at $\alpha = tan^{-1}\ (1/1.2)$, according to Yasuda et al. (1999).

Nappe flow, having nappe and air pockets, has no or little negative pressure at the corner of soil bags. Therefore, suction of the backfill material from the slope behind the soil bags might not occur. Accordingly, no

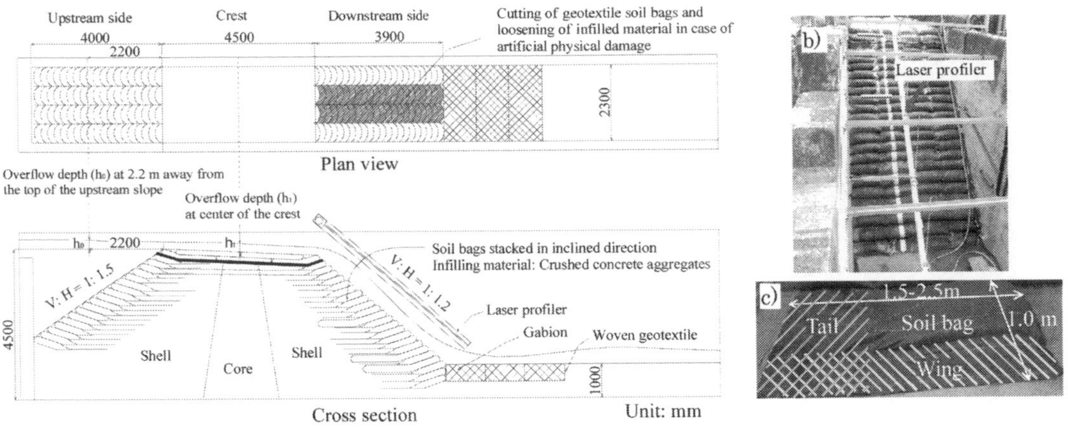

Figure 6. Large-scale overflow-induced collapse test: a) Plan view and cross section of full-scale small earth dam; b) Downstream slope of constructed GSET-spillway model; and c) Large-size soil bag integrated with wing and tail.

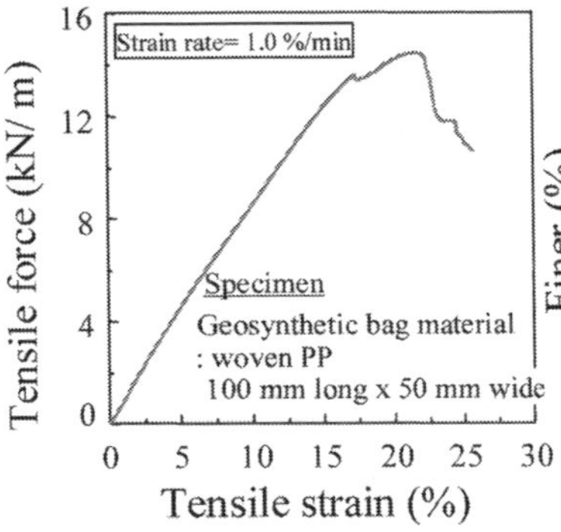

Figure 7. Tensile load-strain relationship for a tensile test performed on the PP woven type geosynthetic used for soil bag.

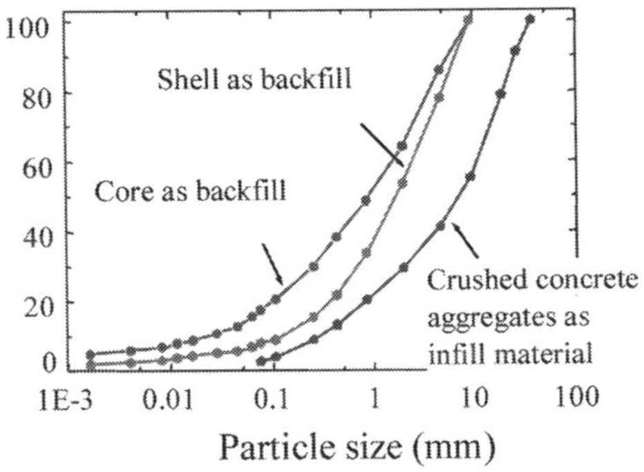

Figure 8. Distribution curves of infill material of soil bags and backfill materials of embankment.

damage on the embankment, such as deformation, suction of backfill material or breakage of soil bag, was observed.

2.4 *Overflow level 2 (minor or moderate damage)*

When the overflow depth, h_0, increased to between 23.8 cm and 32.3 cm (i.e., S/d_c became 0.649–1.248), a thick vein head flow, separated from the top of the downstream slope, started forming, which gave an impact on a limited number of soil bags. As seen from Photo 2b, a heavy flow started entraining air at some distance after having leaped over several soil bag steps below the starting point (Fig. 10b). This stream regime entraining air is basically classified into skimming flow, which is consistent with the upper limit of S/d_c for the formation of skimming flow in the case of a stepped spillway, $S/dc = 1.126$ at $\alpha = tan^{-1}(1/1.2)$, according to Yasuda et al. (1999). Skimming flow having a corner eddy creates negative pressure (i.e., suction) at the corner. Therefore, the damage observed on the slope at this stage included: i) suction of backfill materials from the slope behind the soil bags through the void area between the interfaces between the vertically and/or horizontally adjacent soil bags; ii) attrition of soil bag surfaces; and iii) perforations of soil bags by sharp edges of the crushed concrete aggregates, induced by impact and tractive force. Photo 3a shows the sucked backfill material that remained on the periphery of the void between the soil bags' interfaces after the test. Photo 3b shows the surface attrition and the punching holes at sharp edges of crushed concrete particles.

At the subsequent test stage, artificial physical damage was given to the soil bags repeatedly, four times, by cutting the surface of geotextile soil bag and loosening the infill material, as shown in Photo 4, during overflow testing for a total period of 150 min at a discharge rate of $q = 0.348\,m^3/s/m$ at respective stages between damaging operations. It was found that the

Table 1. Properties of infill material and backfill materials (JIS 1210).

Type	Material	U_c	G_s	F_c(%)	k (cm/s)	Compaction method Ab (Bb in the case of infill material) ρ_{dmax} (g/cm^3)	w_{opt}(%)
Infill (soil bag)	Crushed concrete aggregates	39.2	2.605	2.6	2.67E-04	1.868	12.8
Shell	Kasana sand	20.3	2.650	7.8	1.21E-04	1.935	11.6
Core	Kasama sand and Kanto loam mix	68.3	2.617	17.4	1.27E-06	1.470	24.6

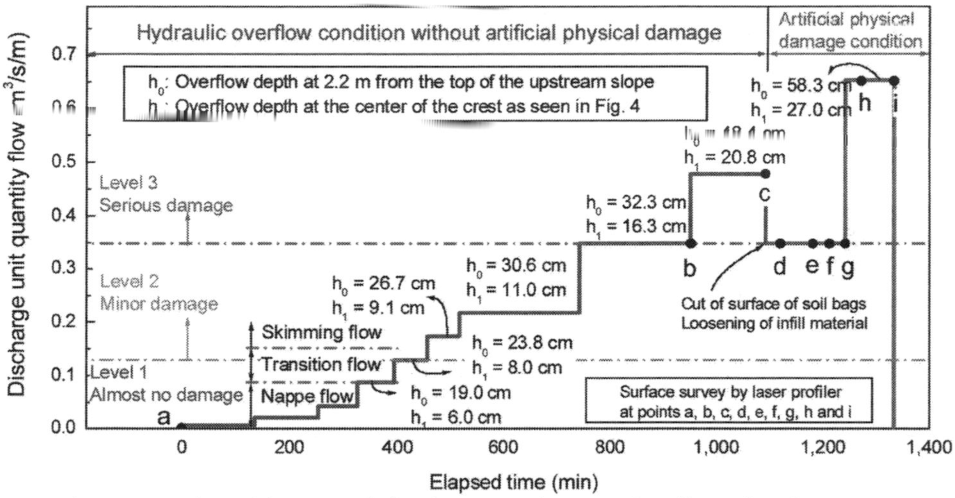

Figure 9. Time history of discharge unit quantity flow for the test per-formed on GSET-spillway model.

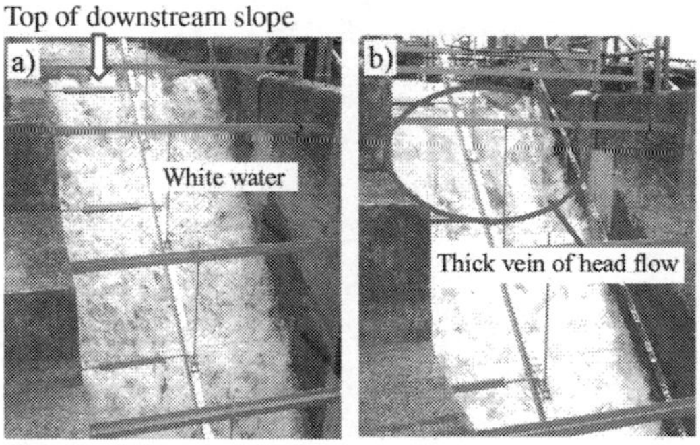

Photo 2. Stream regime on the downstream side: a) Overflow level 1 at q = 0.087 m^3/s/m; and b) Overflow level 2 at q = 0.348 m^3/s/m.

infill material (i.e., crushed concrete aggregate) had been slightly cemented by hydration of remaining cement. Figure 11a shows the profile of the deformed downstream surface at stages d, e, f and g plotted in Fig. 9. It seemed that impact and tractive forces by water flow were strong and infill material was washed out in the vicinity of the places of artificial loosening. Subsequently, as schematically shown in Fig. 12, the torn-off edge part of a geotextile sheet of soil bags became like a flange, covering and sealing the torn-off area while preventing further erosion of infill material from the inside of soil bags. Subsequently at overflow level 2, progressive erosion did not develop. Furthermore, it is likely that the combination of dense arrangement of geotextile sheet inside the slope and slight cementation of the infill material contributed significantly to the high stability and resistance of the slope surface, preventing serious damage.

2.5 *Overflow level 3 (serious damage)*

When the overflow depth, h_0, exceeded 32.3 cm, a very thick vein of head flow with a free fall for a long distance gave severe impact on a limited number of soil bags, resulting in progressive erosion toward the

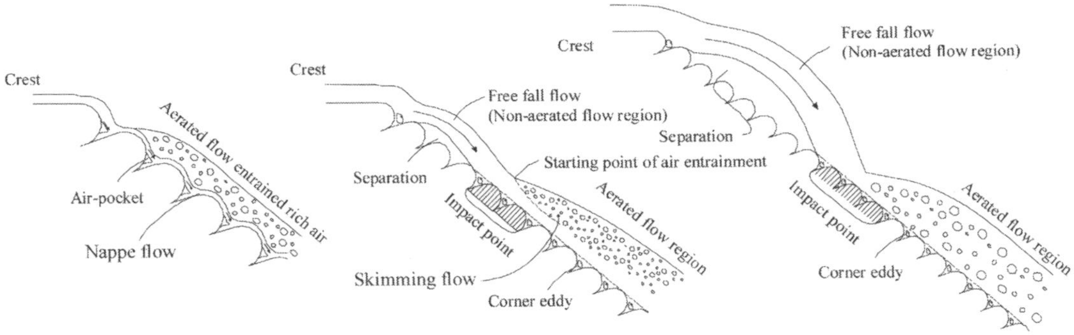

(a) Level 1 (h_0 = ~ 23.8 cm) (b) Level 2 (h_0 = 23.8-32.3 cm) (c) Level 3 (h_0 = 32.3-58.3 cm)

Figure 10. Schematic flow regimes in overflow levels.

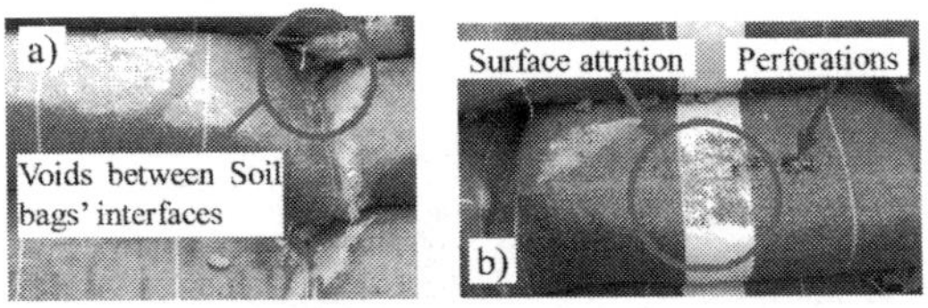

Photo 3. Minor or moderate damages: a) Evidence of sucked backfill material in cell zone behind stacked soil bags; and b) Attrition surface and perforations on soil bag surfaces.

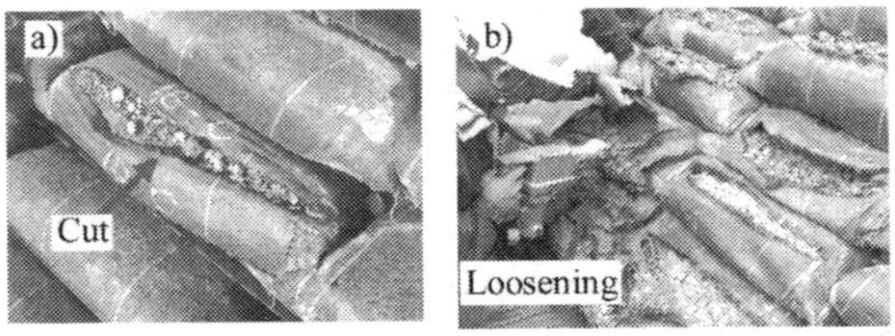

Photo 4. Artificial physical damage: a) Cut of the surface of geotexitle soil bag; and b) Loosening of the infill material.

inside of the slope (Photo 5a and Fig. 10c). Fig. 11b presents the profile of the deformed downstream surface at stages b and c in Fig. 9 (before giving artificial physical damage to the soil bags). It was found that a limited number of soil bags deformed severely due to impacts by over-fall of a very thick vein.

Figure 11c shows the profile of the deformed downstream surface after periods of 30 min and 90 min since the start of flow at a rate of $q = 0.652$ $m^3/s/m$, at stages h and i plotted in Fig. 9, after having given artificial physical damage to the soil bags. It was found that, after 90 minutes, the erosion had reached the foundation of the slope along the axis of waterfall formation. Differently from overflow level 2, a free fall having sufficient force to penetrate the soil bags was formed, which resulted in severer progressive erosion. Photo 5b shows a trace of deep erosion that was formed by a waterfall. This erosion, which reached the foundation, is critical damage to the slope for the stability of the embankment. However, the rate of development of erosion with the reinforced slope was substantially slower than with the unreinforced slope, which collapsed at 5 min (as shown in Photo 6) at a much lower discharge of $q = 0.050$ $m^3/s/m$. It is to be noted that, even after this severer erosion in the slope, the settlement at the crest of the GSET-spillway was negligible due likely to reinforcement effects. This high performance of the GSET-spillway shows that this technology can alleviate serious damage to the downstream slope caused by overflowing of earth dams.

The test results showed that a long free fall creates strong penetration force, which results in severe damage on the downstream slope for a small earth dam. It is therefore important both to reduce the penetration force and to increase the resistance against such force of soil bags. To this end, it is suggested: 1) to arrange a short slope with a gentle gradient from the top corner to the impact point at the upper part of the downstream slope; and 2) to reinforce the surface of soil bags at the impact point to provide sufficient resistance against the penetration force.

3 CONCLUSIONS

The effectiveness of GSET (Geosynthetics soil bags with extended tails) spillway against temporary flooding was evaluated by overflow-induced collapse tests. The damage pattern and erosion development depend on the overflow levels:

Overflow level 1 (nappe flow regime): nearly no damage

Overflow level 2 (skimming flow regime): Some backfill material in the slope may be sucked out by suction created by corner eddies, while attrition and perforations on the soil bag surfaces may be caused by tractive and impact forces. A combination of dense arrangement of geotextile layers in the slope

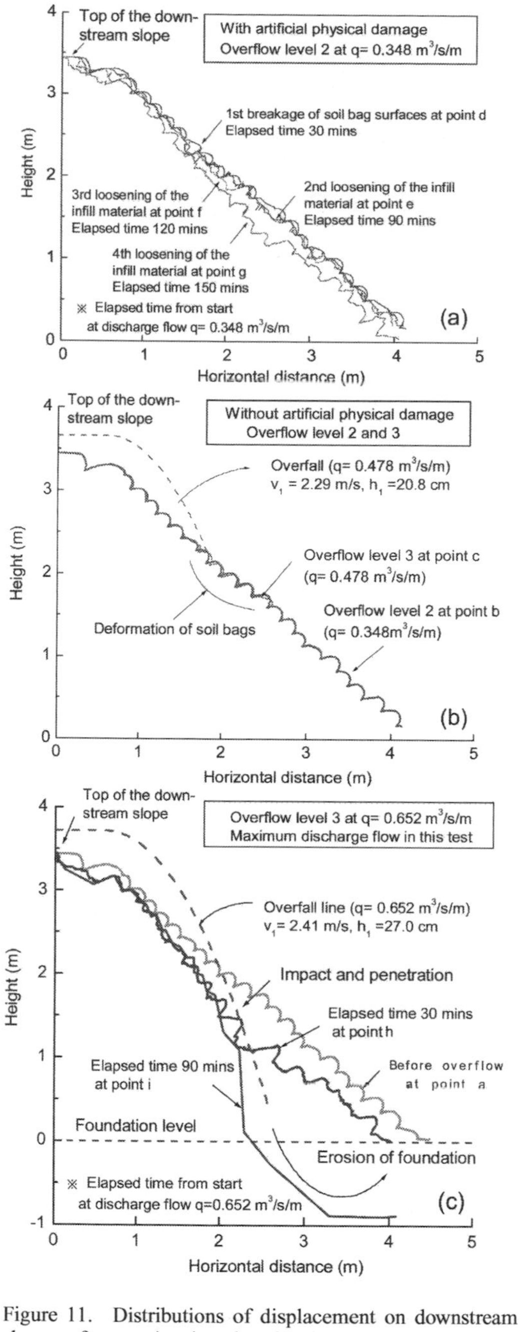

Figure 11. Distributions of displacement on downstream slope surface at: a) points d, e, f and g; b) points b and c; c) points and a, h and i.

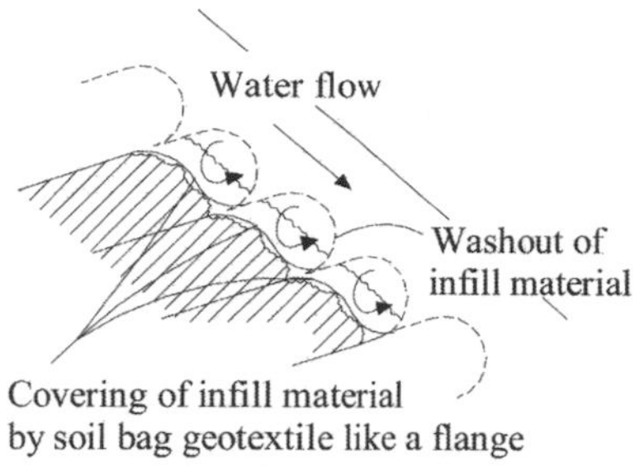

Figure 12. Schematic diagram of erosion on the down-stream slope with densely arranged geotextile layers in case of the ar-tificial physical damage at overflow level 2.

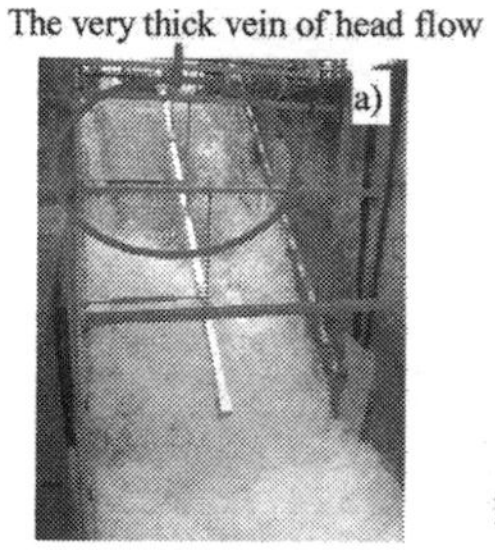

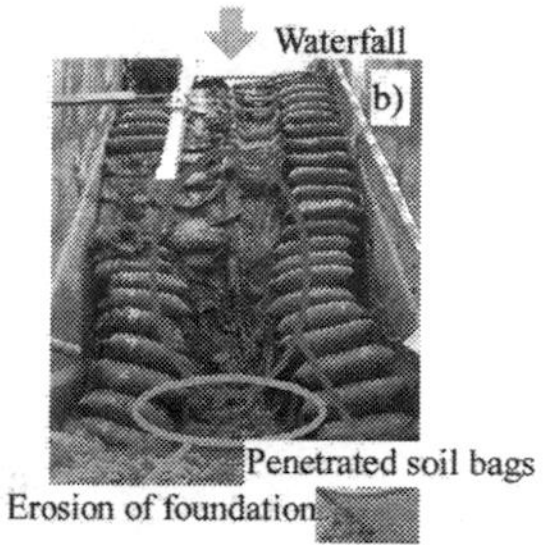

Photo 5. Overflow level 3(serious damage): a) Stream regime on the downstream side at overflow level 3 at $q = 0.652\ m^3/s/m$; and b) Erosion trace like the formation of a waterfall basin af-ter final overtopping.

Photo 6. Erosion trace on un-reinforced soil slope after 5 min-utes at $q = 0.050\ m^3/s/m$.

and slight cementation of the infill material could prevent fast development of erosion. At overflow level 2, which is the design condition in practice, minor or moderate damage to the GSET-spillway can be expected.

Overflow level 3 (formation of a very thick vein of over-fall): A long free fall was formed, which had sufficient energy to penetrate soil bags and then cause fast development of erosion, seriously dam-aging the GSET-spillway. Yet, the development rate of erosion was significantly slower than with the unreinforced slope, while the settlement of the crest was negligible. Therefore, total collapse might not take place even during a very strong flood.

It is concluded that the GSET-spillway is effective technology to prevent the collapse of the downstream slope by overflowing of earth fill dams.

ACKNOWLEDGEMENTS

The authors are grateful to the National Institute for Land and Infrastructure Management (NILIM) and Public Works Research Institute (PWRI) for their cooperation, and also to Dr. Duttie and Dr. Warat for their help, discussions and advice.

REFERENCES

Hubert, C. (1994). Hydraulic design of stepped cascades, channels, weirs and spillways, *Pergamon, First edition*, ISBN 08 041918 6

Matsushima, K., Yamazaki, S., Mohri, Y., and Arangelovski, G. (2005a). Overflow model test of small dam with soil bags, *Proc. of 40th Annual Symposium on Geotechnical Engineering*, pp. 1995–1996(in Japanese).

Matsushima, K., Yamazaki, S., Mohri, Y., and Arangelovski, G. (2005b). Overflow Test of Small Earth Dam with Allowed Overtopping, *Proc. of Annual Symposium on Irrigation, Drainage and Reclamation Engineering*, CD-ROM (in Japanese).

Matsushima, K., et al. (2006). Shear characteristics of geosynthetics soil bags stacked in inclined and horizontal directions, *Proc. of 21st Geosynthetics Symposium (Japan Chapter of IGS)*, pp. 145–152(in Japanese).

Mohri, Y., Matshushima, K., Hori, T., and Tani, S. (2005). Damage to small-size reservoirs and their reconstruction method, *Special Issue on Lessons from the 2004 Niigataken Chu-Etsu Earthquake and Reconstruction, Foundation Engineering and Equipment (Kiso-ko)*, October, pp. 62–65 (in Japanese).

Yasuda, Y. and Ohtsu, I. (1999). Flow resistance of skimming flows in stepped channels, *Proc. of 28th IAHR Congress*, Grz, Austria, Session of B14 (CD-ROM).

New Horizons in Earth Reinforcement – Otani, Miyata & Mukunoki (eds)
© 2008 Taylor & Francis Group, London, ISBN 978-0-415-45775-0

Application of geotextile technology to reduce surface erosion on natural slope

E. Purwanto
Civil Engineering Department, FTSP- Islamic University of Indonesia

ABSTRACT: Continuous erosion can result particle transportation of soil mass which can reduce the strength of soil, it can cause the slope instability. Results of the research indicate soil from Plipir village, Purworejo, Province of Center Java is sandy silt (MH) with plasticity index 16,60%. The use of geotextile can prevent erosion 9,62%–13,10% of natural erosion.

1 INTRODUCTION

Erosion is a natural phenomenon and normal process in the history of shaping the topography in earth surface. When it comes to a rapid step, erosion can be a main process in affecting the soil surface. The erosion process, which is only caused by natural factors or not yet influenced by human activity, is called geological erosion or normal erosion, or natural erosion, while erosion, which is already influenced by human activity, is called accelerated erosion.

There are mounted evidence of natural erosion in Indonesia, where it can be considered that in every hectare of land in highland, there are about 25–65 ton drove away per year by the wiping out of Indonesian forest.(Sabo Technical Center, 1985)[2]

During this erosion process, there is a deliberate change of soil profile, which will affect the arability and slope stability. Some problems appear from erosion is the move of soil mass which can affect soil strength, even more dangerous when in fall of slope dike.

Landslide on natural/human-made slope is caused by the distraction of soil instability. Technically, landslide slope failure is caused by low resistance of soil shear to be able to protect soil shear strength , height, and slanting slope.

In accordance with development of science and technology, especially in civil engineering, erosion treatment system and slope stability started to be developed using synthetic/geosynthetic material, or soil reinforcement as solution alternative, then in this research, this is called geotextile technology.

Based on mechanic and hydraulic role plus its multifunction role (as filter, reinforcement, and separator), the application of geotextile technology to cope with natural disaster (surface erosion and landslide) is considered to be very important to be comprehensively studied.

Erosion prevention has to be linked mainly to root factors. One of the factors is rain. Thereby, it needs to do some research on rain profile which will cause erosion. To prevent further damage, it needs efforts to control erosion and slope stability by using geotextile technology with all the benefits of its physical and mechanical feature. Since the application of geotextile technology to cope with erosion and slope stability problem is still new (not yet developed), it needs to do research to know the affectivity of this material, linked to rain intensity parameter, soil type and the slanting of the slope.

Research was conducted jointly with SABO Technical Center Yogyakarta by observing visually (direct observation) and then studied the performance of surface erosion by learning some factors, which are soil type, slope slanting, rain density, and also the application of geotextile technology as solution alternative prevention. This research took a case study in Plipir village, Purworejo Regency, Province of Center Java.

2 OBJECTIVE

The objective of the research are :

1. To identify the sum and the comparison between eroded soils drifted away in the results of simulated rain to parameters of rain intensity and slope slanting, for an unprotected land slope, and geotextile protected.

2. To measure affectivity of surface erosion prevention system using geotextile technology.

2.1 *Research parameter*

Some parameters were used in this research :

1. Soil used for this research was taken from Plipir village, Purworejo, Province of Center Java.
2. Soil condition was saturated, to make it as the real condition at the time the first flow cause surface erosion.
3. Rain intensity used was 30 mm/hr, 40 mm/hr, and 60 mm/hr based on a consideration that those level of intensity happened frequently in Indonesia.
4. Slope angle used limited only on 30°, 45° , and 60°.
5. Soil thickness was 30 cm, soil density and slope angle were constants during the test.
6. Geosyntethic used was non-woven geotextile (Polyfelt Geosynthetics TS-30) put horizontally.
7. Laboratory test was conducted in Soil Mechanic Laboratory, Civil Engineering Department, Faculty of Civil Engineering and Planning, Islamic University of Indonesia, and *SABO Technical Center* Laboratory, *Yogyakarta.*

3 FUNDAMENTAL THEORY

Erosion happened in two ways, natural erosion, or known as normal erosion and or geological erosion, and accelerated erosion. Normal erosion process causes shapes of topography in the nature. Accelerated erosion is soil transport causing soil damage as result of human activity, which distracts balance of shaping process and soil transport (Hudson, 1977).

The relation between slope failures caused by land surface erosion has been examined by Musgrave (1947, in Kirby, 1980) which informed that there is relationship between rain profile and the quantity of soil eroded which finally causing failure of land slope. Zang (1940, in Kirby,1980) did an experiment using a square of land rain simulation as field condition. This experiment showed that there is relationship between slope slanting angle and erosion, the bigger the angle, erosion will come bigger.

Soil erosion, based on its source, can be classified as sheet erosion, which is land surface erosion caused by surface flow, then develop into rill erosion as erosion caused by small scale of water flow. Rill erosion then shapes gully erosion, which is erosion caused by water flow on a small-gully river, and slowly shaping channel erosion, which is erosion on river channel resulted from degradation of its base.

In accordance with development of geosynthetic technology till the last decade, synthetic material can be used as an alternative solution to prevent erosion and reinforce land, especially on slope (Edy Purwanto, 1996).

Beside that, it is also known that benefit of synthetic material (geotextile) 2 dimension can decrease erosion results. The use of combined geotextile and vegetated grass as a protection eventually can overcome the erosion problem completely. (SABO Technical Center Yogyakarta 1985).

3.1 *Soil profile*

Land profile which can affect erosion is defined by profile affecting infiltration capacity, soil resistance against dispersion, and rainfall erosion drop and flow on land surface (Baver 1958). Those profile includes texture, structure, organic material, etc.

Infiltration capacity is very affected by soil permeability, land surface condition, and soil water capacity (Baver 1958). Rough-textured soil like sand and gravel has high infiltration capacity and resistance on being transported (Arsyad 1982 in Supardi 1989). Fine-textured land commonly resist on splitting, due to its strong cohesion strength. Sand splits into dust and clay, while clay is very easy to be transported. Therefore, sand and clay has same low erodibility, in different way. Clay erodibility is low because it is hard to split, while sand erodibility is low because it's hard to be transported.

3.2 *Rain factor*

Erosion can be defined as an erosion process which is categorized into 3 (three) phase, which is discharge, transportation, and sedimentation (Ellison, 1947 in Hudson 1971).

Erosion caused by rain is an intensity result of two components, which is the rain itself, and the type of the soil the rail fall into. The erosion quantity is affected by those two components, therefore different rain intensity will result different erosion as well for the same soil type.

Rain potential ability to cause erosion is called erosivity. Erosivity is a function of rain physical profile such as rain intensity, rain duration, rain pellet diameter, fall speed, and rain kinetic energy.

Erodibilty is a characteristic of the speed of eroded soil. Erodibilty is a function of soil physical profile and soil treatment management (Hudson, 1971)

A clear relationship between erosivity and erodibility, and erosion is indicated by Universal Set Loss Equation (USLE) as follows :

$$A = (0.224)\ R\ K\ L\ S\ E\ P \qquad (1)$$

where: A = Soil Loss (Kg/m^{2}s); R = Rain erosivity factor; K = Soil erodibilty factor; L = Slope length factor, S = slope slanting factor ; E = Plant treatment factor; P = Erosion handling application factor

The formula includes R as a factor of rain aspect, K as soil aspect, L and S as topography aspect, and E and P as handling technique aspect. (Kirkby 1980)

Rain erosivity factor related to rain physical profile which includes item below substances below.

3.3 *Rain profiles*

Rain profile resulting erosion is as follows: rain intensity, rain duration, rain quantity, size, fall speed, rain pellet, and rain kinetic energy (Kohnke and Bertrand1959, in Supardi 1989).

3.3.1 *Rain intensity*

Rain intensity is a number indicating comparison between rain quantity and rain duration stated in mm/hr and cm/hr.

$$I = \frac{Q}{A.t}\ x\ 600 \qquad (2)$$

where: I = rain intensity (mm/hr); Q = water volume in every container (ml); A = container width (cm^2); t = time (minute)

Erosivity index itself is a function of rain intensity and its application is wider compared to erosivity index which is based on another rain parameter, however, rain intensity is not enough to give us information needed to examine erosion caused by rain. It is because high-intense rain might fall in a short duration and the quantity is little and can't cause erosion. While low-intense rain fall in long duration, can cause erosion.

3.3.2 *Size and distribution of rain pellet*

Rain is distributed into different size of pellets, from slightly bigger than fog, until one with 7 mm diameter. This size becomes important parameter related to erosivity because it's influential on pellet fall speed which determines value of rain kinetic energy.

The first measurement of rain pellet recorded was done by Lowe in 1982 (in Hudson 1971) which contain pellets on a flat stone and scratched with a rectangle pattern so that the rain globule drops can be measured, another method to measure such as "Florir Pellet" etc. From those measuring method can be obtained range of rain pellet size in many countries, and many rain type (SABO Technical Center 1985).

According to Kowal and Kossam 1976, in Setyantono, 1990, in tropical area, average diameter of rain is 3 mm–4 mm. In a research in 1990, in the range of 30 mm/hr; 40 mm/hr; and 60 mm/hr, major pellet quantity was 3 mm and 4 mm.

Beside pellet size, it is also important to examine distribution of pellet size starting from the small to the big one, and how the pellet distribution changes in any rain type. From the experiment it was known that low-intense rain results small pellet, while high-intense rain results big pellet (Laws and Person 1948, in Hudson 1971)

3.3.3 *Rain duration*

High-intense rain with short duration might not cause erosion, while low intense rain with long duration is possible to cause erosion.

3.3.4 *Falling speed of rain pellet*

Free fall pellet will be affected by gravitation and then accelerated to a condition where air shear equals to gravitation, and then pellet will keep falling in that speed. The speed is affected by size and shape of the rain pellet. The bigger it is, the bigger the falling speed will be.

3.3.5 *Rain kinetic energy*

This energy is needed mainly for the release of soil pellet. This energy depends on the size and falling speed of rain pellet. Kinetic energy can be determined if other parameter, falling speed of Rain Pellet and pellet mass is known.

The equation to calculate rain kinetic energy is:

$$Ek = \frac{1}{2}.\ m\ .V^2 \qquad (3)$$

where: Ek = Rain Kinetic Energy (joule); m = Pellet mass (kg) ; v = Falling speed of rain pellet (m/second)

For rainfall simulator, formula used is : (SABO Technical Center 1985)

$$KE = 11.87 + 8.73.Log\ I. \qquad (4)$$

where: KE = Kinetic Energy (joule); I = Rain intensity (mm/hr)

Kinetic energy is the main cause of destruction, transfer, and transport of soil pellet, roughness, and soil profile. Soil with hi-cohesion like clay, has low erodability because of the tight bond of the pellets is hard to release. When the bond is released, clay will be easier to be carried away, so that the erodability increases.

Many researches were done to find the relationship between the measured quantity of eroded soil in a field, and several types of soil profile which can be measured in laboratory. Bouyoucos 1935 (in Hudson 1971) assumed that soil erodability is equal to :

$$Soil\ erodability = \frac{\%\ of\ sand + \%\ of\ silt}{\%\ of\ clay} \qquad (5)$$

While Middleton, 1930 and 1932 (in Hudson 1971) assumed that "Dispersion Ratio" method is used to determine soil erodability, which is based on the change of clay content and silt before and after dispersion in water. Besides that, there are still many techniques used to determine soil erodability ; mud and fine sand percentage (0,002–0,10 mm), organic soi soil content percentage, soil structure, and soil permeability.

Table 1. Distribution of rain pellet size.

Diameter of rain grain (mm)	Quantity of rain pellet		
	I = 30 mm/hr	I = 40 mm/hr	I = 60 mm/hr
1	3	7	19
2	4	9	34
3	18	27	49
4	15	21	46
5	4	6	21
6	2	3	8

Table 2. Speed of soil flow surface.

Slanting angle (°)	Rain intensity (mm/hr)	Average time of flow speed (second)	Flow speed (cm/second)
30	30	60	0.000333
	40	27	0.000740
	60	26.60	0.000751
35	30	60.125	0.000334
	40	26.75	0.000747
	60	18.25	0.001095
40	30	57.50	0.000347
	40	20	0.001000
	60	16.50	0.001212

4 RESULT OF RESEARCH

Soil from Plipir village, Purworejo, Province of Center java is a silt (MH), fine sand, un-organic soil, Gs = 2,58 ; gb = 1,266 gr/cm^3, w = 63,52%, LL = 60,36; PL = 44,76; SL = 2,12.10^{-5} cm/dt; c = 0,205 kg/cm^2, and $\Phi = 8,45°$ This type of soil is soil, which has large characteristic of expendability.

On the distribution test of rain pellet size, which has rain intensity of 30 mm/hr, 40 mm/hr and 60 mm/hr, it is obtained distribution of rain pellet as shown in table 1.

As shown in the table, rain pellet quantity in every intensity per pellet diameter increases, therefore by the addition of pellet diameter cause pellet mass extending, which means increasing of kinetic energy which then accelerate erosion.

Measurement result in the erosivity test can determine the speed of surface flow using big comparison of flow debit on surface of soil flow width as shown in table 2.

From the observation, it is identified that the bigger rain intensity and slanting angle is, the bigger surface flow speed will be, and this will accelerate erosion to occur.

Erosion test on intensity and constant volume mass shows that the bigger slope slanting angle, there will be more soil pellet transported, either with or without geotextile. The test result is shown in table 3.

Table 3. Comparison of dry soil weight eroded (gwet = 12,6 kN/m^3).

Rain intensity (mm/hr)	Eroded dry soil weight (gr)					
	Without geotextile			With geotextile		
	$\alpha =$ 30°	$\alpha =$ 35°	$\alpha =$ 40°	$\alpha =$ 30°	$\alpha =$ 35°	$\alpha =$ 40°
30	0.30	1.25	4.12	0.25	1.10	3.98
40	1.25	1.85	4.94	1.20	1.60	4.65
60	2.45	2.90	1.6	2.25	2.50	5.05

The data above shows that the bigger slope-slanting angle, there will be more soil pellet transported, and the higher rain intensity is, there will be more soil pellet transported also, either with or without geotextile. On a test using non-woven geotextile Polyfet TS-30, shows that the using of geotextile can reduce the quantity of soil pellets eroded until 11,356% without geotextile.

This explain that geotextile surface can hold soil pellets, and can be viewed visually after the test. This also applies on soil density, when it is higger, there will be less soil pellets eroded.

Every erosion test will be ended with a land failure. Big or small, and the speed of the land failure on test is affected by slanting angle and rain intensity. This means, that the bigger the angle is, the failure will run faster. The test result indicates that using geotextile can decelerate failure compared to the test without geotextile. This is because the rough surface of geotextile gives stickiness and high shear coefficient between soil and geotextile so that it will slow down a land failure.

5 CONCLUSION

1. Higher rain intensity will result bigger kinetic energy, which will cause the increase of eroded soil pellets around 0,152 grams. Bigger slanting slope angle will result the increase the speed of water flow on surface which will cause the increase of soil pellets eroded averagely 0,013 grams. In the other hand, bigger soil density results the decrease of soil pellets eroded.
2. The affectivity of the use of non-woven polyfet TS-30 geotextile to decrease soil pellets eroded is around 9,62%–13,10% of naturally eroded soil test. And also, using geotextile can decelerate land failure.

ACKNOWLEDGEMENTS

The authors wish to acknowledge the contributions and cooperations of SABO laboratory on the research of application of geotextile technology to reduce surface erosion on natural slope. The assistance of Sabo Laboratory professional and technical staff is also gratefully appreciated.

REFERENCES

Abramson W.T. (et all). 1996. Slope Stability and Stabilization Methods. John Wiley & Sons, Inc. New York.

American Society for Testing and Material (ASTM), 1977. Annual Book of ASTM Standard, Vol. 04.08.

Baver 1958. Soil Physics, John Wiley & Sons, Inc. New York.

Edy Purwanto. 1996. Etude de Frotement sol – geosinthetique en domain Geotechnique, Ph.D.thesis, University of Joseph FOURRIER, Grenoble, France.

Hayashi, S., Makiuchi, K. and Ochial,H., 1994. Testing Methods for Soil-Geosynthetic Frictional Behavior, Japanese Standard, Proc. Fifth Int. Conf. on Geotextiles, Geomembranes and Related Products, Vol.1, Singapore, pp.411–414

Hudson. 1971. Fabric mats for Stabilizing Embankment and Retaining Structures. Enka Industrial Systems, Netherlands.

Kirby M.J. 1980. Soil Erosion. John Wiley & Sons. Ltd. Norwich. England.

Menacham A. 1996. Soil Erosion Conservation and Rehabilitation. Marcel Dekker Inc. New York.

Peck R.B. 1967. Stability of Natural Slopes. Journal of Soil Mechanics and Foundation Division, ASCE. vol. 93. No. SM4., pp. 403–417.

Sabo Technical Centre, 2002. Water Induced Disaster Engineering II. Geomorphology. Yogyakarta. Indonesia.

New Horizons in Earth Reinforcement – Otani, Miyata & Mukunoki (eds)
© 2008 Taylor & Francis Group, London, ISBN 978-0-415-45775-0

Author Index